Interactive Genetic Problems

For chapters involving inheritance patterns and other quantitative problems, students can set up crosses and attempt to predict the outcome according to genetic principles. They also have the freedom to alter parameters to see how such changes affect the outcome.

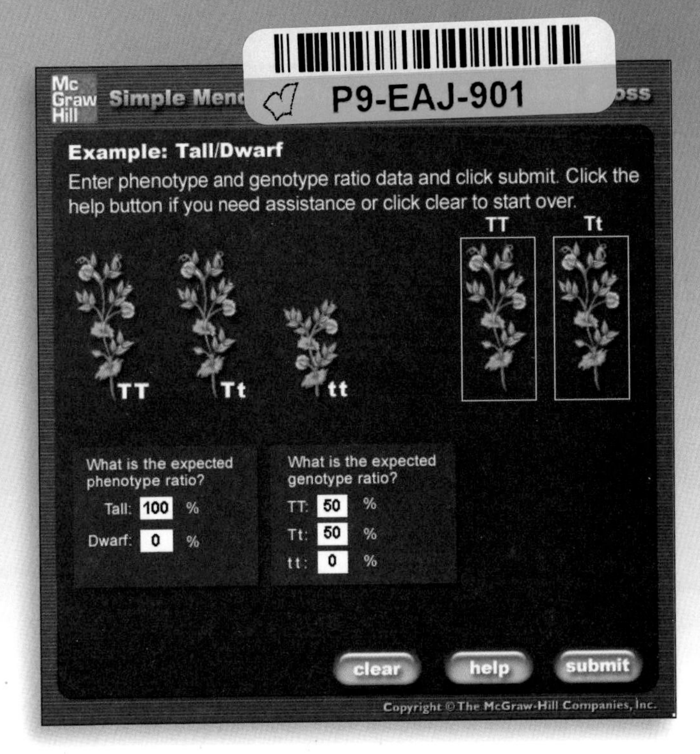

Test Yourself

Students can take a quiz at the Companion Site to gauge their mastery of chapter content. Each chapter quiz is specifically constructed to test comprehension of key concepts. Immediate feedback on responses explains why an answer is correct or incorrect. Students can even e-mail the quiz results to their professor!

Tweaking the Experiment

For the in-depth experiments that are found in each chapter, a self-help quiz is found on the Companion Site. These questions ask students to predict how changes in the experimental parameters may affect the outcome of the experiment.

My Lectures—Tegrity

McGraw-Hill Tegrity Campus™ records and distributes lectures with just a click of a button. Students can view anytime/anywhere via computer, iPod, or mobile device. It indexes as it records PowerPoint® presentations and anything shown on the computer, so students can use keywords to find exactly what they want to study.

Instructor Resources

McGraw-Hill Connect Genetics provides easy access to the following resources:

- Enhanced image PowerPoints with editable labels
- Lecture PowerPoints with animations
- Animation PowerPoints
- JPEG files of art, photos, and tables from the textbook

Powerful Reporting Solutions

McGraw-Hill Connect Genetics offers detailed reporting so instructors can quickly assess how students are doing in regards to overall class performance, individual assignments, and each question. All practice and test bank questions are tagged to the textbook by chapter, section, topic, and Bloom's level of difficulty to aid these reporting features.

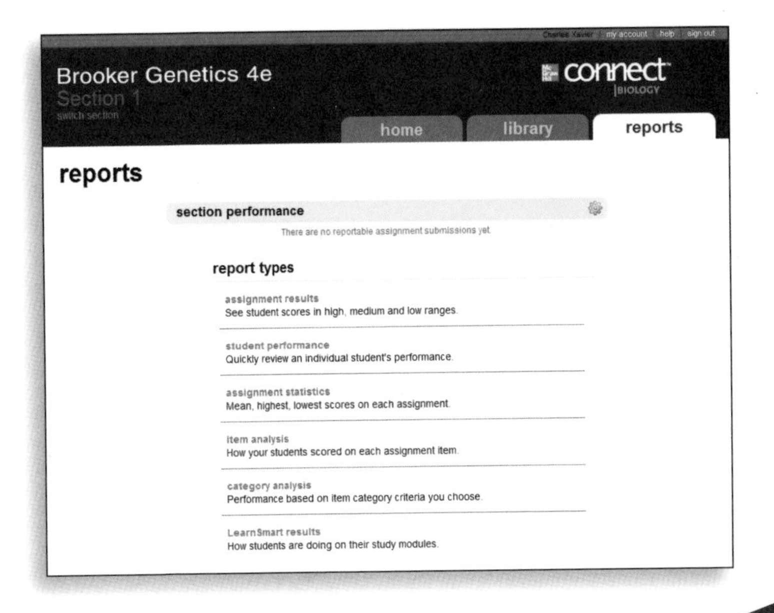

ROBERT J. BROOKER

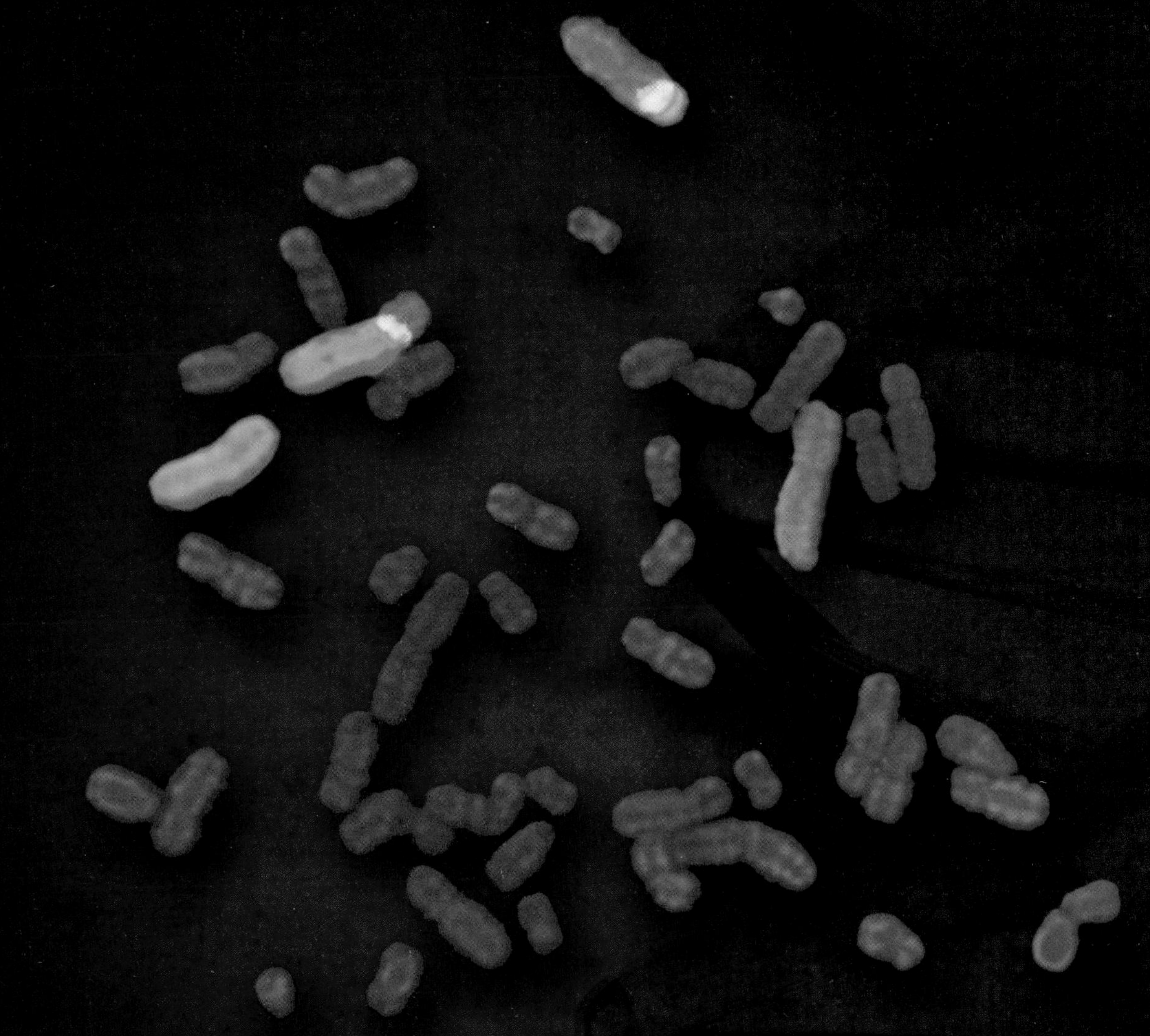

FOURTH EDITION

GENETICS

ANALYSIS & PRINCIPLES

The McGraw-Hill Companies

GENETICS: ANALYSIS & PRINCIPLES, FOURTH EDITION

Published by McGraw-Hill, a business unit of The McGraw-Hill Companies, Inc., 1221 Avenue of the Americas, New York, NY 10020. Copyright © 2012 by The McGraw-Hill Companies, Inc. All rights reserved. Previous editions © 2009, 2005, and 1999. No part of this publication may be reproduced or distributed in any form or by any means, or stored in a database or retrieval system, without the prior written consent of The McGraw-Hill Companies, Inc., including, but not limited to, in any network or other electronic storage or transmission, or broadcast for distance learning.

Some ancillaries, including electronic and print components, may not be available to customers outside the United States.

This book is printed on acid-free paper.

1 2 3 4 5 6 7 8 9 0 DOW/DOW 1 0 9 8 7 6 5 4 3 2 1

ISBN 978-0-07-352528-0
MHID 0-07-352528-6

Vice President, Editor-in-Chief: *Marty Lange*
Vice President, EDP: *Kimberly Meriwether David*
Senior Director of Development: *Kristine Tibbetts*
Publisher: *Janice Roerig-Blong*
Director of Digital Content: *Elizabeth M. Sievers*
Developmental Editor: *Mandy C. Clark*
Executive Marketing Manager: *Patrick Reidy*
Senior Project Manager: *Jayne L. Klein*
Buyer II: *Sherry L. Kane*
Senior Media Project Manager: *Tammy Juran*
Senior Designer: *David W. Hash*
Cover Designer: *John Joran*
Cover Image: *(FISH) micrograph of Chromosomes 2:3 translocation in cancer, ©James King-Holmes/Photo Researchers; DNA structure model, ©Alexander Shirkov/iStock Photo.*
Senior Photo Research Coordinator: *John C. Leland*
Photo Research: *Pronk & Associates, Inc.*
Compositor: *Lachina Publishing Services*
Typeface: *10/12 Minion*
Printer: *R. R. Donnelley*

All credits appearing on page or at the end of the book are considered to be an extension of the copyright page.

Library of Congress Cataloging-in-Publication Data

Brooker, Robert J.
Genetics : analysis & principles / Robert J. Brooker. — 4th ed.
p. cm.
Includes index.
ISBN 978-0-07-352528-0 — ISBN 0-07-352528-6 (hard copy : alk. paper) 1. Genetics. I. Title.
QH430.B766 2012
576.5--dc22

2010015380

www.mhhe.com

BRIEF CONTENTS

TABLE OF CONTENTS

ABOUT THE AUTHOR

Robert J. Brooker is a professor in the Department of Genetics, Cell Biology, and Development at the University of Minnesota–Minneapolis. He received his B.A. in biology from Wittenberg University in 1978 and his Ph.D. in genetics from Yale University in 1983. At Harvard, he conducted postdoctoral studies on the lactose permease, which is the product of the *lacY* gene of the *lac* operon. He continues his work on transporters at the University of Minnesota. Dr. Brooker's laboratory primarily investigates the structure, function, and regulation of iron transporters found in bacteria and *C. elegans*. At the University of Minnesota he teaches undergraduate courses in biology, genetics, and cell biology.

DEDICATION

To my wife, Deborah, and our children, Daniel, Nathan, and Sarah

PREFACE

In the fourth edition of *Genetics: Analysis & Principles,* the content has been updated to reflect current trends in the field. In addition, the presentation of the content has been improved in a way that fosters active learning. As an author, researcher, and teacher, I want a textbook that gets students actively involved in learning genetics. To achieve this goal, I have worked with a talented team of editors, illustrators, and media specialists who have helped me to make the fourth edition of *Genetics: Analysis & Principles* a fun learning tool. The features that we feel are most appealing to students, and which have been added to or improved on in the fourth edition, are the following.

- **Interactive exercises** Education specialists have crafted interactive exercises in which the students can make their own choices in problem-solving activities and predict what the outcomes will be. Previously, these exercises focused on inheritance patterns and human genetic diseases. (For example, see Chapters 4 and 22.) For the fourth edition, we have also added many new interactive exercises for the molecular chapters.
- **Animations** Our media specialists have created over 50 animations for a variety of genetic processes. These animations were made specifically for this textbook and use the art from the textbook. The animations make many of the figures in the textbook "come to life."
- **Experiments** As in the previous editions, each chapter (beginning with Chapter 2) incorporates one or two experiments that are presented according to the scientific method. These experiments are not "boxed off" from the rest of the chapter. Rather, they are integrated within the chapters and flow with the rest of the text. As you are reading the experiments, you will simultaneously explore the scientific method and the genetic principles that have been discovered using this approach. For students, I hope this textbook helps you to see the fundamental connection between scientific analysis and principles. For both students and instructors, I expect that this strategy makes genetics much more fun to explore.
- **Art** The art has been further refined for clarity and completeness. This makes it easier and more fun for students to study the illustrations without having to go back and forth between the art and the text.
- **Engaging text** As in previous editions, a strong effort has been made in the fourth edition to pepper the text with questions. Sometimes these are questions that scientists considered when they were conducting their research. Sometimes they are questions that the students might ask themselves when they are learning about genetics.

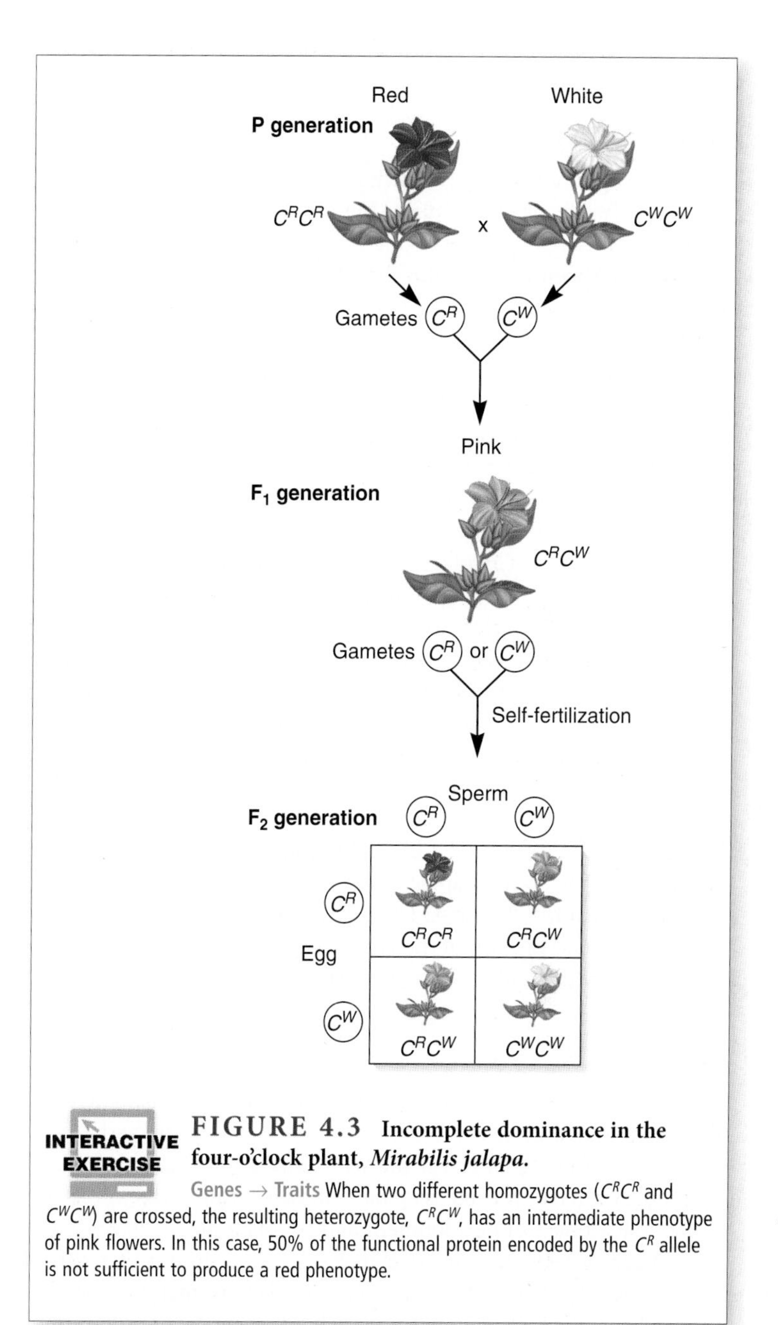

FIGURE 4.3 Incomplete dominance in the four-o'clock plant, *Mirabilis jalapa.*

Genes → Traits When two different homozygotes (C^RC^R and C^WC^W) are crossed, the resulting heterozygote, C^RC^W, has an intermediate phenotype of pink flowers. In this case, 50% of the functional protein encoded by the C^R allele is not sufficient to produce a red phenotype.

Overall, an effective textbook needs to accomplish three goals. First, it needs to provide comprehensive, accurate, and up-to-date content in its field. Second, it needs to expose students to the techniques and skills they will need to become successful in that field. And finally, it should inspire students so they want to pursue that field as a career. The hard work that has gone into the fourth edition of *Genetics: Analysis & Principles* has been aimed at achieving all three of these goals.

HOW WE EVALUATED YOUR NEEDS

ORGANIZATION

In surveying many genetics instructors, it became apparent that most people fall into two camps: **Mendel first** versus **Molecular first.** I have taught genetics both ways. As a teaching tool, this textbook has been written with these different teaching strategies in mind. The organization and content lend themselves to various teaching formats.

Chapters 2 through 8 are largely inheritance chapters, whereas Chapters 24 through 26 examine population and quantitative genetics. The bulk of the molecular genetics is found in Chapters 9 through 23, although I have tried to weave a fair amount of molecular genetics into Chapters 2 through 8 as well. The information in Chapters 9 through 23 *does not assume* that a student has already covered Chapters 2 through 8. Actually, each chapter is written with the perspective that instructors may want to vary the order of their chapters to fit their students' needs.

For those who like to discuss inheritance patterns first, a common strategy would be to cover Chapters 1 through 8 first, and then possibly 24 through 26. (However, many instructors like to cover quantitative and population genetics at the end. Either way works fine.) The more molecular and technical aspects of genetics would then be covered in Chapters 9 through 23. Alternatively, if you like the "Molecular first" approach, you would probably cover Chapter 1, then skip to Chapters 9 through 23, then return to Chapters 2 through 8, and then cover Chapters 24 through 26 at the end of the course. This textbook was written in such a way that either strategy works well.

ACCURACY

Both the publisher and I acknowledge the fact that inaccuracies can be a source of frustration for both the instructor and students. Therefore, throughout the writing and production of this textbook we have worked very hard to catch and correct errors during each phase of development and production.

Each chapter has been reviewed by a minimum of seven people. At least five of these people were faculty members who teach the course or conduct research in genetics or both. In addition, a development editor has gone through the material to check for accuracy in art and consistency between the text and art. With regard to the problem sets, the author personally checked every question and answer when the chapters were completed.

PEDAGOGY

Based on our discussions with instructors from many institutions, some common goals have emerged. Instructors want a broad textbook that clearly explains concepts in a way that is interesting, accurate, concise, and up-to-date. Likewise, most instructors want students to understand the experimentation that revealed these genetic concepts. In this textbook, concepts and experimentation are woven together to provide a story that enables students to learn the important genetic concepts that they will need in their future careers and also to be able to explain the types of experiments that allowed researchers to derive such concepts. The end-of-chapter problem sets are categorized according to their main focus, either conceptual or experimental, although some problems contain a little of both. The problems are meant to strengthen students' abilities in a wide variety of ways.

- By bolstering their understanding of genetic principles
- By enabling students to apply genetic concepts to new situations
- By analyzing scientific data
- By organizing their thoughts regarding a genetic topic
- By improving their writing skills

Finally, since genetics is such a broad discipline, ranging from the molecular to the populational levels, many instructors have told us that it is a challenge for students to see both "the forest and the trees." It is commonly mentioned that students often have trouble connecting the concepts they have learned in molecular genetics with the traits that occur at the level of a whole organism (i.e., What does transcription have to do with blue eyes?). To try to make this connection more meaningful, certain figure legends in each chapter, designated **Genes → Traits,** remind students that molecular and cellular phenomena ultimately lead to the traits that are observed in each species (e.g., see Figure 4.3).

ILLUSTRATIONS

In surveying students whom I teach, I often hear it said that most of their learning comes from studying the figures. Likewise, instructors frequently use the illustrations from a textbook as a central teaching tool. For these reasons, a great amount of effort in improving the fourth edition has gone into the illustrations. The illustrations are created with four goals in mind:

1. **Completeness** For most figures, it should be possible to understand an experiment or genetic concept by looking at the illustration alone. Students have complained that it is difficult to understand the content of an illustration if they have to keep switching back and forth between the figure and text. In cases where an illustration shows the steps in a scientific process, the steps are described in brief statements that allow the students to understand the whole process (e.g., see Figure 11.16). Likewise, such illustrations should make it easier for instructors to explain these processes in the classroom.
2. **Clarity** The figures have been extensively reviewed by students and instructors. This has helped us to avoid drawing things that may be confusing or unclear. I hope

that no one looks at an element in any figure and wonders, "What is that thing?" Aside from being unmistakably drawn, all new elements within each figure are clearly labeled.

3. **Consistency** Before we began to draw the figures for the fourth edition, we generated a style sheet that contained recurring elements that are found in many places in the textbook. Examples include the DNA double helix, DNA polymerase, and fruit flies. We agreed on the best way(s) to draw these elements and also what colors they should be. Therefore, as students and instructors progress through this textbook, they become accustomed to the way things should look.
4. **Realism** An important goal of this and previous editions is to make each figure as realistic as possible. When drawing macroscopic elements (e.g., fruit flies, pea plants), the illustrations are based on real images, not on cartoonlike simplifications. Our most challenging goal, and one that we feel has been achieved most successfully, is the realism of our molecular drawings. Whenever possible, we have tried to depict molecular elements according to their actual structures, if such structures are known. For example, the ways we have drawn RNA polymerase, DNA polymerase, DNA helicase, and ribosomes are based on their crystal structures. When a student sees a figure in this textbook that illustrates an event, for example proofreading DNA, DNA polymerase is depicted in a way that is as realistic as possible (e.g., see Figure 11.16).

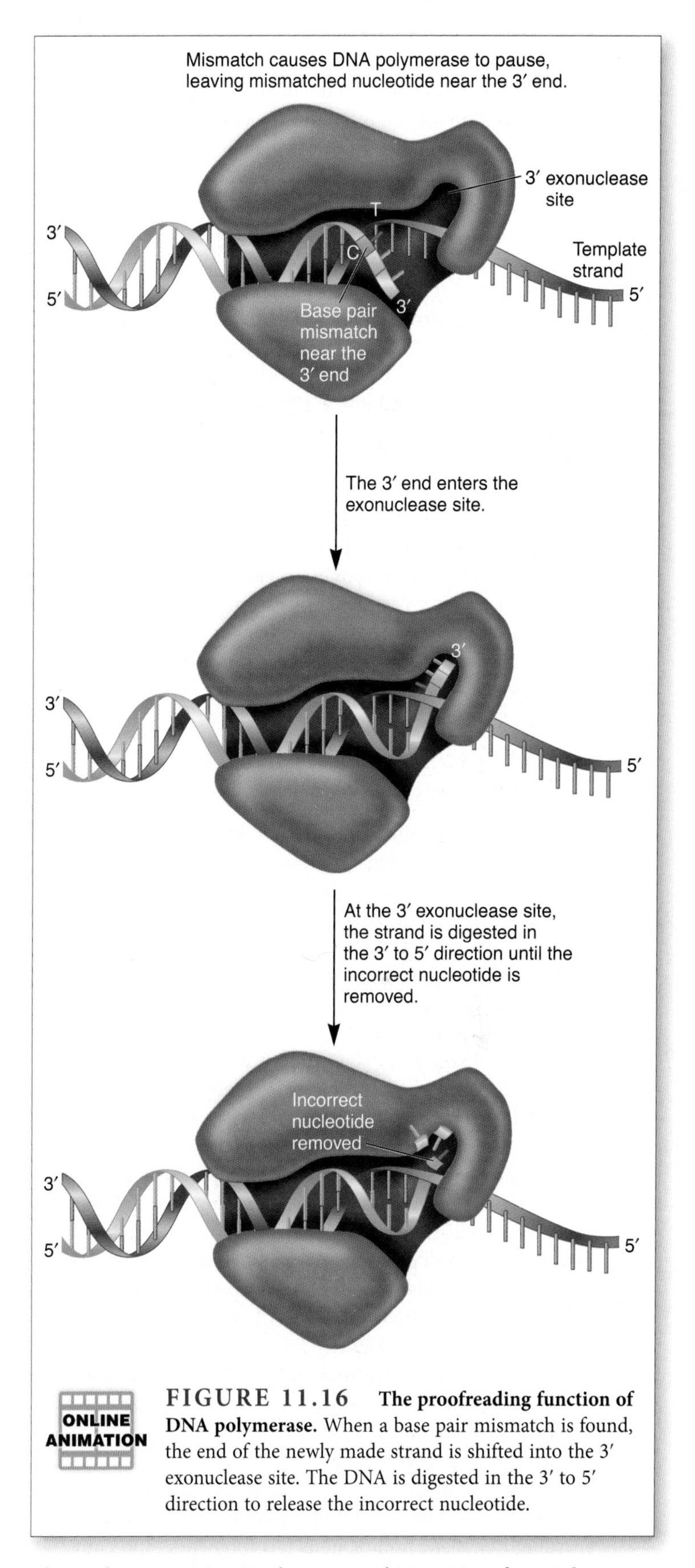

FIGURE 11.16 The proofreading function of DNA polymerase. When a base pair mismatch is found, the end of the newly made strand is shifted into the 3′ exonuclease site. The DNA is digested in the 3′ to 5′ direction to release the incorrect nucleotide.

WRITING STYLE

Motivation in learning often stems from enjoyment. If you enjoy what you're reading, you are more likely to spend longer amounts of time with it and focus your attention more crisply. The writing style of this book is meant to be interesting, down to earth, and easy to follow. Each section of every chapter begins with an overview of the contents of that section, usually with a table or figure that summarizes the broad points. The section then examines how those broad points were discovered experimentally, as well as explaining many of the finer scientific details. Important terms are introduced in a boldface font. These terms are also found in the glossary.

There are various ways to make a genetics book interesting and inspiring. The subject matter itself is pretty amazing, so it's not difficult to build on that. In addition to describing the concepts and experiments in ways that motivate students, it is important to draw on examples that bring the concepts to life. In a genetics book, many of these examples come from the medical realm. This textbook contains lots of examples of human diseases that exemplify some of the underlying principles of genetics. Students often say they remember certain genetic concepts because they remember how defects in certain genes can cause disease. For example, defects in DNA repair genes cause a higher predisposition to develop cancer. In addition, I have tried to be evenhanded in providing examples from the microbial and plant world. Finally, students are often interested in applications of genetics that affect their everyday lives. Because we frequently hear about genetics in the news, it's inspiring for students to learn the underlying basis for such technologies. Chapters 18 to 21 are devoted to genetic technologies, and applications of these and other technologies are found throughout this textbook. By the end of their genetics course, students should come away with a greater appreciation for the influence of genetics in their lives.

SIGNIFICANT CONTENT CHANGES IN THE FOURTH EDITION

- A new feature of the fourth edition is that each chapter ends with a list of key terms. These are the terms in the chapter that are in bold face. The terms are also found in the glossary. This addition was made at the request of students.
- The summary at the end of the chapter has been modified in two ways. First, the key points are found as bulleted lists. Second, the bulleted lists also refer to the figures and tables where the topics can be found. This modification was made at the request of students, who said that it was difficult to easily extract the main points from summaries that were in paragraph form, as they were in previous editions.
- The chapter on Non-Mendelian Inheritance (formerly Chapter 7) is now Chapter 5. This change was made at the request of instructors who often cover the chapters on Mendelian and Non-Mendelian inheritance consecutively.

Examples of Specific Content Changes to Individual Chapters

- Chapter 2 (Mendelian Inheritance) An improved figure on Mendel's law of segregation has been added (Figure 2.6).
- Chapter 3 (Reproduction and Chromosome Transmission) An improved figure emphasizes how chromosomes in a karyotype are pairs of sister chromatids (see Figure 3.6). Also, the stages of mitosis and meiosis are set off as subsections with bold headings, which makes them easier to follow.
- Chapter 5 (Non-Mendelian Inheritance) Information regarding the molecular mechanism of imprinting has been updated, including a descripiton of CTC-binding factor. With regard to human mitochondrial diseases, the topics of heteroplasmy and somatic mutation have been expanded.
- Chapter 6 (Genetic and Linkage Mapping in Eukaryotes) A new figure illustrates the outcome of crossing over between two linked genes in Morgan's classic experiments (see Figure 6.4). This is then followed up with another figure that shows the consequences of crossing over among three linked genes (see Figure 6.5).
- Chapter 7 (Genetic Transfer and Mapping in Bacteria and Bacteriophages) A new figure depicts how F' factors arise by the imprecise excision of F factors from a chromosome (see Figure 7.5b).
- Chapter 8 (Variation in Chromosome Structure and Number) New information and figures have been added regarding nonallelic homologous recombination and copy number variation in populations (see Figures 8.5 and 8.8).
- Chapter 10 (Chromosome Organization and Molecular Structure) New figures have been added on the action of DNA gyrase and the relative amounts of unique and repetitive sequences in the human genome (see Figures 10.9 and 10.12).
- Chapter 11 (DNA Replication) A new figure illustrates DNA replication from a single origin (see Figure 11.11). Also, the topic of how RNA primers are removed by flap endonuclease in eukaryotic cells has been added, which includes a new figure (see Figure 11.23).
- Chapter 12 (Gene Transcription and RNA Modification) The mechanism of transcriptional termination in eukaryotes via the allosteric or torpedo models has been added (see Figure 12.15). Also, RNA editing has been moved to this chapter.
- Chapter 13 (Translation of mRNA) A new figure describes Beadle and Tatum's study of methionine biosynthesis (see Figure 13.2). The topic of the incorporation of selenocysteine and pyrrolysine during translation has been added (see Table 13.3).
- Chapter 14 (Gene Regulation in Bacteria and Bacteriophages) This chapter has a new section on riboswitches (see pp. 377–378).
- Chapter 15 (Gene Regulation in Eukaryotes) A new section has been added on chromatin remodeling, histone variation, and histone modification (see pp. 397–403). A new figure describes the technique of chromatin immunoprecipitation sequencing (see Figure 15.11). A new section has been added on insulators (see pp. 406–407).
- Chapter 16 (Gene Mutation and DNA Repair) The topic of oxidative stress and oxidative DNA damage has been greatly expanded (see pp. 435–437). A new figure depicts the probable mechanism of trinucleotide repeat expansion (see Figure 16.12).
- Chapter 17 (Recombination and Transposition at the Molecular Level) A new figure describes the transposition of non-LTR retrotransposons (see Figure 17.18).
- Chapter 18 (Recombinant DNA Technology) The topic of polymerase chain reaction (PCR) is now expanded to an entire section, which includes several new figures that describe the steps of the PCR cycle, reverse transcriptase PCR, real-time PCR, and the classic experiment that demonstrated the feasibility of real-time PCR (see pp. 491–499).
- Chapter 19 (Biotechnology) A new feature experiment describes the method of gene therapy (see Figure 19.20).
- Chapter 20 (Genomics I: Analysis of DNA) A new subsection has been added on next-generation DNA sequencing methods, including a new figure on pyrosequencing (see pp. 564–566).
- Chapter 21 (Genomics II: Functional Genomics, Proteomics, and Bioinformatics) A new subsection has been added that discusses gene knockout collections.
- Chapter 22 (Medical Genetics and Cancer) Two new subsections have been added on haplotypes and haplotype association studies (see pp. 609–610, Figures 22.5–22.6). The topic of preimplantation genetic diagnosis has also been added. With regard to inherited forms of cancer, a new figure describes how the "loss of heterozygosity" leads to cancer (see Figure 22.22).

- Chapter 23 (Developmental Genetics) A new section has been added at the beginning of the chapter that provides a general overview of animal development (see pp. 638–641). This precedes the two sections on Invertebrate and Vertebrate Development.
- Chapter 24 (Population Genetics) A new figure shows the output from automated DNA fingerprinting (see Figure 24.22).
- Chapter 26 (Evolutionary Genetics) The topic of species concepts is more focused on the factors that are used to distinguish species; the general lineage concept is described (see pp. 734–736). A new example illustrates the concept of a molecular clock (see Figure 26.14).

SUGGESTIONS WELCOME!

It seems very appropriate to use the word *evolution* to describe the continued development of this textbook. I welcome any and all comments. The refinement of any science textbook requires input from instructors and their students. These include comments regarding writing, illustrations, supplements, factual content, and topics that may need greater or less emphasis. You are invited to contact me at:

Dr. Rob Brooker
Dept. of Genetics, Cell Biology, and Development
University of Minnesota
6-160 Jackson Hall
321 Church St.
Minneapolis, MN 55455
brook005@umn.edu

TEACHING AND LEARNING SUPPLEMENTS

www.mhhe.com/brookergenetics4e

McGraw-Hill Connect™ Genetics provides online presentation, assignment, and assessment solutions. It connects your students with the tools and resources they'll need to achieve success.

With Connect™ Genetics you can deliver assignments, quizzes, and tests online. A set of questions and activities are presented for every chapter. As an instructor, you can edit existing questions and author entirely new problems. Track individual student performance—by question, assignment, or in relation to the class overall—with detailed grade reports. Integrate grade reports easily with Learning Management Systems (LMS), such as Blackboard® and WebCT. And much more.

ConnectPlus™ Genetics provides students with all the advantages of Connect™ Genetics, plus 24/7 online access to an eBook.

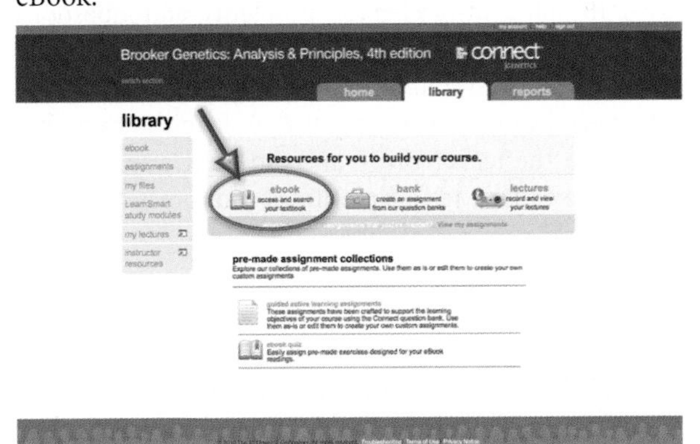

To learn more visit **www.mcgrawhillconnect.com**

PRESENTATION CENTER:

Build instructional materials wherever, whenever, and however you want!

www.mhhe.com/brookergenetics4e

The Presentation Center is an online digital library containing photos, artwork, animations, and other media tools that can be used to create customized lectures, visually enhanced tests and quizzes, compelling course websites, or attractive printed support materials. All assets are copyrighted by McGraw-Hill Higher Education, but can be used by instructors for classroom purposes. The visual resources in this collection include:

- **FlexArt Image PowerPoints®** Full-color digital files of all illustrations in the book with editable labels can be readily incorporated into lecture presentations, exams, or custom-made classroom materials. All files are preinserted into PowerPoint slides for ease of lecture preparation.
- **Photos** The photo collection contains digital files of photographs from the text, which can be reproduced for multiple classroom uses.
- **Tables** Every table that appears in the text has been saved in electronic form for use in classroom presentations or quizzes.
- **Animations** Numerous full-color animations illustrating important processes are also provided. Harness the visual effect of concepts in motion by importing these files into classroom presentations or online course materials.
- **PowerPoint Lecture Outlines** Ready-made presentations that combine art and lecture notes are provided for each chapter of the text.
- **PowerPoint Slides** For instructors who prefer to create their lectures from scratch, all illustrations, photos, tables and animations are preinserted by chapter into blank PowerPoint slides.

FOR THE STUDENT:

Student Study Guide/Solutions Manual

The solutions to the end-of-chapter problems and questions aid the students in developing their problem-solving skills by providing the steps for each solution. The Study Guide follows the order of sections and subsections in the textbook and summarizes the main points in the text, figures, and tables. It also contains concept-building exercises, self-help quizzes, and practice exams.

Companion Website
www.mhhe.com/brookergenetics4e

The Brooker *Genetics: Analysis & Principles* companion website offers an extensive array of learning tools, including a variety of quizzes for each chapter, interactive genetics problems, animations and more.

McGraw-Hill ConnectPlus™ interactive learning platform provides all of the benefits of Connect: online presentation tools, auto-grade assessments, and powerful reporting—all in an easy-to-use interface, as well as a customizable, assignable eBook. This media-rich version of the book is available through the McGraw-Hill Connect™ platform and allows seamless integration of text, media, and assessment.

By choosing ConnectPlus™, instructors are providing their students with a powerful tool for improving academic performance and truly mastering course material. ConnectPlus™ allows students to practice important skills at their own pace and on their own schedule. Students' assessment results and instructors' feedback are saved online—so students can continually review their progress and plot their course to success. Learn more at: **www.mcgrawhillconnect.com**

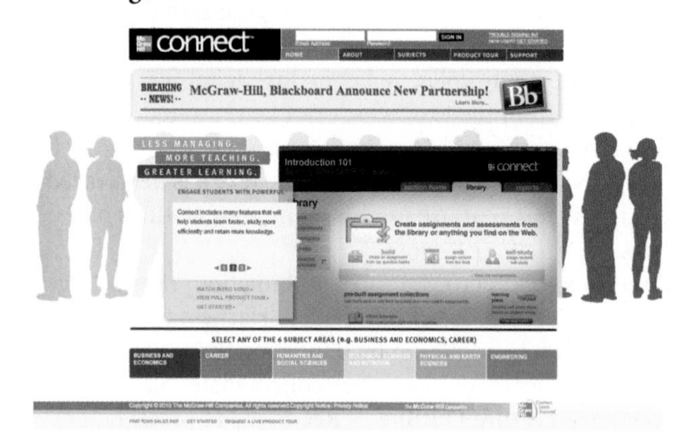

Do More

McGraw-Hill Higher Education and Blackboard® have teamed up.

Blackboard, the Web-based course-management system, has partnered with McGraw-Hill to better allow students and faculty to use online materials and activities to complement face-to-face teaching. Blackboard features exciting social learning and teaching tools that foster more logical, visually impactful and active learning opportunities for students. You'll transform your closed-door classrooms into communities where students remain connected to their educational experience 24 hours a day.

This partnership allows you and your students access to McGraw-Hill's Connect™ and Create™ right from within your Blackboard course—all with one single sign-on.

Not only do you get single sign-on with Connect™ and Create™, you also get deep integration of McGraw-Hill content and content engines right in Blackboard. Whether you're choosing a book for your course or building Connect™ assignments, all the tools you need are right where you want them—inside of Blackboard.

Gradebooks are now seamless. When a student completes an integrated Connect™ assignment, the grade for that assignment automatically (and instantly) feeds your Blackboard grade center.

McGraw-Hill and Blackboard can now offer you easy access to industry leading technology and content, whether your campus hosts it, or we do. Be sure to ask your local McGraw-Hill representative for details.

ACKNOWLEDGMENTS

The production of a textbook is truly a collaborative effort, and I am greatly indebted to a variety of people. All four editions of this textbook went through multiple rounds of rigorous revision that involved the input of faculty, students, editors, and educational and media specialists. Their collective contributions are reflected in the final outcome.

Let me begin by acknowledging the many people at McGraw-Hill whose efforts are amazing. My highest praise goes to Lisa Bruflodt and Mandy Clark (Senior Developmental Editors), who managed and scheduled nearly every aspect of this project. I also would like to thank Janice Roerig-Blong (Publisher) for her patience in overseeing this project. She has the unenviable job of managing the budget for the book and that is not an easy task. Other people at McGraw-Hill have played key roles in producing an actual book and the supplements that go along with it. In particular, Jayne Klein (Project Manager) has done a superb job of managing the components that need to be assembled to produce a book, along with Sherry Kane (Buyer). I would also like to thank John Leland (Photo Research Coordinator), who acted as an interface between me and the photo company. In addition, my gratitude goes to David Hash (Designer), who provided much input into the internal design of the book as well as creating an awesome cover. Finally, I would like to thank Patrick Reidy (Marketing Manager), whose major efforts begin when the fourth edition comes out! I would also like to thank Linda Davoli (Freelance Copy Editor) for making grammatical improvements throughout the text and art, which has significantly improved the text's clarity.

I would also like to extend my thanks to Bonnie Briggle and everyone at Lachina Publishing Services, including the many artists who have played important roles in developing the art for the third and fourth editions. Also, folks at Lachina Publishing Services worked with great care in the paging of the book, making sure that the figures and relevant text are as close to each other as possible. Likewise, the people at Pronk & Associates have done a great job of locating many of the photographs that have been used in the fourth edition.

Finally, I want to thank the many scientists who reviewed the chapters of this textbook. Their broad insights and constructive suggestions were an important factor that shaped its final content and organization. I am truly grateful for their time and effort.

REVIEWERS

Agnes Ayme-Southgate, *College of Charleston*
Diya Banerjee, *Virginia Polytechnic Institute*
Miriam Barlow, *University of California*
Bruce Bejcek, *Western Michigan University*
Michael Benedik, *University of Houston*
Helen Chamberlin, *Ohio State University*
Michael Christoffers, *North Dakota State University*
Craig Coleman, *Brigham Young University–Provo*
Brian Condie, *University of Georgia*
Erin Cram, *Northeastern University*
Mack Crayton, *Xavier University of Louisiana*
Stephen D'Surney, *University of Mississippi*
Sandra Davis, *University of Indianapolis*
Michael Deyholos, *University of Alberta*
Robert Dotson, *Tulane University*
Richard Duhrkopf, *Baylor University*
Aboubaker Elkharroubi, *John Hopkins University*
Matthew Elrod-Erickson, *Middle Tennessee State University*
Rebecca Ferrell, *Metro State College of Denver*
Cedric Feschotte, *The University of Texas–Arlington*
Michael Foster, *Eastern Kentucky University*
Gail Gasparich, *Towson University*
Jayant Ghiara, *University of California–San Diego*
Doreen Glodowski, *Rutgers University*
Richard Gomulkiewicz, *Washington State University - Pullman*
Ernest Hanning, *The University of Texas–Dallas*
Michael Harrington, *University of Alberta*
Jutta Heller, *Loyola University*
Bethany Henderson-Dean, *University of Findlay*
Brett Holland, *California State University–Sacramento*
Margaret Hollingsworth, *SUNY Buffalo*
Dena Johnson, *Tarrant County College NW*
Christopher Korey, *College of Charleston*
Howard Laten, *Loyola University*
Haiying Liang, *Clemson University*
Qingshun Quinn Li, *Miami University*
Dmitri Maslov, *University of California–Riverside*
Debra McDonough, *University of New England–Biddeford*
David McFadyen, *Grant MacEwan College*
Marcie Moehnke, *Baylor University*
Roderick Morgan, *Grand Valley State University*
Sally Pasion, *San Francisco State University*
James Prince, *California State University–Fresno*
Richard Richardson, *University of Texas–Austin*
William Rosche, *Richard Stockton College of NJ*
Mark Rovedo, *Loyola University*
Laurie Russell, *Saint Louis University*
Gwen Sancar, *University of North Carolina–Chapel Hill*
Malcolm Schug, *University of North Carolina–Greensboro*
Julian Kenneth Shull, *Appalachian State University*
Jeffry Shultz, *Louisiana Tech University*
Randall Small, *University of Tennessee–Knoxville*
Terrance Michael Stock, *Grant MacEwan College*
Tin Tin Su, *University of Colorado-Boulder*
John David Swanson, *University of Central Arkansas*
Daniel Yunqiu Wang, *University of Miami–Coral Gables*
Qun-Tian Wang, *University of Illinois–Chicago*
Matthew White, *Ohio University–Athens*
Malcolm Zellars, *Georgia State University*
Robert Zemetra, *University of Idaho*
Chaoyang Zeng, *University of Wisconsin–Milwaukee*

ACCURACY CHECKERS

Agnes Ayme-Southgate, *College of Charleston*
Diya Banerjee, *Virginia Polytechnic Institute*
Miriam Barlow, *University of California*
Bruce Bejcek, *Western Michigan University*
Michael Benedik, *University of Houston*
Helen Chamberlin, *Ohio State University*
Michael Christoffers, *North Dakota State University*
Sandra Davis, *University of Indianapolis*
Michael Deyholos, *University of Alberta*
Aboubaker Elkharroubi, *John Hopkins University*
Michael Foster, *Eastern Kentucky University*
Jutta Heller, *Loyola University*
Bethany Henderson-Dean, *University of Findlay*
Margaret Hollingsworth, *SUNY Buffalo*
Michael Ibba, *Ohio State University*
Dena Johnson, *Tarrant County College NW*
Haiying Liang, *Clemson University*
Qingshun Quinn Li, *Miami University*
Dmitri Maslov, *University of California–Riverside*
Marcie Moehnke, *Baylor University*
Roderick Morgan, *Grand Valley State University*
Laurie Russell, *Saint Louis University*
Tin Tin Su, *University of Colorado-Boulder*
John David Swanson, *University of Central Arkansas*
Matthew White, *Ohio University–Athens*

A Visual Guide to GENETICS: ANALYSIS & PRINCIPLES

Instructional Art

Each figure is carefully designed to follow closely with the text material.

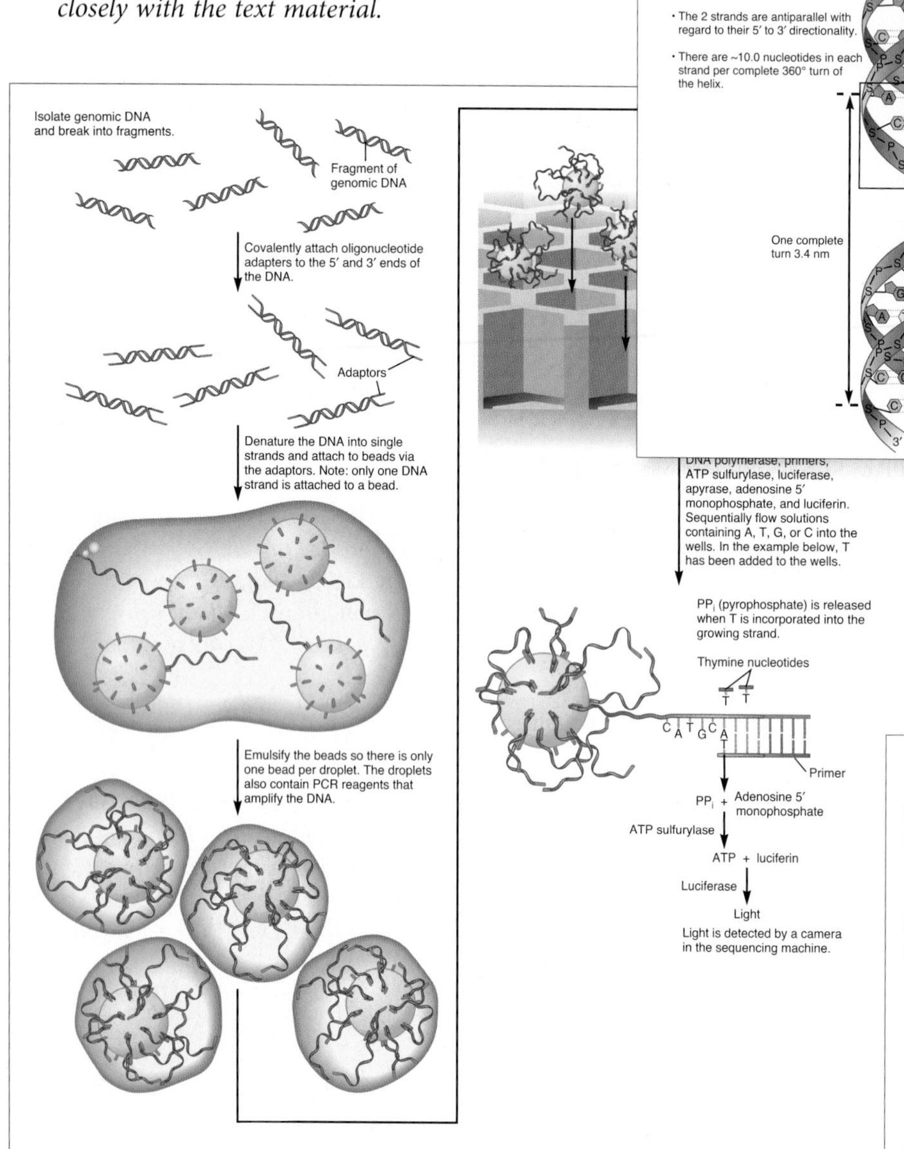

The digitally rendered images have a vivid three-dimensional look that will stimulate a student's interest and enthusiasm.

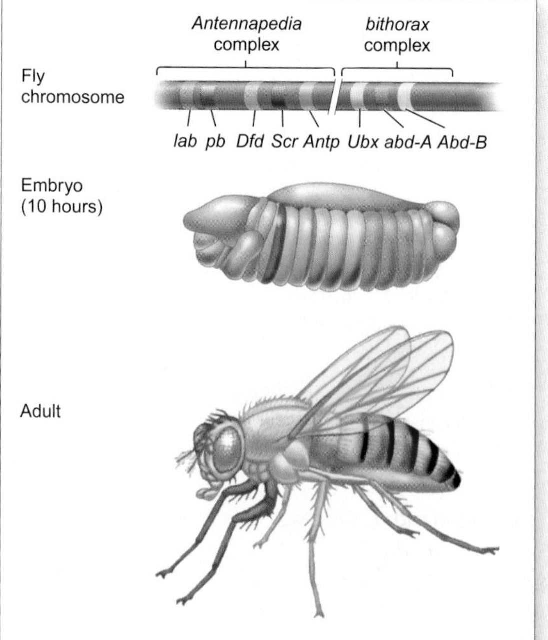

Every illustration was drawn with four goals in mind: completeness, clarity, consistency, and realism.

Learning Through Experimentation

Each chapter (beginning with Chapter 2) incorporates one or two experiments that are presented according to the scientific method. These experiments are integrated within the chapters and flow with the rest of the textbook. As you read the experiments, you will simultaneously explore the scientific method and the genetic principles learned from this approach.

STEP 1: BACKGROUND OBSERVATIONS

Each experiment begins with a description of the information that led researchers to study an experimental problem. Detailed information about the researchers and the experimental challenges they faced help students to understand actual research.

EXPERIMENT 6A

Creighton and McClintock Showed That Crossing Over Produced New Combinations of Alleles and Resulted in the Exchange of Segments Between Homologous Chromosomes

As we have seen, Morgan's studies were consistent with the hypothesis that crossing over occurs between homologous chromosomes to produce new combinations of alleles. To obtain direct evidence that crossing over can result in genetic recombination, Harriet Creighton and Barbara McClintock used an interesting strategy involving parallel observations. In studies conducted in 1931, they first made crosses involving two linked genes to produce parental and recombinant offspring. Second, they used a microscope to view the structures of the chromosomes in the parents and in the offspring. Because the parental chromosomes had some unusual structural features, they could microscopically distinguish the two homologous chromosomes within a pair. As we will see, this enabled them to correlate the occurrence of recombinant offspring with microscopically observable exchanges in segments of homologous chromosomes.

Creighton and McClintock focused much of their attention on the pattern of inheritance of traits in corn. This species has 10 different chromosomes per set, which are named chromosome 1, chromosome 2, chromosome 3, and so on. In previous cytological examinations of corn chromosomes, some strains were found to have an unusual chromosome 9 with a darkly staining knob at one end. In addition, McClintock identified an abnormal version of chromosome 9 that also had an extra piece of chromosome 8 attached at the other end (**Figure 6.6a**). This chromosomal rearrangement is called a translocation.

Creighton and McClintock insightfully realized that this abnormal chromosome could be used to determine if two homologous chromosomes physically exchange segments as a result of crossing over. They knew that a gene was located near the knobbed end of chromosome 9 that provided color to corn kernels. This gene existed in two alleles, the dominant allele *C* (colored) and the recessive allele *c* (colorless). A second gene, located near the translocated piece from chromosome 8, affected the texture of the kernel endosperm. The dominant allele *Wx* caused starchy endosperm, and the recessive *wx* allele caused waxy endosperm. Creighton and McClintock reasoned that a crossover involving a normal chromosome 9 and a knobbed/translocated chromosome 9 would produce a chromosome that had either a knob or a translocation, but not both. These two types of chromosomes would be distinctly different from either of the parental chromosomes (**Figure 6.6b**).

As shown in the experiment of **Figure 6.7**, Creighton and McClintock began with a corn strain that carried an abnormal chromosome that had a knob at one end and a translocation at the other. Genotypically, this chromosome was *C wx*. The cytologically normal chromosome in this strain was *c Wx*. This corn plant, termed parent A, had the genotype *Cc Wx wx*. It was crossed to a strain called parent B that carried two cytologically normal chromosomes and had the genotype *cc Wx wx*.

They then observed the kernels in two ways. First, they examined the phenotypes of the kernels to see if they were colored or colorless, and starchy or waxy. Second, the chromosomes in each kernel were examined under a microscope to determine their cytological appearance. Altogether, they observed a total of 25 kernels (see data of Figure 6.7).

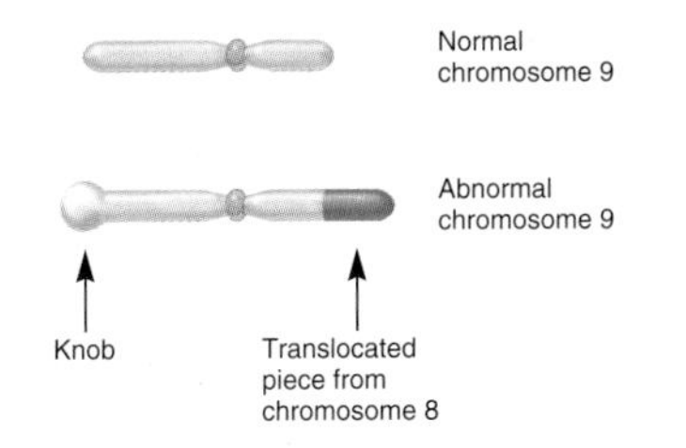

(a) Normal and abnormal chromosome 9

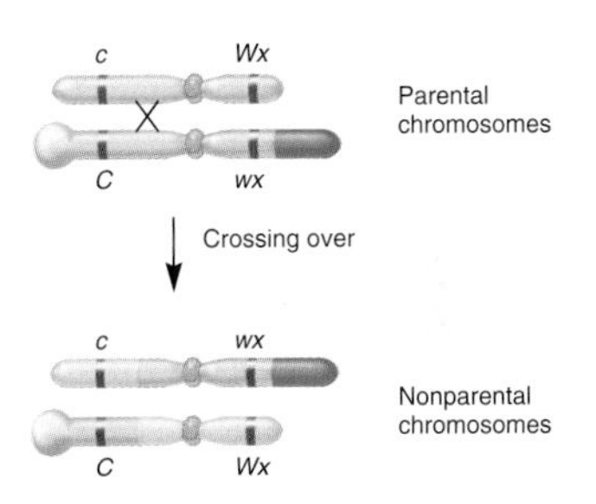

(b) Crossing over between normal and abnormal chromosome 9

FIGURE 6.6 Crossing over between a normal and abnormal chromosome 9 in corn. (a) A normal chromosome 9 in corn is compared to an abnormal chromosome 9 that contains a knob at one end and a translocation at the opposite end. (b) A crossover produces a chromosome that contains only a knob at one end and another chromosome that contains only a translocation at the other end.

THE HYPOTHESIS

Offspring with nonparental phenotypes are the product of a crossover. This crossover should produce nonparental chromosomes via an exchange of chromosomal segments between homologous chromosomes.

STEP 2: HYPOTHESIS

The student is given a statement describing the possible explanation for the observed phenomenon that will be tested. The hypothesis section reinforces the scientific method and allows students to experience the process for themselves.

TESTING THE HYPOTHESIS — FIGURE 6.7 Experimental correlation between genetic recombination and crossing over.

Starting materials: Two different strains of corn. One strain, referred to as parent A, had an abnormal chromosome 9 (knobbed/translocation) with a dominant *C* allele and a recessive *wx* allele. It also had a cytologically normal copy of chromosome 9 that carried the recessive *c* allele and the dominant *Wx* allele. Its genotype was *Cc Wxwx*. The other strain (referred to as parent B) had two normal versions of chromosome 9. The genotype of this strain was *cc Wxwx*.

1. Cross the two strains described. The tassel is the pollen-bearing structure, and the silk (equivalent to the stigma and style) is connected to the ovary. After fertilization, the ovary will develop into an ear of corn.

2. Observe the kernels from this cross.

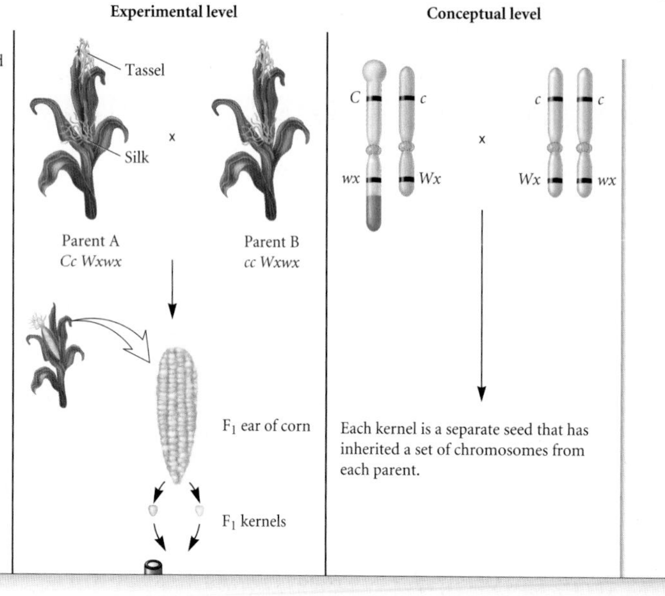

STEP 3: TESTING THE HYPOTHESIS

This section illustrates the experimental process, including the actual steps followed by scientists to test their hypothesis. Science comes alive for students with this detailed look at experimentation.

STEP 4: THE DATA

Actual data from the original research paper help students understand how real-life research results are reported. Each experiment's results are discussed in the context of the larger genetic principle to help students understand the implications and importance of the research.

THE DATA

Phenotype of F_1 Kernel	*Number of Kernels Analyzed*	*Cytological Appearance of Chromosome 9 in F_1 Offspring**		*Did a Crossover Occur During Gamete Formation in Parent A?*
Colored/waxy	3	Knobbed/translocation *C wx*	Normal *c wx*	No
Colorless/starchy	11	Knobless/normal *c Wx*	Normal *c Wx* or *c wx*	No
Colorless/starchy	4	Knobless/translocation *c wx*	Normal *c Wx*	Yes
Colorless/waxy	2	Knobless/translocation *c wx*	Normal *c wx*	Yes
Colored/starchy	5	Knobbed/normal *C Wx*	Normal *c Wx* or *c wx*	Yes
Total	25			

*In this table, the chromosome on the left was inherited from parent A, and the blue chromosome on the right was inherited from parent B.
Data from Harriet B. Creighton and Barbara McClintock (1931) A Correlation of Cytological and Genetical Crossing-Over in *Zea Mays*. *Proc. Natl. Acad. Sci. USA 17*, 492–497.

STEP 5: INTERPRETING THE DATA

This discussion, which examines whether the experimental data supported or refuted the hypothesis, gives students an appreciation for scientific interpretation.

INTERPRETING THE DATA

By combining the gametes in a Punnett square, the following types of offspring can be produced:

Parent B

Parent A ♀ \ ♂	*c Wx*	*c wx*	
C wx	*Cc Wxwx* Colored, starchy	*Cc wxwx* Colored, waxy	Nonrecombinant
c Wx	*cc WxWx* Colorless, starchy	*cc Wxwx* Colorless, starchy	Nonrecombinant

Parent A	*Parent B*
C wx (nonrecombinant)	*c Wx*
c Wx (nonrecombinant)	*c wx*
C Wx (recombinant)	
c wx (recombinant)	

As seen in the Punnett square, two of the phenotypic categories, colored, starchy (*Cc Wxwx* or *Cc WxWx*) and colorless, starchy (*cc WxWx* or *cc Wxwx*), were ambiguous because they could arise from a nonrecombinant and from a recombinant gamete. In other words, these phenotypes could be produced whether or not recombination occurred in parent A. Therefore, let's focus on the two unambiguous phenotypic categories: colored, waxy (*Cc wxwx*) and colorless, waxy (*cc wxwx*). The colored, waxy phenotype could happen only if recombination did not occur in parent A and if parent A passed the knobbed/

End of Chapter Support Materials

These study tools and problems are crafted to aid students in reviewing key information in the text and developing a wide range of skills. They also develop a student's cognitive, writing, analytical, computational, and collaborative abilities.

KEY TERMS

Enhance student development of vital vocabulary necessary for the understanding and application of chapter content. Important terms are boldfaced throughout the chapter and page referenced at the end of each chapter for reflective study.

KEY TERMS

Page 126. genetic map, synteny, genetic linkage
Page 127. linkage groups, dihybrid cross, trihybrid cross, crossing over, bivalent, genetic recombination, nonparental cells, recombinant cells, parental offspring, nonrecombinant offspring.
Page 130. null hypothesis
Page 136. genetic mapping, locus
Page 137. genetic linkage map, testcross, map distance, map units (mu), centiMorgans (cM)
Page 143. positive interference, spores, tetrad, octad, ascus
Page 144. unordered tetrad, unordered octad, ordered tetrad, ordered octad, first-division segregation (FDS)
Page 145. second-division segregation (SDS)
Page 148. parental ditype (PD), tetratype (T), nonparental ditype (NPD)
Page 149. mitotic recombination

CHAPTER SUMMARY

Emphasizes the main concepts from each section of the chapter in a bulleted form to provide students with a thorough review of the main topics covered.

CHAPTER SUMMARY

6.1 Linkage and Crossing Over

- Synteny refers to genes that are located on the same chromosome. Genetic linkage means that the alleles of two or more genes tend to be transmitted as a unit because they are relatively close on the same chromosome.
- Crossing over can change the combination of alleles along a chromosome and produce nonparental, or recombinant, cells and offspring (see Figure 6.1).
- Bateson and Punnett discovered the first example of genetic linkage in sweet peas (see Figure 6.2).
- Morgan also discovered genetic linkage in *Drosophila* and proposed that nonparental offspring are produced by crossing over during meiosis (see Figures 6.3, 6.4).
- When genes are linked, the relative proportions of nonparental offspring depends on the distance between the genes (see Figure 6.5).
- A chi square analysis can be followed to judge whether or not two genes assort independently.
- Creighton and McClintock were able to correlate the forma

- Sturtevant was the first scientist to conduct testcrosses and map the order of a few genes along the X chromosome in *Drosophila* (see Figure 6.10).
- Due to the effects of multiple crossovers, the map distance between two genes obtained from a testcross cannot exceed 50% (see Figure 6.11).
- The data from a trihybrid cross can be used to map genes (see Table 6.1).
- Positive interference refers to the phenomenon that the number of double crossovers in a given region is less than expected based on the frequency of single crossovers.

6.3 Genetic Mapping in Haploid Eukaryotes

- Several haploid eukaryotes have been used in genetic mapping. Ascomycetes have the product of a single meiosis contained with an ascus (see Figure 6.12).
- Certain haploid species may form unordered or ordered tetrads or octads (see Figure 6.13).
- The arrangement of alleles found in spores of an ordered

CONCEPTUAL QUESTIONS

Test the understanding of basic genetic principles. The student is given many questions with a wide range of difficulty. Some require critical thinking skills, and some require the student to write coherent essay questions.

Conceptual Questions

C1. What is the difference in meaning between the terms genetic recombination and crossing over?

C2. When applying a chi square approach in a linkage problem, explain why an independent assortment hypothesis is used.

C3. What is mitotic recombination? A heterozygous individual (*Bb*) with brown eyes has one eye with a small patch of blue. Provide two or more explanations for how the blue patch may have occurred.

C4. Mitotic recombination can occasionally produce a twin spot. Let's suppose an animal species can be heterozygous for two genes that govern fur color and length: One gene affects pigmentation, with dark pigmentation (*A*) dominant to albino (*a*); the other gene affects hair length, with long hair (*L*) dominant to short hair (*l*). The two genes are linked on the same chromosome. Let's assume an animal is *AaLl*; *A* is linked to *l*, and *a* is linked to *L*. Draw the chromosomes labeled with these alleles, and explain how mitotic recombination could produce a twin spot with one spot having albino pigmentation and long fur, the other having dark pigmentation and short fur.

C5. A crossover has occurred in the bivalent shown here.

C7. A diploid organism has a total of 14 chromosomes and about 20,000 genes per haploid genome. Approximately how many genes are in each linkage group?

C8. If you try to throw a basketball into a basket, the likelihood of succeeding depends on the size of the basket. It is more likely that you will get the ball into the basket if the basket is bigger. In your own words, explain how this analogy also applies to the idea that the likelihood of crossing over is greater when two genes are far apart than when they are close together.

C9. By conducting testcrosses, researchers have found that the sweet pea has seven linkage groups. How many chromosomes would you expect to find in leaf cells?

C10. In humans, a rare dominant disorder known as nail-patella syndrome causes abnormalities in the fingernails, toenails, and kneecaps. Researchers have examined family pedigrees with regard to this disorder and, within the same pedigree, also examined the individuals with regard to their blood types. (A description of blood genotypes is found in Chapter 4.) In the following pedigree, individuals affected with nail-patella disorder are shown with filled symbols. The genotype of each individual with regard to their ABO blood type is also shown. Does this pedigree suggest any linkage between the gene that causes nail-patella syndrome and the gene that causes blood type?

EXPERIMENTAL QUESTIONS

Test the ability to analyze data, design experiments, or appreciate the relevance of experimental techniques.

Experimental Questions (Includes Most Mapping Questions)

E1. Figure 6.2 shows the first experimental results that indicated linkage between two different genes. Conduct a chi square analysis to confirm that the genes are really linked and the data could not be explained by independent assortment.

E2. In the experiment of Figure 6.7, the researchers followed the inheritance pattern of chromosomes that were abnormal at both ends to correlate genetic recombination with the physical exchange of chromosome pieces. Is it necessary to use a chromosome that is abnormal at both ends, or could the researchers have used a parental strain with two abnormal versions of chromosome 9, one with a knob at one end and its homologue with a translocation at the other end?

In the heterozygous parent of a testcross, must all of the dominant alleles be linked on the same chromosome and all of the recessive alleles be linked on the homologue?

E6. In your own words, explain why a testcross cannot produce more than 50% recombinant offspring. When a testcross does produce 50% recombinant offspring, what do these results mean?

E7. Explain why the percentage of recombinant offspring in a testcross is a more accurate measure of map distance when two genes are close together. When two genes are far apart, is the percentage of recombinant offspring an underestimate or overestimate of the actual map distance?

STUDENT DISCUSSION/ COLLABORATION QUESTIONS

Encourage students to consider broad concepts and practical problems. Some questions require a substantial amount of computational activities, which can be worked on as a group.

Questions for Student Discussion/Collaboration

1. In mice, a dominant gene that causes a short tail is located on chromosome 2. On chromosome 3, a recessive gene causing droopy ears is 6 mu away from another recessive gene that causes a flaky tail. A recessive gene that causes a jerker (uncoordinated) phenotype is located on chromosome 4. A jerker mouse with droopy ears and a short, flaky tail was crossed to a normal mouse. All F_1 generation mice were phenotypically normal, except they had short tails. These F_1 mice were then testcrossed to jerker mice with droopy ears and long, flaky tails. If this cross produced 400 offspring, what would be the proportions of the 16 possible phenotypic categories?

2. In Chapter 3, we discussed the idea that the X and Y chromosomes have a few genes in common. These genes are inherited in a pseudoautosomal pattern. With this phenomenon in mind, discuss whether or not the X and Y chromosomes are really distinct linkage groups.

3. Mendel studied seven traits in pea plants, and the garden pea happens to have seven different chromosomes. It has been pointed out that Mendel was very lucky not to have conducted crosses involving two traits governed by genes that are closely linked on the same chromosome because the results would have confounded his theory of independent assortment. It has even been suggested that Mendel may not have published data involving traits that were linked! An article by Stig Blixt ("Why Didn't Gregor Mendel Find Linkage?" *Nature 256*:206, 1975) considers this issue. Look up this article and discuss why Mendel did not find linkage.

Note: All answers appear at the Website for this textbook; the answers to even-numbered questions are in the back of the textbook.

www.mhhe.com/brookergenetics4e

Visit the website for practice tests, answer keys, and other learning aids for this chapter. Enhance your understanding of genetics with our interactive exercises, quizzes, animations, and much more.

CHAPTER OUTLINE

Carbon copy, the first cloned pet. *In 2002, the cat shown here, called Carbon copy or Copycat, was produced by cloning, a procedure described in Chapter 19.*

OVERVIEW OF GENETICS

Hardly a week goes by without a major news story involving a genetic breakthrough. The increasing pace of genetic discoveries has become staggering. The Human Genome Project is a case in point. This project began in the United States in 1990, when the National Institutes of Health and the Department of Energy joined forces with international partners to decipher the massive amount of information contained in our **genome**—the DNA found within all of our chromosomes (**Figure 1.1**). Working collectively, a large group of scientists from around the world has produced a detailed series of maps that help geneticists navigate through human DNA. Remarkably, in only a decade, they determined the DNA sequence (read in the bases of A, T, G, and C) covering over 90% of the human genome. The first draft of this sequence, published in 2001, is nearly 3 billion nucleotide base pairs in length. The completed sequence, published in 2003, has an accuracy greater than 99.99%; fewer than one mistake was made in every 10,000 base pairs (bp)!

Studying the human genome allows us to explore fundamental details about ourselves at the molecular level. The results of the Human Genome Project are expected to shed considerable light on basic questions, like how many genes we have, how genes direct the activities of living cells, how species evolve, how single cells develop into complex tissues, and how defective genes cause disease. Furthermore, such understanding may lend itself to improvements in modern medicine by leading to better diagnoses of diseases and the development of new treatments for them.

As scientists have attempted to unravel the mysteries within our genes, this journey has involved the invention of many new technologies. For example, new technologies have made it possible to produce medicines that would otherwise be difficult or impossible to make. An example is human recombinant insulin, sold under the brand name *Humulin.* This medicine is synthesized in strains of *Escherichia coli* bacteria that have been genetically altered by the addition of genes that encode the polypeptides that form human insulin. The bacteria are grown in a laboratory and make large amounts of human insulin. As discussed in Chapter 19, the insulin is purified and administered to many people with insulin-dependent diabetes.

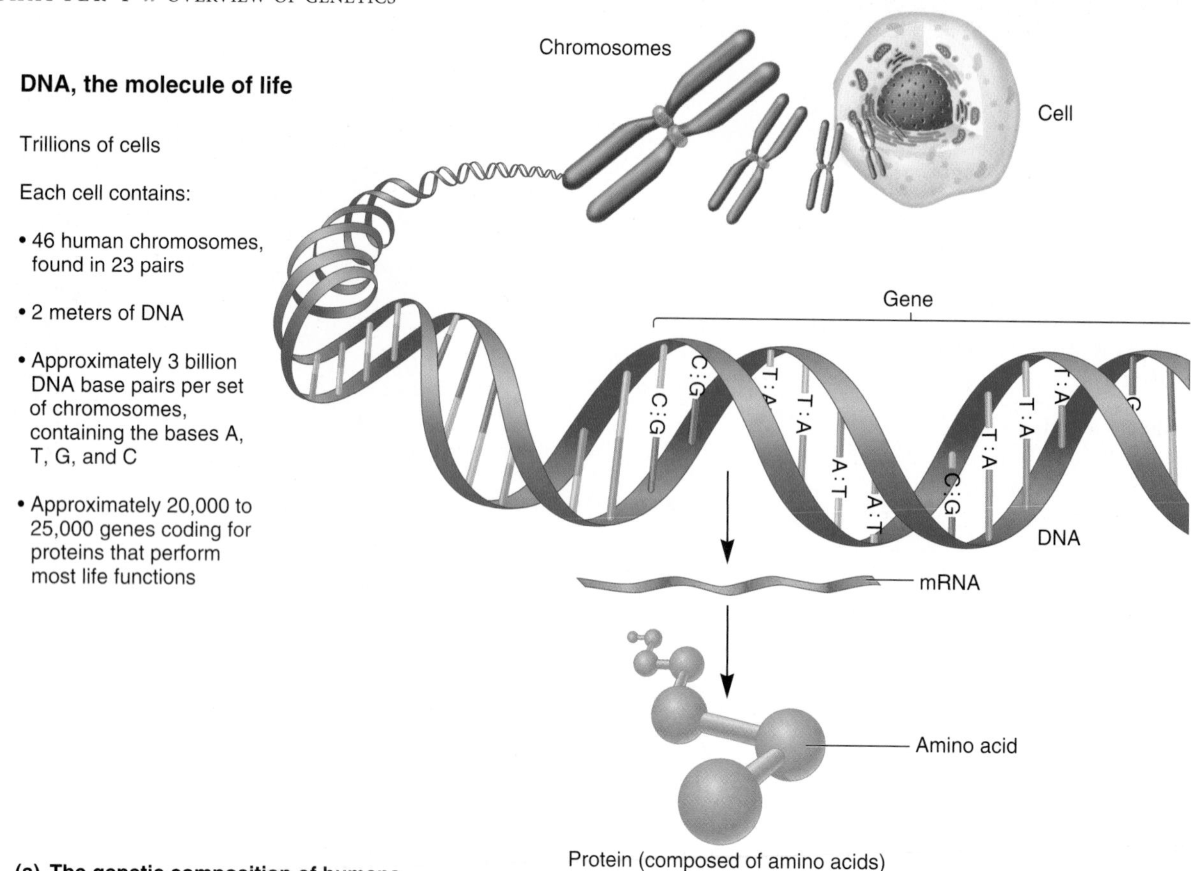

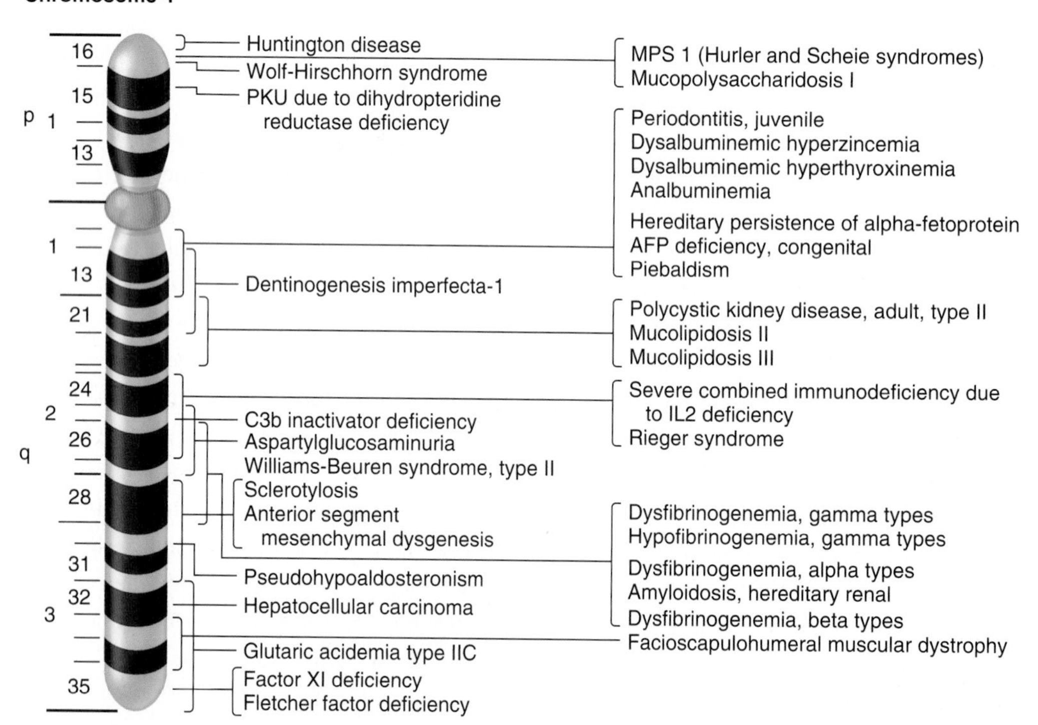

FIGURE 1.1 The Human Genome Project. **(a)** The human genome is a complete set of human chromosomes. People have two sets of chromosomes, one from each parent. Collectively, each set of chromosomes is composed of a DNA sequence that is approximately 3 billion nucleotide base pairs long. Estimates suggest that each set contains about 20,000 to 25,000 different genes. Most genes encode proteins. This figure emphasizes the DNA found in the cell nucleus. Humans also have a small amount of DNA in their mitochondria, which has also been sequenced. **(b)** An important outcome of genetic research is the identification of genes that contribute to human diseases. This illustration depicts a map of a few genes that are located on human chromosome 4. When these genes carry certain rare mutations, they can cause the diseases designated in this figure.

New genetic technologies are often met with skepticism and sometimes even with disdain. An example would be DNA fingerprinting, a molecular method to identify an individual based on a DNA sample (see Chapter 24). Though this technology is now relatively common in the area of forensic science, it was not always universally accepted. High-profile crime cases in the news cause us to realize that not everyone believes in DNA fingerprinting, in spite of its extraordinary ability to uniquely identify individuals. A second controversial example is mammalian cloning. In 1997, Ian Wilmut and his colleagues created clones of sheep, using mammary cells from an adult animal (**Figure 1.2**). More recently, such cloning has been achieved in several mammalian species, including cows, mice, goats, pigs, and cats. In 2002, the first pet was cloned, a cat named Carbon copy, or Copycat (see photo at the beginning of the chapter). The cloning of mammals provides the potential for many practical applications. With regard to livestock, cloning would enable farmers to use cells from their best individuals to create genetically homogeneous herds. This could be advantageous in terms of agricultural yield, although such a genetically homogeneous herd may be more susceptible to certain diseases. However, people have become greatly concerned with the possibility of human cloning. This prospect has raised serious ethical questions. Within the past few years, legislative bills have been introduced that involve bans on human cloning.

Finally, genetic technologies provide the means to modify the traits of animals and plants in ways that would have been unimaginable just a few decades ago. **Figure 1.3a** illustrates a bizarre example in which scientists introduced a gene from jellyfish into mice. Certain species of jellyfish emit a "green glow" produced by a gene that encodes a bioluminescent protein called green fluorescent protein (GFP). When exposed to blue or ultraviolet (UV) light, the protein emits a striking green-colored light. Scientists were able to clone the *GFP* gene from a sample of jellyfish cells and then introduce this gene into laboratory mice. The green fluorescent protein is made throughout the cells of their bodies. As a result, their skin, eyes, and organs give off an eerie green glow when exposed to UV light. Only their fur does not glow.

The expression of green fluorescent protein allows researchers to identify particular proteins in cells or specific body parts. For

FIGURE 1.2 The cloning of a mammal. The lamb on the left is Dolly, the first mammal to be cloned. She was cloned from the cells of a Finn Dorset (a white-faced sheep). The sheep on the right is Dolly's surrogate mother, a Blackface ewe. A description of how Dolly was produced is presented in Chapter 19.

(a) GFP expressed in mice

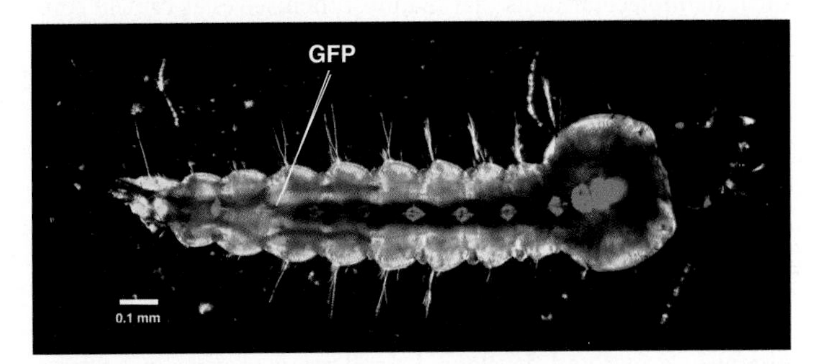

(b) GFP expressed in the gonads of a male mosquito

FIGURE 1.3 The introduction of a jellyfish gene into laboratory mice and mosquitoes. (a) A gene that naturally occurs in the jellyfish encodes a protein called green fluorescent protein (GFP). The GFP gene was cloned and introduced into mice. When these mice are exposed to UV light, GFP emits a bright green color. These mice glow green, just like jellyfish! (b) GFP was introduced next to a gene sequence that causes the expression of GFP only in the gonads of male mosquitoes. This allows researchers to identify and sort males from females.

example, Andrea Crisanti and colleagues have altered mosquitoes to express GFP only in the gonads of males (**Figure 1.3b**). This enables the researchers to identify and sort males from females. Why is this useful? The ability to rapidly sort mosquitoes makes it possible to produce populations of sterile males and then release the sterile males without the risk of releasing additional females. The release of sterile males may be an effective means of controlling mosquito populations because females only breed once before they die. Mating with a sterile male prevents a female from producing offspring. In 2008, Osamu Shimomura, Martin Chalfie, and Roger Tsien received the Nobel Prize in chemistry for the discovery and the development of GFP, which has become a widely used tool in biology.

Overall, as we move forward in the twenty-first century, the excitement level in the field of genetics is high, perhaps higher than it has ever been. Nevertheless, the excitement generated by new genetic knowledge and technologies will also create many ethical and societal challenges. In this chapter, we begin with an overview of genetics and then explore the various fields of genetics and their experimental approaches.

1.1 THE RELATIONSHIP BETWEEN GENES AND TRAITS

Genetics is the branch of biology that deals with heredity and variation. It stands as the unifying discipline in biology by allowing us to understand how life can exist at all levels of complexity, ranging from the molecular to the population level. Genetic variation is the root of the natural diversity that we observe among members of the same species as well as among different species.

Genetics is centered on the study of genes. A gene is classically defined as a unit of heredity, but such a vague definition does not do justice to the exciting characteristics of genes as intricate molecular units that manifest themselves as critical contributors to cell structure and function. At the molecular level, a **gene** is a segment of DNA that produces a functional product. The functional product of most genes is a polypeptide, which is a linear sequence of amino acids that folds into units that constitute proteins. In addition, genes are commonly described according to the way they affect **traits,** which are the characteristics of an organism. In humans, for example, we speak of traits such as eye color, hair texture, and height. The ongoing theme of this textbook is the relationship between genes and traits. As an organism grows and develops, its collection of genes provides a blueprint that determines its characteristics.

In this section of Chapter 1, we examine the general features of life, beginning with the molecular level and ending with populations of organisms. As will become apparent, genetics is the common thread that explains the existence of life and its continuity from generation to generation. For most students, this chapter should serve as a cohesive review of topics they learned in other introductory courses such as General Biology. Even so, it is usually helpful to see the "big picture" of genetics before delving into the finer details that are covered in Chapters 2 through 26.

Living Cells Are Composed of Biochemicals

To fully understand the relationship between genes and traits, we need to begin with an examination of the composition of living organisms. Every cell is constructed from intricately organized chemical substances. Small organic molecules such as glucose and amino acids are produced from the linkage of atoms via chemical bonds. The chemical properties of organic molecules are essential for cell vitality in two key ways. First, the breaking of chemical bonds during the degradation of small molecules provides energy to drive cellular processes. A second important function of these small organic molecules is their role as the building blocks for the synthesis of larger molecules. Four important categories of larger cellular molecules are **nucleic acids** (i.e., DNA and RNA), **proteins, carbohydrates,** and **lipids.** Three of these—nucleic acids, proteins, and carbohydrates—form **macromolecules** that are composed of many repeating units of smaller building blocks. Proteins, RNA, and carbohydrates can be made from hundreds or even thousands of repeating building blocks. DNA is the largest macromolecule found in living cells. A single DNA molecule can be composed of a linear sequence of hundreds of millions of nucleotides!

The formation of cellular structures relies on the interactions of molecules and macromolecules. For example, nucleotides are the building blocks of DNA, which is a constituent of cellular chromosomes (**Figure 1.4**). In addition, the DNA is associated with a myriad of proteins that provide organization to the structure of chromosomes. Within a eukaryotic cell, the chromosomes are contained in a compartment called the cell nucleus. The nucleus is bounded by a double membrane composed of lipids and proteins that shields the chromosomes from the rest of the cell. The organization of chromosomes within a cell nucleus protects the chromosomes from mechanical damage and provides a single compartment for genetic activities such as gene transcription. As a general theme, the formation of large cellular structures arises from interactions among different molecules and macromolecules. These cellular structures, in turn, are organized to make a complete living cell.

Each Cell Contains Many Different Proteins That Determine Cellular Structure and Function

To a great extent, the characteristics of a cell depend on the types of proteins that it makes. All of the proteins that a cell makes at a given time is called its **proteome.** As we will learn throughout this textbook, proteins are the "workhorses" of all living cells. The range of functions among different types of proteins is truly remarkable. Some proteins help determine the shape and structure of a given cell. For example, the protein known as tubulin can assemble into large structures known as microtubules, which provide the cell with internal structure and organization. Other proteins are inserted into cell membranes and aid in the transport of ions and small molecules across the membrane. Proteins may also function as biological motors. An interesting case is the protein known as myosin, which is involved in the contractile properties of muscle cells. Within multicellular organisms, certain proteins also function in cell-to-cell recognition and signaling. For example, hormones such as insulin are secreted by

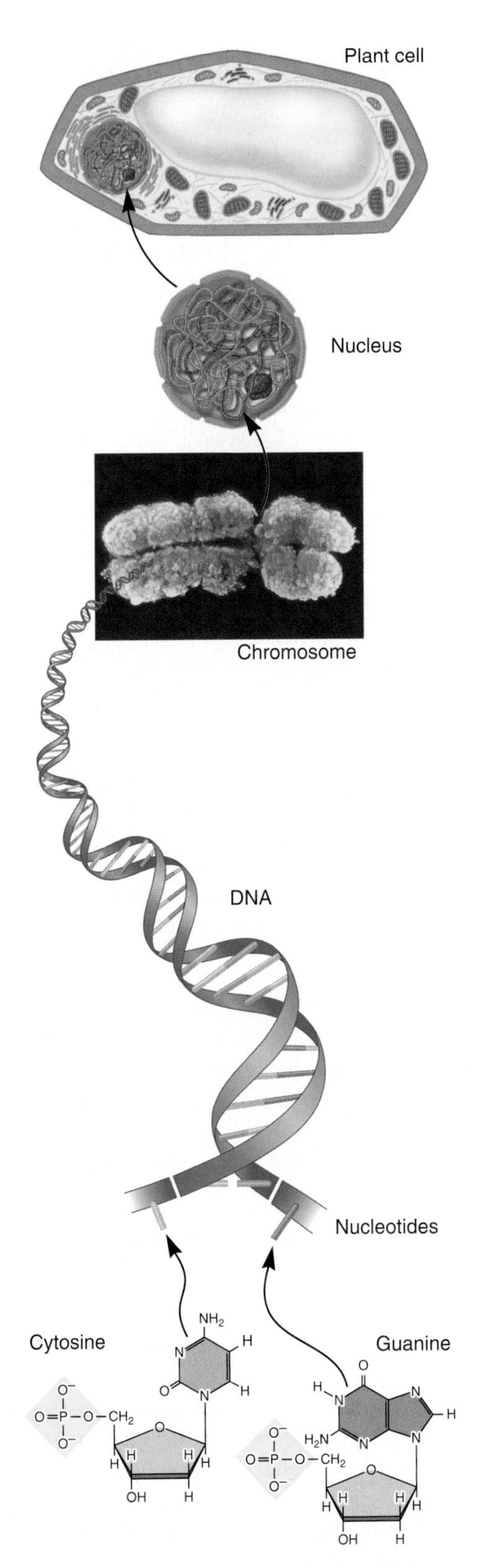

FIGURE 1.4 Molecular organization of a living cell. Cellular structures are constructed from smaller building blocks. In this example, DNA is formed from the linkage of nucleotides to produce a very long macromolecule. The DNA associates with proteins to form a chromosome. The chromosomes are located within a membrane-bound organelle called the nucleus, which, along with many different types of organelles, is found within a complete cell.

endocrine cells and bind to the insulin receptor protein found within the plasma membrane of target cells.

Enzymes, which accelerate chemical reactions, are a particularly important category of proteins. Some enzymes play a role in the breakdown of molecules or macromolecules into smaller units. These are known as catabolic enzymes and are important in the utilization of energy. Alternatively, anabolic enzymes and accessory proteins function in the synthesis of molecules and macromolecules throughout the cell. The construction of a cell greatly depends on its proteins involved in anabolism because these are required to synthesize all cellular macromolecules.

Molecular biologists have come to realize that the functions of proteins underlie the cellular characteristics of every organism. At the molecular level, proteins can be viewed as the active participants in the enterprise of life.

DNA Stores the Information for Protein Synthesis

The genetic material of living organisms is composed of a substance called **deoxyribonucleic acid,** abbreviated **DNA.** The DNA stores the information needed for the synthesis of all cellular proteins. In other words, the main function of the genetic blueprint is to code for the production of cellular proteins in the correct cell, at the proper time, and in suitable amounts. This is an extremely complicated task because living cells make thousands of different proteins. Genetic analyses have shown that a typical bacterium can make a few thousand different proteins, and estimates among higher eukaryotes range in the tens of thousands.

DNA's ability to store information is based on its structure. DNA is composed of a linear sequence of **nucleotides.** Each nucleotide contains one of four nitrogen-containing bases: adenine (A), thymine (T), guanine (G), or cytosine (C). The linear order of these bases along a DNA molecule contains information similar to the way that groups of letters of the alphabet represent words. For example, the "meaning" of the sequence of bases ATGGGCCTTAGC differs from that of TTTAAGCTTGCC. DNA sequences within most genes contain the information to direct the order of amino acids within polypeptides according to the **genetic code.** In the code, a three-base sequence specifies one particular **amino acid** among the 20 possible choices. One or more polypeptides form a functional protein. In this way, the DNA can store the information to specify the proteins made by an organism.

DNA Sequence	Amino Acid Sequence
ATG GGC CTT AGC	Methionine Glycine Leucine Serine
TTT AAG CTT GCC	Phenylalanine Lysine Leucine Alanine

In living cells, the DNA is found within large structures known as **chromosomes.** Figure 1.5 is a micrograph of the 46 chromosomes contained in a cell from a human male. The DNA of an average human chromosome is an extraordinarily long, linear, double-stranded structure that contains well over a hundred

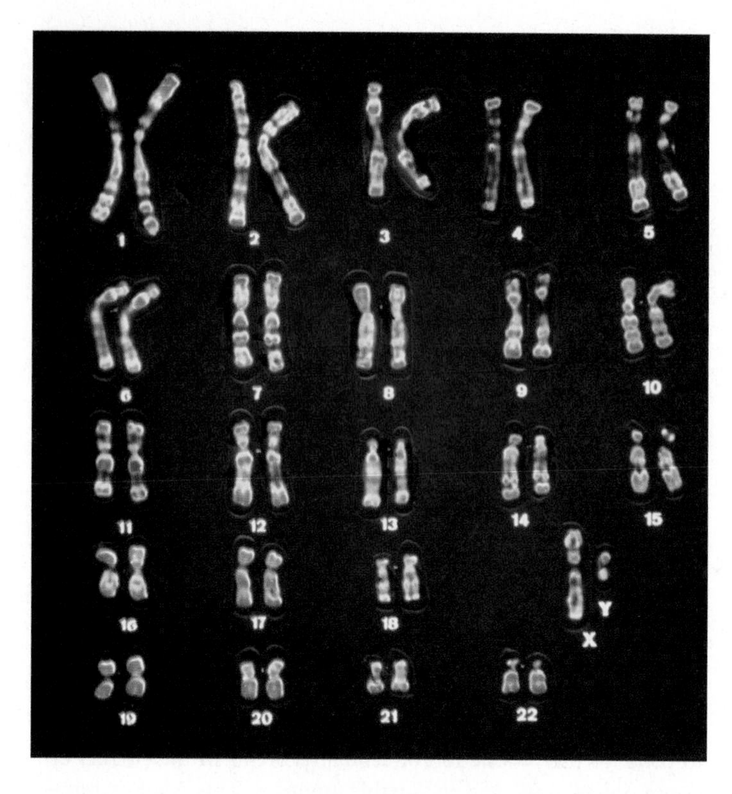

FIGURE 1.5 A micrograph of the 46 chromosomes found in a cell from a human male.

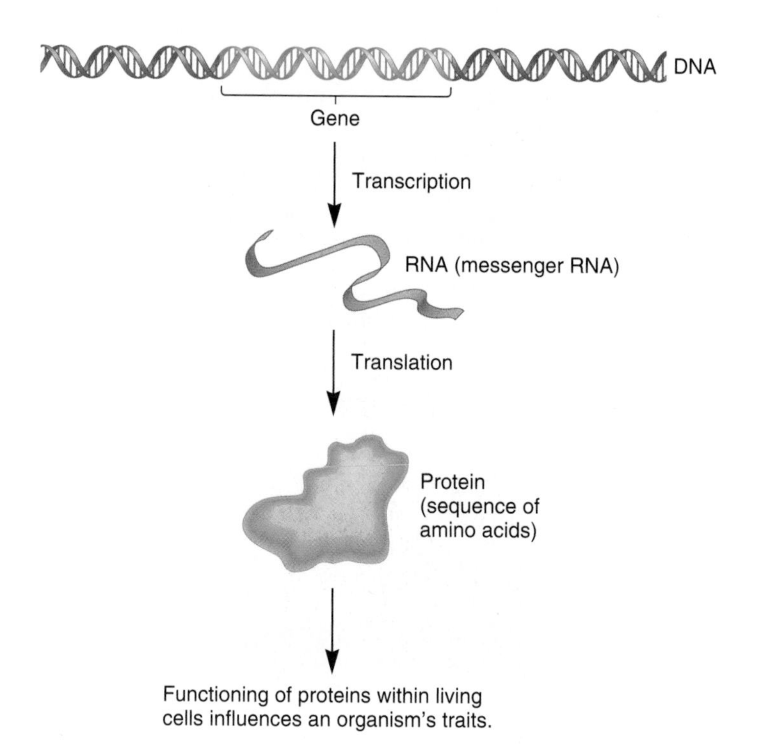

ONLINE ANIMATION

FIGURE 1.6 Gene expression at the molecular level. The expression of a gene is a multistep process. During transcription, one of the DNA strands is used as a template to make an RNA strand. During translation, the RNA strand is used to specify the sequence of amino acids within a polypeptide. One or more polypeptides produce a protein that functions within the cell, thereby influencing an organism's traits.

million nucleotides. Along the immense length of a chromosome, the genetic information is parceled into functional units known as genes. An average-sized human chromosome is expected to contain about 1000 different genes.

The Information in DNA Is Accessed During the Process of Gene Expression

To synthesize its proteins, a cell must be able to access the information that is stored within its DNA. The process of using a gene sequence to affect the characteristics of cells and organisms is referred to as **gene expression.** At the molecular level, the information within genes is accessed in a stepwise process. In the first step, known as **transcription,** the DNA sequence within a gene is copied into a nucleotide sequence of **ribonucleic acid (RNA).** Most genes encode RNAs that contain the information for the synthesis of a particular polypeptide. This type of RNA is called **messenger RNA (mRNA).** For polypeptide synthesis to occur, the sequence of nucleotides transcribed in an mRNA must be **translated** (using the genetic code) into the amino acid sequence of a polypeptide (**Figure 1.6**). After a polypeptide is made, it folds into a three-dimensional structure. As mentioned, a protein is a functional unit. Some proteins are composed of a single polypeptide, and other proteins consist of two or more polypeptides. Some RNA molecules are not mRNA molecules and therefore are not translated into polypeptides. We will consider the functions of these RNA molecules in Chapter 12 (see Table 12.1).

The expression of most genes results in the production of proteins with specific structures and functions. The unique relationship between gene sequences and protein structures is of paramount importance because the distinctive structure of each protein determines its function within a living cell or organism. Mediated by the process of gene expression, therefore, the sequence of nucleotides in DNA stores the information required for synthesizing proteins with specific structures and functions.

The Molecular Expression of Genes Within Cells Leads to an Organism's Traits

A trait is any characteristic that an organism displays. In genetics, we often focus our attention on **morphological traits** that affect the appearance of an organism. The color of a flower and the height of a pea plant are morphological traits. Geneticists frequently study these types of traits because they are easy to evaluate. For example, an experimenter can simply look at a plant and tell if it has red or white flowers. However, not all traits are morphological. **Physiological traits** affect the ability of an organism to function. For example, the rate at which a bacterium metabolizes a sugar such as lactose is a physiological trait. Like morphological traits, physiological traits are controlled, in part, by the expression of genes. **Behavioral traits** also affect the ways that an organism responds to its environment. An example would be the mating calls of bird species. In animals, the nervous system plays a key role in governing such traits.

A complicated, yet very exciting, aspect of genetics is that our observations and theories span four levels of biological organization: molecules, cells, organisms, and populations. This can make it difficult to appreciate the relationship between genes and traits. To understand this connection, we need to relate the following phenomena:

1. Genes are expressed at the **molecular level.** In other words, gene transcription and translation lead to the production of a particular protein, which is a molecular process.
2. Proteins often function at the **cellular level.** The function of a protein within a cell affects the structure and workings of that cell.
3. An organism's traits are determined by the characteristics of its cells. We do not have microscopic vision, yet when we view morphological traits, we are really observing the properties of an individual's cells. For example, a red flower has its color because the flower cells make a red pigment. The trait of red flower color is an observation at the **organism level.** Yet the trait is rooted in the molecular characteristics of the organism's cells.
4. A **species** is a group of organisms that maintains a distinctive set of attributes in nature. The occurrence of a trait within a species is an observation at the **population level.** Along with learning how a trait occurs, we also want to understand why a trait becomes prevalent in a particular species. In many cases, researchers discover that a trait predominates within a population because it promotes the reproductive success of the members of the population. This leads to the evolution of beneficial traits.

As a schematic example to illustrate the four levels of genetics, **Figure 1.7** shows the trait of pigmentation in butterflies. One is light-colored and the other is very dark. Let's consider how we can explain this trait at the molecular, cellular, organism, and population levels.

At the molecular level, we need to understand the nature of the gene or genes that govern this trait. As shown in Figure 1.7a, a gene, which we will call the pigmentation gene, is responsible for the amount of pigment produced. The pigmentation gene can exist in two different forms called **alleles.** In this example, one allele confers a dark pigmentation and one causes a light pigmentation. Each of these alleles encodes a protein that functions as a pigment-synthesizing enzyme. However, the DNA sequences of the two alleles differ slightly from each other. This difference in the DNA sequence leads to a variation in the structure and function of the respective pigmentation enzymes.

At the cellular level (Figure 1.7b), the functional differences between the pigmentation enzymes affect the amount of pigment produced. The allele causing dark pigmentation, which is shown on the left, encodes a protein that functions very well. Therefore, when this gene is expressed in the cells of the wings, a large amount of pigment is made. By comparison, the allele causing light pigmentation encodes an enzyme that functions poorly. Therefore, when this allele is the only pigmentation gene expressed, little pigment is made.

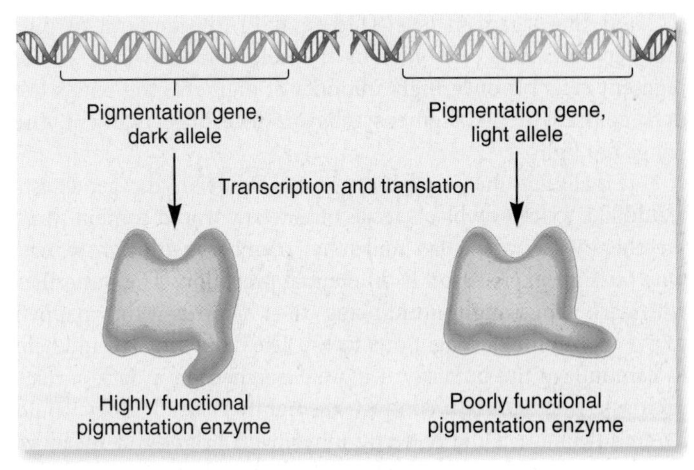

(a) Molecular level

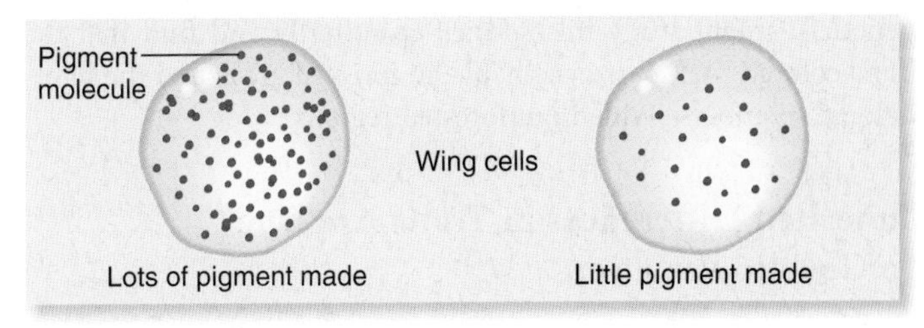

(b) Cellular level

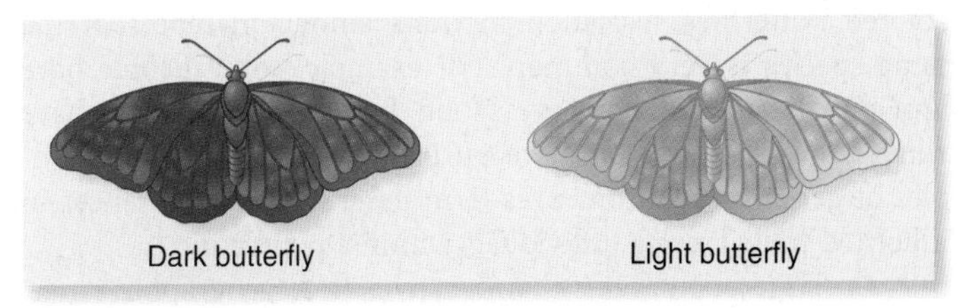

(c) Organism level

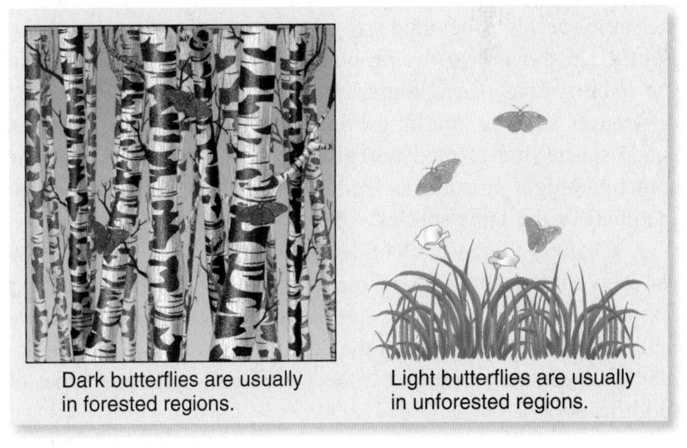

(d) Population level

FIGURE 1.7 The relationship between genes and traits at the (a) molecular, (b) cellular, (c) organism, and (d) population levels.

At the organism level (Figure 1.7c), the amount of pigment in the wing cells governs the color of the wings. If the pigment cells produce high amounts of pigment, the wings are dark-colored; if the pigment cells produce little pigment, the wings are light.

Finally, at the population level (Figure 1.7d), geneticists would like to know why a species of butterfly would contain some members with dark wings and other members with light wings. One possible explanation is differential predation. The butterflies with dark wings might avoid being eaten by birds if they happen to live within the dim light of a forest. The dark wings would help to camouflage the butterfly if it were perched on a dark surface such as a tree trunk. In contrast, the lightly colored wings would be an advantage if the butterfly inhabited a brightly lit meadow. Under these conditions, a bird may be less likely to notice a light-colored butterfly that is perched on a sunlit surface. A population geneticist might study this species of butterfly and find that the dark-colored members usually live in forested areas and the light-colored members reside in unforested regions.

FIGURE 1.8 Two dyeing poison frogs (*Dendrobates tinctorius*) showing different morphs within a single species.

Inherited Differences in Traits Are Due to Genetic Variation

In Figure 1.7, we considered how gene expression could lead to variation in a trait of an organism, such as dark- versus light-colored butterflies. Variation in traits among members of the same species is very common. For example, some people have brown hair, and others have blond hair; some petunias have white flowers, but others have purple flowers. These are examples of **genetic variation.** This term describes the differences in inherited traits among individuals within a population.

In large populations that occupy a wide geographic range, genetic variation can be quite striking. In fact, morphological differences have often led geneticists to misidentify two members of the same species as belonging to separate species. As an example, **Figure 1.8** shows two dyeing poison frogs that are members of the same species, *Dendrobates tinctorius.* They display dramatic differences in their markings. Such contrasting forms within a single species are termed **morphs.** You can easily imagine how someone might mistakenly conclude that these frogs are not members of the same species.

Changes in the nucleotide sequence of DNA underlie the genetic variation that we see among individuals. Throughout this textbook, we will routinely examine how variation in the genetic material results in changes in the outcome of traits. At the molecular level, genetic variation can be attributed to different types of modifications.

1. Small or large differences can occur within gene sequences. When such changes initially occur, they are called **gene mutations.** Mutations result in genetic variation in which a gene is found in two or more alleles, as previously described in Figure 1.7. In many cases, gene mutations alter the expression or function of the protein that the gene specifies.
2. Major alterations can also occur in the structure of a chromosome. A large segment of a chromosome can be lost, rearranged, or reattached to another chromosome.
3. Variation may also occur in the total number of chromosomes. In some cases, an organism may inherit one too many or one too few chromosomes. In other cases, it may inherit an extra set of chromosomes.

Variations within the sequences of genes are a common source of genetic variation among members of the same species. In humans, familiar examples of variation involve genes for eye color, hair texture, and skin pigmentation. Chromosome variation—a change in chromosome structure or number (or both)—is also found, but this type of change is often detrimental. Many human genetic disorders are the result of chromosomal alterations. The most common example is Down syndrome, which is due to the presence of an extra chromosome (**Figure 1.9a**). By comparison, chromosome variation in plants is common and often can lead to plants with superior characteristics, such as increased resistance to disease. Plant breeders have frequently exploited this observation. Cultivated varieties of wheat, for example, have many more chromosomes than the wild species (**Figure 1.9b**).

Traits Are Governed by Genes and by the Environment

In our discussion thus far, we have considered the role that genes play in the outcome of traits. Another critical factor is the **environment**—the surroundings in which an organism exists. A variety of factors in an organism's environment profoundly affect its morphological and physiological features. For example, a person's diet greatly influences many traits such as height, weight, and even intelligence. Likewise, the amount of sunlight a plant receives affects its growth rate and the color of its flowers. The

(a) (b)

FIGURE 1.9 Examples of chromosome variation. (a) A person with Down syndrome competing in the Special Olympics. This person has 47 chromosomes rather than the common number of 46, because she has an extra copy of chromosome 21. **(b)** A wheat plant. Bread wheat is derived from the contributions of three related species with two sets of chromosomes each, producing an organism with six sets of chromosomes.

term **norm of reaction** refers to the effects of environmental variation on an individual's traits.

External influences may dictate the way that genetic variation is manifested in an individual. An interesting example is the human genetic disease **phenylketonuria (PKU).** Humans possess a gene that encodes an enzyme known as phenylalanine hydroxylase. Most people have two functional copies of this gene. People with one or two functional copies of the gene can eat foods containing the amino acid phenylalanine and metabolize it properly.

A rare variation in the sequence of the phenylalanine hydroxylase gene results in a nonfunctional version of this protein. Individuals with two copies of this rare, inactive allele cannot metabolize phenylalanine properly. This occurs in about 1 in 8000 births among Caucasians in the United States. When given a standard diet containing phenylalanine, individuals with this disorder are unable to break down this amino acid. Phenylalanine accumulates and is converted into phenylketones, which are detected in the urine. PKU individuals manifest a variety of detrimental traits, including mental retardation, underdeveloped

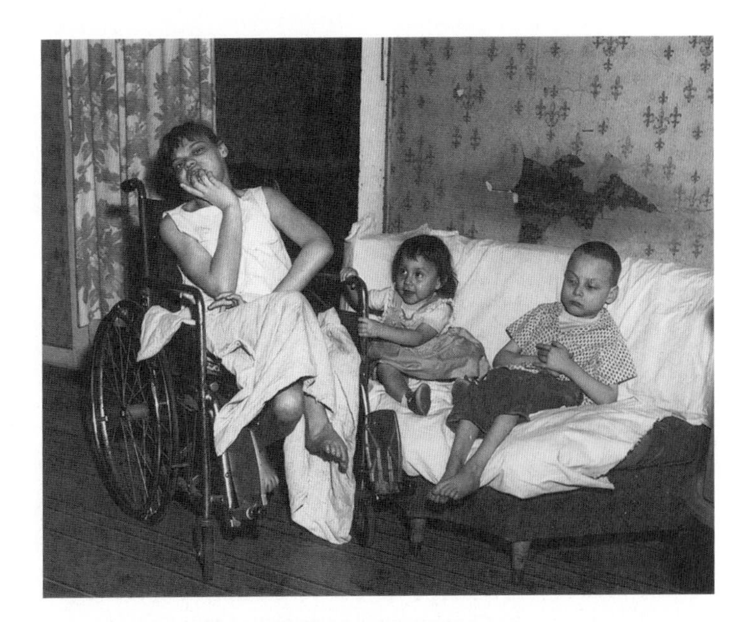

FIGURE 1.10 Environmental influence on the outcome of PKU within a single family. All three children pictured here have inherited the alleles that cause PKU. The child in the middle was raised on a phenylalanine-free diet and developed normally. The other two children were born before the benefits of a phenylalanine-free diet were known and were raised on diets that contained phenylalanine. Therefore, they manifest a variety of symptoms, including mental retardation. People born today with this disorder are usually diagnosed when infants. (*Photo from the March of Dimes Birth Defects Foundation.*)

teeth, and foul-smelling urine. In contrast, when PKU individuals are identified at birth and raised on a restricted diet that is low in phenylalanine, they develop normally (**Figure 1.10**). Fortunately, through routine newborn screening, most affected babies in the United States are now diagnosed and treated early. PKU provides a dramatic example of how the environment and an individual's genes can interact to influence the traits of the organism.

During Reproduction, Genes Are Passed from Parent to Offspring

Now that we have considered how genes and the environment govern the outcome of traits, we can turn to the issue of inheritance. How are traits passed from parents to offspring? The foundation for our understanding of inheritance came from the studies of Gregor Mendel in the nineteenth century. His work revealed that factors that govern traits, which we now call genes, are passed from parent to offspring as discrete units. We can predict the outcome of many genetic crosses based on Mendel's laws of inheritance.

The inheritance patterns identified by Mendel can be explained by the existence of chromosomes and their behavior during cell division. As in Mendel's pea plants, sexually reproducing species are commonly **diploid.** This means they contain two copies of each chromosome, one from each parent. The two

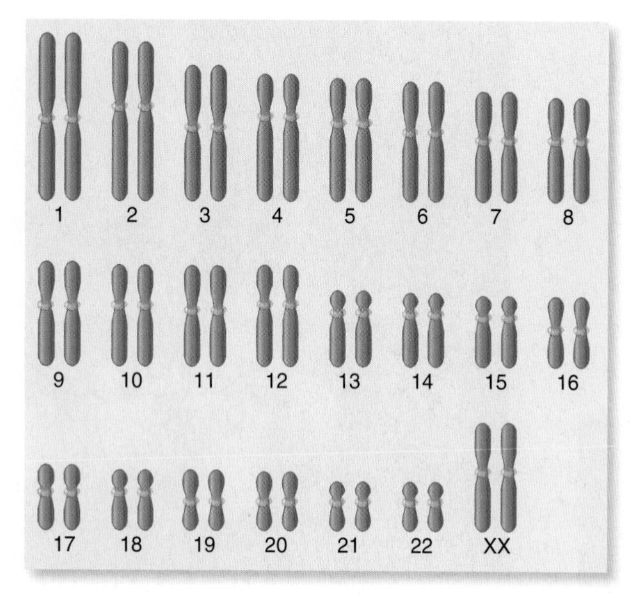

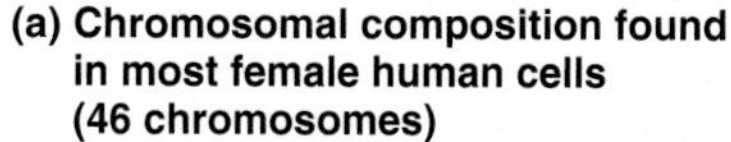
(a) Chromosomal composition found in most female human cells (46 chromosomes)

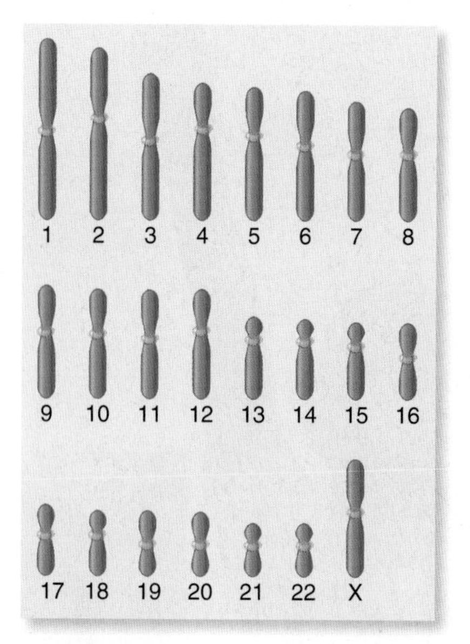

(b) Chromosomal composition found in a human gamete (23 chromosomes)

FIGURE 1.11 The complement of human chromosomes in somatic cells and gametes. **(a)** A schematic drawing of the 46 chromosomes of a human. With the exception of the sex chromosomes, these are always found in homologous pairs. **(b)** The chromosomal composition of a gamete, which contains only 23 chromosomes, one from each pair. This gamete contains an X chromosome. Half of the gametes from human males would contain a Y chromosome instead of the X chromosome.

copies are called **homologs** of each other. Because genes are located within chromosomes, diploid organisms have two copies of most genes. Humans, for example, have 46 chromosomes, which are found in homologous pairs (**Figure 1.11a**). With the exception of the sex chromosomes (X and Y), each homologous pair contains the same kinds of genes. For example, both copies of human chromosome 12 carry the gene that encodes phenylalanine hydroxylase, which was discussed previously. Therefore, an individual has two copies of this gene. The two copies may or may not be identical alleles.

Most cells of the human body that are not directly involved in sexual reproduction contain 46 chromosomes. These cells are called **somatic cells.** In contrast, the **gametes**—sperm and egg cells—contain half that number and are termed **haploid** (**Figure 1.11b**). The union of gametes during fertilization restores the diploid number of chromosomes. The primary advantage of sexual reproduction is that it enhances genetic variation. For example, a tall person with blue eyes and a short person with brown eyes may have short offspring with blue eyes or tall offspring with brown eyes. Therefore, sexual reproduction can result in new combinations of two or more traits that differ from those of either parent.

The Genetic Composition of a Species Evolves over the Course of Many Generations

As we have just seen, sexual reproduction has the potential to enhance genetic variation. This can be an advantage for a population of individuals as they struggle to survive and compete within their natural environment. The term **biological evolution,** or simply, **evolution,** refers to the phenomenon that the genetic makeup of a population can change from one generation to the next.

As suggested by Charles Darwin, the members of a species are in competition with one another for essential resources. Random genetic changes (i.e., mutations) occasionally occur within an individual's genes, and sometimes these changes lead to a modification of traits that promote reproductive success. For example, over the course of many generations, random gene mutations have lengthened the neck of the giraffe, enabling it to feed on leaves that are high in the trees. When a mutation creates a new allele that is beneficial, the allele may become prevalent in future generations because the individuals carrying the allele are more likely to reproduce and pass the beneficial allele to their offspring. This process is known as **natural selection.** In this way, a species becomes better adapted to its environment.

Over a long period of time, the accumulation of many genetic changes may lead to rather striking modifications in a species' characteristics. As an example, **Figure 1.12** depicts the evolution of the modern-day horse. A variety of morphological changes occurred, including an increase in size, fewer toes, and modified jaw structure.

1.2 FIELDS OF GENETICS

Genetics is a broad discipline encompassing molecular, cellular, organism, and population biology. Many scientists who are interested in genetics have been trained in supporting disciplines

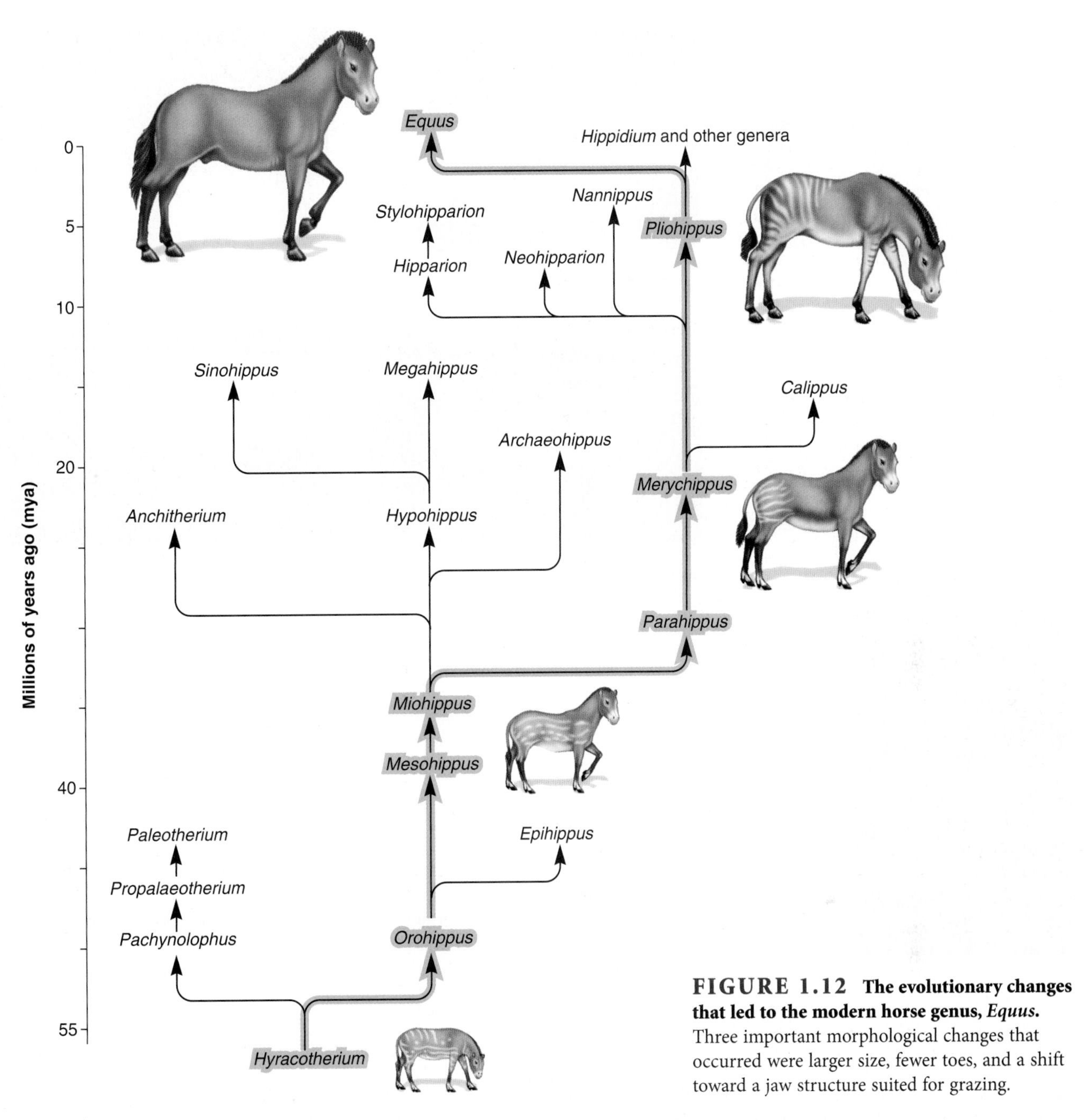

FIGURE 1.12 The evolutionary changes that led to the modern horse genus, *Equus*. Three important morphological changes that occurred were larger size, fewer toes, and a shift toward a jaw structure suited for grazing.

such as biochemistry, biophysics, cell biology, mathematics, microbiology, population biology, ecology, agriculture, and medicine. Experimentally, geneticists often focus their efforts on **model organisms**—organisms studied by many different researchers so they can compare their results and determine scientific principles that apply more broadly to other species. **Figure 1.13** shows some common examples, including *Escherichia coli* (a bacterium), *Saccharomyces cerevisiae* (a yeast), *Drosophila melanogaster* (fruit fly), *Caenorhabditis elegans* (a nematode worm), *Danio rerio* (zebrafish), *Mus musculus* (mouse), and *Arabidopsis thaliana* (a flowering plant). Model organisms offer experimental advantages over other species. For example, *E. coli* is a very simple organism that can be easily grown in the laboratory. By limiting their work to a few such model organisms, researchers can more easily unravel the genetic mechanisms that govern the traits of a given species. Furthermore, the genes found in model organisms often function in a similar way to those found in humans.

The study of genetics has been traditionally divided into three areas—transmission, molecular, and population genetics—although overlap is found among these three fields. In this section, we will examine the general questions that scientists in these areas are attempting to answer.

Transmission Genetics Explores the Inheritance Patterns of Traits as They Are Passed from Parents to Offspring

A scientist working in the field of transmission genetics examines the relationship between the transmission of genes from parent to offspring and the outcome of the offspring's traits. For example, how

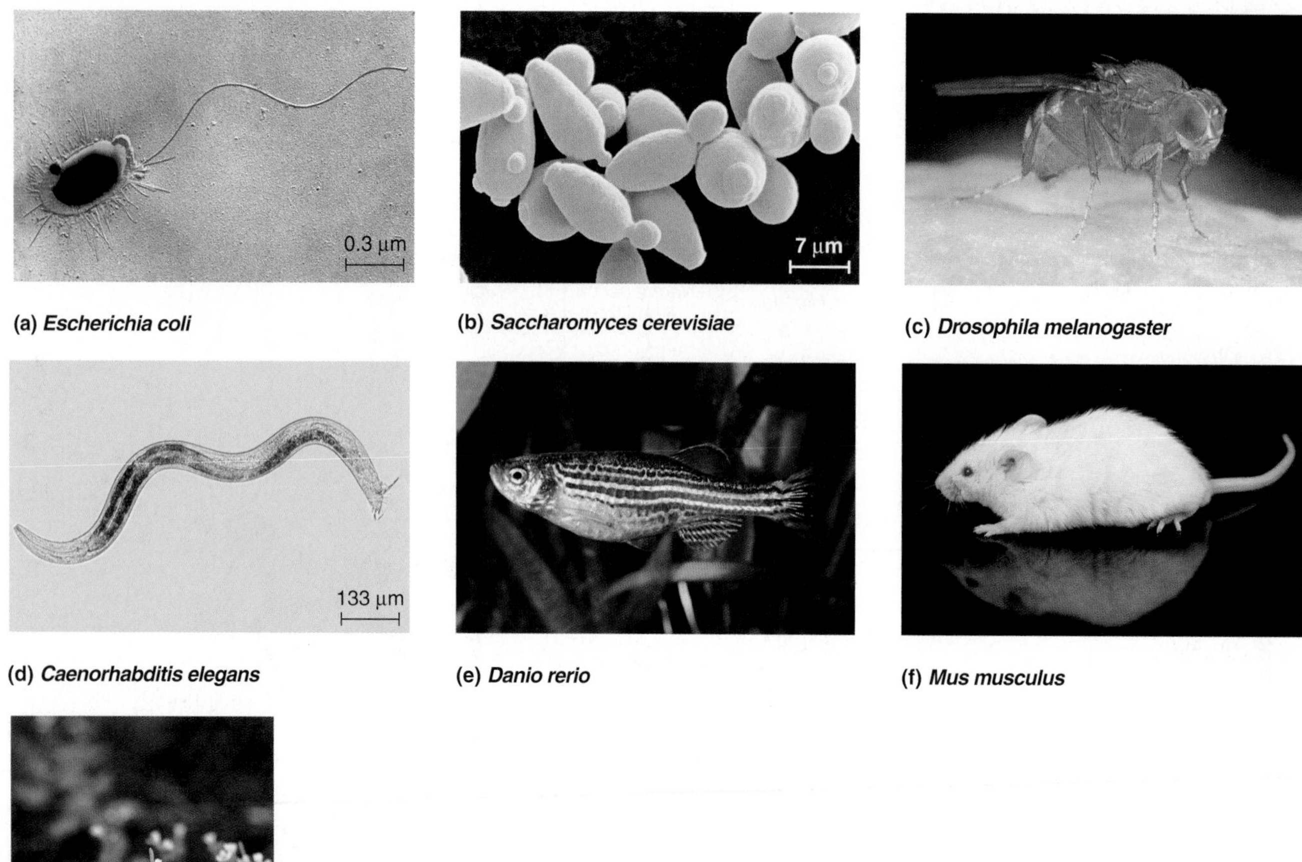

FIGURE 1.13 Examples of model organisms studied by geneticists. **(a)** *Escherichia coli* (a bacterium), **(b)** *Saccharomyces cerevisiae* (a yeast), **(c)** *Drosophila melanogaster* (fruit fly), **(d)** *Caenorhabditis elegans* (a nematode worm), **(e)** *Danio rerio* (zebrafish), **(f)** *Mus musculus* (mouse), and **(g)** *Arabidopsis thaliana* (a flowering plant).

can two brown-eyed parents produce a blue-eyed child? Or why do tall parents tend to produce tall children, but not always? Our modern understanding of transmission genetics began with the studies of Gregor Mendel. His work provided the conceptual framework for transmission genetics. In particular, he originated the idea that factors, which we now call genes, are passed as discrete units from parents to offspring via sperm and egg cells. Since these pioneering studies of the 1860s, our knowledge of genetic transmission has greatly increased. Many patterns of genetic transmission are more complex than the simple Mendelian patterns that are described in Chapter 2. The additional complexities of transmission genetics are examined in Chapters 3 through 8.

Experimentally, the fundamental approach of a transmission geneticist is the **genetic cross.** A genetic cross involves breeding two selected individuals and the subsequent analysis of their offspring in an attempt to understand how traits are passed from parents to offspring. In the case of experimental organisms, the researcher chooses two parents with particular traits and then categorizes the offspring according to the traits they possess. In many cases, this analysis is quantitative in nature. For example, an experimenter may cross two tall pea plants and obtain 100 offspring that fall into two categories: 75 tall and 25 dwarf. As we will see in Chapter 2, the ratio of tall and dwarf offspring provides important information concerning the inheritance pattern of this trait.

Throughout Chapters 2 to 8, we will learn how researchers seek to answer many fundamental questions concerning the passage of traits from parents to offspring. Some of these questions are as follows:

What are the common patterns of inheritance for genes? ***Chapters 2–4***

Are there unusual patterns of inheritance that cannot be explained by the simple transmission of genes located on chromosomes in the cell nucleus? **Chapter 5**

When two or more genes are located on the same chromosome, how does this affect the pattern of inheritance? **Chapters 6, 7**

How do variations in chromosome structure or chromosome number occur, and how are they transmitted from parents to offspring? **Chapter 8**

Molecular Genetics Focuses on a Biochemical Understanding of the Hereditary Material

The goal of molecular genetics, as the name of the field implies, is to understand how the genetic material works at the molecular level. In other words, molecular geneticists want to understand the molecular features of DNA and how these features underlie the expression of genes. The experiments of molecular geneticists are usually conducted within the confines of a laboratory. Their efforts frequently progress to a detailed analysis of DNA, RNA, and proteins, using a variety of techniques that are described throughout Parts III, IV, and V of this textbook.

Molecular geneticists often study mutant genes that have abnormal function. This is called a **genetic approach** to the study of a research question. In many cases, researchers analyze the effects of gene mutations that eliminate the function of a gene. This type of mutation is called a **loss-of-function mutation,** and the resulting gene is called a **loss-of-function allele.** By studying the effects of such mutations, the role of the functional, nonmutant gene is often revealed. For example, let's suppose that a particular plant species produces purple flowers. If a loss-of-function mutation within a given gene causes a plant of that species to produce white flowers, one would suspect the role of the functional gene involves the production of purple pigmentation.

Studies within molecular genetics interface with other disciplines such as biochemistry, biophysics, and cell biology. In addition, advances within molecular genetics have shed considerable light on the areas of transmission and population genetics. Our quest to understand molecular genetics has spawned a variety of modern molecular technologies and computer-based approaches. Furthermore, discoveries within molecular genetics have had widespread applications in agriculture, medicine, and biotechnology.

The following are some general questions within the field of molecular genetics:

What are the molecular structures of DNA and RNA? **Chapters 9, 18**

What is the composition and conformation of chromosomes? **Chapters 10, 20**

How is the genetic material copied? **Chapter 11**

How are genes expressed at the molecular level? **Chapters 12, 13, 18, 19, 21**

How is gene expression regulated so it occurs under the appropriate conditions and in the appropriate cell type? **Chapters 14, 15, 18, 23**

What is the molecular nature of mutations? How are mutations repaired? **Chapter 16**

How does the genetic material become rearranged at the molecular level? **Chapter 17**

What is the underlying relationship between genes and genetic diseases? **Chapter 22**

How do genes govern the development of multicellular organisms? **Chapter 23**

Population Genetics Is Concerned with Genetic Variation and Its Role in Evolution

The foundations of population genetics arose during the first few decades of the twentieth century. Although many scientists of this era did not accept the findings of Mendel or Darwin, the theories of population genetics provided a compelling way to connect the two viewpoints. Mendel's work and that of many succeeding geneticists gave insight into the nature of genes and how they are transmitted from parents to offspring. The work of Darwin provided a natural explanation for the variation in characteristics observed among the members of a species. To relate these two phenomena, population geneticists have developed mathematical theories to explain the prevalence of certain alleles within populations of individuals. The work of population geneticists helps us understand how processes such as natural selection have resulted in the prevalence of individuals that carry particular alleles.

Population geneticists are particularly interested in genetic variation and how that variation is related to an organism's environment. In this field, the frequencies of alleles within a population are of central importance. The following are some general questions in population genetics:

Why are two or more different alleles of a gene maintained in a population? **Chapter 24**

What factors alter the prevalence of alleles within a population? **Chapter 24**

What are the contributions of genetics and environment in the outcome of a trait? **Chapter 25**

How do genetics and the environment influence quantitative traits, such as size and weight? **Chapter 25**

What factors have the most impact on the process of evolution? **Chapter 26**

How does evolution occur at the molecular level? **Chapter 26**

Genetics Is an Experimental Science

Science is a way of knowing about our natural world. The science of genetics allows us to understand how the expression of our genes produces the traits that we possess. Researchers typically follow two general types of scientific approaches: hypothesis testing and discovery-based science. In **hypothesis testing,** also called the **scientific method,** scientists follow a series of steps to reach verifiable conclusions about the world. Although scientists arrive at their theories in different ways, the scientific method provides a way to validate (or invalidate) a particular hypothesis. Alternatively, research may also involve the collection of data without a preconceived hypothesis. For example, researchers might analyze the genes found in cancer cells to identify those genes that have become mutant. In this case, the scientists may not have a hypothesis about which particular genes may be involved. The collection and analysis of data without the need for a preconceived hypothesis is called **discovery-based science** or, simply, discovery science.

In traditional science textbooks, the emphasis often lies on the product of science. Namely, many textbooks are aimed primarily at teaching the student about the observations scientists have made and the hypotheses they have proposed to explain these observations. Along the way, the student is provided with many bits and pieces of experimental techniques and data. Likewise, this textbook also provides you with many observations and hypotheses. However, it attempts to go one step further. Each of the following chapters contains one or two experiments that have been "dissected" into five individual components to help you to understand the entire scientific process:

1. Background information is provided so you can appreciate what previous observations were known prior to conducting the experiment.
2. Most experiments involve hypothesis testing. In those cases, the figure states the hypothesis the scientists were trying to test. In other words, what scientific question was the researcher trying to answer?
3. Next, the figure follows the experimental steps the scientist took to test the hypothesis. The steps necessary to carry out the experiment are listed in the order in which they were conducted. The figure contains two parallel illustrations labeled Experimental Level and Conceptual Level. The illustration shown in the Experimental Level helps you to understand the techniques followed. The Conceptual Level helps you to understand what is actually happening at each step in the procedure.
4. The raw data for each experiment are then presented.
5. Last, an interpretation of the data is offered within the text.

The rationale behind this approach is that it will enable you to see the experimental process from beginning to end. Hopefully, you will find this a more interesting and rewarding way to learn about genetics. As you read through the chapters, the experiments will help you to see the relationship between science and scientific theories.

As a student of genetics, you will be given the opportunity to involve your mind in the experimental process. As you are reading an experiment, you may find yourself thinking about different approaches and alternative hypotheses. Different people can view the same data and arrive at very different conclusions. As you progress through the experiments in this book, you will enjoy genetics far more if you try to develop your own skills at formulating hypotheses, designing experiments, and interpreting data. Also, some of the questions in the problem sets are aimed at refining these skills.

Finally, it is worthwhile to point out that science is a social discipline. As you develop your skills at scrutinizing experiments, it is fun to discuss your ideas with other people, including fellow students and faculty members. Keep in mind that you do not need to "know all the answers" before you enter into a scientific discussion. Instead, it is more rewarding to view science as an ongoing and never-ending dialogue.

KEY TERMS

Page 1. genome
Page 4. genetics, gene, traits, nucleic acids, proteins, carbohydrates, lipids, macromolecules, proteome
Page 5. enzymes, deoxyribonucleic acid (DNA), nucleotides, genetic code, amino acid, chromosomes
Page 6. gene expression, transcription, ribonucleic acid (RNA), messenger RNA (mRNA), translated, morphological traits, physiological traits, behavioral traits
Page 7. molecular level, cellular level, organism level, species, population level, alleles
Page 8. genetic variation, morphs, gene mutations, environment
Page 9. norm of reaction, phenylketonuria (PKU), diploid
Page 10. homologs, somatic cells, gametes, haploid, biological evolution, evolution, natural selection
Page 11. model organisms
Page 12. genetic cross
Page 13. genetic approach, loss-of-function mutation, loss-of-function allele, hypothesis testing, scientific method
Page 14. discovery-based science

CHAPTER SUMMARY

- The complete genetic composition of a cell or organism is called a genome. The genome encodes all of the proteins a cell or organism can make. Many key discoveries in genetics are related to the study of genes and genomes (see Figures 1.1, 1.2, 1.3).

1.1 The Relationship Between Genes and Traits

- Living cells are composed of nucleic acids (DNA and RNA), proteins, carbohydrates, and lipids. The proteome largely determines the structure and function of cells (see Figure 1.4).
- DNA, which is found within chromosomes, stores the information to make proteins (see Figure 1.5).
- Most genes encode polypeptides that are units within functional proteins. Gene expression at the molecular level involves transcription to produce mRNA and translation to produce a polypeptide (see Figure 1.6).

- Genetics, which governs an organism's traits, spans the molecular, cellular, organism, and population levels (see Figure 1.7).
- Genetic variation underlies variation in traits. In addition, the environment plays a key role (see Figures 1.8, 1.9, 1.10).
- During reproduction, genetic material is passed from parents to offspring. In many species, somatic cells are diploid and have two sets of chromosomes whereas gametes are haploid and have a single set (see Figure 1.11).
- Evolution refers to a change in the genetic composition of a population from one generation to the next (see Figure 1.12).

1.2 Fields of Genetics

- Genetics is traditionally divided into transmission genetics, molecular genetics, and population genetics, though overlap occurs among these fields.
- Researchers in genetics carry out hypothesis testing or discovery-based science.

PROBLEM SETS & INSIGHTS

Solved Problems

S1. A human gene called the *CFTR* gene (for cystic fibrosis transmembrane regulator) encodes a protein that functions in the transport of chloride ions across the cell membrane. Most people have two copies of a functional *CFTR* gene and do not have cystic fibrosis. However, a mutant version of the *CFTR* gene is found in some people. If a person has two mutant copies of the gene, he or she develops the disease known as cystic fibrosis. Are the following examples a description of genetics at the molecular, cellular, organism, or population level?

A. People with cystic fibrosis have lung problems due to a buildup of thick mucus in their lungs.

B. The mutant *CFTR* gene encodes a defective chloride transporter.

C. A defect in the chloride transporter causes a salt imbalance in lung cells.

D. Scientists have wondered why the mutant *CFTR* gene is relatively common. In fact, it is the most common mutant gene that causes a severe disease in Caucasians. Usually, mutant genes that cause severe diseases are relatively rare. One possible explanation why CF is so common is that people who have one copy of the functional *CFTR* gene and one copy of the mutant gene may be more resistant to diarrheal diseases such as cholera. Therefore, even though individuals with two mutant copies are very sick, people with one mutant copy and one functional copy might have a survival advantage over people with two functional copies of the gene.

Answer:

A. Organism. This is a description of a trait at the level of an entire individual.

B. Molecular. This is a description of a gene and the protein it encodes.

C. Cellular. This is a description of how protein function affects the cell.

D. Population. This is a possible explanation why two alleles of the gene occur within a population.

S2. Explain the relationship between the following pairs of terms:

A. RNA and DNA

B. RNA and transcription

C. Gene expression and trait

D. Mutation and allele

Answer:

A. DNA is the genetic material. In a cell, DNA is used to make RNA. RNA is then used to specify a sequence of amino acids within a polypeptide.

B. Transcription is a process in which RNA is made using DNA as a template.

C. Genes are expressed at the molecular level to produce functional proteins. The functioning of proteins within living cells ultimately affects an organism's traits.

D. Alleles are alternative forms of the same gene. For example, a particular human gene affects eye color. The gene can exist as a blue allele or a brown allele. The difference between these two alleles is caused by a mutation. Perhaps the brown allele was the first eye color allele in the human population. Within some ancestral person, however, a mutation may have occurred in the eye color gene that converted the brown allele to the blue allele. Now the human population has both the brown allele and the blue allele.

S3. In diploid species that carry out sexual reproduction, how are genes passed from generation to generation?

Answer: When a diploid individual makes haploid cells for sexual reproduction, the cells contain half the number of chromosomes. When two haploid cells (e.g., sperm and egg) combine with each other, a zygote is formed that begins the life of a new individual. This zygote has inherited half of its chromosomes and, therefore, half of its genes from each parent. This is how genes are passed from parents to offspring.

Conceptual Questions

C1. Pick any example of a genetic technology and describe how it has directly affected your life.

C2. At the molecular level, what is a gene? Where are genes located?

C3. Most genes encode proteins. Explain how the structure and function of proteins produce an organism's traits.

C4. Briefly explain how gene expression occurs at the molecular level.

C5. A human gene called the β-globin gene encodes a polypeptide that functions as a subunit of the protein known as hemoglobin. Hemoglobin is found within red blood cells; it carries oxygen. In human populations, the β-globin gene can be found as the common allele called the Hb^A allele, but it can also be found as the Hb^S allele. Individuals who have two copies of the Hb^S allele have the disease called sickle cell anemia. Are the following examples a description of genetics at the molecular, cellular, organism, or population level?

A. The Hb^S allele encodes a polypeptide that functions slightly differently from the polypeptide encoded by the Hb^A allele.

B. If an individual has two copies of the Hb^S allele, that person's red blood cells take on a sickle shape.

C. Individuals who have two copies of the Hb^A allele do not have sickle cell disease, but they are not resistant to malaria. People who have one Hb^A allele and one Hb^S allele do not have sickle cell disease, and they are resistant to malaria. People who have two copies of the Hb^S allele have sickle cell anemia, and this disease may significantly shorten their lives.

D. Individuals with sickle cell disease have anemia because their red blood cells are easily destroyed by the body.

C6. What is meant by the term "genetic variation"? Give two examples of genetic variation not discussed in Chapter 1. What causes genetic variation at the molecular level?

C7. What is the cause of Down syndrome?

C8. Your textbook describes how the trait of phenylketonuria (PKU) is greatly influenced by the environment. Pick a trait in your favorite plant and explain how genetics and the environment may play important roles.

C9. What is meant by the term "diploid"? Which cells of the human body are diploid, and which cells are not?

C10. What is a DNA sequence?

C11. What is the genetic code?

C12. Explain the relationships between the following pairs of genetic terms:

A. Gene and trait

B. Gene and chromosome

C. Allele and gene

D. DNA sequence and amino acid sequence

C13. With regard to biological evolution, which of the following statements is incorrect? Explain why.

A. During its lifetime, an animal evolves to become better adapted to its environment.

B. The process of biological evolution has produced species that are better adapted to their environments.

C. When an animal is better adapted to its environment, the process of natural selection makes it more likely for that animal to reproduce.

C14. What are the primary interests of researchers working in the following fields of genetics?

A. Transmission genetics

B. Molecular genetics

C. Population genetics

Experimental Questions

E1. What is a genetic cross?

E2. The technique known as DNA sequencing (described in Chapter 18) enables researchers to determine the DNA sequence of genes. Would this technique be used primarily by transmission geneticists, molecular geneticists, or population geneticists?

E3. Figure 1.5 shows a micrograph of chromosomes from a normal human cell. If you performed this type of experiment using cells from a person with Down syndrome, what would you expect to see?

E4. Many organisms are studied by geneticists. Of the following species, do you think it would be more likely for them to be studied by a transmission geneticist, a molecular geneticist, or a population geneticist? Explain your answer. Note: More than one answer may be possible.

A. Dogs

B. *E. coli*

C. Fruit flies

D. Leopards

E. Corn

E5. Pick any trait you like in any species of wild plant or animal. The trait must somehow vary among different members of the species. For example, some butterflies have dark wings and others have light wings (see Figure 1.7).

A. Discuss all of the background information that you already have (from personal observations) regarding this trait.

B. Propose a hypothesis that would explain the genetic variation within the species. For example, in the case of the butterflies, your hypothesis might be that the dark butterflies survive better in dark forests, and the light butterflies survive better in sunlit fields.

C. Describe the experimental steps you would follow to test your hypothesis.

D. Describe the possible data you might collect.

E. Interpret your data.

Note: When picking a trait to answer this question, do not pick the trait of wing color in butterflies.

Note: All answers appear at the website for this textbook; the answers to even-numbered questions are in the back of the textbook.

www.mhhe.com/brookergenetics4e

Visit the website for practice tests, answer keys, and other learning aids for this chapter. Enhance your understanding of genetics with our interactive exercises, quizzes, animations, and much more.

The garden pea, studied by Mendel.

CHAPTER OUTLINE

2 MENDELIAN INHERITANCE

An appreciation for the concept of heredity can be traced far back in human history. Hippocrates, a famous Greek physician, was the first person to provide an explanation for hereditary traits (ca. 400 B.C.E.). He suggested that "seeds" are produced by all parts of the body, which are then collected and transmitted to the offspring at the time of conception. Furthermore, he hypothesized that these seeds cause certain traits of the offspring to resemble those of the parents. This idea, known as **pangenesis,** was the first attempt to explain the transmission of hereditary traits from generation to generation.

For the next 2000 years, the ideas of Hippocrates were accepted by some and rejected by many. After the invention of the microscope in the late seventeenth century, some people observed sperm and thought they could see a tiny creature inside, which they termed a homunculus (little man). This homunculus was hypothesized to be a miniature human waiting to develop within the womb of its mother. Those who held that thought, known as spermists, suggested that only the father was responsible for creating future generations and that any resemblance between mother and offspring was due to influences "within the womb." During the same time, an opposite school of thought also developed. According to the ovists, the egg was solely responsible for human characteristics. The only role of the sperm was to stimulate the egg onto its path of development. Of course, neither of these ideas was correct.

The first systematic studies of genetic crosses were carried out by Joseph Kölreuter from 1761 to 1766. In crosses between different strains of tobacco plants, he found that the offspring were usually intermediate in appearance between the two parents. This led Kölreuter to conclude that both parents make equal genetic contributions to their offspring. Furthermore, his observations were consistent with **blending inheritance.** According to this view, the factors that dictate hereditary traits can blend together from generation to generation. The blended traits would then be passed to the next generation. The popular view before the 1860s, which combined the notions of pangenesis and blending inheritance, was that hereditary traits were rather malleable and could change and blend over the course of one or two generations. However, the pioneering work of Gregor Mendel would prove instrumental in refuting this viewpoint.

In Chapter 2, we will first examine the outcome of Mendel's crosses in pea plants. We begin our inquiry into genetics here because the inheritance patterns observed in peas are fundamentally related to inheritance patterns found in other eukaryotic species, such as humans, mice, fruit flies, and corn. We will discover how Mendel's insights into the patterns of inheritance in pea plants revealed some simple rules that govern the process of inheritance. In Chapters 3 through 8, we will explore more complex patterns of inheritance and also consider the role that chromosomes play as the carriers of the genetic material.

In the second part of this chapter, we will become familiar with general concepts in probability and statistics. How are statistical methods useful? First, probability calculations allow us to predict the outcomes of simple genetic crosses, as well as the outcomes of more complicated crosses described in later chapters. In addition, we will learn how to use statistics to test the validity of genetic hypotheses that attempt to explain the inheritance patterns of traits.

2.1 MENDEL'S LAWS OF INHERITANCE

Gregor Johann Mendel, born in 1822, is now remembered as the father of genetics (**Figure 2.1**). He grew up on a small farm in Hynčice (formerly Heinzendorf) in northern Moravia, which was then a part of Austria and is now a part of the Czech Republic. As a young boy, he worked with his father grafting trees to improve the family orchard. Undoubtedly, his success at grafting taught him that precision and attention to detail are important elements of success. These qualities would later be important in his experiments as an adult scientist. Instead of farming, however, Mendel was accepted into the Augustinian monastery of St. Thomas, completed his studies for the priesthood, and was ordained in 1847. Soon after becoming a priest, Mendel worked for a short time as a substitute teacher. To continue that role, he needed to obtain a teaching license from the government. Surprisingly, he failed the licensing exam due to poor answers in the areas of physics and natural history. Therefore, Mendel then enrolled at the University of Vienna to expand his knowledge in these two areas. Mendel's training in physics and mathematics taught him to perceive the world as an orderly place, governed by natural laws. In his studies, Mendel learned that these natural laws could be stated as simple mathematical relationships.

In 1856, Mendel began his historic studies on pea plants. For 8 years, he grew and crossed thousands of pea plants on a small 115- by 23-foot plot. He kept meticulously accurate records that included quantitative data concerning the outcome of his crosses. He published his work, entitled "Experiments on Plant Hybrids," in 1866. This paper was largely ignored by scientists at that time, possibly because of its title. Another reason his work went unrecognized could be tied to a lack of understanding of chromosomes and their transmission, a topic we will discuss in Chapter 3. Nevertheless, Mendel's ground-breaking work allowed him to propose the natural laws that now provide a framework for our understanding of genetics.

Prior to his death in 1884, Mendel reflected, "My scientific work has brought me a great deal of satisfaction and I am convinced that it will be appreciated before long by the whole world." Sixteen years later, in 1900, the work of Mendel was independently rediscovered by three biologists with an interest in plant genetics: Hugo de Vries of Holland, Carl Correns of Germany, and Erich von Tschermak of Austria. Within a few years, the influence of Mendel's studies was felt around the world. In this section, we will examine Mendel's experiments and consider their monumental significance in the field of genetics.

FIGURE 2.1 **Gregor Johann Mendel, the father of genetics.**

Mendel Chose Pea Plants as His Experimental Organism

Mendel's study of genetics grew out of his interest in ornamental flowers. Prior to his work with pea plants, many plant breeders had conducted experiments aimed at obtaining flowers with new varieties of colors. When two distinct individuals with different characteristics are mated, or **crossed,** to each other, this is called a **hybridization** experiment, and the offspring are referred to as **hybrids.** For example, a hybridization experiment could involve a cross between a purple-flowered plant and a white-flowered plant. Mendel was particularly intrigued, in such experiments, by the consistency with which offspring of subsequent generations showed characteristics of one or the other parent. His intellectual foundation in physics and the natural sciences led him to

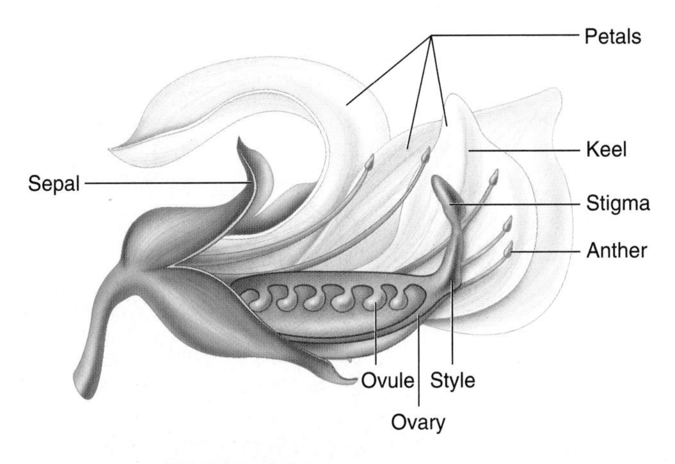

(a) Structure of a pea flower

(b) A flowering pea plant

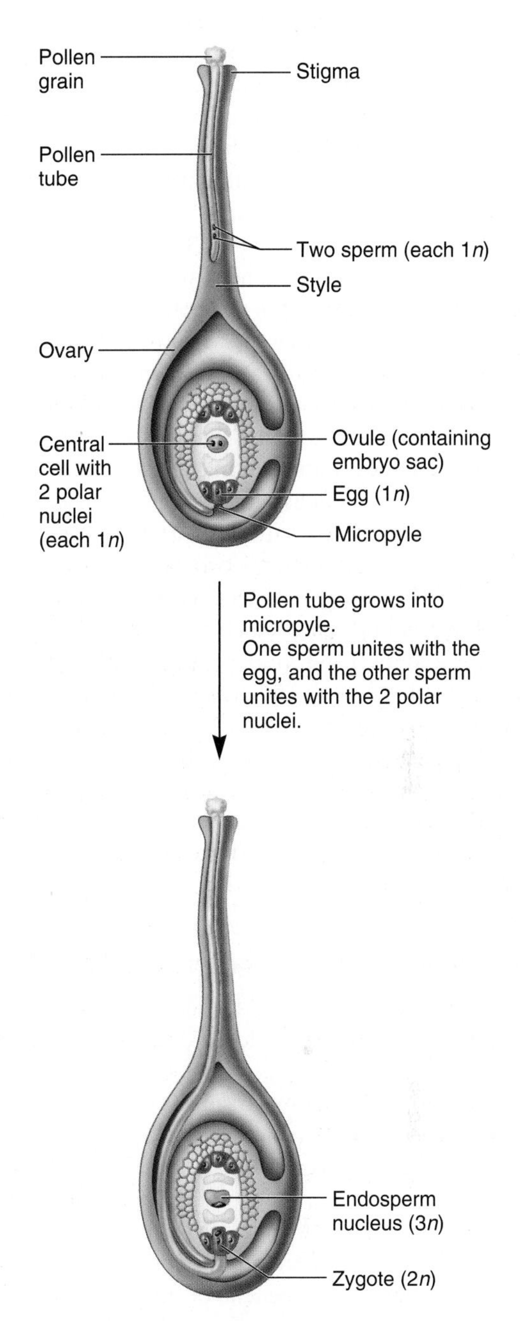

(c) Pollination and fertilization in angiosperms

FIGURE 2.2 Flower structure and pollination in pea plants. **(a)** The pea flower can produce both pollen and egg cells. The pollen grains are produced within the anthers, and the egg cells are produced within the ovules that are contained within the ovary. A modified petal called a keel encloses the anthers and ovaries. **(b)** Photograph of a flowering pea plant. **(c)** A pollen grain must first land on the stigma. After this occurs, the pollen sends out a long tube through which two sperm cells travel toward an ovule to reach an egg cell. The fusion between a sperm and an egg cell results in fertilization and creates a zygote. A second sperm fuses with a central cell containing two polar nuclei to create the endosperm. The endosperm provides a nutritive material for the developing embryo.

consider that this regularity might be rooted in natural laws that could be expressed mathematically. To uncover these laws, he realized that he would need to carry out quantitative experiments in which the numbers of offspring carrying certain traits were carefully recorded and analyzed.

Mendel chose the garden pea, *Pisum sativum*, to investigate the natural laws that govern plant hybrids. The morphological features of this plant are shown in **Figure 2.2a/b**. Several properties of this species were particularly advantageous for studying plant hybridization. First, the species was available in several varieties that had decisively different physical characteristics. Many strains of the garden pea were available that varied in the appearance of their height, flowers, seeds, and pods.

A second important issue is the ease of making crosses. In flowering plants, reproduction occurs by a pollination event (**Figure 2.2c**). Male gametes **(sperm)** are produced within **pollen grains** formed in the **anthers,** and the female gametes **(eggs)** are contained within **ovules** that form in the **ovaries.** For fertilization to occur, a pollen grain lands on the **stigma,** which stimulates the growth of a pollen tube. This enables sperm cells to enter the stigma and migrate toward an ovule. Fertilization occurs when a sperm enters the micropyle, an opening in the ovule wall, and fuses with an egg cell. The term **gamete** is used to describe haploid reproductive cells that can unite to form a zygote. It should be emphasized, however, that the process that produces gametes in animals is quite different from the way that gametes are produced in plants and fungi. These processes are described in greater detail in Chapter 3.

In some experiments, Mendel wanted to carry out **self-fertilization,** which means that the pollen and egg are derived from the same plant. In peas, a modified petal known as the keel covers the reproductive structures of the plant. Because of this covering, pea plants naturally reproduce by self-fertilization. Usually, pollination occurs even before the flower opens. In other experiments, however, Mendel wanted to make crosses between different plants. How did he accomplish this goal? Fortunately, pea plants contain relatively large flowers that are easy to manipulate, making it possible to make crosses between two particular plants and study their outcomes. This process, known as **cross-fertilization,** requires that the pollen from one plant be placed on the stigma of another plant. This procedure is shown in **Figure 2.3.** Mendel was able to pry open immature flowers and remove the anthers before they produced pollen. Therefore, these flowers could not self-fertilize. He would then obtain pollen from another plant by gently touching its mature anthers with a paintbrush. Mendel applied this pollen to the stigma of the flower that already had its anthers removed. In this way, he was able to cross-fertilize his pea plants and thereby obtain any type of hybrid he wanted.

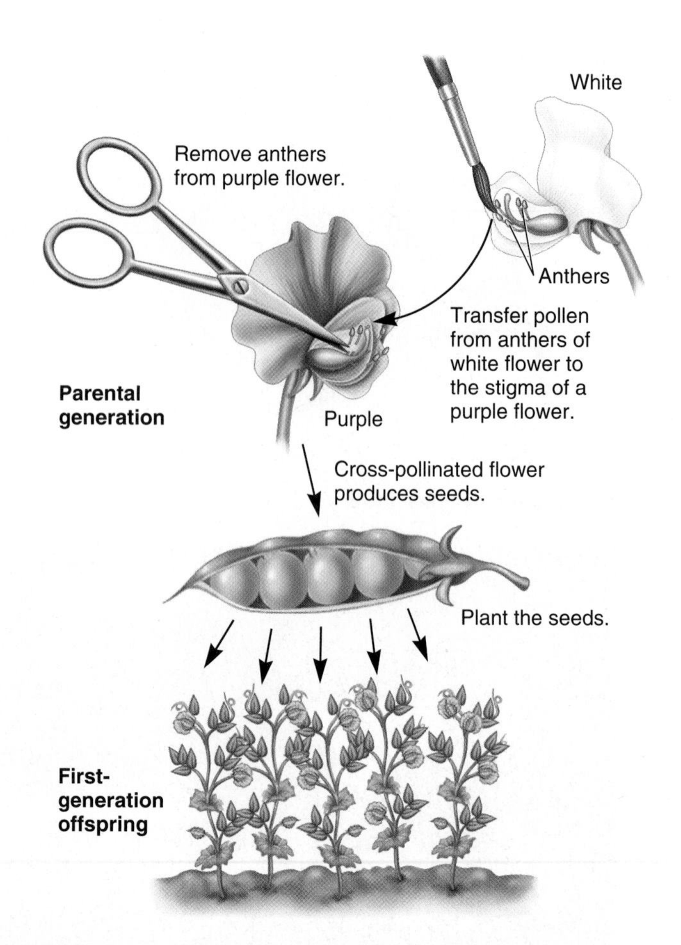

FIGURE 2.3 How Mendel cross-fertilized two different pea plants. This illustration depicts a cross between a plant with purple flowers and another plant with white flowers. The offspring from this cross are the result of pollination of the purple flower using pollen from a white flower.

Mendel Studied Seven Characteristics That Bred True

When he initiated his studies, Mendel obtained several varieties of peas that were considered to be distinct. These plants were different with regard to many morphological characteristics. The general characteristics of an organism are called **characters.** The terms **trait** and **variant** are typically used to describe the specific properties of a character. For example, eye color is a character of humans and blue eyes is a trait (or variant) found in some people. Over the course of 2 years, Mendel tested his pea strains to determine if their characteristics bred true. This means that a trait did not vary in appearance from generation to generation. For example, if the seeds from a pea plant were yellow, the next generation would also produce yellow seeds. Likewise, if these offspring were allowed to self-fertilize, all of their offspring would also produce yellow seeds, and so on. A variety that continues to produce the same trait after several generations of self-fertilization is called a **true-breeding line,** or **strain.**

Mendel next concentrated his efforts on the analysis of characteristics that were clearly distinguishable between different true-breeding lines. **Figure 2.4** illustrates the seven characters that Mendel eventually chose to follow in his breeding experiments. All seven were found in two variants. A variant (or trait) may be found in two or more versions within a single species. For example, one character he followed was height, which was found in two variants: tall and dwarf plants. Mendel studied this character by crossing the variants to each other. A cross in which an experimenter is observing only one character is called a **monohybrid cross,** also called a **single-factor cross.** When the two parents are different variants for a given character, this type of cross produces single-character hybrids, also known as **monohybrids.**

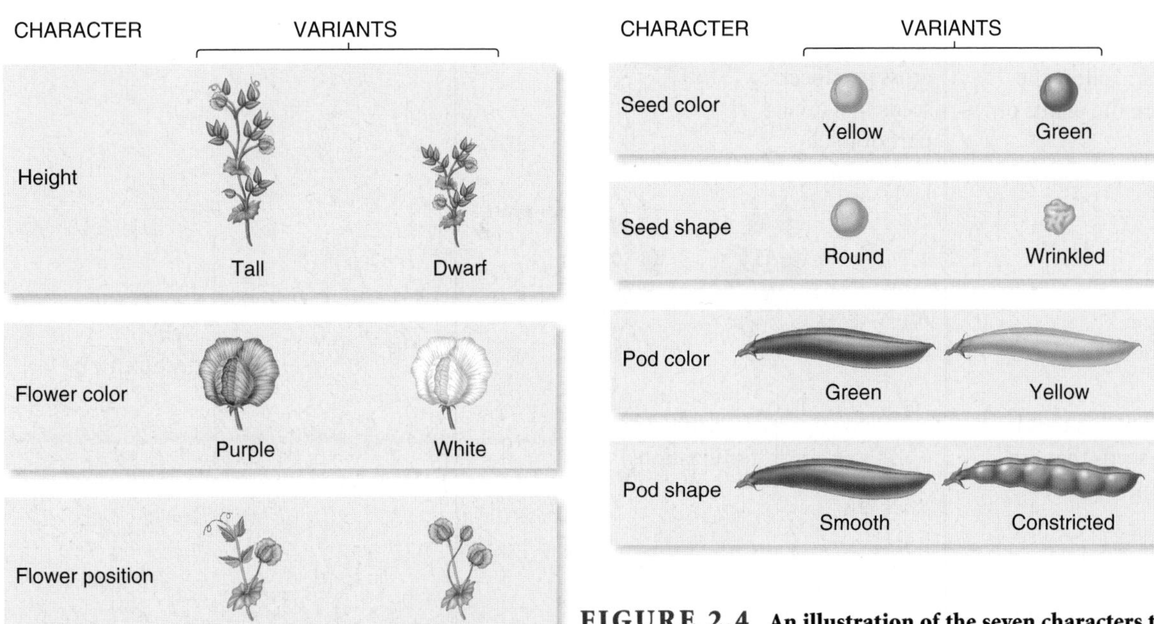

FIGURE 2.4 An illustration of the seven characters that Mendel studied. Each character was found as two variants that were decisively different from each other.

EXPERIMENT 2A

Mendel Followed the Outcome of a Single Character for Two Generations

Prior to conducting his studies, Mendel did not already have a hypothesis to explain the formation of hybrids. However, his educational background caused him to realize that a quantitative analysis of crosses may uncover mathematical relationships that would otherwise be mysterious. His experiments were designed to determine the relationships that govern hereditary traits. This rationale is called an **empirical approach.** Laws that are deduced from an empirical approach are known as empirical laws.

Mendel's experimental procedure is shown in **Figure 2.5**. He began with true-breeding plants that differed with regard to a single character. These are termed the **parental generation,** or **P generation.** When the true-breeding parents were crossed to each other, this is called a P cross, and the offspring constitute the **F_1 generation,** for first filial generation. As seen in the data, all plants of the F_1 generation showed the phenotype of one parent but not the other. This prompted Mendel to follow the transmission of this character for one additional generation. To do so, the plants of the F_1 generation were allowed to self-fertilize to produce a second generation called the **F_2 generation,** for second filial generation.

THE GOAL

Mendel speculated that the inheritance pattern for a single character may follow quantitative natural laws. The goal of this experiment was to uncover such laws.

ACHIEVING THE GOAL — FIGURE 2.5 Mendel's analysis of monohybrid crosses.

Starting material: Mendel began his experiments with true-breeding pea plants that varied with regard to only one of seven different characters (see Figure 2.4).

1. For each of seven characters, Mendel cross-fertilized two different true-breeding lines. Keep in mind that each cross involved two plants that differed in regard to only one of the seven characters studied. The illustration at the right shows one cross between a tall and dwarf plant. This is called a P (parental) cross.

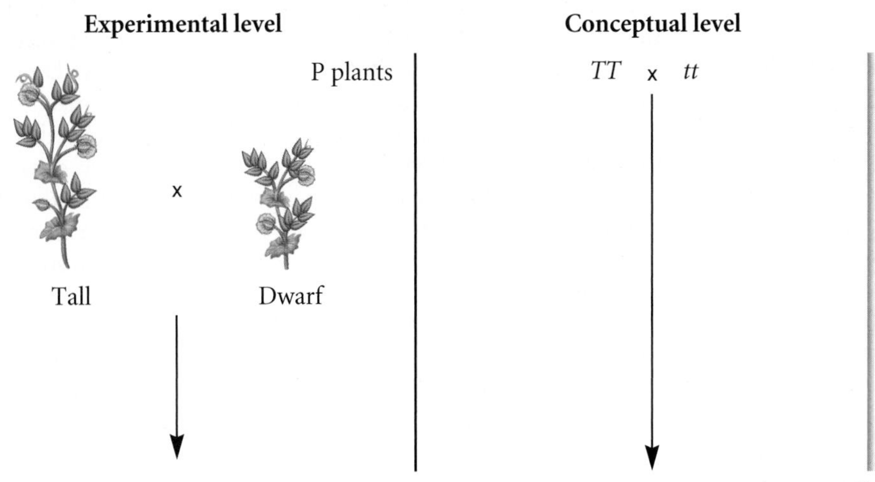

(continued)

2. Collect many seeds. The following spring, plant the seeds and allow the plants to grow. These are the plants of the F_1 generation.

3. Allow the F_1 generation plants to self-fertilize. This produces seeds that are part of the F_2 generation.

4. Collect the seeds and plant them the following spring to obtain the F_2 generation plants.

5. Analyze the characteristics found in each generation.

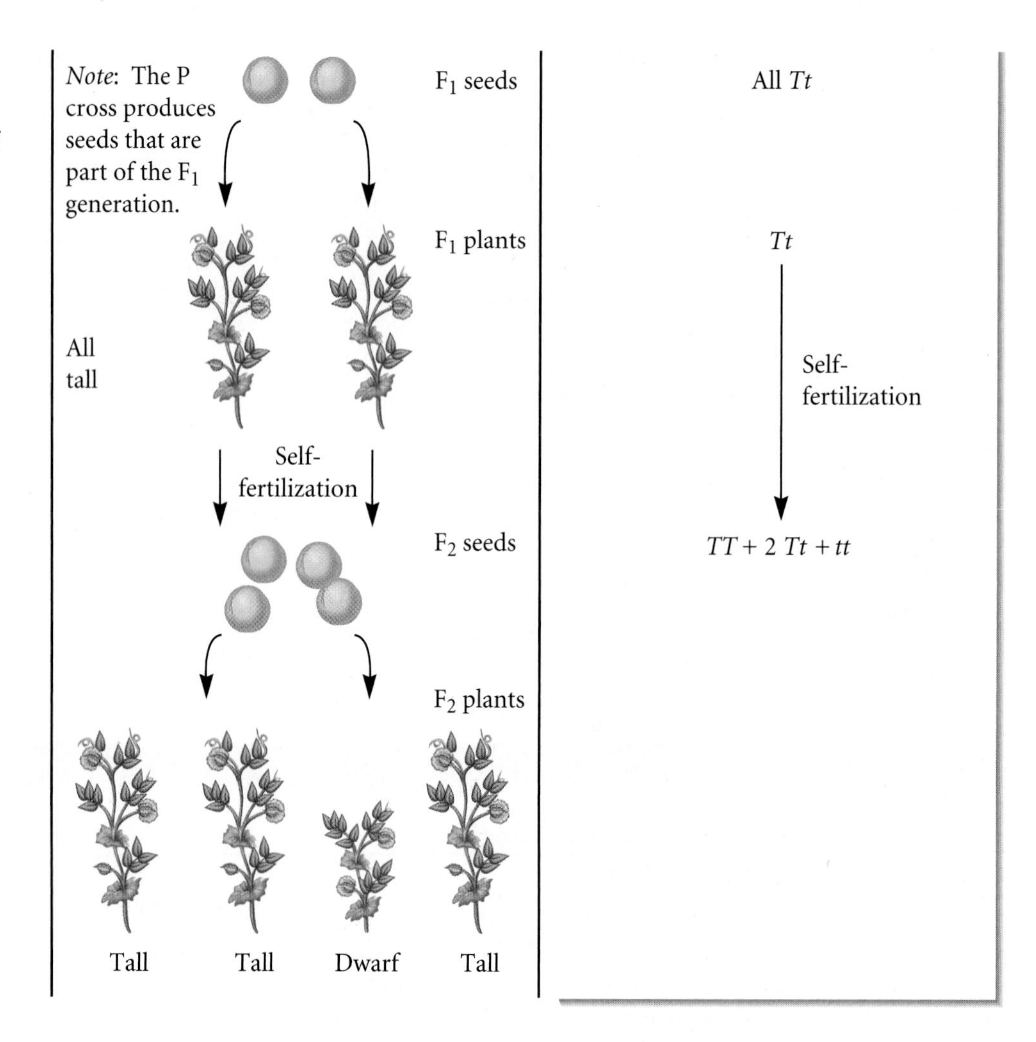

THE DATA

P cross	*F_1 generation*	*F_2 generation*	*Ratio*
Tall × dwarf stem	All tall	787 tall, 277 dwarf	2.84:1
Purple × white flowers	All purple	705 purple, 224 white	3.15:1
Axial × terminal flowers	All axial	651 axial, 207 terminal	3.14:1
Yellow × green seeds	All yellow	6,022 yellow, 2,001 green	3.01:1
Round × wrinkled seeds	All round	5,474 round, 1,850 wrinkled	2.96:1
Green × yellow pods	All green	428 green, 152 yellow	2.82:1
Smooth × constricted pods	All smooth	882 smooth, 299 constricted	2.95:1
Total	All dominant	14,949 dominant, 5,010 recessive	2.98:1

Data from Mendel, Gregor. 1866 Versuche Über Plflanzenhybriden. *Verhandlungen des naturforschenden Vereines in Brünn, Bd IV für das Jahr 1865,* Abhandlungen, 3–47.

INTERPRETING THE DATA

The data shown in Figure 2.5 are the results of producing an F_1 generation via cross-fertilization and an F_2 generation via self-fertilization of the F_1 monohybrids. A quantitative analysis of these data allowed Mendel to propose three important ideas:

1. Mendel's data argued strongly against a blending mechanism of heredity. In all seven cases, the F_1 generation displayed characteristics that were distinctly like one of the two parents rather than traits intermediate in character. His first proposal was that the variant for one character is **dominant** over another variant. For example, the variant of green pods is dominant to that of yellow pods. The term **recessive** is used to describe a variant that is masked by the presence of a dominant trait but reappears in subsequent generations. Yellow pods and dwarf stems are examples of recessive variants. They can also be referred to as recessive traits.
2. When a true-breeding plant with a dominant trait was crossed to a true-breeding plant with a recessive trait, the dominant trait was always observed in the F_1 generation. In the F_2 generation, some offspring displayed the dominant trait, while a smaller proportion showed the recessive trait. How did Mendel explain this observation? Because the

recessive trait appeared in the F_2 generation, he made a second proposal—the genetic determinants of traits are passed along as "unit factors" from generation to generation. His data were consistent with a **particulate theory of inheritance,** in which the genes that govern traits are inherited as discrete units that remain unchanged as they are passed from parent to offspring. Mendel called them unit factors, but we now call them genes.

3. When Mendel compared the numbers of dominant and recessive traits in the F_2 generation, he noticed a recurring pattern. Within experimental variation, he always observed approximately a 3:1 ratio between the dominant trait and the recessive trait. Mendel was the first scientist to apply this type of quantitative analysis in a biological experiment. As described next, this quantitative approach allowed him to make a third proposal—genes **segregate** from each other during the process that gives rise to gametes.

A self-help quiz involving this experiment can be found at www.mhhe.com/brookergenetics4e.

Mendel's 3:1 Phenotypic Ratio Is Consistent with the Law of Segregation

Mendel's research was aimed at understanding the laws that govern the inheritance of traits. At that time, scientists did not understand the molecular composition of the genetic material or its mode of transmission during gamete formation and fertilization. We now know that the genetic material is composed of deoxyribonucleic acid (DNA), a component of chromosomes. Each chromosome contains hundreds or thousands of shorter segments that function as genes—a term that was originally coined by the Danish botanist Wilhelm Johannsen in 1909. A **gene** is defined as a "unit of heredity" that may influence the outcome of an organism's traits. Each of the seven characters that Mendel studied is influenced by a different gene.

Most eukaryotic species, such as pea plants and humans, have their genetic material organized into pairs of chromosomes. For this reason, eukaryotes have two copies of most genes. These copies may be the same or they may differ. The term **allele** refers to different versions of the same gene. With this modern knowledge, the results shown in Figure 2.5 are consistent with the idea that each parent transmits only one copy of each gene (i.e., one allele) to each offspring. **Mendel's law of segregation** states that:

The two copies of a gene segregate (or separate) from each other during transmission from parent to offspring.

Therefore, only one copy of each gene is found in a gamete. At fertilization, two gametes combine randomly, potentially producing different allelic combinations.

Let's use Mendel's cross of tall and dwarf pea plants to illustrate how alleles are passed from parents to offspring (**Figure 2.6**). The letters *T* and *t* are used to represent the alleles of the gene that determines plant height. By convention, the uppercase letter represents the dominant allele (*T* for tall height, in this case), and the recessive allele is represented by the same letter in lowercase (*t*, for dwarf height). For the P cross, both parents are true-breeding plants. Therefore, we know each has identical copies of the height gene. When an individual possesses two identical copies of a gene, the individual is said to be **homozygous**

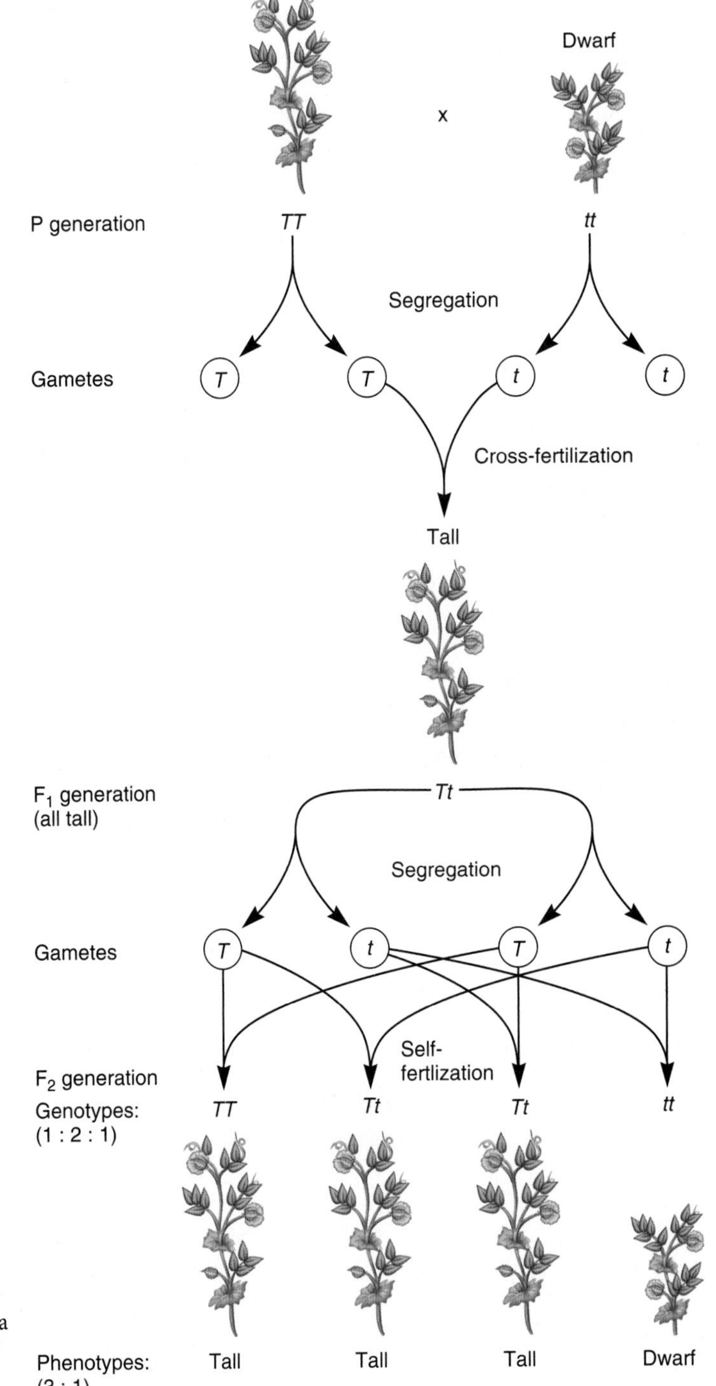

FIGURE 2.6 Mendel's law of segregation. This illustration shows a cross between a true-breeding tall plant and a true-breeding dwarf plant and the subsequent segregation of the tall (*T*) and dwarf (*t*) alleles in the F_1 and F_2 generations.

with respect to that gene. (The prefix *homo-* means like, and the suffix *-zygo* means pair.) In the P cross, the tall plant is homozygous for the tall allele *T*, and the dwarf plant is homozygous for the dwarf allele *t*. The term **genotype** refers to the genetic composition of an individual. *TT* and *tt* are the genotypes of the P generation in this experiment. The term **phenotype** refers to an observable characteristic of an organism. In the P generation, the plants exhibit a phenotype that is either tall or dwarf.

In contrast, the F_1 generation is **heterozygous,** with the genotype *Tt*, because every individual carries one copy of the tall allele and one copy of the dwarf allele. A heterozygous individual carries different alleles of a gene. (The prefix *hetero-* means different.) Although these plants are heterozygous, their phenotypes are tall because they have a copy of the dominant tall allele.

The law of segregation predicts that the phenotypes of the F_2 generation will be tall and dwarf in a ratio of 3:1 (see Figure 2.6). The parents of the F_2 generation are heterozygous. Due to segregation, their gametes can carry either a *T* allele or a *t* allele, but not both. Following self-fertilization, *TT*, *Tt*, and *tt* are the possible genotypes of the F_2 generation (note that the genotype *Tt* is the same as *tT*). By randomly combining these alleles, the genotypes are produced in a 1:2:1 ratio. Because *TT* and *Tt* both produce tall phenotypes, a 3:1 phenotypic ratio is observed in the F_2 generation.

A Punnett Square Can Be Used to Predict the Outcome of Crosses

An easy way to predict the outcome of simple genetic crosses is to use a **Punnett square,** a method originally proposed by Reginald Punnett. To construct a Punnett square, you must know the genotypes of the parents. With this information, the Punnett square enables you to predict the types of offspring the parents are expected to produce and in what proportions. We will follow a step-by-step description of the Punnett square approach using a cross of heterozygous tall plants as an example.

Step 1. *Write down the genotypes of both parents.* In this example, a heterozygous tall plant is crossed to another heterozygous tall plant. The plant providing the pollen is considered the male parent and the plant providing the eggs, the female parent.

Male parent: *Tt*

Female parent: *Tt*

Step 2. *Write down the possible gametes that each parent can make.* Remember that the law of segregation tells us that a gamete can carry only one copy of each gene.

Male gametes: *T* or *t*

Female gametes: *T* or *t*

Step 3. *Create an empty Punnett square.* In the examples shown in this textbook, the number of columns equals the number of male gametes, and the number of rows equals the number of female gametes. Our example has two rows and two columns. Place the male gametes across the top of the Punnett square and the female gametes along the side.

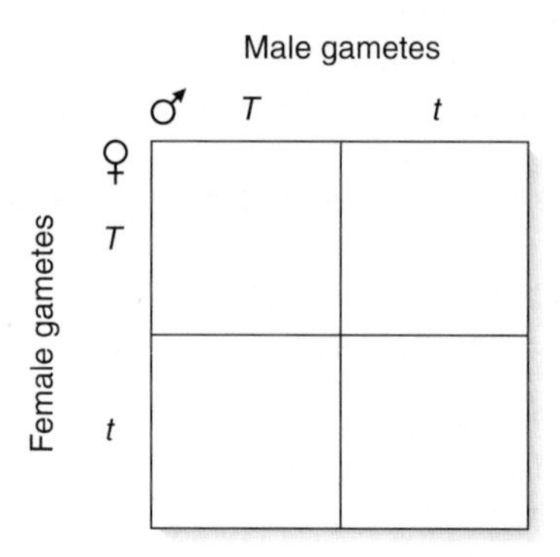

Step 4. *Fill in the possible genotypes of the offspring by combining the alleles of the gametes in the empty boxes.*

Male gametes

♂ / ♀ (Female gametes)	*T*	*t*
T	*TT*	*Tt*
t	*Tt*	*tt*

Step 5. *Determine the relative proportions of genotypes and phenotypes of the offspring.* The genotypes are obtained directly from the Punnett square. They are contained within the boxes that have been filled in. In this example, the genotypes are *TT*, *Tt*, and *tt* in a 1:2:1 ratio. To determine the phenotypes, you must know the dominant/recessive relationship between the alleles. For plant height, we know that *T* (tall) is dominant to *t* (dwarf). The genotypes *TT* and *Tt* are tall, whereas the genotype *tt* is dwarf. Therefore, our Punnett square shows us that the ratio of phenotypes is 3:1, or 3 tall plants : 1 dwarf plant. Additional problems of this type are provided in the Solved Problems at the end of this chapter.

EXPERIMENT 2B

Mendel Also Analyzed Crosses Involving Two Different Characters

Though his experiments described in Figure 2.5 revealed important ideas regarding hereditary laws, Mendel realized that additional insights might be uncovered if he conducted more complicated experiments. In particular, he conducted crosses in which he simultaneously investigated the pattern of inheritance for two different characters. In other words, he carried out **two-factor crosses,** also called **dihybrid crosses,** in which he followed the inheritance of two different characters within the same groups of individuals. For example, let's consider an experiment in which one of the characters was seed shape, found in round or wrinkled variants; the second character was seed color, which existed as yellow and green variants. In this dihybrid cross, Mendel followed the inheritance pattern for both characters simultaneously.

What results are possible from a dihybrid cross? One possibility is that the genetic determinants for two different characters are always linked to each other and inherited as a single unit (**Figure 2.7a**). If this were the case, the F_1 offspring could produce only two types of gametes, *RY* and *ry*. A second possibility is that they are not linked and can assort themselves independently into haploid gametes (**Figure 2.7b**). According to independent assortment, an F_1 offspring could produce four types of gametes, *RY*, *Ry*, *rY*, and *ry*. Keep in mind that the results of Figure 2.5 have already shown us that a gamete carries only one allele for each gene.

The experimental protocol of one of Mendel's two-factor crosses is shown in **Figure 2.8**. He began with two different strains of true-breeding pea plants that were different with regard to two characters: seed shape and seed color. In this example, one plant was produced from seeds that were round and yellow; the other plant from seeds that were wrinkled and green. When these plants were crossed, the seeds, which contain the plant embryo, are considered part of the F_1 generation. As expected, the data revealed that the F_1 seeds displayed a phenotype of round and yellow. This was observed because round and yellow are dominant traits. It is the F_2 generation that supports the independent-assortment model and refutes the linkage model.

THE HYPOTHESES

The inheritance pattern for two different characters follows one or more quantitative natural laws. Two possible hypotheses are described in Figure 2.7.

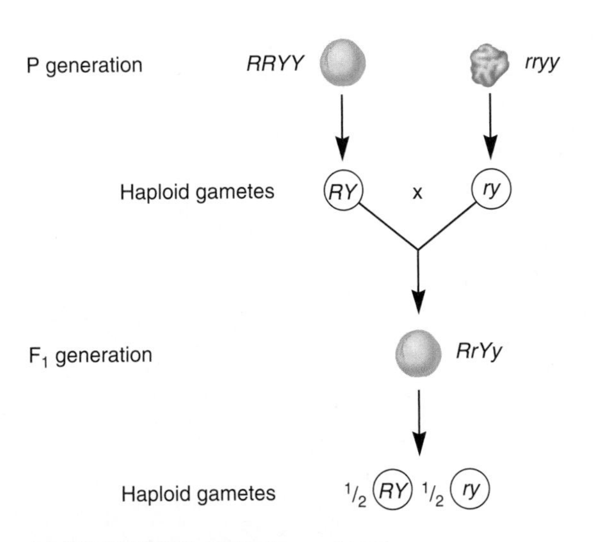

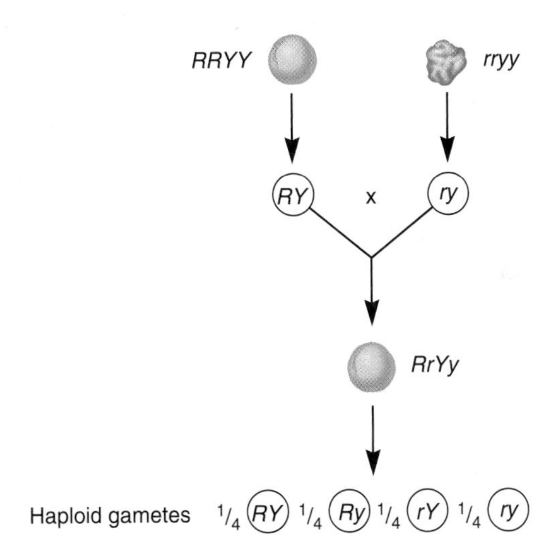

FIGURE 2.7 Two hypotheses to explain how two different genes assort during gamete formation. **(a)** According to the linked hypothesis, the two genes always stay associated with each other. **(b)** In contrast, the independent assortment hypothesis proposes that the two different genes randomly segregate into haploid cells.

TESTING THE HYPOTHESES — FIGURE 2.8 Mendel's analysis of diybrid crosses.

Starting material: In this experiment, Mendel began with two types of true-breeding pea plants that were different with regard to two characters. One plant had round, yellow seeds (*RRYY*); the other plant had wrinkled, green seeds (*rryy*).

1. Cross the two true-breeding plants to each other. This produces F_1 generation seeds.

2. Collect many seeds and record their phenotype.

3. F_1 seeds are planted and grown, and the F_1 plants are allowed to self-fertilize. This produces seeds that are part of the F_2 generation.

4. Analyze the characteristics found in the F_2 generation seeds.

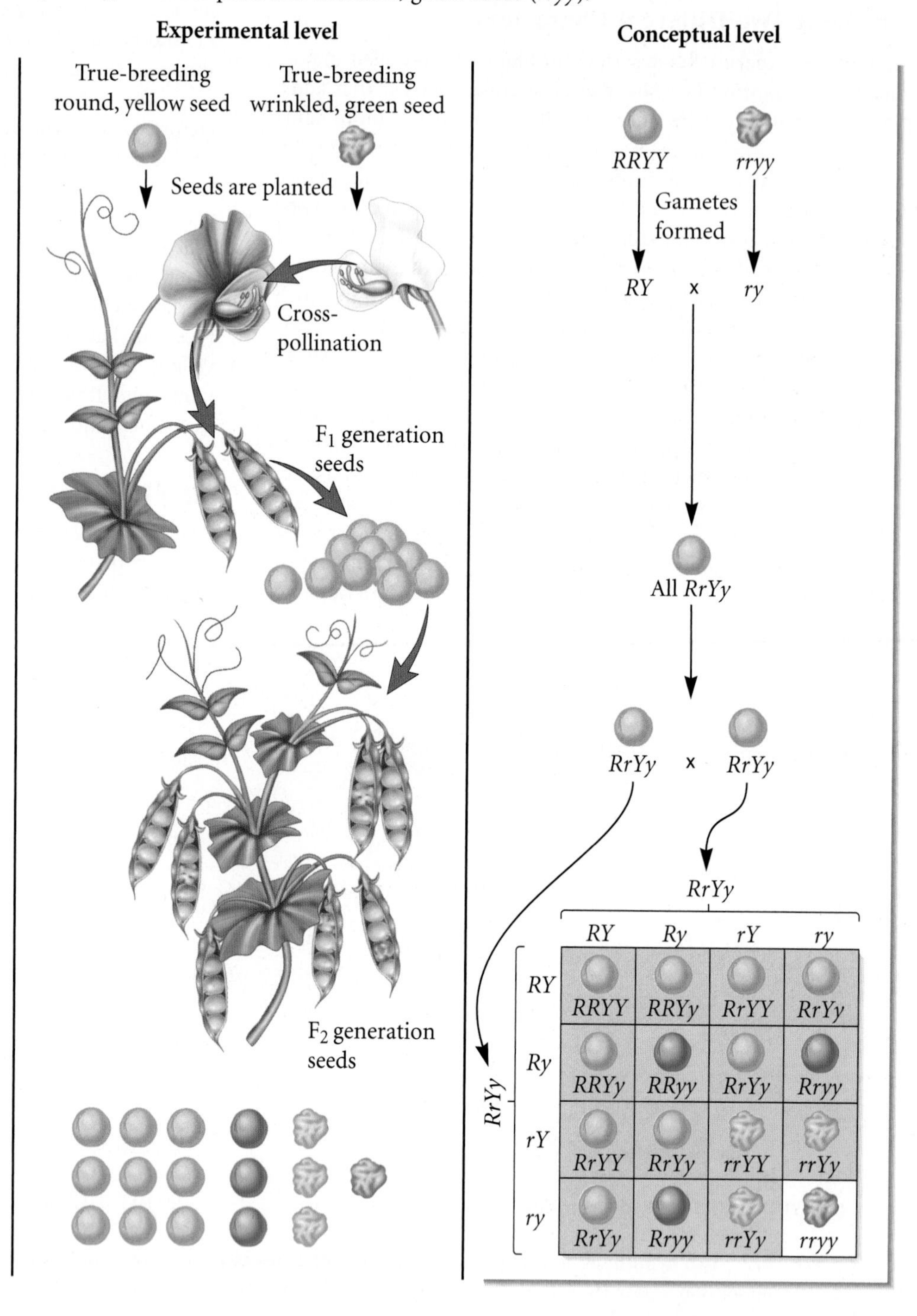

THE DATA

P cross	*F_1 generation*	*F_2 generation*
Round, yellow × wrinkled, green seeds	All round, yellow	315 round, yellow seeds 108 round, green seeds 101 wrinkled, yellow seeds 32 wrinkled, green seeds

INTERPRETING THE DATA

The F_2 generation had seeds that were round and green and seeds that were wrinkled and yellow. These two categories of F_2 seeds are called **nonparentals** because these combinations of traits were not found in the true-breeding plants of the parental generation. The occurrence of nonparental variants contradicts the linkage model. According to the linkage model, the *R* and *Y* alleles should be linked together and so should the *r* and *y* alleles.

If this were the case, the F_1 plants could produce gametes that are only *RY* or *ry*. These would combine to produce *RRYY* (round, yellow), *RrYy* (round, yellow), or *rryy* (wrinkled, green) in a 1:2:1 ratio. Nonparental seeds could not be produced. However, Mendel did not obtain this result. Instead, he observed a phenotypic ratio of 9:3:3:1 in the F_2 generation.

Mendel's results from many dihybrid experiments rejected the hypothesis of linked assortment and, instead, supported the hypothesis that different characters assort themselves independently. Using the modern notion of genes, **Mendel's law of independent assortment** states:

> *Two different genes will randomly assort their alleles during the formation of haploid cells.*

In other words, the allele for one gene will be found within a resulting gamete independently of whether the allele for a different gene is found in the same gamete. Using the example given in Figure 2.8, the round and wrinkled alleles will be assorted into haploid gametes independently of the yellow and green alleles. Therefore, a heterozygous *RrYy* parent can produce four different gametes—*RY*, *Ry*, *rY*, and *ry*—in equal proportions.

In an F_1 self-fertilization experiment, any two gametes can combine randomly during fertilization. This allows for 4^2, or 16, possible offspring, although some offspring will be genetically identical to each other. As shown in **Figure 2.9**, these 16 possible combinations result in seeds with the following phenotypes: 9 round, yellow; 3 round, green; 3 wrinkled, yellow; and 1 wrinkled, green. This 9:3:3:1 ratio is the expected outcome when a dihybrid is allowed to self-fertilize. Mendel was clever enough to realize that the data for his dihybrid experiments were close to a 9:3:3:1 ratio. In Figure 2.8, for example, his F_1 generation produced F_2 seeds with the following characteristics: 315 round, yellow seeds; 108 round, green seeds; 101 wrinkled, yellow seeds; and 32 wrinkled, green seeds. If we divide each of these numbers by 32 (the number of plants with wrinkled, green seeds), the phenotypic ratio of the F_2 generation is 9.8 : 3.2 : 3.4 : 1.0. Within experimental error, Mendel's data approximated the predicted 9:3:3:1 ratio for the F_2 generation.

The law of independent assortment held true for dihybrid crosses involving the traits that Mendel studied in pea plants. However, in other cases, the inheritance pattern of two different genes is consistent with the linkage model described earlier in Figure 2.7a. In Chapter 6, we will examine the inheritance of genes that are linked to each other because they are physically within the same chromosome. As we will see, linked genes do not assort independently.

An important consequence of the law of independent assortment is that a single individual can produce a vast array of genetically different gametes. As mentioned in Chapter 1, diploid species have pairs of homologous chromosomes, which may differ with respect to the alleles they carry. When an offspring receives a combination of alleles that differs from those in the parental generation, this phenomenon is termed **genetic recombination.** One mechanism that accounts for genetic recombination is independent assortment. A second mechanism, discussed in Chapter 6, is crossing over, which can reassort alleles that happen to be linked along the same chromosome.

The phenomenon of independent assortment is rooted in the random pattern by which the homologs assort themselves during the process of meiosis, a topic addressed in Chapter 3. If

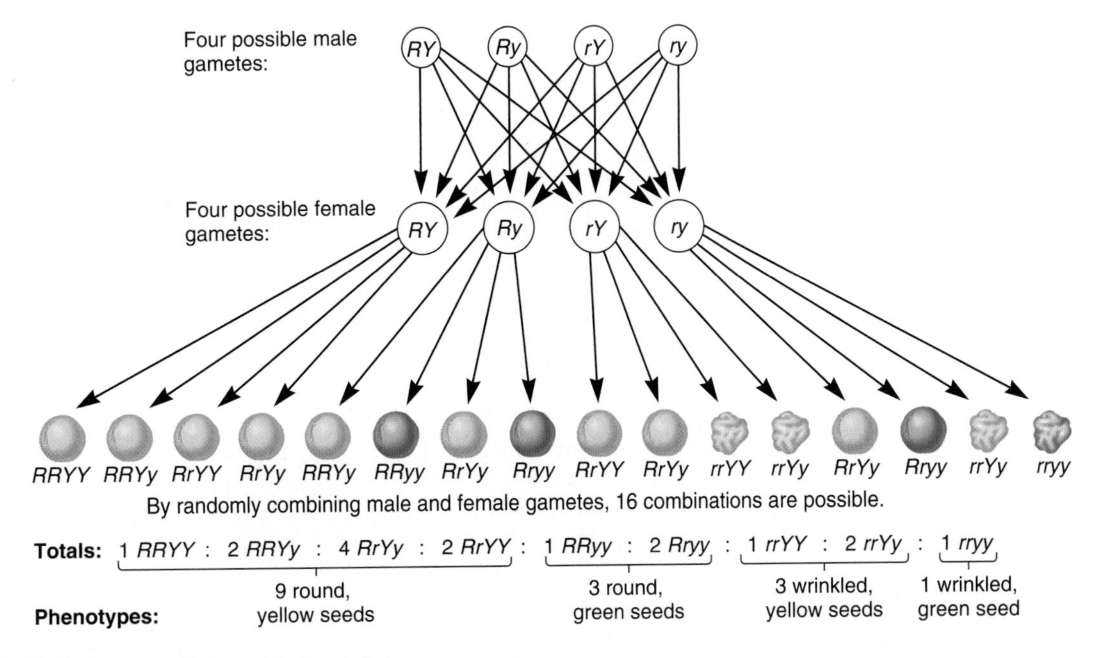

FIGURE 2.9 Mendel's law of independent assortment.

Genes→Traits The cross is between two parents that are heterozygous for seed shape and seed color (*RrYy* × *RrYy*). Four types of male gametes are possible: *RY, Ry, rY,* and *ry*. Likewise, four types of female gametes are possible: *RY, Ry, rY,* and *ry*. These four types of gametes are the result of the independent assortment of the seed shape and seed color alleles relative to each other. During fertilization, any one of the four types of male gametes can combine with any one of the four types of female gametes. This results in 16 types of offspring, each one containing two copies of the seed shape gene and two copies of the seed color gene.

a species contains a large number of homologous chromosomes, this creates the potential for an enormous amount of genetic diversity. For example, human cells contain 23 pairs of chromosomes. These pairs can randomly assort into gametes during meiosis. The number of different gametes an individual can make equals 2^n, where n is the number of pairs of chromosomes. Therefore, humans can make 2^{23}, or over 8 million, possible gametes, due to independent assortment. The capacity to make so many genetically different gametes enables a species to produce individuals with many different combinations of traits. This allows environmental factors to select for those combinations of traits that favor reproductive success.

A self-help quiz involving this experiment can be found at www.mhhe.com/brookergenetics4e.

A Punnett Square Can Also Be Used to Solve Independent Assortment Problems

As already depicted in Figure 2.8, we can make a Punnett square to predict the outcome of crosses involving two or more genes that assort independently. Let's see how such a Punnett square is made by considering a cross between two plants that are heterozygous for height and seed color (**Figure 2.10**). This cross is *TtYy* × *TtYy*. When we construct a Punnett square for this cross, we must keep in mind that each gamete has a single allele for each of two genes. In this example, the four possible gametes from each parent are

TY, *Ty*, *tY*, and *ty*

In this dihybrid experiment, we need to make a Punnett square containing 16 boxes. The phenotypes of the resulting offspring are predicted to occur in a ratio of 9:3:3:1.

In crosses involving three or more genes, the construction of a single large Punnett square to predict the outcome of crosses becomes very unwieldy. For example, in a trihybrid cross between two pea plants that are *Tt Rr Yy*, each parent can make 2^3, or 8, possible gametes. Therefore, the Punnett square must contain $8 \times 8 = 64$ boxes. As a more reasonable alternative, we can consider each gene separately and then algebraically combine them by multiplying together the expected outcomes for each gene. Two such methods, termed the **multiplication method** and the **forked-line method,** are shown in solved problem S3 at the end of this chapter.

Independent assortment is also revealed by a **dihybrid testcross.** In this type of experiment, dihybrid individuals are mated to individuals that are doubly homozygous recessive for the two characters. For example, individuals with a *TtYy* genotype could be crossed to *ttyy* plants. As shown below, independent assortment would predict a 1:1:1:1 ratio among the resulting offspring:

	TY	*Ty*	*tY*	*ty*
ty	*TtYy* Tall, yellow	*Ttyy* Tall, green	*ttYy* Dwarf, yellow	*ttyy* Dwarf, green

Cross: *TtYy* x *TtYy*

♀ \ ♂	*TY*	*Ty*	*tY*	*ty*
TY	*TTYY* Tall, yellow	*TTYy* Tall, yellow	*TtYY* Tall, yellow	*TtYy* Tall, yellow
Ty	*TTYy* Tall, yellow	*TTyy* Tall, green	*TtYy* Tall, yellow	*Ttyy* Tall, green
tY	*TtYY* Tall, yellow	*TtYy* Tall, yellow	*ttYY* Dwarf, yellow	*ttYy* Dwarf, yellow
ty	*TtYy* Tall, yellow	*Ttyy* Tall, green	*ttYy* Dwarf, yellow	*ttyy* Dwarf, green

Genotypes: 1 *TTYY* : 2 *TTYy* : 4 *TtYy* : 2 *TtYY* : 1 *TTyy* : 2 *Ttyy* : 1 *ttYY* : 2 *ttYy* : 1 *ttyy*

Phenotypes: 9 tall plants with yellow seeds; 3 tall plants with green seeds; 3 dwarf plants with yellow seeds; 1 dwarf plant with green seeds

FIGURE 2.10 A Punnett square for a dihybrid cross. The Punnett square shown here involves a cross between two pea plants that are heterozygous for height and seed color. The cross is *TtYy* × *TtYy*.

Modern Geneticists Are Often Interested in the Relationship Between the Molecular Expression of Genes and the Outcome of Traits

Mendel's work with pea plants was critically important because his laws of inheritance pertain to most eukaryotic organisms, such as fruit flies, corn, roundworms, mice, and humans, that transmit their genes through sexual reproduction. During the past several decades, many researchers have focused their attention on the relationship between the phenotypic appearance of traits and the molecular expression of genes. This theme will recur throughout the textbook (and we will draw attention to it by designating certain figure legends with a "Genes → Traits" label). As mentioned in Chapter 1, most genes encode proteins that function within living cells. The specific function of individual proteins affects the outcome of an individual's traits. A genetic approach can help us understand the relationship between a protein's function and its effect on phenotype. Most commonly, a geneticist will try to identify an individual that has a defective copy of a gene to see how that will affect the phenotype of the organism. These defective genes are called **loss-of-function alleles,** and they provide geneticists with a great amount of information. Unknowingly, Gregor Mendel had studied seven loss-of-function alleles among his strains of plants. The recessive characteristics in his pea plants were due to genes that had been rendered defective by a mutation. Such alleles are often inherited in a recessive manner, though this is not always the case.

How are loss-of-function alleles informative? In many cases, such alleles provide critical clues concerning the purpose of the protein's function within the organism. For example, we expect the gene affecting flower color (purple versus white) to encode a protein that is necessary for pigment production. This protein may function as an enzyme that is necessary for the synthesis of purple pigment. Furthermore, a reasonable guess is that the white allele is a loss-of-function allele that is unable to express this protein and therefore cannot make the purple pigment. To confirm this idea, a biochemist could analyze the petals from purple and white flowers and try to identify the protein that is defective or missing in the white petals but functionally active in the purple ones. The identification and characterization of this protein would provide a molecular explanation for this phenotypic characteristic.

Pedigree Analysis Can Be Used to Follow the Mendelian Inheritance of Traits in Humans

Before we end our discussion of simple Mendelian traits, let's address the question of how we can analyze inheritance patterns among humans. In his experiments, Mendel selectively made crosses and then analyzed a large number of offspring. When studying human traits, however, researchers cannot control parental crosses. Instead, they must rely on the information that is contained within family trees. This type of approach, known as a **pedigree analysis,** is aimed at determining the type of inheritance pattern that a gene will follow. Although this method may be less definitive than the results described in Mendel's experiments, a pedigree analysis can often provide important clues concerning the pattern of inheritance of traits within human families. An expanded discussion of human pedigrees is provided in Chapter 22, which concerns the inheritance patterns of many different human diseases.

In order to discuss the applications of pedigree analyses, we need to understand the organization and symbols of a pedigree (**Figure 2.11**). The oldest generation is at the top of the pedigree, and the most recent generation is at the bottom. Vertical

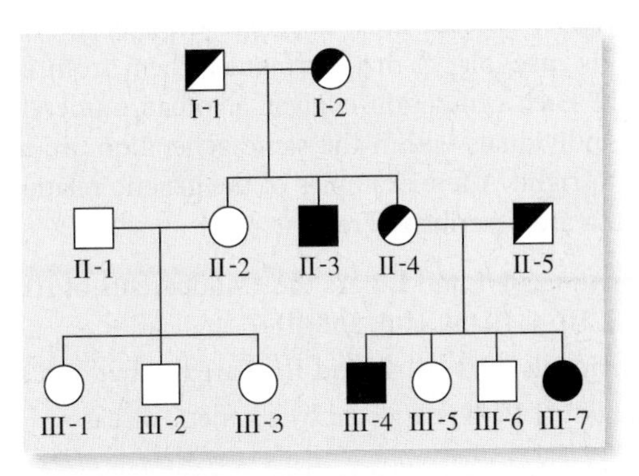

(a) Human pedigree showing cystic fibrosis

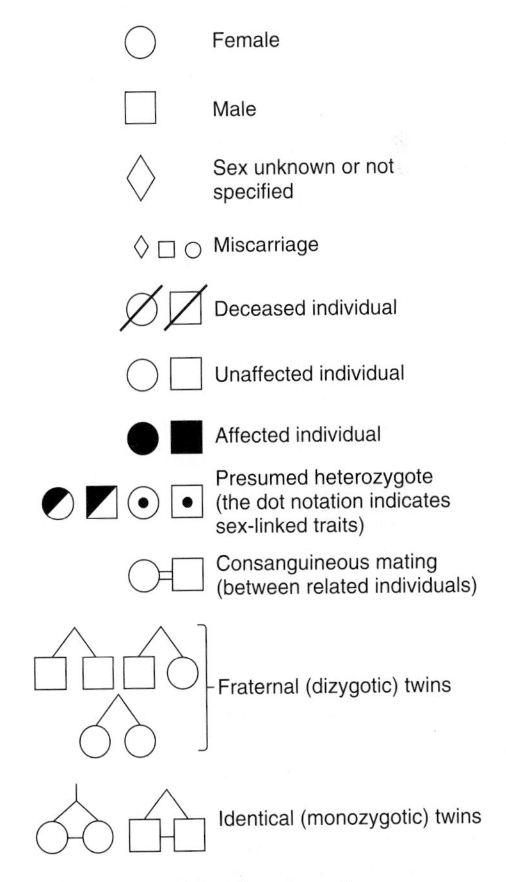

(b) Symbols used in a human pedigree

FIGURE 2.11 **Pedigree analysis.** (a) A family pedigree in which some of the members are affected with cystic fibrosis. Individuals I-1, I-2, II-4, and II-5 are depicted as presumed heterozygotes because they produce affected offspring. (b) The symbols used in a pedigree analysis. Note: In most pedigrees shown in this textbook, such as those found in the problem sets, the heterozygotes are not shown as half-filled symbols. Most pedigrees throughout the book show individuals' phenotypes—open symbols are unaffected individuals and filled (closed) symbols are affected individuals.

lines connect each succeeding generation. A man (square) and woman (circle) who produce one or more offspring are directly connected by a horizontal line. A vertical line connects parents with their offspring. If parents produce two or more offspring, the group of siblings (brothers and sisters) is denoted by two or more individuals projecting from the same horizontal line.

When a pedigree involves the transmission of a human trait or disease, affected individuals are depicted by filled symbols (in this case, black) that distinguish them from unaffected individuals. Each generation is given a roman numeral designation, and individuals within the same generation are numbered from left to right. A few examples of the genetic relationships in Figure 2.11a are described here:

Individuals I-1 and I-2 are the grandparents of III-1, III-2, III-3, III-4, III-5, III-6, and III-7

Individuals III-1, III-2, and III-3 are brother and sisters

Individual III-4 is affected by a genetic disease

The symbols shown in Figure 2.11 depict certain individuals, such as I-1, I-2, II-4, and II-5, as presumed heterozygotes because they are unaffected with a disease but produce homozygous offspring that are affected with a recessive genetic disease. However, in many pedigrees, such as those found in the problem sets at the end of the chapter, the inheritance pattern may not be known, so the symbols reflect only phenotypes. In most pedigrees, affected individuals are shown with closed symbols, and unaffected individuals, including those that might be heterozygous for a recessive disease, are depicted with open symbols.

Pedigree analysis is commonly used to determine the inheritance pattern of human genetic diseases. Human geneticists are routinely interested in knowing whether a genetic disease is inherited as a recessive or dominant trait. One way to discern the dominant/recessive relationship between two alleles is by a pedigree analysis. Genes that play a role in disease may exist as a normal allele or a mutant allele that causes disease symptoms. If the disease follows a simple Mendelian pattern of inheritance and is caused by a recessive allele, an individual must inherit two copies of the mutant allele to exhibit the disease. Therefore, a recessive pattern of inheritance makes two important predictions. First, two heterozygous normal individuals will, on average, have 1/4 of their offspring affected. Second, all offspring of two affected individuals will be affected. Alternatively, a dominant trait predicts that affected individuals will have inherited the gene from at least one affected parent (unless a new mutation has occurred during gamete formation).

The pedigree in Figure 2.11a concerns a human genetic disease known as cystic fibrosis (CF). Among Caucasians, approximately 3% of the population are heterozygous carriers of this recessive allele. In homozygotes, the disease symptoms include abnormalities of the pancreas, intestine, sweat glands, and lungs. These abnormalities are caused by an imbalance of ions across the plasma membrane. In the lungs, this leads to a buildup of thick, sticky mucus. Respiratory problems may lead to early death, although modern treatments have greatly increased the life span of CF patients. In the late 1980s, the gene for CF was identified. The *CF* gene encodes a protein called the cystic fibrosis transmembrane conductance regulator (CFTR). This protein regulates the ion balance across the cell membrane in tissues of the pancreas, intestine, sweat glands, and lungs. The mutant allele causing CF alters the encoded CFTR protein. The altered CFTR protein is not correctly inserted into the plasma membrane, resulting in a decreased function that causes the ionic imbalance. As seen in the pedigree, the pattern of affected and unaffected individuals is consistent with a recessive mode of inheritance. Two unaffected individuals can produce an affected offspring. Although not shown in this pedigree, a recessive mode of inheritance is also characterized by the observation that two affected individuals will produce 100% affected offspring. However, for human genetic diseases that limit survival or fertility (or both), there may never be cases where two affected individuals produce offspring.

2.2 PROBABILITY AND STATISTICS

A powerful application of Mendel's work is that the laws of inheritance can be used to predict the outcome of genetic crosses. In agriculture, for example, plant and animal breeders are concerned with the types of offspring their crosses will produce. This information is used to produce commercially important crops and livestock. In addition, people are often interested in predicting the characteristics of the children they may have. This may be particularly important to individuals who carry alleles that cause inherited diseases. Of course, we cannot see into the future and definitively predict what will happen. Nevertheless, genetic counselors can help couples to predict the likelihood of having an affected child. This probability is one factor that may influence a couple's decision whether to have children.

In this section, we will see how probability calculations are used in genetic problems to predict the outcome of crosses. To compute probability, we will use three mathematical operations known as the sum rule, the product rule, and the binomial expansion equation. These methods allow us to determine the probability that a cross between two individuals will produce a particular outcome. To apply these operations, we must have some knowledge regarding the genotypes of the parents and the pattern of inheritance of a given trait.

Probability calculations can also be used in hypothesis testing. In many situations, a researcher would like to discern the genotypes and patterns of inheritance for traits that are not yet understood. A traditional approach to this problem is to conduct crosses and then analyze their outcomes. The proportions of offspring may provide important clues that allow the experimenter to propose a hypothesis, based on the quantitative laws of inheritance, that explains the transmission of the trait from parent to offspring. Statistical methods, such as the chi square test, can then be used to evaluate how well the observed data from crosses fit the expected data. We will end this chapter with an example that applies the chi square test to a genetic cross.

Probability Is the Likelihood That an Event Will Occur

The chance that an event will occur in the future is called the event's **probability.** For example, if you flip a coin, the probability is 0.50, or 50%, that the head side will be showing when the coin lands. Probability depends on the number of possible outcomes. In this case, two possible outcomes (heads or tails) are equally likely. This allows us to predict a 50% chance that a coin flip will produce heads. The general formula for probability (P) is

$$\text{Probability} = \frac{\text{Number of times an event occurs}}{\text{Total number of events}}$$

$$P_{\text{heads}} = 1 \text{ heads} / (1 \text{ heads} + 1 \text{ tails}) = 1/2 = 50\%$$

In genetic problems, we are often interested in the probability that a particular type of offspring will be produced. Recall that when two heterozygous tall pea plants (*Tt*) are crossed, the phenotypic ratio of the offspring is 3 tall to 1 dwarf. This information can be used to calculate the probability for either type of offspring.

$$\text{Probability} = \frac{\text{Number of individuals with a given phenotype}}{\text{Total number of individuals}}$$

$$P_{\text{tall}} = 3 \text{ tall} / (3 \text{ tall} + 1 \text{ dwarf}) = 3/4 = 75\%$$

$$P_{\text{dwarf}} = 1 \text{ dwarf} / (3 \text{ tall} + 1 \text{ dwarf}) = 1/4 = 25\%$$

The probability is 75% of obtaining a tall plant and 25% of obtaining a dwarf plant. When we add together the probabilities of all possible outcomes (tall and dwarf), we should get a sum of 100% (here, 75% + 25% = 100%).

A probability calculation allows us to predict the likelihood that an event will occur in the future. The accuracy of this prediction, however, depends to a great extent on the size of the sample. For example, if we toss a coin six times, our probability prediction would suggest that 50% of the time we should get heads (i.e., three heads and three tails). In this small sample size, however, we would not be too surprised if we came up with four heads and two tails. Each time we toss a coin, there is a random chance that it will be heads or tails. The deviation between the observed and expected outcomes is called the **random sampling error.** In a small sample, the error between the predicted percentage of heads and the actual percentage observed may be quite large. By comparison, if we flipped a coin 1000 times, the percentage of heads would be fairly close to the predicted 50% value. In a larger sample, we expect the random sampling error to be a much smaller percentage.

The Sum Rule Can Be Used to Predict the Occurrence of Mutually Exclusive Events

Now that we have an understanding of probability, we can see how mathematical operations using probability values allow us to predict the outcome of genetic crosses. Our first genetic problem involves the use of the **sum rule,** which states that

> *The probability that one of two or more mutually exclusive events will occur is equal to the sum of the individual probabilities of the events.*

As an example, let's consider a cross between two mice that are both heterozygous for genes affecting the ears and tail. One gene can be found as an allele designated *de*, which is a recessive allele that causes droopy ears; the normal allele is *De*. An allele of a second gene causes a crinkly tail. This crinkly tail allele (*ct*) is recessive to the normal allele (*Ct*). If a cross is made between two heterozygous mice (*Dede Ctct*), the predicted ratio of offspring is 9 with normal ears and normal tails, 3 with normal ears and crinkly tails, 3 with droopy ears and normal tails, and 1 with droopy ears and a crinkly tail. These four phenotypes are mutually exclusive. For example, a mouse with droopy ears and a normal tail cannot have normal ears and a crinkly tail.

The sum rule allows us to determine the probability that we will obtain any one of two or more different types of offspring. For example, in a cross between two heterozygotes (*Dede Ctct* × *Dede Ctct*), we can ask the following question: What is the probability that an offspring will have normal ears and a normal tail or have droopy ears and a crinkly tail? In other words, if we closed our eyes and picked an offspring out of a litter from this cross, what are the chances that we would be holding a mouse that has normal ears and a normal tail or a mouse with droopy ears and a crinkly tail? In this case, the investigator wants to predict whether one of two mutually exclusive events will occur. A strategy for solving such genetic problems using the sum rule is described here.

The Cross: *Dede Ctct* × *Dede Ctct*

The Question: What is the probability that an offspring will have normal ears and a normal tail or have droopy ears and a crinkly tail?

Step 1. *Calculate the individual probabilities of each phenotype.* This can be accomplished using a Punnett square.

The probability of normal ears and a normal tail is 9/(9 + 3 + 3 + 1) = 9/16

The probability of droopy ears and a crinkly tail is 1/(9 + 3 + 3 + 1) = 1/16

Step 2. *Add together the individual probabilities.*

$$9/16 + 1/16 = 10/16$$

This means that 10/16 is the probability that an offspring will have either normal ears and a normal tail or droopy ears and a crinkly tail. We can convert 10/16 to 0.625, which means that 62.5% of the offspring are predicted to have normal ears and a normal tail or droopy ears and a crinkly tail.

The Product Rule Can Be Used to Predict the Probability of Independent Events

We can use probability to make predictions regarding the likelihood of two or more independent outcomes from a genetic cross. When we say that events are independent, we mean that the occurrence of one event does not affect the probability of another event. As an example, let's consider a rare, recessive human trait known as congenital analgesia. Persons with this trait can distinguish between sharp and dull, and hot and cold, but do not perceive extremes of sensation as being painful. The first case of congenital analgesia, described in 1932, was a man who made his living entertaining the public as a "human pincushion."

For a phenotypically unaffected couple, each being heterozygous for the recessive allele causing congenital analgesia, we can ask the question, What is the probability that the couple's first three offspring will have congenital analgesia? To answer this question, the **product rule** is used. According to this rule,

The probability that two or more independent events will occur is equal to the product of their individual probabilities.

A strategy for solving this type of problem is shown here.

The Cross: *Pp* × *Pp* (where *P* is the common allele and *p* is the recessive congenital analgesia allele)

The Question: What is the probability that the couple's first three offspring will have congenital analgesia?

Step 1. *Calculate the individual probability of this phenotype.* As described previously, this is accomplished using a Punnett square.

The probability of an affected offspring is 1/4 (25%).

Step 2. *Multiply the individual probabilities.* In this case, we are asking about the first three offspring, and so we multiply 1/4 three times.

$$1/4 \times 1/4 \times 1/4 = 1/64 = 0.016$$

Thus, the probability that the first three offspring will have this trait is 0.016. In other words, we predict that 1.6% of the time the first three offspring of a couple, each heterozygous for the recessive allele, will all have congenital analgesia. In this example, the phenotypes of the first, second, and third offspring are independent events. In this case, the phenotype of the first offspring does not have an effect on the phenotype of the second or third offspring.

In the problem described here, we have used the product rule to determine the probability that the first three offspring will all have the same phenotype (congenital analgesia). We can also apply the rule to predict the probability of a sequence of events that involves combinations of different offspring. For example, consider the question, What is the probability that the first offspring will be unaffected, the second offspring will have congenital analgesia, and the third offspring will be unaffected? Again, to solve this problem, begin by calculating the individual probability of each phenotype.

Unaffected = 3/4

Congenital analgesia = 1/4

The probability that these three phenotypes will occur in this specified order is

$$3/4 \times 1/4 \times 3/4 = 9/64 = 0.14, \text{ or } 14\%$$

In other words, this sequence of events is expected to occur only 14% of the time.

The product rule can also be used to predict the outcome of a cross involving two or more genes. Let's suppose an individual with the genotype *Aa Bb CC* was crossed to an individual with the genotype *Aa bb Cc*. We could ask the question, What is the probability that an offspring will have the genotype *AA bb Cc*? If the three genes independently assort, the probability of inheriting alleles for each gene is independent of the other two genes. Therefore, we can separately calculate the probability of the desired outcome for each gene.

Cross: *Aa Bb CC* × *Aa bb Cc*

Probability that an offspring will be *AA* = 1/4, or 0.25

Probability that an offspring will be *bb* = 1/2, or 0.5

Probability that an offspring will be *Cc* = 1/2, or 0.5

We can use the product rule to determine the probability that an offspring will be *AA bb Cc*:

$$P = (0.25)(0.5)(0.5) = 0.0625, \text{ or } 6.25\%$$

The Binomial Expansion Equation Can Be Used to Predict the Probability of an Unordered Combination of Events

A third predictive problem in genetics is to determine the probability that a certain proportion of offspring will be produced with particular characteristics; here they can be produced in an unspecified order. For example, we can consider a group of children produced by two heterozygous brown-eyed (*Bb*) individuals. We can ask the question, What is the probability that two out of five children will have blue eyes?

In this case, we are not concerned with the order in which the offspring are born. Instead, we are only concerned with the final numbers of blue-eyed and brown-eyed offspring. One possible outcome would be the following: firstborn child with blue eyes, second child with blue eyes, and then the next three with brown eyes. Another possible outcome could be firstborn child with brown eyes, second with blue eyes, third with brown eyes, fourth with blue eyes, and fifth with brown eyes. Both of these scenarios would result in two offspring with blue eyes and three with brown eyes. In fact, several other ways to have such a family could occur.

To solve this type of question, the **binomial expansion equation** can be used. This equation represents all of the possibilities for a given set of unordered events.

$$P = \frac{n!}{x!(n-x)!} p^x q^{n-x}$$

where

P = the probability that the unordered outcome will occur

n = total number of events

x = number of events in one category (e.g., blue eyes)

p = individual probability of x

q = individual probability of the other category (e.g., brown eyes)

Note: In this case, $p + q = 1$.

The symbol ! denotes a factorial. $n!$ is the product of all integers from n down to 1. For example, $4! = 4 \times 3 \times 2 \times 1 = 24$. An exception is 0!, which equals 1.

The use of the binomial expansion equation is described next.

The Cross: $Bb \times Bb$

The Question: What is the probability that two out of five offspring will have blue eyes?

Step 1. *Calculate the individual probabilities of the blue-eye and brown-eye phenotypes.* If we constructed a Punnett square, we would find the probability of blue eyes is 1/4 and the probability of brown eyes is 3/4:

$$p = 1/4$$

$$q = 3/4$$

Step 2. *Determine the number of events in category x (in this case, blue eyes) versus the total number of events.* In this example, the number of events in category x is two blue-eyed children among a total number of five.

$$x = 2$$

$$n = 5$$

Step 3. *Substitute the values for p, q, x, and n in the binomial expansion equation.*

$$P = \frac{n!}{x!(n-x)!}\, p^x q^{n-x}$$

$$P = \frac{5!}{2!(5-2)!}\,(1/4)^2(3/4)^{5-2}$$

$$P = \frac{5 \times 4 \times 3 \times 2 \times 1}{(2 \times 1)(3 \times 2 \times 1)}\,(1/16)(27/64)$$

$$P = 0.26 = 26\%$$

Thus, the probability is 0.26 that two out of five offspring will have blue eyes. In other words, 26% of the time we expect a $Bb \times Bb$ cross yielding five offspring to contain two blue-eyed children and three brown-eyed children.

In solved problem S7 at the end of this chapter, we consider an expanded version of this approach that uses a **multinomial expansion equation.** This equation is needed to solve unordered genetic problems that involve three or more phenotypic categories.

The Chi Square Test Can Be Used to Test the Validity of a Genetic Hypothesis

We now look at a different issue in genetic problems, namely **hypothesis testing.** Our goal here is to determine if the data from genetic crosses are consistent with a particular pattern of inheritance. For example, a geneticist may study the inheritance of body color and wing shape in fruit flies over the course of two generations. The following question may be asked about the F_2 generation: Do the observed numbers of offspring agree with the predicted numbers based on Mendel's laws of segregation and independent assortment? As we will see in Chapters 3 through 8, not all traits follow a simple Mendelian pattern of inheritance. Some genes do not segregate and independently assort themselves the same way that Mendel's seven characters did in pea plants.

To distinguish between inheritance patterns that obey Mendel's laws versus those that do not, a conventional strategy is to make crosses and then quantitatively analyze the offspring. Based on the observed outcome, an experimenter may make a tentative hypothesis. For example, it may seem that the data are obeying Mendel's laws. Hypothesis testing provides an objective, statistical method to evaluate whether the observed data really agree with the hypothesis. In other words, we use statistical methods to determine whether the data that have been gathered from crosses are consistent with predictions based on quantitative laws of inheritance.

The rationale behind a statistical approach is to evaluate the **goodness of fit** between the observed data and the data that are predicted from a hypothesis. This is sometimes called a **null hypothesis** because it assumes there is no real difference between the observed and expected values. Any actual differences that occur are presumed to be due to random sampling error. If the observed and predicted data are very similar, we can conclude that the hypothesis is consistent with the observed outcome. In this case, it is reasonable to accept the hypothesis. However, it should be emphasized that this does not prove a hypothesis is correct. Statistical methods can never prove a hypothesis is correct. They can provide insight as to whether or not the observed data seem reasonably consistent with the hypothesis. Alternative hypotheses, perhaps even ones that the experimenter has failed to realize, may also be consistent with the data. In some cases, statistical methods may reveal a poor fit between hypothesis and data. In other words, a high deviation would be found between the observed and expected values. If this occurs, the hypothesis is rejected. Hopefully, the experimenter can subsequently propose an alternative hypothesis that has a better fit with the data.

One commonly used statistical method to determine goodness of fit is the **chi square test** (often written χ^2). We can use the chi square test to analyze population data in which the members of the population fall into different categories. This is the kind of data we have when we evaluate the outcome of genetic crosses, because these usually produce a population of offspring that differ with regard to phenotypes. The general formula for the chi square test is

$$\chi^2 = \Sigma\frac{(O-E)^2}{E}$$

where

O = observed data in each category

E = expected data in each category based on the experimenter's hypothesis

Σ means to sum this calculation for each category. For example, if the population data fell into two categories, the chi square calculation would be

$$\chi^2 = \frac{(O_1 - E_1)^2}{E_1} + \frac{(O_2 - E_2)^2}{E_2}$$

We can use the chi square test to determine if a genetic hypothesis is consistent with the observed outcome of a genetic cross. The strategy described next provides a step-by-step outline

CHAPTER SUMMARY

- Early ideas regarding inheritance included pangenesis and blending inheritance. These ideas were later refuted by the work of Mendel.

2.1 Mendel's Laws of Inheritance

- Mendel chose pea plants as his experimental organism because it was easy to carry out self-fertilization or cross-fertilization experiments with these plants and because pea plants were available in several varieties in which a character existed in two distinct variants (see Figures 2.1, 2.2, 2.3, 2.4).
- By conducting monohybrid crosses, Mendel proposed three key ideas regarding inheritance. (1) Traits may be dominant or recessive. (2) Genes are passed unaltered from generation to generation. (3) The two alleles of a given gene segregate from each other during gamete formation (see Figures 2.5, 2.6).
- A Punnett square can be used to deduce the outcome of crosses.
- By conducting dihybrid crosses, Mendel proposed the law of independent assortment (see Figures 2.8, 2.9).
- A Punnett square can be used to predict the outcome of dihybrid crosses (see Figure 2.10).
- Human inheritance patterns are determined by analyzing family trees known as pedigrees (see Figure 2.11).

2.2 Probability and Statistics

- Probability is the number of times an event occurs divided by the total number of events.
- According to the sum rule, the probability that one of two or more mutually exclusive events will occur is equal to the sum of the individual probabilities of the events.
- According to the product rule, the probability of two or more independent events is equal to the product of their individual probabilities. This rule can be used to predict the outcome of crosses involving two or more genes.
- The binomial expansion is used to predict the probability of an unordered combination of events.
- The chi square test is used to test the validity of a hypothesis (see Table 2.1).

PROBLEM SETS & INSIGHTS

Solved Problems

S1. A heterozygous pea plant that is tall with yellow seeds, *TtYy*, is allowed to self-fertilize. What is the probability that an offspring will be either tall with yellow seeds, tall with green seeds, or dwarf with yellow seeds?

Answer: This problem involves three mutually exclusive events, and so we use the sum rule to solve it. First, we must calculate the individual probabilities for the three phenotypes. The outcome of the cross can be determined using a Punnett square.

Cross: *TtYy* x *TtYy*

♀ \ ♂	*TY*	*Ty*	*tY*	*ty*
TY	*TTYY* Tall, yellow	*TTYy* Tall, yellow	*TtYY* Tall, yellow	*TtYy* Tall, yellow
Ty	*TTYy* Tall, yellow	*TTyy* Tall, green	*TtYy* Tall, yellow	*Ttyy* Tall, green
tY	*TtYY* Tall, yellow	*TtYy* Tall, yellow	*ttYY* Dwarf, yellow	*ttYy* Dwarf, yellow
ty	*TtYy* Tall, yellow	*Ttyy* Tall, green	*ttYy* Dwarf, yellow	*ttyy* Dwarf, green

$$P_{\text{tall with yellow seeds}} = 9/(9 + 3 + 3 + 1) = 9/16$$

$$P_{\text{tall with green seeds}} = 3/(9 + 3 + 3 + 1) = 3/16$$

$$P_{\text{dwarf with yellow seeds}} = 3/(9 + 3 + 3 + 1) = 3/16$$

Sum rule: $9/16 + 3/16 + 3/16 = 15/16 = 0.94 = 94\%$

We expect to get one of these three phenotypes 15/16, or 94%, of the time.

S2. As described in this chapter, a human disease known as cystic fibrosis is inherited as a recessive trait. Two unaffected individuals have a first child with the disease. What is the probability that their next two children will not have the disease?

Answer: An unaffected couple has already produced an affected child. To be affected, the child must be homozygous for the disease allele and thus has inherited one copy from each parent. Therefore, because the parents are unaffected with the disease, we know that both of them must be heterozygous carriers for the recessive disease-causing allele. With this information, we can calculate the probability that they will produce an unaffected offspring. Using a Punnett square, this couple should produce a ratio of 3 unaffected : 1 affected offspring.

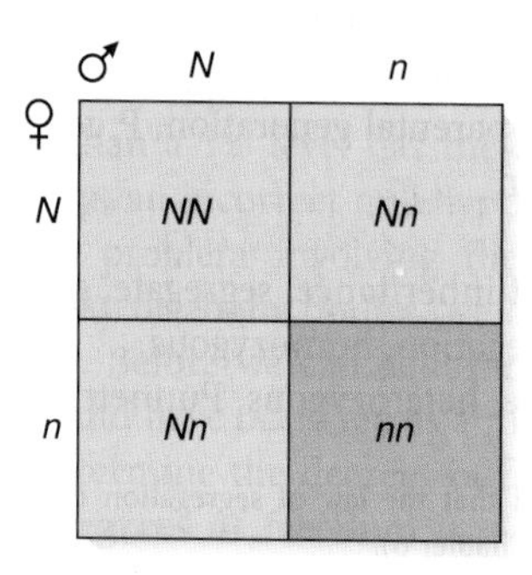

The probability of a single unaffected offspring is

$$P_{\text{unaffected}} = 3/(3 + 1) = 3/4$$

To obtain the probability of getting two unaffected offspring in a row (i.e., in a specified order), we must apply the product rule.

$$3/4 \times 3/4 = 9/16 = 0.56 = 56\%$$

The chance that their next two children will be unaffected is 56%.

S3. A cross was made between two heterozygous pea plants, *TtYy* × *TtYy*. The following Punnett square was constructed:

♀ \ ♂	*TT*	*Tt*	*Tt*	*tt*
YY	*TTYY*	*TtYY*	*TtYY*	*ttYY*
Yy	*TTYy*	*TtYy*	*TtYy*	*ttYy*
Yy	*TTYy*	*TtYy*	*TtYy*	*ttYy*
yy	*TTyy*	*Ttyy*	*Ttyy*	*ttyy*

Phenotypic ratio:

9 tall, yellow seeds : 3 tall, green seeds : 3 dwarf, yellow seeds : 1 dwarf, green seed

What is wrong with this Punnett square?

Answer: The outside of the Punnett square is supposed to contain the possible types of gametes. A gamete should contain one copy of each type of gene. Instead, the outside of this Punnett square contains two copies of one gene and zero copies of the other gene. The outcome happens to be correct (i.e., it yields a 9:3:3:1 ratio), but this is only a coincidence. The outside of the Punnett square must contain one copy of each type of gene. In this example, the correct possible types of gametes are *TY, Ty, tY,* and *ty* for each parent.

S4. A pea plant is heterozygous for three genes (*Tt Rr Yy*), where *T* = tall, *t* = dwarf, *R* = round seeds, *r* = wrinkled seeds, *Y* = yellow seeds, and *y* = green seeds. If this plant is self-fertilized, what are the predicted phenotypes of the offspring, and what fraction of the offspring will occur in each category?

Answer: You could solve this problem by constructing a large Punnett square and filling in the boxes. However, in this case, eight different male gametes and eight different female gametes are possible: *TRY, TRy, TrY, tRY, trY, Try, tRy,* and *try*. It would become rather tiresome to construct and fill in this Punnett square, which would contain 64 boxes. As an alternative, we can consider each gene separately and then algebraically combine them by multiplying together the expected phenotypic outcomes for each gene. In the cross *Tt Rr Yy* × *Tt Rr Yy*, the following Punnett squares can be made for each gene:

♀ \ ♂	*T*	*t*
T	*TT* Tall	*Tt* Tall
t	*Tt* Tall	*tt* Dwarf

3 tall : 1 dwarf

♀ \ ♂	*R*	*r*
R	*RR* Round	*Rr* Round
r	*Rr* Round	*rr* Wrinkled

3 round : 1 wrinkled

♀ \ ♂	*Y*	*y*
Y	*YY* Yellow	*Yy* Yellow
y	*Yy* Yellow	*yy* Green

3 yellow : 1 green

Instead of constructing a large, 64-box Punnett square, we can use two similar ways to determine the phenotypic outcome of this trihybrid cross. In the multiplication method**,** we can simply multiply these three combinations together:

(3 tall +1 dwarf)(3 round + 1 wrinkled)(3 yellow + 1 green)

This multiplication operation can be done in a stepwise manner. First, multiply (3 tall + 1 dwarf) by (3 round + 1 wrinkled).

(3 tall + 1 dwarf)(3 round + 1 wrinkled) = 9 tall, round + 3 tall, wrinkled + 3 dwarf, round, + 1 dwarf, wrinkled

Next, multiply this product by (3 yellow + 1 green).

(9 tall, round + 3 tall, wrinkled + 3 dwarf, round + 1 dwarf, wrinkled) (3 yellow + 1 green) = 27 tall, round, yellow + 9 tall, round, green + 9 tall, wrinkled, yellow + 3 tall, wrinkled, green + 9 dwarf, round, yellow + 3 dwarf, round, green + 3 dwarf, wrinkled, yellow + 1 dwarf, wrinkled, green

Even though the multiplication steps are also somewhat tedious, this approach is much easier than making a Punnett square with 64 boxes, filling them in, deducing each phenotype, and then adding them up!

A second approach that is analogous to the multiplication method is the forked-line method. In this case, the genetic proportions are determined by multiplying together the probabilities of each phenotype.

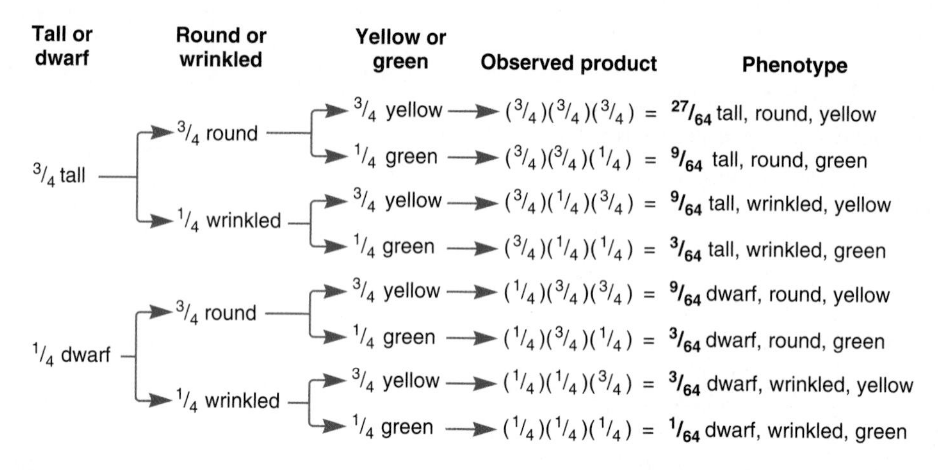

S5. For an individual expressing a dominant trait, how can you tell if it is a heterozygote or a homozygote?

Answer: One way is to conduct a testcross with an individual that expresses the recessive version of the same character. If the individual is heterozygous, half of the offspring will show the recessive trait, but if the individual is homozygous, none of the offspring will express the recessive trait.

$$Dd \times dd \;\rightarrow\; 1\,Dd \text{ (dominant trait)} : 1\,dd \text{ (recessive trait)} \quad \text{or} \quad DD \times dd \;\rightarrow\; \text{All } Dd \text{ (dominant trait)}$$

Another way to determine heterozygosity involves a more careful examination of the individual at the cellular or molecular level. At the cellular level, the heterozygote may not look exactly like the homozygote. This phenomenon is described in Chapter 4. Also, gene cloning methods described in Chapter 18 can be used to distinguish between heterozygotes and homozygotes.

S6. In dogs, black fur color is dominant to white. Two heterozygous black dogs are mated. What would be the probability of the following combinations of offspring?

A. A litter of six pups, four with black fur and two with white fur.

B. A litter of six pups, the firstborn with white fur, and among the remaining five pups, two with white fur and three with black fur.

C. A first litter of six pups, four with black fur and two with white fur, and then a second litter of seven pups, five with black fur and two with white fur.

D. A first litter of five pups, four with black fur and one with white fur, and then a second litter of seven pups in which the firstborn is homozygous, the second born is black, and the remaining five pups are three black and two white.

Answer:

A. Because this is an unordered combination of events, we use the binomial expansion equation, where $n = 6$, $x = 4$, $p = 0.75$ (probability of black), and $q = 0.25$ (probability of white).

The answer is 0.297, or 29.7%, of the time.

B. We use the product rule because the order is specified. The first pup is white and then the remaining five are born later. We also need to use the binomial expansion equation to determine the probability of the remaining five pups.

(probability of a white pup)(binomial expansion for the remaining five pups)

The probability of the white pup is 0.25. In the binomial expansion equation, $n = 5$, $x = 2$, $p = 0.25$, and $q = 0.75$.

The answer is 0.066, or 6.6%, of the time.

C. The order of the two litters is specified, so we need to use the product rule. We multiply the probability of the first litter times the probability of the second litter. We need to use the binomial expansion equation for each litter.

(binomial expansion of the first litter)(binomial expansion of the second litter)

For the first litter, $n = 6$, $x = 4$, $p = 0.75$, $q = 0.25$. For the second litter, $n = 7$, $x = 5$, $p = 0.75$, $q = 0.25$.

The answer is 0.092, or 9.2%, of the time.

D. The order of the litters is specified, so we need to use the product rule to multiply the probability of the first litter times the probability of the second litter. We use the binomial expansion equation to determine the probability of the first litter. The probability of the second litter is a little more complicated. The firstborn is homozygous. There are two mutually exclusive ways to be homozygous, *BB* and *bb*. We use the sum rule to determine the probability of the first pup, which equals $0.25 + 0.25 = 0.5$. The probability of the second pup is 0.75, and we use the binomial expansion equation to determine the probability of the remaining pups.

(binomial expansion of first litter)([0.5][0.75][binomial expansion of second litter])

For the first litter, $n = 5$, $x = 4$, $p = 0.75$, $q = 0.25$. For the last five pups in the second litter, $n = 5$, $x = 3$, $p = 0.75$, $q = 0.25$.

The answer is 0.039, or 3.9%, of the time.

S7. In this chapter, the binomial expansion equation was used in situations where only two phenotypic outcomes are possible. When more than two outcomes are possible, we use a multinomial expansion equation to solve a problem involving an unordered number of events. A general expression for this equation is

$$P = \frac{n!}{a!b!c! \ldots} p^a q^b r^c \ldots$$

where P = the probability that the unordered number of events will occur.

$$n = \text{total number of events}$$

$$a + b + c + \ldots = n$$

$$p + q + r + \ldots = 1$$

(p is the likelihood of a, q is the likelihood of b, r is the likelihood of c, and so on)

The multinomial expansion equation can be useful in many genetic problems where more than two combinations of offspring are possible. For example, this formula can be used to solve problems involving an unordered sequence of events in a dihybrid experiment. This approach is illustrated next.

A cross is made between two heterozygous tall plants with axial flowers (*TtAa*), where tall is dominant to dwarf and axial is dominant to terminal flowers. What is the probability that a group of five offspring will be composed of two tall plants with axial flowers, one tall plant with terminal flowers, one dwarf plant with axial flowers, and one dwarf plant with terminal flowers?

Answer:

Step 1. *Calculate the individual probabilities of each phenotype.* This can be accomplished using a Punnett square.

The phenotypic ratios are 9 tall with axial flowers, 3 tall with terminal flowers, 3 dwarf with axial flowers, and 1 dwarf with terminal flowers.

The probability of a tall plant with axial flowers is $9/(9 + 3 + 3 + 1) = 9/16$.

The probability of a tall plant with terminal flowers is $3/(9 + 3 + 3 + 1) = 3/16$.

The probability of a dwarf plant with axial flowers is $3/(9 + 3 + 3 + 1) = 3/16$.

The probability of a dwarf plant with terminal flowers is $1/(9 + 3 + 3 + 1) = 1/16$.

$$p = 9/16$$

$$q = 3/16$$

$$r = 3/16$$

$$s = 1/16$$

Step 2. *Determine the number of each type of event versus the total number of events.*

$$n = 5$$

$$a = 2$$

$$b = 1$$

$$c = 1$$

$$d = 1$$

Step 3. *Substitute the values in the multinomial expansion equation.*

$$P = \frac{n!}{a!b!c!d!} p^a q^b r^c s^d$$

$$P = \frac{5!}{2!1!1!1!}(9/16)^2(3/16)^1(3/16)^1(1/16)^1$$

$$P = 0.04 = 4\%$$

This means that 4% of the time we would expect to obtain five offspring with the phenotypes described in the question.

Conceptual Questions

C1. Why did Mendel's work refute the idea of blending inheritance?

C2. What is the difference between cross-fertilization and self-fertilization?

C3. Describe the difference between genotype and phenotype. Give three examples. Is it possible for two individuals to have the same phenotype but different genotypes?

C4. With regard to genotypes, what is a true-breeding organism?

C5. How can you determine whether an organism is heterozygous or homozygous for a dominant trait?

C6. In your own words, describe what Mendel's law of segregation means. Do not use the word "segregation" in your answer.

C7. Based on genes in pea plants that we have considered in this chapter, which statement(s) is not correct?

A. The gene causing tall plants is an allele of the gene causing dwarf plants.

B. The gene causing tall plants is an allele of the gene causing purple flowers.

C. The alleles causing tall plants and purple flowers are dominant.

C8. In a cross between a heterozygous tall pea plant and a dwarf plant, predict the ratios of the offspring's genotypes and phenotypes.

C9. Do you know the genotype of an individual with a recessive trait and/or a dominant trait? Explain your answer.

C10. A cross is made between a pea plant that has constricted pods (a recessive trait; smooth is dominant) and is heterozygous for seed color (yellow is dominant to green) and a plant that is heterozygous for both pod texture and seed color. Construct a Punnett square that depicts this cross. What are the predicted outcomes of genotypes and phenotypes of the offspring?

C11. A pea plant that is heterozygous with regard to seed color (yellow is dominant to green) is allowed to self-fertilize. What are the predicted outcomes of genotypes and phenotypes of the offspring?

C12. Describe the significance of nonparentals with regard to the law of independent assortment. In other words, explain how the appearance of nonparentals refutes a linkage hypothesis.

C13. For the following pedigrees, describe what you think is the most likely inheritance pattern (dominant versus recessive). Explain your reasoning. Filled (black) symbols indicate affected individuals.

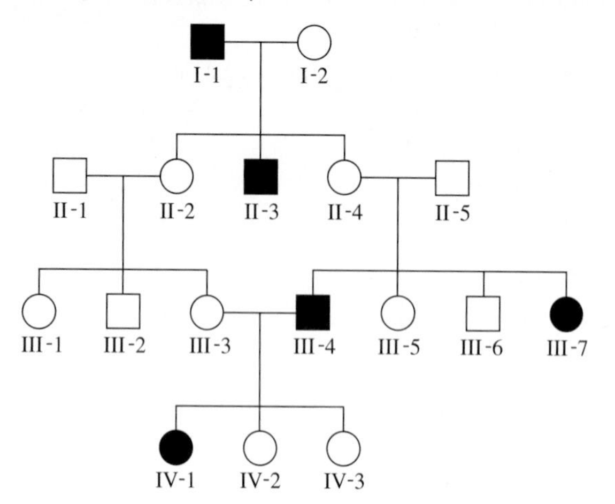

(a)

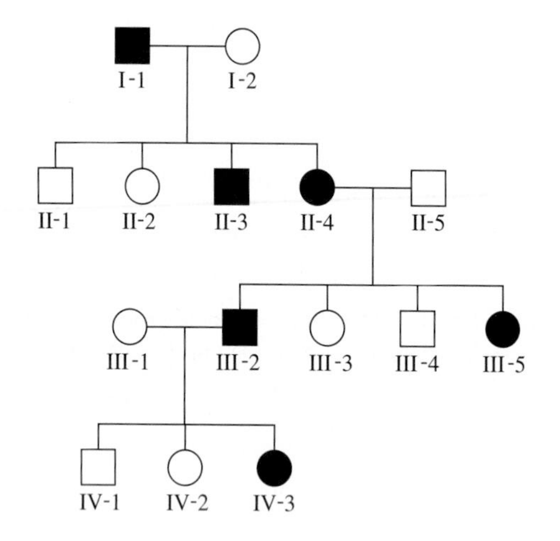

(b)

C14. Ectrodactyly, also known as "lobster claw syndrome," is a recessive disorder in humans. If a phenotypically unaffected couple produces an affected offspring, what are the following probabilities?

A. Both parents are heterozygotes.

B. An offspring is a heterozygote.

C. The next three offspring will be phenotypically unaffected.

D. Any two out of the next three offspring will be phenotypically unaffected.

C15. Identical twins are produced from the same sperm and egg (which splits after the first mitotic division), whereas fraternal twins are produced from separate sperm and separate egg cells. If two parents with brown eyes (a dominant trait) produce one twin boy with blue eyes, what are the following probabilities?

A. If the other twin is identical, he will have blue eyes.

B. If the other twin is fraternal, he or she will have blue eyes.

C. If the other twin is fraternal, he or she will transmit the blue eye allele to his or her offspring.

D. The parents are both heterozygotes.

C16. In cocker spaniels, solid coat color is dominant over spotted coat color. If two heterozygous dogs were crossed to each other, what would be the probability of the following combinations of offspring?

A. A litter of five pups, four with solid fur and one with spotted fur.

B. A first litter of six pups, four with solid fur and two with spotted fur, and then a second litter of five pups, all with solid fur.

C. A first litter of five pups, the firstborn with solid fur, and then among the next four, three with solid fur and one with spotted fur, and then a second litter of seven pups in which the firstborn is spotted, the second born is spotted, and the remaining five are composed of four solid and one spotted animal.

D. A litter of six pups, the firstborn with solid fur, the second born spotted, and among the remaining four pups, two with spotted fur and two with solid fur.

C17. A cross was made between a white male dog and two different black females. The first female gave birth to eight black pups, and the second female gave birth to four white and three black pups. What are the likely genotypes of the male parent and the two female parents? Explain whether you are uncertain about any of the genotypes.

C18. In humans, the allele for brown eye color (*B*) is dominant to blue eye color (*b*). If two heterozygous parents produce children, what are the following probabilities?

A. The first two children have blue eyes.

B. A total of four children, two with blue eyes and the other two with brown eyes.

C. The first child has blue eyes, and the next two have brown eyes.

C19. Albinism, a condition characterized by a partial or total lack of skin pigment, is a recessive human trait. If a phenotypically unaffected couple produced an albino child, what is the probability that their next child will be albino?

C20. A true-breeding tall plant was crossed to a dwarf plant. Tallness is a dominant trait. The F_1 individuals were allowed to self-fertilize. What are the following probabilities for the F_2 generation?

A. The first plant is dwarf.

B. The first plant is dwarf or tall.

C. The first three plants are tall.

D. For any seven plants, three are tall and four are dwarf.

E. The first plant is tall, and then among the next four, two are tall and the other two are dwarf.

C21. For pea plants with the following genotypes, list the possible gametes that the plant can make:

A. *TT Yy Rr*

B. *Tt YY rr*

C. *Tt Yy Rr*

D. *tt Yy rr*

C22. An individual has the genotype *Aa Bb Cc* and makes an abnormal gamete with the genotype *AaBc*. Does this gamete violate the law of independent assortment or the law of segregation (or both)? Explain your answer.

C23. In people with maple syrup urine disease, the body is unable to metabolize the amino acids leucine, isoleucine, and valine. One of the symptoms is that the urine smells like maple syrup. An unaffected couple produced six children in the following order: unaffected daughter, affected daughter, unaffected son, unaffected son, affected son, and unaffected son. The youngest unaffected son marries an unaffected woman and has three children in the following order: affected daughter, unaffected daughter, and unaffected son. Draw a pedigree that describes this family. What type of inheritance (dominant or recessive) would you propose to explain maple syrup urine disease?

C24. Marfan syndrome is a rare inherited human disorder characterized by unusually long limbs and digits plus defects in the heart (especially the aorta) and the eyes, among other symptoms. Following is a pedigree for this disorder. Affected individuals are shown with filled (black) symbols. What type of inheritance pattern do you think is the most likely?

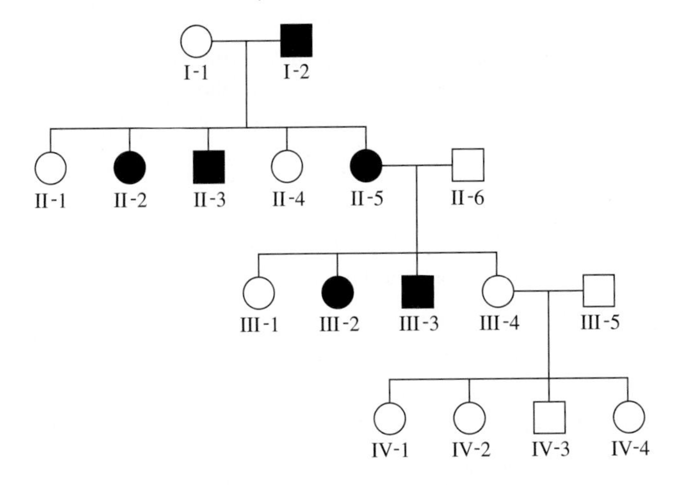

C25. A true-breeding pea plant with round and green seeds was crossed to a true-breeding plant with wrinkled and yellow seeds. Round and yellow seeds are the dominant traits. The F_1 plants were allowed to self-fertilize. What are the following probabilities for the F_2 generation?

A. An F_2 plant with wrinkled, yellow seeds.

B. Three out of three F_2 plants with round, yellow seeds.

C. Five F_2 plants in the following order: two have round, yellow seeds; one has round, green seeds; and two have wrinkled, green seeds.

D. An F_2 plant will not have round, yellow seeds.

C26. A true-breeding tall pea plant was crossed to a true-breeding dwarf plant. What is the probability that an F_1 individual will be true-breeding? What is the probability that an F_1 individual will be a true-breeding tall plant?

C27. What are the expected phenotypic ratios from the following cross: *Tt Rr yy Aa* × *Tt rr YY Aa*, where T = tall, t = dwarf, R = round, r = wrinkled, Y = yellow, y = green, A = axial, a = terminal; T, R, Y, and A are dominant alleles. Note: See solved problem S4 for help in answering this problem.

C28. When an abnormal organism contains three copies of a gene (instead of the normal number of two copies), the alleles for the gene usually segregate so that a gamete will contain one or two copies of the gene. Let's suppose that an abnormal pea plant has three copies of the height gene. Its genotype is *TTt*. The plant is also heterozygous for the seed color gene, *Yy*. How many types of gametes can this plant make, and in what proportions? (Assume that it is equally likely that a gamete will contain one or two copies of the height gene.)

C29. Honeybees are unusual in that male bees (drones) have only one copy of each gene, while female bees have two copies of their genes. That is because drones develop from eggs that have not been fertilized by sperm cells. In bees, the trait of long wings is dominant over short wings, and the trait of black eyes is dominant over white eyes. If a drone with short wings and black eyes was mated to a queen bee that is heterozygous for both genes, what are the predicted genotypes and phenotypes of male and female offspring? What are the phenotypic ratios if we assume an equal number of male and female offspring?

C30. A pea plant that is dwarf with green, wrinkled seeds was crossed to a true-breeding plant that is tall with yellow, round seeds. The F_1 generation was allowed to self-fertilize. What types of gametes, and in what proportions, would the F_1 generation make? What would be the ratios of genotypes and phenotypes of the F_2 generation?

C31. A true-breeding plant with round and green seeds was crossed to a true-breeding plant with wrinkled and yellow seeds. The F_1 plants were allowed to self-fertilize. What is the probability of obtaining the following plants in the F_2 generation: two that have round, yellow seeds; one with round, green seeds; and two with wrinkled, green seeds? (Note: See solved problem S7 for help.)

C32. Wooly hair is a rare dominant trait found in people of Scandinavian descent in which the hair resembles the wool of a sheep. A male with wooly hair, who has a mother with straight hair, moves to an island that is inhabited by people who are not of Scandinavian descent. Assuming that no other Scandinavians immigrate to the island, what is the probability that a great-grandchild of this male will have wooly hair? (Hint: You may want to draw a pedigree to help you figure this out.) If this wooly-haired male has eight great-grandchildren, what is the probability that one out of eight will have wooly hair?

C33. Huntington disease is a rare dominant trait that causes neurodegeneration later in life. A man in his thirties, who already has three children, discovers that his mother has Huntington disease though his father is unaffected. What are the following probabilities?

A. That the man in his thirties will develop Huntington disease.

B. That his first child will develop Huntington disease.

C. That one out of three of his children will develop Huntington disease.

C34. A woman with achondroplasia (a dominant form of dwarfism) and a phenotypically unaffected man have seven children, all of whom have achondroplasia. What is the probability of producing such a family if this woman is a heterozygote? What is the probability that the woman is a heterozygote if her eighth child does not have this disorder?

Experimental Questions

E1. Describe three advantages of using pea plants as an experimental organism.

E2. Explain the technical differences between a cross-fertilization experiment versus a self-fertilization experiment.

E3. How long did it take Mendel to complete the experiment in Figure 2.5?

E4. For all seven characters described in the data of Figure 2.5, Mendel allowed the F_2 plants to self-fertilize. He found that when F_2 plants with recessive traits were crossed to each other, they always bred true. However, when F_2 plants with dominant traits were crossed, some bred true but others did not. A summary of Mendel's results is shown here.

The Ratio of True-Breeding and Non-True-Breeding Parents of the F_2 Generation

F_2 Parents	True-Breeding	Non-True-Breeding	Ratio
Round	193	372	1:1.93
Yellow	166	353	1:2.13
Gray	36	64	1:1.78
Smooth	29	71	1:2.45
Green	40	60	1:1.5
Axial	33	67	1:2.08
Tall	28	72	1:2.57
TOTAL:	525	1059	1:2.02

When considering the data in this table, keep in mind that it describes the characteristics of the F_2 generation parents that had displayed a dominant phenotype. These data were deduced by analyzing the outcome of the F_3 generation. Based on Mendel's laws, explain the 1:2 ratio obtained in these data.

E5. From the point of view of crosses and data collection, what are the experimental differences between a monohybrid and a dihybrid experiment?

E6. As in many animals, albino coat color is a recessive trait in guinea pigs. Researchers removed the ovaries from an albino female guinea pig and then transplanted ovaries from a true-breeding black guinea pig. They then mated this albino female (with the transplanted ovaries) to an albino male. The albino female produced three offspring. What were their coat colors? Explain the results.

E7. The fungus *Melampsora lini* causes a disease known as flax rust. Different strains of *M. lini* cause varying degrees of the rust disease. Conversely, different strains of flax are resistant or sensitive to the various varieties of rust. The Bombay variety of flax is resistant to *M. lini*-strain 22 but sensitive to *M. lini*-strain 24. A strain of flax called 770B is just the opposite; it is resistant to strain 24 but sensitive to strain 22. When 770B was crossed to Bombay, all the F_1 individuals were resistant to both strain 22 and strain 24. When F_1 individuals were self-fertilized, the following data were obtained:

43 resistant to strain 22 but sensitive to strain 24

9 sensitive to strain 22 and strain 24

32 sensitive to strain 22 but resistant to strain 24

110 resistant to strain 22 and strain 24

Explain the inheritance pattern for flax resistance and sensitivity to *M. lini* strains.

E8. For Mendel's data shown in Figure 2.8, conduct a chi square analysis to determine if the data agree with Mendel's law of independent assortment.

E9. Would it be possible to deduce the law of independent assortment from a monohybrid experiment? Explain your answer.

E10. In fruit flies, curved wings are recessive to straight wings, and ebony body is recessive to gray body. A cross was made between true-breeding flies with curved wings and gray bodies to flies with straight wings and ebony bodies. The F_1 offspring were then mated to flies with curved wings and ebony bodies to produce an F_2 generation.

A. Diagram the genotypes of this cross, starting with the parental generation and ending with the F_2 generation.

B. What are the predicted phenotypic ratios of the F_2 generation?

C. Let's suppose the following data were obtained for the F_2 generation:

114 curved wings, ebony body

105 curved wings, gray body

111 straight wings, gray body

114 straight wings, ebony body

Conduct a chi square analysis to determine if the experimental data are consistent with the expected outcome based on Mendel's laws.

E11. A recessive allele in mice results in an abnormally long neck. Sometimes, during early embryonic development, the abnormal neck causes the embryo to die. An experimenter began with a population of true-breeding normal mice and true-breeding mice with long necks. Crosses were made between these two populations to produce an F_1 generation of mice with normal necks. The F_1 mice were then mated to each other to obtain an F_2 generation. For the mice that were born alive, the following data were obtained:

522 mice with normal necks

62 mice with long necks

What percentage of homozygous mice (that would have had long necks if they had survived) died during embryonic development?

E12. The data in Figure 2.5 show the results of the F_2 generation for seven of Mendel's crosses. Conduct a chi square analysis to determine if these data are consistent with the law of segregation.

E13. Let's suppose you conducted an experiment involving genetic crosses and calculated a chi square value of 1.005. There were four categories of offspring (i.e., the degrees of freedom equaled 3). Explain what the 1.005 value means. Your answer should include the phrase "80% of the time."

E14. A tall pea plant with axial flowers was crossed to a dwarf plant with terminal flowers. Tall plants and axial flowers are dominant traits. The following offspring were obtained: 27 tall, axial flowers; 23 tall, terminal flowers; 28 dwarf, axial flowers; and 25 dwarf, terminal flowers. What are the genotypes of the parents?

E15. A cross was made between two strains of plants that are agriculturally important. One strain was disease-resistant but herbicide-sensitive; the other strain was disease-sensitive but herbicide-resistant. A plant breeder crossed the two plants and then allowed the F_1 generation to self-fertilize. The following data were obtained:

F_1 generation: All offspring are disease-sensitive and herbicide-resistant

F_2 generation:

157	disease-sensitive, herbicide-resistant
57	disease-sensitive, herbicide-sensitive
54	disease-resistant, herbicide-resistant
20	disease-resistant, herbicide-sensitive
Total: 288	

Formulate a hypothesis that you think is consistent with the observed data. Test the goodness of fit between the data and your hypothesis using a chi square test. Explain what the chi square results mean.

E16. A cross was made between a plant that has blue flowers and purple seeds to a plant with white flowers and green seeds. The following data were obtained:

F_1 generation: All offspring have blue flowers with purple seeds

F_2 generation:

103	blue flowers, purple seeds
49	blue flowers, green seeds
44	white flowers, purple seeds
104	white flowers, green seeds
Total: 300	

Start with the hypothesis that blue flowers and purple seeds are dominant traits and that the two genes assort independently. Calculate a chi square value. What does this value mean with regard to your hypothesis? If you decide to reject your hypothesis, which aspect of the hypothesis do you think is incorrect (i.e., blue flowers and purple seeds are dominant traits, or the idea that the two genes assort independently)?

Questions for Student Discussion/Collaboration

1. Consider a cross in pea plants: *Tt Rr yy Aa* × *Tt rr Yy Aa*, where *T* = tall, *t* = dwarf, *R* = round, *r* = wrinkled, *Y* = yellow, *y* = green, *A* = axial, *a* = terminal. What is the expected phenotypic outcome of this cross? Have one group of students solve this problem by making one big Punnett square, and have another group solve it by making four single-gene Punnett squares and using the product rule. Time each other to see who gets done first.
2. A cross was made between two pea plants, *TtAa* and *Ttaa*, where *T* = tall, *t* = dwarf, *A* = axial, and *a* = terminal. What is the probability that the first three offspring will be tall with axial flowers or dwarf with terminal flowers and the fourth offspring will be tall with axial flowers. Discuss what operation(s) (e.g., sum rule, product rule, or binomial expansion equation) you used to solve them and in what order they were used.
3. Consider the tetrahybrid cross: *Tt Rr yy Aa* × *Tt RR Yy aa*, where *T* = tall, *t* = dwarf, *R* = round, *r* = wrinkled, *Y* = yellow, *y* = green, *A* = axial, *a* = terminal. What is the probability that the first three plants will have round seeds? What is the easiest way to solve this problem?

Note: All answers appear at the website for this textbook; the answers to even-numbered questions are in the back of the textbook.

www.mhhe.com/brookergenetics4e

Visit the website for practice tests, answer keys, and other learning aids for this chapter. Enhance your understanding of genetics with our interactive exercises, quizzes, animations, and much more.

CHAPTER OUTLINE

Chromosome sorting during cell division. *When eukaryotic cells divide, they replicate and sort their chromosomes (shown in blue), so that each cell receives the correct amount.*

3 REPRODUCTION AND CHROMOSOME TRANSMISSION

In Chapter 2, we considered some patterns of inheritance that explain the passage of traits from parent to offspring. In this chapter, we will survey reproduction at the cellular level and pay close attention to the inheritance of chromosomes. An examination of chromosomes at the microscopic level provides us with insights into understanding the inheritance patterns of traits. To appreciate this relationship, we will first consider how cells distribute their chromosomes during the process of cell division. We will see that in bacteria and most unicellular eukaryotes, simple cell division provides a way to reproduce asexually. Then we will explore a form of cell division called meiosis, which produces cells with half the number of chromosomes. By closely examining this process, we will see how the transmission of chromosomes accounts for the inheritance patterns that were observed by Mendel.

3.1 GENERAL FEATURES OF CHROMOSOMES

The **chromosomes** are structures within living cells that contain the genetic material. Genes are physically located within chromosomes. Biochemically, each chromosome contains a very long segment of DNA, which is the genetic material, and proteins, which are bound to the DNA and provide it with an organized structure. In eukaryotic cells, this complex between DNA and proteins is called **chromatin.** In this chapter, we will focus on the cellular mechanics of chromosome transmission to better understand the patterns of gene transmission that we considered in Chapter 2. In particular, we will examine how chromosomes are copied and sorted into newly made cells. In later chapters, particularly Chapters 8, 10, and 11, we will examine the molecular features of chromosomes in greater detail.

Before we begin a description of chromosome transmission, we need to consider the distinctive cellular differences between bacterial and eukaryotic species. Bacteria and archaea are referred to as **prokaryotes,** from the Greek meaning prenucleus, because their chromosomes are not contained within a membrane-bound nucleus of the cell. Prokaryotes usually have a single type of circular chromosome in a region of the cytoplasm called the **nucleoid** (**Figure 3.1a**). The cytoplasm is enclosed by a plasma membrane that regulates the uptake of nutrients and the excretion of waste products. Outside the plasma membrane is a rigid cell wall that protects the cell from breakage. Certain species of bacteria also have an outer membrane located beyond the cell wall.

Eukaryotes, from the Greek meaning true nucleus, include some simple species, such as single-celled protists and some fungi (such as yeast), and more complex multicellular species, such as plants, animals, and other fungi. The cells of eukaryotic species have internal membranes that enclose highly specialized compartments (**Figure 3.1b**). These compartments form

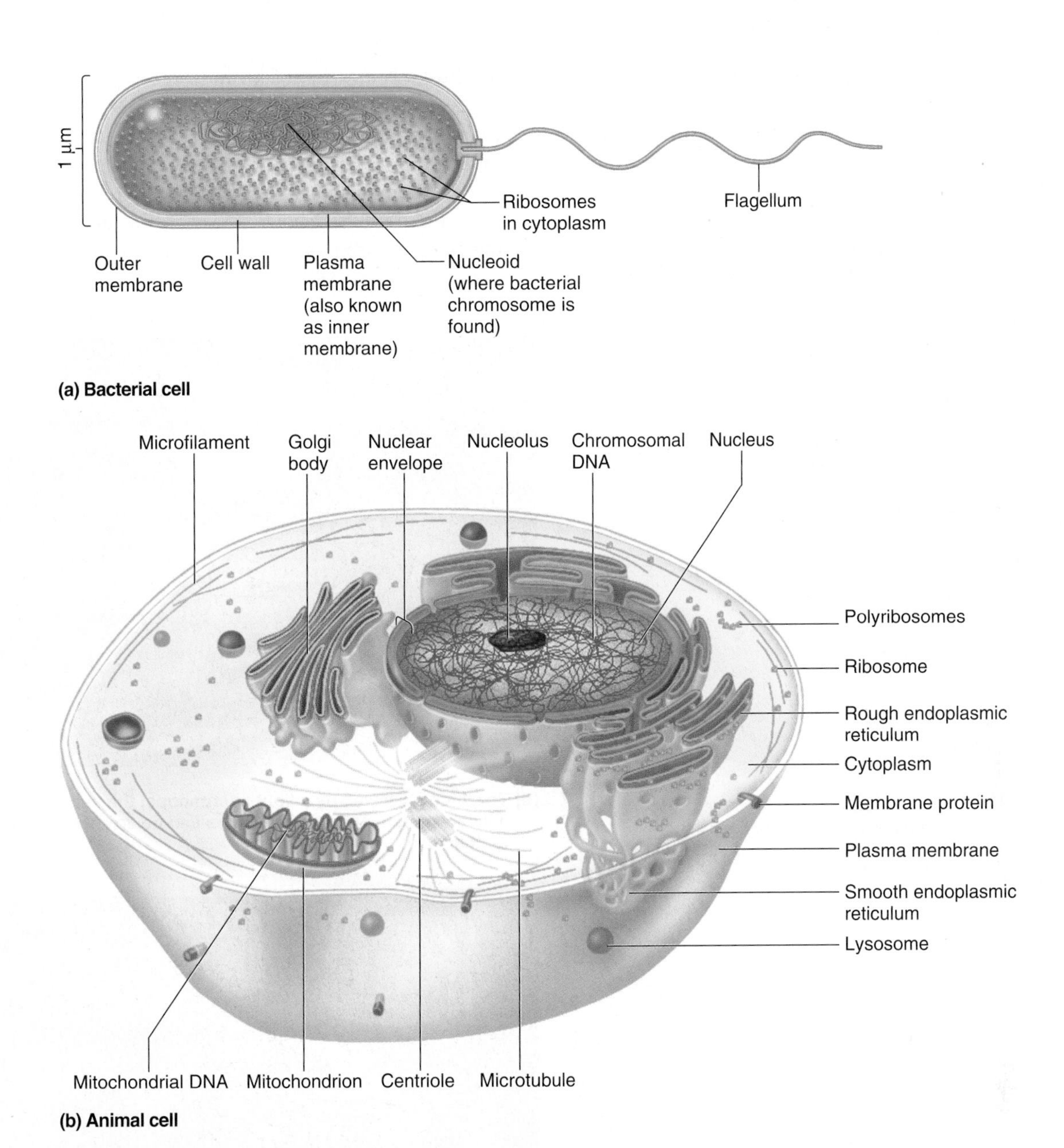

FIGURE 3.1 **The basic organization of cells.** **(a)** A bacterial cell. The example shown here is typical of a bacterium such as *Escherichia coli*, which has an outer membrane. **(b)** A eukaryotic cell. The example shown here is a typical animal cell.

membrane-bound **organelles** with specific functions. For example, the lysosomes play a role in the degradation of macromolecules. The endoplasmic reticulum and Golgi body play a role in protein modification and trafficking. A particularly conspicuous organelle is the **nucleus,** which is bounded by two membranes that constitute the nuclear envelope. Most of the genetic material is found within chromosomes that are located in the nucleus. In addition to the nucleus, certain organelles in eukaryotic cells contain a small amount of their own DNA. These include the mitochondrion, which functions in ATP synthesis, and, in plant cells, the chloroplast, which functions in photosynthesis. The DNA found in these organelles is referred to as extranuclear or extrachromosomal DNA to distinguish it from the DNA that is found in the cell nucleus. We will examine the role of mitochondrial and chloroplast DNA in Chapter 5.

In this section, we will focus on the composition of chromosomes found in the nucleus of eukaryotic cells. As you will learn, eukaryotic species contain genetic material that comes in sets of linear chromosomes.

Eukaryotic Chromosomes Are Examined Cytologically to Yield a Karyotype

Insights into inheritance patterns have been gained by observing chromosomes under the microscope. **Cytogenetics** is the field of genetics that involves the microscopic examination of chromosomes. The most basic observation that a **cytogeneticist** can make is to examine the chromosomal composition of a particular cell. For eukaryotic species, this is usually accomplished by observing the chromosomes as they are found in actively dividing cells. When a cell is preparing to divide, the chromosomes become more tightly coiled, which shortens them and thereby increases their diameter. The consequence of this shortening is that distinctive shapes and numbers of chromosomes become visible with a light microscope. Each species has a particular chromosome composition. For example, most human cells contain 23 pairs of chromosomes, for a total of 46. On rare occasions, some individuals may inherit an abnormal number of chromosomes or a chromosome with an abnormal structure. Such abnormalities can often be detected by a microscopic examination of the chromosomes within actively dividing cells. In addition, a cytogeneticist may examine chromosomes as a way to distinguish between two closely related species.

Figure 3.2a shows the general procedure for preparing human chromosomes to be viewed by microscopy. In this example, the cells were obtained from a sample of human blood;

A sample of blood is collected and treated with drugs that stimulate the cells to divide. Colchicine is added because it disrupts spindle formation and stops cells in mitosis where the chromosomes are highly compacted. The cells are then subjected to centrifugation.

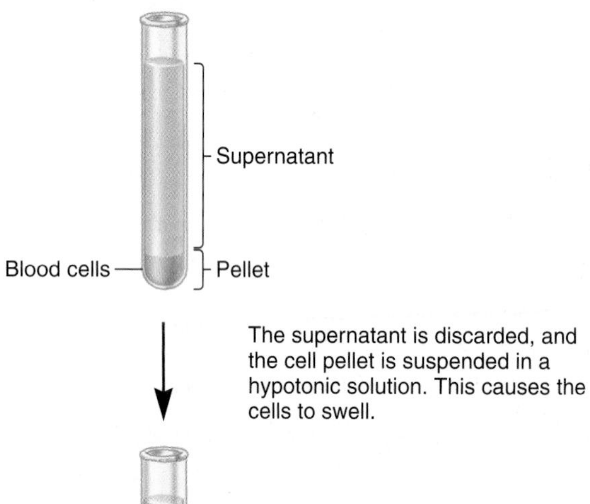

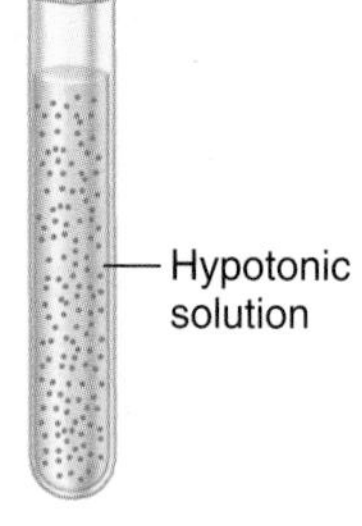

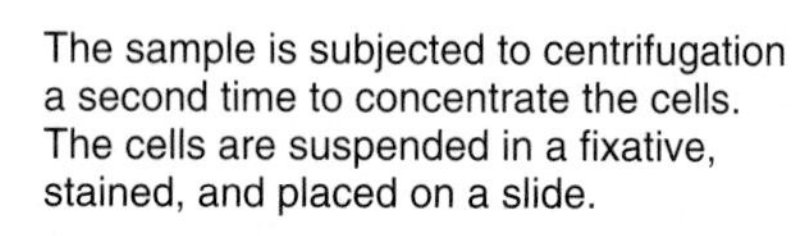

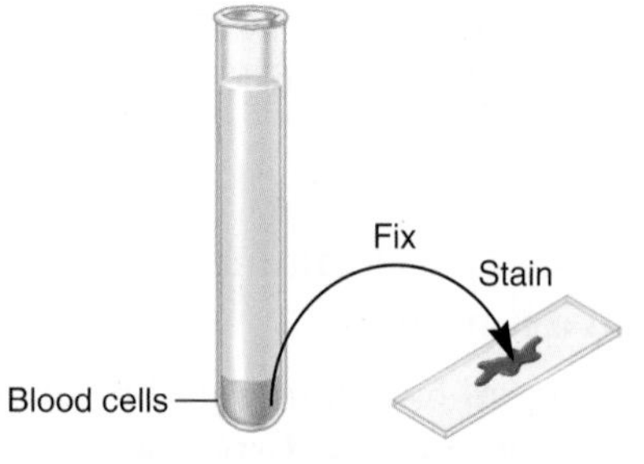

(a) Preparing cells for a karyotype

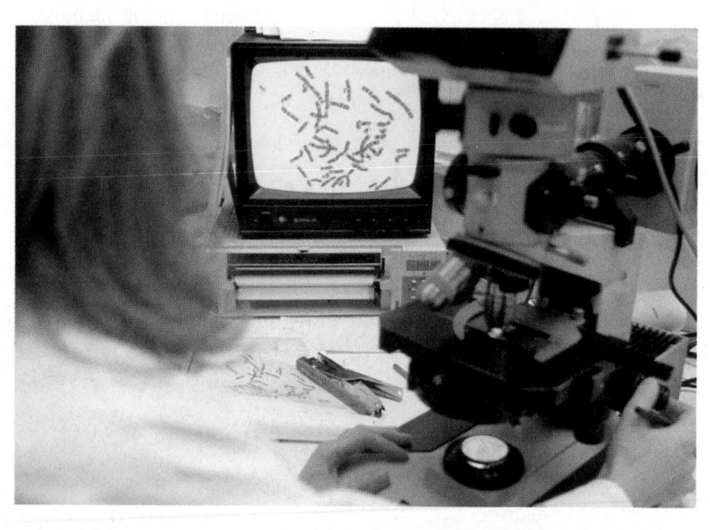

(b) The slide is viewed by a light microscope; the sample is seen on a video screen. The chromosomes can be arranged electronically on the screen.

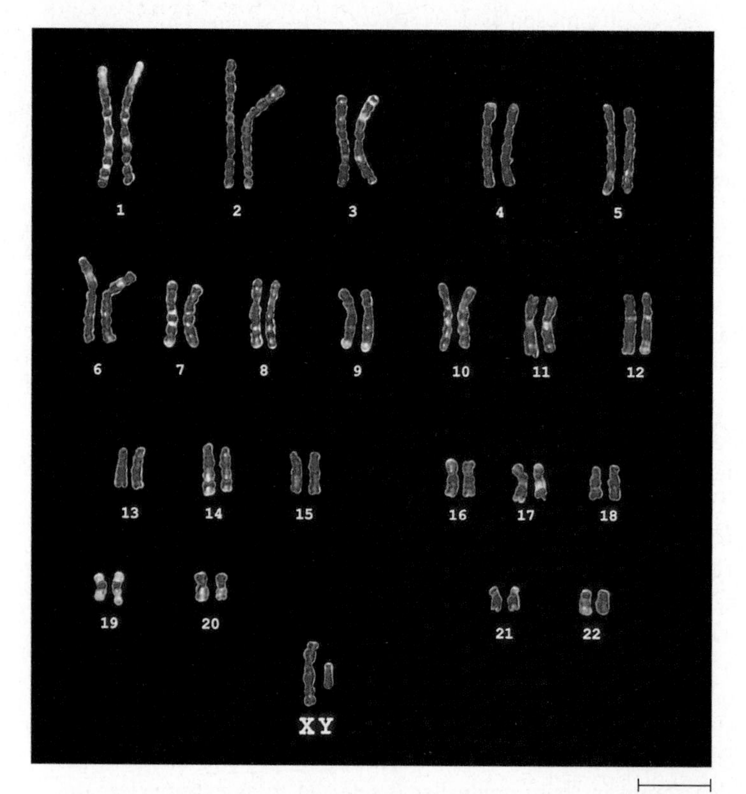

(c) For a diploid human cell, two complete sets of chromosomes from a single cell constitute a karyotype of that cell.

FIGURE 3.2 The procedure for making a human karyotype.

more specifically, the chromosomes within lymphocytes (a type of white blood cell) were examined. Blood cells are a type of **somatic cell.** This term refers to any cell of the body that is not a gamete or a precursor to a gamete. The **gametes** (sperm and egg cells or their precursors) are also called **germ cells.**

After the blood cells have been removed from the body, they are treated with drugs that stimulate them to begin cell division and cause cell division to be halted during mitosis, which is described later in this chapter. As shown in Figure 3.2a, these actively dividing cells are subjected to centrifugation to concentrate them. The concentrated preparation is then mixed with a hypotonic solution that makes the cells swell. This swelling causes the chromosomes to spread out within the cell, thereby making it easier to see each individual chromosome. Next, the cells are treated with a fixative that chemically freezes them so that the chromosomes will no longer move around. The cells are then treated with a chemical dye that binds to the chromosomes and stains them. As discussed in greater detail in Chapter 8, this gives chromosomes a distinctive banding pattern that greatly enhances their visualization and ability to be uniquely identified (also see Figure 8.1c, d). The cells are then placed on a slide and viewed with a light microscope.

In a cytogenetics laboratory, the microscopes are equipped with a camera that can photograph the chromosomes. In recent years, advances in technology have allowed cytogeneticists to view microscopic images on a computer screen (**Figure 3.2b**). On the computer screen, the chromosomes can be organized in a standard way, usually from largest to smallest. As seen in **Figure 3.2c**, the human chromosomes have been lined up, and a number is given to designate each type of chromosome. An exception would be the sex chromosomes, which are designated with the letters X and Y. An organized representation of the chromosomes within a cell is called a **karyotype.** A karyotype reveals how many chromosomes are found within an actively dividing somatic cell.

Eukaryotic Chromosomes Are Inherited in Sets

Most eukaryotic species are **diploid** or have a diploid phase to their life cycle**,** which means that each type of chromosome is a member of a pair. A diploid cell has two sets of chromosomes. In humans, most somatic cells have 46 chromosomes—two sets of 23 each. Other diploid species, however, have different numbers of chromosomes in their somatic cells. For example, the dog has 39 chromosomes per set (78 total), the fruit fly has 4 chromosomes per set (8 total), and the tomato has 12 per set (24 total).

When a species is diploid, the members of a pair of chromosomes are called **homologs;** each type of chromosome is found in a homologous pair. As shown in Figure 3.2c, for example, a human somatic cell has two copies of chromosome 1, two copies of chromosome 2, and so forth. Within each pair, the chromosome on the left is a homolog to the one on the right, and vice versa. In each pair, one chromosome was inherited from the mother and its homolog was inherited from the father. The two chromosomes in a homologous pair are nearly identical in size, have the same banding pattern, and contain a similar composition of genetic material. If a particular gene is found on one copy of a chromosome, it is also found on the other homolog. However, the two homologs may carry different alleles of a given gene. As an example, let's consider a gene in humans, called *OCA2,* which is one of a few different genes that affect eye color. The *OCA2* gene is located on chromosome 15 and comes in variants that result in brown, green, or blue eyes. In a person with brown eyes, one copy of chromosome 15 might carry a dominant brown allele, whereas its homolog could carry a recessive blue allele.

At the molecular level, how similar are homologous chromosomes? The answer is that the sequence of bases of one homolog would usually differ by less than 1% compared to the sequence of the other homolog. For example, the DNA sequence of chromosome 1 that you inherited from your mother would be greater than 99% identical to the sequence of chromosome 1 that you inherited from your father. Nevertheless, it should be emphasized that the sequences are not identical. The slight differences in DNA sequences provide the allelic differences in genes. Again, if we use the eye color gene as an example, a slight difference in DNA sequence distinguishes the brown, green, and blue alleles. It should also be noted that the striking similarities between homologous chromosomes do not apply to the pair of sex chromosomes—X and Y. These chromosomes differ in size and genetic composition. Certain genes that are found on the X chromosome are not found on the Y chromosome, and vice versa. The X and Y chromosomes are not considered homologous chromosomes even though they do have short regions of homology.

Figure 3.3 considers two homologous chromosomes that are labeled with three different genes. An individual carrying these two chromosomes would be homozygous for the dominant allele of gene *A*. The individual would be heterozygous, *Bb*, for the second gene. For the third gene, the individual is homozygous for a recessive allele, *c*. The physical location of a gene is called its **locus** (plural: **loci**). As seen in Figure 3.3, for example, the locus of gene *C* is toward one end of this chromosome, whereas the locus of gene *B* is more in the middle.

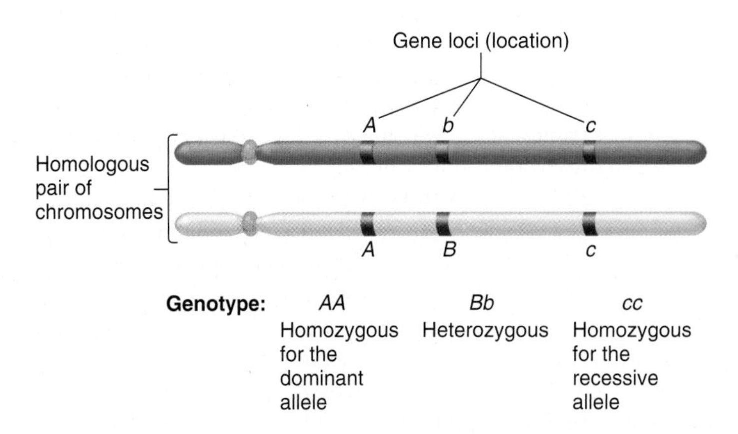

FIGURE 3.3 A comparison of homologous chromosomes. Each pair of homologous chromosomes carries the same types of genes, but, as shown here, the alleles may or may not be different.

3.2 CELL DIVISION

Now that we have an appreciation for the chromosomal composition of living cells, we can consider how chromosomes are copied and transmitted when cells divide. One purpose of cell division is **asexual reproduction.** In this process, a preexisting cell divides to produce two new cells. By convention, the original cell is usually called the mother cell, and the new cells are the two daughter cells. When species are unicellular, the mother cell is judged to be one individual, and the two daughter cells are two new separate organisms. Asexual reproduction is how bacterial cells proliferate. In addition, certain unicellular eukaryotes, such as the amoeba and baker's yeast (*Saccharomyces cerevisiae*), can reproduce asexually.

A second important reason for cell division is multicellularity. Species such as plants, animals, most fungi, and some protists are derived from a single cell that has undergone repeated cellular divisions. Humans, for example, begin as a single fertilized egg; repeated cellular divisions produce an adult with trillions of cells. The precise transmission of chromosomes during every cell division is critical so that all the cells of the body receive the correct amount of genetic material.

In this section, we will consider how the process of cell division requires the duplication, organization, and distribution of the chromosomes. In bacteria, which have a single circular chromosome, the division process is relatively simple. Prior to cell division, bacteria duplicate their circular chromosome; they then distribute a copy into each of the two daughter cells. This process, known as binary fission, is described first. Eukaryotes have multiple numbers of chromosomes that occur as sets. Compared with bacteria, this added complexity requires a more complicated sorting process to ensure that each newly made cell receives the correct number and types of chromosomes. A mechanism known as mitosis entails the organization and distribution of eukaryotic chromosomes during cell division.

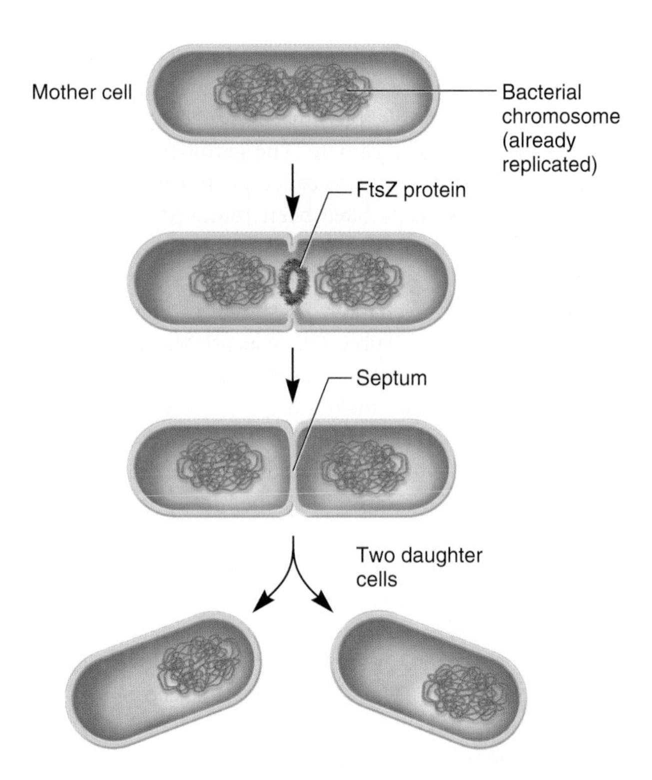

FIGURE 3.4 Binary fission: The process by which bacterial cells divide. Prior to division, the chromosome replicates to produce two identical copies. These two copies segregate from each other, with one copy going to each daughter cell.

Bacteria Reproduce Asexually by Binary Fission

As discussed earlier (see Figure 3.1a), bacterial species are typically unicellular, although individual bacteria may associate with each other to form pairs, chains, or clumps. Unlike eukaryotes, which have their chromosomes in a separate nucleus, the circular chromosomes of bacteria are in direct contact with the cytoplasm. In Chapter 10, we will consider the molecular structure of bacterial chromosomes in greater detail.

The capacity of bacteria to divide is really quite astounding. Some species, such as *Escherichia coli*, a common bacterium of the intestine, can divide every 20 to 30 minutes. Prior to cell division, bacterial cells copy, or replicate, their chromosomal DNA. This produces two identical copies of the genetic material, as shown at the top of **Figure 3.4**. Following DNA replication, a bacterial cell divides into two daughter cells by a process known as **binary fission.** During this event, the two daughter cells become separated from each other by the formation of a septum. As seen in the figure, each cell receives a copy of the chromosomal genetic material. Except when rare mutations occur, the daughter cells are usually genetically identical because they contain exact copies of the genetic material from the mother cell.

Recent evidence has shown that bacterial species produce a protein called FtsZ, which is important in cell division. This protein assembles into a ring at the future site of the septum. FtsZ is thought to be the first protein to move to this division site, and it recruits other proteins that produce a new cell wall between the daughter cells. FtsZ is evolutionarily related to a eukaryotic protein called tubulin. As discussed later in this chapter, tubulin is the main component of microtubules, which play a key role in chromosome sorting in eukaryotes. Both FtsZ and tubulin form structures that provide cells with organization and play key roles in cell division.

Binary fission is an asexual form of reproduction because it does not involve genetic contributions from two different gametes. On occasion, bacteria can exchange small pieces of genetic material with each other. We will consider some interesting mechanisms of genetic exchange in Chapter 7.

Eukaryotic Cells Progress Through a Cell Cycle to Produce Genetically Identical Daughter Cells

The common outcome of eukaryotic cell division is to produce two daughter cells that have the same number and types of chromosomes as the original mother cell. This requires a replication and division process that is more complicated than simple binary

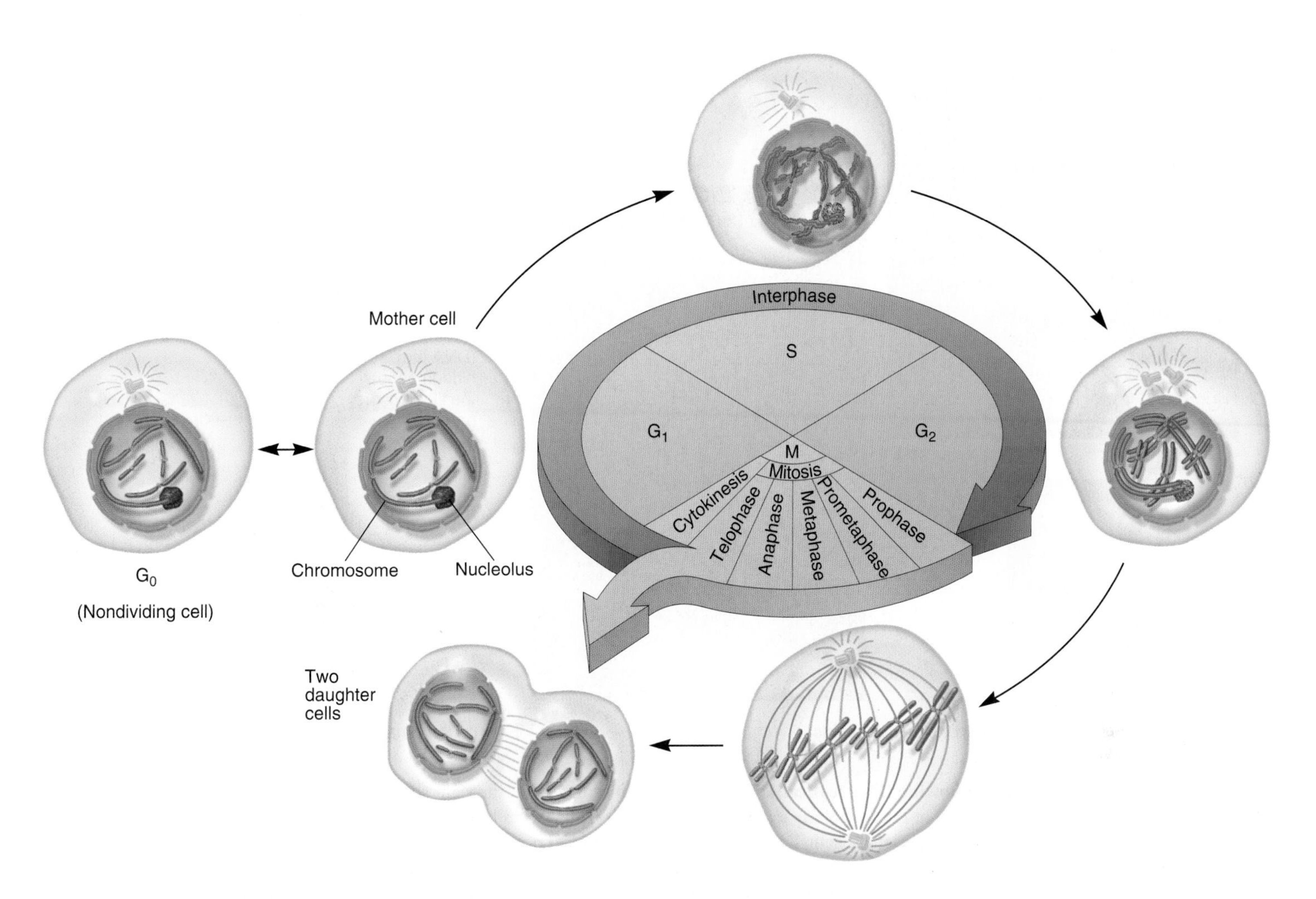

FIGURE 3.5 The eukaryotic cell cycle. Dividing cells progress through a series of phases, denoted G_1, S, G_2, and M phases. This diagram shows the progression of a cell through mitosis to produce two daughter cells. The original diploid cell had three pairs of chromosomes, for a total of six individual chromosomes. During S phase, these have replicated to yield 12 chromatids found in six pairs of sister chromatids. After mitosis and cytokinesis are completed, each of the two daughter cells contains six individual chromosomes, just like the mother cell. Note: The chromosomes in G_0, G_1, S, and G_2 phases are not condensed. In this drawing, they are shown partially condensed so they can be easily counted.

fission. Eukaryotic cells that are destined to divide progress through a series of phases known as the **cell cycle** (**Figure 3.5**). These phases are named G for gap, S for synthesis (of the genetic material), and M for mitosis. There are two G phases: G_1 and G_2. The term "gap" originally described the gaps between S phase and mitosis in which it was not microscopically apparent that significant changes were occurring in the cell. However, we now know that both gap phases are critical periods in the cell cycle that involve many molecular changes. In actively dividing cells, the G_1, S, and G_2 phases are collectively known as **interphase.** In addition, cells may remain permanently, or for long periods of time, in a phase of the cell cycle called G_0. A cell in the G_0 phase is either temporarily not progressing through the cell cycle or, in the case of terminally differentiated cells, such as most nerve cells in an adult mammal, will never divide again.

During the G_1 phase, a cell may prepare to divide. Depending on the cell type and the conditions that it encounters, a cell in the G_1 phase may accumulate molecular changes (e.g., synthesis of proteins) that cause it to progress through the rest of the cell cycle. When this occurs, cell biologists say that a cell has reached a **restriction point** and is committed on a pathway that leads to cell division. Once past the restriction point, the cell will then advance to the S phase, during which the chromosomes are replicated. After replication, the two copies are called **chromatids.** They are joined to each other at a region of DNA called the **centromere** to form a unit known as a pair of **sister chromatids** (**Figure 3.6**). The **kinetochore** is a group of proteins that are bound to the centromere. These proteins help to hold the sister chromatids together and also play a role in chromosome sorting, as discussed later. When S phase is completed, a cell actually has twice as many chromatids as chromosomes in the G_1 phase. For example, a human cell in the G_1 phase has 46 distinct chromosomes, whereas in G_2, it would have 46 pairs of sister chromatids, for a total of 92 chromatids. The term chromosome—meaning colored body—can be a bit confusing because it originally meant a distinct structure that is observable with the microscope. Therefore, the term

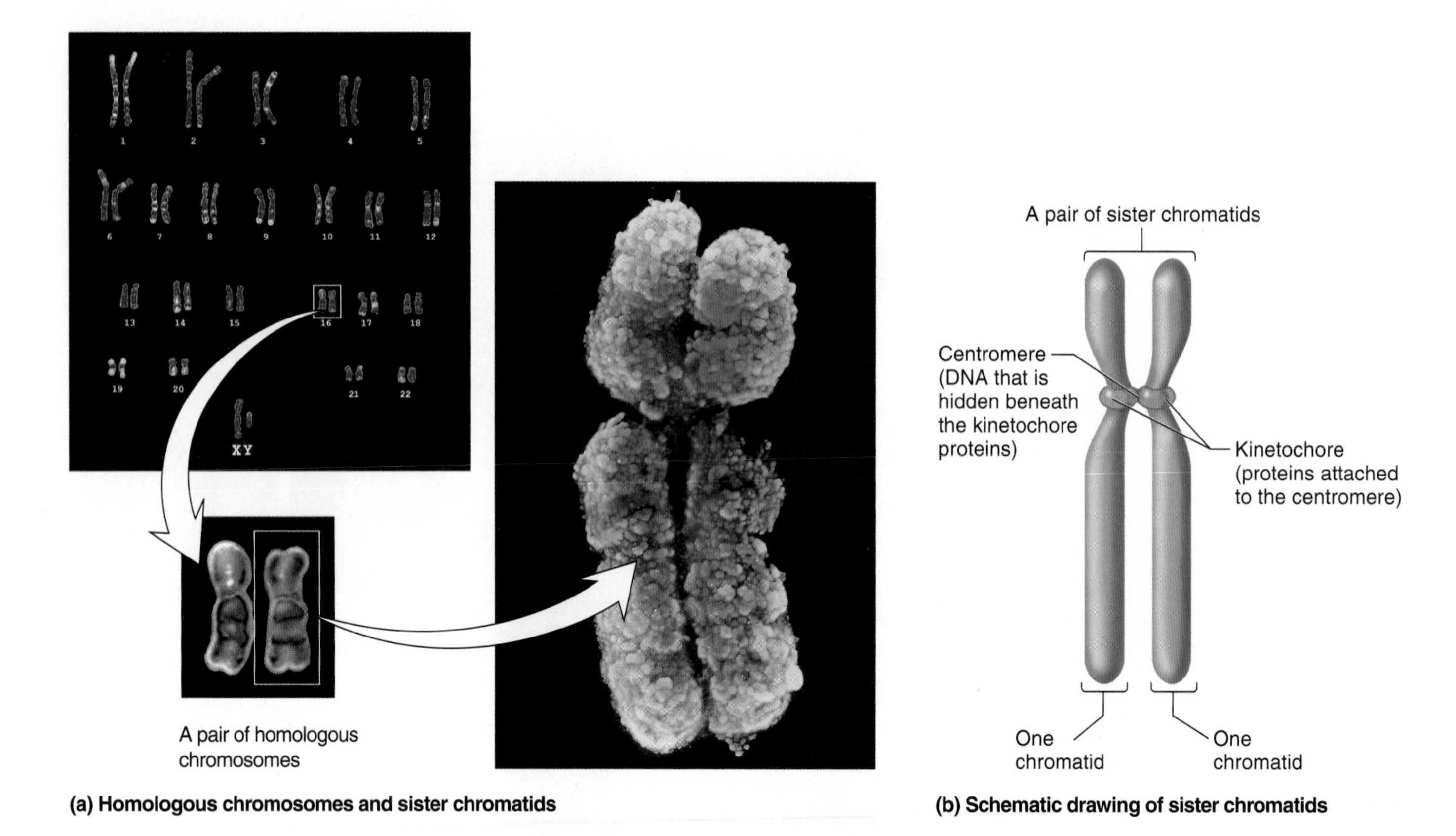

FIGURE 3.6 **Chromosomes following DNA replication.** **(a)** The photomicrograph on the right shows a chromosome in a form called a pair of sister chromatids. This chromosome is in the metaphase stage of mitosis, which is described later in the chapter. Note: Each of the 46 chromosomes that are viewed in a human karyotype (upper left) is actually a pair of sister chromatids. Look closely at the white rectangular boxes in the two insets. **(b)** A schematic drawing of sister chromatids. This structure has two chromatids that lie side by side. As seen here, each chromatid is a distinct unit. The two chromatids are held together by kinetochore proteins that bind to each other and to the centromeres of each chromatid.

chromosome can refer to either a pair of sister chromatids during the G_2 and early stages of M phase or to the structures that are observed at the end of M phase and those during G_1 that contain the equivalent of one chromatid (refer back to Figure 3.5).

During the G_2 phase, the cell accumulates the materials that are necessary for nuclear and cell division. It then progresses into the M phase of the cell cycle, when **mitosis** occurs. The primary purpose of mitosis is to distribute the replicated chromosomes, dividing one cell nucleus into two nuclei, so that each daughter cell receives the same complement of chromosomes. For example, a human cell in the G_2 phase has 92 chromatids, which are found in 46 pairs. During mitosis, these pairs of chromatids are separated and sorted so that each daughter cell receives 46 chromosomes.

Mitosis was first observed microscopically in the 1870s by the German biologist Walter Flemming, who coined the term mitosis (from the Greek mitos, meaning thread). He studied the dividing epithelial cells of salamander larvae and noticed that chromosomes were constructed of two parallel "threads." These threads separated and moved apart, one going to each of the two daughter nuclei. By this mechanism, Flemming pointed out that the two daughter cells received an identical group of threads, of a quantity comparable to the number of threads in the parent cell.

The Mitotic Spindle Apparatus Organizes and Sorts Eukaryotic Chromosomes

Before we discuss the events of mitosis, let's first consider the structure of the **mitotic spindle apparatus** (also known simply as the **mitotic spindle**), which is involved in the organization and sorting of chromosomes (**Figure 3.7**). The spindle apparatus is formed from **microtubule-organizing centers** (MTOCs), which are structures found in eukaryotic cells from which microtubules grow. Microtubules are produced from the rapid polymerization of tubulin proteins. In animal cells, the mitotic spindle is formed from two MTOCs called **centrosomes.** Each centrosome is located at a **spindle pole.** A pair of **centrioles** at right angles to each other is found within each centrosome of animal cells. However, centrosomes and centrioles are not always found in eukaryotic species. For example, plant cells do not have centrosomes. Instead, the nuclear envelope functions as a MTOC for spindle formation.

The mitotic spindle of a typical animal cell has three types of microtubules (see Figure 3.7). The aster microtubules emanate outward from the centrosome toward the plasma membrane. They are important for the positioning of the spindle apparatus within the cell and later in the process of cell division. The polar microtubules project toward the region where the chromosomes

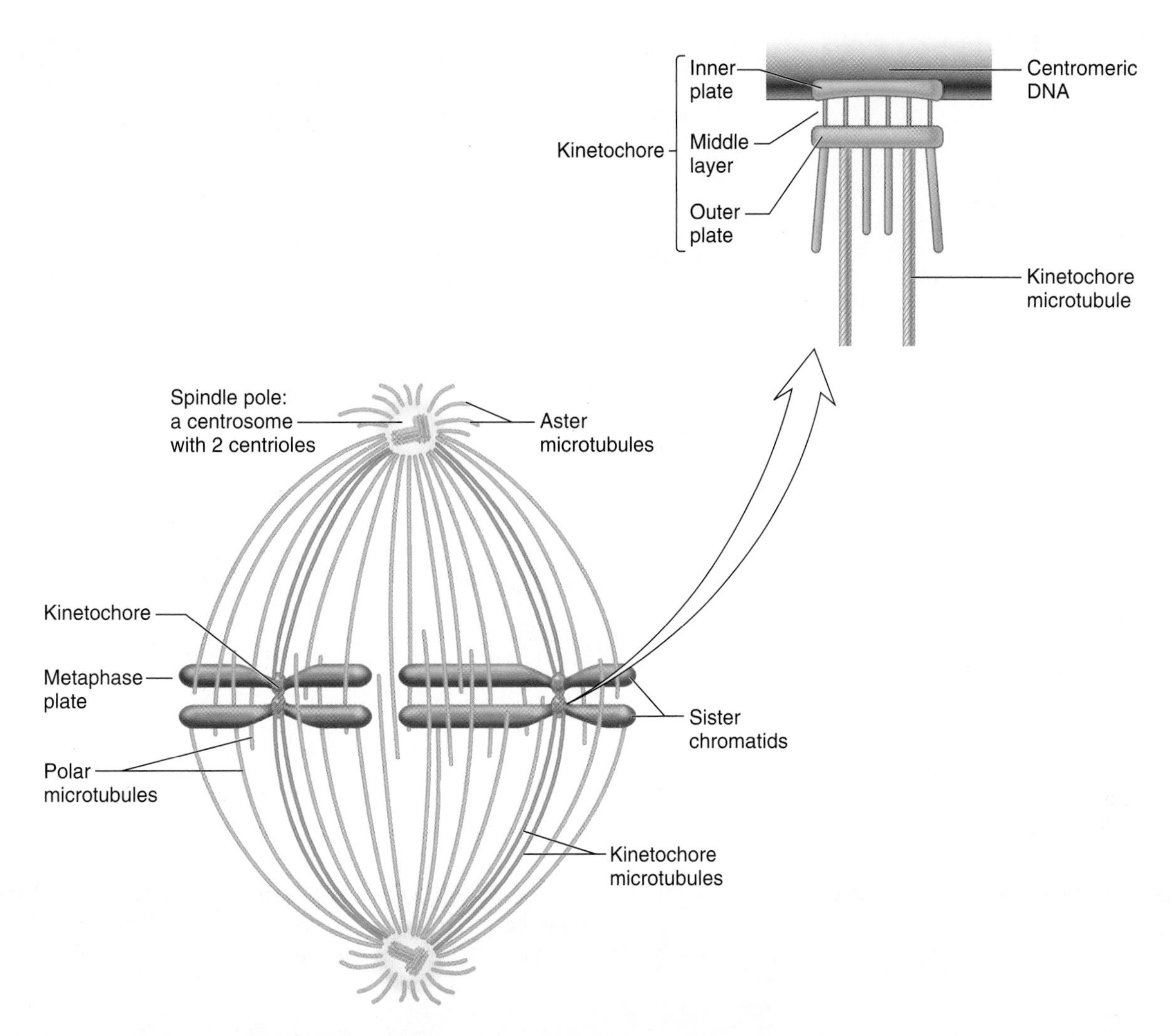

FIGURE 3.7 The structure of the mitotic spindle in a typical animal cell. A single centrosome duplicates during S phase and the two centrosomes separate at the beginning of M phase. The mitotic spindle is formed from microtubules that are rooted in the centrosomes. Each centrosome is located at a spindle pole. The aster microtubules emanate away from the region between the poles. They help to position the spindle within the cell and are used as reference points for cell division. However, astral microtubules are not found in many species, such as plants. The polar microtubules project into the region between the two poles; they play a role in pole separation. The kinetochore microtubules are attached to the kinetochore of sister chromatids. As seen in the inset, the kinetochore is composed of a group of proteins that form three layers: the inner plate, which recognizes the centromere; the outer plate, which recognizes a kinetochore microtubule; and the middle layer, which connects the inner and outer plates.

will be found during mitosis—the region between the two spindle poles. Polar microtubules that overlap with each other play a role in the separation of the two poles. They help to "push" the poles away from each other. Finally, the kinetochore microtubules have attachments to a kinetochore, which is a complex of proteins that is bound to the centromere of individual chromosomes. As seen in the inset to Figure 3.7, the kinetochore proteins form three layers. The proteins of the inner plate make direct contact with the centromeric DNA, whereas the outer plate contacts the kinetochore microtubules. The role of the middle layer is to connect these two regions.

The mitotic spindle allows cells to organize and separate chromosomes so that each daughter cell receives the same complement of chromosomes. This sorting process, known as mitosis, is described next.

The Transmission of Chromosomes During the Division of Eukaryotic Cells Requires a Process Known as Mitosis

In **Figure 3.8**, the process of mitosis is shown for a diploid animal cell. In the simplified diagrams shown along the bottom of this figure, the original mother cell contains six chromosomes; it is diploid ($2n$) and contains three chromosomes per set (n = 3). One set is shown in blue, and the homologous set is shown in red. As discussed next, mitosis is subdivided into phases known as prophase, prometaphase, metaphase, anaphase, and telophase.

Prophase Prior to mitosis, the cells are in interphase, during which the chromosomes are **decondensed**—less tightly

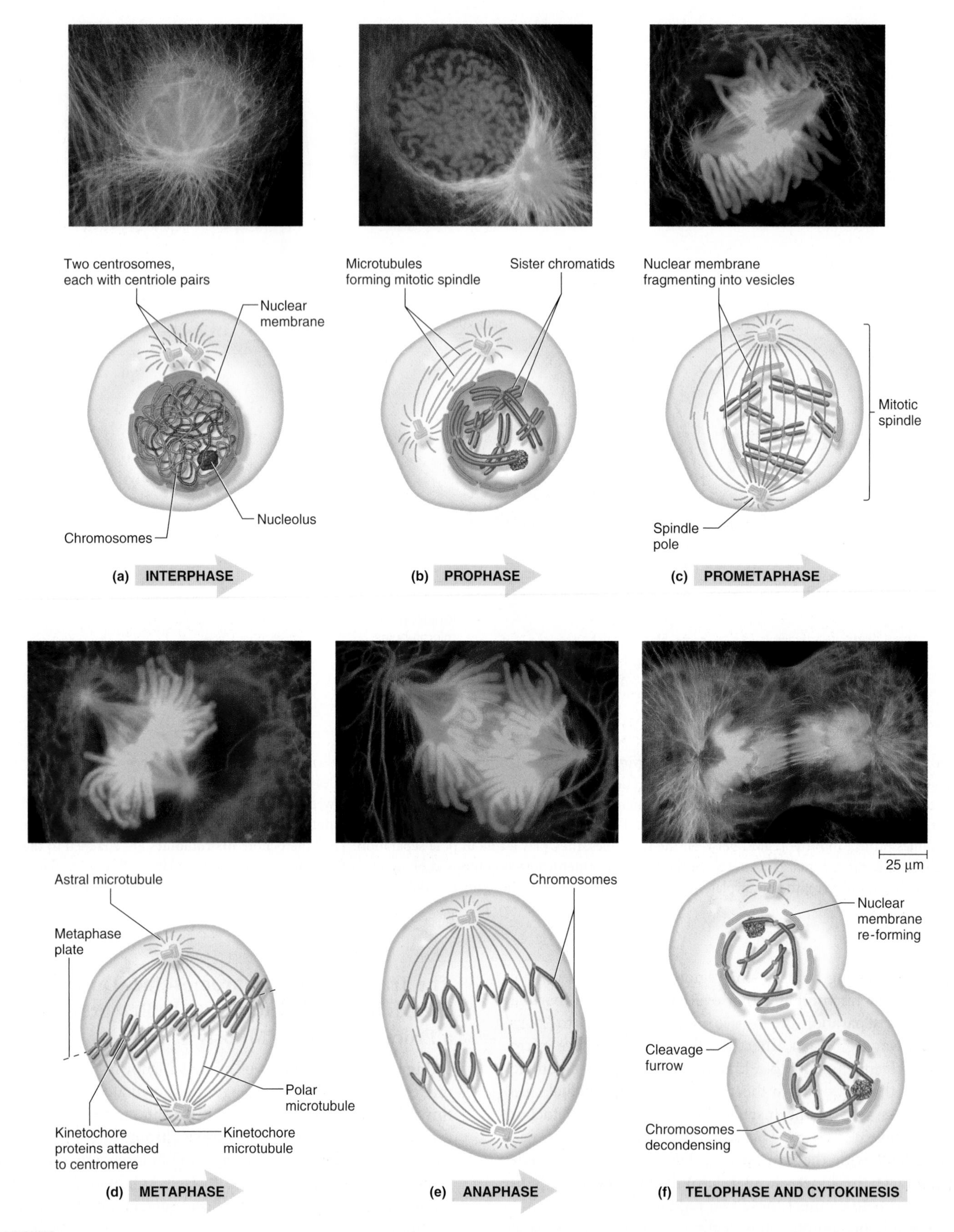

ONLINE ANIMATION

FIGURE 3.8 The process of mitosis in an animal cell. The top panels illustrate cells of a fish embryo progressing through mitosis. The chromosomes are stained in blue and the spindle is green. The bottom panels are schematic drawings that emphasize the sorting and separation of the chromosomes. In this case, the original diploid cell had six chromosomes (three in each set). At the start of mitosis, these have already replicated into 12 chromatids. The final result is two daughter cells each containing six chromosomes.

compacted—and found in the nucleus (Figure 3.8a). At the start of mitosis, in **prophase,** the chromosomes have already replicated to produce 12 chromatids, joined as six pairs of sister chromatids (Figure 3.8b). As prophase proceeds, the nuclear membrane begins to dissociate into small vesicles. At the same time, the chromatids **condense** into more compact structures that are readily visible by light microscopy. The mitotic spindle also begins to form and the nucleolus disappears.

Prometaphase As mitosis progresses from prophase to prometaphase, the centrosomes move to opposite ends of the cell and demarcate two spindle poles, one within each of the future daughter cells. Once the nuclear membrane has dissociated into vesicles, the spindle fibers can interact with the sister chromatids. This interaction occurs in a phase of mitosis called **prometaphase** (Figure 3.8c). How do sister chromatids become attached to the spindle? Initially, microtubules are rapidly formed and can be seen growing out from the two poles. As it grows, if the end of a microtubule happens to make contact with a kinetochore, its end is said to be "captured" and remains firmly attached to the kinetochore. This random process is how sister chromatids become attached to kinetochore microtubules. Alternatively, if the end of a microtubule does not collide with a kinetochore, the microtubule eventually depolymerizes and retracts to the centrosome. As the end of prometaphase nears, the kinetochore on a pair of sister chromatids is attached to kinetochore microtubules from opposite poles. As these events are occurring, the sister chromatids are seen to undergo jerky movements as they are tugged, back and forth, between the two poles. By the end of prometaphase, the mitotic spindle is completely formed.

Metaphase Eventually, the pairs of sister chromatids align themselves along a plane called the **metaphase plate.** As shown in Figure 3.8d, when this alignment is complete, the cell is in **metaphase** of mitosis. At this point, each pair of chromatids is attached to both poles by kinetochore microtubules. The pairs of sister chromatids have become organized into a single row along the metaphase plate. When this organizational process is finished, the chromatids can be equally distributed into two daughter cells.

Anaphase The next step in the division process occurs during **anaphase** (Figure 3.8e). At this stage, the connection that is responsible for holding the pairs of chromatids together is broken. (We will examine the process of sister chromatid cohesion and separation in more detail in Chapter 10.) Each chromatid, now an individual chromosome, is linked to only one of the two poles. As anaphase proceeds, the chromosomes move toward the pole to which they are attached. This involves a shortening of the kinetochore microtubules. In addition, the two poles themselves move farther apart due to the elongation of the polar microtubules, which slide in opposite directions due to the actions of motor proteins.

Telophase During **telophase,** the chromosomes reach their respective poles and decondense. The nuclear membrane now re-forms to produce two separate nuclei. In Figure 3.8f, this has produced two nuclei that contain six chromosomes each. The nucleoli will also reappear.

Cytokinesis In most cases, mitosis is quickly followed by **cytokinesis,** in which the two nuclei are segregated into separate daughter cells. Likewise, cytokinesis also segregates cell organelles such as mitochondria and chloroplasts into daughter cells. In animal cells, cytokinesis begins shortly after anaphase. A contractile ring, composed of **myosin** motor proteins and **actin** filaments, assembles adjacent to the plasma membrane. Myosin hydrolyzes ATP, which shortens the ring and thereby constricts the plasma membrane to form a **cleavage furrow** that ingresses, or moves inward (**Figure 3.9a**). Ingression continues until a midbody structure is formed that physically pinches one cell into two.

In plants, the two daughter cells are separated by the formation of a **cell plate** (**Figure 3.9b**). At the end of anaphase, Golgi-derived vesicles carrying cell wall materials are transported

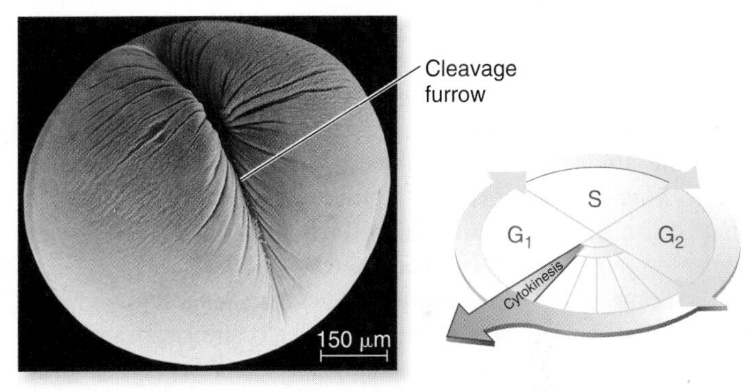

(a) Cleavage of an animal cell

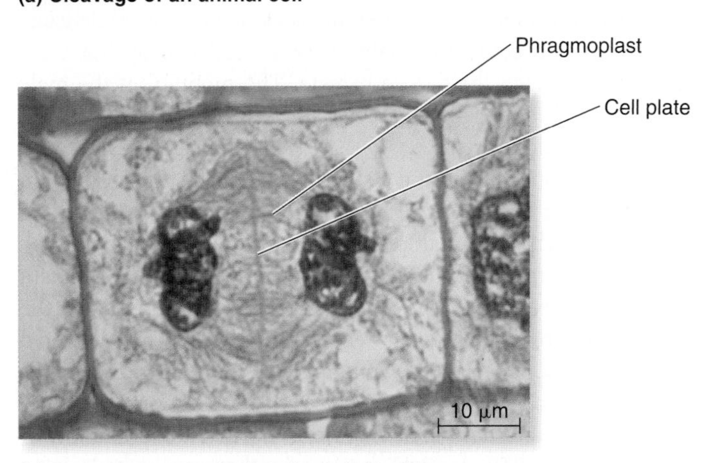

(b) Formation of a cell plate in a plant cell

FIGURE 3.9 Cytokinesis in an animal and plant cell. (a) In an animal cell, cytokinesis involves the formation of a cleavage furrow. (b) In a plant cell, cytokinesis occurs via the formation of a cell plate between the two daughter cells.

to the equator of a dividing cell. These vesicles are directed to their location via the phragmoplast, which is composed of parallel aligned microtubules and actin filaments that serve as tracks for vesicle movement. The fusion of these vesicles gives rise to the cell plate, which is a membrane-bound compartment. The cell plate begins in the middle of the cell and expands until it attaches to the mother cell wall. Once this attachment has taken place, the cell plate undergoes a process of maturation and eventually separates the mother cell into two daughter cells.

Outcome of mitotic cell division Mitosis and cytokinesis ultimately produce two daughter cells having the same number of chromosomes as the mother cell. Barring rare mutations, the two daughter cells are genetically identical to each other and to the mother cell from which they were derived. The critical consequence of this sorting process is to ensure genetic consistency from one somatic cell to the next. The development of multicellularity relies on the repeated process of mitosis and cytokinesis. For diploid organisms that are multicellular, most of the somatic cells are diploid and genetically identical to each other.

3.3 SEXUAL REPRODUCTION

In the previous section, we considered how a cell divides to produce two new cells with identical complements of genetic material. Now we will turn our attention to sexual reproduction, a common way for eukaryotic organisms to produce offspring. During sexual reproduction, gametes are made that contain half the amount of genetic material. These gametes fuse with each other in the process of fertilization to begin the life of a new organism. Gametes are highly specialized cells. The process whereby gametes form is called **gametogenesis.**

Some simple eukaryotic species are **isogamous,** which means that the gametes are morphologically similar. Examples of isogamous organisms include many species of fungi and algae. Most eukaryotic species, however, are **heterogamous**—they produce two morphologically different types of gametes. Male gametes, or **sperm cells,** are relatively small and usually travel far distances to reach the female gamete. The mobility of the male gamete is an important characteristic, making it likely that it will come in close proximity to the female gamete. The sperm of most animal species contain a single flagellum that enables them to swim. The sperm of ferns and nonvascular plants such as bryophytes may have multiple flagella. In flowering plants, however, the sperm are contained within pollen grains. Pollen is a small mobile structure that can be carried by the wind or on the feet or hairs of insects. In flowering plants, sperm are delivered to egg cells via pollen tubes. Compared to sperm cells, the female gamete, known as the **egg cell,** or **ovum,** is usually very large and nonmotile. In animal species, the egg stores a large amount of nutrients that will be available to nourish the growing embryo.

Gametes are typically **haploid,** which means they contain half the number of chromosomes as diploid cells. Haploid cells are represented by $1n$ and diploid cells by $2n$, where n refers to a set of chromosomes. A haploid gamete contains half as many chromosomes (i.e., a single set) as a diploid cell. For example, a diploid human cell contains 46 chromosomes, but a gamete (sperm or egg cell) contains only 23 chromosomes.

During the process known as **meiosis** (from the Greek meaning less), haploid cells are produced from a cell that was originally diploid. For this to occur, the chromosomes must be correctly sorted and distributed in a way that reduces the chromosome number to half its original value. In the case of humans, for example, each gamete must receive half the total number of chromosomes, but not just any 23 chromosomes will do. A gamete must receive one chromosome from each of the 23 pairs. In this section, we will examine the cellular events of gamete development in animal and plant species and how the stages of meiosis lead to the formation of cells with a haploid complement of chromosomes.

Meiosis Produces Cells That Are Haploid

The process of meiosis bears striking similarities to mitosis. Like mitosis, meiosis begins after a cell has progressed through the G_1, S, and G_2 phases of the cell cycle. However, meiosis involves two successive divisions rather than one (as in mitosis). Prior to meiosis, the chromosomes are replicated in S phase to produce pairs of sister chromatids. This single replication event is then followed by two sequential cell divisions called meiosis I and II. As in mitosis, each of these is subdivided into prophase, prometaphase, metaphase, anaphase, and telophase.

Prophase of meiosis I **Figure 3.10** emphasizes some of the important events that occur during prophase of meiosis I, which is further subdivided into stages known as leptotene, zygotene, pachytene, diplotene, and diakinesis. During the **leptotene** stage, the replicated chromosomes begin to condense and become visible with a light microscope. Unlike mitosis, the **zygotene** stage of prophase of meiosis I involves a recognition process known as **synapsis,** in which the homologous chromosomes recognize each other and begin to align themselves. At **pachytene,** the homologs have become completely aligned. The associated chromatids are known as **bivalents.** Each bivalent contains two pairs of sister chromatids, or a total of four chromatids.

In most eukaryotic species, a **synaptonemal complex** is formed between the homologous chromosomes. As shown in **Figure 3.11**, this complex is composed of parallel lateral elements, which are bound to the chromosomal DNA, and a central element, which promotes the binding of the lateral elements to each other via transverse filaments. The synaptonemal complex may not be required for the pairing of homologous chromosomes, because some species, such as *Aspergillus nidulans* and *Schizosaccharomyces pombe,* completely lack such a complex, yet their chromosomes synapse correctly. At present, the precise role of the synaptonemal complex is not clearly understood, and it remains the subject of intense research. It may play more than one role. First, although it may not be required for synapsis, the synaptonemal complex could help to maintain homologous

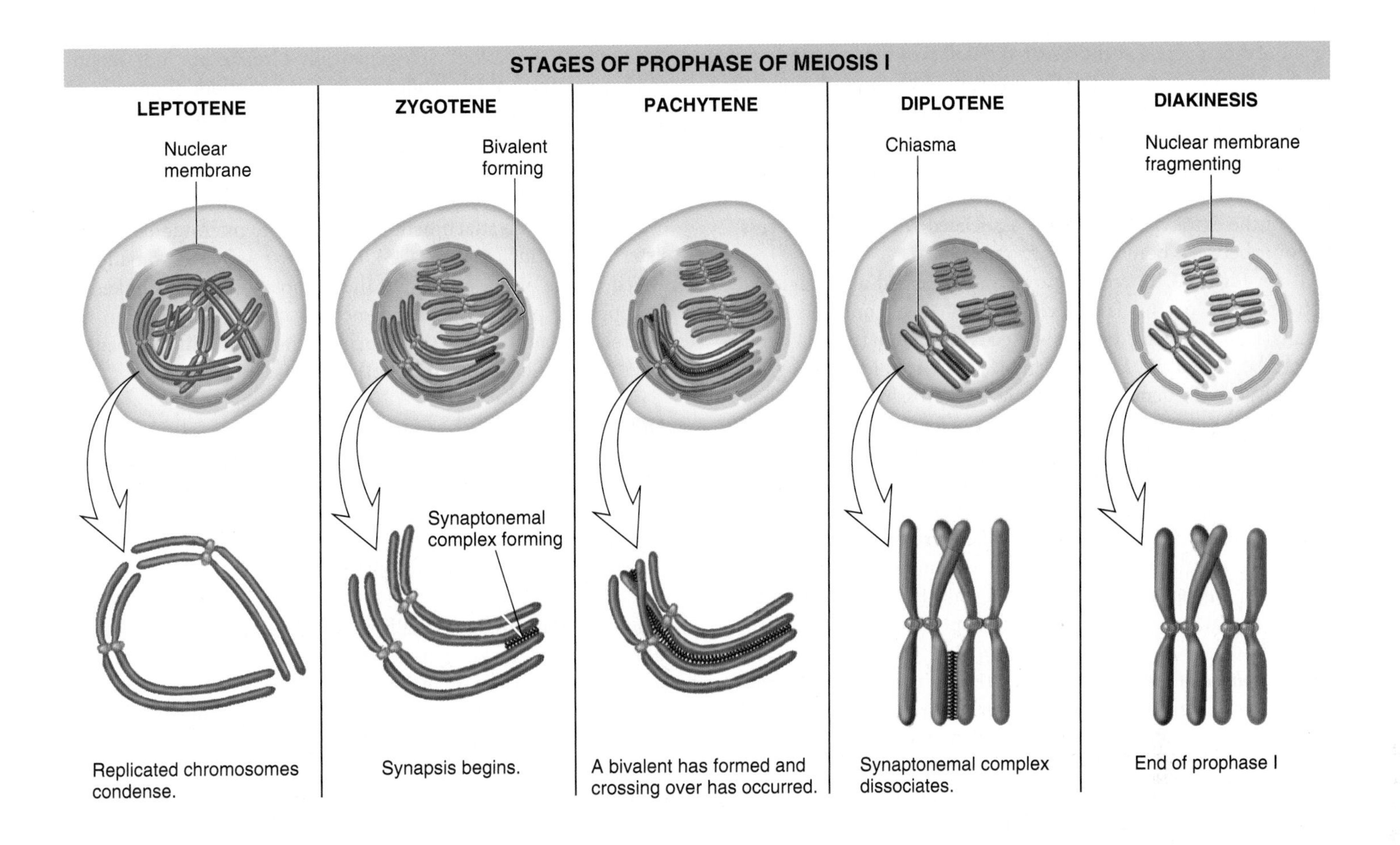

FIGURE 3.10 The events that occur during prophase of meiosis I.

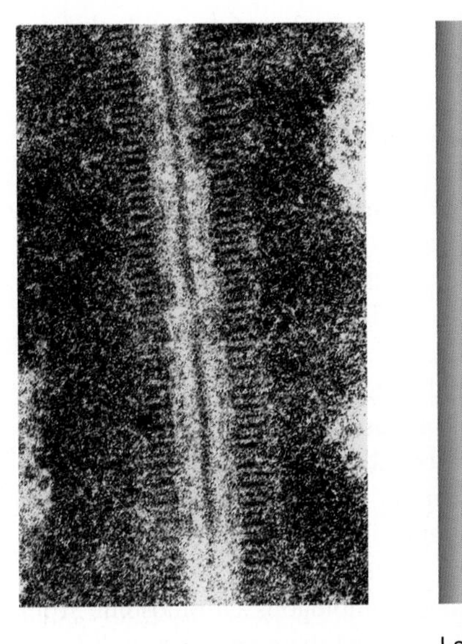

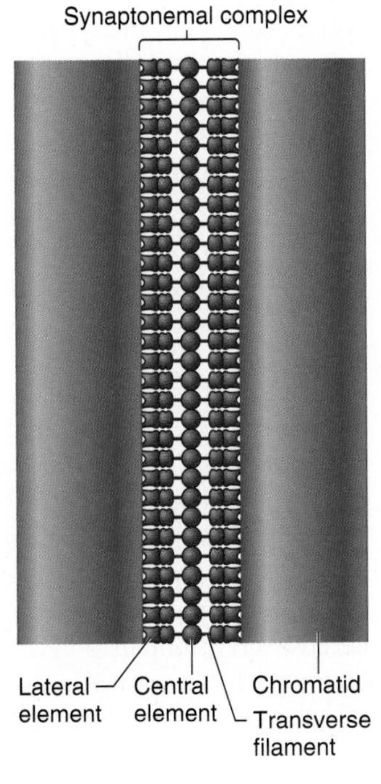

FIGURE 3.11 The synaptonemal complex formed during prophase of meiosis I. The left side is a transmission electron micrograph of a synaptonemal complex. Lateral elements are bound to the chromosomal DNA of homologous chromatids. A central element provides a link between the lateral elements via transverse filaments.

pairing in situations where the normal process has failed. Second, the complex may play a role in meiotic chromosome structure. And third, the synaptonemal complex may serve to regulate the process of crossing over, which is described next.

Prior to the pachytene stage, when synapsis is complete, an event known as **crossing over** usually occurs. Crossing over involves a physical exchange of chromosome pieces. Depending on the size of the chromosome and the species, an average eukaryotic chromosome incurs a couple to a couple dozen crossovers. During spermatogenesis in humans, for example, an average chromosome undergoes slightly more than two crossovers, whereas chromosomes in certain plant species may undergo 20 or more crossovers. Recent research has shown that crossing over is usually critical for the proper segregation of chromosomes. In fact, abnormalities in chromosome segregation may be related to a defect in crossing over. In a high percentage of people with Down syndrome, in which an individual has three copies of chromosome 21 instead of two, research has shown that the presence of the extra chromosome is associated with a lack of crossing over between homologous chromosomes.

In Figure 3.10, crossing over has occurred at a single site between two of the larger chromatids. The connection that results from crossing over is called a **chiasma** (plural: **chiasmata**), because it physically resembles the Greek letter chi, χ. We will consider the genetic consequences of crossing over in Chapter 6 and the molecular process of crossing over in Chapter 17. By the end of the **diplotene** stage, the synaptonemal complex has largely disappeared. The bivalent pulls apart slightly, and microscopically it becomes easier to see that it is actually composed of four chromatids. A bivalent is also called a **tetrad** (from the prefix "tetra-," meaning four) because it is composed of four chromatids. In the last stage of prophase of meiosis I, **diakinesis,** the synaptonemal complex completely disappears.

Prometaphase of meiosis I Figure 3.10 has emphasized the pairing and crossing over that occurs during prophase of meiosis I. In **Figure 3.12**, we turn our attention to the general events in meiosis. Prophase of meiosis I is followed by prometaphase, in which the spindle apparatus is complete, and the chromatids are attached via kinetochore microtubules.

Metaphase of meiosis I At metaphase of meiosis I, the bivalents are organized along the metaphase plate. However, their pattern of alignment is strikingly different from that observed during mitosis (refer back to Figure 3.8d). Before we consider the rest of meiosis I, a particularly critical feature for you to appreciate is how the bivalents are aligned along the metaphase plate. In particular, the pairs of sister chromatids are aligned in a double row rather than a single row, as occurs in mitosis. Furthermore, the arrangement of sister chromatids within this double row is random with regard to the blue and red homologs. In Figure 3.12, one of the blue homologs is above the metaphase plate and the other two are below, whereas one of the red homologs is below the metaphase plate and other two are above.

In an organism that produces many gametes, meiosis in other cells could produce a different arrangement of homologs—three blues above and none below, or none above and three below, and so on. As discussed later in this chapter, the random arrangement of homologs is consistent with Mendel's law of independent assortment. Because most eukaryotic species have several chromosomes per set, the sister chromatids can be randomly aligned along the metaphase plate in many possible ways. For example, consider humans, who have 23 chromosomes per set. The possible number of different, random alignments equals 2^n, where n equals the number of chromosomes per set. Thus, in humans, this would equal 2^{23}, or over 8 million, possibilities! Because the homologs are genetically similar but not identical, we see from this calculation that the random alignment of homologous chromosomes provides a mechanism to promote a vast amount of genetic diversity.

In addition to the random arrangement of homologs within a double row, a second distinctive feature of metaphase of meiosis I is the attachment of kinetochore microtubules to the sister chromatids (**Figure 3.13**). One pair of sister chromatids is linked to one of the poles, and the homologous pair is linked to the opposite pole. This arrangement is quite different from the kinetochore attachment sites during mitosis in which a pair of sister chromatids is linked to both poles (see Figure 3.8).

Anaphase of meiosis I During anaphase of meiosis I, the two pairs of sister chromatids within a bivalent separate from each other (see Figure 3.12). However, the connection that holds sister chromatids together does not break. Instead, each joined pair of chromatids migrates to one pole, and the homologous pair of chromatids moves to the opposite pole.

Telophase of meiosis I Finally, at telophase of meiosis I, the sister chromatids have reached their respective poles, and decondensation occurs in many, but not all, species. The nuclear membrane may re-form to produce two separate nuclei. The end result of meiosis I is two cells, each with three pairs of sister chromatids. It is thus a reduction division. The original diploid cell had its chromosomes in homologous pairs, but the two cells produced at the end of meiosis I are considered to be haploid; they do not have pairs of homologous chromosomes.

Meiosis II The sorting events that occur during meiosis II are similar to those that occur during mitosis, but the starting point is different. For a diploid organism with six chromosomes, mitosis begins with 12 chromatids that are joined as six pairs of sister chromatids (refer back to Figure 3.8). By comparison, the two cells that begin meiosis II each have six chromatids that are joined as three pairs of sister chromatids. Otherwise, the steps that occur during prophase, prometaphase, metaphase, anaphase, and telophase of meiosis II are analogous to a mitotic division.

Meiosis versus mitosis If we compare the outcome of meiosis (see Figure 3.12) to that of mitosis (see Figure 3.8), the results are quite different. (A comparison is also made in solved problem S3 at the end of this chapter.) In these examples, mitosis produced two diploid daughter cells with six chromosomes each, whereas meiosis produced four haploid daughter cells with three chromosomes each. In other words, meiosis has halved the number of chromosomes per cell. With regard to alleles, the results of mitosis and meiosis are also different. The daughter cells produced by mitosis are genetically identical. However, the haploid cells produced by meiosis are not genetically identical to each other because they contain only one homologous chromosome from each pair. Later, we will consider how the gametes may differ in the alleles that they carry on their homologous chromosomes.

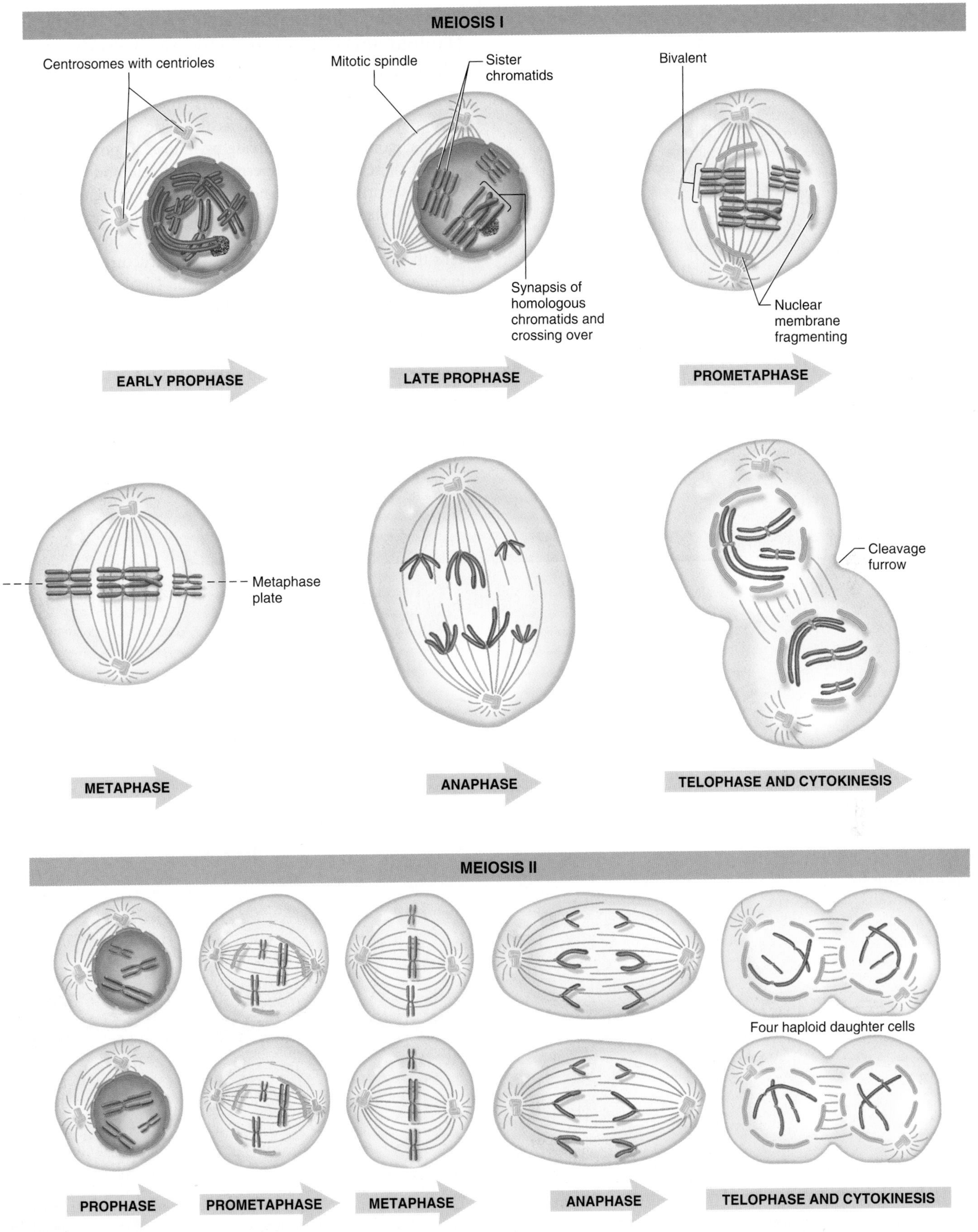

MEIOSIS I

Centrosomes with centrioles

Mitotic spindle

Sister chromatids

Bivalent

Synapsis of homologous chromatids and crossing over

Nuclear membrane fragmenting

EARLY PROPHASE

LATE PROPHASE

PROMETAPHASE

Metaphase plate

Cleavage furrow

METAPHASE

ANAPHASE

TELOPHASE AND CYTOKINESIS

MEIOSIS II

Four haploid daughter cells

PROPHASE

PROMETAPHASE

METAPHASE

ANAPHASE

TELOPHASE AND CYTOKINESIS

ONLINE ANIMATION

FIGURE 3.12 The stages of meiosis in an animal cell. See text for details.

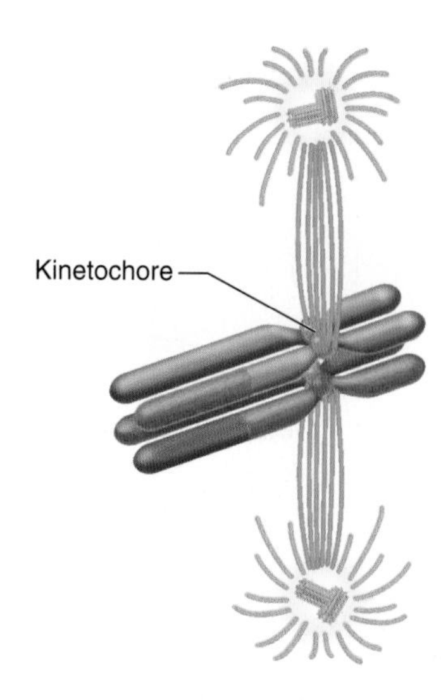

FIGURE 3.13 Attachment of the kinetochore microtubules to replicated chromosomes during meiosis. The kinetochore microtubules from a given pole are attached to one pair of chromatids in a bivalent, but not both. Therefore, each pair of sister chromatids is attached to only one pole.

In Animals, Spermatogenesis Produces Four Haploid Sperm Cells and Oogenesis Produces a Single Haploid Egg Cell

In male animals, **spermatogenesis,** the production of sperm, occurs within glands known as the testes. The testes contain spermatogonial cells that divide by mitosis to produce two cells. One of these remains a spermatogonial cell, and the other cell becomes a primary spermatocyte. As shown in **Figure 3.14a,** the spermatocyte progresses through meiosis I and meiosis II to produce four haploid cells, which are known as spermatids. These cells then mature into sperm cells. The structure of a sperm cell includes a long flagellum and a head. The head of the sperm contains little more than a haploid nucleus and an organelle at its tip, known as an acrosome. The acrosome contains digestive enzymes that are released when a sperm meets an egg cell. These enzymes enable the sperm to penetrate the outer protective layers of the egg and gain entry into the egg cell's cytosol. In animal species without a mating season, sperm production is a continuous process in mature males. A mature human male, for example, produces several hundred million sperm each day.

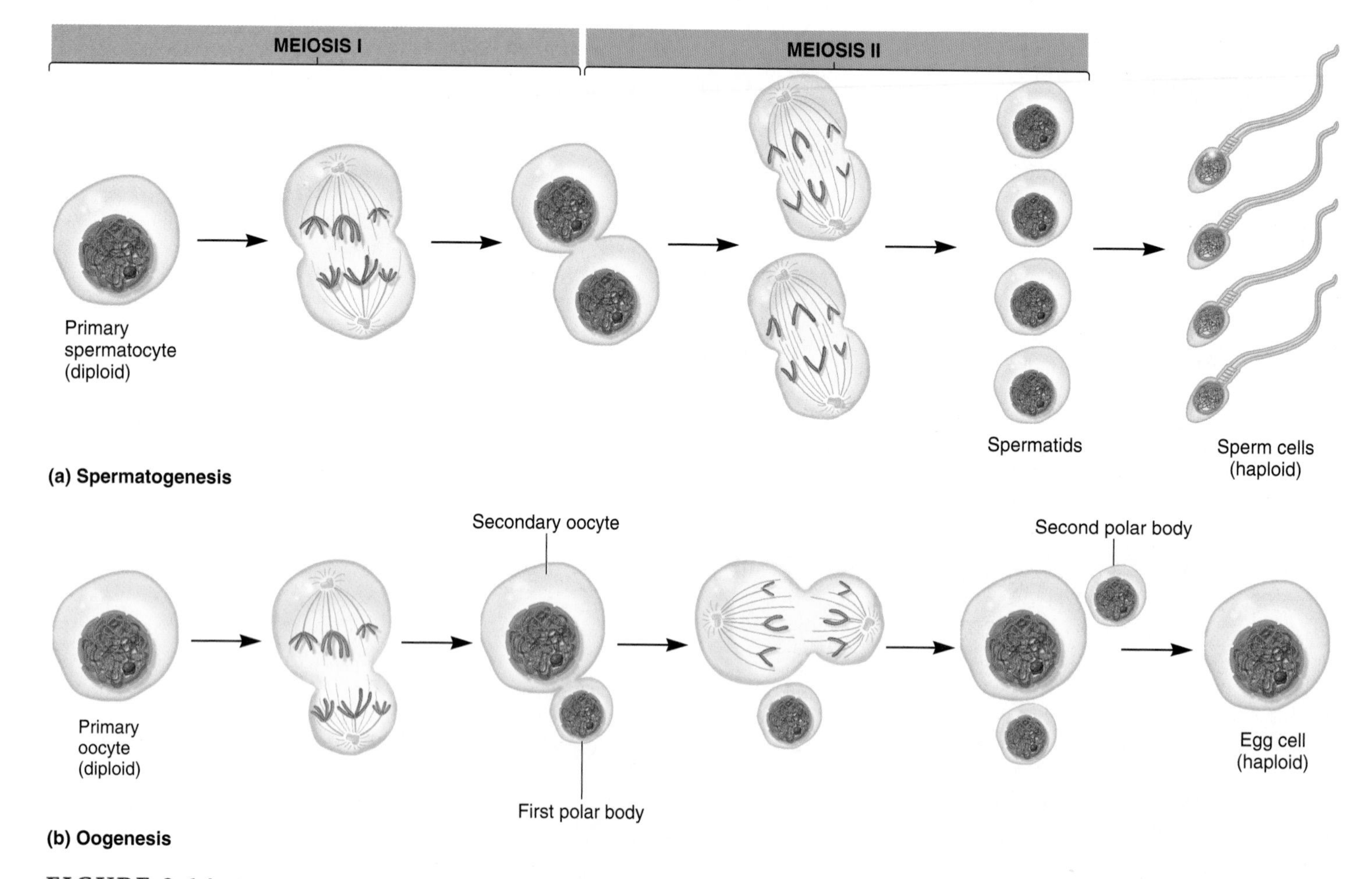

FIGURE 3.14 Gametogenesis in animals. **(a)** Spermatogenesis. A diploid spermatocyte undergoes meiosis to produce four haploid (n) spermatids. These differentiate during spermatogenesis to become mature sperm. **(b)** Oogenesis. A diploid oocyte undergoes meiosis to produce one haploid egg cell and two or three polar bodies. For some species, the first polar body divides; in other species, it does not. Because of asymmetrical cytokinesis, the amount of cytoplasm the egg receives is maximized. The polar bodies degenerate.

In female animals, **oogenesis,** the production of egg cells, occurs within specialized diploid cells of the ovary known as oogonia. Quite early in the development of the ovary, the oogonia initiate meiosis to produce primary oocytes. For example, in humans, approximately 1 million primary oocytes per ovary are produced before birth. These primary oocytes are arrested—enter a dormant phase—at prophase of meiosis I, remaining at this stage until the female becomes sexually mature. Beginning at this stage, primary oocytes are periodically activated to progress through the remaining stages of oocyte development.

During oocyte maturation, meiosis produces only one cell that is destined to become an egg, as opposed to the four gametes produced from each primary spermatocyte during spermatogenesis. How does this occur? As shown in **Figure 3.14b**, the first meiotic division is asymmetrical and produces a secondary oocyte and a much smaller cell, known as a polar body. Most of the cytoplasm is retained by the secondary oocyte and very little by the polar body, allowing the oocyte to become a larger cell with more stored nutrients. The secondary oocyte then begins meiosis II. In mammals, the secondary oocyte is released from the ovary—an event called ovulation—and travels down the oviduct toward the uterus. During this journey, if a sperm cell penetrates the secondary oocyte, it is stimulated to complete meiosis II; the secondary oocyte produces a haploid egg and a second polar body. The haploid egg and sperm nuclei then unite to create the diploid nucleus of a new individual.

Plant Species Alternate Between Haploid (Gametophyte) and Diploid (Sporophyte) Generations

Most species of animals are diploid, and their haploid gametes are considered to be a specialized type of cell. By comparison, the life cycles of plant species alternate between haploid and diploid generations. The haploid generation is called the **gametophyte,** whereas the diploid generation is called the **sporophyte.** Meiosis produces haploid cells called spores, which divide by mitosis to produce the gametophyte. In simpler plants, such as mosses, a haploid spore can produce a large multicellular gametophyte by repeated mitoses and cellular divisions. In flowering plants, however, spores develop into gametophytes that contain only a few cells. In this case, the organism that we think of as a "plant" is the sporophyte, whereas the gametophyte is very inconspicuous. In fact, the gametophytes of most plant species are small structures produced within the much larger sporophyte. Certain cells within the haploid gametophytes then become specialized as haploid gametes.

Figure 3.15 provides an overview of gametophyte development and gametogenesis in flowering plants. Meiosis occurs within two different structures of the sporophyte: the anthers and the ovaries, which produce male and female gametophytes, respectively. This diagram depicts a flower from an angiosperm, which is a plant that produces seeds within an ovary.

In the anther, diploid cells called microsporocytes undergo meiosis to produce four haploid microspores. These separate into individual microspores. In many angiosperms, each microspore

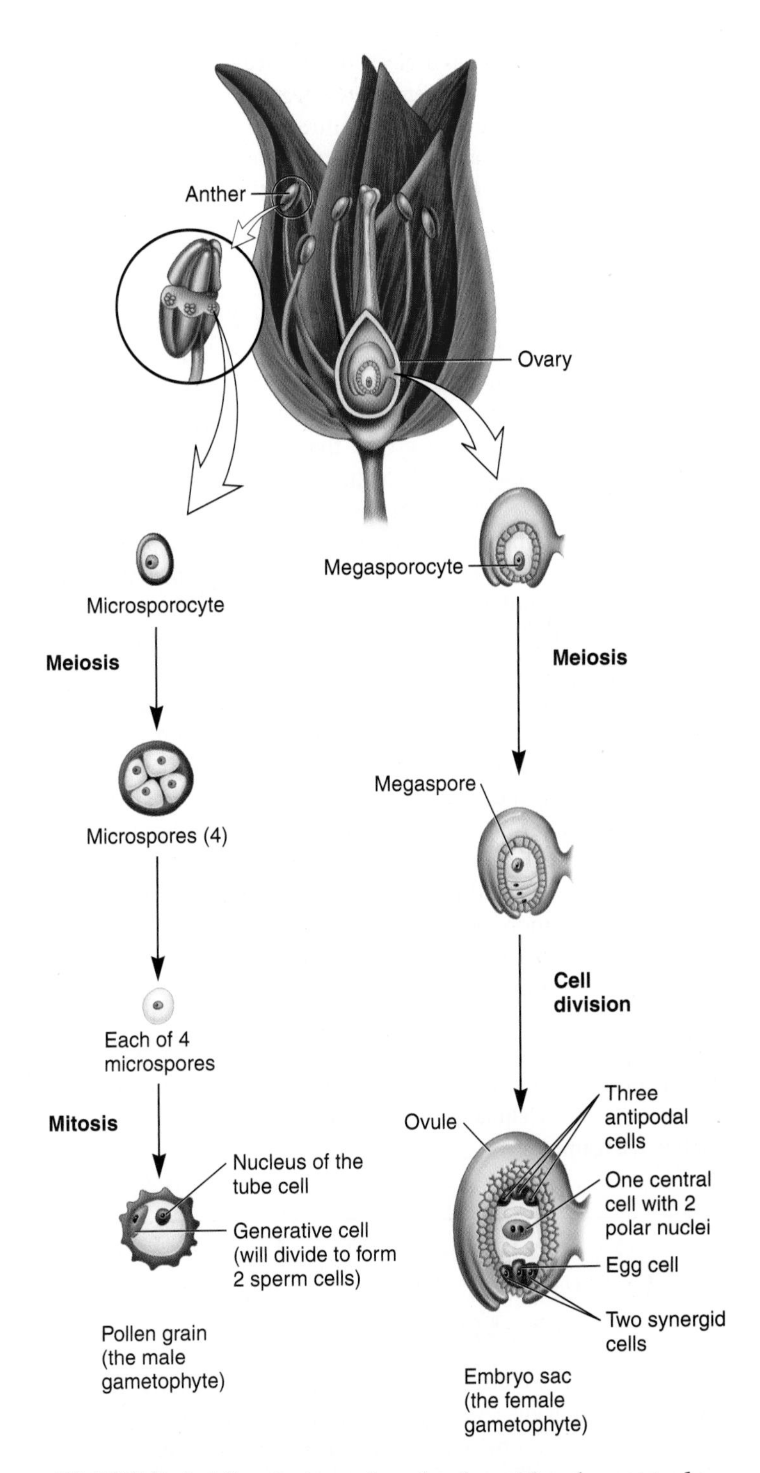

FIGURE 3.15 **The formation of male and female gametes by the gametophytes of angiosperms (flowering plants).**

undergoes mitosis to produce a two-celled structure containing one tube cell and one generative cell, both of which are haploid. This structure differentiates into a **pollen grain,** which is the male gametophyte with a thick cell wall. Later, the generative cell undergoes mitosis to produce two haploid sperm cells. In most plant species, this mitosis occurs only if the pollen grain germinates—if it lands on a stigma and forms a pollen tube (refer back to Figure 2.2c).

By comparison, female gametophytes are produced within ovules found in the plant ovaries. A type of cell known as a megasporocyte undergoes meiosis to produce four haploid megaspores. Three of the four megaspores degenerate. The remaining haploid megaspore then undergoes three successive mitotic divisions accompanied by asymmetrical cytokinesis to produce seven individual cells—one egg, two synergids, three antipodals, and one central cell. This seven-celled structure, also known as the **embryo sac,** is the mature female gametophyte. Each embryo sac is contained within an ovule.

For fertilization to occur, specialized cells within the male and female gametophytes must meet. The steps of plant fertilization were described in Chapter 2. To begin this process, a pollen grain lands on a stigma (refer back to Figure 2.2c). This stimulates the tube cell to sprout a tube that grows through the style and eventually makes contact with an ovule. As this is occurring, the generative cell undergoes mitosis to produce two haploid sperm cells. The sperm cells migrate through the pollen tube and eventually reach the ovule. One of the sperm enters the central cell, which contains the two polar nuclei. This results in a cell that is triploid ($3n$). This cell divides mitotically to produce **endosperm,** which acts as a food-storing tissue. The other sperm enters the egg cell. The egg and sperm nuclei fuse to create a diploid cell, the zygote, which becomes a plant embryo. Therefore, fertilization in flowering plants is actually a double fertilization. The result is that the endosperm, which uses a large amount of plant resources, will develop only when an egg cell has been fertilized. After fertilization is complete, the ovule develops into a seed, and the surrounding ovary develops into the fruit, which encloses one or more seeds.

When comparing animals and plants, it's interesting to consider how gametes are made. Animals produce gametes by meiosis. In contrast, plants produce reproductive cells by mitosis. The gametophyte of plants is a haploid multicellular organism that is produced by mitotic cellular divisions of a haploid spore. Within the multicellular gametophyte, certain cells become specialized as gametes.

3.4 THE CHROMOSOME THEORY OF INHERITANCE AND SEX CHROMOSOMES

Thus far, we have considered how chromosomes are transmitted during cell division and gamete formation. In this section, we will first examine how chromosomal transmission is related to the patterns of inheritance observed by Mendel. This relationship, known as the chromosome theory of inheritance, was a major breakthrough in our understanding of genetics because it established the framework for understanding how chromosomes carry and transmit the genetic determinants that govern the outcome of traits. This theory dramatically unfolded as a result of three lines of scientific inquiry (**Table 3.1**). One avenue concerned Mendel's breeding studies, in which he analyzed the transmission of traits from parent to offspring. A second line of inquiry involved the material basis for heredity. A Swiss botanist, Carl Nägeli, and a German biologist, August Weismann, championed the idea that a substance found in living cells is responsible for the transmission of traits from parents to offspring. Nägeli also suggested that both parents contribute equal amounts of this substance to their offspring. Several scientists, including Oscar Hertwig, Eduard Strasburger, and Walter Flemming, conducted studies suggesting that the chromosomes

TABLE **3.1**

Chronology for the Development and Proof of the Chromosome Theory of Inheritance

1866	Gregor Mendel: Analyzed the transmission of traits from parents to offspring and showed that it follows a pattern of segregation and independent assortment.
1876–77	Oscar Hertwig and Hermann Fol: Observed that the nucleus of the sperm enters the egg during animal cell fertilization.
1877	Eduard Strasburger: Observed that the sperm nucleus of plants (and no detectable cytoplasm) enters the egg during plant fertilization.
1878	Walter Flemming: Described mitosis in careful detail.
1883	Carl Nägeli and August Weismann: Proposed the existence of a genetic material, which Nägeli called idioplasm and Weismann called germ plasm.
1883	Wilhelm Roux: Proposed that the most important event of mitosis is the equal partitioning of "nuclear qualities" to the daughter cells.
1883	Edouard van Beneden: showed that gametes contain half the number of chromosomes and that fertilization restores the normal diploid number.
1884–85	Hertwig, Strasburger and August Weismann: Proposed that chromosomes are carriers of the genetic material.
1889	Theodore Boveri: Showed that enucleated sea urchin eggs that are fertilized by sperm from a different species develop into larva that have characteristics that coincide with the sperm's species.
1900	Hugo de Vries, Carl Correns, and Erich von Tschermak: Rediscovered Mendel's work.
1901	Thomas Montgomery: Determined that maternal and paternal chromosomes pair with each other during meiosis.
1901	C. E. McClung: Discovered that sex determination in insects is related to differences in chromosome composition.
1902	Theodor Boveri: Showed that when sea urchin eggs were fertilized by two sperm, the abnormal development of the embryo was related to an abnormal number of chromosomes.
1903	Walter Sutton: Showed that even though the chromosomes seem to disappear during interphase, they do not actually disintegrate. Instead, he argued that chromosomes must retain their continuity and individuality from one cell division to the next.
1902–03	Theodor Boveri and Walter Sutton: Independently proposed tenets of the chromosome theory of inheritance. Some historians primarily credit this theory to Sutton.
1910	Thomas Hunt Morgan: Showed that a genetic trait (i.e., white-eyed phenotype in *Drosophila*) was linked to a particular chromosome.
1913	E. Eleanor Carothers: Demonstrated that homologous pairs of chromosomes show independent assortment.
1916	Calvin Bridges: Studied chromosomal abnormalities as a way to confirm the chromosome theory of inheritance.

For a description of these experiments, the student is encouraged to read Voeller, B. R. (1968), *The chromosome theory of inheritance. Classic Papers in Development and Heredity.* New York: Appleton-Century-Crofts.

are the carriers of the genetic material. We now know the DNA within the chromosomes is the genetic material.

Finally, the third line of evidence involved the microscopic examination of the processes of fertilization, mitosis, and meiosis. Researchers became increasingly aware that the characteristics of organisms are rooted in the continuity of cells during the life of an organism and from one generation to the next. When the work of Mendel was rediscovered, several scientists noted striking parallels between the segregation and assortment of traits noted by Mendel and the behavior of chromosomes during meiosis. Among them were Theodore Boveri, a German biologist, and Walter Sutton at Columbia University. They independently proposed the chromosome theory of inheritance, which was a milestone in our understanding of genetics. The principles of this theory are described at the beginning of this section.

The remainder of this section focuses on sex chromosomes. The experimental connection between the chromosome theory of inheritance and sex chromosomes is profound. Even though an examination of meiosis provided compelling evidence that Mendel's laws could be explained by chromosome sorting, researchers still needed to correlate chromosome behavior with the inheritance of particular traits. Because sex chromosomes, such as the X and Y chromosome, look very different under the microscope, and because many genes on the X chromosome are not on the Y chromosome, geneticists were able to correlate the inheritance of certain traits with the transmission of specific sex chromosomes. In particular, early studies identified genes on the X chromosome that govern eye color in fruit flies. This phenomenon, which is called **X-linked inheritance,** confirmed the idea that genes are found on chromosomes. In addition, X-linked inheritance showed us that not all traits follow simple Mendelian rules. In later chapters, we will examine a variety of traits that are governed by chromosomal genes yet follow inheritance patterns that are more complex than those observed by Mendel.

The Chromosome Theory of Inheritance Relates the Behavior of Chromosomes to the Mendelian Inheritance of Traits

According to the **chromosome theory of inheritance,** the inheritance patterns of traits can be explained by the transmission patterns of chromosomes during meiosis and fertilization. This theory is based on a few fundamental principles.

1. Chromosomes contain the genetic material that is transmitted from parent to offspring and from cell to cell.
2. Chromosomes are replicated and passed along, generation after generation, from parent to offspring. They are also passed from cell to cell during the development of a multicellular organism. Each type of chromosome retains its individuality during cell division and gamete formation.
3. The nuclei of most eukaryotic cells contain chromosomes that are found in homologous pairs—they are diploid. One member of each pair is inherited from the mother, the other from the father. At meiosis, one of the two members of each pair segregates into one daughter nucleus, and the homolog segregates into the other daughter nucleus. Gametes contain one set of chromosomes—they are haploid.
4. During the formation of haploid cells, different types of (nonhomologous) chromosomes segregate independently of each other.
5. Each parent contributes one set of chromosomes to its offspring. The maternal and paternal sets of homologous chromosomes are functionally equivalent; each set carries a full complement of genes.

The chromosome theory of inheritance allows us to see the relationship between Mendel's laws and chromosomal transmission. As shown in **Figure 3.16**, Mendel's law of segregation can be

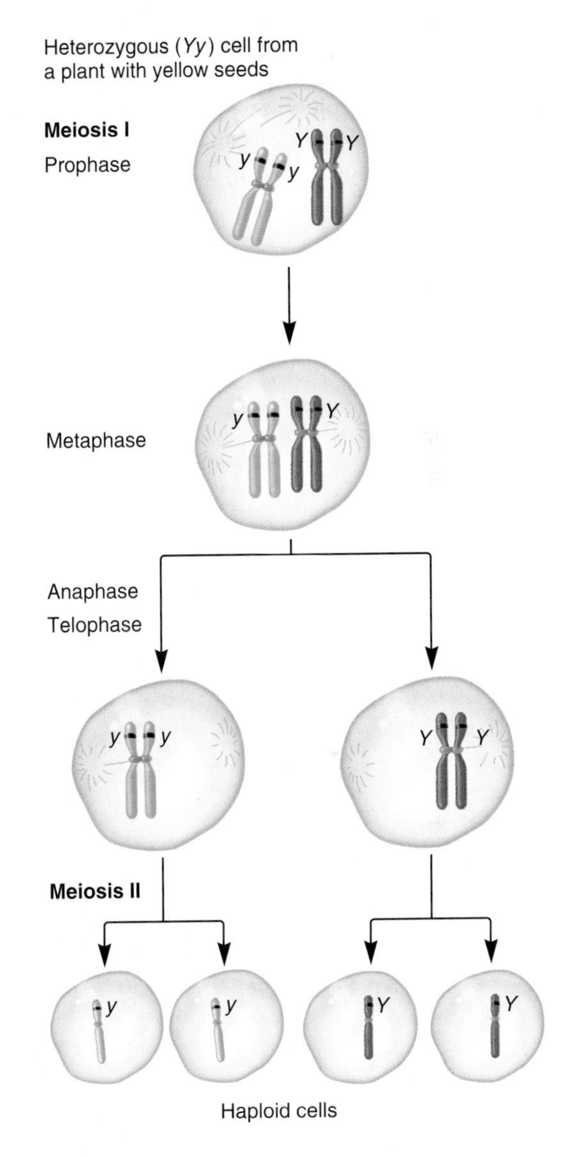

FIGURE 3.16 Mendel's law of segregation can be explained by the segregation of homologs during meiosis. The two copies of a gene are contained on homologous chromosomes. In this example using pea seed color, the two alleles are *Y* (yellow) and *y* (green). During meiosis, the homologous chromosomes segregate from each other, leading to segregation of the two alleles into separate gametes.

Genes→Traits The gene for seed color exists in two alleles, *Y* (yellow) and *y* (green). During meiosis, the homologous chromosomes that carry these alleles segregate from each other. The resulting cells receive the *Y* or *y* allele but not both. When two gametes unite during fertilization, the alleles that they carry determine the traits of the resulting offspring.

explained by the homologous pairing and segregation of chromosomes during meiosis. This figure depicts the behavior of a pair of homologous chromosomes that carry a gene for seed color. One of the chromosomes carries a dominant allele that confers yellow seed color, whereas the homologous chromosome carries a recessive allele that confers green color. A heterozygous individual would pass only one of these alleles to each offspring. In other words, a gamete may contain the yellow allele or the green allele but not both. Because homologous chromosomes segregate from each other, a gamete will contain only one copy of each type of chromosome.

How is the law of independent assortment explained by the behavior of chromosomes? **Figure 3.17** considers the segregation of two types of chromosomes, each carrying a different gene. One pair of chromosomes carries the gene for seed color: the yellow (*Y*) allele is on one chromosome, and the green (*y*) allele is on the homolog. The other pair of (smaller) chromosomes carries the gene for seed shape: one copy has the round (*R*) allele, and the homolog carries the wrinkled (*r*) allele. At metaphase of meiosis I, the different types of chromosomes have randomly aligned along the metaphase plate. As shown in Figure 3.17, this can occur in more than one way. On the left, the *R* allele has sorted with the *y* allele, whereas the *r* allele has sorted with the *Y* allele. On the right, the opposite situation has occurred. Therefore, the random alignment of chromatid pairs during meiosis I can lead to an independent assortment of genes that are found on nonhomologous chromosomes. As we will see in Chapter 6, this law is violated if two different genes are located close to one another on the same chromosome.

Sex Differences Often Correlate with the Presence of Sex Chromosomes

According to the chromosome theory of inheritance, chromosomes carry the genes that determine an organism's traits and are

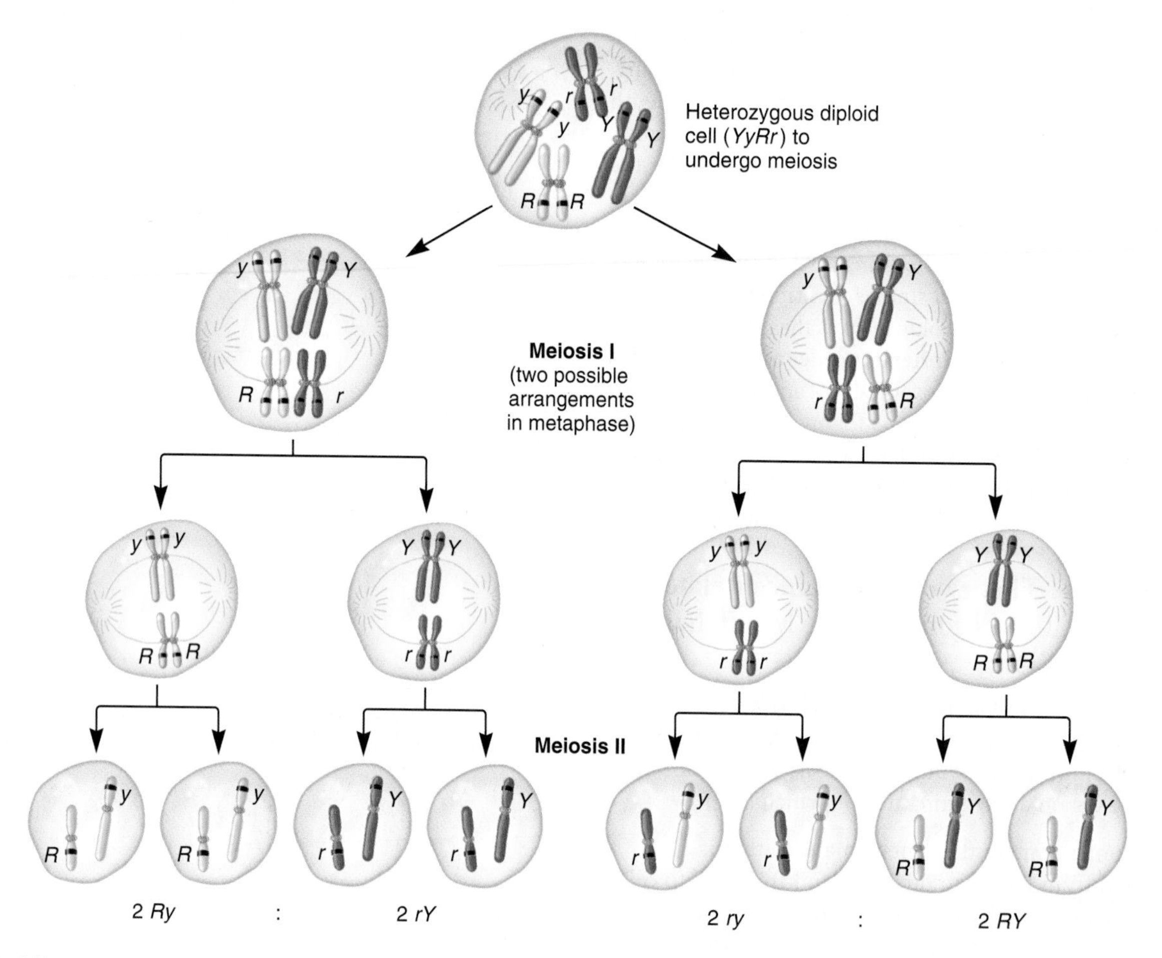

FIGURE 3.17 Mendel's law of independent assortment can be explained by the random alignment of bivalents during metaphase of meiosis I. This figure shows the assortment of two genes located on two different chromosomes, using pea seed color and shape as an example (*YyRr*). During metaphase of meiosis I, different possible arrangements of the homologs within bivalents can lead to different combinations of the alleles in the resulting gametes. For example, on the left, the dominant *R* allele has sorted with the recessive *y* allele; on the right, the dominant *R* allele has sorted with the dominant *Y* allele.

Genes→Traits Most species have several different chromosomes that carry many different genes. In this example, the gene for seed color exists in two alleles, *Y* (yellow) and *y* (green), and the gene for seed shape is found as *R* (round) and *r* (wrinkled) alleles. The two genes are found on different (nonhomologous) chromosomes. During meiosis, the homologous chromosomes that carry these alleles segregate from each other. In addition, the chromosomes carrying the *Y* or *y* alleles will independently assort from the chromosomes carrying the *R* or *r* alleles. As shown here, this provides a reassortment of alleles, potentially creating combinations of alleles that are different from the parental combinations. When two gametes unite during fertilization, the alleles they carry affect the traits of the resulting offspring.

the basis of Mendel's law of segregation and independent assortment. Some early evidence supporting this theory involved the determination of sex. Many species are divided into male and female sexes. In 1901, Clarence McClung, who studied grasshoppers, was the first to suggest that male and female sexes are due to the inheritance of particular chromosomes. Since McClung's initial observations, we now know that a pair of chromosomes, called the **sex chromosomes,** determines sex in many different species. Some examples are described in **Figure 3.18**.

In the X-Y system of sex determination, which operates in mammals, the male contains one X chromosome and one Y chromosome, whereas the female contains two X chromosomes (Figure 3.18a). In this case, the male is called the **heterogametic sex.** Two types of sperm are produced: one that carries only the X chromosome, and another type that carries the Y. In contrast, the female is the **homogametic sex** because all eggs carry a single X chromosome. The 46 chromosomes carried by humans consist of 1 pair of sex chromosomes and 22 pairs of **autosomes**—chromosomes that are not sex chromosomes. In the human male, each of the four sperm produced during gametogenesis contains 23 chromosomes. Two sperm contain an X chromosome, and the other two have a Y chromosome. The sex of the offspring is determined by whether the sperm that fertilizes the egg carries an X or a Y chromosome.

What causes an offspring to develop into a male or female? One possibility is that two X chromosomes are required for female development. A second possibility is that the Y chromosome promotes male development. In the case of mammals, the second possibility is correct. This is known from the analysis of rare individuals who carry chromosomal abnormalities. For example, mistakes that occasionally occur during meiosis may produce an individual who carries two X chromosomes and one Y chromosome. Such an individual develops into a male.

Other mechanisms of sex determination include the X-0, Z-W, and haplodiploid systems. The X-0 system of sex determination operates in many insects (Figure 3.18b). In such species, the male has one sex chromosome (the X) and is designated X0, whereas the female has a pair (two Xs). In other insect species, such as *Drosophila melanogaster*, the male is XY. For both types of insect species (i.e., X0 or XY males, and XX females), the ratio between X chromosomes and the number of autosomal sets determines sex. If a fly has one X chromosome and is diploid for the autosomes ($2n$), the ratio is 1/2, or 0.5. This fly will become a male even if it does not receive a Y chromosome. In contrast to mammals, the Y chromosome in the X-0 system does not determine maleness. If a fly receives two X chromosomes and is diploid, the ratio is 2/2, or 1.0, and the fly becomes a female.

For the Z-W system, which determines sex in birds and some fish, the male is ZZ and the female is ZW (Figure 3.18c). The letters Z and W are used to distinguish these types of sex chromosomes from those found in the X-Y pattern of sex determination of other species. In the Z-W system, the male is the homogametic sex, and the female is heterogametic.

Another interesting mechanism of sex determination, known as the haplodiploid system, is found in bees (Figure 3.18d). The male bee, called the drone, is produced from unfertilized haploid eggs. Female bees, both worker bees and queen bees, are produced from fertilized eggs and therefore are diploid.

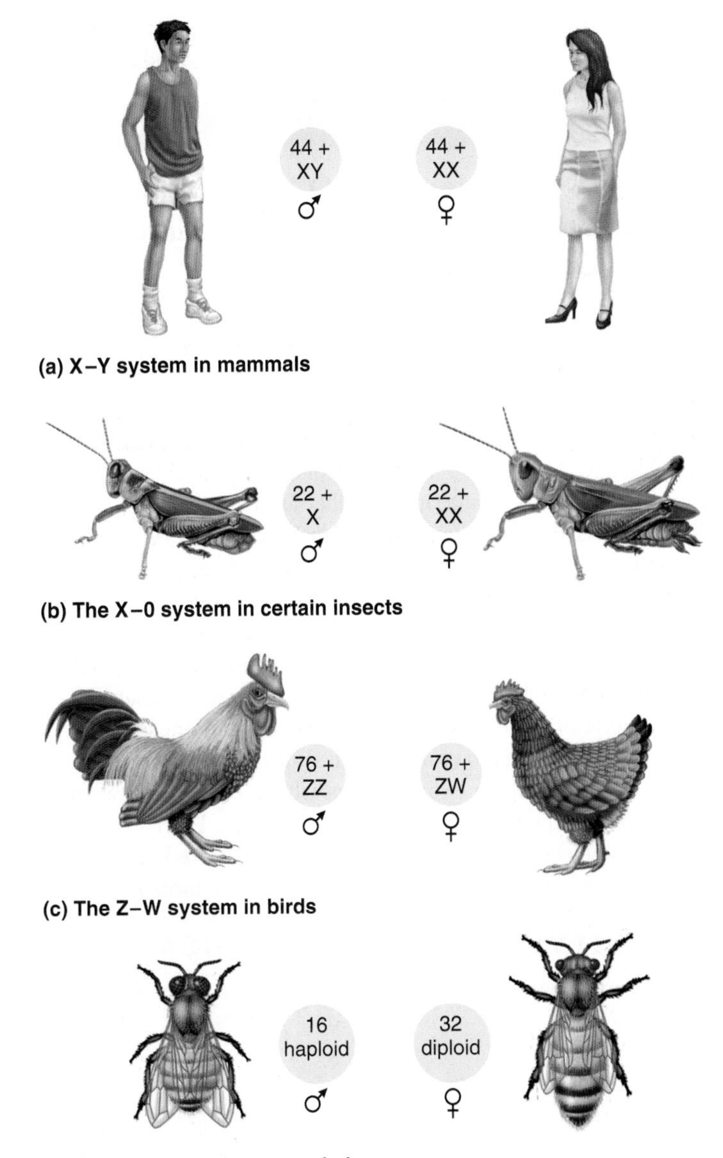

FIGURE 3.18 Different mechanisms of sex determination in animals. See text for a description.

Genes→Traits Certain genes that are found on the sex chromosomes play a key role in the development of sex (male vs. female). For example, in mammals, genes on the Y chromosome initiate male development. In the X-0 system, the ratio of X chromosomes to the sets of autosomes plays a key role in governing the pathway of development toward male or female.

The chromosomal basis for sex determination is rooted in the location of particular genes on the sex chromosomes. The molecular basis for sex determination is described in Chapter 23.

Although sex in many species of animals is determined by chromosomes, other mechanisms are also known. In certain reptiles and fish, sex is controlled by environmental factors such as temperature. For example, in the American alligator (*Alligator mississippiensis*), temperature controls sex development. When fertilized eggs of this alligator are incubated at 33°C, nearly 100%

of them produce male individuals. When the eggs are incubated at a temperature a few degrees below 33°C, they produce nearly all females, whereas at a temperature a few degrees above 33°C, they produce 95% females.

EXPERIMENT 3A

Morgan's Experiments Showed a Connection Between a Genetic Trait and the Inheritance of a Sex Chromosome in *Drosophila*

In the early 1900s, Thomas Hunt Morgan carried out a particularly influential study that confirmed the chromosome theory of inheritance. Morgan was trained as an embryologist, and much of his early research involved descriptive and experimental work in that field. He was particularly interested in ways that organisms change. He wrote, "The most distinctive problem of zoological work is the change in form that animals undergo, both in the course of their development from the egg (embryology) and in their development in time (evolution)." Throughout his life, he usually had dozens of different experiments going on simultaneously, many of them unrelated to each other. He jokingly said there are three kinds of experiments—those that are foolish, those that are damn foolish, and those that are worse than that!

In one of his most famous studies, Morgan engaged one of his graduate students to rear the fruit fly *Drosophila melanogaster* in the dark, hoping to produce flies whose eyes would atrophy from disuse and disappear in future generations. Even after many consecutive generations, however, the flies appeared to have no noticeable changes despite repeated attempts at inducing mutations by treatments with agents such as X-rays and radium. After two years, Morgan finally obtained an interesting result when a true-breeding line of *Drosophila* produced a male fruit fly with white eyes rather than the normal red eyes. Because this had been a true-breeding line of flies, this white-eyed male must have arisen from a new mutation that converted a red-eye allele (denoted w^+) into a white-eye allele (denoted w). Morgan is said to have carried this fly home with him in a jar, put it by his bedside at night while he slept, and then taken it back to the laboratory during the day.

Much like Mendel, Morgan studied the inheritance of this white-eye trait by making crosses and quantitatively analyzing their outcome. In the experiment described in **Figure 3.19**, he began with his white-eyed male and crossed it to a true-breeding red-eyed female. All of the F_1 offspring had red eyes, indicating that red is dominant to white. The F_1 offspring were then mated to each other to obtain an F_2 generation.

THE GOAL

This is an example of discovery-based science rather than hypothesis testing. In this case, a quantitative analysis of genetic crosses may reveal the inheritance pattern for the white-eye allele.

ACHIEVING THE GOAL — FIGURE 3.19 Inheritance pattern of an X-linked trait in fruit flies.

Starting material: A true-breeding line of red-eyed fruit flies plus one white-eyed male fly that was discovered in the culture.

	Experimental level	Conceptual level
1. Cross the white-eyed male to a true-breeding red-eyed female.		X^wY x $X^{w^+}X^{w^+}$
2. Record the results of the F_1 generation. This involves noting the eye color and sexes of many offspring.		$X^{w^+}Y$ male offspring and $X^{w^+}X^w$ female offspring, both with red eyes $X^{w^+}Y$ x $X^{w^+}X^w$
3. Cross F_1 offspring with each other to obtain F_2 offspring. Also record the eye color and sex of the F_2 offspring.		1 $X^{w^+}Y$: 1 X^wY : 1 $X^{w^+}X^{w^+}$: 1 $X^{w^+}X^w$ 1 red-eyed male : 1 white-eyed male : 2 red-eyed females

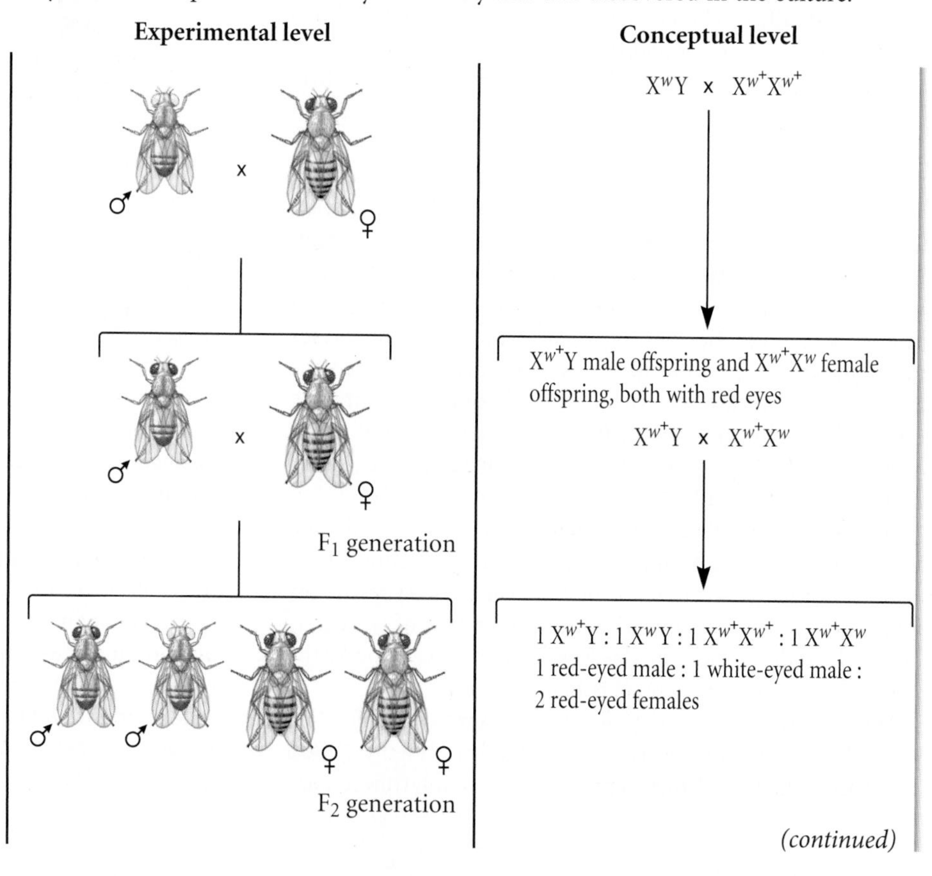

(continued)

4. In a separate experiment, perform a testcross between a white-eyed male and a red-eyed female from the F_1 generation. Record the results.

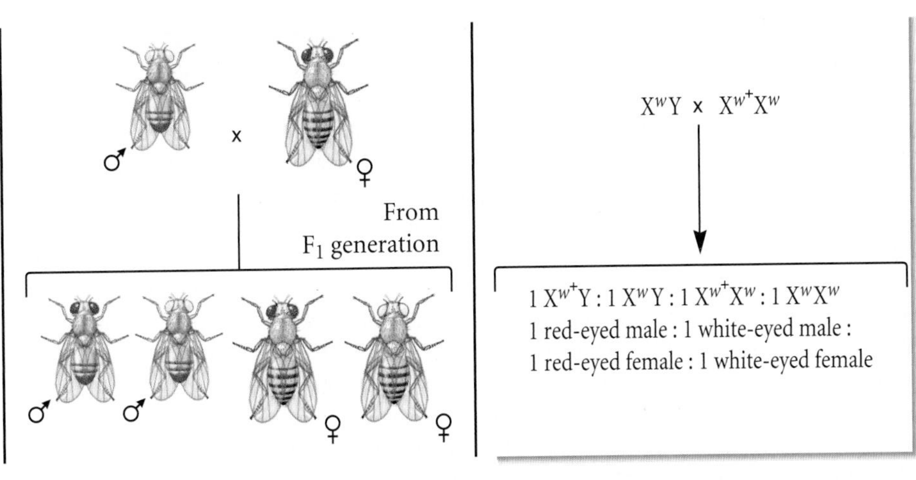

THE DATA

Cross	*Results*	
Original white-eyed male to red-eyed female	F_1 generation:	All red-eyed flies
F_1 male to F_1 female	F_2 generation:	2459 red-eyed females 1011 red-eyed males 0 white-eyed females 782 white-eyed males
White-eyed male to F_1 female	Testcross:	129 red-eyed females 132 red-eyed males 88 white-eyed females 86 white-eyed males

Data from T.H. Morgan (1910) Sex limited inheritance in *Drosophila*. *Science* 32, 120–122.

INTERPRETING THE DATA

As seen in the data table, the F_2 generation consisted of 2459 red-eyed females, 1011 red-eyed males, and 782 white-eyed males. Most notably, no white-eyed female offspring were observed in the F_2 generation. These results suggested that the pattern of transmission from parent to offspring depends on the sex of the offspring and on the alleles that they carry. As shown in the Punnett square below, the data are consistent with the idea that the eye color alleles are located on the X chromosome.

F_1 male is $X^{w^+}Y$
F_1 female is $X^{w^+}X^w$

Female gametes ♀ \ Male gametes ♂	X^{w^+}	Y
X^{w^+}	$X^{w^+}X^{w^+}$ Red, female	$X^{w^+}Y$ Red, male
X^w	$X^{w^+}X^w$ Red, female	X^wY White, male

The Punnett square predicts that the F_2 generation will not have any white-eyed females. This prediction was confirmed experimentally. These results indicated that the eye color alleles are located on the X chromosome. Genes that are physically located within the X chromosome are called **X-linked genes,** or **X-linked alleles.** However, it should also be pointed out that the experimental ratio in the F_2 generation of red eyes to white eyes is (2459 + 1011):782, which equals 4.4:1. This ratio deviates significantly from the predicted ratio of 3:1. How can this discrepancy be explained? Later work revealed that the lower-than-expected number of white-eyed flies is due to their decreased survival rate.

Morgan also conducted a **testcross** (see step 4, Figure 3.19) in which an individual with a dominant phenotype and unknown genotype is crossed to an individual with a recessive phenotype. In this case, he mated an F_1 red-eyed female to a white-eyed male. This cross produced red-eyed males and females, and white-eyed males and females, in approximately equal numbers. The testcross data are also consistent with an X-linked pattern of inheritance. As shown in the following Punnett square, the testcross predicts a 1:1:1:1 ratio:

Testcross:
Male is X^wY
F_1 female is $X^{w^+}X^w$

Female gametes ♀ \ Male gametes ♂	X^w	Y
X^{w^+}	$X^{w^+}X^w$ Red, female	$X^{w^+}Y$ Red, male
X^w	X^wX^w White, female	X^wY White, male

The observed data are 129:132:88:86, which is a ratio of 1.5:1.5:1:1. Again, the lower-than-expected numbers of white-eyed males and females can be explained by a lower survival rate for white-eyed flies. In his own interpretation, Morgan concluded that R (red eye color) and X (a sex factor that is present in two copies in the female) are combined and have never existed apart. In other words, this gene for eye color is on the X chromosome. Morgan was the first geneticist to receive the Nobel Prize.

Calvin Bridges, a graduate student in the laboratory of Morgan, also examined the transmission of X-linked traits. Bridges conducted hundreds of crosses involving several different types of X-linked alleles. In his crosses, he occasionally obtained offspring that had unexpected phenotypes and abnormalities in sex chromosome composition. For example, in a cross between a white-eyed female and a red-eyed male, he occasionally observed a male offspring with red eyes. This event can be explained by nondisjunction, which is described in Chapter 8 (see Figure 8.22). In this example, the rare male offspring with red eyes was produced by a sperm carrying the X-linked red allele and by an egg that underwent nondisjunction and did not receive an X chromosome. The resulting offspring would be a male without a Y chromosome. (As shown earlier in Figure 3.18, the number of X chromosomes determines sex in fruit flies). Bridges observed a parallel between the cytological presence of sex chromosome abnormalities and the occurrence of unexpected traits, which confirmed the idea that the sex chromosomes carry X-linked genes. Together, the work of Morgan and Bridges provided an impressive body of evidence confirming the idea that traits following an X-linked pattern of inheritance are governed by genes that are physically located on the X chromosome. Bridges wrote, "There can be no doubt that the complete parallelism between the unique behavior of chromosomes and the behavior of sex-linked genes and sex in this case means that the sex-linked genes are located in and borne by the X chromosomes." An example of Bridges's work is described in solved problem S5 at the end of this chapter.

A self-help quiz involving this experiment can be found at www.mhhe.com/brookergenetics4e.

KEY TERMS

Page 44. chromosomes, chromatin, prokaryotes, nucleoid, eukaryotes
Page 45. organelles, nucleus, cytogenetics, cytogeneticist
Page 47. somatic cell, gametes, germ cells, karyotype, diploid, homologs, locus, loci
Page 48. asexual reproduction, binary fission
Page 49. cell cycle, interphase, restriction point, chromatids, centromere, sister chromatids, kinetochore
Page 50. mitosis, mitotic spindle apparatus, mitotic spindle, microtubule-organizing centers, centrosomes, spindle pole, centrioles
Page 51. decondensed
Page 53. prophase, condense, prometaphase, metaphase plate, metaphase, anaphase, telophase, cytokinesis, myosin, actin, cleavage furrow, cell plate
Page 54. gametogenesis, isogamous, heterogamous, sperm cells, egg cell, ovum, haploid, meiosis, leptotene, zygotene, synapsis, pachytene, bivalents, synaptonemal complex
Page 55. crossing over
Page 56. chiasma, chiasmata, diplotene, tetrad, diakinesis
Page 58. spermatogenesis
Page 59. oogenesis, gametophyte, sporophyte, pollen grain
Page 60. embryo sac, endosperm
Page 61. X-linked inheritance, chromosome theory of inheritance
Page 63. sex chromosomes, heterogametic sex, homogametic sex, autosomes
Page 65. X-linked genes, X-linked alleles, testcross

CHAPTER SUMMARY

3.1 General Features of Chromosomes

- Chromosomes are structures that contain the genetic material, which is DNA.
- Prokaryotic cells are simple and lack cell compartmentalization, whereas eukaryotic cells contain a cell nucleus and other compartments (see Figure 3.1).
- Chromosomes can be examined under the microscope. An organized representation of the chromosomes from a single cell is called a karyotype (see Figure 3.2).
- In eukaryotic species, the chromosomes are found in sets. Eukaryotic cells are often diploid, which means that each type of chromosome occurs in a homologous pair (see Figure 3.3).

3.2 Cell Division

- Bacteria divide by binary fission (see Figure 3.4).
- To divide, eukaryotic cells progress through a cell cycle (see Figure 3.5).
- Prior to cell division, eukaryotic chromosomes are replicated to form sister chromatids (see Figure 3.6).
- Chromosome sorting in eukaryotes is achieved via a spindle apparatus (see Figure 3.7).
- A common way for eukaryotic cells to divide is by mitosis and cytokinesis. Mitosis is divided into prophase, prometaphase, metaphase, anaphase, and telophase (see Figures 3.8, 3.9).

3.3 Sexual Reproduction

- Another way for eukaryotic cells to divide is via meiosis, which produces four haploid cells. During prophase of meiosis I, homologs synapse and crossing over may occur (see Figures 3.10–3.13).
- Animals produce gametes via spermatogenesis and oogenesis (see Figure 3.14).
- Plants exhibit alternation of generations between a diploid sporophyte and a haploid gametophyte. The gametophyte produces gametes (see Figure 3.15).

3.4 The Chromosome Theory of Inheritance and Sex Chromosomes

- The chromosome theory of inheritance explains how the transmission of chromosomes can explain Mendel's laws.
- Mendel's law of segregation is explained by the separation of homologs during meiosis (see Figure 3.16).
- Mendel's law of independent assortment is explained by the random alignment of different chromosomes during metaphase of meiosis I (see Figure 3.17).
- Mechanisms of sex determination in animals may involve differences in chromosome composition (see Figure 3.18).
- Morgan's work provided strong evidence for the chromosome theory of inheritance by showing that a gene affecting eye color in fruit flies is inherited on the X chromosome (see Figure 3.19).

PROBLEM SETS & INSIGHTS

Solved Problems

S1. A diploid cell has eight chromosomes, four per set. For the following diagram, in what phase of mitosis, meiosis I or meiosis II, is this cell?

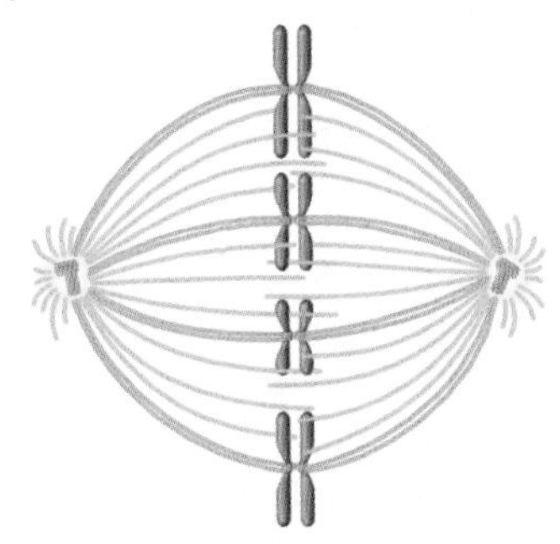

Answer: The cell is in metaphase of meiosis II. You can tell because the chromosomes are lined up in a single row along the metaphase plate, and the cell has only four pairs of sister chromatids. If it were mitosis, the cell would have eight pairs of sister chromatids.

S2. An unaffected woman (i.e., without disease symptoms) who is heterozygous for the X-linked allele causing Duchenne muscular dystrophy has children with a man with a normal allele. What are the probabilities of the following combinations of offspring?

A. An unaffected son

B. An unaffected son or daughter

C. A family of three children, all of whom are affected

Answer: The first thing we must do is construct a Punnett square to determine the outcome of the cross. *N* represents the normal allele, *n* the recessive allele causing Duchenne muscular dystrophy. The mother is heterozygous, and the father has the normal allele.

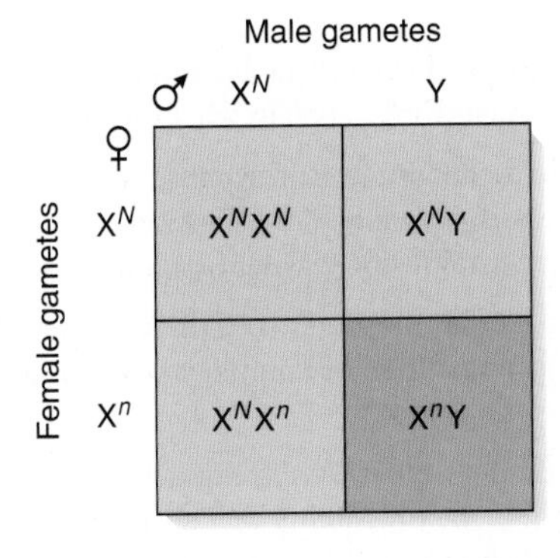

A. There are four possible children, one of whom is an unaffected son. Therefore, the probability of an unaffected son is 1/4.

B. Use the sum rule: 1/4 + 1/2 = 3/4.

C. You could use the product rule because there would be three offspring in a row with the disorder: (1/4)(1/4)(1/4) = 1/64 = 0.016 = 1.6%.

S3. What are the major differences between prophase, metaphase, and anaphase when comparing mitosis, meiosis I, and meiosis II?

Answer: The table summarizes key differences.

A Comparison of Mitosis, Meiosis I, and Meiosis II

Phase	Event	Mitosis	Meiosis I	Meiosis II
Prophase	Synapsis:	No	Yes	No
Prophase	Crossing over:	Rarely	Commonly	Rarely
Metaphase	Alignment along the metaphase plate:	Sister chromatids	Bivalents	Sister chromatids
Anaphase	Separation of:	Sister chromatids	Bivalents	Sister chromatids

S4. Among different plant species, both male and female gametophytes can be produced by single individuals or by separate sexes. In some species, such as the garden pea, a single individual can produce both male and female gametophytes. Fertilization takes place via self-fertilization or cross-fertilization. A plant species that has a single type of flower producing both pollen and eggs is termed a monoclinous plant. In other plant species, two different types of flowers produce either pollen or eggs. When both flower types are on a single individual, such a species is termed monoecious. It is most common for the "male flowers" to be produced near the top of the plant and the "female flowers" toward the bottom. Though less common, some species of plants are dioecious. For dioecious species, one individual makes either male flowers or female flowers, but not both.

Based on your personal observations of plants, try to give examples of monoclinous, monoecious, and dioecious plants. What would be the advantages and disadvantages of each?

Answer: Monoclinous plants—pea plant, tulip, and roses. The same flower produces pollen on the anthers and egg cells within the ovary.

Monoecious plants—corn and pine trees. In corn, the tassels are the male flowers and the ears result from fertilization within the female flowers. In pine trees, pollen is produced in cones near the top of the tree, and eggs cells are found in larger cones nearer the bottom.

Dioecious plants—holly and ginkgo trees. Certain individuals produce only pollen; others produce only eggs.

An advantage of being monoclinous or monoecious is that fertilization is relatively easy because the pollen and egg cells are produced on the same individual. This is particularly true for monoclinous plants. The proximity of the pollen to the egg cells makes it more likely for self-fertilization to occur. This is advantageous if the plant population is relatively sparse. On the other hand, a dioecious species can reproduce only via cross-fertilization. The advantage of cross-fertilization is that it enhances genetic variation. Over the long run, this can be an advantage because cross-fertilization is more likely to produce a varied population of individuals, some of which may possess combinations of traits that promote survival.

S5. To test the chromosome theory of inheritance, Calvin Bridges made crosses involving the inheritance of X-linked traits. One of his experiments concerned two different X-linked genes affecting eye color and wing size. For the eye color gene, the red-eye allele (w^+) is dominant to the white-eye allele (w). A second X-linked trait is wing size; the allele called miniature is recessive to the normal allele. In this case, m represents the miniature allele and m^+ the normal allele. A male fly carrying a miniature allele on its single X chromosome has small (miniature) wings. A female must be homozygous, *mm*, in order to have miniature wings.

Bridges made a cross between $X^{w,m+} X^{w,m+}$ female flies (white eyes and normal wings) to $X^{w+,m}$ Y male flies (red eyes and miniature wings). He then examined the eyes, wings, and sexes of thousands of offspring. Most of the offspring were females with red eyes and normal wings, and males with white eyes and normal wings. On rare occasions (approximately 1 out of 1700 flies), however, he also obtained female offspring with white eyes or males with red eyes. He also noted the wing shape in these flies and then cytologically examined their chromosome composition using a microscope. The following results were obtained:

Offspring	Eye Color	Wing Size	Sex Chromosomes
Expected females	Red	Normal	XX
Expected males	White	Normal	XY
Unexpected females (rare)	White	Normal	XXY
Unexpected males (rare)	Red	Miniature	X0

Data from: Bridges, C. B. (1916) "Non-disjunction as proof of the chromosome theory of heredity," *Genetics 1*, 1–52, 107–163.

Explain these data.

Answer: Remember that in fruit flies, the number of X chromosomes (not the presence of the Y chromosome) determines sex. As seen in the data, the flies with unexpected phenotypes were abnormal in their sex chromosome composition. The white-eyed female flies were due to the union between an abnormal XX female gamete and a normal Y male gamete. Likewise, the unexpected male offspring contained only one X chromosome and no Y. These male offspring were due to the union between an abnormal egg without any X chromosome and a normal sperm containing one X chromosome. The wing size of the unexpected males was a particularly significant result. The red-eyed males showed a miniature wing size. As noted by Bridges, this means they inherited their X chromosome from their father rather than their mother. This observation provided compelling evidence that the inheritance of the X chromosome correlates with the inheritance of particular traits.

At the time of his work, Bridges's results were particularly striking because chromosomal abnormalities had been rarely observed in *Drosophila*. Nevertheless, Bridges first predicted how chromosomal abnormalities would cause certain unexpected phenotypes, and then he actually observed the abnormal number of chromosomes using a microscope. Together, his work provided evidence confirming the idea that traits that follow an X-linked pattern of inheritance are governed by genes physically located on the X chromosome.

Conceptual Questions

C1. The process of binary fission begins with a single mother cell and ends with two daughter cells. Would you expect the mother and daughter cells to be genetically identical? Explain why or why not.

C2. What is a homolog? With regard to genes and alleles, how are homologs similar to and different from each other?

C3. What is a sister chromatid? Are sister chromatids genetically similar or identical? Explain.

C4. With regard to sister chromatids, which phase of mitosis is the organization phase, and which is the separation phase?

C5. A species is diploid containing three chromosomes per set. Draw what the chromosomes would look like in the G_1 and G_2 phases of the cell cycle.

C6. How does the attachment of kinetochore microtubules to the kinetochore differ in metaphase of meiosis I from metaphase of mitosis? Discuss what you think would happen if a sister chromatid was not attached to a kinetochore microtubule.

C7. For the following events, specify whether they occur during mitosis, meiosis I, or meiosis II:

A. Separation of conjoined chromatids within a pair of sister chromatids

B. Pairing of homologous chromosomes

C. Alignment of chromatids along the metaphase plate

D. Attachment of sister chromatids to both poles

C8. Describe the key events during meiosis that result in a 50% reduction in the amount of genetic material per cell.

C9. A cell is diploid and contains three chromosomes per set. Draw the arrangement of chromosomes during metaphase of mitosis and metaphase of meiosis I and II. In your drawing, make one set dark and the other lighter.

C10. The arrangement of homologs during metaphase of meiosis I is a random process. In your own words, explain what this means.

C11. A eukaryotic cell is diploid containing 10 chromosomes (5 in each set). For mitosis and meiosis, how many daughter cells would be produced, and how many chromosomes would each one contain?

C12. If a diploid cell contains six chromosomes (i.e., three per set), how many possible random arrangements of homologs could occur during metaphase of meiosis I?

C13. A cell has four pairs of chromosomes. Assuming that crossing over does not occur, what is the probability that a gamete will contain all of the paternal chromosomes? If n equals the number of chromosomes in a set, which of the following expressions can be used to calculate the probability that a gamete will receive all of the paternal chromosomes: $(1/2)^n$, $(1/2)^{n-1}$, or $n^{1/2}$?

C14. With regard to question C13, how would the phenomenon of crossing over affect the results? In other words, would the probability of a gamete inheriting only paternal chromosomes be higher or lower? Explain your answer.

C15. Eukaryotic cells must sort their chromosomes during mitosis so that each daughter cell receives the correct number of chromosomes. Why don't bacteria need to sort their chromosomes?

C16. Why is it necessary that the chromosomes condense during mitosis and meiosis? What do you think might happen if the chromosomes were not condensed?

C17. Nine-banded armadillos almost always give birth to four offspring that are genetically identical quadruplets. Explain how you think this happens.

C18. A diploid species contains four chromosomes per set for a total of eight chromosomes in its somatic cells. Draw the cell as it would look in late prophase of meiosis II and prophase of mitosis. Discuss how prophase of meiosis II and prophase of mitosis differ from each other, and explain how the difference originates.

C19. Explain why the products of meiosis may not be genetically identical whereas the products of mitosis are.

C20. The period between meiosis I and meiosis II is called interphase II. Does DNA replication take place during interphase II? Explain your answer.

C21. List several ways in which telophase appears to be the reverse of prophase and prometaphase.

C22. Corn has 10 chromosomes per set, and the sporophyte of the species is diploid. If you performed a karyotype, what is the total number of chromosomes you would expect to see in the following types of cells?

A. A leaf cell

B. The sperm nucleus of a pollen grain

C. An endosperm cell after fertilization

D. A root cell

C23. The arctic fox has 50 chromosomes (25 per set), and the common red fox has 38 chromosomes (19 per set). These species can interbreed to produce viable but infertile offspring. How many chromosomes would the offspring have? What problems do you think may occur during meiosis that would explain the offspring's infertility?

C24. Let's suppose that a gene affecting pigmentation is found on the X chromosome (in mammals or insects) or the Z chromosome (in birds) but not on the Y or W chromosome. It is found on an autosome in bees. This gene is found in two alleles, *D* (dark), which is dominant to *d* (light). What would be the phenotypic results of crosses between a true-breeding dark female and true-breeding light male, and the reciprocal crosses involving a true-breeding light female and true-breeding dark male, in the following species? Refer back to Figure 3.18 for the mechanism of sex determination in these species.

A. Birds

B. *Drosophila*

C. Bees

D. Humans

C25. Describe the cellular differences between male and female gametes.

C26. At puberty, the testes contain a finite number of cells and produce an enormous number of sperm cells during the life span of a male. Explain why testes do not run out of spermatogonial cells.

C27. Describe the timing of meiosis I and II during human oogenesis.

C28. Three genes (*A*, *B*, and *C*) are found on three different chromosomes. For the following diploid genotypes, describe all of the possible gamete combinations.

A. *Aa Bb Cc*

B. *AA Bb CC*

C. *Aa BB Cc*

D. *Aa bb cc*

C29. A phenotypically normal woman with an abnormally long chromosome 13 (and a normal homolog of chromosome 13) marries a phenotypically normal man with an abnormally short chromosome 11 (and a normal homolog of chromosome 11). What is the probability of producing an offspring that will have both a long chromosome 13 and a short chromosome 11? If such a child is produced, what is the probability that this child would eventually pass both abnormal chromosomes to one of his or her offspring?

C30. Assuming that such a fly would be viable, what would be the sex of a fruit fly with the following chromosomal composition?

A. One X chromosome and two sets of autosomes

B. Two X chromosomes, one Y chromosome, and two sets of autosomes

C. Two X chromosomes and four sets of autosomes

D. Four X chromosomes, two Y chromosomes, and four sets of autosomes

C31. What would be the sex of a human with the following numbers of sex chromosomes?

A. XXX

B. X (also described as X0)

C. XYY

D. XXY

Experimental Questions

E1. When studying living cells in a laboratory, researchers sometimes use drugs as a way to make cells remain at a particular stage of the cell cycle. For example, aphidicolin inhibits DNA synthesis in eukaryotic cells and causes them to remain in the G_1 phase because they cannot replicate their DNA. In what phase of the cell cycle—G_1, S, G_2, prophase, metaphase, anaphase, or telophase—would you expect somatic cells to stay if the following types of drug were added?

A. A drug that inhibits microtubule formation

B. A drug that allows microtubules to form but prevents them from shortening

C. A drug that inhibits cytokinesis

D. A drug that prevents chromosomal condensation

E2. In Morgan's experiments, which result do you think is the most convincing piece of evidence pointing to X-linkage of the eye color gene? Explain your answer.

E3. In his original studies of Figure 3.19, Morgan first suggested that the original white-eyed male had two copies of the white-eye allele. In this problem, let's assume that he meant the fly was X^wY^w instead of X^wY. Are his data in Figure 3.19 consistent with this hypothesis? What crosses would need to be made to rule out the possibility that the Y chromosome carries a copy of the eye color gene?

E4. How would you set up crosses to determine if a gene was Y linked versus X linked?

E5. Occasionally during meiosis, a mistake can happen whereby a gamete may receive zero or two sex chromosomes rather than one. Calvin Bridges made a cross between white-eyed female flies and red-eyed male flies. As you would expect, most of the offspring were red-eyed females and white-eyed males. On rare occasions, however, he found a white-eyed female or a red-eyed male. These rare flies were not due to new gene mutations but instead were due to mistakes during meiosis in the parent flies. Consider the mechanism of sex determination in fruit flies and propose how this could happen. In your answer, describe the sex chromosome composition of these rare flies.

E6. Let's suppose that you have karyotyped a female fruit fly with red eyes and found that it has three X chromosomes instead of the normal two. Although you do not know its parents, you do know that this fly came from a mixed culture of flies in which some had red eyes, some had white eyes, and some had eosin eyes. Eosin is an allele of the same gene that has white and red alleles. Eosin is a pale orange color. The red allele is dominant and the white allele is recessive. The expression of the eosin allele, however, depends on the number of copies of the allele. When females have two copies of this allele, they have eosin eyes. When females are heterozygous for the eosin allele and white allele, they have light-eosin eyes. When females are heterozygous for the red allele and the eosin allele, they have red eyes. Males that have a single copy of eosin allele have light-eosin eyes.

You cross this female with a white-eyed male and count the number of offspring. You may assume that this unusual female makes half of its gametes with one X chromosome and half of its gametes with two X chromosomes. The following results were obtained:

	Females*	Males
Red eyes	50	11
White eyes	0	0
Eosin	20	0
Light-eosin	21	20

*A female offspring can be XXX, XX, or XXY.

Explain the 3:1 ratio between female and male offspring. What was the genotype of the original mother, which had red eyes and three X chromosomes? Construct a Punnett square that is consistent with these data.

E7. With regard to thickness and length, what do you think the chromosomes would look like if you microscopically examined them during interphase? How would that compare with their appearance during metaphase?

E8. White-eyed flies have a lower survival rate than red-eyed flies. Based on the data in Figure 3.19, what percentage of white-eyed flies survived compared with red-eyed flies, assuming 100% survival of red-eyed flies?

E9. A rare form of dwarfism that also included hearing loss was found to run in a particular family. It is inherited in a dominant manner. It was discovered that an affected individual had one normal copy of chromosome 15 and one abnormal copy of chromosome 15 that was unusually long. How would you determine if the unusually long chromosome 15 was causing this disorder?

E10. Discuss why crosses (i.e., the experiments of Mendel) and the microscopic observations of chromosomes during mitosis and meiosis were both needed to deduce the chromosome theory of inheritance.

E11. A cross was made between female flies with white eyes and miniature wings (both X-linked recessive traits) to male flies with red eyes and normal wings. On rare occasions, female offspring were produced with white eyes. If we assume these females are due to errors in meiosis, what would be the most likely chromosomal composition of such flies? What would be their wing shape?

E12. Experimentally, how do you think researchers were able to determine that the Y chromosome causes maleness in mammals, whereas the ratio of X chromosomes to the sets of autosomes causes sex determination in fruit flies?

Questions for Student Discussion/Collaboration

1. In Figure 3.19, Morgan obtained a white-eyed male fly in a population containing many red-eyed flies that he thought were true-breeding. As mentioned in the experiment, he crossed this fly with several red-eyed sisters, and all the offspring had red eyes. But actually this is not quite true. Morgan observed 1237 red-eyed flies and 3 white-eyed males. Provide two or more explanations why he obtained 3 white-eyed males in the F_1 generation.

2. A diploid eukaryotic cell has 10 chromosomes (5 per set). As a group, take turns having one student draw the cell as it would look during a phase of mitosis, meiosis I, or meiosis II; then have the other students guess which phase it is.

3. Discuss the principles of the chromosome theory of inheritance. Which principles were deduced via light microscopy, and which were deduced from crosses? What modern techniques could be used to support the chromosome theory of inheritance?

Note: All answers appear at the website for this textbook; the answers to even-numbered questions are in the back of the textbook.

www.mhhe.com/brookergenetics4e

Visit the website for practice tests, answer keys, and other learning aids for this chapter. Enhance your understanding of genetics with our interactive exercises, quizzes, animations, and much more.

CHAPTER OUTLINE

Inheritance patterns and alleles. *In the petunia, multiple alleles can result in flowers with several different colors, such as the three shown here.*

4 EXTENSIONS OF MENDELIAN INHERITANCE

The term **Mendelian inheritance** describes inheritance patterns that obey two laws: the law of segregation and the law of independent assortment. Until now, we have mainly considered traits that are affected by a single gene that is found in two different alleles. In these cases, one allele is dominant over the other. This type of inheritance is sometimes called **simple Mendelian inheritance** because the observed ratios in the offspring readily obey Mendel's laws. For example, when two different true-breeding pea plants are crossed (e.g., tall and dwarf) and the F_1 generation is allowed to self-fertilize, the F_2 generation shows a 3:1 phenotypic ratio of tall to dwarf offspring.

In Chapter 4, we will extend our understanding of Mendelian inheritance by first examining the transmission patterns for several traits that do not display a simple dominant/recessive relationship. Geneticists have discovered an amazing diversity of mechanisms by which alleles affect the outcome of traits. Many alleles don't produce the ratios of offspring that are expected from a simple Mendelian relationship. This does not mean that Mendel was wrong. Rather, the inheritance patterns of many traits are more complex and interesting than he had realized. In this chapter, we will examine how the outcome of a trait may be influenced by a variety of factors such as the level of protein expression, the sex of the individual, the presence of multiple alleles of a given gene, and environmental effects. We will also explore how two different genes can contribute to the outcome of a single trait. Later, in Chapters 5 and 6, we will examine eukaryotic inheritance patterns that actually violate the laws of segregation or independent assortment.

4.1 INHERITANCE PATTERNS OF SINGLE GENES

We begin Chapter 4 with the further exploration of traits that are influenced by a single gene. **Table 4.1** describes the general features of several types of Mendelian inheritance patterns that have been observed by researchers. These various patterns occur because the outcome of a trait may be governed by two or more alleles in many different ways. In this section, we will examine these patterns with two goals in mind. First, we want to understand how the molecular expression of genes can account for an individual's phenotype. In other words, we will explore the underlying relationship between molecular genetics—the expression of genes to produce functional proteins—and the traits of individuals that inherit the genes. Our second goal concerns the outcome of crosses. Many of the inheritance patterns described

TABLE 4.1

Types of Mendelian Inheritance Patterns Involving Single Genes

Type	Description
Simple Mendelian	**Inheritance:** This term is commonly applied to the inheritance of alleles that obey Mendel's laws and follow a strict dominant/recessive relationship. In Chapter 4, we will see that some genes can be found in three or more alleles, making the relationship more complex. **Molecular:** 50% of the protein, produced by a single copy of the dominant (functional) allele in the heterozygote, is sufficient to produce the dominant trait.
Incomplete dominance	**Inheritance:** This pattern occurs when the heterozygote has a phenotype that is intermediate between either corresponding homozygote. For example, a cross between homozygous red-flowered and homozygous white-flowered parents will have heterozygous offspring with pink flowers. **Molecular:** 50% of the protein, produced by a single copy of the functional allele in the heterozygote, is not sufficient to produce the same trait as the homozygote making 100%.
Incomplete penetrance	**Inheritance:** This pattern occurs when a dominant phenotype is not expressed even though an individual carries a dominant allele. An example is an individual who carries the polydactyly allele but has a normal number of fingers and toes. **Molecular:** Even though a dominant gene may be present, the protein encoded by the gene may not exert its effects. This can be due to environmental influences or due to other genes that may encode proteins that counteract the effects of the protein encoded by the dominant allele.
Overdominance	**Inheritance:** This pattern occurs when the heterozygote has a trait that is more beneficial than either homozygote. **Molecular:** Three common ways that heterozygotes gain benefits: (1) Their cells may have increased resistance to infection by microorganisms; (2) they may produce more forms of protein dimers, with enhanced function; or (3) they may produce proteins that function under a wider range of conditions.
Codominance	**Inheritance:** This pattern occurs when the heterozygote expresses both alleles simultaneously. For example, in blood typing, an individual carrying the *A* and *B* alleles will have an AB blood type. **Molecular:** The codominant alleles encode proteins that function slightly differently from each other, and the function of each protein in the heterozygote affects the phenotype uniquely.
X-linked	**Inheritance:** This pattern involves the inheritance of genes that are located on the X chromosome. In mammals and fruit flies, males are hemizygous for X-linked genes, whereas females have two copies. **Molecular:** If a pair of X-linked alleles shows a simple dominant/recessive relationship, 50% of the protein, produced by a single copy of the dominant allele in a heterozygous female, is sufficient to produce the dominant trait (in the female).
Sex-influenced inheritance	**Inheritance:** This pattern refers to the effect of sex on the phenotype of the individual. Some alleles are recessive in one sex and dominant in the opposite sex. An example is pattern baldness in humans. **Molecular:** Sex hormones may regulate the molecular expression of genes. This can influence the phenotypic effects of alleles.
Sex-limited inheritance	**Inheritance:** This refers to traits that occur in only one of the two sexes. An example is breast development in mammals. **Molecular:** Sex hormones may regulate the molecular expression of genes. This can influence the phenotypic effects of alleles. In this case, sex hormones that are primarily produced in only one sex are essential to produce a particular phenotype.
Lethal alleles	**Inheritance:** An allele that has the potential of causing the death of an organism. **Molecular:** Lethal alleles are most commonly loss-of-function alleles that encode proteins that are necessary for survival. In some cases, the allele may be due to a mutation in a nonessential gene that changes a protein to function with abnormal and detrimental consequences.

in Table 4.1 do not produce a 3:1 phenotypic ratio when two heterozygotes produce offspring. In this section, we consider how allelic interactions produce ratios that differ from a simple Mendelian pattern. However, as our starting point, we will begin by reconsidering a simple dominant/recessive relationship from a molecular perspective.

Recessive Alleles Often Cause a Reduction in the Amount or Function of the Encoded Proteins

For any given gene, geneticists refer to prevalent alleles in a natural population as **wild-type alleles.** In large populations, more than one wild-type allele may occur—a phenomenon known as **genetic polymorphism.** For example, **Figure 4.1** illustrates a striking example of polymorphism in the elderflower orchid, *Dactylorhiza sambucina.* Throughout the range of this species in Europe, both yellow- and red-flowered individuals are prevalent. Both colors are considered wild type. At the molecular level, a wild-type allele typically encodes a protein that is made in the proper amount and functions normally. As discussed in Chapter 24, wild-type alleles tend to promote the reproductive success of organisms in their native environments.

In addition, random mutations occur in populations and alter preexisting alleles. Geneticists sometimes refer to these kinds of alleles as **mutant alleles** to distinguish them from the more common wild-type alleles. Because random mutations are more likely to disrupt gene function, mutant alleles are often defective in their ability to express a functional protein. Such mutant alleles tend to be rare in natural populations. They are typically, but not always, inherited in a recessive fashion.

Among Mendel's seven traits discussed in Chapter 2, the wild-type alleles are tall plants, purple flowers, axial flowers, yellow seeds, round seeds, green pods, and smooth pods (refer back to Figure 2.4). The mutant alleles are dwarf plants, white flowers, terminal flowers, green seeds, wrinkled seeds, yellow pods, and constricted pods. You may have already noticed that the seven wild-type alleles are dominant over the seven mutant alleles. Likewise, red eyes and normal wings are examples of wild-type alleles in *Drosophila*, and white eyes and miniature wings are recessive mutant alleles.

The idea that recessive alleles usually cause a substantial decrease in the expression of a functional protein is supported by the analysis of many human genetic diseases. Keep in mind that a genetic disease is usually caused by a mutant allele. **Table 4.2** lists several examples of human genetic diseases in which the recessive allele fails to produce a specific cellular protein in its active form. In many cases, molecular techniques have enabled researchers to clone these genes and determine the differences between the wild-type and mutant alleles. They have found that the recessive allele usually contains a mutation that causes a defect in the synthesis of a fully functional protein.

To understand why many defective mutant alleles are inherited recessively, we need to take a quantitative look at protein function. With the exception of sex-linked genes, diploid individuals have two copies of every gene. In a simple dominant/recessive relationship, the recessive allele does not affect the phenotype of the heterozygote. In other words, a single copy of the dominant allele is sufficient to mask the effects of the recessive allele. If the recessive allele cannot produce a functional protein, how do we explain the wild-type phenotype of the heterozygote? As described in **Figure 4.2**, a common explanation is that 50% of the functional protein is adequate to provide the wild-type phenotype. In this example, the *PP* homozygote and *Pp* heterozygote

FIGURE 4.1 An example of genetic polymorphism. Both yellow and red flowers are common in natural populations of the elder-flower orchid, *Dactylorhiza sambucina*, and both are considered wild type.

Dominant (functional) allele: *P* (purple)
Recessive (defective) allele: *p* (white)

Genotype	*PP*	*Pp*	*pp*
Amount of functional protein P	100%	50%	0%

Phenotype	Purple	Purple	White
Simple dominant/recessive relationship			

FIGURE 4.2 A comparison of protein levels among homozygous and heterozygous genotypes *PP*, *Pp*, and *pp*.
Genes →Traits In a simple dominant/recessive relationship, 50% of the protein encoded by one copy of the dominant allele in the heterozygote is sufficient to produce the wild-type phenotype, in this case, purple flowers. A complete lack of the functional protein results in white flowers.

TABLE **4.2**

Examples of Recessive Human Diseases

Disease	Protein That Is Produced by the Normal Gene*	Description
Phenylketonuria	Phenylalanine hydroxylase	Inability to metabolize phenylalanine. The disease can be prevented by following a phenylalanine-free diet. If the diet is not followed early in life, it can lead to severe mental impairment and physical degeneration.
Albinism	Tyrosinase	Lack of pigmentation in the skin, eyes, and hair.
Tay-Sachs disease	Hexosaminidase A	Defect in lipid metabolism. Leads to paralysis, blindness, and early death.
Sandhoff disease	Hexosaminidase B	Defect in lipid metabolism. Muscle weakness in infancy, early blindness, and progressive mental and motor deterioration.
Cystic fibrosis	Chloride transporter	Inability to regulate ion balance across epithelial cells. Leads to production of thick lung mucus and chronic lung infections.
Lesch-Nyhan syndrome	Hypoxanthine-guanine phosphoribosyl transferase	Inability to metabolize purines, which are bases found in DNA and RNA. Leads to self-mutilation behavior, poor motor skills, and usually mental impairment and kidney failure.

*Individuals who exhibit the disease are either homozygous for a recessive allele or hemizygous (for X-linked genes in human males). The disease symptoms result from a defect in the amount or function of the normal protein.

each make sufficient functional protein to yield purple flowers. This means that the homozygous individual makes twice as much of the wild-type protein than it really needs to produce purple flowers. Therefore, if the amount is reduced to 50%, as in the heterozygote, the individual still has plenty of this protein to accomplish whatever cellular function it performs. The phenomenon that "50% of the normal protein is enough" is fairly common among many genes.

A second possible explanation for other genes is that the heterozygote actually produces more than 50% of the functional protein. Due to gene regulation, the expression of the normal gene may be increased or "up-regulated" in the heterozygote to compensate for the lack of function of the defective allele. The topic of gene regulation is discussed in Chapters 14 and 15.

Dominant Mutant Alleles Usually Exert Their Effects in One of Three Ways

Though dominant mutant alleles are much less common than recessive alleles, they do occur in natural populations. How can a mutant allele be dominant over a wild-type allele? Three explanations account for most dominant mutant alleles: a gain-of-function mutation, a dominant-negative mutation, or haploinsufficiency. Some dominant mutant alleles are due to **gain-of-function mutations.** Such mutations change the gene or the protein encoded by a gene so that it gains a new or abnormal function. For example, a mutant gene may be overexpressed and thereby produce too much of the encoded protein. A second category is **dominant-negative mutations** in which the protein encoded by the mutant gene acts antagonistically to the normal protein. In a heterozygote, the mutant protein counteracts the effects of the normal protein and thereby alters the phenotype. Finally, a third way that mutant alleles may affect phenotype is via **haploinsufficiency**. In this case, the mutant allele is a loss-of-function allele. Haploinsufficiency is used to describe patterns of inheritance in which a heterozygote (with one functional allele and one inactive allele) exhibits an abnormal or disease phenotype. An example in humans is a condition called polydactyly in which a heterozygous individual has extra fingers or toes (look ahead to Figure 4.5).

Incomplete Dominance Occurs When Two Alleles Produce an Intermediate Phenotype

Although many alleles display a simple dominant/recessive relationship, geneticists have also identified some cases in which a heterozygote exhibits **incomplete dominance**—a condition in which the phenotype is intermediate between the corresponding homozygous individuals. In 1905, the German botanist Carl Correns first observed this phenomenon in the color of the four-o'clock (*Mirabilis jalapa*). **Figure 4.3** describes Correns' experiment, in which a homozygous red-flowered four-o'clock plant was crossed to a homozygous white-flowered plant. The wild-type allele for red flower color is designated C^R and the white allele is C^W. As shown here, the offspring had pink flowers. If these F_1 offspring were allowed to self-fertilize, the F_2 generation

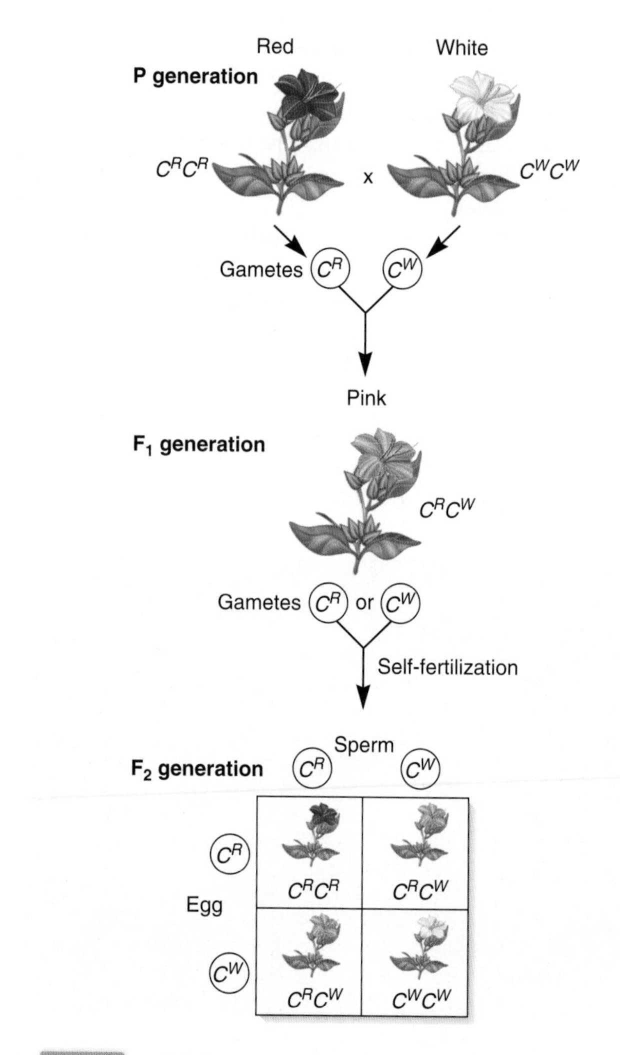

INTERACTIVE EXERCISE

FIGURE 4.3 Incomplete dominance in the four-o'clock plant, *Mirabilis jalapa*.

Genes →Traits When two different homozygotes (C^RC^R and C^WC^W) are crossed, the resulting heterozygote, C^RC^W, has an intermediate phenotype of pink flowers. In this case, 50% of the functional protein encoded by the C^R allele is not sufficient to produce a red phenotype.

consisted of 1/4 red-flowered plants, 1/2 pink-flowered plants, and 1/4 white-flowered plants. The pink plants in the F_2 generation were heterozygotes with an intermediate phenotype. As noted in the Punnett square in Figure 4.3, the F_2 generation displayed a 1:2:1 phenotypic ratio, which is different from the 3:1 ratio observed for simple Mendelian inheritance.

In Figure 4.3, incomplete dominance resulted in a heterozygote with an intermediate phenotype. At the molecular level, the allele that causes a white phenotype is expected to result in a lack of a functional protein required for pigmentation. Depending on the effects of gene regulation, the heterozygotes may produce only 50% of the normal protein, but this amount is not sufficient to produce the same phenotype as the C^RC^R homozygote, which may make twice as much of this protein. In this example, a reasonable explanation is that 50% of the functional protein cannot accomplish the same level of pigment synthesis that 100% of the protein can.

Dominant (functional) allele: *R* (round)
Recessive (defective) allele: *r* (wrinkled)

Genotype	*RR*	*Rr*	*rr*
Amount of functional (starch-producing) protein	100%	50%	0%

Phenotype	Round	Round	Wrinkled
With unaided eye (simple dominant/recessive relationship)			
With microscope (incomplete dominance)			

FIGURE 4.4 A comparison of phenotype at the macroscopic and microscopic levels.

Genes →Traits This illustration shows the effects of a heterozygote having only 50% of the functional protein needed for starch production. This seed appears to be as round as those of the homozygote carrying the *R* allele, but when examined microscopically, it has produced only half the amount of starch.

Finally, our opinion of whether a trait is dominant or incompletely dominant may depend on how closely we examine the trait in the individual. The more closely we look, the more likely we are to discover that the heterozygote is not quite the same as the wild-type homozygote. For example, Mendel studied the characteristic of pea seed shape and visually concluded that the *RR* and *Rr* genotypes produced round seeds and the *rr* genotype produced wrinkled seeds. The peculiar morphology of the wrinkled seed is caused by a large decrease in the amount of starch deposition in the seed due to a defective *r* allele. More recently, other scientists have dissected round and wrinkled seeds and examined their contents under the microscope. They have found that round seeds from heterozygotes actually contain an intermediate number of starch grains compared with seeds from the corresponding homozygotes (**Figure 4.4**). Within the seed, an intermediate amount of the functional protein is not enough to produce as many starch grains as in the homozygote carrying two copies of the *R* allele. Even so, at the level of our unaided eyes, heterozygotes produce seeds that appear to be round. With regard to phenotypes, the *R* allele is dominant to the *r* allele at the level of visual examination, but the *R* and *r* alleles show incomplete dominance at the level of starch biosynthesis.

Traits May Skip a Generation Due to Incomplete Penetrance and Vary in Their Expressivity

As we have seen, dominant alleles are expected to influence the outcome of a trait when they are present in heterozygotes. Occasionally, however, this may not occur. The phenomenon, called **incomplete penetrance,** is a situation in which an allele that is

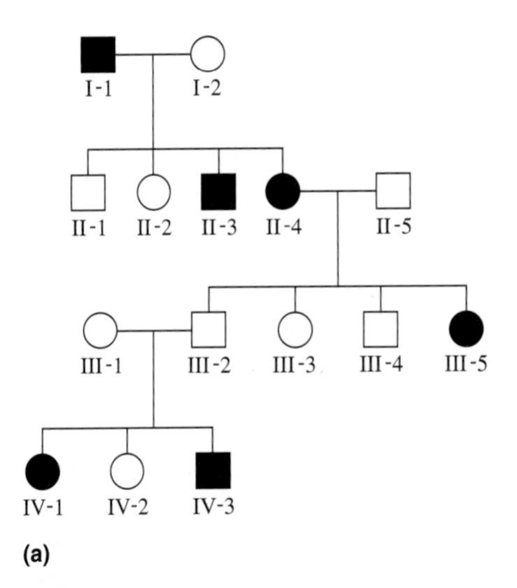

(b)

FIGURE 4.5 Polydactyly, a dominant trait that shows incomplete penetrance. **(a)** A family pedigree. Affected individuals are shown in black. Notice that offspring IV-1 and IV-3 have inherited the trait from a parent, III-2, who is heterozygous but does not exhibit polydactyly. **(b)** Antonio Alfonseca, a baseball player with polydactyly. His extra finger does not give him an advantage when pitching because it is small and does not touch the ball.

expected to cause a particular phenotype does not. **Figure 4.5a** illustrates a human pedigree for a dominant trait known as polydactyly. This trait causes the affected individual to have additional fingers or toes (or both) (**Figure 4.5b**). Polydactyly is due to an autosomal dominant allele—the allele is found in a gene located on an autosome (not a sex chromosome) and a single copy of this allele is sufficient to cause this condition. Sometimes, however, individuals carry the dominant allele but do not exhibit the trait. In Figure 4.5a, individual III-2 has inherited the polydactyly allele from his mother and passed the allele to a daughter and son. However, individual III-2 does not actually exhibit the trait himself, even though he is a heterozygote. In our polydactyly example, the dominant allele does not always "penetrate" into the phenotype of the individual. Alternatively, for recessive traits, incomplete penetrance would occur if a homozygote carrying the recessive allele did not exhibit the recessive trait. The measure of penetrance is described at the populational level. For example, if 60% of the heterozygotes carrying a dominant allele exhibit the trait, we would say that this trait is 60% penetrant. At the individual level, the trait is either present or not.

Another term used to describe the outcome of traits is the degree to which the trait is expressed, or its **expressivity.** In the case of polydactyly, the number of extra digits can vary. For example, one individual may have an extra toe on only one foot, whereas a second individual may have extra digits on both the hands and feet. Using genetic terminology, a person with several extra digits would have high expressivity of this trait, whereas a person with a single extra digit would have low expressivity.

How do we explain incomplete penetrance and variable expressivity? Although the answer may not always be understood, the range of phenotypes is often due to environmental influences and/or due to effects of modifier genes in which one or more genes alter the phenotypic effects of another gene. We will consider the issue of the environment next. The effects of modifier genes will be discussed later in the chapter.

The Outcome of Traits Is Influenced by the Environment

Throughout this book, our study of genetics tends to focus on the roles of genes in the outcome of traits. In addition to genetics, environmental conditions have a great effect on the phenotype of the individual. For example, the arctic fox (*Alopex lagopus*) goes through two color phases. During the cold winter, the arctic fox is primarily white, but in the warmer summer, it is mostly brown (**Figure 4.6a**). As discussed later, such temperature-sensitive alleles affecting fur color are found among many species of mammals.

A dramatic example of the relationship between environment and phenotype can be seen in the human genetic disease known as phenylketonuria (PKU). This autosomal recessive disease is caused by a defect in a gene that encodes the enzyme phenylalanine hydroxylase. Homozygous individuals with this defective allele are unable to metabolize the amino acid phenylalanine properly. When given a standard diet containing

(a) Arctic fox in winter and summer

(b) Healthy person with PKU

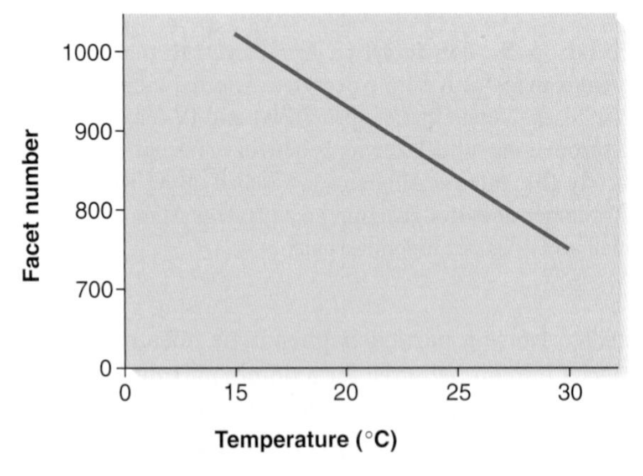

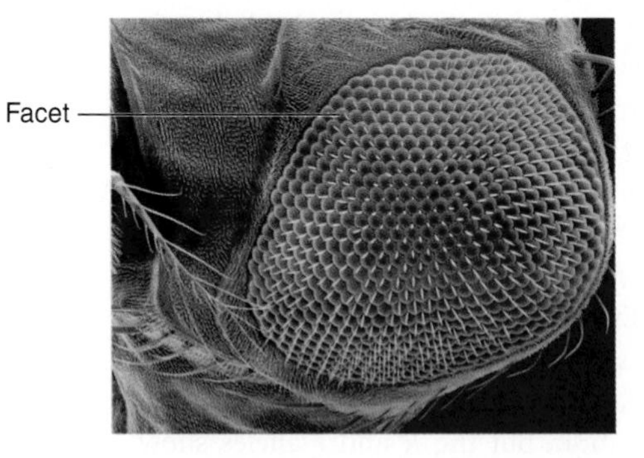

(c) Norm of reaction

FIGURE 4.6 Variation in the expression of traits due to environmental effects. (a) The arctic fox in the winter and summer. **(b)** A person with PKU who has followed a restricted diet and developed normally. **(c)** Norm of reaction. In this experiment, fertilized eggs from a population of genetically identical *Drosophila melanogaster* were allowed to develop into adult flies at different environmental temperatures. This graph shows the relationship between temperature (an environmental factor) and facet number in the eyes of the resulting adult flies. The micrograph shows an eye of *D. melanogaster*.

phenylalanine, which is found in most protein-rich foods, PKU individuals manifest a variety of detrimental traits including mental impairment, underdeveloped teeth, and foul-smelling urine. In contrast, when PKU individuals are diagnosed early and follow a restricted diet free of phenylalanine, they develop normally (**Figure 4.6b**). Since the 1960s, testing methods have been developed that can determine if an individual is lacking the phenylalanine hydroxylase enzyme. These tests permit the identification of infants who have PKU. Their diets can then be modified before the harmful effects of phenylalanine ingestion have occurred. As a result of government legislation, more than 90% of infants born in the United States are now tested for PKU. This test prevents a great deal of human suffering and is also cost-effective. In the United States, the annual cost of PKU testing is estimated to be a few million dollars, whereas the cost of treating severely affected individuals with the disease would be hundreds of millions of dollars.

The examples of the arctic fox and PKU represent dramatic effects of very different environmental conditions. When considering the environment, geneticists often examine a range of conditions, rather than simply observing phenotypes under two different conditions. The term **norm of reaction** refers to the effects of environmental variation on a phenotype. Specifically, it is the phenotypic range seen in individuals with a particular genotype. To evaluate the norm of reaction, researchers begin with true-breeding strains that have the same genotypes and subject them to different environmental conditions. As an example, let's consider facet number in the eyes of fruit flies, *Drosophila melanogaster*. This species has compound eyes composed of many individual facets. **Figure 4.6c** shows the norm of reaction for facet number in genetically identical fruit flies that developed at different temperatures. As shown in the figure, the facet number varies with changes in temperature. At a higher temperature (30°C), the facet number is approximately 750, whereas at a lower temperature (15°C), it is over 1000.

Overdominance Occurs When Heterozygotes Have Superior Traits

As we have just seen, the environment plays a key role in the outcome of traits. For certain genes, heterozygotes may display characteristics that are more beneficial for their survival in a particular environment. Such heterozygotes may be more likely to survive and reproduce. For example, a heterozygote may be larger, disease-resistant, or better able to withstand harsh environmental conditions. The phenomenon in which a heterozygote has greater reproductive success compared with either of the corresponding homozygotes is called **overdominance** or **heterozygote advantage.**

A well-documented example involves a human allele that causes sickle cell disease in homozygous individuals. This disease is an autosomal recessive disorder in which the affected individual produces an altered form of the protein hemoglobin, which carries oxygen within red blood cells. Most people carry the Hb^A allele and make hemoglobin A. Individuals affected with sickle cell anemia are homozygous for the Hb^S allele and produce only hemoglobin S. This causes their red blood cells to deform into a sickle shape under conditions of low oxygen concentration (**Figure 4.7a, b**). The sickling phenomenon causes the life span of these cells to be greatly shortened to only a few weeks compared with a normal span of four months, and therefore, anemia results. In addition, abnormal sickled cells can become clogged in the capillaries throughout the body, leading to localized areas of oxygen depletion. Such an event, called a crisis, causes pain and sometimes tissue and organ damage. For these reasons, the homozygous Hb^SHb^S individual usually has a shortened life span relative to an individual producing hemoglobin A.

In spite of the harmful consequences to homozygotes, the sickle cell allele has been found at a fairly high frequency among human populations that are exposed to malaria. The protozoan genus that causes malaria, *Plasmodium*, spends part of its life

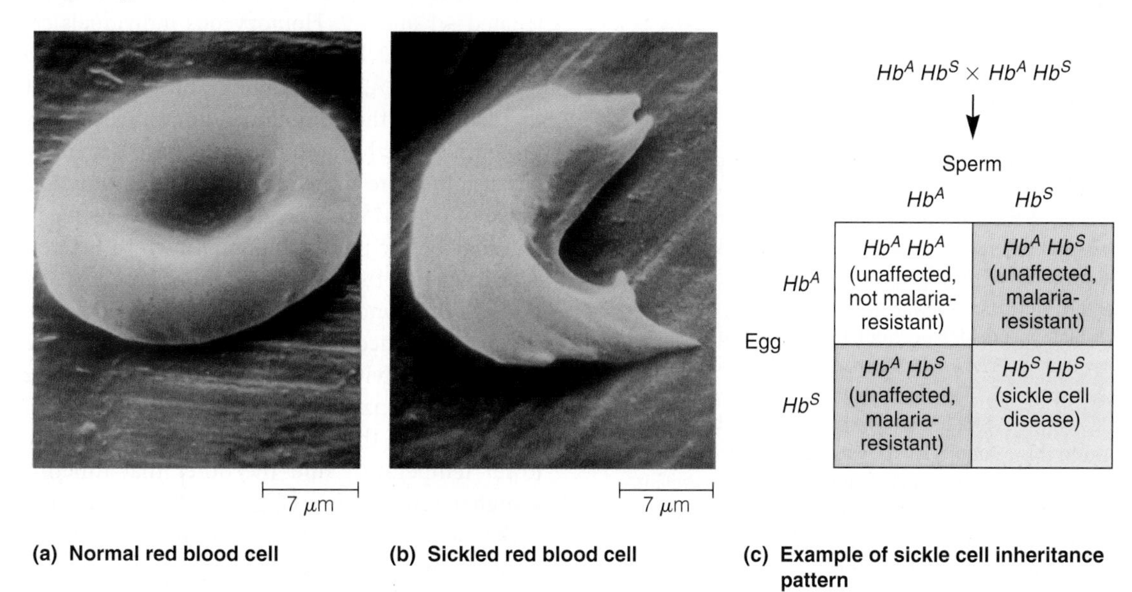

(a) Normal red blood cell **(b) Sickled red blood cell** **(c) Example of sickle cell inheritance pattern**

FIGURE 4.7 Inheritance of sickle cell disease. A comparison of **(a)** normal red blood cells and **(b)** those from a person with sickle cell disease. **(c)** The outcome of a cross between two heterozygous individuals.

cycle within the *Anopheles* mosquito and another part within the red blood cells of humans who have been bitten by an infected mosquito. However, red blood cells of heterozygotes, Hb^AHb^S, are likely to rupture when infected by this parasite, thereby preventing the parasite from propagating. People who are heterozygous have better resistance to malaria than do Hb^AHb^A homozygotes, while not incurring the ill effects of sickle cell disease. Therefore, even though the homozygous Hb^SHb^S condition is detrimental, the greater survival of the heterozygote has selected for the presence of the Hb^S allele within populations where malaria is prevalent. When viewing survival in such a region, overdominance explains the prevalence of the sickle cell allele. In Chapter 24, we will consider the role that natural selection plays in maintaining alleles that are beneficial to the heterozygote but harmful to the homozygote.

Figure 4.7c illustrates the predicted outcome when two heterozygotes have children. In this example, 1/4 of the offspring are Hb^AHb^A (unaffected, not malaria-resistant), 1/2 are Hb^AHb^S (unaffected, malaria-resistant) and 1/4 are Hb^SHb^S (sickle cell disease). This 1:2:1 ratio deviates from a simple Mendelian 3:1 phenotypic ratio.

Overdominance is usually due to two alleles that produce proteins with slightly different amino acid sequences. How can we explain the observation that two protein variants in the heterozygote produce a more favorable phenotype? There are three common explanations. In the case of sickle cell disease, the phenotype is related to the infectivity of *Plasmodium* (**Figure 4.8a**). In the heterozygote, the infectious agent is less likely to propagate within red blood cells. Interestingly, researchers have speculated that other alleles in humans may confer disease resistance in the heterozygous condition but are detrimental in the homozygous state. These include PKU, in which the heterozygous fetus may be resistant to miscarriage caused by a fungal toxin, and Tay-Sachs disease, in which the heterozygote may be resistant to tuberculosis.

A second way to explain overdominance is related to the subunit composition of proteins. In some cases, a protein functions as a complex of multiple subunits; each subunit is composed of one polypeptide. A protein composed of two subunits is called a dimer. When both subunits are encoded by the same gene, the protein is a homodimer. The prefix homo- means that the subunits come from the same type of gene although the gene may exist in different alleles. **Figure 4.8b** considers a situation in which a gene exists in two alleles that encode polypeptides designated A1 and A2. Homozygous individuals can produce only A1A1 or A2A2 homodimers, whereas a heterozygote can also produce an A1A2 homodimer. Thus, heterozygotes can produce three forms of the homodimer, homozygotes only one. For some proteins, A1A2 homodimers may have better functional activity because they are more stable or able to function under a wider range of conditions. The greater activity of the homodimer protein may be the underlying reason why a heterozygote has characteristics superior to either homozygote.

A third molecular explanation of overdominance is that the proteins encoded by each allele exhibit differences in their functional activity. For example, suppose that a gene encodes a metabolic enzyme that can be found in two forms (corresponding to the two alleles), one that functions better at a lower temperature and the other that functions optimally at a higher temperature (**Figure 4.8c**). The heterozygote, which makes a mixture of both enzymes, may be at an advantage under a wider temperature range than either of the corresponding homozygotes.

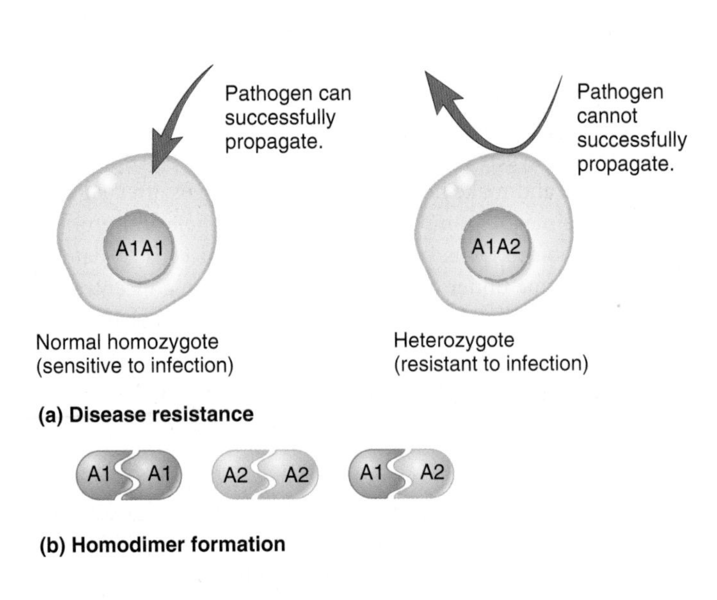

FIGURE 4.8 Three possible explanations for overdominance at the molecular level. **(a)** The successful infection of cells by certain microorganisms depends on the function of particular cellular proteins. In this example, functional differences between A1A1 and A1A2 proteins affect the ability of a pathogen to propagate in the cells. **(b)** Some proteins function as homodimers. In this example, a gene exists in two alleles designated *A1* and *A2*, which encode polypeptides also designated A1 and A2. The homozygotes that are *A1A1* or *A2A2* will make homodimers that are A1A1 and A2A2, respectively. The *A1A2* heterozygote can make A1A1 and A2A2 and can also make A1A2 homodimers, which may have better functional activity. **(c)** In this example, a gene exists in two alleles designated *E1* and *E2*. The *E1* allele encodes an enzyme that functions well in the temperature range of 27° to 32°C. *E2* encodes an enzyme that functions in the range of 30° to 37°C. A heterozygote, *E1E2*, would produce both enzymes and have a broader temperature range (i.e., 27°–37°C) in which the enzyme would function.

Many Genes Exist as Three or More Different Alleles

Thus far, we have considered examples in which a gene exists in two different alleles. As researchers have probed genes at the molecular level within natural populations of organisms, they have discovered that most genes exist in **multiple alleles.** Within a population, genes are typically found in three or more alleles.

An interesting example of multiple alleles involves coat color in rabbits. **Figure 4.9** illustrates the relationship between genotype and phenotype for a combination of four different alleles, which are designated C (full coat color), c^{ch} (chinchilla pattern of coat color), c^{h} (himalayan pattern of coat color), and c (albino). In this case, the gene encodes an enzyme called tyrosinase, which is the first enzyme in a metabolic pathway that leads to the synthesis of melanin from the amino acid tyrosine. This pathway results in the formation of two forms of melanin. Eumelanin, a black pigment, is made first, and then phaeomelanin, an orange/yellow pigment, is made from eumelanin. Alleles of other genes can also influence the relative amounts of eumelanin and phaeomelanin.

Differences in the various alleles are related to the function of tyrosinase. The C allele encodes a fully functional tyrosinase that allows the synthesis of both eumelanin and phaeomelanin, resulting in a full brown coat color. The C allele is dominant to the other three alleles. The chinchilla allele (c^{ch}) is a partial defect in tyrosinase that leads to a slight reduction in black pigment and a greatly diminished amount of orange/yellow pigment, which makes the animal look gray. The albino allele, designated c, is a complete loss of tyrosinase, resulting in white color. The himalayan pattern of coat color, determined by the c^{h} allele, is an example of a **temperature-sensitive allele.** The mutation in this gene has caused a change in the structure of tyrosinase, so it works enzymatically only at low temperature. Because of this property, the enzyme functions only in cooler regions of the body, primarily the tail, the paws, and the tips of the nose and ears. As shown in **Figure 4.10**, similar types of temperature-sensitive alleles have been found in other species of domestic animals, such as the Siamese cat.

Alleles of the ABO Blood Group Can Be Dominant, Recessive, or Codominant

The ABO group of antigens, which determine blood type in humans, is another example of multiple alleles and illustrates yet another allelic relationship called codominance. To understand this concept, we first need to examine the molecular characteristics of human blood types. The plasma membranes of red blood cells have groups of interconnected sugars—oligosaccharides—that act as surface antigens (**Figure 4.11a**). Antigens are molecular

(a) Full coat color CC, Cc^{h}, Cc^{ch}, or Cc.

(b) Chinchilla coat color $c^{ch}c^{ch}$, $c^{ch}c^{h}$, or $c^{ch}c$.

(c) Himalayan coat color $c^{h}c^{h}$ or $c^{h}c$.

(d) Albino coat color cc.

FIGURE 4.9 The relationship between genotype and phenotype in rabbit coat color.

FIGURE 4.10 The expression of a temperature-sensitive conditional allele produces a Siamese pattern of coat color.

Genes →Traits The allele affecting fur pigmentation encodes a pigment-producing protein that functions only at lower temperatures. For this reason, the dark fur is produced only in the cooler parts of the animal, including the tips of the ears, nose, paws, and tail.

structures that are recognized by antibodies produced by the immune system. On red blood cells, two different types of surface antigens, known as A and B, may be found.

The synthesis of these surface antigens is controlled by two alleles, designated I^A and I^B, respectively. The *i* allele is recessive to both I^A and I^B. A person who is homozygous *ii* will have type O blood and does not produce either antigen. A homozygous I^AI^A or heterozygous I^Ai individual will have type A blood. The red blood cells of this individual will contain the surface antigen known as A. Similarly, a homozygous I^BI^B or heterozygous I^Bi individual will produce surface antigen B. As Figure 4.11a indicates, surface antigens A and B have significantly different molecular structures. A person who is I^AI^B will have the blood type AB and express both surface antigens A and B. The phenomenon in which two alleles are both expressed in the heterozygous individual is called **codominance.** In this case, the I^A and I^B alleles are codominant to each other.

As an example of the inheritance of blood type, let's consider the possible offspring between two parents who are I^Ai and I^Bi (**Figure 4.11b**). The I^Ai parent makes I^A and *i* gametes, and the I^Bi parent makes I^B and *i* gametes. These combine to produce I^AI^B, I^Ai, I^Bi, and *ii* offspring in a 1:1:1:1 ratio. The resulting blood types are AB, A, B, and O, respectively.

Biochemists have analyzed the oligosaccharides produced on the surfaces of cells of differing blood types. In type O, the tree is smaller than type A or type B because a sugar has not been attached to a specific site on the tree. This idea is schematically shown in Figure 4.11a. How do we explain this difference at the molecular level? The gene that determines ABO blood type encodes an enzyme called glycosyl transferase that attaches a sugar to the oligosaccharide. The *i* allele carries a mutation that renders this enzyme inactive, which prevents the attachment of an additional sugar. By comparison, the two types of glycosyl transferase encoded by the I^A and I^B alleles have different structures in their active sites. The active site is the part of the protein that recognizes the sugar molecule that will be attached to the oligosaccharide. The glycosyl transferase encoded by the I^A allele recognizes uridine diphosphate *N*-acetylgalactosamine (UDP-GalNAc) and attaches GalNAc

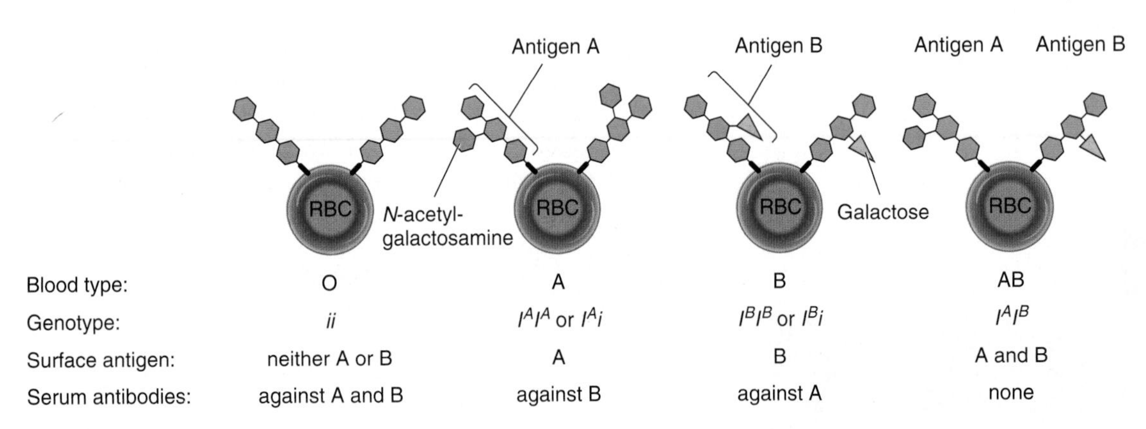

(a) ABO blood type

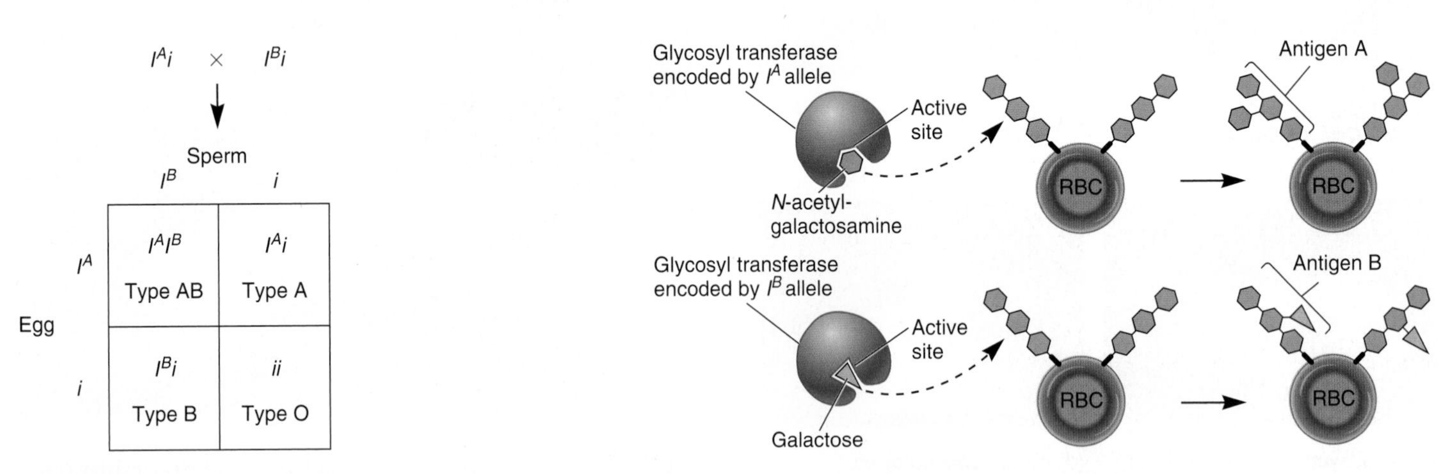

(b) Example of the ABO inheritance pattern

(c) Formation of A and B antigen by glycosyl transferase

FIGURE 4.11 ABO blood type. (a) A schematic representation of blood type at the cellular level. Note: This is not drawn to scale. A red blood cell is much larger than the oligosaccharide on the surface of the cell. (b) The predicted offspring from parents who are I^Ai and I^Bi. (c) The glycosyl transferase encoded by the I^A and I^B alleles recognizes different sugars due to changes in its active site. The *i* allele results in a nonfunctional enzyme.

to the oligosaccharide (**Figure 4.11c**). GalNAc is symbolized as a green hexagon. This produces the structure of surface antigen A. In contrast, the glycosyl transferase encoded by the I^B allele recognizes UDP-galactose and attaches galactose to the oligosaccharide. Galactose is symbolized as an orange triangle. This produces the molecular structure of surface antigen B. A person with type AB blood makes both types of enzymes and thereby has a tree with both types of sugar attached.

A small difference in the structure of the oligosaccharide, namely, a GalNAc in antigen A versus galactose in antigen B, explains why the two antigens are different from each other at the molecular level. These differences enable them to be recognized by different antibodies. A person who has blood type A makes antibodies to blood type B (refer back to Figure 4.11a). The antibodies against blood type B require a galactose in the oligosaccharide for their proper recognition. Their antibodies will not recognize and destroy their own blood cells, but they will recognize and destroy the blood cells from a type B person.

With this in mind, let's consider why blood typing is essential for safe blood transfusions. The donor's blood must be an appropriate match with the recipient's blood. A person with type O blood has the potential to produce antibodies against both A and B antigens if she or he is given type A, type B, or type AB blood. After the antibodies are produced in the recipient, they will react with the donated blood cells and cause them to agglutinate (clump together). This is a life-threatening situation that causes the blood vessels to clog. Other incompatible combinations include a type A person receiving type B or type AB blood, and a type B person receiving type A or type AB blood. Because individuals with type AB blood do not produce antibodies to either A or B antigens, they can receive any type of blood and are known as universal recipients. By comparison, type O persons are universal donors because their blood can be given to type O, A, B, and AB people.

The Inheritance Pattern of X-Linked Genes Can Be Revealed by Reciprocal Crosses

Let's now turn our attention to inheritance patterns of single genes in which the sexes of the parents and offspring play a critical role. As discussed in Chapter 3, many species have males and females that differ in their sex chromosome composition. In mammals, for example, females are XX and males are XY. In such species, certain traits are governed by genes that are located on a sex chromosome. For these traits, the outcome of crosses depends on the genotypes and sexes of the parents and offspring.

As an example, let's consider a human disease known as Duchenne muscular dystrophy (DMD), which was first described by the French neurologist Guillaume Duchenne in the 1860s. Affected individuals show signs of muscle weakness as early as age 3. The disease gradually weakens the skeletal muscles and eventually affects the heart and breathing muscles. Survival is rare beyond the early 30s. The gene for DMD, found on the X chromosome, encodes a protein called dystrophin that is required inside muscle cells for structural support. Dystrophin is thought to strengthen muscle cells by anchoring elements of the internal cytoskeleton to the plasma membrane. Without it, the plasma membrane becomes permeable and may rupture.

DMD is inherited in an **X-linked recessive** pattern—the allele causing the disease is recessive and located on the X chromosome. In the pedigree shown in **Figure 4.12**, several males are affected by this disorder, as indicated by filled squares. The mothers of these males are presumed heterozygotes for this X-linked recessive allele. This recessive disorder is very rare among females because daughters would have to inherit a copy of the mutant allele from their mother and a copy from an affected father.

X-linked muscular dystrophy has also been found in certain breeds of dogs such as golden retrievers (**Figure 4.13a**). Like humans, the mutation occurs in the dystrophin gene, and the symptoms include severe weakness and muscle atrophy that begin at about 6 to 8 weeks of age. Many dogs that inherit this disorder die within the first year of life, though some can live 3 to 5 years and reproduce.

Figure 4.13b (left side) considers a cross between an unaffected female dog with two copies of the wild-type gene and a male dog with muscular dystrophy that carries the mutant allele and has survived to reproductive age. When setting up a Punnett square involving X-linked traits, we must consider the alleles on the X chromosome as well as the observation that males may transmit a Y chromosome instead of the X chromosome. The male makes two types of gametes, one that carries the X chromosome and one that carries the Y. The Punnett square must also include the Y chromosome even though this chromosome does not carry any X-linked genes. The X chromosomes from the female and male are designated with their corresponding alleles. When the Punnett square is filled in, it predicts the X-linked genotypes and sexes of the offspring. As seen on the left side of Figure 4.13b, none of the offspring from this cross are affected with the disorder, although all female offspring are carriers.

The right side of Figure 4.13b shows a **reciprocal cross**—a second cross in which the sexes and phenotypes are reversed. In this case, an affected female animal is crossed to an unaffected

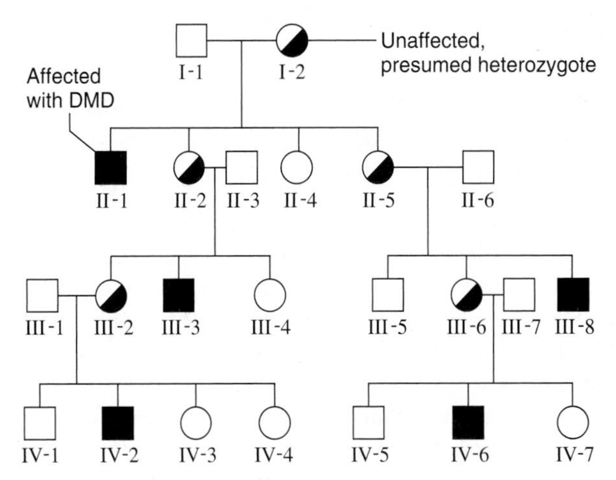

FIGURE 4.12 A human pedigree for Duchenne muscular dystrophy, an X-linked recessive trait. Affected individuals are shown with filled symbols. Females who are unaffected with the disease but have affected sons are presumed to be heterozygous carriers, as shown with half-filled symbols.

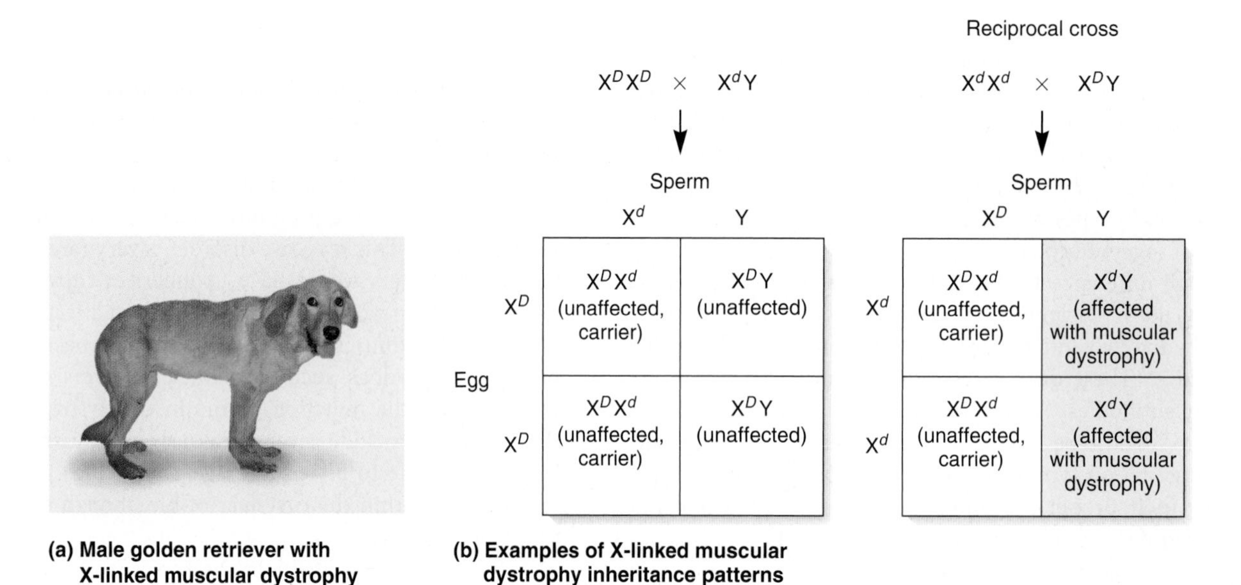

INTERACTIVE EXERCISE

FIGURE 4.13 X-linked muscular dystrophy in dogs. (a) The male golden retriever shown here has the disease. (b) The left side shows a cross between an unaffected female and an affected male. The right shows a reciprocal cross between an affected female and an unaffected male. *D* represents the normal allele for the dystrophin gene, and *d* is the mutant allele that causes a defect in dystrophin function.

male. This cross produces female offspring that are carriers and all male offspring will be affected with muscular dystrophy.

When comparing the two Punnett squares, the outcome of the reciprocal cross yielded different results. This is expected of X-linked genes, because the male transmits the gene only to female offspring, while the female transmits an X chromosome to both male and female offspring. Because the male parent does not transmit the X chromosome to his sons, he does not contribute to their X-linked phenotypes. This explains why X-linked traits do not behave equally in reciprocal crosses. Experimentally, the observation that reciprocal crosses do not yield the same results is an important clue that a trait may be X-linked.

Genes Located on Mammalian Sex Chromosomes Can Be Transmitted in an X-Linked, a Y-Linked, or a Pseudoautosomal Pattern

Our discussion of sex chromosomes has focused on genes that are located on the X chromosome but not on the Y chromosome. The term **sex-linked gene** refers to a gene that is found on one of the two types of sex chromosomes but not on both. Hundreds of X-linked genes have been identified in humans and other mammals.

The inheritance pattern of X-linked genes shows certain distinctive features. For example, males transmit X-linked genes only to their daughters, and sons receive their X-linked genes from their mothers. The term **hemizygous** is used to describe the single copy of an X-linked gene in the male. A male mammal is said to be hemizygous for X-linked genes. Because males of certain species, such as humans, have a single copy of the X chromosome, another distinctive feature of X-linked inheritance is that males are more likely to be affected by rare, recessive X-linked disorders.

By comparison, relatively few genes are located only on the Y chromosome. These few genes are called **holandric genes.** An example of a holandric gene is the *Sry* gene found in mammals. Its expression is necessary for proper male development. A Y-linked inheritance pattern is very distinctive—the gene is transmitted only from fathers to sons.

Besides sex-linked genes, the X and Y chromosomes also contain short regions of homology where the X and Y chromosomes carry the same genes. In addition to several smaller regions, the human sex chromosomes have three homologous regions (**Figure 4.14**). These regions, which are evolutionarily related, promote the necessary pairing of the X and Y chromosomes that occurs during meiosis I of spermatogenesis. Relatively

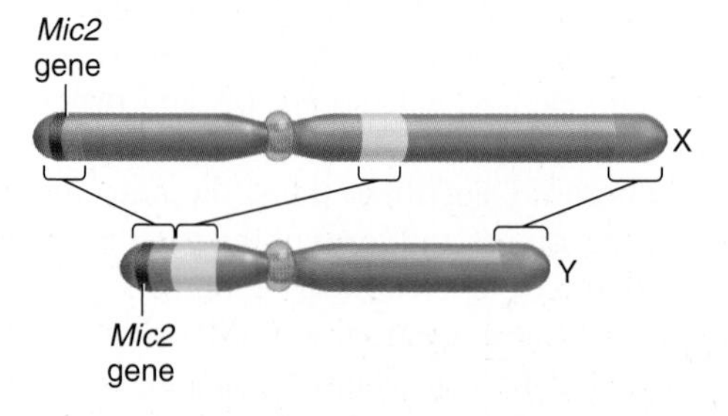

FIGURE 4.14 A comparison of the homologous and nonhomologous regions of the X and Y chromosome in humans. The brackets show three regions of homology between the X and Y chromosome. A few pseudoautosomal genes, such as *Mic2*, are found on both the X and Y chromosomes in these small regions of homology. Researchers estimate that the X chromosome contains between 900 and 1200 genes and the Y chromosome has between 70 and 300 genes.

few genes are located in these homologous regions. One example is a human gene called *Mic2*, which encodes a cell surface antigen. The *Mic2* gene is found on both the X and Y chromosomes. It follows a pattern of inheritance called **pseudoautosomal inheritance.** The term pseudoautosomal refers to the idea that the inheritance pattern of the *Mic2* gene is the same as the inheritance pattern of a gene located on an autosome even though the *Mic2* gene is actually located on the sex chromosomes. As in autosomal inheritance, males have two copies of pseudoautosomally inherited genes, and they can transmit the genes to both daughters and sons.

Some Traits Are Influenced by the Sex of the Individual

As we have just seen, the transmission pattern of sex-linked genes depends on the sex of the parents and offspring. Sex can influence traits in other ways as well. The term **sex-influenced inheritance** refers to the phenomenon in which an allele is dominant in one sex but recessive in the opposite sex. Therefore, sex influence is a phenomenon of heterozygotes. Sex-influenced inheritance should not be confused with sex-linked inheritance. The genes that govern sex-influenced traits are almost always autosomal, not on the X or Y chromosome.

In humans, the common form of pattern baldness provides an example of sex-influenced inheritance. As shown in **Figure 4.15**, the balding pattern is characterized by hair loss on the front and top of the head but not on the sides. This type of pattern baldness is inherited as an autosomal trait. (A common misconception is that this gene is X-linked.) When a male is heterozygous for the baldness allele, he will become bald.

Genotype	Phenotype	
	Males	Females
BB	Bald	Bald
Bb	Bald	Nonbald
bb	Nonbald	Nonbald

In contrast, a heterozygous female will not be bald. Women who are homozygous for the baldness allele will develop the trait, but it is usually characterized by a significant thinning of the hair that occurs relatively late in life.

The sex-influenced nature of pattern baldness is related to the production of the male sex hormone testosterone. The gene that affects pattern baldness encodes an enzyme called 5-α-reductase, which converts testosterone to 5-α-dihydrotestosterone (DHT). DHT binds to cellular receptors and affects the expression of many genes, including those in the cells of the scalp. The allele that causes pattern baldness results in an overexpression of this enzyme. Because mature males normally make more testosterone than females, this allele has a greater phenotypic effect in males. However, a rare tumor of the adrenal gland can cause the secretion of abnormally large amounts of testosterone in females. If this occurs in a woman who is heterozygous *Bb*, she will become bald. If the tumor is removed surgically, her hair will return to its normal condition.

The autosomal nature of pattern baldness has been revealed by the analysis of many human pedigrees. An example is shown in **Figure 4.16a.** A bald male may inherit the bald allele from either parent, and thus a striking observation is that bald fathers can pass this trait to their sons. This could not occur if the trait was X-linked, because fathers do not transmit an X chromosome to their sons. The analyses of many human pedigrees have shown that bald fathers, on average, have at least 50% bald sons. They are expected to produce an even higher percentage of bald male offspring if they are homozygous for the bald allele or the mother also carries one or two copies of the bald allele. For example, a heterozygous bald male and heterozygous (nonbald) female will produce 75% bald sons, whereas a homozygous bald male or homozygous bald female will produce all bald sons.

Figure 4.16b shows the predicted offspring if two heterozygotes produce offspring. In this Punnett square, the phenotypes are designated for both sons and daughters. *BB* offspring are bald, and *bb* offspring are nonbald. *Bb* offspring are bald if they are sons and nonbald if they are daughters. The predicted genotypic

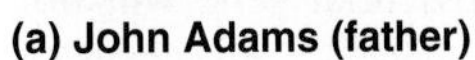

(a) John Adams (father)

(b) John Quincy Adams (son)

(c) Charles Francis Adams (grandson)

(d) Henry Adams (great-grandson)

FIGURE 4.15 **Pattern baldness in the Adams family line.**

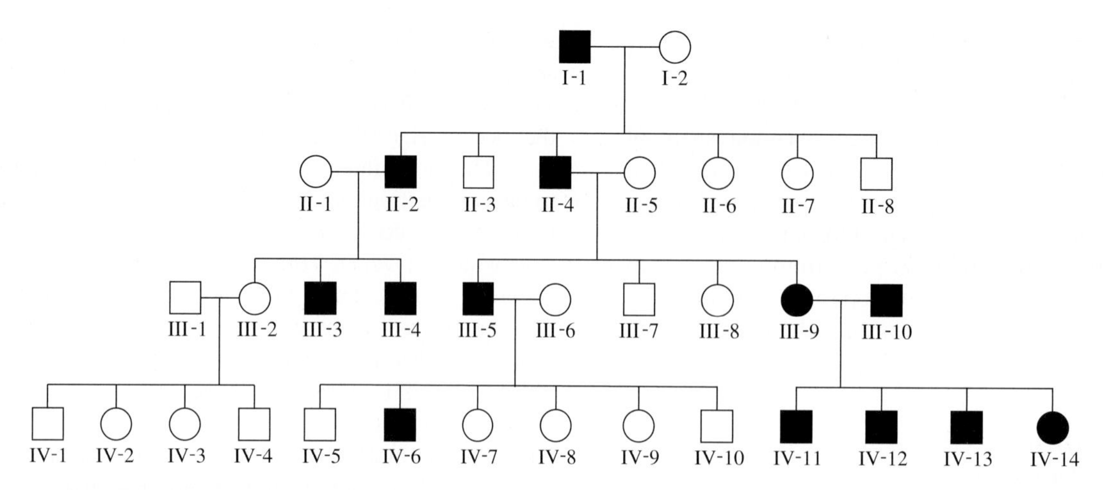

(a) A pedigree for human pattern baldness

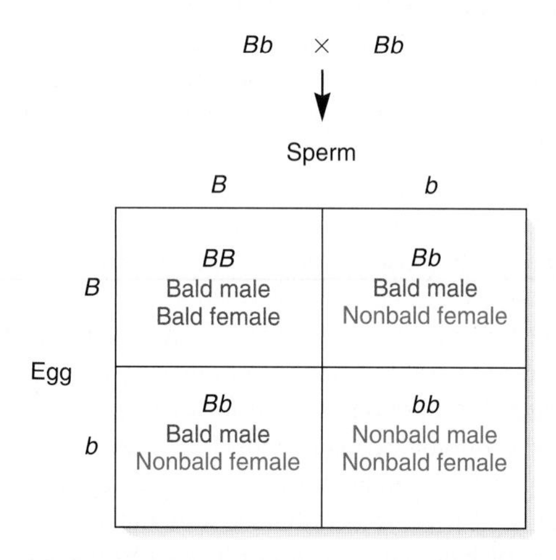

(b) Example of an inheritance pattern involving baldness

FIGURE 4.16 Inheritance of pattern baldness, a sex-influenced trait involving an autosomal gene. **(a)** A family pedigree. Bald individuals are shown in black. **(b)** The predicted offspring from two heterozygous parents.

ratios from this cross would be 1 *BB* bald son to 1 *BB* bald daughter to 2 *Bb* bald sons to 2 *Bb* nonbald daughters to 1 *bb* nonbald son to 1 *bb* nonbald daughter. The predicted phenotypic ratios would be 3 bald sons to 1 bald daughter to 3 nonbald daughters to 1 nonbald son. The ratio of bald to nonbald offspring is 4:4, which is the same as 1:1.

Another example in which sex affects an organism's phenotype is provided by **sex-limited inheritance,** in which a trait occurs in only one of the two sexes. The genes that influence sex-limited traits may be autosomal or X-linked. In humans, examples of sex-limited traits are the presence of ovaries in females and the presence of testes in males. Due to these two sex-limited traits, mature females can only produce eggs, whereas mature males can only produce sperm.

Sex-limited traits are responsible for **sexual dimorphism** in which members of the opposite sex have different morphological features. This phenomenon is common among many animals species and is often striking among various species of birds in which the male has more ornate plumage than the female. As shown in **Figure 4.17**, roosters have a larger comb and wattles and longer neck and tail feathers than do hens. These sex-limited features may be found in roosters but never in normal hens.

Mutations in an Essential Gene May Result in a Lethal Phenotype

Let's now turn our attention to alleles that have the most detrimental effect on phenotype—those that result in death. An allele that has the potential to cause the death of an organism is called a **lethal allele.** These are usually inherited in a recessive manner. When the absence of a specific protein results in a lethal phenotype, the gene that encodes the protein is considered an **essential gene** for survival. Though it varies according to species, researchers estimate that approximately 1/3 of all genes are essential genes. By comparison, **nonessential genes** are not absolutely required for survival, although they are likely to be beneficial to the organism. A loss-of-function mutation in a nonessential gene will not usually cause death. On rare occasions, however, a nonessential gene may acquire a mutation that causes the gene product to be

(a) Hen

(b) Rooster

FIGURE 4.17 Differences in the feathering pattern in female and male chickens, an example of sex-limited inheritance.

abnormally expressed in a way that may interfere with normal cell function and lead to a lethal phenotype. Therefore, not all lethal mutations occur in essential genes, although the great majority do.

Many lethal alleles prevent cell division and thereby cause an organism to die at a very early stage. Others, however, may only exert their effects later in life, or under certain environmental conditions. For example, a human genetic disease known as Huntington disease is caused by a dominant allele. The disease is characterized by a progressive degeneration of the nervous system, dementia, and early death. The age when these symptoms appear, or the **age of onset,** is usually between 30 and 50.

Other lethal alleles may kill an organism only when certain environmental conditions prevail. Such **conditional lethal alleles** have been extensively studied in experimental organisms. For example, some conditional lethals will cause an organism to die only in a particular temperature range. These alleles, called **temperature-sensitive (ts) lethal alleles,** have been observed in many organisms, including *Drosophila.* A ts lethal allele may be fatal for a developing larva at a high temperature (30°C), but the larva will survive if grown at a lower temperature (22°C). Temperature-sensitive lethal alleles are typically caused by mutations that alter the structure of the encoded protein so it does not function correctly at the nonpermissive temperature or becomes unfolded and is rapidly degraded. Conditional lethal alleles may also be identified when an individual is exposed to a particular agent in the environment. For example, people with a defect in the gene that encodes the enzyme glucose-6-phosphate dehydrogenase (G6PD) have a negative reaction to the ingestion of fava beans. This can lead to an acute hemolytic syndrome with 10% mortality if not treated properly.

Finally, it is surprising that certain lethal alleles act only in some individuals. These are called **semilethal alleles.** Of course, any particular individual cannot be semidead. However, within a population, a semilethal allele will cause some individuals to die but not all of them. The reasons for semilethality are not always understood, but environmental conditions and the actions of other genes within the organism may help to prevent the detrimental effects of certain semilethal alleles. An example of a semilethal allele is the X-linked white-eyed allele, which is described in Chapter 3 (see Figure 3.19). Depending on the growth conditions, approximately 1/4 to 1/3 of the flies that would be expected to exhibit this white-eyed trait die prematurely.

In some cases, a lethal allele may produce ratios that seemingly deviate from Mendelian ratios. An example is an allele in a breed of cats known as Manx, which originated on the Isle of Man (**Figure 4.18a**). The Manx cat carries a dominant mutation that affects the spine. This mutation shortens the tail, resulting in a range of tail lengths from normal to tailless. When two Manx cats are crossed to each other, the ratio of offspring is 1 normal to 2 Manx. How do we explain the 1:2 ratio? The answer is that about 1/4 of the offspring die during early embryonic development (**Figure 4.18b**). In this case, the Manx phenotype is dominant, whereas the lethal phenotype occurs only in the homozygous condition.

(a) A Manx cat

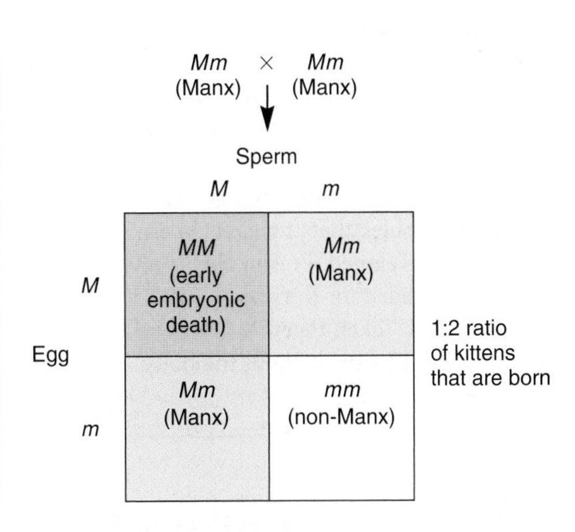

(b) Example of a Manx inheritance pattern

FIGURE 4.18 The Manx cat, which carries a lethal allele. (a) Photo of a Manx cat, which typically has a shortened tail. (b) Outcome of a cross between two Manx cats. Animals that are homozygous for the dominant Manx allele (*M*) die during early embryonic development.

Single Genes Have Pleiotrophic Effects

Before ending our discussion of single-gene inheritance patterns, let's take a broader look at how a single gene may affect phenotype. Although we tend to discuss genes within the context of how they influence a single trait, most genes actually have multiple effects throughout a cell or throughout a multicellular organism. The multiple effects of a single gene on the phenotype of an organism is called **pleiotrophy.** Pleiotrophy occurs for several reasons, including the following:

1. The expression of a single gene can affect cell function in more than one way. For example, a defect in a microtubule protein may affect cell division and cell movement.
2. A gene may be expressed in different cell types in a multicellular organism.
3. A gene may be expressed at different stages of development.

In all or nearly all cases, the expression of a gene is pleiotrophic with regard to the characteristics of an organism. The expression of any given gene influences the expression of many other genes in the genome, and vice versa. Pleiotrophy is revealed when researchers study the effects of gene mutations. As an example of a pleiotrophic mutation, let's consider cystic fibrosis, which is a recessive human disorder. In the late 1980s, the gene for cystic fibrosis was identified. It encodes a protein called the cystic fibrosis transmembrane conductance regulator (CFTR), which regulates ionic balance by allowing the transport of chloride ions (Cl^-) across epithelial cell membranes.

The mutation that causes cystic fibrosis diminishes the function of this Cl^- transporter, affecting several parts of the body in different ways. Because the movement of Cl^- affects water transport across membranes, the most severe symptom of cystic fibrosis is thick mucus in the lungs that occurs because of a water imbalance. In sweat glands, the normal Cl^- transporter has the function of recycling salt out of the glands and back into the skin before it can be lost to the outside world. Persons with cystic fibrosis have excessively salty sweat due to their inability to recycle salt back into their skin cells—a common test for cystic fibrosis is measurement of salt on the skin. Another effect is seen in the reproductive system of males who are homozygous for the cystic fibrosis allele. Males with cystic fibrosis may be infertile because the vas deferens, the tubules that transport sperm from the testes, may be absent or undeveloped. Presumably, a normally functioning Cl^- transporter is needed for the proper development of the vas deferens in the embryo. Taken together, we can see that a defect in CFTR has multiple effects throughout the body.

4.2 GENE INTERACTIONS

In Section 4.1, we considered the effects of a single gene on the outcome of a trait. This approach helps us to understand the various ways that alleles can influence traits. Researchers often examine the effects of a single gene on the outcome of a single trait as a way to simplify the genetic analysis. For example, Mendel studied one gene that affected the height of pea plants—tall versus dwarf alleles. Actually, many other genes in pea plants also affect height, but Mendel did not happen to study variants in those other height genes. How then did Mendel study the effects of a single gene? The answer lies in the genotypes of his strains. Although many genes affect the height of pea plants, Mendel chose true-breeding strains that differed with regard to only one of those genes. As a hypothetical example, let's suppose that pea plants have 10 genes affecting height, which we will call *K, L, M, N, O, P, Q, R, S,* and *T*. The genotypes of two hypothetical strains of pea plants may be

Tall strain: *KK LL MM NN OO PP QQ RR SS TT*

Dwarf strain: *KK LL MM NN OO PP QQ RR SS tt*

In this example, the alleles affecting height may differ at only a single gene. One strain is *TT* and the other is *tt*, and this accounts for the difference in their height. If we make crosses between these tall and dwarf strains, the genotypes of the F_2 offspring will differ with regard to only one gene; the other nine genes will be identical in all of them. This approach allows a researcher to study the effects of a single gene even though many genes may affect a single trait.

Researchers now appreciate that essentially all traits are affected by the contributions of many genes. Morphological features such as height, weight, growth rate, and pigmentation are all affected by the expression of many different genes in combination with environmental factors. In this section, we will further our understanding of genetics by considering how the allelic variants of two different genes affect a single trait. This phenomenon is known as **gene interaction. Table 4.3** considers several examples in which two different genes interact to influence the outcome of particular traits. In this section, we will examine these examples in greater detail.

TABLE **4.3**

Types of Mendelian Inheritance Patterns Involving Two Genes

Type	Description
Epistasis	An inheritance pattern in which the alleles of one gene mask the phenotypic effects of the alleles of a different gene.
Complementation	A phenomenon in which two different parents that express the same or similar recessive phenotypes produce offspring with a wild-type phenotype.
Modifying genes	A phenomenon in which an allele of one gene modifies the phenotypic outcome of the alleles of a different gene.
Gene redundancy	A pattern in which the loss of function in a single gene has no phenotypic effect, but the loss of function of two genes has an effect. Functionality of only one of the two genes is necessary for a normal phenotype; the genes are functionally redundant.
Intergenic suppressors	An inheritance pattern in which the phenotypic effects of one mutation are reversed by a suppressor mutation in another gene.

A Cross Involving a Two-Gene Interaction Can Produce Four Distinct Phenotypes

The first case of two different genes interacting to affect a single trait was discovered by William Bateson and Reginald Punnett in 1906 while they were investigating the inheritance of comb morphology in chickens. Several common varieties of chicken possess combs with different morphologies, as illustrated in **Figure 4.19a**. In their studies, Bateson and Punnett crossed a Wyandotte breed having a rose comb to a Brahma having a pea comb. All F_1 offspring had a walnut comb.

When these F_1 offspring were mated to each other, the F_2 generation consisted of chickens with four types of combs in the following phenotypic ratio: 9 walnut : 3 rose : 3 pea : 1 single comb. As we have seen in Chapter 2, a 9:3:3:1 ratio is obtained in the F_2 generation when the F_1 generation is heterozygous for two different genes and these genes assort independently. However, an important difference here is that we have four distinct categories of a single trait. Based on the 9:3:3:1 ratio, Bateson and Punnett reasoned that a single trait (comb morphology) was determined by two different genes.

R (rose comb) is dominant to *r*.

P (pea comb) is dominant to *p*.

R and *P* (walnut comb) are codominant.

rrpp produces a single comb.

As shown in the Punnett square of **Figure 4.19b**, each of the genes can exist in two alleles, and the two genes show independent assortment.

A Cross Involving a Two-Gene Interaction Can Produce Two Distinct Phenotypes Due to Epistasis

Bateson and Punnett also discovered an unexpected gene interaction when studying crosses involving the sweet pea, *Lathyrus odoratus*. The wild sweet pea has purple flowers. However, they obtained several true-breeding mutant varieties with white flowers. Not surprisingly, when they crossed a true-breeding purple-flowered plant to a true-breeding white-flowered plant, the F_1 generation contained all purple-flowered plants and the F_2 generation (produced by self-fertilization of the F_1 generation) consisted of purple- and white-flowered plants in a 3:1 ratio.

A surprising result came in an experiment where they crossed two different varieties of white-flowered plants (**Figure 4.20**). All of the F_1 generation plants had purple flowers! Bateson and Punnett then allowed the F_1 offspring to self-fertilize. The F_2 generation resulted in purple and white flowers in a ratio of 9 purple to 7 white. From this result, Bateson and Punnett deduced that two different genes were involved, with the following relationship:

C (one purple-color-producing) allele is dominant to *c* (white).

P (another purple-color-producing) allele is dominant to *p* (white).

cc or *pp* masks the *P* or *C* alleles, producing white color.

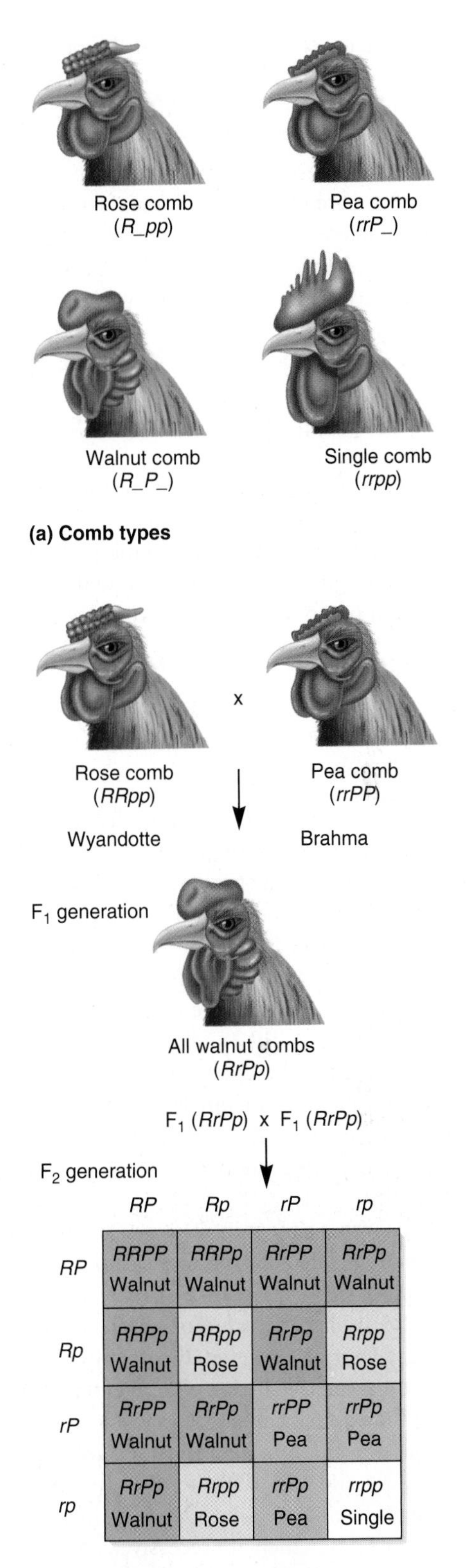

FIGURE 4.19 Inheritance of comb morphology in chickens. This trait is influenced by two different genes, which can each exist in two alleles.
(a) Four phenotypic outcomes are possible. The underline symbol indicates the allele could be either dominant or recessive. **(b)** The crosses of Bateson and Punnett examined the interaction of the two genes.

When the alleles of one gene mask the phenotypic effects of the alleles of another gene, the phenomenon is called **epistasis.** Geneticists consider epistasis relative to a particular phenotype. If possible, geneticists use the wild-type phenotype as their reference phenotype when describing an epistatic interaction. In this case, purple flowers are wild type. Homozygosity for the white allele of one gene masks the expression of the purple-producing allele of another gene. In other words, the *cc* genotype is epistatic to a purple phenotype, and the *pp* genotype is also epistatic to a purple phenotype. At the level of genotypes, *cc* is epistatic to *PP* or *Pp*, and *pp* is epistatic to *CC* or *Cc*. This is an example of **recessive epistasis**. As seen in Figure 4.20, this epistatic interaction produces only two phenotypes—purple or white flowers—in a 9:7 ratio.

Epistatis often occurs because two (or more) different proteins participate in a common function. For example, two or more proteins may be part of an enzymatic pathway leading to the formation of a single product. To illustrate this idea, let's consider the formation of a purple pigment in the sweet pea.

$$\text{Colorless precursor} \xrightarrow{\text{Enzyme C}} \text{Colorless intermediate} \xrightarrow{\text{Enzyme P}} \textbf{Purple pigment}$$

In this example, a colorless precursor molecule must be acted on by two different enzymes to produce the purple pigment. Gene *C* encodes a functional protein called enzyme C, which converts the colorless precursor into a colorless intermediate. Two copies of the recessive allele (*cc*) result in a lack of production of this enzyme in the homozygote. Gene *P* encodes a functional enzyme P, which converts the colorless intermediate into the purple pigment. Like the *c* allele, the recessive *p* allele encodes a defective enzyme P. If an individual is homozygous for either recessive allele (*cc* or *pp*), it will not make any functional enzyme C or enzyme P, respectively. When one of these enzymes is missing, purple pigment cannot be made, and the flowers remain white.

The parental cross shown in Figure 4.20 illustrates another genetic phenomenon called **complementation.** This term refers to the production of offspring with a wild-type phenotype from parents that both display the same or similar recessive phenotype. In this case, purple-flowered F_1 offspring were obtained from two white-flowered parents. Complementation typically occurs because the recessive phenotype in the parents is due to homozgyosity at two different genes. In our sweet pea example, one parent is *CCpp* and the other is *ccPP*. In the F_1 offspring, the *C* and *P* alleles, which are wild-type and dominant, complement the *c* and *p* alleles, which are recessive. The offspring must have one wild-type allele of both genes to display the wild-type phenotype. Why is complementation an important experimental observation? When geneticists observe complementation in a genetic cross, the results suggest that the recessive phenotype in the two parent strains is caused by mutant alleles in two different genes.

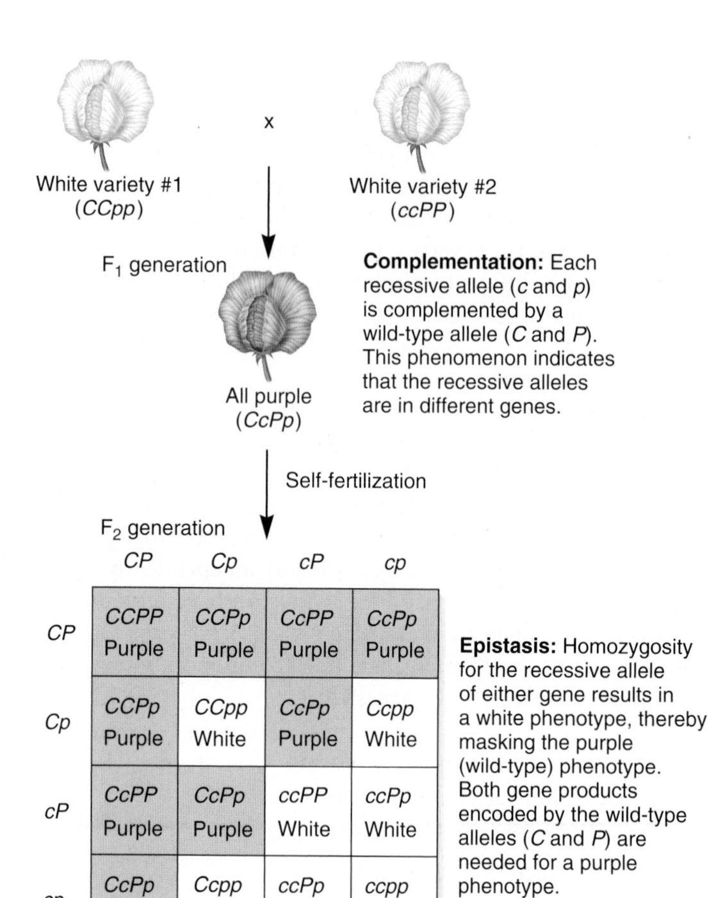

INTERACTIVE EXERCISE

FIGURE 4.20 A cross between two different white varieties of the sweet pea.

Genes →Traits The color of the sweet pea flower is controlled by two genes, which are epistatic to each other and show complementation. Each gene is necessary for the production of an enzyme required for pigment synthesis. The recessive allele of either gene encodes a defective enzyme. If an individual is homozygous recessive for either of the two genes, the purple pigment cannot be synthesized. This results in a white phenotype.

A Cross Involving a Two-Gene Interaction Can Produce Three Distinct Phenotypes Due to Epistasis

Thus far, we have observed two different gene interactions: one producing four phenotypes and the other producing only two. Coat color in rodents provides an example that produces three phenotypes. If a true-breeding black rat is crossed to a true-breeding albino rat, the result is a rat with agouti coat color. Animals with agouti coat color have black pigmentation at the tips of each hair that changes to orange pigmentation near the root. If two agouti animals of the F_1 generation are crossed to each other, they produce agouti, black, and albino offspring in a 9:3:4 ratio (**Figure 4.21**).

How do we explain this ratio? This cross involves two genes that are called *A* (for agouti) and *C* (for colored). The dominant *A* allele of the agouti gene encodes a protein that regulates hair color such that the pigmentation shifts from black (eumelanin) at the tips to orange (phaeomelanin) near the roots. The recessive

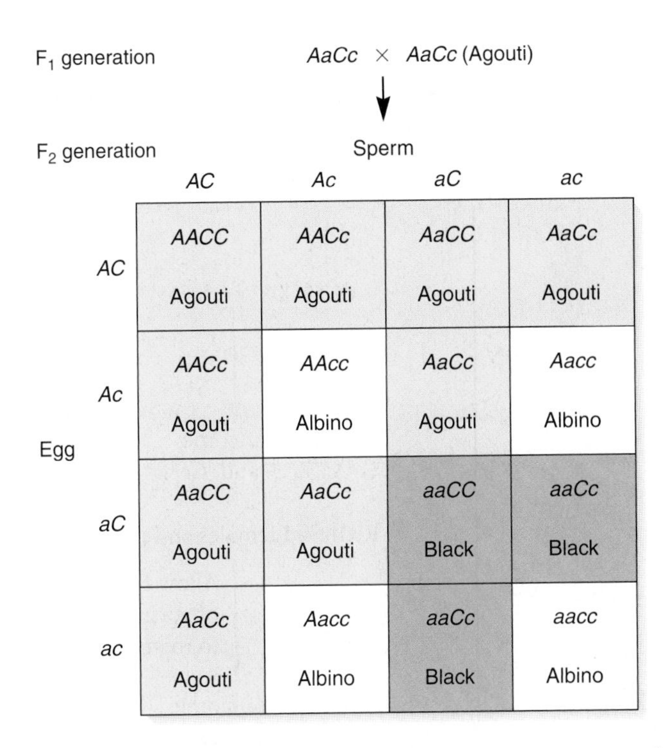

FIGURE 4.21 **Inheritance pattern of coat color in rats involving a gene interaction between the agouti gene (*A* or *a*) and the colored gene (*C* or *c*).**

allele, *a*, inhibits the shift to orange pigmentation and thereby results in black pigment production throughout the entire hair, when an animal is *aa*. As with rabbits, the colored gene encodes tyrosinase, which is needed for the first step in melanin synthesis. The *C* allele allows pigmentation to occur, whereas the *c* allele causes the loss of tyrosinase function. The *C* allele is dominant to the *c* allele; *cc* homozygotes are albino and have white coat color.

As shown at the top of Figure 4.21, the F_1 rats are heterozygous for the two genes. In this case, *C* is dominant to *c*, and *A* is dominant to *a*. If a rat has at least one copy of both dominant alleles, the result is agouti coat color. Let's consider agouti as our reference phenotype. In the F_2 generation, if a rat has a dominant *A* allele but is *cc* homozygous, it will be albino and develop a white coat. The *c* allele is epistatic to *A* and masks pigment production.

By comparison, if an individual has a dominant *C* allele and is homozygous *aa*, the coat color is black. How can we view the effects of the *aa* genotype when an individual carries a *C* allele? Because the *aa* genotype actually masks orange pigmentation, the black phenotype could be viewed as epistasis. However, many geneticists would not view this effect as epistasis but instead would call it a **gene modifier effect**—the alleles of one gene modify the phenotypic effect of the alleles of a different gene. From this alternative viewpoint, the pigmentation is not totally masked, but instead the agouti color is modified to black. Another example of a gene modifer effect is described next.

EXPERIMENT 4A

Bridges Observed an 8:4:3:1 Ratio Because the Cream-Eye Gene Can Modify the X-Linked Eosin Allele But Not the Red or White Alleles

As we have seen, geneticists view epistasis as a situation in which the alleles of a given gene mask the phenotypic effects of the alleles of another gene. In some cases, however, two genes may interact to influence a particular phenotype, but the interaction of particular alleles seems to modify the phenotype, not mask it.

Calvin Bridges discovered an early example in which one gene modifies the phenotypic effects of an X-linked eye color gene in *Drosophila*. As discussed in Chapter 3, the X-linked red allele (w^+) is dominant to the white allele (*w*). Besides these two alleles, Thomas Hunt Morgan and Calvin Bridges found another allele of this gene that they called eosin (*w-e*), which results in eyes that are a pale orange color. The red allele is dominant to the eosin allele. In addition, the expression of the eosin allele depends on the number of copies of the allele. When females have two copies of this allele, they have eosin eyes. When females are heterozygous for the eosin allele and white allele, they have light-eosin eyes. Within true-breeding cultures of flies with eosin eyes, he occasionally found a fly that had a noticeably different eye color. In particular, he identified a rare fly with cream-colored eyes. Bridges reasoned that this new eye color could be explained in two different ways. One possibility is that the cream-colored phenotype could be the result of a new mutation that changed the eosin allele into a cream allele. A second possibility is that a different gene may have incurred a mutation that modified the phenotypic expression of the eosin allele. This second possibility is an example of a gene interaction. To distinguish between these two possibilities, he carried out the crosses described in **Figure 4.22.** He crossed males with cream-colored eyes to wild-type females and then allowed the F_1 generation flies, which all had red eyes, to mate with each other. As shown in the data, all F_2 females had red eyes, but males had red eyes, eosin eyes, or cream eyes.

THE HYPOTHESES

Cream-colored eyes in fruit flies are due to the effect of an allele that is in the same gene as the eosin allele or in a second gene that modifies the expression of the eosin allele.

TESTING THE HYPOTHESES — FIGURE 4.22 A gene interaction between the cream allele and eosin allele.

INTERACTIVE EXERCISE

Starting material: From a culture of flies with eosin eyes, Bridges obtained a fly with cream-colored eyes and used it to produce a true-breeding culture of flies with cream-colored eyes. The allele was called *cream a* (c^a).

Experimental level | **Conceptual level**

1. Cross males with cream-colored eyes to wild-type females.
2. Observe the F_1 offspring and then allow the offspring to mate with each other.
3. Observe and record the eye color and sex of the F_2 generation.

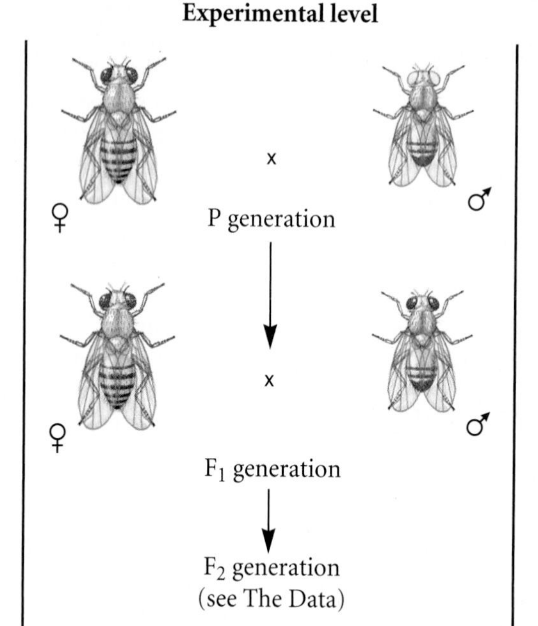

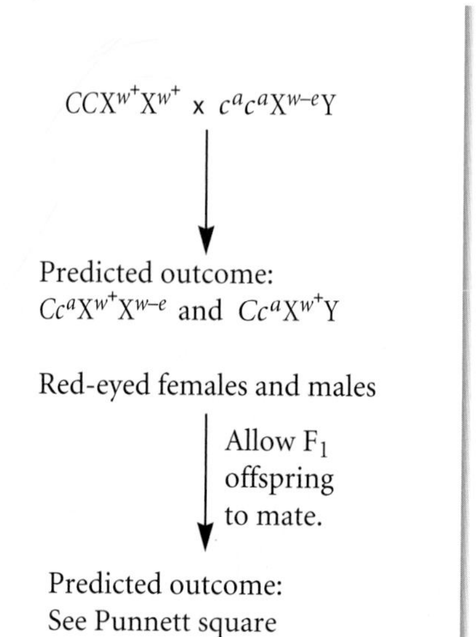

THE DATA

Cross	*Outcome*
P cross:	
Cream-eyed male × wild-type female	F_1: All red eyes
F_1 cross:	
F_1 brother × F_1 sister	F_2: 104 females with red eyes 47 males with red eyes 44 males with eosin eyes 14 males with cream eyes

Data from Calvin Bridges (1919) Specific modifiers of eosin eye color in *Drosophila melanogaster. J. Experimental Zoology 28*, 337–384.

INTERPRETING THE DATA

To interpret these data, keep in mind that Bridges already knew that the eosin allele is X-linked. However, he did not know whether the cream allele was in the same gene as the eosin allele, in a different gene on the X chromosome, or on an autosome. The F_2 generation indicates that the cream allele is not in the same gene as the eosin allele. If the cream allele was in the same gene as the eosin allele, none of the F_2 males would have had eosin eyes; there would have been a 1:1 ratio of red-eyed males and cream-eyed males in the F_2 generation. This result was not obtained. Instead, the actual results are consistent with the idea that the male flies of the parental generation possessed both the eosin and cream alleles. Therefore, Bridges concluded that the cream allele was an allele of a different gene.

One possibility is that the cream allele is an autosomal recessive allele. If so, we can let C represent the dominant allele (which does not modify the eosin phenotype) and c^a represent the cream allele that modifies the eosin color to cream. We already know that the eosin allele is X-linked and recessive to the red allele. The parental cross is expected to produce all red-eyed F_1 flies in which the males are $Cc^aX^{w+}Y$ and the females are $Cc^aX^{w+}X^{w-e}$. When these F_1 offspring are allowed to mate with each other, the Punnett square shown here would predict the following outcome:

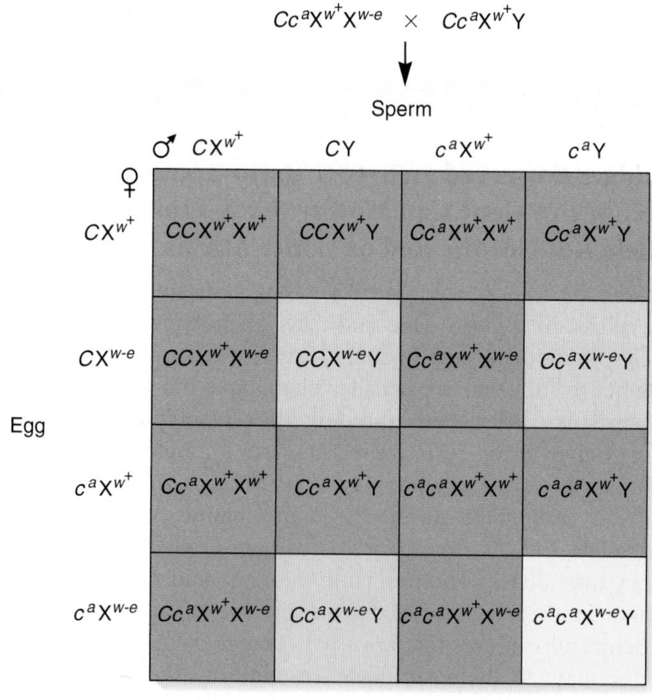

Outcome:

1 $CCX^{w^+}X^{w^+}$: 1 $CCX^{w^+}X^{w-e}$: 2 $Cc^aX^{w^+}X^{w^+}$: 2 $Cc^aX^{w^+}X^{w-e}$: 1 $c^ac^aX^{w^+}X^{w^+}$: 1 $c^ac^aX^{w^+}X^{w-e}$ = **8 red-eyed females**

1 $CCX^{w^+}Y$: 2 $Cc^aX^{w^+}Y$: 1 $c^ac^aX^{w^+}Y$ = **4 red-eyed males**

1 $CCX^{w-e}Y$: 2 $Cc^aX^{w-e}Y$ = **3 light eosin-eyed males**

1 $c^ac^aX^{w-e}Y$ = **1 cream-eyed male**

This phenotypic outcome proposes that the specific modifier allele, c^a, can modify the phenotype of the eosin allele but not the red-eye allele. The eosin allele can be modified only when the c^a allele is homozygous. The predicted 8:4:3:1 ratio agrees reasonably well with Bridges's data.

A self-help quiz involving this experiment can be found at www.mhhe.com/brookergenetics4e.

Due to Gene Redundancy, Loss-of-Function Alleles May Have No Effect on Phenotype

During the past several decades, researchers have discovered new kinds of gene interactions by studying model organisms such as *Escherichia coli* (a bacterium), *Saccharomyces cerevisiae* (baker's yeast), *Arabidopsis thaliana* (a model plant), *Drosophila melanogaster* (fruit fly), *Caenorhabditis elegans* (a nematode worm), and *Mus musculus* (the laboratory mouse). The isolation of mutants that alter the phenotypes of these organisms has become a powerful tool for investigating gene function and has provided ways for researchers to identify new kinds of gene interactions. With the advent of modern molecular techniques (described in Chapters 16, 18, and 19), a common approach for investigating gene function is to intentionally produce loss-of-function alleles in a gene of interest. When a geneticist abolishes gene function by creating an organism that is homozygous for a loss-of-function allele, the resulting organism is said to have undergone a **gene knockout.**

Why are gene knockouts useful? The primary reason for making a gene knockout is to understand how a gene affects the structure and function of cells or the phenotypes of organisms. For example, if a researcher knocked out a particular gene in a mouse and the resulting animal was unable to hear, the researcher would suspect that the role of the functional gene is to promote the formation of ear structures that are vital for hearing.

Interestingly, by studying many gene knockouts in a variety of experimental organisms, geneticists have discovered that many knockouts have no obvious effect on phenotype at the cellular level or the level of discernible traits. To explore gene function further, researchers may make two or more gene knockouts in the same organism. In some cases, gene knockouts in two different genes produce a phenotypic change even though the single knockouts have no effect (**Figure 4.23**). Geneticists may attribute this change to **gene redundancy**—the phenomenon that one gene can compensate for the loss of function of another gene.

Gene redundancy may be due to different underlying causes. One common reason is gene duplication. Certain genes have been duplicated during evolution, so a species may contain two or more copies of similar genes. These copies, which are not identical due to the accumulation of random changes during evolution, are called **paralogs.** When one gene is missing, a paralog may be able to carry out the missing function. For example, genes *A* and *B* in Figure 4.23 could be paralogs of each other. Alternatively, gene redundancy may involve proteins that are involved in a common cellular function. When one of the proteins is missing due to a gene knockout, the function of another protein may be increased to compensate for the missing protein and thereby overcome the defect.

Let's explore the consequences of gene redundancy in a genetic cross. George Shull conducted one of the first studies

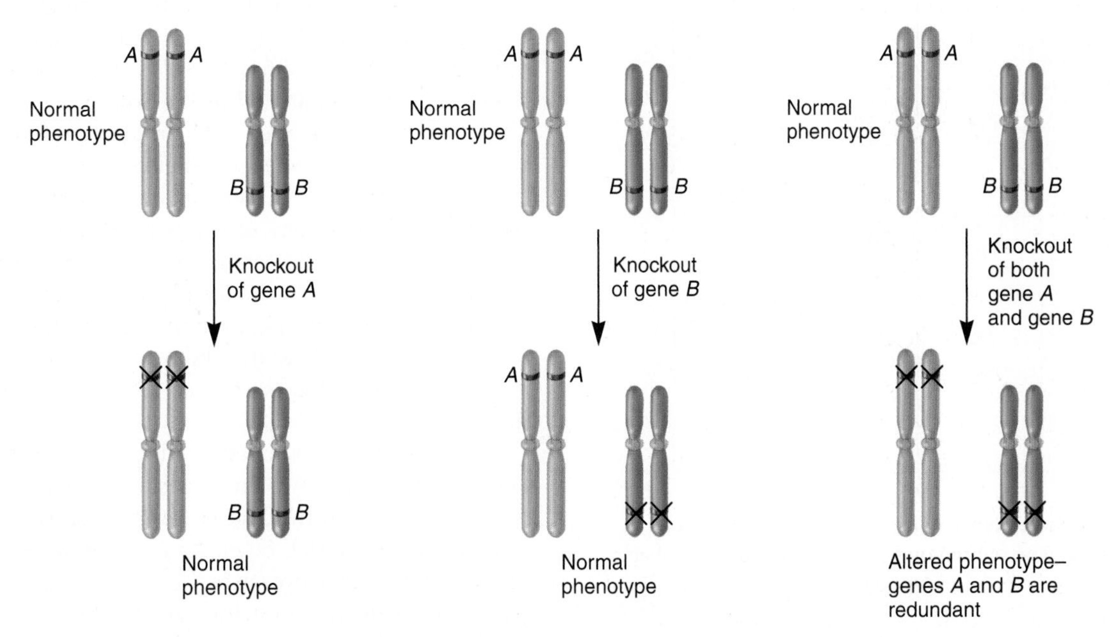

FIGURE 4.23 A molecular explanation for gene redundancy. To have a normal phenotype, an organism must have a functional copy of gene *A* or gene *B*, but not both. If both gene *A* and gene *B* are knocked out, an altered phenotype occurs.

that illustrated the phenomenon of gene redundancy. His work involved a weed known as shepherd's purse, a member of the mustard family. The trait he followed was the shape of the seed capsule, which is commonly triangular (**Figure 4.24**). Strains producing smaller ovate capsules are due to loss-of-function alleles in two different genes (*ttvv*). The ovate strain is an example of a double gene knockout. When Shull crossed a true-breeding plant with triangular capsules to a plant having ovate capsules, the F_1 generation all had triangular capsules. When the F_1 plants were self-fertilized, a surprising result came in the F_2 generation. Shull observed a 15:1 ratio of plants having triangular capsules to ovate capsules. The result can be explained by gene redundancy. Having one functional copy of either gene (*T* or *V*) is sufficient to produce the triangular phenotype. *T* and *V* are functional alleles of redundant genes. Only one of them is necessary for a triangular shape. When the functions of both genes are knocked out, as in the *ttvv* homozygote, the capsule becomes smaller and ovate.

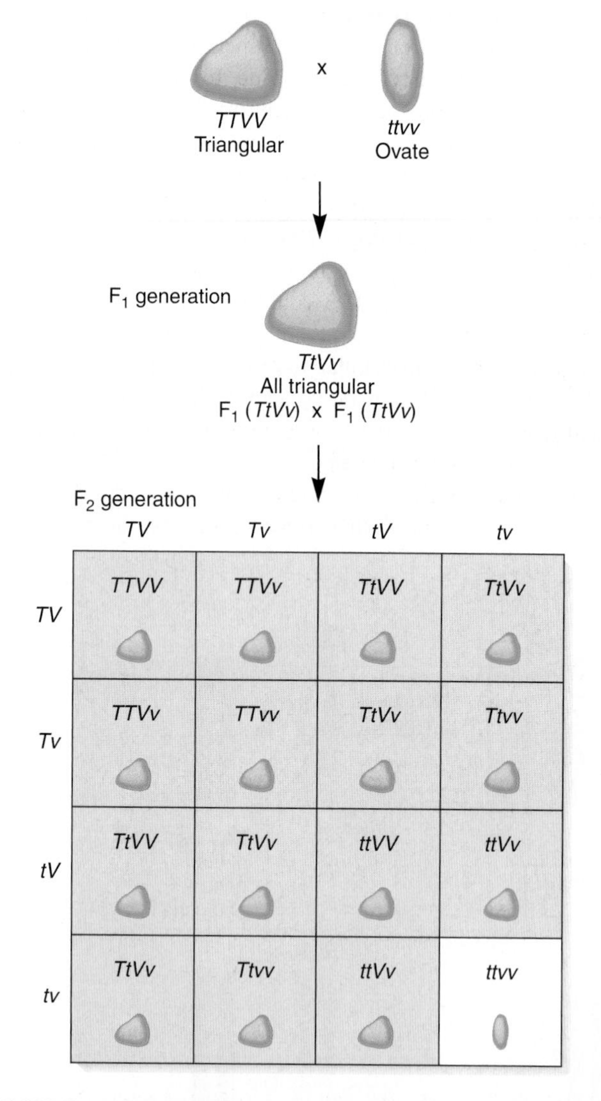

FIGURE 4.24 Inheritance of capsule shape in shepherd's purse, an example of gene redundancy. In this case, triangular shape requires a dominant allele in one of two genes, but not both. The *T* and *V* alleles are redundant.

The Phenotypic Effects of a Mutation Can Be Reversed by a Suppressor Mutation

When studying an experimental organism, a common approach to gain a deeper understanding of gene interaction is the isolation of a **suppressor mutation**—a second mutation that reverses the phenotypic effects of a first mutation. When a suppressor mutation is in a different gene than the first mutation, it is called an **intergenic** (or **extragenic**) **suppressor.**

What type of information might a researcher gain from the analysis of intergenic suppressor mutants? Usually, the primary goal is to identify proteins that participate in a common cellular process that ultimately affects the traits of an organism. In *Drosophila*, several different proteins work together in a signaling pathway that determines whether certain parts of the body contain sensory cells, such as those that make up mechanosensory bristles. Researchers have isolated dominant mutants that result in flies with fewer bristles. The mutated gene was named *Hairless* to reflect this phenotype. In this case, the wild-type allele is designated *h*, and the dominant mutant is *H*. After the *Hairless* mutant was obtained, researchers then isolated mutants that suppressed the hairless phenotype. Such suppressor mutants, which are in a different gene, produced flies that have a wild-type number of bristles. These mutants, which are also dominant, are in a gene that was named *Suppressor of Hairless*. The wild-type allele is designated *soh*, and the dominant mutant allele is *SoH*.

How do we explain the effects of these mutations at the molecular level? Let's first consider the functions of the proteins encoded by the normal (wild-type) genes (**Figure 4.25**). The role of the SoH protein, encoded by the *soh* allele of the *Suppressor of Hairless* gene, is to prevent the formation of sensory structures such as bristles in regions of the body where they should not be made. The Hairless protein is made in regions of the body where bristles should form, and binds to the SoH protein and inhibits its function. When the Hairless protein is properly expressed on the surface of the fly, as in an *hh* homozygote, bristles will form there.

Now let's consider the effects of a single mutation in the *Hairless* gene. In a heterozygote carrying the dominant allele (*H*), only half the amount of functional Hairless protein is made. This is not enough to inhibit all of the SoH proteins that are made. Therefore, the uninhibited SoH proteins prevent bristle formation and result in a hairless (bristleless) phenotype.

What happens in the double mutant? The suppressor mutation eliminates one of the two functional *soh* alleles. The double mutant expresses only one functional *h* allele and one functional *soh* allele. In the double mutant, the reduced amount of Hairless protein is able to inhibit the reduced amount of the SoH protein. Therefore, the ability of the SoH proteins to prevent bristle formation is stopped. Bristles form in the double heterozygote.

The analysis of a mutant and its suppressor often provides key information that two proteins participate in a common function. In some cases, the analysis reveals that two proteins physically interact with each other. As we have just seen, this type of interaction occurs between the Hairless and SoH proteins. Alternatively, two distinct proteins encoded by different genes may

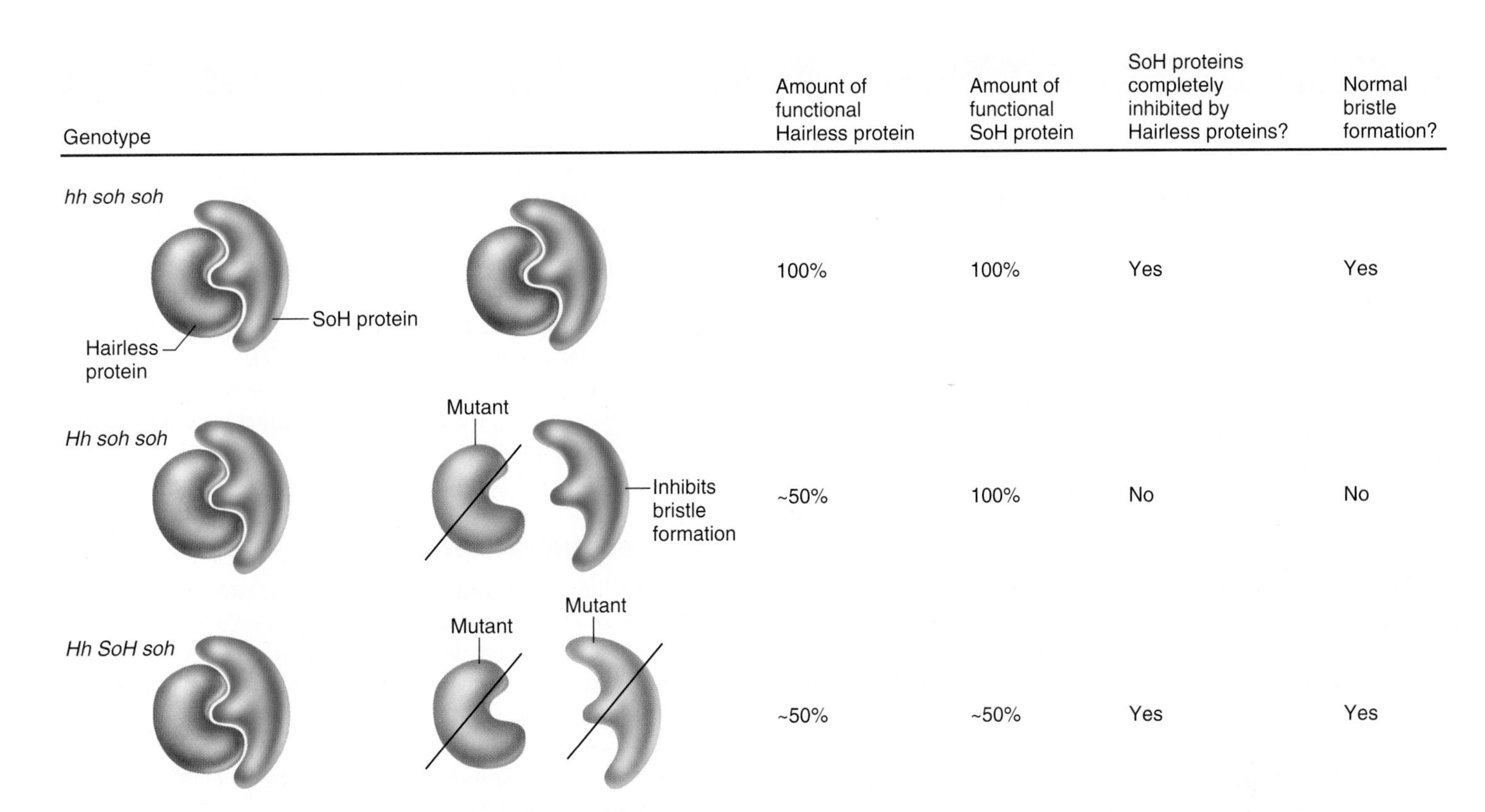

FIGURE 4.25 An example of a gene interaction involving an intergenic suppressor. The *Hairless* mutation, which produces the dominant *H* allele, results in flies with fewer bristles. A dominant suppressor mutation in a second gene restores bristle formation. This dominant allele is designated *SoH*. Examination of the interactions between the mutant and its suppressor reveals that the Hairless and SoH proteins physically interact with each other to determine whether bristles are formed.

participate in a common function, but do not directly interact with each other. For example, two enzymes may be involved in a biochemical pathway that leads to the synthesis of an amino acid. A mutation that greatly decreases the amount of one enzyme may limit the ability of an organism to make the amino acid. If this occurs, the amino acid would have to be supplied to the organism for it to survive. A suppressor mutation could increase the function of another enzyme in the pathway and thereby restore the ability of the organism to make an adequate amount of the amino acid. Such a suppressor would alleviate the need for the organism to have the amino acid supplemented in its diet.

Other suppressors exert their effects by altering the amount of protein encoded by a mutant gene. For example, a mutation may decrease the functional activity of a protein that is needed for sugar metabolism. An organism harboring such a mutation may not be able to metabolize the sugar at a sufficient rate for growth or survival. A suppressor mutation in a different gene could alter genetic regulatory proteins and thereby increase the amount of the protein encoded by the mutant gene. (The proteins involved in gene regulation are described in Chapters 14 and 15.) This suppressor mutation would increase the amount of the defective protein and thereby result in a faster rate of sugar metabolism.

KEY TERMS

Page 71. Mendelian inheritance, simple Mendelian inheritance
Page 72. wild-type alleles, genetic polymorphism, mutant alleles
Page 74. gain-of-function mutations, dominant-negative mutations, haploinsufficiency, incomplete dominance
Page 75. incomplete penetrance
Page 76. expressivity
Page 77. norm of reaction, overdominance, heterozygote advantage
Page 79. multiple alleles, temperature-sensitive allele
Page 80. codominance
Page 81. X-linked recessive, reciprocal cross
Page 82. sex-linked gene, hemizygous, holandric genes
Page 83. pseudoautosomal inheritance, sex-influenced inheritance
Page 84. sex-limited inheritance, sexual dimorphism, lethal allele, essential gene, nonessential genes
Page 85. age of onset, conditional lethal alleles, temperature-sensitive lethal alleles, semilethal alleles
Page 86. pleiotrophy, gene interaction
Page 88. epistasis, recessive epistasis, complementation
Page 89. gene modifier effect
Page 91. gene knockout, gene redundancy, paralogs
Page 92. suppressor mutation, intergenic (extragenic) suppressor

CHAPTER SUMMARY

- Mendelian inheritance patterns obey Mendel's laws.

4.1 Inheritance Patterns of Single Genes

- Several inheritance patterns involving single genes differ from those observed by Mendel (see Table 4.1).
- Wild-type alleles are prevalent in a population. When a gene exists in two or more wild-type alleles, this is a genetic polymorphism (see Figure 4.1).
- Recessive alleles are often due to mutations that result in a reduction or loss of function of the encoded protein (see Figure 4.2 and Table 4.2).
- Dominant alleles are most commonly caused by gain-of-function mutations, dominant negative mutations, or haploinsufficiency.
- Incomplete dominance is an inheritance pattern in which the heterozygote has an intermediate phenotype (see Figure 4.3).
- Whether we judge an allele to be dominant or incompletely dominant may depend on how closely we examine the phenotype (see Figure 4.4).
- Incomplete penetrance is a situation in which an allele that is expected to be expressed is not expressed (see Figure 4.5).
- Traits may vary in their expressivity.
- The outcome of traits is influenced by the environment (see Figure 4.6).
- Overdominance is an inheritance pattern in which the heterozygote has greater reproductive success (see Figures 4.7, 4.8).
- Most genes exist in multiple alleles in a population. Some alleles are temperature-sensitive (see Figures 4.9, 4.10).
- Some alleles, such as those that produce A and B blood antigens, are codominant (see Figure 4.11).
- X-linked inheritance patterns show differences between males and females, and are revealed in reciprocal crosses (see Figures 4.12, 4.13).
- The X and Y chromosomes carry different sets of genes, but they do have regions of short homology that can lead to pseudoautosomal inheritance (see Figure 4.14).
- For sex-influenced traits such as pattern baldness in humans, heterozygous males and females have different phenotypes (see Figures 4.15, 4.16).
- Sex-limited traits are expressed in only one sex, thereby resulting in sexual dimorphism (see Figure 4.17).
- Lethal alleles most commonly occur in essential genes. Lethal alleles may result in inheritance patterns that yield unexpected ratios (see Figure 4.18).
- Single genes have pleiotrophic effects.

4.2 Gene Interactions

- A gene interaction is a situation in which two or more genes affect a single phenotype (see Table 4.3).
- Bateson and Punnett discovered the first case of a gene interaction affecting comb morphology in chickens (see Figure 4.19).
- Epistasis is a situation in which the allele of one gene masks the phenotypic expression of the alleles of a different gene (see Figures 4.20, 4.21).
- A gene modifier effect is a situation in which an allele of one gene modifies (but does not completely mask) the phenotypic effects of the alleles of a different gene. An example is the cream eye color observed by Bridges (see Figure 4.22).
- Two different genes may have redundant functions, which is revealed in a double gene knockout (see Figures 4.23, 4.24).
- An intergenic suppressor mutation reverses the effects of a mutation in a different gene (see Figure 4.25).

PROBLEM SETS & INSIGHTS

Solved Problems

S1. In humans, why are X-linked recessive traits more likely to occur in males than females?

Answer: Because a male is hemizygous for X-linked traits, the phenotypic expression of X-linked traits depends on only a single copy of the gene. When a male inherits a recessive X-linked allele, he will automatically exhibit the trait because he does not have another copy of the gene on the corresponding Y chromosome. This phenomenon is particularly relevant to the inheritance of recessive X-linked alleles that cause human disease. (Some examples will be described in Chapter 22.)

S2. In Ayrshire cattle, the spotting pattern of the animals can be either red and white or mahogany and white. The mahogany and white pattern is caused by the allele *M*. The red and white phenotype is controlled by the allele *m*. When mahogany and white animals are mated to red and white animals, the following results are obtained:

Genotype	**Phenotype**	
	Females	**Males**
MM	Mahogany and white	Mahogany and white
Mm	Red and white	Mahogany and white
mm	Red and white	Red and white

Explain the pattern of inheritance.

Answer: The inheritance pattern for this trait is sex-influenced inheritance. The M allele is dominant in males but recessive in females, whereas the m allele is dominant in females but recessive in males.

S3. The following pedigree involves a single gene causing an inherited disease. If you assume that incomplete penetrance is *not* occurring, indicate which modes of inheritance are *not* possible. (Affected individuals are shown as filled symbols.)

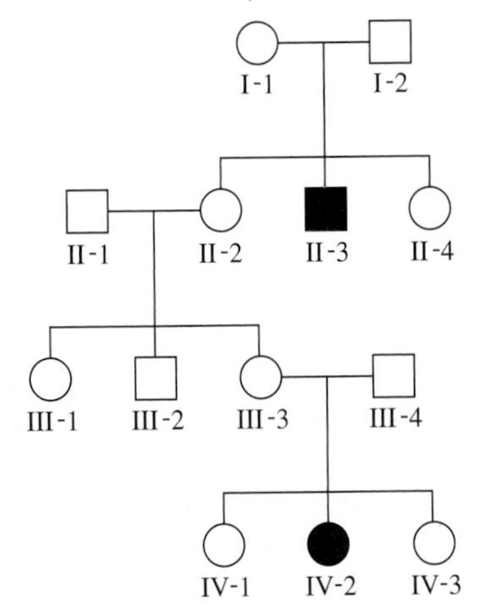

A. Recessive

B. Dominant

C. X-linked, recessive

D. Sex-influenced, dominant in females

E. Sex-limited, recessive in females

Answer:

A. It could be recessive.

B. It is probably not dominant unless it is incompletely penetrant.

C. It could not be X-linked recessive because individual IV-2 does not have an affected father.

D. It could not be sex-influenced, dominant in females because individual II-3 (who would have to be homozygous) has an unaffected mother (who would have to be heterozygous and affected).

E. It is not sex-limited because individual II-3 is an affected male and IV-2 is an affected female.

S4. Red-green color blindness is inherited as a recessive X-linked trait. What are the following probabilities?

A. A woman with phenotypically normal parents and a color-blind brother will have a color-blind son. Assume that she has no previous children.

B. The next child of a phenotypically normal woman, who has already had one color-blind son, will be a color-blind son.

C. The next child of a phenotypically normal woman, who has already had one color-blind son, and who is married to a color-blind man, will have a color-blind daughter.

Answer:

A. The woman's mother must have been a heterozygote. So there is a 50% chance that the woman is a carrier. If she has children, 1/4 (i.e., 25%) will be affected sons if she is a carrier. However, there is only a 50% chance that she is a carrier. We multiply 50% times 25%, which equals $0.5 \times 0.25 = 0.125$, or a 12.5% chance.

B. If she already had a color-blind son, then we know she must be a carrier, so the chance is 25%.

C. The woman is heterozygous and her husband is hemizygous for the color-blind allele. This couple will produce 1/4 offspring that are color-blind daughters. The rest are 1/4 carrier daughters, 1/4 normal sons, and 1/4 color-blind sons. Answer is 25%.

S5. Pattern baldness is an example of a sex-influenced trait that is dominant in males and recessive in females. A couple, neither of whom is bald, produced a bald son. What are the genotypes of the parents?

Answer: Because the father is not bald, we know he must be homozygous, *bb*. Otherwise, he would be bald. A female who is not bald can be either *Bb* or *bb*. Because she has produced a bald son, we know that she must be *Bb* in order to pass the *B* allele to her son.

S6. Two pink-flowered four-o'clocks were crossed to each other. What are the following probabilities for the offspring?

A. A plant will be red-flowered.

B. The first three plants examined will be white.

C. A plant will be either white or pink.

D. A group of six plants contain one pink, two whites, and three reds.

Answer: The first thing we need to do is construct a Punnett square to determine the individual probabilities for each type of offspring.

Because flower color is incompletely dominant, the cross is $Rr \times Rr$.

♀ \ ♂	*R*	*r*
R	*RR* Red	*Rr* Pink
r	*Rr* Pink	*rr* White

The phenotypic ratio is 1 red to 2 pink to 1 white. In other words, 1/4 are expected to be red, 1/2 pink, and 1/4 white.

A. The probability of a red-flowered plant is 1/4, which equals 25%.

B. Use the product rule.

$1/4 \times 1/4 \times 1/4 = 1/64 = 1.6\%$

C. Use the sum rule because these are mutually exclusive events. A given plant cannot be both white and pink.

$1/4 + 1/2 = 3/4 = 75\%$

D. Use the multinomial expansion equation. See solved problem S7 in Chapter 2 for an explanation of the multinomial expansion equation. In this case, three phenotypes are possible.

$$P = \frac{n!}{a!b!c!} p^a q^b r^c$$

where

n = total number of offspring = 6
a = number of reds = 3
p = probability of reds = (1/4)
b = number of pinks = 1
q = probability of pink = (1/2)
c = number of whites = 2
r = probability of whites = (1/4)

If we substitute these values into the equation,

$$P = \frac{6!}{3!1!2!} (1/4)^3(1/2)^1(1/4)^2$$

$$P = 0.029 = 2.9\%$$

This means that 2.9% of the time we would expect to obtain six plants, three with red flowers, one with pink flowers, and two with white flowers.

Conceptual Questions

C1. Describe the differences among dominance, incomplete dominance, codominance, and overdominance.

C2. Discuss the differences among sex-influenced, sex-limited, and sex-linked inheritance. Describe examples.

C3. What is meant by a gene interaction? How can a gene interaction be explained at the molecular level?

C4. Let's suppose a recessive allele encodes a completely defective protein. If the functional allele is dominant, what does that tell you about the amount of the functional protein that is sufficient to cause the phenotype? What if the allele shows incomplete dominance?

C5. A nectarine is a peach without the fuzz. The difference is controlled by a single gene that is found in two alleles, *D* and *d*. At the molecular level, would it make more sense to you that the nectarine is homozygous for a recessive allele or that the peach is homozygous for the recessive allele? Explain your reasoning.

C6. An allele in *Drosophila* produces a "star-eye" trait in the heterozygous individual. However, the star-eye allele is lethal in homozygotes. What would be the ratio and phenotypes of surviving flies if star-eyed flies were crossed to each other?

C7. A seed dealer wants to sell four-o'clock seeds that will produce only red, white, or pink flowers. Explain how this should be done.

C8. The serum from one individual (let's call this person individual 1) is known to agglutinate the red blood cells from a second individual (individual 2). List the pairwise combinations of possible genotypes that individuals 1 and 2 could be. If individual 1 is the parent of individual 2, what are his or her possible genotypes?

C9. Which blood phenotypes (A, B, AB, and/or O) provide an unambiguous genotype? Is it possible for a couple to produce a family of children with all four blood types? If so, what would the genotypes of the parents have to be?

C10. A woman with type B blood has a child with type O blood. What are the possible genotypes and blood types of the father?

C11. A type A woman is the daughter of a type O father and type A mother. If she has children with a type AB man, what are the following probabilities?

A. A type AB child

B. A type O child

C. The first three children with type AB blood

D. A family containing two children with type B blood and one child with type AB

C12. In Shorthorn cattle, coat color is controlled by a single gene that can exist as a red allele (*R*) or white allele (*r*). The heterozygotes (*Rr*) have a color called roan that looks less red than the *RR* homozygotes. However, when examined carefully, the roan phenotype in cattle is actually due to a mixture of completely red hairs and completely white hairs. Should this be called incomplete dominance, codominance, or something else? Explain your reasoning.

C13. In chickens, the Leghorn variety has white feathers due to an autosomal dominant allele. Silkies have white feathers due to a recessive allele in a second (different) gene. If a true-breeding white Leghorn is crossed to a true-breeding white Silkie, what is the expected phenotype of the F_1 generation? If members of the F_1 generation are mated to each other, what is the expected phenotypic outcome of the F_2 generation? Assume the chickens in the parental generation are homozygous for the white allele at one gene and homozygous for the brown allele at the other gene. In subsequent generations, nonwhite birds will be brown.

C14. Propose the most likely mode of inheritance (autosomal dominant, autosomal recessive, or X-linked recessive) for the following pedigrees. Affected individuals are shown with filled (black) symbols.

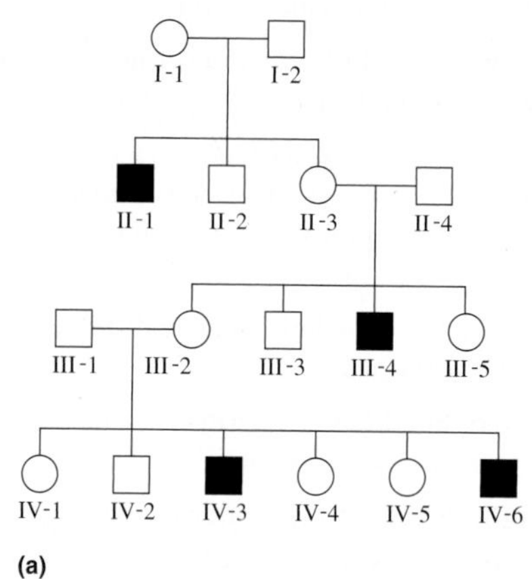

(a)

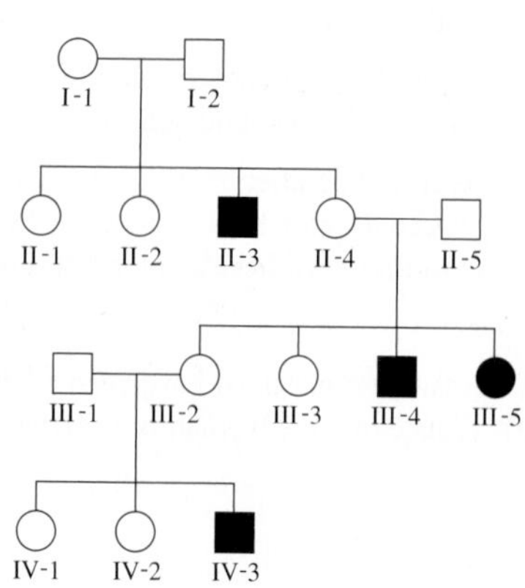

(b)

C15. A human disease known as vitamin D-resistant rickets is inherited as an X-linked dominant trait. If a male with the disease produces children with a female who does not have the disease, what is the expected ratio of affected and unaffected offspring?

C16. Hemophilia is an X-linked recessive trait in humans. If a heterozygous woman has children with an unaffected man, what is the probability of the following combinations of children?

A. An affected son

B. Four unaffected offspring in a row

C. An unaffected daughter or son

D. Two out of five offspring that are affected

C17. Incontinentia pigmenti is a rare, X-linked dominant disorder in humans characterized by swirls of pigment in the skin. If an affected female, who had an unaffected father, has children with an unaffected male, what would be the predicted ratios of affected and unaffected sons and daughters?

C18. With regard to pattern baldness in humans (a sex-influenced trait), a woman who is not bald and whose mother is bald has children with a bald man whose father is not bald. What are their probabilities of having the following types of families?

A. Their first child will not become bald.

B. Their first child will be a male who will not become bald.

C. Their first three children will be females who are not bald.

C19. In rabbits, the color of body fat is controlled by a single gene with two alleles, designated *Y* and *y*. The outcome of this trait is affected by the diet of the rabbit. When raised on a standard vegetarian diet, the dominant *Y* allele confers white body fat, and the *y* allele confers yellow body fat. However, when raised on a xanthophyll-free diet, the homozygote *yy* animal has white body fat. If a heterozygous animal is crossed to a rabbit with yellow body fat, what are the proportions of offspring with white and yellow body fat when raised on a standard vegetarian diet? How do the proportions change if the offspring are raised on a xanthophyll-free diet?

C20. A Siamese cat that spends most of its time outside was accidentally injured in a trap and required several stitches in its right front paw. The veterinarian had to shave the fur from the paw and leg, which originally had rather dark fur. Later, when the fur grew back, it was much lighter than the fur on the other three legs. Do you think this injury occurred in the hot summer or cold winter? Explain your answer.

C21. A true-breeding male fly with eosin eyes is crossed to a white-eyed female that is heterozygous for the wild-type (*C*) and cream alleles (c^a). What are the expected proportions of their offspring?

C22. The trait of hen- versus cock-feathering is a sex-limited trait controlled by a single gene. Females always exhibit hen-featuring as do *HH* and *Hh* males. Only *hh* males show cock-featuring. Starting with two heterozygous fowl that are hen-feathered, explain how you would obtain a true-breeding line that always produced cock-feathered males.

C23. In the pedigree shown here for a trait determined by a single gene (affected individuals are shown in black), state whether it would be possible for the trait to be inherited in each of the following ways:

A. Recessive

B. X-linked recessive

C. Dominant, complete penetrance

D. Sex-influenced, dominant in males

E. Sex-limited

F. Dominant, incomplete penetrance

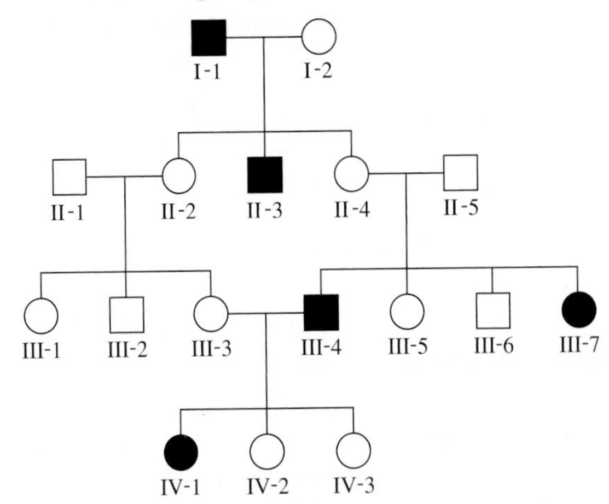

C24. The pedigree shown here also concerns a trait determined by a single gene (affected individuals are shown in black). Which of the following patterns of inheritance are possible?

A. Recessive

B. X-linked recessive

C. Dominant

D. Sex-influenced, recessive in males

E. Sex-limited

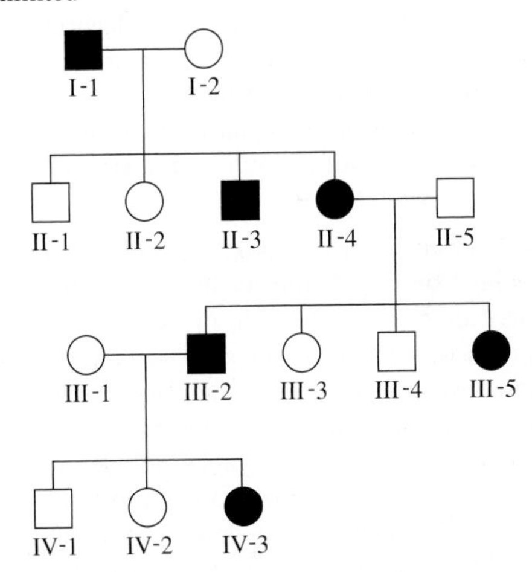

C25. Let's suppose you have pedigree data from thousands of different families involving a particular genetic disease. How would you decide whether the disease is inherited as a recessive trait as opposed to one that is dominant with incomplete penetrance?

C26. Compare phenotypes at the molecular, cellular, and organism levels for individuals who are homozygous for the hemoglobin allele, Hb^AHb^A, and the sickle cell allele, Hb^SHb^S.

C27. A very rare dominant allele that causes the little finger to be crooked has a penetrance value of 80%. In other words, 80% of heterozygotes carrying the allele will have a crooked little finger. If a homozygous unaffected person has children with a heterozygote carrying this mutant allele, what is the probability that an offspring will have little fingers that are crooked?

C28. A sex-influenced trait in humans is one that affects the length of the index finger. A "short" allele is dominant in males and recessive in females. Heterozygous males have an index finger that is significantly shorter than the ring finger. The gene affecting index finger length is located on an autosome. A woman with short index fingers has children with a man who has normal index fingers. They produce five children in the following order: female, male, male, female, male. The oldest female offspring marries a man with normal fingers and then has one daughter. The youngest male among the five children marries a woman with short index fingers, and then they have two sons. Draw the pedigree for this family. Indicate the phenotypes of every individual (filled symbols for individuals with short index fingers and open symbols for individuals with normal index fingers).

C29. In horses, there are three coat-color patterns termed cremello (beige), chestnut (brown), and palomino (golden with light mane and tail). If two palomino horses are mated, they produce about 1/4 cremello, 1/4 chestnut, and 1/2 palomino offspring. In contrast, cremello horses and chestnut horses breed true. (In other words, two cremello horses will produce only cremello offspring and two chestnut horses will produce only chestnut offspring.) Explain this pattern of inheritance.

C30. Briefly describe three explanations for how a suppressor mutation exerts its effects at the molecular level.

Experimental Questions

E1. Mexican hairless dogs have little hair and few teeth. When a Mexican hairless is mated to another breed of dog, about half of the puppies are hairless. When two Mexican hairless dogs are mated to each other, about 1/3 of the surviving puppies have hair, and about 2/3 of the surviving puppies are hairless. However, about two out of eight puppies from this type of cross are born grossly deformed and do not survive. Explain this pattern of inheritance.

E2. In chickens, some varieties have feathered shanks (legs), but others do not. In a cross between a Black Langhans (feathered shanks) and Buff Rocks (unfeathered shanks), the shanks of the F_1 generation are all feathered. When the F_1 generation is crossed, the F_2 generation contains chickens with feathered shanks to unfeathered shanks in a ratio of 15:1. Suggest an explanation for this result.

E3. In sheep, the formation of horns is a sex-influenced trait; the allele that results in horns is dominant in males and recessive in females. Females must be homozygous for the horned allele to have horns. A horned ram was crossed to a polled (unhorned) ewe, and the first offspring they produced was a horned ewe. What are the genotypes of the parents?

E4. A particular breed of dog can have long hair or short hair. When true-breeding long-haired animals were crossed to true-breeding short-haired animals, the offspring all had long hair. The F_2 generation produced a 3:1 ratio of long- to short-haired offspring. A second trait involves the texture of the hair. The two variants are wiry hair and straight hair. F_1 offspring from a cross of these two varieties all had wiry hair, and F_2 offspring showed a 3:1 ratio of wiry-haired to straight-haired puppies. Recently, a breeder of the short-, wiry-haired dogs found a female puppy that was albino. Similarly, another breeder of the long-, straight-haired dogs found a male puppy that was albino. Because the albino trait is always due to a recessive allele, the two breeders got together and mated the two dogs. Surprisingly, all of the puppies in the litter had black hair. How would you explain this result?

E5. In the clover butterfly, males are always yellow, but females can be yellow or white. In females, white is a dominant allele. Two yellow butterflies were crossed to yield an F_1 generation consisting of 50% yellow males, 25% yellow females, and 25% white females. Describe how this trait is inherited and the genotypes of the parents.

E6. The *Mic2* gene in humans is present on both the X and Y chromosome. Let's suppose the *Mic2* gene exists in a dominant *Mic2* allele, which results in normal surface antigen, and a recessive *mic2* allele, which results in defective surface antigen production. Using molecular techniques, it is possible to identify homozygous and heterozygous individuals. By following the transmission of the *Mic2* and *mic2* alleles in a large human pedigree, would it be possible to distinguish between pseudoautosomal inheritance and autosomal inheritance? Explain your answer.

E7. Duroc Jersey pigs are typically red, but a sandy variation is also seen. When two different varieties of true-breeding sandy pigs were crossed to each other, they produced F_1 offspring that were red. When these F_1 offspring were crossed to each other, they produced red, sandy, and white pigs in a 9:6:1 ratio. Explain this pattern of inheritance.

E8. As discussed in this chapter, comb morphology in chickens is governed by a gene interaction. Two walnut comb chickens were crossed to each other. They produced only walnut comb and rose comb offspring, in a ratio of 3:1. What are the genotypes of the parents?

E9. In certain species of summer squash, fruit color is determined by two interacting genes. A dominant allele, *W*, determines white color, and a recessive allele (*w*) allows the fruit to be colored. In a homozygous *ww* individual, a second gene determines fruit color: *G* (green) is dominant to *g* (yellow). A white squash and a yellow squash were crossed, and the F_1 generation yielded approximately 50% white fruit and 50% green fruit. What are the genotypes of the parents?

E10. Certain species of summer squash exist in long, spherical, or disk shapes. When a true-breeding long-shaped strain was crossed to a true-breeding disk-shaped strain, all of the F_1 offspring were disk-shaped. When the F_1 offspring were allowed to self-fertilize, the F_2 generation consisted of a ratio of 9 disk-shaped to 6 round-shaped to 1 long-shaped. Assuming the shape of summer squash is governed by two different genes, with each gene existing in two alleles, propose a mechanism to account for this 9:6:1 ratio.

E11. In a species of plant, two genes control flower color. The red allele (*R*) is dominant to the white allele (*r*); the color-producing allele (*C*) is dominant to the non-color-producing allele (*c*). You suspect that either an *rr* homozygote or a *cc* homozygote will produce white flowers. In other words, *rr* is epistatic to *C*, and *cc* is epistatic to *R*. To test your hypothesis, you allowed heterozygous plants (*RrCc*) to self-fertilize and counted the offspring. You obtained the following data: 201 plants with red flowers and 144 with white flowers. Conduct a chi-square analysis to see if your observed data are consistent with your hypothesis.

E12. In *Drosophila*, red eyes is the wild-type phenotype. Several different genes (with each gene existing in two or more alleles) are known to affect eye color. One allele causes purple eyes, and a different allele causes sepia eyes. Both of these alleles are recessive compared with red eye color. When flies with purple eyes were crossed to flies with sepia eyes, all of the F_1 offspring had red eyes. When the F_1 offspring were allowed to mate with each other, the following data were obtained:

146 purple eyes

151 sepia eyes

50 purplish sepia eyes

444 red eyes

Explain this pattern of inheritance. Conduct a chi-square analysis to see if the experimental data fit your hypothesis.

E13. As mentioned in Experimental Question E12, red eyes is the wild-type phenotype in *Drosophila*, and several different genes (with each gene existing in two or more alleles) affect eye color. One allele causes purple eyes, and a different allele causes vermilion eyes. The purple and vermilion alleles are recessive compared with red eye color. The following crosses were made, and the following data were obtained:

Cross 1: Males with vermilion eyes × females with purple eyes

354 offspring, all with red eyes

Cross 2: Males with purple eyes × females with vermilion eyes

212 male offspring with vermilion eyes

221 female offspring with red eyes

Explain the pattern of inheritance based on these results. What additional crosses might you make to confirm your hypothesis?

E14. Let's suppose you were looking through a vial of fruit flies in your laboratory and noticed a male fly that has pink eyes. What crosses would you make to determine if the pink allele is an X-linked gene? What crosses would you make to determine if the pink allele is an allele of the same X-linked gene that has white and eosin alleles? Note: The white and eosin alleles are discussed in Figure 4.22.

E15. When examining a human pedigree, what features do you look for to distinguish between X-linked recessive inheritance versus autosomal recessive inheritance? How would you distinguish X-linked dominant inheritance from autosomal dominant inheritance in a human pedigree?

E16. The cream allele is a modifier of eosin and the cream allele is autosomal. By comparison, the red and eosin alleles are X-linked. Based on these ideas, conduct a chi-square analysis to determine if Bridges' data of Figure 4.22 agree with the predicted ratio of 8 red-eyed females, 4 red-eyed males, 3 light eosin-eyed males, and 1 cream-eyed male.

Questions for Student Discussion/Collaboration

1. Let's suppose a gene exists as a functional wild-type allele and a nonfunctional mutant allele. At the organism level, the wild-type allele is dominant. In a heterozygote, discuss whether dominance occurs at the cellular or molecular level. Discuss examples in which the issue of dominance depends on the level of examination.
2. A true-breeding rooster with a rose comb, feathered shanks, and cock-feathering was crossed to a hen that is true-breeding for pea comb and unfeathered shanks but is heterozygous for hen-feathering. If you assume these genes can assort independently, what is the expected outcome of the F_1 generation?
3. In oats, the color of the chaff is determined by a two-gene interaction. When a true-breeding black plant was crossed to a true-breeding white plant, the F_1 generation was composed of all black plants. When the F_1 offspring were crossed to each other, the ratio produced was 12 black to 3 gray to 1 white. First, construct a Punnett square that accounts for this pattern of inheritance. Which genotypes produce the gray phenotype? Second, at the level of protein function, how would you explain this type of inheritance?

Note: All answers appear at the website for this textbook; the answers to even-numbered questions are in the back of the textbook.

www.mhhe.com/brookergenetics4e

Visit the website for practice tests, answer keys, and other learning aids for this chapter. Enhance your understanding of genetics with our interactive exercises, quizzes, animations, and much more.

CHAPTER OUTLINE

Shell coiling in the water snail, Lymnaea peregra. *In this species, some snails coil to the left, and others coil to the right. This is due to an inheritance pattern called the maternal effect.*

5 NON-MENDELIAN INHERITANCE

Mendelian inheritance patterns involve genes that directly influence the outcome of an offspring's traits and obey Mendel's laws. To predict phenotype, we must consider several factors. These include the dominant/recessive relationship of alleles, gene interactions that may affect the expression of a single trait, and the roles that sex and the environment play in influencing the individual's phenotype. Once these factors are understood, we can predict the phenotypes of offspring from their genotypes.

Most genes in eukaryotic species follow a Mendelian pattern of inheritance. However, many genes do not. In this chapter, we will examine several additional and even bizarre types of inheritance patterns that deviate from a Mendelian pattern. In the first two sections, we will consider two interesting examples of non-Mendelian inheritance called the maternal effect and epigenetic inheritance. Even though these inheritance patterns involve genes on chromosomes within the cell nucleus, the genotype of the offspring does not directly govern their phenotype in ways predicted by Mendel. We will see how the timing of gene expression and gene inactivation can cause a non-Mendelian pattern of inheritance.

In the third section, we will examine deviations from Mendelian inheritance that arise because some genetic material is not located in the cell nucleus. Certain cellular organelles, such as mitochondria and chloroplasts, contain their own genetic material. We will survey the inheritance of organellar genes and a few other examples in which traits are influenced by genetic material that exists outside of the cell nucleus.

5.1 MATERNAL EFFECT

We will begin by considering genes that have a **maternal effect.** This term refers to an inheritance pattern for certain **nuclear genes**—genes located on chromosomes that are found in the cell nucleus—in which the genotype of the mother directly determines the phenotype of her offspring. Surprisingly, for maternal effect genes, the genotypes of the father and offspring themselves do not affect the phenotype of the offspring. We will see that this phenomenon is explained by the accumulation of gene products that the mother provides to her developing eggs.

The Genotype of the Mother Determines the Phenotype of the Offspring for Maternal Effect Genes

The first example of a maternal effect gene was studied in the 1920s by Arthur Boycott and involved morphological features of the water snail, *Lymnaea peregra*. In this species, the shell and internal organs can be arranged in either a right-handed (dextral) or left-handed (sinistral) direction. The dextral orientation is more common and is dominant to the sinistral orientation. **Figure 5.1** describes the results of a genetic analysis carried out by Boycott. In this experiment, he began with two different true-breeding strains of snails with either a dextral or sinistral morphology. Many combinations of crosses produced results that could not be explained by a Mendelian pattern of inheritance. When a dextral female (*DD*) was crossed to a sinistral male (*dd*), all F_1 offspring were dextral. However, in the **reciprocal cross,** where a sinistral female (*dd*) was crossed to a dextral male (*DD*), all F_1 offspring were sinistral. Taken together, these results contradict a Mendelian pattern of inheritance.

How can we explain the unusual results obtained in Figure 5.1? Alfred Sturtevant proposed the idea that snail coiling is due to a maternal effect gene that exists as a dextral (*D*) or sinistral (*d*) allele. His conclusions were drawn from the inheritance patterns of the F_2 and F_3 generations. In this experiment, the genotype of the F_1 generation is expected to be heterozygous (*Dd*). When these F_1 individuals were crossed to each other, a genotypic ratio of 1 *DD* : 2 *Dd* : 1 *dd* is predicted for the F_2 generation. Because the *D* allele is dominant to the *d* allele, a 3:1 phenotypic ratio of dextral to sinistral snails should be produced according to a Mendelian pattern of inheritance. Instead of this predicted phenotypic ratio, however, the F_2 generation was composed of all dextral snails. This incongruity with Mendelian inheritance is due to the maternal effect. The phenotype of the offspring depended solely on the genotype of the mother. The F_1 mothers were *Dd*. The *D* allele in the mothers is dominant to the *d* allele and caused the offspring to be dextral, even if the offspring's genotype was *dd*. When the members of the F_2 generation were crossed, the F_3 generation exhibited a 3:1 ratio of dextral to sinistral snails. This ratio corresponds to the genotypes of the F_2 females, which were the mothers of the F_3 generation. The ratio of F_2 females was 1 *DD* : 2 *Dd* : 1 *dd*. The *DD* and *Dd* females produced dextral offspring, whereas the *dd* females produced sinistral offspring. This explains the 3:1 ratio of dextral and sinistral offspring in the F_3 generation.

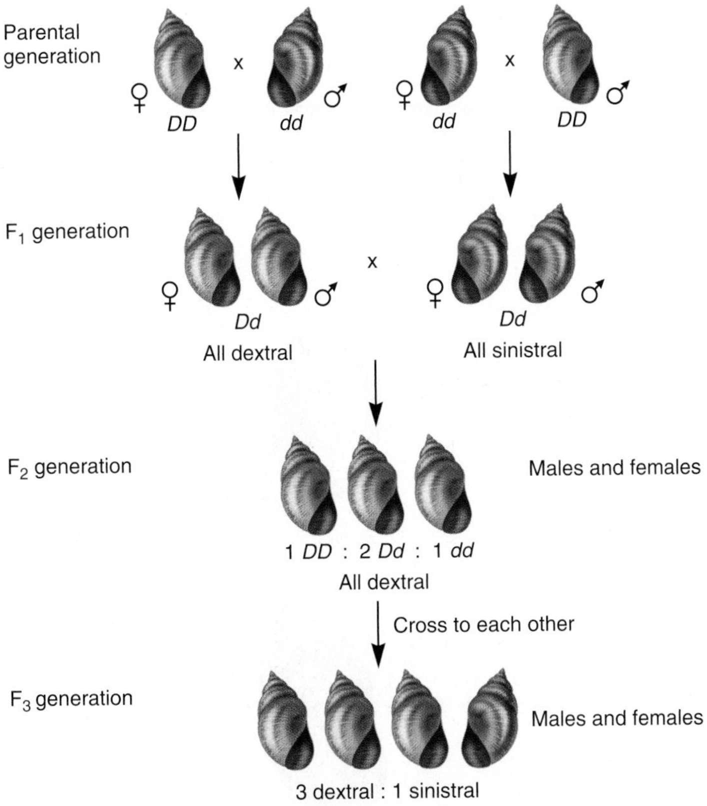

INTERACTIVE EXERCISE

FIGURE 5.1 Experiment showing the inheritance pattern of snail coiling. In this experiment, *D* (dextral) is dominant to *d* (sinistral). The genotype of the mother determines the phenotype of the offspring. This phenomenon is known as the maternal effect. In this case, a *DD* or *Dd* mother produces dextral offspring, and a *dd* mother produces sinistral offspring. The genotypes of the father and offspring do not affect the offspring's phenotype.

Female Gametes Receive Gene Products from the Mother That Affect Early Developmental Stages of the Embryo

At the molecular and cellular level, the non-Mendelian inheritance pattern of maternal effect genes can be explained by the process of oogenesis in female animals (**Figure 5.2a**). As an animal oocyte (egg) matures, many surrounding maternal cells called nurse cells provide the egg with nutrients and other materials. In Figure 5.2a, a female is heterozygous for the snail-coiling maternal effect gene, with the alleles designated *D* and *d*. Depending on the outcome of meiosis, the haploid egg may receive the *D* allele or the *d* allele, but not both. The surrounding nurse cells, however, produce both *D* and *d* gene products (mRNA and proteins). These gene products are then transported into the egg. As shown here, the egg has received both the *D* allele gene products and the *d* allele gene products. These gene products persist for a significant time after the egg has been fertilized and begins its embryonic development. In this way, the gene products of the nurse cells, which reflect the genotype of the mother, influence the early developmental stages of the embryo.

Now that we have an understanding of the relationship between oogenesis and maternal effect genes, let's reconsider the topic of snail coiling. As shown in **Figure 5.2b**, a female snail that is *DD* transmits only the *D* gene products to the egg. During the early stages of embryonic development, these gene products cause the egg cleavage to occur in a way that promotes a right-handed body plan. A heterozygous female transmits both *D* and *d* gene products. Because the *D* allele is dominant, the maternal effect also causes a right-handed body plan. Finally, a *dd* mother contributes only *d* gene products that promote a left-handed body plan, even if the egg is fertilized by a sperm carrying a *D* allele. The sperm's genotype is irrelevant, because the expression of the sperm's gene would occur too late. The origin of dextral and sinistral coiling can be traced to the orientation

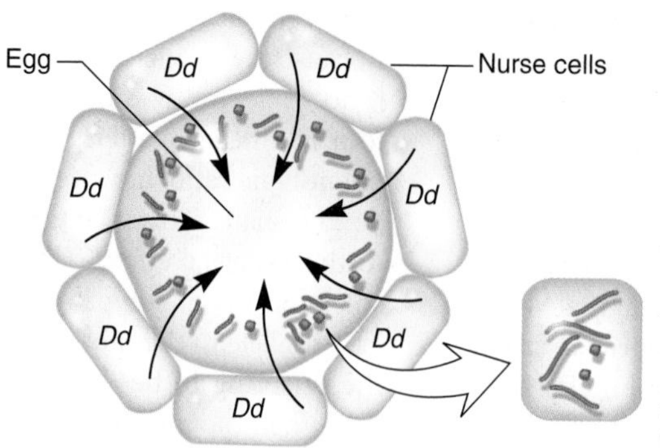

(a) Transfer of gene products from nurse cells to egg

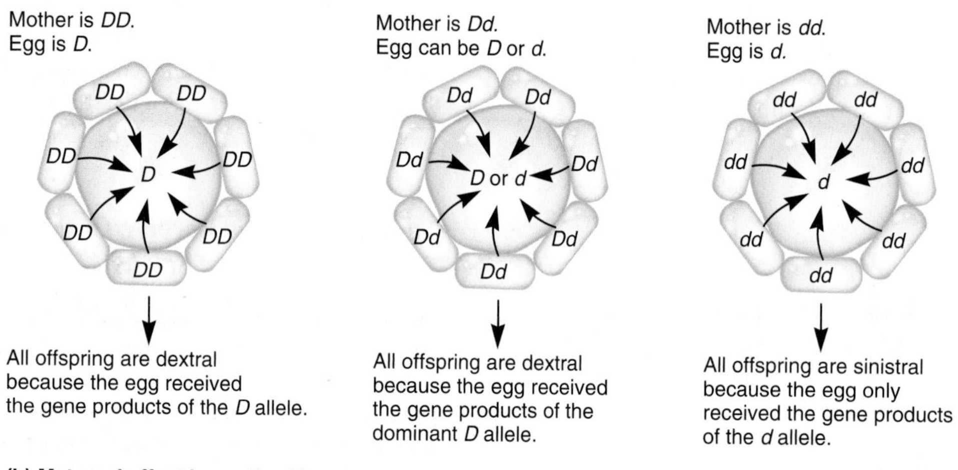

(b) Maternal effect in snail coiling

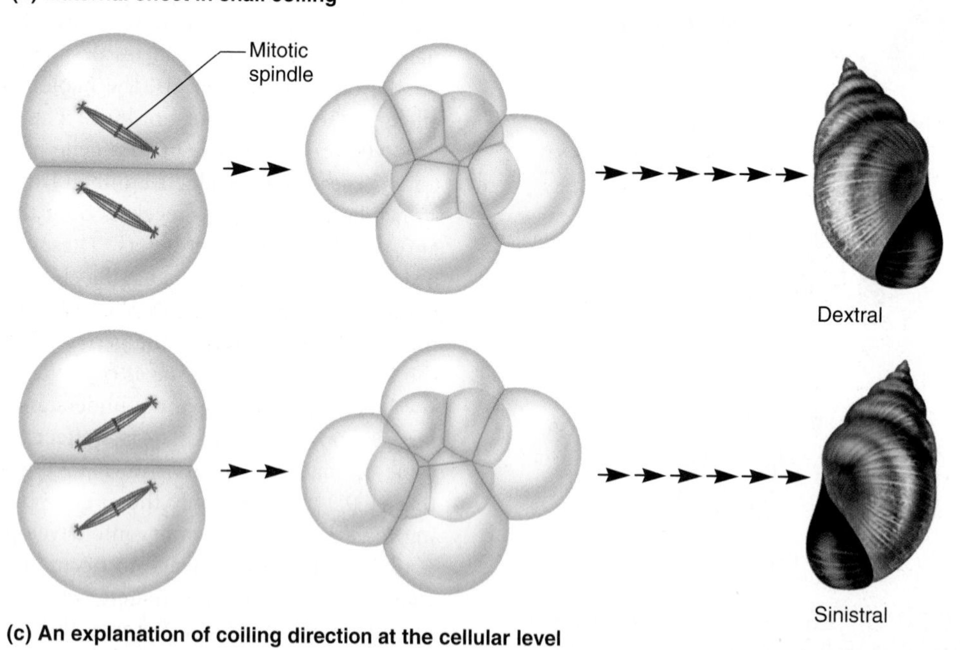

(c) An explanation of coiling direction at the cellular level

FIGURE 5.2 The mechanism of maternal effect in snail coiling. **(a)** Transfer of gene products from nurse cells to an egg. The nurse cells are heterozygous (*Dd*). Both the *D* and *d* alleles are activated in the nurse cells to produce *D* and *d* gene products (mRNA or proteins, or both). These products are transported into the cytoplasm of the egg, where they accumulate to significant amounts. **(b)** Explanation of the maternal effect in snail coiling. **(c)** The direction of snail coiling is determined by differences in the cleavage planes during early embryonic development.

Genes → Traits If the nurse cells are *DD* or *Dd*, they will transfer the *D* gene product to the egg and thereby cause the resulting offspring to be dextral. If the nurse cells are *dd*, only the *d* gene product will be transferred to the egg, so the resulting offspring will be sinistral.

of the mitotic spindle at the two- to four-cell stage of embryonic development. The dextral and sinistral snails develop as mirror images of each other (**Figure 5.2c**).

Since these initial studies, researchers have found that maternal effect genes encode proteins that are important in the early steps of embryogenesis. The accumulation of maternal gene products in the egg allows embryogenesis to proceed quickly after fertilization. Maternal effect genes often play a role in cell division, cleavage pattern, and body axis orientation. Therefore, defective alleles in maternal effect genes tend to have a dramatic effect on the phenotype of the individual, altering major features of morphology, often with dire consequences.

Our understanding of maternal effect genes has been greatly aided by their identification in experimental organisms such as *Drosophila melanogaster*. In such organisms with a short generation time, geneticists have successfully searched for mutant alleles that prevent the normal process of embryonic development. In *Drosophila*, geneticists have identified several maternal effect genes with profound effects on the early stages of development. The pattern of development of a *Drosophila* embryo occurs along axes, such as the anteroposterior axis and the dorsoventral axis. The proper development of each axis requires a distinct set of maternal gene products. For example, the maternal effect gene called *bicoid* produces a gene product that accumulates in a region of the egg that will eventually become anterior structures in the developing embryo. Mutant alleles of maternal effect genes often lead to abnormalities in the anteroposterior or the dorsoventral pattern of development. More recently, several maternal effect genes have been identified in mice and humans that are required for proper embryonic development. Chapter 23 examines the relationships among the actions of several maternal effect genes during embryonic development.

5.2 EPIGENETIC INHERITANCE

Epigenetic inheritance is a pattern in which a modification occurs to a nuclear gene or chromosome that alters gene expression, but is not permanent over the course of many generations. As we will see, epigenetic inheritance patterns are the result of DNA and chromosomal modifications that occur during oogenesis, spermatogenesis, or early stages of embryogenesis. Once they are initiated during these early stages, epigenetic changes alter the expression of particular genes in a way that may be fixed during an individual's lifetime. Therefore, epigenetic changes can permanently affect the phenotype of the individual. However, epigenetic modifications are not permanent over the course of many generations, and they do not change the actual DNA sequence. For example, a gene may undergo an epigenetic change that inactivates it for the lifetime of an individual. However, when this individual makes gametes, the gene may become activated and remain operative during the lifetime of an offspring who inherits the active gene.

In this section, we will examine two examples of epigenetic inheritance called dosage compensation and genomic imprinting. Dosage compensation has the effect of offseting differences in the number of sex chromosomes. One of the sex chromosomes is altered, with the result that males and females have similar levels of gene expression even though they do not possess the same complement of sex chromosomes. In mammals, dosage compensation is initiated during the early stages of embryonic development. By comparison, genomic imprinting happens prior to fertilization; it involves a change in a single gene or chromosome during gamete formation. Depending on whether the modification occurs during spermatogenesis or oogenesis, imprinting governs whether an offspring expresses a gene that has been inherited from its mother or father.

Dosage Compensation Is Necessary to Ensure Genetic Equality Between the Sexes

Dosage compensation refers to the phenomenon that the level of expression of many genes on the sex chromosomes (such as the X chromosome) is similar in both sexes even though males and females have a different complement of sex chromosomes. This term was coined in 1932 by Hermann Muller to explain the effects of eye color mutations in *Drosophila*. Muller observed that female flies homozygous for certain X-linked eye color alleles had a similar phenotype to hemizygous males. He noted that an X-linked gene conferring an apricot eye color produces a very similar phenotype in homozygous females and hemizygous males. In contrast, a female that has one copy of the apricot allele and a deletion of the apricot gene on the other X chromosome has eyes of paler color. Therefore, one copy of the allele in the female is not equivalent to one copy of the allele in the male. Instead, two copies of the allele in the female produce a phenotype that is similar to that produced by one copy in the male. In other words, the difference in gene dosage—two copies in females versus one copy in males—is being compensated for at the level of gene expression.

Since these initial studies, dosage compensation has been studied extensively in mammals, *Drosophila*, and *Caenorhabditis elegans* (a nematode). Depending on the species, dosage compensation occurs via different mechanisms (**Table 5.1**). Female mammals equalize the expression of X-linked genes by turning off one of their two X chromosomes. This process is known as **X inactivation.** In *Drosophila*, the male accomplishes dosage compensation by doubling the expression of most X-linked genes. In *C. elegans*, the XX animal is a hermaphrodite that produces both sperm and egg cells, and an animal carrying a single X chromosome is a male that produces only sperm. The XX hermaphrodite diminishes the expression of X-linked genes on both X chromosomes to approximately 50% of that in the male.

In birds, the Z chromosome is a large chromosome, usually the fourth or fifth largest, which contains almost all of the known sex-linked genes. The W chromosome is generally a much smaller microchromosome containing a high proportion of repeat sequence DNA that does not encode genes. Males are ZZ and females are ZW. Several years ago, researchers studied the level of expression of a Z-linked gene that encodes an enzyme called aconitase. They discovered that males express twice as much aconitase as females do. These results suggested

TABLE 5.1
Mechanisms of Dosage Compensation Among Different Species

Species	Sex Chromosomes in: Females	Males	Mechanism of Compensation
Placental mammals	XX	XY	One of the X chromosomes in the somatic cells of females is inactivated. In certain species, the paternal X chromosome is inactivated, and in other species, such as humans, either the maternal or paternal X chromosome is randomly inactivated throughout the somatic cells of females.
Marsupial mammals	XX	XY	The paternally derived X chromosome is inactivated in the somatic cells of females.
Drosophila melanogaster	XX	XY	The level of expression of genes on the X chromosome in males is increased twofold.
Caenorhabditis elegans	XX*	X0	The level of expression of genes on both X chromosomes in hermaphrodites is decreased to 50% levels compared with males.

*In *C. elegans*, an XX individual is a hermaphrodite, not a female.

that dosage compensation does not occur in birds. More recently, the expression of hundreds of Z-linked genes has been examined in chickens. These newer results also suggest that birds lack a general mechanism of dosage compensation that controls the expression of most Z-linked genes. Even so, the pattern of gene expression between males and females was found to vary a great deal for certain Z-linked genes. Overall, the results suggest that some Z-linked genes may be dosage-compensated, but many of them are not.

Dosage Compensation Occurs in Female Mammals by the Inactivation of One X Chromosome

In 1961, Mary Lyon proposed that dosage compensation in mammals occurs by the inactivation of a single X chromosome in females. Liane Russell also proposed the same idea around the same time. This proposal brought together two lines of study. The first type of evidence came from cytological studies. In 1949, Murray Barr and Ewart Bertram identified a highly condensed structure in the interphase nuclei of somatic cells in female cats that was not found in male cats. This structure became known as the **Barr body** (**Figure 5.3a**). In 1960, Susumu Ohno correctly proposed that the Barr body is a highly condensed X chromosome.

In addition to this cytological evidence, Lyon was also familiar with mammalian examples in which the coat color had a variegated pattern. **Figure 5.3b** is a photo of a calico cat, which is a female that is heterozygous for an X-linked gene that can occur as an orange or a black allele. (The white underside is due to a dominant allele in a different gene.) The orange and black patches are randomly distributed in different female individuals. The calico pattern does not occur in male cats, but similar kinds of mosaic patterns have been identified in the female mouse. Lyon suggested that both the Barr body and the calico pattern are the result of X inactivation in the cells of female mammals.

The mechanism of X inactivation, also known as the **Lyon hypothesis,** is schematically illustrated in **Figure 5.4**. This example involves a white and black variegated coat color found in certain strains of mice. As shown here, a female mouse has inherited an X chromosome from its mother that carries an allele conferring white coat color (X^b). The X chromosome from its father carries a black coat color allele (X^B). How can X inactivation explain a variegated coat pattern? Initially, both X chromosomes are active. However, at an early stage of embryonic development, one of the two X chromosomes is randomly inactivated in each somatic cell and becomes a Barr body. For example, one embryonic cell may have the X^B chromosome inactivated. As the embryo continues

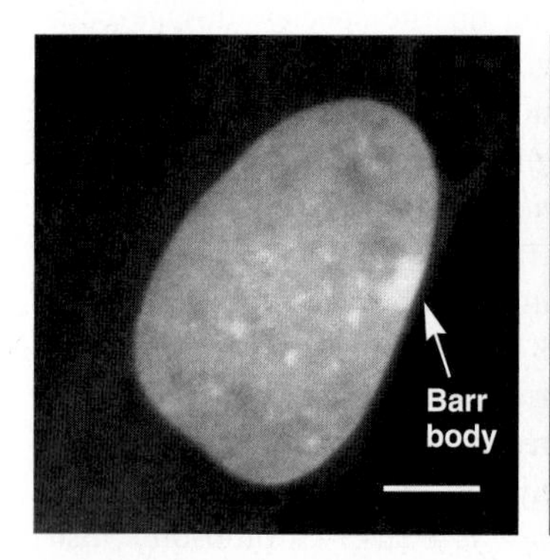

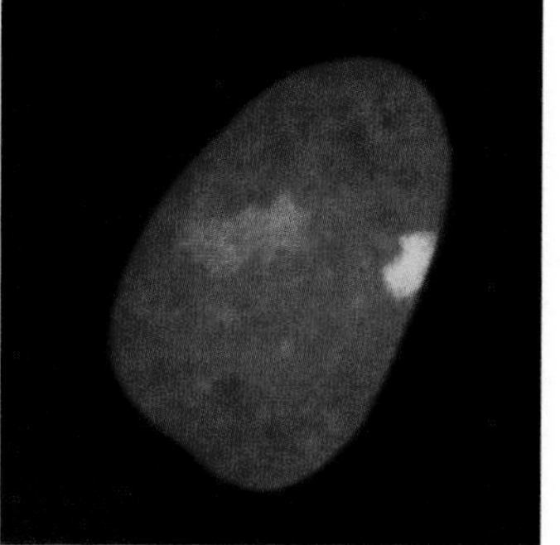

(a) Nucleus with a Barr body

(b) A calico cat

INTERACTIVE EXERCISE

FIGURE 5.3 X chromosome inactivation in female mammals. **(a)** The left micrograph shows the Barr body on the periphery of a human nucleus after staining with a DNA-specific dye. Because it is compact, the Barr body is the most brightly staining. The white scale bar is 5 μm. The right micrograph shows the same nucleus using a yellow fluorescent probe that recognizes the X chromosome. The Barr body is more compact than the active X chromosome, which is to the left of the Barr body. **(b)** The fur pattern of a calico cat.

Genes → Traits The pattern of black and orange fur on this cat is due to random X inactivation during embryonic development. The orange patches of fur are due to the inactivation of the X chromosome that carries a black allele; the black patches are due to the inactivation of the X chromosome that carries the orange allele. In general, only heterozygous female cats can be calico. A rare exception would be a male cat (XXY) that has an abnormal composition of sex chromosomes.

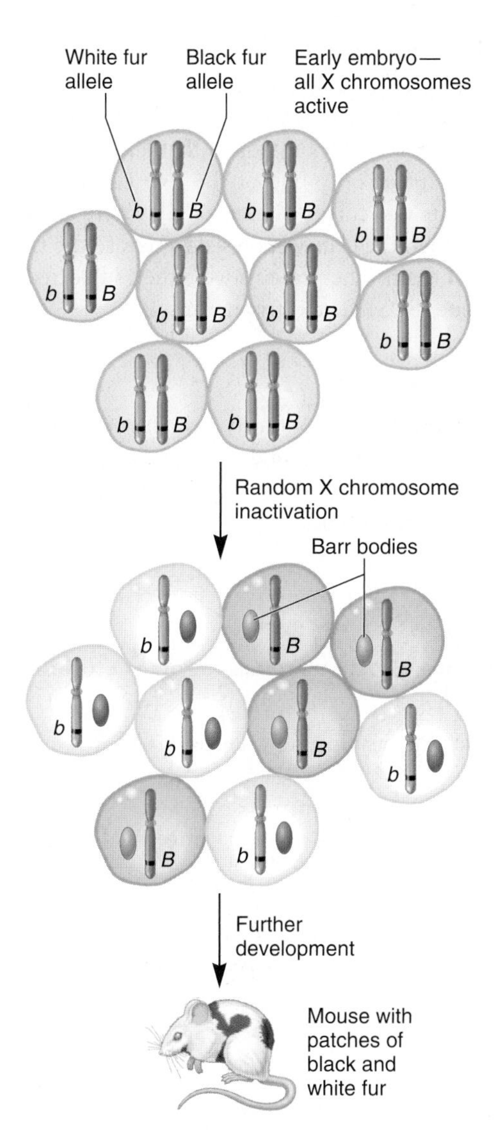

to grow and mature, this embryonic cell will divide and may eventually give rise to billions of cells in the adult animal. The epithelial (skin) cells that are derived from this embryonic cell will produce a patch of white fur because the X^B chromosome has been permanently inactivated. Alternatively, another embryonic cell may have the other X chromosome inactivated (i.e., X^b). The epithelial cells derived from this embryonic cell will produce a patch of black fur. Because the primary event of X inactivation is a random process that occurs at an early stage of development, the result is an animal with some patches of white fur and other patches of black fur. This is the basis of the variegated phenotype.

During inactivation, the chromosomal DNA becomes highly compacted in a Barr body, so most of the genes on the inactivated X chromosome cannot be expressed. When cell division occurs and the inactivated X chromosome is replicated, both copies remain highly compacted and inactive. Likewise, during subsequent cell divisions, X inactivation is passed along to all future somatic cells.

FIGURE 5.4 The mechanism of X chromosome inactivation.

Genes → Traits The top of this figure represents a mass of several cells that compose the early embryo. Initially, both X chromosomes are active. At an early stage of embryonic development, random inactivation of one X chromosome occurs in each cell. This inactivation pattern is maintained as the embryo matures into an adult.

EXPERIMENT 5A

In Adult Female Mammals, One X Chromosome Has Been Permanently Inactivated

According to the Lyon hypothesis, each somatic cell of female mammals expresses the genes on one of the X chromosomes, but not both. If an adult female is heterozygous for an X-linked gene, only one of two alleles will be expressed in any given cell. In 1963, Ronald Davidson, Harold Nitowsky, and Barton Childs set out to test the Lyon hypothesis at the cellular level. To do so, they analyzed the expression of a human X-linked gene that encodes an enzyme involved with sugar metabolism known as glucose-6-phosphate dehydrogenase (G-6-PD).

Prior to the Lyon hypothesis, biochemists had found that individuals vary with regard to the G-6-PD enzyme. This variation can be detected when the enzyme is subjected to gel electrophoresis (see the Appendix for a description of gel electrophoresis). One *G-6-PD* allele encodes a G-6-PD enzyme that migrates very quickly during gel electrophoresis (the "fast" enzyme), whereas another *G-6-PD* allele produces an enzyme that migrates more slowly (the "slow" enzyme). As shown in **Figure 5.5**, a sample of cells from heterozygous adult females produces both types of enzymes, whereas hemizygous males produce either the fast or slow type. The difference in migration between the fast and slow G-6-PD enzymes is due to minor differences in the structures of these enzymes. These minor differences do not significantly affect G-6-PD function, but they do enable geneticists to distinguish the proteins encoded by the two X-linked alleles.

As shown in **Figure 5.6**, Davidson, Nitowsky, and Childs tested the Lyon hypothesis using cell culturing techniques. They removed small samples of epithelial cells from a heterozygous female and grew them in the laboratory. When combined, these samples contained a mixture of both types of enzymes because

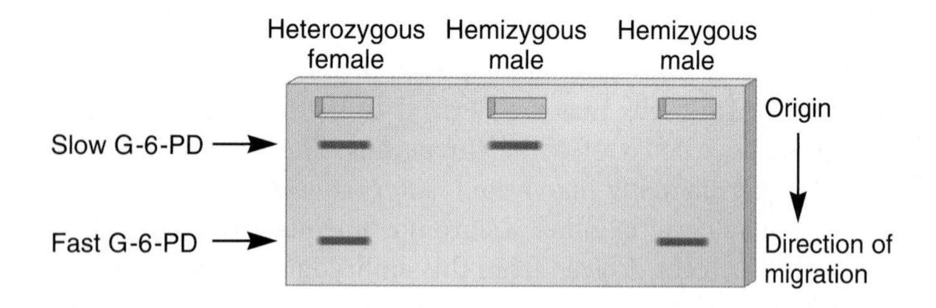

FIGURE 5.5 Mobility of G-6-PD protein on gels. *G-6-PD* can exist as a fast allele that encodes a protein that migrates more quickly to the bottom of the gel and a slow allele that migrates more slowly. The protein encoded by the fast allele is closer to the bottom of the gel.

the adult cells were derived from many different embryonic cells, some that had the slow allele inactivated and some that had the fast allele inactivated. In the experiment of Figure 5.6, these cells were sparsely plated onto solid growth media. After several days, each cell grew and divided to produce a colony, also called a **clone** of cells. All cells within a colony were derived from a single cell. The researchers reasoned that all cells within a single clone would express only one of the two *G-6-PD* alleles if the Lyon hypothesis was correct. Nine colonies were grown in liquid cultures, and then the cells were lysed to release the G-6-PD proteins inside of them. The proteins were then subjected to sodium dodecyl sulfate (SDS) gel electrophoresis.

THE HYPOTHESIS

According to the Lyon hypothesis, an adult female who is heterozygous for the fast and slow *G-6-PD* alleles should express only one of the two alleles in any particular somatic cell and its descendants, but not both.

TESTING THE HYPOTHESIS — FIGURE 5.6 Evidence that adult female mammals contain one X chromosome that has been permanently inactivated.

Starting material: Small skin samples taken from a woman who was heterozygous for the fast and slow alleles of *G-6-PD*.

1. Mince the tissue to separate the individual cells.

2. Grow the cells in a liquid growth medium and then plate (sparsely) onto solid growth medium. The cells then divide to form a clone of many cells.

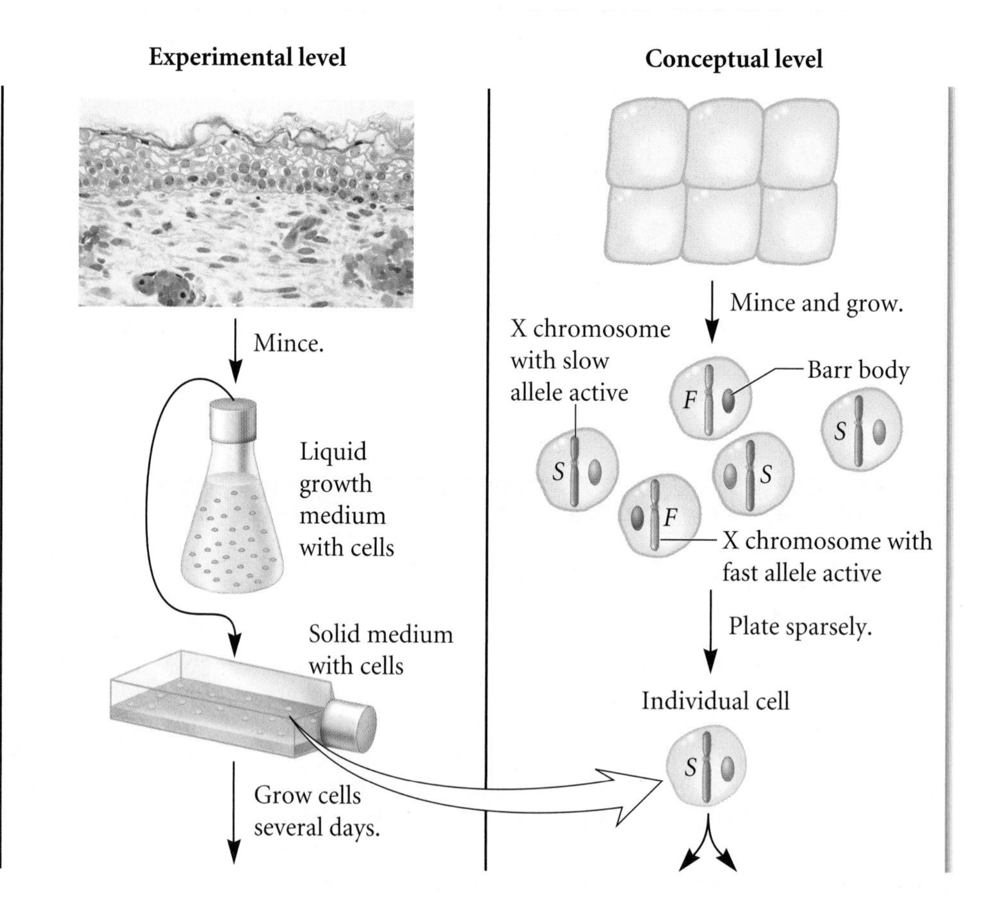

(continued)

3. Take nine isolated clones and grow in liquid cultures. (Only three are shown here.)

4. Take cells from the liquid cultures, lyse cells to obtain proteins, and subject to gel electrophoresis. (This technique is described in the Appendix.)

   Note: As a control, lyse cells from step 1, and subject the proteins to gel electrophoresis. This control sample is not from a clone. It is a mixture of cells derived from a woman's skin sample.

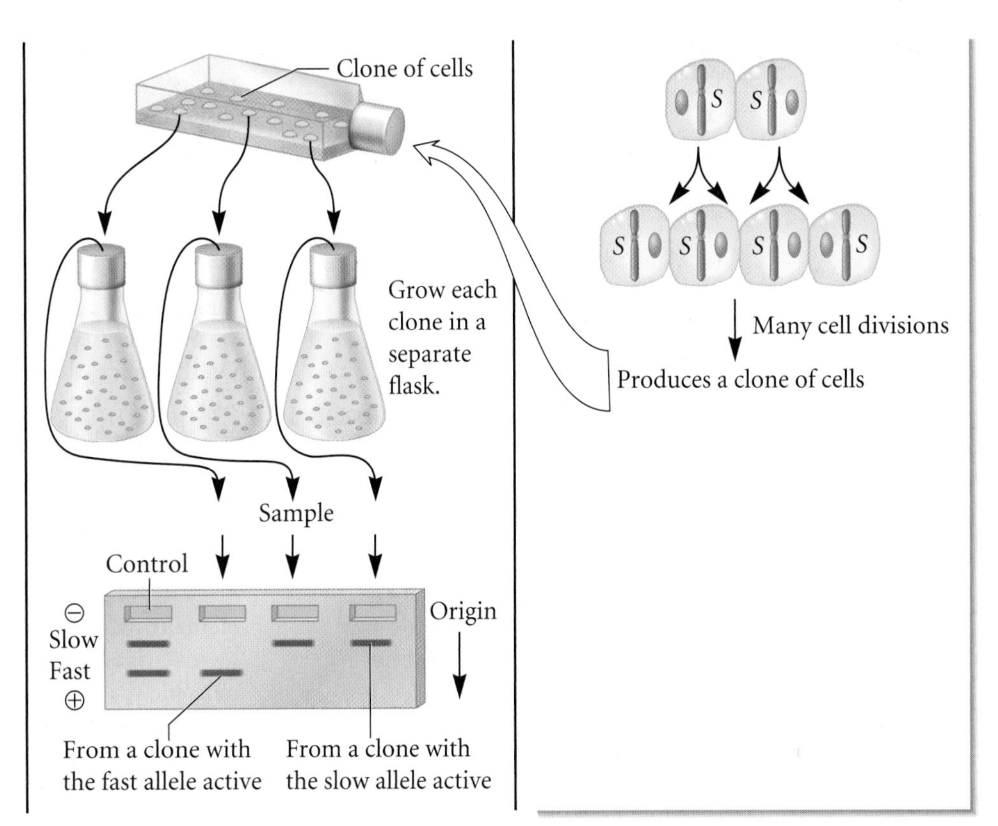

THE DATA

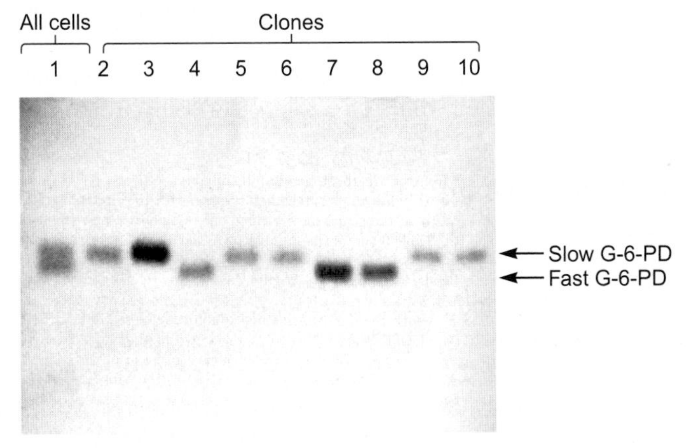

Data from Ronald G. Davidson, Harold M. Nitowsky, and Barton Childs (1963) Demonstration of two population of cells in the human female heterozygous for glucose-6-phosphate dehydrogenase variants. *Proc Natl Acad Sci USA 50*, 481-485.

INTERPRETING THE DATA

In the data shown in Figure 5.6, the control (lane 1) was a protein sample obtained from a mixture of epithelial cells from a heterozygous woman who produced both types of G-6-PD enzymes. Bands corresponding to the fast and slow enzymes were observed in this lane. As described in steps 2 to 4, this mixture of epithelial cells was also used to generate nine clones. The proteins obtained from these clones are shown in lanes 2 through 10. Each clone was a population of cells independently derived from a single epithelial cell. Because the epithelial cells were obtained from an adult female, the Lyon hypothesis predicts that each epithelial cell would already have one of its X chromosomes permanently inactivated and would pass this trait to its progeny cells. For example, suppose that an epithelial cell had inactivated the X chromosome that encoded the fast G-6-PD. If this cell was allowed to form a clone of cells on a plate, all cells in this clonal population would be expected to have the same X chromosome inactivated—the X chromosome encoding the fast G-6-PD. Therefore, this clone of cells should express only the slow G-6-PD. As shown in the data, all nine clones expressed either the fast or slow G-6-PD protein, but not both. These results are consistent with the hypothesis that X inactivation has already occurred in any given epithelial cell and that this pattern of inactivation is passed to all of its progeny cells.

A self-help quiz involving this experiment can be found at www.mhhe.com/brookergenetics4e.

Mammals Maintain One Active X Chromosome in Their Somatic Cells

Since the Lyon hypothesis was confirmed, the genetic control of X inactivation has been investigated further by several laboratories. Research has shown that mammalian cells possess the ability to count their X chromosomes in their somatic cells and allow only one of them to remain active. How was this determined? A key observation came from comparisons of the chromosome composition of people who were born with normal or abnormal numbers of sex chromosomes.

Phenotype	Chromosome Composition	Number of X Chromosomes	Number of Barr Bodies
Normal female	XX	2	1
Normal male	XY	1	0
Turner syndrome (female)	X0	1	0
Triple X syndrome (female)	XXX	3	2
Klinefelter syndrome (male)	XXY	2	1

In normal females, two X chromosomes are counted and one is inactivated, while in males, one X chromosome is counted and none inactivated. If the number of X chromosomes exceeds two, as in triple X syndrome, additional X chromosomes are converted to Barr bodies.

X Inactivation in Mammals Depends on the X-Inactivation Center and the *Xist* Gene

Although the genetic control of inactivation is not entirely understood at the molecular level, a short region on the X chromosome called the **X-inactivation center (Xic)** is known to play a critical role (**Figure 5.7**). Eeva Therman and Klaus Patau identified the Xic from its key role in X inactivation. The counting of human X chromosomes is accomplished by counting the number of Xics. A Xic must be found on an X chromosome for inactivation to occur. Therman and Patau discovered that if one of the two X chromosomes in a female is missing its Xic due to a chromosome mutation, a cell counts only one Xic and X inactivation does not occur. Having two active X chromosomes is a lethal condition for a human female embryo.

Let's consider how the molecular expression of certain genes controls X inactivation. The expression of a specific gene within the Xic is required for the compaction of the X chromosome into a Barr body. This gene, discovered in 1991, is named *Xist* (for X-inactive specific transcript). The *Xist* gene on the inactivated X chromosome is active, which is unusual because most other genes on the inactivated X chromosome are silenced. The *Xist* gene product is an RNA molecule that does not encode a protein. Instead, the role of the *Xist* RNA is to coat the X chromosome and inactivate it. After coating, other proteins associate with the *Xist* RNA and promote chromosomal compaction into a Barr body.

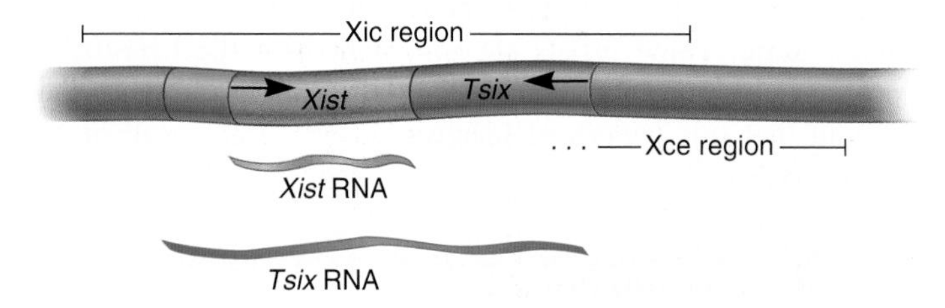

FIGURE 5.7 The X-inactivation center (Xic) of the X chromosome. The *Xist* gene is transcribed into RNA from the inactive X chromosome but not from the active X chromosome. This *Xist* RNA binds to the inactive X chromosome and recruits proteins that promote its compaction. The precise location of the Xce is not known. Some of it is adjacent to Xic and some of it may include all or part of the Xic. Xce may regulate the transcription of the *Xist* or *Tsix* genes and thereby influence the choice of the X chromosome that remains active.

A second gene found within the Xic, designated *Tsix*, plays a role in preventing X inactivation. As shown in Figure 5.7, the *Xist* and *Tsix* genes are overlapping and transcribed in opposite directions. (The name *Tsix* is *Xist* spelled backwards). The expression of the *Tsix* gene inhibits the transcription of the *Xist* gene. On the active X chromosome, the *Tsix* gene is expressed, and the *Xist* gene is not. The opposite situation occurs on the inactive X chromosome—*Xist* is expressed, and *Tsix* is not. Researchers have studied heterozygous females that carry a normal *Tsix* gene on one X chromosome and a defective, mutant *Tsix* gene on the other. The X chromosome carrying the mutant *Tsix* gene is preferentially inactivated.

Another region termed the **X chromosomal controlling element (Xce)** also affects the choice of the X chromosome to be inactivated. Genetic variation occurs in the Xce. An X chromosome that carries a strong Xce is more likely to remain active than an X chromosome that carries a weak Xce, thereby leading to skewed (nonrandom) X inactivation. As shown in Figure 5.7, the Xce is very close to the end of the Xic and may even encompass all or part of the Xic. Although the mechanism by which the Xce exerts its effects are not well understood, some researchers speculate that Xce serves as a binding site for proteins that regulate the expression of genes in the Xic, such as *Xist* or *Tsix*. Genetic variation in Xce that enhances *Xist* expression would tend to promote Barr body formation, whereas Xce variation that enhances *Tsix* expression would tend to prevent X inactivation.

X Inactivation Occurs in Three Phases: Initiation, Spreading, and Maintenance

The process of X inactivation can be divided into three phases: initiation, spreading, and maintenance (**Figure 5.8**). During initiation, which occurs during embryonic development, one of the X chromosomes remains active, and the other is chosen to be inactivated. How is a particular X chromosome chosen for X inactivation? The answer is not well understood, but is thought to involve a complex interplay between *Xist* and *Tsix* gene expression.

During the spreading phase, the chosen X chromosome is inactivated. This spreading requires the expression of the *Xist* gene. The *Xist* RNA coats the inactivated X chromosome and recruits proteins that promote compaction. This compaction involves DNA methylation and also the modification of histone proteins, which are described in Chapter 10. The spreading phase is so named because inactivation begins near the Xic and spreads in both directions along the X chromosome.

Once the initiation and spreading phases occur for a given X chromosome, the inactivated X chromosome is maintained as a Barr body during future cell divisions. When a cell divides, the Barr body is replicated, and both copies remain compacted. This maintenance phase continues from the embryonic stage through adulthood.

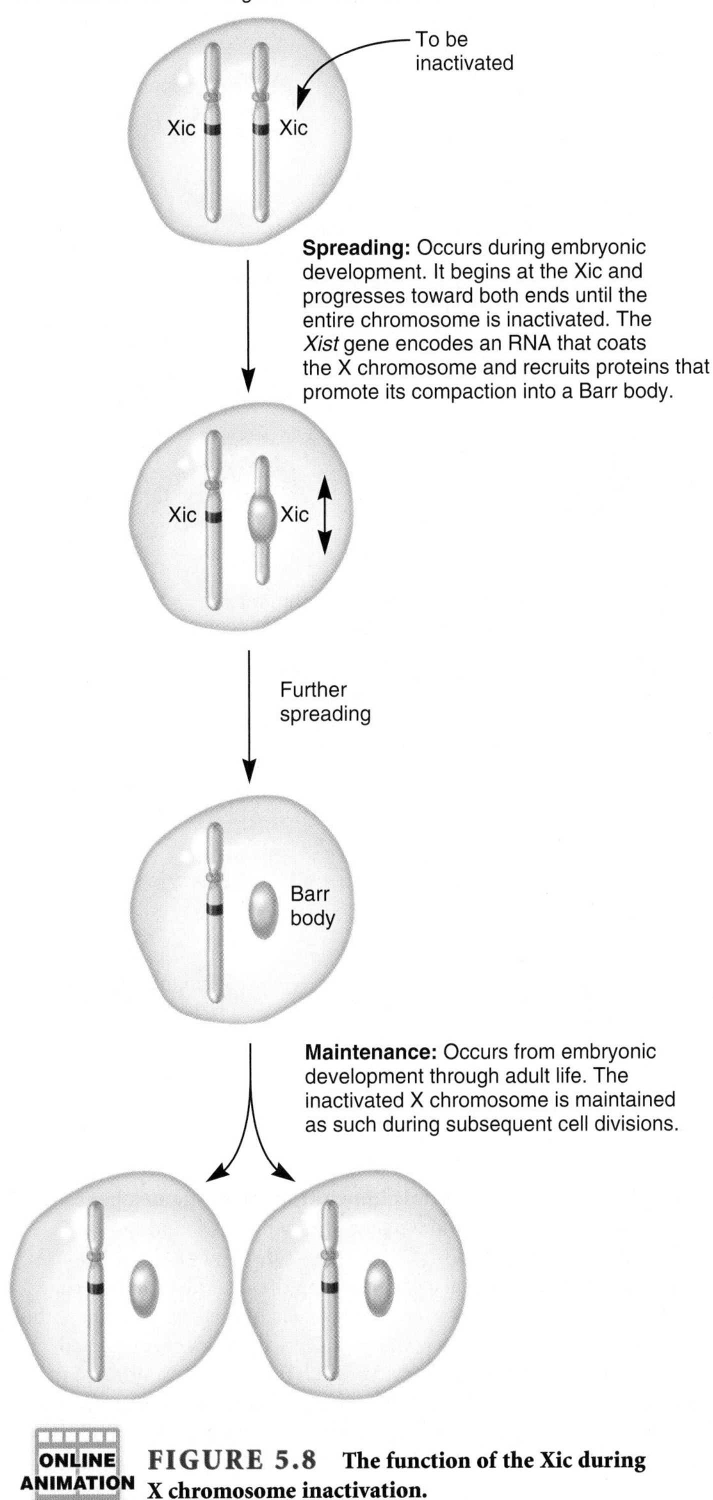

FIGURE 5.8 **The function of the Xic during X chromosome inactivation.**

Some genes on the inactivated X chromosome are expressed in the somatic cells of adult female mammals. These genes are said to escape the effects of X inactivation. As mentioned, *Xist* is an example of a gene that is expressed from the highly condensed Barr body. In humans, up to a quarter of X-linked genes may escape inactivation to some degree. Many of these genes occur in clusters. Among these are the pseudoautosomal genes found on the X and Y chromosomes in the regions of homology described in Chapter 4. Dosage compensation is not necessary for X-linked pseudoautosomal genes because they are located on both the X and Y chromosomes. How are genes on the Barr body expressed? Although the mechanism is not understood, these genes may be found in localized regions where the chromatin is less tightly packed and able to be transcribed.

The Expression of an Imprinted Gene Depends on the Sex of the Parent from Which the Gene Was Inherited

As we have just seen, dosage compensation changes the level of expression of many genes located on the X chromosome. We now turn to another epigenetic phenomenon known as imprinting. The term imprinting implies a type of marking process that has a memory. For example, newly hatched birds identify marks on their parents, which allows them to distinguish their parents from other individuals. The term **genomic imprinting** refers to an analogous situation in which a segment of DNA is marked, and that mark is retained and recognized throughout the life of the organism inheriting the marked DNA. The phenotypes caused by imprinted genes follow a non-Mendelian pattern of inheritance because the marking process causes the offspring to distinguish between maternally and paternally inherited alleles. Depending on how the genes are marked, the offspring expresses only one of the two alleles. This phenomenon is termed **monoallelic expression.**

To understand genomic imprinting, let's consider a specific example. In the mouse, a gene designated *Igf2* encodes a protein growth hormone called insulin-like growth factor 2. Imprinting occurs in a way that results in the expression of the paternal *Igf2* allele but not the maternal allele. The paternal allele is transcribed into RNA, but the maternal allele is transcriptionally silent. With regard to phenotype, a functional *Igf2* gene is necessary for normal size. A loss-of-function allele of this gene, designated *Igf2*$^-$, is defective in the synthesis of a functional Igf2 protein. This may cause a mouse to be a dwarf, but the dwarfism depends on whether the mutant allele is inherited from the male or female parent, as shown in **Figure 5.9**. On the left side, an offspring has inherited the *Igf2* allele from its father and the *Igf2*$^-$ allele from its mother. Due to imprinting, only the *Igf2* allele is expressed in the offspring. Therefore, this mouse grows to a normal size. Alternatively, in the reciprocal cross on the right side, an individual has inherited the *Igf2*$^-$ allele from its father and the *Igf2* allele from its mother. In this case, the *Igf2* allele is not expressed. In this mouse, the *Igf2*$^-$ allele would be transcribed into mRNA, but the mutation renders the Igf2 protein defective. Therefore, the offspring on the right has a dwarf phenotype. As shown here, both offspring have the same genotype; they are heterozygous for the *Igf2* alleles (i.e., *Igf2 Igf2*$^-$). They are phenotypically different, however, because only the paternally inherited allele is expressed.

At the cellular level, imprinting is an epigenetic process that can be divided into three stages: (1) the establishment of the imprint during gametogenesis, (2) the maintenance of the

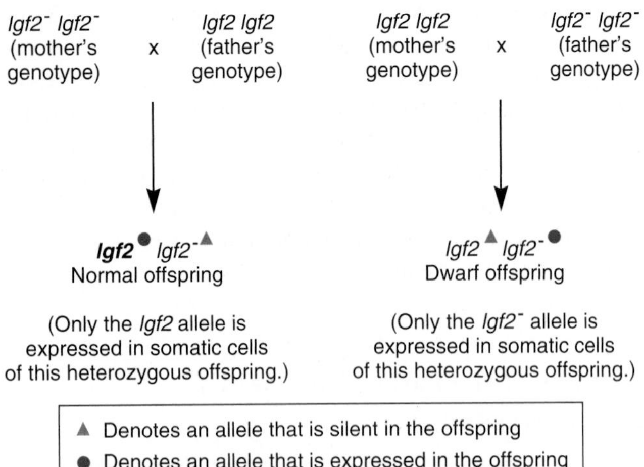

INTERACTIVE EXERCISE

FIGURE 5.9 An example of genomic imprinting in the mouse. In the cross on the left, a homozygous male with the normal *Igf2* allele is crossed to a homozygous female carrying a defective allele, designated *Igf2*$^-$. An offspring is heterozygous and normal because the paternal allele is active. In the reciprocal cross on the right, a homozygous male carrying the defective allele is crossed to a homozygous normal female. In this case, the offspring is heterozygous and dwarf. This is because the paternal allele is defective due to mutation and the maternal allele is not expressed. The photograph shows normal-size (left) and dwarf littermates (right) derived from a cross between a wild-type female and a heterozygous male carrying a loss-of-function *Igf2* allele (courtesy of A. Efstratiadis). The loss-of-function allele was created using gene knockout methods described in Chapter 19 (see Figure 19.7).

imprint during embryogenesis and in adult somatic cells, and (3) the erasure and reestablishment of the imprint in the germ cells. These stages are described in **Figure 5.10**, which shows the imprinting of the *Igf2* gene. The two mice shown here have inherited the *Igf2* allele from their father and the *Igf2*$^-$ allele from their mother. Due to imprinting, both mice express the *Igf2* allele in their somatic cells, and the pattern of imprinting is maintained in the somatic cells throughout development. In the germ cells (i.e., sperm and eggs), the imprint is erased; it will be reestablished according to the sex of the animal. The female mouse on the left will transmit only transcriptionally inactive alleles to her offspring. The male mouse on the right will transmit transcriptionally active alleles. However, because this male is a heterozygote, it will transmit either a functionally active *Igf2* allele or a functionally defective mutant allele (*Igf2*$^-$). An *Igf2*$^-$ allele, which is inherited from a male mouse, can be expressed

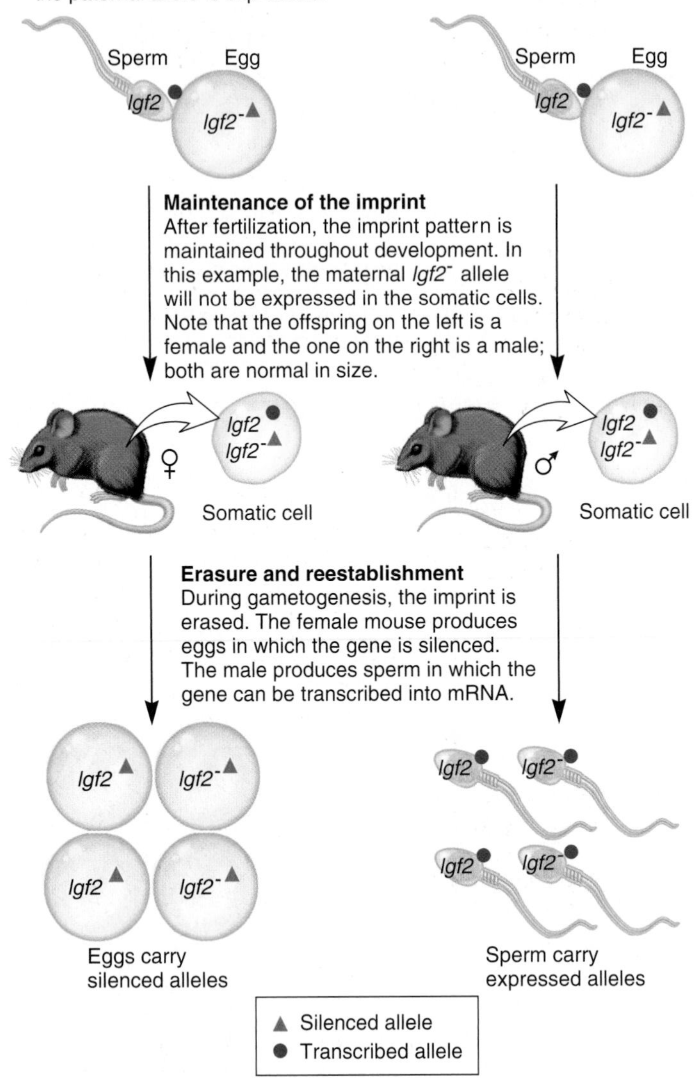

FIGURE 5.10 Genomic imprinting during gametogenesis. This example involves a mouse gene *Igf2*, which is found in two alleles designated *Igf2* and *Igf2*$^-$. The left side shows a female mouse that was produced from a sperm carrying the *Igf2* allele and an egg carrying the *Igf2*$^-$ allele. In the somatic cells of this female animal, the *Igf2* allele is active. However, when this female produces eggs, both alleles are transcriptionally inactive when they are transmitted to offspring. The right side of this figure shows a male mouse that was also produced from a sperm carrying the *Igf2* allele and an egg carrying the *Igf2*$^-$ allele. In the somatic cells of this male animal, the *Igf2* allele is active. However, the sperm from this male contains either a functionally active *Igf2* allele or a functionally defective *Igf2*$^-$ allele.

into mRNA (i.e., it is transcriptionally active), but it will not produce a functional Igf2 protein due to the deleterious mutation that created the *Igf2*$^-$ allele; a dwarf phenotype will result.

As seen in Figure 5.10, genomic imprinting is permanent in the somatic cells of an animal, but the marking of alleles can be altered from generation to generation. For example, the female mouse on the left possesses an active copy of the *Igf2* allele, but any allele this female transmits to its offspring will be transcriptionally inactive.

Genomic imprinting occurs in several species, including numerous insects, mammals, and flowering plants. Imprinting may involve a single gene, a part of a chromosome, an entire chromosome, or even all of the chromosomes from one parent. Helen Crouse discovered the first example of imprinting, which involved an entire chromosome in the housefly, *Sciara coprophila*. In this species, the fly normally inherits three sex chromosomes, rather than two as in most other species. One X chromosome is inherited from the female, and two are inherited from the male. In male flies, both paternal X chromosomes are lost from somatic cells during embryogenesis. In female flies, only one of the paternal X chromosomes is lost. In both sexes, the maternally inherited X chromosome is never lost. These results indicate that the maternal X chromosome is marked to promote its retention or paternal X chromosomes are marked to promote their loss.

Genomic imprinting can also be involved in the process of X inactivation, described previously. In certain species, imprinting plays a role in the choice of the X chromosome that will be inactivated. For example, in marsupials, the paternal X chromosome is marked so that it is the X chromosome that is always inactivated in the somatic cells of females. In marsupials, X inactivation is not random; the maternal X chromosome is always active.

The Imprinting of Genes and Chromosomes Is a Molecular Marking Process That Involves DNA Methylation

As we have seen, genomic imprinting must involve a marking process. A particular gene or chromosome must be marked differently during spermatogenesis versus oogenesis. After fertilization takes place, this differential marking affects the expression of particular genes. What is the molecular explanation for genomic imprinting? As discussed in Chapter 15, **DNA methylation**—the attachment of a methyl group onto a cytosine base—is a common way that eukaryotic genes may be regulated. Research indicates that genomic imprinting involves an **imprinting control region (ICR)** that is located near the imprinted gene. A portion of the DNA in this region is called the differentially methylated domain (DMD). Depending on the particular gene, the DMD is methylated in the egg or the sperm, but not both. The ICR also contains binding sites for one or more proteins that regulate the transcription of the imprinted gene.

For most imprinted genes, methylation causes an inhibition of gene expression. Methylation could enhance the binding of proteins that inhibit transcription or inhibit the binding of proteins that enhance transcription (or both). (The relationship between methylation and gene expression is described in Chapter 15.) For this reason, imprinting is usually described as a marking process that silences gene expression by preventing transcription. However, this is not always the case. Two imprinted human genes, *H19* and *Igf2*, provide an interesting example. These two genes lie close to each other on human chromosome 11 and appear to be controlled by the same ICR, which is a 52,000-base pair (bp) region that lies between the *Igf2* and *H19* genes (**Figure 5.11**). It contains binding sites for proteins that regulate the transcription of the *H19* or *Igf2* genes. This ICR is highly methylated on the paternally inherited chromosome but not on the maternally inherited one.

When the ICR is not methylated, a protein called CTC-binding factor is able to bind to the ICR (Figure 5.11a). The CTC-binding factor gets its name from the observation that it binds to a region of DNA that is rich in CTC (cytosine-thymine-cytosine) sequences. The ICR contains several such CTC sequences. When they are unmethylated, the CTC-binding factor can bind to the ICR. As described in Figure 5.11a, this has two effects. First, it prevents the binding of activator proteins to the *Igf2* gene, thereby shuting off this gene. In contrast, it permits activator proteins to turn on the *H19* gene.

How does methylation affect the transcription of the *Igf2* and *H19* genes? When the cytosines within the ICR become methylated, the CTC-binding factor is unable to bind to the ICR (Figure 5.11b). This permits activator proteins to turn on the *Igf2* gene. The DNA methylation also causes the repression of the *H19* gene so it is not transcribed.

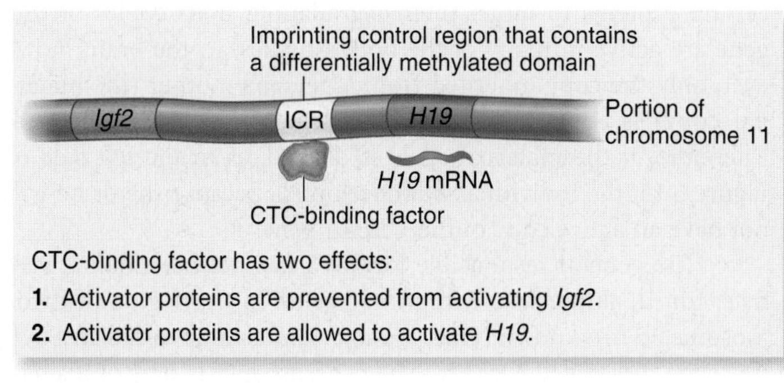

(a) Maternal chromosome

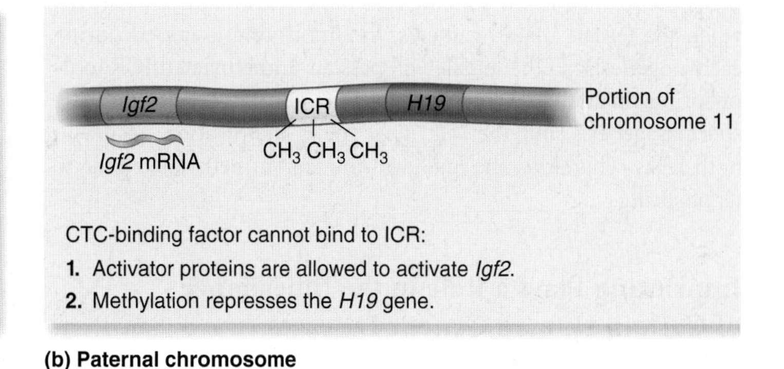

(b) Paternal chromosome

FIGURE 5.11 A simplified scheme for how DNA methylation at the ICR affects the expression of the *Igf2* and *H19* genes. (a) The lack of methylation of the maternal chromosome causes the *Igf2* gene to be turned off and the *H19* gene to be turned on. (b) The methylation of the paternal chromosome has the opposite effect. The *Igf2* gene is turned on and the *H19* gene is turned off.

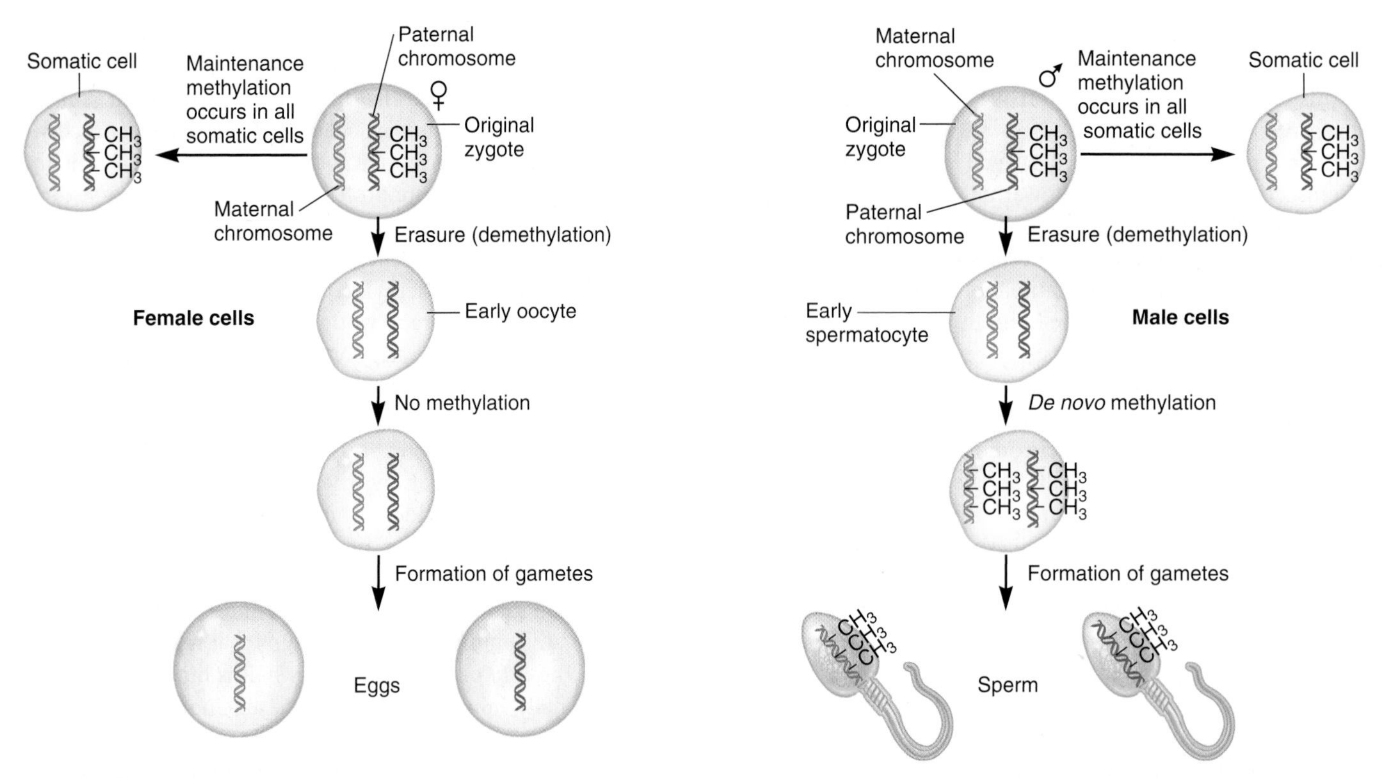

FIGURE 5.12 **The pattern of ICR methylation from one generation to the next.** In this example, a male and a female offspring have inherited a methylated ICR and nonmethylated ICR from their father and mother, respectively. Maintenance methylation retains the imprinting in somatic cells during embryogenesis and in adulthood. Demethylation occurs in cells that are destined to become gametes. In this example, *de novo* methylation occurs only in cells that are destined to become sperm. Haploid male gametes transmit a methylated ICR, whereas haploid female gametes transmit an unmethylated ICR.

Imprinting from Generation to Generation Involves Maintenance, Erasure, and *De Novo* Methylation Steps

Now that we have an understanding of how methylation may affect gene transcription, let's consider the methylation process from one generation to the next. In the example shown in **Figure 5.12**, the paternally inherited allele for a particular gene is methylated but the maternally inherited allele is not. A female (left side) and male (right side) have inherited a methylated ICR from their father and an unmethylated ICR from their mother. This pattern of imprinting is maintained in the somatic cells of both individuals. However, when the female makes gametes, the imprinting is erased during early oogenesis, so the female will pass an unmethylated ICR to its offspring. In the male, the imprinting is also erased during early spermatogenesis, but then *de novo* (new) methylation occurs in both ICRs. Therefore, the male will transmit a methylated gene to its offspring.

Imprinting Plays a Role in the Inheritance of Certain Human Genetic Diseases

Human diseases, such as Prader-Willi syndrome (PWS) and Angelman syndrome (AS), are influenced by imprinting. PWS is characterized by reduced motor function, obesity, and small hands and feet. Individuals with AS are thin and hyperactive, have unusual seizures and repetitive symmetrical muscle movements, and exhibit mental deficiencies. Most commonly, both PWS and AS involve a small deletion in human chromosome 15. If this deletion is inherited from the mother, it leads to Angelman syndrome; if inherited from the father, it leads to Prader-Willi syndrome (**Figure 5.13**).

Researchers have discovered that this region of chromosome 15 contains closely linked but distinct genes that are maternally or paternally imprinted. AS results from the lack of expression of a single gene (*UBE3A*) that codes for a protein called E6-AP, which functions to transfer small ubiquitin molecules to certain proteins to target their degradation. Both copies of this gene are active in many of the body's tissues. In the brain, however, only the copy inherited from a person's mother (the maternal copy) is active. The paternal allele of *UBE3A* is silenced. Therefore, if the maternal allele is deleted, as in the left side of Figure 5.13, the individual will develop AS because she or he will not have an active copy of the *UBE3A* gene.

The gene(s) responsible for PWS has not been definitively determined, although five imprinted genes in this region of chromosome 15 are known. One possible candidate involved in PWS is a gene designated *SNRPN*. The gene product is part of a small nuclear ribonucleoprotein polypeptide N, which is a complex that controls RNA splicing and is necessary for the synthesis of critical proteins in the brain. The maternal allele of *SNRPN* is silenced, and only the paternal copy is active.

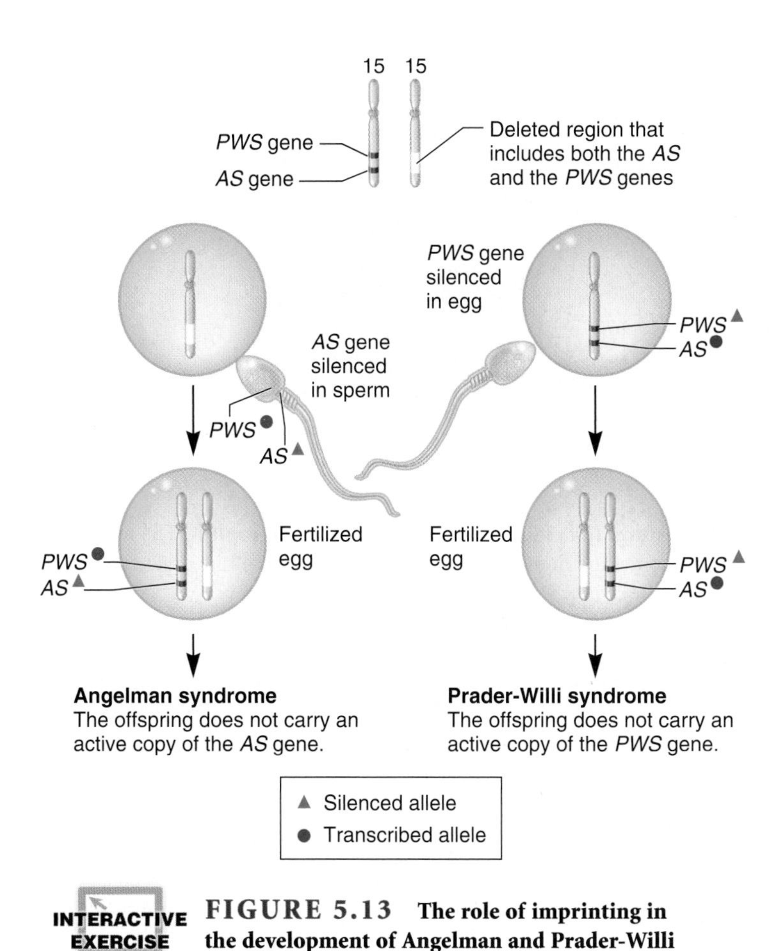

INTERACTIVE EXERCISE

FIGURE 5.13 The role of imprinting in the development of Angelman and Prader-Willi syndromes.

Genes → Traits A small region on chromosome 15 contains two different genes designated the *AS* gene and *PWS* gene in this figure. If a chromosome 15 deletion is inherited from the mother, Angelman syndrome occurs because the offspring does not inherit an active copy of the *AS* gene (left). Alternatively, the chromosome 15 deletion may be inherited from the father, leading to Prader-Willi syndrome. The phenotype of this syndrome occurs because the offspring does not inherit an active copy of the *PWS* gene (right).

The Biological Significance of Imprinting Is Not Well Understood

Genomic imprinting is a fairly new and exciting area of research. Imprinting has been identified in many mammalian genes (**Table 5.2**). In some cases, the female alleles are transcriptionally active in the somatic cells of offspring, whereas in other cases, the male alleles are active.

The biological significance of imprinting is still a matter of much speculation. Several hypotheses have been proposed to explain the potential benefits of genomic imprinting. One example, described by David Haig, involves differences in female versus male reproductive patterns in mammals. As discussed in Chapter 24, natural selection favors types of genetic variation that confer a reproductive advantage. The likelihood that favorable variation will be passed to offspring may differ between the sexes. In many mammalian species, females may mate with multiple males, perhaps generating embryos in the same uterus fathered by different males. For males, silencing genes that inhibit embryonic growth would be an advantage. The embryos of males that silence such genes would grow faster than other embryos in the same uterus, making it more likely for the males to pass their genes to future generations. For females, however, rapid growth of embryos might be a disadvantage because it could drain too many resources from the mother. According to this scenario, the mother would silence genes that cause rapid embryonic growth. From the mother's perspective, she would give all of her offspring an equal chance of survival without sapping her own strength. This would make it more likely for the female to pass her genes to future generations.

TABLE 5.2

Examples of Mammalian Genes and Inherited Human Diseases That Involve Imprinted Genes*

Gene	Allele Expressed	Function
WT1	Maternal	Wilms tumor–suppressor gene; suppresses cell growth
INS	Paternal	Insulin; hormone involved in cell growth and metabolism
Igf2	Paternal	Insulin-like growth factor II; similar to insulin
Igf2R	Maternal	Receptor for insulin-like growth factor II
H19	Maternal	Unknown
SNRPN	Paternal	Splicing factor
Gabrb	Maternal	Neurotransmitter receptor

*Researchers estimate that approximately 1–2% of human genes are subjected to genomic imprinting, but fewer than 100 have actually been demonstrated to be imprinted.

The Haig hypothesis seems to be consistent with the imprinting of several mammalian genes that are involved with growth, such as *Igf2*. Females silence this growth-enhancing gene, whereas males do not. However, several imprinted genes do not seem to play a role in embryonic development. Therefore, an understanding of the biological role(s) of genomic imprinting requires further investigation.

5.3 EXTRANUCLEAR INHERITANCE

Thus far, we have considered several types of non-Mendelian inheritance patterns. These include maternal effect genes, dosage compensation, and genomic imprinting. All of these inheritance patterns involve genes found on chromosomes in the cell nucleus. Another cause of non-Mendelian inheritance patterns involves genes that are not located in the cell nucleus. In eukaryotic species, the most biologically important example of extranuclear inheritance involves genetic material in cellular organelles. In addition to the cell nucleus, the mitochondria and chloroplasts contain their own genetic material. Because these organelles are found within the cytoplasm of the cells, the inheritance of organellar genetic material is called **extranuclear inheritance** (the prefix *extra-* means outside of) or **cytoplasmic inheritance.** In this section, we will examine the genetic composition of mitochondria and chloroplasts and explore the pattern

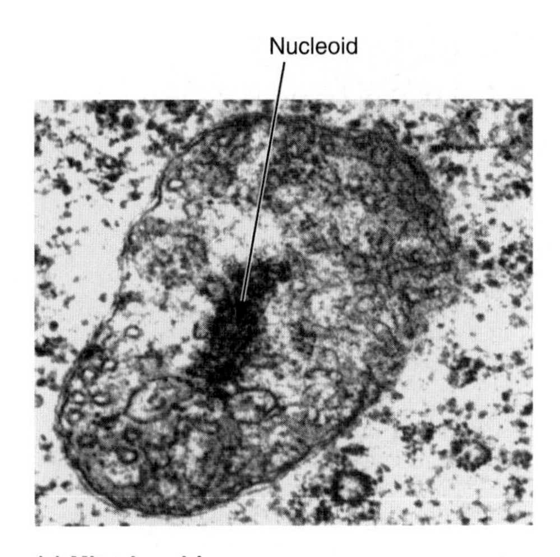

(a) Mitochondrion

nucleoid
nucleoid

(b) Chloroplast

FIGURE 5.14 **Nucleoids within (a) a mitochondrion and (b) a chloroplast.** The mitochondrial and chloroplast chromosomes are found within the nucleoid region of the organelle.

of transmission of these organelles from parent to offspring. We will also consider a few other examples of inheritance patterns that cannot be explained by the transmission of nuclear genes.

Mitochondria and Chloroplasts Contain Circular Chromosomes with Many Genes

In 1951, Yukako Chiba was the first to suggest that chloroplasts contain their own DNA. He based his conclusion on the staining properties of a DNA-specific dye known as Feulgen. Researchers later developed techniques to purify organellar DNA. In addition, electron microscopy studies provided interesting insights into the organization and composition of mitochondrial and chloroplast chromosomes. More recently, the advent of molecular genetic techniques in the 1970s and 1980s has allowed researchers to determine the genome sequences of organellar DNAs. From these types of studies, the chromosomes of mitochondria and chloroplasts were found to resemble smaller versions of bacterial chromosomes.

The genetic material of mitochondria and chloroplasts is located inside the organelle in a region known as the **nucleoid** (**Figure 5.14**). The genome is a single circular chromosome (composed of double-stranded DNA), although a nucleoid contains several copies of this chromosome. In addition, a mitochondrion or chloroplast often has more than one nucleoid. In mice, for example, each mitochondrion has one to three nucleoids, with each nucleoid containing two to six copies of the circular mitochondrial genome. However, this number varies depending on the type of cell and the stage of development. In comparison, the chloroplasts of algae and higher plants tend to have more nucleoids per organelle. **Table 5.3** describes the genetic composition of mitochondria and chloroplasts for a few selected species.

Besides variation in copy number, the sizes of mitochondrial and chloroplast genomes also vary greatly among different species. For example, a 400-fold variation is found in the sizes of mitochondrial chromosomes. In general, the mitochondrial genomes of animal species tend to be fairly small; those of fungi

TABLE **5.3**

Genetic Composition of Mitochondria and Chloroplasts

Species	Organelle	Nucleoids per Organelle	Total Number of Chromosomes per Organelle
Tetrahymena	Mitochondrion	1	6–8
Mouse	Mitochondrion	1–3	5–6
Chlamydomonas	Chloroplast	5–6	~80
Euglena	Chloroplast	20–34	100–300
Higher plants	Chloroplast	12–25	~60

Data from: Gillham, N. W. (1994). *Organelle Genes and Genomes.* Oxford University Press, New York.

and protists are intermediate in size; and those of plant cells tend to be fairly large. Among algae and plants, substantial variation is also found in the sizes of chloroplast chromosomes.

Figure 5.15 illustrates a map of human **mitochondrial DNA (mtDNA).** Each copy of the mitochondrial chromosome consists of a circular DNA molecule that is only 17,000 bp in length. This size is less than 1% of a typical bacterial chromosome. The human mtDNA carries relatively few genes. Thirteen genes encode proteins that function within the mitochondrion. In addition, mtDNA carries genes that encode ribosomal RNA and transfer RNA. These rRNAs and tRNAs are necessary for the synthesis of the 13 polypeptides that are encoded by the mtDNA. The primary role of mitochondria is to provide cells with the bulk of their adenosine triphosphate (ATP), which is used as an energy source to drive cellular reactions. These 13 polypeptides are subunits of proteins that function in a process known as oxidative phosphorylation, in which mitochondria use oxygen and synthesize ATP. However, mitochondria require many additional proteins to carry out oxidative phosphorylation and

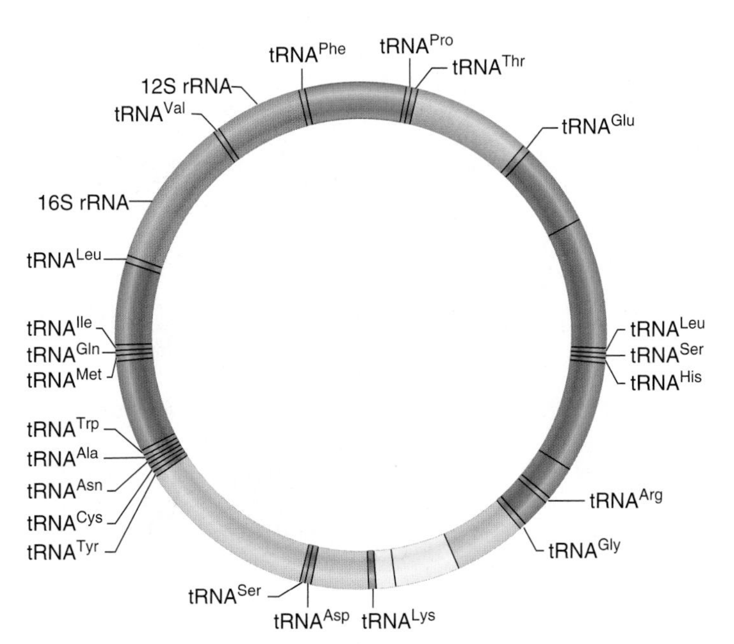

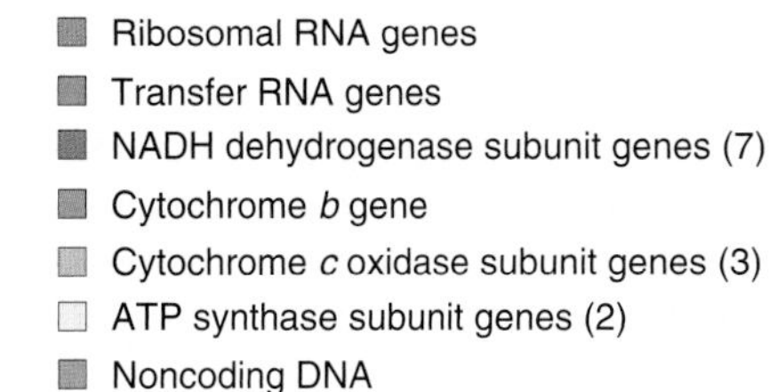

FIGURE 5.15 A genetic map of human mitochondrial DNA (mtDNA). This diagram illustrates the locations of many genes along the circular mitochondrial chromosome. The genes shown in red encode transfer RNAs. For example, $tRNA^{Arg}$ encodes a tRNA that carries arginine. The genes that encode ribosomal RNA are shown in light brown. The remaining genes encode proteins that function within the mitochondrion. The mitochondrial genomes from numerous species have been determined.

other mitochondrial functions. Most mitochondrial proteins are encoded by genes within the cell nucleus. When these nuclear genes are expressed, the mitochondrial polypeptides are first synthesized outside the mitochondria in the cytosol of the cell. They are then transported into the mitochondria where they may associate with other polypeptides and become functional proteins.

Chloroplast genomes tend to be larger than mitochondrial genomes, and they have a correspondingly greater number of genes. A typical chloroplast genome is approximately 100,000 to 200,000 bp in length, which is about 10 times larger than the mitochondrial genome of animal cells. **Figure 5.16** shows the **chloroplast DNA (cpDNA)** of the tobacco plant, which is

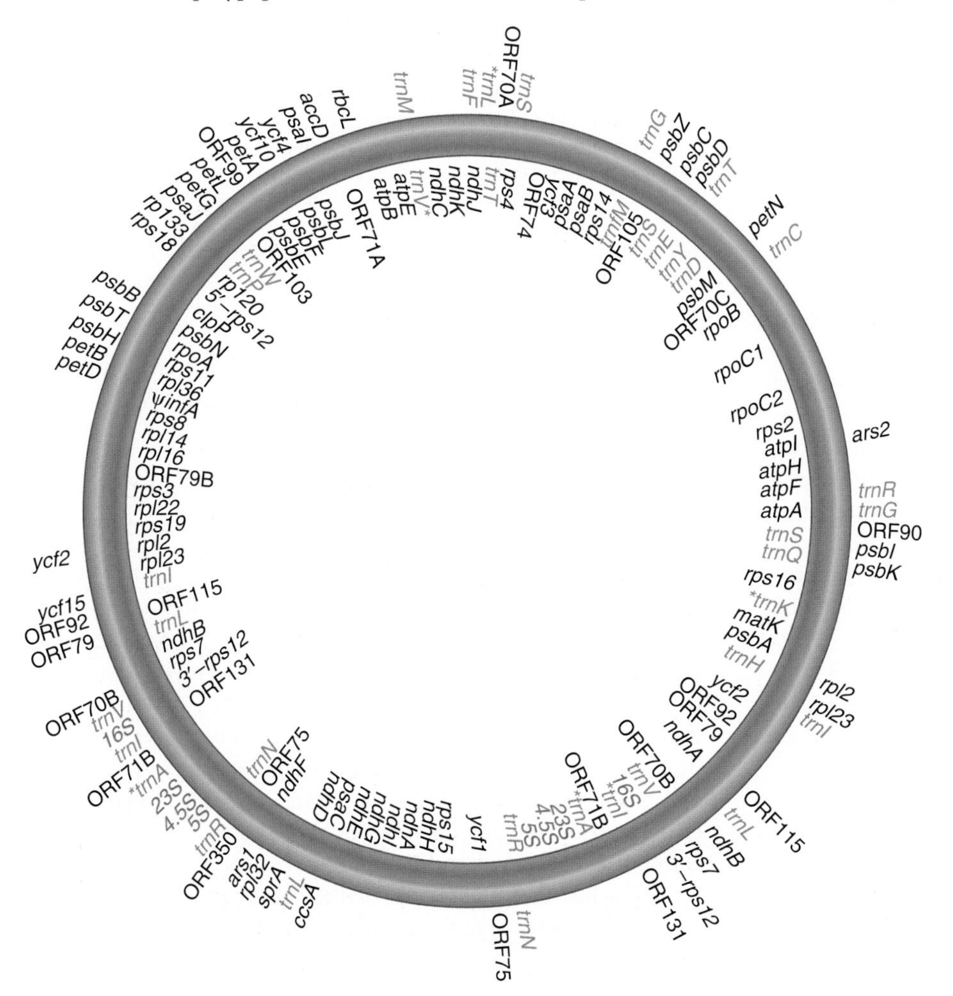

FIGURE 5.16 A genetic map of tobacco chloroplast DNA (cpDNA). This diagram illustrates the locations of many genes along the circular chloroplast chromosome. The gene names shown in blue encode transfer RNAs. The genes that encode ribosomal RNA are shown in red. The remaining genes shown in black encode polypeptides that function within the chloroplast. The genes designated ORF (open reading frame) encode polypeptides with unknown functions.

a circular DNA molecule that contains 156,000 bp of DNA and carries between 110 and 120 different genes. These genes encode ribosomal RNAs, transfer RNAs, and many proteins required for photosynthesis. As with mitochondria, many chloroplast proteins are encoded by genes found in the plant cell nucleus. These proteins contain chloroplast-targeting signals that direct them into the chloroplasts.

Extranuclear Inheritance Produces Non-Mendelian Results in Reciprocal Crosses

In diploid eukaryotic species, most genes within the nucleus obey a Mendelian pattern of inheritance because the homologous pairs of chromosomes segregate during gamete formation. Except for sex-linked traits, offspring inherit one copy of each gene from both the maternal and paternal parents. The sorting of chromosomes during meiosis explains the inheritance patterns of nuclear genes. By comparison, the inheritance of extranuclear genetic material does not display a Mendelian pattern. Mitochondria and chloroplasts are not sorted during meiosis and therefore do not segregate into gametes in the same way as nuclear chromosomes.

In 1909, Carl Correns discovered a trait that showed a non-Mendelian pattern of inheritance involving pigmentation in *Mirabilis jalapa* (the four-o'clock plant). Leaves can be green, white, or variegated with both green and white sectors. Correns demonstrated that the pigmentation of the offspring depended solely on the maternal parent (**Figure 5.17**). If the female parent had white pigmentation, all offspring had white leaves. Similarly, if the female was green, all offspring were green. When the female was variegated, the offspring could be green, white, or variegated.

The pattern of inheritance observed by Correns is a type of extranuclear inheritance called **maternal inheritance** (not to be confused with maternal effect). Chloroplasts are a type of plastid that makes chlorophyll, a green photosynthetic pigment. Maternal inheritance occurs because the chloroplasts are inherited only through the cytoplasm of the egg. The pollen grains of *M. jalapa* do not transmit chloroplasts to the offspring.

The phenotypes of leaves can be explained by the types of chloroplasts within the leaf cells. The green phenotype, which is the wild-type condition, is due to the presence of normal chloroplasts that make green pigment. By comparison, the white phenotype is due to a mutation in a gene within the chloroplast DNA that diminishes the synthesis of green pigment. A cell may contain both types of chloroplasts, a condition known as **heteroplasmy.** A leaf cell containing both types of chloroplasts is green because the normal chloroplasts produce green pigment.

How does a variegated phenotype occur? **Figure 5.18** considers the leaf of a plant that began from a fertilized egg that contained both types of chloroplasts (i.e., a heteroplasmic cell). As a plant grows, the two types of chloroplasts are irregularly distributed to daughter cells. On occasion, a cell may receive only the chloroplasts that have a defect in making green pigment. Such a cell continues to divide and produce a sector of the plant that is entirely white. In this way, the variegated phenotype is produced. Similarly, if we consider the results of Figure 5.17, a female parent that is variegated may transmit green, white, or a mixture

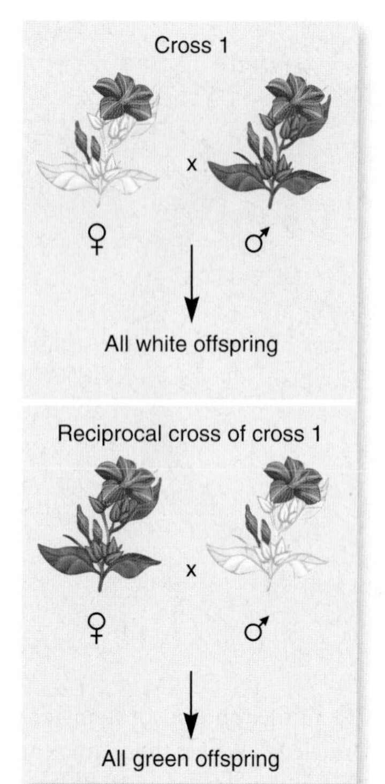

INTERACTIVE EXERCISE

FIGURE 5.17 Maternal inheritance in the four-o'clock plant, *Mirabilis jalapa*. The reciprocal crosses of four-o'clock plants by Carl Correns consisted of a pair of crosses between white-leaved and green-leaved plants, and a pair of crosses between variegated-leaved and green-leaved plants.

Genes → Traits In this example, the white phenotype is due to chloroplasts that carry a mutant allele that diminishes green pigmentation. The variegated phenotype is due to a mixture of chloroplasts, some of which carry the normal (green) allele and some of which carry the white allele. In the crosses shown here, the parent providing the eggs determines the phenotypes of the offspring. This is due to maternal inheritance. The egg contains the chloroplasts that are inherited by the offspring. (Note: The defective chloroplasts that give rise to white sectors are not completely defective in chlorophyll synthesis. Therefore, entirely white plants can survive, though they are smaller than green or variegated plants.)

of these types of chloroplasts to the egg cell, thereby producing green, white, or variegated offspring, respectively.

Studies in Yeast and *Chlamydomonas* Provided Genetic Evidence for Extranuclear Inheritance of Mitochondria and Chloroplasts

The research of Correns and others indicated that some traits, such as leaf pigmentation, are inherited in a non-Mendelian manner. However, such studies did not definitively determine that maternal inheritance is due to genetic material within organelles. Further progress in the investigation of extranuclear inheritance was provided by detailed genetic analyses of eukaryotic microorganisms, such as yeast and algae, by isolating and characterizing mutant phenotypes that specifically affected the chloroplasts or mitochondria.

During the 1940s and 1950s, yeasts and molds became model eukaryotic organisms for investigating the inheritance of mitochondria. Because mitochondria produce energy for cells in

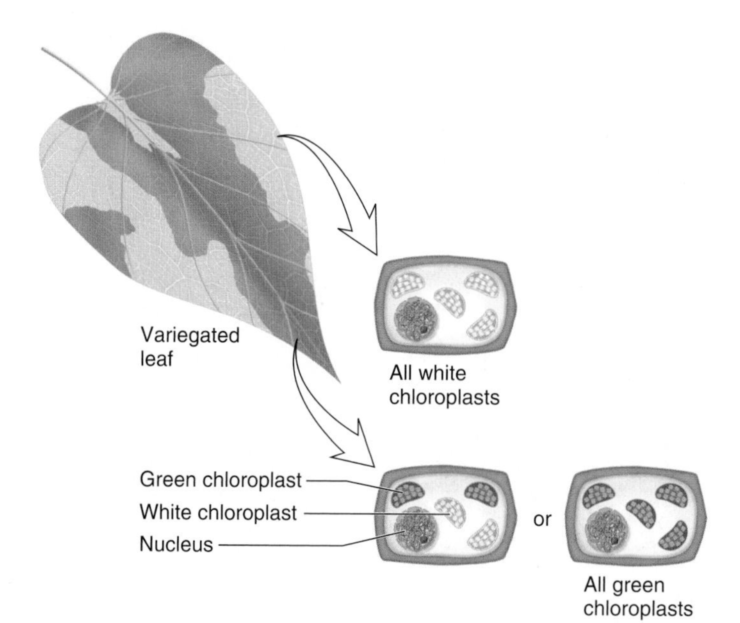

FIGURE 5.18 A cellular explanation of the variegated phenotype in *Mirabilis jalapa*. This plant inherited two types of chloroplasts—those that can produce green pigment and those that are defective. As the plant grows, the two types of chloroplasts are irregularly distributed to daughter cells. On occasion, a leaf cell may receive only the chloroplasts that are defective at making green pigment. Such a cell continues to divide and produces a sector of the leaf that is entirely white. Cells that contain both types of chloroplasts or cells that contain only green chloroplasts produce green tissue, which may be adjacent to a sector of white tissue. This is the basis for the variegated phenotype of the leaves.

the form of ATP, mutations that yield defective mitochondria are expected to make cells grow much more slowly. Boris Ephrussi and his colleagues identified mutations in *Saccharomyces cerevisiae* that had such a phenotype. These mutants were called **petites** to describe their formation of small colonies on agar plates as opposed to wild-type strains that formed larger colonies. Biochemical and physiological evidence indicated that petite mutants had defective mitochondria. The researchers found that petite mutants could not grow when the cells had an energy source requiring only the metabolic activity of mitochondria, but could form small colonies when grown on sugars metabolized by the glycolytic pathway, which occurs outside the mitochondria.

Because yeast cells exist in two mating types, designated a and α, Ephrussi was able to mate a wild-type strain to his petite mutants. Genetic analyses showed that petite mutants were inherited in different ways. When a wild-type strain was crossed to a segregational petite mutant, he obtained a ratio of 2 wild-type cells to 2 petite cells (**Figure 5.19a**). This result is consistent with a Mendelian pattern of inheritance (see the discussion of tetrad analysis in Chapter 6). Therefore, segregational petite mutations cause defects in genes located in the cell nucleus. These genes encode proteins necessary for mitochondrial function. Such proteins are synthesized in the cytosol and are then taken up by the mitochondria, where they perform their functions. Segregational petites get their name because they segregate in a Mendelian manner during meiosis.

By comparison, the second category of petite mutants, known as vegetative petite mutants, did not segregate in a Mendelian manner (**Figure 5.19b**). Ephrussi identified two types of vegetative petites, called neutral petites and suppressive petites. In a cross between a wild-type strain and a neutral petite, all four haploid daughter cells were wild type. This type of inheritance contradicts the normal 2:2 ratio expected for the segregation of Mendelian traits. In comparison, a cross between a wild-type strain and a suppressive petite usually yielded all petite colonies. Thus, both types of vegetative petites are defective in mitochondrial function and show a non-Mendelian pattern of inheritance. These results occurred because vegetative petites carry mutations in the mitochondrial genome itself.

Since these initial studies, researchers have found that neutral petites lack most of their mitochondrial DNA, whereas suppressive petites usually lack small segments of the mitochondrial genetic material. When two yeast cells are mated, the daughter cells inherit mitochondria from both parents. For example, in a cross between a wild-type and a neutral petite strain, the daughter cells inherit both types of mitochondria. Because wild-type

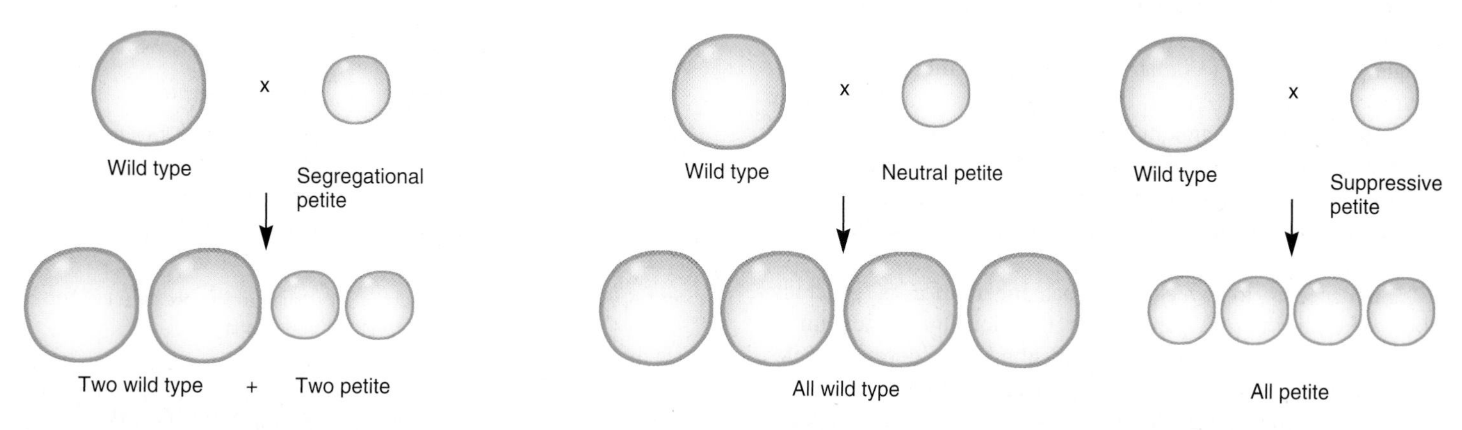

FIGURE 5.19 Transmission of the petite trait in *Saccharomyces cerevisiae*. (**a**) A wild-type strain crossed to a segregational petite. (**b**) A wild-type strain crossed to a neutral vegetative petite and to a suppressive vegetative petite.

mitochondria are inherited, the cells display a normal phenotype. The inheritance pattern of suppressive petites is more difficult to explain because the daughter cells inherit both normal and suppressive petite mitochondria. One possibility is that the suppressive petite mitochondria replicate more rapidly so that the wild-type mitochondria are not maintained in the cytoplasm for many doublings. Alternatively, experimental evidence suggests that genetic exchanges between the mitochondrial genomes of wild-type and suppressive petites may ultimately produce a defective population of mitochondria.

Let's now turn our attention to the inheritance of chloroplasts that are found in eukaryotic species capable of photosynthesis (namely, algae and plants). The unicellular alga *Chlamydomonas reinhardtii* is used as a model organism to investigate the inheritance of chloroplasts. This organism contains a single chloroplast that occupies approximately 40% of the cell volume. Genetic studies of chloroplast inheritance began when Ruth Sager identified a mutant strain of *Chlamydomonas* that is resistant to the antibiotic streptomycin (sm^r). By comparison, most strains are sensitive to killing by streptomycin (sm^s).

Sager conducted crosses to determine the inheritance pattern of the sm^r gene. During mating, two haploid cells unite to form a diploid cell, which then undergoes meiosis to form four haploid cells. Like yeast, *Chlamydomonas* is an organism that can be found in two mating types, in this case, designated mt^+ and mt^-. Mating type is due to nuclear inheritance and segregates in a 1:1 manner. By comparison, Sager and her colleagues discovered that the sm^r gene was inherited from the mt^+ parent but not from the mt^- parent (**Figure 5.20**). Therefore, this sm^r gene was not inherited in a Mendelian manner. This pattern occurred because only the mt^+ parent transmits chloroplasts to daughter cells and the sm^r gene is found in the chloroplast genome.

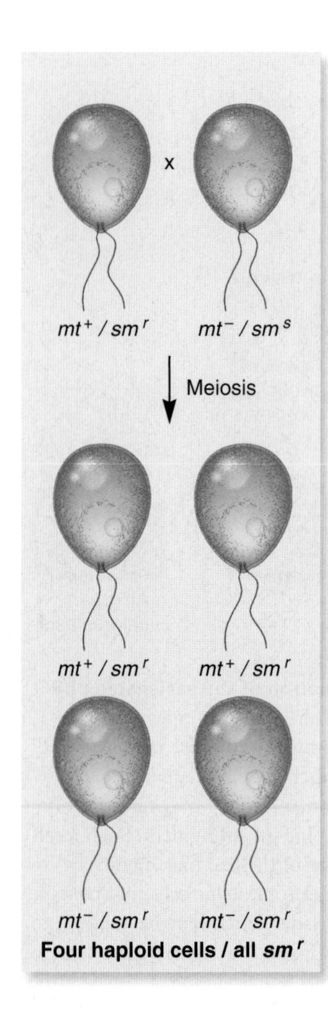

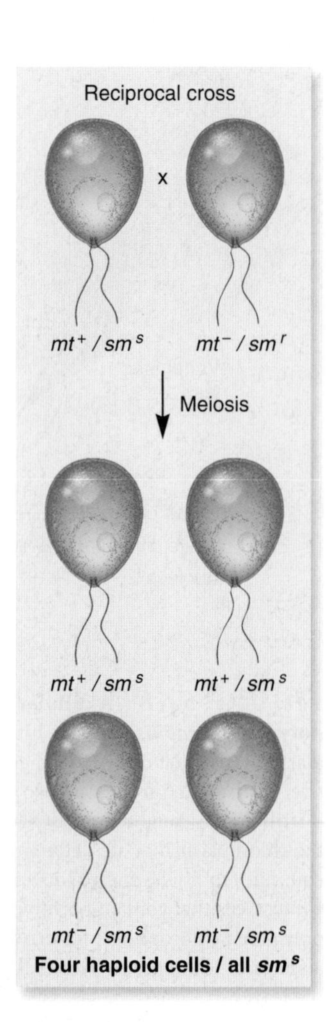

FIGURE 5.20 Chloroplast inheritance in *Chlamydomonas*. The two mating types of the organism are indicated as mt^+ and mt^-. Sm^r indicates streptomycin resistance, whereas sm^s indicates sensitivity to this antibiotic.

The Pattern of Inheritance of Mitochondria and Chloroplasts Varies Among Different Species

The inheritance of traits via genetic material within mitochondria and chloroplasts is now a well-established phenomenon that geneticists have investigated in many different species. In **heterogamous** species, two kinds of gametes are made. The female gamete tends to be large and provides most of the cytoplasm to the zygote, whereas the male gamete is small and often provides little more than a nucleus. Therefore, mitochondria and chloroplasts are most often inherited from the maternal parent. However, this is not always the case. **Table 5.4** describes the inheritance patterns of mitochondria and chloroplasts in several selected species.

In species in which maternal inheritance is generally observed, the paternal parent may occasionally provide mitochondria via the sperm. This phenomenon, called **paternal leakage,** occurs in many species that primarily exhibit maternal inheritance of their organelles. In the mouse, for example, approximately one to four paternal mitochondria are inherited for every 100,000 maternal mitochondria per generation of offspring. Most offspring do not inherit any paternal mitochondria, but a rare individual may inherit a mitochondrion from the sperm.

TABLE **5.4**
Transmission of Organelles Among Different Species

Species	Organelle	Transmission
Mammals	Mitochondria	Maternal inheritance
S. cerevisiae	Mitochondria	Biparental inheritance
Molds	Mitochondria	Usually maternal inheritance; paternal inheritance has been found in the genus *Allomyces*
Chlamydomonas	Mitochondria	Inherited from the parent with the mt^- mating type
Chlamydomonas	Chloroplasts	Inherited from the parent with the mt^+ mating type
Plants		
Angiosperms	Mitochondria and chloroplasts	Often maternal inheritance, although biparental inheritance is found among some species
Gymnosperms	Mitochondria and chloroplasts	Usually paternal inheritance

Many Human Diseases Are Caused by Mitochondrial Mutations

Mitochondrial diseases can occur in two ways. In some cases, mitochondrial mutations that cause disease are transmitted from mother to offspring. Human mtDNA is maternally inherited because it is transmitted from mother to offspring via the cytoplasm of the egg. Therefore, the transmission of inherited human mitochondrial diseases follows a maternal inheritance pattern. In addition, mitochondrial mutations may occur in somatic cells and accumulate as a person ages. Researchers have discovered that mitochondria are particularly susceptible to DNA damage. When more oxygen is consumed than is actually used to make ATP, mitochondria tend to produce free radicals that damage DNA. Unlike nuclear DNA, mitochondrial DNA has very limited repair abilities and almost no protective ability against free radical damage.

Table 5.5 describes several mitochondrial diseases that have been discovered in humans and are caused by mutations in mitochondrial genes. Over 200 diseases associated with defective mitochondria have been discovered. These are usually chronic degenerative disorders that affect cells requiring a high level of ATP, such as nerve and muscle cells. For example, Leber hereditary optic neuropathy (LHON) affects the optic nerve and may lead to the progressive loss of vision in one or both eyes. LHON can be caused by a defective mutation in one of several different mitochondrial genes. Researchers are still investigating how a defect in these mitochondrial genes produces the symptoms of this disease.

An important factor in mitochondrial disease is heteroplasmy, which means that a cell contains a mixed population of mitochondria. Within a single cell, some mitochondria may carry a disease-causing mutation whereas others may not. As cells divide, mutant and normal mitochondria randomly segregate into the resulting daughter cells. Some daughter cells may receive a high ratio of mutant to normal mitochondria, whereas others may receive a low ratio. To cause disease that affects a particular cell or tissue, the ratio of mutant to normal mitochondria must exceed a certain threshold value before disease symptoms are observed.

TABLE **5.5**

Examples of Human Mitochondrial Diseases

Disease	Mitochondrial Gene Mutated
Leber hereditary optic neuropathy	A mutation in one of several mitochondrial genes that encode respiratory chain proteins: *ND1, ND2, CO1, ND4, ND5, ND6,* and *cytb*
Neurogenic muscle weakness	A mutation in the *ATPase6* gene that encodes a subunit of the mitochondrial ATP-synthetase, which is required for ATP synthesis
Mitochondrial encephalomyopathy, lactic acidosis, and strokelike episodes	A mutation in genes that encode tRNAs for leucine and lysine
Mitochondrial myopathy	A mutation in a gene that encodes a tRNA for leucine
Maternal myopathy and cardiomyopathy	A mutation in a gene that encodes a tRNA for leucine
Myoclonic epilepsy with ragged-red muscle fibers	A mutation in a gene that encodes a tRNA for lysine

Extranuclear Genomes of Mitochondria and Chloroplasts Evolved from an Endosymbiotic Relationship

The idea that the nucleus, mitochondria, and chloroplasts contain their own separate genetic material may at first seem puzzling. Wouldn't it be simpler to have all of the genetic material in one place in the cell? The underlying reason for distinct genomes of mitochondria and chloroplasts can be traced back to their evolutionary origin, which is thought to involve a symbiotic association.

A symbiotic relationship occurs when two different species live together in a close association. The symbiont is the smaller of the two species; the host is the larger. The term **endosymbiosis** describes a symbiotic relationship in which the symbiont actually lives inside (*endo-*, inside) the host. In 1883, Andreas Schimper proposed that chloroplasts were descended from an endosymbiotic relationship between cyanobacteria and eukaryotic cells. This idea, now known as the **endosymbiosis theory,** suggested that the ancient origin of chloroplasts was initiated when a cyanobacterium took up residence within a primordial eukaryotic cell (**Figure 5.21**). Over the course of evolution, the

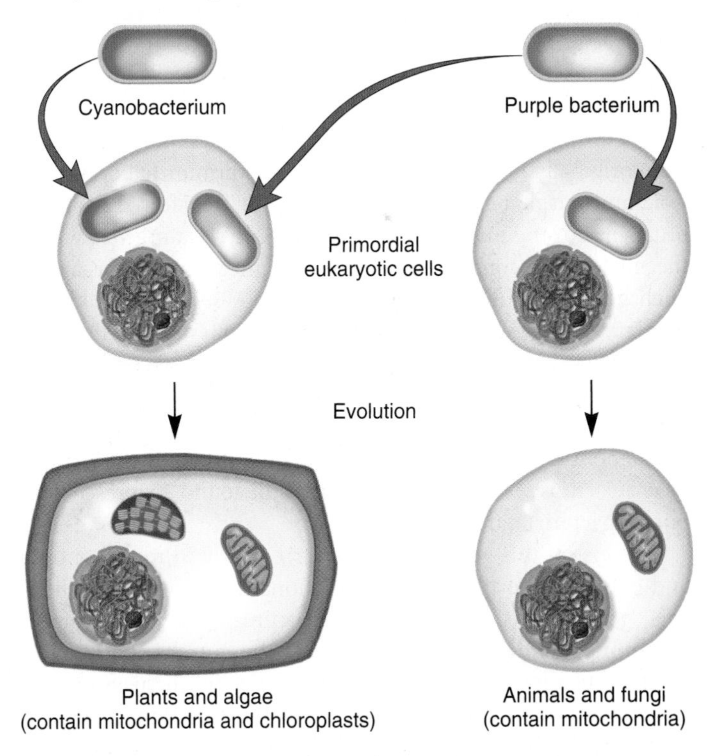

FIGURE 5.21 The endosymbiotic origin of mitochondria and chloroplasts. According to the endosymbiotic theory, chloroplasts descended from an endosymbiotic relationship between cyanobacteria and eukaryotic cells. This arose when a bacterium took up residence within a primordial eukaryotic cell. Over the course of evolution, the intracellular bacterial cell gradually changed its characteristics, eventually becoming a chloroplast. Similarly, mitochondria are derived from an endosymbiotic relationship between purple bacteria and eukaryotic cells.

characteristics of the intracellular bacterial cell gradually changed to those of a chloroplast. In 1922, Ivan Wallin also proposed an endosymbiotic origin for mitochondria.

In spite of these hypotheses, the question of endosymbiosis was largely ignored until researchers in the 1950s discovered that chloroplasts and mitochondria contain their own genetic material. The issue of endosymbiosis was hotly debated after Lynn Margulis published a book entitled *Origin of Eukaryotic Cells* (1970). During the 1970s and 1980s, the advent of molecular genetic techniques allowed researchers to analyze genes from chloroplasts, mitochondria, bacteria, and eukaryotic nuclear genomes. They found that genes in chloroplasts and mitochondria are very similar to bacterial genes but not as similar to those found within the nucleus of eukaryotic cells. This observation provided strong support for the endosymbiotic origin of mitochondria and chloroplasts, which is now widely accepted.

The endosymbiosis theory proposes that the relationship provided eukaryotic cells with useful cellular characteristics. Chloroplasts were derived from cyanobacteria, a bacterial species that is capable of photosynthesis. The ability to carry out photosynthesis enabled algal and plant cells to use the energy from sunlight. By comparison, mitochondria are thought to have been derived from a different type of bacteria known as gram-negative nonsulfur purple bacteria. In this case, the endosymbiotic relationship enabled eukaryotic cells to synthesize greater amounts of ATP. It is less clear how the relationship would have been beneficial to cyanobacteria or purple bacteria, though the cytosol of a eukaryotic cell may have provided a stable environment with an adequate supply of nutrients.

During the evolution of eukaryotic species, most genes that were originally found in the genome of the primordial cyanobacteria and purple bacteria have been lost or transferred from the organelles to the nucleus. The sequences of certain genes within the nucleus are consistent with their origin within an organelle. Such genes are more similar in their DNA sequence to known bacterial genes than to their eukaryotic counterparts. Therefore, researchers have concluded that these genes have been removed from the mitochondrial and chloroplast chromosomes and relocated to the nuclear chromosomes. This has occurred many times throughout evolution, so modern mitochondria and chloroplasts have lost most of the genes that are still found in present-day purple bacteria and cyanobacteria.

Most of this gene transfer occurred early in mitochondrial and chloroplast evolution. The functional transfer of mitochondrial genes seems to have ceased in animals, but gene transfer from mitochondria and chloroplasts to the nucleus continues to occur in plants at a low rate. The molecular mechanism of gene transfer is not entirely understood, but the direction of transfer is well established. During evolution, gene transfer has occurred primarily from the organelles to the nucleus. For example, about 1500 genes have been transferred from the mitochondrial genome to the nuclear genome of eukaryotes. Transfer of genes from the nucleus to the organelles has almost never occurred, although one example is known of a nuclear gene in plants that has been transferred to the mitochondrial genome. This unidirectional gene transfer from organelles to the nucleus partly explains why the organellar genomes now contain relatively few genes. In addition, gene transfer can occur between organelles. It can happen between two mitochondria, between two chloroplasts, and between a chloroplast and mitochondrion. Overall, the transfer of genetic material between the nucleus, chloroplasts, and mitochondria is an established phenomenon, although its biological benefits remain unclear.

Eukaryotic Cells Occasionally Contain Symbiotic Infective Particles

Other unusual endosymbiotic relationships have been identified in eukaryotic organisms. Several examples are known in which infectious particles establish a symbiotic relationship with their host. In some cases, research indicates that symbiotic infectious particles are bacteria that exist within the cytoplasm of eukaryotic cells. Although symbiotic infectious particles are relatively uncommon, they have provided interesting and even bizarre examples of the extranuclear inheritance of traits.

In the 1940s, Tracy Sonneborn studied a phenomenon known as the killer trait in the protozoan *Paramecium aurelia*. Killer paramecia secrete a substance called paramecin, which kills some but not all strains of paramecia. Sonneborn found that killer strains contain particles in their cytoplasm known as kappa particles. Each kappa particle is 0.4 μm long and has its own DNA. Genes within the kappa particle encode the paramecin toxin. In addition, other kappa particle genes provide the killer paramecia with resistance to the toxin.

Nonkiller paramecia are killed when mixed with killer paramecia. However, Sonnenborn found that when nonkiller paramecia were mixed with a cell extract derived from killer paramecia, the kappa particles within the extract are taken up by the nonkiller strains and convert them into killer strains. In other words, the extranuclear particle that determines the killer trait is infectious.

Infectious particles have also been identified in fruit flies. Philippe l'Heritier identified strains of *Drosophila melanogaster* that are highly sensitive to killing by CO_2. Reciprocal crosses between CO_2-sensitive and normal flies revealed that the trait is inherited in a non-Mendelian manner. Furthermore, cell extracts from a sensitive fly can infect a normal fly and make it sensitive to CO_2.

Another example of an infectious particle in fruit flies involves a trait known as sex ratio in which affected flies produce progenies with a large excess of females. Chana Malogolowkin and Donald Poulson discovered one strain of *Drosophila willistoni* in which most of the offspring of female flies were daughters; nearly all the male offspring died. The sex ratio trait is transmitted from mother to offspring. The rare surviving males do not transmit this trait to their male or female offspring. This result indicates a maternal inheritance pattern for the sex ratio trait. The agent in the cytoplasm of female flies responsible for the sex ratio trait was later found to be a symbiotic bacterium, which was named *Spiroplasma poulsonii*. Its presence is usually lethal to males but not to females. This infective agent can be extracted from the tissues of adult females and used to infect the females of a normal strain of flies.

KEY TERMS

CHAPTER SUMMARY

- Non-Mendelian inheritance refers to inheritance patterns that cannot be easily explained by Mendel's experiments.

5.1 Maternal Effect

- Maternal effect is an inheritance pattern in which the genotype of the mother determines the phenotype of the offspring. It occurs because gene products of maternal effect genes are transferred from nurse cells to the oocyte. These gene products affect early stages of development (see Figures 5.1, 5.2).

5.2 Epigenetic Inheritance

- Epigenetic inheritance is a pattern in which a gene or chromosome is modified and gene expression is altered, but the modification is not permanent over the course of many generations.
- Dosage compensation often occurs in species that differ in their sex chromosomes (see Table 5.1).
- In mammals, the process of X inactivation in females compensates for the single X chromosome found in males. The inactivated X chromosome is called a Barr body. The process can lead to a variegated phenotype, such as a calico cat (see Figures 5.3, 5.4).
- After it occurs during embryonic development, the pattern of X inactivation is maintained when cells divide (see Figures 5.5, 5.6).
- X inactivation is controlled by the X-inactivation center that contains the *Xist* and *Tsix* genes. X inactivation occurs as initiation, spreading, and maintenance phases (see Figure 5.7, 5.8).
- Genomic imprinting refers to a marking process in which an offspring expresses a gene that is inherited from one parent but not both (see Figures 5.9, 5.10).
- DNA methylation at imprinting control regions is the marking process that causes imprinting (see Figures 5.11, 5.12).
- Human diseases such as Prader-Willi syndrome and Angelman syndrome are associated with genomic imprinting (see Figure 5.13, Table 5.2).

5.3 Extranuclear Inheritance

- Extranuclear inheritance involves the inheritance of genes that are found in mitochondria or chloroplasts.
- Mitochondria and chloroplasts carry circular chromosomes in a nucleoid region. These circular chromosomes contain relatively few genes compared with the number in the cell nucleus (see Figures 5.14–5.16, Table 5.3).
- Maternal inheritance occurs when organelles, such as mitochondria or chloroplasts, are transmitted via the egg (see Figure 5.17).
- Heteroplasmy for chloroplasts can result in a variegated phenotype (see Figure 5.18).
- Neutral and suppressive petites in yeast are due to defects in mitochondrial DNA and show a non-Mendelian inheritance pattern (see Figure 5.19).
- In the alga *Chlamydomonas*, chloroplasts are transmitted from the *mt*$^+$ parent (see Figure 5.20).
- The transmission patterns of mitochondria and chloroplasts vary among different species (see Table 5.4).
- Many diseases are caused by mutations in mitochondrial DNA (see Table 5.5).
- Mitochondria and chloroplasts were derived from an ancient endosymbiotic relationship (see Figure 5.21).
- On rare occasions, eukaryotic cells may contain infectious particles.

PROBLEM SETS & INSIGHTS

Solved Problems

S1. Our understanding of maternal effect genes has been greatly aided by their identification in experimental organisms such as *Drosophila melanogaster* and *Caenorhabditis elegans*. In experimental organisms with a short generation time, geneticists have successfully searched for mutant alleles that prevent the normal process of embryonic development. In many cases, the offspring die at early embryonic or larval stages. These are called maternal effect lethal alleles. How would a researcher identify a mutation that produced a recessive maternal effect lethal allele?

Answer: A maternal effect lethal allele can be identified when a phenotypically normal mother produces only offspring with gross developmental abnormalities. For example, let's call the normal allele *N* and the maternal effect lethal allele *n*. A cross between two flies that are heterozygous for a maternal effect lethal allele would produce 1/4 of the offspring with a homozygous genotype, *nn*. These flies are viable because of the maternal effect. Their mother would be *Nn* and provide the *n* egg with a sufficient amount of *N* gene product so that the *nn* flies would develop properly. However, homozygous *nn* females cannot provide their eggs with any normal gene product. Therefore, all of their offspring are abnormal and die during early stages.

S2. A maternal effect gene in *Drosophila*, called *torso*, is found as a recessive allele that prevents the correct development of anterior- and posterior-most structures. A wild-type male is crossed to a female of unknown genotype. This mating produces 100% larva that are missing their anterior- and posterior-most structures and therefore die during early development. What is the genotype and phenotype of the female fly in this cross? What are the genotypes and phenotypes of the female fly's parents?

Answer: Because this cross produces 100% abnormal offspring, the female fly must be homozygous for the abnormal *torso* allele. Even so, the female fly must be phenotypically normal in order to reproduce. This female fly had a mother that was heterozygous for a normal and abnormal *torso* allele and a father that was either heterozygous or homozygous for the abnormal *torso* allele.

$torso^+\ torso^-$ (grandmother) × $torso^+\ torso^-$ or $torso^-\ torso^-$ (grandfather)

↓

$torso^-\ torso^-$
(mother of 100% abnormal offspring)

This female fly is phenotypically normal because its mother was heterozygous and provided the gene products of the $torso^+$ allele from the nurse cells. However, this homozygous female will produce only abnormal offspring because it cannot provide them with the normal $torso^+$ gene products.

S3. An individual with Angelman syndrome produced an offspring with Prader-Willi syndrome. Why does this occur? What are the sexes of the parent with Angelman syndrome and the offspring with Prader-Willi syndrome?

Answer: These two different syndromes are most commonly caused by a small deletion in chromosome 15. In addition, genomic imprinting plays a role because genes in this deleted region are differentially imprinted, depending on sex. If this deletion is inherited from the paternal parent, the offspring develops Prader-Willi syndrome. Therefore, in this problem, the person with Angelman syndrome must have been a male because he produced a child with Prader-Willi syndrome. The child could be either a male or female.

S4. In yeast, a haploid petite mutant also carries a mutant gene that requires the amino acid histidine for growth. The petite his^- strain is crossed to a wild-type his^+ strain to yield the following tetrad:

2 cells: petite his^-

2 cells: petite his^+

Explain the inheritance of the petite and his^- mutations.

Answer: The his^- and his^+ alleles are segregating in a 2:2 ratio. This result indicates a nuclear pattern of inheritance. By comparison, all four cells in this tetrad have a petite phenotype. This is a suppressive petite that arises from a mitochondrial mutation.

S5. Suppose that you are a horticulturist who has recently identified an interesting plant with variegated leaves. How would you determine if this trait is nuclearly or cytoplasmically inherited?

Answer: Make crosses and reciprocal crosses involving normal and variegated strains. In many species, chloroplast genomes are inherited maternally, although this is not always the case. In addition, a significant percentage of paternal leakage may occur. Nevertheless, when reciprocal crosses yield different outcomes, an organellar mode of inheritance is possibly at work.

S6. A phenotype that is similar to a yeast suppressive petite was also identified in the mold *Neurospora crassa*. Mary and Herschel Mitchell identified a slow-growing mutant that they called *poky*. Unlike yeast, which are isogamous (i.e., produce one type of gamete), *Neurospora* is sexually dimorphic and produces male and female reproductive structures. When a *poky* strain of *Neurospora* was crossed to a wild-type strain, the results were different between reciprocal crosses. If a *poky* mutant was the female parent, all spores exhibited the *poky* phenotype. By comparison, if the wild-type strain was the female parent, all spores were wild type. Explain these results.

Answer: These genetic studies indicate that the *poky* mutation is maternally inherited. The cytoplasm of the female reproductive cells provides the offspring with their mitochondria. Besides these genetic studies, the Mitchells and their collaborators showed that *poky* mutants are defective in certain cytochromes, which are iron-containing proteins that are known to be located in the mitochondria.

Conceptual Questions

C1. Define the term epigenetic inheritance, and describe two examples.

C2. Describe the inheritance pattern of maternal effect genes. Explain how the maternal effect occurs at the cellular level. What are the expected functional roles of the proteins that are encoded by maternal effect genes?

C3. A maternal effect gene exists in a dominant *N* (normal) allele and a recessive *n* (abnormal) allele. What would be the ratios of genotypes and phenotypes for the offspring of the following crosses?

A. *nn* female × *NN* male

B. *NN* female × *nn* male

C. *Nn* female × *Nn* male

C4. A *Drosophila* embryo dies during early embryogenesis due to a recessive maternal effect allele called *bicoid*. The wild-type allele is designated $bicoid^+$. What are the genotypes and phenotypes of the embryo's mother and maternal grandparents?

C5. For Mendelian traits, the nuclear genotype (i.e., the alleles found on chromosomes in the cell nucleus) directly influences an offspring's traits. In contrast, for non-Mendelian inheritance patterns, the offspring's phenotype cannot be reliably predicted solely from its genotype. For the following traits, what do you need to know to predict the phenotypic outcome?

A. Dwarfism due to a mutant *Igf2* allele

B. Snail coiling direction

C. Leber hereditary optic neuropathy

C6. Suppose a maternal effect gene exists as a normal dominant allele and an abnormal recessive allele. A mother who is phenotypically abnormal produces all normal offspring. Explain the genotype of the mother.

C7. Suppose that a gene affects the anterior morphology in house flies and is inherited as a maternal effect gene. The gene exists in a normal allele, *H*, and a recessive allele, *h*, which causes a small head. A female fly with a normal head is mated to a true-breeding male with a small head. All of the offspring have small heads. What are the genotypes of the mother and offspring? Explain your answer.

C8. Explain why maternal effect genes exert their effects during the early stages of development.

C9. As described in Chapter 19, researchers have been able to "clone" mammals by fusing a cell having a diploid nucleus (i.e., a somatic cell) with an egg that has had its (haploid) nucleus removed.

A. With regard to maternal effect genes, would the phenotype of such a cloned animal be determined by the animal that donated the egg or by the animal that donated the somatic cell? Explain.

B. Would the cloned animal inherit extranuclear traits from the animal that donated the egg or by the animal that donated the somatic cell? Explain.

C. In what ways would you expect this cloned animal to be similar to or different from the animal that donated the somatic cell? Is it fair to call such an animal a "clone" of the animal that donated the diploid nucleus?

C10. With regard to the numbers of sex chromosomes, explain why dosage compensation is necessary.

C11. What is a Barr body? How is its structure different from that of other chromosomes in the cell? How does the structure of a Barr body affect the level of X-linked gene expression?

C12. Among different species, describe three distinct strategies for accomplishing dosage compensation.

C13. Describe when X inactivation occurs and how this leads to phenotypic results at the organism level. In your answer, you should explain why X inactivation causes results such as variegated coat patterns in mammals. Why do two different calico cats have their patches of orange and black fur in different places? Explain whether or not a variegated coat pattern due to X inactivation could occur in marsupials.

C14. Describe the molecular process of X inactivation. This description should include the three phases of inactivation and the role of the Xic. Explain what happens to X chromosomes during embryogenesis, in adult somatic cells, and during oogenesis.

C15. On rare occasions, an abnormal human male is born who is somewhat feminized compared with normal males. Microscopic examination of the cells of one such individual revealed that he has a single Barr body in each cell. What is the chromosomal composition of this individual?

C16. How many Barr bodies would you expect to find in humans with the following abnormal compositions of sex chromosomes?

A. XXY

B. XYY

C. XXX

D. X0 (a person with just a single X chromosome)

C17. Certain forms of human color blindness are inherited as X-linked recessive traits. Hemizygous males are color-blind, but heterozygous females are not. However, heterozygous females sometimes have partial color blindness.

A. Discuss why heterozygous females may have partial color blindness.

B. Doctors identified an unusual case in which a heterozygous female was color-blind in her right eye but had normal color vision in her left eye. Explain how this might have occurred.

C18. A black female cat (X^BX^B) and an orange male cat (X^OY) were mated to each other and produced a male cat that was calico. Which sex chromosomes did this male offspring inherit from its mother and father? Remember that the presence of the Y chromosome determines maleness in mammals.

C19. What is the spreading phase of X inactivation? Why do you think it is called a spreading phase? Discuss the role of the *Xist* gene in the spreading phase of X inactivation.

C20. When does the erasure and reestablishment phase of genomic imprinting occur? Explain why it is necessary to erase an imprint and then reestablish it in order to always maintain imprinting from the same sex of parent.

C21. In what types of cells would you expect *de novo* methylation to occur? In what cell types would it not occur?

C22. On rare occasions, people are born with a condition known as uniparental disomy. It happens when an individual inherits both copies of a chromosome from one parent and no copies from the other parent. This occurs when two abnormal gametes happen to complement each other to produce a diploid zygote. For example, an abnormal sperm that lacks chromosome 15 could fertilize an egg that contains two copies of chromosome 15. In this situation, the individual would be said to have maternal uniparental disomy 15 because both copies of chromosome 15 were inherited from the mother. Alternatively, an abnormal sperm with two copies of chromosome 15 could fertilize an egg with no copies. This is known as paternal uniparental disomy 15. If a female is born with paternal uniparental disomy 15, would you expect her to be phenotypically normal, have Angelman syndrome (AS), or have Prader-Willi syndrome (PWS)? Explain. Would you expect her to produce normal offspring or offspring affected with AS or PWS?

C23. Genes that cause Prader-Willi syndrome and Angelman syndrome are closely linked along chromosome 15. Although people with these syndromes do not usually reproduce, let's suppose that a couple produces two children with Angelman syndrome. The oldest child (named Pat) grows up and has two children with Prader-Willi syndrome. The second child (named Robin) grows up and has one child with Angelman syndrome.

A. Are Pat and Robin's parents both normal or does one of them have Angelman or Prader-Willi syndrome? If one of them has a disorder, explain why it is the mother or the father.

B. What are the sexes of Pat and Robin? Explain.

C24. How is the process of X inactivation similar to genomic imprinting? How is it different?

C25. What is extranuclear inheritance? Describe three examples.

C26. What is a reciprocal cross? Suppose that a gene is found as a wild-type allele and a recessive mutant allele. What would be the expected outcomes of reciprocal crosses if a true-breeding normal individual was crossed to a true-breeding individual carrying the mutant allele? What would be the results if the gene is maternally inherited?

C27. Among different species, does extranuclear inheritance always follow a maternal inheritance pattern? Why or why not?

C28. What is the phenotype of a petite mutant? Where can a petite mutation occur: in nuclear genes, extranuclear genetic material, or both? What is the difference between a neutral and suppressive petite?

C29. Extranuclear inheritance often correlates with maternal inheritance. Even so, paternal leakage is not uncommon. What is paternal leakage? If a cross produced 200 offspring and the rate of mitochondrial paternal leakage was 3%, how many offspring would be expected to contain paternal mitochondria?

C30. Discuss the structure and organization of the mitochondrial and chloroplast genomes. How large are they, how many genes do they contain, and how many copies of the genome are found in each organelle?

C31. Explain the likely evolutionary origin of mitochondrial and chloroplast genomes. How have the sizes of the mitochondrial and chloroplast genomes changed since their origin? How has this occurred?

C32. Which of the following traits or diseases are determined by nuclear genes?

A. Snail coiling pattern

B. Prader-Willi syndrome

C. Streptomycin resistance in *Chlamydomonas*

D. Leber hereditary optic neuropathy

C33. Acute murine leukemia virus (AMLV) causes leukemia in mice. This virus is easily passed from mother to offspring through the mother's milk. (Note: Even though newborn offspring acquire the virus, they may not develop leukemia until much later in life. Testing can determine if an animal carries the virus.) Describe how the formation of leukemia via AMLV resembles a maternal inheritance pattern. How could you determine that this form of leukemia is not caused by extranuclear inheritance?

C34. Describe how a biparental pattern of extranuclear inheritance would resemble a Mendelian pattern of inheritance for a particular gene. How would they differ?

C35. According to the endosymbiosis theory, mitochondria and chloroplasts are derived from bacteria that took up residence within eukaryotic cells. However, at one time, prior to being taken up by eukaryotic cells, these bacteria were free-living organisms. However, we cannot take a mitochondrion or chloroplast out of a living eukaryotic cell and get it to survive and replicate on its own. Why not?

Experimental Questions

E1. Figure 5.1 describes an example of a maternal effect gene. Explain how Sturtevant deduced a maternal effect gene based on the F_2 and F_3 generations.

E2. Discuss the types of experimental observations that Mary Lyon brought together in proposing her hypothesis concerning X inactivation. In your own words, explain how these observations were consistent with her hypothesis.

E3. Chapter 18 describes three blotting methods (i.e., Southern blotting, Northern blotting, and Western blotting) that are used to detect specific genes and gene products. Southern blotting detects DNA, Northern blotting detects RNA, and Western blotting detects proteins. Suppose that a female fruit fly is heterozygous for a maternal effect gene, which we will call gene *B*. The female is *Bb*. The normal allele, *B*, encodes a functional mRNA that is 550 nucleotides long. A recessive allele, *b*, encodes a shorter mRNA that is 375 nucleotides long. (Allele *b* is due to a deletion within this gene.) How could you use one or more of these techniques to show that nurse cells transfer gene products from gene *B* to developing eggs? You may assume that you can dissect the ovaries of fruit flies and isolate eggs separately from nurse cells. In your answer, describe your expected results.

E4. As a hypothetical example, a trait in mice results in mice with very long tails. You initially have a true-breeding strain with normal tails and a true-breeding strain with long tails. You then make the following types of crosses:

Cross 1: When true-breeding females with normal tails are crossed to true-breeding males with long tails, all F_1 offspring have long tails.

Cross 2: When true-breeding females with long tails are crossed to true-breeding males with normal tails, all F_1 offspring have normal tails.

Cross 3: When F_1 females from cross 1 are crossed to true-breeding males with normal tails, all offspring have normal tails.

Cross 4: When F_1 males from cross 1 are crossed to true-breeding females with long tails, half of the offspring have normal tails and half have long tails.

Explain the pattern of inheritance of this trait.

E5. You have a female snail that coils to the right, but you do not know its genotype. You may assume that right coiling (*D*) is dominant to left coiling (*d*). You also have male snails at your disposal of known genotype. How would you determine the genotype of this female snail? In your answer, describe your expected results depending on whether the female is *DD*, *Dd*, or *dd*.

E6. On a recent camping trip, you find one male snail on a deserted island that coils to the right. However, in this same area, you find several shells (not containing living snails) that coil to the left. Therefore, you conclude that you are not certain of the genotype of this male snail. On a different island, you find a large colony of snails of the same species. All of these snails coil to the right, and every snail shell that you find on this second island coils to the right. With regard to the maternal effect gene that determines coiling pattern, how would you determine the genotype of the male snail that you found on the deserted island? In your answer, describe your expected results.

E7. Figure 5.6 describes the results of X inactivation in mammals. If fast and slow alleles of glucose-6-phosphate dehydrogenase (*G-6-PD*) exist in other species, what would be the expected results of gel electrophoresis for a heterozygous female of the following species?

A. Marsupial

B. *Drosophila melanogaster*

C. *Caenorhabditis elegans* (Note: We are considering the hermaphrodite in *C. elegans* to be equivalent to a female.)

E8. Two male mice, which we will call male A and male B, are both phenotypically normal. Male A was from a litter that contained half phenotypically normal mice and half dwarf mice. The mother of male A was known to be homozygous for the normal *Igf2* allele. Male B was from a litter of eight mice that were all phenotypically normal. The parents of male B were a phenotypically normal male and a dwarf female. Male A and male B were put into a cage with two female mice that we will call female A and female B. Female A is dwarf, and female B is phenotypically normal. The parents of these two females were unknown, although it was known that they were from the same litter. The mice were allowed to mate with each other, and the following data were obtained:

Female A gave birth to three dwarf babies and four normal babies.

Female B gave birth to four normal babies and two dwarf babies.

Which male(s) mated with female A and female B? Explain.

E9. In the experiment of Figure 5.6, why does a clone of cells produce only one type of G-6-PD enzyme? What would you expect to happen if a clone was derived from an early embryonic cell? Why does the initial sample of tissue produce both forms of G-6-PD?

E10. Chapter 18 describes a blotting method known as Northern blotting that can be used to determine the amount of mRNA produced by a particular gene. In this method, the amount of a specific mRNA produced by cells is detected as a band on a gel. If one type of cell produces twice as much of a particular mRNA as another cell, the band will appear twice as dark. Also, sometimes mutations affect the length of mRNA that is transcribed from a gene. For example, a small deletion within a gene may shorten an mRNA. Northern blotting also can discern the sizes of mRNAs.

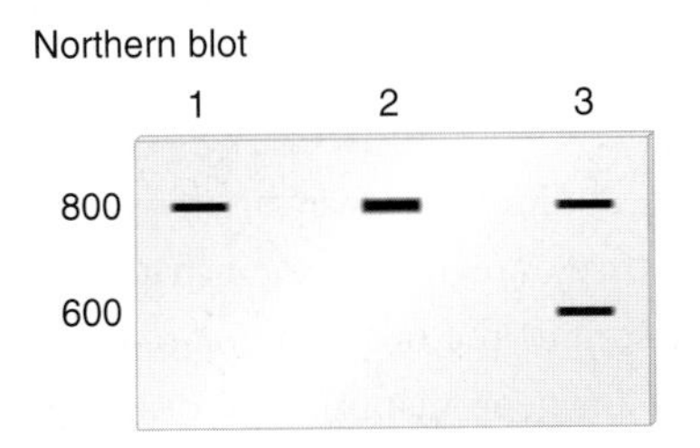

Lane 1 is a Northern blot of mRNA from cell type A that is 800 nucleotides long.

Lane 2 is a Northern blot of the same mRNA from cell type B. (Cell type B produces twice as much of this RNA as cell type A.)

Lane 3 shows a heterozygote in which one of the two genes has a deletion, which shortens the mRNA by 200 nucleotides.

Here is the question. Suppose an X-linked gene exists as two alleles: *B* and *b*. Allele *B* encodes an mRNA that is 750 nucleotides long, and allele *b* encodes a shorter mRNA that is 675 nucleotides long. Draw the expected results of a Northern blot using mRNA isolated from the same type of somatic cells taken from the following individuals:

A. First lane is mRNA from an X^bY male fruit fly.

Second lane is mRNA from an X^bX^b female fruit fly.

Third lane is mRNA from an X^BX^b female fruit fly.

B. First lane is mRNA from an X^BY male mouse.

Second lane is mRNA from an X^BX^b female mouse.*

Third lane is mRNA from an X^BX^B female mouse.*

*The sample is taken from an adult female mouse. It is not a clone of cells. It is a tissue sample, like the one described in the experiment of Figure 5.6.

C. First lane is mRNA from an X^B0 male *C. elegans*.

Second lane is mRNA from an X^BX^b hermaphrodite *C. elegans*.

Third lane is mRNA from an X^BX^B hermaphrodite *C. elegans*.

E11. A variegated trait in plants is analyzed using reciprocal crosses. The following results are obtained:

Variegated female × Normal male	Normal female × Variegated male
↓	↓
1024 variegated + 52 normal	1113 normal + 61 variegated

Explain this pattern of inheritance.

E12. Ruth Sager and her colleagues discovered that the mode of inheritance of streptomycin resistance in *Chlamydomonas* could be altered if the mt^+ cells were exposed to UV irradiation prior to mating. This exposure dramatically increased the frequency of biparental inheritance. What would be the expected outcome of a cross between an mt^+ sm^r and an mt^- sm^s strain in the absence of UV irradiation? How would the result differ if the mt^+ strain was exposed to UV light?

E13. Take a look at Figure 5.19 and describe how you could experimentally distinguish between yeast strains that are neutral petites and those that are suppressive petites.

Questions for Student Discussion/Collaboration

1. Recessive maternal effect genes are identified in flies (for example) when a phenotypically normal mother cannot produce any normal offspring. Because all of the offspring are dead, this female fly cannot be used to produce a strain of heterozygous flies that could be used in future studies. How would you identify heterozygous individuals that are carrying a recessive maternal effect allele? How would you maintain this strain of flies in a laboratory over many generations?

2. What is an infective particle? Discuss the similarities and differences between infective particles and organelles such as mitochondria and chloroplasts. Do you think the existence of infective particles supports the endosymbiosis theory of the origin of mitochondria and chloroplasts?

Note: All answers appear at the website for this textbook; the answers to even-numbered questions are in the back of the textbook.

www.mhhe.com/brookergenetics4e

Visit the website for practice tests, answer keys, and other learning aids for this chapter. Enhance your understanding of genetics with our interactive exercises, quizzes, animations, and much more.

CHAPTER OUTLINE

Crossing over during meiosis. *This event provides a way to reassort the alleles of genes that are located on the same chromosome.*

6 GENETIC LINKAGE AND MAPPING IN EUKARYOTES

In Chapter 2, we were introduced to Mendel's laws of inheritance. According to these principles, we expect that two different genes will segregate and independently assort themselves during the process that creates gametes. After Mendel's work was rediscovered at the turn of the twentieth century, chromosomes were identified as the cellular structures that carry genes. The chromosome theory of inheritance explained how the transmission of chromosomes is responsible for the passage of genes from parents to offspring.

When geneticists first realized that chromosomes contain the genetic material, they began to suspect that a conflict might sometimes occur between the law of independent assortment of genes and the behavior of chromosomes during meiosis. In particular, geneticists assumed that each species of organism must contain thousands of different genes, yet cytological studies revealed that most species have at most a few dozen chromosomes. Therefore, it seemed likely, and turned out to be true, that each chromosome would carry many hundreds or even thousands of different genes. The transmission of genes located close to each other on the same chromosome violates the law of independent assortment.

In this chapter, we will consider the pattern of inheritance that occurs when different genes are situated on the same chromosome. In addition, we will briefly explore how the data from genetic crosses are used to construct a **genetic map**—a diagram that describes the order of genes along a chromosome. Newer strategies for gene mapping are described in Chapter 20. However, an understanding of traditional mapping studies, as described in this chapter, will strengthen our appreciation for these newer molecular approaches. More importantly, traditional mapping studies further illustrate how the location of two or more genes on the same chromosome can affect the transmission patterns from parents to offspring.

6.1 LINKAGE AND CROSSING OVER

In eukaryotic species, each linear chromosome contains a very long segment of DNA. A chromosome contains many individual functional units—called genes—that influence an organism's traits. A typical chromosome is expected to contain many hundreds or perhaps a few thousand different genes. The term **synteny** means that two or more genes are located on the same chromosome. Genes that are syntenic are physically linked to each other, because each eukaryotic chromosome contains a single, continuous, linear molecule of DNA. **Genetic linkage** is the phenomenon in which genes that are close together on the same chromosome tend to be transmitted as a unit. Therefore, genetic linkage has an influence on inheritance patterns.

Chromosomes are sometimes called **linkage groups,** because a chromosome contains a group of genes that are physically linked together. In species that have been characterized genetically, the number of linkage groups equals the number of chromosome types. For example, human somatic cells have 46 chromosomes, which are composed of 22 types of autosomes that come in pairs plus one pair of sex chromosomes, the X and Y. Therefore, humans have 22 autosomal linkage groups, and an X chromosome linkage group, and males have a Y chromosome linkage group. In addition, the human mitochondrial genome is another linkage group.

Geneticists are often interested in the transmission of two or more characters in a genetic cross. When a geneticist follows the variants of two different characters in a cross, this is called a **dihybrid cross;** when three characters are followed, it is a **trihybrid cross;** and so on. The outcome of a dihybrid or trihybrid cross depends on whether or not the genes are linked to each other along the same chromosome. In this section, we will examine how linkage affects the transmission patterns of two or more characters.

Crossing Over May Produce Recombinant Genotypes

Even though the alleles for different genes may be linked along the same chromosome, the linkage can be altered during meiosis. In diploid eukaryotic species, homologous chromosomes can exchange pieces with each other, a phenomenon called **crossing over.** This event occurs during prophase of meiosis I. As discussed in Chapter 3, the replicated chromosomes, known as sister chromatids, associate with the homologous sister chromatids to form a structure known as a **bivalent.** A bivalent is composed of two pairs of sister chromatids. In prophase of meiosis I, it is common for a sister chromatid of one pair to cross over with a sister chromatid from the homologous pair.

Figure 6.1 considers meiosis when two genes are linked on the same chromosome. One of the parental chromosomes carries the *A* and *B* alleles, while the homolog carries the *a* and *b* alleles. In Figure 6.1a, no crossing over has occurred. Therefore, the resulting haploid cells contain the same combination of alleles as the original chromosomes. In this case, two haploid cells carry the dominant *A* and *B* alleles, and the other two carry the recessive *a* and *b* alleles. The arrangement of linked alleles has not been altered.

In contrast, Figure 6.1b illustrates what can happen when crossing over occurs. Two of the haploid cells contain combinations of alleles, namely *A* and *b* or *a* and *B*, which differ from those in the original chromosomes. In these two cells, the grouping of linked alleles has changed. An event such as this, leading to a new combination of alleles, is known as **genetic recombination.** The haploid cells carrying the *A* and *b*, or the *a* and *B*, alleles are called **nonparental cells** or **recombinant cells.** Likewise, if such haploid cells were gametes that participated in fertilization, the resulting offspring are called nonparental offspring or

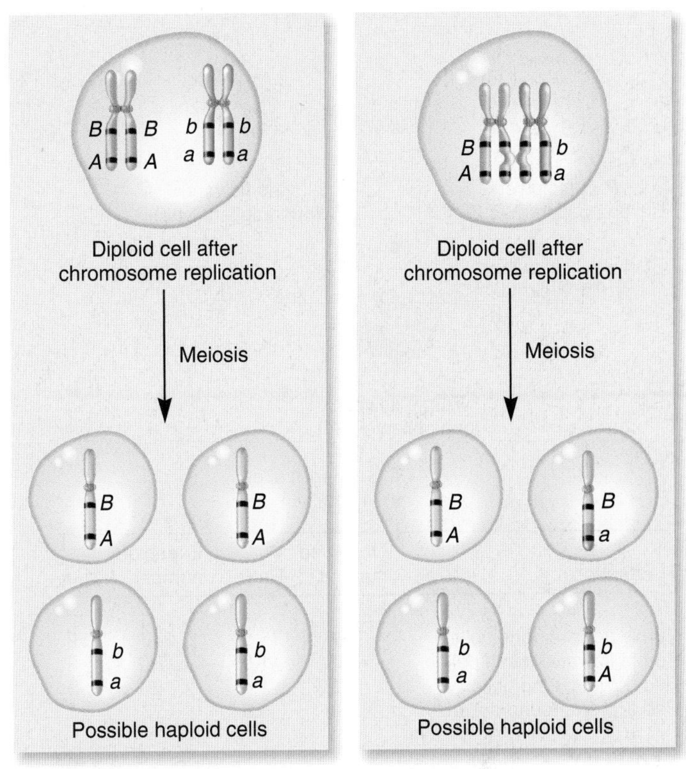

(a) Without crossing over, linked alleles segregate together.

(b) Crossing over can reassort linked alleles.

FIGURE 6.1 Consequences of crossing over during meiosis. **(a)** In the absence of crossing over, the *A* and *B* alleles and the *a* and *b* alleles are maintained in the same arrangement found in the parental chromosomes. **(b)** Crossing over has occurred in the region between the two genes, producing two nonparental haploid cells with a new combination of alleles.

recombinant offspring. These offspring can display combinations of traits that are different from those of either parent. In contrast, offspring that have inherited the same combination of alleles that are found in the chromosomes of their parents are known as **parental offspring** or **nonrecombinant offspring.**

In this section, we will consider how crossing over affects the pattern of inheritance for genes linked on the same chromosome. In Chapter 17, we will consider the molecular events that cause crossing over to occur.

Bateson and Punnett Discovered Two Traits That Did Not Assort Independently

An early study indicating that some traits may not assort independently was carried out by William Bateson and Reginald Punnett in 1905. According to Mendel's law of independent assortment, a dihybrid cross between two individuals, heterozygous for two genes, should yield a 9:3:3:1 phenotypic ratio among the offspring. However, a surprising result occurred when Bateson and Punnett conducted a cross in the sweet pea involving two different traits: flower color and pollen shape.

As seen in **Figure 6.2**, they began by crossing a true-breeding strain with purple flowers (*PP*) and long pollen (*LL*) to

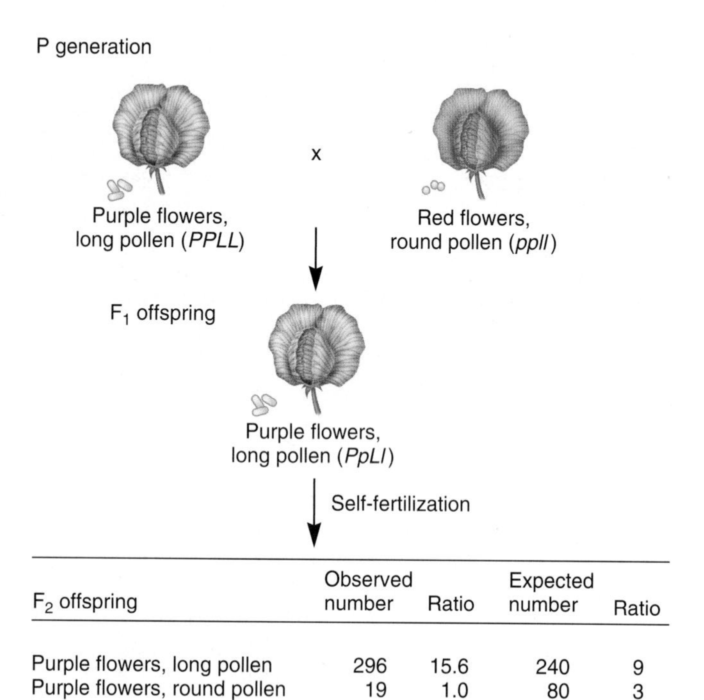

F_2 offspring	Observed number	Ratio	Expected number	Ratio
Purple flowers, long pollen	296	15.6	240	9
Purple flowers, round pollen	19	1.0	80	3
Red flowers, long pollen	27	1.4	80	3
Red flowers, round pollen	85	4.5	27	1

FIGURE 6.2 An experiment of Bateson and Punnett with sweet peas, showing that independent assortment does not always occur. Note: The expected numbers are rounded to the nearest whole number.

Genes →Traits Two genes that govern flower color and pollen shape are found on the same chromosome. Therefore, the offspring tend to inherit the parental combinations of alleles (*PL* or *pl*). Due to occasional crossing over, a lower percentage of offspring inherit nonparental combinations of alleles (*Pl* or *pL*).

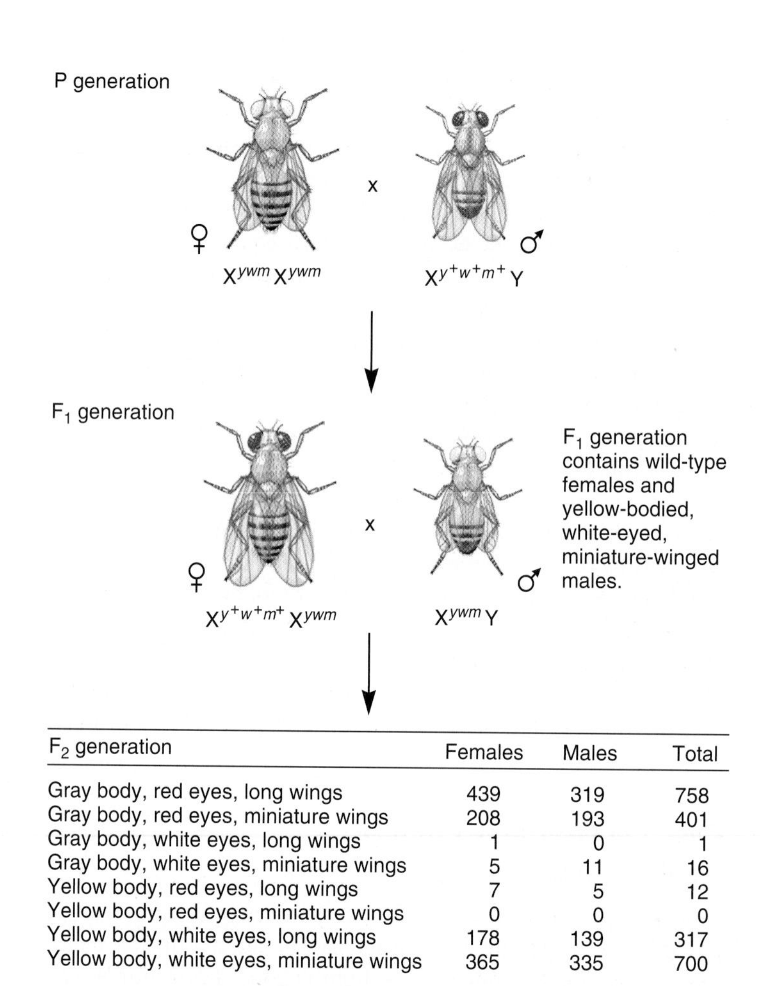

F_2 generation	Females	Males	Total
Gray body, red eyes, long wings	439	319	758
Gray body, red eyes, miniature wings	208	193	401
Gray body, white eyes, long wings	1	0	1
Gray body, white eyes, miniature wings	5	11	16
Yellow body, red eyes, long wings	7	5	12
Yellow body, red eyes, miniature wings	0	0	0
Yellow body, white eyes, long wings	178	139	317
Yellow body, white eyes, miniature wings	365	335	700

FIGURE 6.3 Morgan's trihybrid cross involving three X-linked traits in *Drosophila*.

Genes →Traits Three genes that govern body color, eye color, and wing length are all found on the X chromosome. Therefore, the offspring tend to inherit the parental combinations of alleles (y^+ w^+ m^+ or *y w m*). Figure 6.5 explains how single and double crossovers can create nonparental combinations of alleles.

a strain with red flowers (*pp*) and round pollen (*ll*). This yielded an F_1 generation of plants that all had purple flowers and long pollen (*PpLl*). An unexpected result came from the F_2 generation. Even though the F_2 generation had four different phenotypic categories, the observed numbers of offspring did not conform to a 9:3:3:1 ratio. Bateson and Punnett found that the F_2 generation had a much greater proportion of the two phenotypes found in the parental generation—purple flowers with long pollen and red flowers with round pollen. Therefore, they suggested that the transmission of these two traits from the parental generation to the F_2 generation was somehow coupled and not easily assorted in an independent manner. However, Bateson and Punnett did not realize that this coupling was due to the linkage of the flower color gene and the pollen shape gene on the same chromosome.

Morgan Provided Evidence for the Linkage of X-Linked Genes and Proposed That Crossing Over Between X Chromosomes Can Occur

The first direct evidence that different genes are physically located on the same chromosome came from the studies of Thomas Hunt Morgan in 1911, who investigated the inheritance pattern of different characters that had been shown to follow an X-linked pattern of inheritance. **Figure 6.3** illustrates an experiment involving three characters that Morgan studied. His parental crosses were wild-type male fruit flies mated to females that had yellow bodies (*yy*), white eyes (*ww*), and miniature wings (*mm*). The wild-type alleles for these three genes are designated y^+ (gray body), w^+ (red eyes), and m^+ (long wings). As expected, the phenotypes of the F_1 generation were wild-type females, and males with yellow bodies, white eyes, and miniature wings. The linkage of these genes was revealed when the F_1 flies were mated to each other and the F_2 generation examined.

Instead of equal proportions of the eight possible phenotypes, Morgan observed a much higher proportion of the combinations of traits found in the parental generation. He observed 758 flies with gray bodies, red eyes, and long wings, and 700 flies with yellow bodies, white eyes, and miniature wings. The combination of gray body, red eyes, and long wings was found in the males of the parental generation, and the combination of yellow body, white eyes, and miniature wings was the same as the females of the parental generation. Morgan's explanation for this higher proportion of parental combinations was that all three genes are located on the X chromosome and, therefore, tend to be transmitted together as a unit.

However, to fully account for the data shown in Figure 6.3, Morgan needed to explain why a significant proportion of the F_2 generation had nonparental combinations of alleles. Along with the two parental phenotypes, five other phenotypic combinations appeared that were not found in the parental generation. How did Morgan explain these data? He considered the studies conducted in 1909 of the Belgian cytologist Frans Alfons Janssens, who observed chiasmata under the microscope and proposed that crossing over involves a physical exchange between homologous chromosomes. Morgan shrewdly realized that crossing over between homologous X chromosomes was consistent with his data. He assumed that crossing over did not occur between the X and Y chromosome and that these three genes are not found on the Y chromosome. With these ideas in mind, he hypothesized that the genes for body color, eye color, and wing length are all located on the same chromosome, namely, the X chromosome. Therefore, the alleles for all three characters are most likely to be inherited together. Due to crossing over, Morgan also proposed that the homologous X chromosomes (in the female) can exchange pieces of chromosomes and produce new (nonparental) combinations of alleles and nonparental combinations of traits in the F_2 generation.

To appreciate Morgan's proposals, let's simplify his data and consider only two of the three genes: those that affect body color and eye color. If we use the data from Figure 6.3, the following results were obtained:

Gray body, red eyes	1159	
Yellow body, white eyes	1017	
Gray body, white eyes	17	Nonparental offspring
Yellow body, red eyes	12	
Total	2205	

Figure 6.4 considers how Morgan's proposals could account for these data. The parental offspring with gray bodies and red eyes or yellow body and white eyes were produced when no crossing over had occurred between the two genes (Figure 6.4a). This was the more common situation. By comparison, crossing over could alter the pattern of alleles along each chromosome and account for the nonparental offspring (Figure 6.4b). Why were there relatively few nonparental offspring? These two genes are very close

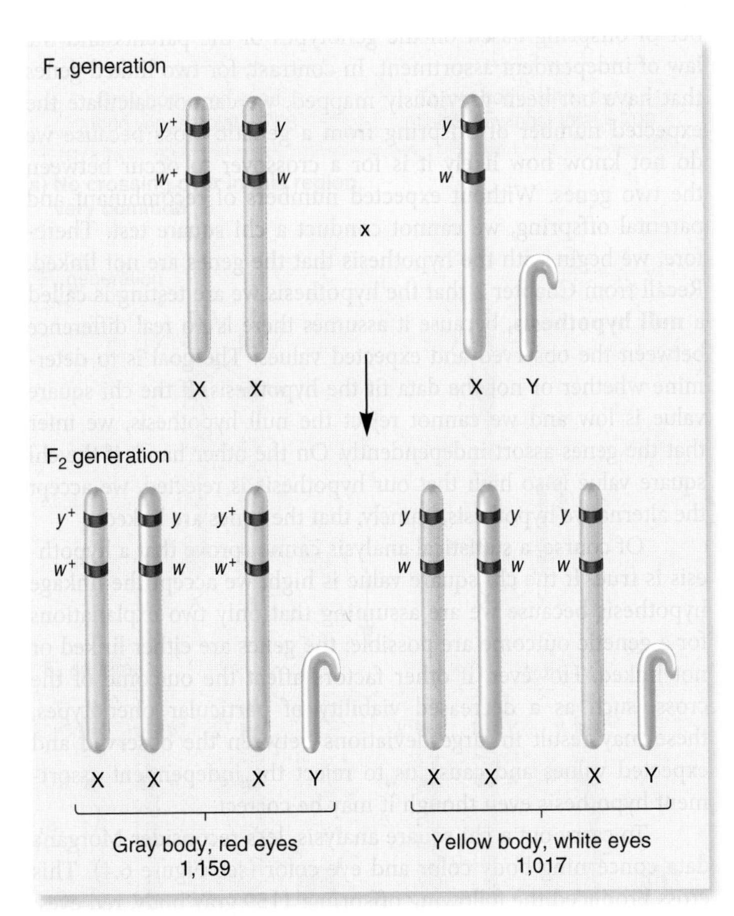

(a) No crossing over, parental offspring

(b) Crossing over, nonparental offspring

FIGURE 6.4 Morgan's explanation for parental and nonparental offspring. As described in Chapter 3, crossing over actually occurs at the bivalent stage, but for simplicity, this figure shows only two X chromosomes (one of each homolog) rather than four chromatids, which would occur during the bivalent stage of meiosis. Also note that this figure shows only a portion of the X chromosome. A map of the entire X chromosome is shown in Figure 6.8.

1017 yellow body, white eyes; 17 gray body, white eyes; and 12 yellow body, red eyes. However, when a heterozygous female ($X^{y+w+} X^{yw}$) is crossed to a hemizygous male ($X^{yw}Y$), the laws of segregation and independent assortment predict the following outcome:

F₁ male gametes

F₁ female gametes ♀ \ ♂	X^{yw}	Y
$X^{y^+w^+}$	$X^{y^+w^+}X^{yw}$ Gray body, red eyes	$X^{y^+w^+}Y$ Gray body, red eyes
X^{y^+w}	$X^{y^+w}X^{yw}$ Gray body, white eyes	$X^{y^+w}Y$ Gray body, white eyes
X^{yw^+}	$X^{yw^+}X^{yw}$ Yellow body, red eyes	$X^{yw^+}Y$ Yellow body, red eyes
X^{yw}	$X^{yw}X^{yw}$ Yellow body, white eyes	$X^{yw}Y$ Yellow body, white eyes

Mendel's laws predict a 1:1:1:1 ratio among the four phenotypes. The observed data obviously seem to conflict with this expected outcome. Nevertheless, we stick to the strategy just discussed. We begin with the hypothesis that the two genes are not linked, and then we conduct a chi square analysis to see if the data fit this hypothesis. If the data do not fit, we reject the idea that the genes assort independently and conclude the genes are linked.

A step-by-step outline for applying the chi square test to distinguish between linkage and independent assortment is described next.

Step 1. *Propose a hypothesis.* Even though the observed data appear inconsistent with this hypothesis, we propose that the two genes for eye color and body color obey Mendel's law of independent assortment. This hypothesis allows us to calculate expected values. Because the data seem to conflict with this hypothesis, we actually anticipate that the chi square analysis will allow us to reject the independent assortment hypothesis in favor of a linkage hypothesis. We are also assuming the alleles follow the law of segregation, and the four phenotypes are equally viable.

Step 2. *Based on the hypothesis, calculate the expected value of each of the four phenotypes.* Each phenotype has an equal probability of occurring (see the Punnett square given previously). Therefore, the probability of each phenotype is 1/4. The observed F_2 generation had a total of 2205 individuals. Our next step is to calculate the expected number of offspring with each phenotype when the total equals 2205; 1/4 of the offspring should be each of the four phenotypes:

$1/4 \times 2205 = 551$ (expected number of each phenotype, rounded to the nearest whole number)

Step 3. *Apply the chi square formula, using the data for the observed values (O) and the expected values (E) that have been calculated in step 2.* In this case, the data consist of four phenotypes.

$$\chi^2 = \frac{(O_1 - E_1)^2}{E_1} + \frac{(O_2 - E_2)^2}{E_2} + \frac{(O_3 - E_3)^2}{E_3} + \frac{(O_4 - E_4)^2}{E_4}$$

$$\chi^2 = \frac{(1159 - 551)^2}{551} + \frac{(17 - 551)^2}{551} + \frac{(12 - 551)^2}{551} + \frac{(1017 - 551)^2}{551}$$

$$\chi^2 = 670.9 + 517.5 + 527.3 + 394.1 = 2109.8$$

Step 4. *Interpret the calculated chi square value.* This is done with a chi square table, as discussed in Chapter 2. The four phenotypes are based on the law of segregation and the law of independent assortment. By itself, the law of independent assortment predicts only two categories, recombinant and nonrecombinant. Therefore, based on a hypothesis of independent assortment, the degree of freedom equals $n-1$, which is $2-1$, or 1.

The calculated chi square value is enormous! This means that the deviation between observed and expected values is very large. With 1 degree of freedom, such a large deviation is expected to occur by chance alone less than 1% of the time (see Table 2.1). Therefore, we reject the hypothesis that the two genes assort independently. As an alternative, we could accept the hypothesis that the genes are linked. Even so, it should be emphasized that rejecting the null hypothesis does not necessarily mean that the linked hypothesis is correct. For example, some of the non-Mendelian inheritance patterns described in Chapter 6 can produce results that do not conform to independent assortment.

EXPERIMENT 6A

Creighton and McClintock Showed That Crossing Over Produced New Combinations of Alleles and Resulted in the Exchange of Segments Between Homologous Chromosomes

As we have seen, Morgan's studies were consistent with the hypothesis that crossing over occurs between homologous chromosomes to produce new combinations of alleles. To obtain direct evidence that crossing over can result in genetic recombination, Harriet Creighton and Barbara McClintock used an interesting strategy involving parallel observations. In studies conducted in 1931, they first made crosses involving two linked genes to produce parental and recombinant offspring. Second, they used a microscope to view the structures of the chromosomes in the parents and in the offspring. Because the parental chromosomes had some unusual structural features, they could microscopically distinguish the two homologous chromosomes within a pair. As we will see, this enabled them to correlate the occurrence of recombinant offspring with microscopically observable exchanges in segments of homologous chromosomes.

Creighton and McClintock focused much of their attention on the pattern of inheritance of traits in corn. This species has 10 different chromosomes per set, which are named chromosome 1, chromosome 2, chromosome 3, and so on. In previous cytological examinations of corn chromosomes, some strains were found to have an unusual chromosome 9 with a darkly staining knob at one end. In addition, McClintock identified an abnormal version of chromosome 9 that also had an extra piece of chromosome 8 attached at the other end (**Figure 6.6a**). This chromosomal rearrangement is called a translocation.

Creighton and McClintock insightfully realized that this abnormal chromosome could be used to determine if two homologous chromosomes physically exchange segments as a result of crossing over. They knew that a gene was located near the knobbed end of chromosome 9 that provided color to corn kernels. This gene existed in two alleles, the dominant allele *C* (colored) and the recessive allele *c* (colorless). A second gene, located near the translocated piece from chromosome 8, affected the texture of the kernel endosperm. The dominant allele *Wx* caused starchy endosperm, and the recessive *wx* allele caused waxy endosperm. Creighton and McClintock reasoned that a crossover involving a normal chromosome 9 and a knobbed/translocated chromosome 9 would produce a chromosome that had either a knob or a translocation, but not both. These two types of chromosomes would be distinctly different from either of the parental chromosomes (**Figure 6.6b**).

As shown in the experiment of **Figure 6.7**, Creighton and McClintock began with a corn strain that carried an abnormal chromosome that had a knob at one end and a translocation at the other. Genotypically, this chromosome was *C wx*. The cytologically normal chromosome in this strain was *c Wx*. This corn plant, termed parent A, had the genotype *Cc Wx wx*. It was crossed to a strain called parent B that carried two cytologically normal chromosomes and had the genotype *cc Wx wx*.

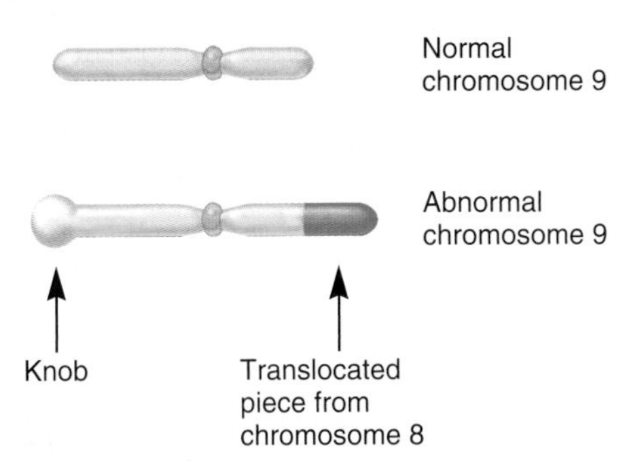

(a) Normal and abnormal chromosome 9

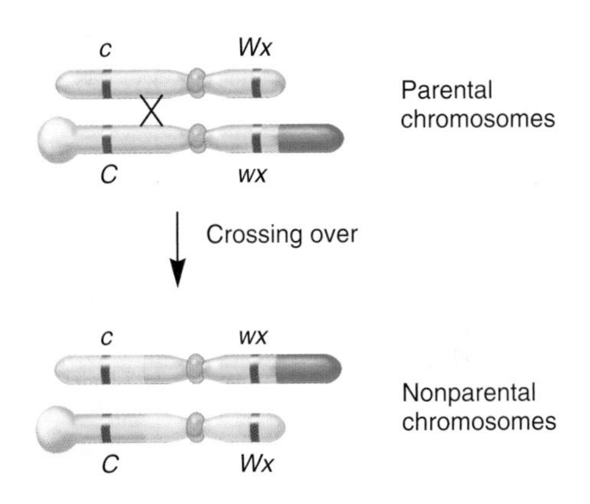

(b) Crossing over between normal and abnormal chromosome 9

FIGURE 6.6 Crossing over between a normal and abnormal chromosome 9 in corn. (a) A normal chromosome 9 in corn is compared to an abnormal chromosome 9 that contains a knob at one end and a translocation at the opposite end. (b) A crossover produces a chromosome that contains only a knob at one end and another chromosome that contains only a translocation at the other end.

They then observed the kernels in two ways. First, they examined the phenotypes of the kernels to see if they were colored or colorless, and starchy or waxy. Second, the chromosomes in each kernel were examined under a microscope to determine their cytological appearance. Altogether, they observed a total of 25 kernels (see data of Figure 6.7).

THE HYPOTHESIS

Offspring with nonparental phenotypes are the product of a crossover. This crossover should produce nonparental chromosomes via an exchange of chromosomal segments between homologous chromosomes.

TESTING THE HYPOTHESIS — FIGURE 6.7 Experimental correlation between genetic recombination and crossing over.

Starting materials: Two different strains of corn. One strain, referred to as parent A, had an abnormal chromosome 9 (knobbed/translocation) with a dominant *C* allele and a recessive *wx* allele. It also had a cytologically normal copy of chromosome 9 that carried the recessive *c* allele and the dominant *Wx* allele. Its genotype was *Cc Wxwx*. The other strain (referred to as parent B) had two normal versions of chromosome 9. The genotype of this strain was *cc Wxwx*.

1. Cross the two strains described. The tassel is the pollen-bearing structure, and the silk (equivalent to the stigma and style) is connected to the ovary. After fertilization, the ovary will develop into an ear of corn.
2. Observe the kernels from this cross.
3. Microscopically examine chromosome 9 in the kernels.

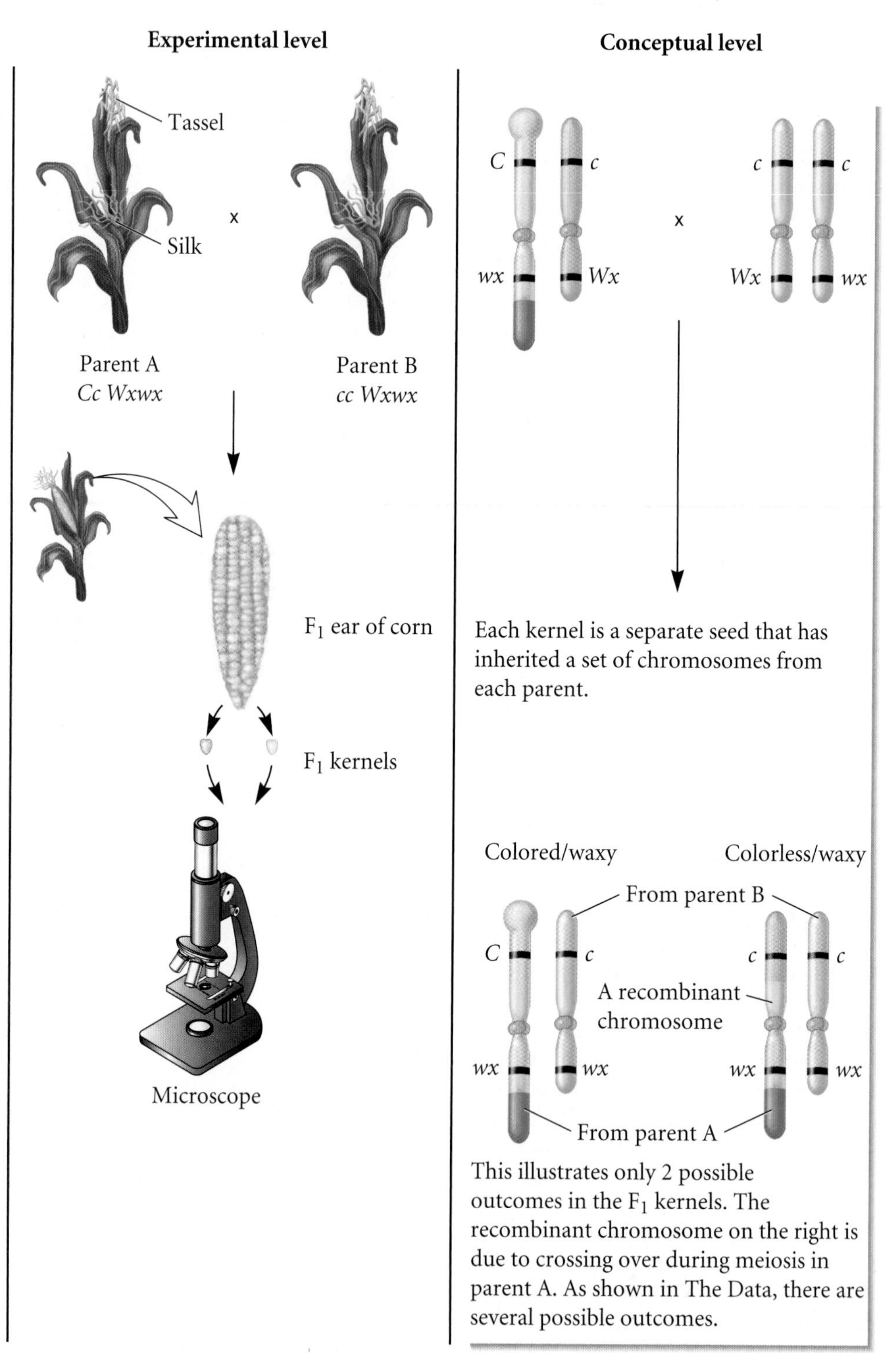

THE DATA

Phenotype of F_1 Kernel	Number of Kernels Analyzed	Cytological Appearance of Chromosome 9 in F_1 Offspring*		Did a Crossover Occur During Gamete Formation in Parent A?
Colored/waxy	3	Knobbed/translocation: *C* *wx*	Normal: *c* *wx*	No
Colorless/starchy	11	Knobless/normal: *c* *Wx*	Normal: *c* *Wx* or *c* *wx*	No
Colorless/starchy	4	Knobless/translocation: *c* *wx*	Normal: *c* *Wx*	Yes
Colorless/waxy	2	Knobless/translocation: *c* *wx*	Normal: *c* *wx*	Yes
Colored/starchy	5	Knobbed/normal: *C* *Wx*	Normal: *c* *Wx* or *c* *wx*	Yes
Total	25			

*In this table, the chromosome on the left was inherited from parent A, and the blue chromosome on the right was inherited from parent B.
Data from Harriet B. Creighton and Barbara McClintock (1931) A Correlation of Cytological and Genetical Crossing-Over in *Zea Mays*. *Proc. Natl. Acad. Sci. USA 17*, 492–497.

INTERPRETING THE DATA

By combining the gametes in a Punnett square, the following types of offspring can be produced:

Parent A (♀) \ Parent B (♂)	*c Wx*	*c wx*	
C wx	*Cc Wxwx* Colored, starchy	*Cc wxwx* Colored, waxy	Nonrecombinant
c Wx	*cc WxWx* Colorless, starchy	*cc Wxwx* Colorless, starchy	Nonrecombinant
C Wx	*Cc WxWx* Colored, starchy	*Cc Wxwx* Colored, starchy	Recombinant
c wx	*cc Wxwx* Colorless, starchy	*cc wxwx* Colorless, waxy	Recombinant

In this experiment, the researchers were interested in whether or not crossing over had occurred in parent A, which was heterozygous for both genes. This parent could produce four types of gametes, but parent B could produce only two types.

Parent A	Parent B
C wx (nonrecombinant)	*c Wx*
c Wx (nonrecombinant)	*c wx*
C Wx (recombinant)	
c wx (recombinant)	

As seen in the Punnett square, two of the phenotypic categories, colored, starchy (*Cc Wxwx* or *Cc WxWx*) and colorless, starchy (*cc WxWx* or *cc Wxwx*), were ambiguous because they could arise from a nonrecombinant and from a recombinant gamete. In other words, these phenotypes could be produced whether or not recombination occurred in parent A. Therefore, let's focus on the two unambiguous phenotypic categories: colored, waxy (*Cc wxwx*) and colorless, waxy (*cc wxwx*). The colored, waxy phenotype could happen only if recombination did not occur in parent A and if parent A passed the knobbed/translocated chromosome to its offspring. As shown in the data, three kernels were obtained with this phenotype, and all of them had the knobbed/translocated chromosome. By comparison, the colorless, waxy phenotype could be obtained only if genetic recombination occurred in parent A and this parent passed a chromosome 9 that had a translocation but was knobless. Two kernels were obtained with this phenotype, and both of them had the expected chromosome that had a translocation but was knobless. Taken together, these results showed a perfect correlation between genetic recombination of alleles and the cytological presence of a chromosome displaying a genetic exchange of chromosomal pieces from parent A.

Overall, the observations described in this experiment were consistent with the idea that a crossover occurred in the region between the *C* and *wx* genes that involved an exchange of segments between two homologous chromosomes. As stated by Creighton and McClintock, "Pairing chromosomes, heteromorphic in two regions, have been shown to exchange parts at the same time they exchange genes assigned to these regions." These results supported the view that genetic recombination involves a physical exchange between homologous chromosomes. This microscopic evidence helped to convince geneticists that recombinant offspring arise from the physical exchange of segments of homologous chromosomes. As shown in the solved problem S4 at the end of this chapter, an experiment by Curt Stern was also consistent with the conclusion that crossing over between homologous chromosomes accounts for the formation of offspring with recombinant phenotypes.

A self-help quiz involving this experiment can be found at www.mhhe.com/brookergenetics4e.

6.2 GENETIC MAPPING IN PLANTS AND ANIMALS

The purpose of **genetic mapping,** also known as gene mapping or chromosome mapping, is to determine the linear order and distance of separation among genes that are linked to each other along the same chromosome. **Figure 6.8** illustrates a simplified genetic map of *Drosophila melanogaster,* depicting the locations of many different genes along the individual chromosomes. As shown here, each gene has its own unique **locus**—the site where the gene is found within a particular chromosome. For example, the gene designated *brown eyes* (*bw*), which affects eye color, is located near one end of chromosome 2. The gene designated *black body* (*b*), which affects body color, is found near the middle of the same chromosome.

Why is genetic mapping useful? First, it allows geneticists to understand the overall complexity and genetic organization of a particular species. The genetic map of a species portrays the underlying basis for the inherited traits that an organism displays. In some cases, the known locus of a gene within a genetic map can help molecular geneticists to clone that gene and thereby obtain greater information about its molecular features. In addition, genetic maps are useful from an evolutionary point of view. A comparison of the genetic maps for different species can improve our understanding of the evolutionary relationships among those species.

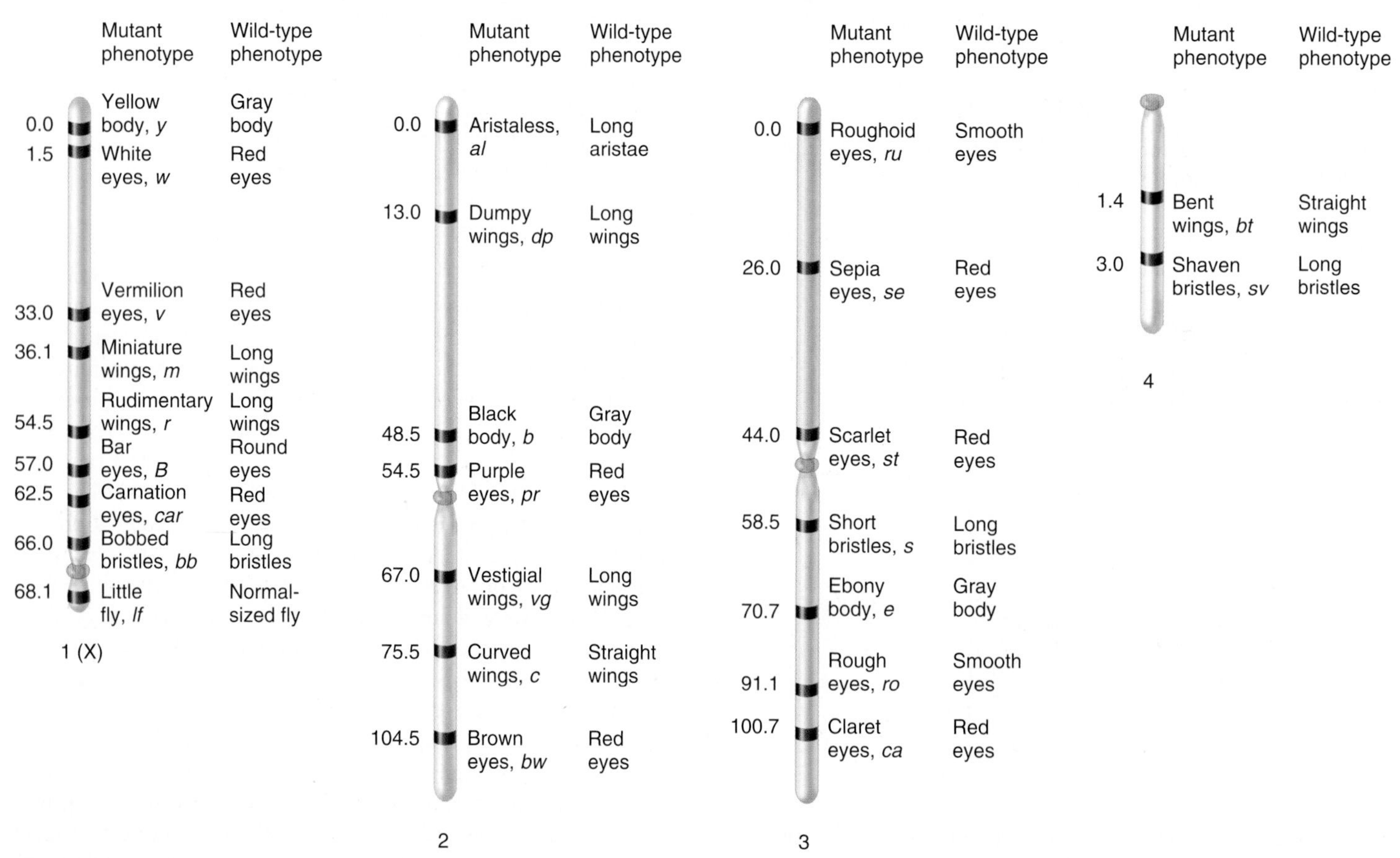

FIGURE 6.8 A simplified genetic linkage map of *Drosophila melanogaster*. This simplified map illustrates a few of the many thousands of genes that have been identified in this organism.

Along with these scientific uses, genetic maps have many practical benefits. For example, many human genes that play a role in human disease have been genetically mapped. This information can be used to diagnose and perhaps someday treat inherited human diseases. It can also help genetic counselors predict the likelihood that a couple will produce children with certain inherited diseases. In addition, genetic maps are gaining increasing importance in agriculture. A genetic map can provide plant and animal breeders with helpful information for improving agriculturally important strains through selective breeding programs.

In this section, we will examine traditional genetic mapping techniques that involve an analysis of crosses of individuals that are heterozygous for two or more genes. The frequency of nonparental offspring due to crossing over provides a way to deduce the linear order of genes along a chromosome. As depicted in Figure 6.8, this linear arrangement of genes is known as a **genetic linkage map.** This approach has been useful for analyzing organisms that are easily crossed and produce a large number of offspring in a short period of time. Genetic linkage maps have been constructed for several plant species and certain species of animals, such as *Drosophila*. For many organisms, however, traditional mapping approaches are difficult due to long generation times or the inability to carry out experimental crosses (as in humans). Fortunately, many alternative methods of gene mapping have been developed to replace the need to carry out crosses. As described in Chapter 20, molecular approaches are increasingly used to map genes.

The Frequency of Recombination Between Two Genes Can Be Correlated with Their Map Distance Along a Chromosome

Genetic mapping allows us to estimate the relative distances between linked genes based on the likelihood that a crossover will occur between them. If two genes are very close together on the same chromosome, a crossover is unlikely to begin in the region between them. However, if two genes are very far apart, a crossover is more likely to be initiated in this region and thereby recombine the alleles of the two genes. Experimentally, the basis for genetic mapping is that the percentage of recombinant offspring is correlated with the distance between two genes. If two genes are far apart, many recombinant offspring will be produced. However, if two genes are close together, very few recombinant offspring will be observed.

To interpret a genetic mapping experiment, the experimenter must know if the characteristics of an offspring are due to crossing over during meiosis in a parent. This is accomplished by conducting a **testcross.** Most testcrosses are between an individual that is heterozygous for two or more genes and an individual that is recessive and homozygous for the same genes. The goal of the testcross is to determine if recombination has occurred during meiosis in the heterozygous parent. Thus, genetic mapping is based on the level of recombination that occurs in just one parent—the heterozygote. In a testcross, new combinations of alleles cannot occur in the gametes of the other parent, which is homozygous for these genes.

Figure 6.9 illustrates how a testcross provides an experimental strategy to distinguish between recombinant and nonrecombinant offspring. This cross concerns two linked genes affecting bristle length and body color in fruit flies. The recessive alleles are *s* (short bristles) and *e* (ebony body), and the dominant (wild-type) alleles are s^+ (long bristles) and e^+ (gray body). One parent displays both recessive traits. Therefore, we know this parent is homozygous for the recessive alleles of the two genes (*ss ee*). The other parent is heterozygous for the linked genes affecting bristle length and body color. This parent was produced from a cross involving a true-breeding wild-type fly and a true-breeding fly with short bristles and an ebony body. Therefore, in this heterozygous parent, we know that the *s* and *e* alleles are located on one chromosome and the corresponding s^+ and e^+ alleles are located on the homologous chromosome.

Now let's take a look at the four possible types of offspring these parents can produce. The offspring's phenotypes are long bristles, gray body; short bristles, ebony body; long bristles, ebony body; and short bristles, gray body. All four types of offspring have inherited a chromosome carrying the *s* and *e* alleles from their homozygous parent (shown on the right in each pair). Focus your attention on the other chromosome. The offspring with long bristles and gray bodies have inherited a chromosome carrying the s^+ and e^+ alleles from the heterozygous parent. This chromosome is not the product of a crossover. The offspring with short bristles and ebony bodies have inherited a chromosome carrying the *s* and *e* alleles from the heterozygous parent. Again, this chromosome is not the product of a crossover.

The other two types of offspring, however, can be produced only if crossing over has occurred in the region between these two genes. Those with long bristles and ebony bodies or short bristles and gray bodies have inherited a chromosome that is the product of a crossover during meiosis in the heterozygous parent. As noted in Figure 6.9, the recombinant offspring are fewer in number than are the nonrecombinant offspring.

The frequency of recombination can be used as an estimate of the physical distance between two genes on the same chromosome. The **map distance** is defined as the number of recombinant offspring divided by the total number of offspring, multiplied by 100. We can calculate the map distance between these two genes using this formula:

$$\text{Map distance} = \frac{\text{Number of recombinant offspring}}{\text{Total number of offspring}} \times 100$$

$$= \frac{76 + 75}{537 + 542 + 76 + 75} \times 100$$

$$= 12.3 \text{ map units}$$

The units of distance are called **map units (mu),** or sometimes **centiMorgans (cM)** in honor of Thomas Hunt Morgan. One map unit is equivalent to a 1% frequency of recombination. In this example, we would conclude that the *s* and *e* alleles are 12.3 mu apart from each other along the same chromosome.

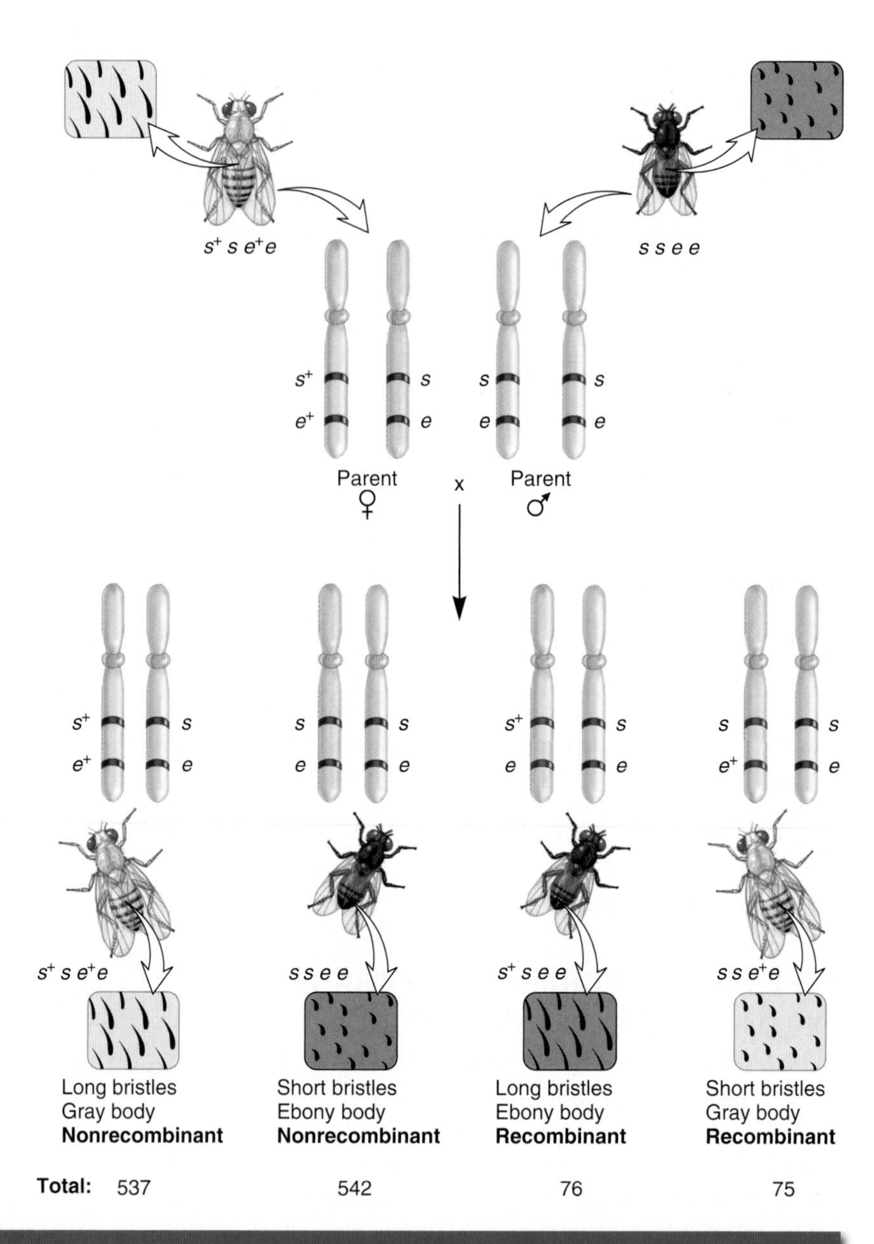

INTERACTIVE EXERCISE

FIGURE 6.9 Use of a testcross to distinguish between recombinant and nonrecombinant offspring. The cross involves one *Drosophila* parent that is homozygous recessive for short bristles (*ss*) and ebony body (*ee*), and one parent heterozygous for both genes ($s^+s\ e^+e$). (Note: *Drosophila* geneticists normally designate the short allele as *ss* and a homozygous fly with short bristles as *ssss*. In this case, the allele causing short bristles is designated with a single *s* to avoid confusion between the allele designation and the genotype of the fly. Also, crossing over does not occur during sperm formation in *Drosophila*, which is unusual among eukaryotes. Therefore, the heterozygote in a testcross involving *Drosophila* must be the female.)

EXPERIMENT 6B

Alfred Sturtevant Used the Frequency of Crossing Over in Dihybrid Crosses to Produce the First Genetic Map

In 1913, the first individual to construct a (very small) genetic map was Alfred Sturtevant, an undergraduate who spent time in the laboratory of Thomas Hunt Morgan. Sturtevant wrote: "In conversation with Morgan . . . I suddenly realized that the variations in the strength of linkage, already attributed by Morgan to differences in the spatial separation of the genes, offered the possibility of determining sequences [of different genes] in the linear dimension of a chromosome. I went home and spent most of the night (to the neglect of my undergraduate homework) in producing the first chromosome map, which included the sex-linked genes, *y*, *w*, *v*, *m*, and *r*, in the order and approximately the relative spacing that they still appear on the standard maps."

In the experiment of **Figure 6.10**, Sturtevant considered the outcome of crosses involving six different mutant alleles that altered the phenotype of flies. All of these alleles were known to be recessive and X-linked. They are *y* (yellow body color), *w* (white eye color), *w-e* (eosin eye color), *v* (vermilion eye color), *m* (miniature wings), and *r* (rudimentary wings). The *w* and *w-e* alleles are alleles of the same gene. In contrast, the *v* allele (vermilion eye color) is an allele of a different gene that also affects eye color. The two alleles that affect wing length, *m* and *r*, are also in different genes. Therefore, Sturtevant studied the inheritance of six recessive alleles, but since *w* and *w-e* are alleles of the same gene, his genetic map contained only five genes. The corresponding wild-type alleles are y^+ (gray body), w^+ (red eyes), v^+ (red eyes), m^+ (long wings), and r^+ (long wings).

THE HYPOTHESIS

When genes are located on the same chromosome, the distance between the genes can be estimated from the proportion of recombinant offspring. This provides a way to map the order of genes along a chromosome.

TESTING THE HYPOTHESIS — FIGURE 6.10 The first genetic mapping experiment.

Starting materials: Sturtevant began with several different strains of *Drosophila* that contained the six alleles already described.

1. Cross a female that is heterozygous for two different genes to a male that is hemizygous recessive for the same two genes. In this example, cross a female that is $X^{y^+w^+}X^{yw}$ to a male that is $X^{yw}Y$.

   This strategy was employed for many dihybrid combinations of the six alleles already described.

2. Observe the outcome of the crosses.

3. Calculate the percentages of offspring that are the result of crossing over (number of nonparental/total).

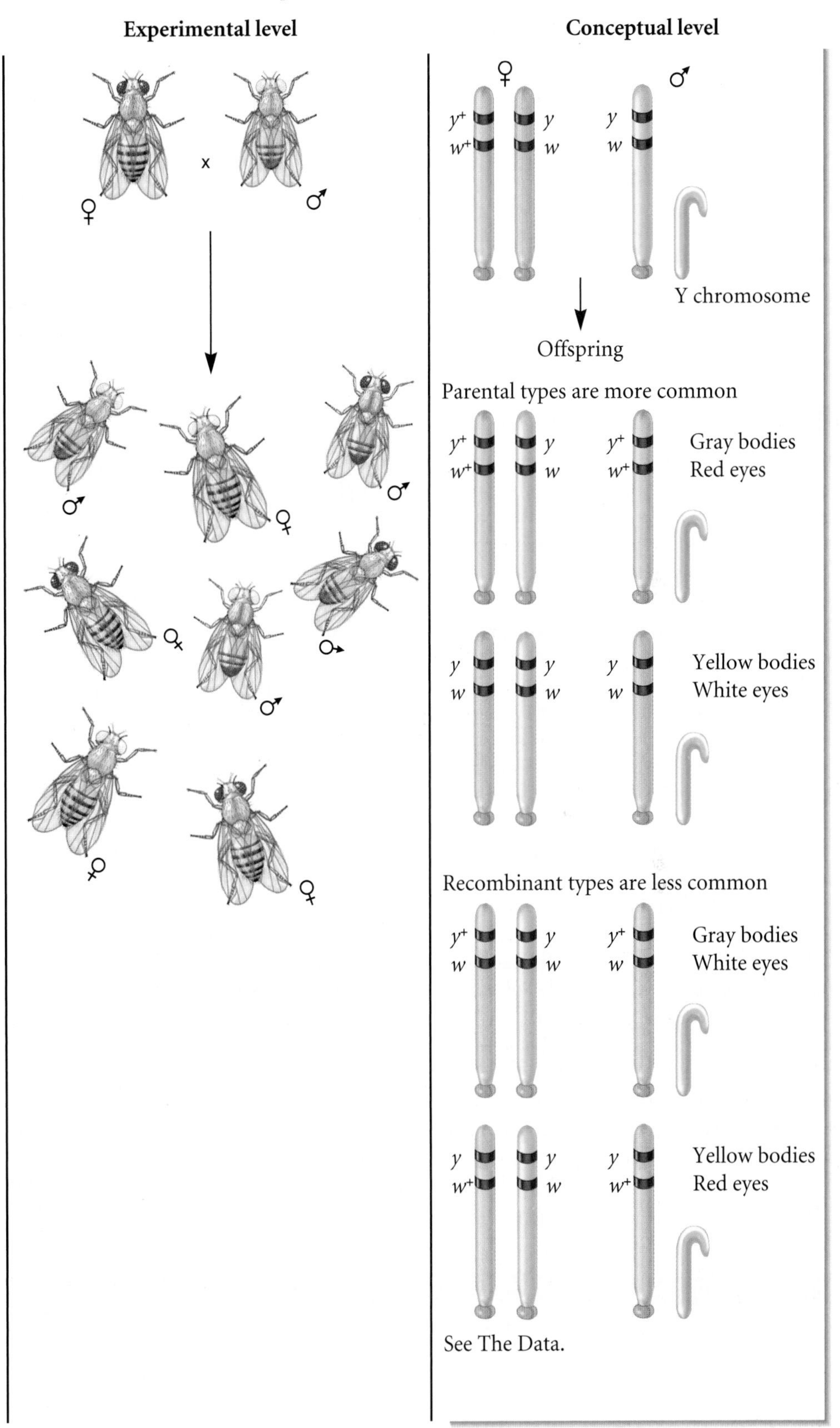

THE DATA

Alleles Concerned	*Number Recombinant/Total Number*	*Percent Recombinant Offspring*
y and *w*/*w-e*	214/21,736	1.0
y and *v*	1464/4551	32.2
y and *m*	115/324	35.5
y and *r*	260/693	37.5
w/*w-e* and *v*	471/1584	29.7
w/*w-e* and *m*	2062/6116	33.7
w/*w-e* and *r*	406/898	45.2
v and *m*	17/573	3.0
v and *r*	109/405	26.9

Data from Alfred H. Sturtevant (1913) The linear arrangement of six sex-linked factors in *Drosophila*, as shown by their mode of association. *J Exp Zool 14*, 43–59.

INTERPRETING THE DATA

As shown in Figure 6.10, Sturtevant made pairwise testcrosses and then counted the number of offspring in the four phenotypic categories. Two of the categories were nonrecombinant and two were recombinant, requiring a crossover between the X chromosomes in the female heterozygote. Let's begin by contrasting the results between particular pairs of genes, shown in the data. In some dihybrid crosses, the percentage of nonparental offspring was rather low. For example, dihybrid crosses involving the *y* allele and the *w* or *w-e* allele yielded 1% recombinant offspring. This result suggested that these two genes are very close together. By comparison, other dihybrid crosses showed a higher percentage of nonparental offspring. For example, crosses involving the *v* and *r* alleles produced 26.9% recombinant offspring. These two genes are expected to be farther apart.

To construct his map, Sturtevant began with the assumption that the map distances would be more accurate between genes that are closely linked. Therefore, his map is based on the distance between *y* and *w* (1.0), *w* and *v* (29.7), *v* and *m* (3.0), and *v* and *r* (26.9). He also considered other features of the data to deduce the order of the genes. For example, the percentage of crossovers between *w* and *m* was 33.7. The percentage of crossovers between *w* and *v* was 29.7, suggesting that *v* is between *w* and *m*, but closer to *m*. The proximity of *v* and *m* is confirmed by the low percentage of crossovers between *v* and *m* (3.0). Sturtevant collectively considered the data and proposed the genetic map shown here.

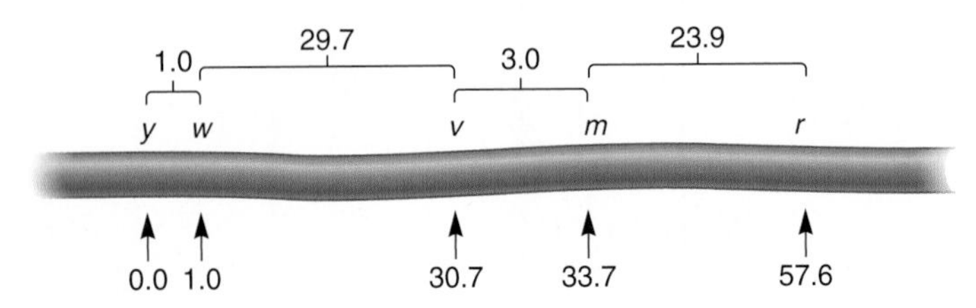

In this genetic map, Sturtevant began at the *y* allele and mapped the genes from left to right. For example, the *y* and *v* alleles are 30.7 mu apart, and the *v* and *m* alleles are 3.0 mu apart. This study by Sturtevant was a major breakthrough, because it showed how to map the locations of genes along chromosomes by making the appropriate crosses.

If you look carefully at Sturtevant's data, you will notice a few observations that do not agree very well with his genetic map. For example, the percentage of recombinant offspring for the *y* and *r* dihybrid cross was 37.5 (but the map distance is 57.6), and the crossover percentage between *w* and *r* was 45.2 (but the map distance is 56.6). As the percentage of recombinant offspring approaches a value of 50%, this value becomes a progressively more inaccurate measure of actual map distance (**Figure 6.11**). What is the basis for this inaccuracy? When the distance between two genes is large, the likelihood of multiple crossovers in the region between them causes the observed number of recombinant offspring to underestimate this distance.

Multiple crossovers set a quantitative limit on the relationship between map distance and the percentage of recombinant offspring. Even though two different genes can be on the same chromosome and more than 50 mu apart, a testcross is expected to yield a maximum of only 50% recombinant offspring. What accounts for this 50% limit? The answer lies in the pattern of multiple crossovers. A single crossover in the region between two genes will produce only 50% recombinant chromosomes (see Figure 6.1b). Therefore, to exceed a 50% recombinant level, it would seem necessary to have multiple crossovers within a tetrad. However, let's consider double crossovers. As shown in the figure to solved problem S5 at the end of the chapter, a double crossover between two genes could involve four, three, or two

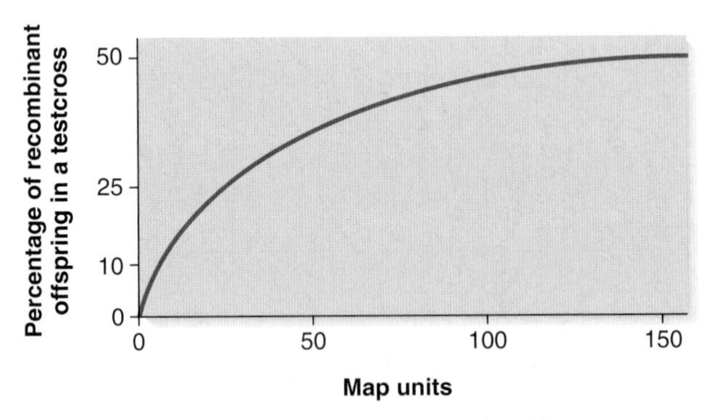

FIGURE 6.11 Relationship between the percentage of recombinant offspring in a testcross and the actual map distance between genes. The *y*-axis depicts the percentage of recombinant offspring that would be observed in a dihybrid testcross. The actual map distance, shown on the *x*-axis, is calculated by analyzing the percentages of recombinant offspring from a series of many dihybrid crosses involving closely linked genes. Even though two genes may be more than 50 mu apart, the percentage of recombinant offspring will not exceed 50%.

chromatids, which would yield 100%, 50%, or 0% recombinants, respectively. Because all of these double crossovers are equally likely, we take the average of them to determine the maximum recombination frequency. This average equals 50%. Therefore, when two different genes are more than 50 mu apart, they follow the law of independent assortment in a testcross and only 50% recombinants are observed.

A self-help quiz involving this experiment can be found at www.mhhe.com/brookergenetics4e.

Trihybrid Crosses Can Be Used to Determine the Order and Distance Between Linked Genes

Thus far, we have considered the construction of genetic maps using dihybrid testcrosses to compute map distance. The data from trihybrid crosses can yield additional information about map distance and gene order. In a trihybrid cross, the experimenter crosses two individuals that differ in three characters. The following experiment outlines a common strategy for using trihybrid crosses to map genes. In this experiment, the parental generation consists of fruit flies that differ in body color, eye color, and wing shape. We must begin with true-breeding lines so that we know which alleles are initially linked to each other on the same chromosome. In this example, all of the dominant alleles are linked on the same chromosome.

Step 1. *Cross two true-breeding strains that differ with regard to three alleles.* In this example, we will cross a fly that has a black body (*bb*), purple eyes (*prpr*), and vestigial wings (*vgvg*) to a homozygous wild-type fly with a gray body (b^+b^+), red eyes (pr^+pr^+), and long wings (vg^+vg^+):

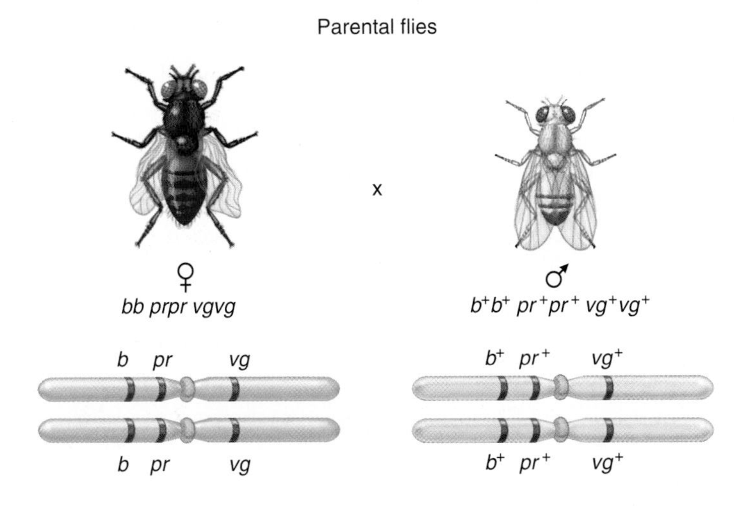

The goal in this step is to obtain F_1 individuals that are heterozygous for all three genes. In the F_1 heterozygotes, all dominant alleles are located on one chromosome, and all recessive alleles are on the other homologous chromosome.

Step 2. *Perform a testcross by mating F_1 female heterozygotes to male flies that are homozygous recessive for all three alleles (bb prpr vgvg).*

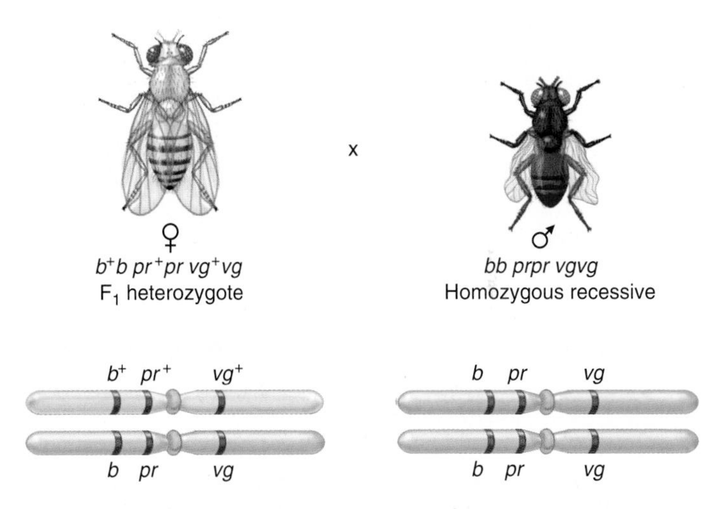

During gametogenesis in the heterozygous female F_1 flies, crossovers may produce new combinations of the three alleles.

Step 3. *Collect data for the F_2 generation.* As shown in **Table 6.1**, eight phenotypic combinations are possible. An analysis of the F_2 generation flies allows us to map these three genes. Because the three genes exist as two alleles each, we have 2^3, or 8, possible combinations of offspring. If these alleles assorted independently, all eight combinations would occur in equal proportions. However, we see that the proportions of the eight phenotypes are far from equal.

The genotypes of the parental generation correspond to the phenotypes gray body, red eyes, and long wings, and black body, purple eyes, and vestigial wings. In crosses involving linked genes, the parental phenotypes occur most frequently in the offspring. The remaining six phenotypes are due to crossing over.

The double crossover is always expected to be the least frequent category of offspring. Two of the phenotypes—gray body, purple eyes, and long wings; and black body, red eyes, and vestigial wings—arose from a double crossover between two pairs of genes.

TABLE 6.1

Data from a Trihybrid Cross (see step 2)

Phenotype	Number of Observed Offspring (males and females)	Chromosome Inherited from F_1 Female
Gray body, red eyes, long wings	411	b^+ pr^+ vg^+
Gray body, red eyes, vestigial wings	61	b^+ pr^+ vg
Gray body, purple eyes, long wings	2	b^+ pr vg^+
Gray body, purple eyes, vestigial wings	30	b^+ pr vg
Black body, red eyes, long wings	28	b pr^+ vg^+
Black body, red eyes, vestigial wings	1	b pr^+ vg
Black body, purple eyes, long wings	60	b pr vg^+
Black body, purple eyes, vestigial wings	412	b pr vg
Total	1005	

Also, the combination of traits in the double crossover tells us which gene is in the middle. When a chromatid undergoes a double crossover, the gene in the middle becomes separated from the other two genes at either end.

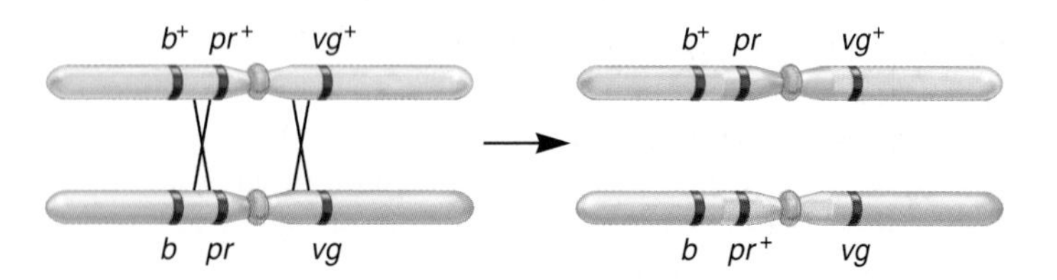

In the double-crossover categories, the recessive purple eye allele is separated from the other two recessive alleles. When mated to a homozygous recessive fly in the testcross, this yields flies with gray bodies, purple eyes, and long wings; or ones with black bodies, red eyes, and vestigial wings. This observation indicates that the gene for eye color lies between the genes for body color and wing shape.

Step 4. *Calculate the map distance between pairs of genes.* To do this, we need to understand which gene combinations are recombinant and which are nonrecombinant. The recombinant offspring are due to crossing over in the heterozygous female parent. If you look back at step 2, you can see the arrangement of alleles in the heterozygous female parent in the absence of crossing over. Let's consider this arrangement with regard to gene pairs:

b^+ is linked to pr^+, and b is linked to pr

pr^+ is linked to vg^+, and pr is linked to vg

b^+ is linked to vg^+, and b is linked to vg

With regard to body color and eye color, the recombinant offspring have gray bodies and purple eyes (2 + 30) or black bodies and red eyes (28 + 1). As shown along the right side of Table 6.1, these offspring were produced by crossovers in the female parents. The total number of these recombinant offspring is 61. The map distance between the body color and eye color genes is

$$\text{Map distance} = \frac{61}{944 + 61} \times 100 = 6.1 \text{ mu}$$

With regard to eye color and wing shape, the recombinant offspring have red eyes and vestigial wings (61 + 1) or purple eyes and long wings (2 + 60). The total number is 124. The map distance between the eye color and wing shape genes is

$$\text{Map distance} = \frac{124}{881 + 124} \times 100 = 12.3 \text{ mu}$$

With regard to body color and wing shape, the recombinant offspring have gray bodies and vestigial wings (61 + 30) or black bodies and long wings (28 + 60). The total number is 179. The map distance between the body color and wing shape genes is

$$\text{Map distance} = \frac{179}{826 + 179} \times 100 = 17.8 \text{ mu}$$

Step 5. *Construct the map.* Based on the map unit calculation, the body color (b) and wing shape (vg) genes are farthest apart. The eye color gene (pr) must lie in the middle. As mentioned earlier, this order of genes is also confirmed by the pattern of traits found in the double crossovers. To construct the map, we use the distances between the genes that are closest together.

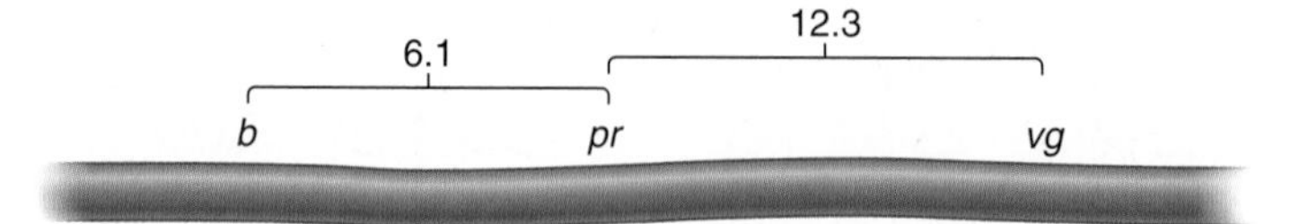

In our example, we have placed the body color gene first and the wing shape gene last. The data also are consistent with a map in which the wing shape gene comes first and the body color gene comes last. In detailed genetic maps, the locations of genes are mapped relative to the centromere.

You may have noticed that our calculations underestimate the distance between the body color and wing shape genes. We obtained a value of 17.8 mu even though the distance seems to be 18.4 mu when we add together the distance between body color and eye color genes (6.1 mu) and the distance between eye color and wing shape genes (12.3 mu). What accounts for this discrepancy? The answer is double crossovers. If you look at the data in Table 6.1, the offspring with gray bodies, purple eyes, and long wings or those with black bodies, red eyes, and vestigial wings are due to a double crossover. From a phenotypic perspective, these offspring are not recombinant with regard to the body color and wing shape alleles. Even so, we know that they arose from a double crossover between these two genes. Therefore, we should consider these crossovers when calculating the distance between the body color and wing shape genes. In this case, three offspring (2 + 1) were due to double crossovers. Because they are double crossovers, we multiply 2 times the number of double crossovers (2 + 1) and add this number to our previous value of recombinant offspring:

$$\text{Map distance} = \frac{179 + 2(2 + 1)}{826 + 179} \times 100 = 18.4 \text{ mu}$$

Interference Can Influence the Number of Double Crossovers That Occur in a Short Region

In Chapter 2, we considered the product rule to determine the probability that two independent events will both occur. The product rule allows us to predict the expected likelihood of a double crossover provided we know the individual probabilities of each single crossover. Let's reconsider the data of the trihybrid testcross just described to see if the frequency of double crossovers is what we would expect based on the product rule. If each crossover is an independent event, we can multiply the likelihood of a single crossover between *b* and *pr* (0.061) times the likelihood of a single crossover between *pr* and *vg* (0.123). The product rule predicts

Expected likelihood of a double crossover = 0.061 × 0.123 = 0.0075 = 0.75%

Expected number of offspring due to a double crossover, based on a total of 1005 offspring produced = 1005 × 0.0075 = 7.5

In other words, we would expect about 7 or 8 offspring to be produced as a result of a double crossover. The observed number of offspring was only 3 (namely, 2 with gray bodies, purple eyes, and long wings, and 1 with a black body, red eyes, and vestigial wings). What accounts for the lower number? This lower-than-expected value is probably not due to random sampling error. Instead, the likely cause is a common genetic phenomenon known as **positive interference,** in which the occurrence of a crossover in one region of a chromosome decreases the probability that a second crossover will occur nearby. In other words, the first crossover interferes with the ability to form a second crossover in the immediate vicinity. To provide interference with a quantitative value, we first calculate the coefficient of coincidence (C), which is the ratio of the observed number of double crossovers to the expected number.

$$C = \frac{\text{Observed number of double crossovers}}{\text{Expected number of double crossovers}}$$

Interference (I) is expressed as

$$I = 1 - C$$

For the data of the trihybrid testcross, the observed number of crossovers is 3 and the expected number is 7.5, so the coefficient of coincidence equals 3/7.5 = 0.40. In other words, only 40% of the expected number of double crossovers were actually observed. The value for interference equals 1 − 0.4 = 0.60, or 60%. This means that 60% of the expected number of crossovers did not occur. Because I has a positive value, this is called positive interference. Rarely, the outcome of a testcross yields a negative value for interference. A negative interference value suggests that a first crossover enhanced the rate of a second crossover in a nearby region. Although the molecular mechanisms that cause interference are not entirely understood, in most organisms the number of crossovers is regulated so that very few occur per chromosome. The reasons for positive and negative interference require further research.

6.3 GENETIC MAPPING IN HAPLOID EUKARYOTES

Before ending our discussion of genetic mapping, let's consider some pioneering studies that involved the genetic mapping of haploid organisms. You may find it surprising that certain species of simple eukaryotes, particularly unicellular algae and fungi, which spend part of their life cycle in the haploid state, have also been used in genetic mapping studies. The sac fungi, called ascomycetes, have been particularly useful to geneticists because of their unique style of sexual reproduction. In fact, much of our earliest understanding of genetic recombination came from the genetic analyses of fungi.

Fungi may be unicellular or multicellular organisms. Fungal cells are typically haploid ($1n$) and can reproduce asexually. In addition, fungi can also reproduce sexually by the fusion of two haploid cells to create a diploid zygote ($2n$) (**Figure 6.12**). The diploid zygote can then proceed through meiosis to produce four haploid cells, which are called **spores.** This group of four spores is known as a **tetrad** (not to be confused with a tetrad of four sister chromatids). In some species, meiosis is followed by a mitotic division to produce eight spores, known as an **octad.**

In ascomycete fungi and certain species of algae, the cells of a tetrad or octad are contained within a sac, which is called an **ascus** (plural: asci) in fungi. In other words, the products of a single meiotic division are contained within one sac. This mode of reproduction does not occur in other eukaryotic groups. Studies of fungi have been pivotal in our fundamental understanding of meiosis and crossing over. By comparison, the products of meiosis are produced differently in animals and plants. For example, in animals, oogenesis produces a single functional egg, and spermatogenesis occurs in the testes, where the resulting sperm become mixed with millions of other sperm.

Using a microscope, researchers can dissect asci and study the traits of each haploid spore. In this way, these organisms offer

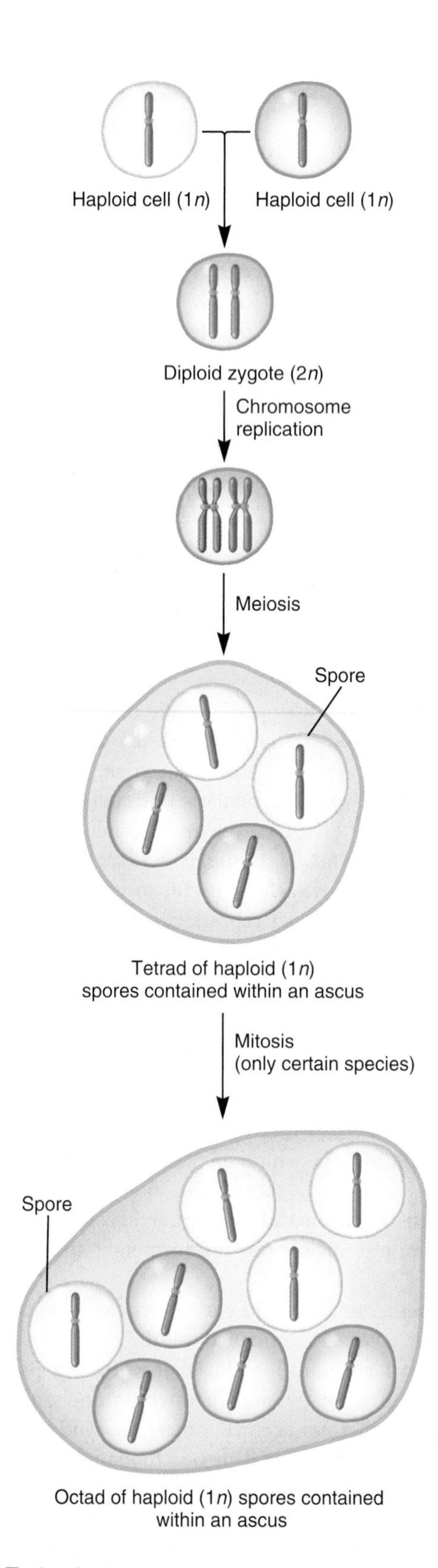

FIGURE 6.12 Sexual reproduction in ascomycetes. For simplicity, this diagram shows each haploid cell as having only one chromosome per haploid set. However, fungal species actually contain several chromosomes per haploid set.

a unique opportunity for geneticists to identify and study all of the cells that are derived from a single meiotic division. In this section, we will consider how the analysis of asci can be used to map genes in fungi.

Ordered Tetrad Analysis Can Be Used to Map the Distance Between a Gene and the Centromere

The arrangement of spores within an ascus varies from species to species (**Figure 6.13a**). In some cases, the ascus provides enough space for the tetrads or octads of spores to randomly mix together. This creates an **unordered tetrad** or **octad.** These occur in fungal species such as *Saccharomyces cerevisiae* and *Aspergillus nidulans* and also in certain unicellular algae (*Chlamydomonas reinhardtii*). By comparison, other species of fungi produce a very tight ascus that prevents spores from randomly moving around, which results in an **ordered tetrad** or **octad. Figure 6.13b** illustrates how an ordered octad is formed in *Neurospora crassa.* In this example, spores that carry the *A* allele have orange pigmentation, and those having the *a* (albino) allele are white.

A key feature of ordered tetrads or octads is that the position and order of spores within the ascus reflect their relationship to each other as they were produced by meiosis and mitosis. This idea is schematically shown in Figure 6.13b. After the original diploid cell has undergone chromosome replication, the first meiotic division produces two cells that are arranged next to each other within the sac. The second meiotic division then produces four cells that are also arranged in a row. Due to the tight enclosure of the sac around the cells, each pair of daughter cells is forced to lie next to each other in a linear fashion. Likewise, when these four cells divide by mitosis, each pair of daughter cells is located next to each other.

In species that make ordered tetrads or octads, experimenters can determine the genotypes of the spores within the asci and map the distance between a single gene and the centromere. Because the location of the centromere can be seen under the microscope, the mapping of a gene relative to the centromere provides a way to correlate a gene's location with the cytological characteristics of a chromosome. This approach has been extensively exploited in *N. crassa.*

Figure 6.14 compares the arrangement of cells within a *Neurospora* ascus depending on whether or not a crossover has occurred between two homologs that differ at a gene with alleles *A* (orange pigmentation) and *a* (albino, which results in a white phenotype). In Figure 6.14a, a crossover has not occurred, so the octad contains a linear arrangement of four haploid cells carrying the *A* allele, which are adjacent to four haploid cells that contain the *a* allele. This 4:4 arrangement of spores within the ascus is called **first-division segregation (FDS),** or an M1 pattern. It is called a first-division segregation pattern because the *A* and *a* alleles have segregated from each other after the first meiotic division.

In contrast, as shown in Figure 6.14b, if a crossover occurs between the centromere and the gene of interest, the ordered octad will deviate from the 4:4 pattern. Depending on the relative locations of the two chromatids that participated in the

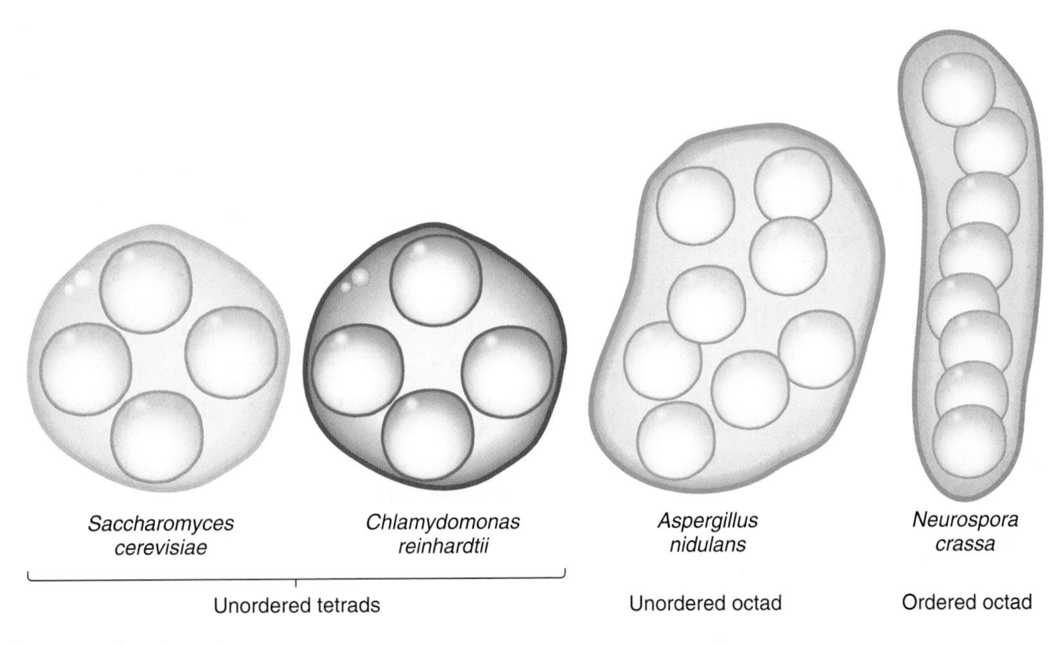

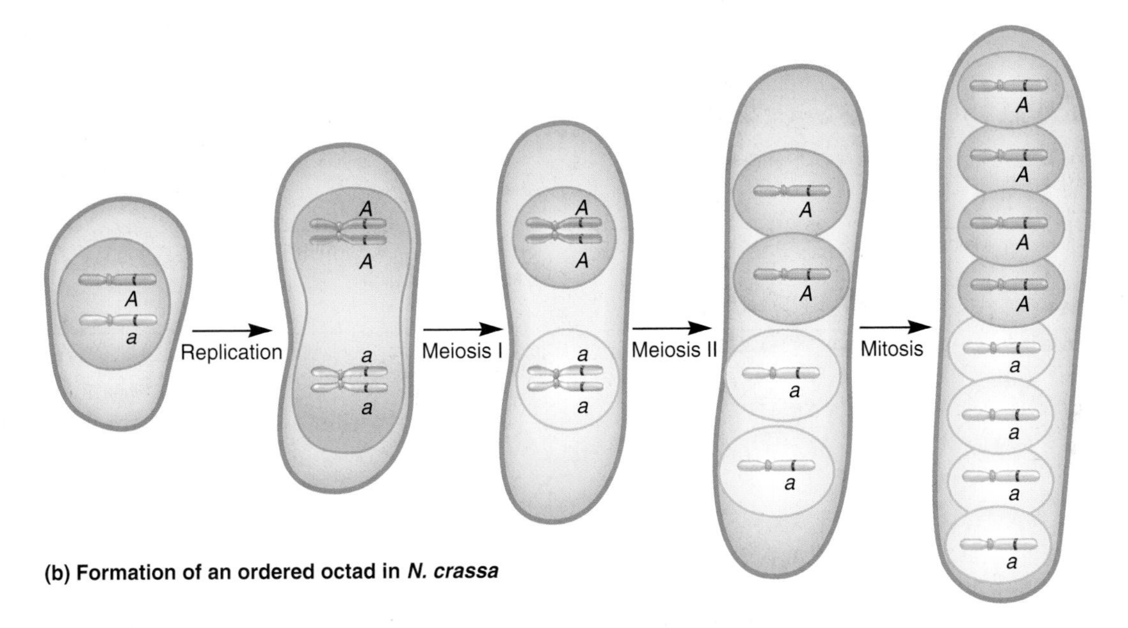

FIGURE 6.13 Arrangement of spores within asci of different species. **(a)** *Saccharomyces cerevisiae* and *Chlamydomonas reinhardtii* (an alga) produce unordered tetrads, *Aspergillus nidulans* produces an unordered octad, and *Neurospora crassa* produces an ordered octad. **(b)** Ordered octads are produced in *N. crassa* by meiosis and mitosis in such a way that the eight resulting cells are arranged linearly.

crossover, the ascus will contain a 2:2:2:2 or 2:4:2 pattern. These patterns are called **second-division segregation (SDS),** or M2 patterns. In this case, the *A* and *a* alleles do not segregate until the second meiotic division is completed.

Because a pattern of second-division segregation is a result of crossing over, the percentage of SDS asci can be used to calculate the map distance between the centromere and the gene of interest. To understand why this is possible, let's consider the relationship between a crossover site and the centromere. As shown in **Figure 6.15**, a crossover will separate a gene from its original centromere only if it begins in the region between the centromere and that gene. Therefore, the chances of getting a 2:2:2:2 or 2:4:2 pattern depend on the distance between the gene of interest and the centromere.

To determine the map distance between the centromere and a gene, the experimenter must count the number of SDS asci and the total number of asci. In SDS asci, only half of the spores are actually the product of a crossover. Therefore, the map distance between the gene of interest and the centromere is calculated as

$$\text{Map distance} = \frac{(1/2)\ (\text{Number of SDS asci})}{\text{Total number of asci}} \times 100$$

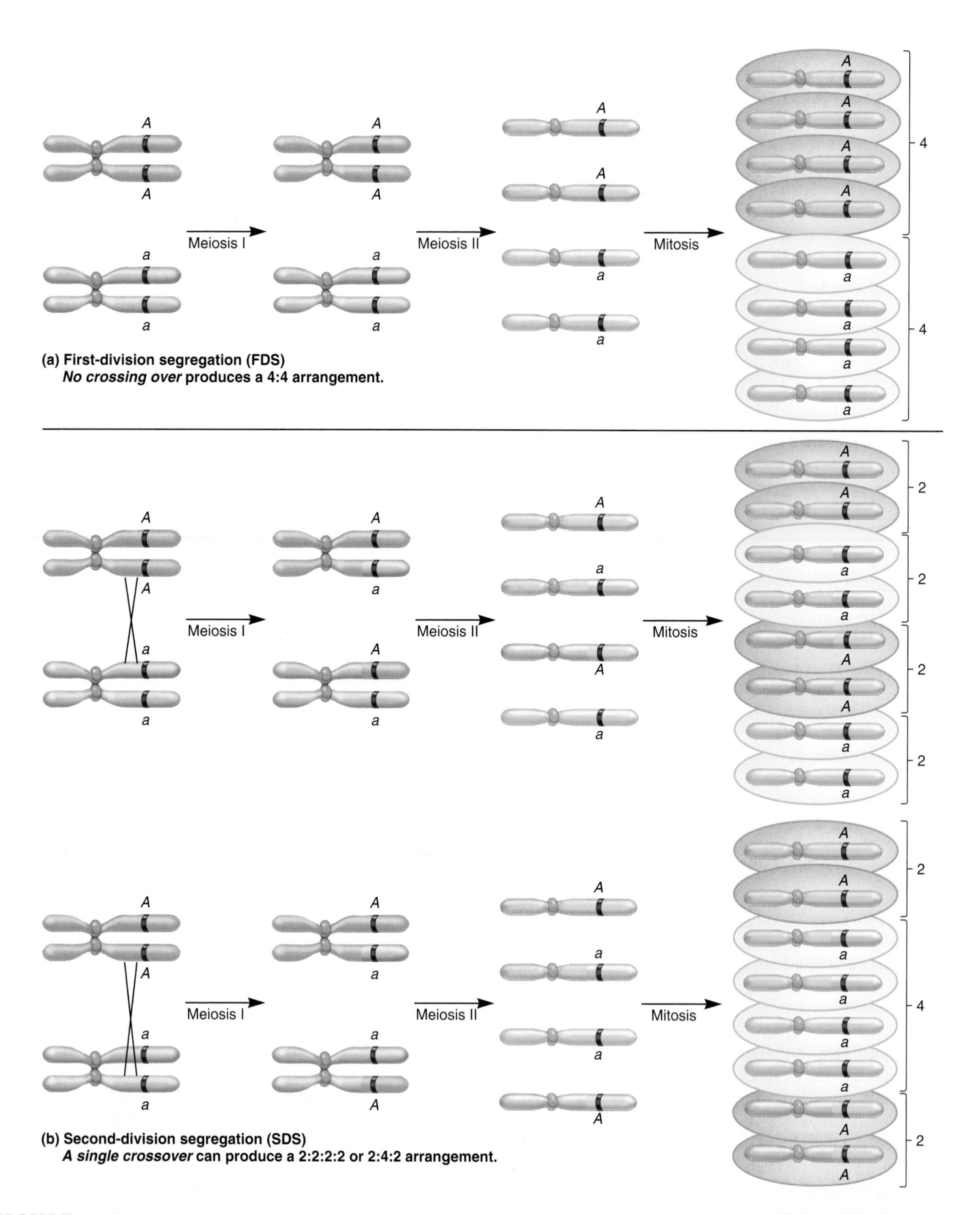

FIGURE 6.14 A comparison of the arrangement of cells within an ordered octad, depending on whether or not crossing over has occurred. **(a)** If no crossing over has occurred, the octad will have a 4:4 arrangement of spores known as an FDS or M1 pattern. **(b)** If a crossover has occurred between the centromere and the gene of interest, a 2:2:2:2 or 2:4:2 pattern, known as an SDS or M2 pattern, is observed.

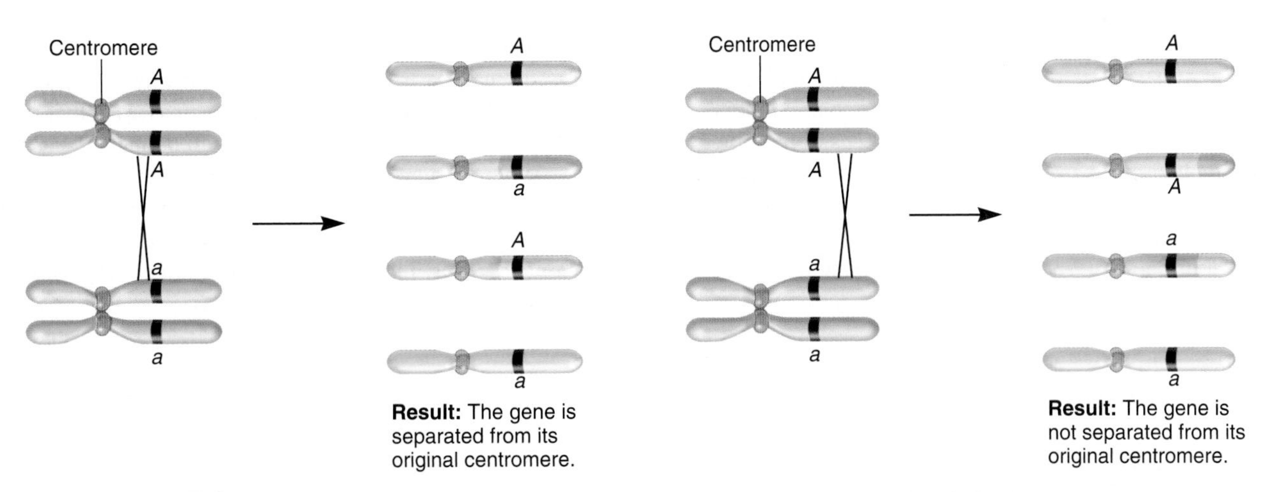

FIGURE 6.15 The relationship between a crossover site and the separation of an allele from its original centromere. (a) If a crossover initially forms between the centromere and the gene of interest, the gene will be separated from its original centromere. (b) If a crossover initiates outside this region, the gene will remain attached to its original centromere.

Unordered Tetrad Analysis Can Be Used to Map Genes in Dihybrid Crosses

Unordered tetrads contain a group of spores that are the product of meiosis and randomly arranged in an ascus. An experimenter can conduct a dihybrid cross, remove the spores from each ascus, and determine the phenotypes of the spores. This analysis can determine if two genes are linked or assort independently. If two genes are linked, a tetrad analysis can also be used to compute map distance.

Figure 6.16 illustrates the possible outcomes starting with two haploid yeast strains. One strain carries the wild-type alleles ura^+ and arg^+, which are required for uracil and arginine biosynthesis,

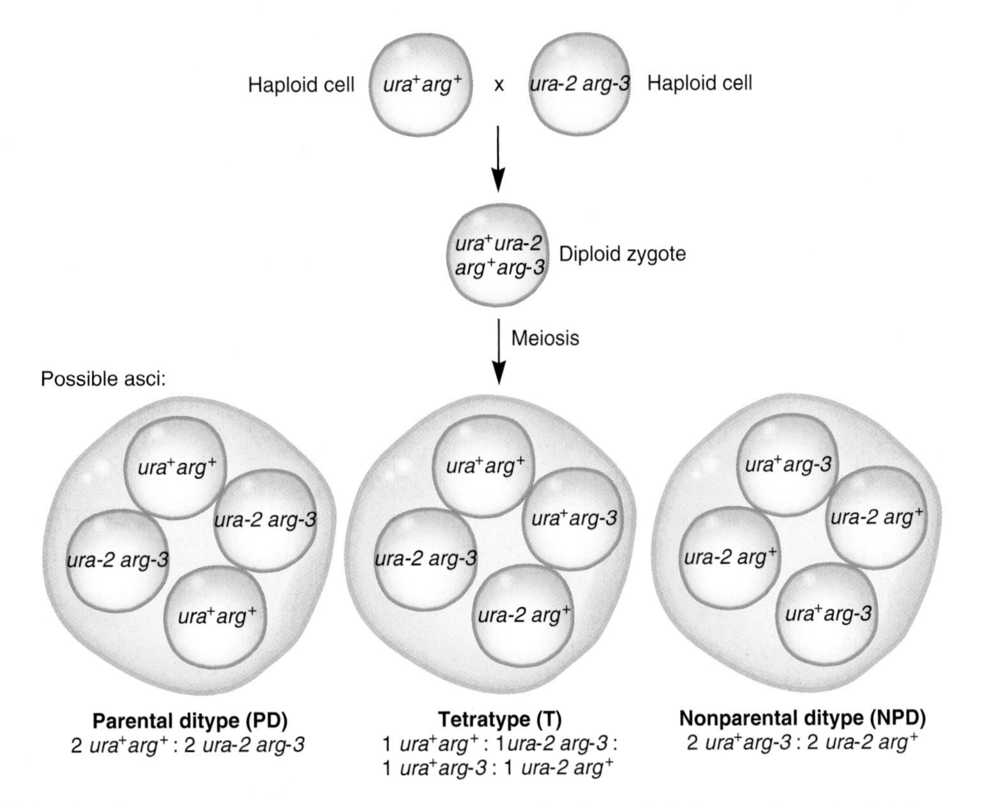

FIGURE 6.16 The assortment of two genes in an unordered tetrad. If the tetrad contains 100% parental cells, this ascus has the parental ditype (PD). If it contains 50% parental and 50% recombinant cells, it is a tetratype (T). Finally, an ascus with 100% recombinant cells is called a nonparental ditype (NPD). This figure does not illustrate the chromosomal locations of the alleles. In this type of experiment, the goal is to determine whether the two genes are linked on the same chromosome.

respectively. The other strain has defective alleles *ura-2* and *arg-3*; these result in yeast strains that require uracil and arginine in the growth medium. A diploid zygote with the genotype $ura^+ ura\text{-}2\ arg^+\ arg\text{-}3$ was produced from the fusion of haploid cells from these two strains. The diploid cell then proceeds through meiosis to produce four haploid cells. After the completion of meiosis, three distinct types of tetrads could be produced. One possibility is that the tetrad will contain four spores with the parental combinations of alleles. This ascus is said to have the **parental ditype (PD).** Alternatively, an ascus may have two parental cells and two nonparental cells, which is called a **tetratype (T).** Finally, an ascus with a **nonparental ditype** (NPD) contains four cells with nonparental genotypes.

When two genes assort independently, the number of asci having a parental ditype is expected to equal the number having a nonparental ditype, thus yielding 50% recombinant spores. For linked genes, **Figure 6.17** illustrates the relationship between

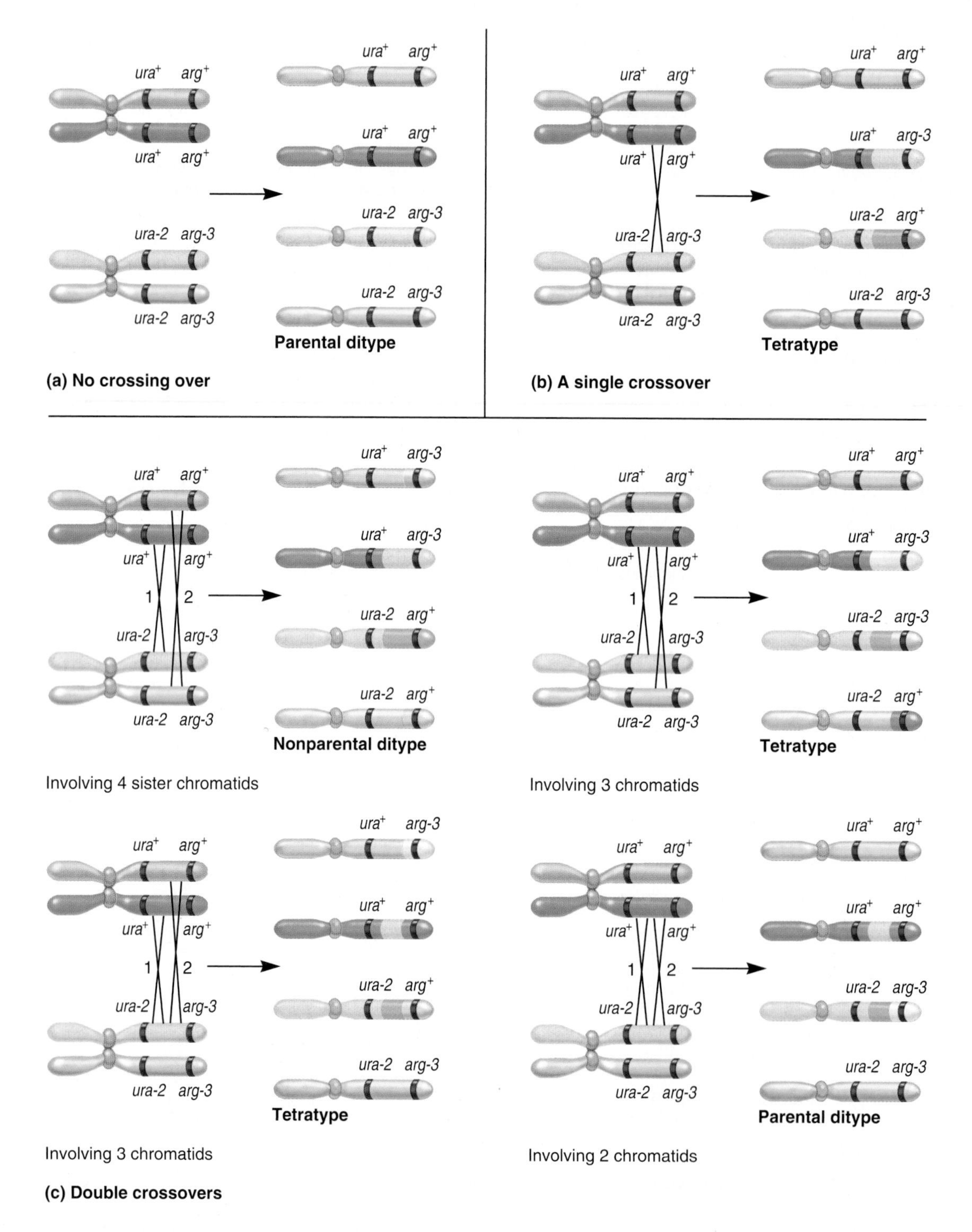

FIGURE 6.17 **Relationship between crossing over and the production of the parental ditype, tetratype, and nonparental ditype for two linked genes.**

crossing over and the type of ascus that will result. If no crossing over occurs in the region between the two genes, the parental ditype will be produced (Figure 6.17a). A single crossover event produces a tetratype (Figure 6.17b). Double crossovers can yield a parental ditype, tetratype, or nonparental ditype, depending on the combination of chromatids that are involved (Figure 6.17c). A nonparental ditype is produced when a double crossover involves all four chromatids. A tetratype results from a three-chromatid crossover. Finally, a double crossover between the same two chromatids produces the parental ditype.

The data from a tetrad analysis can be used to calculate the map distance between two linked genes. As in conventional mapping, the map distance is calculated as the percentage of offspring that carry recombinant chromosomes. As mentioned, a tetratype contains 50% recombinant chromosomes; a nonparental ditype, 100%. Therefore, the map distance is computed as

$$\text{Map distance} = \frac{\text{NPD} + (1/2)\,(\text{T})}{\text{total number of asci}} \times 100$$

Over short map distances, this calculation provides a fairly reliable measure of distance. However, it does not adequately account for double crossovers. When two genes are far apart on the same chromosome, the calculated map distance using this equation underestimates the actual map distance due to double crossovers. Fortunately, a particular strength of tetrad analysis is that we can derive another equation that accounts for double crossovers and thereby provides a more accurate value for map distance. To begin this derivation, let's consider a more precise way to calculate map distance.

$$\text{Map distance} = \frac{\text{Single crossover tetrads} + (2)\,(\text{Double crossover tetrads})}{\text{Total number of asci}} \times 0.5 \times 100$$

This equation includes the number of single and double crossovers in the computation of map distance. The total number of crossovers equals the number of single crossovers plus 2 times the number of double crossovers. Overall, the tetrads that contain single and double crossovers also contain 50% nonrecombinant chromosomes. To calculate map distance, therefore, we divide the total number of crossovers by the total number of asci and multiply by 0.5 and 100.

To be useful, we need to relate this equation to the number of parental ditypes, nonparental ditypes, and tetratypes that are obtained by experimentation. To derive this relationship, we must consider the types of tetrads that are produced from no crossing over, a single crossover, and double crossovers. To do so, let's take another look at Figure 6.17. As shown there, the parental ditype and tetratype are ambiguous. The parental ditype can be derived from no crossovers or a double crossover; the tetratype can be derived from a single crossover or a double crossover. However, the nonparental ditype is unambiguous, because it can be produced only from a double crossover. We can use this observation as a way to determine the actual number of single and double crossovers. As seen in Figure 6.17, 1/4 of all the double crossovers are nonparental ditypes. Therefore, the total number of double crossovers equals four times the number of nonparental ditypes.

Next, we need to know the number of single crossovers. A single crossover yields a tetratype, but double crossovers can also yield a tetratype. Therefore, the total number of tetratypes overestimates the true number of single crossovers. Fortunately, we can compensate for this overestimation. Because two types of tetratypes are due to a double crossover, the actual number of tetratypes arising from a double crossover should equal 2NPD. Therefore, the true number of single crossovers is calculated as T – 2NPD.

Now we have accurate measures of both single and double crossovers. The number of single crossovers equals T – 2NPD, and the number of double crossovers equals 4NPD. We can substitute these values into our previous equation.

$$\text{Map distance} = \frac{(\text{T} - 2\text{NPD}) + (2)\,(4\text{NPD})}{\text{Total number of asci}} \times 0.5 \times 100$$

$$= \frac{\text{T} + 6\text{NPD}}{\text{Total number of asci}} \times 0.5 \times 100$$

This equation provides a more accurate measure of map distance because it considers both single and double crossovers.

6.4 MITOTIC RECOMBINATION

Thus far, we have considered how the arrangement of linked alleles along a chromosome can be rearranged by crossing over. This event can produce cells and offspring with a nonparental combination of traits. In these previous cases, crossing over has occurred during meiosis, when the homologous chromosomes replicate and form bivalents.

In multicellular organisms, the union of egg and sperm is followed by many cellular divisions, which occur in conjunction with mitotic divisions of the cell nuclei. As discussed in Chapter 3, mitosis normally does not involve the homologous pairing of chromosomes to form a bivalent. Therefore, crossing over during mitosis is expected to occur much less frequently than during meiosis. Nevertheless, it does happen on rare occasions. Mitotic crossing-over may produce a pair of recombinant chromosomes that have a new combination of alleles, an event known as **mitotic recombination.** If it occurs during an early stage of embryonic development, the daughter cells containing the recombinant chromosomes continue to divide many times to produce a patch of tissue in the adult. This may result in a portion of tissue with characteristics different from those of the rest of the organism.

In 1936, Curt Stern identified unusual patches on the bodies of certain *Drosophila* strains. He was working with strains carrying X-linked alleles affecting body color and bristle morphology (**Figure 6.18**). A recessive allele confers yellow body color (*y*), and another recessive allele causes shorter body bristles that look singed (*sn*). The corresponding wild-type alleles result in gray body color (y^+) and long bristles (sn^+). Females that are $y^+y\ sn^+sn$ are expected to have gray body color and long bristles. This was generally the case. However, when Stern carefully observed the bodies of these female flies under a low-power microscope, he occasionally noticed places in which two adjacent regions were different from

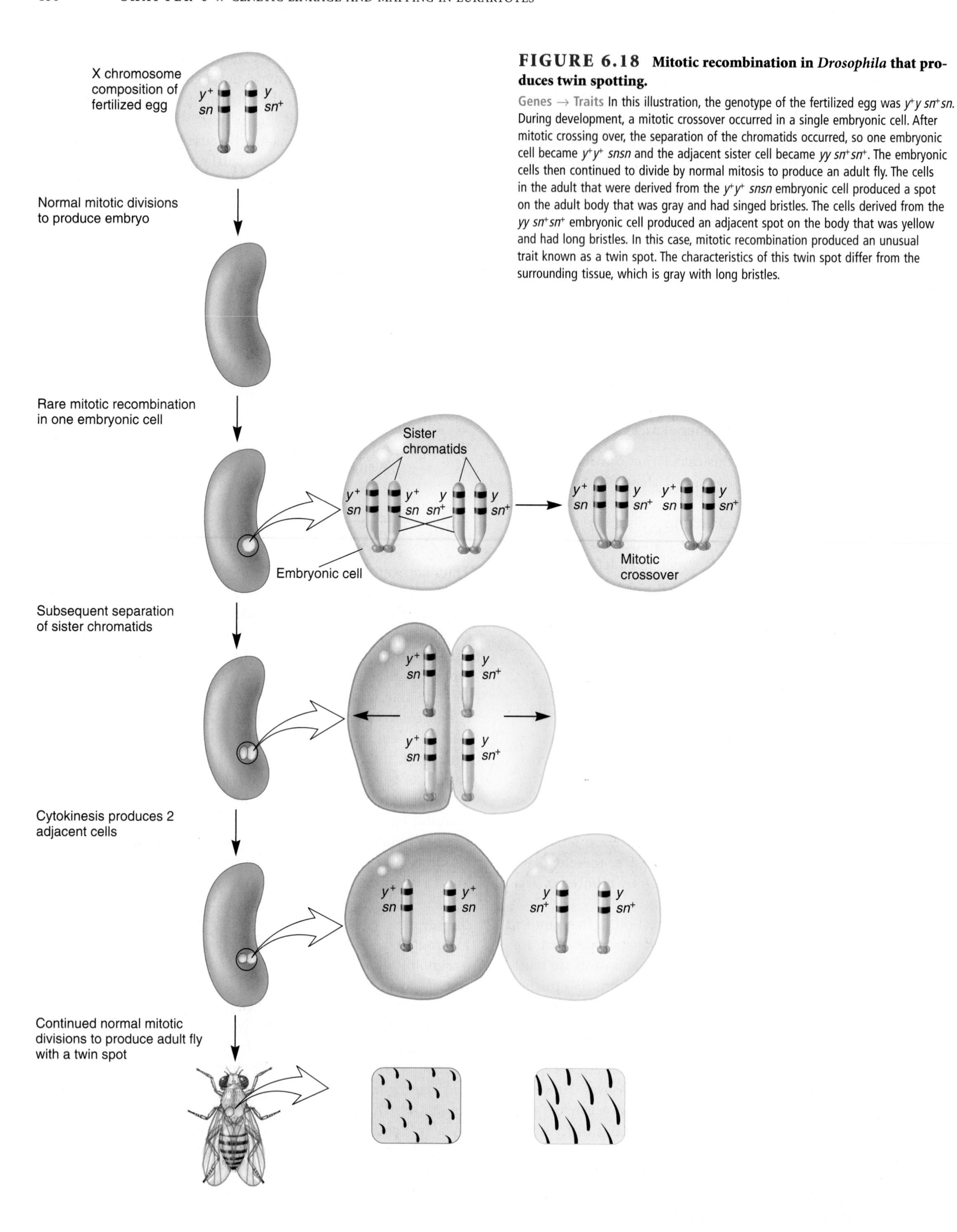

FIGURE 6.18 Mitotic recombination in *Drosophila* that produces twin spotting.

Genes → Traits In this illustration, the genotype of the fertilized egg was $y^+y\ sn^+sn$. During development, a mitotic crossover occurred in a single embryonic cell. After mitotic crossing over, the separation of the chromatids occurred, so one embryonic cell became $y^+y^+\ snsn$ and the adjacent sister cell became $yy\ sn^+sn^+$. The embryonic cells then continued to divide by normal mitosis to produce an adult fly. The cells in the adult that were derived from the $y^+y^+\ snsn$ embryonic cell produced a spot on the adult body that was gray and had singed bristles. The cells derived from the $yy\ sn^+sn^+$ embryonic cell produced an adjacent spot on the body that was yellow and had long bristles. In this case, mitotic recombination produced an unusual trait known as a twin spot. The characteristics of this twin spot differ from the surrounding tissue, which is gray with long bristles.

the rest of the body—a twin spot. He concluded that twin spotting was too frequent to be explained by the random positioning of two independent single spots that happened to occur close together. How then did Stern explain the phenomenon of twin spotting? He proposed that twin spots are due to a single mitotic recombination within one cell during embryonic development.

As shown in Figure 6.18, the X chromosomes of the fertilized egg are y^+ *sn* and *y* sn^+. During development, a rare crossover can occur during mitosis to produce two adjacent daughter cells that are y^+y^+ *snsn* and *yy* sn^+sn^+. As embryonic development proceeds, the cell on the left continues to divide to produce many cells, eventually producing a patch on the body that has gray color with singed bristles. The daughter cell next to it produces a patch of yellow body color with long bristles. These two adjacent patches—a twin spot—are surrounded by cells that are y^+y sn^+sn and have gray color and long bristles. Twin spots provide evidence that mitotic recombination occasionally occurs.

KEY TERMS

Page 126. genetic map, synteny, genetic linkage
Page 127. linkage groups, dihybrid cross, trihybrid cross, crossing over, bivalent, genetic recombination, nonparental cells, recombinant cells, parental offspring, nonrecombinant offspring
Page 130. null hypothesis
Page 136. genetic mapping, locus
Page 137. genetic linkage map, testcross, map distance, map units (mu), centiMorgans (cM)
Page 143. positive interference, spores, tetrad, octad, ascus
Page 144. unordered tetrad, unordered octad, ordered tetrad, ordered octad, first-division segregation (FDS)
Page 145. second-division segregation (SDS)
Page 148. parental ditype (PD), tetratype (T), nonparental ditype (NPD)
Page 149. mitotic recombination

CHAPTER SUMMARY

6.1 Linkage and Crossing Over

- Synteny refers to genes that are located on the same chromosome. Genetic linkage means that the alleles of two or more genes tend to be transmitted as a unit because they are relatively close on the same chromosome.
- Crossing over can change the combination of alleles along a chromosome and produce nonparental, or recombinant, cells and offspring (see Figure 6.1).
- Bateson and Punnett discovered the first example of genetic linkage in sweet peas (see Figure 6.2).
- Morgan also discovered genetic linkage in *Drosophila* and proposed that nonparental offspring are produced by crossing over during meiosis (see Figures 6.3, 6.4).
- When genes are linked, the relative proportions of nonparental offspring depends on the distance between the genes (see Figure 6.5).
- A chi square analysis can be followed to judge whether or not two genes assort independently.
- Creighton and McClintock were able to correlate the formation of nonparental offspring with the presence of chromosomes that had exchanged pieces due to crossing over (see Figures 6.6, 6.7).

6.2 Genetic Mapping in Plants and Animals

- A genetic linkage map is a diagram that portrays the order and relative spacing of genes along one or more chromosomes (see Figure 6.8).
- A testcross can be performed to map the distance between two or more genes (see Figure 6.9).
- Sturtevant was the first scientist to conduct testcrosses and map the order of a few genes along the X chromosome in *Drosophila* (see Figure 6.10).
- Due to the effects of multiple crossovers, the map distance between two genes obtained from a testcross cannot exceed 50% (see Figure 6.11).
- The data from a trihybrid cross can be used to map genes (see Table 6.1).
- Positive interference refers to the phenomenon that the number of double crossovers in a given region is less than expected based on the frequency of single crossovers.

6.3 Genetic Mapping in Haploid Eukaryotes

- Several haploid eukaryotes have been used in genetic mapping. Ascomycetes have the product of a single meiosis contained with an ascus (see Figure 6.12).
- Certain haploid species may form unordered or ordered tetrads or octads (see Figure 6.13).
- The arrangement of alleles found in spores of an ordered octad depends on whether crossing over has occurred. The arrangement can be used to map the distance between a gene and the centromere (see Figures 6.14, 6.15).
- The analysis of unordered tetrads in yeast can be used to map the distance between two linked genes (see Figure 6.16, 6.17).

6.4 Mitotic Recombination

- Mitotic recombination can occur rarely and may lead to twin spots (see Figure 6.18).

PROBLEM SETS & INSIGHTS

Solved Problems

S1. In the garden pea, orange pods (*orp*) are recessive to green pods (*Orp*), and sensitivity to pea mosaic virus (*mo*) is recessive to resistance to the virus (*Mo*). A plant with orange pods and sensitivity to the virus was crossed to a true-breeding plant with green pods and resistance to the virus. The F_1 plants were then testcrossed to plants with orange pods and sensitivity to the virus. The following results were obtained:

160 orange pods, virus sensitive
165 green pods, virus resistant
36 orange pods, virus resistant
39 green pods, virus sensitive
400 total

A. Conduct a chi square analysis to see if these genes are linked.

B. If they are linked, calculate the map distance between the two genes.

Answer:

A. Chi square analysis.

1. Our hypothesis is that the genes are not linked.
2. Calculate the predicted number of offspring based on the hypothesis. The testcross is

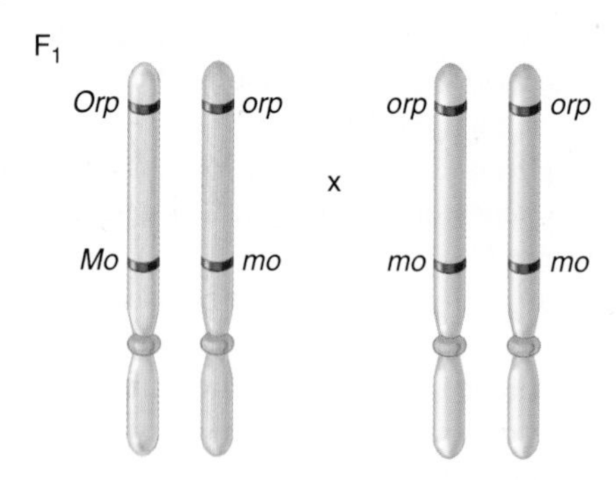

The predicted outcome of this cross under our hypothesis is a 1:1:1:1 ratio of plants with the four possible phenotypes. In other words, 1/4 should have the phenotype orange pods, virus-sensitive; 1/4 should have green pods, virus-resistant; 1/4 should have orange pods, virus-resistant; and 1/4 should have green pods, virus-sensitive. Because a total of 400 offspring were produced, our hypothesis predicts 100 offspring in each category.

3. Calculate the chi square.

$$\chi^2 = \frac{(O_1 - E_1)^2}{E_1} + \frac{(O_2 - E_2)^2}{E_2} + \frac{(O_3 - E_3)^2}{E_3} + \frac{(O_4 - E_4)^2}{E_4}$$

$$\chi^2 = \frac{(160 - 100)^2}{100} + \frac{(165 - 100)^2}{100} + \frac{(36 - 100)^2}{100} + \frac{(39 - 100)^2}{100}$$

$$\chi^2 = 36 + 42.3 + 41 + 37.2 = 156.5$$

4. Interpret the chi square value. The calculated chi square value is quite large. This indicates that the deviation between observed and expected values is very high. For 1 degree of freedom in Table 2.1, such a large deviation is expected to occur by chance alone less than 1% of the time. Therefore, we reject the hypothesis that the genes assort independently. As an alternative, we may infer that the two genes are linked.

B. Calculate the map distance.

$$\text{Map distance} = \frac{(\text{Number of nonparental offspring})}{\text{Total number of offspring}} \times 100$$

$$= \frac{36 + 39}{36 + 39 + 160 + 165} \times 100$$

$$= 18.8 \text{ mu}$$

The genes are approximately 18.8 mu apart.

S2. Two recessive disorders in mice—droopy ears and flaky tail—are caused by genes that are located 6 mu apart on chromosome 3. A true-breeding mouse with normal ears (*De*) and a flaky tail (*ft*) was crossed to a true-breeding mouse with droopy ears (*de*) and a normal tail (*Ft*). The F_1 offspring were then crossed to mice with droopy ears and flaky tails. If this testcross produced 100 offspring, what is the expected outcome?

Answer: The testcross is

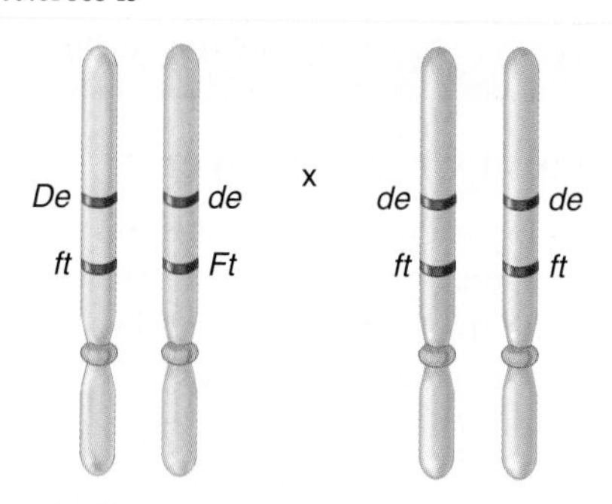

The parental offspring are

Dede ftft Normal ears, flaky tail
dede Ftft Droopy ears, normal tail

The recombinant offspring are

dede ftft Droopy ears, flaky tail
Dede Ftft Normal ears, normal tail

Because the two genes are located 6 mu apart on the same chromosome, 6% of the offspring will be recombinants. Therefore, the expected outcome for 100 offspring is

3 droopy ears, flaky tail
3 normal ears, normal tail
47 normal ears, flaky tail
47 droopy ears, normal tail

S3. The following X-linked recessive traits are found in fruit flies: vermilion eyes are recessive to red eyes, miniature wings are recessive to long wings, and sable body is recessive to gray body. A cross was made between wild-type males with red eyes, long wings, and gray bodies to females with vermilion eyes, miniature wings, and sable bodies. The heterozygous females from this cross, which had red eyes, long wings, and gray bodies, were then crossed to males with vermilion eyes, miniature wings, and sable bodies. The following outcome was obtained:

Males and Females

1320 vermilion eyes, miniature wings, sable body

1346 red eyes, long wings, gray body

102 vermilion eyes, miniature wings, gray body

90 red eyes, long wings, sable body

42 vermilion eyes, long wings, gray body

48 red eyes, miniature wings, sable body

2 vermilion eyes, long wings, sable body

1 red eyes, miniature wings, gray body

A. Calculate the map distance between the three genes.

B. Is positive interference occurring?

Answer:

A. The first step is to determine the order of the three genes. We can do this by evaluating the pattern of inheritance in the double crossovers. The double crossover group occurs with the lowest frequency. Thus, the double crossovers are vermilion eyes, long wings, and sable body, and red eyes, miniature wings, and gray body. Compared with the parental combinations of alleles (vermilion eyes, miniature wings, sable body and red eyes, long wings, gray body), the gene for wing length has been reassorted. Two flies have long wings associated with vermilion eyes and sable body, and one fly has miniature wings associated with red eyes and gray body. Taken together, these results indicate that the wing length gene is found in between the eye color and body color genes.

Eye color——wing length——body color

We now calculate the distance between eye color and wing length, and between wing length and body color. To do this, we consider the data according to gene pairs:

vermilion eyes, miniature wings = 1320 + 102 = 1422

red eyes, long wings = 1346 + 90 = 1436

vermilion eyes, long wings = 42 + 2 = 44

red eyes, miniature wings = 48 + 1 = 49

The recombinants are vermilion eyes, long wings and red eyes, miniature wings. The map distance between these two genes is

(44 + 49)/(1422 + 1436 + 44 + 49) × 100 = 3.2 mu

Likewise, the other gene pair is wing length and body color.

miniature wings, sable body = 1320 + 48 = 1368

long wings, gray body = 1346 + 42 = 1388

miniature wings, gray body = 102 + 1 = 103

long wings, sable body = 90 + 2 = 92

The recombinants are miniature wings, gray body and long wings, sable body. The map distance between these two genes is

(103 + 92)/(1368 + 1388 + 103 + 92) × 100 = 6.6 mu

With these data, we can produce the following genetic map:

3.2　6.6

v　*m*　*s*

B. To calculate the interference value, we must first calculate the coefficient of coincidence.

$$C = \frac{\text{Observed number of double crossovers}}{\text{Expected number of double crossovers}}$$

Based on our calculation of map distances in part A, the percentage of single crossovers equals 3.2% (0.032) and 6.6% (0.066). The expected number of double crossovers equals 0.032 × 0.066, which is 0.002, or 0.2%. A total of 2951 offspring were produced. If we multiply 2951 × 0.002, we get 5.9, which is the expected number of double crossovers. The observed number was 3. Therefore,

$C = 3/5.9 = 0.51$

$I = 1 - C = 1 - 0.51 = 0.49$

In other words, approximately 49% of the expected double crossovers did not occur due to interference.

S4. Around the same time as the study of Creighton and McClintock, described in Figure 6.7, Curt Stern conducted similar experiments with *Drosophila*. He had strains of flies with microscopically detectable abnormalities in the X chromosome. In one case, the X chromosome was shorter than normal due to a deletion at one end. In another case, the X chromosome was longer than normal because an extra piece of the Y chromosome was attached at the other end of the X chromosome, where the centromere is located. He had female flies that had both abnormal chromosomes. On the short X chromosome, a recessive allele (*car*) was located that results in carnation-colored eyes, and a dominant allele (*B*) that causes bar-shaped eyes was also found on this chromosome. On the long X chromosome were located the wild-type alleles for these two genes (designated car^+ and B^+), which confer red eyes and round eyes, respectively. Stern realized that a crossover between the two X chromosomes in such female flies would result in recombinant chromosomes that would be cytologically distinguishable from the parental chromosomes. If a crossover occurred between the *B* and *car* genes on the X chromosome, this is expected to produce a normal-sized X chromosome and an abnormal chromosome with a deletion at one end and an extra piece of the Y chromosome at the other end.

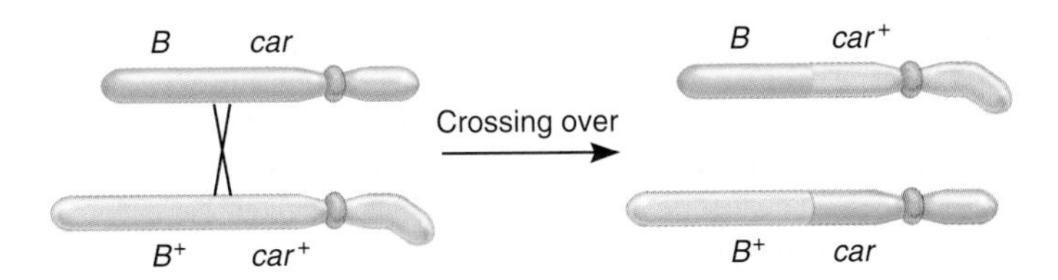

Stern crossed these female flies to male flies that had a normal-length X chromosome with the *car* allele and the allele for round eyes (*car* B^+). Using a microscope, he could discriminate between the morphologies of parental chromosomes—like those contained within the original parental flies—and recombinant chromosomes that may be found in the offspring. What would be the predicted phenotypes and chromosome characteristics in the offspring if crossing over did or did not occur between the X chromosomes in the female flies of this cross?

Answer: To demonstrate that genetic recombination is due to crossing over, Stern needed to correlate recombinant phenotypes (due to genetic recombination) with the inheritance of recombinant chromosomes (due to crossing over). Because he knew the arrangement of alleles in the female flies, he could predict the phenotypes of parental and nonparental offspring. The male flies could contribute the *car* and B^+ alleles (on a cytologically normal X chromosome) or contribute a Y chromosome. In the absence of crossing over, the female flies could contribute a short X chromosome with the *car* and *B* alleles or a long X chromosome with the car^+ and B^+ alleles. If crossing over occurred in the region between these two genes, the female flies would contribute recombinant X chromosomes. One possible recombinant X chromosome would be normal-sized and carry the *car* and B^+ alleles, and the other recombinant X chromosome would be deleted at one end with a piece of the Y chromosome at the other end and carry the car^+ and *B* alleles. When combined with an X or Y chromosome from the males, the parental offspring would have carnation, bar eyes or wild-type eyes; the nonparental offspring would have carnation, round eyes or red, bar eyes.

Female gametes ♀ \ Male gametes ♂	$carB^+$	Y	Phenotype	X chromosome from female
$carB$	$carB$ / $carB^+$	$carB$ / Y	Carnation, bar eyes	Short X chromosome
car^+B^+	car^+B^+ / $carB^+$	car^+B^+ / Y	Red, round eyes	Long X chromosome with a piece of Y
$carB^+$	$carB^+$ / $carB^+$	$carB^+$ / Y	Carnation, round eyes	Normal-sized X chromosome
car^+B	car^+B / $carB^+$	car^+B / Y	Red, bar eyes	Short X chromosome with a piece of Y

The results shown in the Punnett square are the actual results that Stern observed. His interpretation was that crossing over between homologous chromosomes—in this case, the X chromosome—accounts for the formation of offspring with recombinant phenotypes.

S5. Researchers have discovered a limit to the relationship between map distance and the percentage of recombinant offspring. Even though two genes on the same chromosome may be much more than 50 mu apart, we do not expect to obtain greater than 50% recombinant offspring in a testcross. You may be wondering why this is so. The answer lies in the pattern of multiple crossovers. At the pachytene stage of meiosis, a single crossover in the region between two genes produces only 50% recombinant chromosomes (see Figure 6.1b). Therefore, to exceed a 50% recombinant level, it would seem necessary to have multiple crossovers within the tetrad.

Let's suppose that two genes are far apart on the same chromosome. A testcross is made between a heterozygous individual, *AaBb*, and a homozygous individual, *aabb*. In the heterozygous individual, the dominant alleles (*A* and *B*) are linked on the same chromosome, and the recessive alleles (*a* and *b*) are linked on the same chromosome. Draw out all of the possible double crossovers (between two, three, or four chromatids) and determine the average number of recombinant offspring, assuming an equal probability of all of the double crossover possibilities.

Answer: A double crossover between the two genes could involve two chromatids, three chromatids, or four chromatids. The possibilities for all types of double crossovers are shown here:

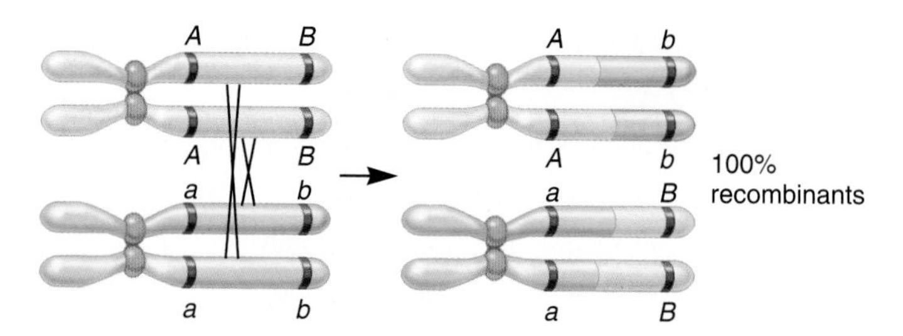

Double crossover (involving 4 chromatids)

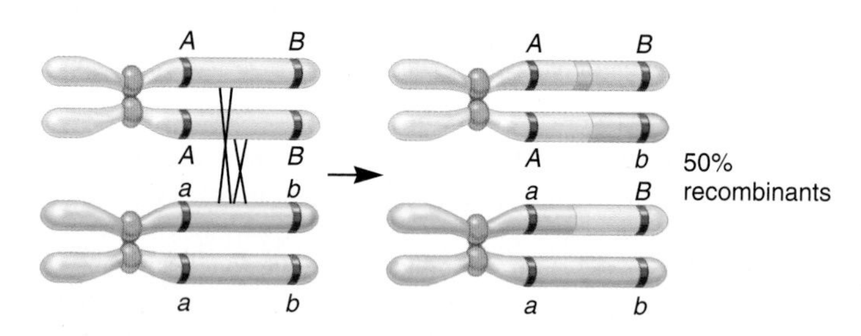

Double crossover (involving 3 chromatids)

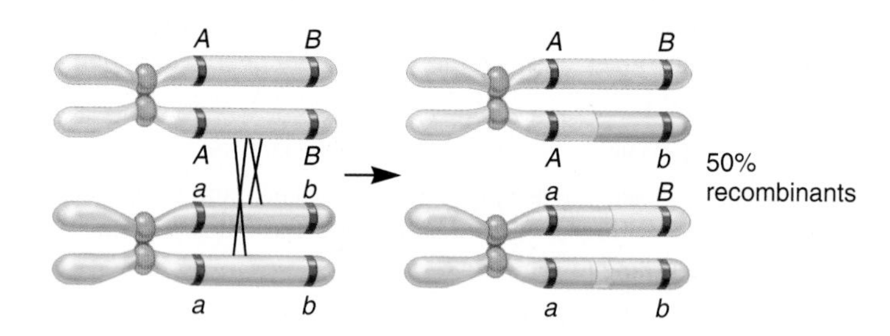

Double crossover (involving 3 chromatids)

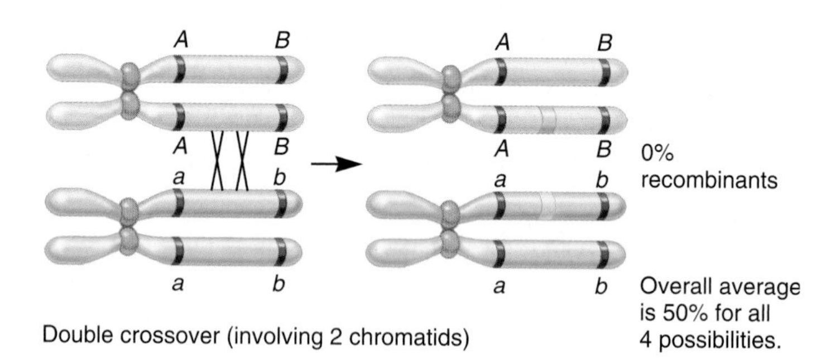

Double crossover (involving 2 chromatids)

This drawing considers the situation where two crossovers are expected to occur in the region between the two genes. Because the tetrad is composed of two pairs of homologs, a double crossover between homologs could occur in several possible ways. In this illustration, the crossover on the right has occurred first. Because all of these double crossing over events are equally probable, we take the average of them to determine the maximum recombination frequency. This average equals 50%.

Conceptual Questions

C1. What is the difference in meaning between the terms genetic recombination and crossing over?

C2. When applying a chi square approach in a linkage problem, explain why an independent assortment hypothesis is used.

C3. What is mitotic recombination? A heterozygous individual (*Bb*) with brown eyes has one eye with a small patch of blue. Provide two or more explanations for how the blue patch may have occurred.

C4. Mitotic recombination can occasionally produce a twin spot. Let's suppose an animal species can be heterozygous for two genes that govern fur color and length: One gene affects pigmentation, with dark pigmentation (*A*) dominant to albino (*a*); the other gene affects hair length, with long hair (*L*) dominant to short hair (*l*). The two genes are linked on the same chromosome. Let's assume an animal is *AaLl*; *A* is linked to *l*, and *a* is linked to *L*. Draw the chromosomes labeled with these alleles, and explain how mitotic recombination could produce a twin spot with one spot having albino pigmentation and long fur, the other having dark pigmentation and short fur.

C5. A crossover has occurred in the bivalent shown here.

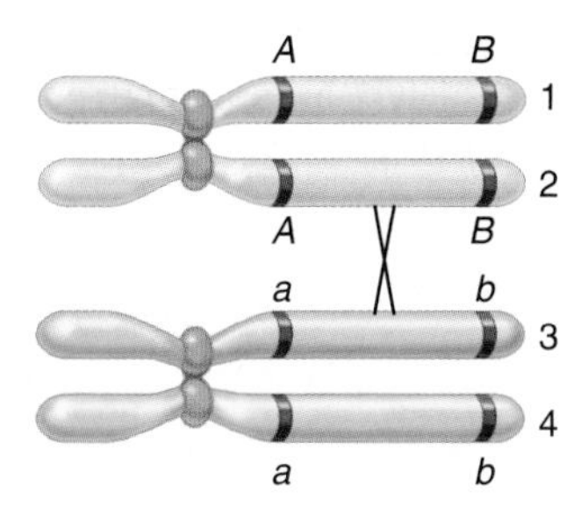

If a second crossover occurs in the same region between these two genes, which two chromatids would be involved to produce the following outcomes?

A. 100% recombinants

B. 0% recombinants

C. 50% recombinants

C6. A crossover has occurred in the bivalent shown here.

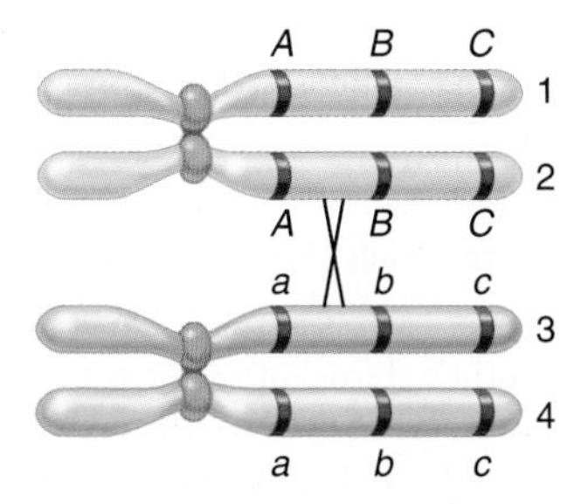

What is the outcome of this single crossover event? If a second crossover occurs somewhere between *A* and *C*, explain which two chromatids it would involve and where it would occur (i.e., between which two genes) to produce the types of chromosomes shown here:

A. *A B C*, *A b C*, *a B c*, and *a b c*

B. *A b c*, *A b c*, *a B C*, and *a B C*

C. *A B c*, *A b c*, *a B C*, and *a b C*

D. *A B C*, *A B C*, *a b c*, and *a b c*

C7. A diploid organism has a total of 14 chromosomes and about 20,000 genes per haploid genome. Approximately how many genes are in each linkage group?

C8. If you try to throw a basketball into a basket, the likelihood of succeeding depends on the size of the basket. It is more likely that you will get the ball into the basket if the basket is bigger. In your own words, explain how this analogy also applies to the idea that the likelihood of crossing over is greater when two genes are far apart than when they are close together.

C9. By conducting testcrosses, researchers have found that the sweet pea has seven linkage groups. How many chromosomes would you expect to find in leaf cells?

C10. In humans, a rare dominant disorder known as nail-patella syndrome causes abnormalities in the fingernails, toenails, and kneecaps. Researchers have examined family pedigrees with regard to this disorder and, within the same pedigree, also examined the individuals with regard to their blood types. (A description of blood genotypes is found in Chapter 4.) In the following pedigree, individuals affected with nail-patella disorder are shown with filled symbols. The genotype of each individual with regard to their ABO blood type is also shown. Does this pedigree suggest any linkage between the gene that causes nail-patella syndrome and the gene that causes blood type?

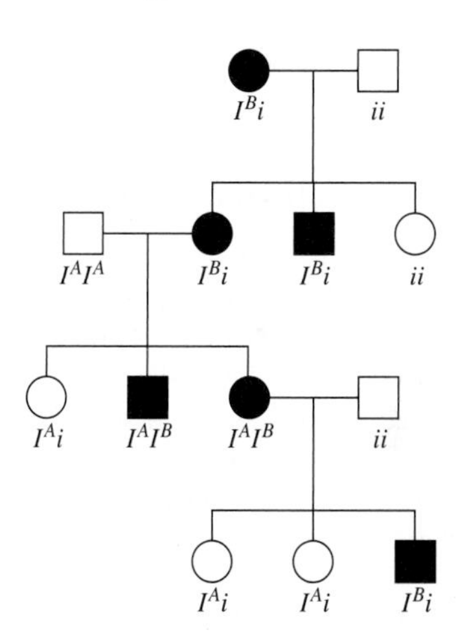

C11. When true-breeding mice with brown fur and short tails (*BBtt*) were crossed to true-breeding mice with white fur and long tails (*bbTT*), all F_1 offspring had brown fur and long tails. The F_1 offspring were crossed to mice with white fur and short tails. What are the possible phenotypes of the F_2 offspring? Which F_2 offspring are recombinant, and which are nonrecombinant? What are the ratios of the F_2 offspring if independent assortment is taking place? How are the ratios affected by linkage?

C12. Though we often think of genes in terms of the phenotypes they produce (e.g., curly leaves, flaky tail, brown eyes), the molecular function of most genes is to encode proteins. Many cellular proteins function as enzymes. The table that follows describes the map distances between six different genes that encode six different enzymes: *Ada*, adenosine deaminase; *Hao-1*, hydroxyacid oxidase-1; *Hdc*, histidine decarboxylase; *Odc-2*, ornithine decarboxylase-2; *Sdh-1*, sorbitol dehydrogenase-1; and *Ass-1*, arginosuccinate synthetase-1.

Map distances between two genes:

	Ada	*Hao-1*	*Hdc*	*Odc-2*	*Sdh-1*	*Ass-1*
Ada		14		8	28	
Hao-1	14		9		14	
Hdc		9		15	5	
Odc-2	8		15			63
Sdh-1	28	14	5			43
Ass-1				63	43	

Construct a genetic map that describes the locations of all six genes.

C13. If the likelihood of a single crossover in a particular chromosomal region is 10%, what is the theoretical likelihood of a double or triple crossover in that same region? How would positive interference affect these theoretical values?

C14. Except for fungi that form asci, in most dihybrid crosses involving linked genes, we cannot tell if a double crossover between the two genes has occurred because the offspring will inherit the parental combination of alleles. How does the inability to detect double crossovers affect the calculation of map distance? Is map distance underestimated or overestimated because of our inability to detect double crossovers? Explain your answer.

C15. Researchers have discovered that some regions of chromosomes are much more likely than others to cross over. We might call such a region a "hot spot" for crossing over. Let's suppose that two genes, gene *A* and gene *B*, are 5,000,000 bp apart on the same chromosome. Genes *A* and *B* are in a hot spot for crossing over. Two other genes, let's call them gene *C* and gene *D*, are also 5,000,000 bp apart but are not in a hot spot for recombination. If we conducted dihybrid crosses to compute the map distance between genes *A* and *B*, and other dihybrid crosses to compute the map distance between genes *C* and *D*, would the map distances be the same between *A* and *B* compared with to *C* and *D*? Explain.

C16. Describe the unique features of ascomycetes that lend themselves to genetic analysis.

C17. In fungi, what is the difference between a tetrad and an octad? What cellular process occurs in an octad that does not occur in a tetrad?

C18. Explain the difference between an unordered versus an ordered octad.

C19. In *Neurospora*, a cross is made between a wild-type and an albino mutant strain, which produce orange and white spores, respectively. Draw two different ways that an octad might look if it was displaying second-division segregation.

C20. One gene in *Neurospora*, let's call it gene *A*, is located close to a centromere, and a second gene, gene *B*, is located more toward the end of the chromosome. Would the percentage of octads exhibiting first-division segregation be higher with respect to gene *A* or gene *B*? Explain your answer.

Experimental Questions (Includes Most Mapping Questions)

E1. Figure 6.2 shows the first experimental results that indicated linkage between two different genes. Conduct a chi square analysis to confirm that the genes are really linked and the data could not be explained by independent assortment.

E2. In the experiment of Figure 6.7, the researchers followed the inheritance pattern of chromosomes that were abnormal at both ends to correlate genetic recombination with the physical exchange of chromosome pieces. Is it necessary to use a chromosome that is abnormal at both ends, or could the researchers have used a parental strain with two abnormal versions of chromosome 9, one with a knob at one end and its homolog with a translocation at the other end?

E3. The experiment of Figure 6.7 is not like a standard testcross, because neither parent is homozygous recessive for both genes. If you were going to carry out this same kind of experiment to verify that crossing over can explain the recombination of alleles of different genes, how would you modify this experiment to make it a standard testcross? For both parents, you should designate which alleles are found on an abnormal chromosome (i.e., knobbed, translocation chromosome 9) and which alleles are found on normal chromosomes.

E4. How would you determine that genes in mammals are located on the Y chromosome linkage group? Is it possible to conduct crosses (let's say in mice) to map the distances between genes along the Y chromosome? Explain.

E5. Explain the rationale behind a testcross. Is it necessary for one of the parents to be homozygous recessive for the genes of interest? In the heterozygous parent of a testcross, must all of the dominant alleles be linked on the same chromosome and all of the recessive alleles be linked on the homolog?

E6. In your own words, explain why a testcross cannot produce more than 50% recombinant offspring. When a testcross does produce 50% recombinant offspring, what do these results mean?

E7. Explain why the percentage of recombinant offspring in a testcross is a more accurate measure of map distance when two genes are close together. When two genes are far apart, is the percentage of recombinant offspring an underestimate or overestimate of the actual map distance?

E8. If two genes are more than 50 mu apart, how would you ever be able to show experimentally that they are located on the same chromosome?

E9. In Morgan's trihybrid testcross of Figure 6.3, he realized that crossing over was more frequent between the eye color and wing length genes than between the body color and eye color genes. Explain how he determined this.

E10. In the experiment of Figure 6.10, list the gene pairs from the particular dihybrid crosses that Sturtevant used to construct his genetic map.

E11. In the tomato, red fruit (*R*) is dominant over yellow fruit (*r*), and yellow flowers (*Wf*) are dominant over white flowers (*wf*). A cross was made between true-breeding plants with red fruit and yellow flowers, and plants with yellow fruit and white flowers. The F_1 generation plants were then crossed to plants with yellow fruit and white flowers. The following results were obtained:

333 red fruit, yellow flowers

64 red fruit, white flowers

58 yellow fruit, yellow flowers

350 yellow fruit, white flowers

Calculate the map distance between the two genes.

E12. Two genes are located on the same chromosome and are known to be 12 mu apart. An *AABB* individual was crossed to an *aabb* individual to produce *AaBb* offspring. The *AaBb* offspring were then crossed to *aabb* individuals.

A. If this cross produces 1000 offspring, what are the predicted numbers of offspring with each of the four genotypes: *AaBb*, *Aabb*, *aaBb*, and *aabb*?

B. What would be the predicted numbers of offspring with these four genotypes if the parental generation had been *AAbb* and *aaBB* instead of *AABB* and *aabb*?

E13. Two genes, designated *A* and *B*, are located 10 mu from each other. A third gene, designated *C*, is located 15 mu from *B* and 5 mu from *A*. The parental generation consisting of *AA bb CC* and *aa BB cc* individuals were crossed to each other. The F_1 heterozygotes were then testcrossed to *aa bb cc* individuals. If we assume no double crossovers occur in this region, what percentage of offspring would you expect with the following genotypes?

A. *Aa Bb Cc*

B. *aa Bb Cc*

C. *Aa bb cc*

E14. Two genes in tomatoes are 61 mu apart; normal fruit (*F*) is dominant to fasciated fruit (*f*), and normal numbers of leaves (*Lf*) is dominant to leafy (*lf*). A true-breeding plant with normal leaves and fruit was crossed to a leafy plant with fasciated fruit. The F_1 offspring were then crossed to leafy plants with fasciated fruit. If this cross produced 600 offspring, what are the expected numbers of plants in each of the four possible categories: normal leaves, normal fruit; normal leaves, fasciated fruit; leafy, normal fruit; and leafy, fasciated fruit?

E15. In the tomato, three genes are linked on the same chromosome. Tall is dominant to dwarf, skin that is smooth is dominant to skin that is peachy, and fruit with a normal tomato shape is dominant to oblate shape. A plant that is true-breeding for the dominant traits was crossed to a dwarf plant with peachy skin and oblate fruit. The F_1 plants were then testcrossed to dwarf plants with peachy skin and oblate fruit. The following results were obtained:

151 tall, smooth, normal

33 tall, smooth, oblate

11 tall, peach, oblate

2 tall, peach, normal

155 dwarf, peach, oblate

29 dwarf, peach, normal

12 dwarf, smooth, normal

0 dwarf, smooth, oblate

Construct a genetic map that describes the order of these three genes and the distances between them.

E16. A trait in garden peas involves the curling of leaves. A dihybrid cross was made involving a plant with yellow pods and curling leaves to a wild-type plant with green pods and normal leaves. All F_1 offspring had green pods and normal leaves. The F_1 plants were then crossed to plants with yellow pods and curling leaves. The following results were obtained:

117 green pods, normal leaves

115 yellow pods, curling leaves

78 green pods, curling leaves

80 yellow pods, normal leaves

A. Conduct a chi square analysis to determine if these two genes are linked.

B. If they are linked, calculate the map distance between the two genes. How accurate do you think this distance is?

E17. In mice, the gene that encodes the enzyme inosine triphosphatase is 12 mu from the gene that encodes the enzyme ornithine decarboxylase. Suppose you have identified a strain of mice homozygous for a defective inosine triphosphatase gene that does not produce any of this enzyme and is also homozygous for a defective ornithine decarboxylase gene. In other words, this strain of mice cannot make either enzyme. You crossed this homozygous recessive strain to a normal strain of mice to produce heterozygotes. The heterozygotes were then backcrossed to the strain that cannot produce either enzyme. What is the probability of obtaining a mouse that cannot make either enzyme?

E18. In the garden pea, several different genes affect pod characteristics. A gene affecting pod color (green is dominant to yellow) is approximately 7 mu away from a gene affecting pod width (wide is dominant to narrow). Both genes are located on chromosome 5. A third gene, located on chromosome 4, affects pod length (long is dominant to short). A true-breeding wild-type plant (green, wide, long pods) was crossed to a plant with yellow, narrow, short pods. The F_1 offspring were then testcrossed to plants with yellow, narrow, short pods. If the testcross produced 800 offspring, what are the expected numbers of the eight possible phenotypic combinations?

E19. A sex-influenced trait is dominant in males and causes bushy tails. The same trait is recessive in females and results in a normal tail. Fur color is not sex influenced. Yellow fur is dominant to white fur. A true-breeding female with a bushy tail and yellow fur was crossed to a white male without a bushy tail. The F_1 females were then crossed to white males without bushy tails. The following results were obtained:

Males	Females
28 normal tails, yellow	102 normal tails, yellow
72 normal tails, white	96 normal tails, white
68 bushy tails, yellow	0 bushy tails, yellow
29 bushy tails, white	0 bushy tails, white

A. Conduct a chi square analysis to determine if these two genes are linked.

B. If the genes are linked, calculate the map distance between them. Explain which data you used in your calculation.

E20. Three recessive traits in garden pea plants are as follows: yellow pods are recessive to green pods, bluish green seedlings are recessive to green seedlings, creeper (a plant that cannot stand up) is recessive to normal. A true-breeding normal plant with green pods and green seedlings was crossed to a creeper with yellow pods and bluish green seedlings. The F_1 plants were then crossed to creepers with yellow pods and bluish green seedlings. The following results were obtained:

2059 green pods, green seedlings, normal

151 green pods, green seedlings, creeper

281 green pods, bluish green seedlings, normal

15 green pods, bluish green seedlings, creeper

2041 yellow pods, bluish green seedlings, creeper

157 yellow pods, bluish green seedlings, normal

282 yellow pods, green seedlings, creeper

11 yellow pods, green seedlings, normal

Construct a genetic map that describes the map distance between these three genes.

E21. In mice, a trait called snubnose is recessive to a wild-type nose, a trait called pintail is dominant to a normal tail, and a trait called jerker (a defect in motor skills) is recessive to a normal gait. Jerker mice with a snubnose and pintail were crossed to normal mice, and then the F_1 mice were crossed to jerker mice that have a snubnose and normal tail. The outcome of this cross was as follows:

560 jerker, snubnose, pintail

548 normal gait, normal nose, normal tail

102 jerker, snubnose, normal tail

104 normal gait, normal nose, pintail

77 jerker, normal nose, normal tail

71 normal gait, snubnose, pintail

11 jerker, normal nose, pintail

9 normal gait, snubnose, normal tail

Construct a genetic map that describes the order and distance between these genes.

E22. In *Drosophila,* an allele causing vestigial wings is 12.5 mu away from another gene that causes purple eyes. A third gene that affects body color has an allele that causes black body color. This third gene is 18.5 mu away from the vestigial wings gene and 6 mu away from the gene causing purple eyes. The alleles causing vestigial wings, purple eyes, and black body are all recessive. The dominant (wild-type) traits are long wings, red eyes, and gray body. A researcher crossed wild-type flies to flies with vestigial wings, purple eyes, and black bodies. All F_1 flies were wild type. F_1 female flies were then crossed to male flies with vestigial wings, purple eyes, and black bodies. If 1000 offspring were observed, what are the expected numbers of the following types of flies?

Long wings, red eyes, gray body

Long wings, purple eyes, gray body

Long wings, red eyes, black body

Long wings, purple eyes, black body

Short wings, red eyes, gray body

Short wings, purple eyes, gray body

Short wings, red eyes, black body

Short wings, purple eyes, black body

Which kinds of flies can be produced only by a double crossover event?

E23. Three autosomal genes are linked along the same chromosome. The distance between gene *A* and *B* is 7 mu, the distance between *B* and *C* is 11 mu, and the distance between *A* and *C* is 4 mu. An individual who is *AA bb CC* was crossed to an individual who is *aa BB cc* to produce heterozygous F_1 offspring. The F_1 offspring were then crossed to homozygous *aa bb cc* individuals to produce F_2 offspring.

A. Draw the arrangement of alleles on the chromosomes in the parents and in the F_1 offspring.

B. Where would a crossover have to occur to produce an F_2 offspring that was heterozygous for all three genes?

C. If we assume that no double crossovers occur in this region, what percentage of F_2 offspring is likely to be homozygous for all three genes?

E24. Let's suppose that two different X-linked genes exist in mice, designated with the letters *N* and *L*. Gene *N* exists in a dominant, normal allele and in a recessive allele, *n*, that is lethal. Similarly, gene *L* exists in a dominant, normal allele and in a recessive allele, *l*, that is lethal. Heterozygous females are normal, but males that carry either recessive allele are born dead. Explain whether or not it would be possible to map the distance between these two genes by making crosses and analyzing the number of living and dead offspring. You may assume that you have strains of mice in which females are heterozygous for one or both genes.

E25. The alleles *his-5* and *lys-1*, found in baker's yeast, result in cells that require histidine and lysine for growth, respectively. A cross was made between two haploid yeast strains that are *his-5 lys-1* and his^+ lys^+. From the analysis of 818 individual tetrads, the following numbers of tetrads were obtained:

2 spores: *his-5* lys^+ + 2 spores: his^+ *lys-1* = 4

2 spores: *his-5 lys-1* + 2 spores: his^+ lys^+ = 502

1 spore: *his-5 lys-1* + 1 spore: *his-5* lys^+ + 1 spore: his^+ *lys-1* + 1 spore: his^+ lys^+ = 312

A. Compute the map distance between these two genes using the method of calculation that considers double crossovers and the one that does not. Which method gives a higher value? Explain why.

B. What is the frequency of single crossovers between these two genes?

C. Based on your answer to part B, how many NPDs are expected from this cross? Explain your answer. Is positive interference occurring?

E26. On chromosome 4 in *Neurospora*, the allele *pyr-1* results in a pyrimidine requirement for growth. A cross was made between a *pyr-1* and a pyr^+ (wild-type) strain, and the following results were obtained:

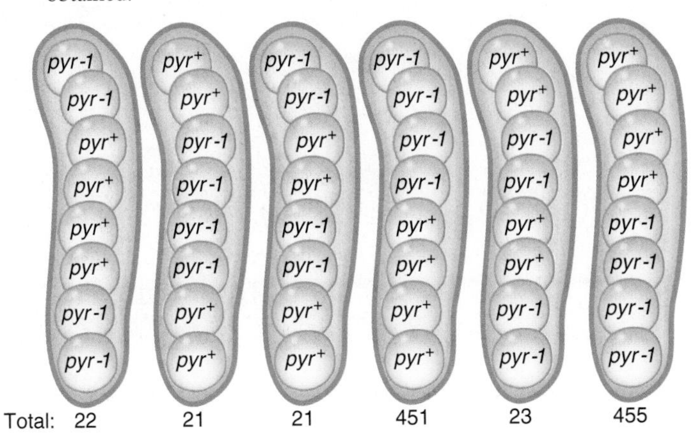

What is the distance between the *pyr-1* gene and the centromere?

E27. On chromosome 3 in *Neurospora*, the *pro-1* allele is located approximately 9.8 mu from the centromere. Let's suppose a cross was made between a *pro-1* and a pro^+ strain and 1000 asci were analyzed.

A. What are the six types of asci that can be produced?

B. What are the expected numbers of each type of ascus?

Questions for Student Discussion/Collaboration

1. In mice, a dominant gene that causes a short tail is located on chromosome 2. On chromosome 3, a recessive gene causing droopy ears is 6 mu away from another recessive gene that causes a flaky tail. A recessive gene that causes a jerker (uncoordinated) phenotype is located on chromosome 4. A jerker mouse with droopy ears and a short, flaky tail was crossed to a normal mouse. All F_1 generation mice were phenotypically normal, except they had short tails. These F_1 mice were then testcrossed to jerker mice with droopy ears and long, flaky tails. If this cross produced 400 offspring, what would be the proportions of the 16 possible phenotypic categories?
2. In Chapter 3, we discussed the idea that the X and Y chromosomes have a few genes in common. These genes are inherited in a pseudoautosomal pattern. With this phenomenon in mind, discuss whether or not the X and Y chromosomes are really distinct linkage groups.
3. Mendel studied seven traits in pea plants, and the garden pea happens to have seven different chromosomes. It has been pointed out that Mendel was very lucky not to have conducted crosses involving two traits governed by genes that are closely linked on the same chromosome because the results would have confounded his theory of independent assortment. It has even been suggested that Mendel may not have published data involving traits that were linked! An article by Stig Blixt ("Why Didn't Gregor Mendel Find Linkage?" *Nature 256*:206, 1975) considers this issue. Look up this article and discuss why Mendel did not find linkage.

Note: All answers appear at the website for this textbook; the answers to even-numbered questions are in the back of the textbook.

www.mhhe.com/brookergenetics4e

Visit the website for practice tests, answer keys, and other learning aids for this chapter. Enhance your understanding of genetics with our interactive exercises, quizzes, animations, and much more.

CHAPTER OUTLINE

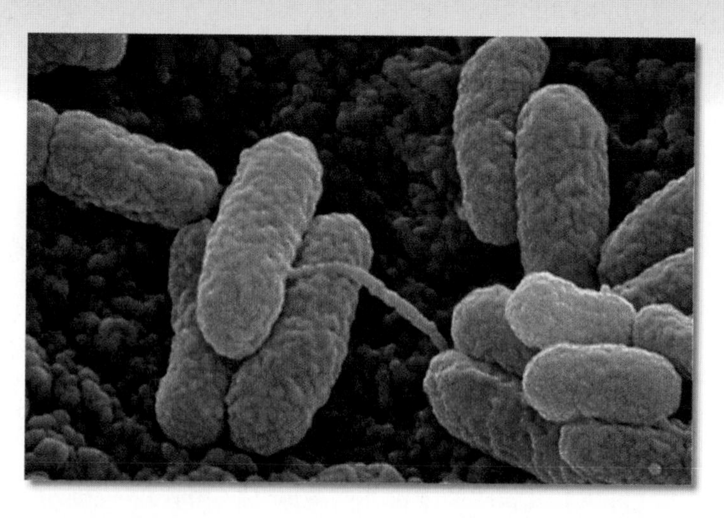

Conjugating bacteria. *The bacteria shown here are transferring genetic material by a process called conjugation.*

7 GENETIC TRANSFER AND MAPPING IN BACTERIA AND BACTERIOPHAGES

One reason researchers are so interested in bacteria and viruses is related to their impact on health. Infectious diseases caused by these agents are a leading cause of human death, accounting for a quarter to a third of deaths worldwide. The spread of infectious diseases results from human behavior, and in recent times has been accelerated by increased trade and travel, and the inappropriate use of antibiotic drugs. Although the incidence of fatal infectious diseases in the United States is relatively low compared with the worldwide average, an alarming increase in more deadly strains of bacteria and viruses has occurred over the past few decades. Since 1980, the number of deaths in the United States due to infectious diseases has approximately doubled.

Thus far, our attention in Part II of this textbook has focused on genetic analyses of eukaryotic species such as fungi, plants, and animals. As we have seen, these organisms are amenable to genetic studies for two reasons. First, allelic differences, such as white versus red eyes in *Drosophila* and tall versus dwarf pea plants, provide readily discernible traits among different individuals. Second, because most eukaryotic species reproduce sexually, crosses can be made, and the pattern of transmission of traits from parent to offspring can be analyzed. The ability to follow allelic differences in a genetic cross is a basic tool in the genetic examination of eukaryotic species.

In Chapter 7, we turn our attention to the genetic analysis of bacteria. Like their eukaryotic counterparts, bacteria often possess allelic differences that affect their cellular traits. Common allelic variations among bacteria that are readily discernible involve traits such as sensitivity to antibiotics and differences in their nutrient requirements for growth. In these cases, the allelic differences are between different strains of bacteria, because any given bacterium is usually haploid for a particular gene. In fact, the haploid nature of bacteria is a feature that makes it easier to identify mutations that produce phenotypes such as altered nutritional requirements. Loss-of-function mutations, which are often recessive in diploid eukaryotes, are not masked by dominant alleles in haploid species. Throughout this chapter, we will consider interesting experiments that examine bacterial strains with allelic differences.

Compared with eukaryotes, another striking difference in prokaryotic species is their mode of reproduction. Because bacteria reproduce asexually, researchers do not use crosses in the genetic analysis of bacterial species. Instead, they rely on a similar mechanism, called genetic transfer, in which a segment of bacterial DNA is transferred from one bacterium to another. In the first part of this chapter, we will explore the different routes of genetic transfer. We will see how researchers have used genetic transfer to map the locations of genes along the chromosome of many bacterial species.

In the second part of this chapter, we will examine **bacteriophages** (also known as **phages**), which are viruses that infect bacteria. Bacteriophages contain their own genetic material that governs the traits of the phage. As we will see, the genetic analysis of phages can yield a highly detailed genetic map of a short region of DNA. These types of analyses have provided researchers with insights regarding the structure and function of genes.

7.1 GENETIC TRANSFER AND MAPPING IN BACTERIA

Genetic transfer is a process by which one bacterium transfers genetic material to another bacterium. Why is genetic transfer an advantage? Like sexual reproduction in eukaryotes, genetic transfer in bacteria is thought to enhance the genetic diversity of bacterial species. For example, a bacterial cell carrying a gene that provides antibiotic resistance may transfer this gene to another bacterial cell, allowing that bacterial cell to survive exposure to the antibiotic.

Bacteria can naturally transfer genetic material in three ways (**Table 7.1**). The first route, known as **conjugation,** involves a direct physical interaction between two bacterial cells. One bacterium acts

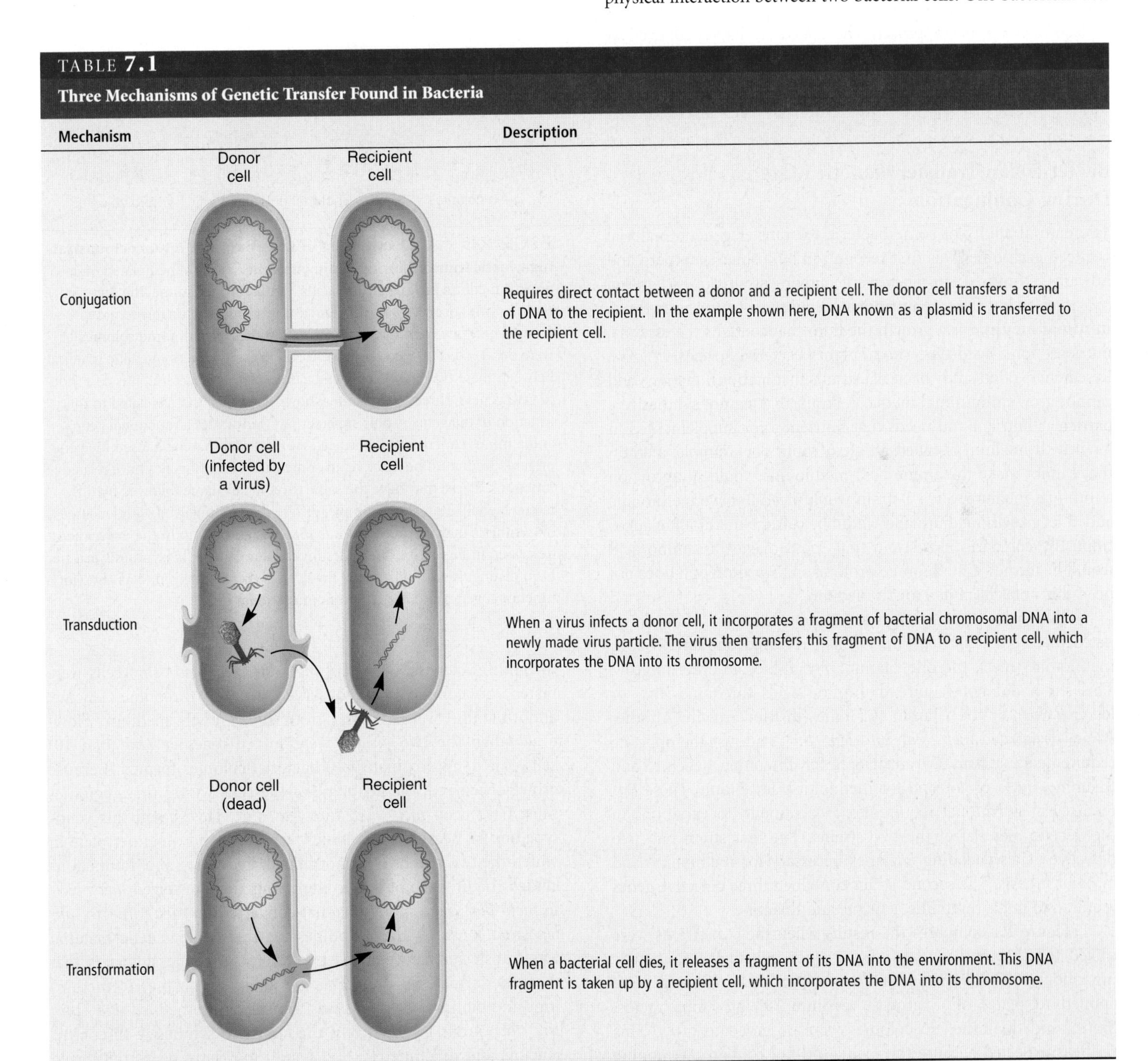

TABLE 7.1
Three Mechanisms of Genetic Transfer Found in Bacteria

Mechanism	Description
Conjugation	Requires direct contact between a donor and a recipient cell. The donor cell transfers a strand of DNA to the recipient. In the example shown here, DNA known as a plasmid is transferred to the recipient cell.
Transduction	When a virus infects a donor cell, it incorporates a fragment of bacterial chromosomal DNA into a newly made virus particle. The virus then transfers this fragment of DNA to a recipient cell, which incorporates the DNA into its chromosome.
Transformation	When a bacterial cell dies, it releases a fragment of its DNA into the environment. This DNA fragment is taken up by a recipient cell, which incorporates the DNA into its chromosome.

as a donor and transfers genetic material to a recipient cell. A second means of transfer is called **transduction.** This occurs when a virus infects a bacterium and then transfers bacterial genetic material from that bacterium to another. The last mode of genetic transfer is **transformation.** In this case, genetic material is released into the environment when a bacterial cell dies. This material then binds to a living bacterial cell, which can take it up. These three mechanisms of genetic transfer have been extensively investigated in research laboratories, and their molecular mechanisms continue to be studied with great interest. In this section, we will examine these three systems of genetic transfer in greater detail.

We will also learn how genetic transfer between bacterial cells has provided unique ways to accurately map bacterial genes. The mapping methods described in this chapter have been largely replaced by molecular approaches described in Chapter 20. Even so, the mapping of bacterial genes serves to illuminate the mechanisms by which genes are transferred between bacterial cells and also helps us to appreciate the strategies of newer mapping approaches.

Bacteria Can Transfer Genetic Material During Conjugation

The natural ability of some bacteria to transfer genetic material between each other was first recognized by Joshua Lederberg and Edward Tatum in 1946. They were studying strains of *Escherichia coli* that had different nutritional requirements for growth. A **minimal medium** is a growth medium that contains the essential nutrients for a wild-type (nonmutant) bacterial species to grow. Researchers often study bacterial strains that harbor mutations and cannot grow on minimal media. A strain that cannot synthesize a particular nutrient and needs that nutrient to be supplemented in its growth medium is called an **auxotroph.** For example, a strain that cannot make the amino acid methionine would not grow on a minimal medium. Such a strain would need to have methionine added its growth medium and would be called a methionine auxotroph. By comparison, a strain that could make this amino acid would be termed a methionine prototroph. A **prototroph** does not need this nutrient in its growth medium.

The experiment in **Figure 7.1** considers one strain, designated met^- bio^- thr^+ leu^+ thi^+, which required one amino acid, methionine (met), and one vitamin, biotin (bio), in order to grow. This strain did not require the amino acids threonine (thr) or leucine (leu), or the vitamin thiamine (thi) for growth. Another strain, designated met^+ bio^+ thr^- leu^- thi^-, had just the opposite requirements. It was an auxotroph for threonine, leucine, and thiamine, but a prototroph for methionine and biotin. These differences in nutritional requirements correspond to variations in the genetic material of the two strains. The first strain had two defective genes encoding enzymes necessary for methionine and biotin synthesis. The second strain contained three defective genes required to make threonine, leucine, and thiamine.

Figure 7.1 compares the results when the two strains were mixed together and when they were not mixed. Without mixing, about 100 million (10^8) met^- bio^- thr^+ leu^+ thi^+ cells were applied to plates on a growth medium lacking amino acids, biotin, and thiamine; no colonies were observed to grow. This result is expected because the media did not contain methionine

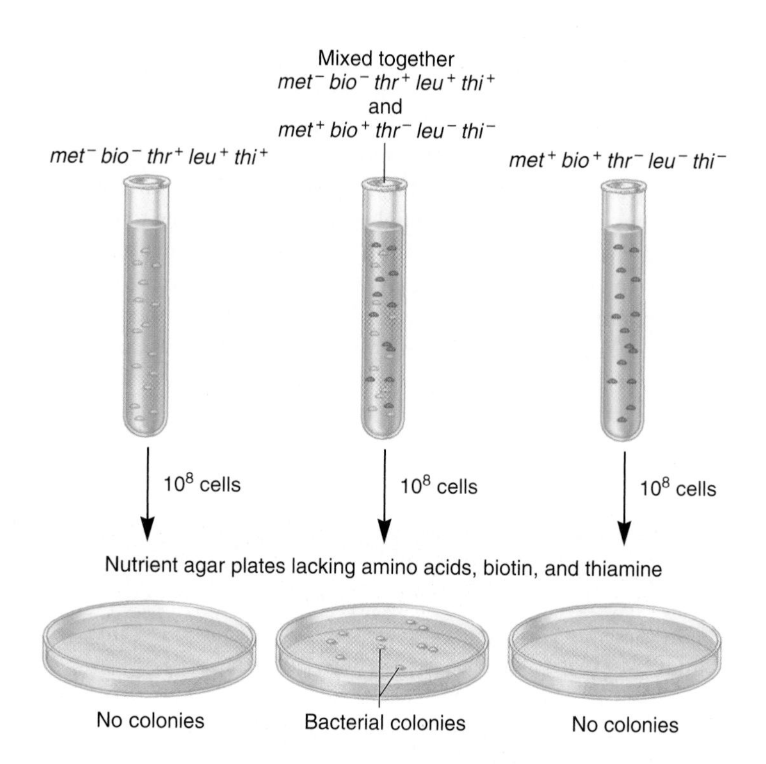

FIGURE 7.1 Experiment of Lederberg and Tatum demonstrating genetic transfer during conjugation in *E. coli*. When plated on a growth medium lacking amino acids, biotin, and thiamine, the met^- bio^- thr^+ leu^+ thi^+ or met^+ bio^+ thr^- leu^- thi^- strains were unable to grow. However, if they were mixed together and then plated, some colonies were observed. These colonies were due to the transfer of genetic material between these two strains by conjugation. Note: In bacteria, it is common to give genes a three-letter name (shown in italics) that is related to the function of the gene. A plus superscript ($^+$) indicates a functional gene, and a minus superscript ($^-$) indicates a mutation that has caused the gene or gene product to be inactive. In some cases, several genes have related functions. These may have the same three-letter name followed by different capital letters. For example, different genes involved with leucine biosynthesis may be called *leuA*, *leuB*, *leuC*, and so on. In the experiment described in Figure 7.1, the genes involved in leucine biosynthesis had not been distinguished, so the gene involved is simply referred to as leu^+ (for a functional gene) and leu^- (for a nonfunctional gene).

or biotin. Likewise, when 10^8 met^+ bio^+ thr^- leu^- thi^- cells were plated, no colonies were observed because threonine, leucine, and thiamine were missing from this growth medium. However, when the two strains were mixed together and then 10^8 cells plated, approximately 10 bacterial colonies formed. Because growth occurred, the genotype of the cells within these colonies must have been met^+ bio^+ thr^+ leu^+ thi^+. How could this genotype occur? Because no colonies were observed on either plate in which the two strains were not mixed, Lederberg and Tatum concluded that it was not due to mutations that converted met^- bio^- to met^+ bio^+ or to mutations that converted thr^- leu^- thi^- to thr^+ leu^+ thi^+. Instead, they hypothesized that some genetic material was transferred between the two strains. One possibility is that the genetic material providing the ability to synthesize methionine and biotin (met^+ bio^+) was transferred to the met^- bio^- thr^+ leu^+ thi^+ strain. Alternatively, the ability to synthesize threonine, leucine, and thiamine (thr^+ leu^+ thi^+) may have been transferred

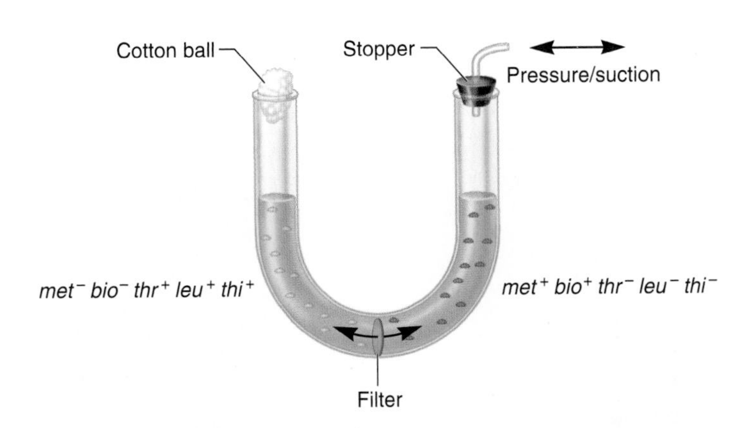

FIGURE 7.2 A U-tube apparatus like that used by Bernard Davis. The fluid in the tube is forced through the filter by alternating suction and pressure. However, the pores in the filter are too small for the passage of bacteria.

to the *met*$^+$ *bio*$^+$ *thr*$^-$ *leu*$^-$ *thi*$^-$ cells. The results of this experiment did not distinguish between these two possibilities.

In 1950, Bernard Davis conducted experiments showing that two strains of bacteria must make physical contact with each other to transfer genetic material. The apparatus he used, known as a U-tube, is shown in **Figure 7.2**. At the bottom of the U-tube is a filter with pores small enough to allow the passage of genetic material (i.e., DNA molecules) but too small to permit the passage of bacterial cells. On one side of the filter, Davis added a bacterial strain with a certain combination of nutritional requirements (the *met*$^-$ *bio*$^-$ *thr*$^+$ *leu*$^+$ *thi*$^+$ strain). On the other side, he added a different bacterial strain (the *met*$^+$ *bio*$^+$ *thr*$^-$ *leu*$^-$ *thi*$^-$ strain). The application of alternating pressure and suction promoted the movement of liquid through the filter. Because the bacteria were too large to pass through the pores, the movement of liquid did not allow the two types of bacterial strains to mix with each other. However, any genetic material that was released from a bacterium could pass through the filter.

After incubation in a U-tube, bacteria from either side of the tube were placed on media that could select for the growth of cells that were *met*$^+$ *bio*$^+$ *thr*$^+$ *leu*$^+$ *thi*$^+$. These selective media lacked methionine, biotin, threonine, leucine, and thiamine, but contained all other nutrients essential for growth. In this case, no bacterial colonies grew on the plates. The experiment showed that, without physical contact, the two bacterial strains did not transfer genetic material to one another.

The term conjugation is now used to describe the natural process of genetic transfer between bacterial cells that requires direct cell-to-cell contact. Many, but not all, species of bacteria can conjugate. Working independently, Joshua and Esther Lederberg, William Hayes, and Luca Cavalli-Sforza discovered in the early 1950s that only certain bacterial strains can act as donors of genetic material. For example, only about 5% of natural isolates of *E. coli* can act as donor strains. Research studies showed that a strain incapable of acting as a donor could subsequently be converted to a donor strain after being mixed with another donor strain. Hayes correctly proposed that donor strains contain a fertility factor that can be transferred to conjugation-defective strains to make them conjugation-proficient.

We now know that certain donor strains of *E. coli* contain a small circular segment of genetic material known as an **F factor** (for fertility factor) in addition to their circular chromosome. Strains of *E. coli* that contain an F factor are designated *F*$^+$, while strains without F factors are termed *F*$^-$. In recent years, the molecular details of the conjugation process have been extensively studied. Though the mechanisms vary somewhat from one bacterial species to another, some general themes have emerged. F factors carry several genes that are required for conjugation to occur. For example, **Figure 7.3** shows the arrangement of genes on the F factor found in certain strains of *E. coli*. The functions

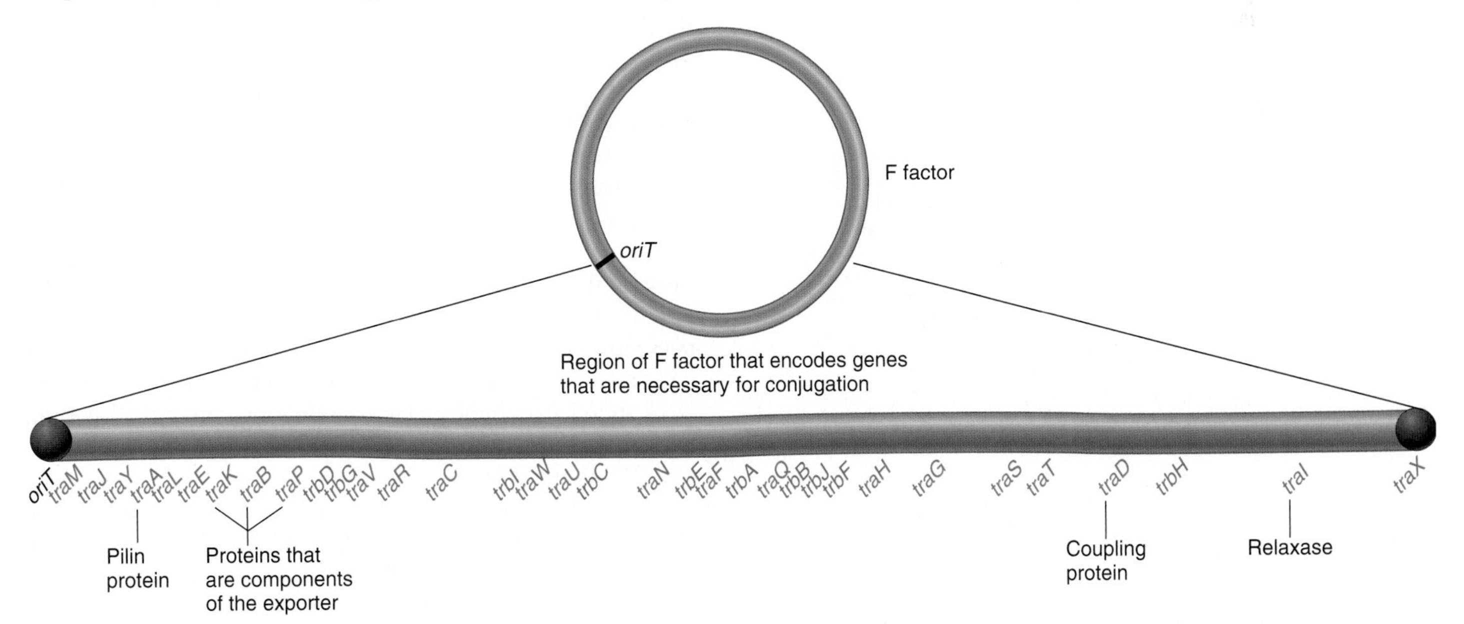

FIGURE 7.3 Genes on the F factor that play a role during conjugation. A region of the F factor carries genes that play a role in the conjugative process. Because they play a role in the transfer of DNA from donor to recipient cell, the genes are designated with the three-letter names of *tra* or *trb*, followed by a capital letter. The *tra* genes are shown in red, and the *trb* genes are shown in blue. The functions of a few examples are indicated. The origin of transfer is designated *oriT*.

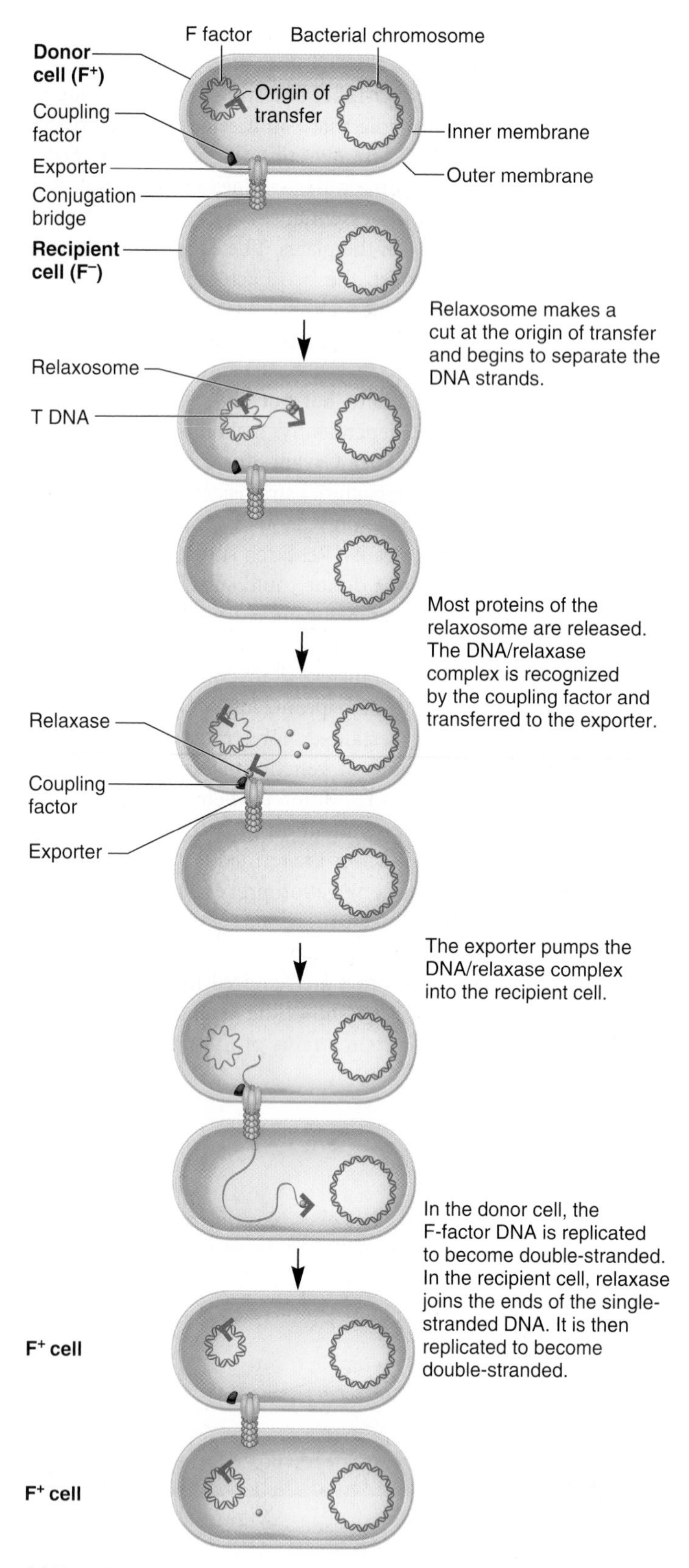

(a) Transfer of an F factor via conjugation

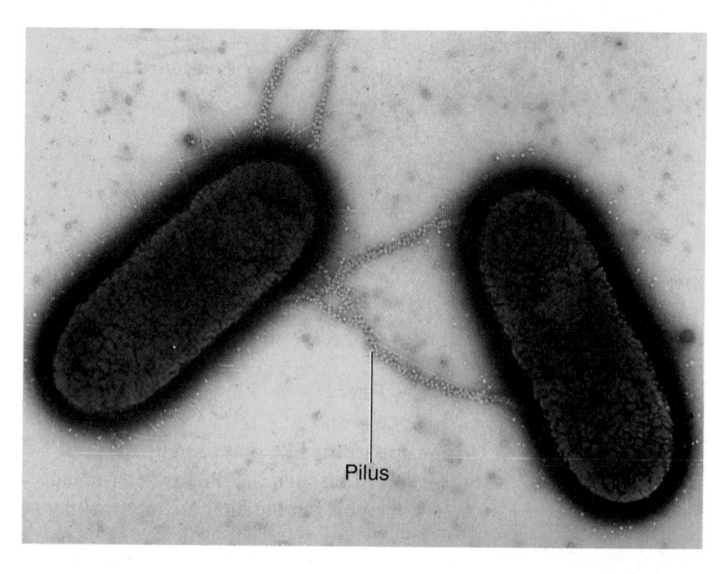

(b) Conjugating *E. coli*

FIGURE 7.4 The transfer of an F factor during bacterial conjugation. **(a)** The mechanism of transfer. The end result is that both cells have an F factor. **(b)** Two *E. coli* cells in the act of conjugation. The cell on the left is F⁺, and the one on the right is F⁻. The two cells make contact with each other via sex pili that are made by the F⁺ cell.

of the proteins encoded by these genes are needed to transfer a strand of DNA from the donor cell to a recipient cell.

Figure 7.4a describes the molecular events that occur during conjugation in *E. coli*. Contact between donor and recipient cells is a key step that initiates the conjugation process. **Sex pili** (singular: **pilus**) are made by F^+ strains (**Figure 7.4b**). The gene encoding the pilin protein (*traA*) is located on the F factor. The pili act as attachment sites that promote the binding of bacteria to each other. In this way, an F^+ strain makes physical contact with an F^- strain. In certain species, such as *E. coli*, long pili project from F^+ cells and attempt to make contact with nearby F^- cells. Once contact is made, the pili shorten and thereby draw the donor and recipient cells closer together. A **conjugation bridge** is then formed between the two cells, which provides a passageway for DNA transfer.

The successful contact between a donor and recipient cell stimulates the donor cell to begin the transfer process. Genes within the F factor encode a protein complex called the **relaxosome.** This complex first recognizes a DNA sequence in the F factor known as the **origin of transfer** (see Figure 7.4.a). Upon recognition, one DNA strand in the site is cut. The relaxosome also catalyzes the separation of the DNA strands, and only the cut DNA strand is transferred to the recipient cell. As the DNA strands separate, most of the proteins within the relaxosome are released, but one protein, called relaxase, remains bound to the end of the cut DNA strand. The complex between the single-stranded DNA and relaxase is called a **nucleoprotein** because it contains both nucleic acid (DNA) and protein (relaxase).

The next phase of conjugation involves the export of the nucleoprotein complex from the donor cell to the recipient cell. To begin this process, the DNA/relaxase complex is recognized by a coupling factor that promotes the entry of the nucleoprotein into the exporter, a complex of proteins that spans both inner and outer membranes of the donor cell. In bacterial species, this complex is formed from 10 to 15 different proteins that are encoded by genes within the F factor.

Once the DNA/relaxase complex is pumped out of the donor cell, it travels through the conjugation bridge and then into the recipient cell. As shown in Figure 7.4a, the other strand of the F factor DNA remains in the donor cell, where DNA replication restores the F factor DNA to its original double-stranded condition. After the recipient cell receives a single strand of the F factor DNA, relaxase catalyzes the joining of the ends of the linear DNA molecule to form a circular molecule. This single-stranded DNA is replicated in the recipient cell to become double-stranded. The result of conjugation is that the recipient cell has acquired an F factor, converting it from an F^- to an F^+ cell. The genetic composition of the donor strain has not changed.

Hfr Strains Contain an F Factor Integrated into the Bacterial Chromosome

Luca Cavalli-Sforza discovered a strain of *E. coli* that was very efficient at transferring many chromosomal genes to recipient F^- strains. Cavalli-Sforza designated this bacterial strain an ***Hfr* strain** (for high frequency of recombination). How is an *Hfr* strain formed? As shown in **Figure 7.5a**, an F factor may align with a similar region found in the bacterial chromosome. Due to recombination, which is described in Chapter 17, the F factor may integrate into the bacterial chromosome. In this example, the F factor has integrated next to a *lac*$^+$ gene. F factors can integrate into several different sites that are scattered around the *E. coli* chromosome.

Occasionally, the integrated F factor in an *Hfr* strain is excised from the bacterial chromosome. This process involves the looping out of the F factor DNA from the chromosome, which is followed by recombination that releases the F factor from the chromosome (**Figure 7.5b**). In the example shown here, the excision is imprecise. This produces an F factor that carries a portion of the bacterial chromosome and leaves behind some of the F factor DNA in the bacterial chromosome. F factors that carry portions of the bacterial chromosome are called **F′ factors** (F prime factors). For example, in the experiment described earlier in Figure 7.1, an F′ factor carrying the *bio*$^+$ and *met*$^+$ genes or an F′ factor carrying the *thr*$^+$, *leu*$^+$, and *thi*$^+$ genes may have been transferred to the recipient strain. Therefore, conjugation may introduce new genes into the recipient strain and thereby alter its genotype. We will also consider F′ factors in Chapter 14 when we discuss mechanisms of bacterial gene regulation.

Hfr Strains Can Transfer a Portion of the Bacterial Chromosome to Recipient Cells

William Hayes, who independently isolated another *Hfr* strain, demonstrated that conjugation between an *Hfr* strain and an F^- strain involves the transfer of a portion of the bacterial

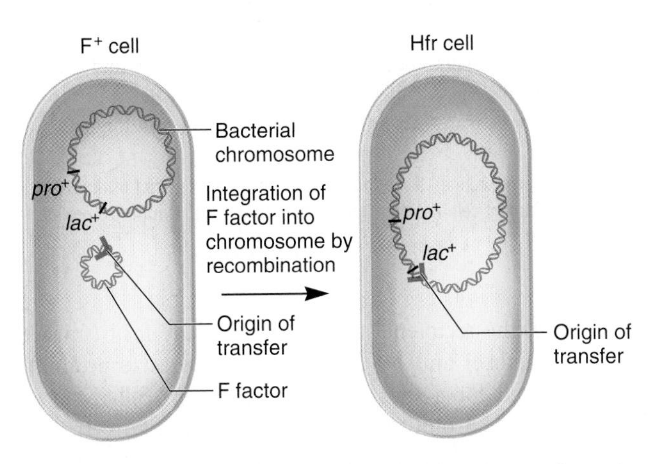

(a) When an F factor integrates into the chromosome, it creates an Hfr cell.

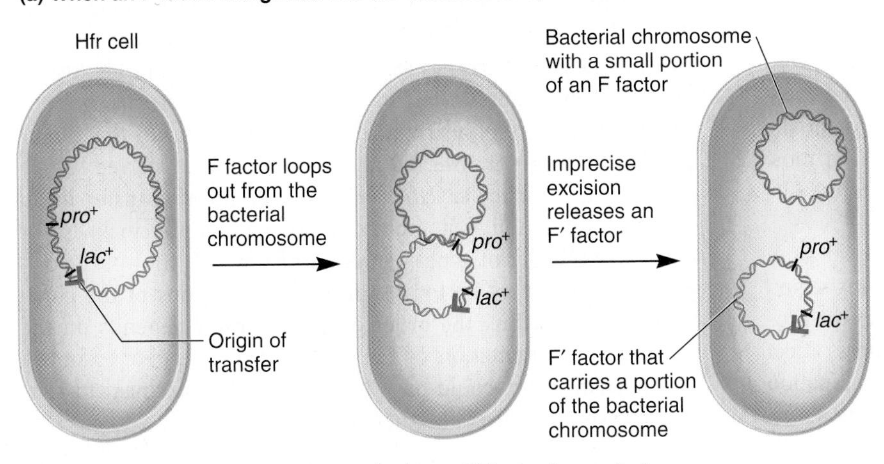

(b) When an F factor excises imprecisely, an F′ factor is created.

FIGURE 7.5 Integration of an F factor to form an *Hfr* cell and its subsequent excision to form an F′ factor. (a) An *Hfr* cell is created when an F factor integrates into the bacterial chromosome. (b) When an F factor is imprecisely excised, an F′ factor is created that carries a portion of the bacterial chromosome.

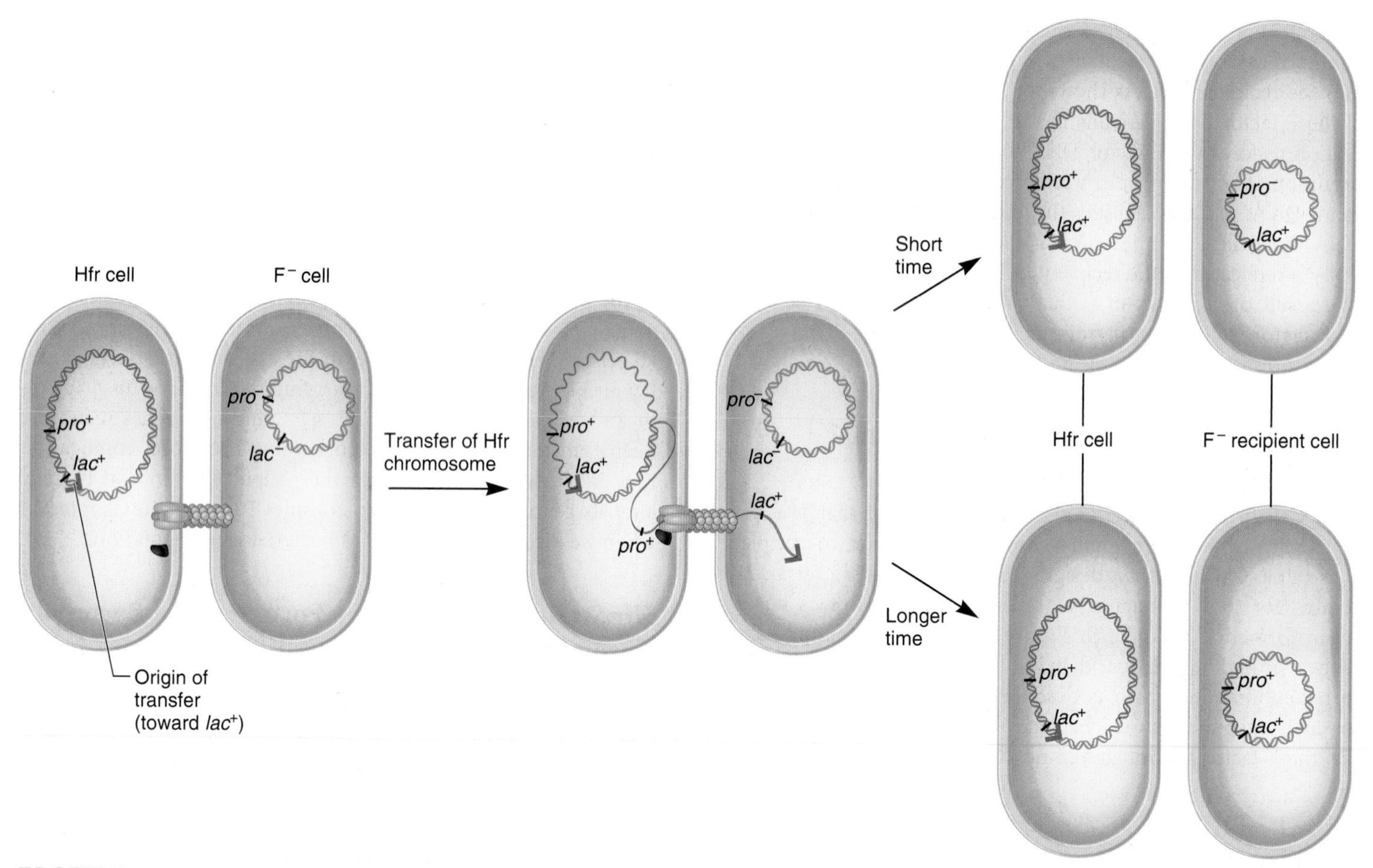

FIGURE 7.6 Transfer of bacterial genes by an *Hfr* strain. The transfer of the bacterial chromosome begins at the origin of transfer and then proceeds around the circular chromosome. After a segment of chromosome has been transferred to the F⁻ recipient cell, it recombines with the recipient cell's chromosome. If mating occurs for a brief period, only a short segment of the chromosome is transferred. If mating is prolonged, a longer segment of the bacterial chromosome is transferred.

Genes → Traits The F⁻ recipient cell was originally *lac*⁻ (unable to metabolize lactose) and *pro*⁻ (unable to synthesize proline). If mating occurred for a short period of time, the recipient cell acquired *lac*⁺, allowing it to metabolize lactose. If mating occurred for a longer period of time, the recipient cell also acquired *pro*⁺, enabling it to synthesize proline.

chromosome from the *Hfr* strain to the F⁻ cell (**Figure 7.6**). The origin of transfer within the integrated F factor determines the starting point and direction of this transfer process. One of the DNA strands is cut at the origin of transfer. This cut, or nicked, site is the starting point at which the Hfr chromosome enters the F⁻ recipient cell. From this starting point, a strand of the DNA of the Hfr chromosome begins to enter the F⁻ cell in a linear manner. The transfer process occurs in conjunction with chromosomal replication, so the Hfr cell retains its original chromosomal composition. About 1.5 to 2 hours is required for the entire Hfr chromosome to pass into the F⁻ cell. Because most matings do not last that long, usually only a portion of the Hfr chromosome is transmitted to the F⁻ cell.

Once inside the F⁻ cell, the chromosomal material from the Hfr cell can swap, or recombine, with the homologous region of the recipient cell's chromosome. (Chapter 17 describes the process of homologous recombination.) How does this process affect the recipient cell? As illustrated in Figure 7.6, this recombination may provide the recipient cell with a new combination of alleles. In this example, the recipient strain was originally *lac*⁻ (unable to metabolize lactose) and *pro*⁻ (unable to synthesize proline). If mating occurred for a short time, the recipient cell received a short segment of chromosomal DNA from the donor. In this case, the recipient cell has become *lac*⁺ but remains *pro*⁻. If the mating is prolonged, the recipient cell will receive a longer segment of chromosomal DNA from the donor. After a longer mating, the recipient becomes *lac*⁺ and *pro*⁺. As shown in Figure 7.6, an important feature of Hfr mating is that the bacterial chromosome is transferred linearly to the recipient strain. In this example, *lac*⁺ is always transferred first, and *pro*⁺ is transferred later.

In any particular *Hfr* strain, the origin of transfer has a specific orientation that promotes either a counterclockwise or clockwise transfer of genes. Among different *Hfr* strains, the origin of transfer may be located in different regions of the chromosome. Therefore, the order of gene transfer depends on the location and orientation of the origin of transfer. For example, another *Hfr* strain could have its origin of transfer next to *pro*⁺ and transfer *pro*⁺ first and then *lac*⁺.

EXPERIMENT 7A

Conjugation Experiments Can Map Genes Along the *E. coli* Chromosome

The first genetic mapping experiments in bacteria were carried out by Elie Wollman and François Jacob in the 1950s. At the time of their studies, not much was known about the organization of bacterial genes along the chromosome. A few key advances made Wollman and Jacob realize that the process of genetic transfer could be used to map the order of genes in *E. coli*. First, the discovery of conjugation by Joshua Lederberg and Edward Tatum and the identification of *Hfr* strains by Cavalli-Sforza and Hayes made it clear that bacteria can transfer genes from donor to recipient cells in a linear fashion. In addition, Wollman and Jacob were aware of previous microbiological studies concerning bacteriophages—viruses that bind to *E. coli* cells and subsequently infect them. These studies showed that bacteriophages can be sheared from the surface of *E. coli* cells if they are spun in a blender. In this treatment, the bacteriophages are detached from the surface of the bacterial cells, but the bacteria themselves remain healthy and viable. Wollman and Jacob reasoned that a blender treatment could also be used to separate bacterial cells that were in the act of conjugation without killing them. This technique is known as an **interrupted mating.**

The rationale behind Wollman and Jacob's mapping strategy is that the time it takes for genes to enter a donor cell is directly related to their order along the bacterial chromosome. They hypothesized that the chromosome of the donor strain in an Hfr mating is transferred in a linear manner to the recipient strain. If so, the order of genes along the chromosome can be deduced by determining the time it takes various genes to enter the recipient strain. Assuming the Hfr chromosome is transferred linearly, they realized that interruptions of mating at different times would lead to various lengths of the Hfr chromosome being transferred to the F^- recipient cell. If two bacterial cells had mated for a short period of time, only a small segment of the Hfr chromosome would be transferred to the recipient bacterium. However, if the bacterial cells were allowed to mate for a longer period before being interrupted, a longer segment of the Hfr chromosome could be transferred (see Figure 7.6). By determining which genes were transferred during short matings and which required longer times, Wollman and Jacob were able to deduce the order of particular genes along the *E. coli* chromosome.

As shown in the experiment of **Figure 7.7**, Wollman and Jacob began with two *E. coli* strains. The donor (*Hfr*) strain had the following genetic composition:

- thr^+: able to synthesize threonine, an essential amino acid for growth
- leu^+: able to synthesize leucine, an essential amino acid for growth
- azi^s: sensitive to killing by azide (a toxic chemical)
- ton^s: sensitive to infection by bacteriophage T1. (As discussed later, when bacteriophages infect bacteria, they may cause lysis, which results in plaque formation.)
- lac^+: able to metabolize lactose and use it for growth
- gal^+: able to metabolize galactose and use it for growth
- str^s: sensitive to killing by streptomycin (an antibiotic)

The recipient (F^-) strain had the opposite genotype: thr^- leu^- azi^r ton^r lac^- gal^- str^r (r = resistant). Before the experiment, Wollman and Jacob already knew the thr^+ gene was transferred first, followed by the leu^+ gene, and both were transferred relatively soon (5–10 minutes) after mating. Their main goal in this experiment was to determine the times at which the other genes (azi^s ton^s lac^+ gal^+) were transferred to the recipient strain. The transfer of the str^s gene was not examined because streptomycin was used to kill the donor strain following conjugation.

Before discussing the conclusions of this experiment, let's consider how Wollman and Jacob monitored gene transfer. To determine if particular genes had been transferred after mating, they took the mated cells and first plated them on growth media that lacked threonine (thr) and leucine (leu) but contained streptomycin (str). On these plates, the original donor and recipient strains could not grow because the donor strain was streptomycin sensitive and the recipient strain required threonine and leucine. However, mated cells in which the donor had transferred chromosomal DNA carrying the thr^+ and leu^+ genes to the recipient cell would be able to grow.

To determine the order of gene transfer of the azi^s, ton^s, lac^+, and gal^+ genes, Wollman and Jacob picked colonies from the first plates and restreaked them on media that contained azide or bacteriophage T1 or on media that contained lactose or galactose as the sole source of energy for growth. The plates were incubated overnight to observe the formation of visible bacterial growth. Whether or not the bacteria could grow depended on their genotypes. For example, a cell that is azi^s cannot grow on media containing azide, and a cell that is lac^- cannot grow on media containing lactose as the carbon source for growth. By comparison, a cell that is azi^r and lac^+ can grow on both types of media.

THE GOAL (DISCOVERY-BASED SCIENCE)

The chromosome of the donor strain in an Hfr mating is transferred in a linear manner to the recipient strain. The order of genes along the chromosome can be deduced by determining the time various genes take to enter the recipient strain.

ACHIEVING THE GOAL — FIGURE 7.7 The use of conjugation to map the order of genes along the *E. coli* chromosome.

Starting materials: The two *E. coli* strains already described, one *Hfr* strain (thr^+ leu^+ azi^s ton^s lac^+ gal^+ str^s) and one F^- (thr^- leu^- azi^r ton^r lac^- gal^- str^r).

Experimental level | **Conceptual level**

1. Mix together a large number of Hfr donor and F⁻ recipient cells.

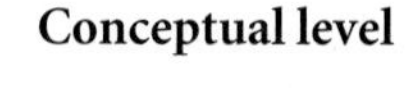

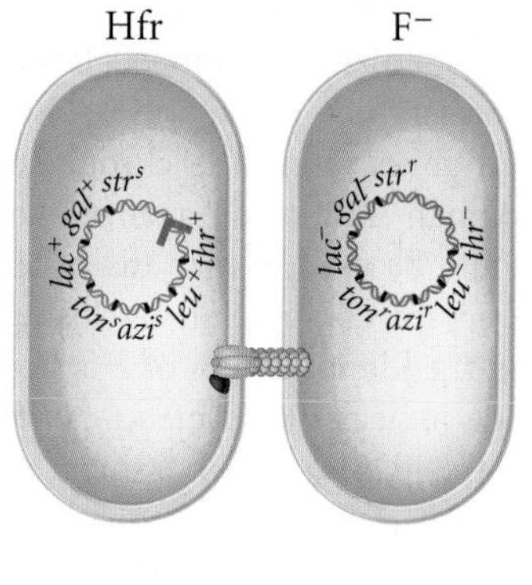

2. After different periods of time, take a sample of cells and interrupt conjugation in a blender.

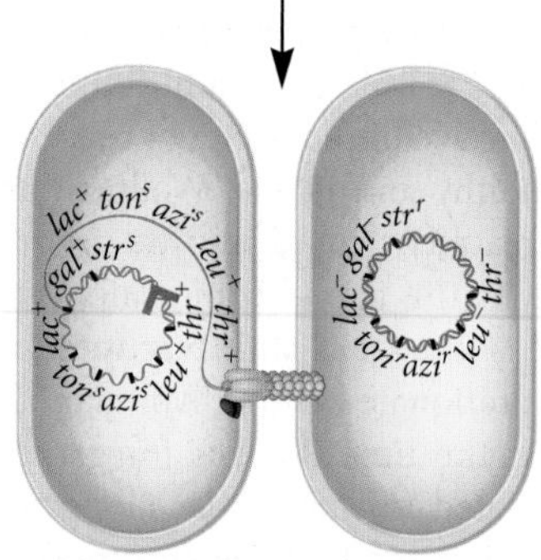

Separate by blending; donor DNA recombines with recipient cell chromosome.

In this conceptual example, the cells have been incubated about 20 minutes.

3. Plate the cells on growth media lacking threonine and leucine but containing streptomycin. Note: The general methods for growing bacteria in a laboratory are described in the Appendix.

4. Pick each surviving colony, which would have to be thr^+ leu^+ str^r, and test to see if it is sensitive to killing by azide, sensitive to infection by T1 bacteriophage, and able to metabolize lactose or galactose.

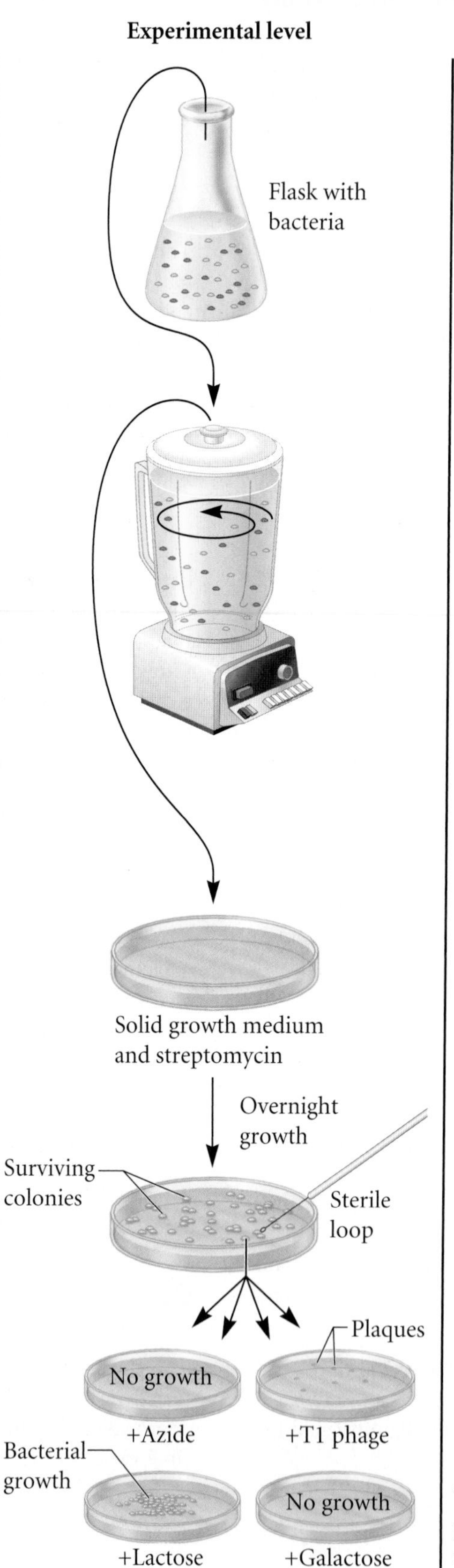

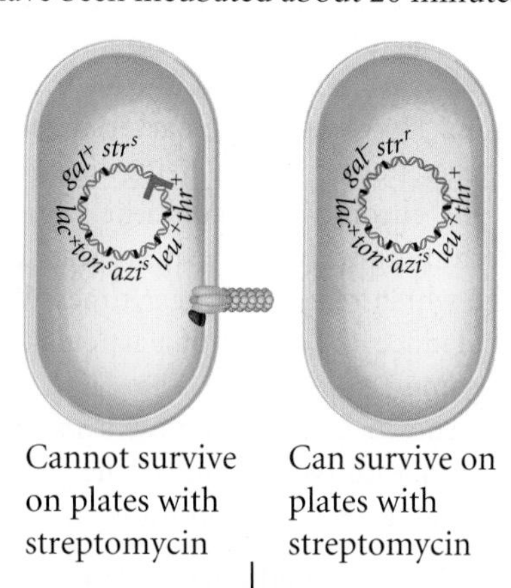

Additional tests

The conclusion is that the colony that was picked contained cells with a genotype of thr^+ leu^+ azi^s ton^s lac^+ gal^- str^r.

THE DATA

Minutes That Bacterial Cells Were Allowed to Mate Before Blender Treatment	*Percent of Surviving Bacterial Colonies with the Following Genotypes:*				
	thr⁺ leu⁺	*azi^s*	*ton^s*	*lac⁺*	*gal⁺*
5	—*	—	—	—	—
10	100	12	3	0	0
15	100	70	31	0	0
20	100	88	71	12	0
25	100	92	80	28	0.6
30	100	90	75	36	5
40	100	90	75	38	20
50	100	91	78	42	27
60	100	91	78	42	27

*There were no surviving colonies within the first 5 minutes of mating.

Data from François Jacob and Elie Wollman (1961) *Sexuality and the Genetics of Bacteria.* Academic Press, New York.

INTERPRETING THE DATA

Now let's discuss the data shown in Figure 7.7. After the first plating, all survivors would be cells in which the thr^+ and leu^+ alleles had been transferred to the F^- recipient strain, which was already streptomycin-resistant. As seen in the data, 5 minutes was not sufficient time to transfer the thr^+ and leu^+ alleles because no surviving colonies were observed. After 10 minutes or longer, however, surviving bacterial colonies with the thr^+ leu^+ genotype were obtained. To determine the order of the remaining genes (azi^s, ton^s, lac^+, and gal^+), each surviving colony was tested to see if it was sensitive to killing by azide, sensitive to infection by T1 bacteriophage, able to use lactose for growth, or able to use galactose for growth. The likelihood of surviving colonies depended on whether the azi^s, ton^s, lac^+, and gal^+ genes were close to the origin of transfer or farther away. For example, when cells were allowed to mate for 25 minutes, 80% carried the ton^s gene, whereas only 0.6% carried the gal^+ gene. These results indicate that the ton^s gene is closer to the origin of transfer compared with the gal^+ gene. When comparing the data in Figure 7.7, a consistent pattern emerged. The gene that conferred sensitivity to azide (azi^s) was transferred first, followed by ton^s, lac^+, and finally, gal^+. From these data, as well as those from other experiments, Wollman and Jacob constructed a genetic map that described the order of these genes along the *E. coli* chromosome.

This work provided the first method for bacterial geneticists to map the order of genes along the bacterial chromosome. Throughout the course of their studies, Wollman and Jacob identified several different *Hfr* strains in which the origin of transfer had been integrated at different places along the bacterial chromosome. When they compared the order of genes among different *Hfr* strains, their results were consistent with the idea that the *E. coli* chromosome is circular (see solved problem S2).

A self-help quiz involving this experiment can be found at www.mhhe.com/brookergenetics4e.

A Genetic Map of the *E. coli* Chromosome Has Been Obtained from Many Conjugation Studies

Conjugation experiments have been used to map more than 1000 genes along the circular *E. coli* chromosome. A map of the *E. coli* chromosome is shown in **Figure 7.8**. This simplified map shows the locations of only a few dozen genes. Because the chromosome is circular, we must arbitrarily assign a starting point on the map, in this case the gene *thrA*. Researchers scale genetic maps from bacterial conjugation studies in units of **minutes.** This unit refers to the relative time it takes for genes to first enter an F^- recipient strain during a conjugation experiment. The *E. coli* genetic map shown in Figure 7.8 is 100 minutes long, which is approximately the time that it takes to transfer the complete chromosome during an Hfr mating.

The distance between two genes is determined by comparing their times of entry during a conjugation experiment. As shown in **Figure 7.9**, the time of entry is found by conducting mating experiments at different time intervals before interruption. We compute the time of entry by extrapolating the data back to the *x*-axis. In this experiment, the time of entry of the *lacZ* gene was approximately 16 minutes, and that of the *galE* gene was 25 minutes. Therefore, these two genes are approximately 9 minutes apart from each other along the *E. coli* chromosome.

Let's look back at Figure 7.8 and consider where the origin of transfer must have been located in the donor strain that was used in the experiment of Figure 7.9. The *lacZ* gene is located at 7 minutes on the chromosome map, and the *galE* is found at 16 minutes. For the donor strain used in Figure 7.9, we can deduce that the origin of transfer was located at approximately 91 minutes on the chromosome map and transferred DNA in the clockwise direction. If we assume the origin was located here, it would take about 16 minutes to transfer *lacZ* and about 25 minutes to transfer *galE*.

Bacteria May Contain Different Types of Plasmids

Thus far, we have considered F factors, which are one type of DNA that can exist independently of the chromosomal DNA. The more general term for this structure is **plasmid.** Most known plasmids are circular, although some are linear. Plasmids occur naturally in many strains of bacteria and in a few types of eukaryotic cells such as yeast. The smallest plasmids consist of just a few thousand base pairs (bp) and carry only a gene or two; the largest are in the range of 100,000 to 500,000 bp and carry several dozen or even hundreds of genes. Some plasmids, such as F factors, can integrate into the chromosome. These are also called **episomes.**

A plasmid has its own origin of replication that allows it to be replicated independently of the bacterial chromosome. The

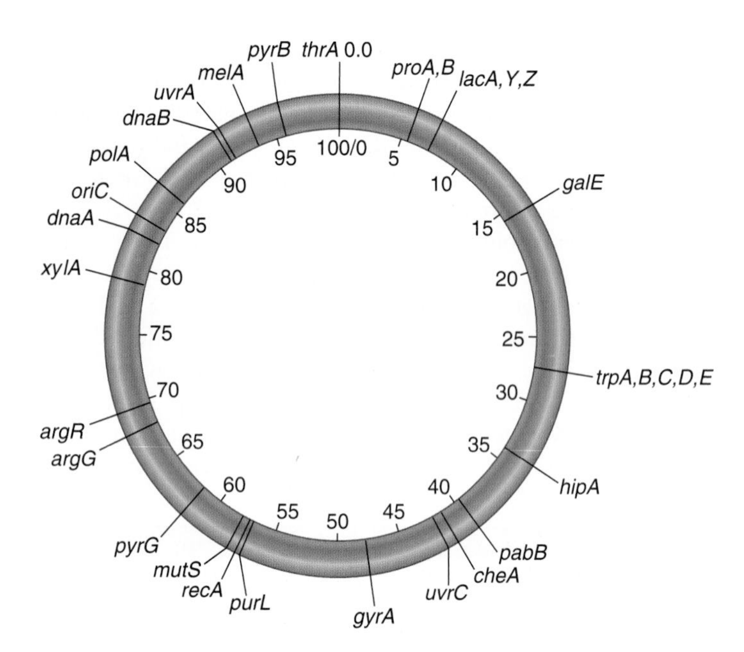

FIGURE 7.8 A simplified genetic map of the *E. coli* chromosome indicating the positions of several genes. *E. coli* has a circular chromosome with about 4300 different genes. This map shows the locations of several of them. The map is scaled in units of minutes, and it proceeds in a clockwise direction. The starting point on the map is the gene *thrA*.

DNA sequence of the origin of replication influences how many copies of the plasmid are found within a cell. Some origins are said to be very strong because they result in many copies of the plasmid, perhaps as many as 100 per cell. Other origins of replication have sequences that are described as much weaker, in that the number of copies created is relatively low, such as one or two per cell.

Why do bacteria have plasmids? Plasmids are not usually necessary for bacterial survival. However, in many cases, certain genes within a plasmid provide some type of growth advantage to the cell. By studying plasmids in many different species, researchers have discovered that most plasmids fall into five different categories:

1. Fertility plasmids, also known as F factors, allow bacteria to mate with each other.
2. Resistance plasmids, also known as R factors, contain genes that confer resistance against antibiotics and other types of toxins.
3. Degradative plasmids carry genes that enable the bacterium to digest and utilize an unusual substance. For example, a degradative plasmid may carry genes that allow a bacterium to digest an organic solvent such as toluene.
4. Col-plasmids contain genes that encode colicines, which are proteins that kill other bacteria.
5. Virulence plasmids carry genes that turn a bacterium into a pathogenic strain.

Bacteriophages Transfer Genetic Material from One Bacterial Cell to Another Via Transduction

We now turn to a second method of genetic transfer, one that involves bacteriophages. Before we discuss the ability of

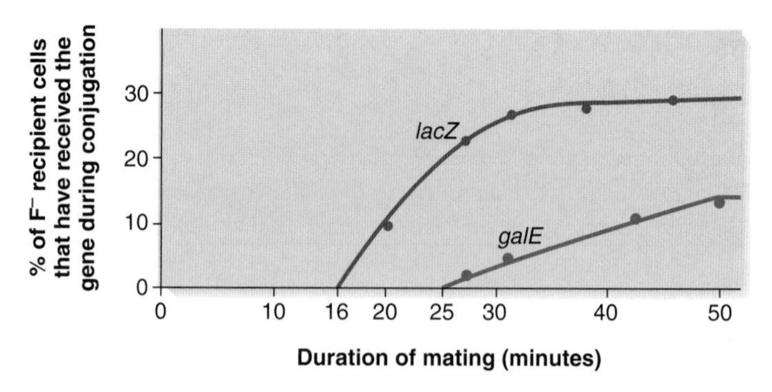

FIGURE 7.9 Time course of an interrupted *E. coli* conjugation experiment. By extrapolating the data back to the origin, the approximate time of entry of the *lacZ* gene is found to be 16 minutes; that of the *galE* gene, 25 minutes. Therefore, the distance between these two genes is 9 minutes.

bacteriophages to transfer genetic material between bacterial cells, let's consider some general features of a phage's reproductive cycle. Bacteriophages are composed of genetic material that is surrounded by a protein coat. As shown in **Figure 7.10**, certain types of bacteriophages bind to the surface of a bacterium and inject their genetic material into the bacterial cytoplasm. At this point, depending on the specific type of virus and its growth conditions, a phage may follow a lytic cycle or a lysogenic cycle. During the **lytic cycle,** the bacteriophage directs the synthesis of many copies of the phage genetic material and coat proteins (see Figure 7.10, left side). These components then assemble to make new phages. When synthesis and assembly are completed, the bacterial host cell is lysed (broken apart), releasing the newly made phages into the environment. **Virulent phages** follow only a lytic cycle, and thus infection results in the death of the host cell.

In other cases, a bacteriophage infects a bacterium and follows the lysogenic cycle (see Figure 7.10, right side). During the **lysogenic cycle,** most types of phages integrate their genetic material into the chromosome of the bacterium. This integrated phage DNA is known as a **prophage.** A prophage can exist in a dormant state for a long time during which no new bacteriophages are made. When a bacterium containing a lysogenic prophage divides to produce two daughter cells, the prophage's genetic material is copied along with the bacterial chromosome. Therefore, both daughter cells inherit the prophage. At some later time, a prophage may become activated to excise itself from the bacterial chromosome and enter the lytic cycle. When this happens, it promotes the synthesis of new phages and eventually lyses the host cell. A bacteriophage that usually exists in the lysogenic cycle is called a **temperate phage.** Under most conditions, temperate phages do not produce new phages and do not kill the host bacterial cell.

With a general understanding of bacteriophage reproductive cycles, we may now examine the ability of phages to transfer genetic material between bacteria. This process is called transduction. Examples of phages that can transfer bacterial chromosomal DNA from one bacterium to another are the P22 and P1 phages, which infect the bacterial species *Salmonella typhimurium* and *E. coli*, respectively. The P22 and P1 phages can follow either the lytic or lysogenic cycle. In the lytic cycle, when the phage infects the bacterial cell, the bacterial chromosome becomes fragmented into small pieces of DNA. The phage DNA directs the synthesis of

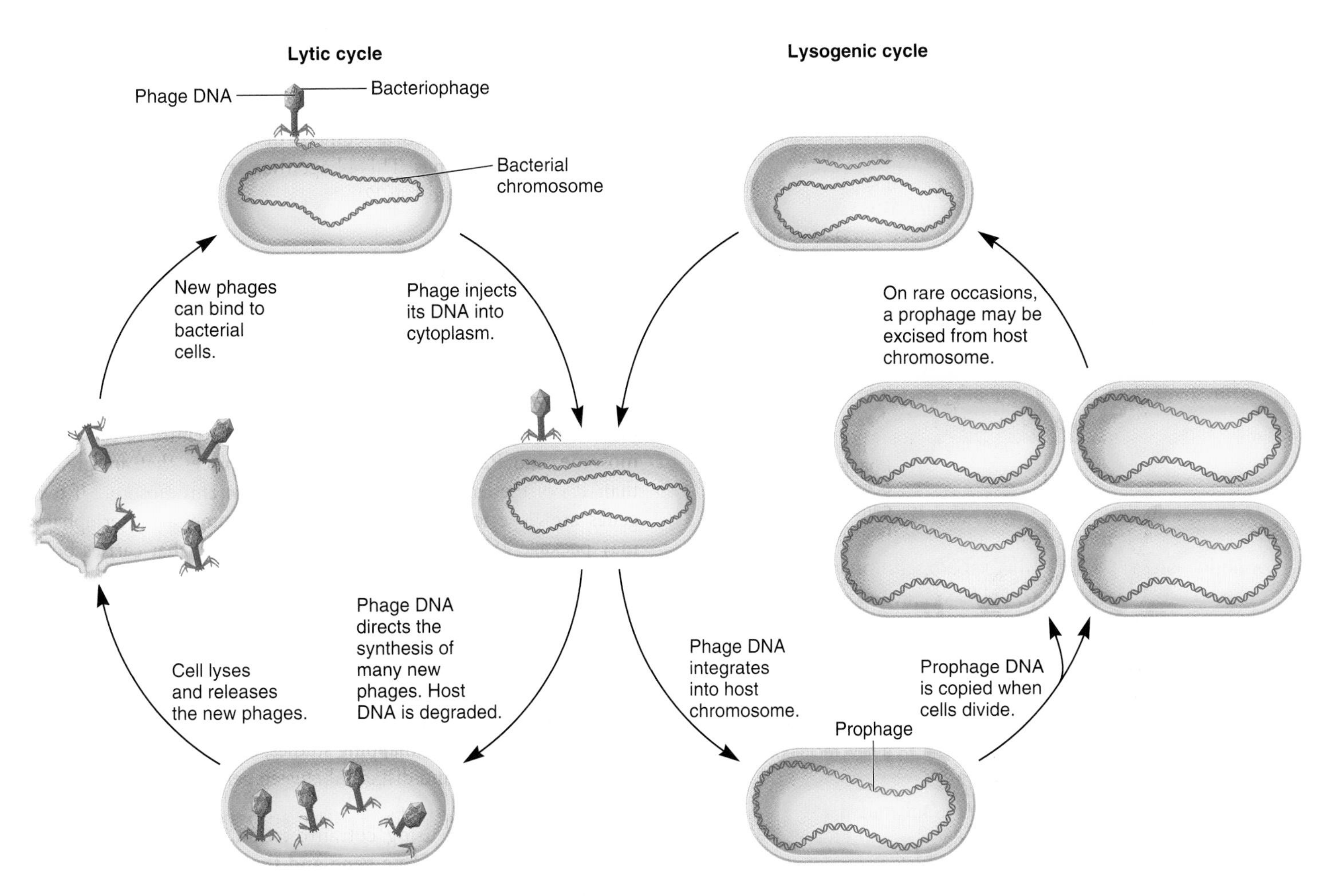

FIGURE 7.10 **The lytic and lysogenic reproductive cycles of certain bacteriophages.** Some bacteriophages, such as temperate phages, can follow both cycles. Other phages, known as virulent phages, can follow only a lytic cycle.

more phage DNA and proteins, which then assemble to make new phages (**Figure 7.11**).

How does a bacteriophage transfer bacterial chromosomal genes from one cell to another? Occasionally, a mistake can happen in which a piece of bacterial DNA assembles with phage proteins. This creates a phage that contains bacterial chromosomal DNA. When phage synthesis is completed, the bacterial cell is lysed and releases the newly made phage into the environment. Following release, this abnormal phage can bind to a living bacterial cell and inject its genetic material into the bacterium. The DNA fragment, which was derived from the chromosomal DNA of the first bacterium, can then recombine with the recipient cell's bacterial chromosome. In this case, the recipient bacterium has been changed from a cell that was *his*$^-$ *lys*$^-$ (unable to synthesize histidine and lysine) to a cell that is *his*$^+$ *lys*$^-$ (able to synthesize histidine but unable to synthesize lysine). In the example shown in Figure 7.11, any piece of the bacterial chromosomal DNA can be incorporated into the phage. This type of transduction is called **generalized transduction.** By comparison, some phages carry out specialized transduction in which only particular bacterial genes are transferred to recipient cells (see solved problem S5 at the end of the chapter).

Transduction was first discovered in 1952 by Joshua Lederberg and Norton Zinder, using an experimental strategy similar to that depicted in Figure 7.1. They mixed together two strains of the bacterium *Salmonella typhimurium*. One strain, designated LA-22, was *phe*$^-$ *trp*$^-$ *met*$^+$ *his*$^+$. This strain was unable to synthesize phenylalanine or tryptophan but was able to synthesize methionine and histidine. The other strain, LA-2, was *phe*$^+$ *trp*$^+$ *met*$^-$ *his*$^-$. It was able to synthesize phenylalanine and tryptophan but not methionine or histidine. When a mixture of these cells was placed on plates with growth media lacking these four amino acids, approximately 1 cell in 100,000 was observed to grow. The genotype of the surviving bacterial cells must have been *phe*$^+$ *trp*$^+$ *met*$^+$ *his*$^+$. Therefore, Lederberg and Zinder concluded that genetic material had been transferred between the two strains.

A novel result occurred when Lederberg and Zinder repeated this experiment using a U-tube apparatus, as previously shown in Figure 7.2. They placed the LA-22 strain (*phe*$^-$ *trp*$^-$ *met*$^+$ *his*$^+$) on one side of the filter and LA-2 (*phe*$^+$ *trp*$^+$ *met*$^-$ *his*$^-$) on the other. After an incubation period of several minutes, they removed samples from either side of the tube and plated the cells on media lacking the four amino acids. Surprisingly, they obtained colonies from the side of the tube that contained LA-22 but not from the side that contained LA-2. From these results, they concluded that some agent, which was small enough to pass through the filter, was being transferred from LA-2 to LA-22 that converted LA-22 to a *phe*$^+$ *trp*$^+$ *met*$^+$ *his*$^+$ genotype. In other

Bacteria Can Also Transfer Genetic Material by Transformation

A third mechanism for the transfer of genetic material from one bacterium to another is known as transformation. This process was first discovered by Frederick Griffith in 1928 while he was working with strains of *Streptococcus pneumoniae* (formerly known as *Diplococcus pneumoniae,* or pneumococcus). During transformation, a living bacterial cell takes up DNA that is released from a dead bacterium. This DNA may then recombine into the living bacterium's chromosome, producing a bacterium with genetic material that it has received from the dead bacterium. (This experiment is discussed in detail in Chapter 9.) Transformation may be either a natural process that has evolved in certain bacteria, in which case it is called **natural transformation,** or an artificial process in which the bacterial cells are forced to take up DNA, an experimental approach termed **artificial transformation.** For example, a technique known as electroporation, in which an electric current causes the uptake of DNA, is used by researchers to promote the transport of DNA into a bacterial cell.

Since the initial studies of Griffith, we have learned a great deal about the events that occur in natural transformation. This form of genetic transfer has been reported in a wide variety of bacterial species. Bacterial cells that are able to take up DNA are known as **competent cells.** Those that can take up DNA naturally carry genes that encode proteins called **competence factors.** These proteins facilitate the binding of DNA fragments to the cell surface, the uptake of DNA into the cytoplasm, and its subsequent incorporation into the bacterial chromosome. Temperature, ionic conditions, and nutrient availability can affect whether or not a bacterium is competent to take up genetic material from its environment. These conditions influence the expression of competence genes.

In recent years, geneticists have unraveled some of the steps that occur when competent bacterial cells are transformed by genetic material in their environment. **Figure 7.13** describes the steps of transformation. First, a large fragment of genetic material binds to the surface of the bacterial cell. Competent cells express DNA receptors that promote such binding. Before entering the cell, however, this large piece of chromosomal DNA must be cut into smaller fragments. This cutting is accomplished by an extracellular bacterial enzyme known as an endonuclease, which makes occasional random cuts in the long piece of chromosomal DNA. At this stage, the DNA fragments are composed of double-stranded DNA.

In the next step, the DNA fragment begins its entry into the bacterial cytoplasm. For this to occur, the double-stranded DNA interacts with proteins in the bacterial membrane. One of the DNA strands is degraded, and the other strand enters the bacterial cytoplasm via an uptake system, which is structurally similar to the one described for conjugation (as shown earlier in Figure 7.4a), but is involved with DNA uptake rather than export.

To be stably inherited, the DNA strand must be incorporated into the bacterial chromosome. If the DNA strand has a sequence that is similar to a region of DNA in the bacterial chromosome, the DNA may be incorporated into the chromosome by a process known as **homologous recombination,** discussed

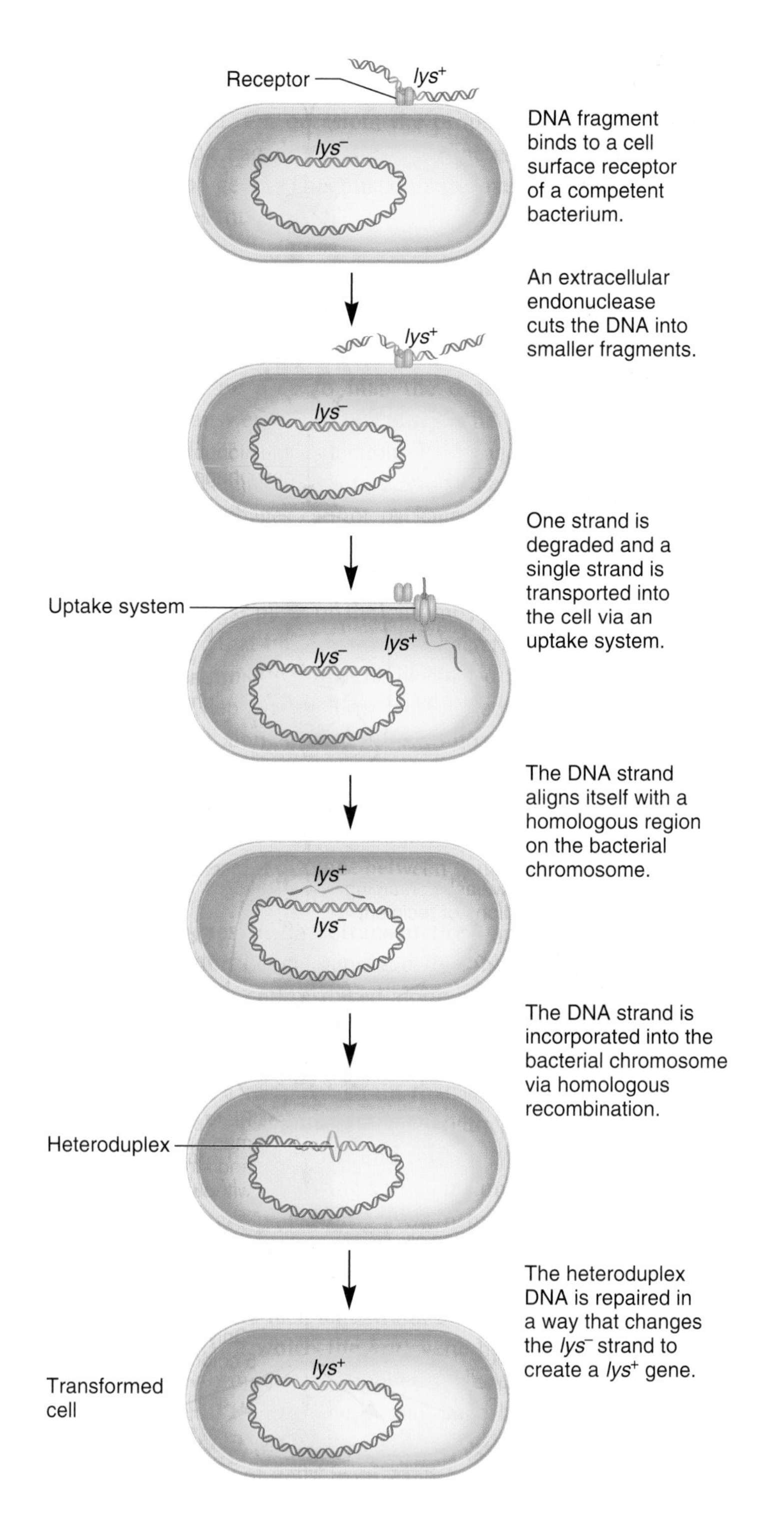

FIGURE 7.13 **The steps of bacterial transformation.** In this example, a fragment of DNA carrying a lys^+ gene enters the competent cell and recombines with the chromosome, transforming the bacterium from lys^- to lys^+.

Genes → Traits Bacterial transformation can also lead to new traits for the recipient cell. The recipient cell was lys^- (unable to synthesize the amino acid lysine). Following transformation, it became lys^+. This result would transform the recipient bacterial cell into a cell that could synthesize lysine and grow on a medium that lacked this amino acid. Before transformation, the recipient lys^- cell would not have been able to grow on a medium lacking lysine.

in detail in Chapter 17. For this to occur, the single-stranded DNA aligns itself with the homologous location on the bacterial chromosome. In the example shown in Figure 7.13, the foreign DNA carries a functional *lys*$^+$ gene that aligns itself with a nonfunctional (mutant) *lys*$^-$ gene already present within the bacterial chromosome. The foreign DNA then recombines with one of the strands in the bacterial chromosome of the competent cell. In other words, the foreign DNA replaces one of the chromosomal strands of DNA, which is subsequently degraded. During homologous recombination, alignment of the *lys*$^-$ and the *lys*$^+$ alleles results in a region of DNA called a **heteroduplex** that contains one or more base sequence mismatches. However, the heteroduplex exists only temporarily. DNA repair enzymes in the recipient cell recognize the heteroduplex and repair it. In this case, the heteroduplex has been repaired by eliminating the mutation that caused the *lys*$^-$ genotype, thereby creating a *lys*$^+$ gene. In this example, the recipient cell has been transformed from a *lys*$^-$ strain to a *lys*$^+$ strain. Alternatively, a DNA fragment that has entered a cell may not be homologous to any genes that are already found in the bacterial chromosome. In this case, the DNA strand may be incorporated at a random site in the chromosome. This process is known as **nonhomologous,** or **illegitimate, recombination.**

Some bacteria preferentially take up DNA fragments from other bacteria of the same species or closely related species. How does this occur? Recent research has shown that the mechanism can vary among different species. In *Streptococcus pneumoniae*, the cells secrete a short peptide called the **competence-stimulating peptide (CSP).** When many *S. pneumoniae* cells are in the vicinity of one another, the concentration of CSP becomes high, which stimulates the cells, via a cell-signaling pathway, to express the competence proteins needed for the uptake of DNA and its incorporation in the chromosome. Because competence requires a high external concentration of CSP, *S. pneuomoniae* cells are more likely to take up DNA from nearby *S. pneumoniae* cells that have died and released their DNA into the environment.

Other bacterial species promote the uptake of DNA among members of their own species via **DNA uptake signal sequences,** which are 9 or 10 bp long. In the human pathogens *Neisseria meningitidis* (a causative agent of meningitis), *N. gonorrhoeae* (a causative agent of gonorrhea), and *Haemophilus influenzae* (a causative agent of ear, sinus, and respiratory infections), these sequences are found at many locations within their respective genomes. For example, *H. influenzae* contains approximately 1500 copies of the sequence 5′-AAGTGCGGT-3′ in its genome, and *N. meningitidis* contains about 1900 copies of the sequence 5′-GCCGTCTGAA-3′. DNA fragments that contain their own uptake signal sequence are preferentially taken up by these species instead of other DNA fragments. For example, *H. influenzae* is much more likely to take up a DNA fragment with the sequence 5′-AAGTGCGGT-3′. For this reason, transformation is more likely to involve DNA uptake between members of the same species.

Transformation has also been used to map many bacterial genes, using methods similar to the cotransduction experiments described earlier. If two genes are close together, the **cotransformation** frequency is expected to be high, whereas genes that are far apart have a cotransformation frequency that is very low or even zero. Like cotransduction, genetic mapping via cotransformation is used only to map genes that are relatively close together.

Bacteria May Acquire New Genes by Horizontal Gene Transfer

The transmission of genes from mother cell to daughter cell or from parent to offspring is called **vertical gene transfer.** By comparison, **horizontal gene transfer** is a process in which an organism incorporates genetic material from another organism without being the offspring of that organism. Horizontal gene transfer can involve the exchange of genetic material between members of the same species or different species. The three mechanisms of genetic transfer that we have considered—conjugation, transduction, and transformation—are important mechanisms for horizontal gene transfer among bacterial species. When analyzing the genomes of bacterial species, researchers have discovered that a sizable fraction of their genes are derived from horizontal gene transfer. For example, over the past 100 million years, *E. coli* and *Salmonella typhimurium* have acquired roughly 17% of their genes via horizontal gene transfer.

The types of genes that bacteria acquire via horizontal gene transfer are quite varied, though they commonly involve functions that are readily acted on by natural selection. These include genes that confer antibiotic resistance, the ability to degrade toxic compounds, and pathogenicity. Geneticists have suggested that much of the speciation that has occurred in prokaryotic species is the result of horizontal gene transfer. In many cases, the acquisition of new genes allows a novel survival strategy that has led to the formation of a new species. These processes are considered in detail in Chapters 24 and 26.

The medical relevance of horizontal gene transfer is quite profound. Antibiotics are commonly prescribed to treat many bacterial illnesses, including infections of the respiratory tract, urinary tract, skin, ears, and eyes. In addition, antibiotics are used in agriculture as a supplement in animal feed and to control certain bacterial diseases of high-value fruits and vegetables. Unfortunately, however, the widespread and uncontrolled use of antibiotics has promoted the prevalence of antibiotic-resistant strains of bacteria. This phenomenon, termed **acquired antibiotic resistance,** may occur via genetic alterations in the bacteria's own genome or by the horizontal transfer of resistance genes from a resistant strain to a sensitive strain. Resistant strains carry genes that counteract the effects of antibiotics. Such resistance genes encode proteins that either break down the drug, pump the drug out of the cell, or prevent the drug from inhibiting cellular processes.

Bacterial resistance to antibiotics in community-acquired respiratory tract infections, such as pneumonia, as well as other medical illnesses, is a serious problem, and it is increasing in prevalence worldwide at an alarming rate. As often mentioned in the news media, antibiotic resistance has increased dramatically over the past few decades, and resistance has been reported in almost all species of bacteria. In many countries, for example, penicillin resistance in *Streptococcus pneumoniae* is found in over 50% of all strains, with resistance to other drugs rising as well. Likewise,

E. coli strains designated *E. coli* K12S and *E. coli* K12(λ). He was growing these two strains to teach his class about the lysogenic cycle. *E. coli* K12(λ) has DNA from another phage, called lambda, integrated into its chromosome, whereas *E. coli* K12S does not. To see if the use of these strains might improve phage yield, *E. coli* B, *E. coli* K12S, and *E. coli* K12(λ) were infected with the *rII* and wild-type T4 phage strains. As expected, the wild-type phage could infect all three bacterial strains. However, the *rII* mutant strains behaved quite differently. In *E. coli* B, the *rII* strains produced large plaques that had poor yields of bacteriophage. In *E. coli* K12S, the *rII* mutants produced normal plaques that gave good yields of phage. Surprisingly, in *E. coli* K12(λ), the *rII* mutants were unable to produce plaques at all, for reasons that were not understood. Nevertheless, as we will see later, this fortuitous observation was a critical feature that allowed intragenic mapping in this bacteriophage.

A Complementation Test Can Reveal If Mutations Are in the Same Gene or in Different Genes

In his experiments, Benzer was interested in a single trait, namely, the ability to form plaques. He had isolated many *rII* mutant strains that could form large plaques in *E. coli* B but could not produce plaques in *E. coli* K12(λ). To attempt gene mapping, he needed to know if the various *rII* mutations were in the same gene or if they involved mutations in different genes. To accomplish this, he conducted a **complementation test.** The goal of this type of approach is to determine if two different mutations that affect the same trait are in the same gene or in two different genes (also see Figure 4.20).

The possible outcomes of complementation tests involving mutations that affect plaque formation are shown in **Figure 7.17.** This example involves four different *rII* mutations in T4 bacteriophage, designated strains 1 through 4, that prevent plaque formation in *E. coli* K12(λ). To conduct this complementation experiment, bacterial cells were coinfected with an excess of two different strains of T4 phage. Two distinct outcomes are possible. In Figure 7.17a, the two *rII* phage strains possess deleterious mutations in the same gene (gene *A*). Because they cannot make a wild-type gene *A* product when coinfected into an *E. coli* K12(λ) cell, plaques do not form. This phenomenon is called **noncomplementation.**

Alternatively, if each *rII* mutation is in a different phage gene (e.g., gene *A* and gene *B*), a bacterial cell that is coinfected by both types of phages will have two mutant genes as well as two wild-type genes (Figure 7.17b). If the mutant phage genes behave in a recessive fashion, the doubly infected cell will have a wild-type phenotype. Why does this phenotype occur? The coinfected cells produce normal proteins that are encoded by the wild-type versions of both genes *A* and *B*. For this reason, coinfected cells are lysed in the same manner as if they were infected by the wild-type strain. Therefore, this coinfection should be able to produce plaques in *E. coli* K12(λ). This result is called **complementation** because the defective genes in each *rII* strain are complemented by the corresponding wild-type genes. It should be noted that, for a variety of reasons, intergenic complementation may not always work. One possibility is that a mutation may behave in a dominant fashion. In addition, mutations that affect regulatory genetic regions rather than the protein-coding region may not show complementation.

By carefully considering the pattern of complementation and noncomplementation, Benzer found that the *rII* mutations occurred in two different genes, which were termed *rIIA* and

rII strain 1
(gene *A* is defective,
gene *B* is normal)
gene *A* gene *B*

rII strain 2
(gene *A* is defective,
gene *B* is normal)
gene *A* gene *B*

Coinfect *E. coli* K12(λ)

Plate and observe if
plaques are formed.

No plaques

No complementation occurs, because the coinfected cell is unable to make the normal product of gene *A*. The coinfected cell will not produce viral particles, thus no bacterial cell lysis and no plaque formation.

(a) Noncomplementation: The phage mutations are in the same gene.

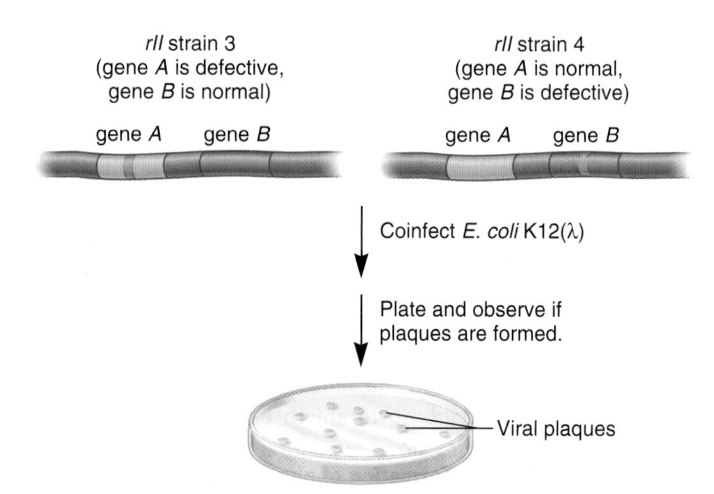

Complementation occurs, because the coinfected cell is able to make normal products of gene *A* and gene *B*. The coinfected bacterial cell will produce viral particles that lyse the cell, resulting in the appearance of clear plaques.

(b) Complementation: The phage mutations are in different genes.

FIGURE 7.17 A comparison of noncomplementation and complementation. Four different T4 phage strains (designated 1 through 4) that carry *rII* mutations were coinfected into *E. coli* K12(λ). **(a)** If two *rII* phage strains possess mutations in the same gene, noncomplementation will occur. **(b)** If the *rII* mutations are in different genes (such as gene *A* and gene *B*), a coinfected cell will have two mutant genes but also two wild-type genes. Doubly infected cells with a wild-type copy of each gene can produce new phages and form plaques. This result is called complementation because the defective genes in each *rII* strain are complemented by the corresponding wild-type genes.

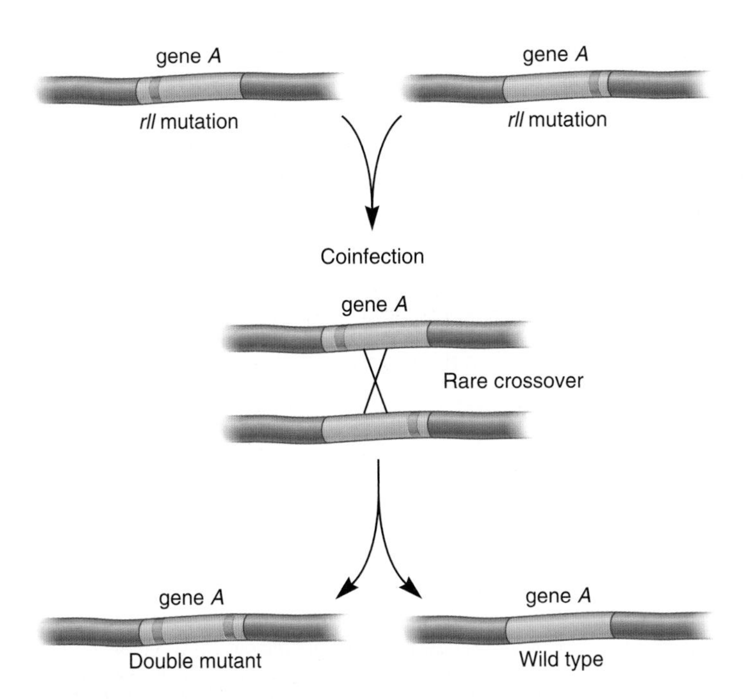

FIGURE 7.18 **Intragenic recombination.** Following coinfection, a rare crossover has occurred between the sites of the two mutations. This produces a wild-type phage with no mutations and a double-mutant phage with both mutations.

rIIB. The identification of two distinct genes affecting plaque formation was a necessary step that preceded his intragenic mapping analysis, which is described next. Benzer coined the term **cistron** to refer to the smallest genetic unit that gives a negative complementation test. In other words, if two mutations occur within the same cistron, they cannot complement each other. Since these studies, researchers have learned that a cistron is equivalent to a gene. In recent decades, the term gene has gained wide popularity but the term cistron is not commonly used. However, the term polycistronic is still used to describe bacterial mRNAs that carry two or more gene sequences, as described in Chapter 14.

Intragenic Maps Were Constructed Using Data from a Recombinational Analysis of Mutants Within the *rII* Region

As we saw in Figure 7.17, the ability of strains with mutations in two different genes to produce viral plaques after coinfection is due to complementation. Noncomplementation occurs when two different strains have mutations in the same gene. However, at an extremely low rate, two noncomplementing strains of viruses can produce an occasional viral plaque if intragenic recombination has taken place. For example, **Figure 7.18** shows a coinfection experiment between two phage strains that both contain *rII* mutations in gene *A*. These mutations are located at different places within the same gene. On rare occasions, a crossover may occur in the very short region between each mutation. This crossover produces a double mutant gene *A* and a wild-type gene *A*. Because this event has produced a wild-type gene *A*, the function of the protein encoded by gene *A* is restored. Therefore, new phages can be made in *E. coli* K12(λ), resulting in the formation of a viral plaque.

Figure 7.19 describes the general strategy for intragenic mapping of *rII* phage mutations. Bacteriophages from two different noncomplementing *rII* phage mutants (here, *r103* and *r104*) were mixed together in equal numbers and then infected into *E. coli* B. In this strain, the *rII* mutants grew and propagated. Recall from Figure 7.18, when two different mutants coinfect the same cell, intragenic recombination can occur, producing wild-type phages and double-mutant phages. However, these intragenic recombinants were produced at a very low rate. Following coinfection and lysis of *E. coli* B, a new population of phages was isolated. This population was expected to contain predominantly nonrecombinant phages. However, due to intragenic recombination, it should also contain a very low percentage of wild-type phages and double-mutant phages (refer back to Figure 7.18).

How could Benzer determine the number of rare phages that were produced by intragenic recombination? The key approach is that *rII* mutant phages cannot grow in *E. coli* K12(λ). Following coinfection, he took this new population of phages and used some of them to infect *E. coli* B and some to infect *E. coli* K12(λ). After plating, the *E. coli* B infection was used to determine the total number of phages, because *rII* mutants as well as wild-type phages can produce plaques in this strain. The overwhelming majority of these phages were expected to be nonrecombinant phages. The *E. coli* K12(λ) infection was used to determine the number of rare intragenic recombinants that produce wild-type phages.

Figure 7.19 illustrates the great advantage of this experimental system in detecting a low percentage of recombinants. In the laboratory, phage preparations containing several billion phages per milliliter are readily made. Among billions of phages, a low percentage (e.g., 1 in every 1000) may be wild-type phages arising from intragenic recombination. The wild-type recombinants can produce plaques in *E. coli* K12(λ), whereas the *rII* mutant strains cannot. In other words, only the tiny fraction of wild-type recombinants would produce plaques in *E. coli* K12(λ).

The frequency of recombinant phages can be determined by comparing the number of wild-type phages, produced by intragenic recombination, and the total number of phages. As shown in Figure 7.19, the total number of phages can be deduced from the number of plaques obtained from the infection of *E. coli* B. In this experiment, the phage preparation was diluted by 10^8 (1:100,000,000), and 1 mL was used to infect *E. coli* B. Because this plate produced 66 plaques, the total number of phages in the original preparation was $66 \times 10^8 = 6.6 \times 10^9$, or 6.6 billion phages per milliliter. By comparison, the phage preparation used to infect *E. coli* K12(λ) was diluted by only 10^6 (1 : 1,000,000). This plate produced 11 plaques. Therefore, the number of wild-type phages was 11×10^6, which equals 11 million wild-type phages per milliliter.

As we have already seen in Chapter 6, genetic mapping distance is computed by dividing the number of recombinants by the total population (nonrecombinants and recombinants) times 100. In this experiment, intragenic recombination produced an equal number of two types of recombinants: wild-type phages and double-mutant phages. Only the wild-type phages are detected

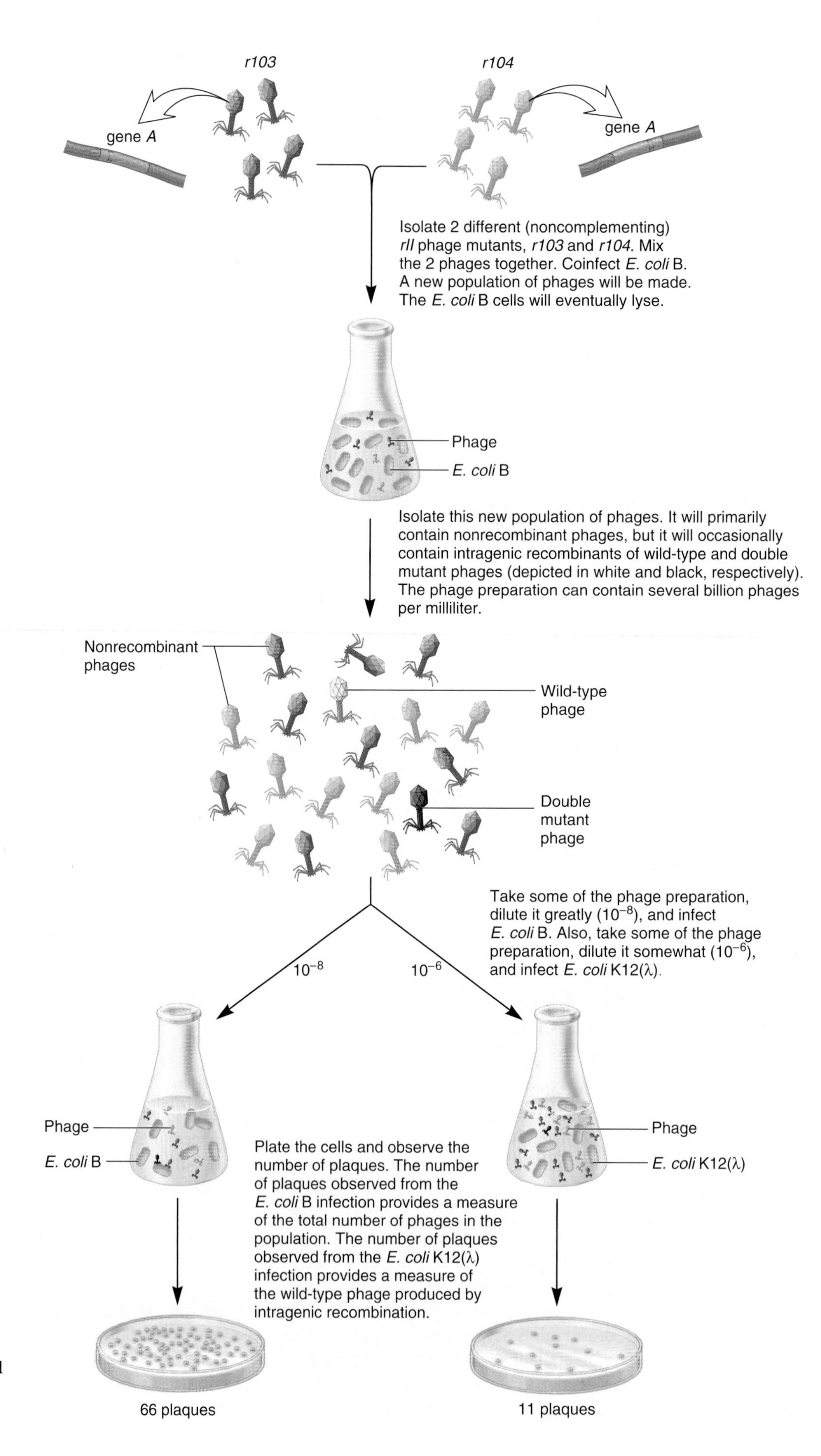

FIGURE 7.19 Benzer's method of intragenic mapping in the *rII* region.

in the infection of *E. coli* K12(λ). Therefore, to obtain the total number of recombinants, the number of wild-type phages must be multiplied by 2. With all this information, we can use the following equation to compute the frequency of recombinants using the experimental approach described in Figure 7.19.

$$\text{Frequency of recombinants} = \frac{2\ [\text{Wild-types plaques obtained in } E.\ coli\ \text{K12}(\lambda)]}{\text{Total number of plaques obtained in } E.\ coli\ \text{B}}$$

$$\text{Frequency of recombinants} = \frac{2\ (11 \times 10^6)}{6.6 \times 10^9}$$

$$= 3.3 \times 10^{-3} = 0.0033$$

In this example, approximately 3.3 recombinants were produced per 1000 phages.

The frequency of recombinants provides a measure of map distance. In eukaryotic mapping studies, we compute the map distance by multiplying the frequency of recombinants by 100 to give a value in map units (also known as centiMorgans). Similarly, in these experiments, the frequency of recombinants can provide a measure of map distance along the bacteriophage DNA. In this case, the map distance is between two mutations within the same gene. Like intergenic mapping, the frequency of intragenic recombinants is correlated with the distance between the two mutations; the farther apart they are, the higher the frequency of recombinants. If two mutations happen to be located at exactly the same site within a gene, coinfection would not be able to produce any wild-type recombinants, and so the map distance would be zero. These are known as **homoallelic** mutations.

Deletion Mapping Can Be Used to Localize Many *rII* Mutations to Specific Regions in the *rIIA* or *rIIB* Genes

Now that we have seen the general approach to intragenic mapping, let's consider a method to efficiently map hundreds of *rII* mutations within the two genes designated *rIIA* and *rIIB*. As you may have realized, the coinfection experiments described in Figure 7.19 are quite similar to Sturtevant's strategy of making dihybrid crosses to map genes along the X chromosome of *Drosophila* (refer back to Figure 6.10). Similarly, Benzer wanted to coinfect different *rII* mutants in order to map the sites of the mutations within the *rIIA* and *rIIB* genes. During the course of his work, he obtained hundreds of different *rII* mutant strains that he wanted to map. However, making all the pairwise combinations would have been an overwhelming task. Instead, Benzer used an approach known as **deletion mapping** as a first step in localizing his *rII* mutations to a fairly short region within gene *A* or gene *B*.

Figure 7.20 describes the general strategy used in deletion mapping. This approach is easier to understand if we use an example. Let's suppose that the goal is to know the approximate location of an *rII* mutation, such as *r103*. To do so, *E. coli* K12(λ) is coinfected with *r103* and a deletion strain. Each deletion strain is a T4 bacteriophage that is missing a known segment of the *rIIA* and/or *rIIB* gene. If the deleted region includes the same

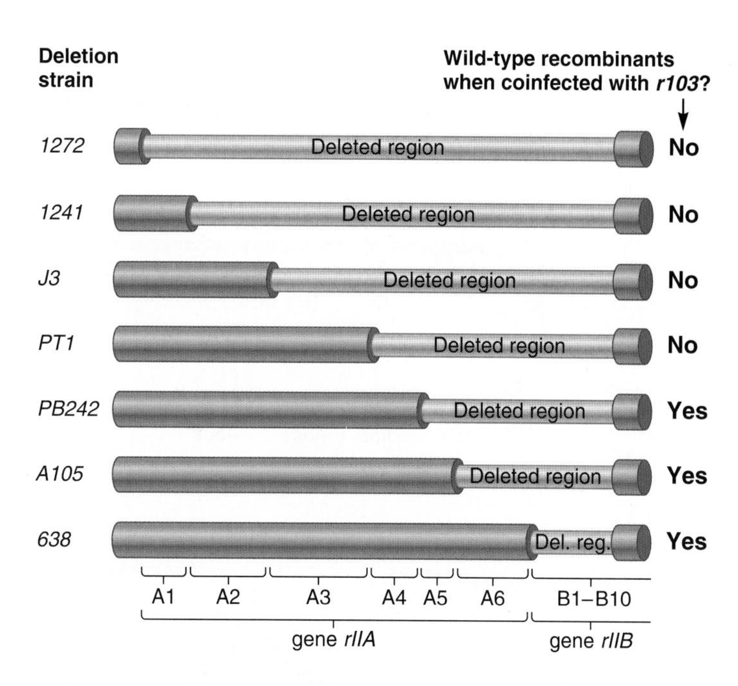

FIGURE 7.20 The use of deletion strains to localize *rII* mutants to short regions within the *rIIA* or *rIIB* gene. The deleted regions are shown in gray.

region that contains the *r103* mutation, a coinfection cannot produce intragenic wild-type recombinants. Therefore, plaques will not form. However, if a deletion strain recombines with *r103* to produce a wild-type phage, the deleted region does not overlap with the *r103* mutation.

In the example shown in Figure 7.20, the *r103* strain produced wild-type recombinants when coinfected with deletion strains *PB242*, *A105*, and *638*. However, coinfection of *r103* with *PT1*, *J3*, *1241*, and *1272* did not produce intragenic wild-type recombinants. Because coinfection with *PB242* produced recombinants and *PT1* did not, the *r103* mutation must be located in the region that is missing in *PT1* but not missing in *PB242*. As shown at the bottom of Figure 7.20, this region is called A4 (the A refers to the *rIIA* gene). In other words, the *r103* mutation is located somewhere within the A4 region, but not in the other six regions (A1, A2, A3, A5, A6, and B).

As described in Figure 7.20, this first step in the deletion mapping strategy localized an *rII* mutation to one of seven regions; six of these were in *rIIA* and one was in *rIIB*. Other deletion strains were used to eventually localize each *rII* mutation to one of 47 short regions; 36 were in *rIIA*, 11 in *rIIB*. At this point, pairwise coinfections were made between mutant strains that had been localized to the same region by deletion mapping. For example, 24 mutations were deletion-mapped to a region called A5d. Pairwise coinfection experiments were conducted among this group of 24 mutants to precisely map their locations relative to each other in the A5d region. Similarly, all mutants in each of the 46 other groups were mapped by pairwise coinfections. In this way, a fine structure map was constructed depicting the locations of hundreds of different *rII* mutations (**Figure 7.21**). As seen in this figure, certain locations contained a relatively high number of mutations compared with other sites. These were termed **hot spots** for mutation.

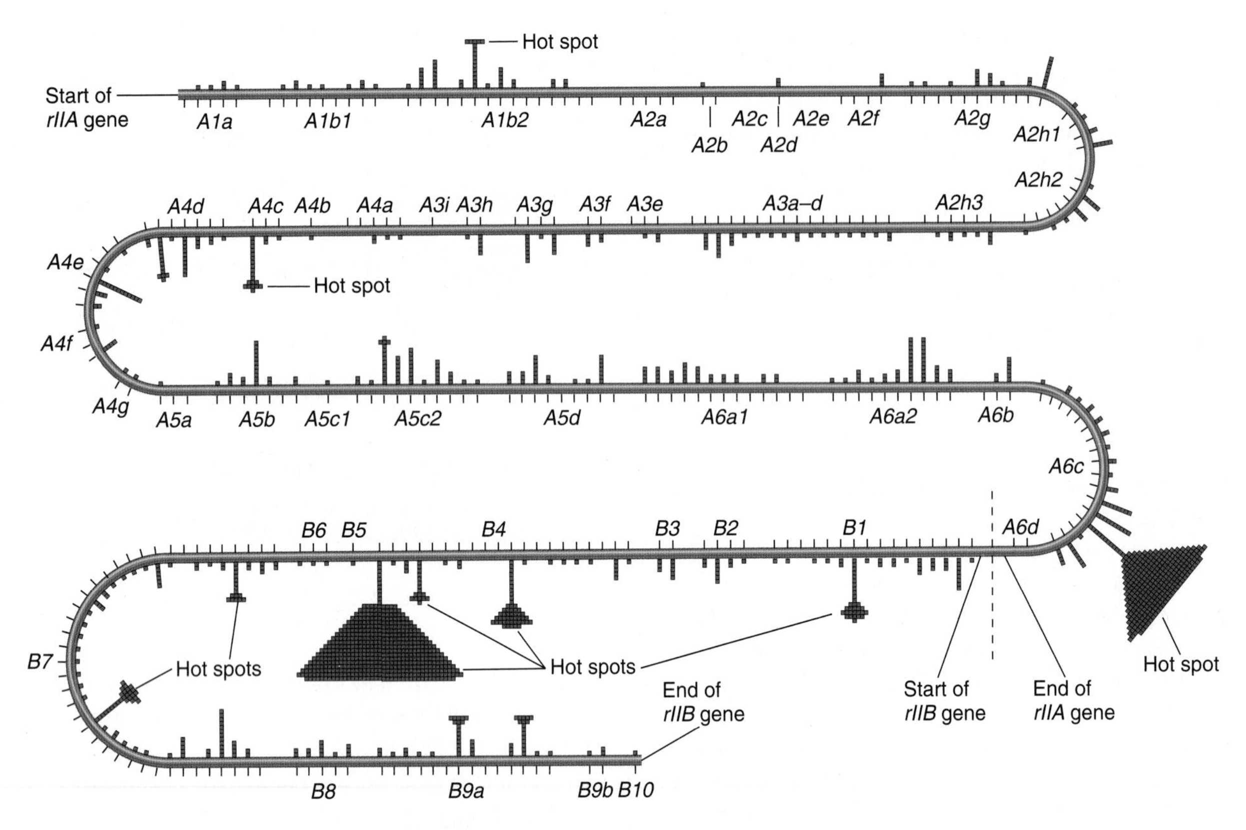

FIGURE 7.21 **The outcome of intragenic mapping of many *rII* mutations.** The blue line represents the linear sequence of the *rIIA* and *rIIB* genes, which are found within the T4 phage's genetic material. Each small purple box attached to the blue line symbolizes a mutation that was mapped by intragenic mapping. Among hundreds of independent mutant phages, several mutations sometimes mapped to the same site. In this figure, mutations at the same site form columns of boxes. Hot spots contain a large number of mutations and are represented as a group of boxes attached to a column of boxes. A hot spot contains many mutations at the same site within the *rIIA* or *rIIB* gene.

Intragenic Mapping Experiments Provided Insight into the Relationship Between Traits and Molecular Genetics

Intragenic mapping studies were a pivotal achievement in our early understanding of gene structure. Since the time of Mendel, geneticists had considered a gene to be the smallest unit of heredity, which provided an organism with its inherited traits. In the late 1950s, however, the molecular nature of the gene was not understood. Because it is a unit of heredity, some scientists envisioned a gene as being a particle-like entity that could not be further subdivided into additional parts. However, intragenic mapping studies revealed, convincingly, that this is not the case. These studies showed that mutations can occur at many different sites within a single gene. Furthermore, intragenic crossing over could recombine these mutations, resulting in wild-type genes. Therefore, rather than being an indivisible particle, a gene must be composed of a large structure that can be subdivided during crossing over.

Benzer's results were published in the late 1950s and early 1960s, not long after the physical structure of DNA had been elucidated by Watson and Crick. We now know that a gene is a segment of DNA that is composed of smaller building blocks called nucleotides. A typical gene is a linear sequence of several hundred to many thousand base pairs. As the genetic map of Figure 7.21 indicates, mutations can occur at many sites along the linear structure of a gene; intragenic crossing over can recombine mutations that are located at different sites within the same gene.

KEY TERMS

Page 161. bacteriophages, phages, genetic transfer, conjugation
Page 162. transduction, transformation, minimal medium, auxotroph, prototroph
Page 163. F factor
Page 164. sex pilus, conjugation bridge, relaxosome, origin of transfer, nucleoprotein
Page 165. *Hfr* strain, F′ factors
Page 167. interrupted mating
Page 169. minutes, plasmid, episomes
Page 170. lytic cycle, virulent phages, lysogenic cycle, prophage, temperate phage
Page 171. generalized transduction
Page 172. cotransduction
Page 174. natural transformation, artificial transformation, competent cells, competence factors, homologous recombination

Page 175. heteroduplex, nonhomologous recombination, illegitimate recombination, competence-stimulating peptide (CSP), DNA uptake signal sequences, cotransformation, vertical gene transfer, horizontal gene transfer, acquired antibiotic resistance
Page 176. viruses, intragenic mapping, fine-structure mapping, plaque
Page 178. complementation test, noncomplementation, complementation
Page 179. cistron
Page 181. homoallelic, deletion mapping, hot spots

CHAPTER SUMMARY

7.1 Genetic Transfer and Mapping in Bacteria

- Three general mechanisms for genetic transfer in various species of bacteria are conjugation, transduction, and transformation (see Table 7.1).
- Joshua Lederberg and Edward Tatum discovered conjugation in *E. coli* by analyzing auxotrophic strains (see Figure 7.1).
- Using a U-tube apparatus, Davis showed that conjugation required cell-to-cell contact (see Figure 7.2).
- Certain strains of bacteria have F factors, which they can transfer via conjugation in a series of steps (see Figures 7.3, 7.4).
- *Hfr* strains are formed when an F factor integrates into the bacterial chromosome. The imprecise excision can produce an F′ factor that carries a portion of the bacterial chromosome (see Figure 7.5).
- *Hfr* strains can transfer a portion of the bacterial chromosome to a recipient cell during conjugation (see Figure 7.6).
- Wollman and Jacob showed that conjugation can be used to map the locations of genes along the bacterial chromosome, thereby creating a genetic map (see Figures 7.7–7.9).
- Bacteriophages may follow a lytic or lysogenic reproductive cycle (see Figure 7.10).
- During transduction, a portion of a bacterial chromosome is transferred to a recipient cell via a bacteriophage (see Figure 7.11).
- A cotransduction experiment can be used to map genes that are close together along a bacterial chromosome (see Figure 7.12).
- During transformation, a segment of DNA is taken up by a bacterial cell and then is incorporated into the bacterial chromosome. (Figure 7.13)

7.2 Intragenic Mapping in Bacteriophages

- Bacteriophages are viruses that infect bacteria (see Figure 7.14).
- Some bacteriophages cause bacteria to lyse. This event can lead to plaque formation when bacteria are plated as a lawn on bacterial media (see Figure 7.15).
- Benzer studied a type of bacteriophage called T4. He identified mutant strains that caused rapid lysis, thereby leading to larger plaques (see Figure 7.16).
- Benzer identified strains of T4 that could not form plaques in *E. coli* K12(λ). He conducted complementation tests to determine if the mutations in these strains were in the same phage gene or in different genes (see Figure 7.17).
- Benzer devised a method to study intragenic recombination. He used deletion mapping to determine the locations of the mutations within two phage genes (see Figures 7.18–7.21).

PROBLEM SETS & INSIGHTS

Solved Problems

S1. In *E. coli*, the gene $bioD^+$ encodes an enzyme involved in biotin synthesis, and $galK^+$ encodes an enzyme involved in galactose utilization. An *E. coli* strain that contained wild-type versions of both genes was infected with P1, and then a P1 lysate was obtained. This lysate was used to transduce (infect) a strain that was $bioD^-$ and $galK^-$. The cells were plated on media containing galactose as the sole carbon source for growth to select for transduction of the $galK^+$ gene. These media also were supplemented with biotin. The colonies were then restreaked on media that lacked biotin to see if the $bioD^+$ gene had been cotransduced. The following results were obtained:

Selected Gene	Nonselected Gene	Number of Colonies That Grew On: Galactose + Biotin	Galactose − Biotin	Cotransduction Frequency
$galK^+$	$bioD^+$	80	10	0.125

How far apart are these two genes?

Answer: We can use the cotransduction frequency to calculate the distance between the two genes (in minutes) using the equation

$$\text{Cotransduction frequency} = (1 - d/2)^3$$

$$0.125 = (1 - d/2)^3$$

$$1 - d/2 = \sqrt[3]{0.125}$$

$$1 - d/2 = 0.5$$

$$d/2 = 1 - 0.5$$

$$d = 1.0 \text{ minute}$$

The two genes are approximately 1 minute apart on the *E. coli* chromosome.

S2. By conducting mating experiments between a single *Hfr* strain and a recipient strain, Wollman and Jacob mapped the order of many bacterial genes. Throughout the course of their studies, they identified several different *Hfr* strains in which the F factor DNA had been integrated at different places along the bacterial chromosome. A sample of their experimental results is shown in the following table:

Hfr strain	Origin	Order of Transfer of Several Different Bacterial Genes: First								Last
H	O	*thr*	*leu*	*azi*	*ton*	*pro*	*lac*	*gal*	*str*	*met*
1	O	*leu*	*thr*	*met*	*str*	*gal*	*lac*	*pro*	*ton*	*azi*
2	O	*pro*	*ton*	*azi*	*leu*	*thr*	*met*	*str*	*gal*	*lac*
3	O	*lac*	*pro*	*ton*	*azi*	*leu*	*thr*	*met*	*str*	*gal*
4	O	*met*	*str*	*gal*	*lac*	*pro*	*ton*	*azi*	*leu*	*thr*
5	O	*met*	*thr*	*leu*	*azi*	*ton*	*pro*	*lac*	*gal*	*str*
6	O	*met*	*thr*	*leu*	*azi*	*ton*	*pro*	*lac*	*gal*	*str*
7	O	*ton*	*azi*	*leu*	*thr*	*met*	*str*	*gal*	*lac*	*pro*

A. Explain how these results are consistent with the idea that the bacterial chromosome is circular.

B. Draw a map that shows the order of genes and the locations of the origins of transfer among these different *Hfr* strains.

Answer:

A. In comparing the data among different *Hfr* strains, the order of the nine genes was always the same or the reverse of the same order. For example, *HfrH* and *Hfr4* transfer the same genes but their orders are reversed relative to each other. In addition, the *Hfr* strains showed an overlapping pattern of transfer with regard to the origin. For example, *Hfr1* and *Hfr2* had the same order of genes, but *Hfr1* began with *leu* and ended with *azi*, whereas *Hfr2* began with *pro* and ended with *lac*. From these findings, Wollman and Jacob concluded that the segment of DNA that was the origin of transfer had been inserted at different points within a circular *E. coli* chromosome in different *Hfr* strains. They also concluded that the origin can be inserted in either orientation, so the direction of gene transfer can be clockwise or counterclockwise around the circular bacterial chromosome.

B. A genetic map consistent with these results is shown here.

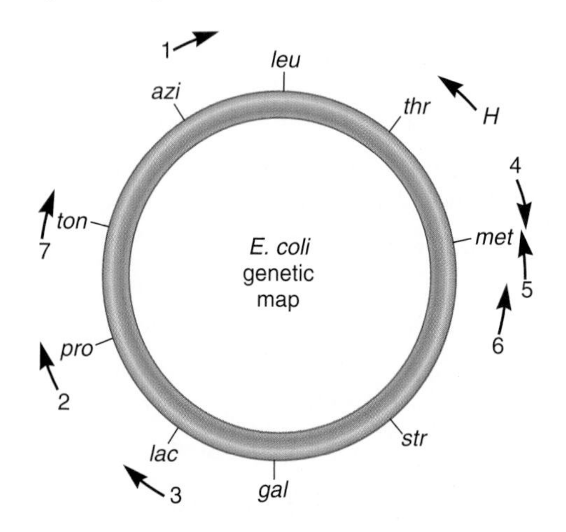

S3. An *Hfr* strain that is *leuA*$^+$ and *thiL*$^+$ was mated to a strain that is *leuA*$^-$ and *thiL*$^-$. In the data points shown here, the mating was interrupted, and the percentage of recombinants for each gene was determined by streaking on media that lacked either leucine or thiamine. The results are shown in the following graph.

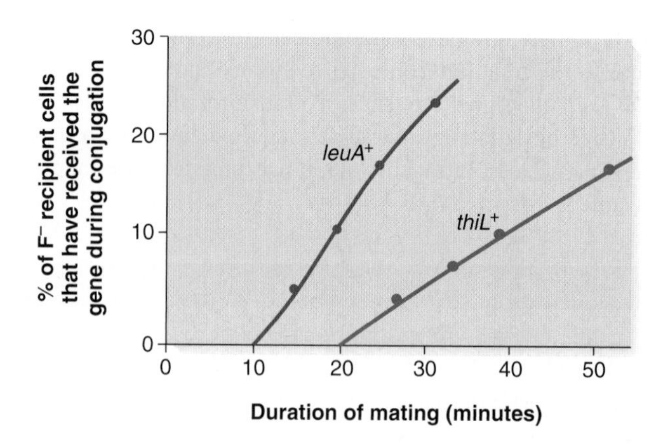

What is the map distance (in minutes) between these two genes?

Answer: This problem is solved by extrapolating the data points to the *x*-axis to determine the time of entry. For *leuA*$^+$, they extrapolate back to 10 minutes. For *thiL*$^+$, they extrapolate back to 20 minutes. Therefore, the distance between the two genes is approximately 10 minutes.

S4. Genetic transfer via transformation can also be used to map genes along the bacterial chromosome. In this approach, fragments of chromosomal DNA are isolated from one bacterial strain and used to transform another strain. The experimenter examines the transformed bacteria to see if they have incorporated two or more different genes. For example, the DNA may be isolated from a donor *E. coli* bacterium that has functional copies of the *araB* and *leuD* genes. Let's call these genes *araB*$^+$ and *leuD*$^+$ to indicate the genes are functional. These two genes are required for arabinose metabolism and leucine synthesis, respectively. To map the distance between these two genes via transformation, a recipient bacterium is used that is *araB*$^-$ and *leuD*$^-$. Following transformation, the recipient bacterium may become *araB*$^+$ and *leuD*$^+$. This phenomenon is called cotransformation because two genes from the donor bacterium have been transferred to the recipient via transformation. In this type of experiment, the recipient cell is exposed to a fairly low concentration of donor DNA, making it unlikely that the recipient bacterium will take up more than one fragment of DNA. Therefore, under these conditions, cotransformation is likely only when two genes are fairly close together and are found on one fragment of DNA.

In a cotransformation experiment, a researcher has isolated DNA from an *araB*$^+$ and *leuD*$^+$ donor strain. This DNA was transformed into a recipient strain that was *araB*$^-$ and *leuD*$^-$. Following transformation, the cells were plated on a medium containing arabinose and leucine. On this medium, only bacteria that are *araB*$^+$ can grow. The bacteria can be either *leuD*$^+$ or *leuD*$^-$ because leucine is provided in the medium. Colonies that grew on this medium were then restreaked on a medium that contained arabinose but lacked leucine. Only *araB*$^+$ and *leuD*$^+$ cells could grow on these secondary plates. Following this protocol, a researcher obtained the following results:

Number of colonies growing on arabinose plus leucine media: 57

Number of colonies that grew when restreaked on an arabinose medium without leucine: 42

What is the map distance between these two genes? Note: This problem can be solved using the strategy of a cotransduction experiment except that the researcher must determine the average size of DNA fragments that are taken up by the bacterial cells. This would correspond to the value of *L* in a cotransduction experiment.

Answer: As mentioned, the basic principle of gene mapping via cotransformation is identical to the method of gene mapping via cotransduction described in this chapter. One way to calculate the map distance is to use the same equation that we used for cotransduction data, except that we substitute cotransformation frequency for cotransduction frequency.

$$\text{Cotransformation frequency} = (1 - d/L)^3$$

(Note: Cotransformation is not quite as accurate as cotransduction because the sizes of chromosomal pieces tend to vary significantly from experiment to experiment, so the value of L is not quite as reliable. Nevertheless, cotransformation has been used extensively to map the order and distance between closely linked genes along the bacterial chromosome.)

The researcher needs to experimentally determine the value of L by running the DNA on a gel and estimating the average size of the DNA fragments. Let's assume they are about 2% of the bacterial chromosome, which, for *E. coli*, would be about 80,000 bp in length. So L equals 2 minutes, which is the same as 2%.

$$\text{Cotransformation frequency} = (1 - d/\text{L})^3$$

$$42/57 = (1 - d/2)^3$$

$$d = 0.2 \text{ minutes}$$

The distance between *araB* and *leuD* is approximately 0.2 minutes.

S5. In our discussion of transduction via P1 or P22, the reproductive cycle of the bacteriophage sometimes resulted in the packaging of many different pieces of the bacterial chromosome. For other bacteriophages, however, transduction may involve the transfer of only a few specific genes from the donor cell to the recipient. This phenomenon is known as specialized transduction. The key event that causes specialized transduction to occur is that the lysogenic phase of the phage reproductive cycle involves the integration of the viral DNA at a single specific site within the bacterial chromosome. The transduction of particular bacterial genes involves an abnormal excision of the phage DNA from this site within the chromosome that carries adjacent bacterial genes. For example, a bacteriophage called lambda (λ) that infects *E. coli* specifically integrates between two genes designated gal^+ and bio^+ (required for galactose utilization and biotin synthesis, respectively). Either of these genes could be packaged into the phage if an abnormal excision event occurred. How would specialized transduction be different from generalized transduction?

Answer: Generalized transduction can involve the transfer of any bacterial gene, but specialized transduction can transfer only genes that are adjacent to the site where the phage integrates. As mentioned, a bacteriophage that infects *E. coli* cells, known as lambda (λ), provides a well-studied example of specialized transduction. In the case of phage lambda, the lysogenic reproductive cycle results in the integration of the phage DNA at a site that is called the attachment site (described further in Chapter 17). The attachment site is located between two bacterial genes, gal^+ and bio^+. An *E. coli* strain that is lysogenic for phage lambda has the lambda DNA integrated between these two bacterial genes. On occasion, the phage may enter the lytic cycle and excise its DNA from the bacterial chromosome. When this occurs normally, the phage excises its entire viral DNA from the bacterial chromosome. The excised phage DNA is then replicated and becomes packaged into newly made phages. However, an abnormal excision does occur at a low rate (i.e., about one in a million). In this abnormal event, the phage DNA is excised in such a way that an adjacent bacterial gene is included and some of the phage DNA is not included in the final product. For example, the abnormal excision may yield a fragment of DNA that includes the gal^+ gene and some of the lambda DNA but is missing part of the lambda DNA. If this DNA fragment is packaged into a virus, it is called a defective phage because it is missing some of the phage DNA. If it carries the gal^+ gene, it is designated λ*dgal* (the letter *d* designates a defective phage). Alternatively, an abnormal excision may carry the bio^+ gene. This phage is designated λ*dbio*. Defective lambda phages can then transduce the gal^+ or bio^+ genes to other *E. coli* cells.

Conceptual Questions

C1. The terms conjugation, transduction, and transformation are used to describe three different natural forms of genetic transfer between bacterial cells. Briefly discuss the similarities and differences among these processes.

C2. Conjugation is sometimes called "bacterial mating." Is it a form of sexual reproduction? Explain.

C3. If you mix together an equal number of F^+ and F^- cells, how would you expect the proportions to change over time? In other words, do you expect an increase in the relative proportions of F^+ or of F^- cells? Explain your answer.

C4. What is the difference between an F^+ and an *Hfr* strain? Which type of strain do you expect to transfer many bacterial genes to recipient cells?

C5. What is the role of the origin of transfer during F^+- and Hfr-mediated conjugation? What is the significance of the direction of transfer in Hfr-mediated conjugation?

C6. What is the role of sex pili during conjugation?

C7. Think about the structure and transmission of F factors and discuss how you think F factors may have originated.

C8. Each species of bacteria has its own distinctive cell surface. The characteristics of the cell surface play an important role in processes such as conjugation and transduction. For example, certain strains of *E. coli* have pili on their cell surface. These pili enable *E. coli* to mate with other *E. coli*, and the pili also enable certain bacteriophages (such as M13) to bind to the surface of *E. coli* and gain entry into the cytoplasm. With these ideas in mind, explain which forms of genetic transfer (i.e., conjugation, transduction, and transformation) are more likely to occur between different species of bacteria. Discuss some of the potential consequences of interspecies genetic transfer.

C9. Briefly describe the lytic and lysogenic cycles of bacteriophages. In your answer, explain what a prophage is.

C10. What is cotransduction? What determines the likelihood that two genes will be cotransduced?

C11. When bacteriophage P1 causes *E. coli* to lyse, the resulting material is called a P1 lysate. What type of genetic material would be found in most of the P1 phages in the lysate? What kind of genetic material is occasionally found within a P1 phage?

C12. As described in Figure 7.11, host DNA is hydrolyzed into small pieces, which are occasionally assembled with phage proteins, creating a phage with bacterial chromosomal DNA. If the breakage of the chromosomal DNA is not random (i.e., it is more likely to break at certain spots as opposed to other spots), how might nonrandom breakage affect cotransduction frequency?

C13. Describe the steps that occur during bacterial transformation. What is a competent cell? What factors may determine whether a cell will be competent?

C14. Which bacterial genetic transfer process does not require recombination with the bacterial chromosome?

C15. Researchers who study the molecular mechanism of transformation have identified many proteins in bacteria that function in the uptake of DNA from the environment and its recombination into the host cell's chromosome. This means that bacteria have evolved molecular mechanisms for the purpose of transformation by extracellular DNA. Of what advantage(s) is it for a bacterium to import DNA from the environment and/or incorporate it into its chromosome?

C16. Antibiotics such as tetracycline, streptomycin, and bacitracin are small organic molecules that are synthesized by particular species of bacteria. Microbiologists have hypothesized that the reason why certain bacteria make antibiotics is to kill other species that occupy the same environment. Bacteria that produce an antibiotic may be able to kill competing species. This provides more resources for the antibiotic-producing bacteria. In addition, bacteria that have the genes necessary for antibiotic biosynthesis contain genes that confer resistance to the same antibiotic. For example, tetracycline is made by the soil bacterium *Streptomyces aureofaciens*. Besides the genes that are needed to make tetracycline, *S. aureofaciens* also contains genes that confer tetracycline resistance; otherwise, it would kill itself when it makes tetracycline. In recent years, however, many other species of bacteria that do not synthesize tetracycline have acquired the genes that confer tetracycline resistance. For example, certain strains of *E. coli* carry tetracycline-resistance genes, even though *E. coli* does not synthesize tetracycline. When these genes were analyzed at the molecular level, it was found that they are evolutionarily related to the genes in *S. aureofaciens*. This observation indicates that the genes from *S. aureofaciens* have been transferred to *E. coli*.

A. What form of genetic transfer (i.e., conjugation, transduction, or transformation) would be the most likely mechanism of interspecies gene transfer?

B. Because *S. aureofaciens* is a nonpathogenic soil bacterium and *E. coli* is an enteric bacterium, do you think it was direct gene transfer, or do you think it may have occurred in multiple steps (i.e., from *S. aureofaciens* to other bacterial species and then to *E. coli*)?

C. How could the widespread use of antibiotics to treat diseases have contributed to the proliferation of many bacterial species that are resistant to antibiotics?

C17. What does the term complementation mean? If two different mutations that produce the same phenotype can complement each other, what can you conclude about the locations of each mutation?

C18. Intragenic mapping is sometimes called interallelic mapping. Explain why the two terms mean the same thing. In your own words, explain what an intragenic map is.

C19. As discussed in Chapter 12, genes are composed of a sequence of nucleotides. A typical gene in a bacteriophage is a few hundred or a few thousand nucleotides in length. If two different strains of bacteriophage T4 have a mutation in the *rIIA* gene that gives a rapid-lysis phenotype, yet they never produce wild-type phages by intragenic recombination when they are coinfected into *E. coli* B, what would you conclude about the locations of the mutations in the two different T4 strains?

Experimental Questions

E1. In the experiment of Figure 7.1, a $met^- \ bio^- \ thr^+ \ leu^+ \ thi^+$ cell could become $met^+ \ bio^+ \ thr^+ \ leu^+ \ thi^+$ by a (rare) double mutation that converts the $met^- \ bio^-$ genetic material into $met^+ \ bio^+$. Likewise, a $met^+ \ bio^+ \ thr^- \ leu^- \ thi^-$ cell could become $met^+ \ bio^+ \ thr^+ \ leu^+ \ thi^+$ by three mutations that convert the $thr^- \ leu^- \ thi^-$ genetic material into $thr^+ \ leu^+ \ thi^+$. From the results of Figure 7.1, how do you know that the occurrence of 10 $met^+ \ bio^+ \ thr^+ \ leu^+ \ thi^+$ colonies is not due to these types of rare double or triple mutations?

E2. In the experiment of Figure 7.1, Joshua Lederberg and Edward Tatum could not discern whether $met^+ \ bio^+$ genetic material was transferred to the $met^- \ bio^- \ thr^+ \ leu^+ \ thi^+$ strain or if $thr^+ \ leu^+ \ thi^+$ genetic material was transferred to the $met^+ \ bio^+ \ thr^- \ leu^- \ thi^-$ strain. Let's suppose that one strain is streptomycin-resistant (say, $met^+ \ bio^+ \ thr^- \ leu^- \ thi^-$) and the other strain is sensitive to streptomycin. Describe an experiment that could determine whether the $met^+ \ bio^+$ genetic material was transferred to the $met^- \ bio^- \ thr^+ \ leu^+ \ thi^+$ strain or if the $thr^+ \ leu^+ \ thi^+$ genetic material was transferred to the $met^+ \ bio^+ \ thr^- \ leu^- \ thi^-$ strain.

E3. Explain how a U-tube apparatus can distinguish between genetic transfer involving conjugation and genetic transfer involving transduction. Do you think a U-tube could be used to distinguish between transduction and transformation?

E4. What is an interrupted mating experiment? What type of experimental information can be obtained from this type of study? Why is it necessary to interrupt mating?

E5. In a conjugation experiment, what is meant by the time of entry? How is the time of entry determined experimentally?

E6. In your laboratory, you have an F^- strain of *E. coli* that is resistant to streptomycin and is unable to metabolize lactose, but it can metabolize glucose. Therefore, this strain can grow on media that contain glucose and streptomycin, but it cannot grow on media containing lactose. A researcher has sent you two *E. coli* strains in two separate tubes. One strain, let's call it strain *A*, has an F factor that carries the genes that are required for lactose metabolism. On its chromosome, it also has the genes that are required for glucose metabolism. However, it is sensitive to streptomycin. This strain can grow on media containing lactose or glucose, but it cannot grow if streptomycin is added to the media. The second strain, let's call it strain *B*, is an F^- strain. On its chromosome, it has the genes that are required for lactose and glucose metabolism. Strain *B* is also sensitive to streptomycin. Unfortunately, when strains *A* and *B* were sent to you, the labels had fallen off the tubes. Describe how you could determine which tubes contain strain *A* and strain *B*.

E7. As mentioned in solved problem S2, origins of transfer can be located in many different locations, and their direction of transfer can be clockwise or counterclockwise. Let's suppose a researcher mated six different *Hfr* strains that were thr^+ leu^+ ton^s str^r azi^s lac^+ gal^+ pro^+ met^+ to an F^- strain that was thr^- leu^- ton^r str^s azi^r lac^- gal^- pro^- met^-, and obtained the following results:

Strain	Order of Gene Transfer
1	ton^s azi^s leu^+ thr^+ met^+ str^r gal^+ lac^+ pro^+
2	leu^+ azi^s ton^s pro^+ lac^+ gal^+ str^r met^+ thr^+
3	lac^+ gal^+ str^r met^+ thr^+ leu^+ azi^s ton^s pro^+
4	leu^+ thr^+ met^+ str^r gal^+ lac^+ pro^+ ton^s azi^s
5	ton^s pro^+ lac^+ gal^+ str^r met^+ thr^+ leu^+ azi^s
6	met^+ str^r gal^+ lac^+ pro^+ ton^s azi^s leu^+ thr^+

Draw a circular map of the *E. coli* chromosome and describe the locations and orientations of the origins of transfer in these six *Hfr* strains.

E8. An *Hfr* strain that is $hisE^+$ and $pheA^+$ was mated to a strain that is $hisE^-$ and $pheA^-$. The mating was interrupted and the percentage of recombinants for each gene was determined by streaking on media that lacked either histidine or phenylalanine. The following results were obtained:

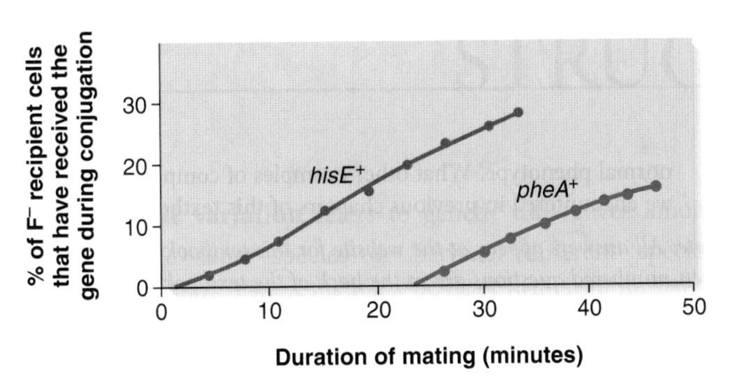

A. Determine the map distance (in minutes) between these two genes.

B. In a previous experiment, it was found that *hisE* is 4 minutes away from the gene *pabB*. *PheA* was shown to be 17 minutes from this gene. Draw a genetic map describing the locations of all three genes.

E9. Acridine orange is a chemical that inhibits the replication of F factor DNA but does not affect the replication of chromosomal DNA, even if the chromosomal DNA contains an Hfr. Let's suppose that you have an *E. coli* strain that is unable to metabolize lactose and has an F factor that carries a streptomycin-resistant gene. You also have an F^- strain of *E. coli* that is sensitive to streptomycin and has the genes that allow the bacterium to metabolize lactose. This second strain can grow on lactose-containing media. How would you generate an *Hfr* strain that is resistant to streptomycin and can metabolize lactose? (Hint: F factors occasionally integrate into the chromosome to become *Hfr* strains, and occasionally *Hfr* strains excise their DNA from the chromosome to become F^+ strains that carry an F′ factor.)

E10. In a P1 transduction experiment, the P1 lysate contains phages that carry pieces of the host chromosomal DNA, but the lysate also contains broken pieces of chromosomal DNA (see Figure 7.11). If a P1 lysate is used to transfer chromosomal DNA to another bacterium, how could you show experimentally that the recombinant bacterium has been transduced (i.e., taken up a P1 phage with a piece of chromosomal DNA inside) versus transformed (i.e., taken up a piece of chromosomal DNA that is not within a P1 phage coat)?

E11. Can you devise an experimental strategy to get P1 phage to transduce the entire lambda genome from one strain of bacterium to another strain? (Note: The general features of phage lambda's reproductive cycle are described in Chapter 14.) Phage lambda has a genome size of 48,502 nucleotides (about 1% of the size of the *E. coli* chromosome) and can follow the lytic or lysogenic reproductive cycle. Growth of *E. coli* on minimal growth medium favors the lysogenic reproductive cycle, whereas growth on rich media and/or under UV light promotes the lytic cycle.

E12. Let's suppose a new strain of P1 has been identified that packages larger pieces of the *E. coli* chromosome. This new P1 strain packages pieces of the *E. coli* chromosome that are 5 minutes long. If two genes are 0.7 minutes apart along the *E. coli* chromosome, what would be the cotransduction frequency using a normal strain of P1 and using this new strain of P1 that packages large pieces? What would be the experimental advantage of using this new P1 strain?

E13. If two bacterial genes are 0.6 minutes apart on the bacterial chromosome, what frequency of cotransductants would you expect to observe in a P1 transduction experiment?

E14. In an experiment involving P1 transduction, the cotransduction frequency was 0.53. How far apart are the two genes?

E15. In a cotransduction experiment, the transfer of one gene is selected for and the presence of the second gene is then determined. If 0 out of 1000 P1 transductants that carry the first gene also carry the second gene, what would you conclude about the minimum distance between the two genes?

E16. In a cotransformation experiment (see solved problem S4), DNA was isolated from a donor strain that was $proA^+$ and $strC^+$ and sensitive to tetracycline. (The *proA* and *strC* genes confer the ability to synthesize proline and confer streptomycin resistance, respectively.) A recipient strain is $proA^-$ and $strC^-$ and is resistant to tetracycline. After transformation, the bacteria were first streaked on a medium containing proline, streptomycin, and tetracycline. Colonies were then restreaked on a medium containing streptomycin and tetracycline. (Note: Both types of media had carbon and nitrogen sources for growth.) The following results were obtained:

> 70 colonies grew on the medium containing proline, streptomycin, and tetracycline, but only 2 of these 70 colonies grew when restreaked on the medium containing streptomycin and tetracycline but lacking proline.

A. If we assume the average size of the DNA fragments is 2 minutes, how far apart are these two genes?

B. What would you expect the cotransformation frequency to be if the average size of the DNA fragments was 4 minutes and the two genes are 1.4 minutes apart?

E17. If you took a pipette tip and removed a phage plaque from a petri plate, what would it contain?

E18. As shown in Figure 7.17, phages with *rII* mutations cannot produce plaques in *E. coli* K12(λ), but wild-type phages can. From an experimental point of view, explain why this observation is so significant.

a given species. In most cases, two phenotypically normal individuals of the same species have the same number and types of chromosomes.

To determine the chromosomal composition of a species, the chromosomes in actively dividing cells are examined microscopically. **Figure 8.1a** shows micrographs of chromosomes from three species: a human, a fruit fly, and a corn plant. As seen here, a human has 46 chromosomes (23 pairs), a fruit fly has 8 chromosomes (4 pairs), and corn has 20 chromosomes (10 pairs). Except for the sex chromosomes, which differ between males and females, most members of the same species have very similar chromosomes. For example, the overwhelming majority of

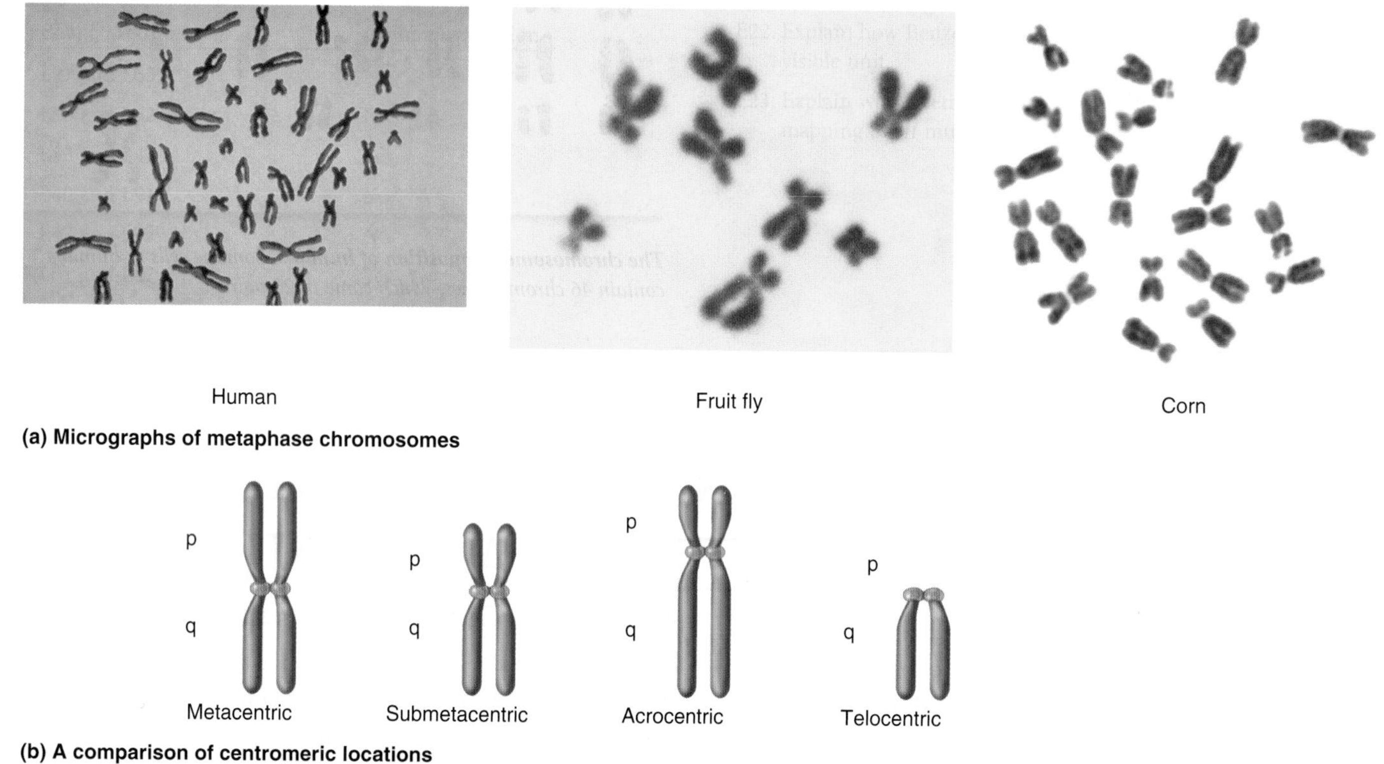

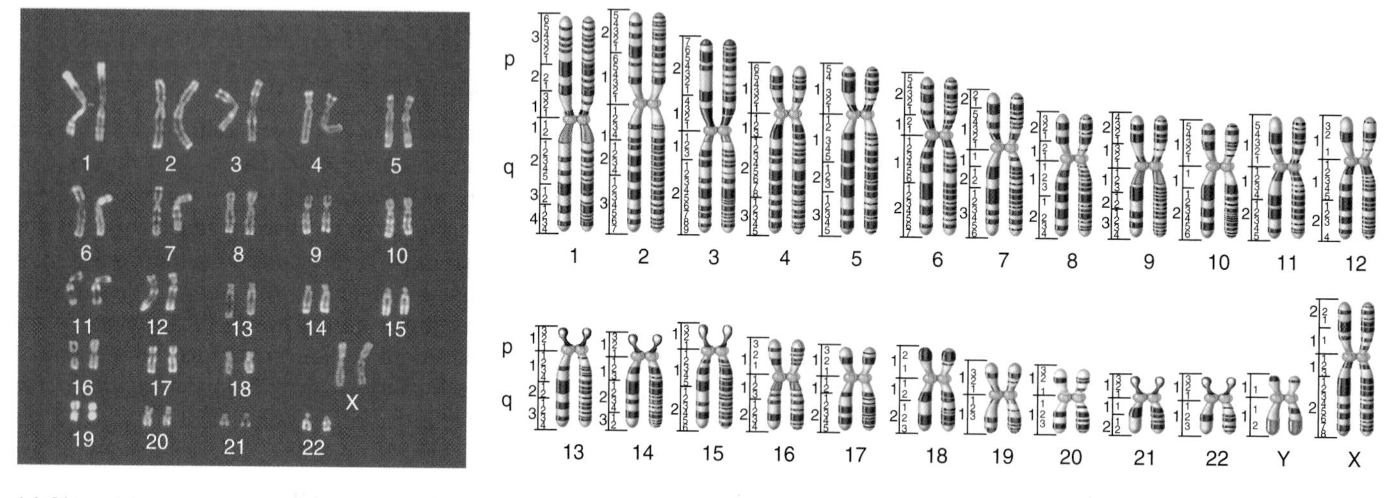

FIGURE 8.1 Features of normal chromosomes. **(a)** Micrographs of chromosomes from a human, a fruit fly, and corn. **(b)** A comparison of centromeric locations. Centromeres can be metacentric, submetacentric, acrocentric (near one end), or telocentric (at the end). **(c)** Human chromosomes that have been stained with Giemsa. **(d)** The conventional numbering of bands in Giemsa-stained human chromosomes. The numbering is divided into broad regions, which then are subdivided into smaller regions. The numbers increase as the region gets farther away from the centromere. For example, if you take a look at the left chromatid of chromosome 1, the uppermost dark band is at a location designated p35. The banding patterns of chromatids change as the chromatids condense. The left chromatid of each pair of sister chromatids shows the banding pattern of a chromatid in metaphase, and the right side shows the banding pattern as it would appear in prometaphase. Note: In prometaphase, the chromatids are more extended than in metaphase.

people have 46 chromosomes in their somatic cells. By comparison, the chromosomal compositions of distantly related species, such as humans and fruit flies, may be very different. A total of 46 chromosomes is normal for humans, whereas 8 chromosomes is the norm for fruit flies.

Cytogeneticists have various ways to classify and identify chromosomes. The three most commonly used features are location of the centromere, size, and banding patterns that are revealed when the chromosomes are treated with stains. As shown in **Figure 8.1b**, chromosomes are classified as **metacentric** (in which the centromere is near the middle), **submetacentric** (in which the centromere is slightly off center), **acrocentric** (in which the centromere is significantly off center but not at the end), and **telocentric** (in which the centromere is at one end). Because the centromere is never exactly in the center of a chromosome, each chromosome has a short arm and a long arm. For human chromosomes, the short arm is designated with the letter p (for the French, petite), and the long arm is designated with the letter q. In the case of telocentric chromosomes, the short arm may be nearly nonexistent.

Figure 8.1c shows a human karyotype. The procedure for making a karyotype is described in Chapter 3 (see Figure 3.2). A **karyotype** is a micrograph in which all of the chromosomes within a single cell have been arranged in a standard fashion. When preparing a karyotype, the chromosomes are aligned with the short arms on top and the long arms on the bottom. By convention, the chromosomes are numbered roughly according to their size, with the largest chromosomes having the smallest numbers. For example, human chromosomes 1, 2, and 3 are relatively large, whereas 21 and 22 are the two smallest. An exception to the numbering system involves the sex chromosomes, which are designated with letters (for humans, X and Y).

Because different chromosomes often have similar sizes and centromeric locations (e.g., compare human chromosomes 8, 9, and 10), geneticists must use additional methods to accurately identify each type of chromosome within a karyotype. For detailed identification, chromosomes are treated with stains to produce characteristic banding patterns. Several different staining procedures are used by cytogeneticists to identify specific chromosomes. An example is **G banding,** which is shown in Figure 8.1c. In this procedure, chromosomes are treated with mild heat or with proteolytic enzymes that partially digest chromosomal proteins. When exposed to the dye called Giemsa, named after its inventor Gustav Giemsa, some chromosomal regions bind the dye heavily and produce a dark band. In other regions, the stain hardly binds at all and a light band results. Though the mechanism of staining is not completely understood, the dark bands are thought to represent regions that are more tightly compacted. As shown in Figure 8.1c and d, the alternating pattern of G bands is a unique feature for each chromosome.

In the case of human chromosomes, approximately 300 G bands can usually be distinguished during metaphase. A larger number of G bands (in the range of 800) can be observed in prometaphase chromosomes because they are more extended than metaphase chromosomes. **Figure 8.1d** shows the conventional numbering system that is used to designate G bands along a set of human chromosomes. The left chromatid in each pair of sister chromatids shows the expected banding pattern during metaphase, and the right chromatid shows the banding pattern as it would appear during prometaphase.

Why is the banding pattern of eukaryotic chromosomes useful? First, when stained, individual chromosomes can be distinguished from each other, even if they have similar sizes and centromeric locations. For example, compare the differences in banding patterns between human chromosomes 8 and 9 (Figure 8.1d). These differences permit us to distinguish these two chromosomes even though their sizes and centromeric locations are very similar. Banding patterns are also used to detect changes in chromosome structure. As discussed next, chromosomal rearrangements or changes in the total amount of genetic material are more easily detected in banded chromosomes. Also, chromosome banding can be used to assess evolutionary relationships between species. Research studies have shown that the similarity of chromosome banding patterns is a good measure of genetic relatedness.

Changes in Chromosome Structure Include Deletions, Duplications, Inversions, and Translocations

With an understanding that chromosomes typically come in a variety of shapes and sizes, let's consider how the structures of normal chromosomes can be modified. In some cases, the total amount of genetic material within a single chromosome can be increased or decreased significantly. Alternatively, the genetic material in one or more chromosomes may be rearranged without affecting the total amount of material. As shown in **Figure 8.2**, these mutations are categorized as deletions, duplications, inversions, and translocations.

Deletions and duplications are changes in the total amount of genetic material within a single chromosome. In Figure 8.2, human chromosomes are labeled according to their normal G-banding patterns. When a **deletion** occurs, a segment of chromosomal material is missing. In other words, the affected chromosome is deficient in a significant amount of genetic material. The term **deficiency** is also used to describe a missing region of a chromosome. In contrast, a **duplication** occurs when a section of a chromosome is repeated compared with the normal parent chromosome.

Inversions and translocations are chromosomal rearrangements. An **inversion** involves a change in the direction of the genetic material along a single chromosome. For example, in Figure 8.2c, a segment of one chromosome has been inverted, so the order of four G bands is opposite to that of the parent chromosome. A **translocation** occurs when one segment of a chromosome becomes attached to a different chromosome or to a different part of the same chromosome. A **simple translocation** occurs when a single piece of chromosome is attached to another chromosome. In a **reciprocal translocation,** two different types of chromosomes exchange pieces, thereby producing two abnormal chromosomes carrying translocations.

Figure 8.2 illustrates the common ways that the structure of chromosomes can be altered. Throughout the rest of this section, we will consider how these changes occur, how the changes are detected experimentally, and how they affect the phenotypes of the individuals who inherit them.

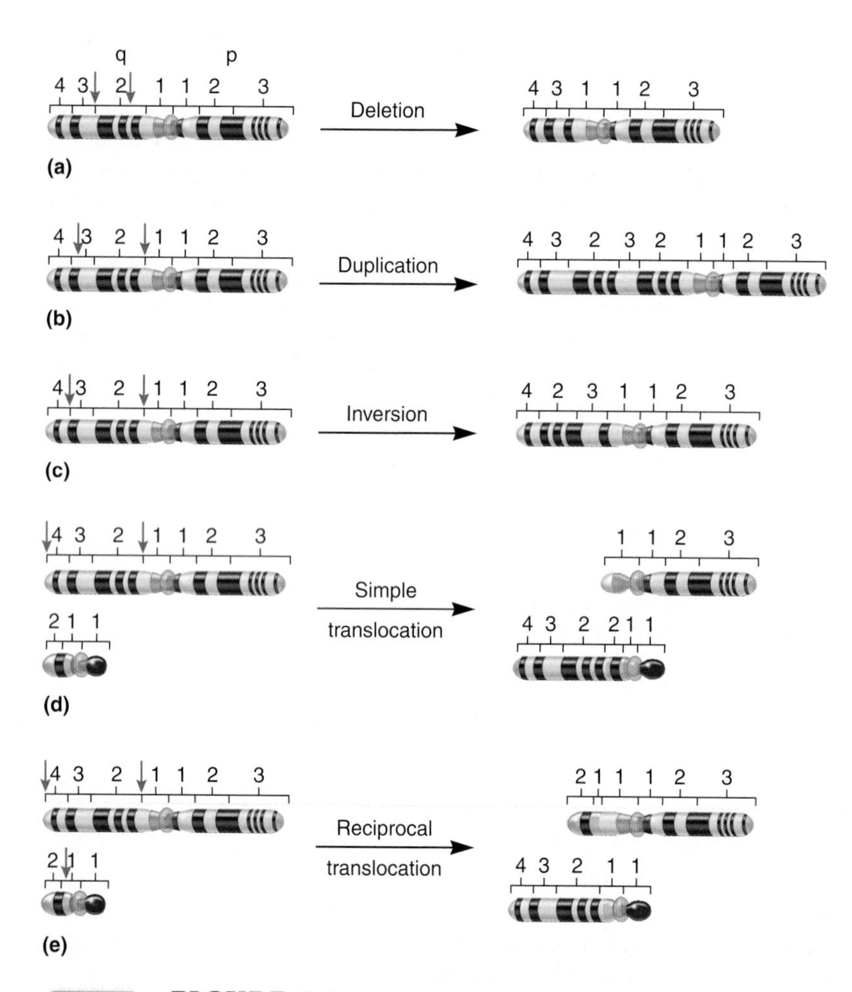

ONLINE ANIMATION

FIGURE 8.2 Types of changes in chromosome structure. The large chromosome shown throughout is human chromosome 1. The smaller chromosome seen in (d) and (e) is human chromosome 21. **(a)** A deletion occurs that removes a large portion of the q2 region, indicated by the red arrows. **(b)** A duplication occurs that doubles the q2–q3 region. **(c)** An inversion occurs that inverts the q2–q3 region. **(d)** The q2–q4 region of chromosome 1 is translocated to chromosome 21. A region of a chromosome cannot be inserted directly to the tip of another chromosome because telomeres at the tips of chromosomes prevent such an event. In this example, a small piece at the end of chromosome 21 must be removed for the q2–q4 region of chromosome 1 to be attached to chromosome 21. **(e)** The q2–q4 region of chromosome 1 is exchanged with most of the q1–q2 region of chromosome 21.

The Loss of Genetic Material in a Deletion Tends to Be Detrimental to an Organism

A chromosomal deletion occurs when a chromosome breaks in one or more places and a fragment of the chromosome is lost. In **Figure 8.3a**, a normal chromosome has broken into two separate pieces. The piece without the centromere is lost and degraded. This event produces a chromosome with a **terminal deletion.** In **Figure 8.3b**, a chromosome has broken in two places to produce three chromosomal fragments. The central fragment is lost, and the two outer pieces reattach to each other. This process has created a chromosome with an **interstitial deletion.** Deletions can also be created when recombination takes place at incorrect locations between two homologous chromosomes. The products of this type of aberrant recombination event are one chromosome with a deletion and another chromosome with a duplication. This process is examined later in this chapter.

The phenotypic consequences of a chromosomal deletion depend on the size of the deletion and whether it includes genes or portions of genes that are vital to the development of the organism. When deletions have a phenotypic effect, they are usually detrimental. Larger deletions tend to be more harmful because more genes are missing. Many examples are known in which deletions have significant phenotypic influences. For example, a human genetic disease known as cri-du-chat, or Lejeune, syndrome is caused by a deletion in a segment of the short arm of human chromosome 5 (**Figure 8.4a**). Individuals who carry a single copy of this abnormal chromosome along with a normal chromosome 5 display an array of abnormalities including mental deficiencies, unique facial anomalies, and an unusual catlike cry in infancy, which is the meaning of the French name for the syndrome (**Figure 8.4b**). Two other human genetic diseases, Angelman syndrome and Prader-Willi syndrome, which are described in Chapter 5, are due to a deletion in chromosome 15.

Duplications Tend to Be Less Harmful Than Deletions

Duplications result in extra genetic material. They are usually caused by abnormal events during recombination. Under normal circumstances, crossing over occurs at analogous sites between homologous chromosomes. On rare occasions, a crossover may occur

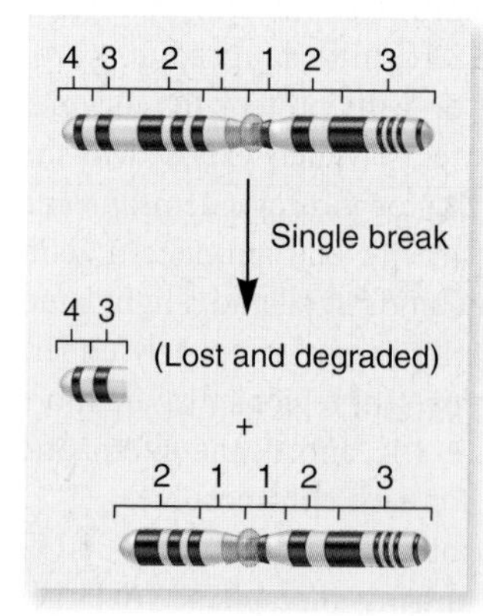

(a) Terminal deletion

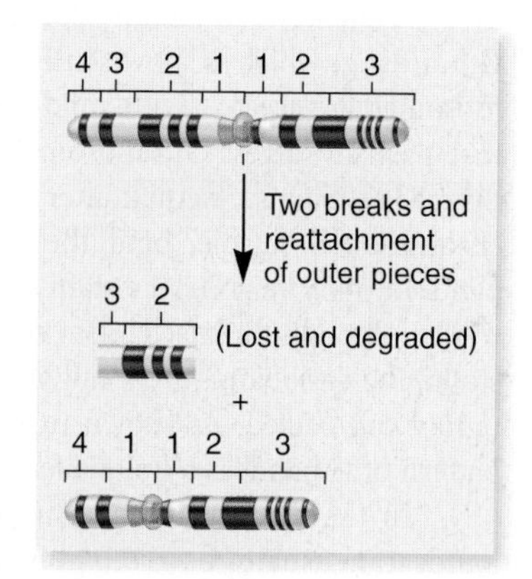

(b) Interstitial deletion

FIGURE 8.3 Production of terminal and interstitial deletions. This illustration shows the production of deletions in human chromosome 1.

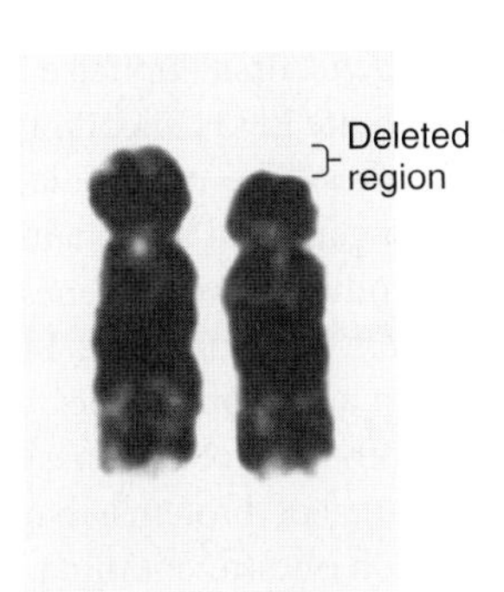

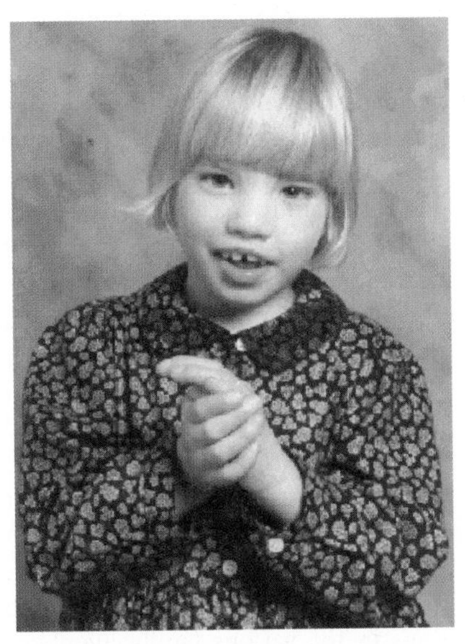

(a) Chromosome 5 (b) A child with cri-du-chat syndrome

FIGURE 8.4 **Cri-du-chat syndrome.** **(a)** Chromosome 5 from the karyotype of an individual with this disorder. A section of the short arm of chromosome 5 is missing. **(b)** An affected individual.

Genes → Traits Compared with an individual who has two copies of each gene on chromosome 5, an individual with cri-du-chat syndrome has only one copy of the genes that are located within the missing segment. This genetic imbalance (one versus two copies of many genes on chromosome 5) causes the phenotypic characteristics of this disorder, which include a catlike cry in infancy, short stature, characteristic facial anomalies (e.g., a triangular face, almond-shaped eyes, broad nasal bridge, and low-set ears), and microencephaly (a smaller than normal brain).

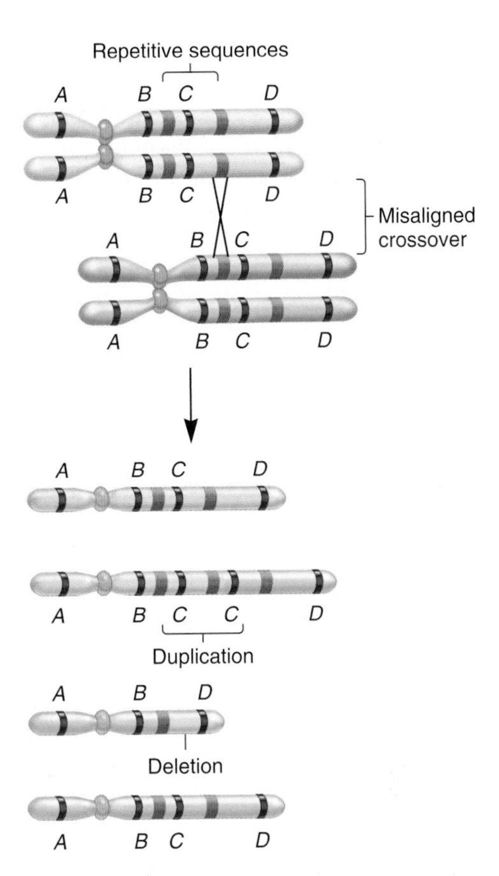

FIGURE 8.5 **Nonallelic homologous recombination, leading to a duplication and a deletion.** A repetitive sequence, shown in red, has promoted the misalignment of homologous chromosomes. A crossover has occurred at sites between genes *C* and *D* in one chromatid and between genes *B* and *C* in another chromatid. After crossing over is completed, one chromatid contains a duplication, and the other contains a deletion.

at misaligned sites on the homologs (**Figure 8.5**). What causes the misalignment? In some cases, a chromosome may carry two or more homologous segments of DNA that have identical or similar sequences. These are called **repetitive sequences** because they occur multiple times. An example of repetitive sequences are transposable elements, which are described in Chapter 17. In Figure 8.5, the repetitive sequence on the right (in the upper chromatid) has lined up with the repetitive sequence on the left (in the lower chromatid). A crossover then occurs. This is called **nonallelic homologous recombination** because it has occurred at homologous sites (i.e., repetitive sequences), but the alleles of neighboring genes are not properly aligned. The result is that one chromatid has an internal duplication and another chromatid has a deletion. In Figure 8.5, the chromosome with the extra genetic material carries a **gene duplication,** because the number of copies of gene *C* has been increased from one to two. In most cases, gene duplications happen as rare, sporadic events during the evolution of species. Later in this section, we will consider how multiple copies of genes can evolve into a family of genes with specialized functions.

Like deletions, the phenotypic consequences of duplications tend to be correlated with size. Duplications are more likely to have phenotypic effects if they involve a large piece of the chromosome. In general, small duplications are less likely to have harmful effects than are deletions of comparable size. This observation suggests that having only one copy of a gene is more harmful than having three copies. In humans, relatively few well-defined syndromes are caused by small chromosomal duplications. An example is Charcot-Marie-Tooth disease (type 1A), a peripheral neuropathy characterized by numbness in the hands and feet that is caused by a small duplication on the short arm of chromosome 17.

Duplications Provide Additional Material for Gene Evolution, Sometimes Leading to the Formation of Gene Families

In contrast to the gene duplication that causes Charcot-Marie-Tooth disease, the majority of small chromosomal duplications have no phenotypic effect. Nevertheless, they are vitally important because they provide raw material for the addition of more genes into a species' chromosomes. Over the course of many generations, this can lead to the formation of a **gene family** consisting of two or more genes that are similar to each other. As shown in **Figure 8.6**, the members of a gene family are derived from the same ancestral gene. Over time, two copies of an ancestral gene can accumulate different mutations. Therefore, after many generations, the two genes will be similar but not identical. During evolution, this type of event can occur several times, creating a family of many similar genes.

When two or more genes are derived from a single ancestral gene, the genes are said to be **homologous.** Homologous genes within a single species are called **paralogs** and constitute a gene family. A well-studied example of a gene family is shown in **Figure 8.7**, which illustrates the evolution of the globin gene family found in humans. The globin genes encode polypeptides that are subunits of proteins that function in oxygen binding. For example, hemoglobin is a protein found in red blood cells; its function is to carry oxygen throughout the body. The globin gene family is composed of 14 paralogs that were originally derived from a single ancestral globin gene. According to an evolutionary analysis, the ancestral globin gene first duplicated about 500 million years ago and became separate genes encoding myoglobin and the hemoglobin group of genes. The primordial hemoglobin gene duplicated into an α-chain gene and a β-chain gene, which subsequently duplicated to produce several genes located on chromosomes 16 and 11, respectively. Currently, 14 globin genes are found on three different human chromosomes.

Why is it advantageous to have a family of globin genes? Although all globin polypeptides are subunits of proteins that play a role in oxygen binding, the accumulation of different mutations in the various family members has produced globins that are more specialized in their function. For example, myoglobin is better at binding and storing oxygen in muscle cells, and the hemoglobins are better at binding and transporting oxygen via the red blood cells. Also, different globin genes are expressed during different stages of human development. The ε- and ζ-globin genes are expressed very early in embryonic life, whereas the α-globin and γ-globin genes are expressed during the second and third trimesters of gestation. Following birth, the α-globin gene remains turned on, but the γ-globin genes are turned off and the β-globin gene is turned on. These differences in the expression of the globin genes reflect the differences in the oxygen transport needs of humans during the embryonic, fetal, and postpartum stages of life.

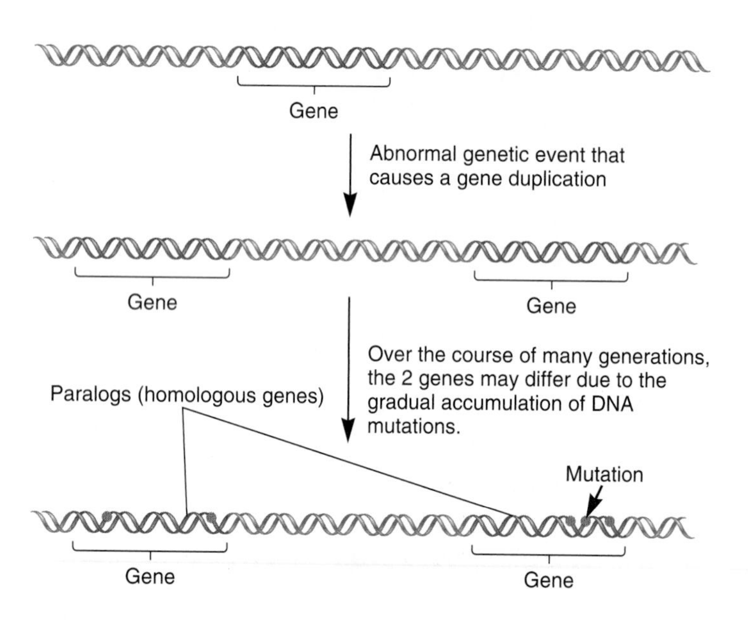

ONLINE ANIMATION

FIGURE 8.6 Gene duplication and the evolution of paralogs. An abnormal crossover event like the one described in Figure 8.5 leads to a gene duplication. Over time, each gene accumulates different mutations.

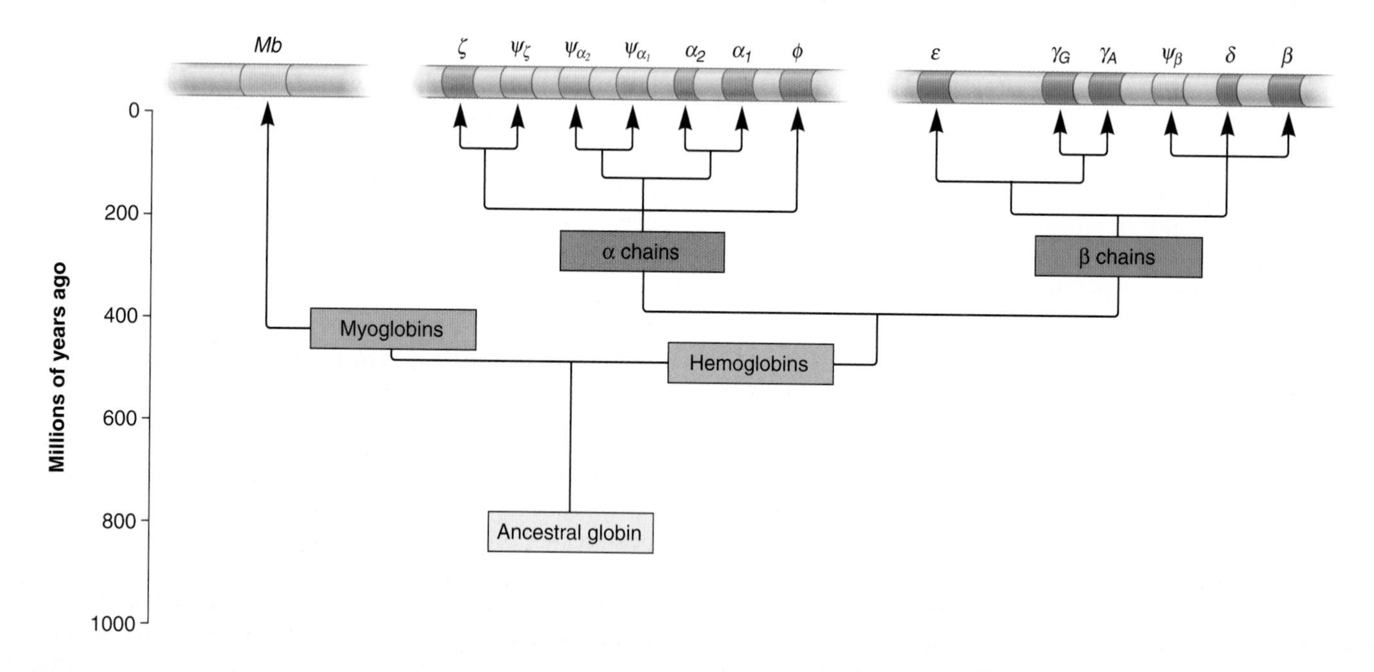

FIGURE 8.7 The evolution of the globin gene family in humans. The globin gene family evolved from a single ancestral globin gene. The first gene duplication produced two genes that accumulated mutations and became the genes encoding myoglobin (on chromosome 22) and the group of hemoglobins. The primordial hemoglobin gene then duplicated to produce several α-chain and β-chain genes, which are found on chromosomes 16 and 11, respectively. The four genes shown in gray are nonfunctional pseudogenes.

Copy Number Variation Is Relatively Common Among Members of the Same Species

The term **copy number variation (CNV)** refers to a type of structural variation in which a segment of DNA, which is 1000 bp or more in length, exhibits copy number differences among members of the same species. One possibility is that some members of a species may carry a chromosome that is missing a particular gene or part of a gene. Alternatively, a CNV may involve a duplication. For example, some members of a diploid species may have one copy of gene *A* on both homologs of a chromosome, and thereby have two copies of the gene (**Figure 8.8**). By comparison, other members of the same species might have one copy of gene *A* on a particular chromosome and two copies on its homolog for a total of three copies. The homolog with two copies of gene *A* is said to have undergone a **segmental duplication.**

In the past 10 years, researchers have discovered that copy number variation is relatively common in animal and plant species. Though the analysis of structural variation is a relatively new area of investigation, researchers estimate that between 1% and 10% of a genome may show CNV within a typical species of animal or plant.

Most CNV is inherited and has happened in the past, but it may also be caused by new mutations. A variety of mechanisms may bring about copy number variation. One common cause is nonallelic homologous recombination, which was described earlier in Figure 8.5. This type of event can produce a chromosome with a duplication or deletion, and thereby alter the copy number of genes. Researchers also speculate that the proliferation of transposable elements, which are described in Chapter 17, may increase the copy number of DNA segments. A third mechanism that underlies CNV may involve errors in DNA replication, which is described in Chapter 11.

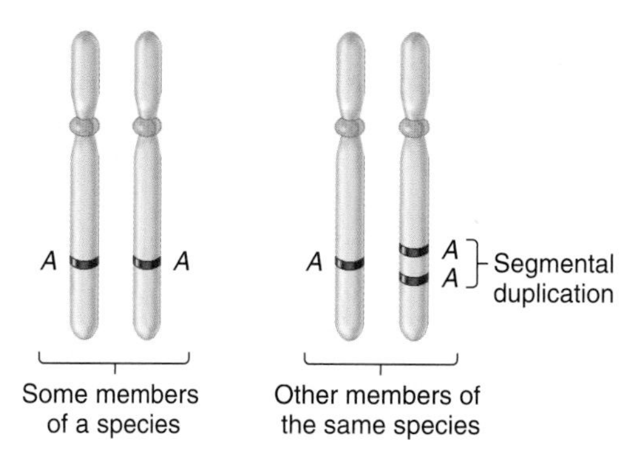

FIGURE 8.8 **An example of copy number variation.** On the left, some individuals have two copies of gene *A*, whereas other individuals, shown on the right, have three copies.

What are the phenotypic consequences of CNV? In many cases, CNV has no obvious phenotypic consequences. However, recent medical research is revealing that some CNV is associated with specific human diseases. For example, particular types of CNV are associated with schizophrenia, autism, and certain forms of learning disabilities. In addition, CNV may affect susceptibility to infectious diseases. An example is the human *CCL3* gene that encodes a chemokine protein, which is involved in immunity. In human populations, the copy number of this gene varies from 1 to 6. In people infected with HIV (human immunodeficiency virus), copy number variation of *CCL3* may affect the progression of AIDS (acquired immune deficiency syndrome). Individuals with a higher copy number of *CCL3* produce more chemokine protein and often show a slower advancement of AIDS. Finally, another reason why researchers are interested in copy number variation is its relationship to cancer, which is described next.

EXPERIMENT 8A

Comparative Genomic Hybridization Is Used to Detect Chromosome Deletions and Duplications

As we have seen, chromosome deletions and duplications may influence the phenotypes of individuals who inherit them. One very important reason why researchers have become interested in these types of chromosomal changes is related to cancer. As discussed in Chapter 22, chromosomal deletions and duplications have been associated with many types of human cancers. Though such changes may be detectable by traditional chromosomal staining and karyotyping methods, small deletions and duplications may be difficult to detect in this manner. Fortunately, researchers have been able to develop more sensitive methods for identifying changes in chromosome structure.

In 1992, Anne Kallioniemi, Daniel Pinkel, and colleagues devised a method called **comparative genomic hybridization (CGH).** This technique is largely used to determine if cancer cells have changes in chromosome structure, such as deletions or duplications. To begin this procedure, DNA is isolated from a test sample, which in this case was a sample of breast cancer cells, and also from a normal reference sample (**Figure 8.9**). The DNA from the breast cancer cells was used as a template to make green fluorescent DNA, and the DNA from normal cells was used to make red fluorescent DNA. These green or red DNA molecules averaged 800 bp in length and were made from sites that were scattered all along each chromosome. The green and red DNA molecules were then denatured by heat treatment. Equal amounts of the two fluorescently labeled DNA samples were mixed together and applied to normal metaphase chromosomes in which the DNA had also been denatured. Because the fluorescently labeled DNA fragments and the metaphase chromosomes had both been denatured, the fluorescently labeled DNA strands can bind to complementary regions on the metaphase chromosomes. This process is called **hybridization** because the DNA from one sample (a green or red DNA strand) forms a double-stranded region with a DNA strand from another sample (an unlabeled metaphase chromosome). Following hybridization, the metaphase chromosomes were visualized using a fluorescence microscope, and the images were analyzed by a computer that can determine the relative intensities of green and red fluorescence. What are the expected results? If a chromosomal region in the breast cancer cells and the normal cells are present in the same amount, the ratio between green and red fluorescence should be 1. If a chromosomal region is deleted in the breast cancer cell line, the ratio will be less than 1, or if a region is duplicated, it will be greater than 1.

THE GOAL

Deletions or duplications in cancer cells can be detected by comparing the ability of fluorescently labeled DNA from cancer cells and normal cells to bind (hybridize) to normal metaphase chromosomes.

ACHIEVING THE GOAL — FIGURE 8.9 **The use of comparative genomic hybridization to detect deletions and duplications in cancer cells.**

Starting materials: Breast cancer cells and normal cells.

1. Isolate DNA from human breast cancer cells and normal cells. This involved breaking open the cells and isolating the DNA by chromatography. (See Appendix for description of chromatography.)

2. Label the breast cancer DNA with a green fluorescent molecule and the normal DNA with a red fluorescent molecule. This was done by using the DNA from step 1 as a template, and incorporating fluorescently labeled nucleotides into newly made DNA strands.

3. The DNA strands were then denatured by heat treatment. Mix together equal amounts of fluorescently labeled DNA and add it to a preparation of metaphase chromosomes from white blood cells. The procedure for preparing metaphase chromosomes is described in Figure 3.2. The metaphase chromosomes were also denatured.

4. Allow the fluorescently labeled DNA to hybridize to the metaphase chromosomes.

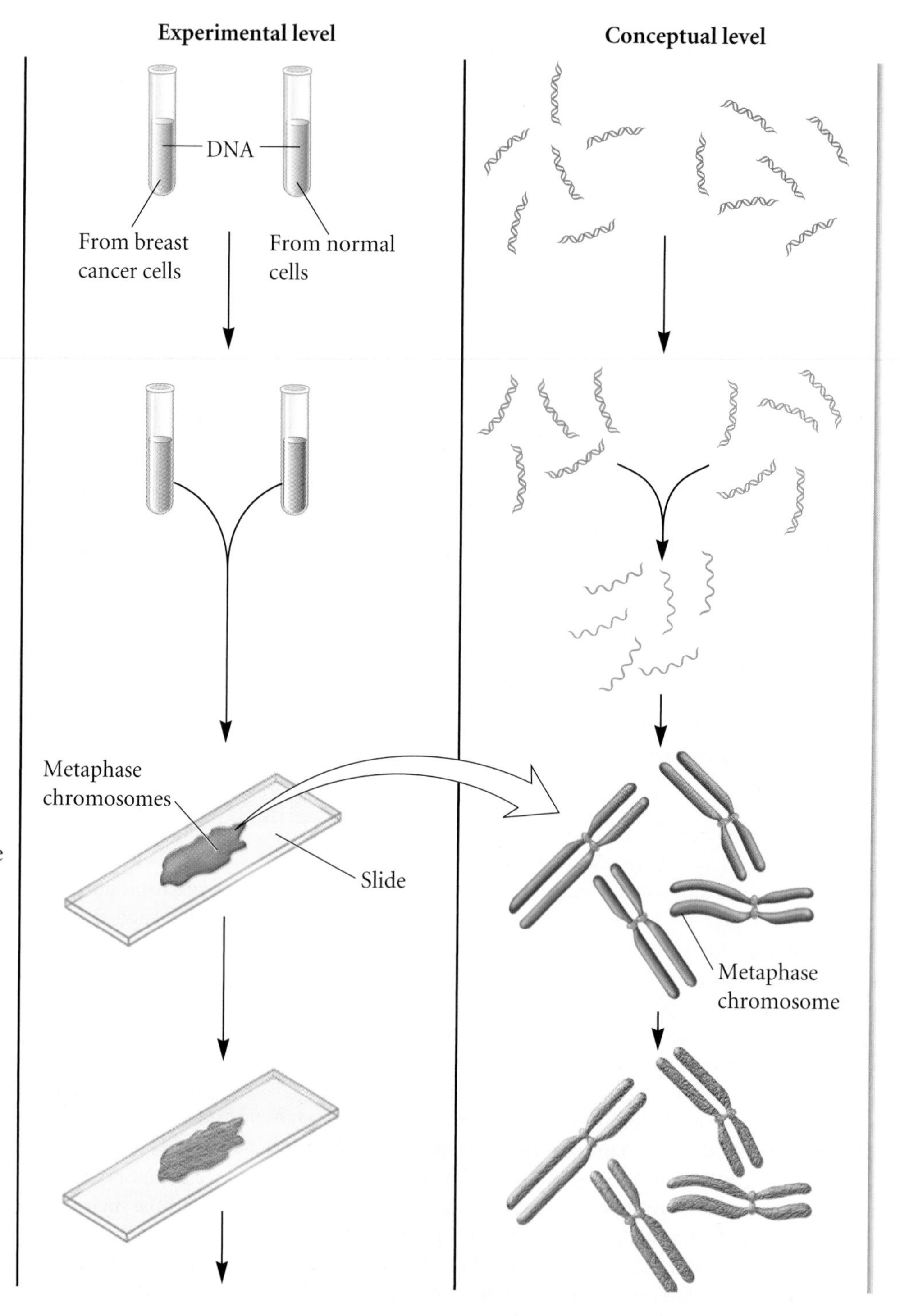

5. Visualize the chromosomes with a fluorescence microscope. Analyze the amount of green and red fluorescence along each chromosome with a computer.

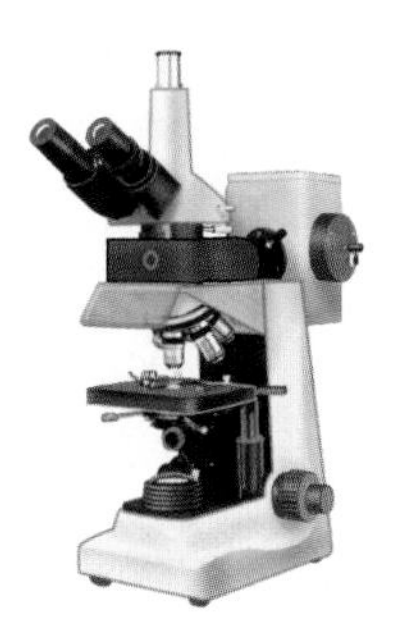

Deletions in the chromosomes of cancer cells show a green to red ratio of less than 1, whereas chromosome duplications show a ratio greater than 1.

THE DATA

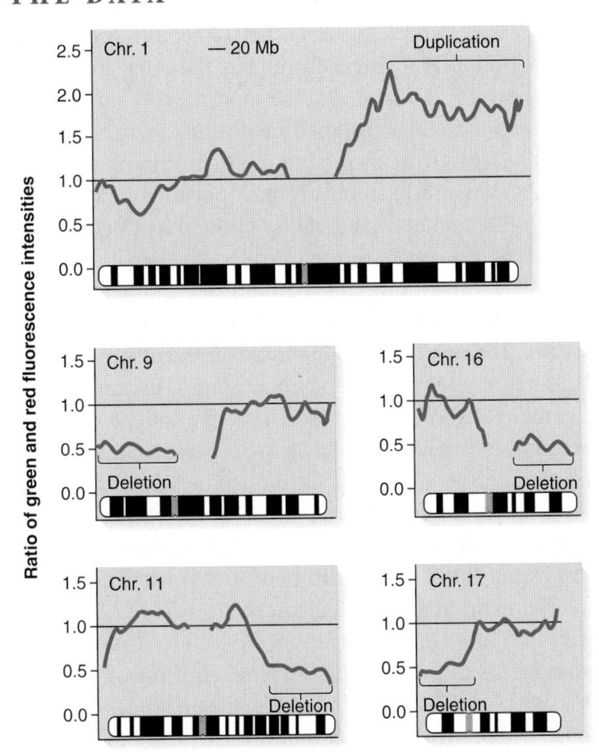

Note: Unlabeled repetitive DNA was also included in this experiment to decrease the level of nonspecific, background labeling. This repetitive DNA also prevents labeling near the centromere. As seen in the data, regions in the chromosomes where the curves are missing are due to the presence of highly repetitive sequences near the centromere. Data from A. Kallioniemi, O. P. Kallioniemi, D. Sudar, et al. (1992) Comparative genomic hybridization for molecular cytogenetic analysis of solid tumors. *Science 258,* 818–821.

INTERPRETING THE DATA

The data of Figure 8.9 show the ratio of green (cancer DNA) to red (normal DNA) fluorescence along five different metaphase chromosomes. Chromosome 1 shows a large duplication, as indicated by the ratio of 2. One interpretation of this observation is that both copies of chromosome 1 carry a duplication. In comparison, chromosomes 9, 11, 16, and 17 have regions with a value of 0.5. This value indicates that one of the two chromosomes of these four types in the cancer cells carries a deletion, but the other chromosome does not. (A value of 0 would indicate both copies of a chromosome had deleted the same region.) Overall, these results illustrate how this technique can be used to map chromosomal duplications and deletions in cancer cells.

This method is named comparative genomic hybridization because a comparison is made between the ability of two DNA samples (cancer versus normal cells) to hybridize to an entire genome. In this case, the entire genome is in the form of metaphase chromosomes. As discussed in Chapter 20, the fluorescently labeled DNAs can be hybridized to a DNA microarray instead of metaphase chromosomes. This newer method, called array comparative genomic hybridization (aCGH), is gaining widespread use in the analysis of cancer cells.

A self-help quiz involving this experiment can be found at www.mhhe.com/brookergenetics4e.

Inversions Often Occur Without Phenotypic Consequences

We now turn our attention to inversions, changes in chromosome structure that involve a rearrangement in the genetic material. A chromosome with an inversion contains a segment that has been flipped to the opposite direction. Geneticists classify inversions according to the location of the centromere. If the centromere lies within the inverted region of the chromosome, the inverted region is known as a **pericentric inversion** (**Figure 8.10b**). Alternatively, if the centromere is found outside the inverted region, the inverted region is called a **paracentric inversion** (**Figure 8.10c**).

When a chromosome contains an inversion, the total amount of genetic material remains the same as in a normal chromosome. Therefore, the great majority of inversions do not have any phenotypic consequences. In rare cases, however, an inversion can alter the phenotype of an individual. Whether or not this occurs is related to the boundaries of the inverted segment. When an inversion occurs, the chromosome is broken in two places, and the center piece flips around to produce the inversion. If either breakpoint occurs within a vital gene, the function of the gene is expected to be disrupted, possibly producing a phenotypic effect. For example, some people with hemophilia (type A)

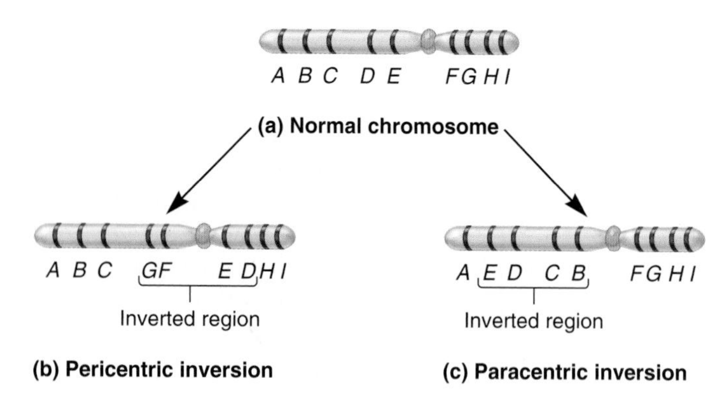

FIGURE 8.10 **Types of inversions.** (a) Depicts a normal chromosome with the genes ordered from *A* through *I*. A pericentric inversion (b) includes the centromere, whereas a paracentric inversion (c) does not.

have inherited an X-linked inversion in which the breakpoint has inactivated the gene for factor VIII—a blood-clotting protein. In other cases, an inversion (or translocation) may reposition a gene on a chromosome in a way that alters its normal level of expression. This is a type of **position effect**—a change in phenotype that occurs when the position of a gene changes from one chromosomal site to a different location. This topic is also discussed in Chapter 16 (see Figures 16.2 and 16.3).

Because inversions seem like an unusual genetic phenomenon, it is perhaps surprising that they are found in human populations in significant numbers. About 2% of the human population carries inversions that are detectable with a light microscope. In most cases, such individuals are phenotypically normal and live their lives without knowing they carry an inversion. In a few cases, however, an individual with an inversion chromosome may produce offspring with phenotypic abnormalities. This event may prompt a physician to request a microscopic examination of the individual's chromosomes. In this way, phenotypically normal individuals may discover they have a chromosome with an inversion. Next, we will examine how an individual carrying an inversion may produce offspring with phenotypic abnormalities.

Inversion Heterozygotes May Produce Abnormal Chromosomes Due to Crossing Over

An individual carrying one copy of a normal chromosome and one copy of an inverted chromosome is known as an **inversion heterozygote.** Such an individual, though possibly phenotypically normal, may have a high probability of producing haploid cells that are abnormal in their total genetic content.

The underlying cause of gamete abnormality is the phenomenon of crossing over within the inverted region. During meiosis I, pairs of homologous sister chromatids synapse with each other. **Figure 8.11** illustrates how this occurs in an inversion heterozygote. For the normal chromosome and inversion chromosome to synapse properly, an **inversion loop** must form to permit the homologous genes on both chromosomes to align next to each other despite the inverted sequence. If a crossover occurs within the inversion loop, highly abnormal chromosomes are produced. A crossover is more likely to occur in this region if the inversion is large. Therefore, individuals carrying large inversions are more likely to produce abnormal gametes.

The consequences of this type of crossover depend on whether the inversion is pericentric or paracentric. Figure 8.11a describes a crossover in the inversion loop when one of the homologs has a pericentric inversion in which the centromere lies within the inverted region of the chromosome. This event consists of a single crossover that involves only two of the four sister chromatids. Following the completion of meiosis, this single crossover yields two abnormal chromosomes. Both of these abnormal chromosomes have a segment that is deleted and a different segment that is duplicated. In this example, one of the abnormal chromosomes is missing genes *H* and *I* and has an extra copy of genes *A*, *B*, and *C*. The other abnormal chromosome has the opposite situation; it is missing genes *A*, *B*, and *C* and has an extra copy of genes *H* and *I*. These abnormal chromosomes may result in gametes that are inviable. Alternatively, if these abnormal chromosomes are passed to offspring, they are likely to produce phenotypic abnormalities, depending on the amount and nature of the duplicated and deleted genetic material. A large deletion is likely to be lethal.

Figure 8.11b shows the outcome of a crossover involving a paracentric inversion in which the centromere lies outside the inverted region. This single crossover event produces a very strange outcome. One chromosome, called a **dicentric** chromosome, contains two centromeres. The region of the chromosome connecting the two centromeres is a **dicentric bridge.** The crossover also produces a piece of chromosome without any centromere—an **acentric fragment,** which is lost and degraded in subsequent cell divisions. The dicentric chromosome is a temporary condition. If the two centromeres try to move toward opposite poles during anaphase, the dicentric bridge will be forced to break at some random location. Therefore, the net result of this crossover is to produce one normal chromosome, one chromosome with an inversion, and two chromosomes that contain deletions. These two chromosomes with deletions result from the breakage of the dicentric chromosome. They are missing the genes that were located on the acentric fragment.

Translocations Involve Exchanges Between Different Chromosomes

Another type of chromosomal rearrangement is a translocation in which a piece from one chromosome is attached to another chromosome. Eukaryotic chromosomes have telomeres, which tend to prevent translocations from occurring. As described in Chapters 10 and 11, **telomeres**—specialized repeated sequences of DNA—are found at the ends of normal chromosomes. Telomeres allow cells to identify where a chromosome ends and prevent the attachment of chromosomal DNA to the natural ends of a chromosome.

If cells are exposed to agents that cause chromosomes to break, the broken ends lack telomeres and are said to be reactive—a reactive end readily binds to another reactive end. If a single chromosome break occurs, DNA repair enzymes will usually recognize the two reactive ends and join them back together;

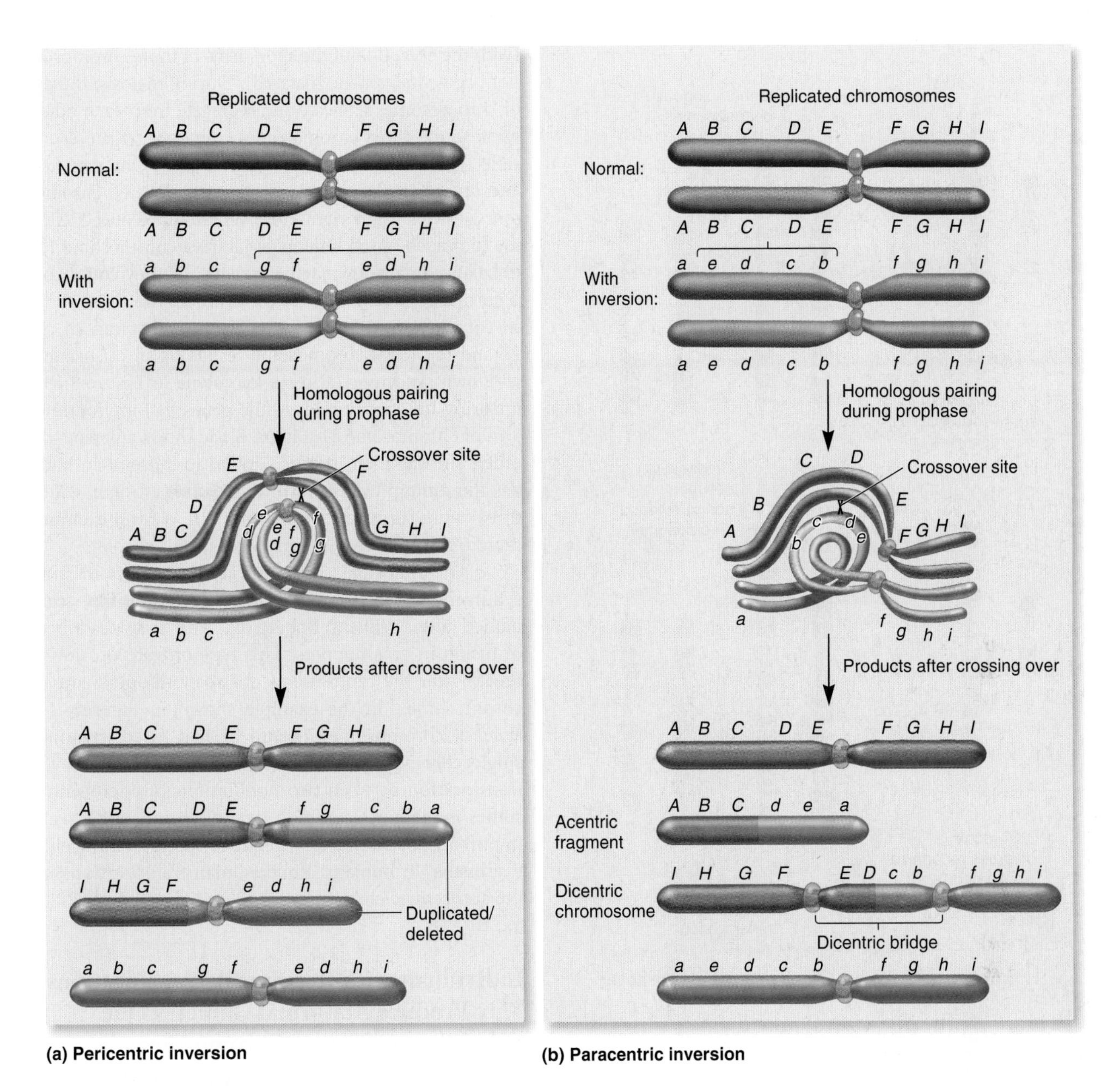

FIGURE 8.11 The consequences of crossing over in the inversion loop. (a) Crossover within a pericentric inversion. (b) Crossover within a paracentric inversion.

the chromosome is repaired properly. However, if multiple chromosomes are broken, the reactive ends may be joined incorrectly to produce abnormal chromosomes (**Figure 8.12a**). This is one mechanism that causes reciprocal translocations to occur.

A second mechanism that can cause a translocation is an abnormal crossover. As shown in **Figure 8.12b**, a reciprocal translocation can be produced when two nonhomologous chromosomes cross over. This type of rare aberrant event results in a rearrangement of the genetic material, though not a change in the total amount of genetic material.

The reciprocal translocations we have considered thus far are also called **balanced translocations** because the total amount of genetic material is not altered. Like inversions, balanced translocations usually occur without any phenotypic consequences because the individual has a normal amount of genetic material. In a few cases, balanced translocations can result in position effects similar to those that can occur in inversions. In addition, carriers of a reciprocal translocation are at risk of having offspring with an **unbalanced translocation,** in which significant portions of genetic material are duplicated and/or deleted. Unbalanced translocations are generally associated with phenotypic abnormalities or even lethality.

Let's consider how a person with a balanced translocation may produce gametes and offspring with an unbalanced translocation. An inherited human syndrome known as familial Down syndrome provides an example. A person with a normal phenotype may have one copy of chromosome 14, one copy of chromosome 21, and one copy of a chromosome that is a fusion between chromosome 14 and 21 (**Figure 8.13a**). The individual has a normal phenotype because the total amount of genetic material is present

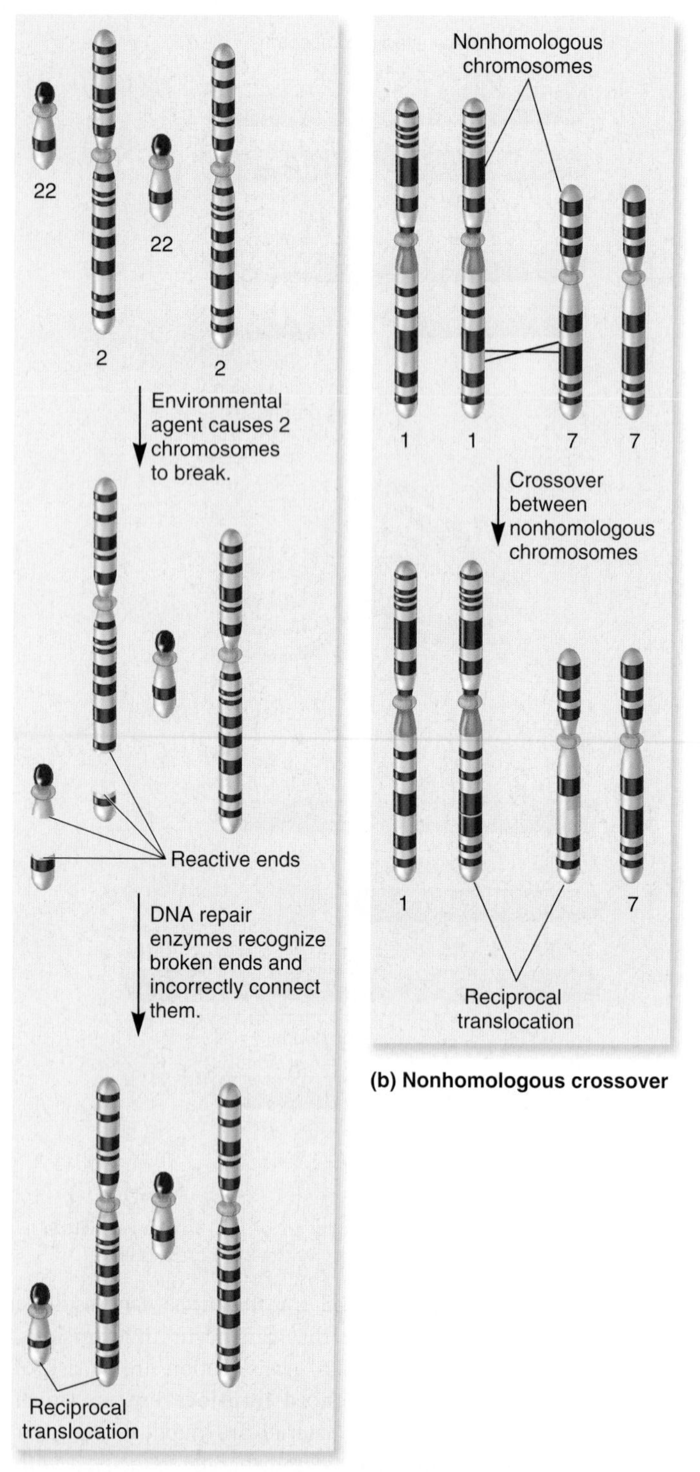

FIGURE 8.12 Two mechanisms that cause a reciprocal translocation. **(a)** When two different chromosomes break, the reactive ends are recognized by DNA repair enzymes, which attempt to reattach them. If two different chromosomes are broken at the same time, the incorrect ends may become attached to each other. **(b)** A nonhomologous crossover has occurred between chromosome 1 and chromosome 7. This crossover yields two chromosomes that carry translocations.

(with the exception of the short arms of these chromosomes that do not carry vital genetic material). During meiosis, these three types of chromosomes replicate and segregate from each other. However, because the three chromosomes cannot segregate evenly, six possible types of gametes may be produced. One gamete is normal, and one is a balanced carrier of a translocated chromosome. The four gametes to the right, however, are unbalanced, either containing too much or too little material from chromosome 14 or 21. The unbalanced gametes may be inviable, or they could combine with a normal gamete. The three offspring on the right will not survive. In comparison, the unbalanced gamete that carries chromosome 21 and the fused chromosome results in an offspring with familial Down syndrome (also see karyotype in **Figure 8.13b**). Such an offspring has three copies of the genes that are found on the long arm of chromosome 21. **Figure 8.13c** shows a person with this disorder. She has characteristics similar to those of an individual who has the more prevalent form of Down syndrome, which is due to three entire copies of chromosome 21. We will examine this common form of Down syndrome later in this chapter.

The abnormal chromosome that occurs in familial Down syndrome is an example of a **Robertsonian translocation,** named after William Robertson, who first described this type of fusion in grasshoppers. This type of translocation arises from breaks near the centromeres of two nonhomologous acrocentric chromosomes. In the example shown in Figure 8.13, the long arms of chromosomes 14 and 21 had fused, creating one large single chromosome; the two short arms are lost. This type of translocation between two nonhomologous acrocentric chromosomes is the most common type of chromosome rearrangement in humans, occurring at a frequency of approximately one in 900 live births. In humans, Robertsonian translocations involve only the acrocentric chromosomes 13, 14, 15, 21, and 22.

Individuals with Reciprocal Translocations May Produce Abnormal Gametes Due to the Segregation of Chromosomes

As we have seen, individuals who carry balanced translocations have a greater risk of producing gametes with unbalanced combinations of chromosomes. Whether or not this occurs depends on the segregation pattern during meiosis I (**Figure 8.14**). In this example, the parent carries a reciprocal translocation and is likely to be phenotypically normal. During meiosis, the homologous chromosomes attempt to synapse with each other. Because of the translocations, the pairing of homologous regions leads to the formation of an unusual structure that contains four pairs of sister chromatids (i.e., eight chromatids), termed a **translocation cross.**

To understand the segregation of translocated chromosomes, pay close attention to the centromeres, which are numbered in Figure 8.14. For these translocated chromosomes, the expected segregation pattern is governed by the centromeres. Each haploid gamete should receive one centromere located on chromosome 1 and one centromere located on chromosome 2. This can occur in two ways. One possibility is alternate segregation. As shown in

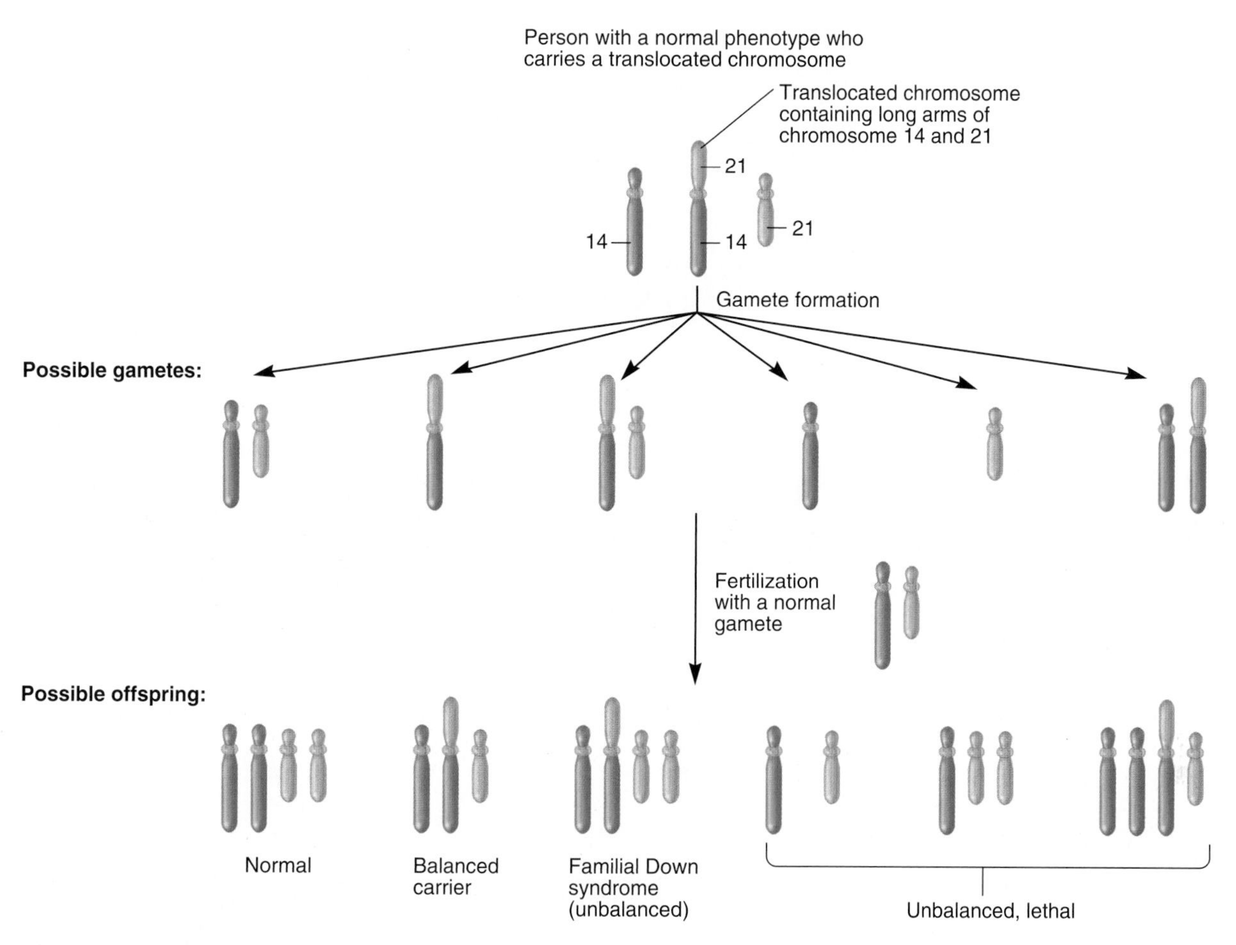

(a) Possible transmission patterns

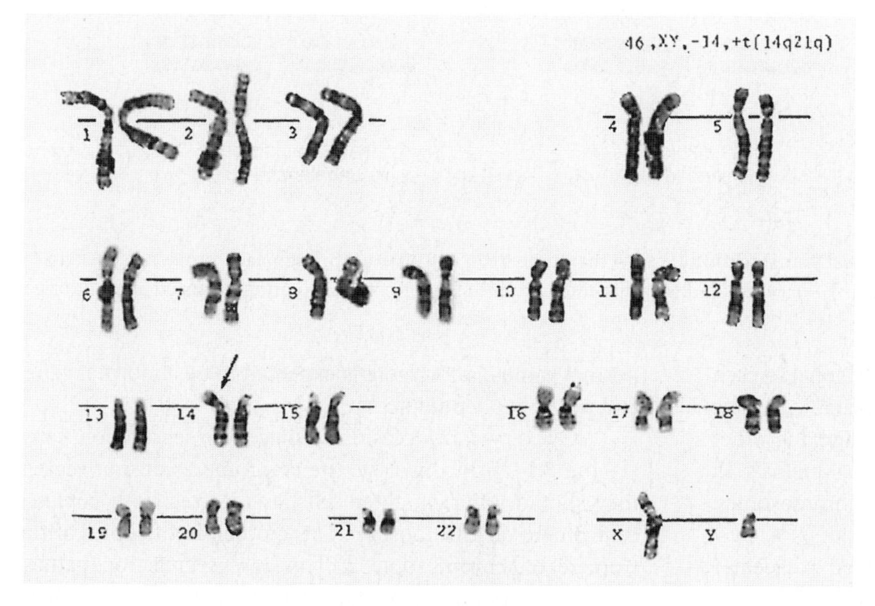

(b) Karyotype of a male with familial Down syndrome

(c) Child with Down syndrome

FIGURE 8.13 **Transmission of familial Down syndrome.** **(a)** Potential transmission of familial Down syndrome. The individual with the chromosome composition shown at the top of this figure may produce a gamete carrying chromosome 21 and a fused chromosome containing the long arms of chromosomes 14 and 21. Such a gamete can give rise to an offspring with familial Down syndrome. **(b)** The karyotype of an individual with familial Down syndrome. This karyotype shows that the long arm of chromosome 21 has been translocated to chromosome 14 (see arrow). In addition, the individual also carries two normal copies of chromosome 21. **(c)** An individual with this disorder.

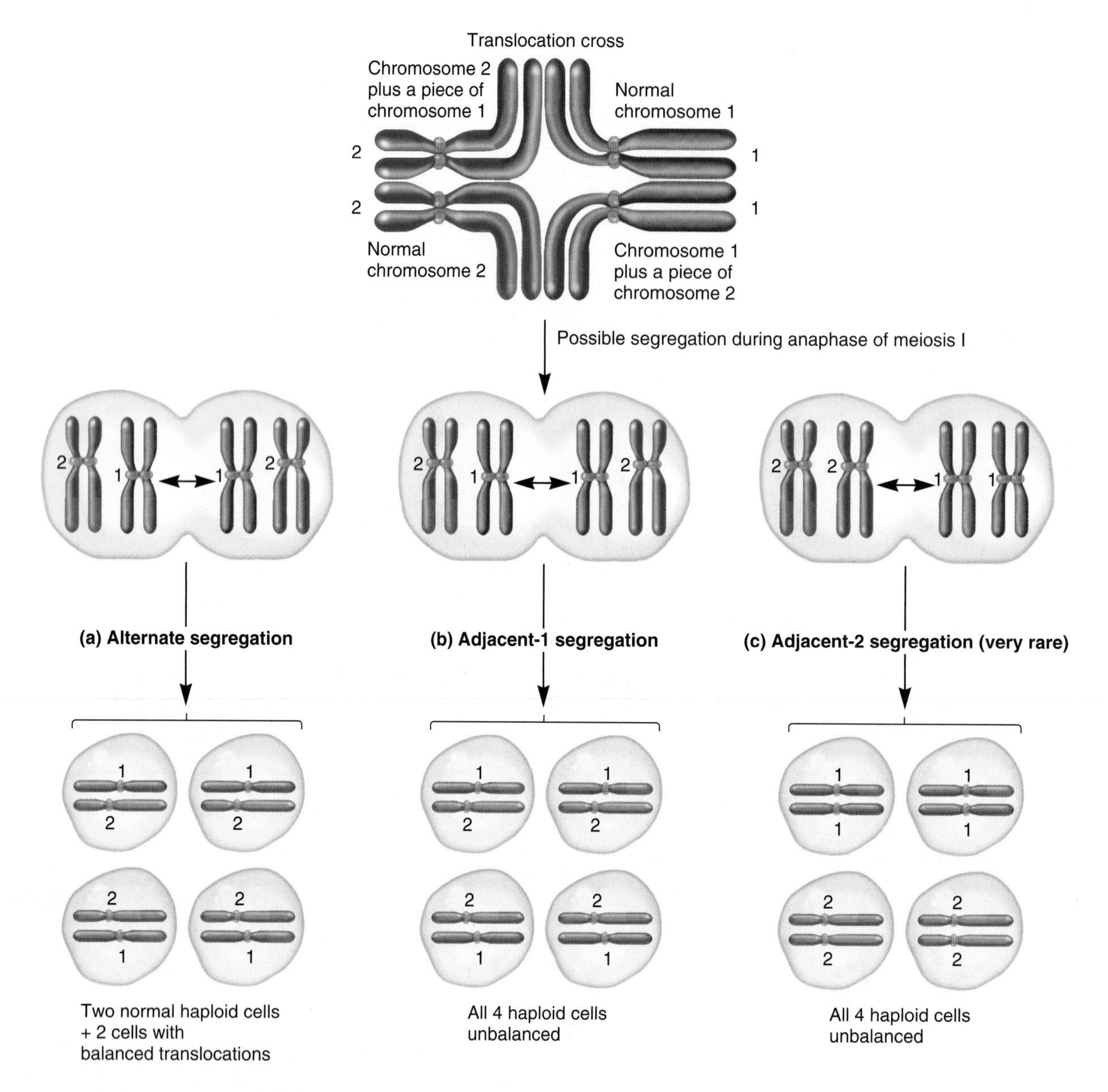

FIGURE 8.14 **Meiotic segregation of a reciprocal translocation.** Follow the numbered centromeres through each process. **(a)** Alternate segregation gives rise to balanced haploid cells, whereas **(b)** adjacent-1 and **(c)** adjacent-2 produce haploid cells with an unbalanced amount of genetic material.

Figure 8.14a, this occurs when the chromosomes diagonal to each other within the translocation cross sort into the same cell. One daughter cell receives two normal chromosomes, and the other cell gets two translocated chromosomes. Following meiosis II, four haploid cells are produced: two have normal chromosomes, and two have reciprocal (balanced) translocations.

Another possible segregation pattern is called adjacent-1 segregation (Figure 8.14b). This occurs when adjacent chromosomes (one of each type of centromere) segregate into the same cell. Following anaphase of meiosis I, each daughter cell receives one normal chromosome and one translocated chromosome. After meiosis II is completed, four haploid cells are produced, all of which are genetically unbalanced because part of one chromosome has been deleted and part of another has been duplicated. If these haploid cells give rise to gametes that unite with a normal gamete, the zygote is expected to be abnormal genetically and possibly phenotypically.

On very rare occasions, adjacent-2 segregation can occur (Figure 8.14c). In this case, the centromeres do not segregate as they should. One daughter cell has received both copies of the centromere on chromosome 1; the other, both copies of the centromere on chromosome 2. This rare segregation pattern also yields four abnormal haploid cells that contain an unbalanced combination of chromosomes.

Alternate and adjacent-1 segregation patterns are the likely outcomes when an individual carries a reciprocal translocation. Depending on the sizes of the translocated segments, both types may be equally likely to occur. In many cases, the haploid cells from adjacent-1 segregation are not viable, thereby lowering the fertility of the parent. This condition is called **semisterility.**

8.2 VARIATION IN CHROMOSOME NUMBER

As we saw in Section 8.1, chromosome structure can be altered in a variety of ways. Likewise, the total number of chromosomes can vary. Eukaryotic species typically contain several chromosomes that are inherited as one or more sets. Variations in chromosome number can be categorized in two ways: variation in the number of sets of chromosomes and variation in the number of particular chromosomes within a set.

Organisms that are **euploid** have a chromosome number that is an exact multiple of a chromosome set. In *Drosophila melanogaster*, for example, a normal individual has 8 chromosomes. The species is diploid, having two sets of 4 chromosomes each (**Figure 8.15a**). A normal fruit fly is euploid because 8 chromosomes divided by 4 chromosomes per set equals two exact sets. On rare occasions, an abnormal fruit fly can be produced with 12 chromosomes, containing three sets of 4 chromosomes each. This alteration in euploidy produces a **triploid** fruit fly with 12 chromosomes. Such a fly is also euploid because it has exactly three sets of chromosomes. Organisms with three or more sets of chromosomes are also called **polyploid** (**Figure 8.15b**). Geneticists use the letter n to represent a set of chromosomes. A diploid organism is referred to as $2n$, a triploid organism as $3n$, a **tetraploid** organism as $4n$, and so on.

A second way in which chromosome number can vary is by **aneuploidy.** Such variation involves an alteration in the number of particular chromosomes, so the total number of chromosomes is not an exact multiple of a set. For example, an abnormal fruit fly could contain nine chromosomes instead of eight because it has three copies of chromosome 2 instead of the normal two copies (**Figure 8.15c**). Such an animal is said to have trisomy 2 or to be **trisomic.** Instead of being perfectly diploid ($2n$), a trisomic animal is $2n + 1$. By comparison, a fruit fly could be lacking a single chromosome, such as chromosome 1, and contain a total of seven chromosomes ($2n - 1$). This animal is **monosomic** and is described as having monosomy 1.

In this section, we will begin by considering several examples of aneuploidy. This is generally regarded as an abnormal condition that usually has a negative effect on phenotype. We will then examine euploid variation that occurs occasionally in animals and quite frequently in plants, and consider how it affects phenotypic variation.

Aneuploidy Causes an Imbalance in Gene Expression That Is Often Detrimental to the Phenotype of the Individual

The phenotype of every eukaryotic species is influenced by thousands of different genes. In humans, for example, a single set of chromosomes contains approximately 20,000 to 25,000 different

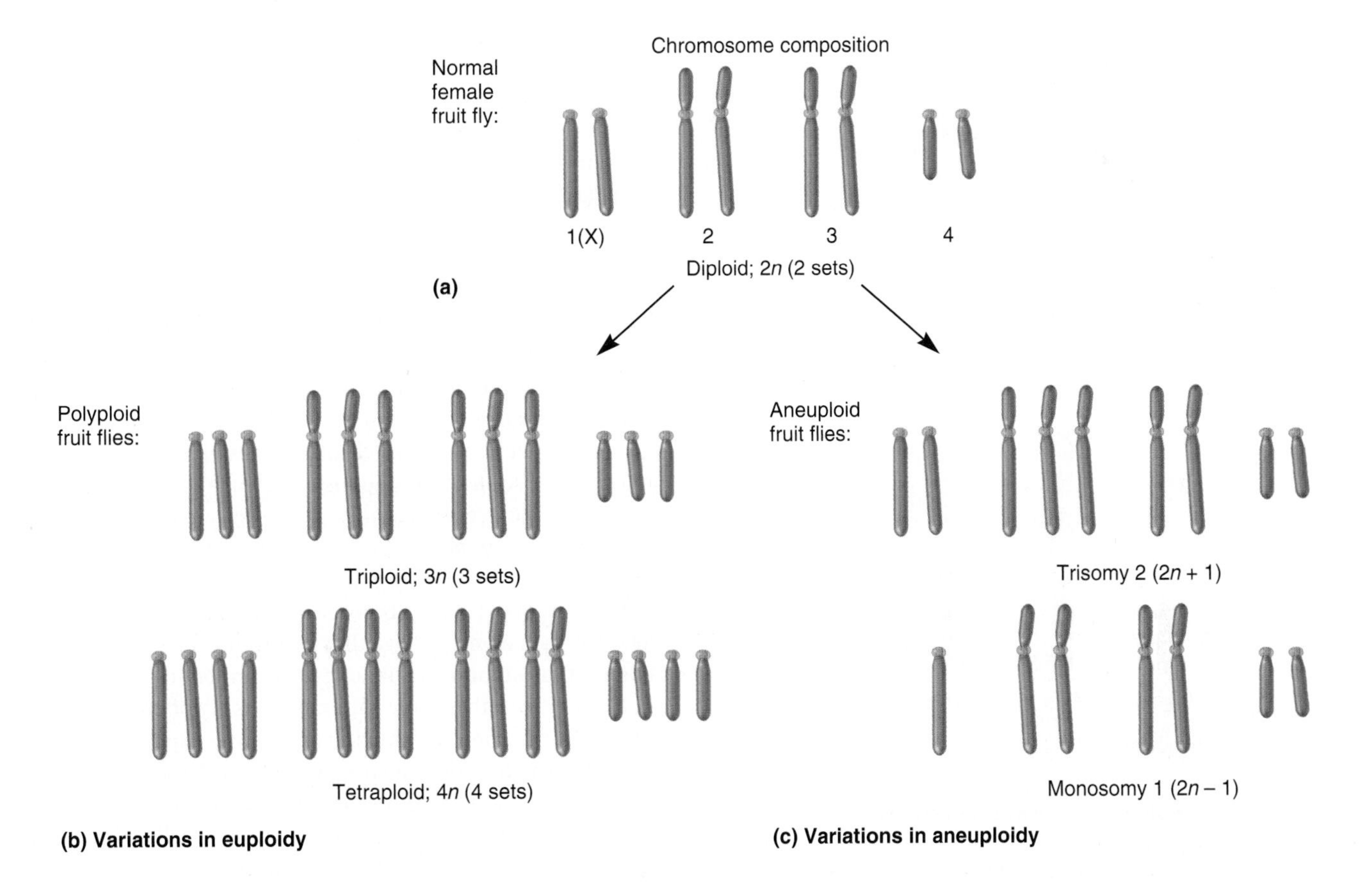

FIGURE 8.15 **Types of variation in chromosome number.** (a) Depicts the normal diploid number of chromosomes in *Drosophila*. (b) Examples of polyploidy. (c) Examples of aneuploidy.

genes. To produce a phenotypically normal individual, intricate coordination has to occur in the expression of thousands of genes. In the case of humans and other diploid species, evolution has resulted in a developmental process that works correctly when somatic cells have two copies of each chromosome. In other words, when a human is diploid, the balance of gene expression among many different genes usually produces a person with a normal phenotype.

Aneuploidy commonly causes an abnormal phenotype. To understand why, let's consider the relationship between gene expression and chromosome number in a species that has three pairs of chromosomes (**Figure 8.16**). The level of gene expression is influenced by the number of genes per cell. Compared with a diploid cell, if a gene is carried on a chromosome that is present in three copies instead of two, more of the gene product is typically made. For example, a gene present in three copies instead of two may produce 150% of the gene product, though that number may vary due to effects of gene regulation. Alternatively, if only one copy of that gene is present due to a missing chromosome, less of the gene product is usually made, perhaps only 50%. Therefore, in trisomic and monosomic individuals, an imbalance occurs between the level of gene expression on the chromosomes found in pairs versus the one type that is not.

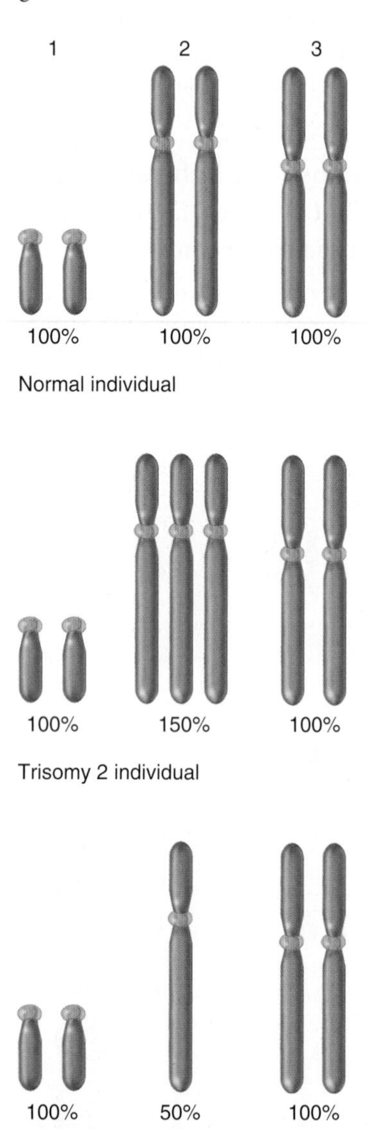

FIGURE 8.16 Imbalance of gene products in trisomic and monosomic individuals. Aneuploidy of chromosome 2 (i.e., trisomy and monosomy) leads to an imbalance in the amount of gene products from chromosome 2 compared with the amounts from chromosomes 1 and 3.

At first glance, the difference in gene expression between euploid and aneuploid individuals may not seem terribly dramatic. Keep in mind, however, that a eukaryotic chromosome carries hundreds or even thousands of different genes. Therefore, when an organism is trisomic or monosomic, many gene products occur in excessive or deficient amounts. This imbalance among many genes appears to underlie the abnormal phenotypic effects that aneuploidy frequently causes. In most cases, these effects are detrimental and produce an individual that is less likely to survive than a euploid individual.

Aneuploidy in Humans Causes Abnormal Phenotypes

A key reason why geneticists are so interested in aneuploidy is its relationship to certain inherited disorders in humans. Even though most people are born with a normal number of chromosomes (i.e., 46), alterations in chromosome number occur fairly frequently during gamete formation. About 5% to 10% of all fertilized human eggs result in an embryo with an abnormality in chromosome number! In most cases, these abnormal embryos do not develop properly and result in a spontaneous abortion very early in pregnancy. Approximately 50% of all spontaneous abortions are due to alterations in chromosome number.

In some cases, an abnormality in chromosome number produces an offspring that survives to birth or longer. Several human disorders involve abnormalities in chromosome number. The most common are trisomies of chromosomes 13, 18, or 21, and abnormalities in the number of the sex chromosomes (**Table 8.1**). Most of the known trisomies involve chromosomes that are relatively small—chromosome 13, 18, or 21—and carry fewer genes compared to larger chromosomes. Trisomies of the other human autosomes and monosomies of all autosomes are presumed to produce a lethal phenotype, and many have been found in spontaneously aborted embryos and fetuses. For example, all possible human trisomies have been found in spontaneously aborted embryos except trisomy 1. It is believed that trisomy 1 is lethal at such an early stage that it prevents the successful implantation of the embryo. Variation in the number of X chromosomes, unlike that of other large chromosomes, is often nonlethal. The survival of trisomy X individuals may be explained by X inactivation, which is described in Chapter 5. In an individual with more than one X chromosome, all additional X chromosomes are converted to Barr bodies in the somatic cells of adult tissues. In an individual with trisomy X, for example, two out of three X chromosomes are converted to inactive Barr bodies. Unlike the level of expression for autosomal genes, the normal level

TABLE 8.1

Aneuploid Conditions in Humans

Condition	Frequency	Syndrome	Characteristics
Autosomal			
Trisomy 13	1/15,000	Patau	Mental and physical deficiencies, wide variety of defects in organs, large triangular nose, early death
Trisomy 18	1/6000	Edward	Mental and physical deficiencies, facial abnormalities, extreme muscle tone, early death
Trisomy 21	1/800	Down	Mental deficiencies, abnormal pattern of palm creases, slanted eyes, flattened face, short stature
Sex Chromosomal			
XXY	1/1000 (males)	Klinefelter	Sexual immaturity (no sperm), breast swelling
XYY	1/1000 (males)	Jacobs	Tall and thin
XXX	1/1500 (females)	Triple X	Tall and thin, menstrual irregularity
X0	1/5000 (females)	Turner	Short stature, webbed neck, sexually undeveloped

of expression for X-linked genes is from a single X chromosome. In other words, the correct level of mammalian gene expression results from two copies of each autosomal gene and one copy of each X-linked gene. This explains how the expression of X-linked genes in males (XY) can be maintained at the same levels as in females (XX). It may also explain why trisomy X is not a lethal condition. The phenotypic effects noted in Table 8.1 involving sex chromosomal abnormalities may be due to the expression of X-linked genes prior to embryonic X inactivation or to the expression of genes on the inactivated X chromosome. As described in Chapter 5, pseudoautosomal genes and some other genes on the inactivated X chromosome are expressed in humans. Having one or three copies of the sex chromosomes results in an under- or overexpression of these X-linked genes, respectively.

Human abnormalities in chromosome number are influenced by the age of the parents. Older parents are more likely to produce children with abnormalities in chromosome number. Down syndrome provides an example. The common form of this disorder is caused by the inheritance of three copies of chromosome 21. The incidence of Down syndrome rises with the age of either parent. In males, however, the rise occurs relatively late in life, usually past the age when most men have children. By comparison, the likelihood of having a child with Down syndrome rises dramatically during a woman's reproductive age (**Figure 8.17**). This syndrome was first described by the English physician John Langdon Down in 1866. The association between maternal age and Down syndrome was later discovered by L. S. Penrose in 1933, even before the chromosomal basis for the disorder was

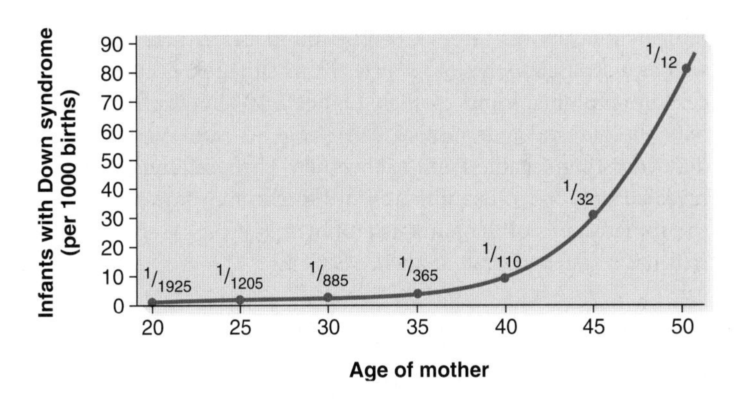

FIGURE 8.17 The incidence of Down syndrome births according to the age of the mother. The *y*-axis shows the number of infants born with Down syndrome per 1000 live births, and the *x*-axis plots the age of the mother at the time of birth. The data points indicate the fraction of live offspring born with Down syndrome.

identified by the French scientist Jérôme Lejeune in 1959. Down syndrome is most commonly caused by **nondisjunction,** which means that the chromosomes do not segregate properly. (Nondisjunction is discussed later in this chapter.) In this case, nondisjunction of chromosome 21 most commonly occurs during meiosis I in the oocyte.

Different hypotheses have been proposed to explain the relationship between maternal age and Down syndrome. One popular idea suggests that it may be due to the age of the oocytes. Human primary oocytes are produced within the ovary of the female fetus prior to birth and are arrested at prophase of meiosis I and remain in this stage until the time of ovulation. Therefore, as a woman ages, her primary oocytes have been in prophase I for a progressively longer period of time. This added length of time may contribute to an increased frequency of nondisjunction. About 5% of the time, Down syndrome is due to an extra paternal chromosome. Prenatal tests can determine if a fetus has Down syndrome and some other genetic abnormalities. The topic of genetic testing is discussed in Chapter 22.

Variations in Euploidy Occur Naturally in a Few Animal Species

We now turn our attention to changes in the number of sets of chromosomes, referred to as variations in euploidy. Most species of animals are diploid. In some cases, changes in euploidy are not well tolerated. For example, polyploidy in mammals is generally a lethal condition. However, many examples of naturally occurring variations in euploidy occur. In **haplodiploid** species, which include many species of bees, wasps, and ants, one of the sexes is haploid, usually the male, and the other is diploid. For example, male bees, which are called drones, contain a single set of chromosomes. They are produced from unfertilized eggs. By comparison, female bees are produced from fertilized eggs and are diploid.

Many examples of vertebrate polyploid animals have been discovered. Interestingly, on several occasions, animals that are morphologically very similar can be found as a diploid species as well as a separate polyploid species. This situation occurs among certain amphibians and reptiles. **Figure 8.18** shows photographs of a diploid and a tetraploid (4*n*) frog. As you can see, they look indistinguishable from each other. Their difference can be revealed only by an examination of the chromosome number in the somatic cells of the animals and by mating calls—*H. chrysoscelis* has a faster trill rate than *H. versicolor*.

Variations in Euploidy Can Occur in Certain Tissues Within an Animal

Thus far, we have considered variations in chromosome number that occur at fertilization, so all somatic cells of an individual contain this variation. In many animals, certain tissues of the body display normal variations in the number of sets of chromosomes. Diploid animals sometimes produce tissues that are polyploid. For example, the cells of the human liver can vary to a great degree in their ploidy. Liver cells contain nuclei that can be triploid, tetraploid, and even octaploid (8*n*). The occurrence of polyploid tissues or cells in organisms that are otherwise diploid is known as **endopolyploidy.** What is the biological significance of endopolyploidy? One possibility is that the increase in chromosome number in certain cells may enhance their ability to produce specific gene products that are needed in great abundance.

An unusual example of natural variation in the ploidy of somatic cells occurs in *Drosophila* and some other insects. Within certain tissues, such as the salivary glands, the chromosomes undergo repeated rounds of chromosome replication without cellular division. For example, in the salivary gland cells of *Drosophila,* the pairs of chromosomes double approximately nine times ($2^9 = 512$). **Figure 8.19a** illustrates how repeated rounds of chromosomal replication produce a bundle of chromosomes that lie together in a parallel fashion. This bundle, termed a **polytene chromosome,** was first observed by E. G. Balbiani in 1881. Later, in the 1930s, Theophilus Painter and colleagues recognized that the size and morphology of polytene chromosomes provided geneticists with unique opportunities to study chromosome structure and gene organization.

Figure 8.19b shows a micrograph of a polytene chromosome. The structure of polytene chromosomes is different from other forms of endopolyploidy because the replicated chromosomes remain attached to each other. Prior to the formation of polytene chromosomes, *Drosophila* cells contain eight chromosomes (two sets of four chromosomes each; see Figure 8.15a). In the salivary gland cells, the homologous chromosomes synapse with each other and replicate to form a polytene structure. During this process, the four types of chromosomes aggregate to form a single structure with several polytene arms. The central point where the chromosomes aggregate is known as the **chromocenter.** Each of the four types of chromosome is attached to the chromocenter near its centromere. The X and Y and chromosome 4 are telocentric, and chromosomes 2 and 3 are metacentric. Therefore, chromosomes 2 and 3 have two arms that radiate from the chromocenter, whereas the X and Y and chromosome 4 have a single arm projecting from the chromocenter (**Figure 8.19c**).

(a) ***Hyla chrysoscelis***

(b) ***Hyla versicolor***

FIGURE 8.18 Differences in euploidy in two closely related frog species. The frog in **(a)** is diploid, whereas the frog in **(b)** is tetraploid.

Genes → Traits Though similar in appearance, these two species differ in their number of chromosome sets. At the level of gene expression, this observation suggests that the number of copies of each gene (two versus four) does not critically affect the phenotype of these two species.

Variations in Euploidy Are Common in Plants

We now turn our attention to variations of euploidy that occur in plants. Compared with animals, plants more commonly exhibit polyploidy. Among ferns and flowering plants, at least 30% to 35% of species are polyploid. Polyploidy is also important in agriculture. Many of the fruits and grains we eat are produced from polyploid plants. For example, the species of wheat that we use to make bread, *Triticum aestivum*, is a hexaploid (6*n*) that arose from the union of diploid genomes from three closely related species (**Figure 8.20a**). Different species of strawberries are diploid, tetraploid, hexaploid, and even octaploid!

In many instances, polyploid strains of plants display outstanding agricultural characteristics. They are often larger in size

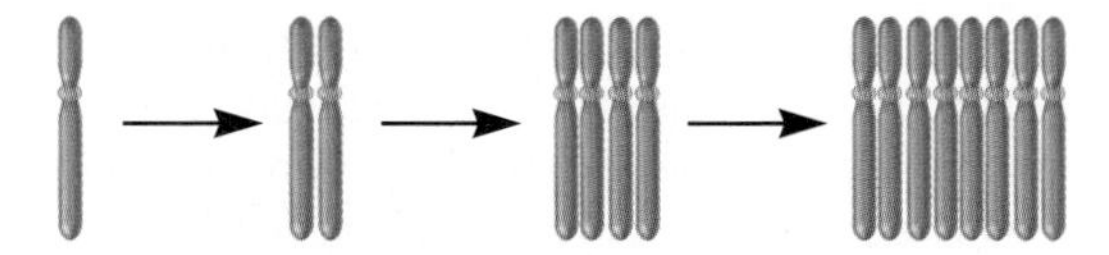

(a) Repeated chromosome replication produces polytene chromosome.

(b) A polytene chromosome

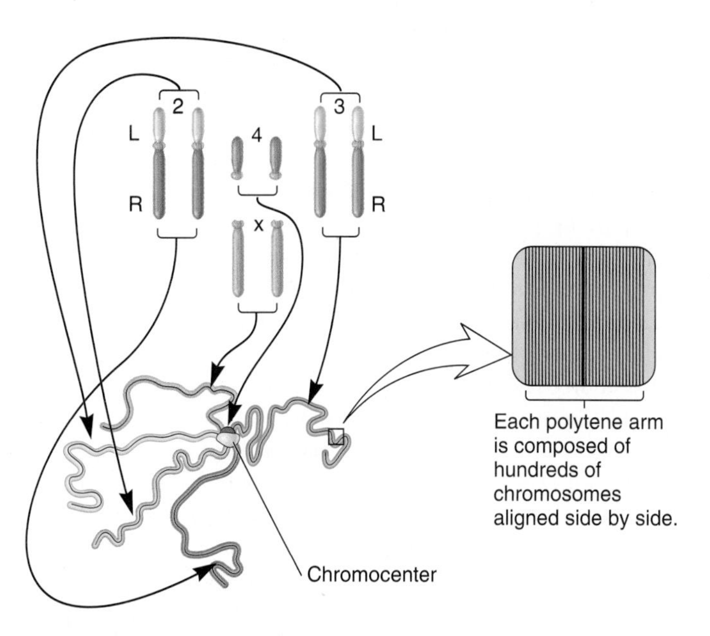

(c) Relationship between a polytene chromosome and regular *Drosophila* chromosomes

FIGURE 8.19 **Polytene chromosomes in *Drosophila*.** (**a**) A schematic illustration of the formation of polytene chromosomes. Several rounds of repeated replication without cellular division result in a bundle of sister chromatids that lie side by side. Both homologs also lie parallel to each other. This replication does not occur in highly condensed, heterochromatic DNA near the centromere. (**b**) A photograph of a polytene chromosome. (**c**) This drawing shows the relationship between the four pairs of chromosomes and the formation of a polytene chromosome in the salivary gland. The heterochromatic regions of the chromosomes aggregate at the chromocenter, and the arms of the chromosomes project outward. In chromosomes with two arms, the short arm is labeled L and the long arm is labeled R.

(a) Cultivated wheat, a hexaploid species

Tetraploid Diploid

(b) A comparison of diploid and tetraploid petunias

FIGURE 8.20 **Examples of polyploid plants.** (**a**) Cultivated wheat, *Triticum aestivum*, is a hexaploid. It was derived from three different diploid species of grasses that originally were found in the Middle East and were cultivated by ancient farmers in that region. (**b**) Differences in euploidy may exist in two closely related petunia species. The larger flower at the left is tetraploid, whereas the smaller one at the right is diploid.

Genes → Traits An increase in chromosome number from diploid to tetraploid or hexaploid affects the phenotype of the individual. In the case of many plant species, a polyploid individual is larger and more robust than its diploid counterpart. This suggests that having additional copies of each gene is somewhat better than having two copies of each gene. This phenomenon in plants is rather different from the situation in animals. Tetraploidy in animals may have little effect (as in Figure 8.18b), but it is also common for polyploidy in animals to be detrimental.

and more robust. These traits are clearly advantageous in the production of food. In addition, polyploid plants tend to exhibit a greater adaptability, which allows them to withstand harsher environmental conditions. Also, polyploid ornamental plants often produce larger flowers than their diploid counterparts (**Figure 8.20b**).

Polyploid plants having an odd number of chromosome sets, such as triploids ($3n$) or pentaploids ($5n$), usually cannot reproduce. Why are they sterile? The sterility arises because they produce highly aneuploid gametes. During prophase of meiosis I, homologous pairs of sister chromatids form bivalents. However, organisms with an odd number of chromosomes, such as three, display an unequal separation of homologous chromosomes during anaphase of meiosis I (**Figure 8.21**). An odd number cannot be divided equally between two daughter cells. For each type of chromosome, a daughter cell randomly gets one or two copies. For example, one daughter cell might receive one copy of chromosome 1, two copies of chromosome 2, two copies of chromosome 3, one copy of chromosome 4, and so forth. For a triploid species containing many different chromosomes in a set, meiosis is very unlikely to produce a daughter cell that is euploid. If we assume that a daughter cell receives either one copy or two copies of each kind of chromosome, the probability that meiosis will produce a cell that is perfectly haploid or diploid is $(1/2)^{n-1}$, where n is the number of chromosomes in a set. As an example, in a triploid organism containing 20 chromosomes per set, the probability of producing a haploid or diploid cell is 0.000001907, or 1 in 524,288. Thus, meiosis is almost certain to produce cells that contain one copy of some chromosomes and two copies of the others. This high probability of aneuploidy underlies the reason for triploid sterility.

Though sterility is generally a detrimental trait, it can be desirable agriculturally because it may result in a seedless fruit.

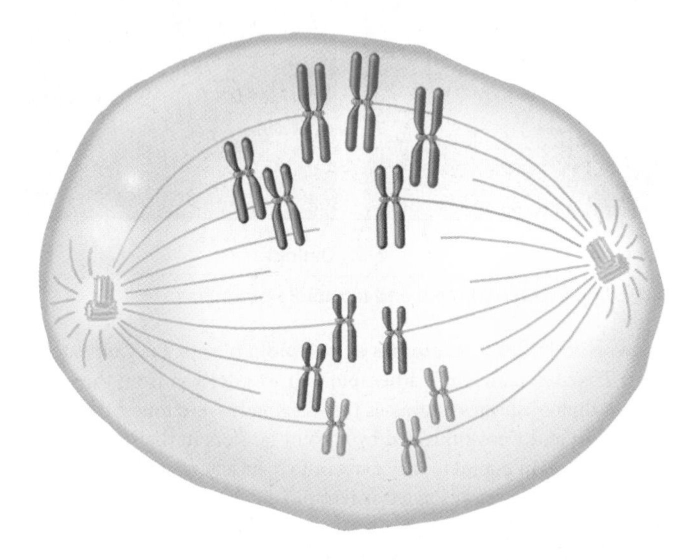

FIGURE 8.21 Schematic representation of anaphase of meiosis I in a triploid organism containing three sets of four chromosomes. In this example, the homologous chromosomes (three each) do not evenly separate during anaphase. Each cell receives one copy of some chromosomes and two copies of other chromosomes. This produces aneuploid gametes.

For example, domestic bananas and seedless watermelons are triploid varieties. The domestic banana was originally derived from a seed-producing diploid species and has been asexually propagated by humans via cuttings. The small black spots in the center of a domestic banana are degenerate seeds. In the case of flowers, the seedless phenotype can also be beneficial. Seed producers such as Burpee have developed triploid varieties of flowering plants such as marigolds. Because the triploid marigolds are sterile and unable to set seed, more of their energy goes into flower production. According to Burpee, "They bloom and bloom, unweakened by seed bearing."

8.3 NATURAL AND EXPERIMENTAL WAYS TO PRODUCE VARIATIONS IN CHROMOSOME NUMBER

As we have seen, variations in chromosome number are fairly widespread and usually have a significant effect on the phenotypes of plants and animals. For these reasons, researchers have wanted to understand the cellular mechanisms that cause variations in chromosome number. In some cases, a change in chromosome number is the result of nondisjunction. The term nondisjunction refers to an event in which the chromosomes do not segregate properly. As we will see, it may be caused by an improper separation of homologous pairs in a bivalent in meiosis or a failure of the centromeres to disconnect during mitosis.

Meiotic nondisjunction can produce haploid cells that have too many or too few chromosomes. If such a cell gives rise to a gamete that fuses with a normal gamete during fertilization, the resulting offspring will have an abnormal chromosome number in all of its cells. An abnormal nondisjunction event also may occur after fertilization in one of the somatic cells of the body. This second mechanism is known as **mitotic nondisjunction.** When this occurs during embryonic stages of development, it may lead to a patch of tissue in the organism that has an altered chromosome number. A third common way in which the chromosome number of an organism can vary is by interspecies crosses. An **alloploid** organism contains sets of chromosomes from two or more different species. This term refers to the occurrence of chromosome sets (ploidy) from the genomes of different (allo) species.

In this section, we will examine these three mechanisms in greater detail. Also, in the past few decades, researchers have devised several methods for manipulating chromosome number in experimentally and agriculturally important species. We will conclude this section by exploring how the experimental manipulation of chromosome number has had an important impact on genetic research and agriculture.

Meiotic Nondisjunction Can Produce Aneuploidy or Polyploidy

Nondisjunction during meiosis can occur during anaphase of meiosis I or meiosis II. If it happens during meiosis I, an entire bivalent migrates to one pole (**Figure 8.22a**). Following the completion of meiosis, the four resulting haploid cells produced from this event are abnormal. If nondisjunction occurs during anaphase of meiosis II (**Figure 8.22b**), the net result is two abnormal and two normal haploid cells. If a gamete that is missing a chromosome is viable and participates in fertilization, the resulting offspring is monosomic for the missing chromosomes. Alternatively, if a gamete carrying an extra chromosome unites with a normal gamete, the offspring will be trisomic.

In rare cases, all of the chromosomes can undergo nondisjunction and migrate to one of the daughter cells. The net result of **complete nondisjunction** is a diploid cell and a cell without any chromosomes. The cell without chromosomes is nonviable, but the diploid cell might participate in fertilization with a normal haploid gamete to produce a triploid individual. Therefore, complete nondisjunction can produce individuals that are polyploid.

Mitotic Nondisjunction or Chromosome Loss Can Produce a Patch of Tissue with an Altered Chromosome Number

Abnormalities in chromosome number occasionally occur after fertilization takes place. In this case, the abnormal event happens during mitosis rather than meiosis. One possibility is that the sister chromatids separate improperly, so one daughter cell has three copies of a chromosome, whereas the other daughter cell has only one (**Figure 8.23a**). Alternatively, the sister chromatids could separate during anaphase of mitosis, but one of the chromosomes could be improperly attached to the spindle, so it does not migrate to a pole (**Figure 8.23b**). A chromosome will be degraded if it is left outside the nucleus when the nuclear membrane re-forms. In this case, one of the daughter cells has two copies of that chromosome, whereas the other has only one.

When genetic abnormalities occur after fertilization, the organism contains a subset of cells that are genetically different from those of the rest of the organism. This condition is referred to as **mosaicism.** The size and location of the mosaic region depend on the timing and location of the original abnormal event. If a genetic alteration happens very early in the embryonic development of an organism, the abnormal cell will be the precursor for a large section of the organism. In the most extreme case, an abnormality can take place at the first mitotic division. As a bizarre example, consider a fertilized *Drosophila* egg that is XX. One of the X chromosomes may be lost during the first mitotic division, producing one daughter cell that is XX and one that is X0. Flies that are XX develop into females, and X0 flies develop into males. Therefore, in this example, one-half of the organism becomes female and one-half becomes male! This peculiar and rare individual is referred to as a **bilateral gynandromorph** (**Figure 8.24**).

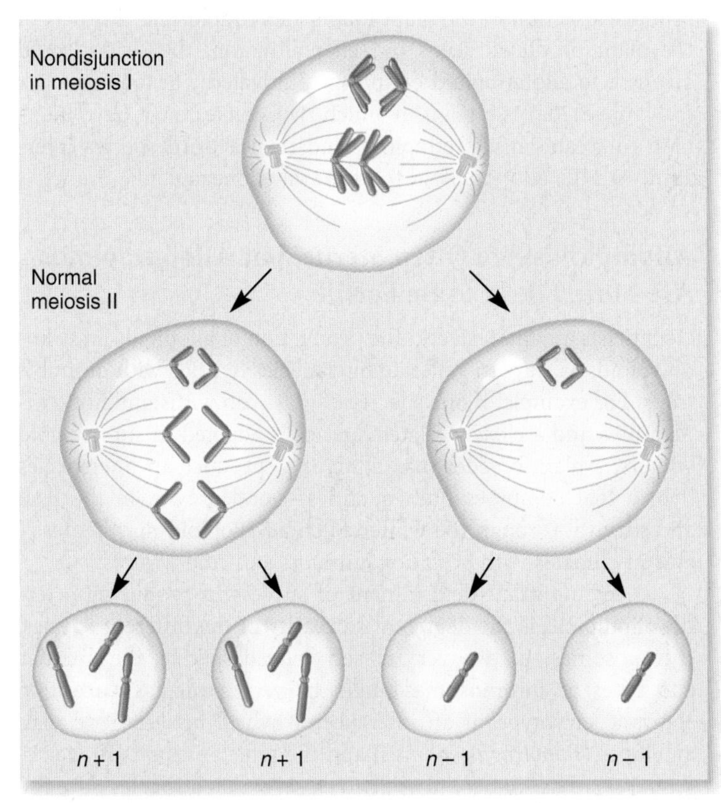

(a) Nondisjunction in meiosis I

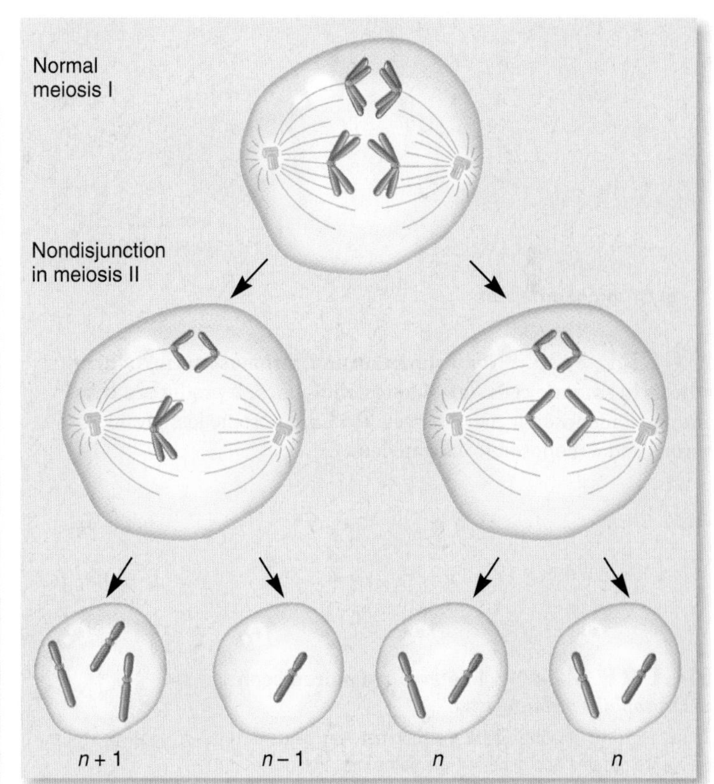

(b) Nondisjunction in meiosis II

FIGURE 8.22 Nondisjunction during meiosis I and II. The chromosomes shown in purple are behaving properly during meiosis I and II, so each cell receives one copy of this chromosome. The chromosomes shown in blue are not disjoining correctly. In (**a**), nondisjunction occurred in meiosis I, so the resulting four cells receive either two copies of the blue chromosome or zero copies. In (**b**), nondisjunction occurred during meiosis II, so one cell has two blue chromosomes and another cell has zero. The remaining two cells are normal.

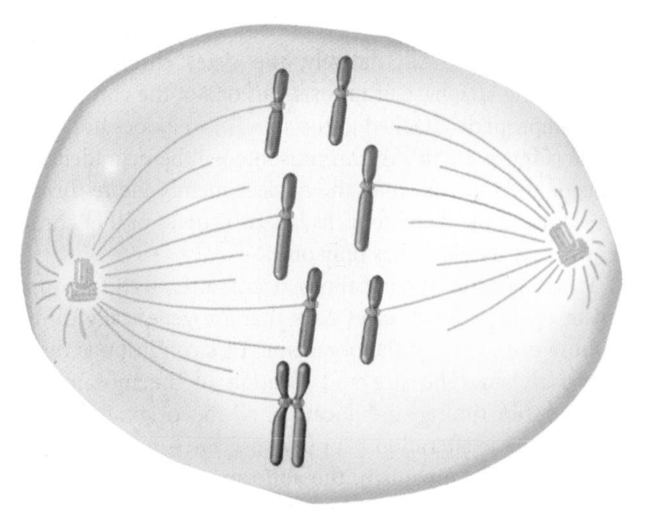

(a) Mitotic nondisjunction

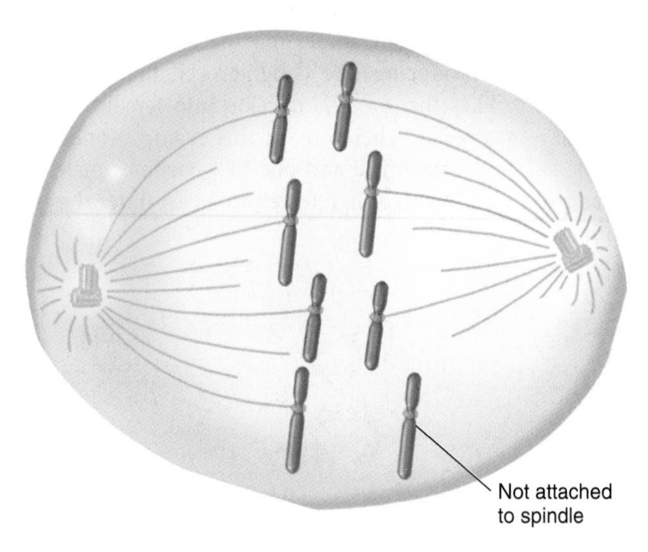

(b) Chromosome loss

FIGURE 8.23 Nondisjunction and chromosome loss during mitosis in somatic cells. (a) Mitotic nondisjunction produces a trisomic and a monosomic daughter cell. (**b**) Chromosome loss produces a normal and a monosomic daughter cell.

FIGURE 8.24 A bilateral gynandromorph of *Drosophila melanogaster.*

Genes → Traits In *Drosophila,* the ratio between genes on the X chromosome and genes on the autosomes determines sex. This fly began as an XX female. One X chromosome carried the recessive white-eye and miniature wing alleles; the other X chromosome carried the wild-type alleles. The X chromosome carrying the wild-type alleles was lost from one of the cells during the first mitotic division, producing one XX cell and one X0 cell. The XX cell became the precursor for the left side of the fly, which developed as female. The X0 cell became the precursor for the other side of the fly, which developed as male with a white eye and a miniature wing.

Changes in Euploidy Can Occur by Autopolyploidy, Alloploidy, and Allopolyploidy

Different mechanisms account for changes in the number of chromosome sets among natural populations of plants and animals (**Figure 8.25**). As previously mentioned, complete nondisjunction, due to a general defect in the spindle apparatus, can produce an individual with one or more extra sets of chromosomes. This individual is known as an **autopolyploid** (Figure 8.25a). The prefix auto- (meaning self) and term polyploid (meaning many sets of chromosomes) refer to an increase in the number of chromosome sets within a single species.

A much more common mechanism for change in chromosome number, called **alloploidy,** is a result of interspecies crosses (Figure 8.25b). An alloploid that has one set of chromosomes from two different species is called an **allodiploid.** This event is most likely to occur between species that are close evolutionary relatives. For example, closely related species of grasses may interbreed to produce allodiploids. As shown in Figure 8.25c, an **allopolyploid** contains two (or more) sets of chromosomes from two (or more) species. In this case, the **allotetraploid** contains two complete sets of chromosomes from two different species, for a total of four sets. In nature, allotetraploids usually arise from allodiploids. This can occur when a somatic cell in an allodiploid undergoes complete nondisjunction to create an allotetraploid cell. In plants, such a cell can continue to grow and produce a section of the plant that is allotetraploid. If this part of the plant produced seeds by self-pollination, the seeds would give rise to allotetraploid offspring. Cultivated wheat (refer back to Figure 8.20a) is a plant in which two species must have interbred to create an allotetraploid, and then a third species interbred with the allotetraploid to create an allohexaploid.

Allodiploids Are Often Sterile, but Allotetraploids Are More Likely to Be Fertile

Geneticists are interested in the production of alloploids and allopolyploids as ways to generate interspecies hybrids with desirable traits. For example, if one species of grass can withstand hot temperatures and a closely related species is adapted to survive cold winters, a plant breeder may attempt to produce an interspecies hybrid that combines both qualities—good growth in the heat and survival through the winter. Such an alloploid may be desirable in climates with both hot summers and cold winters.

An important determinant of success in producing a fertile allodiploid is the degree of similarity of the different species' chromosomes. In two very closely related species, the number and types of chromosomes might be very similar. **Figure 8.26** shows a karyotype of an interspecies hybrid between the roan antelope (*Hippotragus equinus*) and the sable antelope (*Hippotragus niger*). As seen here, these two closely related species have the same number of chromosomes. The sizes and banding patterns of the chromosomes show that they correspond to one another. For example, chromosome 1 from both species is fairly large, has very similar banding patterns, and carries many of the same genes. Evolutionarily related chromosomes from two different species are called **homeologous** chromosomes (not to be confused with

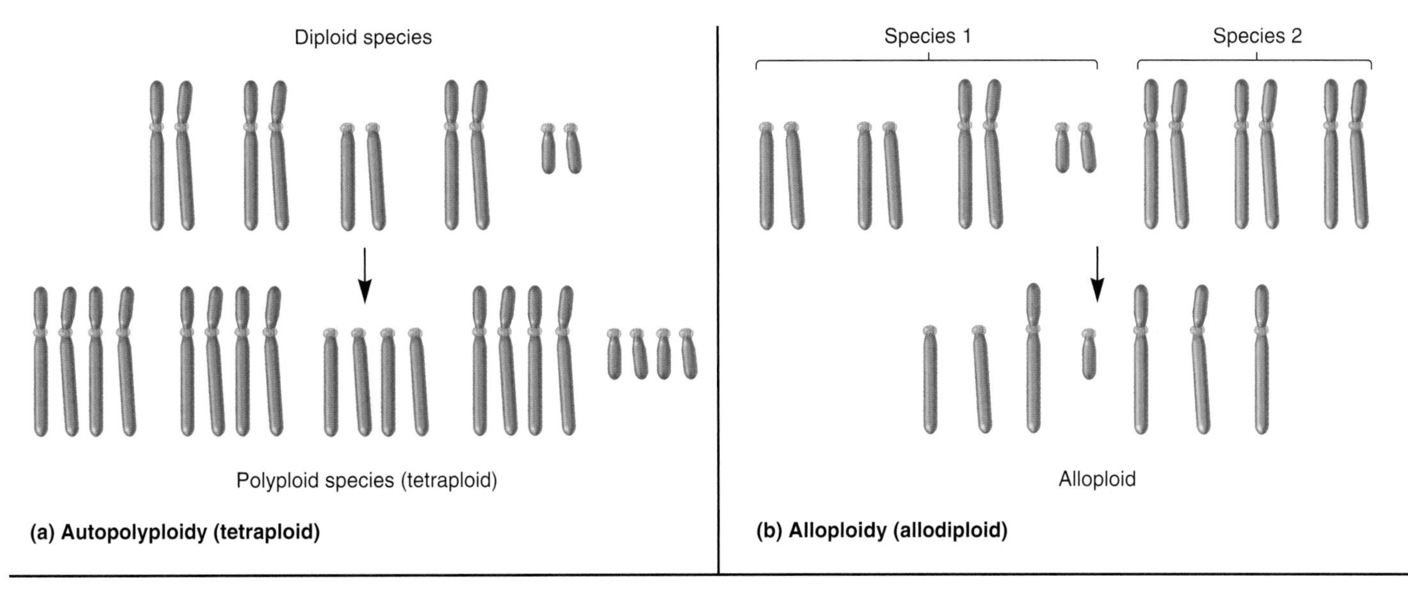

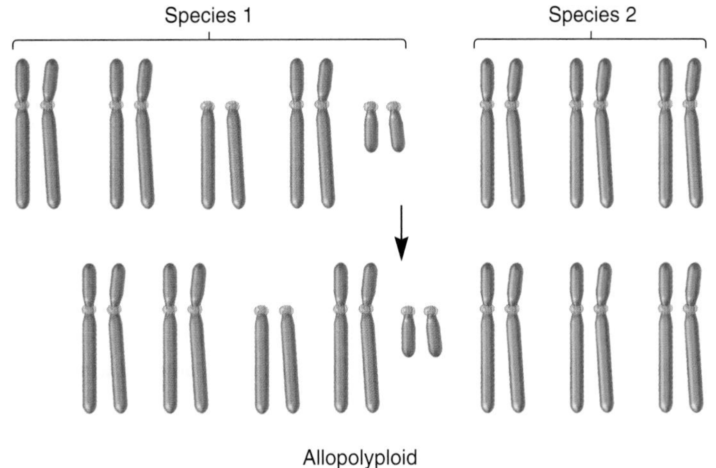

(c) Allopolyploidy (allotetraploid)

FIGURE 8.25 **A comparison of autopolyploidy, alloploidy, and allopolyploidy.**

homologous). This allodiploid is fertile because the homeologous chromosomes can properly synapse during meiosis to produce haploid gametes.

The critical relationship between chromosome pairing and fertility was first recognized by the Russian cytogeneticist Georgi Karpechenko in 1928. He crossed a radish (*Raphanus*) and a cabbage (*Brassica*), both of which are diploid and contain 18 chromosomes. Each of these organisms produces haploid cells containing 9 chromosomes. Therefore, the allodiploid produced from this interspecies cross contains 18 chromosomes. However, because the radish and cabbage are not closely related species, the nine *Raphanus* chromosomes are distinctly different from the nine *Brassica* chromosomes. During meiosis I, the radish and cabbage chromosomes cannot synapse with each other. This prevents the proper chromosome pairing and results in a high degree of aneuploidy (**Figure 8.27a**). Therefore, the radish/cabbage hybrid is sterile.

Among his strains of sterile alloploids, Karpechenko discovered that on rare occasions a plant produced a viable seed. When such seeds were planted and subjected to karyotyping, the plants were found to be allotetraploids with two sets of chromosomes from each of the two species. In the example shown in **Figure 8.27b**, the radish/cabbage allotetraploid contains 36 chromosomes instead of 18. The homologous chromosomes from each of the two species can synapse properly. When anaphase of meiosis I occurs, the pairs of synapsed chromosomes can disjoin equally to produce cells with 18 chromosomes each (a haploid set from the radish plus a haploid set from the cabbage). These cells can give rise to gametes with 18 chromosomes that can combine with each other to produce an allotetraploid containing 36 chromosomes. In this way, the allotetraploid is a fertile organism. Karpechenko's goal was to create a "vegetable for the masses" that would combine the nutritious roots of a radish with the flavorful leaves of a cabbage. Unfortunately, however, the allotetraploid had the leaves of the radish and the roots of the cabbage! Nevertheless, Karpechenko's scientific contribution was still important because he showed that it is possible to artificially produce a new self-perpetuating species of plant by creating an allotetraploid.

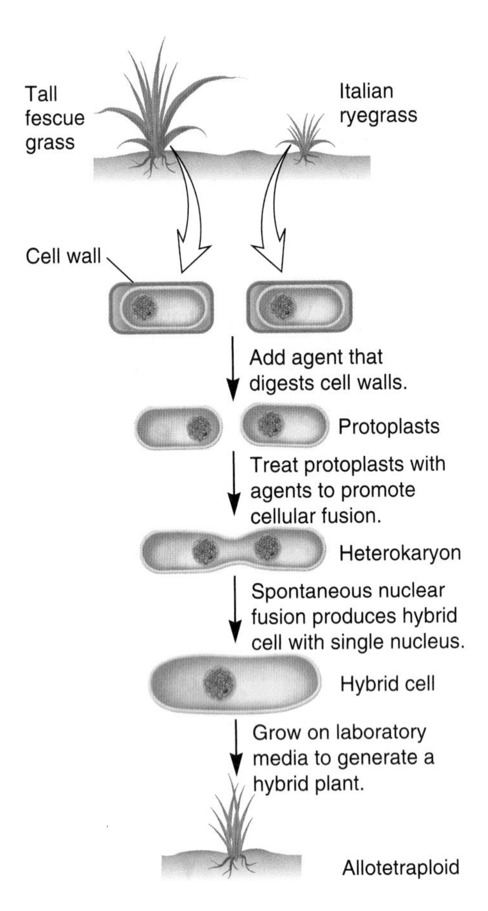

FIGURE 8.29 The technique of cell fusion. This technique is shown here with cells from tall fescue grass (*Festuca arundinacea*) and Italian ryegrass (*Lolium multiflorum*). The resulting hybrid is an allotetraploid.

Genes → Traits The allotetraploid contains two copies of genes from each parent. In this case, the allotetraploid displays characteristics that are intermediate between the tall fescue grass and the Italian ryegrass.

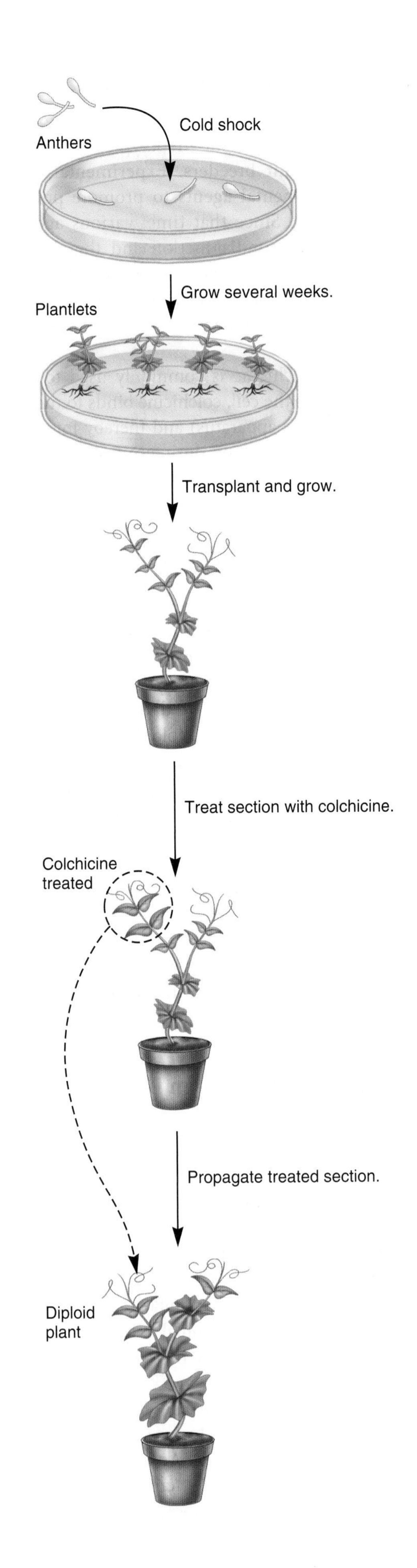

Figure 8.29 has phenotypic characteristics that are intermediate between tall fescue grass and Italian ryegrass.

Monoploids Produced in Agricultural and Genetic Research Can Be Used to Make Homozygous and Hybrid Strains

A goal of some plant breeders is to have diploid strains of crop plants that are homozygous for all of their genes. One true-breeding strain can then be crossed to a different true-breeding strain to produce an F_1 hybrid that is heterozygous for many genes. Such hybrids are often more vigorous than the corresponding homozygous strains. This phenomenon, known as **hybrid vigor,** or **heterosis,** is described in greater detail in Chapter 25. Seed companies often use this strategy to produce hybrid seed for many crops, such as corn and alfalfa. To achieve this goal, the companies must have homozygous parental strains that can be crossed to each other to produce the hybrid seed. One way to obtain these homozygous strains involves inbreeding over many generations. This may be accomplished after several rounds of self-fertilization. As you might imagine, this can be a rather time-consuming endeavor.

FIGURE 8.30 The experimental production of monoploids with anther culture. In the technique of anther culture, the anthers are collected, and the haploid pollen within them is induced to begin development by a cold shock treatment. After several weeks, monoploid plantlets emerge, and these can be grown on an agar medium. Eventually, the plantlets can be transferred to small pots. After the plantlets have grown to a reasonable size, a section of the monoploid plantlet can be treated with colchicine to convert it to diploid tissue. A cutting from this diploid section can then be used to generate a diploid plant.

As an alternative, the production of **monoploids**—organisms that have a single set of chromosomes in their somatic cells—can be used as part of an experimental strategy to develop homozygous diploid strains of plants. Monoploids have been used to improve agricultural crops such as wheat, rice, corn, barley, and potato. In 1964, Sipra Guha-Mukherjee and Satish Maheshwari developed a method to produce monoploid plants directly from pollen grains. **Figure 8.30** describes the experimental technique called **anther culture,** which has been extensively used to produce diploid strains of crop plants that are homozygous for all of their genes. The method involves alternation between monoploid and diploid generations. The parental plant is diploid but not homozygous for all of its genes. The anthers from this diploid plant are collected, and the haploid pollen grains are induced to begin development by a cold shock—an abrupt exposure to cold temperature. After several weeks, monoploid plantlets emerge, which can then be grown on agar media in a laboratory. However, due to the presence of deleterious alleles that are recessive, many of the pollen grains may fail to produce viable plantlets. Therefore, anther culture has been described as a "monoploid sieve" that weeds out individuals that carry deleterious recessive alleles.

Eventually, plantlets that are healthy can be transferred to small pots. After the plants grow to a reasonable size, a section of the monoploid plant can be treated with colchicine to convert it to diploid tissue. A cutting from this diploid section can then be used to generate a separate plant. This diploid plant is homozygous for all of its genes because it was produced by the chromosomal doubling of a monoploid strain.

In certain animal species, monoploids can be produced by experimental treatments that induce the eggs to begin development without fertilization by sperm. This process is known as **parthenogenesis.** In many cases, however, the haploid zygote develops only for a short time before it dies. Nevertheless, a short phase of development can be useful to research scientists. For example, the zebrafish (*Danio rerio*), a common aquarium fish, has gained popularity among researchers interested in vertebrate development. The haploid egg can be induced to begin development by exposure to sperm rendered biologically inactive by UV irradiation.

KEY TERMS

Page 189. genetic variation, allelic variation, cytogeneticist

Page 191. metacentric, submetacentric, acrocentric, telocentric, karyotype, G banding, deletion, deficiency, duplication, inversion, translocation, simple translocation, reciprocal translocation

Page 192. terminal deletion, interstitial deletion

Page 193. repetitive sequences, nonallelic homologous recombination, gene duplication, gene family

Page 194. homologous, paralogs

Page 195. copy number variation (CNV), segmental duplication, comparative genomic hybridization (CGH), hybridization

Page 197. pericentric inversion, paracentric inversion

Page 198. position effect, inversion heterozygote, inversion loop, dicentric, dicentric bridge, acentric fragment, telomeres

Page 199. balanced translocations, unbalanced translocation

Page 200. Robertsonian translocation, translocation cross

Page 202. semisterility

Page 203. euploid, triploid, polyploid, tetraploid, aneuploidy, trisomic, monosomic

Page 205. nondisjunction, haplodiploid

Page 206. endopolyploidy, polytene chromosome, chromocenter

Page 208. meiotic nondisjunction, mitotic nondisjunction, alloploid

Page 209. complete nondisjunction, mosaicism, bilateral gynandromorph

Page 210. autopolyploid, alloploidy, allodiploid, allopolyploid, allotetraploid, homeologous

Page 213. cell fusion, protoplast, heterokaryon, hybrid cell

Page 214. hybrid vigor, heterosis

Page 215. monoploids, anther culture, parthenogenesis

Page 217. pseudodominance

flies with this condition have wings with a notched appearance at their edges. Female flies that are heterozygous for this mutation have notched wings; homozygous females and hemizygous males are unable to survive. The notched phenotype is due to a defect in a single gene called *Notch* (*N*). Geneticists studying fruit flies with this phenotype have discovered that one X chromosome of some of the mutant flies has a small deletion that includes the *Notch* gene as well as a few genes on either side of it. The mutation causing this phenotype in other flies is confined within the *Notch* gene itself. A genetic analysis can distinguish between notched fruit flies carrying a deletion versus those that carry a single-gene mutation. This is possible because the *Notch* gene happens to be located next to the red/white eye color gene on the X chromosome. How would you distinguish between a notched phenotype due to a deletion that included the *Notch* gene and the adjacent eye color gene versus a notch phenotype due to a small mutation only within the *Notch* gene itself?

Answer: To determine if the *Notch* mutation is due to a deletion, red-eyed females with the notched phenotype can be crossed to white-eyed males. Only the daughters with notched wings need to be analyzed. If they have red eyes, this means the *Notch* mutation has not deleted the red-eye allele from the X chromosome. Alternatively, if they have white eyes, this indicates the red-eye allele has been deleted from the X chromosome that carries the *Notch* mutation. In this case, the white-eye allele is expressed because it is present in a single copy in a female fly with two X chromosomes. This phenomenon is pseudodominance.

S6. Albert Blakeslee began using the Jimson weed (*Datura stramonium*) as an experimental organism to teach his students the laws of Mendelian inheritance. Although this plant has not gained widespread use in genetic studies, Blakeslee's work provided a convincing demonstration that changes in chromosome number affect the phenotype of organisms. Blakeslee's assistant, B. T. Avery, identified a Jimson weed mutant that he called "globe" because the capsule is more rounded than normal. In genetic crosses, he found that the globe mutant had a peculiar pattern of inheritance. The globe trait was passed to about 25% of the offspring when the globe plants were allowed to self-fertilize. Unexpectedly, about 25% of the offspring also had the globe phenotype when globe plants were pollinated by a normal plant. In contrast, when pollen from a globe plant was used to pollinate a normal plant, less than 2% of the offspring had the globe phenotype. This non-Mendelian pattern of inheritance caused Blakeslee and his colleagues to investigate the nature of this trait further. We now know the globe phenotype is due to trisomy 11. Can you explain this unusual pattern of inheritance knowing it is due to trisomy 11?

Answer: This unusual pattern of inheritance of these aneuploid strains can be explained by the viability of euploid versus aneuploid gametes or gametophytes. An individual that is trisomic has a 50% chance of producing an egg or sperm that will inherit an extra chromosome and a 50% chance that a gamete will be normal. Blakeslee's results indicate that when a pollen grain inherited an extra copy of chromosome 11, it was almost always nonviable and unable to produce an aneuploid offspring. However, a significant percentage of aneuploid eggs were viable, so some (25%) of the offspring were aneuploid. Because an aneuploid plant should produce a 1:1 ratio between euploid and aneuploid eggs, the observation that only 25% of the offspring were aneuploid also indicates that about half of the female gametophytes or aneuploid eggs from such gametophytes were also nonviable. Overall, these results provide compelling evidence that imbalances in chromosome number can alter reproductive viability and also cause significant phenotypic consequences.

Conceptual Questions

C1. Which changes in chromosome structure cause a change in the total amount of genetic material, and which do not?

C2. Explain why small deletions and duplications are less likely to have a detrimental effect on an individual's phenotype than large ones. If a small deletion within a single chromosome happens to have a phenotypic effect, what would you conclude about the genes in this region?

C3. How does a chromosomal duplication occur?

C4. What is a gene family? How are gene families produced over time? With regard to gene function, what is the biological significance of a gene family?

C5. Following a gene duplication, two genes will accumulate different mutations, causing them to have slightly different sequences. In Figure 8.7, which pair of genes would you expect to have more similar sequences, α_1 and α_2 or $\psi_{\alpha 1}$ and α_2? Explain your answer.

C6. Two chromosomes have the following order of genes:

Normal: *A B C* centromere *D E F G H I*

Abnormal: *A B G F E D* centromere *C H I*

Does the abnormal chromosome have a pericentric or paracentric inversion? Draw a sketch showing how these two chromosomes would pair during prophase of meiosis I.

C7. An inversion heterozygote has the following inverted chromosome:

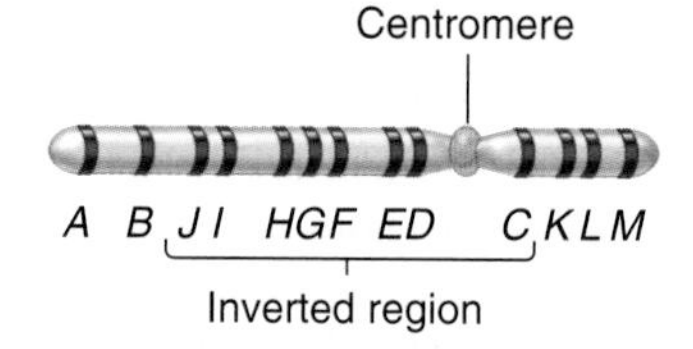

What would be the products if a crossover occurred between genes *H* and *I* on one inverted and one normal chromosome?

C8. An inversion heterozygote has the following inverted chromosome:

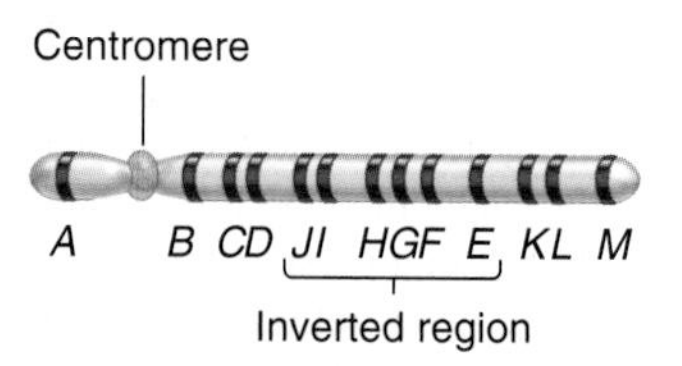

What would be the products if a crossover occurred between genes *H* and *I* on one inverted and one normal chromosome?

C9. Explain why inversions and reciprocal translocations do not usually cause a phenotypic effect. In a few cases, however, they do. Explain how.

C10. An individual has the following reciprocal translocation:

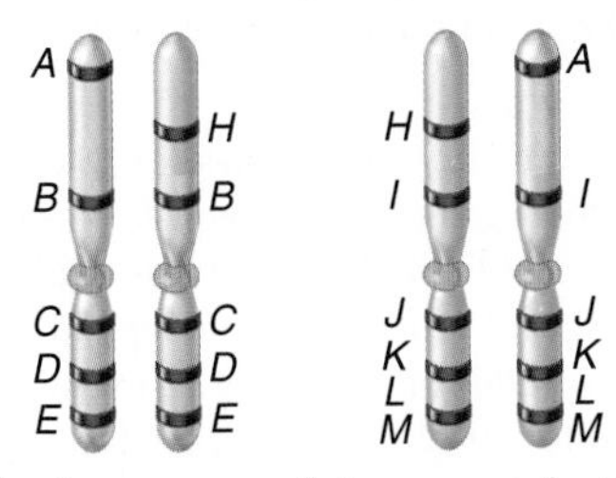

What would be the outcome of alternate and adjacent-1 segregation?

C11. A phenotypically normal individual has the following combinations of abnormal chromosomes:

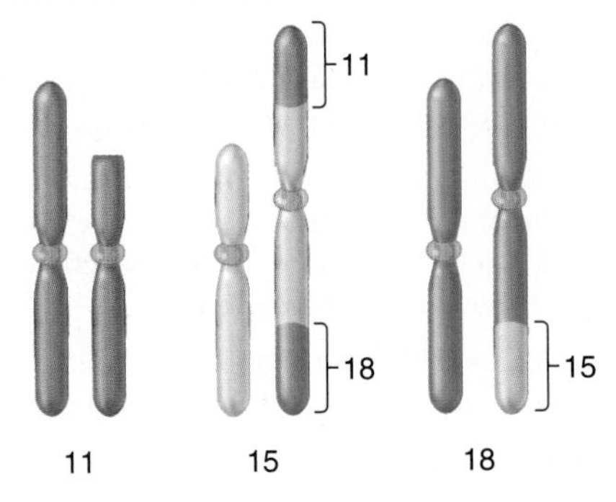

The normal chromosomes are shown on the left of each pair. Suggest a series of events (breaks, translocations, crossovers, etc.) that may have produced this combination of chromosomes.

C12. Two phenotypically normal parents produce a phenotypically abnormal child in which chromosome 5 is missing part of its long arm but has a piece of chromosome 7 attached to it. The child also has one normal copy of chromosome 5 and two normal copies of chromosome 7. With regard to chromosomes 5 and 7, what do you think are the chromosomal compositions of the parents?

C13. In the segregation of centromeres, why is adjacent-2 segregation less frequent than alternate or adjacent-1 segregation?

C14. Which of the following types of chromosomal changes would you expect to have phenotypic consequences? Explain your choices.

A. Pericentric inversion

B. Reciprocal translocation

C. Deletion

D. Unbalanced translocation

C15. Explain why a translocation cross occurs during metaphase of meiosis I when a cell contains a reciprocal translocation.

C16. A phenotypically abnormal individual has a phenotypically normal father with an inversion on one copy of chromosome 7 and a normal mother without any changes in chromosome structure. The order of genes along chromosome 7 in the father is as follows:

R T D M centromere *P U X Z C* (normal chromosome 7)

R T D U P centromere *M X Z C* (inverted chromosome 7)

The phenotypically abnormal offspring has a chromosome 7 with the following order of genes:

R T D M centromere *P U D T R*

With a sketch, explain how this chromosome was formed. In your answer, explain where the crossover occurred (i.e., between which two genes).

C17. A diploid fruit fly has eight chromosomes. How many total chromosomes would be found in the following flies?

A. Tetraploid

B. Trisomy 2

C. Monosomy 3

D. $3n$

E. $4n + 1$

C18. A person is born with one X chromosome, zero Y chromosomes, trisomy 21, and two copies of the other chromosomes. How many chromosomes does this person have altogether? Explain whether this person is euploid or aneuploid.

C19. Two phenotypically unaffected parents produce two children with familial Down syndrome. With regard to chromosomes 14 and 21, what are the chromosomal compositions of the parents?

C20. Aneuploidy is typically detrimental, whereas polyploidy is sometimes beneficial, particularly in plants. Discuss why you think this is the case.

C21. Explain how aneuploidy, deletions, and duplications cause genetic imbalances. Why do you think that deletions and monosomies are more detrimental than duplications and trisomies?

C22. Female fruit flies homozygous for the X-linked white-eye allele are crossed to males with red eyes. On very rare occasions, an offspring is a male with red eyes. Assuming these rare offspring are not due to a new mutation in one of the mother's X chromosomes that converted the white-eye allele into a red-eye allele, explain how this red-eyed male arose.

C23. A cytogeneticist has collected tissue samples from members of the same butterfly species. Some of the butterflies were located in Canada, and others were found in Mexico. Upon karyotyping, the cytogeneticist discovered that chromosome 5 of the Canadian butterflies had a large inversion compared with the Mexican butterflies. The Canadian butterflies were inversion homozygotes, whereas the Mexican butterflies had two normal copies of chromosome 5.

A. Explain whether a mating between the Canadian and Mexican butterflies would produce phenotypically normal offspring.

B. Explain whether the offspring of a cross between Canadian and Mexican butterflies would be fertile.

C24. Why do you think that human trisomies 13, 18, and 21 can survive but the other trisomies are lethal? Even though X chromosomes are large, aneuploidies of this chromosome are also tolerated. Explain why.

C25. A zookeeper has collected a male and female lizard that look like they belong to the same species. They mate with each other and produce phenotypically normal offspring. However, the offspring are sterile. Suggest one or more explanations for their sterility.

C26. What is endopolyploidy? What is its biological significance?

C27. What is mosaicism? How is it produced?

C28. Explain how polytene chromosomes of *Drosophila melanogaster* are produced and how they form a six-armed structure.

C29. Describe some of the advantages of polyploid plants. What are the consequences of having an odd number of chromosome sets?

C30. While conducting field studies on a chain of islands, you decide to karyotype two phenotypically identical groups of turtles, which are found on different islands. The turtles on one island have 24 chromosomes, but those on another island have 48 chromosomes. How would you explain this observation? How do you think the turtles with 48 chromosomes came into being? If you mated the two types of turtles together, would you expect their offspring to be phenotypically normal? Would you expect them to be fertile? Explain.

C31. A diploid fruit fly has eight chromosomes. Which of the following terms should not be used to describe a fruit fly with four sets of chromosomes?

A. Polyploid

B. Aneuploid

C. Euploid

D. Tetraploid

E. $4n$

C32. Which of the following terms should not be used to describe a human with three copies of chromosome 12?

A. Polyploid

B. Triploid

C. Aneuploid

D. Euploid

E. $2n + 1$

F. Trisomy 12

C33. The kidney bean, *Phaseolus vulgaris*, is a diploid species containing a total of 22 chromosomes in somatic cells. How many possible types of trisomic individuals could be produced in this species?

C34. The karyotype of a young girl who is affected with familial Down syndrome revealed a total of 46 chromosomes. Her older brother, however, who is phenotypically unaffected, actually had 45 chromosomes. Explain how this could happen. What would you expect to be the chromosomal number in the parents of these two children?

C35. A triploid plant has 18 chromosomes (i.e., 6 chromosomes per set). If we assume a gamete has an equal probability of receiving one or two copies of each of the six types of chromosome, what are the odds of this plant producing a monoploid or a diploid gamete? What are the odds of producing an aneuploid gamete? If the plant is allowed to self-fertilize, what are the odds of producing a euploid offspring?

C36. Describe three naturally occurring ways that the chromosome number can change.

C37. Meiotic nondisjunction is much more likely than mitotic nondisjunction. Based on this observation, would you conclude that meiotic nondisjunction is usually due to nondisjunction during meiosis I or meiosis II? Explain your reasoning.

C38. A woman who is heterozygous, *Bb*, has brown eyes. *B* (brown) is a dominant allele, and *b* (blue) is recessive. In one of her eyes, however, there is a patch of blue color. Give three different explanations for how this might have occurred.

C39. What is an allodiploid? What factor determines the fertility of an allodiploid? Why are allotetraploids more likely to be fertile?

C40. What are homeologous chromosomes?

C41. Meiotic nondisjunction usually occurs during meiosis I. What is not separating properly: bivalents or sister chromatids? What is not separating properly during mitotic nondisjunction?

C42. Table 8.1 shows that Turner syndrome occurs when an individual inherits one X chromosome but lacks a second sex chromosome. Can Turner syndrome be due to nondisjunction during oogenesis, spermatogenesis, or both? If a phenotypically normal couple has a color-blind child (due to a recessive X-linked allele) with Turner syndrome, did nondisjunction occur during oogenesis or spermatogenesis in this child's parents? Explain your answer.

C43. Male honeybees, which are monoploid, produce sperm by meiosis. Explain what unusual event (compared to other animals) must occur during spermatogenesis in honeybees to produce sperm? Does this unusual event occur during meiosis I or meiosis II?

Experimental Questions

E1. What is the main goal of comparative genome hybridization? Explain how the ratio of green to red fluorescence provides information regarding chromosome structure.

E2. Let's suppose a researcher conducted comparative genomic hybridization (see Figure 8.9) and accidentally added twice as much (red) DNA from normal cells. What green to red ratio would you expect in a region from a chromosome from a cancer cell that carried a duplication on both chromosomal copies? What ratio would be observed for a region that was deleted on just one of the chromosomes from cancer cells?

E3. With regard to the analysis of chromosome structure, explain the experimental advantage that polytene chromosomes offer. Discuss why changes in chromosome structure are more easily detected in polytene chromosomes than in ordinary (nonpolytene) chromosomes.

E4. Describe how colchicine can be used to alter chromosome number.

E5. Describe the steps you would take to produce a tetraploid plant that is homozygous for all of its genes.

E6. In agriculture, what is the primary purpose of anther culture?

E7. What are some experimental advantages of cell fusion techniques as opposed to interbreeding approaches?

E8. It is an exciting time to be a plant breeder because so many options are available for the development of new types of agriculturally useful plants. Let's suppose you wish to develop a seedless tomato that could grow in a very hot climate and is resistant to a viral pathogen that commonly infects tomato plants. At your disposal, you have a seed-bearing tomato strain that is heat-resistant and produces great-tasting tomatoes. You also have a wild strain of tomato plants (which have lousy-tasting tomatoes) that is resistant to the viral pathogen. Suggest a series of steps you might follow to produce a great-tasting, seedless tomato that is resistant to heat and the viral pathogen.

E9. What is a G band? Discuss how G bands are useful in the analysis of chromosome structure.

E10. A female fruit fly contains one normal X chromosome and one X chromosome with a deletion. The deletion is in the middle of the X chromosome and is about 10% of the entire length of the X chromosome. If you stained and observed the chromosomes of this female fly in salivary gland cells, draw a picture of the polytene arm of the X chromosome. Explain your drawing.

E11. Describe two different experimental strategies to create an allotetraploid from two different diploid species of plants.

E12. In the procedure of anther culture (see Figure 8.30), an experimenter may begin with a diploid plant and then cold shock the pollen to get them to grow as haploid plantlets. In some cases, the pollen may come from a phenotypically vigorous plant that is heterozygous for many genes. Even so, many of the haploid plantlets appear rather weak and nonvigorous. In fact, many of them fail to grow at all. In contrast, some of the plantlets are fairly healthy. Explain why some plantlets would be weak, but others are quite healthy.

Questions for Student Discussion/Collaboration

1. A chromosome involved in a reciprocal translocation also has an inversion. In addition, the cell contains two normal chromosomes.

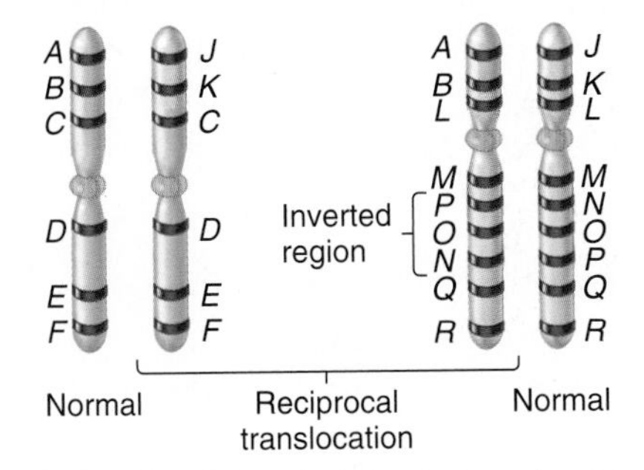

Make a drawing that shows how these chromosomes will pair during metaphase of meiosis I.

2. Besides the ones mentioned in this textbook, look for other examples of variations in euploidy. Perhaps you might look in more advanced textbooks concerning population genetics, ecology, or the like. Discuss the phenotypic consequences of these changes.

3. Cell biology textbooks often discuss cellular proteins encoded by genes that are members of a gene family. Examples of such proteins include myosins and glucose transporters. Look through a cell biology textbook and identify some proteins encoded by members of gene families. Discuss the importance of gene families at the cellular level.

4. Discuss how variation in chromosome number has been useful in agriculture.

Note: All answers appear at the website for this textbook; the answers to even-numbered questions are in the back of the textbook.

www.mhhe.com/brookergenetics4e

Visit the website for practice tests, answer keys, and other learning aids for this chapter. Enhance your understanding of genetics with our interactive exercises, quizzes, animations, and much more.

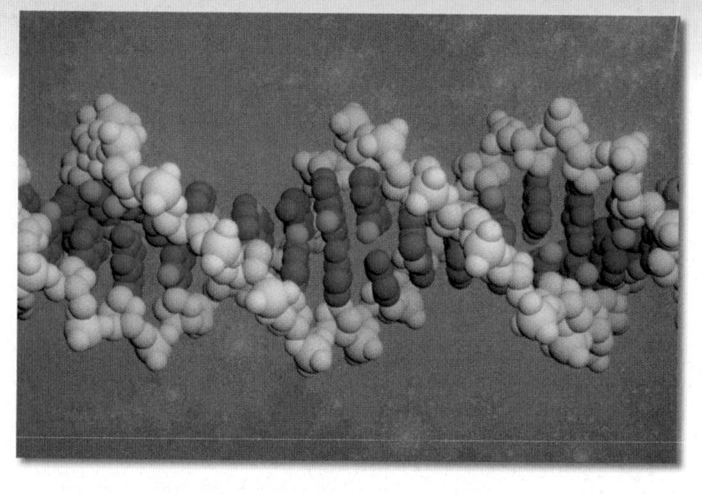

A molecular model showing the structure of the DNA double helix.

CHAPTER OUTLINE

9 MOLECULAR STRUCTURE OF DNA AND RNA

In Chapters 2 through 8, we focused on the relationship between the inheritance of genes and chromosomes and the outcome of an organism's traits. In Chapter 9, we will shift our attention to **molecular genetics**—the study of DNA structure and function at the molecular level. An exciting goal of molecular genetics is to use our knowledge of DNA structure to understand how DNA functions as the genetic material. Using molecular techniques, researchers have determined the organization of many genes. This information, in turn, has helped us understand how the expression of such genes governs the outcome of an individual's inherited traits.

The past several decades have seen dramatic advances in techniques and approaches to investigate and even to alter the genetic material. These advances have greatly expanded our understanding of molecular genetics and also have provided key insights into the mechanisms underlying transmission and population genetics. Molecular genetic technology is also widely used in supporting disciplines such as biochemistry, cell biology, and microbiology.

To a large extent, our understanding of genetics comes from our knowledge of the molecular structure of **DNA (deoxyribonucleic acid)** and **RNA (ribonucleic acid).** In this chapter, we will begin by considering classic experiments which showed that DNA is the genetic material. We will then survey the molecular features of DNA and RNA that underlie their function.

9.1 IDENTIFICATION OF DNA AS THE GENETIC MATERIAL

To fulfill its role, the genetic material must meet four criteria.

1. **Information:** The genetic material must contain the information necessary to construct an entire organism. In other words, it must provide the blueprint to determine the inherited traits of an organism.
2. **Transmission:** During reproduction, the genetic material must be passed from parents to offspring.
3. **Replication:** Because the genetic material is passed from parents to offspring, and from mother cell to daughter cells during cell division, it must be copied.
4. **Variation:** Within any species, a significant amount of phenotypic variability occurs. For example, Mendel studied several characteristics in pea plants that varied among different plants. These included height (tall versus dwarf) and seed color (yellow versus green). Therefore, the genetic material must also vary in ways that can account for the known phenotypic differences within each species.

Along with Mendel's work, the data of many other geneticists in the early 1900s were consistent with these four properties: information, transmission, replication, and variation. However, the experimental study of genetic crosses cannot, by itself, identify the chemical nature of the genetic material.

In the 1880s, August Weismann and Carl Nägeli championed the idea that a chemical substance within living cells is responsible for the transmission of traits from parents to offspring. The chromosome theory of inheritance was developed, and experimentation demonstrated that the chromosomes are the carriers of the genetic material (see Chapter 3). Nevertheless, the story was not complete because chromosomes contain both DNA and proteins. Also, RNA is found in the vicinity of chromosomes. Therefore, further research was needed to precisely identify the genetic material. In this section, we will examine the first experimental approaches to achieve this goal.

Experiments with Pneumococcus Suggested That DNA Is the Genetic Material

Some early work in microbiology was important in developing an experimental strategy to identify the genetic material. Frederick Griffith studied a type of bacterium known then as pneumococci and now classified as *Streptococcus pneumoniae.* Certain strains of *S. pneumoniae* secrete a polysaccharide capsule, whereas other strains do not. When streaked onto petri plates containing a solid growth medium, capsule-secreting strains have a smooth colony morphology, whereas those strains unable to secrete a capsule have a rough appearance.

The different forms of *S. pneumoniae* also affect their virulence, or ability to cause disease. When smooth strains of *S. pneumoniae* infect a mouse, the capsule allows the bacteria to escape attack by the mouse's immune system. As a result, the bacteria can grow and eventually kill the mouse. In contrast, the nonencapsulated (rough) bacteria are destroyed by the animal's immune system.

In 1928, Griffith conducted experiments that involved the injection of live and/or heat-killed bacteria into mice. He then observed whether or not the bacteria caused a lethal infection. Griffith was working with two strains of *S. pneumoniae*, a type S (S for smooth) and a type R (R for rough). When injected into a live mouse, the type S bacteria proliferated within the mouse's bloodstream and ultimately killed the mouse (**Figure 9.1a**). Following the death of the mouse, Griffith found many type S bacteria within the mouse's blood. In contrast, when type R bacteria were injected into a mouse, the mouse lived (**Figure 9.1b**). To verify that the proliferation of the smooth bacteria was causing the death of the mouse, Griffith killed the smooth bacteria with heat treatment before injecting them into the mouse. In this case, the mouse also survived (**Figure 9.1c**).

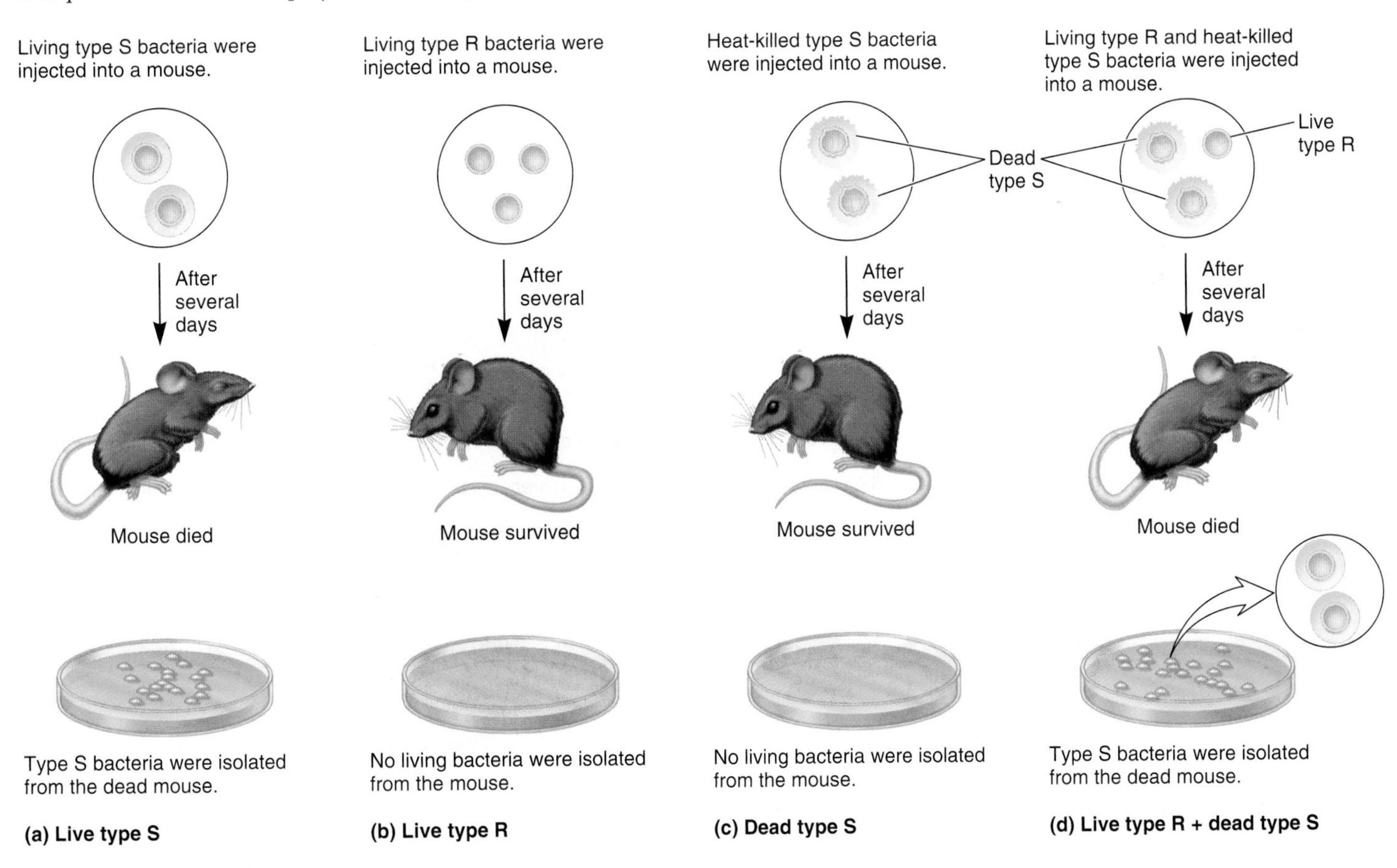

FIGURE 9.1 **Griffith's experiments on genetic transformation in pneumococcus.**

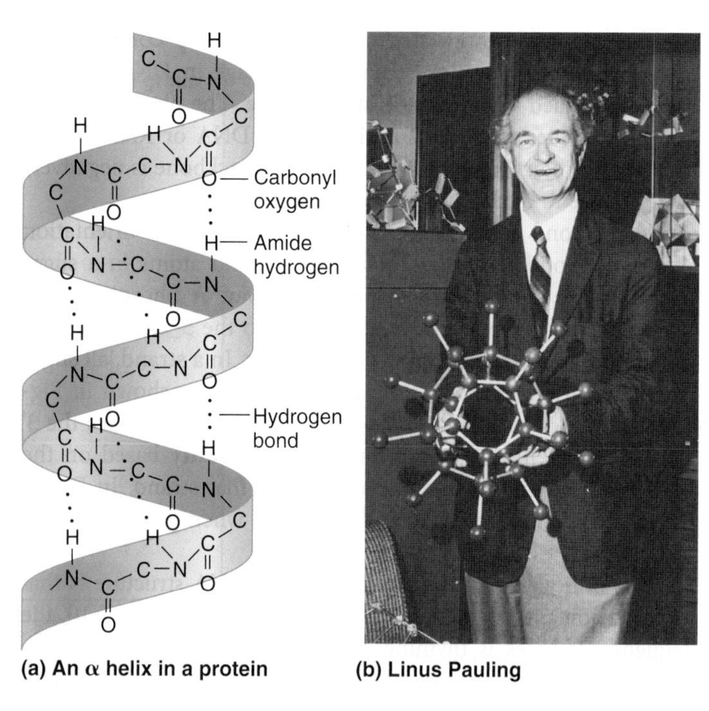

FIGURE 9.11 Linus Pauling and the α-helix protein structure. (a) An α-helix is a secondary structure found in proteins. This structure emphasizes the polypeptide backbone (shown as a tan ribbon), which is composed of amino acids linked together in a linear fashion. Hydrogen bonding between hydrogen and oxygen atoms stabilizes the helical conformation. (b) Linus Pauling with a ball-and-stick model.

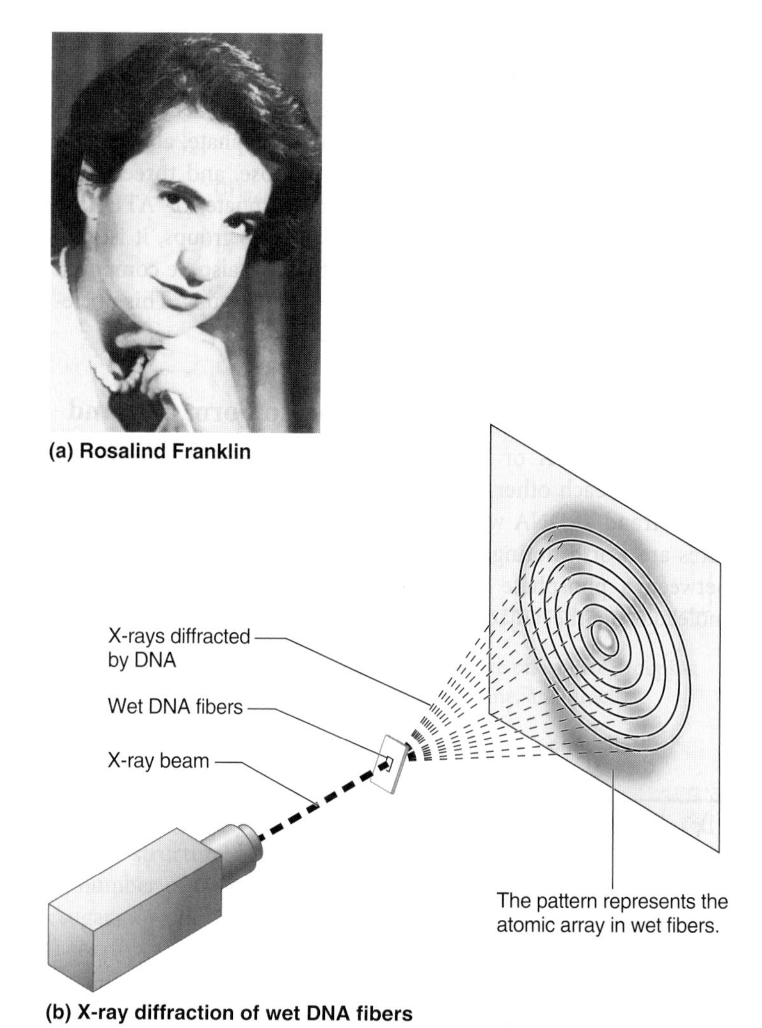

FIGURE 9.12 X-ray diffraction of DNA.

A second important development that led to the elucidation of the double helix was X-ray diffraction data. When a purified substance, such as DNA, is subjected to X-rays, it produces a well-defined diffraction pattern if the molecule is organized into a regular structural pattern. An interpretation of the diffraction pattern (using mathematical theory) can ultimately provide information concerning the structure of the molecule. Rosalind Franklin (**Figure 9.12a**), working in the same laboratory as Maurice Wilkins, used X-ray diffraction to study wet DNA fibers. Franklin made marked advances in X-ray diffraction techniques while working with DNA. She adjusted her equipment to produce an extremely fine beam of X-rays. She extracted finer DNA fibers than ever before and arranged them in parallel bundles. Franklin also studied the fibers' reactions to humid conditions. The diffraction pattern of Franklin's DNA fibers is shown in **Figure 9.12b**. This pattern suggested several structural features of DNA. First, it was consistent with a helical structure. Second, the diameter of the helical structure was too wide to be only a single-stranded helix. Finally, the diffraction pattern indicated that the helix contains about 10 base pairs (bp) per complete turn. These observations were instrumental in solving the structure of DNA.

EXPERIMENT 9B

Chargaff Found That DNA Has a Biochemical Composition in Which the Amount of A Equals T and the Amount of G Equals C

Another piece of information that led to the discovery of the double-helix structure came from the studies of Erwin Chargaff. In the 1940s and 1950s, he pioneered many of the biochemical techniques for the isolation, purification, and measurement of nucleic acids from living cells. This was not a trivial undertaking, because the biochemical composition of living cells is complex. At the time of Chargaff's work, researchers already knew that the building blocks of DNA are nucleotides containing the bases adenine, thymine, guanine, or cytosine. Chargaff analyzed the base composition of DNA, which was isolated from many different species. He expected that the results might provide important clues concerning the structure of DNA.

The experimental protocol of Chargaff is described in **Figure 9.13**. He began with various types of cells as starting material. The chromosomes were extracted from cells and then treated with protease to separate the DNA from chromosomal proteins. The DNA was then subjected to a strong acid treatment that cleaved the bonds between the sugars and bases. Therefore, the strong acid treatment released the individual bases from the DNA strands. This mixture of bases was then subjected to paper chromatography to separate the four types. The amounts of the four bases were determined spectroscopically.

THE GOAL

An analysis of the base composition of DNA in different organisms may reveal important features about the structure of DNA.

ACHIEVING THE GOAL — FIGURE 9.13 An analysis of base composition among different DNA samples.

Starting material: The following types of cells were obtained: *Escherichia coli*, *Streptococcus pneumoniae*, yeast, turtle red blood cells, salmon sperm cells, chicken red blood cells, and human liver cells.

1. For each type of cell, extract the chromosomal material. This can be done in a variety of ways, including the use of high salt, detergent, or mild alkali treatment. Note: The chromosomes contain both DNA and protein.

2. Remove the protein. This can be done in several ways, including treament with protease.

3. Hydrolyze the DNA to release the bases from the DNA strands. A common way to do this is by strong acid treatment.

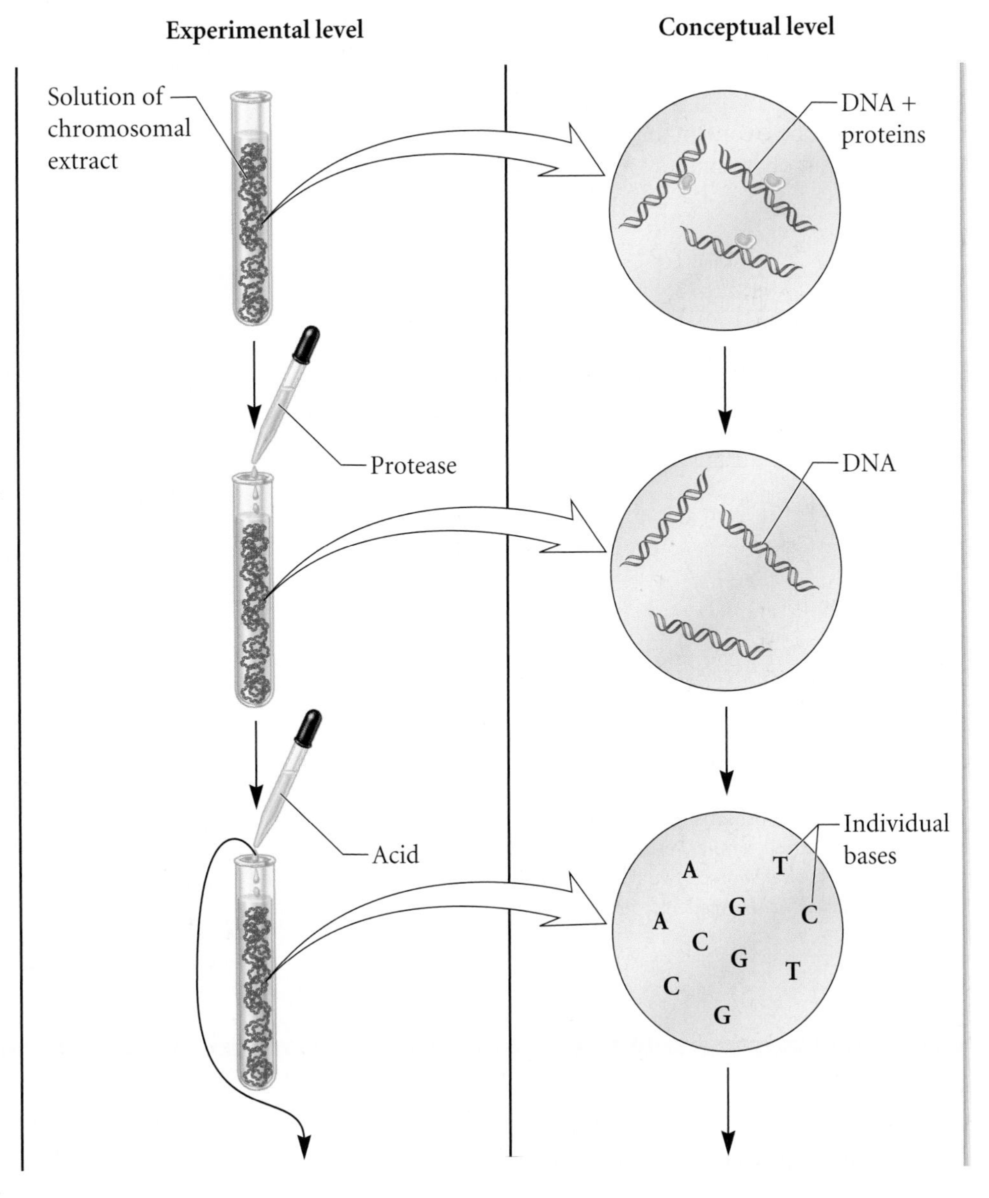

4. Separate the bases by chromatography. Paper chromatography provides an easy way to separate the four types of bases. (The technique of chromatography is described in the Appendix.)

5. Extract bands from paper into solutions and determine the amounts of each base by spectroscopy. Each base will absorb light at a particular wavelength. By examining the absorption profile of a sample of base, it is then possible to calculate the amount of the base. (Spectroscopy is described in the Appendix.)

6. Compare the base content in the DNA from different organisms.

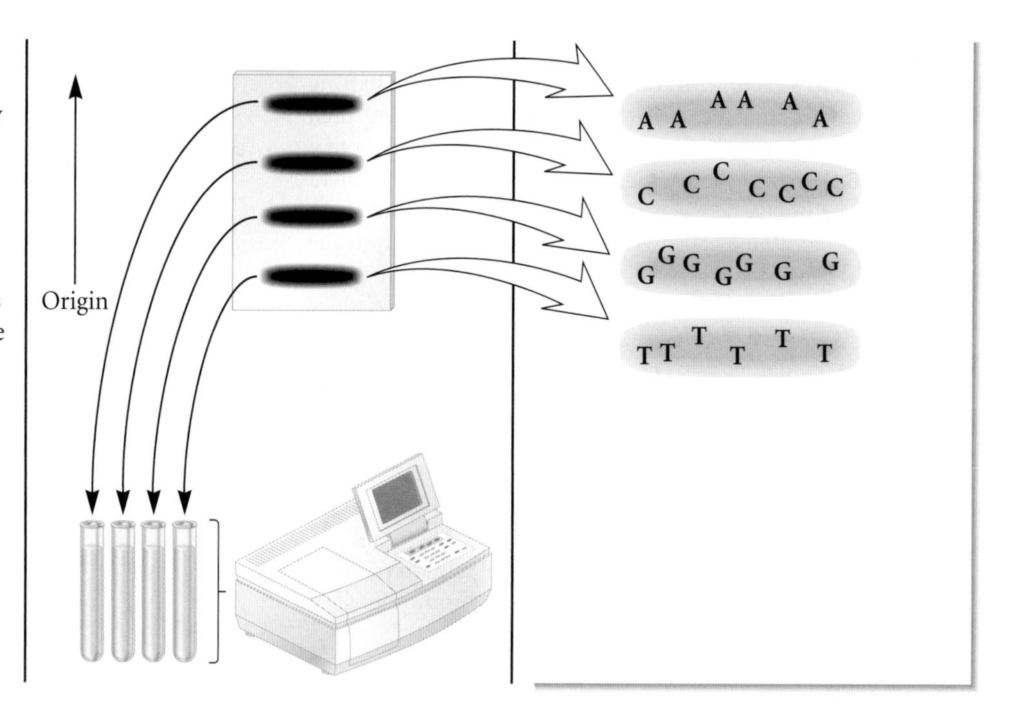

THE DATA

Base Content in the DNA from a Variety of Organisms*

	Percentage of Bases (based on molarity)			
Organism	*Adenine*	*Thymine*	*Guanine*	*Cytosine*
Escherichia coli	26.0	23.9	24.9	25.2
Streptococcus pneumoniae	29.8	31.6	20.5	18.0
Yeast	31.7	32.6	18.3	17.4
Turtle red blood cells	28.7	27.9	22.0	21.3
Salmon sperm	29.7	29.1	20.8	20.4
Chicken red blood cells	28.0	28.4	22.0	21.6
Human liver cells	30.3	30.3	19.5	19.9

*When the base compositions from different tissues within the same species were measured, similar results were obtained. These data were compiled from several sources. See E. Chargaff and J. Davidson, Eds. (1995) *The Nucleic Acids*. Academic Press, New York.

INTERPRETING THE DATA

The data shown in Figure 9.13 are only a small sampling of Chargaff's results. During the late 1940s and early 1950s, Chargaff published many papers concerned with the chemical composition of DNA from biological sources. Hundreds of measurements were made. The compelling observation was that the amount of adenine was similar to that of thymine, and the amount of guanine was similar to cytosine. The idea that the amount of A in DNA equals the amount of T, and the amount of G equals C, is known as **Chargaff's rule.**

These results were not sufficient to propose a model for the structure of DNA. However, they provided the important clue that DNA is structured so that each molecule of adenine interacts with thymine, and each molecule of guanine interacts with cytosine. A DNA structure in which A binds to T, and G to C, would explain the equal amounts of A and T, and G and C observed in Chargaff's experiments. As we will see, this observation became crucial evidence that Watson and Crick used to elucidate the structure of the double helix.

A self-help quiz involving this experiment can be found at www.mhhe.com/brookergenetics4e.

Watson and Crick Deduced the Double-Helical Structure of DNA

Thus far, we have examined key pieces of information used to determine the structure of DNA. In particular, the X-ray diffraction work of Franklin suggested a helical structure composed of two or more strands with 10 bases per turn. In addition, the work of Chargaff indicated that the amount of A equals T, and the amount of G equals C. Furthermore, Watson and Crick were familiar with Pauling's success in using ball-and-stick models to deduce the secondary structure of proteins. With these key observations, they set out to solve the structure of DNA.

Watson and Crick assumed DNA is composed of nucleotides that are linked together in a linear fashion. They also assumed the chemical linkage between two nucleotides is always the same. With these ideas in mind, they tried to build ball-and-stick models that incorporated the known experimental observations. Because the diffraction pattern suggested the helix must have two (or more) strands, a critical question was, How could two strands interact? As discussed in his book, *The Double Helix*, James Watson noted that in an early attempt at model building, they considered the possibility that the negatively charged phosphate groups, together with magnesium ions, were promoting an interaction between the backbones of DNA strands (**Figure 9.14**).

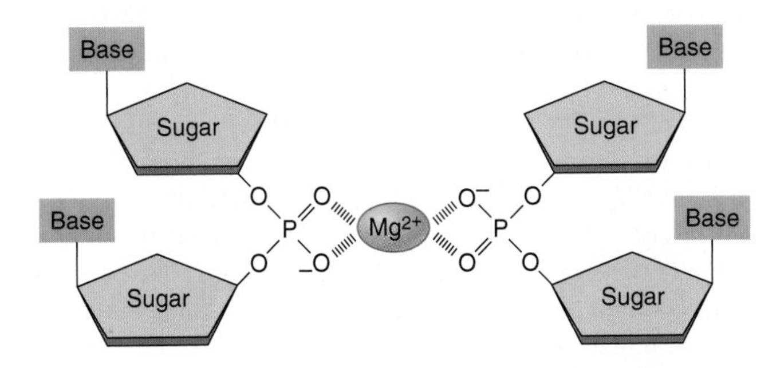

FIGURE 9.14 An incorrect hypothesis for the structure of the DNA double helix. This illustration shows an early hypothesis of Watson and Crick's, suggesting that two DNA strands interact by a cross-link between the negatively charged phosphate groups in the backbone and divalent Mg^{2+} cations.

However, more detailed diffraction data were not consistent with this model.

Because the magnesium hypothesis for DNA structure appeared to be incorrect, it was back to the drawing board (or back to the ball-and-stick units) for Watson and Crick. During this time, Rosalind Franklin had produced even clearer X-ray diffraction patterns, which provided greater detail concerning the relative locations of the bases and backbone of DNA. This major breakthrough suggested a two-strand interaction that was helical. In their model building, Watson and Crick's emphasis shifted to models containing the two backbones on the outside of the model, with the bases projecting toward each other. At first, a structure was considered in which the bases form hydrogen bonds with the identical base in the opposite strand (A to A, T to T, G to G, and C to C). However, the model building revealed that the bases could not fit together this way. The final hurdle was overcome when it was realized that the hydrogen bonding of adenine to thymine was structurally similar to that of guanine to cytosine. With an interaction between A and T and between G and C, the ball-and-stick models showed that the two strands would fit together properly. This ball-and-stick model, shown in **Figure 9.15**, was consistent with all of the known data regarding DNA structure.

(a) Watson and Crick

(b) Original model of the DNA double helix

FIGURE 9.15 Watson and Crick and their model of the DNA double helix. (a) James Watson is shown here on the left and Francis Crick on the right. (b) The molecular model they originally proposed for the double helix. Each strand contains a sugar-phosphate backbone. In opposite strands, A hydrogen bonds to T, and G hydrogen bonds with C.

For their work, Watson, Crick, and Maurice Wilkins were awarded the 1962 Nobel Prize in physiology or medicine. The contribution of Rosalind Franklin to the discovery of the double helix was also critical and has been acknowledged in several books and articles. Franklin was independently trying to solve the structure of DNA. However, Wilkins, who worked in the same laboratory, shared Franklin's X-ray data with Watson and Crick, presumably without her knowledge. This provided important information that helped them solve the structure of DNA, which was published in the journal *Nature* in April 1953. Though she was not given credit in the original publication of the double-helix structure, Franklin's key contribution became known in later years. Unfortunately, however, Rosalind Franklin died in 1958, and the Nobel Prize is not awarded posthumously.

The Molecular Structure of the DNA Double Helix Has Several Key Features

The general structural features of the double helix are shown in **Figure 9.16**. In a DNA double helix, two DNA strands are twisted together around a common axis to form a structure that resembles a spiral staircase. This double-stranded structure is stabilized by **base pairs (bp)**—pairs of bases in opposite strands that are hydrogen bonded to each other. Counting the bases, if you move past 10 bp, you have gone 360° around the backbone. The linear distance of a complete turn is 3.4 nm; each base pair traverses 0.34 nm.

A distinguishing feature of the hydrogen bonding between base pairs is its specificity. An adenine base in one strand hydrogen bonds with a thymine base in the opposite strand, or a guanine base hydrogen bonds with a cytosine. This **AT/GC rule** explained the earlier data of Chargaff showing that the DNA from many organisms contains equal amounts of A and T, and equal amounts of G and C (see Figure 9.13). The AT/GC rule indicates that purines (A and G) always bond with pyrimidines (T and C). This keeps the width of the double helix relatively constant. As noted in Figure 9.16, three hydrogen bonds occur between G and C but only two between A and T. For this reason, DNA sequences that have a high proportion of G and C tend to form more stable double-stranded structures.

The AT/GC rule implies that we can predict the sequence in one DNA strand if the sequence in the opposite strand is known.

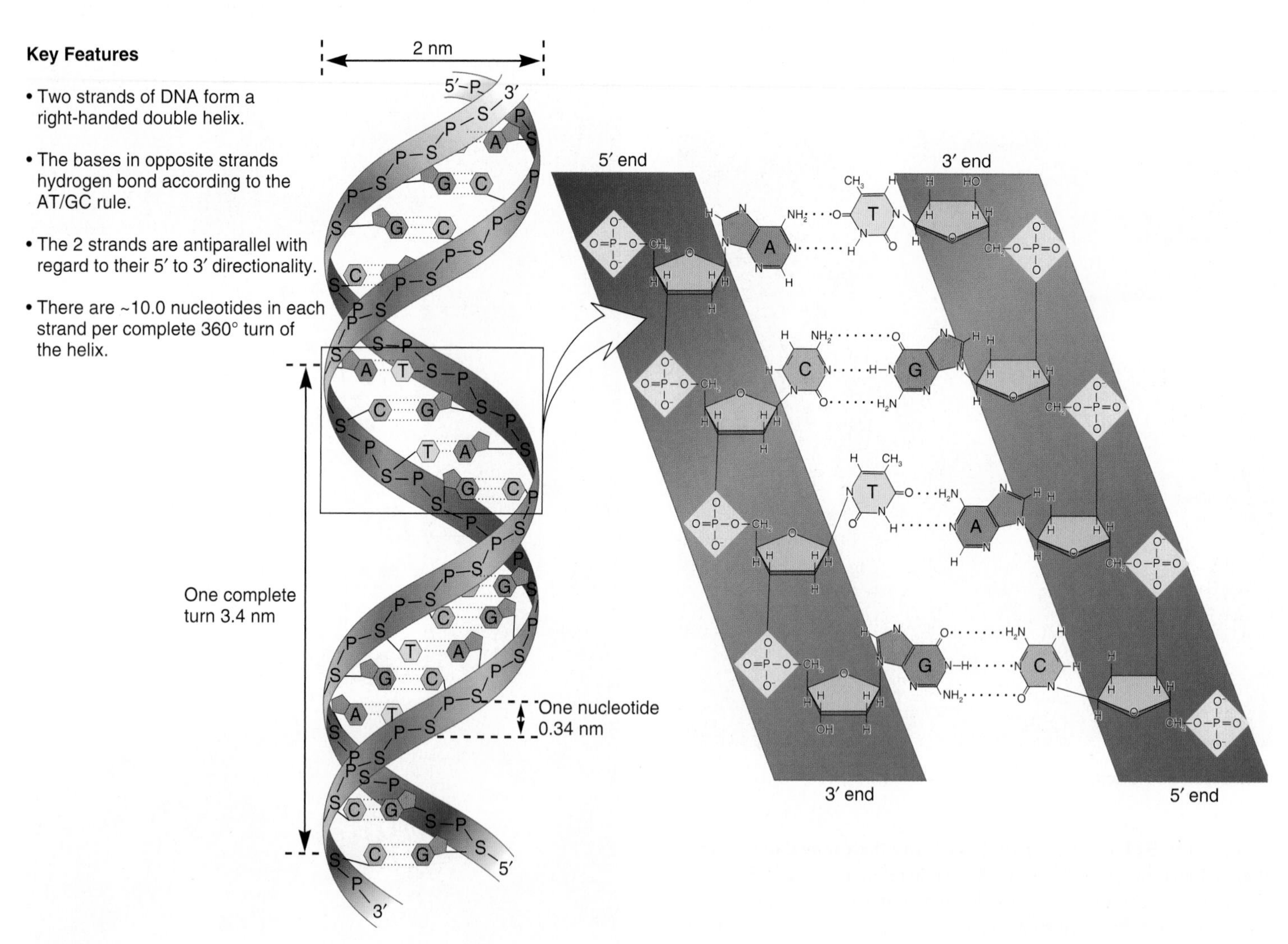

FIGURE 9.16 Key features of the structure of the double helix. Note: In the inset, the planes of the bases and sugars are shown parallel to each other in order to depict the hydrogen bonding between the bases. In an actual DNA molecule, the bases would be rotated about 90° so that the planes of the bases would be facing each other.

For example, let's consider a DNA strand with the sequence of 5′–ATGGCGGATTT–3′. The opposite strand would have to be 3′–TACCGCCTAAA–5′. In genetic terms, we would say that these two sequences are **complementary** to each other or that the two sequences exhibit complementarity. In addition, you may have noticed that the sequences are labeled with 5′ and 3′ ends. These numbers designate the direction of the DNA backbones. The direction of DNA strands is depicted in the inset to Figure 9.16. When going from the top of this figure to the bottom, one strand is running in the 5′ to 3′ direction, and the other strand is 3′ to 5′. This opposite orientation of the two DNA strands is referred to as an **antiparallel** arrangement. An antiparallel structure was initially proposed in the models of Watson and Crick.

Figure 9.17a is a schematic model that emphasizes certain molecular features of DNA structure. The bases in this model are depicted as flat rectangular structures that hydrogen bond in pairs. (The hydrogen bonds are the dotted lines.) Although the bases are not actually rectangular, they do form flattened planar structures. Within DNA, the bases are oriented so that the flattened regions are facing each other, an arrangement referred to as base stacking. In other words, if you think of the bases as flat plates, these plates are stacked on top of each other in the double-stranded DNA structure. Along with hydrogen bonding, base stacking is a structural feature that stabilizes the double helix by excluding water molecules. The helical structure of the DNA backbone depends on the hydrogen bonding between base pairs and also on base stacking.

By convention, the direction of the DNA double helix shown in Figure 9.17a spirals in a direction that is called "right-handed." To understand this terminology, imagine that a double helix is laid on your desk; one end of the helix is close to you, and the other end is at the opposite side of the desk. As it spirals away from you, a right-handed helix turns in a clockwise direction. By comparison, a left-handed helix would spiral in a counterclockwise manner. Both strands in Figure 9.17a spiral in a right-handed direction.

Figure 9.17b is a space-filling model for DNA in which the atoms are represented by spheres. This model emphasizes the surface features of DNA. Note that the backbone—composed of sugar and phosphate groups—is on the outermost surface. In a living cell, the backbone has the most direct contact with water. In contrast, the bases are more internally located within the double-stranded structure. Biochemists use the term **grooves** to describe the indentations where the atoms of the bases are in contact with the surrounding water. As you travel around the DNA helix, the structure of DNA has two grooves: the **major groove** and the **minor groove.**

As discussed in later chapters, proteins can bind to DNA and affect its conformation and function. For example, some proteins can hydrogen bond to the bases within the major groove. This hydrogen bonding can be very precise so that a protein interacts with a particular sequence of bases. In this way, a protein can recognize a specific gene and affect its ability to be transcribed. We will consider such proteins in Chapters 12, 14, and 15. Alternatively, other proteins bind to the DNA backbone. For example, histone proteins, which are discussed in Chapter 10, form ionic interactions with the negatively charged phosphates in the DNA backbone. The histones are important for the proper compaction of DNA in eukaryotic cells and also play a role in gene transcription.

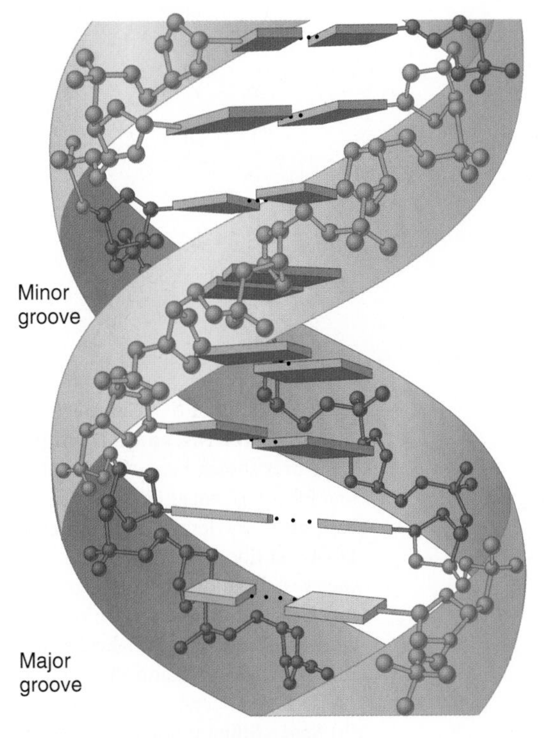

(a) Ball-and-stick model of DNA

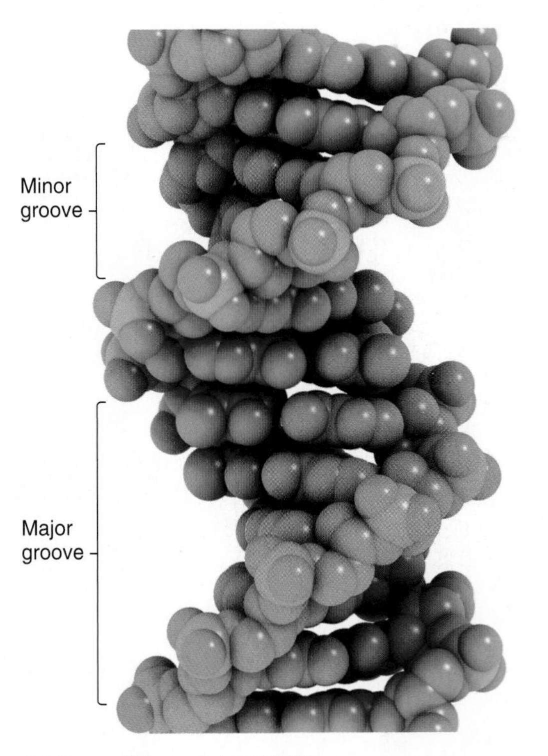

(b) Space-filling model of DNA

FIGURE 9.17 **Two models of the double helix.** **(a)** Ball-and-stick model of the double helix. The deoxyribose-phosphate backbone is shown in detail, whereas the bases are depicted as flattened rectangles. **(b)** Space-filling model of the double helix.

DNA Can Form Alternative Types of Double Helices

The DNA double helix can form different types of structures. **Figure 9.18** compares the structures of **A DNA, B DNA,** and **Z DNA.** The highly detailed structures shown here were deduced by X-ray crystallography on short segments of DNA. B DNA is the predominant form of DNA found in living cells. However, under certain in vitro conditions, the two strands of DNA can twist into A DNA or Z DNA, which differ significantly from B DNA. A and B DNA are right-handed helices; Z DNA has a left-handed conformation. In addition, the helical backbone in Z DNA appears to zigzag slightly as it winds itself around the double-helical structure. The numbers of base pairs per 360° turn

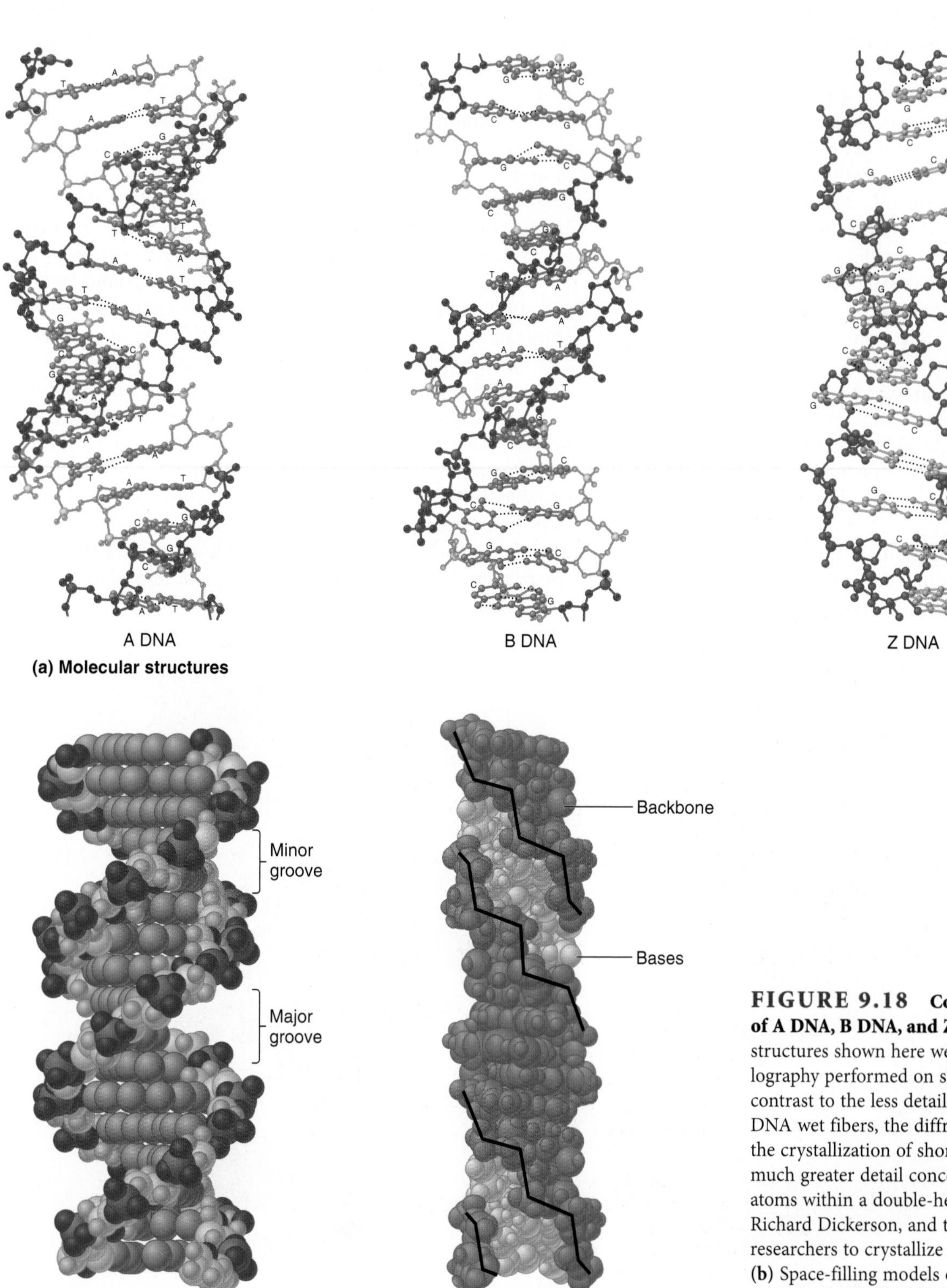

FIGURE 9.18 Comparison of the structures of A DNA, B DNA, and Z DNA. (a) The highly detailed structures shown here were deduced by X-ray crystallography performed on short segments of DNA. In contrast to the less detailed structures obtained from DNA wet fibers, the diffraction pattern obtained from the crystallization of short segments of DNA provides much greater detail concerning the exact placement of atoms within a double-helical structure. Alexander Rich, Richard Dickerson, and their colleagues were the first researchers to crystallize a short piece of DNA.
(b) Space-filling models of the B-DNA and Z-DNA structures. In the case of Z DNA, the black lines connect the phosphate groups in the DNA backbone. As seen here, they travel along the backbone in a zigzag pattern.

are 11.0, 10.0, and 12.0 in A, B, and Z DNA, respectively. In B DNA, the bases tend to be centrally located, and the hydrogen bonds between base pairs occur relatively perpendicular to the central axis. In contrast, the bases in A DNA and Z DNA are substantially tilted relative to the central axis.

The ability of the predominant B DNA to adopt A-DNA and Z-DNA conformations depends on certain conditions. In X-ray diffraction studies, A DNA occurs under conditions of low humidity. The ability of a double helix to adopt a Z-DNA conformation depends on various factors. At high ionic strength (i.e., high salt concentration), formation of a Z-DNA conformation is favored by a sequence of bases that alternates between purines and pyrimidines. One such sequence is

5'–GCGCGCGCG–3'
3'–CGCGCGCGC–5'

At lower ionic strength, the methylation of cytosine bases can favor Z-DNA formation. Cytosine **methylation** occurs when a cellular enzyme attaches a methyl group ($-CH_3$) to the cytosine base. In addition, negative supercoiling (a topic discussed in Chapter 10) favors the Z-DNA conformation.

What is the biological significance of A and Z DNA? Research has not found any biological role for A DNA. However, accumulating evidence suggests a possible biological role for Z DNA in the process of transcription. Recent research has identified cellular proteins that specifically recognize Z DNA. In 2005, Alexander Rich and colleagues reported that the Z-DNA-binding region of one such protein played a role in regulating the transcription of particular genes. In addition, other research has suggested that Z DNA may play a role in chromosome structure by affecting the level of compaction.

DNA Can Form a Triple Helix, Called Triplex DNA

A surprising discovery made in 1957 by Alexander Rich, David Davies, and Gary Felsenfeld was that DNA can form a triple-helical structure called **triplex DNA.** This triplex was formed in vitro using pieces of DNA that were made synthetically. Although this result was interesting, it seemed to have little, if any, biological relevance.

About 30 years later, interest in triplex DNA was renewed by the observation that triplex DNA can form in vitro by mixing natural double-stranded DNA and a third short strand that is synthetically made. The synthetic strand binds into the major groove of the naturally occurring double-stranded DNA (**Figure 9.19**). As shown here, an interesting feature of triplex DNA formation is that it is sequence-specific. In other words, the synthetic third strand incorporates itself into a triple helix due to specific interactions between the synthetic DNA and the biological DNA. The pairing rules are that a thymine in the synthetic DNA hydrogen bonds at an AT pair in the biological DNA and that a cytosine in the synthetic DNA hydrogen bonds at a GC pair.

The formation of triplex DNA has been implicated in several cellular processes, including recombination, which is described in Chapter 17. In addition, researchers are interested in triplex DNA due to its potential as a tool to specifically inhibit particular genes. As shown in Figure 9.19, the synthetic DNA

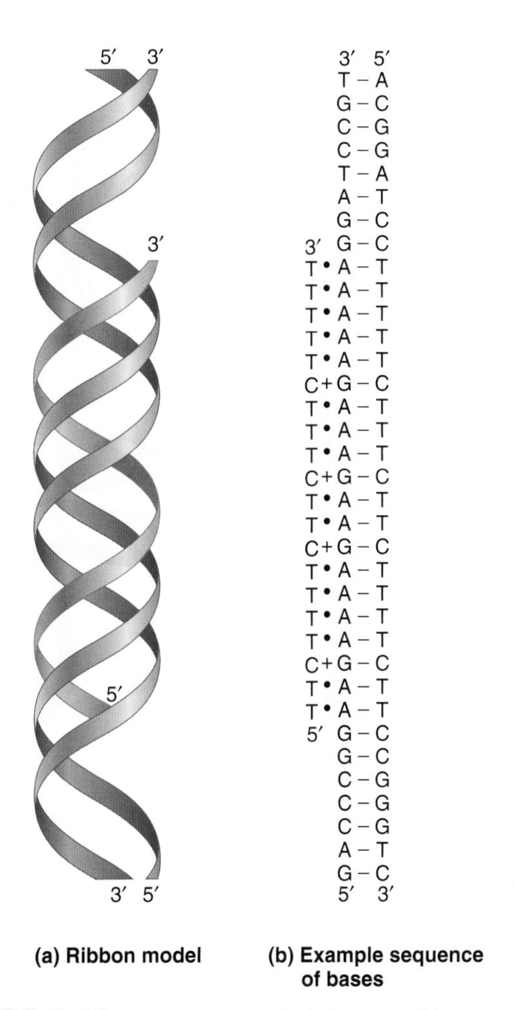

FIGURE 9.19 The structure of triplex DNA. (**a**) As seen in the ribbon model, the third, synthetic strand binds within the major groove of the double-stranded structure. (**b**) Within triplex DNA, the third strand hydrogen bonds according to the rule T to AT, and C to GC. The cytosine bases in the third strand are protonated (i.e., positively charged).

strand binds into the major groove according to specific base-pairing rules. Therefore, researchers can design a synthetic DNA to recognize the base sequence found in a particular gene. When the synthetic DNA binds to a gene, it inhibits transcription. In addition, the synthetic DNA can cause mutations in a gene that inactivate its function. Researchers are excited about the possibility of using such synthetic DNA to silence the expression of particular genes. For example, this approach could be used to silence genes that become overactive in cancer cells. However, further research is needed to develop effective ways to promote the uptake of synthetic DNAs into the appropriate target cells.

The Three-Dimensional Structure of DNA Within Chromosomes Requires Additional Folding and the Association with Proteins

To fit within a living cell, the long double-helical structure of chromosomal DNA must be extensively compacted into a three-dimensional conformation. With the aid of DNA-binding proteins, such as histone proteins, the double helix becomes greatly twisted and folded. **Figure 9.20** depicts the relationship between the DNA

exact copies of DNA. Likewise, the ability to transcribe DNA into RNA is based on complementarity. During transcription, one strand of DNA is used as a template to make a complementary strand of RNA.

In addition to the synthesis of new strands of DNA and RNA, complementarity is important in other ways. As mentioned in this chapter, the folding of RNA into a particular structure is driven by the hydrogen bonding of complementary regions. This event is necessary to produce functionally active tRNA molecules. Likewise, stem-loop structures also occur in other types of RNA. For example, the rapid formation of stem-loop structures is known to occur as RNA is being transcribed and to affect the termination of transcription.

A third way that complementarity can be functionally important is that it can promote the interaction of two separate RNA molecules. During translation, codons in mRNA bind to the anticodons in tRNA (see Chapter 13). This binding is due to complementarity. For example, if a codon is 5′–AGG–3′, the anticodon is 3′–UCC–5′. This type of specific interaction between codons and anticodons is an important step that enables the nucleotide sequence in mRNA to code for an amino acid sequence within a protein. In addition, many other examples of RNA-RNA interactions are known and will be described throughout this textbook.

S5. An important feature of triplex DNA formation is that it is sequence-specific. The synthetic third strand incorporates itself into a triple helix, so a thymine in the synthetic DNA binds near an AT pair in the biological DNA, and a cytosine in the synthetic DNA binds near a GC pair. From a practical point of view, this opens the possibility of synthesizing a short strand of DNA that forms a triple helix at a particular target site. For example, if the sequence of a particular gene is known, researchers can make a synthetic piece of DNA that forms a triple helix somewhere within that gene according to the T to AT, and C to GC rule. Triplex DNA formation is known to inhibit gene transcription. In other words, when the synthetic DNA binds within the DNA of a gene, the formation of triplex DNA prevents that gene from being transcribed into RNA. Discuss how this observation might be used to combat diseases.

Answer: Triplex DNA formation opens the exciting possibility of designing synthetic pieces of DNA to inhibit the expression of particular genes. Theoretically, such a tool could be used to combat viral diseases or to inhibit the growth of cancer cells. To combat a viral disease, a synthetic DNA could be made that specifically binds to an essential viral gene, thereby preventing viral proliferation. To inhibit cancer, a synthetic DNA could be made to bind to an oncogene. (Note: As described in Chapter 22, an oncogene is a gene that promotes cancerous growth.) Inhibition of an oncogene could prevent cancer. At this point, a primary obstacle in applying this approach is devising a method of getting the synthetic DNA into living cells.

Conceptual Questions

C1. What is the meaning of the term genetic material?

C2. After the DNA from type S bacteria is exposed to type R bacteria, list all of the steps that you think must occur for the bacteria to start making a capsule.

C3. Look up the meaning of the word transformation in a dictionary and explain whether it is an appropriate word to describe the transfer of genetic material from one organism to another.

C4. What are the building blocks of a nucleotide? With regard to the 5′ and 3′ positions on a sugar molecule, how are nucleotides linked together to form a strand of DNA?

C5. Draw the structure of guanine, guanosine, and deoxyguanosine triphosphate.

C6. Draw the structure of a phosphodiester linkage.

C7. Describe how bases interact with each other in the double helix. This discussion should address the issues of complementarity, hydrogen bonding, and base stacking.

C8. If one DNA strand is 5′–GGCATTACACTAGGCCT–3′, what is the sequence of the complementary strand?

C9. What is meant by the term DNA sequence?

C10. Make a side-by-side drawing of two DNA helices, one with 10 bp per 360° turn and the other with 15 bp per 360° turn.

C11. Discuss the differences in the structural features of A DNA, B DNA, and Z DNA.

C12. What parts of a nucleotide (namely, phosphate, sugar, and/or bases) occupy the major and minor grooves of double-stranded DNA, and what parts are found in the DNA backbone? If a DNA-binding protein does not recognize a specific nucleotide sequence, do you expect that it recognizes the major groove, the minor groove, or the DNA backbone? Explain.

C13. List the structural differences between DNA and RNA.

C14. Draw the structure of deoxyribose and number the carbon atoms. Describe the numbering of the carbon atoms in deoxyribose with regard to the directionality of a DNA strand. In a DNA double helix, what does the term antiparallel mean?

C15. Write out a sequence of an RNA molecule that could form a stem-loop with 24 nucleotides in the stem and 16 nucleotides in the loop.

C16. Compare the structural features of a double-stranded RNA structure with those of a DNA double helix.

C17. Which of the following DNA double helices would be more difficult to separate into single-stranded molecules by treatment with heat, which breaks hydrogen bonds?

A. GGCGTACCAGCGCAT
   CCGCATGGTCGCGTA

B. ATACGATTTACGAGA
   TATGCTAAATGCTCT

Explain your choice.

C18. What structural feature allows DNA to store information?

C19. Discuss the structural significance of complementarity in DNA and in RNA.

C20. An organism has a G + C content of 64% in its DNA. What are the percentages of A, T, G, and C?

C21. Let's suppose you have recently identified an organism that was scraped from an asteroid that hit the earth. (Fortunately, no one was injured.) When you analyze this organism, you discover that its DNA is a triple helix, composed of six different nucleotides: A, T, G, C, X, and Y. You measure the chemical composition of the bases and find the following amounts of these six bases: A = 24%,

T = 23%, G = 11%, C = 12%, X = 21%, Y = 9%. What rules would you propose that govern triplex DNA formation in this organism? Note: There is more than one possibility.

C22. On further analysis of the DNA described in conceptual question C21, you discover that the triplex DNA in this alien organism is composed of a double helix, with the third helix wound within the major groove (just like the DNA in Figure 9.19). How would you propose that this DNA is able to replicate itself? In your answer, be specific about the base pairing rules within the double helix and which part of the triplex DNA would be replicated first.

C23. A DNA-binding protein recognizes the following double-stranded sequence:

5′–GCCCGGGC–3′
3′–CGGGCCCG–5′

This type of double-stranded structure could also occur within the stem region of an RNA stem-loop molecule. Discuss the structural differences between RNA and DNA that might prevent this DNA-binding protein from recognizing a double-stranded RNA molecule.

C24. Within a protein, certain amino acids are positively charged (e.g., lysine and arginine), some are negatively charged (e.g., glutamate and aspartate), some are polar but uncharged, and some are nonpolar. If you knew that a DNA-binding protein was recognizing the DNA backbone rather than base sequences, which amino acids in the protein would be good candidates for interacting with the DNA?

C25. In what ways are the structures of an α helix in proteins and the DNA double helix similar, and in what ways are they different?

C26. A double-stranded DNA molecule contains 560 nucleotides. How many complete turns would be found in this double helix?

C27. As the minor and major grooves of the DNA wind around a DNA double helix, do they ever intersect each other, or do they always run parallel to each other?

C28. What chemical group (phosphate group, hydroxyl group, or a nitrogenous base) is found at the 3′ end of a DNA strand? What group is found at the 5′ end?

C29. The base composition of an RNA virus was analyzed and found to be 14.1% A, 14.0% U, 36.2% G, and 35.7% C. Would you conclude that the viral genetic material is single-stranded RNA or double-stranded RNA?

C30. The genetic material found within some viruses is single-stranded DNA. Would this genetic material contain equal amounts of A and T and equal amounts of G and C?

C31. A medium-sized human chromosome contains about 100 million bp. If the DNA were stretched out in a linear manner, how long would it be?

C32. A double-stranded DNA molecule is 1 cm long, and the percentage of adenine is 15%. How many cytosines would be found in this DNA molecule?

C33. Could single-stranded DNA form a stem-loop structure? Why or why not?

C34. As described in Chapter 15, the methylation of cytosine bases can have an important effect on gene expression. For example, the methylation of cytosines may inhibit the transcription of genes. A methylated cytosine base has the following structure:

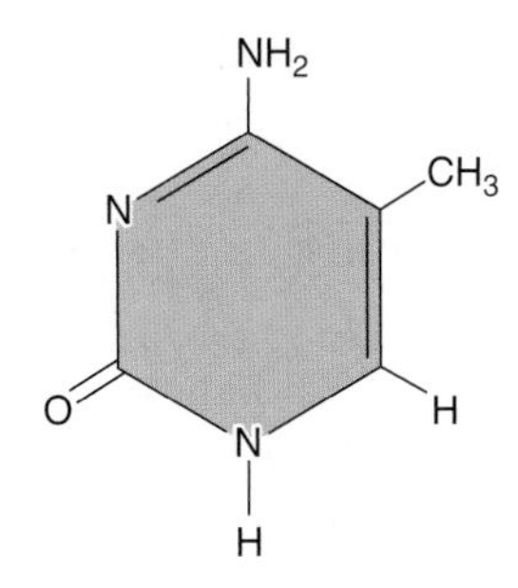

Would you expect the methylation of cytosine to affect the hydrogen bonding between cytosine and guanine in a DNA double helix? Why or why not? (Hint: See Figure 9.16 for help.) Take a look at solved problem S3 and speculate as to how methylation could affect gene expression.

C35. An RNA molecule has the following sequence:

Region 1 Region 2 Region 3
5′-CAUCCAUCCAUUCCCCAUCCGAUAAGGGGAAUGGAUCCGAAUGGAUAAC-3′

Parts of region 1 can form a stem-loop with region 2 and with region 3. Can region 1 form a stem-loop with region 2 and region 3 at the same time? Why or why not? Which stem-loop would you predict to be more stable: a region 1/region 2 interaction or a region 1/region 3 interaction? Explain your choice.

Experimental Questions

E1. Genetic material acts as a blueprint for an organism's traits. Explain how the experiments of Griffith indicated that genetic material was being transferred to the type R bacteria.

E2. With regard to the experiment described in Figure 9.2, answer the following:

A. List several possible reasons why only a small percentage of the type R bacteria was converted to type S.

B. Explain why an antibody must be used to remove the bacteria that are not transformed. What would the results look like, in all five cases, if the antibody/centrifugation step had not been included in the experimental procedure?

C. The DNA extract was treated with DNase, RNase, or protease. Why was this done? (In other words, what were the researchers trying to demonstrate?)

E3. An interesting trait that some bacteria exhibit is resistance to killing by antibiotics. For example, certain strains of bacteria are resistant to tetracycline, whereas other strains are sensitive to tetracycline. Describe an experiment you would carry out to demonstrate that tetracycline resistance is an inherited trait encoded by the DNA of the resistant strain.

E4. With regard to the experiment of Figure 9.5, answer the following:

A. Provide possible explanations why some of the DNA is in the supernatant.

B. Plot the results if the radioactivity in the pellet, rather than in the supernatant, had been measured.

C. Why were ^{32}P and ^{35}S chosen as radioisotopes to label the phages?

D. List possible reasons why less than 100% of the phage protein was removed from the bacterial cells during the shearing process.

E5. Does the experiment of Figure 9.5 rule out the possibility that RNA is the genetic material of T2 phage? Explain your answer. If it does not, could you modify the approach of Hershey and Chase to show that it is DNA and not RNA that is the genetic material of T2 bacteriophage? Note: It is possible to specifically label DNA or RNA by providing bacteria with radiolabeled thymine or uracil, respectively.

E6. In this chapter, we considered two experiments—one by Avery, MacLeod, and McCarty and the second by Hershey and Chase—that indicated DNA is the genetic material. Discuss the strengths and weaknesses of the two approaches. Which experimental approach did you find the most convincing? Why?

E7. The type of model building used by Pauling and Watson and Crick involved the use of ball-and-stick units. Now we can do model building on a computer screen. Even though you may not be familiar with this approach, discuss potential advantages of using computers in molecular model building.

E8. With regard to Chargaff's experiment described in Figure 9.13, answer the following:

A. What is the purpose of paper chromatography?

B. Explain why it is necessary to remove the bases in order to determine the base composition of DNA.

C. Would Chargaff's experiments have been convincing if they had been done on only one species? Discuss.

E9. Gierer and Schramm exposed plant tissue to purified RNA from tobacco mosaic virus, and the plants developed the same types of lesions as if they were exposed to the virus itself. What would be the results if the RNA was treated with DNase, RNase, or protease prior to its exposure to the plant tissue?

Questions for Student Discussion/Collaboration

1. Try to propose structures for a genetic material that are substantially different from the double helix. Remember that the genetic material must have a way to store information and a way to be faithfully replicated.

2. How might you provide evidence that DNA is the genetic material in mice?

Note: All answers appear at the website for this textbook; the answers to even-numbered questions are in the back of the textbook.

www.mhhe.com/brookergenetics4e

Visit the website for practice tests, answer keys, and other learning aids for this chapter. Enhance your understanding of genetics with our interactive exercises, quizzes, animations, and much more.

CHAPTER OUTLINE

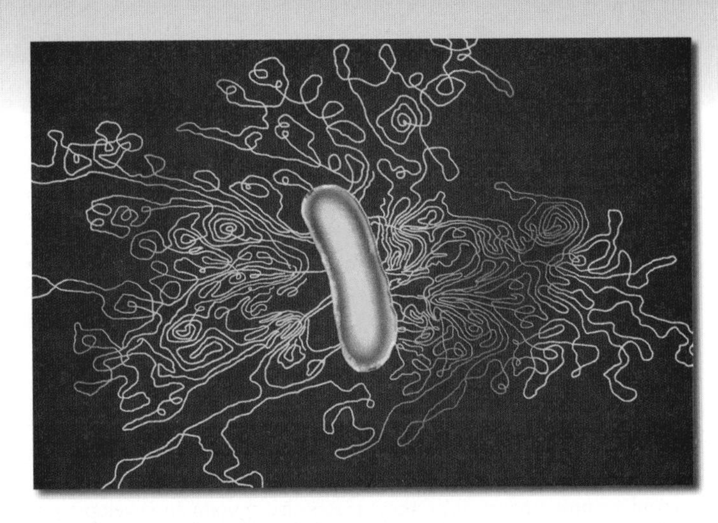

Structure of a bacterial chromosome. *This is an electron micrograph of a bacterial chromosome, which has been released from a bacterial cell.*

10 CHROMOSOME ORGANIZATION AND MOLECULAR STRUCTURE

Chromosomes are the structures within cells that contain the genetic material. The term **genome** refers to a complete set of genetic material in a particular cellular compartment. For bacteria, the genome is typically a single circular chromosome. For eukaryotes, the nuclear genome refers to one complete set of chromosomes that resides in the cell nucleus. In other words, the haploid complement of chromosomes is considered a nuclear genome. Eukaryotes have a mitochondrial genome, and plants also have a chloroplast genome. Unless otherwise noted, the term eukaryotic genome refers to the nuclear genome.

The primary function of the genetic material is to store the information needed to produce the characteristics of an organism. As we saw in Chapter 9, the sequence of bases in a DNA molecule can store information. To fulfill their role at the molecular level, chromosomal sequences facilitate four important processes: (1) the synthesis of RNA and cellular proteins, (2) the replication of chromosomes, (3) the proper segregation of chromosomes, and (4) the compaction of chromosomes so they can fit within living cells. In this chapter, we will examine the general organization of the genetic material within viral, bacterial, and eukaryotic chromosomes. In addition, the molecular mechanisms that account for the packaging of the genetic material in viruses, bacteria, and eukaryotic cells will be described. We begin by considering the comparatively simple genomes of viruses.

10.1 VIRAL GENOMES

Viruses are small infectious particles that contain nucleic acid as their genetic material, surrounded by a protein coat, or capsid (**Figure 10.1a**). The capsid of **bacteriophages,** which are viruses that infect bacteria, may also contain a sheath, base plate, and tail fibers (see Figure 9.3). Certain eukaryotic viruses also have an envelope consisting of a membrane embedded with spike proteins (**Figure 10.1b**). By themselves, viruses are not cellular organisms. They do not contain energy-producing enzymes, ribosomes, or cellular organelles. Instead, viruses rely on their **host cells**—the cells they infect—for making new viruses. In general, most viruses exhibit a limited **host range,** the spectrum of host species that a virus can infect. Many viruses can infect only specific types of cells of one host species. Depending on the life cycle of the virus, the host cell may or may not be destroyed during the process of viral replication and release. In this section, we consider the genetic composition of viruses and examine how viral genomes are packaged into virus particles.

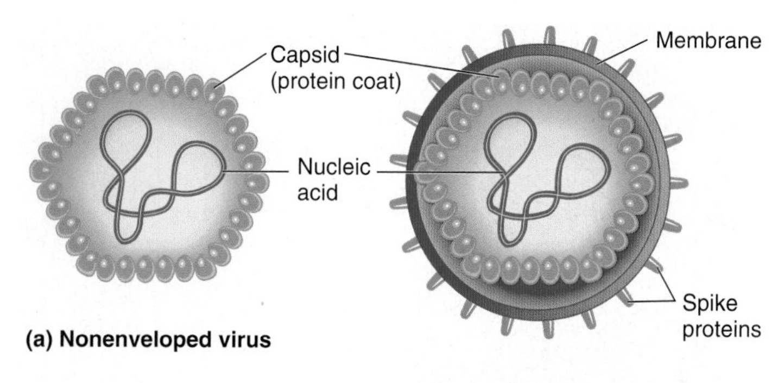

FIGURE 10.1 **General structure of viruses.** (**a**) The simplest viruses contain a nucleic acid molecule (DNA or RNA) surrounded by a capsid, or protein coat. (**b**) Other viruses also contain an envelope composed of a membrane and spike proteins. The membrane is obtained from the host cell when the virus buds through the plasma membrane.

Viral Genomes Are Relatively Small and Are Composed of DNA or RNA

A **viral genome** is the genetic material that a virus contains. The term viral chromosome is also used to describe the viral genome. The nucleic acid composition of viral genomes varies markedly among different types of viruses. **Table 10.1** describes the genome characteristics of a few selected viruses. The genome can be DNA or RNA, but not both. In some cases, it is single-stranded, whereas in others, it is double-stranded. Depending on the type of virus, the genome can be linear or circular.

As shown in Table 10.1, viral genomes vary in size from several thousand to more than a hundred thousand nucleotides in length. For example, the genomes of some simple viruses, such as Qβ virus, are only a few thousand nucleotides in length and contain only a few genes. Other viruses, particularly those with a complex structure, have many more genes. The T-even phages (T2, T4, etc.), discussed in Chapters 7 and 9, are examples of more complex viruses.

TABLE **10.1**

Characteristics of Selected Viral Genomes

Virus	Host	Type of Nucleic Acid*	Size†	Number of Genes‡
Parvovirus	Mammals	ssDNA	5.0	5
Fd	*E. coli*	ssDNA	6.4	10
Lambda	*E. coli*	dsDNA	48.5	71
T4	*E. coli*	dsDNA	169.0	288
Qβ	*E. coli*	ssRNA	4.2	4
TMV	Many plants	ssRNA	6.4	6
Influenza virus	Mammals	ssRNA	13.5	11
Human immunodeficiency virus (HIV)	Primates	ssRNA	9.7	9
Herpes simplex virus, type 2 (genital herpes)	Humans	dsDNA	158.4	77

*ss refers to single-stranded, and ds refers to double-stranded.
†Number of thousands of nucleotides or nucleotide base pairs
‡This number refers to the number of protein-encoding units. In some cases, two or more proteins are made from a single gene due to events such as protein processing.
TMV, tobacco mosaic virus

Viral Genomes Are Packaged into the Capsid in an Assembly Process

In an infected cell, the reproductive cycle of the virus eventually leads to the synthesis of viral nucleic acids and proteins. Newly synthesized viral chromosomes and capsid proteins must then come together and assemble to make mature virus particles. Viruses with a simple structure may self-assemble, which means that the nucleic acid and capsid proteins spontaneously bind to each other to form a mature virus. The structure of one self-assembling virus, the tobacco mosaic virus, is shown in **Figure 10.2**. As shown here, the proteins assemble around the RNA genome, which becomes trapped inside the hollow capsid. This assembly process can occur in vitro if purified capsid proteins and RNA are mixed together.

Some viruses, such as T2 bacteriophage, have more complicated structures that do not self-assemble. The correct assembly of this virus requires the help of proteins not found within the mature virus particle itself. When virus assembly requires the participation of noncapsid proteins, the process is called directed assembly, because the noncapsid proteins direct the proper assembly of the virus. What are the functions of these noncapsid proteins? Some proteins, called scaffolding proteins, catalyze the assembly process and are transiently associated with the capsid. However, as viral assembly nears completion, the scaffolding proteins are expelled from the mature virus. In addition, other noncapsid proteins act as proteases that specifically cleave viral capsid proteins. This cleavage produces a capsid protein that is somewhat smaller and able to assemble correctly. For many viruses, the cleavage of capsid proteins into smaller units is an important event that precedes viral assembly.

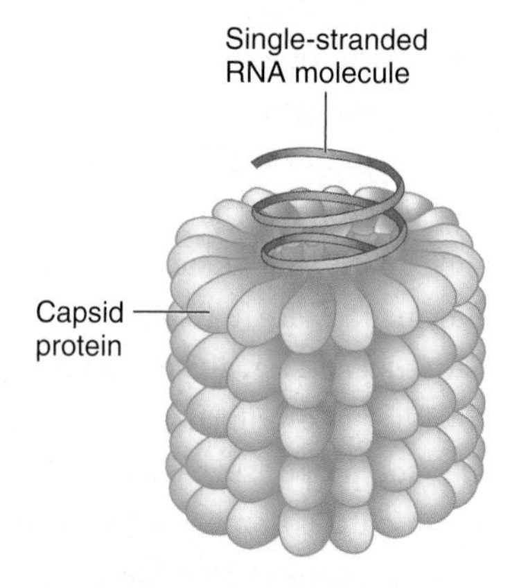

FIGURE 10.2 **Structure of the tobacco mosaic virus.** This self-assembling virus is composed of a coiled RNA molecule surrounded by 2130 identical protein subunits. Only a portion of the tobacco mosaic virus is shown here. Several layers of proteins have been omitted from this illustration to reveal the RNA genome, which is trapped inside the protein coat.

10.2 BACTERIAL CHROMOSOMES

Let's now turn our attention to the organization of chromosomes found in bacterial species. Inside a bacterial cell, the chromosome is highly compacted and found within a region of the cell known as the **nucleoid.** Although bacteria usually contain a single type of chromosome, more than one copy of that chromosome may be found within one bacterial cell. Depending on the growth conditions and phase of the cell cycle, bacteria may have one to four identical chromosomes per cell. In addition, the number of copies varies depending on the bacterial species. As shown in **Figure 10.3**, each chromosome occupies its own distinct nucleoid region within the cell. Unlike the eukaryotic nucleus, the bacterial nucleoid is not a separate cellular compartment bounded by a membrane. Rather, the DNA in a nucleoid is in direct contact with the cytoplasm of the cell.

In this section, we will explore two important features of bacterial chromosomes. First, the organization of DNA sequences along the chromosome is examined. Second, we consider the mechanisms that cause the chromosome to become a compacted structure within a nucleoid of the bacterium.

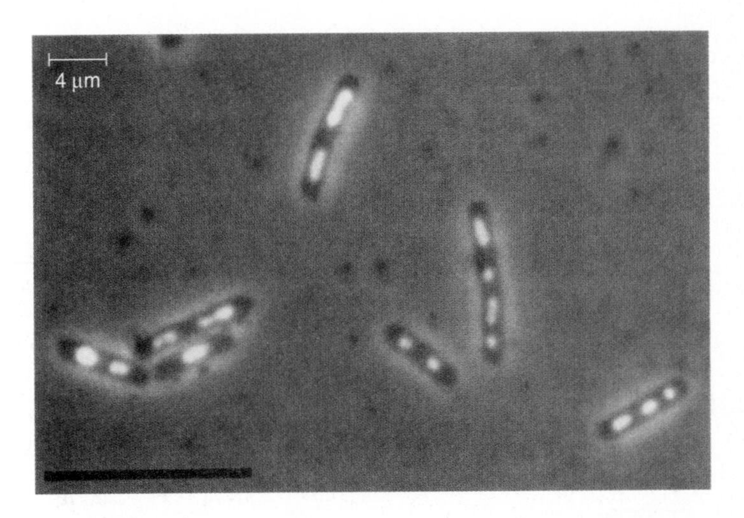

FIGURE 10.3 **The localization of nucleoids within *Bacillus subtilis* bacteria.** The nucleoids are fluorescently labeled and seen as bright, oval-shaped regions within the bacterial cytoplasm. Note that two or more nucleoids are found within each cell. Some of the cells seen here are in the process of dividing.

Bacterial Chromosomes Contain a Few Thousand Gene Sequences Interspersed with Other Functionally Important Sequences

Bacterial chromosomal DNA is usually a circular molecule, though some bacteria have linear chromosomes. A typical chromosome is a few million base pairs (bp) in length. For example, the chromosome of one strain of *Escherichia coli* has approximately 4.6 million bp, and the *Haemophilus influenzae* chromosome has roughly 1.8 million bp. A bacterial chromosome commonly has a few thousand different genes. These genes are interspersed throughout the entire chromosome (**Figure 10.4**). **Structural genes**—nucleotide sequences that encode proteins—account for the majority of bacterial DNA. The nontranscribed regions of DNA located between adjacent genes are termed **intergenic regions.**

Other sequences in chromosomal DNA influence DNA replication, gene transcription, and chromosome structure. For example, bacterial chromosomes have one **origin of replication,** a sequence that is a few hundred nucleotides in length. This nucleotide sequence functions as an initiation site for the assembly of several proteins required for DNA replication. Also, a variety of **repetitive sequences** have been identified in many bacterial species. These sequences are found in multiple copies and are usually interspersed within the intergenic regions throughout the bacterial chromosome. Repetitive sequences may play a role in a variety of genetic processes, including DNA folding, DNA replication, gene regulation, and genetic recombination. As discussed in Chapter 17, some repetitive sequences are transposable elements that can

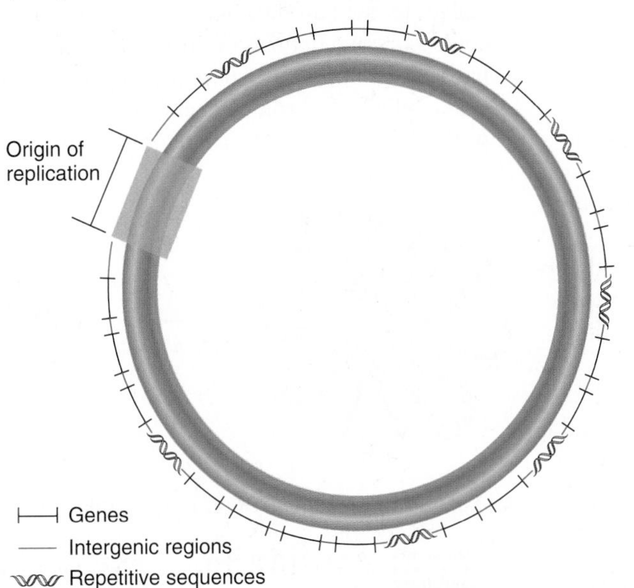

Key features:

- Most, but not all, bacterial species contain circular chromosomal DNA.
- A typical chromosome is a few million base pairs in length.
- Most bacterial species contain a single type of chromosome, but it may be present in multiple copies.
- Several thousand different genes are interspersed throughout the chromosome. The short regions between adjacent genes are called intergenic regions.
- One origin of replication is required to initiate DNA replication.
- Repetitive sequences may be interspersed throughout the chromosome.

FIGURE 10.4 **Organization of sequences in bacterial chromosomal DNA.**

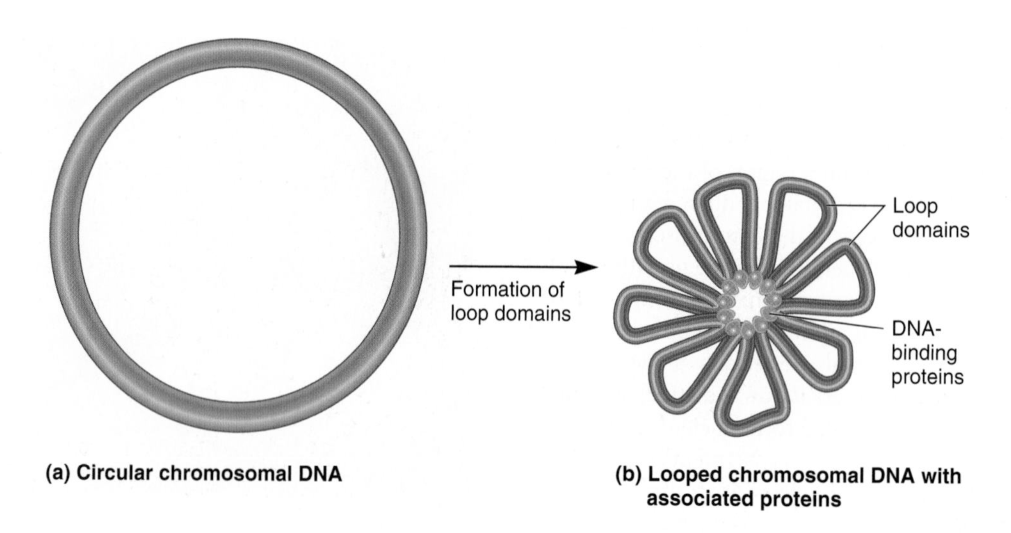

ONLINE ANIMATION

FIGURE 10.5 **The formation of loop domains within the bacterial chromosome.** To promote compaction, **(a)** the large, circular chromosomal DNA is organized into **(b)** smaller, looped chromosomal DNA with loop domains and associated proteins.

move throughout the genome. Figure 10.4 summarizes the key features of sequence organization within bacterial chromosomes.

The Formation of Chromosomal Loops Helps Make the Bacterial Chromosome More Compact

To fit within the bacterial cell, the chromosomal DNA must be compacted about 1000-fold. Part of this compaction process involves the formation of **loop domains** within the bacterial chromosome (**Figure 10.5**). As its name suggests, a loop domain is a segment of chromosomal DNA folded into a structure that resembles a loop. DNA-binding proteins anchor the base of the loops in place. The number of loops varies according to the size of the bacterial chromosome and the species. In *E. coli*, a chromosome has 50 to 100 loop domains with about 40,000 to 80,000 bp of DNA in each loop. This looped structure compacts the circular chromosome about 10-fold.

DNA Supercoiling Further Compacts the Bacterial Chromosome

Because DNA is a long thin molecule, twisting forces can dramatically change its conformation. This effect is similar to twisting a rubber band. If twisted in one direction, a rubber band eventually coils itself into a compact structure as it absorbs the energy applied by the twisting motion. Because the two strands within DNA already coil around each other, the formation of additional coils due to twisting forces is referred to as **DNA supercoiling** (**Figure 10.6**).

How do twisting forces affect DNA structure? **Figure 10.7** illustrates four possibilities. In Figure 10.7a, a double-stranded DNA molecule with five complete turns is anchored between two plates. In this hypothetical example, the ends of the DNA molecule cannot rotate freely. Both underwinding and overwinding of the DNA double helix can induce supercoiling of the helix. Because B DNA is a right-handed helix, underwinding is a left-handed twisting motion, and overwinding is a right-handed twist. Along the left side of Figure 10.7, one of the plates has been given a turn in the direction that tends to unwind the helix. As the helix absorbs this force, two things can happen. The underwinding motion can cause fewer turns (Figure 10.7b) or cause a negative supercoil to form (Figure 10.7c). On the right side of Figure 10.7, one of the plates has been given a right-handed turn, which overwinds the double helix. This can lead to either more turns (Figure 10.7d) or the formation of a positive supercoil (Figure 10.7e). The DNA conformations shown in Figure 10.7a, c, and e differ only with regard to supercoiling. These three DNA conformations are referred to as **topoisomers** of each other. The DNA conformations shown in Figure 10.7b and d are not structurally favorable and do not occur in living cells.

Chromosome Function Is Influenced by DNA Supercoiling

The chromosomal DNA in living bacteria is negatively supercoiled. In the chromosome of *E. coli*, about one negative supercoil occurs per 40 turns of the double helix. Negative supercoiling has several important consequences. As already mentioned, the supercoiling of chromosomal DNA makes it much more compact (see Figure 10.6). Therefore, supercoiling helps to greatly

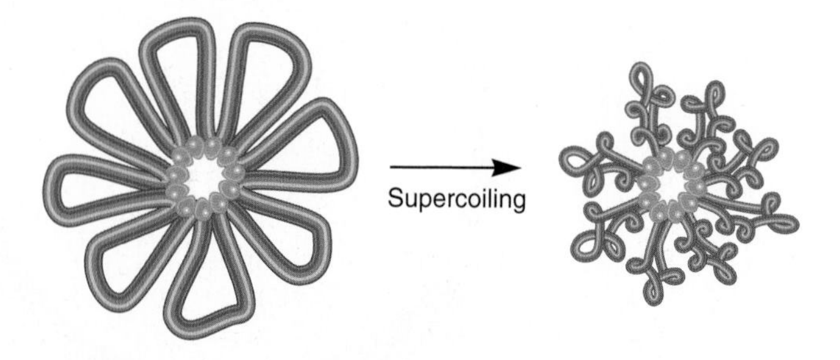

(a) Looped chromosomal DNA (b) Looped and supercoiled DNA

FIGURE 10.6 **DNA supercoiling leads to further compaction of the looped chromosomal DNA.** **(a)** The looped chromosomal DNA becomes much more compacted due to **(b)** supercoiling within the loops.

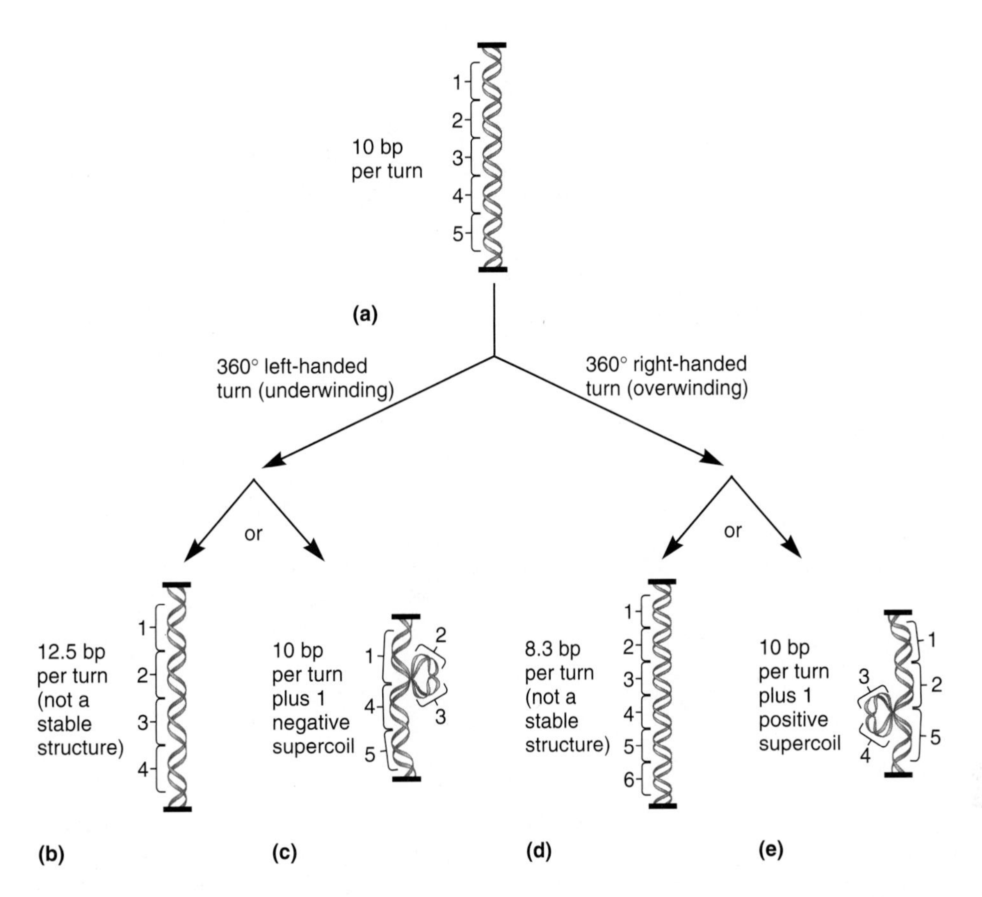

FIGURE 10.7 Schematic representation of DNA supercoiling. In this example, the DNA in **(a)** is anchored between two plates and given a twist as noted by the arrows. A left-handed twist (underwinding) could produce either **(b)** fewer turns or **(c)** a negative supercoil. A right-handed twist (overwinding) produces **(d)** more turns or **(e)** a positive supercoil. The structures shown in (b) and (d) are unstable.

decrease the size of the bacterial chromosome. In addition, negative supercoiling also affects DNA function. To understand how, remember that negative supercoiling is due to an underwinding force on the DNA. Therefore, negative supercoiling creates tension on the DNA strands that may be released by DNA strand separation (**Figure 10.8**). Although most of the chromosomal DNA is negatively supercoiled and compact, the force of negative supercoiling may promote DNA strand separation in small regions. This enhances genetic activities such as replication and transcription that require the DNA strands to be separated.

How does bacterial DNA become supercoiled? In 1976, Martin Gellert and colleagues discovered the enzyme **DNA gyrase,** also known as topoisomerase II. This enzyme, which contains four subunits (two A and two B subunits), introduces negative supercoils (or relaxes positive supercoils) using energy from ATP (**Figure 10.9a**). To alter supercoiling, DNA gyrase has two sets of jaws that allow it to grab onto two regions of DNA. One of the DNA regions is grabbed by the lower jaws and then is wrapped in a right-handed direction around the two A subunits. The upper jaws then clamp onto another region of DNA. The DNA in the lower jaws is cut in both strands, and the other region of DNA is then released from the upper jaws and passed through this double stranded break. The net result is that two negative supercoils have been introduced into the DNA molecule (**Figure 10.9b**). In addition, DNA gyrase in bacteria and topoisomerase II in eukaryotes can untangle DNA molecules. For example, as discussed in Chapter 11, circular DNA molecules are sometimes intertwined following DNA replication (see Figure 11.14). Such interlocked molecules can be separated via topoisomerase II.

A second type of enzyme, **topoisomerase I,** can relax negative supercoils. This enzyme can bind to a negatively supercoiled region and introduce a break in one of the DNA strands. After one DNA strand has been broken, the DNA molecule can rotate to relieve the tension that is caused by negative supercoiling. This rotation relaxes negative supercoiling. The broken strand is then resealed. The competing actions of DNA gyrase and topoisomerase I govern the overall supercoiling of the bacterial DNA.

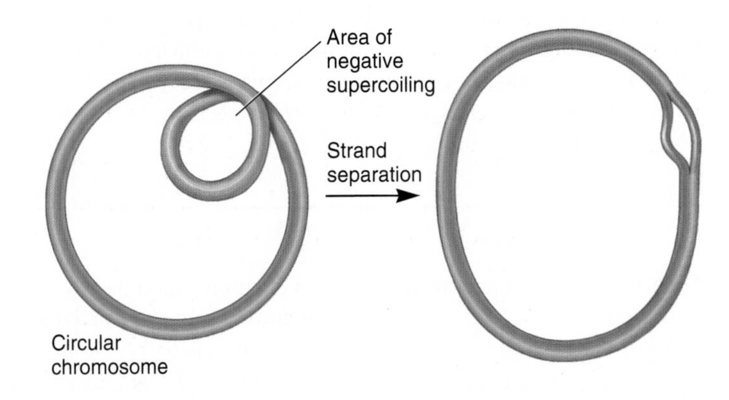

FIGURE 10.8 Negative supercoiling promotes strand separation.

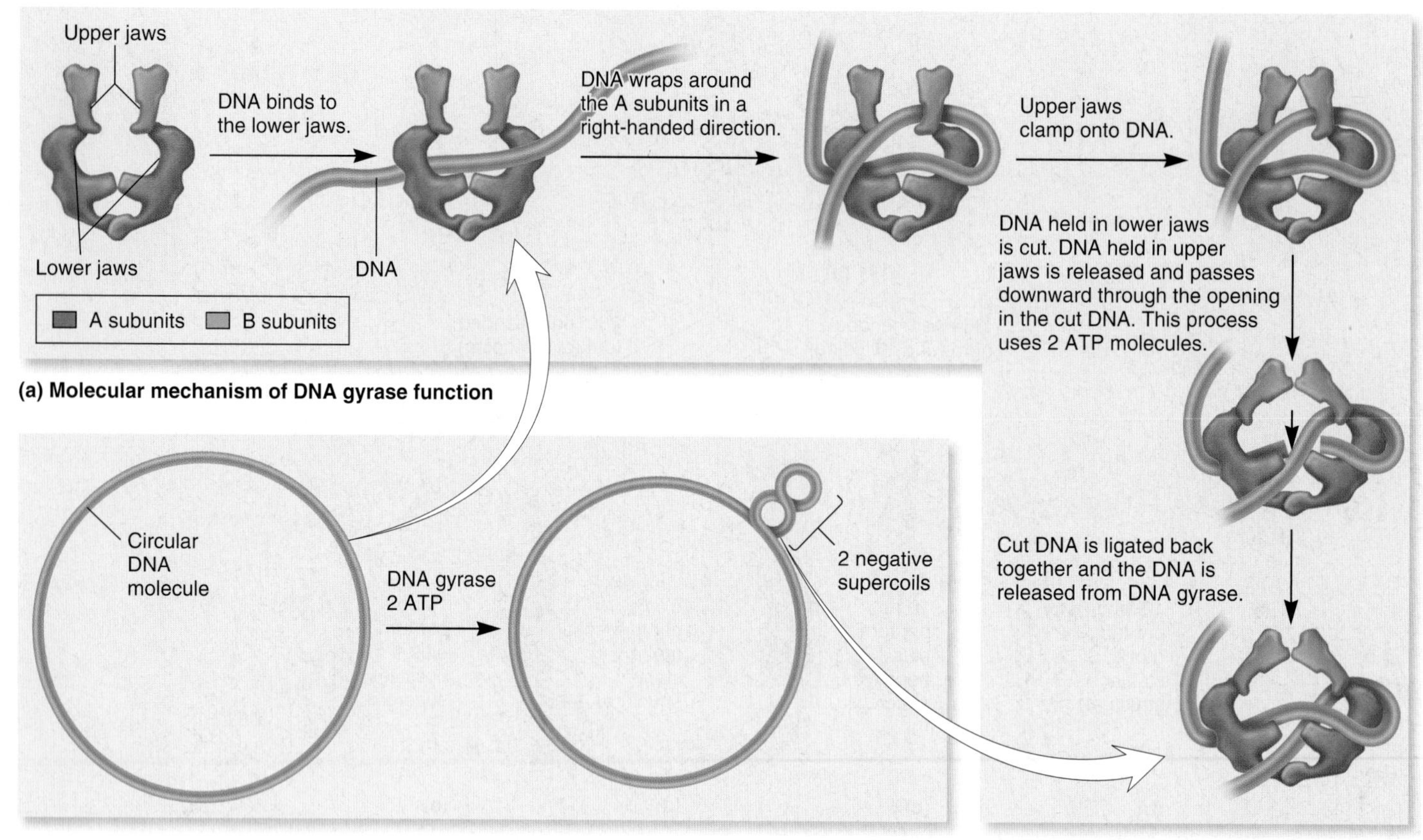

FIGURE 10.9 **The action of DNA gyrase.** **(a)** DNA gyrase, also known as topoisomerase II, is composed of two A and two B subunits. The upper and lower jaws clamp onto two regions of DNA. The lower region is wrapped around the A subunits, which then cleave this DNA. The unbroken segment of DNA is released from the upper jaws and passes through the break. The break is repaired. The B subunits capture the energy from ATP hydrolysis to catalyze this process. **(b)** The result is that two negative turns have been introduced into the DNA molecule.

The ability of gyrase to introduce negative supercoils into DNA is critical for bacteria to survive. For this reason, much research has been aimed at identifying drugs that specifically block bacterial gyrase function as a way to cure or alleviate diseases caused by bacteria. Two main classes—quinolones and coumarins—inhibit gyrase and other bacterial topoisomerases, thereby blocking bacterial cell growth. These drugs do not inhibit eukaryotic topoisomerases, which are structurally different from their bacterial counterparts. This finding has been the basis for the production of many drugs with important antibacterial applications. An example is ciprofloxacin (known also by the brand name Cipro), which is used to treat a wide spectrum of bacterial diseases, including anthrax.

10.3 EUKARYOTIC CHROMOSOMES

Eukaryotic species have one or more sets of chromosomes; each set is composed of several different linear chromosomes (refer back to Figure 8.1). Humans, for example, have two sets of 23 chromosomes each, for a total of 46. The total amount of DNA in cells of eukaryotic species is usually much greater than that in bacterial cells. This enables eukaryotic genomes to contain many more genes than their bacterial counterparts. A distinguishing feature of eukaryotic cells is that their chromosomes are located within a separate cellular compartment known as the **nucleus.** To fit within the nucleus, the length of DNA must be compacted by a remarkable amount. As in bacterial chromosomes, this is accomplished by the binding of the DNA to many different cellular proteins. The term **chromatin** is used to describe the DNA-protein complex found within eukaryotic chromosomes. Chromatin is a dynamic structure that can change its shape and composition during the life of a cell.

In this section, we will examine the sizes of eukaryotic genomes and the organization of DNA sequences along the length of eukaryotic chromosomes. We then examine the levels of compaction of eukaryotic chromosomes during different stages of the cell cycle. Our discussion of chromatin compaction in this chapter largely focuses on structural features between DNA and DNA-binding proteins that occur in eukaryotic chromosomes. By comparison, in Chapter 15, you will learn that chromatin is a dynamic structure that can alternate between loose and compact conformations in a way that regulates gene expression.

The Sizes of Eukaryotic Genomes Vary Substantially

Different eukaryotic species vary dramatically in the size of their genomes (**Figure 10.10a**; note that this is a log scale). In many cases, this variation is not related to the complexity of the species.

For example, two closely related species of salamander, *Plethodon richmondi* and *Plethodon larselli*, differ considerably in genome size (**Figure 10.10b, c**). The genome of *P. larselli* is more than twice as large as the genome of *P. richmondi*. However, the genome of *P. larselli* probably doesn't contain more genes. How do we explain the difference in genome size? The additional DNA in *P. larselli* is due to the accumulation of repetitive DNA sequences present in many copies. In some species, these repetitive sequences can accumulate to enormous levels. Such highly repetitive sequences do not encode proteins, and their function remains a matter of controversy and great interest. The structure and significance of repetitive DNA will be discussed later in this chapter.

Eukaryotic Chromosomes Have Many Functionally Important Regions, Including Genes, Origins of Replication, Centromeres, and Telomeres

Each eukaryotic chromosome contains a long, linear DNA molecule (**Figure 10.11**). Three types of regions are required for chromosomal replication and segregation: origins of replication, centromeres, and telomeres. As mentioned previously, origins of replication are chromosomal sites that are necessary to initiate DNA replication. Unlike most bacterial chromosomes, which contain only one origin of replication, eukaryotic chromosomes contain many origins, interspersed approximately every 100,000 bp apart. The function of origins of replication is discussed in greater detail in Chapter 11.

Centromeres are regions that play a role in the proper segregation of chromosomes during mitosis and meiosis. For most species, each eukaryotic chromosome contains a single centromere, which usually appears as a constricted region of a mitotic chromosome. Centromeres function as a site for the formation of kinetochores, which assemble just before and during the very early stages of mitosis and meiosis. The **kinetochore** is composed of a group of cellular proteins that link the centromere to the spindle apparatus during mitosis and meiosis, ensuring the proper segregation of the chromosomes to each daughter cell. In certain yeast species, such as *Saccharomyces cerevisiae*, centromeres have a defined DNA sequence that is about 125 bp in length. This type of centromere is called a point centromere. By comparison, the centromeres found in more complex eukaryotes are much larger and contain many copies of tandemly repeated DNA sequences. (Tandem repeats are discussed later in this chapter). These are called regional centromeres. They can range in size from several thousand bp in length to over one million bp. The repeated DNA sequences within regional centromeres by themselves are not necessary or sufficient to form a functional centromere with a kinetochore. Instead, other biochemical properties are needed to make a functional centromere. For example, a distinctive feature of all eukaryotic centromeres is that histone H3, which is discussed later in this chapter, is replaced with a histone variant called CENP-A. (Histone variants are described in Chapter 15). However, researchers are still trying to identify all of the biochemical properties that distinguish regional centromeres, and understand how these properties are transmitted during cell division.

At the ends of linear chromosomes are found specialized regions known as **telomeres.** Telomeres serve several important functions in the replication and stability of the chromosome. As discussed in Chapter 8, telomeres prevent chromosomal rearrangements such as translocations. In addition, they prevent chromosome shortening in two ways. First, the telomeres protect

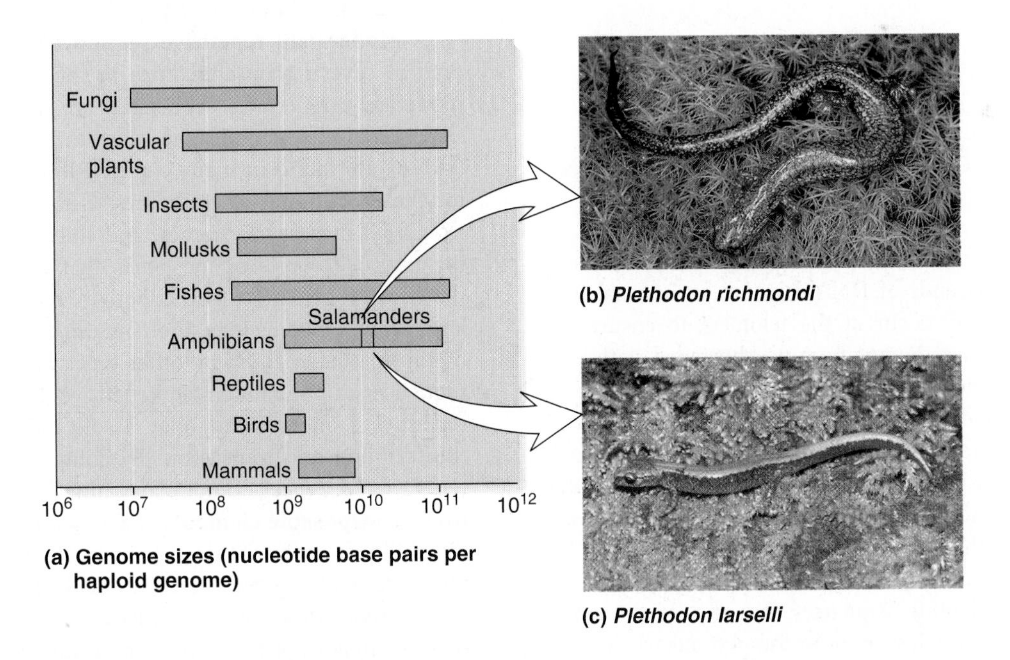

FIGURE 10.10 Haploid genome sizes among groups of eukaryotic species. (a) Ranges of genome sizes among different groups of eukaryotes. Data from T. Ryan Gregory, et al. (2007) Eukaryotic genome size databases. *Nucleic Acids Res.* 35:D332–D338. **(b)** A species of salamander, *Plethodon richmondi*, and **(c)** a close relative, *Plethodon larselli*. The genome of *P. larselli* is over twice as large as that of *P. richmondi*.

Genes→Traits The two species of salamander shown here have very similar traits, even though the genome of *P. larselli* is over twice as large as that of *P. richmondi*. However, the genome of *P. larselli* is not likely to contain more genes. Rather, the additional DNA is due to the accumulation of short repetitive DNA sequences that do not code for genes and are present in many copies.

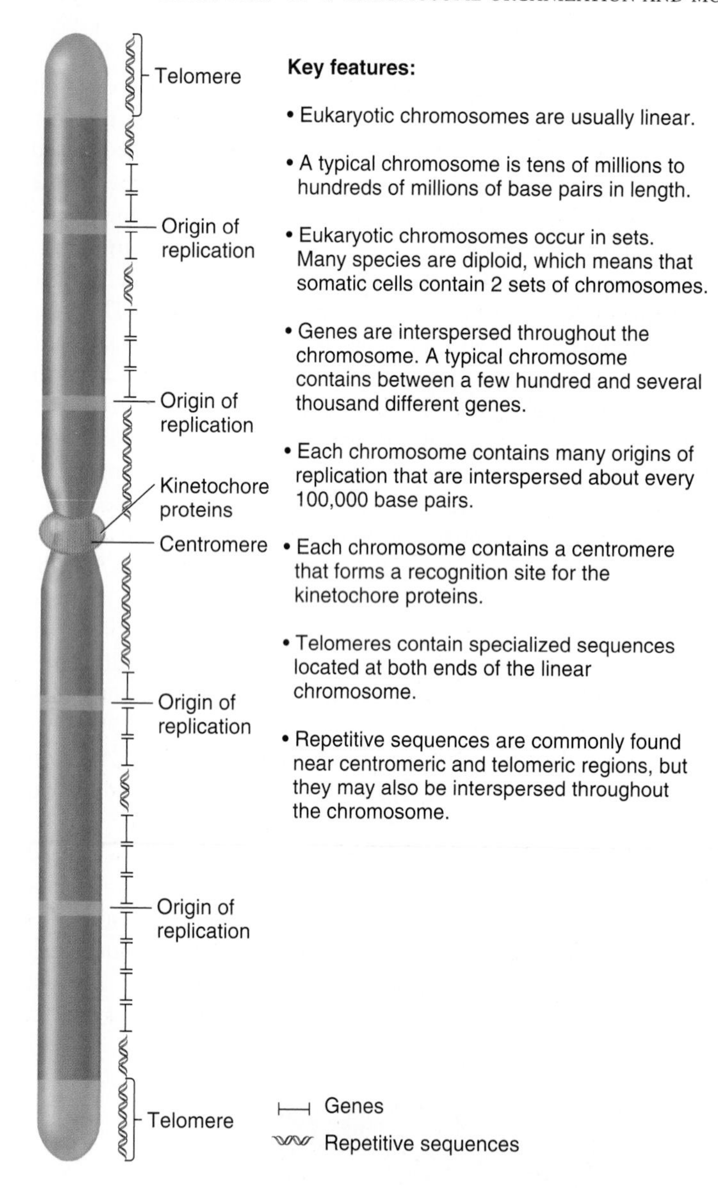

FIGURE 10.11 **Organization of eukaryotic chromosomes.**

chromosomes from digestion via enzymes called exonucleases that recognize the ends of DNA. Second, an unusual form of DNA replication may occur at the telomere to ensure that eukaryotic chromosomes do not become shortened with each round of DNA replication (see Chapter 11).

Genes are located between the centromeric and telomeric regions along the entire eukaryotic chromosome. A single chromosome usually has a few hundred to several thousand different genes. The sequence of a typical eukaryotic gene is several thousand to tens of thousands of base pairs in length. In less complex eukaryotes such as yeast, genes are relatively small and primarily contain nucleotide sequences that encode the amino acid sequences within proteins. In more complex eukaryotes such as mammals and flowering plants, structural genes tend to be much longer due to the presence of **introns**—noncoding intervening sequences. Introns range in size from less than 100 bp to more than 10,000 bp. Therefore, the presence of large introns can greatly increase the lengths of eukaryotic genes.

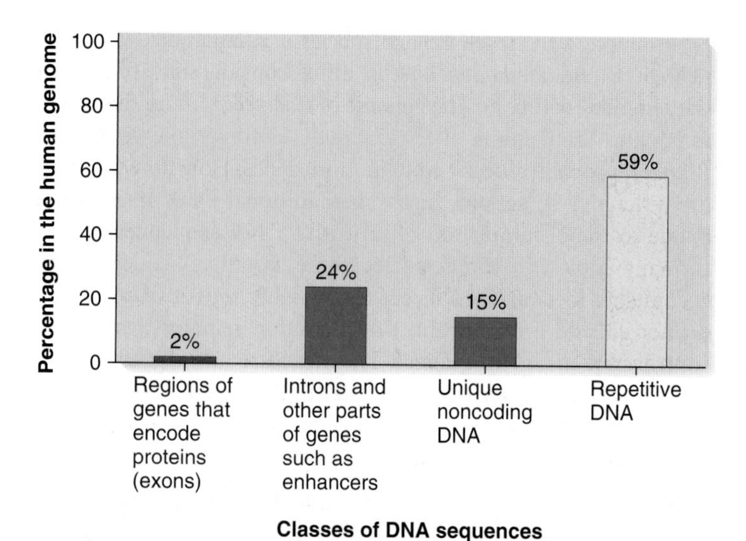

FIGURE 10.12 **Relative amounts of unique and repetitive DNA sequences in the human genome.**

The Genomes of Eukaryotes Contain Sequences That Are Unique, Moderately Repetitive, or Highly Repetitive

The term **sequence complexity** refers to the number of times a particular base sequence appears throughout the genome. Unique or nonrepetitive sequences are those found once or a few times within the genome. Structural genes are typically unique sequences of DNA. The vast majority of proteins in eukaryotic cells are encoded by genes present in one or a few copies. In the case of humans, unique sequences make up roughly 41% of the entire genome (**Figure 10.12**).

Moderately repetitive sequences are found a few hundred to several thousand times in the genome. In a few cases, moderately repetitive sequences are multiple copies of the same gene. For example, the genes that encode ribosomal RNA (rRNA) are found in many copies. Ribosomal RNA is necessary for the functioning of ribosomes. Cells need a large amount of rRNA for making ribosomes, and this is accomplished by having multiple copies of the genes that encode rRNA. Likewise, the histone genes are also found in multiple copies because a large number of histone proteins are needed for the structure of chromatin. In addition, other types of functionally important sequences can be moderately repetitive. For example, moderately repetitive sequences may play a role in the regulation of gene transcription and translation. By comparison, some moderately repetitive sequences do not play a functional role and are derived from **transposable elements (TE)**—segments of DNA that have the ability to move within a genome. This category of repetitive sequences is discussed in greater detail in Chapter 17.

Highly repetitive sequences are found tens of thousands or even millions of times throughout the genome. Each copy of a highly repetitive sequence is relatively short, ranging from a few nucleotides to several hundred in length. A widely studied example is the *Alu* family of sequences found in humans and other primates. The *Alu* sequence is approximately 300 bp long. This sequence derives its name from the observation that it contains a site for cleavage by a restriction enzyme known as *Alu*I. (The function of restriction enzymes is described in Chapter 18.) The *Alu* sequence is present in about 1,000,000 copies in the human genome. It represents about 10% of the total human DNA and occurs approximately every 5000 to 6000 bp! Evolutionary studies suggest that the *Alu* sequence arose 65 million years ago from a section of a single ancestral gene known as the *7SL RNA* gene. Since that time, this gene has become a type of TE called a **retroelement,** which can be transcribed into RNA, copied into DNA, and inserted into the genome. Remarkably, over the course of 65 million years, the *Alu* sequence has been copied and inserted into the human genome to achieve the modern number of about 1,000,000 copies.

Repetitive sequences, like the *Alu* family, are interspersed throughout the genome. However, some moderately and highly repetitive sequences are clustered together in a **tandem array,** also known as tandem repeats. In a tandem array, a very short nucleotide sequence is repeated many times in a row. In *Drosophila*, for example, 19% of the chromosomal DNA is highly repetitive DNA found in tandem arrays. An example is shown here.

```
AATATAATATAATATAATATAATATATAATAT
TTATATTATATTATATTATATTATATATTATA
```

In this particular tandem array, two related sequences, AATAT and AATATAT, are repeated. As mentioned earlier, tandem arrays of short sequences are commonly found in centromeric regions of chromosomes and can be quite long, sometimes more than 1,000,000 bp in length!

What is the functional significance of highly repetitive sequences? Whether highly repetitive sequences have any significant function is controversial. Some experiments in *Drosophila* indicate that highly repetitive sequences may be important in the proper segregation of chromosomes during meiosis. It is not yet clear if highly repetitive DNA plays the same role in other species. The sequences within highly repetitive DNA vary greatly from species to species. Likewise, the amount of highly repetitive DNA can vary a great deal even among closely related species (as noted earlier in Figure 10.10).

Sequence Complexity Can Be Evaluated in a Renaturation Experiment

One approach that has proven useful in understanding genome complexity has come from renaturation studies. These kinds of experiments were first carried out by Roy Britten and David Kohne in 1968. In a renaturation study, the DNA is broken up into pieces containing several hundred base pairs. The double-stranded DNA is then denatured (separated) into single-stranded pieces by heat treatment (**Figure 10.13a**). When the temperature is lowered, the pieces of DNA that are complementary can reassociate, or renature, with each other to form double-stranded molecules.

The rate of renaturation of complementary DNA strands provides a way to distinguish between unique, moderately repetitive, and highly repetitive sequences. For a given category of DNA sequences, the renaturation rate depends on the concentration of its complementary partner. Highly repetitive DNA sequences renature much faster because many copies of the complementary sequences are present. In contrast, unique sequences, such as those

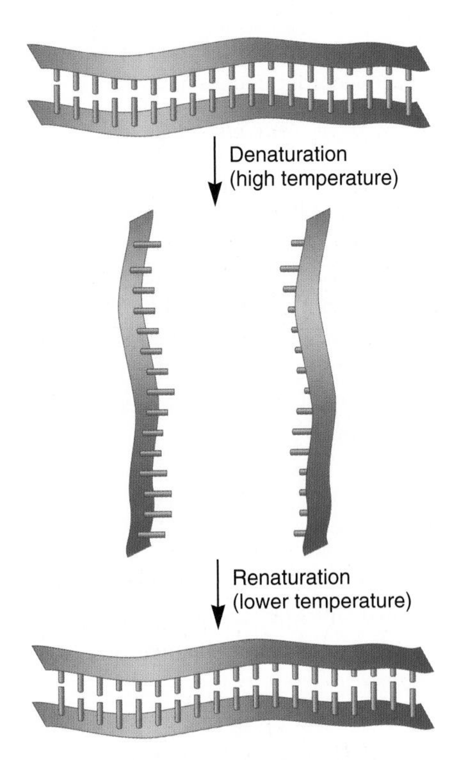

(a) Renaturation of DNA strands

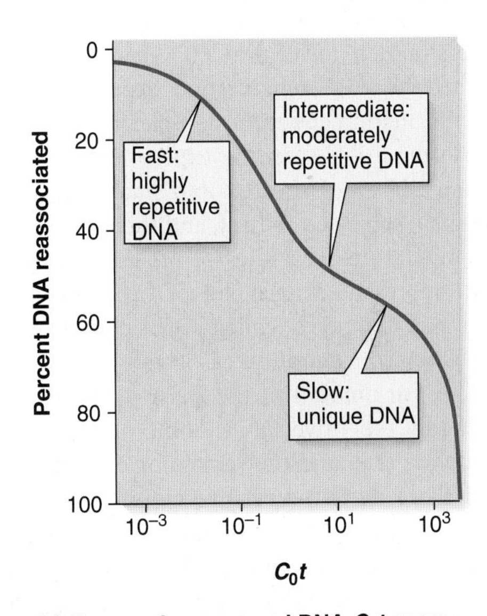

(b) Human chromosomal DNA C_0t curve

FIGURE 10.13 Renaturation and DNA sequence complexity. (a) Denaturation and renaturation (or reassociation) of DNA strands. **(b)** A C_0t curve for human chromosomal DNA.

found within most genes, take longer to renature because of the added time it takes for the unique sequences to find each other.

The renaturation of two DNA strands is a bimolecular reaction that involves the collision of two complementary DNA strands. Its rate is proportional to the product of the concentrations of both strands. If C is the concentration of a single-stranded DNA, then for any DNA derived from a double-stranded fragment, the concentration of one DNA strand (denoted C_1) equals the concentration of its complementary partner (denoted C_2).

Letting $C = C_1 = C_2$, we see the rate of renaturation is represented by the second-order equation

$$-dC/dt = kC^2$$

This is called a second-order equation because the rate depends on the concentration of both reactants—C_1 and C_2. In this case, this product is simplified to C^2 because $C_1 = C_2$.

This equation says that a change in concentration of a single strand ($-dC$) with respect to time (dt) equals a rate constant (k) times the concentration of the single-stranded molecule squared (C^2). This equation can then be integrated to determine how the concentration of the single-stranded DNA changes from time zero to a later time.

$$\frac{C}{C_0} = \frac{1}{1 + k_2 C_0 t}$$

Where

C = the concentration of single-stranded DNA at a later time, t

C_0 = the concentration of single-stranded DNA at time zero

k_2 = the second-order rate constant for renaturation

In this equation, C/C_0 is the fraction of DNA still in the single-stranded form after a given length of time. For example, if C/C_0 equals 0.4 after a certain period of time, 40% of the DNA is still in the single-stranded form, and 60% has renatured into the double-stranded form. A renaturation experiment can provide quantitative information about the complexity of DNA sequences within chromosomal DNA. In the experiment shown in **Figure 10.13b**, human DNA was sheared into small pieces (each about 600 bp in length), subjected to heat, and then allowed to renature at a lower temperature. The rates of renaturation for the DNA pieces can be represented in a plot of C/C_0 versus C_0t, which is referred to as a C_0t curve (called a Cot curve). The term Cot refers to the DNA concentration (C_0) multiplied by the incubation time (t). A fair amount of the DNA renatures very rapidly. This is the highly repetitive DNA. Some of the DNA reassociates at a moderate rate, and the remaining DNA renatures fairly slowly. From these data, the relative amounts of highly repetitive, moderately repetitive, and unique DNA sequences can be approximated. As seen in Figure 10.13b, about 40% of human DNA fragments are unique DNA sequences that renature slowly.

Eukaryotic Chromatin Must Be Compacted to Fit Within the Cell

We now turn our attention to ways that eukaryotic chromosomes are folded to fit within a living cell. A typical eukaryotic chromosome contains a single, linear double-stranded DNA molecule that may be hundreds of millions of base pairs in length. If the DNA from a single set of human chromosomes was stretched from end to end, the length would be over 1 meter! By comparison, most eukaryotic cells are only 10 to 100 μm in diameter, and the cell nucleus is only about 2 to 4 μm in diameter. Therefore, the DNA in a eukaryotic cell must be folded and packaged by a staggering amount to fit inside the nucleus.

The compaction of linear DNA within eukaryotic chromosomes is accomplished through mechanisms that involve interactions between DNA and several different proteins. In recent years, it has become increasingly clear that the proteins bound to chromosomal DNA are subject to change during the life of the cell. These changes in protein composition, in turn, affect the degree of compaction of the chromatin. Chromosomes are very dynamic structures that alternate between tight and loose compaction states in response to changes in protein composition. In the remaining parts of this chapter, we will focus our attention on two issues of chromosome structure. First, we consider how chromosomes are compacted and organized during interphase—the period of the cell cycle that includes the G_1, S, and G_2 phases. Later, we examine the additional compaction that is necessary to produce the highly condensed chromosomes found in M phase.

Linear DNA Wraps Around Histone Proteins to Form Nucleosomes, the Repeating Structural Unit of Chromatin

The repeating structural unit within eukaryotic chromatin is the **nucleosome**—a double-stranded segment of DNA wrapped around an octamer of histone proteins (**Figure 10.14a**). Each octamer contains eight histone subunits, two copies each of four different histone proteins. The DNA lies on the surface and makes 1.65 negative superhelical turns around the histone octamer. The amount of DNA required to wrap around the histone octamer is 146 or 147 bp. At its widest point, a single nucleosome is about 11 nm in diameter.

The chromatin of eukaryotic cells contains a repeating pattern in which the nucleosomes are connected by linker regions of DNA that vary in length from 20 to 100 bp, depending on the species and cell type. It has been suggested that the overall structure of connected nucleosomes resembles beads on a string. This structure shortens the length of the DNA molecule about sevenfold.

Each of the **histone proteins** consists of a globular domain and a flexible, charged amino terminus called an amino terminal tail. Histone proteins are very basic proteins because they contain a large number of positively charged lysine and arginine amino acids. The arginines, in particular, play a major role in binding to the DNA. Arginines within the histone proteins form electrostatic and hydrogen bonding interactions with the phosphate groups along the DNA backbone. The octamer of histones contains two molecules each of four different histone proteins: H2A, H2B, H3, and H4. These are called the core histones. In 1997, Timothy Richmond and colleagues determined the structure of a nucleosome by X-ray crystallography (**Figure 10.14b**).

Another histone, H1, is found in most eukaryotic cells and is called the linker histone. It binds to the DNA in the linker region between nucleosomes and may help to compact adjacent nucleosomes (**Figure 10.14c**). The linker histones are less tightly bound to the DNA than are the core histones. In addition, nonhistone proteins bound to the linker region play a role in the organization and compaction of chromosomes, and their presence may affect the expression of nearby genes.

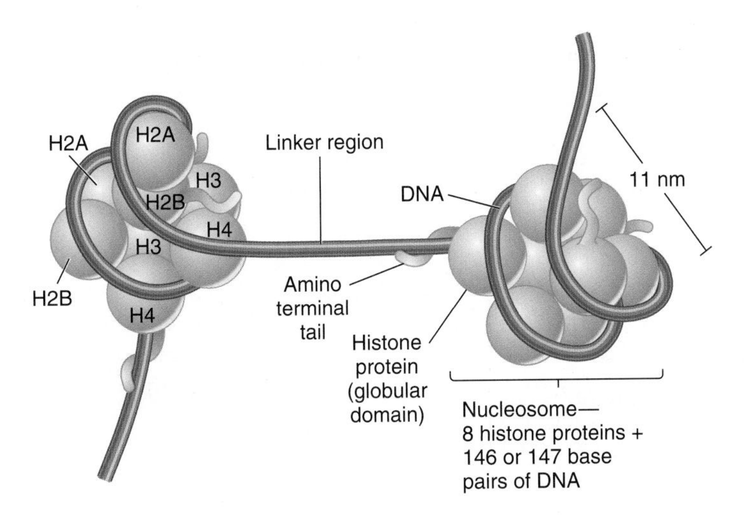

(a) Nucleosomes showing core histone proteins

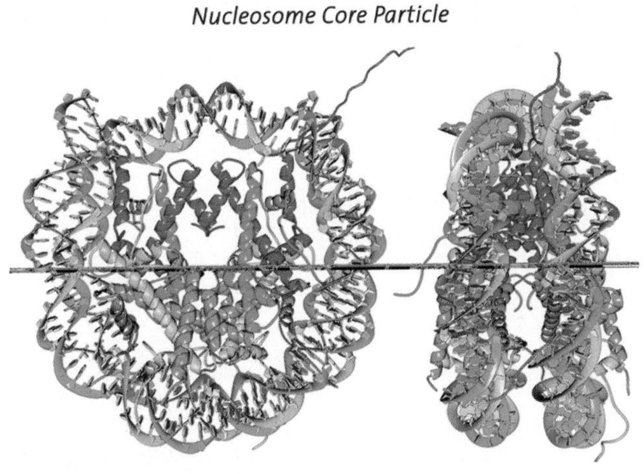

(b) Molecular model for nucleosome structure

(Image courtesy of Timothy J. Richmond. Reprinted by permission from Macmillan Publishers Ltd. Luger K., Mader, AW, Richmond, RK, Sargent, DF, Richmond, TJ [1997] Crystal structure of the nucleosome core particle at 2.8 Å resolution. *Nature 389*:6648, 251–260.) This drawing shows two views of a nucleosome that are at right angles to each other. The horizontal bar passes through the center of the nucleosome in each view.

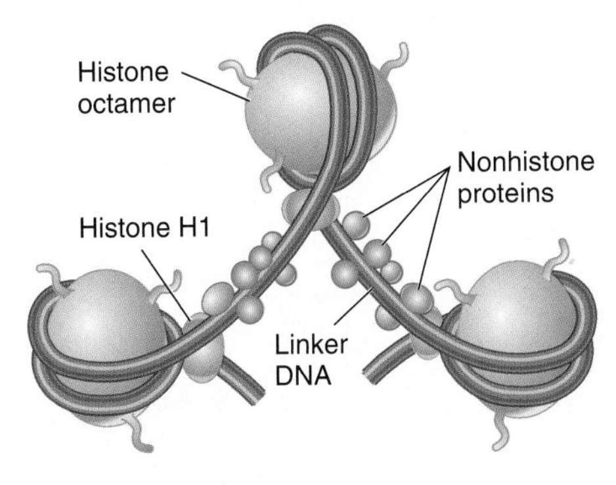

(c) Nucleosomes showing linker histones and nonhistone proteins

FIGURE 10.14 **Nucleosome structure.** **(a)** A nucleosome consists of 146 or 147 bp of DNA wrapped around an octamer of core histone proteins. **(b)** A model for the structure of a nucleosome as determined by X-ray crystallography. **(c)** The linker region of DNA connects adjacent nucleosomes. The linker histone H1 and nonhistone proteins also bind to this linker region.

EXPERIMENT 10A

The Repeating Nucleosome Structure Is Revealed by Digestion of the Linker Region

The model of nucleosome structure was originally proposed by Roger Kornberg in 1974. He based his proposal on several observations. Biochemical experiments had shown that chromatin contains a ratio of one molecule of each of the four core histones (namely, H2A, H2B, H3, and H4) per 100 bp of DNA. Approximately one H1 protein was found per 200 bp of DNA. In addition, purified core histone proteins were observed to bind to each other via specific pairwise interactions. Subsequent X-ray diffraction studies showed that chromatin is composed of a repeating pattern of smaller units. Finally, electron microscopy of chromatin fibers revealed a diameter of approximately 11 nm. Taken together, these observations led Kornberg to propose a model in which the DNA double helix is wrapped around an octamer of core histone proteins. Including the linker region, this involves about 200 bp of DNA.

Markus Noll decided to test Kornberg's model by digesting chromatin with DNase I, an enzyme that cuts the DNA backbone, and then accurately measuring the molecular mass of the DNA fragments by gel electrophoresis. Noll assumed that the linker region of DNA is more accessible to DNase I and, therefore, DNase I is more likely to make cuts in the linker region than in the 146-bp region that is tightly bound to the core histones. If this is correct, incubation with DNase I is expected to make cuts in the linker region and thereby produce DNA pieces approximately 200 bp in length. The size of the DNA fragments may vary somewhat because the linker region is not of constant length and because the cut within the linker region may occur at different sites.

Figure 10.15 describes Noll's experimental protocol. He began with nuclei from rat liver cells and incubated them with low, medium, or high concentrations of DNase I. The DNA was extracted into an aqueous phase and then loaded onto an agarose gel that separated the fragments according to their molecular mass. The DNA fragments within the gel were stained with a UV-sensitive dye, ethidium bromide, which made it possible to view the DNA fragments under UV illumination.

THE HYPOTHESIS

This experiment seeks to test the beads-on-a-string model for chromatin structure. According to this model, DNase I should preferentially cut the DNA in the linker region, thereby producing DNA pieces that are about 200 bp in length.

TESTING THE HYPOTHESIS — FIGURE 10.15 DNase I cuts chromatin into repeating units containing 200 bp of DNA.

Starting material: Nuclei from rat liver cells.

1. Incubate the nuclei with low, medium, and high concentrations of DNase I. The conceptual level illustrates a low DNase I concentration.

2. Extract the DNA. This involves dissolving the nuclear membrane with detergent and extracting with the organic solvent phenol.

3. Load the DNA into a well of an agarose gel and run the gel to separate the DNA pieces according to size. On this gel, also load DNA fragments of known molecular mass (marker lane).

4. Visualize the DNA fragments by staining the DNA with ethidium bromide, a dye that binds to DNA and is fluorescent when excited by UV light.

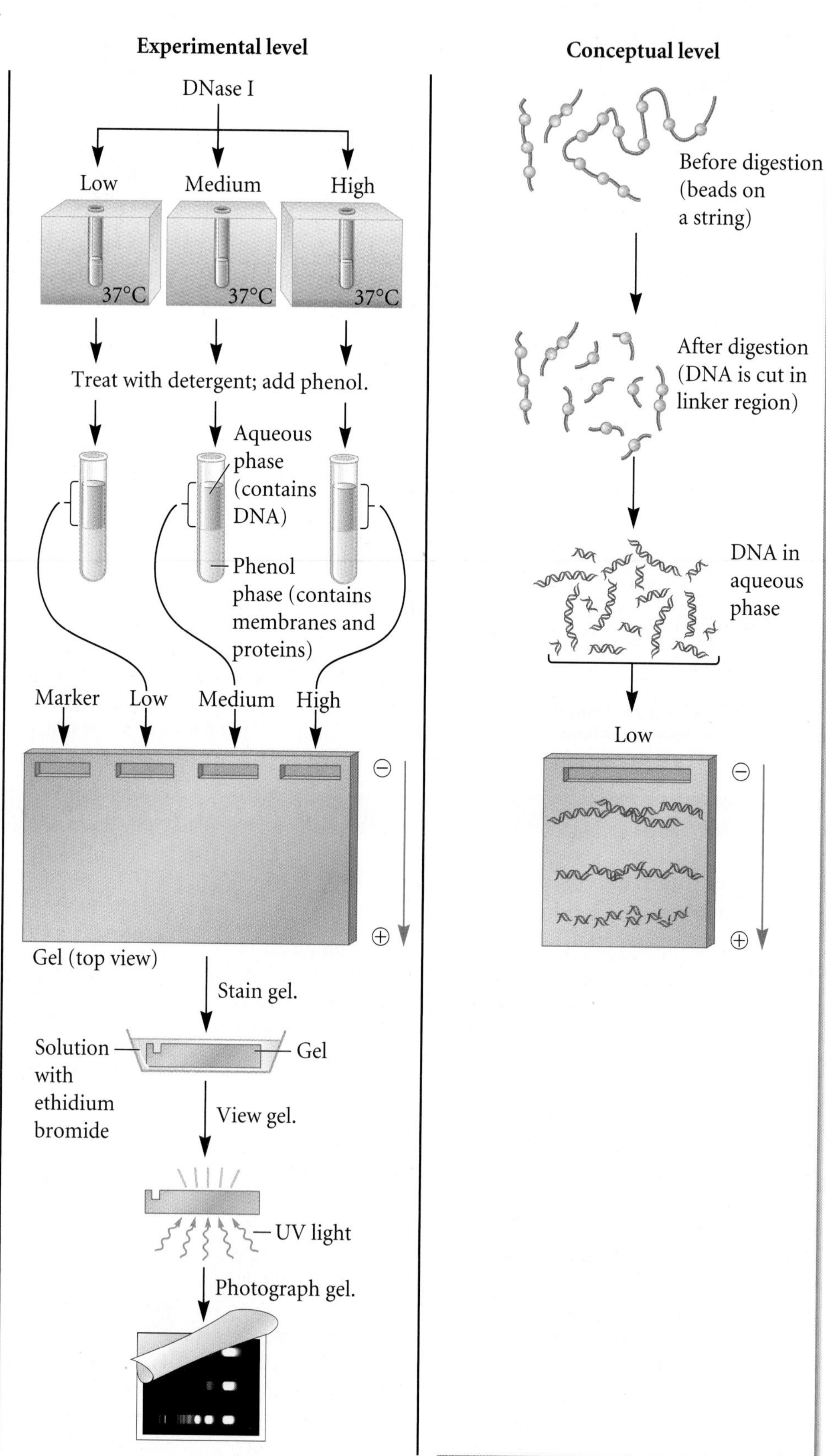

THE DATA

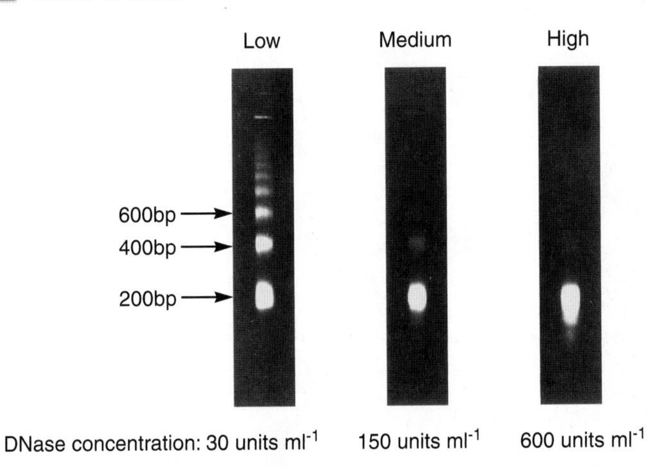

(Reprinted by permission from Macmillan Publishers Ltd. Noll M [1974] Subunit structure of chromatin. *Nature. 251*:249–251.)

INTERPRETING THE DATA

As shown in the data of Figure 10.15, at high DNase I concentrations, the entire sample of chromosomal DNA was digested into fragments of approximately 200 bp in length. This result is predicted by the beads-on-a-string model. Furthermore, at lower DNase I concentrations, longer pieces were observed, and these were in multiples of 200 bp (400, 600, etc.). How do we explain these longer pieces? They occurred because occasional linker regions remained uncut at lower DNase I concentrations. For example, if one linker region was not cut, a DNA piece would contain two nucleosomes and be 400 bp in length. If two consecutive linker regions were not cut, this would produce a piece with three nucleosomes containing about 600 bp of DNA. Taken together, these results strongly supported the nucleosome model for chromatin structure.

A self-help quiz involving this experiment can be found at the Online Learning Center.

Nucleosomes Become Closely Associated to Form a 30-nm Fiber

In eukaryotic chromatin, nucleosomes associate with each other to form a more compact structure that is 30 nm in diameter. Evidence for the packaging of nucleosomes was obtained in the microscopy studies of Fritz Thoma in 1977. Chromatin samples were treated with a resin that removed histone H1, but the removal depended on the salt concentration. A moderate salt solution (100 mM NaCl) removed H1, but a solution with no added NaCl did not remove H1. These samples were then observed with an electron microscope. At moderate salt concentrations (**Figure 10.16a**), the chromatin exhibited the classic beads-on-a-string morphology. Without added NaCl (when H1 is expected to remain bound to the DNA), these "beads" associated with each other into a more compact conformation (**Figure 10.16b**). These results suggest that the nucleosomes are packaged into a more compact unit and that H1 has a role in the packaging and compaction of nucleosomes. However, the precise role of H1 in chromatin compaction remains unclear. Recent data suggest that the core histones also play a key role in the compaction and relaxation of chromatin.

The experiment of Figure 10.16 and other experiments have established that nucleosome units are organized into a more compact structure that is 30 nm in diameter, known as the **30-nm fiber** (**Figure 10.17a**). The 30-nm fiber shortens the total length of DNA another sevenfold. The structure of the 30-nm fiber has proven difficult to determine, because the conformation of the DNA may be substantially altered when it is extracted from living cells. Most models for the 30-nm fiber fall into two main classes. The solenoid model suggests a helical structure in which contact between nucleosomes produces a symmetrically compact structure within the 30-nm fiber (**Figure 10.17b**). This type of model is still favored by some researchers in the field. However, experimental data also suggest that the 30-nm fiber may not form such a regular structure. Instead, an alternative zigzag model, advocated

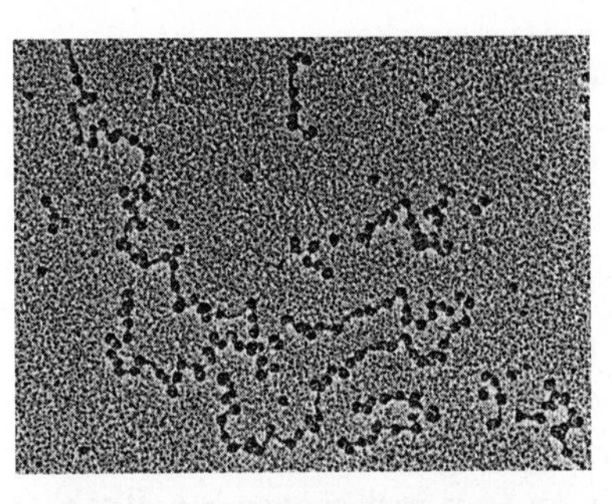

(a) H1 histone not bound—beads on a string

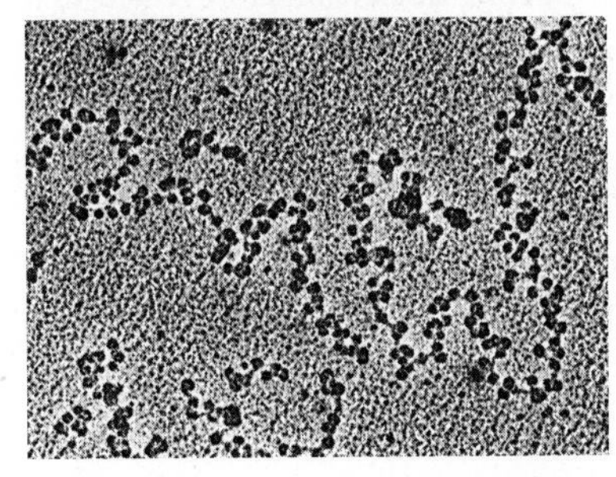

(b) H1 histone bound to linker region—nucleosomes more compact

FIGURE 10.16 The nucleosome structure of eukaryotic chromatin as viewed by electron microscopy. The chromatin in **(a)** has been treated with moderate salt concentrations to remove the linker histone H1. It exhibits the classic beads-on-a-string morphology. The chromatin in **(b)** has been incubated without added NaCl and shows a more compact morphology.

by Rachel Horowitz, Christopher Woodcock, and others, is based on techniques such as cryoelectron microscopy (electron microscopy at low temperature). According to the zigzag model, linker regions within the 30-nm structure are variably bent and twisted, and little face-to-face contact occurs between nucleosomes (**Figure 10.17c**). At this level of compaction, the overall picture of chromatin that emerges is an irregular, fluctuating, three-dimensional zigzag structure with stable nucleosome units

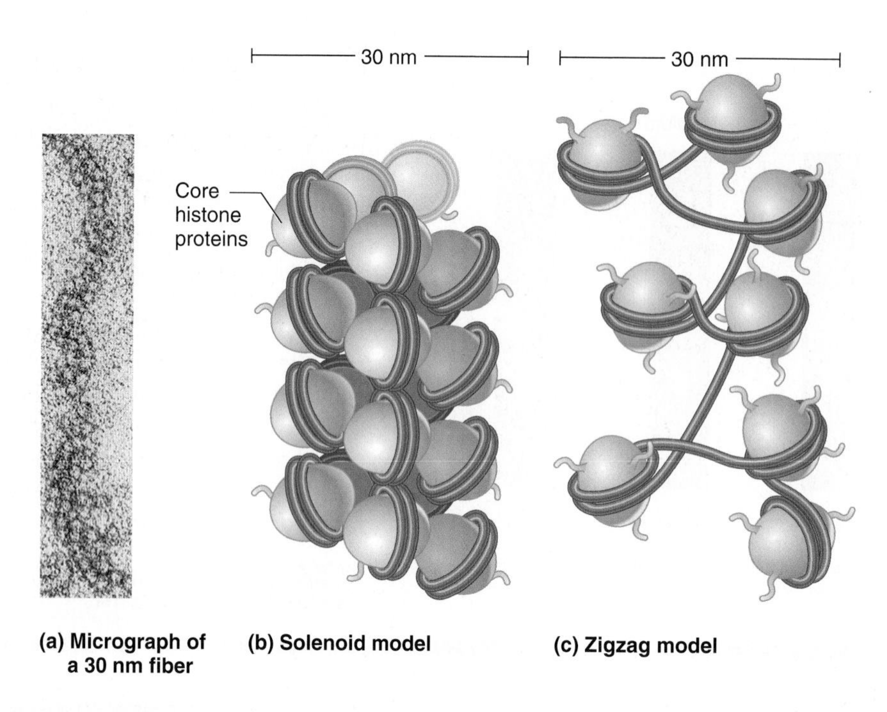

FIGURE 10.17 **The 30-nm fiber.** **(a)** A photomicrograph of the 30-nm fiber. **(b)** In the solenoid model, the nucleosomes are packed in a spiral configuration. **(c)** In the zigzag model, the linker DNA forms a more irregular structure, and less contact occurs between adjacent nucleosomes. The zigzag model is consistent with more recent data regarding chromatin conformation.

connected by deformable linker regions. In 2005, Timothy Richmond and colleagues were the first to solve the crystal structure of a segment of DNA containing multiple nucleosomes, in this case four. The structure with four nucleosomes revealed that the linker DNA zigzags back and forth between each nucleosome, a feature consistent with the zigzag model.

Chromosomes Are Further Compacted by Anchoring the 30-nm Fiber into Radial Loop Domains Along the Nuclear Matrix

Thus far, we have examined two mechanisms that compact eukaryotic DNA. These involve the wrapping of DNA within nucleosomes and the arrangement of nucleosomes to form a 30-nm fiber. Taken together, these two events shorten the DNA nearly 50-fold. A third level of compaction involves interactions between the 30-nm fibers and a filamentous network of proteins in the nucleus called the **nuclear matrix.** As shown in **Figure 10.18a**, the nuclear matrix consists of two parts. The **nuclear lamina** is a collection of fibers that line the inner nuclear membrane. These fibers are composed of intermediate filament proteins. The second part is an **internal nuclear matrix,** which is connected to the nuclear lamina and fills the interior of the nucleus. The internal nuclear matrix, whose structure and functional role remain controversial, is hypothesized to be an intricate fine network of irregular protein fibers plus many other proteins that bind to these fibers. Even when the chromatin is extracted from the nucleus, the internal nuclear matrix may remain intact (**Figure 10.18b** and **c**). However, the matrix should not be considered a static structure. Research indicates that the protein composition of the internal nuclear matrix is very dynamic and complex, consisting of dozens or perhaps hundreds of different proteins. The protein composition varies depending on species, cell type, and environmental conditions. This complexity has made it difficult to propose models regarding its overall organization. Further research is necessary to understand the structure and dynamic nature of the internal nuclear matrix.

The proteins of the nuclear matrix are involved in compacting the DNA into **radial loop domains,** similar to those described for the bacterial chromosome. During interphase, chromatin is organized into loops, often 25,000 to 200,000 bp in size, which are anchored to the nuclear matrix. The chromosomal DNA of eukaryotic species contains sequences called **matrix-attachment regions (MARs)** or **scaffold-attachment regions (SARs),** which are interspersed at regular intervals throughout the genome. The MARs bind to specific proteins in the nuclear matrix, thus forming chromosomal loops (**Figure 10.18d**).

Why is the attachment of radial loops to the nuclear matrix important? In addition to compaction, the nuclear matrix serves to organize the chromosomes within the nucleus. Each chromosome in the cell nucleus is located in a discrete **chromosome territory.** As shown in studies by Thomas Cremer, Christoph Cremer, and others, these territories can be viewed when interphase cells are exposed to multiple fluorescent molecules that recognize specific sequences on particular chromosomes. **Figure 10.19** illustrates an experiment in which chicken cells were exposed to a mixture of probes that recognize multiple sites along several of the larger chromosomes found in this species (*Gallus gallus*). Figure 10.19a shows the chromosomes in metaphase. The probes label each type of metaphase chromosome with a different color. Figure 10.19b shows the use of the same probes during interphase, when the chromosomes are less condensed and found

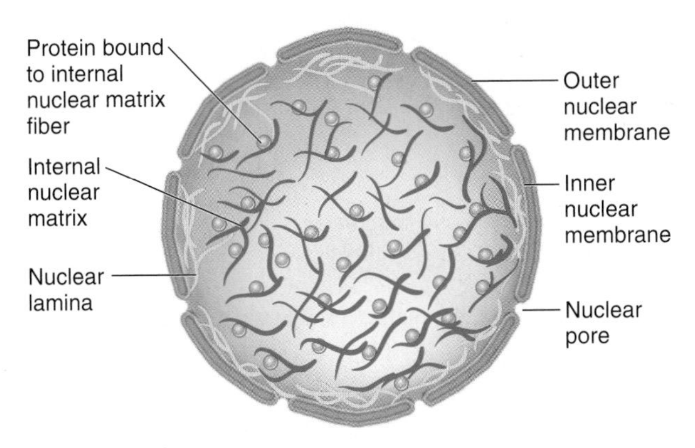

(a) Proteins that form the nuclear matrix

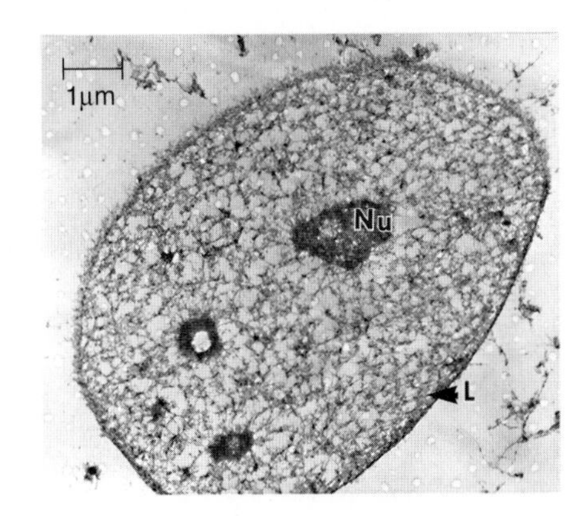

(b) Micrograph of nucleus with chromatin removed

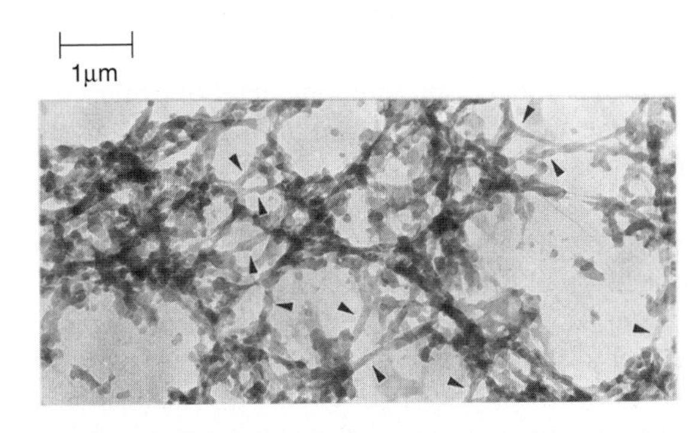

(c) Micrograph showing a close-up of nuclear matrix

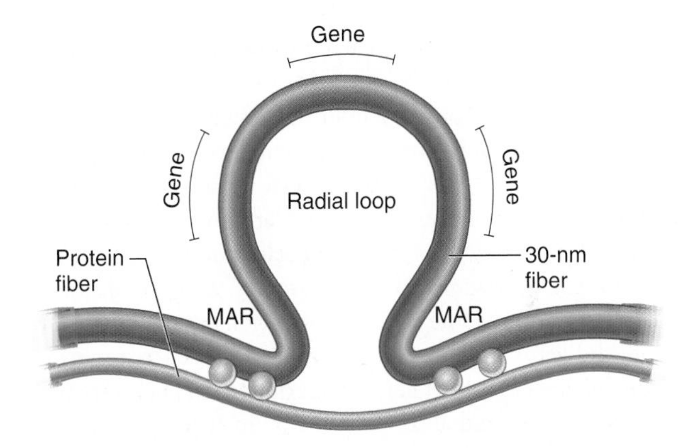

(d) Radial loop bound to a nuclear matrix fiber

FIGURE 10.18 **Structure of the nuclear matrix.** **(a)** This schematic drawing shows the arrangement of the matrix within a cell nucleus. The nuclear lamina (depicted as yellow filaments) is a collection of fibrous proteins that line the inner nuclear membrane. The internal nuclear matrix is composed of protein filaments (depicted in green) that are interconnected. These fibers also have many other proteins associated with them (depicted in orange). **(b)** An electron micrograph of the nuclear matrix during interphase after the chromatin has been removed. The nucleolus is labeled Nu, and the lamina is labeled L. **(c)** At higher magnification, the protein fibers are more easily seen (arrowheads point at fibers). **(d)** The matrix-attachment regions (MARs), which contain a high percentage of A and T bases, bind to the nuclear matrix and create radial loops. This causes a greater compaction of eukaryotic chromosomal DNA.

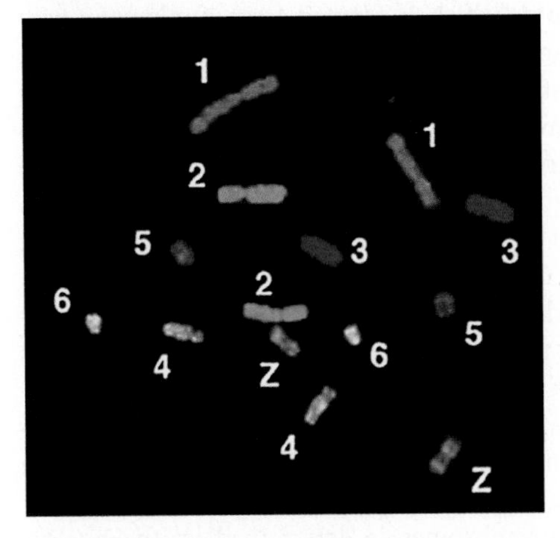

(a) Metaphase chromosomes

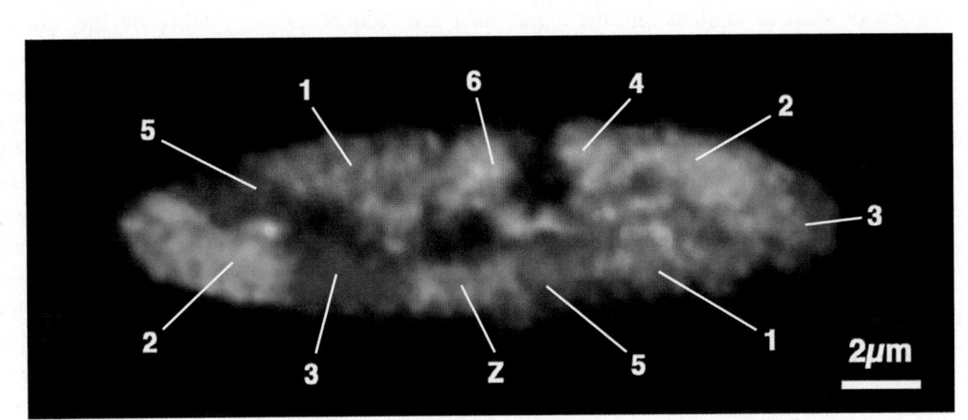

(b) Chromosomes in the cell nucleus during interphase

FIGURE 10.19 **Chromosome territories in the cell nucleus.** **(a)** Several metaphase chromosomes from the chicken were labeled with chromosome-specific probes. Each of seven types of chicken chromosomes (i.e., 1, 2, 3, 4, 5, 6, and Z) is labeled a different color. **(b)** The same probes were used to label interphase chromosomes in the cell nucleus. Each of these chromosomes occupies its own distinct, nonoverlapping territory within the cell nucleus. (Note: Chicken cells are diploid, with two copies of each chromosome.)

(Reprinted by permission from Macmillan Publishers Ltd. Cremer, T. & Cremer, C. [2001] Chromosome territories, nuclear architecture and gene regulation in mammalian cells. *Nature Rev Genet 2*:4, 292–301.)

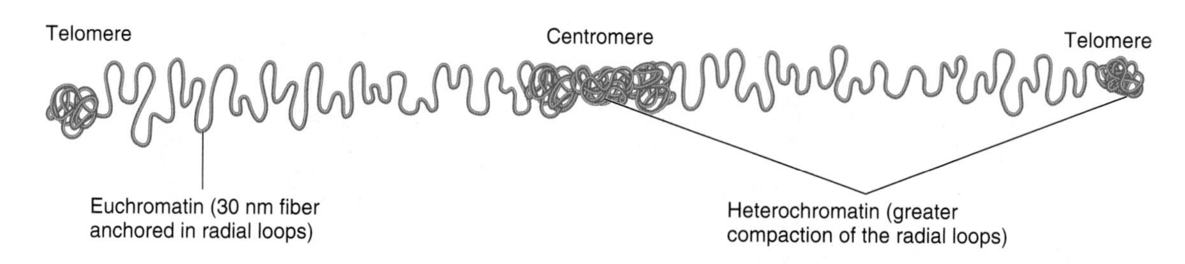

FIGURE 10.20 **Chromatin structure during interphase.** Heterochromatic regions are more highly condensed and tend to be localized in centromeric and telomeric regions.

in the cell nucleus. As seen here, each chromosome occupies its own distinct territory.

Before ending the topic of interphase chromosome compaction, let's consider how the compaction level of interphase chromosomes may vary. This variability can be seen with a light microscope and was first observed by the German cytologist Emil Heitz in 1928. He coined the term **heterochromatin** to describe the tightly compacted regions of chromosomes. In general, these regions of the chromosome are transcriptionally inactive. By comparison, the less condensed regions, known as **euchromatin,** reflect areas that are capable of gene transcription. In euchromatin, the 30-nm fiber forms radial loop domains. In heterochromatin, these radial loop domains become compacted even further.

Figure 10.20 illustrates the distribution of euchromatin and heterochromatin in a typical eukaryotic chromosome during interphase. The chromosome contains regions of both heterochromatin and euchromatin. Heterochromatin is most abundant in the centromeric regions of the chromosomes and, to a lesser extent, in the telomeric regions. The term **constitutive heterochromatin** refers to chromosomal regions that are always heterochromatic and permanently inactive with regard to transcription. Constitutive heterochromatin usually contains highly repetitive DNA sequences, such as tandem repeats, rather than gene sequences. **Facultative heterochromatin** refers to chromatin that can occasionally interconvert between heterochromatin and euchromatin. An example of facultative heterochromatin occurs in female mammals when one of the two X chromosomes is converted to a heterochromatic Barr body. As discussed in Chapter 5, most of the genes on the Barr body are transcriptionally inactive. The conversion of one X chromosome to heterochromatin occurs during embryonic development in the somatic cells of the body.

Condensin and Cohesin Promote the Formation of Metaphase Chromosomes

When cells prepare to divide, the chromosomes become even more condensed. This aids in their proper sorting during metaphase. **Figure 10.21** illustrates the levels of compaction that lead to a metaphase chromosome. During interphase, most of the chromosomal DNA is found in euchromatin, in which the 30-nm fibers form radial loop domains that are attached to a protein scaffold. The average distance that loops radiate from the protein scaffold is approximately 300 nm. This structure can be further compacted via additional folding of the radial loop domains and protein scaffold. This additional level of compaction greatly shortens the overall length of a chromosome and produces a diameter of approximately 700 nm, which is the compaction level found in heterochromatin. During interphase, most chromosomal regions are euchromatic, and some localized regions, such as those near centromeres, are heterochromatic.

As cells enter M phase, the level of compaction changes dramatically. By the end of prophase, sister chromatids are entirely heterochromatic. Two parallel chromatids have a larger diameter of approximately 1400 nm but a much shorter length compared with interphase chromosomes. These highly condensed metaphase chromosomes undergo little gene transcription because it is difficult for transcription proteins to gain access to the compacted DNA. Therefore, most transcriptional activity ceases during M phase, although a few specific genes may be transcribed. M phase is usually a short period of the cell cycle.

In highly condensed chromosomes, such as those found in metaphase, the radial loops are highly compacted and remain anchored to a **scaffold,** which is formed from nonhistone proteins of the nuclear matrix. Experimentally, researchers can delineate the nonhistone proteins of the scaffold that hold the loops in place. **Figure 10.22a** shows a human metaphase chromosome. In this condition, the radial loops of DNA are in a very compact configuration. If this chromosome is treated with a high concentration of salt to remove both the core and linker histones, the highly compact configuration is lost, but the bottoms of the elongated loops remain attached to the scaffold composed of nonhistone proteins. In **Figure 10.22b**, an arrow points to an elongated DNA strand emanating from the darkly staining scaffold. Remarkably, the scaffold retains the shape of the original metaphase chromosome even though the DNA strands have become greatly elongated. These results illustrate that the structure of metaphase chromosomes is determined by the nuclear matrix proteins, which form the scaffold, and by the histones, which are needed to compact the radial loops.

Researchers are trying to understand the steps that lead to the formation and organization of metaphase chromosomes. During the past several years, studies in yeast and frog oocytes

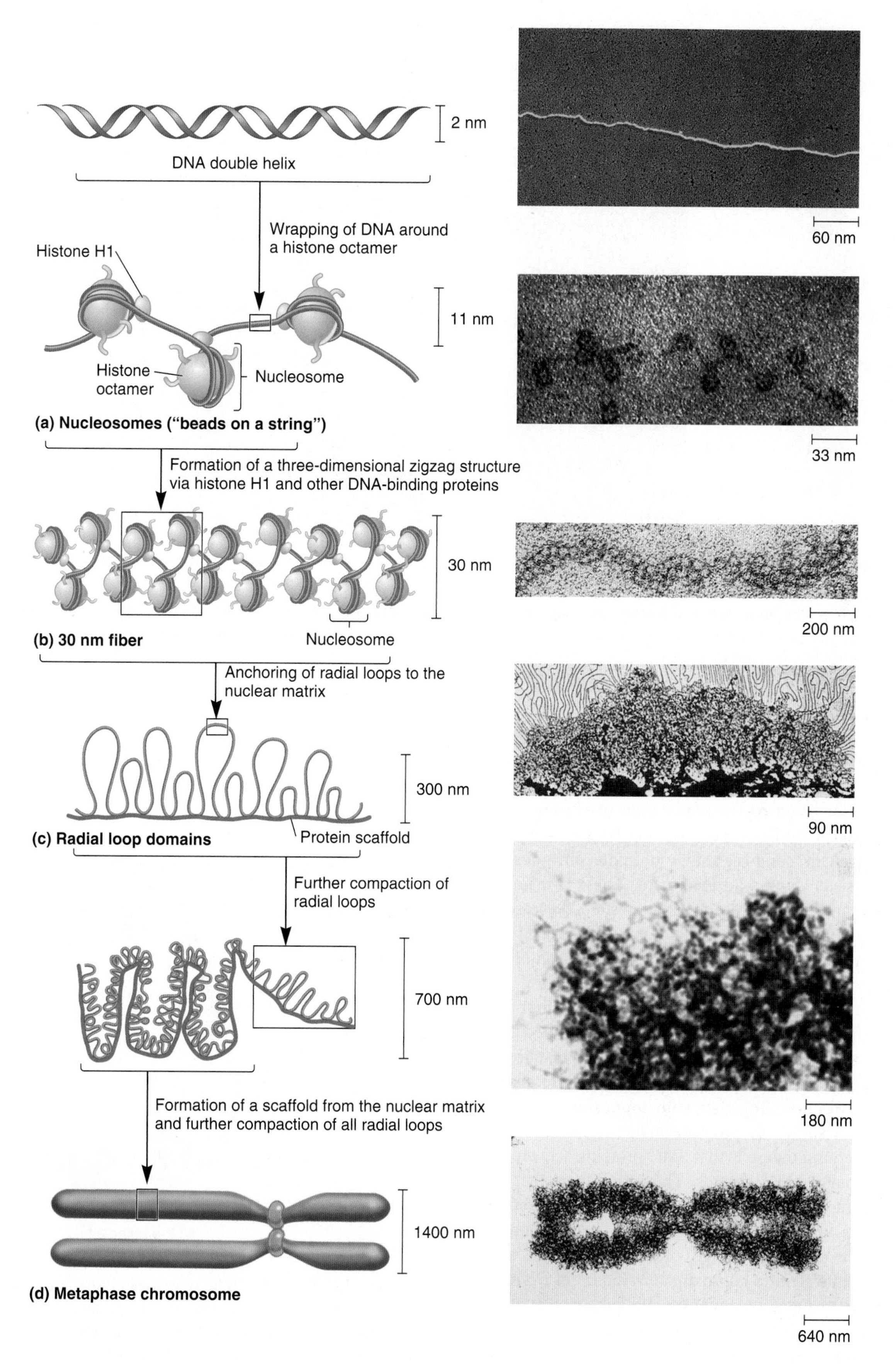

FIGURE 10.21 **The steps in eukaryotic chromosomal compaction leading to the metaphase chromosome.**

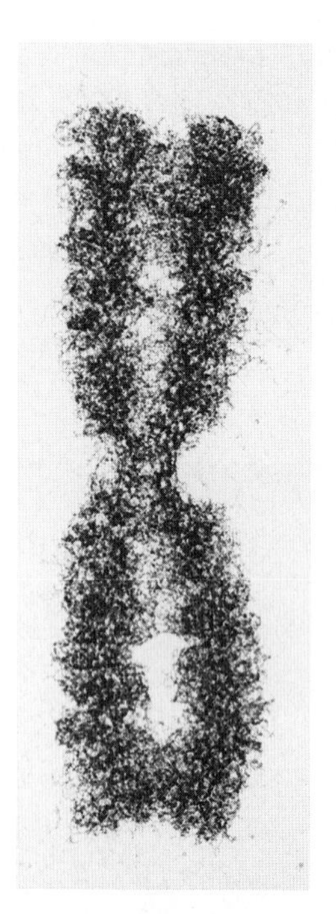

(a) Metaphase chromosome

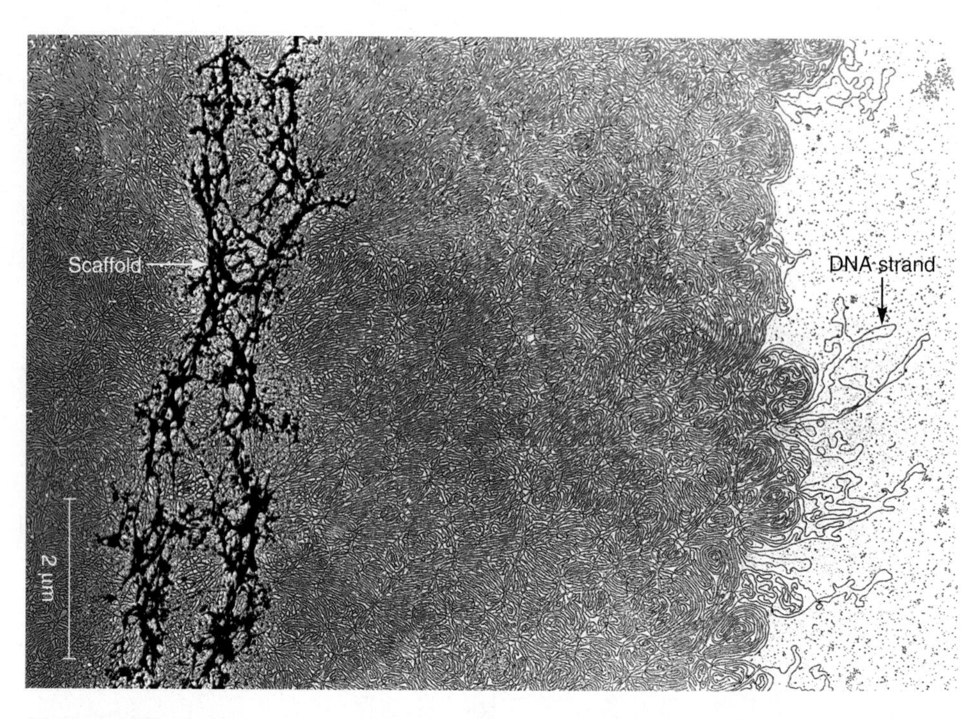

(b) Metaphase chromosome treated with high salt to remove histone proteins

FIGURE 10.22 The importance of histone proteins and scaffolding proteins in the compaction of eukaryotic chromosomes. **(a)** A metaphase chromosome. **(b)** A metaphase chromosome following treatment with high salt concentration to remove the histone proteins. The arrow on the left points to the scaffold (composed of nonhistone proteins), which anchors the bases of the radial loops. The arrow on the right points to an elongated strand of DNA.

have been aimed at the identification of proteins that promote the conversion of interphase chromosomes into metaphase chromosomes. In yeast, mutants have been characterized that have alterations in the condensation or the segregation of chromosomes. Similarly, biochemical studies using frog oocytes resulted in the purification of protein complexes that promote chromosomal condensation or sister chromatid alignment. These two lines of independent research produced the same results. Researchers found that cells contain two multiprotein complexes called **condensin** and **cohesin,** which play a critical role in chromosomal condensation and sister chromatid alignment, respectively.

Condensin and cohesin are two completely distinct complexes, but both contain a category of proteins called **SMC proteins.** SMC stands for structural maintenance of chromosomes. These proteins use energy from ATP to catalyze changes in chromosome structure. Together with topoisomerases, SMC proteins have been shown to promote major changes in DNA structure. An emerging theme is that SMC proteins actively fold, tether, and manipulate DNA strands. They are dimers that have a V-shaped structure. The monomers, which are connected at a hinge region, have two long coiled arms with a head region that binds ATP (**Figure 10.23**). The length of each monomer is about 50 nm, which is equivalent to approximately 150 bp of DNA.

As their names suggest, condensin and cohesin play different roles in metaphase chromosome structure. Prior to M phase, condensin is found outside the nucleus (**Figure 10.24**). However, as M phase begins, condensin is observed to coat the individual chromatids as euchromatin is converted into heterochromatin.

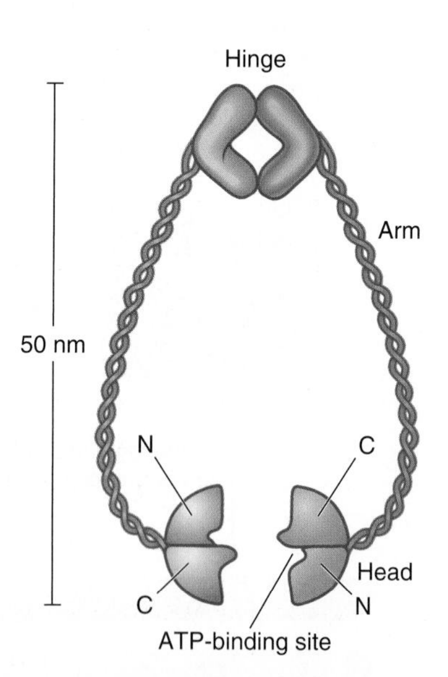

FIGURE 10.23 The structure of SMC proteins. This figure shows the generalized structure of SMC proteins, which are dimers consisting of hinge, arm, and head regions. The head regions bind and hydrolyze ATP. Condensin and cohesin have additional protein subunits not shown here. N indicates the amino terminus. C indicates the carboxyl terminus.

The role of condensin in the compaction process is not well understood. Although condensin is often implicated in the process of chromosomal condensation, researchers have been able to deplete condensin from actively dividing cells, and the chromosomes are still able to condense. However, such condensed chromosomes show abnormalities in their ability to separate from each other during cell division. These results suggest that condensin is important in the proper organization of highly condensed chromosomes, such as those found during metaphase.

In comparison, the function of cohesin is to promote the binding (i.e., cohesion) between sister chromatids. After S phase and until the middle of prophase, sister chromatids remain attached to each other along their length. As shown in **Figure 10.25**, this attachment is promoted by cohesin, which is found along the entire length of each chromatid. In certain species, such as mammals, cohesins located along the chromosome arms are released during prophase, which allows the arms to separate. However, some cohesins remain attached, primarily to the centromeric regions, leaving the centromeric region as the main linkage before anaphase. At anaphase, the cohesins bound to the centromere are rapidly degraded by a protease aptly named separase, thereby allowing sister chromatid separation.

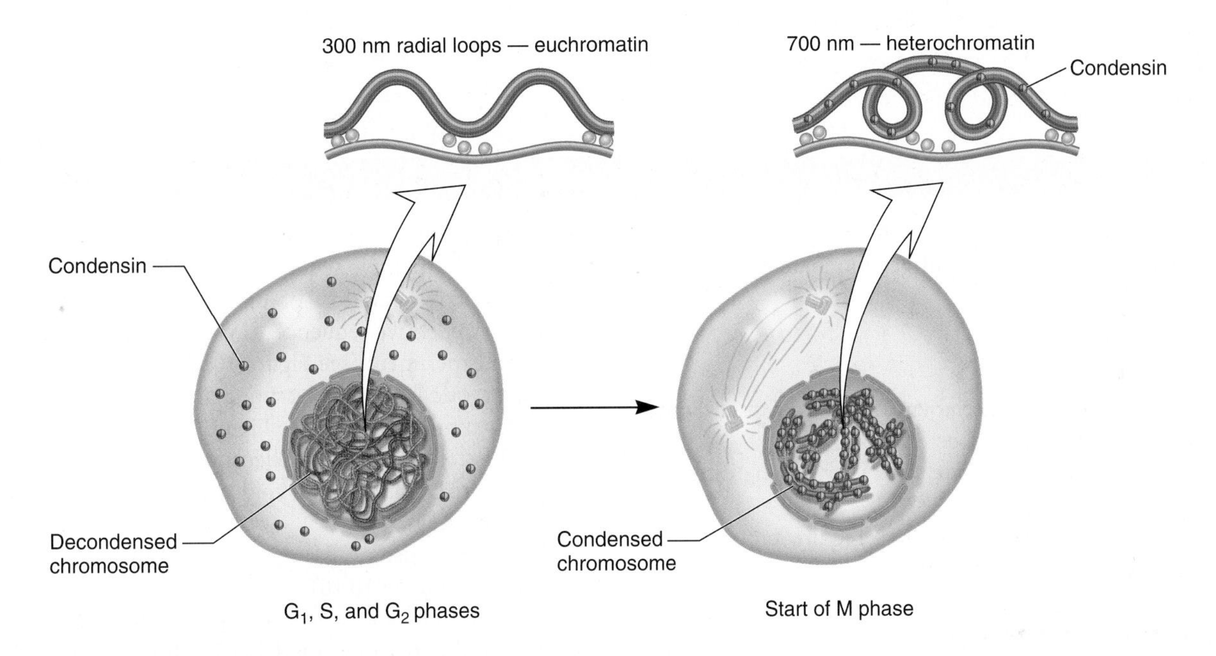

FIGURE 10.24 The localization of condensin during interphase and the start of M phase. During interphase (G_1, S, and G_2), most of the condensin protein is found outside the nucleus. The interphase chromosomes are largely euchromatic. At the start of M phase, condensin travels into the nucleus and binds to the chromosomes, which become heterochromatic due to a greater compaction of the radial loop domains.

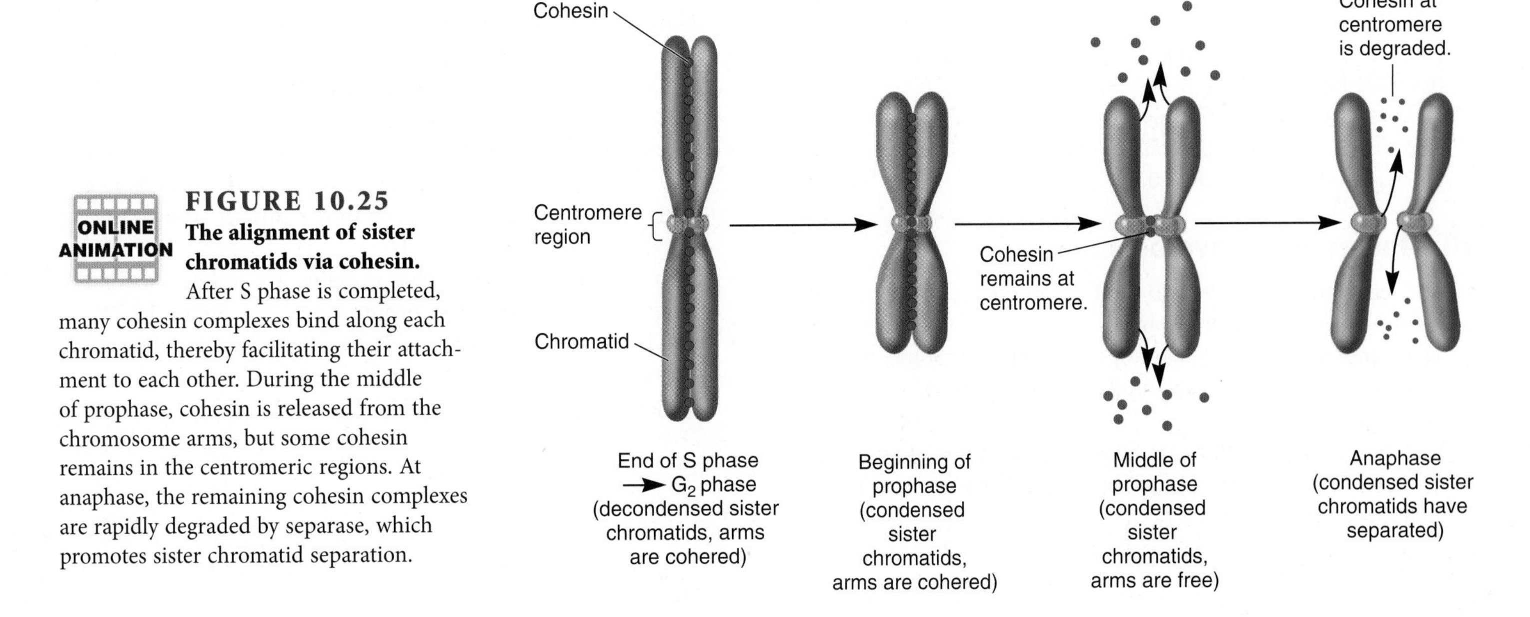

FIGURE 10.25 The alignment of sister chromatids via cohesin. After S phase is completed, many cohesin complexes bind along each chromatid, thereby facilitating their attachment to each other. During the middle of prophase, cohesin is released from the chromosome arms, but some cohesin remains in the centromeric regions. At anaphase, the remaining cohesin complexes are rapidly degraded by separase, which promotes sister chromatid separation.

KEY TERMS

Page 247. chromosomes, genome, viruses, bacteriophages, host cells, host range
Page 248. viral genome
Page 249. nucleoid, structural genes, intergenic regions, origin of replication, repetitive sequences
Page 250. loop domains, DNA supercoiling, topoisomers
Page 251. DNA gyrase, topoisomerase I
Page 252. nucleus, chromatin
Page 253. centromeres, kinetochore, telomeres
Page 254. introns, sequence complexity, moderately repetitive sequences, transposable elements (TEs), highly repetitive sequences
Page 255. retroelement, tandem array
Page 256. nucleosome, histone proteins
Page 259. 30-nm fiber
Page 260. nuclear matrix, nuclear lamina, internal nuclear matrix, radial loop domains, matrix-attachment regions (MARs), scaffold-attachment regions (SARs), chromosome territory
Page 262. heterochromatin, euchromatin, constitutive heterochromatin, facultative heterochromatin, scaffold
Page 264. condensin, cohesin, SMC proteins

CHAPTER SUMMARY

- Chromosomes contain the genetic material, which is DNA. A genome refers to a complete set of genetic material in a particular cellular compartment.

10.1 Viral Genomes

- A virus contains genetic material enclosed in a capsid. Some viruses also have an envelope (see Figure 10.1).
- The genetic material and genome sizes vary among different types of viruses (see Table 10.1).
- Some viruses self-assemble, but others require proteins that direct their assembly (see Figure 10.2).

10.2 Bacterial Chromosomes

- Bacterial chromosomes are found in the nucleoid region of a bacterial cell. They are typically circular and carry an origin of replication and a few thousand genes (see Figures 10.3, 10.4).
- Bacterial chromosomes are made more compact by the formation of loop domains and DNA supercoiling (see Figures 10.5, 10.6, 10.7).
- Negative DNA supercoiling can promote DNA strand separation (see Figure 10.8).
- DNA gyrase (topoisomerase II) is a bacterial enzyme that introduces negative supercoils. Topoisomerase I relaxes negative supercoils (see Figure 10.9).

10.3 Eukaryotic Chromosomes

- The term chromatin refers to the DNA-protein complex found within eukaryotic chromosomes.
- The genome sizes of eukaryotes vary greatly. Some of this variation is due to the accumulation of repetitive sequences (see Figure 10.10).
- Eukaryotic chromosomes are usually linear and contain a centromere, telomeres, multiple origins of replication, and many genes (see Figure 10.11).
- The human genome contains about 41% unique sequences and 59% repetitive sequences (see Figure 10.12).
- The relative amounts of unique, moderately repetitive, and highly repetitive sequences can be determined from a renaturation experiment (see Figure 10.13).
- Eukaryotic DNA wraps around an octamer of histone proteins to form nucleosomes (see Figure 10.14).
- Noll tested Kornberg's nucleosome model by digesting eukaryotic chromatin with varying concentrations of DNase I (see Figure 10.15).
- The linker histone, H1, plays a role in nucleosome compaction (see Figure 10.16).
- Nucleosomes are further compacted to form a 30-nm fiber. Solenoid and zigzag models have been proposed (see Figure 10.17).
- Chromatin is further compacted by the attachment of 30-nm fibers to protein filaments to form radial loop domains (see Figure 10.18).
- With the cell nucleus, each eukaryotic chromosome occupies its own unique chromosome territory (see Figure 10.19).
- In nondividing cells, each chromosome has highly compacted regions called heterochromatin and less compacted regions called euchromatin (see Figure 10.20).
- Chromatin compaction occurs due to nucleosome formation and radial loop formation. A metaphase chromosome is completely heterochromatic due to the further compaction of radial loops (see Figure 10.21).
- Both histone and nonhistone proteins are important for the compaction of metaphase chromosomes (see Figure 10.22).
- Condensin and cohesion are SMC proteins that promote chromosome compaction and sister chromatid cohesion, respectively (see Figures 10.23–10.25).

PROBLEM SETS & INSIGHTS

Solved Problems

S1. Here is a C_0t curve for a hypothetical eukaryotic species:

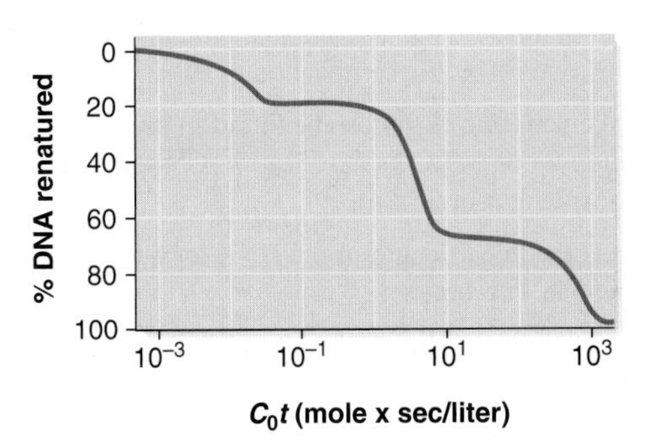

Estimate the amount of highly repetitive DNA, moderately repetitive DNA, and unique DNA.

Answer: About 20% is highly repetitive and renatures quickly, about 50% is moderately repetitive, and about 30% is unique and renatures very slowly.

S2. Let's suppose a bacterial DNA molecule is given a left-handed twist. How does this affect the structure and function of the DNA?

Answer: A left-handed twist is negative supercoiling. Negative supercoiling makes the bacterial chromosome more compact. It also promotes DNA functions that involve strand separation, including gene transcription and DNA replication.

S3. To hold bacterial DNA in a more compact configuration, specific proteins must bind to the DNA and stabilize its conformation (as shown in Figure 10.5). Several different proteins are involved in this process. These proteins have been collectively referred to as "histone-like" due to their possible functional similarity to the histone proteins found in eukaryotes. Based on your knowledge of eukaryotic histone proteins, what biochemical properties would you expect from bacterial histone-like proteins?

Answer: The histone-like proteins have the properties expected for proteins involved in DNA folding. They are all small proteins found in relative abundance within the bacterial cell. In some cases, the histone-like proteins are biochemically similar to eukaryotic histones. For example, they tend to be basic (positively charged) and bind to DNA in a non-sequence-dependent fashion. However, other proteins appear to bind to bacterial DNA at specific sites in order to promote DNA bending.

Conceptual Questions

C1. In viral replication, what is the difference between self-assembly and directed assembly?

C2. Bacterial chromosomes have one origin of replication, whereas eukaryotic chromosomes have several. Would you expect viral chromosomes to have an origin of replication? Why or why not?

C3. What is a bacterial nucleoid? With regard to cellular membranes, what is the difference between a bacterial nucleoid and a eukaryotic nucleus?

C4. In Part II of this textbook, we considered inheritance patterns for diploid eukaryotic species. Bacteria frequently contain two or more nucleoids. With regard to genes and alleles, how is a bacterium that contains two nucleoids similar to a diploid eukaryotic cell, and how is it different?

C5. Describe the two main mechanisms by which the bacterial DNA becomes compacted.

C6. As described in Chapter 9, 1 bp of DNA is approximately 0.34 nm in length. A bacterial chromosome is about 4 million bp in length and is organized into about 100 loops that are about 40,000 bp in length.

A. If it was stretched out linearly, how long (in micrometers) would one loop be?

B. If a bacterial chromosomal loop is circular, what would be its diameter? (Note: Circumference = πD, where D is the diameter of the circle.)

C. Is the diameter of the circular loop calculated in part B small enough to fit inside a bacterium? The dimensions of the bacterial cytoplasm, such as *E. coli*, are roughly 0.5 μm wide and 1.0 μm long.

C7. Why is DNA supercoiling called supercoiling rather than just coiling? Why is positive supercoiling called overwinding and negative supercoiling called underwinding? How would you define the terms positive and negative supercoiling for Z DNA (described in Chapter 9)?

C8. Coumarins and quinolones are two classes of drugs that inhibit bacterial growth by directly inhibiting DNA gyrase. Discuss two reasons why inhibiting DNA gyrase might inhibit bacterial growth.

C9. Take two pieces of string that are approximately 10 inches each, and create a double helix by wrapping them around each other to make 10 complete turns. Tape one end of the strings to a table, and now twist the strings three times (360° each time) in a right-handed direction. Note: As you are looking down at the strings from above, a right-handed twist is in the clockwise direction.

A. Did the three turns create more or fewer turns in your double helix? How many turns are now in your double helix?

B. Is your double helix right-handed or left-handed? Explain your answer.

C. Did the three turns create any supercoils?

D. If you had coated your double helix with rubber cement and allowed the cement to dry before making the three additional right-handed turns, would the rubber cement make it more or less likely for the three turns to create supercoiling? Would a pair of cemented strings be more or less like a real DNA double helix than an uncemented pair of strings? Explain your answer.

C10. Try to explain the function of DNA gyrase with a drawing.

C11. How are two topoisomers different from each other? How are they the same?

C12. On rare occasions, a chromosome can suffer a small deletion that removes the centromere. When this occurs, the chromosome usually is not found within subsequent daughter cells. Explain why a chromosome without a centromere is not transmitted very efficiently from mother to daughter cells. (Note: If a chromosome is located outside the nucleus after telophase, it is degraded.)

C13. What is the function of a centromere? At what stage of the cell cycle would you expect the centromere to be the most important?

C14. Describe the characteristics of highly repetitive DNA.

C15. Describe the structures of a nucleosome and a 30-nm fiber.

C16. Beginning with the G_1 phase of the cell cycle, describe the level of compaction of the eukaryotic chromosome. How does the level of compaction change as the cell progresses through the cell cycle? Why is it necessary to further compact the chromatin during mitosis?

C17. If you assume the average length of linker DNA is 50 bp, approximately how many nucleosomes are found in the haploid human genome, which contains 3 billion bp?

C18. Draw the binding between the nuclear matrix and MARs.

C19. Compare heterochromatin and euchromatin. What are the differences between them?

C20. Compare the structure and cell localization of chromosomes during interphase and M phase.

C21. What types of genetic activities occur during interphase? Explain why these activities cannot occur during M phase.

C22. Let's assume the linker DNA averages 54 bp in length. How many molecules of H2A would you expect to find in a DNA sample that is 46,000 bp in length?

C23. In Figure 10.16, what are we looking at in part b? Is this an 11-nm fiber, a 30-nm fiber, or a 300-nm fiber? Does this DNA come from a cell during M phase or interphase?

C24. What are the roles of the core histone proteins compared with the role of histone H1 in the compaction of eukaryotic DNA?

C25. A typical eukaryotic chromosome found in humans contains about 100 million bp of DNA. As described in Chapter 9, 1 bp of DNA has a linear length of 0.34 nm.

A. What is the linear length of the DNA for a typical human chromosome in micrometers?

B. What is the linear length of a 30-nm fiber of a typical human chromosome?

C. Based on your calculation of part B, would a typical human chromosome fit inside the nucleus (with a diameter of 5 μm) if the 30-nm fiber were stretched out in a linear manner? If not, explain how a typical human chromosome fits inside the nucleus during interphase.

C26. Which of the following terms should not be used to describe a Barr body?

A. Chromatin

B. Euchromatin

C. Heterochromatin

D. Chromosome

E. Genome

C27. Discuss the differences in the compaction levels of metaphase chromosomes compared with interphase chromosomes. When would you expect gene transcription and DNA replication to take place, during M phase or interphase? Explain why.

C28. What is an SMC protein? Describe two examples.

Experimental Questions

E1. Two circular DNA molecules, which we will call molecule A and molecule B, are topoisomers of each other. When viewed under the electron microscope, molecule A appears more compact than molecule B. The level of gene transcription is much lower for molecule A. Which of the following three possibilities could account for these observations?

First possibility: Molecule A has three positive supercoils, and molecule B has three negative supercoils.

Second possibility: Molecule A has four positive supercoils, and molecule B has one negative supercoil.

Third possibility: Molecule A has zero supercoils, and molecule B has three negative supercoils.

E2. Explain how a renaturation experiment can provide quantitative information about genome sequence complexity.

E3. In a renaturation experiment, does the copy number affect only the rate of renaturation, or does it also affect the rate of denaturation? Explain your answer.

E4. Let's suppose that you have isolated DNA from a cell and have viewed it under a microscope. It looks supercoiled. What experiment would you perform to determine if it is positively or negatively supercoiled? In your answer, describe your expected results. You may assume that you have purified topoisomerases at your disposal.

E5. We seem to know more about the structure of eukaryotic chromosomal DNA than bacterial DNA. Discuss why you think this is so, and list several experimental procedures that have yielded important information concerning the compaction of eukaryotic chromatin.

E6. An organism contains 20% highly repetitive DNA, 10% moderately repetitive DNA, and 70% unique sequences. Draw the expected C_0t curve that would be obtained from this organism.

E7. When chromatin is treated with a moderate salt concentration, the linker histone H1 is removed (see Figure 10.16a). Higher salt concentration removes the rest of the histone proteins (see Figure 10.22b). If the experiment of Figure 10.15 were carried out after the DNA was treated with moderate or high salt, what would be the expected results?

E8. Let's suppose you have isolated chromatin from some bizarre eukaryote with a linker region that is usually 300 to 350 bp in length. The nucleosome structure is the same as in other eukaryotes. If you digested this eukaryotic organism's chromatin with a high concentration of DNase I, what would be your expected results?

E9. If you were given a sample of chromosomal DNA and asked to determine if it is bacterial or eukaryotic, what experiment would you perform, and what would be your expected results?

E10. Consider how histone proteins bind to DNA and then explain why a high salt concentration can remove histones from DNA (as shown in Figure 10.22b).

E11. In Chapter 20, the technique of fluorescence in situ hybridization (FISH) is described. This is another method used to examine sequence complexity within a genome. In this method, a particular DNA sequence, such as a particular gene sequence, can be detected within an intact chromosome by using a DNA probe that is complementary to the sequence. For example, let's consider the β-globin gene, which is found on human chromosome 11. A probe complementary to the β-globin gene binds to the β-globin gene and shows up as a brightly colored spot on human chromosome 11. In this way, researchers can detect where the β-globin gene is located within a set of chromosomes. Because the β-globin gene is unique and because human cells are diploid (i.e., have two copies of each chromosome), a FISH experiment shows two bright spots per cell; the probe binds to each copy of chromosome 11. What would you expect to see if you used the following types of probes?

A. A probe complementary to the *Alu*I sequence

B. A probe complementary to a tandemly repeated sequence near the centromere of the X chromosome

Questions for Student Discussion/Collaboration

1. Bacterial and eukaryotic chromosomes are very compact. Discuss the advantages and disadvantages of having a compact structure.
2. The prevalence of highly repetitive sequences seems rather strange to many geneticists. Do they seem strange to you? Why or why not? Discuss whether or not you think they have an important function.
3. Discuss and make a list of the similarities and differences between bacterial and eukaryotic chromosomes.

Note: All answers appear at the website for this textbook; the answers to even-numbered questions are in the back of the textbook.

www.mhhe.com/brookergenetics4e

Visit the website for practice tests, answer keys, and other learning aids for this chapter. Enhance your understanding of genetics with our interactive exercises, quizzes, animations, and much more.

CHAPTER OUTLINE

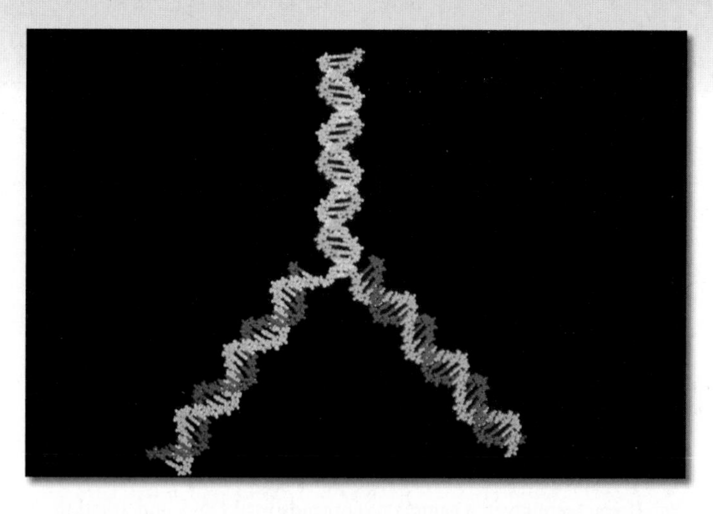

A model for DNA undergoing replication. *This molecular model shows a DNA replication fork, the site where new DNA strands are made. In this model, the original DNA is yellow and blue. The newly made strands are purple.*

DNA REPLICATION

As discussed throughout Chapters 2 to 8, genetic material is transmitted from parent to offspring and from cell to cell. For transmission to occur, the genetic material must be copied. During this process, known as **DNA replication,** the original DNA strands are used as templates for the synthesis of new DNA strands. We will begin Chapter 11 with a consideration of the structural features of the double helix that underlie the replication process. Then we examine how chromosomes are replicated within living cells, addressing the following questions: where does DNA replication begin, how does it proceed, and where does it end? We first consider bacterial DNA replication and examine how DNA replication occurs within living cells, and then we turn our attention to the unique features of the replication of eukaryotic DNA. At the molecular level, it is rather remarkable that the replication of chromosomal DNA occurs very quickly, very accurately, and at the appropriate time in the life of the cell. For this to happen, many cellular proteins play vital roles. In this chapter, we will examine the mechanism of DNA replication and consider the functions of several proteins involved in the process.

11.1 STRUCTURAL OVERVIEW OF DNA REPLICATION

Because they bear directly on the replication process, let's begin by recalling a few important structural features of the double helix from Chapter 9. The double helix is composed of two DNA strands, and the individual building blocks of each strand are nucleotides. The nucleotides contain one of four bases: adenine, thymine, guanine, or cytosine. The double-stranded structure is held together by base stacking and by hydrogen bonding between the bases in opposite strands. A critical feature of the double-helix structure is that adenine hydrogen bonds with thymine, and guanine hydrogen bonds with cytosine. This rule, known as the AT/GC rule, is the basis for the complementarity of the base sequences in double-stranded DNA.

Another feature worth noting is that the strands within a double helix have an antiparallel alignment. This directionality is determined by the orientation of sugar molecules within the sugar-phosphate backbone. If one strand is running in the 5′ to 3′ direction, the complementary strand is running in the 3′ to 5′ direction. The issue of directionality will be important when we consider the function of the enzymes that synthesize new DNA strands. In this section, we will consider how the structure of the DNA double helix provides the basis for DNA replication.

Existing DNA Strands Act as Templates for the Synthesis of New Strands

As shown in **Figure 11.1a**, DNA replication relies on the complementarity of DNA strands according to the AT/GC rule. During the replication process, the two complementary strands of DNA come apart and serve as **template strands,** or **parental strands,** for the synthesis of two new strands of DNA. After the double helix has separated, individual nucleotides have access to the template strands. Hydrogen bonding between individual nucleotides and the template strands must obey the AT/GC rule. To complete the replication process, a covalent bond is formed between the phosphate of one nucleotide and the sugar of the previous nucleotide. The two newly made strands are referred to as the **daughter strands.** Note that the base sequences are identical in both double-stranded molecules after replication (**Figure 11.1b**). Therefore, DNA is replicated so that both copies retain the same information—the same base sequence—as the original molecule.

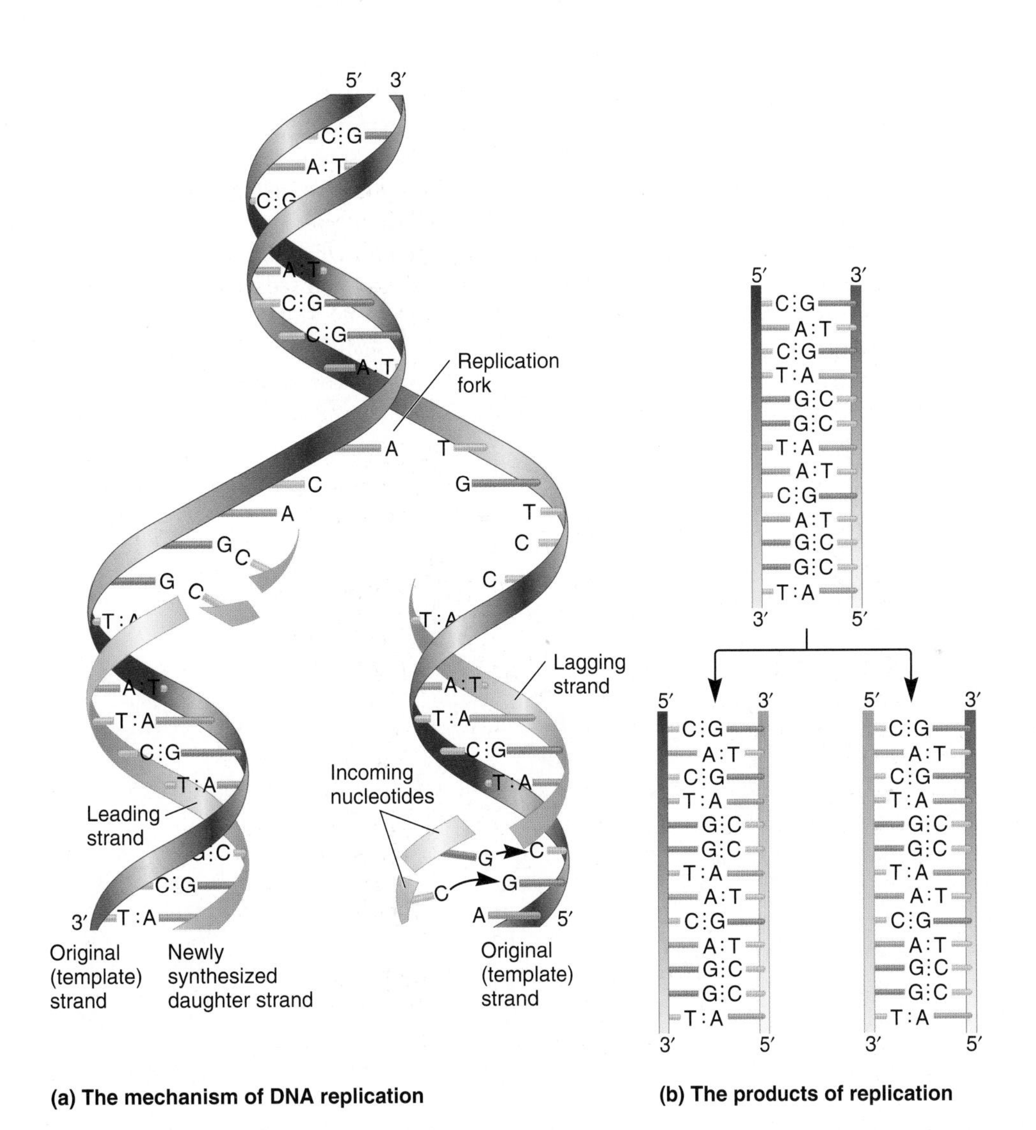

ONLINE ANIMATION

FIGURE 11.1 **The structural basis for DNA replication.** **(a)** The mechanism of DNA replication as originally proposed by Watson and Crick. As we will see, the synthesis of one newly made strand (the leading strand) occurs in the direction toward the replication fork, whereas the synthesis of the other newly made strand (the lagging strand) occurs in small segments away from the replication fork. **(b)** DNA replication produces two copies of DNA with the same sequence as the original DNA molecule.

EXPERIMENT 11A

Three Different Models Were Proposed That Described the Net Result of DNA Replication

Scientists in the late 1950s had considered three different mechanisms to explain the net result of DNA replication. These mechanisms are shown in **Figure 11.2.** The first is referred to as a **conservative model.** According to this hypothesis, both strands of parental DNA remain together following DNA replication. In this model, the original arrangement of parental strands is completely conserved, while the two newly made daughter strands also remain together following replication. The second is called a **semiconservative model.** In this mechanism, the double-stranded DNA is half conserved following the replication process. In other words, the newly made double-stranded DNA contains one parental strand and one daughter strand. The third, called the **dispersive model,** proposes that segments of parental DNA and newly made DNA are interspersed in both strands following the replication process. Only the semiconservative model shown in Figure 11.2b is actually correct.

In 1958, Matthew Meselson and Franklin Stahl devised a method to experimentally distinguish newly made daughter strands from the original parental strands. Their technique involved labeling DNA with a heavy isotope of nitrogen. Nitrogen, which is found within the bases of DNA, occurs in both a heavy (^{15}N) and light (^{14}N) form. Prior to their experiment, they grew *Escherichia coli* cells in the presence of ^{15}N for many generations. This produced a population of cells in which all of the DNA was heavy-labeled. At the start of their experiment, shown in **Figure 11.3** (generation 0), they switched the bacteria to a medium that contained only ^{14}N and then collected samples of cells after various time points. Under the growth conditions they employed, 30 minutes is the time required for one doubling, or one generation time. Because the bacteria were doubling in a medium that contained only ^{14}N, all of the newly made DNA strands are labeled with light nitrogen, but the original strands remain in the heavy form.

Meselson and Stahl then analyzed the density of the DNA by centrifugation, using a cesium chloride (CsCl) gradient. (The procedure of gradient centrifugation is described in the Appendix.) If both DNA strands contained ^{14}N, the DNA would have a light density and sediment near the top of the tube. If one strand contained ^{14}N and the other strand contained ^{15}N, the DNA would be half-heavy and have an intermediate density. Finally, if both strands contained ^{15}N, the DNA would be heavy and would sediment closer to the bottom of the centrifuge tube.

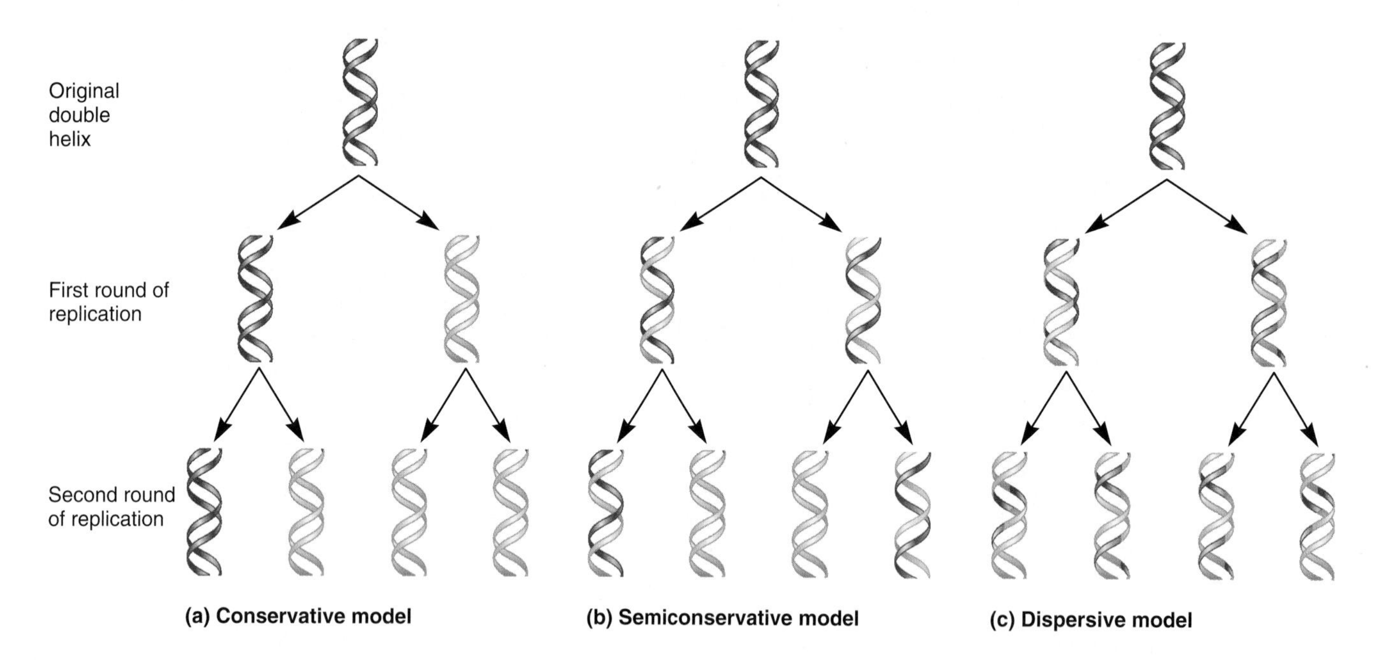

FIGURE 11.2 **Three possible models for DNA replication.** The two original parental DNA strands are shown in purple, and the newly made strands after one and two generations are shown in light blue.

THE HYPOTHESIS

Based on Watson's and Crick's ideas, the hypothesis was that DNA replication is semiconservative. Figure 11.2 also shows two alternative models.

TESTING THE HYPOTHESIS — FIGURE 11.3 Evidence that DNA replication is semiconservative.

Starting material: A strain of *E. coli* that has been grown for many generations in the presence of ^{15}N. All of the nitrogen in the DNA is labeled with ^{15}N.

1. Add an excess of ^{14}N-containing compounds to the bacterial cells so all of the newly made DNA will contain ^{14}N.
2. Incubate the cells for various lengths of time. Note: The ^{15}N-labeled DNA is shown in purple and the ^{14}N-labeled DNA is shown in blue.
3. Lyse the cells by the addition of lysozyme and detergent, which disrupt the bacterial cell wall and cell membrane, respectively.
4. Load a sample of the lysate onto a CsCl gradient. (Note: The average density of DNA is around 1.7 g/cm^3, which is well isolated from other cellular macromolecules.)
5. Centrifuge the gradients until the DNA molecules reach their equilibrium densities.
6. DNA within the gradient can be observed under a UV light.

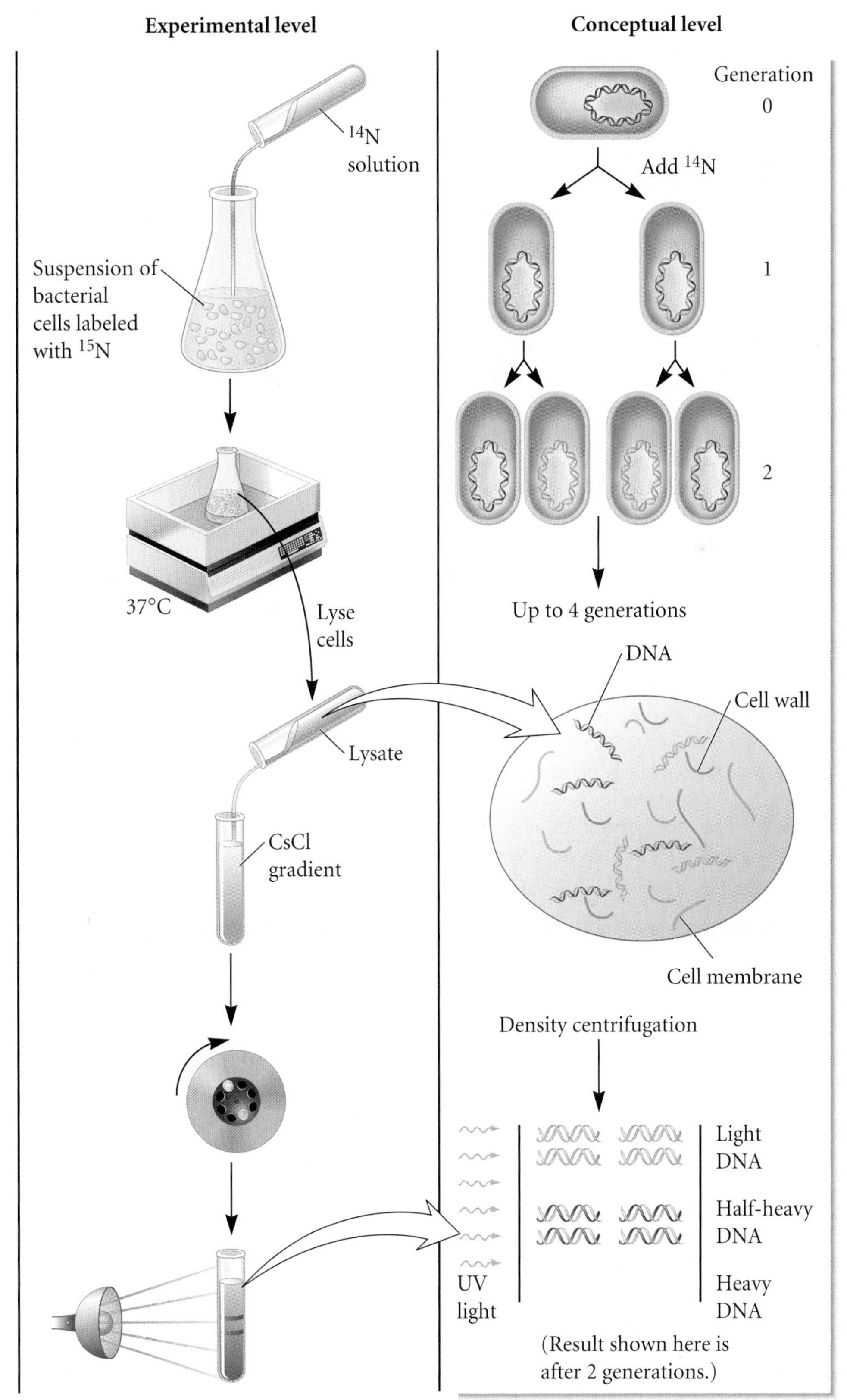

THE DATA

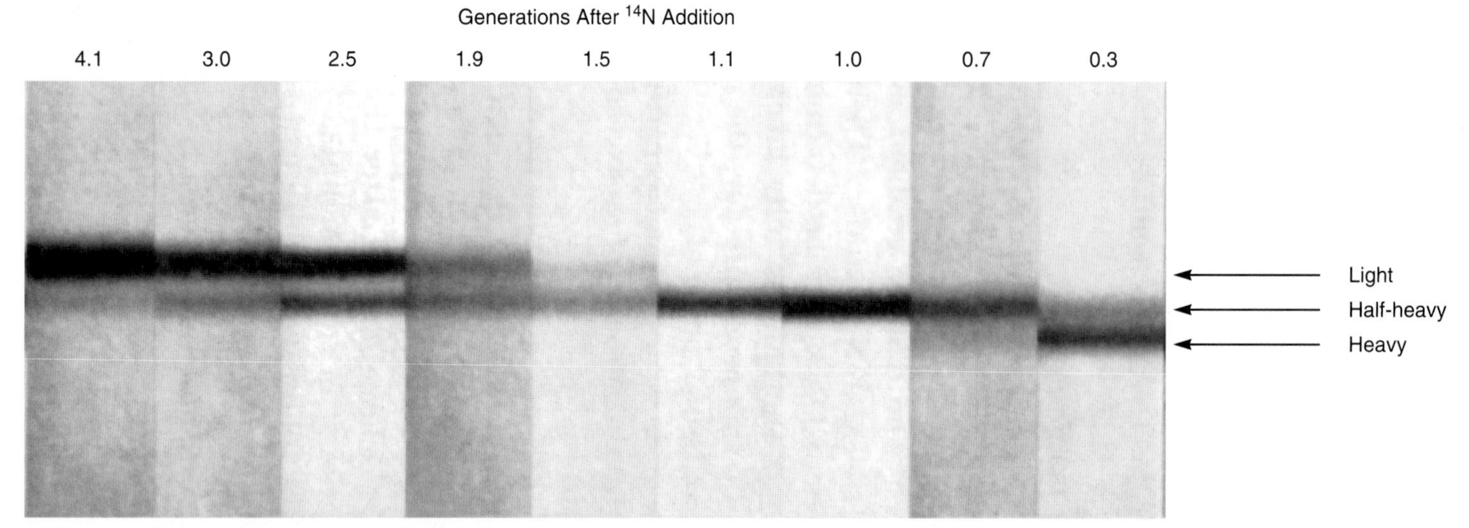

Data from: Meselson, M., and Stahl, F.W. (1958) The Replication of DNA in *Escherichia coli*. *Proc. Natl. Acad. Sci, USA 44*: 671–682.

INTERPRETING THE DATA

As seen in the data following Figure 11.3, after one round of DNA replication (i.e., one generation), all of the DNA sedimented at a density that was half-heavy. Which of the three models is consistent with this result? Both the semiconservative and dispersive models are consistent. In contrast, the conservative model predicts two separate DNA types: a light type and a heavy type. Because all of the DNA had sedimented as a single band, this model was disproved. According to the semiconservative model, the replicated DNA would contain one original strand (a heavy strand) and a newly made daughter strand (a light strand). Likewise, in a dispersive model, all of the DNA should have been half-heavy after one generation as well. To determine which of these two remaining models is correct, therefore, Meselson and Stahl had to investigate future generations.

After approximately two rounds of DNA replication (i.e., 1.9 generations), a mixture of light DNA and half-heavy DNA was observed. This result was consistent with the semiconservative model of DNA replication, because some DNA molecules should contain all light DNA, and other molecules should be half-heavy (see Figure 11.2b). The dispersive model predicts that after two generations, the heavy nitrogen would be evenly dispersed among four strands, each strand containing 1/4 heavy nitrogen and 3/4 light nitrogen (see Figure 11.2c). However, this result was not obtained. Instead, the results of the Meselson and Stahl experiment provided compelling evidence in favor of only the semiconservative model for DNA replication.

A self-help quiz involving this experiment can be found at www.mhhe.com/brookergenetics4e.

11.2 BACTERIAL DNA REPLICATION

Thus far, we have considered how a complementary, double-stranded structure underlies the ability of DNA to be copied. In addition, the experiments of Meselson and Stahl showed that DNA replication results in two double helices, each one containing an original parental strand and a newly made daughter strand. We now turn our attention to how DNA replication actually occurs within living cells. Much research has focused on the bacterium *E. coli*. The results of these studies have provided the foundation for our current molecular understanding of DNA replication. The replication of the bacterial chromosome is a stepwise process in which many cellular proteins participate. In this section, we will follow this process from beginning to end.

Bacterial Chromosomes Contain a Single Origin of Replication

Figure 11.4 presents an overview of the process of bacterial chromosomal replication. The site on the bacterial chromosome where DNA synthesis begins is known as the **origin of replication.** Bacterial chromosomes have a single origin of replication. The synthesis of new daughter strands is initiated within the origin and proceeds in both directions, or **bidirectionally,** around the bacterial chromosome. This means that two **replication forks** move in opposite directions outward from the origin. A replication fork is the site where the parental DNA strands have separated and new daughter strands are being made. Eventually, these replication forks meet each other on the opposite side of the bacterial chromosome to complete the replication process.

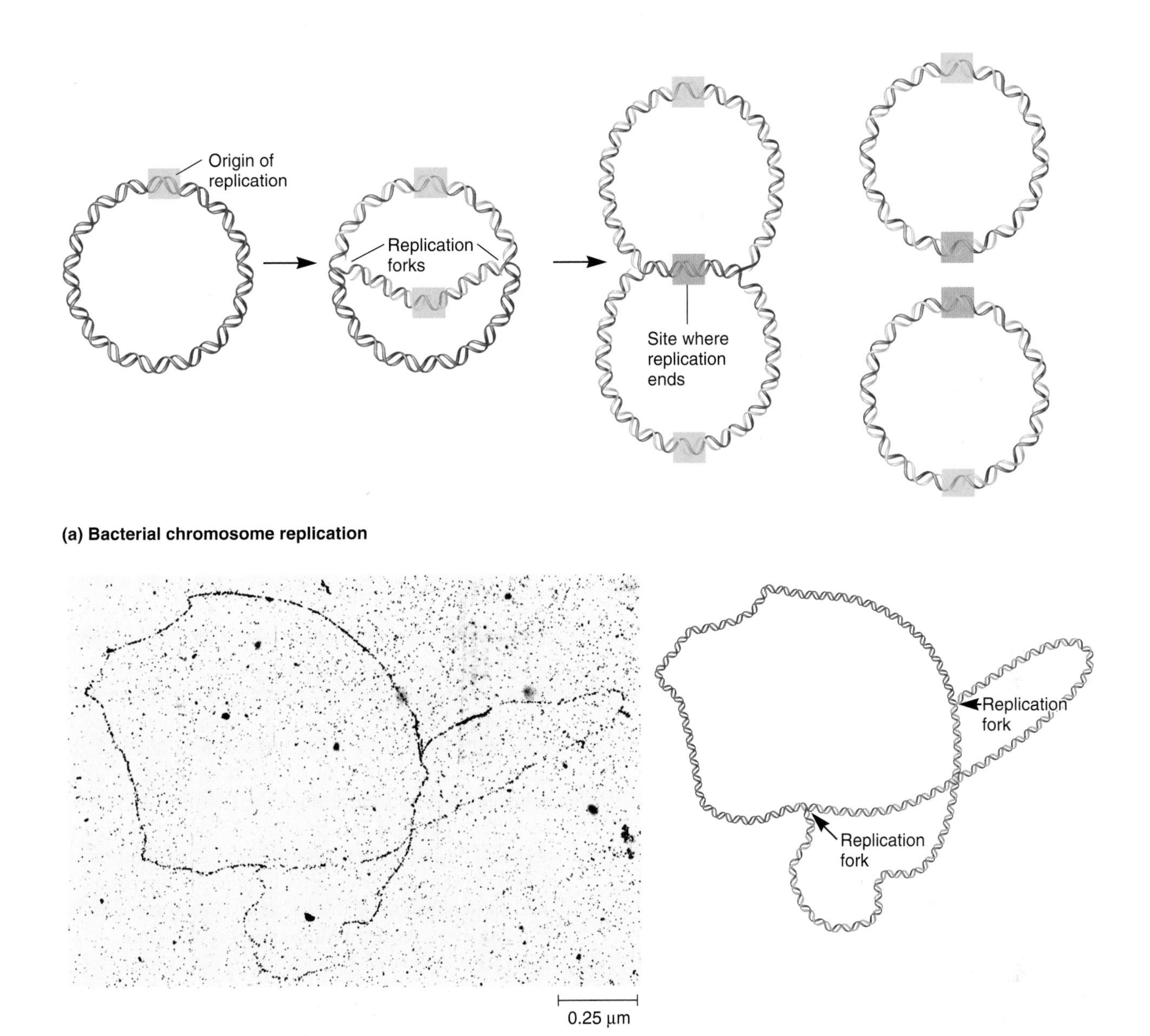

FIGURE 11.4 **The process of bacterial chromosome replication.** **(a)** An overview of the process of bacterial chromosome replication. **(b)** A replicating *E. coli* chromosome visualized by autoradiography and transmission electron microscopy (TEM). This chromosome was radiolabeled by growing bacterial cells in media containing radiolabeled thymidine. The diagram at the right shows the locations of the two replication forks. The chromosome is about one-third replicated. New strands are shown in blue.

Replication Is Initiated by the Binding of DnaA Protein to the Origin of Replication

Considerable research has focused on the origin of replication in *E. coli*. This origin is named *oriC* for origin of Chromosomal replication (**Figure 11.5**). Three types of DNA sequences are found within *oriC*: an AT-rich region, DnaA box sequences, and GATC methylation sites. The GATC methylation sites will be discussed later in this chapter when we consider the regulation of replication.

DNA replication is initiated by the binding of **DnaA proteins** to sequences within the origin known as **DnaA box sequences.** The DnaA box sequences serve as recognition sites for the binding of the DnaA proteins. When DnaA proteins are in their ATP-bound form, they bind to the five DnaA boxes in *oriC* to initiate DNA replication. DnaA proteins also bind to each other to form a complex (**Figure 11.6**). With the aid of other DNA-binding proteins, such as HU and IHF, this causes the DNA to bend around the complex of DnaA proteins and results in the separation of the AT-rich region. Because only two hydrogen bonds form between AT base pairs, whereas three hydrogen bonds occur between G and C, the DNA strands are more easily separated at an AT-rich region.

Following separation of the AT-rich region, the DnaA proteins, with the help of the DnaC protein, recruit **DNA helicase** proteins to this site. DNA helicase is also known as DnaB protein. When a DNA helicase encounters a double-stranded region,

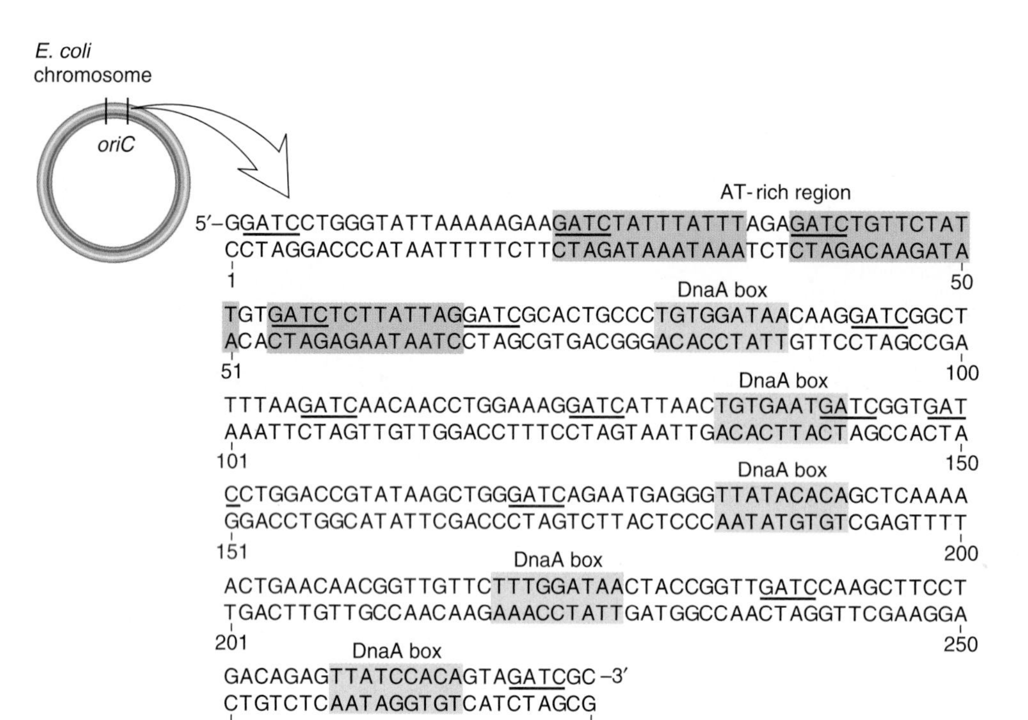

FIGURE 11.5 The sequence of *oriC* in *E. coli*. The AT-rich region is composed of three tandem repeats that are 13 bp long and highlighted in blue. The five DnaA boxes are highlighted in orange. The GATC methylation sites are underlined.

5′ 3′ 3′ 5′
AT-rich region
DnaA boxes
DnaA proteins bind to DnaA boxes and to each other. Additional proteins that cause the DNA to bend also bind (not shown). This causes the region to wrap around the DnaA proteins and separates the AT-rich region.
AT-rich region
DnaA protein
DNA helicase (DnaB protein) binds to the origin. DnaC protein (not shown) assists this process.
Helicase
DNA helicase separates the DNA in both directions, creating 2 replication forks.
Fork
Fork

it breaks the hydrogen bonds between the two strands, thereby generating two single strands. Two DNA helicases begin strand separation within the *oriC* region and continue to separate the DNA strands beyond the origin. These proteins use the energy from ATP hydrolysis to catalyze the separation of the double-stranded parental DNA. In *E. coli,* DNA helicases bind to single-stranded DNA and travel along the DNA in a 5′ to 3′ direction to keep the replication fork moving. As shown in Figure 11.6, the action of DNA helicases promotes the movement of two replication forks outward from *oriC* in opposite directions. This initiates the replication of the bacterial chromosome in both directions, an event termed **bidirectional replication.**

Several Proteins Are Required for DNA Replication at the Replication Fork

Figure 11.7 provides an overview of the molecular events that occur as one of the two replication forks moves around the bacterial chromosome, and **Table 11.1** summarizes the functions of the major proteins involved in *E. coli* DNA replication. Let's begin with strand separation. To act as a template for DNA replication, the strands of a double helix must separate. As mentioned previously, the function of DNA helicase is to break the hydrogen bonds between base pairs and thereby unwind the strands; this action generates positive supercoiling ahead of each replication

FIGURE 11.6 The events that occur at *oriC* to initiate the DNA replication process. To initiate DNA replication, DnaA proteins bind to the five DnaA boxes, which causes the DNA strands to separate at the AT-rich region. DnaA and DnaC proteins then recruit DNA helicase (DnaB) into this region. Each DNA helicase is composed of six subunits, which form a ring around one DNA strand and migrate in the 5′ to 3′ direction. As shown here, the movement of two DNA helicase proteins serves to separate the DNA strands beyond the *oriC* region.

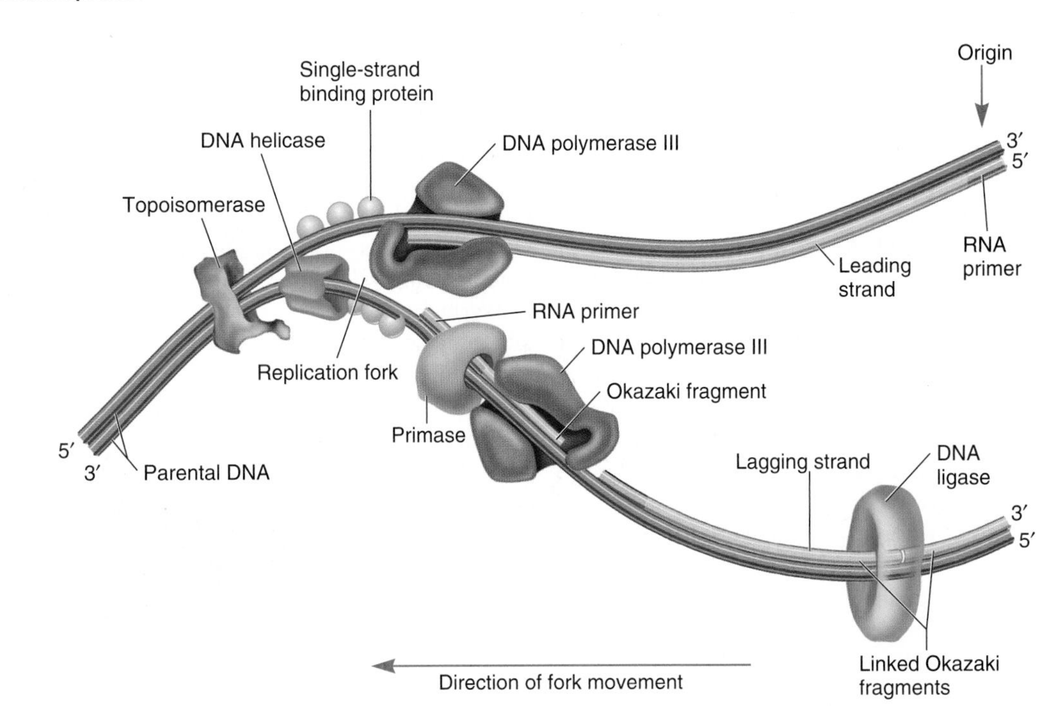

FIGURE 11.7 **The proteins involved with DNA replication.**
Note: The drawing of DNA polymerase III depicts the catalytic subunit that synthesizes DNA.

fork. As shown in Figure 11.7, an enzyme known as a **topoisomerase (type II),** also called **DNA gyrase,** travels in front of DNA helicase and alleviates positive supercoiling.

After the two parental DNA strands have been separated and the supercoiling relaxed, they must be kept that way until the complementary daughter strands have been made. What prevents the DNA strands from coming back together? DNA replication requires **single-strand binding proteins** that bind to the strands of parental DNA and prevent them from re-forming a double helix. In this way, the bases within the parental strands are kept in an exposed condition that enables them to hydrogen bond with individual nucleotides.

The next event in DNA replication involves the synthesis of short strands of RNA (rather than DNA) called **RNA primers.** These strands of RNA are synthesized by the linkage of ribonucleotides via an enzyme known as **primase.** This enzyme synthesizes short strands of RNA, typically 10 to 12 nucleotides in length. These short RNA strands start, or prime, the process of DNA replication. In the **leading strand,** a single primer is made at the origin of replication. In the **lagging strand,** multiple primers are made. As discussed later, the RNA primers are eventually removed.

A type of enzyme known as **DNA polymerase** is responsible for synthesizing the DNA of the leading and lagging strands. This enzyme catalyzes the formation of covalent bonds between adjacent nucleotides and thereby makes the new daughter strands. In *E. coli*, five distinct proteins function as DNA polymerases and are designated polymerase I, II, III, IV, and V. DNA polymerases I and III are involved in normal DNA replication, whereas DNA polymerases II, IV, and V play a role in DNA repair and the replication of damaged DNA.

DNA polymerase III is responsible for most of the DNA replication. It is a large enzyme consisting of 10 different subunits that play various roles in the DNA replication process (**Table 11.2**). The α subunit actually catalyzes the bond formation between adjacent nucleotides, and the remaining nine subunits fulfill other functions. The complex of all 10 subunits

TABLE 11.1
Proteins Involved in *E. coli* DNA Replication

Common Name	Function
DnaA protein	Binds to DnaA boxes within the origin to initiate DNA replication
DnaC protein	Aids DnaA in the recruitment of DNA helicase to the origin
DNA helicase (DnaB)	Separates double-stranded DNA
Topoisomerase	Removes positive supercoiling ahead of the replication fork
Single-strand binding protein	Binds to single-stranded DNA and prevents it from re-forming a double-stranded structure
Primase	Synthesizes short RNA primers
DNA polymerase III	Synthesizes DNA in the leading and lagging strands
DNA polymerase I	Removes RNA primers, fills in gaps with DNA
DNA ligase	Covalently attaches adjacent Okazaki fragments
Tus	Binds to ter sequences and prevents the advancement of the replication fork

together is called DNA polymerase III holoenzyme. By comparison, DNA polymerase I is composed of a single subunit. Its role during DNA replication is to remove the RNA primers and fill in the vacant regions with DNA.

Though the various DNA polymerases in *E. coli* and other bacterial species vary in their subunit composition, several common structural features have emerged. The catalytic subunit of all DNA polymerases has a structure that resembles a human hand. As shown in **Figure 11.8**, the template DNA is threaded through the palm of the hand; the thumb and fingers are wrapped around the DNA. The incoming deoxyribonuleoside triphosphates (dNTPs) enter the catalytic site, bind to the template strand according to the AT/GC rule, and then are covalently attached to the 3′ end of the growing strand. DNA polymerase also contains a 3′ exonuclease site that removes mismatched bases, as described later.

As researchers began to unravel the function of DNA polymerase, two features seemed unusual (**Figure 11.9**). DNA polymerase cannot begin DNA synthesis by linking together the first two individual nucleotides. Rather, this type of enzyme can elongate only a preexisting strand starting with an RNA primer or existing DNA strand (Figure 11.9a). A second unusual feature is the directionality of strand synthesis. DNA polymerase can attach nucleotides only in the 5′ to 3′ direction, not in the 3′ to 5′ direction (Figure 11.9b).

Due to these two unusual features, the synthesis of the leading and lagging strands shows distinctive differences (**Figure 11.10**). The synthesis of RNA primers by primase allows DNA polymerase III to begin the synthesis of complementary daughter strands of DNA. DNA polymerase III catalyzes the attachment of nucleotides to the 3′ end of each primer, in a 5′ to 3′ direction. In the leading strand, one RNA primer is made at the origin, and then DNA polymerase III can attach nucleotides in a 5′ to 3′ direction as it slides toward the opening of the replication fork. The synthesis of the leading strand is therefore continuous.

In the lagging strand, the synthesis of DNA also elongates in a 5′ to 3′ manner, but it does so in the direction away from the replication fork. In the lagging strand, RNA primers must repeatedly initiate the synthesis of short segments of DNA; thus, the synthesis has to be discontinuous. The length of these fragments in bacteria is typically 1000 to 2000 nucleotides. In eukaryotes, the fragments are shorter—100 to 200 nucleotides. Each fragment contains a short RNA primer at the 5′ end, which is made by primase. The remainder of the fragment is a strand of DNA made by DNA polymerase III. The DNA fragments made in this manner are known as **Okazaki fragments,** after Reiji and Tuneko Okazaki, who initially discovered them in the late 1960s.

To complete the synthesis of Okazaki fragments within the lagging strand, three additional events must occur: removal of the RNA primers, synthesis of DNA in the area where the primers have been removed, and the covalent attachment of adjacent

TABLE **11.2**

Subunit Composition of DNA Polymerase III Holoenzyme from *E. coli*

Subunit(s)	Function
α	Synthesizes DNA
ε	3′ to 5′ proofreading (removes mismatched nucleotides)
θ	Accessory protein that stimulates the proofreading function
β	Clamp protein, which allows DNA polymerase to slide along the DNA without falling off
τ, γ, δ, δ′, ψ, and χ	Clamp loader complex, involved with helping the clamp protein bind to the DNA

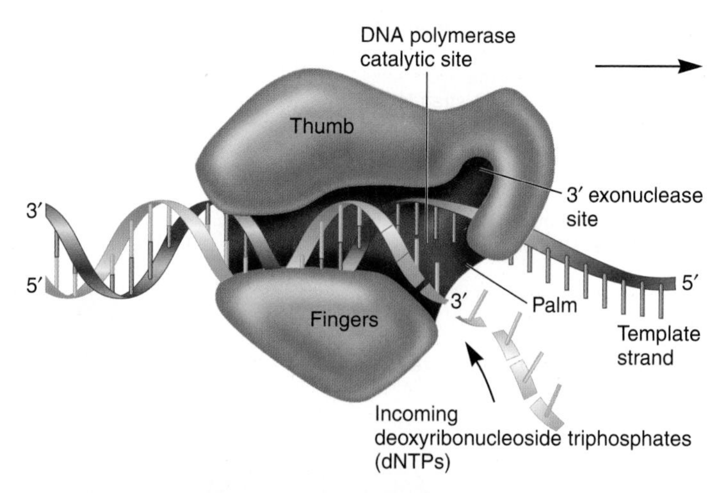

(a) Schematic side view of DNA polymerase III

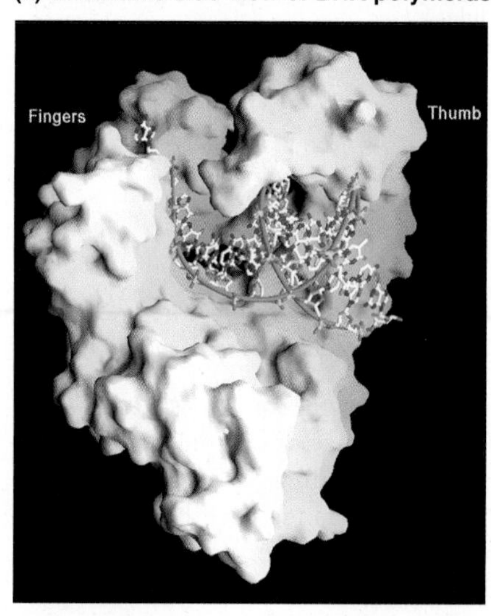

(b) Molecular model for DNA polymerase bound to DNA

(Reprinted by permission from Macmillan Publishers Ltd. Ying Li, et al. [1998] Crystal structures of open and closed forms of binary and ternary complexes of the large fragment of *Thermus aquaticus* DNA polymerase I: Structural basis for nucleotide incorporation. *Embo J 17*:24, 7514–7525.)

FIGURE 11.8 The action of DNA polymerase. **(a)** DNA polymerase slides along the template strand as it synthesizes a new strand by connecting deoxyribonucleoside triphosphates (dNTPs) in a 5′ to 3′ direction. The catalytic subunit of DNA polymerase resembles a hand that is wrapped around the template strand. In this regard, the movement of DNA polymerase along the template strand is similar to a hand that is sliding along a rope. **(b)** The molecular structure of DNA polymerase I from the bacterium *Thermus aquaticus*. This model shows a portion of DNA polymerase I that is bound to DNA. This molecular structure depicts a front view of DNA polymerase; part (a) is a schematic side view.

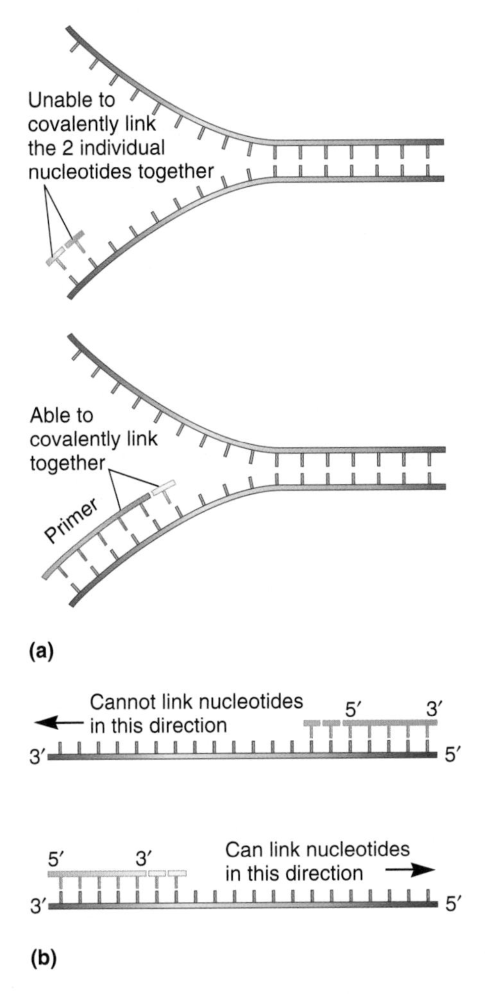

FIGURE 11.9 Unusual features of DNA polymerase function. (a) DNA polymerase can elongate a strand only from an RNA primer or existing DNA strand. (b) DNA polymerase can attach nucleotides only in a 5′ to 3′ direction. Note the template strand is in the opposite, 3′ to 5′, direction.

fragments of DNA (see Figure 11.10 and refer back to Figure 11.7). In *E. coli*, the RNA primers are removed by the action of DNA polymerase I. This enzyme has a 5′ to 3′ exonuclease activity, which means that DNA polymerase I digests away the RNA primers in a 5′ to 3′ direction, leaving a vacant area. DNA polymerase I then synthesizes DNA to fill in this region. It uses the 3′ end of an adjacent Okazaki fragment as a primer. For example, in Figure 11.10, DNA polymerase I would remove the RNA primer from the first Okazaki fragment and then synthesize DNA in the vacant region by attaching nucleotides to the 3′ end of the second Okazaki fragment. After the gap has been completely filled in, a covalent bond is still missing between the last nucleotide added by DNA polymerase I and the adjacent DNA strand that had been previously made by DNA polymerase III. An enzyme known as **DNA ligase** catalyzes a covalent bond between adjacent fragments to complete the replication process in the lagging strand (refer back to Figure 11.7). In *E. coli*, DNA ligase requires NAD^+ to carry out this reaction, whereas the DNA ligases found in archaea and eukaryotes require ATP.

The synthesis of the lagging strand was studied by the Okazakis using radiolabeled nucleotides. They incubated *E. coli* cells

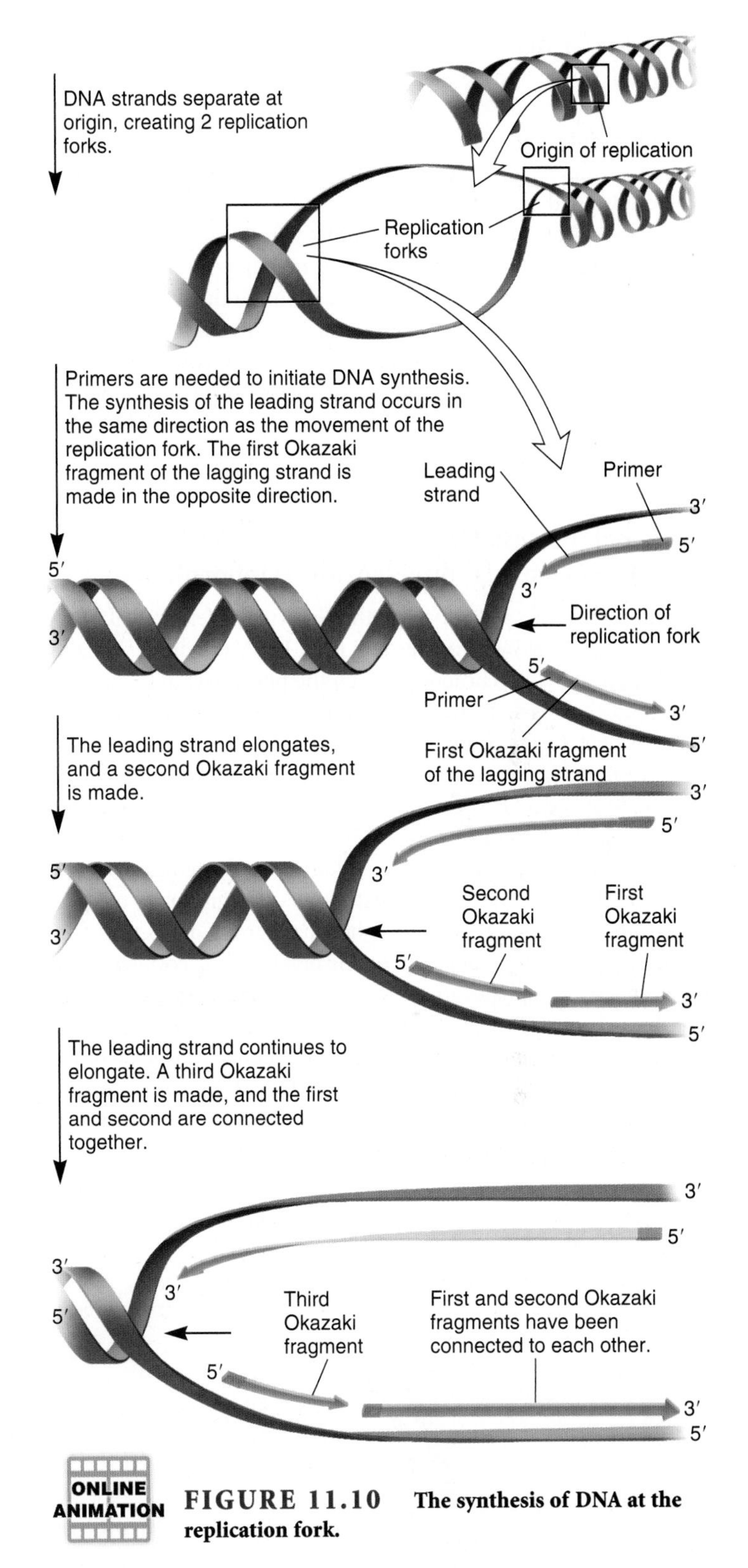

ONLINE ANIMATION **FIGURE 11.10 The synthesis of DNA at the replication fork.**

with radiolabeled thymidine for 15 seconds and then added an excess of nonlabeled thymidine. This is termed a **pulse/chase experiment** because the cells were given the radiolabeled compound for a brief period of time—a pulse—followed by an excess amount of unlabeled compound—a chase. They then isolated DNA from samples of cells at timed intervals after the pulse/chase. The DNA was denatured into single-stranded molecules, and the sizes of the radiolabeled DNA strands were determined by centrifugation. At quick time intervals, such as only a few seconds following the thymidine incubation, the fragments were found to be short, in the range of 1000 to 2000 nucleotides in length. At extended time intervals, the radiolabeled strands became much longer. At these later time points, the adjacent Okazaki fragments would have had enough time to link together.

Now that we understand how the leading and lagging strands are made, **Figure 11.11** shows how new strands are constructed from a single origin of replication. To the left of the origin, the top strand is made continuously, whereas to the right of the origin it is made in Okazaki fragments. By comparison, the synthesis of the bottom strand is just the opposite. To the left of the origin it is made in Okazaki fragments and to the right of the origin the synthesis is continuous.

DNA Polymerase III Is a Processive Enzyme That Uses Deoxyribonucleoside Triphosphates

Let's now turn our attention to other enzymatic features of DNA polymerase. As shown in **Figure 11.12**, DNA polymerases catalyze the covalent attachment between the phosphate in one nucleotide and the sugar in the previous nucleotide. The formation of this covalent (ester) bond requires an input of energy. Prior to bond formation, the nucleotide about to be attached to the growing strand is a dNTP. It contains three phosphate groups attached at the 5′-carbon atom of deoxyribose. The dNTP first enters the catalytic site of DNA polymerase and binds to the template strand according to the AT/GC rule. Next, the 3′-OH group on the previous nucleotide reacts with the phosphate group adjacent to the sugar on the incoming nucleotide. The breakage of a covalent bond between two phosphates in a dNTP is a highly exergonic reaction that provides the energy to form a covalent (ester) bond between the sugar at the 3′ end of the DNA strand and the phosphate of the incoming nucleotide. The formation of this covalent bond causes the newly made strand to grow in the 5′ to 3′ direction. As shown in Figure 11.12, pyrophosphate (PP_i) is released.

As noted in Chapter 9 (Figure 9.10), the term phosphodiester linkage (also called a phosphodiester bond) is used to describe the linkage between a phosphate and two sugar molecules. As its name implies, a phosphodiester linkage involves two ester bonds. In comparison, as a DNA strand grows, a single covalent (ester) bond is formed between adjacent nucleotides (see Figure 11.12). The other ester bond in the phosphodiester linkage—the bond between the 5′-oxygen and phosphorus—is already present in the incoming nucleotide.

DNA polymerase catalyzes the covalent attachment of nucleotides with great speed. In *E. coli*, DNA polymerase III attaches approximately 750 nucleotides per second! DNA polymerase III can catalyze the synthesis of the daughter strands so quickly because it is a **processive enzyme.** This means it does not dissociate from the growing strand after it has catalyzed the covalent joining of two nucleotides. Rather, as depicted in Figure 11.8a, it remains clamped to the DNA template strand and slides along the template as it catalyzes the synthesis of the daughter strand. The β subunit of the holoenzyme, also known as the clamp protein, promotes the association of the holoenzyme with the DNA as it glides along the template strand (refer back to Table 11.2). The β subunit forms a dimer in the shape of a ring; the hole of the ring is large enough to accommodate a double-stranded DNA molecule, and its width is about one turn of DNA. A complex of several subunits functions as a clamp loader that allows the DNA polymerase holoenzyme to initially clamp onto the DNA.

The effects of processivity are really quite remarkable. In the absence of the β subunit, DNA polymerase can synthesize DNA at a rate of approximately only 20 nucleotides per second. On average, it falls off the DNA template after about 10 nucleotides have been linked together. By comparison, when the β subunit is present, as in the holoenzyme, the synthesis rate is approximately 750 nucleotides per second. In the leading strand, DNA polymerase III has been estimated to synthesize a segment of DNA that is over 500,000 nucleotides in length before it inadvertently falls off.

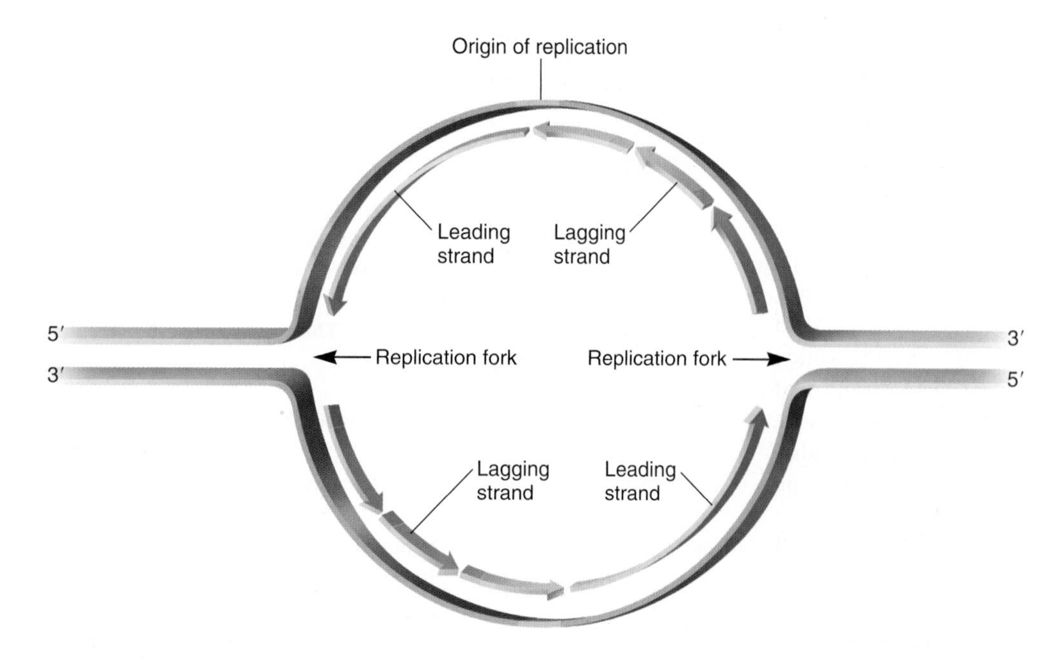

FIGURE 11.11 **The synthesis of leading and lagging strands outward from a single origin of replication.**

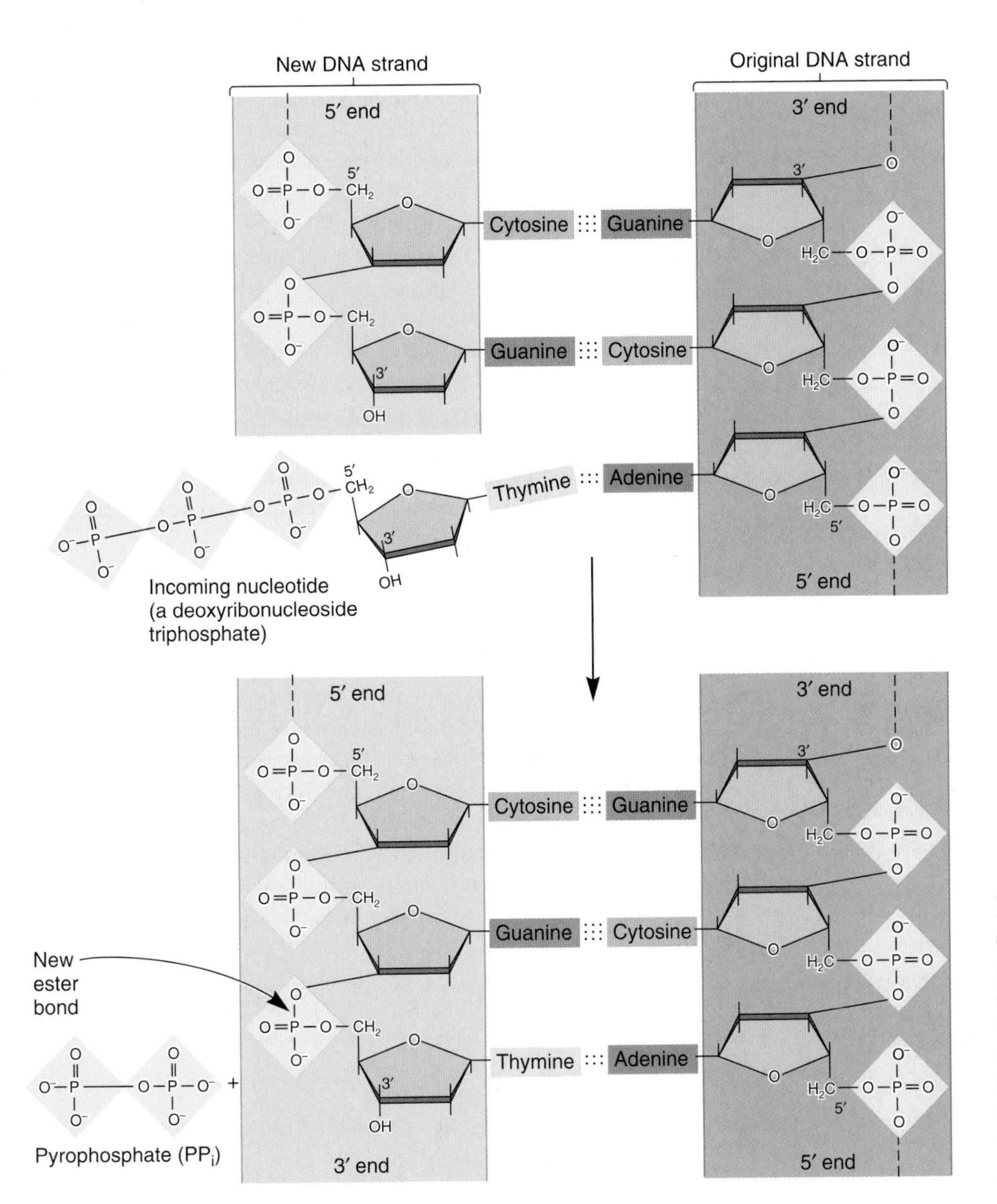

FIGURE 11.12 The enzymatic action of DNA polymerase. An incoming deoxyribonucleoside triphosphate (dNTP) is cleaved to form a nucleoside monophosphate and pyrophosphate (PP$_i$). The energy released from this exergonic reaction allows the nucleoside monophosphate to form a covalent (ester) bond at the 3′ end of the growing strand. This reaction is catalyzed by DNA polymerase. PP$_i$ is released.

Replication Is Terminated When the Replication Forks Meet at the Termination Sequences

On the opposite side of the *E. coli* chromosome from *oriC* is a pair of **termination sequences** called ter sequences. A protein known as the termination utilization substance (Tus) binds to the ter sequences and stops the movement of the replication forks. As shown in **Figure 11.13**, one of the ter sequences designated T1 prevents the advancement of the fork moving left to right, but allows the movement of the other fork (see the inset to Figure 11.13). Alternatively, T2 prevents the advancement of

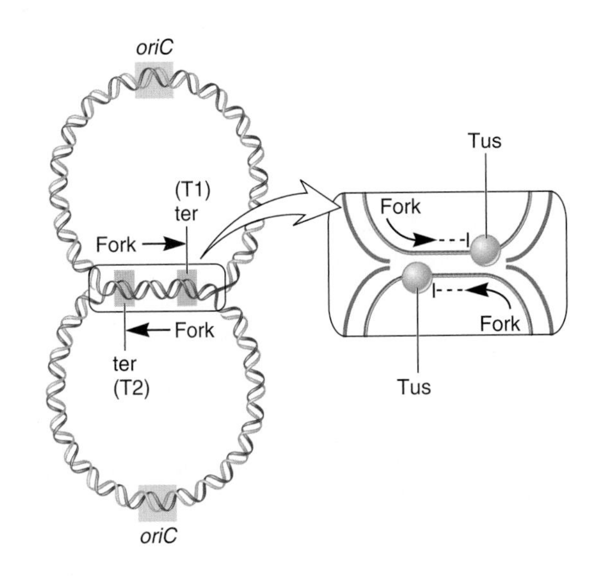

FIGURE 11.13 The termination of DNA replication. Two sites in the bacterial chromosome, shown with rectangles, are ter sequences designated T1 and T2. The T1 site prevents the further advancement of the fork moving left to right, and T2 prevents the advancement of the fork moving right to left. As shown in the inset, the binding of Tus prevents the replication forks from proceeding past the ter sequences in a particular direction.

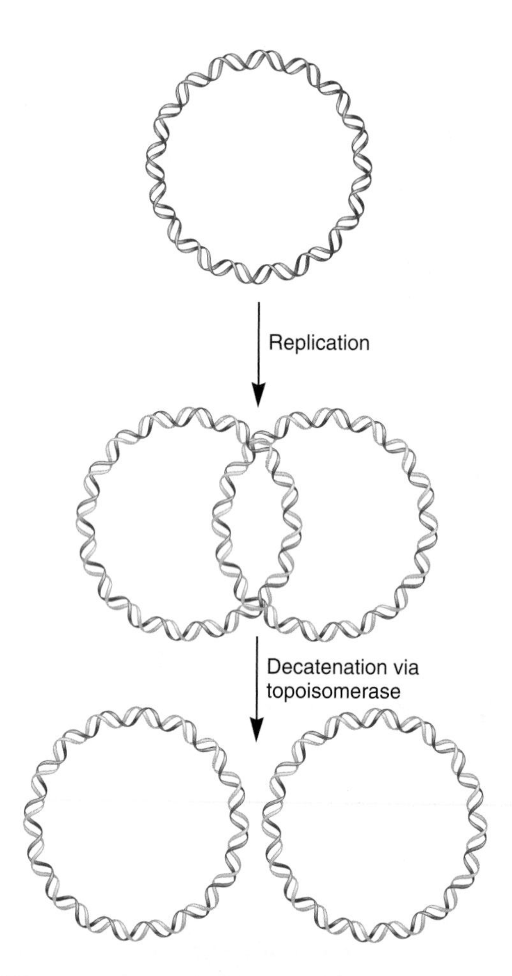

FIGURE 11.14 Separation of catenanes. DNA replication can result in two intertwined chromosomes called catenanes. These catenanes can be separated by the action of topoisomerase.

the fork moving right to left, but allows the advancement of the other fork. In any given cell, only one ter sequence is required to stop the advancement of one replication fork, and then the other fork ends its synthesis of DNA when it reaches the halted replication fork. In other words, DNA replication ends when oppositely advancing forks meet, usually at T1 or T2. Finally, DNA ligase covalently links the two daughter strands, creating two circular, double-stranded molecules.

After DNA replication is completed, one last problem may exist. DNA replication often results in two intertwined DNA molecules known as **catenanes** (**Figure 11.14**). Fortunately, catenanes are only transient structures in DNA replication. In *E. coli*, topoisomerase II introduces a temporary break into the DNA strands and then rejoins them after the strands have become unlocked. This allows the catenanes to be separated into individual circular molecules.

Certain Enzymes of DNA Replication Bind to Each Other to Form a Complex

Figure 11.15 provides a more three-dimensional view of the DNA replication process. DNA helicase and primase are physically bound to each other to form a complex known as a **primosome.** This complex leads the way at the replication fork. The primosome tracks along the DNA, separating the parental strands and synthesizing RNA primers at regular intervals along the lagging strand. By acting within a complex, the actions of DNA helicase and primase can be better coordinated.

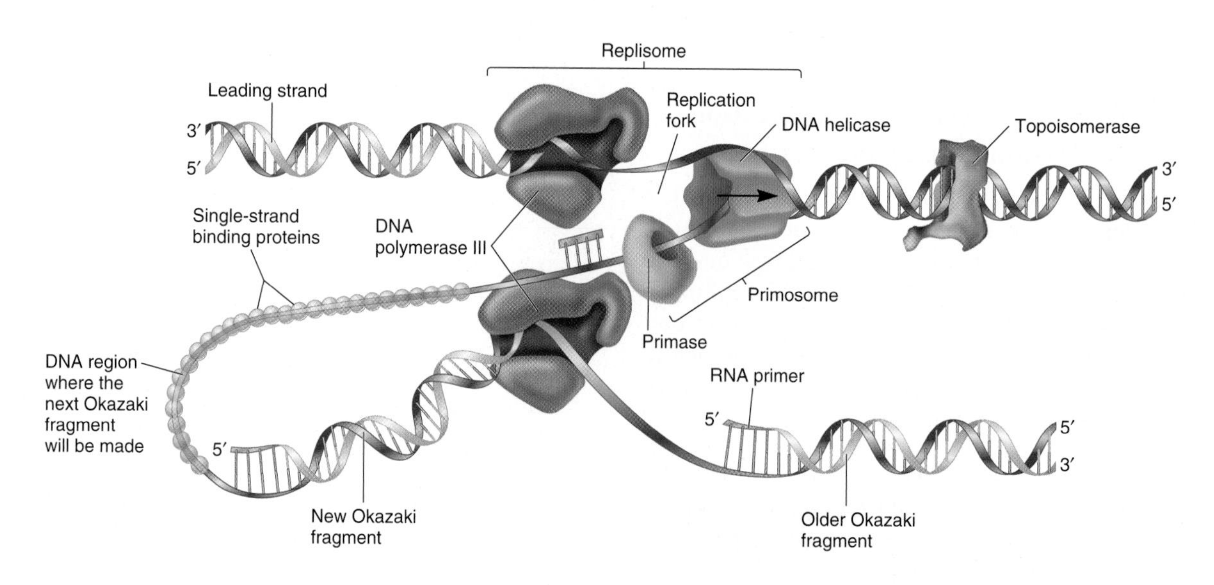

FIGURE 11.15 A three-dimensional view of DNA replication. DNA helicase and primase associate together to form a primosome. The primosome associates with two DNA polymerase enzymes to form a replisome.

The primosome is physically associated with two DNA polymerase holoenzymes to form a **replisome.** As shown in Figure 11.15, two DNA polymerase III proteins act in concert to replicate the leading and lagging strands. The term **dimeric DNA polymerase** is used to describe two DNA polymerase holoenzymes that move as a unit toward the replication fork. For this to occur, the lagging strand is looped out with respect to the DNA polymerase that synthesizes the lagging strand. This loop allows the lagging-strand polymerase to make DNA in a 5′ to 3′ direction yet move toward the opening of the replication fork. Interestingly, when this DNA polymerase reaches the end of an Okazaki fragment, it must be released from the template DNA and "hop" to the RNA primer that is closest to the fork. The clamp loader complex (see Table 11.2), which is part of DNA polymerase holoenzyme, then reloads the enzyme at the site where the next RNA primer has been made. Similarly, after primase synthesizes an RNA primer in the 5′ to 3′ direction, it must hop over the primer and synthesize the next primer closer to the replication fork.

The Fidelity of DNA Replication Is Ensured by Proofreading Mechanisms

With replication occurring so rapidly, one might imagine that mistakes can happen in which the wrong nucleotide is incorporated into the growing daughter strand. Although mistakes can happen during DNA replication, they are extraordinarily rare. In the case of DNA synthesis via DNA polymerase III, only one mistake per 100 million nucleotides is made. Therefore, DNA synthesis occurs with a high degree of accuracy or **fidelity.**

Why is the fidelity so high? First, the hydrogen bonding between G and C or A and T is much more stable than between mismatched pairs. However, this stability accounts for only part of the fidelity, because mismatching due to stability considerations accounts for 1 mistake per 1000 nucleotides.

Two characteristics of DNA polymerase also contribute to the fidelity of DNA replication. First, the active site of DNA polymerase preferentially catalyzes the attachment of nucleotides when the correct bases are located in opposite strands. Helix distortions caused by mispairing usually prevent an incorrect nucleotide from properly occupying the active site of DNA polymerase. By comparison, the correct nucleotide occupies the active site with precision and undergoes induced fit, which is necessary for catalysis. The inability of incorrect nucleotides to undergo induced fit decreases the error rate to a range of 1 in 100,000 to 1 million.

A second way that DNA polymerase decreases the error rate is by the enzymatic removal of mismatched nucleotides. As shown in **Figure 11.16**, DNA polymerase can identify a mismatched nucleotide and remove it from the daughter strand. This occurs by exonuclease cleavage of the bonds between adjacent nucleotides at the 3′ end of the newly made strand. The ability to remove mismatched bases by this mechanism is called the **proofreading function** of DNA polymerase. Proofreading occurs by the removal of nucleotides in the 3′ to 5′ direction at the 3′ exonuclease site. After the mismatched nucleotide is removed, DNA polymerase resumes DNA synthesis in the 5′ to 3′ direction.

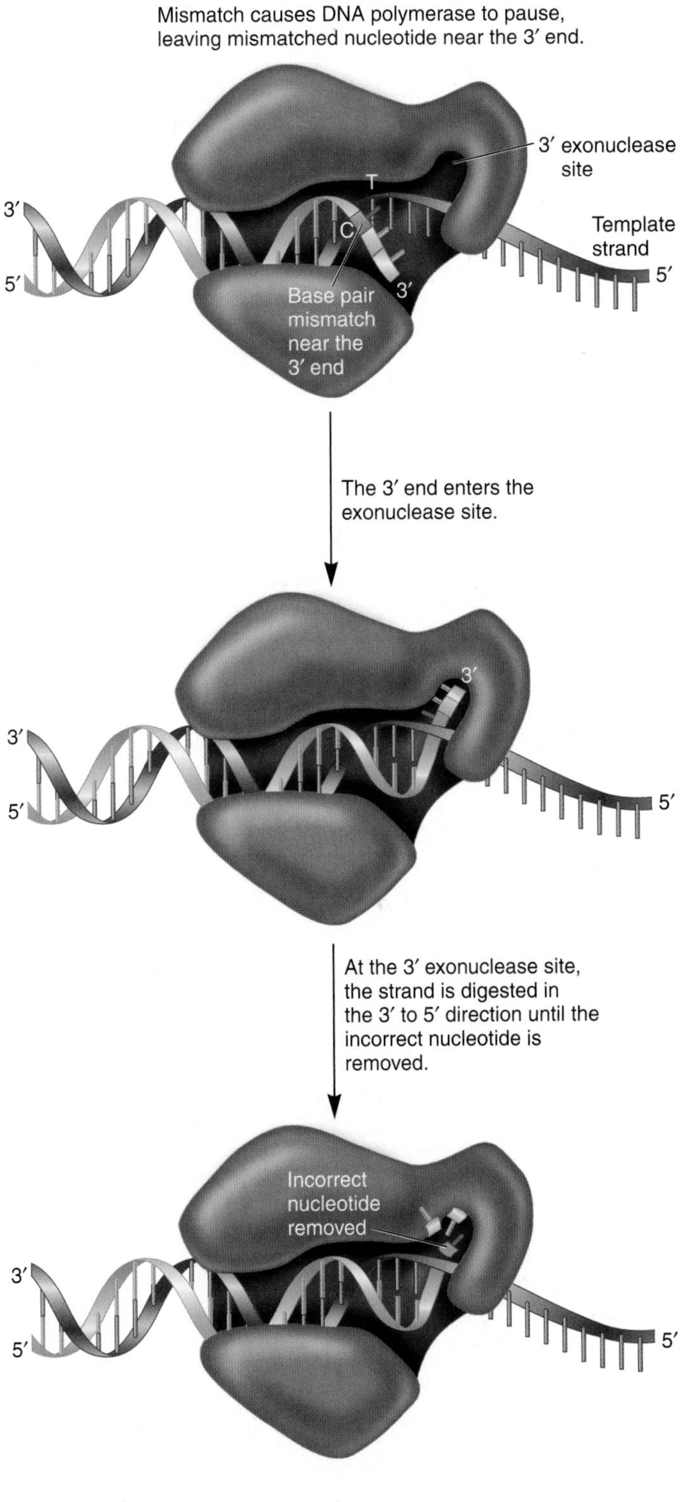

FIGURE 11.16 The proofreading function of DNA polymerase. When a base pair mismatch is found, the end of the newly made strand is shifted into the 3′ exonuclease site. The DNA is digested in the 3′ to 5′ direction to release the incorrect nucleotide.

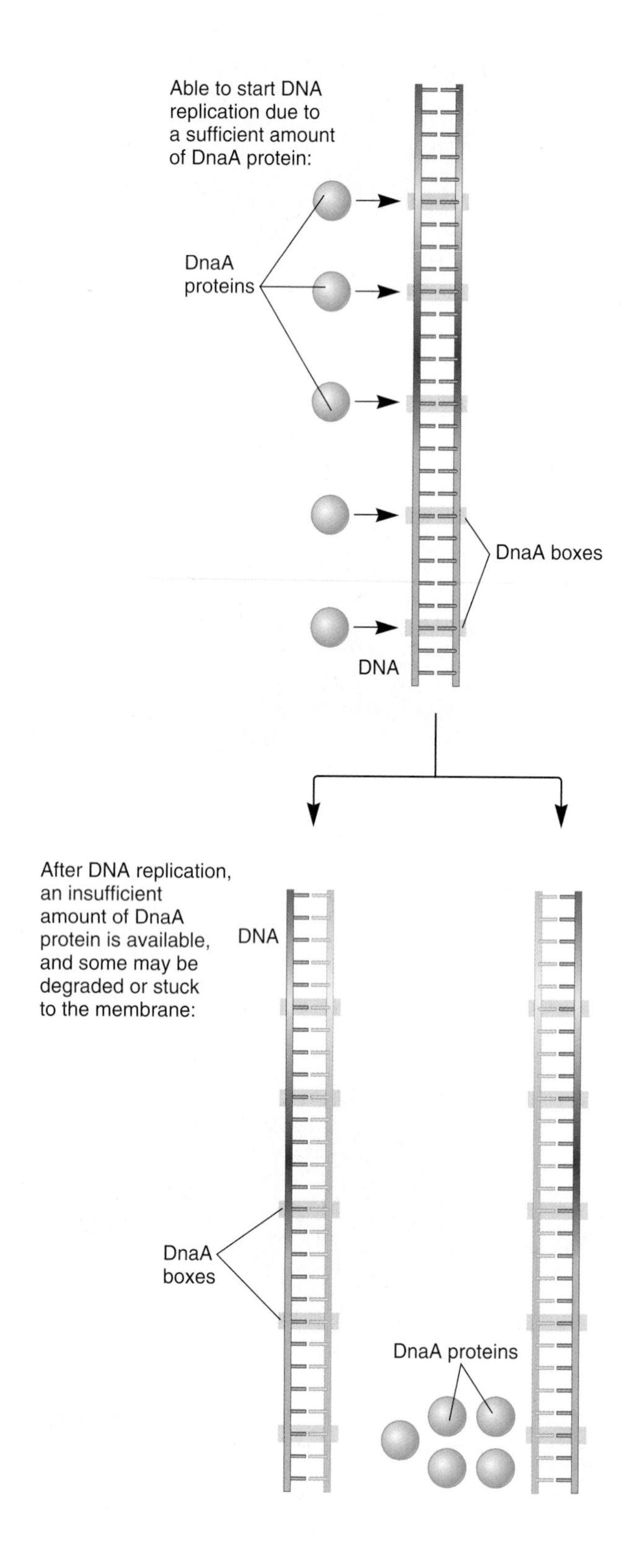

Bacterial DNA Replication Is Coordinated with Cell Division

Bacterial cells can divide into two daughter cells at an amazing rate. Under optimal conditions, certain bacteria such as *E. coli* can divide every 20 to 30 minutes. DNA replication should take place only when a cell is about to divide. If DNA replication occurs too frequently, too many copies of the bacterial chromosome will be found in each cell. Alternatively, if DNA replication does not occur frequently enough, a daughter cell will be left without a chromosome. Therefore, cell division in bacterial cells must be coordinated with DNA replication.

Bacterial cells regulate the DNA replication process by controlling the initiation of replication at *oriC*. This control has been extensively studied in *E. coli*. In this bacterium, several different mechanisms may control DNA replication. In general, the regulation prevents the premature initiation of DNA replication at *oriC*.

After the initiation of DNA replication, DnaA protein hydrolyzes its ATP and therefore switches to an ADP-bound form. DnaA-ADP has a lower affinity for DnaA boxes and does not readily form a complex. This prevents premature initiation. In addition, the initiation of replication is controlled by the amount of the DnaA protein (**Figure 11.17**). To initiate DNA replication, the concentration of the DnaA protein must be high enough so it can bind to all of the DnaA boxes and form a complex. Immediately following DNA replication, the number of DnaA boxes is double, so an insufficient amount of DnaA protein is available to initiate a second round of replication. Also, some of the DnaA protein may be rapidly degraded and some of it may be inactive because it becomes attached to other regions of chromosomal DNA and to the cell membrane during cell division. Because it takes time to accumulate newly made DnaA protein, DNA replication cannot occur until the daughter cells have had time to grow.

Another way to regulate DNA replication involves the GATC methylation sites within *oriC*. These sites can be methylated by an enzyme known as DNA adenine methyltransferase (Dam). The Dam enzyme recognizes the 5′–GATC–3′ sequence, binds there, and attaches a methyl group onto the adenine base, forming methyladenine (**Figure 11.18a**). DNA methylation within

FIGURE 11.17 The amount of DnaA protein provides a way to regulate DNA replication. To begin replication, enough DnaA protein must be present to bind to all of the DnaA boxes. Immediately after DNA replication, insufficient DnaA protein is available to reinitiate a second (premature) round of DNA replication at the two origins of replication. This is because twice as many DnaA boxes are found after DNA replication and because some DnaA proteins may be degraded or stuck to other chromosomal sites and to the cell membrane.

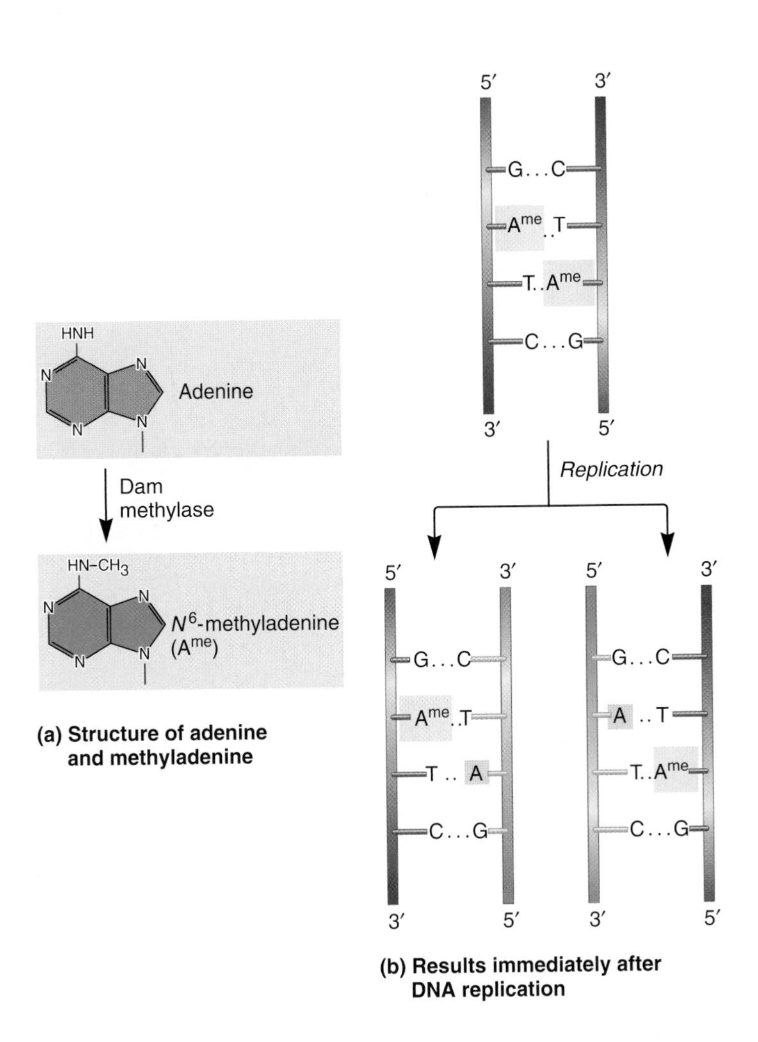

(a) Structure of adenine and methyladenine

(b) Results immediately after DNA replication

oriC helps regulate the replication process. Prior to DNA replication, these sites are methylated in both strands. This full methylation of the 5′–GATC–3′ sites facilitates the initiation of DNA replication at the origin. Following DNA replication, the newly made strands are not methylated, because adenine rather than methyladenine is incorporated into the daughter strands (**Figure 11.18b**). The initiation of DNA replication at the origin does not readily occur until after it has become fully methylated. Because it takes several minutes for Dam to methylate the 5′–GATC–3′ sequences within this region, DNA replication does not occur again too quickly.

FIGURE 11.18 **Methylation of GATC sites in *oriC*.** **(a)** The action of Dam (DNA adenine methyltransferase), which covalently attaches a methyl group to adenine to form methyladenine (A^{me}). **(b)** Prior to DNA replication, the action of Dam causes both adenines within the GATC sites to be methylated. After DNA replication, only the adenines in the original strands are methylated. Several minutes will pass before Dam methylates these unmethylated adenines.

EXPERIMENT 11B

DNA Replication Can Be Studied in Vitro

Much of our understanding of bacterial DNA replication has come from thousands of experiments in which DNA replication has been studied in vitro. This approach was pioneered by Arthur Kornberg in the 1950s, who received the 1959 Nobel Prize in Physiology or Medicine for his efforts.

Figure 11.19 describes Kornberg's approach to monitor DNA replication in vitro. In this experiment, an extract of proteins from *E. coli* was used. Although we do not consider the procedures for purifying replication proteins here, an alternative approach is to purify specific proteins from the extract and study their functions individually. In either case, the proteins are mixed with template DNA and radiolabeled nucleotides. Kornberg correctly hypothesized that dNTPs are the precursors for DNA synthesis. Also, he knew that dNTPs are soluble in an acidic solution, whereas long strands of DNA precipitate out of solution at an acidic pH. This precipitation event provides a method of separating nucleotides—in this case, dNTPs—from strands of DNA. Therefore, after the proteins, template DNA, and nucleotides were incubated for a sufficient time to allow the synthesis of new strands, step 3 of this procedure involved the addition of perchloric acid. This step precipitated strands of DNA, which were then separated from the radiolabeled nucleotides via centrifugation. Newly made strands of DNA, which were radiolabeled, sediment to the pellet, whereas radiolabeled nucleotides that had not been incorporated into new strands remained in the supernatant.

THE HYPOTHESIS

DNA synthesis can occur in vitro if all the necessary components are present.

TESTING THE HYPOTHESIS — FIGURE 11.19 In vitro synthesis of DNA strands.

Starting material: An extract of proteins from *E. coli*.

1. Mix together the extract of *E. coli* proteins, template DNA that is not radiolabeled, and ^{32}P-radiolabeled deoxyribonucleoside triphosphates. This is expected to be a complete system that contains everything necessary for DNA synthesis. As a control, a second sample is made in which the template DNA was omitted from the mixture.

2. Incubate the mixture for 30 minutes at 37°C.

3. Add perchloric acid to precipitate DNA. (It does not precipitate free nucleotides.)

4. Centrifuge the tube.
   Note: The radiolabeled deoxyribonucleoside triphosphates that have not been incorporated into DNA will remain in the supernatant.

5. Collect the pellet, which contains precipitated DNA and proteins. (The control pellet is not expected to contain DNA.)

6. Count the amount of radioactivity in the pellet using a scintillation counter. (See the Appendix.)

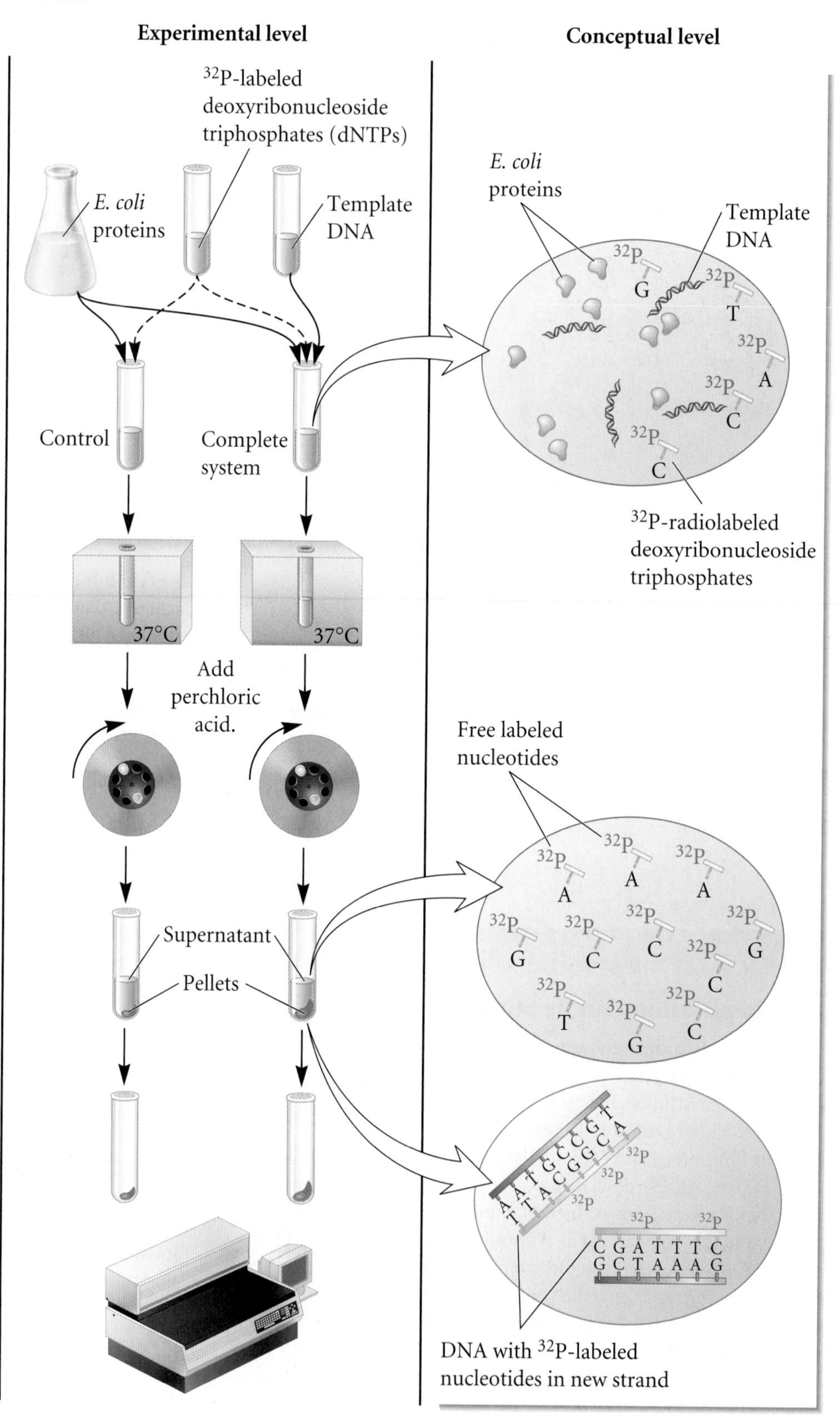

THE DATA

Conditions	*Amount of Radiolabeled DNA**
Complete system	3300
Control (template DNA omitted)	0

*Calculated in picomoles of ^{32}P-labeled DNA. Data from: M.J. Bessman, I.R. Lehman, E.S. Simms, and A. Kornberg (1958) Enzymatic synthesis of deoxyribonucleic acid. II. General properties of the reaction. *J Biol Chem 233*:171–177.

INTERPRETING THE DATA

As shown in the data after Figure 11.19, when the *E. coli* proteins were mixed with nonlabeled template DNA and radiolabeled dNTPs, an acid-precipitable, radiolabeled product was formed. This product was newly synthesized DNA strands. As a control, if nonradiolabeled template DNA was omitted from the assay, no radiolabeled DNA was made. This is the expected result, because the template DNA is necessary to make new daughter strands. Taken together, these results indicate that this technique can be used to measure the synthesis of DNA in vitro.

The in vitro approach has provided the foundation for studying the replication process at the molecular level. A common experimental strategy is to purify proteins from cell extracts and to determine their roles in the replication process. In other words, purified proteins, such as those described in Table 11.1, can be mixed with nucleotides, template DNA, and other substances in a test tube to determine if the synthesis of new DNA strands occurs. This approach still continues, particularly as we try to understand the added complexities of eukaryotic DNA replication.

A self-help quiz involving this experiment can be found at www.mhhe.com/brookergenetics4e.

The Isolation of Mutants Has Been Instrumental to Our Understanding of DNA Replication

In the previous experiment, we considered an experimental strategy for studying DNA synthesis in vitro. In his early experiments, Arthur Kornberg used crude extracts containing *E. coli* proteins and monitored their ability to synthesize DNA. In such extracts, the predominant polymerase is DNA polymerase I. Surprisingly, its activity is so high that it is nearly impossible to detect the activities of the other DNA polymerases. For this reason, researchers in the 1950s and 1960s thought that DNA polymerase I was the only enzyme responsible for DNA replication. This conclusion dramatically changed as a result of mutant isolation.

In 1969, Paula DeLucia and John Cairns identified a mutant *E. coli* strain in which DNA polymerase I lacked its 5′ to 3′ polymerase function but retained its 5′ to 3′ exonuclease function, which is needed to remove RNA primers. This mutant was identified by randomly testing thousands of bacterial colonies that had been subjected to mutagens—agents that cause mutations. Because this mutant strain could grow normally, DeLucia and Cairns concluded that the DNA-synthesizing function of DNA polymerase I is not absolutely required for bacteria to replicate their DNA. How is this possible? The researchers speculated that *E. coli* must have other DNA polymerases. Therefore, DeLucia and Cairns set out to find these seemingly elusive enzymes.

The isolation of mutants was one way that helped researchers identify additional DNA polymerase enzymes, namely DNA polymerase II and III. In addition, mutant isolation played a key role in the identification of other proteins needed to replicate the leading and lagging strands, as well as proteins that recognize the origin of replication and the ter sites. Because DNA replication is vital for cell division, mutations that block DNA replication would be lethal to a growing population of bacterial cells. For this reason, if researchers want to identify loss-of-function mutations in vital genes, they must screen for **conditional mutants.** One type of conditional mutant is a **temperature-sensitive (ts) mutant.** In the case of a vital gene, an organism harboring a ts mutation can survive at the permissive temperature but not at the nonpermissive temperature. For example, a ts mutant might survive and grow at 30°C (the permissive temperature) but fail to grow at 42°C (the nonpermissive temperature). The higher temperature inactivates the function of the protein encoded by the mutant gene.

Figure 11.20 shows a general strategy for the isolation of ts mutants. Researchers expose bacterial cells to a mutagen that increases the likelihood of mutations. The mutagenized cells are plated on growth media and incubated at the permissive temperature. The colonies are then replica plated onto two plates: one incubated at the permissive temperature and one at the nonpermissive temperature. As seen here, this enables researchers to identify ts mutations that are lethal at the nonpermissive temperature.

With regard to the study of DNA replication, researchers analyzed a large number of ts mutants to discover if any of them had a defect in DNA replication. For example, one could expose a ts mutant to radiolabeled thymine (a base that is incorporated into DNA), shift to the nonpermissive temperature, and determine if a mutant strain could make radiolabeled DNA, using procedures that are similar to those described in Figure 11.19. Because *E. coli* has many vital genes not involved with DNA replication, only a small subset of ts mutants would be expected to have mutations in genes that encode proteins that are critical to the replication process. Therefore, researchers had to screen many thousands of ts mutants to identify the few involved in DNA replication. This approach is sometimes called a "brute force" genetic screen.

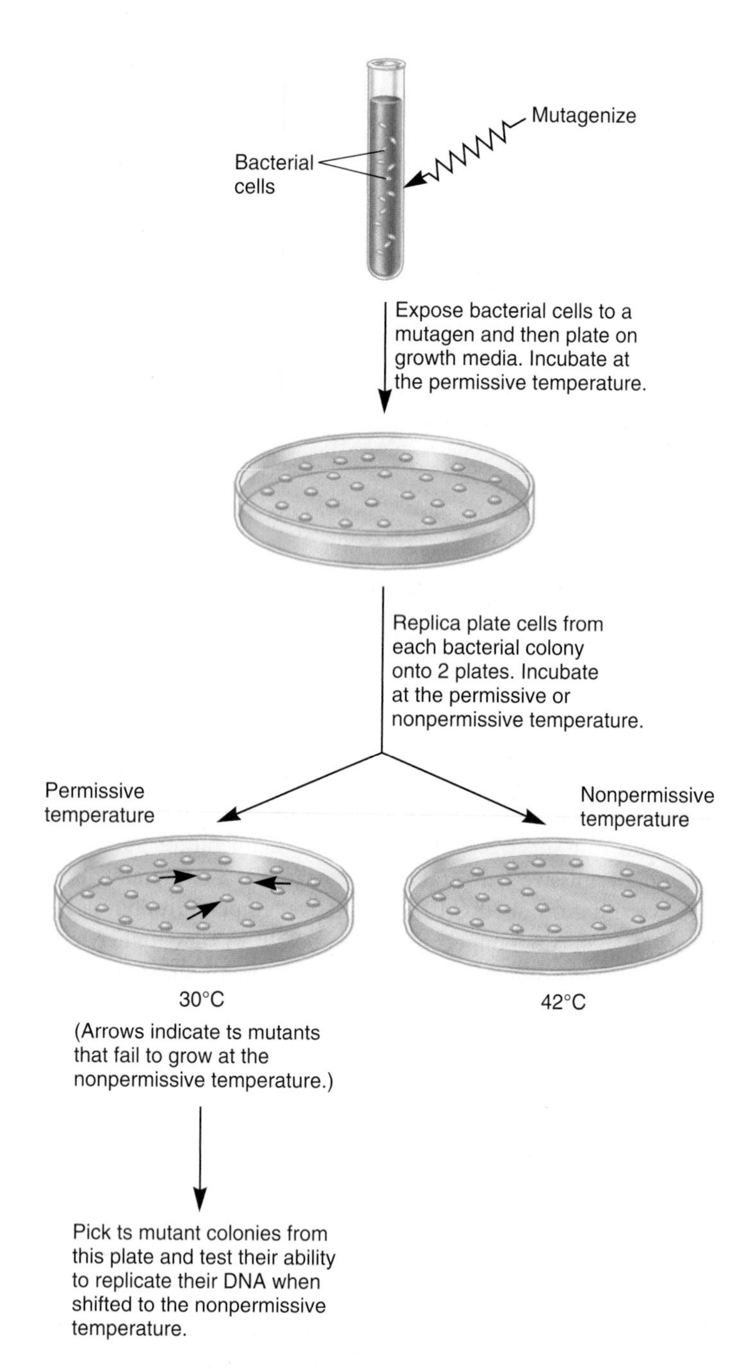

FIGURE 11.20 A strategy to identify ts mutations in vital genes. In this approach, bacteria are mutagenized, which increases the likelihood of mutation, and then grown at the permissive temperature. Colonies are then replica plated and grown at both the permissive and nonpermissive temperatures. (Note: The procedure of replica plating is shown in Chapter 16, Figure 16.7.) Ts mutants fail to grow at the nonpermissive temperature. The appropriate colonies can be picked from the plates, grown at the permissive temperature, and analyzed to see if DNA replication is altered at the nonpermissive temperature.

Table 11.3 summarizes some of the genes that were identified using this type of strategy. The genes were originally designated with the name dna, followed by a capital letter that generally refers to the order in which they were discovered. When shifted to the nonpermissive temperature, certain mutants showed a rapid arrest of DNA synthesis. These so-called rapid-stop mutations inactivated genes that encode enzymes needed for DNA replication. By comparison, other mutants were able to complete their current round of replication but could not start another round. These slow-stop mutants involved genes that encode proteins needed for the initiation of replication at the origin. In later studies, the proteins encoded by these genes were purified, and their functions were studied in vitro. This work contributed greatly to our modern understanding of DNA replication at the molecular level.

TABLE **11.3**

Examples of ts Mutants Involved in DNA Replication in *E. coli*

Gene Name	Protein Function
Rapid-Stop Mutants	
dnaE	α subunit of DNA polymerase III, synthesizes DNA
dnaX	τ subunit of DNA polymerase III, promotes the dimerization of two DNA polymerase III proteins together at the replication fork and stimulates DNA helicase
dnaN	β subunit of DNA polymerase III, functions as a clamp protein that makes DNA polymerase a processive enzyme
dnaZ	γ subunit of DNA polymerase III, helps the β subunit bind to the DNA
dnaG	Primase, needed to make RNA primers
dnaB	Helicase, needed to unwind the DNA strands during replication
Slow-Stop Mutants	
dnaA	DnaA protein that recognizes the DnaA boxes at the origin
dnaC	DnaC protein that recruits DNA helicase to the origin

11.3 EUKARYOTIC DNA REPLICATION

Eukaryotic DNA replication is not as well understood as bacterial replication. Much research has been carried out on a variety of experimental organisms, particularly yeast and mammalian cells. Many of these studies have found extensive similarities between the general features of DNA replication in prokaryotes and eukaryotes. For example, DNA helicases, topoisomerases, single-strand binding proteins, primases, DNA polymerases, and DNA ligases—the types of bacterial enzymes described in Table 11.1—have also been identified in eukaryotes. Nevertheless, at the molecular level, eukaryotic DNA replication appears to be substantially more complex. These additional intricacies of eukaryotic DNA replication are related to several features of eukaryotic cells. In particular, eukaryotic cells have larger, linear chromosomes, the chromatin is tightly packed within nucleosomes,

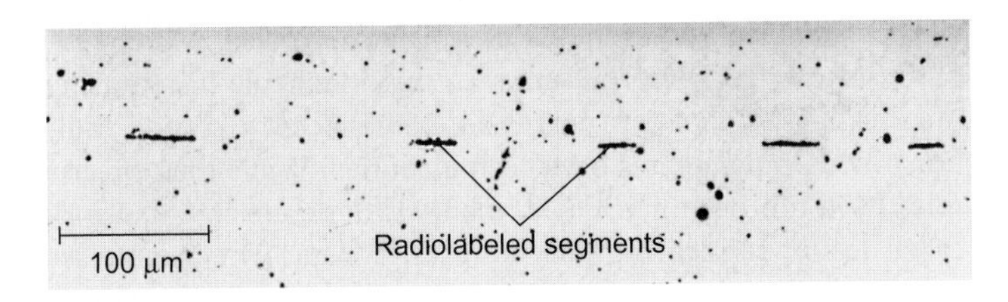

FIGURE 11.21 Evidence for multiple origins of replication in eukaryotic chromosomes. In this experiment, cells were given a pulse/chase of ^{3}H-thymidine and unlabeled thymidine. The chromosomes were isolated and subjected to autoradiography. In this micrograph, radiolabeled segments were interspersed among nonlabeled segments, indicating that eukaryotic chromosomes contain multiple origins of replication.

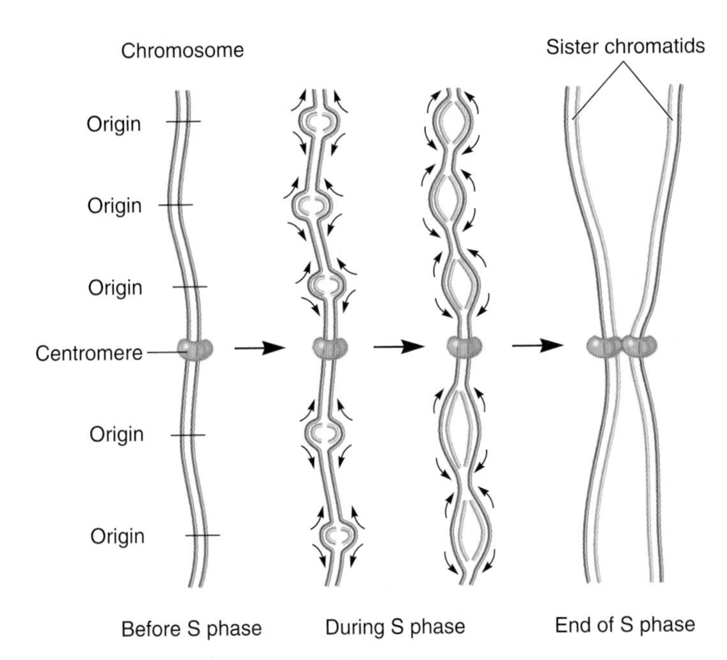

FIGURE 11.22 The replication of eukaryotic chromosomes. At the beginning of the S phase of the cell cycle, eukaryotic chromosome replication begins from multiple origins of replication. As S phase continues, the replication forks move bidirectionally to replicate the DNA. By the end of S phase, all of the replication forks have merged. The net result is two sister chromatids attached to each other at the centromere.

and cell cycle regulation is much more complicated. This section emphasizes some of the unique features of eukaryotic DNA replication.

Initiation Occurs at Multiple Origins of Replication on Linear Eukaryotic Chromosomes

Because eukaryotes have long, linear chromosomes, the chromosomes require multiple origins of replication so the DNA can be replicated in a reasonable length of time. In 1968, Joel Huberman and Arthur Riggs provided evidence for multiple origins of replication by adding a radiolabeled nucleoside (^{3}H-thymidine) to a culture of actively dividing cells, followed by a chase with nonlabeled thymidine. The radiolabeled thymidine was taken up by the cells and incorporated into their newly made DNA strands for a brief period. The chromosomes were then isolated from the cells and subjected to autoradiography. As seen in **Figure 11.21**, radiolabeled segments were interspersed among nonlabeled segments. This result is consistent with the hypothesis that eukaryotic chromosomes contain multiple origins of replication.

As shown schematically in **Figure 11.22**, DNA replication proceeds bidirectionally from many origins of replication during S phase of the cell cycle. The multiple replication forks eventually make contact with each other to complete the replication process.

The molecular features of eukaryotic origins of replication may have some similarities to the origins found in bacteria. At the molecular level, eukaryotic origins of replication have been extensively studied in the yeast *Saccharomyces cerevisiae*. In this organism, several replication origins have been identified and sequenced. They have been named **ARS elements** (for autonomously replicating sequence). ARS elements, which are about 50 bp in length, are necessary to initiate chromosome replication. ARS elements have unique features of their DNA sequences. First, they contain a higher percentage of A and T bases than the rest of the chromosomal DNA. In addition, they contain a copy of the ARS consensus sequence (ACS), ATTTAT(A or G)TTTA, along with additional elements that enhance origin function. This arrangement is similar to bacterial origins.

In *S. cerevisiae*, origins of replication are determined primarily by their DNA sequences. In animals, the critical features that define origins of replication are not completely understood. In many species, origins are not determined by particular DNA sequences but instead occur at specific sites along a chromosome due to chromatin structure and protein modifications.

DNA replication in eukaryotes begins with the assembly of a **prereplication complex (preRC)** consisting of at least 14 different proteins. Part of the preRC is a group of six proteins called the **origin recognition complex (ORC)** that acts as the initiator of eukaryotic DNA replication. The ORC was originally identified in yeast as a protein complex that binds directly to ARS elements. DNA replication at the origin begins with the binding of ORC, which usually occurs during G_1 phase. Other proteins of the preRC then bind, including a group of proteins called **MCM helicase.**[1] The binding of MCM helicase at the origin completes a process called **DNA replication licensing;** only those origins with MCM helicase can initiate DNA synthesis. During S phase, DNA synthesis begins when preRCs are acted on by at least 22 additional proteins that activate MCM helicase and assemble two divergent replication forks at each replication origin. An

[1] MCM is an acronym for minichromosome maintenance. The genes encoding MCM proteins were originally identified in mutant yeast strains that are defective in the maintenance of minichromosomes in the cell. MCM proteins have since been shown to play a role in DNA replication.

important role of these additional proteins is to carefully regulate the initiation of DNA replication so that it happens at the correct time during the cell cycle and occurs only once during the cell cycle. The precise roles of these proteins are under active research investigation.

Eukaryotes Contain Several Different DNA Polymerases

Eukaryotes have many types of DNA polymerases. For example, mammalian cells have well over a dozen different DNA polymerases (**Table 11.4**). Four of these, designated α (alpha), ε (epsilon), δ (delta), and γ (gamma), have the primary function of replicating DNA. DNA polymerase γ functions in the mitochondria to replicate mitochondrial DNA, whereas α, ε, and δ are involved with DNA replication in the cell nucleus during S phase.

DNA polymerase α is the only eukaryotic polymerase that associates with primase. The functional role of the DNA polymerase α/primase complex is to synthesize a short RNA-DNA primer of approximately 10 RNA nucleotides followed by 20 to 30 DNA nucleotides. This short RNA-DNA strand is then used by DNA polymerase ε or δ for the processive elongation of the leading and lagging strands, respectively. For this to happen, the DNA polymerase α/primase complex dissociates from the replication fork and is exchanged for DNA polymerase ε or δ. This exchange is called a **polymerase switch.** Accumulating evidence suggests that DNA polymerase ε is primarily involved with leading-strand synthesis, whereas DNA polymerase δ is responsible for lagging-strand synthesis.

What are the functions of the other DNA polymerases? Several of them also play an important role in DNA repair, a topic that will be examined in Chapter 16. DNA polymerase β, which has been studied for several decades, is not involved in the replication of normal DNA, but plays an important role in removing incorrect bases from damaged DNA. More recently, several additional DNA polymerases have been identified. While their precise roles have not been elucidated, many of these are in a category called **lesion-replicating polymerases.** When DNA polymerase α, δ, and ε encounter abnormalities in DNA structure, such as abnormal bases or cross-links, they may be unable to replicate over the aberration. When this occurs, lesion-replicating polymerases are attracted to the damaged DNA and have special properties that enable them to synthesize a complementary strand over the abnormal region. Each type of lesion-replicating polymerase may be able to replicate over a different kind of DNA damage.

TABLE **11.4**

Eukaryotic DNA Polymerases

Polymerase Types*	Function
α	Initiates DNA replication in conjunction with primase
ε	Replication of the leading strand during S phase
δ	Replication of the lagging strand during S phase
γ	Replication of mitochondrial DNA
η, κ, ι, ξ (lesion-replicating polymerases)	Replication of damaged DNA
α, β, δ, ε, σ, λ, μ, φ, θ, η	DNA repair or other functions†

*The designations are those of mammalian enzymes.

†Many DNA polymerases have dual functions. For example, DNA polymerases α, δ, and ε are involved in the replication of normal DNA and also play a role in DNA repair. In cells of the immune system, certain genes that encode antibodies (i.e., immunoglobulin genes) undergo a phenomenon known as hypermutation. This increases the variation in the kinds of antibodies the cells can make. Certain polymerases in this list, such as η, may play a role in hypermutation of immunoglobulin genes. DNA polymerase σ may play a role in sister chromatid cohesion, a topic discussed in Chapter 10.

Flap Endonuclease Removes RNA Primers During Eukaryotic DNA Replication

Another key difference between bacterial and eukaryotic DNA replication is the way that RNA primers are removed. As discussed earlier in this chapter, bacterial RNA primers are removed by DNA polymerase I. By comparison, a DNA polymerase enzyme does not play this role in eukaryotes. Instead, an enzyme called flap endonuclease is primarily responsible for RNA primer removal.

Flap endonuclease gets its name because it removes small pieces of RNA flaps that are generated by the action of DNA polymerase δ. In the diagram shown in **Figure 11.23**, DNA polymerase δ elongates the left Okazaki fragment until it runs into the RNA primer of the adjacent Okazaki fragment on the right. This causes a portion of the RNA primer to form a short flap, which is removed by the endonuclease function of flap endonuclease. As DNA polymerase δ continues to elongate the DNA, short flaps continue to be generated, which are sequentially removed by flap endonuclease. Eventually, all of the RNA primer is removed, and DNA ligase can seal the DNA fragments together.

Though flap endonuclease is thought to be the primary pathway for RNA primer removal in eukaryotes, it is unable to remove a flap that is too long. In such cases, a long flap is cleaved by the enzyme called Dna2 nuclease/helicase. This enzyme can cut a long flap, thereby generating a short flap. The short flap is then removed via flap endonuclease.

The Ends of Eukaryotic Chromosomes Are Replicated by Telomerase

Linear eukaryotic chromosomes contain **telomeres** at both ends. The term telomere refers to the complex of telomeric sequences within the DNA and the special proteins that are bound to these sequences. Telomeric sequences consist of a moderately repetitive tandem array and a 3′ overhang region that is 12 to 16 nucleotides in length (**Figure 11.24**).

The tandem array that occurs within the telomere has been studied in a wide variety of eukaryotic organisms. A common feature is that the telomeric sequence contains several guanine

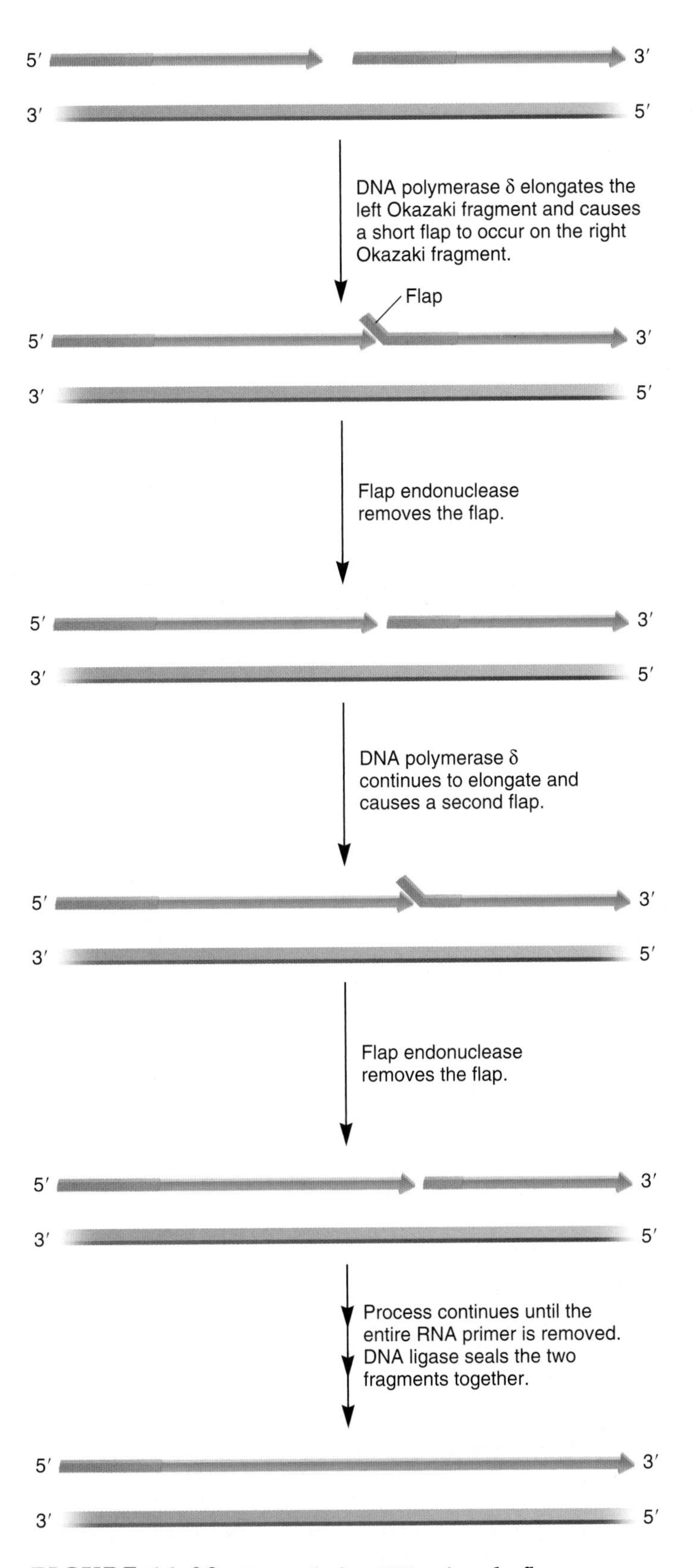

FIGURE 11.23 Removal of an RNA primer by flap endonuclease.

nucleotides and often many thymine nucleotides (**Table 11.5**). Depending on the species and the cell type, this sequence can be tandemly repeated up to several hundred times in the telomere region.

One reason why telomeric repeat sequences are needed is because DNA polymerase is unable to replicate the 3′ ends of DNA strands. Why is DNA polymerase unable to replicate this region? The answer lies in the two unusual enzymatic features of this enzyme. As discussed previously, DNA polymerase synthesizes DNA only in a 5′ to 3′ direction, and it cannot link together the first two individual nucleotides; it can elongate only preexisting strands. These two features of DNA polymerase function pose a problem at the 3′ ends of linear chromosomes. As shown in **Figure 11.25**, the 3′ end of a DNA strand cannot be replicated by DNA polymerase because a primer cannot be made upstream from this point. Therefore, if this problem were not solved, the chromosome would become progressively shorter with each round of DNA replication.

To prevent the loss of genetic information due to chromosome shortening, additional DNA sequences are attached to the ends of telomeres. In 1984, Carol Greider and Elizabeth Blackburn discovered an enzyme called **telomerase** that prevents chromosome shortening. It recognizes the sequences at the ends of eukaryotic chromosomes and synthesizes additional repeats of telomeric sequences. They received the 2009 Nobel Prize in physiology or medicine for their discovery. **Figure 11.26** shows the interesting mechanism by which telomerase works. The telomerase enzyme contains both protein subunits and RNA. The RNA part of telomerase contains a sequence complementary to the DNA sequence found in the telomeric repeat. This allows telomerase to bind to the 3′ overhang region of the telomere. Following binding, the RNA sequence beyond the binding site functions as a template allowing the synthesis of a six-nucleotide sequence at the end of the DNA strand. This is called polymerization, because it is analogous to the function of DNA polymerase. It is catalyzed by two identical protein subunits called **telomerase reverse transcriptase (TERT).** TERT's name indicates that it uses an RNA template to synthesize DNA. Following polymerization, the telomerase can then move—a process called translocation—to the new end of this DNA strand and attach another six nucleotides to the end. This binding-polymerization-translocation cycle occurs many times in a row, thereby greatly lengthening the 3′ end of the DNA strand in the telomeric region. The complementary strand is then synthesized by primase, DNA polymerase, and DNA ligase, as described earlier in this chapter.

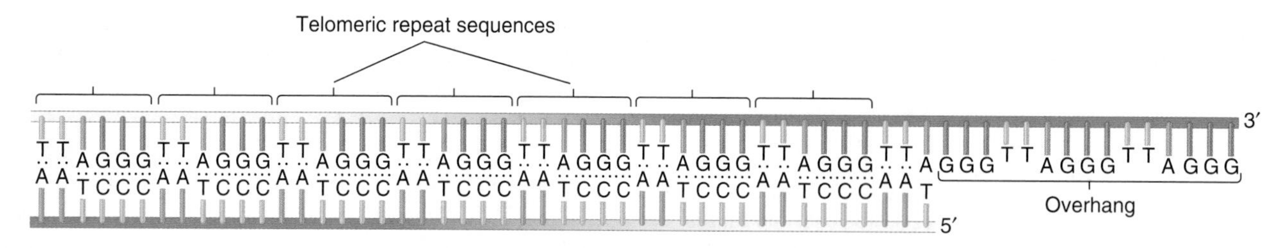

ONLINE ANIMATION

FIGURE 11.24 General structure of telomeric sequences. The telomere DNA consists of a tandemly repeated sequence and a 12- to 16-nucleotide overhang.

TABLE 11.5

Telomeric Repeat Sequences Within Selected Organisms

Group	Examples	Telomeric Repeat Sequence
Mammals	Humans	TTAGGG
Slime molds	*Physarum, Didymium*	TTAGGG
	Dictyostelium	$AG_{(1-8)}$
Filamentous fungi	*Neurospora*	TTAGGG
Budding yeast	*Saccharomyces cerevisiae*	$TG_{(1-3)}$
Ciliates	*Tetrahymena*	TTGGGG
	Paramecium	TTGGG(T/G)
	Euplotes	TTTTGGGG
Higher plants	*Arabidopsis*	TTTAGGG

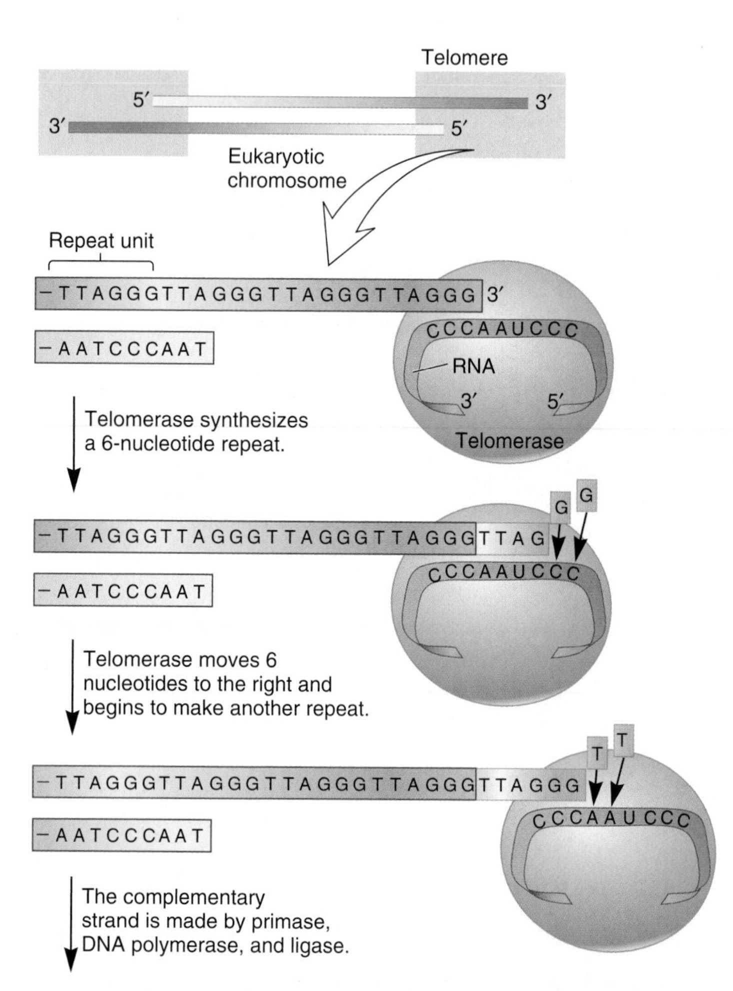

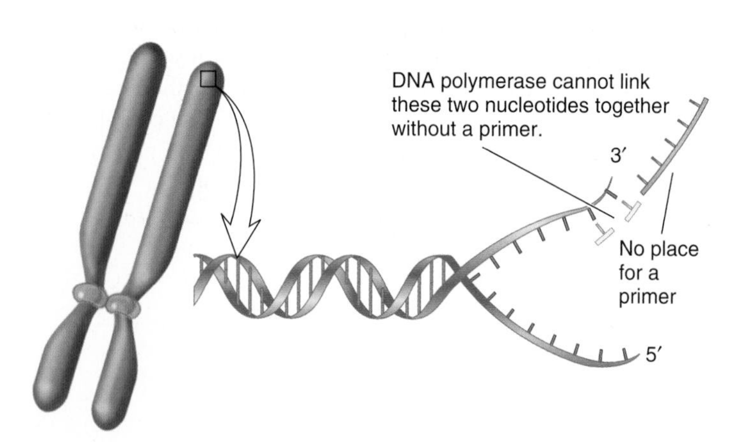

FIGURE 11.25 The replication problem at the ends of linear chromosomes. DNA polymerase cannot synthesize a DNA strand that is complementary to the 3′ end because a primer cannot be made upstream from this site.

ONLINE ANIMATION

FIGURE 11.26 The enzymatic action of telomerase. A short, three-nucleotide segment of RNA within telomerase causes it to bind to the 3′ overhang. The adjacent part of the RNA is used as a template to make a short, six-nucleotide repeat of DNA. After the repeat is made, telomerase moves six nucleotides to the right and then synthesizes another repeat. This process is repeated many times to lengthen the top strand shown in this figure. The bottom strand is made by DNA polymerase, using an RNA primer at the end of the chromosome that is complementary to the telomeric repeat sequence in the top strand. DNA polymerase fills in the region, which is sealed by ligase.

KEY TERMS

Page 270. DNA replication
Page 271. template strands, parental strands, daughter strands
Page 272. conservative model, semiconservative model, dispersive model
Page 274. origin of replication, bidirectionally, replication forks
Page 275. DnaA proteins, DnaA box sequences, DNA helicase
Page 276. bidirectional replication
Page 277. topoisomerase (type II), DNA gyrase, single-strand binding proteins, RNA primers, primase, leading strand, lagging strand, DNA polymerase
Page 278. Okazaki fragments
Page 279. DNA ligase, pulse/chase experiment
Page 280. processive enzyme
Page 281. termination sequences
Page 282. catenanes, primosome
Page 283. replisome, dimeric DNA polymerase, fidelity, proofreading function
Page 287. conditional mutants, temperature-sensitive (ts) mutant
Page 289. ARS elements, prereplication complex (preRC), origin recognition complex (ORC), MCM helicase, DNA replication licensing
Page 290. polymerase switch, lesion-replicating polymerases, flap endonuclease, telomeres
Page 291. telomerase, telomerase reverse transcriptase (TERT)

CHAPTER SUMMARY

- DNA replication is the process in which existing DNA strands are used to make new DNA strands.

11.1 Structural Overview of DNA Replication

- DNA replication occurs when the strands of DNA unwind and each strand is used as a template to make a new strand according to the AT/GC rule. The resulting DNA molecules have the same base sequence as the original DNA (see Figure 11.1).
- By labeling DNA with heavy and light isotopes of nitrogen and using centrifugation, Meselson and Stahl showed that DNA replication is semiconservative (see Figures 11.2, 11.3).

11.2 Bacterial DNA Replication

- Bacterial DNA replication begins at a single origin of replication and proceeds bidirectionally around the circular chromosome (see Figure 11.4).
- In *E. coli*, DNA replication is initiated when DnaA proteins bind to five DnaA boxes at the origin of replication and cause the AT-rich region to unwind. DNA helicases then promote the movement of two forks (see Figures 11.5, 11.6).
- At each replication fork, DNA helicase unwinds the DNA and topisomerase alleviates positive supercoiling. Single-strand binding proteins coat the DNA to prevent the strands from coming back together. Primase synthesizes RNA primers and DNA polymerase synthesizes complementary strands of DNA. DNA ligase seals the gaps between Okazaki fragments (see Figure 11.7, Table 11.1).
- DNA polymerase III in *E. coli* is an enzyme with several subunits that wraps around the DNA like a hand (see Figure 11.8, Table 11.2).
- DNA polymerase enzymes need a primer to synthesize DNA and make new DNA strands in a 5′ to 3′ direction (see Figure 11.9).
- During DNA synthesis, the leading strand is made continuously in the direction of the replication fork, whereas the lagging strand is made as Okazaki fragments in the direction away from the fork (see Figures 11.10, 11.11).
- DNA polymerase III is a processive enzyme that uses deoxynucleoside triphosphates to make new DNA strands (see Figure 11.12).
- In *E. coli*, DNA replication is terminated at ter sequences (see Figure 11.13).
- Following DNA replication, interlocked catenanes sometimes need to be unlocked via topoisomerase II (see Figure 11.14).
- The primosome is a complex between helicase and primase. The replisome is a complex between the primosome and dimeric DNA polymerase (see Figure 11.15).
- The high fidelity of DNA replication is a result of (1) the stability of hydrogen bonding between the correct bases, (2) the phenomenon of induced fit, and (3) the proofreading ability of DNA polymerase (see Figure 11.16).
- Bacterial DNA replication is regulated by the amount of DnaA protein and by the methylation of GATC sites in *oriC* (see Figures 11.17, 11.18).
- Kornberg devised a method to measure DNA replication in vitro (see Figure 11.19).
- The isolation and characterization of temperature-sensitive mutants was a useful strategy for identifying proteins involved with DNA replication (see Figure 11.20, Table 11.3).

11.3 Eukaryotic DNA Replication

- Eukaryotic chromosomes contain multiple origins of replication. Part of the prereplication complex is formed from a group of six proteins called the origin recognition complex. The binding of MCM helicase completes a process called DNA replication licensing (see Figures 11.21, 11.22).
- Eukaryotes have several different DNA polymerases with specialized roles. Different types of DNA polymerases switch with each other during the process of DNA replication (see Table 11.4).
- Flap endonuclease is an enzyme that removes RNA primers from Okazaki fragments (see Figure 11.23).

- The ends of eukaryotic chromosomes contain telomeres, which are composed of short repeat sequences and proteins (see Figure 11.24, Table 11.5).
- DNA polymerase is unable to replicate the very end of a eukaryotic chromosome (see Figure 11.25).
- Telomerase uses a short RNA molecule as a template to add repeat sequences onto telomeres (see Figure 11.26).

PROBLEM SETS & INSIGHTS

Solved Problems

S1. Describe three ways to account for the high fidelity of DNA replication. Discuss the quantitative contributions of each of the three ways.

Answer:

First: AT and GC pairs are preferred in the double-helix structure. This provides fidelity to around 1 mistake per 1000.

Second: Induced fit by DNA polymerase prevents covalent bond formation unless the proper nucleotides are in place. This increases fidelity another 100- to 1000-fold, to about 1 error in 100,000 to 1 million.

Third: Exonuclease proofreading increases fidelity another 100- to 1000-fold, to about 1 error per 100 million nucleotides added.

S2. What do you think would happen if the ter sequences were deleted from the bacterial DNA?

Answer: Instead of meeting at the ter sequences, the two replication forks would meet somewhere else. This would depend on how fast they were moving. For example, if one fork was advancing faster than the other, they would meet closer to where the slower-moving fork started. In fact, researchers have actually conducted this experiment. Interestingly, *E. coli* without the ter sequences seemed to survive just fine.

S3. Summarize the steps that occur in the process of chromosomal DNA replication in *E. coli*.

Answer:

Step 1. DnaA proteins bind to the origin of replication, resulting in the separation of the AT-rich region.

Step 2. DNA helicase breaks the hydrogen bonds between the DNA strands, topoisomerases alleviate positive supercoiling, and single-strand binding proteins hold the parental strands apart.

Step 3. Primase synthesizes one RNA primer in the leading strand and many RNA primers in the lagging strand. DNA polymerase III then synthesizes the daughter strands of DNA. In the lagging strand, many short segments of DNA (Okazaki fragments) are made. DNA polymerase I removes the RNA primers and fills in with DNA, and DNA ligase covalently links the Okazaki fragments together.

Step 4. The processes described in steps 2 and 3 continue until the two replication forks reach the ter sequences on the other side of the circular bacterial chromosome.

Step 5. Topoisomerases unravel the intertwined chromosomes, if necessary.

S4. If a strain of *E. coli* overproduced the Dam enzyme, how would that affect the DNA replication process? Would you expect such a strain to have more or fewer chromosomes per cell compared with a normal strain of *E. coli*? Explain why.

Answer: If a strain overproduced the Dam enzyme, DNA would replicate more rapidly. The GATC methylation sites in the origin of replication have to be fully methylated for DNA replication to occur. Immediately after DNA replication, a delay occurs before the next round of DNA replication because the two copies of newly replicated DNA are hemimethylated. A strain that overproduces Dam would rapidly convert the hemimethylated DNA into fully methylated DNA and more quickly allow the next round of DNA replication to occur. For this reason, the overproducing strain might have more copies of the *E. coli* chromosome because it would not have a long delay in DNA replication.

Conceptual Questions

C1. What key structural features of the DNA molecule underlie its ability to be faithfully replicated?

C2. With regard to DNA replication, define the term bidirectional replication.

C3. Which of the following statements is not true? Explain why.

A. A DNA strand can serve as a template strand on many occasions.

B. Following semiconservative DNA replication, one strand is a newly made daughter strand and the other strand is a parental strand.

C. A DNA double helix may contain two strands of DNA that were made at the same time.

D. A DNA double helix obeys the AT/GC rule.

E. A DNA double helix could contain one strand that is 10 generations older than its complementary strand.

C4. The compound known as nitrous acid is a reactive chemical that replaces amino groups ($-NH_2$) with keto groups ($=O$). When nitrous acid reacts with the bases in DNA, it can change cytosine to uracil and change adenine to hypoxanthine. A DNA double helix has the following sequence:

TTGGATGCTGG
AACCTACGACC

A. What would be the sequence of this double helix immediately after reaction with nitrous acid? Let the letter H represent hypoxanthine and U represent uracil.

B. Let's suppose this DNA was reacted with nitrous acid. The nitrous acid was then removed, and the DNA was replicated for two generations. What would be the sequences of the DNA products after the DNA had replicated twice? Your answer should contain the sequences of four double helices. Note: During DNA replication, uracil hydrogen bonds with adenine, and hypoxanthine hydrogen bonds with cytosine.

C5. One way that bacterial cells regulate DNA replication is by GATC methylation sites within the origin of replication. Would this mechanism work if the DNA was conservatively (rather than semiconservatively) replicated?

C6. The chromosome of *E. coli* contains 4.6 million bp. How long will it take to replicate its DNA? Assuming DNA polymerase III is the primary enzyme involved and this enzyme can actively proofread during DNA synthesis, how many base pair mistakes will be made in one round of DNA replication in a bacterial population containing 1000 bacteria?

C7. Here are two strands of DNA.

——————————————DNA polymerase–>

————————————————————————————

The one on the bottom is a template strand, and the one on the top is being synthesized by DNA polymerase in the direction shown by the arrow. Label the 5′ and 3′ ends of the top and bottom strands.

C8. A DNA strand has the following sequence:

5′–GATCCCGATCCGCATACATTTACCAGATCACCACC–3′

In which direction would DNA polymerase slide along this strand (from left to right or from right to left)? If this strand was used as a template by DNA polymerase, what would be the sequence of the newly made strand? Indicate the 5′ and 3′ ends of the newly made strand.

C9. List and briefly describe the three types of sequences within bacterial origins of replication that are functionally important.

C10. As shown in Figure 11.5, five DnaA boxes are found within the origin of replication in *E. coli*. Take a look at these five sequences carefully.

A. Are the sequences of the five DnaA boxes very similar to each other? (Hint: Remember that DNA is double-stranded; think about these sequences in the forward and reverse direction.)

B. What is the most common sequence for the DnaA box? In other words, what is the most common base in the first position, second position, and so on until the ninth position? The most common sequence is called the consensus sequence.

C. The *E. coli* chromosome is about 4.6 million bp long. Based on random chance, is it likely that the consensus sequence for a DnaA box occurs elsewhere in the *E. coli* chromosome? If so, why aren't there multiple origins of replication in *E. coli*?

C11. Obtain two strings of different colors (e.g., black and white) that are the same length. A length of 20 inches is sufficient. Tie a knot at one end of the black string, and tie a knot at one end of the white string. Each knot designates the 5′ end of your strings. Make a double helix with your two strings. Now tape one end of the double helix to a table so that the tape is covering the knot on the black string.

A. Pretend your hand is DNA helicase and use your hand to unravel the double helix, beginning at the end that is not taped to the table. Should your hand be sliding along the white string or the black string?

B. As in Figure 11.15, imagine that your two hands together form a dimeric replicative DNA polymerase. Unravel your two strings halfway to create a replication fork. Grasp the black string with your left hand and the white string with your right hand. Your thumbs should point toward the 5′ end of each string. You need to loop one of the strings so that one of the DNA polymerases can synthesize the lagging strand. With such a loop, the dimeric replicative DNA polymerase can move toward the replication fork and synthesize both DNA strands in the 5′ to 3′ direction. In other words, with such a loop, your two hands can touch each other with both of your thumbs pointing toward the fork. Should the black string be looped, or should the white string be looped?

C12. Sometimes DNA polymerase makes a mistake, and the wrong nucleotide is added to the growing DNA strand. With regard to pyrimidines and purines, two general types of mistakes are possible. The addition of an incorrect pyrimidine instead of the correct pyrimidine (e.g., adding cytosine where thymine should be added) is called a transition. If a pyrimidine is incorrectly added to the growing strand instead of purine (e.g., adding cytosine where an adenine should be added), this type of mistake is called a transversion. If a transition or transversion is not detected by DNA polymerase, a mutation is created that permanently changes the DNA sequence. Though both types of mutations are rare, transition mutations are more frequent than transversion mutations. Based on your understanding of DNA replication and DNA polymerase, offer three explanations why transition mutations are more common.

C13. A short genetic sequence, which may be recognized by DNA primase, is repeated many times throughout the *E. coli* chromosome. Researchers have hypothesized that DNA primase may recognize this sequence as a site to begin the synthesis of an RNA primer for DNA replication. The *E. coli* chromosome is roughly 4.6 million bp in length. How many copies of the DNA primase recognition sequence would be necessary to replicate the entire *E. coli* chromosome?

C14. Single-strand binding proteins keep the two parental strands of DNA separated from each other until DNA polymerase has an opportunity to replicate the strands. Suggest how single-strand binding proteins keep the strands separated and yet do not impede the ability of DNA polymerase to replicate the strands.

C15. The ability of DNA polymerase to digest a DNA strand from one end is called its exonuclease activity. Exonuclease activity is used to digest RNA primers and also to proofread a newly made DNA strand. Note: DNA polymerase I does not change direction while it is removing an RNA primer and synthesizing new DNA. It does change direction during proofreading.

A. In which direction, 5′ to 3′ or 3′ to 5′, is the exonuclease activity occurring during the removal of RNA primers and during the proofreading and removal of mistakes following DNA replication?

B. Figure 11.16 shows a drawing of the 3′ exonuclease site. Do you think this site would be used by DNA polymerase I to remove RNA primers? Why or why not?

C16. In the following drawing, the top strand is the template DNA, and the bottom strand shows the lagging strand prior to the action of

DNA polymerase I. The lagging strand contains three Okazaki fragments. The RNA primers have not yet been removed.

The top strand is the template DNA

3'————————————————————————————5'

5'*************____**************____*************_____3'

RNA primer ↑ RNA primer ↑ RNA primer

|—————————||——————————||——————————|

Left Okazaki fragment Middle Okazaki fragment Right Okazaki fragment

A. Which Okazaki fragment was made first, the one on the left or the one on the right?

B. Which RNA primer would be the first one to be removed by DNA polymerase I, the primer on the left or the primer on the right? For this primer to be removed by DNA polymerase I and for the gap to be filled in, is it necessary for the Okazaki fragment in the middle to have already been synthesized? Explain why.

C. Let's consider how DNA ligase connects the left Okazaki fragment with the middle Okazaki fragment. After DNA polymerase I removes the middle RNA primer and fills in the gap with DNA, where does DNA ligase function? See the arrows on either side of the middle RNA primer. Is ligase needed at the left arrow, at the right arrow, or both?

D. When connecting two Okazaki fragments, DNA ligase needs to use NAD^+ or ATP as a source of energy to catalyze this reaction. Explain why DNA ligase needs another source of energy to connect two nucleotides, but DNA polymerase needs nothing more than the incoming nucleotide and the existing DNA strand. Note: You may want to refer to Figure 11.12 to answer this question.

C17. What is DNA methylation? Why is DNA in a hemimethylated condition immediately after DNA replication? What are the functional consequences of methylation in the regulation of DNA replication?

C18. Describe the three important functions of the DnaA protein.

C19. If a strain of bacteria was making too much DnaA protein, how would you expect this to affect its ability to regulate DNA replication? With regard to the number of chromosomes per cell, how might this strain differ from a normal bacterial strain?

C20. Draw a picture that illustrates how DNA helicase works.

C21. What is an Okazaki fragment? In which strand of DNA are Okazaki fragments found? Based on the properties of DNA polymerase, why is it necessary to make these fragments?

C22. Discuss the similarities and differences in the synthesis of DNA in the lagging and leading strands. What is the advantage of a primosome and a replisome as opposed to having all replication enzymes functioning independently of each other?

C23. Explain the proofreading function of DNA polymerase.

C24. What is a processive enzyme? Explain why this is an important feature of DNA polymerase.

C25. Why is it important for living organisms to regulate DNA replication?

C26. What enzymatic features of DNA polymerase prevent it from replicating one of the DNA strands at the ends of linear chromosomes? Compared with DNA polymerase, how is telomerase different in its ability to synthesize a DNA strand? What does telomerase use as its template for the synthesis of a DNA strand? How does the use of this template result in a telomere sequence that is tandemly repetitive?

C27. As shown in Figure 11.26, telomerase attaches additional DNA, six nucleotides at a time, to the ends of eukaryotic chromosomes. However, it works in only one DNA strand. Describe how the opposite strand is replicated.

C28. If a eukaryotic chromosome has 25 origins of replication, how many replication forks does it have at the beginning of DNA replication?

C29. A diagram of a linear chromosome is shown here. The end of each strand is labeled with an A, B, C, or D. Which ends could not be replicated by DNA polymerase? Why not?

5'–A————————————————————B–3'

3'–C————————————————————D–5'

C30. As discussed in Chapter 10, some viruses contain RNA as their genetic material. Certain RNA viruses can exist as a provirus in which the viral genetic material has been inserted into the chromosomal DNA of the host cell. For this to happen, the viral RNA must be copied into a strand of DNA. An enzyme called reverse transcriptase, encoded by the viral genome, copies the viral RNA into a complementary strand of DNA. The strand of DNA is then used as a template to make a double-stranded DNA molecule. This double-stranded DNA molecule is then inserted into the chromosomal DNA, where it may exist as a provirus for a long period of time.

A. How is the function of reverse transcriptase similar to the function of telomerase?

B. Unlike DNA polymerase, reverse transcriptase does not have a proofreading function. How might this affect the proliferation of the virus?

C31. Telomeres contain a 3' overhang region, as shown in Figure 11.24. Does telomerase require a 3' overhang to replicate the telomere region? Explain.

Experimental Questions

E1. Answer the following questions that pertain to the experiment of Figure 11.3.

A. What would be the expected results if the Meselson and Stahl experiment were carried out for four or five generations?

B. What would be the expected results of the Meselson and Stahl experiment after three generations if the mechanism of DNA replication was dispersive?

C. As shown in the data, explain why three different bands (i.e., light, half-heavy, and heavy) can be observed in the CsCl gradient.

E2. An absentminded researcher follows the steps of Figure 11.3, and when the gradient is viewed under UV light, the researcher does not see any bands at all. Which of the following mistakes could account for this observation? Explain how.

A. The researcher forgot to add ^{14}N-containing compounds.

B. The researcher forgot to add lysozyme.

C. The researcher forgot to turn on the UV lamp.

E3. Figure 11.4b shows an autoradiograph of a replicating bacterial chromosome. If you analyzed many replicating chromosomes, what types of information could you learn about the mechanism of DNA replication?

E4. The experiment of Figure 11.19 described a method for determining the amount of DNA made during replication. Let's suppose that you can purify all of the proteins required for DNA replication. You then want to "reconstitute" DNA synthesis by mixing together all of the purified components necessary to synthesize a complementary strand of DNA. If you started with single-stranded DNA as a template, what additional proteins and molecules would you have to add for DNA synthesis to occur? What additional proteins would be necessary if you started with a double-stranded DNA molecule?

E5. Using the reconstitution strategy described in experimental question E4, what components would you have to add to measure the ability of telomerase to synthesize DNA? Be specific about the type of template DNA that you would add to your mixture.

E6. As described in Figure 11.19, perchloric acid precipitates strands of DNA, but it does not precipitate free nucleotides. (Note: The term free nucleotide means nucleotides that are not connected covalently to other nucleotides.) Explain why this is a critical step in the experimental procedure. If a researcher used a different reagent that precipitated DNA strands and free nucleotides instead of using perchloric acid (which precipitates only DNA strands), how would that affect the results?

E7. Would the experiment of Figure 11.19 work if the ^{32}P-labeled nucleotides were deoxyribonucleoside monophosphates instead of dNTPs? Explain why or why not.

E8. To synthesize DNA in vitro, single-stranded DNA can be used as a template. As described in Figure 11.19, you also need to add DNA polymerase, dNTPs, and a primer in order to synthesize a complementary strand of DNA. The primer can be a short sequence of DNA or RNA. The primer must be complementary to the template DNA. Let's suppose a single-stranded DNA molecule is 46 nucleotides long and has the following sequence:

GCCCCGGTACCCCGTAATATACGGGACTAGGCCGGAGGTCCGGGCG

This template DNA is mixed with a primer with the sequence 5′–CGCCCGGACC–3′, DNA polymerase, and dNTPs. In this case, a double-stranded DNA molecule is made. However, if the researcher substitutes a primer with the sequence 5′–CCAGGCCCGC–3′, a double-stranded DNA molecule is not made.

A. Which is the 5′ end of the DNA molecule shown, the left end or the right end?

B. If you added a primer that was 10 nucleotides long and complementary to the left end of the single-stranded DNA, what would be the sequence of the primer? You should designate the 5′ and 3′ ends of the primer. Could this primer be used to replicate the single-stranded DNA?

E9. The technique of dideoxy sequencing of DNA is described in Chapter 18. The technique relies on the use of dideoxyribonucleotides (shown in Figures 18.18 and 18.19). A dideoxyribonucleotide has a hydrogen atom attached to the 3′-carbon atom instead of an –OH group. When a dideoxyribonucleotide is incorporated into a newly made strand, the strand cannot grow any longer. Explain why.

E10. Another technique described in Chapter 18 is the polymerase chain reaction (PCR) (see Figure 18.6). This method is based on our understanding of DNA replication. In this method, a small amount of double-stranded template DNA is mixed with a high concentration of primers. Nucleotides and DNA polymerase are also added. The template DNA strands are separated by heat treatment, and when the temperature is lowered, the primers can bind to the single-stranded DNA, and then DNA polymerase replicates the DNA. This increases the amount of DNA made from the primers. This cycle of steps (i.e., heat treatment, lower temperature, allow DNA replication to occur) is repeated again and again and again. Because the cycles are repeated many times, this method is called a chain reaction. It is called a polymerase chain reaction because DNA polymerase is the enzyme needed to increase the amount of DNA with each cycle. In a PCR experiment, the template DNA is placed in a tube, and the primers, nucleotides, and DNA polymerase are added to the tube. The tube is then placed in a machine called a thermocycler, which raises and lowers the temperature. During one cycle, the temperature is raised (e.g., to 95°C) for a brief period and then lowered (e.g., to 60°C) to allow the primers to bind. The sample is then incubated at a slightly higher temperature for a few minutes to allow DNA replication to proceed. In a typical PCR experiment, the tube may be left in the thermocycler for 25 to 30 cycles. The total time for a PCR experiment is a few hours.

A. Why is DNA helicase not needed in a PCR experiment?

B. How is the sequence of each primer important in a PCR experiment? Do the two primers recognize the same strand or opposite strands?

C. The DNA polymerase used in PCR experiments is a DNA polymerase isolated from thermophilic bacteria. Why is this kind of polymerase used?

D. If a tube initially contained 10 copies of double-stranded DNA, how many copies of double-stranded DNA (in the region flanked by the two primers) would be obtained after 27 cycles?

Questions for Student Discussion/Collaboration

1. The complementarity of double-stranded DNA is the underlying reason that DNA can be faithfully copied. Propose alternative chemical structures that could be faithfully copied.
2. The technique described in Figure 11.19 makes it possible to measure DNA synthesis in vitro. Let's suppose you have purified the following proteins: DNA polymerase, DNA helicase, ligase, primase, single-strand binding protein, and topoisomerase. You also have the following reagents available:

   A. Radiolabeled nucleotides (labeled with ^{32}P, a radioisotope of phosphorus)

   B. Nonlabeled double-stranded DNA

   C. Nonlabeled single-stranded DNA

   D. An RNA primer that binds to one end of the nonlabeled single-stranded DNA

   With these reagents, how could you show that DNA helicase is necessary for strand separation and primase is necessary for the synthesis of an RNA primer? Note: In this question, think about conditions where DNA helicase or primase would be necessary to allow DNA replication and other conditions where they would be unnecessary.
3. DNA replication is fast, virtually error-free, and coordinated with cell division. Discuss which of these three features you think is the most important.

Note: All answers appear at the website for this textbook; the answers to even-numbered questions are in the back of the textbook.

www.mhhe.com/brookergenetics4e

Visit the website for practice tests, answer keys, and other learning aids for this chapter. Enhance your understanding of genetics with our interactive exercises, quizzes, animations, and much more.

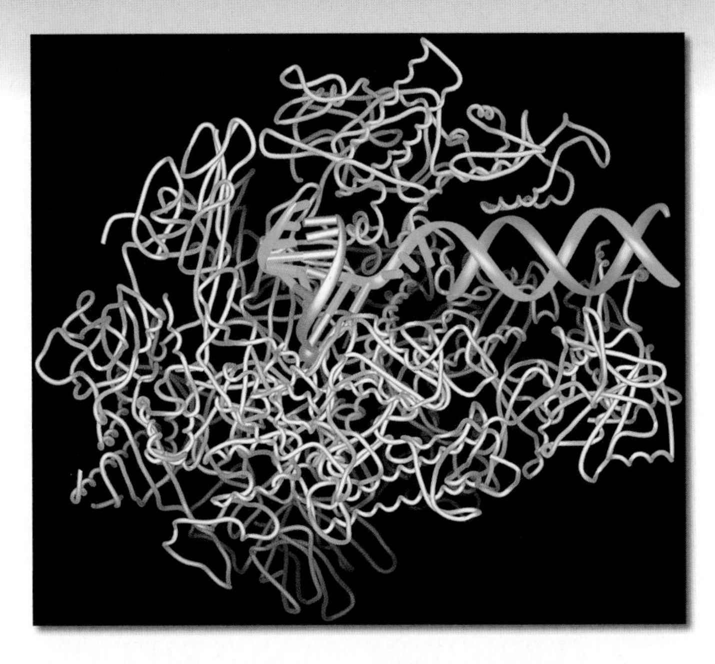

A molecular model showing the enzyme RNA polymerase in the act of sliding along the DNA and synthesizing a copy of RNA.

CHAPTER OUTLINE

12 GENE TRANSCRIPTION AND RNA MODIFICATION

The function of the genetic material is that of a blueprint. It stores the information necessary to create a living organism. The information is contained within units called genes. At the molecular level, a **gene** is defined as a segment of DNA that is used to make a functional product, either an RNA molecule or a polypeptide. How is the information within a gene accessed? The first step in this process is called **transcription,** which literally means the act or process of making a copy. In genetics, this term refers to the process of synthesizing RNA from a DNA sequence (**Figure 12.1**). The structure of DNA is not altered as a result of transcription. Rather, the DNA base sequence has only been accessed to make a copy in the form of RNA. Therefore, the same DNA can continue to store information. DNA replication, which was discussed in Chapter 11, provides a mechanism for copying that information so it can be transmitted from cell to cell and from parent to offspring.

Structural genes encode the amino acid sequence of a polypeptide. When a structural gene is transcribed, the first product is an RNA molecule known as **messenger RNA (mRNA).**

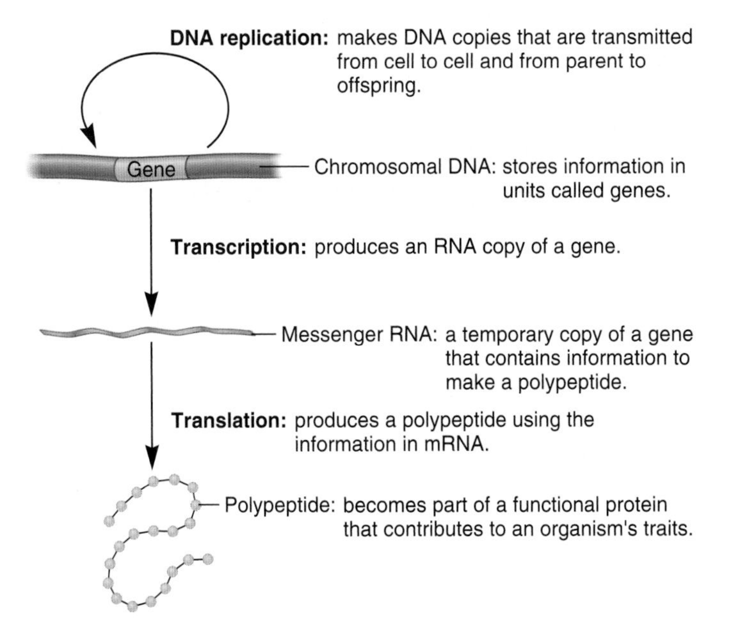

FIGURE 12.1 The central dogma of genetics. The usual flow of genetic information is from DNA to mRNA to polypeptide. Note: The direction of informational flow shown in this figure is the most common direction found in living organisms, but exceptions occur. For example, RNA viruses and certain transposable elements use an enzyme called reverse transcriptase to make a copy of DNA from RNA.

During polypeptide synthesis—a process called **translation**—the sequence of nucleotides within the mRNA determines the sequence of amino acids in a polypeptide. One or more polypeptides then assemble into a functional protein. The synthesis of functional proteins ultimately determines an organism's traits. The model depicted in Figure 12.1, which is called the **central dogma of genetics** (also called the central dogma of molecular biology), was first enunciated by Francis Crick in 1958. It forms a cornerstone of our understanding of genetics at the molecular level. The flow of genetic information occurs from DNA to mRNA to polypeptide.

In this chapter, we begin to study the molecular steps in gene expression, with an emphasis on transcription and the modifications that may occur to an RNA transcript after it has been made. Chapter 13 will examine the process of translation, and Chapters 14 and 15 will focus on how the level of gene expression is regulated at the molecular level.

12.1 OVERVIEW OF TRANSCRIPTION

One key concept important in the process of transcription is that short base sequences define the beginning and ending of a gene and also play a role in regulating the level of RNA synthesis. In this section, we begin by examining the sequences that determine where transcription starts and ends, and also briefly consider DNA sequences, called regulatory sites, that influence whether a gene is turned on or off. The functions of regulatory sites will be examined in greater detail in Chapters 14 and 15. A second important concept is the role of proteins in transcription. DNA sequences, in and of themselves, just exist. For genes to be actively transcribed, proteins must recognize particular DNA sequences and act on them in a way that affects the transcription process. In the later part of this section, we will consider how proteins participate in the general steps of transcription and the types of RNA transcripts that can be made.

Gene Expression Requires Base Sequences That Perform Different Functional Roles

At the molecular level, **gene expression** is the overall process by which the information within a gene is used to produce a functional product, such as a polypeptide. Along with environmental factors, the molecular expression of genes determines an organism's traits. For a gene to be expressed, a few different types of base sequences perform specific roles. **Figure 12.2** shows a common organization of base sequences needed to create a structural gene that functions in a bacterium such as *E. coli*. Each type of base sequence performs its role during a particular stage of gene expression. For example, the promoter and terminator are base sequences used during gene transcription. Specifically, the **promoter** provides a site to begin transcription, and the **terminator** specifies the end of transcription. These two sequences cause RNA synthesis to occur within a defined location. As shown in Figure 12.2, the DNA is transcribed into RNA from the end of the promoter to the

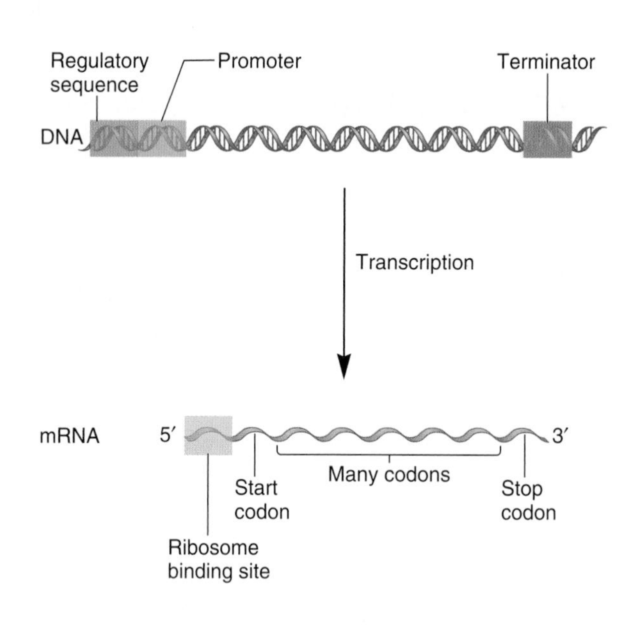

DNA:

- **Regulatory sequences:** site for the binding of regulatory proteins; the role of regulatory proteins is to influence the rate of transcription. Regulatory sequences can be found in a variety of locations.
- **Promoter:** site for RNA polymerase binding; signals the beginning of transcription.
- **Terminator:** signals the end of transcription.

mRNA:

- **Ribosomal binding site:** site for ribosome binding; translation begins near this site in the mRNA. In eukaryotes, the ribosome scans the mRNA for a start codon.
- **Start codon:** specifies the first amino acid in a polypeptide sequence, usually a formylmethionine (in bacteria) or a methionine (in eukaryotes).
- **Codons:** 3-nucleotide sequences within the mRNA that specify particular amino acids. The sequence of codons within mRNA determines the sequence of amino acids within a polypeptide.
- **Stop codon:** specifies the end of polypeptide synthesis.
- Bacterial mRNA may be polycistronic, which means it encodes two or more polypeptides.

FIGURE 12.2 Organization of sequences of a bacterial gene and its mRNA transcript. This figure depicts the general organization of sequences that are needed to create a functional gene that encodes an mRNA.

terminator. As described later, the base sequence in the RNA transcript is complementary to the **template strand** of DNA. The opposite strand is the **nontemplate strand.** For structural genes, the nontemplate strand is also called the **coding strand** because its sequence is the same as the transcribed mRNA that encodes a polypeptide, except that the DNA has T's in places where the mRNA contains U's.

A category of proteins called **transcription factors** recognizes base sequences in the DNA and controls transcription. Some transcription factors bind directly to the promoter and facilitate transcription. Other transcription factors recognize **regulatory sequences,** or **regulatory elements**—short stretches of DNA involved in the regulation of transcription. Certain transcription factors bind to such regulatory sequences and increase the rate of transcription while others inhibit transcription.

Base sequences within an mRNA are used during the translation process. In bacteria, a short sequence within the mRNA, the **ribosome-binding site,** provides a location for the ribosome to bind and begin translation. The bacterial ribosome recognizes this site because it is complementary to a sequence in ribosomal RNA. In addition, mRNA contains a series of **codons,** read as groups of three nucleotides, which contain the information for a polypeptide's sequence. The first codon, which is very close to the ribosome-binding site, is the **start codon.** This is followed by many more codons that dictate the sequence of amino acids within the synthesized polypeptide. Finally, a **stop codon** signals the end of translation. Chapter 13 will examine the process of translation in greater detail.

The Three Stages of Transcription Are Initiation, Elongation, and Termination

Transcription occurs in three stages: **initiation; elongation,** or synthesis of the RNA transcript; and **termination** (**Figure 12.3**). These steps involve protein-DNA interactions in which proteins such as **RNA polymerase,** the enzyme that synthesizes RNA, interact with DNA sequences. What causes transcription to begin? The initiation stage in the transcription process is a recognition step. The sequence of bases within the promoter region is recognized by transcription factors. The specific binding of transcription factors to the promoter sequence identifies the starting site for transcription.

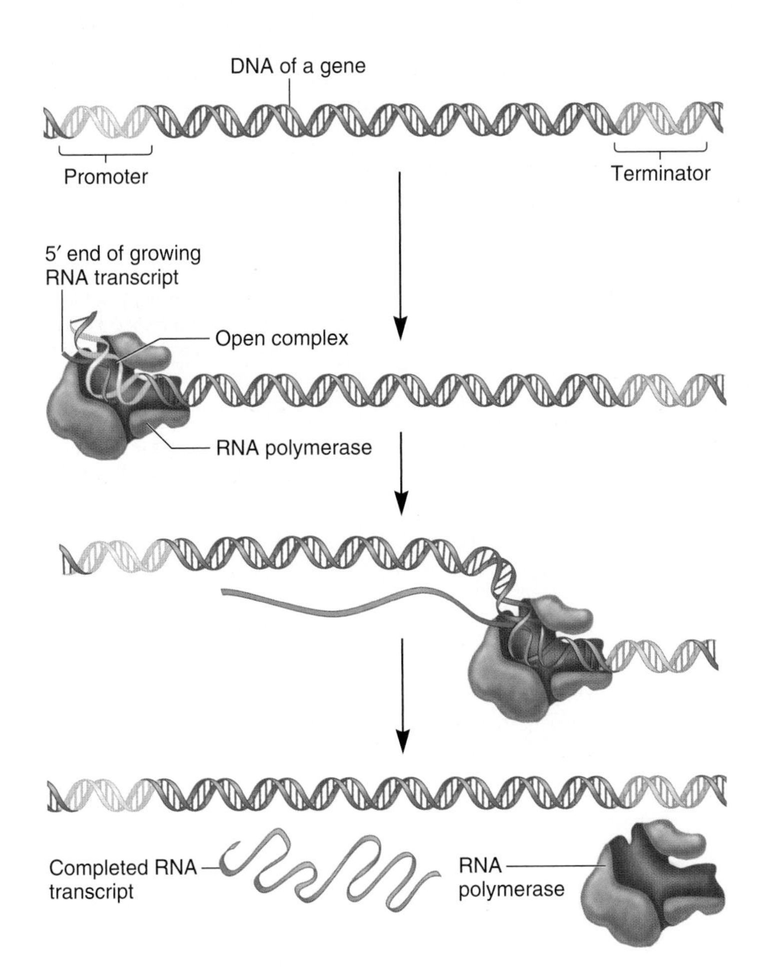

Initiation: The promoter functions as a recognition site for transcription factors (not shown). The transcription factor(s) enables RNA polymerase to bind to the promoter. Following binding, the DNA is denatured into a bubble known as the open complex.

Elongation/synthesis of the RNA transcript: RNA polymerase slides along the DNA in an open complex to synthesize RNA.

Termination: A terminator is reached that causes RNA polymerase and the RNA transcript to dissociate from the DNA.

ONLINE ANIMATION

FIGURE 12.3 Stages of transcription.

Genes→Traits The ability of genes to produce an organism's traits relies on the molecular process of gene expression. Transcription is the first step in gene expression. During transcription, the gene's sequence within the DNA is used as a template to make a complementary copy of RNA. In Chapter 13, we will examine how the sequence in mRNA is translated into a polypeptide chain. After polypeptides are made within a living cell, they fold into functional proteins that govern an organism's traits.

Transcription factors and RNA polymerase first bind to the promoter region when the DNA is in the form of a double helix. For transcription to occur, the DNA strands must be separated. This allows one of the two strands to be used as a template for the synthesis of a complementary strand of RNA. This synthesis occurs as RNA polymerase slides along the DNA, forming a small bubble-like structure known as the open promoter complex, or simply as the **open complex.** Eventually, RNA polymerase reaches a terminator, which causes both RNA polymerase and the newly made RNA transcript to dissociate from the DNA.

RNA Transcripts Have Different Functions

Once they are made, RNA transcripts play different functional roles (**Table 12.1**). Well over 90% of all genes are structural genes, which are transcribed into mRNA. For structural genes, mRNAs are made first, but the final, functional products are polypeptides that are components of proteins. The remaining types of RNAs described in Table 12.1 are never translated. The RNA transcripts from such nonstructural genes have various important cellular functions. For nonstructural genes, the functional product is the RNA. In some cases, the RNA transcript becomes part of a complex that contains both protein subunits and one or more RNA molecules. Examples of protein-RNA complexes include ribosomes, signal recognition particles, RNaseP, spliceosomes, and telomerase.

TABLE 12.1
Functions of RNA Molecules

Type of RNA	Description
mRNA	Messenger RNA (mRNA) encodes the sequence of amino acids within a polypeptide. In bacteria, some mRNAs encode a single polypeptide. Other mRNAs are polycistronic—a single mRNA encodes two or more polypeptides. In most species of eukaryotes, each mRNA usually encodes a single polypeptide. However, in some species, such as *Caenorhabditis elegans* (a nematode worm), polycistronic mRNAs are relatively common.
tRNA	Transfer RNA (tRNA) is necessary for the translation of mRNA. The structure and function of transfer RNA are outlined in Chapter 13.
rRNA	Ribosomal RNA (rRNA) is necessary for the translation of mRNA. Ribosomes are composed of both rRNAs and protein subunits. The structure and function of ribosomes are examined in Chapter 13.
MicroRNA	MicroRNAs (miRNAs) are short RNA molecules that are involved in gene regulation in eukaryotes (see Chapter 15).
scRNA	Small cytoplasmic RNA (scRNA) is found in the cytoplasm of bacteria and eukaryotes. In bacteria, scRNA is needed for protein secretion. An example in eukaryotes is 7S RNA, which is necessary in the targeting of proteins to the endoplasmic reticulum. It is a component of a complex known as signal recognition particle (SRP), which is composed of 7S RNA and six different protein subunits.
RNA of RNaseP	RNaseP is a catalyst necessary in the processing of tRNA molecules. The RNA is the catalytic component. RNaseP is composed of a 350- to 410-nucleotide RNA and one protein subunit.
snRNA	Small nuclear RNA (snRNA) is necessary in the splicing of eukaryotic pre-mRNA. snRNAs are components of a spliceosome, which is composed of both snRNAs and protein subunits. The structure and function of spliceosomes are examined later in this chapter.
Telomerase RNA	The enzyme telomerase, which is involved in the replication of eukaryotic telomeres, is composed of an RNA molecule and protein subunits.
snoRNA	Small nucleolar RNA (snoRNA) is necessary in the processing of eukaryotic rRNA transcripts. snoRNAs are also associated with protein subunits. In eukaryotes, snoRNAs are found in the nucleolus, where rRNA processing and ribosome assembly occur.
Viral RNAs	Some types of viruses use RNA as their genome, which is packaged within the viral capsid.

12.2 TRANSCRIPTION IN BACTERIA

Our molecular understanding of gene transcription initially came from studies involving bacteria and bacteriophages. Several early investigations focused on the production of viral RNA after bacteriophage infection. The first suggestion that RNA is derived from the transcription of DNA was made by Elliot Volkin and Lazarus Astrachan in 1956. When the researchers exposed *E. coli* cells to T2 bacteriophage, they observed that the RNA made immediately after infection had a base composition substantially different from the base composition of RNA prior to infection. Furthermore, the base composition after infection was very similar to the base composition in the T2 DNA, except that the RNA contained uracil instead of thymine. These results were consistent with the idea that the bacteriophage DNA is used as a template for the synthesis of bacteriophage RNA.

In 1960, Matthew Meselson and François Jacob found that proteins are synthesized on ribosomes. One year later, Jacob and his colleague Jacques Monod proposed that a certain type of RNA acts as a genetic messenger (from the DNA to the ribosome) to provide the information for protein synthesis. They hypothesized that this RNA, which they called messenger RNA (mRNA), is transcribed from the sequence within DNA and then directs the synthesis of particular polypeptides. In the early 1960s, this proposal was remarkable, considering that it was made before the actual isolation and characterization of the mRNA molecules in vitro. In 1961, the hypothesis was confirmed by Sydney Brenner in collaboration with Jacob and Meselson. They found that when a virus infects a bacterial cell, a virus-specific RNA is made that rapidly associates with preexisting ribosomes in the cell.

Since these pioneering studies, a great deal has been learned about the molecular features of bacterial gene transcription. Much of our knowledge comes from studies of *E. coli*. In this section, we will examine the three steps in the gene transcription process as they occur in bacteria.

A Promoter Is a Short Sequence of DNA That Is Necessary to Initiate Transcription

The type of DNA sequence known as the promoter gets its name from the idea that it "promotes" gene expression. More precisely, this sequence of bases directs the exact location for the initiation of RNA transcription. Most of the promoter region is located just ahead of or upstream from the site where transcription of a gene actually begins. By convention, the bases in a promoter sequence are numbered in relation to the **transcriptional start site** (**Figure 12.4**). This site is the first base used as a template for RNA transcription and is denoted +1. The bases preceding this site are numbered in a negative direction. No base is numbered zero. Therefore, most of the promoter region is labeled with negative numbers that describe the number of bases preceding the beginning of transcription.

Although the promoter may encompass a region several dozen nucleotides in length, short **sequence elements** are particularly critical for promoter recognition. By comparing the sequence of DNA bases within many promoters, researchers have learned that certain sequences of bases are necessary to create a functional promoter. In many promoters found in *E. coli* and similar species, two sequence elements are important. These are located at approximately the −35 and −10 sites in the promoter region (see Figure 12.4). The sequence in the top DNA strand at the −35 region is 5′–TTGACA–3′, and the one at the −10 region is 5′–TATAAT–3′. The TATAAT sequence is called the **Pribnow box** after David Pribnow, who initially discovered it in 1975.

The sequences at the −35 and −10 sites can vary among different genes. For example, **Figure 12.5** illustrates the sequences found in several different *E. coli* promoters. The most commonly occurring bases within a sequence element form the **consensus sequence.** This sequence is efficiently recognized by proteins that initiate transcription. For many bacterial genes, a strong correlation is found between the maximal rate of RNA transcription and the degree to which the −35 and −10 regions agree with their consensus sequences.

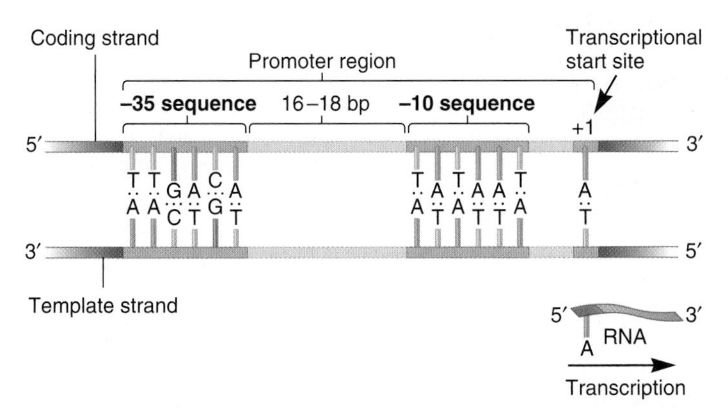

FIGURE 12.4 The conventional numbering system of promoters. The first nucleotide that acts as a template for transcription is designated +1. The numbering of nucleotides to the left of this spot is in a negative direction, whereas the numbering to the right is in a positive direction. For example, the nucleotide that is immediately to the left of the +1 nucleotide is numbered −1, and the nucleotide to the right of the +1 nucleotide is numbered +2. There is no zero nucleotide in this numbering system. In many bacterial promoters, sequence elements at the −35 and −10 regions play a key role in promoting transcription.

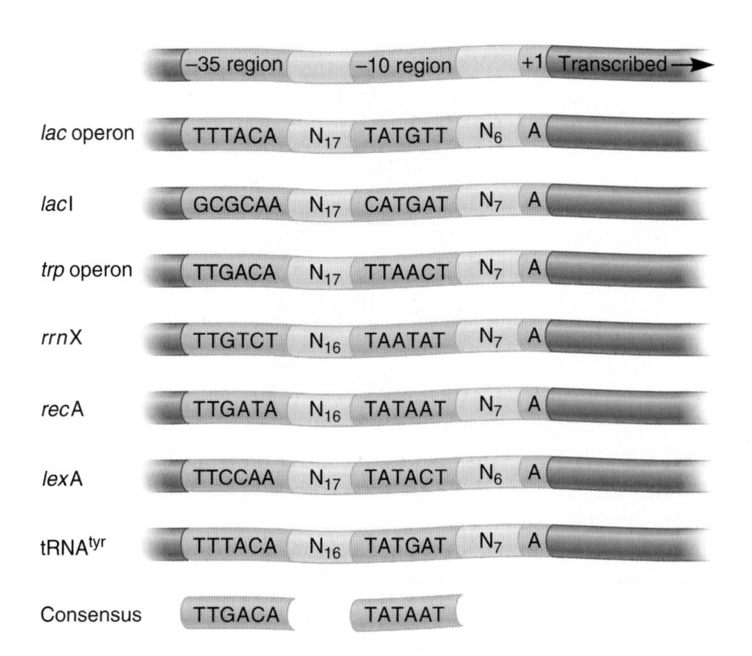

FIGURE 12.5 Examples of −35 and −10 sequences within a variety of bacterial promoters. This figure shows the −35 and −10 sequences for one DNA strand found in seven different bacterial and bacteriophage promoters. The consensus sequence is shown at the bottom. The spacer regions contain the designated number of nucleotides between the −35 and −10 region or between the −10 region and the transcriptional start site. For example, N_{17} means there are 17 nucleotides between the end of the −35 region and the beginning of the −10 region.

Bacterial Transcription Is Initiated When RNA Polymerase Holoenzyme Binds at a Promoter Sequence

Thus far, we have considered the DNA sequences that constitute a functional promoter. Let's now turn our attention to the proteins that recognize those sequences and carry out the transcription process. The enzyme that catalyzes the synthesis of RNA is RNA polymerase. In *E. coli*, the **core enzyme** is composed of five subunits, $\alpha_2\beta\beta'\omega$. The association of a sixth subunit, **sigma (σ) factor,** with the core enzyme is referred to as RNA polymerase **holoenzyme.** The different subunits within the holoenzyme play distinct functional roles. The two α subunits are important in the proper assembly of the holoenzyme and in the process of binding to DNA. The β and β' subunits are also needed for binding to the DNA and carry out the catalytic synthesis of RNA. The ω (omega) subunit is important for the proper assembly of the core enzyme. The holoenzyme is required to initiate transcription; the primary role of σ factor is to recognize the promoter. Proteins, such as σ factor, that influence the function of RNA polymerase are types of transcription factors.

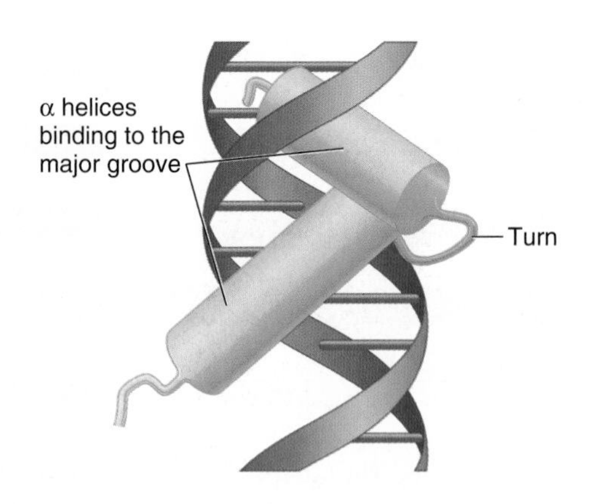

FIGURE 12.6 The binding of σ factor protein to the DNA double helix. In this example, the protein contains two α helices connected by a turn, termed a helix-turn-helix motif. Two α helices of the protein can fit within the major groove of the DNA. Amino acids within the α helices form hydrogen bonds with the bases in the DNA.

After RNA polymerase holoenzyme is assembled into its six subunits, it binds loosely to the DNA and then slides along the DNA, much as a train rolls down the tracks. How is a promoter identified? When the holoenzyme encounters a promoter sequence, σ factor recognizes the bases at both the -35 and -10 regions. σ factor protein contains a structure called a **helix-turn-helix motif** that can bind tightly to these regions. Alpha (α) helices within the protein fit into the major groove of the DNA double helix and form hydrogen bonds with the bases. This phenomenon of molecular recognition is shown in **Figure 12.6**. Hydrogen bonding occurs between nucleotides in the -35 and -10 regions of the promoter and amino acid side chains in the helix-turn-helix structure of σ factor.

As shown in **Figure 12.7**, the process of transcription is initiated when σ factor within the holoenzyme has bound to the promoter region to form the **closed complex.** For transcription to begin, the double-stranded DNA must then be unwound into an open complex. This unwinding first occurs at the TATAAT sequence in the -10 region, which contains only AT base pairs, as shown in Figure 12.4. AT base pairs form only two hydrogen bonds, whereas GC pairs form three. Therefore, DNA in an AT-rich region is more easily separated because fewer hydrogen bonds must be broken. A short strand of RNA is made within the open complex, and then σ factor is released from the core enzyme. The release of σ factor marks the transition to the elongation phase of transcription. The core enzyme may now slide down the DNA to synthesize a strand of RNA.

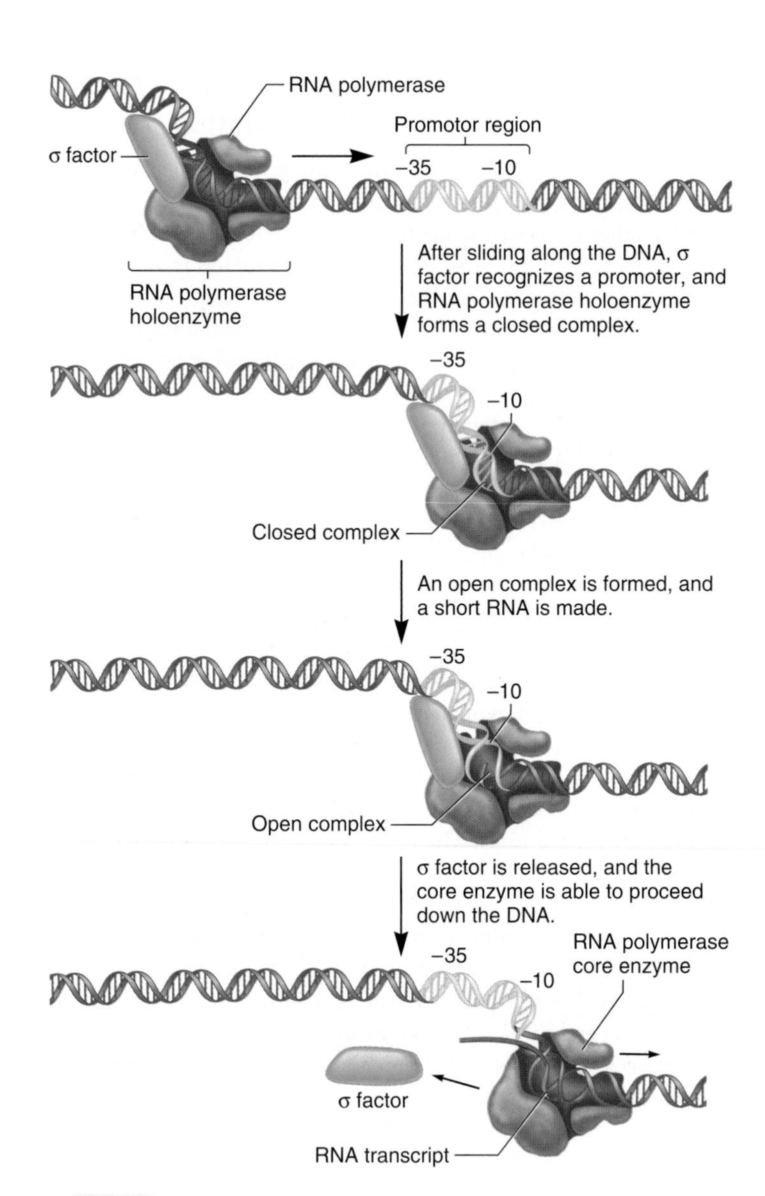

ONLINE ANIMATION **FIGURE 12.7 The initiation stage of transcription in bacteria.** The σ factor subunit of the RNA polymerase holoenzyme recognizes the -35 and -10 regions of the promoter. The DNA unwinds in the -10 region to form an open complex, and a short RNA is made. σ factor then dissociates from the holoenzyme, and the RNA polymerase core enzyme can proceed down the DNA to transcribe RNA, forming an open complex as it goes.

The RNA Transcript Is Synthesized During the Elongation Stage

After the initiation stage of transcription is completed, the RNA transcript is made during the elongation stage. During the synthesis of the RNA transcript, RNA polymerase moves along the DNA, causing it to unwind (**Figure 12.8**). As previously mentioned, the DNA strand used as a template for RNA synthesis is called the template, or antisense, strand. The opposite DNA strand is the coding, or sense, strand; it has the same sequence as the RNA transcript except that T in the DNA corresponds to U in the RNA. Within a given gene, only the template strand is used for RNA synthesis, whereas the coding strand is never used. As it moves along the DNA, the open complex formed by the action of RNA polymerase is approximately 17 bp long. On average, the rate of RNA synthesis is about 43 nucleotides per second! Behind the open complex, the DNA rewinds back into a double helix.

As described in Figure 12.8, the chemistry of transcription by RNA polymerase is similar to the synthesis of DNA via DNA polymerase, which is discussed in Chapter 11. RNA polymerase

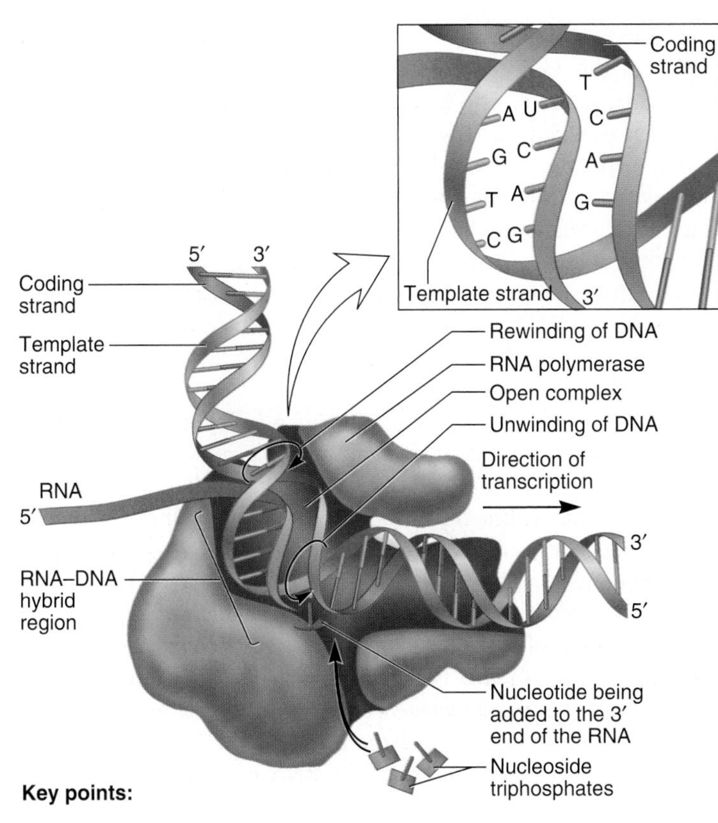

Key points:

- RNA polymerase slides along the DNA, creating an open complex as it moves.
- The DNA strand known as the template strand is used to make a complementary copy of RNA as an RNA–DNA hybrid.
- RNA polymerase moves along the template strand in a 3′ to 5′ direction, and RNA is synthesized in a 5′ to 3′ direction using nucleoside triphosphates as precursors. Pyrophosphate is released (not shown).
- The complementarity rule is the same as the AT/GC rule except that U is substituted for T in the RNA.

ONLINE ANIMATION **FIGURE 12.8** **Synthesis of the RNA transcript.**

always connects nucleotides in the 5′ to 3′ direction. During this process, RNA polymerase catalyzes the formation of a bond between the 5′ phosphate group on one nucleotide and the 3′-OH group on the previous nucleotide. The complementarity rule is similar to the AT/GC rule, except that uracil substitutes for thymine in the RNA. In other words, RNA synthesis obeys an $A_{DNA}U_{RNA}$/ $T_{DNA}A_{RNA}$/$G_{DNA}C_{RNA}$/$C_{DNA}G_{RNA}$ rule.

When considering the transcription of multiple genes within a chromosome, the direction of transcription and the DNA strand used as a template varies among different genes. **Figure 12.9** shows three genes adjacent to each other within a chromosome. Genes *A* and *B* are transcribed from left to right, using the bottom DNA strand as a template. By comparison, gene *C* is transcribed from right to left and uses the top DNA strand as a template. Note that in all three cases, the template strand is read in the 3′ to 5′ direction, and the synthesis of the RNA transcript occurs in a 5′ to 3′ direction.

Transcription Is Terminated by Either an RNA-Binding Protein or an Intrinsic Terminator

The end of RNA synthesis is referred to as termination. Prior to termination, the hydrogen bonding between the DNA and RNA within the open complex is of central importance in preventing dissociation of RNA polymerase from the template strand. Termination occurs when this short RNA-DNA hybrid region is forced to separate, thereby releasing RNA polymerase as well as the newly made RNA transcript. In *E. coli*, two different mechanisms for termination have been identified. For certain genes, an RNA-binding protein known as **ρ (rho)** is responsible for terminating transcription, in a mechanism called ρ-dependent termination. For other genes, termination does not require the involvement of the ρ protein. This is referred to as ρ-independent termination.

In **ρ-dependent termination,** the termination process requires two components. First, a sequence upstream from the

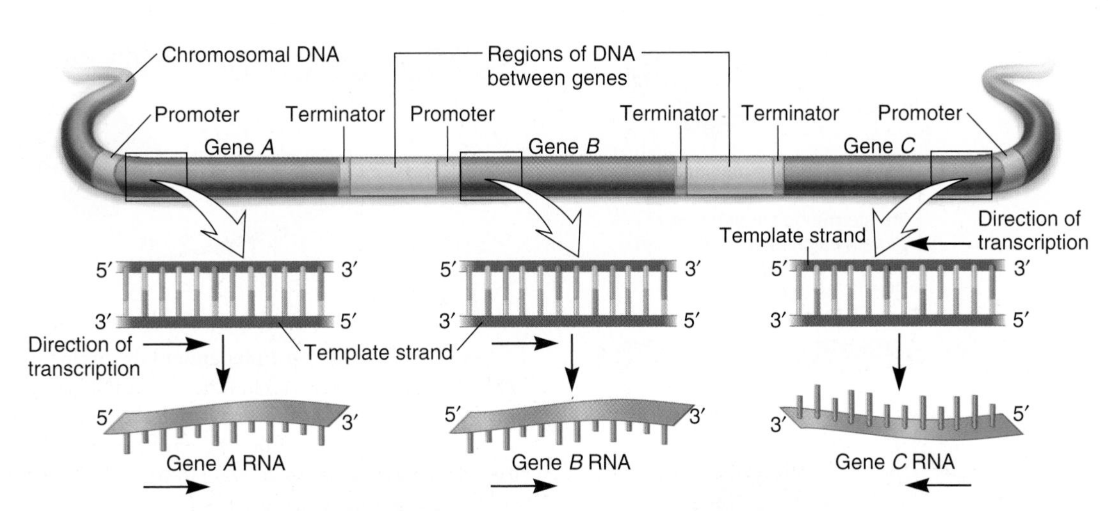

FIGURE 12.9 **The transcription of three different genes found in the same chromosome.** RNA polymerase synthesizes each RNA transcript in a 5′ to 3′ direction, sliding along a DNA template strand in a 3′ to 5′ direction. However, the use of the template strand varies from gene to gene. For example, genes *A* and *B* use the bottom strand, but gene *C* uses the top strand.

terminator, called the *rut* site for rho utilization site, acts as a recognition site for the binding of the ρ protein (**Figure 12.10**). How does ρ protein facilitate termination? The ρ protein functions as a helicase, an enzyme that can separate RNA-DNA hybrid regions. After the *rut* site is synthesized in the RNA, ρ protein binds to the RNA and moves in the direction of RNA polymerase. The second component of ρ-dependent termination is the site where termination actually takes place. At this terminator site, the DNA encodes an RNA sequence containing several GC base pairs that form a stem-loop structure. RNA synthesis terminates several nucleotides beyond this stem-loop. As discussed in Chapter 9, a stem-loop structure, also called a hairpin, can form due to complementary sequences within the RNA (refer back to Figure 9.22). This stem-loop forms almost immediately after the RNA sequence is synthesized and quickly binds to RNA polymerase. This binding results in a conformational change that causes RNA polymerase to pause in its synthesis of RNA. The pause allows ρ protein to catch up to the stem-loop, pass through it, and break the hydrogen bonds between the DNA and RNA within the open complex. When this occurs, the completed RNA strand is separated from the DNA along with RNA polymerase.

Let's now turn our attention to **ρ-independent termination,** a process that does not require the ρ protein. In this case, the terminator is composed of two adjacent nucleotide sequences that function within the RNA (**Figure 12.11**). One is a uracil-rich sequence located at the 3′ end of the RNA. The second sequence is adjacent to the uracil-rich sequence and promotes the formation of a stem-loop structure. As shown in Figure 12.11, the formation of the stem-loop causes RNA polymerase to pause in its synthesis of RNA. This pausing is stabilized by other proteins that bind to RNA polymerase. For example, a protein called NusA, which is bound to RNA polymerase, promotes pausing at stem-loop sequences. At the precise time RNA polymerase pauses, the

FIGURE 12.10 ρ-Dependent termination.

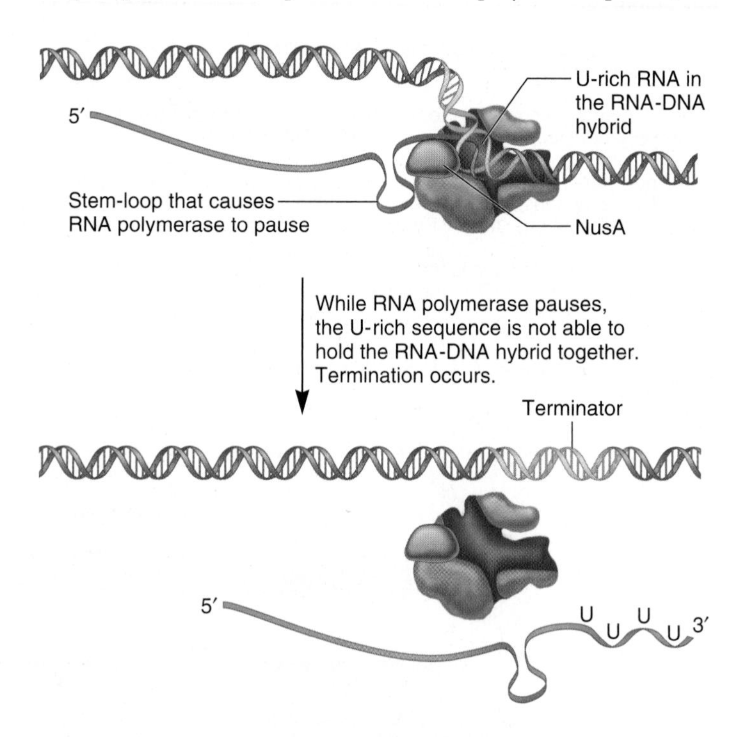

FIGURE 12.11 ρ-Independent or intrinsic termination. When RNA polymerase reaches the end of the gene, it transcribes a uracil-rich sequence. As this uracil-rich sequence is transcribed, a stem-loop forms just upstream from the open complex. The formation of this stem-loop causes RNA polymerase to pause in its synthesis of the transcript. This pausing is stabilized by NusA, which binds near the region where RNA exits the open complex. While it is pausing, the RNA in the RNA-DNA hybrid is a uracil-rich sequence. Because hydrogen bonds between U and A are relatively weak interactions, the transcript and RNA polymerase dissociate from the DNA.

uracil-rich sequence in the RNA transcript is bound to the DNA template strand. As previously mentioned, the hydrogen bonding of RNA to DNA keeps RNA polymerase clamped onto the DNA. However, the binding of this uracil-rich sequence to the DNA template strand is relatively weak, causing the RNA transcript to spontaneously dissociate from the DNA and cease further transcription. Because this process does not require a protein (the ρ protein) to physically remove the RNA transcript from the DNA, it is also referred to as **intrinsic termination.** In *E. coli*, about half of the genes show intrinsic termination, and the other half are terminated by ρ protein.

12.3 TRANSCRIPTION IN EUKARYOTES

Many of the basic features of gene transcription are very similar in bacterial and eukaryotic species. Much of our understanding of transcription has come from studies in *Saccharomyces cerevisiae* (baker's yeast) and other eukaryotic species, including mammals. In general, gene transcription in eukaryotes is more complex than that of their bacterial counterparts. Eukaryotic cells are larger and contain a variety of compartments known as organelles. This added level of cellular complexity dictates that eukaryotes contain many more genes encoding cellular proteins. In addition, most eukaryotic species are multicellular, being composed of many different cell types. Multicellularity adds the requirement that genes be transcribed in the correct type of cell and during the proper stage of development. Therefore, in any given species, the transcription of the thousands of different genes that an organism possesses requires appropriate timing and coordination. In this section, we will examine the basic features of gene transcription in eukaryotes. We will focus on the proteins that are needed to make an RNA transcript. In addition, an important factor that affects eukaryotic gene transcription is chromatin structure. Eukaryotic gene transcription requires changes in the positions and structures of nucleosomes. However, because these changes are important for regulating transcription, they are described in Chapter 15 rather than this chapter.

Eukaryotes Have Multiple RNA Polymerases That Are Structurally Similar to the Bacterial Enzyme

The genetic material within the nucleus of a eukaryotic cell is transcribed by three different RNA polymerase enzymes, designated RNA polymerase I, II, and III. What are the roles of these enzymes? Each of the three RNA polymerases transcribes different categories of genes. RNA polymerase I transcribes all of the genes that encode ribosomal RNA (rRNA) except for the 5S rRNA. RNA polymerase II plays a major role in cellular transcription because it transcribes all of the structural genes. It is responsible for the synthesis of all mRNA and also transcribes certain snRNA genes, which are needed for pre-mRNA splicing. RNA polymerase III transcribes all tRNA genes and the 5S rRNA gene.

All three RNA polymerases are structurally very similar and are composed of many subunits. They contain two large catalytic subunits similar to the β and β′ subunits of bacterial RNA polymerase. The structures of RNA polymerase from a few different species have been determined by X-ray crystallography. A remarkable similarity exists between the bacterial enzyme and its eukaryotic counterparts. **Figure 12.12a** compares the

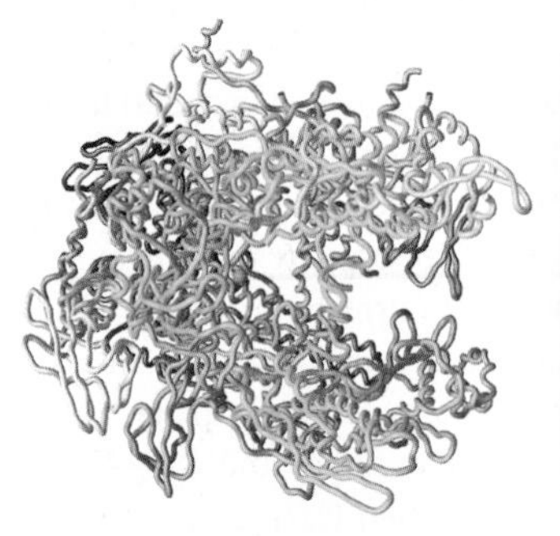

(a) Structure of a bacterial RNA polymerase

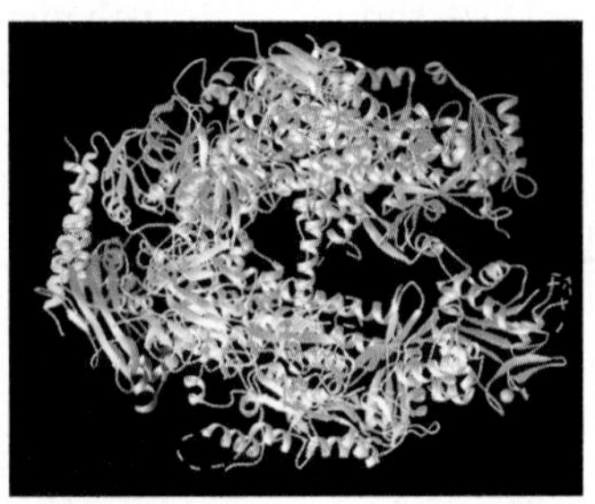

Structure of a eukaryotic RNA polymerase II (yeast)

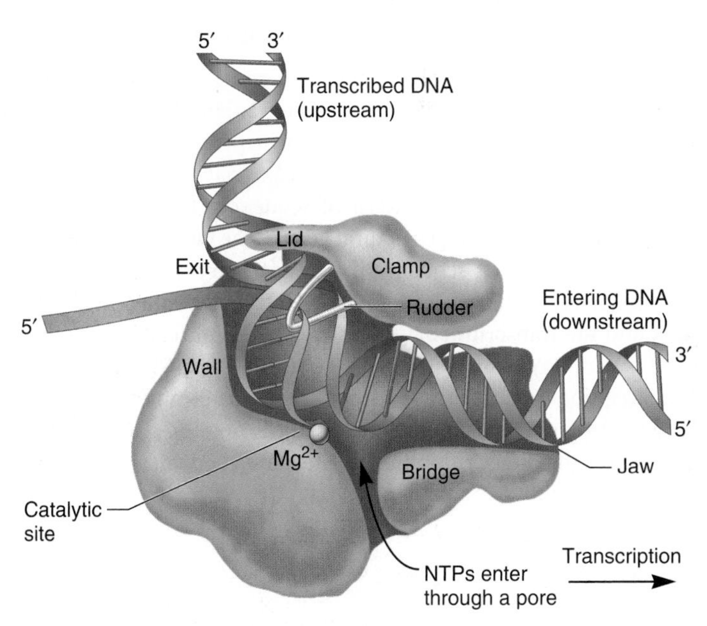

(b) Schematic structure of RNA polymerase

FIGURE 12.12 Structure and molecular function of RNA polymerase. **(a)** A comparison of the crystal structures of a bacterial RNA polymerase (left) to a eukaryotic RNA polymerase II (right). The bacterial enzyme is from *Thermus aquaticus*. The eukaryotic enzyme is from *Saccharomyces cerevisiae.* **(b)** A mechanism for transcription based on the crystal structure. In this diagram, the direction of transcription is from left to right. The double-stranded DNA enters the polymerase along a bridge surface that is between the jaw and clamp. At a region termed the wall, the RNA-DNA hybrid is forced to make a right-angle turn, which enables nucleotides to bind to the template strand. Mg^{2+} is located at the catalytic site. Nucleoside triphosphates (NTPs) enter the catalytic site via a pore region and bind to the template DNA. At the catalytic site, the nucleotides are covalently attached to the 3′ end of the RNA. As RNA polymerase slides down the template, a small region of the protein termed the rudder separates the RNA-DNA hybrid. The single-stranded RNA then exits under a small lid.

structures of a bacterial RNA polymerase with RNA polymerase II from yeast. As seen here, both enzymes have a very similar structure. Also, it is very exciting that this structure provides a way to envision how the transcription process works. As seen in **Figure 12.12b**, DNA enters the enzyme through the jaw and lies on a surface within RNA polymerase termed the bridge. The part of the enzyme called the clamp is thought to control the movement of the DNA through RNA polymerase. A wall in the enzyme forces the RNA-DNA hybrid to make a right-angle turn. This bend facilitates the ability of nucleotides to bind to the template strand. Mg^{2+} is located at the catalytic site, which is precisely at the 3′ end of the growing RNA strand. Nucleoside triphosphates (NTPs) enter the catalytic site via a pore region. The correct nucleotide binds to the template DNA and is covalently attached to the 3′ end. As RNA polymerase slides down the template, a rudder, which is about 9 bp away from the 3′ end of the RNA, forces the RNA-DNA hybrid apart. The single-stranded RNA then exits under a small lid.

Eukaryotic Structural Genes Have a Core Promoter and Regulatory Elements

In eukaryotes, the promoter sequence is more variable and often more complex than that found in bacteria. For structural genes, at least three features are found in most promoters: regulatory elements, a TATA box, and a transcriptional start site. **Figure 12.13** shows a common pattern of sequences found within the promoters of eukaryotic structural genes. The **core promoter** is a relatively short DNA sequence that is necessary for transcription to take place. It consists of a TATAAA sequence called the **TATA box** and the transcriptional start site, where transcription begins. The TATA box, which is usually about 25 bp upstream from a transcriptional start site, is important in determining the precise starting point for transcription. If it is missing from the core promoter, the transcription start site point becomes undefined, and transcription may start at a variety of different locations. The core promoter, by itself, produces a low level of transcription. This is termed **basal transcription.**

Regulatory elements are short DNA sequences that affect the ability of RNA polymerase to recognize the core promoter and begin the process of transcription. These elements are recognized by transcription factors—proteins that bind to regulatory elements and influence the rate of transcription. There are two categories of regulatory elements. Activating sequences, known as **enhancers,** are needed to stimulate transcription. In the absence of enhancer sequences, most eukaryotic genes have very low levels of basal transcription. Under certain conditions, it may also be necessary to prevent transcription of a given gene. This occurs via **silencers**—DNA sequences that are recognized by transcription factors that inhibit transcription. As seen in Figure 12.13, a common location for regulatory elements is the −50 to −100 region. However, the locations of regulatory elements vary considerably among different eukaryotic genes. These elements can be far away from the core promoter yet strongly influence the ability of RNA polymerase to initiate transcription.

DNA sequences such as the TATA box, enhancers, and silencers exert their effects only over a particular gene. They are called ***cis*-acting elements.** The term *cis* comes from chemistry nomenclature meaning "next to." *Cis*-acting elements, though possibly far away from the core promoter, are always found within the same chromosome as the genes they regulate. By comparison, the regulatory transcription factors that bind to such elements are called ***trans*-acting factors** (the term *trans* means "across from"). The transcription factors that control the expression of a gene are themselves encoded by genes; regulatory genes that encode transcription factors may be far away from the genes they control. When a gene encoding a *trans*-acting factor is expressed, the transcription factor protein that is made can diffuse throughout the cell and bind to its appropriate *cis*-acting element. Let's now turn our attention to the function of such proteins.

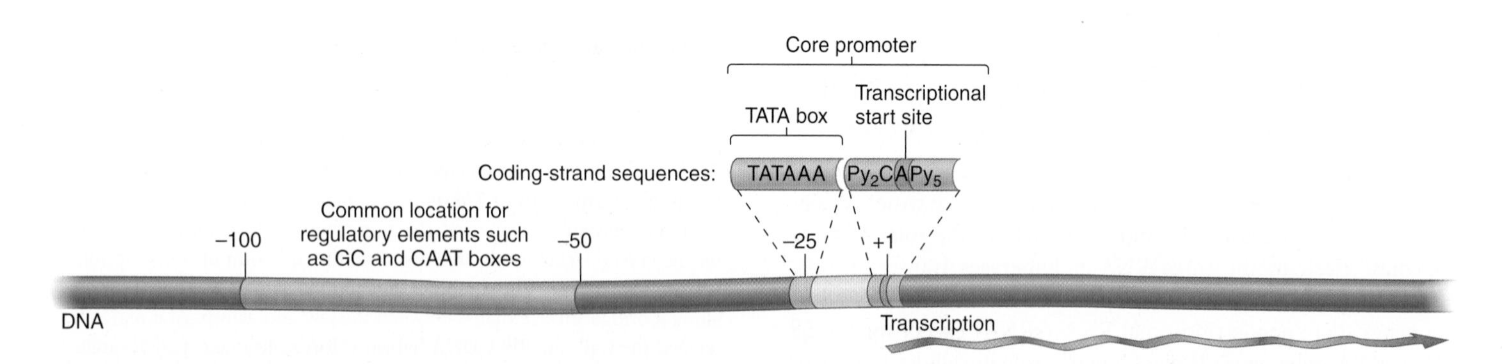

FIGURE 12.13 A common pattern found for the promoter of structural genes recognized by RNA polymerase II. The start site usually occurs at adenine; two pyrimidines (Py: cytosine or thymine) and a cytosine precede this adenine, and five pyrimidines (Py) follow it. A TATA box is approximately 25 bp upstream. However, the sequences that constitute eukaryotic promoters are quite diverse, and not all structural genes have a TATA box. Regulatory elements, such as GC or CAAT boxes, vary in their locations but are often found in the −50 to −100 region. The core promoters for RNA polymerase I and III are quite different. A single upstream regulatory element is involved in the binding of RNA polymerase I to its promoter, whereas two regulatory elements, called A and B boxes, facilitate the binding of RNA polymerase III.

Transcription of Eukaryotic Structural Genes Is Initiated When RNA Polymerase II and General Transcription Factors Bind to a Promoter Sequence

Thus far, we have considered the DNA sequences that play a role in the promoter region of eukaryotic structural genes. By studying transcription in a variety of eukaryotic species, researchers have discovered that three categories of proteins are needed for basal transcription at the core promoter: RNA polymerase II, general transcription factors, and mediator (**Table 12.2**).

Five different proteins called **general transcription factors (GTFs)** are always needed for RNA polymerase II to initiate transcription of structural genes. **Figure 12.14** describes the assembly of GTFs and RNA polymerase II at the TATA box. As shown here, a series of interactions leads to the formation of the open complex. Transcription factor IID (TFIID) first binds to the TATA box and thereby plays a critical role in the recognition of the core promoter. TFIID is composed of several subunits, including TATA-binding protein (TBP), which directly binds to the TATA box, and several other proteins called TBP-associated factors (TAFs). After TFIID binds to the TATA box, it associates with TFIIB. TFIIB promotes the binding of RNA polymerase II and TFIIF to the core promoter. Lastly, TFIIE and TFIIH bind to the complex. This completes the assembly of proteins to form a closed complex, also known as a **preinitiation complex.**

TFIIH plays a major role in the formation of the open complex. TFIIH has several subunits that perform different functions.

TABLE **12.2**

Proteins Needed for Transcription via the Core Promoter of Eukaryotic Structural Genes

RNA polymerase II: The enzyme that catalyzes the linkage of ribonucleotides in the 5′ to 3′ direction, using DNA as a template. Essentially all eukaryotic RNA polymerase II proteins are composed of 12 subunits. The two largest subunits are structurally similar to the β and β′ subunits found in *E. coli* RNA polymerase.

General transcription factors:

TFIID: Composed of TATA-binding protein (TBP) and other TBP-associated factors (TAFs). Recognizes the TATA box of eukaryotic structural gene promoters.

TFIIB: Binds to TFIID and then enables RNA polymerase II to bind to the core promoter. Also promotes TFIIF binding.

TFIIF: Binds to RNA polymerase II and plays a role in its ability to bind to TFIIB and the core promoter. Also plays a role in the ability of TFIIE and TFIIH to bind to RNA polymerase II.

TFIIE: Plays a role in the formation or the maintenance (or both) of the open complex. It may exert its effects by facilitating the binding of TFIIH to RNA polymerase II and regulating the activity of TFIIH.

TFIIH: A multisubunit protein that has multiple roles. First, certain subunits act as helicases and promote the formation of the open complex. Other subunits phosphorylate the carboxyl terminal domain (CTD) of RNA polymerase II, which releases its interaction with TFIIB, thereby allowing RNA polymerase II to proceed to the elongation phase.

Mediator: A multisubunit complex that mediates the effects of regulatory transcription factors on the function of RNA polymerase II. Though mediator typically has certain core subunits, many of its subunits vary, depending on the cell type and environmental conditions. The ability of mediator to affect the function of RNA polymerase II is thought to occur via the CTD of RNA polymerase II. Mediator can influence the ability of TFIIH to phosphorylate CTD, and subunits within mediator itself have the ability to phosphorylate CTD. Because CTD phosphorylation is needed to release RNA polymerase II from TFIIB, mediator plays a key role in the ability of RNA polymerase II to switch from the initiation to the elongation stage of transcription.

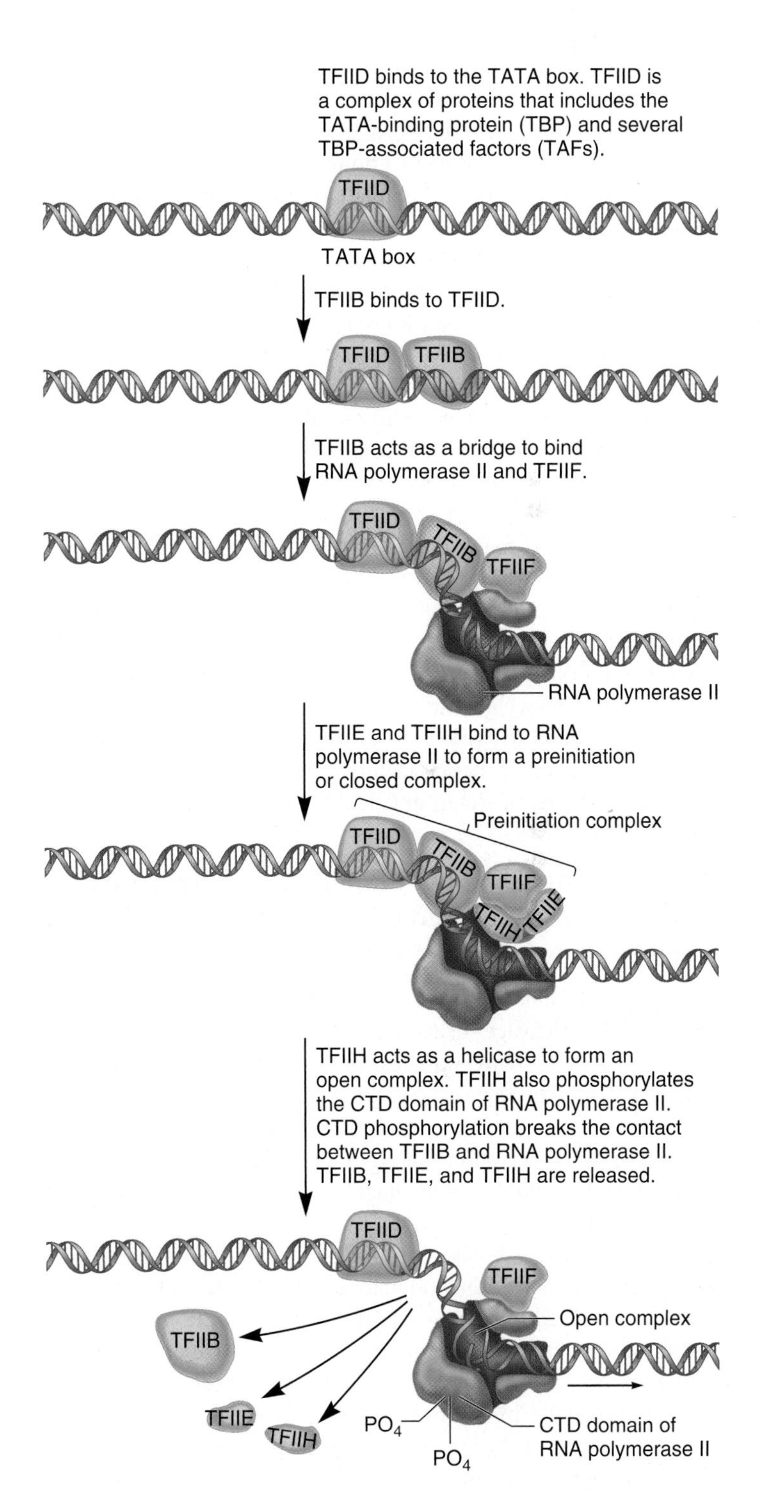

FIGURE 12.14 **Steps leading to the formation of the open complex.**

Certain subunits act as helicases, which break the hydrogen bonding between the double-stranded DNA and thereby promote the formation of the open complex. Another subunit hydrolyzes ATP and phosphorylates a domain in RNA polymerase II known as the carboxyl terminal domain (CTD). Phosphorylation of the CTD releases the contact between RNA polymerase II and TFIIB. Next, TFIIB, TFIIE, and TFIIH dissociate, and RNA polymerase II is free to proceed to the elongation stage of transcription.

In vitro, when researchers mix together TFIID, TFIIB, TFIIF, TFIIE, TFIIH, RNA polymerase II, and a DNA sequence containing a TATA box and transcriptional start site, the DNA is transcribed into RNA. Therefore, these components are referred to as the **basal transcription apparatus.** In a living cell, however, additional components regulate transcription and allow it to proceed at a reasonable rate.

In addition to GTFs and RNA polymerase II, another component required for transcription is a large protein complex termed mediator. This complex was discovered by Roger Kornberg and colleagues in 1990. In 2006, Kornberg was awarded the Nobel Prize in chemistry for his studies regarding the molecular basis of eukaryotic transcription. **Mediator** derives its name from the observation that it mediates interactions between RNA polymerase II and regulatory transcription factors that bind to enhancers or silencers. It serves as an interface between RNA polymerase II and many diverse regulatory signals. The subunit composition of mediator is quite complex and variable. The core subunits form an elliptically shaped complex that partially wraps around RNA polymerase II. Mediator itself may phosphorylate the CTD of RNA polymerase II, and it may regulate the ability of TFIIH to phosphorylate the CTD. Therefore, it can play a pivotal role in the switch between transcriptional initiation and elongation. The function of mediator during eukaryotic gene regulation is explored in greater detail in Chapter 15.

Transcriptional Termination of RNA Polymerase II Occurs After the 3′ End of the Transcript Is Cleaved Near the PolyA Signal Sequence

As discussed later in this chapter, eukaryotic pre-mRNAs are modified by cleavage near their 3′ end and the subsequent attachment of a string of adenine nucleotides (look ahead at Figure 12.24). This processing, which is called polyadenylation, requires a polyA signal sequence that directs the cleavage of the pre-mRNA. Transcription via RNA polymerase II typically terminates about 500 to 2000 nucleotides downstream from the polyA signal.

Figure 12.15 shows a simplified scheme for the transcriptional termination of RNA polymerase II. After RNA polymerase II has transcribed the polyA signal sequence, the RNA is cleaved just downstream from this sequence. This cleavage occurs before transcriptional termination. Two models have been proposed for transcriptional termination. According to the allosteric model, RNA polymerase II becomes destabilized after it has transcribed the polyA signal sequence, and it eventually dissociates from the DNA. This destabilization may be caused by the loss of proteins that function as elongation factors or by the binding of proteins that function as termination factors. A second model, called the torpedo model, suggests that RNA polymerase II is physically removed from the DNA. According to this model, the region of RNA that is downstream from the polyA signal sequence is cleaved by an exonuclease that degrades the transcript in the 5′ to 3′ direction. When the exonuclease catches up to RNA polymerase II, this causes RNA polymerase II to dissociate from the DNA.

Which of these two models is correct? Additional research is needed, but the results of studies over the past few years have provided evidence that the two models are not mutually exclusive. Therefore, both mechanisms may play a role in transcriptional termination.

12.4 RNA MODIFICATION

During the 1960s and 1970s, studies in bacteria established the physical structure of the gene. The analysis of bacterial genes showed that the sequence of DNA within the coding strand corresponds to the sequence of nucleotides in the mRNA, except that T is replaced with U. During translation, the sequence of codons in the mRNA is then read, providing the instructions for the correct amino acid sequence in a polypeptide. The one-to-one correspondence between the sequence of codons in the DNA coding strand and the amino acid sequence of the polypeptide has been termed the **colinearity** of gene expression.

The situation dramatically changed in the late 1970s, when the tools became available to study eukaryotic genes at the molecular level. The scientific community was astonished by the discovery that eukaryotic structural genes are not always colinear with their functional mRNAs. Instead, the coding sequences within many eukaryotic genes are separated by DNA sequences that are not translated into protein. The coding sequences are found within **exons,** which are regions that are contained within mature RNA. By comparison, the sequences that are found between the exons are called **intervening sequences,** or **introns.** During transcription, an RNA is made corresponding to the entire gene sequence. Subsequently, as it matures, the sequences in the RNA that correspond to the introns are removed and the exons are connected, or spliced, together. This process is called **RNA splicing.** Since the 1970s, research has revealed that splicing is a common genetic phenomenon in eukaryotic species. Splicing occurs occasionally in bacteria as well.

Aside from splicing, research has also shown that RNA transcripts can be modified in several other ways. **Table 12.3** describes the general types of RNA modifications. For example, rRNAs and tRNAs are synthesized as long transcripts that are processed into smaller functional pieces. In addition, most eukaryotic mRNAs have a cap attached to their 5′ end and a tail attached at their 3′ end. In this section, we will examine the molecular mechanisms that account for several types of RNA modifications and consider why they are functionally important.

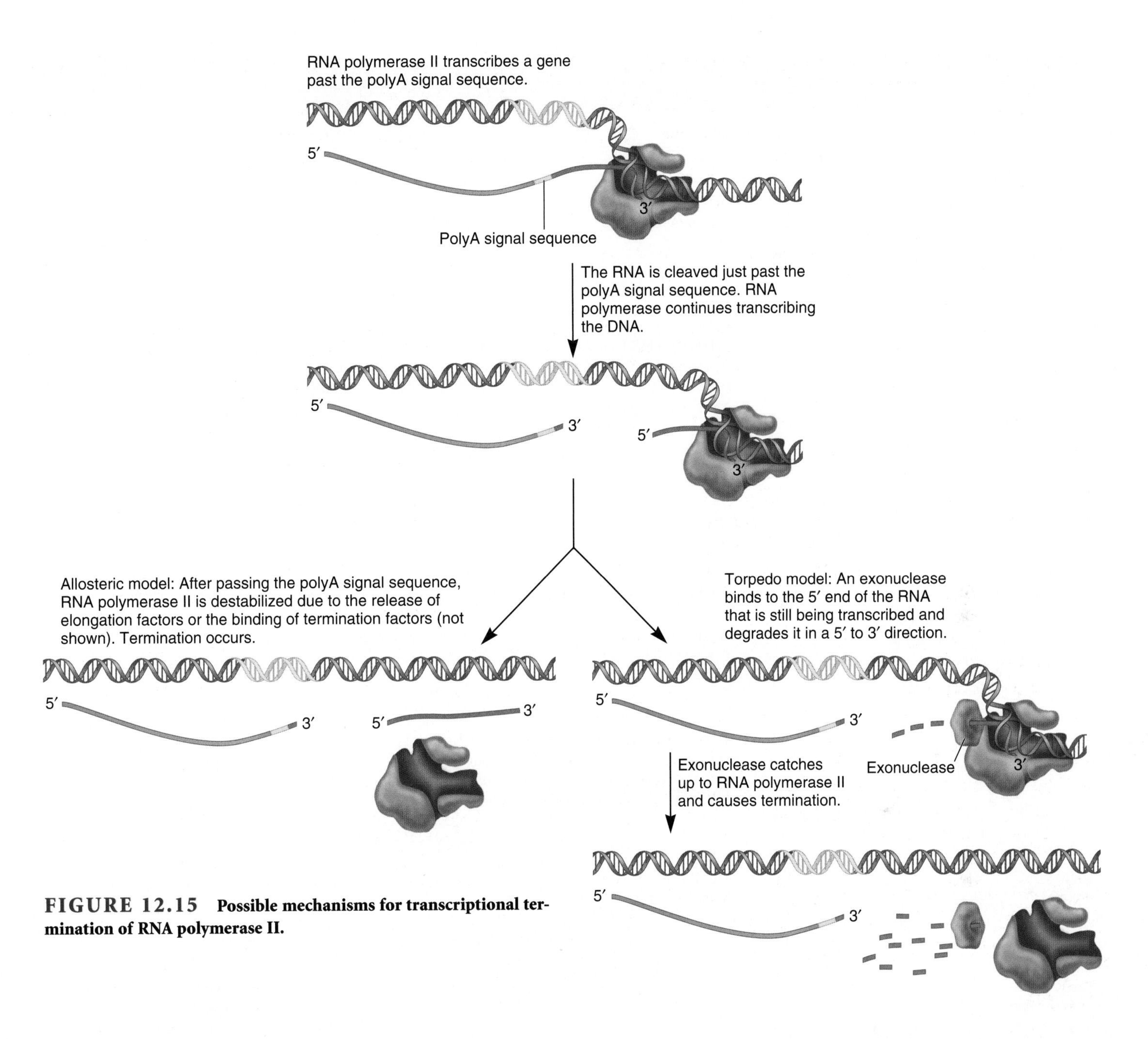

FIGURE 12.15 **Possible mechanisms for transcriptional termination of RNA polymerase II.**

Some Large RNA Transcripts Are Cleaved into Smaller Functional Transcripts

For many nonstructural genes, the RNA transcript initially made during gene transcription is processed or cleaved into smaller pieces. As an example, **Figure 12.16** shows the processing of mammalian ribosomal RNA. The ribosomal RNA gene is transcribed by RNA polymerase I to make a long primary transcript, known as 45S rRNA. The term 45S refers to the sedimentation characteristics of this transcript in Svedberg units. Following the synthesis of the 45S rRNA, cleavage occurs at several points to produce three fragments, termed 18S, 5.8S, and 28S rRNA. These are functional rRNA molecules that play a key role in forming the structure of the ribosome. In eukaryotes, the cleavage of 45S rRNA into smaller rRNAs and the assembly of ribosomal subunits occur in a structure within the cell nucleus known as the **nucleolus.**

The production of tRNA molecules requires processing via exonucleases and endonucleases. An **exonuclease** is a type of enzyme that cleaves a covalent bond between two nucleotides at one end of a strand. Starting at one end, an exonuclease can digest a strand, one nucleotide at a time. Some exonucleases can begin this digestion only from the 3′ end, traveling in the 3′ to 5′ direction, whereas others can begin only at the 5′ end and digest in the 5′ to 3′ direction. By comparison, an **endonuclease** can cleave the bond between two adjacent nucleotides within a strand.

Like ribosomal RNA, tRNAs are synthesized as large precursor tRNAs that must be cleaved to produce mature, functional tRNAs that bind to amino acids. This processing has

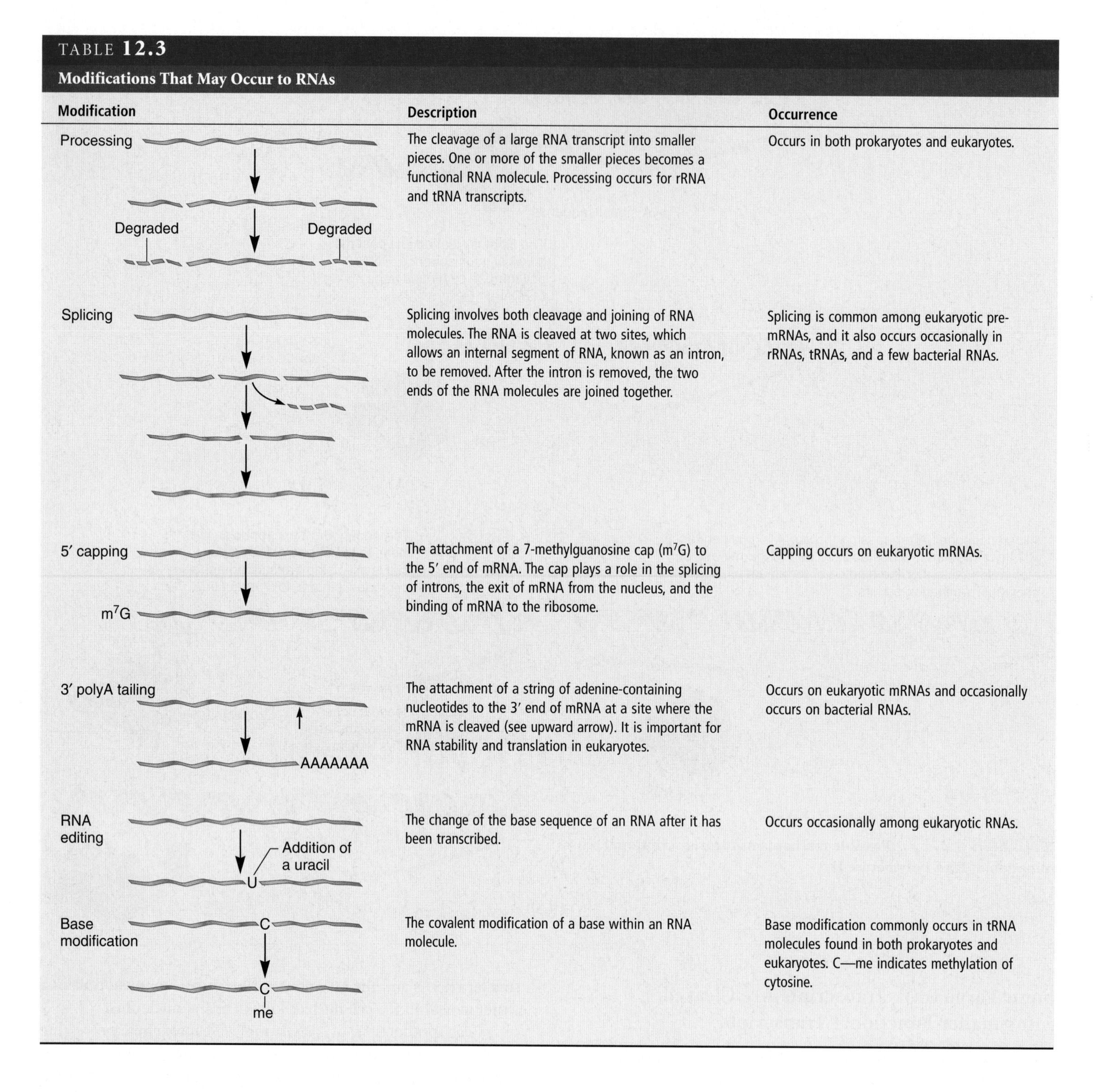

TABLE **12.3**

Modifications That May Occur to RNAs

Modification	Description	Occurrence
Processing (Degraded, Degraded)	The cleavage of a large RNA transcript into smaller pieces. One or more of the smaller pieces becomes a functional RNA molecule. Processing occurs for rRNA and tRNA transcripts.	Occurs in both prokaryotes and eukaryotes.
Splicing	Splicing involves both cleavage and joining of RNA molecules. The RNA is cleaved at two sites, which allows an internal segment of RNA, known as an intron, to be removed. After the intron is removed, the two ends of the RNA molecules are joined together.	Splicing is common among eukaryotic pre-mRNAs, and it also occurs occasionally in rRNAs, tRNAs, and a few bacterial RNAs.
5′ capping (m^{7}G)	The attachment of a 7-methylguanosine cap (m^{7}G) to the 5′ end of mRNA. The cap plays a role in the splicing of introns, the exit of mRNA from the nucleus, and the binding of mRNA to the ribosome.	Capping occurs on eukaryotic mRNAs.
3′ polyA tailing (AAAAAAA)	The attachment of a string of adenine-containing nucleotides to the 3′ end of mRNA at a site where the mRNA is cleaved (see upward arrow). It is important for RNA stability and translation in eukaryotes.	Occurs on eukaryotic mRNAs and occasionally occurs on bacterial RNAs.
RNA editing (Addition of a uracil; U)	The change of the base sequence of an RNA after it has been transcribed.	Occurs occasionally among eukaryotic RNAs.
Base modification (C; C—me)	The covalent modification of a base within an RNA molecule.	Base modification commonly occurs in tRNA molecules found in both prokaryotes and eukaryotes. C—me indicates methylation of cytosine.

been studied extensively in *E. coli*. **Figure 12.17** shows the processing of a precursor tRNA, which involves the action of two endonucleases and one exonuclease. The precursor tRNA is recognized by RNaseP, which is an endonuclease that cuts the precursor tRNA. The action of RNaseP produces the correct 5′ end of the mature tRNA. A different endonuclease cleaves the precursor tRNA to remove a 170-nucleotide segment from the 3′ end. Next, an exonuclease, called RNaseD, binds to the 3′ end and digests the RNA in the 3′ to 5′ direction. When it reaches an ACC sequence, the exonuclease stops digesting the precursor tRNA molecule. Therefore, all tRNAs in *E. coli* have an ACC sequence at their 3′ ends. Finally, certain bases in tRNA molecules may be covalently modified to alter their structure. The functional importance of modified bases in tRNAs is discussed in Chapter 13.

As researchers studied tRNA processing, they discovered certain features that were very unusual and exciting, changing the way biologists view the actions of catalysts. RNaseP has been found to be a catalyst that contains both RNA and protein subunits. In 1983, Sidney Altman and colleagues made the surprising

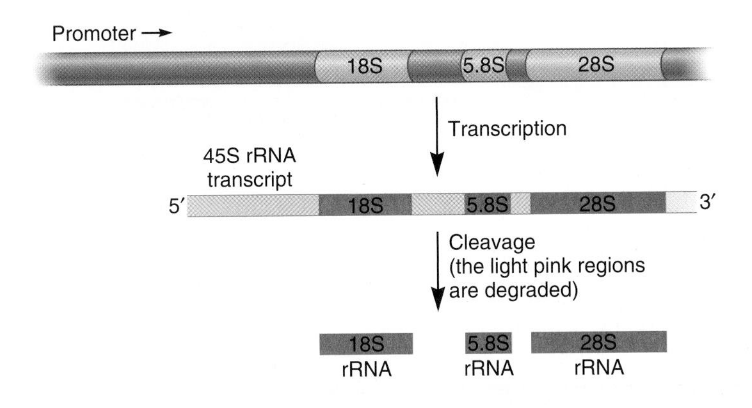

FIGURE 12.16 The processing of ribosomal RNA in eukaryotes. The large ribosomal RNA gene is transcribed into a long 45S rRNA primary transcript. This transcript is cleaved to produce 18S, 5.8S, and 28S rRNA molecules, which become associated with protein subunits in the ribosome. This processing occurs within the nucleolus of the cell.

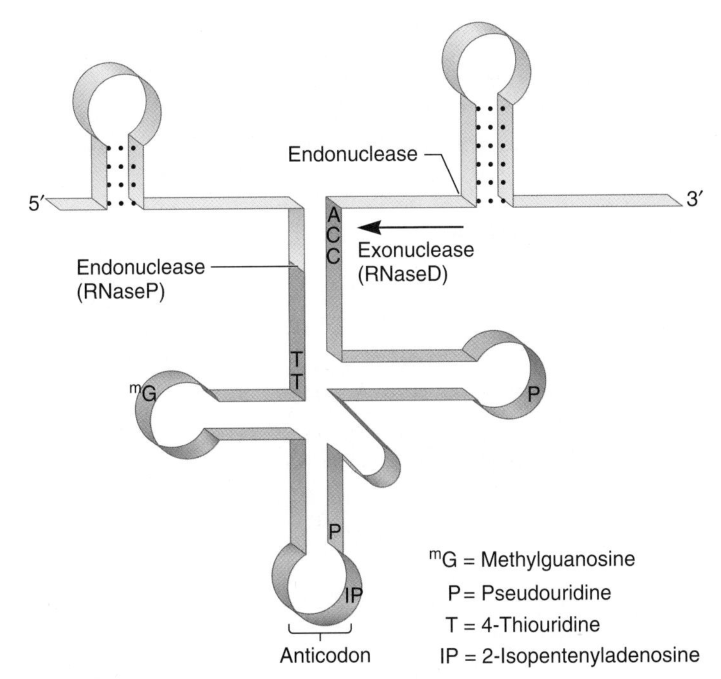

FIGURE 12.17 The processing of a precursor tRNA molecule in *E. coli*. RNaseP is an endonuclease that makes a cut that creates the 5′ end of the mature tRNA. To produce the 3′ end of mature tRNA, an endonuclease makes a cut, and then the exonuclease RNaseD removes nine nucleotides at the 3′ end. In addition to these cleavage steps, several bases within the tRNA molecule are modified to other bases as schematically indicated. A similar type of precursor tRNA processing occurs in eukaryotes.

discovery that the RNA portion of RNaseP, not the protein subunit, contains the catalytic ability to cleave the precursor tRNA. RNaseP is an example of a **ribozyme,** an RNA molecule with catalytic activity. Prior to the study of RNaseP and the identification of self-splicing RNAs (discussed later), biochemists had staunchly believed that only protein molecules could function as biological catalysts.

EXPERIMENT 12A

Introns Were Experimentally Identified via Microscopy

Although the discovery of ribozymes was very surprising, the observation that tRNA and rRNA transcripts are processed to a smaller form did not seem unusual to geneticists and biochemists, because the cleavage of RNA was similar to the cleavage that can occur for other macromolecules such as DNA and proteins. In sharp contrast, when splicing was detected in the 1970s, it was a novel concept. Splicing involves cleavage at two sites. An intron is removed, and—in a unique step—the remaining fragments are hooked back together again.

Eukaryotic introns were first detected by comparing the base sequence of viral genes and their mRNA transcripts during viral infection of mammalian cells by adenovirus. This research was carried out in 1977 by two groups headed by Philip Sharp and Richard Roberts. This pioneering observation led to the next question: Are introns a peculiar phenomenon that occurs only in viral genes, or are they found in eukaryotic genes as well?

In the late 1970s, several research groups, including those of Pierre Chambon, Bert O'Malley, and Philip Leder, investigated the presence of introns in eukaryotic structural genes. The experiments of Leder used electron microscopy to identify introns in the β-globin gene. β globin is a polypeptide that is a subunit of hemoglobin, the protein that carries oxygen in red blood cells. To detect introns within the gene, Leder considered the possible effects of mRNA binding to a gene. **Figure 12.18a** considers the situation in which a gene does not contain an intron. In this experiment, a segment of double-stranded chromosomal DNA containing a gene was first denatured and mixed with mature mRNA encoded by that gene. Because the mRNA is complementary to the template strand of the DNA, the template strand and the mRNA bind to each other to form a hybrid molecule. This event is called **hybridization.** Later, when the DNA is allowed to renature, the binding of the mRNA to the template strand of DNA prevents the two strands of DNA from forming a double helix. In the absence of any introns, the single-stranded DNA forms a loop. Because the RNA has displaced one of the DNA strands, this structure is known as an RNA displacement loop, or **R loop,** as shown in Figure 12.18a.

In contrast, Leder and colleagues realized that a different type of R loop structure would form if a gene contained an intron (**Figure 12.18b**). When mRNA is hybridized to a region of a gene containing one intron and then the other DNA strand

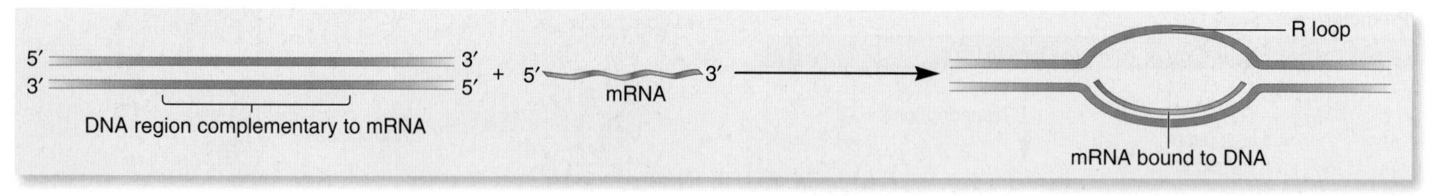

(a) No introns in the DNA

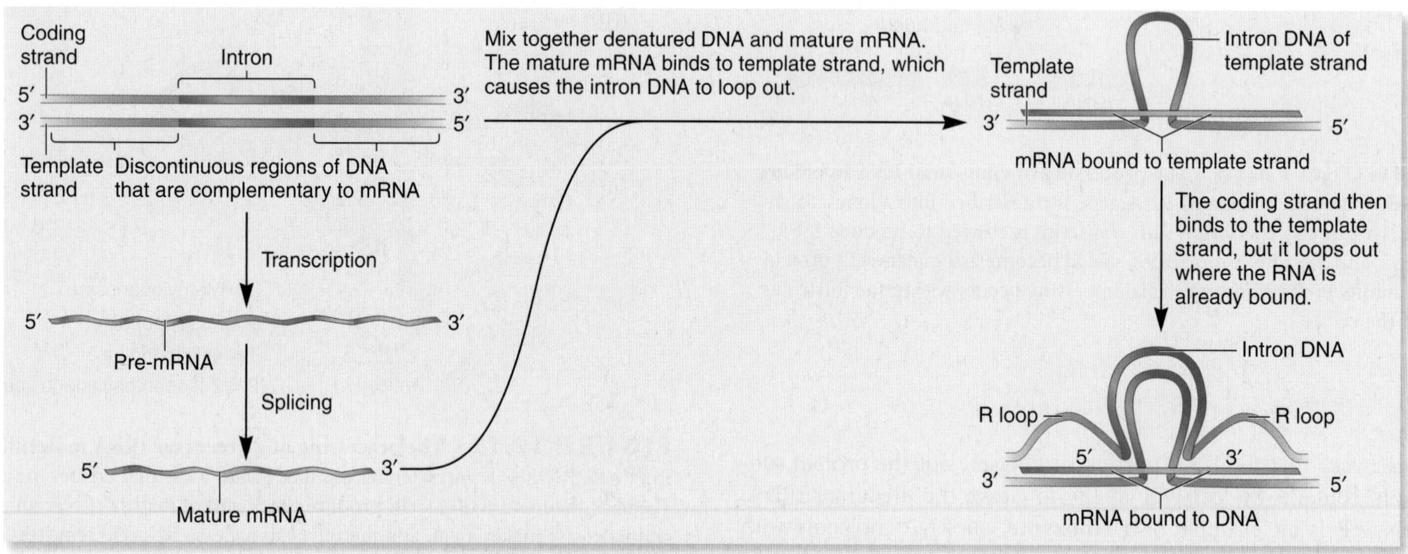

(b) One intron in the DNA. The intron in the pre-mRNA is spliced out.

FIGURE 12.18 Hybridization of mRNA to double-stranded DNA. In this experiment, the DNA is denatured and then allowed to renature under conditions that favor an RNA-DNA hybrid. **(a)** If the DNA does not contain an intron, the binding of the mRNA to the template strand of DNA prevents the two strands of DNA from forming a double helix. The single-stranded region of DNA will form an R loop. **(b)** When mRNA hybridizes to a gene containing one intron, two single-stranded R loops will form that are separated by a double-stranded DNA region. The intervening double-stranded region occurs because an intron has been spliced out of the mRNA and the mRNA cannot hybridize to this segment of the gene.

is allowed to renature, two single-stranded R loops form that are separated by a double-stranded DNA region. The intervening double-stranded region occurs because an intron has been spliced out of the mature mRNA, so the mRNA cannot hybridize to this segment of the gene.

As shown in steps 1 through 4 of **Figure 12.19**, this hybridization approach was used to identify introns within the β-globin gene. Following hybridization, the samples were placed on a microscopy grid, shadowed with heavy metal, and then observed by electron microscopy.

THE HYPOTHESIS

The mouse β-globin gene contains one or more introns.

TESTING THE HYPOTHESIS — FIGURE 12.19 RNA hybridization to the β-globin gene reveals an intron.

Starting material: A cloned fragment of chromosomal DNA that contains the mouse β-globin gene.

1. Isolate mature mRNA for the mouse β-globin gene. Note: Globin mRNA is abundant in reticulocytes, which are immature red blood cells.

2. Mix together the β-globin mRNA and cloned DNA of the β-globin gene.

3. Separate the double-stranded DNA and allow the mRNA to hybridize. This is done using 70% formamide, at 52°C, for 16 hours.

4. Dilute the sample to decrease the formamide concentration. This allows the DNA to re-form a double-stranded structure. Note: The DNA cannot form a double-stranded structure in regions where the mRNA has already hybridized.

5. Spread the sample onto a microscopy grid.

6. Stain with uranyl acetate and shadow with heavy metal. Note: The technique of electron microscopy is described in the Appendix.

7. View the sample under the electron microscope.

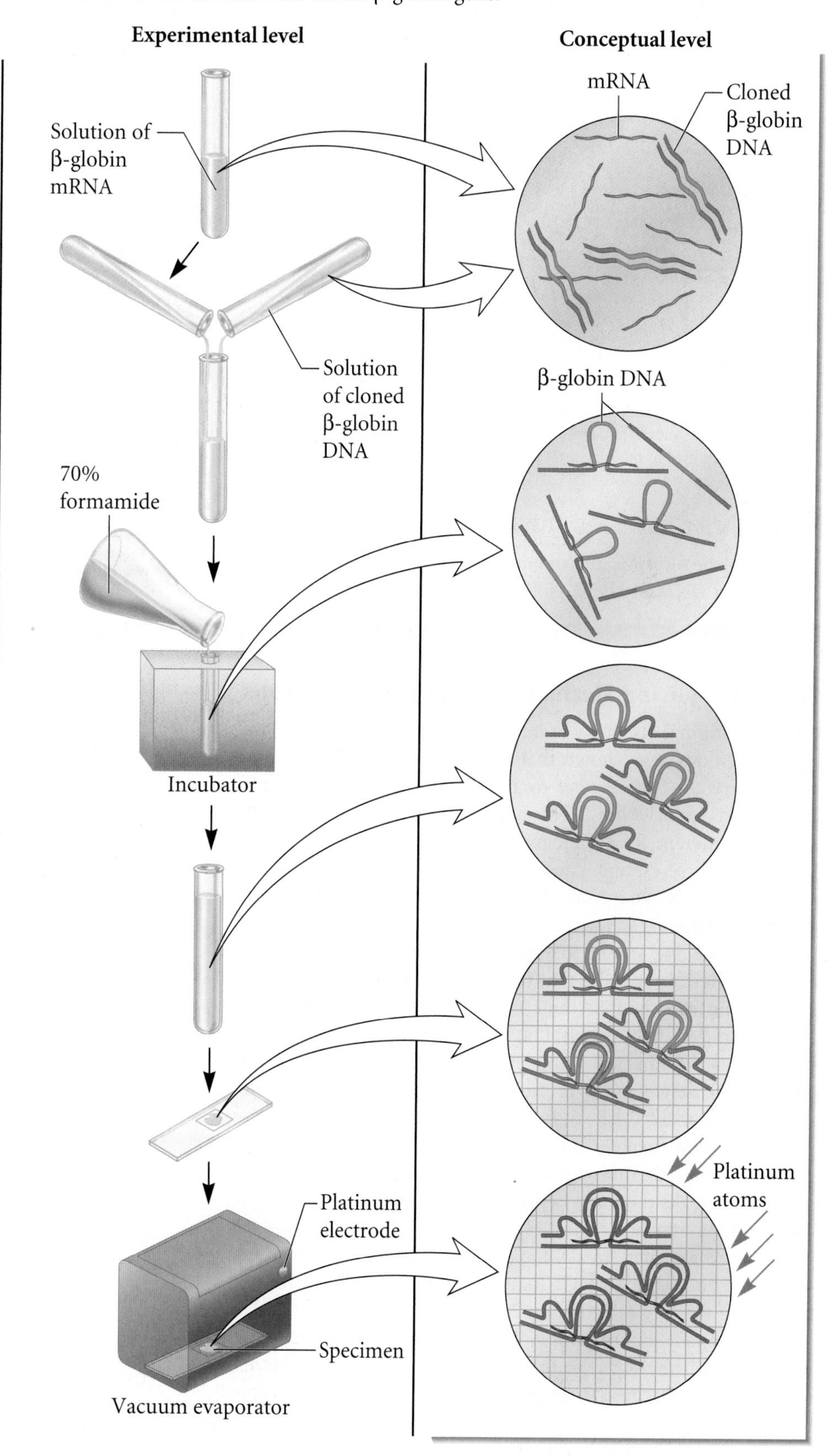

THE DATA

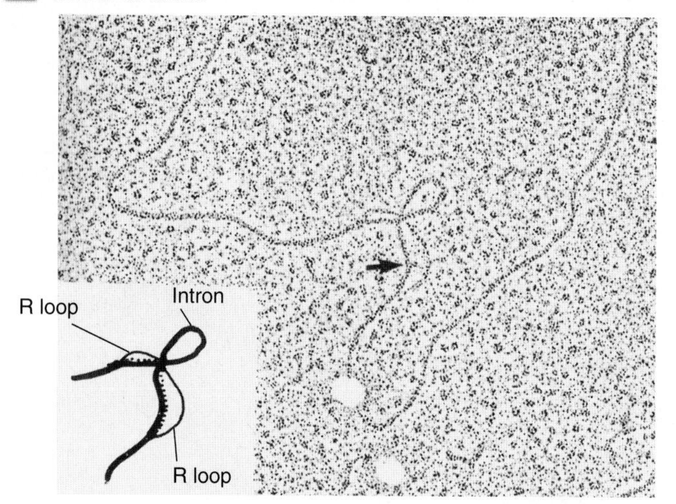

Data from: Tilghman, S.M., Tiemeier, D.C., Seidman, J.G., Peterlin, B.M., Sullivan, M., Maizel, J.V., and Leder, P. (1978) Intervening sequence of DNA identified in the structural portion of a mouse beta-globin gene. *Proc. Natl. Acad. Sci. USA* 75:725–729.

INTERPRETING THE DATA

As seen in the electron micrograph, the β-globin mRNA hybridized to the DNA of the β-globin gene, which resulted in the formation of two R loops separated by a double-stranded DNA region. These data were consistent with the idea that the DNA of the β-globin gene contains an intron. Similar results were obtained by Chambon and O'Malley for other structural genes. Since these initial discoveries, introns have been found in many eukaryotic genes. The prevalence and biological significance of introns are discussed later in this chapter.

Since the late 1970s, DNA sequencing methods have permitted an easier and more precise way of detecting introns. Researchers can clone a fragment of chromosomal DNA that contains a particular gene. This is called a **genomic clone.** In addition, mRNA can be used as a starting material to make a copy of DNA known as **complementary DNA (cDNA).** The cDNA does not contain introns, because the introns have been previously removed during RNA splicing. In contrast, if a gene contains introns, a genomic clone for a eukaryotic gene also contains introns. Therefore, a comparison of the DNA sequences from genomic and cDNA clones can provide direct evidence that a particular gene contains introns. Compared with genomic DNA, the cDNA is missing base sequences that were removed during splicing.

A self-help quiz involving this experiment can be found at www.mhhe.com/brookergenetics4e.

Different Splicing Mechanisms Can Remove Introns

Since the original discovery of introns, the investigations of many research groups have shown that most structural genes in complex eukaryotes contain one or more introns. Less commonly, introns can occur within tRNA and rRNA genes. At the molecular level, different RNA splicing mechanisms have been identified. In the three examples shown in **Figure 12.20**, splicing leads to the removal of the intron RNA and the covalent connection of the exon RNA.

The splicing among **group I** and **group II introns** occurs via **self-splicing**—splicing that does not require the aid of other catalysts. Instead, the RNA functions as its own ribozyme. Group I and II differ in the ways that the introns are removed and the exons are connected. Group I introns that occur within the rRNA of *Tetrahymena* (a protozoan) have been studied extensively by Thomas Cech and colleagues. In this organism, the splicing process involves the binding of a single guanosine to a guanosine-binding site within the intron (Figure 12.20a). This guanosine breaks the bond between the first exon and the intron and becomes attached to the 5′ end of the intron. The 3′—OH group of exon 1 then breaks the bond next to a different nucleotide (in this example, a G) that lies at the boundary between the end of the intron and exon 2; exon 1 forms a phosphoester bond with the 5′ end of exon 2. The intron RNA is subsequently degraded. In this example, the RNA molecule functions as its own ribozyme, because it splices itself without the aid of a catalytic protein.

In group II introns, a similar splicing mechanism occurs, except the 2′—OH group on ribose found in an adenine nucleotide already within the intron strand begins the catalytic process (Figure 12.20b). Experimentally, group I and II self-splicing can occur in vitro without the addition of any proteins. However, in a living cell, proteins known as **maturases** often enhance the rate of splicing of group I and II introns.

In eukaryotes, the transcription of structural genes produces a long transcript known as **pre-mRNA,** which is located within the nucleus. This pre-mRNA is usually altered by splicing and other modifications before it exits the nucleus. Unlike group I and II introns, which may undergo self-splicing, pre-mRNA splicing requires the aid of a multicomponent structure known as the **spliceosome.** As discussed shortly, this is needed to recognize the intron boundaries and to properly remove it.

Table 12.4 describes the occurrence of introns among the genes of different species. The biological significance of group I and II introns is not understood. By comparison, pre-mRNA splicing is a widespread phenomenon among complex eukaryotes. In mammals and flowering plants, most structural genes have at least one intron that can be located anywhere within the gene. For example, an average human gene has about eight introns. In some cases, a single gene can have many introns. As an extreme example, the human dystrophin gene, which, when mutated, causes Duchenne muscular dystrophy, has 79 exons punctuated by 78 introns.

Pre-mRNA Splicing Occurs by the Action of a Spliceosome

As noted previously, the spliceosome is a large complex that splices pre-mRNA in eukaryotes. It is composed of several

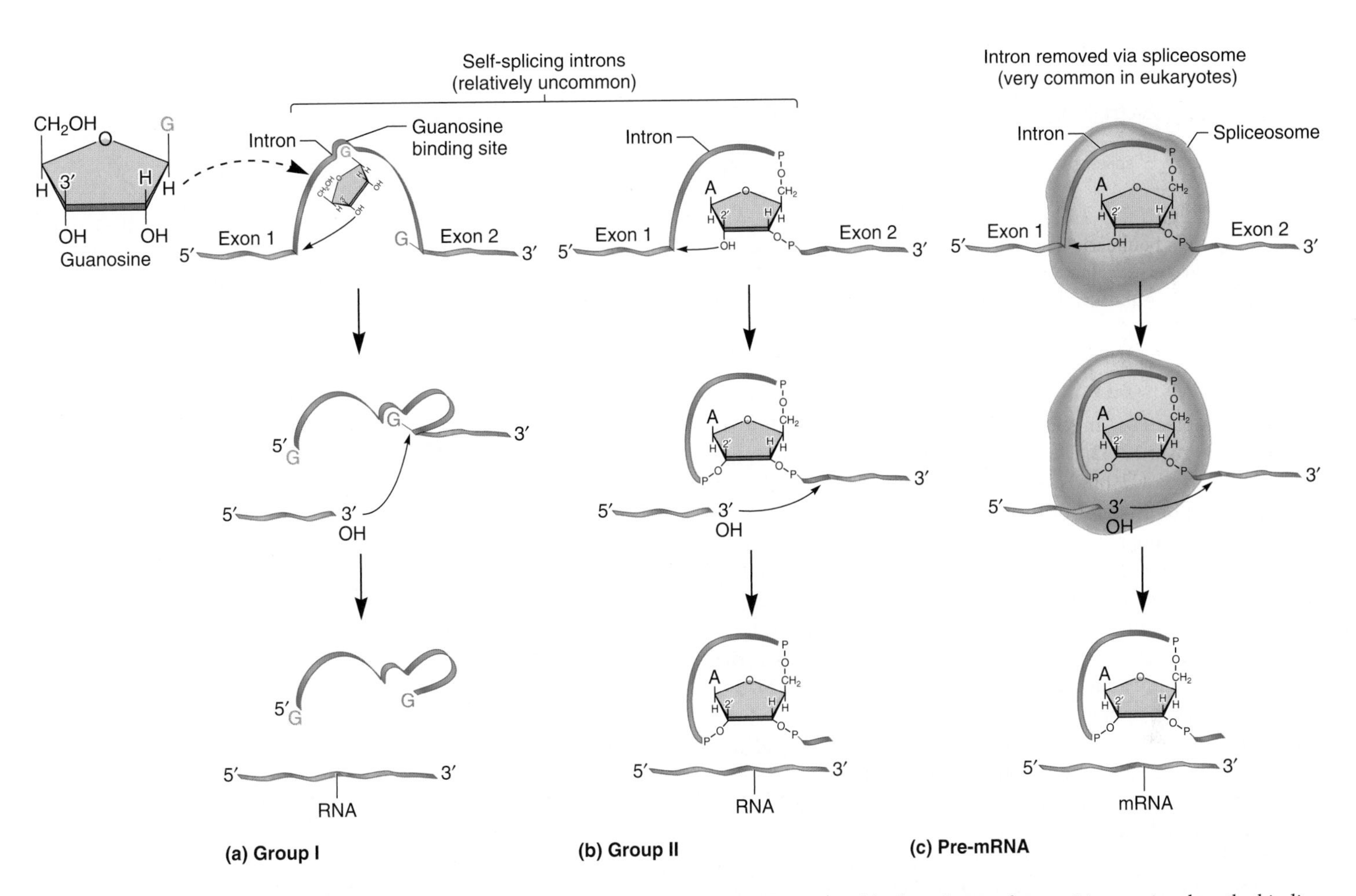

FIGURE 12.20 **Mechanisms of RNA splicing.** Group I and II introns are self-splicing. **(a)** The splicing of group I introns involves the binding of a free guanosine to a site within the intron, leading to the cleavage of RNA at the 3′ end of exon 1. The bond between a different guanine nucleotide (in the intron strand) and the 5′ end of exon 2 is cleaved. The 3′ end of exon 1 then forms a covalent bond with the 5′ end of exon 2. **(b)** In group II introns, a similar splicing mechanism occurs, except that the 2′—OH group on an adenine nucleotide (already within the intron) begins the catalytic process. **(c)** Pre-mRNA splicing requires the aid of a multicomponent structure known as the spliceosome.

TABLE 12.4

Occurrence of Introns

Type of Intron	Mechanism of Removal	Occurrence
Group I	Self-splicing	Found in rRNA genes within the nucleus of *Tetrahymena* and other simple eukaryotes. Found in a few structural, tRNA, and rRNA genes within the mitochondrial DNA (fungi and plants) and in chloroplast DNA. Found very rarely in tRNA genes within bacteria.
Group II	Self-splicing	Found in a few structural, tRNA, and rRNA genes within the mitochondrial DNA (fungi and plants) and in chloroplast DNA. Also found rarely in bacterial genes.
Pre-mRNA	Spliceosome	Very commonly found in structural genes within the nucleus of eukaryotes.

subunits known as **snRNPs** (pronounced "snurps"). Each snRNP contains small nuclear RNA and a set of proteins. During splicing, the subunits of a spliceosome carry out several functions. First, spliceosome subunits bind to an intron sequence and precisely recognize the intron-exon boundaries. In addition, the spliceosome must hold the pre-mRNA in the correct configuration to ensure the splicing together of the exons. And finally, the spliceosome catalyzes the chemical reactions that cause the introns to be removed and the exons to be covalently linked.

Intron RNA is defined by particular sequences within the intron and at the intron-exon boundaries. The consensus sequences for the splicing of mammalian pre-mRNA are shown in **Figure 12.21**. These sequences serve as recognition sites for the binding of the spliceosome. The bases most commonly found at these sites—those that are highly conserved evolutionarily—are shown in bold. The 5′ and 3′ splice sites occur at the ends of the intron, whereas the branch site is somewhere in the middle. These sites are recognized by components of the spliceosome.

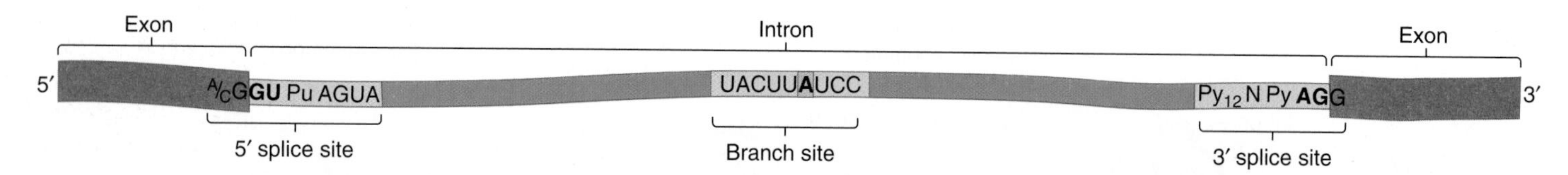

ONLINE ANIMATION **FIGURE 12.21 Consensus sequences for pre-mRNA splicing in complex eukaryotes.** Consensus sequences exist at the intron-exon boundaries and at a branch site found within the intron itself. The adenine nucleotide shown in blue in this figure corresponds to the adenine nucleotide at the branch site in Figure 12.22. The nucleotides shown in bold are highly conserved. Designations: A/C = A or C, Pu = purine, Py = pyrimidine, N = any of the four bases.

The molecular mechanism of pre-mRNA splicing is depicted in **Figure 12.22**. The snRNP designated U1 binds to the 5′ splice site, and U2 binds to the branch site. This is followed by the binding of a trimer of three snRNPs: a U4/U6 dimer plus U5. The intron loops outward, and the two exons are brought closer together. The 5′ splice site is then cut, and the 5′ end of the intron becomes covalently attached to the 2′—OH group of a specific adenine nucleotide in the branch site. U1 and U4 are released. In the final step, the 3′ splice site is cut, and then the exons are covalently attached to each other. The three snRNPs—U2, U5, and U6—remain attached to the intron, which is in a lariat configuration. Eventually, the intron is degraded, and the snRNPs are used again to splice other pre-mRNAs.

The chemical reactions that occur during pre-mRNA splicing are not completely understood. Though further research is needed, evidence is accumulating that certain snRNA molecules within the spliceosome may play a catalytic role in the removal of introns and the connection of exons. In other words, snRNAs may function as ribozymes that cleave the RNA at the exon-intron boundaries and connect the remaining exons. Researchers have speculated that RNA molecules within U2 and U6 may have this catalytic function.

Before ending our discussion of pre-mRNA splicing, let's consider why it may be an advantage for a species to have genes that contain introns. One benefit is a phenomenon called **alternative splicing.** When a pre-mRNA has multiple introns, variation may occur in the pattern of splicing, so the resulting mRNAs contain alternative combinations of exons. The variation in splicing may happen in different cell types or during different stages of development. The biological advantage of alternative splicing is that two or more different proteins can be derived from a single gene. This allows an organism to carry fewer genes in its genome. The molecular mechanism of alternative splicing is examined in Chapter 15. It involves the actions of proteins (not shown in Figure 12.22) that influence whether or not U1 and U2 can begin the splicing process.

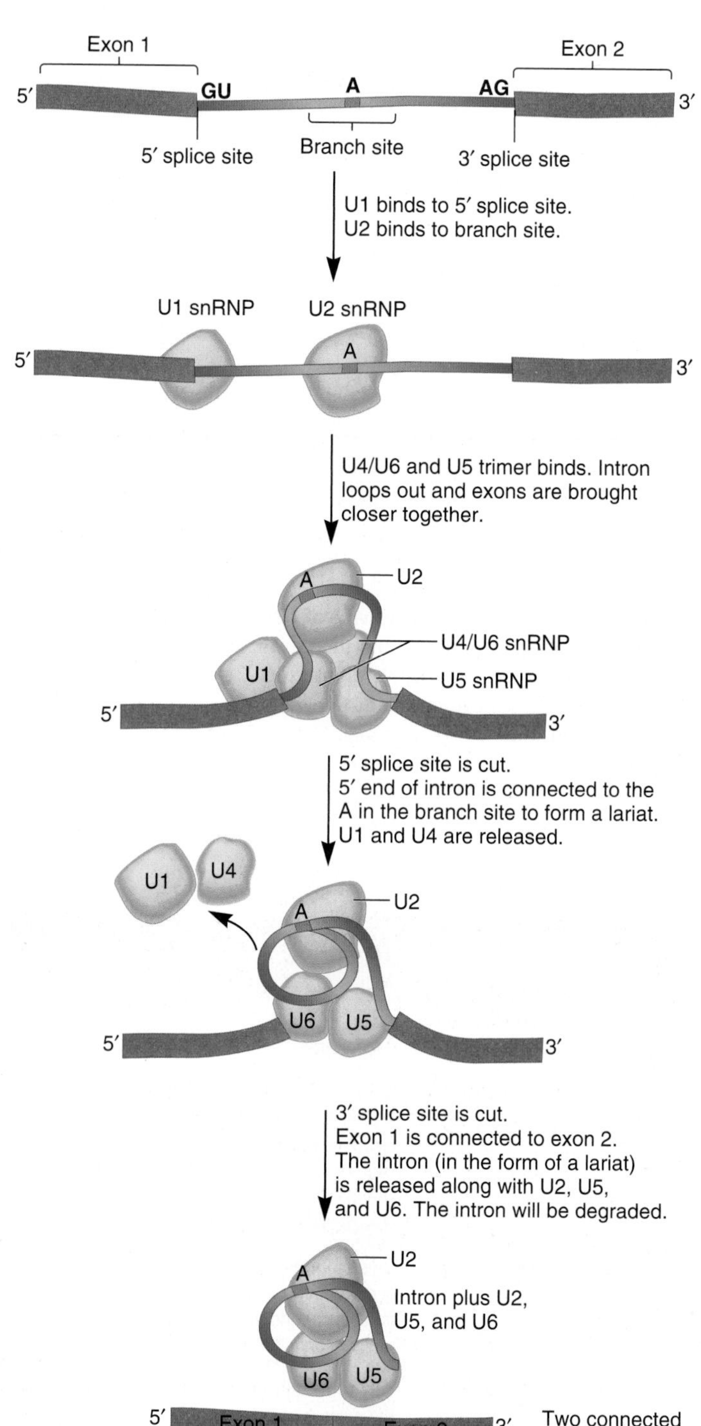

FIGURE 12.22 Splicing of pre-mRNA via a spliceosome.

The Ends of Eukaryotic Pre-mRNAs Have a 5′ Cap and a 3′ Tail

In addition to splicing, pre-mRNAs in eukaryotes are also subjected to modifications at their 5′ and 3′ ends. At their 5′ end, most mature mRNAs have a 7-methylguanosine covalently attached—an event known as **capping.** Capping occurs while the pre-mRNA is being made by RNA polymerase II, usually when the transcript is only 20 to 25 nucleotides in length. As shown in **Figure 12.23**, it is a three-step process. The nucleotide at the 5′ end of the transcript has three phosphate groups. First, an enzyme called RNA 5′-triphosphatase removes one of the phosphates, and then a second enzyme, guanylyltransferase, uses guanosine triphosphate (GTP) to attach a guanosine monophosphate (GMP) to the 5′ end. Finally, a methyltransferase attaches a methyl group to the guanine base.

What are the functions of the 7-methylguanosine cap? The cap structure is recognized by cap-binding proteins, which perform various roles. For example, cap-binding proteins are required for the proper exit of most mRNAs from the nucleus. Also, the cap structure is recognized by initiation factors that are needed during the early stages of translation. Finally, the cap structure may be important in the efficient splicing of introns, particularly the first intron located nearest the 5′ end.

Let's now turn our attention to the 3′ end of the RNA molecule. Most mature mRNAs have a string of adenine nucleotides, referred to as a **polyA tail,** which is important for mRNA stability and in the synthesis of polypeptides. The polyA tail is not encoded in the gene sequence. Instead, it is added enzymatically after the pre-mRNA has been completely transcribed—a process termed **polyadenylation.**

The steps required to synthesize a polyA tail are shown in **Figure 12.24**. To acquire a polyA tail, the pre-mRNA contains a polyadenylation signal sequence near its 3′ end. In mammals, the consensus sequence is AAUAAA. This sequence is downstream (toward the 3′ end) from the stop codon in the pre-mRNA. An endonuclease recognizes the signal sequence and cleaves the pre-mRNA at a location that is about 20 nucleotides beyond the 3′ end of the AAUAAA sequence. The fragment beyond the 3′ cut is degraded. Next, an enzyme known as polyA-polymerase attaches many adenine-containing nucleotides. The length of the polyA tail varies among different mRNAs, from a few dozen to several hundred adenine nucleotides. As discussed in Chapter 15, a long polyA tail increases the stability of mRNA and plays a role during translation.

The Nucleotide Sequence of RNA Can Be Modified by RNA Editing

The term **RNA editing** refers to a change in the nucleotide sequence of an RNA molecule that involves additions or deletions of particular bases, or a conversion of one type of base to another, such as a cytosine to a uracil. In the case of mRNAs, editing can have various effects, such as generating start codons, generating stop codons, and changing the coding sequence for a polypeptide.

The phenomenon of RNA editing was first discovered in trypanosomes, the protists that cause sleeping sickness. As with the

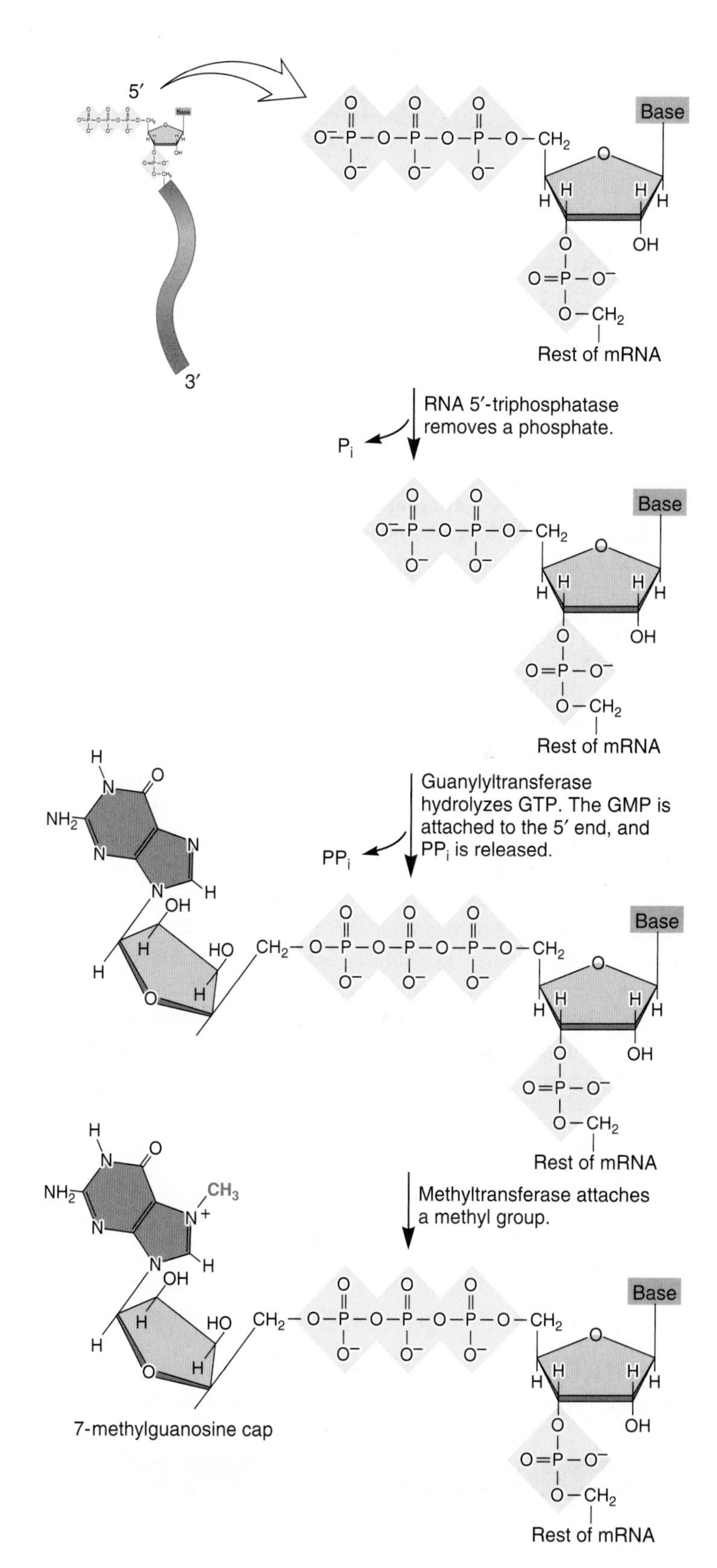

FIGURE 12.23 Attachment of a 7-methylguanosine cap to the 5′ end of mRNA. When the transcript is about 20 to 25 nucleotides in length, RNA 5′-triphosphatase removes one of the three phosphates, and then a second enzyme, guanylyltransferase, attaches GMP to the 5′ end. Finally, a methyltransferase attaches a methyl group to the guanine base.

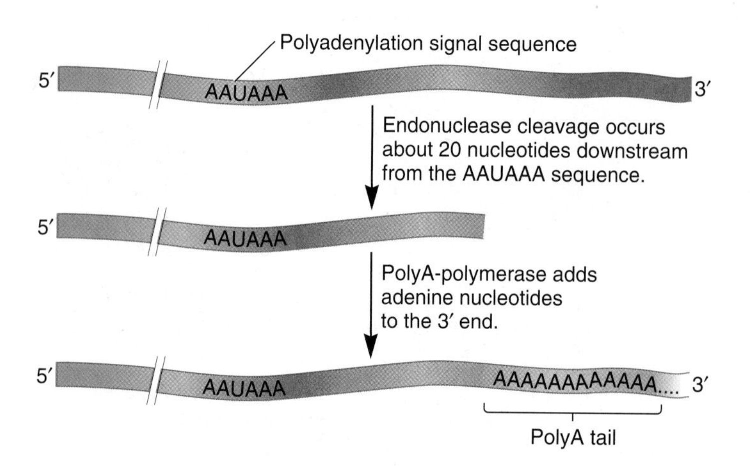

FIGURE 12.24 **Attachment of a polyA tail.** First, an endonuclease cuts the RNA at a location that is 11 to 30 nucleotides after the AAUAAA polyadenylation sequence, making the RNA shorter at its 3′ end. Adenine-containing nucleotides are then attached, one at a time, to the 3′ end by the enzyme polyA-polymerase.

discovery of RNA splicing, the initial finding of RNA editing was met with great skepticism. Since that time, however, RNA editing has been shown to occur in various organisms and in a variety of ways, although its functional significance is slowly emerging (**Table 12.5**). In the specific case of trypanosomes, the editing process involves the addition or deletion of one or more uracil nucleotides.

A more widespread mechanism for RNA editing involves changes of one type of base to another. In this form of editing, a base in the RNA is deaminated—an amino group is removed from the base. When cytosine is deaminated, uracil is formed, and when adenine is deaminated, inosine is formed (**Figure 12.25**). Inosine is recognized as guanine during translation.

An example of RNA editing occurs in mammals involving an mRNA that encodes a protein called apolipoprotein B. In the liver, the RNA editing process produces apolipoprotein B-100, a protein that is essential for the transport of cholesterol in the blood. In intestinal cells, the mRNA may be edited so that a single C is changed to a U. What is the significance of this base substitution? This change converts a glutamine codon (CAA) to a stop codon (UAA) and thereby results in a shorter apolipoprotein. In this case, RNA editing produces an apolipoprotein B with an altered structure. Therefore, RNA editing can produce two proteins from the same gene, much like the phenomenon of alternative splicing.

How widespread is RNA editing that involves C to U and A to I substitutions? In invertebrates such as *Drosophila*, researchers estimate that 50 to 100 RNAs are edited for the purpose of changing the RNA coding sequence. In mammals, the RNAs from less than 25 genes are known to be edited.

TABLE **12.5**

Examples of RNA Editing

Organism	Type of Editing	Found In
Trypanosomes (protozoa)	Primarily additions but occasionally deletions of uracil nucleotides	Many mitochondrial mRNAs
Land plants	C-to-U conversion	Many mitochondrial and chloroplast mRNAs, tRNAs, and rRNAs
Slime mold	C additions	Many mitochondrial mRNAs
Mammals	C-to-U conversion	Apoliproprotein B mRNA, and NFI mRNA, which encodes a tumor-suppressor protein
	A-to-I conversion	Glutamate receptor mRNA, many tRNAs
Drosophila	A-to-I conversion	mRNA for calcium and sodium channels

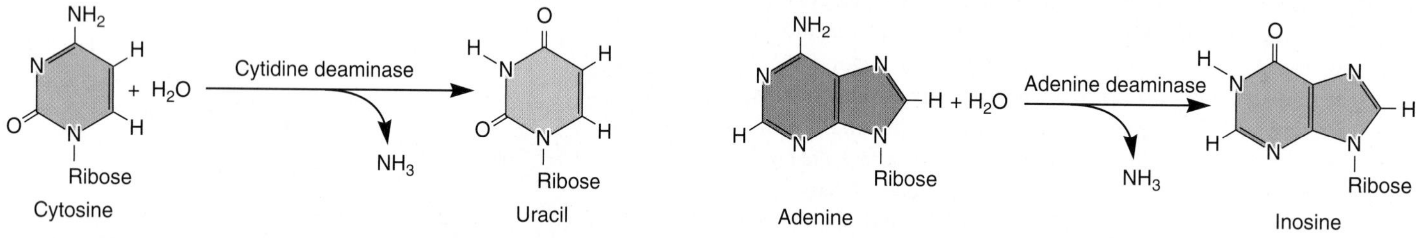

FIGURE 12.25 **RNA editing by deamination.** A cytidine deaminase can remove an amino group from cytosine, thereby creating uracil. An adenine deaminase can remove an amino group from adenine to make inosine.

KEY TERMS

Page 299. gene, transcription, structural genes, messenger RNA (mRNA)

Page 300. translation, central dogma of genetics, gene expression, promoter, terminator

Page 301. template strand, nontemplate strand, coding strand, transcription factors, regulatory sequences, regulatory elements, ribosome-binding site, codons, start codon, stop codon, initiation, elongation, termination, RNA polymerase

Page 302. open complex

Page 303. transcriptional start site, sequence elements, Pribnow box, consensus sequence, core enzyme, sigma (σ) factor, holoenzyme

Page 304. helix-turn-helix motif, closed complex

Page 305. ρ (rho), ρ-dependent termination

Page 306. ρ-independent termination

Page 307. intrinsic termination

Page 308. core promoter, TATA box, basal transcription, enhancers, silencers, *cis*-acting elements, *trans*-acting factors
Page 309. general transcription factors (GTFs), preinitiation complex
Page 310. basal transcription apparatus, mediator, colinearity, exons, intervening sequences, introns, RNA splicing
Page 311. nucleolus, exonuclease, endonuclease
Page 313. ribozyme, hybridization, R loop
Page 316. genomic clone, complementary DNA (cDNA), group I introns, group II introns, self-splicing, maturases, pre-mRNA, spliceosome
Page 317. snRNPs
Page 318. alternative splicing
Page 319. capping, polyA tail, polyadenylation, RNA editing

CHAPTER SUMMARY

- According to the central dogma of genetics, DNA is transcribed into mRNA, and mRNA is translated into a polypeptide. DNA replication allows the DNA to be passed from cell to cell and from parent to offspring (see Figure 12.1).

12.1 Overview of Transcription

- A gene is an organization of DNA sequences. A promoter signals the start of transcription, and a terminator signals the end. Regulatory sequences control the rate of transcription. For genes that encode polypeptides, the gene sequence also specifies a start codon, a stop codon, and many codons in between. Bacterial genes also specify a ribosomal binding sequence (see Figure 12.2).
- Transcription occurs in three phases called initiation, elongation, and termination (see Figure 12.3).
- RNA transcripts have several different functions (see Table 12.1).

12.2 Transcription in Bacteria

- Many bacterial promoters have sequence elements at the −35 and −10 regions. The transcriptional start site is at +1 (see Figures 12.4, 12.5).
- During the initiation phase of transcription in *E. coli*, sigma (σ) factor, which is bound to RNA polymerase, binds into the major groove of DNA and recognizes sequence elements at the promoter. This process forms a closed complex. Following the formation of an open complex, σ factor is released (see Figures 12.6, 12.7).
- During the elongation phase of transcription, RNA polymerase slides along the DNA and maintains an open complex as it goes. RNA is made in the 5′ to 3′ direction according to complementary base pairing (see Figure 12.8).
- In a given chromosome, the use of the template strand varies from gene to gene (see Figure 12.9).
- Transcriptional termination in *E. coli* occurs by a rho (ρ)-dependent or ρ-independent mechanism (see Figures 12.10, 12.11).

12.3 Transcription in Eukaryotes

- Eukaryotes use RNA polymerase I, II, and III to transcribe different categories of genes. Prokaryotic and eukaryotic RNA polymerases have similar structures (see Figure 12.12).
- Eukaryotic promoters consist of a core promoter and regulatory elements such as enhancers and silencers (see Figure 12.13).
- Transcription of structural genes in eukaryotes requires RNA polymerase II, five general transcription factors, and mediator. The five general transcription factors and RNA polymerase assemble together to form an open complex (see Table 12.2, Figure 12.14).
- Transcriptional termination of RNA polymerase II occurs while the 3′ end of the transcript is being processed (see Figure 12.15).

12.4 RNA Modification

- RNA transcripts can be modified in a variety of ways, which include processing, splicing, 5′ capping, 3′ polyA tailing, RNA editing, and base modification (see Table 12.3).
- Certain RNA molecules such as ribosomal RNAs and precursor tRNAs are processed to smaller, functional molecules via cleavage steps (see Figures 12.16, 12.17).
- Leder and colleagues identified introns in a globin gene using microscopy (see Figures 12.18, 12.19).
- Group I and II introns are removed by self-splicing. Pre-mRNA introns are removed via a spliceosome (see Table 12.4, Figure 12.20).
- The spliceosome is a multicomponent structure that recognizes intron sequences and removes them from pre-mRNA (see Figures 12.21, 12.22).
- In eukaryotes, mRNA is given a methylguanosine cap at the 5′ end and polyA tail at the 3′ end (see Figures 12.23, 12.24).
- RNA editing changes the base sequence of an RNA after it has been synthesized (see Table 12.5, Figure 12.25).

PROBLEM SETS & INSIGHTS

Solved Problems

S1. Describe the important events that occur during the three stages of gene transcription in bacteria. What proteins play critical roles in the three stages?

Answer: The three stages are initiation, elongation, and termination.

Initiation: RNA polymerase holoenzyme slides along the DNA until σ factor recognizes a promoter. σ factor binds tightly to this sequence, forming a closed complex. The DNA is then denatured to form a bubble-like structure known as the open complex.

Elongation: RNA polymerase core enzyme slides along the DNA, synthesizing RNA as it goes. The α subunits of RNA polymerase keep the enzyme bound to the DNA, while the β subunits are responsible for binding and for the catalytic synthesis of RNA. The ω (omega) subunit is also important for the proper assembly of the core enzyme. During elongation, RNA is made according to the AU/GC rule, with nucleotides being added in the 5′ to 3′ direction.

Termination: RNA polymerase eventually reaches a sequence at the end of the gene that signals the end of transcription. In ρ-independent termination, the properties of the termination sequences in the DNA are sufficient to cause termination. In ρ-dependent termination, the ρ (rho) protein recognizes a sequence within the RNA, binds there, and travels toward RNA polymerase. When the formation of a stem-loop structure causes RNA polymerase to pause, ρ catches up and separates the RNA-DNA hybrid region, releasing the RNA polymerase.

S2. What is the difference between a structural gene and a nonstructural gene?

Answer: Structural genes encode mRNAs that are translated into polypeptide sequences. Nonstructural genes encode RNAs that are never translated. Products of nonstructural genes include tRNA and rRNA, which function during translation; microRNA, which is involved in gene regulation; 7S RNA, which is part of a complex known as SRP; scRNA, small cytoplasmic RNA found in bacteria; the RNA of RNaseP; telomerase RNA, which is involved in telomere replication; snoRNA, which is involved in rRNA trimming; and snRNA, which is a component of spliceosome. In many cases, the RNA from nonstructural genes becomes part of a complex composed of RNA molecules and protein subunits.

S3. When RNA polymerase transcribes DNA, only one of the two DNA strands is used as a template. Take a look at Figure 12.4 and explain how RNA polymerase determines which DNA strand is the template strand.

Answer: The binding of σ factor and RNA polymerase depends on the sequence of the promoter. RNA polymerase binds to the promoter in such a way that the −35 sequence TTGACA and the −10 sequence TATAAT are within the coding strand, whereas the −35 sequence AACTGT and the −10 sequence ATATTA are within the template strand.

S4. The process of transcriptional termination is not as well understood in eukaryotes as it is in bacteria. Nevertheless, current evidence suggests several different mechanisms exist for eukaryotic termination. The termination of structural genes appears to occur via the release of elongation factors and/or an RNA-binding protein that functions as an exonuclease. Another type of mechanism is found for the termination of rRNA genes by RNA polymerase I. In this case, a protein known as TTFI (transcription termination factor I) binds to the DNA downstream from the termination site. Discuss how the binding of a protein downstream from the termination site could promote transcriptional termination.

Answer: First, the binding of TTFI could act as a roadblock to the movement of RNA polymerase I. Second, TTFI could promote the dissociation of the RNA transcript and RNA polymerase I from the DNA; it may act like a helicase. Third, it could cause a change in the structure of the DNA that prevents RNA polymerase from moving past the termination site. Though multiple effects are possible, the third effect seems the most likely because TTFI is known to cause a bend in the DNA when it binds to the termination sequence.

Conceptual Questions

C1. Genes may be structural genes that encode polypeptides, or they may be nonstructural genes.

A. Describe three examples of genes that are not structural genes.

B. For structural genes, one DNA strand is called the template strand, and the complementary strand is called the coding strand. Are these two terms appropriate for nonstructural genes? Explain.

C. Do nonstructural genes have a promoter and terminator?

C2. In bacteria, what event marks the end of the initiation stage of transcription?

C3. What is the meaning of the term consensus sequence? Give an example. Describe the locations of consensus sequences within bacterial promoters. What are their functions?

C4. What is the consensus sequence of the following six DNA sequences?

GGCATTGACT

GCCATTGTCA

CGCATAGTCA

GGAAATGGGA

GGCTTTGTCA

GGCATAGTCA

C5. Mutations in bacterial promoters may increase or decrease the level of gene transcription. Promoter mutations that increase transcription are termed up-promoter mutations, and those that decrease transcription are termed down-promoter mutations. As shown in Figure 12.5, the sequence of the −10 region of the promoter for the *lac* operon is TATGTT. Would you expect the following mutations to be up-promoter or down-promoter mutations?

A. TATGTT to TATATT

B. TATGTT to TTTGTT

C. TATGTT to TATGAT

C6. According to the examples shown in Figure 12.5, which positions of the −35 sequence (i.e., first, second, third, fourth, fifth, or sixth) are more tolerant of changes? Do you think that these positions play a more or less important role in the binding of σ factor? Explain why.

C7. In Chapter 9, we considered the dimensions of the double helix (see Figure 9.16). In an α helix of a protein, there are 3.6 amino acids per complete turn. Each amino acid advances the α helix by 0.15 nm; a complete turn of an α helix is 0.54 nm in length. As shown in Figure 12.6, two α helices of a transcription factor occupy the major groove of the DNA. According to Figure 12.6, estimate the number of amino acids that bind to this region. How many complete turns of the α helices occupy the major groove of DNA?

C8. A mutation within a gene sequence changes the start codon to a stop codon. How will this mutation affect the transcription of this gene?

C9. What is the subunit composition of bacterial RNA polymerase holoenzyme? What are the functional roles of the different subunits?

C10. At the molecular level, describe how σ factor recognizes bacterial promoters. Be specific about the structure of σ factor and the type of chemical bonding.

C11. Let's suppose a DNA mutation changes the consensus sequence at the −35 location so that σ factor does not bind there as well. Explain how a mutation could inhibit the binding of σ factor to the DNA. Look at Figure 12.5 and describe two specific base substitutions you think would inhibit the binding of σ factor. Explain why you think your base substitutions would have this effect.

C12. What is the complementarity rule that governs the synthesis of an RNA molecule during transcription? An RNA transcript has the following sequence:

5′–GGCAUGCAUUACGGCAUCACACUAGGGAUC–3′

What is the sequence of the template and coding strands of the DNA that encodes this RNA? On which side (5′ or 3′) of the template strand is the promoter located?

C13. Describe the movement of the open complex along the DNA.

C14. Describe what happens to the chemical bonding interactions when transcriptional termination occurs. Be specific about the type of chemical bonding.

C15. Discuss the differences between ρ-dependent and ρ-independent termination.

C16. In Chapter 11, we discussed the function of DNA helicase, which is involved in DNA replication. The structure and function of DNA helicase and ρ protein are rather similar to each other. Explain how the functions of these two proteins are similar and how they are different.

C17. Discuss the similarities and differences between RNA polymerase (described in this chapter) and DNA polymerase (described in Chapter 11).

C18. Mutations that occur at the end of a gene may alter the sequence of the gene and prevent transcriptional termination.

A. What types of mutations would prevent ρ-independent termination?

B. What types of mutations would prevent ρ-dependent termination?

C. If a mutation prevented transcriptional termination at the end of a gene, where would gene transcription end? Or would it end?

C19. If the following RNA polymerases were missing from a eukaryotic cell, what types of genes would not be transcribed?

A. RNA polymerase I

B. RNA polymerase II

C. RNA polymerase III

C20. What sequence elements are found within the core promoter of structural genes in eukaryotes? Describe their locations and specific functions.

C21. For each of the following transcription factors, how would eukaryotic transcriptional initiation be affected if it were missing?

A. TFIIB

B. TFIID

C. TFIIH

C22. Describe the allosteric and torpedo models for transcriptional termination of RNA polymerase II. Which model is more similar to ρ-dependent termination in bacteria and which model is more similar to ρ-independent termination?

C23. Which eukaryotic transcription factor(s) shown in Figure 12.14 plays an equivalent role to σ factor found in bacterial cells?

C24. The initiation phase of eukaryotic transcription via RNA polymerase II is considered an assembly and disassembly process. Which types of biochemical interactions—hydrogen bonding, ionic bonding, covalent bonding, and/or hydrophobic interactions—would you expect to drive the assembly and disassembly process? How would temperature and salt concentration affect assembly and disassembly?

C25. A eukaryotic structural gene contains two introns and three exons: exon 1–intron 1–exon 2–intron 2–exon 3. The 5′ splice site at the boundary between exon 2 and intron 2 has been eliminated by a small deletion in the gene. Describe how the pre-mRNA encoded by this mutant gene would be spliced. Indicate which introns and exons would be found in the mRNA after splicing occurs.

C26. Describe the processing events that occur during the production of tRNA in *E. coli.*

C27. Describe the structure and function of a spliceosome. Speculate why the spliceosome subunits contain snRNA. In other words, what do you think is/are the functional role(s) of snRNA during splicing?

C28. What is the unique feature of ribozyme function? Give two examples described in this chapter.

C29. What does it mean to say that gene expression is colinear?

C30. What is meant by the term self-splicing? What types of introns are self-splicing?

C31. In eukaryotes, what types of modification occur to pre-mRNA?

C32. What is alternative splicing? What is its biological significance?

C33. The processing of ribosomal RNA in eukaryotes is shown in Figure 12.16. Why is this called cleavage or processing but not splicing?

C34. In the splicing of group I introns shown in Figure 12.20, does the 5′ end of the intron have a phosphate group? Explain.

C35. According to the mechanism shown in Figure 12.22, several snRNPs play different roles in the splicing of pre-mRNA. Identify the snRNP that recognizes the following sites:

A. 5′ splice site

B. 3′ splice site

C. Branch site

C36. After the intron (which is in a lariat configuration) is released during pre-mRNA splicing, a brief moment occurs before the two exons are connected to each other. Which snRNP(s) holds the exons in place so they can be covalently connected to each other?

C37. A lariat contains a closed loop and a linear end. An intron has the following sequence: 5′–GUPuAGUA–60 nucleotides–UACUUAUCC–100 nucleotides–Py_{12}NPyAG–3′. Which sequence would be found within the closed loop of the lariat, the 60-nucleotide sequence or the 100-nucleotide sequence?

Experimental Questions

E1. A research group has sequenced the cDNA and genomic DNA from a particular gene. The cDNA is derived from mRNA, so it does not contain introns. Here are the DNA sequences.

cDNA:

5′-ATTGCATCCAGCGTATACTATCTCGGGCCCAATTAATGCCA-
GCGGCCAGACTATCACCCAACTCGGTTACCTACTAGTATATC-
CCATATACTAGCATATATTTTACCCATAATTTGTGTGTGGGTATA-
CAGTATAATCATATA-3′

Genomic DNA (contains one intron):

5′-ATTGCATCCAGCGTATACTATCTCGGGCCCAATTAATGC-
CAGCGGCCAGACTATCACCCAACTCGGCCCACCCCCCAGGTTTA-
CACAGTCATACCATACATACAAAAATCGCAGTTACTTATCCCA-
AAAAAACCTAGATACCCCACATACTATTAACTCTTTCTTTCTAG-
GTTACCTACTAGTATATCCCATATACTAGCATATATTTTAC-
CCATAATTTGTGTGTGGGTATACAGTATAATCATATA-3′

Indicate where the intron is located. Does the intron contain the normal consensus splice site sequences based on those described in Figure 12.21? Underline the splice site sequences, and indicate whether or not they fit the consensus sequence.

E2. What is an R loop? In an R loop experiment, to which strand of DNA does the mRNA bind, the coding strand or the template strand?

E3. If a gene contains three introns, draw what it would look like in an R loop experiment.

E4. Chapter 18 describes a technique known as Northern blotting that can be used to detect RNA transcribed from a particular gene. In this method, a specific RNA is detected using a short segment of cloned DNA as a probe. The DNA probe, which is radioactive, is complementary to the RNA that the researcher wishes to detect. After the radioactive probe DNA binds to the RNA, the RNA is visualized as a dark (radioactive) band on an X-ray film. As shown here, the method of Northern blotting can be used to determine the amount of a particular RNA transcribed in a given cell type. If one type of cell produces twice as much of a particular mRNA as occurs in another cell, the band will appear twice as intense. Also, the method can distinguish if alternative RNA splicing has occurred to produce an RNA that has a different molecular mass.

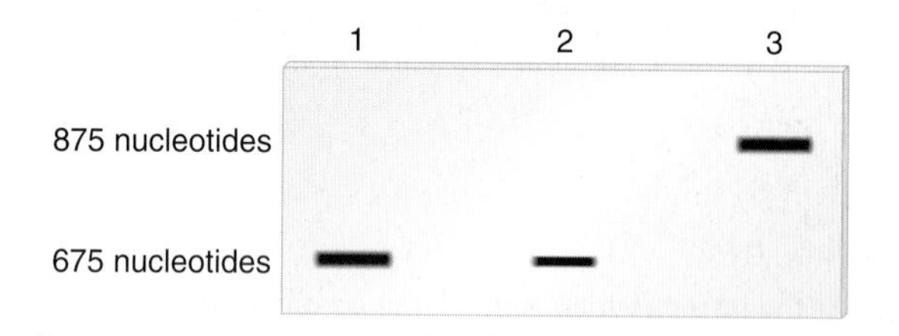

Lane 1 is a sample of RNA isolated from nerve cells.

Lane 2 is a sample of RNA isolated from kidney cells. Nerve cells produce twice as much of this RNA as do kidney cells.

Lane 3 is a sample of RNA isolated from spleen cells. Spleen cells produce an alternatively spliced version of this RNA that is about 200 nucleotides longer than the RNA produced in nerve and kidney cells.

Let's suppose a researcher was interested in the effects of mutations on the expression of a particular structural gene in eukaryotes. The gene has one intron that is 450 nucleotides long. After this intron is removed from the pre-mRNA, the mRNA transcript is 1100 nucleotides in length. Diploid somatic cells have two copies of this gene. Make a drawing that shows the expected results of a Northern blot using mRNA from the cytosol of somatic cells, which were obtained from the following individuals:

Lane 1: A normal individual

Lane 2: An individual homozygous for a deletion that removes the −50 to −100 region of the gene that encodes this mRNA

Lane 3: An individual heterozygous in which one gene is normal and the other gene had a deletion that removes the −50 to −100 region

Lane 4: An individual homozygous for a mutation that introduces an early stop codon into the middle of the coding sequence of the gene

Lane 5: An individual homozygous for a three-nucleotide deletion that removes the AG sequence at the 3′ splice site

E5. A gel retardation assay can be used to study the binding of proteins to a segment of DNA. This method is described in Chapter 18. When a protein binds to a segment of DNA, it retards the movement of the DNA through a gel, so the DNA appears at a higher point in the gel (see the following).

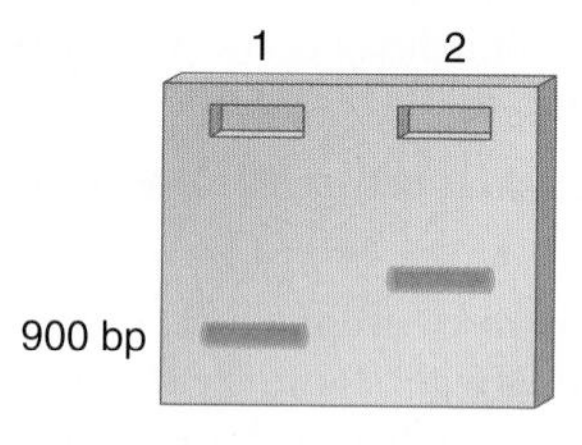

Lane 1: 900-bp fragment alone

Lane 2: 900-bp fragment plus a protein that binds to the 900-bp fragment

In this example, the segment of DNA is 900 bp in length, and the binding of a protein causes the DNA to appear at a higher point in the gel. If this 900-bp fragment of DNA contains a eukaryotic promoter for a structural gene, draw a gel that shows the relative locations of the 900-bp fragment under the following conditions:

Lane 1: 900 bp plus TFIID

Lane 2: 900 bp plus TFIIB

Lane 3: 900 bp plus TFIID and TFIIB

Lane 4: 900 bp plus TFIIB and RNA polymerase II

Lane 5: 900 bp plus TFIID, TFIIB, and RNA polymerase II/TFIIF

E6. As described in Chapter 18 and in experimental question E5, a gel retardation assay can be used to determine if a protein binds to DNA. This method can also determine if a protein binds to RNA. In the combinations described here, would you expect the migration of the RNA to be retarded due to the binding of a protein?

A. mRNA from a gene that is terminated in a ρ-independent manner plus ρ protein

B. mRNA from a gene that is terminated in a ρ-dependent manner plus ρ protein

C. pre-mRNA from a structural gene that contains two introns plus the snRNP called U1

D. Mature mRNA from a structural gene that contains two introns plus the snRNP called U1

E7. The technique of DNA footprinting is described in Chapter 18. If a protein binds over a region of DNA, it will protect chromatin in that region from digestion by DNase I. To carry out a DNA footprinting experiment, a researcher has a sample of a cloned DNA fragment. The fragments are exposed to DNase I in the presence and absence of a DNA-binding protein. Regions of the DNA fragment not covered by the DNA-binding protein will be digested by DNase I, and this will produce a series of bands on a gel. Regions of the DNA fragment not digested by DNase I (because a DNA-binding protein is preventing DNase I from gaining access to the DNA) will be revealed, because a region of the gel will not contain any bands.

In the DNA footprinting experiment shown here, a researcher began with a sample of cloned DNA 300 bp in length. This DNA contained a eukaryotic promoter for RNA polymerase II. For the sample loaded in lane 1, no proteins were added. For the sample loaded in lane 2, the 300-bp fragment was mixed with RNA polymerase II plus TFIID and TFIIB.

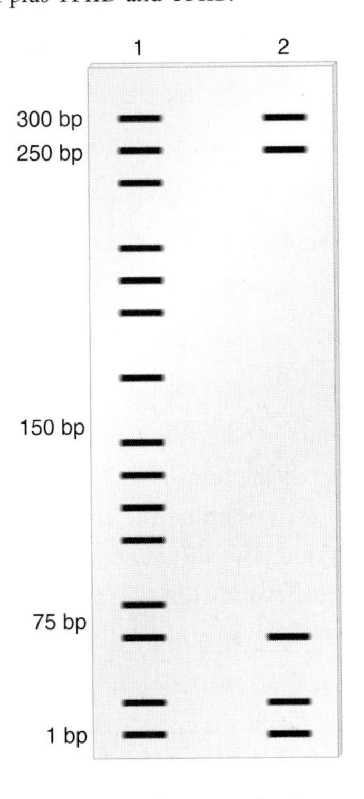

A. How long of a region of DNA is "covered up" by the binding of RNA polymerase II and the transcription factors?

B. Describe how this binding would occur if the DNA was within a nucleosome structure. (Note: The structure of nucleosomes is described in Chapter 10.) Do you think that the DNA is in a nucleosome structure when RNA polymerase and transcription factors are bound to the promoter? Explain why or why not.

E8. As described in Table 12.1, several different types of RNA are made, especially in eukaryotic cells. Researchers are sometimes interested in focusing their attention on the transcription of structural genes in eukaryotes. Such researchers want to study mRNA. One method that is used to isolate mRNA is column chromatography. (Note: See the Appendix for a general description of chromatography.) Researchers can covalently attach short pieces of DNA that contain stretches of thymine (i.e., TTTTTTTTTTTT) to the column matrix. This is called a poly-dT column. When a cell extract is poured over the column, mRNA binds to the column, but other types of RNA do not.

A. Explain how you would use a poly-dT column to obtain a purified preparation of mRNA from eukaryotic cells. In your description, explain why mRNA binds to this column and what you would do to release the mRNA from the column.

B. Can you think of ways to purify other types of RNA, such as tRNA or rRNA?

Questions for Student Discussion/Collaboration

1. Based on your knowledge of introns and pre-mRNA splicing, discuss whether or not you think alternative splicing fully explains the existence of introns. Can you think of other possible reasons to explain the existence of introns?

2. Discuss the types of RNA transcripts and the functional roles they play. Why do you think some RNAs form complexes with protein subunits?

Note: All answers appear at the website for this textbook; the answers to even-numbered questions are in the back of the textbook.

www.mhhe.com/brookergenetics4e

Visit the website for practice tests, answer keys, and other learning aids for this chapter. Enhance your understanding of genetics with our interactive exercises, quizzes, animations, and much more.

CHAPTER OUTLINE

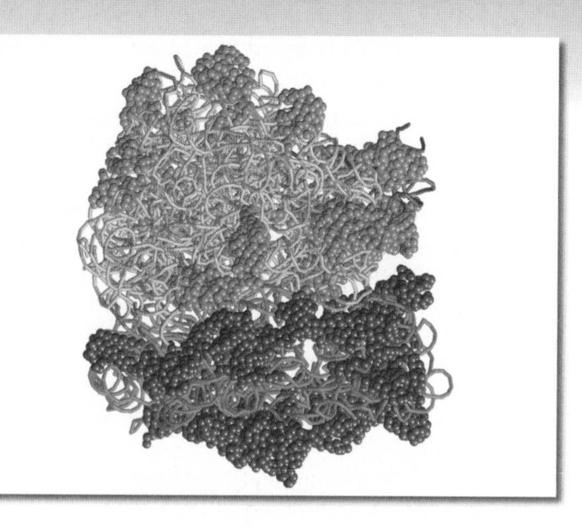

A molecular model for the structure of a ribosome. *This is a model of ribosome structure based on X-ray crystallography. Ribosomes are needed to synthesize polypeptides, using mRNA as a template. A detailed description of this model is discussed later in Figure 13.15.*

13 TRANSLATION OF mRNA

The synthesis of cellular proteins occurs via the translation of the sequence of codons within mRNA into a sequence of amino acids that constitute a polypeptide. The general steps that occur in this process were already outlined in Chapter 1. In this chapter, we will explore the current state of knowledge regarding translation, with an eye toward the specific molecular interactions responsible for this process. During the past few decades, the concerted efforts of geneticists, cell biologists, and biochemists have profoundly advanced our understanding of translation. Even so, many questions remain unanswered, and this topic continues to be an exciting area of investigation.

We will begin by considering classic experiments that revealed the purpose of some genes is to encode proteins that function as enzymes. Next, we examine how the genetic code is used to decipher the information within mRNA to produce a polypeptide with a specific amino acid sequence. The rest of this chapter is devoted to an understanding of translation at the molecular level as it occurs in living cells. This will involve an examination of the cellular components—including many different proteins, RNAs, and small molecules—needed for the translation process. We will consider the structure and function of tRNA molecules, which act as the translators of the genetic information within mRNA, and then examine the composition of ribosomes. Finally, we will explore the differences between translation in bacterial cells and eukaryotic cells.

13.1 THE GENETIC BASIS FOR PROTEIN SYNTHESIS

Proteins are critically important as active participants in cell structure and function. The primary role of DNA is to store the information needed for the synthesis of all the proteins that an organism makes. As we discussed in Chapter 12, genes that encode an amino acid sequence are known as **structural genes.** The RNA transcribed from structural genes is called **messenger RNA (mRNA).** The main function of the genetic material is to encode the production of cellular proteins in the correct cell, at the proper time, and in suitable amounts. This is an extremely complicated task because living cells make thousands of different proteins. Genetic analyses have shown that a typical bacterium can make a few thousand different proteins, and estimates for eukaryotes range from several thousand in simple eukaryotic organisms, such as yeast, to tens of thousands in plants and animals.

In this section, we will begin by considering early experiments that showed the role of genes is to encode proteins. We then examine the general features of the genetic code—the sequence of bases in a codon that specifies an amino acid—and explore the experiments through which the code was deciphered, or "cracked." Finally, we will look at the biochemistry of polypeptide synthesis to see how this determines the structure and function of proteins, which are ultimately responsible for the characteristics of living cells and an organism's traits.

Archibald Garrod Proposed That Some Genes Code for the Production of a Single Enzyme

The idea that a relationship exists between genes and the production of proteins was first suggested at the beginning of the twentieth century by Archibald Garrod, a British physician. Prior to Garrod's studies, biochemists had studied many metabolic pathways within living cells. These pathways consist of a series of metabolic conversions of one molecule to another, each step catalyzed by a specific enzyme. Each enzyme is a distinctly different protein that catalyzes a particular chemical reaction. **Figure 13.1** illustrates part of the metabolic pathway for the degradation of phenylalanine, an amino acid commonly found in human diets. The enzyme phenylalanine hydroxylase catalyzes the conversion of phenylalanine to tyrosine, and a different enzyme, tyrosine aminotransferase, converts tyrosine into *p*-hydroxyphenylpyruvic acid, and so on. In all of the steps shown in Figure 13.1, a specific enzyme catalyzes a single type of chemical reaction.

Garrod studied patients who had defects in their ability to metabolize certain compounds. He was particularly interested in the inherited disease known as **alkaptonuria.** In this disorder, the patient's body accumulates abnormal levels of homogentisic acid (also called alkapton), which is excreted in the urine, causing it to appear black on exposure to air. In addition, the disease is characterized by bluish black discoloration of cartilage and skin (ochronosis). Garrod proposed that the accumulation of homogentisic acid in these patients is due to a missing enzyme, namely, homogentisic acid oxidase (see Figure 13.1).

How did Garrod realize that certain genes encode enzymes? He already knew that alkaptonuria is an inherited trait that follows an autosomal recessive pattern of inheritance. Therefore, an individual with alkaptonuria must have inherited the mutant (defective) gene that causes this disorder from both parents. From these observations, Garrod proposed that a relationship exists between the inheritance of the trait and the inheritance of a defective enzyme. Namely, if an individual inherited the mutant gene (which causes a loss of enzyme function), she or he would not produce any normal enzyme and would be unable to metabolize homogentisic acid. Garrod described alkaptonuria as an **inborn error of metabolism.** This hypothesis was the first suggestion that a connection exists between the function of genes and the production of enzymes. At the turn of the century, this idea was particularly insightful, because the structure and function of the genetic material were completely unknown.

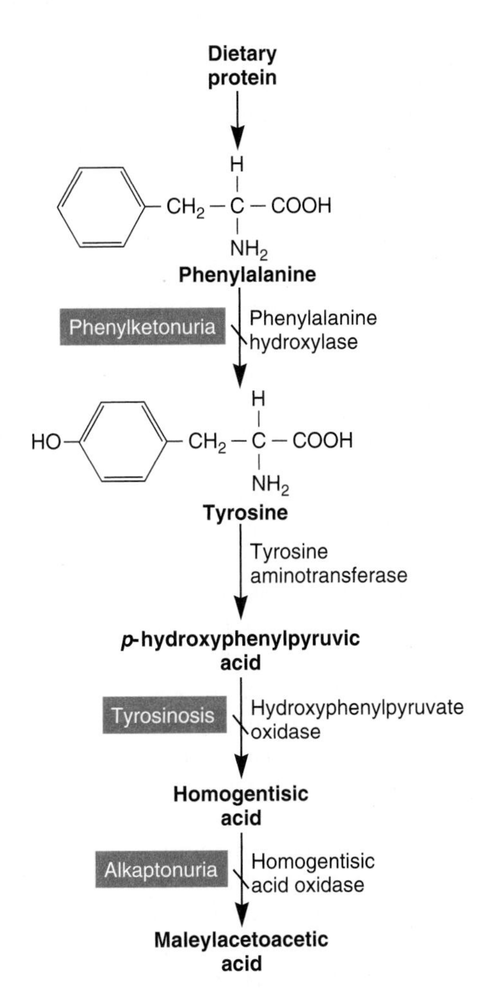

FIGURE 13.1 The metabolic pathway of phenylalanine breakdown. This diagram shows part of the pathway of phenylalanine metabolism, which consists of enzymes that successively convert one molecule to another. Certain human genetic diseases (shown in red boxes) are caused when enzymes in this pathway are missing or defective.

Genes → Traits When a person inherits two defective copies of the gene that encodes homogentisic acid oxidase, he or she cannot convert homogentisic acid into maleylacetoacetic acid. Such a person accumulates large amounts of homogentisic acid in the urine and has other symptoms of the disease known as alkaptonuria. Similarly, if a person has two defective copies of the gene encoding phenylalanine hydroxylase, he or she is unable to synthesize the enzyme phenylalanine hydroxylase and has the disease called phenylketonuria (PKU).

Beadle and Tatum's Experiments with *Neurospora* Led Them to Propose the One-Gene/One-Enzyme Hypothesis

In the early 1940s, George Beadle and Edward Tatum were also interested in the relationship among genes, enzymes, and traits. They developed an experimental system for investigating the connection between genes and the production of particular enzymes. Consistent with the ideas of Garrod, the underlying assumption behind their approach was that a relationship exists between genes and the production of enzymes. However, the quantitative nature of this relationship was unclear. In particular, they asked the question, Does one gene control the production of one enzyme, or does one gene control the synthesis of many enzymes involved in a complex biochemical pathway?

At the time of their studies, many geneticists were trying to understand the nature of the gene by studying morphological traits. However, Beadle and Tatum realized that morphological traits are likely to be based on systems of biochemical reactions so complex as to make analysis exceedingly difficult. Therefore, they turned their genetic studies to the analysis of simple nutritional requirements in *Neurospora crassa*, a common bread mold. *Neurospora* can be easily grown in the laboratory and has few nutritional requirements: a carbon source (sugar), inorganic salts, and the vitamin biotin. Normal *Neurospora* cells produce many different enzymes that can synthesize the organic molecules, such as amino acids and other vitamins, which are essential for growth.

Beadle and Tatum wanted to understand how enzymes are controlled by genes. They reasoned that a mutation in a gene, causing a defect in an enzyme needed for the cellular synthesis of an essential molecule, would prevent that mutant strain from growing on minimal medium, which contains only a carbon source, inorganic salts, and biotin. In their original study of 1941, Beadle and Tatum exposed *Neurospora* cells to X-rays, which caused mutations to occur, and studied the growth of the resulting cells. By plating the cells on media with or without vitamins, they were able to identify mutant strains that required vitamins for growth. In each case, a single mutation resulted in the requirement for a single type of vitamin in the growth media.

This early study by Beadle and Tatum led to additional research to study enzymes involved with the synthesis of other substances, including the amino acid methionine. They first isolated several different mutant strains that required methionine for growth. They hypothesized that each mutant strain might be blocked at only a single step in the consecutive series of reactions that lead to methionine synthesis. To test this hypothesis, the mutant strains were examined for their ability to grow in the presence of *O*-acetylhomoserine, cystathionine, homocysteine, or methionine. *O*-Acetylhomoserine, cystathionine, and homocysteine are intermediates in the synthesis of methionine from homoserine. A simplified depiction of the results is shown in **Figure 13.2a**. The wild-type strain could grow on minimal growth media that contained the minimum set of nutrients that is required for growth. The minimal media did not contain *O*-acetylhomoserine, cystathionine, homocysteine, or methionine. Based on their growth properties, the mutant strains that had been originally identified as requiring methionine for growth could be placed into four groups designated strains 1, 2, 3, and 4 in this figure. A strain 1 mutant was missing enzyme 1, needed for the conversion of homoserine into *O*-acetylhomoserine. The cells could grow only if *O*-acetylhomoserine, cystathionine, homocysteine, or methionine was added to the growth medium. A strain 2 mutant was missing the second enzyme in this pathway that is needed for the conversion of *O*-acetylhomoserine into cystathionine and a strain 3 mutant was unable to convert cystathionine into homocysteine. Finally, a strain 4 mutant could not make methionine from homocysteine. Based on these results, the researchers could order the enzymes into a biochemical pathway as depicted in **Figure 13.2b**. Taken together, the analysis of these mutants allowed Beadle and Tatum to conclude that a single gene controlled the synthesis of a single enzyme. This was referred to as the **one-gene/one-enzyme hypothesis.**

In later decades, this hypothesis had to be modified in four ways. First, enzymes are only one category of cellular proteins. All proteins are encoded by genes, and many of them do not function as enzymes. Second, some proteins are composed of two or more different polypeptides. Therefore, it is more accurate

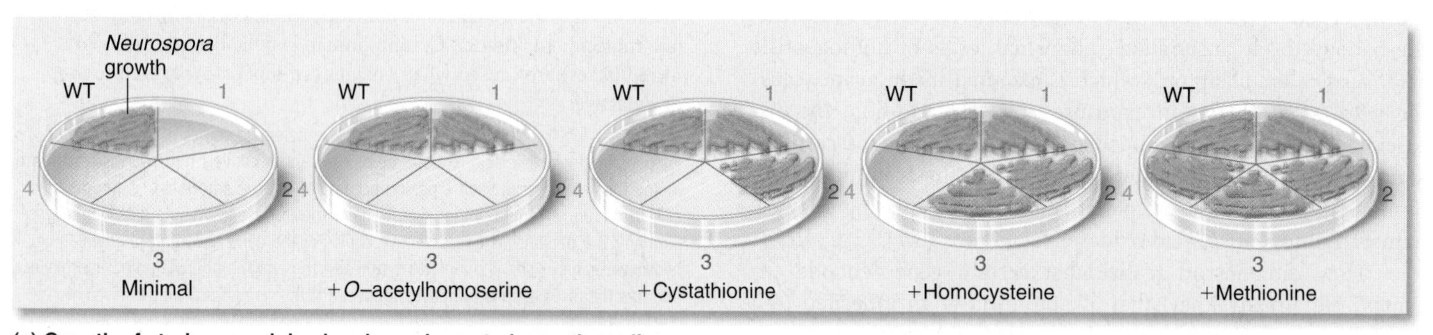

(a) Growth of strains on minimal and supplemented growth media

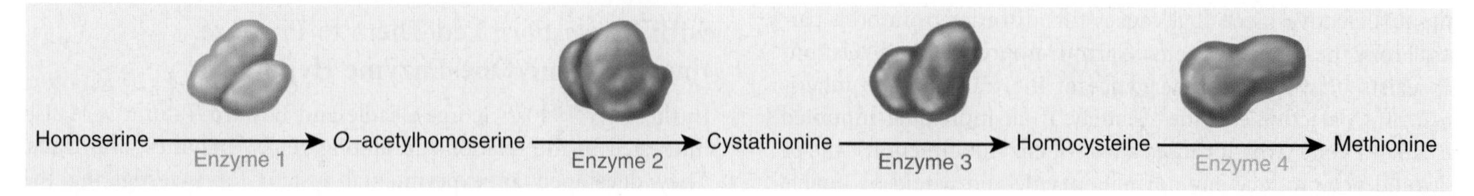

(b) Simplified pathway for methionine biosynthesis

FIGURE 13.2 An example of an experiment that supported Beadle and Tatum's one-gene/one-enzyme hypothesis. (a) Growth of wild-type (WT) and mutant strains on minimal media or in the presence of *O*-acetylhomoserine, cystathionine, homocysteine, or methionine. (b) A simplified pathway for methionine biosynthesis. Note: Homoserine is made by *Neurospora* via enzymes and precursor molecules not discussed in this experiment.

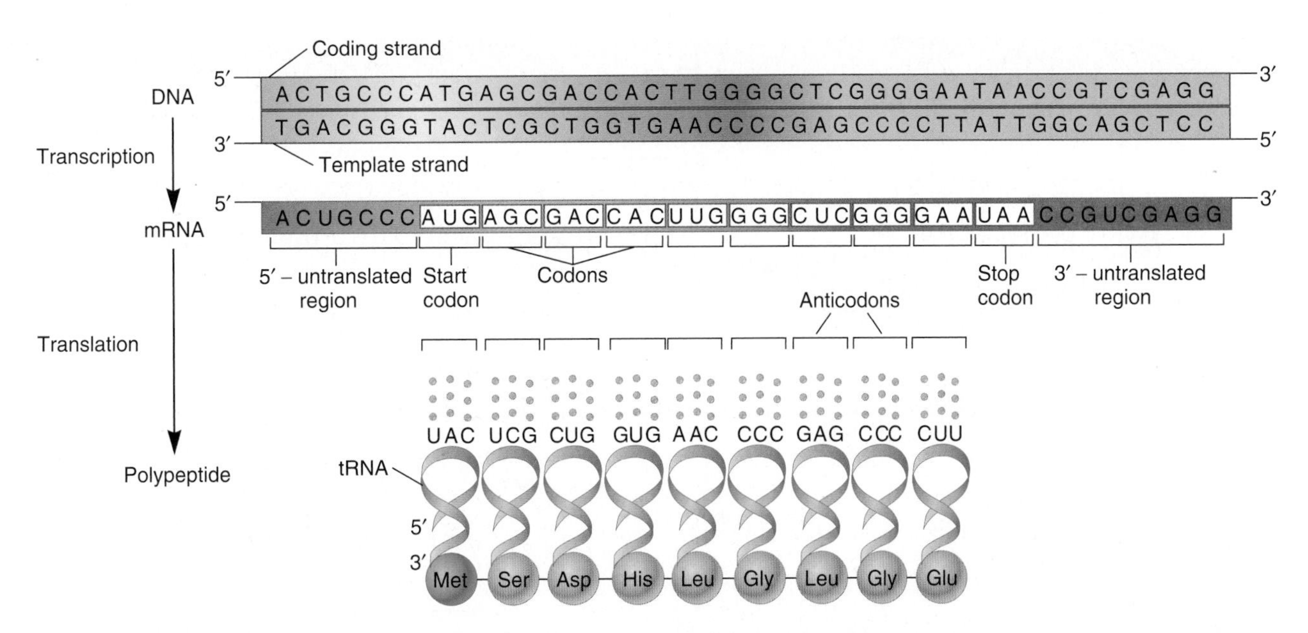

FIGURE 13.3 The relationships among the DNA coding sequence, mRNA codons, tRNA anticodons, and amino acids in a polypeptide. The sequence of nucleotides within DNA is transcribed to make a complementary sequence of nucleotides within mRNA. This sequence of nucleotides in mRNA is translated into a sequence of amino acids of a polypeptide. tRNA molecules act as intermediates in this translation process.

to say that a structural gene encodes a polypeptide. The term **polypeptide** refers to a structure; it is a linear sequence of amino acids. A structural gene encodes a polypeptide. By comparison, the term **protein** denotes function. Some proteins are composed of one polypeptide. In such cases, a single gene does encode a single protein. In other cases, however, a functional protein is composed of two or more different polypeptides. An example is hemoglobin, which is composed of two α-globin and two β-globin polypeptides. In this case, the expression of two genes—the α-globin and β-globin genes—is needed to create one functional protein. A third reason why the one-gene/one-polypeptide hypothesis needed revision is that we now know that many genes do not encode polypeptides. As discussed in Chapter 12, several types of genes specify functional RNA molecules that are not used to encode polypeptides (refer back to Table 12.1). Finally, as discussed in Chapter 15, one gene can encode multiple polypeptides due to alternative splicing and RNA editing.

During Translation, the Genetic Code Within mRNA Is Used to Make a Polypeptide with a Specific Amino Acid Sequence

Let's now turn to a general description of translation. Why have researchers named this process translation? At the molecular level, **translation** involves an interpretation of one language—the language of mRNA, a nucleotide sequence—into the language of proteins—an amino acid sequence. The ability of mRNA to be translated into a specific sequence of amino acids relies on the **genetic code.** The sequence of bases within an mRNA molecule provides coded information that is read in groups of three nucleotides known as codons (**Figure 13.3**). The sequence of three bases in most codons specifies a particular amino acid. These codons are termed **sense codons.** For example, the codon AGC specifies the amino acid serine, whereas the codon GGG encodes the amino acid glycine. The codon AUG, which specifies methionine, is used as a **start codon;** it is usually the first codon that begins a polypeptide sequence. The AUG codon can also be used to specify additional methionines within the coding sequence. Finally, three codons are used to end the process of translation. These are UAA, UAG, and UGA, which are known as **stop codons.** They are also known as **termination** or **nonsense codons.**

The codons in mRNA are recognized by the anticodons in transfer RNA (tRNA) molecules (see Figure 13.3). **Anticodons** are three-nucleotide sequences that are complementary to codons in mRNA. The tRNA molecules carry the amino acids that correspond to the codons in the mRNA. In this way, the order of codons in mRNA dictates the order of amino acids within a polypeptide.

The details of the genetic code are shown in **Table 13.1.** Because polypeptides are composed of 20 different kinds of amino acids, a minimum of 20 codons is needed in order to specify each type. With four types of bases in mRNA (A, U, G, and C), a genetic code containing two bases in a codon would not be sufficient because it would only have 4^2, or 16, possible types. By comparison, a three-base codon system can specify 4^3, or 64, different codons. Because the number of possible codons exceeds 20—which is the number of different types of amino acids—the genetic code is termed **degenerate.** This means that more than one codon can specify the same amino acid. For example, the codons GGU, GGC, GGA, and GGG all specify the amino acid glycine. Such codons are termed **synonymous codons.** In most instances, the third base in the codon is the base that varies. The third base is sometimes referred to as the **wobble base.** This term is derived from the idea that the complementary base in the tRNA can "wobble" a bit during the recognition of the third base of the codon in mRNA. The significance of the wobble base will be discussed later in this chapter.

The start codon (AUG) defines the **reading frame** of an mRNA—a sequence of codons determined by reading bases in

TABLE **13.1**

The Genetic Code

First base	Second base: U	Second base: C	Second base: A	Second base: G	Third base
U	UUU UUC Phenylalanine (Phe) UUA UUG Leucine (Leu)	UCU UCC UCA UCG Serine (Ser)	UAU UAC Tyrosine (Tyr) UAA Stop codon UAG Stop codon	UGU UGC Cysteine (Cys) UGA Stop codon UGG Tryptophan (Trp)	U C A G
C	CUU CUC CUA CUG Leucine (Leu)	CCU CCC CCA CCG Proline (Pro)	CAU CAC Histidine (His) CAA CAG Glutamine (Gln)	CGU CGC CGA CGG Arginine (Arg)	U C A G
A	AUU AUC AUA Isoleucine (Ile) AUG Methionine (Met); start codon	ACU ACC ACA ACG Threonine (Thr)	AAU AAC Asparagine (Asn) AAA AAG Lysine (Lys)	AGU AGC Serine (Ser) AGA AGG Arginine (Arg)	U C A G
G	GUU GUC GUA GUG Valine (Val)	GCU GCC GCA GCG Alanine (Ala)	GAU GAC Aspartic acid (Asp) GAA GAG Glutamic acid (Glu)	GGU GGC GGA GGG Glycine (Gly)	U C A G

groups of three, beginning with the start codon. This concept is best understood with a few examples. The mRNA sequence shown below encodes a short polypeptide with 7 amino acids:

5′–AUGCCCGGAGGCACCGUCCAAU–3′

Met–Pro–Gly–Gly–Thr–Val–Gln

If we remove one base (C) adjacent to the start codon, this changes the reading frame to produce a different polypeptide sequence:

5′–AUGCCGGAGGCACCGUCCAAU–3′

Met–Pro–Glu–Ala–Pro–Ser–Asn

Alternatively, if we remove three bases (CCC) next to the start codon, the resulting polypeptide has the same reading frame as the first polypeptide, though one amino acid (Pro, proline) has been deleted:

5′–AUGGGAGGCACCGUCCAAU–3′

Met–Gly–Gly–Thr–Val–Gln

How did researchers discover that the genetic code is read in triplets? The first evidence came from studies of Francis Crick and his colleagues in 1961. These experiments involved the isolation of mutants in a bacteriophage called T4. As described in Chapter 7, mutations in T4 genes that affect plaque morphology are easily identified (see Figure 7.16). In particular, loss-of-function mutations within certain T4 genes, designated *rII*, resulted in plaques that were larger and had a clear boundary. In comparison, wild-type phages, designated r^+, produced smaller plaques with a fuzzy boundary. Crick and colleagues exposed T4 phages to a chemical called proflavin, which causes single-nucleotide additions or deletions in gene sequences. The mutagenized phages were plated to identify large (*rII*) plaques. Though proflavin can cause either single-nucleotide additions or deletions, the first mutant strain that the researchers identified was arbitrarily called a (+) mutation. Many years later, when methods of DNA sequencing became available, it was determined that the (+) mutation is a single-nucleotide addition. **Table 13.2** shows a hypothetical wild-type sequence in a phage gene (first line) and considers how nucleotide additions and/or deletions could affect the resulting amino acid sequence. A single-nucleotide addition (+) would alter the reading frame beyond the point of insertion, thereby abolishing the proper function of the encoded protein. This is called a **frameshift mutation,** because it has changed the reading frame. This mutation resulted in a loss of function for the protein encoded by this viral gene and thereby produced an *rII* plaque phenotype.

The (+) mutant strain was then subjected to a second round of mutagenesis via proflavin. Several plaques were identified that had reverted to a wild-type (r^+) phenotype. By analyzing these strains using methods described in Chapter 7, it was determined that each one contained a second mutation that was close to the original (+) mutation. These second mutations were designated (–) mutations. Three different (–) mutations, designated a, b, and c, were identified. Each of these (–) mutations was a single-nucleotide deletion that was close to the original (+) mutation. Therefore, it restored the reading frame and produced a protein with a nearly normal amino acid sequence.

The critical experiment that suggested the genetic code is read in triplets came by combining different (–) mutations together. Mutations in different phages can be brought together into the same phage via crossing over, as described in Chapter 7. Using such an approach, the researchers constructed strains containing one, two, or three (–) mutations. The results showed that

TABLE **13.2**

Evidence That the Genetic Code Is Read in Triplets*

Strain	Plaque Phenotype	DNA Coding Sequence/Polypeptide Sequence§	Downstream Sequence‡
Wild type	r^+	ATG GGG CCC GTC CAT CCG TAC GCC GGA ATT ATA Met Gly Pro Val His Pro Tyr Ala Gly Ile Ile---------	In frame
(+)	*rII*	↓A ATG GGG ACC CGT CCA TCC GTA CGC CGG AAT TAT A Met Gly Thr Arg Pro Ser Val Arg Arg Asn Tyr--------	Out of frame
$(+)(-)_a$	r^+	↓A ↑C ATG GGG ACC GTC CAT CCG TAC GCC GGA ATT ATA Met Gly Thr Val His Pro Tyr Ala Gly Ile Ile---------	In frame
$(+)(-)_b$	r^+	↓A ↑T ATG GGG ACC CGC CAT CCG TAC GCC GGA ATT ATA Met Gly Thr Arg His Pro Tyr Ala Gly Ile Ile---------	In frame
$(+)(-)_c$	r^+	↓A ↑G ATG GGG ACC CTC CAT CCG TAC GCC GGA ATT ATA Met Gly Thr Leu His Pro Tyr Ala Gly Ile Ile---------	In frame
$(-)_a$	*rII*	↑C ATG GGG CCG TCC ATC CGT ACG CCG GAA TTA TA Met Gly Pro Ser Ile Arg Thr Pro Glu Leu---------------	Out of frame
$(-)_a(-)_b$	*rII*	↑C ↑T ATG GGG CCG CCA TCC GTA CGC CGG AAT TAT A Met Gly Pro Pro Ser Val Arg Arg Asn Tyr----------------	Out of frame
$(-)_a(-)_b(-)_c$	r^+	↑↑↑GTC ATG GGG CCC CAT CCG TAC GCC GGA ATT ATA Met Gly Pro His Pro Tyr Ala Gly Ile Ile-------------	In frame [Only Val is missing]

*This table shows only a small portion of a hypothetical coding sequence.
§A down arrow (↓) indicates the location of a single nucleotide addition; an up arrow (↑) indicates the location of a single nucleotide deletion.
‡The term downstream sequence refers to the remaining part of the sequence that is not shown in this figure. It could include hundreds of codons. An "in-frame" sequence is wild type, whereas an "out-of-frame" sequence (caused by the addition or deletion of one or two base pairs) is not.

a wild-type plaque morphology was obtained only when three (−) mutations were combined in the same phage (see Table 13.2). The three (−) mutations in the same phage restored the normal reading frame. These results were consistent with the hypothesis that the genetic code is read in multiples of three nucleotides.

Exceptions to the Genetic Code Are Known to Occur, Which Includes the Incorporation of Selenocysteine and Pyrrolysine into Polypeptides

From the analysis of many different species, including bacteria, protists, fungi, plants, and animals, researchers have found that the genetic code is nearly universal. However, a few exceptions to the genetic code have been noted (**Table 13.3**). The eukaryotic organelles known as mitochondria have their own DNA, which encodes a few structural genes. In mammals, the mitochondrial genetic code contains differences such as AUA = methionine and UGA = tryptophan. Also, in mitochondria and certain ciliated protists, AGA and AGG specify stop codons instead of arginine.

Selenocysteine (Sec) and **pyrrolysine** (Pyl) are sometimes called the 21st and 22nd amino acids in polypeptides. Their structures are shown later in Figure 13.7f. Selenocysteine is found in several enzymes involved in oxidation-reduction reactions in bacteria, archaea, and eukaryotes. Pyrrolysine is found in a few enzymes of methane-producing archaea. Selenocysteine and pyrrolysine are encoded by UGA and UAG codons, respectively, which normally function as stop codons. Like the standard 20 amino acids, selenocysteine and pyrrolysine are bound to tRNAs that specifically carry them to the ribosome for their incorporation into polypeptides. The anticodon of the tRNA that carries selenocysteine is complementary to a UGA codon, and the tRNA that carries pyrrolysine has an anticodon that is complementary to UAG.

TABLE **13.3**

Examples of Exceptions to the Genetic Code*

Codon	Universal Meaning	Exception
AUA	Isoleucine	Methionine in yeast and mammalian mitochondria
UGA	Stop	Tryptophan in mammalian mitochondria
CUU, CUA, CUC, CUG	Leucine	Threonine in yeast mitochondria
AGA, AGG	Arginine	Stop codon in ciliated protozoa and in yeast and mammalian mitochondria
UAA, UAG	Stop	Glutamine in ciliated protozoa
UGA	Stop	Encodes selenocysteine in certain genes found in bacteria, archaea, and eukaryotes
UAC	Stop	Encodes pyrrolysine in certain genes found in methane-producing archaea

*Several other exceptions, sporadically found among various species, are also known.

How are UGA and UAG codons "recoded" to specify the incorporation of selenocysteine or pyrrolysine, respectively? In the case of selenocysteine, a UGA codon is followed by a sequence called the selenocysteine insertion sequence (SECIS), which forms a stem-loop structure. In bacteria, a SECIS may be located immediately following the UGA codon, whereas a SECIS may be further downstream in the 3′-untranslated region of the mRNA in archaea and eukaryotes. The SECIS is recognized by proteins that favor the binding of a UGA codon to a tRNA carrying selenocysteine instead of the binding of release factors that are needed for polypeptide termination. Similarly, pyrrolysine incorporation may involve sequences downstream from a UAG codon that form a stem-loop structure.

EXPERIMENT 13A

Synthetic RNA Helped to Decipher the Genetic Code

Having determined that the genetic code is read in triplets, how did scientists determine the functions of the 64 codons of the genetic code? During the early 1960s, three research groups headed by Marshall Nirenberg, Severo Ochoa, and H. Gobind Khorana set out to decipher the genetic code. Though they used different methods, all of these groups used synthetic mRNA in their experimental approaches to "crack the code." We first consider the work of Nirenberg and his colleagues. Prior to their studies, several laboratories had already determined that extracts from bacterial cells, containing a mixture of components including ribosomes, tRNAs, and other factors required for translation, are able to synthesize polypeptides if mRNA and amino acids are added. This mixture is termed an in vitro, or **cell-free translation system.** If radiolabeled amino acids are added to a cell-free translation system, the synthesized polypeptides are radiolabeled and easy to detect.

To decipher the genetic code, Nirenberg and colleagues needed to gather information regarding the relationship between mRNA composition and polypeptide composition. To accomplish this goal, they made mRNA molecules of a known base composition, added them to a cell-free translation system, and then analyzed the amino acid composition of the resultant polypeptides. For example, if an mRNA molecule consisted of a string of adenine-containing nucleotides (e.g., 5′–AAAAAAAAAAAAAAAA–3′), researchers could add this polyA mRNA to a cell-free translation system and ask the question, Which amino acid is specified by a codon that contains only adenine nucleotides? (As Table 13.1 shows, it is lysine.)

Before discussing the details of this type of experiment, let's consider how the synthetic mRNA molecules were made. To synthesize mRNA, an enzyme known as polynucleotide phosphorylase was used. In the presence of excess ribonucleoside diphosphates, also called nucleoside diphosphates (NDPs), this enzyme catalyzes the covalent linkage of nucleotides to make a polymer of RNA. Because it does not use a template, the order of the nucleotides is random. For example, if only uracil-containing diphosphates (UDPs) are added, then a polyU mRNA (5′–UUUUUUUUUUUUUUUU–3′) is made. If nucleotides containing two different bases, such as uracil and guanine, are added, then the phosphorylase makes a random polymer containing both nucleotides (5′–GGGUGUGUGGUGGGUG–3′). An experimenter can control the amounts of the nucleotides that are added. For example, if 70% G and 30% U are mixed together with polynucleotide phosphorylase, the predicted amounts of the codons within the random polymer are as follows:

Codon Possibilities	*Percentage in the Random Polymer*
GGG	0.7 × 0.7 × 0.7 = 0.34 = 34%
GGU	0.7 × 0.7 × 0.3 = 0.15 = 15%
GUU	0.7 × 0.3 × 0.3 = 0.06 = 6%
UUU	0.3 × 0.3 × 0.3 = 0.03 = 3%
UUG	0.3 × 0.3 × 0.7 = 0.06 = 6%
UGG	0.3 × 0.7 × 0.7 = 0.15 = 15%
UGU	0.3 × 0.7 × 0.3 = 0.06 = 6%
GUG	0.7 × 0.3 × 0.7 = 0.15 = 15%
	100%

By controlling the amounts of the NDPs in the phosphorylase reaction, the relative amounts of the possible codons can be predicted.

The first experiment that demonstrated the ability to synthesize polypeptides from synthetic mRNA was performed by Marshall Nirenberg and J. Heinrich Matthaei in 1961. As shown in **Figure 13.4**, a cell-free translation system was added to 20 different tubes. An mRNA template made via polynucleotide phosphorylase was then added to each tube. In this example, the mRNA was made using 70% G and 30% U. Next, the 20 amino acids were added to each tube, but each tube differed with regard to the type of radiolabeled amino acid. For example, radiolabeled glycine would be found in only 1 of the 20 tubes. The tubes were incubated a sufficient length of time to allow translation to occur. The newly made polypeptides were then precipitated onto a filter by treatment with trichloroacetic acid. This step precipitates polypeptides but not amino acids. A washing step caused amino acids that had not been incorporated into polypeptides to pass through the filter. Finally, the amount of radioactivity trapped on the filter was determined by liquid scintillation counting.

THE GOAL

The researchers assumed that the sequence of bases in mRNA determines the incorporation of specific amino acids into a polypeptide. The purpose of this experiment was to provide information that would help to decipher the relationship between base composition and particular amino acids.

ACHIEVING THE GOAL — FIGURE 13.4 Elucidation of the genetic code.

Starting material: A cell-free translation system that can synthesize polypeptides if mRNA and amino acids are added.

1. Add the cell-free translation system to each of 20 tubes.
2. To each tube, add random mRNA polymers of G and U made via polynucleotide phosphorylase using 70% G and 30% U.
3. Add a different radiolabeled amino acid to each tube, and add the other 19 non-radiolabeled amino acids. The translation system contained enzymes (discussed later) that attach amino acids to the appropriate tRNAs.
4. Incubate for 60 minutes to allow translation to occur.
5. Add 15% trichloroacetic acid (TCA), which precipitates polypeptides but not amino acids.

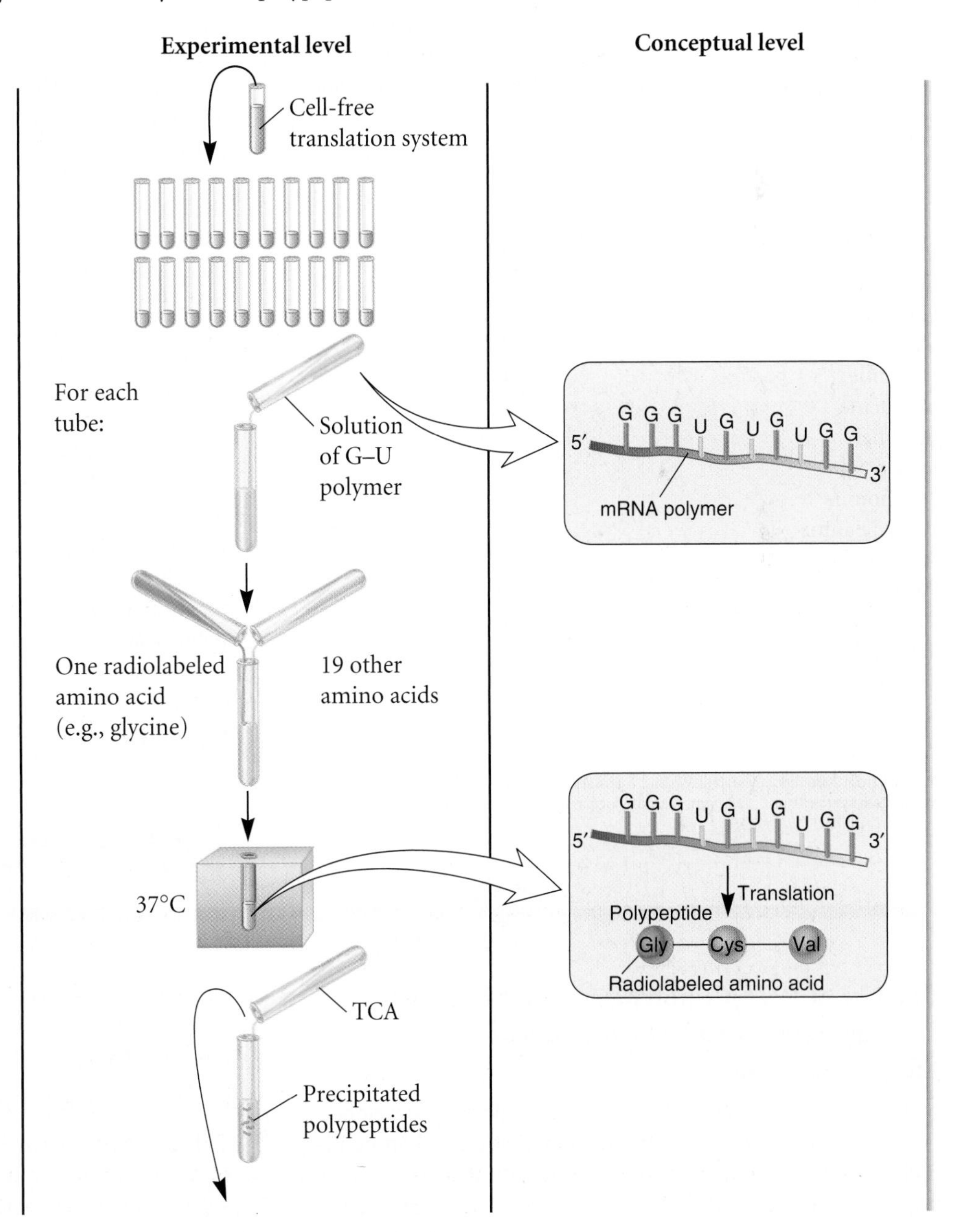

6. Place the precipitate onto a filter and wash to remove unused amino acids.

7. Count the radioactivity on the filter in a scintillation counter (see the Appendix for a description).

8. Calculate the amount of radiolabeled amino acids in the precipitated polypeptides.

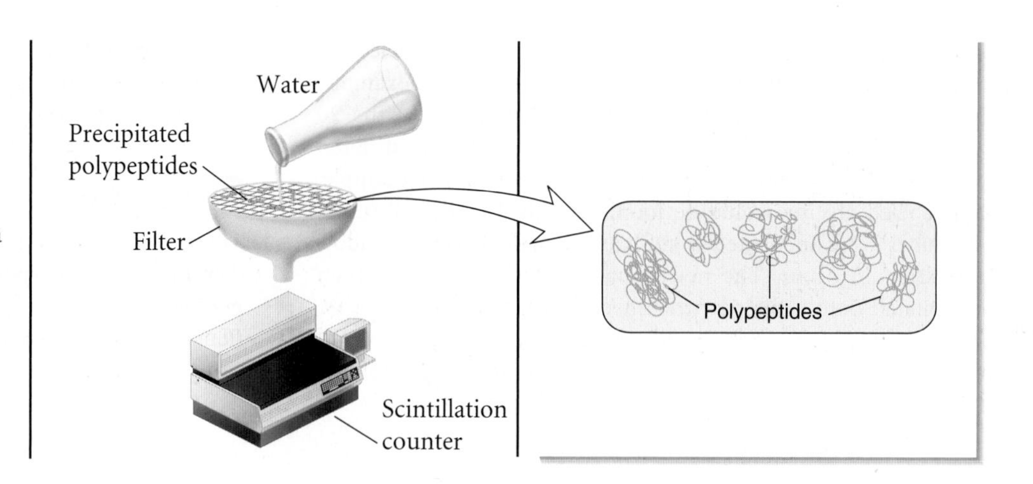

THE DATA

Radiolabeled Amino Acid Added	Relative Amount of Radiolabeled Amino Acid Incorporated into Translated Polypeptides (% of total)
Alanine	0
Arginine	0
Asparagine	0
Aspartic acid	0
Cysteine	6
Glutamic acid	0
Glutamine	0
Glycine	49
Histidine	0
Isoleucine	0
Leucine	6
Lysine	0
Methionine	0
Phenylalanine	3
Proline	0
Serine	0
Threonine	0
Tryptophan	15
Tyrosine	0
Valine	21

Adapted from Nirenberg, Marshall W., and Matthaei, J.H. (1961) The dependence of cell-free protein synthesis in *E. coli* upon naturally occurring or synthetic polyribonucleotides. *Proc Natl Acad Sci USA 47*, 1588–1602.

INTERPRETING THE DATA

According to the calculation previously described, codons should occur in the following percentages: 34% GGG, 15% GGU, 6% GUU, 3% UUU, 6% UUG, 15% UGG, 6% UGU, and 15% GUG. In the data shown in Figure 13.4, the value of 49% for glycine is due to two codons: GGG (34%) and GGU (15%). The 6% cysteine is due to UGU, and so on. It is important to realize that the genetic code was not deciphered in a single experiment such as the one described here. Furthermore, this kind of experiment yields information regarding only the nucleotide content of codons, not the specific order of bases within a single codon. For example, this experiment indicates that a cysteine codon contains two U's and one G. However, it does not tell us that a cysteine codon is UGU. Based on these data alone, a cysteine codon could be UUG, GUU, or UGU. However, by comparing many different RNA polymers, the laboratories of Nirenberg and Ochoa established patterns between the specific base sequences of codons and the amino acids they encode. In their first experiments, Nirenberg and Matthaei showed that a random polymer containing only uracil produced a polypeptide containing only phenylalanine. From this result, they inferred that UUU specifies phenylalanine. This idea is consistent with the results shown in the data table. In the random 70% G and 30% U polymer, 3% of the codons will be UUU. Likewise, 3% of the amino acids within the polypeptides were found to be phenylalanine.

A self-help quiz involving this experiment can be found at www.mhhe.com/brookergenetics4e.

The Use of RNA Copolymers and the Triplet-Binding Assay Also Helped to Crack the Genetic Code

In the 1960s, H. Gobind Khorana and colleagues developed a novel method to synthesize RNA. They first created short RNA molecules, two to four nucleotides in length, that had a defined sequence. For example, RNA molecules with the sequence 5′–AUC–3′ were synthesized chemically. These short RNAs were then linked together enzymatically, in a 5′ to 3′ manner, to create long copolymers with the sequence

5′–AUCAUCAUCAUCAUCAUCAUCAUCAUCAUC–3′

This is called a copolymer, because it is made from the linkage of several smaller molecules. Depending on the reading frame, such a copolymer would contain three different codons: AUC,

TABLE **13.4**

Examples of Copolymers That Were Analyzed by Khorana and Colleagues

Synthetic RNA*	Codon Possibilities	Amino Acids Incorporated into Polypeptides
UC	UCU, CUC	Serine, leucine
AG	AGA, GAG	Arginine, glutamic acid
UG	UGU, GUG	Cysteine, valine
AC	ACA, CAC	Threonine, histidine
UUC	UUC, UCU, CUU	Phenylalanine, serine, leucine
AAG	AAG, AGA, GAA	Lysine, arginine, glutamic acid
UUG	UUG, UGU, GUU	Leucine, cysteine, valine
CAA	CAA, AAC, ACA	Glutamine, asparagine, threonine
UAUC	UAU, AUC, UCU, CUA	Tyrosine, isoleucine, serine, leucine
UUAC	UUA, UAC, ACU, CUU	Leucine, tyrosine, threonine

*The synthetic RNAs were linked together to make copolymers.

UCA, and CAU. Using a cell-free translation system like the one described in Figure 13.4, such a copolymer produced polypeptides containing isoleucine, serine, and histidine. **Table 13.4** summarizes some of the copolymers that were made using this approach and the amino acids that were incorporated into polypeptides.

Finally, another method that helped to decipher the genetic code also involved the chemical synthesis of short RNA molecules. In 1964, Marshall Nirenberg and Philip Leder discovered that RNA molecules containing three nucleotides—a triplet—could stimulate ribosomes to bind a tRNA. In other words, the RNA triplet acted like a codon. Ribosomes were able to bind RNA triplets, and then a tRNA with the appropriate anticodon could subsequently bind to the ribosome. To establish the relationship between triplet sequences and specific amino acids, samples containing ribosomes and a particular triplet were exposed to tRNAs with different radiolabeled amino acids.

As an example, in one experiment the researchers began with a sample of ribosomes that were mixed with 5′–CCC–3′ triplets. Portions of this sample were then added to 20 different tubes that had tRNAs with different radiolabeled amino acids. For example, one tube contained radiolabeled histidine, a second tube had radiolabeled proline, a third tube contained radiolabeled glycine, and so on. Only one radiolabeled amino acid was added to each tube. After allowing sufficient time for tRNAs to bind to the ribosomes, the samples were filtered; only the large ribosomes and anything bound to them were trapped on the filter (**Figure 13.5**). Unbound tRNAs passed through the filter. Next, the researchers determined the amount of radioactivity trapped on each filter. If the filter contained a large amount of radioactivity, the results indicated that the added triplet encoded the amino acid that was radiolabeled.

Using the triplet-binding assay, Nirenberg and Leder were able to establish relationships between particular triplet sequences and the binding of tRNAs carrying specific (radiolabeled) amino acids. In the case of the 5′–CCC–3′ triplet, they determined that tRNAs carrying radiolabeled proline were bound to the ribosomes. Unfortunately, in some cases, a triplet could not promote sufficient tRNA binding to yield unambiguous results. Nevertheless, the triplet-binding assay was an important tool in the identification of the majority of codons.

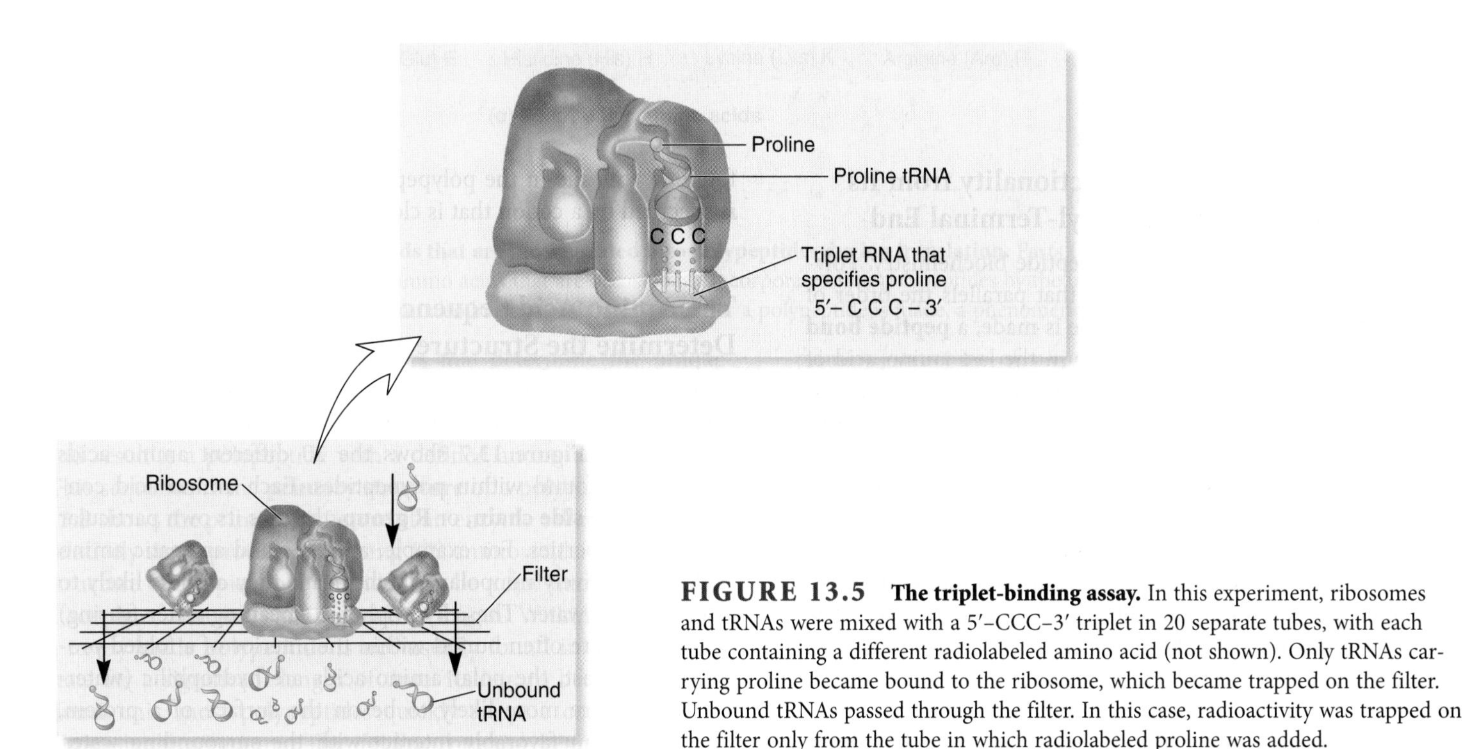

FIGURE 13.5 The triplet-binding assay. In this experiment, ribosomes and tRNAs were mixed with a 5′–CCC–3′ triplet in 20 separate tubes, with each tube containing a different radiolabeled amino acid (not shown). Only tRNAs carrying proline became bound to the ribosome, which became trapped on the filter. Unbound tRNAs passed through the filter. In this case, radioactivity was trapped on the filter only from the tube in which radiolabeled proline was added.

FIGURE 13.8 **An example of a protein's primary structure.** This is the amino acid sequence of the enzyme lysozyme, which contains 129 amino acids in its primary structure. As you may have noticed, the first amino acid is not methionine; instead, it is lysine. The first methionine residue in this polypeptide sequence is removed after or during translation. The removal of the first methionine occurs in many (but not all) proteins.

secondary structures, certain amino acids, such as glutamic acid, alanine, and methionine, are good candidates to form an α helix. Other amino acids, such as valine, isoleucine, and tyrosine, are more likely to be found in a β-sheet conformation. Secondary structures within polypeptides are primarily stabilized by the formation of hydrogen bonds between atoms that are located in the polypeptide backbone. In addition, some regions do not form a repeating secondary structure. Such regions have shapes that look very irregular in their structure because they do not follow a repeating folding pattern.

The short regions of secondary structure within a polypeptide are folded relative to each other to make the **tertiary structure** of a polypeptide. As shown in Figure 13.9c, α-helical regions and β-sheet regions are connected by irregularly shaped segments to determine the tertiary structure of the polypeptide. The folding of a polypeptide into its secondary and then tertiary conformation can usually occur spontaneously because it is a thermodynamically favorable process. The structure is determined by various interactions, including the tendency of hydrophobic amino acids to avoid water, ionic interactions among charged amino acids, hydrogen bonding among amino acids in the folded polypeptide, and weak bonding known as van der Waals interactions.

A protein is a functional unit that can be composed of one or more polypeptides. Some proteins are composed of a single polypeptide. Many proteins, however, are composed of two or more polypeptides that associate with each other to make a functional protein with a **quaternary structure** (Figure 13.9d). The individual polypeptides are called **subunits** of the protein, each of which has its own tertiary structure. The association of multiple subunits is the quaternary structure of a protein.

Cellular Proteins Are Primarily Responsible for the Characteristics of Living Cells and an Organism's Traits

Why is the genetic material largely devoted to storing the information to make proteins? To a great extent, the characteristics of a cell depend on the types of proteins that it makes. In turn, the traits of multicellular organisms are determined by the properties of their cells. Proteins perform a variety of functions critical to the life of cells and to the morphology and function of organisms (**Table 13.5**). Some proteins are important in determining the shape and structure of a given cell. For example, the protein tubulin assembles into large cytoskeletal structures known as microtubules, which provide eukaryotic cells with internal structure and organization. Some proteins are inserted into the cell membrane and aid in the transport of ions and small molecules across the membrane. An example is a sodium channel that transports sodium ions into nerve cells. Another interesting category of proteins are those that function as biological motors, such as myosin, which is involved in the contractile properties of muscle cells. Within multicellular organisms, certain proteins function in cell signaling and cell surface recognition. For example, proteins, such as the hormone insulin, are secreted by endocrine cells and

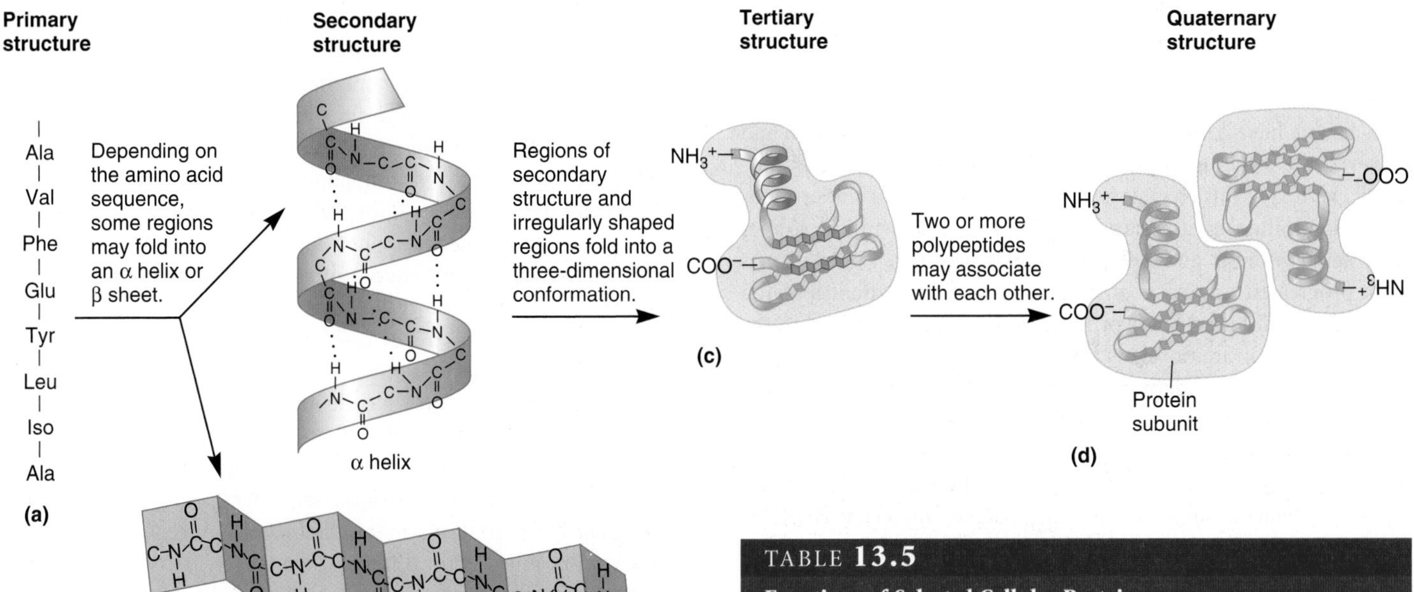

ONLINE ANIMATION

FIGURE 13.9 Levels of structures formed in proteins. **(a)** The primary structure of a polypeptide within a protein is its amino acid sequence. **(b)** Certain regions of a primary structure will fold into a secondary structure; the two types of secondary structures are called α helices and β sheets. **(c)** Both of these secondary structures can be found within the tertiary structure of a polypeptide. **(d)** Some polypeptides associate with each other to form a protein with a quaternary structure.

bind to the insulin receptor proteins found within the plasma membrane of target cells.

Many proteins are **enzymes,** which function to accelerate chemical reactions within the cell. Some enzymes assist in the breakdown of molecules or macromolecules into smaller units. These are known as catabolic enzymes and are important in utilizing cellular energy. In contrast, anabolic enzymes function in the synthesis of molecules and macromolecules. Several anabolic enzymes are listed in Table 13.5, including DNA polymerase, which is required for the synthesis of DNA from nucleotide building blocks. Throughout the cell, the synthesis of molecules and macromolecules relies on enzymes and accessory proteins. Ultimately, then, the construction of a cell greatly depends on its anabolic enzymes because these are required to synthesize all cellular macromolecules.

TABLE 13.5
Functions of Selected Cellular Proteins

Function	Examples
Cell shape and organization	Tubulin: Forms cytoskeletal structures known as microtubules
	Ankyrin: Anchors cytoskeletal proteins to the plasma membrane
Transport	Sodium channels: Transport sodium ions across the nerve cell membrane
	Lactose permease: Transports lactose across the bacterial cell membrane
	Hemoglobin: Transports oxygen in red blood cells
Movement	Myosin: Involved in muscle cell contraction
	Kinesin: Involved in the movement of chromosomes during cell division
Cell signaling	Insulin: A hormone that influences target cell metabolism and growth
	Epidermal growth factor: A growth factor that promotes cell division
	Insulin receptor: Recognizes insulin and initiates a cell response
Cell surface recognition	Integrins: Bind to large extracellular proteins
Enzymes	Hexokinase: Phosphorylates glucose during the first step in glycolysis
	β-Galactosidase: Cleaves lactose into glucose and galactose
	Glycogen synthetase: Uses glucose molecules as building blocks to synthesize a large carbohydrate known as glycogen
	Acyl transferase: Links together fatty acids and glycerol phosphate during the synthesis of phospholipids
	RNA polymerase: Uses ribonucleotides as building blocks to synthesize RNA
	DNA polymerase: Uses deoxyribonucleotides as building blocks to synthesize DNA

13.2 STRUCTURE AND FUNCTION OF tRNA

Thus far, we have considered the general features of translation and surveyed the structure and functional significance of cellular proteins. The rest of this chapter is devoted to a molecular understanding of translation as it occurs in living cells. Biochemical studies of protein synthesis and tRNA molecules began in the 1950s. As work progressed toward an understanding of translation, research revealed that different kinds of RNA molecules are involved in the incorporation of amino acids into growing polypeptides. Francis Crick proposed the **adaptor hypothesis.** According to this idea, the position of an amino acid within a polypeptide chain is determined by the binding between the mRNA and an adaptor molecule carrying a specific amino acid. Later, work by Paul Zamecnik and Mahlon Hoagland suggested that the adaptor molecule is tRNA. During translation, a tRNA has two functions: (1) It recognizes a three-base codon sequence in mRNA, and (2) it carries an amino acid specific for that codon. In this section, we will examine the general function of tRNA molecules. We begin by considering an experiment that was critical in supporting the adaptor hypothesis and then explore some of the important structural features that underlie tRNA function.

The Function of a tRNA Depends on the Specificity Between the Amino Acid It Carries and Its Anticodon

The adaptor hypothesis proposes that tRNA molecules recognize the codons within mRNA and carry the correct amino acids to the site of polypeptide synthesis. During mRNA-tRNA recognition, the anticodon in a tRNA molecule binds to a codon in mRNA due to their complementary sequences (**Figure 13.10**). Importantly, the anticodon in the tRNA corresponds to the amino acid that it carries. For example, if the anticodon in the tRNA is 3′–AAG–5′, it is complementary to a 5′–UUC–3′ codon. According to the genetic code, described earlier in this chapter, the UUC codon specifies phenylalanine. Therefore, the tRNA with a 3′–AAG–5′ anticodon must carry a phenylalanine. As another example, if the tRNA has a 3′–GGC–5′ anticodon, it is complementary to a 5′–CCG–3′ codon that specifies proline. This tRNA must carry proline.

Recall that the genetic code has 64 codons. Of these, 61 are sense codons that specify the 20 amino acids. Therefore, to synthesize proteins, a cell must produce many different tRNA molecules having specific anticodon sequences. To do so, the chromosomal DNA contains many distinct tRNA genes that encode tRNA molecules with different sequences. According to the adaptor hypothesis, the anticodon in a tRNA specifies the type of amino acid that it carries. Due to this specificity, tRNA molecules are named according to the type of amino acid they carry. For example, a tRNA that attaches to phenylalanine is described as $tRNA^{Phe}$, whereas a tRNA that carries proline is $tRNA^{Pro}$.

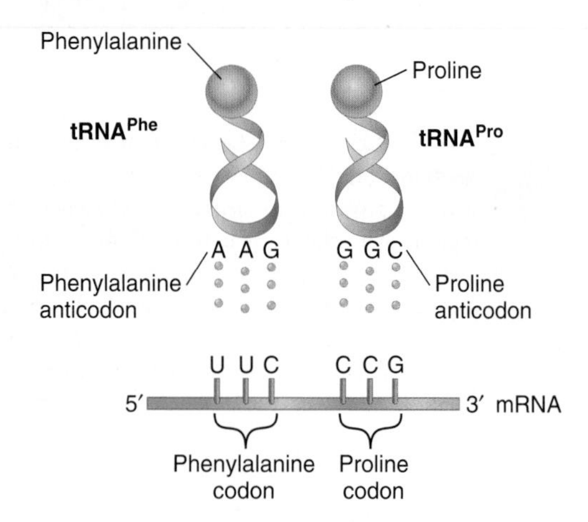

FIGURE 13.10 Recognition between tRNAs and mRNA. The anticodon in the tRNA binds to a complementary sequence in the mRNA. At its other end, the tRNA carries the amino acid that corresponds to the codon in the mRNA via the genetic code.

EXPERIMENT 13B

tRNA Functions as the Adaptor Molecule Involved in Codon Recognition

In 1962, François Chapeville and his colleagues conducted experiments aimed at testing the adaptor hypothesis. Their technical strategy was similar to that of the Nirenberg experiments that helped to decipher the genetic code (see Experiment 13A). In this approach, a cell-free translation system was made from cell extracts that contained the components necessary for translation. These components include ribosomes, tRNAs, and other translation factors. A cell-free translation system can synthesize polypeptides in vitro if mRNA and amino acids are added. Such a translation system can be used to investigate the role of specific factors by adding a particular mRNA template and varying individual components required for translation.

According to the adaptor hypothesis, the amino acid attached to a tRNA is not directly involved in codon recognition. Chapeville reasoned that if this were true, the alteration of an amino acid already attached to a tRNA should cause that altered amino acid to be incorporated into the polypeptide instead of the normal amino acid. For example, consider a $tRNA^{Cys}$ that carries the amino acid cysteine. If the attached cysteine were changed to an alanine, this $tRNA^{Cys}$ should insert an alanine into a polypeptide where it would normally put a cysteine. Fortunately, Chapeville could carry out this strategy because he had a reagent, known as Raney nickel, that can chemically convert cysteine to alanine.

A key aspect of the experimental design was the choice of the mRNA template. Chapeville and his colleagues synthesized an mRNA template that contained only U and G. Therefore, this template contained only the following codons (refer back to the genetic code in Table 13.1):

UUU = phenylalanine	GGU = glycine
UUG = leucine	GUU = valine
UGG = tryptophan	GUG = valine
GGG = glycine	UGU = cysteine

Among the eight possible codons, one cysteine codon occurs, but no alanine codons can be formed from a polyUG template.

As shown in the experiment of **Figure 13.11**, Chapeville began with a cell-free translation system that contained tRNA molecules. Amino acids, which would become attached to tRNAs, were added to this mixture. Of the 20 amino acids, only cysteine was radiolabeled. After allowing sufficient time for the amino acids to become attached to the correct tRNAs, the sample was divided into two tubes. One tube was treated with Raney nickel, whereas the control tube was not. As mentioned, Raney nickel converts cysteine into alanine by removing the –SH (sulfhydryl) group. However, it did not remove the radiolabel, which was a ^{14}C-label within the cysteine amino acid. Next, the polyUG mRNA was added as a template, and the samples were incubated to allow the translation of the mRNA into a polypeptide. In the control tube, we would expect the polypeptide to contain phenylalanine, leucine, tryptophan, glycine, valine, and cysteine, because these are the codons that contain only U and G. However, in the Raney nickel-treated sample, if the $tRNA^{Cys}$ was using its anticodon region to recognize the mRNA, we would expect to see alanine instead of cysteine.

Following translation, the polypeptides were isolated and hydrolyzed via a strong acid treatment, and then the individual amino acids were separated by column chromatography. The column separated cysteine from alanine; alanine eluted in a later fraction. The amount of radioactivity in each fraction was determined by liquid scintillation counting.

THE HYPOTHESIS

Codon recognition is dictated only by the tRNA anticodon; the chemical structure of the amino acid attached to the tRNA does not play a role.

TESTING THE HYPOTHESIS — FIGURE 13.11 **Evidence that tRNA uses its anticodon sequence to recognize mRNA.**

Starting material: A cell-free translation system that can synthesize polypeptides if mRNA and amino acids are added.

1. Place cell-free translation system into a tube. Note: This drawing emphasizes only $tRNA^{Cys}$, even though the cell-free translation system contains all types of tRNAs, and other components, such as ribosomes. In the translation system, a substantial proportion of the tRNAs do not have an attached amino acid. The translation system also contains enzymes that attach amino acids to tRNAs. (These enzymes will be described later in the chapter.)

2. Add amino acids, including radiolabeled cysteine. An enzyme within the translation system will specifically attach the radiolabeled cysteine to $tRNA^{Cys}$. The other tRNAs will have unlabeled amino acids attached to them. Incubate and divide into two tubes.

3. In one tube, treat the tRNAs with Raney nickel. This removes the –SH group from cysteine, converting it to alanine. In the control tube, do not add Raney nickel.

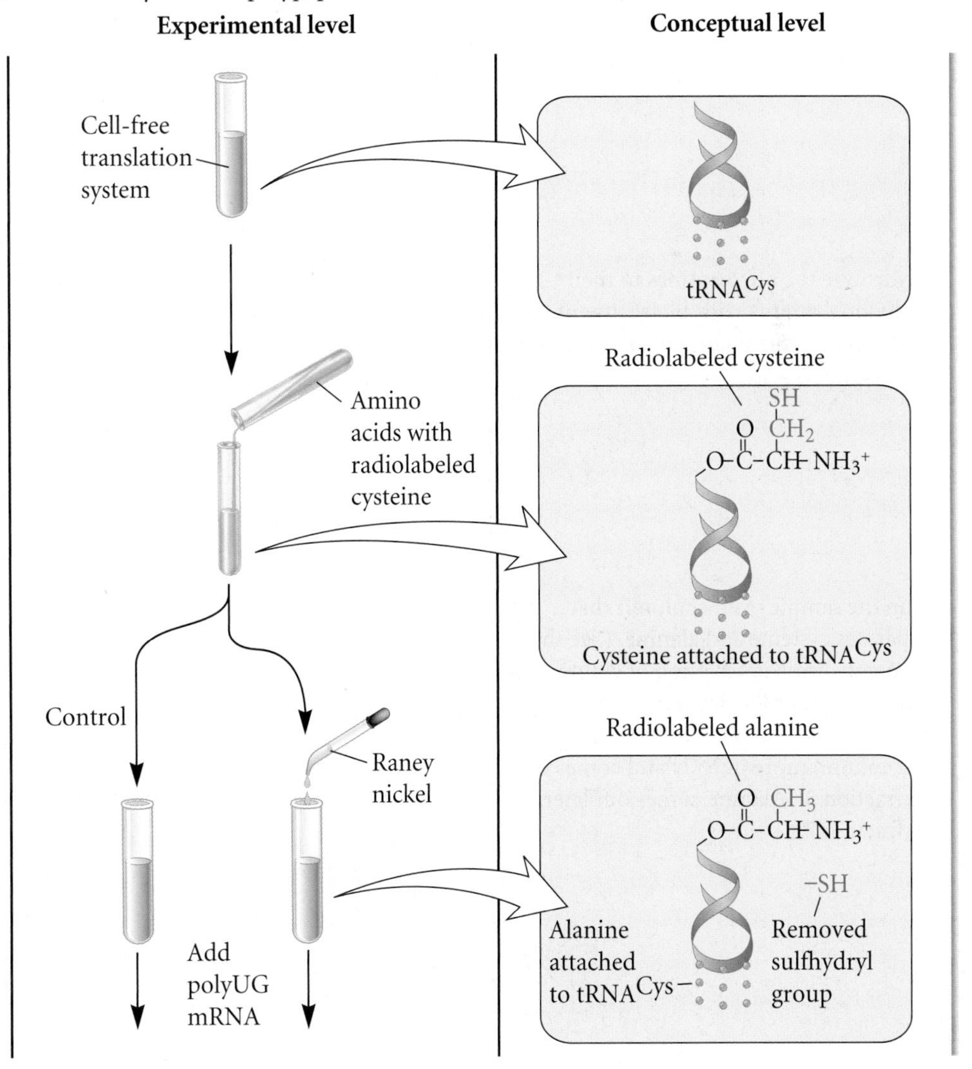

4. Add polyUG mRNA made via polynucleotide phosphorylase as a template. A polyUG mRNA contains cysteine codons but no alanine codons.

5. Allow translation to proceed.

6. Precipitate the newly made polypeptides with trichloroacetic acid and then isolate the polypeptides on a filter.

7. Hydrolyze the polypeptides to their individual amino acids by treatment with a solution containing concentrated hydrochloric acid.

8. Run the sample over a column that separates cysteine and alanine. (See the Appendix for a description of column chromatography.) Separate into fractions. Note: Cysteine runs through the column more quickly and comes out in fraction 3. Alanine comes out later, in fraction 7.

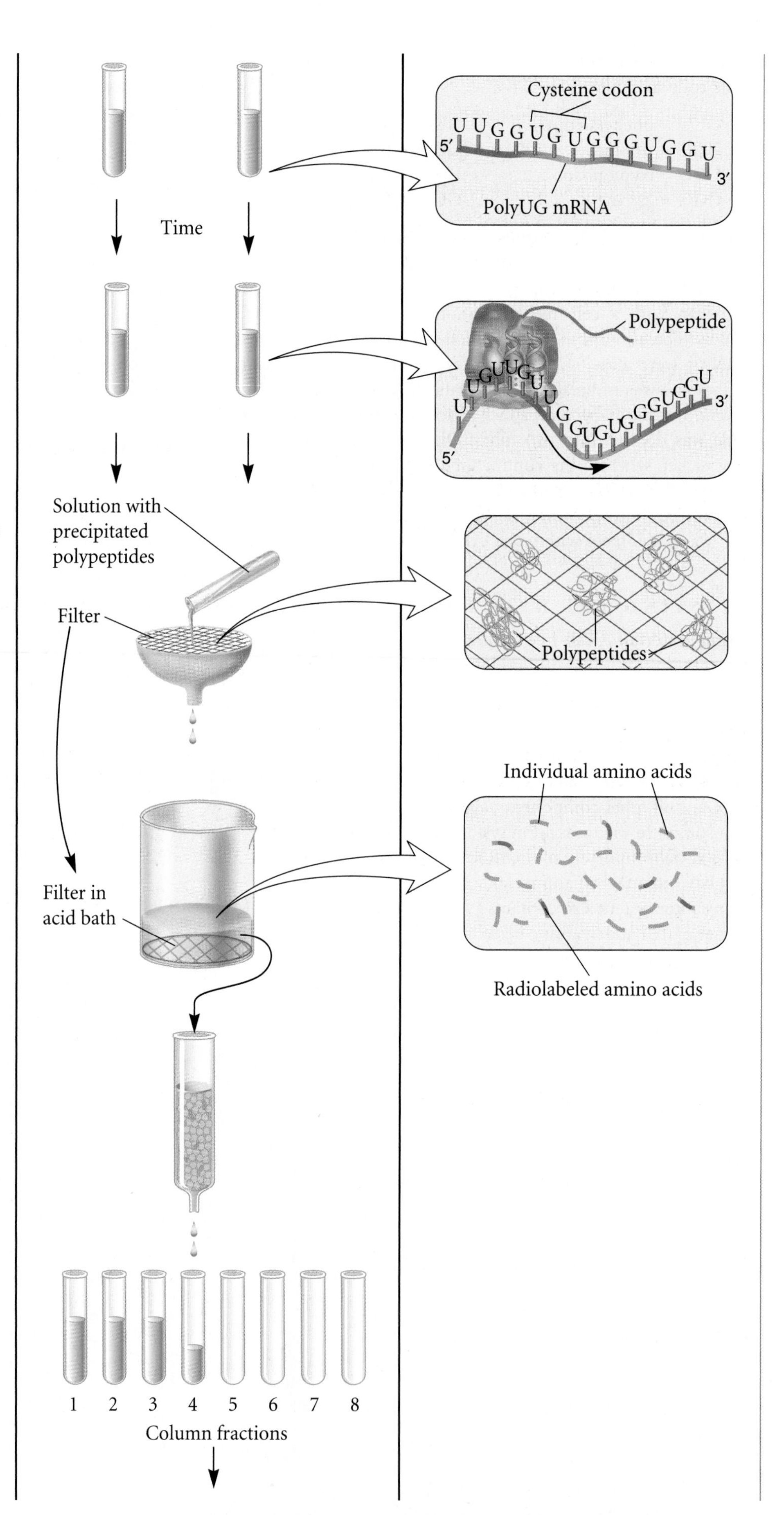

9. Determine the amount of radioactivity in the fractions that contain alanine and cysteine.

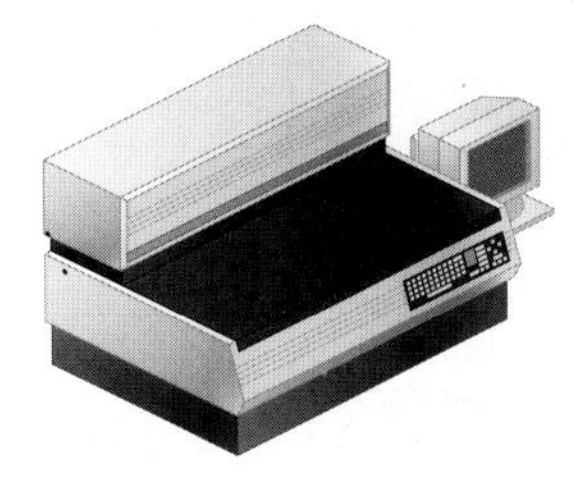

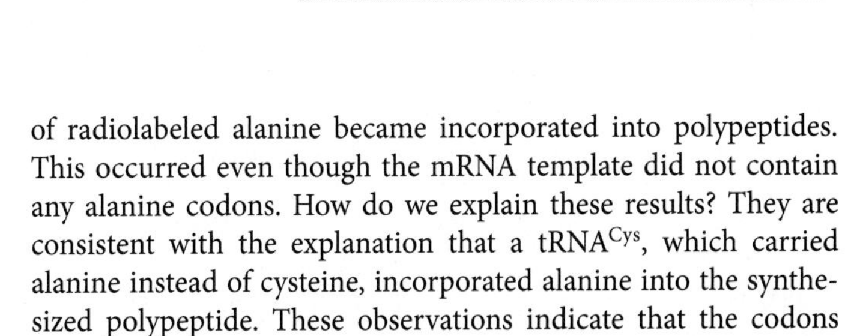

THE DATA

Conditions	Amount of Radiolabeled Amino Acids Incorporated into Polypeptide (cpm)*		
	Cysteine	*Alanine*	*Total*
Control, untreated tRNA	2835	83	2918
Raney nickel-treated tRNA	990	2020	3010

*cpm is the counts per minute of radioactivity in the sample.
Adapted from Chapeville F., Lipmann F., von Ehrenstein G., et al. (1962) On the role of soluble ribonucleic acid in coding of amino acids. *Proc Natl Acad Sci USA 48*, 1086–1092.

INTERPRETING THE DATA

In the control sample, nearly all of the radioactivity was found in the fraction containing cysteine. This result was expected because the only radiolabeled amino acid attached to tRNAs was cysteine. The low radioactivity in the alanine fraction (83 counts per minute [cpm]) probably represents contamination of this fraction by a small amount of cysteine. By comparison, when the tRNAs were treated with Raney nickel, a substantial amount of radiolabeled alanine became incorporated into polypeptides. This occurred even though the mRNA template did not contain any alanine codons. How do we explain these results? They are consistent with the explanation that a $tRNA^{Cys}$, which carried alanine instead of cysteine, incorporated alanine into the synthesized polypeptide. These observations indicate that the codons in mRNA are identified directly by the tRNA rather than the attached amino acid.

As seen in the data of Figure 13.11, the Raney nickel-treated sample still had 990 cpm of cysteine incorporated into polypeptides. This is about one-third of the total amount of radioactivity (namely, 990/3010). In other experiments conducted in this study, the researchers showed that the Raney nickel did not react with about one-third of the $tRNA^{Cys}$. Therefore, this proportion of the Raney nickel-treated $tRNA^{Cys}$ would still carry cysteine. This observation was consistent with the data shown here. Overall, the results of this experiment supported the adaptor hypothesis, indicating that tRNAs act as adaptors to carry the correct amino acid to the ribosome based on their anticodon sequence.

A self-help quiz involving this experiment can be found at www.mhhe.com/brookergenetics4e.

Common Structural Features Are Shared by All tRNAs

To understand how tRNAs act as carriers of the correct amino acids during translation, researchers have examined the structural characteristics of these molecules in great detail. Though a cell makes many different tRNAs, all tRNAs share common structural features. As originally proposed by Robert W. Holley in 1965, the secondary structure of tRNAs exhibits a cloverleaf pattern. A tRNA has three stem-loop structures, a few variable sites, and an acceptor stem with a 3′ single-stranded region (**Figure 13.12**). The acceptor stem is where an amino acid becomes attached to a tRNA (see inset). A conventional numbering system for the nucleotides within a tRNA molecule begins at the 5′ end and proceeds toward the 3′ end. Among different types of tRNA molecules, the variable sites (shown in blue) can differ in the number of nucleotides they contain. The anticodon is located in the second loop region.

The actual three-dimensional, or tertiary, structure of tRNA molecules involves additional folding of the secondary structure. In the tertiary structure of tRNA, the stem-loop regions are folded into a much more compact molecule. The ability of RNA molecules to form stem-loop structures and the tertiary folding of tRNA molecules are described in Chapter 9 (see Figure 9.23). Interestingly, in addition to the normal A, U, G, and C nucleotides, tRNA molecules commonly contain modified nucleotides within their primary structures. For example, Figure 13.12 illustrates a tRNA that contains several modified bases. Among many different species, researchers have found that more than 80 different nucleotide modifications can occur in tRNA molecules. We will explore the significance of modified bases in codon recognition later in this chapter.

Aminoacyl-tRNA Synthetases Charge tRNAs by Attaching the Appropriate Amino Acid

To function correctly, each type of tRNA must have the appropriate amino acid attached to its 3′ end. How does an amino acid get attached to a tRNA with the correct anticodon? Enzymes

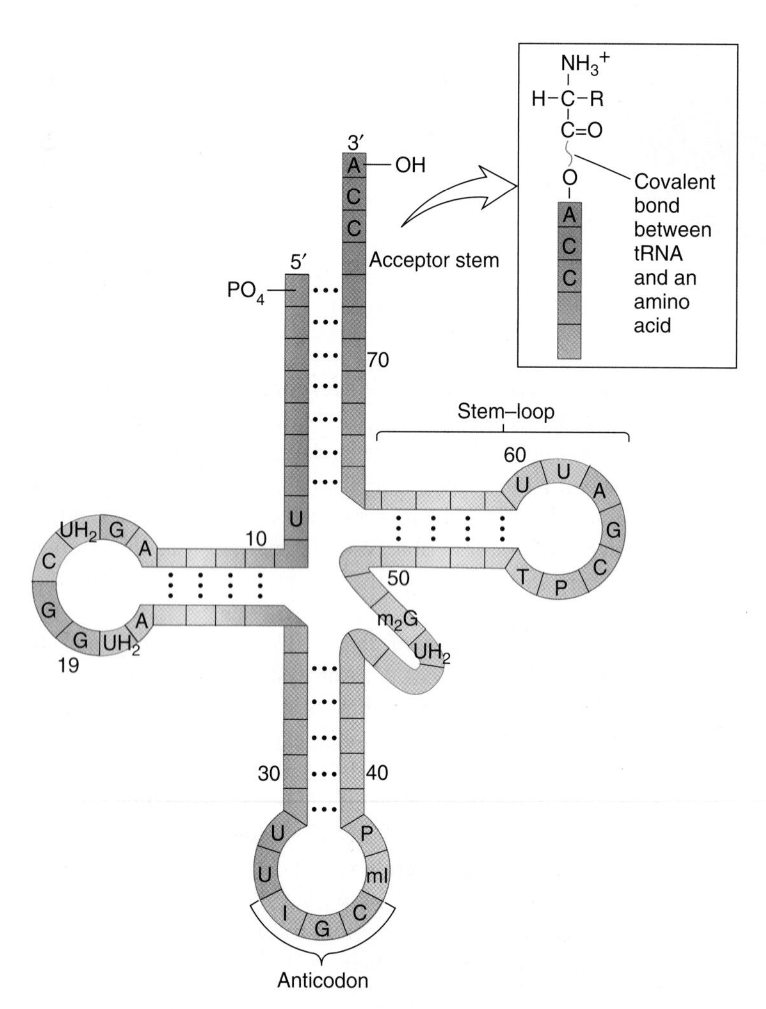

FIGURE 13.12 Secondary structure of tRNA. The conventional numbering of nucleotides begins at the 5′ end and proceeds toward the 3′ end. In all tRNAs, the nucleotides at the 3′ end contain the sequence CCA. Certain locations can have additional nucleotides not found in all tRNA molecules. These variable sites are shown in blue. The figure also shows the locations of a few modified bases specifically found in a yeast tRNA that carries alanine. The modified bases are as follows: I = inosine, mI = methylinosine, T = ribothymidine, UH_2 = dihydrouridine, m_2G = dimethylguanosine, and P = pseudouridine. The inset shows an amino acid covalently attached to the 3′ end of a tRNA.

in the cell known as **aminoacyl-tRNA synthetases** catalyze the attachment of amino acids to tRNA molecules. Cells produce 20 different aminoacyl-tRNA synthetase enzymes, 1 for each of the 20 distinct amino acids. Each aminoacyl-tRNA synthetase is named for the specific amino acid it attaches to tRNA. For example, alanyl-tRNA synthetase recognizes a tRNA with an alanine anticodon—$tRNA^{Ala}$—and attaches an alanine to it.

Aminoacyl-tRNA synthetases catalyze a chemical reaction involving three different molecules: an amino acid, a tRNA molecule, and ATP. In the first step of the reaction, a synthetase recognizes a specific amino acid and also ATP (**Figure 13.13**). The ATP is hydrolyzed, and AMP becomes attached to the amino acid; pyrophosphate is released. During the second step, the correct tRNA binds to the synthetase. The amino acid becomes covalently attached to the 3′ end of the tRNA molecule at the

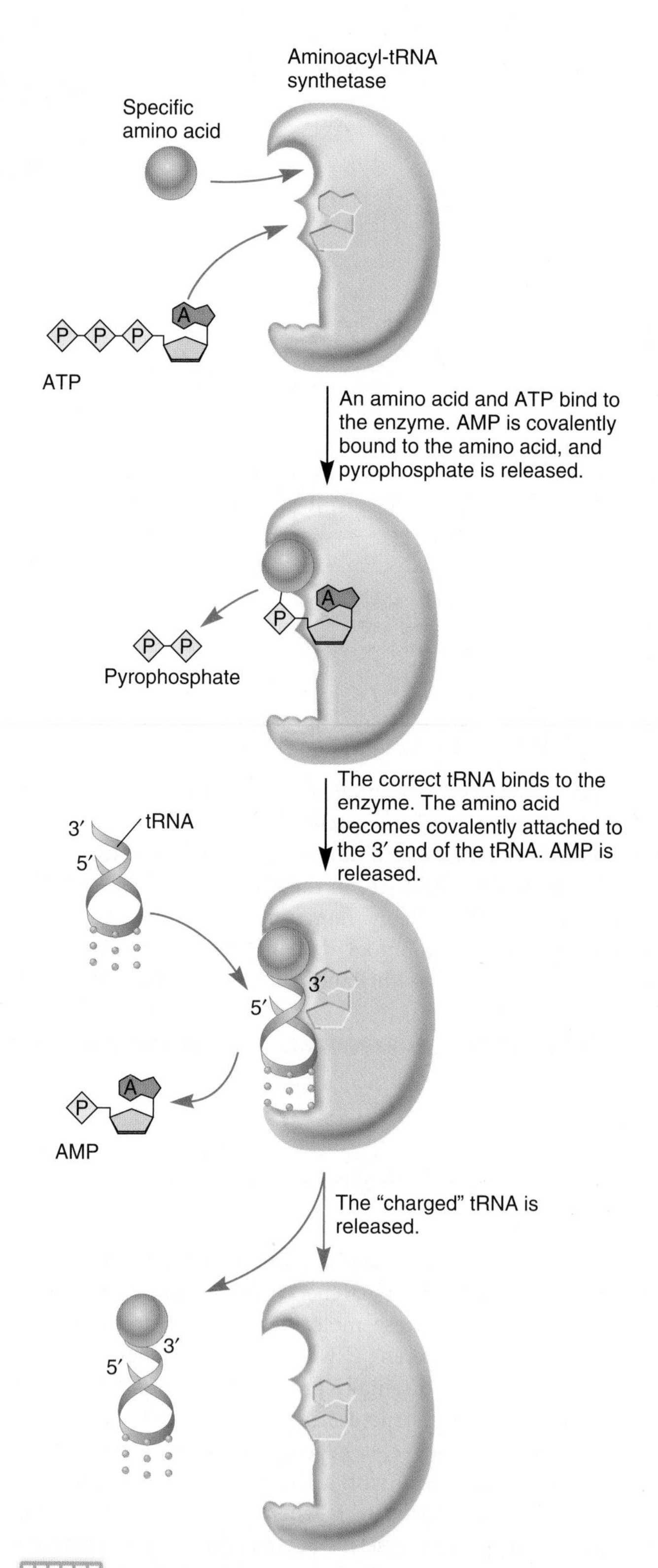

ONLINE ANIMATION **FIGURE 13.13 Catalytic function of aminoacyl-tRNA synthetase.** Aminoacyl-tRNA synthetase has binding sites for a specific amino acid, ATP, and a particular tRNA. In the first step, the enzyme catalyzes the covalent attachment of AMP to an amino acid, yielding an activated amino acid. In the second step, the activated amino acid is attached to the appropriate tRNA.

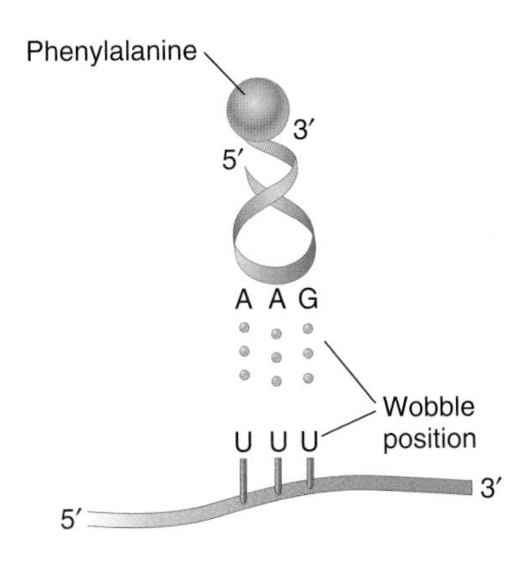

(a) Location of wobble position

Nucleotide of tRNA anticodon	Third nucleotide of mRNA codon
G	C, U
C	G
A	U, C, G, (A)
U	A, U, G, (C)
I	U, C, A
xm^5s^2U, xm^5Um, Um, xm^5U	A, (G)
xo^5U	U, A, G
k^2C	A

(b) Revised wobble rules

FIGURE 13.14 Wobble position and base-pairing rules. **(a)** The wobble position occurs between the first base (meaning the first base in the 5′ to 3′ direction) in the anticodon and the third base in the mRNA codon. **(b)** The revised wobble rules are slightly different from those originally proposed by Crick. The standard bases found in RNA are G, C, A, and U. In addition, the structures of bases in tRNAs may be modified. Some modified bases that may occur in the wobble position in tRNA are I = inosine; xm^5s^2U = 5-methyl-2-thiouridine; xm^5Um = 5-methyl-2′-*O*-methyluridine; Um = 2′-*O*-methyluridine; xm^5U = 5-methyluridine; xo^5U = 5-hydroxyuridine; k^2C = lysidine (a cytosine derivative). The mRNA bases in parentheses are recognized very poorly by the tRNA.

acceptor stem, and AMP is released. Finally, the tRNA with its attached amino acid is released from the enzyme. At this stage, the tRNA is called a **charged tRNA** or an **aminoacyl-tRNA.** In a charged tRNA molecule, the amino acid is attached to the 3′ end of the tRNA by a covalent bond (see Figure 13.12 inset).

The ability of the aminoacyl-tRNA synthetases to recognize tRNAs has sometimes been called the "second genetic code." This recognition process is necessary to maintain the fidelity of genetic information. The frequency of error for aminoacyl-tRNA synthetases is less than 10^{-5}. In other words, the wrong amino acid is attached to a tRNA less than once in 100,000 times! As you might expect, the anticodon region of the tRNA is usually important for precise recognition by the correct aminoacyl-tRNA synthetase. In studies of *Escherichia coli* synthetases, 17 of the 20 types of aminoacyl-tRNA synthetases recognize the anticodon region of the tRNA. However, other regions of the tRNA are also important recognition sites. These include the acceptor stem and bases in the stem-loop regions.

As mentioned previously, tRNA molecules frequently contain bases within their structure that have been chemically modified. These modified bases can have important effects on tRNA function. For example, modified bases within tRNA molecules affect the rate of translation and the recognition of tRNAs by aminoacyl-tRNA synthetases. Positions 34 and 37 contain the largest variety of modified nucleotides; position 34 is the first base in the anticodon that matches the third base in the codon of mRNA. As discussed next, a modified base at position 34 can have important effects on codon-anticodon recognition.

Mismatches That Follow the Wobble Rule Can Occur at the Third Position in Codon-Anticodon Pairing

After considering the structure and function of tRNA molecules, let's reexamine some subtle features of the genetic code. As discussed earlier, the genetic code is degenerate, which means that more than one codon can specify the same amino acid. Degeneracy usually occurs at the third position in the codon. For example, valine is specified by GUU, GUC, GUA, and GUG. In all four cases, the first two bases are G and U. The third base, however, can be U, C, A, or G. To explain this pattern of degeneracy, Francis Crick proposed in 1966 that it is due to "wobble" at the third position in the codon-anticodon recognition process. According to the **wobble rules,** the first two positions pair strictly according to the AU/GC rule. However, the third position can tolerate certain types of mismatches (**Figure 13.14**). This proposal suggested that the base at the third position in the codon does not have to hydrogen bond as precisely with the corresponding base in the anticodon.

Because of the wobble rules, some flexibility is observed in the recognition between a codon and anticodon during the process of translation. When two or more tRNAs that differ at the wobble base are able to recognize the same codon, these are termed **isoacceptor tRNAs.** As an example, tRNAs with an anticodon of 3′–CCA–5′ or 3′–CCG–5′ can recognize a codon with the sequence of 5′–GGU–3′. In addition, the wobble rules enable a single type of tRNA to recognize more than one codon. For example, a tRNA with an anticodon sequence of 3′–AAG–5′ can recognize a 5′–UUC–3′ and a 5′–UUU–3′ codon. The 5′–UUC–3′ codon is a perfect match with this tRNA. The 5′–UUU–3′ codon is mismatched according to the standard RNA-RNA hybridization rules (namely, G in the anticodon is mismatched to U in the codon), but the two can fit according to the wobble rules described in Figure 13.14. Likewise, the modification of the wobble base to an inosine allows a tRNA to recognize three different codons. At the cellular level, the ability of a single tRNA to recognize more than one codon makes it unnecessary for a cell to make 61 different tRNA molecules with anticodons that are complementary to the 61 possible sense codons. *E. coli* cells, for example, make a population of tRNA molecules that have just 40 different anticodon sequences.

13.3 RIBOSOME STRUCTURE AND ASSEMBLY

In Section 13.2, we examined how the structure and function of tRNA molecules are important in translation. According to the adaptor hypothesis, tRNAs bind to mRNA due to complementarity between the anticodons and codons. Concurrently, the tRNA molecules have the correct amino acid attached to their 3′ ends.

To synthesize a polypeptide, additional events must occur. In particular, the bond between the 3′ end of the tRNA and the amino acid must be broken, and a peptide bond must be formed between the adjacent amino acids. To facilitate these events, translation occurs on the surface of a macromolecular complex known as the **ribosome.** The ribosome can be thought of as the macromolecular arena where translation takes place.

In this section, we will begin by outlining the biochemical compositions of ribosomes in bacterial and eukaryotic cells. We will then examine the key functional sites on ribosomes for the translation process.

Bacterial and Eukaryotic Ribosomes Are Assembled from rRNA and Proteins

Bacterial cells have one type of ribosome that is found within the cytoplasm. Eukaryotic cells contain biochemically distinct ribosomes in different cellular locations. The most abundant type of ribosome functions in the cytosol, which is the region of the eukaryotic cell that is inside the plasma membrane but outside the membrane-bound organelles. Besides the cytosolic ribosomes, all eukaryotic cells have ribosomes within the mitochondria. In addition, plant cells and algae have ribosomes in their chloroplasts. The compositions of mitochondrial and chloroplast ribosomes are quite different from that of the cytosolic ribosomes. Unless other-wise noted, the term eukaryotic ribosome refers to ribosomes in the cytosol, not to those found within organelles. Likewise, the description of eukaryotic translation refers to translation via cytosolic ribosomes.

Each ribosome is composed of structures called the large and small subunits. This term is perhaps misleading because each ribosomal subunit itself is formed from the assembly of many different proteins and RNA molecules called ribosomal RNA or rRNA. In bacterial ribosomes, the 30S subunit is formed from the assembly of 21 different ribosomal proteins and a 16S rRNA molecule; the 50S subunit contains 34 different proteins and 5S and 23S rRNA molecules (**Table 13.6**). The designations 30S and 50S refer to the rate that these subunits sediment when subjected to a centrifugal force. This rate is described as a sedimentation coefficient in Svedberg units (S), in honor of Theodor Svedberg, who invented the ultracentrifuge. Together, the 30S and 50S subunits form a 70S ribosome. (Note: Svedberg units do not add up linearly.) In bacteria, the ribosomal proteins and rRNA molecules are synthesized in the cytoplasm, and the ribosomal subunits are assembled there.

The synthesis of eukaryotic rRNA occurs within the nucleus, and the ribosomal proteins are made in the cytosol, where translation takes place. The 40S subunit is composed of 33 proteins and an 18S rRNA; the 60S subunit is made of 49 proteins and 5S, 5.8S, and 28S rRNAs (see Table 13.6). The assembly of the rRNAs and ribosomal proteins to make the 40S and 60S subunits occurs within the **nucleolus,** a region of the nucleus specialized for this purpose. The 40S and 60S subunits are then exported into the cytosol, where they associate to form an 80S ribosome during translation.

Components of Ribosomal Subunits Form Functional Sites for Translation

To understand the structure and function of the ribosome at the molecular level, researchers must determine the locations and functional roles of the individual ribosomal proteins and rRNAs. In recent years, many advances have been made toward a molecular understanding of ribosomes. Microscopic and biophysical

TABLE **13.6**

Composition of Bacterial and Eukaryotic Ribosomes

	Small subunit	Large subunit	Assembled ribosome
Bacterial			
Sedimentation coefficient	30S	50S	70S
Number of proteins	21	34	55
rRNA	16S rRNA	5S rRNA, 23S rRNA	16S rRNA, 5S rRNA, 23S rRNA
Eukaryotic			
Sedimentation coefficient	40S	60S	80S
Number of proteins	33	49	82
rRNA	18S rRNA	5S rRNA, 5.8S rRNA, 28S rRNA	18S rRNA, 5S rRNA, 5.8S rRNA, 28S rRNA

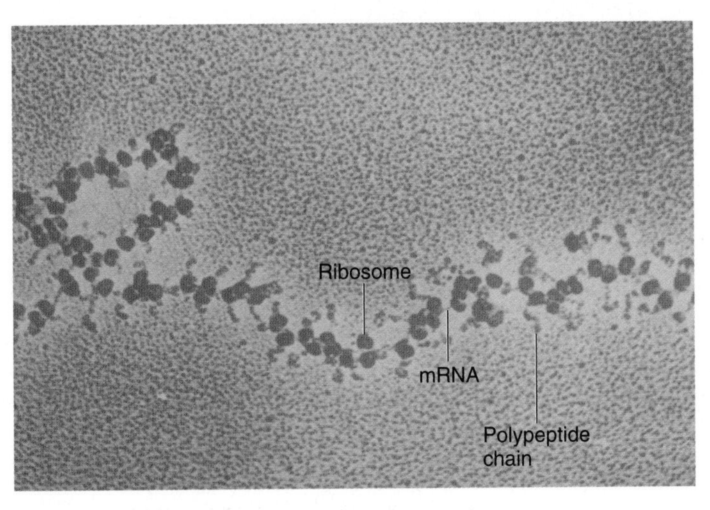

(a) Ribosomes as seen with electron microscope

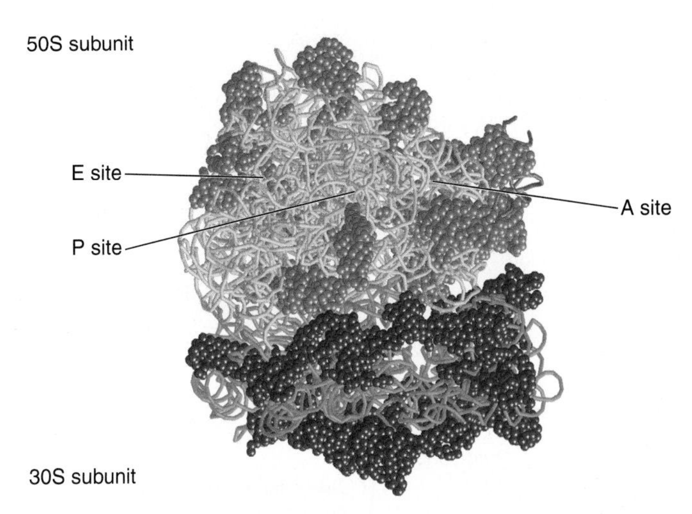

(b) Bacterial ribosome model based on X-ray diffraction studies

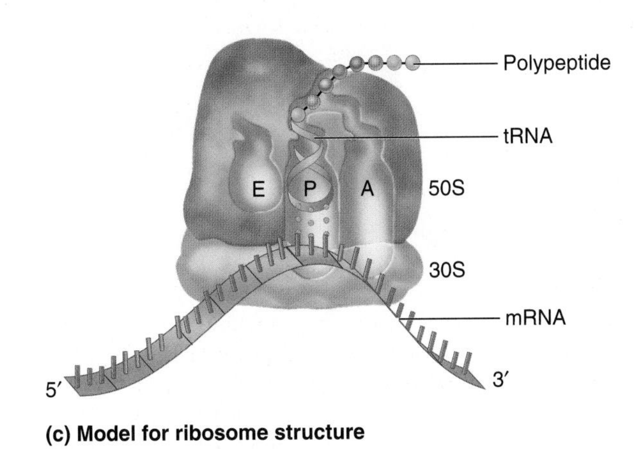

(c) Model for ribosome structure

FIGURE 13.15 Ribosomal structure. **(a)** Electron micrograph of ribosomes attached to a bacterial mRNA molecule. **(b)** Crystal structure of the 50S and 30S subunits in bacteria. This model shows the interface between the two subunits. The rRNA is shown in gray strands (50S subunit) and turquoise strands (30S subunit), and proteins are shown in violet (50S subunit) and navy blue (30S subunit). **(c)** A model depicting the sites where tRNA and mRNA bind to an intact ribosome. The mRNA lies on the surface of the 30S subunit. The E, P, and A sites are formed at the interface between the large and small subunits. The growing polypeptide chain exits through a hole in the 50S subunit.

methods have been used to study ribosome structure. An electron micrograph of bacterial ribosomes is shown in **Figure 13.15a.** More recently, a few research groups have succeeded in crystallizing ribosomal subunits, and even intact ribosomes. This is an amazing technical feat, because it is difficult to find the right conditions under which large macromolecules will form highly ordered crystals. **Figure 13.15b** shows the crystal structure of bacterial ribosomal subunits. The overall shape of each subunit is largely determined by the structure of the rRNAs, which constitute most of the mass of the ribosome. The interface between the 30S and 50S subunits is primarily composed of rRNA. Ribosomal proteins cluster on the outer surface of the ribosome and on the periphery of the interface.

During bacterial translation, the mRNA lies on the surface of the 30S subunit within a space between the 30S and 50S subunits. As the polypeptide is being synthesized, it exits through a channel within the 50S subunit (**Figure 13.15c**). Ribosomes contain discrete sites where tRNAs bind and the polypeptide is synthesized. In 1964, James Watson was the first to propose a two-site model for tRNA binding to the ribosome. These sites are known as the **peptidyl site (P site)** and **aminoacyl site (A site).** In 1981, Knud Nierhaus, Hans Sternbach, and Hans-Jörg Rheinberger proposed a three-site model. This model incorporated the observation that uncharged tRNA molecules can bind to a site on the ribosome that is distinct from the P and A sites. This third site is now known as the **exit site (E site).** The locations of the E, P, and A sites are shown in Figure 13.15c. Next, we will examine the roles of these sites during the three stages of translation.

13.4 STAGES OF TRANSLATION

Like transcription, the process of translation can be viewed as occurring in three stages: initiation, elongation, and termination. **Figure 13.16** presents an overview of these stages. During **initiation,** the ribosomal subunits, mRNA, and the first tRNA assemble to form a complex. After the initiation complex is formed, the ribosome slides along the mRNA in the 5′ to 3′ direction, moving over the codons. This is the **elongation** stage of translation. As the ribosome moves, tRNA molecules sequentially bind to the mRNA at the A site in the ribosome, bringing with them the appropriate amino acids. Therefore, amino acids are linked in

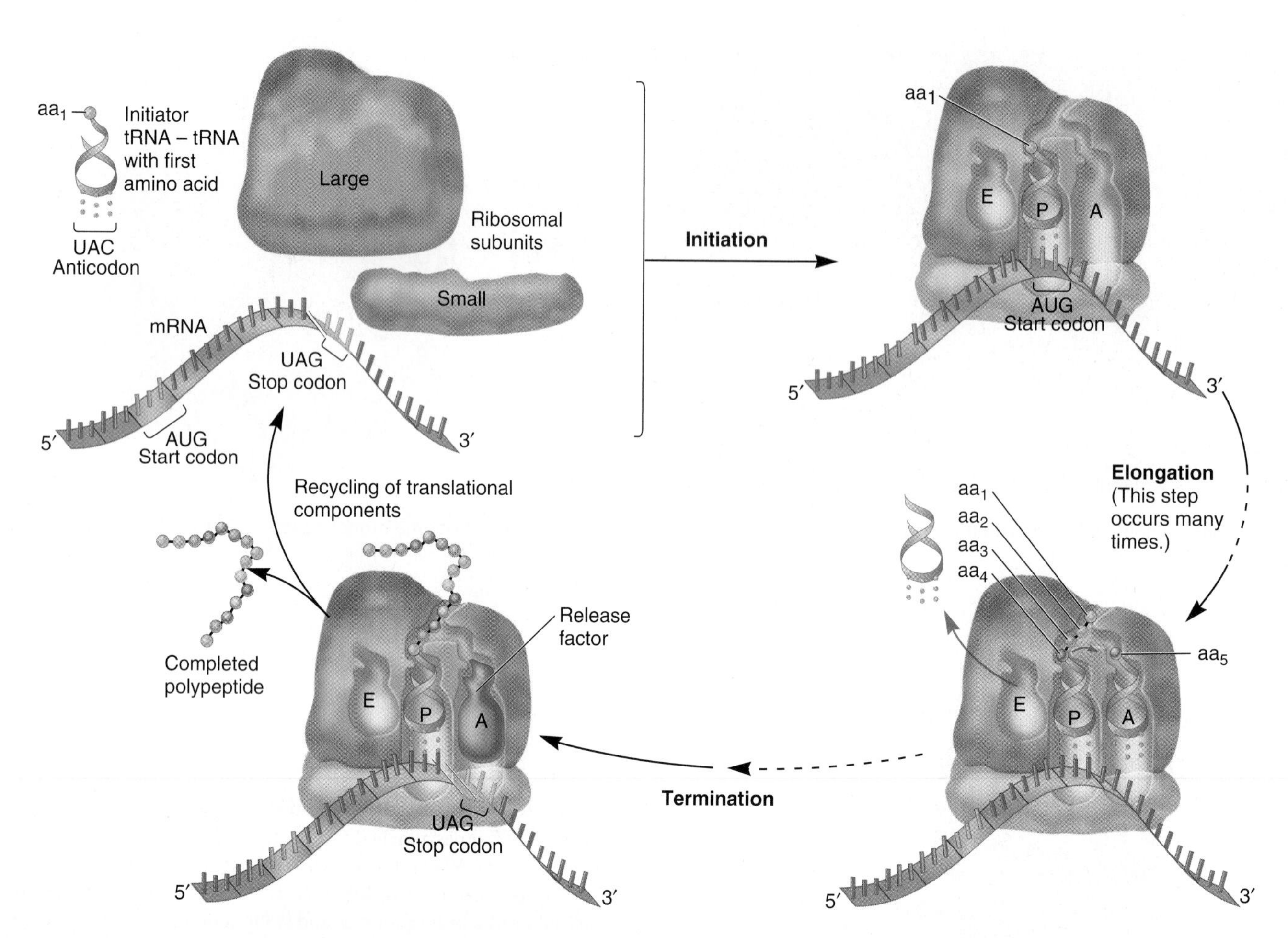

FIGURE 13.16 Overview of the stages of translation. The initiation stage involves the assembly of the ribosomal subunits, mRNA, and the initiator tRNA carrying the first amino acid. During elongation, the ribosome slides along the mRNA and synthesizes a polypeptide chain. Translation ends when a stop codon is reached and the polypeptide is released from the ribosome. (Note: In this and succeeding figures in this chapter, the ribosomes are drawn schematically to emphasize different aspects of the translation process. The structures of ribosomes are described in Figure 13.15.)

Genes → Traits The ability of genes to produce an organism's traits relies on the molecular process of gene expression. During translation, the codon sequence within mRNA (which is derived from a gene sequence during transcription) is translated into a polypeptide sequence. After polypeptides are made within a living cell, they function as proteins to govern an organism's traits. For example, once the β-globin polypeptide is made, it functions within the hemoglobin protein and provides red blood cells with the ability to carry oxygen, a vital trait for survival. Translation allows functional proteins to be made within living cells.

the order dictated by the codon sequence in the mRNA. Finally, a stop codon is reached, signaling the **termination** of translation. At this point, disassembly occurs, and the newly made polypeptide is released. In this section, we will examine the components required for the translation process and consider their functional roles during the three stages of translation.

The Initiation Stage Involves the Binding of mRNA and the Initiator tRNA to the Ribosomal Subunits

During initiation, an mRNA and the first tRNA bind to the ribosomal subunits. A specific tRNA functions as the **initiator tRNA,** which recognizes the start codon in the mRNA. In bacteria, the initiator tRNA, which is also designated $tRNA^{fMet}$, carries a methionine that has been covalently modified to *N*-formylmethionine. In this modification, a formyl group (—CHO) is attached to the nitrogen atom in methionine after the methionine has been attached to the tRNA.

Figure 13.17 describes the initiation stage of translation in bacteria during which the mRNA, $tRNA^{fMet}$, and ribosomal subunits associate with each other to form an initiation complex. The formation of this complex requires the participation of three initiation factors: IF1, IF2, and IF3. First, IF1 and IF3 bind to the 30S subunit. IF1 and IF3 prevent the association of the 50S subunit. Next, the mRNA binds to the 30S subunit. This binding is facilitated by a nine-nucleotide sequence within the bacterial mRNA called the **Shine-Dalgarno sequence.** The location of this sequence is shown in Figure 13.17 and in more detail in **Figure 13.18**. How does the Shine-Dalgarno sequence facilitate the binding of mRNA to the ribosome? The Shine-Dalgarno sequence is complementary to a short sequence within the 16S rRNA, which promotes the hydrogen bonding of the mRNA to the 30S subunit.

Next, $tRNA^{fMet}$ binds to the mRNA that is already attached to the 30S subunit (see Figure 13.17). This step requires the function of IF2, which uses GTP. The $tRNA^{fMet}$ binds to the start

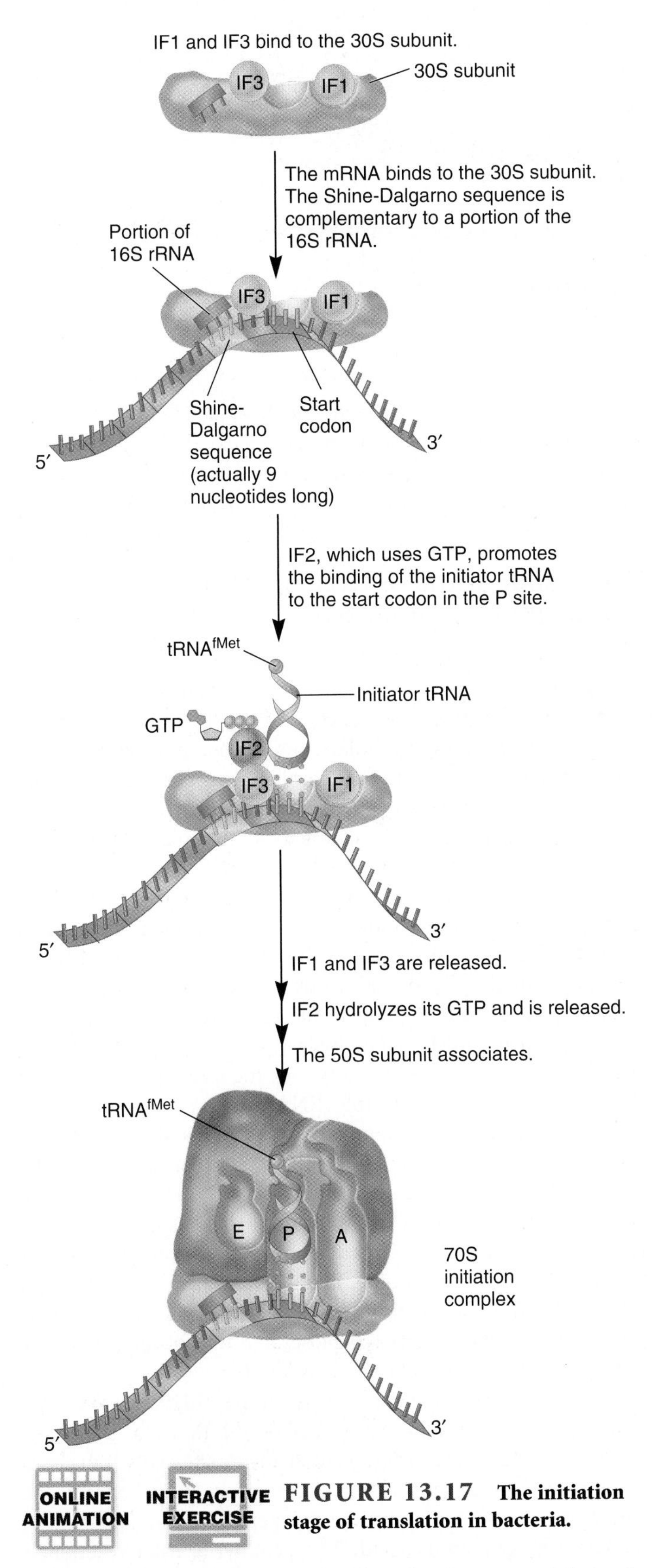

FIGURE 13.17 The initiation stage of translation in bacteria.

codon, which is typically a few nucleotides downstream from the Shine-Dalgarno sequence. The start codon is usually AUG, but in some cases it can be GUG or UUG. Even when the start codon is GUG (which normally encodes valine) or UUG (which normally encodes leucine), the first amino acid in the polypeptide is still a formylmethionine because only a tRNA^{fMet} can initiate translation. During or after translation of the entire polypeptide, the formyl group or the entire formylmethionine may be removed. Therefore, some polypeptides may not have formylmethionine or methionine as their first amino acid. As noted in Figure 13.17, the tRNA^{fMet} binds to the P site on the ribosome. IF1 is thought to occupy a portion of the A site, thereby preventing the binding of tRNA^{fMet} to the A site during initiation. By comparison, during the elongation stage that is discussed later, all of the other tRNAs initially bind to the A site.

After the mRNA and tRNA^{fMet} have become bound to the 30S subunit, IF1 and IF3 are released, and then IF2 hydrolyzes its GTP and is also released. This allows the 50S ribosomal subunit to associate with the 30S subunit. Much later, after translation is completed, IF1 binding is necessary to dissociate the 50S and 30S ribosomal subunits so that the 30S subunit can reinitiate with another mRNA molecule.

In eukaryotes, the assembly of the initiation complex bears similarities to that in bacteria. However, as described in **Table 13.7**, additional factors are required for the initiation process. Note that the initiation factors are designated eIF (for eukaryotic Initiation Factor) to distinguish them from bacterial initiation factors. The initiator tRNA in eukaryotes carries methionine rather than formylmethionine, as in bacteria. A eukaryotic initiation factor, eIF2, binds directly to tRNA^{Met} to recruit it to the 40S subunit. Eukaryotic mRNAs do not have a Shine-Dalgarno sequence. How then are eukaryotic mRNAs recognized by the ribosome? The mRNA is recognized by eIF4, which is a multiprotein complex that recognizes the 7-methylguanosine cap and facilitates the binding of the mRNA to the 40S subunit.

The identification of the correct AUG start codon in eukaryotes differs greatly from that in bacteria. After the initial binding of mRNA to the ribosome, the next step is to locate an AUG start codon that is somewhere downstream from the 5′ cap structure. In 1986, Marilyn Kozak proposed that the ribosome begins at the 5′ end and then scans along the mRNA in the 3′ direction in search of an AUG start codon. In many, but not all, cases, the ribosome uses the first AUG codon that it encounters as a start codon. When a start codon is identified, the 60S subunit assembles onto the 40S subunit with the aid of eIF5.

By analyzing the sequences of many eukaryotic mRNAs, researchers have found that not all AUG codons near the 5′ end of mRNA can function as start codons. In some cases, the scanning ribosome passes over the first AUG codon and chooses an AUG farther down the mRNA. The sequence of bases around the AUG codon plays an important role in determining whether or not it is selected as the start codon by a scanning ribosome. The consensus sequence for optimal start codon recognition in more complex eukaryotes, such as vertebrates and vascular plants, is shown here.

						Start Codon			
G	C	C	(A/G)	C	C	A	U	G	G
−6	−5	−4	−3	−2	−1	+1	+2	+3	+4

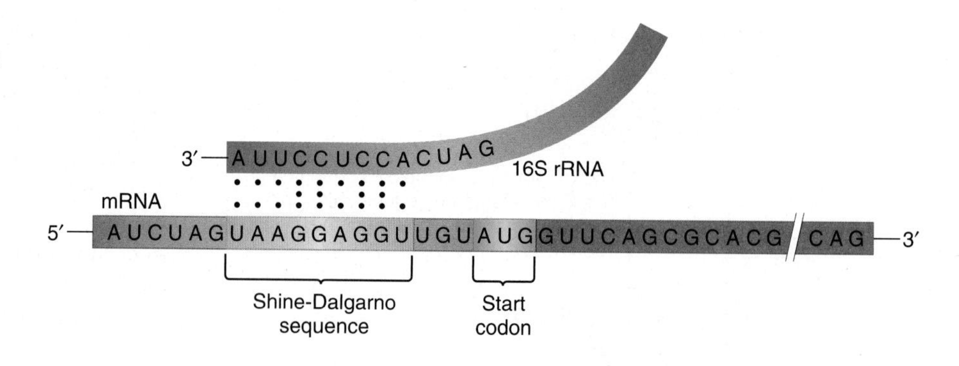

FIGURE 13.18 The locations of the Shine-Dalgarno sequence and the start codon in bacterial mRNA. The Shine-Dalgarno sequence is complementary to a sequence in the 16S rRNA. It hydrogen bonds with the 16S rRNA to promote initiation. The start codon is typically a few nucleotides downstream from the Shine-Dalgarno sequence.

Aside from an AUG codon itself, a guanine at the +4 position and a purine, preferably an adenine, at the −3 position are the most important sites for start codon selection. These rules for optimal translation initiation are called **Kozak's rules.**

TABLE **13.7**

A Simplified Comparison of Translational Protein Factors in Bacteria and Eukaryotes

Bacterial Factors	Eukaryotic Factors*	Function
Initiation Factors		
	eIF4	Involved with the recognition of the 7-methylguanosine cap and the binding of the mRNA to the small ribosomal subunit
IF1, IF3	eIF1, eIF3, eIF6	Prevent the association between the small and large ribosomal subunits and favor their disassociation
IF2	eIF2	Promotes the binding of the initiator tRNA to the small ribosomal subunit
	eIF5	Helps to dissociate the other elongation factors, which allows the large ribosomal subunit to bind
Elongation Factors		
EF-Tu	eEF1α	Involved in the binding of tRNAs to the A site
EF-Ts	eEF1βγ	Nucleotide exchange factors required for the functioning of EF-Tu and eEF1α, respectively
EF-G	eEF2	Required for translocation
Release Factors		
RF1, RF2	eRF1	Recognize a stop codon and trigger the cleavage of the polypeptide from the tRNA
RF3	eRF3	GTPases that are also involved in termination

*Eukaryotic translation factors are typically composed of multiple proteins.

Polypeptide Synthesis Occurs During the Elongation Stage

During the elongation stage of translation, amino acids are added, one at a time, to the polypeptide chain (**Figure 13.19**). Even though this process is rather complex, it occurs at a remarkable rate. Under normal cellular conditions, a polypeptide chain can elongate at a rate of 15 to 20 amino acids per second in bacteria and 2 to 6 amino acids per second in eukaryotes!

To begin elongation, a charged tRNA brings a new amino acid to the ribosome so it can be attached to the end of the growing polypeptide chain. At the top of Figure 13.19, which describes bacterial translation, a short polypeptide is attached to the tRNA located at the P site of the ribosome. A charged tRNA carrying a single amino acid binds to the A site. This binding occurs because the anticodon in the tRNA is complementary to the codon in the mRNA. The hydrolysis of GTP by the elongation factor, EF-Tu, provides energy for the binding of a tRNA to the A site. In addition, the 16S rRNA, which is a component of the small 30S ribosomal subunit, plays a key role that ensures the proper recognition between the mRNA and correct tRNA. The 16S rRNA can detect when an incorrect tRNA is bound at the A site and will prevent elongation until the mispaired tRNA is released from the A site. This phenomenon, termed the **decoding function** of the ribosome, is important in maintaining high fidelity of mRNA translation. An incorrect amino acid is incorporated into a growing polypeptide at a rate of approximately one mistake per 10,000 amino acids, or 10^{-4}.

The next step of elongation is the **peptidyl transfer** reaction—the polypeptide is removed from the tRNA in the P site and transferred to the amino acid at the A site. This transfer is accompanied by the formation of a peptide bond between the amino acid at the A site and the polypeptide chain, lengthening the chain by one amino acid. The peptidyl transfer reaction is catalyzed by a component of the 50S subunit known as **peptidyl transferase,** which is composed of several proteins and rRNA. Interestingly, based on the crystal structure of the 50S subunit,

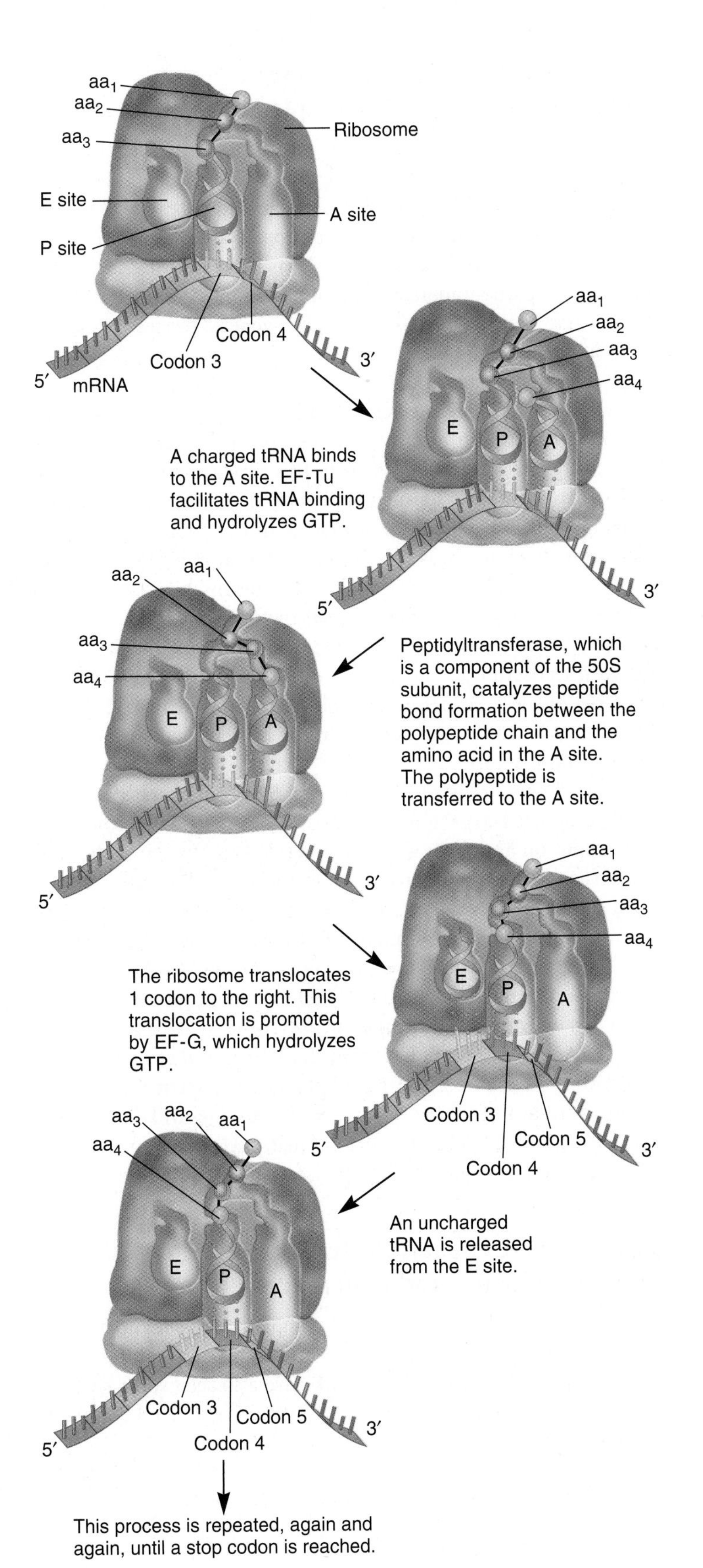

FIGURE 13.19 The elongation stage of translation in bacteria. This process begins with the binding of an incoming tRNA. The hydrolysis of GTP by EF-Tu provides the energy for the binding of the tRNA to the A site. A peptide bond is then formed between the incoming amino acid and the last amino acid in the growing polypeptide chain. This moves the polypeptide chain to the A site. The ribosome then translocates in the 3′ direction so that the two tRNAs are moved to the E and P sites. The tRNA carrying the polypeptide is now back in the P site. This translocation requires the hydrolysis of GTP via EF-G. The uncharged tRNA in the E site is released from the ribosome. Now the process is ready to begin again. Each cycle of elongation causes the polypeptide chain to grow by one amino acid.

Thomas Steitz, Peter Moore, and their colleagues concluded that the 23S rRNA—not the ribosomal protein—catalyzes bond formation between adjacent amino acids. In other words, the ribosome is a ribozyme!

After the peptidyl transfer reaction is complete, the ribosome moves, or translocates, to the next codon in the mRNA. This moves the tRNAs at the P and A sites to the E and P sites, respectively. Finally, the uncharged tRNA exits the E site. You should notice that the next codon in the mRNA is now exposed in the unoccupied A site. At this point, a charged tRNA can enter the empty A site, and the same series of steps can add the next amino acid to the polypeptide chain. As you may have realized, the A, P, and E sites are named for the role of the tRNA that is usually found there. The A site binds an aminoacyl-tRNA (also called a charged tRNA), the P site usually contains the peptidyl-tRNA (a tRNA with an attached peptide), and the E site is where the uncharged tRNA exits.

Termination Occurs When a Stop Codon Is Reached in the mRNA

The final stage of translation, known as termination, occurs when a stop codon is reached in the mRNA. In most species, the three stop codons are UAA, UAG, and UGA. The stop codons are not recognized by a tRNA with a complementary sequence. Instead, they are recognized by proteins known as **release factors** (see Table 13.7). Interestingly, the three-dimensional structures of release factor proteins are "molecular mimics" that resemble the structure of tRNAs. Such proteins can specifically bind to a stop codon sequence. In bacteria, RF1 recognizes UAA and UAG, and RF2 recognizes UGA and UAA. A third release factor, RF3, is also required. In eukaryotes, a single release factor, eRF, recognizes all three stop codons and eRF3 is also required for termination.

Figure 13.20 illustrates the termination stage of translation in bacteria. At the top of this figure, the completed polypeptide chain is attached to a tRNA in the P site. A stop codon is located at the A site. In the first step, RF1 or RF2 binds to the stop codon at the A site and RF3 (not shown) binds at a different location on the ribosome. After RF1 (or RF2) and RF3 have bound, the bond between the polypeptide and the tRNA is hydrolyzed. The polypeptide and tRNA are then released from the ribosome.

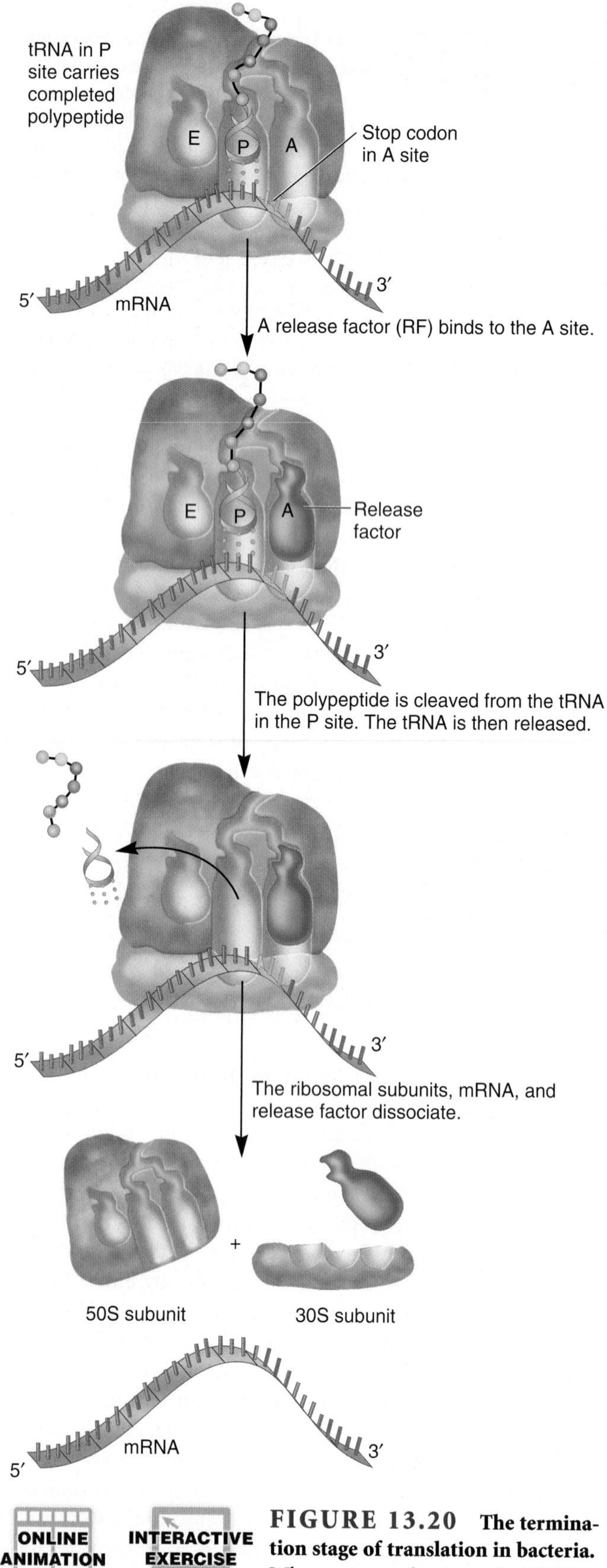

FIGURE 13.20 The termination stage of translation in bacteria. When a stop codon is reached, RF1 or RF2 binds to the A site. (RF3 binds elsewhere and uses GTP to facilitate the termination process.) The polypeptide is cleaved from the tRNA in the P site and released. The tRNA is released, and the rest of the components disassemble.

The final step in translational termination is the disassembly of ribosomal subunits, mRNA, and the release factors.

Bacterial Translation Can Begin Before Transcription Is Completed

Although most of our knowledge concerning transcription and translation has come from genetic and biochemical studies, electron microscopy (EM) has also been an important tool in elucidating the mechanisms of transcription and translation. As described earlier in this chapter, EM has been a critical technique in facilitating our understanding of ribosome structure. In addition, it has been employed to visualize genetic processes such as translation.

The first success in the EM observation of gene expression was achieved by Oscar Miller, Jr., and his colleagues in 1967. **Figure 13.21** shows an electron micrograph of a bacterial gene in the act of gene expression. Prior to this experiment, biochemical and genetic studies had suggested that the translation of a bacterial structural gene begins before the mRNA transcript is completed. In other words, as soon as an mRNA strand is long enough, a ribosome attaches to the 5′ end and begins translation, even before RNA polymerase has reached the transcriptional termination site within the gene. This phenomenon is termed the coupling between transcription and translation in bacterial cells. Note that coupling of these processes does not usually occur in eukaryotes, because transcription takes place in the nucleus while translation occurs in the cytosol.

As shown in Figure 13.21, several RNA polymerase enzymes have recognized a gene and begun to transcribe it. Because the transcripts on the right side are longer than those on the left, Miller concluded that transcription was proceeding from left to right in the micrograph. This EM image also shows the process of translation. Relatively small mRNA transcripts, near the left side of the figure, have a few ribosomes attached to them. As the transcripts become longer, additional ribosomes are attached to them. The term **polyribosome,** or **polysome,** is used to describe an mRNA transcript that has many bound ribosomes in the act of translation. In this electron micrograph, the nascent polypeptide chains were too small for researchers to observe. In later studies, as EM techniques became more refined, the polypeptide chains emerging from the ribosome were also visible (see Figure 13.15a).

Bacterial and Eukaryotic Translation Show Similarities and Differences

Throughout this chapter, we have compared translation in both bacteria and eukaryotic organisms. The general steps of translation are similar in all forms of life, but we have also seen some striking differences between bacteria and eukaryotes. **Table 13.8** compares translation between these groups.

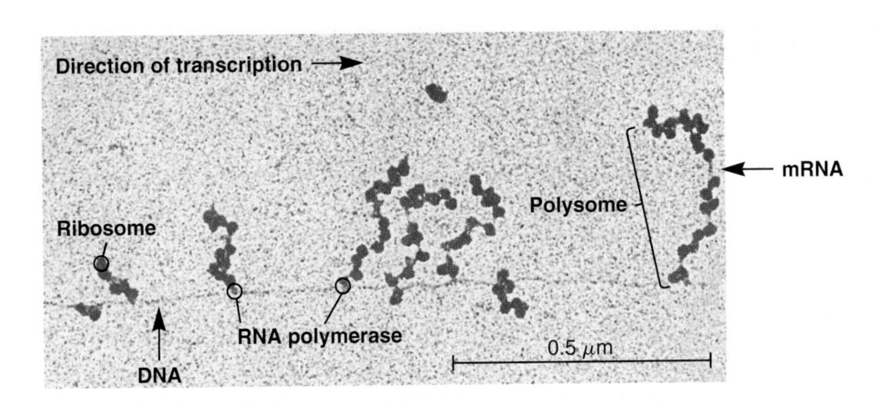

FIGURE 13.21 Coupling between transcription and translation in bacteria. An electron micrograph showing the simultaneous transcription and translation processes. The DNA is transcribed by many RNA polymerases that move along the DNA from left to right. Note that the RNA transcripts are getting longer as you go from left to right. Ribosomes attach to the mRNA, even before transcription is completed. The complex of many ribosomes bound to the same mRNA is called a polyribosome or a polysome. Several polyribosomes are seen here.

TABLE **13.8**

A Comparison of Bacterial and Eukaryotic Translation

	Bacterial	**Eukaryotic**
Ribosome composition:	70S ribosomes: 30S subunit— 21 proteins + 1 rRNA 50S subunit— 34 proteins + 2 rRNAs	80S ribosomes: 40S subunit— 33 proteins + 1 rRNA 60S subunit— 49 proteins + 3 rRNAs
Initiator tRNA:	$tRNA^{fmet}$	$tRNA^{Met}$
Formation of the initiation complex:	Requires IF1, IF2, and IF3	Requires more initiation factors compared to bacterial initiation
Initial binding of mRNA to the ribosome:	Requires a Shine-Dalgarno sequence	Requires a 7-methylguanosine cap
Selection of a start codon:	AUG, GUG, or UUG located just downstream from the Shine-Dalgarno sequence	According to Kozak's rules
Elongation rate	Typically 15 to 20 amino acids per second	Typically 2 to 6 amino acids per second
Termination:	Requires RF1, RF2, and RF3	Requires eRF1 and eRF3
Coupled to transcription:	Yes	No

KEY TERMS

Page 326. structural genes, messenger RNA (mRNA)
Page 327. alkaptonuria, inborn error of metabolism
Page 328. one-gene/one-enzyme hypothesis
Page 329. polypeptide, protein, translation, genetic code, sense codons, start codon, stop codons, termination codons, nonsense codons, anticodons, degenerate, synonymous codons, wobble base, reading frame
Page 330. frameshift mutation
Page 331. selenocysteine, pyrrolysine
Page 332. cell-free translation system
Page 336. peptide bond, N-terminus, amino-terminal end, C-terminus, carboxyl-terminal end, side chain, R group
Page 337. primary structure, chaperones, secondary structure, α helix, β sheet
Page 338. tertiary structure, quaternary structure, subunits
Page 339. enzymes
Page 340. adaptor hypothesis
Page 344. aminoacyl-tRNA synthetases
Page 345. charged tRNA, aminoacyl-tRNA, wobble rules, isoacceptor tRNAs
Page 346. ribosome, nucleolus
Page 347. peptidyl site (P site), aminoacyl site (A site), exit site (E site), initiation, elongation
Page 348. termination, initiator tRNA, Shine-Dalgarno sequence
Page 350. Kozak's rules, decoding function, peptidyl transfer, peptidyl transferase
Page 351. release factors
Page 352. polyribosome, polysome

CHAPTER SUMMARY

- Cellular proteins are made via the translation of mRNA.

13.1 The Genetic Basis for Protein Synthesis

- Garrod studied the disease called alkaptonuria and suggested that some genes may encode enzymes (see Figure 13.1).
- Beadle and Tatum studied *Neurospora* mutants that were altered in their nutritional requirements and hypothesized that one gene encodes one enzyme. This hypothesis was later modified because: (1) some proteins are not enzymes; (2) some proteins are composed of two or more different polypeptides; and (3) some genes encode RNAs that are not translated into polypeptides (see Figure 13.2).
- During translation, the sequence of codons in mRNA is used via tRNA molecules to make a polypeptide with a specific amino acid sequence (see Figure 13.3).
- The genetic code refers to the relationship between three-base codons in the mRNA and the amino acids that are incorporated into a polypeptide. One codon (AUG) is a start codon, which determines the reading frame of the mRNA. Three codons (UAA, UAG, and UGA) can function as stop codons (see Table 13.1).
- Crick studied mutations in T4 phage and determined that the genetic code is read in triplets (see Table 13.2).
- The genetic code is largely universal but some exceptions are known to occur (see Table 13.3).
- Nirenberg and colleagues used synthetic RNA and a cell-free translation system to decipher the genetic code (see Figure 13.4).
- Other methods to decipher the genetic code included the synthesis of copolymers by Khorana and the triplet-binding assay of Nirenberg and Leder (see Table 13.4, Figure 13.5).
- A polypeptide is made by the formation of peptide bonds between adjacent amino acids. Each polypeptide has a directionality from its amino terminus to its carboxyl terminus that parallels the arrangement of codons in mRNA in the 5′ to 3′ direction (see Figure 13.6).
- Amino acids differ in their side chain structure (see Figure 13.7).
- Cellular proteins carry out a variety of functions. The structures and functions of proteins are largely responsible for an organism's traits (see Table 13.5).
- Protein structure can be viewed at different levels, which include primary structure (sequence of amino acids), secondary structure (repeating folding patterns such as the α helix and the β sheet), tertiary structure (additional folding), and quaternary structure (the binding of multiple subunits to each other) (see Figures 13.8, 13.9).

13.2 Structure and Function of tRNA

- The anticodon in a tRNA is complementary to a codon in mRNA. The tRNA carries a specific amino acid that corresponds to the codon in the mRNA according to the genetic code (see Figure 13.10).
- Chapeville used Raney nickel to determine that the recognition between tRNA and mRNA is not due to the chemistry of the amino acid that is attached to the tRNA (see Figure 13.11).
- The secondary structure of tRNA resembles a cloverleaf. The anticodon is in the second loop and the amino acid is attached to the 3′ end (see Figure 13.12).
- Aminoacyl-tRNA synthetases are a group of enzymes that attach the correct amino acid to a tRNA. The resulting tRNA is called a charged tRNA or an aminoacyl-tRNA (see Figure 13.13).
- Mismatches are allowed between the pairing of tRNAs and mRNA according to the wobble rules (see Figure 13.14).

13.3 Ribosome Structure and Assembly

- Ribosomes are the site of polypeptide synthesis. The small and large subunit of ribosomes are composed of rRNAs and multiple proteins (see Table 13.6).
- A ribosome contains an A (aminoacyl), P (peptidyl) and E (exit) site, which are occupied by tRNA molecules (see Figure 13.15).

13.4 Stages of Translation

- The three stages of translation are initiation, elongation, and termination (see Figure 13.16).
- During the initiation stage of translation, the mRNA, initiator tRNA, and ribosomal subunits assemble. Initiation factors are involved in the process. In bacteria, the Shine-Dalgarno sequence promotes the binding of the mRNA to the small ribosomal subunit (see Figures 13.17, 13.18, Table 13.7).
- During elongation, tRNAs bring amino acids to the A site and a series of peptidyl transferase reactions creates a polypeptide. At each step, the polypeptide is transferred from the P site to the A site. The tRNAs are released from the E site. Elongation factors are involved in this process (see Figure 13.19).
- Start codon selection in complex eukaryotes follows Kozak's rules.
- During termination, a release factor binds to a stop codon in the A site. This promotes the cleavage of the polypeptide from the tRNA and the subsequent disassembly of the tRNA, mRNA, and ribosomal subunits (see Figure 13.20).
- Bacterial translation can begin before transcription is completed (see Figure 13.21).
- Bacterial and eukaryotic translation show many similarities and differences (see Table 13.8).

PROBLEM SETS & INSIGHTS

Solved Problems

S1. The first amino acid in a certain bacterial polypeptide is methionine. The start codon in the mRNA is GUG, which codes for valine. Why isn't the first amino acid formylmethionine or valine?

Answer: The first amino acid in a polypeptide is carried by the initiator tRNA, which always carries formylmethionine. This occurs even when the start codon is GUG (valine) or UUG (leucine). The formyl group can be later removed to yield methionine as the first amino acid.

S2. A tRNA has the anticodon sequence 3′–CAG–5′. What amino acid does it carry?

Answer: Because the anticodon is 3′–CAG–5′, it would be complementary to a codon with the sequence 5′–GUC–3′. According to the genetic code, this codon specifies the amino acid valine. Therefore, this tRNA must carry valine at its acceptor stem.

S3. In eukaryotic cells, the assembly of ribosomal subunits occurs in the nucleolus. As discussed in Chapter 12 (see Figure 12.16), a single 45S rRNA transcript is cleaved to produce the three rRNA fragments—18S, 5.8S, and 28S rRNA—that play a key role in creating the structure of the ribosome. The genes that encode the 45S precursor are found in multiple copies (i.e., they are moderately repetitive). The segments of chromosomes that contain the 45S rRNA genes align themselves at the center of the nucleolus. This site is called the nucleolar-organizing center. In this region, active transcription of the 45S gene takes place. Briefly explain how the assembly of the ribosomal subunits occurs.

Answer: In the nucleolar-organizing center, the 45S RNA is cleaved to the 18S, 5.8S, and 28S rRNAs. The other components of the ribosomal subunits, 5S rRNA and ribosomal proteins, must also be imported into the nucleolar region. Because proteins are made in the cytosol, they must enter the nucleus through the nuclear pores. When all the components are present, they assemble into 40S and 60S ribosomal subunits. Following assembly, the ribosomal subunits exit the nucleus through the nuclear pores and enter the cytosol.

S4. An antibiotic is a drug that kills or inhibits the growth of microorganisms. The use of antibiotics has been of great importance in the battle against many infectious diseases caused by microorganisms. For many antibiotics, their mode of action is to inhibit the translation process within bacterial cells. Certain antibiotics selectively bind to bacterial (70S) ribosomes but do not inhibit eukaryotic (80S) ribosomes. Their ability to inhibit translation can occur at different steps in the translation process. For example, tetracycline prevents the attachment of tRNA to the ribosome, whereas erythromycin inhibits the translocation of the ribosome along the mRNA. Why would an antibiotic bind to a bacterial ribosome but not to a eukaryotic ribosome? How does inhibition of translation by antibiotics such as tetracycline prevent bacterial growth?

Answer: Because bacterial ribosomes have a different protein and rRNA composition than eukaryotic ribosomes, certain drugs can recognize these different components, bind specifically to bacterial ribosomes, thereby interferimg with the process of translation. In other words, the surface of a bacterial ribosome must be somewhat different from the surface of a eukaryotic ribosome so that the drugs bind to the surface of only bacterial ribosomes. If a bacterial cell is exposed to tetracycline or other antibiotics, it cannot synthesize new polypeptides. Because polypeptides form functional proteins needed for processes such as cell division, the bacterium is unable to grow and proliferate.

Conceptual Questions

C1. An mRNA has the following sequence:

5′–GGCGAUGGGCAAUAAACCGGGCCAGUAAGC–3′

Identify the start codon and determine the complete amino acid sequence that would be translated from this mRNA.

C2. What does it mean when we say that the genetic code is degenerate? Discuss the universality of the genetic code.

C3. According to the adaptor hypothesis, are the following statements true or false?

A. The sequence of anticodons in tRNA directly recognizes codon sequences in mRNA, with some room for wobble.

B. The amino acid attached to the tRNA directly recognizes codon sequences in mRNA.

C. The amino acid attached to the tRNA affects the binding of the tRNA to a codon sequence in mRNA.

C4. In bacteria, researchers have isolated strains that carry mutations within tRNA genes. These mutations can change the sequence of the anticodon. For example, a normal $tRNA^{Trp}$ gene would encode a tRNA with the anticodon 3′–ACC–5′. A mutation could change this sequence to 3′–CCC–5′. When this mutation occurs, the tRNA still carries a tryptophan at its 3′ acceptor stem, even though the anticodon sequence has been altered.

A. How would this mutation affect the synthesis of polypeptides within the bacterium?

B. What does this mutation tell you about the recognition between tryptophanyl-tRNA synthetase and $tRNA^{Trp}$? Does the enzyme primarily recognize the anticodon or not?

C5. The covalent attachment of an amino acid to a tRNA is an endergonic reaction. In other words, it requires an input of energy for the reaction to proceed. Where does the energy come from to attach amino acids to tRNA molecules?

C6. The wobble rules for tRNA-mRNA pairing are shown in Figure 13.14. If we assume that the tRNAs do not contain modified bases, what is the minimum number of tRNAs needed to efficiently recognize the codons for the following types of amino acids?

A. Leucine

B. Methionine

C. Serine

C7. How many different sequences of mRNA could encode a peptide with the sequence proline-glycine-methionine-serine?

C8. If a tRNA molecule carries a glutamic acid, what are the two possible anticodon sequences that it could contain? Be specific about the 5′ and 3′ ends.

C9. A tRNA has an anticodon sequence 3′–GGU–5′. What amino acid does it carry?

C10. If a tRNA has an anticodon sequence 3′–CCI–5′, what codon(s) can it recognize?

C11. Describe the anticodon of a single tRNA that could recognize the codons 5′–AAC–3′ and 5′–AAU–3′. How would this tRNA need to be modified for it to also recognize 5′–AAA–3′?

C12. Describe the structural features that all tRNA molecules have in common.

C13. In the tertiary structure of tRNA, where is the anticodon region relative to the attachment site for the amino acid? Are they adjacent to each other?

C14. What is the role of aminoacyl-tRNA synthetase? The ability of the aminoacyl-tRNA synthetases to recognize tRNAs has sometimes been called the "second genetic code." Why has the function of this type of enzyme been described this way?

C15. What is an activated amino acid?

C16. Discuss the significance of modified bases within tRNA molecules.

C17. How and when does formylmethionine become attached to the initiator tRNA in bacteria?

C18. Is it necessary for a cell to make 61 different tRNA molecules, corresponding to the 61 codons for amino acids? Explain your answer.

C19. List the components required for translation. Describe the relative sizes of these different components. In other words, which components are small molecules, macromolecules, or assemblies of macromolecules?

C20. Describe the components of eukaryotic ribosomal subunits and where the assembly of the subunits occurs within living cells.

C21. The term subunit can be used in a variety of ways. Compare the use of the term subunit in proteins versus ribosomal subunit.

C22. Do the following events during bacterial translation occur primarily within the 30S subunit, within the 50S subunit, or at the interface between these two ribosomal subunits?

A. mRNA-tRNA recognition

B. Peptidyl transfer reaction

C. Exit of the polypeptide chain from the ribosome

D. Binding of initiation factors IF1, IF2, and IF3

C23. What are the three stages of translation? Discuss the main events that occur during these three stages.

C24. Describe the sequence in bacterial mRNA that promotes recognition by the 30S subunit.

C25. For each of the following initiation factors, how would eukaryotic initiation of translation be affected if it were missing?

A. eIF2

B. eIF4

C. eIF5

C26. How does a eukaryotic ribosome select its start codon? Describe the sequences in eukaryotic mRNA that provide an optimal context for a start codon.

C27. For each of the following sequences, rank them in order (from best to worst) as sequences that could be used to initiate translation according to Kozak's rules.

GACGCCAUGG

GCCUCCAUGC

GCCAUCAAGG

GCCACCAUGG

C28. Explain the functional roles of the A, P, and E sites during translation.

C29. An mRNA has the following sequence: 5′–AUG UAC UAU GGG GCG UAA–3′

Describe the amino acid sequence of the polypeptide that would be encoded by this mRNA. Be specific about the amino and carboxyl terminal ends.

C30. Which steps during the translation of bacterial mRNA involve an interaction between complementary strands of RNA?

C31. What is the function of the nucleolus?

C32. In which of the ribosomal sites, the A site, P site, and/or E site, could the following be found?

A. A tRNA without an amino acid attached

B. A tRNA with a polypeptide attached

C. A tRNA with a single amino acid attached

C33. What is a polysome?

C34. According to Figure 13.19, explain why the ribosome translocates along the mRNA in a 5′ to 3′ direction rather than a 3′ to 5′ direction.

C35. The lactose permease of *E. coli* is a protein composed of a single polypeptide that is 417 amino acids in length. By convention, the amino acids within a polypeptide are numbered from the amino-terminal end to the carboxyl-terminal end. Are the following questions about the lactose permease true or false?

A. Because the 64th amino acid is glycine and the 68th amino acid is aspartic acid, the codon for glycine, 64, is closer to the 3′ end of the mRNA than the codon for aspartic acid, 68.

B. The mRNA that encodes the lactose permease must be greater than 1241 nucleotides in length.

C36. An mRNA encodes a polypeptide that is 312 amino acids in length. The 53rd codon in this polypeptide is a tryptophan codon. A mutation in the gene that encodes this polypeptide changes this tryptophan codon into a stop codon. How many amino acids would be in the resulting polypeptide: 52, 53, 259, or 260?

C37. Explain what is meant by the coupling of transcription and translation in bacteria. Does coupling occur in bacterial and/or eukaryotic cells? Explain.

Experimental Questions

E1. In the experiment of Figure 13.4, what would be the predicted amounts of amino acids incorporated into polypeptides if the RNA was a random polymer containing 50% C and 50% G?

E2. With regard to the experiment described in Figure 13.11, answer the following questions:

A. Why was a polyUG mRNA template used?

B. Would you radiolabel the cysteine with the isotope ^{14}C or ^{35}S? Explain your choice.

C. What would be the expected results if the experiment was followed in the same way except that a polyGC template was used? Note: A polyGC template could contain two different alanine codons (GCC and GCG), but it could not contain any cysteine codons.

E3. An experimenter has a chemical reagent that modifies threonine to another amino acid. Following the protocol described in Figure 13.11, an mRNA is made composed of 50% C and 50% A. The amino acid composition of the resultant polypeptides is 12.5% lysine, 12.5% asparagine, 25% serine, 12.5% glutamine, 12.5% histidine, and 25% proline. One of the amino acids present in this polypeptide is due to the modification of threonine. Which amino acid is it? Based on the structure of the amino acid side chains, explain how the structure of threonine has been modified.

E4. Polypeptides can be translated in vitro. Would a bacterial mRNA be translated in vitro by eukaryotic ribosomes? Would a eukaryotic mRNA be translated in vitro by bacterial ribosomes? Why or why not?

E5. Discuss how the elucidation of the structure of the ribosome can help us to understand its function.

E6. Figure 13.21 shows an electron micrograph of a bacterial gene as it is being transcribed and translated. In this figure, label the 5′ and 3′ ends of the DNA and RNA strands. Place an arrow where you think the start codons are found in the mRNA transcripts.

E7. Chapter 18 describes a blotting method known as Western blotting that can be used to detect the production of a polypeptide that is translated from a particular mRNA. In this method, a protein is detected with an antibody that specifically recognizes and binds to its amino acid sequence. The antibody acts as a probe to detect the presence of the protein. In a Western blotting experiment, a mixture of cellular proteins is separated using gel electrophoresis according to their molecular masses. After the antibody has bound to the protein of interest within a blot of a gel, the protein is visualized as a dark band. For example, an antibody that recognizes the β-globin polypeptide could be used to specifically detect the β-globin polypeptide in a blot. As shown here, the method of Western blotting can be used to determine the amount and relative size of a particular protein that is produced in a given cell type.

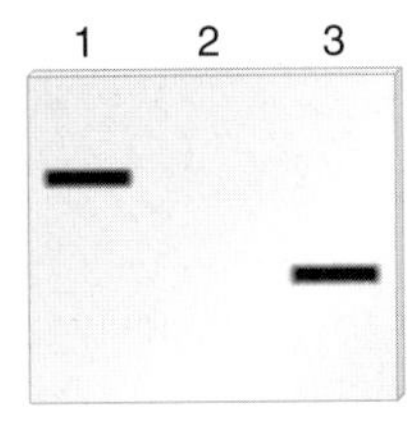

Lane 1 is a sample of proteins isolated from normal red blood cells.

Lane 2 is a sample of proteins isolated from kidney cells. Kidney cells do not produce β globin.

Lane 3 is a sample of proteins isolated from red blood cells from a patient with β-thalassemia. This patient is homozygous for a mutation that results in the shortening of the β-globin polypeptide.

Now here is the question. A protein called troponin contains 334 amino acids. Because each amino acid weighs 120 daltons (Da) (on average), the molecular mass of this protein is about 40,000 Da, or 40 kDa. Troponin functions in muscle cells, and it is not expressed in nerve cells. Draw the expected results of a Western blot for the following samples:

Lane 1: Proteins isolated from muscle cells

Lane 2: Proteins isolated from nerve cells

Lane 3: Proteins isolated from the muscle cells of an individual who is homozygous for a mutation that introduces a stop codon at codon 177

E8. The technique of Western blotting can be used to detect proteins that are translated from a particular mRNA. This method is described in Chapter 18 and also in experimental question E7. Let's suppose a researcher was interested in the effects of mutations on the expression of a structural gene that encodes a protein we will call protein X. This protein is expressed in skin cells and contains 572 amino acids. Its molecular mass is approximately 68,600 Da, or 68.6 kDa. Make a drawing that shows the expected results of a Western blot using proteins isolated from the skin cells obtained from the following individuals:

Lane 1: A normal individual

Lane 2: An individual who is homozygous for a deletion, which removes the promoter for this gene

Lane 3: An individual who is heterozygous in which one gene is normal and the other gene has a mutation that introduces an early stop codon at codon 421

Lane 4: An individual who is homozygous for a mutation that introduces an early stop codon at codon 421

Lane 5: An individual who is homozygous for a mutation that changes codon 198 from a valine codon into a leucine codon

E9. The protein known as tyrosinase is needed to make certain types of pigments. Tyrosinase is composed of a single polypeptide with 511 amino acids. Because each amino acid weighs 120 Da (on average), the molecular mass of this protein is approximately 61,300 Da, or 61.3 kDa. People who carry two defective copies of the tyrosinase gene have the condition known as albinism. They are unable to make pigment in the skin, eyes, and hair. Western blotting can be used to detect proteins that are translated from a particular mRNA. This method is described in Chapter 18 and also in experimental question E7. Skin samples were collected from a pigmented individual (lane 1) and from three unrelated albino individuals (lanes 2, 3, and 4) and subjected to a Western blot analysis using an antibody that recognizes tyrosinase. Explain the possible cause of albinism in the three albino individuals.

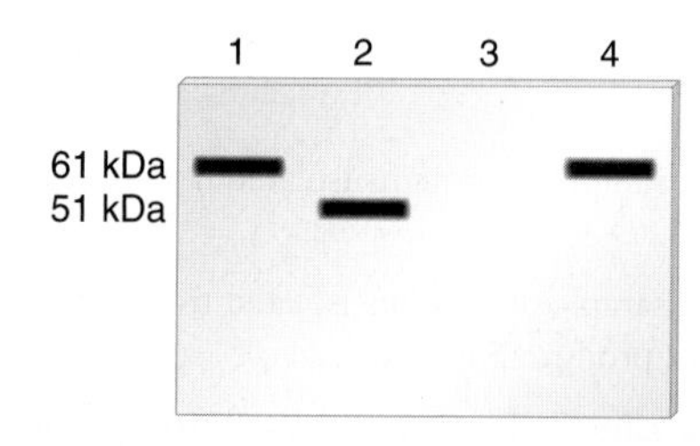

E10. Although 61 codons specify the 20 amino acids, most species display a codon bias. This means that certain codons are used much more frequently than other codons. For example, UUA, UUG, CUU, CUC, CUA, and CUG all specify leucine. In yeast, however, the UUG codon is used to specify leucine approximately 80% of the time.

A. The experiment of Figure 13.4 shows the use of an in vitro or cell-free translation system. In this experiment, the RNA, which was used for translation, was chemically synthesized. Instead of using a chemically synthesized RNA, researchers can isolate mRNA from living cells and then add the mRNA to the cell-free translation system. If a researcher isolated mRNA from kangaroo cells and then added it to a cell-free translation system that came from yeast cells, how might the phenomenon of codon bias affect the production of proteins?

B. Discuss potential advantages and disadvantages of codon bias for translation.

Questions for Student Discussion/Collaboration

1. Discuss why you think the ribosome needs to contain so many proteins and rRNA molecules. Does it seem like a waste of cellular energy to make such a large structure so that translation can occur?
2. Discuss and make a list of the similarities and differences in the events that occur during the initiation, elongation, and termination stages of transcription (see Chapter 12) and translation (Chapter 13).
3. Which events during translation involve molecular recognition of a nucleotide base sequence within RNA? Which events involve recognition between different protein molecules?

Note: All answers appear at the website for this textbook; the answers to even-numbered questions are in the back of the textbook.

www.mhhe.com/brookergenetics4e

Visit the website for practice tests, answer keys, and other learning aids for this chapter. Enhance your understanding of genetics with our interactive exercises, quizzes, animations, and much more.

CHAPTER OUTLINE

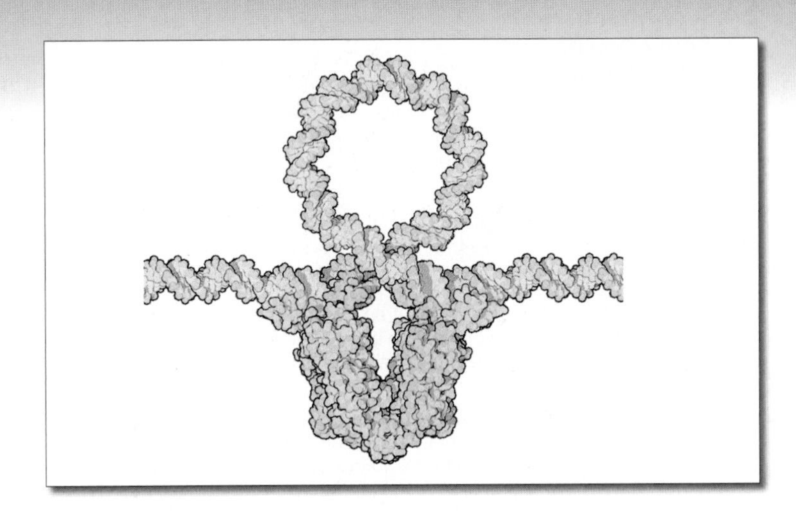

A model showing the binding of a genetic regulatory protein to DNA, which results in a DNA loop. *The model shown here involves the lac repressor protein found in E. coli binding to the operator site in the lac operon.*

GENE REGULATION IN BACTERIA AND BACTERIOPHAGES

Chromosomes of bacteria, such as *Escherichia coli*, contain a few thousand different genes. **Gene regulation** is the phenomenon in which the level of gene expression can vary under different conditions. In comparison, unregulated genes have essentially constant levels of expression in all conditions over time. Unregulated genes are also called **constitutive genes.** Frequently, constitutive genes encode proteins that are continuously needed for the survival of the bacterium. In contrast, the majority of genes are regulated so that the proteins they encode can be produced at the proper times and in the proper amounts.

A key benefit of gene regulation is that the encoded proteins are produced only when they are required. Therefore, the cell avoids wasting valuable energy making proteins it does not need. From the viewpoint of natural selection, this enables an organism such as a bacterium to compete as efficiently as possible for limited resources. Gene regulation is particularly important because bacteria exist in an environment that is frequently changing with regard to temperature, nutrients, and many other factors. The following are a few common processes regulated at the genetic level:

1. **Metabolism:** Some proteins function in the metabolism of small molecules. For example, certain enzymes are needed for a bacterium to metabolize particular sugars. These enzymes are required only when the bacterium is exposed to such sugars in its environment.
2. **Response to environmental stress:** Certain proteins help a bacterium to survive environmental stress such as osmotic shock or heat shock. These proteins are required only when the bacterium is confronted with the stress.
3. **Cell division:** Some proteins are needed for cell division. These are necessary only when the bacterial cell is getting ready to divide.

The expression of structural genes, which encode polypeptides, ultimately leads to the production of functional cellular proteins. As we saw in Chapters 12 and 13, gene expression is a multistep process that proceeds from transcription to translation, and it may involve posttranslational effects on protein structure and function. As shown in **Figure 14.1**, gene regulation can occur at any of these steps in the pathway of gene expression. In this chapter, we will examine the molecular mechanisms that account for these types of gene regulation.

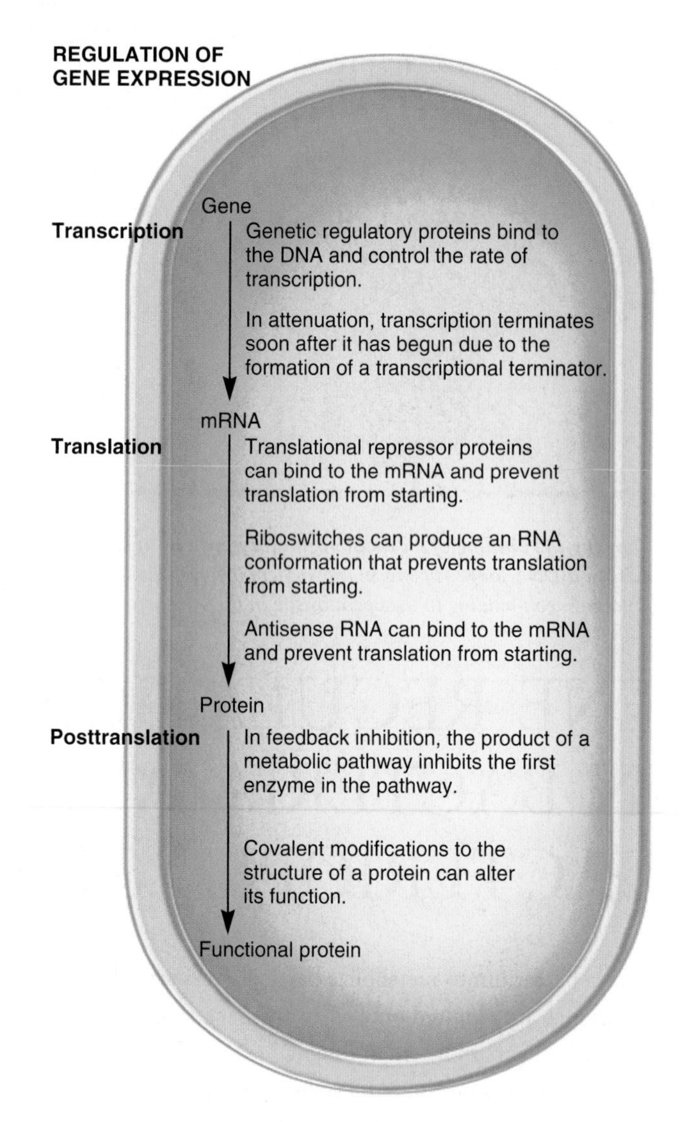

FIGURE 14.1 **Common points where regulation of gene expression occurs in bacteria.**

14.1 TRANSCRIPTIONAL REGULATION

In bacteria, the most common way to regulate gene expression is by influencing the rate at which transcription is initiated. Although we frequently refer to genes as being "turned on or off," it is more accurate to say that the level of gene expression is increased or decreased. At the level of transcription, this means that the rate of RNA synthesis can be increased or decreased.

In most cases, transcriptional regulation involves the actions of regulatory proteins that can bind to the DNA and affect the rate of transcription of one or more nearby genes. Two types of regulatory proteins are common. A **repressor** is a regulatory protein that binds to the DNA and inhibits transcription, whereas an **activator** is a regulatory protein that increases the rate of transcription. Transcriptional regulation by a repressor protein is termed **negative control,** and regulation by an activator protein is considered to be **positive control.**

In conjunction with regulatory proteins, small effector molecules often play a critical role in transcriptional regulation. However, small effector molecules do not bind directly to the DNA to alter transcription. Rather, an effector molecule exerts its effects by binding to an activator or repressor. The binding of the effector molecule causes a conformational change in the regulatory protein and thereby influences whether or not the protein can bind to the DNA. Genetic regulatory proteins that respond to small effector molecules have two functional domains. One domain is a site where the protein binds to the DNA; the other domain is the binding site for the effector molecule.

Regulatory proteins are given names describing how they affect transcription when they are bound to the DNA (repressor or activator). In contrast, small effector molecules are given names that describe how they affect transcription when they are present in the cell at a sufficient concentration to exert their effect (**Figure 14.2**). An **inducer** is a small effector molecule that causes transcription to increase. An inducer may accomplish this in two ways: It could bind to a repressor protein and prevent it from binding to the DNA, or it could bind to an activator protein and cause it to bind to the DNA. In either case, the transcription rate is increased. Genes that are regulated in this manner are called **inducible genes.**

Alternatively, the presence of a small effector molecule may inhibit transcription. This can also occur in two ways. A **corepressor** is a small molecule that binds to a repressor protein, thereby causing the protein to bind to the DNA. An **inhibitor** binds to an activator protein and prevents it from binding to the DNA. Both corepressors and inhibitors act to reduce the rate of transcription. Therefore, the genes they regulate are termed **repressible genes.** Unfortunately, this terminology can be confusing because a repressible system could involve an activator protein, or an inducible system could involve a repressor protein.

In this section, we will examine several examples in which genes are regulated by the actions of genetic regulatory proteins that influence the rate of transcription. We will see how gene regulation provides an efficient way for bacteria to synthesize proteins.

The Phenomenon of Enzyme Adaptation Is Due to the Synthesis of Cellular Proteins

Our initial understanding of gene regulation can be traced back to the creative minds of François Jacob and Jacques Monod at the Pasteur Institute in Paris, France. Their research into genes and gene regulation stemmed from an interest in the phenomenon known as **enzyme adaptation,** which had been identified at the turn of the twentieth century. Enzyme adaptation refers to the observation that a particular enzyme appears within a living cell only after the cell has been exposed to the substrate for that enzyme. When a bacterium is not exposed to a particular substance, it does not make the enzymes needed to metabolize that substance.

To investigate this phenomenon, Jacob and Monod focused their attention on lactose metabolism in *E. coli*. Key experimental observations that led to an understanding of this genetic system are listed here.

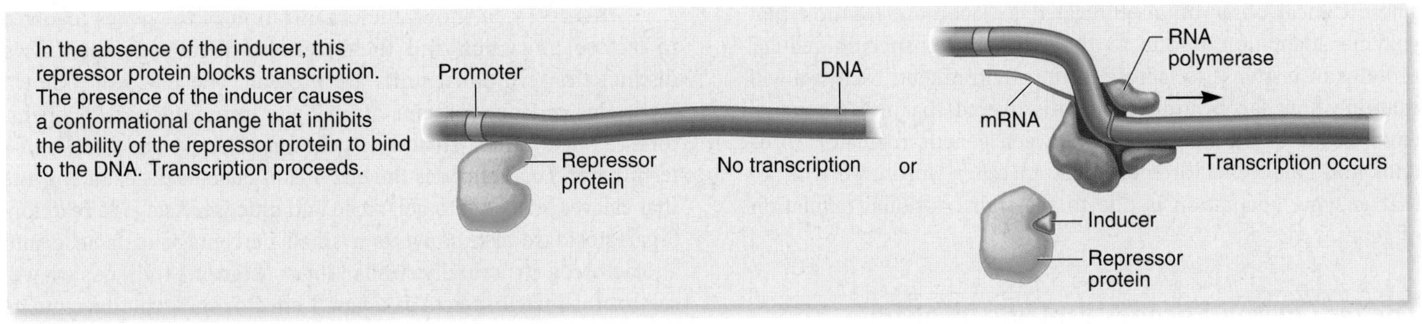

(a) Repressor protein, inducer molecule, inducible gene

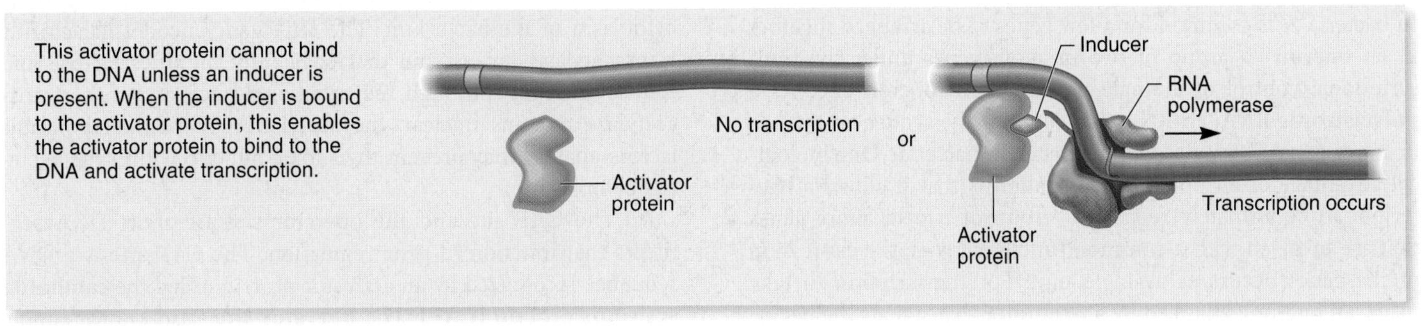

(b) Activator protein, inducer molecule, inducible gene

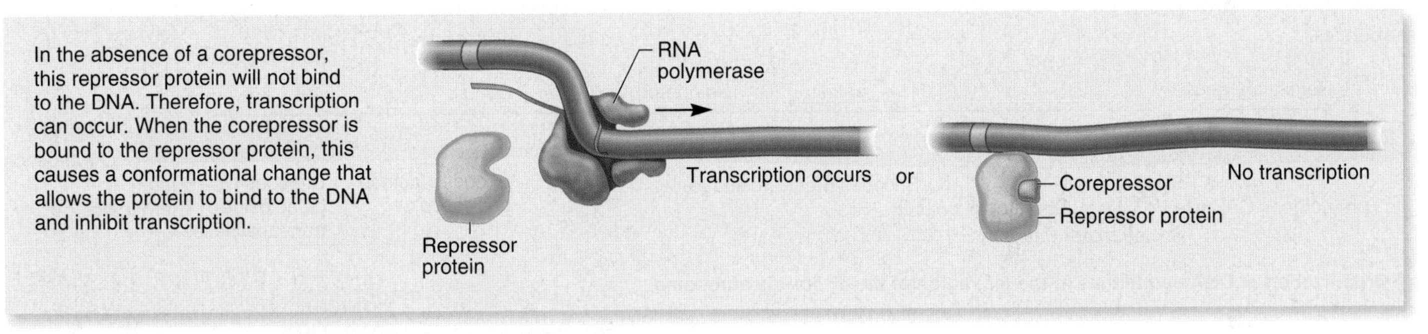

(c) Repressor protein, corepressor molecule, repressible gene

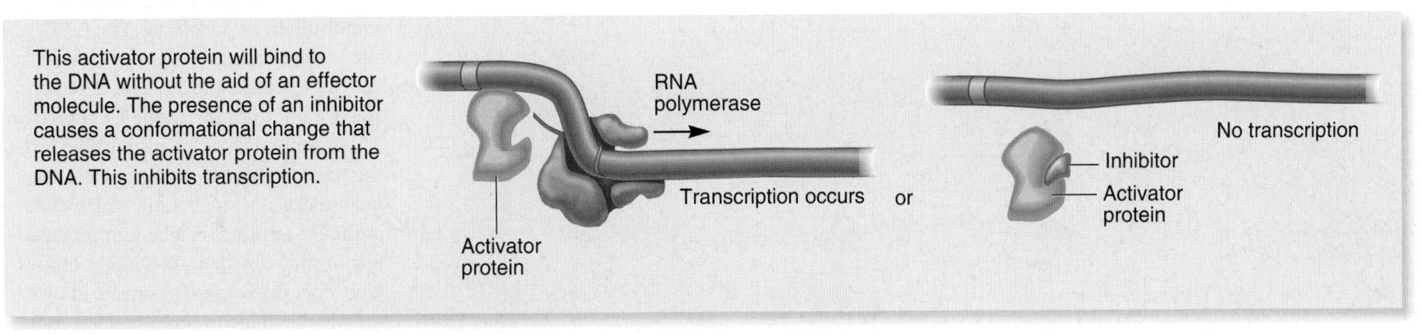

(d) Activator protein, inhibitor molecule, repressible gene

FIGURE 14.2 Binding sites on a genetic regulatory protein. In these examples, a regulatory protein has two binding sites: one for a small effector molecule and one for DNA. The binding of the small effector molecule changes the conformation of the regulatory protein, which alters the DNA-binding site structure, thereby influencing whether the protein can bind to the DNA.

1. The exposure of bacterial cells to lactose increased the levels of lactose-utilizing enzymes by 1000- to 10,000-fold.
2. Antibody and labeling techniques revealed that the increase in the activity of these enzymes was due to the increased synthesis of the enzymes.
3. The removal of lactose from the environment caused an abrupt termination in the synthesis of the enzymes.
4. Mutations that prevented the synthesis of a particular protein involved in lactose utilization showed that a separate gene encoded each protein.

These critical observations indicated to Jacob and Monod that enzyme adaptation is due to the synthesis of specific cellular proteins in response to lactose in the environment. Next, we will examine how Jacob and Monod discovered that this phenomenon is due to the interactions between genetic regulatory proteins and small effector molecules. In other words, we will see that enzyme adaptation is due to the transcriptional regulation of genes.

The *lac* Operon Encodes Proteins Involved in Lactose Metabolism

In bacteria, it is common for a few genes to be arranged together in an **operon**—a group of two or more genes under the transcriptional control of a single promoter. An operon encodes a **polycistronic RNA,** an RNA that contains the sequences for two or more genes. Why do operons occur in bacteria? One biological advantage of an operon organization is that it allows a bacterium to coordinately regulate a group of two or more genes that are involved with a common functional goal; the expression of the genes occurs as a single unit. For transcription to take place, an operon is flanked by a **promoter** that signals the beginning of transcription and a **terminator** that specifies the end of transcription. Two or more genes are found between these two sequences.

Figure 14.3a shows the organization of the genes involved in lactose utilization and their transcriptional regulation. Two distinct transcriptional units are present. The first unit, known as the *lac* operon, contains a CAP site; promoter (*lacP*); operator site (*lacO*); three structural genes, *lacZ, lacY,* and *lacA*; and a terminator. *LacZ* encodes the enzyme β-galactosidase, an enzyme that cleaves lactose into galactose and glucose. As a side reaction, β-galactosidase also converts a small percentage of lactose into allolactose, a structurally similar sugar (**Figure 14.3b**). As we will see later, allolactose acts as a small effector molecule to regulate the *lac* operon. The *lacY* gene encodes lactose permease, a membrane protein required for the active transport of lactose into the cytoplasm of the bacterium. The *lacA* gene encodes galactoside transacetylase, an enzyme that covalently modifies lactose and lactose analogs. Although the functional necessity of the transacetylase remains unclear, the acetylation of nonmetabolizable lactose analogs may prevent their toxic buildup within the bacterial cytoplasm.

The CAP site and the operator site are short DNA segments that function in gene regulation. The **CAP site** is a DNA sequence recognized by an activator protein called the **catabolite activator protein (CAP).** The **operator site** (also known simply as the **operator**) is a sequence of bases that provides a binding site for a repressor protein.

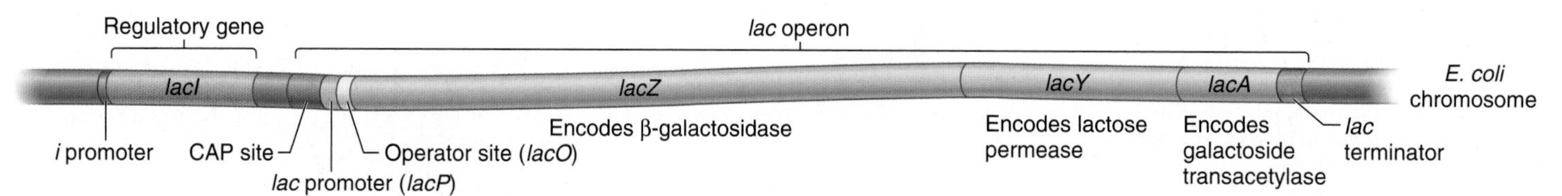

(a) Organization of DNA sequences in the *lac* region of the *E. coli* chromosome

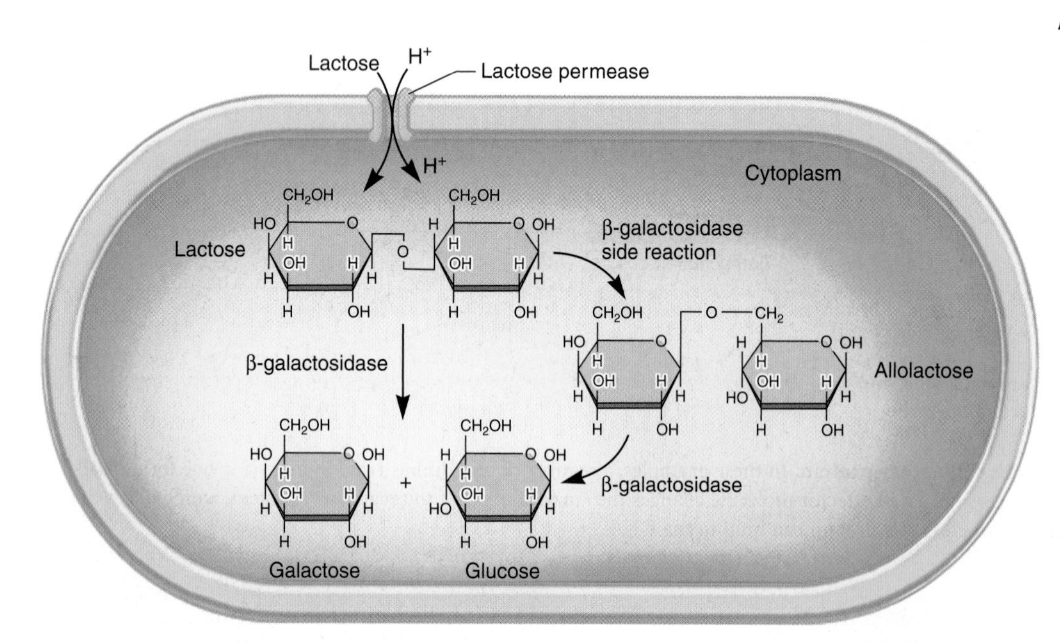

(b) Functions of lactose permease and β-galactosidase

ONLINE ANIMATION

FIGURE 14.3 Organization of the *lac* operon and other genes involved with lactose metabolism in *E. coli.* **(a)** The CAP site is the binding site for the catabolite activator protein (CAP). The operator site is a binding site for the lac repressor. The promoter (*lacP*) is responsible for the transcription of the *lacZ, lacY,* and *lacA* genes as a single unit, which ends at the *lac* terminator. The *i* promoter is responsible for the transcription of the *lacI* gene. **(b)** Lactose permease allows the uptake of lactose into the bacterial cytoplasm. It cotransports lactose with H^+. Because bacteria maintain an H^+ gradient across their cytoplasmic membrane, this cotransport permits the active accumulation of lactose against a gradient. β-galactosidase is a cytoplasmic enzyme that cleaves lactose and related compounds into galactose and glucose. As a minor side reaction, β-galactosidase also converts lactose into allolactose. Allolactose can also be broken down into galactose and glucose.

A second transcriptional unit involved in genetic regulation is the *lacI* gene (see Figure 14.3a), which is not part of the *lac* operon. The *lacI* gene, which is constitutively expressed at fairly low levels, has its own promoter, the *i* promoter. The *lacI* gene encodes the **lac repressor,** a protein that is important for the regulation of the *lac* operon. The lac repressor functions as a homotetramer, a protein composed of four identical subunits. The amount of lac repressor made is approximately 10 homotetramer proteins per cell. Only a small amount of the lac repressor protein is needed to repress the *lac* operon.

The *lac* Operon Is Regulated by a Repressor Protein

The *lac* operon can be transcriptionally regulated in more than one way. The first mechanism that we will examine is one that is inducible and under negative control. As shown in **Figure 14.4**, this form of regulation involves the lac repressor protein, which binds to the sequence of nucleotides found within the *lac* operator site. Once bound, the lac repressor prevents RNA polymerase from transcribing the *lacZ*, *lacY*, and *lacA* genes (Figure 14.4a). The binding of the repressor to the operator site is a reversible process. In the absence of allolactose, the lac repressor is bound to the operator site most of the time.

The ability of the lac repressor to bind to the operator site depends on whether or not allolactose is bound to it. Each of the repressor protein's four subunits has a single binding site for allolactose, the inducer. How does a small molecule like allolactose exert its effects? When allolactose binds to the repressor, a conformational change occurs in the lac repressor protein that prevents it from binding to the operator site. Under these conditions, RNA polymerase is now free to transcribe the operon (Figure 14.4b). In genetic terms, we would say that the operon has been **induced.** The action of a small effector molecule, such as allolactose, is called **allosteric regulation.** Allosteric proteins have at least two binding sites. The effector molecule binds to the protein's **allosteric site,** which is a site other than the protein's active site. In the case of the lac repressor, the active site is the part of the protein that binds to the DNA.

Rare mutations in the *lacI* gene that alter the regulation of the *lac* operon reveal that the lac repressor is composed of a protein domain that binds to the DNA and another domain that contains the allolactose-binding site. As discussed later in Figure 14.7, researchers have identified *lacI*$^-$ mutations that result in the constitutive expression of the *lac* operon, which means that it is expressed in the presence and absence of lactose. Such mutations may result in an inability to synthesize any repressor protein, or they may produce a repressor protein that is unable to bind to the DNA at the *lac* operator site. If the lac repressor is unable to bind to the DNA, the *lac* operon cannot be repressed. By comparison, *lacI*^{S} mutations have the opposite effect: the *lac* operon cannot be induced even in the presence of lactose. These mutations, which are called super-repressor mutations, are typically located in the domain that binds allolactose. The mutation usually results in a lac repressor protein that cannot bind allolactose. If the lac repressor is unable to bind allolactose, it will remain bound to the *lac* operator site and therefore induction cannot occur.

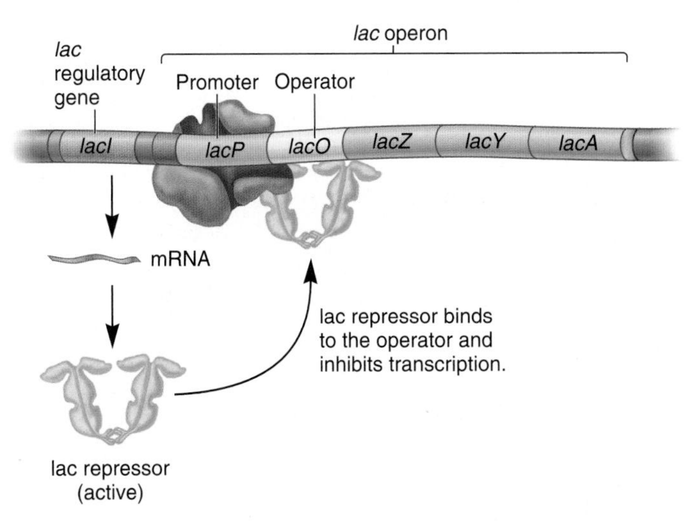

(a) No lactose in the environment

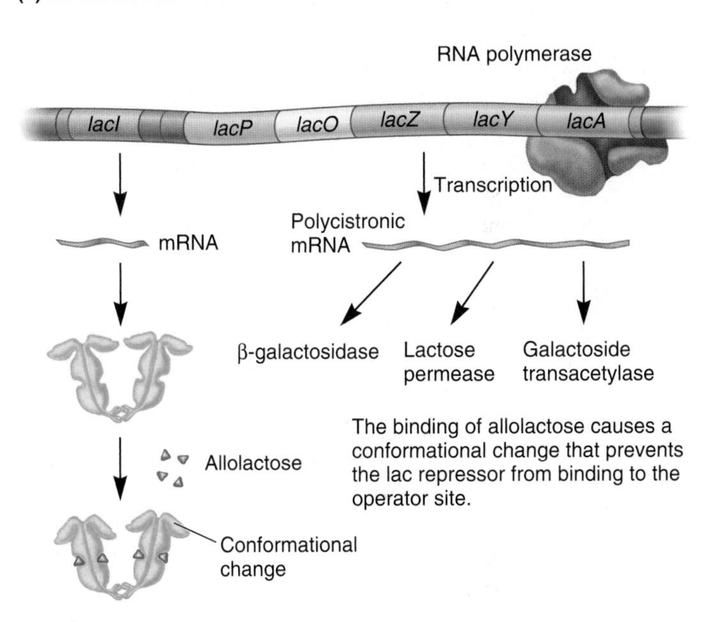

(b) Lactose present

ONLINE ANIMATION **FIGURE 14.4 Mechanism of induction of the *lac* operon.** **(a)** In the absence of the inducer allolactose, the repressor protein is tightly bound to the operator site, thereby inhibiting the ability of RNA polymerase to transcribe the operon. **(b)** When allolactose is available, it binds to the repressor. This alters the conformation of the repressor protein that prevents it from binding to the operator site. Therefore, RNA polymerase can transcribe the operon. (The CAP site is not labeled in this drawing.)

The Regulation of the *lac* Operon Allows a Bacterium to Respond to Environmental Change

To better appreciate *lac* operon regulation at the cellular level, let's consider the process as it occurs over time. **Figure 14.5** illustrates the effects of external lactose on the regulation of the *lac* operon. In the absence of lactose, no inducer is available to bind to the lac repressor. Therefore, the lac repressor binds to the

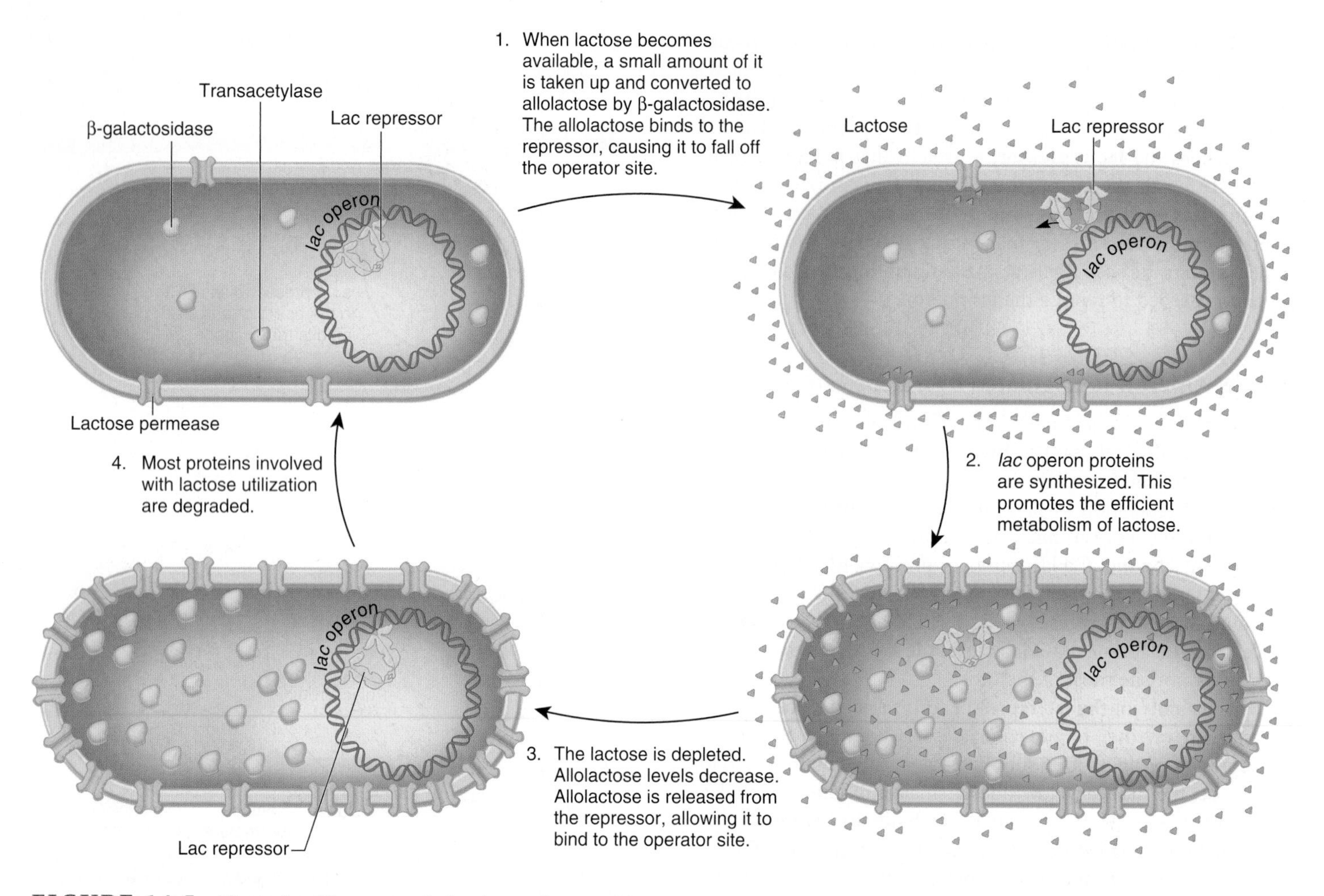

FIGURE 14.5 The cycle of *lac* operon induction and repression.

Genes → Traits The genes and the genetic regulation of the *lac* operon provide the bacterium with the trait of being able to metabolize lactose in the environment. When lactose is present, the genes of the *lac* operon are induced, and the bacterial cell can efficiently metabolize this sugar. When lactose is absent, these genes are repressed so the bacterium does not waste its energy expressing them. Note: The proteins involved with lactose utilization are fairly stable, but they will eventually be degraded.

operator site and inhibits transcription. In reality, the repressor does not completely inhibit transcription, so very small amounts of β-galactosidase, lactose permease, and transacetylase are made. However, the levels are far too low for the bacterium to readily use lactose. When the bacterium is exposed to lactose, a small amount can be transported into the cytoplasm via lactose permease, and β-galactosidase converts some of it to allolactose. As this occurs, the cytoplasmic level of allolactose gradually rises; eventually, allolactose binds to the lac repressor. The binding of allolactose promotes a conformational change that prevents the repressor from binding to the *lac* operator site and thereby allows transcription of the *lacZ*, *lacY*, and *lacA* genes to occur. Translation of the encoded polypeptides produces the proteins needed for lactose uptake and metabolism.

To understand how the induction process is shut off in a lactose-depleted environment, let's consider the interaction between allolactose and the lac repressor. The lac repressor has a measurable affinity for allolactose. The binding of allolactose to the lac repressor is reversible. The likelihood that allolactose will bind to the repressor depends on the allolactose concentration. During induction of the operon, the concentration of allolactose rises and approaches the affinity for the repressor protein. This makes it likely that allolactose will bind to the lac repressor, thereby causing it to be released from the operator site. Later on, however, the bacterial cell metabolizes the sugars, thereby lowering the concentration of allolactose below its affinity for the repressor. At this point, the lac repressor is unlikely to be bound to allolactose. When allolactose is released, the lac repressor returns to the conformation that binds to the operator site. In this way, the binding of the repressor shuts down the *lac* operon when lactose is depleted from the environment. After repression occurs, the mRNA and proteins encoded by the *lac* operon are eventually degraded (see Figure 14.5).

EXPERIMENT 14A

The *lacI* Gene Encodes a Diffusible Repressor Protein

Now that we have an understanding of the *lac* operon, let's consider one of the experimental approaches that was used to elucidate its regulation. In the 1950s, Jacob, Monod, and their colleague Arthur Pardee had identified a few rare mutant strains of bacteria that had abnormal lactose adaptation. As mentioned earlier, one type of mutant, designated *lacI*$^-$, resulted in the constitutive expression of the *lac* operon even in the absence of lactose. As shown in **Figure 14.6a**, the correct explanation is that a loss-of-function mutation in the *lacI* gene prevented the lac repressor protein from binding to the *lac* operator site and inhibiting transcription. At the time of their work, however, the function of the lac repressor was not yet known. Instead, the researchers incorrectly hypothesized that the *lacI*$^-$ mutation resulted in the synthesis of an internal inducer, making it unnecessary for cells to be exposed to lactose for induction (**Figure 14.6b**).

To further explore the nature of this mutation, Jacob, Monod, and Pardee applied a genetic approach. In order to understand their approach, let's briefly consider the process of bacterial conjugation (described in Chapter 7). The earliest studies of Jacob, Monod, and Pardee in 1959 involved matings between recipient cells, termed F$^-$, and donor cells, which were *Hfr* strains that transferred a portion of the bacterial chromosome. Later experiments in 1961 involved the transfer of circular segments of DNA known as F factors. We consider the latter type of experiment here. Sometimes an F factor also carries genes that were originally found within the bacterial chromosome. These types of F factors are called F′ factors (F prime factors). In their studies, Jacob, Monod, and Pardee identified F′ factors that carried the *lacI* gene and portions of the *lac* operon. These F′ factors can be transferred from one cell to another by bacterial conjugation. A strain of bacteria containing F′ factor genes is called a **merozygote,** or partial diploid.

The production of merozygotes was instrumental in allowing Jacob, Monod, and Pardee to elucidate the function of the *lacI* gene. This experimental approach has two key points. First, the two *lacI* genes in a merozygote may be different alleles. For example, the *lacI* gene on the chromosome may be a *lacI*$^-$ allele that causes constitutive expression, whereas the *lacI* gene on the F′ factor may be normal. Second, the genes on the F′ factor and the genes on the bacterial chromosome are not physically adjacent to each other. As we now know, the expression of the *lacI* gene on an F′ factor should produce repressor proteins that can diffuse within the cell and eventually bind to the operator site of the *lac* operon located on the chromosome and also to the operator site on an F′ factor.

Figure 14.7 shows one experiment of Jacob, Monod, and Pardee in which they analyzed a *lacI*$^-$ mutant strain that was already known to constitutively express the *lac* operon and compared it to the corresponding merozygote. The merozygote had a *lacI*$^-$ mutant gene on the chromosome and a normal *lacI* gene on an F′ factor. These two strains were grown and then divided into two tubes each. In half of the tubes, lactose was omitted. In the other tubes, the strains were incubated with lactose to determine if lactose was needed to induce the expression of the operon. The cells were lysed by sonication, and then a lactose analog, β-*o*-nitrophenylgalactoside (β-ONPG), was added. This molecule is colorless, but β-galactosidase cleaves it into a product that has a yellow color. Therefore, the amount of yellow color produced in a given amount of time is a measure of the amount of β-galactosidase that is being expressed from the *lac* operon.

THE HYPOTHESIS

The *lacI*$^-$ mutation results in the synthesis of an internal inducer.

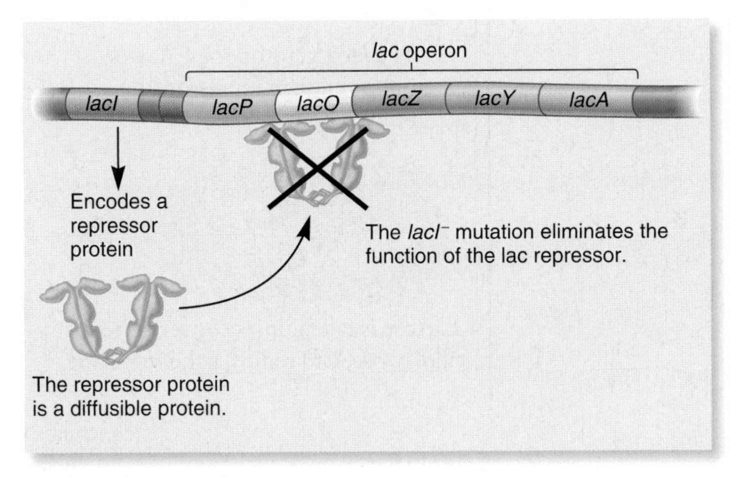

(a) Correct explanation

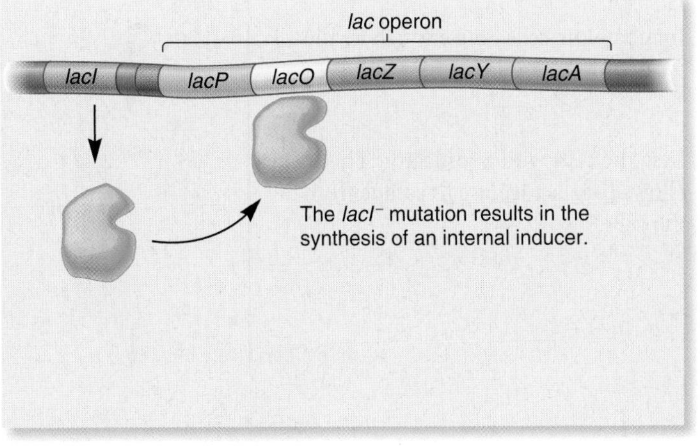

(b) Internal inducer hypothesis

FIGURE 14.6 Alternative hypotheses to explain how a *lacI*$^-$ mutation could cause the constitutive expression of the *lac* operon. (a) The correct explanation in which the *lacI*$^-$ mutation eliminates the function of the lac repressor protein, which prevents it from repressing the *lac* operon. (b) The hypothesis of Jacob, Monod, and Pardee. In this case, the *lacI*$^-$ mutation would result in the synthesis of an internal inducer that turns on the *lac* operon.

TESTING THE HYPOTHESIS — FIGURE 14.7 **Evidence that the *lacI* gene encodes a diffusible repressor protein.**

Starting material: The genotype of the mutant strain was $lacI^-$ $lacZ^+$ $lacY^+$ $lacA^+$. The merozygote strain had an F′ factor that was $lacI^+$ $lacZ^+$ $lacY^+$ $lacA^+$, which had been introduced into the mutant strain via conjugation.

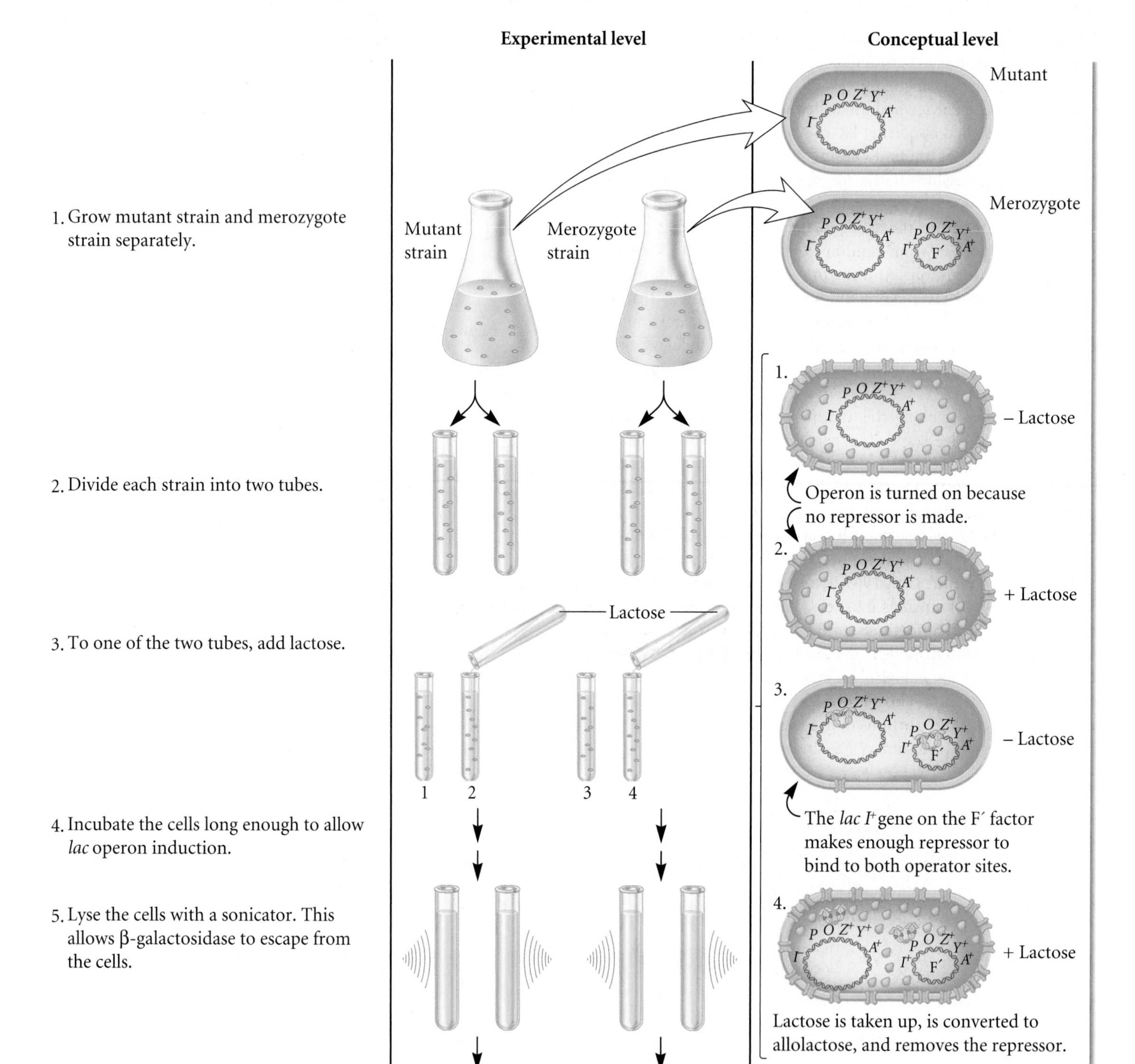

(continued)

6. Add β-*o*-nitrophenylgalactoside (β-ONPG). This is a colorless compound. β-galactosidase will cleave the compound to produce galactose and *o*-nitrophenol (O-NP). *O*-nitrophenol has a yellow color. The deeper the yellow color, the more β-galactosidase was produced.

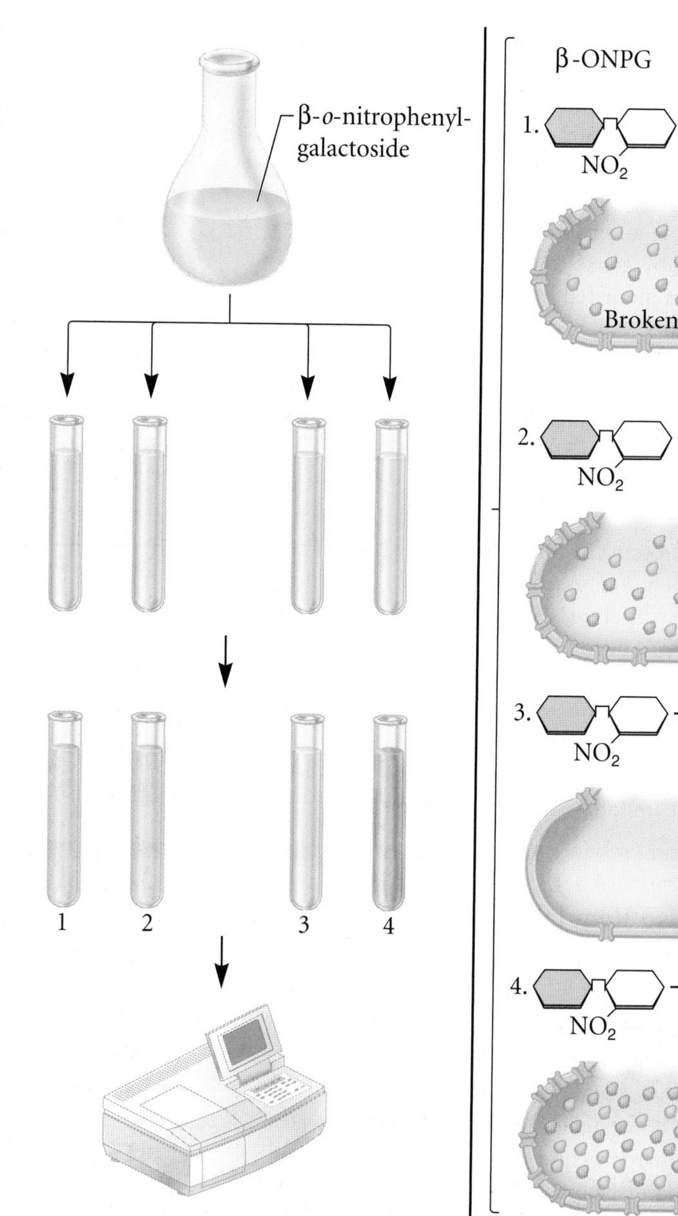

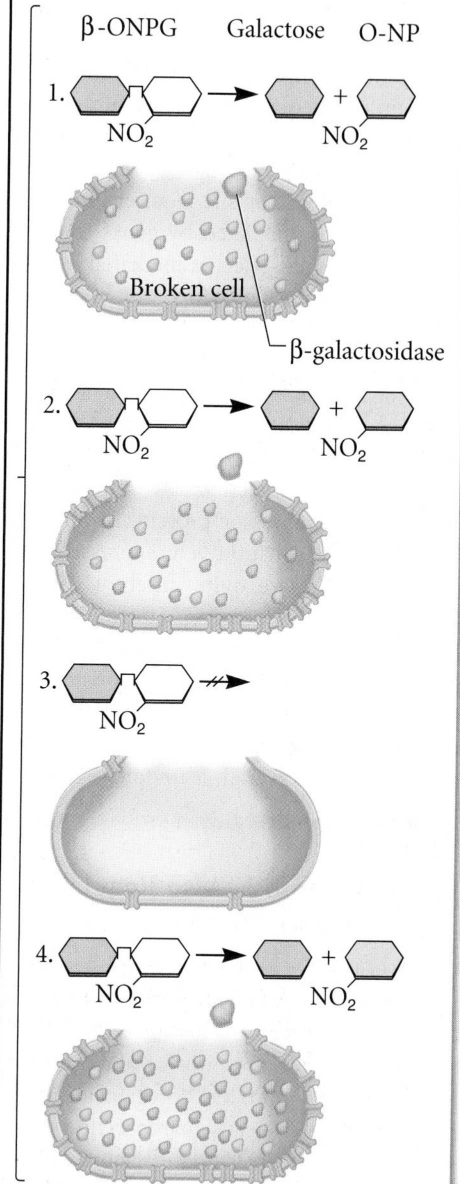

7. Incubate the sonicated cells to allow β-galactosidase time to cleave β-*o*-nitrophenylgalactoside.

8. Measure the yellow color produced with a spectrophotometer. (See the Appendix for a description of spectrophotometry.)

THE DATA

Strain	*Addition of Lactose*	*Amount of β-Galactosidase (percentage of parent strain)*
Mutant	No	100%
Mutant	Yes	100%
Merozygote	No	<1%
Merozygote	Yes	220%

Data from: F. Jacob, and J. Monod (1961) Genetic regulatory mechanisms in the synthesis of proteins. *J Mol Biol.* 3, 318–356.

INTERPRETING THE DATA

As seen in the data, the yellow production in the original mutant strain was the same in the presence or absence of lactose. This result is expected because the expression of β-galactosidase in the *lacI*$^-$ mutant strain was already known to be constitutive. In other words, the presence of lactose was not needed to induce the operon due to a defective *lacI* gene. In the merozygote strain, however, a different result was obtained. In the absence of lactose, the *lac* operons were repressed—even the operon on the bacterial chromosome. How do we explain these results? Because the normal *lacI* gene on the F′ factor was not physically located next to the chromosomal *lac* operon, this result is consistent with the idea that the *lacI* gene codes for a repressor protein that can diffuse throughout the cell and bind to any *lac* operon. The hypothesis that the *lacI*$^-$ mutation resulted in the synthesis of an internal inducer was rejected. If that hypothesis had been correct, the merozygote strain would have still made an internal inducer, and the *lac* operons in the merozygote would have been expressed in the absence of lactose. This result was not obtained.

The interactions between regulatory proteins and DNA sequences illustrated in this experiment have led to the definition of two genetic terms. A ***trans*-effect** is a form of genetic regulation that can occur even though two DNA segments are not physically adjacent. The action of the lac repressor on the *lac* operon is a *trans*-effect. A regulatory protein, such as the lac repressor, is called a ***trans*-acting factor.** In contrast, a ***cis*-acting element** is a DNA segment that must be adjacent to the gene(s) that it regulates, and it is said to have a ***cis*-effect** on gene expression. The *lac* operator site is an example of a *cis*-acting element. A *trans*-effect is mediated by genes that encode regulatory proteins, whereas a *cis*-effect is mediated by DNA sequences that are bound by regulatory proteins.

Jacob and Monod also isolated constitutive mutants that affected the operator site, *lacO*. **Table 14.1** summarizes the effects of mutations based on their locations in the *lacI* regulatory gene versus *lacO* and their analysis in merozygotes. As seen here, a loss-of-function mutation in a gene encoding a repressor protein has the same effect as a mutation in an operator site that cannot bind a repressor protein. In both cases, the genes of the *lac* operon are constitutively expressed. In a merozygote, however, the results are quite different. When a normal *lacI* gene and a normal *lac* operon are introduced into a cell harboring a defective *lacI* gene, the normal *lacI* gene can regulate both operons. In contrast, when a *lac* operon with a normal operator site is introduced into a cell with a defective operator site, the operon with the defective operator site continues to be expressed without lactose present. Overall, a mutation in a *trans*-acting factor can be complemented by the introduction of a second gene with normal function. However, a mutation in a *cis*-acting element is not affected by the introduction of another *cis*-acting element with normal function into the cell.

TABLE **14.1**

A Comparison of Loss-of-Function Mutations in the *lacI* Gene Versus the Operator Site

		Expression of the *lac* Operon (%)	
Chromosome	**F′ factor**	**With Lactose**	**Without Lactose**
Wild type	None	100	<1
lacI⁻	None	100	100
lacO⁻	None	100	100
lacI⁻	*lacI*⁺ and a normal *lac* operon	200	<1
lacO⁻	*lacI*⁺ and a normal *lac* operon	200	100

A self-help quiz involving this experiment can be found at www.mhhe.com/brookergenetics4e.

The *lac* Operon Is Also Regulated by an Activator Protein

The *lac* operon can be transcriptionally regulated in a second way, known as **catabolite repression** (as we shall see, a somewhat imprecise term). This form of transcriptional regulation is influenced by the presence of glucose, which is a catabolite—a substance that is broken down inside the cell. The presence of glucose ultimately leads to repression of the *lac* operon. When exposed to both glucose and lactose, *E. coli* cells first use glucose, and catabolite repression prevents the use of lactose. Why is this an advantage? The explanation is efficiency. The bacterium does not have to express all of the genes necessary for both glucose and lactose metabolism. If the glucose is used up, catabolite repression is alleviated, and the bacterium then expresses the *lac* operon. The sequential use of two sugars by a bacterium, known as diauxic growth, is a common phenomenon among many bacterial species. Glucose, a more commonly encountered sugar, is metabolized preferentially, and then a second sugar is metabolized after glucose is depleted from the environment.

Glucose, however, is not itself the small effector molecule that binds directly to a genetic regulatory protein. Instead, this form of regulation involves a small effector molecule, **cyclic AMP (cAMP),** which is produced from ATP via an enzyme known as adenylyl cyclase. When a bacterium is exposed to glucose, the transport of glucose into the cell stimulates a signaling pathway that causes the intracellular concentration of cAMP to decrease because the pathway inhibits adenylyl cyclase, the enzyme needed for cAMP synthesis. The effect of cAMP on the *lac* operon is mediated by an activator protein called the catabolite activator protein (CAP). CAP is composed of two subunits, each of which binds one molecule of cAMP.

Figure 14.8 considers how the interplay between the lac repressor and CAP determines whether the *lac* operon is expressed in the presence or absence of lactose and/or glucose. When only lactose is present, cAMP levels are high (Figure 14.8a). The cAMP binds to CAP, and then CAP binds to the CAP site and stimulates the ability of RNA polymerase to begin transcription. In the presence of lactose, the lac repressor is not bound to the operator site, so transcription can proceed at a high rate. In the absence of both lactose and glucose, cAMP levels are also high (Figure 14.8b). Under these conditions, however, the binding of the lac repressor inhibits transcription even though CAP is bound to the DNA. Therefore, the transcription rate is very low.

Figure 14.8c considers the situation in which both sugars are present. The presence of lactose causes the lac repressor to be inactive, which prevents it from binding to the operator site. Even so, the presence of glucose decreases cAMP levels so that cAMP is released from CAP, which prevents CAP from binding to the CAP site. Because CAP is not bound to the CAP site, the transcription of the *lac* operon is low in the presence of both sugars. Finally, Figure 14.8d illustrates what happens when only glucose is present. The transcription of the *lac* operon is very low because the lac repressor is bound to the operator site and CAP is not bound to the CAP site due to low cAMP levels.

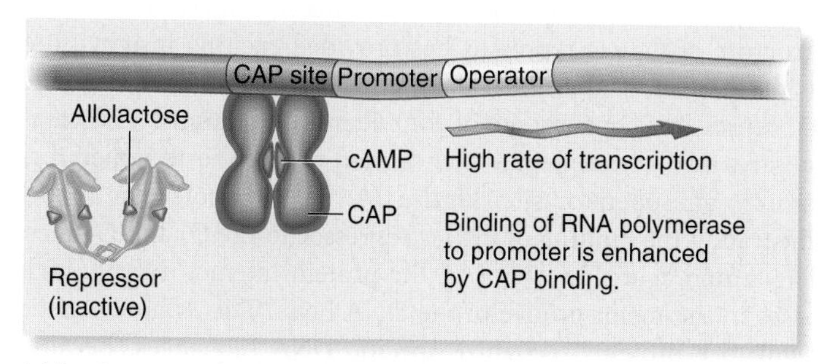

(a) Lactose, no glucose (high cAMP)

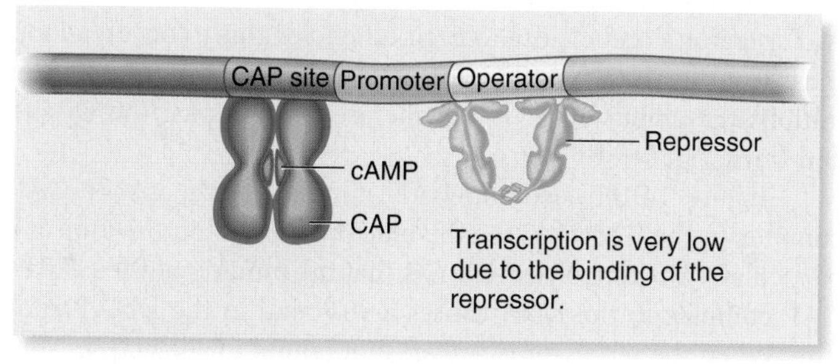

(b) No lactose or glucose (high cAMP)

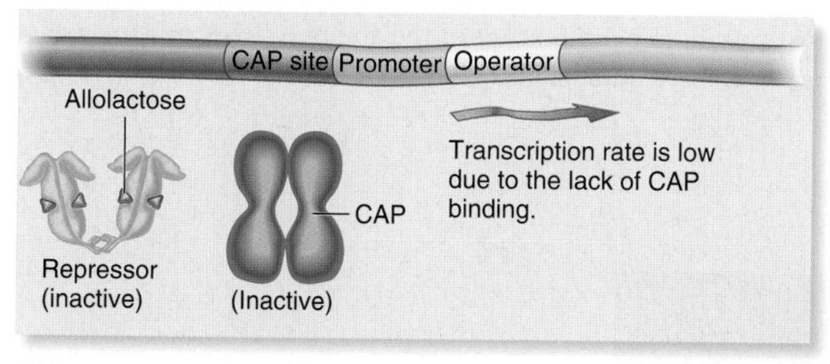

(c) Lactose and glucose (low cAMP)

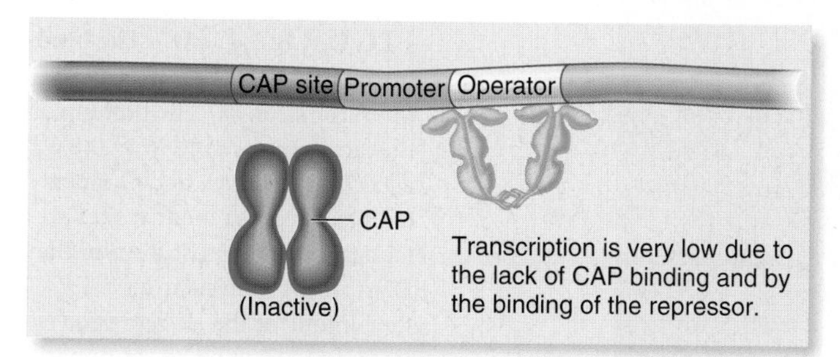

(d) Glucose, no lactose (low cAMP)

INTERACTIVE EXERCISE **FIGURE 14.8 The roles of the lac repressor and catabolite activator protein (CAP) in the regulation of the *lac* operon.** This figure illustrates how the *lac* operon is regulated depending on its exposure to lactose or glucose.

Genes → Traits The mechanism of catabolite repression provides the bacterium with the trait of being able to choose between two sugars. When exposed to both glucose and lactose, the bacterium chooses glucose first. After the glucose is used up, it then expresses the genes necessary for lactose metabolism. This trait allows the bacterium to more efficiently use sugars from its environment.

The effect of glucose, called catabolite repression, may seem like a puzzling way to describe this process because this regulation involves the action of an inducer (cAMP) and an activator protein (CAP), not a repressor. The term was coined before the action of the cAMP-CAP complex was understood. At that time, the primary observation was that glucose (a catabolite) inhibited (repressed) lactose metabolism.

Many bacterial promoters that transcribe genes involved in the breakdown of other sugars, such as maltose, arabinose, and melibiose, also have binding sites for CAP. Therefore, when glucose levels are high, these operons are inhibited. This promotes diauxic growth.

Further Studies Have Revealed That the *lac* Operon Has Three Operator Sites for the lac Repressor

Our traditional view of the regulation of the *lac* operon has been modified as we gain a greater understanding of the molecular process. In particular, detailed genetic and crystallographic studies have shown that the binding of the lac repressor is more complex than originally realized. The site in the *lac* operon commonly called the operator site was first identified by mutations that prevented lac repressor binding. These mutations, called $lacO^-$ or $lacO^C$ mutants, resulted in the constitutive expression of the *lac* operon even in strains that make a normal lac repressor protein. $LacO^C$ mutations were localized in the *lac* operator site, which is now known as O_1. This led to the view that a single operator site was bound by the lac repressor to inhibit transcription, as in Figure 14.4.

In the late 1970s and 1980s, two additional operator sites were identified. As shown at the top of **Figure 14.9**, these sites are called O_2 and O_3. O_1 is the operator site slightly downstream from the promoter. O_2 is located farther downstream in the *lacZ* coding sequence, and O_3 is located slightly upstream from the promoter. The O_2 and O_3 operator sites were initially called pseudo-operators, because substantial repression occurred in the absence of either one of them. However, studies by Benno Müller-Hill and his colleagues revealed a surprising result. As shown in Figure 14.9 (fourth example down), if both O_2 and O_3 are missing, repression is dramatically reduced even when O_1 is present. When O_1 is missing, even in the presence of one of the other operator sites, repression is nearly abolished.

How were these results interpreted? The data of Figure 14.9 supported a hypothesis that the lac repressor must bind to O_1 and either O_2 or O_3 to cause full repression. According to this view, the lac repressor can readily bind to O_1 and O_2, or to O_1 and O_3, but not to O_2 and O_3. If either O_2 or O_3 were missing, maximal repression is not achieved because it is less likely for the repressor to bind when only two operator sites are present. Look at Figure 14.9 and you will notice that the operator sites are a fair distance from each other. For this reason, it was proposed that the binding of the lac repressor to two operator sites requires the DNA to form a loop. A loop in the DNA would bring the operator sites closer together, thereby facilitating the binding of the repressor protein (**Figure 14.10a**).

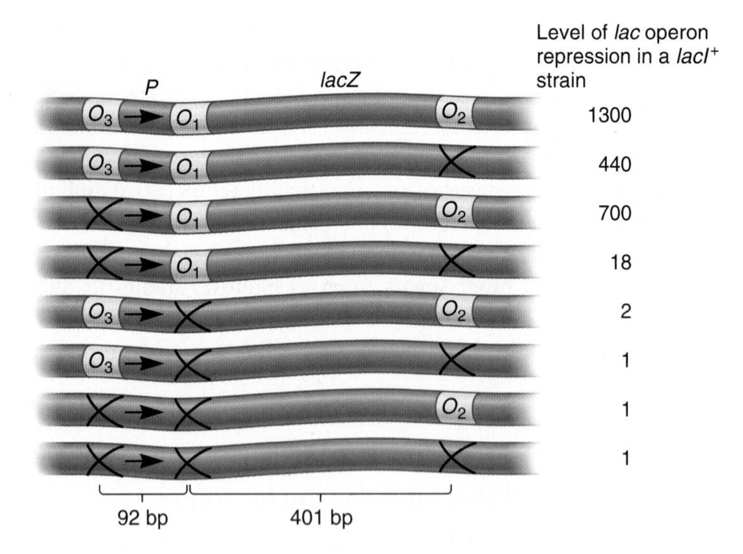

FIGURE 14.9 The identification of three *lac* operator sites. The top of this figure shows the locations of three *lac* operator sites, designated O_1, O_2, and O_3. O_1 is the *lac* operator site shown in previous figures. The arrows depict the starting site for transcription. Defective operator sites are indicated with an X. When all three operator sites are present, the repression of the *lac* operon is 1300-fold; this means there is 1/1300 the level of expression as compared to when lactose is present. This figure also shows the amount of repression when one or more operator sites are removed. A repression value of 1.0 indicates that no repression is occurring. In other words, a value of 1.0 indicates constitutive expression.

In 1996, the proposal that the lac repressor binds to two operator sites was confirmed by studies in which the lac repressor was crystallized by Mitchell Lewis and his colleagues. The crystal structure of the lac repressor has provided exciting insights into its mechanism of action. As mentioned earlier in this chapter, the lac repressor is a tetramer of four identical subunits. The crystal structure revealed that each dimer within the tetramer recognizes one operator site. **Figure 14.10b** is a molecular model illustrating the binding of the lac repressor to the O_1 and O_3 sites. The amino acid side chains in the protein interact directly with bases in the major groove of the DNA helix. This is how genetic regulatory proteins recognize specific DNA sequences. Because each dimer within the tetramer recognizes a single operator site, the association of two dimers to form a tetramer requires that the two operator sites be close to each other. For this to occur, a loop must form in the DNA. The formation of this loop dramatically inhibits the ability of RNA polymerase to slide past the O_1 site and transcribe the operon.

Figure 14.10b also shows the binding of the cAMP-CAP complex to the CAP site (see the blue protein within the loop). A particularly striking observation is that the binding of the cAMP-CAP complex to the DNA causes a 90° bend in the DNA structure. When the repressor is active—not bound to allolactose—the cAMP-CAP complex facilitates the binding of the lac repressor to the O_1 and O_3 sites. When the repressor is inactive, this bending also appears to be important in the ability of RNA polymerase to initiate transcription slightly downstream from the bend.

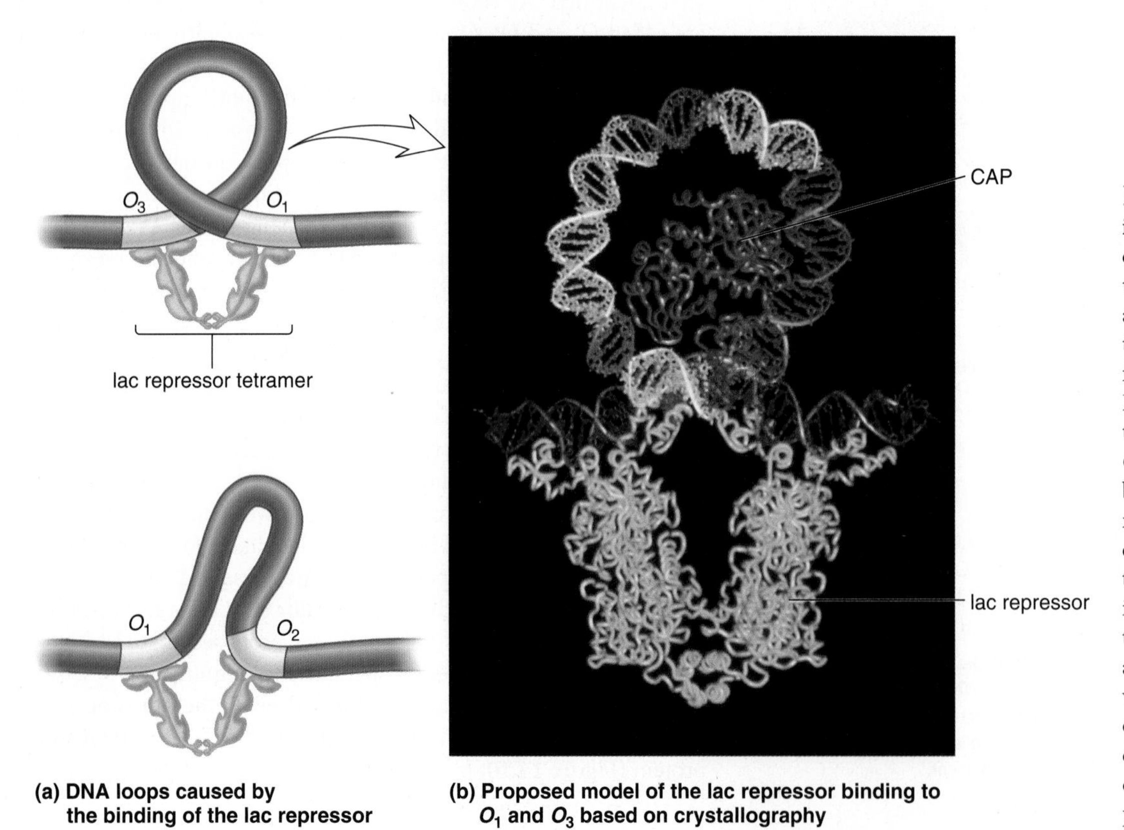

FIGURE 14.10 The binding of the lac repressor to two operator sites. **(a)** The binding of the lac repressor protein to the O_1 and O_3 or to the O_1 and O_2 operator sites. Because the two sites are far apart, a loop must form in the DNA. **(b)** A molecular model for the binding of the lac repressor to O_1 and O_3. Each repressor dimer binds to one operator site, so the repressor tetramer brings the two operator sites together. This causes the formation of a DNA loop in the intervening region. Note that the DNA loop contains the –35 and –10 regions (shown in green), which are recognized by σ factor of RNA polymerase. This loop also contains the binding site for the cAMP-CAP complex, which is the protein within the loop.

The *ara* Operon Can Be Regulated Positively or Negatively by the Same Regulatory Protein

Now that we have considered the regulation of the *lac* operon, let's compare its regulation to that of other genes in the bacterial chromosome. Another operon in *E. coli* involved in sugar metabolism is the *ara* (arabinose) operon. The sugar arabinose is a constituent of the cell walls of a few types of plants. As shown in **Figure 14.11**, the *ara* operon contains three structural genes, *araB*, *araA*, and *araD*, encoding a polycistronic mRNA for the three enzymes involved in arabinose metabolism. The actions of the three enzymes metabolize arabinose into D-xylulose-5-phosphate.

Like the *lac* operon, the *ara* operon contains a single promoter, designated P_{BAD}. The operon also contains a CAP site for the binding of the catabolite activator protein. The *araC* gene, which has its own promoter (P_C), is adjacent to the *ara* operon. *AraC* encodes a regulatory protein, called the AraC protein, that can bind to sites designated *araI*, $araO_1$, and $araO_2$.

As we have seen with the *lac* operon, some regulatory proteins such as the lac repressor inhibit transcription, whereas others such as CAP turn on transcription. AraC is a rather interesting protein because it can act as either a negative or positive regulator of transcription, depending on whether or not arabinose is present. How can a protein function in both ways? In the absence of arabinose, the AraC protein binds to the *araI*, $araO_1$, and $araO_2$ sites (**Figure 14.12a**). An AraC protein dimer is bound to $araO_1$, whereas monomers are bound at $araO_2$ and *araI*. The binding of the AraC dimer to the $araO_1$ site inhibits the transcription of the *araC* gene. In other words, the AraC protein is a repressor of the *araC* gene. This keeps AraC protein levels fairly low. The AraC proteins bound at $araO_2$ and *araI* repress the *ara* operon, but only in the absence of arabinose. As shown in Figure 14.12a, the AraC proteins at $araO_2$ and *araI* can bind to each other by causing a loop in the DNA, as originally proposed by Robert Schleif and his colleagues in 1990. This DNA loop prevents RNA polymerase from binding to the DNA and transcribing the *ara* operon via P_{BAD}. Therefore, in the absence of arabinose, the *ara* operon is turned off.

Figure 14.12b illustrates the activation of the *ara* operon in the presence of arabinose. When arabinose is bound to the AraC protein, the interaction between the AraC proteins at the $araO_2$ and *araI* sites is broken. This opens the DNA loop. In addition, a second AraC protein binds at the *araI* site. This AraC dimer at the *araI* operator site activates transcription by directly interacting with RNA polymerase. This activation can occur in conjunction with the activation of the *ara* operon by CAP and cAMP if glucose levels are low. When the *ara* operon is activated, the bacterial cell can efficiently metabolize arabinose.

The *trp* Operon Is Regulated by a Repressor Protein and Also by Attenuation

The *trp* operon (pronounced "trip") encodes enzymes that are needed for the biosynthesis of the amino acid tryptophan. The *trpE*, *trpD*, *trpC*, *trpB*, and *trpA* genes encode enzymes involved in tryptophan biosynthesis. The *trpR* and *trpL* genes are involved in regulating the *trp* operon in two different ways. The *trpR* gene encodes the **trp repressor** protein. When tryptophan levels within the cell are very low, the trp repressor cannot bind to the operator site. Under these conditions, RNA polymerase transcribes the *trp* operon (**Figure 14.13a**). In this way, the cell expresses the genes required for the synthesis of tryptophan. When the tryptophan levels within the cell become high, tryptophan acts as a corepressor that binds to the trp repressor protein. This causes a conformational change in the trp repressor that allows it to bind to the *trp* operator site (**Figure 14.13b**). This inhibits the ability of RNA polymerase to transcribe the operon. Therefore, when a high level of tryptophan is present within the cell—when the cell does not need to make more tryptophan—the *trp* operon is turned off.

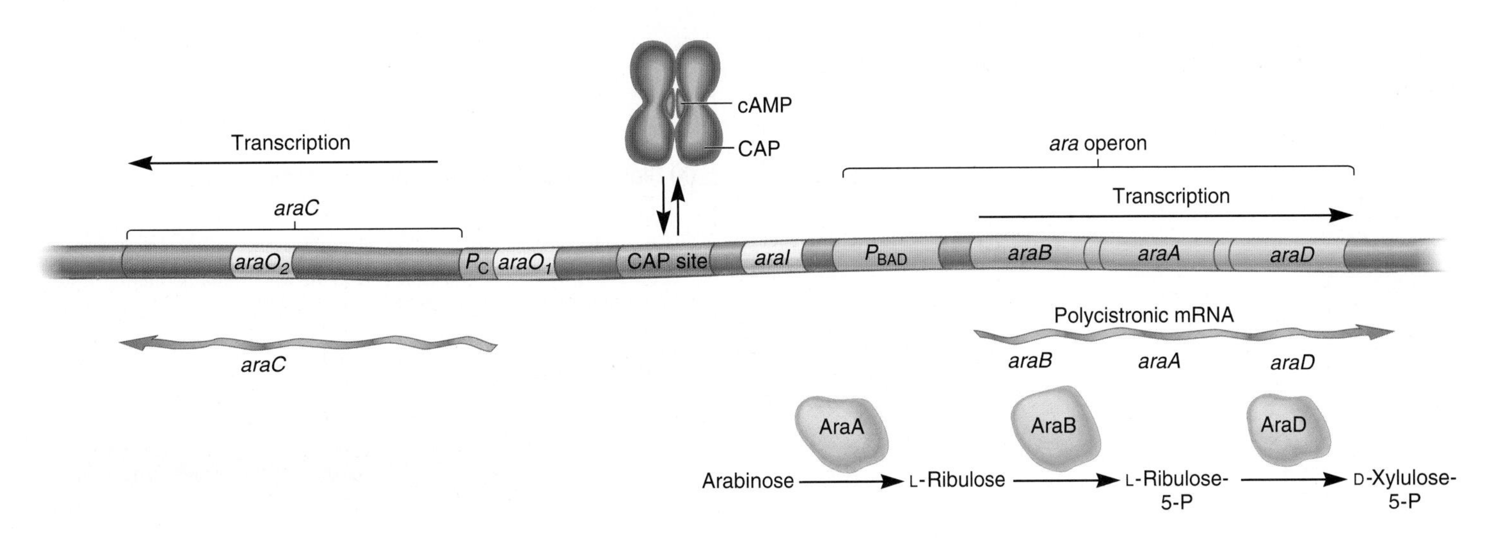

FIGURE 14.11 **Organization of the *ara* operon and other genes involved in arabinose metabolism.** *AraC* encodes a genetic regulatory protein called the AraC protein. This protein can bind to three different regulatory sites, called *araI*, $araO_1$, and $araO_2$. P_C is the promoter for the *araC* gene; P_{BAD} is the promoter for the *ara* operon, which contains three genes (*araB*, *araA*, and *araD*) encoding proteins involved in arabinose metabolism.

In the absence of arabinose, an AraC protein binds to the *araO*$_2$ operator site and another to the *araI* site. These two AraC proteins interact to promote a loop in the DNA. This loop prevents RNA polymerase from transcribing the *ara* operon from the P_{BAD} promoter.

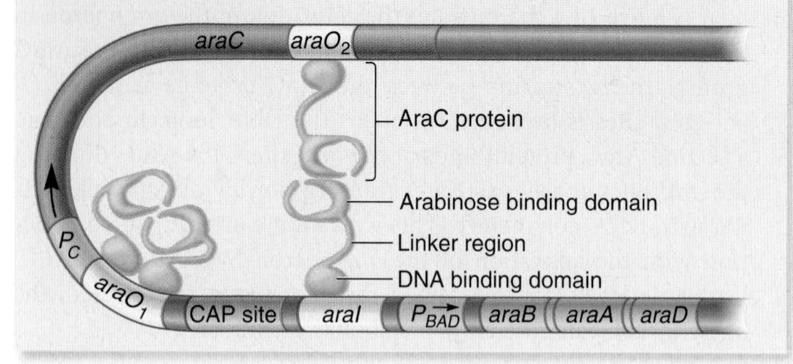

(a) Operon inhibited in the absence of arabinose

In the presence of arabinose, arabinose binds to the AraC proteins and causes a conformational change that breaks the interaction between the AraC proteins bound at *araO*$_2$ and *araI*. This causes the DNA loop to be broken. Under these conditions, an additional AraC protein binds to *araI*, and RNA polymerase can transcribe the *ara* operon.

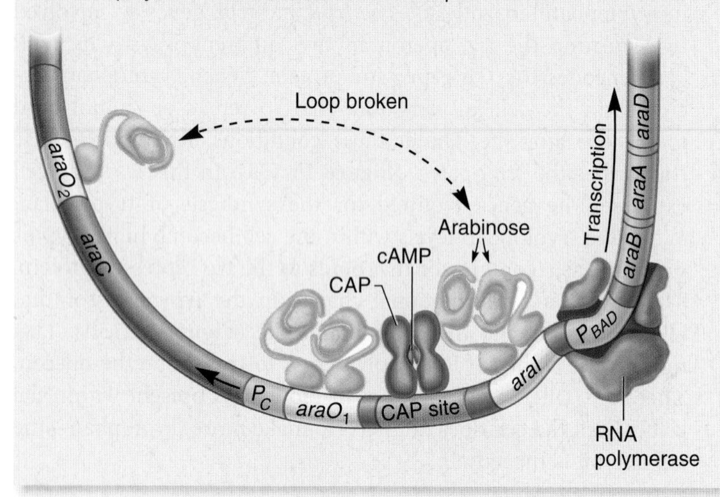

(b) Operon activated in the presence of arabinose

FIGURE 14.12 **DNA looping and unlooping via the AraC protein.**

When tryptophan levels are low, tryptophan does not bind to the trp repressor protein, which prevents the repressor protein from binding to the operator site. Under these conditions, RNA polymerase can transcribe the operon, which leads to the expression of the *trpE*, *trpD*, *trpC*, *trpB*, and *trpA* genes. These genes encode enzymes involved in tryptophan biosynthesis.

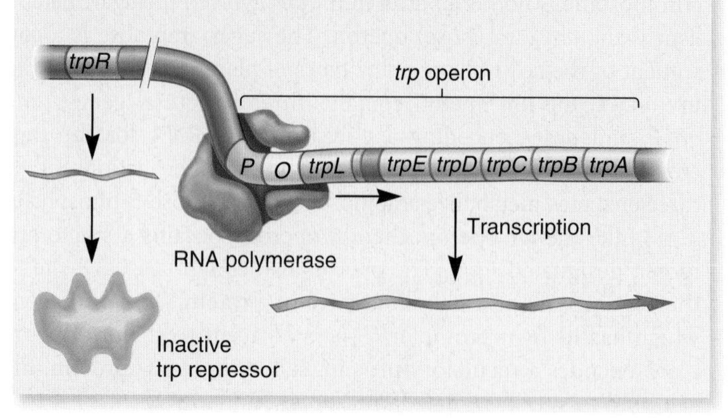

(a) Low tryptophan levels, transcription of the entire *trp* operon occurs

When tryptophan levels are high, tryptophan acts as a corepressor that binds to the trp repressor protein. The tryptophan-trp repressor complex then binds to the operator site to inhibit transcription.

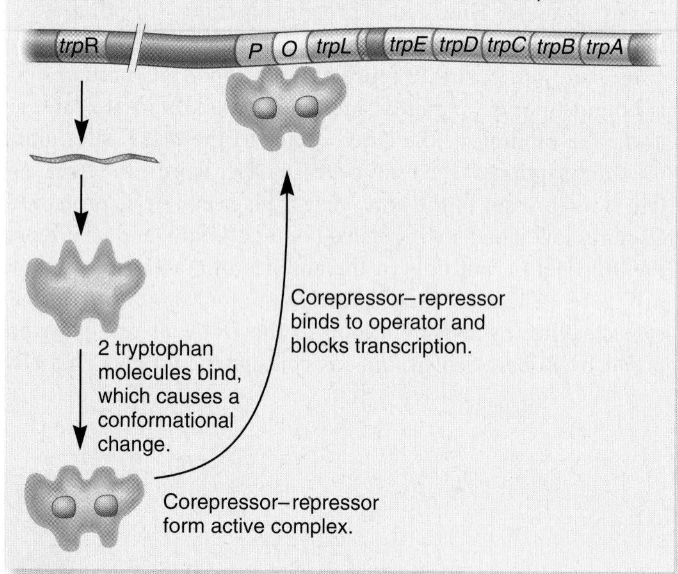

(b) High tryptophan levels, repression occurs

Another mechanism of regulation is attenuation. When attenuation occurs, the RNA is transcribed only to the attenuator sequence, and then transcription is terminated.

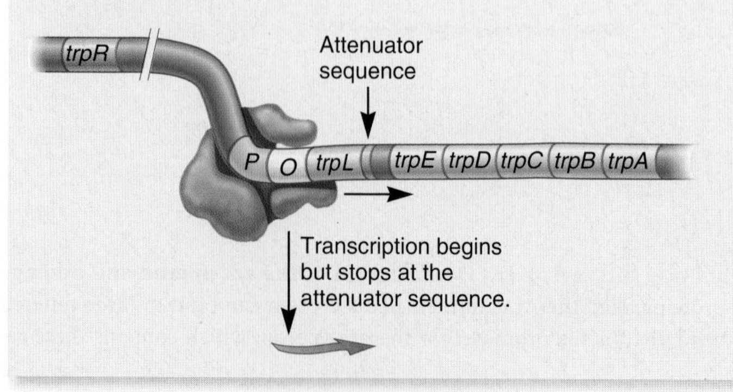

(c) High tryptophan levels, attenuation occurs

FIGURE 14.13 **Organization of the *trp* operon and regulation via the trp repressor protein and attenuation.**

In the 1970s, after the action of the trp repressor was elucidated, Charles Yanofsky and coworkers made a few unexpected observations. Mutant strains were found that lacked the trp repressor protein. Surprisingly, these mutant strains still could inhibit the expression of the *trp* operon in the presence of tryptophan. In addition, *trp* operon mutations were identified in which a region including the *trpL* gene was missing from the operon. These mutations resulted in higher levels of expression of the other genes in the *trp* operon. As is often the case, unusual observations can lead scientists into productive avenues of study. By pursuing this research further, Yanofsky discovered a second regulatory mechanism in the *trp* operon, called **attenuation,** that is mediated by the region that includes the *trpL* gene (**Figure 14.13c**).

Attenuation can occur in bacteria because the processes of transcription and translation are coupled. As described in Chapter 13, bacterial ribosomes quickly attach to the 5′ end of mRNA soon after its synthesis begins via RNA polymerase. During attenuation, transcription actually begins, but it is terminated before the entire mRNA is made. A segment of DNA, termed the **attenuator sequence,** is important in facilitating this termination. When attenuation occurs, the mRNA from the *trp* operon is made as a short piece that terminates shortly past the *trpL* gene (see Figure 14.13c). Because this short mRNA has been terminated before RNA polymerase has transcribed the *trpE*, *trpD*, *trpC*, *trpB*, and *trpA* genes, it will not encode the proteins required for tryptophan biosynthesis. In this way, attenuation inhibits the further production of tryptophan in the cell.

The segment of the *trp* operon immediately downstream from the operator site plays a critical role during attenuation. The first gene in the *trp* operon is the *trpL* gene, which encodes a peptide containing 14 amino acids called the leader peptide. As shown in **Figure 14.14**, two features are key in the attenuation mechanism. First, two tryptophan (Trp) codons are found within the mRNA that encodes the *trp* leader peptide. What is the role of these codons? As we will see later, these two codons provide a way to sense whether or not the bacterium has sufficient tryptophan to synthesize its proteins. Second, the mRNA can form stem-loop structures. The type of stem-loop structure that forms underlies attenuation.

Different combinations of stem-loop structures are possible due to interactions among four regions within the RNA transcript (see the color key in Figure 14.14). Region 2 is complementary to region 1 and also to region 3. Region 3 is complementary to region 2 as well as to region 4. Therefore, three stem-loop structures are possible: 1–2, 2–3, and 3–4. Even so, keep in mind that a particular segment of RNA can participate in the formation of only one stem-loop structure. For example, if region 2 forms a stem-loop with region 1, it cannot (at the same time) form a stem-loop with region 3. Alternatively, if region 2 forms a stem-loop with region 3, then region 3 cannot form a stem-loop with region 4. Though three stem-loop structures are possible, the 3–4 stem-loop structure is functionally unique. The 3–4 stem-loop together with the U-rich attenuator sequence acts as an intrinsic terminator—a ρ-independent terminator, as described in Chapter 12. Therefore, the formation of the 3–4 stem-loop causes RNA polymerase to

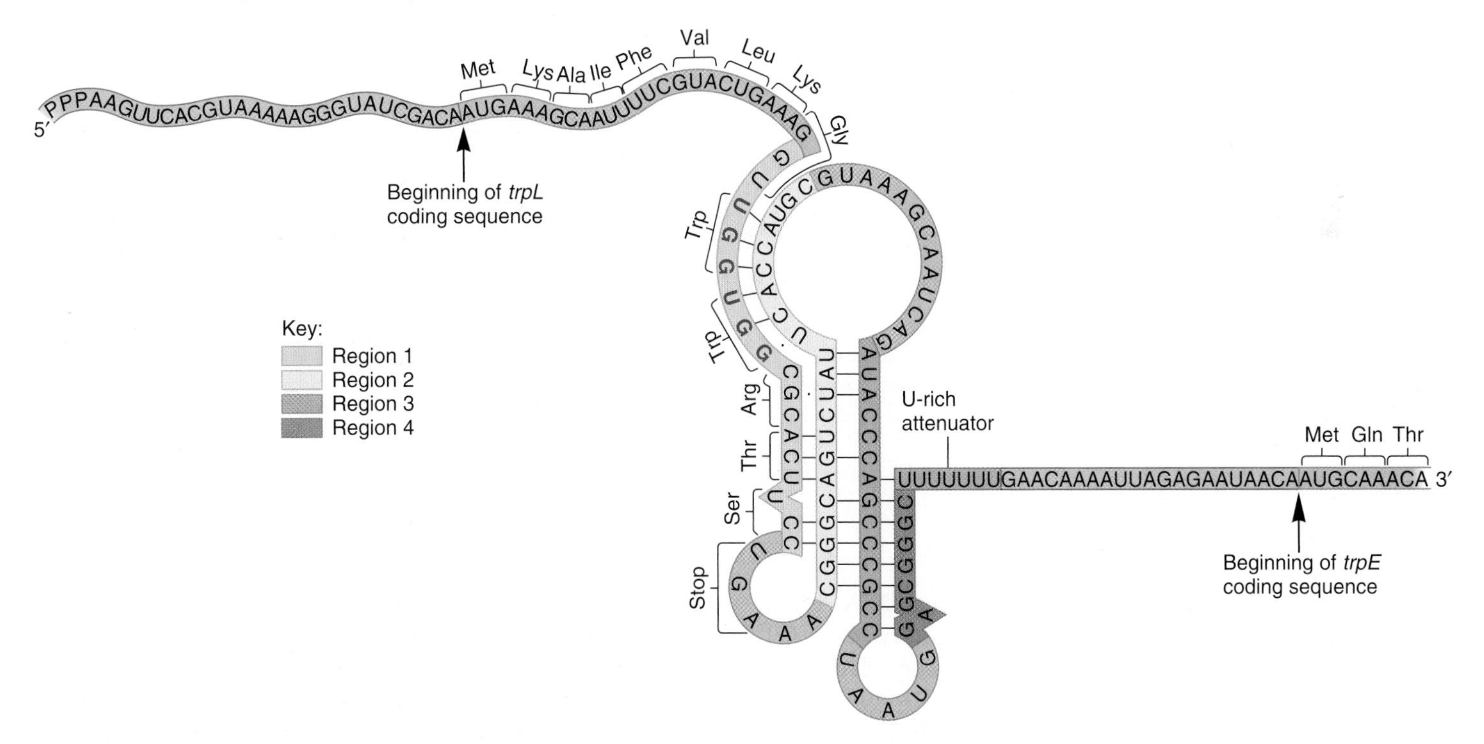

FIGURE 14.14 Sequence of the *trpL* mRNA produced during attenuation. A second method of regulation of the *trp* operon is attenuation, which occurs when tryptophan levels are sufficient for protein synthesis. A short mRNA is made that includes the *trpL* region. As shown here, this mRNA has several regions that are complementary to each other. The possible hydrogen bonding between regions 1 and 2, 2 and 3, and 3 and 4 is also shown. The last U in the purple attenuator sequence is the last nucleotide that would be transcribed during attenuation. At very low tryptophan concentrations, however, transcription occurs beyond the end of *trpL* and proceeds through the *trpE* gene and the rest of the *trp* operon.

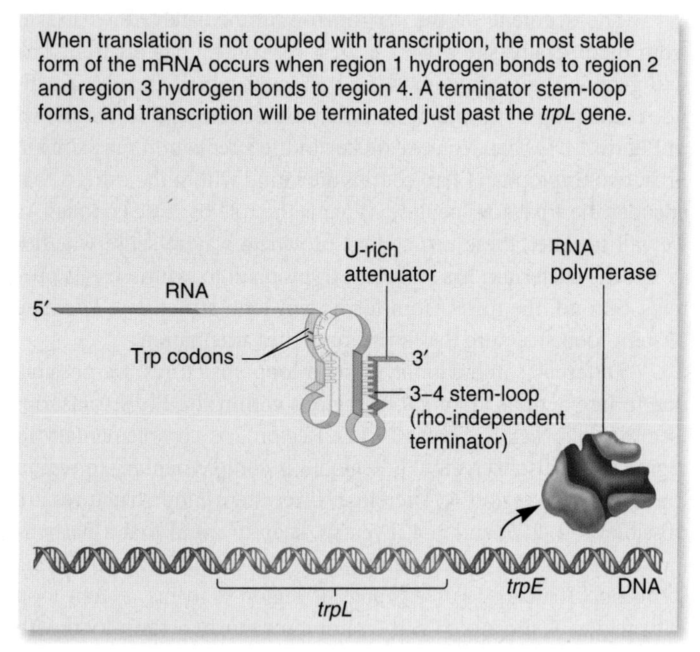

(a) No translation

Coupled transcription and translation occur under conditions in which the tryptophan concentration is very low. The ribosome pauses at the Trp codons in the *trpL* gene because insufficient amounts of charged tRNA^{Trp} are present. This pause blocks region 1 of the mRNA, so region 2 can hydrogen bond only with region 3. When this happens, the 3–4 stem-loop structure cannot form. Transcriptional termination does not occur, and RNA polymerase transcribes the rest of the operon.

Leader peptide
Trp codon
Ribosome stalls
5′
1
2
3
4
mRNA for *trpE*
trpL
DNA
trpE

(b) Low tryptophan levels, 2–3 stem-loop forms

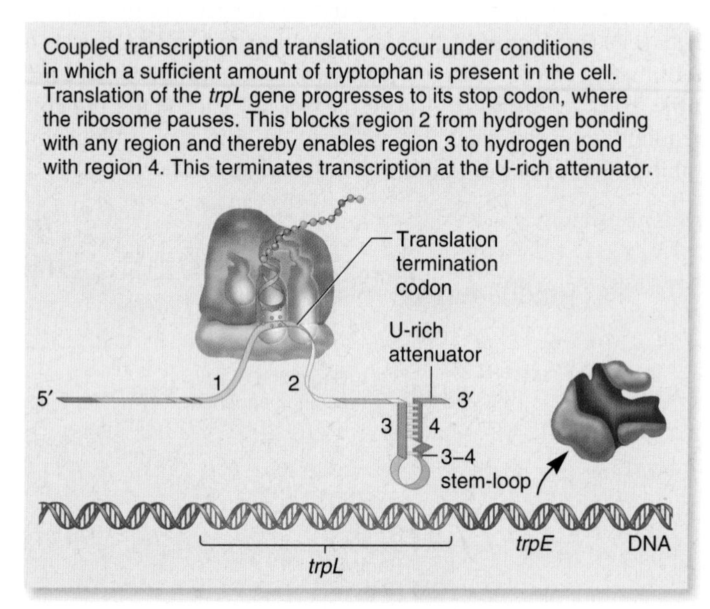

(c) High tryptophan levels, 3–4 stem-loop forms

FIGURE 14.15 Possible stem-loop structures formed from *trpL mRNA* under different conditions of translation. Attenuation occurs in parts (a) and (c) due to the formation of a 3-4 stem-loop.

pause, and the U-rich sequence dissociates from the DNA. This terminates transcription at the U-rich attenuator. In comparison, if region 3 forms a stem-loop with region 2, transcription will not be terminated because a 3–4 stem-loop cannot form.

Conditions that favor the formation of the 3–4 stem-loop ultimately rely on the translation of the *trpL* gene. As shown in **Figure 14.15**, three scenarios are possible. In Figure 14.15a, translation is not coupled with transcription. Region 1 rapidly hydrogen bonds to region 2, and region 3 is left to hydrogen bond to region 4. Therefore, the terminator stem-loop forms, and transcription is terminated just past the *trpL* gene at the U-rich attenuator.

In Figure 14.15b, coupled transcription and translation occur under conditions in which the tryptophan concentration is low. When tryptophan levels are low, the cell cannot make a sufficient amount of charged tRNA^{Trp}. As we see in Figure 14.15b, the ribosome pauses at the Trp codons in the *trpL* mRNA because it is waiting for charged tRNA^{Trp}. This pause occurs in such a way that the ribosome shields region 1 of the mRNA. This sterically prevents region 1 from hydrogen bonding to region 2. As an alternative, region 2 hydrogen bonds to region 3. Therefore, because region 3 is already hydrogen bonded to region 2, the 3–4 stem-loop structure cannot form. Under these conditions, transcriptional termination does not occur, and RNA polymerase transcribes the rest of the operon. This ultimately enables the bacterium to make more tryptophan.

Finally, in Figure 14.15c, coupled transcription and translation occur under conditions in which a sufficient amount of tryptophan is present in the cell. In this case, translation of the *trpL* mRNA progresses to its stop codon, where the ribosome pauses. The pausing at the stop codon prevents region 2 from hydrogen bonding with any region and thereby enables region 3 to hydrogen bond with region 4. As in Figure 14.15a, this terminates transcription. Of course, keep in mind that the *trpL* gene contains two tryptophan codons. For the ribosome to smoothly progress to the *trpL* stop codon, enough charged $tRNA^{Trp}$ must be available to translate this mRNA. It follows that the bacterium must have a sufficient amount of tryptophan. Under these conditions, the rest of the transcription of the operon is terminated.

Attenuation is a mechanism to regulate transcription that is found in several other operons involved with amino acid biosynthesis. In all cases, the mRNAs that encode the leader peptides are rich in codons for the particular amino acid that is synthesized by the enzymes encoded by the particular operon. For example, the mRNA that encodes the leader peptide of the histidine operon has seven histidine codons in its sequence, and the mRNA for the leader peptide of the leucine operon has four leucine codons. Like the *trp* operon, these other operons have alternative stem-loop structures, one of which is a transcriptional terminator.

Inducible Operons Encode Catabolic Enzymes, and Repressible Operons Usually Encode Anabolic Enzymes

Thus far, we have seen that bacterial genes can be transcriptionally regulated in a positive or negative way—and sometimes both. The *lac* operon and *ara* operon are inducible systems regulated by sugar molecules that activate transcription of these operons. By comparison, the *trp* operon is a repressible operon regulated by tryptophan, a corepressor that binds to the repressor and turns the operon off. In addition, an abundance of charged $tRNA^{Trp}$ in the cytoplasm can turn the *trp* operon off via attenuation.

By studying the genetic regulation of many operons, geneticists have discovered a general trend concerning inducible versus repressible regulation. When the genes in an operon encode proteins that function in the catabolism or breakdown of a substance, they are usually regulated in an inducible manner. The substance to be broken down or a related compound often acts as the inducer. For example, allolactose and arabinose act as inducers of the *lac* and *ara* operons, respectively. An inducible form of regulation allows the bacterium to express the appropriate genes only when they are needed to catabolize these sugars.

In contrast, other enzymes are important for the anabolism or synthesis of small molecules. The genes that encode these anabolic enzymes tend to be regulated by a repressible mechanism. The corepressor or inhibitor is commonly the small molecule that is the product of the enzymes' biosynthetic activities. For example, tryptophan is produced by the sequential action of several enzymes that are encoded by the *trp* operon. Tryptophan itself acts as a corepressor that can bind to the trp repressor protein when the intracellular levels of tryptophan become relatively high. This mechanism turns off the genes required for tryptophan biosynthesis when enough of this amino acid has been made. Therefore, genetic regulation via repression provides the bacterium with a way to prevent the overproduction of the product of a biosynthetic pathway.

14.2 TRANSLATIONAL AND POSTTRANSLATIONAL REGULATION

Though genetic regulation in bacteria occurs predominantly at transcription, many examples are known in which regulation is exerted at a later stage in gene expression. In some cases, specialized mechanisms have evolved to regulate the translation of certain mRNAs. Recall that the translation of mRNA occurs in three stages: initiation, elongation, and termination. Genetic regulation of translation is usually aimed at preventing the initiation step. In addition, as described later in Section 14.3, translation can be regulated by riboswitches.

The net result of translation is the synthesis of a protein. The activities of proteins within living cells and organisms ultimately determine an individual's traits. Therefore, to fully understand how proteins influence an organism's traits, researchers have investigated how the functions of proteins are regulated. The term **posttranslational** regulation refers to the functional control of proteins that are already present in the cell rather than regulation of transcription or translation. Posttranslational control can either activate or inhibit the function of a protein. Compared with transcriptional or translational regulation, posttranslational control can be relatively fast, occurring in a matter of seconds, which is an important advantage. In contrast, transcriptional and translational regulation typically require several minutes or even hours to take effect because these two mechanisms involve the synthesis and turnover of mRNA and polypeptides. In this section, we will examine some of the ways that bacteria can regulate the initiation of translation, as well as ways that protein function can be regulated posttranslationally.

Repressor Proteins and Antisense RNA Can Inhibit Translation

For some bacterial genes, the translation of mRNA is regulated by the binding of proteins or other RNA molecules that influence the ability of ribosomes to translate the mRNA into a polypeptide. A **translational regulatory protein** recognizes sequences within the mRNA, much as transcription factors recognize DNA sequences. In most cases, translational regulatory proteins act to inhibit translation. These are known as **translational repressors.** When a translational repressor protein binds to the mRNA, it can inhibit translational initiation in one of two ways. One possibility is that it can bind in the vicinity of the Shine-Dalgarno sequence and/or the start codon and thereby sterically block the ribosome's ability to initiate translation in this region. Alternatively, the repressor protein may bind outside the

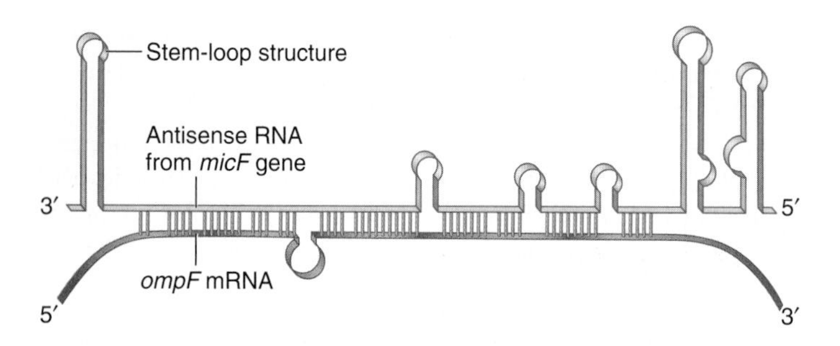

FIGURE 14.16 The double-stranded RNA structure formed between the *micF* antisense RNA and the *ompF* mRNA. Because they have regions that are complementary to each other, the *micF* antisense RNA binds to the *ompF* mRNA to form a double-stranded structure in which the *ompF* mRNA cannot be translated.

Shine-Dalgarno/start codon region but stabilize the mRNA secondary structure that prevents initiation. Translational repression is also a form of genetic regulation found in eukaryotic species, and we will consider specific examples in Chapter 15.

A second way to regulate translation is via the synthesis of **antisense RNA,** an RNA strand that is complementary to a strand of mRNA. (The mRNA strand has the same sequence as the DNA sense strand.) To understand this form of genetic regulation, let's consider a trait known as osmoregulation, which is essential for the survival of most bacteria. Osmoregulation refers to the ability to control the amount of water inside the cell. Because the solute concentrations in the external environment may rapidly change between hypotonic and hypertonic conditions, bacteria must have an osmoregulation mechanism to maintain their internal cell volume. Otherwise, bacteria would be susceptible to the harmful effects of lysis or shrinking.

In *E. coli,* an outer membrane protein encoded by the *ompF* gene is important in osmoregulation. At low osmolarity, the ompF protein is preferentially produced, whereas at high osmolarity, its synthesis is decreased. The expression of another gene, known as *micF,* is responsible for inhibiting the expression of the *ompF* gene at high osmolarity. As shown in **Figure 14.16**, the inhibition occurs because the *micF* RNA is complementary to the *ompF* mRNA; it is an antisense strand of RNA. When the *micF* gene is transcribed, its RNA product binds to the *ompF* mRNA via hydrogen bonding between their complementary regions. The binding of the *micF* RNA to the *ompF* mRNA prevents the *ompF* mRNA from being translated. The RNA transcribed from the *micF* gene is called antisense RNA because it is complementary to the *ompF* mRNA, which is a sense strand of mRNA that encodes a protein. The *micF* RNA does not encode a protein.

Posttranslational Regulation Can Occur Via Feedback Inhibition and Covalent Modifications

Let's now turn our attention to ways that protein function is regulated posttranslationally. A common mechanism to regulate the activity of metabolic enzymes is **feedback inhibition.** The synthesis of many cellular molecules such as amino acids, vitamins, and nucleotides occurs via the action of a series of enzymes that convert precursor molecules to a particular product. The final product in a metabolic pathway then inhibits an enzyme that acts early in the pathway.

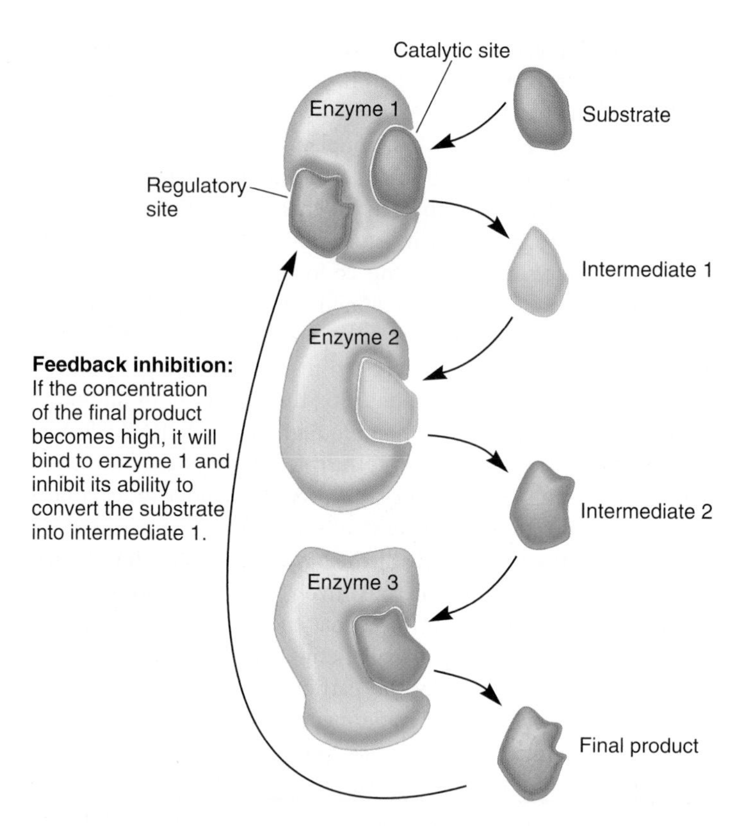

FIGURE 14.17 Feedback inhibition in a metabolic pathway. The substrate is converted to a product by the sequential action of three different enzymes. Enzyme 1 has a catalytic site that recognizes the substrate, and it also has a regulatory site that recognizes the final product. When the final product binds to the regulatory site, it inhibits enzyme 1.

Figure 14.17 depicts feedback inhibition in a metabolic pathway. Enzyme 1 is an example of an **allosteric enzyme,** an enzyme that contains two different binding sites. (The lac repressor is an allosteric protein, but not an enzyme.) The catalytic site is responsible for the binding of the substrate and its conversion to intermediate 1. The second site is a regulatory or allosteric site. This site binds the final product of the metabolic pathway. When bound to the regulatory site, the final product inhibits the catalytic ability of enzyme 1.

To appreciate feedback inhibition at the cellular level, we can consider the relationship between the product concentration and the regulatory site on enzyme 1. As the final product is made within the cell, its concentration gradually increases. Once the final product concentration has reached a level that is similar to its affinity for enzyme 1, the final product is likely to bind to the regulatory site on enzyme 1 and inhibit its function. In this way, the net result is that the final product of a metabolic pathway inhibits the further synthesis of more product. Under these conditions, the concentration of the final product has reached a level sufficient for the purpose of the bacterium.

A second strategy to control the function of proteins is by the covalent modification of their structure, a process called

posttranslational covalent modification (look ahead to Figure 21.4). Certain types of modifications are involved primarily in the assembly and construction of a functional protein. These alterations include proteolytic processing; disulfide bond formation; and the attachment of prosthetic groups, sugars, or lipids. These are typically irreversible changes required to produce a functional protein. In contrast, other types of modifications, such as phosphorylation ($—PO_4$), acetylation ($—COCH_3$), and methylation ($—CH_3$), are often reversible modifications that transiently affect the function of a protein.

14.3 RIBOSWITCHES

In 2001 and 2002, researchers in a few different laboratories discovered a mechanism of gene regulation called a **riboswitch.** In this form of regulation, an RNA molecule can exist in two different secondary conformations. The conversion from one conformation to the other is due to the binding of a small molecule. As described in **Table 14.2**, a riboswitch can regulate transcription, translation, RNA stability, and splicing.

TABLE **14.2**
Types of Riboswitches

Type of Regulation	Description
Transcription	The 5′ region of an mRNA may exist in one conformation that forms a ρ-independent terminator, which causes attenuation of transcription. The other conformation does not form a terminator and is completely transcribed.
Translation	The 5′ region of an mRNA may in exist in one conformation in which the Shine-Dalgarno sequence cannot be recognized by the ribosome, whereas the other conformation has an accessible Shine-Dalgarno sequence that allows the mRNA to be translated.
RNA stability	One mRNA conformation may be stable, whereas the other conformation acts as ribozyme that causes self-degradation.
Splicing	In eukaryotes, one pre-mRNA conformation may be spliced in one way, whereas another conformation is spliced in a different way.

Riboswitches are widespread in bacteria. Researchers estimate that 3 to 5% of all bacterial genes may be regulated by riboswitches. In bacteria, they are involved in regulating genes associated with the biosynthesis of purines, amino acids, vitamins, and other essential molecules. Riboswitches are also found in archaea, algae, fungi, and plants. In this section, we will examine two examples of riboswitches in bacteria. The first example shows how riboswitches can regulate transcription and the second example involves translational regulation.

A Riboswitch Can Regulate Transcription

Thiamin, also called vitamin B_1, is an important organic molecule for bacteria, archaea, and eukaryotes. The active form of this vitamin is thiamin pyrophosphate (TPP). TPP is an essential coenzyme for the functioning of a variety of enzymes such as certain enzymes in the citric acid cycle. In bacteria, TPP is made via biosynthetic enzymes that are encoded in the bacterial genome. For example, in *Bacillus subtilis*, the majority of genes involved in TPP synthesis are found within the *thi* operon that contains seven genes.

Certain genes that encode TPP biosynthetic enzymes, such as those of the *thi* operon in *B. subtilis*, are regulated by a riboswitch that controls transcription (**Figure 14.18**). As the polycistronic mRNA for the *thi* operon is being made, the 5′ end quickly folds into a secondary structure. When TPP levels are low, the secondary structure has a stem-loop called an antiterminator, which prevents the formation of the terminator stem-loop. Therefore, under these conditions, transcription of the entire *thi* operon occurs. In this way, the bacterium is able to make more TPP, which is in short supply. By comparison, when TPP levels are high, TPP binds to the RNA and causes a change in its secondary structure. As shown on the right side of Figure 14.18, a ρ-independent terminator forms instead of the antiterminator stem-loop. The ρ-independent terminator abruptly stops transcription, thereby inhibiting the production of the enzymes that are needed to make more TPP. Similar to the *trp* operon discussed earlier in this chapter, this is an example of attenuation.

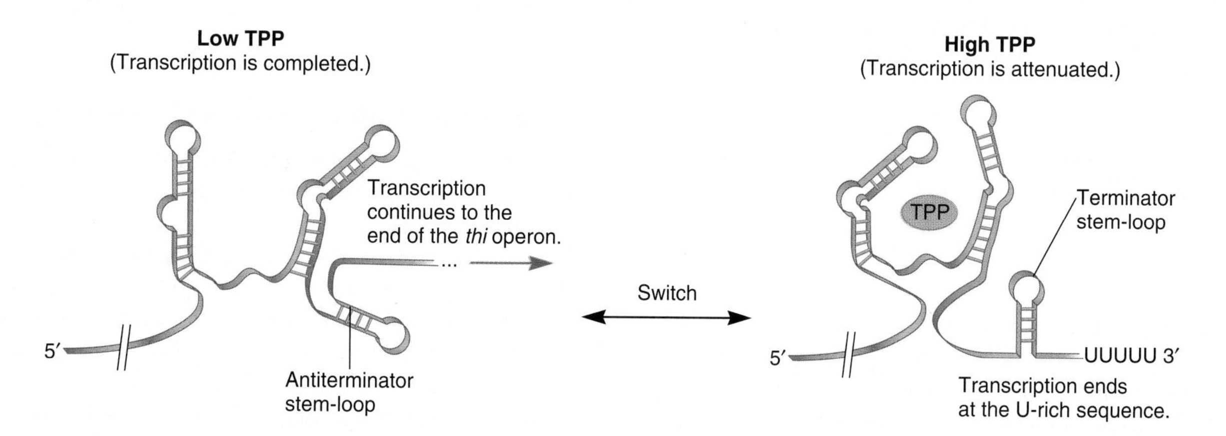

FIGURE 14.18 **A TPP riboswitch in *B. subtilis* that regulates transcription.**

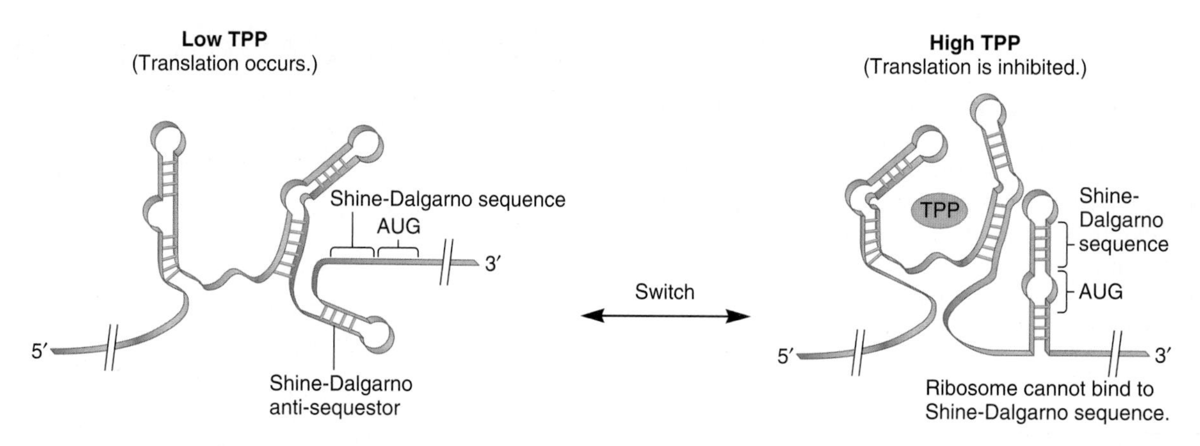

FIGURE 14.19 A TPP riboswitch in *E. coli* that regulates translation. Note: In both cases, the mRNA for the *thiMD* operon is completely transcribed.

A Riboswitch Can Regulate Translation

Let's now consider an example in which a riboswitch regulates translation. In gram-positive bacteria, such as *B. subtilis*, the regulation of TPP biosynthetic enzymes occurs via a riboswitch that controls transcription. However, in gram-negative bacteria, such as *Escherichia coli*, a structurally similar riboswitch regulates translation.

Figure 14.19 shows how a riboswitch can regulate translation. In *E. coli*, the *thiMD* operon encodes two enzymes involved with TPP biosynthesis. When TPP levels are low, the 5′ end of the mRNA folds into a structure that contains a stem-loop called the Shine-Dalgarno anti-sequestor. When this stem-loop forms, the Shine-Dalgarno sequence is accessible to ribosome binding. Therefore, the mRNA is translated when TPP is in short supply. By comparison, when TPP levels are high, TPP binds to the RNA and causes a change in its secondary structure. As shown on the right side of Figure 14.19, a stem-loop forms that contains the Shine-Dalgarno sequence and the start codon. The formation of this stem-loop sequesters the Shine-Dalgarno sequence, thereby preventing ribosomal binding. This blocks the translation of the enzymes that are needed to make more TPP.

When comparing Figures 14.18 and 14.19, it is worth noting that the 5′ end of the *thiMD* mRNA of *E. coli* can exist in two different secondary structures that greatly resemble those that occur at the 5′ end of the *thi* operon of *B. subtilis*. However, the effects on regulation are quite different. The TPP riboswitch in *E. coli* controls translation, whereas the TPP riboswitch in *B. subtilis* controls transcription.

14.4 GENE REGULATION IN THE BACTERIOPHAGE REPRODUCTIVE CYCLE

Viruses are small particles that contain genetic material surrounded by a protein coat. They can infect a living cell and then propagate themselves by using the energy and metabolic machinery of the host cell. For this to occur, the genetic material of viruses orchestrates an intricate series of steps, involving the expression of many viral genes. During the past several decades, the reproduction of viruses has presented an interesting and challenging problem for geneticists to investigate. The study of bacteriophages, which are viruses that infect bacteria, has greatly advanced our basic knowledge of how genetic regulatory proteins work. In addition, the study of viruses has been instrumental in our ability to devise medical strategies aimed at combating viral diseases. For example, our knowledge of the reproductive cycles of human viruses has led to the development of drugs that inhibit viral growth. Azidothymidine (AZT), which is used to combat human immunodeficiency virus (HIV), suppresses the production of viral DNA by inhibiting a viral gene product called reverse transcriptase that is involved in viral DNA synthesis.

In this section, we will focus on the function of bacteriophage genes that encode genetic regulatory proteins. The structural genes of bacteriophages are often arranged in operons. This enables all of the genes within an operon to be controlled by regulatory proteins that bind to operator sites and influence the function of nearby promoters. Like bacterial operons, phage operons can be controlled by repressor proteins or activator proteins. To understand how this works, we will carefully examine the two reproductive cycles—lytic and lysogenic—of a virus called **phage λ** (lambda), which was discovered by Esther Lederberg in 1951. Since its discovery, phage λ has been investigated extensively and has provided geneticists with a model on which to base our understanding of viral proliferation.

Phage λ Can Follow a Lytic or Lysogenic Cycle

Phage λ can bind to the surface of a bacterium and inject its genetic material into the bacterial cytoplasm. After this occurs, the phage proceeds along only one of two alternative cycles, known as the **lytic cycle** and the **lysogenic cycle.** This topic is also discussed in Chapter 7 (see Figure 7.10). During the lytic cycle, the genetic instructions of the bacteriophage direct the synthesis of many copies of the phage genetic material and coat proteins that are then assembled to make new phages. When synthesis and assembly are completed, the bacterial host cell is lysed, and the newly made phages are released into the environment.

Alternatively, in the lysogenic cycle, the phage can act as a **temperate phage,** which usually does not produce new phages and does not kill the bacterial cell that acts as its host. During the lysogenic cycle, phage λ integrates its genetic material into the chromosome of the bacterium. This integrated phage DNA is known as a **prophage.** A prophage can exist in a dormant state for a long time, during which no new bacteriophages are made. When a bacterium containing a lysogenic prophage divides to produce two daughter cells, it copies the prophage's genetic material along with its own chromosome. Therefore, both daughter cells inherit the prophage. At some later time, in a process called induction, a prophage may become activated to excise itself from the bacterial chromosome and enter the lytic cycle. During induction, the phage promotes the synthesis of new phages and eventually lyses the host cell.

Inside the virus particle, phage λ DNA is linear. After injection into the bacterium, the two ends of the DNA become covalently attached to each other to form a circular piece of DNA. **Figure 14.20** shows the genome of phage λ. The organization of the genes within this circular structure reflects the two alternative cycles of this virus. The genes in the top center of the figure are transcribed very soon after infection. This occurs at the beginning of either reproductive cycle. As we will see, the expression pattern of these early genes determines whether the lytic or lysogenic cycle prevails.

If the lysogenic cycle prevails, the integrase (*int*) gene is subsequently turned on. The integrase gene encodes an enzyme that integrates the λ DNA into the bacterial chromosome. (The mechanism of phage λ integration is described in Chapter 17.) If the lysogenic cycle is not chosen, the genes on the right side and bottom of the figure are transcribed. These genes are necessary for the synthesis of new phages. They encode replication proteins, coat proteins, proteins involved in coat assembly, proteins involved in packaging the DNA into the phage head, and enzymes that cause the bacterium to lyse.

The Choice Between the Lytic or Lysogenic Cycle Depends on the Relative Levels of the cII and Cro Proteins

Now that we understand the phage λ reproductive cycles and genome organization, let's examine how the decision is made between the lytic and lysogenic cycles. This choice depends on the actions of certain genetic regulatory proteins.

Soon after the λ DNA enters the bacterial cell, two promoters, designated P_L and P_R, are used for transcription. This initiates a competition between the lytic and lysogenic cycles (**Figure 14.21**). Initially, transcription from P_L and P_R results in the synthesis of two short RNA transcripts that encode two proteins called the N protein and the cro protein, both of which are genetic regulatory proteins. The N protein is a genetic regulatory protein with an interesting function that we have not yet considered. Its function, known as **antitermination,** is to prevent

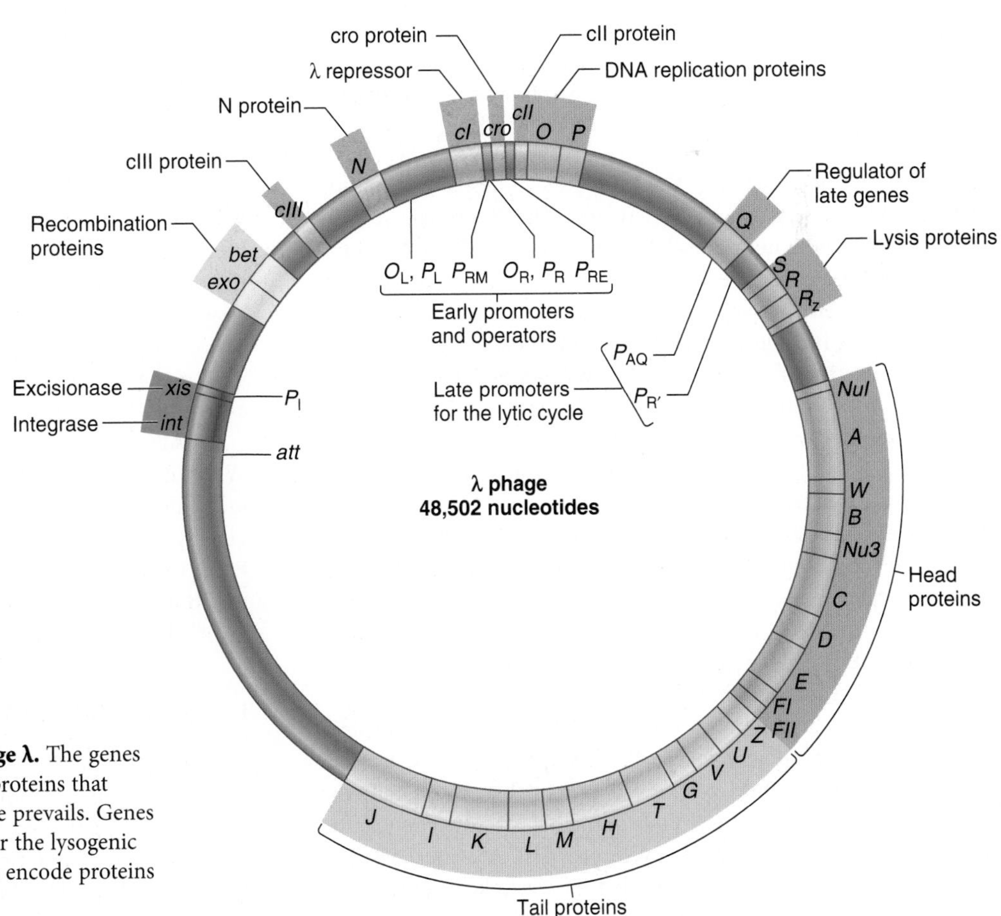

FIGURE 14.20 **The genome of phage λ.** The genes shaded in orange encode genetic regulatory proteins that determine whether the lysogenic or lytic cycle prevails. Genes shaded in green encode proteins necessary for the lysogenic cycle. The genes shaded in dark and light tan encode proteins required for the lytic cycle.

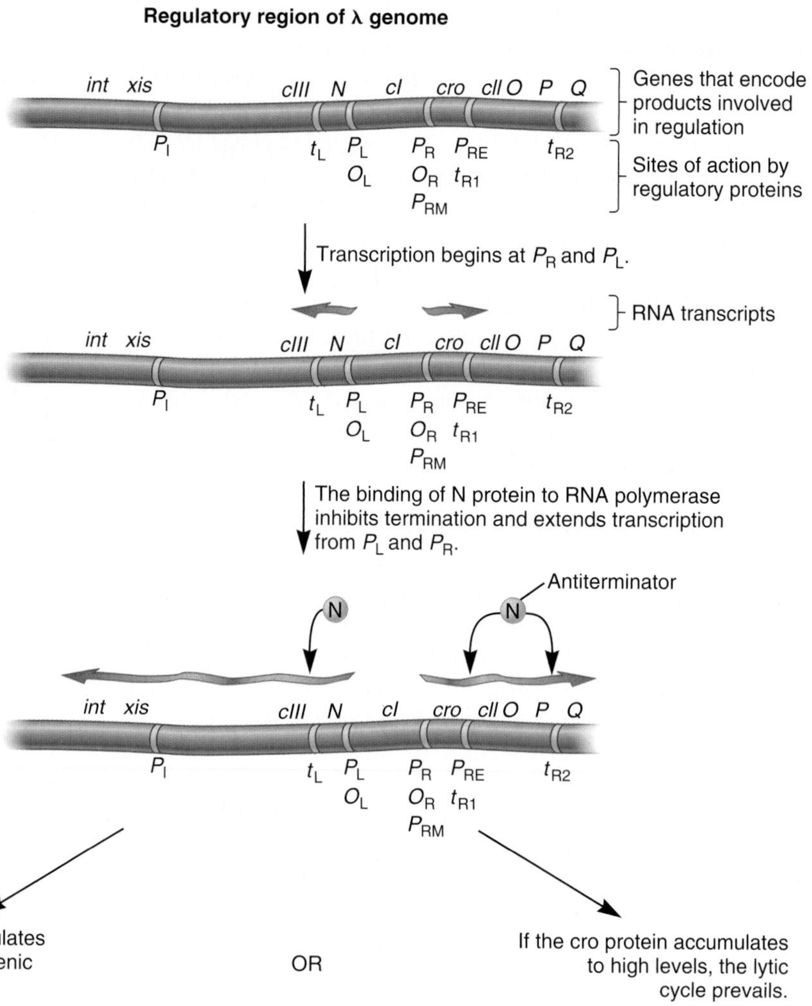

FIGURE 14.21 The events that lead to the beginning of the lysogenic and lytic cycles of phage λ. This figure shows the region of the phage λ genome that regulates the choice between the lytic and lysogenic cycles. DNA is shown in blue. The names of genes that encode proteins are shown above the DNA. Promoters and operator sites are shown below the DNA. In this figure and Figure 14.22, RNA transcripts are shown in red or green. The key regulatory proteins are indicated as spheres.

Immediately after infection, P_L and P_R are used to make two short mRNAs. These mRNAs encode two early proteins, designated N and cro. The N protein prevents transcriptional termination at three sites in the RNA (t_L, t_{R1}, and t_{R2}). This allows the transcription of several genes, which include *cIII*, *cII*, O, *P*, and Q. The expression of the O, *P*, and Q genes is necessary only for the lytic cycle. If the lysogenic cycle is chosen, transcription of these genes is abruptly inhibited.

transcriptional termination. The N protein inhibits termination at three sites, designated t_L, t_{R1}, and t_{R2}. The N protein actually binds to RNA polymerase and prevents transcriptional termination when these sites are being transcribed. When the N protein prevents termination at t_{R1} and t_{R2}, the transcript from P_R is extended to include the *cII*, *O*, *P*, and *Q* genes. The *cII* gene encodes an activator protein, the *O* and *P* genes encode enzymes needed for the initiation of λ DNA synthesis, and the *Q* gene encodes another antiterminator that is required for the lytic cycle. When the N protein prevents termination at t_L, the transcript from P_L is extended to include the *int*, *xis*, and *cIII* genes. The *int* gene encodes integrase, which is involved with integrating λ DNA into the *E. coli* chromosome, and the *xis* gene encodes excisionase, which can excise the λ DNA if a switch is made from the lysogenic to the lytic cycle. The *cIII* gene encodes the cIII protein, which inhibits a cellular protease and thereby makes cII less vulnerable to protease digestion.

As shown at the bottom of Figure 14.21, if the cII protein accumulates to sufficient levels, the lysogenic cycle is favored. Alternatively, if the level of the cro protein becomes high, the lytic cycle prevails. What environmental factors determine whether the lytic or lysogenic cycle prevails? A critical issue is that the cII protein is easily degraded by a cellular protease that is produced by *E. coli*. Whether or not this protease is made at high levels depends on the environmental conditions. If the growth conditions are very favorable, such as a rich growth medium, the intracellular protease levels are relatively high, and the cII protein tends to be degraded. In this case, the cro protein slowly accumulates to sufficient levels. Therefore, environmental conditions that are favorable for growth promote the lytic cycle. This makes sense, because a sufficient supply of nutrients is necessary to synthesize new bacteriophages.

Alternatively, starvation conditions favor the lysogenic cycle. When nutrients are limited, the cellular protease level is relatively low. Under these conditions, the cII protein builds up much more quickly than the cro protein. This event favors the lysogenic cycle. From the perspective of the bacteriophage, lysogeny may be a better choice under starvation conditions because nutrients may be insufficient for the production of new λ phages.

Integrase and the λ Repressor Control the Lysogenic Cycle

Let's now consider how the lysogenic cycle is followed (**Figure 14.22a**). If the level of cII-cIII complex becomes high, the cII protein activates two different promoters in the λ genome. When the cII protein binds to the promoter P_{RE}, it turns on the transcription of *cI*, a gene that encodes the λ repressor. The cII

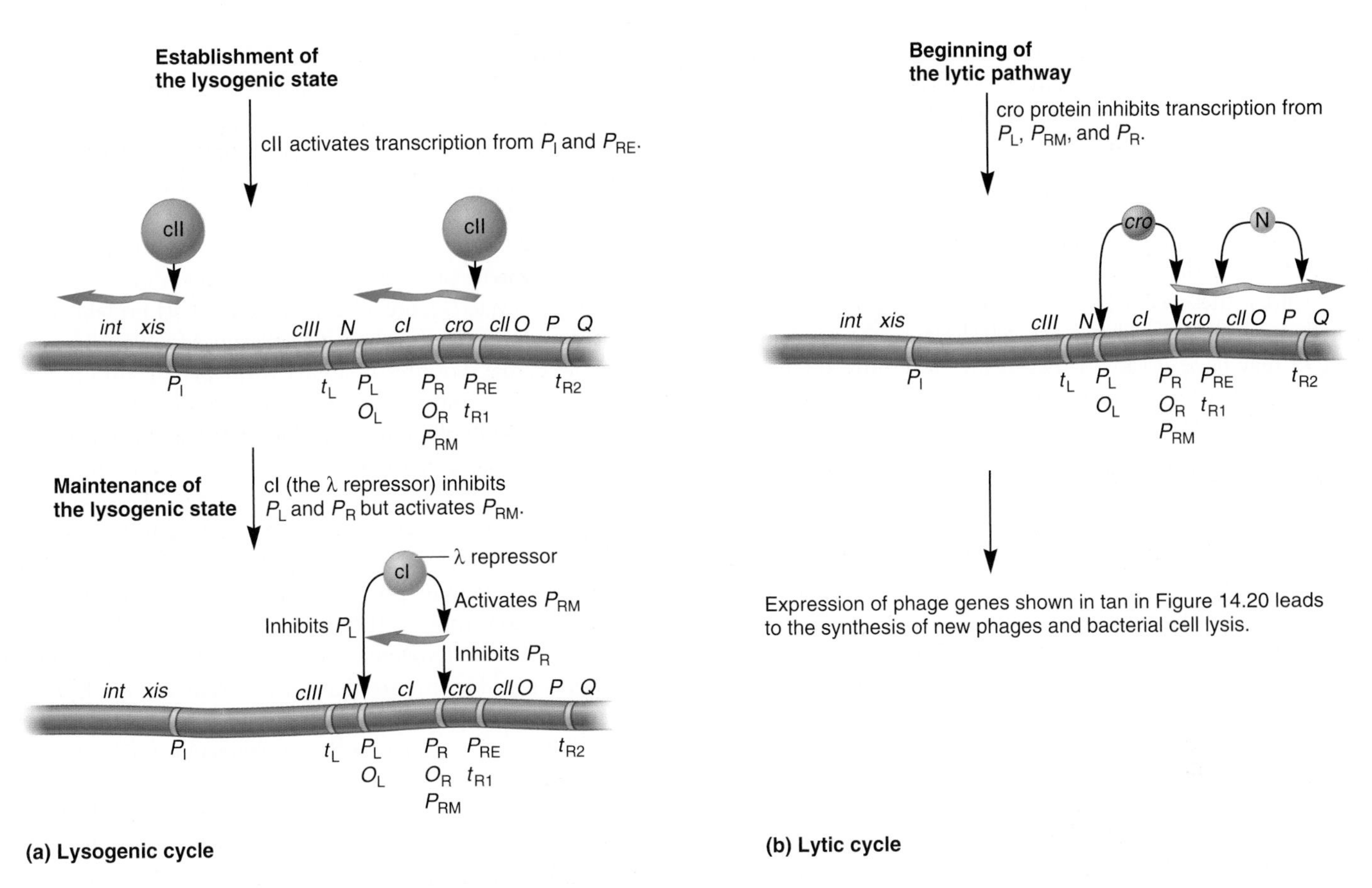

FIGURE 14.22 Phage λ gene regulation during the lysogenic and lytic cycles. (**a**) Gene regulation pattern that leads to the lysogenic cycle. The cII protein binds to P_{RE}. This activates the transcription of the *cI* gene, which encodes the λ repressor. The cII protein also binds to P_I to activate the transcription of the *int* gene. The λ repressor binds to O_L and O_R to inhibit transcription from P_L and P_R. This prevents the lytic cycle. The integrase protein catalyzes the integration of the λ DNA into the *E. coli* chromosome. At a later stage in the lysogenic cycle, the *cI* gene is transcribed from the P_{RM} promoter. This results in a low but steady synthesis of the λ repressor, which is necessary to further maintain the lysogenic state. (**b**) Steps that lead to the lytic cycle. The cro protein binds to O_L and blocks transcription from P_L. Cro also binds to O_R and blocks transcription from P_{RM}. This prevents the synthesis of the λ repressor. Transcription from P_R is also inhibited, but this occurs after the transcription of the O, P, and Q genes. The O and P proteins catalyze the replication of additional λ DNA. The Q protein allows transcription through $P_{R'}$ (not shown in this figure, but see Figure 14.20). This leads to the synthesis of many proteins that are necessary to make new λ phages and to lyse the cell.

Genes → Traits The ability to choose between two alternative reproductive cycles can be viewed as a trait of this bacteriophage. As described here, the choice between the two cycles depends on the pattern of gene regulation.

protein also activates the *int* gene by binding to the promoter P_I. The λ repressor and integrase proteins play central roles in promoting the lysogenic cycle. When the λ repressor is made in sufficient quantities, it binds to operator sites (O_L and O_R) that are adjacent to P_L and P_R. When the λ repressor is bound to O_R, it inhibits the expression of the genes required for the lytic cycle.

Notice in Figure 14.22a that the binding of the λ repressor to O_R will inhibit the expression of *cII*. This may seem counterintuitive, because the cII protein was initially required to activate the *cI* gene that encodes the λ repressor. You may be thinking that the inhibition of the *cII* gene eventually prevents the expression of the *cI* gene and ultimately stops the synthesis of the λ repressor protein. What prevents the inhibition of *cI* gene expression? The explanation is that the *cI* gene actually has two promoters: P_{RE}, which is activated by the cII protein, and a second promoter called P_{RM}. Transcription from P_{RE} occurs early in the lysogenic cycle. P_{RE} gets its name because the use of this promoter results in the expression of the λ Repressor during the Establishment of the lysogenic cycle. The transcript made from the use of P_{RE} is very stable and quickly leads to a buildup of the λ repressor protein. This causes an abrupt inhibition of the lytic cycle because the binding of the λ repressor protein to O_R blocks the P_R promoter. Later in the lysogenic cycle, it is no longer necessary to make a large amount of the λ repressor. At this point, the use of the P_{RM} promoter is sufficient to make enough Repressor protein to Maintain the lysogenic cycle. Interestingly, the P_{RM} promoter is activated by the λ repressor protein. The λ repressor was named when it was understood that it repressed the lytic cycle. Later studies revealed that it also activates its own transcription from P_{RM}.

After the lysogenic cycle is established, certain environmental conditions can also favor induction to the lytic cycle at some later time. For example, exposure to UV light promotes induction. This also is caused by the activation of cellular proteases. In this case, a cellular protein known as recA (a protein ordinarily involved in facilitating recombination between DNA molecules) detects the DNA damage from UV light and is activated to become a mediator of protein cleavage. RecA protein mediates the cleavage of the λ repressor, thereby inactivating it. This allows transcription from P_R and eventually leads to the accumulation of the cro protein, which thereby favors the lytic cycle. Under these conditions, it may be advantageous for λ to make new phages and lyse the cell, because the exposure to UV light may have already damaged the bacterium to the point where further bacterial growth and division are prevented.

The Lytic Cycle Depends on the Action of the Cro Protein

If the activity of the cro protein exceeds the activity of the cII protein, the lytic cycle prevails (**Figure 14.22b**). As was mentioned, an early step in the expression of λ genes is the transcription from P_R to produce the cro protein. If the concentration of the cro protein builds to sufficient levels, it will bind to two operator regions: O_L and O_R. The binding of cro to O_L inhibits transcription from P_L; the binding of cro to O_R has several effects. When the cro protein binds to O_R, it inhibits transcription from P_{RM} in the leftward direction. This inhibition prevents the expression of the *cI* gene, which encodes the λ repressor; the λ repressor is needed to maintain the lysogenic state. Therefore, the λ repressor cannot successfully shut down transcription from P_R.

Although the binding of the cro protein to O_R inhibits transcription from P_R in the rightward direction, this inhibition occurs after the transcription of the *O*, *P*, and *Q* genes. The O and P proteins are necessary for the replication of the λ DNA. The Q protein is an antiterminator protein that permits transcription through another promoter, designated $P_{R'}$ (not to be confused with P_R). The $P_{R'}$ promoter controls a very large operon that encodes the proteins necessary for the phage coat, the assembly of the coat proteins, the packaging of the λ DNA, and the lysis of the bacterial cell (refer back to Figure 14.20). These proteins are made toward the end of the lytic cycle. The expression of these late genes leads to the synthesis and assembly of many new λ phages that are released from the bacterial cell when it lyses.

The O_R Region Provides a Genetic Switch Between the Lytic and Lysogenic Cycles

Before we end this section on the λ reproductive cycles, let's consider how the O_R region acts as a genetic switch between the two cycles. Depending on the binding of genetic regulatory proteins to this region, the switch can be turned to favor the lytic or lysogenic cycle. How does this switch work? To understand the mechanism, we need to take a closer look at the O_R region (**Figure 14.23**).

The O_R region contains three operator sites, designated O_{R1}, O_{R2}, and O_{R3}. These operator sites control two promoters called P_{RM} and P_R that transcribe in opposite directions. The λ repressor protein or the cro protein can bind to any or all of the three operator sites. The binding of these two proteins at these sites governs the switch between the lysogenic and lytic cycles. Two critical issues influence this binding event. The first is the relative affinities that the regulatory proteins have for these operator sites. The second is the concentrations of the λ repressor protein and the cro protein within the cell.

Let's consider how an increasing concentration of the λ repressor protein can switch on the lysogenic cycle and switch off the lytic cycle (Figure 14.23, left side). This protein was first isolated by Mark Ptashne and his colleagues in 1967. Their studies showed that λ repressor binds with highest affinity to O_{R1}, followed by O_{R2} and O_{R3}. As the concentration of the λ repressor builds within the cell, a dimer of the λ repressor protein first binds to O_{R1} because it has the highest affinity for this site. Next, a second λ repressor dimer binds to O_{R2}. This occurs very rapidly, because the binding of the first dimer to O_{R1} favors the binding of a second dimer to O_{R2}. This is called a cooperative interaction. The binding of the λ repressor to O_{R1} and O_{R2} inhibits transcription from P_R, thereby switching off the lytic cycle.

Early in the lysogenic cycle, the λ repressor protein concentration may become so high that it occupies O_{R3}. Eventually, however, the λ repressor concentration begins to drop, because the inhibition of P_R decreases the synthesis of cII, which activates the λ repressor gene from P_{RE}. As the λ repressor concentration gradually falls, it is first removed from O_{R3}. This allows transcription from P_{RM}. As mentioned earlier, the term λ repressor is somewhat misleading because the binding of the λ repressor at only O_{R1} and O_{R2} acts as an activator of P_{RM}. The ability of the λ repressor to activate its own transcription allows the switch to the lysogenic cycle to be maintained.

In the lytic cycle (Figure 14.23, right side), the binding of the cro protein controls the switch. The cro protein has its highest affinity for O_{R3} and has a similar affinity for O_{R2} and O_{R1}. Under conditions that favor the lytic cycle, the cro protein accumulates, and a cro dimer first binds to O_{R3}. This blocks transcription from P_{RM}, thereby switching off the lysogenic cycle. Later in the lytic cycle, the cro protein concentration continues to rise, so eventually it binds to O_{R2} and O_{R1}. This inhibits transcription from P_R, which is not needed in the later stages of the lytic cycle.

Genetic switches, like the one just described for phage λ, represent an important form of genetic regulation. As we have just seen, a genetic switch can be used to control two alternative reproductive cycles of a bacteriophage. In addition, genetic switches are also important in the developmental pathways of bacteria and eukaryotes. For example, certain species of bacteria can grow in a vegetative state when nutrients are abundant but produce spores when conditions are unfavorable for growth. The choice between vegetative growth and sporulation involves genetic switches. Likewise, genetic switches operate in the developmental pathways in eukaryotes. As we will discover in Chapter 23, they are key events in the initiation of cell differentiation during development. Studies of the phage λ life cycle have provided fundamental information with which to understand how these other switches can operate at the molecular level.

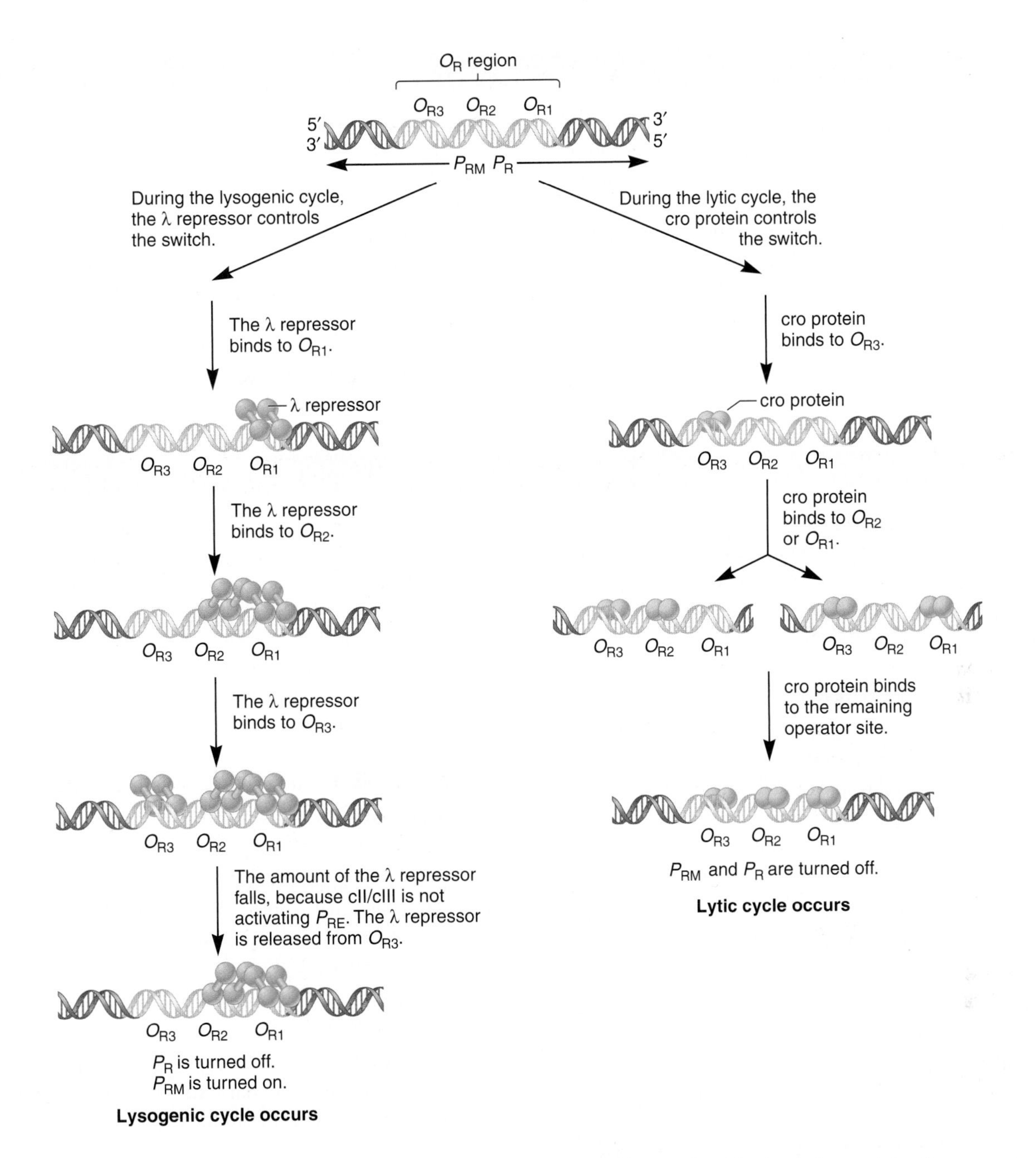

FIGURE 14.23 The O_R region, the genetic switch between the lysogenic and lytic cycles. The O_R region contains three operator sites and the P_{RM} and P_R promoters, which transcribe in opposite directions. The left side of the figure depicts the events that promote the lysogenic cycle. The λ repressor protein is a dimer; each subunit contains two globular domains connected by a short link. The λ repressor protein dimer first binds to O_{R1} because it has the highest affinity for this site. A second λ repressor dimer binds to O_{R2}. This occurs very rapidly, because the binding of the first dimer to O_{R1} favors the binding of a second dimer to O_{R2}. The binding of the λ repressor to O_{R1} and O_{R2} inhibits transcription from P_R and thereby switches off the lytic cycle. Early in the lysogenic cycle, the λ repressor protein concentration may become so high that it occupies O_{R3}. Later, the λ repressor concentration begins to drop, because the inhibition of P_R decreases the synthesis of cII protein, which activates the λ repressor gene (from P_{RE}). As the λ repressor concentration gradually falls, it is first removed from O_{R3}. This allows transcription from P_{RM} and maintains the lysogenic cycle.

The right side depicts events at the O_R region that promote the lytic cycle. The cro repressor protein is a small globular protein that also binds to each operator site as a dimer. Because the cro protein has its highest affinity for O_{R3}, it binds there first. This blocks transcription from P_{RM} and thereby switches off the lysogenic cycle. The cro protein has a similar affinity for O_{R2} and O_{R1}, and so it may occupy either of these sites next. Later in the lytic cycle, the cro protein concentration continues to rise, so eventually it binds to both O_{R2} and O_{R1}. This inhibits transcription from the P_R promoter, which is not needed in the later stages of the lytic cycle.

KEY TERMS

Page 359. gene regulation, constitutive genes
Page 360. repressor, activator, negative control, positive control, inducer, inducible genes, corepressor, inhibitor, repressible genes, enzyme adaptation
Page 362. operon, polycistronic RNA, promoter, terminator, CAP site, catabolite activator protein (CAP), operator site, operator
Page 363. lac repressor, induced, allosteric regulation, allosteric site
Page 365. merozygote
Page 367. *trans*-effect
Page 368. *trans*-acting factor, *cis*-acting element, *cis*-effect, catabolite repression, cyclic AMP (cAMP)
Page 371. trp repressor
Page 373. attenuation, attenuator sequence
Page 375. posttranslational, translational regulatory protein, translational repressors
Page 376. antisense RNA, feedback inhibition, allosteric enzyme
Page 377. posttranslational covalent modification, riboswitch
Page 378. phage λ, lytic cycle, lysogenic cycle
Page 379. temperate phage, prophage, antitermination

CHAPTER SUMMARY

- Gene regulation is the phenomenon in which the level of gene expression can vary under different conditions. By comparison, constitutive genes are expressed at relatively constant levels. Gene regulation can occur at transcription, translation, or posttranslationally (see Figure 14.1).

14.1 Transcriptional Regulation

- Repressors and activators are proteins that exert negative control or positive control, respectively. Small effector molecules control the function of repressors and activators. These include inducers, inhibitors, and corepressors (see Figure 14.2).
- Enzyme adaptation is the phenomenon in which the enzymes that metabolize a molecule are made only when the molecule is present in the environment.
- The *lac* operon encodes a polycistronic mRNA for proteins that are involved with the uptake and metabolism of lactose (see Figure 14.3).
- The lac repressor protein binds to the *lac* operator and inhibits its transcription. Allolactose binds to the repressor and causes a conformational change that prevents the repressor from binding to the operator. This event induces transcription (see Figure 14.4).
- The regulation of the *lac* operon enables the bacterium *E. coli* to respond to changes in the level of lactose in its environment (see Figure 14.5).
- Jacob and Monod constructed merozygotes to show that the *lacI* gene encodes a diffusible repressor protein (see Figures 14.6, 14.7).
- Mutations in the *lacI* gene and the *lac* operator site may have different effects in a merozygote (see Table 14.1).
- CAP is an activator protein of the *lac* operon. It binds to the CAP site when cAMP levels are high. Glucose inhibits cAMP levels, thereby inhibiting the *lac* operon (see Figure 14.8).
- The lac repressor, which is a tetramer, binds to two operator sites, either to O_1 and O_2, or to O_1 and O_3 (see Figures 14.9, 14.10).
- The *ara* operon is regulated by the AraC protein that can act as a repressor in the absence of arabinose or an activator in the presence of arabinose (see Figures 14.11, 14.12).
- The *trp* operon is regulated by the trp repressor that binds to the *trp* operator when tryptophan, a corepressor, is present (see Figure 14.13).
- A second way that the *trp* operon is regulated is via attenuation in which the formation of a terminator stem-loop causes early termination of transcription (see Figures 14.14, 14.15).
- Inducible operons typically encode catabolic enzymes, whereas repressible operons often encode anabolic enzymes.

14.2 Translational and Posttranslational Regulation

- Translation can be regulated by translational regulatory proteins and by antisense RNA (see Figure 14.16).
- Posttranslational control of protein function may involve feedback inhibition and posttranslational covalent modifications (see Figure 14.17).

14.3 Riboswitches

- A riboswitch is a form of regulation in which an RNA exists in two different secondary structures. The conversion from one secondary structure to the other is due to the binding of a small molecule.
- A riboswitch can regulate transcription, translation, RNA stability, and splicing (see Table 14.2, Figures 14.18, 14.19).

14.4 Gene Regulation in the Bacteriophage Reproductive Cycle

- Phage λ has different sets of genes that allow it to follow a lytic or lysogenic cycle (see Figure 14.20).
- The expression of the cII protein favors the lysogenic cycle, whereas the expression of cro protein favors the lytic cycle (see Figures 14.21, 14.22).
- The O_R region has binding sites for the λ repressor and cro protein. The O_R region acts as a genetic switch between the lytic and lysogenic cycles (see Figure 14.23).

PROBLEM SETS & INSIGHTS

Solved Problems

S1. Researchers have identified mutations in the promoter region of the *lacI* gene that make it more difficult for the *lac* operon to be induced. These are called *lacI*^{Q} mutants, because a greater Quantity of lac repressor protein is made. Explain why an increased transcription of the *lacI* gene makes it more difficult to induce the *lac* operon.

Answer: An increase in the amount of lac repressor protein makes it easier to repress the *lac* operon. When the cells become exposed to lactose, allolactose levels slowly rise. Some of the allolactose binds to the lac repressor protein and causes it to be released from the operator site. If many more lac repressor proteins are found within the cell, more allolactose is needed to ensure that no unoccupied repressor proteins can repress the operon.

S2. Explain how the pausing of the ribosome in the presence or absence of tryptophan affects the formation of a terminator stem-loop.

Answer: The key issue is the location where the ribosome stalls. In the absence of tryptophan, it stalls over the Trp codons in the *trpL* mRNA. Stalling at this site shields region 1 in the attenuator region. Because region 1 is unavailable to hydrogen bond with region 2, region 2 hydrogen bonds with region 3. Therefore, region 3 cannot form a terminator stem-loop with region 4. Alternatively, if tryptophan levels in the cell are sufficient, the ribosome pauses over the stop codon in the *trpL* RNA. In this case, the ribosome shields region 2. Therefore, regions 3 and 4 hydrogen bond with each other to form a terminator stem-loop, which abruptly halts the continued transcription of the *trp* operon.

S3. Which key protein(s) that affect the choice between the lytic and lysogenic cycles may be degraded by cellular proteases?

Answer: After infection, the key protein affected by cellular proteases is cII. If protease levels are high, as under good growth conditions, cII is degraded. This promotes the lytic cycle. Under starvation conditions, the protease levels are low. This prevents the degradation of cII, thereby promoting the lysogenic cycle. After lysogeny has been established, the key protein affected by cellular proteases is the λ repressor. Agents such as UV light activate cellular proteases that digest the λ repressor. This permits induction of the lytic cycle.

S4. In bacteria, it is common for two or more genes to be arranged together in an operon. Discuss the arrangement of genetic sequences within an operon. What is the biological advantage of an operon organization?

Answer: An operon contains several different DNA sequences that play specific roles. For transcription to take place, an operon is flanked by a promoter to signal the beginning of transcription and a terminator to signal the end of transcription. Two or more structural genes that encode different proteins are found between these two sequences. A key feature of an operon is that the expression of the structural genes occurs as a unit. When transcription takes place, a polycistronic mRNA is made that encodes all of the structural genes.

In order to control the ability of RNA polymerase to transcribe an operon, an additional DNA sequence, known as the operator site, is usually present. The base sequence within the operator site can serve as a binding site for genetic regulatory proteins called activators or repressors. The advantage of an operon organization is that it allows a bacterium to coordinately regulate a group of genes whose encoded proteins have a common function. For example, an operon may contain a group of genes involved in lactose breakdown or a group of genes involved in tryptophan synthesis. The genes within an operon usually encode proteins within a common metabolic pathway or cellular function.

S5. The sequential use of two sugars by a bacterium is known as diauxic growth. It is a common phenomenon among many bacterial species. When glucose is one of the two sugars available, it is typical that the bacterium metabolizes glucose first, and then a second sugar after the glucose has been used up. Among *E. coli* and related species, diauxic growth is regulated by intracellular cAMP levels and the catabolite activator protein (CAP). Summarize the effects of glucose and lactose on the ability of the lac repressor and the cAMP-CAP complex to regulate the *lac* operon.

Answer: In the absence of lactose, the lac repressor has the dominating effect of shutting off the *lac* operon. Even when glucose is also absent and the cAMP-CAP complex is formed, the presence of the bound lac repressor prevents the expression of the *lac* operon. The effects of the cAMP-CAP complex are exerted only in the presence of lactose. When lactose is present and glucose is absent, the cAMP-CAP complex acts to enhance the rate of transcription. However, when both lactose and glucose are present, the inability of CAP to bind to the *lac* operon decreases the rate of transcription. The table shown here summarizes these effects.

***lac* Operon Regulation**

Sugar Present	Transcription of the *lac* Operon
None	The operon is turned off due to the dominating effect of the lac repressor protein.
Lactose	The operon is maximally turned on. The repressor protein is removed from the operator site, and the cAMP-CAP complex is bound to the CAP site.
Glucose	The operon is turned off due to the dominating effect of the lac repressor protein.
Lactose and glucose	The expression of the *lac* operon is greatly decreased. The lac repressor is removed from the operator site, and the catabolite activator protein is not bound to the CAP site. The absence of CAP at the CAP site makes it difficult for RNA polymerase to begin transcription. However, a little more transcription occurs under these conditions than in the absence of lactose, when the repressor is bound.

S6. The ability of DNA-binding proteins to promote a loop in DNA structure is an interesting phenomenon that is important in the structure and function of DNA. Besides the regulation of genes and operons, DNA looping is required in the compaction of DNA within the nucleoid of a bacterium and the nucleus of a eukaryotic cell (see Chapter 10). In addition, DNA looping is frequently involved in the expression of eukaryotic genes (see Chapter 15). In this solved problem, we examine an experimental approach that made it possible for Robert Schleif and his colleagues to determine that the AraC protein causes a loop to form in the DNA. This work relied on the mobility of DNA in an acrylamide gel. A segment of DNA that contains a loop is more compact than the same

DNA segment without a loop. Therefore, when these two alternative structures (looped versus unlooped) are run through a gel, the looped structure migrates more quickly to the bottom of the gel because it can more easily penetrate the gel matrix. As a starting material, Schleif had a sample of DNA that contained a portion of the *ara* operon including both the *araI* and *araO*$_2$ sites. In the gel shown here, this segment of DNA was exposed to the following conditions before it was run on the gel:

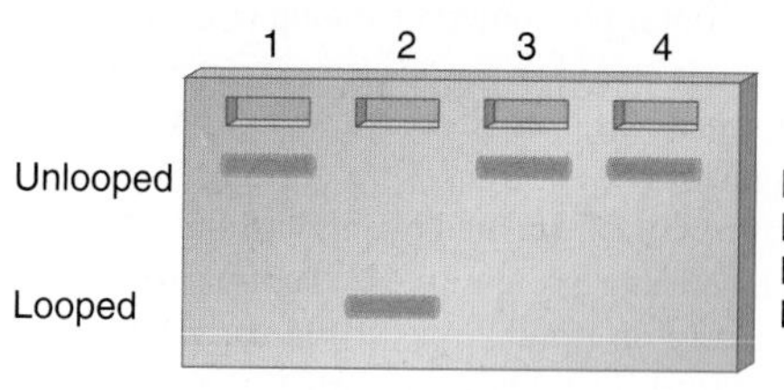

Lane 1. No further additions
Lane 2. Add AraC protein
Lane 3. Add arabinose
Lane 4. Add AraC protein and arabinose

Explain these results.

Answer: In lane 2, AraC protein was added and arabinose was not. As expected from the DNA looping hypothesis, a DNA loop was formed as evidenced from the faster mobility on the acrylamide gel. In lane 1, no AraC protein was added, so no DNA loop was able to form. These results confirm the idea that the AraC protein causes a loop to form in the sample loaded into lane 2. In lane 3, only the sugar arabinose is present but not the AraC protein. Because the DNA remains unlooped, the sugar by itself has no effect. In the sample loaded into lane 4, AraC protein was added, and arabinose was added as well. In this case, no DNA loop is observed. These results are consistent with the idea that arabinose binds to the AraC protein and breaks the DNA loop that is promoted by the AraC protein.

Conceptual Questions

C1. What is the difference between a constitutive gene and a regulated gene?

C2. In general, why is it important to regulate genes? Discuss examples of situations in which it would be advantageous for a bacterial cell to regulate genes.

C3. If a gene is repressible and under positive control, describe what kind of effector molecule and regulatory protein are involved. Explain how the binding of the effector molecule affects the regulatory protein.

C4. Transcriptional regulation often involves a regulatory protein that binds to a segment of DNA and a small effector molecule that binds to the regulatory protein. Do the following terms apply to a regulatory protein, a segment of DNA, or a small effector molecule?

A. Repressor
B. Inducer
C. Operator site
D. Corepressor
E. Activator
F. Attenuator
G. Inhibitor

C5. An operon is repressible—a small effector molecule turns off transcription. Which combinations of small effector molecules and regulatory proteins could be involved?

A. An inducer plus a repressor

B. A corepressor plus a repressor

C. An inhibitor plus an activator

D. An inducer plus an activator

C6. Some mutations have a *cis*-effect, whereas others have a *trans*-effect. Explain the molecular differences between *cis*- and *trans*-mutations. Which type of mutation (*cis* or *trans*) can be complemented in a merozygote experiment?

C7. What is enzyme adaptation? From a genetic point of view, how does it occur?

C8. In the *lac* operon, how would gene expression be affected if one of the following segments was missing?

A. *lac* operon promoter

B. Operator site

C. *lacA* gene

C9. If an abnormal repressor protein could still bind allolactose but the binding of allolactose did not alter the conformation of the repressor protein, how would this affect the expression of the *lac* operon?

C10. What is diauxic growth? Explain the roles of cAMP and the catabolite activator protein in this process.

C11. Mutations may have an effect on the expression of the *lac* operon, the *ara* operon, and the *trp* operon. Would the following mutations have a *cis*- or *trans*-effect on the expression of the structural genes in the operon?

A. A mutation in the operator site that prevents the lac repressor from binding to it

B. A mutation in the *lacI* gene that prevents the lac repressor from binding to DNA

C. A mutation in the *araC* gene that prevents two AraC proteins from binding to each other and forming a loop

D. A mutation in *trpL* that prevents attenuation

C12. Would a mutation that inactivated the lac repressor and prevented it from binding to the *lac* operator site result in the constitutive expression of the *lac* operon under all conditions? Explain. What is the disadvantage to the bacterium of having a constitutive *lac* operon?

C13. Describe the function of the AraC protein. How does it positively and negatively regulate the *ara* operon?

C14. Explain how a mutation would affect the regulation of the *ara* operon if the mutation prevented AraC protein from binding to the following sites:

A. *araO*$_2$

B. *araO*$_1$

C. *araI*

D. *araO*$_2$ and *araI*

C15. What is meant by the term attenuation? Is it an example of gene regulation at the level of transcription or translation? Explain your answer.

C16. As described in Figure 14.14, four regions within the *trpL* gene can form stem-loop structures. Let's suppose that mutations have been previously identified that prevent the ability of a particular region to form a stem-loop structure with a complementary region. For example, a region 1 mutant cannot form a 1–2 stem-loop structure, but it can still form a 2–3 or 3–4 structure. Likewise, a region 4 mutant can form a 1–2 or 2–3 stem-loop but not a 3–4 stem-loop. Under the following conditions, would attenuation occur?

A. Region 1 is mutant, tryptophan is high, and translation is not occurring.

B. Region 2 is mutant, tryptophan is low, and translation is occurring.

C. Region 3 is mutant, tryptophan is high, and translation is not occurring.

D. Region 4 is mutant, tryptophan is low, and translation is not occurring.

C17. As described in Chapter 13, enzymes known as aminoacyl-tRNA synthetases are responsible for attaching amino acids to tRNAs. Let's suppose that tryptophanyl-tRNA synthetase was partially defective at attaching tryptophan to tRNA; its activity was only 10% of that found in a normal bacterium. How would that affect attenuation of the *trp* operon? Would it be more or less likely to be attenuated? Explain your answer.

C18. The 3–4 stem-loop and U-rich attenuator found in the *trp* operon (see Figure 14.14) is an example of ρ-independent termination. The function of ρ-independent terminators is described in Chapter 12. Would you expect attenuation to occur if the tryptophan levels were high and mutations at the end of the *trpL* gene changed the UUUUUUUU sequence to UGGUUGUC? Explain why or why not.

C19. Mutations in tRNA genes can create tRNAs that recognize stop codons. Because stop codons are sometimes called nonsense codons, these types of mutations that affect tRNAs are called nonsense suppressors. For example, a normal $tRNA^{Gly}$ has an anticodon sequence CCU that recognizes a glycine codon in mRNA (GGA) and puts in a glycine during translation. However, a mutation in the gene that encodes $tRNA^{Gly}$ could change the anticodon to ACU. This mutant $tRNA^{Gly}$ would still carry glycine, but it would recognize the stop codon UGA. Would this mutation affect attenuation of the *trp* operon? Explain why or why not. Note: To answer this question, you need to look carefully at Figure 14.14 and see if you can identify any stop codons that may exist beyond the UGA stop codon that is found after region 1.

C20. Translational control is usually aimed at preventing the initiation of translation. With regard to cellular efficiency, why do you think this is the case?

C21. What is antisense RNA? How does it affect the translation of a complementary mRNA?

C22. A species of bacteria can synthesize the amino acid histidine so it does not require histidine in its growth medium. A key enzyme, which we will call histidine synthetase, is necessary for histidine biosynthesis. When these bacteria are given histidine in their growth media, they stop synthesizing histidine intracellularly. Based on this observation alone, propose three different regulatory mechanisms to explain why histidine biosynthesis ceases when histidine is in the growth medium. To explore this phenomenon further, you measure the amount of intracellular histidine synthetase protein when cells are grown in the presence and absence of histidine. In both conditions, the amount of this protein is identical. Which mechanism of regulation would be consistent with this observation?

C23. Using three examples, describe how allosteric sites are important in the function of genetic regulatory proteins.

C24. In what ways are the actions of the lac repressor and trp repressor similar and how are they different with regard to their binding to operator sites, their effects on transcription, and the influences of small effector molecules?

C25. Transcriptional repressor proteins (e.g., lac repressor), antisense RNA, and feedback inhibition are three different mechanisms that turn off the expression of genes and gene products. Which of these three mechanisms would be most effective in each of the following situations?

A. Shutting down the synthesis of a polypeptide

B. Shutting down the synthesis of mRNA

C. Shutting off the function of a protein

For your answers in parts A–C that have more than one mechanism, which mechanism would be the fastest or the most efficient?

C26. What key features distinguish the lytic and lysogenic cycles?

C27. With regard to promoting the lytic or lysogenic cycle, what would happen if the following genes were missing from the λ genome?

A. *cro*

B. *cI*

C. *cII*

D. *int*

E. *cII* and *cro*

C28. How do the λ repressor and the cro protein affect the transcription from P_R and P_{RM}? Explain where these proteins are binding to cause their effects.

C29. In your own words, explain why it is necessary for the *cI* gene to have two promoters. What would happen if it had only P_{RE}?

C30. A mutation in P_R causes its transcription rate to be increased 10-fold. Do you think this mutation would favor the lytic or lysogenic cycle? Explain your answer.

C31. When an *E. coli* bacterium already has a λ prophage integrated into its chromosome, another λ phage cannot usually infect the cell and establish the lysogenic or lytic cycle. Based on your understanding of the genetic regulation of the λ life cycles, why do you think the other phage would be unsuccessful?

C32. If a bacterium were exposed to a drug that inhibited the N protein, what would you expect to happen if the bacterium was later infected by phage λ? Would phage λ follow the lytic cycle, the lysogenic cycle, or neither? Explain your answer.

C33. Figure 14.23 shows a genetic switch that controls the choice between the lytic and lysogenic cycles of phage λ. What is a genetic switch? Compare the roles of a genetic switch and a simple operator site (like the one found in the *lac* operon) in gene regulation.

C34. This question combines your knowledge of conjugation (described in Chapter 7) and the genetic regulation that directs the phage λ reproductive cycles. When donor *Hfr* strains conjugate with recipient F^- bacteria that are lysogenic for phage λ, the conjugated cells survive normally. However, if donor *Hfr* strains that are lysogenic for phage λ conjugate with recipient F^- bacteria that do not contain any phage λ, the conjugated cells often lyse, due to the induction of λ into the lytic cycle. Based on your knowledge of the regulation of the two λ cycles, explain this observation.

Experimental Questions

E1. Answer the following questions that pertain to the experiment of Figure 14.7.

A. Why was β-ONPG used? Why was no yellow color observed in one of the four tubes? Can you propose alternative methods to measure the level of expression of the *lac* operon?

B. The optical density values were twice as high for the mated strain as for the parent strain. Why was this result obtained?

E2. Chapter 18 describes a blotting method known as Northern blotting, which can be used to detect RNA transcribed from a particular gene or a particular operon. In this method, a specific RNA is detected by using a short segment of cloned DNA as a probe. The DNA probe, which is radioactive, is complementary to the RNA that the researcher wishes to detect. After the radioactive probe DNA binds to the RNA within a blot of a gel, the RNA is visualized as a dark (radioactive) band on an X-ray film. For example, a DNA probe complementary to the mRNA of the *lac* operon could be used to specifically detect the *lac* operon mRNA on a gel blot. As shown here, the method of Northern blotting can be used to determine the amount of a particular RNA transcribed under different types of growth conditions. In this Northern blot, bacteria containing a normal *lac* operon were grown under different types of conditions, and then the mRNA was isolated from the cells and subjected to a Northern blot, using a probe that is complementary to the mRNA of the *lac* operon.

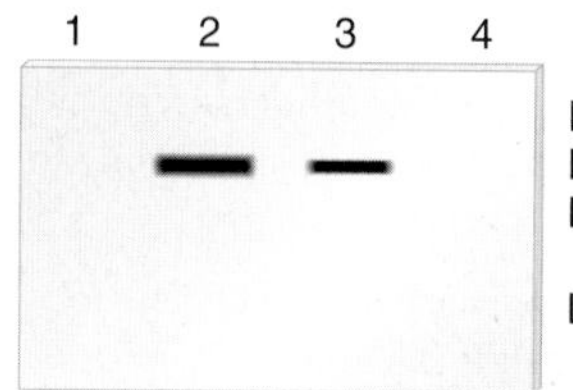

Lane 1. Growth in media containing glucose
Lane 2. Growth in media containing lactose
Lane 3. Growth in media containing glucose and lactose
Lane 4. Growth in media that doesn't contain glucose or lactose

Based on your understanding of the regulation of the *lac* operon, explain these results. Which is more effective at shutting down the *lac* operon, the binding of the *lac* repressor or the removal of CAP? Explain your answer based on the results shown in the Northern blot.

E3. As described in experimental question E2 and also in Chapter 18, the technique of Northern blotting can be used to detect the transcription of RNA. Draw the results you would expect from a Northern blot if bacteria were grown in media containing lactose (and no glucose) but had the following mutations:

Lane 1. Normal strain

Lane 2. Strain with a mutation that inactivates the lac repressor

Lane 3. Strain with a mutation that prevents allolactose from binding to the lac repressor

Lane 4. Strain with a mutation that inactivates CAP

How would your results differ if these bacterial strains were grown in media that did not contain lactose or glucose?

E4. An absentminded researcher follows the protocol described in Figure 14.7 and (at the end of the experiment) does not observe any yellow color in any of the tubes. Yikes! Which of the following mistakes could account for this observation?

A. Forgot to sonicate the cells

B. Forgot to add lactose to two of the tubes

C. Forgot to add β-ONPG to the four tubes

E5. Explain how the data shown in Figure 14.9 indicate that two operator sites are necessary for repression of the *lac* operon. What would the results have been if all three operator sites were required for the binding of the lac repressor?

E6. A mutant strain has a defective *lac* operator site that results in the constitutive expression of the *lac* operon. Outline an experiment you would carry out to demonstrate that the operator site must be physically adjacent to the genes that it influences. Based on your knowledge of the *lac* operon, describe the results you would expect.

E7. Let's suppose you have isolated a mutant strain of *E. coli* in which the *lac* operon is constitutively expressed. To understand the nature of this defect, you create a merozygote in which the mutant strain contains an F′ factor with a normal *lac* operon and a normal *lacI* gene. You then compare the mutant strain and the merozygote with regard to their β-galactosidase activities in the presence and absence of lactose. You obtain the following results:

	Addition of Lactose	Amount of β-Galactosidase (percentage of mutant strain in the presence of lactose)
Mutant	No	100
Mutant	Yes	100
Merozygote	No	100
Merozygote	Yes	200

Explain the nature of the defect in the mutant strain.

E8. In the experiment of Figure 14.7, a *lacI*⁻ mutant was conjugated to a strain that had a functional *lacI* gene on an F′ factor. The results of this experiment were important in determining the action of the lac repressor protein. What results would you expect if you used the same approach to investigate the regulation of the *ara* operon via AraC? In other words, what would be the level of expression of the *ara* operon in an araC⁻ strain in the presence and absence of arabinose, and how would the level of expression change when a functional *araC* gene was introduced into the strain via conjugation?

E9. A segment of DNA that contains a loop is more compact than the same DNA segment without a loop. Therefore, when these two alternative structures (looped versus unlooped) are electrophoresed through a gel, the looped structure migrates more quickly to the bottom of the gel because it can more easily penetrate the gel matrix (see solved problem S6). Let's suppose that a mutant *E. coli* strain has been identified in which the AraC protein represses the *ara* operon, even in the presence of arabinose. In the experiment shown here, mutant or normal AraC protein was mixed with a segment of DNA containing the *ara* operon in the absence or presence of arabinose and then run on a gel.

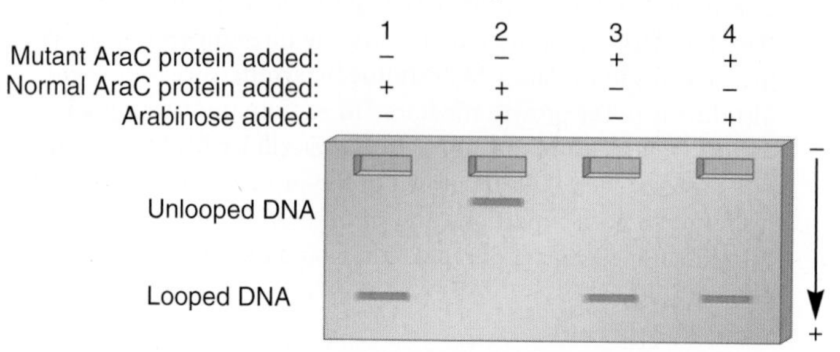

Describe the defect in this mutant AraC protein.

Questions for Student Discussion/Collaboration

1. Discuss the advantages and disadvantages of genetic regulation at the different levels described in Figure 14.1.
2. As you look at Figure 14.10, discuss possible "molecular ways" that the cAMP-CAP complex and lac repressor may influence RNA polymerase function. In other words, try to explain how the bending and looping in DNA may affect the ability of RNA polymerase to initiate transcription.
3. Certain environmental conditions such as UV light are known to activate lysogenic λ prophages and cause them to progress into the lytic cycle. UV light initially causes the repressor protein to be proteolytically degraded. Make a flow diagram that describes the subsequent events that would lead to the lytic cycle. Note: The *xis* gene codes for an enzyme that is necessary to excise the λ prophage from the *E. coli* chromosome. The integrase enzyme is also necessary to excise the λ prophage.

Note: All answers appear at the website for this textbook; the answers to even-numbered questions are in the back of the textbook.

www.mhhe.com/brookergenetics4e

Visit the website for practice tests, answer keys, and other learning aids for this chapter. Enhance your understanding of genetics with our interactive exercises, quizzes, animations, and much more.

CHAPTER OUTLINE

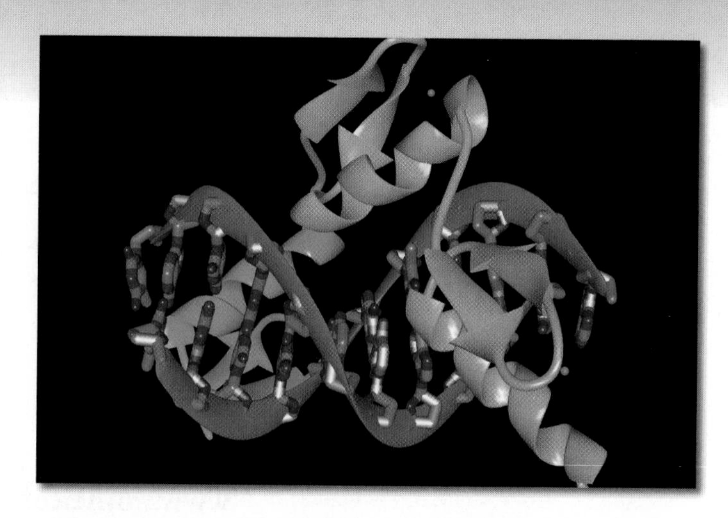

Certain proteins, known as regulatory transcription factors, have the ability to bind into the major groove of DNA and regulate gene transcription.

GENE REGULATION IN EUKARYOTES

Gene regulation refers to the phenomenon that the level of gene expression can be controlled so that genes can be expressed at high or low levels. The ability to regulate genes provides many benefits to eukaryotic organisms, a category that includes protists, fungi, plants, and animals. Like their prokaryotic counterparts, eukaryotic cells need to adapt to changes in their environment. For example, eukaryotic cells can respond to changes in nutrient availability by enzyme adaptation, much as prokaryotic cells do. Eukaryotic cells also respond to environmental stresses such as ultraviolet (UV) radiation by inducing genes that provide protection against harmful environmental agents. An example is the ability of humans to develop a tan. The tanning response helps to protect a person's skin cells against the damaging effects of UV rays.

Among plants and animals, multicellularity and a more complex cell structure also demand a much greater level of gene regulation. The life cycle of complex eukaryotic organisms involves the progression through developmental stages to achieve a mature organism. Some genes are expressed only during early stages of development, such as the embryonic stage, whereas others are expressed in the adult. In addition, complex eukaryotic species are composed of many different tissues that contain a variety of cell types. Gene regulation is necessary to ensure the differences in structure and function among distinct cell types. It is amazing that the various cells within a multicellular organism usually contain the same genetic material, yet phenotypically may look quite different. For example, the appearance of a human nerve cell seems about as similar to a muscle cell as an amoeba is to a paramecium. In spite of these phenotypic differences, a human nerve cell and muscle cell actually contain the same complement of human chromosomes. Nerve and muscle cells look strikingly different because of gene regulation rather than differences in DNA content. Many genes are expressed in the nerve cell and not the muscle cell, and vice versa.

The molecular mechanisms that underlie gene regulation in eukaryotes bear many similarities to the ways that bacteria regulate their genes. As in bacteria, regulation in eukaryotes can occur at any step in the pathway of gene expression (**Figure 15.1**). We will begin this chapter with an exploration of gene regulation at the level of transcription, an important form of control. In addition, research in the past few decades has revealed that eukary-

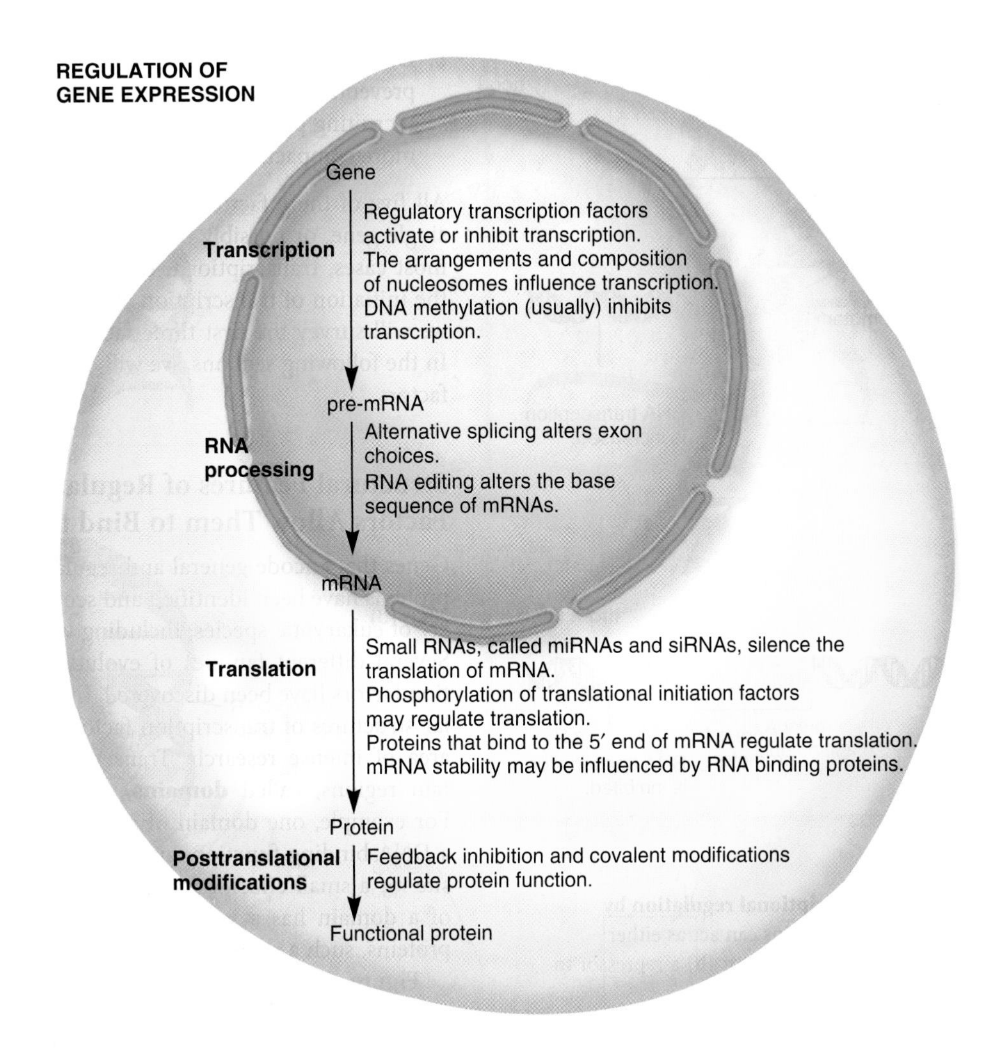

FIGURE 15.1 **Levels of gene expression commonly subject to regulation.**

otic organisms frequently regulate gene expression at points other than transcription. We will discuss some well-studied examples in which genes are regulated at many of these other control points.

15.1 REGULATORY TRANSCRIPTION FACTORS

The term **transcription factor** is broadly used to describe proteins that influence the ability of RNA polymerase to transcribe a given gene. We will focus our attention on transcription factors that affect the ability of RNA polymerase to begin the transcription process. Such transcription factors can regulate the binding of the transcriptional apparatus to the core promoter and/or control the switch from the initiation to the elongation stage of transcription. Two categories of transcription factors play a key role in these processes. In Chapter 12, we considered **general transcription factors,** which are required for the binding of RNA polymerase to the core promoter and its progression to the elongation stage. General transcription factors are necessary for a basal level of transcription. In addition, eukaryotic cells possess a diverse array of **regulatory transcription factors** that serve to regulate the rate of transcription of target genes. The importance of transcription factors is underscored by the number of genes that encode this category of proteins. A sizeable portion of the genomes of complex organisms such as plants and animals is devoted to the process of gene regulation. For example, in *Arabidopsis thaliana,* a plant that is used as a model organism by plant geneticists, approximately 5% of its genome encodes transcription factors. This species has more than 1500 different genes that encode proteins that regulate the transcription of other genes. Similarly, 2 to 3% of the genes in the human genome encode transcription factors.

As discussed in this section, regulatory transcription factors exert their effects by influencing the ability of RNA polymerase to begin transcription of a particular gene. They typically recognize *cis*-acting elements that are located in the vicinity of the core promoter. These DNA sequences are analogous to the operator sites found near bacterial promoters. In eukaryotes, these DNA sequences are generally known as **control elements,** or **regulatory elements.** When a regulatory transcription factor binds to a regulatory element, it affects the transcription of an associated gene. For example, the binding of regulatory transcription factors may

can stimulate transcription 10- to 1000-fold, a phenomenon known as **up regulation.** Alternatively, regulatory elements that serve to inhibit transcription are called silencers, and their action is called **down regulation.**

Many regulatory elements are **orientation-independent,** or **bidirectional.** This means that the regulatory element can function in the forward or reverse direction. For example, if the forward orientation of an enhancer is

5′–GATA–3′
3′–CTAT–5′

this enhancer is also bound by a regulatory transcription factor and enhances transcription even when it is oriented in the reverse direction:

5′–TATC–3′
3′–ATAG–5′

Striking variation is also found in the location of regulatory elements relative to a gene's promoter. Regulatory elements are often located in a region within a few hundred base pairs (bp) upstream from the promoter site. However, they can be quite distant from the promoter, even 100,000 bp away, yet exert strong effects on the ability of RNA polymerase to initiate transcription at the core promoter! Regulatory elements were first discovered by Susumu Tonegawa and coworkers in the 1980s. While studying genes that play a role in immunity, they identified a region that is far away from the core promoter, but is needed for high levels of transcription to take place. In some cases, regulatory elements are located downstream from the promoter site and may even be found within introns, the noncoding parts of genes. As you may imagine, the variation in regulatory element orientation and location profoundly complicates the efforts of geneticists to identify the regulatory elements that affect the expression of any given gene.

Regulatory Transcription Factors May Exert Their Effects Through TFIID and Mediator

Different mechanisms have been discovered that explain how a regulatory transcription factor can bind to a regulatory element and thereby affect gene transcription. For most genes, more than one mechanism is involved. The net effect of a regulatory transcription factor is to influence the ability of RNA polymerase to transcribe a given gene. However, most regulatory transcription factors do not bind directly to RNA polymerase. How then do most regulatory transcription factors exert their effects? For structural genes in eukaryotes, regulatory transcription factors commonly influence the function of RNA polymerase II by interacting with other proteins that directly bind to RNA polymerase II. Two protein complexes that communicate the effects of regulatory transcription factors are TFIID and mediator.

Some regulatory transcription factors bind to a regulatory element and then influence the function of TFIID. As discussed in Chapter 12, **TFIID** is a general transcription factor that binds to the TATA box and is needed to recruit RNA polymerase II to the core promoter. Activator proteins are expected to enhance the ability of TFIID to initiate transcription. One possibility is that activator proteins could help recruit TFIID to the TATA box or they could enhance the function of TFIID in a way that facilitates its ability to bind RNA polymerase II. In some cases, activator proteins exert their effects by interacting with **coactivators**—proteins that increase the rate of transcription but do not directly bind to the DNA itself. This type of activation is shown in **Figure 15.4a**. In contrast, repressors inhibit the function of TFIID. They could exert their effects by preventing the binding of TFIID to the TATA box (**Figure 15.4b**) or by inhibiting the ability of TFIID to recruit RNA polymerase II to the core promoter.

A second way that regulatory transcription factors control RNA polymerase II is via mediator—a protein complex discovered by Roger Kornberg and colleagues in 1990. The term **mediator** refers to the observation that it mediates the interaction between RNA polymerase II and regulatory transcription factors. As discussed in Chapter 12, mediator controls the ability of RNA polymerase II to progress to the elongation stage of transcription. Transcriptional activators stimulate the ability of mediator to facilitate the switch between the initiation and elongation stages, whereas repressors have the opposite effect. In the example shown in **Figure 15.5**, an activator binds to a distant enhancer element. The activator protein and mediator are brought together by the formation of a loop within the intervening DNA.

A third way that regulatory transcription factors can influence transcription is by recruiting proteins to the promoter region that affect nucleosome positions and compositions. For example, certain transcriptional activators can recruit proteins to the promoter region that promote the conversion of chromatin from a closed to an open conformation. We will return to this topic later in this chapter.

The Function of Regulatory Transcription Factor Proteins Can Be Modulated in Three Ways

Thus far, we have considered the structures of regulatory transcription factors and the molecular mechanisms that account for their abilities to control transcription. The functions of the regulatory transcription factors themselves must also be modulated. Why is this necessary? The answer is that the genes they control must be turned on at the proper time, in the correct cell type, and under the appropriate environmental conditions. Therefore, eukaryotes have evolved different ways to modulate the functions of these proteins.

The functions of regulatory transcription factor proteins are controlled in three common ways: through (1) the binding of a small effector molecule, (2) protein–protein interactions, and (3) covalent modifications. **Figure 15.6** depicts these three mechanisms of modulating regulatory transcription factor function. Usually, one or more of these modulating effects are important in determining whether a transcription factor can bind to the DNA or influence transcription by RNA polymerase. For example, a small effector molecule may bind to a regulatory transcription factor and promote its binding to DNA (Figure 15.6a). We will see that steroid hormones function in this manner. Another important way is via protein–protein interactions (Figure 15.6b). The formation of homodimers and heterodimers is a fairly common means of controlling transcription. Finally, the function of a regulatory transcription factor can be affected by covalent

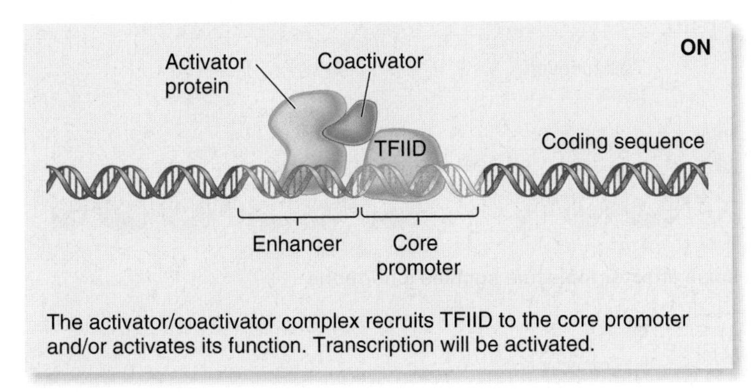

(a) Transcriptional activation via TFIID

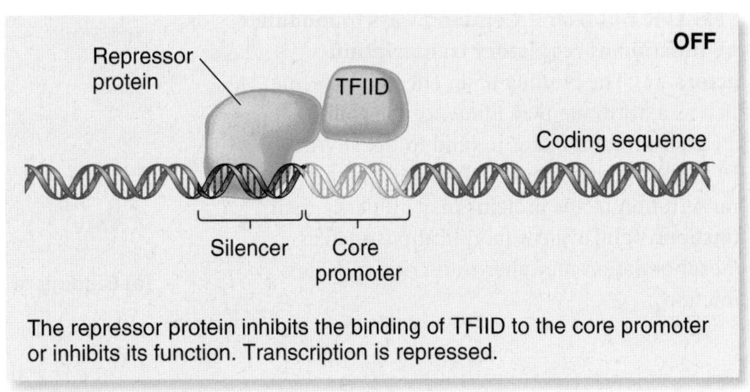

(b) Transcriptional repression via TFIID

FIGURE 15.4 **Effects of regulatory transcription factors on TFIID.** **(a)** Some activators stimulate the function of TFIID, thereby enhancing transcription. In this example, the activator interacts with a coactivator that directly binds to TFIID and stimulates its function. **(b)** Regulatory transcription factors may also functions as repressors. In the example shown here, the repressor binds to a silencer and inhibits the ability of TFIID to bind to the TATA box at the core promoter.

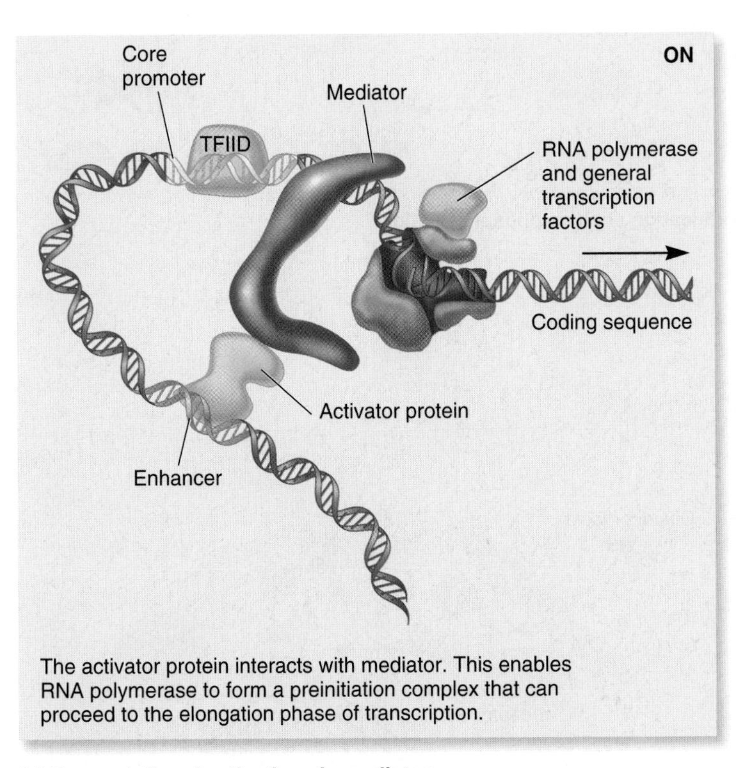

(a) Transcriptional activation via mediator

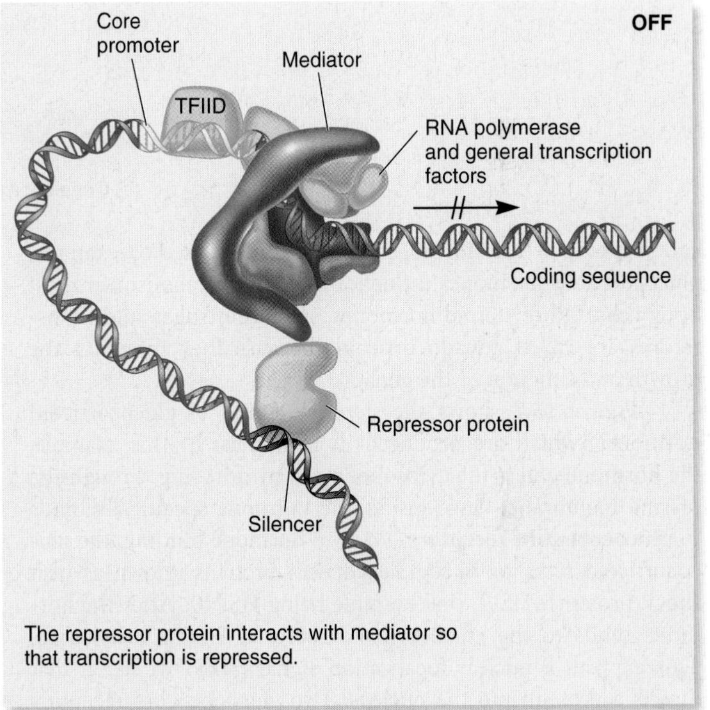

(b) Transcriptional repression via mediator

FIGURE 15.5 **Effects of an activator protein on mediator.** Some activators interact with mediator in a way that causes RNA polymerase to proceed to the elongation phase of transcription. **(a)** In this example, the enhancer is relatively far away from the core promoter. Therefore, a loop must form in the DNA so the activator can interact with mediator. **(b)** Alternatively, repressors can interact with mediator, thereby preventing RNA polymerase from proceeding to the elongation phase of transcription.

modifications such as the attachment of a phosphate group (Figure 15.6c). As discussed later, the phosphorylation of activators can control their ability to stimulate transcription.

Steroid Hormones Exert Their Effects by Binding to a Regulatory Transcription Factor

Now that we have a general understanding of the structure and function of transcription factors, let's turn our attention to specific examples that illustrate how regulatory transcription factors carry out their roles within living cells. Our first example is a category that responds to steroid hormones. This type of regulatory transcription factor is known as a **steroid receptor,** because the steroid hormone binds directly to the protein.

The ultimate action of a steroid hormone is to affect gene transcription. In animals, steroid hormones act as signaling molecules that are synthesized by endocrine glands and secreted into the bloodstream. The hormones are then taken up by cells that

FIGURE 15.6 Common ways to modulate the function of regulatory transcription factors. (a) The binding of an effector molecule such as a hormone may influence the ability of a transcription factor to bind to the DNA. (b) Protein–protein interactions among transcription factor proteins may influence their functions. (c) Covalent modifications such as phosphorylation may alter transcription factor function.

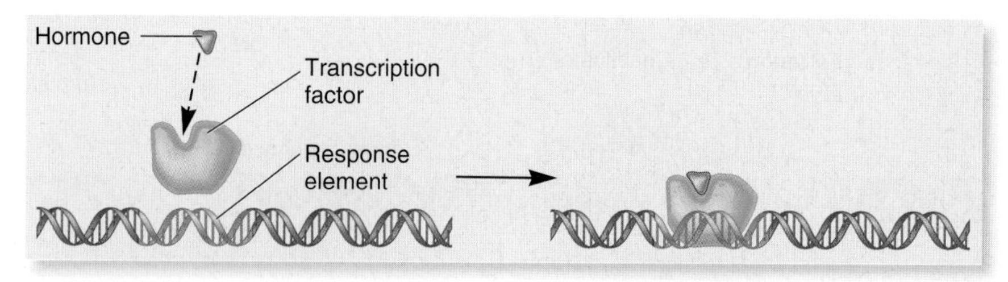

(a) Binding of a small effector molecule such as a hormone

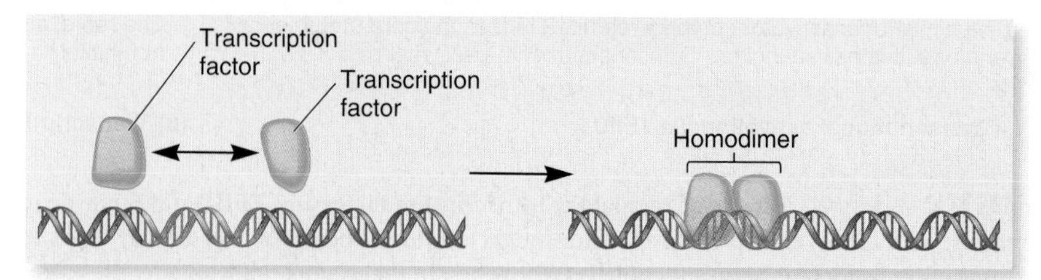

(b) Protein–protein interaction

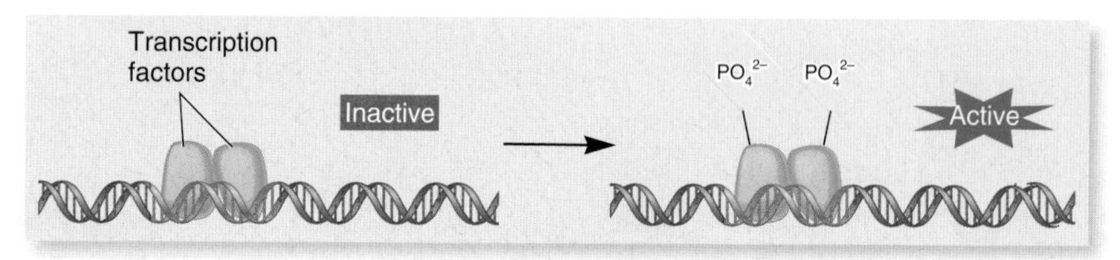

(c) Covalent modification such as phosphorylation

can respond to the hormones in different ways. For example, glucocorticoid hormones influence nutrient metabolism in most body cells. Other steroid hormones, such as estrogen and testosterone, are called gonadocorticoids because they influence the growth and function of the gonads.

Figure 15.7 shows the stepwise action of glucocorticoid hormones, which are produced in mammals. In this example, the hormone enters the cytosol of a cell by diffusing through the plasma membrane. Once inside, the hormone specifically binds to **glucocorticoid receptors.** Prior to hormone binding, the glucocorticoid receptor is complexed with proteins known as heat shock proteins (HSP), one example being HSP90. After the hormone binds to the glucocorticoid receptor, HSP90 is released. This exposes a nuclear localization signal (NLS)—a signal that directs a protein into the nucleus. Two glucocorticoid receptors form a homodimer and then travel through a nuclear pore into the nucleus.

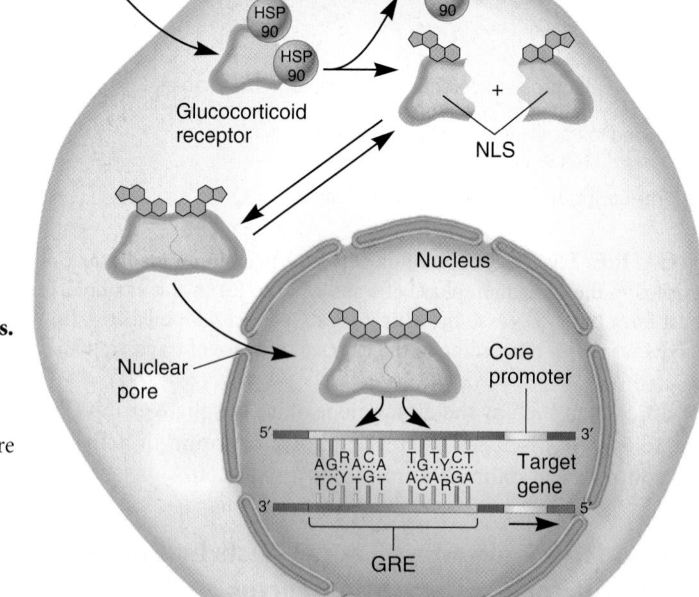

ONLINE ANIMATION

FIGURE 15.7 The action of glucocorticoid hormones. Once inside the cell, the glucocorticoid hormone binds to the glucocorticoid receptor, releasing it from a protein known as HSP90. This exposes a nuclear localization signal (NLS). Two glucocorticoid receptors then form a dimer and travel into the nucleus, where the dimer binds to a glucocorticoid response element (GRE) that is next to a particular gene. The binding of the glucocorticoid receptors to the GRE activates the transcription of the adjacent target gene.

Genes → Traits Glucocorticoid hormones are produced by the endocrine glands in response to fasting and activity. They enable the body to regulate its metabolism properly. When glucocorticoids are produced, they are taken into cells and bind to the glucocorticoid receptors. This eventually leads to the activation of genes that encode proteins involved in the synthesis of glucose, the breakdown of proteins, and the mobilization of fats.

How does the glucocorticoid receptor regulate the expression of particular genes? In the nucleus, the glucocorticoid receptor homodimer binds to DNA sites with the following consensus sequence:

5′–AGRACA–3′
3′–TCYTGT–5′

where R is a purine and Y is pyrimidine. A glucocorticoid response element (GRE) contains two of these sequences in opposite directions (Figure 15.7). A GRE is found next to many genes and functions as an enhancer. The binding of the glucocorticoid receptor homodimer to a GRE activates the transcription of the nearby gene, eventually leading to the synthesis of the encoded protein.

Mammalian cells usually have a large number of glucocorticoid receptors within the cytoplasm. Because GREs are located near dozens of different genes, the uptake of many hormone molecules can activate many glucocorticoid receptors, thereby stimulating the transcription of many different genes. For this reason, a cell can respond to the presence of the hormone in a very complex way. Glucocorticoid hormones stimulate many genes that encode proteins involved in several different cellular processes, including the synthesis of glucose, the breakdown of proteins, and the mobilization of fats. Although the genes are not physically adjacent to each other, the regulation of multiple genes via glucocorticoid hormones is much like the ability of bacterial operons to simultaneously control the expression of several genes.

The CREB Protein Is an Example of a Regulatory Transcription Factor Modulated by Covalent Modification

As we have just seen, steroid hormones function as signaling molecules that bind directly to regulatory transcription factors to alter their function. This enables a cell to respond to a hormone by up-regulating a particular set of genes. Most extracellular signaling molecules, however, do not enter the cell or bind directly to transcription factors. Instead, most signaling molecules must bind to receptors in the plasma membrane. This binding activates the receptor and may lead to the synthesis of an intracellular signal that causes a cellular response. One type of cellular response is to affect the transcription of particular genes within the cell.

As our second example of regulatory transcription factor function within living cells, we will examine the **cAMP response element-binding protein (CREB protein).** The CREB protein is a regulatory transcription factor that becomes activated in response to cell-signaling molecules that cause an increase in the cytoplasmic concentration of the molecule cyclic adenosine monophosphate (cAMP). This transcription factor recognizes a DNA site that has two adjacent copies of the following consensus sequence.

5′–TGACGTCA–3′
3′–ACTGCAGT–5′

This response element, which is found near many different genes, has been termed a **cAMP response element (CRE).**

Figure 15.8 shows the steps leading to the activation of the CREB protein. A wide variety of hormones, growth factors,

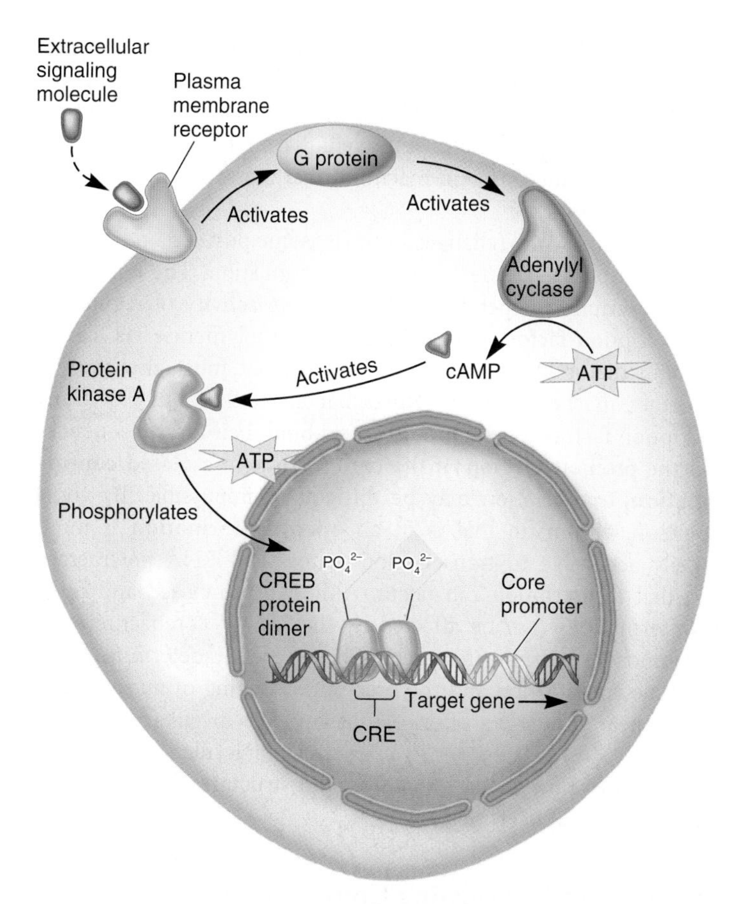

FIGURE 15.8 The activity of the CREB protein. When an extracellular signaling molecule binds to a receptor in the plasma membrane, this activates a G protein, which activates adenylyl cyclase, leading to the synthesis of cAMP. Next, cAMP binds to protein kinase A, which activates it. Protein kinase A then travels into the nucleus and phosphorylates the CREB protein. Once phosphorylated, the CREB protein acts as a transcriptional activator.

neurotransmitters, and other signaling molecules can bind to plasma membrane receptors to initiate an intracellular response. In this case, the response involves the production of a second messenger, cAMP. The extracellular signaling molecule itself is considered the primary messenger. When the signaling molecule binds to the receptor, it activates a G protein that subsequently activates the enzyme adenylyl cyclase. The activated adenylyl cyclase catalyzes the synthesis of cAMP. The cAMP molecule then binds to a second enzyme, protein kinase A, and activates it. This enzyme travels into the nucleus and phosphorylates several different cellular proteins, including CREB proteins. When phosphorylated, CREB proteins stimulate transcription. In contrast, unphosphorylated CREB proteins can still bind to CREs but do not activate RNA polymerase.

15.2 CHROMATIN REMODELING, HISTONE VARIATION, AND HISTONE MODIFICATION

The term ATP-dependent chromatin remodeling, or simply **chromatin remodeling,** refers to dynamic changes in the structure of chromatin that occur during the life of a cell. These changes range from local alterations in the positioning of one or a few nucleosomes to larger changes that affect chromatin structure over a longer distance. Chromatin remodeling is carried out by ATP-dependent chromatin-remodeling complexes, which are a set of diverse multiprotein machines that reposition and restructure nucleosomes.

In eukaryotes, changes in nucleosome position and histone composition are key features of gene regulation. The regulation of transcription depends not only on the activity of regulatory transcription factors that influence RNA polymerase via TFIID and mediator to turn genes on or off, but must also involve changes in chromatin structure that affect the ability of transcription factors to gain access to and bind their target sequences in the promoter region. If the chromatin is in a **closed conformation,** transcription may be difficult or impossible. By comparison, chromatin that is in an **open conformation** is more easily accessible to transcription factors and RNA polymerase so that transcription can occur. Although the closed and open conformations may be affected by the relative compaction of a chromosomal region, researchers have recently determined that histone composition and the precise positioning of nucleosomes at or near promoters often play a key role in eukaryotic gene regulation. In this section, we examine molecular mechanisms that explain how changes in chromatin structure can control the regulation of eukaryotic genes.

Chromatin Remodeling Complexes Alter the Positions and Compositions of Nucleosomes

In recent years, geneticists have been trying to identify the steps that promote the interconversion between the closed and open conformations of chromatin. Nucleosomes have been shown to change position in cells that normally express a particular gene compared with cells in which the gene is inactive. For example, in reticulocytes that express the β-globin gene, an alteration in nucleosome positioning occurs in the promoter region from nucleotide −500 to +200. This alteration is thought to be an important step in gene activation. Based on the analysis of many genes, researchers have discovered that a key role of some transcriptional activators is to orchestrate changes in chromatin structure from the closed to the open conformation by altering nucleosomes.

One way to change chromatin structure is through **ATP-dependent chromatin remodeling.** In this process, the energy of ATP hydrolysis is used to drive a change in the locations and/or compositions of nucleosomes, thereby making the DNA more or less amenable to transcription. Therefore, chromatin remodeling is important for both the activation and repression of transcription.

The remodeling process is carried out by a protein complex that recognizes nucleosomes and uses ATP to alter their configuration. All remodeling complexes have a catalytic ATPase subunit that is similar to other motor proteins, called **DNA translocases,** that move along the DNA. Eukaryotes have multiple families of chromatin remodelers. Though their names may differ depending on the species, common chromatin-remodeling complex families are referred to as the SWI/SNF-family, the ISWI-family, the INO80-family, and the Mi-2-family. The names of these remodelers sometimes refer to the effects of mutations in genes that encode remodeling proteins. For example, the abbreviations SWI and SNF refer to the effects that occur in yeast when these remodeling enzymes are defective. SWI mutants are defective in mating-type switching, and SNF mutations create a sucrose nonfermenting phenotype.

How do chromatin remodelers change chromatin structure? Three effects are possible. One result of ATP-dependent chromatin remodeling is a change in the location of nucleosomes (**Figure 15.9a**). This may involve a shift in nucleosomes to a new location or a change in the relative spacing of nucleosomes over a long stretch of DNA. A second effect is that remodelers may evict histones from the DNA, thereby creating gaps where nucleosomes are not found (**Figure 15.9b**). A third possibility is that remodelers may change the composition of nucleosomes by removing standard histones and replacing them with histone variants (**Figure 15.9c**). The functions of histone variants are described next.

Histone Variants Play Specialized Roles in Chromatin Structure and Function

As discussed in Chapter 10, the five histone genes, which encode histones H1, H2A, H2B, H3, and H4, are moderately repetitive. The total number of histone genes varies from species to species. As an example, the human genome contains over 70 histone genes that have been produced by gene duplication events during evolution. Most of the histone genes encode standard histone proteins. However, a few have accumulated mutations that change the amino acid sequence of the histone proteins. These are called **histone variants.** Among eukaryotic species, histone variants have been identified for histone H1, H2A, H2B, and H3 genes, but not for the H4 gene.

What are the consequences of histone variation? Research over the past two decades has shown that certain histone variants play specialized roles in chromatin structure and function. In all eukaryotes, histone variants are incorporated into a subset of nucleosomes to create functionally specialized regions of chromatin. In most cases, the standard histones are incorporated into the nucleosomes while new DNA is synthesized during S phase of the cell cycle. Later, some of the standard histones are replaced by histone variants via chromatin-remodeling complexes.

Table 15.1 describes the standard histones and a few histone variants that are found in humans. A key role of many histone variants is to regulate the structure of chromatin, thereby influencing gene transcription. Such variants can have opposite effects. The incorporation of histone H2A.Bbd into a chromosomal region where a particular gene is found favors gene activation. In contrast, the incorporation of histone $H1^0$ represses gene expression.

Although our focus in this chapter is on gene regulation, it is worth noting that histone variants play other important roles. For example, histone cenH3, which is a variant of histone H3, is found at the centromeres of each chromosome and functions in the binding of kinetochore proteins. Histone cenH3 is

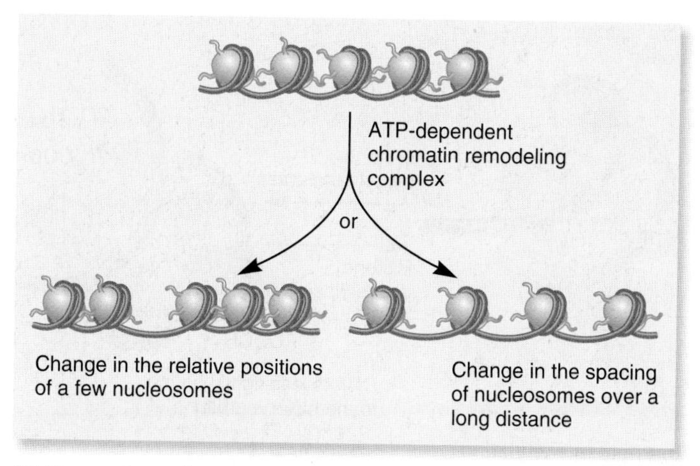

(a) Change in nucleosome position

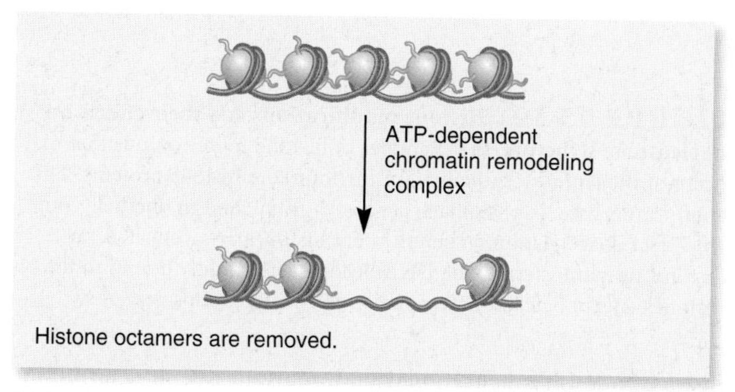

(b) Histone eviction

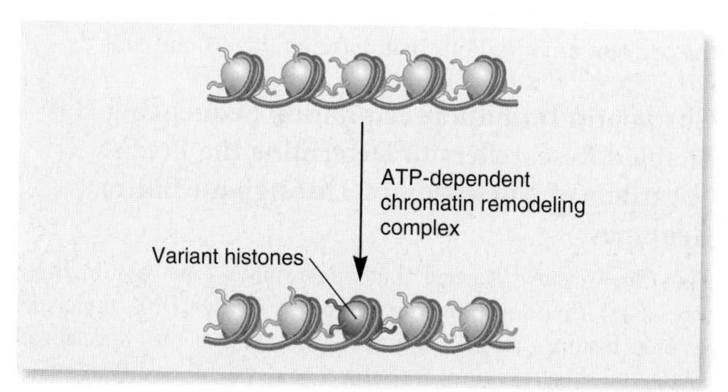

(c) Replacement with variant histones

FIGURE 15.9 ATP-dependent chromatin remodeling. Chromatin-remodeling complexes may **(a)** change the locations of nucleosomes, **(b)** remove histones from the DNA, or **(c)** replace core histones with variant histones.

required for the proper segregation of eukaryotic chromosomes. Other histone variants are primarily found at specialized sites in certain cells. Histone macroH2A is found along the inactivated X chromosome in female mammals, whereas spH2B is found at the telomeres in sperm cells. Finally, certain histone variants appear to play a role in DNA repair. For example, histone H2A.X becomes phosphorylated where a double-stranded DNA break occurs. This phosphorylation is thought to be important for the proper repair of that break.

TABLE 15.1

Standard Human Histones and Examples of Histone Variants

Histone	Type	Number of Genes in Humans	Function
H1	Standard	11	Standard linker histone*
$H1^0$	Variant	1	Linker histone associated with chromatin compaction and gene repression
H2A	Standard	15	Standard core histone
MacroH2A	Variant	1	Core histone that is abundant on the inactivated X chromosome in female mammals. Plays a role in chromatin compaction
H2A.Z	Variant	1	Core histone that is usually found in nucleosomes that flank the transcriptional start site of promoters. Plays a role in gene transcription
H2A.Bbd	Variant	1	Core histone that promotes open chromatin. Plays a role in gene activation
H2A.X	Variant	1	Plays a role in DNA repair
H2B	Standard	17	Standard core histone
spH2B	Variant	1	Core histone found in the telomeres of sperm cells
H3	Standard	10	Standard core histone
cenH3	Variant	1	Core histone found at centromeres. Involved with the binding of kinetochore proteins
H3.3	Variant	2	Core histone that promotes open chromatin. Plays a role in gene activation
H4	Standard	14	Standard core histone

*H1 in mammals is found in five subtypes.

The Histone Code Also Controls Gene Transcription

As described in Chapter 10, each of the core histone proteins consists of a globular domain and a flexible, charged amino terminus called an amino-terminal tail (refer back to Figure 10.14a). The DNA wraps around the globular domains, and the amino-terminal tails protrude from the chromatin. In recent years, researchers have discovered that particular amino acids in the amino-terminal tails of standard histones and histone variants are subject to several types of covalent modifications, including acetylation, methylation, and phosphorylation. Over 50 different enzymes have been identified in mammals that selectively modify amino-terminal tails of histones. **Figure 15.10a** shows examples of sites in the tails of H2A, H2B, H3, and H4 that can be modified.

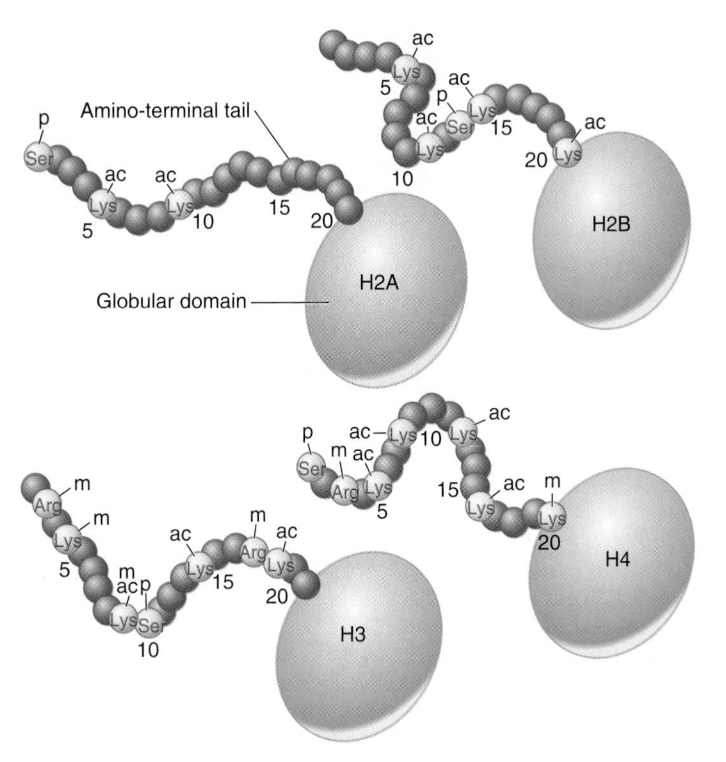

(a) Examples of histone modifications

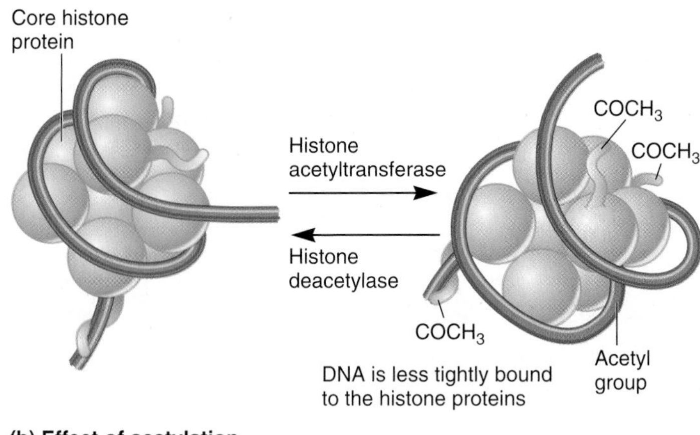

(b) Effect of acetylation

FIGURE 15.10 Histone modifications and their effects on nucleosome structure. **(a)** Examples of histone modifications that may occur at the amino-terminal tail of the four core histone proteins. The abbreviations are: p, phosphate; ac, acetyl group; and m, methyl group. **(b)** Effect of acetylation. When the core histones are acetylated via histone acetyltransferase, the DNA becomes less tightly bound to the histones. Histone deacetylase removes the acetyl groups.

How do histone modifications affect the level of transcription? First, they may directly influence interactions between nucleosomes. For example, positively charged lysines within the core histone proteins can be acetylated by enzymes called **histone acetyltransferases.** The attachment of the acetyl group ($—COCH_3$) eliminates the positive charge on the lysine side chain, thereby disrupting the electrostatic attraction between the histone protein and the negatively charged DNA backbone (**Figure 15.10b**).

In addition, histone modifications occur in patterns that are recognized by proteins. According to the **histone code hypothesis,** proposed by Brian Strahl, C. David Allis, and Bryan Turner in 2000, the pattern of histone modification acts much like a language or code in specifying alterations in chromatin structure. For example, one pattern might involve phosphorylation of the serine at the first position in H2A and acetylation of the fifth and eighth amino acids in H4, which are lysines. A different pattern could involve acetylation of the fifth amino acid, a lysine, in H2B and methylation of the third amino acid in H4, which is an arginine.

The pattern of covalent modifications to the amino-terminal tails provides binding sites for proteins that subsequently affect the degree of transcription. One pattern of histone modification may attract proteins that inhibit transcription, which would silence the transcription of genes in the region. A different combination of histone modifications may attract proteins, such as chromatin-remodeling enzymes, which would serve to alter the positions of nucleosomes in a way that promotes gene transcription. For example, the acetylation of histones attracts certain chromatin remodelers that can shift or evict nucleosomes, thereby aiding in the transcription of genes. Overall, the histone code is thought to play an important role in determining whether the information within the genomes of eukaryotic species is accessed. Researchers are trying to decipher the effects of the covalent modifications that make up the histone code.

Chromatin Immunoprecipitation Sequencing Has Enabled Researchers to Determine the Precise Locations of Nucleosomes Throughout Entire Genomes

Thus far, we have learned that nucleosomes can vary in three ways. First, their locations can be altered along a DNA molecule. Second, histones exist in different variants that play specialized roles. And third, histones are subject to covalent modifications at their amino-terminal tails.

Within the last 10 years researchers have been able to map the locations of specific nucleosomes within a genome. This allows for the determination of (1) where nucleosomes are located, (2) where histone variants are found, and (3) where covalent modifications of histones occur. This amazing achievement has been possible in large part by using an approach called **chromatin immunoprecipitation sequencing (ChIP-Seq).**

As shown in **Figure 15.11**, ChIP-Seq begins with living cells. The cells are treated with formaldehyde, which covalently crosslinks proteins to the DNA. Next, the cells are broken open, and the chromatin is exposed to a high concentration of micrococcal nuclease (MNase), which is an enzyme that digests DNA. MNase digests the linker regions between nucleosomes, but it is unable to cleave DNA that is attached to the core histone proteins. Following this digestion, what remains is millions of nucleosomes that have DNA attached to them.

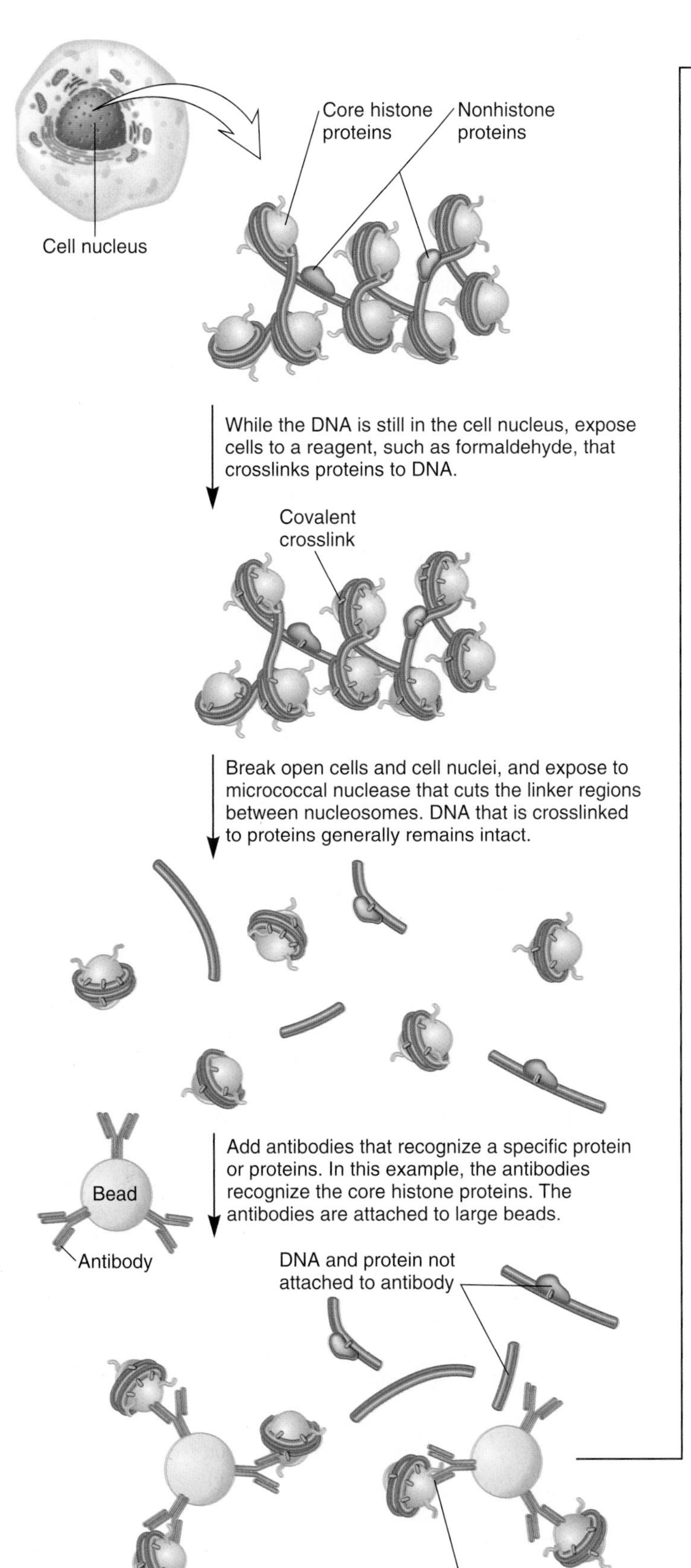

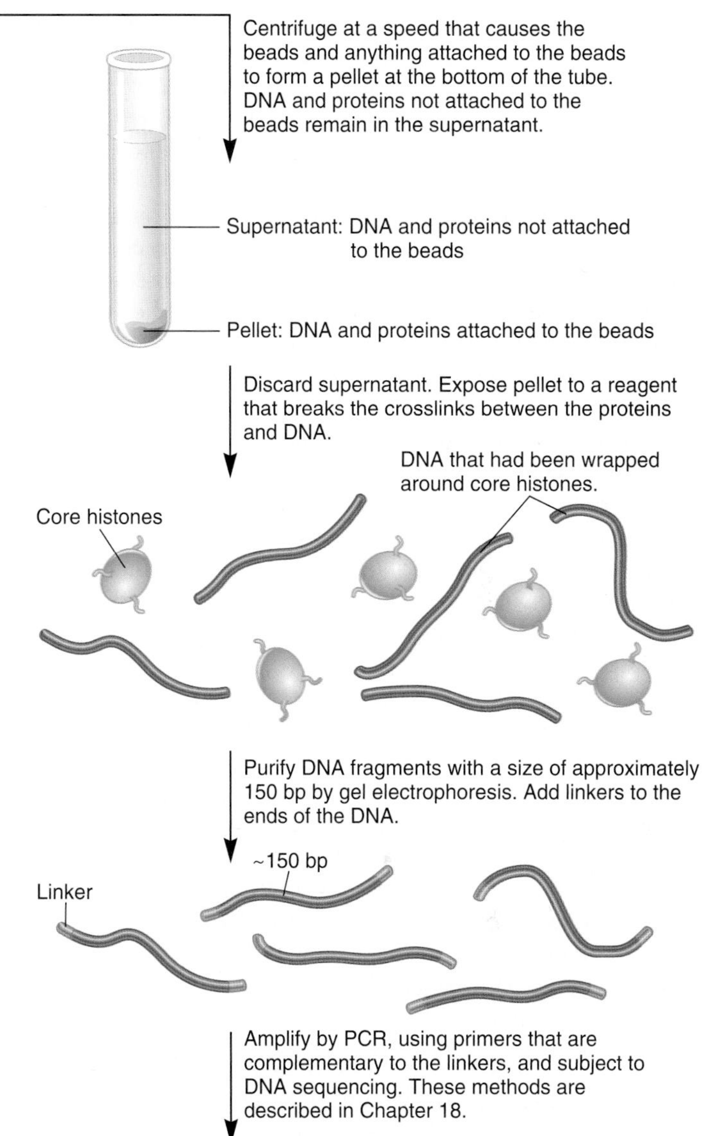

FIGURE 15.11 The use of chromatin immunoprecipitation sequencing (ChIP-Seq) to determine the locations of nucleosomes in the genome of an organism. ChIP-Seq is used to determine where in the genome particular proteins, such as histones, are bound to the DNA. ChIp-Seq may also be used to determine whether other DNA-binding proteins bind within a genome. For example, it may be used to determine where particular regulatory transcription factors bind to specific genes. Note: In this example, micrococcal nuclease is used to separate nucleosomes. An alternative way to break up the DNA is to use shearing forces.

The nucleosomes can be removed from this mixture using antibodies that specifically recognize histone proteins. All nucleosomes can be removed if the antibody recognizes a histone such as H4, which is found in all nucleosomes and does not exist as a histone variant. Alternatively, specific types of nucleosomes can be removed from this mixture using antibodies that recognize particular histones. For example, an antibody could be added that recognizes a variant histone. It is also possible to use antibodies that recognize a histone that has been covalently modified in a certain way. For example, a researcher could use an antibody that recognizes an acetylated histone.

After the addition of a particular type of antibody, the antibodies recognize specific nucleosomes and cause them to aggregate and precipitate out of solution. This process is called **immunoprecipitation.** Following this step, the crosslinks between the histones and DNA are broken, and the DNA is separated from the core histones. The DNA is then subjected to gel electrophoresis. Only those DNA fragments that are about 150 bp in length are saved. This is the length of DNA that wraps around one nucleosome. At this stage, the researcher has thousands or millions of DNA fragments that are about 150 bp in length. Short oligonucleotides, also called linkers, are added to the ends of the DNA fragments, which facilitates their ability to be amplified by the process of polymerase chain reaction (PCR) and subjected to DNA sequencing. (PCR and DNA sequencing are described in Chapter 18).

How are the DNA sequences analyzed? The ChIP-Seq method is used on species in which the entire genome has already been sequenced, such as yeast, *Drosophila*, humans, and *Arabidopsis*. A genome map describes the locations of DNA sequences along each chromosome in a genome. Using computer software, the sequences obtained via ChIP-Seq can be matched to identical sequences on a genome map. Because the genome map also shows the locations of genes, this method allows researchers to determine the relative positions of nucleosomes from the beginning of a gene to the end, as described next.

Eukaryotic Genes Are Flanked by Nucleosome-Free Regions and Well-Positioned Nucleosomes

Over the last 10 years or so, researchers have used ChIP-Seq to analyze the genomes of various species, including yeast, *Drosophila*, humans, and other eukaryotic species. These studies have revealed that many eukaryotic genes show a common pattern of nucleosome organization (**Figure 15.12**). For active genes or those genes that can be activated, the core promoter is found at a **nucleosome-free region (NFR),** which is a site that is missing histones. The NFR is typically 150 bp in length. Although the NFR may be required for transcription, it is not, by itself, sufficient for gene activation. At any given time in the life of a cell, many genes that contain an NFR are not being actively transcribed.

The NFR is flanked by two well-positioned nucleosomes that are termed the −1 and +1 nucleosomes. In yeast, the transcriptional start site (TSS) is usually at the boundary between the NFR and the +1 nucleosome. However, in animals, the TSS is about 60 bp farther upstream into the NFR. The +1 nucleosome typically contains histone variants H2A.Z and H3.3. Depending on the species and the gene, these variants may also be found in the −1 nucleosome and in some of the nucleosomes that immediately follow the +1 nucleosome in the transcribed region. For example, the +2 nucleosome is likely to contain H2A.Z but not as likely as the +1 nucleosome. Similarly, the +3 nucleosome is likely to contain H2A.Z, but not as likely as the +2 nucleosome.

The nucleosomes downstream from the +1 nucleosome tend to be evenly spaced near the beginning of a eukaryotic gene, but their spacing becomes less regular farther downstream. The ends of many eukaryotic genes appear to have a well-positioned nucleosome that is followed by an NFR. This arrangement at the ends of genes may be important for transcriptional termination.

Transcriptional Activation Involves Changes in Nucleosome Locations, Composition, and Histone Modifications

A key role of certain transcriptional activators is to recruit chromatin-remodeling enzymes and histone-modifying enzymes to the promoter region. Though the order of recruitment may differ among specific transcriptional activators, this appears to be critical for transcriptional initiation and elongation. **Figure 15.13** presents a general scheme for how transcriptional activators may facilitate transcription in a eukaryotic gene, such as a gene found in yeast.

As discussed earlier, a gene that is able to be activated has an NFR at the transcriptional start site. In the scenario shown in Figure 15.13, a transcriptional activator binds to the NFR. The

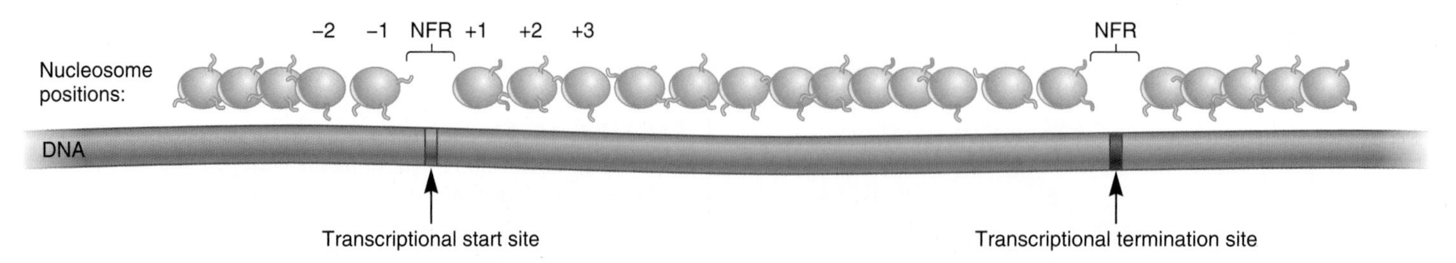

FIGURE 15.12 **Nucleosome arrangements and composition in the vicinity of a structural gene.**

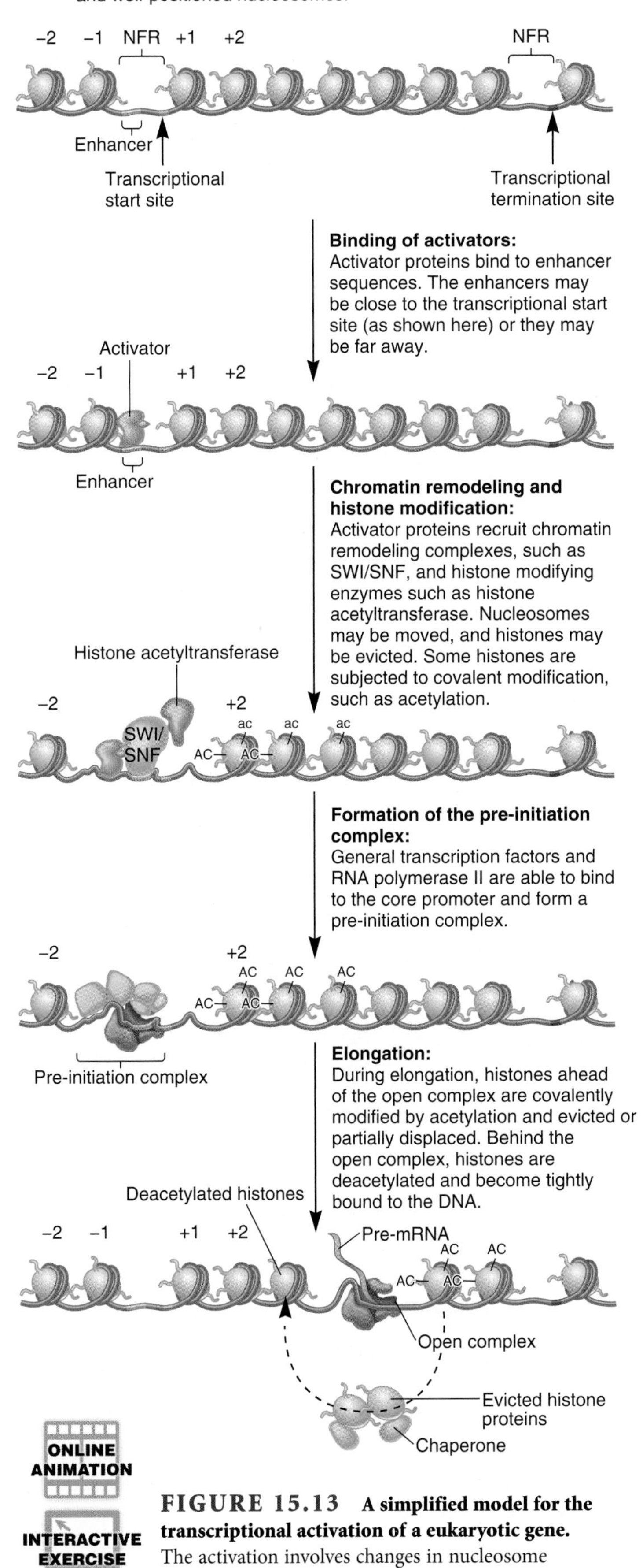

FIGURE 15.13 A simplified model for the transcriptional activation of a eukaryotic gene. The activation involves changes in nucleosome arrangements and histone modifications.

activator then recruits chromatin-remodeling complexes and histone-modifying enzymes to this region. The chromatin remodelers may shift nucleosomes or temporarily evict nucleosomes from the promoter region. Nucleosomes containing the histone variant H2A.Z, which are typically found at the +1 nucleosome, are thought to be more easily removed from the DNA than those containing the standard histone H2A. Histone-modifying enzymes, such as histone acetyltransferase, covalently modify histone proteins and may affect nucleosome contact with the DNA. The actions of chromatin remodelers and histone-modifying enzymes facilitate the binding of general transcription factors and RNA polymerase II to the core promoter, thereby allowing the formation of a preinitiation complex (see Figure 15.13).

Further changes in chromatin structure are necessary for elongation to occur. RNA polymerase II cannot transcribe DNA that is tightly wrapped in nucleosomes. For transcription to occur, histones are evicted, partially displaced, or destabilized so that RNA polymerase II can pass. Evicted histones are transferred to histone chaperones, which are proteins that bind histones and aid in the assembly of histones. Assembled histones are then placed back on the DNA behind the moving RNA polymerase II (see Figure 15.13).

Histone-modifying enzymes also play a key role in histone removal and replacement during the elongation phase of transcription. Histone-modifying enzymes have been found to travel with RNA polymerase II during the elongation phase of transcription. These include enzymes that carry out histone acetylation, H3 methylation, and H2B ubiquitination. These modifications facilitate histone removal ahead of the traveling RNA polymerase II. Behind RNA polymerase II, histone deacetylase removes the acetyl groups, thereby favoring the binding of histones to the DNA to form nucleosomes. The re-formation of nucleosomes behind RNA polymerase II is thought to be critical to maintaining the fidelity of transcription. If nucleosome re-formation did not occur properly, transcriptional initiation might occur at multiple points in a gene, thereby producing faulty transcripts.

Changes in the relative amounts of histone variants also occur within actively transcribed genes. For example, histone variant H3.3 is often found in the transcribed regions of genes, but is less common in silent genes. H3.3 may facilitate the eviction and replacement of nucleosomes ahead and behind RNA polymerase II, respectively. Also, genes with very high levels of transcription may be largely devoid of nucleosomes, because multiple RNA polymerases are transcribing the gene.

15.3 DNA METHYLATION

We now turn our attention to a regulatory mechanism that usually silences gene expression. DNA structure can be modified by the covalent attachment of methyl groups, a mechanism called **DNA methylation.** This process is common in some but not all eukaryotic species. For example, yeast and *Drosophila* have little or no detectable methylation of their DNA, whereas DNA methylation in vertebrates and plants is relatively abundant. In mammals, approximately 2 to 7% of the DNA is methylated. In this section, we will examine how DNA methylation occurs and how it can control gene expression.

DNA Methylation Occurs on the Cytosine Base and Usually Inhibits Gene Transcription

As shown in **Figure 15.14**, eukaryotic DNA methylation occurs via an enzyme called **DNA methyltransferase,** which attaches a methyl group to the number 5 position of the cytosine base, forming 5-methylcytosine. The sequence that is methylated is shown here.

$$\begin{array}{c} CH_3 \\ | \\ 5'\text{-CG-}3' \\ 3'\text{-GC-}5' \\ | \\ CH_3 \end{array}$$

Note that this sequence contains cytosines in both strands. Methylation of the cytosine in both strands is termed full methylation, whereas methylation of only one strand is called hemimethylation.

DNA methylation usually inhibits the transcription of eukaryotic genes, particularly when it occurs in the vicinity of the promoter. In vertebrates and plants, **CpG islands** occur near many promoters of genes. (Note: CpG refers to a dinucleotide of C and G in DNA that is connected by a phosphodiester linkage.) These CpG islands are commonly 1000 to 2000 bp in length and contain a high number of CpG sites. In the case of **housekeeping genes**—genes that encode proteins required in most cells of a multicellular organism—the cytosine bases in the CpG islands are unmethylated. Therefore, housekeeping genes tend to be expressed in most cell types. By comparison, other genes are highly regulated and may be expressed only in a particular cell type. These are **tissue-specific genes.** In some cases, it has been found that the expression of such genes may be silenced by the methylation of CpG islands. Unmethylated CpG islands are correlated with active genes, whereas suppressed genes contain methylated CpG islands. In this way, DNA methylation is thought to play an important role in the silencing of tissue-specific genes to prevent them from being expressed in the wrong tissue.

Methylation can affect transcription in two general ways. First, methylation of CpG islands may prevent or enhance the binding of regulatory transcription factors to the promoter region. For example, methylated CG sequences could prevent the binding of an activator protein to an enhancer element, presumably by the methyl group protruding into the major groove of the DNA (**Figure 15.15a**). The inability of an activator protein to bind to the DNA would inhibit the initiation of transcription. However, CG methylation does not slow down the movement of RNA polymerase along a gene. In vertebrates and plants, coding regions downstream from the core promoter usually contain methylated CG sequences, but these do not hinder the elongation phase of transcription. This suggests that methylation must occur in the vicinity of the promoter to have an effect on transcription.

A second way that methylation inhibits transcription is via proteins known as **methyl-CpG-binding proteins,** which bind methylated sequences (**Figure 15.15b**). These proteins contain a domain called the methyl-binding domain that specifically recognizes a methylated CG sequence. Once bound to the DNA, the methyl-CpG-binding protein recruits other proteins to the region that inhibit transcription. For example, methyl-CpG-binding proteins may recruit histone deacetylase to a methylated CpG island near a promoter. Histone deacetylation removes acetyl groups from the histone proteins, which makes it more difficult for nucleosomes to be removed from the DNA. In this way, deacetylation tends to inhibit transcription.

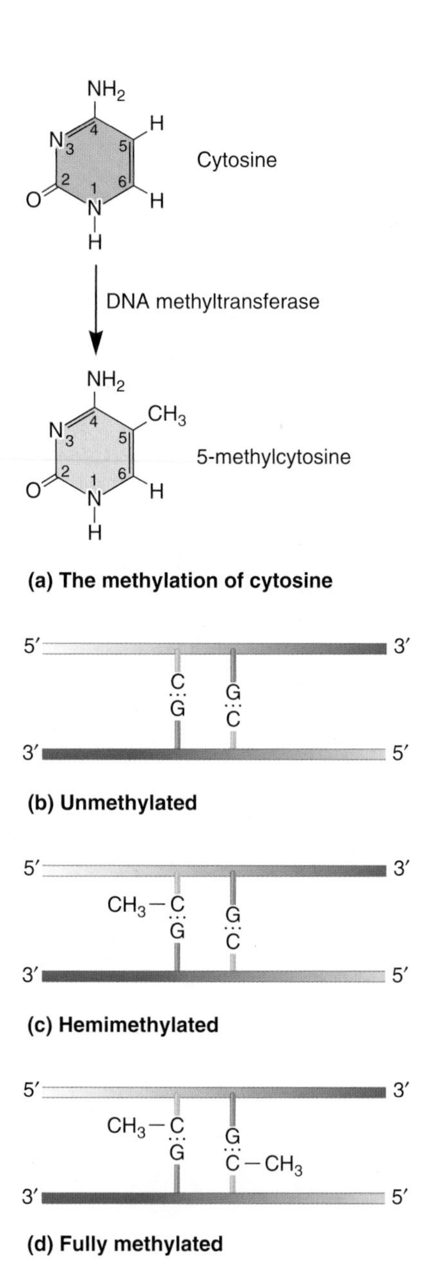

FIGURE 15.14 DNA methylation on cytosine bases. (**a**) Methylation occurs via an enzyme known as DNA methyltransferase, which attaches a methyl group to the number 5 carbon on cytosine. The CG sequence can be (**b**) unmethylated, (**c**) hemimethylated, or (**d**) fully methylated.

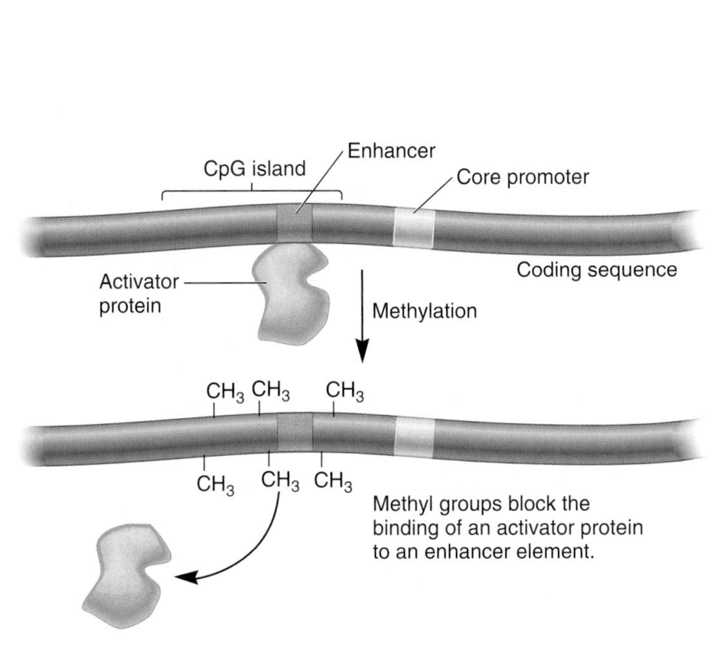

(a) Methylation inhibits the binding of an activator protein.

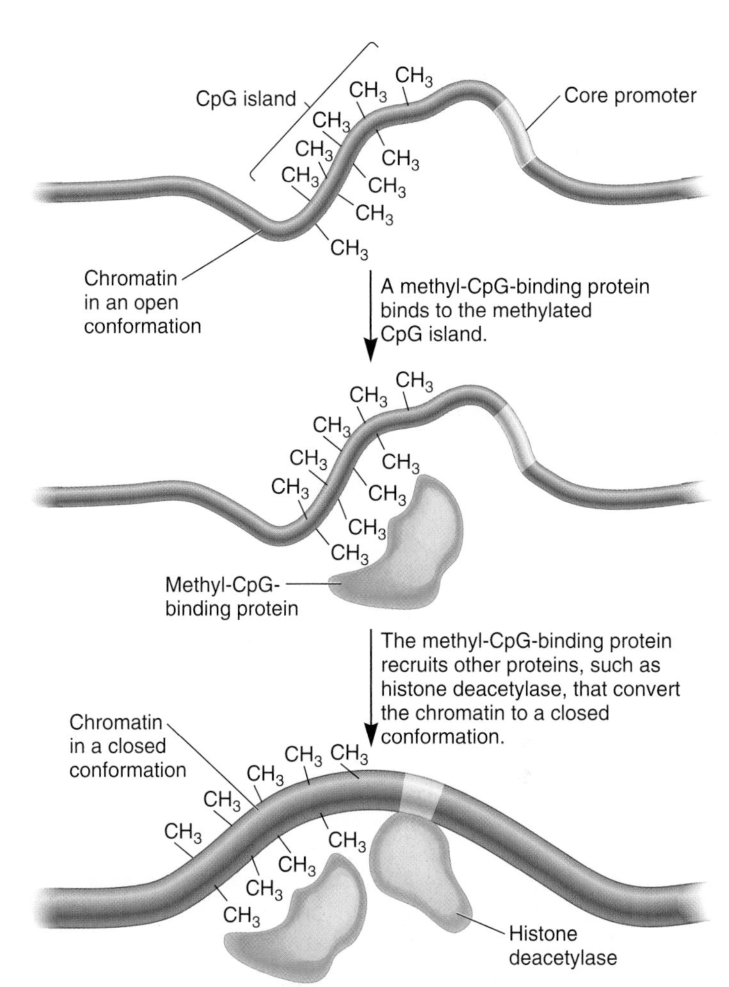

(b) Methyl-CpG-binding protein recruits other proteins that change the chromatin to a closed conformation.

FIGURE 15.15 **Transcriptional silencing via methylation.** (a) The methylation of a CpG island may inhibit the binding of transcriptional activators to the promoter region. (b) The binding of a methyl-CpG-binding protein to a CpG island may lead to the recruitment of proteins, such as histone deacetylase, that convert chromatin to a closed conformation and thus suppress transcription.

DNA Methylation Is Heritable

Methylated DNA sequences are inherited during cell division. Experimentally, if fully methylated DNA is introduced into a plant or vertebrate cell, the DNA will remain fully methylated even in subsequently produced daughter cells. However, if the same sequence of nonmethylated DNA is introduced into a cell, it will remain nonmethylated in the daughter cells. These observations indicate that the pattern of methylation is retained following DNA replication and, therefore, is inherited in future daughter cells.

How can methylation be inherited from cell to cell? **Figure 15.16** illustrates a molecular model that explains this process, which was originally proposed by Arthur Riggs, Robin Holliday, and J. E. Pugh. The DNA in a particular cell may become methylated by ***de novo* methylation**—the methylation of DNA that was previously unmethylated. When a fully methylated segment of DNA replicates in preparation for cell division, the newly made daughter strands contain unmethylated cytosines. Because only one strand is methylated, such DNA is said to be hemimethylated. This hemimethylated DNA is efficiently recognized by DNA methyltransferase, which makes it fully methylated. This process is called **maintenance methylation,** because it preserves the methylated condition in future cells. However, maintenance methylation does not act on unmethylated DNA. Overall, maintenance methylation appears to be an efficient process that routinely occurs within vertebrate and plant cells. By comparison, *de novo* methylation and demethylation are infrequent and highly regulated events. According to this view, the initial methylation or demethylation of a given gene can be regulated so that it occurs in a specific cell type or stage of development. Once methylation has occurred, it can then be transmitted from mother to daughter cells via maintenance methylation.

The methylation mechanism shown in Figure 15.16 can explain the phenomenon of genomic imprinting, which is described in Chapter 7. In this case, specific genes are methylated during oogenesis or spermatogenesis, but not both. Following fertilization, the pattern of methylation is maintained in the offspring. For example, if a gene is methylated only during spermatogenesis, the allele that is inherited from the father will be methylated in the somatic cells of the offspring, but the maternal allele will remain unmethylated. Along these lines, geneticists are also eager to determine how variations in DNA methylation

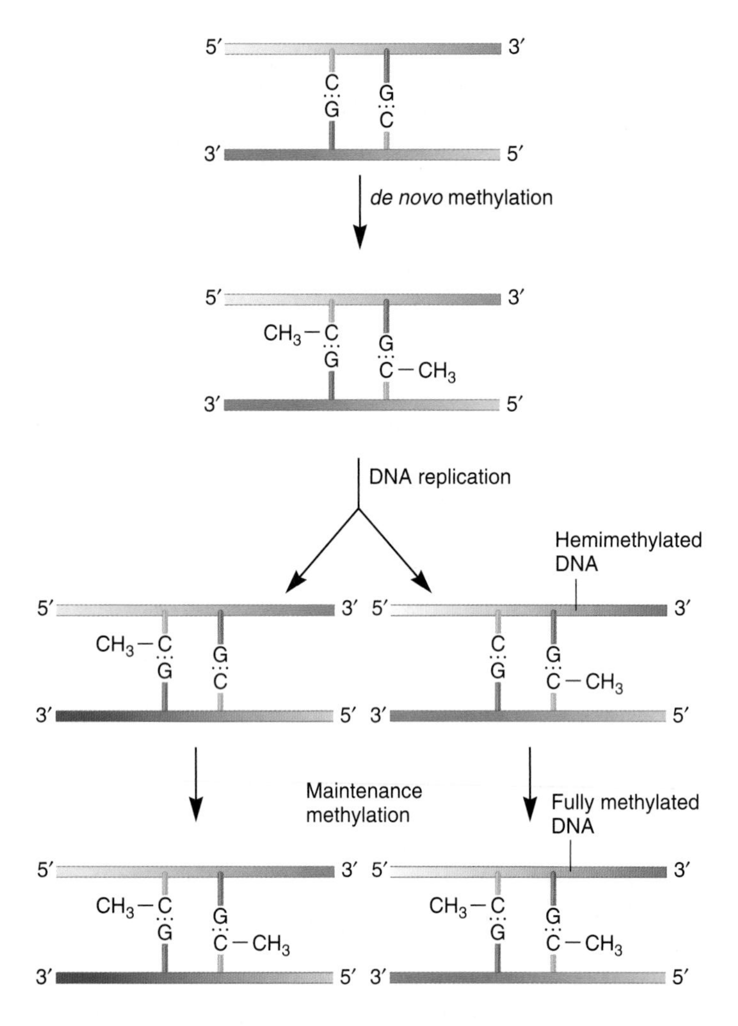

FIGURE 15.16 A molecular model for the inheritance of DNA methylation. The DNA initially undergoes *de novo* methylation, which is a rare, highly regulated event. Once this occurs, DNA replication produces hemimethylated DNA molecules, which are then fully methylated by DNA methyltransferase. This process, called maintenance methylation, is a routine event that is expected to occur for all hemimethylated DNA.

patterns may be important for cell differentiation. It may be a key way to silence genes in different cell types. However, additional research is necessary to understand how specific genes may be targeted for *de novo* methylation or demethylation during different developmental stages or in specific cell types.

15.4 INSULATORS

Thus far, we have considered how regulatory transcription factors, chromatin remodeling, and DNA methylation can regulate gene transcription. As we have seen, eukaryotic gene regulation involves changes in chromatin structure that may occur over a relatively long distance. In addition, gene regulation often involves regulatory elements such as enhancers that are far away from the promoters they regulate.

In eukaryotes, adjacent genes usually exhibit different patterns of gene regulation. Because eukaryotic gene regulation can occur over long distances, a bewildering aspect of such regulation is its ability to control a particular gene but not affect neighboring genes. How are the effects of chromatin remodeling and histone modifications constrained so they only affect a particular gene and not its neighbors? How is an enhancer prevented from controlling multiple genes?

An **insulator** is a segment of DNA that functions as a boundary between two genes. An insulator is so named because it protects, or "insulates," a gene from the regulatory effects of a neighboring gene. In this section, we will consider how insulators are able to carry out this critical function.

Insulators May Act as a Barrier to Changes in Chromatin Structure or Block the Effects of Neighboring Enhancers

Insulators typically perform two roles. Some insulators act as a barrier to chromatin-remodeling or histone-modifying enzymes. As an example, let's consider the effects of histone deacetylase, which removes acetyl groups from core histones, thereby favoring a closed chromatin conformation that is transcriptionally silent. Histone deacetylase may act over a long region of chromatin. How is the action of histone deacetylase controlled so that certain genes in a chromosomal region are not silenced? In the example of **Figure 15.17a**, a gene is found in a chromosomal region in which most of the histones are not acetylated due to

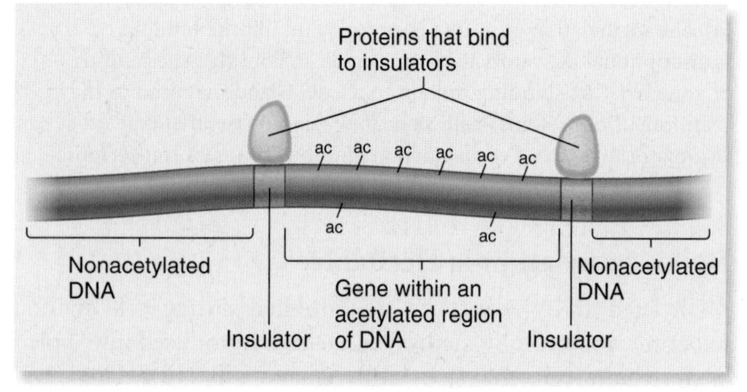

(a) Insulators as a barrier to changes in chromatin structure

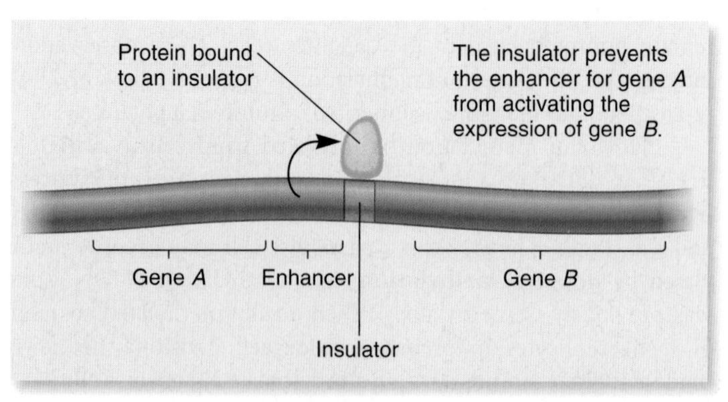

(b) Insulator that blocks the effects of a neighboring enhancer

FIGURE 15.17 Potential effects of insulators. (**a**) Some insulators act as a barrier to enzymes that move along the DNA and change chromatin structure or composition. (**b**) Other insulators prevent enhancers of an adjacent gene from exerting their effects.

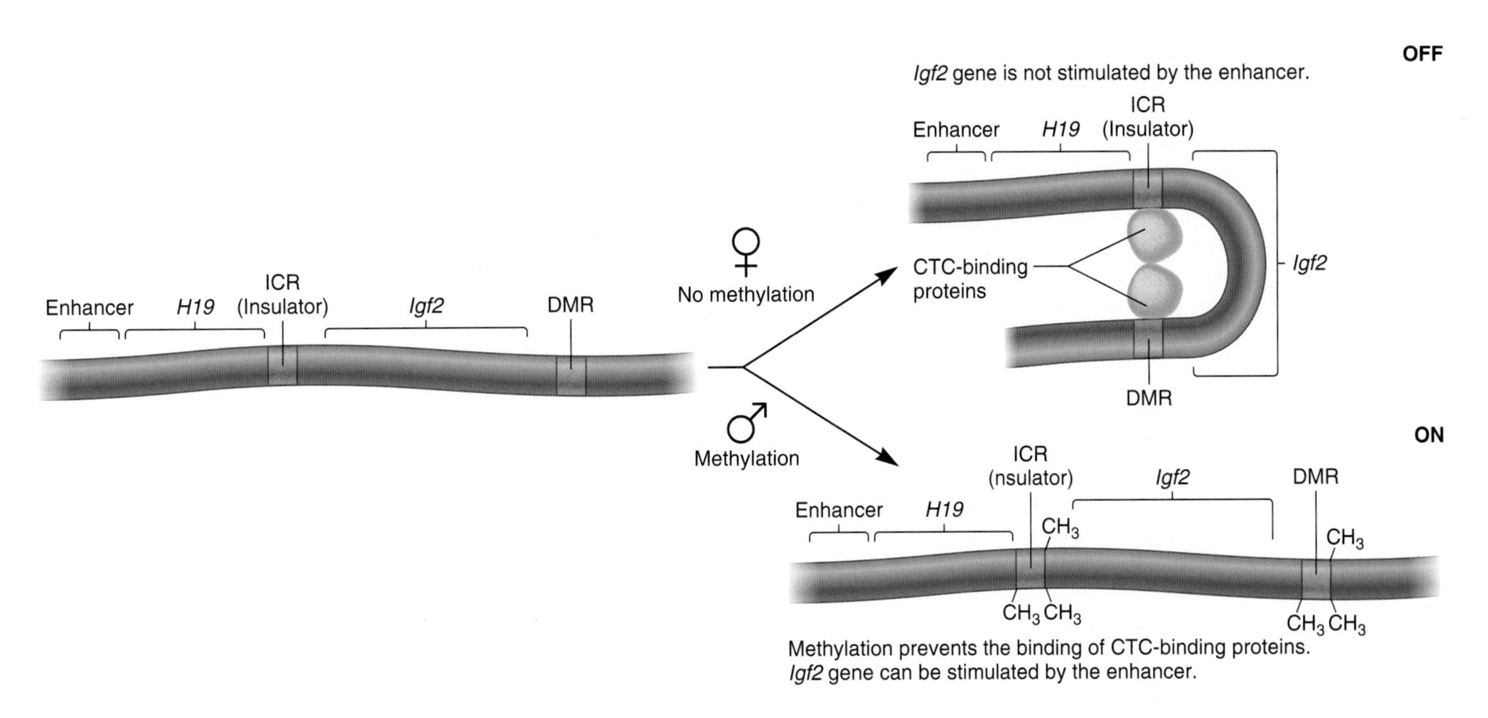

FIGURE 15.18 A looping mechanism to explain the function of an insulator. In this simplified example, the formation of a DNA loop prevents an enhancer of the *H19* gene from exerting its effects on the *Igf2* gene.

the action of histone deacetylase. However, this gene is flanked by two insulators that allow the region where the gene is located to be highly acetylated and transcriptionally activated. In this example, the insulators act as barriers to the action of histone deacetylase. The mechanisms that enable insulators to act as barriers are not well understood. However, such insulators often bind proteins that recruit histone-modifying enzymes or chromatin-remodeling enzymes to the region. For example, an insulator could bind a protein that recruits histone acetyltransferase, which would favor the acetylation of core histones.

A second role of insulators is to block the effects of enhancers that exert their effects on neighboring genes. As discussed earlier in this chapter, enhancers may stimulate the expression of genes that are relatively far away. To prevent an enhancer of one gene from activating the expression of an adjacent gene, an insulator may be located in between them. In the example shown in **Figure 15.17b**, the enhancer can activate the expression of gene *A* but the insulator prevents it from exerting its effects on gene *B*. In other words, gene *B* is insulated from the effects of this enhancer. How does this occur? Although the mechanism of enhancer blocking may vary among different genes, one mechanism is chromosome looping, which is described next.

Insulators May Promote the Formation of Loops That Block the Effects of Nearby Enhancers

How do insulators block the regulatory effects of enhancers? The mechanisms of insulator function are still being actively investigated, but one way they may block the activity of enhancers is by a looping mechanim. This type of mechanism is thought to operate with regard to an insulator that affects the *Igf2* gene, which is discussed in Chapter 5 (see Figure 5.11). Furthermore, the insulator plays a role in imprinting. **Figure 15.18** shows a simplified mechanism for the effects of an insulator (also called the imprinting control region, or ICR) that is located between the *H19* and *Igf2* genes. An enhancer is found next to the *H19* gene. To prevent this enhancer from also stimulating the *Igf2* gene, a loop is formed. For this to occur, a protein called the CTC-binding factor, or CTCF, binds to both the insulator and to a site that follows the *Igf2* gene called a differentially methylated region (DMR). The CTCFs bound to these sites then bind to each other to form a loop in the DNA. This loop blocks the ability of the enhancer to stimulate the *Igf2* gene.

Alternatively, if the insulator and DMR are methylated, which occurs during sperm formation, CTCF is unable to bind to the DNA (see bottom right of Figure 15.18). This prevents loop formation, which allows the enhancer to stimulate the *Igf2* gene. Therefore, the paternally inherited *Igf2* gene is transcriptionally activated. By comparison, the maternally inherited *Igf2* gene, which is not methylated, is transcriptionally silent due to the looping mechanism shown in Figure 15.18.

15.5 REGULATION OF RNA PROCESSING, RNA STABILITY, AND TRANSLATION

Thus far, we have considered a variety of mechanisms that regulate the level of gene transcription. These mechanisms control the amount of RNA transcribed from a given gene. In addition, eukaryotic gene expression is commonly regulated after the RNA is made (**Table 15.2**). One mechanism is pre-mRNA processing. Following transcription, a pre-mRNA transcript is processed before it becomes a functional mRNA. These processing events, which include splicing, capping, polyA tailing, and RNA editing, were described in Chapter 12. We begin this section by

TABLE 15.2

Gene Regulation via RNA Processing and Translation

Effect	Description
Alternative splicing	Certain pre-mRNAs can be spliced in more than one way, leading to polypeptides that have different amino acid sequences. Alternative splicing is often cell-specific so that a protein can be fine-tuned to function in a particular cell type. It is an important form of gene regulation in multicellular eukaryotic species.
RNA stability	The amount of RNA is greatly influenced by the half-life of RNA transcripts. A long polyA tail on mRNAs promotes their stability due to the binding of polyA-binding protein. Some RNAs with a relatively short half-life contain sequences that target them for rapid destruction. Some RNAs are stabilized by specific RNA-binding proteins that usually bind near the 3′ end.
RNA interference	Double-stranded RNA can mediate the degradation of homologous mRNAs in the cell. This is a mechanism of gene regulation. Also, it probably provides eukaryotic cells with protection from invasion by certain types of viruses and may prevent the movement of transposable elements.
General regulation of translation	The function of translational initiation factors may be regulated to permit or inhibit translation. This regulation affects the translation of all cellular mRNAs. Inhibition of translation is desirable if a cell has been exposed to a virus or to toxic materials.
Translational regulation of specific mRNAs	Some mRNAs are regulated via binding proteins that inhibit the ability of the ribosomes to initiate translation. These proteins usually bind at the 5′ end of the mRNA, thereby preventing the ribosome from binding.

examining how the process of alternative splicing is regulated at the RNA level.

Another strategy for regulating gene expression is to influence the concentration of mRNA. This can be accomplished by regulating the rate of transcription. When the transcription of a gene is increased, a higher concentration of the corresponding RNA results. In addition, RNA concentration is greatly affected by the stability or half-life of a particular RNA. Factors that increase RNA stability are expected to raise the concentration of that RNA molecule. We will explore how sequences within mRNA molecules greatly affect their stability.

We also consider a newly discovered mechanism of mRNA silencing, known as RNA interference, which involves double-stranded RNAs that may direct the breakdown of specific RNAs or inhibit their translation. In addition to RNA interference, the process of mRNA translation may be regulated at the level of the translational machinery such as ribosomes. In eukaryotes, this can occur by directly controlling the function of translational initiation factors. Also, RNA-binding proteins can prevent ribosomes from initiating the translation process for specific mRNAs. In this section, we will examine these interesting mechanisms for regulating mRNA translation.

Alternative Splicing Regulates Which Exons Occur in an RNA Transcript, Allowing Different Polypeptides to Be Made from the Same Structural Gene

When it was first discovered, the phenomenon of splicing seemed like a rather wasteful process. During transcription, energy is used to synthesize intron sequences. Likewise, energy is also used to remove introns via a large spliceosome complex. This observation intrigued many geneticists, because natural selection tends to eliminate wasteful processes. Therefore, instead of simply viewing splicing as a wasteful process, many geneticists expected to find that pre-mRNA splicing has one or more important biological roles. In recent years, one very important biological advantage has become apparent. This is **alternative splicing,** which refers to the phenomenon that a pre-mRNA can be spliced in more than one way.

What is the advantage of alternative splicing? To understand the biological effects of alternative splicing, remember that the sequence of amino acids within a polypeptide determines the structure and function of a protein. Alternative splicing produces two or more polypeptides with differences in their amino acid sequences, leading to possible changes in their functions. In most cases, the alternative versions of the protein have similar functions, because most of their amino acid sequences are identical to each other. Nevertheless, alternative splicing produces differences in amino acid sequences that provide each polypeptide with its own unique characteristics. Because alternative splicing allows two or more different polypeptide sequences to be derived from a single gene, some geneticists have speculated that an important advantage of this process is that it allows an organism to carry fewer genes in its genome.

The degree of splicing and alternative splicing varies greatly among different species. Baker's yeast (*Saccharomyces cerevisiae*), for example, contains about 6300 genes and approximately 300 (i.e., approximately 5%) encode pre-mRNAs that are spliced. Of these, only a few have been shown to be alternatively spliced. Therefore, in this unicellular eukaryote, alternative splicing is not a major mechanism for generating protein diversity. In comparison, complex multicellular organisms seem to rely on alternative splicing to a great extent. Humans have approximately 20,000 to 25,000 different genes, and most of these contain one or more introns. Recent estimates suggest that about 70% of all human pre-mRNAs are alternatively spliced. Furthermore, certain pre-mRNAs are alternatively spliced to an extraordinary extent. Some pre-mRNAs can be alternatively spliced to produce dozens of different mRNAs. This provides a much greater potential for human cells to create protein diversity.

Figure 15.19 considers an example of alternative splicing for a gene that encodes a protein known as α-tropomyosin. This protein functions in the regulation of cell contraction. It is located along the thin filaments found in smooth muscle cells, such as those in the uterus and small intestine, and in striated

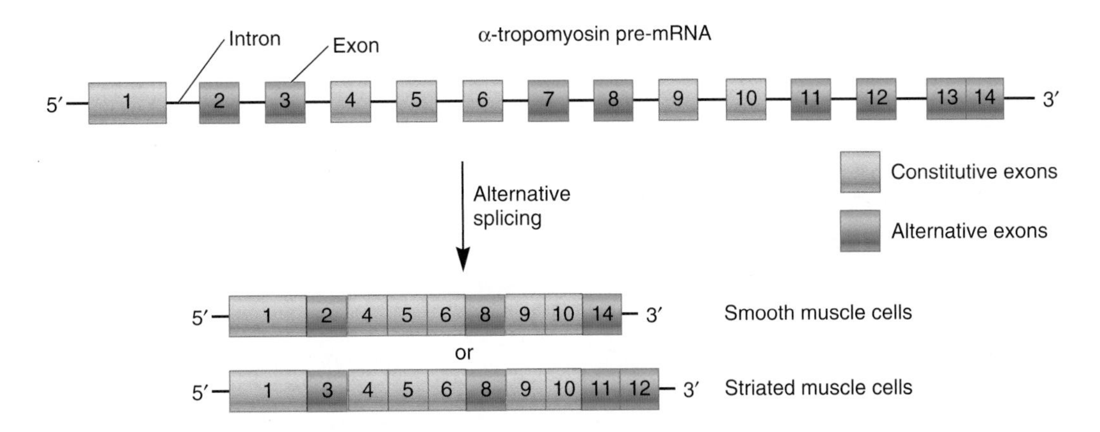

FIGURE 15.19 Alternative ways that the rat α-tropomyosin pre-mRNA can be spliced. The top part of this figure depicts the structure of the rat α-tropomyosin pre-mRNA. Exons are shown as colored boxes, and introns are illustrated as connecting black lines. The lower part of the figure describes the final mRNA products in smooth and striated muscle cells. Note: Exon 8 is found in the final mRNA of smooth and striated muscle cells, but not in the mRNA of certain other cell types. The junction between exons 13 and 14 contains a 3′ splice site so that exon 13 can be separated from exon 14.

Genes → Traits Alpha-tropomyosin functions in the regulation of cell contraction in muscle and nonmuscle cells. Alternative splicing of the pre-mRNA provides a way to vary contractibility in different types of cells by modifying the function of α-tropomyosin. As shown here, the alternatively spliced versions of the pre-mRNA produce α-tropomyosin proteins that differ slightly from each other in their structure (i.e., amino acid sequence). These alternatively spliced versions vary in function to meet the needs of the cell type in which they are found. For example, the sequence of exons 1–2–4–5–6–8–9–10–14 produces an α-tropomyosin protein that functions suitably in smooth muscle cells. Overall, alternative splicing affects the traits of an organism by allowing a single gene to encode several versions of a protein, each optimally suited to the cell type in which it is made.

muscle cells that are found in cardiac and skeletal muscle. Alpha-tropomyosin is also synthesized in many types of nonmuscle cells but in lower amounts. Within a multicellular organism, different types of cells must regulate their contractibility in subtly different ways. One way this may be accomplished is by the production of different forms of α-tropomyosin.

The intron–exon structure of the rat α-tropomyosin pre-mRNA and two alternative ways that the pre-mRNA can be spliced are described in Figure 15.19. The pre-mRNA contains 14 exons, 6 of which are **constitutive exons** (shown in red), which are always found in the mature mRNA from all cell types. Presumably, constitutive exons encode polypeptide segments of the α-tropomyosin protein that are necessary for its general structure and function. By comparison, **alternative exons** (shown in green) are not always found in the mRNA after splicing has occurred. The polypeptide sequences encoded by alternative exons may subtly change the function of α-tropomyosin to meet the needs of the cell type in which it is found. For example, Figure 15.19 shows the predominant splicing products found in smooth muscle cells and striated muscle cells. Exon 2 encodes a segment of the α-tropomyosin protein that alters its function to make it suitable for smooth muscle cells. By comparison, the α-tropomyosin mRNA found in striated muscle cells does not include exon 2. Instead, this mRNA contains exon 3, which is more suitable for that cell type.

Alternative splicing is not a random event. Rather, the specific pattern of splicing is regulated in any given cell. The molecular mechanism for the regulation of alternative splicing involves proteins known as **splicing factors.** Such splicing factors play a key role in the choice of particular splice sites. **SR proteins** are an example of a type of splicing factor. SR proteins contain a domain at their carboxyl-terminal end that is rich in serines (S) and arginines (R) and is involved in protein–protein recognition. They also contain an RNA-binding domain at their amino-terminal end.

As discussed in Chapter 12, components of the spliceosome recognize the 5′ and 3′ splice sites and then remove the intervening intron. The key effect of splicing factors is to modulate the ability of the spliceosome to choose 5′ and 3′ splice sites. This can occur in two ways. Some splicing factors act as repressors that inhibit the ability of the spliceosome to recognize a splice site. In **Figure 15.20a**, a splicing repressor binds to a 3′-splice site and prevents the spliceosome from recognizing the site. Instead, the spliceosome binds to the next available 3′-splice site. The splicing repressor causes exon 2 to be spliced out of the mature mRNA, an event called **exon skipping.** Alternatively, other splicing factors enhance the ability of the spliceosome to recognize particular splice sites. In **Figure 15.20b**, splicing enhancers bind to the 3′- and 5′-splice sites that flank exon 3, which results in the inclusion of exon 3 in the mature mRNA.

Alternative splicing in different tissues is thought to occur because each cell type has its own characteristic concentration of many kinds of splicing factors. Furthermore, much like transcription factors, splicing factors may be regulated by the binding of small effector molecules, protein–protein interactions, and covalent modifications. Overall, the differences in the composition of splicing factors and the regulation of their activities form the basis for alternative splicing decisions.

The Stability of mRNA Influences mRNA Concentration

In eukaryotes, the stability of mRNAs can vary considerably. Certain mRNAs have very short half-lives, such as several minutes, whereas others can persist for many days or even months. In some cases, the stability of an mRNA can be regulated so that

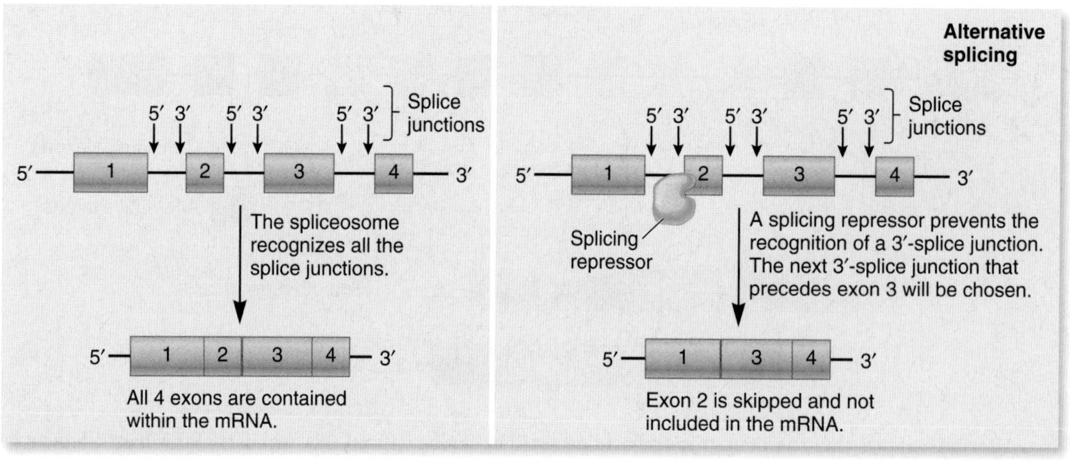

(a) Splicing repressors

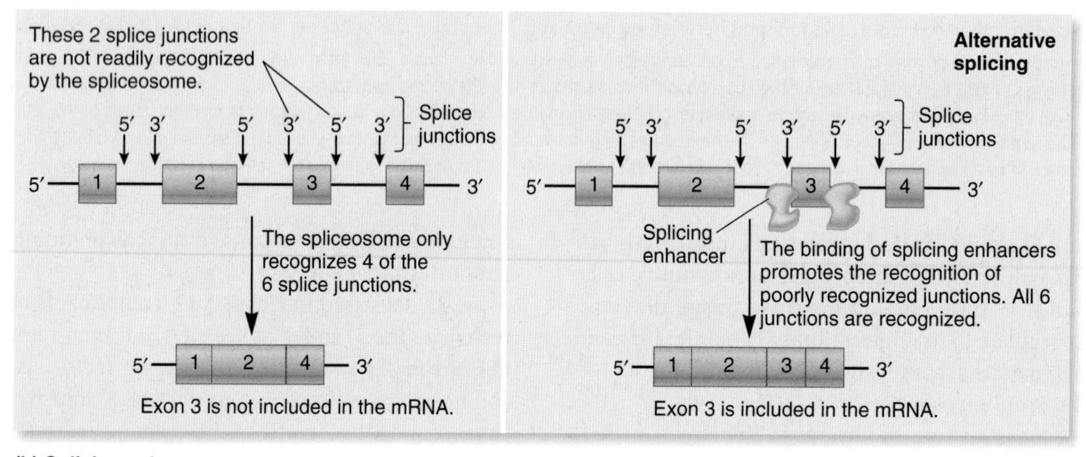

(b) Splicing enhancers

FIGURE 15.20 **The roles of splicing factors during alternative splicing.** **(a)** Splicing factors can act as repressors to prevent the recognition of splice sites. In this example, the presence of the splicing repressor causes exon 2 to be skipped and thus not included in the mRNA. **(b)** Other splicing factors can enhance the recognition of splice sites. In this example, the splicing enhancers promote the recognition of sites that flank exon 3, thereby causing its inclusion in the mRNA.

its half-life is shortened or lengthened. A change in the stability of mRNA can greatly influence the cellular concentration of that mRNA molecule. In this way, mechanisms that control mRNA stability can dramatically affect gene expression.

Various factors play a role in mRNA stability. One important structural feature is the length of the polyA tail. As you may recall from Chapter 12, most newly made mRNAs contain a polyA tail that averages 200 nucleotides in length. The polyA tail is recognized by the **polyA-binding protein.** The binding of this protein enhances RNA stability. However, as an mRNA ages, its polyA tail tends to be shortened by the action of cellular exonucleases. Once it becomes less than 10 to 30 adenosines in length, the polyA-binding protein can no longer bind, and the mRNA is rapidly degraded by exo- and endonucleases.

Certain mRNAs, particularly those with short half-lives, contain sequences that act as destabilizing elements. Although these destabilizing elements can be located anywhere within the mRNA, they are most commonly located near the 3′ end between the stop codon and the polyA tail. This region of the mRNA is known as the **3′-untranslated region (3′-UTR).** An example of a destabilizing element is the **AU-rich element (ARE)** that is found in many short-lived mRNAs (**Figure 15.21**). This element, which contains the consensus sequence AUUUA, is recognized by cellular proteins that bind to the ARE, thereby influencing whether or not the mRNA is rapidly degraded.

Double-Stranded RNA Can Silence the Expression of mRNA

Another way that specific RNAs can be targeted for degradation or translational inhibition is via a recently discovered mechanism involving double-stranded RNA. Research in plants and the nematode *Caenorhabditis elegans* led to the discovery that double-stranded RNA can silence the expression of particular genes. Studies of plant viruses were one avenue of research that identified this mechanism of gene regulation. Certain plant viruses produce double-stranded RNA as part of their life cycle. In addition, these plant viruses may carry genes very similar to genes

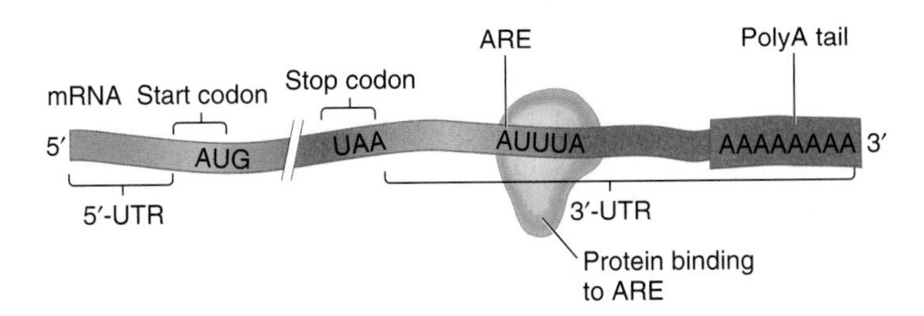

FIGURE 15.21 The location of AU-rich elements (AREs) within mRNAs. One or more AREs are commonly found within the 3′-UTRs (untranslated regions) of mRNAs with short half-lives. The 5′-UTR is the untranslated region of the mRNA that precedes the start codon. AREs are recognized by cellular proteins, which influence whether or not the mRNA is rapidly degraded.

that already exist in the genome of the plant cell. When such viruses infect plant cells, they silence the expression of the plant gene that is similar to the viral gene.

Plant research that involved the production of transgenic plants also supported the idea that double-stranded RNA can cause mRNA to be degraded or inhibited. Using molecular techniques described in Chapter 19, cloned genes can be introduced into the genome of plants. Surprisingly, researchers observed that when cloned genes were introduced in multiple copies, the expression of the gene was often silenced. How were these results explained? We now know that this may be due to the formation of double-stranded RNA. When a cloned gene randomly inserts into a genome, it may, as a matter of random chance, happen to insert itself next to a promoter for a plant gene that is already present in the genome (**Figure 15.22**). The cloned gene itself has a promoter to make a copy of the sense strand. If it randomly inserts next to a plant gene promoter that is oriented in the opposite direction, both strands of the cloned gene would be transcribed, thereby generating double-stranded RNA. As more copies of a cloned gene are introduced into a genome, the likelihood is greater that the scenario described in Figure 15.22 will occur. This event silences the expression of the cloned gene even though it is present in multiple copies. Furthermore, if the cloned gene is homologous to a plant gene that is already present in the plant cell, this phenomenon will also silence the endogenous plant gene. Therefore, the curious observation made by researchers was that increasing the number of copies of a cloned gene frequently led to gene silencing.

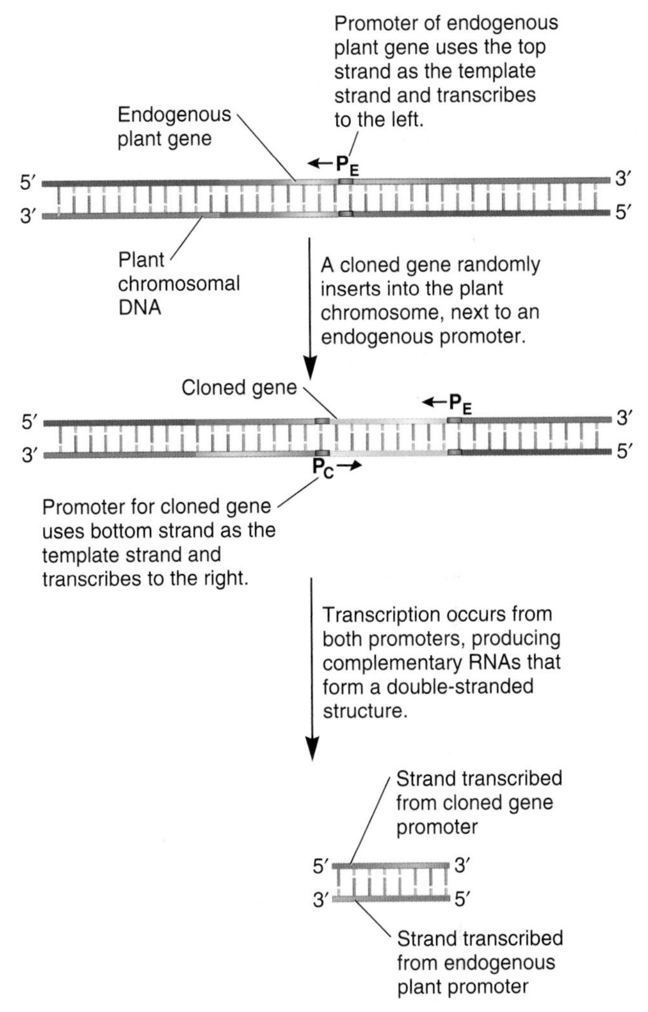

FIGURE 15.22 Gene insertion leading to the production of double-stranded RNA. When a cloned gene with its own promoter is introduced into a plant cell, it randomly inserts into the chromosomal DNA. In some cases, the cloned gene may insert next to a promoter for an endogenous plant gene, so the promoter for the endogenous gene transcribes the antisense strand of the cloned gene. Therefore, the two promoters transcribe the sense strand and the antisense strand to produce complementary RNAs that will form double-stranded RNA.

EXPERIMENT 15A

Fire and Mello Show That Double-Stranded RNA Is More Potent Than Antisense RNA at Silencing mRNA

The discovery that double-stranded RNA inhibits mRNA function came about as a result of studies involving gene expression. As a research tool to investigate the function of particular genes, researchers had often introduced antisense RNA (RNA that is complementary to mRNA) into cells as a way to inhibit mRNA translation. Because antisense RNA is complementary to mRNA, the antisense RNA binds to the mRNA, thereby preventing translation. Oddly, in some experiments, researchers introduced sense RNA (RNA with the same sequence as mRNA) into cells, and this also inhibited mRNA translation. Another curious observation was that the effects of antisense RNA often persisted for a very long time, much longer than would have been predicted by the relatively short half-lives of most RNA molecules in the cell. These two unusual observations caused Andrew Fire, Craig

Mello, and colleagues to investigate how the injection of RNA into cells inhibits mRNA.

They used *C. elegans* as their experimental organism because it was relatively easy to inject with RNA and the expression of many genes had already been established. In 1998, Fire and Mello investigated the effects of several injected RNAs known to be complementary to cellular mRNAs. In the investigation of **Figure 15.23**, we focus on one of their experiments involving a gene called *mex-3*, which had already been shown to be highly expressed in early embryos. They started with the cloned gene for *mex-3*. This gene was genetically engineered using techniques described in Chapter 18 so that one version had a promoter that would result in the synthesis of the normal mRNA, which is the sense RNA. They made another version of *mex-3* in which a promoter directed the synthesis of the opposite strand, which produced antisense RNA. To make RNA in vitro, they added RNA polymerase and nucleotides to these cloned genes to make sense or antisense RNA.

Next, they injected RNA into the gonads of *C. elegans* and observed the effects in developing embryos. They either injected antisense RNA or they mixed sense and antisense RNA and injected double-stranded RNA. They also used uninjected worms as controls. To determine the expression of *mex-3*, they incubated the resulting embryos with a probe complementary to the *mex-3* mRNA. The probe was labeled so it could be observed under the microscope. After this incubation step, any probe that was not bound to this mRNA was washed away.

THE GOAL

The goal was to further understand how the experimental injection of RNA was responsible for the silencing of particular mRNAs.

ACHIEVING THE GOAL — FIGURE 15.23 Injection of antisense and double-stranded RNA into *C. elegans* to compare their effects on mRNA silencing.

Starting material: The researchers used *C. elegans* as their model organism. They also had the cloned *mex-3* gene, which had been previously shown to be highly expressed in the embryo.

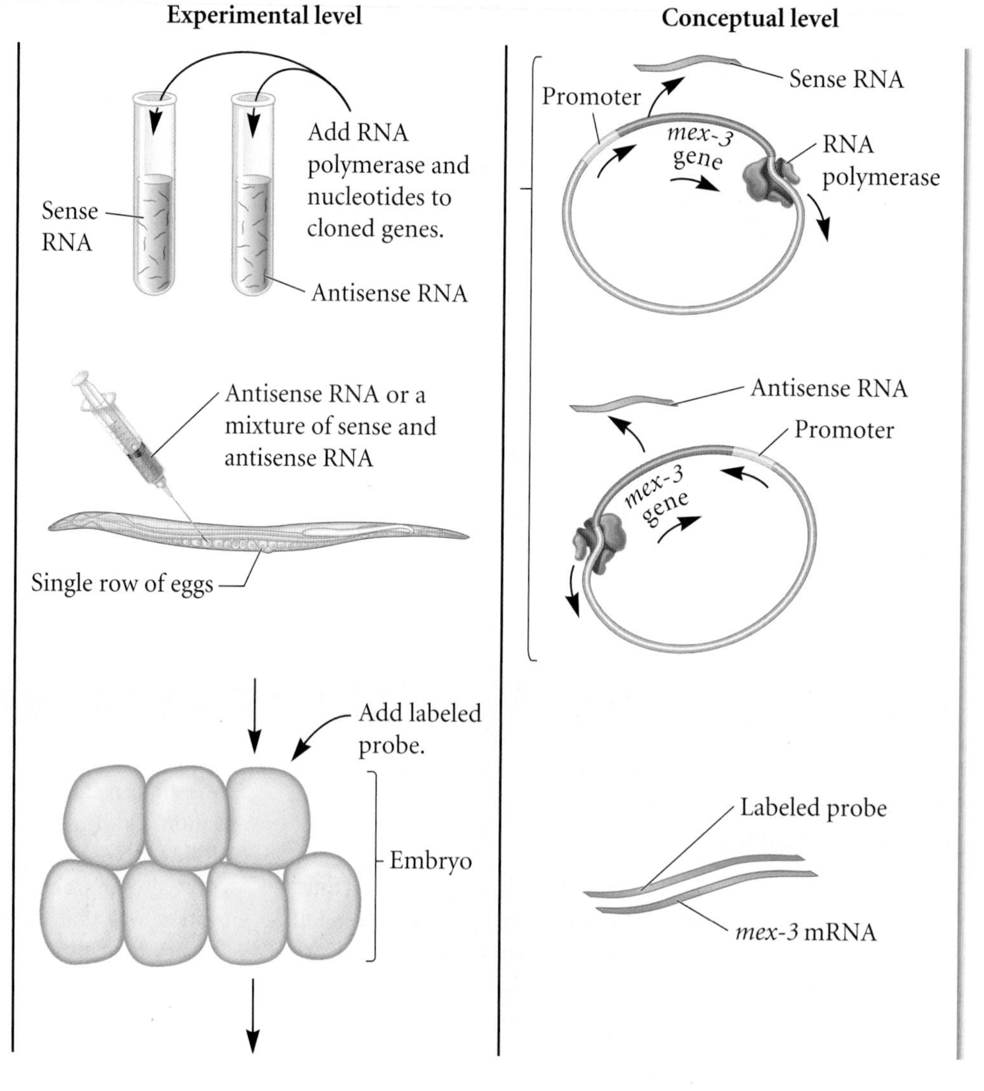

1. Make sense and antisense *mex-3* RNA in vitro using cloned genes for *mex-3* with promoters on either side of the gene. RNA polymerase and nucleotides are added to synthesize the RNAs.

2. Inject either *mex-3* antisense RNA or a mixture of *mex-3* sense and antisense RNA into the gonads of *C. elegans.* This RNA is taken up by the eggs and early embryos. As a control, do not inject any RNA.

3. Incubate and then subject early embryos to *in situ* hybridization. In this method, a labeled probe is added that is complementary to *mex-3* mRNA. If cells express *mex-3*, the mRNA in the cells will bind to the probe and become labeled. After incubation with a labeled probe, the cells are washed to remove unbound probe.

4. Observe embryos under the microscope.

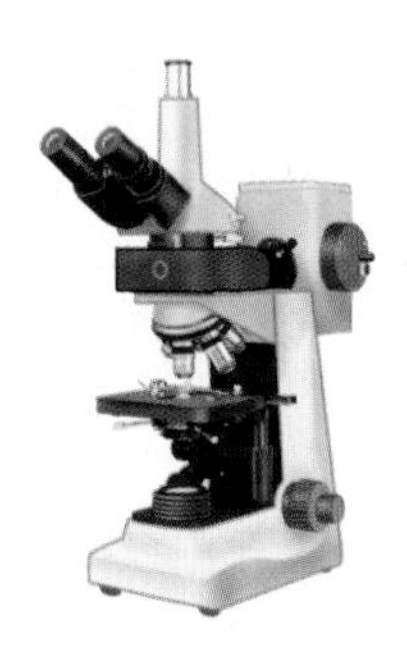

THE DATA

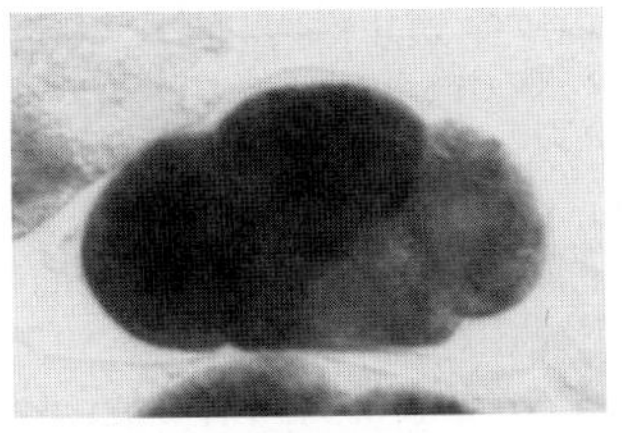

Control

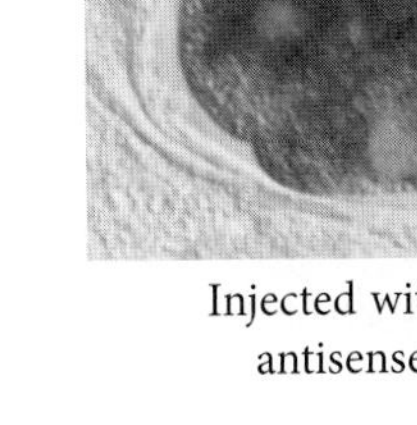

Injected with *mex-3* antisense RNA

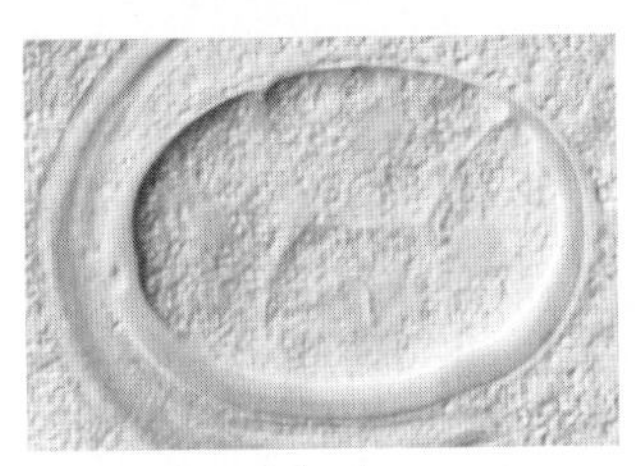

Injected with both *mex-3* sense and antisense RNA

Photos reprinted by permission from Macmillan Publishers Ltd. A. Fire, S. Xu, M.K. Montgomery, et al. (1998) Potent and specific genetic interference by double-stranded RNA in *Caenorhabditis elegans*. *Nature*. 391:6669, 806–811.

INTERPRETING THE DATA

As seen in the data of Figure 15.23, the control embryos were very darkly staining. These results indicated that the *mex-3* gene is highly expressed, which was known from previous research. In the embryos injected with antisense RNA, *mex-3* mRNA levels were decreased, but detectable. Remarkably, in samples injected with double-stranded RNA, no *mex-3* mRNA was detected! These results indicated that double-stranded RNA is more potent at silencing mRNA than is antisense RNA. In this case, the double-stranded RNA caused the *mex-3* mRNA to be degraded. They used the term **RNA interference (RNAi)** to describe the phenomenon in which double-stranded RNA causes the silencing of mRNA. This surprising observation led researchers to investigate the underlying molecular mechanism that accounts for this phenomenon, as described next.

A self-help quiz involving this experiment can be found at www.mhhe.com/brookergenetics4e.

RNA Interference Is Mediated by MicroRNAs or Short-Interfering RNAs via the RNA-Induced Silencing Complex

RNA interference is widely found in eukaryotic species. Double-stranded RNA can come from a variety of sources. As mentioned earlier, viruses sometimes produce double-stranded RNA as part of their reproductive cycle, and the insertion of multiple copies of a gene into the genome can also result in double-stranded RNA. In addition, the genomes of eukaryotic organisms have been found to contain genes that encode RNA molecules that cause RNA interference. These RNA molecules, called **microRNAs (miRNAs)** or **short-interfering RNAs (siRNAs),** are small RNA molecules, typically 21 to 23 nucleotides in length, that silence the expression of specific mRNAs. MicroRNAs are partially complementary to certain cellular mRNAs and inhibit their translation, whereas short-interfering RNAs are usually a perfect match to specific mRNAs and cause the mRNAs to be degraded.

In 1993, Victor Ambros and his colleagues, who were interested in the developmental stages that occur in the worm *C. elegans*, determined that the transcription of a particular gene produced a small RNA, now called a microRNA, that does not encode a protein. Instead, this miRNA was found to be complementary to an mRNA, and it inhibited the translation of the mRNA. Since this study, researchers have discovered that genes encoding miRNAs are widely found in animals and plants. In humans, for example, approximately 200 different genes encode miRNAs.

MiRNAs and siRNAs represent an important mechanism of mRNA silencing. How do miRNAs and siRNAs cause the silencing of specific mRNAs? **Figure 15.24** shows how an miRNA or siRNA, encoded by a gene, leads to RNA interference. In this example, the miRNA or siRNA is first synthesized as a pre-miRNA or pre-siRNA, which forms a hairpin structure that is cut by an endonuclease called **Dicer.** This releases a double-stranded RNA molecule that associates with cellular proteins to form a complex called the **RNA-induced silencing complex (RISC).** One of the RNA strands is degraded. The remaining single-stranded miRNA or siRNA, which is complementary to specific cellular mRNAs, allows the RISC to specifically recognize and bind to those mRNA molecules.

After binding to the mRNA, two different things may happen. In some cases, the RISC may direct the degradation of the mRNA. This usually occurs when the siRNA and mRNA are a perfect match or are highly complementary. Alternatively, the RISC may inhibit translation. This is more common when the miRNA and mRNA are not a perfect match or are only partially complementary. In either case, the expression of the mRNA is silenced. The effect is termed RNA interference because the miRNA or siRNA interferes with the proper expression of an mRNA. Likewise, Dicer and RISC may act on double-stranded RNA that comes from viruses or from the insertion of multiple copies of a gene into the genome (as shown at the bottom of Figure 15.22). In 2006, Fire and Mello received the Nobel Prize in physiology or medicine for their discovery of this phenomenon.

RNA interference is believed to have at least three benefits. First, this phenomenon represents a newly identified form of gene regulation. When genes encoding pre-miRNAs or pre-siRNAs are turned on, the production of miRNAs or siRNAs silences the expression of other genes. Second, RNAi may offer a defense mechanism against certain viruses. In particular, this mechanism may inhibit the proliferation of viruses that have a double-stranded RNA genome and viruses that produce double-stranded RNAs as part of their reproductive cycles. After entering the host cell, the viral RNA would be degraded, as in Figure 15.24, and the cell would survive the infection. Third, researchers have speculated that RNAi may also play a role in the silencing of certain transposable elements. As discussed in Chapter 17, transposable elements are DNA segments that have the capacity to move throughout the genome, an event termed transposition. Transposable elements carry genes, such as transposase, that are needed in the transposition process. The random insertion of many transposable elements in a cell may ultimately lead to gene silencing due to the scenario described in Figure 15.22. The silencing of genes within transposable elements via RNAi would protect the organism against the potentially harmful effects of transposition.

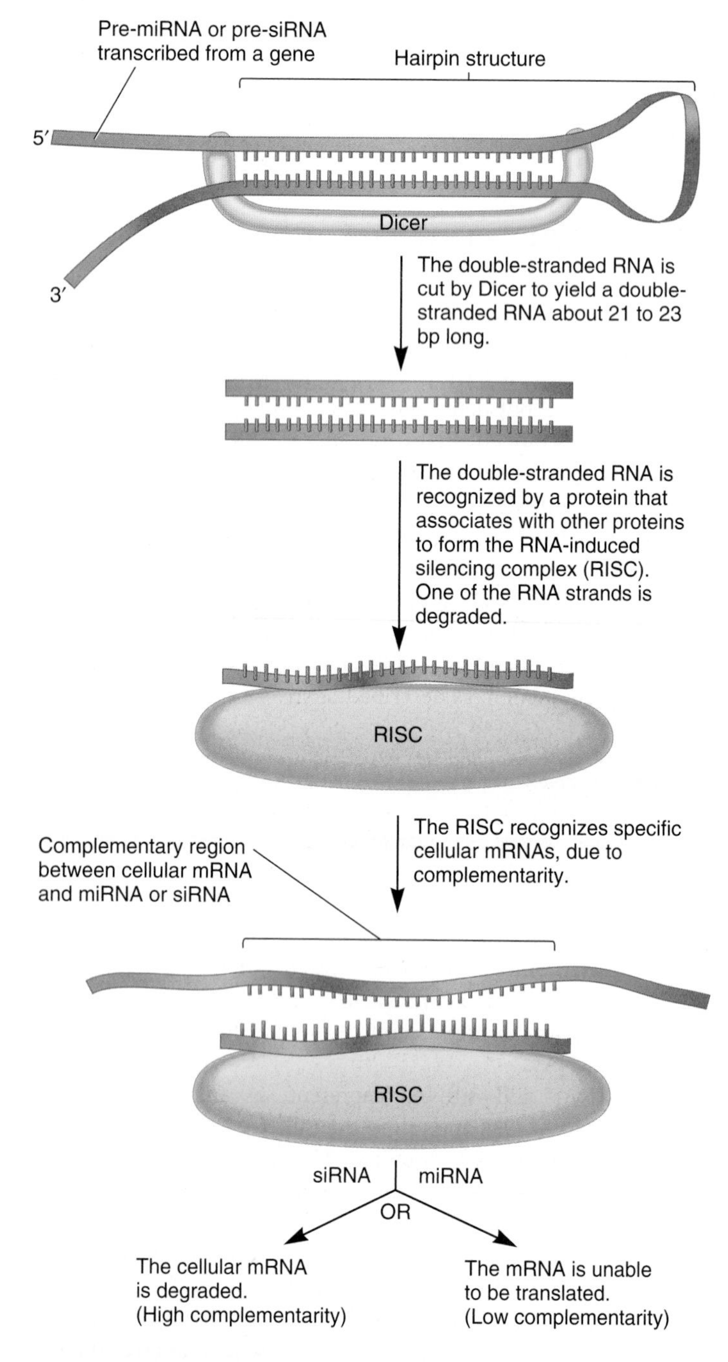

FIGURE 15.24 Mechanism of RNA interference. A double-stranded region of RNA within a pre-miRNA is recognized by an endonuclease called Dicer that digests the RNA to produce a small piece, 21 to 23 bp in length. This small RNA binds with proteins to form an RNA-induced silencing complex (RISC). One strand of the RNA is degraded. The remaining RNA strand in RISC binds to complementary mRNAs. After RISC binding, endonucleases within the complex may cut the mRNA, thereby inactivating it. It should be emphasized that the miRNAs and siRNAs are mRNA-specific. For example, if an miRNA within RISC was complementary to the β-globin mRNA, only the β-globin mRNA would be silenced.

Phosphorylation of Ribosomal Initiation Factors Can Alter the Rate of Translation

Let's now turn our attention to another regulatory mechanism that affects translation. Modulation of translational initiation factors is widely used to control fundamental cellular processes. Under certain conditions, it is advantageous for a cell to stop synthesizing proteins. For example, if a virus infects a cell, the inhibition of protein synthesis can prevent viral proliferation by inhibiting the production of new viral proteins. Likewise, if critical nutrients are in short supply, it is beneficial for a cell to conserve its resources by inhibiting unnecessary protein synthesis.

Translational initiation factors are required to begin protein synthesis. The phosphorylation of many different initiation factors has been found to affect translation in eukaryotic cells. Two factors, eIF2 and eIF4F, appear to play a central role in controlling the initiation of translation. The functions of these two translational initiation factors are modulated by phosphorylation in opposite ways. When the α subunit of eIF2 (known as eIF2α) is phosphorylated, translation is inhibited, whereas the phosphorylation of eIF4F increases the rate of translation.

Figure 15.25 shows the events leading to translational inhibition by eIF2α. A variety of conditions can lead to a shutdown of protein synthesis, including viral infection, nutrient deprivation, heat shock, and the presence of toxic heavy metals. These conditions promote the activation of protein kinases known as eIF2α protein kinases, several of which have been identified. Once activated, an eIF2α protein kinase can phosphorylate eIF2α. The

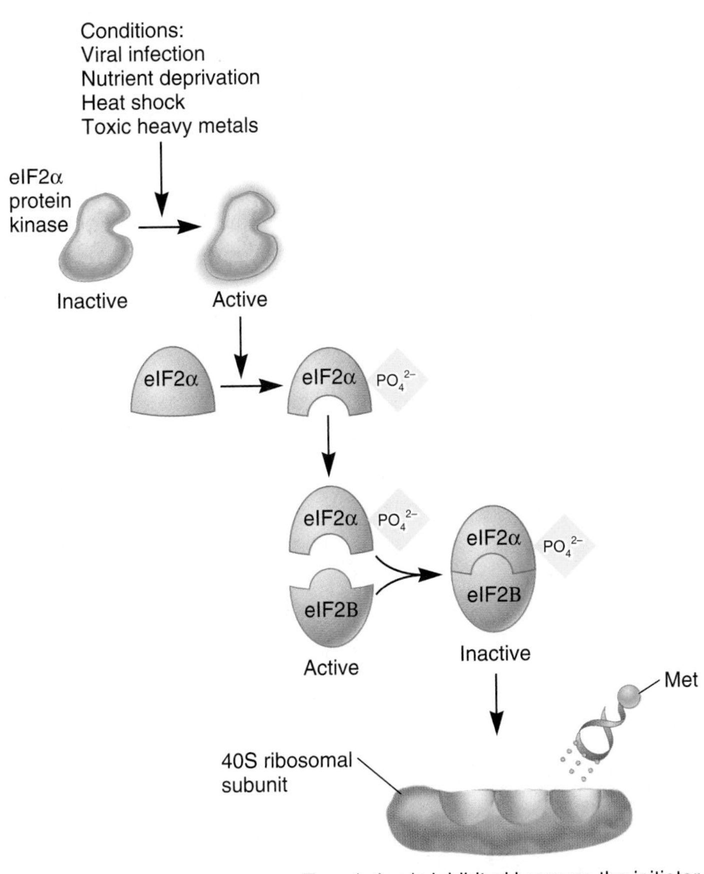

FIGURE 15.25 The pathway that leads to the phosphorylation of eIF2α (eukaryotic initiation factor α) and the inhibition of translation.

phosphorylation of eIF2α causes it to bind tightly to another initiation factor subunit called eIF2B. Functional eIF2B is necessary so that eIF2 can promote the binding of the initiator tRNA^{Met} to the 40S ribosomal subunit. However, when the phosphorylated eIF2α binds to eIF2B, it prevents eIF2B from functioning. Therefore, the initiator tRNA^{Met} does not bind to the 40S subunit, and translation is inhibited.

A second important way to control translation is via the eIF4F translation factor that modulates the binding of mRNA to the ribosomal initiation complex. The function of eIF4F is stimulated by phosphorylation. A variety of conditions have been shown to cause eIF4F to become phosphorylated. These include the presence of growth factors, insulin, and other signaling molecules that promote cell proliferation. Conversely, conditions such as heat shock and viral infection decrease the level of eIF4F phosphorylation, thereby inhibiting translation.

The Regulation of Iron Assimilation Is an Example of the Regulatory Effect of RNA-Binding Proteins on Translation

As we have just seen, the phosphorylation of translational initiation factors can modulate the translation of mRNA. Because these initiation factors are necessary to translate all of a cell's mRNA, this form of regulation affects the expression of many mRNAs. By comparison, specific mRNAs are sometimes regulated by RNA-binding proteins that directly affect translational initiation or RNA stability. The regulation of iron assimilation provides a well-studied example in which both of these phenomena occur. Before discussing this form of translational control, let's consider the biology of iron metabolism.

Iron is an essential element for the survival of living organisms because it is required for the function of many different enzymes. The pathway by which mammalian cells take up iron is depicted in **Figure 15.26**. Iron ingested by an animal is absorbed into the bloodstream and becomes bound to transferrin, a protein that carries iron through the bloodstream. The transferrin–Fe^{3+} complex is recognized by a transferrin receptor on the surface of cells; the complex binds to the receptor and then is transported into the cytosol by endocytosis. Once inside, the iron is then released from transferrin. At this stage, Fe^{3+} may bind to cellular enzymes that require iron for their activity. Alternatively, if too much iron is present, the excess iron is stored within a hollow, spherical protein known as ferritin. The storage of excess iron within ferritin helps to prevent the toxic buildup of too much iron within the cell.

Because iron is a vital yet potentially toxic substance, mammalian cells have evolved an interesting way to regulate its assimilation. The two mRNAs that encode ferritin and the transferrin receptor are both influenced by an RNA-binding protein known as the **iron regulatory protein (IRP)**. How does IRP exert its

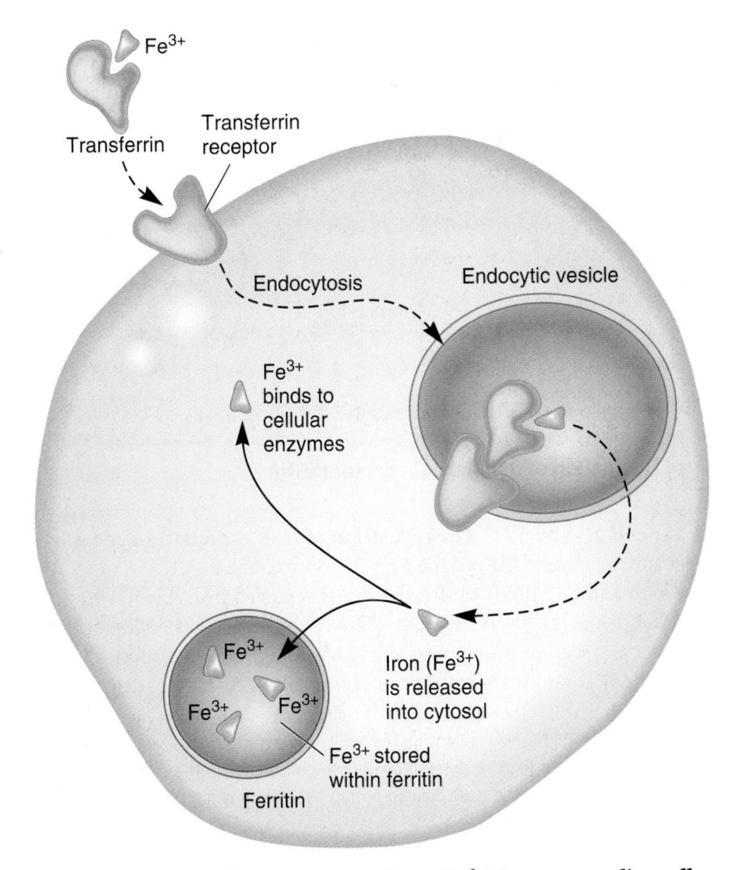

FIGURE 15.26 The uptake of iron (Fe^{3+}) into mammalian cells.

effects? This protein binds to a regulatory element within these two mRNAs known as the **iron response element (IRE).** The ferritin mRNA has an IRE in its 5′-untranslated region (5′-UTR). When IRP binds to this IRE, it inhibits the translation of the ferritin mRNA (**Figure 15.27a**, left). However, when iron is abundant in the cytosol, the iron binds directly to IRP and prevents it from binding to the IRE. Under these conditions, the ferritin mRNA is translated to make more ferritin protein (Figure 15.27a, right), which prevents the toxic buildup of iron within the cytosol.

The transferrin receptor mRNA also contains an iron response element, but it is located in the 3′-UTR. When IRP binds to this IRE, it does not inhibit translation. Instead, the binding of IRP increases the stability of the mRNA by blocking the action of RNA-degrading enzymes. This leads to increased amounts of transferrin receptor mRNA within the cell when the cytosolic levels of iron are very low (**Figure 15.27b**, left). Under these conditions, more transferrin receptor is made. This promotes the uptake of iron when it is in short supply. In contrast, when iron is abundant within the cytosol, IRP is removed from the transferrin receptor mRNA, and the mRNA becomes rapidly degraded (Figure 15.27b, right). This leads to a decrease in the amount of transferrin receptor, thereby helping to prevent the uptake of too much iron into the cell.

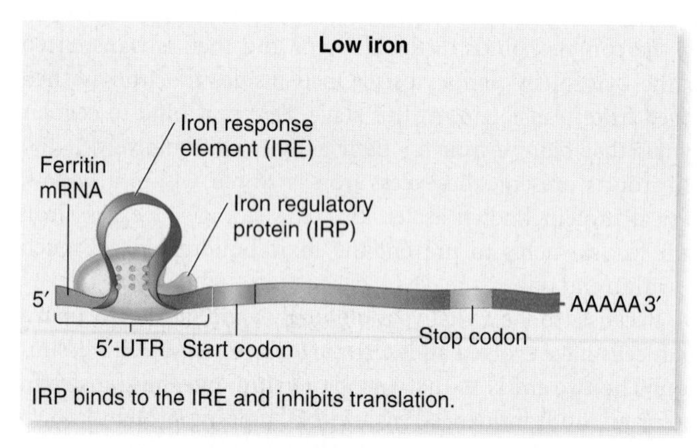

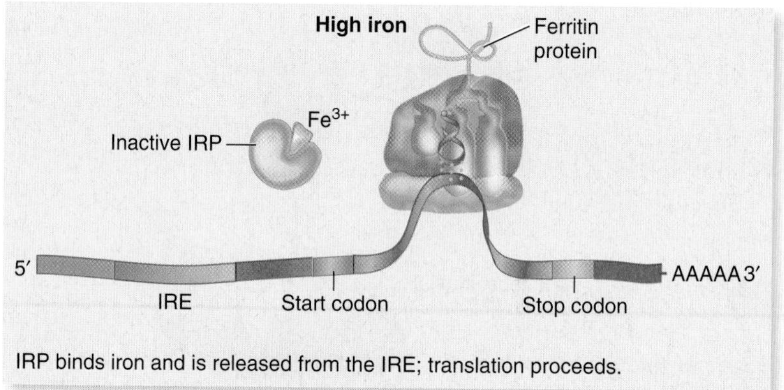

(a) Regulation of ferritin mRNA

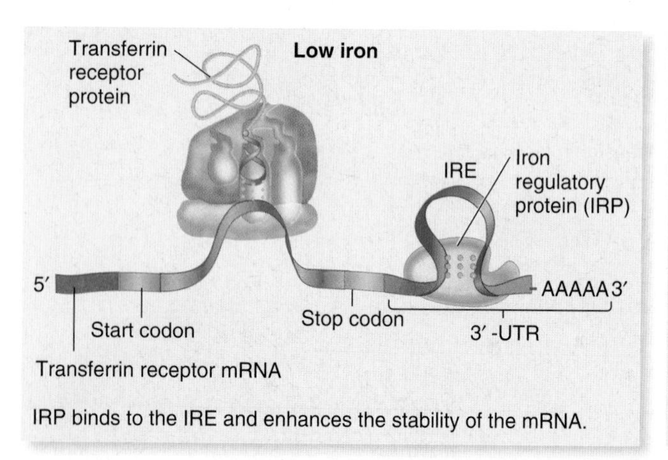

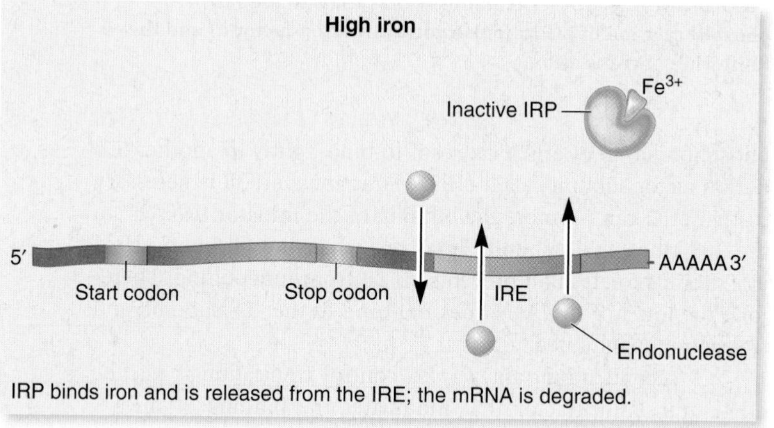

(b) Regulation of transferrin receptor mRNA

FIGURE 15.27 The regulation of iron assimilation genes by IRP and IRE. (a) When Fe^{3+} concentrations are low, the binding of the iron regulatory protein (IRP) to the iron response element (IRE) in the 5′-UTR of ferritin mRNA inhibits translation (left). When Fe^{3+} concentrations are high and Fe^{3+} binds to IRP, the IRP is removed from the ferritin mRNA, so translation can proceed (right). **(b)** The binding of IRP to IRE in the 3′-UTR of the transferrin receptor mRNA enhances the stability of the mRNA and leads to a higher concentration of this mRNA. Therefore, more transferrin receptor proteins are made when Fe^{3+} concentration is low (left). When Fe^{3+} levels are high and this metal binds to IRP, the IRP dissociates from the IRE, and the transferrin receptor mRNA is rapidly degraded (right).

Genes → Traits This form of translational control allows cells to use iron appropriately. When the cellular concentration of iron is low, the stability of the transferrin receptor mRNA is increased, thereby enhancing the ability of cells to synthesize more receptors and thereby take up more iron. Also, the translation of ferritin mRNA is inhibited, because ferritin is not needed to store excess iron. By comparison, when the cellular concentration of iron is high, the translation of ferritin mRNA is enhanced. This leads to the synthesis of ferritin and prevents the toxic buildup of iron. When iron is high, the transferrin receptor mRNA is degraded, which decreases further uptake of iron.

KEY TERMS

Page 390. gene regulation
Page 391. transcription factor, general transcription factors, regulatory transcription factors, control elements, regulatory elements
Page 392. activator, enhancer, repressors, silencers, combinatorial control, domains, motif
Page 393. homodimer, heterodimer
Page 394. up regulation, down regulation, orientation-independent, bidirectional, TFIID, coactivators, mediator
Page 395. steroid receptor
Page 396. glucocorticoid receptors
Page 397. cAMP response element-binding protein (CREB protein), cAMP response element (CRE), chromatin remodeling
Page 398. closed conformation, open conformation, ATP-dependent chromatin remodeling, DNA translocases, histone variants
Page 400. histone acetyltransferases, histone code hypothesis, chromatin immunoprecipitation sequencing (ChIP-Seq)
Page 402. immunoprecipitation, nucleosome-free region (NFR)
Page 403. DNA methylation
Page 404. DNA methyltransferase, CpG islands, housekeeping genes, tissue-specific genes, methyl-CpG-binding proteins
Page 405. *de novo* methylation, maintenance methylation
Page 406. insulator
Page 408. alternative splicing
Page 409. constitutive exons, alternative exons, splicing factors, SR proteins, exon skipping
Page 410. polyA-binding protein, 3′-untranslated region (3′-UTR), AU-rich element (ARE)
Page 413. RNA interference (RNAi), microRNAs (miRNAs), short-interfering RNAs (siRNAs), Dicer, RNA-induced silencing complex (RISC)
Page 415. iron regulatory protein (IRP)
Page 416. iron response element (IRE)
Page 419. promoter bashing

CHAPTER SUMMARY

- Gene regulation refers to the phenomenon that the level of gene expression can be controlled so that genes can be expressed at high or low levels. Gene regulation can occur at any step in the pathway of gene expression (see Figure 15.1).

15.1 Regulatory Transcription Factors

- Regulatory transcription factors can be activators that bind to enhancers and increase transcription, or they can be repressors that bind to silencers and inhibit transcription (see Figure 15.2).
- Combinatorial control means that the transcription of a gene is controlled by a variety of factors.
- Transcription factor proteins contain specific domains that may be involved in a variety of processes such as DNA binding and protein dimerization (see Figure 15.3).
- Control or response elements are usually orientation-independent.
- Regulatory transcription factors may exert their effects by interacting with TFIID (a general transcription factor) or mediator (see Figures 15.4, 15.5).
- The functions of regulatory transcription factors can be modulated by small effector molecules, protein–protein interactions, and covalent modifications (see Figure 15.6).
- Steroid hormones, such as glucocorticoids, bind to receptors that function as transcripitonal activators (see Figure 15.7).
- The CREB protein activates transcription when it has become phosphorylated by protein kinase A (see Figure 15.8).

15.2 Chromatin Remodeling, Histone Variation, and Histone Modification

- Chromatin remodeling occurs via ATP-dependent chromatin-remodeling complexes that alter the positions and compositions of nucleosomes (see Figure 15.9).
- Histone variants, which have amino acid sequences that differ slightly from the standard histones, play specialized roles in chromatin structure and function (see Table 15.1).
- Amino-terminal histone tails are subject to covalent modifications that act as a histone code for the binding of proteins that affect chromatin structure and gene expression (see Figure 15.10).
- Chromatin immunoprecipitation sequencing (ChIP-Seq) is a method that can be used to determine the precise locations of nucleosomes throughout an entire genome (see Figure 15.11).
- Many eukaryotic genes are flanked by a nucleosome-free region (NFR) with well-positioned nucleosomes (see Figure 15.12).
- Gene transcription involves changes in nucleosome locations, composition, and histone modifications (see Figure 15.13).

15.3 DNA Methylation

- DNA methylation is the attachment of a methyl group to a cytosine base via DNA methyltransferase (see Figure 15.14).
- The methylation of CpG islands near promoters usually silences transcription. Methylation may affect the binding of regulatory transcription factors or it may inhibit transcription via methyl-CpG-binding proteins (see Figure 15.15).

- Maintenance methylation can preserve a methylation pattern following cell division (see Figure 15.16).

15.4 Insulators

- An insulator is a segment of DNA that functions as a boundary that protects one gene from the regulatory effects of a neighboring gene. Insulators may act as a barrier to changes in chromatin structure or block the effects of neighboring enhancers (see Figure 15.17).
- Insulators may block the effects of enhancers by promoting the formation of a loop in the DNA (see Figure 15.18).

15.5 Regulation of RNA Processing, RNA Stability, and Translation

- Eukaryotic gene regulation commonly occurs after the mRNA is made. This includes the regulation of RNA processing and translation (see Table 15.2).
- During alternative splicing, proteins called splicing factors regulate which exons are included in a resulting mRNA (see Figures 15.19, 15.20).
- RNA stability may be regulated by an AU-rich element (ARE) found in the 3′-untranslated region (see Figure 15.21).
- The insertion of multiple copies of the same cloned gene into a genome may result in the production of double-stranded RNA that causes mRNA silencing (see Figure 15.22).
- Fire and Mello showed that double-stranded RNA is more potent at inhibiting mRNA than is antisense RNA (see Figure 15.23).
- RNA interference is a mechanism of RNA silencing in which miRNA or siRNA become part of an RNA-induced silencing complex (RISC) that inhibits the translation of mRNA or causes its degradation (see Figure 15.24).
- The phosphorylation of ribosomal elongation factor eIF2α inhibits translation, whereas the phosphorylation of eIF4F stimulates translation (see Figure 15.25).
- The regulation of iron uptake and storage is needed to prevent the toxic buildup of iron. When iron levels are low, the binding of iron regulatory protein (IRP) to an iron regulatory element (IRE) found in the 5′-untranslated region of ferritin mRNA inhibits translation. By comparison, the binding of IRP to an IRE in the 3′-untranslated region of the transferrin receptor mRNA promotes its stability when iron levels are low (see Figures 15.26, 15.27).

PROBLEM SETS & INSIGHTS

Solved Problems

S1. Describe how the arrangements and compositions of nucleosomes may result in a closed conformation that prevents gene transcription.

Answer: Nucleosome composition and arrangements are important for transcriptional initiation and elongation. The locations of nucleosomes may physically prevent general transcription factors or RNA polymerase from binding to the DNA. Also, nucleosome composition and certain types of covalent modifications may prevent RNA polymerase from forming an open complex, which is necessary to begin transcription. These types of changes are also important for transcription to proceed to the elongation phase.

S2. What are the two alternative ways that IRP can affect gene expression at the RNA level?

Answer: The ferritin mRNA has an IRE in its 5′-UTR. When IRP binds to this IRE, it inhibits the translation of the ferritin mRNA. This decreases the amount of ferritin protein, which is not needed when iron levels are low. However, when iron is abundant in the cytosol, the iron binds directly to IRP and prevents it from binding to the IRE. This allows the ferritin mRNA to be translated, producing more ferritin protein. An IRE is also located in the transferrin receptor mRNA in the 3′-UTR. When IRP binds to this IRE, it increases the stability of the mRNA, which leads to an increase in the amount of transferrin receptor mRNA within the cell when the cytosolic levels of iron are very low. Under these conditions, more transferrin receptor is made to promote the uptake of iron, which is in short supply. When the iron is found in abundance within the cytosol, IRP is removed from the mRNA, and the mRNA becomes rapidly degraded. This leads to a decrease in the amount of the transferrin receptor.

S3. Eukaryotic regulatory elements are often orientation-independent and can function in a variety of locations. Explain the meaning of this statement.

Answer: Orientation independence means that the regulatory element can function in the forward or reverse direction. In addition, regulatory elements can function at a variety of locations that may be upstream or downstream from the core promoter. A loop in the DNA must form to bring the regulatory transcription factors bound at the regulatory elements and the core promoter in close proximity with one another.

S4. To gain a molecular understanding of how the glucocorticoid receptor works, geneticists have attempted to "dissect" the protein to identify smaller domains that play specific functional roles. Using recombinant DNA techniques described in Chapter 18, particular segments in the coding region of the glucocorticoid receptor gene can be removed. The altered gene can then be expressed in a living cell to see if functional aspects of the receptor have been changed or lost. For example, the removal of the portion of the gene encoding the carboxyl-terminal half of the protein causes a loss of glucocorticoid binding. These results indicate that the carboxyl-terminal portion contains a domain that functions as a glucocorticoid-binding site. The following figure (see next page) illustrates the locations of several functional domains within the glucocorticoid receptor relative to the entire amino acid sequence.

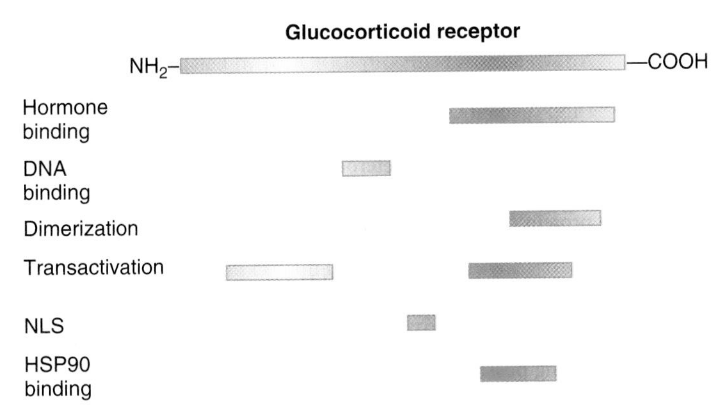

Based on your understanding of the mechanism of glucocorticoid receptor function described in Figure 15.7, explain the functional roles of the different domains in the glucocorticoid receptor.

Answer: The hormone-binding domain is located in the carboxyl-terminal half of the protein. This part of the protein also contains the region necessary for HSP90 binding and receptor dimerization. A nuclear localization sequence (NLS) is located near the center of the protein. After hormone binding, the NLS is exposed on the surface of the protein and allows it to be targeted to the nucleus. The DNA-binding domain, which contains zinc fingers, is also centrally located in the primary amino acid sequence. Zinc fingers promote DNA binding to the major groove. Finally, two separate regions of the protein, one in the amino-terminal half and one in the carboxyl-terminal half, are necessary for the transactivation of RNA polymerase. If these domains are removed from the receptor, it can still bind to the DNA, but it cannot activate transcription.

S5. A common approach to identify genetic sequences that play a role in the transcriptional regulation of a gene is the strategy sometimes called **promoter bashing.** This approach requires gene cloning methods, which are described in Chapter 18. A clone is obtained that has the coding region for a structural gene as well as the region that is upstream from the core promoter. This upstream region is likely to contain genetic regulatory elements such as enhancers and silencers. The diagram shown here depicts a cloned DNA region that contains the upstream region, the core promoter, and the coding sequence for a protein that is expressed in human liver cells. The upstream region may be several thousand base pairs in length.

To determine if promoter bashing has an effect on transcription, it is helpful to have an easy way to measure the level of gene expression. One way to accomplish this is to swap the coding sequence of the gene of interest with the coding sequence of another gene. For example, the coding sequence of the *lacZ* gene, which encodes β-galactosidase, is frequently swapped because it is easy to measure the activity of β-galactosidase using an assay for its enzymatic activity. The *lacZ* gene is called a "reporter gene" because its activity is easy to measure. As shown here, the coding sequence of the *lacZ* gene has been swapped with the coding sequence of the liver-specific gene. In this new genetic construct, the expression and transcriptional regulation of the *lacZ* gene are under the control of the core promoter and upstream region of the liver-specific gene.

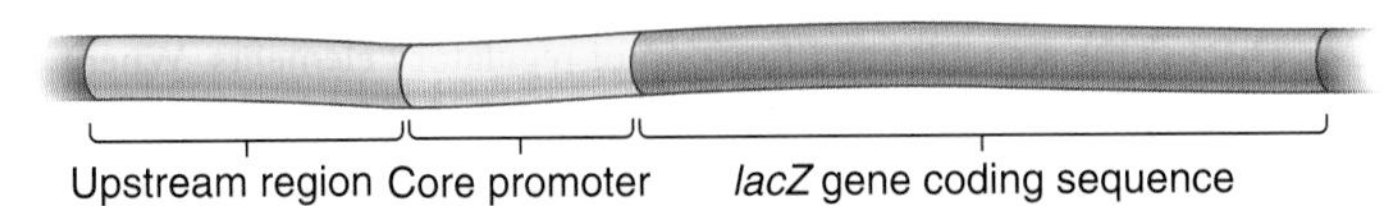

Now comes the "bashing" part of the experiment. Different segments of the upstream region are deleted, and then the DNA is transformed into living cells. In this case, the researcher would probably transform the DNA into liver cells, because those are the cells where the gene is normally expressed. The last step is to measure the β-galactosidase activity in the transformed liver cells.

In the diagram shown here, the upstream region and the core promoter have been divided into five regions, labeled A–E.

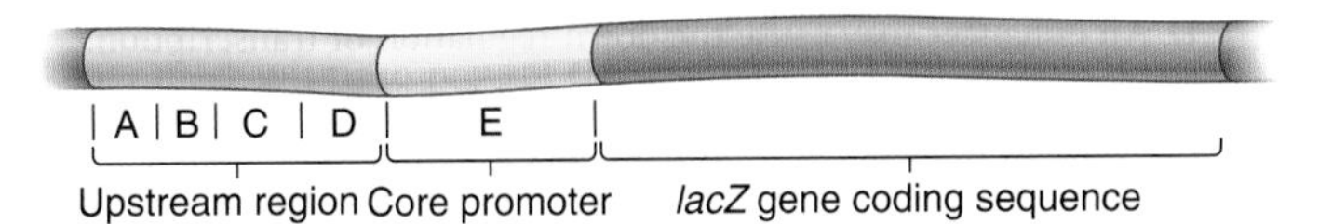

One of these regions was deleted (i.e., bashed out), and the rest of the DNA segment was transformed into liver cells. The data shown here are from this experiment.

Region Deleted	Percentage of β-Galactosidase Activity*
None	100
A	100
B	330
C	100
D	5
E	<1

*The amount of β-galactosidase activity in the cells carrying an undeleted upstream and promoter region was assigned a value of 100%. The amounts of activity in the cells carrying a deletion were expressed relative to this 100% value.

Explain what these results mean.

Answer: The amount of β-galactosidase activity found in liver cells that do not carry a deletion reflects the amount of expression under normal circumstances. If the core promoter (region E) is deleted, very little expression is observed. This is expected because a core promoter is needed for transcription. If enhancers are deleted, the activity should be less than 100%. It appears that one or more enhancers are found in region D. If a silencer is deleted, the activity should be above 100%. From the data shown it appears that one or more silencers are found in region B. Finally, if a deletion has no effect, there may not be any regulatory elements there. This was observed for regions A and C.

Note: The deletion of an enhancer has an effect on β-galactosidase activity only if the cell is expressing the regulatory transcription factor that binds to the enhancer and activates transcription. Likewise, the deletion of a silencer has an effect only if the cell is expressing the repressor protein that binds to the silencer and inhibits transcription. In the problem described, the liver cells must be expressing the activator(s) and repressor(s) that recognize the regulatory elements found in regions D and B, respectively.

Suppressor mutations are classified according to their relative locations with regard to the mutation they suppress (**Table 16.3**). When the second mutant site is within the same gene as the first, the mutation is termed an **intragenic suppressor.** This type of suppressor often involves a change in protein structure that compensates for an abnormality in protein structure caused by the first mutation. Researchers often isolate suppressor mutations to obtain information about protein structure and function. For example, Robert Brooker and colleagues have isolated many intragenic suppressors in the *lacY* gene of *E. coli*, which encodes the lactose permease described in Chapter 15. This protein must undergo conformational changes to transport lactose across the cell membrane. They began with single mutations that altered amino acids on transmembrane regions, which inhibited this conformational change, thereby preventing growth on media containing lactose. Suppressor mutations were then isolated that allowed growth on lactose by restoring transport function. By analyzing the locations of these suppressor mutations, the researchers were able to determine that certain transmembrane regions in the protein are critical for conformational changes required for lactose transport.

Alternatively, a suppressor mutation can be in a different gene from the first mutation—an **intergenic suppressor.** Researchers often study intergenic suppressors as a way to gain information about proteins that have similar or redundant functional roles, proteins that participate in a common pathway, multimeric proteins with two or more subunits, and the regulation of protein expression by transcription factors. An example of an intergenic suppressor analysis in *Drosophila* is described in Chapter 4 (see Figure 4.25).

How do intergenic suppressors work? These suppressor mutations usually involve a change in the expression of one gene that compensates for a loss-of-function mutation affecting another gene (see Table 16.3). For example, a first mutation may cause one protein to be partially or completely defective. An intergenic suppressor mutation in a different structural gene might overcome this defect by altering the structure of a second protein so that it could take over the functional role the first protein cannot perform. Alternatively, intergenic suppressors may involve proteins that participate in a common cellular pathway. When a first mutation affects the activity of a protein, a suppressor mutation could alter the function of a second protein involved in this pathway, thereby overcoming the defect in the first.

In some cases, intergenic suppressors involve multimeric proteins, with each subunit encoded by a different gene. A mutation in one subunit that inhibits function may be compensated for by a mutation in another subunit. Another type of intergenic suppressor is one that involves mutations in genetic regulatory proteins such as transcription factors. When a first mutation causes a protein to be defective, a suppressor mutation may occur in a gene that encodes a transcription factor. The mutant transcription factor transcriptionally activates another gene that can compensate for the loss-of-function mutation in the first gene.

Less commonly, intergenic suppression involves mutations in nonstructural genes that alter the translation of particular codons. Examples are suppressor tRNA mutants, which have been identified in microorganisms (see solved problem S1). Suppressor tRNA mutants have a change in the anticodon region that causes the tRNA to behave contrary to the genetic code. Nonsense suppressors are mutant tRNAs that recognize a stop codon, and instead of stopping translation, put an amino acid into the growing polypeptide chain. This type of mutant tRNA can suppress a nonsense mutation in the mRNA from a structural gene. However, such bacterial strains grow poorly because they may also suppress stop codons in the mRNAs from normal genes.

Changes in Chromosome Structure Can Affect the Expression of a Gene

Thus far, we have considered small changes in the DNA sequence of particular genes. A change in chromosome structure can also be associated with an alteration in the expression of specific genes. Quite commonly, an inversion or translocation has no obvious phenotypic consequence. However, in 1925, Alfred Sturtevant was the first to recognize that chromosomal rearrangements in *Drosophila melanogaster* can influence phenotypic expression (namely, eye morphology). In some cases, a chromosomal rearrangement may affect a gene because a chromosomal **breakpoint**—the region where two chromosome pieces break and rejoin with other chromosome pieces—occurs within a gene. A breakpoint within the middle of a gene is very likely to inhibit gene function because it separates the gene into two pieces.

In other cases, a gene may be left intact, but its expression may be altered when it is moved to a new location. When this occurs, the change in gene location is said to have a **position effect.** How do position effects alter gene expression? Researchers have discovered two common explanations. **Figure 16.2** depicts a schematic example in which a piece of one chromosome has been inverted or translocated to a different chromosome. One possibility is that a gene may be moved next to regulatory sequences for a different gene, such as silencers or enhancers, that influence the expression of the relocated gene (Figure 16.2a). Alternatively, a chromosomal rearrangement may reposition a gene from a less condensed or euchromatic chromosome to a very highly condensed or heterochromatic chromosome. When the gene is moved to a heterochromatic region, its expression may be turned off (Figure 16.2b). This second type of position effect may produce a variegated phenotype in which the expression of the gene is variable. For genes that affect pigmentation, this produces a mottled appearance rather than an even color. **Figure 16.3** shows a position effect that alters eye color in *Drosophila*. Figure 16.3a depicts a normal red-eyed fruit fly, and Figure 16.3b shows a mutant fly that has inherited a chromosomal rearrangement in which a gene affecting eye color has been relocated to a heterochromatic chromosome. The variegated appearance of the eye occurs because the degree of heterochromatin formation varies across different regions of the eye. In cells where heterochromatin formation has turned off the eye color gene, a white phenotype occurs, but other cells allow this same region to remain euchromatic and produce a red phenotype.

TABLE 16.3

Examples of Suppressor Mutations

Type	No Mutation	First Mutation	Second Mutation	Description
Intragenic	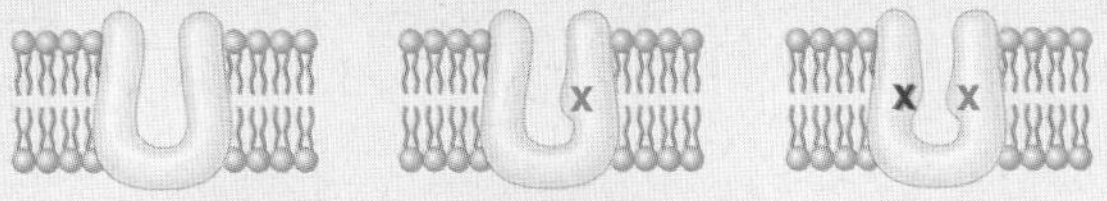 Transport can occur	Transport inhibited	Transport can occur	A first mutation disrupts normal protein function, and a suppressor mutation affecting the same protein restores function. In this example, the first mutation inhibits lactose transport function, and the second mutation restores it.
Intergenic Redundant function	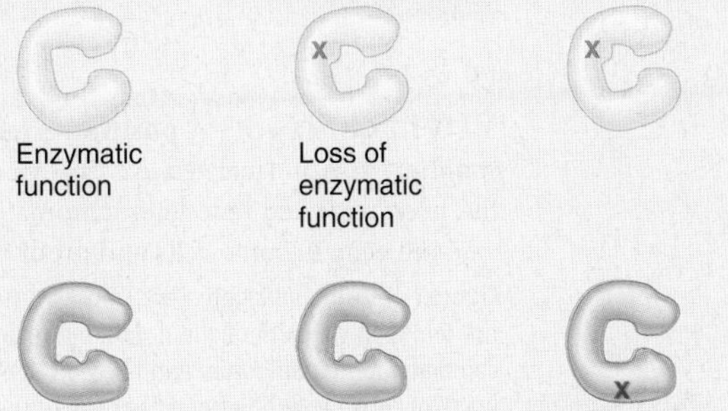 Enzymatic function	Loss of enzymatic function	Gain of a new enzymatic function	A first mutation inhibits the function of a protein, and a second mutation alters a different protein to carry out that function. In this example, the proteins function as enzymes.
Common pathway	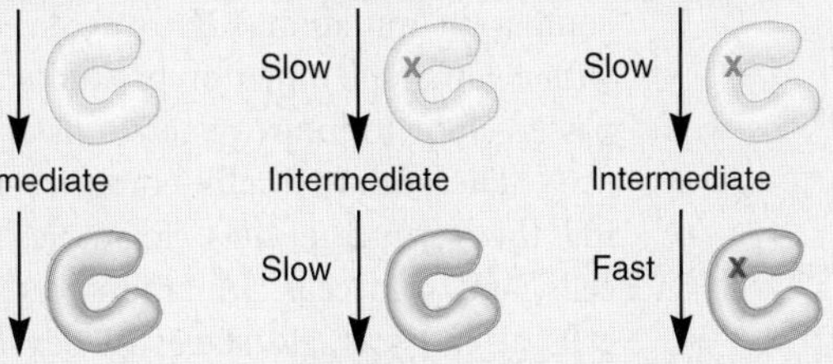 Precursor Fast Intermediate Slow Product	Precursor Slow Intermediate Slow Little product	Precursor Slow Intermediate Fast Product	Two or more different proteins may be involved in a common pathway. A mutation that causes a defect in one protein may be compensated for by a mutation that alters the function of a different protein in the same pathway.
Multimeric protein	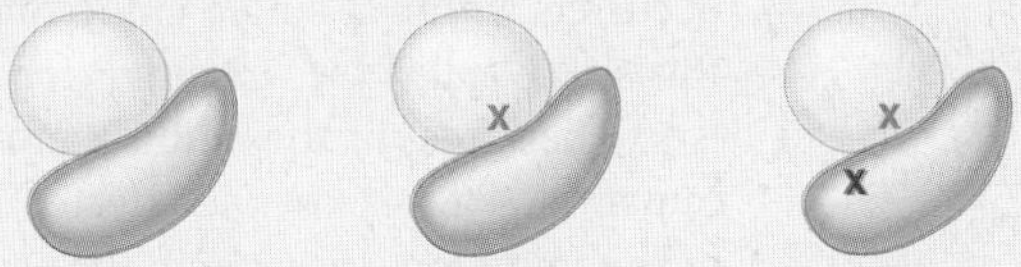 Active	Inactive	Active	A mutation in a gene encoding one protein subunit that inhibits function may be suppressed by a mutation in a gene that encodes a different subunit. The double mutant has restored function.
Transcription factor	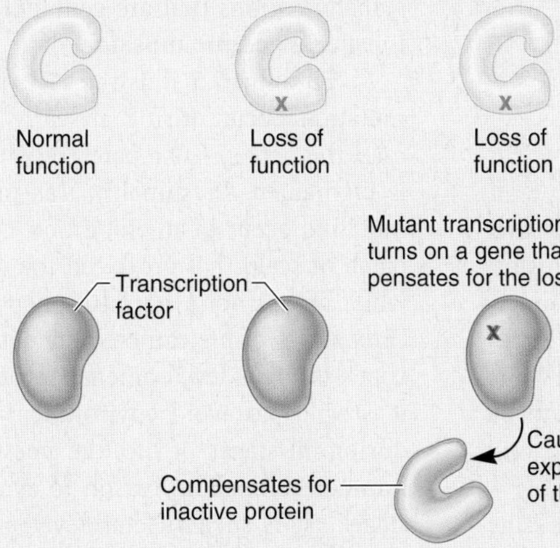 Normal function Transcription factor	Loss of function	Loss of function Mutant transcription factor turns on a gene that compensates for the loss of function Causes expression of this protein. Compensates for inactive protein	A first mutation causes loss of function of a particular protein. A second mutation may alter a transcription factor and cause it to activate the expression of another gene. This other gene encodes a protein that can compensate for the loss of function caused by the first mutation.

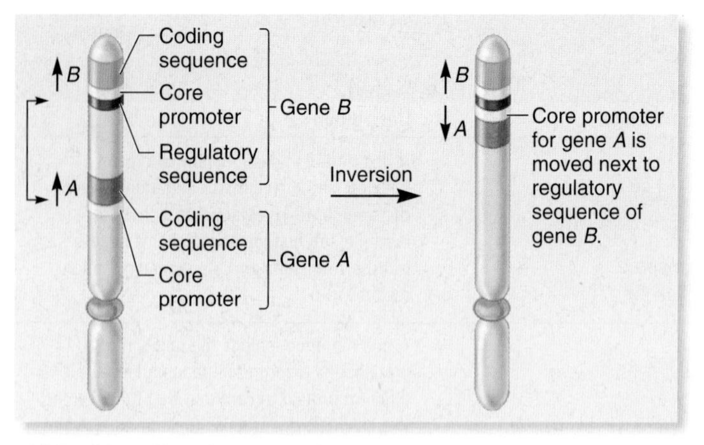

(a) Position effect due to regulatory sequences

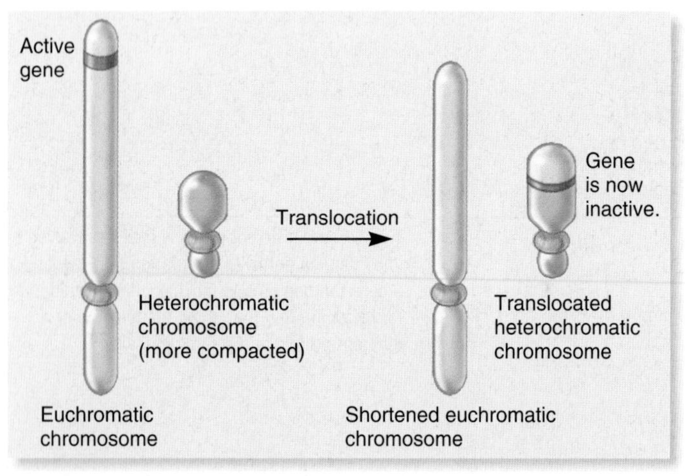

(b) Position effect due to translocation to a heterochromatic chromosome

FIGURE 16.2 **Causes of position effects.** (a) A chromosomal inversion has repositioned the core promoter of gene *A* next to the regulatory sequences for gene *B*. Because regulatory sequences are often bidirectional, the regulatory sequences for gene *B* may regulate the transcription of gene *A*. (b) A translocation has moved a gene from a euchromatic to a heterochromatic chromosome. This type of position effect prevents the expression of the relocated gene.

Mutations Can Occur in Germ-Line or Somatic Cells

In this section, we have considered many different ways that mutations affect gene expression. For multicellular organisms, the timing of mutations also plays an important role. A mutation can occur very early in life, such as in a gamete or a fertilized egg, or it may occur later in life, such as in the embryonic or adult stages. The exact time when mutations occur can be important with regard to the severity of the genetic effect and whether they are passed from parent to offspring.

Geneticists classify the cells of animals into two types—the germ line and the somatic cells. The term **germ line** refers to cells that give rise to the gametes such as eggs and sperm. A **germ-line mutation** can occur directly in a sperm or egg cell, or it can occur in a precursor cell that produces the gametes. If a mutant gamete participates in fertilization, all cells of the resulting offspring will contain the mutation (**Figure 16.4a**). Likewise, when an individual with a germ-line mutation produces gametes, the mutation may be passed along to future generations of offspring.

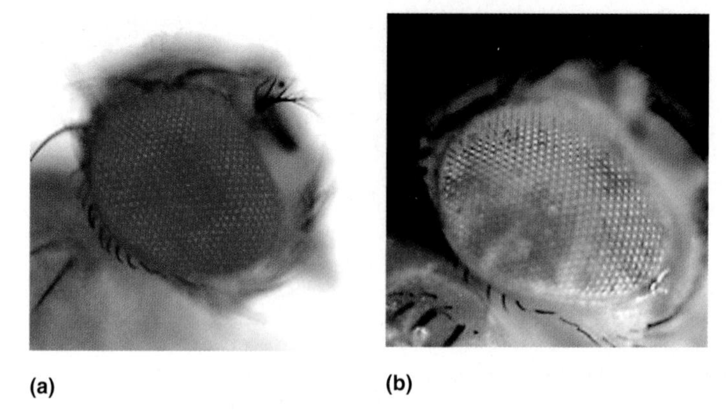

(a) (b)

FIGURE 16.3 **A position effect that alters eye color in *Drosophila*.** (a) A normal red eye. (b) An eye in which an eye color gene has been relocated to a heterochromatic chromosome. This can inactivate the gene in some cells and produces a variegated phenotype.

Genes → Traits Variegated eye color occurs because the degree of heterochromatin formation varies throughout different regions of the eye. In some cells, heterochromatin formation occurs and turns off the eye color gene, thereby leading to the white phenotype. In other patches of cells, the region containing the eye color allele remains euchromatic, yielding a red phenotype.

The **somatic cells** comprise all cells of the body excluding the germ-line cells. Examples include muscle cells, nerve cells, and skin cells. Mutations can also happen within somatic cells at early or late stages of development. **Figure 16.4b** illustrates the consequences of a mutation that took place during the embryonic stage. In this example, a **somatic mutation** has occurred within a single embryonic cell. As the embryo grows, this single cell is the precursor for many cells of the adult organism. Therefore, in the adult, a portion of the body contains the mutation. The size of the affected region depends on the timing of the mutation. In general, the earlier the mutation occurs during development, the larger the region. An individual that has somatic regions that are genotypically different from each other is called a **genetic mosaic.**

Figure 16.5 illustrates an individual who had a somatic mutation occur during an early stage of development. In this case, the person has a patch of white hair, but the rest of the hair is pigmented. Presumably, this individual initially had a single mutation occur in an embryonic cell that ultimately gave rise to a patch of scalp that produced the white hair. Although a patch of white hair is not a harmful phenotypic effect, mutations during early stages of life can be quite detrimental, especially if they disrupt essential developmental processes. Therefore, even though it is smart to avoid environmental agents that cause mutations during all stages of life, the possibility of somatic mutations is a rather compelling reason to avoid them during the very early stages of life such as fetal development, infancy, and early childhood. For example, the possibility of somatic mutations in an embryo is a reason why women are advised to avoid getting X-rays during pregnancy.

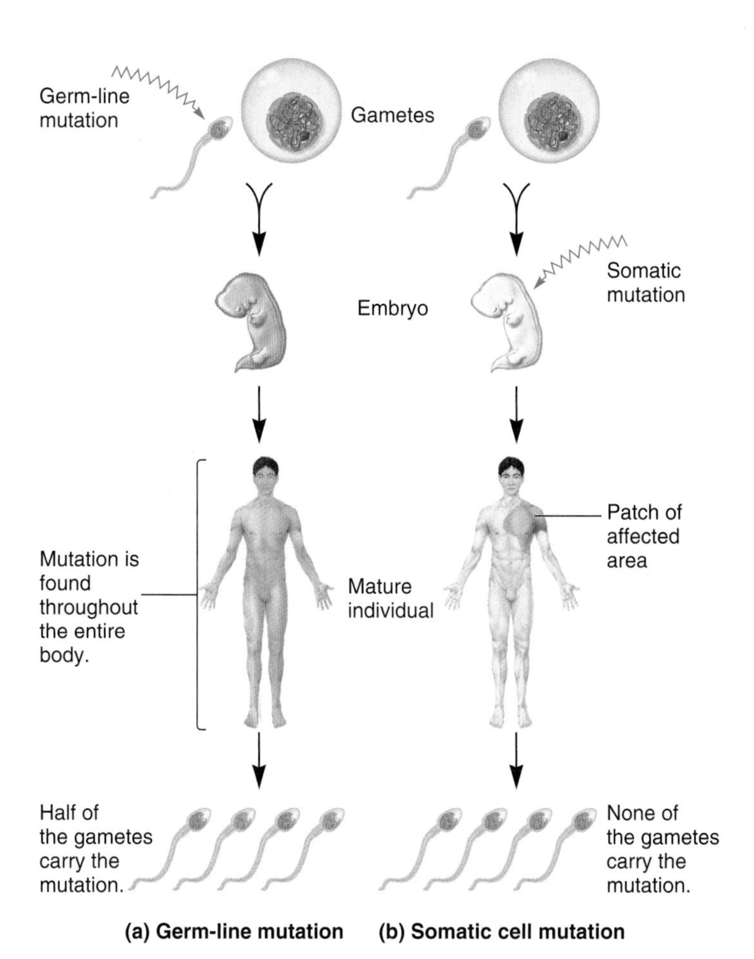

FIGURE 16.4 The effects of germ-line versus somatic mutations.

FIGURE 16.5 Example of a somatic mutation.
Genes → Traits This person has a patch of white hair because a somatic mutation occurred in a single cell during embryonic development that prevented pigmentation of the hair. This cell continued to divide to produce a patch of white hair.

16.2 OCCURRENCE AND CAUSES OF MUTATION

As we have seen, mutations can have a wide variety of effects on the phenotypic expression of genes. For this reason, geneticists have expended a great deal of effort identifying the causes of mutations. This task has been truly challenging, because a staggering number of agents can alter the structure of DNA and thereby cause mutation. Geneticists categorize the cause of mutation in one of two ways. **Spontaneous mutations** are changes in DNA structure that result from abnormalities in biological processes, whereas **induced mutations** are caused by environmental agents (**Table 16.4**).

Many causes of spontaneous mutations are examined in other chapters throughout this textbook. As discussed in Chapter 8, abnormalities in crossing over can produce mutations such as deletions, duplications, translocations, and inversions. In addition, aberrant segregation of chromosomes during meiosis can cause changes in chromosome number. In Chapter 11, we learned that DNA polymerase can make a mistake during DNA replication by

TABLE 16.4
Causes of Mutations

Common Causes of Mutations	Description
Spontaneous	
Aberrant recombination	Abnormal crossing over may cause deletions, duplications, translocations, and inversions (see Chapter 8).
Aberrant segregation	Abnormal chromosomal segregation may cause aneuploidy or polyploidy (see Chapter 8).
Errors in DNA replication	A mistake by DNA polymerase may cause a point mutation (see Chapter 11).
Transposable elements	Transposable elements can insert themselves into the sequence of a gene (see Chapter 17).
Depurination	On rare occasions, the linkage between purines (i.e., adenine and guanine) and deoxyribose can spontaneously break. If not repaired, it can lead to mutation.
Deamination	Cytosine and 5-methylcytosine can spontaneously deaminate to create uracil or thymine.
Tautomeric shifts	Spontaneous changes in base structure can cause mutations if they occur immediately prior to DNA replication.
Toxic metabolic products	The products of normal metabolic processes, such as reactive oxygen species, may be chemically reactive agents that can alter the structure of DNA.
Induced	
Chemical agents	Chemical substances may cause changes in the structure of DNA.
Physical agents	Physical phenomena such as UV light and X-rays can damage the DNA.

putting the wrong base in a newly synthesized daughter strand. Errors in DNA replication are usually infrequent except in certain viruses, such as HIV, that have relatively high rates of spontaneous mutations. Also, normal metabolic processes may produce chemicals within the cell that can react directly with the DNA and alter its structure. As discussed in Chapter 17, transposable genetic elements can alter gene sequences by inserting themselves into genes. In this section, we will examine how spontaneous changes in nucleotide structure, such as depurination, deamination, and tautomeric shifts, can cause mutation. Overall, a distinguishing feature of spontaneous mutations is that their underlying cause originates within the cell. By comparison, the cause of induced mutations originates outside the cell. Induced mutations are produced by environmental agents, either chemical or physical, that enter the cell and lead to changes in DNA structure. Agents known to alter the structure of DNA which lead to mutations are called **mutagens.**

We begin by examining the random nature of spontaneous mutations and general features of the mutation rate. Then we will explore several mechanisms by which mutagens can alter the structure of DNA. Finally, laboratory tests that can identify potential mutagens will be described.

Spontaneous Mutations Are Random Events

For a couple of centuries, biologists had questioned whether heritable changes occur purposefully as a result of behavior or exposure to particular environmental conditions or whether they are spontaneous events that may happen randomly. In the nineteenth century, the naturalist Jean-Baptiste Lamarck proposed that physiological events—such as the use or disuse of muscles—determine whether traits are passed along to offspring. For example, his hypothesis suggested that an individual who practiced and became adept at a physical activity and developed muscular legs would pass that characteristic on to the next generation. The alternative point of view is that genetic variation exists in a population as a matter of random chance, and natural selection results in the differential reproductive success of organisms that are better adapted to their environments. Those individuals who, by chance, happen to have beneficial mutations will be more likely to survive and pass these genes to their offspring. These opposing ideas of the nineteenth century—one termed physiological adaptation and the other termed random mutation—were tested in bacterial studies in the 1940s and 1950s, two of which are described here.

Salvadore Luria and Max Delbrück were interested in the ability of bacteria to become resistant to infection by a bacteriophage called T1. When a population of *E. coli* cells is exposed to T1, a small percentage of bacteria are found to be resistant to T1 infection and pass this trait to their progeny. Luria and Delbrück were interested in whether such resistance, called *ton*^{r} (T one resistance), is due to the occurrence of random mutations or whether it is a physiological adaptation that occurs at a low rate within the bacterial population.

According to the physiological adaptation hypothesis, the rate of mutation should be a relatively constant value and depend on exposure to the bacteriophage. Therefore, when comparing different populations of bacteria, the number of *ton*^{r} bacteria should be an essentially constant proportion of the total population. In contrast, the random mutation hypothesis depends on the timing of mutation. If a *ton*^{r} mutation occurs early within the proliferation of a bacterial population, many *ton*^{r} bacteria will be found within that population because mutant bacteria will have time to grow and multiply. However, if it occurs much later in population growth, fewer *ton*^{r} bacteria will be observed. In general, a random mutation hypothesis predicts a much greater fluctuation in the number of *ton*^{r} bacteria among different populations.

To distinguish between the physiological adaptation and random mutation hypotheses, Luria and Delbrück inoculated one large flask and 20 individual tubes with *E. coli* cells and grew them in the absence of T1 phage. The flask was grown to produce a very large population of cells, while each individual culture was grown to a smaller population of approximately 20 million cells. They then plated the individual cultures onto media containing T1 phage. Likewise, 10 subsamples, each consisting of 20 million bacteria, were removed from the large flask and plated onto media with T1 phage.

The results of the Luria and Delbrück experiment are shown in **Figure 16.6**. Within the smaller individual cultures, a great fluctuation was observed in the number of *ton*^{r} mutants. This test, therefore, has become known as the **fluctuation test.** Which hypothesis do these results support? The data are consistent with a random mutation hypothesis in which the timing of a mutation during the growth of a culture greatly affects the number of mutant cells. For example, in tube 14, many *ton*^{r} bacteria were observed. Luria and Delbrück reasoned that a mutation occurred randomly in one bacterium at an early stage of the population growth, before the bacteria were exposed to T1 on plates. This mutant bacterium then divided to produce many daughter cells that inherited the *ton*^{r} trait. In other tubes, such as 1 and 3, this spontaneous mutation did not occur, so none of the bacteria had a *ton*^{r} phenotype. By comparison, the cells plated from the large flask tended toward a relatively constant and intermediate number of *ton*^{r} bacteria. Because the large flask had so many cells, several independent *ton*^{r} mutations were likely to have occurred during different stages of its growth. In a single flask, however, these independent events would be mixed together to give an average value of *ton*^{r} cells.

Several years later, Joshua and Esther Lederberg were also interested in the relationship between mutation and the environmental conditions that select for mutation. To distinguish between the physiological adaptation and random mutation hypotheses, they developed a technique known as **replica plating** in the 1950s. As shown in **Figure 16.7**, they plated a large number of bacteria onto a master plate that did not contain any selective agent (namely, the T1 phage). A sterile piece of velvet cloth was lightly touched to this plate in order to pick up a few bacterial cells from each colony. This replica was then transferred to two secondary plates that contained an agent that selected for the growth of bacterial cells with a particular genotype.

In the example shown in Figure 16.7, the secondary plates contained T1 bacteriophages. On these plates, only those mutant cells that are *ton*^{r} could grow. On the secondary plates, a few colonies were observed. Strikingly, they occupied the same location on each plate. These results indicated that the mutations conferring *ton*^{r} occurred randomly while the cells were growing

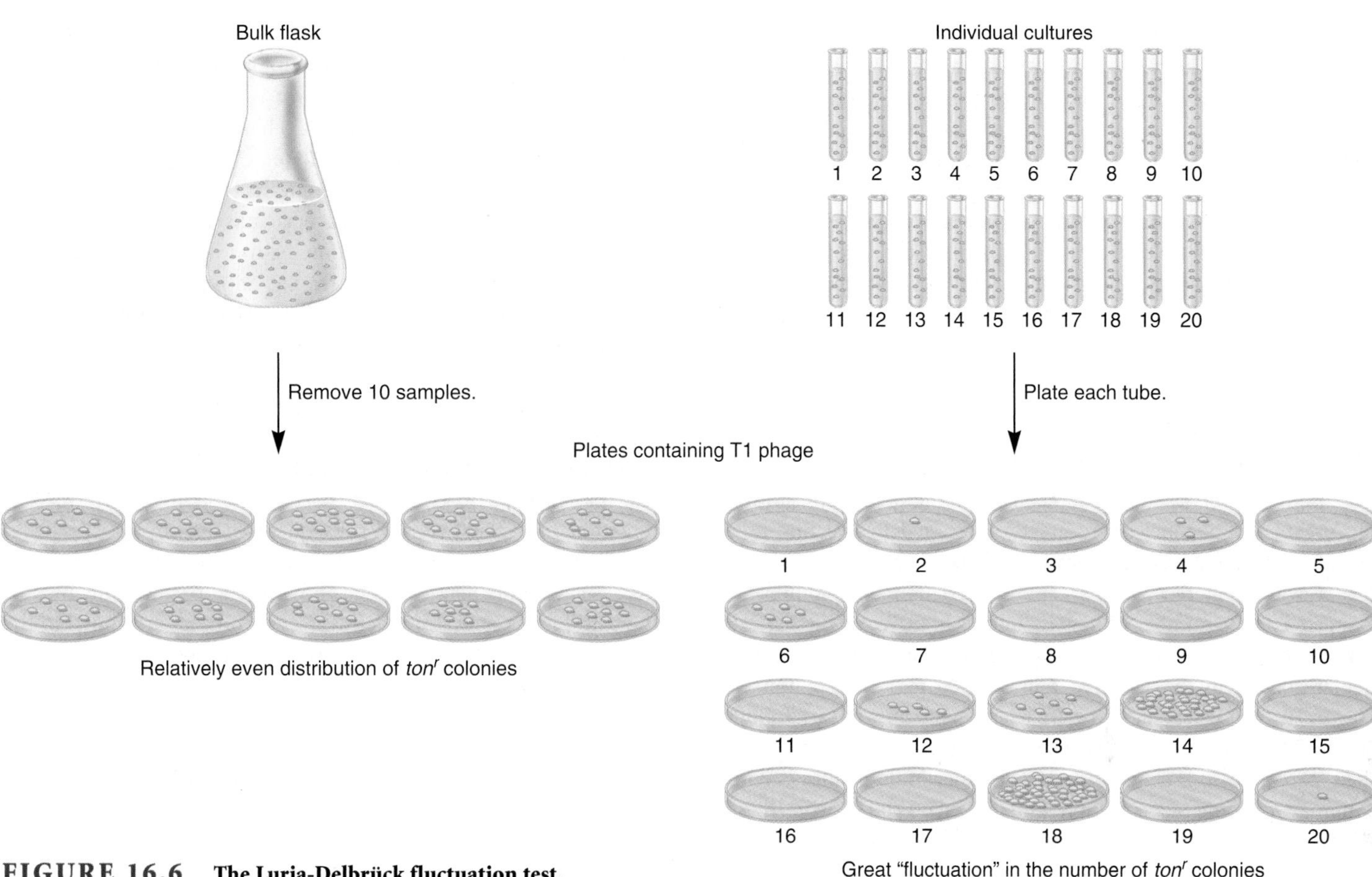

FIGURE 16.6 **The Luria-Delbrück fluctuation test.**

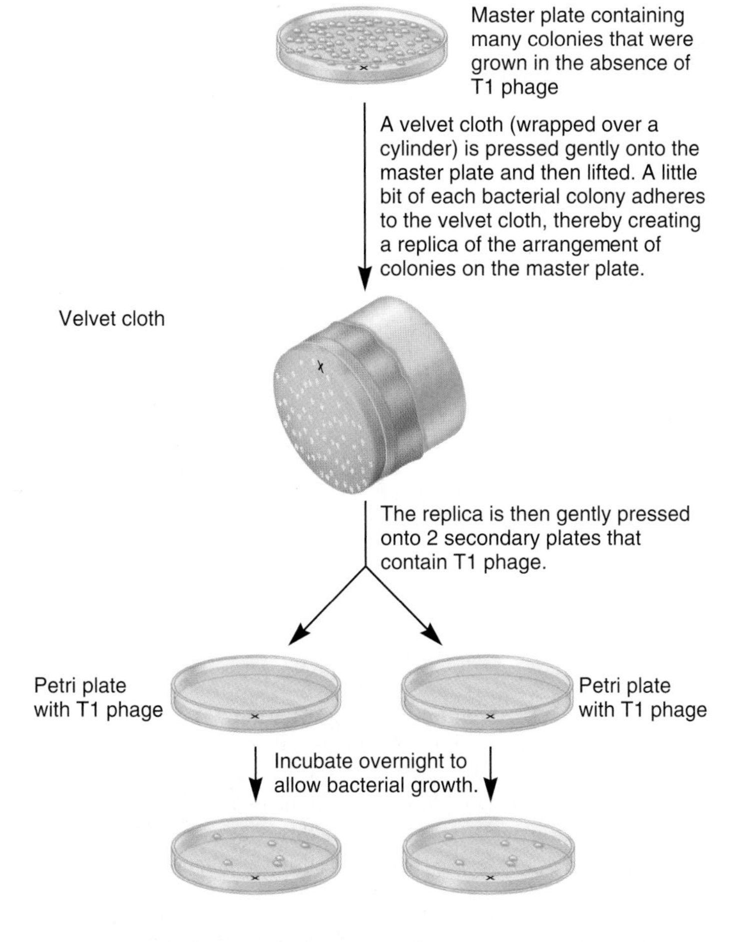

on the nonselective master plate. The presence of the T1 phage in the secondary plates simply selected for the growth of previously occurring *ton*^{r} mutants. These results supported the random mutation hypothesis. In contrast, the physiological adaptation hypothesis would have predicted that *ton*^{r} bacterial mutants would occur after exposure to the selective agent. If that had been the case, the colonies would not be expected to arise in identical locations on different secondary plates but rather in random patterns.

Taken together, the results of Luria and Delbrück and those of the Lederbergs supported the random mutation hypothesis, now known as the **random mutation theory.** According to this theory, mutations are a random process—they can occur in any gene and do not involve exposure of an organism to a particular condition that causes specific types of mutations to happen. In some cases, a random mutation may provide a mutant organism with an advantage, such as resistance to T1 phage. Although such mutations occur as a matter of random chance, growth conditions may select for organisms that happen to carry them.

FIGURE 16.7 **Replica plating.** Bacteria were first plated on a master plate under nonselective conditions. A sterile velvet cloth was used to make a replica of the master plate. This replica was gently pressed onto two secondary plates that contained a selective agent. In this case, the two secondary plates contained T1 bacteriophage. Only those mutant cells that are *ton*^{r} (resistant to T1) could grow to form visible colonies. Note: The black × indicates the alignment of the velvet and the plates. Data adapted from: Lederberg, J., and Lederberg, E.M. (1952) Replica plating and indirect selection of bacterial mutants. *J. Bacteriol. 63,* 399–406.

As researchers have learned more about mutation at the molecular level, the view that mutations are a totally random process has required some modification. Within the same individual, some genes mutate at a much higher rate than other genes. Why does this happen? Some genes are larger than others, which provides a greater chance for mutation. Also, the relative locations of genes within a chromosome may cause some genes to be more susceptible to mutation than others. Even within a single gene, **hot spots** are usually found—certain regions of a gene that are more likely to mutate than other regions (refer back to Chapter 7, Figure 7.21).

Mutation Rates and Frequencies Are Ways to Quantitatively Assess Mutation in a Population

Because mutations occur spontaneously among populations of living organisms, geneticists are greatly interested in learning how prevalent they are. The term **mutation rate** is the likelihood that a gene will be altered by a new mutation. This rate is commonly expressed as the number of new mutations in a given gene per cell generation. In general, the spontaneous mutation rate for a particular gene is in the range of 1 in 100,000 to 1 in 1 billion, or 10^{-5} to 10^{-9} per cell generation. An average person will carry about 100 to 200 new mutations in their entire genome compared with their parents. Most of these are single nucleotide changes. With a human genome size of approximately 3,200,000,000 bp, these numbers tell us that a mutation is a relatively infrequent event. However, the mutation rate is not a constant number. The presence of certain environmental agents, such as X-rays, can increase the rate of induced mutations to a much higher value than the spontaneous mutation rate. In addition, mutation rates vary substantially from species to species and even within different strains of the same species. One explanation for this variation is that there are many different causes of mutations (refer back to Table 16.4).

Before we end our discussion of mutation rate, let's distinguish the rate of new mutation from the concept of **mutation frequency.** The mutation frequency for a gene is the number of mutant genes divided by the total number of genes within a population. If 1 million bacteria were plated and 10 were found to be mutant, the mutation frequency would be 1 in 100,000, or 10^{-5}. As we have seen, Luria and Delbrück showed that among the bacteria in the 20 tubes, the timing of mutations influenced the mutation frequency within any particular tube. Some tubes had a high frequency of mutation, but others did not. Therefore, the mutation frequency depends not only on the mutation rate but also on the timing of mutation, and on the likelihood that a mutation will be passed to future generations. The mutation frequency is an important genetic concept, particularly in the field of population genetics. As we will see in Chapter 24, mutation frequencies may rise above the mutation rate due to evolutionary factors such as natural selection and genetic drift.

Spontaneous Mutations Can Arise by Depurination, Deamination, and Tautomeric Shifts

Thus far, we have considered the random nature of mutation and the quantitative difference between mutation rate and mutation frequency. We now turn our attention to the molecular changes in DNA structure that can cause mutation. Our first examples concern changes that can occur spontaneously, albeit at a low rate. The most common type of chemical change that occurs naturally is **depurination,** which involves the removal of a purine (adenine or guanine) from the DNA. The covalent bond between deoxyribose and a purine base is somewhat unstable and occasionally undergoes a spontaneous reaction with water that releases the base from the sugar, thereby creating an **apurinic site** (**Figure 16.8a**). In a typical mammalian cell, approximately 10,000 purines are lost from the DNA in a 24-hour period at 37°C. The rate of loss is higher if the DNA is exposed to agents that cause certain types of base modifications such as the attachment of alkyl groups (methyl or ethyl groups). Fortunately, as discussed later in this chapter, apurinic sites are recognized by DNA base excision repair enzymes that repair the site. If the repair system fails, however, a mutation may result during subsequent rounds of DNA replication. What happens at an apurinic site during DNA replication? Because a complementary base is not present to specify the incoming base for the new strand, any of the four bases are added to the new strand in the region that is opposite the apurinic site (**Figure 16.8b**). This may lead to a new mutation.

A second spontaneous lesion that may occur in DNA is the **deamination** of cytosines. The other bases are not readily deaminated. As shown in **Figure 16.9a**, deamination involves the removal of an amino group from the cytosine base. This produces uracil. As discussed later, DNA repair enzymes can recognize uracil as an inappropriate base within DNA and subsequently

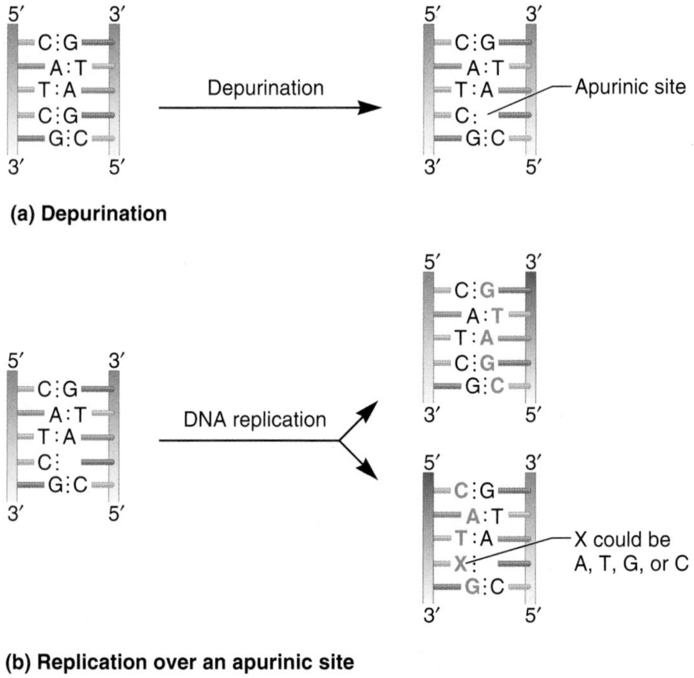

FIGURE 16.8 **Spontaneous depurination.** (**a**) The bond between guanine and deoxyribose is broken, thereby releasing the base. This leaves an apurinic site in the DNA. (**b**) If an apurinic site remains in the DNA as it is being replicated, any of the four nucleotides can be added to the newly made strand. Because three out of four (A, T, and G) are the incorrect base, the chance of causing a mutation is 75%.

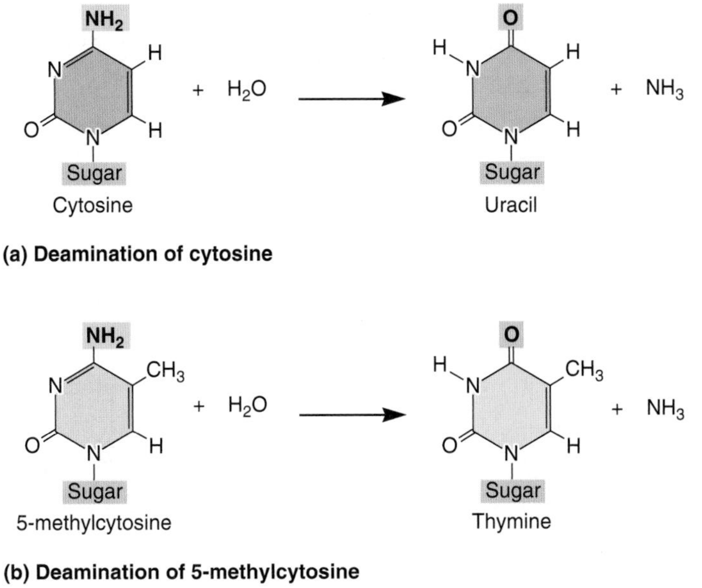

Cytosine + H_2O → Uracil + NH_3

(a) Deamination of cytosine

5-methylcytosine + H_2O → Thymine + NH_3

(b) Deamination of 5-methylcytosine

FIGURE 16.9 Spontaneous deamination of cytosine and 5-methylcytosine. (a) The deamination of cytosine produces uracil. (b) The deamination of 5-methylcytosine produces thymine.

remove it. However, if such repair does not take place, a mutation may result because uracil hydrogen bonds with adenine during DNA replication. Therefore, if a DNA template strand has uracil instead of cytosine, a newly made strand will incorporate adenine into the daughter strand instead of guanine.

Figure 16.9b shows the deamination of 5-methylcytosine. As discussed in Chapter 15, the methylation of cytosine occurs in many eukaryotic species. It also occurs in prokaryotes. If 5-methylcytosine is deaminated, the resulting base is thymine, which is a normal constituent of DNA. Therefore, this poses a problem for DNA repair because DNA repair proteins cannot distinguish which is the incorrect base—the thymine that was produced by deamination or the guanine in the opposite strand that originally base-paired with the methylated cytosine. For this reason, methylated cytosine bases tend to produce hot spots for mutation. As an example, researchers analyzed 55 spontaneous mutations that occurred within the *lacI* gene of *E. coli* and determined that 44 of them involved changes at sites that were originally occupied by a methylated cytosine base.

A third way that mutations may arise spontaneously involves a temporary change in base structure called a **tautomeric shift.** In this case, the **tautomers** are bases, which exist in keto and enol or amino and imino forms. These forms can interconvert by a chemical reaction that involves the migration of a hydrogen atom and a switch of a single bond and an adjacent double bond. The common, stable form of guanine and thymine is the keto form; the common form of adenine and cytosine is the amino form (**Figure 16.10a**). At a low rate, G and T can interconvert to an enol form, and A and C can change to an imino form. Though the relative amounts of the enol and imino forms of these bases are relatively small, they can cause a mutation because these rare forms of the bases do not conform to the AT/GC rule of base pairing. Instead, if one of the bases is in the enol or imino form, hydrogen bonding will promote TG and CA base pairs, as shown in **Figure 16.10b**.

How does a tautomeric shift cause a mutation? The answer is that it must occur immediately prior to DNA replication. When DNA is in a double-stranded condition, the base pairing usually holds the bases in their more stable forms. After the strands unwind, however, a tautomeric shift may occur. In the example shown in **Figure 16.10c**, a thymine base in the template strand has undergone a tautomeric shift just prior to the replication of the complementary daughter strand. During replication, the daughter strand incorporates a guanine opposite this thymine, creating a base mismatch. This mismatch could be repaired via the proofreading function of DNA polymerase or via a mismatch repair system (discussed later in this chapter). However, if these repair mechanisms fail, the next round of DNA replication will produce a double helix with a CG base pair, whereas the correct base pair should be TA. As shown in the right side of Figure 16.10c, one of four daughter cells inherits this CG mutation.

Oxidative Stress May Also Lead to DNA Damage and Mutation

Aerobic organisms use oxygen as a terminal acceptor of their electron transport chains. **Reactive oxygen species (ROS),** such as hydrogen peroxide, superoxide, and hydroxyl radical, are products of oxygen metabolism in all aerobic organisms. In eukaryotes, ROS are naturally produced as unwanted by-products of energy production in mitochondria. They may also be produced during certain types of immune responses and from a variety of detoxification reactions in the cell. If ROS accumulate, they can damage cellular molecules, including DNA, proteins, and lipids. To prevent this from happening, cells use a variety of enzymes, such as superoxide dismutase and catalase, to prevent the buildup of ROS. In addition, small molecules, such as vitamin C, may act as antioxidants. Likewise, certain foods contain chemicals that act as antioxidants. Colorful fruits and vegetables, including grapes, blueberries, cranberries, citrus fruits, spinach, broccoli, beets, beans, red peppers, carrots, and strawberries, are usually high in antioxidants. In humans, the overaccumulation of ROS has been implicated in a wide variety of medical conditions, including cardiovascular disease, Alzheimer disease, chronic fatigue syndrome, and aging. However, the production of ROS is not always harmful. ROS are produced by the immune system as a means of killing pathogens. In addition, some ROS are used in cell signaling.

Oxidative stress refers to an imbalance between the production of ROS and an organism's ability to break them down. If ROS overaccumulate, one particularly harmful consequence is **oxidative DNA damage,** which refers to changes in DNA structure that are caused by ROS. DNA bases are very susceptible to oxidation. Guanine bases are particularly vulnerable to oxidation, which can lead to several different oxidized guanine products. The most thoroughly studied guanine oxidation product is 7,8-dihydro-8-oxoguanine, which is commonly known as 8-oxoguanine (8-oxoG) (**Figure 16.11**). Researchers often measure the amount of 8-oxoG in a sample of DNA to determine the extent of oxidative stress. Why are oxidized bases harmful? In the case of 8-oxoG, it base pairs with adenine during DNA

Common — Rare
Keto form → Tautomeric shift (Guanine) → Enol form
Keto form → Tautomeric shift (Thymine) → Enol form
Amino form → Tautomeric shift (Adenine) → Imino form
Amino form → Tautomeric shift (Cytosine) → Imino form

(a) Tautomeric shifts that occur in the 4 bases found in DNA

Thymine (enol) Guanine (keto) Cytosine (imino) Adenine (amino)

(b) Mis–base pairing due to tautomeric shifts

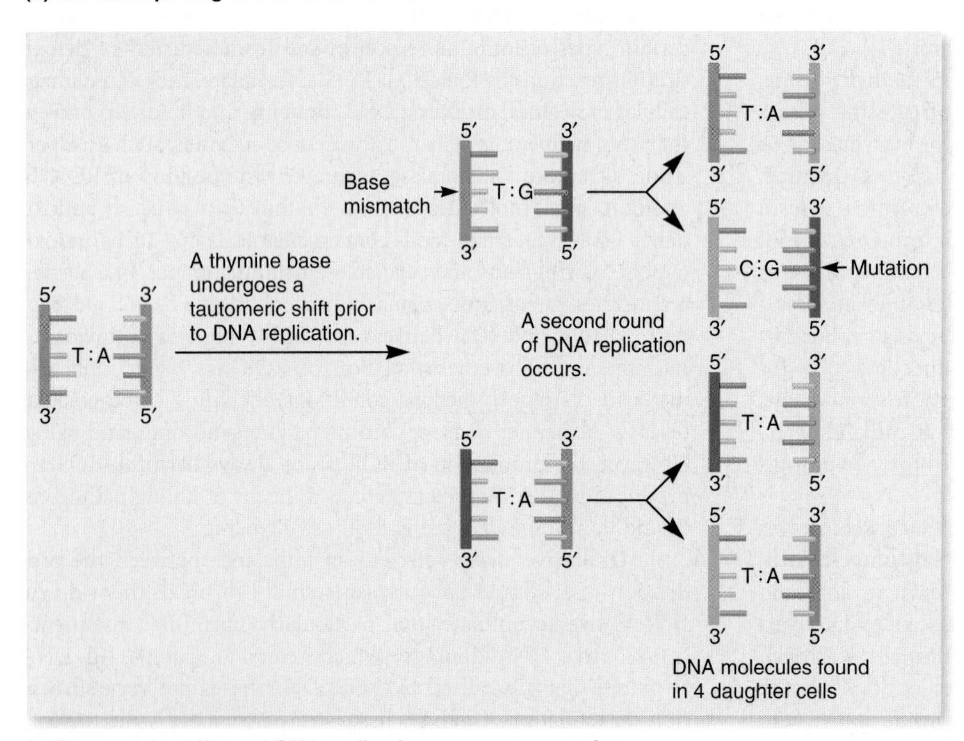

(c) Tautomeric shifts and DNA replication can cause mutation

FIGURE 16.10 Tautomeric shifts and their ability to cause mutation. (a) The common forms of the bases are shown on the left, and the rare forms produced by a tautomeric shift are shown on the right. **(b)** On the left, the rare enol form of thymine pairs with the common keto form of guanine (instead of adenine); on the right, the rare imino form of cytosine pairs with the common amino form of adenine (instead of guanine). **(c)** A tautomeric shift occurred in a thymine base just prior to replication, causing the formation of a TG base pair. If not repaired, a second round of replication will lead to the formation of a permanent CG mutation. Note: A tautomeric shift is a very temporary situation. During the second round of replication, the thymine base that shifted prior to the first round of DNA replication is likely to have shifted back to its normal form. Therefore, during the second round of replication, an adenine base is found opposite this thymine.

replication, causing mutations in which a GC base pair becomes a TA base pair. This is a transversion mutation.

Although oxidative DNA damage can occur spontaneously, it also results from environmental agents, such as ultraviolet light, X-rays, and many chemicals including those found in cigarette smoke. Later in this section, we will discuss the discovery of such environment agents and their abilities to cause mutation.

DNA Sequences Known as Trinucleotide Repeats Are Hotspots for Mutation

Researchers have discovered that several human genetic diseases are caused by an unusual form of mutation known as **trinucleotide repeat expansion (TNRE).** The term refers to the phenomenon in which a repeated sequence of three nucleotides can

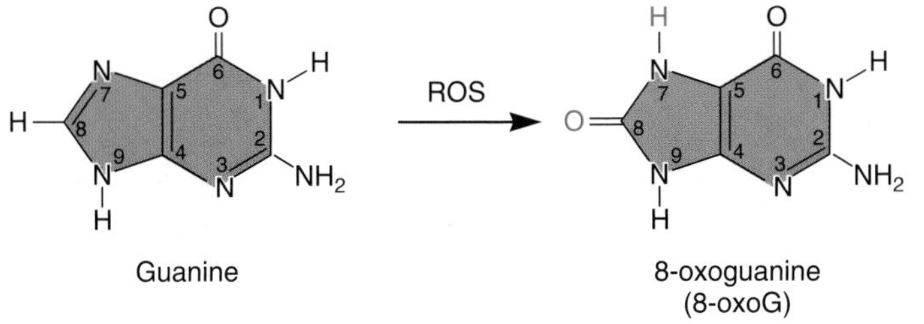

FIGURE 16.11 Oxidation of guanine to 8-oxoguanine by reactive oxygen species (ROS).

readily increase in number from one generation to the next. In humans and other species, certain genes and chromosomal locations contain regions where trinucleotide sequences are repeated in tandem. These sequences are usually transmitted normally from parent to offspring without mutation. However, in persons with TNRE disorders, the length of a trinucleotide repeat has increased above a certain critical size and thereby causes disease symptoms.

Table 16.5 describes several human diseases that involve these types of expansions, including spinal and bulbar muscular atrophy (SBMA), Huntington disease (HD), spinocerebellar ataxia (SCA1), fragile X syndromes (FRAXA and FRAXE), and myotonic muscular dystrophy (DM). In some cases, the expansion is within the coding sequence of the gene. Typically, such an expansion is a CAG repeat. Because CAG encodes a glutamine codon, these repeats cause the encoded proteins to contain long tracts of glutamine. The presence of glutamine tracts causes the proteins to aggregate. This aggregation of proteins or protein fragments carrying glutamine repeats is correlated with the progression of the disease. In other TNRE disorders, the expansions are located in the noncoding regions of genes. In the case of the two fragile X syndromes, the repeat produces CpG islands that become methylated. As discussed in Chapter 15, methylation can silence gene transcription. For DM, it has been hypothesized that these expansions cause abnormal changes in RNA structure, which produce disease symptoms.

Some TNRE disorders have the unusual feature of a progressively worsening severity in future generations—a phenomenon called **anticipation** or **dynamic mutation.** An example is depicted here, where the trinucleotide repeat of CAG has expanded from 11 tandem copies to 18.

CAGCAGCAGCAGCAGCAGCAGCAGCAGCAGCAG $n = 11$

to

CAGCAGCAGCAGCAGCAGCAGCAGCAGCAGCAGCAGCAG-
CAGCAGCAGCAGCAG $n = 18$

However, anticipation does not occur with all TNRE disorders and usually depends on whether the disease is inherited from the mother or father. In the case of HD, anticipation is likely to occur if the mutant gene is inherited from the father. In contrast, DM is more likely to get worse if the gene is inherited from the mother. These results suggest that TNRE can happen more frequently during oogenesis or spermatogenesis, depending on the particular gene involved. The phenomenon of anticipation makes it particularly difficult for genetic counselors to advise couples about the likely severity of these diseases if they are passed to their children.

How does TNRE occur? Though it may occur in more than one way, researchers have determined that a key aspect of TNRE is that the triplet repeat can form a hairpin, also called a stem-loop. A consistent feature of the triplet sequences associated with TNRE is they contain at least one C and one G (see Table 16.5). As shown in **Figure 16.12a**, such a sequence can form a hairpin due to the formation of CG base pairs. The formation of a hairpin during DNA replication can lead to an increase in the length of a DNA region if it occurs in the newly made daughter strand (**Figure 16.12b**). In such a scenario, DNA polymerase slips off the DNA after the repeat sequence is synthesized. A hairpin quickly forms, and then DNA polymerase hops back onto the DNA and continues with DNA replication. When this occurs, DNA polymerase is synthesizing most of the hairpin region twice. Depending on how the DNA is repaired, this may result in a trinucleotide

TABLE **16.5**

TNRE Disorders

Disease	SBMA	HD	SCA1	FRAXA	FRAXE	DM
Repeat Sequence	CAG	CAG	CAG	CGG	GCC	CTG
Location of Repeat	Coding sequence	Coding sequence	Coding sequence	5′-UTR	5′-UTR	3′-UTR
Number of Repeats in Unaffected Individuals	11–33	6–37	6–44	6–53	6–35	5–37
Number of Repeats in Affected Individuals	36–62	27–121	43–81	>200	>200	>200
Pattern of Inheritance	X-linked	Autosomal dominant	Autosomal dominant	X-linked	X-linked	Autosomal dominant
Disease Symptoms	Neuro-degenerative	Neuro-degenerative	Neuro-degenerative	Mental impairment	Mental impairment	Muscle disease
Anticipation*	None	Male	Male	Female	None	Female

*Indicates the parent in which anticipation occurs most prevalently.

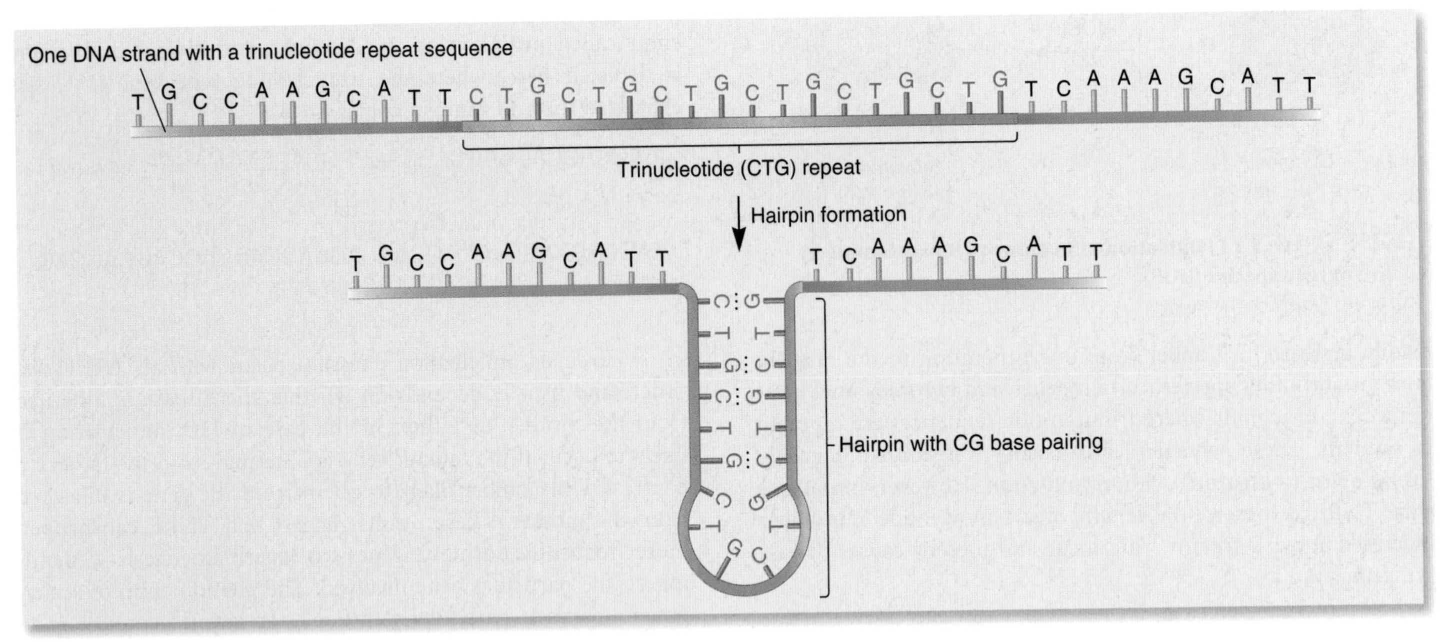

(a) Formation of a hairpin with a trinucleotide (CTG) repeat sequence

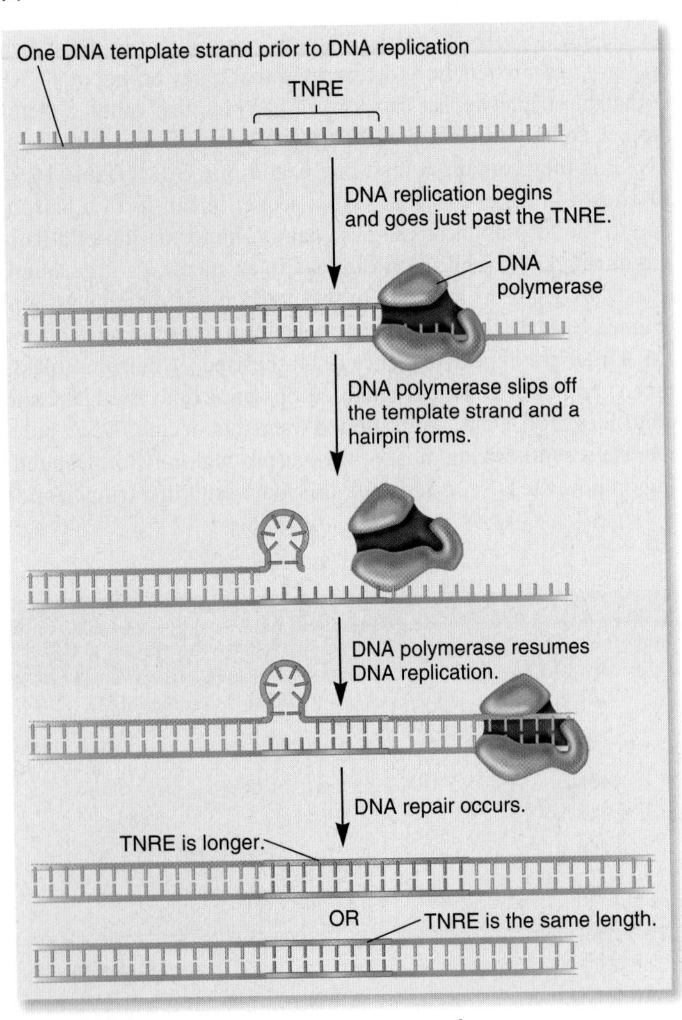

(b) Mechanism of trinucleotide repeat expansion

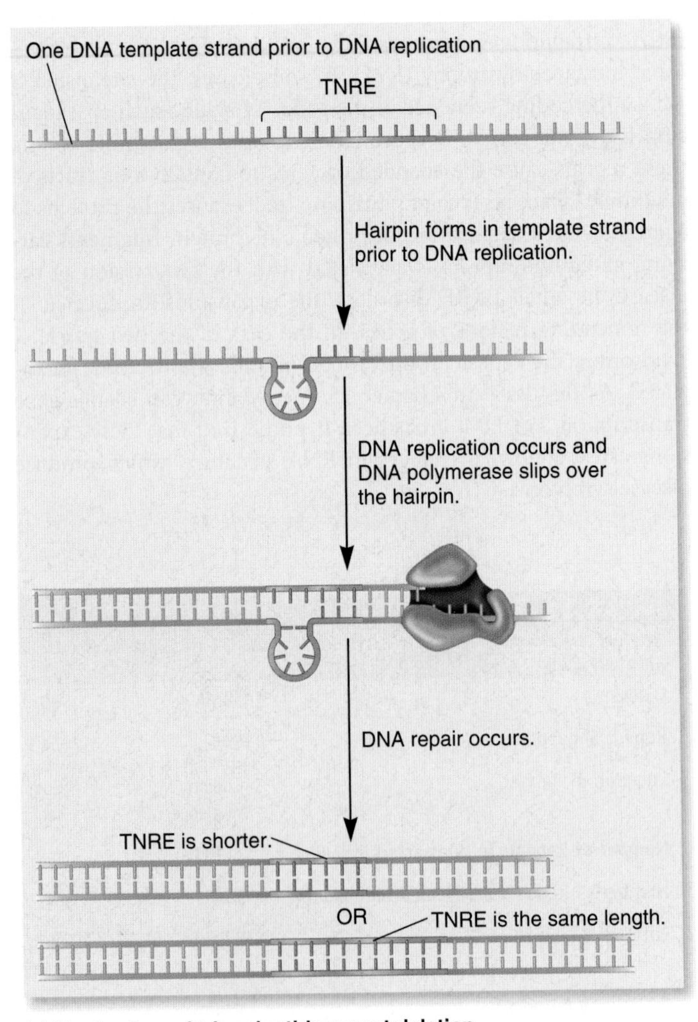

(c) Mechanism of trinucleotide repeat deletion

FIGURE 16.12 **Proposed mechanism of trinucleotide repeat expansion (TNRE).** **(a)** Trinucleotide repeats can form hairpin structures due to CG base pairing. **(b)** Formation of a trinucleotide repeat expansion. **(c)** Formation of a deletion.

repeat expansion. Alternatively, if a hairpin forms in the template strand, a deletion may result in the trinucleotide repeat, which is also common (**Figure 16.12c**). When the trinucleotide repeat sequence is abnormally long, such changes may frequently occur during gamete formation, and therefore future generations may have trinucleotide repeat sequences that are longer (or shorter) than their parents. In addition, as discussed later in this chapter, certain forms of DNA repair involve short regions where the DNA is replicated. Such DNA repair may occur in the somatic cells of individuals who already have abnormally long trinucleotide repeats in a particular gene, which may frequently cause those repeats to become even longer and more harmful. In this way, the severity of some TNRE disorders may get worse with age.

EXPERIMENT 16A

X-Rays Were the First Environmental Agent Shown to Cause Induced Mutations

As shown earlier in Table 16.4, changes in DNA structure can also be caused by environmental agents, either chemical or physical agents. These agents are called mutagens, and the mutations they cause are referred to as induced mutations. In 1927, Hermann Muller devised an approach to show that X-rays can cause induced mutations in *Drosophila melanogaster*. Muller reasoned that a mutagenic agent might cause some genes to become defective. His experimental approach focused on the ability of a mutagen to cause defects in X-linked genes that result in a recessive lethal phenotype.

To determine if X-rays increase the likelihood of recessive, X-linked lethal mutations, Muller sought an easy way to detect the occurrence of such mutations. He cleverly realized that he had a laboratory strain of fruit flies that could make this possible. In particular, he conducted his crosses in such a way that a female fly that inherited a new mutation causing a recessive X-linked lethal allele would not be able to produce any male offspring. This made it very easy for him to detect lethal mutations; he had to determine the relative number of female flies that could not produce sons compared with those that could. To understand Muller's crosses, we need to take a closer look at a peculiar version of one of the X chromosomes in a strain of flies that he used in his crosses. This X chromosome, designated *ClB*, had three important genetic alterations.

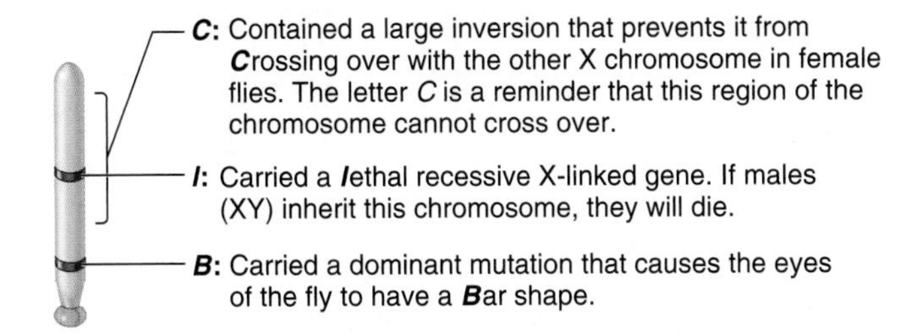

(Note: *C* and *B* are uppercase letters because they are inherited in a dominant manner, and *l* is lowercase because it is a recessive allele.)

A female fly that has one copy of this X chromosome would have bar-shaped eyes, because bar is a dominant allele. Even though this X chromosome has a recessive lethal allele, a female fly can survive if the corresponding gene on the other X chromosome is a normal allele. In Muller's experiments, the goal was to determine if exposure to X-rays caused a mutation on the normal X chromosome (not the *ClB* chromosome) that produced a recessive lethal allele in any essential X-linked gene except for the gene that already had a lethal allele on the *ClB* chromosome

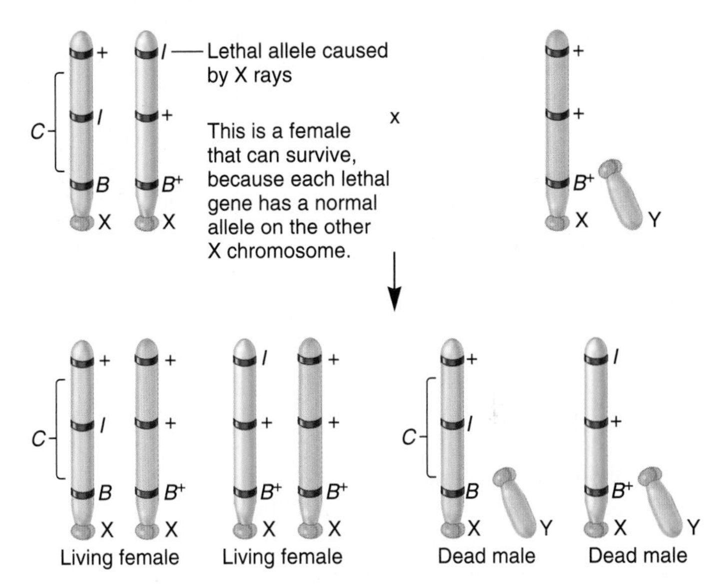

FIGURE 16.13 A strategy to detect the presence of lethal X-linked mutations. X-rays may cause a recessive lethal mutation to occur in the normal X chromosome. This female also contains another lethal allele in the *ClB* chromosome. Nevertheless, this female could survive because it would be heterozygous for recessive lethal mutations in two different genes. Because each X chromosome would have a lethal mutation, this female would not be able to produce any living sons.

(**Figure 16.13**). If a recessive lethal mutation occurred on the normal X chromosome, this female could survive because it would be heterozygous for recessive lethal mutations in two different genes. However, because each X chromosome would have a lethal mutation, this female would not be able to produce any living sons.

The steps in Muller's protocol are shown in **Figure 16.14**. He began with wild-type males and exposed them to X-rays. These X-rays may mutate the X chromosome in sperm cells, resulting in a recessive lethal allele. These males, and a control group of males that were not exposed to X-rays, were then mated to females carrying the *ClB* chromosome. Daughters with bar eyes were saved from this cross and mated to nonirradiated males. You should look carefully at this cross and realize that if these daughters also contained a lethal allele on the X chromosome they inherited from their father (e.g., an irradiated male in step 1), they would not be able to produce living sons.

THE HYPOTHESIS

The exposure of flies to X-rays will increase the rate of mutation.

TESTING THE HYPOTHESIS — FIGURE 16.14 Evidence that X-rays cause mutation.

Starting material: The female flies used in this study had one normal X chromosome and a *ClB* X chromosome. The male flies had a normal X chromosome.

1. Expose male flies to X-rays. Also, have a control group that is not exposed to X-rays.
2. Mate the male flies to female flies carrying one normal X chromosome and one *ClB* X chromosome.
3. Save about 1000 daughters with bar eyes. Note: These females contain a *ClB* X chromosome from their mothers and an X chromosome from their fathers that may (or may not) have a recessive lethal mutation.
4. Mate each bar-eyed daughter with normal (nonirradiated) males. Note: This is done in (1000) individual tubes. (Only six are shown.)
5. Count the number of crosses that do not contain any male offspring. These crosses indicate that the bar-eyed female parent contained an X-linked lethal recessive mutation on the non-*ClB* X chromosome.

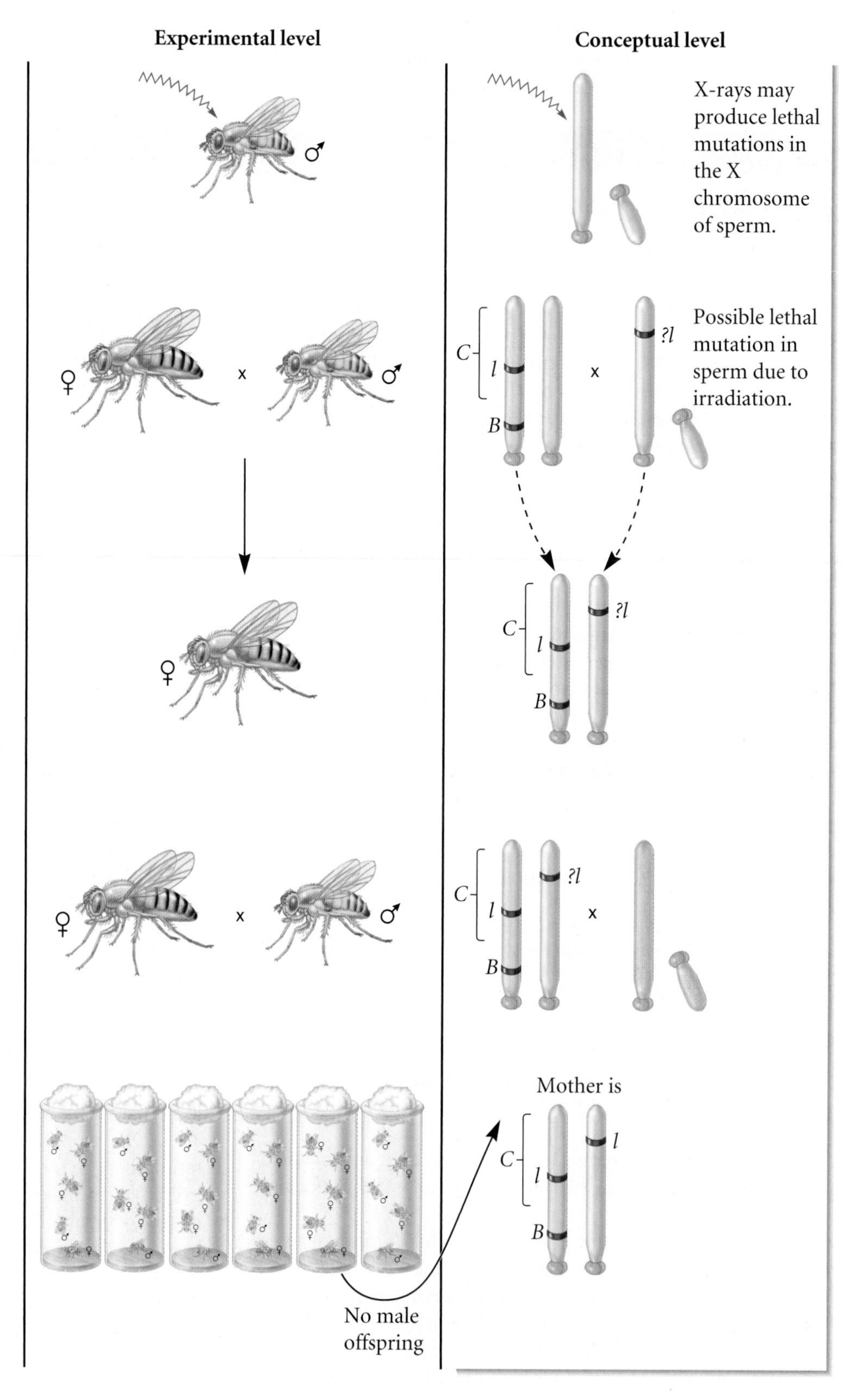

THE DATA

Treatment of Fathers of the *ClB* Daughters	Number of *ClB* Daughters Crossed to Normal Males*	Number of Tubes Containing Any Offspring†	Number of Tubes with Female Offspring but Lacking Male Offspring
Control	1011	947	1
X-ray treated	1015	783	91

*See step 4.

†These values are less than the numbers of mated *ClB* daughters because some crosses did not produce living offspring.

Data from: Muller, Hermann J. (1927) Artificial transmutation of the gene. *Science 66*, 84–87.

INTERPRETING THE DATA

As shown in the data of Figure 16.14, in the absence of X-ray treatment, only 1 cross in approximately 1000 was unable to produce male offspring. This means that the spontaneous rate for any X-linked lethal mutation was relatively low. By comparison, X-ray treatment of the fathers that gave rise to these *ClB* females resulted in 91 crosses without male offspring. Because these females inherited their non-*ClB* chromosome from irradiated fathers, these results indicate that X-rays greatly increase the rate of X-linked, recessive lethal mutations. This conclusion has been confirmed in many subsequent studies, which have shown that the increase in mutation rate is correlated with the amount of exposure to X-rays.

A self-help quiz involving this experiment can be found at www.mhhe.com/brookergenetics4e.

Mutagens Alter DNA Structure in Different Ways

Since this pioneering study of Muller, researchers have found that an enormous array of agents can act as mutagens that permanently alter the structure of DNA. We often hear in the news media that we should avoid these agents in our foods and living environment. For example, we use products such as sunscreens to help us avoid the mutagenic effects of ultraviolet (UV) rays. The public is concerned about mutagens for two important reasons. First, mutagenic agents are often involved in the development of human cancers. In addition, because new mutations may be deleterious, people want to avoid mutagens to prevent gene mutations that may have harmful effects on their future offspring.

Mutagenic agents are usually classified as chemical or physical mutagens. Examples of both types of agents are listed in **Table 16.6**. In some cases, chemicals that are not mutagenic can be altered to a mutagenically active form after being ingested. Cellular enzymes such as oxidases have been shown to activate some mutagens.

Mutagens can alter the structure of DNA in various ways. Some mutagens act by covalently modifying the structure of bases. For example, **nitrous acid** (HNO_2) replaces amino groups with keto groups ($=NH_2$ to $=O$), a process called deamination. This can change cytosine to uracil and adenine to hypoxanthine. When this altered DNA replicates, the modified bases do not pair with the appropriate bases in the newly made strand. Instead, uracil pairs with adenine, and hypoxanthine pairs with cytosine (**Figure 16.15**).

Other chemical mutagens can also disrupt the appropriate pairing between nucleotides by alkylating bases within the DNA. During alkylation, methyl or ethyl groups are covalently attached to the bases. Examples of alkylating agents include **nitrogen mustard** (a type of mustard gas) and **ethyl methanesulfonate (EMS).** Mustard gas was used as a chemical weapon during World War I. Such agents damage the skin, eyes, mucous membranes, lungs, and blood-forming organs.

TABLE 16.6

Examples of Mutagens

Mutagen	Effect(s) on DNA Structure
Chemical	
Nitrous acid	Deaminates bases
Nitrogen mustard	Alkylating agent
Ethyl methanesulfonate	Alkylating agent
Proflavin	Intercalates within DNA helix
5-Bromouracil	Base analogue
2-Aminopurine	Base analogue
Physical	
X-rays	Cause base deletions, single-stranded nicks in DNA, cross-linking, and chromosomal breaks
UV light	Promotes pyrimidine dimer formation, such as thymine dimers

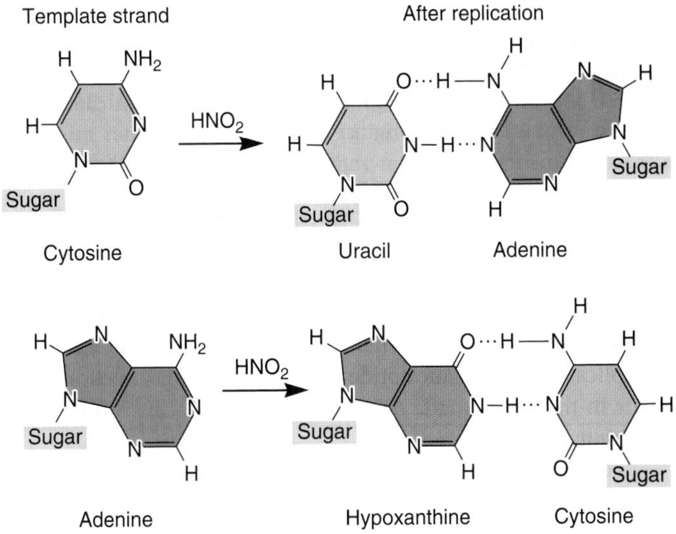

FIGURE 16.15 Mispairing of modified bases that have been deaminated by nitrous acid. Nitrous acid converts cytosine to uracil, and adenine to hypoxanthine. During DNA replication, uracil pairs with adenine, and hypoxanthine pairs with cytosine. This creates mutations in the newly replicated strand during DNA replication.

systems. Yet, when even a single system is absent, the bacteria have a much higher rate of mutation. In fact, the rate of mutation is so high that these bacterial strains are sometimes called mutator strains. Likewise, in humans, an individual who is defective in only a single DNA repair system may manifest various disease symptoms, including a higher risk of skin cancer. This increased risk is due to the inability to repair UV-induced mutations.

Living cells contain several DNA repair systems that can fix different types of DNA alterations (**Table 16.7**). Each repair system is composed of one or more proteins that play specific roles in the repair mechanism. In most cases, DNA repair is a multistep process. First, one or more proteins in the DNA repair system detect an irregularity in DNA structure. Next, the abnormality is removed by the action of DNA repair enzymes. Finally, normal DNA is synthesized via DNA replication enzymes. In this section, we will examine several different repair systems that have been characterized in bacteria, yeast, mammals, and plants. Their diverse ways of repairing DNA underscore the extreme necessity for the structure of DNA to be maintained properly.

Damaged Bases Can Be Directly Repaired

In a few cases, the covalent modification of nucleotides by mutagens can be reversed by specific cellular enzymes. As discussed earlier in this chapter, UV light causes the formation of thymine dimers. Bacteria, fungi, most plants, and some animals produce an enzyme called **photolyase** that recognizes thymine dimers and splits them, which returns the DNA to its original condition (**Figure 16.19a**). Photolyases are flavoproteins that contain two light-sensitive cofactors. The repair mechanism itself requires light and is known as **photoreactivation.** This process directly restores the structure of DNA. Because plants are exposed to sunlight throughout the day, photolyase is a critical DNA repair enzyme for many plant species.

TABLE 16.7

Common Types of DNA Repair Systems

System	Description
Direct repair	An enzyme recognizes an incorrect alteration in DNA structure and directly converts it back to a correct structure.
Base excision repair and nucleotide excision repair	An abnormal base or nucleotide is first recognized and removed from the DNA. A segment of DNA is excised, and then the complementary DNA strand is used as a template to synthesize normal DNA.
Mismatch repair	Similar to excision repair except the DNA defect is a base pair mismatch, not an abnormal nucleotide. The mismatch is recognized, and a segment of DNA is removed. The parental strand is used to synthesize a normal daughter strand of DNA.
Homologous recombination repair	Occurs at double-strand breaks or when DNA damage causes a gap in synthesis during DNA replication. The strands of a normal chromatid are used to repair a damaged chromatid.
Nonhomologous end joining	Occurs at double-strand breaks. The broken ends are recognized by proteins that keep the ends together; the broken ends are eventually rejoined.

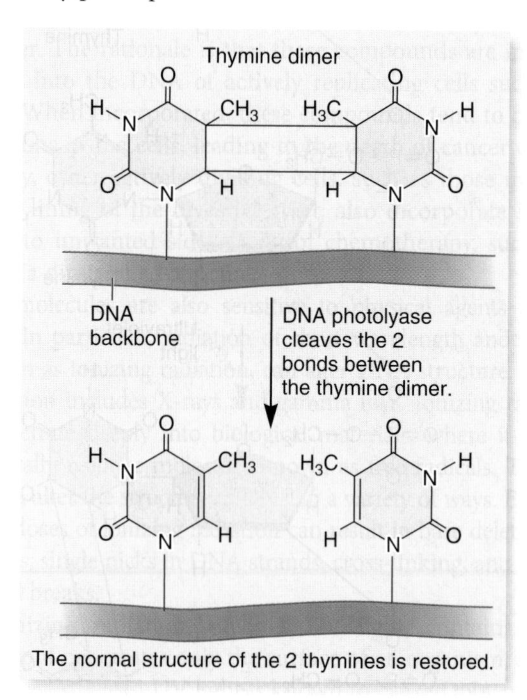

(a) Direct repair of a thymine dimer

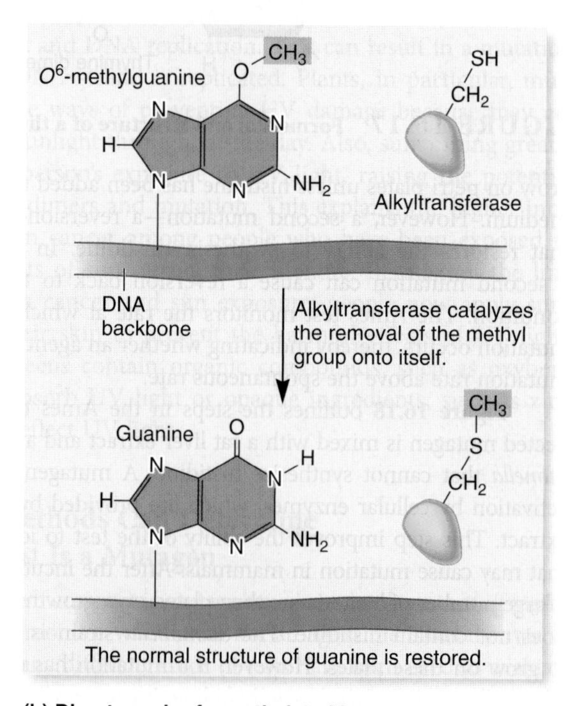

(b) Direct repair of a methylated base

FIGURE 16.19 Direct repair of damaged bases in DNA. **(a)** The repair of thymine dimers by photolyase. **(b)** The repair of methylguanine by the transfer of the methyl group to alkyltransferase.

A protein known as **alkyltransferase** can remove methyl or ethyl groups from guanine bases that have been mutagenized by alkylating agents such as nitrogen mustard and EMS. This protein is called alkyltransferase because it transfers the methyl or ethyl group from the base to a cysteine side chain within the alkyltransferase protein (**Figure 16.19b**). Surprisingly, this permanently inactivates alkyltransferase, which means it can be used only once!

Base Excision Repair Removes a Damaged Base

A second type of repair system, called **base excision repair (BER),** involves the function of a category of enzymes known as **DNA *N*-glycosylases.** This type of enzyme can recognize an abnormal base and cleave the bond between it and the sugar in the DNA backbone, creating an apurinic or apyrimidinic site (**Figure 16.20**). Living organisms produce multiple types of DNA *N*-glycosylases, each recognizing particular types of abnormal base structures. Depending on the DNA *N*-glycosylase, this repair system can eliminate abnormal bases such as uracil, 3-methyladenine, 7-methylguanine, and pyrimidine dimers. Base excision repair is particularly important for the repair of oxidative DNA damage.

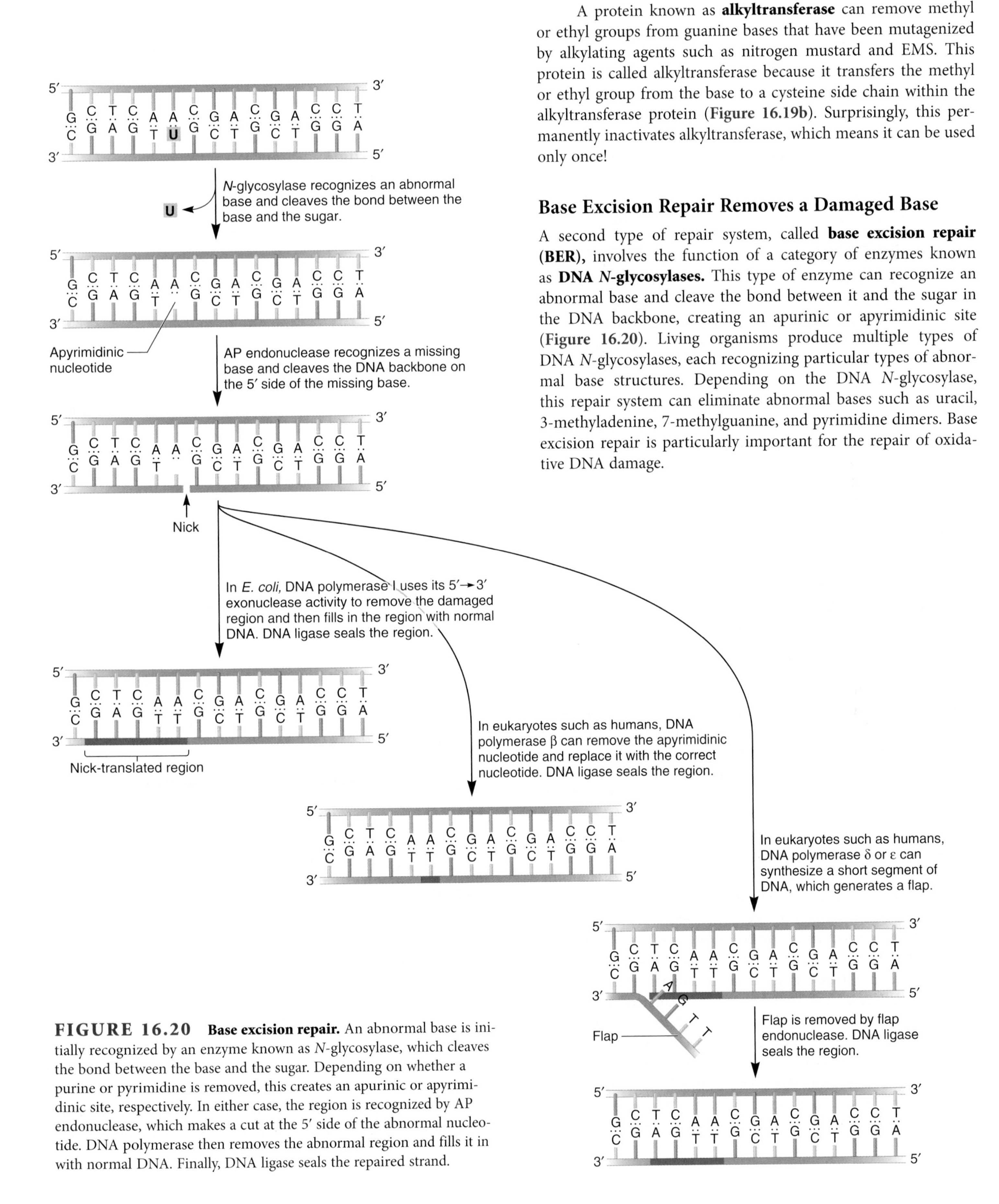

FIGURE 16.20 **Base excision repair.** An abnormal base is initially recognized by an enzyme known as *N*-glycosylase, which cleaves the bond between the base and the sugar. Depending on whether a purine or pyrimidine is removed, this creates an apurinic or apyrimidinic site, respectively. In either case, the region is recognized by AP endonuclease, which makes a cut at the 5′ side of the abnormal nucleotide. DNA polymerase then removes the abnormal region and fills it in with normal DNA. Finally, DNA ligase seals the repaired strand.

this textbook, *E. coli* has other pathways to carry out homologous recombination.

RecBCD is a protein complex composed of the RecB, RecC, and RecD proteins. (The term Rec indicates that these proteins are involved with recombination.) The RecBCD complex plays an important role in the initiation of recombination involving double-strand breaks, as outlined in Figure 17.5. In this process, RecBCD recognizes a double-strand break within DNA and catalyzes DNA unwinding and strand degradation. The action of RecBCD produces single-stranded DNA ends that can participate in strand invasion and exchange. The single-stranded DNA ends are coated with single-stranded binding protein to prevent their further degradation. The RecBCD complex can also create breaks in the DNA at sites known as chi sequences. In *E. coli*, the chi sequence is 5′–GCTGGTGG–3′. As RecBCD tracks along the DNA, when it encounters a chi sequence, it cuts one of the DNA strands to the 3′ side of this sequence. This generates a single nick that can also initiate homologous recombination.

The function of the RecA protein is to promote strand invasion. To accomplish this task, it binds to the single-stranded ends of DNA molecules generated from the activity of RecBCD. A large number of RecA proteins bind to single-stranded DNA, forming a structure called a filament. During strand invasion, this filament makes contact with the unbroken chromosome. Initially, this contact is most likely to occur at nonhomologous regions. The contact point slides along the DNA until it reaches a homologous region. Once a homologous site is located, RecA catalyzes the displacement of one DNA strand, and the invading single-stranded DNA quickly forms a double helix with the other strand. This results in a D-loop, as shown in Figure 17.5. RecA proteins mediate the movement of the invading strand and the displacement of the complementary strand. This occurs in such a way that the displaced strand invades the vacant region of the broken chromosome.

Proteins that bind specifically to Holliday junctions have also been identified. These include a complex of proteins termed RuvABC and a protein called RecG. RuvA is a tetramer that forms a platform on which the Holliday junction is held in a square planar configuration. This platform also contains two hexameric rings of RuvB. Together, RuvA and RuvB act as a helicase that catalyzes an ATP-dependent migration of a four-way branched DNA junction in either the 5′ → 3′ or 3′ → 5′ direction. RuvC is an endonuclease that binds as a dimer to Holliday junctions and resolves these structures by making cuts in the DNA (see bottom of Figure 17.4). RecG is a helicase that also catalyzes branch migration of Holliday junctions.

Before ending this discussion regarding the molecular mechanism of homologous recombination, let's consider recombinational events during meiosis in eukaryotic cells. As described in Chapter 3, crossing over between homologous chromosomes is an important event during prophase of meiosis I. An intriguing question is, How are crossover sites chosen between two homologous chromosomes? Although the answer is not entirely understood, molecular studies in two different yeast species, *Saccharomyces cerevisiae* and *Schizosaccharomyces pombe*, suggest that double-strand breaks initiate the homologous recombination that occurs during meiosis. In other words, double-strand breaks create sites where a crossover will occur. In *S. cerevisiae*, the formation of DNA double-strand breaks that initiate meiotic recombination requires at least 10 different proteins. One particular protein, termed Spo11 protein, is thought to be instrumental in cleaving the DNA, thereby creating a double-strand break. However, the roles of the other proteins, and the interactions among them, are not completely understood. Once a double-strand break is made, homologous recombination can then occur according to the model described in Figure 17.5.

Gene Conversion May Result from DNA Mismatch Repair or DNA Gap Repair

As mentioned earlier, homologous recombination can lead to gene conversion, in which one allele is converted to the allele on the homologous chromosome. The original Holliday model was based on this phenomenon.

How can homologous recombination account for gene conversion? Researchers have identified two possible ways this can occur. One mechanism involves DNA mismatch repair, a topic that was described in Chapter 16. To understand how this works, let's take a closer look at the heteroduplexes formed during branch migration of a Holliday junction (see Figure 17.4a). A heteroduplex contains a DNA strand from each of the two original parental chromosomes. The two parental chromosomes may contain an allelic difference within this region. In other words, this short region may contain DNA sequence differences. If this is the case, the heteroduplex formed after branch migration will contain an area of base mismatch. Gene conversion occurs when recombinant chromosomes are repaired and result in two copies of the same allele.

As shown in **Figure 17.6**, mismatch repair of a heteroduplex may result in gene conversion. In this example, the two parental chromosomes had different alleles due to a single base-pair difference in their DNA sequences, as shown at the top of the figure. During recombination, branch migration has occurred across this region, thereby creating two heteroduplexes with base mismatches. As described in Chapter 16, DNA mismatches will be recognized by DNA repair systems and repaired to a double helix that obeys the AT/GC rule. These two mismatches can be repaired in four possible ways. As shown here, two possibilities produce no gene conversion, whereas the other two lead to gene conversion.

A second mechanism for gene conversion occurs via DNA gap repair synthesis. **Figure 17.7** illustrates how gap repair synthesis can lead to gene conversion according to the double-strand break model. The top chromosome, which carries the recessive *b* allele, has suffered a double-strand break in this gene. A gap is created by the digestion of the DNA in the double helix. This digestion eliminates the *b* allele. The two template strands used in gap repair synthesis are from one homologous chromatid. This helix carries the dominant *B* allele. Therefore, after gap repair synthesis takes place, the top chromosome carries the *B* allele, as does the bottom chromosome. Gene conversion has changed the recessive *b* allele to a dominant *B* allele.

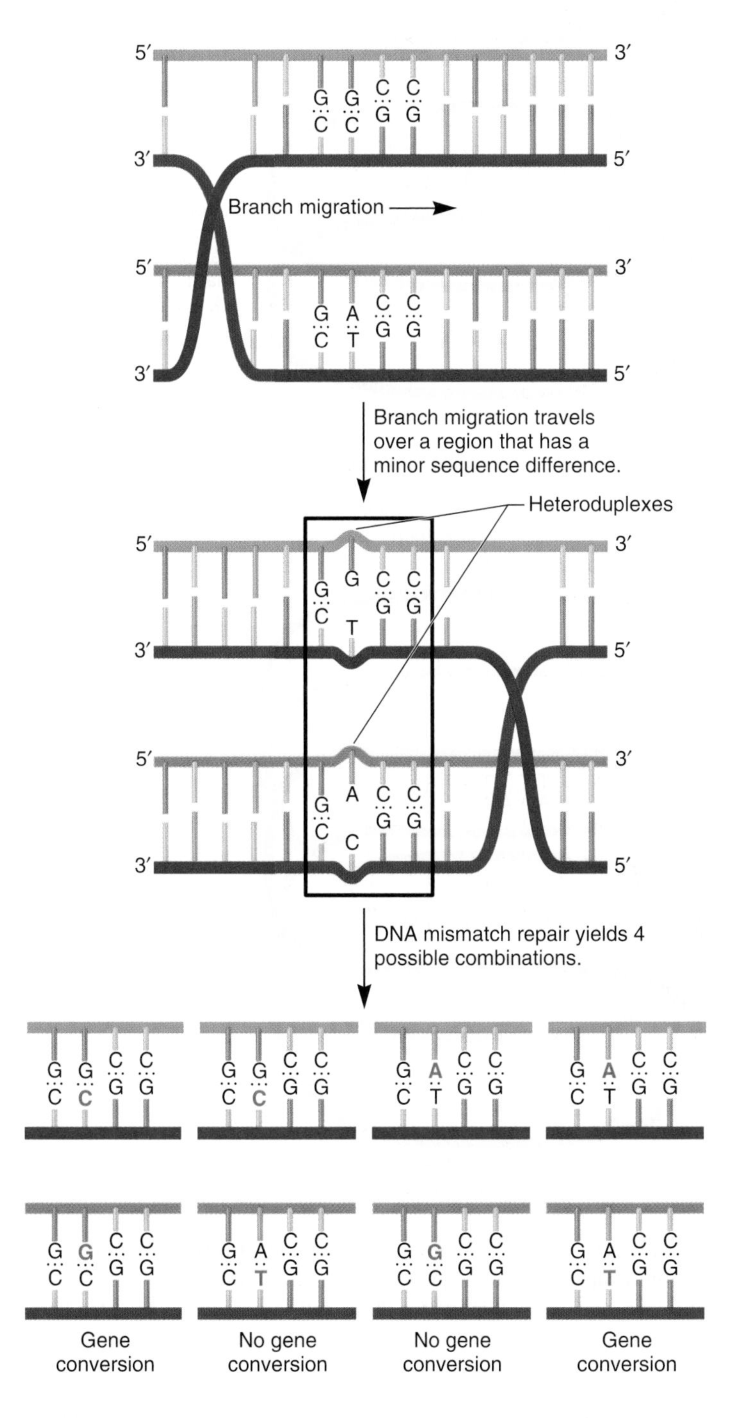

FIGURE 17.6 Gene conversion by DNA mismatch repair. A branch migrates past a homologous region that contains slightly different DNA sequences. This produces two heteroduplexes: DNA double helices with mismatches. The mismatches can be repaired in four possible ways by the mismatch repair system described in Chapter 16. Two of these ways result in gene conversion. The repaired base is shown in red.

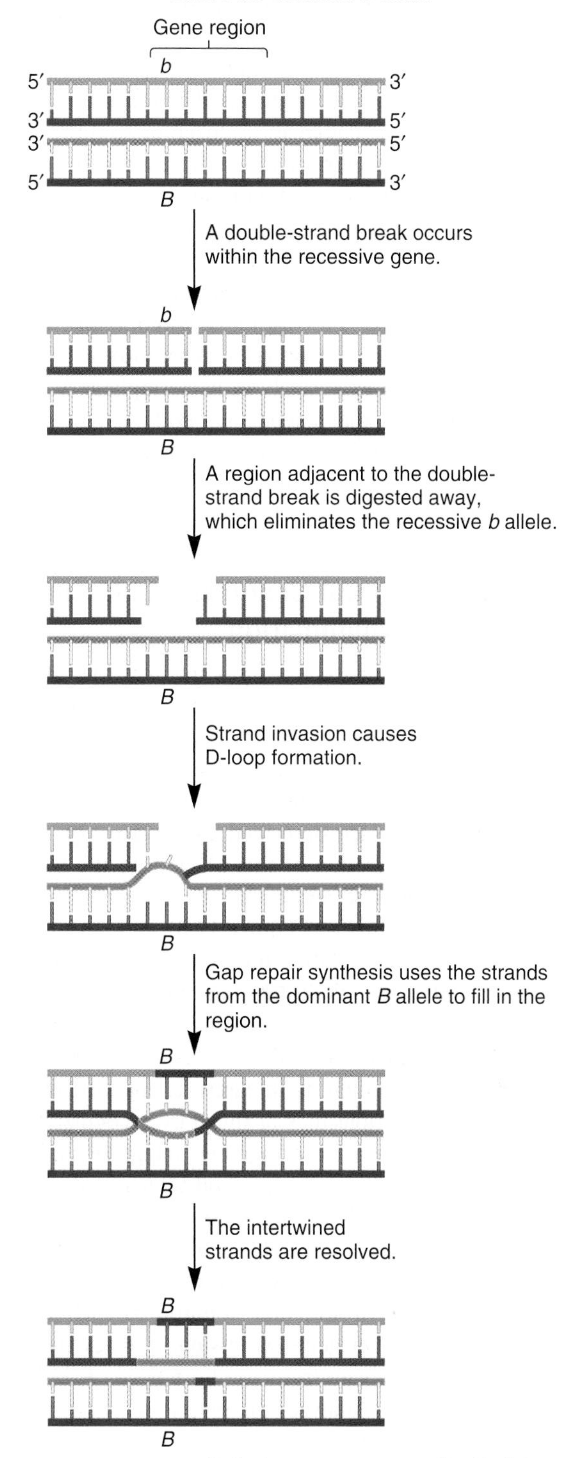

FIGURE 17.7 Gene conversion by gap repair synthesis in the double-strand break model. A gene is found in two alleles, designated *B* and *b*. A double-strand break occurs in the DNA encoding the *b* allele. Both of these DNA strands are digested away, thereby eliminating the *b* allele. A complementary DNA strand encoding the *B* allele migrates to this region and provides the template to synthesize a double-stranded region. Following resolution, both DNA double helices carry the *B* allele.

generate two double-strand breaks: one at the end of a V domain and one at the beginning of a J domain. For example, in the recombination event shown in Figure 17.9, RAG1 and RAG2 have made cuts at the end of variable domain number 78 and the beginning of joining domain number 2. The intervening region is lost, and the two ends are then joined to each other. The connection phase of this process is catalyzed by a group of proteins termed **non-homologous end-joining (NHEJ) proteins.** The fusion process may not be entirely precise, so a few nucleotides can be added or lost at the junction between the variable and joining domains. This imprecision further accentuates the diversity in antibody genes.

Following transcription, the fused VJ region is contained within a pre-mRNA transcript that is then spliced to connect the J and C domains. After this has occurred, each B cell produces only the particular κ light chain encoded by the specific VJ fusion domain and the constant domain.

The heavy-chain polypeptides are produced by a similar recombination mechanism. Though mammals have multiple copies of heavy-chain genes, a typical gene in mice may have about 500 variable domains and four joining domains. In addition, a heavy-chain gene encodes several diversity (D) domains, which are found between the variable and joining domains. The recombination first involves the connection of a D and J domain, followed by the connection of a V and DJ domain. The same proteins that catalyze VJ fusion are involved in the recombination with the heavy-chain gene. Collectively, this process is called **V(D)J recombination.** The D is in parentheses because this type of domain is found only in the heavy-chain genes, not in the light-chain genes.

The recombination process within immunoglobulin genes produces an enormous diversity in polypeptides. Even though it occurs at specific junctions within the antibody gene, the recombination is fairly random with regard to the particular V and J domains that can be joined.

Overall, the possible number of functional antibodies that can be produced by V(D)J recombination is rather staggering. For example, if we assume that any of the 300 different variable sequences can be spliced next to any of the four joining sequences, this results in 1200 possible light-chain combinations. If we also assume that a heavy chain gene has 500 variable regions, 12 diversity domains, and 4 joining regions, the number of heavy-chain possibilities is $500 \times 12 \times 4 = 24{,}000$. Because any light-chain–heavy-chain combination is possible, this yields $1200 \times 24{,}000 = 28{,}800{,}000$ possible antibody molecules from the random recombination within two precursor genes! The diversity is actually higher because mammals have multiple copies of nonidentical precursor genes for the light and heavy chains of antibodies.

17.3 TRANSPOSITION

The last form of recombination we will consider is transposition. In some ways, transposition resembles the site-specific recombination we examined for phage λ. In that case, a segment of λ DNA was able to integrate itself into the *E. coli* chromosome. Transposition also involves the integration of small segments of DNA into the chromosome. Transposition, though, can occur at many different locations within the genome. The DNA segments that transpose themselves are known as **transposable elements (TEs).** TEs have sometimes been referred to as "jumping genes" because they are inherently mobile.

Transposable elements were first identified by Barbara McClintock in the early 1950s from her classic studies with corn plants. Since that time, geneticists have discovered many different types of TEs in organisms as diverse as bacteria, fungi, plants, and animals. The advent of molecular technology has allowed scientists to understand more about the characteristics of TEs that enable them to be mobile. In this section, we examine the characteristics of TEs and explore the mechanisms that explain how they move. We will also discuss the biological significance of TEs and their uses as experimental tools.

EXPERIMENT 17B

McClintock Found That Chromosomes of Corn Plants Contain Loci That Can Move

Barbara McClintock began her scientific career as a student at Cornell University. Her interests quickly became focused on the structure and function of the chromosomes of corn plants, an interest that continued for the rest of her life. She spent countless hours examining corn chromosomes under the microscope. She was technically gifted and had a theoretical mind that could propose ideas that conflicted with conventional wisdom.

During her long career as a scientist, McClintock identified many unusual features of corn chromosomes. She noticed that chromosome 9 in one strain of corn had the strange characteristic of tending to break at a fairly high rate at the same site. McClintock termed this a **mutable site,** or locus. This observation initiated a 6-year study concerned with highly unstable chromosomal locations. In 1951, at the end of her study, McClintock proposed that these sites are actually locations where TEs have been inserted into the chromosomes. At the time of McClintock's studies, such an idea was entirely unorthodox.

McClintock focused her efforts on the relationship between a mutable locus and its phenotypic effects on corn kernels. Chromosome 9 with a mutable locus also carried several genes that affected the phenotype of corn kernels. Each gene existed in (at least) one dominant and one recessive allele. The mutable locus was termed *Ds* (for dissociation), because the locus was known to frequently cause chromosomal breaks. In the chromosome shown here, the *Ds* locus is located next to several genes affecting kernel traits.

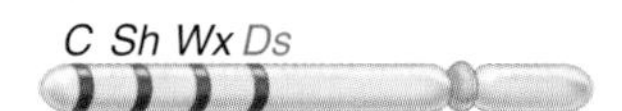

In this case, there are three genes that exist as two or more alleles:

1. *C* is an allele for normal kernel color (dark red), *c* is a recessive allele of the same gene that causes a colorless kernel, and C^I is a third allele of this gene that is dominant to both *C* and *c* and causes a colorless kernel.

2. *Sh* is an allele that produces normal endosperm, whereas *sh* is a recessive allele that causes shrunken endosperm. (Note: The endosperm is the storage material in the kernel that is used by the plant embryo to provide energy for growth.)
3. *Wx* is the allele that produces normal starch in the endosperm, and *wx* is a recessive allele that produces a waxy-appearing phenotype.

During her intensive work, in which she studied corn chromosomes under the microscope, McClintock identified strains of corn in which the *Ds* locus was found in different locations within the corn genome. She could determine the location of *Ds* because the movement of *Ds* occasionally causes chromosome breakage (**Figure 17.10**). Keep in mind that the endosperm of a kernel is triploid because it is derived from the fusion of two maternal haploid nuclei and one paternal haploid nucleus (refer back to Chapter 2, Figure 2.2c). The kernel shown in Figure 17.10 was produced by a cross in which the pollen carried the top chromosome—C^I *Sh Wx Ds*—whereas the two maternal chromosomes are *C sh wx*. This kernel is expected to be colorless, because the C^I allele is dominant and causes a colorless phenotype. However, as the kernel grows by cell division, the movement of the *Ds* locus out of its original location may occasionally cause a chromosome to break, and the distal part of this chromosome is lost. This chromosome breakage may happen in several cells, which continue to divide and grow as the kernel becomes larger. This process produces a sectoring phenotype—patches of cells occur in the kernel that are red, shrunken, and waxy.

By analyzing many kernels, McClintock was also able to identify cases in which *Ds* had moved to a new location. For example, if *Ds* had moved out of its original location and inserted

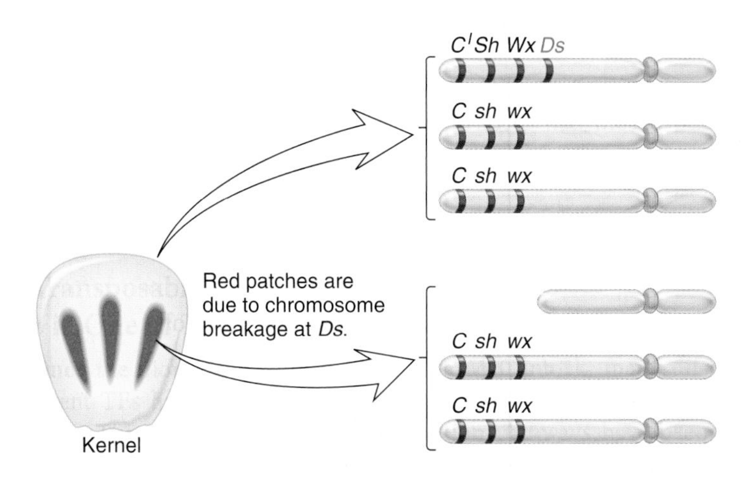

FIGURE 17.10 The sectoring trait in corn kernels.

Genes → Traits This kernel is expected to be colorless, because the C^I allele is dominant. On occasion, though, the movement of *Ds* may cause a chromosome to break, thereby losing the C^I, *Sh*, and *Wx* alleles. As such a cell continues to divide, it produces a patch of daughter cells that are red, shrunken, and waxy. Therefore, this sectoring trait arises from the loss of genes that occurs when the movement of *Ds* causes chromosome breakage. Note: The movement of *Ds* does not usually cause chromosome breakage. In most cases, the movement of *Ds* out of a site is followed by nonhomologous end joining (described in Chapter 16) in which the chromosome pieces are connected together.

between *Sh* and *Wx*, a break at *Ds* would produce the following combination:

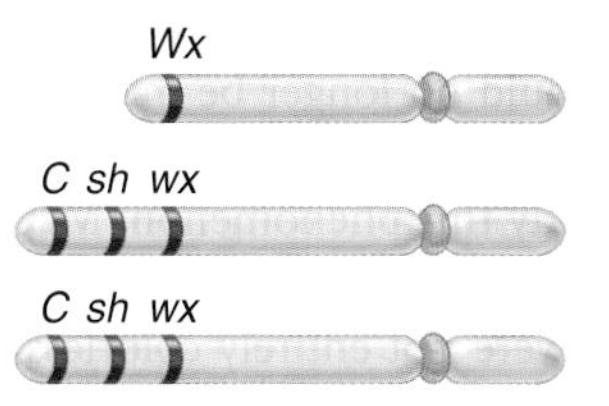

This genotype would produce patches on the kernel that are red and shrunken but not waxy. In this way, McClintock identified 20 independent cases in which the *Ds* locus had moved to a new location within this chromosome. Overall, the results from many crosses were consistent with the idea that *Ds* can transpose itself throughout the corn genome. McClintock also found that a second locus, termed *Ac* (for activator), was necessary for the *Ds* locus to move. Researchers later discovered that the *Ac* locus contains a gene that encodes an enzyme called transposase, which is necessary for *Ds* to move. We will discuss the function of transposase later in this chapter. Some strains of McClintock's corn contained the *Ac* locus, but others did not.

During her studies, McClintock noticed a particularly exciting and unusual event. By making the appropriate cross, she sought to produce kernels with the following genotype:

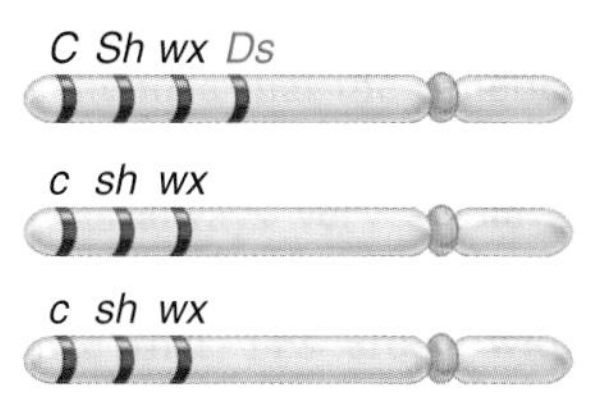

The kernels were expected to be red. Because the strain also contained the *Ac* locus, breakage would occur occasionally at the *Ds* locus to produce colorless patches. Among 4000 kernels, she noticed 1 kernel with the opposite phenotype—a colorless kernel with red patches. This observation suggested that the inherited genotype in this case, which produced a colorless phenotype, was mutable to become *C*.

How did McClintock explain these results? She postulated that the colorless phenotype was due to a transposition of *Ds* into the *C* gene:

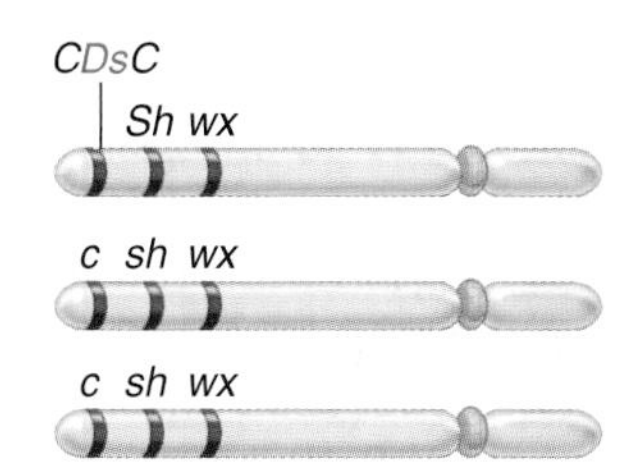

When *Ds* was located within the *C* gene, it inactivated the *C* gene, thereby resulting in the colorless phenotype. However, McClintock proposed that when *Ds* occasionally transposed out of the gene during kernel growth, the *C* allele would be restored and a red patch would result. In this case, the formation of patches, or sectoring, was due to the movement of *Ds* out of its location within the *C* gene and the rejoining of the two ends, not

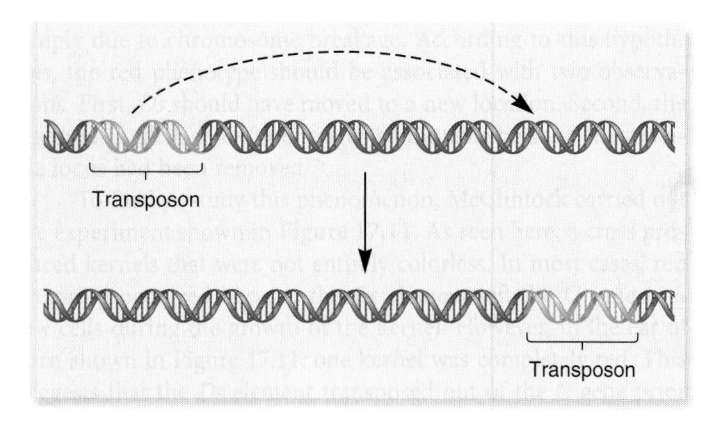

(a) Simple transposition

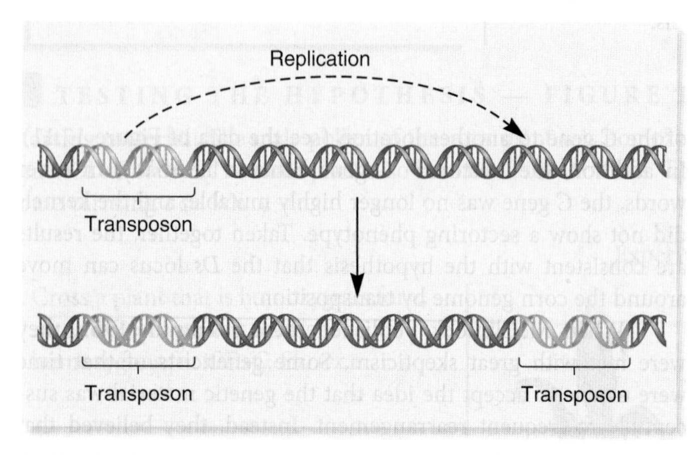

(b) Replicative transposition

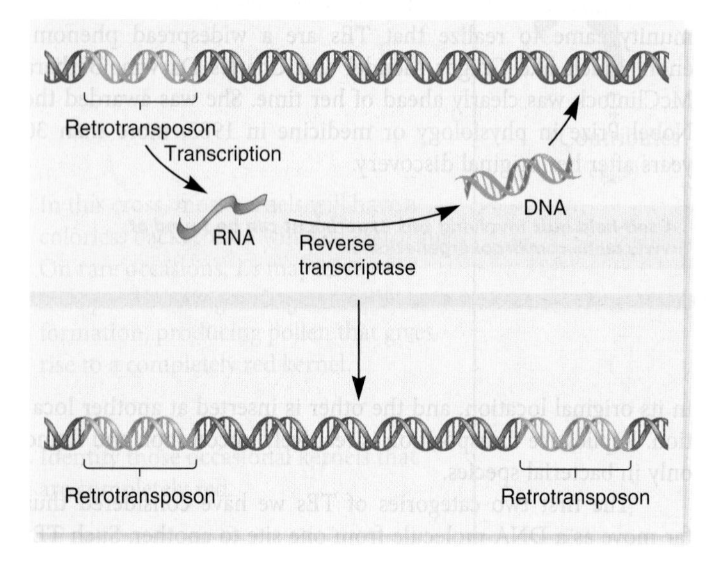

(c) Retrotransposition

FIGURE 17.12 Three mechanisms of transposition.

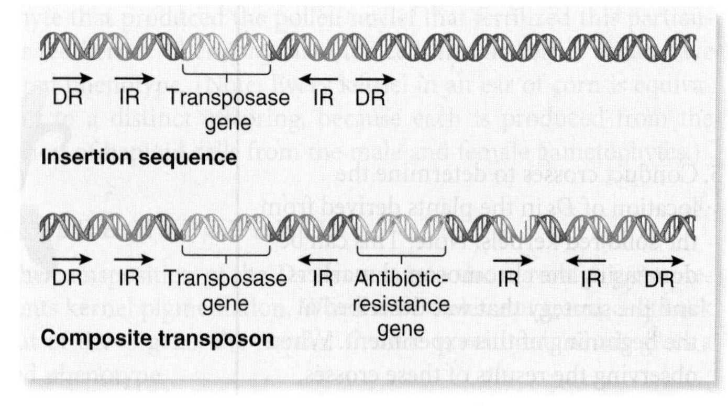

(a) Elements that move by simple transposition

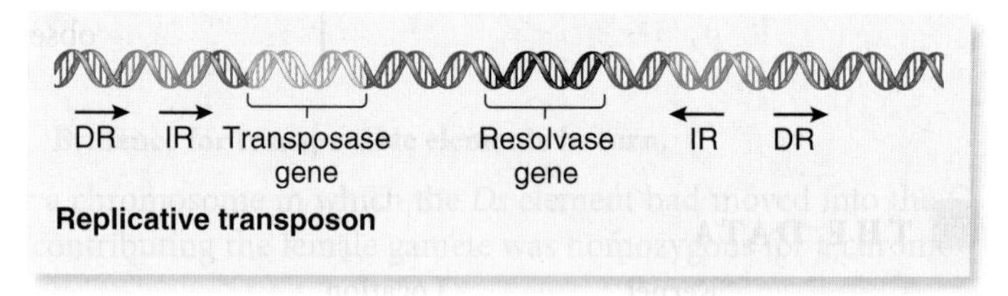

(b) An element that moves by replicative transposition

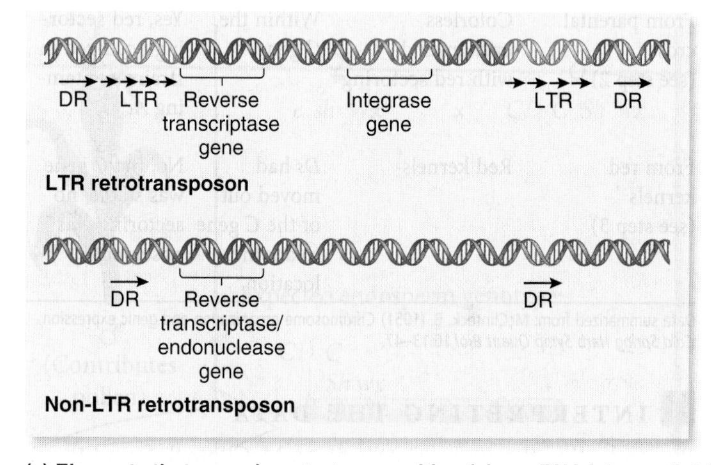

(c) Elements that move by retrotransposition (via an RNA intermediate)

FIGURE 17.13 Common organizations of transposable elements. Direct repeats (DRs) are found within the host DNA. Inverted repeats (IRs) are at the ends of most transposable elements. Long terminal repeats (LTRs) are regions containing a large number of tandem repeats.

Each Type of Transposable Element Has a Characteristic Pattern of DNA Sequences

Research on TEs from many species has established that DNA sequences within them are organized in several different ways. **Figure 17.13** describes a few of those ways, although many variations are possible. All TEs are flanked by **direct repeats (DRs),** also called **target-site duplications,** which are identical nucleotide sequences that are oriented in the same direction and repeated. Direct repeats are adjacent to both ends of the element. The simplest TE, which is commonly found in bacteria, is known as an **insertion sequence.** As shown in Figure 17.13a, an insertion sequence has two important characteristics. First, both ends of the insertion sequence contain **inverted repeats (IRs).** Inverted repeats are DNA sequences that are identical (or very similar) but run in opposite directions, such as the following:

```
5'-CTGACTCTT-3'     and     5'-AAGAGTCAG-3'
3'-GACTGAGAA-5'             3'-TTCTCAGTC-5'
```

Depending on the particular element, the lengths of inverted repeats range from 9 to 40 bp in length. In addition, insertion sequences may contain a central region that encodes the enzyme **transposase,** which catalyzes the transposition event.

Composite transposons contain additional genes that are not necessary for transposition per se. They commonly contain genes that confer a selective advantage to the organism under certain growth conditions. Composite transposons are prevalent in bacteria, where they often contain genes that provide resistance to antibiotics or toxic heavy metals. For example, the composite transposon shown in Figure 17.13a contains two insertion sequences flanking a gene that confers antibiotic resistance. During transposition of a composite transposon, only the inverted repeats at the ends of the transposon are involved in the transpositional event. Both insertion sequences and composite transposons are elements that move via simple transposition, also called cut-and-paste transposition.

Replicative transposons, elements that move by replicative transposition, have a sequence organization that is similar to insertion sequences except that replicative transposons have a resolvase gene that is found between the inverted repeats (Figure 17.13b). As discussed later in this chapter, both transposase and resolvase are needed to catalyze the transposition of replicative transposons.

The organization of retrotransposons can vary greatly, and they are categorized based on their evolutionary relationship to retroviral sequences. Retroviruses are RNA viruses that make a DNA copy that integrates into the host's genome. The **LTR retrotransposons** are evolutionarily related to known retroviruses. These TEs have retained the ability to move around the genome, though, in most cases, they do not produce mature viral particles. LTR retrotransposons are so named because they contain **long terminal repeats (LTRs)** at both ends of the element (Figure 17.13c). The LTRs are typically a few hundred nucleotides in length. Like their viral counterparts, LTR retrotransposons encode virally related proteins such as reverse transcriptase and integrase that are needed for the retrotransposition process.

By comparison, **non-LTR retrotransposons** appear less like retroviruses in their sequence. They may contain a gene that encodes a protein that has both reverse transcriptase and endonuclease function (see Figure 17.13c). As discussed later, these functions are needed for retrotransposition. Some non-LTR retrotransposons are evolutionarily derived from normal eukaryotic genes. For example, the *Alu* family of repetitive sequences found in humans is derived from a single ancestral gene known as the *7SL RNA* gene (a component of the complex called signal recognition particle, which targets newly made proteins to the endoplasmic reticulum). This gene sequence has been copied by retrotransposition to achieve the current number of approximately 1,000,000 copies.

Transposable elements are considered to be complete or **autonomous elements** when they contain all of the information necessary for transposition or retrotransposition to take place. However, TEs are often incomplete or nonautonomous. A **nonautonomous element** typically lacks a gene such as transposase or reverse transcriptase that is necessary for transposition. The *Ds* locus described in the experiment of Figure 17.11 is a non-autonomous element, because it lacks a transposase gene. An element that is similar to *Ds* but contains a functional transposase gene is called the *Ac* locus or *Ac* element, which stands for Activator element. As mentioned earlier, an *Ac* locus provides a transposase gene that enables *Ds* to transpose. Therefore, nonautonomous TEs such as *Ds* can transpose only when the *Ac* locus is present at another region in the genome.

Transposase Catalyzes the Excision and Insertion of Transposons

Now that we understand the typical organization of TEs, let's examine the steps of the transposition process. The enzyme transposase catalyzes the removal of a TE from its original site in the chromosome and its subsequent insertion at another location. A general scheme for simple transposition is shown in **Figure 17.14**. Transposase binds to the inverted repeat sequences at the ends of the TE and brings them close together. The DNA is cleaved at the ends of the TE, excising it from its original site within the chromosome. The transposase carries the TE to a new site and cleaves the target DNA sequence at staggered recognition sites. The TE is then inserted and ligated to the target DNA.

As noted in Figure 17.14, the ligation of the transposable element into its new site initially leaves short gaps in the target DNA. Notice that the DNA sequences in these gaps are complementary to each other (in this case, ATGCT and TACGA). Therefore, when they are filled in by DNA gap repair synthesis, this produces direct repeats that flank both ends of the TE. These direct repeats are common features found adjacent to all TEs (see Figure 17.13).

Although the transposition process depicted in Figure 17.14 does not directly alter the number of TEs, simple transposition is known to increase their number in genomes, in some cases to fairly high levels. How can this happen? The answer is that transposition often occurs around the time of DNA replication (**Figure 17.15**). After a replication fork has passed a region containing a TE, two TEs will be found behind the fork—one in each of the replicated regions. One of these TEs could then transpose from its original location into a region ahead of the replication fork. After the replication fork has passed this second region and DNA replication is completed, two TEs will be found in one of the chromosomes and one TE in the other chromosome. In this way, simple transposition can lead to an increase in TEs. We will discuss the biological significance of transposon proliferation later in this chapter.

Replicative Transposition Requires Both Transposase and Resolvase

Replicative transposition has been studied in several bacterial transposons and in bacteriophage μ (mu), which behaves like a transposon. The net result of replicative transposition is that a

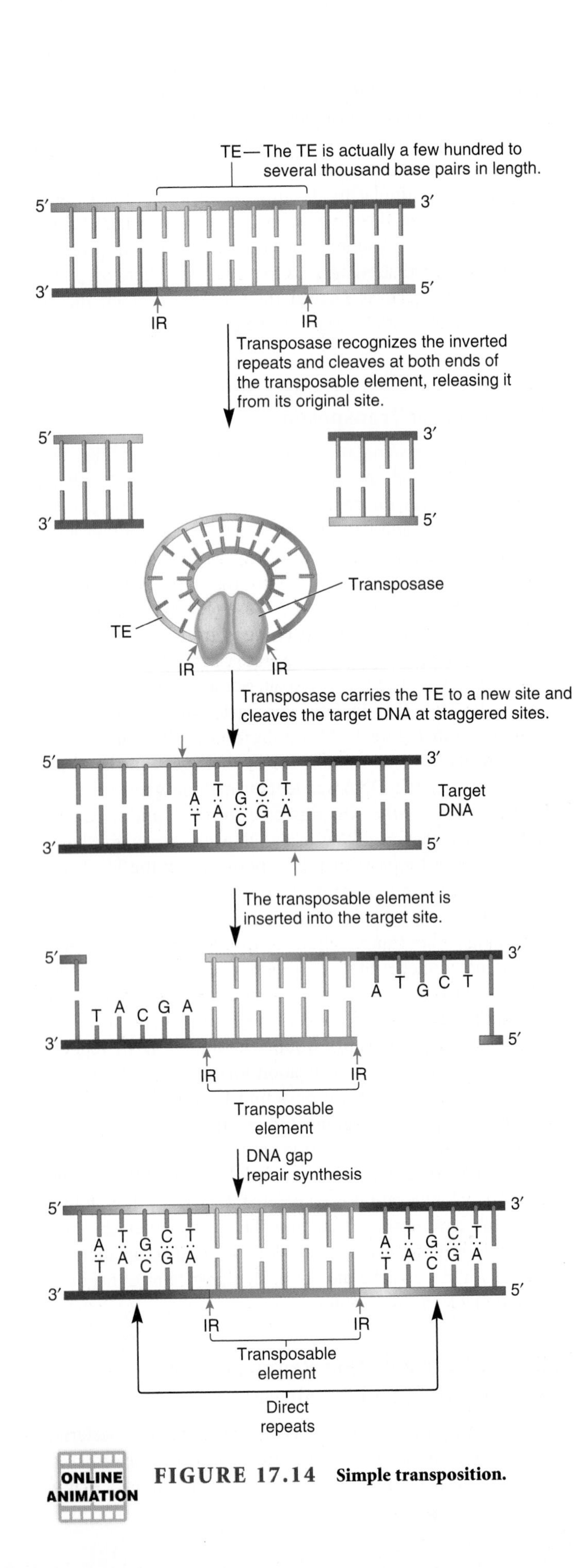

FIGURE 17.14 Simple transposition.

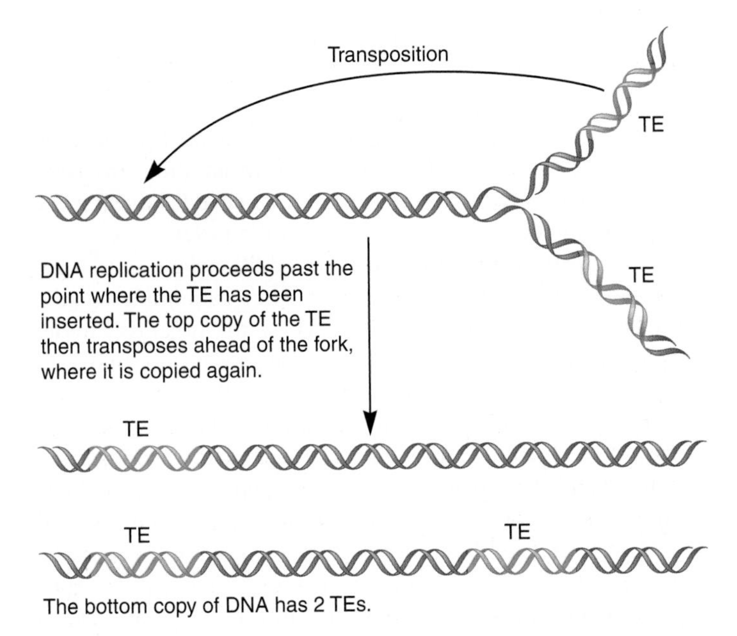

FIGURE 17.15 Increase in transposable element (TE) copy number via simple transposition. In this example, a TE that has already been replicated transposes to a new site that has not yet replicated. Following the completion of DNA replication, the TE has increased in number.

TE occurs at a new site, and one also remains in its original location. **Figure 17.16** describes a model for replicative transposition between two circular DNA molecules. One DNA molecule already has a TE, whereas the other does not. In this mechanism, transposase initially makes one cut at each end of the TE and two cuts in the target DNA. Note that this differs from simple transposition, in which the transposase makes four cuts and completely removes the TE from its original site (refer back to Figure 17.14). In replicative transposition, the TE is left at its original location.

Following ligation, both the target DNA and the TE have a long gap. DNA gap repair synthesis copies the target DNA gap as well as the TE. This creates two copies of the TE within a large circular molecule known as a cointegrant. The enzyme resolvase catalyzes homologous recombination within the TEs so that the cointegrant can be resolved into two separate DNA molecules. One of these molecules contains the TE in its original location, and the other has a TE at a new location.

Retrotransposons Use Reverse Transcriptase for Retrotransposition

Thus far, we have considered how DNA elements—transposons—can move throughout the genome. By comparison, retrotransposons use an RNA intermediate in their transposition mechanism. Let's begin with LTR retrotransposons. As shown in **Figure 17.17**, the movement of LTR retrotransposons requires two key enzymes: reverse transcriptase and integrase. In this example, the cell already contains a retrotransposon known as *Ty* within its genome. This retrotransposon is transcribed into RNA. In a series of steps, **reverse transcriptase** uses this RNA as a template to synthesize a double-stranded DNA molecule. The LTRs at the ends of the

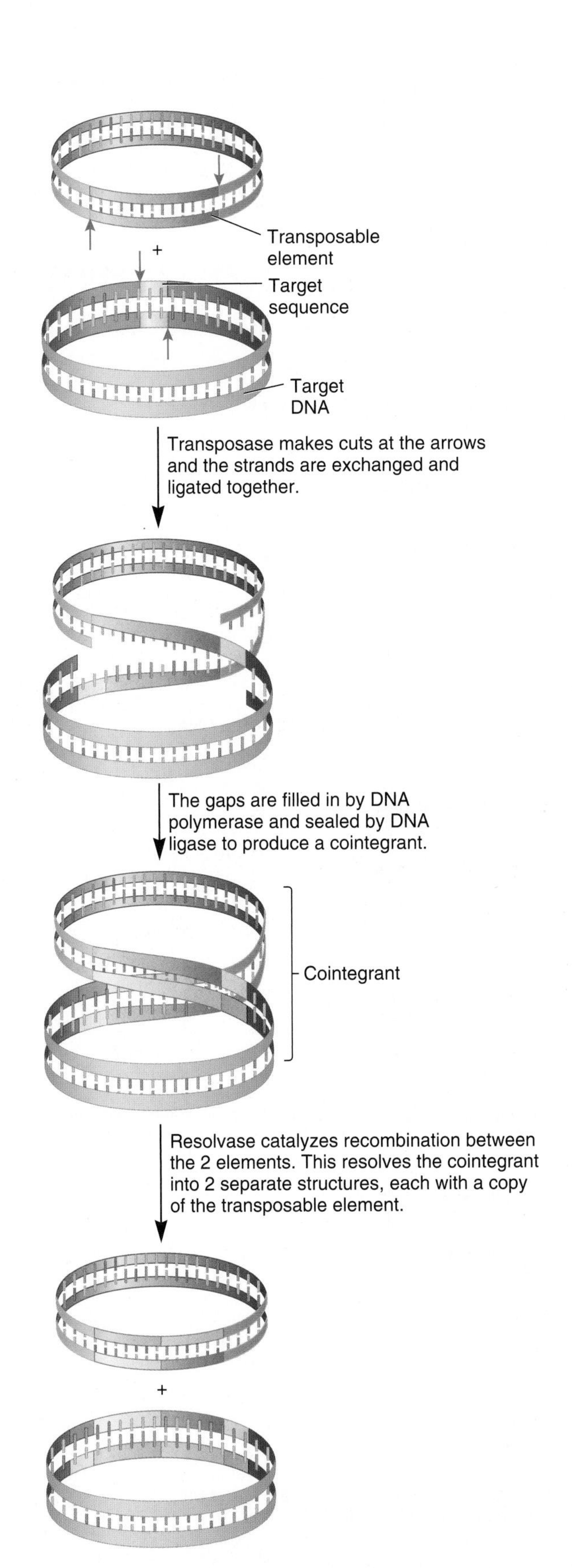

FIGURE 17.16 Replicative transposition. The end result of this process is that one TE remains in its original site and another TE is inserted at a new site.

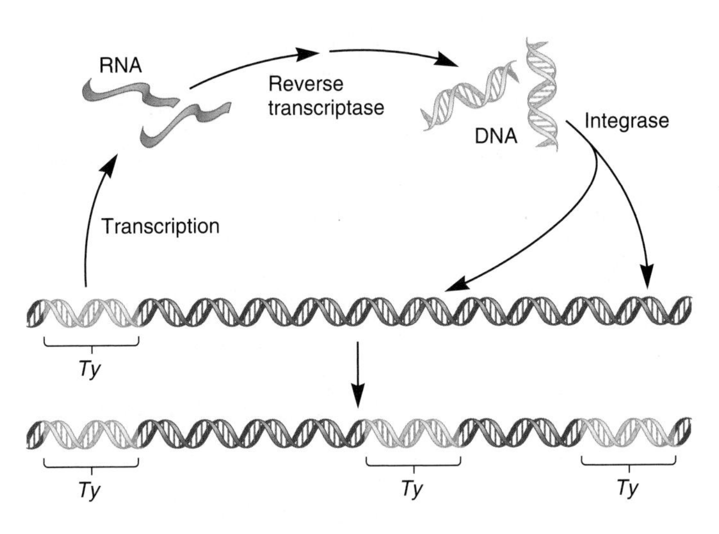

FIGURE 17.17 Retrotransposition of an LTR retrotransposon.

double-stranded DNA are then recognized by **integrase,** which catalyzes the insertion of the DNA into the target chromosomal DNA. The integration of retrotransposons can occur at many locations within the genome. Furthermore, because a single retrotransposon can be copied into many RNA transcripts, retrotransposons may accumulate rapidly within a genome.

The currently accepted model for the replication and integration of non-LTR retrotransposons is called **target-site primed reverse transcription (TPRT).** As shown in **Figure 17.18**, the retrotransposon is first copied into RNA with a polyA tail. The target DNA site is recognized by an endonuclease, which may be encoded by the retrotransposon. This endonuclease recognizes a consensus sequence of 5′-TTTTA-3′, and initially cuts just one of the DNA strands. The 3′-polyA tail of the retrotransposon RNA can bind to this nicked site due to AT base pairing. Reverse transcriptase then uses the target DNA as a primer and makes a DNA copy of the RNA. This is the process of target-site primed reverse transcription. To be fully integrated into the target DNA, the endonuclease makes a cut in the other DNA strand usually about 7 to 20 nucleotides away from the first cut. The retrotransposon DNA is then ligated into the target site within a chromosome, perhaps by nonhomologous end joining. The mechanism for synthesis of the other DNA strand of the retrotransposon is not completely understood. It could occur via DNA repair mechanisms described in Chapter 16.

Transposable Elements May Have Important Influences on Mutation and Evolution

Over the past few decades, researchers have found that TEs probably occur in the genomes of all species. **Table 17.2** describes a few TEs that have been studied in great detail. As discussed in Chapter 10, the genomes of eukaryotic species typically contain moderately and highly repetitive sequences. In some cases, these repetitive sequences are due to the proliferation of TEs. In mammals, for example, **LINEs** are long interspersed elements that are usually 1000 to 10,000 bp in length and found in a few thousand to several hundred thousands of copies. In humans, a particular family

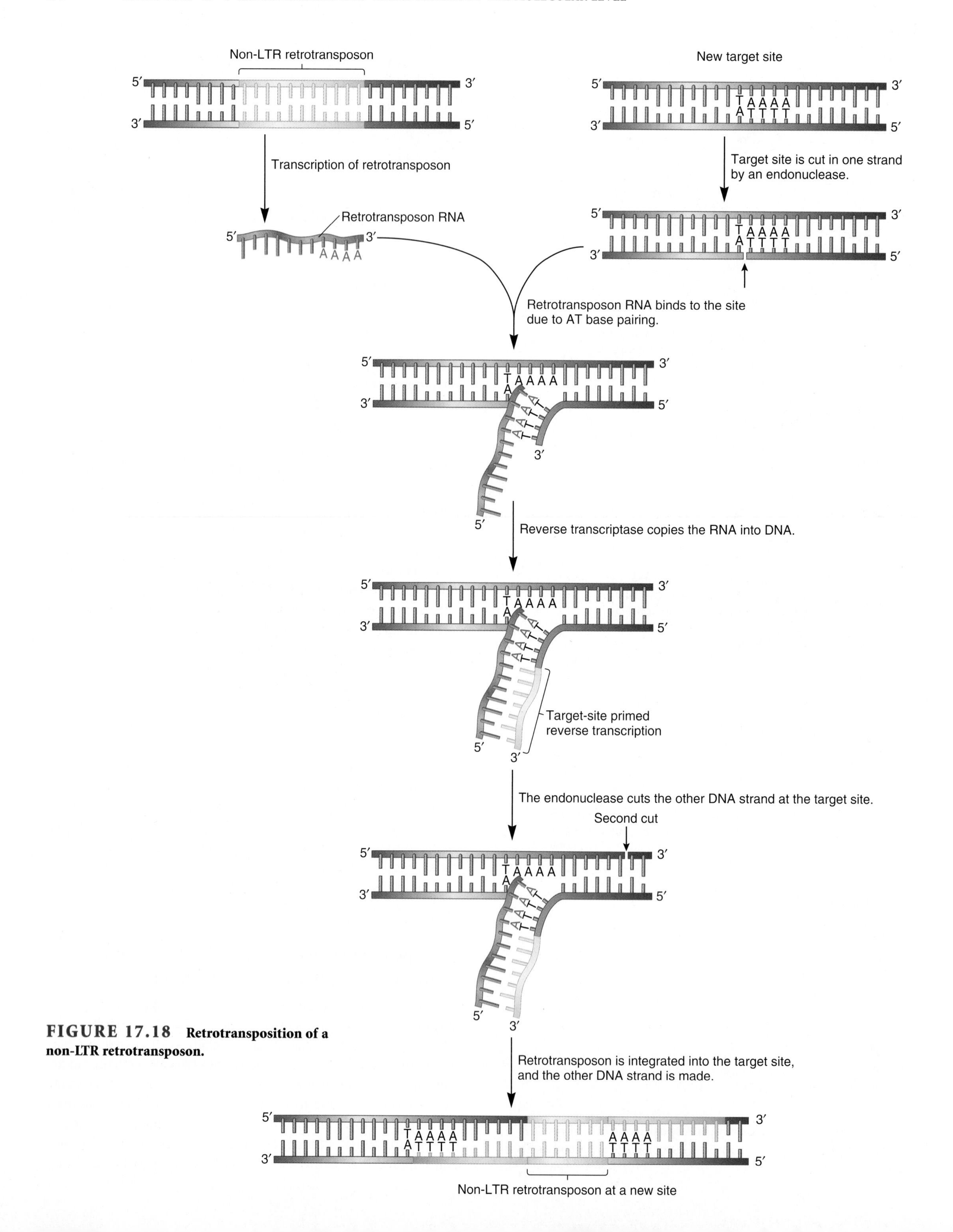

FIGURE 17.18 Retrotransposition of a non-LTR retrotransposon.

TABLE 17.2
Examples of Transposable Elements

Element	Type	Approximate Length (bp)	Description
Bacterial			
IS1	Insertion sequence	768	An insertion sequence that is commonly found in five to eight copies in *E. coli.*
Mu	Replicative transposon	36,000	A true virus that can insert itself anywhere in the *E. coli* chromosome. Its name, *Mu,* is derived from its ability to insert into genes and mutate them.
Tn10	Composite transposon	9300	One of many different bacterial transposons that carries antibiotic resistance.
Tn951	Composite transposon	16,600	A transposon that provides bacteria with genes that allow them to metabolize lactose.
Yeast			
Ty elements	LTR retrotransposon	6300	A retroelement found in *S. cerevisiae* at about 35 copies per genome.
Drosophila			
P elements	Simple transposon	500–3000	A transposon that may be found in 30–50 copies in P strains of *Drosophila.* It is absent from M strains.
Copia-like elements	LTR retrotransposon	5000–8000	A family of *copia*-like elements found in *Drosophila,* which vary slightly in their lengths and sequences. Typically, each family member is found at about 5–100 copies per genome.
Humans			
Alu sequence	Non-LTR retrotransposon	300	A SINE that is abundantly interspersed throughout the human genome.
L1	Non-LTR retrotransposon	6500	A LINE found in about 500,000 copies in the human genome.
Plants			
Ac/Ds	Simple transposon	4500	*Ac* is an autonomous transposon found in corn and other plant species. It carries a transposase gene. *Ds* is a nonautonomous version that lacks a functional transposase gene.
Opie	LTR retrotransposon	9000	An LTR retrotransposon found in plants that is related to the *copia* element found in animals.

of related lines called line-1 or l1 is found in about 500,000 copies and represents about 17% of the total human DNA! By comparison, **SINEs** are short interspersed elements that are less than 500 bp in length. A specific example of a SINE is the *Alu* sequence, present in about 1 million copies in the human genome. About 10% of the human genome is composed of this particular TE.

LINEs and SINEs continue to proliferate in the human genome but at a fairly low rate. In about 1 live birth in 100, an *Alu* or an l1 (or both) sequence has been inserted into a new site in the human genome. On rare occasions, a new insertion can disrupt a gene and cause phenotypic abnormalities. For example, new insertions of l1 or *Alu* sequences into particular genes have been shown, on occasion, to be associated with diseases such as hemophilia, muscular dystrophy, and breast and colon cancer.

The relative abundance of TEs varies widely among different species. As shown in **Table 17.3**, TEs can be quite prevalent in amphibians, mammals, and flowering plants, but tend to be less abundant in simpler organisms such as bacteria and yeast. The biological significance of TEs in the evolution of prokaryotic and eukaryotic species remains a matter of debate. According to the **selfish DNA hypothesis,** TEs exist because they contain characteristics that allow them to multiply within the chromosomal DNA of living cells. In other words, they resemble parasites in the sense that they inhabit a cell without offering any selective advantage to the organism. They can proliferate as long as they do not harm the organism to the extent that they significantly disrupt survival.

TABLE 17.3
Abundance of TEs in the Genomes of Selected Species

Species	Percentage of the Total Genome Composed of Transposable Elements*
Frog (*Xenopus laevis*)	77
Corn (*Zea mays*)	60
Human (*Homo sapiens*)	45
Mouse (*Mus musculus*)	40
Fruit fly (*Drosophila melanogaster*)	20
Nematode (*Caenorhabditis elegans*)	12
Yeast (*Saccharomyces cerevisiae*)	4
Bacterium (*Escherichia coli*)	0.3

*In some cases, the abundance of TEs may vary somewhat among different strains of the same species. The values reported here are typical values.

Alternatively, other geneticists have argued that most transpositional events are deleterious. Therefore, TEs would be eliminated

- Antibody precursor genes are rearranged via site-specific recombination to produce a vast array of different antibodies (see Figure 17.9).

17.3 Transposition

- Barbara McClintock discovered the phenomenon of transposition in which a segment of DNA called a transposable element can move to multiple sites in a genome (see Figures 17.10, 17.11).
- Transposable elements can move via three general mechanisms called simple transposition (cut-and-paste), replicative transposition, and retrotransposition (see Figure 17.12).
- Each type of transposable element has its own pattern of DNA sequences, which always includes direct repeats (see Figure 17.13).
- Transposase catalyzes the excision and insertion of transposons, which move via a DNA intermediate (see Figure 17.14).
- Simple transposition can increase the copy number of a transposon if it occurs just after a transposon has been replicated (see Figure 17.15).
- During replicative transposition, a transposon remains in its original site and a copy is also inserted into a new site (see Figure 17.16).
- Retrotransposition of LTR retrotransposons occurs via reverse transcriptase and integrase, whereas retrotransposition of non-LTR retrotransposons occurs via target-site primed reverse transcription (see Figures 17.17, 17.18).
- Many different transposons are found among living organisms. There abundance varies among different species (see Tables 17.2, 17.3).
- Transposition can have a variety of effects on chromosome structure and gene expression (see Table 17.4).
- Transposon tagging is a method to identify a particular gene in a genome (see Figure 17.19).

PROBLEM SETS & INSIGHTS

Solved Problems

S1. Zickler was the first person to demonstrate gene conversion by observing unusual ratios in *Neurospora* octads. At first, it was difficult for geneticists to believe these results because they seemed to contradict the Mendelian concept that alleles do not physically interact with each other. However, work by Mary Mitchell provided convincing evidence that gene conversion actually takes place. She investigated three different genes in *Neurospora*. One *Neurospora* strain had three mutant alleles: *pdx-1* (pyridoxine-requiring), *pyr-1* (pyrimidine-requiring), and *col-4* (a mutation that affected growth morphology). The *pdx-l* gene had been previously shown to map between the *pyr-1* and *col-4* genes. As shown here, Mitchell crossed this strain to a wild-type *Neurospora* strain:

pyr-1 pdx-1 col-4 × *pyr-1*$^+$ *pdx-1*$^+$ *col-4*$^+$

She first analyzed many octads with regard to their requirement for pyridoxine. Out of 246 octads, 2 had an aberrant ratio in which two spores were *pdx-1,* and six were *pdx-1*$^+$. These same spores were then analyzed with regard to the other two genes. In both cases, the aberrant asci gave a normal 4:4 ratio of *pyr-1:pyr-1*$^+$ and *col-4:col-4*$^+$. Explain these results.

Answer: These results can be explained by gene conversion. The gene conversion took place in a limited region of the chromosome (within the *pdx-1* gene), but it did not affect the flanking genes (*pyr-1* and *col-4*) located on either side of the *pdx-1* gene. In the asci containing two *pdx-1* alleles and six *pdx-1*$^+$ alleles, a crossover occurred during meiosis I in the region of the *pdx-1* gene. Gene conversion changed the *pdx-1* allele into the *pdx-1*$^+$ allele. This gene conversion could have occurred by two mechanisms. If branch migration occurred across the *pdx-1* gene, a heteroduplex may have formed, which could be repaired by DNA mismatch repair, as described in Figure 17.6. In the aberrant asci with two *pdx-1* and six *pdx-1*$^+$ alleles, the *pdx-1* allele was converted to *pdx-1*$^+$. Alternatively, gene conversion of *pdx-1* into *pdx-1*$^+$ could have taken place via gap repair synthesis, as described in Figure 17.7. In this case, the *pdx-1* allele would have been digested away, and the DNA encoding the *pdx-1*$^+$ allele would have migrated into the digested region and provided a template to make a copy of the *pdx-1*$^+$ allele. (Note: Since this pioneering work, additional studies have shed considerable light concerning the phenomenon of gene conversion. It occurs at a fairly high rate in fungi, approximately 0.1 to 1% of the time. It is not due to new mutations occurring during meiosis.)

S2. Recombination involves the pairing of identical or similar sequences, followed by crossing over and the resolution of the intertwined helices. On rare occasions, the direct repeats or the inverted repeats within a single TE can align and undergo homologous recombination. What are the consequences when the direct repeats recombine? What are the consequences when the inverted repeats recombine?

Answer:

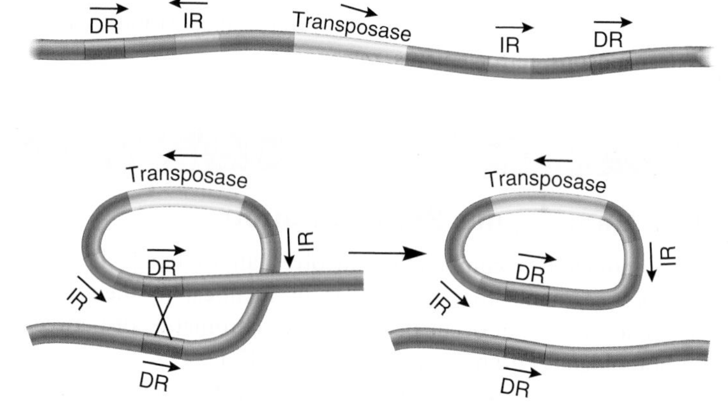

Most of the transposable element has been excised.

(a) Recombination between direct repeats

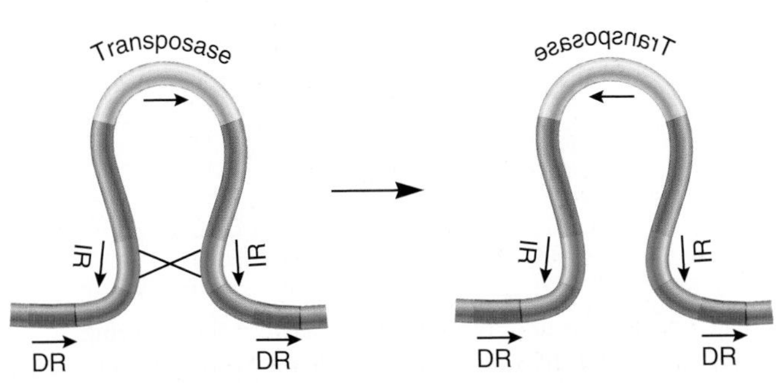

The sequence within the transposable element has been inverted.
Note that the transposase gene has changed to the opposite direction.

(b) Recombination between inverted repeats

S3. A schematic drawing of an uncrossed Holliday junction is shown here. One chromatid is shown in red, and the homologous chromatid is shown in blue. The red chromatid carries a dominant allele labeled *A* and a recessive allele labeled *b*, whereas the blue chromatid carries a recessive allele labeled *a* and a dominant allele labeled *B*.

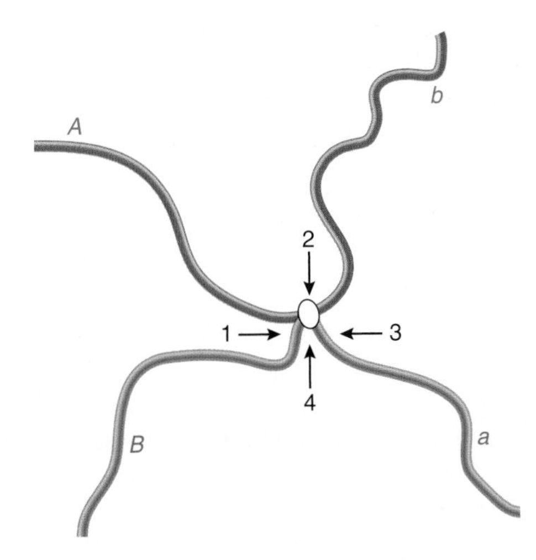

Where would the breakage of the crossed strands have to occur to get recombinant chromosomes? Would it have to occur at sites 1 and 3, or at sites 2 and 4? What would be the genotypes of the two recombinant chromosomes?

Answer:

Breakage would have to occur at the arrows labeled 2 and 4. This would connect the *A* allele with the *B* allele. The *a* allele in the other homologue would become connected with the *b* allele. In other words, one chromosome would be *AB*, and the homolog would be *ab*.

Conceptual Questions

C1. Describe the similarities and differences between homologous recombination involving sister chromatid exchange (SCE) and that involving homologs. Would you expect the same types of proteins to be involved in both processes? Explain.

C2. The molecular mechanism of SCE is similar to homologous recombination between homologs except that the two segments of DNA are sister chromatids instead of homologous chromatids. If branch migration occurs during SCE, will a heteroduplex be formed? Explain why or why not. Can gene conversion occur during sister chromatid exchange?

C3. Which steps in the double-strand break model for recombination would be inhibited if the following proteins were missing? Explain the function of each protein required for the step that is inhibited.

A. RecBCD

B. RecA

C. RecG

D. RuvABC

C4. What two molecular mechanisms can explain the phenomenon of gene conversion? Would both occur in the double-strand break model?

C5. Is homologous recombination an example of mutation? Explain.

C6. What are recombinant chromosomes? How do they differ from the original parental chromosomes from which they are derived?

C7. In the Holliday model for homologous recombination (see Figure 17.4), the resolution steps can produce recombinant or nonrecombinant chromosomes. Explain how this can occur.

C8. What is gene conversion?

C9. Make a list of the differences between the Holliday model and the double-strand break model.

C10. In recombinant chromosomes, where is gene conversion likely to have taken place: near the breakpoint or far away from the breakpoint? Explain.

C11. What events does the RecA protein facilitate?

C12. According to the double-strand break model, does gene conversion necessarily involve DNA mismatch repair? Explain.

C13. What type of DNA structure is recognized by RecG and RuvABC? Do you think these proteins recognize DNA sequences? Be specific about what type(s) of molecular recognition these proteins can perform.

C14. Briefly describe three ways that antibody diversity is produced.

C15. Describe the function of RAG1 and RAG2 proteins and NHEJ proteins.

C16. According to the scenario shown in Figure 17.9, how many segments of DNA (one, two, or three) are removed during site-specific recombination within the gene that encodes the κ (kappa) light chain for IgG proteins? How many segments are spliced out of the pre-mRNA?

C17. Describe the role that integrase plays during the insertion of λ DNA into the host chromosome.

C18. If you were examining a sequence of chromosomal DNA, what characteristics would cause you to believe that the DNA contained a transposable element?

C19. According to the model for replicative transposition shown in Figure 17.16, does the TE replicate before or after it transposes? Explain your answer.

C20. Why does transposition always produce direct repeats in the chromosomal DNA?

C21. Which types of TEs have the greatest potential for proliferation: insertion sequences, replicative TEs, or retrotransposons? Explain your choice.

C22. Do you consider TEs to be mutagens? Explain.

C23. Let's suppose that a species of mosquito has two different types of simple transposons that we will call X elements and Z elements. The X elements appear quite stable. When analyzing a population of 100 mosquitoes, every mosquito has six X elements, and they are always located in the same chromosomal locations among different individuals. In contrast, the Z elements seem to "move around" quite a bit. Within the same 100 mosquitoes, the number of Z elements ranges from 2 to 14, and the locations of the Z elements tend to vary considerably among different individuals. Explain how one simple transposon can be stable and another simple transposon can be mobile, within the same group of individuals.

C24. This chapter describes five different types of TEs, including insertion sequences, composite transposons, replicative transposons, LTR retrotransposons, and non-LTR retrotransposons. Which of these five types of TEs would have the following features?

A. Require reverse transcriptase to transpose

B. Require transposase to transpose

C. Are flanked by direct repeats

D. Have inverted repeats

C25. What features distinguish a transposon from a retrotransposon? How are their sequences different, and how are their mechanisms of transposition different?

C26. Solved problem S2 illustrates the consequences of crossing over between the direct and inverted repeats within a single TE. The drawing here shows the locations of two copies of the same TE within a single chromosome. The chromosome is depicted according to its G banding pattern. (Note: G bands are illustrated in Figure 8.1.) The direct and inverted repeats are labeled 1, 2, 3, and 4, from left to right.

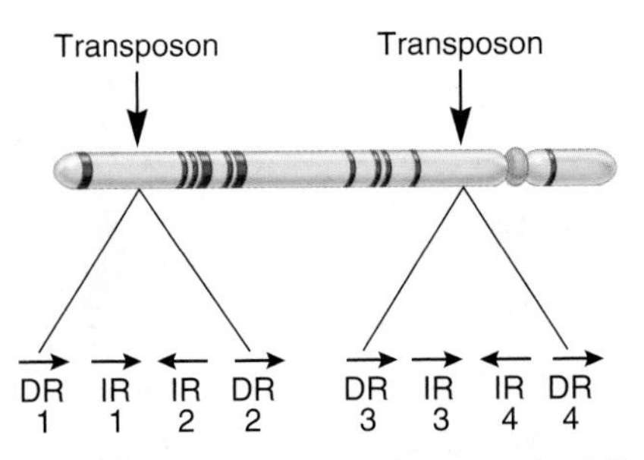

Draw the end result of recombination between the following sequences. Your drawing should include the banding pattern of the resulting chromosome.

A. DR-1 and DR-4

B. IR-1 and IR-4

C27. What is the difference between an autonomous and a nonautonomous TE? Is it possible for nonautonomous TEs to move? If yes, explain how.

C28. An operon in the bacterium *Salmonella typhimurium* has the following arrangement:

DR IR IR DR

↓ ↓ promoter→ ↓ ↓ *H2 rH1*

The promoter for this operon is contained within a TE. The *H2* gene encodes a protein that is part of the bacterial flagellum. The *rH1* gene encodes a repressor protein that represses the *H1* gene, which is found at another location in the bacterial chromosome. The *H1* gene also encodes a flagellar protein. When the promoter is found in the arrangement shown here, the *H2*/*rH1* operon is turned on. This results in flagella that contain the *H2* protein. The *H1* protein is not made because the *rH1* repressor prevents the transcription of the *H1* gene. At a frequency of approximately 1 in 10,000 (which is much higher than the spontaneous mutation rate), this strain of bacterium can "switch" its expression so that *H2* is turned off and *H1* is turned on. Bacteria that have *H1* turned on and *H2* turned off can also switch back to having *H2* turned on and *H1* turned off. This switch also occurs at a frequency of about 1 in 10,000. Based on your understanding of transposons and recombination, explain how switching occurs in *Salmonella typhimurium*. Hint: Take a look at solved problem S2.

C29. The occurrence of multiple transposons within the genome of organisms has been suggested as a possible cause of chromosomal rearrangements such as deletions, translocations, and inversions. How could the occurrence of transposons promote these kinds of structural rearrangements?

Experimental Questions

E1. With the harlequin staining technique, one sister chromatid appears to fluoresce more brightly than the other. Why?

E2. In the data shown here, harlequin staining was used to determine the frequency of sister chromatid exchanges (SCEs) in the presence of a suspected mutagen.

	Frequency of SCEs/Chromosome
No mutagen	0.67
With suspected mutagen	14.7

Would you conclude that this substance is a mutagen?

E3. In the experiment described in experimental question E2, at what point would you need to add the mutagen: before the first round of DNA replication, after the first round but before the second round, or after the second round?

E4. Let's suppose that a researcher followed the protocol described in the experiment of Figure 17.3 but exposed the cells to BrdU for three cell generations instead of two. Near the end of growth, the cells were exposed to colcemid to prevent them from completing mitosis following the third round of DNA replication. What would be the expected results if a parental cell contained a total of four chromosomes (two homologous pairs) and a single sister chromatid exchange occurred after the second replication? Your drawing should show four cells that contain four condensed chromosomes in each cell.

E5. Based on your understanding of the experiment of Figure 17.3, does BrdU enhance the binding of Giemsa to the chromatids or inhibit the binding of Giemsa? Explain.

E6. Briefly explain how McClintock determined that *Ds* was occasionally moving from one chromosomal location to another. Discuss the type of data she examined to arrive at this conclusion.

E7. In the data of Figure 17.11, is the solid red phenotype due to chromosome breakage or the excision of a TE? Explain how you have arrived at your conclusion.

E8. As discussed in the experiment of Figure 17.11, the presence of a transposon can create a mutable site or locus that is subject to frequent chromosome breakage. Why do you think a transposon creates a mutable site? If chromosome breakage occurs, do you think the transposon has moved somewhere else? How would you experimentally determine if it has?

E9. In your own words, explain the term transposon tagging.

E10. Tumor-suppressor genes are normal human genes that prevent uncontrollable cell growth. Starting with a normal laboratory human cell line, describe how you could use transposon tagging to identify tumor-suppressor genes. Note: When a TE hops into a tumor-suppressor gene, it may cause uncontrolled cell growth. This is detected as a large clump of cells among a normal monolayer of cells.

E11. Gerald Rubin and Allan Spradling devised a method of introducing a transposon into *Drosophila*. This approach has been important for the transposon tagging of many *Drosophila* genes. They began with a P element that had been cloned on a plasmid. (Note: Methods of cloning are described in Chapter 18.) Using cloning methods, they inserted the wild-type allele for the *rosy* gene into the P element in this plasmid. The recessive allele, *rosy*, results in a rosy eye color, while the wild-type allele, *rosy*$^+$, produces red eyes. This plasmid also has an intact transposase gene. The cloned DNA is shown here.

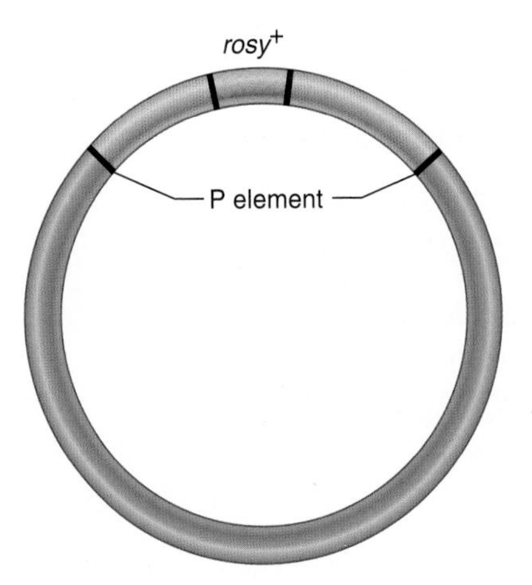

They used a micropipette to inject this DNA into regions of embryos that would later become reproductive cells. These embryos were originally homozygous for the recessive *rosy* allele. However, the P element carrying the *rosy*$^+$ allele could "hop" out of the plasmid and into a chromosome of the cells that were destined to become germ cells (i.e., sperm or egg cells). After they had matured to adults, these flies were then mated to flies that were homozygous for the recessive *rosy* allele. If offspring inherited a chromosome carrying the P element with the *rosy*$^+$ gene, such offspring would have red eyes. Therefore, the phenotype of red eyes provided a way to identify offspring that had a P element insertion.

Now here is the question. Let's suppose you were interested in identifying genes that play a role in wing development. Outline the experimental steps you would follow, using the plasmid with the P element containing the *rosy*$^+$ gene, as a way to transposon tag genes that play a role in wing development. Note: You should assume that the inactivation of a gene involved in wing development would cause an abnormality in wing shape. Also keep in mind that most P element insertions inactivate genes and may be inherited in a recessive manner.

Questions for Student Discussion/Collaboration

1. Make a list of the similarities and differences among homologous recombination, site-specific recombination, and transposition.
2. If no homologous recombination of any kind could occur, what would be the harmful and beneficial consequences?
3. Based on your current knowledge of genetics, discuss whether or not you think the selfish DNA hypothesis is correct.

Note: All answers appear at the website for this textbook; the answers to even-numbered questions are in the back of the textbook.

www.mhhe.com/brookergenetics4e

Visit the website for practice tests, answer keys, and other learning aids for this chapter. Enhance your understanding of genetics with our interactive exercises, quizzes, animations, and much more.

CHAPTER OUTLINE

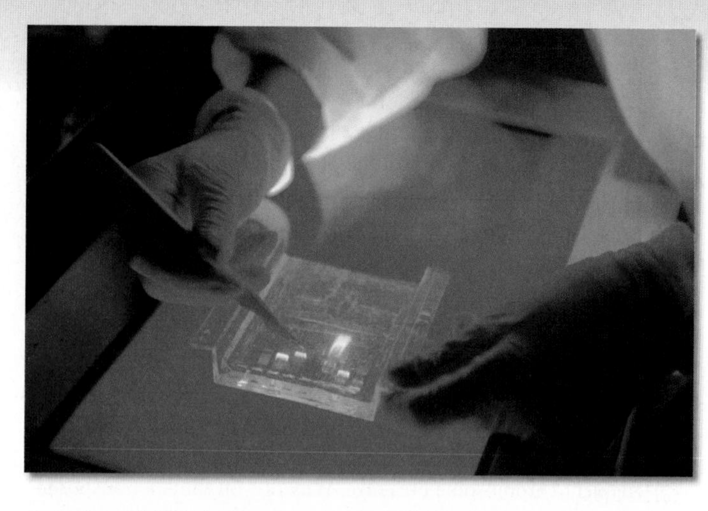

Detection of DNA bands on a gel. *DNA can be cut into fragments that can be separated via gel electrophoresis and then observed by staining with a dye called ethidium bromide. The cutting and pasting of DNA fragments allow researchers to clone genes.*

18 RECOMBINANT DNA TECHNOLOGY

As we learned in Chapter 17, recombination involves the cutting and pasting of DNA segments to produce new arrangements. **Recombinant DNA technology** is the use of in vitro molecular techniques to manipulate fragments of DNA and produce new arrangements. In the early 1970s, the first successes in making recombinant DNA molecules were accomplished independently by two groups at Stanford University: David Jackson, Robert Symons, and Paul Berg; and Peter Lobban and A. Dale Kaiser. Both groups were able to isolate and purify pieces of DNA in a test tube and then covalently link DNA fragments from two different sources. In other words, they constructed molecules called **recombinant DNA molecules.** Shortly thereafter, researchers were able to introduce such recombinant DNA molecules into living cells. Once inside a host cell, the recombinant molecules can be replicated to produce many identical copies of a gene—a process called gene cloning.

Recombinant DNA technology and gene cloning have enabled geneticists to probe relationships between gene sequences and phenotypic consequences, and thereby have been fundamental to our understanding of gene structure and function. Most researchers in molecular genetics are familiar with recombinant DNA technology and apply it frequently in their work. Significant practical applications of recombinant DNA technology also have been developed, including exciting advances such as gene therapy, screening for human diseases, recombinant vaccines, and the production of transgenic plants and animals in agriculture, in which a cloned gene from one species is transferred to some other species. Transgenic organisms have also been important in basic research.

In this chapter, we will focus primarily on the use of recombinant DNA technology as a way to further our understanding of gene structure and function. We will look at the materials and molecular techniques used in gene cloning, and explore polymerase chain reaction (PCR), which can make many copies of DNA from a defined region. We will then explore techniques for identifying specific genes or gene products, as well as methods for detecting the binding of proteins to DNA or RNA sequences. Finally, we will examine how scientists analyze and alter DNA sequences through the techniques of DNA sequencing (a method that enables researchers to determine the base sequence of a DNA strand) and site-directed mutagenesis (a procedure that allows researchers to make mutations within a cloned segment of DNA).

In Chapter 19, we will consider many of the practical applications that have arisen as a result of these technologies. Chapters 20 and 21 are devoted to genomics, the molecular analysis of many genes and even the entire genome of a species.

18.1 GENE CLONING USING VECTORS

Molecular biologists want to understand how the molecules within living cells contribute to cell structure and function. Because proteins are the workhorses of cells and because they are the products of genes, many molecular biologists focus their attention on the structure and function of proteins or the genes that encode them. Researchers may focus their efforts on the study of just one or perhaps a few different genes or proteins. At the molecular level, this poses a daunting task. In eukaryotic species, any given cell can express thousands of different proteins, making the study of any single gene or protein akin to a "needle in a haystack" exploration. To overcome this truly formidable obstacle, researchers frequently take the approach of cloning the genes that encode their proteins of interest. The term **gene cloning** refers to the phenomenon of making many copies of a gene. The laboratory methods necessary to clone a gene were devised during the early 1970s. Since then, many technical advances have enabled gene cloning to become a widely used procedure among scientists, including geneticists, cell biologists, biochemists, plant biologists, microbiologists, evolutionary biologists, clinicians, and biotechnologists.

Table 18.1 summarizes some of the common uses of gene cloning. In modern molecular biology, the diversity of uses for gene cloning is remarkable. For this reason, gene cloning has provided the foundation for critical technical advances in a variety of disciplines, including molecular biology, genetics, cell biology, biochemistry, and medicine. In this section and the following section, we will examine the two general strategies used to make copies of a gene—the insertion of a gene into a vector that is then propagated in living cells, and cloning via polymerase chain reaction. Later sections in this chapter, as well as Chapters 19 through 21, will consider many of the uses of gene cloning that are described in Table 18.1.

Cloning Experiments May Involve Two Kinds of DNA Molecules: Chromosomal DNA and Vector DNA

If a scientist wants to clone a particular gene, a common source of the gene is the chromosomal DNA of the species that carries the gene. For example, if the goal is to clone the rat β-globin gene, this gene is found within the chromosomal DNA of rat cells. In this case, therefore, the rat's chromosomal DNA is one type of DNA needed in a cloning experiment. To prepare chromosomal DNA, an experimenter first obtains cellular tissue from the organism of interest. The preparation of chromosomal DNA then involves the breaking open of cells and the extraction and purification of the DNA using biochemical techniques such as chromatography and centrifugation (see the Appendix for a description of these techniques).

Let's begin our discussion of gene cloning by considering a recombinant DNA technology in which a gene is removed from

TABLE 18.1
Some Uses of Gene Cloning

Technique	Description
Gene sequencing	Cloned genes provide enough DNA to subject the gene to DNA sequencing (described later in this chapter). The sequence of the gene can reveal the gene's promoter, regulatory sequences, and coding sequence. Gene sequencing is also important in the identification of alleles that cause cancer and inherited human diseases.
Site-directed mutagenesis	A cloned gene can be manipulated to change its DNA sequence. Mutations within genes can help to identify gene sequences such as promoters and regulatory elements. The study of a mutant gene can also help to elucidate its normal function and how its expression may affect the roles of other genes. Mutations in the coding sequence can reveal which amino acids are important for a protein's structure and function.
Gene probes	Labeled DNA strands from a cloned gene can be used as probes for identifying similar or identical genes or RNA. These methods, known as Southern and Northern blot analysis, are described later in this chapter. Probes can also be used to localize genes within intact chromosomes (see Chapter 20). DNA probes are used in techniques such as DNA fingerprinting for identifying criminals (see Chapter 24).
Expression of cloned genes	Cloned genes can be introduced into a different cell type or different species. The expression of cloned genes has many uses: *Research* 1. The expression of a cloned gene can help to elucidate its cellular function. 2. The coding sequence of a gene can be placed next to an active promoter and then introduced into a culture of cells that will express a large amount of the protein. This greatly aids in the purification of large amounts of protein that may be needed for biochemical or biophysical studies. *Biotechnology* 1. Cloned genes can be introduced into bacteria to make pharmaceutical products such as insulin (see Chapter 19). 2. Cloned genes can be introduced into plants and animals to make transgenic species with desirable traits (see Chapter 19). *Clinical trials* 1. Cloned genes have been used in clinical trials involving gene therapy (see Chapter 19).

its native site within a chromosome and inserted into a smaller segment of DNA known as a **vector**—a small DNA molecule that can replicate independently of host cell chromosomal DNA and produce many identical copies of an inserted gene. The purpose of vector DNA is to act as a carrier of the DNA segment that is to be cloned. (The term vector comes from a Latin term meaning carrier.) In cloning experiments, a vector may carry a small segment of chromosomal DNA, perhaps only a single gene. By comparison, a chromosome carries many more genes, perhaps a few hundred or thousand. Like a chromosome, a vector is replicated when it resides within a living cell; a cell that harbors a vector is called a **host cell.** When a vector is replicated within a host cell, the DNA that it carries is also replicated.

The vectors commonly used in gene-cloning experiments were derived originally from two natural sources: plasmids or viruses. Most vectors are **plasmids,** which are small circular pieces of DNA. As discussed in Chapter 7, plasmids are found naturally in many strains of bacteria and occasionally in eukaryotic cells. Many naturally occurring plasmids carry genes that confer resistance to antibiotics or other toxic substances. These plasmids are called **R factors.** Some of the plasmids used in modern cloning experiments were derived from R factors.

Plasmids also contain a DNA sequence, known as an **origin of replication,** that is recognized by the replication enzymes of the host cell and allow it to be replicated. The sequence of the origin of replication determines whether or not the vector can replicate in a particular type of host cell. Some plasmids have origins of replications with a broad host range. Such a plasmid can replicate in the cells of many different species. Alternatively, many vectors used in cloning experiments have a limited host cell range. In cloning experiments, researchers must choose a vector that replicates in the appropriate cell type(s) for their experiments. For example, if researchers want a cloned gene to be propagated in *Escherichia coli,* the vector they employ must have an origin of replication that is recognized by this species of bacterium. The origin of replication also determines the copy number of a plasmid. Some plasmids are said to have strong origins because they can achieve high copy number—perhaps 100 to 200 copies of the plasmid per cell. Others have weaker origins whereby only one or two plasmids are found per cell.

Commercially available plasmids have been genetically engineered for effective use in cloning experiments. They contain unique sites where geneticists can easily insert pieces of DNA. Another useful feature of cloning vectors is that they often contain resistance genes that provide host cells with the ability to grow in the presence of a toxic substance. Such a gene is called a **selectable marker,** because the expression of the gene selects for the growth of the host cells. Many selectable markers are genes that confer antibiotic resistance to the host cell. For example, the gene amp^R encodes an enzyme known as β-lactamase. This enzyme degrades ampicillin, an antibiotic that normally kills bacteria. Bacteria containing the amp^R gene can grow on media containing ampicillin because they can degrade it. In a cloning experiment where the amp^R gene is found within the plasmid, the growth of cells in the presence of ampicillin identifies bacteria that contain the plasmid. These bacteria can grow and form visible colonies on solid growth media. In contrast, those cells that do not contain the plasmid are ampicillin-sensitive and do not grow.

An alternative type of vector used in cloning experiments is a viral vector. As discussed in Chapter 7, viruses can infect living cells and propagate themselves by taking control of the host cell's metabolic machinery. When a chromosomal gene is inserted into a viral genome, the gene is replicated whenever the viral DNA is replicated. Therefore, viruses can be used as vectors to carry other pieces of DNA. When a virus is used as a vector, the researcher may analyze viral plaques rather than bacterial colonies. The characteristics of viral plaques are described in Chapter 7 (see Figure 7.15).

Molecular biologists may chose from hundreds of different vectors to use in their cloning experiments. **Table 18.2** provides a general description of several different types of vectors that are commonly used to clone small segments of DNA. In addition, other types of vectors, such as cosmids, bacterial artificial chromosomes (BACs), and yeast artificial chromosomes (YACs), are used to clone large pieces of DNA. These vectors are described in detail in Chapter 20. Vectors designed to introduce genes into plants and animals are discussed in Chapter 19.

TABLE **18.2**

Some Vectors Used in Cloning Experiments

Example	Type	Description
pBluescript	Plasmid	A type of vector like the one shown in Figure 18.2. It is used to clone small segments of DNA and propagate them in *E. coli.*
YEp24	Plasmid	This plasmid is an example of a **shuttle vector,** which can replicate in two different host species, *E. coli* and *Saccharomyces cerevisiae.* It carries an origin of replication for both species.
λgt11	Viral	This vector is derived from the bacteriophage λ, which is described in Chapter 14. λgt11 also contains a promoter from the *lac* operon. When fragments of DNA are cloned next to this promoter, the DNA is expressed in *E. coli.* This is an example of an **expression vector.** An expression vector is designed to clone the coding sequence of genes so they are transcribed and translated correctly.
SV40	Viral	This virus naturally infects mammalian cells. Genetically altered derivatives of the SV40 viral DNA are used as vectors for the cloning and expression of genes in mammalian cells that are grown in the laboratory.
Baculovirus	Viral	This virus naturally infects insect cells. In a laboratory, insect cells can be grown in liquid media. Unlike many other types of cells, insect cells often express large amounts of proteins that are encoded by cloned genes. When researchers want to make a large amount of a protein, they can clone the gene that encodes the protein into baculovirus and then purify the protein from insect cells.

Enzymes Are Used to Cut DNA into Pieces and Join the Pieces Together

A key step in a cloning experiment is the insertion of chromosomal DNA into a plasmid or viral vector. This requires the cutting and pasting of DNA fragments. To cut DNA, researchers use enzymes known as **restriction endonucleases,** or **restriction enzymes.** The restriction enzymes used in cloning experiments bind to a specific base sequence and then cleave the DNA backbone at two defined locations, one in each strand. Proposed by Werner Arber in the 1960s and discovered by Hamilton Smith and Daniel Nathans in the 1970s, restriction enzymes are made naturally by many different species of bacteria and protect bacterial cells from invasion by foreign DNA, particularly that of bacteriophages. Researchers have isolated and purified restriction enzymes from many bacterial species and now use them in their cloning experiments.

Figure 18.1 shows the role of a restriction enzyme, called *Eco*RI, in producing a recombinant DNA molecule. Certain types of restriction enzymes are useful in cloning because they digest DNA into fragments with "sticky ends." As shown in Figure 18.1, the sticky ends are single-stranded regions of DNA that can hydrogen bond to a complementary sequence of DNA from a different source. The ends of two different DNA pieces hydrogen bond to each other because of their complementary sticky ends.

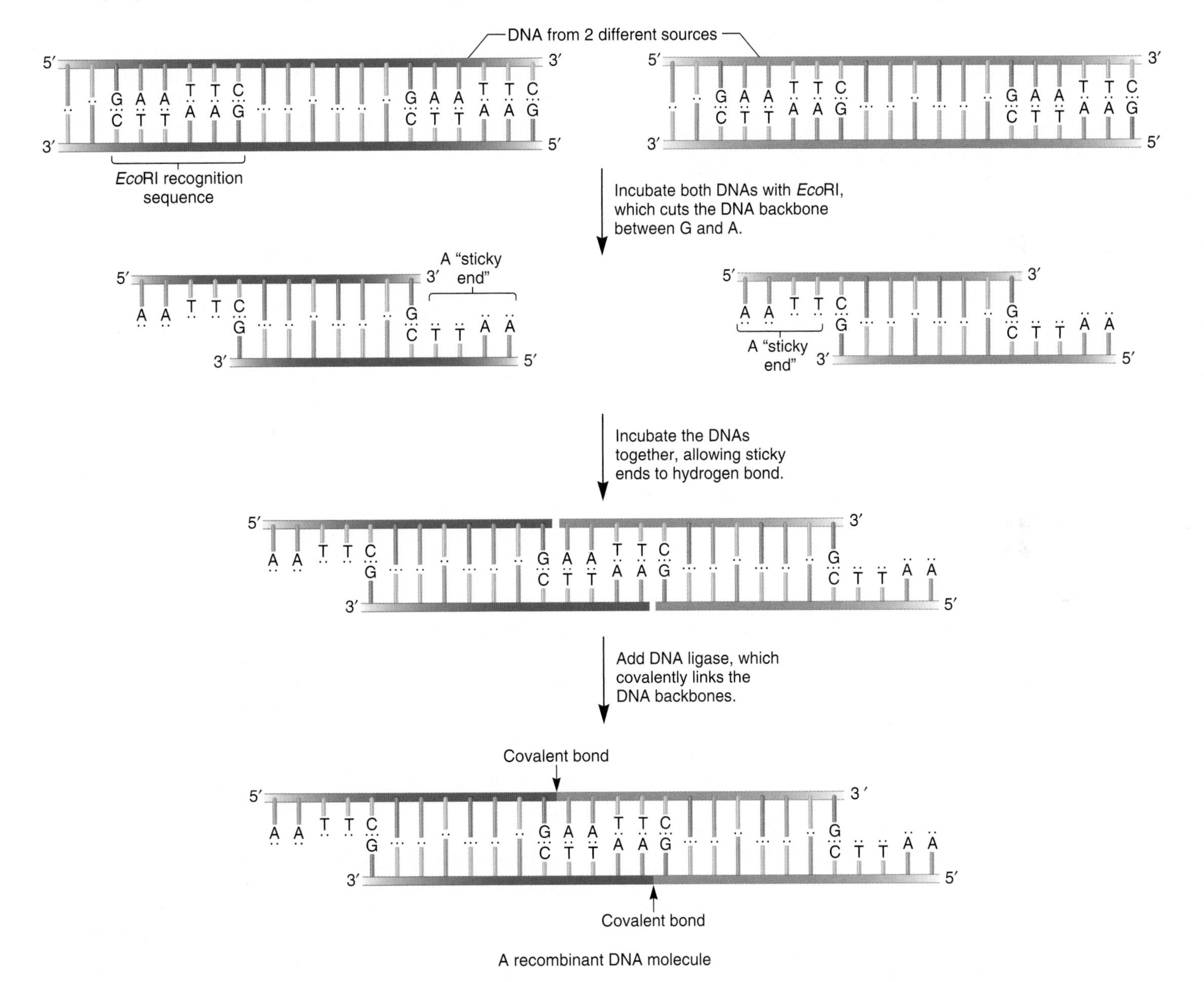

FIGURE 18.1 The action of a restriction enzyme and the production of recombinant DNA. The restriction enzyme *Eco*RI binds to a specific sequence, in this case 5′–GAATTC–3′. It then cleaves the DNA backbone between G and A, producing DNA fragments. The single-stranded ends of the DNA fragments can hydrogen bond with each other, because they have complementary sequences. The enzyme DNA ligase then catalyzes the formation of covalent bonds in the DNA backbones of the fragments.

The hydrogen bonding between the sticky ends of DNA fragments promotes a temporary interaction between the two fragments. However, this interaction is not stable because it involves only a few hydrogen bonds between complementary bases. How can this interaction be made more permanent? The answer is that the sugar-phosphate backbones within the DNA strands must be covalently linked together. This linkage is catalyzed by **DNA ligase.** Figure 18.1 illustrates the action of DNA ligase, which catalyzes covalent bond formation in the sugar-phosphate backbones of both DNA strands after the sticky ends have hydrogen bonded with each other.

Currently, several hundred different restriction enzymes from various bacterial species have been identified and are available commercially to molecular biologists. **Table 18.3** gives a few examples. Restriction enzymes usually recognize sequences that are **palindromic;** that is, the sequence in one strand is identical when read in the opposite direction in the complementary strand. For example, the sequence recognized by *Eco*RI is 5′–GAATTC–3′ in the top strand. Read in the opposite direction in the bottom strand, this sequence is also 5′–GAATTC–3′.

TABLE **18.3**

Some Restriction Enzymes Used in Gene Cloning

Restriction Enzyme*	Bacterial Source	Sequence Recognized†
*Bam*HI	*Bacillus amyloliquefaciens* H	↓ 5′–GGATCC–3′ 3′–CCTAGG–5′ ↑
*Cla*I	*Caryophanon latum*	↓ 5′–ATCGAT–3′ 3′–TAGCTA–5′ ↑
*Eco*RI	*E. coli* RY13	↓ 5′–GAATTC–3′ 3′–CTTAAG–5′ ↑
*Nae*I	*Nocardia aerocolonigenes*	↓ 5′–GCCGGC–3′ 3′–CGGCCG–5′ ↑
*Pst*I	*Providencia stuartii*	↓ 5′–CTGCAG–3′ 3′–GACGTC–5′ ↑
*Sac*I	*Streptomyces achromogenes*	↓ 5′–GAGCTC–3′ 3′–CTCGAG–5′ ↑

*Restriction enzymes are named according to the species in which they are found. The first three letters are italicized because they indicate the genus and species names. Because a species may produce more than one restriction enzyme, the enzymes are designated I, II, III, and so on, to indicate the order in which they were discovered in a given species. Some restriction enzymes, like *Eco*RI, produce a sticky end with a 5′ overhang (see Figure 18.1), whereas others, such as *Pst*I, produce a 3′ overhang. However, not all restriction enzymes cut DNA to produce sticky ends. For example, the enzyme *Nae*I cuts DNA to produce blunt ends.

†The arrows show the locations in the upper and lower DNA strands where the restriction enzymes cleave the DNA backbone. A complete list of restriction enzymes can be found at http://rebase.neb.com/rebase/rebase.html.

Gene Cloning Involves the Insertion of DNA Fragments into Vectors, Which Are Then Propagated Within Host Cells

Now that we are familiar with the materials, let's outline the general strategy that is followed in a typical cloning experiment. In the procedure shown in **Figure 18.2**, the goal is to clone a chromosomal gene of interest into a plasmid vector that already carries the amp^R gene. To begin this experiment, the chromosomal DNA is isolated and digested with a restriction enzyme. This enzyme cuts the chromosomes into many small fragments. The plasmid DNA is also cut with the same restriction enzyme. However, the plasmid has only one unique site for the restriction enzyme. After cutting, the plasmid has two ends that are complementary to the sticky ends of the chromosomal DNA fragments. The digested chromosomal DNA and plasmid DNA are mixed together and incubated under conditions that promote the binding of these complementary sticky ends.

DNA ligase is then added to catalyze the covalent linkage between DNA fragments. In some cases, the two ends of the vector simply ligate back together, restoring the vector to its original structure. This is called a recircularized vector. In other cases, a fragment of chromosomal DNA may become ligated to both ends of the vector. In this way, a segment of chromosomal DNA has been inserted into the vector. The vector containing a piece of chromosomal DNA is referred to as a **recombinant vector.**

Following ligation, the DNA is introduced into living cells treated with agents that render them permeable to DNA molecules. Cells that can take up DNA from the extracellular medium are called **competent cells.** This step in the procedure is commonly called **transformation** (see Chapter 7). In the experiment shown in Figure 18.2, a plasmid is introduced into bacterial cells that were originally sensitive to ampicillin. The bacteria are then streaked onto plates containing bacterial growth media and ampicillin. A bacterium that has taken up a plasmid carrying the amp^R gene continues to divide and forms a bacterial colony containing tens of millions of cells. Because each cell within a single colony is derived from the same original cell, all cells within a colony contain the same type of plasmid DNA.

In the experiment shown in Figure 18.2, how can the experimenter distinguish between bacterial colonies that contain a recircularized vector versus those with a recombinant vector carrying a piece of chromosomal DNA? As shown here, the chromosomal DNA has been inserted into a region of the vector that contains the *lacZ* gene, which encodes the enzyme β-galactosidase (see Chapter 14). The insertion of chromosomal DNA into the vector disrupts the *lacZ* gene so it is no longer able to produce a functional enzyme. By comparison, a recircularized vector has a functional *lacZ* gene. The functionality of *lacZ* can be determined by providing the growth medium with a colorless compound,

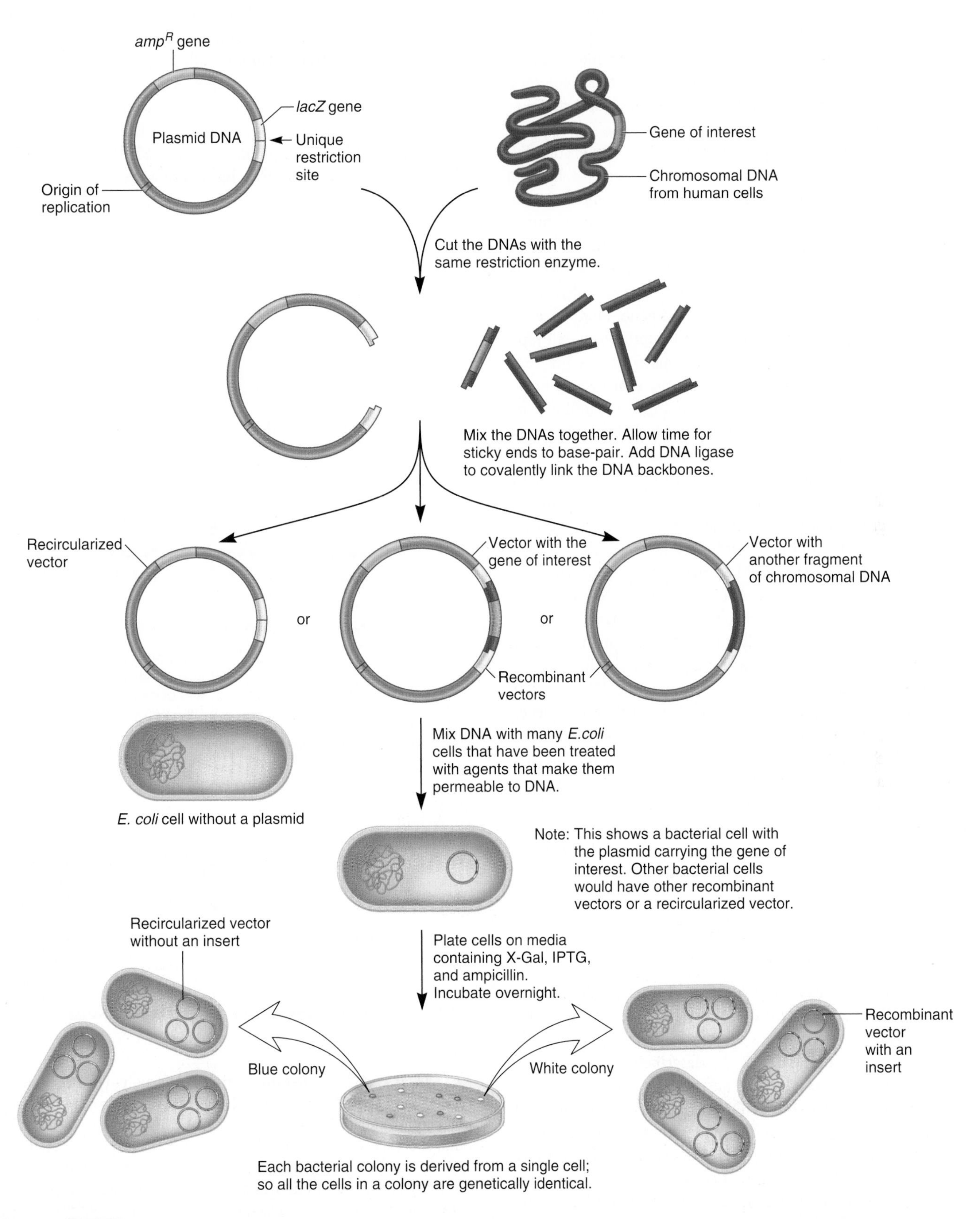

FIGURE 18.2 **The steps in gene cloning.** Note: X-Gal refers to the colorless compound 5-bromo-4-chloro-3-indolyl-β-D-galactoside. IPTG is an acronym for isopropyl-β-D-thiogalactopyranoside, which is a nonmetabolizable lactose analogue that can induce the *lac* promoter.

X-Gal (5-bromo-4-chloro-3-indolyl-β-D-galactopyranoside), which is cleaved by β-galactosidase into a blue dye. Bacteria grown in the presence of X-Gal and IPTG (an inducer of the *lacZ* gene) form blue colonies if they have a functional *lacZ* gene and white colonies if they do not. In this experiment, therefore, bacterial colonies containing recircularized vectors form blue colonies, whereas colonies containing recombinant vectors are white.

In the example of Figure 18.2, one of the white colonies contains cells with a recombinant vector that carries a human gene of interest; the segment containing the human gene is shown in red. The goal of gene cloning is to produce an enormous number of copies of a recombinant vector that carry the gene of interest. During transformation, a single bacterial cell usually takes up a single copy of a recombinant vector. However, two subsequent events lead to the amplification of the cloned gene. First, because the vector has an origin of replication, the bacterial host cell replicates the recombinant vector to produce many copies per cell. Second, the bacterial cells divide approximately every 20 minutes. Following overnight growth, a population of many millions of bacteria is obtained. Each of these bacterial cells contains many copies of the cloned gene. For example, a bacterial colony may be composed of 10 million cells, with each cell containing 50 copies of the recombinant vector. Therefore, this bacterial colony would contain 500 million copies of the cloned gene! As described earlier in Table 18.1, gene cloning has several different uses.

The preceding description has acquainted you with the steps required to clone a gene. A misleading aspect of Figure 18.2 is that the digestion of the chromosomal DNA with restriction enzymes appears to yield only a few DNA fragments, one of which contains the gene of interest. In an actual cloning experiment, however, the digestion of the chromosomal DNA with a restriction enzyme produces tens of thousands of different pieces of chromosomal DNA, not just a few. Later, we will consider methods of identifying bacterial colonies containing the specific gene that a researcher wants to clone.

Recombinant DNA technology can also be used to clone fragments of DNA that do not code for genes. For example, sequences such as telomeres, centromeres, and highly repetitive sequences have been cloned by this procedure.

cDNA Can Be Made from mRNA via Reverse Transcriptase

In the discussion of gene cloning described earlier in Figure 18.2, chromosomal DNA and plasmid DNA were used as the material to clone genes. Alternatively, a sample of RNA can provide a starting point for cloning DNA. As described in Chapter 17, the enzyme **reverse transcriptase** can use RNA as a template to make a complementary strand of DNA. This enzyme is encoded in the genome of retroviruses and provides a way for retroviruses to copy their RNA genome into DNA molecules that integrate into the host cell's chromosomes. Likewise, reverse transcriptase is encoded in some retrotransposons and is needed in the retrotransposition of such elements.

Researchers can use purified reverse transcriptase in a strategy for cloning genes, using mRNA as the starting material (**Figure 18.3**). To begin this experiment, mRNA is purified from a sample of cells. The mRNA is mixed with primers composed of a string of thymine-containing nucleotides. This short strand of DNA, or **oligonucleotide,** is called a poly-dT primer. Because eukaryotic mRNAs contain a polyA tail, poly-dT primers are complementary to the 3′ end of mRNAs. Reverse transcriptase and deoxyribonucleotides (dNTPs) are then added to make a DNA strand that is complementary to the mRNA. Next, RNaseH, DNA polymerase, and DNA ligase are added. One way to make

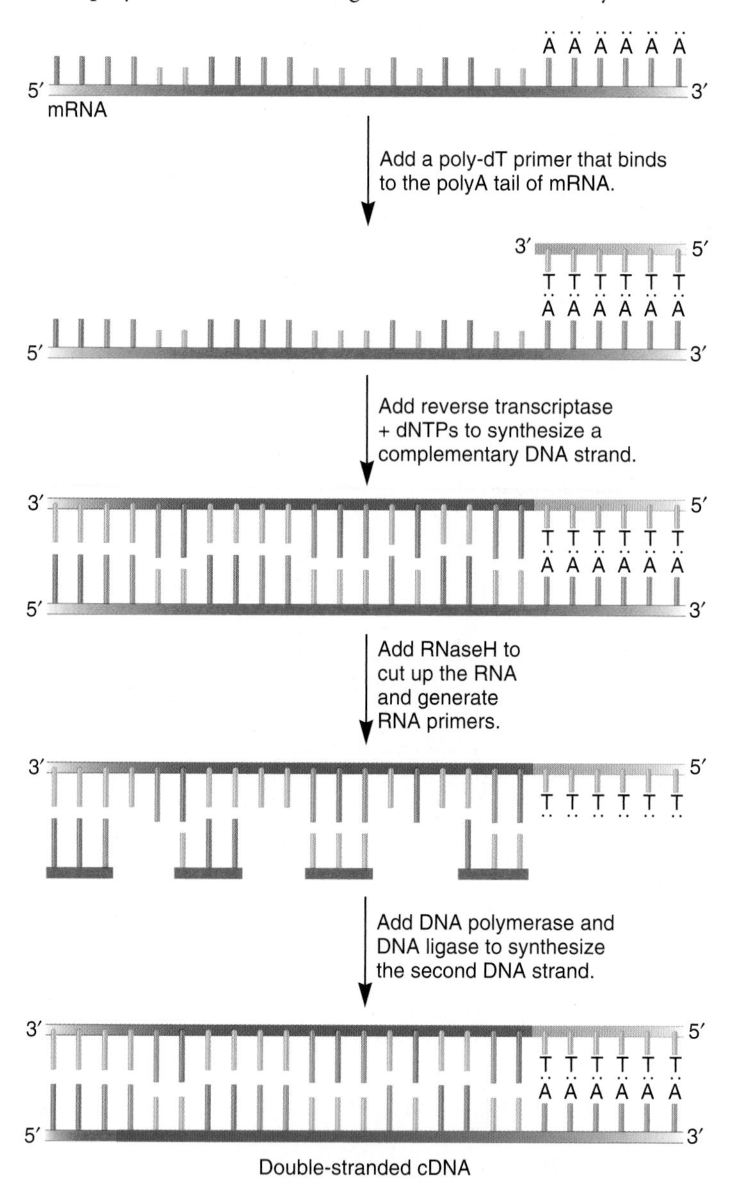

FIGURE 18.3 **Synthesis of cDNA.** A poly-dT primer anneals to the 3′ end of mRNAs. Reverse transcriptase then catalyzes the synthesis of a complementary DNA strand (cDNA). RNaseH digests the mRNA into short pieces that are used as primers by DNA polymerase I to synthesize the second DNA strand. The 5′ to 3′ exonuclease function of DNA polymerase I removes all of the RNA primers except the one at the 5′ end (because there is no primer upstream from this site). This RNA primer can be removed by the subsequent addition of an RNase. After the double-stranded cDNA is made, it can then be inserted into a vector as described later in Figure 18.11b.

the other DNA strand is to use RNaseH, which partially digests the RNA, generating short RNAs that are used as primers by DNA polymerase to make a second DNA strand that is complementary to the strand made by reverse transcriptase. Finally, DNA ligase seals any nicks in this second DNA strand. When DNA is made from RNA as the starting material, the DNA is called **complementary DNA (cDNA).** The term originally referred to the single strand of DNA that is complementary to the RNA template. However, cDNA now refers to any DNA, whether it is single- or double-stranded, that is made using RNA as the starting material.

Why is cDNA cloning useful? From a research perspective, an important advantage of cDNA is that it lacks introns, which are often found in eukaryotic genes. Because introns can be quite large, it is much simpler to insert cDNAs into vectors if researchers want to focus their attention on the coding sequence of a gene. For example, if the primary goal is to determine the coding sequence of a structural gene, a researcher inserts cDNA into a vector and then determines the DNA sequence of the insert, as described later in this chapter. Similarly, if a scientist wants to express an encoded protein of interest in a cell that does not splice out the introns properly (e.g., in a bacterial cell), it is necessary to make cDNA clones of the respective gene.

Restriction Mapping Is Used to Locate the Restriction Sites Within a Vector

As we have seen, DNA or gene cloning involves the digestion of vector and chromosomal DNA with restriction enzymes and the subsequent ligation of DNA fragments into vectors. In this type of procedure, the locations of restriction enzyme sites, or simply restriction sites, are important for the design of experiments. In the vector, for example, it is desirable to have unique restriction sites for the insertion of chromosomal DNA.

A common approach for determining the locations of restriction sites is known as **restriction mapping. Figure 18.4** outlines the restriction mapping of a bacterial plasmid. To begin this experiment, the small circular plasmid DNA is isolated and purified from host cells. Samples of the purified DNA are then placed in separate test tubes that contain a particular restriction enzyme or combination of enzymes. The plasmid DNA is incubated with the restriction enzymes long enough for digestion to occur. The DNA fragments are then separated by gel electrophoresis. (See Appendix for a description of gel electrophoresis.) In lane 1 of this experiment, a DNA sample called a set of molecular markers is also subjected to gel electrophoresis. This sample contains a mixture of DNA fragments with molecular masses that are known from previous experiments. (These markers are obtained from commercial sources or can be prepared in the laboratory.) To determine the sizes of the fragments obtained by digesting the plasmid, the fragments in lanes 2 through 8 are compared with the known markers in lane 1.

The restriction map shown in Figure 18.4 was deduced by comparing the sizes of fragments obtained from digestions with one, two, or all three of the restriction enzymes. The starting plasmid is a circular molecule 4363 bp in length. A single digestion with any of the three enzymes yields a single linear fragment of size 4363 bp. This means that *Eco*RI, *Bam*HI, and *Pst*I cut the plasmid at a single site. The double digestion with *Eco*RI and *Bam*HI yields two fragments of about 380 bp and 3980 bp. This result indicates that the *Eco*RI and *Bam*HI sites are approximately 380 bp apart in one direction along the circle and 3980 bp apart along the circle in the opposite direction. Likewise, the pairwise combinations of *Eco*RI/*Pst*I and *Bam*HI/*Pst*I indicate how far apart these sites are along the circular plasmid. Finally, the triple digestion confirms the locations of the single sites for *Eco*RI, *Bam*HI, and *Pst*I. Taken together, these results provide a map of the restriction sites within this plasmid. A similar approach can be used on a recombinant vector to determine the locations of restriction sites within a fragment of DNA that has been inserted into a plasmid.

Alternatively, another way to obtain a restriction map is via DNA sequencing, a technique described later in the chapter. If the DNA sequence of a recombinant vector has been determined, computer programs can scan the sequence and identify sites that are recognized by particular restriction enzymes. For example, *Eco*RI recognizes the sequence 5′–GAATTC–3′. If a recombinant vector contains this sequence, a computer program would identify this as an *Eco*RI site and place it on a map. Such a program would scan the DNA sequence for the recognition sequences of many different restriction enzymes and generate a detailed restriction map of a recombinant vector.

A second use of restriction enzymes is gene mapping. The technique known as restriction fragment length polymorphism (RFLP) enables researchers to map particular genes within a species' genome. This approach is discussed in Chapter 20.

18.2 POLYMERASE CHAIN REACTION

In our previous discussions of gene cloning, the DNA of interest was inserted into a vector, which then was introduced into a host cell. The replication of the vector within the host cell, and the proliferation of the host cells, led to the production of many copies of the DNA. Another way to copy DNA, without the aid of vectors and host cells, is a technique called **polymerase chain reaction (PCR),** which was developed by Kary Mullis in 1985. In this section, we begin with a general description of PCR and then examine how it can be used to quantitate the amount of DNA or RNA in a biological sample.

Each Cycle of PCR Involves Three Steps: Denaturation, Primer Annealing, and Primer Extension

The PCR method is used to make large of amounts of DNA in a defined region that is flanked by two **primers.** The primers are oligonucleotides, which are short segments of DNA, usually about 15 to 20 nucleotides in length. As shown in **Figure 18.5a**, the starting material of a PCR experiment can be a complex mixture of DNA. The two primers bind to specific sites in the DNA because their bases are complementary at these sites. The end result of PCR is that the region that is flanked by the primers, which contains the gene of interest, is amplified. The term amplification means that

primers have annealed, the temperature is raised slightly, and *Taq* polymerase catalyzes the synthesis of complementary DNA strands in the 5′ to 3′ direction, starting at the primers. This process, which is called **primer extension,** doubles the amount of the template DNA. The three steps shown in Figure 18.5b consititute one cycle of a PCR reaction. The sequential process of denaturation—primer annealing—primer extension is then repeated for many cycles to double the amount of template DNA many times in a row. This method is called a chain reaction because the products of each previous reaction (the newly made DNA strands) are used as reactants (the template strands) in subsequent reactions.

Figure 18.6 follows a PCR experiment through four cycles. Because the region of interest is doubled with each cycle, the end

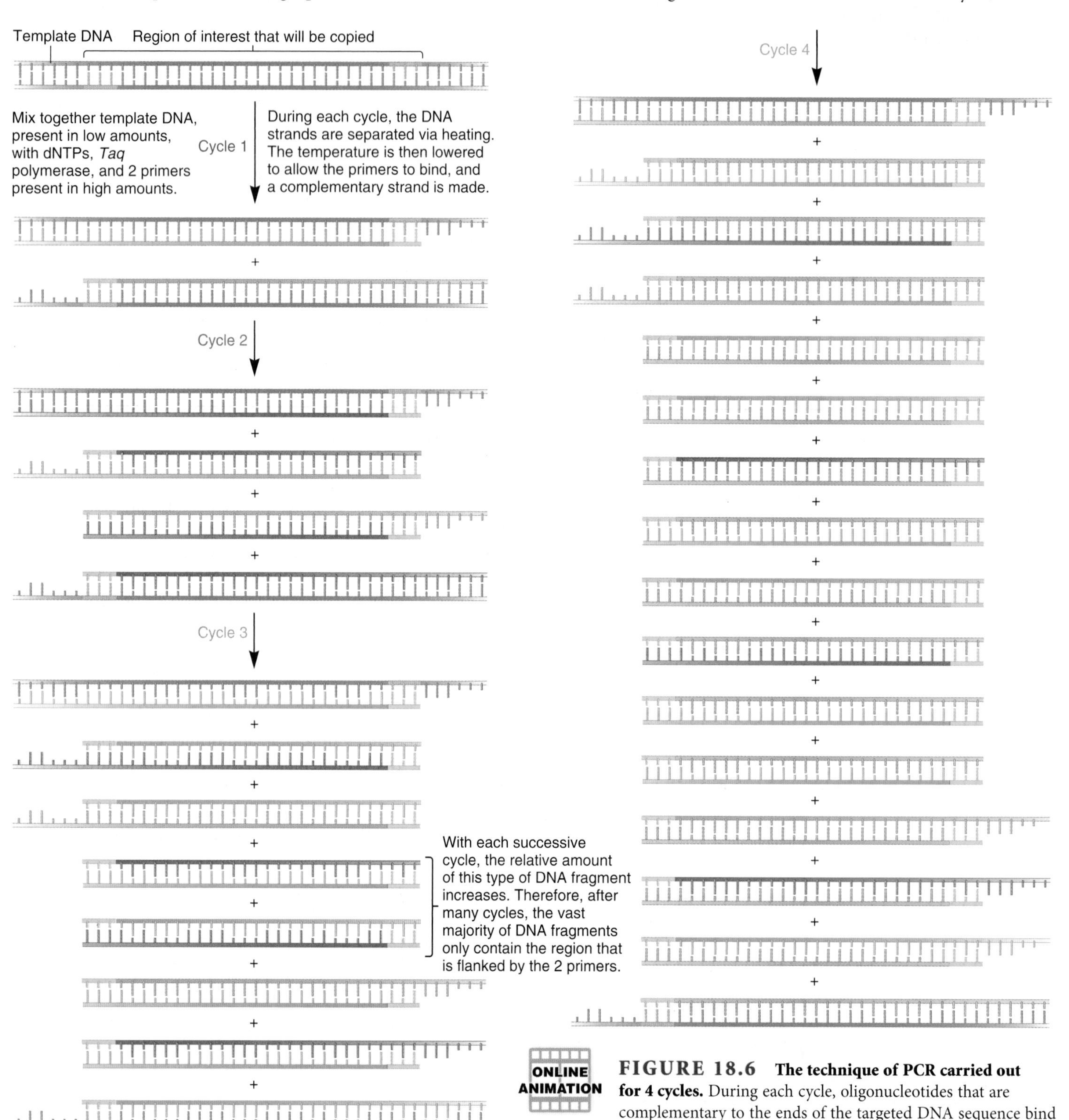

FIGURE 18.6 The technique of PCR carried out for 4 cycles. During each cycle, oligonucleotides that are complementary to the ends of the targeted DNA sequence bind to the DNA and act as primers for the synthesis of this DNA region. Note: The original DNA template strands are purple, and those strands made during the first, second, third, and fourth cycles are blue, red, gray, and orange, respectively.

result of four cycles is 2^4 or 16 copies of the region of interest. Because the starting material contains long strands of chromosomal DNA, some of the products of a PCR experiment contain the region of interest plus some additional DNA at either end. However, after many cycles, the products that contain only the region of interest greatly predominate in the mixture.

PCR is carried out in a thermal reactor, known as a **thermocycler,** that automates the timing of each cycle. The experimenter mixes the DNA sample, dNTPs, *Taq* polymerase, and an excess amount of primers together in a single tube. The tube is placed in a thermocycler, and the experimenter sets the machine to operate within a defined temperature range and number of cycles. During each cycle, the thermocycler increases the temperature to denature the DNA strands and then lowers the temperature to allow annealing and extension to take place. Typically, each cycle lasts 2 to 3 minutes and is then repeated. A typical PCR run is likely to involve 20 to 30 cycles of replication and takes a couple of hours to complete. The PCR technique can amplify the amount of DNA by a staggering amount. Assuming 100% efficiency, the intervening region between the two primers increases 2^{20}-fold after 20 cycles. This is approximately a million-fold!

An important advantage of PCR is that it can amplify a particular region of DNA from a very complex mixture of template DNA. For example, if a researcher uses two primers that anneal to the human β-globin gene, PCR can amplify just the β-globin gene from a DNA sample that contains all of the human chromosomes!

Alternatively, PCR can be used to amplify a sample of chromosomal DNA semispecifically or nonspecifically. As discussed in Chapter 24, this approach is used in DNA fingerprinting analysis. In a semispecific PCR experiment, the primers recognize a known repetitive DNA sequence found at several sites within the genome. Using chromosomal DNA as a template amplifies many different DNA fragments. A nonspecific approach uses a mixture of short PCR primers with many different random sequences. These primers anneal randomly throughout the genome and amplify most of the chromosomal DNA. Nonspecific DNA amplification is used to increase the total amount of DNA in very small samples, such as blood stains found at crime scenes.

Reverse Transcriptase PCR Can Be Used to Amplify RNA

PCR is also used to detect and quantitate the amount of specific RNAs in living cells. To accomplish this goal, RNA is isolated from a sample and mixed with deoxynucleotides, reverse transcriptase, and a primer that binds near the 3′ end of the RNA of interest (**Figure 18.7**). This generates a single-stranded cDNA, which then can be used as template DNA in a conventional PCR reaction. The end result is that the RNA has been amplified to produce many copies of DNA.

This method, called **reverse transcriptase PCR,** is extraordinarily sensitive. Reverse transcriptase PCR can detect the expression of small amounts of RNA from a single cell! As discussed next, certain modifications to PCR can allow researchers to observe the accumulation of PCR products and to quantitate the amount of DNA or RNA in a biological sample.

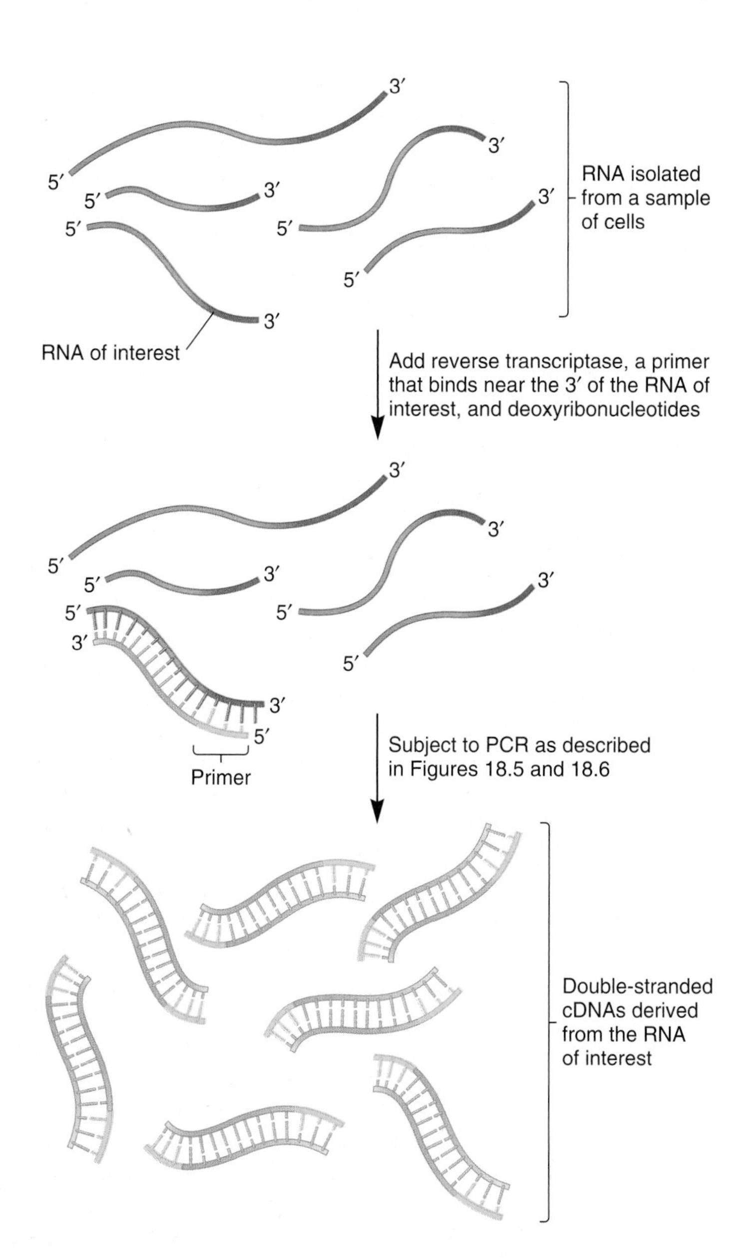

FIGURE 18.7 **The technique of reverse transcriptase PCR.**

Real-time PCR Can Be Used to Quantitate the Amount of a Specific Gene or a Specific mRNA in a Sample

In some applications of PCR, the goal is to obtain large amounts of a DNA region. To determine if PCR is successful, a researcher typically runs a sample of DNA on a gel, stains the gel with ethidium bromide (EtBr) that binds to the DNA, and then observes the gel under UV light, which causes EtBr to fluoresce. If a band of the correct size is seen, the experiment is likely to have been successful. This PCR approach is sometimes called endpoint analysis, because the success of the experiment is judged after PCR is completed.

By comparison, **real-time PCR** technology allows quantitation of specific PCR products in "real time" as PCR is taking place in a thermocycler. Because the PCR products are ultimately derived from the template DNA that was initially added to the reaction, this approach allows researchers to determine how much DNA, such as the DNA that encodes a specific gene, was originally in the sample before PCR was conducted. Similarly, if the starting material is mRNA that is reverse transcribed into DNA, real-time PCR can be used to determine how much mRNA from a specific gene was in a sample. This provides a way to quantitatively measure gene expression.

How do researchers determine the amount of a PCR product during real-time PCR? The procedure is carried out in a thermocycler that has the capacity to measure changes in the level of fluorescence that is emitted from detector molecules that are added to the PCR mixture. The fluorescence given off by the detector molecules depends on the amount of the PCR product. Several detector molecules have been developed. We will consider one type called TaqMan.

The TaqMan detector is an oligonucleotide that has a reporter molecule at one end and a quencher molecule at the other (**Figure 18.8a**). The oligonucleotide is complementary to a site within the PCR product of interest. The reporter molecule emits fluorescence at a certain wavelength, but that fluorescence is largely absorbed by the nearby quencher. Therefore, the close proximity of the reporter molecule to the quencher molecule prevents the detection of fluorescence from the reporter molecule.

As shown in **Figure 18.8b**, a primer and TaqMan detector both anneal to the template DNA after the denaturation step. During the primer extension step, the 5′ to 3′ exonuclease activity of *Taq* polymerase cleaves the oligonucleotide in the TaqMan detector into individual nucleotides, thereby separating the reporter from the quencher. This allows the reporter to emit (unquenched) fluorescence that can be measured within the thermocycler. As PCR products accumulate, more and more of the TaqMan detectors are digested, and therefore the level of fluorescence increases.

Figure 18.9a considers a real-time PCR experiment as it occurs over the course of many cycles, such as 20 to 40. Real-time PCR goes through three main phases. Initially, when the amount of PCR product is small and reagents are not limiting, the PCR product is made exponentially and the reaction is close to 100% efficiency. This exponential accumulation is difficult to detect in the earliest cycles because the amount of PCR product is small. The second phase is relatively linear as PCR products continue to accumulate, but the reaction efficiency falls as reagents become limiting. Finally, in the third phase, the accumulation of PCR products reaches a plateau as one or more reagents are used up.

During the exponential phase of PCR, the amount of PCR product is proportional to the amount of starting template that was initially added. Therefore, the exponential phase of PCR is analyzed to quantitate the initial template concentration. The commonly used method is called the **cycle threshold method (C_t method).**

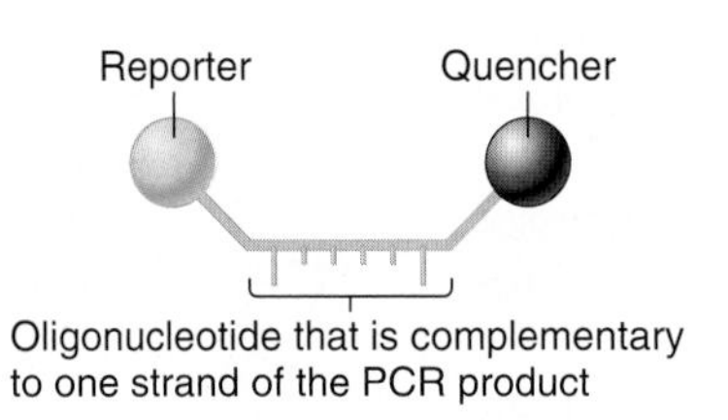

(a) TaqMan detector

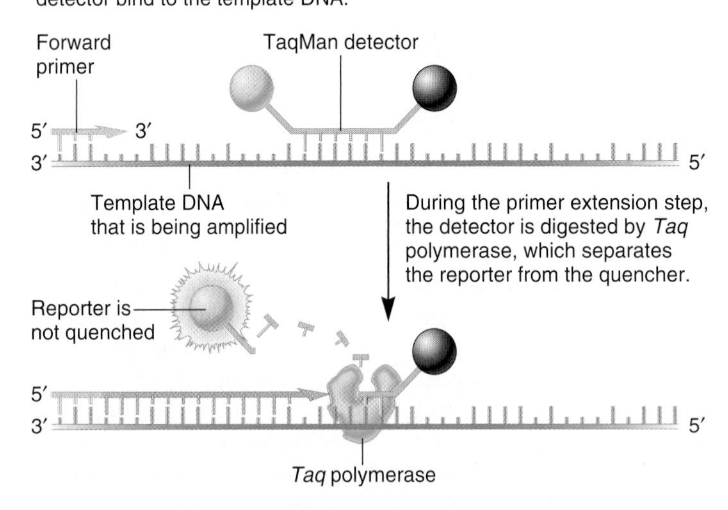

(b) Use of a TaqMan detector in real-time PCR

FIGURE 18.8 **An example of a detector, called TaqMan, which is used in a real-time PCR experiment.**

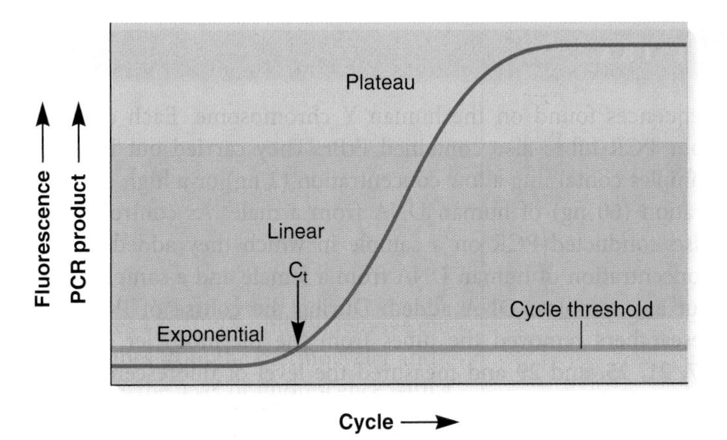

(a) Phases of PCR

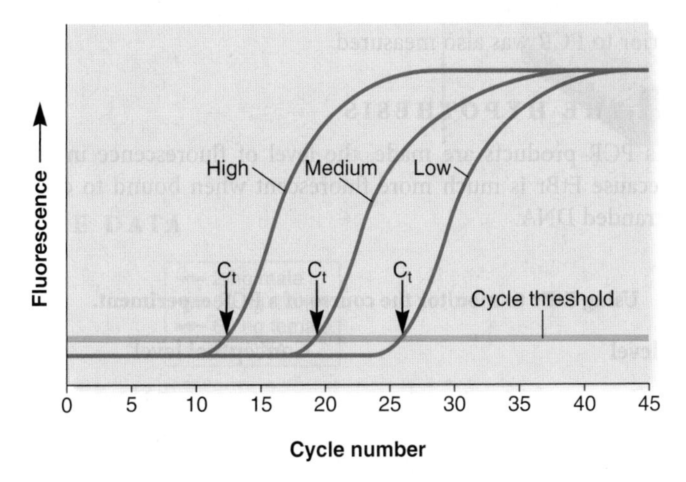

(b) Real-time PCR at high, medium, and low concentrations of the starting template DNA

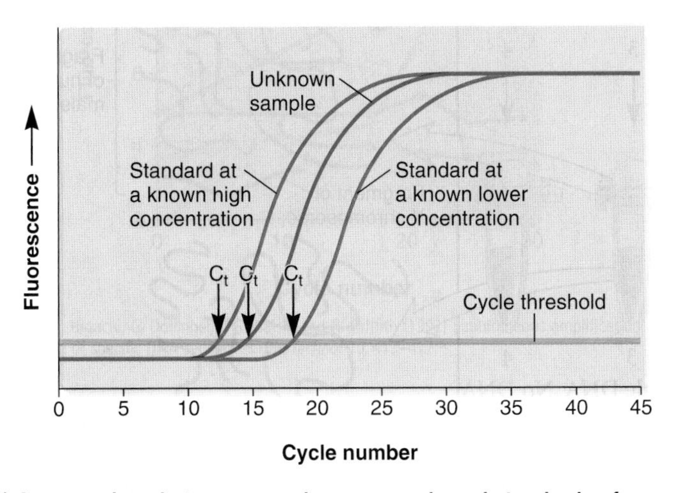

(c) A comparison between an unknown sample and standards of known concentrations

The cycle threshold (C_t) is reached when the accumulation of fluorescence is significantly greater than the background level. During the early cycles, the fluorescence signal due to the background level of fluorescence is greater than that derived from the amplification of the PCR product. Once the C_t value is exceeded, the exponential accumulation of product can be measured. **Figure 18.9b** considers three PCR runs in which the template DNA was initially added at a high, medium, or low concentration. When the initial concentration of the template DNA is higher, the C_t is reached at an earlier amplification cycle.

To determine the amount of starting template DNA, the sample of interest that has an unknown amount of starting template DNA is compared with some type of standard. Real-time PCR involves the coamplification of two templates: the sample of interest and the standard. The two types of PCR products can be monitored simultaneously with different colored fluorescent molecules.

Different types of standards may be used. One possibility is that a standard of known concentration is added to the PCR mixture. For example, plasmid DNA carrying a specific gene can be added to the PCR mixture in known amounts, and the amplification of this plasmid-encoded gene would provide a standard. By comparing the C_t values of the standard and the unknown sample, researchers can determine the concentration of the unknown sample. This is schematically shown in **Figure 18.9c**, but is actually done via computer software. Alternatively, researchers may use an internal standard in which another gene that is already present in the sample is also amplified. This relative quantitation method is somewhat simpler. The amount of unknown template DNA of interest is expressed relative to the internal standard.

FIGURE 18.9 Examples of data that are obtained from a real-time PCR experiment. (**a**) The three phases that occur in a typical PCR experiment. (**b**) PCR carried out at three different starting template concentrations. (**c**) A comparison between a sample with an unknown amount of template DNA and a standard.

RNA that is transcribed from a particular gene or the protein that is encoded by an mRNA. In this section, we will consider the methodology and uses of several common techniques used to detect DNA, RNA, and proteins.

A DNA Library May Be Constructed Using Genomic DNA or cDNA

In a typical cloning experiment that involves the use of vectors (refer back to Figure 18.2), the treatment of the chromosomal DNA with restriction enzymes yields tens of thousands of different DNA fragments. Therefore, after the DNA fragments are ligated individually to vectors, the researcher has a collection of recombinant vectors, with each vector containing a particular fragment of chromosomal DNA. A collection of recombinant vectors is known as a **DNA library.** When the starting material is chromosomal DNA, the library is called a **genomic library** (**Figure 18.11a**). The library shown here uses a plasmid vector. Alternatively, a viral vector could be used, which would result in viral plaques rather than bacterial colonies.

It is also common for researchers to make a **cDNA library** that contains recombinant vectors with cDNA inserts. Because cDNA is produced from mature mRNA via reverse transcriptase, it lacks any introns. A cDNA library could be made because a researcher wanted to express the encoded protein of interest in a cell that would not splice out the introns properly.

Figure 18.11b illustrates how cDNAs are made and inserted into vectors. As described earlier in Figure 18.3, cDNA is first made via reverse transcriptase. To insert the cDNAs into vectors, short oligonucleotides called linkers are attached to the cDNAs via DNA ligase. The linkers contain DNA sequences with a unique site for a restriction enzyme. After the linkers are attached to the cDNAs, the cDNAs and the vectors are cut with restriction enzymes and then ligated to each other. This produces a cDNA library.

A DNA Library May Be Screened by Colony Hybridization to Identify a Cloned Gene

In many cases, the goal of a gene cloning experiment is to obtain many copies of a specific gene or region of a gene. In cloning via PCR, this is achieved by selecting primers that flank the region of interest. This allows a researcher to specifically clone a gene of interest without cloning other segments of chromosomal DNA. However, when cloning is done using vectors, many different segments of chromosomal DNA are cloned during the construction of a DNA library. For example, let's suppose that a geneticist wishes to clone the rat β-globin gene. To begin a cloning experiment, chromosomal DNA is isolated from rat cells. This chromosomal DNA is then digested with a restriction enzyme, yielding thousands of DNA fragments. The chromosomal fragments are then ligated to vector DNA and transformed into bacterial cells to make a DNA library. Unfortunately, only a small percentage of the recombinant vectors, perhaps one in a few thousand, actually contains the rat β-globin gene. For this reason, researchers must have some way of identifying those rare bacterial colonies within a DNA library that happen to contain the cloned gene of interest, in this case, a colony that contains the rat β-globin gene.

Figure 18.12 describes one method for identifying a bacterial colony that contains the gene of interest. This procedure is referred to as **colony hybridization.** The master plate shown at the top of Figure 18.12 has many bacterial colonies. Each bacterial colony is composed of bacterial cells containing a recombinant vector with a different piece of rat chromosomal DNA. The goal is to identify a colony that contains the gene of interest, in this case, the rat β-globin gene. To do so, a nylon membrane is laid gently onto the master plate containing many bacterial colonies. After the membrane is lifted, some cells from each colony are attached to it. In this way, the membrane contains a replica of the colonies on the master plate.

In this procedure, cells are treated with detergent, which dissolves the cell membrane and makes the DNA accessible. The DNA within the cells is then fixed (i.e., adhered) to the nylon membrane and denatured with NaOH. The membrane is then submerged in a solution containing a **probe,** which can be an oligonucleotide, a fragment of DNA from a cloned gene, or a specific RNA. In this case, the probe is single-stranded DNA with a base sequence that is complementary to one of the DNA strands of the rat β-globin gene. The probe is labeled in some way for ease of detection. It could be radiolabeled using a radioisotope such as ^{32}P, or it could be fluorescently labeled. The example in Figure 18.12 uses a radiolabeled probe. The probe is given time to hybridize to the DNA on the membrane, which has already been denatured into single strands. If a bacterial colony contains the rat β-globin gene, the probe will hybridize to the DNA in this colony. Most bacterial colonies are not expected to contain this gene. The unbound probe is then washed away, and the membrane is placed next to X-ray film. If the DNA within a bacterial colony did hybridize to the probe, a dark spot will appear on the film in the corresponding location. The developed film is compared with the master plate to identify bacterial colonies carrying the rat β-globin gene. Following the identification of labeled colonies, the experimenter can pick them from the plate, and grow bacteria containing the cloned gene.

A scientist might follow several different strategies to obtain a probe for a colony hybridization experiment. In some cases, the gene of interest may have already been cloned. A piece of the cloned gene then can be used as a radiolabeled probe. However, in many circumstances, a scientist may attempt to clone a gene that has never been cloned before. In such a case, one strategy is to use a probe that likely has a sequence similar to that of the gene of interest. For example, if the goal is to clone the β-globin gene from a South American rodent, it is expected that this gene is similar to the rat β-globin gene, which has already been cloned. Therefore, the cloned rat β-globin gene can be used as a DNA probe to "fish out" the β-globin gene from a similar species.

Can a researcher design a DNA probe for a novel type of gene that no one else has ever cloned before from any species? If the protein of interest has been previously isolated from living cells, fragments of the protein may be subjected to amino acid

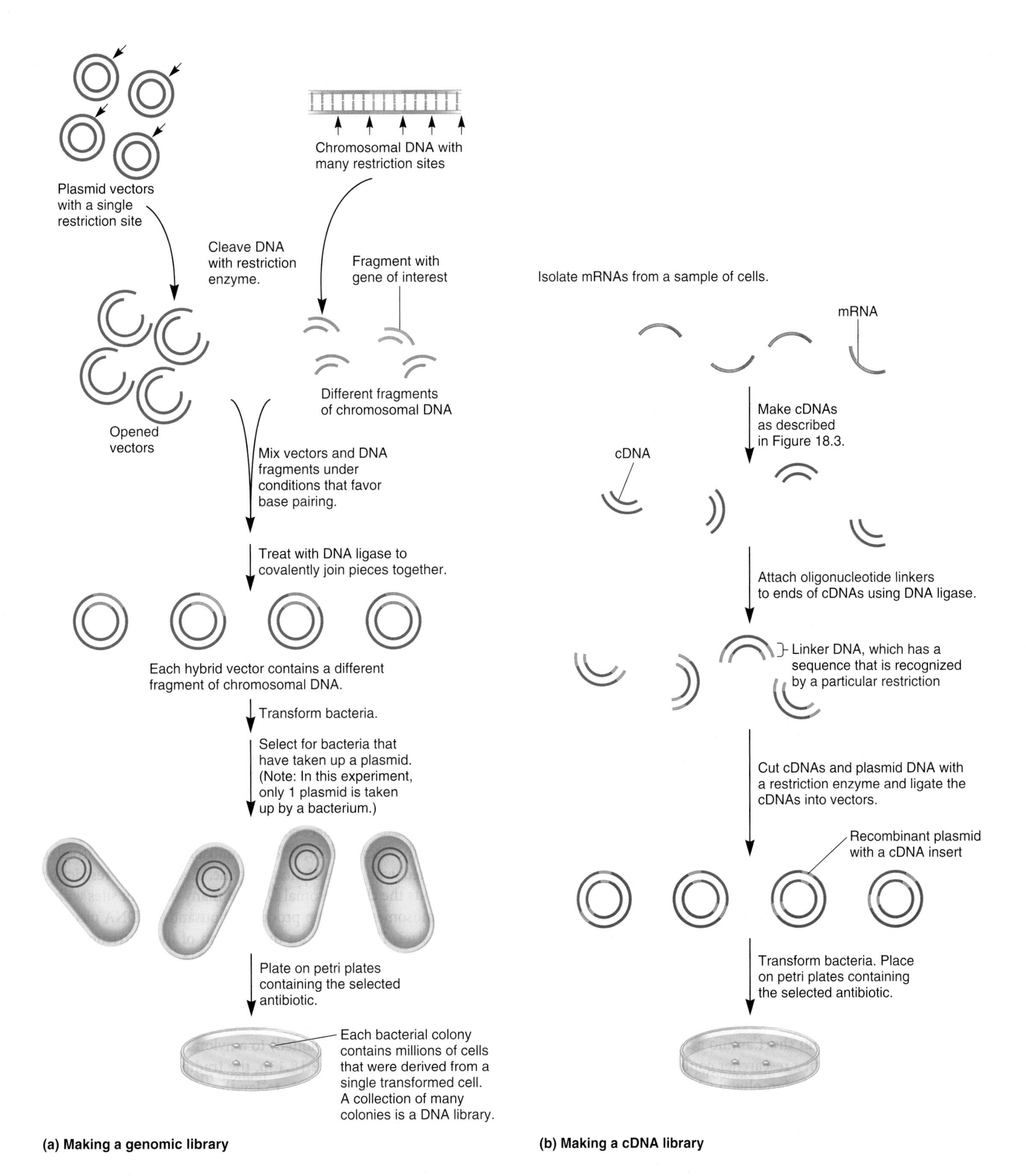

FIGURE 18.11 **The construction of a DNA library.** (**a**) The digestion of chromosomal DNA produces many fragments. The fragment containing the gene of interest is highlighted in red. Following ligation, each vector contains a different piece of chromosomal DNA. (**b**) To make a cDNA library, oligonucleotide linkers that contain a restriction site are attached to the cDNAs, so they can be inserted into vectors.

which are used to study protein-DNA interactions. Gel retardation is also used to study protein-RNA interactions.

The Gel Retardation Assay Can Be Used to Determine If a Protein Binds to a Specific DNA Fragment or RNA Molecule

A technically simple, widely used method for identifying DNA- or RNA-binding proteins is the **gel retardation assay,** also known as the **gel mobility shift assay.** This technique was used originally to study interactions between specific proteins and rRNA molecules and quickly became popular after its success in studying protein-DNA interactions in the *lac* operon. Now it is commonly used as a technique for detecting interactions between mRNAs and RNA-binding proteins and between eukaryotic transcription factors and DNA regulatory elements.

In the case of DNA-binding proteins, the technical basis for a gel retardation assay is that the binding of a protein to a DNA fragment retards the fragment's ability to move within a polyacrylamide or agarose gel. During electrophoresis, DNA fragments are pulled through the gel matrix toward the bottom of the gel by a voltage gradient. Smaller fragments of DNA migrate more quickly through a gel matrix than do larger ones. As you might expect, therefore, the binding of a protein to a DNA fragment retards the DNA's rate of movement through the gel matrix, because a protein-DNA complex has a higher mass. When comparing a DNA fragment and a protein-DNA complex after electrophoresis, the complex is shifted to a higher band than that for the DNA alone, because the complex migrates more slowly to the bottom of the gel (**Figure 18.16**). The bands can be visualized by staining the DNA with a dye such as EtBr. To increase the sensitivity of the gel retardation assay, the DNA can also be radiolabeled or labeled with a fluorescent molecule.

A gel retardation assay must be carried out under nondenaturing conditions. This means that the buffers and gel cannot cause the unfolding of proteins or the separation of the DNA double helix. This is necessary so the proteins and DNA retain their proper structure and are thereby able to bind to each other. The nondenaturing conditions of a gel retardation assay differ from the more common SDS gel electrophoresis, in which the proteins are denatured by the detergent SDS.

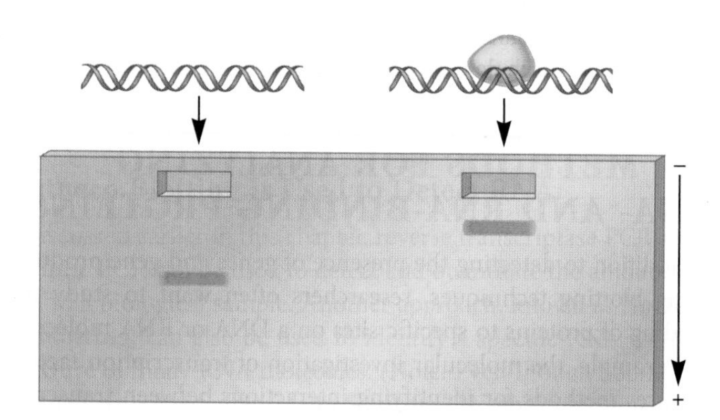

FIGURE 18.16 The results of a gel retardation assay. The binding of protein to a labeled fragment of DNA retards its rate of movement through a gel. For the results shown in the lane on the right, if the concentration of the DNA fragment was higher than the concentration of the protein, there would be two bands: one band with protein bound (at a higher molecular mass) and one band without protein bound (corresponding to the band found in the left lane).

DNase I Footprinting Shows the Detailed Interactions Between a Protein and DNA

Another method for studying protein-DNA interactions is **DNase I footprinting,** a technique described by David Galas and Albert Schmitz in 1978. A DNase I footprinting experiment attempts to identify one or more regions of DNA that interact with a DNA-binding protein. In their original study, Galas and Schmitz identified a site in the *lac* operon, known as the *lac* operator site, that is bound by a DNA-binding protein called the lac repressor. Compared with a gel retardation assay, DNase I footprinting provides more detailed information about the interactions between a protein and DNA. One drawback is that DNase I footprinting is a more complicated technique compared to a gel retardation assay.

To understand the basis of a DNase I footprinting experiment, we need to consider the interactions among three types of molecules: a fragment of DNA, DNA-binding proteins, and agents that can alter DNA structure. As an example, let's examine the binding of RNA polymerase to a bacterial promoter, a topic discussed in Chapter 12. When RNA polymerase holoenzyme binds to the promoter to form a closed complex, it binds tightly to the –35 and –10 promoter regions, but the protein covers up an even larger region of the DNA. Therefore, holoenzyme bound at the promoter prevents other molecules from gaining access to this region of the DNA. The enzyme DNase I, which can cleave covalent bonds in the DNA backbone, is used as a reagent for determining if a DNA region has a protein bound to it. Galas and Schmitz reasoned that DNase I cannot cleave the DNA at locations where a protein is bound. In this example, it is expected that RNA polymerase holoenzyme will bind to a promoter and protect this DNA region from DNase I cleavage.

Figure 18.17 shows the results of a DNase I footprinting experiment. In this experiment, a sample of many identical DNA fragments, all of which are 150 bp in length, were radiolabeled at only one end. The sample of fragments was then divided into two tubes: tube A, which did not contain any holoenzyme, and tube B, which contained RNA polymerase holoenzyme. DNase I was then added to both tubes. The tubes were incubated long enough for DNase I to cleave the DNA at a single site in each DNA fragment. Each tube contained many 150-bp DNA fragments, and the cutting in any DNA strand by DNase I occurred randomly. Therefore, the DNase I treatment should produce a mixture of many smaller DNA fragments. A key point, however, is that DNase I cannot cleave the DNA in a region where RNA polymerase holoenzyme is bound. After DNase I treatment, the DNA fragments within the two tubes were separated by gel electrophoresis, and DNA fragments containing the labeled end were detected using autoradiography.

In the absence of RNA polymerase holoenzyme (tube A), DNase I should cleave the 150-bp fragments randomly at any

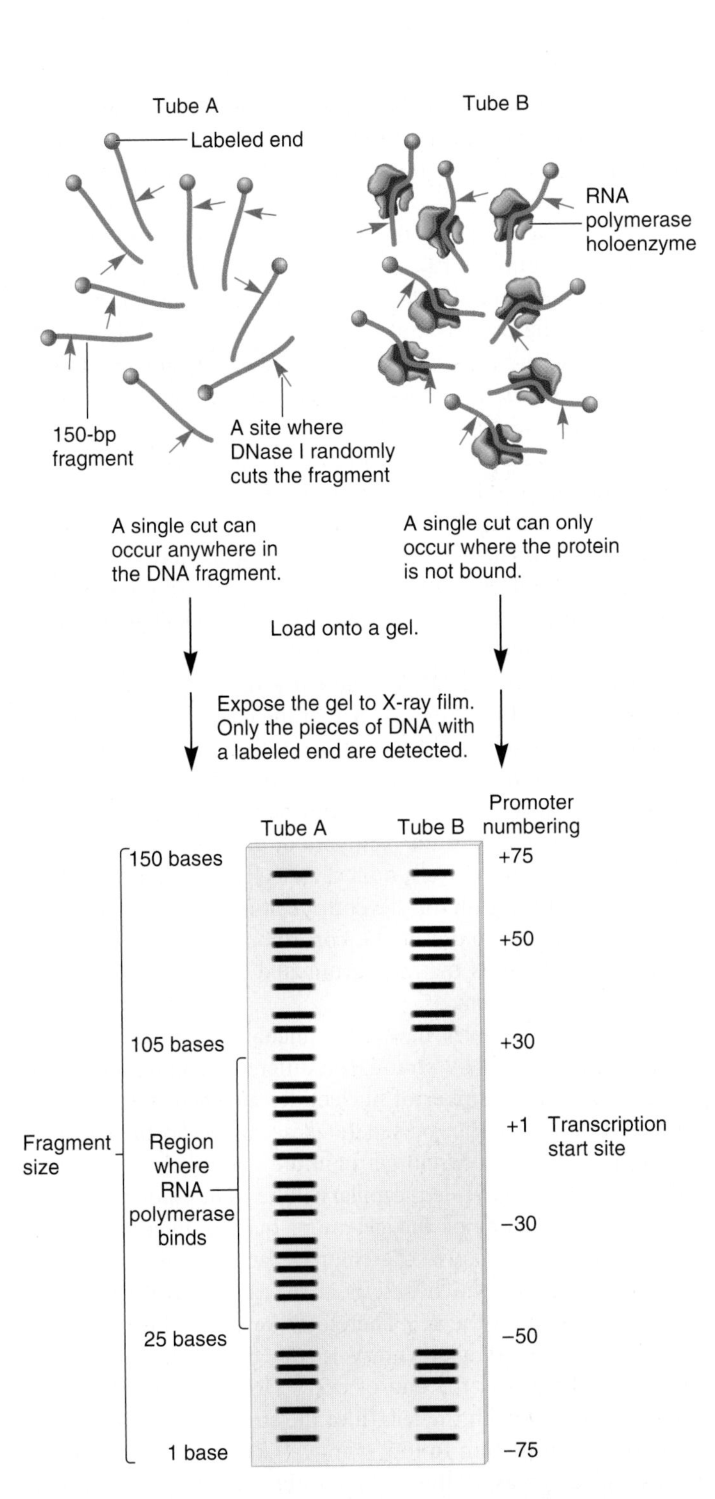

FIGURE 18.17 A DNase I footprinting experiment. Both tubes contained 150-bp fragments of DNA that were incubated with DNase I. Tube B also contained RNA polymerase holoenzyme. The binding of RNA polymerase holoenzyme protected a region of about 80 base pairs (namely, the −50 region to the +30 region) from DNase I digestion. Note: The promoter numbering convention shown here is the same as described previously in Figure 12.4.

single location. Therefore, a continuous range of sizes occurs (see Figure 18.17). However, if we look at the gel lane from tube B, no bands are observed in the size range from 25 to 105 nucleotides. Why are these bands missing? The answer is that DNase I cannot cleave the DNA within the region where the holoenzyme is bound. The middle portion of the 150-bp fragment contains a promoter sequence that binds the RNA polymerase holoenzyme. Along the right side of the gel, the bases are numbered according to their position within the gene. (The site labeled +1 is where transcription begins.) As seen here, RNA polymerase covers up a fairly large region (its "footprint") of about 80 base pairs, from the −50 region to the +30 region.

As illustrated in this experiment, DNase I footprinting can identify the DNA region that interacts with a DNA-binding protein. In addition to RNA polymerase-promoter binding, DNase I footprinting has been used to identify the binding sites for many other types of DNA-binding proteins, such as eukaryotic transcription factors and histones. This technique has greatly facilitated our understanding of protein-DNA interactions.

18.5 DNA SEQUENCING AND SITE-DIRECTED MUTAGENESIS

As we have seen throughout this textbook, our knowledge of genetics can be largely attributed to an understanding of DNA structure and function. The feature that underlies all aspects of inherited traits is the DNA sequence. For this reason, analyzing and altering DNA sequences is a powerful approach for understanding genetics. In this last section, we will begin by examining a technique called **DNA sequencing.** This method enables researchers to determine the base sequence of DNA found in genes and other chromosomal regions. It is one of the most important tools for exploring genetics at the molecular level. DNA sequencing is practiced by scientists around the world, and the amount of scientific information contained within experimentally determined DNA sequences has become enormous. In Chapter 20, we will learn how researchers can determine the complete DNA sequence of entire genomes. In Chapter 21, we will consider how computers play an essential role in the storage and analysis of genetic sequences.

Not only can researchers determine DNA sequences, they can also use another technique, known as site-directed mutagenesis, to change the sequence of cloned DNA segments. At the end of this section, we will examine how site-directed mutagenesis is conducted and how it provides information regarding the function of genes.

The Dideoxy Method of DNA Sequencing Is Based on Our Knowledge of DNA Replication

Molecular geneticists often want to determine DNA base sequences as a first step toward understanding the function and expression of genes. For example, the investigation of genetic sequences has been vital to our understanding of promoters, regulatory elements, and the genetic code itself. Likewise, an examination of sequences has facilitated our understanding of origins of replication, centromeres, telomeres, and transposable elements.

During the 1970s, two methods for DNA sequencing were devised. One method, developed by Allan Maxam and Walter

Gilbert, involves the base-specific chemical cleavage of DNA. Another method, developed by Frederick Sanger and colleagues, is known as **dideoxy sequencing.** Because it has become the more popular method of DNA sequencing, we consider the dideoxy method here. In addition, Chapter 20 considers some newer methods of sequencing DNA that are not based on the Sanger dideoxy method. These newer methods are commonly used in projects aimed at determining the DNA sequence of entire genomes.

The dideoxy procedure of DNA sequencing is based on our knowledge of DNA replication but uses a clever twist. As described in Chapter 11, DNA polymerase connects adjacent deoxyribonucleotides by catalyzing a covalent bond between the 5′ phosphate on one nucleotide and the 3′ –OH group on the previous nucleotide (refer back to Figure 11.12). Chemists, though, can synthesize deoxyribonucleotides that are missing the –OH group at the 3′ position (**Figure 18.18**). These synthetic nucleotides are called **dideoxyribonucleotides (ddNTPs).** (Note: The prefix dideoxy- indicates that two (di) oxygens (oxy) are removed (de) from this sugar compared with ribose; ribose has –OH groups at both the 2′ and 3′ positions.) Sanger reasoned that if a dideoxyribonucleotide is added to a growing DNA strand, the strand can no longer grow because the dideoxyribonucleotide is missing the 3′ –OH group. The incorporation of a dideoxyribonucleotide into a growing strand is therefore referred to as **chain termination.**

To detect the incorporation of dideoxynucleotides during DNA replication, the newly made DNA strands must be labeled in some way. When dideoxy sequencing was first invented, researchers labeled the newly made DNA with radioisotopes, which allowed them to be detected via autoradiography. This traditional DNA sequencing method has been largely replaced by **automated DNA sequencing** in which each type of dideoxynucleotide is labeled with a different colored fluorescent molecule. For example, a common way to fluorescently label ddNTPs is the following: ddA is green, ddT is red, ddG is yellow, and ddC is blue.

Prior to automated DNA sequencing, the segment of DNA to be sequenced must usually be obtained in large amounts. This is accomplished using gene cloning, which was described earlier in this chapter. In the example in **Figure 18.19**, the segment of DNA to be sequenced, which we will call the target DNA, was cloned into a vector at a defined location. The target DNA was inserted next to a site in the vector where a primer will bind, which is called the primer-annealing site. The aim of the experiment is to determine the base sequence of the target DNA. In

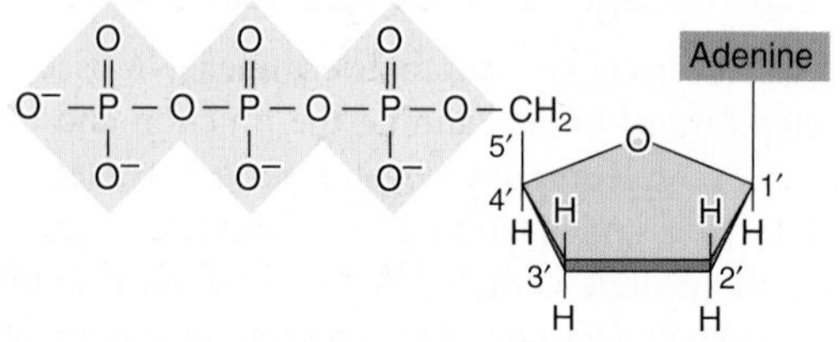

FIGURE 18.18 The structure of a dideoxyribonucleotide. Note that the 3′ group is a hydrogen rather than an –OH group. For this reason, another nucleotide cannot be attached at the 3′ position.

the experiment shown in Figure 18.19, the recombinant vector DNA has been previously denatured into single strands, usually via heat treatment. Only the strand needed for DNA sequencing is shown here.

Let's now consider the steps that are involved in DNA sequencing (Figure 18.19). First, a sample containing many copies of the single-stranded DNA is mixed with many primers that will bind to the primer-annealing site. The primer binds to the DNA because the primer and primer-annealing site are complementary to each other. All four types of deoxyribonucleotides and DNA polymerase are then added to the annealed DNA fragments. The tube also has a low concentration of each dideoxyribonucleotide (ddG, ddA, ddT, or ddC), which are fluorescently labeled. The tube is then incubated to allow DNA polymerase to make strands complementary to the target DNA sequence.

DNA synthesis continues until a dideoxynucleotide is incorporated into a growing strand. For example, chain termination can occasionally occur at the sixth or thirteenth position of the newly synthesized DNA strand if a ddT becomes incorporated at either of these sites. Note that the complementary A base is found at the sixth and thirteenth position in the target DNA. Therefore, we expect to make DNA strands that terminate at the sixth or thirteenth positions and have a ddT at their ends. Because these DNA strands contain a ddT, they are fluorescently labeled in red. Alternatively, ddA causes chain termination at the second, seventh, eighth, or eleventh positions because a complementary T base is found at the corresponding positions in the target strand. Strands that are terminated with ddA are fluorescently labeled in green.

After the samples have been incubated for several minutes, mixtures of DNA strands of different lengths are made, depending on the number of nucleotides attached to the primer. These DNA strands can be separated according to their lengths by running them on a slab gel or more commonly by running them through a gel-filled capillary tube. The shorter strands move to the bottom of the gel more quickly than the longer strands. Because we know the color of each dideoxyribonucleotide, we also know which base is at the very end of each DNA strand separated on the gel. Therefore, we can deduce the DNA sequence that is complementary to the target DNA by "reading" which base is at the end of every DNA strand and matching this sequence with the length of the strand. Reading the base sequence, from bottom to top, is much like climbing a ladder of bands. For this reason, the sequence obtained by this method is referred to as a **sequencing ladder.**

Theoretically, it is possible to read this sequence directly from the gel. From a practical perspective, however, it is faster and more efficient to automate the procedure using a laser and fluorescence detector. As the gel is running, each band passes the laser and the laser beam excites the fluorescent dye. The fluorescence detector records the amount of fluorescence emission from the excited dye. The detector reads the level of fluorescence at four wavelengths, corresponding to the four different colored dyes. An example of the printout from the fluorescence detector is shown in **Figure 18.19b**. As seen here, the peaks of fluorescence correspond to the DNA sequence that is complementary to the target DNA. Note that ddG is usually labeled with a yellow

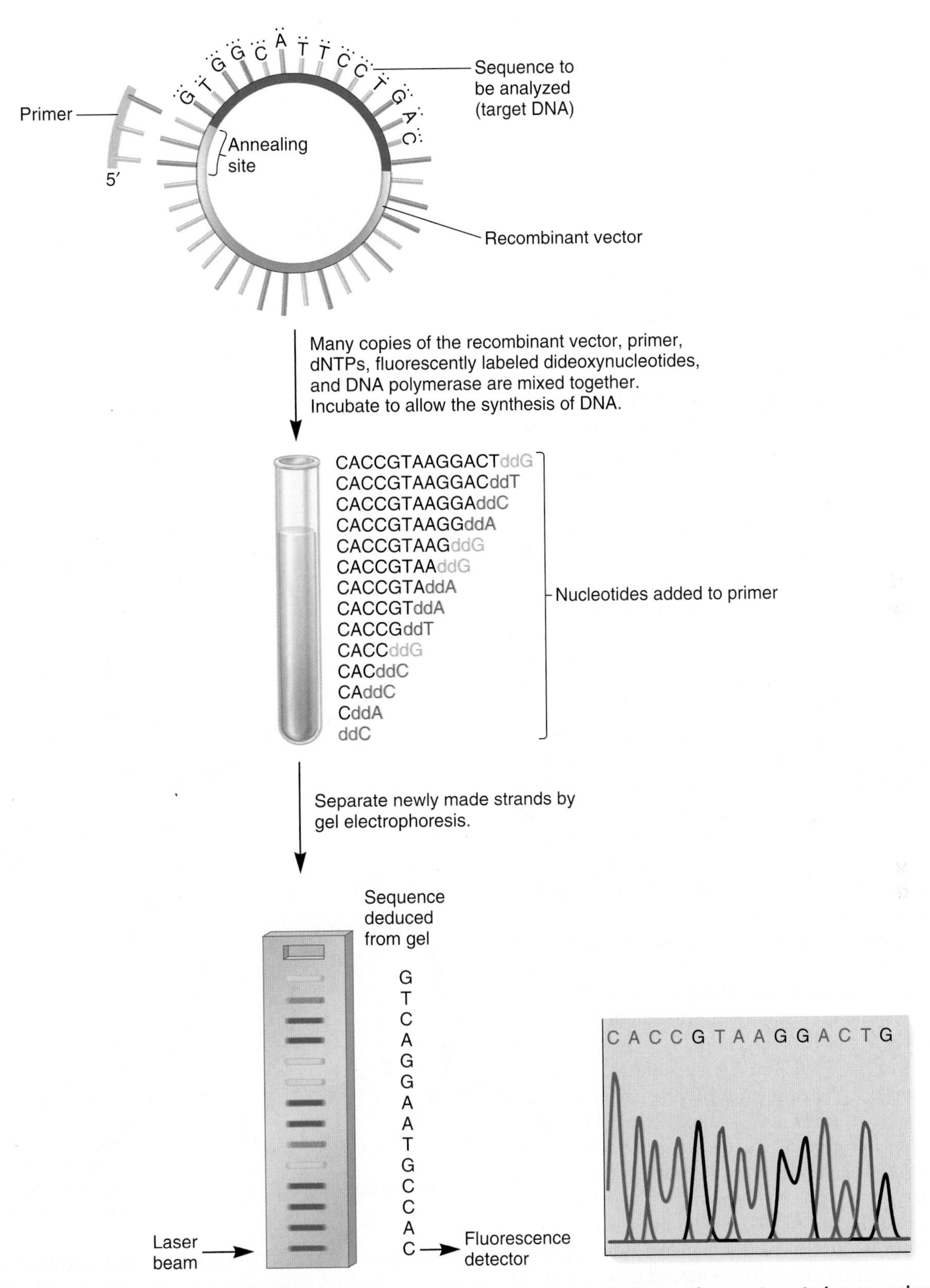

FIGURE 18.19 The protocol for DNA sequencing by the dideoxy method. **(a)** The method shown here begins with single-stranded DNA in which the target DNA has been inserted into a vector. This diagram schematically depicts a series of bands on a gel; the four colors of the bands occur because each type of dideoxynucleotide is labeled with a different colored fluorescent molecule. As each band passes a laser, the fluorescent dye is excited by the laser beam, and the fluorescence emission is recorded by a fluorescence detector. The detector reads the level of fluorescence at four wavelengths, corresponding to the four dyes. **(b)** As shown in the printout, the peaks of fluorescence correspond to the DNA sequence that is complementary to the target DNA.

dye, but it is converted to black ink on the printout shown in Figure 18.19b for ease of reading. Though improvements in automated sequencing continue to be made, a typical sequencing run can provide a DNA sequence that is approximately 700 to 900 bases long, and perhaps even longer.

Site-Directed Mutagenesis Is a Technique for Altering DNA Sequences

As we have seen, dideoxy sequencing provides a way of determining the base sequence of DNA. To understand how the genetic material functions, researchers often analyze mutations that alter the normal DNA sequence, thereby affecting the expression of genes and the outcome of traits. For example, geneticists have discovered that many inherited human diseases, such as sickle cell disease and hemophilia, involve mutations within specific genes. These mutations provide insight into the function of the genes in unaffected individuals. Hemophilia, for example, involves deleterious mutations in genes that encode blood clotting factors.

Because the analysis of mutations can provide important information about normal genetic processes, researchers often wish to obtain mutant organisms. As we discussed in Chapter 16, mutations can arise spontaneously or can be induced by environmental agents. Mendel's pea plants are a classic example of allelic strains with different phenotypes that arose from spontaneous mutations. X-rays and UV light are physical agents that can cause induced mutations. In addition, experimental organisms can be treated with chemical mutagens that increase the rate of mutations.

More recently, researchers have developed molecular techniques for making mutations within cloned genes or other DNA segments. One widely used method, known as **site-directed mutagenesis,** allows a researcher to produce a mutation at a specific site within a cloned DNA segment. For example, if a DNA sequence is 5′–AAATTTCTTTAAA–3′, a researcher can use site-directed mutagenesis to change it to 5′–AAATTTGTTTAAA–3′. In this case, the researcher deliberately changed the seventh base from a C to a G. Why is this method useful? The site-directed mutant can then be introduced into a living organism to see how the mutation affects the expression of a gene, the function of a protein, and the phenotype of an organism.

The first successful attempts at site-directed mutagenesis involved changes in the sequences of viral genomes. These studies were conducted in the 1970s. Mark Zoller and Michael Smith also developed a protocol for the site-directed mutagenesis of DNA that has been cloned into a viral vector. Since these early studies, many approaches have been used to achieve site-directed mutagenesis. **Figure 18.20** describes the general steps in the procedure. Prior to this experiment, the DNA was denatured into single strands; only the single strand needed for site-directed mutagenesis is shown. As in PCR, this single-stranded DNA is referred to as the template DNA, because it is used as a template to synthesize a complementary strand.

As shown in Figure 18.20, an oligonucleotide primer is allowed to hybridize or anneal to the template DNA. The primer, typically 20 or so nucleotides in length, is synthesized chemically.

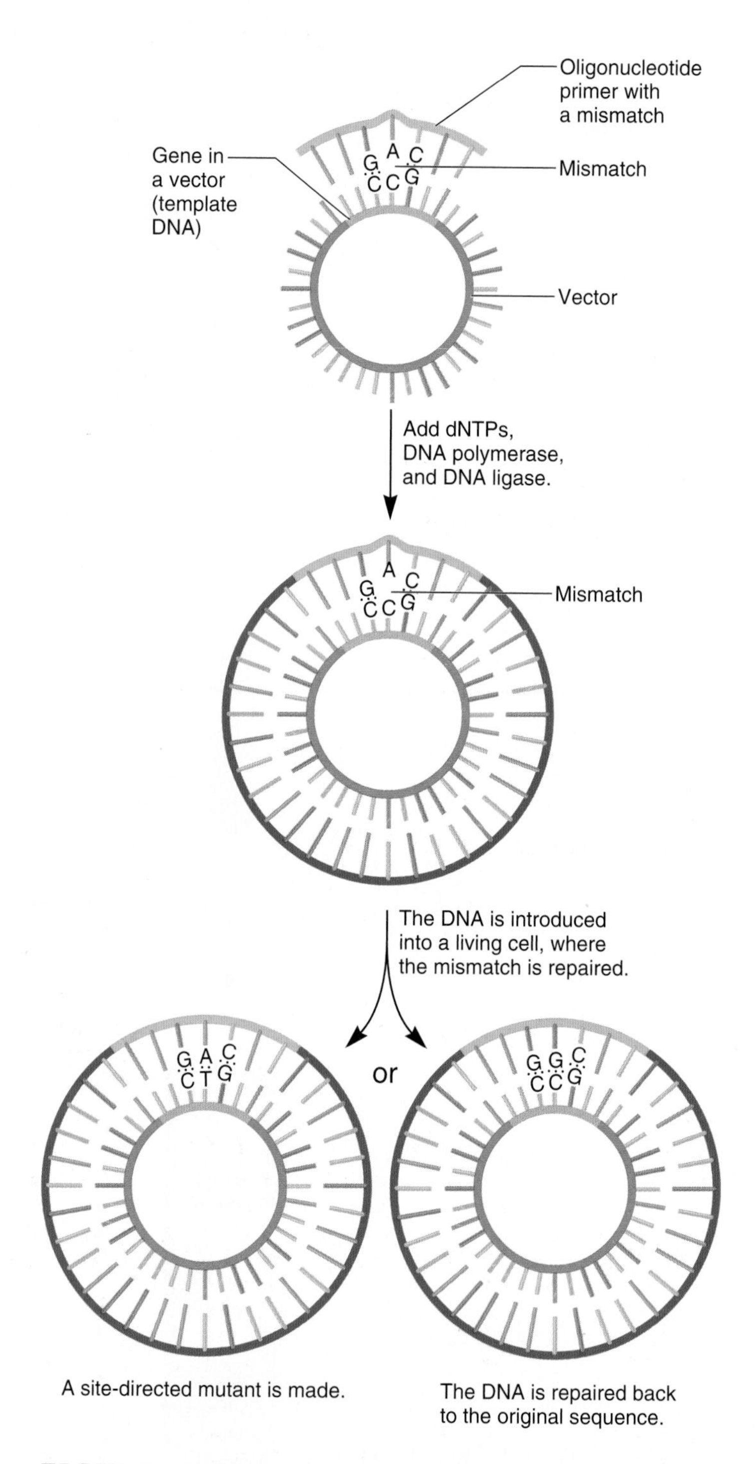

FIGURE 18.20 The method of site-directed mutagenesis.
Genes → Traits To examine the relationship between genes and traits, researchers can alter gene sequences via site-directed mutagenesis. The altered gene can then be introduced into a living organism to examine how the mutation affects the organism's traits. For example, a researcher could introduce a nonsense mutation into the middle of the *lacY* gene in the *lac* operon. If this site-directed mutant was introduced into an *E. coli* bacterium that did not a have a normal copy of the *lacY* gene, the bacterium would be unable to use lactose. These results indicate that a functional *lacY* gene is necessary for bacteria to have the trait of lactose utilization.

(A shorter version of the primer is shown in Figure 18.20 for simplicity.) The scientist designs the base sequence of the primer. The primer has two important characteristics. First, most of the sequence of the primer is complementary to the site in the DNA where the mutation is to be made. However, a second feature is that the primer contains a region of mismatch where the primer and template DNA are not complementary. The mutation occurs in this mismatched region. For this reason, site-directed mutagenesis is sometimes referred to as oligonucleotide-directed mutagenesis.

After the primer and template have annealed, the complementary strand is synthesized by adding deoxyribonucleoside triphosphates (dNTPs), DNA polymerase, and DNA ligase. This yields a double-stranded molecule that contains a mismatch only at the desired location. This double-stranded DNA can then be introduced into a bacterial cell. Within the cell, the DNA mismatch is likely to be repaired (see Chapter 16). Depending on which base is replaced, this may produce the mutant sequence or the original sequence. Clones containing the desired mutation can be identified by DNA sequencing and used for further studies.

After a site-directed mutation has been made within a cloned gene, its consequences are analyzed by introducing the mutant gene into a living cell or organism. As described earlier, recombinant vectors containing cloned genes can be introduced into bacterial cells. Following transformation into a bacterium, a researcher can study the differences in function between the mutant and wild-type genes and the proteins they encode. Similarly, mutant genes made via site-directed mutagenesis can be introduced into plants and animals, as discussed in Chapter 19.

KEY TERMS

Page 484. recombinant DNA technology, recombinant DNA molecules
Page 485. gene cloning
Page 486. vector, host cell, plasmids, R factors, origin of replication, selectable marker, shuttle vector, expression vector
Page 487. restriction endonucleases, restriction enzymes
Page 488. DNA ligase, palindromic, recombinant vector, competent cells, transformation
Page 490. reverse transcriptase, oligonucleotide
Page 491. complementary DNA (cDNA), restriction mapping, polymerase chain reaction (PCR), primers
Page 493. template DNA, *Taq* polymerase, annealing
Page 494. primer extension
Page 495. thermocycler, reverse transcriptase PCR, real-time PCR
Page 496. cycle threshold method (C_t method)
Page 500. DNA library, genomic library, cDNA library, colony hybridization, probe
Page 502. Southern blotting
Page 504. high stringency, low stringency, Northern blotting, Western blotting
Page 505. antibody, epitopes, antigen
Page 506. gel retardation assay, gel mobility shift assay, DNase I footprinting
Page 507. DNA sequencing
Page 508. dideoxy sequencing, dideoxyribonucleotides (ddNTPs), chain termination, automated DNA sequencing, sequencing ladder
Page 510. site-directed mutagenesis

CHAPTER SUMMARY

- Recombinant DNA technology is the use of in vitro molecular techniques to manipulate fragments of DNA and produce new arrangements.

18.1 Gene Cloning Using Vectors

- Gene cloning has many uses including gene sequencing, site-directed mutagenesis, the use of genes as probes, and the expression of cloned genes (see Table 18.1).
- Cloning vectors can be derived from plasmids or viruses (see Table 18.2).
- Restriction enzymes, also called restriction endonucleases, cut chromosomal DNA and vector DNA to produce sticky ends that will hydrogen bond with each other. DNA ligase is needed to make a covalent link in the DNA backbone (see Figure 18.1, Table 18.3).
- Gene cloning using vectors involves the insertion of the gene into a vector and then its propagation in a living cell such as *E. coli* (see Figure 18.2).
- Complementary DNA (cDNA) can be made via reverse transcriptase by starting with mRNA (see Figure 18.3).
- Restriction mapping can be used to determine the locations of restriction sites within a vector (see Figure 18.4).

18.2 Polymerase Chain Reaction

- Polymerase chain reaction (PCR) uses oligonucleotide primers to copy a specific region of DNA. Each cycle of PCR involves three steps: denaturation, primer annealing, and primer extension (see Figures 18.5, 18.6).
- Reverse transcriptase PCR begins with mRNA to study gene expression (see Figure 18.7).
- Real-time PCR monitors PCR as it occurs in a thermocycler. It can be used to quantitatively measure the amount of starting DNA or mRNA in a sample. Real-time PCR uses fluorescent detectors to follow the PCR reaction. A sample that has an unknown amount of DNA is compared with a standard (see Figures 18.8. 18.9).

- Higuchi and colleagues carried out the first real-time PCR experiment, using ethidium bromide as a fluorescent detector (see Figure 18.10).

18.3 DNA Libraries and Blotting Methods

- A DNA library is a collection of recombinant vectors. The inserts can be chromosomal DNA or cDNA (see Figure 18.11).
- Colony hybridization can be used to identify colonies that carry a gene or cDNA of interest (see Figure 18.12).
- Southern blotting uses a labeled DNA probe to detect the presence of a particular gene sequence within a mixture of many gene sequences. At low stringency, gene families may be detected (see Figure 18.13).
- Northern blotting uses a labeled DNA probe to detect a specific RNA within a mixture of many different RNAs (see Figure 18.14).
- Western blotting uses antibodies to detect a specific protein within a mixture of many different proteins (see Figure 18.15).

18.4 Methods for Analyzing DNA- and RNA-Binding Proteins

- A gel retardation assay can determine if a protein binds to a specific DNA fragment or RNA molecule because the binding of the protein slows down the movement of the DNA through a gel (see Figure 18.16).
- DNase I footprinting can determine the regions of a DNA molecule that are bound by a protein (see Figure 18.17).

18.5 DNA Sequencing and Site-Directed Mutagenesis

- The most commonly used method of DNA sequencing, called dideoxy sequencing, uses fluorescently labeled dideoxynucleotides that cause chain termination (see Figures 18.18, 18.19).
- In the method of site-directed mutagenesis, an oligonucleotide with one or more mismatches directs a mutation into a specific region of DNA (see Figure 18.20).

PROBLEM SETS & INSIGHTS

Solved Problems

S1. RNA was isolated from four different cell types and probed with radiolabeled DNA strands from a cloned gene that is called gene *X*. The results are shown here.

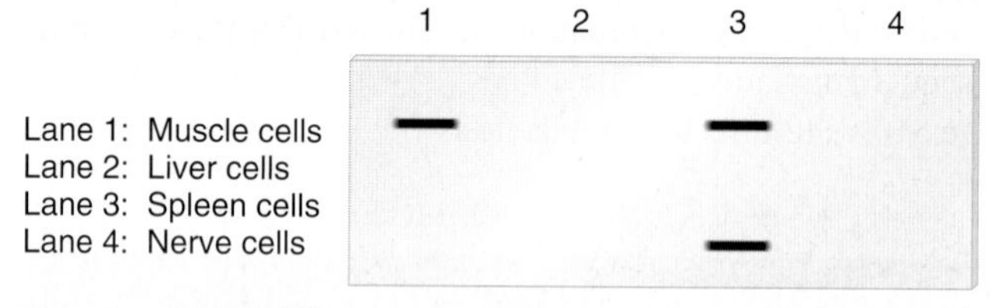

Explain the results of this experiment.

Answer: In this Northern blot, a dark band appears in those lanes where RNA was isolated from muscle and spleen cells but not from liver and nerve cells. These results indicate that the muscle and spleen cells contain a significant amount of RNA from gene *X*, but the liver and nerve cells do not. The muscle cells show a single band, whereas the spleen cells show this band plus a second band of lower molecular mass. An interpretation of these results is that the spleen cells can alternatively splice the RNA to produce a second RNA containing fewer exons.

S2. In the Western blotting experiment shown here, proteins were extracted from red blood cells obtained from tissue samples at different stages of human development. An equal amount of total cellular proteins was added to each lane. The primary antibody recognizes the β-globin polypeptide that is found in the hemoglobin protein.

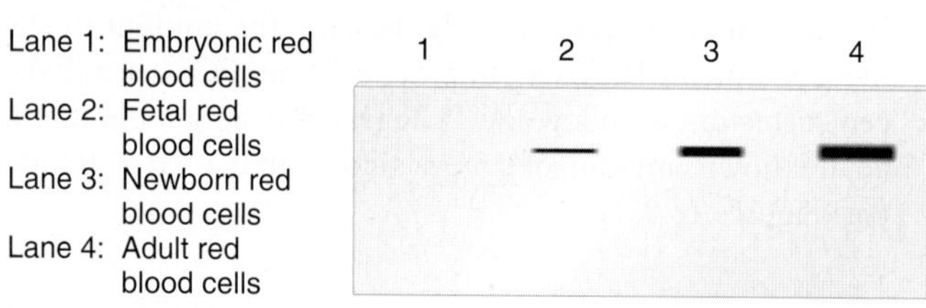

Explain these results.

Answer: As shown here, the amount of β globin increases during development. Little detectable β globin is produced during embryonic development. The amount increases significantly during fetal development and becomes maximal in the adult. These results indicate that the β-globin gene is "turned on" in later stages of development, leading to the synthesis of the β-globin polypeptide. This experiment illustrates how a Western blot can provide information concerning the relative amount of a specific protein within living cells.

S3. A DNA strand is 3′–ATACGACTAGTCGGGACCATATC–5′. If the primer in a dideoxy sequencing reaction anneals just to the left of this sequence, draw what the sequencing ladder would look like.

Answer:

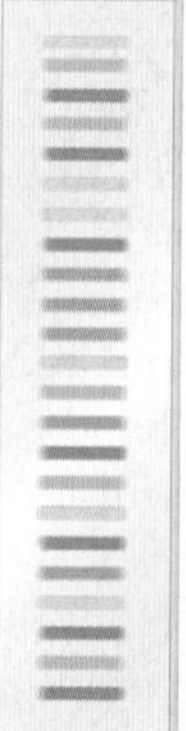

S4. The human genetic disease phenylketonuria (PKU) involves a defect in a gene that encodes the enzyme phenylalanine hydroxylase. It is inherited as a recessive autosomal disorder. Using a strand of DNA from the nonmutant phenylalanine hydroxylase gene as a probe, a Southern blot was carried out on a PKU patient, one of her parents, and an unaffected, unrelated person. In this example, the DNA fragments were subjected to acrylamide gel electrophoresis, rather than agarose gel electrophoresis, because acrylamide gel electrophoresis is better able to detect small deletions within genes. The following results were obtained:

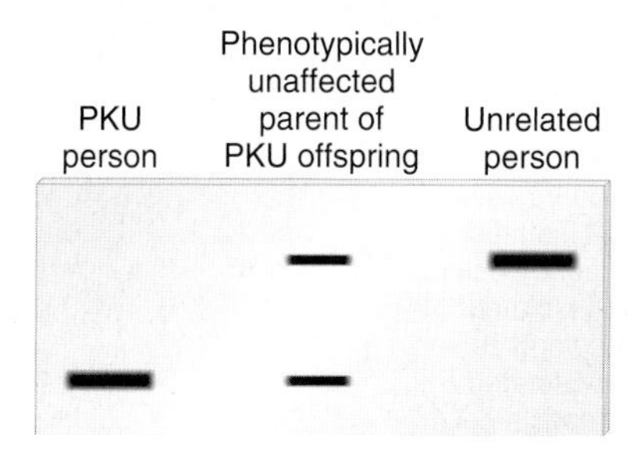

Suggest an explanation for these results.

Answer: In the affected person, the PKU defect is caused by a small deletion within the PKU gene. The unaffected parent is heterozygous for the nonmutant gene and the deletion. The PKU-affected person carries only the deletion, which runs at a lower molecular mass than the nonmutant gene. The unrelated person carries two copies of the nonmutant (nondeleted) gene.

Conceptual Questions

C1. Discuss three important advances that have resulted from gene cloning.

C2. What is a restriction enzyme? What structure does it recognize? What type of chemical bond does it cleave? Be as specific as possible.

C3. Write a double-stranded sequence that is 20 bases long and is palindromic.

C4. What is cDNA? In eukaryotes, how does cDNA differ from genomic DNA?

C5. Explain and draw the structural feature of a dideoxyribonucleotide that causes chain termination.

Experimental Questions

E1. What is the functional significance of sticky ends in a cloning experiment? What type of bonding makes the ends sticky?

E2. Table 18.3 describes the cleavage sites of six different restriction enzymes. After these restriction enzymes have cleaved the DNA, five of them produce sticky ends that can hydrogen bond with complementary sticky ends, as shown in Figure 18.1. The efficiency of sticky ends binding together depends on the number of hydrogen bonds; more hydrogen bonds makes the ends "stickier" and more likely to stay attached. Rank these five restriction enzymes in Table 18.3 (from best to worst) with regard to the efficiency of their sticky ends binding to each other.

E3. Describe the important features of cloning vectors. Explain the purpose of selectable marker genes in cloning experiments.

E4. How does gene cloning produce many copies of a gene?

E5. In your own words, describe the series of steps necessary to clone a gene. Your answer should include the use of a probe to identify a bacterial colony that contains the cloned gene of interest.

E6. What is a recombinant vector? How is a recombinant vector constructed? Explain how X-Gal can be used in a method of identifying recombinant vectors that contain segments of chromosomal DNA.

E7. A circular plasmid was digested with one or more restriction enzymes, run on a gel, and the following results were obtained:

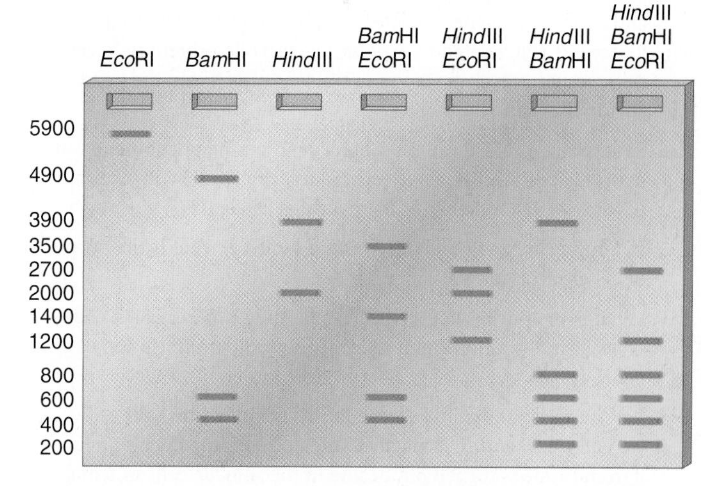

Construct a restriction map for this plasmid.

E8. If a researcher began with a sample that contained three copies of double-stranded DNA, how many copies would be present after 27 cycles of PCR, assuming 100% efficiency?

E9. Why is a thermostable form of DNA polymerase (e.g., *Taq* polymerase) used in PCR? Is it necessary to use a thermostable form of DNA polymerase in the techniques of dideoxy DNA sequencing or site-directed mutagenesis?

E10. Reverse transcriptase is an enzyme that uses RNA as template to make a complementary strand of DNA. Experimentally, it is used to make cDNA. Reverse transcriptase can also be used in conjunction with PCR to amplify RNAs. This method is called reverse transcriptase PCR. In other words, it is possible to make many copies of double-stranded DNA using RNA as a template. Starting with a sample of RNA that contained the mRNA for the β-globin gene, explain how you could create many copies of the β-globin cDNA using reverse transcriptase PCR.

E11. What type of detector is used for real-time PCR? Explain how the level of fluorescence correlates with the level of PCR product.

E12. What phase of PCR (exponential, linear, or stationary) is analyzed to quantitate the amount of DNA or RNA in a sample? Explain why this phase is chosen.

E13. Let's suppose you have recently cloned a gene, which we will call gene *X*, from corn. You use a labeled DNA strand from this cloned gene to probe genomic DNA from corn in a Southern blot experiment under conditions of low and high stringency. The following results were obtained:

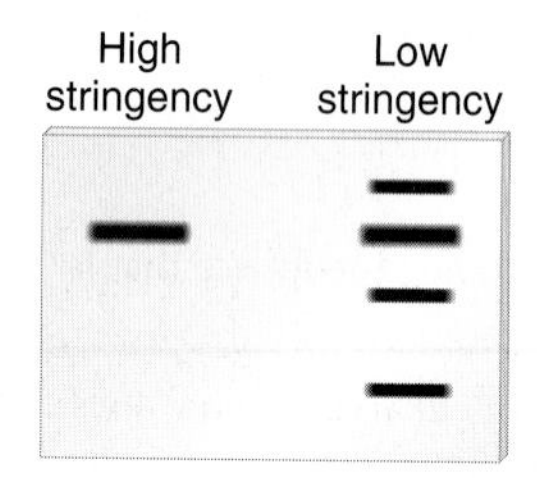

What do these results mean?

E14. What is a DNA library? Do you think this is an appropriate name?

E15. Some vectors used in cloning experiments contain bacterial promoters that are adjacent to unique cloning sites. This makes it possible to insert a gene sequence next to the bacterial promoter and express the gene in bacterial cells. These are called expression vectors. If you wanted to express a eukaryotic protein in bacterial cells, would you clone genomic DNA or cDNA into the expression vector? Explain your choice.

E16. Southern and Northern blotting depend on the phenomenon of hybridization. In these two techniques, explain why hybridization occurs. Which member of the hybrid is labeled?

E17. In Southern, Northern, and Western blotting, what is the purpose of gel electrophoresis?

E18. What is the purpose of a Northern blotting experiment? What types of information can it tell you about the transcription of a gene?

E19. Let's suppose an X-linked gene in mice exists as two alleles, which we will call *B* and *b*. X inactivation, a process in which one X chromosome is turned off, occurs in the somatic cells of female mammals (see Chapter 5). Allele *B* encodes an mRNA that is 900 nucleotides long, whereas allele *b* contains a small deletion that shortens the mRNA to a length of 825 nucleotides. Draw the expected results of a Northern blot using mRNA isolated from somatic tissue of the following mice:

Lane 1. mRNA from an X^bY male mouse

Lane 2. mRNA from an X^bX^b female mouse

Lane 3. mRNA from an X^BX^b female mouse. Note: The sample taken from the female mouse is not from a clone of cells. It is from a tissue sample, like the one shown at the beginning of the experiment of Figure 5.6.

E20. The method of Northern blotting can be used to determine the amount and size of a particular RNA transcribed in a given cell type. Alternative splicing (discussed in Chapter 15) can produce mRNAs from the same gene that have different lengths. A Northern blot is shown here using a DNA probe that is complementary to the mRNA encoded by a particular gene. The mRNA in lanes 1 through 4 was isolated from different cell types, and equal amounts of total cellular mRNA were added to each lane.

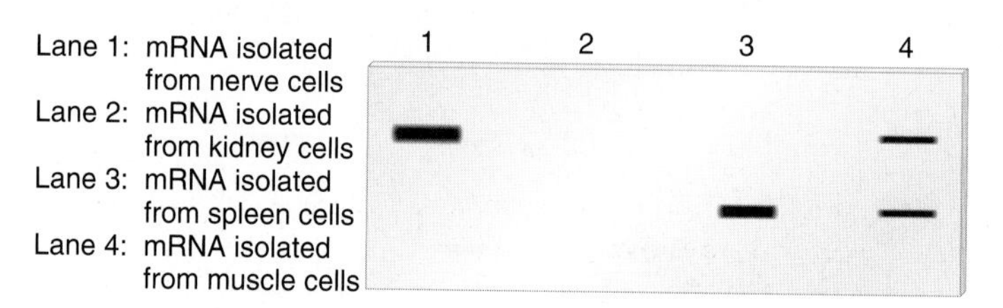

Explain these results.

E21. Southern blotting can be used to detect the presence of repetitive sequences, such as transposable elements, that are present in multiple copies within the chromosomal DNA of an organism. (Note: Transposable elements are described in Chapter 17.) In the Southern blot shown here, chromosomal DNA was isolated from three different strains of baker's yeast, digested with a restriction enzyme, run on a gel, blotted, and then probed with a radioactive DNA probe that is complementary to a transposable element called the *Ty* element.

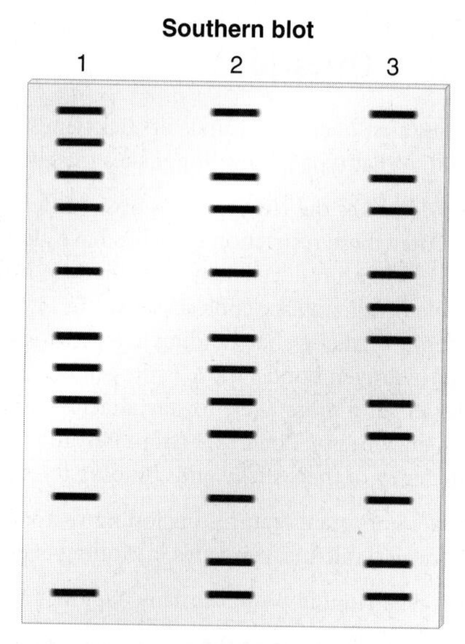

Explain, in a general way, why the banding patterns are not the same in lanes 1, 2, and 3.

E22. In Chapter 8, Figure 8.7 describes the evolution of the globin gene family. All of the genes in this family are homologous to each other, though the degree of sequence similarity varies depending on the time of divergence. Genes that have diverged more recently have sequences that are more similar. For example, the α_1 and α_2

genes have DNA sequences that are more similar to each other compared to the α_1 and ξ genes. In a Southern blotting experiment, the degree of sequence similarity can be discerned by varying the stringency of hybridization. At high temperature (i.e., high stringency), the probe recognizes genes that are only a perfect or very close match. At a lower temperature, however, homologous genes with lower degrees of similarities can be detected because slight mismatches are tolerated. If a Southern blot was conducted on a sample of human chromosomal DNA and a probe was used that was a perfect match to the β-globin gene, rank the following genes (from those that are detected at high stringency down to those that are only detected at low stringency) as they would appear in a Southern blot experiment: *Mb*, α_1, β, γ_A, δ, and ε.

E23. In the Western blot shown here, polypeptides were isolated from red blood cells and muscle cells from two different individuals. One individual was unaffected, and the other individual suffered from a disease known as thalassemia, which involves a defect in hemoglobin. In the Western blot, the gel blot was exposed to an antibody that recognizes β globin, which is one of the polypeptides that constitute hemoglobin. Equal amounts of total cellular proteins were added to each lane.

Lane 1: Proteins isolated from normal red blood cells
Lane 2: Proteins isolated from the red blood cells of a thalassemia patient
Lane 3: Proteins isolated from normal muscle cells
Lane 4: Proteins isolated from the muscle cells of a thalassemia patient

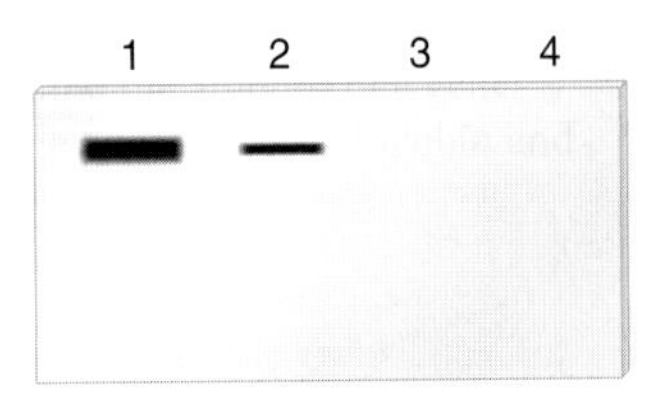

Explain these results.

E24. Let's suppose a researcher was interested in the effects of mutations on the expression of a structural gene that encodes a polypeptide 472 amino acids in length. This polypeptide is expressed in leaf cells of *Arabidopsis thaliana*. Because the average molecular mass of an amino acid is 120 Da, this protein has a molecular mass of approximately 56,640 Da. Make a drawing that shows the expected results of a Western blot using polypeptides isolated from the leaf cells that were obtained from the following individuals:

Lane 1. A plant homozygous for a nonmutant gene

Lane 2. A plant homozygous for a deletion that removes the promoter for this gene

Lane 3. A heterozygous plant in which one gene is nonmutant and the other gene has a mutation that introduces an early stop codon at codon 112

Lane 4. A plant homozygous for a mutation that introduces an early stop codon at codon 112

Lane 5. A plant homozygous for a mutation that changes codon 108 from a phenylalanine codon into a leucine codon

E25. If you wanted to know if a protein was made during a particular stage of development, what technique would you choose?

E26. Explain the basis for using an antibody as a probe in a Western blotting experiment.

E27. Starting with pig cells and a probe that is a labeled DNA strand from the human β-globin gene, describe how you would clone the β-globin gene from pigs. You may assume that you have available all of the materials needed in a cloning experiment. How would you confirm that a putative clone really contained the β-globin gene?

E28. A cloned gene fragment contains a response element that is recognized by a regulatory transcription factor. Previous experiments have shown that the presence of a hormone results in transcriptional activation by this transcription factor. To study this effect, you conduct a gel retardation assay and obtain the following results:

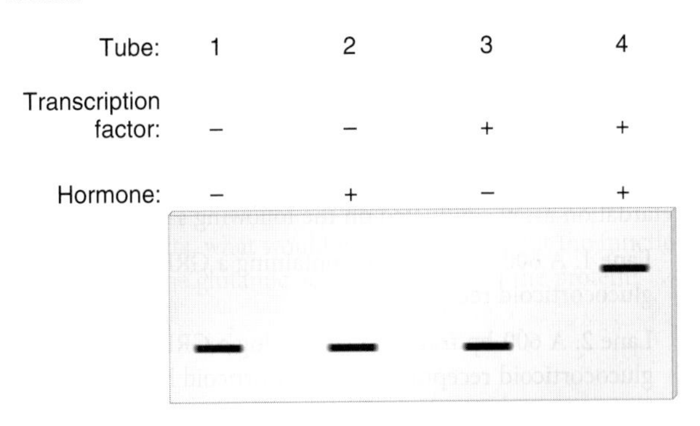

Explain the action of the hormone.

E29. Describe the rationale behind a gel retardation assay.

E30. Certain hormones, such as epinephrine, can increase the levels of cAMP within cells. Let's suppose you can pretreat cells with or without epinephrine and then prepare a cell extract that contains the CREB protein (see Chapter 15 for a description of the CREB protein). You then use a gel retardation assay to analyze the ability of the CREB protein to bind to a DNA fragment containing a cAMP response element (CRE). Describe what the expected results would be.

E31. A gel retardation assay can be used to study the binding of proteins to a segment of DNA. In the experiment shown here, a gel retardation assay was used to examine the requirements for the binding of RNA polymerase II (from eukaryotic cells) to the promoter of a structural gene. The assembly of general transcription factors and RNA polymerase II at the core promoter is described in Chapter 12 (Figure 12.14). In this experiment, the segment of DNA containing a promoter sequence was 1100 bp in length. The fragment was mixed with various combinations of proteins and then subjected to a gel retardation assay.

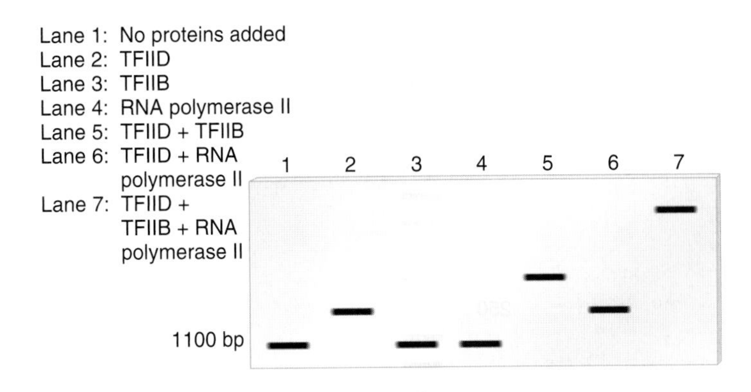

Explain which proteins (TFIID, TFIIB, or RNA polymerase II) are able to bind to this DNA fragment by themselves. Which transcription factors (i.e., TFIID or TFIIB) are needed for the binding of RNA polymerase II?

CHAPTER OUTLINE

The sheep named Dolly, which was cloned using genetic material from a somatic cell.

BIOTECHNOLOGY

Biotechnology is broadly defined as the application of technologies that involve the use of living organisms, or products from living organisms, for the development of products that benefit humans. Biotechnology is not a new topic. It began several thousand years ago when humans began to domesticate animals and plants for the production of food. Since that time, many species of microorganisms, animals, and plants have become routinely used by people. More recently, the term biotechnology has become associated with molecular genetics. Since the 1970s, molecular genetic tools have provided novel ways to make use of living organisms for products and services. As discussed in Chapter 18, recombinant DNA techniques can be used to genetically engineer microorganisms. In addition, recombinant methods enable the introduction of genetic material into animals and plants. **Genetically modified organisms (GMOs)** have received genetic material via recombinant DNA technology. If an organism has received genetic material from a different species, it is called a **transgenic organism.** A gene from one species that is introduced into another species is called a transgene.

In the 1980s, court rulings made it possible to patent recombinant organisms such as transgenic animals and plants. This was one factor that contributed to the growth of many biotechnology industries. In this chapter, we will examine how molecular techniques have expanded our knowledge of the genetic characteristics of commercially important species. We will also discuss examples in which recombinant microorganisms and transgenic animals and plants have been given characteristics that are useful in the treatment of disease or in agricultural production. These include recombinant bacteria that make human insulin, transgenic livestock that produce human proteins in their milk, and transgenic tomatoes with a longer shelf life. In addition, the topics of mammalian cloning and stem cell research are examined from a technical point of view. Likewise, the current and potential use of human gene therapy—the introduction of cloned genes into living cells in the treatment of a disease—will be addressed. In the process, we will also touch upon some of the ethical issues associated with these technologies.

19.1 USES OF MICROORGANISMS IN BIOTECHNOLOGY

Microorganisms are used to benefit humans in various ways (**Table 19.1**). In this section, we will examine how molecular genetic tools have become increasingly important for improving our use of microorganisms. Such tools can produce recombinant microorganisms with genes that have been manipulated in vitro. Why are recombinant organisms useful? Recombinant techniques can improve strains of microorganisms and have even yielded strains that make products not normally produced by microorganisms. For example, human genes have been introduced into

TABLE **19.1**

Common Uses of Microorganisms

Application	Examples
Production of medicines	Antibiotics
	Synthesis of human insulin in recombinant *E. coli*
Food fermentation	Cheese, yogurt, vinegar, wine, and beer
Biological control	Control of plant diseases, insect pests, and weeds
	Symbiotic nitrogen fixation
	Prevention of frost formation
Bioremediation	Cleanup of environmental pollutants such as petroleum hydrocarbons and synthetics that are difficult to degrade

bacteria to produce medically important products such as insulin and human growth hormone. As discussed in this section, several recombinant strains are in widespread use. However, in some areas of biotechnology and in some parts of the world, the commercialization of recombinant strains has proceeded very slowly. This is particularly true for applications in which recombinant microorganisms may be used to produce food products or where they are released into the environment. In such cases, safety and environmental concerns along with negative public perceptions have slowed or even halted the commercial use of recombinant microorganisms. Nevertheless, molecular genetic research continues, and many biotechnologists expect an expanding use of recombinant microbes in the future.

Many Important Medicines Are Produced by Recombinant Microorganisms

During the 1970s, geneticists became aware of the great potential of recombinant DNA technology to produce therapeutic agents for treating certain human diseases. Healthy individuals possess many different genes that encode short peptide and longer polypeptide hormones. Diseases can result when an individual is unable to produce these hormones.

In 1976, Robert Swanson and Herbert Boyer formed Genentech Inc. The aspiration of this company was to engineer bacteria to synthesize useful products, particularly peptide and polypeptide hormones. Their first contract was with researchers Keiichi Itakura and Arthur Riggs. They were able to engineer a bacterial strain that produced somatostatin, a human hormone that inhibits the secretion of a number of other hormones, including growth hormone, insulin, and glucagon. Somatostatin was not chosen for its commercial potential. Instead, it was chosen because the researchers thought it would be technically less difficult to produce than other hormones. Somatostatin is very small (only 14 amino acids long), which requires a short coding sequence, and it can be detected easily. Since this pioneering work, recombinant DNA technology has been used to develop bacterial strains that synthesize several other medical agents, a few of which are described in **Table 19.2**.

TABLE **19.2**

Examples of Medical Agents Produced by Recombinant Microorganisms

Drug	Action	Treatment
Insulin	A hormone that promotes glucose uptake	For diabetes
Tissue plasminogen activator (TPA)	Dissolves blood clots	For heart attacks and other arterial occlusions
Superoxide dismutase	Antioxidant	For heart attacks and to minimize tissue damage
Factor VIII	Blood-clotting factor	For certain types of hemophilias
Renin inhibitor	Lowers blood pressure	For hypertension
Erythropoietin	Stimulates the production of red blood cells	For anemia

In 1982, the U.S. Food and Drug Administration approved the sale of the first genetically engineered drug, human insulin, which was produced by Genentech and marketed by Eli Lilly. In nondiabetic individuals, insulin is produced by the β cells of the pancreas. Insulin functions to regulate several physiological processes, particularly the uptake of glucose into fat and muscle cells. Persons with insulin-dependent diabetes cannot synthesize an adequate amount of insulin due to a loss of their β cells. Prior to 1982, insulin was isolated from the pancreases removed from cattle and pigs. Unfortunately, in some cases, diabetic individuals became allergic to such insulin and had to use expensive combinations of insulin from other animals and human cadavers. Today, people with diabetes can use genetically engineered human insulin to treat their disease.

Insulin is a hormone composed of two polypeptide chains, called the A and B chains. To make this hormone using bacteria, the coding sequences of the A and B chains are placed next to the coding sequence of a native *E. coli* protein, β-galactosidase (**Figure 19.1**). This creates a fusion protein comprising β-galactosidase and the A or B chain. This step is necessary because the A and B chains are rapidly degraded when expressed in bacterial cells by themselves. The fusion proteins, however, are not. How are the two fusion proteins used to make human insulin? After the fusion proteins are expressed in bacteria, they can be purified and then treated with cyanogen bromide (CNBr), which cleaves after a methionine that is found at the junction between β-galactosidase and the A or B chain. This cleavage step separates β-galactosidase from the A or B chain. The A and B chains are then purified and mixed together under conditions in which they refold and form disulfide bonds with each other to make an active insulin hormone.

Bacterial Species Can Be Used as Biological Control Agents

The term **biological control** refers to the use of living organisms or their products to alleviate plant diseases or damage from

The gene of interest has been cloned. A neomycin resistance gene is inserted into the center of this gene, and a thymidine kinase gene is inserted next to the gene.

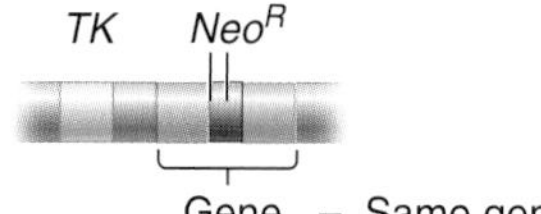

This cloned DNA is then introduced into embryonic stem cells. In this case, the cells were derived from a mouse with dark fur color. The cells are grown in the presence of neomycin and gancyclovir. Only those cells that contain the *Neo^R* gene but are lacking the *TK* gene will survive.

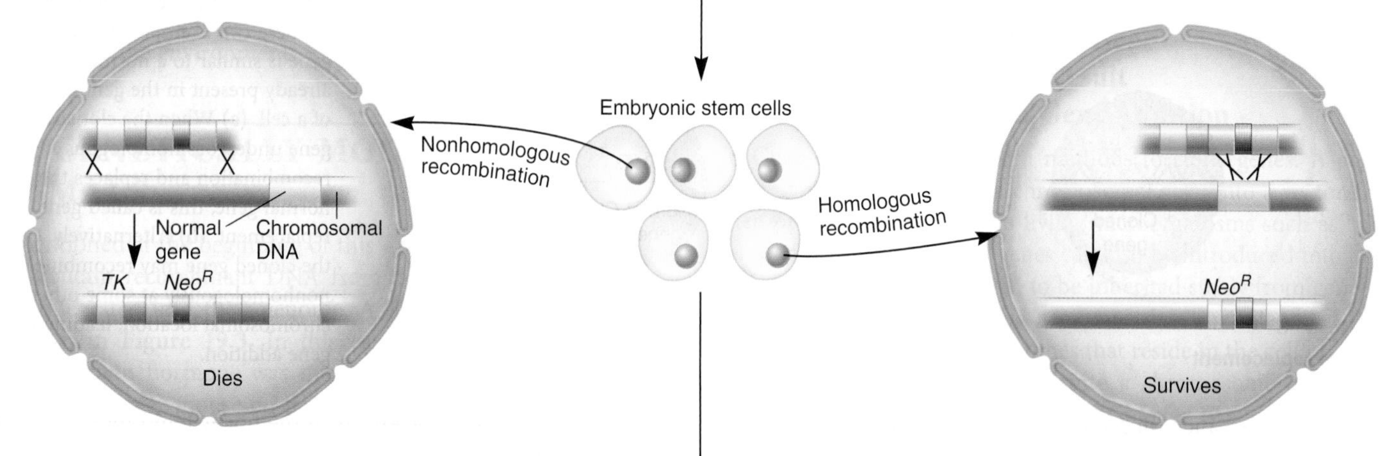

Surviving cells are injected into embryonic blastocysts derived from a mouse with white coat color. The injected blastocysts are reimplanted into the uterus of a female mouse.

Blastocyst

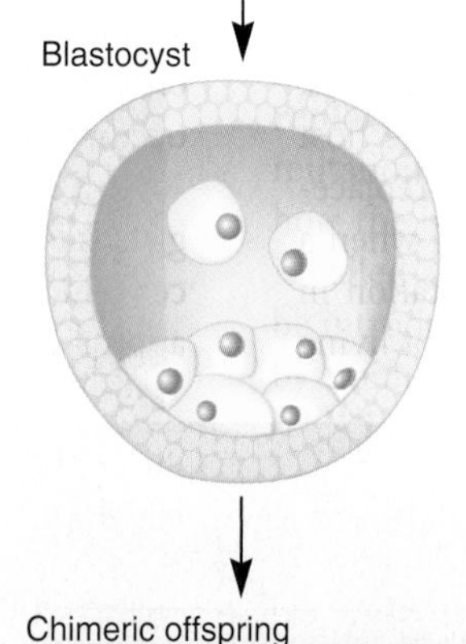

Following birth, chimeric mice are identified as those that contain a coat with both dark and white fur. The appropriate crosses are made in order to produce mice that have two copies of the target gene.

Chimeric offspring

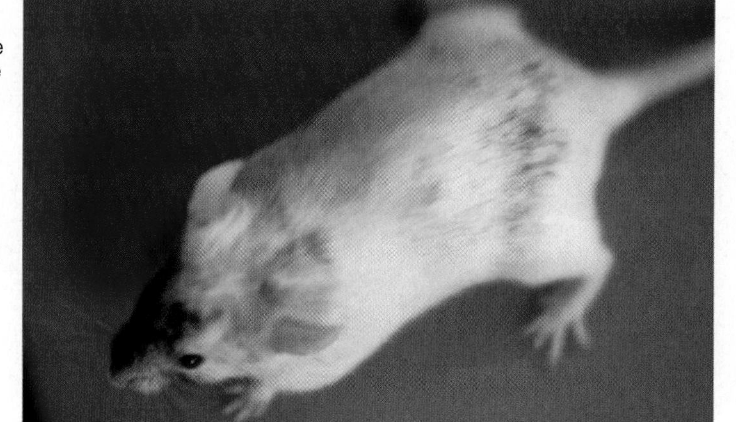

FIGURE 19.6 Producing a gene replacement in mice. The bottom of this figure shows a photograph of a chimeric mouse. Note the patches of black and white fur.

in a mouse chromosome and is cloned so that it can be manipulated in vitro. The cloned gene is shown at the top of Figure 19.6. The cloning procedure involves two selectable marker genes that influence whether or not mouse embryonic cells can grow in the presence of certain drugs. First, the gene of interest is inactivated by inserting a neomycin-resistance gene (called *Neo*^{R}) into the center of its coding sequence. *Neo*^{R} provides cells with resistance to neomycin. Next, a thymidine kinase gene, designated *TK*, is inserted adjacent to the gene of interest but not within the gene itself. The *TK* gene renders cells sensitive to killing by a drug called gancyclovir.

After the cloned gene has been modified in vitro, it is introduced into mouse embryonic stem cells. How do researchers identify the cells in which homologous recombination has occurred? When the cells are grown in the presence of neomycin and gancyclovir, most nonhomologous recombinants are killed because they also carry the *TK* gene. In contrast, homologous recombinants in which the normal gene has been partially replaced with the cloned gene contain only the *Neo*^{R} gene, so they are resistant to both drugs. The surviving embryonic cells can then be injected into blastocysts, early embryos that are obtained from a pregnant mouse. In the example shown in Figure 19.6, the embryonic cells are from a mouse with dark fur, and the blastocysts are from a mouse with white fur. The embryonic cells can mix with the blastocyst cells to create a **chimera,** an organism that contains cells from two different individuals. To identify chimeras, the injected blastocysts are reimplanted into the uterus of a female mouse and allowed to develop. When this mouse gives birth, chimeras are easily identified because they contain patches of white and dark fur (see Figure 19.6).

Chimeric animals that contain a single-gene replacement can then be mated to other mice to produce offspring that carry the mutant gene. Because mice are diploid, researchers must make two or more subsequent crosses to create a strain of mice that contains both copies of the mutant target gene. However, homozygous strains for a gene knockout cannot be produced if the mutant gene is lethal in the homozygous state.

An alternative method for producing genetically modified mice is to inject the desired gene into a fertilized egg. To conduct this type of experiment, researchers obtain mouse eggs and fertilize them in vitro. Immediately following fertilization, the cloned gene is injected into the sperm pronucleus—the haploid nucleus that has not yet fused with the egg nucleus. The cloned DNA then integrates into the genome, and the two pronuclei fuse to form the diploid nucleus of a zygote. The zygote begins to divide and is introduced into the uterus of a female mouse, where it becomes implanted and grows. As discussed next, genetically modified mice are used in basic research and to study human diseases.

Gene Knockouts and Knockins Are Produced in Mice to Understand Gene Function and Human Disease

As we have seen, researchers can replace a normal mouse gene with one that has been inactivated by the insertion of an antibiotic-resistance gene. As mentioned, when a mouse is homozygous for an inactivated gene, this condition is called a gene knockout. The inactive mutant gene has replaced both copies of the normal gene. In other words, the function of the normal gene has been "knocked out." By creating gene knockouts, researchers can study how the loss of normal gene function affects the organism. Gene knockouts frequently have specific effects on the phenotype of a mouse, which helps researchers to determine that the function of a gene is critical within a particular tissue or during a specific stage of development. In many cases, however, a gene knockout produces no obvious phenotypic effect. One explanation is that a single gene may make such a small contribution to an organism's phenotype that its loss may be difficult to detect. Alternatively, another possible explanation for a lack of observable phenotypic change in a knockout mouse may involve **gene redundancy.** This means that when one type of gene is inactivated, another gene with a similar function may be able to compensate for the inactive gene.

A particularly exciting avenue of gene knockout research is its application in the study and treatment of human disease. How is this useful? Knocking out the function of a gene may provide clues about what that gene normally does. Because humans share many genes with mice, observing the characteristics of knockout mice gives researchers information that can be used to better understand how a similar gene may cause or contribute to a disease in humans. Examples of research areas in which knockout mice have been useful include cancer, obesity, heart disease, diabetes, and many inherited disorders. The use of knockouts in the area of functional genomics is also discussed in Chapter 21.

In contrast to knockouts, researchers may introduce genes into the mouse genome to study the effects of gene overexpression or to examine the effects of particular alleles, such as those that may cause disease in humans. To accomplish this, researchers can produce a **gene knockin.** A gene knockin is a gene addition in which a gene of interest has been added to a particular site in the mouse genome (**Figure 19.7**). In this example, the cloned gene is inserted into the middle of a segment of DNA from a noncritical site in the mouse genome. The noncritical segment is very long, which allows the transgene to be targeted to that specific, noncritical integration site by homologous recombination after it is introduced into mouse cells. Such gene knockins tend to result in a more consistent level of expression of the transgene compared to gene additions that may occur randomly in another place in the genome. Also, because a targeted transgene is not interfering with a critical locus, the researcher can be more certain that any resulting phenotypic effect is due to the expression of the transgene.

To study human diseases, researchers have produced strains of transgenic mice that harbor both gene knockouts and gene knockins. A strain of mice engineered to carry a mutation that is analogous to a disease-causing mutation in a human gene is termed a mouse model. As an example, let's consider sickle cell disease, which is due to a mutation in the human β-globin gene (refer back to Figure 4.7). This gene encodes a polypeptide called β globin; adult hemoglobin is composed of both α-globin and β-globin polypeptides. When researchers produced a gene knockin by introducing the mutant human β-globin gene into mice, the resulting mice showed only mild symptoms of the disease. However, Chris Pászty and Edward Rubin produced a mouse model with multiple gene knockins

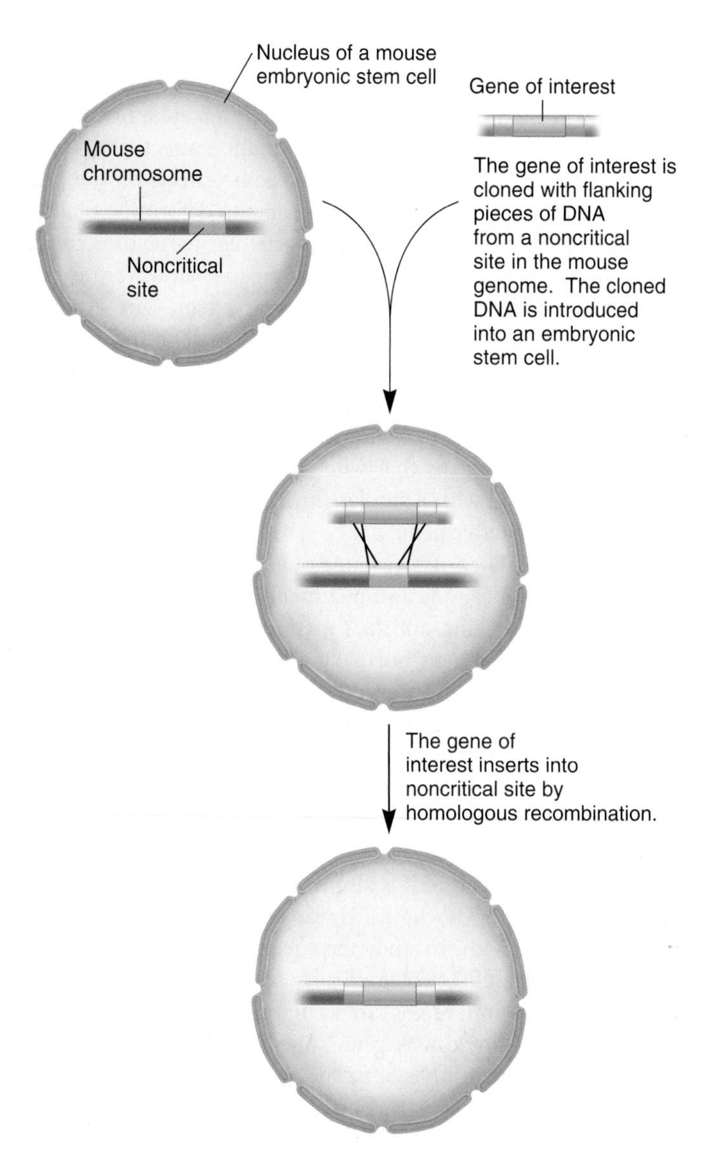

FIGURE 19.7 Producing a gene knockin in mice. This simplified diagram illustrates the strategy of producing a gene knockin. The gene of interest is cloned, and the cloned gene is flanked by DNA sequences that are homologous to a noncritical site in the mouse genome. When introduced into a mouse embryonic stem cell, the gene inserts into the genome by homologous recombination, which is an example of a targeted gene addition.

and gene knockouts. In particular, the mice had gene knockins for the normal human α-globin gene and the mutant β-globin gene from patients with sickle cell anemia. The strain also had gene knockouts of the mouse α-globin gene and β-globin gene:

normal human α-globin gene knockin;
mutant human β-globin gene knockin;
mouse α-globin gene knockout;
mouse β-globin gene knockout.

Therefore, these mice made adult hemoglobin just like people with sickle cell disease, but they did not produce any normal mouse hemoglobin. These transgenic mice exhibit the major features of sickle cell disease—sickled red blood cells, anemia, and multiorgan pathology. They have been useful as a model for studying the disease and testing potential therapies.

Biotechnology Holds Promise in Producing Transgenic Livestock

The technology for creating transgenic mice has been extended to other animals, and much research is under way to develop transgenic species of livestock, including fish, sheep, pigs, goats, and cattle. For some farmers, the ability to modify the characteristics of livestock via the introduction of cloned genes is an exciting prospect. In addition, work is currently under way to produce genetically modified pigs that are expected to be resistant to rejection mechanisms that occur following organ transplantation to humans. These strains may in time become a source of organs or cells for patients.

A novel avenue of research involves the production of medically important proteins in the mammary glands of livestock. This approach is sometimes called **molecular pharming.** (The term is also used to describe the manufacture of medical products by agricultural plants.) As shown in **Table 19.3**, several human proteins have been successfully produced in the milk of domestic livestock. Compared with the production of proteins in bacteria, one advantage is that certain proteins are more likely to function properly when expressed in mammals. This may be due to covalent modifications, such as the attachment of carbohydrate groups, which occur in eukaryotes but not in bacteria. In addition, certain proteins may be degraded rapidly or folded improperly when expressed in bacteria. Furthermore, the yield of recombinant proteins in milk can be quite large. Dairy cows, for example, produce about 10,000 liters of milk per year per cow. In some cases, a transgenic cow can produce approximately 1 g of the transgenic protein per liter of milk.

To introduce a human gene into an animal so the encoded protein will be secreted into its milk, the strategy is to insert the gene next to a milk-specific promoter. Eukaryotic genes often are expressed in a tissue-specific fashion. In mammals, certain

TABLE 19.3
Proteins That Can Be Produced in the Milk of Domestic Animals

Protein	Host	Use
Lactoferrin	Cattle	Used as an iron supplement in infant formula
Tissue plasminogen activator (TPA)	Goat	Dissolves blood clots
Antibodies	Cattle	Used to combat specific infectious diseases
α_1-Antitrypsin	Sheep	Treatment of emphysema
Factor IX	Sheep	Treatment of certain inherited forms of hemophilia
Insulin-like growth factor	Cattle	Treatment of diabetes

genes are expressed specifically within the mammary gland so their protein product is secreted into the milk. Examples of milk-specific genes include genes that encode milk proteins such as β-lactoglobulin, casein, and whey acidic protein. To express a human gene that encodes a protein hormone into a domestic animal's milk, the promoter for a milk-specific gene is linked to the coding sequence for the human gene (**Figure 19.8**). The DNA is then injected into an oocyte, where it is integrated into the genome. The fertilized oocyte is then implanted into the uterus of a female animal, which later gives birth to a transgenic offspring. If the offspring is a female, the protein hormone encoded by the human gene is expressed within the mammary gland and secreted into the milk. The milk can then be obtained from the animal, and the human hormone isolated.

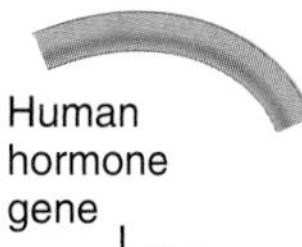

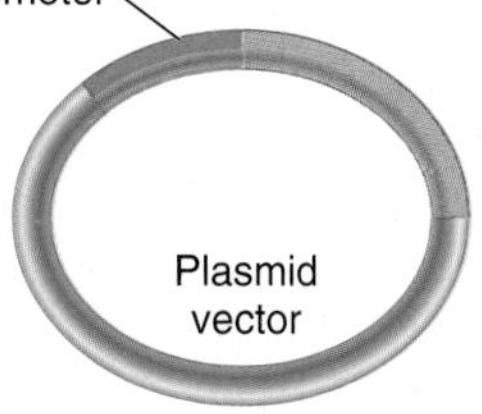

FIGURE 19.8 Strategy for expressing human genes in a domestic animal's milk. The β-lactoglobulin gene is normally expressed in mammary cells, whereas the human hormone gene is not. To express the human hormone gene in milk, the promoter from the milk-specific gene in sheep is linked to the coding sequence of the human hormone gene. In addition to the promoter, a short signal sequence may also be necessary so the protein is secreted from the mammary cells and into the milk.

Genes → Traits By using genetic engineering, researchers can give sheep the trait of producing a human hormone in their milk. This hormone can be purified from the milk and used to treat humans.

19.3 REPRODUCTIVE CLONING AND STEM CELLS

The previous section focused on the area of biotechnology in which cloned genes are introduced into animals. Another aspect of biotechnology involves the cloning of whole organisms or the manipulation of stem cells. In this section, we will consider mammalian cloning and stem cell research. These topics have received enormous public attention due to the complex ethical issues they raise.

Researchers Have Succeeded in Cloning Mammals from Somatic Cells

The term "cloning" has several different meanings. In Chapter 18, we discussed gene cloning, which involves methods that produce many copies of a gene. The cloning of an entire organism is a different matter. **Reproductive cloning** refers to methods that produce two or more genetically identical individuals. This happens occasionally in nature; identical twins are genetic clones that began from the same fertilized egg. Similarly, researchers can take mammalian embryos at an early stage of development (e.g., the two-cell to eight-cell stage), separate the cells, implant them into the uterus, and obtain multiple births of genetically identical individuals.

In the case of plants, cloning is an easier undertaking, as we will explore later in the chapter. Plants can be cloned from somatic cells. In most cases, it is relatively easy to take a cutting from a plant, expose it to growth hormones, and obtain a separate plant that is genetically identical to the original. However, this approach has not been possible with mammals. For several decades, scientists believed that chromosomes within the somatic cells of mammals had incurred irreversible genetic changes that render them unsuitable for cloning. However, this hypothesis has proven to be incorrect. In 1997, Ian Wilmut and his colleagues at the Roslin Institute in Scotland announced that a sheep, named Dolly, had been cloned using the genetic material from somatic cells.

How was Dolly produced? As shown in **Figure 19.9**, the researchers removed mammary cells from an adult female sheep

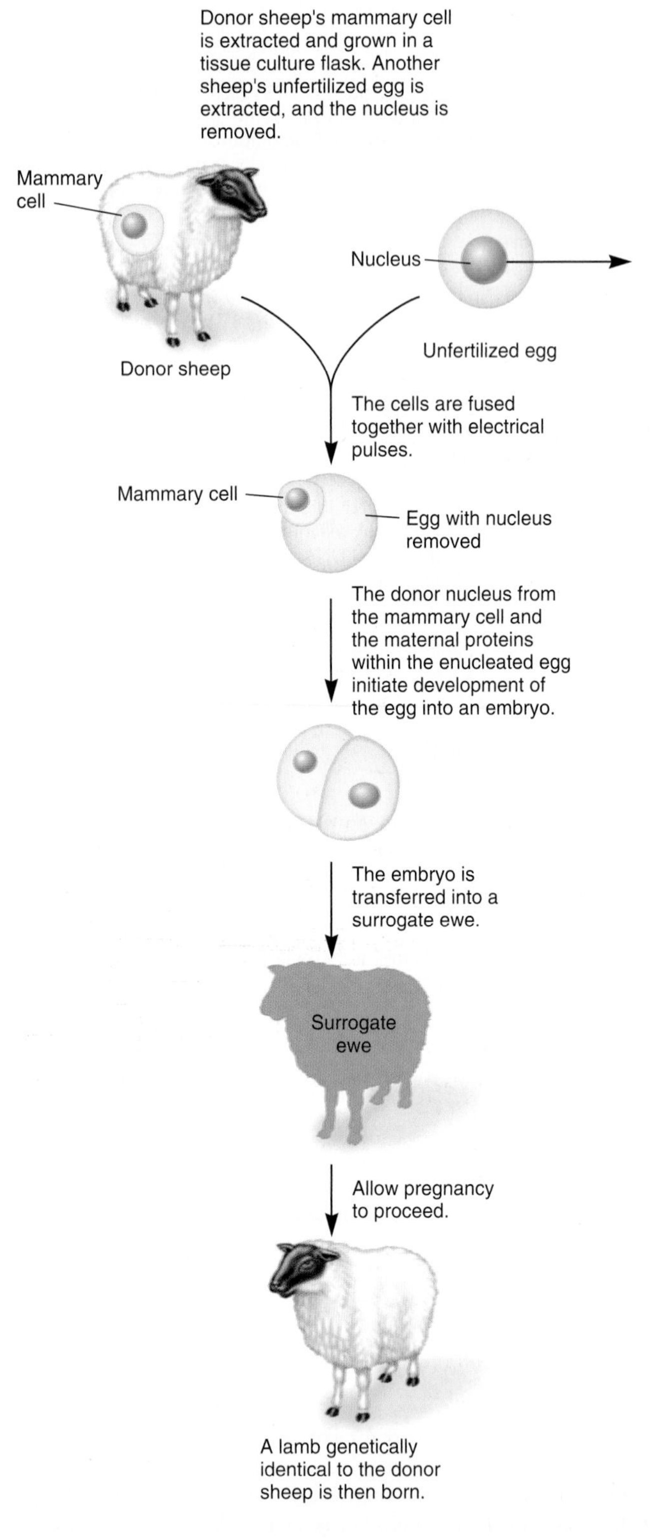

FIGURE 19.9 Protocol for the successful cloning of sheep.

Genes → Traits Dolly was (almost) genetically identical to the sheep that donated a mammary cell to create her. Dolly and the donor sheep were (almost) genetically identical in the same way that identical twins are; they carried the same set of genes and looked remarkably similar. However, they may have had minor genetic differences due to possible variation in their mitochondrial DNA and may have exhibited some phenotypic differences due to maternal effect or imprinted genes.

and grew them in the laboratory. The researchers then extracted the nucleus from an egg cell of a different sheep and used electrical pulses to fuse the diploid mammary cell with the enucleated egg cell. After fusion, the zygote began embryonic development, and the resulting embryo was implanted into the uterus of a surrogate mother sheep. One hundred and forty-eight days later, Dolly was born.

Although Dolly was clearly a clone of the initial adult female sheep, tests conducted when she was 3 years old suggested that she was "genetically older" than her actual age indicated. As mammals age, chromosomes in somatic cells tend to shorten from the telomeres—the ends of eukaryotic chromosomes. Therefore, older individuals have shorter chromosomes in their somatic cells than younger ones do. This shortening does not seem to occur in the cells of the germ line, however. When researchers analyzed the chromosomes in Dolly's somatic cells when she was about 3 years old, the lengths of her chromosomes were consistent with a sheep that was significantly older, say, 9 or 10 years old. The sheep that donated the somatic cell that produced Dolly was 6 years old, and her mammary cells had been grown in culture for several cell doublings before a mammary cell was fused with an oocyte. This led researchers to postulate that Dolly's shorter telomeres were a result of chromosome shortening in the somatic cells of the sheep that donated the nucleus. In 2003, the Roslin Institute announced the decision to euthanize 6-year-old Dolly after an examination showed progressive lung disease. Her death has raised concerns among experts that the techniques used to produce Dolly could have caused premature aging.

With regard to telomere length, research in mice and cattle has shown different results; the telomeres of these cloned animals appear to be the correct length. For example, cloning was conducted on mice via the method described in Figure 19.9 for six consecutive generations. The cloned mice of the sixth generation had normal telomeres. Further research is necessary to determine if cloning via somatic cells has an effect on the length of telomeres in subsequent generations. However, other studies in mice point to various types of genetic flaws in cloned animals. For example, Rudolf Jaenisch and his colleagues used DNA microarray technology (described in Chapter 21) to analyze the transcription patterns of over 10,000 genes in cloned mice. As much as 4% of those genes were not expressed normally. Furthermore, research has shown that cloned mice die at a younger age than their naturally bred counterparts.

Mammalian cloning is still at an early stage of development. Nevertheless, the breakthrough of creating Dolly has shown that it is technically possible. In recent years, cloning from somatic cells has been achieved in several mammalian species, including sheep, cattle, mice, goats, and pigs. In 2002, the first pet was cloned. She was named Carbon Copy, also called Copy Cat (**Figure 19.10**). Mammalian cloning may potentially have many practical applications. Cloning livestock would enable farmers to use the somatic cells from their best individuals to create genetically homogeneous herds, which could increase agricultural yield. However, such a genetically homogeneous herd may be more susceptible to rare diseases.

FIGURE 19.10 Carbon Copy, the first cloned pet. The animal shown here was produced using a procedure similar to the one shown in Figure 19.9.

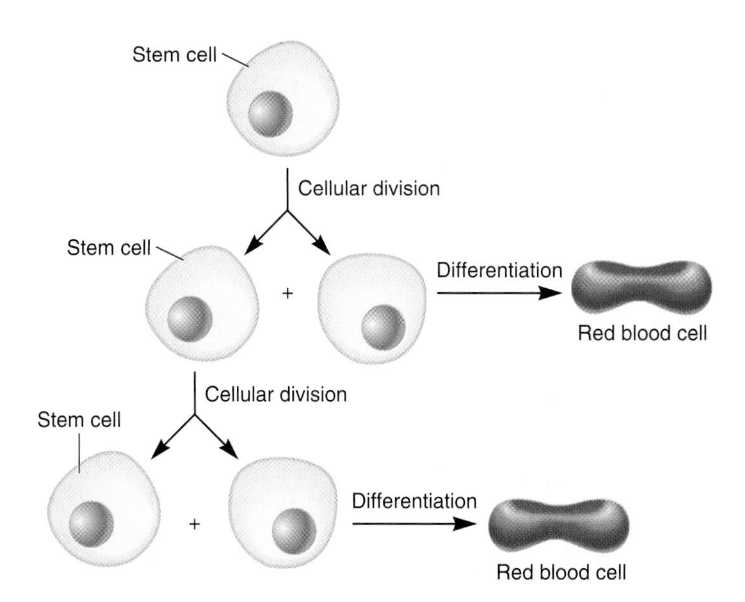

FIGURE 19.11 Growth pattern of stem cells. The two main traits that stem cells exhibit are an ability to divide and an ability to differentiate. When a stem cell divides, one of the two cells remains a stem cell, and the other daughter cell differentiates into a specialized cell type.

Though some people are concerned about the practical uses of cloning agricultural species, a majority have become very concerned with the possibility of human cloning. This prospect has raised a host of serious ethical questions. For example, some people feel that it is morally wrong and threatens the basic fabric of parenthood and family. Others feel that it is a technology that could offer a new avenue for reproduction, one that could be offered to infertile couples, for example. In the public sector, the sentiment toward human cloning has been generally negative. Many countries have issued an all-out ban on human cloning, but others permit limited research in this area. Because the technology for cloning exists, our society will continue to wrestle with the legal and ethical aspects of cloning as it applies not only to animals but also to people.

Stem Cells Have the Ability to Divide and Differentiate into Different Cell Types

Stem cells supply the cells that construct our bodies from a fertilized egg. In adults, stem cells also replenish worn-out or damaged cells. To accomplish this task, stem cells have two common characteristics. First, they have the capacity to divide, and second, they can differentiate into one or more specialized cell types. As shown in **Figure 19.11**, the two daughter cells produced from the division of a stem cell can have different fates. One of the cells may remain an undifferentiated stem cell, while the other daughter cell can differentiate into a specialized cell type. With this type of asymmetrical division/differentiation pattern, the population of stem cells remains relatively constant, yet the stem cells provide a population of specialized cells. In the adult, this type of mechanism is needed to replenish cells that have a finite life span, such as skin epithelial cells and red blood cells.

In mammals, stem cells are commonly categorized according to their developmental stage and their ability to differentiate (**Figure 19.12**). The ultimate stem cell is the fertilized egg, which, via multiple cellular divisions, can give rise to an entire organism. A fertilized egg is considered **totipotent,** because it can give rise to all the cell types in the adult organism. The early mammalian embryo contains **embryonic stem cells (ES cells),** which are found in the inner cell mass of the blastocyst. The blastocyst is the stage of embryonic development prior to uterine implantation—the preimplantation embryo. ES cells are **pluripotent,** which means they can differentiate into almost every cell type of the body. However, a single ES cell has lost the ability to produce an entire, intact individual.

During the early fetal stage of development, the germ-line cells found in the gonads also are pluripotent. These cells are called **embryonic germ cells (EG cells).** Interestingly, certain types of human cancers called teratocarcinomas arise from cells that are pluripotent. These bizarre tumors contain a variety of tissues including cartilage, neuroectoderm, muscle, bone, skin, ganglionic structures, and primitive glands. Due to the seemingly embryonic origin of teratocarcinoma cells, these cells are termed **embryonic carcinoma cells (EC cells).**

As mentioned, adults also contain stem cells, but these are thought to be multipotent or unipotent. A **multipotent** stem cell can differentiate into several cell types but far fewer than an ES cell. For example, hematopoietic stem cells (HSCs) found in the

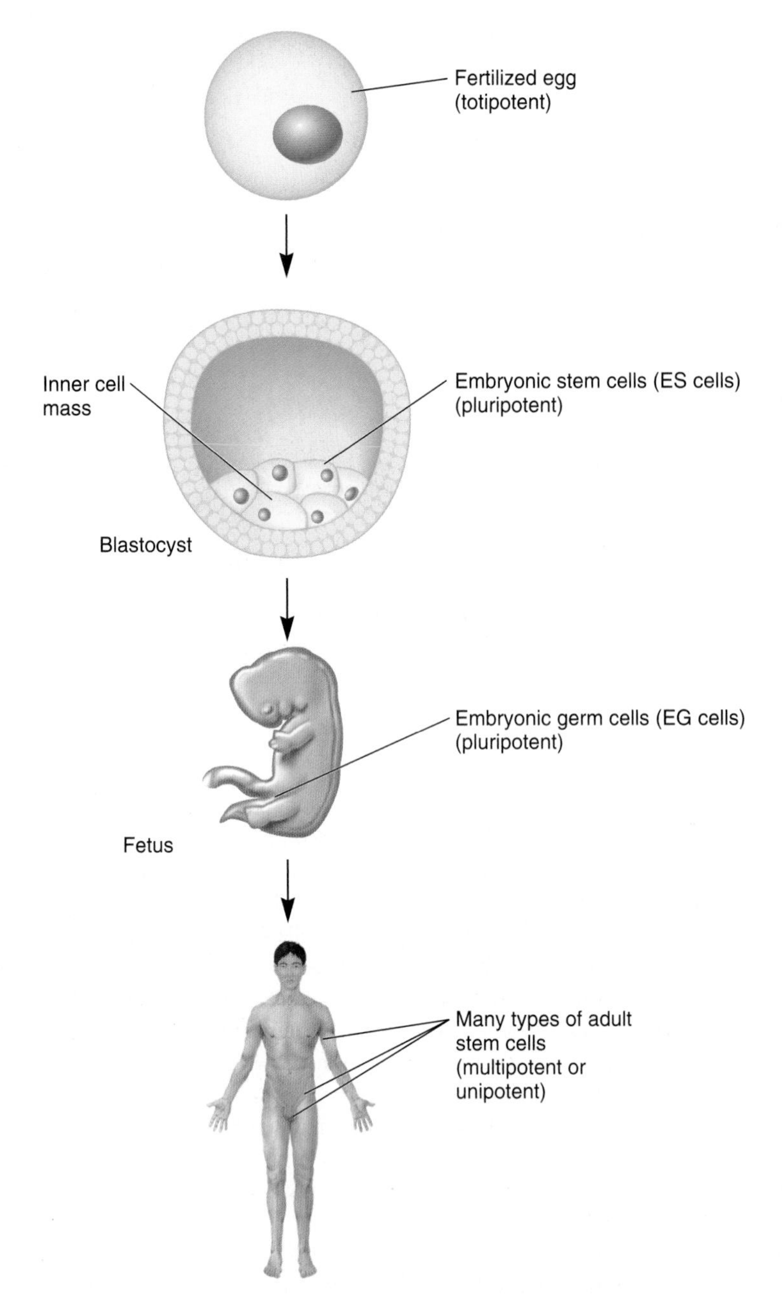

FIGURE 19.12 **Occurrence of stem cells at different stages of development.**

bone marrow can supply cells that populate two different tissues, namely, the blood and lymphoid tissues (**Figure 19.13**). Furthermore, each of these tissues contains several cell types. Multipotent HSCs can follow a pathway in which cell division produces a myeloid progenitor cell, which can then differentiate into a red blood cell, megakaryocyte, basophil, monocyte, eosinophil, neutrophil, or dendritic cell. Alternatively, an HSC can follow a path in which it becomes a lymphoid progenitor cell, which then differentiates into a T cell, B cell, natural killer cell, or dendritic cell. Other stem cells found in the adult seem to be **unipotent.** For example, primordial germ cells in the testis differentiate only into a single cell type, the sperm.

Stem Cells Have the Potential to Treat a Variety of Diseases

What are the potential uses of stem cells? Interest in stem cells centers on two main areas. Because stem cells have the capacity to differentiate into multiple cell types, the study of stem cells may help us to understand basic genetic mechanisms that underlie the process of development, the details of which are described in Chapter 23. A second compelling reason why people have become interested in stem cells is their potential to treat human diseases or injuries that cause cell and tissue damage. This application has already become a reality in certain cases. For example, bone marrow transplantation is used to treat patients with certain forms of cancers. Such patients may be given radiation treatments that destroy their immune systems. When these patients are injected with bone marrow from a healthy person, the stem cells within the transplanted marrow have the ability to proliferate and differentiate within their bodies and provide them with a functioning immune system.

Renewed interest in the use of stem cells in the potential treatment of many other diseases was fostered in 1998 by studies of two separate teams, headed by James Thomson and John Gearhart, showing that embryonic cells, either ES or EG cells, can be successfully propagated in the laboratory. As mentioned, ES and EG cells are pluripotent and therefore have the capacity to produce many different kinds of tissue. As shown in **Table 19.4**, embryonic cells could potentially be used to treat a wide variety of diseases associated with cell and tissue damage. By comparison, it would be difficult, based on our modern knowledge, to treat these diseases with adult stem cells because of the inability to locate most types of adult stem cells within the body and successfully grow them in the laboratory. Even HSCs are elusive. In the bone marrow, about 1 cell in 10,000 is a stem cell, yet that is enough to populate all of the blood and lymphoid cells of the body. The stem cells of most other adult tissues are equally difficult to locate, if not more so. In addition, with the exception of stem cells in the blood, other types of stem cells in the adult body are difficult to remove in sufficient numbers for transplantation. By comparison, ES and EG cells are easy to identify and have the great advantage of rapid growth in the laboratory. For these reasons, ES and EG cells offer a greater potential for transplantation, based on our current knowledge of stem cell biology.

For ES or EG cells to be used in transplantation, researchers need to derive methods that cause them to differentiate into the appropriate type of tissue. For example, if the goal was to repair a spinal cord injury, ES or EG cells would need the appropriate cues that cause them to differentiate into neural tissue. At present, much research is needed to understand and potentially control the fate of ES or EG cells. Currently, researchers speculate that a complex variety of factors determines the developmental fates of stem cells. These include internal factors within the stem cells themselves, as well as external factors such as the properties of neighboring cells and the presence of hormones and growth factors in the environment.

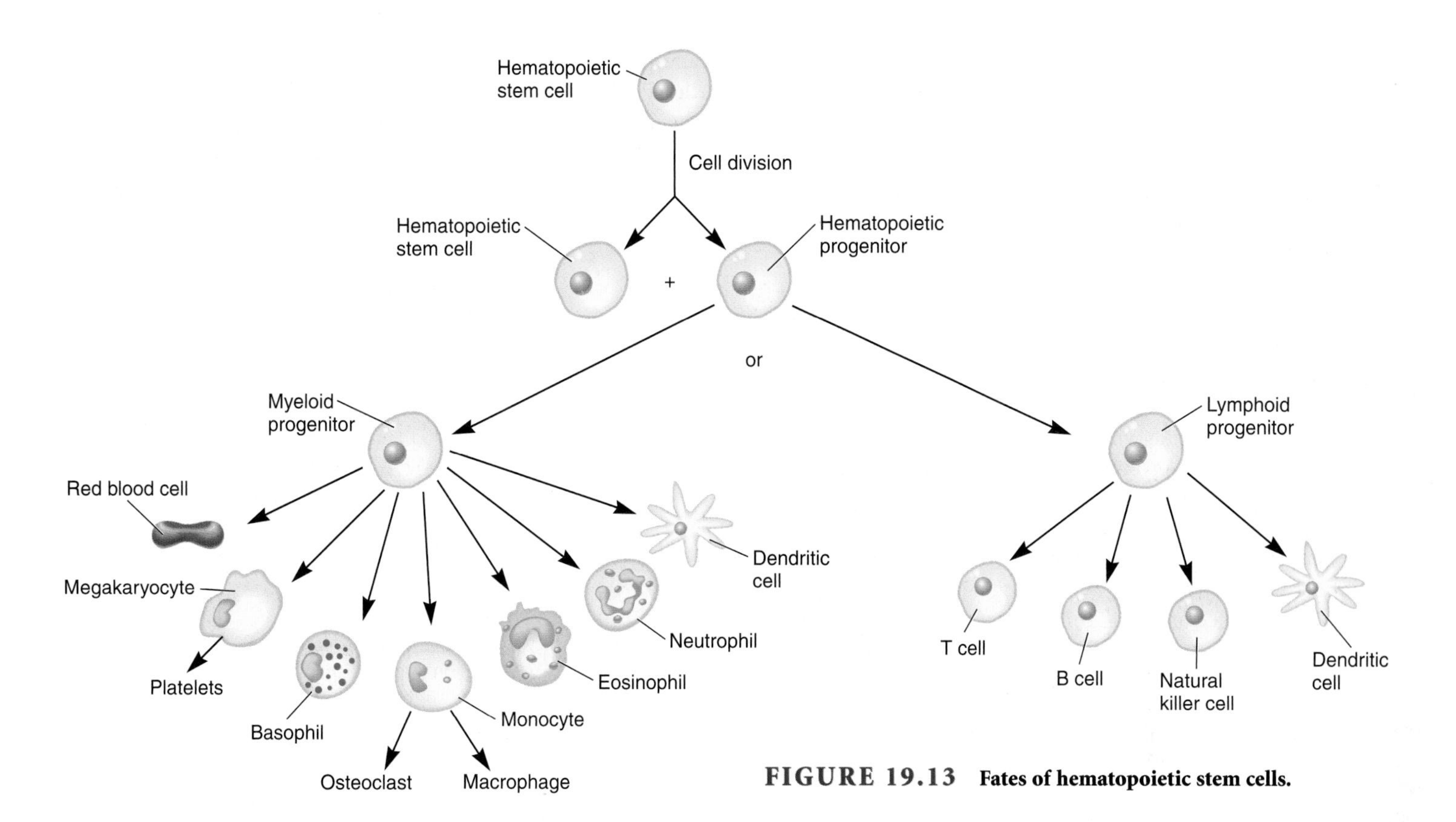

FIGURE 19.13 **Fates of hematopoietic stem cells.**

From an ethical perspective, the primary issue that raises debate is the source of the stem cells for research and potential treatments. Most ES cells have been derived from human embryos that were produced from in vitro fertilization and were subsequently not used. Most EG cells are obtained from aborted fetuses. Some feel that it is morally wrong to use such tissue in research and/or the treatment of disease or they fear that this technology could lead to intentional abortions for the sole purpose of obtaining fetal tissues for transplantation. Alternatively, others feel that the embryos and fetuses that provide the ES and EG cells are not going to become living individuals, and therefore, it is beneficial to study these cells and use them in a positive way to treat human diseases and injury. It is not clear whether these two opposing viewpoints can reach a common ground.

As a compromise, some governments have enacted laws that limit or prohibit the use of embryos or fetuses to obtain stem cells, yet permit the use of stem cell lines that are already available in research laboratories.

If stem cells could be obtained from adult cells and propagated in the laboratory, an ethical dilemma may be avoided because most people do not have serious moral objections to current procedures such as bone marrow transplantation. In 2006, work by Shinya Yamanaka and colleagues showed that adult mouse fibroblasts (a type of connective tissue cell) could become pluripotent via the injection of four different genes that encode transcription factors. In 2007, Yamanaka's laboratory and two other research groups showed that such induced pluripotent stem cells (iPS) can differentiate into all cell types when injected into mouse blastocysts and grown into baby mice. Though further research is still needed, these recent results indicate that adult cells can be reprogrammed to become embryonic stem cells.

TABLE **19.4**
Potential Uses of Stem Cells to Treat Diseases

Cell/Tissue Type	Disease Treatment
Neural	Implantation of cells into the brain to treat Parkinson disease
	Treatment of injuries such as spinal cord injuries
Skin	Treatment of burn victims and other types of skin disorders
Cardiac	Repair of heart damage associated with heart attacks
Cartilage	Repair of joints damaged by injury or arthritis
Bone	Repair of damaged bone or replacement with new bone
Liver	Repair or replacement of liver tissue that has been damaged by injury or disease
Skeletal muscle	Repair or replacement of damaged muscle

19.4 GENETICALLY MODIFIED PLANTS

As we have seen, researchers have succeeded in making genetically modified animals for a variety of reasons. In this section, we will examine the methods that scientists follow to make transgenic plants.

A large amount of research has been aimed at the use of transgenic species in agriculture. For centuries, agriculture has relied on selective breeding programs to produce plants and animals with desirable characteristics. For agriculturally important species, this often means the production of strains that are larger, are disease-resistant, and yield high-quality food. Agricultural scientists can now complement traditional breeding strategies with modern molecular genetic approaches. In the mid-1990s, genetically modified crops first became commercialized. Since that time, their use has progressively increased. In 2009, roughly 25% of all agricultural crops were transgenic. Worldwide, more than 100 million hectares (247 million acres) of transgenic crops were planted. In this section, we will discuss some current and potential uses of transgenic plants in agriculture.

Agrobacterium tumefaciens and Other Methods Can Be Used to Make Transgenic Plants

As we have seen, the introduction of cloned genes into embryonic cells can produce transgenic animals. The production of transgenic plants is somewhat easier, because some somatic cells are totipotent, which means they are capable of developing into an entire organism. Therefore, a transgenic plant can be made by the introduction of cloned genes into somatic tissue, such as the tissue of a leaf. After the cells of a leaf have become transgenic, an entire plant can be regenerated by the treatment of the leaf with plant growth hormones, which cause it to form roots and shoots.

Molecular biologists can use the bacterium *Agrobacterium tumefaciens*, which naturally infects plant cells, to produce transgenic plants. A plasmid from the bacterium, known as the Ti plasmid (Tumor-inducing plasmid), naturally induces tumor formation after a plant has been infected (**Figure 19.14a**). A segment of the plasmid DNA, known as **T DNA** (for transferred DNA), is transferred from the bacterium to the infected plant cells. The T DNA from the Ti plasmid is integrated into the chromosomal DNA of the plant cell by recombination. After this occurs, genes within the T DNA that encode plant growth hormones cause uncontrolled plant cell growth. This produces a cancerous plant growth known as a crown gall tumor (**Figure 19.14b**).

Because *A. tumefaciens* inserts its T DNA into the chromosomal DNA of plant cells, it can be used as a vector to introduce cloned genes into plants. Molecular geneticists have been able to modify the Ti plasmid to make this an efficient process. Such vectors are called **T-DNA vectors**. The T DNA genes that cause the development of a gall have been identified. Fortunately for genetic engineers, when these genes are deleted, the T DNA is still taken up into plant cells and integrated within the plant chromosomal DNA. However, a gall does not form. In addition, geneticists have inserted selectable marker genes into the T DNA to allow selection of plant cells that have taken up the T DNA. A gene that provides resistance to the antibiotic kanamycin (kan^R) is a commonly used selectable marker. The T-DNA vectors used in cloning experiments are also modified

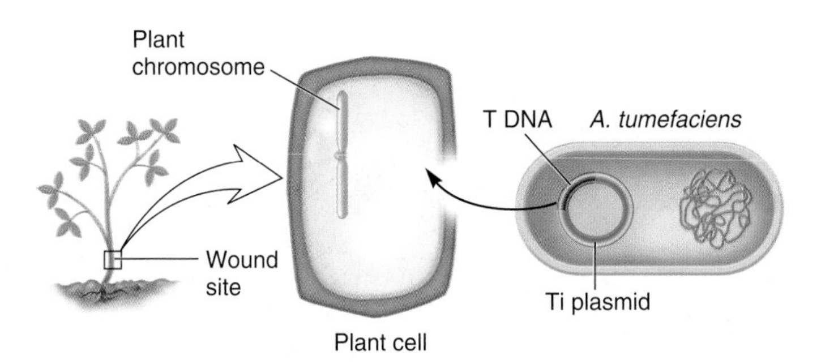

Agrobacterium tumefaciens is found within the soil. A wound on the plant enables the bacterium to infect the plant cells.

During infection, the T DNA within the Ti plasmid is transferred to the plant cell. The T DNA becomes integrated into the plant cell's DNA. Genes within the T DNA promote uncontrolled plant cell growth.

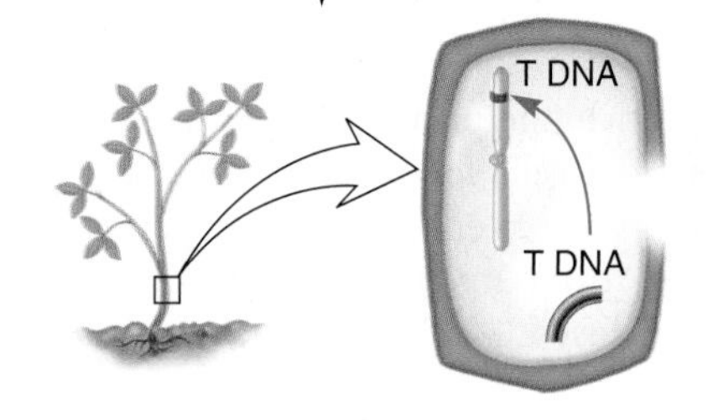

The growth of the recombinant plant cells produces a crown gall tumor.

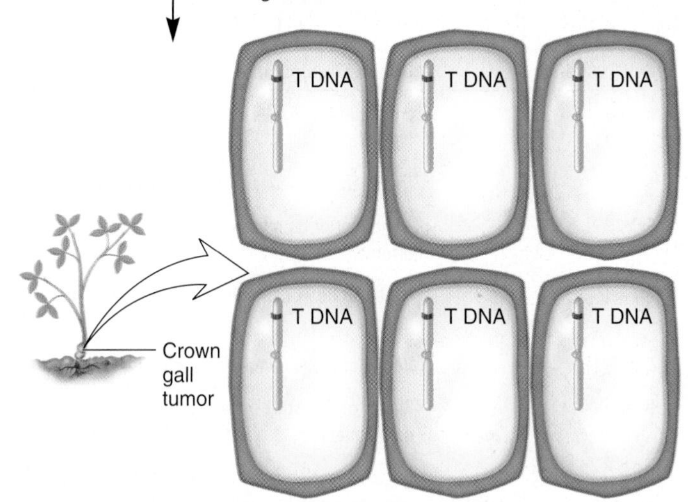

(a) The production of a crown gall tumor by *A. tumefaciens* infection

FIGURE 19.14 ***Agrobacterium tumefaciens*** **infecting a plant and causing a crown gall tumor.**

(b) A crown gall tumor on a pecan tree

FIGURE 19.14 *cont.*

to contain unique restriction sites for the convenient insertion of any gene.

Figure 19.15 shows the general strategy for producing transgenic plants via T DNA-mediated gene transfer. A gene of interest is inserted into a genetically engineered T-DNA vector and then transformed into *A. tumefaciens*. Plant cells are exposed to the transformed *A. tumefaciens*. After allowing time for infection, the plant cells are exposed to the antibiotics kanamycin and carbenicillin. Carbenicillin kills *A. tumefaciens*, and kanamycin kills any plant cells that have not taken up the T DNA with the antibiotic-resistance gene. Therefore, the only surviving cells are those plant cells that have integrated the T DNA into their genome. Because the T DNA also contains the cloned gene of interest, the selected plant cells are expected to have received this cloned gene as well. The cells are then transferred to a medium that contains the plant growth hormones necessary for the regeneration of entire plants. These plants can then be analyzed to verify that they are transgenic plants containing the cloned gene.

A. tumefaciens infects a wide range of plant species, including most dicotyledonous plants, most gymnosperms, and some monocotyledonous plants. However, not all plant species are infected by this bacterium. Fortunately, other methods are available for introducing genes into plant cells. Another common way to produce transgenic plants is an approach known as **biolistic gene transfer.** In this method, plant cells are bombarded with high-velocity microprojectiles coated with DNA. When fired upon by this "gene gun," the microprojectiles penetrate the cell wall and membrane, thereby entering the plant cell. The cells that take up the DNA are identified with a selectable marker and regenerated into new plants.

Other methods are also available for introducing DNA into plant cells (and also animal cells). For example, DNA can

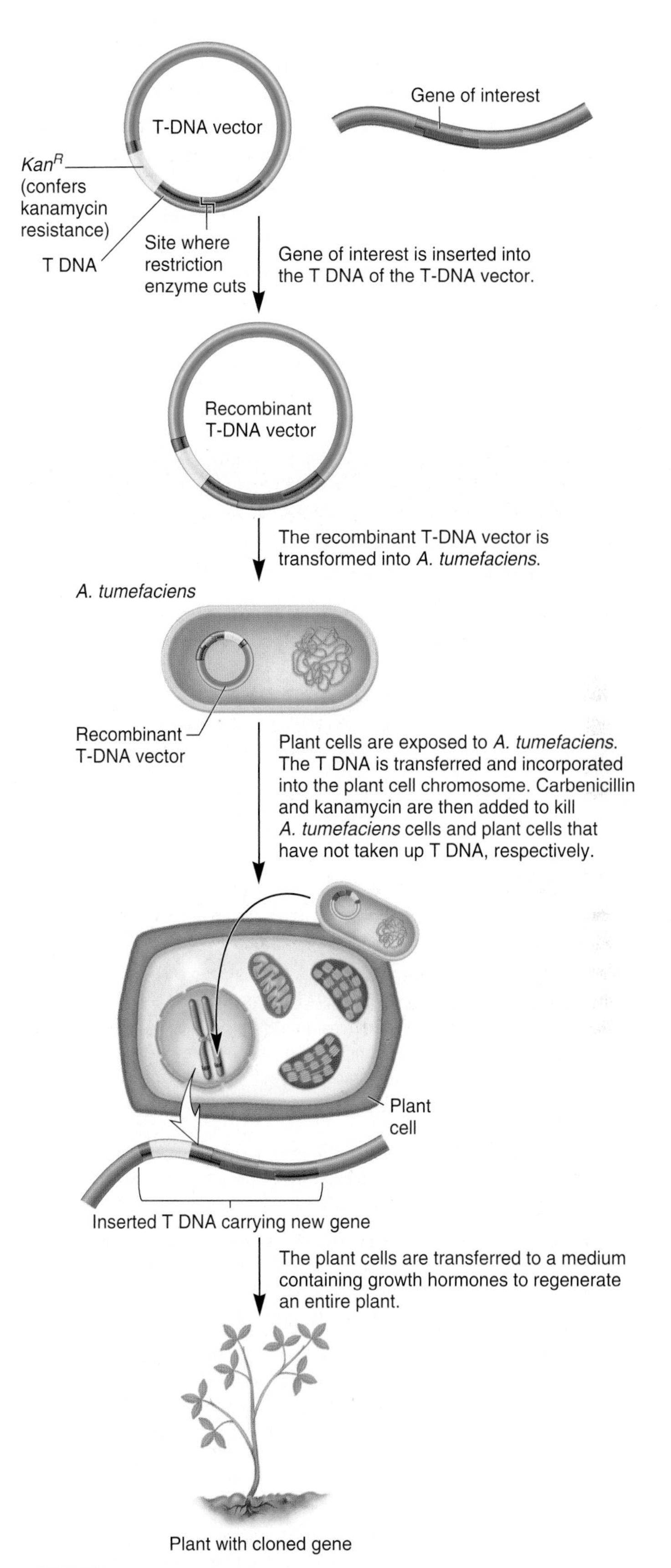

FIGURE 19.15 The transfer of genes into plants using a T-DNA vector from *A. tumefaciens*.

enter plant cells by **microinjection**—the use of microscopic-sized needles—or by **electroporation**—the use of electrical current to create temporary pores in the plasma membrane. Because the rigid plant cell wall is a difficult barrier for DNA entry, other approaches involve the use of protoplasts, which are plant cells that have had their cell walls removed. DNA can be introduced into protoplasts using a variety of methods, including treatment with polyethylene glycol and calcium phosphate.

The production of transgenic plants has been achieved for many agriculturally important plant species, including alfalfa, corn, cotton, soybean, tobacco, and tomato. Some of the applications of transgenic plants are described next.

TABLE **19.5**

Traits That Have Been Modified in Transgenic Plants

Trait	Examples
Plant Protection	
Resistance to viral, bacterial, and fungal pathogens	Transgenic plants that express the pokeweed antiviral protein are resistant to a variety of viral pathogens.
Resistance to insects	Transgenic plants that express the CryIA protein from *Bacillus thuringiensis* are resistant to a variety of insects (see Figure 19.18).
Resistance to herbicides	Transgenic plants can express proteins that render them resistant to particular herbicides (see Figure 19.16).
Plant Quality	
Improvement in storage	Transgenic plants can express antisense RNA that silences a gene involved in fruit softening (see Figure 19.17).
Change in plant composition	Transgenic strains of canola have been altered with regard to oil composition; the seeds of the Brazil nut have been rendered methionine-rich via transgenic technology.
New Products	
Biodegradable plastics	Transgenic plants have been made that can synthesize polyhydroxyalkanoates, which are used as biodegradable plastics.
Vaccines	Transgenic plants have been modified to produce vaccines in their leaves against many human and animal diseases, including hepatitis B, cholera, and malaria.
Pharmaceuticals	Transgenic plants have been made that produce a variety of medicines, including human interferon-α (to fight viral diseases and cancer), human epidermal growth factor (for wound repair), and human aprotinin (for reducing blood loss during transplantation surgery).
Antibodies	Human antibodies have been made in transgenic plants to battle various diseases such as non-Hodgkin lymphoma.

Transgenic Plants Can Be Given Characteristics That Are Agriculturally Useful

Various traits can be modified in transgenic plants (**Table 19.5**). Frequently, transgenic research has sought to produce plant strains resistant to insects, disease, and herbicides. For example, transgenic plants highly tolerant of particular herbicides have been made. The Monsanto Company has produced transgenic plant strains tolerant of glyphosate, the active agent in the herbicide Roundup. The herbicide remains effective against weeds, but the herbicide-resistant crop is spared (**Figure 19.16**).

Another important approach is to make plant strains that are disease-resistant. In many cases, virus-resistant plants have been developed by introducing a gene that encodes a viral coat protein. When the plant cells express the viral coat protein, they become resistant to infection by that pathogenic virus.

Many transgenic plants have been approved for human consumption. The first example was the Flavr Savr tomato (**Figure 19.17**), which was developed by Calgene Inc., which is now part of Monsanto. In this technique, a tomato plant was given a gene that encodes an antisense RNA complementary to the mRNA that encodes the enzyme polygalacturonase. This enzyme, which is expressed during ripening, digests sugar linkages within the pectin found in plant cell walls and thus softens the tomato. The antisense RNA binds to the polygalacturonase mRNA, preventing it from being translated. In addition, the double-stranded RNA is targeted for degradation (RNA interference), as discussed in Chapter 15 (Figure 15.24). Silencing the expression of polygalacturonase has the practical advantage of preventing the tomatoes from softening as quickly as unmodified tomatoes. Therefore, these tomatoes can be allowed to ripen on the vine

FIGURE 19.16 Transgenic plants that are resistant to glyphosate.

Genes → Traits This field of soybean plants has been treated with glyphosate. The plants on the left have been genetically engineered to contain a herbicide-resistance gene. They are resistant to killing by glyphosate. By comparison, the dead plants in the row with the orange stick do not contain this gene.

FIGURE 19.17 The genetically engineered Flavr Savr tomato.

Genes → Traits This transgenic tomato plant has been genetically engineered to contain an artificial gene that encodes an antisense RNA that binds to the mRNA that encodes polygalacturonase. This prevents the mRNA from being translated, which inhibits the synthesis of polygalacturonase. Without this enzyme, the Flavr Savr tomato can be left on the vine to ripen longer than traditional tomatoes, thereby improving its flavor.

for a longer period of time, enhancing their flavor and extending their shelf life. Improved taste is an important consideration in the $5-billion annual U.S. tomato market. By comparison, other commercial tomatoes commonly are picked when green and allowed to ripen later in order to maintain their firmness longer.

The Flavr Savr tomato was not a commercial success, however, and its sales were eventually discontinued. The failure of the Flavr Savr has been attributed to a variety of issues. In particular, the variety of tomato that was genetically engineered may have not been the best choice, and the antirotting trait was not as helpful for the tomato business as Calgene had anticipated. Another factor was that the ripe Flavr Savr tomatoes were more delicate and required the use of expensive handling equipment that wasn't needed for unmodified, green tomatoes.

A much more successful example of the use of transgenic plants has involved the introduction of genes from *Bacillus thuringiensis* (Bt). As discussed earlier, this bacterium produces toxins that kill certain types of caterpillars and beetles and has been widely used as a biological control agent for several decades. These toxins are proteins encoded in the genome of *B. thuringiensis*. Researchers have succeeded in cloning toxin genes from *B. thuringiensis* and transferring those genes into plants. Such Bt varieties of plants produce the toxins themselves and therefore are resistant to many types of caterpillars and beetles. Examples of commercialized crops include Bt corn (**Figure 19.18a**) and Bt cotton. Since their introduction in 1996, the commercial use of these two Bt crops has steadily increased (**Figure 19.18b**).

The introduction of transgenic plants into agriculture has been strongly opposed by some people. What are the perceived risks? One potential risk is that transgenes in commercial crops could endanger native species. For example, Bt crops may kill pollinators of native species. Another worry is that the planting of transgenic crops could potentially lead to the proliferation of resistant insects. To prevent this from happening, researchers are producing transgenic strains that carry more than one toxin gene, which makes it more difficult for insect resistance to arise. Despite these and other concerns, many farmers are embracing transgenic crops, and their use continues to rise.

(a) A field of Bt corn

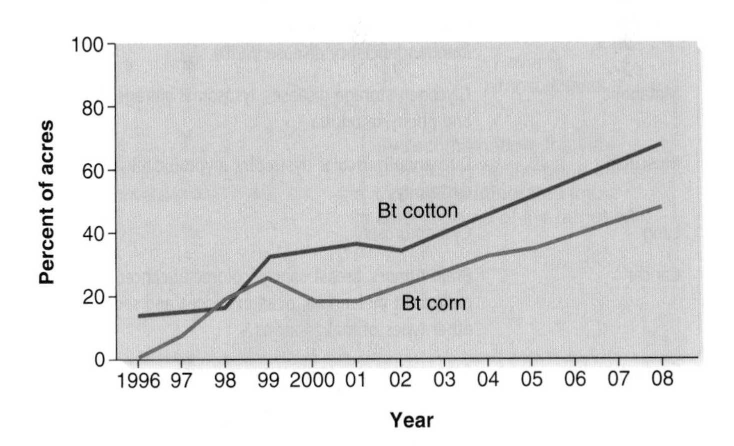

(b) Bt corn and Bt cotton usage since 1996

FIGURE 19.18 The production of Bt crops. (a) A field of Bt corn. These corn plants carry an endotoxin gene from *Bacillus thuringiensis* that provides them with resistance to insects such as corn borers, which are a major pest of corn plants. (b) A graph showing the increase in usage of Bt corn and Bt cotton in the United States since their commercial introduction in 1996.

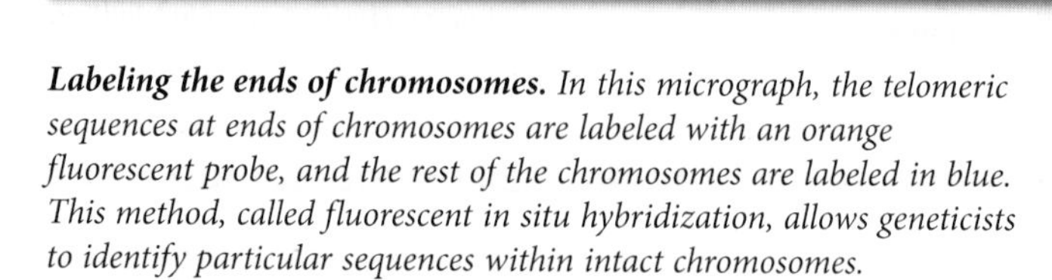

Labeling the ends of chromosomes. *In this micrograph, the telomeric sequences at ends of chromosomes are labeled with an orange fluorescent probe, and the rest of the chromosomes are labeled in blue. This method, called fluorescent in situ hybridization, allows geneticists to identify particular sequences within intact chromosomes.*

20 GENOMICS I: ANALYSIS OF DNA

The term **genome** refers to the total genetic composition of an organism or species. For example, the nuclear genome of humans is composed of 22 different autosomes and an X chromosome and (in males) a Y chromosome. In addition, humans have a mitochondrial genome composed of a single circular chromosome.

As genetic technology has progressed over the past few decades, researchers have gained an increasing ability to analyze the composition of genomes as a whole unit. The term **genomics** refers to the molecular analysis of the entire genome of a species. Genome analysis is a molecular dissection process applied to a complete set of chromosomes. Segments of chromosomes are analyzed in progressively smaller pieces, the locations of which are known on the intact chromosomes. This is the mapping phase of genome analysis. The mapping of the genome ultimately progresses to the determination of the complete DNA sequence, which provides the most detailed description of an organism's genome at the molecular level.

In 1995, a team of researchers headed by Craig Venter and Hamilton Smith obtained the first complete DNA sequence of the bacterial genome from *Haemophilus influenzae*. As discussed later in this chapter, its genome is composed of a single circular chromosome 1.83 million base pairs (bp) in length and contains approximately 1743 genes. In 1996, the first entire DNA sequence of a eukaryote, *Saccharomyces cerevisiae* (baker's yeast), was completed. This work was carried out by a European-led consortium of more than 100 laboratories, including some in the United States, Japan, and Canada. The overall coordinator was André Goffeau in Belgium. The yeast genome contains 16 linear chromosomes, which have a combined length of about 12.1 million bp and contain approximately 6300 genes. Since that time, genome sequences from many prokaryotes and eukaryotes have been completed.

In this chapter, we will focus on methods aimed at elucidating the organization of the sequences within a species' genome. This process may begin with the mapping of regions along a species' chromosomes. The process is finished when the complete DNA sequence has been determined. We will consider three mapping strategies—cytogenetic, linkage, and physical mapping—and the approaches used to carry them out. Then we explore genome-sequencing projects—research endeavors that have the ultimate goal of determining the sequence of DNA bases of the entire genome of a given species. We will examine the methods, goals, and results of these large undertakings, which include the Human Genome Project.

Once a genome sequence is known, researchers can examine, at the level of many genes, how the components of a genome interact to produce the traits of an organism. This approach is called **functional genomics.** Ultimately, a long-term goal of researchers is to determine the roles of all cellular proteins, as well as the interactions that these proteins experience, to produce the characteristics of particular cell types and the traits of complete organisms. This is the research area known as **proteomics.** In this chapter, we explore genomics at the level of DNA segments and sequences. In Chapter 21, we will consider the exciting prospects that functional genomics and proteomics have to offer regarding our current and future understanding of genetics.

20.1 OVERVIEW OF CHROMOSOME MAPPING

In genetics, the term **mapping** refers to the experimental process of determining the relative locations of genes or other segments of DNA along individual chromosomes. Researchers may follow three general approaches to mapping a chromosome: cytogenetic, linkage, and physical mapping strategies. Before we discuss these approaches in detail, let's compare them. **Cytogenetic mapping** (also called cytological mapping) relies on the localization of gene sequences within chromosomes that are viewed microscopically. When stained, each chromosome of a given species has a characteristic banding pattern, and genes are mapped cytogenetically relative to a band location. By comparison, in Chapter 6, we considered how genetic crosses are conducted to map the relative locations of genes within a chromosome. Such genetic studies, which are called **linkage mapping** or genetic mapping, use the frequency of genetic recombination between different genes to determine their relative spacing and order along a chromosome. In eukaryotes, linkage mapping involves crosses among organisms that are heterozygous for two or more genes. The number of recombinant offspring provides a relative measure of the distance between genes, which is computed in map units (or centiMorgans, cM). Finally, a third approach is **physical mapping** in which DNA-cloning techniques are used to determine the location of and distance between genes and other DNA regions. In a physical map, the distances are computed as the number of nucleotide base pairs between genes.

A **genetic map,** or **chromosome map,** is a chart that describes the relative locations of genes or other DNA segments along a chromosome. The term **locus** (plural, **loci**) refers to the site within a genetic map where a specific gene or other DNA segment is found. **Figure 20.1** compares genetic maps that show the loci for two X-linked genes, *sc* (scute, a gene affecting bristle morphology) and *w* (a gene affecting eye color), in *Drosophila melanogaster.* In the cytogenetic map (top), the *sc* gene is located at band 1A8, and the *w* gene is located at band 3B6. In the linkage map, genetic crosses indicate that the two genes are approximately 1.5 map units (mu) apart. The physical map shows that the two genes are approximately 2.4×10^6 bp apart along the X chromosome. Correlations between cytogenetic, linkage, and physical maps often vary from species to species and from one region of the chromosome to another. For example, a distance of 1 mu may correspond to 1 to 2 million bp in one region of the chromosome, but other regions may recombine at a much lower rate, so a distance of 1 mu may be a much longer physical segment of DNA.

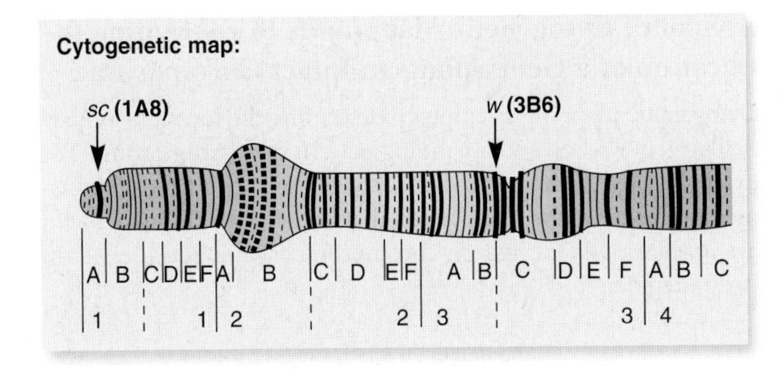

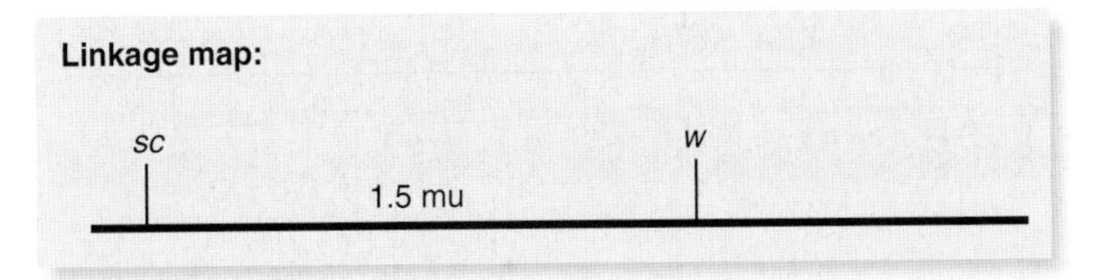

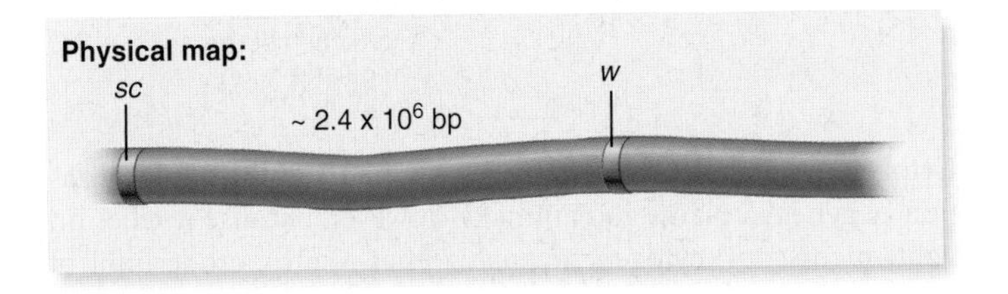

FIGURE 20.1 A comparison of cytogenetic, linkage, and physical maps. Each of these maps shows the distance between the *sc* and *w* genes along the X chromosome in *Drosophila melanogaster*. The cytogenetic map is that of the polytene chromosome.

20.2 CYTOGENETIC MAPPING VIA MICROSCOPY

As we have seen, one way to determine the relative locations of genes along a chromosome is via cytogenetic mapping. This approach relies on the localization of gene sequences within chromosomes that are viewed microscopically. Cytogenetic mapping is commonly used in eukaryotes, which have very large chromosomes relative to those of bacteria. Microscopically, eukaryotic chromosomes can be distinguished from one another by their size, centromeric location, and banding patterns (refer to Chapter 8, Figure 8.1). By treating chromosomal preparations with particular dyes, a discrete banding pattern is obtained for each chromosome. Cytogeneticists use this banding pattern as a way to describe specific regions along a chromosome. In this section, we will explore techniques that are aimed at producing cytogenetic maps.

A Goal of Cytogenetic Mapping Is to Determine the Location of a Gene Along an Intact Chromosome

Cytogenetic mapping attempts to determine the locations of particular genes relative to a banding pattern of a chromosome. For example, the human gene that encodes the cystic fibrosis transmembrane regulator, the protein that is defective in people with cystic fibrosis, is located on chromosome 7, at a specific site in the q3 region.

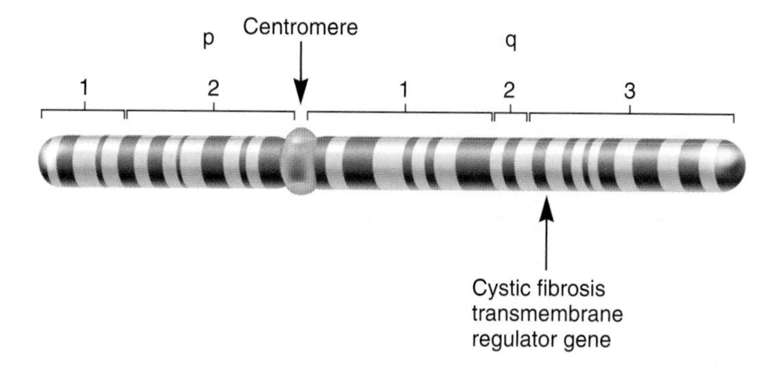

Cytogenetic mapping may be used as a first step in the localization of genes in plants and animals. However, because it relies on light microscopy, cytogenetic analysis has a fairly crude limit of resolution. In most species, cytogenetic mapping is accurate only within limits of approximately 5 million bp along a chromosome. In species that have large polytene chromosomes, such as *Drosophila*, the resolution is much better. A common strategy used by geneticists is to roughly locate a gene by cytogenetic analysis and then determine its location more precisely by the physical mapping methods described later.

In situ Hybridization Can Localize Genes Along Particular Chromosomes

The technique of **in situ hybridization** is widely used to cytogenetically map the locations of genes or other DNA sequences within large eukaryotic chromosomes. The term *in situ* (from the Latin for "in place") indicates that the procedure is conducted on chromosomes that are being held in place—adhered to a surface.

To map a gene via in situ hybridization, researchers use a probe to detect the location of the gene within a set of chromosomes. If the gene of interest has been cloned previously, as described in Chapter 18, the DNA of the cloned gene can be used as a probe. Because a DNA strand from a cloned gene, which is a very small piece of DNA relative to a chromosome, hybridizes only to its complementary sequence on a particular chromosome, this technique provides the ability to localize the gene of interest. For example, let's consider the gene that causes the white-eye phenotype in *Drosophila* when it carries a loss-of-function mutation. This gene has already been cloned. If a single-stranded piece of this cloned DNA is mixed with *Drosophila* chromosomes in which the DNA has been denatured, it will bind only to the X chromosome at the location corresponding to the site of the eye color gene.

The most common method of in situ hybridization uses fluorescently labeled DNA probes and is referred to as **fluorescence in situ hybridization (FISH).** **Figure 20.2** describes the steps of the FISH procedure. The cells are prepared using a technique that keeps the chromosomes intact. The cells are treated with agents that cause them to swell, and their contents are fixed to the slide. The chromosomal DNA is then denatured, and a DNA probe is added. For example, the added DNA probe might be single-stranded DNA that is complementary to a specific gene. In this case, the goal of a FISH experiment is to determine the location of the gene within a set of chromosomes. The probe binds to a site in the chromosomes where the gene is located because the probe and chromosomal gene line up and hydrogen bond with each other. To detect where the probe has bound to a chromosome, the probe is subsequently tagged with a fluorescent molecule. This is usually accomplished by first incorporating biotin-labeled nucleotides into the probe. Biotin, a small, nonfluorescent molecule, has a very high affinity for a protein called avidin. Fluorescently labeled avidin is added, which binds tightly to the biotin and thereby labels the probe as well.

How is the fluorescently labeled probe detected? A fluorescent molecule is one that absorbs light at a particular wavelength and then emits light at a longer wavelength. To detect the light emitted by a fluorescently labeled probe, a fluorescence microscope is used. Such a microscope contains filters that allow the passage of light only within a defined wavelength range. The sample is illuminated at the wavelength of light that is absorbed by the fluorescent molecule. The fluorescent molecule then emits light at a longer wavelength. The fluorescence microscope has a filter that allows the transmission of the emitted light. Because only the emitted light is viewed, the background of the sample is dark, and the fluorescence is seen as a brightly glowing color on a dark background. For most FISH experiments, chromosomes are generally counterstained by a fluorescent dye that is specific for DNA. A commonly used dye is DAPI (4′,6-diamidino-2-phenylindol) that is excited by UV light. This provides all of the DNA with a blue background. The results of a FISH experiment are then compared with a sample of chromosomes that have been stained with Giemsa to produce banding, so the location of a probe can be mapped relative to the banding pattern.

Figure 20.3 illustrates the results of an experiment involving six different DNA probes. The six probes were strands of DNA corresponding to six different DNA segments of human chromosome 5. In this experiment, each probe was labeled with a different fluorescent molecule. This enabled researchers to distinguish the probes when they became bound to their corresponding locations on chromosome 5. In this experiment, computer-imaging methods were used to assign each fluorescently labeled probe a different color. This method is called **chromosome painting.** In this way, FISH discerns the sites along chromosome 5 corresponding to the six different probes. In a visual, colorful way, FISH was used here to determine the order and relative distances between several specific sites along a single chromosome. FISH is commonly used in genetics and cell biology research, and its use has become more widespread in clinical applications. For example, clinicians may use FISH to detect changes in chromosome structure such as deletions, duplications, and translocations, which may occur in patients with genetic disorders (see book cover).

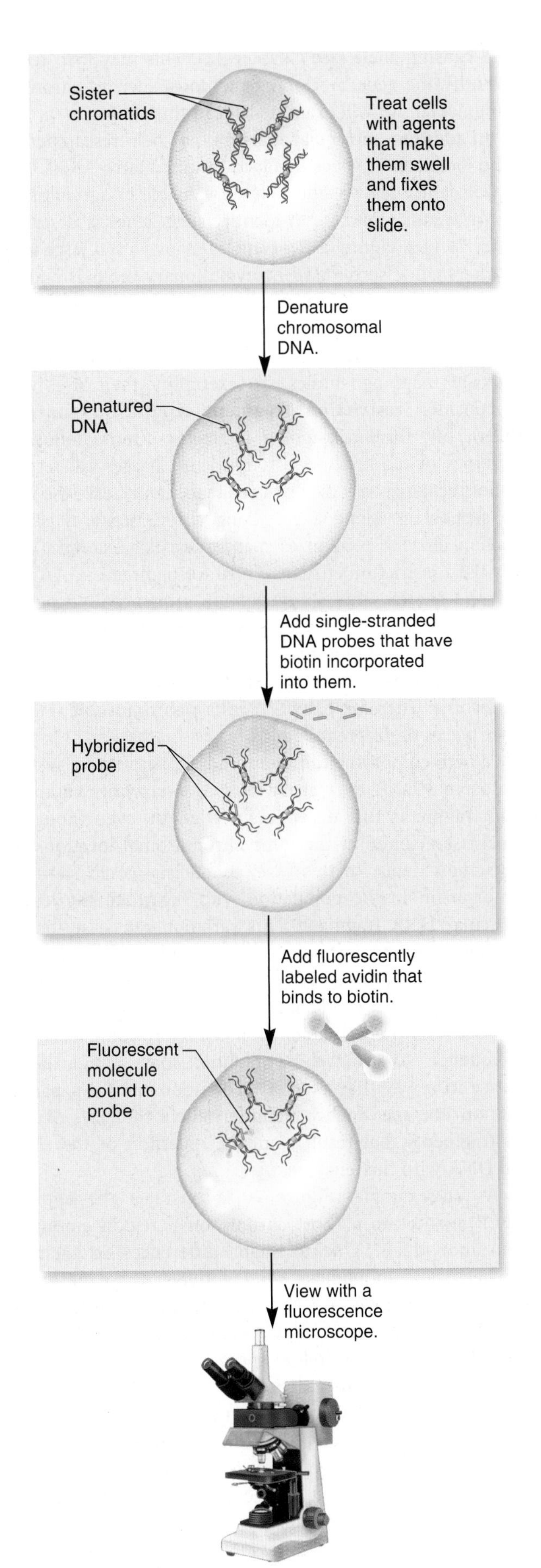

FIGURE 20.2 The technique of fluorescence in situ hybridization (FISH). The probe hybridizes to the denatured chromosomal DNA only at specific, complementary sites in the genome. Note that the chromosomes are highly condensed metaphase chromosomes that have already replicated. These are sister chromatids. Therefore, each X-shaped chromosome actually contains two copies of a particular gene. Because the sister chromatids are identical, a probe that recognizes one sister chromatid will also bind to the other.

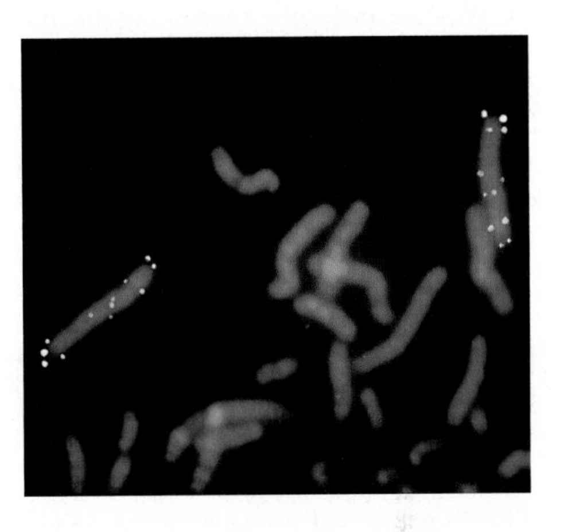

FIGURE 20.3 The results of a fluorescence in situ hybridization experiment. In this experiment, six different probes were used to locate six different sites along chromosome 5. The colors are due to computer imaging of fluorescence emission; they are not the actual colors of the fluorescent labels. Two spots are usually seen at each site because the probe binds to both sister chromatids.

20.3 LINKAGE MAPPING VIA CROSSES

Let's now turn to linkage mapping, which relies on the frequency of recombinant offspring to determine the distance between sites located along the same chromosome. We already considered linkage mapping methods in Chapter 6, where allelic differences between genes were used to map the relative locations of those genes along a chromosome by conducting testcrosses. In this section, we will focus on the use of molecular markers to map genes.

Linkage Mapping Can Use Molecular Markers

As an alternative to relying on allelic differences between genes, geneticists have realized that regions of DNA that do not encode genes can be used as markers along a chromosome. A **molecular marker** is a segment of DNA found at a specific site along a chromosome and has properties that enable it to be uniquely recognized using molecular tools such as polymerase chain reaction (PCR) and gel electrophoresis. As with alleles, the molecular markers may be **polymorphic;** that is, within a population, they may vary from individual to individual. Therefore, the distances between linked molecular markers can be determined from the outcomes of crosses. Using molecular techniques, researchers have found it easier to identify many molecular markers within a given species' genome rather than identifying many allelic differences among individuals. For this reason, geneticists have increasingly turned to molecular markers as points of reference along genetic maps. As described in **Table 20.1**, many different kinds of molecular markers are used by geneticists.

Researchers have constructed detailed genetic maps in which a series of many molecular markers have been identified along each chromosome of certain species. These species include humans, model organisms, agricultural species, and many others. Why are molecular markers useful? One key reason is that molecular markers can be used to determine the approximate location of an unknown gene that causes a human disease. Clinical geneticists sometimes follow the transmission patterns of polymorphic molecular markers in family pedigrees to identify genes that cause human disease when they are mutant. The discovery of a particular marker in those who have the disease can indicate that the marker is close to the disease-causing allele (see Chapter 22). This may help researchers identify the gene by cloning methods, such as chromosome walking, which we will examine later in this chapter.

In addition, molecular markers may help researchers identify the locations of genes involved in quantitative traits, such as fruit yield and meat weight, that are valuable in agriculture. The use of molecular markers to identify such genes is described in Chapter 25 (see Figure 25.7). Genetic maps with a large number of markers can also be used by evolutionary biologists to determine patterns of genetic variation within a species and the evolutionary relatedness of different species.

All of the markers described in Table 20.1 have been used in linkage mapping studies. To exemplify their use, we will first consider **restriction fragment length polymorphisms (RFLPs),** but similar mapping strategies can be followed for other types of markers. As discussed in Chapter 18, restriction enzymes recognize specific DNA sequences and cleave the DNA at those sequences. Along a very long chromosome, a particular restriction enzyme recognizes many sites. For example, a commonly used restriction enzyme, *Eco*RI, recognizes 5′-GAATTC-3′. Simply by chance, this six-nucleotide sequence is expected to occur (on average) every 4^{-6} bases, or once in every 4096 bases. A chromosome composed of millions of nucleotides contains many *Eco*RI sites, which are randomly distributed along the chromosome. Therefore, *Eco*RI digests a chromosome into many smaller pieces of different lengths.

When comparing different individuals, the digestion of chromosomal DNA by a given restriction enzyme may produce certain fragments that differ in their length, even though such fragments are found at the same chromosomal locations in the various individuals. In this case, geneticists would say there is polymorphism in the population with regard to the length of a particular DNA fragment. This variation can arise for different reasons. Among a population of individuals, genetic changes such as small deletions and duplications may subtract or add segments of DNA to a particular region of the chromosome. This tends to occur more frequently if a region contains a repetitive sequence. Alternatively, a mutation may change the DNA sequence in a way that alters a recognition site for a particular restriction enzyme. Such a mutation alters the sizes of certain DNA fragments that result from the digestion of the chromosomal DNA with that enzyme.

As an example, **Figure 20.4** considers the application of RFLP analysis to a short chromosomal region among three diploid individuals. Due to slight differences in their DNA sequences, these individuals vary with regard to the existence of an *Eco*RI site shown in red. In certain individuals, a mutation has changed the sequence at this site so that it is no longer recognized by *Eco*RI. In a diploid species, each individual has two copies of this region. In individual 1, both of these *Eco*RI sites are present, but in individual 2, both are absent. By comparison, individual 3 is a heterozygote. In one of the chromosomes, the site is present, but in the other, it is absent. In this experiment, the DNA was isolated and then cut by *Eco*RI. The DNA from individual 1 is cut at six sites by *Eco*RI, whereas individual 2 has only five locations where the DNA is cut. In individual 2, an

TABLE **20.1**

Common Types of Molecular Markers

Marker	Description
Restriction fragment length polymorphism (RFLP)	A site in a genome where the distance between two restriction sites varies among different individuals. These sites are identified by restriction enzyme digestion of chromosomal DNA and the use of Southern blotting.
Amplified restriction fragment length polymorphism (AFLP)	The same as an RFLP except that the fragment is amplified via PCR instead of isolating the chromosomal DNA
Minisatellite, also called a site with a variable number of tandem repeats (VNTR)	A site in the genome that contains many repeat sequences. The total length is usually in the size range of several hundred to a few thousand base pairs. Because their total lengths are usually polymorphic, they were once used in DNA fingerprinting but have been largely superseded by microsatellites (see Chapter 24).
Microsatellite, also called a short tandem repeat (STR)	A site in the genome that contains many short tandem repeat sequences. The total length is usually in the size range of 100–500 bp, and their lengths may be polymorphic within a population. They are isolated via PCR.
Single-nucleotide polymorphism (SNP)	A site in the genome where a single nucleotide is polymorphic among different individuals. These sites occur commonly in all genomes, and they are gaining greater use in the mapping of disease-causing alleles and in the mapping of genes that contribute to quantitative traits that are valuable in agriculture (see Chapter 24).
Sequence-tagged site (STS)	This is a general term to describe any molecular marker that is found at a unique site in the genome and is amplified by PCR. AFLPs, microsatellites, and SNPs can provide sequence-tagged sites within a genome.

*Eco*RI site is missing because the sequence of DNA in this region has been changed so that it is no longer GAATTC and is not recognized by *Eco*RI.

When subjected to gel electrophoresis, the change in the DNA sequence at a single *Eco*RI restriction site produces a variation in the sizes of homologous DNA fragments. The term *polymorphism,* meaning many forms, refers to the idea that the individuals within a population differ with regard to these particular DNA fragments. As noted in Figure 20.4, not all restriction sites are polymorphic. The three individuals share many DNA fragments that are identical in size. When a DNA segment is identical among all members of a population, it is said to be **monomorphic** (meaning one form). As a practical rule of thumb, a DNA segment is considered monomorphic when over 99% of the individuals in the population have identical sequences at that segment.

The preceding discussion was meant as an overview of the phenomenon of RFLPs. However, the experiment of Figure 20.4 is a technical oversimplification, because it considers only a very short region of chromosomal DNA. In an actual RFLP analysis, DNA samples containing all of the chromosomal DNA would be isolated from these three individuals. The digestion of the chromosomal DNA with *Eco*RI would then yield so many fragments that the results would be very difficult to analyze. To circumvent this problem, Southern blotting is used to identify specific RFLPs.

Figure 20.5 illustrates such an experiment that reconsiders the chromosomes from the three individuals described in Figure 20.4. A DNA strand from a cloned piece of DNA that corresponds to the region near the fourth *Eco*RI site is used as a radiolabeled probe in a Southern blot of the chromosomal DNA. When the blot is exposed to X-ray film, we see only the DNA bands that can hybridize to the radiolabeled

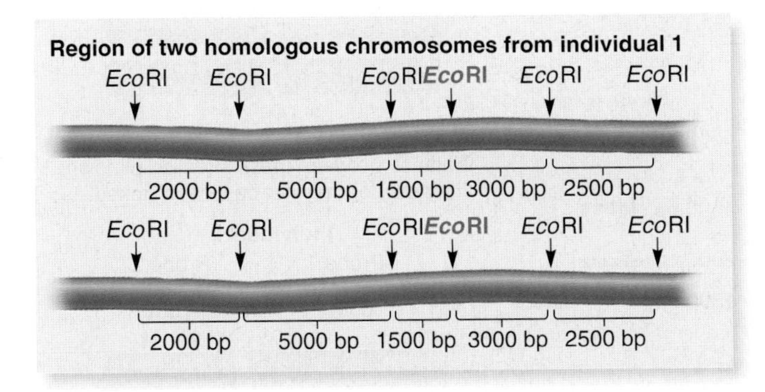

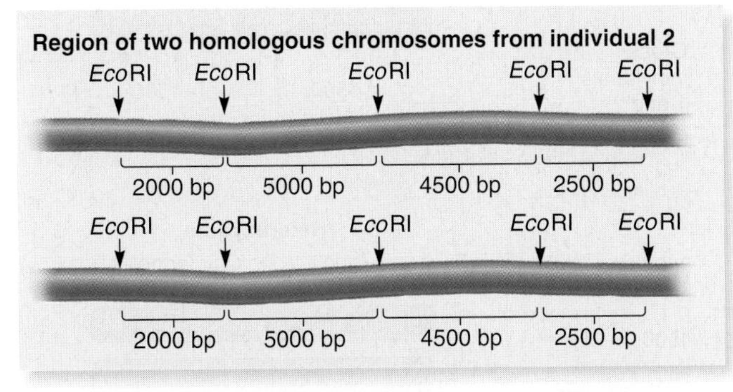

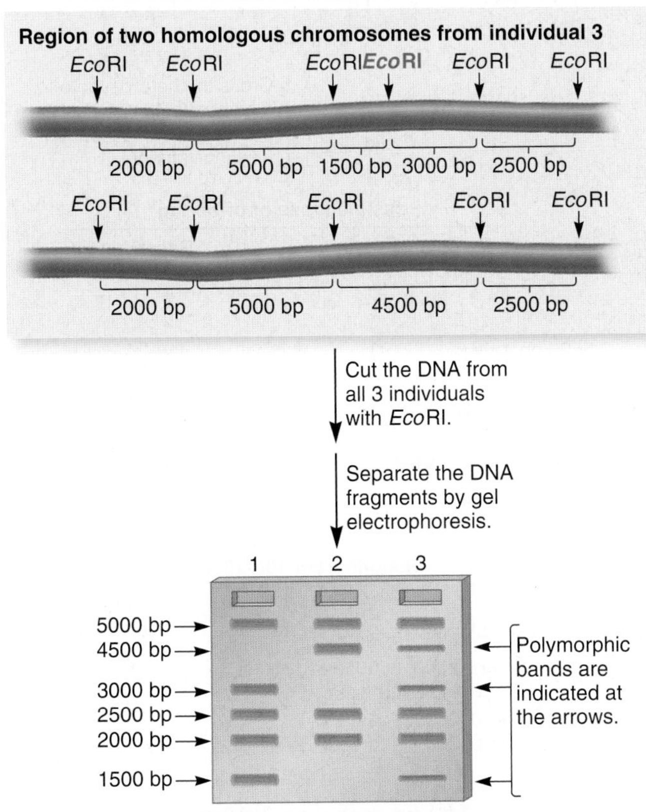

FIGURE 20.4 An RFLP analysis of chromosomal DNA from three different individuals. The *Eco*RI site shown in red is sometimes missing because a mutation has changed the DNA sequence in this region.

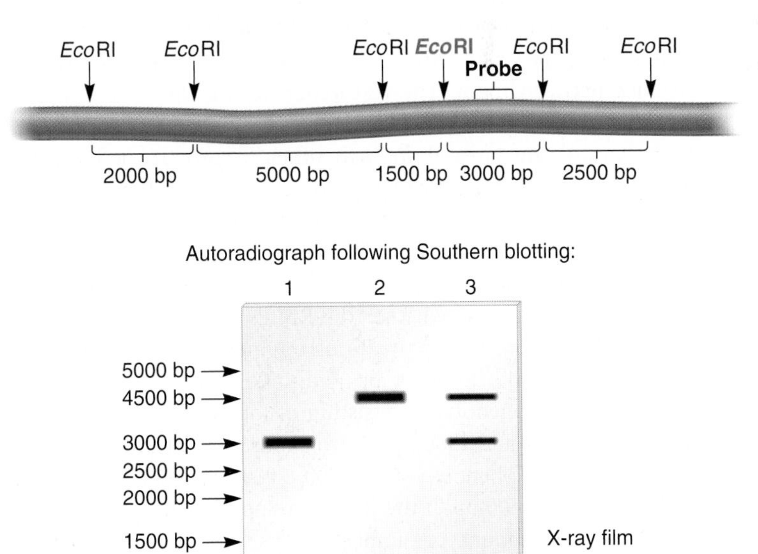

FIGURE 20.5 Southern blotting of a specific RFLP. In this experiment, a Southern blot was conducted on the chromosomal DNA from the same three individuals described in Figure 20.4. A labeled strand of DNA, just to the right of the fourth *Eco*RI site, was used as a probe.

probe. As shown in Figure 20.5, individual 1 shows one band of 3000 bp, and individual 2 has a band 4500 bp long. As discussed in regard to Figure 20.4, this difference is due to the absence of an *Eco*RI site in individual 2. Individual 3 has both bands because of heterozygosity for this *Eco*RI site. In other words, the heterozygote has two thinner bands of different lengths, whereas homozygotes display only one band, because the RFLP procedure detects the molecular products of each chromosome. In an actual RFLP analysis, researchers analyze hundreds or thousands of RFLPs by using a collection of probes that recognize RFLPs throughout the genome.

The Distance Between Two Linked RFLPs Can Be Determined

Now that we understand what an RFLP is, let's consider how RFLPs are mapped and used as chromosomal markers. In the linkage mapping experiments described in Chapter 6, we learned how genetic crosses are used to map the distance between two genes. In that type of analysis, the proportions of offspring with recombinant phenotypes were used to calculate the map distance between genes. Likewise, we can map the distance between two RFLPs by making crosses and analyzing the offspring. In an RFLP analysis, however, we do not look at the phenotypic characteristics of the offspring (e.g., white eyes or miniature wings). Instead, we isolate DNA from a tissue sample and look at the DNA bands on a gel.

Figure 20.6 shows an example of how RFLP mapping is done. Let's suppose two RFLPs are located at different sites in the genome. One RFLP is detected with probe 1 and yields either a 4500-bp band or a 6500-bp band. A second RFLP is detected with probe 2 and yields either a 2000-bp band or a 1500-bp band. In the experiment of Figure 20.6, the goal is to determine whether or not the 4500/6500 and 2000/1500 sites are linked along the same chromosome and, if so, to find the map distance between them. The researcher begins with two strains that are homozygous: 4500 and 2000 versus 6500 and 1500. These strains are crossed to each other to produce a heterozygote. The heterozygote can then be crossed to a homozygote that carries only the 4500 and 2000 bands. This is analogous to a dihybrid testcross described in Chapter 6, in which an F_1 heterozygote is crossed to a homozygote (see Figure 6.9). The results of this cross depend on whether the RFLPs are closely linked or not. If an offspring inherits pairs of fragments like the parents (4500 bp and 2000 bp; 6500 bp and 1500 bp), such offspring show a parental phenotype consisting of four bands of 4500 bp, 2000 bp, 6500 bp, and 1500 bp or two bands of 4500 and 2000 bp. The recombinant phenotypes are 4500, 2000 bp, and 6500 bp or 4500, 2000, and 1500 bp. If the RFLPs are not linked, we expect a 1:1:1:1 ratio of the four types among the offspring. If the RFLPs are linked, we expect a higher percentage of offspring with a parental combination of fragments. Figure 20.6 describes the results of 100 offspring. Many more parental offspring were observed, which indicates that the RFLPs are closely linked.

In an actual analysis of RFLP data, researchers simultaneously analyze many RFLPs and decide on the likelihood of linkage between two RFLPs by using a statistical test called the **lod** (logarithm of the odds) **score method.** This method was devised by Newton Morton in 1955. Although the theoretical basis of this approach is beyond the scope of this textbook, computer programs analyze pooled data from a large number of pedigrees

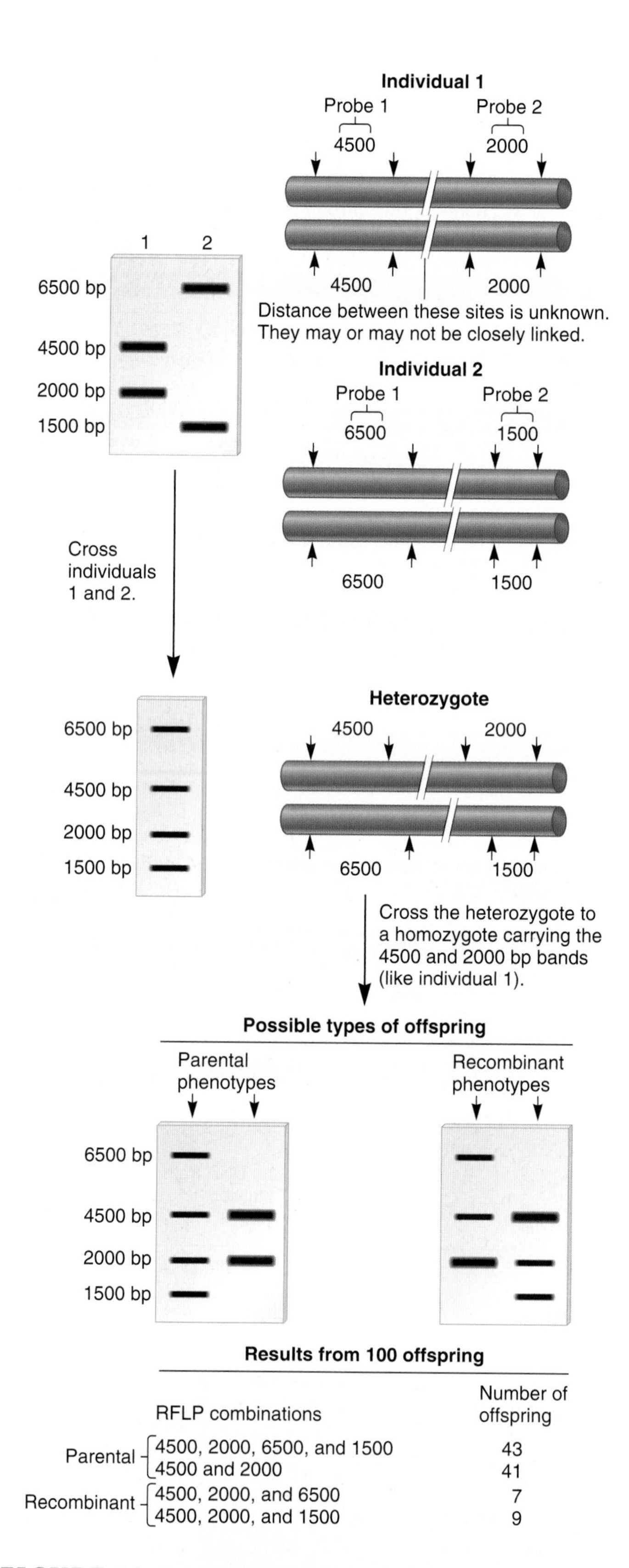

FIGURE 20.6 **Linkage analysis of RFLP markers.**

or crosses involving many RFLPs. The programs determine the probability, or odds, that any two markers exhibit a certain degree of linkage (i.e., are within a particular number of map units apart) and also the probability that the data would have been obtained if the two markers were unlinked. The lod score is then calculated as the ratio of these values:

$$\text{Lod score} = \log_{10} \frac{\text{Probability of a certain degree of linkage}}{\text{Probability of independent assortment}}$$

For example, a lod score of 3 reflects a $\log_{10}$ of 1000. This means the probability is 1000-fold greater that the two markers are linked rather than assort independently. Traditionally, geneticists accept that two markers are linked if the lod score is 3 or greater.

Figure 20.6 illustrates the general type of approach that researchers can follow to map RFLP markers. If a lod score suggested that these two markers were linked, they would calculate the map distance by dividing the recombinant offspring $(7 + 9 = 16)$ by the total number (100) and multiplying by 100, to obtain a map distance of 16 mu.

An RFLP Map Describes the Locations of Many Different RFLPs Throughout a Genome

As we have seen, the map distance between two RFLPs can be determined by analyzing the transmission of RFLP markers from parents to offspring. Such an analysis can be conducted on many different RFLPs to determine their relative locations throughout a genome. RFLPs are quite common in virtually all species, so geneticists can easily map the locations of many RFLPs within a genome. A linkage map composed of many RFLP markers is called an **RFLP map.** Figure 20.7 shows a simplified RFLP map of the plant *Arabidopsis thaliana* (a small plant in the mustard

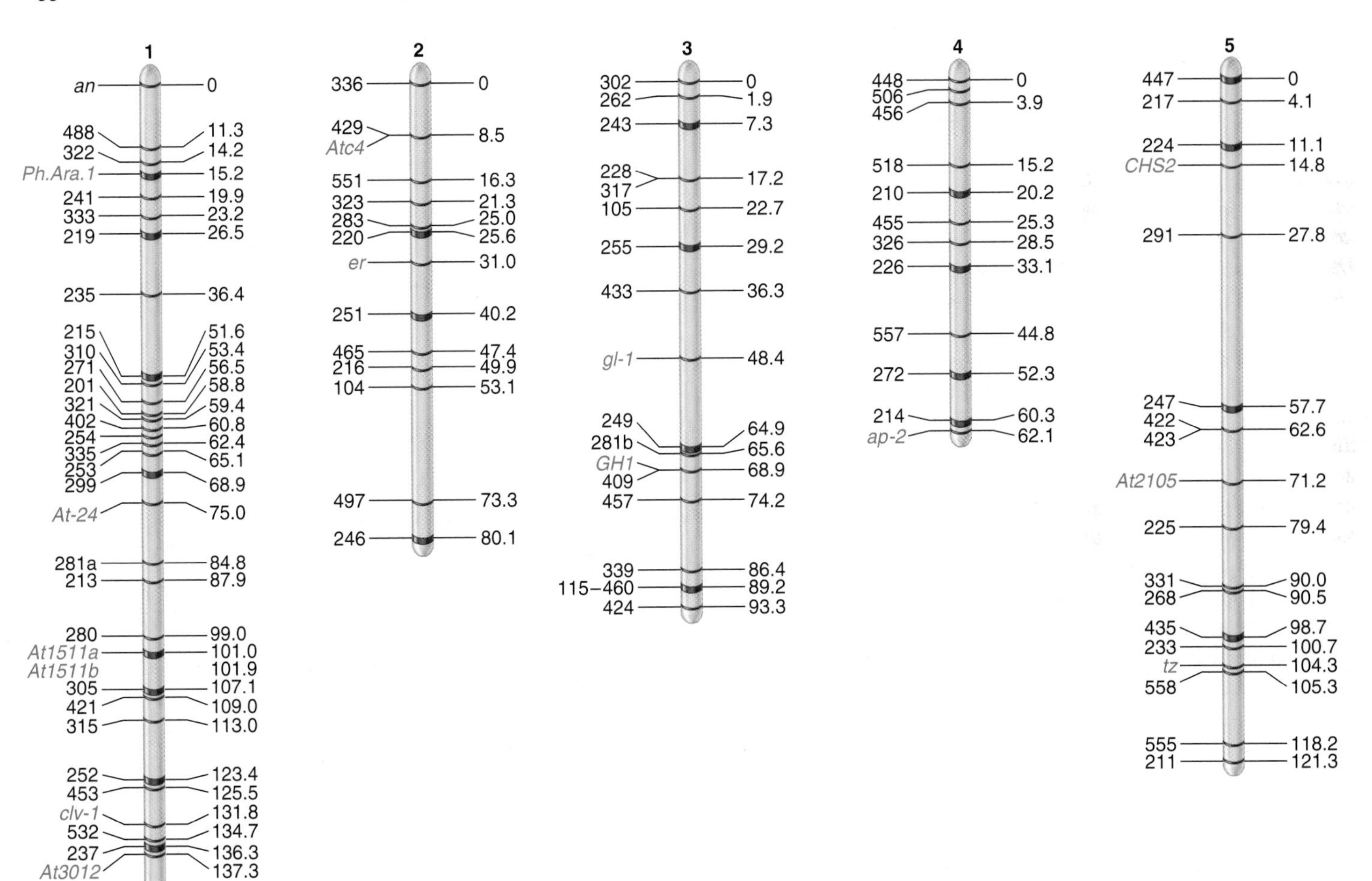

FIGURE 20.7 An RFLP linkage map of *Arabidopsis thaliana*. This plant has five different chromosomes. The left side of each chromosome describes the designations and points to the locations of RFLP markers. The numbers along the right side of each chromosome are the map distances in map units. For example, 219 and 235 are RFLPs located 9.9 mu apart on chromosome 1. The top marker at the end of each chromosome was arbitrarily assigned as the starting point (zero) for each chromosome. In addition, the map shows the locations of a few known genes (shown in red): *Ph.Ara.1* = phytochrome, *At-24* = nitrate reductase, *At1511a/b* = small RNA found in seeds, *clv-1* = clavata-1, *At3012* = alcohol dehydrogenase, *Atc4* = actin, *er* = erecta, *gl-1* = glabra-1, *GH1* = acetolactate synthase, *ap-2* = apelata-2, *CHS2* = chalcone synthase, *At2105* = 12S seed storage protein, and *tz* = thiazole-requiring allele.

Genes → Traits As a first step in mapping the locations of an organism's genes, researchers may initially determine the sites of RFLPs along the chromosomes. In this case, researchers determined the locations of many of these sites along the *Arabidopsis* chromosomes. By mapping these RFLP sites, it becomes easier to locate genes within the *Arabidopsis* genome. The identification of genes helps researchers to elucidate the relationship between genes and traits.

family), which is one of the favorite model organisms of plant molecular geneticists. Many RFLPs have been mapped to different locations along the five *Arabidopsis* chromosomes.

An RFLP map, like the one shown in Figure 20.7, can be used to locate functional genes within the genome. As an example, solved problem S2 illustrates how RFLP analysis can be used to map a gene that confers herbicide resistance.

Linkage Mapping Commonly Uses Molecular Markers Called Microsatellites

To make a highly refined map of a genome, many different polymorphic sites must be identified and their transmission followed from parent to offspring over many generations. RFLPs were among the first molecular markers studied by geneticists. More recently, other molecular markers have been used more commonly because they are easier to generate via PCR. As an example, let's consider **microsatellites,** also called **short tandem repeats (STRs),** which are short repetitive sequences that are abundantly interspersed throughout a species' genome and vary considerably in length among different individuals. Such sequences contain di-, tri-, tetra-, or pentanucleotide sequences that are repeated many times in a row. For example, the most common microsatellite encountered in humans is a dinucleotide sequence $(CA)_n$, where n may range from 5 to more than 50. In other words, this dinucleotide sequence can be tandemly repeated 5 to 50 or more times. The $(CA)_n$ microsatellite is found, on average, about every 10,000 bases in the human genome. Researchers have identified thousands of different DNA segments that contain $(CA)_n$ microsatellites, located at many distinct sites within the human genome. Using primers complementary to the unique DNA sequences that flank a specific $(CA)_n$ region, a particular microsatellite can be amplified by PCR. In other words, the PCR primers copy only a particular microsatellite, but not the thousands of others that are interspersed throughout the genome (**Figure 20.8**).

When a pair of PCR primers copies a single site within a set of chromosomes, the amplified region is called a **sequence-tagged site (STS).** When DNA is collected from a haploid cell, an STS produces only a single band on a gel. In a diploid species, an individual has two copies of a given STS. When an STS contains a microsatellite, the two PCR products may be identical and result in a single band on a gel if the region is the same length in both copies (i.e., if the individual is homozygous for the microsatellite). However, if an individual has two copies that differ in the number of repeats in the microsatellite sequence (i.e., if the individual is heterozygous for the microsatellite), the two PCR products obtained will be different in length (as in Figure 20.8). The two bands observed in this figure differ in their masses because the same site on the two homologous chromosomes contains different numbers of repeat sequences.

Like RFLPs, microsatellites that have length polymorphisms allow researchers to follow their transmission from parent to offspring. PCR amplification of particular microsatellites provides an important strategy in the genetic analysis of human pedigrees, as shown in **Figure 20.9**. Prior to this analysis, a unique

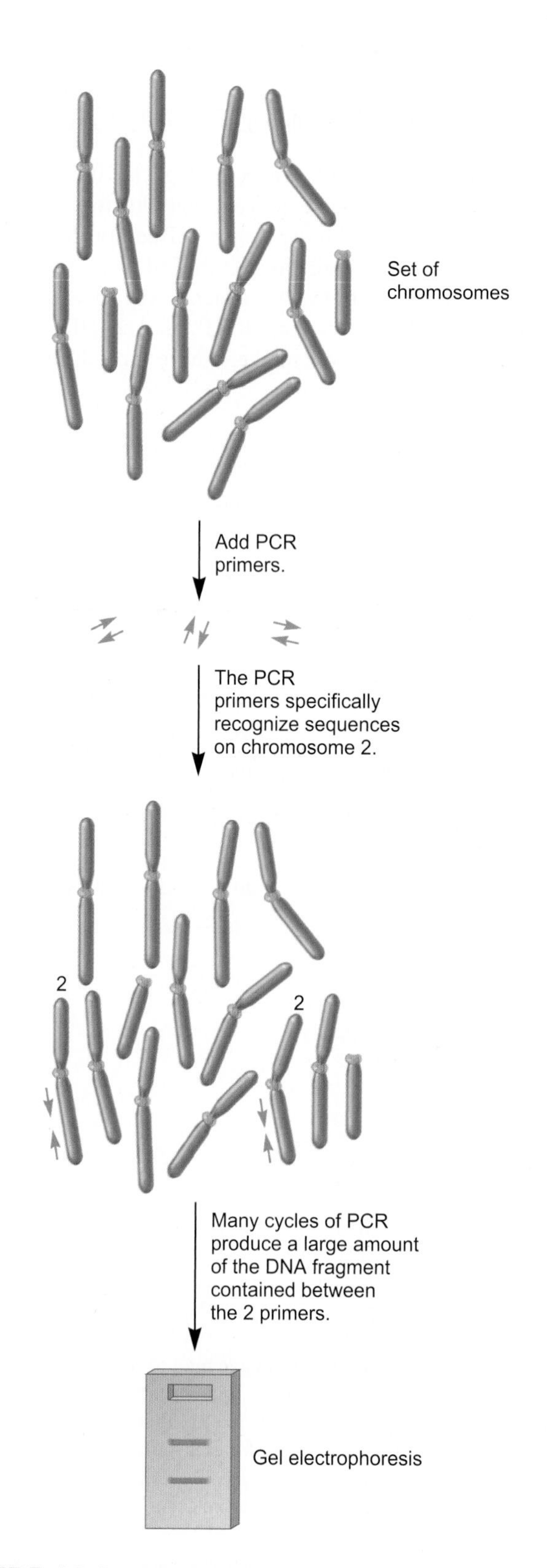

FIGURE 20.8 **Identifying a microsatellite using PCR primers.**

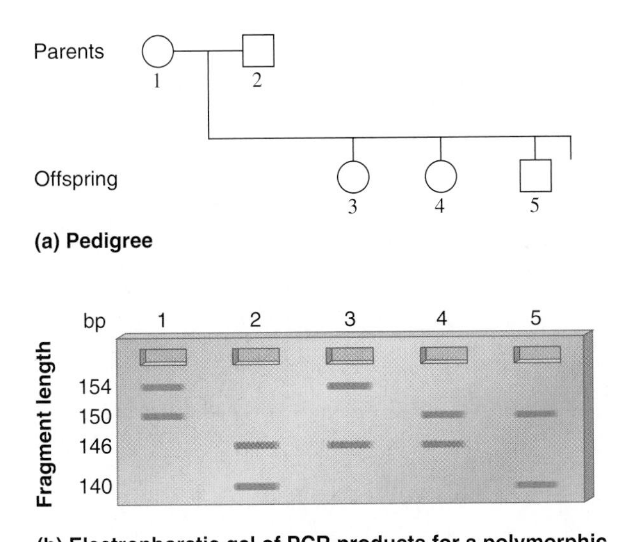

FIGURE 20.9 Inheritance pattern of a polymorphic microsatellite in a human pedigree.

segment of DNA containing a microsatellite had been identified. Using PCR primers complementary to this microsatellite's unique flanking segments, two parents and their three offspring were tested for the inheritance of this microsatellite. A small sample of cells was obtained from each individual and subjected to PCR amplification, as described earlier in Figure 20.8. The amplified PCR products were then analyzed by high-resolution gel electrophoresis, which can detect small differences in the lengths of DNA fragments. The mother's PCR products were 154 and 150 bp in length; the father's were 146 and 140 bp. Their first offspring inherited the 154-bp product from the mother and the 146-bp one from the father, the second inherited the 150 bp from the mother and the 146 bp from the father, and the third inherited the 150 bp from the mother and the 140 bp from the father. As shown in the figure, the transmission of polymorphic microsatellites is relatively easy to follow from generation to generation.

The simple pedigree analysis shown in Figure 20.9 illustrates the general method used to follow the transmission of a single microsatellite that is polymorphic in length. In linkage studies, the goal is to follow the transmission of many different microsatellites to determine those that are linked along the same chromosome versus those that are not. Those that are not linked will independently assort from generation to generation. Those that are linked tend to be transmitted together to the same offspring. In a large pedigree, it is possible to identify cases in which linked microsatellites have segregated due to crossing over. The frequency of crossing over provides a measure of the map distance, in this case between different microsatellites. This approach can help researchers obtain a finely detailed genetic linkage map of the human chromosomes.

In addition, pedigree analysis involving STSs, such as polymorphic microsatellites, can help researchers identify the location of disease-causing alleles. The assumption behind this approach is that a disease-causing allele had its origin in a single individual known as a **founder,** who lived many generations ago. Since that time, the allele has spread throughout portions of the human population. In the case of recessive disorders, descendants of the founder who are affected with the disease have inherited two copies of the mutant allele. A second assumption is that the founder is likely to have had a polymorphic molecular marker that lies somewhere near the mutant allele. This is a reasonable assumption, because all people carry many polymorphic markers throughout their genomes. If a polymorphic marker lies very close to the disease-causing allele, it is unlikely that a crossover will occur in the intervening region. Therefore, such a polymorphic marker may be linked to the disease-causing allele for many generations. For this reason, an association between a particular polymorphic marker and a disease-causing allele may help to predict whether a person is heterozygous for a recessive disease-causing allele. By following the transmission of many polymorphic markers within large family pedigrees, it may be possible to determine that particular markers are found in people who carry specific disease-causing alleles. (An example is shown in Chapter 22, Figure 22.7.) After the identification of a closely linked marker, a disease-causing allele can be identified using a technique called chromosome walking, which is described later in this chapter.

20.4 PHYSICAL MAPPING VIA CLONING

We now turn our attention to methods aimed at establishing a physical map of a species' genome. Physical mapping typically involves the cloning of many pieces of chromosomal DNA. The cloned DNA fragments are then characterized by size (i.e., their length in base pairs), as well as the genes they contain and their relative locations along a chromosome.

As mentioned in Chapter 10, eukaryotic genomes are very large; the *Drosophila* genome is roughly 175 million bp long, and the human genome is approximately 3 billion bp in length. When making a physical map of a genome, researchers must characterize many DNA clones that contain much smaller pieces of the genome. In this section, we examine the general strategies used in creating a physical map of a species' genome. We will also consider how physical mapping information can be used to clone genes.

A Physical Map of a Chromosome Is Constructed by Creating a Contiguous Series of Clones That Span a Chromosome

As discussed in Chapter 18, a DNA library contains a collection of recombinant vectors in which each vector contains a particular fragment of chromosomal DNA. In physical mapping studies, the goal is to determine the relative locations of the cloned chromosomal fragments from a DNA library, as they would occur in an intact chromosome. In other words, the members of the library must be organized according to their actual locations along a

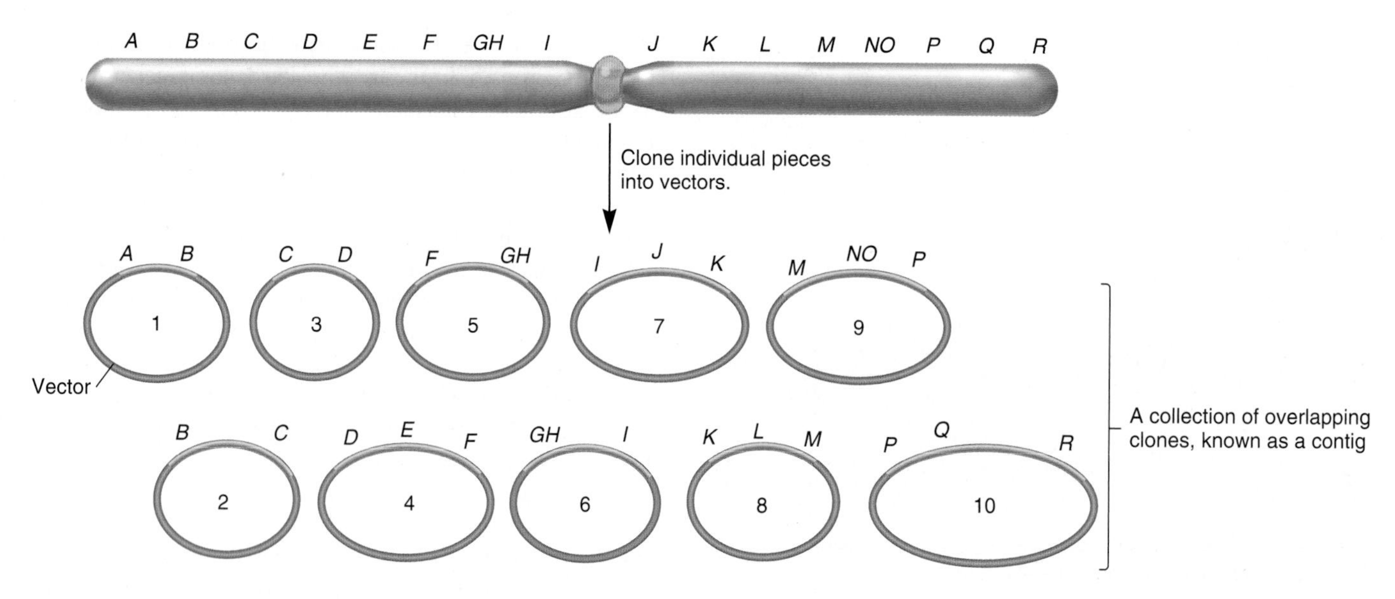

FIGURE 20.10 The construction of a contig. Large pieces of chromosomal DNA are cloned into vectors, and their order is determined by the identification of overlapping regions. For example, clone 3 ends with gene *D*, and clone 4 begins with gene *D*; clone 4 ends with gene *F*, and clone 5 begins with gene *F*.

chromosome. To obtain a complete physical map of a chromosome, researchers need a series of clones that contain contiguous, overlapping pieces of chromosomal DNA. Such a collection of clones is known as a **contig** (**Figure 20.10**). A contig represents a physical map of a chromosome. As discussed later, cloning vectors known as BACs and cosmids are commonly used in the construction of a contig.

Different experimental strategies can be used to align the members of a contig. The general approach is to identify clones that contain overlapping regions. Historically, Southern blotting was used to determine if two different clones contain an overlapping region. In the example shown in Figure 20.10, the DNA from clone 1 could be labeled in vitro and then used as a probe in a Southern blot of the other clones shown in this figure. Clone 1 would hybridize to clone 2, because they share identical DNA sequences in the overlapping region. Similarly, clone 2 could be used as a probe to show that it hybridizes to clone 1 and clone 3. By conducting Southern blots between many combinations of clones, researchers can determine which clones have common overlapping regions and thereby order them as they would occur along the chromosome.

Alternatively, other methods for ordering the members of a contig involve the use of molecular markers. For example, if many STSs have already been identified along a chromosome via linkage or cytogenetic analysis, the STSs can be used as probes to order the members of a contig. Another approach is to subject the members of a library to a restriction enzyme to produce a restriction digestion pattern for each member. The patterns of fragments are then analyzed by computer programs, which can identify regions that are potentially overlapping.

An ultimate goal of physical mapping procedures is to obtain a complete contig for each type of chromosome within a full set. For example, in the case of humans, a complete physical map requires a contig for each of the 22 autosomes and for the X and Y chromosomes. Geneticists can also correlate cloned DNA fragments in a contig with locations along a chromosome obtained from linkage or cytogenetic mapping. For example, a member of a contig may contain a gene, RFLP, or STS that was previously mapped by linkage analysis. **Figure 20.11** considers a situation in which two members of a contig carry genes previously mapped by linkage analysis to be approximately 1.5 mu apart on chromosome 11. In this example, clone 2 has an insert that carries gene *A*, and clone 7 has an insert that carries gene *B*. Because a contig is composed of overlapping members, a researcher can align the contig along chromosome 11, starting with gene *A* and gene *B* as reference points. In this example, genes *A* and *B* serve as markers that identify the location of specific clones within the contig.

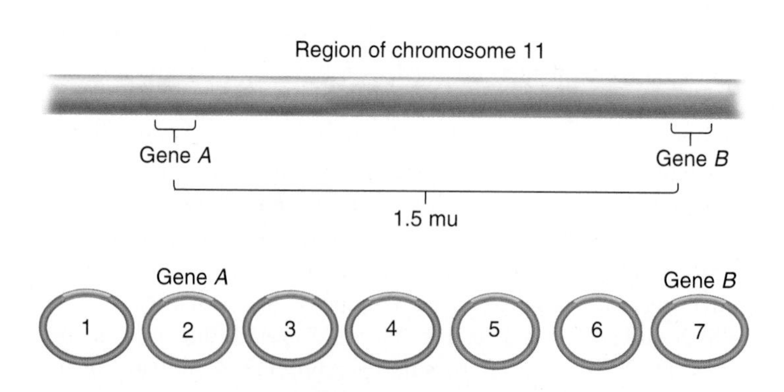

FIGURE 20.11 The use of genetic markers to align a contig. In this example, gene *A* and gene *B* had been mapped previously to specific regions of chromosome 11. Gene *A* was found within the insert of clone 2, gene *B* within the insert of clone 7. This made it possible to align the contig using gene *A* and gene *B* as genetic markers (i.e., reference points) along chromosome 11.

YAC, BAC, and PAC Vectors Are Used to Clone Large Segments of DNA

For large eukaryotic genomes, researchers often begin with vectors that can accept chromosomal DNA inserts of very large size. By having large insert sizes, a contig is more easily constructed and aligned because fewer recombinant vectors are needed. In general, most plasmid and viral vectors can accommodate inserts only a few thousand to perhaps tens of thousands of nucleotides in length. If a plasmid or viral vector has a DNA insert that is too large, it will have difficulty with DNA replication and is likely to suffer deletions in the insert.

By comparison, other cloning vectors, known as **artificial chromosomes,** can accommodate much larger sizes of DNA inserts. As their name suggests, they behave like chromosomes when inside of living cells. The first type to be made was the **yeast artificial chromosome (YAC),** which was developed by David Burke, Georges Carle, and Maynard Olson in 1987. An insert within a YAC can be several hundred thousand to perhaps 2 million bp in length. For an average human chromosome, a few hundred YACs are sufficient to create a contig with fragments that span the entire length of the chromosome. By comparison, it would take thousands or even tens of thousands of recombinant plasmid vectors to create such a contig.

At the molecular level, YACs have structural similarities to normal eukaryotic chromosomes, yet have characteristics that make them suitable for cloning. The general structure of a YAC vector is shown at the top left of **Figure 20.12**. The YAC vector contains two telomeres (*TEL*), a centromere (*CEN*), a bacterial origin of replication (*ORI*), a yeast origin of replication (known as an *ARS,* for autonomous replication sequence), selectable markers, and unique cloning sites that are each recognized by a single restriction enzyme. Without an insert, the circular form of this vector can replicate in *E. coli*. After a large fragment has been inserted, the linear form of the vector can replicate in yeast.

In the experiment shown in Figure 20.12, chromosomal DNA is digested with the restriction enzyme *Eco*RI at a low concentration so that only some of the restriction sites are cut. This partial digestion results in only occasional cleavage of the chromosomal DNA to yield very large DNA fragments.

The circular YAC vector is also digested with *Eco*RI and a second restriction enzyme, *Bam*HI, to yield two arms of the YAC. The YAC arms are then ligated to the large fragments of chromosomal DNA and transformed into yeast cells. When ligation occurs

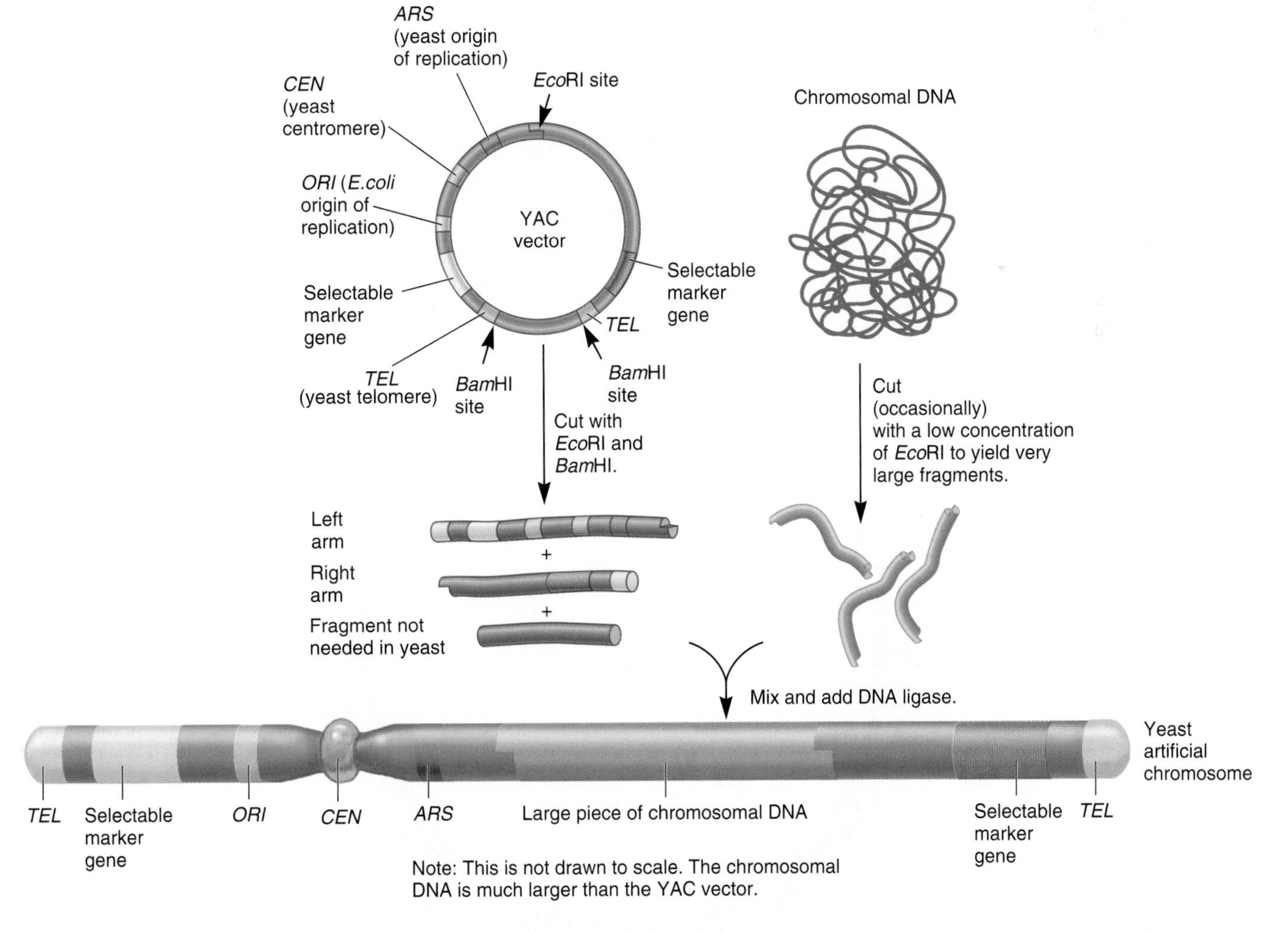

FIGURE 20.12 The use of a YAC vector in DNA cloning. Note: *Eco*RI and *Bam*HI both cut DNA to yield sticky ends. In this diagram, only the *Eco*RI ends are shown with a notch to emphasize that these are the sticky ends that bind to the chromosomal DNA.

in the desired way, a large piece of chromosomal DNA becomes ligated to both arms of the YAC. Because each arm contains a different selectable marker, it is possible to select for the growth of yeast cells that carry a YAC construct having both arms.

Newer types of cloning vectors called **bacterial artificial chromosomes (BACs)** and **P1 artificial chromosomes (PACs)** have been constructed. BACs were developed from bacterial F factors, which are described in Chapter 7, and PACs were developed from P1 bacteriophage chromosomes. BACs and PACs typically can contain inserts up to 300,000 bp and sometimes larger. These vectors are somewhat easier to use than YACs because the DNA is inserted into a circular molecule and transformed into *E. coli*. BACs and PACs are more commonly used than YACs for the cloning of large DNA fragments.

Figure 20.13 shows a simplified drawing of a BAC cloning vector. The vector contains several genes that function in vector replication and segregation. The origin of replication is designated *oriS*, and the *repE* gene encodes a protein essential for replication at *oriS*. The *parA*, *parB*, and *parC* genes encode proteins required for the proper segregation of the vector into daughter cells. A chloramphenicol resistance gene, cm^R, provides a way to select for cells that have taken up the vector based on their ability to grow in the presence of the antibiotic chloramphenicol. The vector also contains unique restriction enzyme sites, such as *Hin*dIII, *Bam*HI, and *Sph*I, for the insertion of large fragments of DNA. These sites are located within the *lacZ* gene, which encodes the enzyme β-galactosidase. Vectors with DNA inserts can be determined by plating cells on media containing the compound X-Gal (as described for plasmid vectors in Chapter 18, Figure 18.2).

YAC, BAC, and PAC cloning vectors have been very useful in the construction of contigs that span long segments of chromosomes. They have been used as the first step in creating a rough physical map of a genome. Although this is an important step in physical mapping, the large insert sizes make them difficult to use in gene-cloning and sequencing experiments. Therefore, libraries containing hybrid vectors with smaller insert sizes are needed. In many cases, a type of cloning vector called a cosmid is used. A **cosmid** is a hybrid between a plasmid vector and phage λ; its DNA can replicate in a cell like a plasmid or be packaged into a protein coat like a phage. Cosmid vectors typically can accept DNA fragments that are tens of thousands of base pairs in length.

Researchers Can Make Genetic Maps of Large Eukaryotic Chromosomes

Figure 20.14 illustrates a comparison between cytogenetic, linkage, and physical maps of human chromosome 16. Actually, this is a very simplified map of chromosome 16. A much more detailed map is available, although it would take well over 10 pages of this textbook to print it! The top of this figure shows the banding pattern of this chromosome. Underneath the banded chromosome are molecular markers that have been mapped by linkage analysis. These same markers have been mapped cytogenetically. A complete contig of this chromosome has also been produced by generating a series of overlapping YACs. However, Figure 20.14 shows the location of only one YAC within the contig. Cosmids within this region are shown below the YAC. In addition, an STS is found within the region and provides a molecular marker for cosmids N16Y1-19 and N16Y1-10, as well as YAC N16Y1.

Positional Cloning Can Be Achieved by Chromosome Walking

The creation of a contig bears many similarities to a gene-cloning strategy known as **positional cloning,** a strategy in which a gene is cloned based on its mapped position along a chromosome.

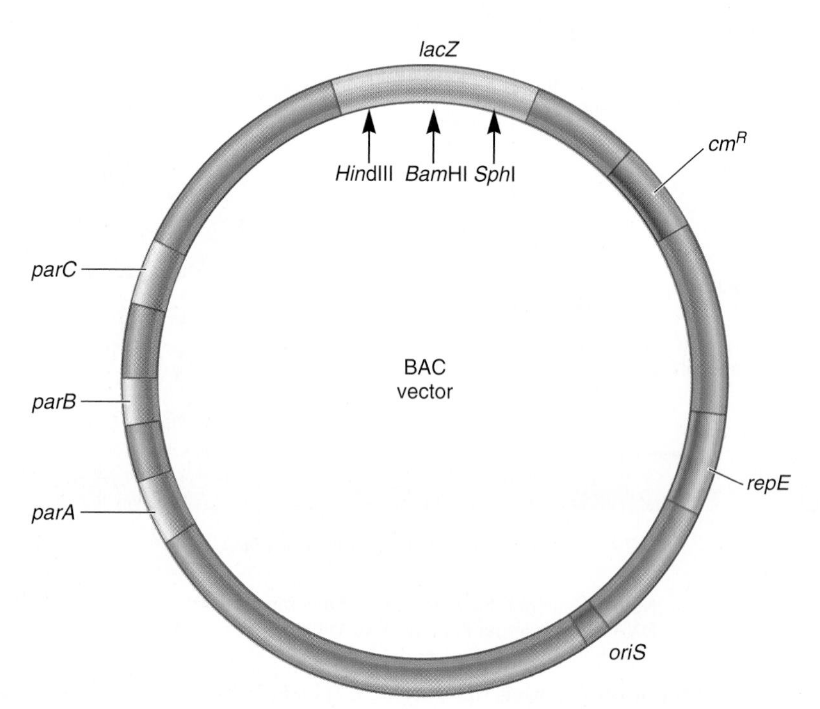

FIGURE 20.13 **A BAC vector.**

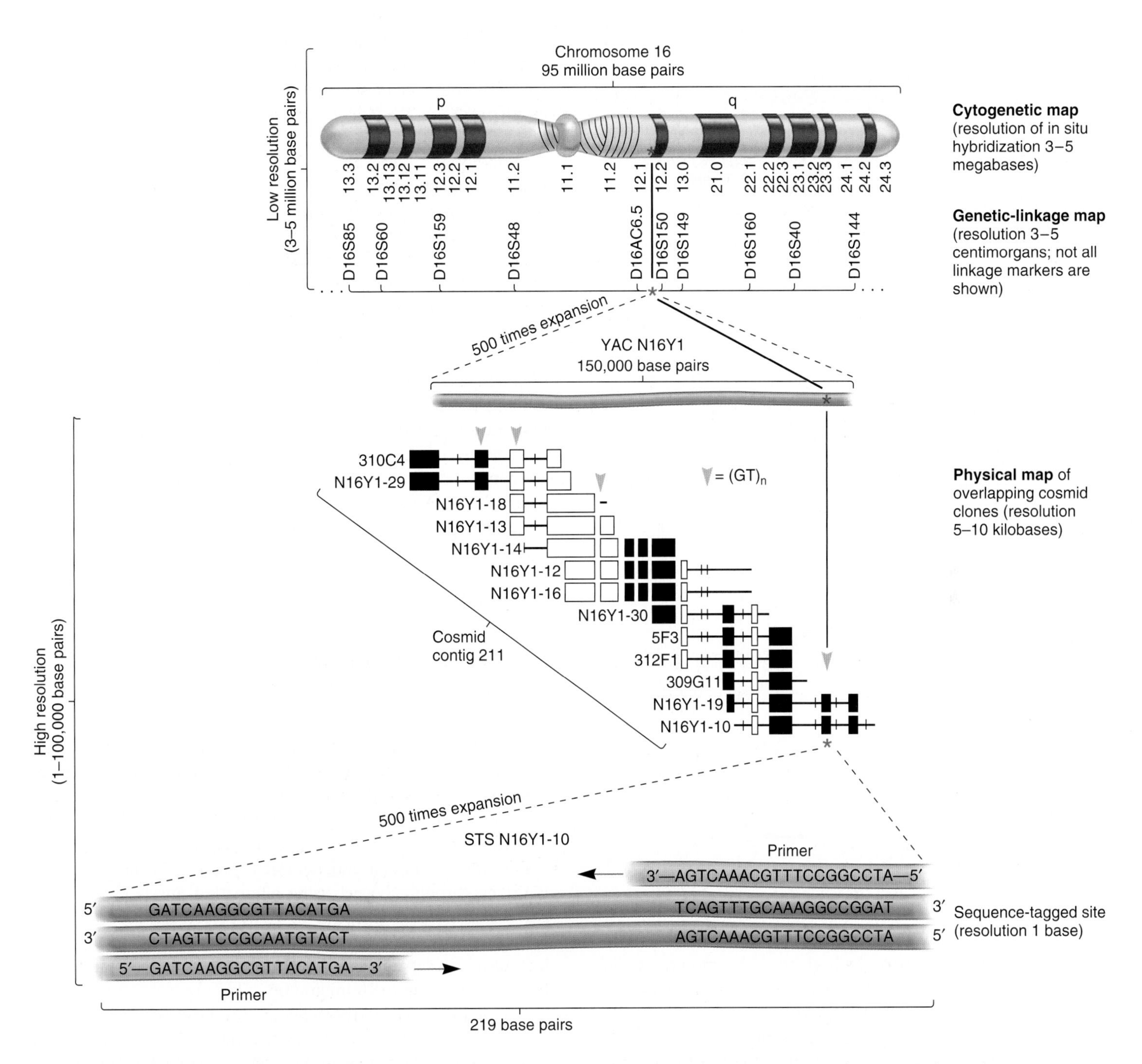

FIGURE 20.14 A correlation of cytogenetic, linkage, and physical maps of human chromosome 16. The top part shows the cytogenetic map of human chromosome 16 according to its G banding pattern. A very simple linkage map of molecular markers (D16S85, D16S60, etc.) is aligned below the cytogenetic map. A correlation between the linkage map and a segment of the physical map is shown below the linkage map. A YAC clone designated YAC N16Y1 is located between markers D16AC6.5 and D16S150 on the linkage map. Pieces of DNA from this YAC were subcloned into cosmid vectors, and the cosmids (310C4, N16Y1-29, N16Y1-18, etc.) were aligned relative to each other. One of the cosmids (N16Y1-10) was sequenced, and this sequence was used to generate an STS shown at the bottom of the figure.

This approach has been successful in the cloning of many human genes, particularly those that cause genetic diseases when mutated. These include genes involved in cystic fibrosis, Huntington disease, and Duchenne muscular dystrophy.

One method of positional cloning is known as **chromosome walking.** To initiate this type of experiment, a gene's position relative to a marker must be known from mapping studies. For example, a gene may be known to be fairly close to a previously mapped gene or RFLP marker. This provides a starting point to molecularly "walk" toward the gene of interest.

Figure 20.15 considers a chromosome walk in which the goal is to locate a gene that we will call gene *A*. In this example, linkage mapping studies have revealed that gene *A* is relatively close to another gene, called gene *B*, that was previously cloned.

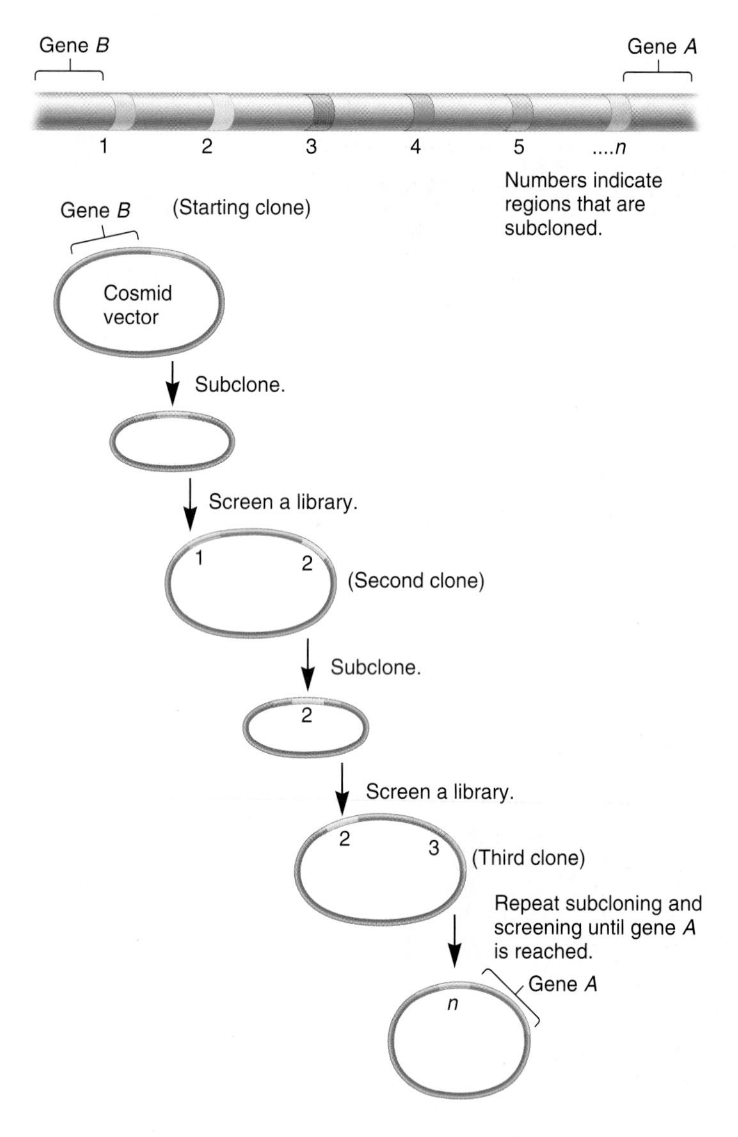

FIGURE 20.15 **The technique of chromosome walking.**

Gene *A* and gene *B* have been deduced from genetic crosses to be approximately 1 mu apart. To begin this chromosome walk, a cloned DNA fragment that contains gene *B* and flanking sequences can be used as a starting point.

To walk from gene *B* to gene *A*, a series of library screening methods are followed. In this example, the starting materials are a cosmid library and a clone containing gene *B*. A small piece of DNA from the first cosmid vector containing gene *B* is inserted into another vector. This is called **subcloning.** The subcloned DNA is labeled and used as a probe to screen a cosmid library. This enables the researchers to identify a second clone that extends into the region that is closer to gene *A*. A subclone from this second clone is then used to screen the library a second time. This allows the researchers to identify a clone that is even closer to gene *A*. This repeated pattern of subcloning and library screening is used to reach gene *A*. The term chromosome walking is an appropriate description of this technique, because each clone takes you a step closer to the gene of interest. When starting at gene *B* in Figure 20.15, researchers also want to have markers to the left of gene *B* to ensure they were not walking in the wrong direction.

The number of steps required to reach the gene of interest depends on the distance between the starting and ending points and on the sizes of the DNA inserts in the library. If the two points are 1 mu apart, they are expected to be approximately 1 million bp apart, although the correlation between map units and physical distances can vary greatly. In a typical walking experiment, each clone might have an average insert size of 50,000 bp. Therefore, it takes about 20 walking steps to reach the gene of interest. Researchers want to locate starting points in a chromosome walking experiment that are as close as possible to the gene they wish to identify.

How do researchers know when they have reached a gene of interest? In the case of a gene that causes a disease when mutant, researchers conduct their walking steps on DNA from both an unaffected and an affected individual. Each set of clones is subjected to DNA sequencing, and those DNA sequences are compared to each other. When the researchers reach a spot where the DNA sequences differ between the unaffected and the affected individual, such a site may be within the gene of interest. However, this has to be confirmed by sequencing the region from several unaffected and affected individuals to be certain the change in DNA sequence is correlated with the disease.

Genomes Can Be Compared by Pulsed-Field Gel Electrophoresis

Researchers may also compare genomes using techniques that separate large DNA fragments and/or small chromosomes. One technique that can accomplish this goal is **pulsed-field gel electrophoresis (PFGE),** developed in 1984 by David Schwartz and Charles Cantor. Using this method, large pieces of chromosomes or entire small chromosomes can be separated and identified. PFGE differs from conventional electrophoresis in that the DNA is driven through the gel using alternating pulses of electrical current at different angles. This can be achieved using electrophoresis devices that have two sets of electrodes. The alternating pulses can separate larger DNA fragments that vary in length from 50,000 bp up to 10 million bp.

For PFGE, large fragments of chromosomes or individual chromosomes must be isolated very carefully. To avoid mechanical breakage of the chromosomes, cells are embedded in agarose blocks (also called plugs). The agarose protects the chromosomal DNA from mechanical breakage. Once inside the plug, the cells are lysed, and if desired, the chromosomal DNA can be subjected to restriction digestion to obtain large DNA fragments. This can be achieved using a restriction enzyme that cuts very infrequently. To separate the large pieces of DNA, the plug is inserted into a well in an agarose gel and then subjected to PFGE.

Figure 20.16a illustrates a separation of yeast chromosomes that have not been subjected to restriction digestion. **Figure 20.16b** shows fragments of DNA that have been obtained by digesting a bacterial chromosome with three different restriction enzymes that cut the chromosome infrequently.

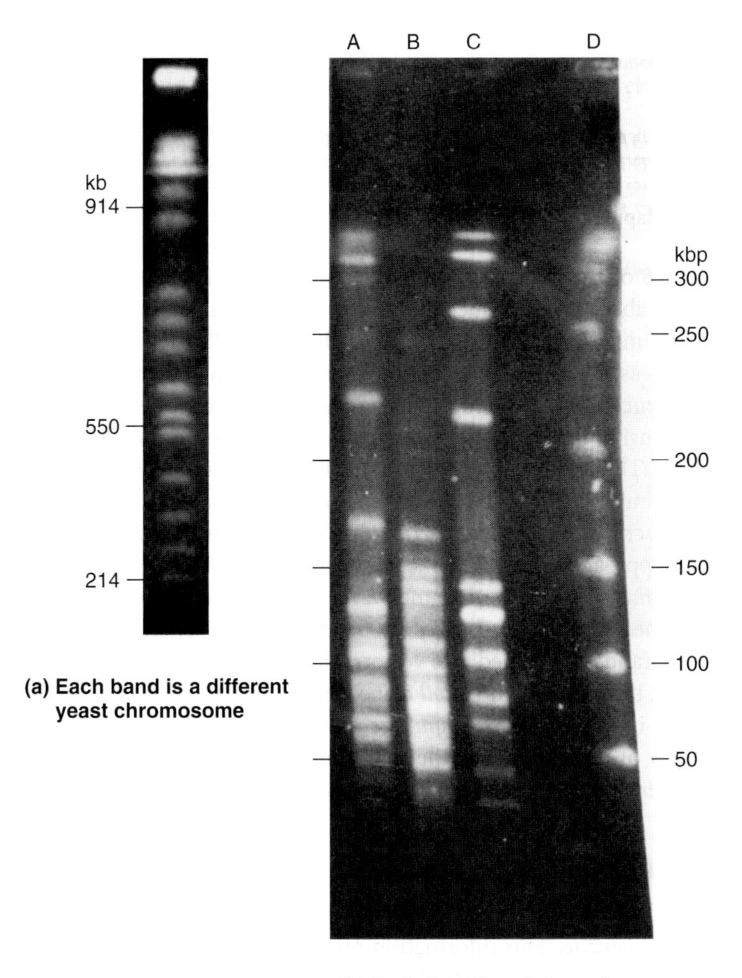

FIGURE 20.16 Pulsed-field gel electrophoresis (PFGE). (**a**) Yeast chromosomes that have not been digested with a restriction enzyme. The numbers at the left indicate the size of the chromosomes in kilobase pairs (kb). (**b**) The *Haemophilus influenzae* chromosome digested with *Eag*I (lane A), *Nae*I (lane B), and *Sma*I (lane C). Lane D shows molecular weight markers.

What are the applications of PFGE? In the past, PFGE was often used to isolate large pieces of chromosomes to construct contigs. However, in recent years, PFGE has been most commonly used to identify species of microorganisms or strains of the same species. For example, some strains of *E. coli* are pathogenic, whereas others are not. When subjected to PFGE, such strains may differ in their banding patterns. Therefore, this method has gained widespread use to distinguish different strains of microorganisms.

20.5 GENOME-SEQUENCING PROJECTS

Genome-sequencing projects are research endeavors that have the ultimate goal of determining the sequence of DNA bases of the entire genome of a given species. Such projects involve many participants, including scientists who isolate DNA and perform DNA-sequencing reactions, as well as theoreticians who gather the DNA sequence information and assemble it into a long DNA sequence for each chromosome. For bacteria and archaea, which usually have just one chromosome, the genome sequence is that of a single chromosome. For eukaryotes, each chromosome must be sequenced. Thus, for humans, the genome sequence includes sequences of 22 autosomes, 2 sex chromosomes, and the mitochondrial genome.

In just a couple of decades, our ability to map and sequence genomes has improved dramatically. As of 2011, the complete genome sequences have been obtained from hundreds of different species, including over 300 prokaryotes and 100 eukaryotes. Considering that the first genome sequence was generated in 1995, the progress of genome sequencing projects since then has been truly remarkable! In this section, we examine the approaches that researchers follow when tackling such a large project. We will also survey some of the general goals of the Human Genome Project, the largest of its kind, and compare the results from the genome sequencing of various species.

EXPERIMENT 20A

Venter, Smith, and Colleagues Sequenced the First Genome in 1995

The first genome to be entirely sequenced was that of the bacterium *Haemophilus influenzae*. This bacterium causes a variety of diseases in humans, including respiratory illnesses and bacterial meningitis. *H. influenzae* has a relatively small genome consisting of approximately 1.8 Mb of DNA in a single circular chromosome.

When sequencing an entire genome, researchers must consider factors such as genome size, the efficiency of the methods used to sequence DNA, and the costs of the project. Since genome-sequencing projects began in the 1990s, researchers have learned that the most efficient and inexpensive way to sequence genomes is via an approach called **shotgun sequencing,** in which DNA fragments to be sequenced are randomly generated from larger DNA fragments. In this method, genomic DNA is isolated and broken into smaller DNA fragments, typically 1500 bp or longer in length. Until recently, the researchers then used the technique of dideoxy sequencing, described in Chapter 18, to randomly sequence fragments from the genome. (Newer methods of DNA sequencing are described later in this chapter.) As a matter of chance, some of the fragments overlap, as shown schematically in **Figure 20.17.** The DNA sequences in two different fragments are identical in the overlapping region. This allows researchers to order them as they genome provides researchers with insight into the types of proteins encoded by these genes. The cloning and sequencing of disease-causing alleles is expected to play an increasingly important role in the diagnosis and treatment of disease.

Innovations in DNA Sequencing Have Made It Faster and Less Expensive

Since it was invented in the 1980s, technological advances in DNA sequencing have been aimed at making it faster and less expensive. The Human Genome Project, which began in the early 1990s, was originally estimated to cost about $3 billion to sequence a single genome. However, cost reductions due to innovations in DNA-sequencing technology drove the actual cost down to about $300 million. By the end of the project, researchers estimated that if they were starting again, they could have sequenced the genome for less than $50 million. The project took about 13 years to complete. In 2007, researchers undertook the sequencing of James Watson's genome, which cost less than $1 million. By 2010, the cost of sequencing of a human genome had been reduced to approximately $10,000 to $100,000. By 2015, some scientists predict that the sequencing of a single human genome will cost less than $1000 and take only a few days to complete. Such innovation will make it feasible to sequence an individual's genome as a routine diagnostic procedure.

The ability to rapidly sequence large amounts of DNA is often referred to as **high-throughput sequencing.** Different types of technological advances have made this possible. First, different aspects of DNA sequencing have become automated so that samples can be processed rapidly in a machine. For example, in Chapter 18, we considered how fluorescently labeled nucleotides can automate the ability to read a DNA sequence by using a fluorescence detector. A second advance from high-throughput sequencing technologies is they enable parallel sequencing. This means multiple samples can be processed at once. The first sequencing machines, also called platforms, could simultaneously perform many sequencing runs via multiple capillary gels. For example, DNA-sequencing machines produced by Applied Biosystems, which rely on Sanger dideoxy sequencing, run 96 capillaries in parallel. Each capillary gel is capable of producing between 500 and 1000 bases of DNA sequence.

Although the Sanger dideoxy sequencing method is still in use, newer high-throughput platforms are based on different methods of sequencing DNA. **Table 20.2** describes a few of these methods, which are often referred to as **next-generation sequencing technologies** because they have superseded the Sanger dideoxy method for large sequencing projects. What sets next-generation sequencers apart from conventional capillary-based sequencing? One key technological advance is the ability to process thousands or even millions of sequence reads in parallel rather than only 96 at a time. This massively parallel throughput may require only one or two instrument runs to complete the sequencing of an entire prokaryotic genome. Also, next-generation sequencers are able to use samples that contain mixtures of DNA fragments that have not been subjected to the conventional vector-based cloning. By comparison, shotgun methods of DNA sequencing using capillary sequencing involve DNA-cloning steps (see Figure 20.19). The elimination of such cloning steps saves a great deal of time and money.

The newer sequencing platforms employ a complex interplay of enzymology, chemistry, high-resolution optics, and new approaches to processing the data. These instruments allow for easy sample preparation steps prior to DNA sequencing. Most of them involve strategies in which fragmented DNA is immobilized in a fixed position and repeatedly exposed to reagents. The second-generation sequencing platforms use PCR to amplify the DNA, whereas third-generation sequencers actually read single DNA molecules.

As an example of next-generation sequencing, **Figure 20.20** considers the technology called **pyrosequencing,** which was developed by Pal Nyren and Mostafa Ronaghi in 1996 and is the basis for the Roche 454/FLX Pyrosequencer. Samples, such as

TABLE **20.2**

Examples of Next-Generation DNA-Sequencing Technologies

Technology*	DNA Preparation	Enzyme(s) Used	Detection
Second-Generation DNA Sequencers			
Roche 454/FLX Pyrosequencer	DNA fragments are bound to small beads, which are dropped into tiny wells in a fiber optic chip.	DNA polymerase, ATP sulfurylase, luciferase, apyrase	Pyrophosphate release activates luciferase, which gives off light.
Illumina/Solexa Genome Analyzer	DNA fragments are bound to a flow cell surface.	DNA polymerase	Four different fluorescently labeled nucleotides are detected.
Third-Generation DNA Sequencers			
Pacific Biosciences SMRT	DNA fragments and DNA polymerase are trapped within tiny holes on a thin metal film.	DNA polymerase	The growth of individual DNA molecules is monitored by fluorescence imaging
Helicos Biosciences tSMS	DNA fragments are bound to a flow cell surface.	DNA polymerase	The growth of individual DNA molecules is monitored by fluorescence imaging.
ZS Genetics TEM	DNA is labeled with heavier elements, such as iodine or bromine.	None	The DNA sequence of a single molecule is read via transmission electron microscopy (TEM).

*Also includes the company name associated with the technology.

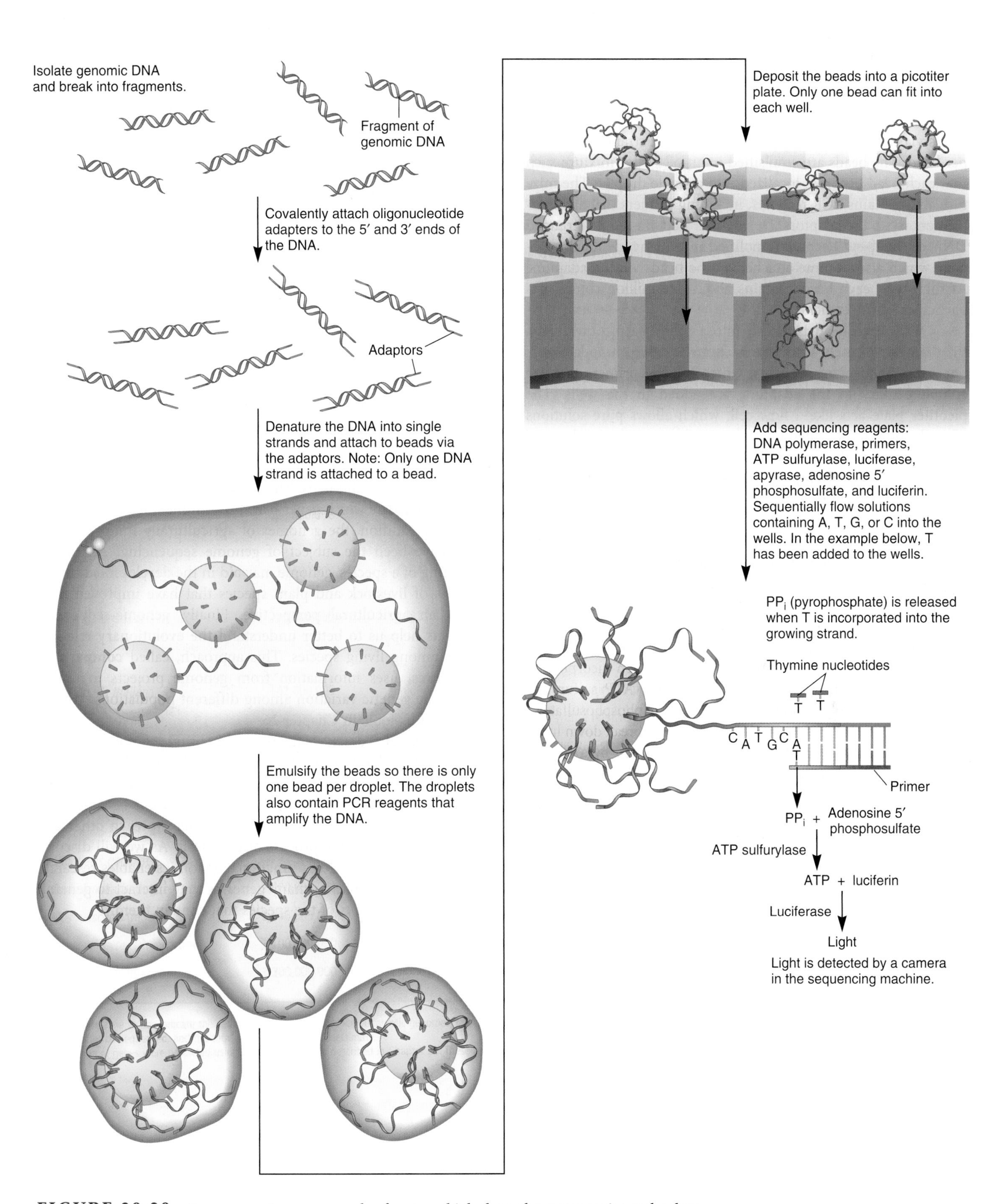

FIGURE 20.20 **Pyrosequencing, an example of a newer high-throughput sequencing technology.**

KEY TERMS

Page 544. genome, genomics
Page 545. functional genomics, proteomics, mapping, cytogenetic mapping, linkage mapping, physical mapping, genetic map, chromosome map, locus, loci
Page 546. in situ hybridization, fluorescence in situ hybridization (FISH), chromosome painting
Page 547. molecular marker, polymorphic
Page 548. restriction fragment length polymorphisms (RFLPs)
Page 549. monomorphic
Page 550. lod score method
Page 551. RFLP map
Page 552. microsatellites, short tandem repeats (STRs), sequence-tagged site (STS)
Page 553. founder
Page 554. contig
Page 555. artificial chromosomes, yeast artificial chromosome (YAC)
Page 556. bacterial artificial chromosomes (BACs), P1 artificial chromosomes (PACs), cosmid, positional cloning
Page 557. chromosome walking
Page 558. subcloning, pulsed-field gel electrophoresis (PFGE)
Page 559. genome-sequencing projects, shotgun sequencing
Page 562. hierarchical shotgun sequencing, whole-genome shotgun sequencing, double-barrel shotgun sequencing, Human Genome Project
Page 564. high-throughput sequencing, next-generation sequencing technologies, pyrosequencing
Page 566. sequencing by synthesis (SBS), comparative genomics

CHAPTER SUMMARY

- The genome is the total genetic composition of an organism or species. Genomics refers to the molecular analysis of a genome as a whole unit. Functional genomics is aimed at studying the expression of genes, whereas proteomics is focused on the structures and functions of proteins.

20.1 Overview of Chromosome Mapping

- The term mapping refers to the process of determining the relative locations of genes or other DNA segments along a chromosome, thereby producing a genetic or chromosome map. The process may involve cytogenetic, linkage, or physical mapping techniques (see Figure 20.1).

20.2 Cytogenetic Mapping Via Microscopy

- Cytogenetic mapping attempts to determine the locations of particular genes relative to a banding pattern of a chromosome.
- Fluorescence in situ hybridization is a commonly used method to map genes and other segments along a chromosome (see Figures 20.2, 20.3).

20.3 Linkage Mapping Via Crosses

- Linkage mapping often relies on molecular markers to map the locations of genes along chromosomes (see Table 20.1).
- Restriction fragment length polymorphisms and microsatellites are used as molecular markers (see Figures 20.4–20.9).

20.4 Physical Mapping Via Cloning

- A physical map of a chromosome is made by producing a contiguous collection of clones called a contig, which spans an entire chromosome (see Figures 20.10, 20.11).
- YACs, BACs, PACs, and cosmids are vectors that can carry a long segment of DNA and are used to make contigs (see Figures 20.12–20.14).
- Chromosome walking is a technique for cloning a gene that is close to a known marker (see Figure 20.15).
- The genomes of similar species and strains within a species may be compared by pulsed-field gel electrophoresis (see Figure 20.16).

20.5 Genome-Sequencing Projects

- Shotgun DNA sequencing has been commonly used to sequence the genomes of many species. Overlapping regions allow researchers to determine the contiguous DNA sequence. Shotgun sequencing can follow a hierarchical or whole-genome shotgun sequencing approach (see Figures 20.17–20.19).
- The Human Genome Project resulted in the successful sequencing of the entire human genome.
- Newer methods of DNA sequencing can process many samples of DNA simultaneously and are superseding the Sanger dideoxy method. An example is pyrosequencing (see Table 20.2, Figure 20.20).
- Since 1995, the genomes of hundreds of species have been sequenced (see Table 20.3).

PROBLEM SETS & INSIGHTS

Solved Problems

S1. An RFLP marker is located 1 million bp away from a gene of interest. Your goal is to start at this RFLP marker and walk to this gene. The average insert size in the library is 55,000 bp, and the average overlap at each end is 5000 bp. Approximately how many steps will it take to get there?

Answer: Each step is only 50,000 bp (i.e., 55,000 minus 5000), because you have to subtract the overlap between adjacent fragments, which is 5000 bp, from the average inset size. Therefore, it will take about 20 steps to go 1 million bp.

S2. When many RFLPs have been mapped within the genome of a plant, an RFLP analysis can be used to map a herbicide-resistance gene. For example, let's suppose that an agricultural geneticist has two strains: one that is herbicide-resistant and one that is herbicide-sensitive. The two strains differ with regard to many RFLPs. The sensitive and resistant strains are crossed, and the F_1 offspring are allowed to self-fertilize. The F_2 offspring are then analyzed with regard to their herbicide sensitivity and RFLP markers. The following results are obtained:

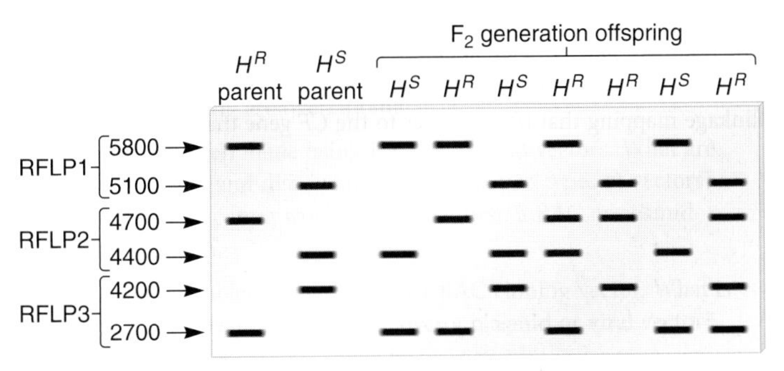

To which RFLP might the herbicide-resistance gene be linked?

Answer: The aim of this experiment is to correlate the presence of a particular RFLP with the herbicide-resistant phenotype. As shown in the data, the herbicide-resistant parent and all of the herbicide-resistant offspring have an RFLP that is 4700 bp in length. In an actual experiment, a thorough lod analysis would be conducted to determine if linkage is considered likely. If so, the 4700-bp RFLP may either contain the gene that confers herbicide resistance, or, as is more likely, the two may be linked. If the 4700-bp RFLP has already been mapped to a particular site in the plant's genome, the herbicide-resistance gene also maps to the same site or very close to it. This information may then be used by a plant breeder when making future crosses to produce herbicide-resistant plant strains.

S3. Does a molecular marker have to be polymorphic to be useful in mapping studies? Does a molecular marker have to be polymorphic to be useful in linkage mapping (i.e., involving family pedigree studies or genetic crosses)? Explain why or why not.

Answer: A molecular marker does not have to be polymorphic to be useful in mapping studies. Many sequence-tagged sites (STSs) that are used in physical or cytogenetic studies are monomorphic. Monomorphic markers can provide landmarks in mapping studies.

In linkage mapping studies, a marker must be polymorphic to be useful. Polymorphic molecular markers can be RFLPs, microsatellites, or SNPs. To compute map distances in linkage analysis, individuals must be heterozygous for two or more markers (or genes). For experimental organisms, heterozygotes are testcrossed to homozygotes, and then the number of recombinant offspring and nonrecombinant offspring are determined. For markers that do not assort independently (i.e., linked markers), the map distance is computed as the number of recombinant offspring divided by the total number of offspring times 100.

S4. The distance between two molecular markers that are linked along the same chromosome can be determined by analyzing the outcome of crosses. This can be done in humans by analyzing the members of a pedigree. However, the accuracy of linkage mapping in human pedigrees is fairly limited because the number of people in most families is relatively small. As an alternative, researchers can analyze a population of sperm, produced from a single male, and compute linkage distance in this manner. As an example, let's suppose a male is heterozygous for two polymorphic STSs. STS-1 exists in two sizes: 234 bp and 198 bp. STS-2 also exists in two sizes: 423 bp and 322 bp. A sample of sperm was collected from this man, and individual sperm were placed into 40 separate tubes. In other words, there was one sperm in each tube. Believe it or not, PCR is sensitive enough to allow analysis of DNA in a single sperm! Into each of the 40 tubes were added the primers that amplify STS-1 and STS-2, and then the samples were subjected to PCR. The following results were obtained.

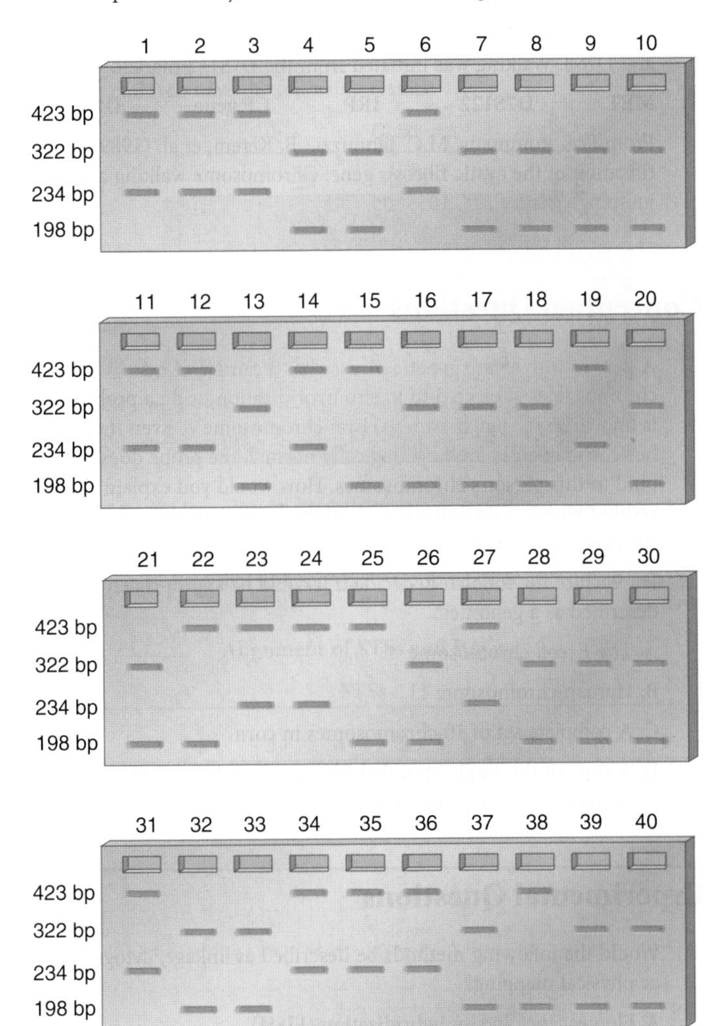

E17. In the Human Genome Project, researchers have collected linkage data from many crosses in which the male was heterozygous for markers and many crosses where the female was heterozygous for markers. The distance between the same two markers, computed in map units or centiMorgans, is different between males and females. In other words, the linkage maps for human males and females are not the same. Propose an explanation for this discrepancy. Do you think the sizes of chromosomes (excluding the Y chromosome) in human males and females are different? How could physical mapping resolve this discrepancy?

E18. Take a look at solved problem S4. Let's suppose a male is heterozygous for two polymorphic sequence-tagged sites. STS-1 exists in two sizes: 211 bp and 289 bp. STS-2 also exists in two sizes: 115 bp and 422 bp. A sample of sperm was collected from this man, and individual sperm were placed into 30 separate tubes. Into each of the 30 tubes were added the primers that amplify STS-1 and STS-2, and then the samples were subjected to PCR. The following results were obtained:

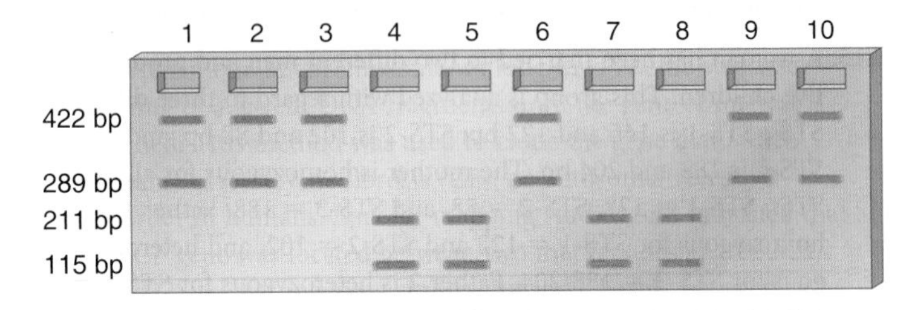

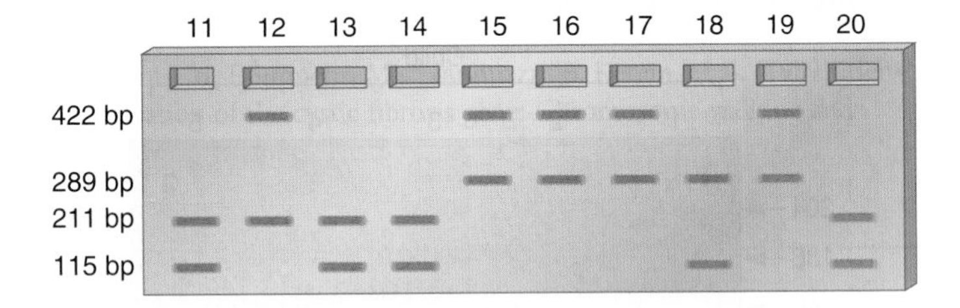

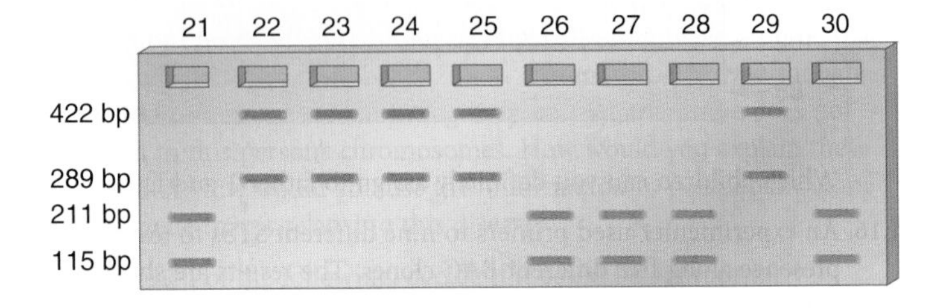

A. What is the arrangement of these STSs in this individual?

B. What is the linkage distance between STS-1 and STS-2?

C. Could this approach of analyzing a population of sperm be applied to RFLPs?

E19. A gene affecting flower color in petunias is closely linked to an RFLP. A red allele for this gene is associated with a 4000-bp RFLP, and a purple allele is linked to a 3400-bp version of this same RFLP. A second gene in petunias affects flower size. An allele causing big flowers is linked to a 7200-bp RFLP, and a small-flower allele is linked to the same RFLP that is 1600 bp. A true-breeding strain with small, red flowers was crossed to a true-breeding strain with big, purple flowers. All of the F_1 offspring had big, purple flowers. These F_1 offspring were then crossed to true-breeding petunias with small, red flowers. The following results were obtained:

Red, small	725
Red, big	111
Purple, small	109
Purple, big	729

Are these two genes linked to each other? If so, compute the map distance. What would be the expected outcome regarding the inheritance of the RFLPs among the offspring?

E20. An agricultural geneticist has studied a gene in alfalfa that affects pesticide resistance. It exists in three alleles that confer low, medium, and high levels of resistance. This gene significantly affects the yield of alfalfa, depending on seasonal variation in pest problems. This geneticist has followed the basic protocol in Figure 20.4, using *Eco*RI as the enzyme to digest the chromosomal DNA. Unfortunately, after tireless efforts and the analysis of thousands of offspring, it has not been possible to identify an RFLP that is associated with the three alleles of this pesticide-resistance gene. What should the geneticist do next? In other words, discuss ways to vary the RFLP method or propose alternative approaches for identifying molecular markers that may be linked to the pesticide-resistance gene.

E21. Figure 20.6 describes the transmission of two RFLPs that were linked and 16 mu apart. If these two RFLPs had not been linked and 100 offspring had been analyzed, what would have been the expected results?

E22. Explain why it is necessary to use the technique of Southern blotting for RFLP mapping.

E23. Compared with a conventional plasmid, what additional sequences are required in a YAC vector so it can behave like an artificial chromosome? Describe the importance of each required sequence.

E24. Explain the technique of pulsed-field gel electrophoresis (PFGE). What special precautions are needed to prevent the mechanical breakage of the chromosomes? What are the uses of PFGE?

E25. When conducting physical mapping studies, place the following methods in their most logical order:

A. Clone large fragments of DNA to make a BAC library.

B. Determine the DNA sequence of subclones from a cosmid library.

C. Subclone BAC fragments to make a cosmid library.

D. Subclone cosmid fragments for DNA sequencing.

E26. Four cosmid clones, which we will call cosmid A, B, C, and D, were subjected to a Southern blot in pairwise combinations. The insert size of each cosmid was also analyzed. The following results were obtained:

Cosmid	Insert Size (bp)	Hybridized to?
A	6000	C
B	2200	C, D
C	11,500	A, B, D
D	7000	B, C

Draw a map that shows the order of the inserts within these four cosmids.

E27. What is an STS? How are STSs generated experimentally? What are the uses of STSs? Explain how a microsatellite can produce a polymorphic STS.

E28. A human gene, which we will call gene *X*, is located on chromosome 11 and is found as a normal allele and a recessive disease-causing allele. The location of gene *X* has been approximated on the map shown here that contains four STSs, labeled STS-1, STS-2, STS-3, and STS-4.

STS-1 STS-2 STS-3 Gene *X* STS-4

A. Explain the general strategy of positional cloning.

B. If you applied the approach of positional cloning to clone gene *X*, where would you begin? As you progressed in your cloning efforts, how would you know if you were walking toward or away from gene *X*?

C. How would you know you had reached gene *X*? (Keep in mind that gene *X* exists as a normal allele and a disease-causing allele.)

E29. Describe how you would clone a gene by positional cloning. Explain how a (previously made) contig would make this task much easier.

E30. A bacterium has a genome size of 4.4 Mb. If a researcher carries out shotgun DNA sequencing and sequences a total of 19 Mb, what is the probability that a base will be left unsequenced? What percentage of the total genome will be left unsequenced?

E31. Discuss the general differences between hierarchical shotgun sequencing and whole-genome shotgun sequencing.

E32. Discuss the advantages of next-generation sequencing technologies.

E33. What is meant by "sequence by synthesis"?

Questions for Student Discussion/Collaboration

1. How is it possible to obtain an RFLP linkage map? What kind of experiments would you conduct to correlate the RFLP linkage map with the positions of known genes that had already been cloned? Discuss the uses of RFLPs in genetic analyses.

2. What is a molecular marker? Give two examples. Discuss why it is easier to locate and map many molecular markers rather than functional genes.

3. Which goals of the Human Genome Project do you think are the most important? Why? Discuss the types of ethical problems that might arise as a result of identifying all of our genes.

Note: All answers appear at the website for this textbook; the answers to even-numbered questions are in the back of the textbook.

www.mhhe.com/brookergenetics4e

Visit the website for practice tests, answer keys, and other learning aids for this chapter. Enhance your understanding of genetics with our interactive exercises, quizzes, animations, and much more.

THE DATA

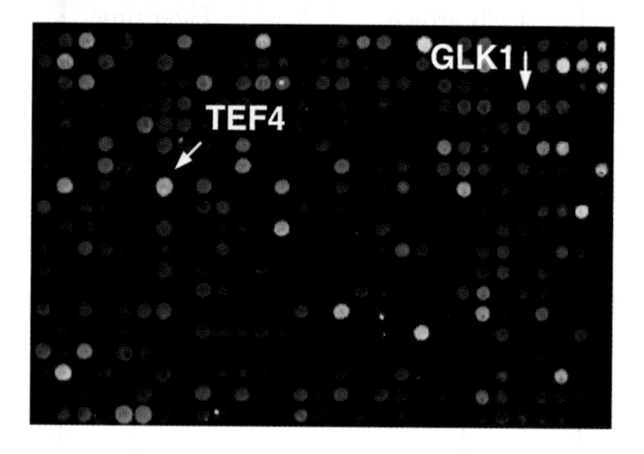

Green spots are genes expressed early in growth, whereas red spots are expressed later. Yellow spots are expressed more evenly. Spots that are barely visible indicate genes that are not substantially expressed under these growth conditions. [Data from J.L. DeRisi, V.R. Iyer, and P.O. Brown (1997) Exploring the metabolic and genetic control of gene expression on a genomic scale. *Science 278*, 680–686.]

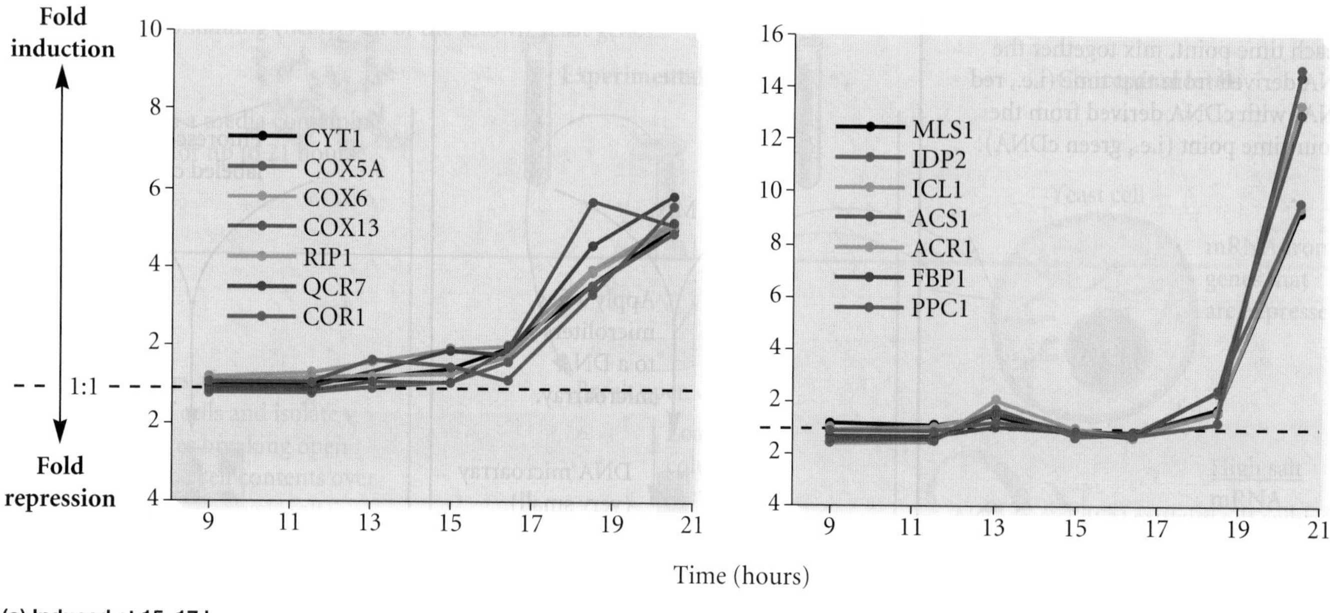

(a) Induced at 15–17 hours

(b) Induced at 19 hours

(c) Repressed at 15–17 hours

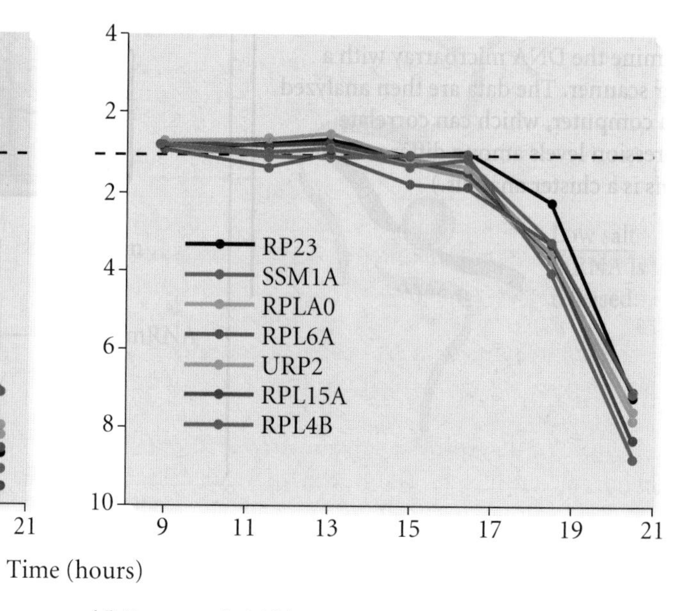

(d) Repressed at 19 hours

INTERPRETING THE DATA

A portion of one microarray is shown at the beginning of the data of Figure 21.2. As seen here, the array shows many spots, some of which are relatively green and some of which are relatively red. The green spots indicate genes that are expressed at higher levels during early stages of growth when glucose levels are high. An example is a gene designated *TEF4*, which is involved with protein synthesis. Red spots identify genes that are expressed when the glucose is depleted. The gene *GLK1* encodes an enzyme that phosphorylates glucose but is expressed only when glucose levels are low. In addition, many spots are not fluorescent, indicating there was not much cDNA in the sample to hybridize to the DNA strands in those locations. These spots correspond to genes that are not greatly expressed under either condition. Judging from the numbers of red and green spots, the shift from glycolysis to the TCA cycle involved a great amount of gene induction and repression. By determining the red:green ratio at each spot, it was found that 710 genes were induced by at least a factor of 2, whereas 1030 genes were repressed by a factor of 2 or more. Therefore, it was found that a diauxic shift involves a staggering amount of gene regulation. Of the 6200 yeast genes analyzed, 1740, or roughly 28%, appeared to be regulated as a result of a diauxic shift.

Because the gene sequence in each spot was known, the next step was to relate the levels of gene expression based on the microarray data to specific genes. A common goal is to identify genes whose pattern of expression seems to strongly correlate with each other. This statistical technique is called a **cluster analysis.** The graphs shown in the data of Figure 21.2 illustrate how microarray data can be used to make this type of comparison. Though the sequences of these genes were known from the yeast genome-sequencing project, the functions of some of these genes were not known. Each time point involved the measurement of the red:green ratio at a particular spot. The data of Figure 21.2 shows the analysis of 28 genes (7 in each panel) over the course of 9 to 21 hours following the addition of glucose. The diauxic shift occurred at approximately 15 hours.

Let's examine the data shown in Figure 21.2. In part (a), the genes were induced at 15 to 17 hours of growth. Due to their coordinate regulation, these genes may be controlled by the same transcription factor(s) and may participate in a common metabolic response to the induction of the TCA cycle. In fact, several of the genes in part (a) have already been studied, and they are known to play a role in the ability of the mitochondria to run the TCA cycle. This makes sense because the diauxic shift occurred around 15 hours of growth, which correlates with the time when the genes were induced. By comparison, the genes in part (b) were induced later, after the TCA cycle had operated for a few hours and when the carbon sources were becoming depleted. Therefore, it would seem that these genes were induced as a response to the operation of the TCA cycle or because the carbon sources in the media were low. Why is this information useful? Though further work needs to be done to elucidate the functions of some of the genes shown in parts (a) and (b), the data suggest a common regulation and metabolic function for particular groups of genes. It seems more likely that the transcription factors that regulate the genes in part (a) are different from those that regulate the genes in part (b). Likewise, the proteins encoded by these genes may work in different cellular pathways.

The genes shown in parts (c) and (d) illustrate a similar phenomenon, except that the switch to the TCA cycle and the operation of the TCA cycle represses those genes. The genes shown in part (c) were active during glycolysis and then were repressed at the time of the diauxic shift, whereas those shown in part (d) were repressed after the TCA cycle had operated for a few hours. Most of the genes shown in part (d) were already known to function as ribosomal proteins. Therefore, as the carbon sources in the media were depleted, these results suggest that one of the cellular responses is to diminish the synthesis of ribosomes, which, in turn, slows down the rate of protein synthesis. It seems that the yeast cells were trying to conserve energy at this late stage of growth.

Overall, the data shown in the experiment of Figure 21.2 illustrate how a microarray analysis can shed light on gene function at the genomic level. It provides great insight regarding gene regulation and may help to identify groups of proteins (i.e., clusters) that share a common cellular function.

A self-help quiz involving this experiment can be found at www.mhhe.com/brookergenetics4e.

DNA Microarrays Can Be Used to Identify DNA-Protein Binding at the Genome Level

As discussed throughout this textbook, the binding of proteins to specific DNA sites is critical for a variety of molecular processes, including gene transcription and DNA replication. To study these processes, researchers have devised a variety of techniques for identifying whether or not specific proteins bind to particular sites in the DNA. For example, the techniques of gel retardation and DNaseI footprinting, which are described in Chapter 18, are used for this purpose (see Figures 18.16 and 18.17).

More recently, a newer approach called **chromatin immunoprecipitation (ChIP)** has gained widespread use in the analysis of DNA–protein interactions. This method can determine whether proteins can bind to a particular region of DNA. A distinguishing feature of this method is that it analyzes DNA–protein interactions as they occur in the chromatin of living cells. In contrast, gel retardation and DNaseI footprinting are in vitro techniques, which typically use cloned DNA and purified proteins.

In Chapter 15, we considered how chromatin immunoprecipitation can be used in conjunction with DNA sequencing to determine the sites in the genome where nucleosomes are found (refer back to Figure 15.11). **Figure 21.3** describes the steps of the chromatin immunoprecipitation when it is used in conjunction with microarrays. Proteins in living cells, which are noncovalently bound to DNA, can be more tightly attached to sequence. If we ran this program, we would obtain the following result:

GJTRLLAMAQLHEOGYLTOBWENTMNMTORXXXTGOODNTHEQ ALLRTLSTORE

In this case, a computer program has identified locations where the sequence of letters forms a word. Several words (which are underlined) have been located within this sequence.

A second computer program could be aimed at locating a series of words that are organized in the correct order to form a grammatically logical English sentence. If we used our sequence file and ran this program, we would obtain the following result:

GJTRLLAMAQLHEOGYLTOBWENTMNMTORXXXTGOODNTHEQ ALLRTLSTORE

The second program has identified five words that form a logical sentence.

Finally, a computer program might be used to identify patterns of letters, rather than words or sentences. For example, a computer program could locate a pattern of five letters that occurs in both the forward and reverse directions. If we applied this program to our sequence file, we would obtain the following:

GJTRLLAMAQLHEOGYLTOBWENTMNMTORXXXTGOODNTHEQ ALLRTLSTORE

In this case, the program has identified a pattern where five letters are found in both the forward and reverse directions.

In the three previous examples, we can distinguish between **sequence recognition** (as in our first example) and **pattern recognition** (as in our third example). In sequence recognition, the program has the information that a specific sequence of symbols has a specialized meaning. This information must be supplied to the computer program. For example, the first program has access to the information from a dictionary with all known English words. With this information, the first program can identify sequences of letters that make words. By comparison, the third program does not rely on specialized sequence information. Rather, it is looking for a pattern of symbols that can occur within any group of symbol arrangements.

Overall, the simple programs we have considered illustrate three general types of identification strategies:

1. *Locate specialized sequences within a very long sequence.* A specialized sequence with a particular meaning or function is called a **sequence element** or **motif.** The computer program has a list of predefined sequence elements and can identify such elements within a sequence of interest.
2. *Locate an organization of sequences.* As shown in the second program, this could be an organization of sequence elements. Alternatively, it could be an organization of a pattern of symbols.
3. *Locate a pattern of symbols.* The third program is an example of locating a pattern of symbols.

The great power of computer analysis is that these types of operations can be performed with great speed and accuracy on sequences that may be enormously long.

Now that we understand the general ways that computer programs identify sequences, let's consider specific examples. As we have discussed throughout this textbook, many short nucleotide sequences play specialized roles in the structure or function of genetic material. A geneticist may want to locate a short sequence element within a longer nucleotide sequence in a data file. For example, a sequence of chromosomal DNA might be tens of thousands of nucleotides in length, and a geneticist may want to know whether a sequence element, such as a TATA box, is found at one or more sites within the chromosomal DNA. To do so, a researcher could visually examine the long chromosomal DNA sequence in search of a TATA box. Of course, this process would be tedious and prone to error. By comparison, the appropriate computer program can locate a sequence element in seconds. Therefore, computers are very useful for this type of application. **Table 21.4** lists some examples of sequence elements that can be identified by computer analysis.

By comparing the amino acid sequences and known functions of proteins in thousands of cases, researchers have also found amino acid sequence motifs that carry out specialized functions within proteins. For example, researchers have determined that the amino acid sequence motif asparagine–X–serine (where X is any amino acid except proline) within eukaryotic proteins is a glycosylation site (i.e., it may have a carbohydrate attached to it). The Prosite database (refer back to Table 21.3) contains a collection of all amino acid sequence motifs known to be functionally important. Researchers can use computer programs to determine whether an amino acid sequence contains any of the motifs found in the Prosite database. This may help them to understand the role of a newly found protein of unknown function.

TABLE 21.4

Short Sequence Elements That Can Be Identified by Computer Analysis

Type of Sequence	Examples*
Promoter	Many *E. coli* promoters contain TTGACA (−35 site) and TATAAT (−10 site). Eukaryotic core promoters may contain CAAT boxes, GC boxes, TATA boxes, etc.
Response elements	Glucocorticoid response element (AGRACA), cAMP response element (GTGACGTRA)
Start codon	ATG
Stop codons	TAA, TAG, TGA
Splice site	GTRAGT————YNYTRAC$(Y)_n$AG
Polyadenylation signal	AATAAA
Highly repetitive sequences	Relatively short sequences that are repeated many times throughout a genome
Transposable elements	Often characterized by a pattern in which direct repeats flank inverted repeats

*The sequences shown in this table would be found in the DNA. For gene sequences, only the coding strand is shown. R = purine (A or G); Y = pyrimidine (T or C); N = A, T, G, or C; U in RNA = T in DNA.

Several Computer-Based Approaches Can Identify Genes Within a Nucleotide Sequence

A typical gene that encodes a protein contains a promoter, followed by a start codon, a coding sequence, a stop codon, and a transcriptional termination site. In addition, most genes contain regulatory sequences (e.g., eukaryotic response elements or prokaryotic operator sites), and eukaryotic genes are likely to contain introns. After researchers have sequenced a long segment of chromosomal DNA, they frequently want to know if the sequence contains any genes. In an attempt to answer this question, geneticists can use computer programs that are aimed at identifying genes in long genomic DNA sequences.

How do computer programs identify a gene within a long genetic sequence? Such programs employ different strategies. A **search-by-signal** approach relies on known sequences such as promoters, start and stop codons, and splice sites to help predict whether or not a DNA sequence contains a structural gene. The program tries to locate an organization of known sequence elements that normally are found within a gene. It tries to locate a region that contains a promoter sequence, followed by a start codon, a coding sequence, a stop codon, and a transcriptional terminator.

A second strategy is a **search-by-content** approach. The goal here is to identify sequences with a nucleotide content that differs significantly from a random distribution. Within protein-coding genes, this occurs primarily due to codon usage. Although there are 64 codons, most organisms display a **codon bias** within the coding regions of genes. This means that certain codons are used much more frequently than others. For example, UUA, UUG, CUU, CUC, CUA, and CUG all specify leucine. In yeast, however, 80% of the leucine codons are UUG. Codon bias allows organisms to more efficiently rely on a smaller population of tRNA molecules. A search by content strategy, therefore, attempts to locate coding regions by identifying regions where the nucleotide content displays a known codon bias.

A third way to locate coding regions within a DNA sequence is to examine translational reading frames. Recall that the reading frame is a sequence of codons determined by reading bases in groups of three. In a new DNA sequence, researchers must consider that the reading of codons (in groups of three nucleotides) could begin with the first nucleotide (reading frame 1), the second nucleotide (reading frame 2), or the third nucleotide (reading frame 3). An **open reading frame (ORF)** is a region of a nucleotide sequence that does not contain any stop codons. Because most proteins are several hundred amino acids in length, a relatively long reading frame is required to encode them. In prokaryotic species, long ORFs are contained within the chromosomal gene sequences. In eukaryotic genes, however, the coding sequence may be interrupted by introns. As described earlier, one way to determine eukaryotic ORFs is to clone and sequence cDNA, which is complementary to mRNA. Alternatively, a computer program can translate a genomic DNA sequence in all three reading frames, seeking to identify a long ORF. In **Figure 21.7**, a DNA sequence has been translated in all three reading frames. Only one of the three reading frames (3) contains a very long ORF without any stop codons, suggesting that this DNA sequence encodes a protein. In Figure 21.7, it was assumed that the reading frame was from left to right. In an uncharacterized genetic sequence, however, a reading frame could proceed from right to left. Therefore, in a newly discovered genetic sequence, six reading frames are possible—three in the forward direction and three in the reverse direction.

DNA sequence

Reading frame

1 S S S S S S S S

2 S S S S S S S S S

3 S

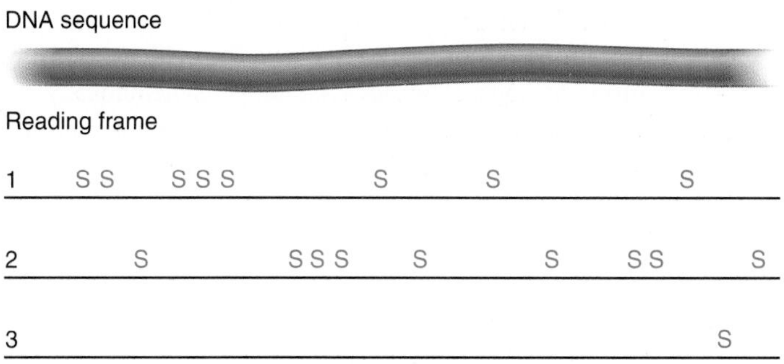

FIGURE 21.7 Translation of a DNA sequence in all three reading frames. The three lines represent the translation of a gene sequence in each of three forward reading frames; the reading frames proceed from left to right. The letter S indicates the location of a stop codon. Reading frame 3 has a very long open reading frame, suggesting that it may be the reading frame for a structural gene. Reading frames 1 and 2 are not likely to be the reading frames for a structural gene, because they contain many stop codons. During the cloning of DNA, the orientation of a gene may become flipped so that the coding sequence is inverted. Therefore, when analyzing many cloned DNA fragments, six reading frames (i.e., three forward and three reverse) are evaluated. Only the three forward frames are shown here.

Even though computer programs are a valuable tool, they are not always accurate in their prediction of gene sequences. In particular, it is often difficult for programs to predict the correct start codon and the precise intron–exon boundaries. In some cases, computer programs may even suggest that a region encodes a gene when it does not. Therefore, although a bioinformatic approach is a relatively easy tool for identifying potential genes, it should not be viewed as a definitive method. The confirmation that a DNA region encodes an actual gene requires laboratory experimentation to show that it is truly transcribed into RNA.

Computer Programs Can Identify Homologous Sequences

Let's now turn our attention to the uses of computer technology to identify genes that are evolutionarily related. The ability to sequence DNA allows geneticists to examine evolutionary relationships at the molecular level. This has become an extremely powerful tool in the field of genomics. When comparing genetic sequences, researchers frequently find two or more similar sequences. For example, the sequence of the *lacY* gene that encodes the lactose permease in *E. coli* is similar to that of the *lacY* gene that encodes the lactose permease in another bacterium, *Klebsiella pneumoniae*. As shown here, when segments of the two *lacY* genes are lined up, approximately 78% of their bases are a perfect match (**Figure 21.8a**).

In this case, the two sequences are similar because the genes are **homologous,** meaning they have been derived from the same ancestral gene. This idea is shown schematically in **Figure 21.8b**. An ancestral *lacY* gene was located in a bacterium that preceded the evolutionary divergence between *E. coli* and *K. pneumoniae*. After these two species diverged from each other, their *lacY* genes accumulated distinct mutations that produced somewhat different base sequences for this gene. Therefore, in these two species of bacteria, the *lacY* genes are similar but not identical. When two homologous genes are found in different species, they are termed **orthologs.**

Two or more homologous genes can also be found within a single species. These are termed paralogous genes, or **paralogs.** As discussed in Chapter 8, abnormal gene duplications may happen several times during evolution, which results in multiple copies of a gene and ultimately leads to the formation of a gene family. A **gene family** consists of two or more paralogs within the genome of a single organism. When a gene family occurs, the concept of orthologs becomes more complex. For example, let's consider the globin gene family found in mammals (see Chapter 8, Figure 8.7). Researchers would say that the β-globin gene in humans is an ortholog to the β-globin gene found in mice. Likewise, α-globin genes found in both species are considered orthologs. However, the α-globin gene in humans is not called an ortholog of the β-globin gene in mice, though they could be called homologous. The most closely related genes in two different species are considered orthologs.

Homologous genes, whether they are orthologs or paralogs, have similar sequences. What is the distinction between homology and sequence similarity? **Homology** implies a common ancestry. **Similarity** means that two sequences are similar to each other. In many cases, such as the *lacY* example, similarity is due to homology. However, this is not always the case. Short genetic sequences may be similar to each other even though two genes are not related evolutionarily. For example, many nonhomologous bacterial genes contain similar promoter sequences at the −35 and −10 regions.

```
              151                                                    200
E. coli       TCTTTTTCTT TTACTTTTTT ATCATGGGAG CCTACTTCCC GTTTTTCCCG
K. pneumoniae TCTTTTTCTT TTACTATTTC ATTATGTCAG CCTACTTTCC TTTTTTTCCG

              201                                                    250
              ATTTGGCTAC ATGACATCAA CCATATCAGC AAAAGTGATA CGGGTATTAT
              GTGTGGCTGG CGGAAGTTAA CCATTTAACC AAAACCGAGA CGGGTATTAT
```

(a) A comparison of a portion of the *lacY* gene from *E. coli* and *K. pneumoniae*

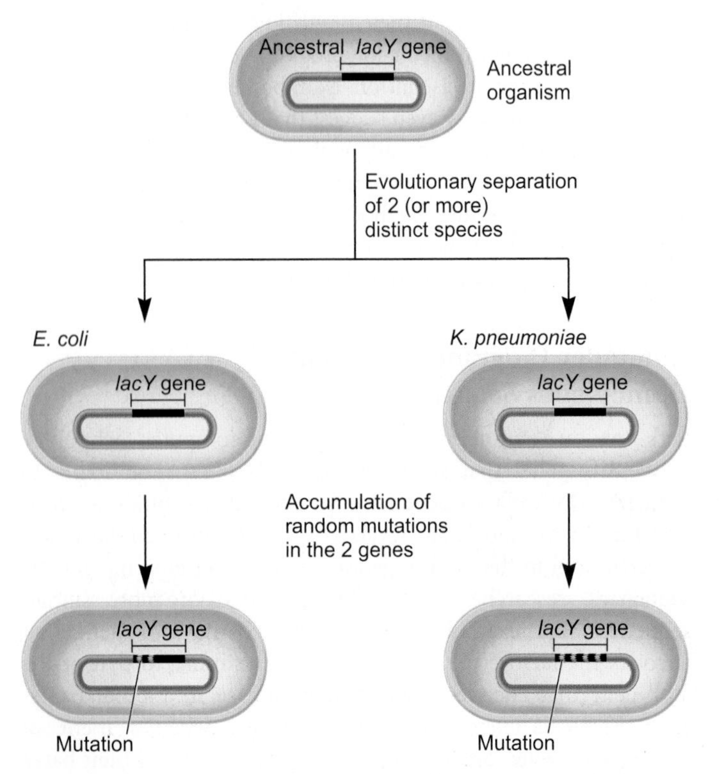

(b) The formation of homologous genes, also known as orthologs

FIGURE 21.8 The origin of homologous *lacY* genes in *Escherichia coli* and *Klebsiella pneumoniae*. (**a**) A comparison of a short region of the *lacY* gene from *E. coli* and *K. pneumoniae*. (**b**) This figure emphasizes a single gene within an ancestral organism. During evolution, the ancestral organism diverged into two different species: *E. coli* and *K. pneumoniae*. After this divergence, the *lacY* gene in the two separate species accumulated mutations, yielding *lacY* genes with somewhat different sequences.

Genes → Traits After two species diverge evolutionarily, their genes accumulate different random mutations. This example concerns the *lacY* gene, which encodes lactose permease. In both species, the function of lactose permease is to transport lactose into the cell. The *lacY* gene in these two species has accumulated different mutations that alter the amino acid sequence of the protein. Researchers have determined that these two species transport lactose at significantly different rates. Therefore, the changes in gene sequences have affected the ability of these two species to transport lactose.

A Simple Dot Matrix Can Compare the Degree of Similarity Between Two Sequences

To evaluate the similarity between two sequences, a matrix can be constructed. In a general way, **Figure 21.9** illustrates the use of a simple dot matrix. As shown in Figure 21.9a, the sequence GENETICSISCOOL is compared with itself. Each point in the grid corresponds to one position of each sequence. The matrix allows all such pairs to be compared simultaneously. Dots are placed where the same letter occurs in both sequences. The key observation is that regions of similarity are distinguished by the occurrence of many dots along a diagonal line within the dot matrix. In contrast, Figure 21.9b compares two unrelated sequences: GENETICSISCOOL and THECOURSEISFUN. In this comparison, no diagonal line is seen. In some cases, two sequences may be related to each other but differ in length. Figure 21.9c compares the sequences GENETICSISCOOL and GENETICSISVERYCOOL. In this example, the second sequence is four letters longer than the first. In the dot matrix, two diagonal lines occur. To align these two lines, a gap must be created in the first sequence. Figure 21.9d shows the insertion of a gap that aligns the two sequences so that the two diagonal lines fall along the same line.

Overall, Figure 21.9 illustrates two important features of dot matrix methods. First, regions of homology are recognized by a series of dots that lie along a diagonal line. Second, gaps can be inserted to align sequences of unequal length but that still have some similarity. These same concepts hold true when genetic sequences are compared to each other. Unfortunately, for anything but the shortest genetic sequences, a simple dot matrix approach is not adequate. Instead, dynamic programming methods are used to identify similarities between genetic sequences. These methods, originally proposed by Saul Needleman and Christian Wunsch, are theoretically similar to a dot matrix, but they involve complex mathematical operations beyond the scope of this textbook.

In their original work of 1970, Needleman and Wunsch demonstrated that whale myoglobin and human β hemoglobin have similar sequences. Since then, this approach has been extended to compare more than two genetic sequences. Newer computer programs can align several genetic sequences

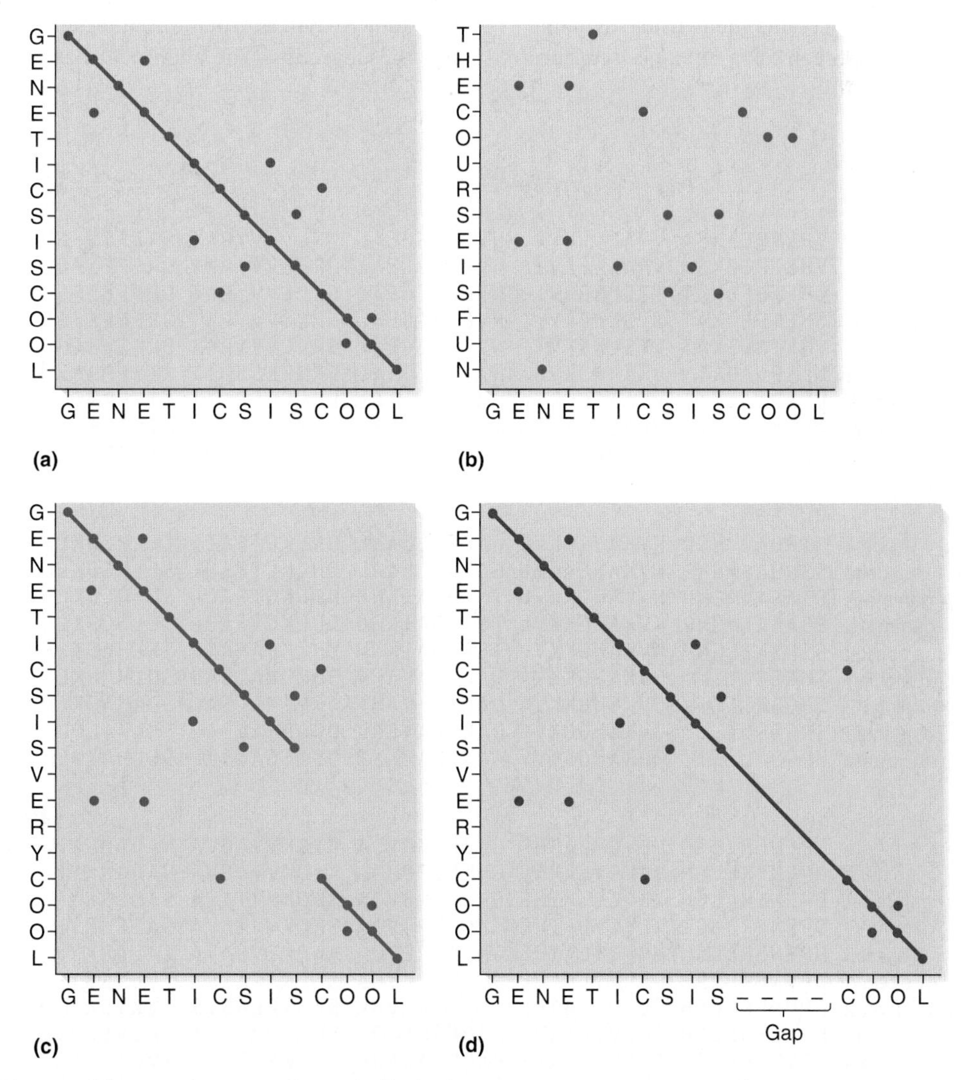

FIGURE 21.9 The use of dot matrices to evaluate similarity between two sequences. On a dot matrix, many points that fall on a diagonal line indicate regions of similarity.

and sensibly put in gaps. This produces a **multiple sequence alignment.**

To illustrate the usefulness of a multiple sequence alignment, let's use the general methods of Needleman and Wunsch and apply them to the globin gene family. Hemoglobin, a protein found in red blood cells, is responsible for carrying oxygen through the bloodstream. In humans, nine paralogous globin genes are functionally expressed. (There are also four pseudogenes that are not expressed and one myoglobin gene.) The nine globin genes fall into two categories: the α chains and the β chains. The α-chain genes are α_1, α_2, θ, and ξ; the β-chain genes are β, δ, γ_A, γ_G, and ε. Each hemoglobin protein is composed of two α chains and two β chains.

Because the globin genes are expressed at different stages of human development, the composition of hemoglobin changes during the course of growth. For example, the ξ and ε genes are expressed during early embryonic development, whereas the α and β genes are expressed in the adult.

Insights into the structure and function of the hemoglobin polypeptide chains can be gained by comparing their sequences. In **Figure 21.10**, the sequences of the human globin polypeptides are compared in a multiple sequence alignment using dynamic programming methods. An inspection of a multiple sequence alignment may reveal important features concerning the similarities and differences within a gene family. In this alignment, dots are shown where it is necessary to create gaps to keep the amino acid sequences aligned. As we can see, the sequence similarity is very high between α_1, α_2, θ, and ξ. In fact, the amino acid sequences encoded by the α_1 and α_2 genes are identical. This suggests that the four types of α chains likely carry out very similar functions. Likewise, the β chains encoded by the β, δ, γ_A, γ_G, and ε genes are very similar to each other. In the globin gene family, the α chains are much more similar to each other than they are to the β chains, and vice versa.

In general, amino acids that are highly conserved within a gene family are more likely to be important functionally. The arrows in the multiple sequence alignment point to histidine amino acids that are conserved in all nine members of the hemoglobin gene family. These histidines, which are highlighted in red, are involved in the necessary function of binding the heme molecule to the globin polypeptides.

Overall, the alignment shown in Figure 21.10 illustrates the type of information that can be derived from a multiple sequence alignment. In this case, multiple sequence alignment has shown that a group of nine genes falls into two closely related subgroups. The alignment has also identified particular

```
                                         Gap
               1                                                      50
    β beta     VHLTPEEKSA VTALWGKV.. NVDEVGGEAL GRLLVVYPWT QRFFESFGDL
    δ delta    VHLTPEEKSA VNALWGKV.. NVDAVGGEAL GRLLVVYPWT QRFFESFGDL
  γA gamma-A   GHFTEEDKAT ITSLWGKV.. NVEDAGGEAL GRLLVVYPWT QRFFESFGDL
  γG gamma-G   GHFTEEDKAT ITSLWGKV.. NVEDAGGEAL GRLLVVYPWT QRFFESFGDL
  ε epsilon    VHFTAEEKAA VTSLWSKM.. NVEEAGGEAL GRLLVVYPWT QRFFESFGDL
  α1 alpha-1   VLSP.ADKTN VKAAWGKVGA HAGEGAEAL  ERMFLSFPTT KTYFPHF.DL
  α2 alpha-2   VLSP.ADKTN VKAAWGKVGA HAGEGAEAL  ERMFLSFPTT KTYFPHF.DL
    θ theta    ALSA.EDRAL VRALWKKLGS NVGVYTTEAL ERTFLAFPAT KTYFSHL.DL
    ξ zeta     SLTK.TERTI IVSMWAKIST QADTIGTETL ERLFLSHPQT KTYFPHF.DL
                   Gap
               51                                                    100
    β beta     STPDAVMGNP KVKAHGKKVL GAFSDGLAHL DNLKGTFATL SELHCDKLHV
    δ delta    SSPDAVMGNP KVKAHGKKVL GAFSDGLAHL DNLKGTFSQL SELHCDKLHV
  γA gamma-A   SSASAIMGNP KVKAHGKKVL TSLGDAIKHL DDLKGTFAQL SELHCDKLHV
  γG gamma-G   SSASAIMGNP KVKAHGKKVL TSLGDAIKHL DDLKGTFAQL SELHCDKLHV
  ε epsilon    SSPSAILGNP KVKAHGKKVL TSFGDAIKNM DNLKPAFAKL SELHCDKLHV
  α1 alpha-1   SHGSA..... QVKGHGKKVA DALTNAVAHV DDMPNALSAL SDLHAHKLRV
  α2 alpha-2   SHGSA..... QVKGHGKKVA DALTNAVAHV DDMPNALSAL SDLHAHKLRV
    θ theta    SPGSS..... QVKAHGQKVA DALSLAVERL DDLPHALSAL SHLHACQLRV
    ξ zeta     HPGSA..... QLRAHGSKVV AAVGDAVKSI DDIGGALSKL SELHAYQLRV
                    Gap
               101                                                   148
    β beta     DPENFRLLGN VLVCVLAHHF GKEFTPPVQA AYQKVVAGVA NALAHKYH
    δ delta    DPENFRLLGN VLVCVLARNF GKEFTPQMQA AYQKVVAGVA NALAHKYH
  γA gamma-A   DPENFRLLGN VLVCVLAIHF GKEFTPEVQA SWQKMVTAVA SALSSRYH
  γG gamma-G   DPENFRLLGN VLVCVLAIHF GKEFTPEVQA SWQKMVTAVA SALSSRYH
  ε epsilon    DPENFRLLGN VMVIILATHF GKEFTPEVQA AWQKLVSAVA IALAHKYH
  α1 alpha-1   DPVNFKLLSH CLLVTLAAHL PAEFTPAVHA SLDKFLASVS TVLTSKYR
  α2 alpha-2   DPVNFKLLSH CLLVTLAAHL PAEFTPAVHA SLDKFLASVS TVLTSKYR
    θ theta    DPASFQLLGH CLLVTLARHL PGDFSPALQA SLDKFLSHSVI SALVSEYR
    ξ zeta     DPVNFKLLSH CLLVTLAARF PADFTAEAHA AWDKFLSVVS SVLTEKYR
```

FIGURE 21.10 **A multiple sequence alignment among selected members of the globin gene family in humans.**

amino acids within the proteins' sequences that are highly conserved. This conservation is consistent with an important role in protein function.

A Database Can Be Searched to Identify Homologous Sequences

Homologous genes usually encode proteins that carry out similar or identical functions. As we have just considered, the members of the globin gene family are all involved with carrying and transporting oxygen. Likewise, the *lacY* genes in *E. coli* and *K. pneumoniae* both encode lactose permease, a protein that transports lactose across the bacterial cell membrane.

A strong correlation is typically found between homology and function. How is this relationship useful with regard to bioinformatics? In many cases, the first indication of the function of a newly determined sequence is through homology to known sequences in a database. An example is the *CFTR* gene that is altered in cystic fibrosis patients. After the *CFTR* gene was identified in humans, a database search revealed it is homologous to several genes found in other species. Moreover, a few of the homologous genes were already known to encode proteins that function in the transport of ions and small molecules across the plasma membrane. This observation provided an important clue that cystic fibrosis involves a defect in ion transport.

The ability of computer programs to identify homology between genetic sequences provides a powerful tool for predicting the function of genetic sequences. In 1990, Stephen Altschul, David Lipman, and colleagues developed a program called **BLAST** (for **basic local alignment search tool**). The BLAST program has been described by many geneticists as the single most important bioinformatic tool. This computer program starts with a particular genetic sequence and then locates homologous sequences within a large database. Because there are 20 amino acids but only four bases, homology among protein sequences is easier to identify than DNA sequence homology. Among proteins, sequences that diverged more than 2.5 billion years ago can still be correlated. By comparison, it becomes difficult to identify homologous DNA sequences that diverged more than 100 million years ago.

To see how the BLAST program works, let's consider the human enzyme phenylalanine hydroxylase, which functions in the metabolism of phenylalanine, an amino acid. Recessive mutations in the gene that encodes this enzyme are responsible for the disease called phenylketonuria (PKU). The computational experiment shown in **Table 21.5** started with the amino acid sequence of this protein and used the BLAST program to search the Swissprot database, which contains millions of different protein sequences. The BLAST program can determine which sequences in this database are the closest matches to the amino acid sequence of human phenylalanine hydroxylase. Table 21.5 shows a portion of the results—10 selected matches to human phenylalanine hydroxylase that were identified by the program. Because this enzyme is found in nearly all eukaryotic species, the program identified phenylalanine hydroxylase from many different species. The column to the right of the match number shows the percentage of amino acids that are identical between the species indicated and the human sequence. Because the human phenylalanine hydroxylase sequence is already in the Swissprot database, the closest match of human phenylalanine hydroxylase is to itself (100%). The next nine sequences are in order of similarity. The next most similar sequence is from the orangutan (99%), a close relative of humans. This is followed by two mammals, the mouse and rat, and then four vertebrates that are not mammals.

TABLE **21.5**

Results from a BLAST Program Comparing Human Phenylalanine Hydroxylase with Database Sequences

Match	% of Identical Amino Acids*	Species	Function of Sequence†	E-value
1	100	Human (*Homo sapiens*)	Phenylalanine hydroxylase	0
2	99	Orangutan (*Pongo pygmaeus*)	Phenylalanine hydroxylase	0
3	92	Mouse (*Mus musculus*)	Phenylalanine hydroxylase	0
4	92	Rat (*Rattus norvegicus*)	Phenylalanine hydroxylase	0
5	83	Chicken (*Gallus gallus*)	Phenylalanine hydroxylase	0
6	78	Western clawed frog (*Xenopus tropicalis*)	Phenylalanine hydroxylase	0
7	75	Zebrafish (*Danio rerio*)	Phenylalanine hydroxylase	0
8	72	Japanese pufferfish (*Takifugu rubripes*)	Phenylalanine hydroxylase	0
9	62	Fruit fly (*Drosophila melanogaster*)	Phenylalanine hydroxylase	10^{-154}
10	57	Nematode (*Caenorhabditis elegans*)	Phenylalanine hydroxylase	10^{-141}

*The number indicates the percentage of amino acids that are identical to the amino acid sequence of human phenylalanine hydroxylase.

†In some cases, the function of the sequence was determined by biochemical assay. In other cases, the function was inferred due to the high degree of sequence similarity with other species.

The ninth and tenth best matches are from *Drosophila* and *C. elegans*, which are invertebrates.

As shown in the right column of Table 21.5, the relationship between the query sequence and each matching sequence is given an E-value (Expect value). The **E-value** represents the number of times that the match or a better one would be expected to occur purely by random chance in a search of the entire database. An E-value that is very small indicates that the similarity between the query sequence and the matching sequence is unlikely to have occurred by random chance. Instead, researchers would accept the hypothesis that the two sequences are homologous, which means they are derived from the same ancestral sequence.

E-values are very dependent on different parameters such as the length of the query sequence, the number of gaps between the query sequence and the matching sequence, and the database size. As a general rule, if the E-value is less than 1×10^{-50}, the match is very similar to the query sequence and is likely to be homologous. (Values that are much less than 10^{-100} are reported as zero by the BLAST program). If the value lies between 1×10^{-50} and 1×10^{-10}, the match, or part of it, is likely to be homologous. If the value is between 1×10^{-10} and 1×10^{-2}, the match has a significant chance of being related to the query sequence, whereas values between 1 and 1×10^{-2} have a relatively low probability of being homologous. Values above 1 are usually not evolutionarily related. As seen in Table 21.5, all of the E-values are below 1×10^{-140}, which suggests that all of these matches are homologous to the query sequence.

You can see two trends in Table 21.5. First, the order of the matches follows the evolutionary relatedness of the various species to humans. The similarity between any two sequences is related to the time that has passed since they diverged from a common ancestor. Among the species listed in this table, the human sequence is most similar to the orangutan, a closely related primate. The next most similar sequences are found in other mammals, followed by other vertebrates, and finally invertebrates. A second trend you may have noticed is that several of the matches involve species that are important from a research, medical, or agricultural perspective. Currently, our genetic databases are biased toward organisms that are of interest to humans, particularly model organisms such as mice and *Drosophila*. Over the next several decades, the sequencing of genomes from many different species will tend to lessen this bias.

The results shown in Table 21.5 illustrate the remarkable computational abilities of current computer technology. In minutes, the human phenylalanine hydroxylase sequence can be compared with millions of different sequences.

Genetic Sequences Can Be Used to Predict the Structure of RNA and Proteins

Another topic in which bioinformatics has influenced functional genomics and proteomics is the area of structure prediction. The function of macromolecules such as DNA, RNA, and proteins relies on their three-dimensional structure, which, in turn, depends on the linear sequences of their building blocks. In the case of DNA and RNA, this means a linear sequence of nucleotides; proteins are composed of a linear sequence of amino acids. Currently, the three-dimensional structure of macromolecules is determined primarily through the use of biophysical techniques such as X-ray crystallography and nuclear magnetic resonance (NMR). These methods are technically difficult and very time-consuming. DNA sequencing, by comparison, requires much less effort. Therefore, because the three-dimensional structure of macromolecules depends ultimately on the linear sequence of their building blocks, it would be far easier if we could predict the structure (and function) of DNAs, RNAs, and proteins from their sequence of building blocks.

RNA molecules typically are folded into a secondary structure, which commonly contains double-stranded regions. This secondary structure is further folded and twisted to adopt a tertiary conformation. Such structural features of RNA molecules are functionally important. For example, the folding of RNA into secondary structures, such as stem loops, affects transcriptional termination and other regulatory events. Therefore, geneticists are interested in the secondary and tertiary structures that RNA molecules can adopt.

Many approaches are available for investigating RNA structure. In addition to biophysical and biochemical techniques, computer modeling of RNA structure has become an important tool. Modeling programs can consider different types of information. For example, the known characteristics of RNA secondary structure, such as the ability to form double-stranded regions, can provide parameters for use in a modeling program.

A comparative approach can also be used in RNA structure prediction. This method assumes that RNAs of similar function and sequence have a similar structure. For example, the genes that encode certain types of RNAs, such as the 16S rRNA that makes up most of the small ribosomal subunit, have been sequenced from many different species. Among different species, the 16S rRNAs have similar but not identical sequences. Computer programs can compare many different 16S rRNA sequences to aid in the prediction of secondary structure. **Figure 21.11** illustrates a secondary structural model for 16S rRNA from *E. coli* based on a comparative sequence analysis of many bacterial 16S rRNA sequences. This large RNA contains 45 stem-loops. As you can imagine, it would be rather difficult to deduce such a model without the aid of a computer! In addition, RNA secondary structure prediction may be aimed at predicting the lowest energy state of a folded molecule. This approach, called free energy minimization, is also related to the base sequence of an RNA molecule.

Structure prediction is also used in the area of proteomics. As described in Chapter 12, proteins contain repeating secondary structural patterns known as α helices and β sheets. Several computer-based approaches attempt to predict secondary structure from the primary amino acid sequence. These programs base their predictions on different types of parameters. Some programs rely on the physical and energetic properties of the amino acids and the polypeptide backbone. More commonly, however, secondary structure predictions are based on the statistical frequency of amino acids within secondary structures that have already been crystallized.

For example, Peter Chou and Gerald Fasman compiled X-ray crystallographic data to calculate the likelihood that an

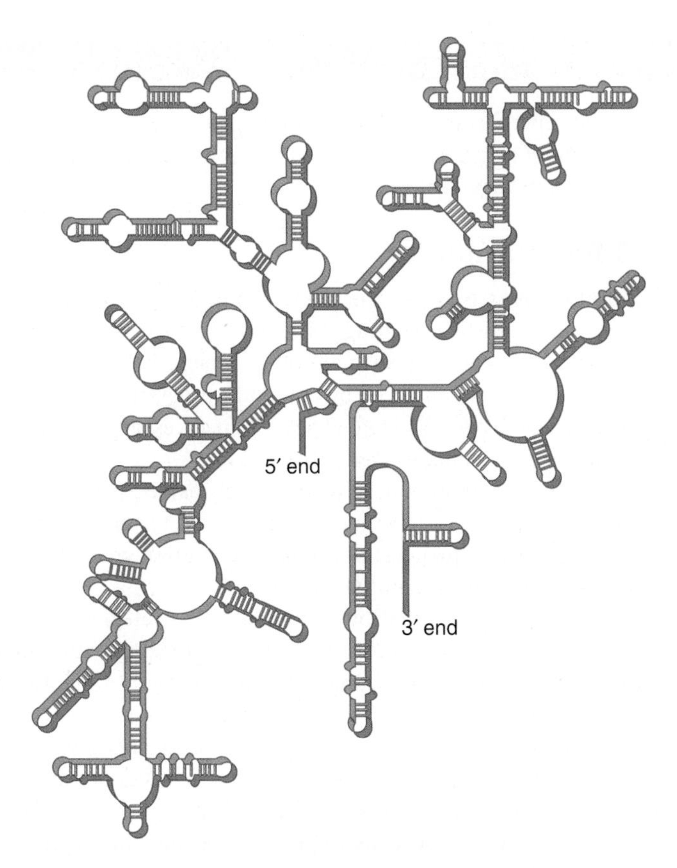

FIGURE 21.11 A secondary structural model for *E. coli* 16S rRNA.

amino acid will be found in an α helix or a β sheet. Certain amino acids, such as glutamic acid and alanine, are likely to be found in an α helix; others, such as valine and isoleucine, are more likely to be found in a β sheet structure. Such information can be used to predict whether a sequence of amino acids within a protein is likely to be folded in an α helix or β sheet conformation. Secondary structure prediction is correct for approximately 60 to 70% of all sequences. Although this degree of accuracy is promising, it is generally not sufficient to predict protein secondary structure reliably. Therefore, one must be cautious in interpreting the results of a secondary structure prediction program.

In recent years, an exciting computer methodology known as a neural network has been applied to protein secondary structure prediction. A computer neural network is a large number of calculation units organized into interconnected layers; this structure is reminiscent of the organization of neurons in the brain. The input layer receives data and may (or may not) transmit that information to the next layer. A neural network can adjust the parameters that define the interconnections among its units in response to data; the network can thus be trained to identify complex patterns coming from the input data. For example, the amino acid sequences of proteins with known crystal structures can be used to train a network (i.e., adjust its parameters) to predict secondary structures for new amino acid sequences. Thus far, neural networks have yielded small improvements in the prediction of protein secondary structure. In the future, a combination of innovative predictive approaches and increased information concerning the biophysical properties of amino acids may make secondary structure prediction a reliable strategy.

Although the three-dimensional structure of a protein is extremely difficult to predict solely from its amino acid sequence, researchers have had some success in predicting tertiary structure using a comparative approach. This strategy requires that the protein of interest be homologous to another protein, the tertiary structure of which already has been solved by X-ray crystallography. In this situation, the crystal structure of the known protein can be used as a starting point to model the three-dimensional structure of the protein of interest. This approach is known as **homology-based modeling,** knowledge-based modeling, or comparative homology. As an example, two research groups independently predicted the three-dimensional structure of a protein, now known as HIV protease, that is encoded by a gene in HIV. The protein is homologous to other proteases with structures that had been solved by X-ray crystallography. Two similar models of the HIV protease were predicted before its actual crystal structure was determined. Both models turned out to be fairly accurate representations of the actual structure, which was later solved by X-ray crystallography. Knowledge of the structure of HIV protease helped scientists synthesize compounds known as HIV protease inhibitors, which are an important part of drug therapy to treat HIV/AIDS.

KEY TERMS

Page 574. functional genomics, proteome, proteomics, bioinformatics
Page 575. DNA microarray, gene chip
Page 581. cluster analysis, chromatin immunoprecipitation (ChIP)
Page 582. ChIP-chip assay
Page 583. gene knockout, alternative splicing, RNA editing
Page 584. posttranslational covalent modification, two-dimensional gel electrophoresis, isoelectric focusing
Page 585. mass spectrometry
Page 586. tandem mass spectrometry, protein microarrays
Page 587. antibody microarray, functional protein microarray, computer programs, computer data file
Page 589. database, annotated, genome databases
Page 590. sequence recognition, pattern recognition, sequence element, motif
Page 591. search-by-signal, search-by-content, codon bias, open reading frame (ORF)
Page 592. homologous, orthologs, paralogs, gene family, homology, similarity
Page 594. multiple-sequence alignment
Page 595. BLAST (basic local alignment search tool)
Page 596. E-value
Page 597. homology-based modeling

CHAPTER SUMMARY

21.1 Functional Genomics

- The goal of functional genomics is to understand the role of genetic sequences in a given species.
- A microarray is a slide dotted with many DNA sequences. It can be used to study the expression of many genes simultaneously and also has other uses (see Figure 21.1, Table 21.1).
- A microarray can be used to study the coordinate regulation of groups of genes, which is called a cluster analysis (see Figure 21.2).
- In a ChIP-chip assay, chromatin immunoprecipitation is used in conjunction with a microarray to study DNA–protein interactions at the genomic level (see Figure 21.3).
- Researchers are producing gene knockout collections for certain species such as mice to determine the functions of genes at the genomic level.

21.2 Proteomics

- The proteome is the entire collection of proteins that a cell or organism makes. The study of the function and interactions of many proteins is called proteomics.
- The proteome is much larger than the genome due to alternative splicing, RNA editing, and posttranslational covalent modifications (see Figure 21.4).
- Two-dimensional gel electrophoresis is used to separate a complex mixture of proteins (see Figure 21.5).
- Mass spectrometry and tandem mass spectrometry are used to identify short amino acids sequences within purified proteins. These short sequences can be used to identify a protein (see Figure 21.6).
- Protein microarrays are used to study protein expression, protein function, protein–protein interactions, and protein–drug interactions (see Table 21.2).

21.3 Bioinformatics

- Bioinformatics involves the use of computers, mathematical tools, and statistical techniques to record, store, and analyze biological information, such as DNA sequences.
- Sequence files are analyzed by computer programs.
- Researchers have collected genetic sequences and compiled them in large databases (see Table 21.3).
- Different computational strategies, such as sequence recognition and pattern recognition, can be used to identify functional genetic sequences. Sequence recognition may identify sequence elements or motifs (see Table 21.4).
- Genes may be identified by computational strategies such as search-by-signal or search-by-content approaches. Searching for a long open reading frame may also be used to identify a gene (see Figure 21.7).
- Homologous genes are derived from the same ancestral gene. They can be orthologs (genes in different species) or paralogs (genes in the same species) (see Figure 21.8).
- A dot matrix can be constructed to evaluate the degree of similarity between short sequences. Researchers may produce a multiple sequence alignment to compare the sequences of several homologous genes (see Figures 21.9, 21.10).
- The BLAST program is used to identify homologous sequences that are found within a database (see Table 21.5).
- Computer programs may also use genetic sequences and try to predict the structure of DNA, RNA, or proteins (see Figure 21.11).

PROBLEM SETS & INSIGHTS

Solved Problems

S1. To answer this question, you will need to look back at the evolution of the globin gene family, which is shown in Chapter 8, Figure 8.7. Throughout the evolution of this gene family, mutations have occurred that have resulted in globin polypeptides with similar but significantly different amino acid sequences. If we look at the sequence alignment in Figure 21.10, we can make logical guesses regarding the timing of mutations, based on a comparison of the amino acid sequences of family members. What is/are the most probable time(s) that mutations occurred to produce the following amino acid differences? Note: You need to examine the alignment in Figure 21.10 and the evolutionary timescale in Figure 8.7 to answer this question.

A. Val-111 and Cys-111

B. Met-112 and Leu-112

C. Ser-141, Asn-141, Ile-141, and Thr-141

Answer:

A. We do not know if the original globin gene encoded a cysteine or valine at codon 111. The mutation could have changed cysteine to valine or valine to cysteine. The mutation probably occurred after the duplication that produced the α-globin family and β-globin family (about 300 million years ago) but before the gene duplications that occurred in the last 200 millions years to produce the multiple copies of the globin genes on chromosome 11 and chromosome 16. Therefore, all of the globin genes on chromosome 11 have a valine at codon 111, and all of the globin genes on chromosome 16 have a cysteine.

B. Met-112 occurs only in the ε-globin polypeptide; all of the other globin polypeptides contain a leucine at position 112. Therefore, the primordial globin gene probably contained a leucine codon at position 112. After the gene duplication that produced the ε-globin gene, a mutation occurred that changed this leucine codon into a methionine codon. This would have occurred since the evolution of primates (i.e., within the last 10 or 20 million years).

C. When we look at the possible codons at position 141 (i.e., Ser-141, Asn-141, Ile-141, and Thr-141), we notice that a serine codon is found in θ globin, ξ globin, and γ globin. Because the θ- and ξ-globin genes are found on chromosome 16 and the γ-globin genes are found on chromosome 11, it is probable that serine is the primordial codon and that the other codons (asparagine, isoleucine, and threonine) arose later by mutation of the serine codon. If this is correct, the Thr-141 codon arose before the gene duplication that produced the α-globin genes. The Asn-141 and Ile-141 mutations arose after the gene duplications that produced the γ-globin genes. Therefore, the Thr-141, Asn-141, and Ile-141 arose since the evolution of primates (i.e., within the last 10 or 20 million years).

S2. Using a comparative sequence analysis, the secondary structures of rRNAs have been predicted. Among many homologous rRNAs, one stem-loop usually has the following structure:

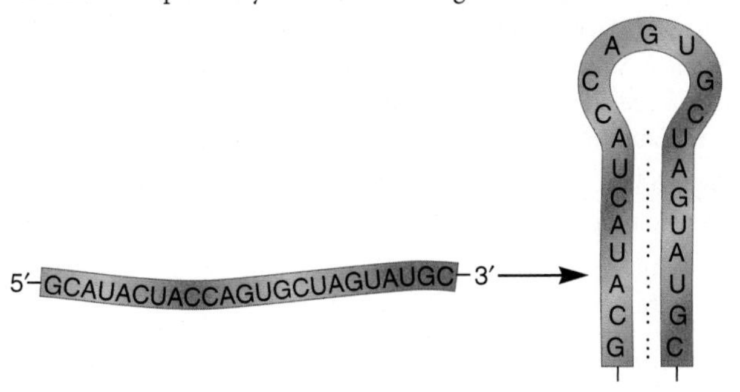

You have sequenced a homologous rRNA from a new species and have obtained most of its sequence, but you cannot read the last five bases on your sequencing gel.

5′–GCATTCTACCAGTGCTAG?????–3′

Of course, you will eventually repeat this experiment to determine the last five bases. However, before you get around to doing this, what do you expect will be the sequence of the last five bases?

Answer: AATGC–3′

This will also form a similar stem-loop structure.

S3. How can codon bias be used to search for structural genes within uncharacterized genetic sequences?

Answer: Most species exhibit a bias in the codons they use within the coding sequence of structural genes. This causes the base content within coding sequences to differ significantly from that of noncoding DNA regions. By knowing the codon bias for a particular species, researchers can use a computer to locate regions that display this bias and thereby identify what are likely to be the coding regions of structural genes.

Conceptual Questions

C1. Discuss the meaning of the following terms: genomics, functional genomics, and proteomics.

C2. Discuss the reasons why the proteome is larger than the genome of a given species.

C3. What is a database? What types of information are stored within a database? Where does the information come from? Discuss the objectives of a genome database.

C4. Besides the examples listed in Table 21.4, list five types of short sequences that a geneticist might want to locate within a DNA sequence.

C5. Discuss the distinction between sequence recognition and pattern recognition.

C6. A multiple sequence alignment of five homologous proteins is shown here:

```
  1                                                50
1 MLAFLNQVRK PTLDLPLEVR RKMWFKPFM. QSYLVVFIGY LTMYLIRKNF
2 MLAFLNQVRK PTLDLALDVR RKMWFKPFM. QSYLVVFIGY LTMYLIRKNF
3 MLPFLKAPAD APL.MTDKYE IDARYRYWRR HILLTIWLGY ALFYFTRKSF
4 MLSFLKAPAN APL.ITDKHE VDARYRYWRR HILITIWLGY ALFYFTRKSF
5 MLSIFKPAPH KAR.LPAA.E IDPTYRRLRW QIFLGIFFGY AAYYLVRKNF

  51                                              100
1 NIAQNDMIST YGLSMTQLGM IGLGFSITYG VGKTLVSYYA DGKNTKQFLP
2 NIAQNDMIST YGLSMTELGM IGLGFSITYG VGKTLVSYYA DGKNTKQFLP
3 NAAVPEILAN GVLSRSDIGL LATLFYITYG VSKFVSGIVS DRSNARYFMG
4 NAAAPEILAS GILTRSDIGL LATLFYITYG VSKFVSGIVS DRSNARYFMG
5 ALAMPYLVEQ .GFSRGDLGF ALSGISIAYG FSKFIMGSVS DRSNPRVFLP
```

Discuss some of the interesting features that this alignment reveals.

C7. What is the difference between similarity and homology?

C8. Which of the following statements uses the term homologous correctly?

A. The two X chromosomes in female mammalian cells are homologous to each other.

B. The α-tubulin gene in *Saccharomyces cerevisiae* is homologous to the α-tubulin gene in *Arabidopsis thaliana.*

C. The promoter of the *lac* operon is homologous to the promoter of the *trp* operon.

D. The *lacY* gene of *E. coli* and *Klebsiella pneumoniae* are approximately 60% homologous to each other.

C9. When comparing (i.e., aligning) two or more genetic sequences, it is sometimes necessary to put in gaps. Explain why. Discuss two changes (i.e., two types of mutations) that could happen during the evolution of homologous genes that would explain the occurrence of gaps in multiple sequence alignments.

Experimental Questions

E1. With regard to DNA microarrays, answer the following questions:

A. What is attached to the slide? Be specific about the number of spots, the lengths of DNA fragments, and the origin of the DNA fragments.

B. What is hybridized to the microarray?

C. How is hybridization detected?

E2. In the experiment of Figure 21.2, explain how the ratio of red:green fluorescence provides information regarding gene regulation.

E3. What is meant by the term cluster analysis? How is this approach useful?

E4. For two-dimensional gel electrophoresis, what physical properties of proteins promote their separation in the first dimension and the second dimension?

E5. Can two-dimensional gel electrophoresis be used as a purification technique? Explain.

E6. Explain how tandem mass spectroscopy can be used to determine the sequence of a peptide. Once a peptide sequence is known, how is this information used to determine the sequence of the entire protein?

E7. Describe the two general types of protein microarrays. What are their possible applications?

E8. Discuss the strategies that can be used to identify a protein-encoding gene using bioinformatics.

E9. What is a motif? Why is it useful for computer programs to identify functional motifs within amino acid sequences?

E10. Discuss why it is useful to search a database to identify sequences that are homologous to a newly determined sequence.

E11. The secondary structure of 16S rRNA has been predicted using a computer-based sequence analysis. In general terms, discuss what type of information is used in a comparative sequence analysis, and explain what assumptions are made concerning the structure of homologous RNAs.

E12. Discuss the basis for secondary structure prediction in proteins. How reliable is it?

E13. To reliably predict the tertiary structure of a protein based on its amino acid sequence, what type of information must be available?

E14. In this chapter, we considered a computer program that can translate a DNA sequence into a polypeptide sequence. A researcher has a sequence file that contains the amino acid sequence of a polypeptide and runs a program that is opposite to this program. This other program is called BACKTRANSLATE. It can take an amino acid sequence file and determine the sequence of DNA that would encode such a polypeptide. How does this program work? In other words, what are the genetic principles that underlie this program? What type of sequence file would this program generate: a nucleotide sequence or an amino acid sequence? Would the BACKTRANSLATE program produce only a single sequence file? Explain why or why not.

E15. In this chapter, we considered a computer program that can translate a DNA sequence into a polypeptide sequence. Instead of running this program, a researcher could simply look the codons up in a genetic code table and determine the sequence by hand. What are the advantages of running this program rather than doing it the old-fashioned way by hand?

E16. To identify the following types of genetic occurrences, would a program use sequence recognition, pattern recognition, or both?

A. Whether a segment of *Drosophila* DNA contains a P element (which is a specific type of transposable element)

B. Whether a segment of DNA contains a stop codon

C. In a comparison of two DNA segments, whether there is an inversion in one segment compared to the other segment

D. Whether a long segment of bacterial DNA contains one or more genes

E17. The goal of many computer programs is to identify "sequence elements" within a long segment of DNA. What is a sequence element? Give two examples. How is the specific sequence of a sequence element determined? In other words, is it determined by the computer program or by genetic studies? Explain.

E18. Take a look at the multiple sequence alignment in Figure 21.10 of the globin polypeptides from amino acids 101 to 148.

A. Which of these amino acids are likely to be most important for globin structure and function? Explain why.

B. Which are likely to be least important?

E19. See solved problem S1 before answering this question. Based on the sequence alignment in Figure 21.10, what is/are the most probable time(s) that mutations occurred in the human globin gene family to produce the following amino acid differences?

A. His-119 and Arg-119

B. Gly-121 and Pro-121

C. Glu-103, Val-103, and Ala-103

E20. Here is a short nucleotide sequence within a gene. Via the Internet (e.g., see www.ncbi.nlm.nih.gov/Tools), determine what gene this sequence is found within. Also, determine the species in which this gene sequence is found.

5′-GGGCGCAATTACTTAACGCCTCGATT
ATCTTCTTGCGCCACTGATCATTA-3′

E21. Take a look at solved problem S1 and the codon table found in Chapter 13 (Table 13.1). Assuming that a mutation involving a single-base change is more likely than a double-base change, propose how the Asn-141, Ile-141, and Thr-141 codons arose. In your answer, describe which of the six possible serine codons is/are likely to be the primordial serine codon of the globin gene family and how that codon changed to produce the Asn-141, Ile-141, and Thr-141 codons.

E22. Membrane proteins often have transmembrane regions that span the membrane in an α-helical conformation. These transmembrane segments are about 20 amino acids long and usually contain amino acids with nonpolar (i.e., hydrophobic) amino acid side chains. Researchers can predict whether a polypeptide sequence has transmembrane segments based on the occurrence of segments that contain 20 nonpolar amino acids. To do so, each amino acid is assigned a hydropathy value, based on the chemistry of its amino acid side chain. Amino acids with very nonpolar side chains are given a high (positive) value, whereas amino acids that are charged and/or polar are given low (negative) values. The hydropathy values usually range from about +4 to −4.

Computer programs have been devised that scan the amino acid sequence of a polypeptide and calculate values based on the hydropathy values of the amino acid side chains. The program usually scans a window of seven amino acids and assigns an average hydropathy value. For example, the program would scan amino acids 1 through 7 and give an average value, then it would scan 2 through 8 and give a value, then it would scan 3 through 9 and give a value, and so on, until it reached the end of the polypeptide sequence (i.e., until it reached the carboxyl terminus).

The program then produces a figure, known as a hydropathy plot, which describes the average hydropathy values throughout the entire polypeptide sequence. An example of a hydropathy plot is shown here.

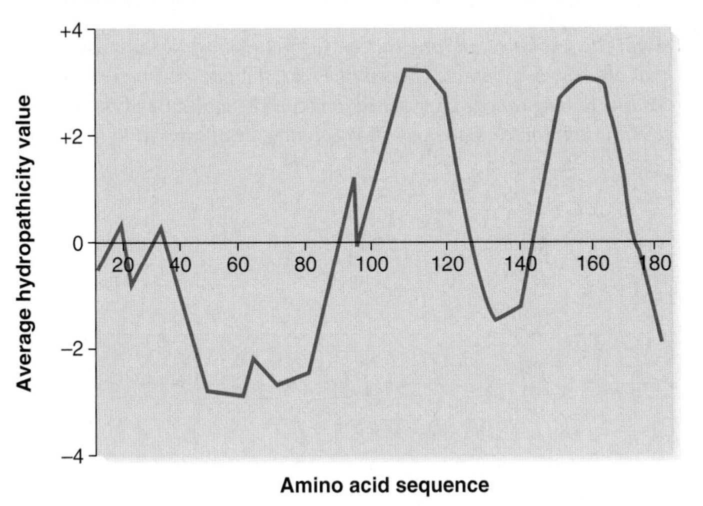

A. How many transmembrane segments are likely in this polypeptide?

B. Draw the structure of this polypeptide if it were embedded in the plasma membrane. Assume that the amino terminus is found in the cytoplasm of the cell.

E23. Explain how a computer program can predict RNA secondary structure. What is the underlying genetic concept used by the program to predict secondary structure?

E24. Are the following statements about protein structure prediction true or false?

A. The prediction of secondary structure relies on information regarding the known occurrence of amino acid residues in α helices or β sheets from X-ray crystallographic data.

B. The prediction of secondary structure is highly accurate, nearly 100% correct.

C. To predict the tertiary structure of a protein based on its amino acid sequence, it is necessary that the protein of interest is homologous to another protein whose tertiary structure is already known.

Questions for Student Discussion/Collaboration

1. Let's suppose you are in charge of organizing and publicizing a genomic database for the mouse genome. Make a list of innovative strategies you would initiate to make the mouse genome database useful and effective.

2. Let's suppose a 5-year-old told you that she was interested in pursuing a career studying the three-dimensional structure of proteins. (Okay, so she's a bit precocious.) Would you advise her to become a geneticist, a mathematical theoretician, or a biophysicist?

3. If you have access to the necessary computer software, make a sequence file and analyze it in the following ways: What is the translated sequence in all three reading frames? What is the longest open reading frame? Is the sequence homologous to any known sequences? If so, does this provide any clues about the function of the sequence?

Note: All answers appear at the website for this textbook; the answers to even-numbered questions are in the back of the textbook.

www.mhhe.com/brookergenetics4e

Visit the website for practice tests, answer keys, and other learning aids for this chapter. Enhance your understanding of genetics with our interactive exercises, quizzes, animations, and much more.

CHAPTER OUTLINE

Cigarette smoking and lung cancer. *Cigarette smoke contains chemicals that are known to mutate genes in the cells of a person's lungs, thereby leading to lung cancer. Lung cancer remains the top cause of cancer death in the United States, with 87% of those deaths linked to smoking.*

22 MEDICAL GENETICS AND CANCER

Genetic information is highly personal and unique. Our genes underlie every aspect of human health, both in function and dysfunction. Obtaining a detailed understanding of how genes work together and interact with environmental factors ultimately allows us to appreciate the differences between the events in normal cellular processes and those that occur in disease pathogenesis. Such knowledge profoundly affects the way many diseases are defined, diagnosed, treated, and prevented. Genetic insight is expected to bring about revolutionary changes in medical practices. In fact, changes are already beginning. Currently, several hundred genetic tests are in clinical use, with many more under development. Most of these tests detect mutations associated with rare genetic disorders that follow Mendelian inheritance patterns. These include Duchenne muscular dystrophy, cystic fibrosis, sickle cell disease, and Huntington disease. In addition, genetic tests are available to detect the predisposition to develop certain forms of cancer.

Approximately 12,000 genetic diseases are known to afflict people, but this is almost certainly an underestimate. The website Online Mendelian Inheritance in Man (OMIM) is a compendium of human genes and phenotypes (see www.ncbi.nlm.nih.gov/omim/). OMIM contains information on all known disorders that are inherited in a Mendelian manner.

Most of the genetic disorders discussed in the first part of this chapter are the direct result of a mutation in one gene. However, many diseases have a complex pattern of inheritance involving several genes. These include common medical disorders such as diabetes, asthma, and mental illness. In these cases, a single mutant gene does not determine whether a person has a disease. Instead, a number of genes may each make a subtle contribution to a person's susceptibility to a disease. Unraveling these complexities will be a challenge for some time to come. The availability of the human genome sequence, discussed in Chapter 20, will be of great help.

In this chapter, we will focus our attention on ways that mutant genes contribute to human disease. In the first part of the chapter, we explore the molecular basis of several genetic disorders and their patterns of inheritance. We also examine how genetic testing can determine if an individual carries a defective allele. The last part of the chapter concerns cancer, a disease that involves the uncontrolled growth of somatic cells. We will examine the underlying genetic basis for cancer and discuss the roles that many different genes may play in the development of this disease.

22.1 INHERITANCE PATTERNS OF GENETIC DISEASES

Human genetics is a topic that is hard to resist. Almost everyone who looks at a newborn is tempted to speculate whether the baby resembles the mother, the father, or perhaps a distant relative. In this section, we will focus primarily on the inheritance of human genetic diseases rather than common traits found in the general population. Even so, the study of human genetic diseases provides insights regarding our traits. The disease hemophilia illustrates this point. Hemophilia (also spelled haemophilia) is a condition in which the blood does not clot properly. By analyzing people with this disorder, researchers have identified genes that participate in the process of blood clotting. The study of hemophilia has helped to elucidate a clotting pathway involving several different proteins. Therefore, as with the study of mutants in model organisms such as *Drosophila,* mice, and yeast, when we study the inheritance of genetic diseases, we often learn a great deal about the genetic basis for normal physiological processes as well.

Because thousands of human diseases have an underlying genetic basis, human genetic analysis is of great medical importance. In this section, we will examine the causes and inheritance patterns of human genetic diseases that result from defects in single genes. As you will learn, the mutant genes that cause these diseases often follow simple Mendelian inheritance patterns.

A Genetic Basis for a Human Disease May Be Suggested from a Variety of Observations

When we view the characteristics of people, we usually think that some traits are inherited, whereas others are caused by environmental factors. For example, when the facial features of two related individuals look strikingly similar, we think that this similarity has a genetic basis. The profound resemblance between identical twins is an obvious example. By comparison, other traits are governed by the environment. If we see a person with purple hair, we likely suspect that he or she has used hair dye as opposed to showing an unusual genetic trait.

For human diseases, geneticists would like to know the relative contributions from genetics and the environment. Is a disease caused by a pathogenic microorganism, a toxic agent in the environment, or a faulty gene? Unlike the case with experimental organisms, we cannot conduct human crosses to determine the genetic basis for diseases. Instead, we must rely on analyzing the occurrence of a disease in families that already exist. As described in the following list, several observations are consistent with the idea that a disease is caused, at least in part, by the inheritance of mutant genes. When the occurrence of a disease correlates with several of these observations, a geneticist becomes increasingly confident that it has a genetic basis.

1. *When an individual exhibits a disease, this disorder is more likely to occur in genetic relatives than in the general population.* For example, someone with cystic fibrosis is more likely to have relatives with this disease than would a randomly chosen member of the general population.
2. *Identical twins share the disease more often than nonidentical twins.* Identical twins, also called **monozygotic (MZ) twins,** are genetically identical to each other, because they were formed from the same sperm and egg. By comparison, nonidentical twins, also called fraternal, or **dizygotic (DZ) twins,** are formed from separate pairs of sperm and egg cells. Fraternal twins share, on average, 50% of their genetic material, the same as any two siblings. When a disorder has a genetic component, a pair of identical twins is more likely to exhibit the disorder than are fraternal twins.

   Geneticists evaluate a disorder's **concordance,** the degree to which it is inherited, by calculating the percentage of twin pairs in which both twins exhibit the disorder relative to pairs where only one twin shows the disorder. Theoretically, for diseases caused by a single gene, concordance among identical twins should be 100%. For fraternal twins, concordance for dominant disorders is expected to be 50%, assuming only one parent is heterozygous for the disease. For recessive diseases, concordance among fraternal twins would be 25% if we assume both parents are heterozygous carriers. However, the actual concordance values observed for most single-gene disorders are usually less than such theoretical values for a variety of reasons. Some disorders are not completely penetrant, meaning that the symptoms associated with the disorder are not always produced. Also, one twin may have a disorder due to a new mutation that occurred after fertilization; it would be very unlikely for the other twin to have the same mutation.
3. *The disease does not spread to individuals sharing similar environmental situations.* Inherited disorders cannot spread from person to person. The only way genetic diseases can be transmitted is from parent to offspring during sexual reproduction.
4. *Different populations tend to have different frequencies of the disease.* Due to evolutionary factors, the frequencies of traits usually vary among different populations of humans. For example, the frequency of the disease sickle cell anemia is highest among certain African and Asian populations and relatively low in other parts of the world (see Chapter 24, Figure 24.14).
5. *The disease tends to develop at a characteristic age.* Many genetic disorders exhibit a characteristic **age of onset** at which the disease appears. Some mutant genes exert their effects during embryonic and fetal development, so their effects are apparent at birth. Other genetic disorders tend to develop much later in life.
6. *The human disorder may resemble a disorder that is already known to have a genetic basis in an animal.* In animals, where we can conduct experiments, various traits are known to be governed by genes. For example, the albino phenotype is found in humans as well as in many animals (**Figure 22.1**).

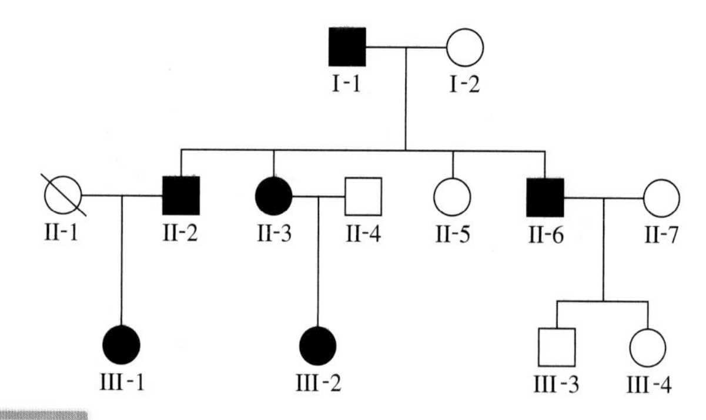

INTERACTIVE EXERCISE **FIGURE 22.3 A family pedigree of Huntington disease, indicating dominant inheritance.**

dominant mutation may occur during gametogenesis, so two unaffected parents may produce an affected offspring.

2. *An affected individual with only one affected parent is expected to produce 50% affected offspring (on average).*
3. *Two affected, heterozygous individuals have (on average) 25% unaffected offspring.*
4. *The trait occurs with the same frequency in both sexes.*
5. *For most dominant, disease-causing alleles, the homozygote is more severely affected with the disorder. In some cases, a dominant allele may be lethal in the homozygous condition.*

Numerous autosomal dominant diseases have been identified in humans (**Table 22.2**). The three common explanations for dominant disorders are haploinsufficiency, a gain-of-function mutation, or a dominant-negative mutation. Let's consider examples of all three types.

The term **haploinsufficiency** refers to the phenomenon in which a person has only a single functional copy of a gene, and that single functional copy does not produce a normal phenotype. In these disorders, 50% of the normal protein is not sufficient to produce a normal phenotype. Haploinsufficiency shows a dominant pattern of inheritance because a heterozygote (with one normal allele and one inactive allele) has the disease. An example is aniridia, which is a rare disorder that results in an absence of the iris of the eye. Aniridia leads to visual impairment and blindness in severe cases.

A second category of dominant disorders involves **gain-of-function mutations.** Such mutations change the gene product so that it "gains" a new or abnormal function. An example is achondroplasia, which is characterized by abnormal bone growth that results in short stature with relatively short arms and legs. This disorder is caused by a point mutation that occurs in the fibroblast growth factor receptor-3 gene. In achondroplasia, the mutant form of the receptor is overactive. This overactivity disrupts the normal signaling pathway and leads to severely shortened bones.

A third category of dominant disorders is characterized by **dominant-negative mutations** in which the altered gene product acts antagonistically to the normal gene product. In humans, Marfan syndrome, which is due to a mutation in the *fibrillin-1* gene, is an example. The *fibrillin-1* gene encodes a glycoprotein that is a structural component of the extracellular matrix that provides structure and elasticity to tissues. The mutant gene encodes a fibrillin-1 protein that antagonizes the effects of the normal protein, thereby weakening the elasticity of certain body parts. For example, the walls of the major arteries such as the aorta, the large artery that leaves the heart, are often affected.

X-Linked Recessive Inheritance Let's now turn to another inheritance pattern common in humans that is called X-linked recessive inheritance (**Table 22.3**). Recessive X-linked inherited diseases pose a special problem for males. Why are males more

TABLE 22.2

Examples of Human Disorders Inherited in an Autosomal Dominant Manner

Disorder	Chromosomal Location of Gene	Gene Product	Effects of Disease-Causing Allele
Aniridia	11p	Pax6 transcription factor	An absence of the iris of the eye, leading to visual impairment and sometimes blindness
Achondroplasia	4p	Fibroblast growth factor receptor-3	A common form of dwarfism associated with a defect in the growth of long bones
Marfan syndrome	15q	Fibrillin-1	Tall and thin individuals with abnormalities in the skeletal, ocular, and cardiovascular systems due to a weakening in the elasticity of certain body parts
Osteoporosis	7q	Collagen (type $1_{\alpha2}$)	Brittle, weakened bones
Familial hypercholesterolemia	19p	LDL receptor	Very high serum levels of low-density lipoprotein (LDL), a predisposing factor in heart disease
Huntington disease	4p	Huntingtin	Neurodegeneration that occurs relatively late in life, usually in middle age
Neurofibromatosis I	17q	Neurofibromin	Individuals may exhibit spots of abnormal pigmentation (café-au-lait spots) and growth of noncancerous tumors in the nervous system

TABLE **22.3**

Examples of Human Disorders Inherited in an X-linked Recessive Manner

Disorder	Gene Product	Effects of Disease-Causing Allele
Duchenne muscular dystrophy	Dystrophin	Progressive degeneration of muscles that begins in early childhood
Hemophilia A	Clotting factor VIII	Defect in blood clotting
Hemophilia B	Clotting factor IX	Defect in blood clotting
Androgen insensitivity syndrome	Androgen receptor	Missing male steroid hormone receptor; XY individuals have external features that are feminine but internally have undescended testes and no uterus

likely to be affected? Most X-linked genes lack a counterpart on the Y chromosome. Males are hemizygous—have a single copy—for these genes. Therefore, a female heterozygous for an X-linked recessive gene passes this trait on to 50% of her sons, as shown in the following Punnett square for hemophilia. In this example, $X^{h\text{-}A}$ is the chromosome that carries a mutant allele causing hemophilia, whereas X^H carries the wild-type allele.

♀ \ ♂	X^H	Y
X^H	X^HX^H Unaffected female	X^HY Unaffected male
$X^{h\text{-}A}$	$X^HX^{h\text{-}A}$ Carrier female	$X^{h\text{-}A}Y$ Male with hemophilia

As mentioned previously, hemophilia is a disorder in which the blood cannot clot properly when a wound occurs. For individuals with this trait, a minor cut may bleed for a very long time, and small injuries can lead to large bruises, because internal broken capillaries may leak blood profusely before they are repaired. For hemophiliacs, common injuries pose a threat of severe internal or external bleeding. Hemophilia A, also called classical hemophilia, is caused by a defect in an X-linked gene that encodes the protein clotting factor VIII. This disease has also been called the "royal disease," because it affected many members of European royal families. The pedigree shown in **Figure 22.4** illustrates the prevalence of hemophilia A among the descendants of Queen Victoria of England. The pattern of X-linked recessive inheritance is revealed by the following observations:

1. *Males are much more likely to exhibit the trait.*
2. *The mothers of affected males often have brothers or fathers who are affected with the same trait.*
3. *The daughters of affected males produce, on average, 50% affected sons.*

X-Linked Dominant Inheritance Relatively few genetic disorders in humans follow an X-linked dominant inheritance pattern. In most cases, males are more severely affected than females, probably because females carry an X chromosome with a normal copy of the gene in question. In most of the X-linked dominant disorders listed in **Table 22.4**, male embryos die at an early stage of development so that most individuals exhibiting the disorder are females. Also, due to their dominant nature and severity, persons with some of the disorders listed in Table 22.4 do not reproduce. Therefore, these dominant disorders, which include Rett syndrome and Aicardi syndrome, are not passed from parent to offspring. Instead, they are caused by new mutations that occur during gamete formation or early embryogenesis. For those X-linked dominant disorders in which the offspring can reproduce, the following pattern is often observed:

1. *Females are much more likely to exhibit the trait when it is lethal to males.*
2. *Affected mothers have a 50% chance of passing the trait to daughters. Note: Affected mothers also have a 50% chance of passing the trait to sons, but for many of these disorders, affected sons are not observed because of lethality.*

Many Genetic Disorders Exhibit Locus Heterogeneity

Hemophilia, which we considered earlier in this chapter, can be used to illustrate another concept in genetics called **locus heterogeneity.** This term refers to the phenomenon in which a particular type of disease may be caused by mutations in two or more different genes. For example, blood clotting involves the participation of several different proteins that take part in a cellular cascade that leads to the formation of a clot. Hemophilia is usually caused by a defect in one of three different clotting factors. In hemophilia A, also called classic hemophilia, a protein called factor VIII is missing. Hemophilia B is a deficiency in a different clotting factor, called factor IX. Both factor VIII and factor IX are encoded by different genes on the X chromosome. These two types of hemophilia show an X-linked recessive pattern of inheritance. By comparison, hemophilia C is due to a factor XI deficiency. The gene encoding factor XI is found on chromosome 4, and this form of hemophilia follows an autosomal recessive pattern of inheritance.

In hemophilia, locus heterogeneity arises from the participation of several proteins in a common cellular process. Another mechanism that may lead to locus heterogeneity occurs when proteins are composed of two or more different subunits, with each subunit being encoded by a different gene. The disease thalassemia is an example of locus heterogeneity caused by a mutation in a protein composed of multiple subunits. This potentially life-threatening disease involves defects in the ability of the red blood cells to transport oxygen. The underlying cause is an alteration in hemoglobin. In adults, hemoglobin is a tetrameric protein composed of two α-globin and two β-globin subunits; α globin and β globin are encoded by separate genes (namely, the α-globin and β-globin genes). Two main types of thalassemia have been discovered in human populations: α thalassemia, in which the α-globin subunit of hemoglobin is defective; and β thalassemia, in which the β-globin subunit is defective.

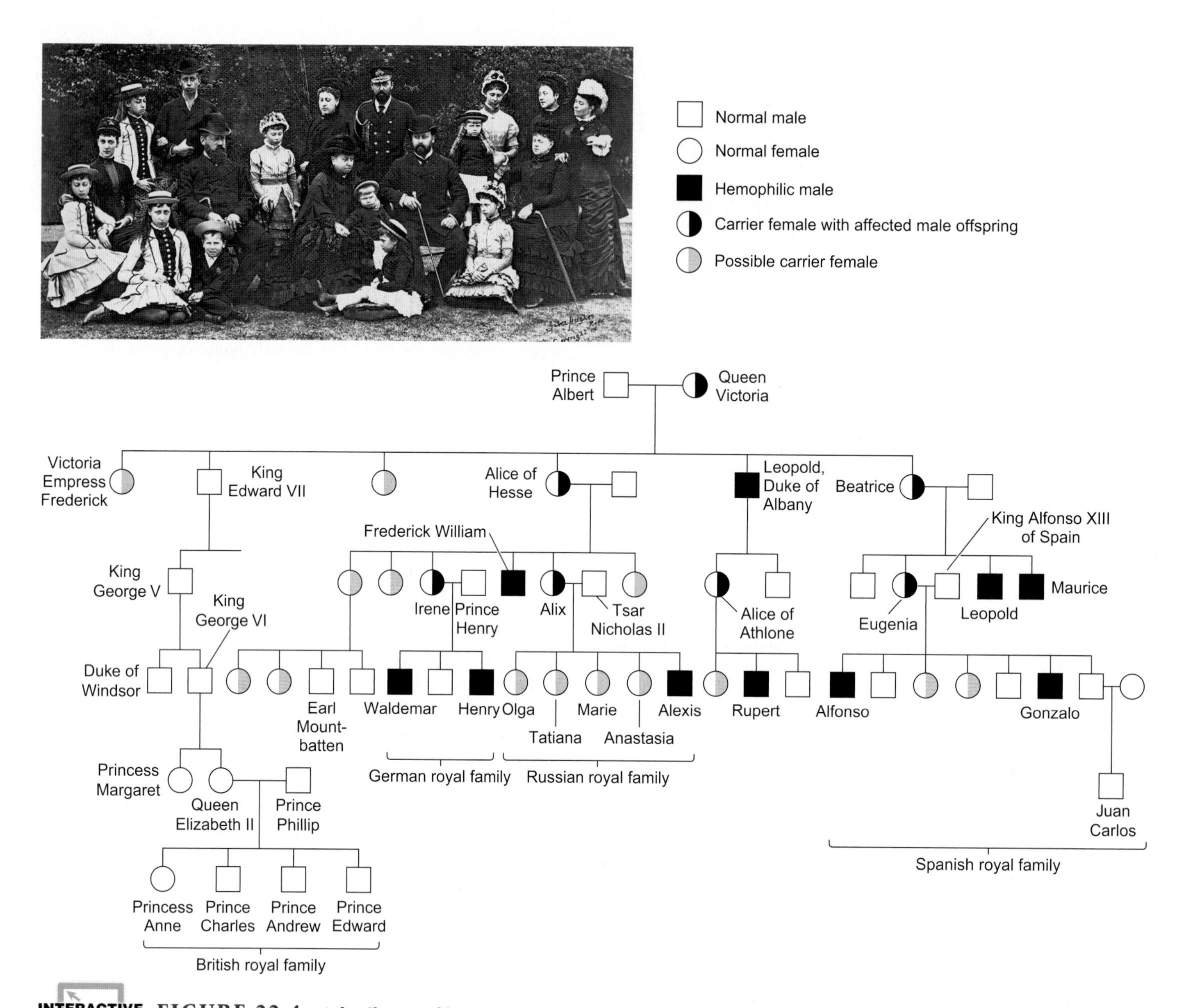

INTERACTIVE EXERCISE

FIGURE 22.4 A family tree of hemophilia A in the royal families of Europe, indicating an X-linked recessive inheritance. Pictured are Queen Victoria and Prince Albert of Great Britain with some of their descendants.

TABLE 22.4

Examples of Human Disorders Inherited in an X-linked Dominant Manner

Disorder	Gene Product	Effects of Disease-Causing Allele
Vitamin D-resistant rickets	Metallopeptidase	Defects in bone mineralization at the sites of bone growth or remodeling, leading to bone deformity and stunted growth in children
Rett syndrome	Methyl-CpG-binding protein-2	A neurodevelopmental disorder that includes a deceleration of head growth and small hands and feet; fatal in males
Aicardi syndrome	Unknown	Characterized by the partial or complete absence of a key structure in the brain called the corpus callosum, and the presence of retinal abnormalities; fatal in males
Incontinentia pigmenti	NFκβ essential modulator	Characterized by morphological and pigmentation abnormalities in the skin, hair, teeth, and nails; fatal in males

Unfortunately, locus heterogeneity may greatly confound pedigree analysis. For example, a human pedigree might contain individuals with X-linked hemophilia and other individuals with hemophilia C. A geneticist who assumed all affected individuals had defects in the same gene would be unable to explain the resulting pattern of inheritance. For disorders such as hemophilia and thalassemia, pedigree analysis is not a major problem because the biochemical basis for these diseases is well understood. However, for rare diseases that are poorly understood at the molecular level, locus heterogeneity may profoundly obscure the pattern of inheritance.

22.2 DETECTION OF DISEASE-CAUSING ALLELES

Because mutant genes are known to play a role in thousands of diseases, researchers have devoted great effort to identifying alleles associated with genetic diseases. In this section, we will explore various approaches used to identify mutant alleles that cause disease and also consider methods that can determine if an individual carries a disease-causing allele.

Haplotypes Exhibit Genetic Variation

To identify disease-causing alleles, researchers often rely on the known locations of genes and molecular markers along chromosomes that have been characterized in human populations. A disease-causing allele may be identified due to its proximity to another known gene or to its proximity to molecular markers.

As discussed in Chapter 20, researchers can characterize chromosomes at the molecular level and determine the precise locations of genes and molecular markers along each chromosome. During the course of evolution, new mutations arise that alter the DNA sequences of genes and molecular markers. For this reason, homologous chromosomes exhibit gene differences (i.e., allelic variation) and show variation in their molecular markers.

As an example, **Figure 22.5** considers a pair of homologous chromosomes from two different individuals and focuses on four sites (called 1, 2, 3, and 4) that occur at particular locations along those chromosomes. These sites could be within particular genes or they could be molecular markers used in mapping studies. In this drawing, each site is also given a letter designation (A, B, or C) depending on the variation in the DNA sequence at the site. In individual 1, sites 1, 2, and 4 differ at one base pair between the homologs. In individual 2, all four sites differ at one base pair. Also note that individuals 1 and 2 differ with regard to some of these sites.

The term **haplotype,** which is a contraction for haploid genotype, refers to the linkage of alleles or molecular markers along a single chromosome. In Figure 22.5, the haplotype for these four sites are shown at the bottom of each chromosome. For example, the haplotype of the left homolog in individual 2 is 1A 2B 3B 4C.

Because mutations are rare events, haplotypes do not dramatically change from one generation to the next due to new mutations. By comparison, haplotypes are more likely to change over the course of a few generations due to crossing over. However, the likelihood of changing a haplotype depends on the distance between the alleles or molecular markers. If two sites are far apart, a crossover is more likely to alter their pattern than if they are close together. If sites 1, 2, 3, and 4 were very close together along this chromosome, the haplotypes shown in this figure would be likely to stay the same after a few generations. For example, a great-great-great grandchild of individual 2 may inherit the haplotype 1A 2B 3B 4C or 1C 2C 3C 4A. In contrast, the inheritance of either haplotype would be much less likely if the sites are far apart and could recombine by crossing over.

Haplotype Association Studies Are Conducted to Identify Disease-Causing Alleles

How do geneticists identify genes that cause disease when they are mutant? Although a variety of approaches may be followed, the hunt often begins with family pedigrees. The goal is to localize a disease-causing allele to a small region on a chromosome that is distinguished by its haplotype. This approach is based on two assumptions:

1. The disease-causing allele had its origin in a single individual known as a **founder,** who lived many generations ago. Since that time, the allele has spread throughout portions of the human population.

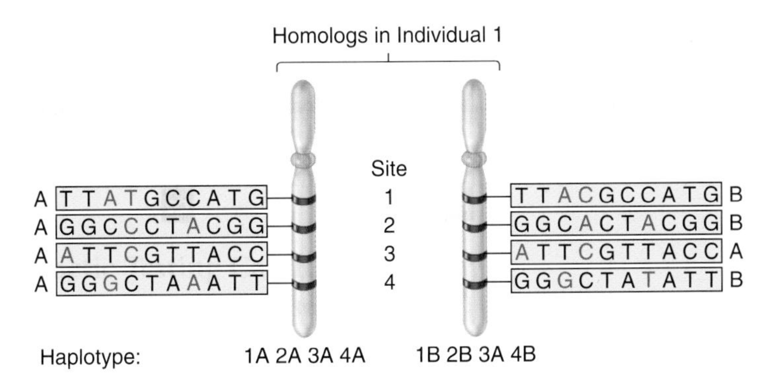

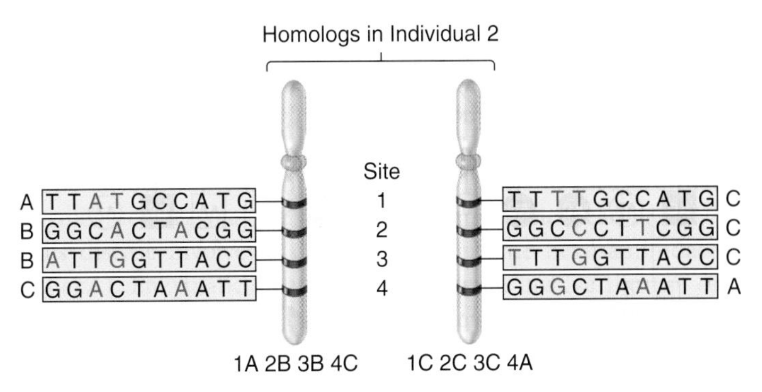

FIGURE 22.5 A schematic representation of haplotypes along a human chromosome. This example considers four sites along a chromosome that may exist in different versions designated A, B, or C. Bases that differ are shown in red.

To perform this assay, a small sample of cells is collected and incubated with MU–GlcNAc, and the fluorescence is measured with a device called a fluorometer. Individuals affected with Tay-Sachs, who do not produce the hexA enzyme, produce little or no fluorescence, whereas individuals who are homozygous for the normal *hexA* allele produce a high level of fluorescence. Heterozygotes, who have 50% hexA activity, produce intermediate levels of fluorescence.

An alternative and more common approach is to detect single-gene mutations at the DNA level. To apply this testing strategy, researchers must have previously identified the mutant gene using molecular techniques. The identification of many human genes, such as those involved in Duchenne muscular dystrophy, cystic fibrosis, and Huntington disease, has made it possible to test for affected individuals or those who may be carriers of these diseases. Table 22.5 describes several ways to test for gene mutations. These laboratory techniques are described in Chapters 18 and 20.

Many human genetic abnormalities involve changes in chromosome number and/or structure. In fact, changes in chromosome number are a common class of human genetic abnormality. Most of these result in spontaneous abortions. However, approximately 1 in 200 live births are aneuploid—have an abnormal number of chromosomes (see Chapter 8, Table 8.1). About 5% of infant and childhood deaths are related to such genetic abnormalities. Changes in chromosome number and many changes in chromosome structure can be detected by karyotyping the chromosomes with a light microscope.

In the United States, genetic screening for certain disorders has become common medical practice. For example, pregnant women older than 35 often have tests conducted to see if their fetuses are carrying chromosomal abnormalities. As discussed in Chapter 8, these tests are indicated because the rate of such defects increases with the age of the mother. Another example is the widespread screening for phenylketonuria (PKU). An inexpensive test can determine if newborns have this disease. Those who test positive can then be given a low-phenylalanine diet to avoid PKU's devastating effects.

Genetic screening also has been conducted on specific populations in which a genetic disease is prevalent. For example, in 1971, community-based screening for heterozygous carriers of Tay-Sachs disease was begun among specific Ashkenazi Jewish populations. With the use of this screening, over the course of one generation, the incidence of TSD births was reduced by 90%. For most rare genetic abnormalities, however, genetic screening is not routine practice. Rather, genetic testing is performed only when a family history reveals a strong likelihood that a couple may produce an affected child. This typically involves a couple that already has an affected child or has other relatives with a genetic disease.

Genetic testing and screening are medical practices with many social and ethical dimensions. For example, people must decide whether or not they want to make use of available tests, particularly when the disease in question has no cure. For example, Huntington disease typically does not affect people until their 50s and can last 20 years. People who learn they are carriers of genetic diseases such as Huntington disease can be devastated by the news. Some argue that people have a right to know about their genetic makeup; others assert that it does more harm than good. Another issue is privacy. Who should have access to personal genetic information, and how could it be used? Could routine genetic testing lead to discrimination by employers or medical insurance companies? In the coming years, we will gain an ever-increasing awareness of our genetic makeup and the underlying causes of genetic diseases. As a society, establishing guidelines for the uses of genetic testing will be a necessary, yet very difficult, task.

Genetic Testing Can Be Performed Prior to Birth

Genetic testing can be performed during pregnancy, which may affect a woman's decision to terminate that pregnancy. The two common ways of obtaining cellular material from a fetus for the purpose of genetic testing are **amniocentesis** and **chorionic villus sampling.** In amniocentesis, a doctor removes amniotic fluid containing fetal cells, using a needle that is passed through the abdominal wall (**Figure 22.8**). The fetal cells are cultured for several weeks and then karyotyped to determine the number of chromosomes per cell and whether changes in chromosome structure have occurred. In chorionic villus sampling, a small piece of the chorion (the fetal part of the placenta) is removed, and a karyotype is prepared directly from the collected cells. Chorionic villus sampling can be performed earlier during pregnancy than amniocentesis, usually around the eighth to tenth week, compared to the fourteenth to sixteenth week for amniocentesis, and results are available sooner. Weighed against these advantages, however, is that chorionic villus sampling may pose a slightly greater risk of causing a miscarriage.

Another method of genetic screening prior to birth is called **preimplantation genetic diagnosis (PGD).** This approach, which is conducted before pregnancy even occurs, involves the genetic testing of embryos that have been produced by **in vitro fertilization (IVF)**—a process in which sperm and egg are combined outside of the mother's body. The testing is typically done to check for a specific genetic abnormality, such as the allele that causes Huntington disease. PGD can also determine if an embryo contains the correct number of chromosomes (also called aneuploidy screening).

PGD is done by removing one or two cells usually at about the eight-cell stage, which occurs 3 days after fertilization. This process is called embryo biopsy or blastomere biopsy. Molecular techniques described in Table 22.5 are then conducted on the removed cell(s) to either check for a particular genetic disease or determine the chromosome composition. The testing can usually be completed in a day or so. Depending on the outcome of the results, a decision can be made whether or not to transfer the embryo into the uterus of the prospective mother in hopes of implantation and the eventual birth of a baby. In most cases, only embryos that do not harbor genetic abnormalites are used. As with the genetic screening of adults, the screening of embryos and fetuses raises many ethical questions.

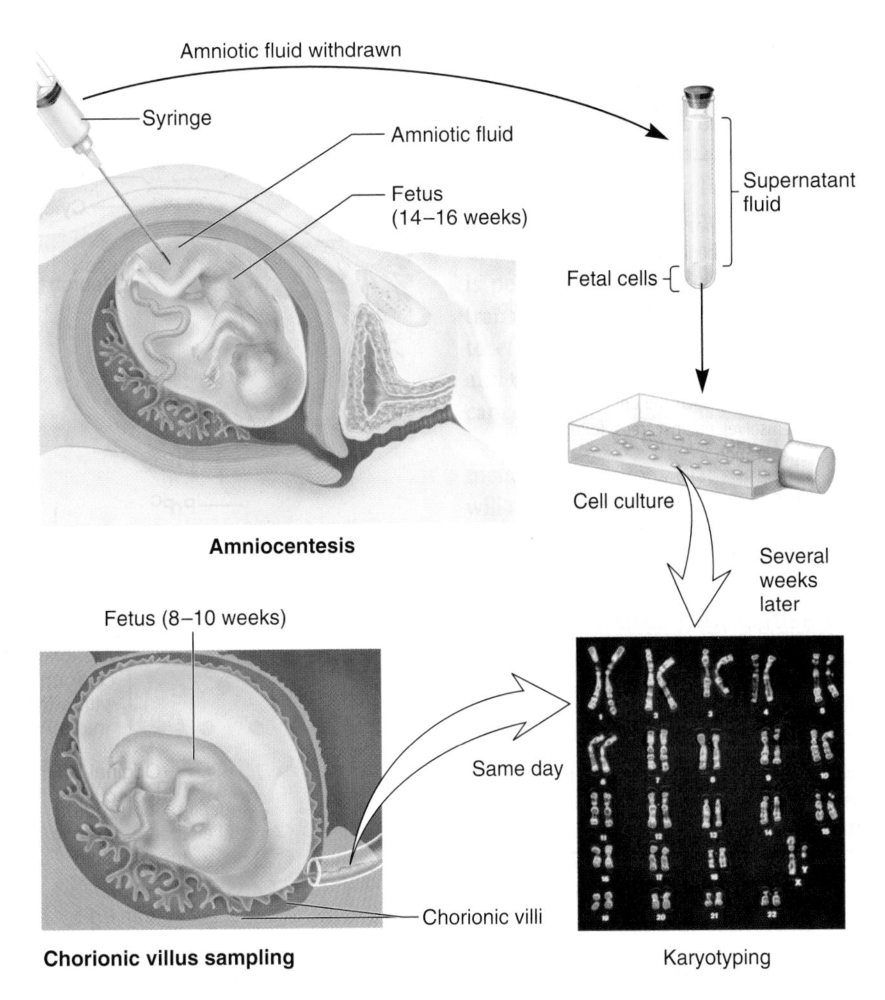

FIGURE 22.8 Techniques to determine genetic abnormalities during pregnancy. In amniocentesis, amniotic fluid is withdrawn, and fetal cells are collected by centrifugation. The cells are then allowed to grow in a laboratory culture medium for several weeks prior to karyotyping. In chorionic villus sampling, a small piece of the chorion is removed. These cells can be prepared directly for karyotyping.

22.3 PRIONS

We now turn to an unusual mechanism in which agents known as prions cause disease. As shown in **Table 22.6**, prions cause several types of neurodegenerative diseases affecting humans and livestock, including mad cow disease. Recent evidence has shown that prions also exist in yeast. In the 1960s, British researchers Tikvah Alper and John Stanley Griffith discovered that preparations from animals with certain neurodegenerative diseases remained infectious even after exposure to radiation that would destroy any DNA or RNA. They suggested that the infectious agent was a protein. Furthermore, Alper and Griffith speculated that the protein usually preferred one folding pattern but could sometimes misfold and then catalyze other proteins to do the same. In the early 1970s, Stanley Prusiner, moved by the death of a patient from a neurodegenerative disease, began to search for the causative agent. In 1982, he isolated a disease-causing agent composed entirely of protein, which he called a **prion.** The term emphasizes the prion's unusual character as a proteinaceous infectious agent. Before the discovery of prions, all known infectious agents such as viruses and bacteria contained their own genetic material (either DNA or RNA).

Prion-related diseases arise from the ability of the prion protein to exist in two conformational states: a normal form PrP^{C}, which does not cause disease, and an abnormal form, PrP^{Sc}, which does. (Note: The superscript C refers to the normal conformation, while the superscript Sc refers to the abnormal conformation, such as the one found in the disease called scrapie.) The gene encoding the prion protein (*PrP*) is found in humans and other mammals, and the protein is expressed at low levels in certain types of cells such as nerve cells. The abnormal conformation of the prion protein can come from two sources. An individual can be infected with the abnormal protein by taking the abnormal protein into their bodies. For example, someone may eat products from an animal that had the disease. Alternatively, some people carry alleles of the *PrP* gene that cause their prion protein to convert spontaneously to the abnormal conformation at a very low rate. These individuals have an inherited predisposition to develop a prion-related disease. An example of an inherited prion disease is familial fatal insomnia (Table 22.6).

What is the molecular mechanism through which prions cause disease? As noted, the prion protein can exist in two conformations, PrP^{C} and PrP^{Sc}. As shown in **Figure 22.9**, the abnormal conformation, PrP^{Sc}, acts as a catalyst to convert normal

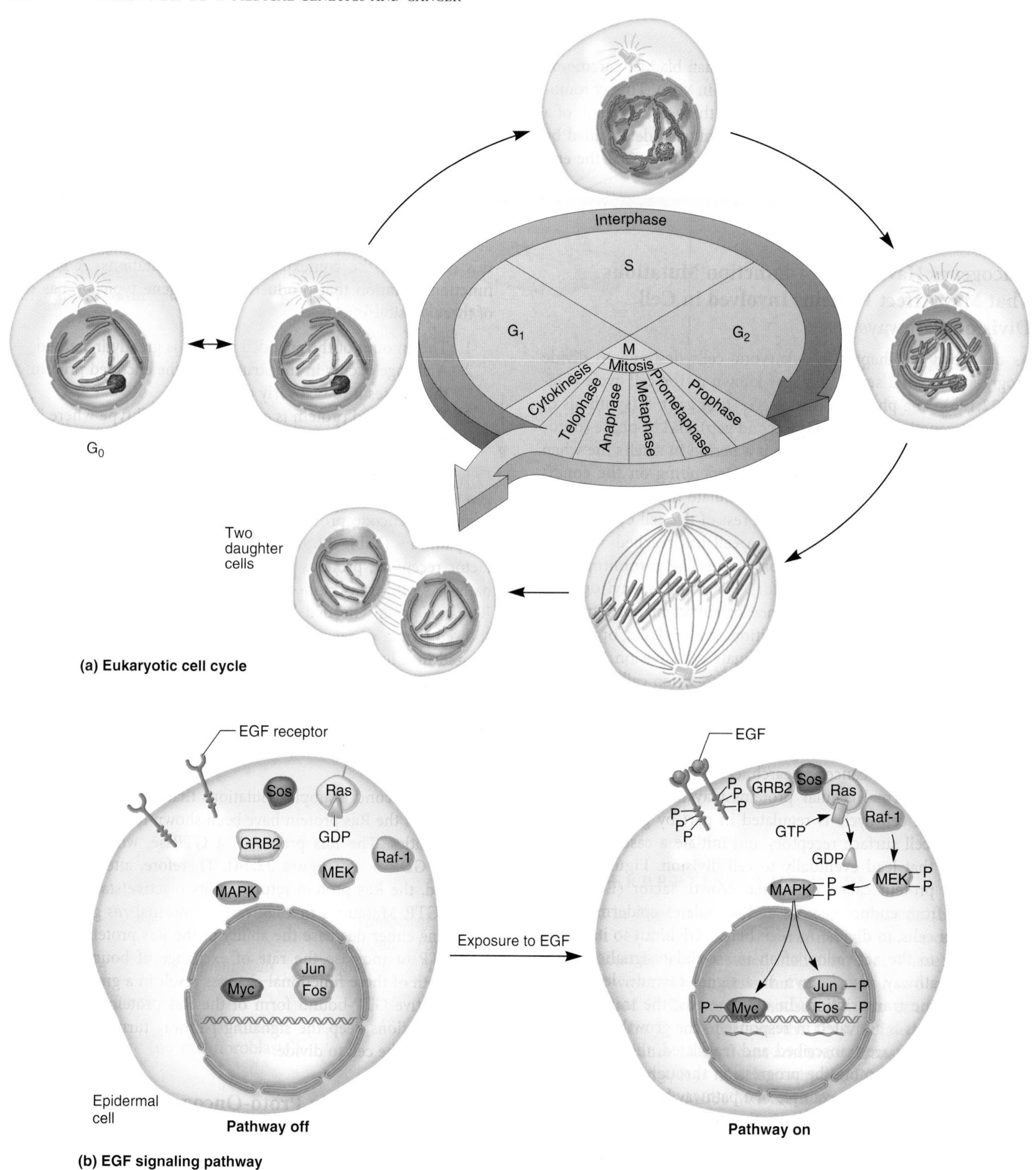

FIGURE 22.13 **The eukaryotic cell cycle and activation of a cell-signaling pathway by a growth factor.** **(a)** The cell cycle involves a progression through the G_1, S, G_2, and M phases. Progression through the cell cycle is often stimulated by growth factors. (Note: Chromosomes are not condensed during G_0, G_1, S, and G_2 phases. They are shown that way in this figure so they can be counted.) **(b)** In this example, epidermal growth factor (EGF) binds to two EGF receptors, causing them to dimerize and phosphorylate each other. An intracellular protein called GRB2 is attracted to the phosphorylated EGF receptor, and it is subsequently bound by another protein called Sos. The binding of Sos to GRB2 enables Sos to activate a protein called Ras. This activation involves the release of GDP and the binding of GTP. The activated Ras/GTP complex then activates Raf-1, which is a protein kinase. Raf-1 phosphorylates MEK, and then MEK phosphorylates MAPK. More than one MAPK may be involved. Finally, the phosphorylated form of MAPK activates transcription factors, such as Myc, Jun, and Fos. This leads to the transcription of genes, which encode proteins that promote cell division.

TABLE 22.8

Examples of Proto-Oncogenes That Can Mutate into Oncogenes

Gene	Cellular Function
Growth Factors*	
sis	Platelet-derived growth factor
int-2	Fibroblast growth factor
Growth Factor Receptors	
erbB	Growth factor receptor for EGF (epidermal growth factor)
trk	Growth factor receptor for CSF-1 (cytostatic factor that inhibits cell division)
fms	Growth factor receptor for NGF (nerve growth factor)
K-sam	Growth factor receptor for FGF (fibroblast growth factor)
Intracellular Signaling Proteins	
ras	GTP/GDP-binding protein
raf	Serine/threonine kinase
src	Tyrosine kinase
abl	Tyrosine kinase
gsp	G-protein α subunit
Transcription Factors	
myc	Transcription factor
jun	Transcription factor
fos	Transcription factor
gli	Transcription factor
erbA	Steroid receptor (which functions as a transcription factor)

*The genes described in this table are found in humans as well as other vertebrate species. Many of these genes were initially identified in retroviruses. Most of the genes have been given three-letter names that are abbreviations for the type of cancer the oncogene causes or the type of virus in which the gene was first identified.

TABLE 22.9

Examples of Genetic Changes That Convert Proto-Oncogenes into Oncogenes

Type of Change	Description and Examples
Missense mutation	A change in the amino acid sequence of a proto-oncogene protein may cause it to function in an abnormal way. Missense mutations can convert *ras* genes into oncogenes.
Gene amplification	The copy number of a proto-oncogene may be increased by gene duplication. *Myc* genes have been amplified in human leukemias; breast, stomach, lung, and colon carcinomas; and neuroblastomas and glioblastomas. *ErbB* genes have been amplified in glioblastomas, squamous cell carcinomas, and breast, salivary gland, and ovarian carcinomas.
Chromosomal translocations	A piece of chromosome may be translocated to another chromosome and affect the expression of genes at the breakpoint site. In Burkitt lymphoma, a region of chromosome 8 is translocated to either chromosome 2, 14, or 22. The breakpoint in chromosome 8 causes the overexpression of the c-*myc* gene.
Viral integration	When a virus integrates into a chromosome, it may enhance the expression of nearby proto-oncogenes. In avian lymphomas, the integration of the avian leukosis virus can enhance the transcription of the c-*myc* gene.

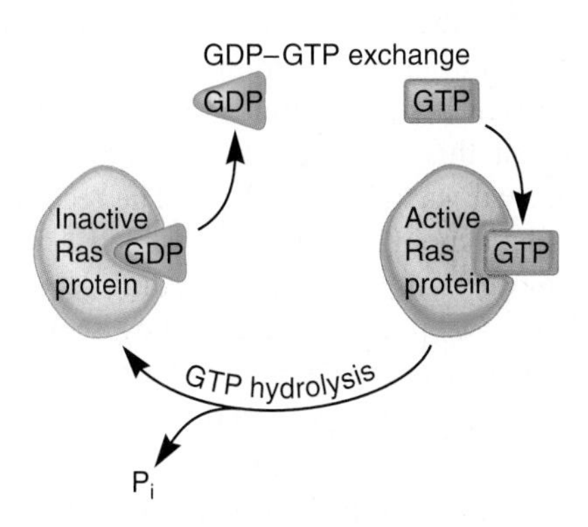

FIGURE 22.14 Functional cycle of the Ras protein. The binding of GTP to Ras activates the function of Ras and promotes cell division. The hydrolysis of GTP to GDP and P_i converts the active form of Ras to an inactive form.

Missense Mutations As mentioned previously, changes in the structure of the Ras protein can cause it to become permanently activated. These changes are caused by a missense mutation in the *ras* gene. The human genome contains four different but evolutionarily related *ras* genes: *ras*H, *ras*N, *ras*K-4a, and *ras*K-4b. All four homologous genes encode proteins with very similar amino acid sequences containing a total of 188 or 189 amino acids. Missense mutants in these normal *ras* genes are associated with particular forms of cancer. For example, a missense mutation in *ras*H that changes a glycine to a valine is responsible for the conversion of *ras*H into an oncogene:

	1	2	3	4	5	6	7	8	9	10	11	12	13		188	189
Normal	Met	Thr	Glu	Tyr	Lys	Leu	Val	Val	Val	Gly	Ala	Gly	Gly		Leu	Ser
Human rasH	ATG	ACG	GAA	TAT	AAG	CTG	GTG	GTG	GTG	GGC	GCC	GGC	GGT	....	CTC	TCC
												↓				
Oncogenic												GTC				
rasH	Met	Thr	Glu	Tyr	Lys	Leu	Val	Val	Val	Gly	Ala	Val	Gly	....	Leu	Ser

Experimentally, chemical carcinogens have been shown to cause these missense mutations and thereby lead to cancer.

Gene Amplification Another genetic event that may occur in cancer cells is gene amplification, or an abnormal increase in the copy number of a proto-oncogene. An increase in gene copy number is expected to increase the amount of the encoded protein, thereby contributing to malignancy. Gene amplification does not normally happen in mammalian cells, but it is a common occurrence in cancer cells. As mentioned previously, Gallo and Groudine discovered that c-*myc* was amplified in a human leukemia cell line. Many human cancers are associated with the amplification of particular oncogenes. In such cases, the extent of oncogene amplification may be correlated with the progression of tumors to increasing malignancy. These include the amplification of N-*myc* in neuroblastomas and *erbB-2* in breast carcinomas. In other types of malignancies, gene amplification is more random and may be a secondary event that increases the expression of oncogenes previously activated by other genetic changes.

Chromosomal Translocation A third type of genetic alteration that can lead to cancer is a chromosomal translocation. Although structural abnormalities are common in cancer cells, very specific types of chromosomal translocations have been identified in certain types of tumors. In 1960, Peter Nowell and David Hungerford discovered that chronic myelogenous leukemia (CML) is correlated with the presence of a shortened version of chromosome 22, which they called the Philadelphia chromosome after the city where it was discovered. Rather than a deletion, this shortened chromosome is the result of a reciprocal translocation between chromosomes 9 and 22. Later studies revealed that this translocation activates a proto-oncogene, *abl*, in an unusual way (**Figure 22.15**). The reciprocal translocation involves breakpoints within the *abl* and *bcr* genes. Following the reciprocal translocation, the coding sequence of the *abl* gene fuses with the promoter and coding sequence of the *bcr* gene. This yields an oncogene that encodes an abnormal fusion protein, which contains the polypeptide sequences encoded from both genes. The *abl* gene encodes a tyrosine kinase enzyme, which uses ATP to attach phosphate groups onto target proteins. This phosphorylation activates certain proteins involved with cell division. Normally, the *abl* gene is highly regulated. However, in the Philadelphia chromosome, the fusion gene is controlled by the *bcr* promoter, which is active in white blood cells. This leads to an overexpression of the tyrosine kinase function in such cells. This explains why this fusion causes a type of cancer called a leukemia, which involves a proliferation of white blood cells.

Interestingly, the study of the *abl* gene has led to an effective treatment for CML. Until recently, the only successful treatment was to destroy the patient's bone marrow and then restore blood-cell production by infusing stem cells from the bone marrow of a healthy donor. With knowledge about the function of the ABL protein, researchers have developed the drug imatinib mesylate (Gleevec) that appears to dramatically improve survival. This molecule fits into the active site of the ABL protein, preventing ATP from binding there. Without ATP, the ABL protein cannot phosphorylate its target proteins. This prevents the ABL

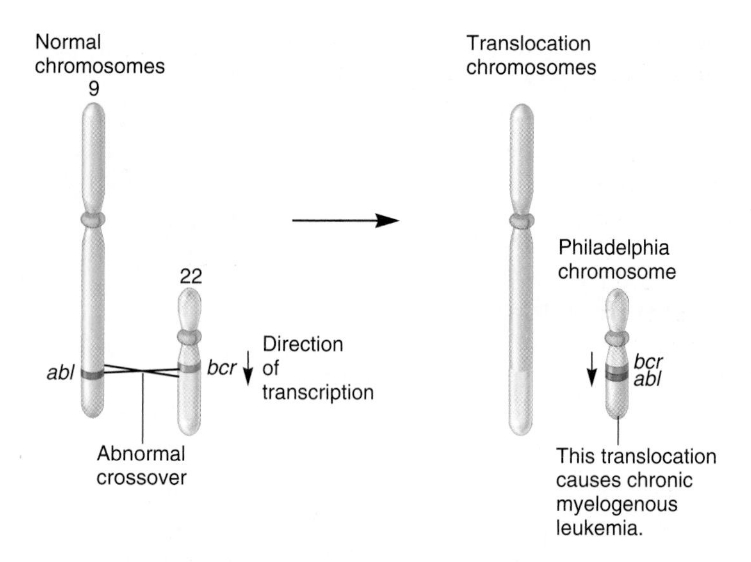

FIGURE 22.15 The reciprocal translocation commonly found in people with chronic myelogenous leukemia.

Genes → Traits In healthy individuals, the *abl* gene is located on chromosome 9, and the *bcr* gene is on chromosome 22. In certain forms of myelogenous leukemia, a reciprocal translocation causes the *abl* gene to fuse with the *bcr* gene. This combined gene, under the control of the *bcr* promoter, encodes an abnormal fusion protein that overexpresses the tyrosine kinase function of the ABL protein and leads to leukemia.

protein from stimulating cell division. In a clinical trial, almost 90% of the CML patients treated with the drug showed no further progression of their disease!

Other forms of cancer also involve chromosomal translocations that cause an overexpression of an oncogene. In Burkitt lymphoma, for example, a region of chromosome 8 is translocated to chromosome 2, 14, or 22. The breakpoint in chromosome 8 is near the c-*myc* gene, and the sites on chromosomes 2, 14, and 22 correspond to locations of different immunoglobulin genes that are normally expressed in lymphocytes. The translocation of the c-*myc* gene near the immunoglobulin genes leads to the overexpression of the c-*myc* gene, thereby promoting malignancy in lymphocytes.

Viral Integration A fourth way that oncogenes can occur is via viral integration. As part of their reproductive cycle, certain viruses integrate their genomes into the chromosomal DNA of their host cell. If the integration occurs next to a proto-oncogene, a viral promoter or enhancer sequence may cause the proto-oncogene to be overexpressed. For example, in certain lymphomas that occur in birds, the genome of the avian leukosis virus has been found to be integrated next to the c-*myc* gene and enhances its level of transcription.

Tumor-Suppressor Genes Play a Role in Preventing the Proliferation of Cancer Cells

Thus far, we have considered how oncogenes promote cancer resulting from gain-of-function mutations. An oncogene is an abnormally activated gene that leads to uncontrolled cell growth. We now turn our attention to a second category of genes called **tumor-suppressor genes.** As the name suggests, the role of a tumor-suppressor gene is to prevent cancerous growth. Therefore, when a tumor-suppressor gene becomes inactivated by mutation, it becomes more likely that cancer will occur. It is a loss-of-function mutation in a tumor-suppressor gene that promotes cancer.

The first identification of a human tumor-suppressor gene involved studies of retinoblastoma, a tumor that occurs in the retina of the eye. Some people have inherited a predisposition to develop this disease within the first few years of life. By comparison, the noninherited form of retinoblastoma, which is caused by environmental agents, tends to occur later in life but only rarely.

Based on these differences, in 1971, Alfred Knudson proposed a "two-hit" model for retinoblastoma. According to this model, retinoblastoma requires two mutations to occur. People with the hereditary form already have received one mutant gene from one of their parents. They need only one additional mutation in the other copy of this tumor-suppressor gene to develop the disease. Because the retina has more than 1 million cells, it is relatively likely that a mutation may occur in one of these cells at an early age, leading to the disease. However, people with the noninherited form of the disease must have two mutations in the same retinal cell to cause the disease. Because two rare events are much less likely to occur than a single such event, the noninherited form of this disease is expected to occur much later in life and only rarely. Therefore, this hypothesis explains the different populations typically affected by the inherited and noninherited forms of retinoblastoma.

Since Knudson's original hypothesis, molecular studies have confirmed the two-hit hypothesis for retinoblastoma. In this case, the gene in which mutations occur is designated *rb* (for retinoblastoma). This tumor suppressor gene is found on the long arm of chromosome 13. Most people have two normal copies of the *rb* gene. Persons with hereditary retinoblastoma have inherited one normal and one defective copy. In nontumorous cells throughout the body, they have one functional copy and one defective copy of *rb*. However, in retinal tumor cells, the normal *rb* gene has also suffered the second hit (i.e., a mutation), which renders it defective. Without the tumor-suppressor ability, cells are allowed to grow and divide in an unregulated manner, which ultimately leads to cancer. (In contrast, as discussed later, most other forms of cancer involve mutations in several genes.)

More recent studies have revealed how the Rb protein suppresses the proliferation of cancer cells (**Figure 22.16**). The Rb protein regulates a transcription factor called E2F, which activates genes required for cell cycle progression. (The eukaryotic cell cycle is described earlier in Figure 22.13a.) The binding of the Rb protein to E2F inhibits its activity and prevents the cell from progressing through the cell cycle. As discussed later in this chapter, when a normal cell is supposed to divide, cellular proteins called cyclins bind to cyclin-dependent protein kinases (CDKs). This activates the kinases, which then leads to the phosphorylation of the Rb protein. The phosphorylated form of the Rb protein is released from E2F, thereby allowing E2F to activate genes needed to progress through the cell cycle. What happens when both copies of the *rb* gene are rendered inactive by mutation? The answer is that the E2F protein is always active, which explains why uncontrolled cell division occurs.

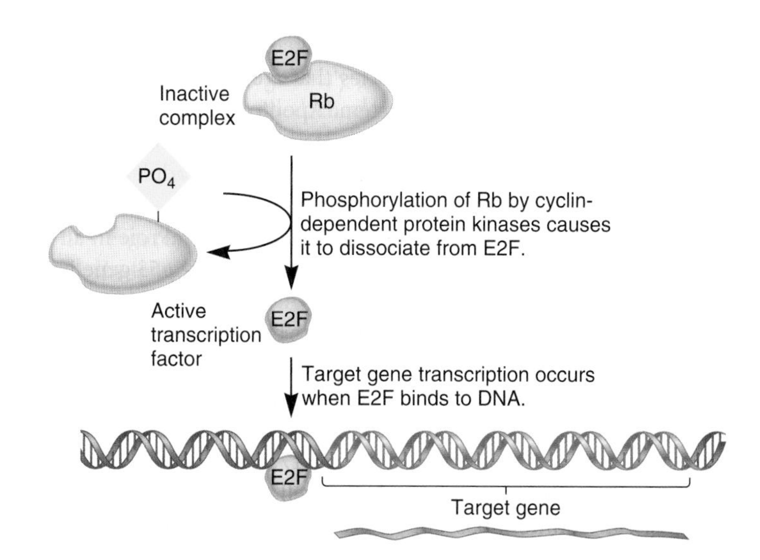

FIGURE 22.16 Interactions between the Rb and E2F proteins. The binding of the Rb protein to the transcription factor E2F inhibits the ability of E2F to function. This prevents cell division. For cell division to occur, cyclins bind to cyclin-dependent protein kinases, which then phosphorylate the Rb protein. The phosphorylated Rb protein is released from E2F. The free form of E2F can activate target genes needed to progress through the cell cycle.

The Vertebrate *p53* Gene Is a Master Tumor-Suppressor Gene That Senses DNA Damage

After the *rb* gene, the second tumor-suppressor gene discovered was the *p53* gene. The *p53* gene is the most commonly altered gene in human cancers. About 50% of all human cancers are associated with defects in *p53*. These include malignant tumors of the lung, breast, esophagus, liver, bladder, and brain as well as sarcomas, lymphomas, and leukemias. For this reason, an enormous amount of research has been aimed at elucidating the function of the p53 protein.

A primary role of the p53 protein is to determine if a cell has incurred DNA damage. If damage is detected, p53 can promote three types of cellular pathways aimed at preventing the proliferation of cells with damaged DNA. First, when confronted with DNA damage, the cell can try to repair its DNA. This may prevent the accumulation of mutations that activate oncogenes or inactivate tumor-suppressor genes. Second, if a cell is in the process of dividing, it can arrest itself in the cell cycle. By stopping the cell cycle, a cell has more time to repair its DNA and avoid producing two mutant daughter cells. For this to happen, p53 stimulates the expression of another gene termed *p21*. The p21 protein inhibits cyclin/CDK protein complexes that are needed to progress from the G_1 phase of the cell cycle to the S phase.

The third, and most drastic, event is that a cell can initiate a series of events called **apoptosis,** or programmed cell death. In response to DNA-damaging agents, a cell may self-destruct. Apoptosis is an active process that involves cell shrinkage, chromatin condensation, and DNA degradation. This process is facilitated by proteases known as **caspases.** These types of proteases are sometimes called the "executioners" of the cell. Caspases digest selected cellular proteins such as microfilaments, which

Molecular Profiling Is Increasingly Used to Classify Tumors

As we have just seen, cancer cells are usually the result of multiple genetic alterations that cause the activation of oncogenes and the loss of function of tumor-suppressor genes. Therefore, each type of tumor is characterized by a particular set of gene and chromosome alterations. Traditionally, different types of tumors have not been identified on the basis of genetic changes but instead have been classified according to their appearance under a microscope. Although this approach is useful, a major drawback is that two tumors may have a very similar microscopic appearance but yet have very different underlying genetic changes and clinical outcomes. For this reason, researchers and clinicians are turning to methods that enable them to understand the molecular changes that occur in diseases such as cancer. This general approach is called **molecular profiling.**

In cancer biology, molecular profiling involves the identification of the genes that play a role in the development of cancer. Why is this useful? First, molecular profiling can distinguish between tumors that look very similar under the microscope. Second, researchers are optimistic that molecular profiling may lead to improved treatment options. As we gain a better understanding of the genetic changes associated with particular types of cancers, researchers may be able to develop drugs that specifically target the proteins that are encoded by cancer-causing gene mutations. As discussed earlier, the drug imatinib mesylate, which is used to treat chronic myelogenous leukemia, was developed in this way.

A third benefit of molecular profiling is that it may affect treatment options and patient outcome. For example, about 70% of all breast cancers exhibit an overexpression of the estrogen receptor. These types of breast cancer are better treated with drugs that either block the estrogen receptor or block the synthesis of estrogen. An example is tamoxifen, which is an antagonist of the estrogen receptor. The application of genetic or molecular data in the treatment of disease is called **personalized medicine.** As we gain a better understanding of human genes and disease states, researchers expect that personalized medicine will become an increasingly important aspect of health care.

DNA Microarrays Are Used in the Molecular Profiling of Tumors

DNA microarrays, which are described in Chapter 21, are often used as a tool in the molecular profiling of tumors. The goal is to identify those genes whose pattern of expression correlates with each other—an approach called cluster analysis (see Figure 21.2). In the study of cancer, researchers can compare cancer cells to normal cells and identify groups (clusters) of genes that are turned on in the cancer cells and off in the normal cells, and other groups of genes that are turned off in the cancer cells and on in the normal cells. Likewise, researchers can compare two different types of tumors and identify groups of genes that show different patterns of expression.

As an example, **Figure 22.21a** shows a computer-generated image that illustrates the results of a microarray analysis of 47 samples, most of which came from the tumors of patients with a type of cancer called diffuse large B-cell lymphoma (DLBCL). Each column represents the expression pattern of a set of genes from a particular sample. Genes that are expressed are shown in red; those that are not expressed are shown in green. During the course of these studies, the researchers identified two different patterns of gene expression. The tumor samples on the left side showed a set of genes (next to the orange bar) that tended to be turned on in the tumor and another set of genes (next to the blue bar) that tended to be turned off in the tumor. This pattern of gene expression was similar to the pattern found in a type of B cell called germinal center B cells. In contrast, the tumors on the right side showed the opposite pattern. The upper genes tended to be turned off in these patients, and the lower genes were turned on. These samples showed a gene expression pattern found in normal activated peripheral blood B cells. These results suggest that the two groups of tumors may have originated in B cells at different stages of development—those on the left originated in germinal center B cells, whereas those on the right originated in activated B cells. Furthermore, the patients from whom these tumors were derived also appeared to have very different clinical outcomes (**Figure 22.21b**). The patients whose tumors had a pattern of gene expression similar to activated B cells had a significantly lower overall survival rate than did the other patients.

Inherited Forms of Cancers May Be Caused by Defects in Tumor-Suppressor Genes and DNA Repair Genes

Before we end our discussion of the genetic basis of cancer, let's consider which genes are most likely to be affected in inherited forms of the disease. As mentioned previously, about 5 to 10% of all cases of cancer involve inherited (germ-line) mutations. These familial forms of cancer occur because people have inherited mutations from one or both parents that give them an increased susceptibility to developing cancer. This does not mean they will definitely get cancer, but they are more likely to develop the disease than are individuals in the general population. When individuals have family members who have developed certain forms of cancer, they may be tested to determine if they also carry a mutant gene. For example, von Hippel-Lindau disease and familial adenomatous polyposis are examples of syndromes for which genetic testing to identify at-risk family members is considered the standard of care.

What types of genes are mutant in familial cancers? Most inherited forms of cancer involve a defect in a tumor-suppressor gene (**Table 22.11**). In these cases, the individual is heterozygous, with one normal and one inactive allele.

At the phenotypic level, a predisposition for developing cancer is inherited in a dominant fashion because a heterozygote exhibits this predisposition. **Figure 22.22a** shows a pedigree for familial breast cancer. In this case, individuals with the disorder have inherited a loss-of-function mutation in the *BRCA-1* gene. As seen in the pedigree, the development of breast cancer shows a dominant pattern of inheritance with incomplete penetrance. Most affected individuals have an affected parent. However, the actual development of cancer is recessive, because it initially relies on the loss of function of the normal copy of the *BRCA-1* gene from a somatic cell (**Figure 22.22b**). This phenomenon is called **loss of heterozygosity (LOH)**—the loss of function of a normal allele when the other allele was already inactivated.

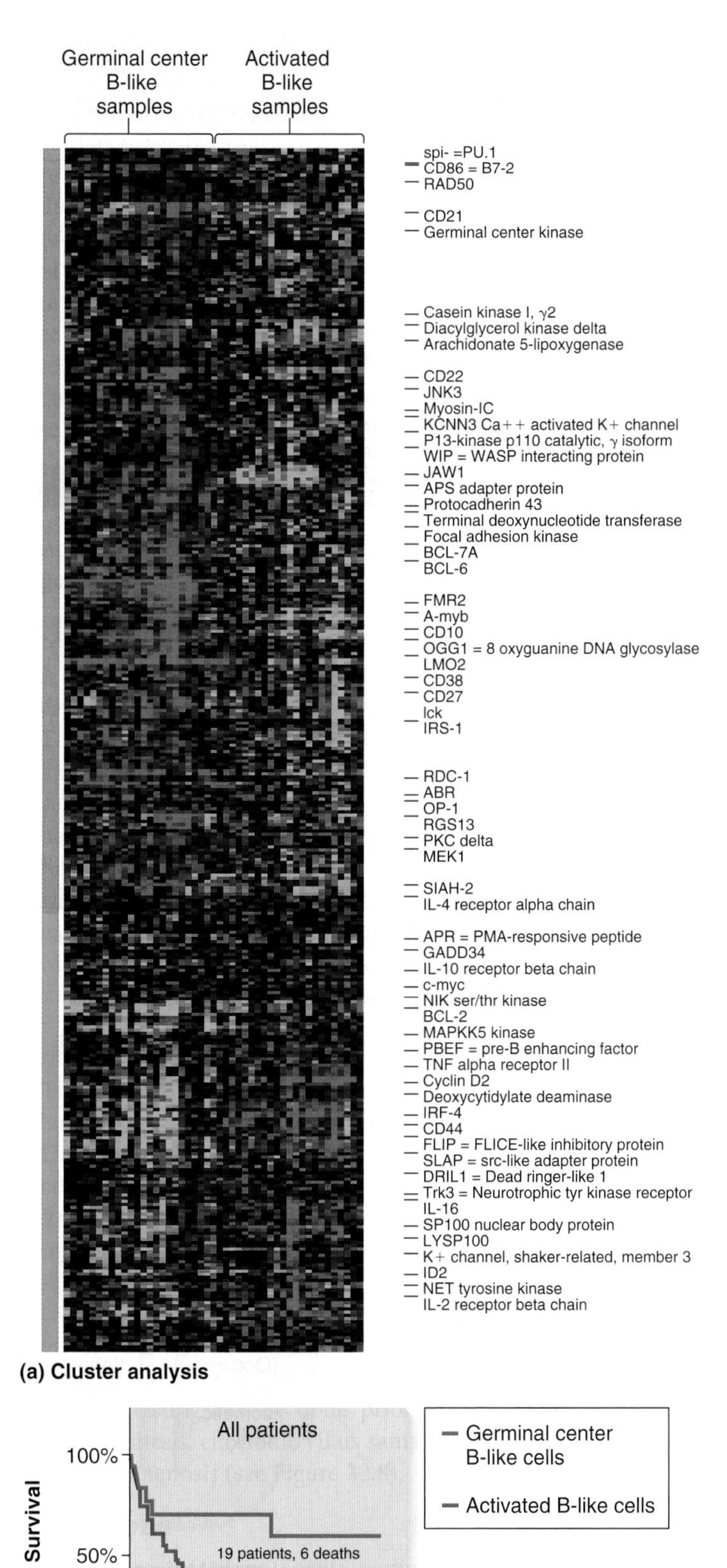

FIGURE 22.21 The use of DNA microarrays to classify types of tumors. **(a)** Forty-seven samples, mostly from patients with diffuse large B-cell lymphoma (DLBCL), were subjected to a DNA microarray analysis. The DNA microarray data were then subjected to a cluster analysis to identify genes that are coordinately expressed. The figure shown here is a graphical illustration of a cluster analysis. Each column represents one sample; each row represents the expression of a particular gene. The names of some of the genes are shown along the right side. (Note: The rows and columns are not easily resolved in this illustration.) Genes highly expressed are shown in red; those not expressed are shown in green. One group of samples had an expression pattern similar to that found in germinal center B cells; the other group had an expression pattern typical of activated B cells in the peripheral blood. **(b)** Survival of patients with DLBCL. (Reprinted by permission from Macmillan Publishers Ltd. A.A. Alizadeh, M.B. Eisen, R.E. Davis, et al. (2000) Distinct types of diffuse large B-cell lymphoma identified by gene expression profiling. *Nature. 403,* 6769, 503–511. Image courtesy of Ash Alizadeh.)

Loss of heterozygosity occurs in different ways. For example, the normal *BRCA-1* gene could suffer a point mutation or a deletion that inactivates its function. Alternatively, the chromosome carrying the normal *BRCA-1* gene could be lost during cell division. Another mechanism is mitotic recombination, which is described in Chapter 6. When this occurs in a heterozygote, a daughter cell may receive two copies of the chromosome carrying the normal allele or two copies of the chromosome carrying the mutant allele (see below). The cell carrying two copies of the mutant allele may continue to divide and accumulate additional genetic changes that result in breast cancer.

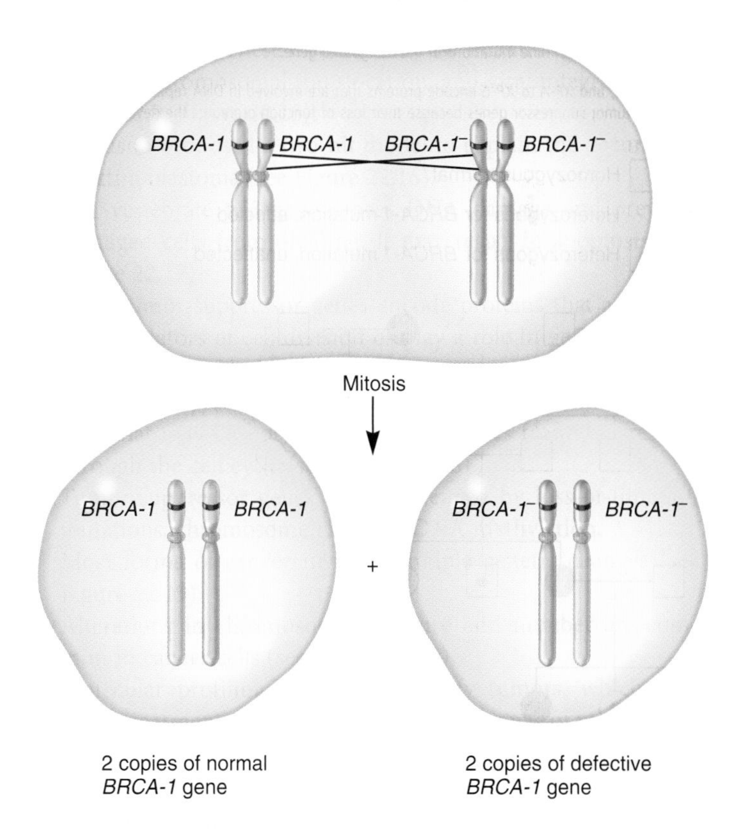

As noted in Table 22.11, not all hereditary forms of cancer are due to defective tumor-suppressor genes. For example, multiple endocrine neoplasia type 2 is due to the activation of an oncogene.

PROBLEM SETS & INSIGHTS

Solved Problems

S1. The pedigree shown below concerns a human disease known as familial hypercholesterolemia.

This disorder is characterized by an elevation of serum cholesterol in the blood. Though relatively rare, this genetic abnormality can be a contributing factor to heart attacks. At the molecular level, this disease is caused by a defective gene that encodes a protein called low-density lipoprotein receptor (LDLR). In the bloodstream, serum cholesterol is bound to a carrier protein known as low-density lipoprotein (LDL). LDL binds to LDLR so that cells can absorb cholesterol. When LDLR is defective, it becomes more difficult for the cells to absorb cholesterol. This explains why the levels of LDL blood cholesterol remain high. Based on the pedigree, what is the most likely pattern of inheritance of this disorder?

Answer: The pedigree is consistent with a dominant pattern of inheritance. An affected individual always has an affected parent. Also, individuals III-8 and III-9, who are both affected, produced unaffected offspring. If this trait was recessive, two affected parents should always produce affected offspring. However, because the trait is dominant, two heterozygous parents can produce homozygous unaffected offspring. The ability of two affected parents to have unaffected offspring is a striking characteristic of dominant inheritance. On average, we would expect that two heterozygous parents should produce 25% unaffected offspring. In the family containing IV-4, IV-5, IV-6, and IV-7, three out of four offspring are actually unaffected. This higher-than-expected proportion of unaffected offspring is not too surprising because the family is a very small group and may deviate substantially from the expected value due to random sampling error.

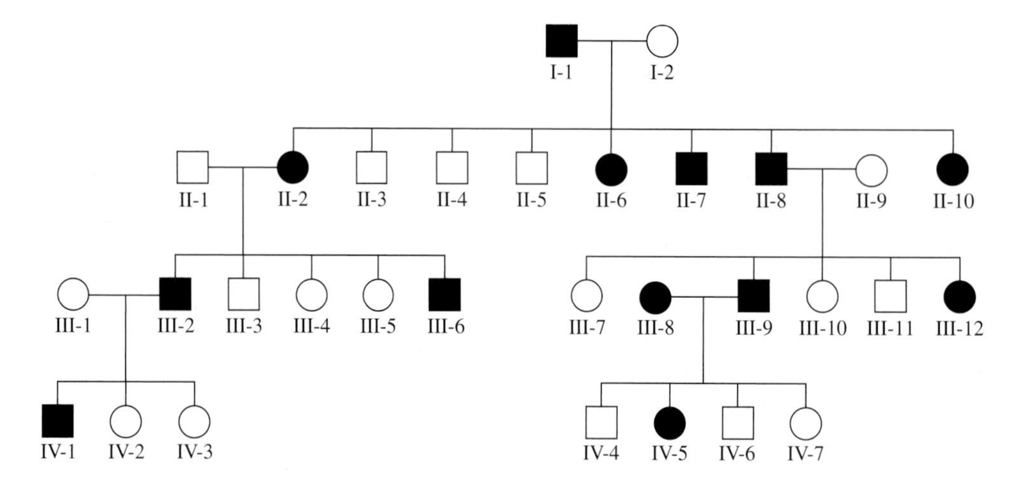

S2. One way to identify a human cellular oncogene is to use human DNA from a malignant cell to transform a mouse cell. A mouse cell that has been transformed by DNA from a malignant human cell may have a human oncogene incorporated into its genome; it also is likely to have *Alu* sequences that are closely linked to this human oncogene. (Note: As discussed in Chapter 10, the human genome contains *Alu* sequences interspersed every 5000 to 6000 bp. *Alu* sequences are not found in the mouse genome.) Discuss how the *Alu* sequence can provide a way to clone human oncogenes.

Answer: One approach is transposon tagging, described in Chapter 17. When a mouse cell is transformed with a human DNA fragment containing an oncogene, that fragment is likely to contain an *Alu* sequence as well. To clone the human oncogene, the chromosomal DNA can be isolated from the transformed mouse cells, digested with a restriction enzyme, and cloned into vectors to create a library of DNA fragments. The members of the library that carry the human oncogene can be identified using a probe complementary to the *Alu* sequence because this sequence is not found in the mouse genome. Using this strategy, researchers have identified several human cellular oncogenes. In human bladder carcinoma, for example, the human cellular oncogene called *ras* was identified this way.

S3. Oncogenes sometimes result from genetic rearrangements (e.g., translocations) that produce gene fusions. An example is the Philadelphia chromosome, in which a reciprocal translocation between chromosomes 9 and 22 leads to fusion of the first part of the *bcr* gene with the *abl* gene. Suggest two different reasons why a gene fusion could create an oncogene.

Answer: An oncogene is derived from a genetic change that abnormally activates the expression of a gene that plays a role in cell division. When a genetic change creates a gene fusion, this can abnormally activate the expression of the gene in two ways.

The first way is at the level of transcription. The promoter and part of the coding sequence of one gene may become fused with the coding sequence of another gene. For example, the promoter and part of the coding sequence of the *bcr* gene may fuse with the coding sequence of the *abl* gene. After this has occurred, the *abl* gene is now under the control of the *bcr* promoter, rather than its own normal promoter. Because the *bcr* promoter is turned on in different cells compared to the *abl* promoter, overexpression of the abl protein occurs in certain cell types compared with its normal level of expression.

A second way that a gene fusion can cause abnormal activation is at the level of protein structure. A fusion protein has parts of two different polypeptides. The first portion of a fusion protein may affect the structure of the second portion of the polypeptide in such a way that the second portion becomes abnormally active, or vice versa.

Conceptual Questions

C1. With regard to pedigree analysis, make a list of the patterns that distinguish recessive, dominant, and X-linked genetic diseases from each other.

C2. Explain, at the molecular level, why human genetic diseases often follow a simple Mendelian pattern of inheritance, whereas most normal traits, such as the shape of your nose or the size of your head, are governed by multiple gene interactions.

C3. Many genetic disorders exhibit locus heterogeneity. Define and give two examples of locus heterogeneity. How does locus heterogeneity confound a pedigree analysis?

C4. In general, why do changes in chromosome structure or number tend to affect an individual's phenotype? Explain why some changes in chromosome structure, such as reciprocal translocations, do not.

C5. We often speak of diseases such as phenylketonuria (PKU) and achondroplasia as having a "genetic basis." Explain whether the following statements are accurate with regard to the genetic basis of any human disease (not just PKU and achondroplasia).

A. An individual must inherit two copies of a mutant allele to have disease symptoms.

B. A genetic predisposition means that an individual has inherited one or more alleles that make it more likely that he or she will develop disease symptoms than other individuals in a population will.

C. A genetic predisposition to develop a disease may be passed from parents to offspring.

D. The genetic basis for a disease is always more important than the environment.

C6. Figure 22.1 illustrates albinism in different species. Describe two other genetic disorders found in both humans and animals.

C7. Discuss why a genetic disease might have a particular age of onset. Would an infectious disease have an age of onset? Explain why or why not.

C8. Gaucher disease (type I) is due to a defect in a gene that encodes a protein called acid β glucosidase. This enzyme plays a role in carbohydrate metabolism within the lysosome. The gene is located on the long arm of chromosome 1. Persons who inherit two defective copies of this gene exhibit Gaucher disease, the major symptoms of which include an enlarged spleen, bone lesions, and changes in skin pigmentation. Let's suppose a phenotypically unaffected woman, whose father had Gaucher disease, has a child with a phenotypically unaffected man, whose mother had Gaucher disease.

A. What is the probability that this child will have the disease?

B. What is the probability that this child will have two normal copies of this gene?

C. If this couple has five children, what is the probability that one of them will have Gaucher disease and four will be phenotypically unaffected?

C9. Ehler-Danlos syndrome is a relatively rare disorder caused by a mutation in a gene that encodes a protein called collagen (type 3 A1). Collagen is a protein found in the extracellular matrix that plays an important role in the formation of skin, joints, and other connective tissues. Persons with this syndrome have extraordinarily flexible skin and very loose joints. The pedigree shown below contains several members affected with Ehler-Danlos syndrome, shown with black symbols. Based on this pedigree, does this syndrome appear to be an autosomal recessive, autosomal dominant, X-linked recessive, or X-linked dominant trait? Explain your reasoning.

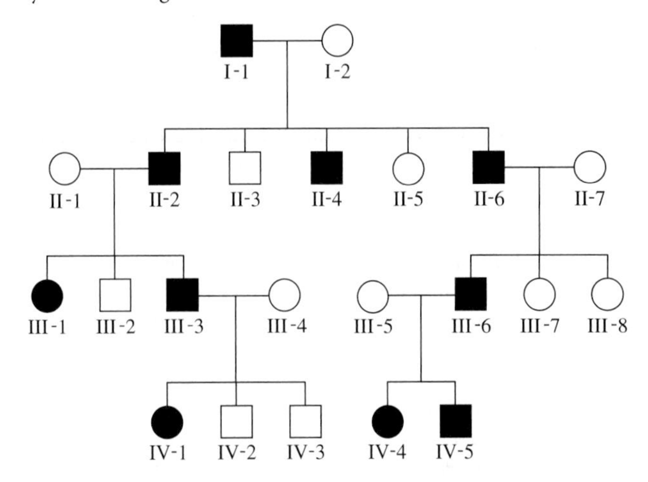

C10. Hurler syndrome is due to a mutation in a gene that encodes a protein called α-L-iduronidase. This protein functions within the lysosome as an enzyme that breaks down mucopolysaccharides (a type of polysaccharide that has many acidic groups attached). When this enzyme is defective, excessive amounts of the mucopolysaccharides dermatan sulfate and heparin sulfate accumulate within the lysosomes, especially in liver cells and connective tissue cells. This leads to symptoms such as an enlarged liver and spleen, bone abnormalities, corneal clouding, heart problems, and severe neurological problems. The pedigree shown below contains three members affected with Hurler syndrome, indicated with black symbols. Based on this pedigree, does this syndrome appear to be an autosomal recessive, autosomal dominant, X-linked recessive, or X-linked dominant trait? Explain your reasoning.

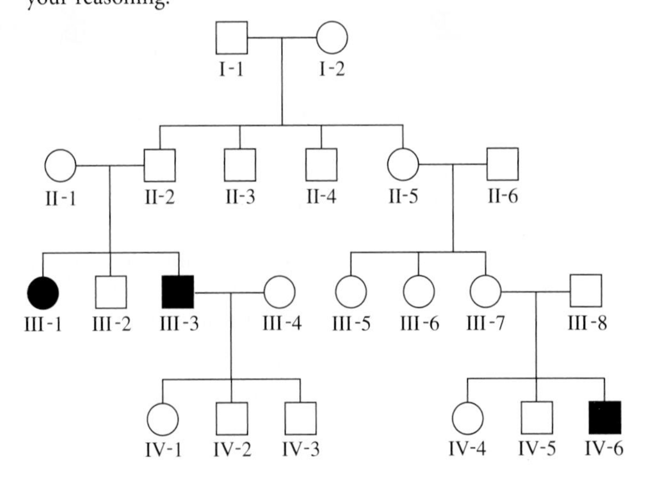

C11. Like Hurler syndrome, Fabry disease involves an abnormal accumulation of substances within lysosomes. However, the lysosomes of individuals with Fabry disease show an abnormal accumulation of lipids. The defective enzyme is α-galactosidase A, which is a

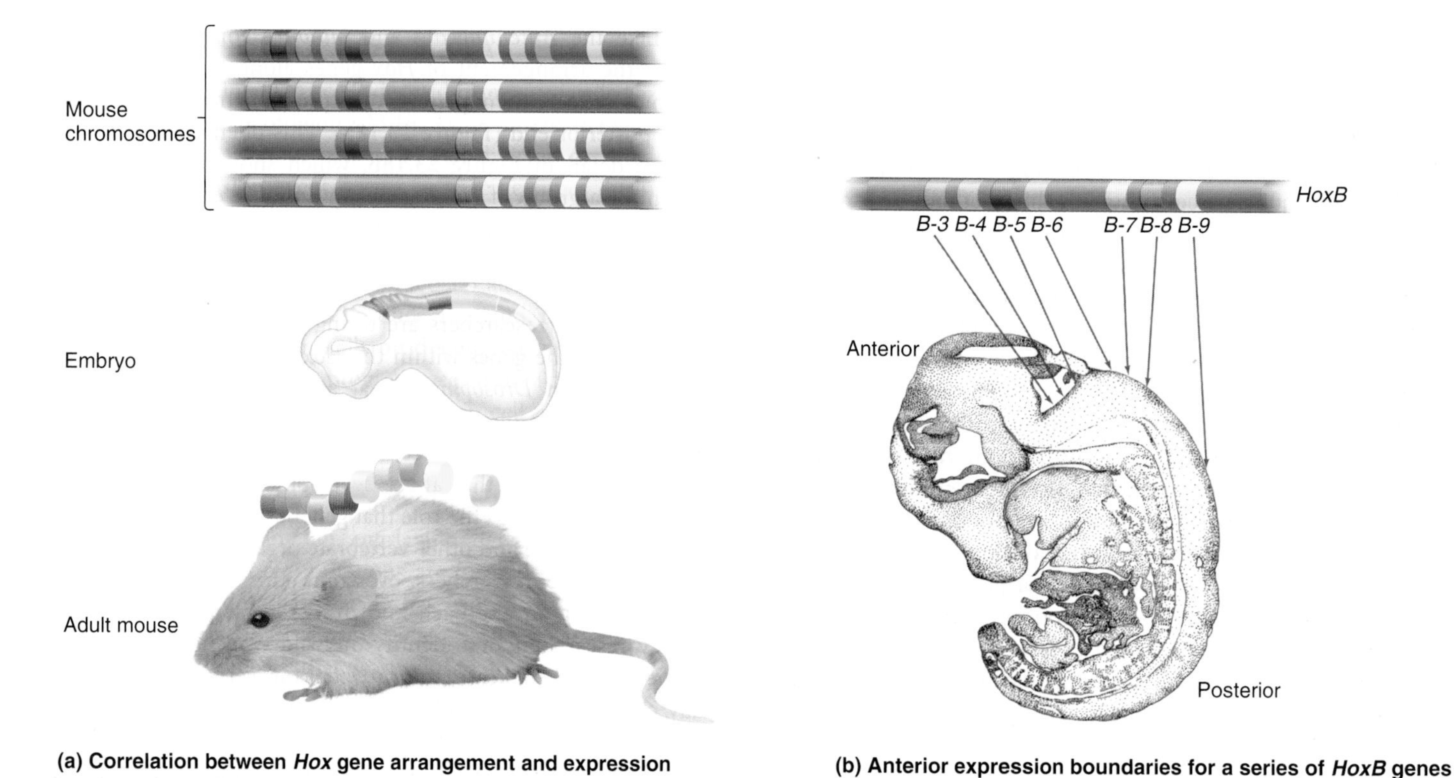

FIGURE 23.18 **Expression pattern of *Hox* genes in the mouse.** **(a)** A schematic illustration of the *Hox* gene expression in the embryo and the corresponding regions in the adult. **(b)** A more-detailed description of *HoxB* expression in a mouse embryo. The arrows indicate the anterior-most boundaries for the expression of *HoxB-3* to *HoxB-9*. The order of *Hox* gene expression, from the anterior end to posterior of the embryo, is in the same order as the genes are found along the chromosome.

consider the expression of the *HoxC-6* gene. Changes in its pattern of expression among vertebrate species are associated with changes in the boundary between cervical (neck) vertebrae and thoracic (chest) vertebrae. The *HoxC-6* gene is expressed during embryonic development prior to vertebrae formation. Differences in the relative position of its expression correlate with the number of neck vertebrae produced (**Figure 23.19**). In the mouse, which has a relatively short neck, *HoxC-6* expression begins in the region of the early embryo that later develops into vertebrae 7 and 8. In contrast, *HoxC-6* expression in the chicken and goose begins much farther back, between vertebrae 14 and 15, or 17 and 18, respectively. The forelimbs also arise at this boundary in all vertebrates. However, snakes, which have no neck or forelimbs, do not have such a boundary because *HoxC-6* expression begins toward their heads.

Genes That Encode Transcription Factors Also Play a Key Role in Cell Differentiation

Thus far, we have focused our attention on patterns of gene expression that occur during the very early stages of development. These genes control the basic body plan of the organism. As this process occurs, cells become **determined.** As mentioned earlier, this refers to the phenomenon that a cell is destined to become a particular cell type. In other words, its fate has been predetermined to eventually become a particular type of cell such as a nerve cell. This occurs long before a cell becomes **differentiated.** This later term means that a cell's morphology and function have changed, usually permanently, into a highly specialized cell type. For example, an undifferentiated mesodermal cell may differentiate into a specialized muscle cell, or an ectodermal cell may differentiate into a nerve cell.

At the molecular level, the profound morphological differences between muscle cells and nerve cells arise from gene regulation. Though muscle and nerve cells contain the same set of genes, they regulate the expression of those genes in very different ways. Certain genes that are transcriptionally active in muscle cells are inactive in nerve cells, and vice versa. Therefore, muscle and nerve cells express different proteins, which affect the morphological and physiological characteristics of the respective cells in distinct ways. In this manner, differential gene regulation underlies cell differentiation.

We learned earlier that a hierarchy of gene regulation is responsible for establishing the body pattern in *Drosophila.* Maternal effect genes control the expression of gap genes, which control the expression of pair-rule genes, and so forth. A similar type of hierarchy is thought to underlie cell differentiation in vertebrates. Researchers have identified specific genes that cause

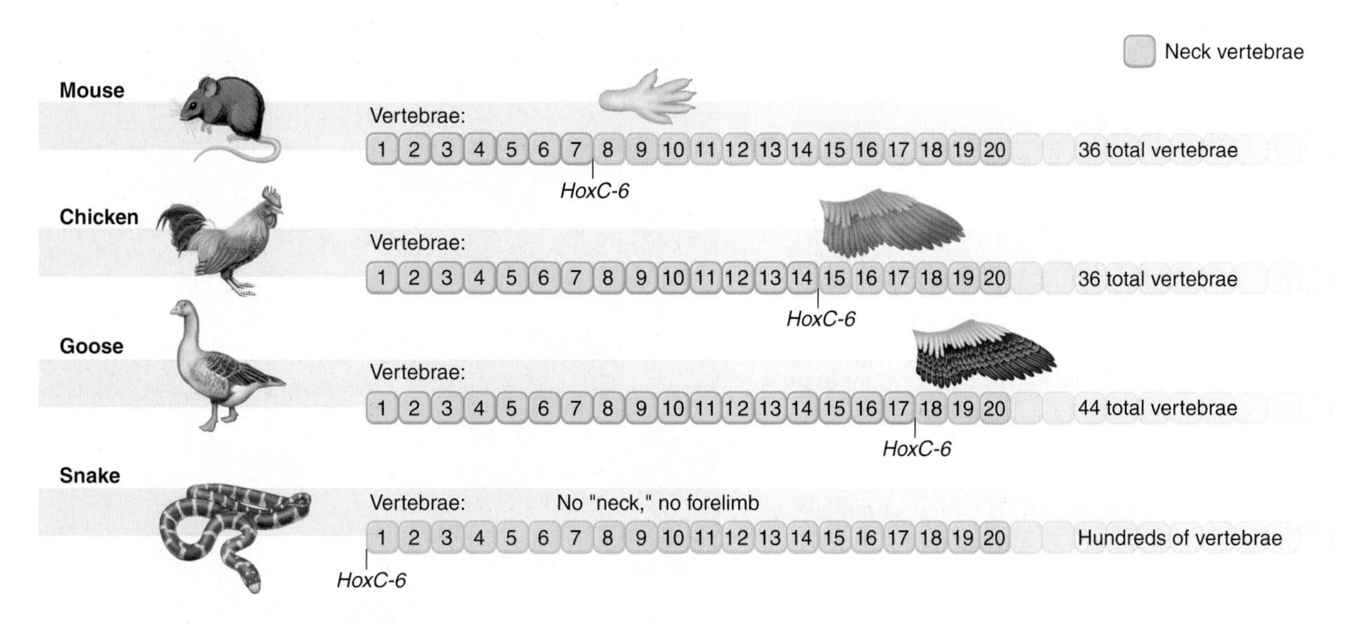

FIGURE 23.19 Expression of the *HoxC-6* gene in different species of vertebrates. This figure indicates the anterior boundary where the *HoxC-6* gene is expressed in the mouse, chicken, goose, and snake.

cells to differentiate into particular cell types. These genes trigger undifferentiated cells to differentiate and follow their proper cell fates.

In 1987, Harold Weintraub and colleagues identified a gene, which they called *MyoD*. This gene plays a key role in skeletal muscle cell differentiation. Experimentally, when the cloned *MyoD* gene was introduced into fibroblast cells in a laboratory, the fibroblasts differentiated into skeletal muscle cells. This result was particularly remarkable because fibroblasts normally differentiate into osteoblasts (bone cells), chondrocytes (cartilage cells), adipocytes (fat cells), and smooth muscle cells, but in vivo, they never differentiate into skeletal muscle cells.

Since this initial discovery, researchers have found that *MyoD* belongs to a small group of genes that initiate muscle development. Besides *MyoD*, these include *Myogenin*, *Myf5*, and *Mrf4*. All four of these genes encode transcription factors that contain a **basic domain** and a **helix-loop-helix domain (bHLH).** The basic domain is responsible for DNA binding and the activation of skeletal muscle-cell-specific genes. The helix-loop-helix domain is necessary for dimer formation between transcription factor proteins. Because of their common structural features and their role in muscle differentiation, MyoD, Myogenin, Myf5, and Mrf4 constitute a family of proteins called **myogenic bHLH proteins.** They are found in all vertebrates and have been identified in several invertebrates, such as *Drosophila* and *C. elegans*. In all cases, the myogenic bHLH genes are activated during skeletal muscle cell development.

At the molecular level, certain key features enable myogenic bHLH proteins to promote muscle cell differentiation. The basic domain binds to a muscle-cell-specific enhancer sequence; this sequence is adjacent to genes that are expressed only in muscle cells (**Figure 23.20**). Therefore, when myogenic bHLH proteins are activated, they can bind to these enhancers and activate the expression of many different muscle-cell-specific genes. They may exert their effects via alterations in chromatin structure or via the activation of RNA polymerase to a transcriptionally active state. In this way, myogenic bHLH proteins function as master switches that activate the expression of many muscle-specific genes. When the encoded proteins are synthesized, they

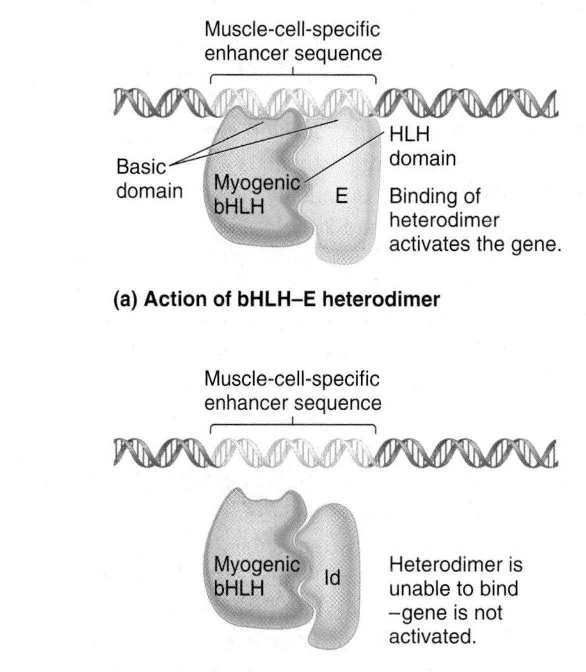

FIGURE 23.20 Regulation of muscle-cell-specific genes by myogenic bHLH proteins. (**a**) A heterodimer formed from a myogenic bHLH protein and an E protein can bind to a muscle-cell-specific enhancer sequence and activate gene expression. (**b**) When a myogenic bHLH protein forms a heterodimer with an Id protein, it cannot bind to the DNA and therefore does not activate gene transcription.

change the characteristics of an undifferentiated cell into those of a highly specialized skeletal muscle cell.

Another important aspect of myogenic bHLH proteins is that their activity is regulated by dimerization, which occurs via the helix-loop-helix domain. As shown in Figure 23.20, heterodimers—dimers formed from two different proteins—may be activating or inhibitory. When a heterodimer forms between a myogenic bHLH protein and an E protein, which also contains a basic domain, the heterodimer binds to the DNA and activates gene expression (Figure 23.20a). However, when a heterodimer forms between a myogenic bHLH protein and a protein called Id (for inhibitor of differentiation), the heterodimer cannot bind to DNA because the Id protein lacks a basic domain (Figure 23.20b). Two basic domains are needed for the heterodimer to bind to the DNA. The Id protein is produced during early stages of development and prevents myogenic bHLH proteins from promoting muscle differentiation too soon. At later stages of development, the amount of Id protein falls, and myogenic bHLH proteins can then combine with E proteins to induce muscle-cell differentiation.

23.4 PLANT DEVELOPMENT

In developmental plant biology, the model organism for genetic analysis is *Arabidopsis thaliana* (**Figure 23.21**). Unlike most flowering plants, which have long generation times and large genomes, *Arabidopsis* has a generation time of about 2 months and a genome size of 14×10^7 bp, which is similar to *Drosophila* and *C. elegans*. A flowering *Arabidopsis* plant produces a large number of seeds and is small enough to be grown in the laboratory. Like *Drosophila*, *Arabidopsis* can be subjected to mutagens to generate mutations that alter developmental processes. The small genome size of this organism makes it relatively easy to map these mutant alleles and eventually clone the relevant genes (as described in Chapters 18 and 20).

FIGURE 23.21 **The model organism *Arabidopsis*.** The plant is relatively small, making it easy to grow many of them in the laboratory.

The morphological patterns of growth are markedly different between animals and plants. As described previously, animal embryos become organized along anteroposterior, dorsoventral, and left-right axes, and then they subdivide into segments. By comparison, the form of plants has two key features. The first is the root-shoot axis. Most plant growth occurs via cell division near the tips of the shoots and the bottoms of the roots.

Second, this growth occurs in a well-defined radial pattern. For example, early in *Arabidopsis* growth, a rosette of leaves or flowers is produced from buds that emanate in a spiral pattern directly from the main shoot (see Figure 23.21). Later, the shoot generates branches that also produce leaf buds as they grow. Overall, the radial pattern in which a plant shoot gives off the buds that produce leaves, flowers, and branches is an important mechanism that determines much of the general morphology of the plant.

At the cellular level, too, plant development differs markedly from animal development. For example, cell migration does not occur during plant development. In addition, the development of a plant does not rely on morphogens that are deposited asymmetrically in the oocyte. In plants, an entirely new individual can be regenerated from many types of somatic cells. In other words, many plant cells are **totipotent,** meaning that they have the ability to differentiate into every cell type and to produce an entire individual. By comparison, animal development typically relies on the organization within an oocyte as a starting point for development.

In spite of these apparent differences, the underlying molecular mechanisms of pattern development in plants still share some similarities with those in animals. In this section, we will consider a few examples in which the genes encoding transcription factors play a key role in plant development.

Plant Growth Occurs from Meristems Formed During Embryonic Development

Figure 23.22 illustrates a common sequence of events that takes place in the development of seed plants such as *Arabidopsis*. After fertilization, the first cellular division is asymmetrical and produces a smaller cell, called the apical cell, and a larger basal cell

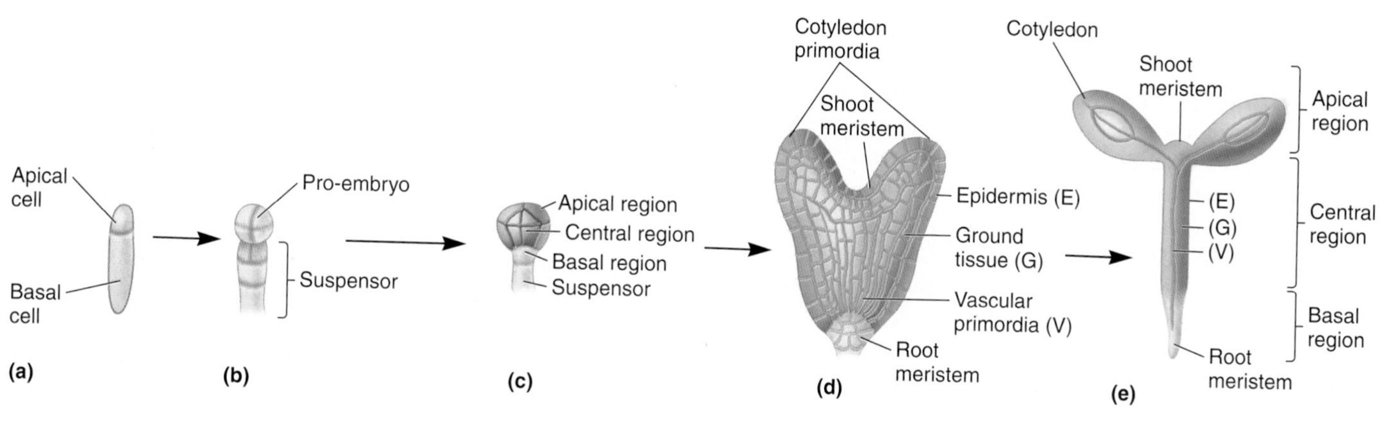

FIGURE 23.22 Developmental steps in the formation of a plant embryo. **(a)** The two-cell stage consists of the apical cell and basal cell. **(b)** The eight-cell stage consists of a proembryo and a suspensor. The suspensor gives rise to extraembryonic tissue, which is needed for seed formation. **(c)** At this stage of embryonic development, the three main regions of the embryo (i.e., apical, central, and basal) have been determined. **(d)** At the heart stage, all of the plant tissues have begun to form. Note that the shoot meristem is located between the future cotyledons, and the root meristem is on the opposite side. **(e)** A seedling.

(Figure 23.22a). The apical cell gives rise to most of the embryo, and it later develops into the shoot of the plant. The basal cell gives rise to the root, along with the suspensor that produces extraembryonic tissue required for seed formation. At the heart stage, which is composed of only about 100 cells, the basic organization of the plant has been established (Figure 23.22d). The **shoot meristem** arises from a group of cells located between the cotyledons. These cells are the precursors that will produce the shoot of the plant, along with lateral structures such as leaves and flowers. The **root meristem** is located at the opposite side and creates the root.

A meristem contains an organized group of actively dividing stem cells. As discussed in Chapter 19, stem cells retain the ability to divide and differentiate into multiple cell types. As they grow, meristems produce offshoots of proliferating cells. On a shoot meristem, for example, these offshoots or buds give rise to structures such as leaves and flowers. The organization of a shoot meristem is shown in **Figure 23.23**. It is organized into three areas called the **organizing center,** the **central zone,** and the **peripheral zone.** The role of the organizing center is to ensure the proper organization of the meristem and preserve the correct number of actively dividing stem cells. The central zone is an area where undifferentiated stem cells are always maintained. The peripheral zone contains dividing cells that eventually differentiate into plant structures. For example, the peripheral zone may form a bud, which will produce a leaf or flower.

In *Arabidopsis*, the organization of a shoot meristem is controlled by two critical genes termed *WUS* and *CLV3*. The *WUS* gene encodes a transcription factor that is expressed in the organizing center (Figure 23.23). The expression of the *WUS* gene induces the adjacent cells in the central zone to become undifferentiated stem cells. These stem cells then turn on the *CLV3* gene, which encodes a secreted protein. The CLV3 protein binds to receptors in the cells of the peripheral zone, preventing them from expressing the *WUS* gene. This limits the area of *WUS* gene expression to the underlying organizing center, thereby maintaining a small population of stem cells at the growing tip. A shoot meristem in *Arabidopsis* contains only about 100 cells. The

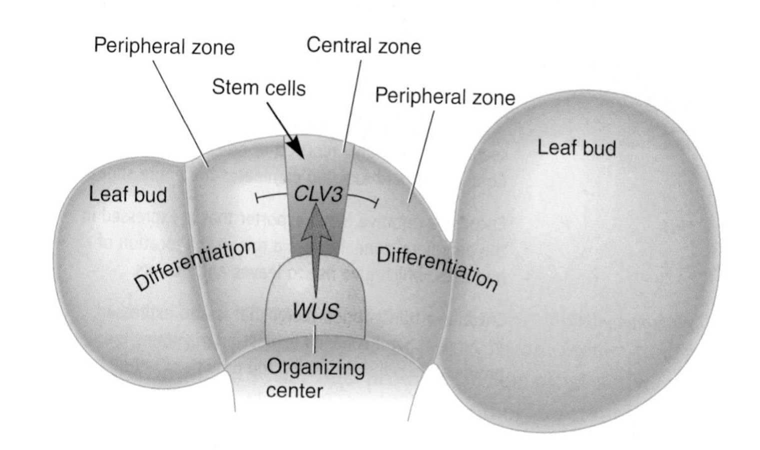

FIGURE 23.23 Organization of a shoot meristem. The organization of a shoot meristem is controlled by the *WUS* and *CLV3* genes, which are abbreviations for *Wuschel* and *clavata*, respectively. The *WUS* gene is expressed in the organizing center and induces the cells in the central zone to become undifferentiated stem cells. The red arrow indicates that the WUS protein induces these stem cells to turn on the *CLV3* gene, which encodes a secreted protein that binds to receptors in the cells of the peripheral zone. The black lines with a vertical slash indicate that the CLV3 protein prevents the cells in the peripheral zone from expressing the *WUS* gene. This limits the area of *WUS* gene expression to the underlying organizing center, thereby maintaining a small population of stem cells at the growing tip. The cells in the peripheral zone are allowed to grow and differentiate into lateral structures such as leaves.

inhibition of *WUS* expression in the peripheral cells also allows them to embark on a path of cell differentiation so they can produce structures such as leaves and flowers.

In the seedling shown earlier in Figure 23.22e, three main regions are observed. The **apical region** produces the leaves and flowers of the plant. The **central region** (not to be confused with the central zone) creates the stem. The radial pattern of cells in the central region causes the radial growth observed in plants. Finally, the **basal region** produces the roots. Each of these three

TABLE 23.3
Mechanisms of Sex Determination in Selected Species

Species	Mechanism
Drosophila, C. elegans	Ratio of the number of X chromosomes to the number of sets of autosomes determines sex. An X:A ratio of 1.0 results in females (or hermaphrodites in *C. elegans*), and an X:A ratio of 0.5 produces males.
Mammals	The presence of the *SRY* gene on the Y chromosome causes maleness.
Birds	Females are ZW, and males are ZZ.
Bees	Males are haploid, and females (workers and queen bees) are diploid.
Certain species of reptiles, fish, and turtles	Environmental conditions, usually temperature, influence the ratio of male and female offspring.
Plants	The male plant produces pollen. In some sexually dimorphic species, the males are XY. In other species, there are not distinct sex chromosomes, but the male plant is heterozygous for a dominant allele that suppresses the female pathway.

the male gametophyte has two types of sex chromosomes, similar to the XY system in mammals. In birds, the female has heteromorphic sex chromosomes: ZW birds are female, and ZZ birds are male. Finally, sex determination is not always caused by two types of sex chromosomes. In bees, for example, males are haploid, and females are diploid.

In many animal species, variation in chromosome composition is not the underlying factor that distinguishes female from male development. Sex determination in many species of reptiles and fish is controlled by environmental factors such as temperature. For example, in the American alligator (*Alligator mississippiensis*), temperature controls sex determination. Eggs incubated at 33°C typically result in ~100% male individuals, whereas eggs incubated at lower or higher temperatures such as 30°C or 34.5°C result in ~100% or 95% females, respectively.

The adoption of one of two sexual fates is an event that has been studied in great detail in several species. Researchers have discovered that sex determination is a process controlled genetically by a hierarchy of genes that exert their effects in early embryonic development. In this section, we will consider features of these hierarchies in *Drosophila*, *C. elegans*, mammals, and plants.

In *Drosophila*, Sex Determination Involves a Regulatory Cascade That Includes Alternative Splicing

In diploid fruit flies, XX flies develop into females, and X0 flies become males. The ratio of the number of X chromosomes to the number of sets of autosomes is the determining factor. In a diploid fly that carries two sets of chromosomes, this ratio is 1.0 in females versus 0.5 in males. Although male fruit flies usually carry a Y chromosome, it is not necessary for male development. The mechanism of sex determination begins in early embryonic development and involves a regulatory cascade composed of several genes. Females and males follow one of two alternative pathways. Simplified versions of these pathways are depicted in **Figure 23.26**.

Let's begin with the pathway that produces female flies. In females, the higher ratio of X chromosomes results in the embryonic expression of a gene designated *Sxl* (Figure 23.26a). The *Sxl* gene product is a protein that functions in the splicing of pre-mRNA. In female embryos, the Sxl protein enhances its own expression by splicing its own pre-mRNA, an event termed an **autoregulatory loop.** In addition, it splices the pre-mRNA from two other genes called *msl-2* and *tra*. The Sxl protein promotes the splicing of the *msl-2* pre-mRNA in a way that introduces an early stop codon in the coding sequence, thereby producing a shortened version of the msl-2 protein that is functionally inactive. By comparison, the Sxl protein promotes the splicing of *tra* pre-mRNA to produce an mRNA that is translated into a functional protein. Therefore, *Sxl* activates *tra*.

The *tra* gene product and a constitutively expressed product from a gene called *tra-2* are also splicing factors. In the female, they cause the alternative splicing of the pre-mRNAs that are expressed from the *fru* and *dsx* genes. The tra and tra-2 proteins cause these pre-mRNAs to be spliced into mRNAs designated fru^F and dsx^F, respectively. The female-specific fru^F mRNA is not translated into a sex-specific gene product. However, the dsx^F mRNA, together with two other gene products from the *ix* and *her* genes, promotes female sexual development and controls some aspects of female-specific behavior via the central nervous system. The dsx^F protein is known to be a transcription factor that regulates certain genes that promote these changes.

How are males produced? In X0 or XY flies, the *Sxl* gene is transcriptionally activated, but it is spliced in a way that places an early stop codon in the coding sequence. Therefore, a functional Sxl protein is not made (Figure 23.26b). This permits the expression of *msl-2*, which promotes dosage compensation. In fruit flies, dosage compensation is accomplished by turning up the expression of X-linked genes in the male to a level that is twofold higher. Therefore, even though the male has only one X chromosome, the expression of X-linked genes in the male and female is approximately equal. The absence of *Sxl* expression in male embryos also promotes the development of maleness. Without *Sxl*, the *tra* mRNA is not properly spliced, so *tra* is not expressed. Without the tra protein, the *fru* and *dsx* mRNAs are spliced in a different way to produce mRNAs designated fru^M and dsx^M. Along with *ix* and *her* gene products, the dsx^M protein promotes male development as well as male-specific behavior. Like the dsx^F protein, the dsx^M protein is a transcription factor that regulates certain genes. Because dsx^M is spliced differently from dsx^F, the dsx^M protein's structure is different from the dsx^F protein, and this difference alters the regulation pattern of dsx^M. In addition, the fru^M gene product is necessary for the regulation of genes involved in male-specific behaviors.

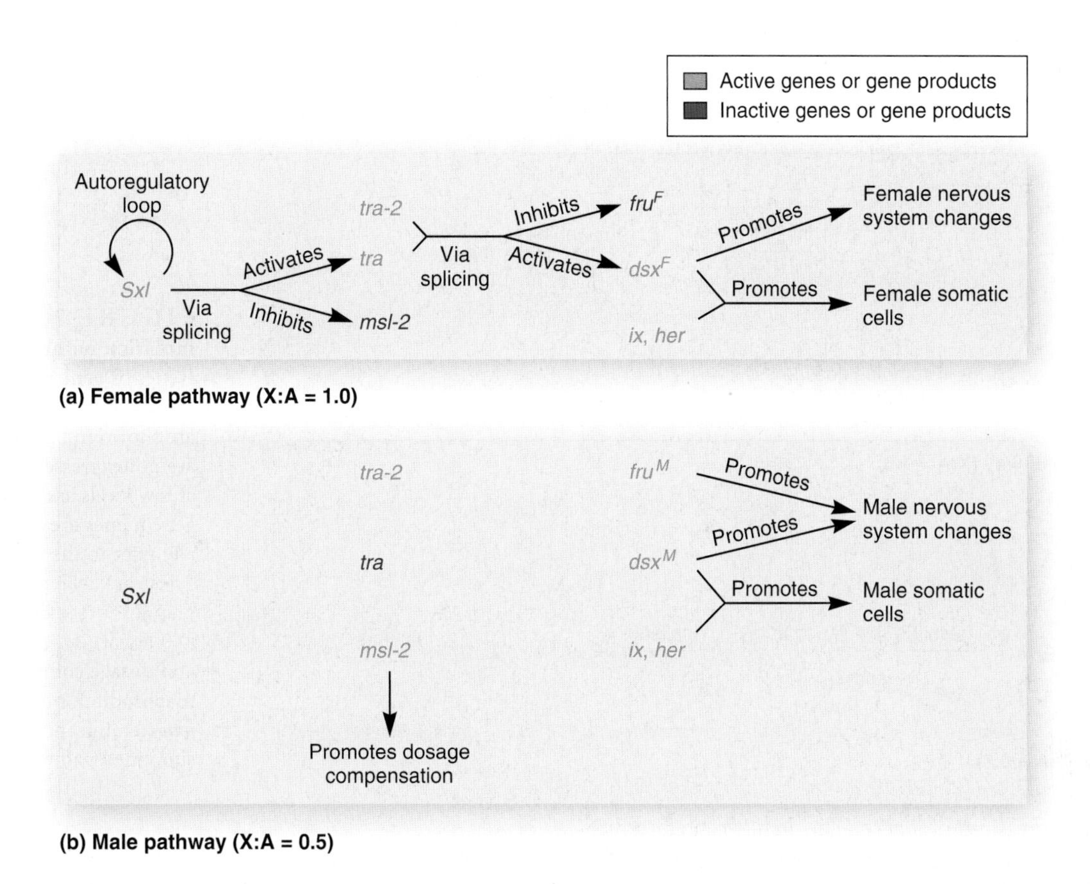

FIGURE 23.26 **Sex determination pathway for *Drosophila melanogaster*.** Genes or gene products that are functionally expressed are shown in light orange; those that are not expressed are shown in blue. The gene names are acronyms for the phenotypes that result from mutations that cause loss-of-function or aberrant expression. These are as follows: *Sxl* (sex lethal), *msl* (male sex lethal), *tra* (transformer), *dsx* (double sex), *fru* (fruitless), *ix* (intersex), and *her* (hermaphrodite). Note: This is a simplified pathway. More gene products are involved than shown here.

In *C. elegans* the Ratio of X Chromosomes to Sets of Autosomes Initiates a Regulatory Cascade That Determines Sex

C. elegans has two sexes: hermaphrodites and males. During larval development, the hermaphrodites produce sperm that are stored in a structure called the spermathecae. In adulthood, hermaphrodites are anatomically female. They produce oocytes that are fertilized when they are forced through the spermathecae by muscular contractions. Fertilization can take place in either of two ways. One possibility is that a sperm can fertilize an oocyte from the same worm. In other words, the hermaphrodite is capable of self-fertilization. Alternatively, a hermaphrodite can mate with a male worm, which produces only sperm.

The sexual identity of *C. elegans*, hermaphrodite versus male, is a trait determined very early in embryonic development. As in *Drosophila*, sexual identity is controlled by the activities of many genes that interact in a regulatory cascade that determines the sex of the worm and controls dosage compensation. In *C. elegans*, the expression of genes on the X chromosome in the hermaphrodite, which carries two X chromosomes, is decreased to 50% of that in males, which carry one X chromosome.

As in *Drosophila*, the ratio of the number of X chromosomes to the number of sets of autosomes (the X:A ratio) is the factor that causes the regulatory cascade to follow one of two alternative pathways (**Figure 23.27**). Let's first consider the events that occur during early embryonic development in the hermaphrodite. Because hermaphrodites have two X chromosomes, they have an X:A ratio of 1.0, compared with the ratio of 0.5 for males. This higher X:A ratio results in an enhanced expression of *fox-1* and *sex-1*, which are located on the X chromosome. In the XX hermaphrodite, the expression of *fox-1* and *sex-1* inhibits the expression of *xol-1*. The protein product of the *xol-1* gene inhibits the expression of three genes called *sdc-1*, *sdc-2*, and *sdc-3*. The inhibition of *xol-1* activity in hermaphrodites permits the expression of the *sdc* genes. The expression of *sdc-1*, *sdc-2*, and *sdc-3* has two effects. First, these three genes are necessary for dosage compensation, and second, they are necessary to permit the downstream expression of *tra-1*, which is needed to promote hermaphrodite development. This second effect occurs via a chain of events that are also shown in Figure 23.27a. The *sdc* gene products inhibit the expression of *her-1*. When *her-1* is inhibited, the expression of *tra-2* is permitted. When *tra-2* is active, the activities of *fem-1*, *fem-2*, and *fem-3* are inhibited. When the *fem* gene products are inhibited, the expression of *tra-1* is permitted. It is the *tra-1* gene product that promotes hermaphrodite development and inhibits male development.

Figure 23.27b describes the pathway for male development. Because males contain a single X chromosome, the expression of *fox-1* and *sex-1* is insufficient to inhibit the expression of *xol-1*. When *xol-1* is expressed, its protein product inhibits the expression of the *sdc* genes. This prevents dosage compensation

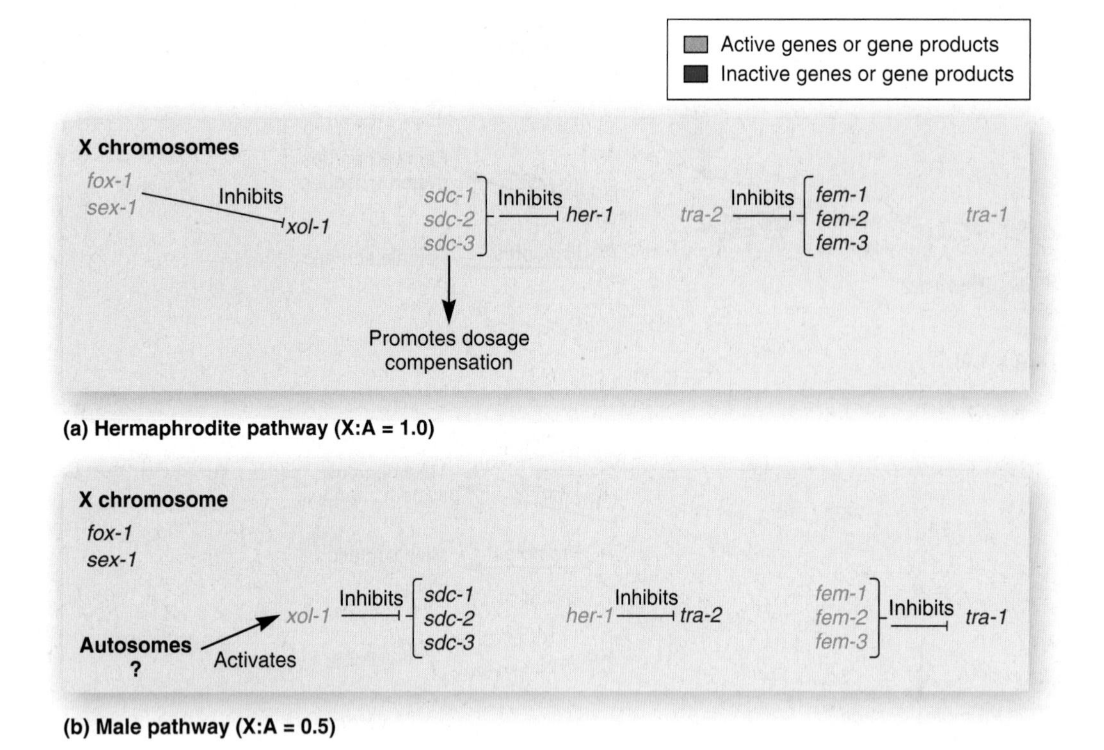

FIGURE 23.27 **Sex determination pathway for *C. elegans*.** Genes or gene products that are expressed at relatively high levels are shown in light orange; those that are not expressed or are expressed at low levels are shown in blue. The gene names are acronyms that usually refer to the effects of mutations. These are as follows: *fox* (feminizing on X), *sex* (sex determination), *xol* (X0 lethal), *sdc* (sex determination and dosage compensation), *her* (hermaphrodite), *tra* (transformer), and *fem* (feminization). Note: This is a simplified pathway.

and permits the expression of *her-1*. The *her-1* gene product then inhibits *tra-2*. This, in turn, permits the expression of the *fem* genes, which inhibit the expression of *tra-1*. Without *tra-1* expression, the worm develops into a male instead of a hermaphrodite. Experimentally, it is has been shown that XX worms lacking a functional *tra-1* gene develop into males.

In Mammals, the *SRY* Gene on the Y Chromosome Determines Maleness

In most mammals, such as humans, mice, and marsupials, the presence of the *SRY* gene on the Y chromosome determines maleness. In cases of abnormal sex chromosome composition such as XXY, an individual develops into a male. The *SRY* gene, which is located on the Y chromosome, causes the sex determination pathway to follow a male developmental scheme. The *SRY* gene encodes a protein that contains a DNA-binding domain called an HMG box, which is found in a broad category of DNA-binding proteins known as the high-mobility group. In several of these proteins, the HMG box is known to cause DNA bends. Thus far, the ability of the SRY protein to promote male sex determination is not well understood, but it may act as a transcription factor and/or promote changes in chromatin architecture, much like the chromatin remodeling proteins discussed in Chapters 10 and 15.

Sex determination in mammals, like that of fruit flies and worms, involves a cascade that is initiated in early embryonic development. However, the details of the pathway are not completely elucidated. **Figure 23.28** illustrates a simplified pathway. Several genes in mammals have been identified that are expressed very early in embryonic development and may be directly or indirectly involved in turning on the *SRY* gene. For example, two genes, designated *SFI* and *WTI*, are expressed in the early embryo prior to sexual differentiation and may regulate *SRY* expression. Once activated,

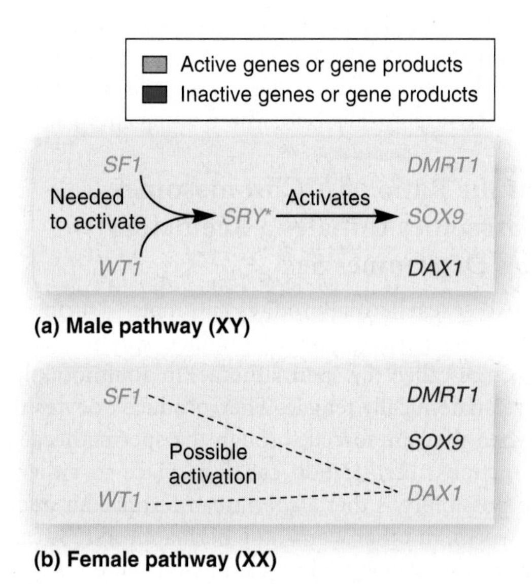

FIGURE 23.28 **Sex determination pathway for mammals.** Genes or gene products that are expressed at relatively high levels are shown in light orange; those that are not expressed or are inactive are shown in blue. The regulation of the *DAX1* gene is not well understood, but it may involve activation via *SF1* and/or *WT1* gene products. The gene names are acronyms that usually refer to the effects of mutations or their relatedness to other genes or chromosomal regions. These are as follows: *SF1* (steroidogenic factor-1), *WT1* (Wilms tumor gene), *SRY* (sex-determining region of the Y chromosome), *DMRT1* (doublesex- and mab-3-related transcription factor), *SOX9* (SRY-like HMG box), and *DAX1* (dosage-sensitive sex-reversal, adrenal hypoplasia congenital, X chromosome). Note: This is a simplified pathway.

researchers postulate that *SRY* expression causes the expression of other genes that promote male development. However, the target genes that are turned on by the SRY protein have not been definitively identified. One likely candidate is a gene called *SOX9*. Like the *SRY* gene, *SOX9* encodes a protein with an HMG domain. In addition, the SOX9 protein is larger than the SRY protein because it also contains two transcriptional activation domains, indicating that it is a transcriptional activator. *SOX9* expression is necessary for male development, because a loss-of-function allele of *SOX9* results in female development in XY animals. Furthermore, when researchers produced XX mice with three copies of the *SOX9* gene, such individuals developed into males.

Two other genes, designated *DAX1* and *DMRT1*, also play a role in sex determination. The *DAX1* gene is X-linked, and its gene product is thought to prevent male development. In XX animals, its expression remains high, in contrast to XY males. XY animals with two copies of the *DAX1* gene develop into females when *SRY* gene expression is low due to a weak allele in the *SRY* gene. This suggests that *DAX1* can inhibit the effects of the *SRY* gene. However, the *DAX1* gene is not needed for female development because XX mice lacking the *DAX1* gene develop as normal females. *DAX1* encodes a hormone-receptor protein. Further research is needed to understand the pathways that are activated via this receptor. Finally, the *DMRT1* gene in mammals is evolutionarily related to the *mab-3* gene in *C. elegans* and the *dsx* gene in *Drosophila*. These genes encode transcription factors involved in the differentiation of the gonads. In mammals, *DMRT1* expression is gonad specific, and it is expressed at higher levels in the testes. XY mice that are lacking the *DMRT1* gene develop as males, but they are infertile due to severe defects in testis structure and the inability to produce sperm cells.

The mechanism of dosage compensation in mammals involves the process of X inactivation. In normal females, one of the two X chromosomes in females is compacted into a Barr body. This topic is described in Chapter 5.

In Sexually Dimorphic Plants, the Male Plant Is Usually Heteromorphic

As mentioned earlier, most species of plants, about 95%, are sexually monomorphic. Only a single type of individual is produced that can make both male and female gametophytes. By comparison, sexually dimorphic species have two separate types of individuals. The most common types are **dioecious,** in which one kind of individual produces male gametophytes, and the other produces female gametophytes. A few plants species are gynodioecious, which produce hermaphrodite plants and female plants. Alternatively, a few species are androdioecious and produce hermaphrodite plants and male plants.

The genetics of sex determination in dioecious plant species is beginning to emerge. Because *Arabidopsis* is sexually monomorphic, researchers have had to turn to other species to study sex determination in plants. For example, the white campion, *Silene latifolia*, has been the subject of numerous investigations. In this species, sex chromosomes, designated X and Y, are responsible for sex determination. The male plant has heteromorphic sex chromosomes, XY; the female plant is XX. In other dioecious species, cytological examination of the chromosomes does not always reveal distinct types of sex chromosomes. Nevertheless, the male plants appear to be the heterozygote. This has been determined because male plants are often "inconstant," which means that they produce an occasional fruit. When a male plant undergoes such self-fertilization, the seeds from the fruit produce a 3:1 ratio of male to female plants. This is the expected result if the male parent was heterozygous for a dominant gene that promotes maleness.

Researchers are beginning to identify genes that are important for sex determination in plants, though further studies are needed to unravel the genetic hierarchy that promotes sexual development. A chromosomal site designated Su^F plays a role in this process. The Su^F locus is believed to contain a gene that acts as a dominant suppressor of the female pathway and prevents carpel development. It is responsible for producing the 3:1 ratio when male plants undergo self-fertilization. In addition, other loci that appear to control early and late anther development have been discovered. Along with Su^F, these loci are linked on the Y chromosome in *S. latifolia*. Eventually, the discovery of more genes that control sexual development will allow researchers to propose alternative pathways that promote male and female development in dioecious plant species.

KEY TERMS

Page 637. developmental genetics
Page 638. pattern, positional information, apoptosis, morphogen
Page 639. induction, cell adhesion, cell adhesion molecules (CAMs), homeotic gene
Page 640. bilaterians, determination
Page 641. parasegments, segments, embryogenesis, anteroposterior axis, dorsoventral axis, left-right axis, proximodistal axis
Page 645. segmentation genes
Page 646. zygotic genes
Page 647. cell fate, homeotic
Page 648. gain-of-function mutation, homeobox, homeodomain
Page 649. lineage diagram, cell lineage
Page 652. heterochronic mutations, homologous genes
Page 653. orthologs, *Hox* complexes, gene knockout, reverse genetics
Page 654. determined, differentiated
Page 655. basic domain, helix-loop-helix domain (bHLH), myogenic bHLH proteins
Page 656. totipotent
Page 657. shoot meristem, root meristem, organizing center, central zone, peripheral zone, apical region, central region, basal region

Page 658. apical-basal-patterning genes, ABC model
Page 659. sex determination
Page 660. autoregulatory loop
Page 663. dioecious

CHAPTER SUMMARY

- Developmental genetics is concerned with the roles that genes play in orchestrating the changes that occur during development.

23.1 Overview of Animal Development

- Positional information in animals may cause a cell to divide, migrate, differentiate, or undergo apoptosis (see Figure 23.1).
- A morphogen is a molecule that conveys positional information.
- Three molecular mechanisms that convey positional information include asymmetrical distribution of morphogens in an oocyte, asymmetrical synthesis and extracellular distribution of a morphogen in an embryo, and cell-to-cell contact (see Figure 23.2).
- The study of mutants that disrupt development has identified genes that control development (see Figure 23.3).
- Development in animals involves four overlapping phases, which include the formation of axes, segmentation, determination, and cell differentiation (see Figure 23.4).

23.2 Invertebrate Development

- *Drosophila* proceeds through different developmental stages from a fertilized egg to an adult. Various sets of genes are responsible for developmental changes (see Figure 23.5, Table 23.1).
- Maternal effect genes establish the anteroposterior and dorsoventral axes due to their asymmetrical distribution. An example is *bicoid*, which promotes the formation of anterior structures (see Figures 23.6–23.8).
- The *Drosophila* embryo is divided into segments (see Figure 23.9).
- Three categories of segmentation genes have been identified based on their effects on development when mutant (see Figure 23.10).
- A hierarchy of gene expression, which includes maternal effect genes, gap genes, pair-rule genes, and segment-polarity genes, gives rise to a segmented embryo (see Figure 23.11).
- Homeotic genes control the developmental fate of particular segments (see Figures 23.12, 23.13).
- Homeotic proteins contain a DNA-binding domain and a transcriptional activation domain (see Figure 23.14).
- In *C. elegans*, a cell lineage diagram describes the cell fate of each cell of the worm's body (see Figure 23.15).
- Heterochronic mutations in *C. elegans* disrupt the timing of developmental changes (see Figure 23.16).

23.3 Vertebrate Development

- Homeotic genes in vertebrates are found in *Hox* complexes (see Figure 23.17).
- *Hox* genes control the fate of regions along the anteroposterior axis (see Figures 23.18, 23.19).
- Transcription factors also control cell differentiation. An example is *myoD*, which causes cells to differentiate into skeletal muscle cells (see Figure 23.20).

23.4 Plant Development

- *Arabidopsis thaliana* is a model organism for studying plant development (see Figure 23.21).
- Plant growth occurs from shoot and root meristems (see Figure 23.21).
- The expression of *WUS* and *CLV3* controls the correct number of stem cells in the central zone of a shoot meristem (see Figure 23.23).
- Researchers are identifying genes that affect apical-basal patterning in plants (see Table 23.2).
- The ABC model describes how homeotic gene control flower development in plants (see Figures 23.24, 23.25).

23.5 Sex Determination in Animals and Plants

- Mechanisms of sex determination differ among various species (see Table 23.3).
- At the molecular level, sex determination is controlled by pathways that activate specific genes or proteins and inactivate others (see Figures 23.26–23.28).

PROBLEM SETS & INSIGHTS

Solved Problems

S1. Discuss and distinguish the functional roles of the maternal effect genes, gap genes, pair-rule genes, and segment-polarity genes in *Drosophila*.

Answer: These genes are involved in pattern formation of the *Drosophila* embryo. The asymmetrical distribution of maternal effect gene products in the oocyte establishes the anteroposterior and dorsoventral axes. These gene products also control the expression of the gap genes, which are expressed as broad bands in certain regions of the embryo. The overlapping expression of maternal effect genes and gap genes controls the pair-rule genes, which are expressed in alternating stripes. A stripe corresponds to a parasegment. Within each parasegment, the expression of segment-polarity genes defines an anterior and posterior compartment. With regard to morphology, an anterior compartment of one parasegment and the posterior compartment of an adjacent parasegment form a segment of the fly.

S2. With regard to genes affecting development, what are the phenotypic effects of gain-of-function mutations versus loss-of-function mutations?

Answer: Gain-of-function mutations cause a gene to be expressed in the wrong place, at the wrong time, or at an abnormal level. When the gene is expressed in the wrong place, that region may develop into an inappropriate structure. For example, when *Antp* is abnormally expressed in an anterior segment, this segment develops legs in place of antennae. When gain-of-function mutations cause a gene to be expressed at the wrong time, this can also disrupt the development process. Gain-of-function heterochronic mutations cause cell lineages to be reiterated, thereby altering the course of development. By comparison, loss-of-function mutations result in a defect in the expression of a gene with the result that protein function is reduced or abolished. This usually disrupts the developmental process, because the cells in the region where the gene is normally expressed are not directed to develop along the correct pathway.

S3. Mutations in genes that control the early stages of development are often lethal (e.g., see Figure 23.7b). To circumvent this problem, developmental geneticists may try to isolate temperature-sensitive developmental mutants, or *ts* alleles. If an embryo carries a *ts* allele, it will develop correctly at the permissive temperature (e.g., 25°C) but will fail to develop if incubated at the nonpermissive temperature (e.g., 30°C). In most cases, *ts* alleles have missense mutations that slightly alter the amino acid sequence of a protein, causing a change in its structure that prevents it from working properly at the nonpermissive temperature. *Ts* alleles are particularly useful because they can provide insight regarding the stage of development when the protein is necessary. Researchers can take groups of embryos that carry a *ts* allele and expose them to the permissive and nonpermissive temperature at different stages of development. In the experiment described next, embryos were divided into five groups and exposed to the permissive or nonpermissive temperature at different times after fertilization.

Time After			**Group**		
Fertilization (hours):	**1**	**2**	**3**	**4**	**5**
0–1	25°C	25°C	25°C	25°C	25°C
1–2	25°C	30°C	25°C	25°C	25°C
2–3	25°C	25°C	30°C	25°C	25°C
3–4	25°C	25°C	25°C	30°C	25°C
4–5	25°C	25°C	25°C	25°C	30°C
5–6	25°C	25°C	25°C	25°C	25°C
SURVIVAL:	Yes	Yes	Yes	No	Yes

Explain these results.

Answer: By varying the temperature during different stages of development, researchers can pinpoint the stage when the function of the protein encoded by this *ts* allele is critical. As shown, embryos fail to survive if they are subjected to the nonpermissive temperature between 3 and 4 hours after fertilization, but they do survive if subjected to the nonpermissive temperature at other times of development. These results indicate that this protein plays a crucial role at the 3- to 4-hour stage of development.

S4. An intriguing question in developmental genetics is, how can a particular gene, such as *even-skipped*, be expressed in a multiple banding pattern as seen in Figure 23.11? Another way of asking this question is, how is the positional information within the broad bands of the gap genes able to be deciphered in a way that causes the pair-rule genes to be expressed in this alternating banding pattern? The answer lies in a complex mechanism of genetic regulation. Certain pair-rule genes have several stripe-specific enhancers that are controlled by multiple transcription factors. A stripe-specific enhancer is typically a short segment of DNA, 300 to 500 bp in length, that contains binding sequences recognized by several different transcription factors. This term is a bit misleading because a stripe-specific enhancer is a regulatory region that contains both enhancer and silencer elements.

In 1992, Michael Levine and his colleagues investigated stripe-specific enhancers that are located near the promoter of the *even-skipped* gene. A segment of DNA, termed the stripe 2 enhancer, controls the expression of the *even-skipped* gene; this enhancer is responsible for the expression of the *even-skipped* gene in stripe 2, which corresponds to parasegment 3 of the embryo. The stripe 2 enhancer is a segment of DNA that contains binding sites for four transcription factors that are the products of the *Krüppel*, *bicoid*, *hunchback*, and *giant* genes. The Hunchback and Bicoid transcription factors bind to this enhancer and activate the transcription of the *even-skipped* gene. In contrast, the transcription factors encoded by the *Krüppel* and *giant* genes bind to the stripe 2 enhancer and repress transcription. The figure shown here describes the concentrations of these four transcription factor proteins in the region of parasegment 3 (i.e., stripe 2) in the *Drosophila* embryo.

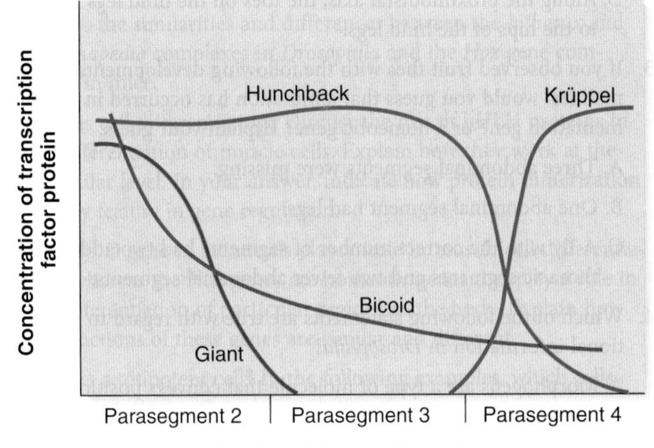

To study stripe-specific enhancers, researchers have constructed artificial genes in which the enhancer is linked to a reporter gene, the expression of which is easy to detect. The next figure shows the results of an experiment in which an artificial gene was made by putting the stripe 2 enhancer next to the β-galactosidase gene. This artificial gene was introduced into *Drosophila*, and then embryos containing this gene were analyzed for β-galactosidase activity. If a region of the embryo is expressing β galactosidase, the region will stain darkly because β galactosidase converts a colorless compound into a dark blue compound.

of being eliminated from the population due to random genetic drift. On the other hand, when N is small, the probability of new mutations is also small, but if they occur, the likelihood of fixation is relatively large.

Now that we have an appreciation for the phenomenon of genetic drift, we can ask a third question:

3. If fixation of a new allele does occur, how many generations is it likely to take?

The formula for calculating this also depends on the number of individuals in the population:

$$\bar{t} = 4N$$

where

$\bar{t}$ equals the average number of generations to achieve fixation

N equals the number of individuals in the population, assuming that males and females contribute equally to each succeeding generation

As you may have expected, allele fixation takes much longer in large populations. If a population has 1 million breeding members, it takes, on average, 4 million generations, perhaps an insurmountable period of time, to reach fixation. In a small group of 100 individuals, however, fixation takes only 400 generations, on average. As discussed in Chapter 26, the drifting of neutral alleles among different populations and species provides a way to measure the rate of evolution and can be used to determine evolutionary relationships.

The preceding discussion of random genetic drift has emphasized two important points. First, genetic drift ultimately operates in a random manner with regard to allele frequency and, over the long run, leads to either allele fixation or elimination. The process is random with regard to particular alleles. Genetic drift can lead to the fixation of deleterious, neutral, or beneficial alleles. A second important feature of genetic drift is that its influence is greatly affected by population size. Genetic drift may lead more quickly to allele loss or fixation in a small population.

In nature, there are different ways that geography and population size influence how genetic drift affects the genetic composition of a species. Some species occupy wide ranges in which small, local populations become geographically isolated from the rest of the species. The allele frequencies within these small populations are more susceptible to genetic drift. Because this is a random process, small isolated populations tend to be more genetically disparate in relation to other populations. This may occur in the following two ways.

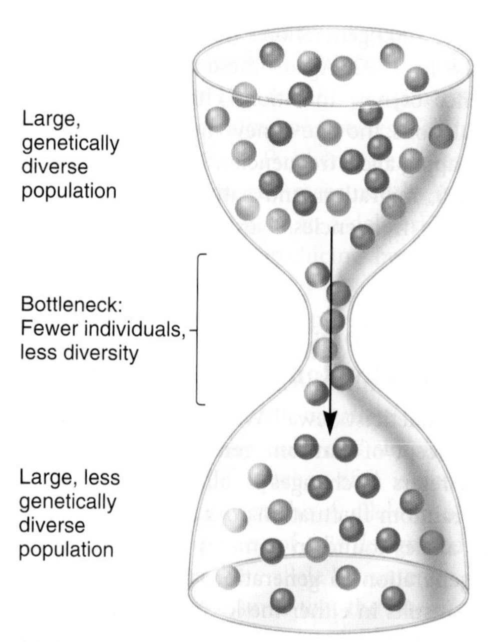

(b) An African cheetah

FIGURE 24.7 The bottleneck effect, an example of genetic drift. (a) A representation of the bottleneck effect. Note that the genetic variation denoted by the green balls has been lost. (b) The African cheetah. The modern species has low genetic variation due to a genetic bottleneck that is thought to have occurred about 10,000 to 12,000 years ago.

Bottleneck Effect Changes in population size may influence genetic drift via the **bottleneck effect.** In nature, a population can be reduced dramatically in size by events such as earthquakes, floods, drought, or human destruction of habitat. Such events may randomly eliminate most of the members of the population without regard to genetic composition. The initial bottleneck may be greatly influenced by genetic drift because the surviving members may have allele frequencies that differ from those of the original population. In addition, allele frequencies are expected to drift substantially during the generations when the population size is small. In extreme cases, alleles may even be eliminated. Eventually, the population with the bottleneck may regain its original size (**Figure 24.7**). However, the new population has less genetic variation than the original large population. As an example, the African cheetah population lost a substantial amount of its genetic variation due to a bottleneck effect. DNA analysis by population geneticists has suggested that a severe bottleneck occurred approximately 10,000 to 12,000 years ago when the population size was dramatically reduced. The population eventually rebounded, but the bottleneck significantly decreased the genetic variation.

Founder Effect Geography and population size may also influence genetic drift via the **founder effect.** Compared to the bottleneck effect, the key difference is that the founder effect involves migration; a small group of individuals separates from a larger population and establishes a colony in a new location. For example, a few individuals may migrate from a large continental population and become the founders of an island population. The founder effect has two important consequences. First, the founding population is expected to have less genetic variation than the original population from which it was derived. Second, as a matter of chance, the allele frequencies in the founding population may differ markedly from those of the original population.

Population geneticists have studied many examples of isolated populations that were started from a few members of another population. In the 1960s, Victor McKusick studied allele frequencies in the Old Order Amish of Lancaster County, Pennsylvania. At that time, this was a group of about 8000 people, descended from just three couples that immigrated to the United States in 1770. Among this population of 8000, a genetic disease known as the Ellis-van Creveld syndrome (a recessive form of dwarfism) was found at a frequency of 0.07, or 7%. By comparison, this disorder is extremely rare in other human populations, even the population from which the founding members had originated. The high frequency of dwarfism in the Lancaster County population is a chance occurrence due to the founder effect. Perhaps one of the original founders carried the recessive allele for dwarfism, or a new mutation may have occurred in one of their early descendants.

Migrations Between Two Populations Can Alter Allele Frequencies

We have just seen how migration to a new location by a relatively small group can result in a population with an altered genetic composition. In addition, migration between two different established populations can alter allele frequencies. For example, a species of birds may occupy two geographic regions that are separated by a large body of water. On rare occasions, the prevailing winds may allow birds from the western population to fly over this body of water and become members of the eastern population. If the two populations have different allele frequencies and if migration occurs in sufficient numbers, this migration may alter the allele frequencies in the eastern population.

After migration has occurred, the new (eastern) population is called a **conglomerate.** To calculate the allele frequencies in the conglomerate, we need two kinds of information. First, we must know the original allele frequencies in the donor and recipient populations. Second, we must know the proportion of the conglomerate population that is due to migrants. With these data, we can calculate the change in allele frequency in the conglomerate population using the following equation:

$$\Delta p_C = m(p_D - p_R)$$

where

Δp_C is the change in allele frequency in the conglomerate population

p_D is the allele frequency in the donor population

p_R is the allele frequency in the original recipient population

m is the proportion of migrants that make up the conglomerate population

$$m = \frac{\text{Number of migrants in the conglomerate population}}{\text{Total number of individuals in the conglomerate population}}$$

As an example, let's suppose the allele frequency of *A* is 0.7 in the donor population and 0.3 in the recipient population. A group of 20 individuals migrates and joins the recipient population, which originally had 80 members. Thus,

$$m = \frac{20}{20 + 80}$$
$$= 0.2$$
$$\Delta p_C = m(p_D - p_R)$$
$$= 0.2(0.7 - 0.3)$$
$$= 0.08$$

We can now calculate the allele frequency in the conglomerate:

$$p_C = p_R + \Delta p_C$$
$$= 0.3 + 0.08 = 0.38$$

Therefore, in the conglomerate population, the allele frequency of *A* has changed from 0.3 (its value before migration) to 0.38. This increase in allele frequency arises from the higher allele frequency of *A* in the donor population. **Gene flow** occurs whenever individuals migrate between populations and the migrants are able to breed successfully with the members of the recipient population. Gene flow depends not only on migration, but also on the ability of the migrants' alleles to be passed to subsequent generations.

In our previous example, we considered the consequences of a unidirectional migration from a donor to a recipient population. In nature, it is common for individuals to migrate in both directions. What are the main consequences of bidirectional migration? Depending on its rate, migration tends to reduce differences in allele frequencies between neighboring populations. In fact, population geneticists can analyze allele frequencies in two different populations to evaluate the rate of migration between them. Populations that frequently mix their gene pools via migration tend to have similar allele frequencies, whereas isolated populations are expected to be more disparate. By comparison, genetic drift, which we considered earlier, tends to make local populations more disparate from each other.

Migration can also enhance genetic diversity within a population. As mentioned, new mutations are relatively rare events. Therefore, a particular mutation may arise only in one population. Migration may then introduce this new allele into neighboring populations.

Natural Selection Is Based on the Relative Reproductive Success of Genotypes

In the 1850s, Charles Darwin and Alfred Russel Wallace independently proposed the theory of evolution by **natural selection.** According to this theory, the conditions found in nature result in the selective survival and reproduction of individuals whose characteristics make them better adapted to their environment. These surviving individuals are more likely to reproduce and contribute offspring to the next generation. Natural selection can be related not only to differential survival but also to mating efficiency and fertility.

A modern restatement of the principles of natural selection can relate our knowledge of molecular genetics to the phenotypes of individuals.

1. Within a population, allelic variation arises in various ways, such as through random mutations that cause differences in DNA sequences. A mutation that creates a new allele may alter the amino acid sequence of the encoded protein, which, in turn, may alter the function of the protein.
2. Some alleles may encode proteins that enhance an individual's survival or reproductive capability compared with that of other members of the population. For example, an allele may produce a protein that is more efficient at a higher temperature, conferring on the individual a greater probability of survival in a hot climate.
3. Individuals with beneficial alleles are more likely to survive and contribute to the gene pool of the next generation.
4. Over the course of many generations, allele frequencies of many different genes may change through this process, thereby significantly altering the characteristics of a species. The net result of natural selection is a population that is better adapted to its environment and more successful at reproduction. Even so, it should be emphasized that species are not perfectly adapted to their environments, because mutations are random events and because the environment tends to change from generation to generation.

As mentioned at the beginning of the chapter, Fisher, Wright, and Haldane developed mathematical relationships to explain the theory of natural selection. As our knowledge of the process of natural selection has increased, it has become apparent that it operates in many different ways. In this chapter, we will consider a few examples of natural selection involving a single trait or a single gene that exists in two alleles. In reality, however, natural selection acts on populations of individuals in which many genes are polymorphic and each individual contains thousands or tens of thousands of different genes.

To begin our quantitative discussion of natural selection, we must examine the concept of **Darwinian fitness**—the relative likelihood that one genotype will contribute to the gene pool of the next generation rather than other genotypes. Natural selection acts on phenotypes that are derived from individuals' genotypes. Although Darwinian fitness often correlates with physical fitness, the two concepts are not identical. Darwinian fitness is a measure of reproductive success. An extremely fertile genotype may have a higher Darwinian fitness than a less fertile genotype that appears more physically fit.

To consider Darwinian fitness, let's use our example of a gene existing in the *A* and *a* alleles. If the three genotypes have the same level of mating success and fertility, we can assign fitness values to each genotype class based on their likelihood of surviving to reproductive age. For example, let's suppose that the relative survival to adulthood of each of the three genotype classes is as follows: For every five *AA* individuals that survive, four *Aa* individuals survive, and one *aa* individual survives. By convention, the genotype with the highest reproductive ability is given a fitness value of 1.0. Relative fitness values are denoted by the variable *W*. The fitness values of the other genotypes are assigned values relative to this 1.0 value:

$$\text{Fitness of } AA\text{: } W_{AA} = 1.0$$

$$\text{Fitness of } Aa\text{: } W_{Aa} = 4/5 = 0.8$$

$$\text{Fitness of } aa\text{: } W_{aa} = 1/5 = 0.2$$

Keep in mind that differences in reproductive success among genotypes may stem from various reasons. In this case, the fittest genotype is more likely to survive to reproductive age. In other situations, the most fit genotype is more likely to mate. For example, a bird with brightly colored feathers may have an easier time attracting a mate than a bird with duller plumage. Finally, a third possibility is that the fittest genotype may be more fertile. It may produce a higher number of gametes or gametes that are more successful at fertilization.

By studying species in their native environments, population geneticists have discovered that natural selection can occur in several ways. The patterns of natural selection depend on the relative fitness values of the different genotypes and on the variation of environmental effects. The four patterns of natural selection that we will consider are called directional, stabilizing, disruptive, and balancing selection. In most of the examples described next, natural selection leads to adaptation so that a species is better able to survive to reproductive age.

Directional Selection Favors the Extreme Phenotype

Directional selection favors individuals at one extreme of a phenotypic distribution that are more likely to survive and reproduce in a particular environment. Different phenomena may initiate the process of directional selection. One way that directional selection may arise is that a new allele may be introduced into a population by mutation, and the new allele may promote a higher fitness in individuals that carry it (**Figure 24.8**). If the homozygote carrying the favored allele has the highest fitness value, directional selection may cause this favored allele to eventually become the predominant allele in the population, perhaps even becoming a monomorphic allele.

Another possibility is that a population may be exposed to a prolonged change in its living environment. Under the new environmental conditions, the relative fitness values may change to favor one genotype, which will promote the elimination of other genotypes. As an example, let's suppose a population of finches on the mainland already has genetic variation in beak size. A small number of birds migrate to an island where the seeds are generally larger than they are on the mainland. In this new environment, birds with larger beaks have a higher fitness because they are better able to crack open the larger seeds and thereby survive to reproductive age. Over the course of many

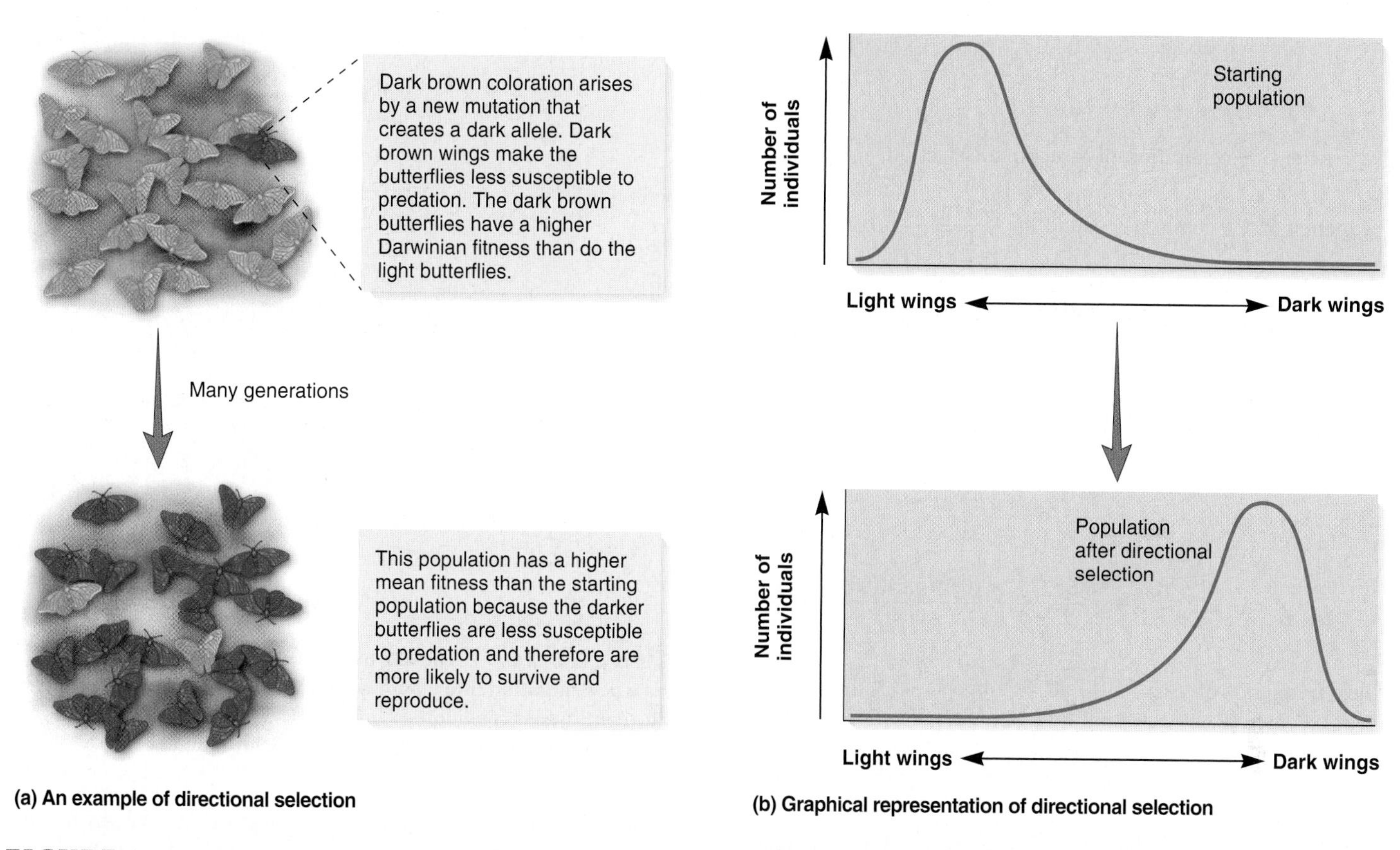

FIGURE 24.8 **Directional selection.** **(a)** A new mutation arises in a population that confers higher Darwinian fitness. In this example, butterflies with dark wings are more likely to survive and reproduce. Over many generations, directional selection favors the prevalence of darker individuals. **(b)** A graphical representation of directional selection.

generations, directional selection produces a population of birds carrying alleles that promote larger beak size.

In the case of directional selection, allele frequencies may change in a step-by-step, generation-per-generation way. To appreciate how this occurs, let's take a look at how fitness affects the Hardy-Weinberg equilibrium and allele frequencies. Again, let's suppose a gene exists in two alleles: *A* and *a*. The three fitness values, which are based on relative survival levels, are

$$W_{AA} = 1.0$$
$$W_{Aa} = 0.8$$
$$W_{aa} = 0.2$$

In the next generation, we expect that the Hardy-Weinberg equilibrium is modified in the following way due to directional selection:

Frequency of *AA*: p^2W_{AA}

Frequency of *Aa*: $2pqW_{Aa}$

Frequency of *aa*: q^2W_{aa}

In a population that is changing due to natural selection, these three terms may not add up to 1.0, as they do in the Hardy-Weinberg equilibrium. Instead, the three terms sum to a value known as the **mean fitness of the population ($\overline{W}$):**

$$p^2W_{AA} + 2pqW_{Aa} + q^2W_{aa} = \overline{W}$$

Dividing both sides of the equation by the mean fitness of the population,

$$\frac{p^2W_{AA}}{\overline{W}} + \frac{2pqW_{Aa}}{\overline{W}} + \frac{q^2W_{aa}}{\overline{W}} = 1$$

Using this equation, we can calculate the expected genotype and allele frequencies after one generation of directional selection:

Frequency of *AA* genotype: $\frac{p^2W_{AA}}{\overline{W}}$

Frequency of *Aa* genotype: $\frac{2pqW_{Aa}}{\overline{W}}$

Frequency of *aa* genotype: $\frac{q^2W_{aa}}{\overline{W}}$

Allele frequency of *A*: $p_A = \frac{p^2W_{AA}}{\overline{W}} + \frac{pqW_{Aa}}{\overline{W}}$

Allele frequency of *a*: $q_a = \frac{q^2W_{aa}}{\overline{W}} + \frac{pqW_{Aa}}{\overline{W}}$

As an example, let's suppose that the starting allele frequencies are $A = 0.5$ and $a = 0.5$, and use fitness values of 1.0, 0.8, and 0.2 for the three genotypes, *AA*, *Aa*, and *aa*, respectively. We begin by calculating the mean fitness of the population:

$$p^2W_{AA} + 2pqW_{Aa} + q^2W_{aa} = \overline{W}$$

$$\overline{W} = (0.5)^2(1) + 2(0.5)(0.5)(0.8) + (0.5)^2(0.2)$$

$$\overline{W} = 0.25 + 0.4 + 0.05 = 0.7$$

After one generation of directional selection,

Frequency of AA genotype: $\dfrac{p^2W_{AA}}{\overline{W}} = \dfrac{(0.5)^2(1)}{0.7} = 0.36$

Frequency of Aa genotype: $\dfrac{2pqW_{Aa}}{\overline{W}} = \dfrac{2(0.5)(0.5)(0.8)}{0.7} = 0.57$

Frequency of aa genotype: $\dfrac{q^2W_{aa}}{\overline{W}} = \dfrac{(0.5)^2(0.2)}{0.7} = 0.07$

Allele frequency of A: $p_A = \dfrac{p^2W_{AA}}{\overline{W}} + \dfrac{pqW_{Aa}}{\overline{W}}$

$$= \frac{(0.5)^2(1)}{0.7} + \frac{(0.5)(0.5)(0.8)}{0.7} = 0.64$$

Allele frequency of a: $q_a = \dfrac{q^2W_{aa}}{\overline{W}} + \dfrac{pqW_{Aa}}{\overline{W}}$

$$= \frac{(0.5)^2(0.2)}{0.7} + \frac{(0.5)(0.5)(0.8)}{0.7} = 0.36$$

After one generation, the allele frequency of A has increased from 0.5 to 0.64, and the frequency of a has decreased from 0.5 to 0.36. This has occurred because the AA genotype has the highest fitness, whereas the Aa and aa genotypes have lower fitness values. Another interesting feature of natural selection is that it raises the mean fitness of the population. If we assume the individual fitness values are constant, the mean fitness of this next generation is

$$\overline{W} = p^2W_{AA} + 2pqW_{Aa} + q^2W_{aa}$$
$$= (0.64)^2(1) + 2(0.64)(0.36)(0.8) + (0.36)^2(0.2)$$
$$= 0.80$$

The mean fitness of the population has increased from 0.7 to 0.8.

What are the consequences of natural selection at the population level? This population is better adapted to its environment than the previous one. Another way of viewing this calculation is that the subsequent population has a greater reproductive potential than the previous one. We could perform the same types of calculations to find the allele frequencies and mean fitness value in the next generation. If we assume the individual fitness values remain constant, the frequencies of A and a in the next generation are 0.85 and 0.15, respectively, and the mean fitness increases to 0.931. As we can see, the general trend is to increase A, decrease a, and increase the mean fitness of the population.

In the previous example, we considered the effects of natural selection by beginning with allele frequencies at intermediate levels (namely, $A = 0.5$ and $a = 0.5$). **Figure 24.9** illustrates what would happen if a new mutation introduced the A allele into a population that was originally monomorphic for the a allele. As

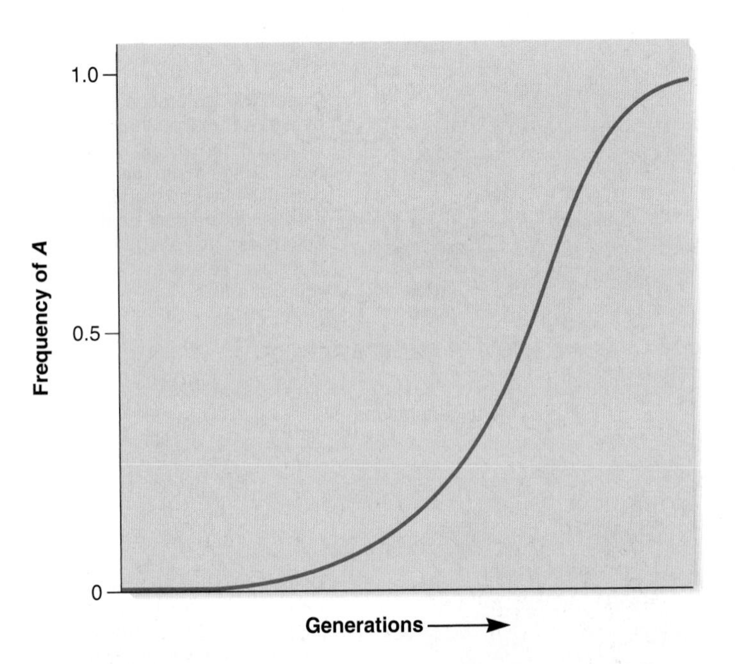

FIGURE 24.9 The fate of a beneficial allele that is introduced as a new mutation into a population. A new allele (A) is beneficial in the homozygous condition: $W_{AA} = 1.0$. The heterozygote, Aa ($W_{Aa} = 0.8$), and homozygote, aa ($W_{aa} = 0.2$), have lower fitness values.

before, the AA homozygote has a fitness of 1.0, the Aa heterozygote 0.8, and the recessive aa homozygote 0.2. Initially, the A allele is at a very low frequency in the population. If it is not lost initially due to genetic drift, its frequency slowly begins to rise and then, at intermediate values, rises much more rapidly.

Eventually, this type of natural selection may lead to fixation of a beneficial allele. However, a new beneficial allele is in a precarious situation when its frequency is very low. As we have seen, random genetic drift is likely to eliminate new mutations, even beneficial ones, due to chance fluctuations.

Researchers have identified many examples of directional selection in nature. As mentioned in Chapter 7, resistance to antibiotics is a growing concern in the treatment of infection. The selection of bacterial strains resistant to one or more antibiotics typically occurs in a directional manner. Similarly, the resistance of insects to pesticides, such as DDT (dichlorodiphenyltrichloroethane), occurs in a directional manner. DDT usage began in the 1940s as a way to decrease the populations of mosquitoes and other insects. However, certain insect species can become resistant to DDT by a dominant mutation in a single enzyme-encoding gene. The mutant enzyme detoxifies DDT, making it harmless to the insect. **Figure 24.10** shows the results of an experiment in which mosquito larvae (*Aedes aegypti*) were exposed to DDT over the course of seven generations. The starting population showed a low level of DDT resistance, as evidenced by the low percentage of survivors after exposure to DDT. By comparison, in seven generations, nearly 100% of the population was DDT-resistant. These results illustrate the power of directional selection in promoting change in a population. Since the 1950s, resistance to nearly every known insecticide has evolved within 10 years of its commercial introduction!

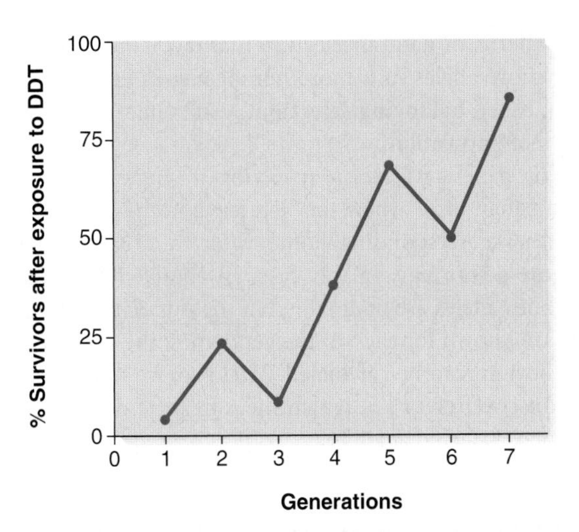

FIGURE 24.10 Directional selection for DDT resistance in a mosquito population. In this experiment, mosquito larvae (*Aedes aegypti*) were exposed to 10 mg/L of DDT. The percentage of survivors was recorded, and then the survivors of each generation were used as parents for the next generation.

Stabilizing Selection Favors Individuals with Intermediate Phenotypes

In **stabilizing selection,** the extreme phenotypes for a trait are selected against, and those individuals with the intermediate phenotypes have the highest fitness values. Stabilizing selection tends to decrease genetic diversity for a particular gene because it eliminates alleles that cause a greater variation in phenotypes. An example of stabilizing selection involves clutch size in birds, which was first proposed by British biologist David Lack in 1947. Under stabilizing selection, birds that lay too many or too few eggs have lower fitness values than those that lay an intermediate value (**Figure 24.11**). Laying too many eggs may cause many offspring to die due to inadequate parental care and food. In addition, the strain on the parents themselves may decrease their likelihood of survival and therefore their ability to produce more offspring. Having too few offspring, on the other hand, does not contribute many individuals to the next generation. Therefore, the most successful parents are those that produce an intermediate clutch size. In the 1980s, Swedish evolutionary biologist Lars Gustafsson and colleagues examined the phenomenon of stabilizing selection in the collared flycatcher, *Ficedula albicollis*, on the island of Gotland, which is southeast of the mainland of Sweden. They discovered that Lack's hypothesis that clutch size is subject to the action of stabilizing selection also appears to be true for this species.

Disruptive Selection Favors Multiple Phenotypes

Disruptive selection, also known as diversifying selection, favors the survival of two or more different genotypes that produce different phenotypes (**Figure 24.12**). In disruptive selection, the fitness values of a particular genotype are higher in one environment and lower in a different one. Disruptive selection is likely to occur in populations that occupy diverse environments so that some members of the species survive in each type of environmental condition.

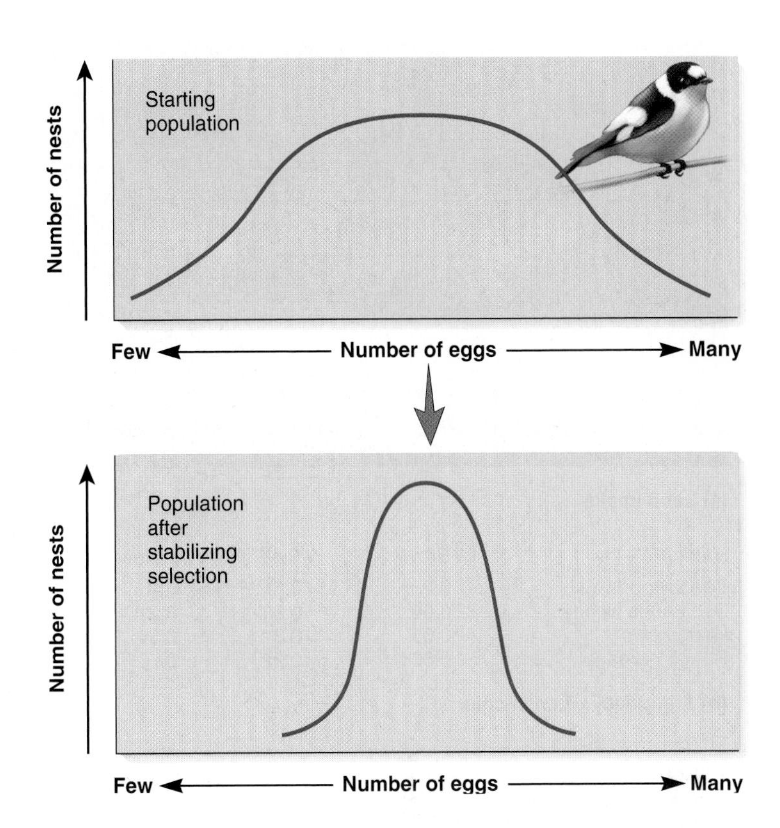

FIGURE 24.11 Stabilizing selection. In this pattern of natural selection, the extremes of a phenotypic distribution are selected against. Those individuals with intermediate traits have the highest fitness. This results in a population with less diversity and more uniform traits.

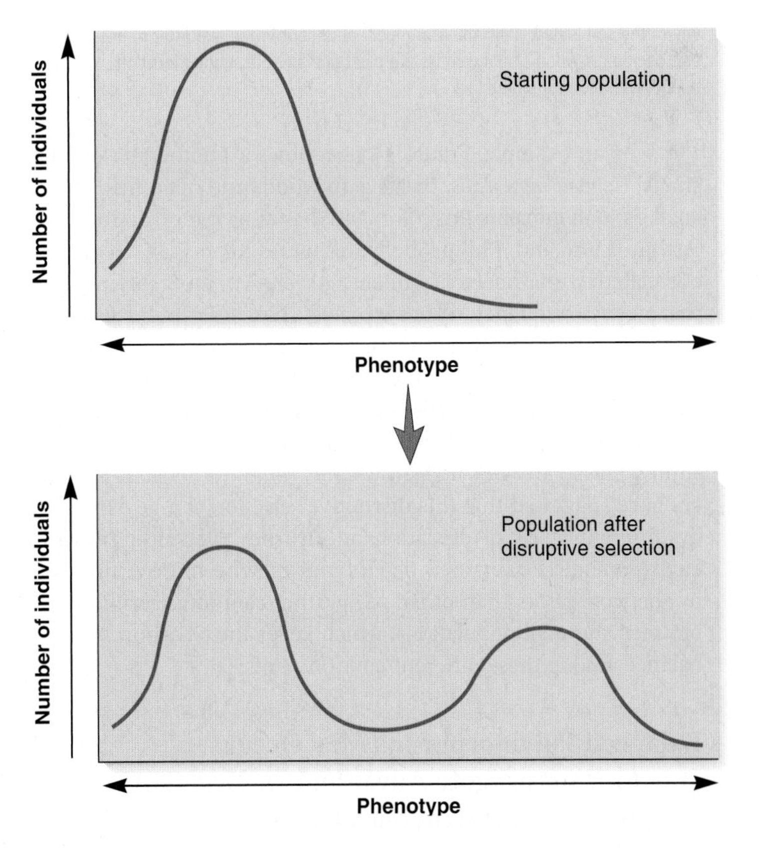

FIGURE 24.12 Disruptive selection. Over time, this form of selection favors two or more phenotypes due to heterogeneous environments.

(a) Land snails

Habitat	*Brown*	*Pink*	*Yellow*
Beechwoods	0.23	0.61	0.16
Deciduous woods	0.05	0.68	0.27
Hedgerows	0.05	0.31	0.64
Rough herbage	0.004	0.22	0.78

(b) Frequency of snail color

FIGURE 24.13 Polymorphism in the land snail, *Cepaea nemoralis.* **(a)** This species of snail can exist in several different colors and banding patterns. **(b)** Coloration of the snails is correlated with the specific environments where they are located.

Genes → Traits Snail coloration is an example of genetic polymorphism due to heterogeneous environments; the genes governing shell coloration are polymorphic. The predation of snails is correlated with their ability to be camouflaged in their natural environment. Snails with brown shells are most prevalent in beechwoods, where the soil is dark. Pink snails are most abundant in the leaf litter of beechwoods and deciduous woods. Yellow snails are most prevalent in sunnier locations, such as hedgerows and rough herbage.

As an example, **Figure 24.13a** shows a photograph of land snails, *Cepaea nemoralis*, that live in woods and open fields. This snail is polymorphic in color and banding patterns. In 1954, Arthur Cain and Philip Sheppard found that snail color was correlated with the environment. As shown in **Figure 24.13b**, the highest frequency of brown shell color was found in snails in the beechwoods, where there are wide expanses of dark soil. Their frequency was substantially less in other environments. By comparison, pink snails are most common in the leaf litter of forest floors, and the yellow snails are most abundant in the sunny, grassy areas of hedgerows and rough herbage. Researchers have suggested that this disruptive selection can be explained by different levels of predation by thrushes. Depending on the environment, certain snail phenotypes may be more easily seen by their predators than others. Migration can occasionally occur between the snail populations, which keeps the polymorphism in balance among these different environments.

Balanced Polymorphisms May Occur Due to Heterozygote Advantage or Negative Frequency-Dependent Selection

As we have just seen, polymorphisms may occur when a species occupies a diverse environment. Researchers have discovered other patterns of natural selection that favor the maintenance of two or more alleles in a more homogeneous environment. This pattern, called **balancing selection,** results in a genetic polymorphism in a population.

For genetic variation involving a single gene, balancing selection may arise when the heterozygote has a higher fitness than either corresponding homozygote, a situation called **heterozygote advantage.** In this case, an equilibrium is reached in which both alleles are maintained in the population. If the fitness values are known for each of the genotypes, the allele frequencies at equilibrium can be calculated. To do so, we must consider the **selection coefficient (*s*),** which measures the degree to which a genotype is selected against.

$$s = 1 - W$$

By convention, the genotype with the highest fitness has an *s* value of zero. Genotypes at a selective disadvantage have *s* values that are greater than 0 but less than or equal to 1.0. An extreme case is a recessive lethal allele. It would have an *s* value of 1.0 in the homozygote, while the *s* value in the heterozygote could be 0.

Let's consider the following case of relative fitness, where

$$W_{AA} = 0.7$$

$$W_{Aa} = 1.0$$

$$W_{aa} = 0.4$$

The selection coefficients are

$$s_{AA} = 1 - 0.7 = 0.3$$

$$s_{Aa} = 1 - 1.0 = 0$$

$$s_{aa} = 1 - 0.4 = 0.6$$

The population reaches an equilibrium when

$$s_{AA}p = s_{aa}q$$

If we take this equation, let $q = 1 - p$, and then solve for p:

$$p = \text{Allele frequency of } A = \frac{s_{aa}}{s_{AA} + s_{aa}}$$

$$= \frac{0.6}{0.3 + 0.6} = 0.67$$

If we let $p = 1 - q$ and then solve for q:

$$q = \text{Allele frequency of } a = \frac{s_{AA}}{s_{AA} + s_{aa}}$$

$$= \frac{0.3}{0.3 + 0.6} = 0.33$$

In this example, balancing selection maintains the two alleles in the population at frequencies in which *A* equals 0.67 and *a* equals 0.33.

Heterozygote advantage can sometimes explain the high frequency of alleles that are deleterious in a homozygous condition. A classic example is the Hb^S allele of the human β-globin gene. A homozygous Hb^SHb^S individual displays sickle cell disease, a disorder that leads to the sickling of the red blood cells. The Hb^SHb^S homozygote has a lower fitness than a homozygote with two copies of the more common β-globin allele, Hb^AHb^A. However, the heterozygote, Hb^AHb^S, has a higher level of fitness than either homozygote in areas where malaria is endemic (**Figure 24.14**).

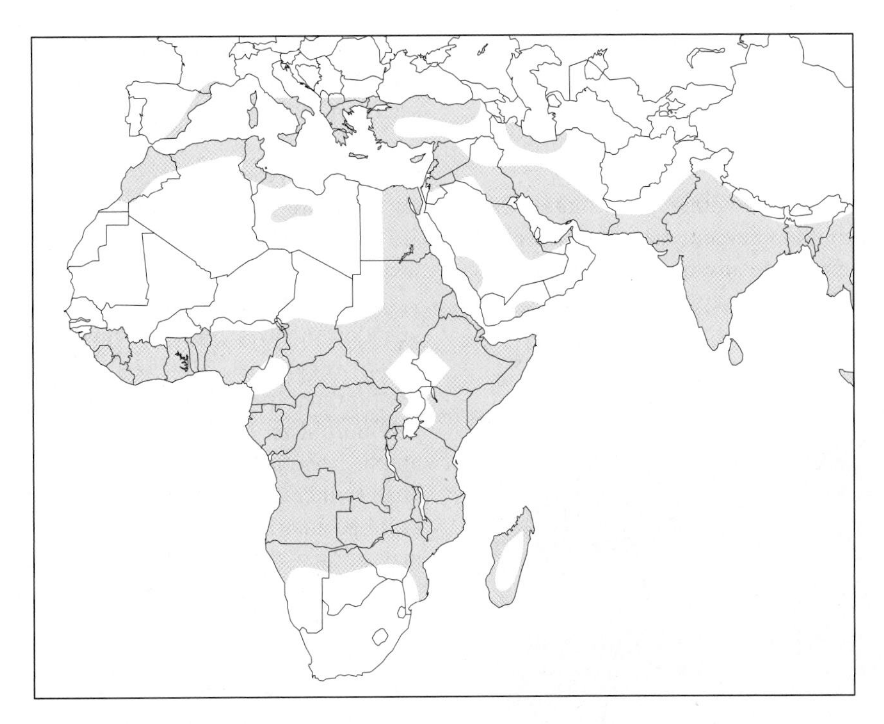

(a) Malaria prevalence

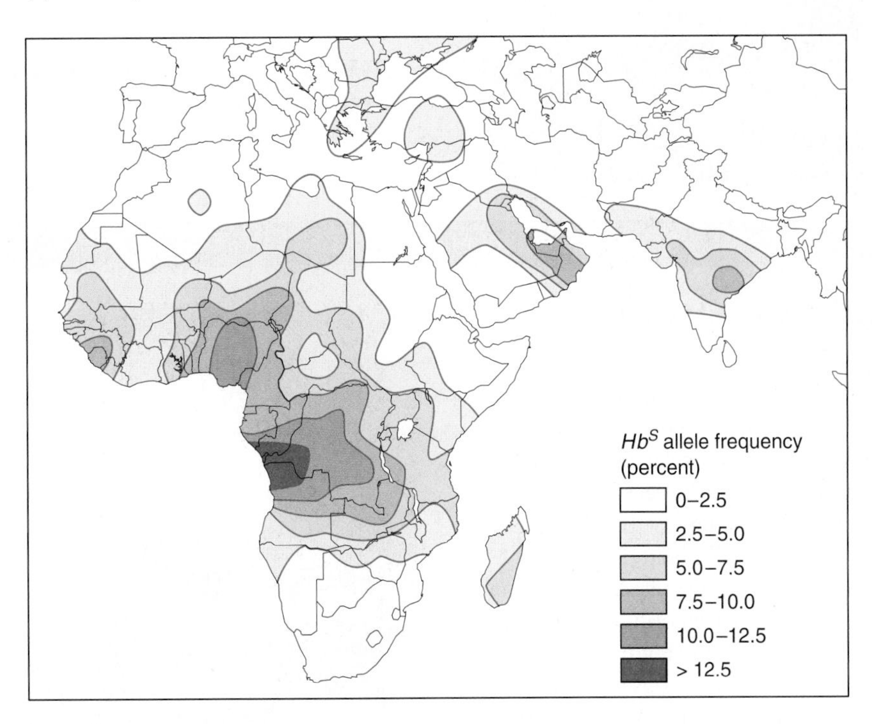

(b) Hb^S allele frequency

FIGURE 24.14 The geographic relationship between malaria and the frequency of the sickle cell allele in human populations. **(a)** The geographic prevalence of malaria in Africa and surrounding areas. **(b)** The frequency of the Hb^S allele in the same areas.

Genes → Traits The sickle cell allele of the β-globin gene is maintained in human populations as a balanced polymorphism. In areas where malaria is prevalent, the heterozygote carrying one copy of the Hb^S allele has a greater fitness than either of the corresponding homozygotes (Hb^AHb^A and Hb^SHb^S). Therefore, even though the Hb^SHb^S homozygotes suffer the detrimental consequences of sickle cell disease, this negative aspect is balanced by the beneficial effects of malarial resistance in the heterozygotes.

Compared with Hb^AHb^A homozygotes, heterozygotes have a 10 to 15% better chance of survival if infected by the malarial parasite, *Plasmodium falciparum*. Therefore, the Hb^S allele is maintained in populations where malaria is prevalent, even though the allele is detrimental in the homozygous state.

In addition to sickle cell disease, other gene mutations that cause human disease in the homozygous state are thought to be prevalent because of heterozygote advantage. For example, the high prevalence of the allele causing cystic fibrosis may be related to this phenomenon, but the advantage that a heterozygote may possess is not understood.

Negative frequency-dependent selection is a second mechanism of balancing selection. In this pattern of natural selection, the fitness of a genotype decreases when its frequency becomes higher. In other words, rare individuals have a higher fitness than more common individuals. Therefore, rare individuals are more likely to reproduce, whereas common individuals are less likely, thereby producing a balanced polymorphism in which no genotype becomes too rare or too common.

An interesting example of negative frequency-dependent selection involves the elder-flowered orchid, *Dactylorhiza sambucina* (**Figure 24.15**). Throughout its range, both yellow- and red-flowered individuals are prevalent. The explanation for this polymorphism is related to its pollinators, which are mainly bumblebees such as *Bombus lapidarius* and *B. terrestris*. The pollinators increase their visits to the flower color of *D. sambucina* as it becomes less common in a given area. One reason why this may occur is because *D. sambucina* is a rewardless flower; that is, it does not provide its pollinators with any reward for visiting, such as sweet nectar. Pollinators learn that the more common color of *D. sambucina* in a given area does not offer a reward, and they increase their visits to the less-common flower. Thus, the relative fitness of the less-common flower increases.

FIGURE 24.15 The two color variations found in the elder-flowered orchid, *Dactylorhiza sambucina*. The two colors are maintained in the population due to negative frequency-dependent selection.

EXPERIMENT 24A

The Grants Have Observed Natural Selection in Galápagos Finches

Let's now turn to a study that demonstrates natural selection in action. Since 1973, Peter Grant, Rosemary Grant, and their colleagues have studied the process of natural selection in finches found on the Galápagos Islands. For over 30 years, the Grants have focused much of their research on one of the Galápagos Islands known as Daphne Major. This small island (0.34 km^2) has a moderate degree of isolation (8 km from the nearest island), an undisturbed habitat, and a resident population of finches, including the medium ground finch, *Geospiza fortis* (**Figure 24.16**).

To study natural selection, the Grants observed various traits in the medium ground finch, including beak size, over the course of many years. The medium ground finch has a relatively small crushing beak, suitable for breaking open small, tender seeds. The Grants quantified beak size among the medium ground finches of Daphne Major by carefully measuring beak depth (a measurement of the beak from top to bottom, at its base) on individual birds. During the course of their studies, they compared the beak sizes of parents and offspring by examining many broods over several years. The depth of the beak was transmitted from parents to offspring, regardless of environmental conditions, indicating that differences in beak sizes are due to genetic differences in the population. In other words, they found that beak depth is a heritable trait.

FIGURE 24.16 The medium ground finch (*Geospiza fortis*), which is found on Daphne Major.

By measuring many birds every year, the Grants were able to assemble a detailed portrait of natural selection from generation to generation. In the study shown in **Figure 24.17**, they measured beak depth in 1976 and 1978.

THE HYPOTHESIS

Beak size will be influenced by natural selection. Environments that produce larger seeds will select for birds with large beaks.

TESTING THE HYPOTHESIS — FIGURE 24.17 Natural selection in medium ground finches of Daphne Major.

	Experimental level	Conceptual level
1. In 1976, measure beak depth in parents and offspring of the species *G. fortis*.	Capture birds and measure beak depth. 	This is a way to measure a trait that may be subject to natural selection.
2. Repeat the procedure on offspring that were born in 1978 and had reached mature size. A drought had occurred in 1977 that caused plants on the island to produce mostly larger seeds and relatively few small seeds.	Capture birds and measure beak depth. 	This is a way to measure a trait that may be subject to natural selection.

THE DATA

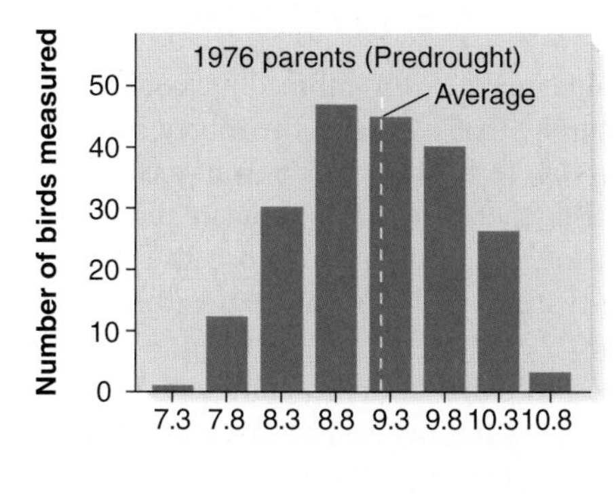

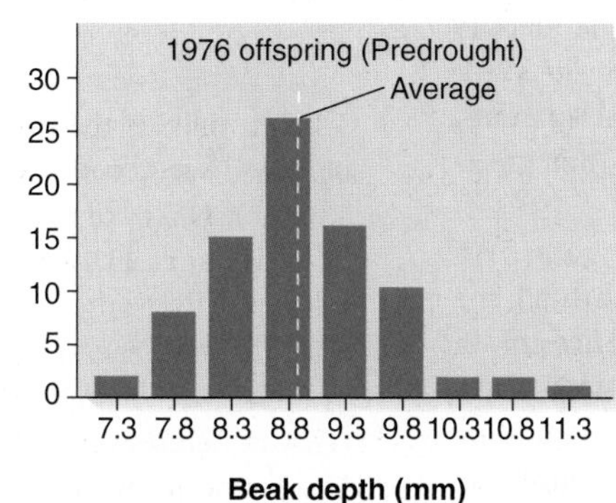

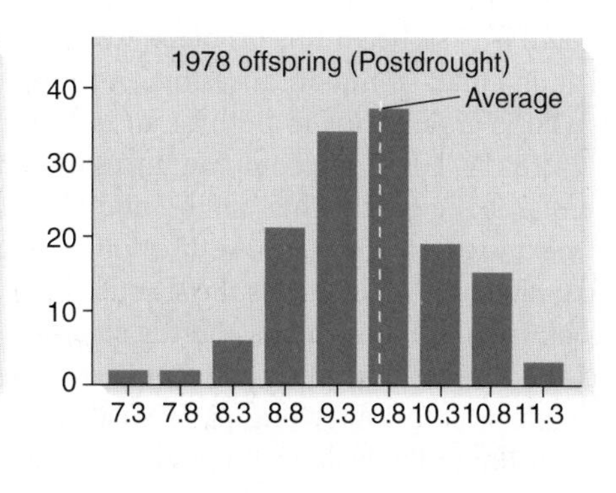

Beak depth (mm)

Data from: B. Rosemary Grant and Peter R. Grant (2003) What Darwin's finches can teach us about the evolutionary origin and regulation of biodiversity. *Bioscience 53*, 965–975.

INTERPRETING THE DATA

In the wet year of 1976, the plants of Daphne Major produced the small seeds that these finches were able to eat in abundance. However, a drought occurred in 1977. During this year, the plants on Daphne Major tended to produce few of the smaller seeds, which the finches rapidly consumed. To survive, the finches resorted to eating larger, drier seeds, which were harder to crush. As a result, the birds that survived tended to have larger beaks, because they were better able to break open these large seeds. In the year after the drought, the average beak depth of offspring in the population increased to approximately 9.8 mm because the surviving birds with larger beaks passed this trait on to their offspring. This is likely to be due to directional selection (see Figure 24.8), although genetic drift could also contribute to these data. Overall, these results illustrate the power of natural selection to alter the nature of a trait, in this case, beak depth, in a given population.

A self-help quiz involving this experiment can be found at www.mhhe.com/brookergenetics4e.

Nonrandom Mating May Occur in Populations

As mentioned earlier, one of the conditions required to establish the Hardy-Weinberg equilibrium is random mating. This means that individuals choose their mates irrespective of their genotypes and phenotypes. In many cases, particularly in human populations, this condition is violated frequently.

When mating is nonrandom in a population, the process is called **assortative mating.** Positive assortative mating occurs when individuals with similar phenotypes choose each other as mates. The opposite situation, where dissimilar phenotypes mate preferentially, is called negative assortative mating. In addition, individuals may choose a mate that is part of the same genetic lineage. The mating of two genetically related individuals, such as cousins, is called **inbreeding.** This is also termed consanguinity. Inbreeding sometimes occurs in human societies and is more likely to take place in nature when population size becomes very limited. In Chapter 25, we will examine how inbreeding is a useful strategy for developing agricultural breeds or strains with desirable characteristics. Conversely, **outbreeding,** which involves mating between unrelated individuals, can create hybrids that are heterozygous for many genes.

In the absence of other evolutionary processes, inbreeding and outbreeding do not affect allele frequencies in a population. However, these patterns of mating do disrupt the balance of genotypes that is predicted by the Hardy-Weinberg equation. Let's first consider inbreeding in a family pedigree. **Figure 24.18** illustrates a human pedigree involving a mating between cousins. Individuals III-2 and III-3 are cousins and have produced the daughter labeled IV-1. She is said to be inbred, because her parents are genetically related to each other.

During inbreeding, the gene pool is smaller, because the parents are related genetically. In the 1940s, Gustave Malécot developed methods to quantify the degree of inbreeding. The **inbreeding coefficient** is the probability that two alleles in a particular individual will be identical for a given gene because both copies are due to descent from a common ancestor. An inbreeding coefficient (F) can be computed by analyzing the degree of relatedness within a pedigree.

As an example, let's determine the inbreeding coefficient for individual IV-1. To begin this problem, we must first identify all of this individual's common ancestors. A common ancestor is anyone who is an ancestor to both of an individual's parents. In Figure 24.18, IV-1 has one common ancestor, I-2, her great-grandfather. I-2 is the grandfather of III-2 and III-3.

Our next step is to determine the inbreeding paths. An inbreeding path for an individual is the shortest path through the pedigree that includes both parents and the common ancestor. In a pedigree, there is an inbreeding path for each common ancestor. The length of each inbreeding path is calculated by adding together all of the individuals in the path except the individual of interest. In this case, there is only one path because IV-1 has only one common ancestor. To add the members of the path, we begin with individual IV-1, but we do not count her. We then move to her father (III-2); to her grandfather (II-2); to I-2, her great-grandfather (the common ancestor); back down to her other grandmother (II-3); and finally to her mother (III-3). This

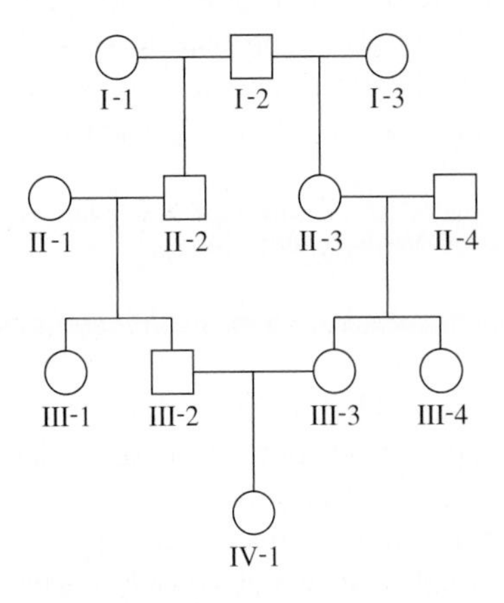

FIGURE 24.18 **A human pedigree containing inbreeding.** Individual IV-1 is the result of inbreeding because her parents are related.

path has five members. Finally, to calculate the inbreeding coefficient, we use the following formula:

$$F = \Sigma(1/2)^n(1 + F_A)$$

where

- F is the inbreeding coefficient of the individual of interest
- n is the number of individuals in the inbreeding path, excluding the inbred offspring
- F_A is the inbreeding coefficient of the common ancestor
- Σ indicates that we add together $(1/2)^n(1 + F_A)$ for each inbreeding path

In this case, there is only one common ancestor and, therefore, only one inbreeding path. Also, we do not know anything about the heritage of the common ancestor, so we assume that F_A is zero. Thus, in our example of Figure 24.18,

$$F = \Sigma(1/2)^n(1 + 0)$$
$$= (1/2)^5 = 1/32 = 3.125\%$$

What does this value mean? Our inbreeding coefficient, 3.125%, tells us the probability that a gene in the inbred individual (IV-1) is homozygous due to its inheritance from a common ancestor (I-2). In this case, therefore, each gene in individual IV-1 has a 3.125% chance of being homozygous because she has inherited the same allele twice from her great-grandfather (I-2), once through each parent.

As an example, let's suppose that the common ancestor (I-2) is heterozygous for the gene involved with cystic fibrosis. His genotype would be *Cc*, where *c* is the recessive allele that causes cystic fibrosis. There is a 3.125% probability that the inbred individual (IV-1) is homozygous (*CC* or *cc*) for this gene because she has inherited both copies from her great-grandfather. She has a 1.56% probability of inheriting both normal alleles (*CC*) and a 1.56% probability of inheriting both mutant alleles (*cc*). The inbreeding coefficient is denoted by the letter F (for Fixation) because it is the probability that an allele will be fixed in the homozygous condition. The term fixation signifies that the homozygous individual can pass only one type of allele to their offspring.

In other pedigrees, an individual may have two or more common ancestors. In this case, the inbreeding coefficient F is calculated as the sum of the inbreeding paths. Such an example is described in solved problem S2 at the end of the chapter.

In addition to pedigree analysis, the effects of inbreeding and outbreeding can also be considered within a population. For example, let's consider the situation in which the frequency of $A = p$ and the frequency of $a = q$. In a given population, the genotype frequencies are determined in the following way:

$p^2 + fpq$ equals the frequency of *AA*

$2pq(1 - f)$ equals the frequency of *Aa*

$q^2 + fpq$ equals the frequency of *aa*

where f is a measure of how much the genotype frequencies deviate from Hardy-Weinberg equilibrium due to nonrandom mating. The value of f ranges from -1 to $+1$. When inbreeding occurs, the value is greater than zero. When outbreeding occurs, the value is less than zero.

As an example, let's suppose that $p = 0.8$, $q = 0.2$, and $f = 0.25$. We can calculate the frequencies of the *AA*, *Aa*, and *aa* genotypes under these conditions as follows:

$$AA = p^2 + fpq = (0.8)^2 + (0.25)(0.8)(0.2) = 0.68$$

$$Aa = 2pq(1 - f) = 2(0.8)(0.2)(1 - 0.25) = 0.24$$

$$aa = q^2 + fpq = (0.2)^2 + (0.25)(0.8)(0.2) = 0.08$$

There will be 68% *AA* homozygotes, 24% heterozygotes, and 8% *aa* homozygotes. If mating had been random (i.e., $f = 0$), the genotype frequencies of *AA* would be p^2, which equals 64%, and *aa* would be q^2, which equals 4%. The frequency of heterozygotes would be $2pq$, which equals 32%. When comparing these numbers, we see that inbreeding raises the proportions of homozygotes and decreases the proportion of heterozygotes. In natural populations, the value of f tends to become larger as a population becomes smaller, because each individual has a more limited choice in mate selection.

What are the consequences of inbreeding in a population? From an agricultural viewpoint, it results in a higher proportion of homozygotes, which may exhibit a desirable trait. For example, an animal breeder may use inbreeding to produce animals that are larger because they have become homozygous for alleles promoting larger size. On the negative side, many genetic diseases are inherited in a recessive manner (see Chapter 22). For these disorders, inbreeding increases the likelihood that an individual will be homozygous and therefore afflicted with the disease. Also, in natural populations, inbreeding lowers the mean fitness of the population if homozygous offspring have a lower fitness value. This can be a serious problem as natural populations become smaller due to human habitat destruction. As the population shrinks, inbreeding becomes more likely because individuals have fewer potential mates from which to choose. The inbreeding, in turn, produces homozygotes that are less fit, thereby decreasing the reproductive success of the population. This phenomenon is called **inbreeding depression.** Conservation biologists sometimes try to circumvent this problem by introducing individuals from one population into another. For example, the endangered Florida panther (*Felis concolor coryi*) suffers from inbreeding-related defects, which include poor sperm quality and quantity and morphological abnormalities. To help alleviate these effects, panthers of the same species from Texas have been introduced into the Florida population.

24.3 SOURCES OF NEW GENETIC VARIATION

In the previous section, we primarily focused on genetic variation in which a single gene exists in two or more alleles. This simplified scenario allows us to appreciate the general principles behind evolutionary mechanisms. As researchers have analyzed genetic variation at the molecular, cellular, and population level, however, they have come to understand that new genetic variation occurs in many ways (**Table 24.2**). Among eukaryotic species, sexual reproduction is an important way that new genetic variation occurs among offspring. In Chapters 3 and 6, we considered how independent assortment and crossing over during sexual reproduction may produce new combinations of alleles among different genes, thereby producing new genetic variation in the resulting offspring. Similarly, in Chapter 26, we will consider

TABLE 24.2

Sources of New Genetic Variation That Occur in Populations

Type	Description
Independent assortment	The independent segregation of different homologous chromosomes may give rise to new combinations of alleles in offspring (see Chapter 3).
Crossing over	Recombination (crossing over) between homologous chromosomes can also produce new combinations of alleles that are located on the same chromosome (see Chapter 6).
Interspecies crosses	On occasion, members of different species may breed with each other to produce hybrid offspring. This topic is discussed in Chapter 26.
Prokaryotic gene transfer	Prokaryotic species possess mechanisms of genetic transfer such as conjugation, transduction, and transformation (see Chapter 7).
New alleles	Point mutations can occur within a gene to create single-nucleotide polymorphisms (SNPs). In addition, genes can be altered by small deletions and additions. Gene mutations are also discussed in Chapter 16.
Gene duplications	Events, such as misaligned crossovers, can add additional copies of a gene into a genome and lead to the formation of gene families. This topic is discussed in Chapter 8.
Chromosome structure and number	Chromosome structure may be changed by deletions, duplications, inversions, and translocations. Changes in chromosome number result in aneuploid, polyploid, and alloploid offspring. These mechanisms are discussed in Chapters 8 and 26.
Exon shuffling	New genes can be created when exons of preexisting genes are rearranged to make a gene that encodes a protein with a new combination of protein domains.
Horizontal gene transfer	Genes from one species can be introduced into another species and become incorporated into that species' genome.
Changes in repetitive sequences	Short repetitive sequences are common in genomes due to the occurrence of transposable elements and due to tandem arrays. The number and lengths of repetitive sequences tend to show considerable variation in natural populations.

A comparison of the DNA fingerprints among different individuals has found two applications. First, DNA fingerprinting can be used as a method of identification. In forensics, DNA fingerprinting can identify a crime suspect. In medicine, the technique can identify the type of bacterium that is causing an infection in a particular patient. A second use of DNA fingerprinting is relationship testing. Closely related individuals have more similar fingerprints than do distantly related ones (see solved problem S6). In humans, this can be used in paternity testing. In population genetics, DNA fingerprinting can provide evidence regarding the degree of relatedness among members of a population. Such information may help geneticists determine if a population is likely to be suffering from inbreeding depression.

The development of DNA fingerprinting has relied on the identification of DNA sites that vary greatly in length among members of a population. This naturally occurring variation causes each individual to have a unique DNA fingerprint. In the 1980s, Alec Jeffreys and his colleagues found that certain minisatellites within human chromosomes are particularly variable in their lengths. As discussed earlier, minisatellites tend to vary within populations due to changes in the number of tandem repeats at each site.

In the past decade, the technique of DNA fingerprinting has become automated, much like the automation that changed the procedure of DNA sequencing described in Chapter 18. DNA fingerprinting is now done using the technique of polymerase chain reaction (PCR), which amplifies microsatellites. Like minisatellites, microsatellites are found in multiple sites in the genome of humans and other species and vary in length among different individuals. In this procedure, the microsatellites from a sample of DNA are amplified by PCR using primers that flank the repetitive region and then separated by gel electrophoresis according to their molecular masses. As in automated DNA sequencing, the amplified microsatellite fragments are fluorescently labeled. A laser excites the fluorescent molecule within a microsatellite, and a detector records the amount of fluorescence emission for each microsatellite. As shown in **Figure 24.22**, this type of DNA fingerprint yields a series of peaks, each peak having a characteristic molecular mass. In this automated approach, the pattern of peaks rather than bands constitutes an individual's DNA fingerprint.

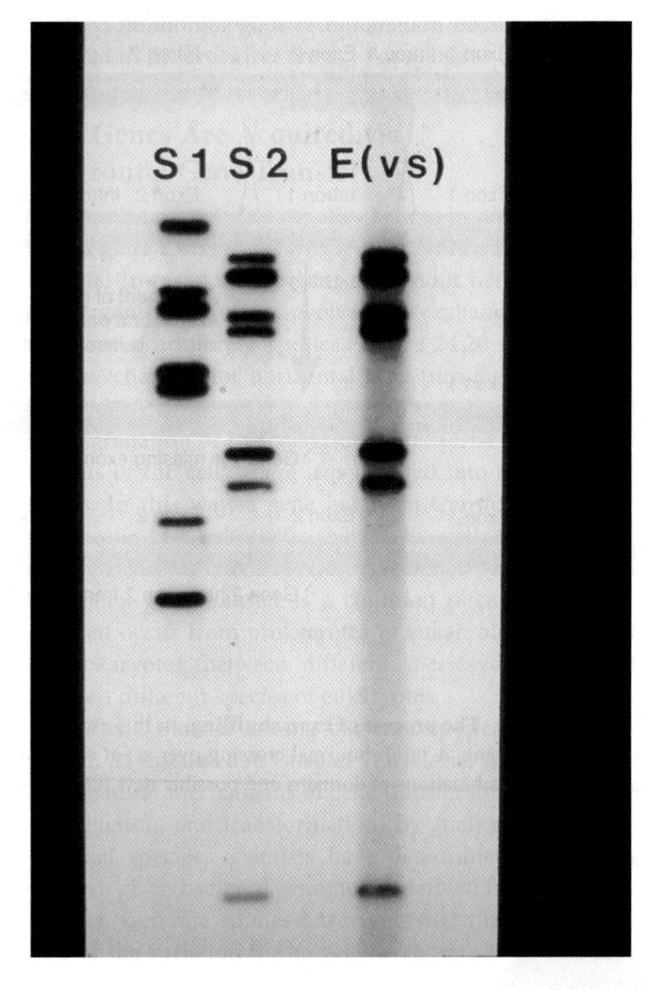

FIGURE 24.21 A comparison of two DNA fingerprints. The chromosomal DNA from two different individuals (Suspect 1—S1, and Suspect 2—S2) was subjected to DNA fingerprinting. The DNA evidence at a crime scene, E(vs), was also subjected to DNA fingerprinting. Following the hybridization of a radiolabeled probe, the DNA appears as a series of bands on a gel. The dissimilarity in the pattern of these bands distinguishes different individuals, much as the differences in physical fingerprint patterns can be used for identification. As seen here, S2 matches the DNA found at the crime scene.

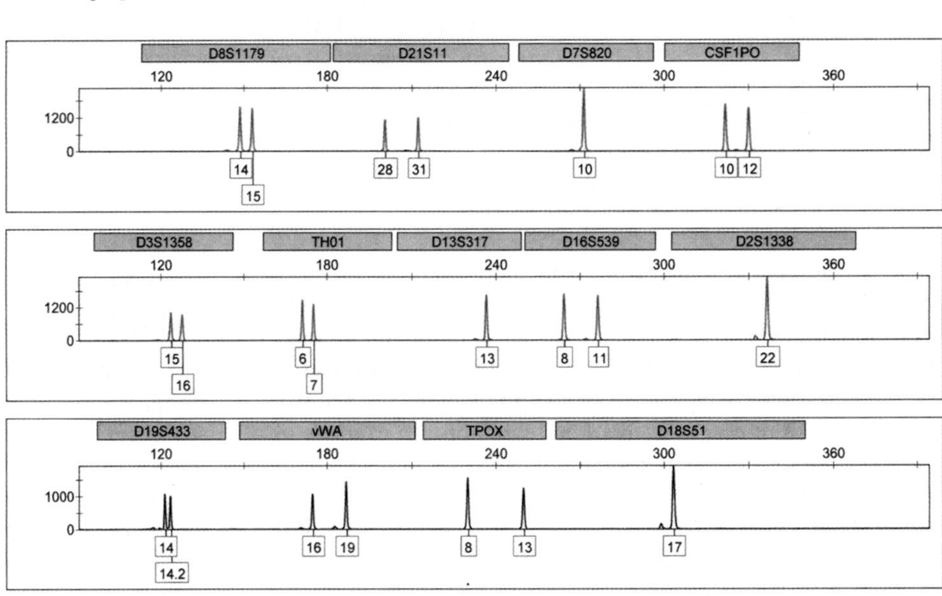

ONLINE ANIMATION

FIGURE 24.22 Automated DNA fingerprinting. In automated DNA fingerprinting, a sample of DNA is amplified, using primers that recognize the ends of microsatellites. The microsatellite fragments are fluorescently labeled and then separated by gel electrophoresis. The fluorescent molecules within each microsatellite are excited with a laser, and the amount of fluorescence is measured via a fluorescence detector. A printout from the detector is shown here. The gray boxes indicate the names of specific microsatellites. The peaks show the relative amounts of each microsatellite. The boxes beneath each peak indicate the number of tandem repeats in a given microsatellite. In this example, the individual is heterozygous for certain microsatellites (e.g., D8S1179) and homozygous for others (e.g., D7S820).

KEY TERMS

Page 670. population genetics, gene pool
Page 671. population, local populations, demes, polymorphism, polymorphic, monomorphic, single-nucleotide polymorphism (SNP)
Page 672. allele frequencies, genotype frequencies
Page 673. Hardy-Weinberg equilibrium, Hardy-Weinberg equation
Page 675. disequilibrium, microevolution
Page 676. mutation rate
Page 677. random genetic drift, genetic drift
Page 678. bottleneck effect, founder effect
Page 679. conglomerate, gene flow, natural selection
Page 680. Darwinian fitness, directional selection
Page 681. mean fitness of the population ($\overline{W}$)
Page 683. stabilizing selection, disruptive selection
Page 684. balancing selection, heterozygote advantage, selection coefficient (s)
Page 686. negative frequency-dependent selection
Page 687. assortative mating, inbreeding
Page 688. outbreeding, inbreeding coefficient
Page 689. inbreeding depression
Page 690. exon shuffling, horizontal gene transfer, repetitive sequences, microsatellite
Page 691. minisatellite, DNA fingerprinting, DNA profiling

CHAPTER SUMMARY

- Population genetics is concerned with changes in genetic variation within a population of individuals over time.

24.1 Genes in Populations and the Hardy-Weinberg Equation

- All of the alleles of every gene in a population constitute the gene pool.
- For sexually reproducing organisms, a population is a group of individuals of the same species that occupy the same region and can interbreed with one another (see Figure 24.1).
- Polymorphism refers to traits or genes that exhibit variation in a population (see Figure 24.2).
- Single-nucleotide polymorphisms are the most common type of variation in genes (see Figure 24.3).
- Geneticists analyze genetic variation by determining allele and genotype frequencies.
- The Hardy-Weinberg equation can be used to calculate genotype frequencies based on allele frequencies (see Figures 24.4, 24.5).
- Deviation from a Hardy-Weinberg equilibrium indicates that evolutionary change is occurring.

24.2 Factors That Change Allele and Genotype Frequencies in Populations

- Mutations are the source of new genetic variation. However, the occurrence of new mutations does not greatly change allele frequencies because they are relatively rare (see Table 24.1).
- Random genetic drift, migration, natural selection, and nonrandom mating are mechanisms that may alter allele or genotype frequencies in a population over time.
- Genetic drift is a change in allele frequency due to chance fluctuations. Over the long run, it often results in allele fixation or loss. The effect of genetic drift is greater in small populations (see Figure 24.6).
- Two examples of genetic drift are the bottleneck effect and the founder effect (see Figure 24.7).
- Migration can alter allele frequencies and tends to make the allele frequencies in neighboring populations more similar.
- Natural selection is the process that changes allele frequencies from one generation to the next based on fitness, which is the relative reproductive successes of different genotypes.
- Directional selection favors the extreme phenotype (see Figures 24.8–24.10).
- Stabilizing selection favors individuals with intermediate phenotypes (see Figure 24.11).
- Disruptive selection favors multiple phenotypes. This may occur due to heterogeneous environments (see Figures 24.12, 24.13).
- Balancing selection results in a stable polymorphism. Examples include heterozygote advantage and negative-frequency dependent selection (see Figures 24.14, 24.15).
- The Grants observed natural selection in a finch population. The selection involved a change in beak size due to drought conditions (see Figures 24.16, 24.17).
- Nonrandom mating may alter the genotype frequencies that would be based on a Hardy-Weinberg equilibrium. Inbreeding promotes homozygosity (see Figure 24.18).

24.3 Sources of New Genetic Variation

- A variety of different mechanisms can bring about genetic variation (see Table 24.2).
- New genes in eukaryotes are produced by exon shuffling (see Figure 24.19).
- A species may acquire a new gene from another species via horizontal gene transfer (see Figure 24.20).
- A common source of genetic variation in populations involves changes in repetitive sequences, such as microsatellites.
- DNA fingerprinting is a technique that relies on variation in repetitive sequences within a population. It is used as a means of identification and in relationship testing (see Figures 24.21, 24.22).

PROBLEM SETS & INSIGHTS

Solved Problems

S1. The phenotypic frequency of people who cannot taste phenylthiocarbamide (PTC) is approximately 0.3. The inability to taste this bitter substance is due to a recessive allele. If we assume there are only two alleles in the population (namely, tasters, *T*, and nontasters, *t*) and that the population is in Hardy-Weinberg equilibrium, calculate the frequencies of these two alleles.

Answer: Let p = allele frequency of the taster allele and q = the allele frequency of the nontaster allele. The frequency of nontasters is 0.3.

This is the frequency of the genotype *tt*, which in this case is equal to q^2:

$$q^2 = 0.3$$

To determine the frequency q of the nontaster allele, we take the square root of both sides of this equation:

$$q = 0.55$$

With this value, we can calculate the frequency p of the taster allele:

$$p = 1 - q$$
$$= 1 - 0.55 = 0.45$$

S2. In the pedigree shown here, answer the following questions with regard to individual VII-1:

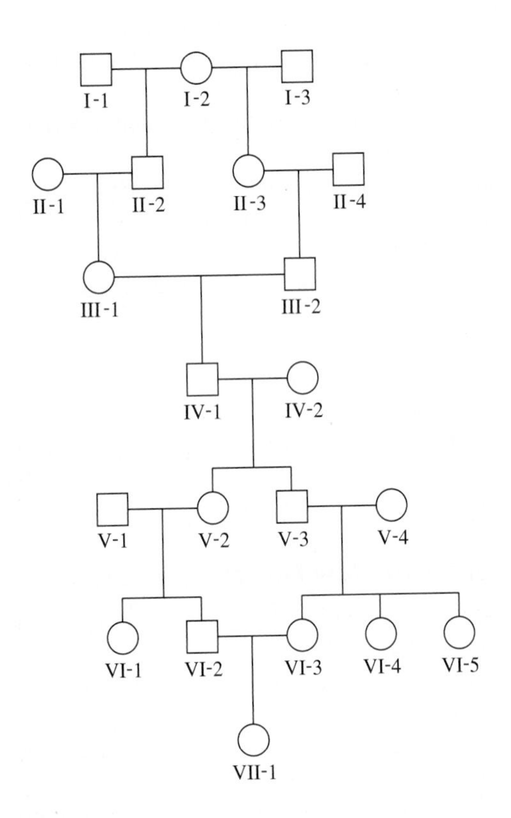

A. Who are the common ancestors of her parents?

B. What is the inbreeding coefficient?

Answer:

A. The common ancestors are IV-1 and IV-2. They are the grandparents of VI-2 and VI-3, who are the parents of VII-1.

B. The inbreeding coefficient is calculated using the formula

$$F = \Sigma(1/2)^n(1 + F_A)$$

In this case, there are two common ancestors, IV-1 and IV-2. Also, IV-1 is inbred, because I-2 is a common ancestor to both of IV-1's parents. The first step is to calculate F_A, the inbreeding coefficient for this common ancestor. The inbreeding path for IV-1 contains five people: III-1, II-2, I-2, II-3, and III-2. Therefore,

$$n = 5$$
$$F_A = (1/2)^5 = 0.03$$

Now we can calculate the inbreeding coefficient for VII-1. Each inbreeding path contains five people: VI-2, V-2, IV-1, V-3, and VI-3; and VI-2, V-2, IV-2, V-3, and VI-3. Thus,

$$F = (1/2)^5 (1 + 0.03) + (1/2)^5 (1 + 0)$$
$$= 0.032 + 0.031 = 0.063$$

S3. The Hardy-Weinberg equation provides a way to predict genotype frequency based on allele frequency. In the case of mammals, males are hemizygous for X-linked genes, whereas females have two copies. Among males, the frequency of any X-linked trait equals the frequency of males with the trait. For example, if an allele frequency for an X-linked disease-causing allele was 5%, then 5% of all males would be affected with the disorder. Female genotype frequencies are computed using the Hardy-Weinberg equation.

As a specific example, let's consider the human X-linked trait known as hemophilia A (see Chapter 22 for a description of this disorder). In human populations, the allele frequency of the hemophilia A allele is approximately 1 in 10,000, or 0.0001. The other allele for this gene is the normal allele. Males can be affected or unaffected, whereas females can be affected, unaffected carriers, or unaffected noncarriers.

A. What are the allele frequencies for the mutant and normal allele in the human population?

B. Among males, what is the frequency of affected individuals?

C. Among females, what is the frequency of affected individuals and heterozygous carriers?

D. Within a population of 100,000 people, what is the expected number of affected males? In this same population, what is the expected number of carrier females?

Answer: Let p represent the normal allele and q represent the allele that causes hemophilia.

A. X^H normal allele, frequency $= 0.9999 = p$

X^h hemophilia allele, frequency $= 0.0001 = q$

B. X^hY genotype frequency of affected males $= q = 0.0001$

C. X^hX^h genotype frequency of affected females $= q^2 = (0.0001)^2 = 0.00000001$

X^HX^h genotype frequency of carrier females $= 2pq = 2(0.9999)(0.0001) = 0.0002$

D. We assume this population is composed of 50% males and 50% females.

Number of affected males $= 50{,}000 \times 0.0001 = 5$

Number of carrier females $= 50{,}000 \times 0.0002 = 10$

S4. The Hardy-Weinberg equation can be modified to include situations of three or more alleles. In its standard (two-allele) form, the Hardy-Weinberg equation reflects the Mendelian notion that each individual inherits two copies of each allele, one from each parent. For a two-allele situation, it can also be written as

$(p + q)^2 = 1$ (Note: The number 2 in this equation reflects the idea that the genotype is due to the inheritance of two alleles, one from each parent.)

This equation can be expanded to include three or more alleles. For example, let's consider a situation in which a gene exists as three alleles: *A1*, *A2*, and *A3*. The allele frequency of *A1* is designated by the letter *p*, *A2* by the letter *q*, and *A3* by the letter *r*. Under these circumstances, the Hardy-Weinberg equation becomes

$$(p + q + r)^2 = 1$$

$$p^2 + q^2 + r^2 + 2pq + 2pr + 2qr = 1$$

where

p^2 is the genotype frequency of *A1A1*

q^2 is the genotype frequency of *A2A2*

r^2 is the genotype frequency of *A3A3*

$2pq$ is the genotype frequency of *A1A2*

$2pr$ is the genotype frequency of *A1A3*

$2qr$ is the genotype frequency of *A2A3*

Now here is the question. As discussed in Chapter 4, the gene that affects human blood type can exist in three alleles. In a Japanese population, the allele frequencies are

$I^A = 0.28$

$I^B = 0.17$

$i = 0.55$

Based on these allele frequencies, calculate the different possible genotype frequencies and blood type frequencies.

Answer: If we let *p* represent I^A, *q* represent I^B, and *r* represent *i*, then

p^2 is the genotype frequency of I^AI^A,
which is type A blood $= (0.28)^2 = 0.08$

q^2 is the genotype frequency of I^BI^B,
which is type B blood $= (0.17)^2 = 0.03$

r^2 is the genotype frequency of *ii*,
which is type O blood $= (0.55)^2 = 0.30$

$2pq$ is the genotype frequency of I^AI^B,
which is type AB blood $= 2(0.28)(0.17) = 0.09$

$2pr$ is the genotype frequency of I^Ai,
which is type A blood $= 2(0.28)(0.55) = 0.31$

$2qr$ is the genotype frequency of I^Bi,
which is type B blood $= 2(0.17)(0.55) = 0.19$

Type A $= 0.08 + 0.31 = 0.39$, or 39%

Type B $= 0.03 + 0.19 = 0.22$, or 22%

Type O $= 0.30$, or 30%

Type AB $= 0.09$, or 9%

S5. Let's suppose that pigmentation in a species of insect is controlled by a single gene existing in two alleles, *D* for dark and *d* for light. The heterozygote *Dd* is intermediate in color. In a heterogeneous environment, the allele frequencies are $D = 0.7$ and $d = 0.3$. This polymorphism is maintained because the environment contains some dimly lit forested areas and some sunny fields. During a hurricane, a group of 1000 insects is blown to a completely sunny area. In this environment, the fitness values are $DD = 0.3$, $Dd = 0.7$, and $dd = 1.0$. Calculate the allele frequencies in the next generation.

Answer: The first step is to calculate the mean fitness of the population:

$$p^2W_{DD} + 2pqW_{Dd} + q^2W_{dd} = \overline{W}$$

$$\overline{W} = (0.7)^2(0.3) + 2(0.7)(0.3)(0.7) + (0.3)^2(1.0)$$
$$= 0.15 + 0.29 + 0.09 = 0.53$$

After one generation of selection, we get

$$\text{Allele frequency of } D\text{: } p_D = \frac{p^2W_{DD}}{\overline{W}} + \frac{pqW_{Dd}}{\overline{W}}$$
$$= \frac{(0.7)^2(0.3)}{0.53} + \frac{(0.7)(0.3)(0.7)}{0.53}$$
$$= 0.55$$

$$\text{Allele frequency of } d\text{: } q_d = \frac{q^2W_{dd}}{\overline{W}} + \frac{pqW_{Dd}}{\overline{W}}$$
$$= \frac{(0.3)^2(1.0)}{0.53} + \frac{(0.7)(0.3)(0.7)}{0.53}$$
$$= 0.45$$

After one generation, the allele frequency of *D* has decreased from 0.7 to 0.55, while the frequency of *d* has increased from 0.3 to 0.45.

S6. An important application of DNA fingerprinting is relationship testing. Persons who are related genetically have some bands or peaks in common. The number they share depends on the closeness of their genetic relationship. For example, an offspring is expected to receive half of his or her minisatellites from one parent and the rest from the other. The diagram on the next page schematically shows a traditional DNA fingerprint of an offspring, mother, and two potential fathers.

In paternity testing, the offspring's DNA fingerprint is first compared with that of the mother. The bands that the offspring have in common with the mother are depicted in purple. The bands that are not similar between the offspring and the mother must have been inherited from the father. These bands are depicted in red. Which male could be the father?

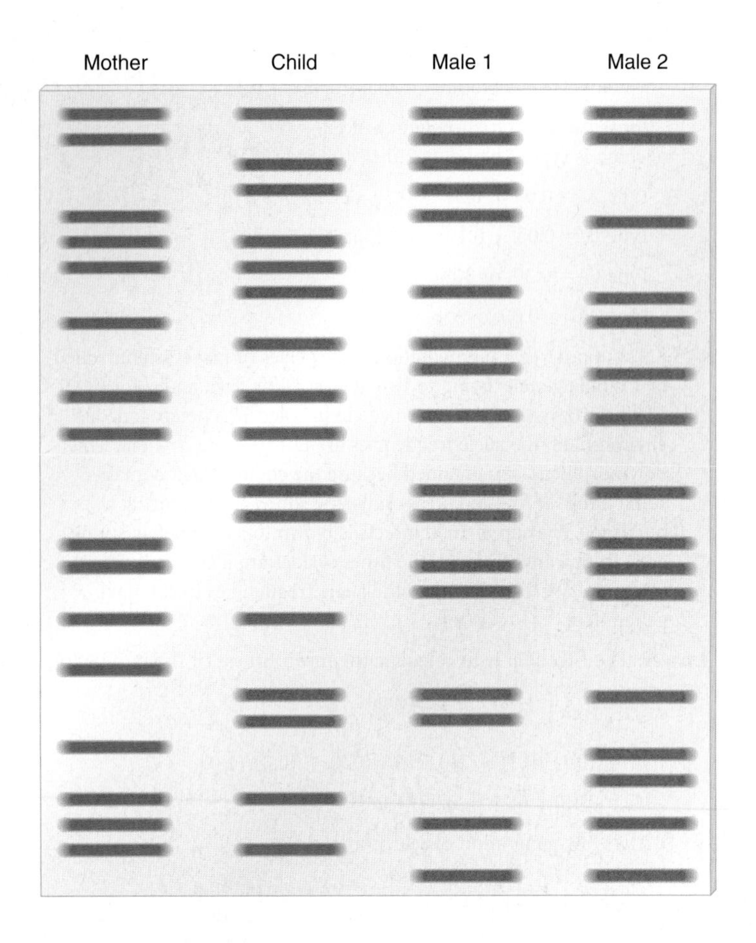

Answer: Male 2 does not have many of the paternal bands. Therefore, he can be excluded as being the father of this child. However, male 1 has all of the paternal bands. He is very likely to be the father.

Geneticists can calculate the likelihood that the matching bands between the offspring and a prospective father could occur just as a matter of random chance. To do so, they must analyze the frequency of each band in a reference population (e.g., Caucasians living in the United States). For example, let's suppose that DNA fingerprinting analyzed 40 bands. Of these, 20 bands matched with the mother and 20 bands matched with a prospective father. If the probability of each of these bands in a reference population was 1/4, the likelihood of such a match occurring by random chance would be $(1/4)^{20}$, or roughly 1 in 1 trillion. Therefore, a match between two samples is rarely a matter of random chance.

Conceptual Questions

C1. What is the gene pool? How is a gene pool described in a quantitative way?

C2. In genetics, what does the term population mean? Pick any species you like and describe how its population might change over the course of many generations.

C3. What is a genetic polymorphism? What is the source of genetic variation?

C4. State for each of the following whether it is an example of an allele, genotype, and/or phenotype frequency:

A. Approximately 1 in 2500 Caucasians is born with cystic fibrosis.

B. The percentage of carriers of the sickle cell allele in West Africa is approximately 13%.

C. The number of new mutations for achondroplasia, a genetic disorder, is approximately 5×10^{-5}.

C5. The term polymorphism can refer to both genes and traits. Explain the meaning of a polymorphic gene and a polymorphic trait. If a gene is polymorphic, does the trait that the gene affects also have to be polymorphic? Explain why or why not.

C6. Cystic fibrosis (CF) is a recessive autosomal trait. In certain Caucasian populations, the number of people born with this disorder is about 1 in 2500. Assuming Hardy-Weinberg equilibrium for this trait,

A. What are the frequencies for the normal and CF alleles?

B. What are the genotype frequencies of homozygous normal, heterozygous, and homozygous affected individuals?

C. Assuming random mating, what is the probability that two phenotypically unaffected heterozygous carriers will choose each other as mates?

C7. Does inbreeding affect allele frequencies? Why or why not? How does it affect genotype frequencies? With regard to rare recessive diseases, what are the consequences of inbreeding in human populations?

C8. For a gene existing in two alleles, what are the allele frequencies when the heterozygote frequency is at its maximum value? What if there are three alleles?

C9. In a population, the frequencies of two alleles are $B = 0.67$ and $b = 0.33$. The genotype frequencies are $BB = 0.50$, $Bb = 0.37$, and $bb = 0.13$. Do these numbers suggest inbreeding? Explain why or why not.

C10. The ability to roll your tongue is inherited as a recessive trait. The frequency of the rolling allele is approximately 0.6, and the dominant (nonrolling) allele is 0.4. What is the frequency of individuals who can roll their tongues?

C11. Using the pedigree shown here, answer the following questions for individual VI-1:

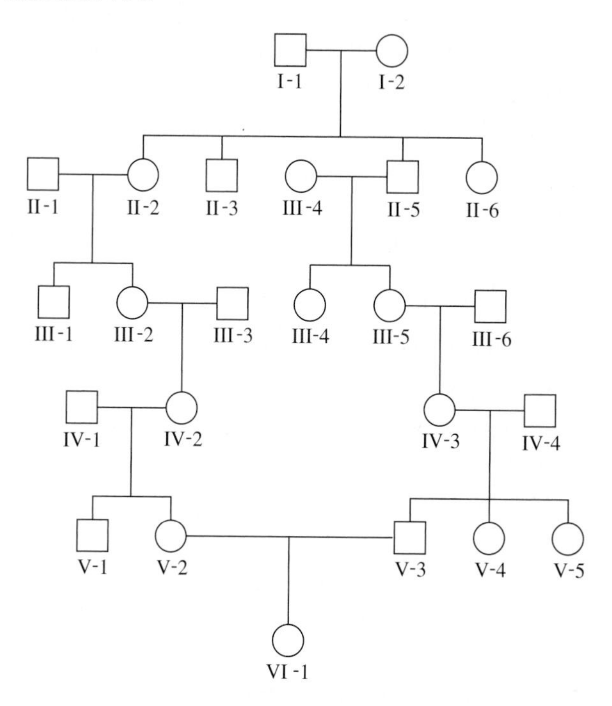

A. Is this individual inbred?

B. If so, who are her common ancestor(s)?

C. Calculate the inbreeding coefficient for VI-1.

D. Are the parents of VI-1 inbred?

C12. A family pedigree is shown here.

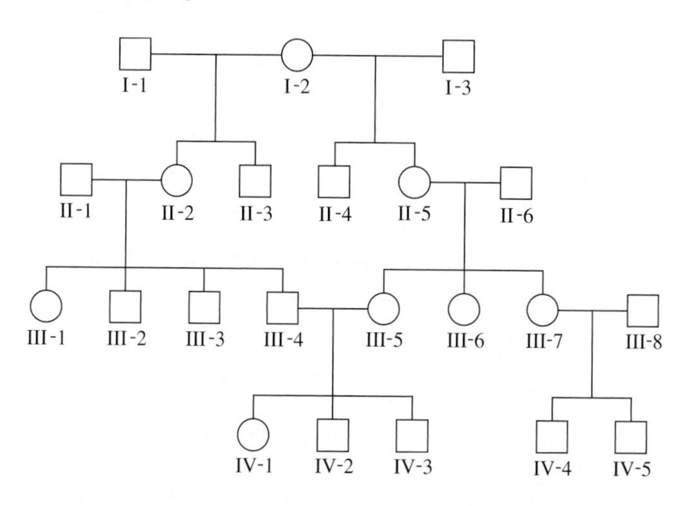

A. What is the inbreeding coefficient for individual IV-3?

B. Based on the data shown in this pedigree, is individual IV-4 inbred?

C13. A family pedigree is shown here.

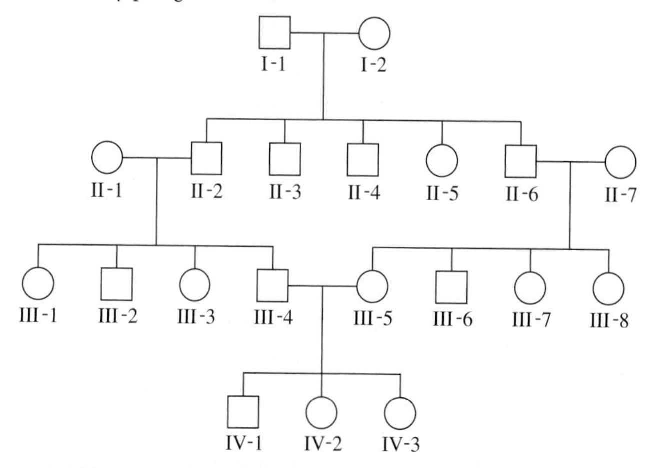

A. What is the inbreeding coefficient for individual IV-2? Who is/are her common ancestors?

B. Based on the data shown in this pedigree, is individual III-4 inbred?

C14. What evolutionary factors can cause allele frequencies to change and possibly lead to a genetic polymorphism? Discuss the relative importance of each type of process.

C15. In the term genetic drift, what is drifting? Why is this an appropriate term to describe this phenomenon?

C16. What is the difference between a random and an adaptive evolutionary process? Describe two or more examples of each. At the molecular level, explain how mutations can be random or adaptive.

C17. Let's suppose the mutation rate for converting a *B* allele into a *b* allele is 10^{-4}. The current allele frequencies are $B = 0.6$ and $b = 0.4$. How long will it take for the allele frequencies to equal each other, assuming that no genetic drift is taking place?

C18. Why is genetic drift more significant in small populations? Why does it take longer for genetic drift to cause allele fixation in large populations than in small ones?

C19. A group of four birds flies to a new location and initiates the formation of a new colony. Three of the birds are homozygous *AA*, and one bird is heterozygous *Aa*.

A. What is the probability that the *a* allele will become fixed in the population?

B. If fixation of the *a* allele occurs, how long will it take?

C. How will the growth of the population, from generation to generation, affect the answers to parts A and B? Explain.

C20. Describe what happens to allele frequencies as a result of the bottleneck effect. Discuss the relevance of this effect with regard to species that are approaching extinction.

C21. When two populations frequently intermix due to migration, what are the long-term consequences with regard to allele frequencies and genetic variation?

C22. Discuss the similarities and differences among directional, disruptive, balancing, and stabilizing selection.

C23. What is Darwinian fitness? What types of characteristics can promote high fitness values? Give several examples.

C24. What is the intuitive meaning of the mean fitness of a population? How does its value change in response to natural selection?

C25. Antibiotics are commonly used to combat bacterial and fungal infections. During the past several decades, however, antibiotic-resistant strains of microorganisms have become alarmingly prevalent. This has undermined the effectiveness of antibiotics in treating many types of infectious disease. Discuss how the following processes that alter allele frequencies may have contributed to the emergence of antibiotic-resistant strains:

A. Random mutation

B. Genetic drift

C. Natural selection

C26. With regard to genetic drift, are the following statements true or false? If a statement is false, explain why.

A. Over the long run, genetic drift leads to allele fixation or loss.

B. When a new mutation occurs within a population, genetic drift is more likely to cause the loss of the new allele rather than the fixation of the new allele.

C. Genetic drift promotes genetic diversity in large populations.

D. Genetic drift is more significant in small populations.

C27. Two populations of antelope are separated by a mountain range. The antelope are known to occasionally migrate from one population to the other. Migration can occur in either direction. Explain how migration affects the following phenomena:

A. Genetic diversity in the two populations

B. Allele frequencies in the two populations

C. Genetic drift in the two populations

C28. Do the following examples describe directional, disruptive, balancing, or stabilizing selection?

A. Polymorphisms in snail color and banding pattern as described in Figure 24.13

B. Thick fur among mammals exposed to cold climates

C. Birth weight in humans

D. Sturdy stems and leaves among plants exposed to windy climates

Experimental Questions

E1. You will need to be familiar with the techniques described in Chapter 18 to answer this question. Gene polymorphisms can be detected using a variety of cellular and molecular techniques. Which techniques would you use to detect gene polymorphisms at the following levels?

A. DNA level

B. RNA level

C. Polypeptide level

E2. You will need to understand solved problem S4 to answer this question. The gene for coat color in rabbits can exist in four alleles termed C (full coat color), c^{ch} (chinchilla), c^h (Himalayan), and c (albino). In a population of rabbits in Hardy-Weinberg equilibrium, the allele frequencies are

$C = 0.34$

$c^{ch} = 0.17$

$c^h = 0.44$

$c = 0.05$

Assume that C is dominant to the other three alleles. c^{ch} is dominant to c^h and c, and c^h is dominant to c.

A. What is the frequency of albino rabbits?

B. Among 1000 rabbits, how many would you expect to have a Himalayan coat color?

C. Among 1000 rabbits, how many would be heterozygotes with a chinchilla coat color?

E3. In a large herd of 5468 sheep, 76 animals have yellow fat, compared with the rest of the members of the herd, which have white fat. Yellow fat is inherited as a recessive trait. This herd is assumed to be in Hardy-Weinberg equilibrium.

A. What are the frequencies of the white and yellow fat alleles in this population?

B. Approximately how many sheep with white fat are heterozygous carriers of the yellow allele?

E4. The human MN blood group is determined by two codominant alleles, M and N. The following data were obtained from various human populations:

		Percentages		
Population	**Place**	***MM***	***MN***	***NN***
Inuit	East Greenland	83.5	15.6	0.9
Navajo Indians	New Mexico	84.5	14.4	1.1
Finns	Karajala	45.7	43.1	11.2
Russians	Moscow	39.9	44.0	16.1
Aborigines	Queensland	2.4	30.4	67.2

(Data from E.B. Speiss (1990). *Genes in Populations*, 2d ed. Wiley-Liss, New York.)

A. Calculate the allele frequencies in these five populations.

B. Which populations appear to be in Hardy-Weinberg equilibrium?

C. Which populations do you think have had significant intermixing due to migration?

E5. You will need to understand solved problem S4 before answering this question. In an island population, the following data were obtained for the numbers of people with each of the four blood types:

Type O	721
Type A	932
Type B	235
Type AB	112

Is this population in Hardy-Weinberg equilibrium? Explain your answer.

E6. In a donor population, the allele frequencies for the common (Hb^A) and sickle cell alleles (Hb^S) are 0.9 and 0.1, respectively. A group of 550 individuals migrates to a new population containing 10,000 individuals; in the recipient population, the allele frequencies are $Hb^A = 0.99$ and $Hb^S = 0.01$.

A. Calculate the allele frequencies in the conglomerate population.

B. Assuming the donor and recipient populations are each in Hardy-Weinberg equilibrium, calculate the genotype frequencies in the conglomerate population prior to further mating between the donor and recipient populations.

C. What will be the genotype frequencies of the conglomerate population in the next generation, assuming it achieves Hardy-Weinberg equilibrium in one generation?

E7. A recessive lethal allele has achieved a frequency of 0.22 due to genetic drift in a very small population. Based on natural selection, how would you expect the allele frequencies to change in the next three generations? Note: Your calculation can assume that genetic drift is not altering allele frequencies in either direction.

E8. Among a large population of 2 million gray mosquitoes, one mosquito is heterozygous for a body color gene; this mosquito has one gray allele and one blue allele. There is no selective advantage or disadvantage between gray and blue body color. All of the other mosquitoes carry the gray allele.

A. What is the probability of fixation of the blue allele?

B. If fixation happens to occur, how many generations is it likely to take?

C. Qualitatively, how would the answers to parts A and B be affected if the blue allele conferred a slight survival advantage?

E9. Resistance to the poison warfarin is a genetically determined trait in rats. Homozygotes carrying the resistance allele (*WW*) have a lower fitness because they suffer from vitamin K deficiency, but heterozygotes (*Ww*) do not. However, the heterozygotes are still resistant to warfarin. In an area where warfarin is applied, the heterozygote has a survival advantage. Due to warfarin resistance, the heterozygote is also more fit than the normal homozygote (*ww*). If the relative fitness values for *Ww*, *WW*, and *ww* individuals are 1.0, 0.37, and 0.19, respectively, in areas where warfarin is applied, calculate the allele frequencies at equilibrium. How would this equilibrium be affected if the rats were no longer exposed to warfarin?

E10. Describe, in as much experimental detail as possible, how you would test the hypothesis that snail color distribution is due to predation.

E11. In the Grants' study of the medium ground finch, do you think the pattern of natural selection was directional, stabilizing, disruptive, or balancing? Explain your answer. If the environment remained dry indefinitely (for many years), what do you think would be the long-term outcome?

E12. Here are traditional DNA fingerprints of five people: a child, mother, and three potential fathers:

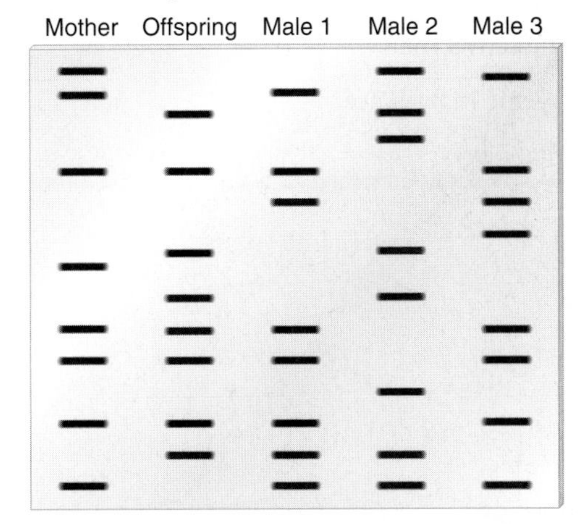

Which males can be ruled out as being the father? Explain your answer. If one of the males could be the father, explain the general strategy for calculating the likelihood that he could match the offspring's DNA fingerprint by chance alone. (See solved problem S6 before answering this question.)

E13. What is DNA fingerprinting? How can it be used in human identification?

E14. When analyzing the automated DNA fingerprints of a father and his biological daughter, a technician examined 50 peaks and found that 30 of them were a perfect match. In other words, 30 out of 50 peaks, or 60%, were a perfect match. Is this percentage too high, or would you expect a value of only 50%? Explain why or why not.

E15. What would you expect to be the minimum percentage of matching peaks in an automated DNA fingerprint for the following pairs of individuals?

A. Mother and son

B. Sister and brother

C. Uncle and niece

D. Grandfather and grandson

Questions for Student Discussion/Collaboration

1. Discuss examples of positive and negative assortive mating in natural populations, human populations, and agriculturally important species.

2. Discuss the role of mutation in the origin of genetic polymorphisms. Suppose that a genetic polymorphism has two alleles at frequencies of 0.45 and 0.55. Describe three different scenarios to explain these observed allele frequencies. You can propose that the alleles are neutral, beneficial, or deleterious.

3. Most new mutations are detrimental, yet rare beneficial mutations can be adaptive. With regard to the fate of new mutations, discuss whether you think it is more important for natural selection to select against detrimental alleles or to select in favor of beneficial ones. Which do you think is more significant in human populations?

Note: All answers appear at the website for this textbook; the answers to even-numbered questions are in the back of the textbook.

www.mhhe.com/brookergenetics4e

Visit the website for practice tests, answer keys, and other learning aids for this chapter. Enhance your understanding of genetics with our interactive exercises, quizzes, animations, and much more.

CHAPTER OUTLINE

Domesticated wheat. *The color of wheat ranges from a dark red to white, which is an example of a complex or quantitative trait.*

25 QUANTITATIVE GENETICS

In this chapter, we will examine **complex traits**—characteristics that are determined by several genes and are significantly influenced by environmental factors. Many complex traits are viewed as **quantitative traits** because they can be described numerically. In humans, quantitative traits include height, the shape of our noses, and the rate at which we metabolize food, to name a few examples. The field of genetics that studies the mode of inheritance of complex or quantitative traits is called **quantitative genetics.** Quantitative genetics is an important branch of genetics for several reasons. In agriculture, most of the key characteristics of interest to plant and animal breeders are quantitative traits. These include traits such as weight, fruit size, resistance to disease, and the ability to withstand harsh environmental conditions. As we will see later in this chapter, genetic techniques have improved our ability to develop strains of agriculturally important species with desirable quantitative traits. In addition, many human diseases are viewed as complex traits that are influenced by several genes.

Quantitative genetics is also important in the study of evolution. Many of the traits that allow a species to adapt to its environment are quantitative. Examples include the swift speed of the cheetah and the sturdiness of tree branches in windy climates. The importance of quantitative traits in the evolution of species will be discussed in Chapter 26. In this chapter, we examine how genes and the environment contribute to the phenotypic expression of complex or quantitative traits. We will begin with an examination of quantitative traits and how to analyze them using statistical techniques. We then look at the inheritance of polygenic traits and at quantitative trait loci—locations on chromosomes containing genes that affect the outcome of quantitative traits. Advances in genetic mapping strategies have enabled researchers to identify these genes. Last, we look at heritability and consider various ways of calculating and modifying the genetic variation that affects phenotype.

25.1 QUANTITATIVE TRAITS

When we compare characteristics among members of the same species, the differences are often quantitative rather than qualitative. Humans, for example, all have the same basic anatomical features (two eyes, two ears, and so on), but they differ in quantitative ways. People vary with regard to height, weight, the shape of facial features, pigmentation, and many other characteristics. As shown in **Table 25.1**, quantitative traits can be categorized as anatomical, physiological, and behavioral. In addition, many human diseases exhibit characteristics and inheritance patterns analogous to those of quantitative traits. Three of the leading

TABLE 25.1

Types of Quantitative Traits

Trait	Examples
Anatomical traits	Height, weight, number of bristles in *Drosophila*, ear length in corn, and the degree of pigmentation in flowers and skin
Physiological traits	Metabolic traits, speed of running and flight, ability to withstand harsh temperatures, and milk production in mammals
Behavioral traits	Mating calls, courtship rituals, ability to learn a maze, and the ability to grow or move toward light
Diseases	Atherosclerosis, hypertension, cancer, diabetes, and arthritis

causes of death worldwide—heart disease, cancer, and diabetes—are considered complex traits.

In many cases, quantitative traits are easily measured and described numerically. Height and weight can be measured in centimeters (or inches) and kilograms (or pounds), respectively. Speed can be measured in kilometers per hour, and metabolic rate can be assessed as the grams of glucose burned per minute. Behavioral traits can also be quantified. A mating call can be evaluated with regard to its duration, sound level, and pattern. The ability to learn a maze can be described as the time and/or repetitions it takes to learn the skill. Finally, complex diseases such as diabetes can also be studied and described via numerical parameters. For example, the severity of the disease can be assessed by the age of onset or by the amount of insulin needed to prevent adverse symptoms.

From a scientific viewpoint, the measurement of quantitative traits is essential when comparing individuals or evaluating groups of individuals. It is not very informative to say that two people are tall. Instead, we are better informed if we know that one person is 5 feet 7 inches and the other is 5 feet 10 inches. In this branch of genetics, the measurement of a quantitative trait is how we describe the phenotype.

In the early 1900s, Francis Galton in England and his student Karl Pearson showed that many traits in humans and domesticated animals are quantitative in nature. To understand the underlying genetic basis of these traits, they founded what became known as the **biometric field** of genetics, which involved the statistical study of biological traits. During this period, Galton and Pearson developed various statistical tools for studying the variation of quantitative traits within groups of individuals; many of these tools are still in use today. In this section, we will examine how quantitative traits are measured and how statistical tools are used to analyze their variation within groups.

Quantitative Traits Exhibit a Continuum of Phenotypic Variation That May Follow a Normal Distribution

In Part II of this textbook, we discussed many traits that fall into discrete categories. For example, fruit flies might have white eyes or red eyes, and pea plants might have wrinkled seeds or smooth seeds. The alleles that govern these traits affect the phenotype in a qualitative way. In analyzing crosses involving these types of traits, each offspring can be put into a particular phenotypic category. Such attributes are called **discontinuous traits.**

In contrast, quantitative traits show a continuum of phenotypic variation within a group of individuals. For such traits, it is often impossible to place organisms into a discrete phenotypic class. For example, **Figure 25.1a** is a classic photograph from 1914 showing the range of heights of 175 students at the Connecticut Agricultural College. Though height is found at minimum and maximum values, the range of heights between these values is fairly continuous.

How do geneticists describe traits that show a continuum of phenotypes? Because quantitative traits do not naturally fall into a small number of discrete categories, an alternative way to describe them is a **frequency distribution.** To construct a frequency distribution, the trait is divided arbitrarily into a number of convenient, discrete phenotypic categories. For example, in Figure 25.1, the range of heights is partitioned into 1-inch intervals. Then a graph is made that shows the numbers of individuals found in each of the categories.

Figure 25.1b shows a frequency distribution for the heights of students pictured in Figure 25.1a. The measurement of height is plotted along the *x*-axis, and the number of individuals who exhibit that phenotype is plotted on the *y*-axis. The values along the *x*-axis are divided into the discrete 1-inch intervals that define the phenotypic categories, even though height is essentially continuous within a group of individuals. For example, in Figure 25.1a, 22 students were between 64.5 and 65.5 inches in height, which is plotted as the point (65 inches, 22 students) on the graph in Figure 25.1b. This type of analysis can be conducted on any group of individuals who vary with regard to a quantitative trait.

The line in the frequency distribution depicts a **normal distribution,** a distribution for a large sample in which the trait of interest varies in a symmetrical way around an average value. The distribution of measurements of many biological characteristics is approximated by a symmetrical bell curve like that in Figure 25.1b. Normal distributions are common when the phenotype is determined by the cumulative effect of many small independent factors. We will consider the significance of this type of distribution next.

Statistical Methods Are Used to Evaluate a Frequency Distribution Quantitatively

Statistical tools are used to analyze a normal distribution in a number of ways. One measure you are probably familiar with is a parameter called the **mean,** which is the sum of all the values in the group divided by the number of individuals in the group. The mean is computed using the following formula:

$$\overline{X} = \frac{\Sigma X}{N}$$

where

$\overline{X}$	is the mean
ΣX	is the sum of all the values in the group
N	is the number of individuals in the group

Number of students	1	0	0	1	5	7	7	22	25	26	27	17	11	17	4	4	1
Height (inches)	58	59	60	61	62	63	64	65	66	67	68	69	70	71	72	73	74

(a)

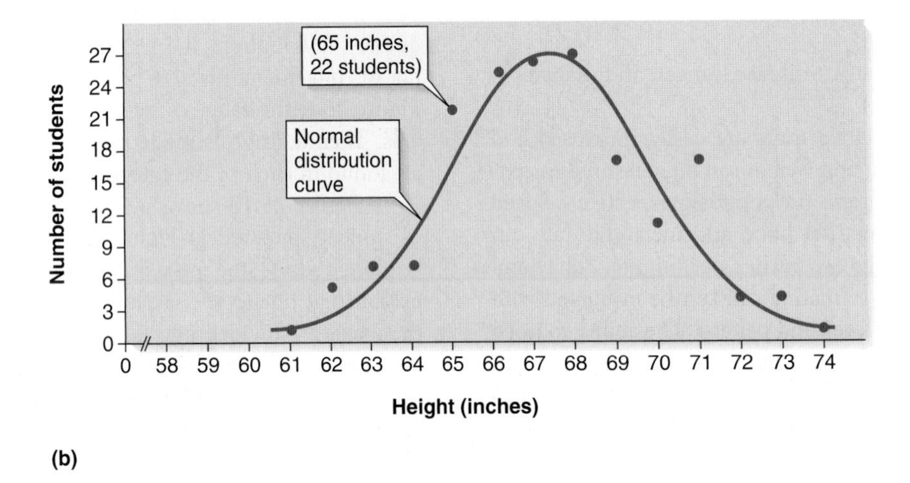

FIGURE 25.1 **Normal distribution of a quantitative trait.** **(a)** The distribution of heights in 175 students at the Connecticut Agricultural College in 1914. **(b)** A frequency distribution for the heights of students shown in (a).

A more generalized form of this equation can be used:

$$\overline{X} = \frac{\Sigma f_i X_i}{N}$$

where

$\overline{X}$ is the mean

$\Sigma f_i X_i$ is the sum of all the values in the group; each value in the group is multiplied by its frequency (f_i) in the group

N is the number of individuals in the group

For example, suppose a bushel of corn had ears with the following lengths (rounded to the nearest centimeter): 15, 14, 13, 14, 15, 16, 16, 17, 15, and 15. Then

$$\overline{X} = \frac{4(15) + 2(14) + 13 + 2(16) + 17}{10}$$

$$\overline{X} = 15 \text{ cm}$$

In genetics, we are often interested in the amount of phenotypic variation that exists in a group. As we will see later in this chapter and in Chapter 26, variation lies at the heart of breeding experiments and of evolution. Without variation, selective breeding is not possible, and natural selection cannot favor one phenotype over another. A common way to evaluate variation within a population is with a statistic called the **variance,** which is a measure of the variation around the mean. The variance is the sum of the squared deviations from the mean divided by the degrees of freedom (*df* equals $N - 1$; see Chapter 2 for a review of degrees of freedom).

$$V_x = \frac{\Sigma f_i (X_i - \overline{X})^2}{N - 1}$$

where

V_X is the variance

$X_i - \overline{X}$ is the difference between each value and the mean

N equals the number of observations

For example, if we use the values given previously for the lengths of ears of corn, the variance in this group is calculated as follows:

$$\Sigma f_i (X_i - \overline{X})^2 = 4(15 - 15)^2 + 2(14 - 15)^2 + (13 - 15)^2 + 2(16 - 15)^2 + (17 - 15)^2$$

$$\Sigma f_i (X_i - \overline{X})^2 = 0 + 2 + 4 + 2 + 4$$

$$\Sigma f_i (X_i - \overline{X})^2 = 12 \text{ cm}^2$$

$$V_X = \frac{\Sigma f_i (X_i - \overline{X})^2}{N - 1}$$

$$V_X = \frac{12 \text{ cm}^2}{9}$$

$$V_X = 1.33 \text{ cm}^2$$

Although variance is a measure of the variation around the mean, it is a statistic that may be difficult to understand intuitively because the variance is computed from squared deviations. For example, weight can be measured in grams; the corresponding variance is measured in square grams. Even so, variances are centrally important in the analysis of quantitative traits because they are additive under certain conditions. This means that the variances for different factors that contribute to a quantitative trait, such as genetic and environmental factors, can be added together to predict the total variance for that trait. Later, we will examine how this property is useful in predicting the outcome of genetic crosses.

To gain a more intuitive grasp of variation, we can take the square root of the variance. This statistic is called the **standard deviation (SD).** Again, using the same values for length, the standard deviation is

$$SD = \sqrt{V_X} = \sqrt{1.33}$$

$$SD = 1.15 \text{ cm}$$

If the values in a population follow a normal distribution, it is easier to appreciate the amount of variation by considering the standard deviation. **Figure 25.2** illustrates the relationship between the standard deviation and the percentages of individuals that deviate from the mean. Approximately 68% of all individuals have values within one standard deviation from the mean, either in the positive or negative direction. About 95% are within two standard deviations, and 99.7% are within three standard deviations. When a quantitative characteristic follows a normal distribution, less than 0.3% of the individuals have values that are more or less than three standard deviations from the mean of the population. In our corn example, three standard deviations equal 3.45 cm. Therefore, we expect that approximately 0.3% of the ears of corn would be less than 11.55 cm or greater than 18.45 cm, assuming that length follows a normal distribution.

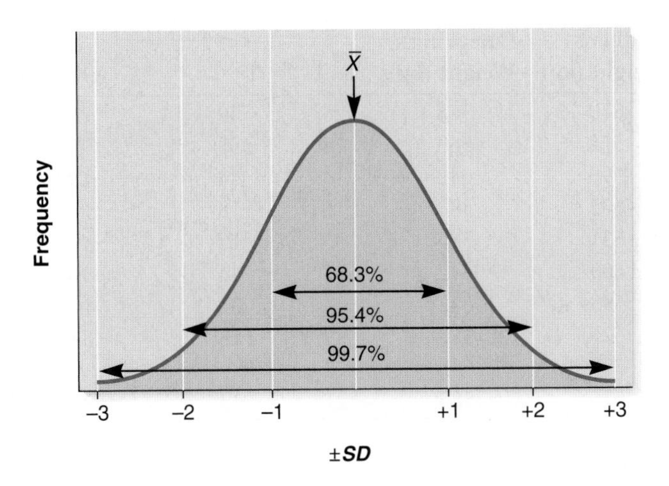

FIGURE 25.2 The relationship between the standard deviation and the proportions of individuals in a normal distribution. For example, approximately 68% of the individuals in a population are between the mean and one standard deviation (1 *SD*) above or below the mean.

Some Statistical Methods Compare Two Variables to Each Other

In many biological problems, it is useful to compare two different variables. For example, we may wish to compare the occurrence of two different phenotypic traits. Do obese animals have larger hearts? Are brown eyes more likely to occur in people with dark skin pigmentation? A second type of comparison is between traits and environmental factors. Does insecticide resistance occur more frequently in areas that have been exposed to insecticides? Is heavy body weight more prevalent in colder climates? Finally, a third type of comparison is between traits and genetic relationships. Do tall parents tend to produce tall offspring? Do women with diabetes tend to have brothers with diabetes?

To gain insight into such questions, a statistic known as the correlation coefficient is often applied. To calculate this statistic, we first need to determine the **covariance,** which describes the relationship between two variables within a group. The covariance is similar to the variance, except that we multiply together the deviations of two different variables rather than squaring the deviations from a single factor.

$$CoV_{(X,Y)} = \frac{\Sigma f_i[(X_i - \overline{X})(Y_i - \overline{Y})]}{N - 1}$$

where

$CoV_{(X,Y)}$ is the covariance between X and Y values

X_i represents the values for one variable, and $\overline{X}$ is the mean value in the group

Y_i represents the values for another variable, and $\overline{Y}$ is the mean value in that group

N is the total number of pairs of observations

As an example, let's compare the weight of cows and that of their adult female offspring. A farmer might be interested in this relationship to determine if genetic variation plays a role in the weight of cattle. The data (see next page) describe the weights at 5 years of age for 10 different cows and their female offspring.

Mother's Weight (kg)	Offspring's Weight (kg)	$X-\overline{X}$	$Y-\overline{Y}$	$(X-\overline{X})(Y-\overline{Y})$
570	568	−26	−30	780
572	560	−24	−38	912
599	642	3	44	132
602	580	6	−18	−108
631	586	35	−12	−420
603	642	7	44	308
599	632	3	34	102
625	580	29	−18	−522
584	605	−12	7	−84
575	585	−21	−13	273
$\overline{X} = 596$	$\overline{Y} = 598$			$\Sigma = 1373$
$SD_X = 21.1$	$SD_Y = 30.5$			

$$CoV_{(X,Y)} = \frac{\Sigma f_i[(X_i - \overline{X})(Y_i - \overline{Y})]}{N - 1}$$

$$CoV_{(X,Y)} = \frac{1373}{10 - 1}$$

$$CoV_{(X,Y)} = 152.6$$

After we have calculated the covariance, we can evaluate the strength of the association between the two variables by calculating a **correlation coefficient (*r*)**. This value, which ranges between −1 and +1, indicates how two factors vary in relation to each other. The correlation coefficient is calculated as

$$r_{(X,Y)} = \frac{CoV_{(X,Y)}}{SD_X SD_Y}$$

A positive *r* value means that two factors tend to vary in the same way relative to each other; as one factor increases, the other increases with it. A value of zero indicates that the two factors do not vary in a consistent way relative to each other; the values of the two factors are not related. Finally, a negative correlation, in which the correlation coefficient is negative, indicates that the two factors tend to vary in opposite ways to each other; as one factor increases, the other decreases.

Let's use the data of 5-year weights for mother and offspring to calculate a correlation coefficient.

$$r_{(X,Y)} = \frac{152.6}{(21.1)(30.5)}$$

$$r_{(X,Y)} = 0.237$$

The result is a positive correlation between the 5-year weights of mother and offspring. In other words, the positive correlation value suggests that heavy mothers tend to have heavy offspring and that lighter mothers have lighter offspring.

How do we evaluate the value of *r*? After a correlation coefficient has been calculated, one must consider whether the *r* value represents a true association between the two variables or whether it could be simply due to chance. To accomplish this, we can test the hypothesis that there is no real correlation (i.e., the null hypothesis, $r = 0$). The null hypothesis is that the observed *r* value differs from zero due only to random sampling error. We followed a similar approach in the chi square analysis described in Chapter 2. Like the chi square value, the significance of the correlation coefficient is directly related to sample size and the degrees of freedom (*df*). In testing the significance of correlation coefficients, *df* equals $N - 2$, because two variables are involved. *N* equals the number of paired observations. **Table 25.2** shows the relationship between the *r* values and degrees of freedom at the 5% and 1% significance levels. (Note: Significance levels are discussed in Chapter 2.)

The use of Table 25.2 is valid only if several assumptions are met. First, the values of *X* and *Y* in the study must have been obtained by an unbiased sampling of the entire population. In addition, this approach assumes that the values of *X* and *Y* follow a normal distribution, like that of Figure 25.1, and that the relationship between *X* and *Y* is linear.

To illustrate the use of Table 25.2, let's consider the correlation we have just calculated for 5-year weights of cows and their

TABLE **25.2**

Values of *r* at the 5% and 1% Significance Levels

Degrees of Freedom (*df*)	5%	1%	Degrees of Freedom (*df*)	5%	1%
1	.997	1.000	24	.388	.496
2	.950	.990	25	.381	.487
3	.878	.959	26	.374	.478
4	.811	.917	27	.367	.470
5	.754	.874	28	.361	.463
6	.707	.834	29	.355	.456
7	.666	.798	30	.349	.449
8	.632	.765	35	.325	.418
9	.602	.735	40	.304	.393
10	.576	.708	45	.288	.372
11	.553	.684	50	.273	.354
12	.532	.661	60	.250	.325
13	.514	.641	70	.232	.302
14	.497	.623	80	.217	.283
15	.482	.606	90	.205	.267
16	.468	.590	100	.195	.254
17	.456	.575	125	.174	.228
18	.444	.561	150	.159	.208
19	.433	.549	200	.138	.181
20	.423	.537	300	.113	.148
21	.413	.526	400	.098	.128
22	.404	.515	500	.088	.115
23	.396	.505	1000	.062	.081

Note: *df* equals $N - 2$.
From J. T. Spence, B. J. Underwood (1976). *Elementary Statistics*. Prentice-Hall, Englewood Cliffs, New Jersey.

female offspring. In this case, we obtained a value of 0.237 for r, and the value of N was 10. Under these conditions, df equals 8. To be valid at a 5% confidence interval, the value of r would have to be 0.632 or higher. Because the value that we obtained is much less than this, it is fairly likely that this value could have occurred as a matter of random sampling error. In this case, we cannot reject the null hypothesis, and, therefore, we cannot conclude the positive correlation is due to a true association between the weights of mothers and offspring.

In an actual experiment, however, a researcher examines many more pairs of cows and offspring, perhaps 500 to 1000. If a correlation of 0.237 was observed for $N = 1000$, the value would be significant at the 1% level. We would reject the null hypothesis that weights are not associated with each other. Instead, we would conclude that a real association occurs between the weights of mothers and their offspring. In fact, these kinds of experiments have been done for cattle weights, and the correlations between mothers and offspring have often been found to be significant.

If a statistically significant correlation is obtained, how do we interpret its meaning? An r value that is statistically significant suggests a true association, but it does not necessarily imply a cause-and-effect relationship. When parents and offspring display a significant correlation for a trait, we should not jump to the conclusion that genetics is the underlying cause of the positive association. In many cases, parents and offspring share similar environments, so the positive association might be rooted in environmental factors. In general, correlations are quite useful in identifying positive or negative associations between two variables. We should use caution, however, because this statistic, by itself, cannot prove that the association is due to cause and effect.

A **regression analysis** may be used when researchers suspect, or when their experimentation has shown, that two variables are related due to cause and effect—that one variable (the independent variable) affects the outcome of another (the dependent variable). Researchers use a regression analysis to predict how much the dependent variable changes in response to the independent variable. This approach is described in solved problem S4 at the end of the chapter.

25.2 POLYGENIC INHERITANCE

In Section 25.1, we saw that quantitative traits tend to show a continuum of variation and can be analyzed with various statistical tools. At the beginning of the 1900s, a great debate focused on the inheritance of quantitative traits. The biometric school, founded by Francis Galton and Karl Pearson, argued that these types of traits are not controlled by discrete genes that affect phenotypes in a predictable way. To some extent, the biometric school favored the idea of blending inheritance, which had been proposed many years earlier (see Chapter 2).

Alternatively, the followers of Mendel, led by William Bateson in England and William Castle in the United States, held firmly to the idea that traits are governed by genes, which are inherited as discrete units. As we know now, Bateson and Castle were correct. However, as we will see in this section, studying quantitative traits is difficult because these traits are controlled by multiple genes and substantially influenced by environmental factors.

Most quantitative traits are polygenic and exhibit a continuum of phenotypic variation. The term **polygenic inheritance** refers to the transmission of a trait governed by two or more different genes. The location on a chromosome that harbors one or more genes that affect the outcome of a quantitative trait is called a **quantitative trait locus (QTL).** As discussed later, QTLs are chromosomal regions that are identified by genetic mapping. Because such mapping usually locates the QTL to a relatively large chromosomal region, a QTL may contain a single gene or two or more closely linked genes that affect a quantitative trait.

Just a few years ago, it was extremely difficult for geneticists to determine the inheritance patterns for genes underlying polygenic traits, particularly those determined by three or more genes having multiple alleles for each gene. Recently, however, molecular genetic tools (described in Chapters 19 and 20) have greatly enhanced our ability to find regions in the genome where QTLs are likely to reside. This has been a particularly exciting advance in the field of quantitative genetics. In some cases, the identification of QTLs may allow the improvement of quantitative traits in agriculturally important species.

Polygenic Inheritance and Environmental Factors May Produce a Continuum of Phenotypes

The first experiment demonstrating that continuous variation is related to polygenic inheritance was conducted by the Swedish geneticist Herman Nilsson-Ehle in 1909. He studied the inheritance of red pigment in the hull of bread wheat, *Triticum aestivum* (**Figure 25.3a**). When true-breeding plants with white hulls were crossed to a variety with red hulls, the F_1 generation had an intermediate color. When the F_1 generation was allowed to self-fertilize, great variation in redness was observed in the F_2 generation, ranging from white, light red, intermediate red, medium red, and dark red. An unsuspecting observer might conclude that this F_2 generation displayed a continuous variation in hull color. However, as shown in **Figure 25.3b**, Nilsson-Ehle carefully categorized the colors of the hulls and discovered that they followed a 1:4:6:4:1 ratio. He concluded that this species is diploid for two different genes that control hull color, each gene existing in a red or white allelic form. He hypothesized that these two loci must contribute additively to the color of the hull; the contribution of each red allele to the color of the hull is additive.

Later, researchers discovered a third gene that also affects hull color. The two strains that Nilsson-Ehle had used in his original experiments must have been homozygous for the white allele of this third gene. It makes sense that wheat would have two copies of three genes that affect hull color because we now know that *T. aestivum* is a hexaploid derived from three closely related diploid species, as discussed in Chapter 8. Therefore, *T. aestivum* has six copies of many genes.

As we have just seen, Nilsson-Ehle categorized wheat hull colors into several discrete genotypic categories. However, for many polygenic traits, this is difficult or impossible. In general, as the number of genes controlling a trait increases and

(a) Red and white hulls of wheat

♀ \ ♂	R1R2	R1r2	r1R2	r1r2
R1R2	R1R1R2R2 Dark red	R1R1R2r2 Medium red	R1r1R2R2 Medium red	R1r1R2r2 Intermediate red
R1r2	R1R1R2r2 Medium red	R1R1r2r2 Intermediate red	R1r1R2r2 Intermediate red	R1r1r2r2 Light red
r1R2	R1r1R2R2 Medium red	R1r1R2r2 Intermediate red	r1r1R2R2 Intermediate red	r1r1R2r2 Light red
r1r2	R1r1R2r2 Intermediate red	R1r1r2r2 Light red	r1r1R2r2 Light red	r1r1r2r2 White

(b) *R1r1R2r2* x *R1r1R2r2*

FIGURE 25.3 The Nilsson-Ehle experiment studying how continuous variation is related to polygenic inheritance in wheat. **(a)** Red (top) and white (bottom) varieties of wheat, *Triticum aestivum*. **(b)** Nilsson-Ehle carefully categorized the colors of the hulls in the F_2 generation and discovered that they followed a 1:4:6:4:1 ratio. This occurs because the contributions of the red alleles are additive.

Genes → Traits In this example, two genes, with two alleles each (red and white), govern hull color. Offspring can display a range of colors, depending on how many copies of the red allele they inherit. If an offspring is homozygous for the red allele of both genes, it will have very dark red hulls. By comparison, if it carries three red alleles and one white allele, it will be medium red (which is not quite as deep in color). In this way, this polygenic trait can exhibit a range of phenotypes from dark red to white.

the influence of the environment increases, the categorization of phenotypes into discrete genotypic classes becomes increasingly difficult, if not impossible. Therefore, a Punnett square cannot be used to analyze most quantitative traits. Instead, statistical methods, which are described later, must be employed.

Figure 25.4 illustrates how genotypes and phenotypes may overlap for polygenic traits. In this example, the environment (sunlight, soil conditions, and so forth) may affect the phenotypic outcome of a trait in plants (namely, seed weight). Figure 25.4a considers a situation in which seed weight is controlled by one gene with light (*w*) and heavy (*W*) alleles. A heterozygous plant (*Ww*) is allowed to self-fertilize. When the weight is only slightly influenced by variation in the environment, as seen on the left, the light, intermediate, and heavy seeds fall into separate, well-defined categories. When the environmental variation has a greater effect on seed weight, as shown on the right, more phenotypic variation is found in seed weight within each genotypic class. The variance in the frequency distribution on the right is much higher. Even so, most individuals can be classified into the three main categories.

By comparison, Figure 25.4b illustrates a situation in which seed weight is governed by three genes instead of one, each existing in light and heavy alleles. When the environmental variation is low and/or plays a minor role in the outcome of this trait, a cross between two heterozygotes is expected to produce a 1:6:15:20:15:6:1 ratio. As shown in the upper illustration in Figure 25.4b, nearly all individuals fall within a phenotypic category that corresponds to their genotype. When the environment has a more variable effect on phenotype, as shown in the lower illustration, the situation becomes more ambiguous. For example, individuals with one *w* allele and five *W* alleles have a phenotype that overlaps with that of individuals having six *W* alleles or two *w* alleles and four *W* alleles. Therefore, it becomes difficult to categorize each phenotype into a unique genotypic class. Instead, the trait displays a continuum ranging from light to heavy seed weight.

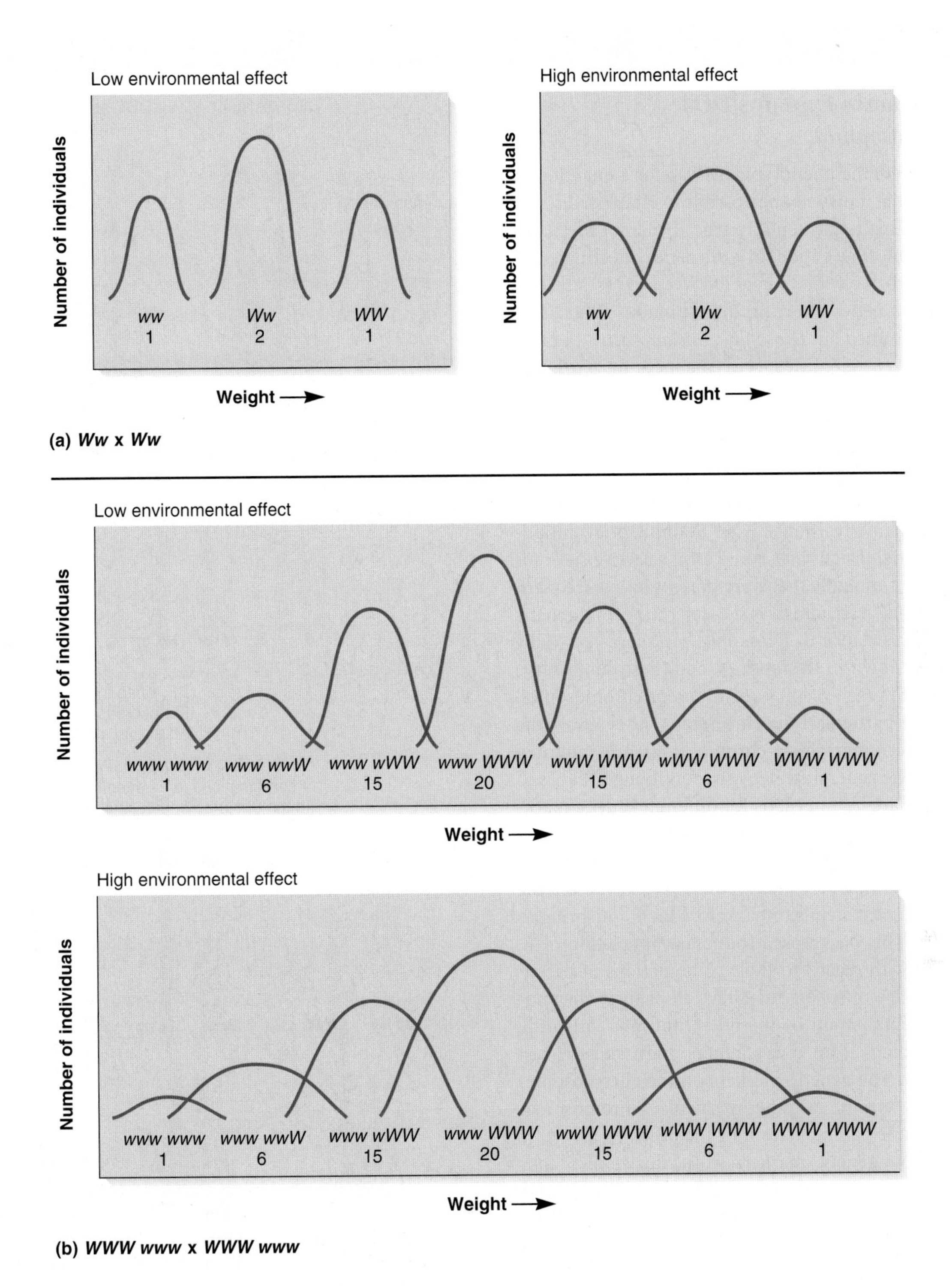

FIGURE 25.4 How genotypes and phenotypes may overlap for polygenic traits. **(a)** Situations in which seed weight is controlled by one gene, existing in light (*w*) and heavy (*W*) alleles. **(b)** Situations in which seed weight is governed by three genes instead of one, each existing in light and heavy alleles. Note: The 1:2:1 and 1:6:15:20:15:6:1 ratios were derived by using a Punnett square and assuming a cross between individuals that are both heterozygous for three different genes.

Genes → Traits The ability of geneticists to correlate genotype and phenotype depends on how many genes are involved and how much the environment causes the phenotype to vary. In (a), a single gene influences weight. In the graph on the left side, the environment does not cause much variation in weight. This makes it easy to distinguish the three genotypes. There is no overlap in the weights of *ww*, *Ww*, and *WW* individuals. In the graph on the right side, the environment causes more variation in weight. In this case, a few individuals with *ww* genotypes have the same weight as a few individuals with *Ww* genotypes; and a few *Ww* genotypes have the same weight as *WW* genotypes. As shown in (b), it becomes even more difficult to distinguish genotype based on phenotype when three genes are involved. The overlaps are minor when the environment does not cause much weight variation. However, when the environment causes substantial phenotypic variation, the overlaps between genotypes and phenotypes are very pronounced and greatly confound genetic analysis.

EXPERIMENT 25A

Polygenic Inheritance Explains DDT Resistance in *Drosophila*

As we have just learned, the phenotypic overlap for a quantitative trait may be so great that it may not be possible to establish discrete phenotypic classes. This is particularly true if many genes contribute to the trait. One way to identify the genes affecting polygenic inheritance is to look for linkage between genes affecting quantitative traits and genes affecting discontinuous traits. This approach was first studied in *Drosophila melanogaster* because many alleles had been identified and mapped to particular chromosomes.

In 1957, James Crow conducted one of the earliest studies to show linkage between genes affecting quantitative traits and genes affecting discontinuous traits. Crow, who was interested in evolution, spent time studying insecticide resistance in *Drosophila*. He noted, "Insecticide resistance is an example of evolutionary change, the insecticide acting as a powerful selective sieve for concentrating resistant mutants that were present in low frequencies in the population." His aim was to determine the genetic basis for insecticide resistance in *Drosophila melanogaster*. Many alleles were already known in this species, and these could serve as **genetic markers** for each of the four different chromosomes. Dominant alleles are particularly useful because they allow the experimenter to determine which chromosomes are inherited from either parent. The general strategy in identifying QTLs is to cross two strains that are homozygous for different genetic markers and also differ with regard to the quantitative trait of interest. This produces an F_1 generation that is heterozygous for the markers and usually exhibits an intermediate phenotype for the quantitative trait. The next step is to backcross the F_1 offspring to the parental strains. This backcross produces a population of F_2 offspring that differ with regard to their combinations of parental chromosomes. A few offspring may have all of their chromosomes from one parental strain or the other, but most offspring have a few chromosomes from one parental strain and the rest from the other strain. The genetic markers on the chromosomes provide a way to determine whether particular chromosomes were inherited from one parental strain or the other.

To illustrate how genetic markers work, **Figure 25.5** considers a situation in which two strains differ in a quantitative trait—resistance to DDT—and also differ in dominant alleles on chromosome 3. The dominant alleles serve as markers for this chromosome. One strain is resistant to DDT and carries a dominant allele that causes minute bristles (*M*), whereas another strain is sensitive to DDT and carries a dominant allele that causes a rough eye (*R*). The wild-type alleles, which are recessive, produce long bristles (*m*) and smooth eyes (*r*). At the start of this experiment, it is not known if alleles affecting DDT resistance are located on this chromosome. If offspring from a backcross inherit both copies of chromosome 3 from the DDT-resistant strain, they will have smooth eyes and minute bristles. If they inherit both copies from the DDT-sensitive strain, they will have rough eyes and long bristles. By comparison, a fly with

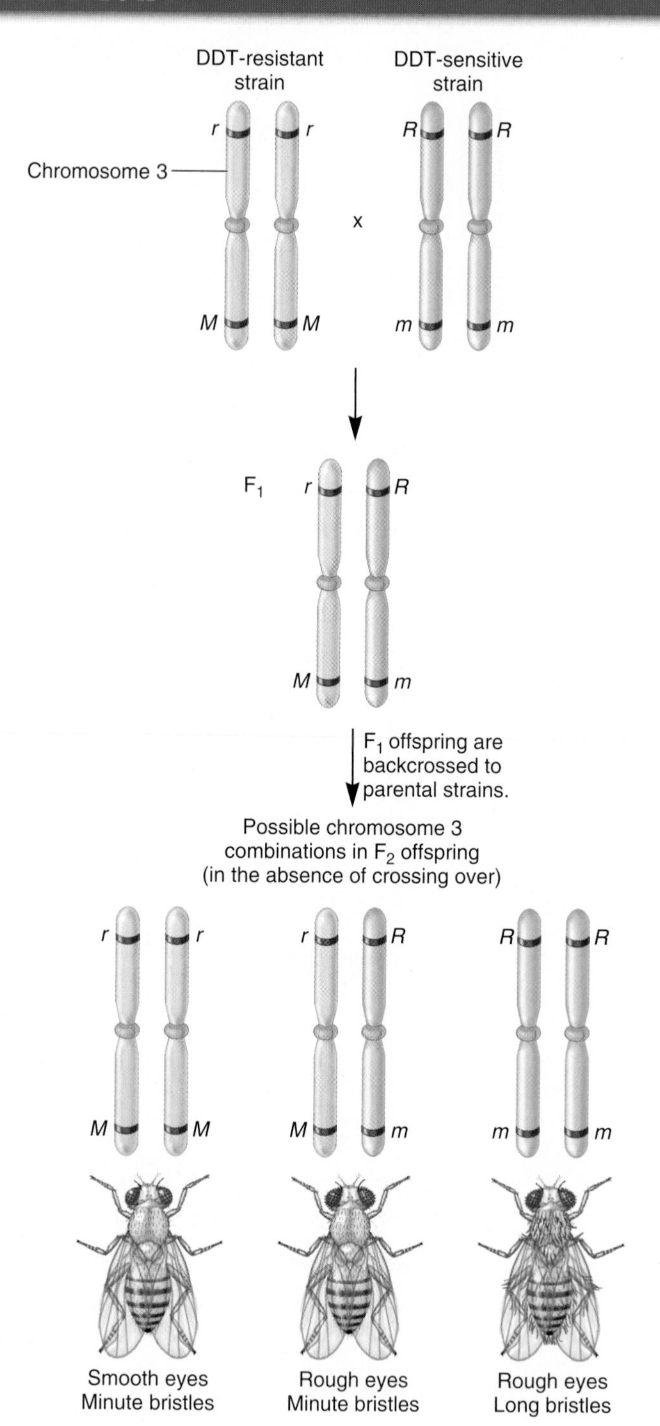

FIGURE 25.5 The use of genetic markers to map a QTL affecting DDT resistance. One strain is DDT-resistant. On chromosome 3, it also carries a dominant allele that causes minute bristles (*M*). The other strain is DDT-sensitive and carries a dominant allele that causes a roughness to the eye (*R*). The wild-type alleles, which are recessive, produce long bristles (*m*) and smooth eyes (*r*). F_2 offspring can have either both copies of chromosome 3 from the DDT-resistant strain, both from the sensitive strain, or one of each. This can be discerned by the phenotypes of the F_2 offspring.

rough eyes (*R*) and minute bristles (*M*) inherited one copy of chromosome 3 from the DDT-resistant strain and one copy from the DDT-sensitive strain. The transmission of the other *Drosophila* chromosomes can also be followed in a similar way. Therefore, the phenotypes of the offspring from the backcross provide a way to discern whether particular chromosomes were inherited from the DDT-resistant or DDT-sensitive strain.

Figure 25.6 shows the protocol followed by James Crow. He began with a DDT-resistant strain that had been produced by exposing flies to DDT for many generations. This DDT-resistant strain was crossed to a sensitive strain. As described previously in Figure 25.5, the two strains had allelic markers that made it possible to determine the origins of the different *Drosophila* chromosomes. Recall that *Drosophila* has four chromosomes. In this study, only chromosomes X, 2, and 3 were marked with alleles. Chromosome 4 was neglected due to its very small size. The F_1 flies were backcrossed to both parental strains, and then the F_2 female progeny were examined in two ways. First, their phenotypes were examined to determine whether particular chromosomes were inherited from the DDT-resistant or DDT-sensitive strain. Next, the female flies were exposed to filter paper impregnated with DDT. It was then determined if the flies survived this exposure for 18 to 24 hours.

THE HYPOTHESIS

DDT resistance is a polygenic trait.

TESTING THE HYPOTHESIS — FIGURE 25.6 Polygenic inheritance of DDT-resistance alleles in *Drosophila melanogaster.*

Starting material: DDT-resistant and DDT-sensitive strains of fruit flies.

1. Cross the DDT-resistant strain to the sensitive strain. In each strain, chromosomes X, 2, and 3 were marked with alleles that provided easily discernible phenotypes.
2. Take the F_1 flies and backcross to both parental strains.
3. Identify the origin of the chromosomes in the F_2 flies according to their phenotypes.
4. Expose the F_2 female flies to DDT on a filter paper for 18–24 hours.

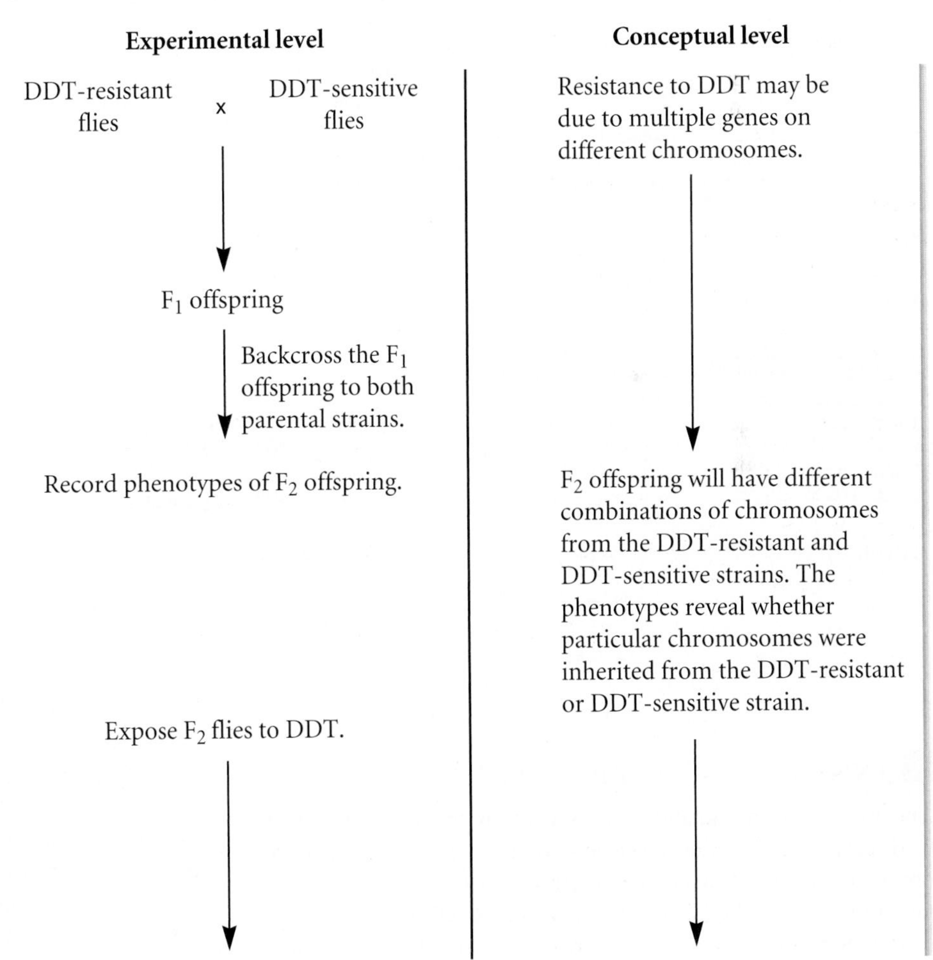

5. Record the number of survivors.

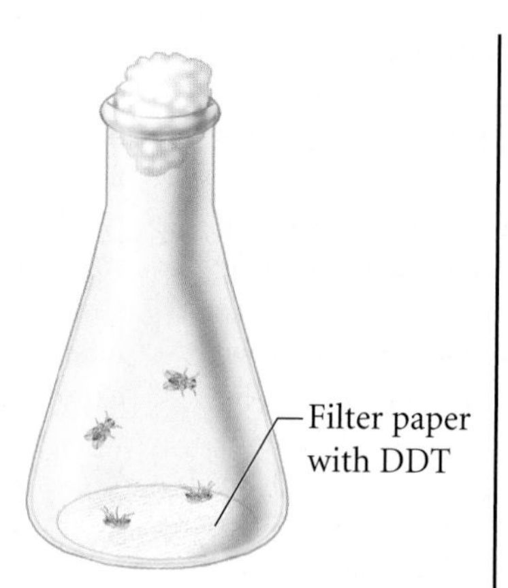

The goal is to determine if survival is correlated with the inheritance of specific chromosomes from the DDT-resistant strain.

THE DATA

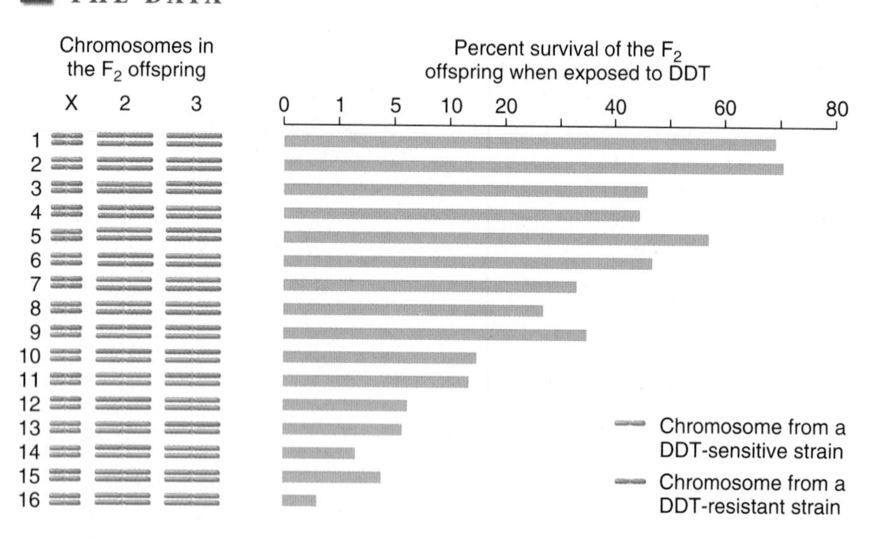

Data from: J.F. Crow (1957) Genetics of insect resistance to chemicals. *Ann Rev Entomol* 2: 227–246.

INTERPRETING THE DATA

The results of this analysis are shown in the data of Figure 25.6. Based on the inheritance of markers, some offspring were observed to inherit all of their chromosomes from one parental strain or the other, but most offspring contained a few chromosomes from one parental strain and the rest from the other. (Note: The illustrations along the left side are simplified and do not take into account the phenomenon of crossing over.) The data in Figure 25.6 suggest that each copy of the X chromosome and chromosomes 2 and 3 from the DDT-resistant strain confer a significant amount of insecticide resistance. A general trend was observed in which flies inheriting more chromosomes from the DDT-resistant strain had greater levels of resistance. (However, exceptions to the trend did occur; compare examples 1 and 2.) Overall, the results are consistent with the hypothesis that insecticide resistance is a polygenic trait involving multiple genes that reside on the X chromosome and on chromosomes 2 and 3.

A self-help quiz involving this experiment can be found at www.mhhe.com/brookergenetics4e.

Quantitative Trait Loci (QTLs) Are Now Mapped by Linkage to Molecular Markers

In the previous experiment, we saw how the locations of genes affecting a quantitative trait, such as DDT resistance, were determined by the linkage of such unknown genes to known genes on the *Drosophila* chromosomes. In the past few years, newer research techniques have identified molecular markers, such as RFLPs and microsatellites, that serve as reference points along chromosomes. This topic is discussed in Chapter 20. These markers have been used to construct detailed genetic maps of several species' genomes. Once a genome map is obtained, it becomes much easier to determine the locations of genes that affect a quantitative trait. In addition to model organisms such as *Drosophila*, *Arabidopsis*, *Caenorhabditis elegans*, and mice, detailed molecular maps have been obtained for many species of agricultural importance. These include crops such as corn, rice, and tomatoes, as well as livestock such as cattle, pigs, sheep, and chickens.

To map the genes in a eukaryotic species, researchers now determine their locations by identifying molecular markers that are close to such genes. This approach is described in Chapter 22 (see Figures 22.5–22.7). In 1989, Eric Lander and David Botstein extended this technique to identify QTLs that govern a quantitative trait. The basis of **QTL mapping** is the association between genetically determined phenotypes for quantitative traits and molecular markers such as RFLPs, microsatellites, and single nucleotide polymorphisms (SNPs).

The general strategy for QTL mapping is shown in **Figure 25.7**. This figure depicts two different strains of a diploid species with four chromosomes per set. The strains are highly inbred, which means they are homozygous for most molecular markers and genes. They differ in two important ways. First, the two strains differ with regard to many molecular markers. These markers are designated 1A and 1B, 2A and 2B, and so forth. The markers 1A and 1B mark the same chromosomal location in this species, namely, the upper tip of chromosome 1. However, the two markers are distinguishable in the two strains at the molecular level. For example, 1A might be a microsatellite that is 148 bp, whereas 1B might be 212 bp. Second, the two strains differ in a quantitative trait of interest. In this example, the strain on the left produces large fruit, whereas the strain on the right produces small fruit. The unknown genes affecting this trait are designated with the letter X. A black X indicates a QTL that harbors alleles that promote large fruit, and a blue X is the same site that carries alleles that promote small fruit. Prior to conducting their crosses, researchers would not know the chromosomal locations of the QTLs shown in this figure. The purpose of the experiment is to determine their locations.

With these ideas in mind, the protocol shown in Figure 25.7 begins by mating the two inbred strains to each other and then backcrossing the F_1 offspring to both parental strains. This produces an F_2 generation with a great degree of variation. The F_2 offspring are then characterized in two ways. First, they are examined for their fruit size, and second, a sample of cells is analyzed to determine which molecular markers are found in their chromosomes. The goal is to find an association between particular molecular markers and fruit size. For example, 2A is strongly associated with large size, whereas 2B is strongly associated with small size. By comparison, 9A and 9B are not associated with large or small size, because a QTL affecting this trait is not found on this chromosome. Also, markers such as 14A and 14B, which are fairly far away from a QTL, are not strongly associated with a particular QTL. Markers that are on the same chromosome but far away from a QTL are often separated from the QTL during meiosis in the F_1 heterozygote due to crossing over. Only closely linked markers are strongly associated with a particular QTL.

Overall, QTL mapping involves the analysis of a large number of markers and offspring. The data are analyzed by computer programs that can statistically associate the phenotype (e.g., fruit size) with particular markers. Markers found throughout the genome of a species provide a way to identify the locations of several different genes that possess allelic differences that may affect the outcome of a quantitative trait.

As an example of QTL mapping, in 1988, Andrew Paterson and his colleagues examined quantitative trait inheritance in the tomato. They studied a domestic strain of tomato and a South American green-fruited variety. These two strains differed in their RFLPs, and they also exhibited dramatic differences in three agriculturally important characteristics: fruit mass, soluble solids content, and fruit pH. The researchers crossed the two strains and then backcrossed the offspring to the domestic tomato. A total of 237 plants was then examined with regard to 70 known RFLP markers. In addition, between 5 and 20 tomatoes from each plant were analyzed with regard to fruit mass, soluble solids content, and fruit pH. Using this approach, the researchers were able to map genes contributing much of the variation in these traits to particular intervals along the tomato chromosomes. They identified six loci causing variation in fruit mass, four affecting soluble solids content, and five with effects on fruit pH.

More recently, the DNA sequence of the entire genome of many species has been determined. In such cases, the mapping of QTLs to a defined chromosomal region may allow researchers to analyze the DNA sequence in that region and to the identify one or more genes that influence the trait of interest.

25.3 HERITABILITY

As we have just seen, recent approaches in molecular mapping have enabled researchers to identify the genes that contribute to a quantitative trait. The other key factor that affects the phenotypic outcomes of quantitative traits is the environment. All traits of organisms are influenced by genetics and the environment, and this is particularly pertinent in the study of quantitative traits. Researchers want to understand how variation, both genetic and environmental, affects the phenotypic results.

The term **heritability** refers to the amount of phenotypic variation within a group of individuals that is due to genetic variation. Genes play a role in the development of essentially all of an

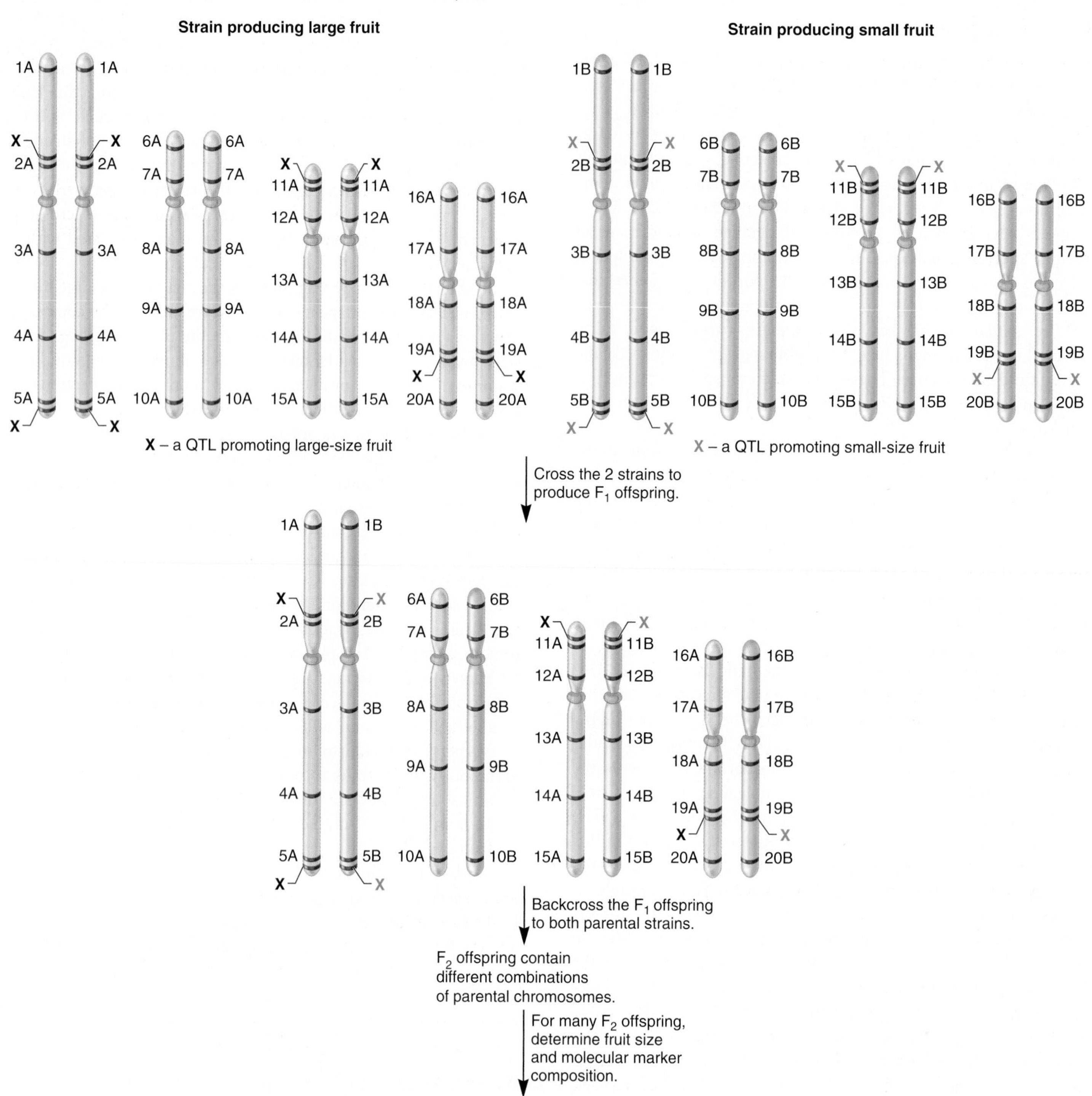

FIGURE 25.7 The general strategy for QTL mapping via molecular markers. Two different inbred strains have four chromosomes per set. The strain on the left produces large fruit, and the strain on the right produces small fruit. The goal of this mapping strategy is to locate the unknown genes affecting this trait, which are designated with the letter X. A black X indicates a site promoting large fruit, and a blue X is a site promoting small fruit. The two strains differ with regard to many molecular markers designated 1A and 1B, 2A and 2B, and so forth. The two strains are mated, and then the F_1 offspring are backcrossed to the parental strains. Many F_2 offspring are then examined for their fruit size and to determine which molecular markers are found in their chromosomes. The data are analyzed by computer programs that can statistically associate the phenotype (e.g., fruit size) with particular markers. Markers found throughout the genome of this species provide a way to locate many different genes that may affect the outcome of a single quantitative trait. In this case, the analysis predicts four QTLs promoting heavier fruit weight that are linked to regions of the chromosomes containing the following markers: 2A, 5A, 11A, and 19A.

organism's traits. Even so, variation of a trait in a population may be due entirely to environmental variation, entirely to genetic variation, or to a combination of the two. If all of the phenotypic variation in a group is due to genetic variation, the heritability would have a value of 1. If all of the variation is due to environmental effects, the heritability would equal 0. For most groups of organisms, the heritability for a given trait lies between these two extremes. For example, both genes and diet affect the size that an individual will attain. Some individuals inherit alleles that tend to make them large, and a proper diet also promotes larger size. Other individuals inherit alleles that make them small, and an inadequate diet may contribute to small size. Taken together, both genetics and the environment influence the phenotypic results.

In the study of quantitative traits, a primary goal is to determine how much of the phenotypic variation arises from genetic variation and how much comes from environmental variation. In this section, we examine how geneticists analyze the genetic and environmental components that affect quantitative traits. As we will see, this approach has been applied with great success in breeding strategies to produce domesticated species with desirable and commercially valuable characteristics.

Genetic Variance and Environmental Variance: Both May Contribute to Phenotypic Variance

Earlier, we examined the amount of phenotypic variation within a group by calculating the variance. Geneticists partition quantitative trait variation into components that are attributable to the following different causes:

Genetic variation (V_G)
Environmental variation (V_E)
Variation due to interactions between genetic and environmental factors ($V_{G \times E}$)
Variation due to associations between genetic and environmental factors ($V_{G \leftrightarrow E}$)

Let's begin by considering a simple situation in which V_G and V_E are the only factors that determine phenotypic variance and the genetic and environmental factors are independent of each other. If so, then the total variance for a trait in a group of individuals is

$$V_T = V_G + V_E$$

where

V_T is the total variance. It reflects the amount of variation that is measured at the phenotypic level.

V_G is the relative amount of variance due to genetic variation.

V_E is the relative amount of variance due to environmental variation.

Why is this equation useful? The partitioning of variance into genetic and environmental components allows us to estimate their relative importance in influencing the variation within a group. If V_G is very high and V_E is very low, genetics plays a greater role in promoting variation within a group. Alternatively, if V_G is low and V_E is high, environmental factors underlie much of the phenotypic variation. As described later in this chapter, a livestock breeder might want to apply selective breeding if V_G for an important (quantitative) trait is high. In this way, the characteristics of the herd may be improved. Alternatively, if V_G is negligible, it would make more sense to investigate (and manipulate) the environmental causes of phenotypic variation.

With experimental and domesticated species, one possible way to determine V_G and V_E is by comparing the variation in traits between genetically identical and genetically disparate groups. For example, researchers have followed the practice of **inbreeding** to develop genetically homogeneous strains of mice. Inbreeding in mice involves many generations of brother-sister matings, which eventually produces strains that are **monomorphic** for all or nearly all of their genes. The term monomorphic means that all the members of a population are homozygous for the same allele of a given gene. Within such an inbred strain of mice, V_G equals zero. Therefore, all phenotypic variation is due to V_E. When studying quantitative traits such as weight, an experimenter might want to know the genetic and environmental variance for a different, genetically heterogeneous group of mice. To do so, the genetically homogeneous and heterogeneous mice could be raised under the same environmental conditions and their weights measured. The phenotypic variance for weight could then be calculated as described earlier. Let's suppose we obtained the following results:

$V_T = 15\ g^2$ for the group of genetically homogeneous mice

$V_T = 22\ g^2$ for the group of genetically heterogeneous mice

In the case of the homogeneous mice, $V_T = V_E$, because V_G equals 0. Therefore, V_E equals 15 g^2. To estimate V_G for the heterogeneous group of mice, we could assume that V_E (i.e., the environmentally produced variance) is the same for them as it is for the homogeneous mice, because the two groups were raised in identical environments. This assumption allows us to calculate the genetic variance for the heterogeneous mice.

$$V_T = V_G + V_E$$

$$22\ g^2 = V_G + 15\ g^2$$

$$V_G = 7\ g^2$$

This result tells us that some of the phenotypic variance in the genetically heterogeneous group is due to the environment (namely, 15 g^2) and some (7 g^2) is due to genetic variation in alleles that affect weight.

Phenotypic Variation May Also Be Influenced by Interactions and Associations Between Genotype and the Environment

Thus far, we have considered the simple situation in which genetic variation and environmental variation are independent of each other and affect the phenotypic variation in an additive way. As another example, let's suppose that three genotypes,

TT, *Tt*, and *tt*, affect height, producing tall, medium, and short plants, respectively. Greater sunlight makes the plants grow taller regardless of their genotypes. In this case, our assumption that $V_T = V_G + V_E$ would be reasonably valid.

However, let's consider a different environmental factor such as minerals in the soil. As a hypothetical example, let's suppose the *t* allele is a loss-of-function allele that eliminates the function of a protein involved with mineral uptake from the soil. In this case, the *Tt* and *tt* plants are shorter because they cannot take up a sufficient supply of certain minerals to support maximal growth, whereas the *TT* plants are not limited by mineral uptake. According to this hypothetical scenario, adding minerals to the soil enhances the growth rate of *tt* plants by a large amount and the *Tt* plants by a smaller amount (**Figure 25.8**). The height of *TT* plants is not affected by mineral supplementation. When the environmental effects on phenotype differ according to genotype, this phenomenon is called a **genotype-environment interaction.** Variation due to interactions between genetic and environmental factors is termed $V_{G \times E}$ as noted earlier.

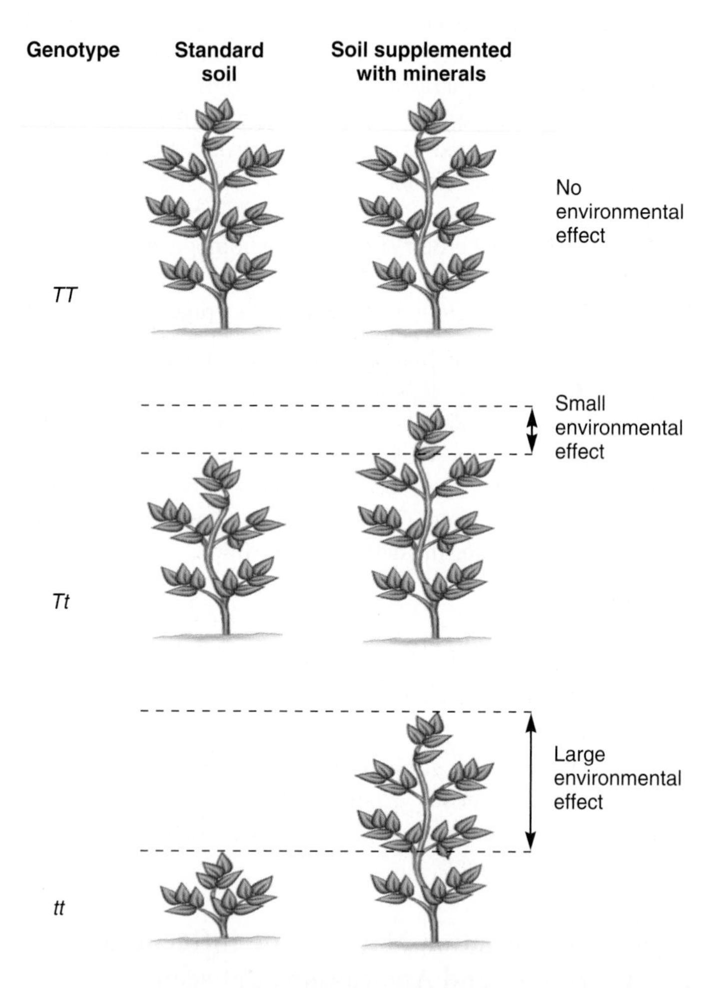

FIGURE 25.8 A schematic example of genotype-environment interaction. When grown in standard soil, the three genotypes *TT*, *Tt*, and *tt* show large, medium, and small heights, respectively. When the soil is supplemented with minerals, a great effect is seen on the *tt* genotype and a smaller effect on the *Tt* genotype. The *TT* genotype is unaffected by the environmental change.

TABLE 25.3

Longevity of Two Strains of *Drosophila melanogaster**

	Strain A		Strain B	
Temperature	Male	Female	Male	Female
Standard	33.6	39.5	37.5	28.9
High	36.3	33.9	23.2	28.6
Low	77.5	48.3	45.8	77.0

*Longevity was measured in the mean number of days of survival. Strains A and B were inbred strains of *D. melanogaster* called Oregon and 2b, respectively. The standard, high, and low temperature conditions were 25°C, 29°C, and 14°C, respectively.

Interactions of genetic and environmental factors are common. As an example, **Table 25.3** shows results from a study conducted in 2000 by Cristina Vieira, Trudy Mackay, and colleagues in which they investigated genotype-environment interaction for quantitative trait loci affecting life span in *Drosophila melanogaster*. The data seen in the table compare the life span in days of male and female flies from two different strains of *D. melanogaster* raised at different temperatures. Because males and females differ in their sex chromosomes and gene expression patterns, they can be viewed as having different genotypes. The effects of environmental changes depended greatly on the strain and the sex of the flies. Under standard culture conditions, the females of strain A had the longest life span, whereas females of strain B had the shortest. In strain A, high temperature increased the longevity of males and decreased the longevity of females. In contrast, under hotter conditions, the longevity of males of strain B was dramatically reduced, whereas females of this same strain were not significantly affected. Lower growth temperature also had different effects in these two strains. Although low temperature increased the longevity of both strains, the effects were most dramatic in the males of strain A and the females of strain B. Taken together, these results illustrate the potential complexity of genotype-environmental interaction when measuring a quantitative trait such as life span.

Another issue confronting geneticists is that genotypes may not be randomly distributed in all possible environments. When certain genotypes are preferentially found in particular environments, this phenomenon is called a **genotype-environment association** ($V_{G \leftrightarrow E}$). When such an association occurs, the effects of genotype and environment are not independent of each other, and the association needs to be considered when determining the effects of genetic and environmental variation on the total phenotypic variation. Genotype-environment associations are very common in the study of human genetics, in which large families tend to have more similar environments than the population as a whole. One way to evaluate this effect is to compare different genetic relationships, such as identical versus fraternal twins. We will examine this approach later in the chapter. Another strategy that geneticists might follow is to analyze siblings that have been adopted by different parents at birth. Their environmental conditions tend to be more disparate, and this may help to minimize the effects of genotype-environment association.

Heritability Is the Relative Amount of Phenotypic Variation That Is Due to Genetic Variation

Another way to view variance is to focus our attention on the genetic contribution to phenotypic variation. Heritability is the proportion of the phenotypic variance that is attributable to genetic variation. If we assume again that environment and genetics are independent and the only two factors affecting phenotype, then

$$h_B^2 = V_G/V_T$$

where

h_B^2 is the heritability in the broad sense

V_G is the variance due to genetics

V_T is the total phenotypic variance, which equals $V_G + V_E$

The heritability defined here, h_B^2, called the **broad-sense heritability,** takes into account different types of genetic variation that may affect the phenotype. As we have seen throughout this textbook, genes can affect phenotypes in various ways. As described earlier, the Nilsson-Ehle experiment showed that the alleles determining hull color in wheat affect the phenotype in an additive way. Alternatively, alleles affecting other traits may show a dominant/recessive relationship. In this case, the alleles are not strictly additive, because the heterozygote has a phenotype closer to, or perhaps the same as, the homozygote containing two copies of the dominant allele. For example, both *TT* and *Tt* pea plants show a tall phenotype. In addition, another complicating factor is epistasis (described in Chapter 4), in which the alleles for one gene can mask the phenotypic expression of the alleles of another gene. To account for these differences, geneticists usually subdivide V_G into these three different genetic categories:

$$V_G = V_A + V_D + V_I$$

where

V_A is the variance due to the additive effects of alleles. A heterozygote shows a phenotype that is intermediate between the respective homozygotes.

V_D is the variance due to the effects of alleles that follow a dominant/recessive pattern of inheritance.

V_I is the variance due to the effects of alleles that interact in an epistatic manner.

In analyzing quantitative traits, geneticists may focus on V_A and neglect the contributions of V_D and V_I. They do this for scientific as well as practical reasons. For some quantitative traits, the additive effects of alleles may play a primary role in the phenotypic outcome. In addition, when the alleles behave additively, we can predict the outcomes of crosses based on the quantitative characteristics of the parents. The heritability of a trait due to the additive effects of alleles is called the **narrow-sense heritability:**

$$h_N^2 = V_A/V_T$$

For many quantitative traits, the value of V_A may be relatively large compared with V_D and V_I. In such cases, the determination of the narrow-sense heritability provides an estimate of the broad-sense heritability.

How can the narrow-sense heritability be determined? In this chapter, we will consider two common ways. As discussed later, one way to calculate the narrow-sense heritability involves selective breeding practices, which are done with agricultural species. A second common strategy to determine h_N^2 involves the measurement of a quantitative trait among groups of genetically related individuals. For example, agriculturally important traits, such as egg weight in poultry, can be analyzed in this way. To calculate the heritability, a researcher determines the observed egg weights between individuals whose genetic relationships are known, such as a mother and her female offspring. These data can then be used to compute a correlation between the parent and offspring, using the methods described earlier. The narrow-sense heritability is then calculated as

$$h_N^2 = r_{obs}/r_{exp}$$

where

r_{obs} is the observed phenotypic correlation between related individuals

r_{exp} is the expected correlation based on the known genetic relationship

In our example, r_{obs} is the observed phenotypic correlation between parent and offspring. In particular research studies, the observed phenotypic correlation for egg weights between mothers and daughters has been found to be about 0.25 (although this varies among strains). The expected correlation, r_{exp}, is based on the known genetic relationship. A parent and child share 50% of their genetic material, so r_{exp} equals 0.50. So,

$$\begin{aligned} h_N^2 &= r_{obs}/r_{exp} \\ &= 0.25/0.50 \\ &= 0.50 \end{aligned}$$

(Note: For siblings, r_{exp} = 0.50; for identical twins, r_{exp} = 1.0; and for an aunt-niece relationship, r_{exp} = 0.25.)

According to this calculation, about 50% of the phenotypic variation in egg weight is due to additive genetic variation; the other half is due to the environment.

When calculating heritabilities from correlation coefficients, keep in mind that this computation assumes that genetics and the environment are independent variables. However, this is not always the case. The environments of parents and offspring are often more similar to each other than they are to those of unrelated individuals. As mentioned earlier, there are several ways to minimize this confounding factor. First, in human studies, one may analyze the heritabilities from correlations between adopted children and their biological parents. Alternatively, one can examine a variety of relationships (aunt-niece, identical twins versus fraternal twins, and so on) and see if the heritability values are roughly the same in all cases. This approach was applied in the study that is described next.

INTERPRETING THE DATA

As seen in the data table, the results indicate that genetics plays the major role in explaining the variation in this trait. Genetically unrelated individuals (namely, parent-parent relationships) have a negligible correlation for this trait. By comparison, individuals who are genetically related have a substantially higher correlation. When the observed correlation coefficient is divided by the expected correlation coefficient based on the known genetic relationships, the average heritability value is 0.97, which is very close to 1.0.

What do these high heritability values mean? They indicate that nearly all of the variation in fingerprint pattern is due to genetic variation. Significantly, fraternal and identical twins have substantially different observed correlation coefficients, even though we expect that they have been raised in very similar environments. These results support the idea that genetics is playing the major role in promoting variation and that the results are not biased heavily by environmental similarities that may be associated with genetically related individuals. From an experimental viewpoint, the results show us how the determination of correlation coefficients between related and unrelated individuals can provide insight regarding the relative contributions of genetics and environment to the variation of a quantitative trait.

A self-help quiz involving this experiment can be found at www.mhhe.com/brookergenetics4e.

Heritability Values Are Relevant Only to Particular Groups Raised in a Particular Environment

Table 25.4 describes heritability values that have been calculated for traits in particular populations. Unfortunately, heritability is a widely misunderstood concept. Heritability describes the amount of phenotypic variation due to genetic variation for a particular population raised in a particular environment. The words *variation*, *particular population*, and *particular environment* cannot be overemphasized. For example, in one population of cattle, the heritability for milk production may be 0.35, whereas in another group (with less genetic variation), the heritability may be 0.1.

TABLE 25.4
Examples of Heritabilities for Quantitative Traits

Trait	Heritability Value*
Humans	
Stature	0.65
IQ testing ability	0.60
Cattle	
Body weight	0.65
Butterfat, %	0.40
Milk yield	0.35
Mice	
Tail length	0.40
Body weight	0.35
Litter size	0.20
Poultry	
Body weight	0.55
Egg weight	0.50
Egg production	0.10

*As emphasized in this chapter, these values apply to particular populations raised in particular environments. The value for IQ testing ability is an average value from many independent studies. The other values were taken from D. S. Falconer (1989). *Introduction to Quantitative Genetics*, 3rd ed. Longman, Essex, England.

Second, if a group displays a heritability of 1.0 for a particular trait, this does not mean that the environment is unimportant in affecting the outcome of the trait. A heritability value of 1.0 only means that the amount of variation within this group is due to genetics. Perhaps the group has been raised in a relatively homogeneous environment, so the environment has not caused a significant amount of variation. Nevertheless, the environment may be quite important. It just is not causing much variation within this particular group.

As a hypothetical example, let's suppose that we take a species of rodent and raise a group on a poor diet; we find their weights range from 1.5 to 2.5 pounds, with a mean weight of 2 pounds. We allow them to mate and then raise their offspring on a healthy diet of rodent chow. The weights of the offspring range from 2.5 to 3.5 pounds, with a mean weight of 3 pounds. In this hypothetical experiment, we might find a positive correlation in which the small parents tended to produce small offspring, and the large parents produce large offspring. The correlation of weights between parent and offspring might be, say, 0.5. In this case, the heritability for weight would be calculated as r_{obs}/r_{exp}, which equals 0.5/0.5, or 1.0. The value of 1.0 means that the total amount of phenotypic variation within this group is due to genetic variation among the individuals. The offspring vary from 2.5 to 3.5 pounds because of genetic variation, and also the parents range from 1.5 to 2.5 pounds because of genetics. However, as we see here, environment has played an important role. Presumably, the mean weight of the offspring is higher because of their better diet. This example is meant to emphasize the point that heritability tells us only the relative contributions of genetic variation and environment in influencing phenotypic *variation* in a *particular population* in a *particular environment*. Heritability does not describe the relative importance of these two factors in determining the outcomes of traits. When a heritability value is high, it does not mean that a change in the environment cannot have a major impact on the outcome of the trait.

With regard to the roles of genetics and environment (sometimes referred to as nature versus nurture), the topic of human intelligence has been hotly debated. As a trait, intelligence is difficult to define or to measure. Nevertheless, performance on an IQ test has been taken by some people as a reflection of intelligence ever since 1916 when Alfred Binet's test was used in the

United States. Even though such tests may have inherent bias and consider only a limited subset of human cognitive abilities, IQ tests still remain a method of assessing intelligence. By comparing IQ scores among related and unrelated individuals, various studies have attempted to estimate heritability values in selected human populations. These values have ranged from 0.3 to 0.8. A heritability value of around 0.6 is fairly common among many studies. Such a value indicates that over half of the heritability for IQ testing ability is due to genetic factors.

Let's consider what a value of 0.6 means, and what it does not mean. It means that 60% of the variation in IQ testing ability is due to genetic variation in a selected population raised in a particular environment. It does not mean that 60% of an individual's IQ testing ability is due to genetics and 40% is due to the environment. Heritability is meaningless at the level of a single individual. Furthermore, even at the population level, a heritability value of 0.6 does not mean that 60% of the IQ testing ability is due to genetics and 40% is due to the environment. Rather, it means that in the selected population that was examined, 60% of the <u>variation</u> in IQ testing ability is due to genetics, whereas 40% of the <u>variation</u> is due to the environment. Heritability is strictly a population value that pertains to variation.

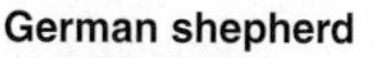

FIGURE 25.11 Some common breeds of dogs that have been obtained by selective breeding.

Genes → Traits By selecting parents carrying the alleles that influence certain quantitative traits in a desired way, dog breeders have produced breeds with distinctive sets of traits. For example, the bulldog has alleles that give it short legs and a flat face. By comparison, the corresponding genes in a German shepherd are found in alleles that produce longer legs and a more pointy snout. All the dogs shown in this figure carry the same kinds of genes (e.g., many genes that affect their sizes, shapes, and fur color). However, the alleles for many of these genes are different among these dogs, thereby producing breeds with strikingly different phenotypes.

Selective Breeding of Species Can Alter Quantitative Traits Dramatically

The term **selective breeding** refers to programs and procedures designed to modify phenotypes in species of economically important plants and animals. This phenomenon, also called **artificial selection,** is related to natural selection, discussed in Chapter 24. In forming his theory of natural selection, Charles Darwin was influenced by his observations of selective breeding by pigeon fanciers and other breeders. The primary difference between artificial and natural selection is how the parents are chosen. Natural selection is due to natural variation in reproductive success. In artificial selection, the breeder chooses individuals that possess traits that are desirable from a human perspective.

For centuries, humans have been practicing selective breeding to obtain domestic species with interesting or agriculturally useful characteristics. The common breeds of dogs and cats have been obtained by selective breeding strategies (**Figure 25.11**). As shown here, it is very striking how selective breeding can modify the quantitative traits in a species. When comparing a greyhound with a bulldog, the magnitude of the differences is fairly amazing. They hardly look like members of the same species. Recent work in 2007 by Nathan Sutter and colleagues indicates that the size of dogs is often determined by alleles in the *Igf1* gene that encodes a growth hormone called insulin-like growth factor 1. A particular allele of this gene was found to be common to all small breeds of dogs and nearly absent from very large breeds, suggesting that this allele is a major contributor to body size in small breeds of dogs.

Likewise, most of the food we eat is obtained from species that have been modified profoundly by selective breeding strategies. This includes products such as grains, fruits, vegetables, meat, milk, and juices. **Figure 25.12** illustrates how certain characteristics in the wild mustard plant (*Brassica oleracea*) have been modified by selective breeding to create several varieties of important domesticated crops. This plant is native to Europe and Asia, and plant breeders began to modify its traits approximately 4000 years ago. As seen here, certain quantitative traits in the domestic strains, such as stems and lateral buds, differ considerably from those of the original wild species.

The phenomenon that underlies selective breeding is variation. Within a group of individuals, allelic variation may affect the outcome of quantitative traits. The fundamental strategy of the selective breeder is to choose parents that will pass on alleles to their offspring that produce desirable phenotypic characteristics. For example, if a breeder wants large cattle, the largest members of the herd are chosen as parents for the next generation. These large cattle will transmit an array of alleles to their offspring that confer large size. The breeder often chooses genetically related individuals (e.g., brothers and sisters) as the parental stock. As mentioned previously, the practice of mating between genetically related individuals is known as inbreeding. Some of the consequences of inbreeding are also described in Chapter 24.

What is the outcome when selective breeding is conducted for a quantitative trait? **Figure 25.13a** shows the results of a program begun at the Illinois Agricultural Experiment Station in 1896, even before the rediscovery of Mendel's laws. This experiment began with 163 ears of corn with an oil content ranging from 4 to 6%. In each of 80 succeeding generations, corn plants

Wild mustard plant

Strain	Modified trait
Kohlrabi	Stem
Kale	Leaves
Broccoli	Flower buds and stem
Brussels sprouts	Lateral leaf buds
Cabbage	Terminal leaf bud
Cauliflower	Flower buds

FIGURE 25.12 Crop plants developed by selective breeding of the wild mustard plant (*Brassica oleracea*).

Genes → Traits The wild mustard plant carries a large amount of genetic (i.e., allelic) variation, which was used by plant breeders to produce modern strains that are agriculturally desirable and economically important. For example, by selecting for alleles that promote the formation of large lateral leaf buds, the strain of Brussels sprouts was created. By selecting for alleles that alter the leaf morphology, kale was developed. Although these six agricultural plants look quite different from each other, they carry many of the same alleles as the wild mustard. However, they differ in alleles affecting the formation of stems, leaves, flower buds, and leaf buds.

were divided into two separate groups. In one group, members with the highest oil content were chosen as parents of the next generation. In the other group, members with the lowest oil content were chosen. After 80 generations, the oil content in the first group rose to over 18%; in the other group, it dropped to less than 1%. These results show that selective breeding can modify quantitative traits in a very directed manner.

Similar results have been obtained for many other quantitative traits. **Figure 25.13b** shows an experiment by Kenneth Mather conducted in the 1940s, in which flies were selected on the basis of their bristle number. The starting group had an average of 40 bristles for females and 35 bristles for males. After eight generations, the group selected for high bristle number had an average of 46 bristles for females and 40 for males, whereas the group selected for low bristle number had an average of 36 bristles for females and 30 for males.

When comparing the curves in Figure 25.13, keep in mind that quantitative traits are often at an intermediate value in unselected populations. Therefore, artificial selection can increase or decrease the magnitude of the trait. Oil content can go up or down, and bristle number can increase or decrease. Artificial selection tends to be the most rapid and effective in changing the frequency of alleles that are at intermediate range in a starting population, such as 0.2 to 0.8.

Figure 25.13 also shows the phenomenon known as a **selection limit**—after several generations a plateau is reached where artificial selection is no longer effective. A selection limit may occur for two reasons. Presumably, the starting population possesses a large amount of genetic variation, which contributes to the diversity in phenotypes. By carefully choosing the parents, each succeeding generation has a higher proportion of the desirable alleles. However, after many generations, the population may be nearly monomorphic for all or most of the desirable alleles that affect the trait of interest. At this point, additional selective breeding will have no effect. When this occurs, the heritability for the trait is near zero, because nearly all genetic variation for the trait of interest has been eliminated from the population. Without the introduction of new mutations into the population, further selection is not possible. A second reason for a selection limit is related to fitness. Some alleles that accumulate in a population due to artificial selection may have a negative influence on the population's overall fitness. A selection limit is reached in which the desired effects of artificial selection are balanced by the negative effects on fitness.

Using artificial selection experiments, the response to selection is a common way to estimate the narrow-sense heritability in

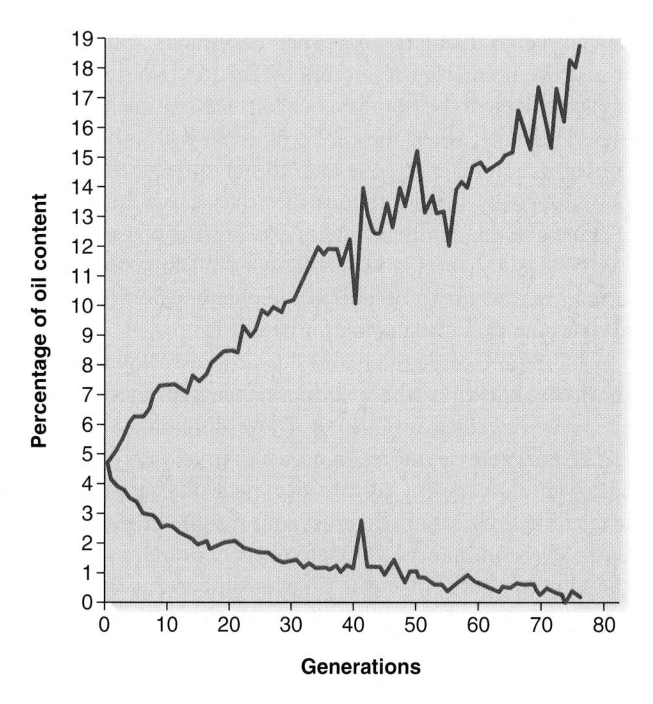

(a) Results of selective breeding for high and low oil content in corn

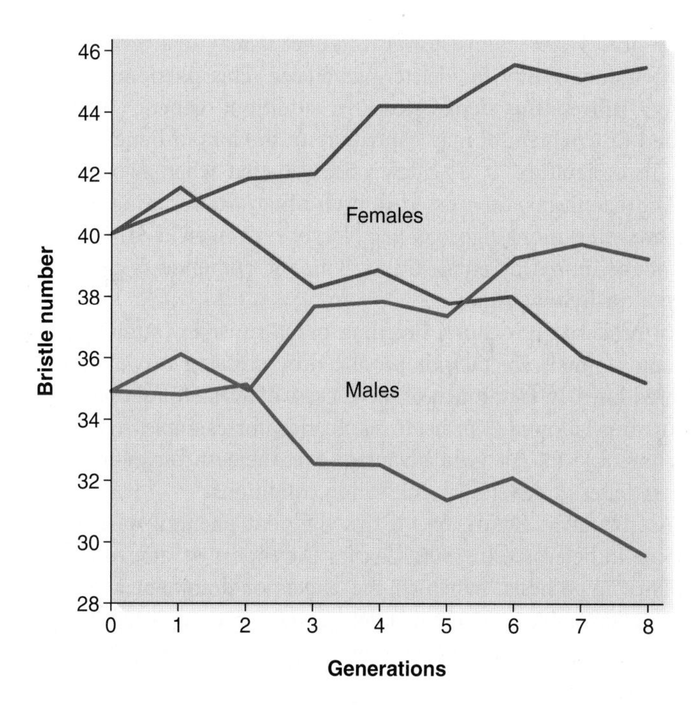

(b) Results of selective breeding for high and low bristle number in flies

FIGURE 25.13 **Common results of selective breeding for a quantitative trait.**

a starting population. The narrow-sense heritability measured in this way is also called the **realized heritability.** It is calculated as

$$h_N^2 = \frac{R}{S}$$

where

- R is the response in the offspring to selection, or the difference between the mean of the offspring and the mean of the population of the starting population.
- S is the selection differential in the parents, or the difference between the mean of the parents and the mean of the starting population.

Here,

$$R = \overline{X}_O - \overline{X}$$

$$S = \overline{X}_P - \overline{X}$$

where

$\overline{X}$ is the mean of the starting population
$\overline{X}_O$ is the mean of the offspring
$\overline{X}_P$ is the mean of the parents

So,

$$h_N^2 = \frac{\overline{X}_O - \overline{X}}{\overline{X}_P - \overline{X}}$$

The narrow-sense heritability is the proportion of the variance in phenotype that can be used to predict changes in the population mean when selection is practiced.

As an example, let's suppose we began with a population of fruit flies in which the average bristle number for both sexes was 37.5. The parents chosen from this population had an average bristle number of 40. The offspring of the next generation had an average bristle number of 38.7. With these values, the realized heritability is

$$h_N^2 = \frac{38.7 - 37.5}{40 - 37.5}$$

$$h_N^2 = \frac{1.2}{2.5}$$

$$h_N^2 = 0.48$$

This result tells us that about 48% of the phenotypic variation is due to the additive effects of alleles.

An important aspect of narrow-sense heritabilities is their ability to predict the outcome of selective breeding. In this case, the goal is to predict the mean phenotypes of offspring. If we rearrange our realized heritability equation

$$R = h_N^2 S$$

$$\overline{X}_O - \overline{X} = h_N^2 (\overline{X}_P - \overline{X})$$

This equation is referred to the breeder's equation, because it is used to calculate the mean phenotypes of offspring based on the mean weights of the parents, the mean weights of the starting population, and the heritability. Solved problem S1 at the end of the chapter illustrates the use of this equation.

Heterosis May Be Explained by Dominance or Overdominance

As we have just seen, selective breeding can alter the phenotypes of domesticated species in a highly directed way. An unfortunate consequence of inbreeding, however, is that it may inadvertently promote homozygosity for deleterious alleles. This phenomenon is called **inbreeding depression.** In addition, genetic drift, described in Chapter 24, may contribute to the loss of beneficial alleles. In agriculture, it is widely observed that when two different inbred strains are crossed to each other, the resulting offspring are often more vigorous (e.g., larger or longer-lived) than either of the inbred parental strains. This phenomenon is called **heterosis,** or **hybrid vigor.**

In modern agricultural breeding practices, many strains of plants and animals are hybrids produced by crossing two different inbred lines. In fact, much of the success of agricultural breeding programs is founded in heterosis. In rice, for example, hybrid strains have a 15 to 20% yield advantage over the best conventional inbred varieties under similar cultivation conditions.

As shown in **Figure 25.14**, two different phenomena may contribute to heterosis. In 1908, Charles Davenport developed the dominance hypothesis, in which the effects of dominant alleles explain the favorable outcome in a heterozygote. He suggested that highly inbred strains have become homozygous for one or more

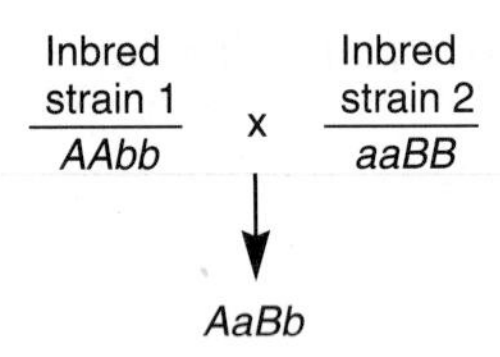

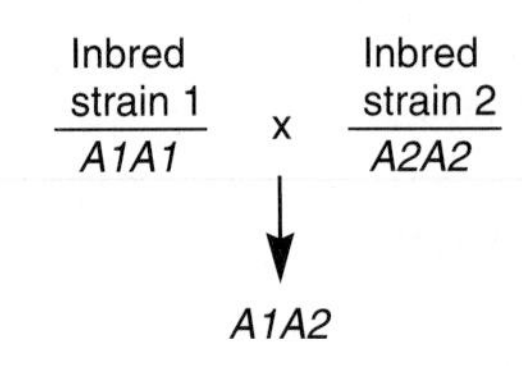

The recessive alleles (*a* and *b*) are slightly harmful in the homozygous condition.

The hybrid offspring is more vigorous, because the harmful effects of the recessive alleles are masked by the dominant alleles.

The Dominance Hypothesis

Neither the *A1* nor *A2* allele is recessive.

The hybrid offspring is more vigorous, because the heterozygous combination of alleles exhibits overdominance. This means that the *A1A2* heterozygote is more vigorous than either the *A1A1* or *A2A2* homozygote.

The Overdominance Hypothesis

FIGURE 25.14 Mechanisms to explain heterosis. The two common explanations are the dominance hypothesis and the overdominance hypothesis.

recessive genes that are somewhat deleterious (but not lethal). Because the homozygosity occurs by chance, two different inbred strains are likely to be homozygous for recessive alleles in different genes. Therefore, when they are crossed to each other, the resulting hybrids are heterozygous and do not suffer the consequences of homozygosity for deleterious recessive alleles. In other words, the benefit of the dominant alleles explains the observed heterosis. Steven Tanksley, working with colleagues in China, found that heterosis in rice seems to be due to the phenomenon of dominance. This is a common explanation for heterosis.

In 1908, George Shull and Edward East proposed a second hypothesis, known as the overdominance hypothesis (see Figure 25.14). As described in Chapter 4, overdominance occurs when the heterozygote is more vigorous than either corresponding homozygote. According to this idea, heterosis can occur because the resulting hybrids are heterozygous for one or more genes that display overdominance. The heterozygote is more vigorous than either homozygote. In corn, Charles Stuber and his colleagues have found that several QTLs for grain yield support the overdominance hypothesis.

Finally, it should be pointed out that overdominance is very difficult to distinguish from **pseudo-overdominance,** a phenomenon initially suggested by James Crow. Pseudo-overdominance is really the same as dominance, except that the chromosomal region contains two or more genes that are very closely linked. For example, a QTL may be identified in a mapping experiment to be close to a particular molecular marker. However, a QTL could contain two genes, both affecting the same quantitative trait. For example, at a single QTL, the alleles of two genes, *a* and *B*, may be closely linked in one strain, whereas *A* and *b* are closely linked in another strain. The hybrid is really heterozygous (*AaBb*) for two different genes, but this may be difficult to discern in mapping experiments because the genes are so close together. If a researcher assumed there was only one gene at a QTL, the overdominance hypothesis would be favored, whereas if two genes were actually present, the dominance hypothesis would be correct. Therefore, without very fine mapping, which is rarely done for QTLs, it is hard to distinguish between overdominance and pseudo-overdominance. However, if the genome sequence is available, the identification of candidate genes in the mapped region may be able to discern between overdominance and pseudo-overdominance.

KEY TERMS

Page 700. complex traits, quantitative traits, quantitative genetics
Page 701. biometric field, discontinuous traits, frequency distribution, normal distribution, mean
Page 702. variance
Page 703. standard deviation (SD), covariance
Page 704. correlation coefficient (*r*)
Page 705. regression analysis, polygenic inheritance, quantitative trait locus (QTL)
Page 708. genetic markers
Page 711. QTL mapping, heritability
Page 713. inbreeding, monomorphic
Page 714. genotype-environment interaction, genotype-environment association
Page 715. broad-sense heritability, narrow-sense heritability
Page 719. selective breeding, artificial selection
Page 720. selection limit
Page 721. realized heritability
Page 722. inbreeding depression, heterosis, hybrid vigor, pseudo-overdominance
Page 724. regression coefficient

CHAPTER SUMMARY

- Quantitative genetics is the field of genetics concerned with complex and quantitative traits.

25.1 Quantitative Traits

- Quantitative traits can be categorized as anatomical, physiological, or behavioral (see Table 25.1).
- Quantitative traits often exhibit a continuum and may follow a normal distribution (see Figure 25.1).
- Statistical methods are used to analyze quantitative traits. These include the mean, variance, standard deviation, covariance, and correlation (see Figure 25.2, Table 25.2).

25.2 Polygenic Inheritance

- Polygenic inheritance refers to an inheritance pattern in which multiple genes affect a single trait.
- The locations along a chromosome that contain genes affecting a quantitative trait are called quantitative trait loci (QTLs).
- Polygenic inheritance and environmental factors may cause a quantitative trait to fall along a continuum (see Figures 25.3, 25.4).
- By following the transmission of genetic markers, Crow determined that DDT resistance in fruit flies is explained by polygenic inheritance (see Figures 25.5, 25.6).
- Quantitative trait loci are mapped by their linkage to molecular markers (see Figure 25.7).

25.3 Heritability

- Heritability is the amount of phenotypic variation within a group of individuals that is due to genetic variation.
- Genetic variance and environmental variance may contribute additively to phenotypic variance.
- Genetic variance and environmental variance may exhibit interactions and associations (see Figure 25.8, Table 25.3).
- Broad-sense heritability refers to all genetic factors affecting heritability, which includes the additive effects of alleles, effects due to dominant/recessive relationships, and effects due to epistatic interactions.
- Narrow-sense heritability is heritability that is due to the additive effects of alleles.
- Holt determined that dermal ridge count has a very high heritability value in humans (See Figures 25.9, 25.10).
- Heritability values refer only to particular groups raised in a particular environment (see Table 25.4).
- Selective breeding refers to programs and procedures designed to modify phenotypes in commercially important plants and animals (see Figures 25.11, 25.12).
- When starting with a genetically diverse population, selective breeding can usually modify a trait in different directions until a selection limit is reached (see Figure 25.13).
- Heterosis is the phenomenon in which the crossing of different inbred strains produces hybrids that are more vigorous than the inbred strains. This may be due to dominance or overdominance (see Figure 25.14).

PROBLEM SETS & INSIGHTS

Solved Problems

S1. The narrow-sense heritability for potato weight in a starting population of potatoes is 0.42, and the mean weight is 1.4 lb. If a breeder crosses two strains with average potato weights of 1.9 and 2.1 lb, respectively, what is the predicted average weight of potatoes in the offspring?

Answer: The mean weight of the parental strains is 2.0 lb. To solve for the mean weight of the offspring:

$$R = h_N^2 S$$

$$\overline{X}_O - \overline{X} = h_N^2 (\overline{X}_P - \overline{X})$$

$$\overline{X}_O - 1.4 = 0.42\,(2.0 - 1.4)$$

$$\overline{X}_O = 1.65 \text{ lb}$$

S2. A farmer wants to increase the average body weight in a herd of cattle. She begins with a herd having a mean weight of 595 kg and chooses individuals to breed that have a mean weight of 625 kg. Twenty offspring were obtained, having the following weights in kilograms: 612, 587, 604, 589, 615, 641, 575, 611, 610, 598, 589, 620, 617, 577, 609, 633, 588, 599, 601, and 611. Calculate the realized heritability for body weight in this herd.

Answer:

$$h_N^2 = \frac{R}{S} = \frac{\overline{X}_O - \overline{X}}{\overline{X}_P - \overline{X}}$$

We already know the mean weight of the starting herd (595 kg) and the mean weight of the parents (625 kg). The only calculation missing is the mean weight of the offspring, $\overline{X}_O$.

$$\overline{X}_O = \frac{\text{Sum of the offsprings' weights}}{\text{Number of offspring}}$$

$$\overline{X}_O = 604 \text{ kg}$$

$$\overline{X}_O = \frac{604 - 595}{625 - 595} = 0.3$$

S3. The following data describe the 6-week weights (in grams) of mice and their offspring of the same sex:

Parent (g)	Offspring (g)
24	26
21	24
24	22
27	25
23	21
25	26
22	24
25	24
22	24
27	24

Calculate the correlation coefficient.

Answer: To calculate the correlation coefficient, we first need to calculate the means and standard deviations for each group:

$$\overline{X}_{parents} = \frac{24 + 21 + 24 + 27 + 23 + 25 + 22 + 25 + 22 + 27}{10} = 24$$

$$\overline{X}_{offspring} = \frac{26 + 24 + 22 + 25 + 21 + 26 + 24 + 24 + 24 + 24}{10} = 24$$

$$SD_{parents} = \sqrt{\frac{0 + 9 + 0 + 9 + 1 + 1 + 4 + 1 + 4 + 9}{9}} = 2.1$$

$$SD_{offspring} = \sqrt{\frac{4 + 0 + 4 + 1 + 9 + 4 + 0 + 0 + 0 + 0}{9}} = 1.6$$

Next, we need to calculate the covariance.

$$C_O V_{(parents,\ offspring)} = \frac{\Sigma\left[(X_P - \overline{X}_P)(X_O - \overline{X}_O)\right]}{N - 1}$$

$$= \frac{0 + 0 + 0 + 3 + 3 + 2 + 0 + 0 + 0 + 0}{9}$$

$$= 0.9$$

Finally, we calculate the correlation coefficient:

$$V_{(parents,\ offspring)} = \frac{C_O V_{(P,\ O)}}{SD_P SD_O}$$

$$V_{(parents,\ offspring)} = \frac{0.9}{(2.1)(1.6)}$$

$$r_{(parent,\ offspring)} = 0.27$$

S4. As described in this chapter, the correlation coefficient provides a way to determine the strength of association between two variables. When the variables are related due to cause and effect (i.e., one variable affects the outcome of another), researchers may use a regression analysis to predict how much one variable changes in response to the other. This is easier to understand if we plot the data for two variables. The graph shown here compares mothers' and offsprings' body weights in cattle. The line running through the data points is called a regression line. It is a line that minimizes the squared vertical distances to all of the points.

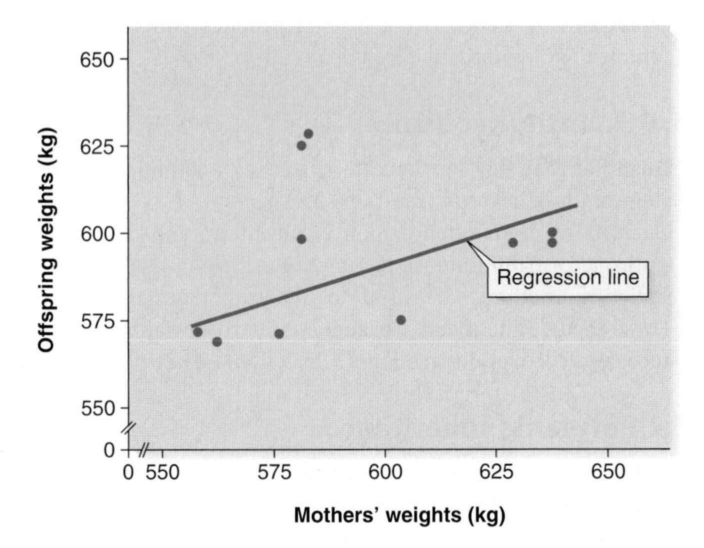

For many types of data, particularly those involving quantitative traits, the regression line can be represented by the equation

$$Y = bX + a$$

where

b is the regression coefficient

a is a constant

In this example, X is the value of a mother's weight, and Y is the value of its offspring's weight. The value of b, known as the **regression coefficient,** represents the slope of the regression line. The value of a is the y-intercept (i.e., the value of Y when X equals zero). The equation shown is very useful because it allows us to predict the value of Y at any given value of X, and vice versa. To do so, we first need to determine the values of b and a. This can be accomplished in the following manner:

$$b = \frac{C_O V_{(X,Y)}}{V_X}$$

$$a = \overline{Y} - b\overline{X}$$

Once the values of a and b have been computed, we can use them to predict the values of Y or X by using the equation

$$Y = bX - a$$

For example, if $b = 0.5$, $a = 2$, and $X = 58$, this equation can be used to compute a value of Y that equals 31. It is important to keep in mind that this equation predicts the average value of Y. As we see in the preceding figure, the data points tend to be scattered around the regression line. Deviations may occur between the data points and the line. The equation predicts the values that are the most likely to occur. In an actual experiment, however, some deviation will occur between the predicted values and the experimental values due to random sampling error.

Now here is the question. Using the data found in this chapter regarding weight in cattle, what is the predicted weight of an offspring if its mother weighed 660 lb?

Answer: We first need to calculate a and b.

$$b = \frac{C_O V_{(X,Y)}}{V_X}$$

We need to use the data on page 704 to calculate V_X, which is the variance for the mothers' weights. The variance equals 445.1. The covariance is already calculated on page 704; it equals 152.6.

$$b = \frac{152.6}{445.1}$$

$$b = 0.34$$

$$a = \overline{Y} - b\overline{X}$$

$$a = 598 - (0.34)(596) = 395.4$$

Now we are ready to calculate the predicted weight of the offspring using the equation

$$Y = bX + a$$

In this problem, $X = 660$ pounds.

$$Y = 0.34(660) + 395.4$$

$$Y = 619.8 \text{ lb}$$

The average weight of the offspring is predicted to be 619.8 lb.

S5. Genetic variance can be used to estimate the number of genes affecting a quantitative trait by using the following equation:

$$n = \frac{D^2}{8V_G}$$

where

n is the number of genes affecting the trait

D is the difference between the mean values of the trait in two strains that have allelic differences at every gene that influences the trait

V_G is the genetic variance for the trait; it is calculated using data from both strains

For this method to be valid, several assumptions must be met. In particular, the alleles of each gene must be additive, each gene must contribute equally to the trait, all of the genes must assort independently, and the two strains must be homozygous for alternative alleles of each gene. For example, if three genes affecting a quantitative trait exist in two alleles each, one strain could be *AA bb CC* and the other would be *aa BB cc*. In addition, the strains must be raised under the same environmental conditions. Unfortunately, these assumptions are not typically met with regard to most quantitative traits. Even so, when one or more assumptions are invalid, the calculated value of n is smaller than the actual number. Therefore, this calculation can be used to estimate the minimum number of genes that affect a quantitative trait.

Now here is the question. The average bristle number in two strains of flies was 35 and 42. The genetic variance for bristle number calculated for both strains was 0.8. What is the minimum number of genes that affect bristle number?

Answer: We apply the equation described previously.

$$n = \frac{D^2}{8V_G}$$

$$n = \frac{(35 - 42)^2}{8(0.8)}$$

$$n = 7.7 \text{ genes}$$

Because genes must come in whole numbers and because this calculation is a minimum estimate, we conclude that at least eight genes affect bristle number.

S6. Are the following statements regarding heritability true or false?

A. Heritability applies to a specific population raised in a particular environment.

B. Heritability in the narrow sense takes into account all types of genetic variance.

C. Heritability is a measure of the amount that genetics contributes to the outcome of a trait.

Answer:

A. True

B. False. Narrow-sense heritability considers only the effects of additive alleles.

C. False. Heritability is a measure of the amount of phenotypic variation that is due to genetic variation; it applies to the variation of a specific population raised in a particular environment.

Conceptual Questions

C1. Give several examples of quantitative traits. How are these quantitative traits described within groups of individuals?

C2. At the molecular level, explain why quantitative traits often exhibit a continuum of phenotypes within a population. How does the environment help produce this continuum?

C3. What is a normal distribution? Discuss this curve with regard to quantitative traits within a population. What is the relationship between the standard deviation and the normal distribution?

C4. Explain the difference between a continuous trait and a discontinuous trait. Give two examples of each. Are quantitative traits likely to be continuous or discontinuous? Explain why.

C5. What is a frequency distribution? Explain how the graph is made for a quantitative trait that is continuous.

C6. The variance for weight in a particular herd of cattle is 484 lb^2. The mean weight is 562 lb. How heavy would an animal have to be if it was in the top 2.5% of the herd? The bottom 0.13%?

C7. Two different varieties of potatoes both have the same mean weight of 1.5 lb. One group has a very low variance, and the other has a much higher variance.

A. Discuss the possible reasons for the differences in variance.

B. If you were a potato farmer, would you rather raise a variety with a low or high variance? Explain your answer from a practical point of view.

C. If you were a potato breeder and you wanted to develop potatoes with a heavier weight, would you choose the variety with a low or high variance? Explain your answer.

C8. If an r value equals 0.5 and $N = 4$, would you conclude a positive correlation is found between the two variables? Explain your answer. What if $N = 500$?

C9. What does it mean when a correlation coefficient is negative? Can you think of examples?

C10. When a correlation coefficient is statistically significant, what do you conclude about the two variables? What do the results mean with regard to cause and effect?

C11. What is polygenic inheritance? Discuss the issues that make polygenic inheritance difficult to study.

C12. What is a quantitative trait locus (QTL)? Does a QTL contain one gene or multiple genes? What technique is commonly used to identify QTLs?

C13. Let's suppose that weight in a species of mammal is polygenic, and each gene exists as a heavy and light allele. If the allele frequencies in the population were equal for both types of allele (i.e., 50% heavy alleles and 50% light alleles), what percentage of individuals would be homozygous for the light alleles at all of the genes affecting this trait, if the trait was determined by the following number of genes?

A. Two

B. Three

C. Four

C14. The broad-sense heritability for a trait equals 1.0. In your own words, explain what this value means. Would you conclude that the environment is unimportant in the outcome of this trait? Explain your answer.

C15. Compare and contrast the dominance and overdominance hypotheses. Based on your knowledge of mutations and genetics, which do you think tends to be the more common explanation for heterosis?

C16. What is hybrid vigor (also known as heterosis)? Give examples that you might find in a vegetable garden.

C17. From an agricultural point of view, discuss the advantages and disadvantages of selective breeding. It is common for plant breeders to take two different, highly inbred strains, which are the product of many generations of selective breeding, and cross them to make hybrids. How does this approach overcome some of the disadvantages of selective breeding?

C18. Many beautiful varieties of roses have been produced, particularly in the last few decades. These newer varieties often have very striking and showy flowers, making them desirable as horticultural specimens. However, breeders and novices alike have noticed that some of these newer varieties do not have very fragrant flowers compared with the older, more traditional varieties. From a genetic point of view, suggest an explanation why some of these newer varieties with superb flowers are not as fragrant.

C19. In your own words, explain the meaning of the term heritability. Why is a heritability value valid only for a particular population of individuals raised in a particular environment?

C20. What is the difference between broad-sense heritability and narrow-sense heritability? Why is narrow-sense heritability such a useful concept in the field of agricultural genetics?

C21. The heritability for egg weight in a group of chickens on a farm in Maine is 0.95. Are the following statements regarding heritability true or false? If a statement is false, explain why.

A. The environment in Maine has very little effect on the outcome of this trait.

B. Nearly all of the phenotypic variation for this trait in this group of chickens is due to genetic variation.

C. The trait is polygenic and likely to involve a large number of genes.

D. Based on the observation of the heritability in the Maine chickens, it is reasonable to conclude that the heritability for egg weight in a group of chickens on a farm in Montana is also very high.

C22. In a fairly large population of people living in a commune in the southern United States, everyone cares about good nutrition. All of the members of this population eat very nutritious foods, and their diets are very similar to each other. How do you think the height of individuals in this commune population would compare with that of the general population in the following categories?

A. Mean height

B. Heritability for height

C. Genetic variation for alleles that affect height

C23. When artificial selection is practiced over many generations, it is common for the trait to reach a plateau in which further selection has little effect on the outcome of the trait. This phenomenon is illustrated in Figure 25.13. Explain why.

C24. Discuss whether a natural population of wolves or a domesticated population of German shepherds is more likely to have a higher heritability for the trait of size.

C25. With regard to heterosis, would the following statements be consistent with the dominance hypothesis, the overdominance hypothesis, or both?

A. Strains that have been highly inbred have become monomorphic for one or more recessive alleles that are somewhat detrimental to the organism.

B. Hybrid vigor occurs because highly inbred strains are monomorphic for many genes, whereas hybrids are more likely to be heterozygous for those same genes.

C. If a gene exists in two alleles, hybrids are more vigorous because heterozygosity for the gene is more beneficial than homozygosity of either allele.

Experimental Questions

E1. Here are data for height and weight among 10 male college students.

Height (cm)	Weight (kg)
159	48
162	50
161	52
175	60
174	64
198	81
172	58
180	74
161	50
173	54

A. Calculate the correlation coefficients for this group.

B. Is the correlation coefficient statistically significant? Explain.

E2. The abdomen length (in millimeters) was measured in 15 male *Drosophila,* and the following data were obtained: 1.9, 2.4, 2.1, 2.0, 2.2, 2.4, 1.7, 1.8, 2.0, 2.0, 2.3, 2.1, 1.6, 2.3, and 2.2. Calculate the mean, standard deviation, and variance for this population of male fruit flies.

E3. You need to understand solved problem S5 before answering this question. The average weights for two varieties of cattle were 514 kg and 621 kg. The genetic variance for weight calculated for both strains was 382 kg^2. What is the minimum number of genes that affect weight variation in these two varieties of cattle?

E4. Using the same strategy as the experiment of Figure 25.6, the following data are the survival of F_2 offspring obtained from backcrosses to insecticide-resistant and control strains:

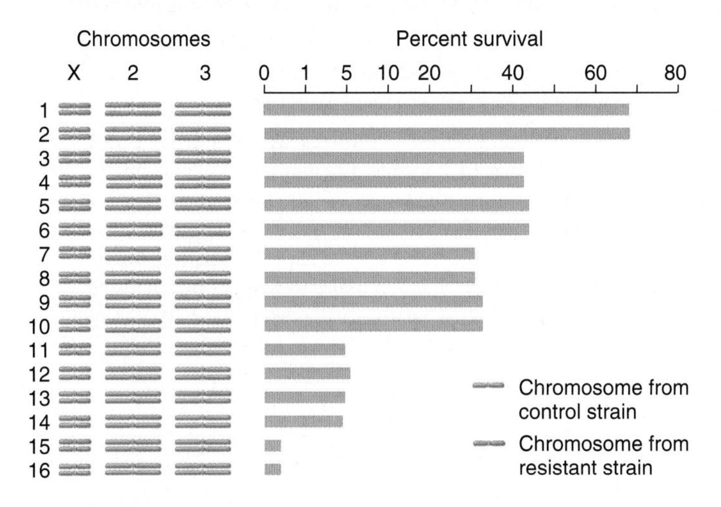

Interpret these results with regard to the locations of QTLs.

E5. In one strain of cabbage, you conduct an RFLP analysis of head weight; you determine that seven QTLs affect this trait. In another strain of cabbage, you find that only four QTLs affect this trait. Note that both strains of cabbage are from the same species, although they may have been subjected to different degrees of inbreeding. Explain how one strain can have seven QTLs and another strain four QTLs for exactly the same trait. Is the second strain missing three genes?

E6. From an experimental viewpoint, what does it mean to say that an RFLP is associated with a trait? Let's suppose that two strains of pea plants differ in two RFLPs that are linked to two genes governing pea size. RFLP-1 is found in 2000-bp and 2700-bp bands, and RFLP-2 is found in 3000-bp and 4000-bp bands. The plants producing large peas have RFLP-1 (2000 bp) and RFLP-2 (3000 bp); those producing small peas have RFLP-1 (2700 bp) and RFLP-2 (4000 bp). A cross is made between these two strains, and the F_1 offspring are allowed to self-fertilize. Five phenotypic classes are observed: small peas, small-medium peas, medium peas, medium-large peas, and large peas. We assume that each of the two genes makes an equal contribution to pea size and that the genetic variance is additive. Draw a gel and explain what RFLP banding patterns you would expect to observe for these five phenotypic categories. Note: Certain phenotypic categories may have more than one possible banding pattern.

E7. Let's suppose that two strains of pigs differ in 500 RFLPs. One strain is much larger than the other. The pigs are crossed to each other, and the members of the F_1 generation are also crossed among themselves to produce an F_2 generation. Three distinct RFLPs are associated with F_2 pigs that are larger. How would you interpret these results?

E8. Outline the steps you would follow to determine the number of genes that influence the yield of rice. Describe the results you might get if rice yield is governed by variation in six different genes.

E9. A researcher has two highly inbred strains of mice. One strain is susceptible to infection by a mouse leukemia virus, whereas the other strain is resistant. Susceptibility/resistance is a polygenic trait. The two strains were crossed together, and all of the F_1 mice were resistant. The F_1 mice were then allowed to interbreed, and 120 F_2 mice were obtained. Among these 120 mice, 118 were resistant to the viral pathogen, and 2 were sensitive. Discuss how many different genes may be involved in this trait. How would your answer differ if none of the F_2 mice had been susceptible to the leukemia virus? Hint: You should assume that the inheritance of one viral-resistance allele is sufficient to confer resistance.

E10. In a wild strain of tomato plants, the phenotypic variance for tomato weight is 3.2 g^2. In another strain of highly inbred tomatoes raised under the same environmental conditions, the phenotypic variance is 2.2 g^2. With regard to the wild strain,

A. Estimate V_G.

B. What is h_B^2?

C. Assuming that all of the genetic variance is additive, what is h_N^2?

E11. The average thorax length in a *Drosophila* population is 1.01 mm. You want to practice selective breeding to make larger *Drosophila.* To do so, you choose 10 parents (5 males and 5 females) of the following sizes: 0.97, 0.99, 1.05, 1.06, 1.03, 1.21, 1.22, 1.17, 1.19, and 1.20. You mate them and then analyze the thorax sizes of 30 offspring (half male and half female):

0.99, 1.15, 1.20, 1.33, 1.07, 1.11, 1.21, 0.94, 1.07, 1.11, 1.20, 1.01, 1.02, 1.05, 1.21, 1.22, 1.03, 0.99, 1.20, 1.10, 0.91, 0.94, 1.13, 1.14, 1.20, 0.89, 1.10, 1.04, 1.01, 1.26

Calculate the realized heritability in this group of flies.

E12. In a strain of mice, the average 6-week body weight is 25 g and the narrow-sense heritability for this trait is 0.21.

A. What would be the average weight of the offspring if parents with a mean weight of 27 g were chosen?

B. What weight of parents would you have to choose to obtain offspring with an average weight of 26.5 g?

E13. Two tomato strains, A and B, both produce fruit that weighs, on average, 1 lb each. All of the variance is due to V_G. When these two strains are crossed to each other, the F_1 offspring display heterosis with regard to fruit weight, with an average weight of 2 lb. You take these F_1 offspring and backcross them to strain A. You then grow several plants from this cross and measure the weights of their fruit. What would be the expected results for each of the following scenarios?

A. Heterosis is due to a single overdominant gene.

B. Heterosis is due to two dominant genes, one in each strain.

C. Heterosis is due to two overdominant genes.

D. Heterosis is due to dominance of several genes each from strains A and B.

E14. You need to understand solved problem S4 before answering this question. The variance in height for fathers (in square inches) was 112, the variance for sons was 122, and the covariance was 144. The mean height for fathers was 68 in., and the mean height for sons was 69 in. If a father had a height of 70 in., what is the most probable height of his son?

E15. A danger in computing heritability values from studies involving genetically related individuals is the possibility that these individuals share more similar environments than do unrelated individuals. In the experiment of Figure 25.10, which data are the most compelling evidence that ridge count is not caused by genetically related individuals sharing common environments? Explain.

E16. A large, genetically heterogeneous group of tomato plants was used as the original breeding stock by two different breeders, named Mary and Hector. Each breeder was given 50 seeds and began an artificial selection strategy, much like the one described in Figure 25.13. The seeds were planted, and the breeders selected the 10 plants with the highest mean tomato weights as the breeding stock for the next generation. This process was repeated over the course of 12 growing seasons, and the following data were obtained:

	Mean Weight of Tomatoes (lb)	
Year	Mary's Tomatoes	Hector's Tomatoes
1	0.7	0.8
2	0.9	0.9
3	1.1	1.2
4	1.2	1.3
5	1.3	1.3
6	1.4	1.4
7	1.4	1.5
8	1.5	1.5
9	1.5	1.5
10	1.5	1.5
11	1.5	1.5
12	1.5	1.5

A. Explain these results.

B. Another tomato breeder, named Martin, got some seeds from Mary's and Hector's tomato strains (after 12 generations), grew the plants, and then crossed them to each other. The mean weight of the tomatoes in these hybrids was about 1.7 lb. For a period of 5 years, Martin subjected these hybrids to the same experimental strategy that Mary and Hector had followed, and he obtained the following results:

	Mean Weight of Tomatoes (lb)
Year	Martin's Tomatoes
1	1.7
2	1.8
3	1.9
4	2.0
5	2.0

Explain Martin's data. Is heterosis occurring? Why was Martin able to obtain tomatoes heavier than 1.5 lb, whereas Mary's and Hector's strains appeared to plateau at this weight?

E17. The correlations for height were determined for 15 pairs of individuals with the following genetic relationships:

Mother/daughter: 0.36

Mother/granddaughter: 0.17

Sister/sister: 0.39

Sister/sister (fraternal twins): 0.40

Sister/sister (identical twins): 0.77

What is the average heritability for height in this group of females?

E18. An animal breeder had a herd of sheep with a mean weight of 254 lb at 3 years of age. He chose animals with mean weights of 281 lb as parents for the next generation. When these offspring reached 3 years of age, their mean weights were 269 lb.

A. Calculate the narrow-sense heritability for weight in this herd.

B. Using the heritability value that you calculated in part A, what weight of animals would you have to choose to get offspring that weigh 275 lb (at 3 years of age)?

E19. The trait of blood pressure in humans has a frequency distribution that is similar to a normal distribution. The following graph (see next page) shows the ranges of blood pressures for a selected population of people. The red line depicts the frequency distribution of the systolic pressures for the entire population. Several individuals with high blood pressure were identified, and the blood pressures of their relatives were determined. This frequency distribution is depicted with a blue line. (Note: The blue line does not include the people who were identified with high blood pressure; it includes only their relatives.)

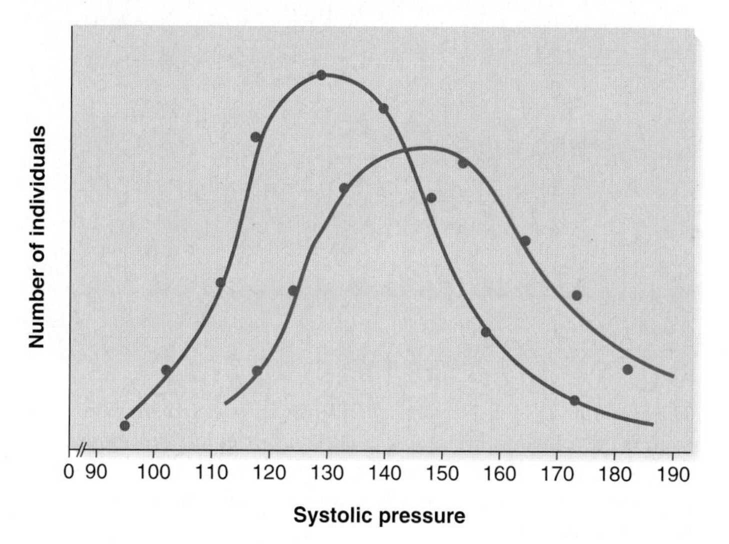

What do these data suggest with regard to a genetic basis for high blood pressure? What statistical approach could you use to determine the heritability for this trait?

Questions for Student Discussion/Collaboration

1. Discuss why heritability is an important phenomenon in agriculture. Discuss how it is misunderstood.
2. From a biological viewpoint, speculate as to why many traits seem to fit a normal distribution. Students with a strong background in math and statistics may want to explain how a normal distribution is generated, and what it means. Can you think of biological examples that do not fit a normal distribution?
3. What is heterosis? Discuss whether it is caused by a single gene or several genes. Discuss the two major hypotheses proposed to explain heterosis. Which do you think is more likely to be correct?

Note: All answers appear at the website for this textbook; the answers to even-numbered questions are in the back of the textbook.

www.mhhe.com/brookergenetics4e

Visit the website for practice tests, answer keys, and other learning aids for this chapter. Enhance your understanding of genetics with our interactive exercises, quizzes, animations, and much more.

CHAPTER OUTLINE

The evolution of eyes. *Developmental biologists have recently discovered that the eyes of many diverse species, including fruit flies, frogs, mice, and people, are under the control of the homologous gene called Pax6, suggesting that animal eyes may have originated once during the evolution of animals.*

26 EVOLUTIONARY GENETICS

Biological evolution, or simply **evolution,** is a heritable change in one or more characteristics of a population or species from one generation to the next. Evolution can be viewed on a small scale as it relates to a single gene or it can be viewed on a larger scale as it relates to the formation of new species. In Chapter 24, we examined several factors that cause allele frequencies to change in populations. This process, also known as **microevolution,** concerns the changing composition of gene pools with regard to particular alleles over measurable periods of time. As we have seen, several evolutionary mechanisms, such as mutation, genetic drift, migration, natural selection, and inbreeding, affect the allele and genotypic frequencies within natural populations. On a microevolutionary scale, evolution can be viewed as a change in allele frequency over time.

A goal of this chapter is to relate phenotypic changes that occur during evolution to the underlying genetic changes that cause them to happen. In the first part of the chapter, we will be concerned with evolution on a large scale, which leads to the origin of new species. The question of how species form has been central to the development of evolutionary theory. The term **macroevolution** refers to large-scale evolutionary changes that create new species and higher taxa. It concerns the establishment of the diversity of organisms over long periods of time through the accumulated evolution and extinction of many species.

In the second part of this chapter, we will link molecular genetics to the evolution of species. Techniques for analyzing chromosomes and DNA sequences have greatly enhanced our understanding of evolutionary processes at the molecular level. The term **molecular evolution** refers to patterns and processes associated with evolutionary change at the molecular level. Such changes may be phenotypically neutral or they may underlie the phenotypic changes associated with evolution. In this chapter, we examine how molecular data can provide information about the phylogenetic relationships among different organisms. Finally, as discussed in the last part of the chapter, we will see how genes that affect embryonic development can have a dramatic effect on the phenotypes of organisms. The field of evolutionary developmental biology (evo-devo) focuses on the role of developmental genes in the formation of traits that are important in the evolution of new species.

The topics of molecular evolution and evo-devo are a fitting way to end our discussion of genetics because they integrate the ongoing theme of this textbook—the relationship between molecular genetics and traits—in the broadest and most profound ways. Theodosius Dobzhansky, an influential evolutionary scientist, once said, "Nothing in biology makes sense except in the light of evolution." The extraordinarily diverse and seemingly bizarre array of species on our planet can be explained naturally within the context of evolution. An examination of molecular evolution allows us to make sense of the existence of these species at both the population and the molecular levels.

26.1 ORIGIN OF SPECIES

Charles Darwin, a British naturalist born in 1809, proposed the theory of evolution and provided evidence that existing species have evolved from preexisting ones. Like many great scientists, Darwin had a broad background in science, which enabled him to see connections among different disciplines. His thinking was influenced by the field of geology. According to Charles Lyell, the processes that alter the Earth are uniform through time. This view, which was known as uniformitarianism, suggested that the Earth is very old and that slow geological processes can lead eventually to substantial changes in the Earth's characteristics.

Darwin's own experimental observations also greatly affected his thinking. His famous voyage on the *HMS Beagle*, which lasted from 1832 to 1836, involved a careful examination of many different species. He observed the similarities among many discrete species, yet noted the differences that enabled them to be adapted to their environmental conditions. He was particularly struck by the distinctive adaptations of island species. For example, the finches found on the Galápagos Islands had unique phenotypic characteristics compared with those of similar finches found on the mainland.

A third important influence on Darwin was a paper published in 1798, "Essay on the Principle of Population," by Thomas Malthus, an English economist. Malthus asserted that the population size of humans can, at best, increase arithmetically due to increased land usage and improvements in agriculture, whereas the reproductive potential of humans can increase geometrically. He argued that famine, war, and disease work to limit population growth, especially among the poor.

With these three ideas in mind, Darwin had largely formulated his theory of evolution by natural selection by the mid-1840s. He then spent several years studying barnacles without having published his ideas. The geologist Charles Lyell, who had greatly influenced Darwin's thinking, strongly encouraged Darwin to publish his theory of evolution. In 1856, Darwin began to write a long book to explain his ideas. In 1858, however, Alfred Wallace, a naturalist working in the East Indies, sent Darwin an unpublished manuscript to read prior to its publication. In it, Wallace proposed the same ideas concerning evolution. Darwin therefore quickly excerpted some of his own writings on this subject, and two papers, one by Darwin and one by Wallace, were published in the *Proceedings of the Linnaean Society of London*. These papers were not widely recognized. A short time later, however, Darwin finished his book, *On the Origin of Species by Means of Natural Selection*, which expounded his ideas in greater detail and with experimental support. This book, which received high praise from many scientists and scorn from others, started a great debate concerning evolution. Although some of his ideas were incomplete because the genetic basis of traits was not understood at that time, Darwin's work represents one of the most important contributions to our understanding of biology.

Darwin called evolution "the theory of descent with modification through variation and natural selection." As its name suggests, evolution is based on two fundamental principles: genetic variation and natural selection. A modern interpretation of evolution can view these two principles at the species level (macroevolution) and at the level of genes in populations (microevolution).

1. *Genetic variation at the species level:* As we have seen in Chapter 24, genetic variation is a consistent feature of natural populations. Darwin observed that many species exhibit a great amount of phenotypic variation. Although the theory of evolution preceded Mendel's pioneering work in genetics, Darwin (as well as many other people before him) observed that offspring resemble their parents more than they do unrelated individuals. Therefore, he assumed that traits are passed from parent to offspring. However, the genetic basis for the inheritance of traits was not understood at that time.

   At the gene level: Genetic variation can involve allelic differences in genes. These differences are caused by random mutations. Alternative alleles may affect the functions of the proteins they encode, thereby affecting the phenotype of the organism. Likewise, changes in chromosome structure and number may affect gene expression, thereby influencing the phenotype of the individual.
2. *Natural selection at the species level:* Darwin agreed with Malthus that most species produce many more offspring than will survive and reproduce, resulting in an ever-present struggle for existence. Over the course of many generations, those individuals who happen to possess the most favorable traits will dominate the composition of the population. The result of natural selection is to make a species better adapted to its environment and more successful at reproduction.

   At the gene level: Some alleles encode proteins that provide the individual with a selective advantage. Over time, natural selection may change the allele frequencies of genes, thereby leading to the fixation of beneficial alleles and the elimination of detrimental alleles.

In this section, we will consider the different characteristics that biologists examine when deciding if two groups of organisms constitute different species and explore the features of evolution as it occurs in natural populations over time.

Each Species Is Established Using Characteristics and Histories That Distinguish It from Other Species

Before we begin to consider how biologists study the evolution of new species, we need to consider how species are defined and identified. A **species** refers to a group of organisms that maintains a distinctive set of attributes in nature. How many different species are on Earth? The number is astounding. A study done by biologist Edward O. Wilson and colleagues in 1990 estimated the known number of species at approximately 1.4 million. Currently, about 1.75 million species have been identified. However, a vast number of species have yet to be established. This is particularly true among prokaryotic organisms, which are difficult to categorize into distinct species. Common estimates of the total number of species range from 5 to 50 million!

When studying natural populations, evolutionary biologists are often confronted with situations where some differences between two populations are apparent, but it is difficult to decide whether the two populations truly represent separate species. When two or more geographically restricted groups of the same species display one or more traits that are somewhat different but not enough to warrant their placement into different species, biologists sometimes classify such groups as **subspecies.** Similarly, many bacterial species are subdivided into **ecotypes.** Each ecotype is a genetically distinct population adapted to its local environment.

Members of the same species share an evolutionary history that is distinct from other species. Although this may seem like a reasonable way to characterize a given species, evolutionary biologists would agree that the identification of many species is a difficult undertaking. What criteria do we use to distinguish species? How many differences must exist between two populations to classify them as distinct species? Such questions are often difficult to answer.

The characteristics that a biologist uses to identify a species depend, in large part, on the species in question. For example, the traits used to distinguish insect species are quite different from those used to identify different bacterial species. The relatively high level of horizontal gene transfer among bacteria presents special challenges in the grouping of bacterial species. Among bacteria, it is sometimes very difficult and perhaps arbitrary to divide closely related organisms into separate species. The most commonly used characteristics to identify species are morphological traits, the ability to interbreed, molecular features, ecological factors, and evolutionary relationships. A comparison of these concepts will help you to appreciate the various approaches that biologists use to identify the bewildering array of species on our planet.

Morphological Traits One way to establish that a population constitutes a unique species is based on their physical characteristics. Organisms are classified as the same species if their anatomical traits appear to be very similar. Likewise, microorganisms can be classified according to morphological traits at the cellular level. By comparing many different morphological traits, biologists may decide that certain populations constitute a unique species.

Although an analysis of morphological traits is a common way for biologists to establish that a particular group constitutes a species, it has a few drawbacks. First, it may be difficult to decide how many traits to consider. In addition, it is difficult to analyze quantitative traits, such as size and weight, which vary in a continuous way among members of the same species. Another drawback is that the degree of dissimilarity that distinguishes different species may not show a simple relationship; the members of the same species sometimes look very different, and conversely, members of different species sometimes look remarkably similar to each other. For example, **Figure 26.1a** shows two different frogs of the species *Dendrobates tinctorius*, commonly called the dyeing poison frog. This species exists in many different-colored morphs, which are individuals of the same species that have noticeably dissimilar appearances. In contrast, **Figure 26.1b** shows two different species of meadowlarks, the western meadowlark (*Sturnella neglecta*) and eastern meadowlark (*Sturnella magna*). Both species are nearly

(a) Frogs of the same species

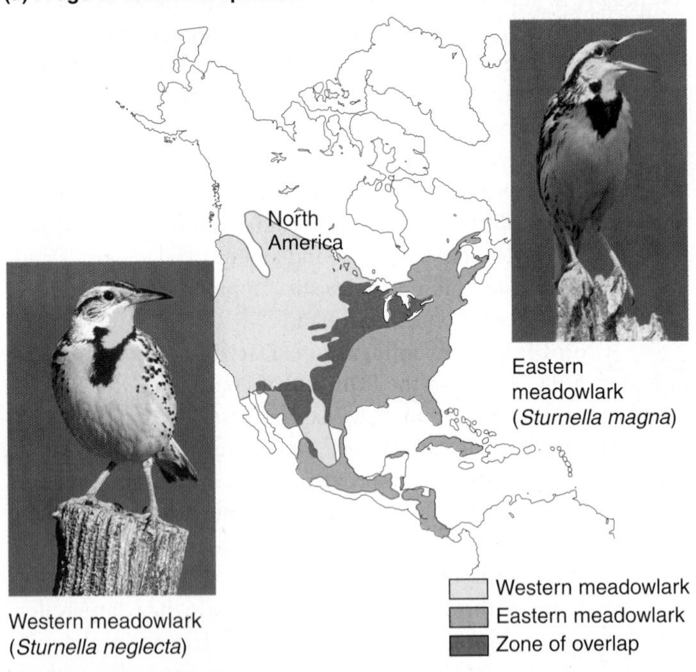

(b) Birds of different species

FIGURE 26.1 Morphological dissimilarities and similarities in species. **(a)** In some cases, two members of the same species can look quite different. These photographs show two members of the same species, the dyeing poison frog (*Dendrobates tinctorius*). **(b)** In comparison, members of different species can look quite similar as illustrated by the western meadowlark (*Sturnella neglecta*) and eastern meadowlark (*Sturnella magna*).

identical in shape, coloration, and habitat, and their ranges overlap in the central United States.

Reproductive Isolation Why do biologists describe two species, such as the western and eastern meadowlarks, as being different if they are morphologically similar? One reason is that biologists have discovered that they are unable to breed with each other in nature. In the zone of overlap, very little interspecies mating takes place between western and eastern meadowlarks, largely due to differences in their songs. The song of the western meadowlark is a long series of flutelike gurgling notes that go down the scale. By comparison, the eastern meadowlark's song is a simple series of whistles, typically about four or five notes. These differences in songs enable meadowlarks to recognize potential mates as members of their own species.

Therefore, a second way to identify a species is by the ability to interbreed. In the late 1920s, geneticist Theodosius Dobzhansky proposed that each species is reproductively isolated from other species. Such **reproductive isolation** prevents one species from successfully interbreeding with other species. In 1942, evolutionary biologist Ernst Mayr expanded on the ideas of Dobzhansky to provide a reproductive definition of a species. According to Mayr, a key feature of sexually reproducing species is that, in nature, the members of one species have the potential to interbreed with one another to produce viable, fertile offspring but cannot successfully interbreed with members of other species.

TABLE **26.1**

Types of Reproductive Isolation Among Different Species

Prezygotic Isolating Mechanisms	
Habitat isolation	Species may occupy different habitats so they never come in contact with each other.
Temporal isolation	Species have different mating or flowering seasons, mate at different times of day, or become sexually active at different times of the year.
Sexual isolation	Sexual attraction between males and females of different animal species is limited due to differences in behavior, physiology, or morphology.
Mechanical isolation	The anatomical structures of genitalia prevent mating between different species.
Gametic isolation	Gametic transfer takes place, but the gametes fail to unite with each other. This can occur because the male and female gametes fail to attract, because they are unable to fuse, or because the male gametes are inviable in the female reproductive tract of another species.
Postzygotic Isolating Mechanisms	
Hybrid inviability	The egg of one species is fertilized by the sperm from another species, but the fertilized egg fails to develop past early embryonic stages.
Hybrid sterility	The interspecies hybrid survives, but it is sterile. For example, the mule, which is sterile, is a cross between a female horse (*Equus caballus*) and a male donkey (*Equus asinus*).
Hybrid breakdown	The F_1 interspecies hybrid is viable and fertile, but succeeding generations (i.e., F_2, etc.) become increasingly inviable. This is usually due to the formation of less fit genotypes by genetic recombination.

Reproductive isolation has been used to distinguish many plant and animal species, especially those that look alike but do not interbreed. How does reproductive isolation occur? **Table 26.1** describes several ways. These are classified as **prezygotic isolating mechanisms,** which prevent the formation of a zygote, and **postzygotic isolating mechanisms,** which prevent the development of a viable and fertile individual after fertilization has taken place. Reproductive isolation in nature may be circumvented when species are kept in captivity. For example, different species of the genus *Drosophila* rarely mate with each other in nature. In the laboratory, however, it is fairly easy to produce interspecies hybrids.

Although reproductive isolation has been commonly used to classify species, it suffers from four main problems. First, in nature, it may be difficult to determine if two populations are reproductively isolated, particularly if they are populations with nonoverlapping geographical ranges. Second, biologists have noted many cases in which two different species can interbreed in nature yet consistently maintain themselves as separate species. For example, different species of yucca plants, such as *Yucca pallida* and *Yucca constricta*, do interbreed in nature yet typically maintain populations with distinct characteristics. For this reason, they are viewed as distinct species. A third drawback of reproductive isolation is that it does not apply to asexual species such as bacteria. Likewise, some species of plants and fungi only reproduce asexually. Finally, a fourth drawback is that it cannot be applied to extinct species. For these reasons, reproductive isolation has been primarily used to distinguish closely related species of modern animals and plants that reproduce sexually.

Molecular Features Molecular features are now commonly used to determine if two different populations are different species. Evolutionary biologists often compare DNA sequences within genes, gene order along chromosomes, chromosome structure, and chromosome number as features to identify similarities and differences among different populations. DNA sequence differences are often used to compare populations. For example, researchers may compare the DNA sequence of the 16S rRNA gene between different bacterial populations as a way to decide if the two populations represent different species. When the sequences are very similar, such populations are probably judged as the same species. However, it may be difficult to draw the line when separating groups into different species. Is a 2% difference in their genome sequences sufficient to warrant placement into two different species, or do we need a 5% difference?

Ecological Factors A variety of factors related to an organism's habitat can be used to distinguish one species from another. For example, certain species of warblers can be distinguished by the habitat in which they forage for food. Some species search the ground for food, others forage in bushes or small trees, and some species primarily forage in tall trees. Such habitat differences can be used to distinguish different species that look morphologically similar.

Many bacterial species have been categorized as distinct species based on ecological factors. Bacterial cells of the same species are likely to use the same types of resources (e.g., sugars and vitamins) and grow under the same types of conditions (e.g., temperature and pH). However, a drawback of this approach is that different groups of bacteria sometimes display very similar growth characteristics, and even the same species may show great variation in the growth conditions it will tolerate.

Evolutionary Relationships Later in this chapter, we will examine the methods used to produce evolutionary trees that describe the evolutionary relationships among different species. In some cases, such relationships are based on an analysis of the fossil record. Alternatively, another way to establish evolutionary relationships is by the analysis of DNA sequences. Researchers can obtain samples of cells from different individuals and compare the genes within those cells to see how similar or different they are.

Species Concepts Thus far, we have considered multiple ways to identify species. The most commonly used characteristics to distinguish species are morphological traits, the ability to interbreed, molecular features, ecological factors, and evolutionary relationships.

A **species concept** is a way to define the concept of a species and/or provide an approach for distinguishing one species from another. In 1942, Ernst Mayr proposed an early species concept called the **biological species concept.** According to this idea, a species is a group of individuals whose members have the potential to interbreed with one another in nature to produce viable, fertile offspring but cannot successfully interbreed with members of other species. The biological species concept emphasizes reproductive isolation as the most important criterion for delimiting species. Since then, over 20 different species concepts have been proposed by a variety of evolutionary biologists. Another example is the **evolutionary lineage concept** proposed by American paleontologist George Gaylord Simpson in 1961. According this idea, a species should be defined based on the separate evolution of lineages. A third example is the ecological species concept, described by American evolutionary biologist Leigh Van Valen in 1976. According to this viewpoint, each species occupies an ecological niche, which is the unique set of habitat resources that a species requires, as well as its influence on the environment and other species.

Most evolutionary biologists agree that different methods are needed to distinguish the vast array of species on Earth. Even so, some evolutionary biologists have questioned whether it is valid to have many different species concepts. In 1998, Kevin de Queiroz suggested that there is only a single general species concept, which concurs with Simpson's evolutionary lineage concept and includes all previous concepts. According to de Queiroz's **general lineage concept,** each species is a population of an independently evolving lineage. Each species has evolved from a specific series of ancestors and, as a consequence, forms a group of organisms with a particular set of characteristics. Multiple criteria are used to determine if a population is part of an independent evolutionary lineage, and thus a species, which is distinct from others. Typically, researchers use analyses of morphology, reproductive isolation, DNA sequences, and ecology to determine if a population or group of populations is distinct from others. Because of its generality, the general lineage concept has received significant support.

Speciation Usually Occurs via a Branching Process Called Cladogenesis

Speciation is the process by which new species are formed via evolution. One way it can occur is by **anagenesis** (from the Greek *ana*, up, and *genesis*, origin) in which a single species evolves into a different species over the course of many generations (**Figure 26.2a**). However, most evolutionary biologists would argue that anagenesis is not a common mechanism for speciation, though it could occur if an entire species was confined to a single environment for long periods of time.

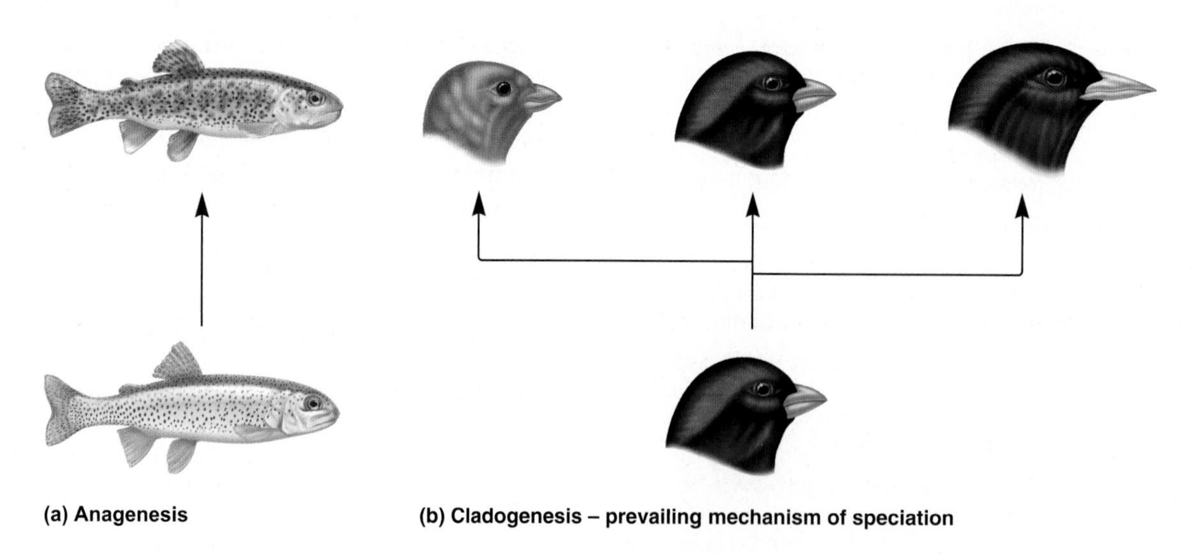

FIGURE 26.2 A comparison between anagenesis and cladogenesis, two patterns of speciation.
(a) Anagenesis is the change of one species into another. (b) Cladogenesis involves a process in which one original species is separated into two or more different species. Within a group of related species, both of these processes may occur, but cladogenesis is far more common.

By comparison, speciation primarily occurs via **cladogenesis** (from the Greek *clados,* branch), which involves the division of a species into two or more species (**Figure 26.2b**). This form of speciation increases species diversity. Although cladogenesis is usually thought of as a splitting process, it commonly occurs as a budding process, which results in the original species plus one or more new species with different characteristics. If we view evolution as a tree, the new species bud from the original species and develop characteristics that prevent them from breeding with the original one.

Cladogenesis Can Be Allopatric, Parapatric, or Sympatric

Depending on the geographic locations of the evolving population(s) and the environment that a species occupies, cladogenesis is categorized as allopatric, parapatric, or sympatric (**Table 26.2**).

Allopatric Speciation **Allopatric speciation** (from the Greek *allos,* other, and Latin *patria,* homeland) is thought to be the most prevalent way for a species to diverge. It happens when members of a species become geographically separated from the other members. This form of speciation can occur by the geographic subdivision of large populations via geological processes. For example, a mountain range may emerge and split a species that occupies the lowland regions, or a creeping glacier may divide a population. **Figure 26.3** shows an interesting example in which geological separation promoted speciation. Two species of antelope squirrels occupy opposite rims of the Grand Canyon. On the south rim is Harris's antelope squirrel (*Ammospermophilus harrisi*), whereas a closely related white-tailed antelope squirrel (*Ammospermophilus leucurus*) is found on the north rim. Presumably, these two species evolved from a common species that existed before the canyon was formed. Over time, the accumulation of genetic changes in the two separated populations led to the formation of two morphologically distinct species. Interestingly, birds that can easily fly across the canyon have not diverged into different species on the opposite rims.

Allopatric speciation can also occur via a second mechanism, known as the **founder effect,** which is thought to be more rapid and frequent than allopatric speciation caused by geological events. The founder effect, which was discussed in Chapter

A. harrisi

A. leucurus

FIGURE 26.3 An example of allopatric speciation: two closely related species of antelope squirrels that occupy opposite rims of the Grand Canyon.

Genes → Traits Harris's antelope squirrel (*Ammospermophilus harrisi*) is found on the south rim of the Grand Canyon, whereas the white-tailed antelope squirrel (*Ammospermophilus leucurus*) is found on the north rim. These two species evolved from a common species that existed before the canyon was formed. After the canyon was formed, the two separated populations accumulated genetic changes due to mutation, genetic drift, and natural selection that eventually led to the formation of two distinct species.

24, occurs when a small group migrates to a new location that is geographically separated from the main population. For example, a storm may force a small group of birds from the mainland to a distant island. In this case, the migration of individuals between the island and the mainland is a very infrequent event. In a relatively short time, the founding population on the island may evolve into a new species. Two evolutionary mechanisms may contribute to this rapid evolution. First, genetic drift may quickly lead to the random fixation of certain alleles and the elimination of other alleles from the population. Another factor is natural selection. The environment on an island may differ significantly from the mainland environment. For this reason, natural selection on the island may favor different types of alleles.

Parapatric Speciation **Parapatric speciation** (from the Greek *para,* beside) occurs when members of a species are partially separated. In other words, the geographic separation is not complete. For example, members of a given species may invade a new ecological niche at the periphery of an existing population. Alternatively, a mountain range may divide a species into two populations but have breaks in the range where the two groups are connected physically. In these zones of contact, the members

TABLE **26.2**

Common Genetic Mechanisms That Underlie Allopatric, Parapatric, and Sympatric Speciation

Type of Speciation	Common Genetic Mechanisms Responsible for Speciation
Allopatric—two large populations are separated by geographic barriers	Many small genetic differences may accumulate over a long period, leading to reproductive isolation. Some of these genetic differences may be adaptive, whereas others are neutral.
Allopatric—a small founding population separates from the main population	Genetic drift may lead to the rapid formation of a new species. If the group has moved to an environment that is different from its previous one, natural selection is expected to favor beneficial alleles and eliminate harmful alleles.
Parapatric—two populations occupy overlapping ranges, so a limited amount of interbreeding occurs	A new combination of alleles or chromosomal rearrangement may rapidly limit the amount of gene flow between neighboring populations because hybrid offspring have a very low fitness.
Sympatric—within a population occupying a single habitat in a continuous range, a small group evolves into a reproductively isolated species	An abrupt genetic change leads to reproductive isolation. For example, a mutation may affect gamete recognition. In plants, the formation of a tetraploid often leads to the formation of a new species because the interspecies hybrid (e.g., diploid × tetraploid) is triploid and sterile.

of two populations can interbreed, although this tends to occur infrequently. In addition, parapatric speciation may occur among very sedentary species even though no geographic isolation exists. Certain organisms are so sedentary that as little as 100 to 1000 m may be sufficient to limit the interbreeding between neighboring groups. Plants, terrestrial snails, rodents, grasshoppers, lizards, and many flightless insects may speciate in a parapatric manner.

During parapatric speciation, **hybrid zones** exist where two populations can interbreed. For speciation to occur, the amount of gene flow within the hybrid zones must become very limited. In other words, there must be selection against the offspring produced in the hybrid zone. This can happen if each of the two parapatric populations accumulates different chromosomal rearrangements, such as inversions and balanced translocations. How do chromosomal rearrangements, such as inversions, prevent interbreeding? As discussed in Chapter 8, if a hybrid individual has one chromosome with a large inversion and one that does not carry the inversion, crossing over during meiosis can lead to the production of grossly abnormal chromosomes. Therefore, such a hybrid individual is substantially less fertile. By comparison, an individual homozygous for two normal chromosomes or for two chromosomes carrying the same inversion is fertile, because crossing over can proceed normally.

Sympatric speciation **Sympatric speciation** (from the Greek *sym*, together) occurs when a new species arises in the same geographic area as the species from which it was derived. In plants, a common way for sympatric speciation to occur is the formation of polyploids. As discussed in Chapter 8, complete nondisjunction of chromosomes during gamete formation can increase the number of chromosome sets within a single species (autopolyploidy) or between different species (allopolyploidy). Polyploidy is so frequent in plants that it is a major form of speciation. In ferns and flowering plants, at least 30% of the species are polyploid. By comparison, polyploidy is much less common in animals, but it can occur. For example, roughly 30 species of reptiles and amphibians have been identified that are polyploids derived from diploid relatives.

The formation of a polyploid can abruptly lead to reproductive isolation. As an example, let's consider the probable events that led to the formation of a natural species of common hemp nettle known as *Galeopsis tetrahit*. This species is thought to be an allotetraploid derived from two diploid species: *Galeopsis pubescens* and *Galeopsis speciosa*. As shown in **Figure 26.4a**, *G. tetrahit* has 32 chromosomes, whereas the two diploid species contain 16 chromosomes each ($2n = 16$). **Figure 26.4b** illustrates what would happen in crosses between

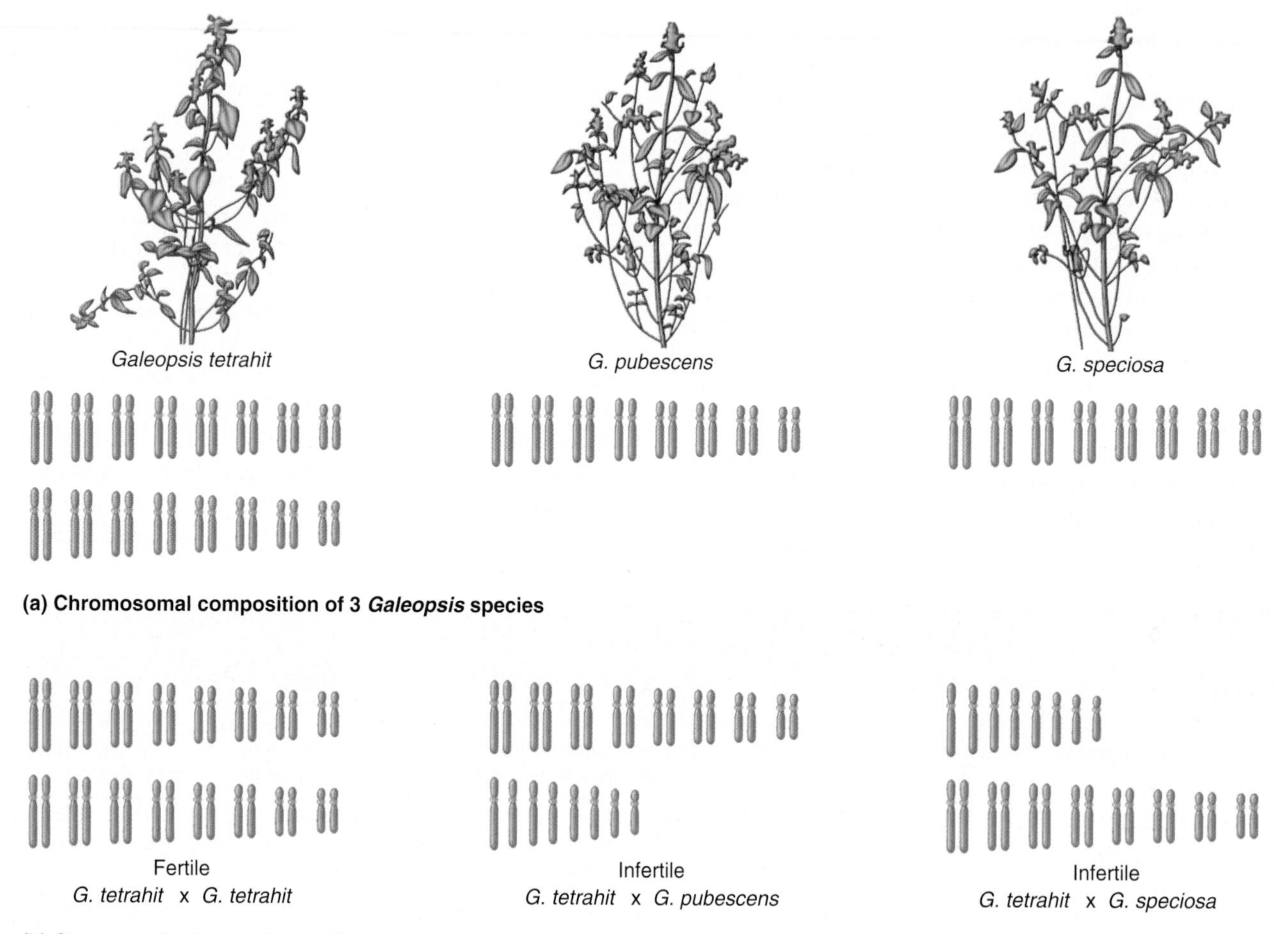

FIGURE 26.4 A comparison of crosses between three natural species of hemp nettle with different ploidy levels. **(a)** *Galeopsis tetrahit* is an allotetraploid that is thought to be derived from *Galeopsis pubescens* and *Galeopsis speciosa*. **(b)** If *G. tetrahit* is mated with the other two species, the F_1 hybrid offspring will be monoploid for one chromosome set and diploid for the other. The F_1 offspring are likely to be sterile, because they will produce highly aneuploid gametes.

the allotetraploid and the diploid species. The allotetraploid crossed to another allotetraploid produces an allotetraploid. The allotetraploid is fertile, because all of its chromosomes occur in homologous pairs that can segregate evenly during meiosis. However, a cross between an allotetraploid and a diploid produces an offspring that is monoploid for one chromosome set and diploid for the other. These offspring are expected to be sterile, because they produce highly aneuploid gametes that have incomplete sets of chromosomes. This hybrid sterility renders the allotetraploid reproductively isolated from the diploid species.

In the 1930s, Arne Müntzing first proposed that *Galeopsis tetrahit* is an allotetraploid that arose from an interspecies cross between *G. pubescens* and *G. speciosa*. To test this hypothesis, he performed a series of crosses between *G. pubescens* and *G. speciosa* and succeeded in producing an allotetraploid that had two sets of chromosomes from both species. This artificial *G. tetrahit* had traits similar to the natural *G. tetrahit* species but different from *G. pubescens* and *G. speciosa*. Furthermore, the artificial and natural *G. tetrahit* strains could be mated to each other to produce fertile offspring. Overall, Müntzing came to the conclusion that "not only the artificial tetraploid but probably also natural *G. tetrahit* represents a synthesis of *pubescens*- and *speciosa*-genomes."

Another way that sympatric speciation may occur is when members of a population occupy different local environments that are continuous with each other. An example of this type of sympatric speciation was described by Jeffrey Feder, Guy Bush, and colleagues. They studied the North American apple maggot fly (*Rhagoletis pomonella*). This fly originally fed on native hawthorn trees. However, the introduction of apple trees approximately 200 years ago provided a new local environment for this species. The apple-feeding populations of this species develop more rapidly because apples mature more quickly than hawthorne fruit. The result is partial temporal isolation in reproduction (see Table 26.1). Although the two populations—those that feed on apple trees and those that feed on hawthorne trees—are considered subspecies, evolutionary biologists speculate they may eventually become distinct species due to reproductive isolation and the accumulation of independent mutations in the two populations.

Speciation Can Be Gradual or Punctuated by Periods of Rapid Change

As we have seen, many different genetic mechanisms give rise to new species. For this reason, the rates of evolutionary change are not constant, although the degree of inconstancy has been debated since the time of Darwin. Even Darwin himself suggested that evolution can occur at fast and slow paces. **Figure 26.5** illustrates contrasting views about the rates of evolutionary change. These views are not mutually exclusive but represent two different ways to consider the tempo of evolution. The concept

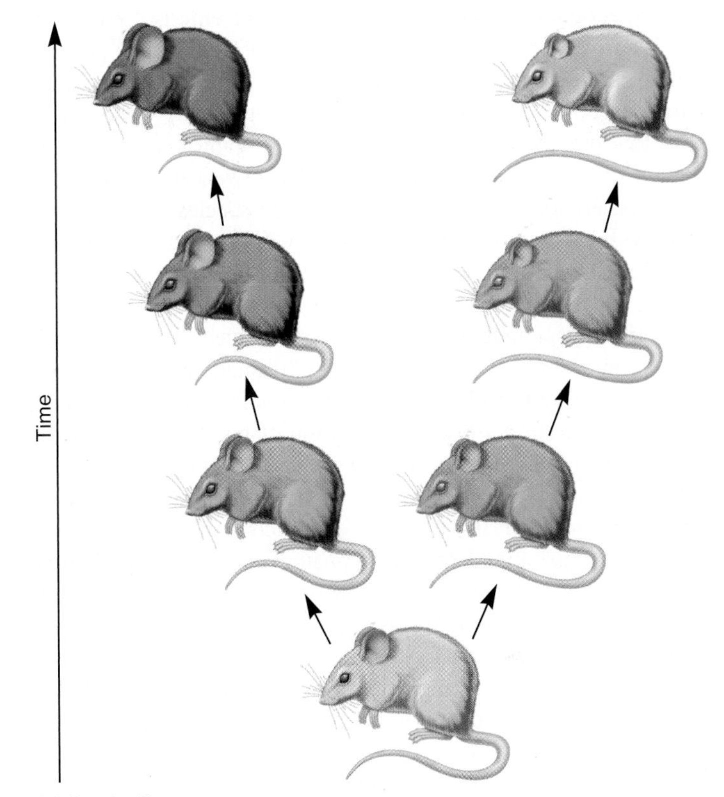

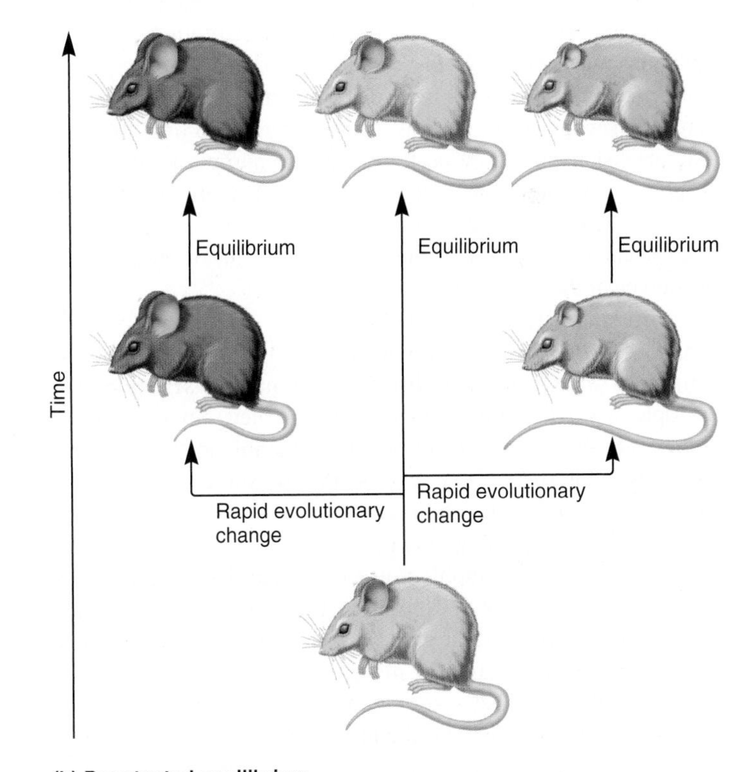

FIGURE 26.5 A comparison of gradualism and punctuated equilibrium. (**a**) During gradualism, a species gradually changes due to the accumulation of small genetic changes. (**b**) During punctuated equilibrium, a species exists essentially unchanged for long periods of time, during which it is in equilibrium with its environment. These equilibrium periods are punctuated by short periods of evolutionary change, during which its features may change rapidly.

of **gradualism** suggests that each new species evolves gradually over long spans of time (Figure 26.5a). The principal idea is that large phenotypic differences that produce the emergence of new species are due to the slow accumulation of many small genetic changes. By comparison, the concept of **punctuated equilibrium,** proposed by Niles Eldredge and Stephen Jay Gould, suggests that the tempo of evolution is more sporadic (Figure 26.5b). According to this model, species exist relatively unchanged for many generations. During this period, the species is in equilibrium with its environment. These long periods of equilibrium are punctuated by relatively short periods (on a geological timescale) during which evolution occurs at a far more rapid rate.

In reality, neither of the views presented in Figure 26.5 fully account for evolutionary change. The phenomenon of punctuated equilibrium is often supported by the fossil record. Paleontologists rarely find a gradual transition of fossil forms. Instead, it is much more common to observe species appearing as new forms rather suddenly in a layer of rocks, persisting relatively unchanged for a very long time, and then suddenly becoming extinct. It is presumed that the transition period during which a previous species evolved into a new, morphologically distinct species was so short that few, if any, of the transitional members were preserved as fossils. Even so, these rapid periods of change were likely followed by long periods of equilibrium that involved the additional accumulation of many small genetic changes, consistent with gradualism.

As discussed earlier, rapid evolutionary change can be explained by genetic phenomena. As we have seen throughout this textbook, single-gene mutations can have dramatic effects on phenotypic characteristics. Therefore, only a small number of new mutations may be required to alter phenotypic characteristics, eventually producing a group of individuals that make up a new species. Likewise, genetic events such as changes in chromosome structure (e.g., inversions and translocations) or chromosome number may abruptly create individuals with new phenotypic traits. On an evolutionary timescale, these types of events can be rather rapid because one or only a few genetic changes can have a major effect on the phenotype of the organism.

In conjunction with genetic changes, species may also be subjected to sudden environmental shifts that quickly drive the gene pool in a particular direction via natural selection. For example, a small group may migrate to a new environment in which certain alleles provide better adaptation to the new surroundings. Alternatively, a species may be subjected to a relatively sudden environmental event that has a major influence on survival. There may be a change in climate, or a new predator may infiltrate the geographic range of the species. Natural selection may lead to rapid evolution of the gene pool by favoring those genetic changes that allow members of the population to survive the climatic change or more easily avoid the predator.

Overall, the fossil record and known genetic phenomena tend to support the idea that the tempo of evolution can be quite variable. In some cases, rapid evolutionary change has taken place and led to the formation of new species. During other periods, smaller phenotypic changes may occur over a longer timescale.

26.2 PHYLOGENETIC TREES

Thus far, we have considered the various factors that play a role in the formation of new species. In this section, we will examine **phylogeny**—the sequence of events involved in the evolutionary development of a species or group of species. A systematic approach is followed to produce a **phylogenetic tree,** which is a diagram that describes a phylogeny. Such a tree is a hypothesis of the evolutionary relationships among various species, based on the information available to and gathered by biologists termed **systematists.**

In the 1950s, the German entomologist Willi Hennig proposed that evolutionary relationships should be inferred from new features shared by descendants of a common ancestor. Phylogenetic trees are now based on **homology,** which refers to similarities among various species that occur because the species are derived from a common ancestor. Attributes that are the result of homology are said to be **homologous.** For example, the wing of a bat, the arm of a human, and the front leg of a cat are homologous structures. By comparison, a bat wing and insect wing are not homologous; they arose independently of each other. When constructing phylogenetic trees, researchers identify homologous features that are shared by some species but not by others. This allows them to group species based on their shared characteristics. Researchers typically study homology at the level of morphological traits or at the level of genes.

Historically, comparisons of morphological similarities and differences have been used to construct evolutionary trees. In this approach, species that share certain characteristics (i.e., homologous traits) tend to be placed closer together on the tree. In addition, species have been categorized based on physiology, biochemistry, and even behavior. Although these approaches continue to be used, systematists are increasingly using molecular data to infer evolutionary relationships. In 1963, Linus Pauling and Emile Zuckerkandl were the first to suggest the use of molecular data to establish evolutionary relationships. When comparing homologous genes in different species, the DNA sequences from closely related species are more similar to each other than are the sequences from distantly related species. In this section, we will examine the general features of phylogentic trees, how they can be constructed, and the types of information they reveal.

A Phylogenetic Tree Depicts the Evolutionary Relationships Among Different Species

Let's first take a look at what information is found within a phylogenetic tree and the form in which it is presented. **Figure 26.6** shows a hypothetical phylogenetic tree of the relationships between various insect species in which the species (butterflies) are labeled A through J. The vertical axis represents time, with the oldest species at the bottom.

The prevailing mechanism of speciation is through cladogenesis, in which a species diverges into two or more species. The nodes or branch points in a phylogenetic tree illustrate times when cladogenesis has occurred. For example, approximately 12 million years ago (mya), species A diverged into species A and

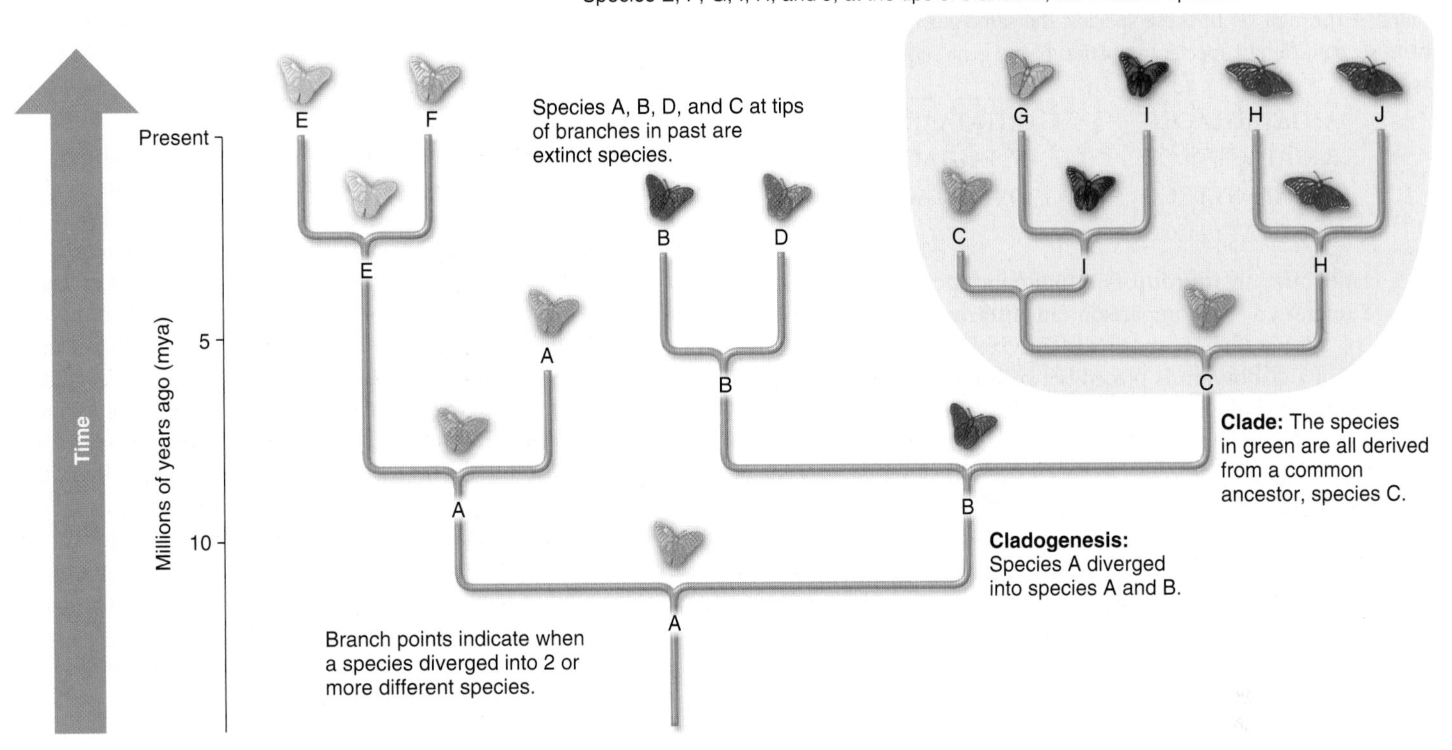

FIGURE 26.6 **How to read a phylogenetic tree.** This hypothetical tree shows the proposed relationships among various butterfly species.

species B by cladogenesis. The tips of branches represent either species that became extinct in the past, such as species A, B, D and C, or modern species, such as E, F, G, I, H, and J, that are at the top of the tree.

By studying the branch points of a phylogenetic tree, researchers can group species according to common ancestry. A **monophyletic group,** also known as a **clade,** is a group of species consisting of all descendents of the group's most common ancestor. For example, the group highlighted in light green is a clade derived from the common ancestor labeled C. The present-day descendents of a common ancestor can also be called a clade. In this case, species G, I, H, and J form a modern clade. Likewise, the entire tree shown in Figure 26.6 forms a clade, with species A as a common ancestor. As we see in this figure, smaller and more recent clades are subsets of larger ones.

The phylogenetic tree in Figure 26.6 includes ancestral species. This type of tree could be obtained by examining the fossil record. However, many phylogenetic trees do not include ancestral species, but instead focus on the relationships among modern species. An example of this type of phylogenetic tree is a cladogram, which is described next.

A Phylogenetic Tree Can Be Constructed Using Different Approaches

Now that we appreciate the concept of a phylogenetic tree, let's turn our attention to how evolutionary biologists actually construct them. One method, known as a **cladistic approach,** reconstructs a phylogenetic tree, also called a **cladogram,** by considering the various possible pathways of evolution and then choosing the most plausible tree. Cladistics is a commonly used method for the construction of phylogenetic trees.

A cladistic approach compares traits, also called characters, that are either shared or not shared by different species. These can be morphological traits, such as the shapes of birds' beaks, or molecular traits, such as sequences of homologous genes. Such characters may come in different versions called character states. Those that are shared with a distant ancestor are called **ancestral characters** (also called primitive characters). Such characters are viewed as being older—ones that arose earlier in evolution. In contrast, a **shared derived character,** or **synapomorphy,** is a trait that is shared by a group of organisms but not by a distant common ancestor. Compared with ancestral characters, shared derived characters are more recent traits on an evolutionary timescale. For example, among mammals, only some species have flippers, such as whales and dolphins. In this case, flippers were derived from the two front limbs of an ancestral species. The word "derived" refers to the observation that evolution involves the modification of traits in preexisting species. In other words, the features of newer populations of organisms are derived from changes in preexisting populations. The basis of the cladistic approach is to analyze many shared derived characters among groups of species to deduce the pathway that gave rise to the species.

To understand the concept of ancestral versus shared derived characters, let's consider how a cladogram can be constructed based on molecular data such as a sequence of a gene. Our example uses molecular data obtained from seven different hypothetical species called A through G. In these species, the same gene was sequenced from seven different species; a portion of the gene sequence is shown as follows:

A: GATAGTACCC	E: GGTATAACCC
B: GATAGTTCCC	F: GGTAGTACCA
C: GATAGTTCCG	G: GGTAGTACCC
D: GGTATTACCC	

In a cladogram, an **ingroup** is a monophyletic group in which we are interested. By comparison, an **outgroup** is a species or group of species that is more distantly related to an ingroup. The root of a cladogram is placed between the outgroup and the ingroup. In the cladogram of **Figure 26.7**, the outgroup is species E. This may have been inferred because the other species may share traits that are not found in species E. The other species (A, B, C, D, F, and G) form the ingroup. For these data, a mutation that changes the DNA sequence is analogous to a modification of a characteristic. Species that share such genetic changes possess shared derived characters because the new genetic sequence was derived from a more ancestral sequence.

Now that we understand some of the general principles of cladistics, let's consider the steps a researcher would follow to construct a cladogram using this approach.

1. **Choose the species in whose evolutionary relationships you are interested.** In a simple cladogram, individual species are compared with each other. In more complex cladograms, species may be grouped into larger taxa (e.g., families) and compared with each other. If such grouping is done, the groups must be clades for the results to be reliable.
2. **Choose characters for comparing different species.** As mentioned, a character is a general feature of an organism. Characters may come in different versions called character states. For example, a base at a particular location in a gene can be considered a character, and this character could exist in different character states, such as A, T, G, or C, due to mutations.
3. **Determine the polarity of character states.** In other words, determine if a character state is ancestral or derived. In the case of morphological traits, this information may be available by examining the fossil record. If possible, identify an outgroup.
4. **Group species (or higher taxa) based on shared derived characters.**
5. **Build a cladogram based on the following principles:**
   - All species (or higher taxa) are placed on tips in the phylogenetic tree, not at branch points. A cladogram does not include ancestral species at branch points.
   - Each cladogram branch point should have a list of one or more shared derived characters that are common to all species above the branch point unless the character is later modified.
   - All shared derived characters appear together only once in a cladogram unless they independently arose during evolution more than once in the ancestors of different clades.
6. **Choose the best cladogram among possible options.** When grouping species (or higher taxa), more than one cladogram may be possible. Therefore, analyzing the data and producing the best possible cladogram is a key aspect of this process. As described next, different theoretical approaches can be followed to achieve this goal.

The greatest challenge in a cladistic approach is to determine the correct order of events. It may not always be obvious which traits are ancestral and came earlier and which are derived and came later in evolution. Different approaches can be used to deduce the correct order. First, for morphological traits, a common way to deduce the order of events is to analyze fossils and determine the relative dates that certain traits arose. A second strategy assumes that the best hypothesis is the one that requires the fewest number of evolutionary changes. This concept, called the **principle of parsimony,** states that the preferred hypothesis is the one that is the simplest. For example, if two species possess a tail, we would initially assume that a tail arose once during evolution and that both species have descended from a common ancestor with a tail. Such a hypothesis is simpler than assuming that tails arose twice during evolution and that the tails in the two species are not due to descent from a common ancestor.

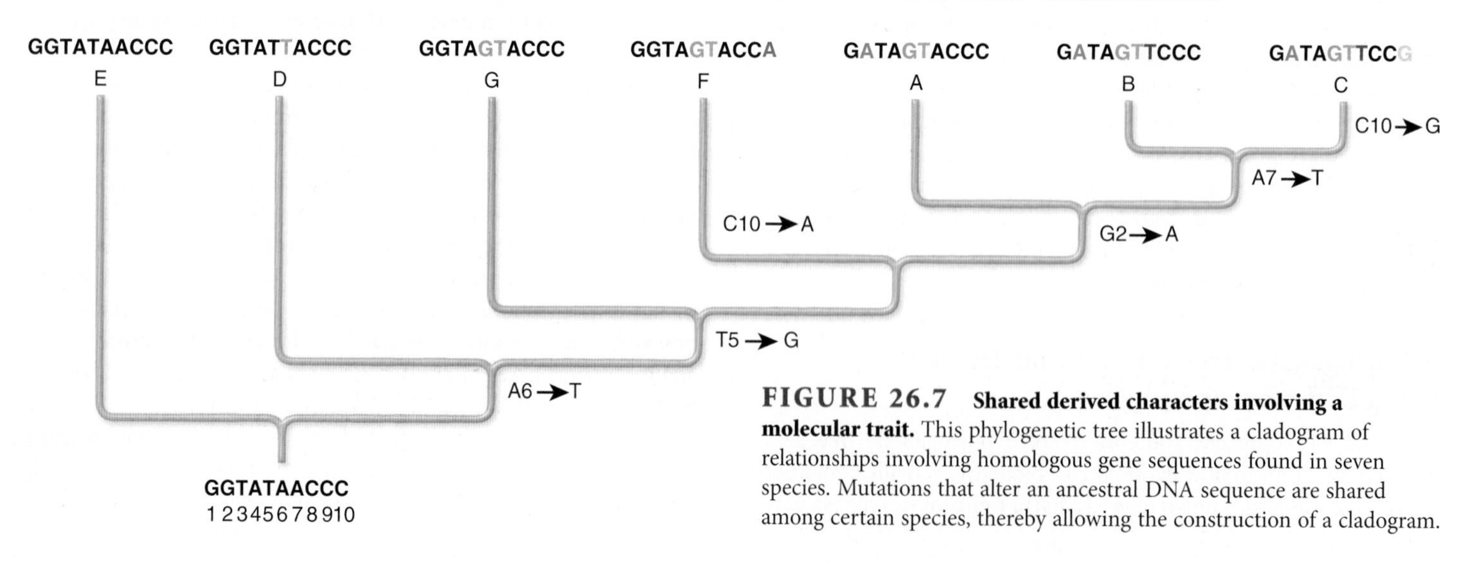

FIGURE 26.7 Shared derived characters involving a molecular trait. This phylogenetic tree illustrates a cladogram of relationships involving homologous gene sequences found in seven species. Mutations that alter an ancestral DNA sequence are shared among certain species, thereby allowing the construction of a cladogram.

Let's consider a simple example to illustrate how the principle of parsimony can be used. Our example involves molecular data obtained from four different hypothetical species called A through D. In these species, a homologous region of DNA was sequenced as shown here.

1 2 3 4 5
A: GTACA
B: GACAG
C: GTCAA
D: GACCG

Given this information, three different trees are shown in **Figure 26.8,** although more are possible. In these examples, tree 1 requires seven mutations, tree 2 requires six, and tree 3 requires only five. Because tree 3 requires the fewest number of mutations, it is considered the most parsimonious. Based on the principle of parsimony, tree 3 is the more likely choice.

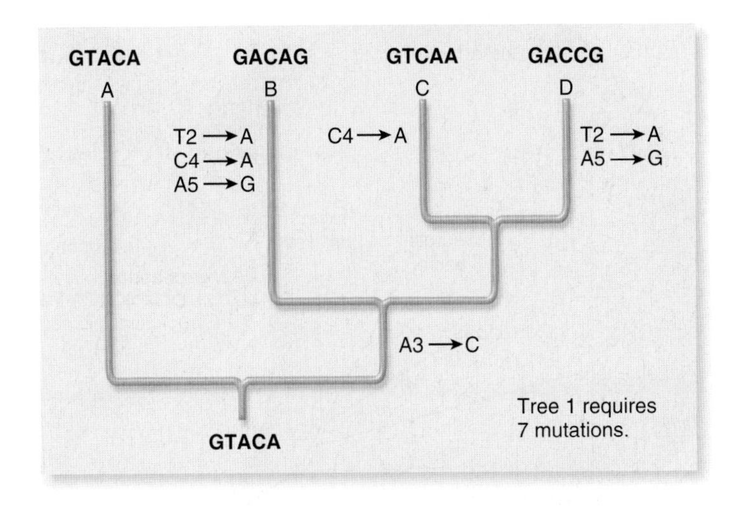

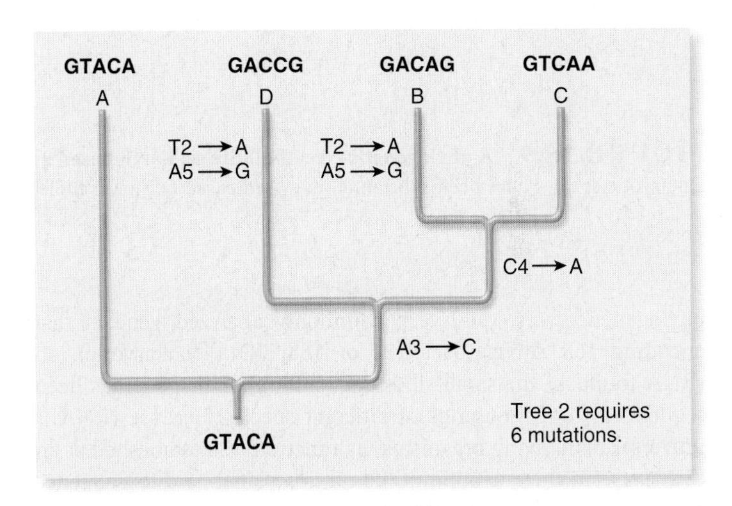

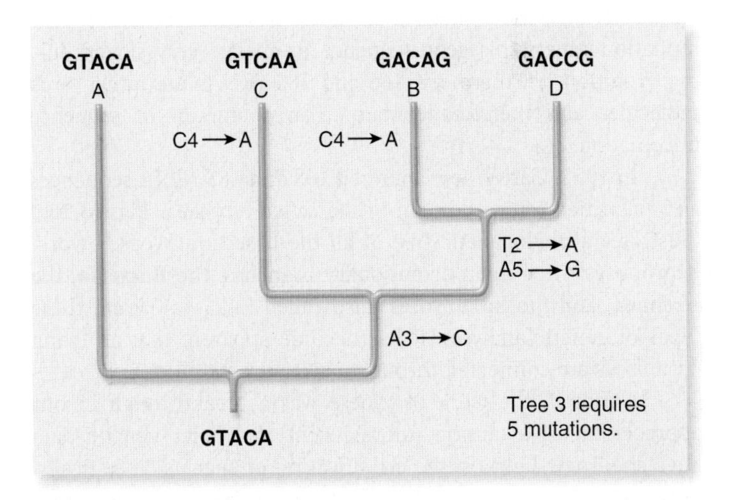

FIGURE 26.8 The cladistic approach: Choosing a cladogram from molecular genetic data. This figure shows three different phylogenetic trees for the evolution of a short DNA sequence, but many more are possible. According to the principle of parsimony, the cladogram shown in tree 3 is the more plausible choice because it requires only five mutations. When constructing cladograms based on long genetic sequences, researchers use computers to generate trees with the fewest possible genetic changes.

Maximum Likelihood and Bayesian Methods Are Used to Discriminate Among Possible Phylogenetic Trees

In addition to the fossil record and the principle of parsimony, evolutionary biologists also apply other approaches, such as maximum likelihood and Bayesian methods, when proposing and evaluating phylogenetic trees. These methods involve the use of an evolutionary model—a set of assumptions about how evolution is likely to happen. For example, mutations affecting the third base in a codon are often neutral because they don't affect the amino acid sequence of the encoded protein and therefore don't affect the fitness of an organism. As discussed later in this chapter, such neutral mutations are more likely to become prevalent in a population than are mutations in the first or second base. Therefore, one possible assumption of an evolutionary model is that neutral mutations are more likely than nonneutral ones.

According to an approach called **maximum likelihood,** researchers ask the question: What is the probability that an evolutionary model and a proposed phylogenetic tree would give rise to the observed data? The rationale is that a phylogenetic tree that gives a higher probability of producing the observed data is preferred to any trees that give a lower probability. By comparison, **Bayesian methods** ask the question: What is the probability that a particular phylogenetic tree is correct, given the observed data and a particular evolutionary model. Though the computational strategies of maximum likelihood and Bayesian methods are different (and beyond the scope of this textbook), the goal of both approaches is to identify one or more trees that are most likely to be correct based on an evolutionary model and the available data.

Phylogenetic Trees Refine Our Understanding of Evolutionary Relationships

For molecular evolutionary studies, the DNA sequences of many genes have been obtained from a wide range of sources. Several different types of gene sequences have been used to construct

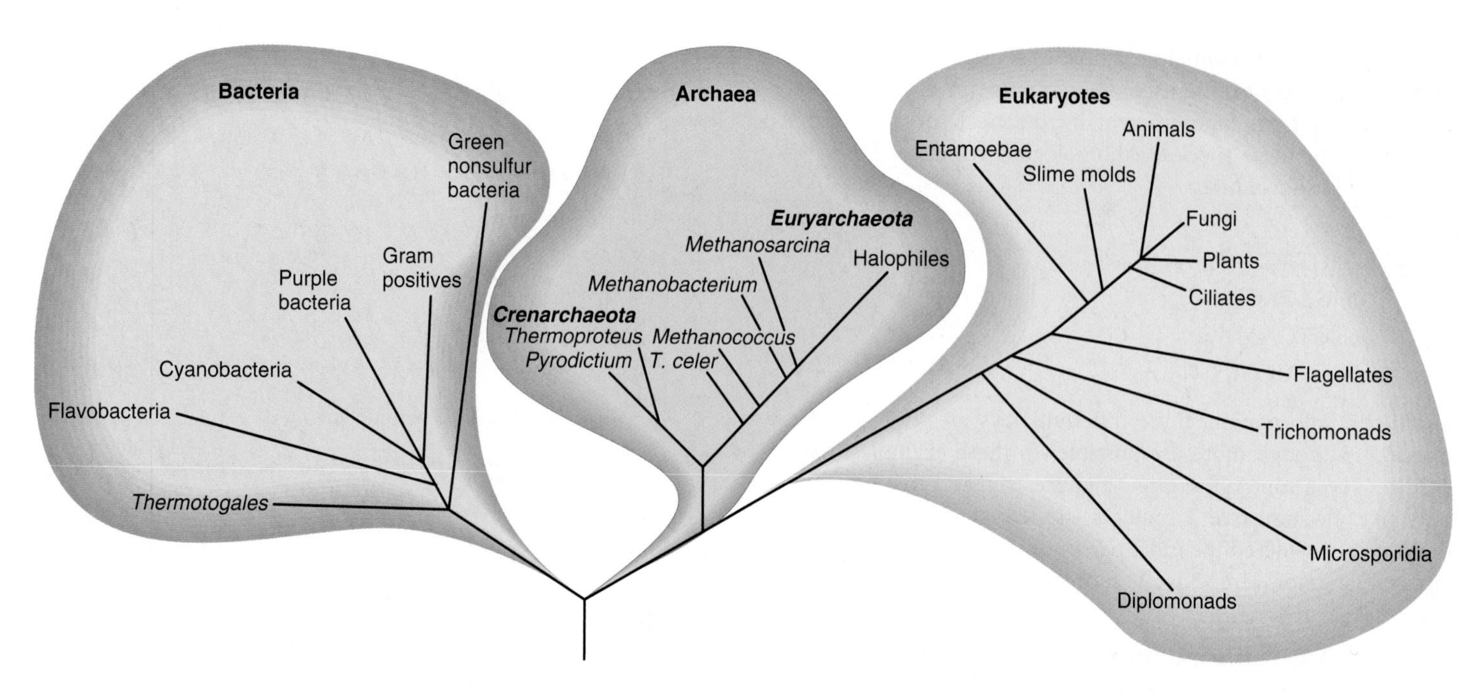

FIGURE 26.9 A phylogenetic tree of all life on Earth based on 16S and 18S rRNA sequences. The relationships in this figure among eukaryotic species have been substantially revised based on newer molecular data (see Figure 26.10).

phylogenetic trees. One very commonly analyzed gene is that encoding 16S rRNA (bacterial) or 18S rRNA (eukaryotic), an rRNA found in the small ribosomal subunit. This gene has been sequenced from thousands of different species. Because rRNA is universal in all living organisms, its function was established at an early stage in the evolution of life on this planet, and its sequence has changed fairly slowly. Presumably, most mutations in this gene are deleterious, so few neutral or beneficial alleles can occur. This limitation causes this gene sequence to change very slowly during evolution. Furthermore, 16S and 18S rRNAs are rather large molecules and therefore contain a large amount of sequence information.

In 1977, Carl Woese analyzed 16S and 18S rRNA sequences and identified a new domain of life called Archaea. **Figure 26.9** illustrates a phylogenetic tree of all life based on Woese's work. It proposes three main evolutionary branches: the **Bacteria,** the **Archaea,** and the **Eukaryotes** (also called Eukarya). From these types of genetic analyses, it has become apparent that all living organisms are connected through a complex evolutionary tree.

Although the work of Woese was a breakthrough in our appreciation of evolution, more recent molecular genetic data have shed new light on the classification of species. Specifically, biologists once categorized eukaryotic species into four kingdoms: protists (Protista), fungi (Fungi), animals (Animalia), and plants (Plantae). However, recent models propose several major groups, called **supergroups,** as a way to organize eukaryotes into monophyletic groups. **Figure 26.10** shows a diagram that hypothesizes several supergroups. Of the four traditional kingdoms, both Fungi and Animalia are within the Opisthokonta supergroup, and Plantae is found within land plants and algal relatives. The remaining branches and supergroups used to be classified within the single kingdom Protista. As seen in this figure, molecular data and newer ways of building trees reveal that protists played a key role in the evolution of many diverse groups of eukaryotic species, producing several large monophyletic supergroups.

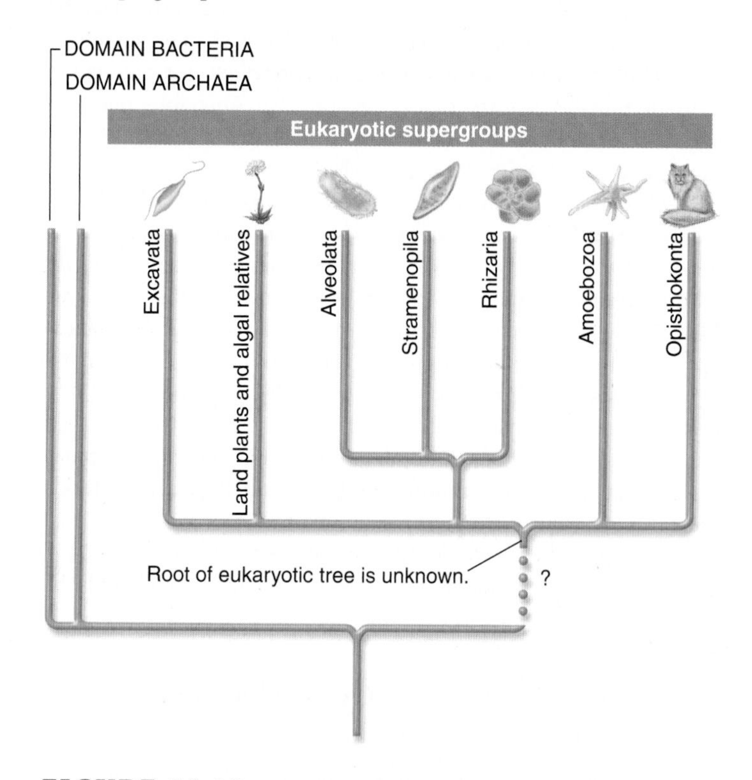

FIGURE 26.10 A modern cladogram for eukaryotes. Each of the supergroups shown here are hypothesized to be monophyletic. This drawing should be considered a working hypothesis. The arrangement of these supergroups relative to each other is not entirely certain.

Horizontal Gene Transfer Also Contributes to the Evolution of Species

The types of phylogenetic trees considered thus far are examples of **vertical evolution,** in which species evolve from preexisting species by the accumulation of gene mutations and by changes in chromosome structure and number. Vertical evolution involves genetic changes in a series of ancestors that form a lineage. In addition to vertical evolution, however, species accumulate genetic changes by another process called **horizontal gene transfer.** This refers to a process in which an organism incorporates genetic material from another organism without being the offspring of that organism. It often involves the exchange of genetic material between different species.

An analysis of many genomes suggests that horizontal gene transfer was prevalent during the early stages of evolution, when all organisms were unicellular, but continued even after the divergence of the three major domains of life. With regard to modern organisms, horizontal gene transfer remains prevalent among prokaryotic species. By comparison, this process is less common among eukaryotes, though it does occur. Researchers have speculated that multicellularity and sexual reproduction have presented barriers to horizontal gene transfer in many eukaryotes. For a gene to be transmitted to eukaryotic offspring, it must be transferred into a eukaryotic gamete or a cell that gives rise to gametes.

How has horizontal gene transfer affected evolution? In the past few decades, scientists have debated the role of horizontal gene transfer in the earliest stages of evolution, prior to the emergence of the two prokaryotic domains. The traditional viewpoint was that the three domains of life arose from a single type of prokaryotic cell (or pre-prokaryotic cell) called the universal ancestor. However, genomic research has suggested that horizontal gene transfer may have been particularly common during the early stages of evolution on Earth, when all species were unicellular. Rather than proposing that all life arose from a single type of prokaryotic cell, horizontal gene transfer may have been so prevalent that the universal ancestor may have actually been an ancestral community of cell lineages that evolved as a whole. If that were the case, the tree of life cannot be traced back to a single universal ancestor.

Figure 26.11 illustrates a schematic scenario for the evolution of life on Earth that includes the roles of both vertical evolution and horizontal gene transfer. This has been described as

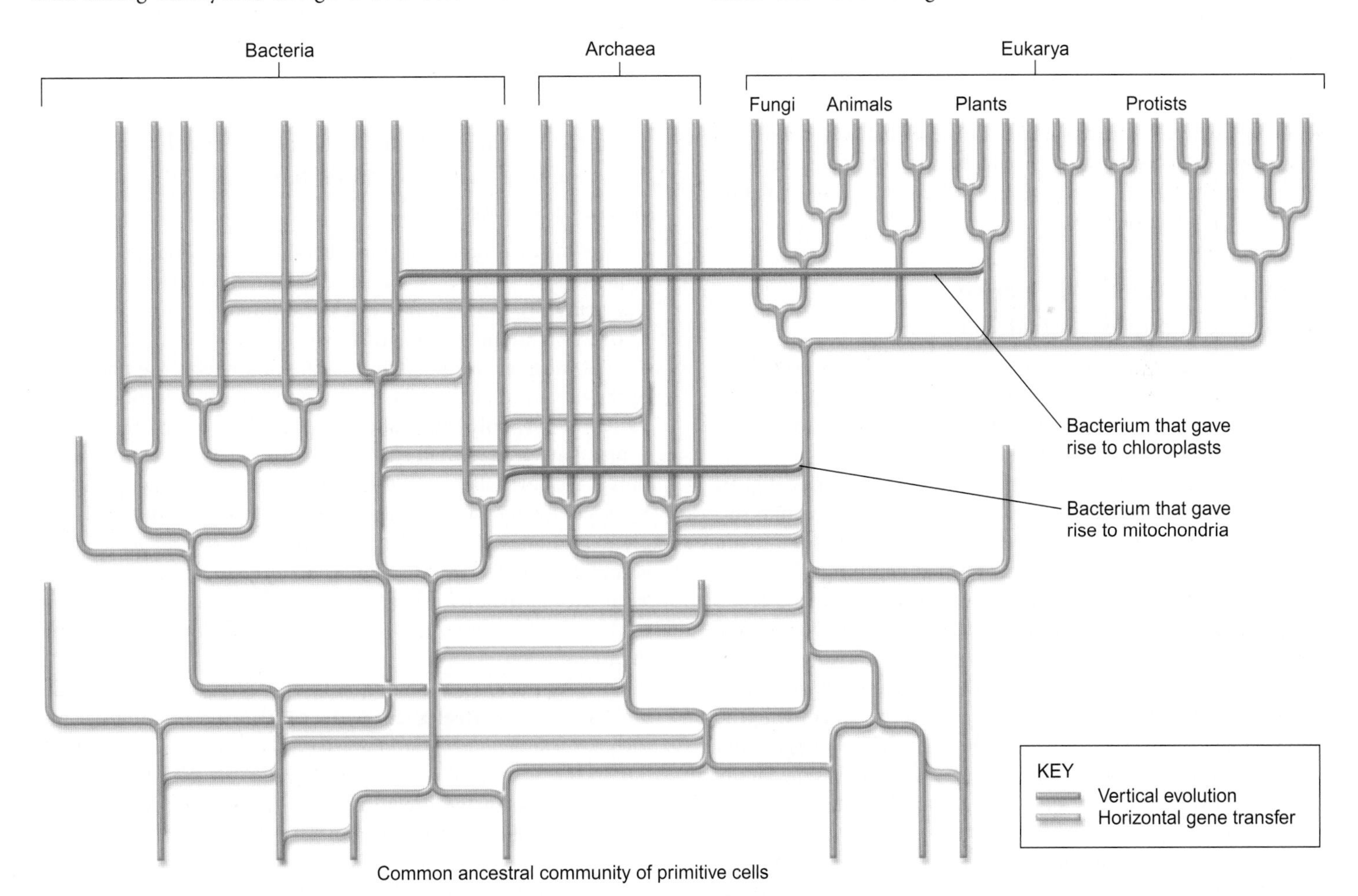

FIGURE 26.11 A revised view regarding the evolution of life, incorporating the concept of horizontal gene transfer. This phylogenetic tree shows a classification of life on Earth that includes the contribution of horizontal gene transfer in the evolution of species on our planet. This phenomenon was prevalent during the early stages of evolution when all organisms were unicellular. Horizontal gene transfer continues to be a prominent factor in the speciation of bacteria and archaea. Note: This tree is meant to be schematic. Figure 26.10 is a more realistic representation of the evolutionary relationships among modern species. In this simplified drawing, the horizontal gene transfer that gave rise to chloroplasts is shown as a single event, though multiple transfers of plastids have occurred during the evolution of life on Earth.

a "web of life" rather than a "tree of life." Instead of a universal ancestor, a web of life began with a community of primitive cells that transferred genetic material in a horizontal fashion. Horizontal gene transfer was also prevalent during the early evolution of bacteria and archaea and when eukaryotes first emerged as unicellular species. Although horizontal gene transfer remains a prominent way to foster evolutionary change in modern bacteria and archaea, the region of the diagram that contains eukaryotic species has a more treelike structure, because horizontal gene transfer has become much less common in these species.

26.3 MOLECULAR EVOLUTION

Molecular evolution refers to patterns and processes associated with evolutionary change at the molecular level. Such changes may be phenotypically neutral or they may underlie the phenotypic changes associated with evolution. Differences in nucleotide sequences are quantitative and can be analyzed using mathematical principles in conjunction with computer programs. Evolutionary changes at the DNA level can be objectively compared among different species to establish evolutionary relationships. Furthermore, this approach can be used to compare any two existing organisms, no matter how greatly they differ in their morphological traits. For example, we can compare DNA sequences between humans and bacteria, or between plants and fruit flies. Such comparisons would be very difficult at a morphological level. In this section, we will examine how evolution occurs at the molecular level.

Homologous Genes Are Derived from a Common Ancestral Gene

As we have seen, a phylogenetic tree is based on homology, which refers to similarities among various species that occur because the species are derived from a common ancestor. In this section, we will focus on genetic homology. Two genes are said to be homologous if they are derived from the same ancestral gene. During evolution, a single species may become divided into two or more different ones. When two homologous genes are found in different species, these genes are termed **orthologs.** In Chapter 23, we considered orthologs of *Hox* genes. In that case, several homologous genes were identified in the fruit fly and the mouse. In addition, two or more homologous genes can be found within a single species. These are termed paralogous genes, or **paralogs.** As discussed in Chapter 8, this can occur because abnormal gene duplication events can produce multiple copies of a gene and ultimately lead to the formation of a **gene family.** A gene family consists of two or more paralogs within the genome of a particular species.

Figure 26.12a shows examples of both orthologs and paralogs in the globin gene family. Hemoglobin is an oxygen-carrying protein found in all vertebrate species. It is composed of two different subunits, encoded by the α-globin and β-globin genes. Figure 26.12a shows an alignment of the deduced amino acid sequences encoded by these genes. The sequences are homologous between humans and horses because of their evolutionary relationship. We say that the human and horse α-globin genes are orthologs of each other, as are the human and horse β-globin genes. The α-globin and β-globin genes in humans are paralogs of each other.

As shown in **Figure 26.12b**, the sequences of the orthologs are more similar to each other than they are to the paralogs. For example, the sequences of human and horse β globins show 25 differences, whereas the human β globin and human α globins show 84 differences. What do these results mean? They indicate that the gene duplications that created the α-globin and β-globin genes occurred long before the evolutionary divergence that produced different species of mammals. For this reason, there was a greater amount of time for the α- and β-globin genes to accumulate changes compared to the amount of time that has elapsed since the evolutionary divergence of mammalian species. This idea is schematically shown in **Figure 26.13.**

Based on the analysis of genetic sequences, evolutionary biologists have estimated that the gene duplication that produced the α-globin and β-globin gene lineages occurred approximately 400 mya, whereas the speciation events that resulted in different species of mammals occurred less than 200 mya. Therefore, the α-globin and β-globin genes have had much more time to accumulate changes relative to each other.

Genetic Variation at the Molecular Level Is Associated with Neutral Changes in Gene Sequences

As we have seen, the globin genes exhibit variation in their sequences. Researchers have asked the question, Is such variation due primarily to mutations that are favored by natural selection or due to random genetic drift?

A **nonneutral mutation** is one that affects the phenotype of the organism and can be acted on by natural selection. Such a mutation may only subtly alter the phenotype of an organism, or it may have a major effect. According to Darwin, natural selection is the agent that leads to evolutionary change in populations. It selects for individuals with the highest Darwinian fitness and often promotes the establishment of beneficial alleles and the elimination of deleterious ones. Therefore, many geneticists have assumed that natural selection is the dominant factor in changing the genetic composition of natural populations, thereby leading to variation.

In opposition to this viewpoint, in 1968, Motoo Kimura proposed the **neutral theory of evolution.** According to this theory, most genetic variation observed in natural populations is due to the accumulation of neutral mutations that do not affect the phenotype of the organism and are not acted on by natural selection. For example, a mutation within a structural gene that changes a glycine codon from GGG to GGC does not affect the amino acid sequence of the encoded protein. Because neutral mutations do not affect phenotype, they spread throughout a population according to their frequency of appearance and to random genetic drift. This theory has been called the "survival of the luckiest" and also **non-Darwinian evolution** to contrast

Hemoglobin

β	1	2	3	4	5	6	7	8	9	10	11	12	13	14	15	16	17	18	—	—	19	20	21	22	23	24	25	26	27	28	29
Human	Val	His	Leu	Thr	Pro	Glu	Glu	Lys	Ser	Ala	Val	Thr	Ala	Leu	Trp	Gly	Lys	Val	—	—	Asn	Val	Asp	Glu	Val	Gly	Gly	Glu	Ala	Leu	Gly
Horse	Val	Gln	Leu	Ser	Gly	Glu	Glu	Lys	Ala	Ala	Val	Leu	Ala	Leu	Trp	Asp	Lys	Val	—	—	Asn	Glu	Glu	Glu	Val	Gly	Gly	Glu	Ala	Leu	Gly
α	1	—	2	3	4	5	6	7	8	9	10	11	12	13	14	15	16	17	18	19	20	21	22	23	24	25	26	27	28	29	30
Human	Val	—	Leu	Ser	Pro	Ala	Asp	Lys	Thr	Asn	Val	Lys	Ala	Ala	Trp	Gly	Lys	Val	Gly	Ala	His	Ala	Gly	Glu	Tyr	Gly	Ala	Glu	Ala	Leu	Glu
Horse	Val	—	Leu	Ser	Ala	Ala	Asp	Lys	Thr	Asn	Val	Lys	Ala	Ala	Trp	Ser	Lys	Val	Gly	Gly	His	Ala	Gly	Glu	Val	Gly	Ala	Glu	Ala	Leu	Glu

β	30	31	32	33	34	35	36	37	38	39	40	41	42	43	44	45	46	47	48	49	50	51	52	53	54	55	56	57	58	59	60
Human	Arg	Leu	Leu	Val	Val	Tyr	Pro	Trp	Thr	Gln	Arg	Phe	Phe	Glu	Ser	Phe	Gly	Asp	Leu	Ser	Thr	Pro	Asp	Ala	Val	Met	Gly	Asn	Pro	Lys	Val
Horse	Arg	Leu	Leu	Val	Val	Tyr	Pro	Trp	Thr	Gln	Arg	Phe	Phe	Asp	Ser	Phe	Gly	Asp	Leu	Ser	Asn	Pro	Gly	Ala	Val	Met	Gly	Asn	Pro	Lys	Val
α	31	32	33	34	35	36	37	38	39	40	41	42	43	44	45	46	—	47	48	49	50	—	—	—	—	—	51	52	53	54	55
Human	Arg	Met	Phe	Leu	Ser	Phe	Pro	Thr	Thr	Lys	Thr	Tyr	Phe	Pro	His	Phe	—	Asp	Leu	Ser	His	—	—	—	—	—	Gly	Ser	Ala	Gln	Val
Horse	Arg	Met	Phe	Leu	Gly	Phe	Pro	Thr	Thr	Lys	Thr	Tyr	Phe	Pro	His	Phe	—	Asp	Leu	Ser	His	—	—	—	—	—	Gly	Ser	Ala	Gln	Val

β	61	62	63	64	65	66	67	68	69	70	71	72	73	74	75	76	77	78	79	80	81	82	83	84	85	86	87	88	89	90	91
Human	Lys	Ala	His	Gly	Lys	Lys	Val	Leu	Gly	Ala	Phe	Ser	Asp	Gly	Leu	Ala	His	Leu	Asp	Asn	Leu	Lys	Gly	Thr	Phe	Ala	Thr	Leu	Ser	Glu	Leu
Horse	Lys	Ala	His	Gly	Lys	Lys	Val	Leu	His	Ser	Phe	Gly	Glu	Gly	Val	His	His	Leu	Asp	Asn	Leu	Lys	Gly	Thr	Phe	Ala	Ala	Leu	Ser	Glu	Leu
α	56	57	58	59	60	61	62	63	64	65	66	67	68	69	70	71	72	73	74	75	76	77	78	79	80	81	82	83	84	85	86
Human	Lys	Gly	His	Gly	Lys	Lys	Val	Ala	Asp	Ala	Leu	Thr	Asn	Ala	Val	Ala	His	Val	Asp	Asp	Met	Pro	Asn	Ala	Leu	Ser	Ala	Leu	Ser	Asp	Leu
Horse	Lys	Ala	His	Gly	Lys	Lys	Val	Gly	Asp	Ala	Leu	Thr	Leu	Ala	Val	Gly	His	Leu	Asp	Asp	Leu	Pro	Gly	Ala	Leu	Ser	Asp	Leu	Ser	Asn	Leu

β	92	93	94	95	96	97	98	99	100	101	102	103	104	105	106	107	108	109	110	111	112	113	114	115	116	117	118	119	120	121	122
Human	His	Cys	Asp	Lys	Leu	His	Val	Asp	Pro	Glu	Asn	Phe	Arg	Leu	Leu	Gly	Asn	Val	Leu	Val	Cys	Val	Leu	Ala	His	His	Phe	Gly	Lys	Glu	Phe
Horse	His	Cys	Asp	Lys	Leu	His	Val	Asp	Pro	Glu	Asn	Phe	Arg	Leu	Leu	Gly	Asn	Val	Leu	Ala	Val	Val	Leu	Ala	Arg	His	Phe	Gly	Lys	Asp	Phe
α	87	88	89	90	91	92	93	94	95	96	97	98	99	100	101	102	103	104	105	106	107	108	109	110	111	112	113	114	115	116	117
Human	His	Ala	His	Lys	Leu	Arg	Val	Asp	Pro	Val	Asn	Phe	Lys	Leu	Leu	Ser	His	Cys	Leu	Leu	Val	Thr	Leu	Ala	Ala	His	Leu	Pro	Ala	Glu	Phe
Horse	His	Ala	His	Lys	Leu	Arg	Val	Asp	Pro	Val	Asn	Phe	Lys	Leu	Leu	Ser	His	Cys	Leu	Leu	Ser	Thr	Leu	Ala	Val	His	Leu	Pro	Asn	Asp	Phe

β	123	124	125	126	127	128	129	130	131	132	133	134	135	136	137	138	139	140	141	142	143	144	145	146
Human	Thr	Pro	Pro	Val	Gln	Ala	Ala	Tyr	Gln	Lys	Val	Val	Ala	Gly	Val	Ala	Asn	Ala	Leu	Ala	His	Lys	Tyr	His
Horse	Thr	Pro	Glu	Leu	Gln	Ala	Ser	Tyr	Gln	Lys	Val	Val	Ala	Gly	Val	Ala	Asn	Ala	Leu	Ala	His	Lys	Tyr	His
α	118	119	120	121	122	123	124	125	126	127	128	129	130	131	132	133	134	135	136	137	138	139	140	141
Human	Thr	Pro	Ala	Val	His	Ala	Ser	Leu	Asp	Lys	Phe	Leu	Ala	Ser	Val	Ser	Thr	Val	Leu	Thr	Ser	Lys	Tyr	Arg
Horse	Thr	Pro	Ala	Val	His	Ala	Ser	Leu	Asp	Lys	Phe	Leu	Ser	Ser	Val	Ser	Thr	Val	Leu	Thr	Ser	Lys	Tyr	Arg

(a) Alignment of human and horse globin polypeptides

Orthologs	Number of Amino Acid Differences
Human β globin vs horse β globin	25 out of 146
Human α globin vs horse α globin	18 out of 141
Paralogs	
Human β globin vs human α globin	84 out of 146
Horse β globin vs horse α globin	81 out of 146

(b) A comparison of amino acid differences between orthologs and paralogs

FIGURE 26.12 A comparison of the α- and β-globin polypeptides from humans and horses. **(a)** An alignment of the deduced amino acid sequences obtained by sequencing the exon portions of the corresponding genes. The gaps indicate where additional amino acids are found in the sequence of myoglobin, another member of this gene family. **(b)** A comparison of amino acid differences between orthologs and paralogs.

it with Darwin's theory, which focuses on fitness. Kimura agreed with Darwin that natural selection is responsible for adaptive changes in a species during evolution. His main argument is that most modern variation in gene sequences is neutral with respect to natural selection.

In support of the theory, Kimura and his colleague Tomoko Ohta outlined five principles that govern the evolution of genes at the molecular level:

1. For each protein, the rate of evolution, in terms of amino acid substitutions, is approximately constant with regard to neutral substitutions that do not affect protein structure or function.

   Evidence: As an example, the amount of genetic variation between the coding sequence of the human α-globin and β-globin genes is approximately the same as the difference between the α-globin and β-globin genes in the horse (shown earlier in Figure 26.12b). This type of comparison holds true in many different genes compared among many different species.
2. Proteins that are functionally less important for the survival of an organism, or parts of a protein that are less important for its function, tend to evolve faster than more important proteins or regions of a protein. In other words, during evolution, less important proteins or protein domains accumulate amino acid substitutions more rapidly than important ones do.

   Evidence: Certain proteins are critical for survival, and their structure is precisely suited to their function. Examples are the histone proteins necessary for nucleosome formation in eukaryotes. Histone genes tolerate very few mutations and have evolved extremely slowly. By comparison, fibrinopeptides, which bind to fibrinogen to form a blood clot, evolve very rapidly.

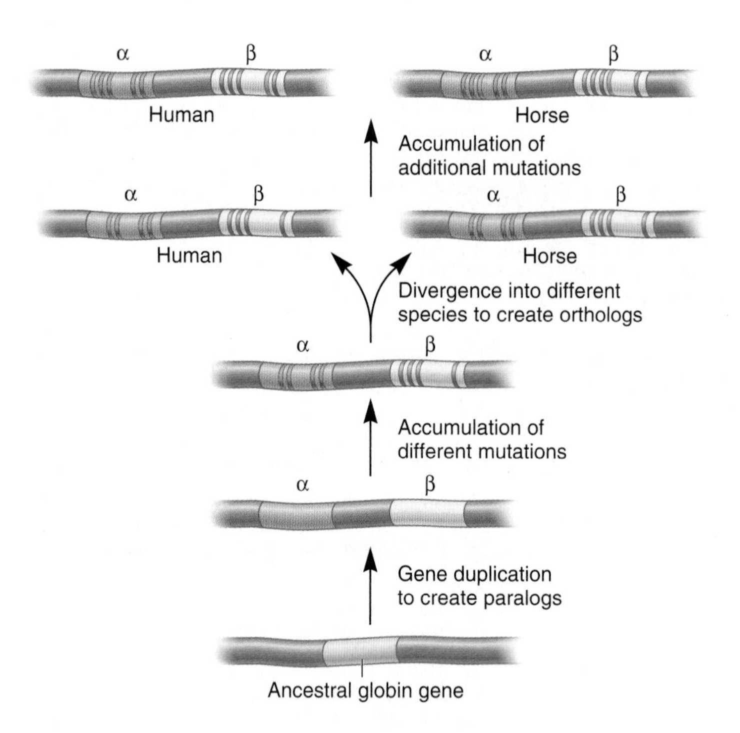

FIGURE 26.13 Evolution of paralogs and orthologs. In this schematic example, the ancestral globin gene duplicated to create the α- and β-globin genes, which are paralogous genes. (Note: Actually, gene duplications occurred several times to produce a large globin gene family, but this is a simplified example that shows only one gene duplication event. Also, chromosomal rearrangements have placed the α- and β-globin genes on different chromosomes.) Over time, these two paralogs accumulated different mutations, designated with red and blue lines. At a later point in evolution, a species divergence occurred to create different mammalian species including humans and horses. Over time, the orthologs also accumulated different mutations, designated with green lines. As noted here, the orthologs have fewer differences than the paralogs because the gene duplication occurred prior to the species divergence.

Presumably, the sequence of amino acids in this polypeptide is not very important for allowing it to aggregate and form a clot. Another example concerns the amino acid sequences of enzymes. Amino acid substitutions are very rare within the active site, which is critical for function, but are more frequent in other parts of the protein.

3. Amino acid substitutions that do not significantly alter the existing structure and function of a protein are found more commonly than disruptive amino acid changes.

   Evidence: When comparing the coding sequences within homologous genes of modern species, nucleotide differences are more likely to be observed in the wobble base than in the first or second base within a codon. Mutations in the wobble base are often silent because they do not change the amino acid sequence of the protein. In addition, conservative substitutions (i.e., a substitution with a similar amino acid, such as a nonpolar amino acid for another nonpolar amino acid) are fairly common. By comparison, nonconservative substitutions—those that significantly alter the structure and function of a protein—are less frequent. Nonsense and frameshift mutations are very rare within the coding sequences of genes. Also, intron sequences evolve more rapidly than exon sequences.

4. Gene duplication often precedes the emergence of a gene having a new function.

   Evidence: When a single copy of a gene exists in a species, it usually plays a functional role similar to that of the homologous gene found in another species. Gene duplications have produced gene families in which each family member can evolve somewhat different functional roles. Examples include the globin family described in Chapter 8 and the *Hox* genes described in Chapter 23.

5. Selective elimination of definitely deleterious mutations and the random fixation of selectively neutral or very slightly deleterious alleles occur far more frequently in evolution than selection of advantageous mutants.

   Evidence: As mentioned in principle 3, silent and conservative mutations are much more common than nonconservative substitutions. Presumably these nonconservative mutations usually have a negative effect on the phenotype of the organism, so they are effectively eliminated from the population by natural selection. On rare occasions, however, an amino acid substitution due to a mutation may have a beneficial effect on the phenotype. For example, a nonconservative mutation in the β-globin gene produced the Hb^S allele, which gives an individual resistance to malaria in the heterozygous condition.

In general, the DNA sequencing of hundreds of thousands of different genes from thousands of species has provided compelling support for these five principles of gene evolution at the molecular level. When it was first proposed, the neutral theory sparked a great debate. Some geneticists, called selectionists, strongly opposed the neutralist theory of evolution. However, the debate largely cooled after Ohta incorporated the concept of nearly neutral mutations into the theory. Nearly neutral mutations have a minimal effect on phenotype—they may be slightly beneficial or slightly detrimental. Ohta suggested that the prevalence of such alleles can depend mostly on natural selection or mostly on genetic drift, depending on the population size.

Why do evolutionary biologists care about neutral or nearly neutral mutations? One reason is that their prevalence is used as a tool to add a time scale to phylogenetic trees. This topic is discussed next.

Molecular Clocks Can Be Used to Date the Divergence of Species

According to the neutral theory of evolution, most of the observed variation is due to neutral mutations. In a sense, the relatively constant rate of neutral or nearly neutral mutations acts as a **molecular clock** on which to measure evolutionary time. According to this idea, neutral mutations become fixed in a population at a rate that is proportional to the rate of mutation per generation. On this basis, the genetic divergence between species that is due to neutral mutations reflects the time elapsed since their last common ancestor.

Figure 26.14a shows an example of a molecular clock involving a study of superoxide dismutase found in 27 different fruit fly species. (Superoxide dismutase protects cells against harmful free radicals). Twenty-three species were in the genus *Drosophila*, two in the genus *Chymomyza*, one in the genus *Scaptodrosophila*, and one in the genus *Ceratitis*. The genus *Ceratitis* is in the family Tephritidae and is more distantly related to the other 26 species, which are in the family Drosophitidae. In this figure, the *y*-axis is a measure of the average number of amino acid differences in superoxide dismutase between pairs of species or between groups of species. The *x*-axis plots the amount of time that has elapsed since a pair or two different groups shared a common ancestor. For example, the yellow dot represents the average number of amino acid differences between *Ceratitis* (in family Tephritidae) versus the other species (in family Drosophitidae). Approximately 30 amino acid differences were observed. By comparison, each of the two red dots compares one of the *Chymomyza* species with the *Drosophila* species. The blue dots are pairwise comparisons of *Drosophila* species with each other or with the *Scaptodrosophila* species. As a general trend, species that diverged a longer time ago show a greater number of amino acid differences than species that diverged more recently. The explanation for this phenomenon is that species accumulate independent mutations (e.g., nearly neutral mutations) after they have diverged from each other. A longer period of time since their divergence allows for a greater accumulation of mutations that makes their sequences different.

To further understand the concept of a molecular clock, let's consider how the molecular clock data are related to a phylogenetic tree (**Figure 26.14b**). In this diagram, the *Drosophila* genus is divided into five subgenera. The divergence between the genus *Ceratitis* and the other genera occurred a long time ago, nearly 100 mya. The molecular clock data described in Figure 26.14a showed a relatively large number of amino acid differences (about 30) between *Ceratitis* and the other genera. By comparison, the divergence between *Chymomyza* species and *Drosophila* species occurred more recently—about 65 mya as shown in Figure 26.14b. Likewise the molecular clock data revealed fewer amino acid differences of about 20 between *Chromomyza* and *Drosophila* species (see Figure 26.14a). Finally, the five subgenera of *Drosophila* diverged the most recently and had the fewest number of amino acids differences.

Figure 26.14a suggests a linear relationship between the number of sequence changes and the time of divergence. Such a relationship indicates that the observed rate of neutral or nearly neutral mutations remains constant over millions of years. For example, a linear relationship predicts that a pair of species showing 20 nucleotide differences in a given genetic sequence would have a (most recent) common ancestor that existed twice as long ago as another pair showing 10 nucleotide differences. Although actual data, such as the data shown in Figure 16.14a, sometimes show a relatively linear relationship over a defined period, evolutionary biologists have discovered that molecular clocks are often not linear. This is particularly true over very long periods or when comparing distantly related taxa. When comparing different

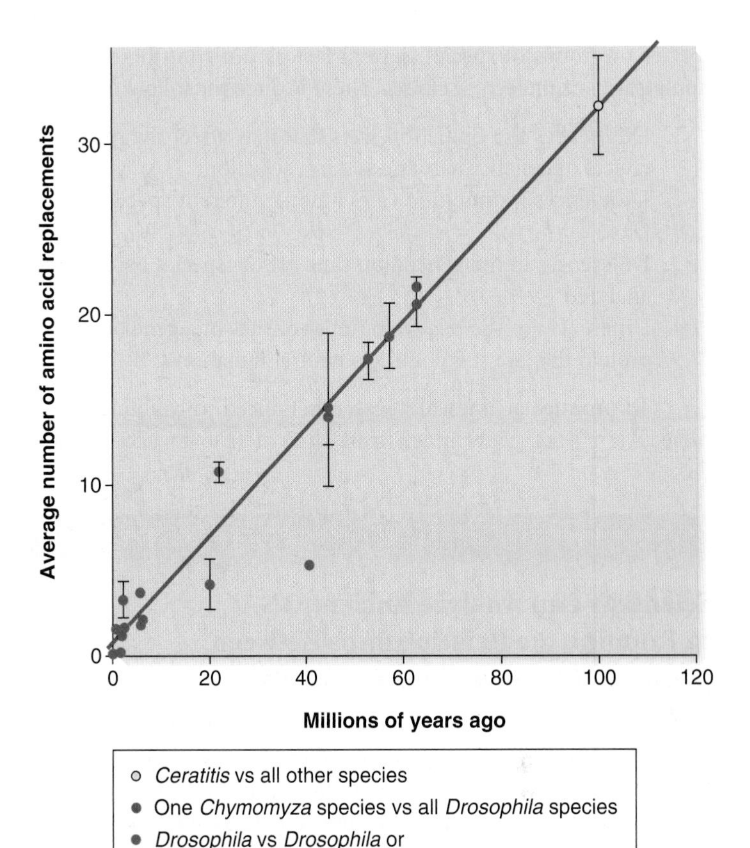

(a) An example of a molecular clock

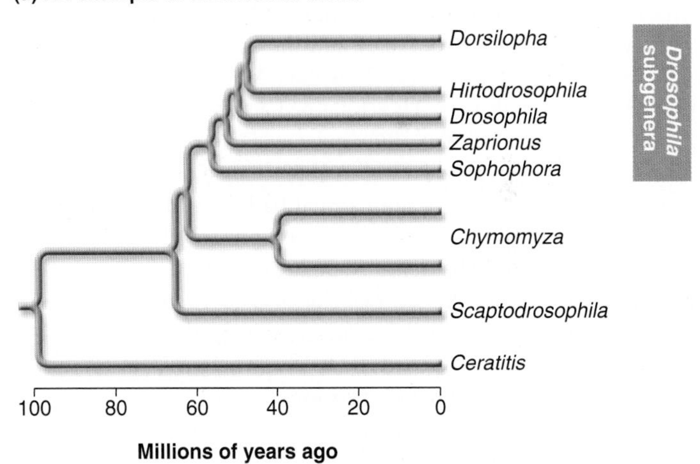

(b) A phylogenetic tree with a timescale

FIGURE 26.14 A molecular clock. According to the concept of a molecular clock, neutral or nearly neutral mutations accumulate over evolutionary time at a fairly constant rate. (**a**) The clock shown here is based on an analysis of superoxide dismutase found in 27 different fruit fly species. Twenty-three species were in the genus *Drosophila*, two in the genus *Chymomyza*, one in the genus *Scaptodrosophila*, and one in the genus *Ceratitis*. When comparing the amino acid sequence of this enzyme between species or groups of species, those that diverged more recently tend to have fewer differences than those whose common ancestor occurred in the very distant past. (**b**) A phylogenetic tree for the species shown in part (a). Parts (a) and (b) are modified from F. J. Ayala (1997) Vagaries of the molecular clock. *Proc Natl Acad Sci USA 94*, 7776–7783.

species or groups of species, several factors can contribute to the nonlinearity of molecular clocks. These include the following:

1. Differences in population sizes that may affect the relative effects of genetic drift and natural selection
2. Differences in mutation rates among different species and taxa
3. Differences in the generation times of the species being analyzed
4. Differences in the relative number of sites in a gene or protein that are susceptible to neutral mutations

To produce a timescale, researchers need to calibrate their molecular clocks. How much time does it take to accumulate a certain percentage of nucleotide changes? To perform such a calibration, researchers must have information regarding the date when two species shared a common ancestor. Such information could come from the fossil record, for instance. The genetic differences between those species are then divided by the time elapsed since their last common ancestor to calculate a rate of change. For example, research suggests that humans and chimpanzees diverged from a common ancestor approximately 6 mya. The percentage of nucleotide differences in mitochondrial DNA between humans and chimpanzees is 12%. From these data, the molecular clock for changes in mitochondrial DNA sequences of primates is calibrated at roughly 2% nucleotide changes per million years.

EXPERIMENT 26A

Scientists Can Analyze Ancient DNA to Examine the Relationships Between Living and Extinct Flightless Birds

The majority of phylogenetic trees have been constructed from molecular data using DNA samples collected from living species. With this approach, we can infer the prehistoric changes that gave rise to present-day DNA sequences. As an alternative, scientists have discovered that it is occasionally possible to obtain DNA sequence information from species that lived in the past. In 1984, the first successful attempt at determining DNA sequences from an extinct species was accomplished by groups at the University of California at Berkeley and the San Diego Zoo, including Russell Higuchi, Barbara Bowman, Mary Freiberger, Oliver Ryder, and Allan Wilson. They obtained a sample of dried muscle from a museum specimen of the quagga (*Equus quagga*), a zebra-like species that became extinct in 1883. This piece of muscle tissue was obtained from an animal that had died 140 years ago. A sample of its skin and muscle had been preserved in salt in the Museum of Natural History at Mainz, Germany. The researchers extracted DNA from the sample, cloned pieces of it into vectors, and then sequenced hybrid vectors containing the quagga DNA. This pioneering study opened the field of **ancient DNA analysis,** also known as **molecular paleontology.**

Since the mid-1980s, many researchers have become excited about the information that might be derived from sequencing DNA obtained from older specimens. Currently there is debate about how long DNA can remain significantly intact after an organism has died. Over time, the structure of DNA is degraded by hydrolysis and the loss of purines. Nevertheless, under certain conditions (e.g., cold temperature, low oxygen), DNA samples may remain stable for as long as 50,000 to 100,000 years, and perhaps longer.

In most studies involving prehistoric specimens (in particular, those that are much older than the salt-preserved quagga sample), the ancient DNA is extracted from bone, dried muscle, or preserved skin. These samples are often obtained from museum specimens that have been gathered by archaeologists. However, it is unlikely that enough DNA can be extracted to enable a researcher to directly clone the DNA into a vector. Since 1985, however, the advent of PCR technology, described in Chapter 18, has made it possible to amplify very small amounts of DNA using PCR primers that flank a region within the 12S rRNA gene, a mitochondrial gene. In recent years, this approach has been used to elucidate the phylogenetic relationships between modern and extinct species.

In the experiment described in **Figure 26.15**, Alan Cooper, Cécile Mourer-Chauviré, Geoffrey Chambers, Arndt von Haeseler, Allan Wilson, and Svante Pääbo investigated the evolutionary relationships among some extinct and modern species of flightless birds. Two groups of flightless birds, the moas and the kiwis, existed in New Zealand during the Pleistocene era. The moas are now extinct, although 11 species were formerly present. In this study, the researchers investigated the phylogenetic relationships among four extinct species of moas that were available as museum samples, three kiwis of New Zealand, and several other (nonextinct) species of flightless birds. These included the emu and the cassowary (found in Australia and New Guinea), the ostrich (found in Africa and formerly Asia), and two rheas (found in South America).

The samples from the various species were subjected to PCR to amplify the 12S rRNA gene. This provided enough DNA to subject the gene to DNA sequencing. The sequences of the genes were aligned using computer programs described in Chapter 21.

THE GOAL

Because DNA is a relatively stable molecule, it can be amplified by PCR from a preserved sample of a deceased organism and subjected to DNA sequencing. A comparison of these DNA sequences with modern species may help elucidate the phylogenetic relationships between extinct and modern species.

ACHIEVING THE GOAL — FIGURE 26.15 DNA analysis reveals phylogenetic relationships among extinct and modern flightless birds.

Starting material: Tissue samples from four extinct species of moas were obtained from museum specimens. Tissue samples were also obtained from three species of kiwis, one emu, one cassowary, one ostrich, and two species of rhea.

1. For soft tissue samples, treat with proteinase K (which digests protein) and a detergent that dissolves cell membranes. This releases the DNA from the cells.

2. Individually, mix the DNA samples with a pair of PCR primers that are complementary to the 12S rRNA gene.

3. Subject the samples to PCR. See Chapter 18 (Figure 18.6) for a description of PCR.

4. Subject the amplified DNA fragments to DNA sequencing. See Chapter 18 for a description of DNA sequencing.

5. Align the DNA sequences to each other. Methods of DNA sequence alignment are described in Chapter 21.

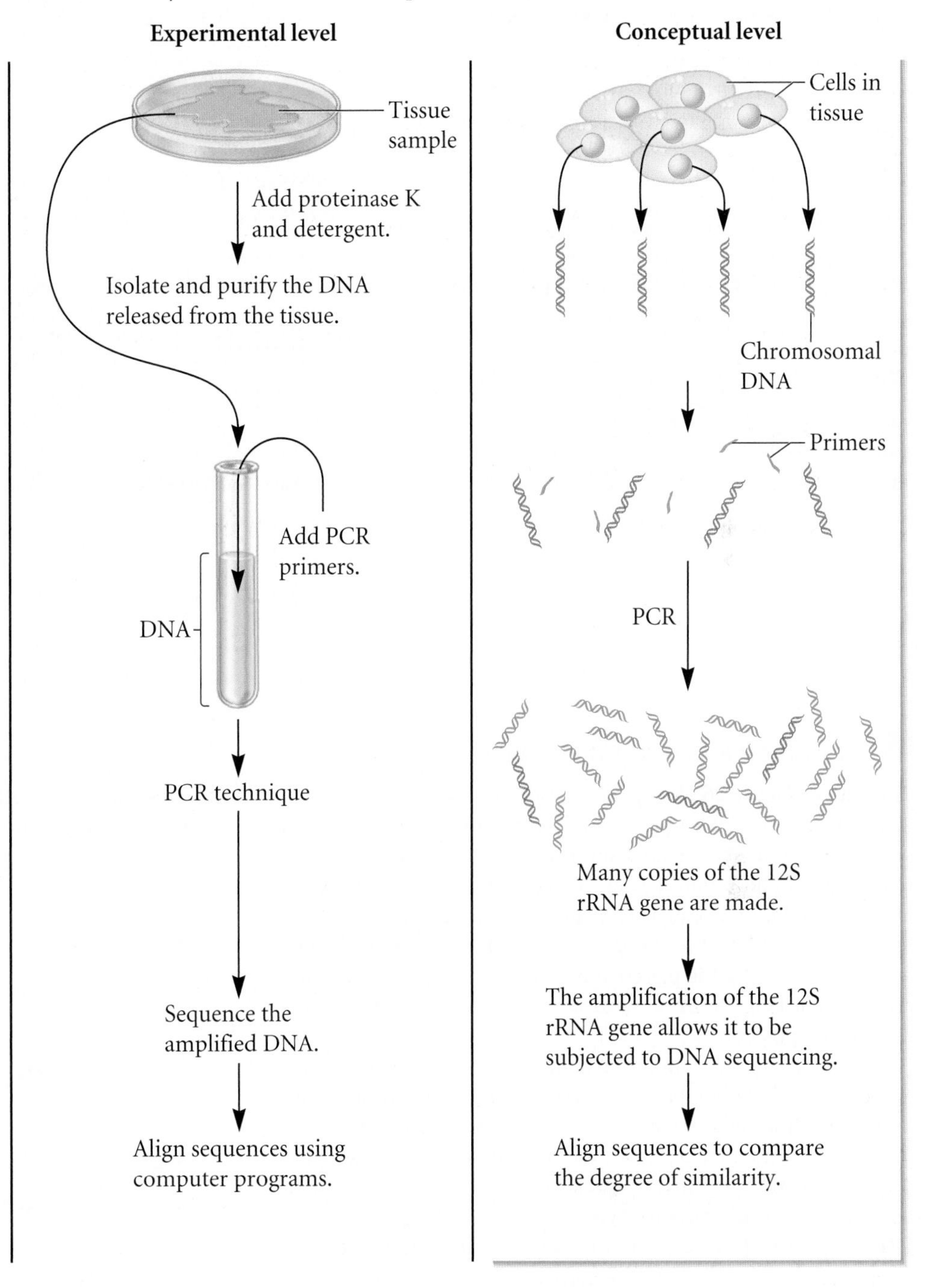

THE DATA

```
MOA 1      GCTTAGCCCTAAATCCAGATACTTACCCTACACAAGTATCCGCCCGAGAACTACGAGCACAAACGCTTAAAACTCTAAGGACTTGGCGGTGCCCCAAACCCACCTAGAGGAGCCTGTTCTATAATCGATAATCCACGATA
MOA 2      ············································································································································································
MOA 3      ····G··········T·······························T································································································
MOA 4      ····························C······································································································C·····T··
KIWI 1     ················T·G·······GT···CT·····C·······················································T····································C········
KIWI 2     ················T·G·G·····AT···CT·····C·······················································T····································C········
KIWI 3     ················T·G·G···G·AT···C······C·······················································T····································C········
EMU        ···············TT······C···T···CAG··C······T··················································T····································C········
CASSOWARY  ···············TT······CG·TA···CTG···························································T····································C········
OSTRICH    ········T······AT···········C··CT·····························································T·············································T
RHEA 1     ················T···········C··CT·····························································T····································C········
RHEA 2     ················C···········C··C······························································T····································C········

MOA 1      CACCCGACCATCCCTCGCCCGT-GCAGCCTACATACCGCCGTCCCCAGCCCGCCT--AATGAAAG-AACAATAGCGAGCACAACAGCCCTCCCCCGCTAACAAGACAGGTCAAGGTATAGCATATGAGATGGAAGAAATG
MOA 2      ····················A·–·······························––········–······················TCA–················································
MOA 3      ············T··T····A·–·······························––········–·····················TA·––·T··············································
MOA 4      ············T··T····A·–·······························––········–·················T····AC––················································
KIWI 1     ·····A······T··T···AAC–A·······T···········G·····T·····AA·····G··–·····C·····A·········TA··–··A··························C··················
KIWI 2     ·····A······T··T···AAC–A·······T···········G·····T·····AA·····G··–·····C·····A·········TA··–··A··························C··················
KIWI 3     ·····A······T··T···AAC–A···················G···········AA·········–····GC···············TACA–··A··························CC·C·······G·······
EMU        ·····AG·····T··T···AA·–A···················G···········––········–·················T····AC––TT·····························G··················
CASSOWARY  ·····A······T··T···AA·TA···················G···········––·G···G··–··············T·······AC––T······························G··················
OSTRICH    ·····A····C····T···A––T····················G··········C––·····G··–·················T·····A––··································GAG···········
RHEA 1     ············T··T····A·–································TA·G······–·····C···AG···T··T···TA–––··································G·············
RHEA 2     ············T··T····A·–································TA·····G··–·····C···A····T··T···TA–––·······G·········································

MOA 1      GGCTACATTTTCTAACATAGAACACCC-------------ACGAAAGAGAAGGTGAAACCCTCCTCAAAAGGCGGATTTAGCAGTAAAATAGAACAAGAATGCCTATTTTAAGCCCGGCCCTGGGGC
MOA 2      ···························–––––––––––––············A······T·······G···············································T···········
MOA 3      ··························T–––––––––––––···························G························G·······C·····C········T···········
MOA 4      ···························–––––––––––––············A··············G························G·······C···C··········T···········
KIWI 1     ···············A······T··T––––––––––––––······A··GGT······T·–C···T·G·····················C····T····GA·T··········–·T······A····
KIWI 2     ···············A······T··T––––––––––––––······A··GGT······T··C···T·G·····················C····T····GA·T··········–·T······A····
KIWI 3     ···············A······T··T––––––––––––––······A··GGTA·····T··C···T·G····A················C····T·····A·T··········–········A····
EMU        ······················T·T·––––––––––––––········AG·T······T·AC·T···G·····················C····T····GA·T·········A–·T···T··A····
CASSOWARY  ······················T···––––––––––––––······A··G·T······T·A····T·G·····················C·········GA·T·········A–········A····
OSTRICH    ······················T··A––––––––––––––·········G·TA·····T·A······G··························T····GA·T··········–T····T··A····
RHEA 1     ···············TC·······A·––––––––––––––···G·····GGCA······–AC····CG····················G···G·TC····A···C·C······–········A····
RHEA 2     ··············GTC·······G·––––––––––––––·········GGCA······–AC····CG····················G·G·G·TC····A···C·C······–········A····
```

Data from: A. Cooper, C. Mourer-Chauviré, G. K. Chambers, et al. (1992) Independent origins of New Zealand moas and kiwis. *Proc Natl Acad Sci USA 89*, 8741–8744.

INTERPRETING THE DATA

The data of Figure 26.15 illustrate a multiple sequence alignment of the amplified DNA sequences. The first line shows the DNA sequence of one extinct moa species, and underneath it are the sequences of the other species. When the other sequences are identical to the first sequence, a dot is placed in the corresponding position. When the sequences are different, the nucleotide base (A, T, G, or C) is placed there. In a few regions, the genes are different lengths. In these cases, a dash is placed at the corresponding position.

As you can see from the large number of dots, the sequences among these flightless birds are very similar. To establish evolutionary relationships, researchers focus on sites where the gene sequences are not identical. At these sites, nucleotide changes have occurred, thereby identifying species with shared derived characters. Some surprising results were obtained. Certain sites in the DNA sequences from the kiwis (a New Zealand species) are the same as the sequence from the ostrich (an African species), but different from those of the moas, which were once found in New Zealand. Likewise, several sites in the DNA sequences of the kiwis are the same as the emu and cassowary (found in Australia and New Guinea), but different from the moas. Contrary to their original expectations, the authors concluded that the kiwis are more closely related to Australian and African flightless birds than they are to the moas. They proposed that New Zealand was colonized twice by ancestors of flightless birds. As shown in **Figure 26.16**, the researchers constructed a new evolutionary tree to illustrate the relationships among these modern and extinct species.

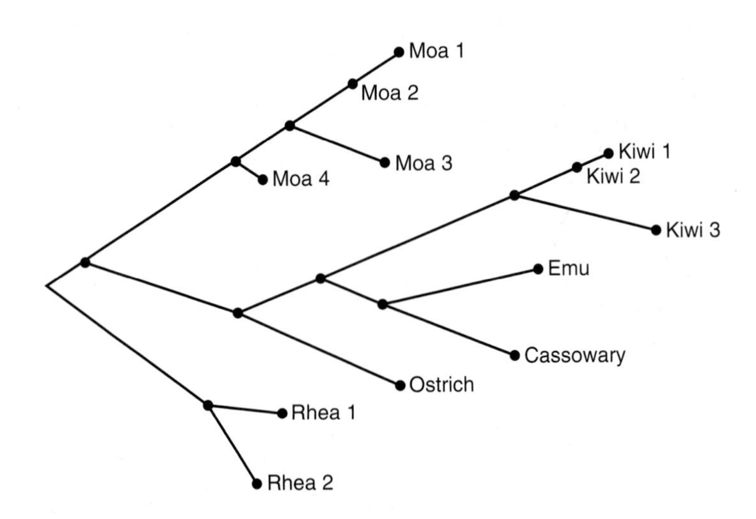

FIGURE 26.16 A revised phylogenetic tree of moas, kiwis, emus, cassowaries, ostriches, and rheas.

FIGURE 26.17 Extinct organisms from which DNA sequences have been obtained. Right panel, from bottom right to top right: quagga, marsupial wolf, sabre-toothed cat, moa, mammoth, cave bear, blue antelope, giant ground sloth, and Aurochs. Left panel, from bottom left to top right: mastodon, New Zealand coot, South Island piopio, Steller's sea cow, Neanderthal man, South Island adzebill (*Aptornis defossor*), Shasta ground sloth, pig-footed bandicoot, moa-nalo, and Balearic Islands cave goat (*Myotragus balearicus*). (Adapted from M. Hofreiter, D. Serre, H.N. Poinar, et al. [2001], Ancient DNA, *Nat Rev Genet 2*, p. 357.)

Since these early studies, sequences of ancient DNA have been derived from a variety of species. **Figure 26.17** shows some extinct organisms from which DNA sequences have been determined. Many of these samples were tens of thousands of years old. For example, the sample of a Neanderthal man was approximately 30,000 years old. The oldest samples are likely to be in the range of 50,000 to 100,000 years old.

A self-help quiz involving this experiment can be found at www.mhhe.com/brookergenetics4e.

Evolution Is Associated with Changes in Chromosome Structure and Number

In this section, we focused on mutations that alter the DNA sequences within genes. In addition to gene mutations, other types of changes, such as gene duplications, transpositions, inversions, translocations, and changes in chromosome number, are important features of evolution.

As discussed earlier, changes in chromosome structure and/or number may not always be adaptive, but they can lead to reproductive isolation and the origin of new species. As an example of variation in chromosome structure among closely related species, **Figure 26.18** compares the banding pattern of the three largest chromosomes in humans and the corresponding chromosomes in chimpanzees, gorillas, and orangutans. The banding patterns are strikingly similar because these species are closely related evolutionarily. However, some interesting differences are observed. Humans have one large chromosome 2, but this chromosome is divided into two separate chromosomes in the other three species. This explains why humans have 23 types of chromosomes and the other species have 24. This may have occurred by a fusion of the two smaller chromosomes during the development of the human lineage. Another interesting change in chromosome structure is seen in chromosome 3. The banding patterns among humans, chimpanzees, and gorillas are very similar, but the orangutan has a large inversion that flips the arrangement of bands in the centromeric region.

Synteny Groups Contain the Same Group of Linked Genes

With the advent of molecular techniques, researchers can analyze the chromosomes of two or more different species and identify regions that contain the same groups of linked genes, which are called **synteny groups.** Within a particular synteny group, the same types of genes are found in the same order. In 1995, Graham Moore and colleagues analyzed the locations of molecular markers along the chromosomes of several cereal grasses including rice (*Oryza sativa*), wheat (*Triticum aestivum*), maize (*Zea mays*), foxtail millet (*Setaria italica*), sugarcane (*Saccharum officinarum*), and sorghum (*Sorghum vulgare*). From this analysis, they were able to identify several large synteny groups that are common to most of these species (**Figure 26.19**). As an example, let's compare rice (12 chromosomes per set) with wheat (7 chromosomes per set). In rice, chromosome 6 contains two synteny groups designated R6a and R6b, whereas chromosome 8 contains a single synteny group, R8. In wheat, chromosome 7 consists of R6a and R6b at either end, while R8 is sandwiched in the middle. One possible explanation for this difference could be a chromosomal rearrangement during the evolution of wheat in which R8 became inserted into the middle of a chromosome containing R6a and R6b. Overall, the evolution of cereal grass species has maintained most of the same types of genes, but many chromosomal rearrangements have occurred. These types of rearrangements promote reproductive isolation.

How do these observations relate to bird evolution? As we have seen, variation in the expression of these genes determines whether or not a bird's feet are webbed. At some point in the evolution of birds, mutations occurred that provided variation in the expression of the *BMP4* and *gremlin* genes, which resulted in nonwebbed or webbed feet. In terrestrial settings, having nonwebbed feet is an advantage because these are more effective at holding on to perches, running along the ground, and snatching prey. Therefore, natural selection maintains nonwebbed feet in terrestrial environments. This process explains the occurrence of nonwebbed feet in chickens, hawks, crows, and many other terrestrial birds. In aquatic environments, webbed feet are an advantage because they act as paddles for swimming, so genetic variation that produced webbed feet has been promoted by natural selection. Over the course of many generations, this gave rise to webbed feet that are now found in ducks, geese, penguins, and other aquatic birds.

How does having webbed or nonwebbed feet influence speciation? This trait may not directly affect the ability of two individuals to mate. However, due to natural selection, birds with webbed feet became more prevalent in aquatic environments, whereas birds with nonwebbed feet are common in terrestrial locations. Therefore, reproductive isolation occurs because the populations occupy different environments.

The Evolution of Animal Body Plans Is Related to Changes in *Hox* Gene Number and Expression

Hox genes, which are discussed in Chapter 23, are found in all animals. Developmental biologists have speculated that genetic variation in the *Hox* genes may have spawned the formation of many new body types, yielding many different animal species. As shown in **Figure 26.21**, the number and arrangement of *Hox* genes varies considerably among different types of animals. Sponges, the simplest of animals, have at least one *Hox* gene, whereas insects typically have nine or more. In most cases, multiple *Hox* genes occur in a cluster in which the genes are close to each other along a chromosome. In mammals, such *Hox* gene clusters have been duplicated twice during the course of evolution to form four clusters for a total of 38 genes.

How would an increase in *Hox* genes enable more complex body forms to evolve? Part of the answer lies in the spatial expression of the *Hox* genes. In fruit flies, for example, different *Hox* genes are expressed in different segments of the body along the anteroposterior axis (see Figure 23.17). Therefore, an increase in the number of *Hox* genes allows each of these master control genes to become more specialized in the region it controls. One

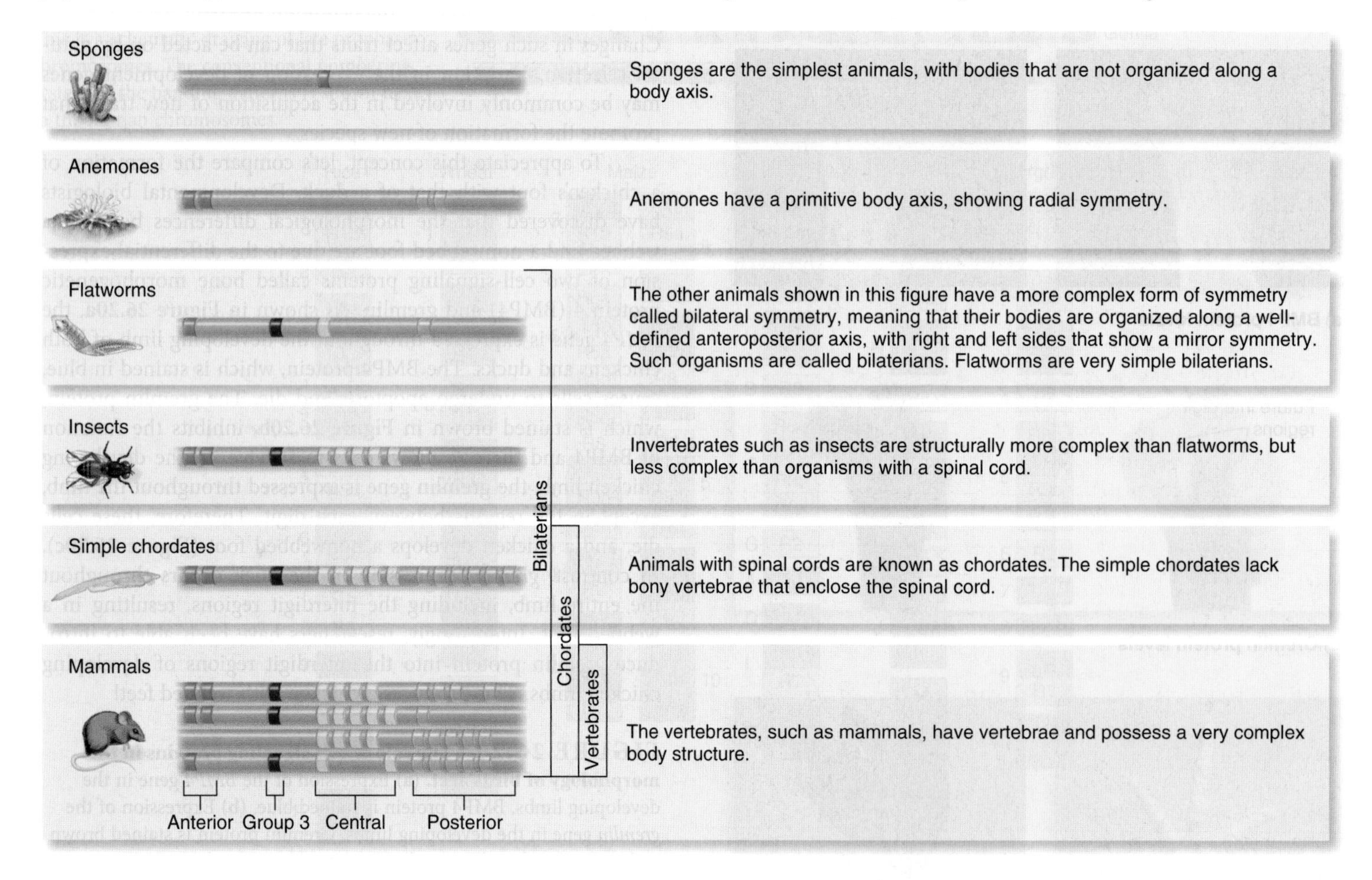

FIGURE 26.21 ***Hox* gene composition in different types of animals.** Researchers speculate that the duplication of *Hox* genes and *Hox* gene clusters played a key role in the evolution of more complex body plans in animals. The *Hox* genes are divided into four groups, called anterior, group 3, central, and posterior, based on their relative similarities. Each group is represented by a different color in this figure.

segment in the middle of the fruit fly body can be controlled by a particular *Hox* gene and form wings and legs, whereas a segment in the head region can be controlled by a different *Hox* gene and develop antennae. Therefore, research suggests that one way that new, more complex body forms evolved was through an increase in the number of *Hox* genes, thereby making it possible to form many specialized parts of the body that are organized along a body axis.

Three lines of evidence support the idea that *Hox* gene complexity has been instrumental in the evolution and speciation of animals with different body patterns. First, *Hox* genes are known to control body development. Second, as described in Figure 26.21, a general trend is observed in which animals with a more complex body structure tend to have more *Hox* genes and more *Hox* clusters in their genomes than do simpler animals. Finally, a comparison of *Hox* gene evolution and animal evolution bears striking parallels. Researchers can analyze *Hox* gene sequences among modern species and estimate the timing of past events via molecular clock data. Geneticists can estimate when the first *Hox* gene arose by gene innovation; the date is difficult to pinpoint but is well over 600 mya. The single *Hox* gene found in the sponge has descended from this primordial *Hox* gene. In addition, gene duplications of this primordial gene produced clusters of *Hox* genes in other species. Clusters, such as those found in modern insects, were likely to have arisen approximately 600 mya. A duplication of that cluster is estimated to have occurred around 520 mya. Remarkably, these estimates of *Hox* gene origins correlate with major speciation events in the history of animals. The Cambrian explosion, which occurred from 533 to 525 mya, saw a phenomenal diversification in the body plan of invertebrate species. This diversification occurred after the *Hox* cluster was formed and was possibly undergoing its first duplication that resulted in two *Hox* clusters. Also, approximately 420 mya, a second duplication produced species with four *Hox* clusters. This event precedes the proliferation of tetrapods—vertebrates with four limbs—that occurred approximately 417 to 354 mya. Modern tetrapods, such as mammals, have four *Hox* clusters. This second duplication may have been a critical event that led to the evolution of complex terrestrial vertebrates with four limbs, such as lizards, bears, and humans.

The Study of the *Pax6* Gene Indicates That Different Types of Eyes May Have Evolved from a Simpler Form

Explaining how a complex organ comes into existence is another major challenge for evolutionary biologists. Although it is relatively easy to understand how a limb could undergo evolutionary modifications to become a wing, flipper, or arm, it is more difficult to understand how a body structure comes into being in the first place. In his book *The Origin of Species*, Charles Darwin addressed this question and pointed out that the existence of complex organs, such as the eye, was difficult to understand. As Darwin noted, the eye of vertebrate species is exceedingly complex, being able to adjust focus, let in different amounts of light, and detect a spectrum of colors. He suggested that such a complex eye must have evolved from a simpler structure through the process of descent with modification. With amazing insight, he speculated that a very simple eye could be composed of just two cells, a photoreceptor cell and an adjacent pigment cell. The photoreceptor cell, which is a type of nerve cell, is able to absorb light and respond to it. The function of the pigment cell is to stop the light from reaching one side of the photoreceptor cell. This primitive, two-cell arrangement would allow an organism to sense both light and the direction from which the light comes.

How would natural selection play a role in the evolution of eyes? A primitive eye would provide an additional way for a mobile organism to sense its environment, possibly allowing it to avoid predators or locate food. Vision is nearly universal among animals, which indicates that there must be a strong selective advantage to better eyesight. Over time, eyes could become more complex by enhancing the ability to absorb different amounts and wavelengths of light and also by refinements in structures such as the addition of lenses that detect the direction of light.

Since the time of Darwin, many evolutionary biologists have wrestled with the question of eye evolution. From an anatomical point of view, researchers have discovered many different types of eyes. For example, the eyes of fruit flies, squid, and humans are quite different from each other. Furthermore, species that are closely related evolutionarily sometimes have different types of eyes. This observation led evolutionary biologists Luitfried von Salvini-Plawen and Ernst Mayr to propose that eyes may have independently arisen many different times during evolution. Based solely on morphology, such a hypothesis seemed reasonable, and for many years was accepted by the scientific community.

The situation took a dramatic turn when geneticists began to study eye development. Researchers identified a gene, *Pax6*[1], that influences eye development in both rodents and humans. In mice and rats, a mutation in this gene results in small eyes. In humans, a mutation in the *Pax6* gene results in an eye disorder called aniridia. In heterozygotes carrying one defective copy of the gene, the iris does not form. Researchers subsequently discovered a gene in *Drosophila*, named *eyeless*, that also causes a defect in eye development when mutant. DNA sequencing revealed the *eyeless* and *Pax6* genes are homologous; they are derived from the same ancestral gene.

In 1995, Walter Gehring and his colleagues were able to show experimentally that the abnormal expression of the *eyeless* gene in other parts of the fruit fly body could promote the formation of additional eyes. For example, using genetic engineering techniques, they were able to express the *eyeless* gene in the region where antennae should form. As seen in **Figure 26.22a**, this resulted in the formation of fruit fly eyes where antennae are normally found! Remarkably, the expression of the mouse *Pax6* gene in *Drosophila* can also cause the formation of eyes in unusual places. **Figure 26.22b** shows the formation of an eye on the leg of a fruit fly. This eye was caused by the expression of the mouse *Pax6* gene in this region.

The mouse *Pax6* gene switches on eye formation in *Drosophila*, but the eye produced is a *Drosophila* eye, not a mouse eye.

[1] *Pax* is an acronym for paired box. The protein encoded by this gene contains a domain called a paired box.

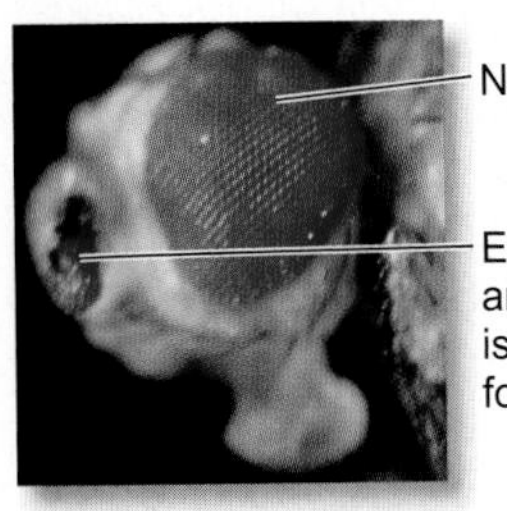

(a) Abnormal expression of *Drosophila eyeless* gene

(b) Abnormal expression of mouse *Pax6* gene in a fruit fly leg

FIGURE 26.22 Formation of additional eyes in *Drosophila* due to the abnormal expression of a master control gene for eye morphogenesis. **(a)** When the *Drosophila eyeless* gene is expressed in the antenna region, eyes are formed where antennae should be located. **(b)** When the mouse *Pax6* gene is expressed in the leg region of *Drosophila*, a small eye is formed there.

This happens because the genes activated by the *Pax6* gene are all from the *Drosophila* genome. In *Drosophila*, the *eyeless* gene switches on a cascade involving 2500 different genes required for eye morphogenesis. The *Pax6* gene and its *Drosophila* homolog, *eyeless,* are master control genes that promote the formation of an eye. Based on these observations, Gehring has suggested that the eyes of *Drosophila* and mammals are evolutionarily derived from the modification of an eye that arose once during evolution. If *Drosophila* and mammalian eyes had arisen independently, the *Pax6* gene from mice would not be expected to induce the formation of eyes in *Drosophila*. Alternatively, other researchers argue that *Pax6* may control only certain features of eye development and that different types of eyes may have evolved independently. Future research will be needed to resolve this controversy.

Since the initial discovery of the *Pax6* and *eyeless* genes, homologs of this gene have been discovered in many different species. In all cases where it has been tested, this gene directs eye development. The *Pax6* gene and its homologs encode a transcription factor protein that controls the expression of many different genes. Gehring and colleagues have hypothesized that the eyes from many different species all evolved from a common ancestral form consisting of, as proposed by Darwin, one photoreceptor cell and one pigment cell (**Figure 26.23**). As mentioned

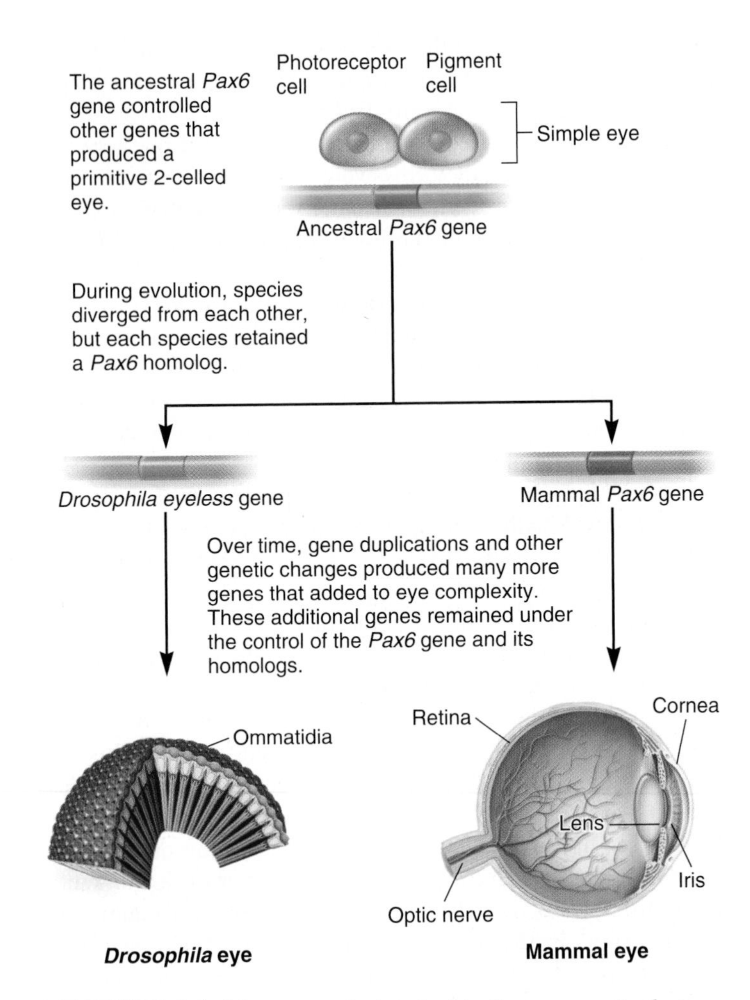

FIGURE 26.23 Eye evolution. In this diagram, genetic changes, under control of the ancestral *Pax6* gene, led to the evolution of different types of eyes. Ommatidia are units of the insect compound eye.

previously, such a very simple eye can accomplish some rudimentary form of vision by detecting light and its direction. Eyes such as these are still found in modern species such as the larvae of certain types of mollusks. Over the course of evolution, simple eyes were transformed into more complex types of eyes by modifications that resulted in the addition of more types of cells such as lens and muscle cells.

KEY TERMS

Page 730. biological evolution, evolution, microevolution, macroevolution, molecular evolution

Page 732. species, subspecies, ecotypes

Page 733. reproductive isolation, prezygotic isolating mechanisms, postzygotic isolating mechanisms

Page 734. species concept, biological species concept, evolutionary lineage concept, general lineage concept, speciation, anagenesis

Page 735. cladogenesis, allopatric speciation, founder effect, parapatric speciation

Page 736. hybrid zones, sympatric speciation

Page 738. gradualism, punctuated equilibrium, phylogeny, phylogenetic tree, systematists, homology, homologous

Page 739. monophyletic group, clade, cladistic approach, cladogram, ancestral characters, shared derived characters, synaptomorphy

Page 740. ingroup, outgroup, principle of parsimony

Page 741. maximum likelihood, Bayesian methods

Page 742. Bacteria, Archaea, Eukaryotes, supergroups

Page 743. vertical evolution, horizontal gene transfer

Page 744. molecular evolution, orthologs, paralogs, gene family, nonneutral mutation, neutral theory of evolution, non-Darwinian evolution

CHAPTER SUMMARY

- Evolution is a heritable change in one or more characteristics of a population or species from one generation to the next.

26.1 Origin of Species

- The most commonly used characteristics to distinguish species are morphological traits, the ability to interbreed, molecular features, ecological factors, and evolutionary relationships. However, each approach has its drawbacks (see Figure 26.1, Table 26.1).
- A species concept is a way to define the concept of a species and/or provide an approach for distinguishing one species from another. The general lineage concept is receiving wide support.
- Speciation is the process by which new species are formed via evolution. It may occur by anagenesis, when one species evolves into a new one, or more commonly by cladogenesis, in which a species diverges into two or more different ones (see Figure 26.2).
- Cladogenesis may occur via allopatric, parapatric, or sympatric speciation (see Table 26.2, Figures 26.3, 26.4).
- The tempo of evolution may exhibit gradualism and/or punctuated equilibrium (see Figure 26.5).

26.2 Phylogenetic Trees

- A phylogenetic tree is a diagram that describes a phylogeny—the sequence of events involved in the evolutionary development of a species or group of species (see Figure 26.6).
- Different methods, including cladistics, are used to construct phylogenetic trees. One way to evaluate the validity of possible trees is by the principle of parsimony (see Figures 26.7, 26.8).
- Maximum likelihood and Bayesian methods are also used to discriminate among possible phylogenetic trees.
- Phylogenetic trees help us to understand the relationships among different taxa such as domains and supergroups (see Figures 26.9, 26.10).
- In addition to vertical evolution, horizontal gene transfer has played an important role in the evolution of life on Earth (see Figure 26.11).

26.3 Molecular Evolution

- Molecular evolution refers to patterns and processes associated with evolutionary change at the molecular level.
- Homologous genes are derived from the same ancestral gene. They can be orthologs or paralogs (see Figures 26.12, 26.13).
- The neutral theory of evolution suggests that most variation in gene sequences is due to mutations that are neutral or nearly neutral with regard to phenotype. Such variation accumulates in populations largely due to genetic drift.
- The relatively constant rate of neutral or nearly mutations acts as a molecular clock on which to measure evolutionary time (see Figure 26.14).
- Phylogenetic trees can be constructed from ancient DNA isolated from extinct organisms (see Figures 26.15–26.17).
- Evolution also involves changes in chromosome structure and number (see Figures 26.18, 26.19).

26.4 Evo-Devo: Evolutionary Developmental Biology

- Evolutionary developmental biology is a relatively new field of biology that compares the development of species in an attempt to understand ancestral relationships between different species and the developmental mechanisms that bring about evolutionary change.
- Variation in the pattern of expression of developmentally important genes, such as those involved with cell signaling, can have a dramatic effect on morphology (see Figure 26.20).
- The evolution of animal body plans has involved changes in the number and expression patterns of *Hox* genes (see Figure 26.21).
- The *Pax6* gene and its *Drosophila* homolog, *eyeless,* are master control genes that promote the formation of an eye. This observation suggests that the eyes of all animals may be evolutionarily derived from the modification of an eye that arose once during evolution (see Figures 26.22, 26.23).

PROBLEM SETS & INSIGHTS

Solved Problems

S1. A codon for leucine is UUA. A mutation causing a single-base substitution in a gene can change this codon in the transcribed mRNA into GUA (valine), AUA (isoleucine), CUA (leucine), UGA (stop), UAA (stop), UCA (serine), UUG (leucine), UUC (phenylalanine), or UUU (phenylalanine). According to the neutral theory, which of these mutations would you expect to see within the genetic variation of a natural population? Explain.

Answer: The neutral theory proposes that neutral mutations accumulate to the greatest extent in a population. Leucine is a nonpolar amino acid. For a UUA codon, single-base changes of CUA and UUG are silent, so they would be the most likely to occur in a natural population. Likewise, conservative substitutions to other nonpolar amino acids such as isoleucine (AUA), valine (GUA), and phenylalanine (UUC and UUU) may not affect protein structure and function, so they may also occur and not be eliminated rapidly by natural selection. The polar amino acid serine (UCA) is a nonconservative substitution; one would predict that it is more likely to disrupt protein function. Therefore, it may be less likely to be found. Finally, the stop codons, UGA and UAA, would be expected to diminish or eliminate protein function, particularly if they occur early in the coding sequence. These types of mutations are selected against and, therefore, are not usually found in natural populations.

S2. Explain why homologous genes have sequences that are similar but not identical.

Answer: Homologous genes are derived from the same ancestral gene. Therefore, as a starting point, they had identical sequences. Over time, however, each gene accumulates random mutations that the other homologous genes did not acquire. These random mutations change the gene from its original sequence. Therefore, much of the sequence between homologous genes remains identical, but some of the sequence is altered due to the accumulation of independent random mutations.

S3. Explain why plants are more likely to evolve by sympatric speciation compared to animals.

Answer: A common way for sympatric speciation to occur is by the formation of polyploids. For example, if one species is diploid ($2n$), nondisjunction could produce an individual that is tetraploid ($4n$). If the tetraploid individual was a plant and if it was monoecious (i.e., produces both pollen and egg cells), the plant could multiply to produce many tetraploid offspring. These offspring would be reproductively isolated from the diploid plants that are found in the same geographic area. This isolation occurs because the offspring of a cross between a diploid and tetraploid plant would be infertile. For example, if the pollen from a diploid plant fertilized the egg from a tetraploid plant, the offspring would be triploid. The triploid offspring might be viable, but it is likely that triploids would be sterile because they would produce highly aneuploid gametes. The gametes would be aneuploid because there are an odd number of homologous chromosomes. In this case, there would be three copies of each homologous chromosome, and these could not be equally distributed into gametes. Therefore, when a tetraploid is produced, it can immediately become its own unique species that is reproductively isolated from other species.

In contrast, polyploidy rarely occurs in animals. Perhaps the main reason is because animals usually cannot tolerate polyploidy. Tetraploid animals typically die during early stages of development. In addition, most species of animals have male and female sexes. To develop a tetraploid species from a diploid species, nondisjunction would have to produce both a male and female offspring that could reproduce with each other. The chance of producing one tetraploid offspring is relatively rare. The chances of producing two tetraploid offspring that happen to be male and female and happen to mate with each other to produce many offspring would be extremely rare.

S4. As described in Figure 26.18, evolution is associated with changes in chromosome structure and number. As seen there, chromosome 2 in humans is divided into two distinct chromosomes in chimpanzees, gorillas, and orangutans. In addition, chromosome 3 in the orangutan has a large inversion not found in the other three primates. Discuss the potential role of these types of changes in the evolution of these primate species.

Answer: As discussed in Chapter 8, changes in chromosome structure, such as inversions and balanced translocations, may not have any phenotypic effects. Likewise, the division of a single chromosome into two distinct chromosomes may not have any phenotypic effect as long as the total amount of genetic material remains the same. Overall, the types of changes in chromosome structure and number shown in Figure 26.18 may not have caused any changes in the phenotypes of primates. However, the changes would be expected to promote reproductive isolation. For example, if a gorilla mated with an orangutan, the offspring would be an inversion heterozygote for chromosome 3. As shown in Figure 8.10, crossing over during gamete formation in an inversion heterozygote may produce chromosomes that have too much or too little genetic material. This is particularly likely if the inversion is fairly large (e.g., the one shown in Figure 26.18). The inheritance of too much or too little genetic material is likely to be detrimental or even lethal. For this reason, the hybrid offspring of a gorilla and orangutan would probably not be fertile. (Note: In reality, there are several other reasons why interspecies matings between gorillas and orangutans do not produce viable offspring.)

Overall, the primary effect of changes in chromosome structure and number, like the ones shown in Figure 26.18, is to promote reproductive isolation. Once two populations become reproductively isolated, they accumulate different mutations, which over the course of many generations leads to two different species with distinct characteristics.

It should be noted that changes in chromosome number in plants are more likely to have effects that abruptly lead to the formation of new species. This idea is discussed in solved problem S3.

Conceptual Questions

C1. Discuss the two principles on which evolution is based.

C2. Evolution, which involves genetic change in a population of organisms over time, is often described as the unifying theme in biology. Discuss how evolution is unifying at the molecular and cellular levels.

C3. What is a species? What types of observations do researchers analyze when trying to identify species?

C4. What is meant by the term reproductive isolation? Give several examples.

C5. Would the following examples of reproductive isolation be considered a prezygotic or postzygotic mechanism?

A. Horses and donkeys can interbreed to produce mules, but the mules are infertile.

B. Three species of the orchid genus *Dendrobium* produce flowers 8 days, 9 days, and 11 days after a rainstorm. The flowers remain open for 1 day.

C. Two species of fish release sperm and eggs into seawater at the same time, but the sperm of one species do not fertilize the eggs of the other species.

D. Two tree frogs, *Hyla chrysoscelis* (diploid) and *H. versicolor* (tetraploid), can produce viable offspring, but the offspring are sterile.

C6. Distinguish between anagenesis and cladogenesis. Which type of speciation is more prevalent? Why?

C7. Describe three or more genetic mechanisms that may lead to the rapid evolution of a new species. Which of these genetic mechanisms are influenced by natural selection, and which are not?

C8. Explain the type of speciation (allopatric, parapatric, or sympatric) most likely to occur under each of the following conditions:

A. A pregnant female rat is transported by an ocean liner to a new continent.

B. A meadow containing several species of grasses is exposed to a pesticide that promotes nondisjunction.

C. In a very large lake containing several species of fish, the water level gradually falls over the course of several years. Eventually, the large lake becomes subdivided into smaller lakes, some of which are connected by narrow streams.

C9. Alloploids are produced by crosses involving two different species. Explain why alloploids are reproductively isolated from the two original species from which they were derived. Explain why alloploids are usually sterile, whereas allotetraploids (containing a diploid set from each species) are commonly fertile.

C10. Discuss the evidence in favor of the punctuated equilibrium model of evolution. What mechanisms could account for this pattern of evolution? In contrast, what type of genetic changes are consistent with gradualism?

C11. Discuss whether the phenomenon of reproductive isolation applies to bacteria, which reproduce asexually. How would a geneticist divide bacteria into separate species?

C12. Discuss the major differences among allopatric, parapatric, and sympatric speciation.

C13. The following are two DNA sequences from homologous genes:

TTGCATAGGCATACCGTATGATATCGAAAACTAGAAAAATAGGGCGATAGCTA

GTATGTTATCGAAAAGTAGCAAAATAGGGCGATAGCTACCCAGACTACCGGAT

The two sequences, however, do not begin and end at the same location. Try to line them up according to their homologous regions.

C14. What is meant by the term molecular clock? How is this concept related to the neutral theory of evolution?

C15. Would the rate of deleterious or beneficial mutations be a good molecular clock? Why or why not?

C16. Which would you expect to exhibit a faster rate of evolutionary change, the nucleotide sequence of a gene or the amino acid sequence of the encoded polypeptide of the same gene? Explain your answer.

C17. When comparing the coding region of structural genes among closely related species, it is commonly found that certain regions of the gene have evolved more rapidly (i.e., have tolerated more changes in sequence) than other regions of the gene. Explain why different regions of a structural gene evolve at different rates.

C18. Plant seeds contain storage proteins that are encoded by plant genes. When the seed germinates, these proteins are rapidly hydrolyzed (i.e., the covalent bonds between amino acids within the polypeptides are broken), which releases amino acids for the developing seedling. Would you expect the genes that encode plant storage proteins to evolve slowly or rapidly compared with genes that encode enzymes? Explain your answer.

C19. Figure 26.9 shows a phylogenetic tree of all life on Earth based on 16S rRNA data. Based on your understanding of molecular genetics (in this and other chapters), describe three or more observations that suggest that all life-forms on Earth evolved from a common ancestor or group of ancestors.

C20. Take a look at the α-globin and β-globin sequences in Figure 26.12. Which sequences are more similar, the α globin in humans and the α globin in horses, or the α globin in humans and the β globin in humans? Based on your answer, would you conclude that the gene duplication that gave rise to the α-globin and β-globin genes occurred before or after the divergence of humans and horses? Explain your reasoning.

C21. Compare and contrast the neutral theory of evolution versus the Darwinian (i.e., selectionist) theory of evolution. Explain why the neutral theory of evolution is sometimes called non-Darwinian evolution.

C22. For each of the following examples, discuss whether it would be the result of neutral mutation or mutation that has been acted on by natural selection, or both:

A. When comparing sequences of homologous genes, differences in the coding sequence are most common at the wobble base (i.e., the third base in each codon).

B. For a structural gene, the regions that encode portions of the polypeptide that are vital for structure and function are less likely to incur mutations compared to other regions of the gene.

C. When comparing the sequences of homologous genes, introns usually have more sequence differences compared to exons.

C23. As discussed in Chapter 24, genetic variation is prevalent in natural populations. This variation is revealed in the DNA sequencing of genes. According to the neutral theory of evolution, discuss the relative importance of natural selection against detrimental mutations, natural selection in favor of beneficial mutations, and neutral mutations in accounting for the genetic variation we see in natural populations.

C24. If you were comparing the karyotypes of species that are closely related evolutionarily, what types of similarities and differences would you expect to find?

C25. In the developing bud that gives rise to a hand in a human embryo, where would you expect the *gremlin* gene to be expressed?

C26. Discuss how *Hox* gene number is related to body complexity.

Experimental Questions

E1. Two populations of snakes are separated by a river. The snakes cross the river only on rare occasions. The snakes in the two populations look very similar to each other except that the members of the population on the eastern bank of the river have a yellow spot on the top of their head, whereas the members of the western population have an orange spot on the top of their head. Discuss two experimental methods that you might follow to determine whether the two populations are members of the same species or members of different ones.

E2. Sympatric speciation by allotetraploidy has been proposed as a common mechanism for speciation. Let's suppose you were interested in the origin of certain grass species in southern California. Experimentally, how would you go about determining if some of the grass species are the result of allotetraploidy?

E3. Two diploid species of closely related frogs, which we will call species A and species B, were analyzed with regard to genes that encode an enzyme called hexokinase. Species A has two distinct copies of this gene: *A1* and *A2*. In other words, this diploid species is *A1A1 A2A2*. The other species has three copies of the hexokinase gene, which we will call *B1*, *B2*, and *B3*. A diploid individual of species B would be *B1B1 B2B2 B3B3*. These hexokinase genes from the two species were subjected to DNA sequencing, and the percentage of sequence identity was compared among these genes. The results are shown here:

Percentage of DNA Sequence Identity

	A1	*A2*	*B1*	*B2*	*B3*
A1	100	62	54	94	53
A2	62	100	91	49	92
B1	54	91	100	67	90
B2	94	49	67	100	64
B3	53	92	90	64	100

If we assume that hexokinase genes were never lost in the evolution of these frog species, how many distinct hexokinase genes do you think there were in the most recent ancestor that preceded the divergence of these two species? Explain your answer. Also explain why species B has three distinct copies of this gene, whereas species A has only two.

E4. A researcher sequenced a portion of a bacterial gene and obtained the following sequence, beginning with the start codon, which is underlined:

<u>ATG</u> CCG GAT TAC CCG GTC CCA AAC AAA ATG ATC GGC CGC CGA ATC TAT CCC

The bacterial strain that contained this gene has been maintained in the laboratory and grown serially for many generations. Recently, another person working in the laboratory isolated DNA from the bacterial strain and sequenced the same region. The following results were obtained.

<u>ATG</u> CCG GAT TAT CCG GTC CCA AAT AAA ATG ATC GGC CGC CGA ATC TAC CCC

Explain why these sequencing differences may have occurred.

E5. F_1 hybrids between two species of cotton, *Gossypium barbadense* and *G. hirsutum*, are very vigorous plants. However, F_1 crosses produce many seeds that do not germinate and a low percentage of very weak F_2 offspring. Suggest two reasons for these observations.

E6. A species of antelope contains 20 chromosomes per set. The species is divided by a mountain range into two separate populations, which we will call the eastern and western population. When comparing the karyotypes of these two populations, it was discovered that the members of the eastern population are homozygous for a large inversion within chromosome 14. How would this inversion affect the interbreeding between the two populations? Could such an inversion play an important role in speciation?

E7. Explain why molecular techniques were needed as a way to provide evidence for the neutral theory of evolution.

E8. Prehistoric specimens often contain minute amounts of ancient DNA. What technique can be used to increase the amount of DNA in an older sample? Explain how this technique is performed and how it increases the amount of a specific region of DNA.

E9. In the experiment of Figure 26.15, explain how we know that the kiwis are more closely related to the emu and cassowary than to the moas. Cite particular regions in the sequences that support your answer.

E10. In Chapter 20, we learned about a technique called fluorescence in situ hybridization (FISH), during which a labeled piece of DNA is hybridized to a set of chromosomes. Let's suppose that we cloned a piece of DNA from *G. pubescens* (see Figure 26.4) and used it as a labeled probe for in situ hybridization. What would you expect to happen if we hybridized it to the *G. speciosa*, the natural *G. tetrahit*, or the artificial *G. tetrahit* strains? Describe your expected results.

E11. A team of researchers has obtained a dinosaur bone (*Tyrannosaurus rex*) and has attempted to extract ancient DNA from it. Using primers to the 12S rRNA mitochondrial gene, they have used PCR and obtained a DNA segment that yields a sequence homologous to crocodile DNA. Other scientists are skeptical that this sequence is really from the dinosaur. Instead, they believe that it may be due to contamination from more recent DNA, such as the remains of a reptile that lived much more recently. What criteria might you use to establish the credibility of the dinosaur sequence?

E12. Discuss how the principle of parsimony can be used in a cladistics approach of constructing a phylogenetic tree.

E13. As discussed in this chapter and Chapter 24, genes are sometimes transferred between different species via horizontal gene transfer. Discuss how horizontal gene transfer might give misleading results when constructing a phylogenetic tree. How could you overcome this problem?

E14. If a researcher used genetic engineering techniques to express the *Drosophila eyeless* gene in the embryo at the region that will become the tip of the mouse's tail, what results would you expect in the resulting offspring?

Questions for Student Discussion/Collaboration

1. The raw material for evolution is random mutation. Discuss whether or not you view evolution as a random process.
2. Compare the forms of speciation that are slow with those that occur more rapidly. Make a list of the slow and fast forms. With regard to mechanisms of genetic change, what features do slow and rapid speciation have in common? What features are different?
3. Do you think that Darwin would object to the neutral theory of evolution?

Note: All answers appear at the website for this textbook; the answers to even-numbered questions are in the back of the textbook.

www.mhhe.com/brookergenetics4e

Visit the website for practice tests, answer keys, and other learning aids for this chapter. Enhance your understanding of genetics with our interactive exercises, quizzes, animations, and much more.

APPENDIX
EXPERIMENTAL TECHNIQUES

OUTLINE

A.1 METHODS FOR GROWING CELLS

Researchers often grow cells in a laboratory as a way to study their properties. This is known as a **cell culture.** Cell culturing offers several technical advantages. The primary advantage is that the growth medium is defined and can be controlled. Minimal growth medium contains the bare essentials for cell growth: salts, a carbon source, an energy source, essential vitamins, amino acids, and trace elements. In their experiments, geneticists often compare strains that can grow in minimal media and mutant strains that cannot grow unless the medium is supplemented with additional components. A rich growth medium contains many more components than are required for growth.

Researchers also add substances to the culture medium for other experimental reasons. For example, radioactive isotopes can be added to the culture medium to radiolabel cellular macromolecules. In addition, an experimenter could add a hormone to the growth medium and then monitor the cells' response to the hormone. In all of these cases, cell culturing is advantageous because the experimenter can control and vary the composition of the growth medium.

The first step in creating a cell culture is the isolation of a cell population that the researchers wish to study. For bacteria, such as *Escherichia coli,* and eukaryotic microorganisms, such as yeast and *Neurospora,* the researchers simply obtain a sample of cells from a colleague or a stock center. For animal or plant tissues, the procedure is a bit more complicated. When cells are contained within a complex tissue, they must first be dispersed by treating the tissue with agents that separate it into individual cells to create a cell suspension.

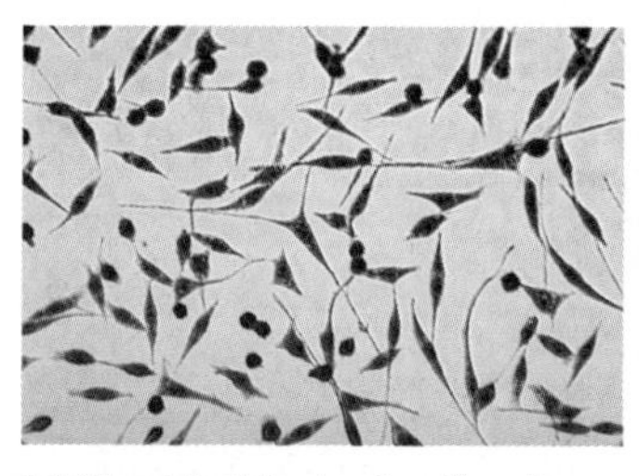

(a) Fibroblast (animal cell) culture

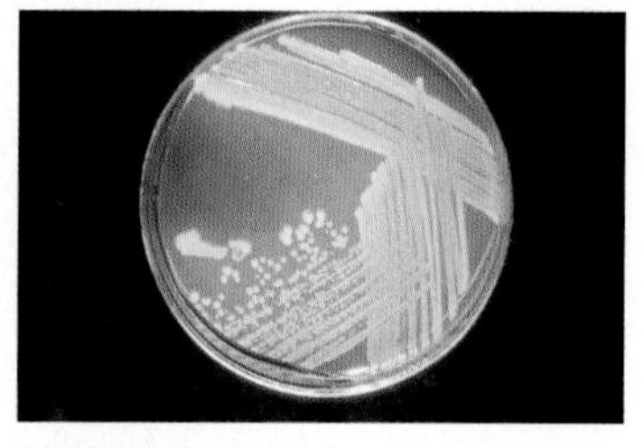

(b) Bacterial colonies

FIGURE A.1 **Growth of cells on solid growth media.** **(a)** This micrograph shows fibroblasts growing as a monolayer on a solid growth medium. **(b)** Bacterial cells form colonies that are a clonal population of cells derived from a single cell.

Once a desired population of cells has been obtained, researchers can grow them in a laboratory (i.e., in vitro) either suspended in a liquid growth medium or attached to a solid surface such as agar. Both methods have been commonly used in the experiments considered throughout this textbook. Liquid culture is often used when researchers want to obtain a large quantity of cells and isolate individual cellular components, such as nuclei or DNA. By comparison, **Figure A.1** shows animal cells and bacteria cells that are grown on solid growth media. As shown in Figure 22.11, solid media are used to study cancer cells, because such cells can be distinguished by the formation of foci in which malignant cells pile up on top of each other. In gene-cloning experiments with bacteria and yeast, solid media are also used. Each colony of cells is a clone of cells that is derived from a single cell that divided to produce many cells (Figure A.1b). As discussed in Chapter 18, a solid medium is used in the isolation of individual clones that contain a desired gene.

A.2 MICROSCOPY

Microscopy is a technique for observing things that cannot be seen (or can hardly be seen) with the naked eye. A key concept in microscopy is **resolution,** which is the minimum distance between two objects that enables them to be seen as separate

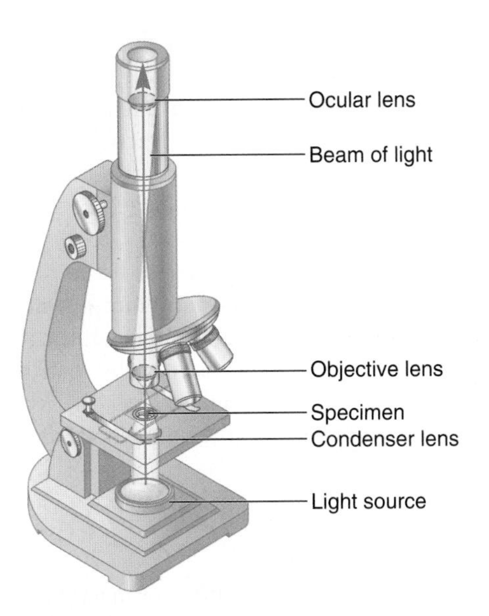

(a) Light microscope

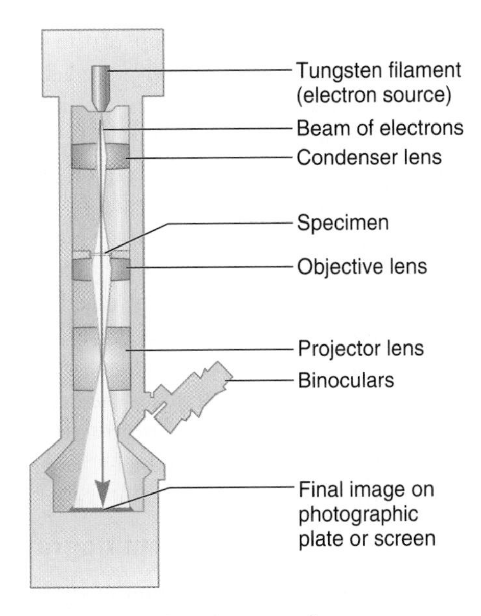

(b) Transmission electron microscope

FIGURE A.2 **Design of (a) optical (light) and (b) transmission electron microscopes.**

from each other. The ability to resolve two points as being separate depends on several factors, including the wavelength of the illumination source (light or electron beam), the medium in which the sample is immersed, and the structural features of the microscope (which are beyond the scope of this textbook).

As shown in **Figure A.2**, two widely used kinds of microscopes are the optical (light) microscope and the transmission electron microscope (TEM). The light microscope is used to resolve cellular structures to a limit of approximately 0.3 μm. (For comparison, a typical bacterium is about 1 μm long.) At this resolution, the individual cell organelles in eukaryotic cells can be discerned easily, and chromosomes are also visible. Karyotyping is accomplished via light microscopy after the chromosomes have been treated with stains. A variation of light microscopy known as fluorescence microscopy is often used to highlight a particular feature of a chromosome or cellular structure. The technique of fluorescence in situ hybridization (FISH; see Chapter 20) makes use of this type of microscope. Also, optical modifications in certain light microscopes (e.g., phase contrast and differential interference) can be used to exaggerate the differences in densities between neighboring cells or cell structures. These kinds of light microscopes are useful in monitoring cell division in living (unstained) cells or in transparent worms (as in Figure 23.16).

The structural details of large macromolecules such as DNA and ribosomes are not observable by light microscopy. The coarse topology of these macromolecules can be determined by electron microscopy. Electron microscopes have a limit of resolution of about 2 nm, which is about 100 times finer than the best light microscopes. The primary advantage of electron microscopy over light microscopy is its better resolution. Disadvantages include a much higher expense and more extensive sample preparation. In transmission electron microscopy, the sample is bombarded with an electron beam. This requires that the sample be dried, fixed, and usually coated with a heavy metal that absorbs electrons.

A.3 SEPARATION METHODS

Biologists often wish to take complex systems and separate them into less complex components. For example, the cells within a complex tissue can be separated into individual cells, or the macromolecules within cells can be separated from the other cellular components. In this section, we focus primarily on methods aimed at separating and purifying macromolecules.

Disruption of Cellular Components

In many experiments described in this textbook, researchers have obtained a sample of cells and then wish to isolate particular components from the cells. For example, a researcher may want to purify a protein that functions as a transcription factor. To do so, he or she would begin with a sample of cells that synthesize this protein and then break open the cells using one of the methods described in **Table A.1**. In eukaryotes, the breakage of cells releases the soluble proteins from the cell; it also dissociates the cell organelles that are bounded by membranes. This mixture of proteins and cell organelles can then be isolated and purified by centrifugation and chromatographic methods, which are described next.

Centrifugation

Centrifugation is a method commonly used to separate cell organelles and macromolecules. A **centrifuge** contains a motor that causes a rotor holding centrifuge tubes to spin very rapidly. As the rotor spins, particles move toward the bottom of the cen-

TABLE **A.1**
Common Methods of Cell Disruption

Method	Description
Sonication	The exposure of cells to intense sound waves, which breaks the cell membranes.
French press	The passage of cells through a small aperture under high pressure, which breaks the cell membranes and cell wall.
Homogenization	Cells are placed in a tube that contains a pestle. When the pestle is spun, the cells are squeezed through the small space between the pestle and the glass wall of the tube, thereby breaking them.
Osmotic shock	The transfer of cells into a hypo-osmotic medium. The cells take up water and eventually burst.

trifuge tube; the rate at which they move depends on several factors, including their densities, sizes, and shapes and the viscosity of the medium. The rate at which a macromolecule or cell organelle sediments to the bottom of a centrifuge tube is called its **sedimentation coefficient,** which is normally expressed in Svedberg units (S). A sedimentation coefficient has the unit of seconds: $1 \text{ S} = 1 \times 10^{-13}$ seconds.

When a sample contains a mixture of macromolecules or cell organelles, it is likely that different components will sediment at different rates. This phenomenon, known as **differential centrifugation,** is shown in **Figure A.3**. As seen here, particles with large sedimentation coefficients reach the bottom of the tube more quickly than those with smaller coefficients. Researchers can follow two different strategies that use differential centrifugation as a separation technique. One way is to separate the **supernatant** from the **pellet** following centrifugation. The pellet is a collection of particles found at the bottom of the tube, and the supernatant is the liquid found above the pellet. In Figure A.3, when the experimenter had subjected the sample to a low-speed spin, most of the particles with large sedimentation coefficients are found in the pellet, whereas most of the particles with small and intermediate coefficients are found in the supernatant. A high-speed spin of the supernatant then separates the small and intermediate particles. Therefore, differential centrifugation provides a way of segregating these three types of particles.

A second way to separate particles using centrifugation is to collect fractions. A **fraction** is a portion of the liquid contained within a centrifuge tube. The collection of fractions is done when the solution within the centrifuge tube contains a gradient. For example, as shown in **Figure A.4**, the solution at the top of the tube has a lower concentration of cesium chloride (CsCl) than that at the bottom. In this experiment, a sample is layered on the top of the gradient and then centrifuged. In this example, the DNA and RNA separate from each other, because they have different sedimentation coefficients. The experimenter then punctures the bottom of the tube and collects fractions. The DNA fragments, which are heavier, come out of the tube in the earlier fractions; the RNA molecules are collected in later fractions.

A type of gradient centrifugation that may also be used to separate macromolecules and organelles is **equilibrium density centrifugation.** In this method, the particles sediment through the gradient, reaching a position where the density of the particle matches the density of the solution. At this point, the particle is at equilibrium and does not move any farther toward the bottom of the tube.

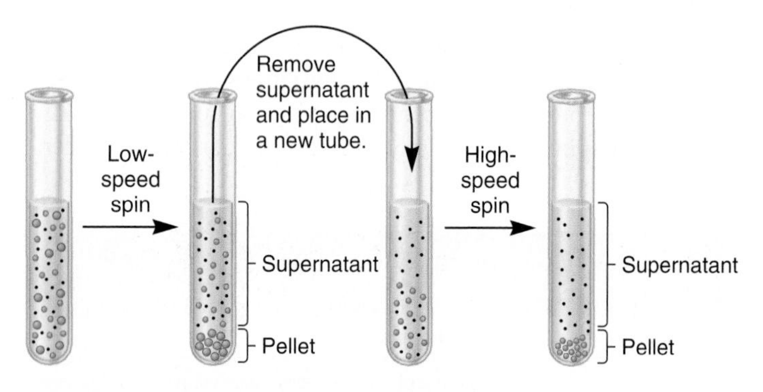

FIGURE A.3 The method of differential centrifugation. A sample containing a mixture of particles with different sedimentation coefficients is placed in a centrifuge tube. The tube is subjected to a low-speed spin that pellets the particles with large sedimentation coefficients. After a high-speed spin of the supernatant, the particles with an intermediate sedimentation coefficient are found in the pellet, and those with a small sedimentation coefficient are in the liquid supernatant.

Chromatography and Gel Electrophoresis

Chromatography is a method of separating different macromolecules and small molecules based on their chemical and physical

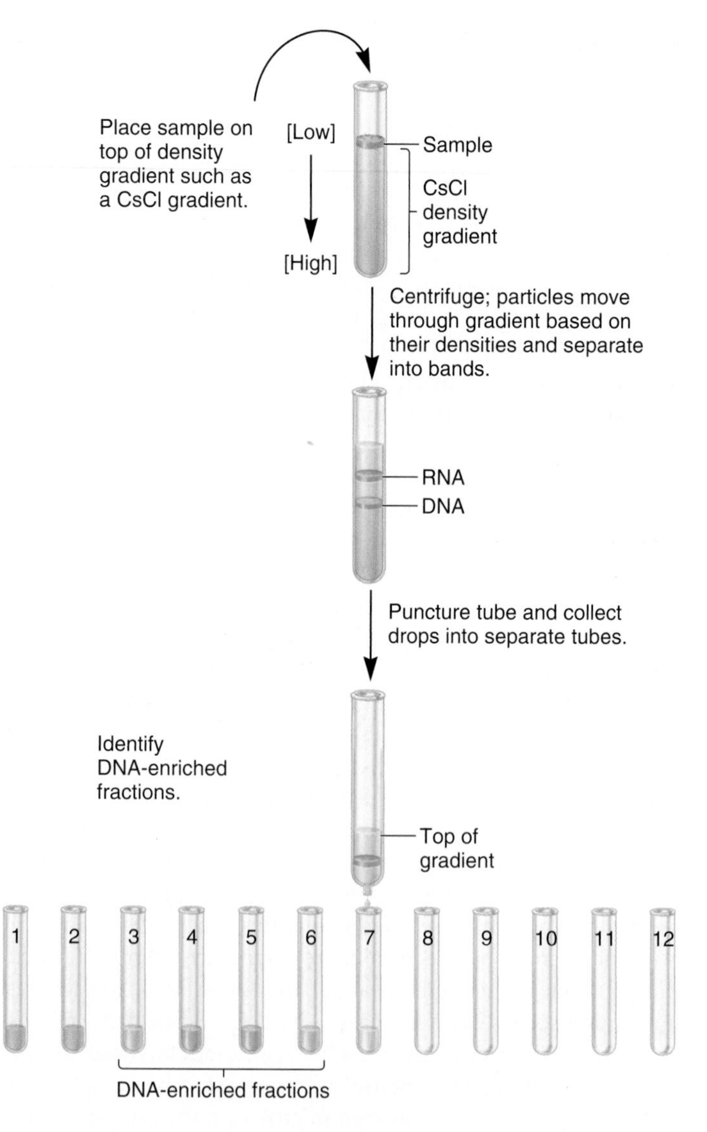

FIGURE A.4 Gradient centrifugation and the collection of fractions.

properties. In this method, a sample is dissolved in a liquid solvent and exposed to some type of matrix, such as a column containing beads or a thin strip of paper. The degree to which the molecules interact with the matrix depends on their chemical and physical characteristics. For example, a positively charged molecule binds tightly to a negatively charged matrix, but a neutral molecule does not.

Figure A.5 illustrates how column chromatography can be used to separate molecules that differ with respect to charge. Prior to this experiment, a column is packed with beads that are positively charged. There is plenty of space between the beads for molecules to flow from the top of the column to the bottom. However, if the molecules are negatively charged, they will spend some of their time binding to the positive charges on the surface of the beads. In the example shown in Figure A.5, the red proteins are positively charged and, therefore, flow rapidly from the top of the column to the bottom. They emerge in the fractions that are collected early in this experiment. The blue proteins, however, are negatively charged and tend to bind to the beads. The binding of the blue proteins to the beads can be disrupted by changing the ionic strength or pH of the solution that is added to the column. Eventually, the blue proteins will be eluted (i.e., leave the column) in later fractions.

Researchers use many variations of chromatography to separate molecules and macromolecules. The type shown in Figure A.5 is called ion-exchange chromatography, because its basis for separation depends on the charge of the molecules. In another type of column chromatography, known as gel filtration chromatography, the beads are porous. Small molecules are temporarily trapped within the beads, whereas large molecules flow between the beads. In this way, gel filtration separates molecules on the basis of size. To separate different types of macromolecules, such as proteins, researchers may use another type of bead; this bead has a preattached molecule that binds specifically to the protein they want to purify. For example, if a transcription factor binds a particular DNA sequence as part of its function, the beads within a column may have this DNA sequence preattached to them. Therefore, the transcription factor binds tightly to the DNA attached to these beads, whereas all other proteins are eluted rapidly from the column. This form of chromatography is called affinity chromatography, because the beads have a special affinity for the macromolecule of interest.

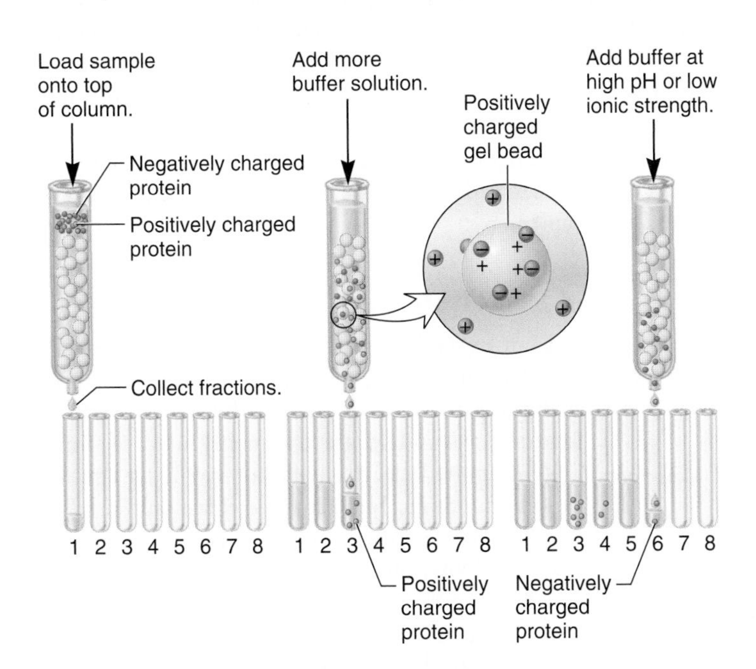

FIGURE A.5 Ion-exchange chromatography.

Besides column chromatography, in which beads are packed into a column, a matrix can be made in other ways. In paper chromatography, molecules pass through a matrix composed of paper. The rate of movement of molecules through the paper depends on their degree of interaction with the solvent and paper. In thin-layer chromatography, a matrix is spread out as a very thin layer on a rigid support such as a glass plate. In general, paper and thin-layer chromatography are effective at separating small molecules, whereas column chromatography is used to separate macromolecules such as DNA fragments or proteins.

Gel electrophoresis combines chromatography and electrophoresis to separate molecules and macromolecules. As its name suggests, the matrix used in gel electrophoresis is composed of a gel. As shown in **Figure A.6**, samples are loaded into wells at one end of the gel, and an electric field is applied across the gel. This electric field causes charged molecules to migrate from one side of the gel to the other. The migration of molecules in response to an electric field is called **electrophoresis.** In the examples of gel electrophoresis found in this textbook, the macromolecules within the sample migrate toward the positive end of the gel. In most forms of gel electrophoresis, a mixture of macromolecules is separated according to their molecular masses. Small proteins or DNA fragments move to the bottom of the gel more quickly than larger ones. Because the samples are loaded in rectangular wells at the top of the gel, the molecules within the sample are separated into bands within the gel. These bands of separated macromolecules can be visualized with stains. For example, ethidium bromide is a stain that binds to DNA and RNA and can be seen under ultraviolet light.

The two most commonly used gels are polymers made from acrylamide or agarose. Proteins typically are separated on polyacrylamide gels, whereas DNA fragments are separated on agarose gels. Occasionally, researchers use polyacrylamide gels to separate DNA fragments that are relatively small (namely, less than 1000 bp in length).

A.4 METHODS FOR MEASURING CONCENTRATIONS OF MOLECULES AND DETECTING RADIOISOTOPES AND ANTIGENS

To understand the structure and function of cells, researchers often need to detect the presence of molecules and macromolecules and to measure their concentrations. In this section, we consider a variety of methods for detecting and measuring the concentrations of biological molecules and macromolecules.

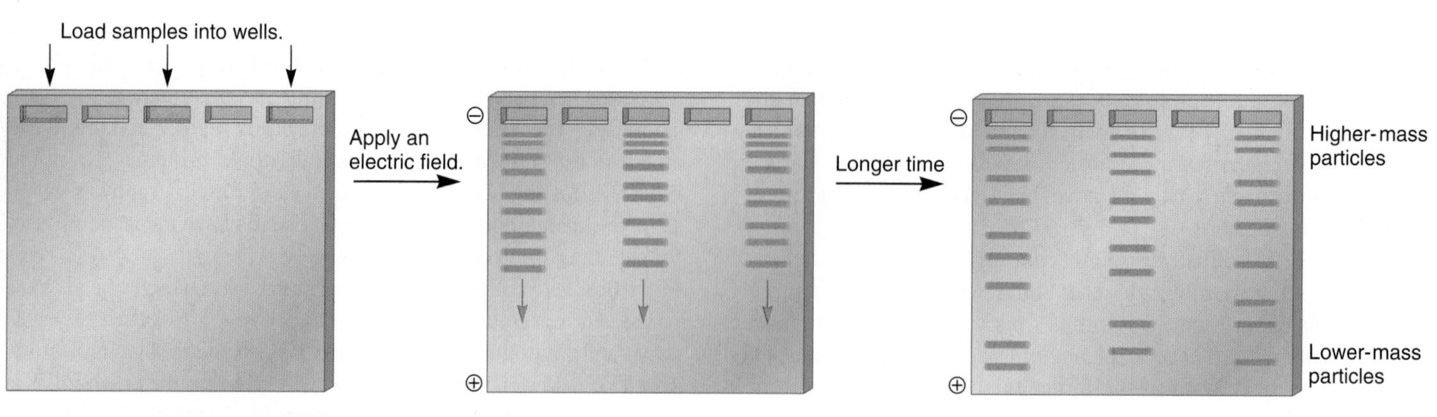

(a) Separation of a mixture of particles by gel electrophoresis

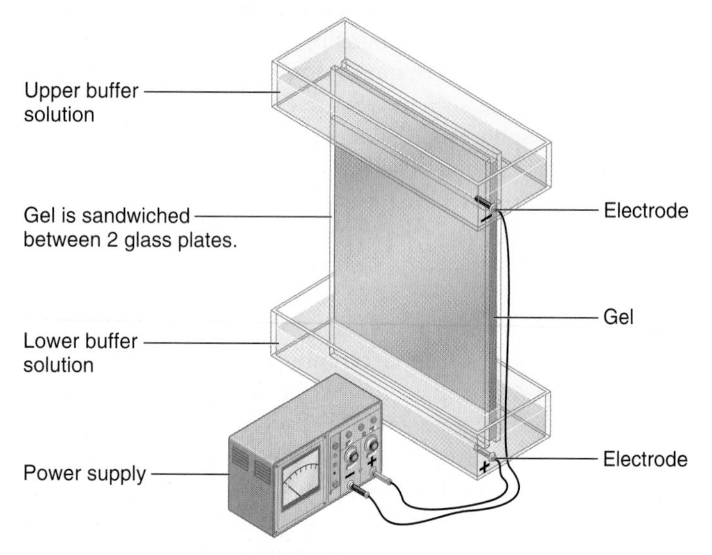

(b) Apparatus used in gel electrophoresis

FIGURE A.6 Acrylamide gel electrophoresis of DNA fragments.

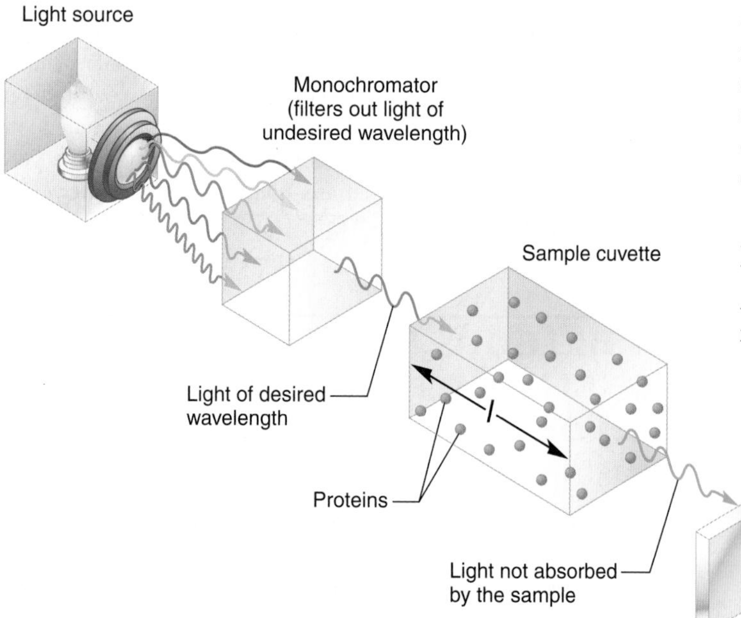

FIGURE A.7 Design of a spectrophotometer.

Spectroscopy

Macromolecules found in living cells, such as proteins, DNA, and RNA, are fairly complex molecules that can absorb radiation (e.g., light). Likewise, small molecules such as amino acids and nucleotides can also absorb light. A device known as a **spectrophotometer** is used by researchers to determine how much radiation at various wavelengths a sample can absorb. The amount of absorption can be used to determine the concentration of particular molecules within a sample, because each type of molecule or macromolecule has its own characteristic wavelength(s) of absorption, called its absorption spectrum.

A spectrophotometer typically has two light sources, which can emit ultraviolet or visible light. As shown in **Figure A.7**, the light source is passed through a monochromator, which emits the light at a desired wavelength. This incident light then strikes a sample contained within a cuvette. Some of the incident light is absorbed, and some is not. The amount and wavelengths of light that are absorbed depend on the concentration and structures of the molecules and macromolecules in the cuvette. The unabsorbed light passes through the sample and is detected by the spectrophotometer. The amount of light that strikes the detector is subtracted from the amount of incident light, yielding the measure of absorption. In this way, the spectrophotometer provides an absorption reading for the sample. This reading can be used to calculate the concentration of particular molecules or macromolecules in a sample.

Detection of Radioisotopes

A radioisotope is an unstable form of an atom that decays to a more stable form by emitting α-, β-, or γ-rays, which are types of ionizing radiation. In research, radioisotopes that are β and/or γ emitters are commonly used. A β-ray is an emitted electron, and a γ-ray is an emitted photon. Some radioisotopes commonly used in biological experiments are shown in **Table A.2**.

Experimentally, radioisotopes are used because they are easy to detect. Therefore, if a particular compound is radiolabeled, its presence can be detected specifically throughout the course of the experiment. For example, if a

TABLE **A.2**
Some Useful Isotopes in Genetics

Isotope	Stable or Radioactive	Emission	Half-life
^{2}H	Stable		
^{3}H	Radioactive	β	12.3 years
^{13}C	Stable		
^{14}C	Radioactive	β	5730 years
^{15}N	Stable		
^{18}O	Stable		
^{24}Na	Radioactive	β (and γ)	15 hours
^{32}P	Radioactive	β	14.3 days
^{35}S	Radioactive	β	87.4 days
^{45}Ca	Radioactive	β	164 days
^{59}Fe	Radioactive	β (and γ)	45 days
^{131}I	Radioactive	β (and γ)	8.1 days

nucleotide is radiolabeled with ^{32}P, a researcher can determine whether the isotope becomes incorporated into newly made DNA or whether it remains as the free nucleotide. Researchers commonly use two different methods of detecting radioisotopes: scintillation counting and autoradiography.

The technique of **scintillation counting** permits a researcher to count the number of radioactive emissions from a sample containing a population of radioisotopes. In this approach, the sample is dissolved in a solution (called the scintillant) that contains organic solvents and one or more compounds known as fluors. When radioisotopes emit ionizing radiation, the energy is absorbed by the fluors in the solvent. This excites the fluor molecules, causing their electrons to be boosted to higher energy levels. The excited electrons return to lower, more stable energy levels by releasing photons of light. When a fluor is struck by ionizing radiation, it also absorbs the energy and then releases a photon of light within a particular wavelength range. The role of a device known as a scintillation counter is to count the photons of light emitted by the fluor. **Figure A.8** shows a scintillation counter. To use this device, a researcher dissolves her or his sample in a scintillant and then places the sample in a scintillation vial. The vial is then placed in the scintillation counter, which detects the amount of radioactivity. The scintillation counter has a digital meter that displays the amount of radioactivity in the sample and provides a printout of the amount of radioactivity in counts per minute. A scintillation counter contains several rows for the loading and analysis of many scintillation vials. After they have been loaded, the scintillation counter counts the amount of radioactivity in each vial and provides the researcher with a printout of the amount of radioactivity in each vial.

A second way of detecting radioisotopes is via **autoradiography.** This technique is not as quantitative as scintillation counting, because it does not provide the experimenter with a precise measure of the amount of radioactivity in counts per minute.

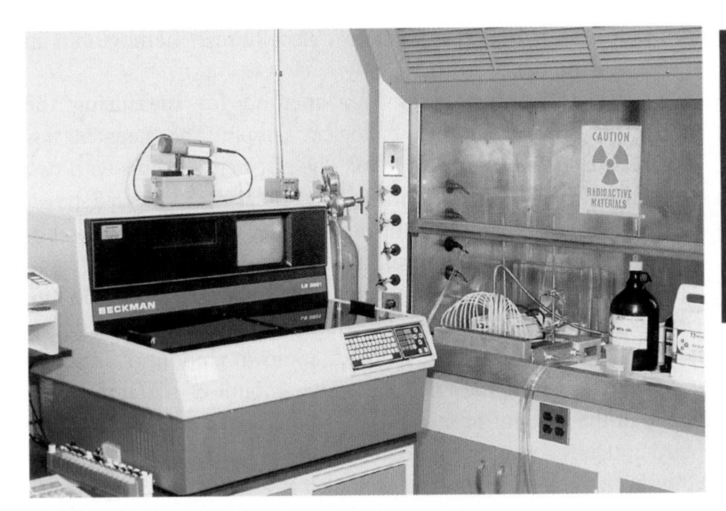

FIGURE A.8 **A scintillation counter.**

However, autoradiography has the great advantage that it can detect the location of radioisotopes as they are found in macromolecules or cells. For example, autoradiography is used to detect a particular band on a gel or to map the location of a gene within an intact chromosome.

To conduct autoradiography, a sample containing a radioisotope is fixed and usually dried. If it is a cellular sample, it also may be thin-sectioned. The sample is then pressed next to X-ray film (in the dark) and placed in a lightproof cassette. When a radioisotope decays, it emits a β- or γ-ray, which may strike a thin layer of photoemulsion next to the film. The photoemulsion contains silver salts such as AgBr. When a radioactive particle is emitted and strikes the photoemulsion, a silver grain is deposited on the film. This produces a dark spot on the film, which correlates with the original location of the radioisotope in the sample. In this way, the dark image on the film reveals the location(s) of the radioisotopes in the sample. Figure 11.4b shows how autoradiography can be used to visualize the process of bacterial chromosome replication. In this case, radiolabeled nucleotides were incorporated into the DNA, making it possible to picture the topology of the chromosome as two replication forks pass around the circular chromosome.

Detection of Antigens by Radioimmunoassay

Antibodies, also known as **immunoglobulins,** are proteins that are used to ward off infection by foreign substances; they are produced by cells of the immune system. Antibodies bind to structures on the surface of foreign substances known as **epitopes;** the foreign substance is called an **antigen.** A particular antibody binds to a particular antigen with a very high degree of specificity. For this reason, antibodies have been used extensively by researchers to detect particular antigens. For example, a human protein such as hemoglobin can be injected into a rabbit. Human hemoglobin is a foreign substance in the rabbit's bloodstream. Therefore, the rabbit makes antibodies that specifically recognize human hemoglobin and are designed to destroy it. Researchers can isolate and purify these antibodies from a sample of the rabbit's blood and then use them to detect human hemoglobin in their experiments.

A **radioimmunoassay** is a method for measuring the amount of an antigen in a biological sample. The steps in this method are shown in **Figure A.9a.** The researcher begins with two tubes that have a known amount of radiolabeled antigen (shown in blue). An unknown amount of the same antigen, which is not radiolabeled (shown in orange), is added to the tube on the right. The nonradiolabeled antigen comes from a biological sample; the goal of this experiment is to determine how much of this antigen is contained within the sample. Next, a known amount of antibody is added to each of the two tubes. The amount of the antibody is less than the amount of the antigen, so the nonlabeled and radiolabeled antigens compete with each other for binding to the antibody. After binding, a precipitating agent such as an anti-immunoglobulin antibody is added, and the precipitate is centrifuged to the bottom of the tube. The radioactivity in the precipitate is then determined by scintillation counting.

To calculate the amount of antigen in the sample being assayed, the researcher must determine the percentage of antibody that has bound to nonlabeled antigen. To do so, a second component of the experiment is to develop a standard curve in which a fixed amount of radiolabeled antigen is mixed with varying amounts of unlabeled antigen (**Figure A.9b**). Using this standard curve, a researcher can determine how much antigen is found in the unknown sample. For example, as shown in the dashed line, if the unknown sample had about 45% of the antibody bound, then the concentration of antigen in the sample is between 50 and 75 nanomolar.

Radioimmunoassays are used to determine the concentrations of many different kinds of antigens. This includes small molecules such as hormones) or macromolecules such as proteins.

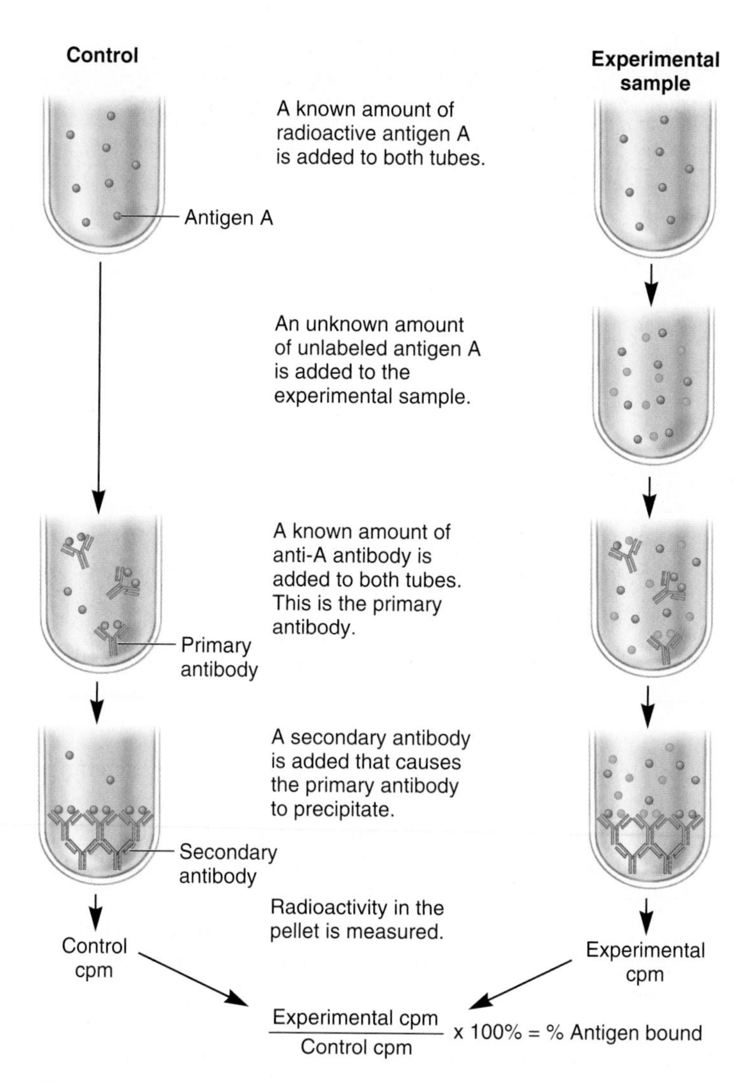

(a) A radioimmunoassay

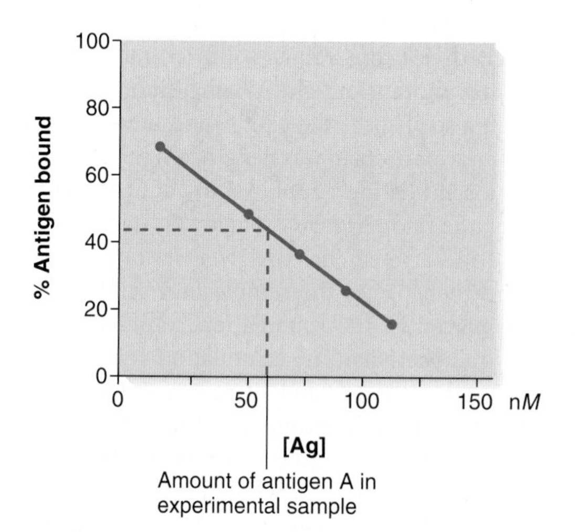

(b) Standard curve

FIGURE A.9 The method of radioimmunoassay (a) and the construction of a standard curve (b). In (**b**), the dashed line corresponds to the amount of antigen (Ag) bound by an unknown sample. This amounts to a concentration between 50 and 75 nanomolar of antigen.

APPENDIX SOLUTIONS TO EVEN-NUMBERED PROBLEMS

CHAPTER 1

Conceptual Questions

C2. A chromosome is a very long polymer of DNA. A gene is a specific sequence of DNA within that polymer; the sequence of bases creates a gene and distinguishes it from other genes. Genes are located in chromosomes, which are found within living cells.

C4. At the molecular level, a gene (a sequence of DNA) is first transcribed into RNA. The genetic code within the RNA is used to synthesize a protein with a particular amino acid sequence. This second process is called translation.

C6. Genetic variation involves the occurrence of genetic differences within members of the same species or different species. Within any population, variation may occur in the genetic material. Variation may occur in particular genes so that some individuals carry one allele and other individuals carry a different allele. An example would be differences in coat color among mammals. There also may be variation in chromosome structure and number. In plants, differences in chromosome number can affect disease resistance.

C8. You could pick almost any trait. For example, flower color in petunias would be an interesting choice. Some petunias are red and others are purple. There must be different alleles in a flower color gene that affect this trait in petunias. In addition, the amount of sunlight, fertilizer, and water also affects the intensity of flower color.

C10. A DNA sequence is a sequence of nucleotides. Each nucleotide may have one of four different bases (i.e., A, T, G, or C). When we speak of a DNA sequence, we focus on the sequence of bases.

C12. A. A gene is a segment of DNA. For most genes, the expression of the gene results in the production of a functional protein. The functioning of proteins within living cells affects the traits of an organism.

B. A gene is a segment of DNA that usually encodes the information for the production of a specific protein. Genes are found within chromosomes. Many genes are found within a single chromosome.

C. An allele is an alternative version of a particular gene. For example, suppose a plant has a flower color gene. One allele could produce a white flower, while a different allele could produce an orange flower. The white allele and orange allele are alleles of the flower color gene.

D. A DNA sequence is a sequence of nucleotides. The information within a DNA sequence (which is transcribed into an RNA sequence) specifies the amino acid sequence within a protein.

C14. A. How genes and traits are transmitted from parents to offspring

B. How the genetic material functions at the molecular and cellular levels

C. Why genetic variation exists in populations, and how it changes over the course of many generations

Experimental Questions

E2. This would be used primarily by molecular geneticists. The sequence of DNA is a molecular characteristic of DNA. In addition, as we will learn throughout this textbook, the sequence of DNA is interesting to transmission and population geneticists as well.

E4. A. Transmission geneticists. Dog breeders are interested in how genetic crosses affect the traits of dogs.

B. Molecular geneticists. This is a good model organism to study genetics at the molecular level.

C. Both transmission geneticists and molecular geneticists. Fruit flies are easy to cross and study the transmission of genes and traits from parents to offspring. Molecular geneticists have also studied many genes in fruit flies to see how they function at the molecular level.

D. Population geneticists. Most wild animals and plants would be the subject of population geneticists. In the wild, you cannot make controlled crosses. But you can study genetic variation within populations and try to understand its relationship to the environment.

E. Transmission geneticists. Agricultural breeders are interested in how genetic crosses affect the outcome of traits.

CHAPTER 2

Conceptual Questions

C2. In the case of plants, cross-fertilization occurs when the pollen and eggs come from different plants, whereas in self-fertilization, they come from the same plant.

C4. A homozygote that has two copies of the same allele

C6. Diploid organisms contain two copies of each type of gene. When they make gametes, only one copy of each gene is found in a gamete. Two alleles cannot stay together within the same gamete.

C8. Genotypes: 1:1 *Tt* and *tt*

Phenotypes: 1:1 Tall and dwarf

C10. *c* is the recessive allele for constricted pods; *Y* is the dominant allele for yellow color. The cross is *ccYy* × *CcYy*. Follow the directions for setting up a Punnett square, as described in Chapter 2. The genotypic ratio is 2 *CcYY* : 4 *CcYy* : 2 *Ccyy* : 2 *ccYY* : 4 *ccYy* : 2 *ccyy*. This 2:4:2:2:4:2 ratio could be reduced to a 1:2:1:1:2:1 ratio.

The phenotypic ratio is 6 smooth pods, yellow seeds : 2 smooth pods, green seeds : 6 constricted pods, yellow seeds : 2 constricted pods, green seeds. This 6:2:6:2 ratio could be reduced to a 3:1:3:1 ratio.

C12. Offspring with a nonparental phenotype are consistent with the idea of independent assortment. If two different traits were always transmitted together as unit, it would not be possible to get nonparental phenotypic combinations. For example, if a true-breeding parent had two dominant traits and was crossed to a true-breeding parent having the two recessive traits, the F_2 generation could not have offspring with one recessive and one dominant phenotype. However, because independent assortment can occur, it is possible for F_2 offspring to have one dominant and one recessive trait.

C14. A. Barring a new mutation during gamete formation, the chance is 100% because they must be heterozygotes in order to produce a child with a recessive disorder.

B. Construct a Punnett square. There is a 50% chance of heterozygous children.

C. Use the product rule. The chance of being phenotypically normal is 0.75 (i.e., 75%), so the answer is $0.75 \times 0.75 \times 0.75 = 0.422$, which is 42.2%.

D. Use the binomial expansion equation, where $n = 3$, $x = 2$, $p = 0.75$, $q = 0.25$. The answer is 0.422, or 42.2%.

C16. First construct a Punnett square. The chances are 75% of producing a solid pup and 25% of producing a spotted pup.

A. Use the binomial expansion equation, where $n = 5$, $x = 4$, $p = 0.75$, $q = 0.25$. The answer is $0.396 = 39.6\%$ of the time.

B. You can use the binomial expansion equation for each litter. For the first litter, $n = 6$, $x = 4$, $p = 0.75$, $q = 0.25$; for the second litter, $n = 5$, $x = 5$, $p = 0.75$, $q = 0.25$. Because the litters are in a specified order, we use the product rule and multiply the probability of the first litter times the probability of the second litter. The answer is 0.070, or 7.0%.

C. To calculate the probability of the first litter, we use the product rule and multiply the probability of the first pup (0.75) times the probability of the remaining four. We use the binomial expansion equation to calculate the probability of the remaining four, where $n = 4$, $x = 3$, $p = 0.75$, $q = 0.25$. The probability of the first litter is 0.316. To calculate the probability of the second litter, we use the product rule and multiply the probability of the first pup (0.25) times the probability of the second pup (0.25) times the probability of the remaining five. To calculate the probability of the remaining five, we use the binomial expansion equation, where $n = 5$, $x = 4$, $p = 0.75$, $q = 0.25$. The probability of the second litter is 0.025. To get the probability of these two litters occurring in this order, we use the product rule and multiply the probability of the first litter (0.316) times the probability of the second litter (0.025). The answer is 0.008, or 0.8%.

D. Because this is a specified order, we use the product rule and multiply the probability of the firstborn (0.75) times the probability of the second born (0.25) times the probability of the remaining four. We use the binomial expansion equation to calculate the probability of the remaining four pups, where $n = 4$, $x = 2$, $p = 0.75$, $q = 0.25$. The answer is 0.040, or 4.0%.

C18. A. Use the product rule:

$(^1/_4)(^1/_4) = {^1/_{16}}$

B. Use the binomial expansion equation:

$n = 4, p = {^1/_4}, q = {^3/_4}, x = 2$

$P = 0.21 = 21\%$

C. Use the product rule:

$(^1/_4)(^3/_4)(^3/_4) = 0.14$, or 14%

C20. A. $^1/_4$

B. 1, or 100%

C. $(^3/_4)(^3/_4)(^3/_4) = {^{27}/_{64}} = 0.42$, or 42%

D. Use the binomial expansion equation, where

$n = 7, p = {^3/_4}, q = {^1/_4}, x = 3$

$P = 0.058$, or 5.8%

E. The probability that the first plant is tall is $^3/_4$. To calculate the probability that among the next four, any two will be tall, we use the binomial expansion equation, where $n = 4$, $p = {^3/_4}$, $q = {^1/_4}$, and $x = 2$.

The probability P equals 0.21.

To calculate the overall probability of these two events:

$(^3/_4)(0.21) = 0.16$, or 16%

C22. It violates the law of segregation because two copies of one gene are in the gamete. The two alleles for the *A* gene did not segregate from each other.

C24. Based on this pedigree, it is likely to be dominant inheritance because an affected child always has an affected parent. In fact, it is a dominant disorder.

C26. It is impossible for the F_1 individuals to be true-breeding because they are all heterozygotes.

C28. 2 *TY*, *tY*, 2 *Ty*, *ty*, *TTY*, *TTy*, 2 *TtY*, 2 *Tty*

It may be tricky to think about, but you get 2 *TY* and 2 *Ty* because either of the two *T* alleles could combine with *Y* or *y*. Also, you get 2 *TtY* and 2 *Tty* because either of the two *T* alleles could combine with *t* and then combine with *Y* or *y*.

C30. The genotype of the F_1 plants is *Tt Yy Rr*. According to the laws of segregation and independent assortment, the alleles of each gene will segregate from each other, and the alleles of different genes will randomly assort into gametes. A *Tt Yy Rr* individual could make eight types of gametes: *TYR*, *TyR*, *Tyr*, *TYr*, *tYR*, *tyR*, *tYr*, and *tyr*, in equal proportions (i.e., $^1/_8$ of each type of gamete). To determine genotypes and phenotypes, you could make a large Punnett square that would contain 64 boxes. You would need to line up the eight possible gametes across the top and along the side and then fill in the 64 boxes. Alternatively, you could use one of the two approaches described in solved problem S4. The genotypes and phenotypes would be

1 *TT YY RR*
2 *TT Yy RR*
2 *TT YY Rr*
2 *Tt YY RR*
4 *TT Yy Rr*
4 *Tt Yy RR*
4 *Tt YY Rr*
8 *Tt Yy Rr* = 27 tall, yellow, round
1 *TT yy RR*
2 *Tt yy RR*
2 *TT yy Rr*
4 *Tt yy Rr* = 9 tall, green, round
1 *TT YY rr*
2 *TT Yy rr*
2 *Tt YY rr*
4 *Tt Yy rr* = 9 tall, yellow, wrinkled
1 *tt YY RR*
2 *tt Yy RR*
2 *tt YY Rr*
4 *tt Yy Rr* = 9 dwarf, yellow, round
1 *TT yy rr*
2 *Tt yy rr* = 3 tall, green, wrinkled
1 *tt yy RR*
2 *tt yy Rr* = 3 dwarf, green, round
1 *tt YY rr*
2 *tt Yy rr* = 3 dwarf, yellow, wrinkled
1 *tt yy rr* = 1 dwarf, green, wrinkled

C32. The wooly-haired male is a heterozygote, because he has the trait and his mother did not. (He must have inherited the normal allele from his mother.) Therefore, he has a 50% chance of passing the wooly allele to his offspring; his offspring have a 50% of passing the allele to their offspring; and these grandchildren have a 50% chance of passing the allele to their offspring (the wooly-haired man's great-grandchildren). Because this is an ordered sequence of independent events, we use the product rule: $0.5 \times 0.5 \times 0.5 = 0.125$, or 12.5%. Because no other Scandinavians are on the island, the chance is 87.5% for the offspring being normal (because they could not inherit the wooly hair allele from anyone else). We use the binomial expansion equation to determine the likelihood that one out of eight great-grandchildren will have wooly hair, where $n = 8$, $x = 1$, $p = 0.125$, $q = 0.875$. The answer is 0.393, or 39.3%, of the time.

C34. Use the product rule. If the woman is heterozygous, there is a 50% chance of having an affected offspring: $(0.5)^7 = 0.0078$, or 0.78%, of the time. This is a pretty small probability. If the woman has an eighth child who is unaffected, however, she has to be a heterozygote, because it is a dominant trait. She would have to pass a normal allele to an unaffected offspring. The answer is 100%.

Experimental Questions

E2. The experimental difference depends on where the pollen comes from. In self-fertilization, the pollen and eggs come from the same plant. In cross-fertilization, they come from different plants.

E4. According to Mendel's law of segregation, the genotypic ratio should be 1 homozygote dominant : 2 heterozygotes : 1 homozygote recessive. This data table considers only the plants with a dominant phenotype. The genotypic ratio should be 1 homozygote dominant : 2 heterozygotes. The homozygote dominants would be true-breeding, while the heterozygotes would not be true-breeding. This 1:2 ratio is very close to what Mendel observed.

E6. All three offspring had black fur. The ovaries from the albino female could produce eggs with only the dominant black allele (because they were obtained from a true-breeding black female). The actual phenotype of the albino mother does not matter. Therefore, all offspring would be heterozygotes (*Bb*) and have black fur.

E8. If we construct a Punnett square according to Mendel's laws, we expect a 9:3:3:1 ratio. Because a total of 556 offspring were observed, the expected number of offspring are

$556 \times 9/16 = 313$ round, yellow

$556 \times 3/16 = 104$ wrinkled, yellow

$556 \times 3/16 = 104$ round, green

$556 \times 1/16 = 35$ wrinkled, green

If we plug the observed and expected values into the chi square equation, we get a value of 0.51. With four categories, our degrees of freedom equal $n - 1$, or 3. If we look up the value of 0.51 in the chi square table (see Table 2.1), we see that it falls between the *P* values of 0.80 and 0.95. This means that the probability is 80% to 95% that any deviation between observed results and expected results was caused by random sampling error. Therefore, we accept the hypothesis. In other words, the results are consistent with the law of independent assortment.

E10. A. If we let c^+ represent normal wings and c represent curved wings, and e^+ represents gray body and e represents ebony body,

Parental Cross: $cce^+e^+ \times c^+c^+ee$.

F_1 generation is heterozygous c^+ce^+e

An F_1 offspring crossed to flies with curved wings and ebony bodies is

$c^+ce^+e \times ccee$

The F_2 offspring would be a 1:1:1:1 ratio of flies:

$c^+ce^+e : c^+cee : cce^+e : ccee$

B. The phenotypic ratio of the F_2 flies would be a 1:1:1:1 ratio of flies:

normal wings, gray body : normal wings, ebony bodies : curved wings, gray bodies : curved wings, ebony bodies

C. From part B, we expect 1/4 of each category. There are a total of 444 offspring. The expected number of each category is $1/4 \times 444$, which equals 111.

$$\chi^2 = \frac{(114-111)^2}{111} + \frac{(105-111)^2}{111} + \frac{(111-111)^2}{111} + \frac{(114-111)^2}{111}$$

$$\chi^2 = 0.49$$

With 3 degrees of freedom, a value of 0.49 or greater is likely to occur between 80% and 95% of the time. Therefore, we accept our hypothesis.

E12. Follow through the same basic chi square strategy as before. We expect a 3:1 ratio, or 3/4 of the dominant phenotype and 1/4 of the recessive phenotype.

The observed and expected values are as follows (rounded to the nearest whole number):

Observed*	Expected	$\frac{(O-E)^2}{E}$
5474	5493	0.066
1850	1831	0.197
6022	6017	0.004
2001	2006	0.012
705	697	0.092
224	232	0.276
882	886	0.018
299	295	0.054
428	435	0.113
152	145	0.338
651	644	0.076
207	215	0.298
787	798	0.152
277	266	0.455
		$\chi^2 = 2.15$

*Due to rounding, the observed and expected values may not add up to precisely the same number.

Because $n = 14$, there are 13 degrees of freedom. If we look up this value in the chi square table, we have to look between 10 and 15 degrees of freedom. In either case, we would expect the value of 2.15 or greater to occur more than 99% of the time. Therefore, we accept the hypothesis.

E14. The dwarf parent with terminal flowers must be homozygous for both genes, because it is expressing these two recessive traits: *ttaa*, where *t* is the recessive dwarf allele, and *a* is the recessive allele for terminal flowers. The phenotype of the other parent is dominant for both traits. However, because this parent was able to produce dwarf offspring with axial flowers, it must have been heterozygous for both genes: *TtAa*.

E16. Our hypothesis is that blue flowers and purple seeds are dominant traits and they are governed by two genes that assort independently. According to this hypothesis, the F_2 generation should yield a ratio of 9 blue flowers, purple seeds : 3 blue flowers, green seeds : 3 white flowers, purple seeds : 1 white flower, green seeds. Because a total of 300 offspring are produced, the expected numbers would be

$9/16 \times 300 = 169$ blue flowers, purple seeds

$3/16 \times 300 = 56$ blue flowers, green seeds

$3/16 \times 300 = 56$ white flowers, purple seeds

$1/16 \times 300 = 19$ white flowers, green seeds

$$\chi^2 = \frac{(103-169)^2}{169} + \frac{(49-56)^2}{56} + \frac{(44-56)^2}{56} + \frac{(104-19)^2}{19}$$

$$\chi^2 = 409.5$$

If we look up this value in the chi square table under 3 degrees of freedom, the value is much higher than would be expected 1% of the time by chance alone. Therefore, we reject the hypothesis. The idea that the two genes are assorting independently seems to be incorrect. The F_1 generation supports the idea that blue flowers and purple seeds are dominant traits.

Questions for Student Discussion/Collaboration

2. If we construct a Punnett square, the following probabilities will be obtained:

tall with axial flowers, 3/8

dwarf with terminal flowers, 1/8

To calculate the probability of being tall with axial flowers or dwarf with terminal flowers, we use the sum rule:

$$^3/_8 + {^1/_8} = {^4/_8} = {^1/_2}$$

We use the product rule to calculate the ordered events of the first three offspring being tall/axial or dwarf/terminal, and the fourth offspring being tall axial:

$$(^1/_2)(^1/_2)(^1/_2)(^3/_8) = {^3/_{64}} = 0.047 = 4.7\%$$

CHAPTER 3

Conceptual Questions

C2. The term homolog refers to the members of a chromosome pair. Homologs are usually the same size and carry the same types and order of genes. They may differ in that the genes they carry may be different alleles.

C4. Metaphase is the organization phase, and anaphase is the separation phase.

C6. In metaphase I of meiosis, each pair of chromatids is attached to only one pole via the kinetochore microtubules. In metaphase of mitosis, there are two attachments (i.e., to both poles). If the attachment was lost, a chromosome would not migrate to a pole and may not become enclosed in a nuclear membrane after telophase. If left out in the cytoplasm, it would eventually be degraded.

C8. The reduction occurs because there is a single DNA replication event but two cell divisions. Because of the nature of separation during anaphase I, each cell receives one copy of each type of chromosome.

C10. It means that the maternally derived and paternally derived chromosomes are randomly aligned along the metaphase plate during metaphase I. Refer to Figure 3.17.

C12. There are three pairs of chromosomes. The number of different, random alignments equals 2^n, where n equals the number of chromosomes per set. So the possible number of arrangements equals 2^3, which is 8.

C14. The probability would be much lower because pieces of maternal chromosomes would be mixed with the paternal chromosomes. Therefore, it is unlikely to inherit a chromosome that was completely paternally derived.

C16. During interphase, the chromosomes are greatly extended. In this conformation, they might get tangled up with each other and not sort properly during meiosis and mitosis. The condensation process probably occurs so that the chromosomes easily align along the equatorial plate during metaphase without getting tangled up.

C18. During prophase II, your drawing should show four replicated chromosomes (i.e., four structures that look like Xs). Each chromosome is one homolog. During prophase of mitosis, there should be eight replicated chromosomes (i.e., eight Xs). During prophase of mitosis, there are pairs of homologs. The main difference is that prophase II has a single copy of each of the four chromosomes, whereas prophase of mitosis has four pairs of homologs. At the end of meiosis I, each daughter cell has received only one copy of a homologous pair, not both. This is due to the alignment of homologs during metaphase I and their separation during anaphase I.

C20. DNA replication does not take place during interphase II. The chromosomes at the end of telophase I have already replicated (i.e., they are found in pairs of sister chromatids). During meiosis II, the sister chromatids separate from each other, yielding individual chromosomes.

C22. A. 20   C. 30

B. 10   D. 20

C24. A. Dark males and light females; reciprocal: all dark offspring

B. All dark offspring; reciprocal: dark females and light males

C. All dark offspring; reciprocal: dark females and light males

D. All dark offspring; reciprocal: dark females and light males

C26. To produce sperm, a spermatogonial cell first goes through mitosis to produce two cells. One of these remains a spermatogonial cell and the other progresses through meiosis. In this way, the testes continue to maintain a population of spermatogonial cells.

C28. A. *A B C, A B c, A b C, A b c, a B C, a b C, a B c, a b c*

B. *A B C, A b C*

C. *A B C, A B c, a B C, a B c*

D. *A b c, a b c*

C30. A. The fly is a male because the ratio of X chromosomes to sets of autosomes is $^1/_2$, or 0.5.

B. The fly is female because the ratio of X chromosomes to sets of autosomes is 1.0.

C. The fly is male because the ratio of X chromosomes to sets of autosomes is 0.5.

D. The fly is female because the ratio of X chromosomes to sets of autosomes is 1.0.

Experimental Questions

E2. Perhaps the most convincing observation was that all of the white-eyed flies of the F_2 generation were males. This suggests a link between sex determination and the inheritance of this trait. Because sex determination in fruit flies is determined by the number of X chromosomes, this suggests a relationship between the inheritance of the X chromosome and the inheritance of this trait.

E4. The basic strategy is to set up a pair of reciprocal crosses. The phenotype of sons is usually the easiest way to discern the two patterns. If it is Y linked, the trait will be passed only from father to son. If it is X linked, the trait will be passed from mother to son.

E6. The 3:1 sex ratio occurs because the female produces 50% gametes that are XX (and must produce female offspring) and 50% that are X (and produce half male and half female offspring). The original female had one X chromosome carrying the red allele and two other X chromosomes carrying the eosin allele. Set up a Punnett square assuming that this female produces the following six types of gametes: $X^{w+}X^{w-e}$, $X^{w+}X^{w-e}$, $X^{w-e}X^{w-e}$, X^{w+}, X^{w-e}, X^{w-e}. The male of this cross is X^wY.

Male gametes

Female gametes ♀ \ ♂	X^w	Y
$X^{w^+}X^{w-e}$	$X^{w^+}X^{w-e}X^w$ Red, female	$X^{w^+}X^{w-e}Y$ Red, female
$X^{w^+}X^{w-e}$	$X^{w^+}X^{w-e}X^w$ Red, female	$X^{w^+}X^{w-e}Y$ Red, female
$X^{w-e}X^{w-e}$	$X^{w-e}X^{w-e}X^w$ Eosin female	$X^{w-e}X^{w-e}Y$ Eosin female
X^{w^+}	$X^{w^+}X^w$ Red, female	$X^{w^+}Y$ Red, male
X^{w-e}	$X^{w-e}X^w$ Light-eosin female	$X^{w-e}Y$ Light-eosin male
X^{w-e}	$X^{w-e}X^w$ Light-eosin female	$X^{w-e}Y$ Light-eosin male

E8. If we use the data from the F_1 mating (i.e., F_2 results), there were 3470 red-eyed flies. We would expect a 3:1 ratio between red- and white-eyed flies. Therefore, assuming that all red-eyed offspring survived, there should have been about 1157 (i.e., $^{3470}/_3$) white-eyed flies. However, there were only 782. If we divide 782 by 1157, we get a value of 0.676, or a 67.6% survival rate.

E10. You need to make crosses to understand the pattern of inheritance of traits (determined by genes) from parents to offspring. And you need to microscopically examine cells to understand the pattern of transmission of chromosomes. The correlation between the pattern of transmission of chromosomes during meiosis, and Mendel's laws of segregation and independent assortment is what led to the chromosome theory of inheritance.

E12. Originally, individuals who had abnormalities in their composition of sex chromosomes provided important information. In mammals, X0 individuals are females, whereas in flies, X0 individuals are males. In mammals, XXY individuals are males, while in flies, XXY individuals are females. These results indicate that the presence of the Y chromosome causes maleness in mammals, but it does not in flies. A further analysis of flies with abnormalities in the number of sets of autosomes revealed that it is the ratio between the number of X chromosomes and the number of sets of autosomes that determines sex in flies.

Questions for Student Discussion/Collaboration

2. It's not possible to give a direct answer, but the point is for students to be able to draw chromosomes in different configurations and understand the various phases. The chromosomes may or may not be

    1. In homologous pairs
    2. Connected as sister chromatids
    3. Associated in bivalents
    4. Lined up in metaphase
    5. Moving toward the poles.

    And so on.

CHAPTER 4

Conceptual Questions

C2. Sex-influenced traits are influenced by the sex of the individual even though the gene that governs the trait may be autosomally inherited. Pattern baldness in people is an example. Sex-limited traits are an extreme example of sex influence. The expression of a sex-limited trait is limited to one sex. For example, colorful plumage in certain species of birds is limited to the male sex. Sex-linked traits involve traits whose genes are found on the sex chromosomes. Examples in humans include hemophilia and color blindness.

C4. If the normal allele is dominant, it tells you that one copy of the gene produces a sufficient amount of the protein encoded by the gene. Having twice as much of this protein, as in the normal homozygote, does not alter the phenotype. If the allele is incompletely dominant, this means that one copy of the normal allele does not produce the same trait as the homozygote.

C6. The ratio would be 1 normal : 2 star-eyed individuals.

C8. If individual 1 is *ii*, individual 2 could be I^Ai, I^AI^A, I^Bi, I^BI^B, or I^AI^B.

If individual 1 is I^Ai or I^AI^A, individual 2 could be I^Bi, I^BI^B, or I^AI^B.

If individual 1 is I^Bi or I^BI^B, individual 2 could be I^Ai, I^AI^A, or I^AI^B.

Assuming individual 1 is the parent of individual 2:

If individual 1 is *ii*, individual 2 could be I^Ai or I^Bi.

If individual 1 is I^Ai, individual 2 could be I^Bi or I^AI^B.

If individual 1 is I^AI^A, individual 2 could be I^AI^B.

If individual 1 is I^Bi, individual 2 could be I^Ai or I^AI^B.

If individual 1 is I^BI^B, individual 2 could be I^AI^B.

C10. The father could not be I^AI^B, I^BI^B, or I^AI^A. He is contributing the O allele to his offspring. Genotypically, he could be I^Ai, I^Bi, or *ii* and have type A, B, or O blood, respectively.

C12. Perhaps it should be called codominant at the "hair level" because one or the other allele is dominant with regard to a single hair. However, this is not the same as codominance in blood types, in which every cell can express both alleles.

C14. A. X-linked recessive (unaffected mothers transmit the trait to sons)

B. Autosomal recessive (affected daughters and sons are produced from unaffected parents)

C16. First set up the following Punnett square:

Male gametes

♀ \ ♂	X^H	Y
X^H	X^HX^H	X^HY
X^h	X^HX^h	X^hY

Female gametes

There is a $^1/_4$ probability of each type of offspring.

A. $^1/_4$

B. $(^3/_4)(^3/_4)(^3/_4)(^3/_4) = {}^{81}/_{256}$

C. $^3/_4$

D. The probability of an affected offspring is $^1/_4$, and the probability of an unaffected offspring is $^3/_4$. For this problem, you use the binomial expansion equation where $x = 2$, $n = 5$, $p = {}^1/_4$, and $q = {}^3/_4$. The answer is 0.26, or 26%, of the time.

C18. We know that the parents must be heterozygotes for both genes.

The genotypic ratio of their children is 1 *BB* : 2 *Bb* : 1 *bb*

The phenotypic ratio depends on sex: 1 *BB* bald male : 1 *BB* bald female : 2 *Bb* bald males : 2 *Bb* nonbald females : 1 *bb* nonbald male : 1 *bb* nonbald female

A. 50%

B. 1/8, or 12.5%

C. (3/8)(3/8)(3/8)= 27/512 = 0.05, or 5%

C20. It probably occurred in the summer. In the Siamese cat, dark fur occurs in cooler regions of the body. If the fur grows during the summer, these regions are likely to be somewhat warmer, and therefore the fur will be lighter.

C22. First, you would cross heterozygous birds to each other. This would yield an F_1 generation consisting of a ratio of 1 *HH* : 2 *Hh* : 1 *hh*. The male offspring that are *hh* would have cock-feathering. All the female offspring would have hen-feathering. You would then take cock-feathered males and cross them to F_1 females. If all of the males within a brood were cock-feathered, it is likely that the mother was *hh*. If so, all of the offspring would be *hh*. The offspring from such a brood could be crossed to each other. If they are truly *hh*, the males of the F_3 generation should all be cock-feathered.

C24. A. Could be.

B. No, because an affected female has an unaffected son.

C. Could be.

D. No, because an affected male has an unaffected daughter.

E. No, because it affects both sexes.

C26. Molecular: The β-globin gene for Hb^A homozygotes encodes a β-globin polypeptide with a normal amino acid sequence compared to Hb^S homozygotes, whose β-globin genes encode a polypeptide that has an abnormal structure. The abnormal structure affects the ability of hemoglobin to carry oxygen.

Cellular: Under conditions of low oxygen, Hb^SHb^S cells form a sickle shape compared to the normal biconcave disk shape of Hb^AHb^A cells.

Organism: In Hb^SHb^S individuals, the sickle shape decreases the life span of the red blood cell, which causes anemia. Also, the clogging of red blood cells in the capillaries causes tissue damage and painful crises. This does not occur in Hb^AHb^A individuals.

C28.

I-1 I-2

II-1 II-2 II-3 II-4 II-5 II-6 II-7

III-1 III-2 III-3

C30. In some cases, two proteins, one encoded by the first mutant and one by the suppressor mutation, physically interact with each other. The two mutations compensate for each other. Second, two distinct proteins encoded by different genes may participate in a common function, but not directly interact with each other. A mutation that greatly decreases the function of one protein may be compensated by a suppressor mutation that increases the function of another protein. Third, some suppressors affect the amount of protein encoded by a mutant gene. A first mutation may decrease the amount or functional activity of a protein. A suppressor mutation involving a genetic regulatory protein could increase the amount of the first protein and thereby overcome its defect in amount or functional activity.

Experimental Questions

E2. Two redundant genes are involved in feathering. The unfeathered Buff Rocks are homozygous recessive for the two genes. The Black Langhans are homozygous dominant for both genes. In the F_2 generation (which is a double heterozygote crossed to another double heterozygote), 1 out of 16 offspring will be doubly homozygous for both recessive genes. All the others will have at least one dominant allele for one of the two (redundant) genes.

E4. The reason why all the puppies have black hair is because albino alleles are found in two different genes. If we let the letters *A* and *B* represent the two different pigmentation genes, then one of the dogs is *AAbb*, and the other is *aaBB*. Their offspring are *AaBb* and therefore are not albinos because they have one dominant copy of each gene.

E6. In general, you cannot distinguish between autosomal and pseudoautosomal inheritance from a pedigree analysis. Mothers and fathers have an equal probability of passing the alleles to sons and daughters. However, if an offspring had a chromosomal abnormality, you might be able to tell. For example, in a family tree involving the *Mic2* allele, an offspring that was X0 would have less of the gene product, and an offspring that was XXX or XYY or XXY would have extra amounts of the gene products. This may lead you to suspect that the gene is located on the sex chromosomes.

E8. One parent must be *RRPp*. The other parent could be *RRPp* or *RrPp*. All the offspring would inherit (at least) one dominant *R* allele. With regard to the other gene, $^3/_4$ would inherit at least one copy of the dominant *P* allele. These offspring would have a walnut comb. The other $^1/_4$ would be homozygous *pp* and have a rose comb (because they would also have a dominant *R* allele).

E10. Let's use the letters *A* and *B* for these two genes. Gene *A* exists in two alleles, which we will call *A* and *a*. Gene *B* exists in two alleles, *B* and *b*. The uppercase alleles are dominant to the lowercase alleles. The true-breeding long-shaped squash is *aabb*, and the true-breeding disk-shaped squash is *AABB*. The F_1 offspring are *AaBb*. You can construct a Punnett square, with 16 boxes, to determine the outcome of self-fertilization of the F_1 plants.

To get the disk-shaped phenotype, an offspring must inherit at least one dominant allele from both genes.

1 *AABB* + 2 *AaBB* + 2 *AABb* + 4 *AaBb* = 9 disk-shaped offspring

To get the round phenotype, an offspring must inherit at least one dominant allele for one of the two genes but must be homozygous recessive for only one of the two genes.

1 *aaBB* + 1 *AAbb* + 2 *aaBb* + 2 *Aabb* = 6 round-shaped offspring

To get the long phenotype, an offspring must inherit all recessive alleles:

1 *aabb* = 1 long-shaped offspring

E12. The results obtained when crossing two F_1 offspring appear to yield a 9:3:3:1 ratio, which would be expected if eye color is affected by two different genes that exist in dominant and recessive alleles. Neither gene is X linked. Let pr^+ represent the red allele of the first gene and *pr* the purple allele. Let sep^+ represent the red allele of the second gene and *sep* the sepia allele.

The first cross is $prpr\ sep^+sep^+ \times pr^+pr^+sep\ sep$

All the F_1 offspring would be $pr^+pr\ sep^+sep$. They have red eyes because they have a dominant red allele for each gene. When the F_1 offspring are crossed to each other, the following results would be obtained:

♀ / ♂	pr^+sep^+	pr^+sep	$pr\ sep^+$	$pr\ sep$
pr^+sep^+	pr^+pr^+ sep^+sep^+ Red	pr^+pr^+ sep^+sep Red	pr^+pr sep^+sep^+ Red	pr^+pr sep^+sep Red
pr^+sep	pr^+pr^+ sep^+sep Red	pr^+pr^+ $sep\ sep$ Sepia	pr^+pr sep^+sep Red	pr^+pr $sep\ sep$ Sepia
$pr\ sep^+$	pr^+pr sep^+sep^+ Red	pr^+pr sep^+sep Red	$pr\ pr$ sep^+sep^+ Purple	$pr\ pr$ sep^+sep Purple
$pr\ sep$	pr^+pr sep^+sep Red	pr^+pr $sep\ sep$ Sepia	$pr\ pr$ sep^+sep Purple	$pr\ pr$ $sep\ sep$ Pur/Sepia

In this case, one gene exists as the red (dominant) or purple (recessive) allele, and the second gene exists as the red (dominant) or sepia (recessive) allele. If an offspring is homozygous for the purple allele, it will have purple eyes. Similarly, if an offspring is homozygous for the sepia allele, it will have sepia eyes. An offspring homozygous for both recessive alleles has purplish sepia eyes. To have red eyes, it must have at least one copy of the dominant red allele for both genes. Based on an expected 9 red : 3 purple : 3 sepia : 1 purplish sepia, the observed and expected numbers of offspring are as follows:

Observed	*Expected*
146 purple eyes	148 purple eyes (791 × $^3/_{16}$)
151 sepia eyes	148 sepia eyes (791 × $^3/_{16}$)
50 purplish sepia eyes	49 purplish sepia eyes (791 × $^1/_{16}$)
444 red eyes	445 red eyes (791 × $^9/_{16}$)
791 total offspring	

If we plug the observed and expected values into our chi square formula, we obtain a chi square value of about 0.11. With 3 degrees of freedom, this is well within our expected range of values, so we cannot reject our hypothesis that purple and sepia alleles are in two different genes and that these recessive alleles are epistatic to each other.

E14. To see if the allele is X linked, the pink-eyed male could be crossed to a red-eyed female. All the offspring would have red eyes, assuming that the pink allele is recessive. When crossed to red-eyed males, the F_1 females will produce $1/2$ red-eyed daughters, $1/4$ red-eyed sons, and $1/4$ pink-eyed sons if the pink allele is X linked.

If the pink allele is X linked, then one could determine if it is in the same X-linked gene as the white and eosin alleles by crossing pink-eyed males to white-eyed females. (Note: We already know that white and eosin are alleles of the same gene.) If the pink and white alleles are in the same gene, the F_1 female offspring should have pink eyes (assuming that the pink allele is dominant over white). However, if the pink and white alleles are in different genes, the F_1 females will have red eyes (assuming that pink is recessive to red). This is because the F_1 females will be heterozygous for two genes, X^{w+p} X^{wp+}, in which the X^{w+} and X^{p+} alleles are the dominant wild-type alleles that produce red eyes, and the X^{w} and X^{p} alleles are recessive alleles for these two different genes, which produce white eyes and pink eyes, respectively.

E16. The total number of offspring shown in the data was 209. Rounded to the nearest whole number, the expected numbers would be as follows:

red-eyed females = 8/16 x 209 = 105

red-eyed males = 4/16 x 209 = 52

light eosin-eyed males = 3/16 x 209 = 39

cream-eyed males = 1/16 x 209 = 13

If we plug the observed and expected values into a chi square equation (see Chapter 2), the chi square value equals 1.2. If we look this value up in the chi square table (Table 2.1) with three degrees of freedom, the value lies between a probability of 0.8 and 0.5. Therefore, we cannot reject the hypothesis that cream is autosomal and modifier of eosin, and eosin and red are X-linked genes.

Questions for Student Discussion/Collaboration

2. Perhaps the easiest way to solve this problem is to take one trait at a time. With regard to combs, all the F_1 generation would be *RrPp* or walnut comb. With regard to shanks, they would all be feathered, because they would inherit one dominant copy of a feathered allele. With regard to hen- or cock-feathering: 1 male cock-feathered : 1 male hen-feathered : 2 females hen-feathered. Overall then, we would have a 1:1:2 ratio of

walnut comb/feathered shanks/cock-feathered males

walnut comb/feathered shanks/hen-feathered males

walnut comb/feathered shanks/hen-feathered females

CHAPTER 5

Conceptual Questions

C2. A maternal effect gene is one in which the genotype of the mother determines the phenotype of the offspring. At the cellular level, this happens because maternal effect genes are expressed in diploid nurse cells and then the gene products are transported into the egg. These gene products play key roles in the early steps of embryonic development.

C4. The genotype of the mother must be bic^- bic^-. That is why it produces abnormal offspring. Because the mother is alive and able to produce offspring, its mother (the maternal grandmother) must have been bic^+ bic^- and passed the bic^- allele to its daughter (the mother in this problem). The maternal grandfather also must have passed the bic^- allele to its daughter. The maternal grandfather could be either bic^+ bic^- or bic^- bic^-.

C6. The mother must be heterozygous. She is phenotypically abnormal because her mother must have been homozygous for the abnormal recessive allele. However, because she produces all normal offspring, she must have inherited the normal dominant allele from her father. She produces all normal offspring because this is a maternal effect gene, and the gene product of the normal dominant allele is transferred to the egg.

C8. Maternal effect genes exert their effects because the gene products are transferred from nurse cells to eggs. The gene products, mRNA and proteins, do not last a very long time before they are eventually degraded. Therefore, they can exert their effects only during early stages of embryonic development.

C10. Dosage compensation refers to the phenomenon that the level of expression of genes on the sex chromosomes is the same in males and females, even though they have different numbers of sex chromosomes. In many species it seems necessary so that the balance of gene expression between the autosomes and sex chromosomes is similar between the two sexes.

C12. In mammals, one of the X chromosomes is inactivated in females; in *Drosophila*, the level of transcription on the X chromosome in males is doubled; in *C. elegans*, the level of transcription of the X chromosome in hermaphrodites is decreased by 50% of that of males.

C14. X inactivation begins with the counting of Xics. If there are two X chromosomes, in the process of initiation, one is targeted for inactivation. During embryogenesis, this inactivation begins at the Xic locus and spreads to both ends of the X chromosome until it becomes a highly condensed Barr body. The *Tsix* gene may play a role in the choice of the X chromosome that remains active. The *Xist* gene, which is located in the Xic region, remains transcriptionally active on the inactivated X chromosome. It is thought to play an important role in X inactivation by coating the inactive X chromosome. After X inactivation is established, it is maintained in the same X chromosome in somatic cells during subsequent cell divisions. In germ cells, however, the X chromosomes are not inactivated, so an egg can transmit either copy of an active (noncondensed) X chromosome.

C16. A. One C. Two

B. Zero D. Zero

C18. The offspring inherited X^B from its mother and X^O and Y from its father. It is an XXY animal, which is male (but somewhat feminized).

C20. Erasure and reestablishment of the imprint occurs during gametogenesis. It is necessary to erase the imprint because each sex will transmit either inactive or active alleles of a gene. In somatic cells, the two alleles for a gene are imprinted according to the sex of the parent from which the allele was inherited.

C22. A person born with paternal uniparental disomy 15 would have Angelman syndrome, because this individual would not have an active copy of the *AS* gene; the paternally inherited copies of the *AS* gene are silenced. This individual would have normal offspring, because she does not have a deletion in either copy of chromosome 15.

C24. In some species, such as marsupials, X inactivation depends on the sex. This is similar to imprinting. Also, once X inactivation occurs during embryonic development, it is remembered throughout the rest of the life of the organism, which is also similar to imprinting. X inactivation in mammals is different from genomic imprinting, in that it is not sex dependent. The X chromosome that is inactivated could be inherited from the mother or the father. There was no marking process on the X chromosome that occurred during gametogenesis. In contrast, genomic imprinting always involves a marking process during gametogenesis.

C26. The term reciprocal cross refers to two parallel crosses that involve the same genotypes of the two parents, but their sexes are opposite in the two crosses. For example, the reciprocal cross of female *BB* × male *bb* is the cross female *bb* × male *BB*. Autosomal inheritance gives the same result because the autosomes are transmitted from parent to offspring in the same way for both sexes. However, for extranuclear inheritance, the mitochondria and plastids are not transmitted via the gametes in the same way for both sexes. For maternal inheritance, the reciprocal crosses would show that the gene is always inherited from the mother.

C28. The phenotype of a petite mutant is that it forms small colonies on growth media, as opposed to wild-type strains that formed larger colonies. These mutants are unable to grow when the cells only have an energy source that requires mitochondrial function. Because nuclear and mitochondrial genes are necessary for mitochondrial function, it is possible for a petite mutation to involve a gene in the nucleus or in the mitochondrial genome. The difference between neutral and suppressive petites is that neutral petites lack most of their mitochondrial DNA, whereas suppressive petites usually lack small segments of the mitochondrial genetic material.

C30. The mitochondrial and chloroplast genomes are composed of a circular chromosome found in one or more copies in a region of the organelle known as the nucleoid. The number of genes per chromosome varies from species to species. Chloroplasts genomes tend to be larger than mitochondria genomes. See Table 5.3 for examples of the variation among mitochondrial and chloroplast genomes.

C32. A. Yes

B. Yes

C. No, it is determined by a gene in the chloroplast genome.

D. No, it is determined by a gene in the mitochondria.

C34. Biparental extranuclear inheritance would resemble Mendelian inheritance in that offspring could inherit alleles of a given gene from both parents. It differs, however, when you think about it from the perspective of heterozygotes. For a Mendelian trait, the law of segregation tells us that a heterozygote passes one allele for a given gene to an offspring, but not both. In contrast, if a parent has a mixed population of mitochondria (e.g., some carrying a mutant gene and some carrying a normal gene), that parent could pass both types of genes (mutant and normal) to a single offspring, because more than one mitochondrion could be contained within a sperm or egg cell.

Experimental Questions

E2. The first type of observation was based on cytological studies. The presence of the Barr body in female cells was consistent with the idea that one of the X chromosomes was highly condensed. The second type of observation was based on genetic mutations. A variegated phenotype that is found only in females is consistent with the idea that certain patches express one allele and other patches express the other allele. This variegated phenotype would occur only if the inactivation of one X chromosome happened at an early stage of embryonic development and was inherited permanently thereafter.

E4. The pattern of inheritance is consistent with imprinting. In every cross, the allele that is inherited from the father is expressed in the offspring, but the allele inherited from the mother is not.

E6. We assume that the snails in the large colony on the second island are true-breeding, *DD*. Let the male snail from the deserted island mate with a female snail from the large colony. Then let the F_1 snails mate with each other to produce an F_2 generation. Then let the F_2 generation mate with each other to produce an F_3 generation. Here are the expected results:

Female *DD* × Male *DD*

All F_1 snails coil to the right.

All F_2 snails coil to the right.

All F_3 snails coil to the right.

Female *DD* × Male *Dd*

All F_1 snails coil to the right.

All F_2 snails coil to the right because all of the F_1 females are *DD* or *Dd*.

$^{15}/_{16}$ of F_3 snails coil to the right; $^{1}/_{16}$ of F_3 snails coil to the left (because $^{1}/_{16}$ of the F_2 females are *dd*).

Female *DD* × Male *dd*

All F_1 snails coil to the right.

All F_2 snails coil to the right because all of the F_1 females are *Dd*.

$^{3}/_{4}$ of F_3 snails coil to the right; $^{1}/_{4}$ of F_3 snails coil to the left (because $^{1}/_{4}$ of the F_2 females are *dd*).

E8. Let's first consider the genotypes of male A and male B. Male A must have two normal copies of the *Igf2* gene. We know this because male A's mother was *Igf2 Igf2*; the father of male A must have been a heterozygote *Igf2 Igf2*$^-$ because half of the litter that contained male A also contained dwarf offspring. But because male A was not dwarf, it must have inherited the normal allele from its father. Therefore, male A must be *Igf2 Igf2*. We cannot be completely sure of the genotype of male B. It must have inherited the normal *Igf2* allele from its father because male B is phenotypically normal. We do not know the genotype of male B's mother, but she could be either *Igf2*$^-$ *Igf2*$^-$ or *Igf2 Igf2*$^-$. In either case, the mother of male B could pass the *Igf2*$^-$ allele to an offspring, but we do not know for sure if she did. So, male B could be either *Igf2 Igf2*$^-$ or *Igf2 Igf2*.

For the *Igf2* gene, we know that the maternal allele is inactivated. Therefore, the genotypes and phenotypes of females A and B are irrelevant. The phenotype of the offspring is determined only by the allele that is inherited from the father. Because we know that male A has to be *Igf2 Igf2*, we know that it can produce only normal offspring. Because both females A and B both produced dwarf offspring, male A cannot be the father. In contrast, male B could be either *Igf2 Igf2* or *Igf2 Igf2*$^-$. Because both females gave birth to dwarf babies (and because male A and male B were the only two male mice in the cage), we conclude that male B must be *Igf2 Igf2*$^-$ and is the father of both litters.

E10. In fruit flies, the expression of a male's X-linked genes is turned up twofold. In mice, one of the two X chromosomes is inactivated; that is why females and males produce the same total amount of mRNA for most X-linked genes. In *C. elegans*, the expression of hermaphrodite X-linked genes is turned down twofold. Overall, the total amount of expression of X-linked genes is the same in males and females (or hermaphrodites) of these three species. In fruit flies and *C. elegans*, heterozygous females and hermaphrodites express 50% of each allele compared to a homozygous male, so that heterozygous females and hermaphrodites produce the same total amount of mRNA from X-linked genes compared to males. Note: In heterozygous females of fruit flies, mice, and *C. elegans*, there is 50% of each gene product (compared to hemizygous males and homozygous females).

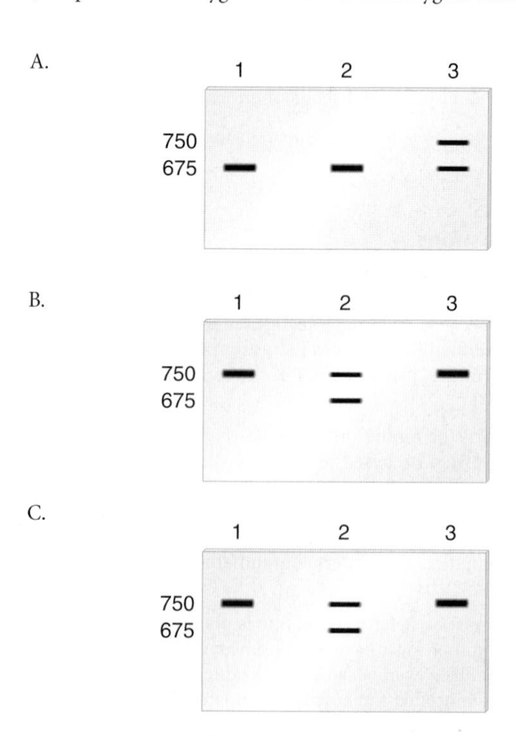

E12. In the absence of UV light, we would expect all sm^r offspring. With UV light, we would expect a greater percentage of sm^s offspring.

Questions for Student Discussion/Collaboration

2. An infective particle is something in the cytoplasm that contains its own genetic material and isn't an organelle. Some symbiotic infective particles, such as those found in killer paramecia, are similar to mitochondria and chloroplasts in that they contain their own genomes and are known to be bacterial in origin. The observation that these endosymbiotic relationships can initiate in modern species tells us that endosymbiosis can spontaneously happen. Therefore, it is reasonable that it happened a long time ago and led to the evolution of mitochondria and chloroplasts.

CHAPTER 6

Conceptual Questions

C2. An independent assortment hypothesis is used because it enables us to calculate the expected values based on Mendel's ratios. Using the observed and expected values, we can calculate whether or not the deviations between the observed and expected values are too large to occur as a matter of chance. If the deviations are very large, we reject the hypothesis of independent assortment.

C4. If the chromosomes (on the right side) labeled 2 and 4 move into one daughter cell, that will lead to a patch that is albino and has long fur. The other cell will receive chromosomes 1 and 3, which will produce a patch that has dark, short fur.

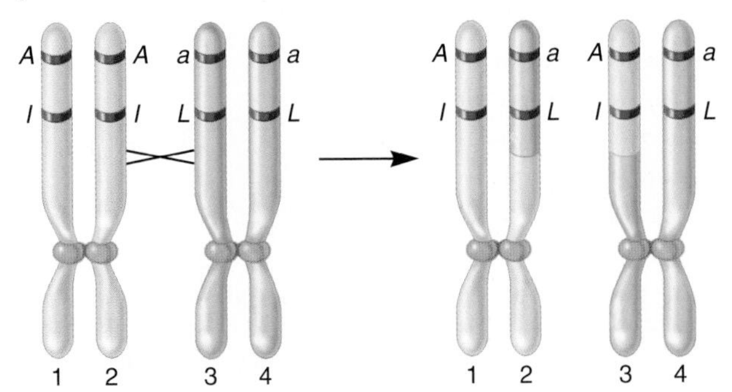

C6. A single crossover produces *A B C*, *A b c*, *a B C*, and *a b c*.

A. Between 2 and 3, between genes *B* and *C*

B. Between 1 and 4, between genes *A* and *B*

C. Between 1 and 4, between genes *B* and *C*

D. Between 2 and 3, between genes *A* and *B*

C8. The likelihood of scoring a basket would be greater if the basket was larger. Similarly, the chances of a crossover initiating in a region between two genes is proportional to the size of the region between the two genes. There are a finite number (usually a few) of crossovers that occur between homologous chromosomes during meiosis, and the likelihood that a crossover will occur in a region between two genes depends on how big that region is.

C10. The pedigree suggests a linkage between the dominant allele causing nail-patella syndrome and the I^B allele of the ABO blood type gene. In every case, the individual who inherits the I^B allele also inherits this disorder.

C12. *Ass-1* 43 *Sdh-1* 5 *Hdc* 9 *Hao-1* 6 *Odc-2* 8 *Ada-1*

C14. The inability to detect double crossovers causes the map distance to be underestimated. In other words, more crossovers occur in the region than we realize. When we have a double crossover, we do not get a recombinant offspring (in a dihybrid cross). Therefore, the second crossover cancels out the effects of the first crossover.

C16. The key feature is that all the products of a single meiosis are contained within a single sac. The spores in this sac can be dissected, and then their genetic traits can be analyzed individually.

C18. In an unordered ascus, the products of meiosis are free to move around. In an ordered octad (or tetrad), they are lined up according to their relationship to each other during meiosis and mitosis. An ordered octad can be used to map the distance between a single gene and its centromere.

C20. The percentage would be higher with respect to gene *A*. First-division segregation patterns occur when there is not a crossover between the centromere and the gene of interest. Because gene *A* is closer to the centromere compared to gene *B*, it would be less likely to have a crossover between gene *A* and the centromere. This would make it more likely to observe first-division segregation.

Experimental Questions (Includes Most Mapping Questions)

E2. They could have used a strain with two abnormal chromosomes. In this case, the recombinant chromosomes would either look normal or have abnormalities at both ends.

E4. A gene on the Y chromosome in mammals would be transmitted only from father to son. It would be difficult to genetically map Y-linked genes because a normal male has only one copy of the Y chromosome, so you do not get any crossing over between two Y chromosomes. Occasionally, abnormal males (XYY) are born with two Y chromosomes. If such males were heterozygous for alleles of Y-linked genes, one could examine the normal male offspring of XYY fathers and determine if crossing over has occurred.

E6. The answer is explained in solved problem S5. We cannot get more than 50% recombinant offspring because the pattern of multiple crossovers can yield an average maximum value of only 50%. When a testcross does yield a value of 50% recombinant offspring, it can mean two different things. Either the two genes are on different chromosomes or the two genes are on the same chromosome but at least 50 mu apart.

E8. If two genes are at least 50 mu apart, you would need to map genes between them to show that the two genes were actually in the same linkage group. For example, if gene *A* was 55 mu from gene *B*, there might be a third gene (e.g., gene *C*) that was 20 mu from *A* and 35 mu from *B*. These results would indicate that *A* and *B* are 55 mu apart, assuming dihybrid testcrosses between genes *A* and *B* yielded 50% recombinant offspring.

E10. Sturtevant used the data involving the following pairs: *y* and *w*, *w* and *v*, *v* and *r*, and *v* and *m*.

E12. A. Because they are 12 mu apart, we expect 12% (or 120) recombinant offspring. This would be approximately 60 *Aabb* and 60 *aaBb* plus 440 *AaBb* and 440 *aabb*.

B. We would expect 60 *AaBb*, 60 *aabb*, 440 *Aabb*, and 440 *aaBb*.

E14. Due to the large distance between the two genes, they will assort independently even though they are actually on the same chromosome. According to independent assortment, we expect 50% parental and 50% recombinant offspring. Therefore, this cross will produce 150 offspring in each of the four phenotypic categories.

E16. A. If we hypothesize two genes independently assorting, then the predicted ratio is 1:1:1:1. There are a total of 390 offspring. The expected number of offspring in each category is about 98. Plugging the figures into our chi square formula,

$$\chi^2 = \frac{(117 - 98)^2}{98} + \frac{(115 - 98)^2}{98} + \frac{(78 - 98)^2}{98} + \frac{(80 - 98)^2}{98}$$

$$\chi^2 = 3.68 + 2.95 + 4.08 + 3.31$$

$$\chi^2 = 14.02$$

Looking up this value in the chi square table under 1 degree of freedom, we reject our hypothesis, because the chi square value is above 7.815.

B. Map distance:

$$\text{Map Distance} = \frac{78 + 80}{117 + 115 + 78 + 80}$$

$$= 40.5 \text{ mu}$$

APPENDIX B

Because the value is relatively close to 50 mu, it is probably a significant underestimate of the true distance between these two genes.

E18. The percentage of recombinants for the green, yellow and wide, narrow is 7%, or 0.07; there will be 3.5% of the green, narrow and 3.5% of the yellow, wide. The remaining 93% parentals will be 46.5% green, wide and 46.5% yellow, narrow. The third gene assorts independently. There will be 50% long and 50% short with respect to each of the other two genes. To calculate the number of offspring out of a total of 800, we multiply 800 by the percentages in each category.

(0.465 green, wide)(0.5 long)(800) = 186 green, wide, long

(0.465 yellow, narrow)(0.5 long)(800) = 186 yellow, narrow, long

(0.465 green, wide)(0.5 short)(800) = 186 green, wide, short

(0.465 yellow, narrow)(0.5 short)(800) = 186 yellow, narrow, short

(0.035 green, narrow)(0.5 long)(800) = 14 green, narrow, long

(0.035 yellow, wide)(0.5 long)(800) = 14 yellow, wide, long

(0.035 green, narrow)(0.5 short)(800) = 14 green, narrow, short

(0.035 yellow, wide)(0.5 short)(800) = 14 yellow, wide, short

E20. Let's use the following symbols: *G* for green pods, *g* for yellow pods, *S* for green seedlings, *s* for bluish green seedlings, *C* for normal plants, *c* for creepers. The parental cross is *GG SS CC* crossed to *gg ss cc*.

The F_1 plants would all be *Gg Ss Cc*. If the genes are linked, the alleles *G*, *S*, and *C* would be linked on one chromosome, and the alleles *g*, *s*, and *c* would be linked on the homologous chromosome.

The testcross is F_1 plants, which are *Gg Ss Cc* crossed to *gg ss cc*.

To measure the distances between the genes, we can separate the data into gene pairs.

Pod color, seedling color

2210 green pods, green seedlings—nonrecombinant

296 green pods, bluish green seedlings—recombinant

2198 yellow pods, bluish green seedlings—nonrecombinant

293 yellow pods, green seedlings—recombinant

$$\text{Map Distance} = \frac{296 + 293}{2210 + 296 + 2198 + 293} \times 100 = 6.7 \text{ mu}$$

Pod color, plant stature

2340 green pods, normal—nonrecombinant

166 green pods, creeper—recombinant

2323 yellow pods, creeper—nonrecombinant

168 yellow pods, normal—recombinant

$$\text{Map Distance} = \frac{166 + 168}{2340 + 166 + 2323 + 168} \times 100 = 6.7 \text{ mu}$$

Seedling color, plant stature

2070 green seedlings, normal—nonrecombinant

433 green seedlings, creeper—recombinant

2056 bluish green seedlings, creeper—nonrecombinant

438 bluish green seedlings, normal—recombinant

$$\text{Map Distance} = \frac{433 + 438}{2070 + 433 + 2056 + 438} \times 100 = 17.4 \text{ mu}$$

The order of the genes is seedling color, pod color, and plant stature (or you could say the opposite order). Pod color is in the middle. If we use the two shortest distances to construct our map:

S ——— 11.8 ——— G ——— 6.7 ——— C

E22. To answer this question, we can consider genes in pairs. Let's consider the two gene pairs that are closest together. The distance between the wing length and eye color genes is 12.5 mu. From this cross, we expect 87.5% to have long wings and red eyes or short wings and purple eyes, and 12.5% to have long wings and purple eyes or short wings and red eyes. Therefore, we expect 43.75% to have long wings and red eyes, 43.75% to have short wings and purple eyes, 6.25% to have long wings and purple eyes, and 6.25% to have short wings and red eyes. If we have 1000 flies, we expect 438 to have long wings and red eyes, 438 to have short wings and purple eyes, 62 to have long wings and purple eyes, and 62 to have short wings and red eyes (rounding to the nearest whole number).

The distance between the eye color and body color genes is 6 mu. From this cross, we expect 94% to have a parental combination (red eyes and gray body or purple eyes and black body) and 6% to have a nonparental combination (red eyes and black body or purple eyes and gray body). Therefore, of our 438 flies with long wings and red eyes, we expect 94% of them (or about 412) to have long wings, red eyes, and gray body and 6% of them (or about 26) to have long wings, red eyes, and black bodies. Of our 438 flies with short wings and purple eyes, we expect about 412 to have short wings, purple eyes, and black bodies and 26 to have short wings, purple eyes, and gray bodies.

Of the 62 flies with long wings and purple eyes, we expect 94% of them (or about 58) to have long wings, purple eyes, and black bodies and 6% of them (or about 4) to have long wings, purple eyes, and gray bodies. Of the 62 flies with short wings and red eyes, we expect 94% (or about 58) to have short wings, red eyes, and gray bodies and 6% (or about 4) to have short wings, red eyes, and black bodies.

In summary,

Long wings, red eyes, gray body	412
Long wings, purple eyes, gray body	4
Long wings, red eyes, black body	26
Long wings, purple eyes, black body	58
Short wings, red eyes, gray body	58
Short wings, purple eyes, gray body	26
Short wings, red eyes, black body	4
Short wings, purple eyes, black body	412

The flies with long wings, purple eyes, and gray bodies, or short wings, red eyes, and black bodies, are produced by a double-crossover event.

E24. Yes. Begin with females that have one X chromosome that is X^{Nl} and the other X chromosome that is X^{nL}. These females have to be mated to $X^{NL}Y$ males because a living male cannot carry the *n* or *l* allele. In the absence of crossing over, a mating between $X^{Nl}X^{nL}$ females to $X^{NL}Y$ males should not produce any surviving male offspring. However, during oogenesis in these heterozygous female mice, there could be a crossover in the region between the two genes, which would produce an X^{NL} chromosome and an X^{nl} chromosome. Male offspring inheriting these recombinant chromosomes will be either $X^{NL}Y$ or $X^{nl}Y$ (whereas nonrecombinant males will be $X^{nL}Y$ or $X^{Nl}Y$). Only the male mice that inherit $X^{NL}Y$ will live. The living males represent only half of the recombinant offspring. (The other half are $X^{nl}Y$, which are born dead.)

To compute map distance:

$$\text{Map Distance} = \frac{2(\text{Number of male living offspring})}{\text{Number of males born dead} + \text{Number of males born alive}} \times 100$$

E26. $$\text{Map Distance} = \frac{(^1/_2)(\text{SDS})}{\text{Total}} \times 100$$

$$= \frac{(^1/_2)(22 + 21 + 21 + 23)}{22 + 21 + 21 + 21 + 451 + 23 + 455} \times 100$$

$$= 4.4 \text{ mu}$$

Questions for Student Discussion/Collaboration

2. The X and Y chromosomes are not completely distinct linkage groups. One might describe them as overlapping linkage groups having some genes in common, but most genes are not common to both.

CHAPTER 7

Conceptual Questions

C2. It is not a form of sexual reproduction, in which two distinct parents produce gametes that unite to form a new individual. However, conjugation is similar to sexual reproduction in the sense that the genetic material from two cells are somewhat mixed. In conjugation, there is not the mixing of two genomes, one from each gamete. Instead, there is a transfer of genetic material from one cell to another. This transfer can alter the combination of genetic traits in the recipient cell.

C4. An F^+ strain contains a separate, circular piece of DNA that has its own origin of transfer. An *Hfr* strain has its origin of transfer integrated into the bacterial chromosome. An F^+ strain can transfer only the DNA contained on the F factor. If given enough time, an *Hfr* strain can actually transfer the entire bacterial chromosome to the recipient cell.

C6. Sex pili promote the binding of donor and recipient cells.

C8. Though exceptions are common, interspecies genetic transfer via conjugation is not as likely because the cell surfaces do not interact correctly. Interspecies genetic transfer via transduction is also not very likely because each species of bacteria is sensitive to particular bacteriophages. The correct answer is transformation. A consequence of interspecies genetic transfer is that new genes can be introduced into a bacterial species from another species. For example, interspecies genetic transfer could provide the recipient bacterium with a new trait, such as resistance to an antibiotic. Evolutionary biologists call this horizontal gene transfer, while the passage of genes from parents to offspring is termed vertical gene transfer.

C10. Cotransduction is the transduction of two or more genes. The distance between the genes determines the frequency of cotransduction. When two genes are close together, the cotransduction frequency would be higher compared to two genes that are relatively farther apart.

C12. If a site that frequently incurred a breakpoint was between two genes, the cotransduction frequency of these two genes would be much less than expected. This is because the site where the breakage occurred would separate the two genes from each other.

C14. The transfer of conjugative plasmids such as F factor DNA

C16. A. If it occurred in a single step, transformation is the most likely mechanism because conjugation does not usually occur between different species, particularly distantly related species, and different species are not usually infected by the same bacteriophages.

B. It could occur in a single step, but it may be more likely to have involved multiple steps.

C. The use of antibiotics selects for the survival of bacteria that have resistance genes. If a population of bacteria is exposed to an antibiotic, those carrying resistance genes will survive, and their relative numbers will increase in subsequent generations.

C18. The term allele means alternative forms of the same gene. Therefore, mutations in the same gene among different phages are alleles of each other; the mutations may be at different positions within the same gene. When we map the distance between mutations in the same gene, we are mapping the distance between the mutations that create different alleles of the same gene. An intragenic map describes the locations of mutations within the same gene.

Experimental Questions

E2. Mix the two strains together and then put some of them on plates containing streptomycin and some of them on plates without streptomycin. If mated colonies are present on both types of plates, then the thr^+, leu^+, and thi^+ genes were transferred to the met^+ bio^+ thr^- leu^- thi^- strain. If colonies are found only on the plates that lack streptomycin, then the met^+ and bio^+ genes are being transferred to the met^- bio^- thr^+ leu^+ thi^+ strain. This answer assumes a one-way transfer of genes from a donor to a recipient strain.

E4. An interrupted mating experiment is a procedure in which two bacterial strains are allowed to mate, and then the mating is interrupted at various time points. The interruption occurs by agitation of the solution in which the bacteria are found. This type of study is used to map the locations of genes. It is necessary to interrupt mating so that you can vary the time and obtain information about the order of transfer; which gene transferred first, second, and so on.

E6. Mate unknown strains *A* and *B* to the F^- strain in your lab that is resistant to streptomycin and cannot use lactose. This is done in two separate tubes (i.e., strain *A* plus your F^- strain in one tube, and strain *B* plus your F^- strain in the other tube). Plate the mated cells on growth media containing lactose plus streptomycin. If you get growth of colonies, the unknown strain had to be strain *A*, the F^+ strain that had lactose utilization genes on its F factor.

E8. A. If we extrapolate these lines back to the *x*-axis, the *hisE* intersects at about 3 minutes, and the *pheA* intersects at about 24 minutes. These are the values for the times of entry. Therefore, the distance between these two genes is 21 minutes (i.e., 24 minus 3).

B. ↑ 4 ↑ 17 ↑

hisE ***pabB*** ***pheA***

E10. One possibility is that you could treat the P1 lysate with DNase I, an enzyme that digests DNA. (Note: If DNA were digested with DNase I, the function of any genes within the DNA would be destroyed.) If the DNA were within a P1 phage, it would be protected from DNase I digestion. This would allow you to distinguish between transformation (which would be inhibited by DNase I) versus transduction (which would not be inhibited by DNase I). Another possibility is that you could try to fractionate the P1 lysate. Naked DNA would be smaller than a P1 phage carrying DNA. You could try to filter the lysate to remove naked DNA, or you could subject the lysate to centrifugation and remove the lighter fractions that contain naked DNA.

E12. Cotransduction frequency $= (1 - d/L)^3$

For the normal strain,

Cotransduction frequency $= (1 - {}^{0.7}/_{2})^3 = 0.275$, or 27.5%

For the new strain,

Cotransduction frequency $= (1 - {}^{0.7}/_{5})^3 = 0.64$, or 64%

The experimental advantage is that you could map genes that are farther than 2 minutes apart. You could map genes that are up to 2 minutes apart.

E14. Cotransduction frequency $= (1 - {}^{d}/_{L})^3$

$$0.53 = (1 - {}^{d}/_{2}\text{ minutes})^3$$

$$(1 - {}^{d}/_{2}\text{ minutes}) = \sqrt[3]{0.53}$$

$$(1 - {}^{d}/_{2}\text{ minutes}) = 0.81$$

$$d = 0.38\text{ minutes}$$

E16. A. We first need to calculate the cotransformation frequency, which equals 2/70, or 0.029.

$$\text{Cotransformation frequency} = (1 - {}^{d}/_{L})^3$$

$$0.029 = (1 - {}^{d}/_{2}\text{ minutes})^3$$

$$d = 1.4\text{ minutes}$$

B. Cotransformation frequency $= (1 - {}^{d}/_{L})^3$

$$= (1 - {}^{1.4}/_{4})^3$$

$$= 0.27$$

As you may have expected, the cotransformation frequency is much higher when the transformation involves larger pieces of DNA.

E18. Benzer could use this observation as a way to evaluate if intragenic recombination had occurred. If two *rII* mutations recombined to make a wild-type gene, the phage would produce plaques in this *E. coli* K12(λ) strain.

C22. One possibility is a sequential mechanism. First, the double helix could unwind and replicate itself as described in Chapter 11. This would produce two double helices. Next, the third strand (bound in the major groove) could replicate itself via a semiconservative mechanism. This new strand could be copied to make a copy that is identical to the strand that lies in the major groove. At this point, you would have two double helices and two strands that could lie in the major groove. These could assemble to make two triple helices.

C24. Lysines and arginines, and also polar amino acids

C26. This DNA molecule contains 280 bp. There are 10 base pairs per turn, so there are 28 complete turns.

C28. A hydroxyl group is at the 3' end, and a phosphate group is at the 5' end.

C30. Not necessarily. The AT/GC rule is required only of double-stranded DNA molecules.

C32. The first thing we need to do is to determine how many base pairs are in this DNA molecule. The linear length of 1 base pair is 0.34 nm, which equals 0.34×10^{-9} m. One centimeter equals 10^{-2} meters.

$$\frac{10^{-2}}{0.34 \times 10^{-9}} = 2.9 \times 10^{7} \text{ bp}$$

There are approximately 2.9×10^7 bp in this DNA molecule, which equals 5.8×10^7 nucleotides. If 15% are adenine, then 15% must also be thymine. This leaves 70% for cytosine and guanine. Because cytosine and guanine bind to each other, there must be 35% cytosine and 35% guanine. If we multiply 5.8×10^7 times 0.35, we get

$(5.8 \times 10^7)(0.35) = 2.0 \times 10^7$ cytosines, or about 20 million cytosines

C34. The methyl group is not attached to one of the atoms that hydrogen bonds with guanine, so methylation would not directly affect hydrogen bonding. It could indirectly affect hydrogen bonding if it perturbed the structure of DNA. Methylation may affect gene expression because it could alter the ability of proteins to recognize DNA sequences. For example, a protein might bind into the major groove by interacting with a sequence of bases that includes one or more cytosines. If the cytosines are methylated, this may prevent a protein from binding into the major groove properly. Alternatively, methylation could enhance protein binding. In Chapter 5, we considered DNA-binding proteins that were influenced by the methylation of DMRs (differentially methylated regions) that occur during genomic imprinting.

Experimental Questions

E2. A. There are different possible reasons why most of the cells were not transformed.

1. Most of the cells did not take up any of the type S DNA.
2. The type S DNA was usually degraded after it entered the type R bacteria.
3. The type S DNA was usually not expressed in the type R bacteria.

B. The antibody/centrifugation steps were used to remove the bacteria that had not been transformed. It enabled the researchers to determine the phenotype of the bacteria that had been transformed. If this step was omitted, there would have been so many colonies on the plate it would have been difficult to identify any transformed bacterial colonies, because they would have represented a very small proportion of the total number of bacterial colonies.

C. They were trying to demonstrate that it was really the DNA in their DNA extract that was the genetic material. It was possible that the extract was not entirely pure and could contain contaminating RNA or protein. However, treatment with RNase and protease did not prevent transformation, indicating that RNA and protein were not the genetic material. In contrast, treatment with DNase blocked transformation, confirming that DNA is the genetic material.

E4. A. There are several possible explanations why about 35% of the DNA is in the supernatant. One possibility is that not all of the DNA was injected into the bacterial cells. Alternatively, some of the cells may have been broken during the shearing procedure, thereby releasing the DNA.

B. If the radioactivity in the pellet had been counted instead of the supernatant, the following figure would be produced:

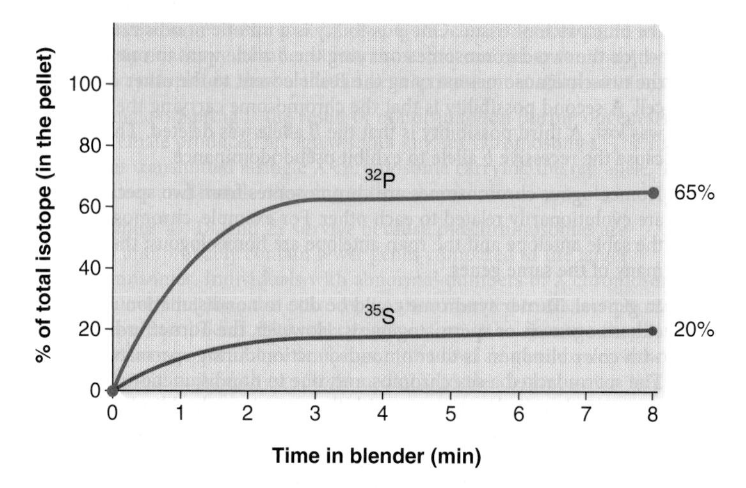

C. ^{32}P and ^{35}S were chosen as radioisotopes to label the phages because phosphorous is found in nucleic acids, while sulfur is found only in proteins.

D. There are multiple reasons why less than 100% of the phage protein is removed from the bacterial cells during the shearing process. For example, perhaps the shearing just is not strong enough to remove all of the phages, or perhaps the tail fibers remain embedded in the bacterium and only the head region is sheared off.

E6. This is really a matter of opinion. The Avery, MacLeod, and McCarty experiment seems to indicate directly that DNA is the genetic material, because DNase prevented transformation and RNase and protease did not. However, one could argue that the DNA is required for the rough bacteria to take up some other contaminant in the DNA preparation. It would seem that the other contaminant would not be RNA or protein. The Hershey and Chase experiments indicate that DNA is being injected into bacteria, although quantitatively the results are not entirely convincing. Some ^{35}S-labeled protein was not sheared off, so the results do not definitely rule out the possibility that protein could be the genetic material. But the results do indicate that DNA is the more likely candidate.

E8. A. The purpose of chromatography was to separate the different types of bases.

B. It was necessary to separate the bases and determine the total amount of each type of base. In a DNA strand, all the bases are found within a single molecule, so it is difficult to measure the total amount of each type of base. When the bases are removed from the strand, each type can be purified, and then the total amount of each type of base can be measured by spectroscopy.

C. Chargaff's results would probably not be very convincing if done on a single species. The strength of his data was that all species appeared to conform to the AT/GC rule, suggesting that this is a consistent feature of DNA structure. In a single species, the observation is that A = T and G = C could occur as a matter of chance.

Questions for Student Discussion/Collaboration

2. There are many possibilities. You could use a DNA-specific chemical and show that it causes heritable mutations. Perhaps you could inject an oocyte with a piece of DNA and produce a mouse with a new trait.

CHAPTER 10

Conceptual Questions

C2. Viruses also need sequences that enable them to be replicated. These sequences are equivalent to the origins of replication found in bacterial and eukaryotic chromosomes.

C4. A bacterium with two nucleoids is similar to a diploid eukaryotic cell because it would have two copies of each gene. The bacterium is different, however, with regard to alleles. A eukaryotic cell can have two different alleles for the same gene. For example, a cell from a pea plant could be heterozygous, *Tt*, for the gene that affects height. By comparison, a bacterium with two nucleoids has two identical chromosomes. Therefore, a bacterium with two nucleoids is homozygous for its chromosomal genes. Note: As discussed in Chapter 7, a bacterium can contain another piece of DNA, called an F' factor, that can carry a few genes. The alleles on an F' factor can be different from the alleles on the bacterial chromosome.

C6. A. One loop is 40,000 bp. One base pair is 0.34 nm, which equals 0.34×10^{-3} µm. If we multiply the two together:

$$(40{,}000)(0.34 \times 10^{-3}) = 13.6\ \mu m$$

B. Circumference = πD

$$13.6\ \mu m = \pi D$$

$$D = 4.3\ \mu m$$

C. No, it is too big to fit inside of *E. coli*. Supercoiling is needed to make the loops more compact.

C8. These drugs would diminish the amount of negative supercoiling in DNA. Negative supercoiling is needed to compact the chromosomal DNA, and it also aids in strand separation. Bacteria might not be able to survive and/or transmit their chromosomes to daughter cells if their DNA was not compacted properly. Also, because negative supercoiling aids in strand separation, these drugs would make it more difficult for the DNA strands to separate. Therefore, the bacteria would have a difficult time transcribing their genes and replicating their DNA, because both processes require strand separation. As discussed in Chapter 11, DNA replication is needed to make new copies of the genetic material to transmit from mother to daughter cells. If DNA replication was inhibited, the bacteria could not grow and divide into new daughter cells. As discussed in Chapters 12–14, gene transcription is necessary for bacterial cells to make proteins. If gene transcription was inhibited, the bacteria could not make many proteins that are necessary for survival.

C10.

Gyrase
Negative supercoil

C12. The centromere is the attachment site for the kinetochore, which attaches to the spindle. If a chromosome is not attached to the spindle, it is free to "float around" within the cell, and it may not be near a pole when the nuclear membrane re-forms during telophase. If a chromosome is left outside of the nucleus, it is degraded during interphase. That is why the chromosome without a centromere may not be found in daughter cells.

C14. Highly repetitive DNA, as its name suggests, is a DNA sequence that is repeated many times, from tens of thousands to millions of times throughout the genome. It can be interspersed in the genome or found clustered in a tandem array, in which a short nucleotide sequence is repeated many times in a row. In DNA renaturation studies, highly repetitive DNA renatures at a much faster rate because there are many copies of the complementary sequences.

C16. During interphase (i.e., G_1, S, and G_2), the euchromatin is found primarily as a 30-nm fiber in a radial loop configuration. Most interphase chromosomes also have some heterochromatic regions where the radial loops are more highly compacted. During M phase, each chromosome becomes entirely heterochromatic, which is needed for the proper sorting of the chromosomes during nuclear division.

C18.

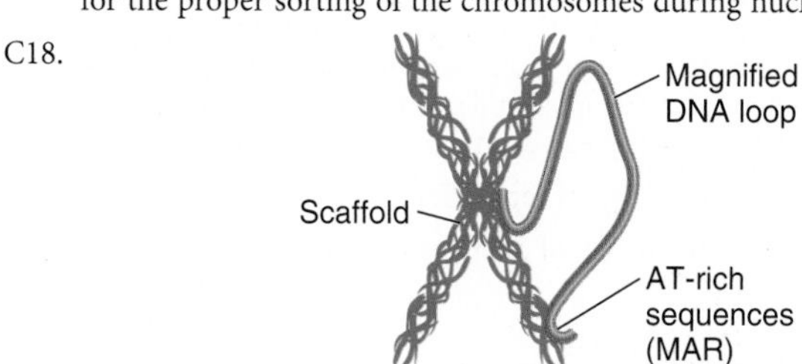

C20. During interphase, the chromosomes are found within the cell nucleus. They are less tightly packed and are transcriptionally active. Segments of chromosomes are anchored to the nuclear matrix. During M phase, the chromosomes become highly condensed, and the nuclear membrane is fragmented into vesicles. The chromosomal DNA remains anchored to a scaffold, formed from the nuclear matrix. The chromosomes eventually become attached to the spindle apparatus via microtubules that attach to the kinetochore, which is attached to the centromere.

C22. There are 146 bp around the core histones. If the linker region is 54 bp, we expect 200 bp of DNA (i.e., 146 + 54) for each nucleosome and linker region. If we divide 46,000 bp by 200 bp, we get 230. Because there are two molecules of H2A for each nucleosome, there would be 460 molecules of H2A in a 46,000-bp sample of DNA.

C24. The role of the core histones is to form the nucleosomes. In a nucleosome, the DNA is wrapped 1.65 times around the core histones. Histone H1 binds to the linker region. It may play a role in compacting the DNA into a 30-nm fiber.

C26. The answer is B and E. A Barr body is composed of a type of highly compacted chromatin called heterochromatin. Euchromatin is not so compacted. A Barr body is not composed of euchromatin. A Barr body is one chromosome, the X chromosome. The term genome refers to all the types of chromosomes that make up the genetic composition of an individual.

C28. SMC stands for structural maintenance of chromosomes. SMC proteins use energy from ATP to catalyze changes in chromosome structure. Together with topoisomerases, SMC proteins have been shown to promote major changes in DNA structure. Two examples of SMC proteins are condensin and cohesin, which play different roles in metaphase chromosome structure. The function of condensin is to promote the proper compaction of metaphase chromosomes, while the function of cohesin is to promote the binding (i.e., cohesion) between sister chromatids.

Experimental Questions

E2. This type of experiment gives the relative proportions of highly repetitive, moderately repetitive, and unique DNA sequences within the genome. The highly repetitive sequences renature at a fast rate, the moderately repetitive sequences renature at an intermediate rate, and the unique sequences renature at a slow rate.

E4. Supercoiled DNA would look curled up into a relatively compact structure. You could add different purified topoisomerases and see how they affect the structure via microscopy. For example, DNA gyrase relaxes positive supercoils, and topoisomerase I relaxes negative supercoils. If you added topoisomerase I to a DNA preparation and it became less compacted, then the DNA was negatively supercoiled.

E6.

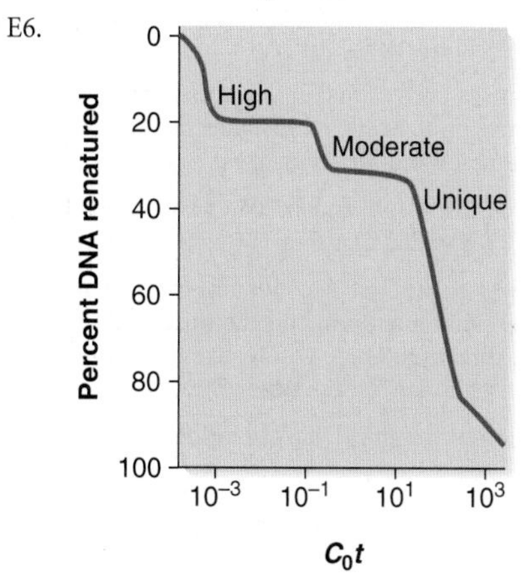

E8. You would get DNA fragments of about 446 to 496 bp (i.e., 146 bp plus 300 to 350 bp).

E10. Histones are positively charged, and DNA is negatively charged. They bind to each other by these ionic interactions. Salt is composed of positively charged ions and negatively charged ions. For example, when dissolved in water, NaCl becomes individual ions of Na^+ and Cl^-. When chromatin is exposed to a salt such as NaCl, the positively charged Na^+ could bind to the DNA, and the negatively charged Cl^- could bind to the histones. This would prevent the histones and DNA from binding to each other.

Questions for Student Discussion/Collaboration

2. This is a matter of opinion. It seems strange to have so much DNA that seems to have no obvious function. It's a waste of energy. Perhaps it has a function that we don't know about yet. On the other hand, evolution does allow bad things to accumulate within genomes, such as genes that cause diseases, etc. Perhaps this is just another example of the negative consequences of evolution.

CHAPTER 11

Conceptual Questions

C2. Bidirectional replication refers to DNA replication in both directions starting from one origin.

C4. A.

```
TTGGHTGUTGG
HHUUTHUGHUU
```

B.

```
TTGGHTGUTGG
HHUUTHUGHUU
     ↓
TTGGHTGUTGG CCAAACACCAA
AACCCACAACC HHUUTHUGHUU
     ↓
TTGGHTGUTGG TTGGGTGTTGG CCAAACACCAA CCAAACACCAA
AACCCACAACC AACCCACAACC GGTTTGTGGTT HHUUTHUGHUU
```

C6. Let's assume there are 4,600,000 bp of DNA and that DNA replication is bidirectional at a rate of 750 nucleotides per second.

If there were just a single replication fork,

4,600,000/750 = 6133 seconds, or 102.2 minutes.

Because replication is bidirectional, $^{102.2}/_{2}$ = 51.1 minutes.

Actually, this is an average value based on a variety of growth conditions. Under optimal growth conditions, replication can occur substantially faster.

With regard to errors, if we assume an error rate of one mistake per 100,000,000 nucleotides,

4,600,000 × 1000 bacteria = 4,600,000,000 nucleotides of replicated DNA.

4,600,000,000/100,000,000 = 46 mistakes.

When you think about it, this is pretty amazing. In this population, DNA polymerase would cause only 46 single mistakes in a total of 1000 bacteria, each containing 4.6 million bp of DNA.

C8. DNA polymerase would slide from right to left. The new strand would be 3'-CTAGGGCTAGGCGTATGTAAATGGTCTAGTGGTGG-5'

C10. A. When looking at Figure 11.5, the first, second, and fourth DnaA boxes are running in the same direction, and the third and fifth are running in the opposite direction. Once you realize that, you can see the sequences are very similar to each other.

B. According to the direction of the first DnaA box, the consensus sequence is

```
TGTGGATAA
ACACCTATT
```

C. This sequence is nine nucleotides long. Because there are four kinds of nucleotides (i.e., A, T, G, and C), the chance of this sequence occurring by random chance is 4^{-9}, which equals once every 262,144 nucleotides. Because the *E. coli* chromosome is more than 10 times longer than this, it is fairly likely that this consensus sequence occurs elsewhere. The reason why there are not multiple origins, however, is because the origin has five copies of the consensus sequence very close together. The chance of having five copies of this consensus sequence occurring close together (as a matter of random chance) is very small.

C12. 1. According to the AT/GC rule, a pyrimidine always hydrogen bonds with a purine. A transition still involves a pyrimidine hydrogen bonding to a purine, but a transversion causes a purine to hydrogen bond with a purine or a pyrimidine to hydrogen bond with a pyrimidine. The structure of the double helix makes it much more difficult for this latter type of hydrogen bonding to occur.

2. The induced-fit phenomenon of the active site of DNA polymerase makes it unlikely for DNA polymerase to catalyze covalent bond formation if the wrong nucleotide is bound to the template strand. A transition mutation creates a somewhat bad interaction between the bases in opposite strands, but it is not as bad as the fit caused by a transversion mutation. In a transversion, a purine is opposite another purine, or a pyrimidine is opposite a pyrimidine. This is a very bad fit.

3. The proofreading function of DNA polymerase is able to detect and remove an incorrect nucleotide that has been incorporated into the growing strand. A transversion is going to cause a larger distortion in the structure of the double helix and make it more likely to be detected by the proofreading function of DNA polymerase.

C14. Primase and DNA polymerase are able to knock the single-strand binding proteins off the template DNA.

C16. A. The right Okazaki fragment was made first. It is farthest away from the replication fork. The fork (not seen in this diagram) would be to the left of the three Okazaki fragments, and moving from right to left.

B. The RNA primer in the right Okazaki fragment would be removed first. DNA polymerase would begin by elongating the DNA strand of the middle Okazaki fragment and removing the right RNA primer with its 5' to 3' exonuclease activity. DNA polymerase I would use the 3' end of the DNA of the middle Okazaki fragment as a primer to synthesize DNA in the region where the right RNA primer is removed. If the middle fragment was not present, DNA polymerase could not fill in this DNA (because it needs a primer).

C. You need DNA ligase only at the right arrow. DNA polymerase I begins at the end of the left Okazaki fragment and synthesizes DNA to fill in the region as it removes the middle RNA primer. At the left arrow, DNA polymerase I is simply extending the length of the left Okazaki fragment. No ligase is needed here. When DNA polymerase I has extended the left Okazaki fragment through the entire region where the RNA primer has been removed, it hits the DNA of the middle Okazaki fragment. This occurs at the right arrow. At this point, the DNA of the middle Okazaki fragment has a 5' end that is a monophosphate. DNA ligase is needed to connect this monophosphate with the 3' end of the region where the middle RNA primer has been removed.

D. As mentioned in the answer to part C, the 5' end of the DNA in the middle Okazaki fragment is a monophosphate. It is a monophosphate because it was previously connected to the RNA primer by a phosphoester bond. At the location of the right arrow, there was only one phosphate connecting this deoxyribonucleotide to the last ribonucleotide in the RNA primer. For DNA polymerase to function, the energy to connect two nucleotides comes from the hydrolysis of the incoming triphosphate. In this location shown at the right arrow, however, the nucleotide is already present at the 5' end of the DNA, and it is a monophosphate. DNA ligase needs energy to connect this nucleotide with the left Okazaki fragment. It obtains energy from the hydrolysis of ATP or NAD^+.

C18. 1. It recognizes the origin of replication.

2. It initiates the formation of a replication bubble.

3. It recruits helicase to the region.

C20. The picture would depict a ring of helicase proteins traveling along a DNA strand and separating the two helices, as shown in Figure 11.6.

C22. The leading strand is primed once, at the origin, and then DNA polymerase III synthesizes DNA continuously in the direction of the replication fork. In the lagging strand, many short pieces of DNA (Okazaki fragments) are made. This requires many RNA primers. The primers are removed by DNA polymerase I, which then fills in the gaps with DNA. DNA ligase then covalently connects the Okazaki fragments. Having the enzymes within a complex such as a primosome or replisome provides coordination among the different steps in the replication process and thereby allows it to proceed faster and more efficiently.

C24. A processive enzyme is one that remains clamped to one of its substrates. In the case of DNA polymerase, it remains clamped to the template strand as it makes a new daughter strand. This is important to ensure a fast rate of DNA synthesis.

C26. The reason is the inability to synthesize DNA in the 3' to 5' direction and the need for a primer prevent replication at the 3' end of the DNA strands. Telomerase is different than DNA polymerase in that it uses a short RNA sequence, which is part of its structure, as a template for DNA synthesis. Because it uses this sequence many times in row, it produces a tandemly repeated sequence in the telomere at the 3' ends of linear chromosomes.

C28. Fifty, because two replication forks emanate from each origin of replication. DNA replication is bidirectional.

C30. A. Both reverse transcriptase and telomerase use an RNA template to make a complementary strand of DNA.

B. Because reverse transcriptase does not have a proofreading function, it is more likely for mistakes to occur. This creates many mutant strains of the virus. Some mutations might prevent the virus from proliferating. However, other mutations might prevent the immune system from battling the virus. These kinds of mutations would enhance the proliferation of the virus.

Experimental Questions

E2. A. You would probably still see a band of DNA, but you would only see a heavy band.

B. You would probably not see a band because the DNA would not be released from the bacteria. The bacteria would sediment to the bottom of the tube.

C. You would not see a band. UV light is needed to see the DNA, which absorbs light in the UV region.

E4. If you started with single-stranded DNA, you would need to add a primer (or primase), dNTPs, and DNA polymerase. If you started with double-stranded DNA, you would also need helicase. Adding single-strand binding protein and topoisomerase may also help.

E6. This is a critical step because you need to separate the radioactivity in the free nucleotides from the radioactivity in the newly made DNA strands. If you used an acid that precipitated free nucleotides and DNA strands, all of the radioactivity would be in the pellet. You would get the same amount of radioactivity in the pellet no matter how much DNA was synthesized into newly made strands. For this reason, the perchloric acid step is very critical. It separates radioactivity in the free nucleotides from radioactivity in the newly made strands.

E8. A. The left end is the 5' end. If you flip the sequence of the first primer around, you will notice it is complementary to the right end of the template DNA. The 5' end of the first primer binds to the 3' end of the template DNA.

B. The sequence would be 3'–CGGGGCCATG–5'. It could not be used because the 3' end of the primer is at the end of the template DNA. There wouldn't be any place for nucleotides to be added to the 3' end of the primer and bind to the template DNA strand.

E10. A. Heat is used to separate the DNA strands, so you do not need helicase.

B. Each primer must be a sequence that is complementary to one of the DNA strands. There are two types of primers, and each type binds to one of the two complementary strands.

C. A thermophilic DNA polymerase is used because DNA polymerases isolated from nonthermophilic species would be permanently inactivated during the heating phase of the PCR cycle. Remember that DNA polymerase is a protein, and most proteins are denatured by heating. However, proteins from thermophilic organisms have evolved to withstand heat, which is how thermophilic organisms survive at high temperatures.

D. With each cycle, the amount of DNA is doubled. Because there are initially 10 copies of the DNA, there will be 10×2^{27} copies after 27 cycles. $10 \times 2^{27} = 1.34 \times 10^9 = 1.34$ billion copies of DNA. As you can see, PCR can amplify the amount of DNA by a staggering amount!

Questions for Student Discussion/Collaboration

2. Basically, the idea is to add certain combinations of enzymes and see what happens. You could add helicase and primase to double-stranded DNA, along with radiolabeled ribonucleotides. Under these conditions, you would make short (radiolabeled) primers. If you also added DNA polymerase and deoxyribonucleotides, you would make DNA strands. Alternatively, if you added DNA polymerase plus an RNA primer, you wouldn't need to add primase. Or, if you used single-stranded DNA as a template rather than double-stranded DNA, you wouldn't need to add helicase.

CHAPTER 12

Conceptual Questions

C2. The release of sigma factor marks the transition to the elongation stage of transcription.

C4. GGCATTGTCA

C6. The most highly conserved positions are the first, second, and sixth. In general, when promoter sequences are conserved, they are more likely to be important for binding. That explains why changes are not found at these positions; if a mutation altered a conserved position, the promoter would probably not work very well. By comparison, changes are occasionally tolerated at the fourth position and frequently at the third and fifth positions. The positions that tolerate changes are less important for binding by sigma factor.

C8. This will not affect transcription. However, it will affect translation by preventing the initiation of polypeptide synthesis.

C10. Sigma factor can slide along the major groove of the DNA. In this way, it is able to recognize base sequences that are exposed in the groove. When it encounters a promoter sequence, hydrogen bonding between the bases and the sigma factor protein can promote a tight and specific interaction.

C12. DNA-G/RNA-C

DNA-C/RNA-G

DNA-A/RNA-U

DNA-T/RNA-A

The template strand is 3'–CCGTACGTAATGCCGTAGTGTGATCCCTAG–5' and the coding strand is 5'–GGCATGCATTACGGCATCACACTAGGGATC–3'. The promoter would be to the left (in the 3' direction) of the template strand.

C14. Transcriptional termination occurs when the hydrogen bonding is broken between the DNA and the part of the newly made RNA transcript that is located in the open complex.

C16. DNA helicase and ρ protein bind to a nucleic acid strand and travel in the 5' to 3' direction. When they encounter a double-stranded region, they break the hydrogen bonds between complementary strands. ρ protein is different from DNA helicase in that it moves along an RNA strand, while DNA helicase moves along a DNA strand. The purpose of DNA helicase function is to promote DNA replication; the purpose of ρ protein function is to promote transcriptional termination.

C18. A. Mutations that alter the uracil-rich region by introducing guanines and cytosines, and mutations that prevent the formation of the stem-loop structure.

B. Mutations that alter the termination sequence, and mutations that alter the ρ recognition site

C. Eventually, somewhere downstream from the gene, another transcriptional termination sequence would be found, and transcription would terminate there. This second termination sequence might be found randomly, or it might be at the end of an adjacent gene.

C20. Eukaryotic promoters are somewhat variable with regard to the pattern of sequence elements that may be found. In the case of structural genes that are transcribed by RNA polymerase II, it is common to have a TATA box, which is about 25 bp upstream from a transcriptional start site. The TATA box is important in the identification of the transcriptional start site and the assembly of RNA polymerase and various transcription factors. The transcriptional start site defines where transcription actually begins.

C22. The two models are described in Figure 12.15. The allosteric model is more like ρ-independent termination whereas the torpedo model is more like ρ-dependent termination. In the torpedo model, a protein knocks RNA polymerase off the DNA, much like the effects of ρ protein.

C24. Hydrogen bonding is usually the predominant type of interaction when proteins and DNA follow an assembly and disassembly process. In addition, ionic bonding and hydrophobic interactions could occur. Covalent interactions would not occur. High temperature and low salt concentrations tend to break hydrogen bonds. Therefore, high temperature and low salt would inhibit assembly and stimulate disassembly.

C26. In bacteria, the 5' end of the tRNA is cleaved by RNase P. The 3' end is cleaved by a different endonuclease, and then a few nucleotides are digested away by an exonuclease that removes nucleotides until it reaches a CCA sequence.

C28. A ribozyme is an enzyme whose catalytic part is composed of RNA. Examples are RNase P and self-splicing group I and II introns. It is thought that the spliceosome may contain catalytic RNAs as well.

C30. Self-splicing means that an RNA molecule can splice itself without the aid of a protein. Group I and II introns can be self-splicing, although proteins can also enhance the rate of splicing.

C32. In alternative splicing, variation occurs in the pattern of splicing, so the resulting mRNAs contain alternative combinations of exons. The biological significance is that two or more different proteins can be produced from a single gene. This is a more efficient use of the genetic material. In multicellular organisms, alternative splicing is often used in a cell-specific manner.

C34. As shown at the left side of Figure 12.20, the guanosine, which binds to the guanosine-binding site, does not have a phosphate group attached to it. This guanosine is the nucleoside that winds up at the 5' end of the intron. Therefore, the intron does not have a phosphate group at its 5' end.

C36. U5

Experimental Questions

E2. An R loop is a loop of DNA that occurs when RNA is hybridized to double-stranded DNA. While the RNA is hydrogen bonding to one of the DNA strands, the other strand does not have a partner to hydrogen bond with, so it bubbles out as a loop. RNA is complementary to the template strand, so that is the strand it binds to.

E4. The 1100-nucleotide band would be observed from a normal individual (lane 1). A deletion that removed the −50 to −100 region would greatly diminish transcription, so the homozygote would produce hardly any of the transcript (just a faint amount, as shown in lane 2), and the heterozygote would produce roughly half as much of the 1100-nucleotide transcript (lane 3) compared to a normal individual. A nonsense codon would not have an effect on transcription; it affects only translation. So the individual with this mutation would produce a normal amount of the 1100-nucleotide transcript (lane 4). A mutation that removed the splice acceptor site would prevent splicing. Therefore, this individual would produce a 1550-nucleotide transcript (actually, 1547 to be precise, 1550 minus 3). The Northern blot is shown here:

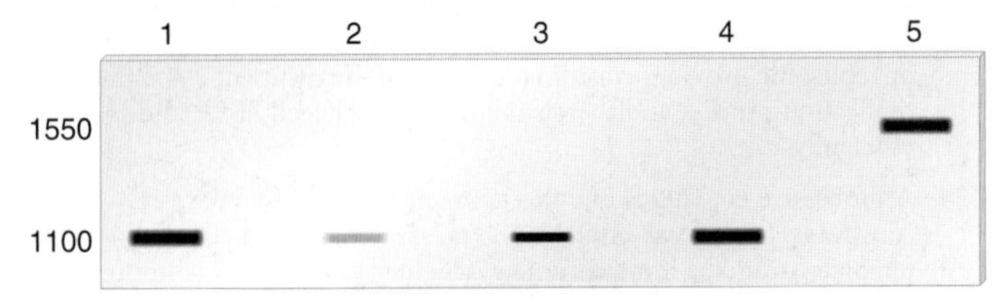

E6. A. It would not be retarded because ρ protein would not bind to the mRNA encoded by a gene that is terminated in a ρ-independent manner. The mRNA from such genes does not contain the sequence near the 3' end that acts as a recognition site for the binding of ρ protein.

B. It would be retarded because ρ protein would bind to the mRNA.

C. It would be retarded because U1 would bind to the pre-mRNA.

D. It would not be retarded because U1 would not bind to mRNA that has already had its introns removed. U1 binds only to pre-mRNA.

E8. A. mRNA molecules would bind to this column because they have a polyA tail. The string of adenine nucleotides in the polyA tail is complementary to stretch of thymine in the poly-dT column, so the two would hydrogen bond to each other. To purify mRNAs, one begins with a sample of cells; the cells need to be broken open by some technique such as homogenization or sonication. This would release the RNAs and other cellular macromolecules. The large cellular structures (organelles, membranes, etc.) could be removed from the cell extract by a centrifugation step. The large cellular structures would be found in the pellet, while soluble molecules such as RNA and proteins would stay in the supernatant. At this point, you would want the supernatant to contain a high salt concentration and neutral pH. The supernatant would then be poured over the poly-dT column. The mRNAs would bind to the poly-dT column and other molecules (i.e., other types of RNAs and proteins) would flow through the column. Because the mRNAs would bind to the poly-dT column via hydrogen bonds, to break the bonds, you could add a solution that contains a low salt concentration and/or a high pH. This would release the mRNAs, which would then be collected in a low salt/high pH solution as it dripped from the column.

B. The basic strategy is to attach a short stretch of DNA nucleotides to the column matrix that is complementary to the type of RNA that you want to purify. For example, if an rRNA contained a sequence 5'–AUUCCUCCA–3', a researcher could chemically synthesize an oligonucleotide with the sequence 3'–TAAGGAGGT–5' and attach it to the column matrix. To purify rRNA, one would use this 3'–TAAGGAGGT–5' column and follow the general strategy described in part A.

Questions for Student Discussion/Collaboration

2. RNA transcripts come in two basic types: those that function as RNA (e.g., tRNA, rRNA, etc.) versus those that are translated (i.e., mRNA). As described in this chapter, they play a myriad of functional roles. RNAs that form complexes with proteins carry out some interesting roles. In some cases, the role is to bind other types of RNA molecules. For example, rRNA in bacteria plays a role in binding mRNA. In other cases, the RNA plays a catalytic role. An example is RNaseP. The structure and function of RNA molecules may be enhanced by forming a complex with proteins.

CHAPTER 13

Conceptual Questions

C2. When we say the genetic code is degenerate, it means that more than one codon can specify the same amino acid. For example, GGG, GGC, GGA, and GGU all specify glycine. In general, the genetic code is nearly universal, because it is used in the same way by viruses, prokaryotes, fungi, plants, and animals. As discussed in Table 13.3, there are a few exceptions, which occur primarily in protists and yeast and mammalian mitochondria.

C4. A. This mutant tRNA would recognize glycine codons in the mRNA but would put in tryptophan amino acids where glycine amino acids are supposed to be in the polypeptide chain.

B. This mutation tells us that the aminoacyl-tRNA synthetase is primarily recognizing other regions of the tRNA molecule besides the anticodon region. In other words, tryptophanyl-tRNA synthetase (the aminoacyl-tRNA synthetase that attaches tryptophan) primarily recognizes other regions of the $tRNA^{Trp}$ sequence (that is, other than the anticodon region), such as the T- and D-loops. If aminoacyl-tRNA synthetases recognized only the anticodon region, we would expect glycyl-tRNA synthetase to recognize this mutant tRNA and attach glycine. That is not what happens.

C6. A. The answer is three. There are six leucine codons: UUA, UUG, CUU, CUC, CUA, and CUG. The anticodon AAU would recognize UUA and UUG. You would need two other tRNAs to efficiently recognize the other four leucine codons. These could be GAG and GAU or GAA and GAU.

B. The answer is one. There is only one codon, AUG, so you need only one tRNA with the anticodon UAC.

C. The answer is three. There are six serine codons: AGU, AGC, UCU, UCC, UCA, and UCG. You would need only one tRNA to recognize AGU and AGC. This tRNA could have the anticodon UCG or UCA. You would need two tRNAs to efficiently recognize the other four tRNAs. These could be AGG and AGU or AGA and AGU.

C8. 3'–CUU–5' or 3'–CUC–5'

C10. It can recognize 5'–GGU–3', 5'–GGC–3', and 5'–GGA–3'. All of these specify glycine.

C12. All tRNA molecules have some basic features in common. They all have a cloverleaf structure with three stem-loop structures. The second stem-loop contains the anticodon sequence that recognizes the codon sequence in mRNA. At the 3' end, there is an acceptor stem, with the sequence CCA, that serves as an attachment site for an amino acid. Most tRNAs also have base modifications that occur within their nucleotide sequences.

C14. The role of aminoacyl-tRNA synthetase is to specifically recognize tRNA molecules and attach the correct amino acid to them. This ability is sometimes described as the second genetic code because the specificity of the attachment is a critical step in deciphering the genetic code. For example, if a tRNA has a 3'–GGG–5' anticodon, it will recognize a 5'–CCC–3' codon, which should specify proline. It is essential that the aminoacyl-tRNA synthetase known as prolyl-tRNA-synthetase recognizes this tRNA and attaches proline to the 3' end. The other aminoacyl-tRNA synthetases should not recognize this tRNA.

C16. Bases that have been chemically modified can occur at various locations throughout the tRNA molecule. The significance of all of these modifications is not entirely known. However, within the anticodon region, base modification may alter base pairing to allow the anticodon to recognize two or more different bases within the codon.

C18. No, it is not. Due to the wobble rules, the 5' base in the anticodon of a tRNA can recognize two or more bases in the third (3') position of the mRNA. Therefore, any given cell type synthesizes far fewer than 61 types of tRNAs.

C20. The assembly process is very complex at the molecular level. In eukaryotes, 33 proteins and one rRNA assemble to form a 40S subunit, and 49 proteins and three rRNAs assemble to form a 60S subunit. This assembly occurs within the nucleolus.

C22. A. On the surface of the 30S subunit and at the interface between the two subunits

B. Within the 50S subunit

C. From the 50S subunit

D. To the 30S subunit

C24. Most bacterial mRNAs contain a Shine-Dalgarno sequence, which is necessary for the binding of the mRNA to the small ribosomal subunit. This sequence, UUAGGAGGU, is complementary to a sequence in the 16S rRNA. Due to this complementarity, these sequences will hydrogen bond to each other during the initiation stage of translation.

C26. The ribosome binds at the 5' end of the mRNA and then scans in the 3' direction in search of an AUG start codon. If it finds one that reasonably obeys Kozak's rules, it will begin translation at that site. Aside from an AUG start codon, two other important features are a guanosine at the +4 position and a purine at the −3 position.

C28. The A (aminoacyl) site is the location where a tRNA carrying a single amino acid initially binds. The only exception is the initiator tRNA, which binds to the P (peptidyl) site. The growing polypeptide chain is removed from the tRNA in the P site and transferred to the amino acid attached to the tRNA in the A site. The ribosome translocates in the 3' direction, with the result that the two tRNAs in the P and A sites are moved to the E (exit) and P sites, and the uncharged tRNA in the E site is released.

C30. The initiation phase involves the binding of the Shine-Dalgarno sequence to the rRNA in the 30S subunit. The elongation phase involves the binding of anticodons in tRNA to codons in mRNA.

C32. A. The E site and P sites. (Note: A tRNA without an amino acid attached is only briefly found in the P site, just before translocation occurs.)

B. P site and A site. (Note: A tRNA with a polypeptide chain attached is only briefly found in the A site, just before translocation occurs.)

C. Usually the A site, except the initiator tRNA, which can be found in the P site.

C34. The tRNAs bind to the mRNA because their anticodon and codon sequences are complementary. When the ribosome translocates in the 5' to 3' direction, the tRNAs remain bound to their complementary codons, and the two tRNAs shift from the A site and P site to the P site and E site. If the ribosome moved in the 3' direction, it would have to dislodge the tRNAs and drag them to a new position where they would not (necessarily) be complementary to the mRNA.

C36. 52

Experimental Questions

E2. A. There could have been other choices, but this template would be predicted to contain a cysteine codon, UGU, but would not contain any alanine codons.

B. You do not want to use ^{35}S because the radiolabel would be removed during treatment with Raney nickel.

C. There would not be a significant amount of radioactivity incorporated into newly made polypeptides with or without Raney nickel treatment. The only radiolabeled amino acid in this experiment was cysteine, which became attached to $tRNA^{Cys}$. When exposed to Raney nickel, these cysteines were converted to alanine but only after they were already attached to $tRNA^{Cys}$. If there were not any cysteine codons in the mRNA template, the $tRNA^{Cys}$ would not recognize this mRNA. Therefore, we would not expect to see much radioactivity in the newly made polypeptides.

E4. The initiation phase of translation is very different in bacteria and in eukaryotes, so they would not be translated very efficiently. A bacterial mRNA would not be translated very efficiently in a eukaryotic translation system, because it lacks a cap structure attached to its 5' end. A eukaryotic mRNA would not have a Shine-Dalgarno sequence near its 5' end, so it would not be translated very efficiently in a bacterial translation system.

E6. Looking at the figure, the 5' end of the template DNA strand is toward the right side. The 5' ends of the mRNAs are farthest from the DNA, and the 3' ends of the mRNAs are closest to the DNA. The start codons are slightly downstream from the 5' ends of the RNAs.

E8.

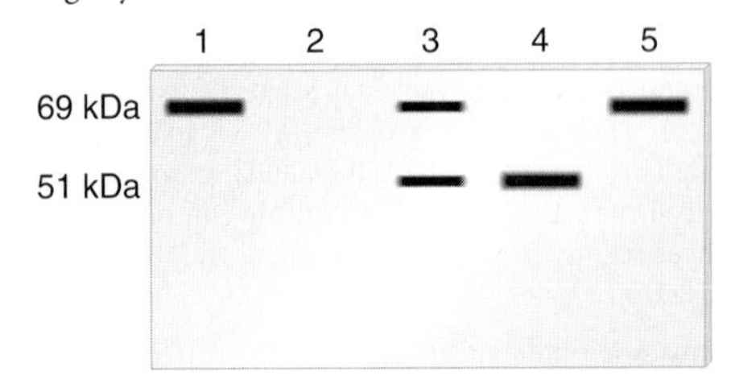

E10. A. If codon usage were significantly different between kangaroo and yeast cells, this would inhibit the translation process. For example, if the preferred leucine codon in kangaroos was CUU, translation would probably be slow in a yeast translation system. We would expect the cell-free translation system from yeast cells to primarily contain leucine tRNAs with an anticodon sequence that is AAC, because this $tRNA^{Leu}$ would match the preferred yeast leucine codon, which is UUG. In a yeast translation system, there probably would not be a large amount of tRNA with an anticodon of GAA, which would match the preferred leucine codon, CUU, of kangaroos. For this reason, kangaroo mRNA would not be translated very well in a yeast translation system, but it probably would be translated to some degree.

B. The advantage of codon bias is that a cell can rely on a smaller population of tRNA molecules to efficiently translate its proteins. A disadvantage is that mutations, which do not change the amino acid sequence but do change a codon (e.g., UUG to UUA), may inhibit the production of a polypeptide if a preferred codon is changed to a nonpreferred codon.

Questions for Student Discussion/Collaboration

2. This could be a very long list. There are similarities along several lines:

1. There is a lot of molecular recognition going on, either between two nucleic acid molecules or between proteins and nucleic acid molecules. Students may see these as similarities or differences, depending on their point of view.
2. There is biosynthesis going on in both processes. Small building blocks are being connected together. This requires an input of energy.
3. There are genetic signals that determine the beginning and ending of these processes.

There are also many differences:

1. Transcription produces an RNA molecule with a similar structure to the DNA, whereas translation produces a polypeptide with a structure that is very different from RNA.
2. Depending on your point of view, it seems that translation is more biochemically complex, requiring more proteins and RNA molecules to accomplish the task.

CHAPTER 14

Conceptual Questions

C2. In bacteria, gene regulation greatly enhances the efficiency of cell growth. It takes a lot of energy to transcribe and translate genes. Therefore, a cell is much more efficient and better at competing in its environment if it expresses genes only when the gene product is needed. For example, a bacterium will express only the genes that are necessary for lactose metabolism when a bacterium is exposed to lactose. When the environment is missing lactose, these genes are turned off. Similarly, when tryptophan levels are high within the cytoplasm, the genes required for tryptophan biosynthesis are repressed.

C4. A. Regulatory protein
B. Effector molecule
C. DNA segment
D. Effector molecule
E. Regulatory protein
F. DNA segment
G. Effector molecule

C6. A mutation that has a *cis*-effect is within a genetic regulatory sequence, such as an operator site, that affects the binding of a genetic regulatory protein. A *cis*-effect mutation affects only the adjacent genes that the genetic regulatory sequence controls. A mutation having a *trans*-effect is usually in a gene that encodes a genetic regulatory protein. A *trans*-effect mutation can be complemented in a merozygote experiment by the introduction of a normal gene that encodes the regulatory protein.

C8. A. No transcription would take place. The *lac* operon could not be expressed.

B. No regulation would take place. The operon would be continuously turned on.

C. The rest of the operon would function normally, but none of the transacetylase would be made.

C10. Diauxic growth refers to the phenomenon in which a cell first uses up one type of sugar (such as glucose) before it begins to metabolize a second sugar (such as lactose). In this case, it is caused by gene regulation. When a bacterial cell is exposed to both sugars, the uptake of glucose causes the cAMP levels in the cell to fall. When this occurs, the catabolite activator protein (CAP) is removed from the *lac* operon, so it is not able to be activated by CAP.

C12. A mutation that prevented the lac repressor from binding to the operator would make the *lac* operon constitutive only in the absence of glucose. However, this mutation would not be entirely constitutive because transcription would be inhibited in the presence of glucose. The disadvantage of constitutive expression of the *lac* operon is that the bacterial cell would waste a lot of energy transcribing the genes and translating the mRNA when lactose was not present.

C14. A. Without *ara*O_2, the repression of the *ara* operon could not occur. The operon would be constitutively expressed at high levels because AraC protein could still activate transcription of the *ara* operon by binding to *araI*. The presence of arabinose would have no effect. Note: The binding of arabinose to AraC is not needed to form an AraC dimer at *araI*. The dimer is able to form because the loop has been broken. This point may be figured out if you notice that an AraC dimer is bound to *ara*O_1 in the presence and absence of arabinose (see Figure 14.12).

B. Without *ara*O_1, the AraC protein would be overexpressed. It would probably require more arabinose to alleviate repression. In addition, activation might be higher because there would be more AraC protein available.

C. Without *araI*, transcription of the *ara* operon cannot be activated. You might get a very low level of constitutive transcription.

D. Without *ara*O_2, the repression of the *ara* operon could not occur. However, without *araI*, transcription of the *ara* operon cannot be activated. You might get a very low level of constitutive transcription.

C16. A. Attenuation will not occur because loop 2–3 will form.

B. Attenuation will occur because 2–3 cannot form, so 3–4 will form.

C. Attenuation will not occur because 3–4 cannot form.

D. Attenuation will not occur because 3–4 cannot form.

C18. The addition of Gs and Cs into the U-rich sequence would prevent attenuation. The U-rich sequence promotes the dissociation of the mRNA from the DNA, when the terminator stem-loop forms. This causes RNA polymerase to dissociate from the DNA and thereby causes transcriptional termination. The UGGUUGUC sequence would probably not dissociate because of the Gs and Cs. Remember that GC base pairs have three hydrogen bonds and are more stable than AU base pairs, which have only two hydrogen bonds.

C20. It takes a lot of cellular energy to translate mRNA into a protein. A cell wastes less energy if it prevents the initiation of translation rather than a later stage such as elongation or termination.

C22. One mechanism is that histidine could act as corepressor that shuts down the transcription of the histidine synthetase gene. A second mechanism would be that histidine could act as an inhibitor via feedback inhibition. A third possibility is that histidine inhibits the ability of the mRNA encoding histidine synthetase to be translated. Perhaps it induces a gene that encodes an antisense RNA. If the amount of histidine synthetase protein was identical in the presence and absence of extracellular histidine, a feedback inhibition mechanism is favored, because this affects only the activity of the histidine synthetase enzyme, not the amount of the enzyme. The other two mechanisms would diminish the amount of this protein.

C24. The two proteins are similar in that both bind to a segment of DNA and repress transcription. They are different in three ways. (1) They recognize different effector molecules (i.e., the lac repressor recognizes allolactose, and the trp repressor recognizes tryptophan). (2) Allolactose causes the lac repressor to release from the operator, while tryptophan causes the trp repressor to bind to its operator. (3) The sequences of the operator sites that these two proteins recognize are different from each other. Otherwise, the lac repressor could bind to the *trp* operator, and the trp repressor could bind to the *lac* operator.

C26. In the lytic cycle, the virus directs the bacterial cell to make more virus particles until eventually the cell lyses and releases them. In the lysogenic cycle, the viral genome is incorporated into the host cell's genome as a prophage. It remains there in a dormant state until some stimulus causes it to excise itself from the bacterial chromosome and enter the lytic cycle.

C28. The O_R region contains three operator sites and two promoters. P_{RM} and P_R transcribe in opposite directions. The λ repressor will first bind to O_{R1} and then O_{R2}. The binding of the λ repressor to O_{R1} and O_{R2} inhibits transcription from P_R and thereby switches off the lytic cycle. Early in the lysogenic cycle, the λ repressor protein concentration may become so high that it will occupy O_{R3}. Later, when the λ repressor concentration begins to drop, it will first be removed from O_{R3}. This allows transcription from P_{RM} and maintains the lysogenic cycle. By comparison, the cro protein has its highest affinity for O_{R3}, and so it binds there first. This blocks transcription from P_{RM} and thereby switches off the lysogenic cycle. The cro protein has a similar affinity for O_{R2} and O_{R1}, and so it may occupy either of these sites next. It will bind to both O_{R2} and O_{R1}. This turns down the expression from P_R, which is not needed in the later stages of the lytic cycle.

C30. It would first increase the amount of cro protein, so the lytic cycle would be favored.

C32. Neither cycle could be followed. As shown in Figure 14.21, N protein is needed to make a longer transcript from P_L for the lysogenic cycle and also to make a longer transcript from P_R for the lytic cycle.

C34. If the F^- strain is lysogenic for phage λ, the λ repressor is already being made in that cell. If the F^- strain receives genetic material from an *Hfr* strain, you would not expect it to have an effect on the lysogenic cycle, which is already established in the F^- cell. However, if the *Hfr* strain is lysogenic for λ and the F^- strain is not, the *Hfr* strain could transfer the integrated λ DNA (i.e., the prophage) to the F^- strain. The cytoplasm of the F^- strain would not contain any λ repressor. Therefore, this λ DNA could choose between the lytic and lysogenic cycle. If it follows the lytic cycle, the F^- recipient bacterium will lyse.

Experimental Questions

E2. In samples loaded in lanes 1 and 4, we expect the repressor to bind to the operator because no lactose is present. In the sample loaded into lane 4, the CAP protein could still bind cAMP because there is no glucose. However, there really is no difference between lanes 1 and 4, so it does not look like the CAP can activate transcription when the lac repressor is bound. If we compare samples loaded into lanes 2 and 3, the lac repressor would not be bound in either case, and the CAP would not be bound in the sample loaded into lane 3. There is less transcription in lane 3 compared to lane 2, but because there is some transcription seen in lane 3, we can conclude that the removal of the CAP (because cAMP levels are low) is not entirely effective at preventing transcription. Overall, the results indicate that the binding of the lac repressor is more effective at preventing transcription of the *lac* operon compared to the removal of the CAP.

E4. A. Yes, if you do not sonicate, then β-galactosidase will not be released from the cell, and not much yellow color will be observed. (Note: You may observe a little yellow color because some β-ONPG may be taken into the cell.)

B. No, you should still get yellow color in the first two tubes even if you forgot to add lactose because the unmated strain does not have a functional lac repressor.

C. Yes, if you forgot to add β-ONPG, you could not get yellow color because the cleavage of β-ONPG by β-galactosidase is what produces the yellow color.

E6. You could mate a strain that has an F' factor carrying a normal *lac* operon and a normal *lacI* gene to this mutant strain. Because the mutation is in the operator site, you would still continue to get expression of β-galactosidase, even in the absence of lactose.

E8. In this case, things are more complex because AraC acts as a repressor and an activator protein. If AraC were missing due to mutation, there would not be repression or activation of the *ara* operon in the presence or absence of arabinose. It would be expressed constitutively at low levels. The introduction of a normal *araC* gene into the bacterium on an F' factor would restore normal regulation (i.e., a *trans*-effect).

Questions for Student Discussion/Collaboration

2. A DNA loop may inhibit transcription by preventing RNA polymerase from recognizing the promoter. Or, it may inhibit transcription by preventing the formation of the open complex. Alternatively, a bend may enhance transcription by exposing the base sequence that the sigma factor of RNA polymerase recognizes. The bend may expose the major groove in such a way that this base sequence is more accessible to binding by sigma factor and RNA polymerase.

CHAPTER 15

Conceptual Questions

C2. Regulatory elements are relatively short genetic sequences that are recognized by regulatory transcription factors. After the regulatory transcription factor has bound to the regulatory element, it will affect the rate of transcription, either activating it or repressing it, depending on the action of the regulatory protein. Regulatory elements are typically located in the upstream region near the promoter, but they can be located almost anywhere (i.e., upstream and downstream) and even quite far from the promoter.

C4. Transcriptional activation occurs when a regulatory transcription factor binds to a response element and activates transcription. Such proteins, called activators, may interact with TFIID and/or mediator to promote the assembly of RNA polymerase and general transcription factors at the promoter region. They also could alter the structure of chromatin so that RNA polymerase and transcription factors are able to gain access to the promoter. Transcriptional inhibition occurs when a regulatory transcription factor inhibits transcription. Such repressors also may interact with TFIID and/or mediator to inhibit RNA polymerase.

C6. A. DNA binding

B. DNA binding

C. Protein dimerization

C8. For the glucocorticoid receptor to bind to a GRE, a steroid hormone must first enter the cell. The hormone then binds to the glucocorticoid receptor, which releases HSP90. The release of HSP90 exposes a nuclear localization signal (NLS) within the receptor, which enables it to dimerize and then enter the nucleus. Once inside the nucleus, the dimer binds to a GRE, which activates transcription of the adjacent genes.

C10. Phosphorylation of the CREB protein causes it to act as a transcriptional activator. The unphosphorylated CREB protein can still bind to CREs, but it does not stimulate transcription.

C12. A. Eventually, the glucocorticoid hormone will be degraded by the cell. The glucocorticoid receptor binds the hormone with a certain affinity. The binding is a reversible process. Once the concentration of the hormone falls below the affinity of the hormone for the receptor, the receptor will no longer have the glucocorticoid hormone bound to it. When the hormone is released, the glucocorticoid receptor will change its conformation, and it will no longer bind to the DNA.

B. An enzyme known as a phosphatase will eventually cleave the phosphate groups from the CREB protein. When the phosphates are removed, the CREB protein will stop activating transcription.

C14. The enhancer found in A would work, but the ones found in B and C would not. The sequence that is recognized by the transcriptional activator is 5'–GTAG–3' in one strand and 3'–CATC–5' in the opposite strand. This is the same arrangement found in A. In B and C, however, the arrangement is 5'–GATG–3' and 3'–CATC–5'. In the arrangement found in B and C, the two middle bases (i.e., A and T) are not in the correct order.

C16.

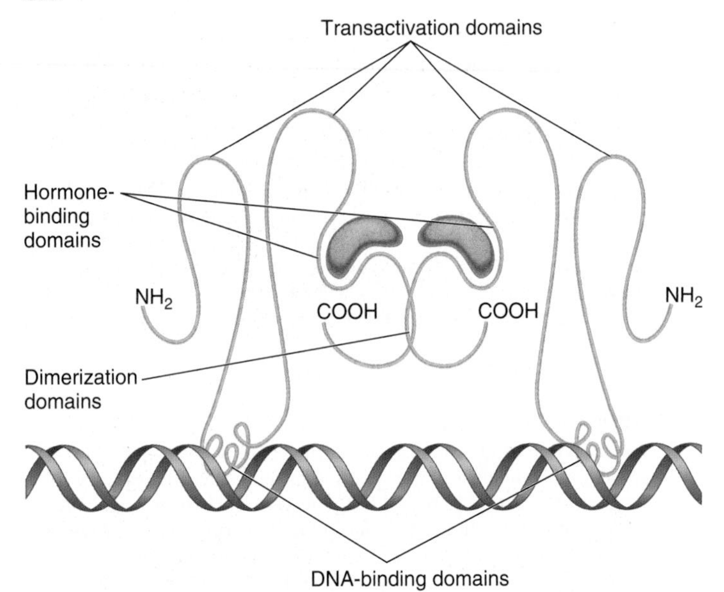

This is one hypothetical drawing of the glucocorticoid receptor. As discussed in your textbook, the glucocorticoid receptor forms a homodimer. The dimer shown here has rotational symmetry. If you flip the left side around, it has the same shape as the right side. The hormone-binding, DNA-binding, dimerization, and transactivation domains are labeled. The glucocorticoid hormone is shown in orange.

C18. ATP-dependent chromatin remodeling complexes may change the positions of nucleosomes, evict histones, and/or replace histones with histone variants.

C20. The attraction between DNA and histones occurs because the histones are positively charged and the DNA is negatively charged. The covalent attachment of acetyl groups decreases the amount of positive charge on the histone proteins and thereby may decrease the binding of the DNA. In addition, histone acetylation may attract proteins to the region that loosen chromatin compaction.

C22. A nucleosome-free region (NFR) is a location in the genome where nucleosomes are missing. They are typically found at the beginning and ends of genes. An NFR at the beginning of a gene is thought to be important so that genes can be activated. The NFR at the end of a gene may be important for its proper termination.

C24. An insulator is a segment of DNA that functions as a boundary between two adjacent genes. An insulator may act as a barrier to changes in chromatin structure or block the effects of a neighboring enhancer.

C26. Perhaps the methyltransferase is responsible for methylating and inhibiting a gene that causes a cell to become a muscle cell. The methyltransferase is inactivated by the mutation.

C28. The function of splicing factors is to influence the selection of splice sites in RNA. In certain cell types, the concentration of particular splicing factors is higher than in other tissues. The high concentration of particular splicing factors, and the regulation of their activities, may promote the selection of particular splice sites and thereby lead to tissue-specific splicing.

C30. This person would be unable to make ferritin, because the IRP would always be bound to the IRE. The amount of transferrin receptor mRNA would be high, even in the presence of high amounts of iron, because the IRP would always remain bound to the IRE and stabilize the transferrin receptor mRNA. Such a person would not have any problem taking up iron into his/her cells. In fact, this person would take up a lot of iron via the transferrin receptor, even when the iron concentrations were high. Therefore, this person would not need more iron in the diet. However, excess iron in the diet would be very toxic for two reasons. First, the person cannot make ferritin, which prevents the toxic buildup of iron in the cytosol. Second, when iron levels are high, the person would continue to synthesize the transferrin receptor, which functions in the uptake of iron.

C32. A disadvantage of mRNAs with a short half-life is that the cells probably waste a lot of energy making them. If a cell needs the protein encoded by a short-lived mRNA, the cell has to keep transcribing the gene that encodes the mRNA because the mRNAs are quickly degraded. An advantage of short-lived mRNAs is that the cell can rapidly turn off protein synthesis. If a cell no longer needs the polypeptide encoded by a short-lived mRNA, it can stop transcribing the gene, and the mRNA will be quickly degraded. This will shut off the synthesis of more proteins rather quickly. With most long-lived mRNAs, it will take much longer to shut off protein synthesis after transcription has been terminated.

C34. First, the double-stranded miRNA could come from the transcription of a gene, as pre-miRNA. Second, it could come from a virus. Third, multiple insertions of the same gene in the genome may lead to both DNA strands being transcribed, thereby forming double-stranded RNA.

C36. If mRNA stability is low, this means it is degraded more rapidly. Therefore, low stability results in a low mRNA concentration. The length of the polyA tail is one factor that affects stability. A longer tail makes mRNA more stable. Certain mRNAs have sequences that affect their half-lives. For example, AU-rich elements (AREs) are found in many short-lived mRNAs. The AREs are recognized by cellular proteins that cause the mRNAs to be rapidly degraded.

Experimental Questions

E2. These results indicate that the fibroblasts undergo maintenance methylation because they can replicate and methylate DNA if it has already been methylated. However, the cells do not undergo de novo methylation, because if the donor DNA was unmethylated, the DNA in the daughter cells remains unmethylated.

E4. Based on these results, there are enhancers that are located in regions A, D, and E. When these enhancers are deleted, the level of transcription is decreased. There also appears to be a silencer in region B, because a deletion of this region increases the rate of transcription. There do not seem to be any response elements in region C, or at least not any that function in muscle cells. Region F contains the core promoter, so the deletion of this region inhibits transcription.

E6. The results indicate that protein X binds to the DNA fragment and retards its mobility (lanes 3 and 4). However, the hormone is not required for DNA binding. Because we already know that the hormone is needed for transcriptional activation, it must play some other role. Perhaps the hormone activates a signaling pathway that leads to the phosphorylation of the transcription factor, and phosphorylation is necessary for translational activation. This situation would be similar to the CREB protein, which is activated by phosphorylation. The CREB protein can bind to the DNA whether or not it is phosphorylated.

E8. The gel shown in (a) is correct for ferritin mRNA regulation. The presence and absence of iron does not affect the amount of mRNA; it affects the ability of the mRNA to be translated. The gel shown in (b) is correct for transferrin receptor mRNA. In the absence of iron, the IRP binds to the IRE in the mRNA and stabilizes it, so the mRNA level is high. In the presence of iron, the iron binds to IRP and IRP is released from the IRE, and the transferrin receptor mRNA is degraded.

Questions for Student Discussion/Collaboration

2. Probably the most efficient method would be to systematically make deletions of progressively smaller sizes. For example, you could begin by deleting 20,000 bp on either side of the gene and see if that affects transcription. If you found that only the deletion on the 5' end of the gene had an effect, you could then start making deletions from the 5' end, perhaps in 10,000-bp or 5000-bp increments until you localized response elements. You would then make smaller deletions in the putative region until it was down to a hundred or a few dozen nucleotides. At this point, you might conduct site-directed mutagenesis, as described in Chapter 18, as a way to specifically identify the regulatory element sequence.

CHAPTER 16

Conceptual Questions

C2. It is a gene mutation, a point mutation, a base substitution, a transition mutation, a deleterious mutation, a mutant allele, a nonsense mutation, a conditional mutation, and a temperature-sensitive lethal mutation.

C4. A. It would probably inhibit protein function, particularly if it was not near the end of the coding sequence.

B. It may or may not affect protein function, depending on the nature of the amino acid substitution and whether the substitution is in a critical region of the protein.

C. It would increase the amount of functional protein.

D. It may affect protein function if the alteration in splicing changes an exon in the mRNA that results in a protein with a perturbed structure.

C6. A. Not appropriate, because the second mutation is at a different codon

B. Appropriate

C. Not appropriate, because the second mutation is in the same gene as the first mutation

D. Appropriate

C8. A. Silent, because the same amino acid (glycine) is encoded by GGA and GGT

B. Missense, because a different amino acid is encoded by CGA compared to GGA

C. Missense, because a different amino acid is encoded by GTT compared to GAT

D. Frameshift, because an extra base is inserted into the sequence

C10. One possibility is that a translocation may move a gene next to a heterochromatic region of another chromosome and thereby diminish its expression, or it could be moved next to a euchromatic region and increase its expression. Another possibility is that the translocation breakpoint may move the gene next to a new promoter or regulatory sequences that may now influence the gene's expression.

C12. A. No; the position (i.e., chromosomal location) of a gene has not been altered.

B. Yes; the expression of a gene has been altered because it has been moved to a new chromosomal location.

C. Yes; the expression of a gene has been altered because it has been moved to a new chromosomal location.

C14. If a mutation within the germ line is passed to an offspring, all of the cells of the offspring's body will carry the mutation. A somatic mutation affects only the somatic cell in which it originated and all of the daughter cells that the somatic cell produces. If a somatic mutation occurs early during embryonic development, it may affect a fairly large region of the organism. Because germ-line mutations affect the entire organism, they are potentially more harmful (or beneficial), but this is not always the case. Somatic mutations can cause quite harmful effects, such as cancer.

C16. A thymine dimer can interfere with DNA replication because DNA polymerase cannot slide past the dimer and add bases to the newly growing strand. Alkylating mutagens such as nitrous acid will cause DNA replication to make mistakes in the base pairing. For example, an alkylated cytosine will base-pair with adenine during DNA replication, thereby creating a mutation in the newly made strand. A third example is 5-bromouracil, which is a thymine analogue. It may base-pair with guanine instead of adenine during DNA replication.

C18. During TNRE, a trinucleotide repeat sequence gets longer. If someone was mildly affected with a TNRE disorder, he or she might be concerned that an expansion of the repeat might occur during gamete formation, yielding offspring more severely affected with the disorder, a phenomenon called anticipation. This phenomenon may depend on the sex of the parent with the TNRE.

C20. According to the random mutation theory, spontaneous mutations can occur in any gene and do not involve exposure of the organism to a particular environment that selects for specific types of mutation. However, the structure of chromatin may cause certain regions of the DNA to be more susceptible to random mutations. For example, DNA in an open conformation may be more accessible to mutagens and more likely to incur mutations. Similarly, hot spots—certain regions of a gene that are more likely to mutate than other regions—can occur within a single gene. Also, another reason that some genes mutate at a higher rate is that some genes are larger than others, which provides a greater chance for mutation.

C22. Excision repair systems could fix this damage. Also, homologous recombination repair could fix the damage.

C24. Anticipation means that the TNRE expands even further in future generations. Anticipation may depend on the sex of the parent with the TNRE.

C26. The mutation frequency is the total number of mutant alleles divided by the total number of alleles in the population. If there are 1,422,000 babies, there are 2,844,000 copies of this gene (because each baby has two copies). The mutation frequency is $^{31}/_{2,844,000}$, which equals 1.09×10^{-5}. The mutation rate is the number of new mutations per generation. There are 13 babies who did not have a parent with achondroplasia; thus, 13 is the number of new mutations. If we calculate the mutation rate as the number of new mutations in a given gene per generation, then we should divide 13 by 2,844,000. In this case, the mutation rate would be 4.6×10^{-6}.

C28. The effects of mutations are cumulative. If one mutation occurs in a cell, this mutation will be passed to the daughter cells. If a mutation occurs in the daughter cell, now there will be two mutations. These two mutations will be passed to the next generation of daughter cells, and so forth. The accumulation of many mutations eventually kills the cells. That is why mutagens are more effective at killing dividing cells compared to nondividing cells. It is because the number of mutations accumulates to a lethal level.

There are two main side effects to this treatment. First, some normal (noncancerous) cells of the body, particularly skin cells and intestinal cells, are actively dividing. These cells are also killed by chemotherapy and radiation therapy. Secondly, it is possible that the therapy may produce mutations that will cause noncancerous cells to become cancerous. For these reasons, there is a maximal dose of chemotherapy or radiation therapy that is recommended.

C30. A. Yes

B. No; the albino trait affects the entire individual.

C. No; the early apple-producing trait affects the entire tree.

D. Yes

C32. Mismatch repair is aimed at eliminating mismatches that may have occurred during DNA replication. In this case, the wrong base is in the newly made strand. The binding of MutH, which occurs on a hemimethylated sequence, provides a sensing mechanism to distinguish between the unmethylated and methylated strands. In other words, MutH binds to the hemimethylated DNA in a way that allows the mismatch repair system to distinguish which strand is methylated and which is not.

C34. Because sister chromatids are genetically identical, an advantage of homologous recombination is that it can be an error-free mechanism to repair a DSB. A disadvantage, however, is that it occurs only during the S and G_2 phase of the cell cycle in eukaryotes or following DNA replication in bacteria. An advantage of NHEJ is that it doesn't involve the participation of a sister chromatid, so it can occur at any stage of the cell cycle. However, a disadvantage is that NHEJ can result in small deletions in the region that has been repaired. Overall, NHEJ is a quick but error-prone repair mechanism, while HR is a more accurate method of repair that is limited to certain stages of the cell cycle.

C36. In *E. coli*, the TRCF recognizes when RNA polymerase is stalled on the DNA. This stalling may be due to DNA damage such as a thymine dimer. The TRCF removes RNA polymerase and recruits the excision DNA repair system to the region, thereby promoting the repair of the template strand of DNA. It is beneficial to preferentially repair actively transcribed DNA because it is functionally important. It is a DNA region that encodes a gene.

C38. The underlying genetic defect that causes xeroderma pigmentosum is a defect in one of the genes that encode a polypeptide involved with nucleotide excision repair. These individuals are defective in repairing DNA abnormalities such as thymine dimers and abnormal bases. Therefore, they are very sensitive to environmental agents such as UV light. Because they are defective at repair, UV light is more likely to cause mutations in these people compared to unaffected individuals. For this reason, people with XP develop pigmentation abnormalities and premalignant lesions and have a high predisposition to skin cancer.

C40. Both types of repair systems recognize an abnormality in the DNA and excise the abnormal strand. The normal strand is then used as a template to synthesize a complementary strand of DNA. The systems differ in the types of abnormalities they detect. The mismatch repair system detects base pair mismatches, while the excision repair system recognizes thymine dimers, chemically modified bases, missing bases, and certain types of cross-links. The mismatch repair system operates immediately after DNA replication, allowing it to distinguish between the daughter strand (which contains the wrong base) and the parental strand. The excision repair system can operate at any time in the cell cycle.

Experimental Questions

E2. When cells from a master plate were replica plated onto two plates containing selective media with the T1 phage, T1-resistant colonies were observed at the same locations on both plates. These results indicate that the mutations occurred randomly while on the master plate (in the absence of T1) rather than occurring as a result of exposure to T1. In other words, mutations are random events, and selective conditions may promote the survival of mutant strains that occur randomly.

To show that antibiotic resistance is due to random mutation, one could follow the same basic strategy except the secondary plates would contain the antibiotic instead of T1 phage. If the antibiotic resistance arose as a result of random mutation on the master plate, one would expect the antibiotic-resistant colonies to appear at the same locations on two different secondary plates.

E4. Perhaps the X-rays also produce mutations that make the *ClB* daughters infertile. Many different types of mutations could occur in the irradiated males and be passed to the *ClB* daughters. Some of these mutations could prevent the *ClB* daughters from being fertile. These mutations could interfere with oogenesis, etc. Such *ClB* daughters would be unable to have any offspring.

E6. You would conclude that chemical A is not a mutagen. The percentage of *ClB* daughters (whose fathers had been exposed to chemical A) that did not produce sons was similar to the control (compare 3 out of 2108 with 2 out of 1402). In contrast, chemical B appears to be a mutagen. The percentage of *ClB* daughters (whose fathers had been exposed to chemical B) that did not produce sons was much higher than the control (compare 3 out of 2108 with 77 out of 4203).

E8. You would expose the bacteria to the physical agent. You could also expose the bacteria to the rat liver extract, but it is probably not necessary for two reasons. First, a physical mutagen is not something that a person would eat. Therefore, the actions of digestion via the liver are probably irrelevant if you are concerned that the agent might be a mutagen. Second, the rat liver extract would not be expected to alter the properties of a physical mutagen.

E10. The results suggest that the strain is defective in excision repair. If we compare the normal and mutant strains that have been incubated for 2 hours at 37°C, much of the radioactivity in the normal strain has been transferred to the soluble fraction because it has been excised. In the mutant strain, however, less of the radioactivity has been transferred to the soluble fraction, suggesting that it is not as efficient at removing thymine dimers.

Questions for Student Discussion/Collaboration

2. The worst time to be exposed to mutagens would be at very early stages of embryonic development. An early embryo is most sensitive to mutation because it will affect a large region of the body. Adults must also worry about mutagens for several reasons. Mutations in somatic cells can cause cancer, a topic discussed in Chapter 22. Also, adults should be careful to avoid mutagens that may affect the ovaries or testes because these mutations could be passed to offspring.

CHAPTER 17

Conceptual Questions

C2. Branch migration will not create a heteroduplex during SCE because the sister chromatids are genetically identical. There should not be any mismatches between the complementary strands. Gene conversion cannot take place because the sister chromatids carry alleles that are already identical to each other.

C4. The two molecular mechanisms that can explain the phenomenon of gene conversion are mismatch DNA repair and gap repair synthesis. Both mechanisms could occur in the double-strand break model.

C6. A recombinant chromosome is one that has been derived from a crossover and contains a combination of alleles that is different from the parental chromosomes. A recombinant chromosome is a hybrid of the parental chromosomes.

C8. Gene conversion occurs when a pair of different alleles is converted to a pair of identical alleles. For example, a pair of *Bb* alleles could be converted to *BB* or *bb*.

C10. Gene conversion is likely to take place near the breakpoint. According to the double-strand break model, a gap may be created by the digestion of one DNA strand in the double helix. Gap repair synthesis may result in gene conversion. A second way that gene conversion can occur is by mismatch repair. A heteroduplex may be created after DNA strand migration and may be repaired in such a way as to cause gene conversion.

C12. No, it does not necessarily involve DNA mismatch repair; it could also involve gap repair synthesis. The double-strand break model involves a migration of DNA strands and the digestion of a gap. Therefore, in this gap region, only one chromatid is providing the DNA strands. As seen in Figure 17.7, the top chromosome is using the top DNA strand from the bottom chromosome in the gap region. The bottom chromosome is using the bottom DNA strand from the bottom chromosome. After DNA synthesis, both chromosomes may have the same allele.

C14. First, gene rearrangement of V, D, and J domains occurs within the light- and heavy-chain genes. Second, within a given B cell, different combinations of light and heavy chains are possible. And third, imprecise fusion may occur between the V, D, and J domains.

C16. One segment, which includes some variable (V) domains and perhaps one or more joining (J) domain, of DNA is removed from the κ light-chain gene. One segment, which may include one or more J domains and the region between the J domain and C domain, is removed during pre-mRNA splicing.

C18. The ends of a short region would be flanked by direct repeats. This is a universal characteristic of all transposable elements. In addition, many elements contain IRs or LTRs that are involved in the transposition process. One might also look for the presence of a transposase or reverse transcriptase gene, although this is not an absolute requirement, because nonautonomous transposable elements typically lack transposase or reverse transcriptase.

C20. Direct repeats occur because transposase or integrase produces staggered cuts in the two strands of chromosomal DNA. The transposable element is then inserted into this site, which temporarily leaves two gaps. The gaps are filled in by DNA polymerase. Because this gap filling is due to complementarity of the base sequences, the two gaps end up with the exact same sequence.

C22. Transposable elements are mutagens, because they alter (disrupt) the sequences of chromosomes and genes within chromosomes. They do this by inserting themselves into genes.

C24. A. Viral-like retroelements and nonviral-like retroelements

B. Insertion sequences, composite transposons, and replicative transposons

C. All five types have direct repeats

D. Insertion sequences, composite transposons, and replicative transposons

C26. A. As shown in solved problem S2a, a crossover between direct repeats (DRs) will excise the region between the repeats. In this case, it will excise a large chromosomal region, which does not contain a centromere, and will be lost. The remaining portion of the chromosome is shown here. It has one direct repeat that is formed partly from DR-1 and partly from DR-4. It is designated DR-$^1/_4$.

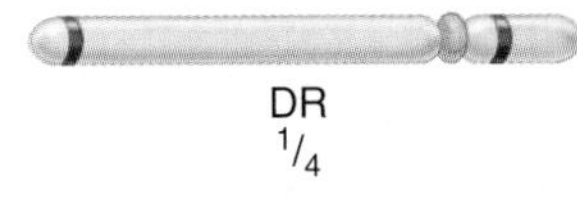

B. A crossover between IR-1 and IR-4 will cause an inversion between the intervening region. This is a similar effect to the crossover within a single transposable element, as described in solved problem S2b. The difference is that the crossover causes a large chromosomal region to be inverted.

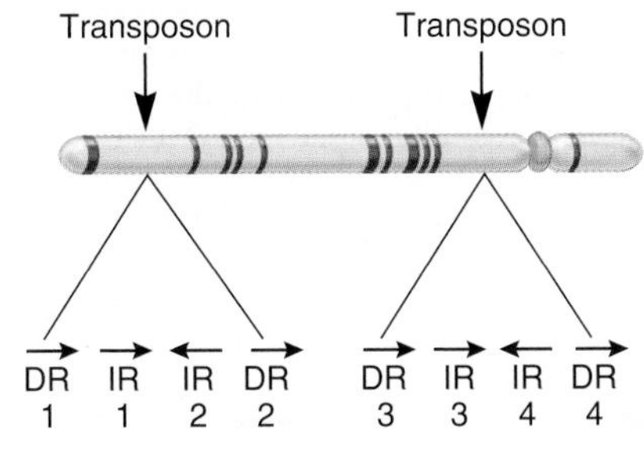

C28. At a frequency of about 1 in 10,000, recombination occurs between the inverted repeats within this transposon. As shown in solved problem S2b, recombination between inverted repeats causes the sequence within the transposon to be reversed. If this occurred in a strain that expressed the *H2/rH1* operon, the promoter would be flipped in the opposite direction, so the *H2* gene and the *rH1* gene would not be expressed. The *H2* flagellar protein would not be made, and the *H1* repressor protein would not be made. The *H1* flagellar protein would be made because the expression of the *H1* gene would not be repressed. A strain expressing the *H1* gene could also "switch" at a frequency of about 1 in 10,000 by the same mechanism. If a crossover occurred between the inverted repeats within the transposable element, the promoter would be flipped again, and the *H2* and *rH1* genes would be expressed again.

Experimental Questions

E2. You would conclude that the substance is a mutagen. Substances that damage DNA tend to increase the level of genetic exchange such as SCE.

E4. The drawing here shows the progression through three rounds of BrdU exposure. After one round, all of the chromosomes would be dark. After two rounds, all of the chromosomes would be harlequin. After three rounds, the number of light sister chromatids would be twice as much as the number of light sister chromatids found after two rounds of replication.

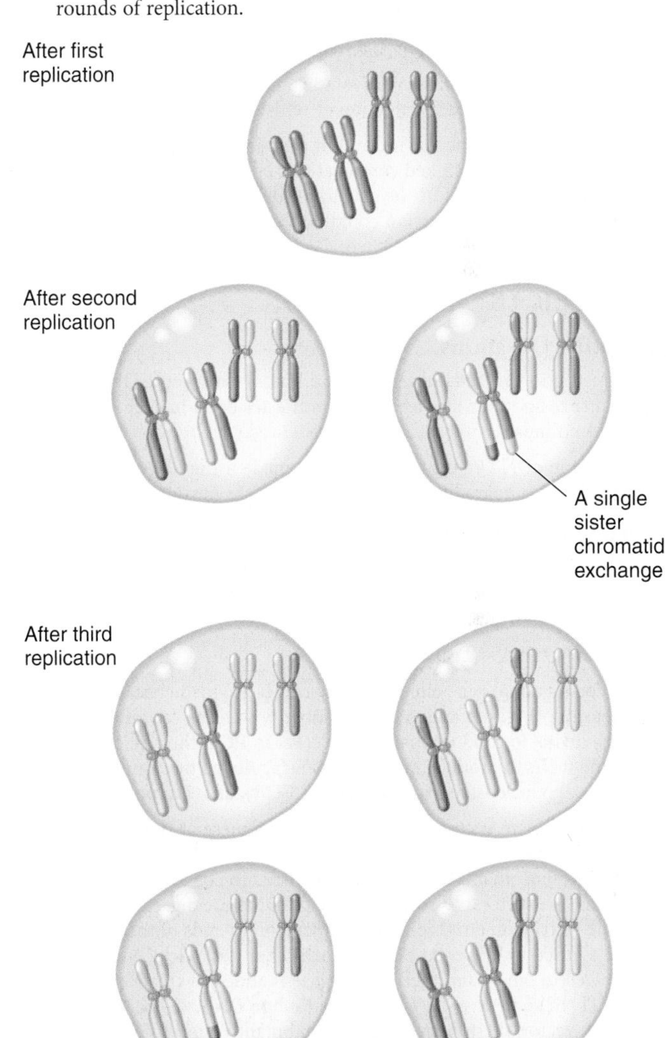

E6. When McClintock started with a colorless strain containing *Ds*, she identified 20 cases where *Ds* had moved to a new location to produce red and shrunken kernels. This identification was possible because the 20 strains had a higher frequency of chromosomal breaks at a specific site and because of the mutability of particular genes. She also had found a strain where *Ds* had inserted into the red-color-producing gene, resulting in the colorless phenotype, so its transposition out of the gene would produce a red patch. Overall, her analysis of the data showed that the sectoring (i.e., mutability) phenotype was consistent with the transposition of *Ds*.

E8. A transposon creates a mutable site because the excision of a transposon causes chromosomal breakage if the two ends are not reconnected. After it has moved out of its original locus (causing chromosomal breakage), it may be inserted into a new locus somewhere else. You could experimentally determine this by examining a strain that has incurred a chromosomal breakage at the first locus. You could microscopically examine many cells that had such a broken chromosome and see if a new locus had been formed. This new locus would be the site into which the transposon had moved. On occasion, there would be chromosomal breakage at this new locus, which could be observed microscopically.

E10. You could begin with the assumption that the inactivation of a tumor-suppressor gene would cause cancerous cell growth. If so, you could begin with a normal human line and introduce a transposon. The next step would be to identify cells that have become immortal. This may be possible by identifying clumps of cells that have lost contact inhibition. You could then grow these cells and make a genomic DNA library from them. The library would be screened using the transposon as a labeled probe.

Questions for Student Discussion/Collaboration

2. Beneficial consequences: You wouldn't get (as many) translocations, inversions, and the accumulation of selfish DNA.

   Harmful consequences: The level of genetic diversity would be decreased, because linked combinations of alleles would not be able to recombine. You wouldn't be able to produce antibody diversity in the same way. Gene duplication could not occur, so the evolution of new genes would be greatly inhibited.

CHAPTER 18

Conceptual Questions

C2. A restriction enzyme recognizes and binds to a specific DNA sequence and then cleaves a (covalent) ester bond in each of two DNA strands.

C4. The term cDNA refers to DNA that is made using RNA as the starting material. Compared to genomic DNA, it lacks introns.

Experimental Questions

E2. Remember that AT base pairs form two hydrogen bonds, while GC base pairs form three hydrogen bonds. The order (from stickiest to least sticky) would be

*Bam*HI = *Pst*I = *Sac*I > *Eco*RI > *Cla*I

E4. In conventional gene cloning, many copies are made because the vector replicates to a high copy number within the cell, and the cells divide to produce many more cells. In PCR, the replication of the DNA to produce many copies is facilitated by primers, deoxyribonucleoside triphosphates (dNTPs), and *Taq* polymerase.

E6. A recombinant vector is a vector that has a piece of "foreign" DNA inserted into it. The foreign DNA came from somewhere else, such as the chromosomal DNA of some organism. To construct a recombinant vector, the vector and source of foreign DNA are digested with the same restriction enzyme. The complementary ends of the fragments are allowed to hydrogen bond to each other (i.e., sticky ends are allowed to bind), and then DNA ligase is added to create covalent bonds. In some cases, a piece of the foreign DNA will become ligated to the vector, thereby creating a recombinant vector. In other cases, the two ends of the vector ligate back together, restoring the vector to its original structure.

As described in Figure 18.2, the insertion of foreign DNA can be detected using X-Gal. As seen here, the insertion of the foreign DNA causes the inactivation of the *lacZ* gene. The *lacZ* gene encodes the enzyme β-galactosidase, which converts the colorless compound X-Gal to a blue compound. If the *lacZ* gene is inactivated by the insertion of foreign DNA, the enzyme will not be produced, and the bacterial colonies will be white. If the vector has simply recircularized and the *lacZ* gene remains intact, the enzyme will be produced, and the colonies will be blue.

E8. 3×2^{27}, which equals 4.0×10^{8}, or about 400 million copies.

E10. Initially, the mRNA would be mixed with reverse transcriptase and nucleotides to create a complementary strand of DNA. Reverse transcriptase also needs a primer. This could be a primer that is known to be complementary to the β-globin mRNA. Alternatively, mature mRNAs have a polyA tail, so one could add a primer that consists of many Ts, called a poly-dT primer. After the complementary DNA strand has been made, the sample would then be mixed with primers, *Taq* polymerase, and nucleotides and subjected to the standard PCR protocol. Note: The PCR reaction would have two kinds of primers. One primer would be complementary to the 5' end of the mRNA and would be unique to the β-globin sequence. The other primer would be complementary to the 3' end. This second primer could be a poly-dT primer, or it could be a unique primer that would bind slightly upstream from the polyA-tail region.

E12. The exponential phase of PCR is chosen because it is during this phase that the amount of PCR product is proportional to the amount of the original DNA in the sample.

E14. A DNA library is a collection of recombinant vectors that contain different pieces of DNA from a source of chromosomal DNA. Because it is a diverse collection of many different DNA pieces, the name library seems appropriate.

E16. Hybridization occurs due to the hydrogen bonding of complementary sequences. Due to the chemical properties of DNA and RNA strands, they form double-stranded regions when the base sequences are complementary. In a Southern and Northern experiment, the cloned DNA is labeled and used as a probe.

E18. The purpose of a Northern blotting experiment is to identify a specific RNA within a mixture of many RNA molecules, using a fragment of cloned DNA as a probe. It can tell you if a gene is transcribed in a particular cell or at a particular stage of development. It can also tell you if a pre-mRNA is alternatively spliced into two or more mRNAs of different sizes.

E20. It appears that this mRNA is alternatively spliced to create a high molecular mass and a lower molecular mass product. Nerve cells produce a very large amount of the larger mRNA, whereas spleen cells produce a moderate amount of the smaller mRNA. Both types are produced in small amounts by the muscle cells. It appears that kidney cells do not transcribe this gene.

E22. 1. β (detected at the highest stringency)
   2. δ
   3. γ_A and ϵ
   4. α_1
   5. *Mb* (detected only at the lowest stringency)

E24. The Western blot is shown here. The sample in lane 2 came from a plant that was homozygous for a mutation that prevented the expression of this polypeptide. Therefore, no protein was observed in this lane. The sample in lane 4 came from a plant that is homozygous for a mutation that introduces an early stop codon into the coding sequence; thus, the polypeptide is shorter than normal (13.3 kDa). The sample in lane 3 was from a heterozygote that expresses about 50% of each type of polypeptide. Finally, the sample in lane 5 came from a plant homozygous for a mutation that changed one amino acid to another amino acid. This type of mutation, termed a missense mutation, may not be detectable on a gel. However, a single amino acid substitution could affect polypeptide function.

E26. The products of structural genes are proteins with a particular amino acid sequence. Antibodies can specifically recognize proteins due to their amino acid sequence. Therefore, an antibody can detect whether or not a cell is making a particular type of protein.

E28. In this case, the transcription factor binds to the response element when the hormone is present. Therefore, the hormone promotes the binding of the transcription factor to the DNA and thereby promotes transcriptional activation.

E30. The levels of cAMP affect the phosphorylation of CREB, and this affects whether or not it can activate transcription. However, CREB can bind to CREs whether or not it is phosphorylated. Therefore, in a gel retardation assay, we would expect CREB to bind to CREs and retard their mobility whether or not the cell extracts were pretreated with epinephrine.

E32. The glucocorticoid receptor will bind to GREs if glucocorticoid hormone is also present (lane 2). The glucocorticoid receptor does not bind without hormone (lane 1), and it does not bind to CREs (lane 6). The CREB protein will bind to CREs with or without hormone (lanes 4 and 5), but it will not bind to GREs (lane 3). The expected results are shown here. In this drawing, the binding of CREB protein to the 700-bp fragment results in a complex with a higher mass compared to the glucocorticoid receptor binding to the 600-bp fragment.

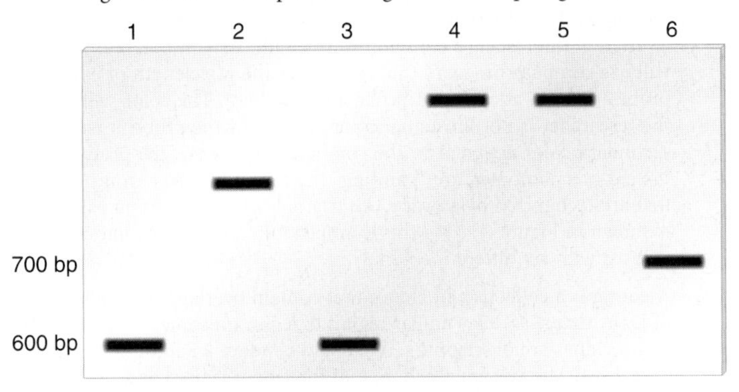

E34. The rationale behind a footprinting experiment has to do with accessibility. If a protein is bound to the DNA, it will cover up the part of the DNA where it is bound. This region of the DNA will be inaccessible to the actions of chemicals or enzymes that cleave the DNA, such as DNase I.

E36. A. AGGTCGGTTGCCATCGCAATAATTTCTGCCTGAACCCAATA

B. Automated sequencing has several advantages. First, the reactions are done in a single tube as opposed to four tubes. Second, the detector can "read" the sequence and provide the researcher with a printout of the sequence. This is much easier than looking at an X-ray film and writing the sequence out by hand. It also avoids human error.

E38. There are lots of different strategies one could follow. For example, you could mutate every other base and see what happens. It would be best to make very nonconservative mutations, such as a purine for a pyrimidine or a pyrimidine for a purine. If the mutation prevents protein binding in a gel retardation assay, then the mutation is probably within the response element. If the mutation has no effect on protein binding, it probably is outside the response element.

Questions for Student Discussion/Collaboration

2. A. Does a particular amino acid within a protein sequence play a critical role in the protein's structure or function?

B. Does a DNA sequence function as a promoter?

C. Does a DNA sequence function as a regulatory site?

D. Does a DNA sequence function as a splicing junction?

E. Is a sequence important for correct translation?

F. Is a sequence important for RNA stability?

And many others . . .

CHAPTER 19

Conceptual Questions

C2. *A. radiobacter* synthesizes an antibiotic that kills *A. tumefaciens*. The genes, which are necessary for antibiotic biosynthesis and resistance, are plasmid encoded and can be transferred during interspecies conjugation. If *A. tumefaciens* received this plasmid during conjugation, it would be resistant to killing. Therefore, the conjugation-deficient strain prevents the occurrence of *A. tumefaciens*–resistant strains.

C4. A biological control agent is an organism that prevents the harmful effects of some other agent in the environment. Examples include *Bacillus thuringiensis*, a bacterium that synthesizes compounds that act as toxins to kill insects, *Ice*$^-$ bacteria that inhibit the proliferation of *Ice*$^+$ bacteria, and the use of *Agrobacterium radiobacter* to prevent crown gall disease caused by *Agrobacterium tumefaciens*.

C6. A mouse model is a strain of mice that carries a mutation in a mouse gene that is analogous to a mutation in a human gene that causes disease. These mice can be used to study the disease and to test potential therapeutic agents.

C8. The T DNA gets transferred to the plant cell; it then is incorporated into the plant cell's genome.

C10. A. With regard to maternal effect genes, the phenotype would depend on the animal that donated the egg. It is the cytoplasm of the egg that accumulates the gene products of maternal effect genes.

B. The extranuclear traits depend on the mitochondrial genome. Mitochondria are found in the egg and in the somatic cell. So, theoretically, both cells could contribute extranuclear traits. In reality, however, researchers have found that the mitochondria in Dolly were from the animal that donated the egg. It is not clear why she had no mitochondria from the mammary cell.

C. The cloned animal would be genetically identical to the animal that donated the nucleus with regard to traits that are determined by nuclear genes, which are expressed during the lifetime of the organism. The cloned animal would/could differ from the animal that donated the nucleus with regard to traits that are determined by maternal effect genes and mitochondrial genes. Such an animal is not a true clone, but it is likely that it would greatly resemble the animal that donated the nucleus, because the vast majority of genes are found in the cell nucleus.

C12. Some people are concerned with the release of genetically engineered microorganisms into the environment. The fear is that such organisms may continue to proliferate and it may not be possible to "stop them." A second concern involves the use of genetically engineered organisms in the food we eat. Some people are worried that genetically engineered organisms may pose an unknown health risk. A third issue is ethics. Some people feel that it is morally wrong to tamper with the genetics of organisms. This opinion may also apply to genetic techniques such as cloning, stem cell research, and gene therapy.

Experimental Questions

E2. One possibility is to clone the toxin-producing genes from *B. thuringiensis* and introduce them into *P. syringae*. This bacterial strain would have the advantage of not needing repeated applications. However, it would be a recombinant strain and might be viewed in a negative light by people who are hesitant to use recombinant organisms in the field. By comparison, *B. thuringiensis* is a naturally occurring species.

E4. Basically, one can follow the strategy described in Figure 19.6. If homologous recombination occurs, only the *Neo*^{R} gene is incorporated into the genome. The cells will be neomycin resistant and also resistant to gancyclovir. If gene addition occurs, the cells will usually be sensitive to gancyclovir. By growing the cells in the presence of neomycin and gancyclovir, one can select for homologous recombinants. The chimeras are identified by the observation that they have a mixture of light and dark fur.

E6. 1. *A. tumefaciens* (T DNA-mediated gene transfer): Genes are cloned into the T DNA of the Ti plasmid. The T DNA from this plasmid is transferred to the plant cell when it is infected by the bacterium.

2. Biolistic gene transfer: DNA is coated on microprojectiles that are "shot" into the plant cell.
3. Microinjection: A microscopic needle containing a solution of DNA is used to inject DNA into plant cells.
4. Electroporation: An electrical current is used to introduce DNA into plant cells.
5. Also, DNA can be introduced into protoplasts by treatment with polyethylene glycol and calcium phosphate. This method is similar to bacterial transformation procedures.

E8. In Mendel's work, and the work of many classical geneticists, an altered (mutant) phenotype is the initial way to identify a gene. For example, Mendel recognized a gene that affects plant height by the identification of tall and dwarf plants. The transmission of this gene could be followed in genetic crosses, and eventually, the gene could be cloned using molecular techniques. Reverse genetics uses the opposite sequence of steps. The gene is cloned first, and a phenotype for the gene is discovered later, by making a transgenic animal with a gene knockout.

E10. A chimera is an organism that is composed of cells from two different individuals (usually of the same species). Chimeras are made by mixing together embryonic cells from two individuals and allowing the cells to organize themselves and develop into a single individual.

E12. You would use Southern blotting to determine the number of gene copies. As described in Chapter 18, you will observe multiple bands in a Southern blot if there are multiple copies of a gene. You need to know this information to predict the outcome of crosses. For example, if there are four integrated copies at different sites in the genome, an offspring could inherit anywhere from zero to four copies of the gene. You would want to understand this so you could predict the phenotypes of the offspring.

You would use Northern blotting or Western blotting to monitor the level of gene expression. A Northern blot will indicate if the gene is transcribed into mRNA, and a Western blot will indicate if the mRNA is translated into protein.

E14. Reproductive cloning means the cloning of entire multicellular organisms. In plants, this is easy. Most species of plants can be cloned by asexual cuttings. In animals, cloning occurs naturally, as in identical twins. Identical twins are genetic replicas of each other because they begin from the same fertilized egg. (Note: There could be some somatic mutations that occur in identical twins that would make them slightly different.) Recently, as in the case of Dolly, reproductive cloning has become possible by fusing somatic cells with enucleated eggs. The advantage, from an agricultural point of view, is that reproductive cloning could allow one to choose the best animal in a herd and make many clones from it. Breeding would no longer be necessary. Also, breeding may be less reliable because the offspring inherit traits from both the mother and father.

E16. Ex vivo therapy involves the removal of living cells from the body and their modification after they have been removed. The modified cells are then reintroduced back into a person's body. This approach works well for cells such as blood cells that are easily removed and replaced. By comparison, this approach would not work very well for many cell types. For example, lung cells cannot be removed and put back again. In this case, in vivo approaches must be sought.

E18. It is the gene product (i.e., the polypeptide) of an oncogene that causes cancerous cell growth. The antisense RNA from the gene introduced via gene therapy would bind to the mRNA from an oncogene. This would prevent the translation of the mRNA into polypeptides and thereby prevent cancerous cell growth.

Questions for Student Discussion/Collaboration

2. From a genetic viewpoint, the recombinant and nonrecombinant strains are very similar. The main difference is their history. The recombinant strain has been subjected to molecular techniques to eliminate a particular gene. The nonrecombinant strain has had the same gene eliminated by a spontaneous mutation or via mutagens. The nonrecombinant strain has the advantage of a better public perception. People are less worried about releasing nonrecombinant strains into the environment.

CHAPTER 20

Conceptual Questions

C2. A. Yes
B. No; this is only one chromosome in the genome.
C. Yes
D. Yes

Experimental Questions

E2. They are complementary to each other.

E4. Because normal cells contain two copies of chromosome 14, one would expect that a probe would bind to complementary DNA sequences on both of these chromosomes. If a probe recognized only one of two chromosomes, this means that one of the copies of chromosome 14 has been lost, or it has suffered a deletion in the region where the probe binds. With regard to cancer, the loss of this genetic material may be related to the uncontrollable cell growth.

E6. After the cells and chromosomes have been fixed to the slide, it is possible to add two or more different probes that recognize different sequences (i.e., different sites) within the genome. Each probe has a different fluorescence emission wavelength. Usually, a researcher will use computer imagery that recognizes the wavelength of each probe and then assigns that probe a bright color. The color seen by the researcher is not the actual color emitted by the probe; it is a secondary color assigned by the computer. In a sense, the probes, with the aid of a computer, are "painting" the regions of the chromosomes that are recognized by a probe. An example of chromosome painting is shown in Figure 20.3. In this example, human chromosome 5 is painted with six different colors.

E8. A contig is a collection of clones that contain overlapping segments of DNA that span a particular region of a chromosome. To determine if two clones are overlapping, one could conduct a Southern blotting experiment. In this approach, one of the clones is used as a probe. If it is overlapping with the second clone, it will bind to it in a Southern blot. Therefore, the second clone is run on a gel, and the first clone is used as a probe. If the band corresponding to the second clone is labeled, this means that the two clones are overlapping.

E10. BAC cloning vectors have the replication properties of a bacterial chromosome and the cloning properties of a plasmid. To replicate like a chromosome, the BAC vector contains an origin of replication from an F factor. Therefore, in a bacterial cell, a BAC can behave as a chromosome. Like a plasmid, BACs also contain selectable markers and convenient cloning sites for the insertion of large segments of DNA. The primary advantage is the ability to clone very large pieces of DNA.

E12. The resistance gene appears to be linked to RFLP 4B.

E14. For most organisms, it is usually easy to locate many RFLPs throughout the genome. The RFLPs can be used as molecular markers to make a map of the genome. This is done using the strategy described in Figure 20.6. To map a functional gene, one would also follow the same general strategy described in Figure 20.6, except that the two strains would also have an allelic difference in the gene of interest. The experimenter would make crosses, such as dihybrid crosses, and determine the number of parental and recombinant offspring based on the alleles and RFLPs they had inherited. If an allele and RFLP are linked, there will be a much lower percentage (i.e., less than 50%) of the recombinant offspring.

E16. *Deduced Outcome*

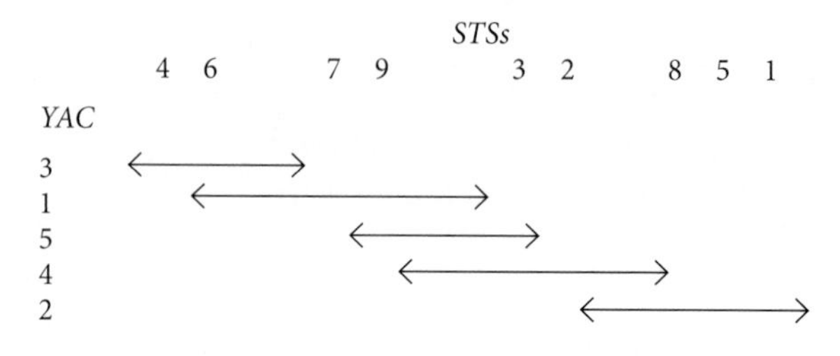

E18. A. One homolog contains the STS-1 that is 289 bp and STS-2 that is 422 bp, while the other homolog contains STS-1 that is 211 bp and STS-2 that is 115 bp. This is based on the observation that 28 of the sperm have either the 289 bp and 422 bp bands or the 211 bp and 115 bp bands.

B. There are two recombinant sperm; see lanes 12 and 18. Because there are two recombinant sperm out of a total of thirty,

$$\text{Map Distance} = \frac{2}{30} \times 100$$
$$= 6.7 \text{ mu}$$

C. In theory, this method could be used. However, there is not enough DNA in one sperm to carry out an RFLP analysis unless the DNA is amplified by PCR.

E20. One possibility is that the geneticist could try a different restriction enzyme. Perhaps there is sequence variation in the vicinity of the pesticide-resistance gene that affects the digestion pattern of a restriction enzyme other than *Eco*RI. There are hundreds of different restriction enzymes that recognize a myriad of different sequences.

Alternatively, the geneticist could give up on the RFLP approach and try to identify one or more sequence-tagged sites that are in the vicinity of the pesticide-resistance gene. In this case, the geneticist would want to identify STSs that are also microsatellites. As shown in Figure 20.9, the transmission of microsatellites can be followed in genetic crosses. Therefore, if the geneticist could identify microsatellites in the vicinity of the pesticide-resistance gene, this would make it possible to predict the outcome of crosses. For example, let's suppose a microsatellite linked to the pesticide-resistance gene existed in three forms: 234 bp, 255 bp, and 311 bp. And let's also suppose that the 234-bp form was linked to the high-resistance allele, the 255-bp form was linked to the moderate-resistance allele, and the 311-bp form was linked to the low-resistance allele. According to this hypothetical example, the geneticist could predict the level of resistance in an alfalfa plant by analyzing the inheritance of these microsatellites.

E22. When chromosomal DNA is isolated and digested with a restriction enzyme, this produces thousands of DNA fragments of different sizes. This makes it impossible to see any particular band on a gel if you simply stained the gel for DNA. Southern blotting allows you to detect one or more RFLPs that are complementary to the radioactive probe that is used.

E24. PFGE is a method of electrophoresis that is used to separate small chromosomes and large DNA fragments. The electrophoresis devices used in PFGE have two sets of electrodes. The two sets of electrodes produce alternating pulses of current, and this facilitates the separation of large DNA fragments.

It is important to handle the sample gently to prevent the breakage of the DNA due to mechanical forces. Cells are first embedded in agarose blocks, and then the blocks are loaded into the wells of the gel. The agarose keeps the sample very stable and prevents shear forces that might mechanically break the DNA. After the blocks are in the gel, the cells within the blocks are lysed, and, if desired, restriction enzymes can be added to digest the DNA. For PFGE, a restriction enzyme that cuts very infrequently might be used.

PFGE can be used as a preparative technique to isolate and purify individual chromosomes or large DNA fragments. PFGE can also be used, in conjunction with Southern blotting, as a mapping technique.

E26. Note that the insert of cosmid B is contained completely within the insert of cosmid C.

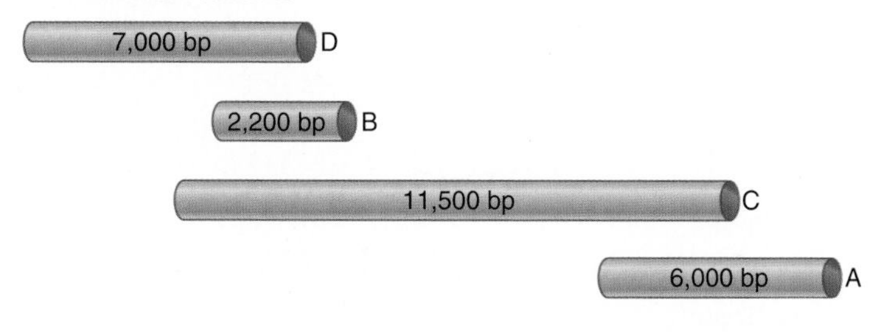

E28. A. The general strategy is shown in Figure 20.15. The researcher begins at a certain location and then walks toward the gene of interest. You begin with a clone that has a marker that is known to map relatively close to the gene of interest. A piece of DNA at the end of the insert is subcloned and then used in a Southern blot to identify an adjacent clone in a cosmid DNA library. This is the first "step." The end of this clone is subcloned to make the next step. And so on. Eventually, after many steps, you will arrive at your gene of interest.

B. In this example, you would begin at STS-3. If you walked a few steps and happened upon STS-2, you would know that you were walking in the wrong direction.

C. This is a difficult aspect of chromosome walking. Basically, you would walk toward gene *X* using DNA from a normal individual and DNA from an individual with a mutant gene *X*. When you have found a site where the sequences are different between the normal and mutant individual, you may have found gene *X*. You would eventually have to confirm this by analyzing the DNA sequence of this region and determining that it encodes a functional gene.

E30. We can calculate the probability that a base will not be sequenced using this approach with the following equation:

$$P = e^{-m}$$

where

P is the probability that a base will be left unsequenced

e is the base of the natural logarithm; $e = 2.72$

m is the number of bases sequenced divided by the total genome size

In this case, $m = 19$ divided by 4.4, which equals 4.3

$$P = e^{-m} = e^{-4.3} = 0.0136 = 1.36\%$$

This means that if we randomly sequence 19.0 Mb, we are likely to miss only 1.36% of the genome. With a genome size of 4.4 Mb, we would miss about 57,840 base pairs out of approximately 4,400,000.

E32. The overall advantage of next-generation sequencing is that a large amount of DNA sequence can be obtained in a short period of time, and at a lower cost. One common innovation is that the subcloning of fragments of DNA into vectors is no longer necessary. Also, some of the methods involve the parallel analysis of an enormous number of samples simultaneously.

Questions for Student Discussion/Collaboration

2. A molecular marker is a segment of DNA, not usually encoding a gene, that has a known location within a particular chromosome. It marks the location of a site along a chromosome. RFLPs and STSs are examples. It is easier to use these types of markers because they can be readily identified by molecular techniques such as restriction digestion analysis and PCR. The locations of functional genes are usually more difficult to use because this relies on conventional genetic mapping approaches whereby allelic differences in the gene are mapped by making crosses or following a pedigree. For monomorphic genes, this approach doesn't work.

CHAPTER 21

Conceptual Questions

C2. There are two main reasons why the proteome is larger than the genome. The first reason involves the processing of pre-mRNA, a phenomenon that occurs primarily in eukaryotic species. RNA splicing and editing can alter the codon sequence of mRNA and thereby produce alternative forms of proteins that have different amino acid sequences. The second reason for protein diversity is posttranslational modifications. There are many ways that a given protein's structure can be covalently modified by cellular enzymes. These include proteolytic processing, disulfide bond formation, glycosylation, attachment of lipids, phosphorylation, methylation, and acetylation, to name a few.

C4. Centromeric sequences, origins of replication, telomeric sequences, repetitive sequences, and enhancers. Other examples are possible.

C6. There are a few interesting trends. Sequences 1 and 2 are similar to each other, as are sequences 3 and 4. There are a few places where amino acid residues are conserved among all five sequences. These amino acids may be particularly important with regard to function.

C8. A. Correct

B. Correct

C. This is not correct. These are short genetic sequences that happen to be similar to each other. The *lac* operon and *trp* operon are not derived from the same primordial operon.

D. This is not correct. The two genes are homologous to each other. It is correct to say that their sequences are 60% identical.

Experimental Questions

E2. The cDNA labeled with a green dye is derived from mRNA obtained from cells at an early time point, when glucose levels were high. The other samples of cDNA were derived from cells collected at later time points, when glucose levels were falling and when the diauxic shift was occurring. These were labeled with a red dye. The green fluorescence provides a baseline for gene expression when glucose is high. At later time points, if the red : green ratio is high (i.e., greater than one), this means that a gene is induced as glucose levels fall, because there is more red cDNA compared to green cDNA. If the ratio is low (i.e., less than one), this means that a gene is being repressed.

E4. In the first dimension (i.e., in the tube gel), proteins migrate to the point in the gel where their net charge is zero. In the second dimension (i.e., the slab gel), proteins are coated with SDS and separated according to their molecular mass.

E6. In tandem mass spectroscopy, the first spectrometer determines the mass of a peptide fragment from a protein of interest. The second spectrometer determines the masses of progressively smaller fragments derived from that peptide. Because the masses of each amino acid are known, the molecular masses of these smaller fragments reveal the amino acid sequence of the peptide. With peptide sequence information, it is possible to use the genetic code and produce DNA sequences that could encode such a peptide. More than one sequence is possible due to the degeneracy of the genetic code. These sequences are used as query sequences to search a genomic database. This program will (hopefully) locate a match. The genomic sequence can then be analyzed to determine the entire coding sequence for the protein of interest.

E8. One strategy is a search by signal approach, which relies on known sequences such as promoters, start and stop codons, and splice sites to help predict whether or not a DNA sequence contains a structural gene. It attempts to identify a region that contains a promoter sequence, then a start codon, a coding sequence, and a stop codon. A second strategy is a search-by-content approach. The goal is to identify sequences whose nucleotide content differs significantly from a random distribution, which is usually due to codon bias. A search-by-content approach attempts to locate coding regions by identifying regions where the nucleotide content displays a bias. A third method to locate structural genes is to search for long open reading frames within a DNA sequence. An open reading frame is a sequence that does not contain any stop codons.

E10. By searching a database, one can identify genetic sequences that are homologous to a newly determined sequence. In most cases, homologous sequences carry out identical or very similar functions. Therefore, if one identifies a homologous member of a database whose function is already understood, this provides an important clue regarding the function of the newly determined sequence.

E12. The basis for secondary structure prediction is that certain amino acids tend to be found more frequently in α helices or β sheets. This information is derived from the statistical frequency of amino acids within secondary structures that have already been crystallized. Predictive methods are perhaps 60 to 70% accurate, which is not very good.

E14. The BACKTRANSLATE program works by using the genetic code. Each amino acid has one or more codons (i.e., three-base sequences) that are specified by the genetic code. This program would produce a sequence file that was a nucleotide base sequence. The BACKTRANSLATE program would produce a degenerate base sequence because the genetic code is degenerate. For example, lysine can be specified by AAA or AAG. The program would probably store a single file that had degeneracy at particular positions. For example, if the amino acid sequence was lysine–methionine–glycine–glutamine, the program would produce the following sequence:

```
5'-AA(A/G)ATGGG(T/C/A/G)CA(A/G)
```

The bases found in parentheses are the possible bases due to the degeneracy of the genetic code.

E16. A. To identify a specific transposable element, a program would use sequence recognition. The sequence of P elements is already known. The program would be supplied with this information and scan a sequence file looking for a match.

B. To identify a stop codon, a program would use sequence recognition. There are three stop codons that are specific three-base sequences. The program would be supplied with these three sequences and scan a sequence file to identify a perfect match.

C. To identify an inversion of any kind, a program would use pattern recognition. In this case, the program would be looking for a pattern in which the same sequence was running in opposite directions in a comparison of the two sequence files.

D. A search by signal approach uses both sequence recognition and pattern recognition as a means to identify genes. It looks for an organization of sequence elements that would form a functional gene. A search-by-content approach identifies genes based on patterns, not on specific sequence elements. This approach looks for a pattern in which the nucleotide content is different from a random distribution. The third approach to identify a gene is to scan a genetic sequence for long open reading frames. This approach is a combination of sequence recognition and pattern recognition. The program is looking for specific sequence elements (i.e., stop codons), but it is also looking for a pattern in which the stop codons are far apart.

E18. A. The amino acids that are most conserved (i.e., the same in all of the family members) are most likely to be important for structure and/or function. This is because a mutation that changed the amino acid might disrupt structure and function, and these kinds of mutations would be selected against. Completely conserved amino acids are found at the following positions: 101, 102, 105, 107, 108, 116, 117, 124, 130, 134, 139, 143, and 147.

B. The amino acids that are least conserved are probably not very important because changes in the amino acid does not seem to inhibit function. (If it did inhibit function, natural selection would eliminate such a mutation.) At one location, position 118, there are five different amino acids.

E20. This sequence is within the *lacY* gene of the *lac* operon of *E. coli*. It is found on page 588, nucleotides 801–850.

E22. A. This sequence has two regions that are about 20 amino acids long and very hydrophobic. Therefore, it is probable that this polypeptide has two transmembrane segments.

B.

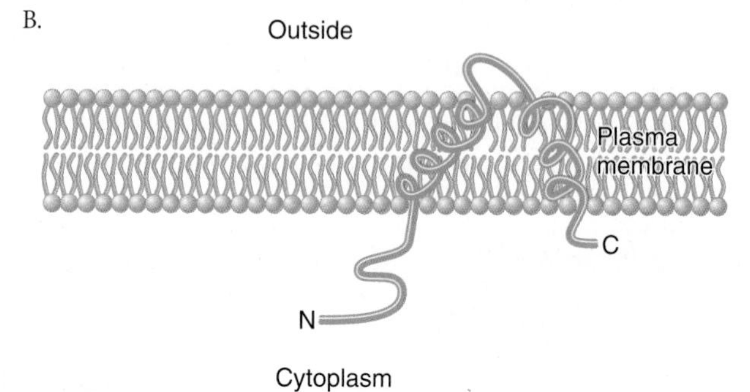

E24. A. True

B. False. The programs are only about 60 to 70% accurate.

C. True

Questions for Student Discussion/Collaboration

2. This is a very difficult question. In 20 years, we may have enough predictive information so that the structure of macromolecules can be predicted from their genetic sequences. If so, it would be better to be a mathematical theoretician with some genetics background. If not, it is probably better to be a biophysicist with some genetics background.

CHAPTER 22

Conceptual Questions

C2. When a disease-causing allele affects a trait, it is causing a deviation from normality, but the gene involved is not usually the only gene that governs the trait. For example, an allele causing hemophilia prevents the normal blood clotting pathway from operating correctly. It follows a simple Mendelian pattern because a single gene affects the phenotype. Even so, it is known that normal blood clotting is due to the actions of many genes.

C4. Changes in chromosome number and unbalanced changes in chromosome structure tend to affect phenotype because they create an imbalance of gene expression. For example, in Down syndrome, there are three copies of chromosome 21 and, therefore, three copies of all the genes on chromosome 21. This leads to a relative overexpression of genes that are located on chromosome 21 compared to the other chromosomes. Balanced translocations and inversions often are without phenotypic consequences because the total amount of genetic material is not altered, and the level of gene expression is not significantly changed.

C6. There are lots of possible answers; here are a few. Dwarfism occurs in people and dogs. Breeds like the dachshund and basset hound are types of dwarfism in dogs. There are diabetic people and mice. There are forms of inherited obesity in people and mice. Hip dysplasia is found in people and dogs.

C8. A. Because a person must inherit two defective copies of this gene and it is known to be on chromosome 1, the mode of transmission is autosomal recessive. Both members of this couple must be heterozygous, because they have one affected parent (who had to transmit the mutant allele to them) and their phenotypes are unaffected (so they must have received the normal allele from their other parent). Because both parents are heterozygotes, there is a $^1/_4$ chance of producing an affected child (a homozygote) with Gaucher disease. If we let *G* represent the nonmutant allele and *g* the mutant allele:

♀ \ ♂	*G*	*g*
G	*GG* Normal	*Gg* Normal
g	*Gg* Normal	*gg* Gaucher

B. From this Punnett square, we can also see that there is a $^1/_4$ chance of producing a homozygote with both normal copies of the gene.

C. We need to apply the binomial expansion equation to solve this problem (see Chapter 2 for a description of this equation). In this problem, $n = 5$, $x = 1$, $p = 0.25$, $q = 0.75$. The answer is 0.396, or 39.6%.

C10. The mode of transmission is autosomal recessive. All of the affected individuals do not have affected parents. Also, the disorder is found in both males and females. If it were X-linked recessive, individual III-1 would have to have an affected father, which she does not.

C12. The 13 babies have acquired a new mutation. In other words, during spermatogenesis or oogenesis, or after the egg was fertilized, a new mutation occurred in the fibroblast growth factor gene. These 13 individuals have the same chances of passing the mutant allele to their offspring as the 18 individuals who inherited the mutant allele from a parent. The chance is 50%.

C14. Because this is a dominant trait, the mother must have two normal copies of the gene, and the father (who is affected) is most likely to be a heterozygote. (Note: The father could be a homozygote, but this is extremely unlikely because the dominant allele is very rare.) If we let *M* represent the mutant Marfan allele and *m* the normal allele, the following Punnett square can be constructed:

♀ \ ♂	*M*	*m*
m	*Mm* Marfan	*mm* Normal
m	*Mm* Marfan	*mm* Normal

A. There is a 50% chance that this couple will have an affected child.

B. We use the product rule. The odds of having an unaffected child are 50%. So if we multiply $0.5 \times 0.5 \times 0.5$, this equals 0.125, or a 12.5% chance of having three unaffected offspring.

C16. A prion is a protein that behaves like an infectious agent. The infectious form of the prion protein has an abnormal conformation. This abnormal conformation is termed PrP^{Sc}, and the normal conformation of the protein is termed PrP^{C}. An individual can be "infected" with the abnormal conformation of the protein by eating products from another animal that had the disease, or the prion protein may convert spontaneously to the abnormal conformation. A prion protein in the PrP^{Sc} conformation can bind to a prion protein in the PrP^{C} conformation and convert it to the PrP^{Sc} form. An accumulation of prions in the PrP^{Sc} form is what causes the disease symptoms.

C18. A. Keep in mind that a conformational change is a stepwise process. It begins in one region of a protein and involves a series of small changes in protein structure. The entire conformational change from PrP^{C} to PrP^{Sc} probably involves many small changes in protein structure that occur in a stepwise manner. Perhaps the conformational change (from PrP^{C} to PrP^{Sc}) begins in the vicinity of position 178 and then proceeds throughout the rest of the protein. If there is a methionine at position 129, the complete conformational change (from PrP^{C} to PrP^{Sc}) can take place. However, perhaps the valine at position 129 somehow blocks one of the steps needed to complete the conformational change.

B. Once the PrP^{Sc} conformational change is completed, a PrP^{Sc} protein can bind to another prion protein in the PrP^{C} conformation. Perhaps it begins to convert it (to the PrP^{Sc} conformation) by initiating a small change in protein structure in the vicinity of position 178. The conversion would then proceed in a stepwise manner until the PrP^{Sc} conformation has been achieved. If a valine is at position 129, this could somehow inhibit one of the steps that are needed to complete the conformational change. If an individual had Val-129 in the polypeptide encoded by the second *PrP* gene, half of their prion proteins would be less sensitive to conversion by PrP^{Sc}, compared to individuals who had Met-129. This would explain why individuals with Val-129 in half of the prion proteins would have disease symptoms that would progress more slowly.

C20. A proto-oncogene is a normal cellular gene that typically plays a role in cell division. It can be altered by mutation to become an oncogene and thereby cause cancer. At the level of protein function, a proto-oncogene can become an oncogene by synthesizing too much of a protein or synthesizing the same amount of a protein that is abnormally active.

C22. The predisposition to develop cancer is inherited in a dominant fashion because the heterozygote has the higher predisposition. The mutant allele is actually recessive at the cellular level. But because we have so many cells in our bodies, it becomes relatively likely that a defective mutation will occur in the single normal gene and lead to a cancerous cell. Some heterozygous individuals may not develop the disease as a matter of chance. They may be lucky and not get a defective mutation in the normal gene, or perhaps their immune system is better at destroying cancerous cells once they arise.

C24. If an oncogene was inherited, it may cause uncontrollable cell growth at an early stage of development and thereby cause an embryo to develop improperly. This could lead to an early spontaneous abortion and thereby explain why we do not observe individuals with inherited oncogenes. Another possibility is that inherited oncogenes may adversely affect gamete survival, which would make it difficult for them to be passed from parent to offspring. A third possibility would be that oncogenes could affect the fertilized zygote in a way that would prevent the correct implantation in the uterus.

C26. The role of *p53* is to sense DNA damage and prevent damaged cells from proliferating. Perhaps, prior to birth, the fetus is in a protected environment, so DNA damage may be minimal. In other words, the fetus may not really need *p53*. After birth, agents such as UV light may cause DNA damage. At this point, *p53* is important. A *p53* knockout is more sensitive to UV light because it cannot repair its DNA properly in response to this DNA-damaging agent, and it cannot kill cells that have become irreversibly damaged.

C28. The p53 protein is a regulatory transcription factor; it binds to DNA and influences the transcription rate of nearby genes. This transcription factor (1) activates genes that promote DNA repair; (2) activates a protein that inhibits cyclin/CDK protein complexes that are required for cell division; and (3) activates genes that promote apoptosis. If a cell is exposed to DNA damage, it has a greater potential to become malignant. Therefore, an organism wants to avoid the proliferation of such a cell. When exposed to an agent that causes DNA damage, a cell will try to repair the damage. However, if the damage is too extensive, the p53 protein will stop the cell from dividing and program it to die. This helps to prevent the proliferation of cancer cells in the body.

Experimental Questions

E2. Perhaps the least convincing is the higher incidence of the disease in particular populations. Because populations living in specific geographic locations are exposed to their own unique environment, it is difficult to distinguish genetic versus environmental causes for a particular disease. The most convincing evidence might be the higher incidence of a disease in related individuals and/or the ability to correlate a disease with the presence of a mutant gene. Overall, however, the reliability that a disease has a genetic component should be based on as many observations as possible.

E4. You would probably conclude that it is less likely to have a genetic component. If it were rooted primarily in genetics, it would be likely to be found in the Central American population. Of course, there is a chance that very few or none of the people who migrated to Central America were carriers of the mutant gene, which is somewhat unlikely for a large migrating population. By comparison, one might suspect that an environmental agent present in South America but not present in Central America may underlie the disease. Researchers could try to search for this environmental agent (e.g., a pathogenic organism).

E6. Males I-1, II-1, II-4, II-6, III-3, III-8, and IV-5 have a normal copy of the gene. Males II-3, III-2, and IV-4 are hemizygous for an inactive mutant allele. Females III-4, III-6, IV-1, IV-2, and IV-3 have two normal copies of the gene, whereas females I-2, II-2, II-5, III-1, III-5, and III-7 are heterozygous carriers of a mutant allele.

E8. A transformed cell is one that has become malignant. In a laboratory, this can be done in three ways. First, the cells could be treated with a mutagen that would convert a proto-oncogene into an oncogene. Second, cells could be exposed to the DNA from a malignant cell line. Under the appropriate conditions, this DNA can be taken up by the cells and integrated into their genome so that they become malignant. A third way to transform cells is by exposure to an oncogenic virus.

E10. By comparing oncogenic viruses with strains that have lost their oncogenicity, researchers have been able to identify particular genes that cause cancer. This has led to the identification of many oncogenes. From this work, researchers have also learned that normal cells contain proto-oncogenes that usually play a role in cell division. This suggests that oncogenes exert their effects by upsetting the cell division process. In particular, it appears that oncogenes are abnormally active and keep the cell division cycle in a permanent "on" position.

E12. One possible category of drugs would be GDP analogues (i.e., compounds that resemble the structure of GDP). Perhaps one could find a GDP analogue that binds to the Ras protein and locks it in the inactive conformation.

One way to test the efficacy of such a drug would be to incubate the drug with a type of cancer cell that is known to have an overactive Ras protein and then plate the cells on solid media. If the drug locked the Ras protein in the inactive conformation, it should inhibit the formation of malignant growth or malignant foci.

There are possible side effects of such drugs. First, they might block the growth of normal cells, because Ras protein plays a role in normal cell proliferation. Second (if you have taken a cell biology course), there are many GTP/GDP-binding proteins in cells, and the drugs could somehow inhibit cell growth and function by interacting with these proteins.

Questions for Student Discussion/Collaboration

2. There isn't a clearly correct answer to this question, but it should stimulate a large amount of discussion.

CHAPTER 23

Conceptual Questions

C2. A. False; the head is anterior to the tail.

B. True

C. False; the feet are ventral to the hips.

D. True

C4. A. True

B. False; because gradients are also established after fertilization during embryonic development.

C. True

C6. A. This is a mutation in a pair-rule gene (*runt*).

B. This is a mutation in a gap gene (*knirps*).

C. This is a mutation in a segment-polarity gene (*patched*).

C8. Positional information refers to the phenomenon whereby the spatial locations of morphogens and cell adhesion molecules (CAMs) provide a cell with information regarding its position relative to other cells. In *Drosophila*, the formation of a segmented body pattern relies initially on the spatial location of maternal gene products. These gene products lead to the sequential activation of the segmentation genes. Positional information can come from morphogens that are found within the oocyte, from morphogens secreted from cells during development, and from cell-to-cell contact. Although all three are important, morphogens in the oocyte have the greatest impact on the overall body structure.

C10. The anterior portion of the antero-posterior axis is established by the action of Bicoid. During oogenesis, the mRNA for Bicoid enters the anterior end of the oocyte and is sequestered there to establish an anterior (high) to posterior (low) gradient. Later, when the mRNA is translated, the Bicoid protein in the anterior region establishes a genetic hierarchy that leads to the formation of anterior structures. If Bicoid was not trapped in the anterior end, it is likely that anterior structures would not form.

C12. Maternal effect gene products influence the formation of the main body axes, including the antero-posterior, dorso-ventral, and terminal regions. They are expressed during oogenesis and needed very early in development. Zygotic genes, particularly the three classes of the segmentation genes, are necessary after the axes have been established. The segmentation genes are expressed after fertilization.

C14. The coding sequence of homeotic genes contains a 180-bp consensus sequence known as a homeobox. The protein domain encoded by the homeobox is called a homeodomain. The homeodomain contains three conserved sequences that are folded into α-helical conformations. The arrangement of these α helices promotes the binding of the protein to the major groove of the DNA. Helix III is called the recognition helix because it recognizes a particular nucleotide sequence within the major groove. In this way, homeotic proteins are able to bind to DNA in a sequence-specific manner and thereby activate particular genes.

C16. It would normally be expressed in the three thoracic segments that have legs (T1, T2, and T3).

C18. A. When a mutation inactivates a gap gene, a contiguous section of the larva is missing.

B. When a mutation inactivates a pair-rule gene, some regions that are derived from alternating parasegments are missing.

C. When a mutation inactivates a segment-polarity gene, portions are missing at either the anterior or posterior end of the segments.

C20. Proper development in mammals is likely to require the products of maternal effect genes that play a key role in initiating embryonic development. The adult body plan is merely an expansion of the embryonic body plan, which is established in the oocyte. Because the starting point for the development of an embryo is the oocyte, this explains why an enucleated oocyte is needed to clone mammals.

C22. A heterochronic mutation is one that alters the timing when a gene involved in development is normally expressed. The gene may be expressed too early or too late, which causes certain cell lineages to be out of sync with the rest of the animal. If a heterochronic mutation affected the intestine, the animal may end up with too many intestinal cells if it is a gain-of-function mutation or too few if it is a loss-of-function mutation. In either case, the effects might be detrimental because the growth of the intestine must be coordinated with the growth of the rest of the animal.

C24. Cell differentiation is the specialization of a cell into a particular cell type. In the case of skeletal muscle cells, the bHLH proteins play a key role in the initiation of cell differentiation. When bHLH proteins are activated, they are able to bind to enhancers and activate the expression of many different muscle-specific genes. In this way, myogenic bHLH proteins turn on the expression of many muscle-specific proteins. When these proteins are synthesized, they change the characteristics of the cell into those of a muscle cell. Myogenic bHLH proteins are regulated by dimerization. When a heterodimer forms between a myogenic bHLH protein and an E protein, it activates gene expression. However, when a heterodimer forms between myogenic bHLH proteins and a protein called Id, the heterodimer is unable to bind to DNA. The Id protein is produced during early stages of development and prevents myogenic bHLH proteins from promoting muscle differentiation too soon. At later stages of development, the amount of Id protein falls, and myogenic bHLH proteins can combine with E proteins to induce muscle differentiation.

C26. A totipotent cell is a cell that has the potential to create a complete organism.

A. In humans, a fertilized egg is totipotent, and the cells during the first few embryonic divisions are totipotent. However, after several divisions, embryonic cells lose their totipotency and, instead, are determined to become particular tissues within the body.

B. In plants, many living cells are totipotent.

C. Because yeast are unicellular, one cell is a complete individual. Therefore, yeast cells are totipotent; they can produce new individuals by cell division.

D. Because bacteria are unicellular, one cell is a complete individual. Therefore, bacteria are totipotent; they can produce new individuals by cell division.

C28. Animals begin their development from an egg and then form antero-posterior and dorso-ventral axes. The formation of an adult organism is an expansion of the embryonic body plan. Plants grow primarily from two meristems: shoot and root meristems. At the cellular level, plant development is different in that it does not involve cell migration, and most plant cells are totipotent. Animals require the organization within an oocyte to begin development. At the genetic level, however, animal and plant development are similar in that a genetic hierarchy of transcription factors governs pattern formation and cell specialization.

Experimental Questions

E2. *Drosophila* has an advantage in that researchers have identified many more mutant alleles that alter development in specific ways. The hierarchy of gene regulation is particularly well understood in the fruit fly. *C. elegans* has the advantage of simplicity and a complete knowledge of cell fate. This enables researchers to explore how the timing of gene expression is critical to the developmental process.

E4. To determine that a mutation is affecting the timing of developmental decisions, a researcher needs to know the normal time or stage of development when cells are supposed to divide and what types of cells will be produced. A lineage diagram provides this. With this information, one can then determine if particular mutations alter the timing when cell division occurs.

E6. Mutant 1 is a gain-of-function allele; it keeps reiterating the L1 pattern of division. Mutant 2 is a loss-of-function allele; it skips the L1 pattern and immediately follows an L2 pattern.

E8. As discussed in Chapter 15, most eukaryotic genes have a core promoter that is adjacent to the coding sequence; regulatory elements that control the transcription rate at the promoter are typically upstream from the core promoter. Therefore, to get the *Antp* gene product expressed where the *abd-A* gene product is normally expressed, you would link the upstream genetic regulatory region of the *abd-A* gene to the coding sequence of the *Antp* gene. This construct would be inserted into the middle of a P element (see next). The construct shown here would then be introduced into an embryo by P element transformation.

P element	*abd-A* regulatory region	/	*Antp* coding sequence	P element

The *Antp* gene product is normally expressed in the thoracic region and produces segments with legs, as illustrated in Figure 23.12. Therefore, because the *abd-A* gene product is normally expressed in the anterior abdominal segments, one might predict that the genetic construct shown above would produce a fly with legs attached to the segments that are supposed to be the anterior abdominal segments. In other words, the anterior abdominal segments might resemble thoracic segments with legs.

E10. A. The female flies must have had mothers that were heterozygous for a (dominant) normal allele and the mutant allele. Their fathers were either homozygous for the mutant allele or heterozygous. The female flies inherited a mutant allele from both their father and mother. Nevertheless, because their mother was heterozygous for the normal (dominant) allele and mutant allele, and because this is a maternal effect gene, their phenotype is based on the genotype of their mother. The normal allele is dominant, so they have a normal phenotype.

B. *Bicoid-A* appears to have a deletion that removes part of the sequence of the gene and thereby results in a shorter mRNA. *Bicoid-B* could also have a deletion that removes all of the sequence of the *bicoid* gene, or it could have a promoter mutation that prevents the expression of the bicoid gene. *Bicoid-C* seems to have a point mutation that does not affect the amount of the bicoid mRNA.

C. With regard to function, all three mutations are known to be loss-of-function mutations. *Bicoid-A* probably eliminates function by truncating the Bicoid protein. The Bicoid protein is a transcription factor. The *bicoid-A* mutation probably shortens this protein and thereby inhibits its function. The *bicoid-B* mutation prevents expression of the bicoid mRNA. Therefore, none of the Bicoid protein would be made, and this would explain the loss of function. The *bicoid-C* mutation seems to prevent the proper localization of the *bicoid* mRNA in the oocyte. There must be proteins within the oocyte that recognize specific sequences in the *bicoid* mRNA and trap it in the anterior end of the oocyte. This mutation must change these sequences and prevent these proteins from recognizing the bicoid mRNA.

E12. An egg-laying defect is somehow related to an abnormal anatomy. The *n540* strain has fewer neurons compared to a normal worm. Perhaps the *n540* strain is unable to lay eggs because it is missing neurons that are needed for egg laying. The *n536* and *n355* strains have an abnormal abundance of neurons. Perhaps this overabundance also interferes with the proper neural signals needed for egg laying.

E14. Geneticists interested in mammalian development have used reverse genetics because it has been difficult for them to identify mutations in developmental genes based on phenotypic effects in the embryo. This is because it is difficult to screen a large number of mammalian embryos in search of abnormal ones that carry mutant genes. It is easy to have thousands of flies in a laboratory, but it is not easy to have thousands of mice. Instead, it is easier to clone the normal gene based on its homology to invertebrate genes and then make mutations in vitro. These mutations can be introduced into a mouse to create a gene knockout. This strategy is opposite to that of Mendel, who characterized genes by first identifying phenotypic variants (e.g., tall versus dwarf, green seeds versus yellow seeds, etc.).

Questions for Student Discussion/Collaboration

2. In this problem, the students should try to make a flow diagram that begins with maternal effect genes, then gap genes, pair-rule genes, and segment polarity genes. These genes then lead to homeotic genes and finally genes that promote cell differentiation. It's almost impossible to make an accurate flow diagram because there are so many gene interactions, but it is instructive to think about developmental genetics in this way. It is probably easier to identify mutant phenotypes that affect later stages of development because they are less likely to be lethal. However, modern methods can screen for conditional mutants as described in solved problem S3. To identify all of the genes necessary for chicken development, you may begin with early genes, but this assumes you have some way to identify them. If they had been identified, you would then try to identify the genes that they stimulate or repress. This could be done using molecular methods described in Chapters 14, 15, and 18.

CHAPTER 24

Conceptual Questions

C2. A population is a group of interbreeding individuals. Let's consider a squirrel population in a forested area. Over the course of many generations, several things could happen to this population. A forest fire, for example, could dramatically decrease the number of individuals and thereby cause a bottleneck. This would decrease the genetic diversity of the population. A new predator may enter the region and natural selection may select for the survival of squirrels that are best able to evade the predator. Another possibility is that a group of squirrels within the population may migrate to a new region and found a new squirrel population.

C4. A. Phenotype frequency and genotype frequency

B. Genotype frequency

C. Allele frequency

C6. A. The genotype frequency for the *CF* homozygote is $^1/_{2500}$, or 0.004. This would equal q^2. The allele frequency is the square root of this value, which equals 0.02. The frequency of the corresponding normal allele equals $1 - 0.02 = 0.98$.

B. The frequency for the *CF* homozygote is 0.004; for the unaffected homozygote, $(098)^2 = 0.96$; and for the heterozygote, 2(0.98)(0.02), which equals 0.039.

C. If a person is known to be a heterozygous carrier, the chances that this particular person will happen to choose another as a mate is equal to the frequency of heterozygous carriers in the population, which equals 0.039, or 3.9%. The chances that two randomly chosen individuals will choose each other as mates equals $0.039 \times 0.039 = 0.0015$, or 0.15%.

C8. For two alleles, the heterozygote is at a maximum when they are 0.5 each. For three alleles, the two heterozygotes are at a maximum when each allele is 0.33.

C10. Because this is a recessive trait, only the homozygotes for the rolling allele will be able to roll their tongues. If p equals the rolling allele and q equals the nonrolling allele, the Hardy-Weinberg equation predicts that the frequency of homozygotes who can roll their tongues would be p^2. In this case, $p^2 = (0.6)^2 = 0.36$, or 36%.

C12. A. The inbreeding coefficient is calculated using the formula

$$F = \Sigma\, (^1/_2)^n\, (1 + F_A)$$

In this case, there is one common ancestor, I-2. Because we have no prior history on I-2, we assume she is not inbred, which makes $F_A = 0$. The inbreeding loop for IV-3 contains five people, III-4, II-2, I-2, II-5, and III-5. Therefore, $n = 5$.

$$F = (^1/_2)^5(1 + 0) = {}^1/_{32} = 0.031$$

B. Based on the data shown in this pedigree, individual IV-4 is not inbred.

C14. Migration, genetic drift, and natural selection are the driving forces that alter allele frequencies within a population. Natural selection acts to eliminate harmful alleles and promote beneficial alleles. Genetic drift involves random changes in allele frequencies that may eventually lead to elimination or fixation of alleles. It is thought to be important in the establishment of neutral alleles in a population. Migration is important because it introduces new alleles into neighboring populations. According to the neutral theory, genetic drift is largely responsible for the variation seen in natural populations.

C16. A random process is one that alters allele frequencies without any regard to whether the changes are beneficial or not. Genetic drift and migration are the two main ways this can occur. An example is the founder effect, when a group of individuals migrate to a new location such as an island. Adaptive processes increase the reproductive success of a species. Natural selection is the adaptive process that tends to eliminate harmful alleles from a population and increase the frequency of beneficial alleles. An example would be the long neck of the giraffe, which enables it to feed in tall trees. At the molecular level, beneficial mutations may alter the coding sequence of a gene and change the structure and function of the protein in a way that is beneficial. For example, the sickle cell disease allele alters the structure of hemoglobin, and in the heterozygous condition, this inhibits the sensitivity of red blood cells to the malaria pathogen.

C18. Genetic drift is due to sampling error, and the degree of sampling error depends on the population size. In small populations, the relative proportion of sampling error is much larger. If genetic drift is moving an allele toward fixation, it will take longer in a large population because the degree of sampling error is much smaller.

C20. During the bottleneck effect, allele frequencies are dramatically altered due to genetic drift. In extreme cases, some alleles are lost, while others may become fixed at 100%. The overall effect is to decrease genetic diversity within the population. This may make it more difficult for the species to respond in a positive way to changes in the environment. Species that are approaching extinction also face a bottleneck as their numbers decrease. The loss of genetic diversity may make it even more difficult for the species to rebound.

C22. In all cases, these forms of natural selection favor one or more phenotypes because such phenotypes have a reproductive advantage. However, the patterns differ with regard to whether a single phenotype or multiple phenotypes are favored and whether the favored phenotype is in the middle of the phenotypic range or at one or both extremes. Directional selection favors one phenotype at a phenotypic extreme. Over time, natural selection is expected to favor the fixation of alleles that cause these phenotypic characteristics. Disruptive selection favors two or more phenotypic categories. It will lead to a population with a balanced polymorphism for the trait. Examples of balancing selection are heterozygote advantage and negative-frequency dependent selection. These promote a stable polymorphism in a population. Stabilizing selection favors individuals with intermediate phenotypes. It tends to decrease genetic diversity because alleles that favor extreme phenotypes are eliminated.

C24. The intuitive meaning of the mean fitness of a population is the average likelihood that members of a population will reproduce. If the mean fitness is high, it is likely that an average member will survive and produce offspring. Natural selection increases the mean fitness of a population.

C26. A. True

B. True

C. False; it causes allele loss or fixation, which results in less diversity.

D. True

C28. A. Disruptive. There are multiple environments that favor different phenotypes.

B. Directional. The thicker the fur, the more likely that survival will occur.

C. Stabilizing. Low birth weight is selected against because it results in low survival. Also, very high birth weight is selected against because it could cause problems in delivery, which also could decrease the survival rate.

D. Directional. Sturdy stems and leaves will promote survival in windy climates.

Experimental Questions

E2. Solved problem S4 shows how the Hardy-Weinberg equation can be modified to include situations of three or more alleles. In this case

$(p + q + r + s)^2 = 1$

$p^2 + q^2 + r^2 + s^2 + 2pq + 2qr + 2qs + 2rp + 2rs + 2sp = 1$

Let $p = C$, $q = c^{ch}$, $r = c^h$, and $s = c$.

A. The frequency of albino rabbits is s^2:

$s^2 = (0.05) = 0.0025 = 0.25\%$

B. Himalayan is dominant to albino but recessive to full and chinchilla. Therefore, Himalayan rabbits would be represented by r^2 and by $2rs$:

$r^2 + 2rs = (0.44)^2 + 2(0.44)(0.05) = 0.24 = 24\%$

Among 1000 rabbits, about 240 would have a Himalayan coat color.

C. Chinchilla is dominant to Himalayan and albino but recessive to full coat color. Therefore, heterozygotes with chinchilla coat color would be represented by $2qr$ and by $2qs$.

$2qr + 2qs = 2(0.17)(0.44) + 2(0.17)(0.05) = 0.17$, or 17%

Among 1000 rabbits, about 170 would be heterozygotes with chinchilla fur.

E4. A.

Inuit	$M = 0.913$	$N = 0.087$
Navajo	$M = 0.917$	$N = 0.083$
Finns	$M = 0.673$	$N = 0.327$
Russians	$M = 0.619$	$N = 0.381$
Aborigines	$M = 0.176$	$N = 0.824$

B. To determine if these populations are in equilibrium, we can use the Hardy-Weinberg equation and calculate the expected number of individuals with each genotype. For example

Inuit $MM = (0.913)^2 = 0.833 = 83.3\%$

$MN = 2(0.913)(0.087) = 0.159 = 15.9\%$

$NN = (0.087)^2 = 0.0076 = 0.76\%$

In general, the values agree pretty well with an equilibrium. The same is true for the other four populations.

C. Based on similar allele frequencies, the Inuit and Navajo Indians seem to have interbred, as well as the Finns and Russians.

E6. A. $\Delta p_C = m(p_D - p_R)$

With regard to the sickle cell allele,

$\Delta p_C = (550/10{,}550)(0.1 - 0.01) = 0.0047$

$p_C = p_R + \Delta p_C = 0.01 + 0.0047 = 0.0147$

B. We need to calculate the genotypes separately:

For the 550 migrating individuals,

$Hb^AHb^A = (0.9)2 = 0.81$, or 81%
We expect $(0.81)550 = 445.5$ individuals to have this genotype.

$Hb^AHb^S = 2(0.9)(0.1) = 0.18$
We expect $(0.18)550 = 99$ heterozygotes.

$Hb^SHb^S = (0.1)^2 = 0.01$
We expect $(0.01)550 = 5.5$ Hb^SHb^S.

For the original recipient population,

$Hb^AHb^A = (0.99)^2 = 0.98$
We expect 9801 individuals to have this genotype.

$Hb^AHb^S = 2(0.99)(0.01) = 0.0198$
We expect 198 with this genotype.

$Hb^SHb^S = (0.01)^2 = 0.0001$
We expect 1 with this genotype.

To calculate the overall population,

$(445.5 + 9801)/10{,}550 = 0.971$ Hb^AHb^A homozygotes

$(99 + 198)/10{,}550 = 0.028$ heterozygotes

$(5.5 + 1)/10{,}550 = 0.00062$ Hb^SHb^S homozygotes

C. After one round of mating, the allele frequencies in the conglomerate (calculated in part A), should yield the expected genotype frequencies according to the Hardy-Weinberg equilibrium.

Allele frequency of $Hb^S = 0.0147$, so $Hb^A = 0.985$.

$Hb^AHb^A = (0.985)^2 = 0.97$

$Hb^AHb^S = 2(0.985)(0.0147) = 0.029$

$Hb^SHb^S = (0.0147)^2 = 0.0002$

E8. A. Probability of fixation $= {}^1/_2N$ (Assuming equal numbers of males and females contributing to the next generation)

Probability of fixation $= {}^1/_2(2{,}000{,}000)$

$= 1$ in 4,000,000 chance

B. $\bar{t} = 4N$

Where $\bar{t}$ = the average number of generations to achieve fixation

N = the number of individuals in population, assuming that males and females contribute equally to each succeeding generation

$\bar{t} = 4(2 \text{ million}) = 8$ million generations

C. If the blue allele had a selective advantage, the value calculated in part A would be slightly larger; there would be a higher chance of allele fixation. The value calculated in part B would be smaller; it would take a shorter period of time to reach fixation.

E10. You could mark snails with a dye and release equal numbers of dark and light snails into dimly lit forested regions and sunny fields. At a later time, recapture the snails and count them. It would be important to have a method of unbiased recapture because the experimenter would have an easier time locating the light snails in a forest and the dark snails in a field. Perhaps one could bait the region with something that the snails like to eat and only collect snails that are at the bait. In addition to this type of experiment, one could also observe predation as it occurs.

E12. Male 2 is the potential father, because he contains the bands found in the offspring but not found in the mother. To calculate the probability, one would have to know the probability of having each of the types of bands that match. In this case, for example, male 2 and the offspring have four bands in common. As a simple calculation, we could eliminate the four bands the offspring shares with the mother. If the probability of having each paternal band is $^1/_4$, the chances that this person is not the father are $(^1/_4)^4$.

E14. This percentage is not too high. Based on their genetic relationship, we expect that a father and daughter must share at least 50% of the same bands in a DNA fingerprint. However, the value can be higher than that because the mother and father may have some bands in common, even though they are not genetically related. For example, at one site in the genome, the father may be heterozygous for a 4100-bp and 5200-bp minisatellite, and the mother may also be heterozygous in this same region and have 4100-bp and 4700-bp minisatellites. The father could pass the 5200-bp band to his daughter, and the mother could pass the 4100-bp band. Thus, the daughter would inherit the 4100-bp and 5200-bp bands. This would be a perfect match to both of the father's bands, even though the father transmitted only the 5200-bp band to his daughter. The 4100-bp band matches because the father and mother happened to have a minisatellite in common. Therefore, the 50% estimate of matching bands in a DNA fingerprint based on genetic relationships is a minimum estimate. The value can be higher than that.

Questions for Student Discussion/Collaboration

2. Mutation is responsible for creating new alleles, but the rate of new mutations is so low that it cannot explain allele frequencies in this range. Let's call the two alleles *B* and *b* and assume that *B* was the original allele and *b* is a more recent allele that arose as a result of mutation. Three scenarios to explain the allele frequencies:

   1. The *b* allele is neutral and reached its present frequency by genetic drift. It hasn't reached elimination or fixation yet.
   2. The *b* allele is beneficial, and its frequency is increasing due to natural selection. However, there hasn't been enough time to reach fixation.
   3. The *Bb* heterozygote is at a selective advantage, leading to a balanced polymorphism.

CHAPTER 25

Conceptual Questions

C2. At the molecular level, quantitative traits often exhibit a continuum of phenotypic variation because they are usually influenced by multiple genes that exist as multiple alleles. A large amount of environmental variation will also increase the overlap between genotypes and phenotypes for polygenic traits.

C4. A discontinuous trait is one that falls into discrete categories. Examples include brown eyes versus blue eyes in humans and purple versus white flowers in pea plants. A continuous trait is one that does not fall into discrete categories. Examples include height in humans and fruit weight in tomatoes. Most quantitative traits are continuous; the trait falls within a range of values. The reason why quantitative traits are continuous is because they are usually polygenic and greatly influenced by the environment. As shown in Figure 25.4b, this tends to create ambiguities between genotypes and a continuum of phenotypes.

C6. To be in the top 2.5% is about two standard deviation units. If we take the square root of the variance, the standard deviation would be 22 lb. To be in the top 2.5%, an animal would have to weigh at least 44 lb heavier than the mean, which equals 606 lb. To be in the bottom 0.13%, an animal would have to be three standard deviations lighter, which would be at least 66 lb lighter than the mean, or 496 lb.

C8. There is a positive correlation, but it could have occurred as a matter of chance alone. According to Table 25.2, this value could have occurred through random sampling error. You would need to conduct more experimentation to determine if there is a significant correlation, such as examining a greater number of pairs of individuals. If $N = 500$, the correlation would be statistically significant, and you would conclude that the correlation did not occur as a matter of random chance. However, you could not conclude cause and effect.

C10. When a correlation coefficient is statistically significant, it means that the association is likely to have occurred for reasons other than random sampling error. It may indicate cause and effect but not necessarily. For example, large parents may have large offspring due to genetics (cause and effect). However, the correlation may be related to the sharing of similar environments rather than cause and effect.

C12. Quantitative trait loci are sites within chromosomes that contain genes that affect a quantitative trait. It is possible for a QTL to contain one gene, or it may contain two or more closely linked genes. QTL mapping, which involves linkage to known molecular markers, is commonly used to determine the locations of QTLs.

C14. If the broad sense heritability equals 1.0, it means that all of the variation in the population is due to genetic variation rather than environmental variation. It does not mean that the environment is unimportant in the outcome of the trait. Under another set of environmental conditions, the trait may have turned out quite differently.

C16. Hybrid vigor is the phenomenon in which an offspring produced from two inbred strains is more vigorous than the corresponding parents. Tomatoes and corn are often the products of hybrids.

C18. When a species is subjected to selective breeding, the breeder is focusing his/her attention on improving one particular trait. In this case, the rose breeder is focused on the size and quality of the flowers. Because the breeder usually selects a small number of individuals (e.g., the ones with best flowers) as the breeding stock for the next generation, this may lead to a decrease in the allelic diversity at other genes. For example, several genes affect flower fragrance. In an unselected population, these genes may exist as "fragrant alleles" and "nonfragrant alleles." After many generations of breeding for large flowers, the fragrant alleles may be lost from the population, just as a matter of random chance. This is a common problem of selective breeding. As you select for an improvement in one trait, you may inadvertently diminish the quality of an unselected trait.

Others have suggested that the lack of fragrance may be related to flower structure and function. Perhaps the amount of energy that a flower uses to make beautiful petals somehow diminishes its capacity to make fragrance.

C20. Broad-sense heritability takes into account all genetic factors that affect the phenotypic variation in a trait. Narrow-sense heritability considers only alleles that behave in an additive fashion. In many cases, the alleles affecting quantitative traits appear to behave additively. More importantly, if a breeder assumes that the heritability of a trait is due to the additive effects of alleles, it is possible to predict the outcome of selective breeding. This is also termed the *realized heritability.*

C22. A. Because of their good nutrition, you may speculate that they would grow to be taller.

B. If the environment is rather homogeneous, then heritability values tend to be higher because the environment contributes less to the amount of variation in the trait. Therefore, in the commune, the heritability might be higher, because they uniformly practice good nutrition. On the other hand, because the commune is a smaller size than the general population, the amount of genetic variation might be less, so this would make the heritability lower. However, because the problem states that the commune population is large, we would probably assume that the amount of genetic variation is similar to that in the general population. Overall, the best guess would be that the heritability in the commune population is higher because of the uniform nutrition standards.

C. As stated in part B, the amount of variation would probably be similar, because the commune population is large. As a general answer, larger populations tend to have more genetic variation. Therefore, the general population probably has a bit more variation.

C24. A natural population of animals is more likely to have a higher genetic diversity compared to a domesticated population. This is because domesticated populations have been subjected to many generations of selective breeding, which decreases the genetic diversity. Therefore, V_G is likely to be higher for the natural population. The other issue is the environment. It is difficult to say which group would have a more homogeneous environment. In general, natural populations tend to have a more heterogeneous environment, but not always. If the environment is more heterogeneous, this tends to cause more phenotypic variation, which makes V_E higher.

Heritability $= V_G/V_T$

$= V_G/(V_G + V_E)$

When V_G is high, heritability increases. When V_E is high, heritability decreases. In the natural wolf population, we would expect that V_G would be high. In addition, we would guess that V_E might be high as well (but that is less certain). Nevertheless, if this were the case, the heritability of the wolf population might be similar to the domestic population. This is because the high V_G in the wolf population is balanced by its high V_E. On the other hand, if V_E is not that high in the wolf population, or if it is fairly high in the domestic population, then the wolf population would have a higher heritability for this trait.

Experimental Questions

E2. To calculate the mean, we add the values together and divide by the total number.

$$\text{Mean} = \frac{1.9 + 2(2.4) + 2(2.1) + 3(2.0) + 2(2.2) + 1.7 + 1.8 + 2(2.3) + 1.6}{15}$$

Mean = 2.1

The variance is the sum of the squared deviations from the mean divided by $N - 1$. The mean value of 2.1 must be subtracted from each value, and then the square is taken. These 15 values are added together and then divided by 14 (which is $N - 1$).

$$\text{Variance} = \frac{0.85}{14}$$

$= 0.061$

The standard deviation is the square root of the variance.

Standard deviation = 0.25

E4. The results are consistent with the idea that there are QTLs for this trait on chromosomes 2 and 3 but not on the X chromosome.

E6. When we say an RFLP is associated with a trait, we mean that a gene that influences a trait is closely linked to an RFLP. At the chromosomal level, the gene of interest is so closely linked to the RFLP that a crossover almost never occurs between them.

Note: Each plant inherits four RFLPs, but it may be homozygous for one or two of them.

Small: 2700 and 4000 (homozygous for both)

Small–medium: 2700 (homozygous), 3000, and 4000; or 2000, 2700, and 4000 (homozygous)

Medium: 2000 and 4000 (homozygous for both); or 2700 and 3000 (homozygous for both); or 2000, 2700, 3000, and 4000

Medium–large: 2000 (homozygous), 3000, and 4000; or 2000, 2700, and 3000 (homozygous)

Large: 2000 and 3000 (homozygous for both)

E8. Let's assume there is an extensive molecular marker map for the rice genome. We would begin with two strains of rice, one with a high yield and one with a low yield, that greatly differ with regard to the molecular markers they carry. We would make a cross between these two strains to get F_1 hybrids. We would then backcross the F_1 hybrids to either of the parental strains and then examine hundreds of offspring with regard to their rice yields and molecular markers. In this case, our expected results would be that six different markers in the high-producing strain would be correlated with offspring that produce higher yields. We might get fewer than six bands if some of these genes are closely linked and associate with the same marker. We also might get fewer than six if the two parental strains have the same marker that is associated with one or more of the genes that affect yield.

E10. A. If we assume that the highly inbred strain has no genetic variance,

V_G (for the wild strain) $= 3.2\ g^2 - 2.2\ g^2 = 1.0\ g^2$

B. $h_B^2 = 1.0\ g^2/3.2\ g^2 = 0.31$

C. It is the same as h_B^2, so it also equals 0.31.

E12. A. $h_N^2 = \dfrac{\overline{X}_O - \overline{X}}{\overline{X}_P - \overline{X}}$

$0.21 = (26.5\ g - 25\ g)/(27\ g - 25\ g)$

$\overline{X}_O - 25\ g = 2\ g\ (0.21)$

$\overline{X}_O = 25.42\ g$

B. $0.21 = (26.5\ g - 25\ g)/(\overline{X}_P - 25\ g)$

$(\overline{X}_P - 25\ g)(0.21) = 1.5\ g$

$\overline{X}_P = 32.14$ g parents

However, because this value is so far from the mean, there may not be 32.14 g parents in the population of mice that you have available.

E14. We first need to calculate a and b. In this calculation, X represents the height of fathers, and Y represents the height of sons.

$b = \dfrac{144}{112} = 1.29$

$a = 69 - (1.29)(68) = -18.7$

For a father who is 70 in. tall,

$Y = (1.29)(70) + (-18.7) = 71.6$

The most likely height of the son would be 71.6 in.

E16. A. After six or seven generations, the selective breeding seems to have reached a plateau. This suggests that the tomato plants have become monomorphic for the alleles that affect tomato weight.

B. There does seem to be heterosis, because the first generation has a weight of 1.7 lb, which is heavier than either Mary's or Hector's tomatoes. This partially explains why Martin has obtained tomatoes heavier than 1.5 lb. However, heterosis is not the whole story; it does not explain why Martin obtained tomatoes that weigh 2 lb. Even though Mary's and Hector's tomatoes were selected for heavier weight, they may not have all of the "heavy alleles" for each gene that controls weight. For example, let's suppose there are 20 genes that affect weight, with each gene existing in a light and heavy allele. During the early stages of selective breeding, when Mary and Hector picked their 10 plants as seed producers for the next generation, as a matter of random chance, some of these plants may have been homozygous for the light alleles at a few of the 20 genes that control weight. Therefore, just as a matter of chance, they probably "lost" a few of the heavy alleles that affect weight. So, after 12 generations of breeding, they have predominantly heavy alleles but also have light alleles for some of the genes. If we represent heavy alleles with a capital letter and light alleles with a lowercase letter, Mary's and Hector's strains could be the following:

Mary's strain: *AA BB cc DD EE FF gg hh II JJ KK LL mm NN OO PP QQ RR ss TT*

Hector's strain: *AA bb CC DD EE ff GG HH II jj kk LL MM NN oo PP QQ RR SS TT*

As we see here, Mary's strain is homozygous for the heavy allele at 15 of the genes but carries the light allele at the other 5. Similarly, Hector's strain is homozygous for the heavy allele at 15 genes and carries the light allele at the other 5. It is important to note, however, that the light alleles in Mary's and Hector's strains are not in the same genes. Therefore, when Martin crosses them together, he will initially get the following:

APPENDIX B

Martin's F_1 offspring: *AA Bb Cc DD EE Ff Gg Hh II Jj Kk LL Mm NN Oo PP QQ RR Ss* TT

If the alleles are additive and contribute equally to the trait, we would expect about the same weight (1.5 lb), because this hybrid has a total of 10 light alleles. However, if heterosis is occurring, genes (which were homozygous recessive in Mary's and Hector's strains) will become heterozygous in the F_1 offspring, and this may make the plants healthier and contribute to a higher weight. If Martin's F_1 strain is subjected to selective breeding, the 10 genes that are heterozygous in the F_1 offspring may eventually become homozygous for the heavy allele. This would explain why Martin's tomatoes achieved a weight of 2.0 pounds after five generations of selective breeding.

E18. A. $h_N^2 = \dfrac{\overline{X}_O - \overline{X}}{\overline{X}_P - \overline{X}}$

$h_N^2 = \dfrac{269 - 254}{281 - 254} = 0.56$

B. $0.56 = \dfrac{275 - 254}{\overline{X}_P - 254}$

$\overline{X}_P = 291.5$ lb

Questions for Student Discussion/Collaboration

2. Most traits depend on the influence of many genes. Also, genetic variation is a common phenomenon in most populations. Therefore, most individuals have a variety of alleles that contribute to a given trait. For quantitative traits, some alleles may make the trait bigger, and other alleles may make the trait turn out smaller. If a population contains many different genes and alleles that govern a quantitative trait, most individuals will have an intermediate phenotype because they will have inherited some large and some small alleles. Fewer individuals will inherit a predominance of large alleles or a predominance of small alleles. An example of a quantitative trait that does not fit a normal distribution is snail pigmentation. The dark snails and light snails are favored rather than the intermediate colors because they are less susceptible to predation.

CHAPTER 26

Conceptual Questions

C2. Evolution is unifying because all living organisms on this planet evolved from an interrelated group of common ancestors. At the molecular level, all organisms have a great deal in common. With the exception of some viruses, they all use DNA as their genetic material. This DNA is found within chromosomes, and the sequence of the DNA is organized into units called genes. Most genes are structural genes that encode the amino acid sequence of polypeptides. Polypeptides fold to form functional units called proteins. At the cellular level, all living organisms also share many similarities. For example, living cells share many of the same basic features including a plasma membrane, ribosomes, enzymatic pathways, and so on. In addition, as discussed in Chapter 5, the mitochondria and chloroplasts of eukaryotic cells are evolutionarily derived from bacterial cells.

C4. Reproductive isolation occurs when two species are unable to mate and produce viable offspring. As discussed in Table 26.1, several prezygotic and postzygotic mechanisms can prevent interspecies matings.

C6. Anagenesis is the evolution of one species into another, whereas cladogenesis is the divergence of one species into two or more species. Of the two, cladogenesis is more prevalent. There may be many reasons why. It is common for an abrupt genetic change such as alloploidy to produce a new species from a preexisting one. Also, migrations of a few members of species into a new region may lead to the formation of a new species in the region (i.e., allopatric speciation).

C8. A. Allopatric

B. Sympatric

C. At first, it may involve parapatric speciation with a low level of intermixing. Eventually, when smaller lakes are formed, allopatric speciation will occur.

C10. The main evidence in favor of punctuated equilibrium is the fossil record. Paleontologists rarely find a gradual transition of fossil forms. The transition period in which environment pressure and genetic changes cause a previous species to evolve into a new species is thought to be so short that few, if any, of the transitional members would be preserved as fossils. Therefore, the fossil record primarily contains representatives from the long equilibrium periods. Also, rapid evolutionary change is consistent with known genetic phenomena, including single-gene mutations that have dramatic effects on phenotypic characteristics, the founder effect, and genetic events such as changes in chromosome structure (e.g., inversions and translocations) or chromosome number, which may abruptly create individuals with new phenotypic traits. In some cases, however, gradual changes are observed in certain species over long periods of time. In addition, the gradual accumulation of mutations is known to occur from the molecular analyses of DNA.

C12. Allopatric speciation involves a physical separation of a species into two or more separate populations. Over time, each population accumulates mutations that alter the characteristics of each population. Because the populations are separated, each will evolve different characteristics and eventually become distinct species. In parapatric speciation, there is some physical separation of two or more populations, but the separation is not absolute. On occasion, members of different populations can interbreed. Even so, the (somewhat) separated populations will tend to accumulate different genetic changes (e.g., inversions) that will ultimately lead to reproductive isolation among the different populations. In sympatric speciation, members of a population are not physically separated, but something happens (e.g., polyploidy) that abruptly results in reproductive isolation between members of the population. For example, a species could be diploid and a member of the population could become tetraploid. The tetraploid member would be reproductively isolated from the diploid members because hybrid offspring would be triploid and sterile. Therefore, the tetraploid individual has become a separate species.

C14. The relatively constant rate of neutral mutations acts as a molecular clock on which to measure evolutionary time. Neutral mutations occur at a rate proportional to the rate of mutation per generation. Therefore, the genetic divergence between species that is due to neutral mutations reflects the time elapsed since their last common ancestor. The concept is related to the neutral theory of evolution because it assumes that most genetic variation is due to the accumulation of neutral or nearly neutral mutations.

C16. A gene sequence can evolve more rapidly. The purpose of structural genes is to encode a polypeptide with a defined amino acid sequence. Many nucleotide changes will have no effect on the amino acid sequence of the polypeptide. For example, mutations in intron sequences and mutations at the wobble base may not affect the amino acid sequence of the encoded polypeptide. These neutral mutations will happen rather rapidly on an evolutionary timescale because natural selection will not remove them from the population. In contrast, changes in the amino acid sequence may alter the structure and function of the polypeptide. Most random mutations that affect the polypeptide sequence are more likely to be detrimental than beneficial, and detrimental mutations will be eliminated by natural selection. This makes it more difficult for the amino acid sequence of the polypeptide to evolve. Only neutral changes and beneficial changes will happen rapidly, and these are less likely to occur in the amino acid sequence compared to the gene sequence.

C18. You would expect the sequences of plant storage proteins to evolve rapidly. The polypeptide sequence is not particularly important for the structure or function of the protein. The purpose of the protein is to provide nutrients to the developing embryo. Changing the sequence would likely be tolerated. However, major changes in the amino acid composition (not the sequence) may be selected against. For example, the storage protein would have to contain some cysteine in its amino acid sequence because the embryo would need some cysteine to grow. However, the location of cysteine codons within the amino acid sequence would not be important; it would only be important that the gene sequence have some cysteine codons.

C20. The α-globin sequences in humans and horses are more similar to each other, compared to the α-globin in humans and the β-globin in humans. This suggests that the gene duplication that produced the α-globin and β-globin genes occurred first. After this gene duplication occurred, each gene accumulated several different mutations that caused the sequences of the two genes to diverge. At a much later time, during the evolution of mammals, a split occurred that produced different branches in the evolutionary tree of mammals. One branch eventually led to the formation of horses and a different branch led to the formation of humans. During the formation of these mammalian branches (which has been more recent), some additional mutations occurred in the α- and β-globin genes. This explains why the α-globin gene in humans and horses is not exactly the same. However, it is more similar than the α- and β-globin genes within humans because the divergence of humans and horses occurred much more recently than the gene duplication that produced the α- and β-globin genes. In other words, there has been much less time for the α-globin gene in humans to diverge from the α-globin gene in horses.

C22. A. This is an example of neutral mutation. Mutations in the wobble base are neutral when they do not affect the amino acid sequence.

B. This is an example of natural selection. Random mutations that occur in vital regions of a polypeptide sequence are likely to inhibit function. Therefore, these types of mutations are eliminated by natural selection. That is why they are relatively rare.

C. This is a combination of neutral mutation and natural selection. The prevalence of mutations in introns is due to the accumulation of neutral mutations. Most mutations within introns do not have any effect on the expression of the exons, which contain the polypeptide sequence. In contrast, mutations within the exons are more likely to be affected by natural selection. As mentioned in the answer to part B, mutations in vital regions are likely to inhibit function. Natural selection tends to eliminate these mutations. Therefore, mutations within exons are less likely than mutations within introns.

C24. Generally, one would expect a similar number of chromosomes with very similar banding patterns. However, there may be a few notable differences. An occasional translocation could change the size or chromosomal number between two different species. Also, an occasional inversion may alter the banding pattern between two species.

C26. Animals with very simple body plans, such as the sponge, have relatively few *Hox* genes. In contrast, animals with more complicated bodies, such as mammals, have many. Researchers speculate that each *Hox* gene can govern the morphological features of a particular region of the body. By having multiple *Hox* genes, different regions of the body can become more specialized, and therefore, more complex.

Experimental Questions

E2. Perhaps the easiest way to determine allotetraploidy is by the chromosomal examination of closely related species. A researcher could karyotype the chromosomes from many different species and look for homologous chromosomes that have similar banding patterns. This may enable them to identify allotetraploids that contain a diploid set of chromosomes from two different species.

E4. The mutations that have occurred in this sequence are neutral mutations. In all cases, the wobble base has changed, and this change would not affect the amino acid sequence of the encoded polypeptide. Therefore, a reasonable explanation is that the gene has accumulated random neutral mutations over the course of many generations. This observation would be consistent with the neutral theory of evolution. A second explanation would be that one of these two researchers made a few experimental mistakes when determining the sequence of this region.

E6. Inversions do not affect the total amount of genetic material. Usually, inversions do not affect the phenotype of the organism. Therefore, if members of the two populations were to interbreed, the offspring would probably be viable because they would have inherited a normal amount of genetic material from each parent. However, such offspring would be inversion heterozygotes. As described in Chapter 8 (see Figure 8.11), crossing over during meiosis may create chromosomes that have too much or too little genetic material. If these unbalanced chromosomes are passed to the next generation of offspring, the offspring may not survive. For this reason, inversion heterozygotes (that are phenotypically normal) may not be very fertile because many of their offspring will die. Because inversion heterozygotes are less fertile, this would tend to keep the eastern and western populations reproductively isolated. Over time, this would aid in the independent evolution of the two populations and would ultimately promote the evolution of the two populations into separate species.

E8. The technique of PCR is used to amplify the amount of DNA in a sample. To accomplish this, one must use oligonucleotide primers that are complementary to the region that is to be amplified. For example, as described in the experiment of Figure 26.15, PCR primers that were complementary to and flank the 12S rRNA gene can be used to amplify the 12S rRNA gene. The technique of PCR is described in Chapter 18.

E10. We would expect the probe to hybridize to the natural *G. tetrahit* and also the artificial *G. tetrahit*, because both of these strains contain two sets of chromosomes from *G. pubescens*. We would expect two bright spots in the *in situ* experiment. Depending on how closely related *G. pubescens* and *G. speciosa* are, the probe may also hybridize to two sites in the *G. speciosa* genome, but this is difficult to predict *a priori*. If so, the *G. tetrahit* species would show four spots.

E12. The principle of parsimony chooses a phylogenetic tree that requires the fewest number of evolutionary changes. When using molecular data, researchers can use computer programs that compare DNA sequences from homologous genes of different species and construct a tree that requires the fewest numbers of mutations. Such a tree is the most likely pathway for the evolution of such species.

E14. Possibly, the mouse would have an eye at the tip of its tail!

Questions for Student Discussion/Collaboration

2. The founder effect and allotetraploidy are examples of rapid forms of evolution. In addition, some single gene mutations may have a great impact on phenotype and lead to the rapid evolution of new species by cladogenesis. Geological processes may promote the slower accumulation of alleles and alter a species' characteristics more gradually. In this case, it is the accumulation of many phenotypically minor genetic changes that ultimately leads to reproductive isolation. Slow and fast mechanisms of evolution have the common theme that they result in reproductive isolation. This is a prerequisite for the evolution of new species. Fast mechanisms tend to involve small populations and a few number of genetic changes. Slower mechanisms may involve larger populations and involve the accumulation of a large number of genetic changes that each contributes in a small way.

GLOSSARY

A

A an abbreviation for adenine.
ABC model a model for flower development.
acentric describes a chromosome without a centromere.
acentric fragment a fragment of a chromosome that lacks a centromere.
acquired antibiotic resistance the acquisition of antibiotic resistance because a bacterium has taken up a gene or plasmid from another bacterial strain.
acridine dye a type of chemical mutagen that causes frameshift mutations.
acrocentric a chromosome with the centromere significantly off center, but not at the very end.
activator a transcriptional regulatory protein that increases the rate of transcription.
acutely transforming virus (ACT) a virus that readily transforms normal cells into malignant cells, when grown in a laboratory.
adaptor hypothesis a hypothesis that proposes a tRNA has two functions: recognizing a three-base codon sequence in mRNA and carrying an amino acid that is specific for that codon.
adenine a purine base found in DNA and RNA. It base-pairs with thymine in DNA.
A DNA a right-handed DNA double helix with 11 bp per turn. Does not occur in living cells.
AFLP see *amplified restriction fragment length polymorphism.*
age of onset for alleles that cause genetic diseases, the time of life at which disease symptoms appear.
alkaptonuria a human genetic disorder involving the accumulation of homogentisic acid due to a defect in homogentisic acid oxidase.
alkyltransferase an enzyme that can remove methyl or ethyl groups from guanine bases.
allele an alternative form of a specific gene.
allele frequency the number of copies of a particular allele in a population divided by the total number of all alleles for that gene in the population.
allelic variation genetic variation in a population that involves the occurrence of two or more different alleles for a particular gene.
allodiploid an organism that contains one set of chromosomes from two different species.
allopatric speciation (Greek, *allos*, "other"; Latin, *patria*, "homeland") an evolutionary phenomenon in which speciation occurs when members of a species become geographically separated from the other members.
alloploid an organism that contains chromosomes from two or more different species.
alloploidy the phenomenon in which a cell or organism contains sets of chromosomes from two or more different species.
allopolyploid an organism that contains two (or more) sets of chromosomes from two (or more) species.
allosteric enzyme an enzyme that contains two binding sites: a catalytic site and a regulatory site.
allosteric regulation the phenomenon in which an effector molecule binds to a noncatalytic site on a protein and causes a conformational change that regulates its function.
allosteric site the site on a protein where a small effector molecule binds to regulate the function of the protein.
allotetraploid an organism that contains two sets of chromosomes from two different species.
allozymes two or more enzymes (encoded by the same type of gene) with alterations in their amino acid sequences, which may affect their gel mobilities.
α helix a type of secondary structure found in proteins.
alternative exon an exon that is not always found in mRNA. It is only found in certain types of alternatively spliced mRNAs.
alternative splicing refers to the phenomenon in which a pre-mRNA can be spliced in more than one way.
amber a stop codon with the sequence UAG.
Ames test a test using strains of a bacterium, *Salmonella typhimurium,* to determine if a substance is a mutagen.
amino acid a building block of polypeptides and proteins. It contains an amino group, a carboxyl group, and a side chain.
aminoacyl site (A site) a site on the ribosome where a charged tRNA initially binds.
aminoacyl tRNA a tRNA molecule that has an amino acid covalently attached to its 3′ end.
aminoacyl-tRNA synthetase an enzyme that catalyzes the attachment of a specific amino acid to the correct tRNA.
2-aminopurine a base analog that acts as a chemical mutagen.
amino terminal end see *amino terminus.*
amino terminus the location of the first amino acid in a polypeptide chain. The amino acid at the amino terminus still retains a free amino group that is not covalently attached to the second amino acid.
amniocentesis a method of obtaining cellular material from a fetus for the purpose of genetic testing.
amplified restriction fragment length polymorphism (AFLP) an RFLP that is amplified via PCR.
anabolic enzyme an enzyme involved in connecting organic molecules to create larger molecules.
anagenesis (Greek, *ana*, "up," and *genesis*, "origin") the evolutionary phenomenon in which a single species is transformed into a different species over the course of many generations.
anaphase the fourth stage of M phase. As anaphase proceeds, half of the chromosomes move to one pole, and the other half move to the other pole.
ancestral character see *primitive character.*
ancient DNA analysis analysis of DNA that is extracted from the remains of extinct species.
aneuploid not euploid. Refers to a variation in chromosome number such that the total number of chromosomes is not an exact multiple of a set or *n* number.
aneuploidy a cell or organism that is aneuploid.
annealing the process in which two complementary segments of DNA bind to each other.
annotated in files involving genetic sequences, annotation is a description of the known function and features of the sequence, as well as other pertinent information.
anteroposterior axis in animals, the axis that runs from the head (anterior) to the tail or base of the spine (posterior).
anther the structure in flowering plants that gives rise to pollen grains.
anther culture the generation of monoploid plants by cold-shock treatment of anthers.
antibodies proteins produced by the B cells of the immune system that recognize foreign substances (namely, viruses, bacteria, and so forth) and target them for destruction.
antibody microarray a small silica, glass, or plastic slide that is dotted with many different antibodies, which recognize particular amino acid sequences within proteins.
anticipation the phenomenon in which the severity of an inherited disease tends to get worse in future generations.
anticodon a three-nucleotide sequence in tRNA that is complementary to a codon in mRNA.
antigens foreign substances that are recognized by antibodies.
antiparallel an arrangement in a double helix in which one strand is running in the 5′ to 3′ direction, while the other strand is 3′ to 5′.
antisense RNA an RNA strand that is complementary to a strand of mRNA.
antisense strand also called the template strand. It is the strand of DNA that is used as a template for RNA synthesis.
antitermination the function of certain proteins, such as N protein in bacteria, that prevents transcriptional termination.
AP endonuclease a DNA repair enzyme that recognizes a DNA region that is missing a base and makes a cut in the DNA backbone near that site.
apical–basal-patterning gene one of several plant genes that play a role in embryonic development.
apical region in plants, the region that produces the leaves and flowers.
apoptosis programmed cell death.
apurinic site a site in DNA that is missing a purine base.
Archaea also called archaebacteria; one of the three domains of life. Archaea are prokaryotic species. They tend to live in extreme environments and are less common than bacteria.
ARS elements DNA sequences found in yeast that function as origins of replication.
artificial chromosomes cloning vectors that can accommodate large DNA inserts and behave like chromosomes when inside of living cells.
artificial selection see *selective breeding.*
artificial transformation transformation of bacteria that occurs via experimental treatments.
ascus (pl. **asci)** a sac that contains haploid spores of fungi (i.e., yeast or molds).

asexual reproduction a form of reproduction that does not involve the union of gametes; at the cellular level, a preexisting cell divides to produce two new cells.
A site see *aminoacyl site.*
assortative mating breeding in which individuals preferentially mate with each other based on their phenotypes.
AT/GC rule in DNA, the phenomenon in which an adenine base in one strand always hydrogen bonds with a thymine base in the opposite strand, and a guanine base always hydrogen bonds with a cytosine.
ATP-dependent chromatin remodeling see *chromatin remodeling.*
attachment site a site in a host cell chromosome where a virus integrates during site-specific recombination.
attenuation a mechanism of genetic regulation, seen in the *trp* operon, in which a short RNA is made but its synthesis is terminated before RNA polymerase can transcribe the rest of the operon.
attenuator sequence a sequence found in certain operons (e.g., *trp* operon) in bacteria that stops transcription soon after it has begun.
AU-rich element (ARE) a sequence found in many short-lived mRNAs that contains the consensus sequence AUUUA.
automated DNA sequencing the use of fluorescently labeled dideoxyribonucleotides and a fluorescence detector to sequence DNA.
autonomous transposable element a transposable element that contains all of the information necessary for transposition or retroposition to take place.
autopolyploid a polyploid produced within a single species due to nondisjunction.
autoradiography a technique that involves the use of X-ray film to detect the location of radioisotopes as they are found in macromolecules or cells. It is used to detect a particular band on a gel or to map the location of a gene within an intact chromosome.
autoregulatory loop a form of gene regulation in which a protein, such as a splicing factor or a transcription factor, regulates its own expression.
autosomes chromosomes that are not sex chromosomes.
auxotroph a strain that cannot synthesize a particular nutrient and needs that nutrient supplemented in its growth medium or diet.

B

BAC see *bacterial artificial chromosome.*
backbone the portion of a DNA or RNA strand that is composed of the repeated covalent linkage of the phosphates and sugar molecules.
backcross in genetics, this usually refers to a cross of F_1 hybrids to individuals that have genotypes of the parental generation.
Bacteria one of the three domains of living organisms. Also called eubacteria. Bacteria are prokaryotic species.
bacterial artificial chromosome (BAC) a cloning vector that propagates in bacteria and is used to clone large fragments of DNA.
bacteriophages (or **phages**) viruses that infect bacteria.
balanced polymorphism when natural selection favors the maintenance of two or more alleles in a population.
balanced translocation a translocation, such as a reciprocal translocation, in which the total amount of genetic material is normal or nearly normal.
balancing selection a pattern of natural selection that favors the maintenance of two or more alleles. Examples include heterozygote advantage and negative frequency-dependent selection.
band shift assay see *gel retardation assay.*
Barr body a structure in the interphase nuclei of somatic cells of female mammals that is a highly condensed X chromosome.
basal region in plants, the region that produces the roots.
basal transcription in eukaryotes, a low level of transcription via the core promoter. The binding of transcription factors to enhancer elements may increase transcription above the basal level.
basal transcription apparatus the minimum number of proteins needed to transcribe a gene.
base a nitrogen-containing molecule that is a portion of a nucleotide in DNA or RNA. Examples of bases are adenine, thymine, guanine, cytosine, and uracil.
base excision repair a type of DNA repair in which a modified base is removed from a DNA strand. Following base removal, a short region of the DNA strand is removed, which is then resynthesized using the complementary strand as a template.
base mismatch when two bases opposite each other in a double helix do not conform to the AT/GC rule. For example, if A is opposite C, that would be a base mismatch.
base pair the structure in which two nucleotides in opposite strands of DNA hydrogen bond with each other. For example, an AT base pair is a structure in which an adenine-containing nucleotide in one DNA strand hydrogen bonds with a thymine-containing nucleotide in the complementary strand.
base substitution a point mutation in which one base is substituted for another.
basic domain a protein domain containing several basic amino acids, which is often involved in binding to DNA.
Bayesian methods with regard to evolutionary trees, this method asks the question: What is the probability that a particular phylogenetic tree is correct given the observed data and a particular evolutionary model?
B DNA the predominant form of DNA in living cells. It is a right-handed DNA helix with 10 bp per turn.
behavioral trait a trait that involves behavior. An example is the ability to learn a maze.
beneficial mutation a mutation that has a beneficial effect on phenotype.
benign a noncancerous tumor that is not invasive and cannot metastasize.
β sheet a type of secondary structure found in proteins.
bHLH a structure found in transcription factor proteins with a basic domain involved in DNA binding and a helix-loop-helix domain involved in dimerization.
bidirectionally the phenomenon in which two replication forks move in opposite directions outward from the origin.
bidirectional replication the phenomenon in which two DNA replication forks emanate in both directions from an origin of replication.
bilateral gynandromorph an animal in which one side is phenotypically male and the other side is female.
bilateralism an animal with right-left symmetry.
binary fission the physical process whereby a bacterial cell divides into two daughter cells. During this event, the two daughter cells become divided by the formation of a septum.
binomial expansion equation an equation used to solve genetic problems involving two types of unordered events.
biodegradation the breakdown of a larger molecule into a smaller molecule via cellular enzymes.
bioinformatics the study of biological information. Recently, this term has been associated with the analysis of genetic sequences, using computers and computer programs.
biolistic gene transfer the use of microprojectiles to introduce DNA into plant cells.
biological control the use of microorganisms or products from microorganisms to alleviate plant diseases or damage from undesirable environmental conditions (e.g., frost damage).
biological evolution the accumulation of genetic changes in a species or population over the course of many generations.
biological species concept definition of a species as a group of individuals whose members have the potential to interbreed with one another in nature to produce viable, fertile offspring, but that cannot interbreed successfully with members of other species.
biometric field a field of genetics that involves the statistical study of biological traits.
bioremediation the use of microorganisms to decrease pollutants in the environment.
biotechnology technologies that involve the use of living organisms, or products from living organisms, as a way to benefit humans.
biotransformation the conversion of one molecule into another via cellular enzymes. This term is often used to describe the conversion of a toxic molecule into a nontoxic one.
bivalent a structure in which two pairs of homologous sister chromatids have synapsed (i.e., aligned) with each other.
BLAST (basic local alignment search tool) a computer program that can start with a particular genetic sequence and then locate homologous sequences within a large database.
blending hypothesis of inheritance an early, incorrect hypothesis of heredity. According to this view, the seeds that dictate hereditary traits are able to blend together from generation to generation. The blended traits would then be passed to the next generation.
bottleneck effect a type of genetic drift that occurs when most members of a population are eliminated without any regard to their genetic composition.
box in genetics, a term used to describe a sequence with a specialized function.
branch migration the lateral movement of a Holliday junction.
breakpoint the region where two chromosome pieces break and rejoin with other chromosome pieces.
broad-sense heritability heritability that takes into account all genetic factors.
5-bromodeoxyuridine (BrdU) a base analog that can be incorporated into chromosomes during DNA replication. The presence of this analog can affect the ability of the chromosomes to absorb certain dyes. This is the basis for the staining of harlequin chromosomes.
5-bromouracil a base analog that acts as a chemical mutagen.

C

C an abbreviation for cytosine.
cAMP see *cyclic AMP.*

cAMP response element (CRE) a short DNA sequence found next to certain eukaryotic genes that is recognized by the cAMP response element-binding (CREB) protein.
cancer a disease characterized by uncontrolled cell division.
cancer cell a cell that has lost its normal growth control. Cancer cells are invasive (i.e., they can invade normal tissues) and metastatic (i.e., they can migrate to other parts of the body).
CAP an abbreviation for the catabolite activator protein, a genetic regulatory protein found in bacteria.
capping the covalent attachment of a 7-methylguanosine nucleotide to the 5′ end of mRNA in eukaryotes.
CAP site the sequence of DNA that is recognized by CAP.
carbohydrate organic molecules with the general formula $C(H_2O)$. An example of a simple carbohydrate is the sugar glucose. Large carbohydrates are composed of multiple sugar units.
carboxyl terminus the location of the last amino acid in a polypeptide chain. The amino acid at the carboxyl terminus still retains a free carboxyl group that is not covalently attached to another amino acid.
carcinogen an agent that can cause cancer.
caspases proteolytic enzymes that play a role in apoptosis.
catabolic enzyme an enzyme that is involved in the breakdown of organic molecules into smaller units.
catabolite activator protein see *CAP*.
catabolite repression the phenomenon in which a catabolite (e.g., glucose) represses the expression of certain genes (e.g., the *lac* operon).
catenane interlocked circular molecules.
cDNA see *complementary DNA*.
cDNA library a DNA library made from a collection of cDNAs.
cell adhesion when the surfaces of cells bind to each other or to the extracellular matrix.
cell adhesion molecule (CAM) a molecule (e.g., surface protein or carbohydrate) that plays a role in cell adhesion.
cell culture refers to the growth of cells in a laboratory.
cell cycle in eukaryotic cells, a series of stages through which a cell progresses in order to divide. The phases are G for growth, S for synthesis (of the genetic material), and M for mitosis. There are two G phases, G_1 and G_2.
cell fate the final morphological features that a cell or group of cells will adopt.
cell-free translation system an experimental mixture that can synthesize polypeptides.
cell fusion the process in which individual cells are mixed together and made to fuse with each other.
cell lineage a series of cells that are descended from a cell or group of cells by cell division.
cell plate the structure that forms between two daughter plant cells that leads to the separation of the cells by the formation of an intervening cell wall.
cellular trait a trait that is observed at the cellular level. An example is the shape of a cell.
centiMorgans (cM) (same as a map unit) a unit of map distance obtained from genetic crosses. Named in honor of Thomas Hunt Morgan.
central dogma of genetics the idea that the usual flow of genetic information is from DNA to RNA to polypeptide (protein). In addition, DNA replication serves to copy the information so that it can be transmitted from cell to cell and from parent to offspring.
central region in plants, it is the region that creates the stem. It is the radial pattern of cells in the central region that causes the radial growth observed in plants.
central zone in plants, an area in the meristem where undifferentiated stem cells are always maintained.
centrifugation a method to separate cell organelles and macromolecules in which samples are placed in tubes and spun very rapidly. The rate at which particles move toward the bottom of the tube depends on their densities, sizes, shapes, and the viscosity of the medium.
centrifuge a machine that contains a motor, which causes a rotor holding centrifuge tubes to spin very rapidly.
centromere a segment of eukaryotic chromosomal DNA that provides an attachment site for the kinetochore.
centrosome a cellular structure from which microtubules emanate.
chain termination the stoppage of growth of a DNA strand, RNA strand, or polypeptide sequence.
chaperone a protein that aids in the folding of polypeptides.
character in genetics, a general characteristic such as eye color.
Chargaff's rule the observation that in DNA the amount of A equals T and the amount of G equals C.
charged tRNA a tRNA that has an amino acid attached to its 3′ end by an ester bond.
checkpoint protein a protein that monitors the conditions of DNA and chromosomes and may prevent a cell from progressing through the cell cycle if an abnormality is detected.
chiasma (pl. **chiasmata**) the site where crossing over occurs between two chromosomes. It resembles the Greek letter chi, χ.
chimera an organism composed of cells that are embryonically derived from two different individuals.
ChIP chip assay a form of chromatin immunoprecipitation that utilizes a microarray to determine where in the genome a particular protein binds.
chi square (χ^2) test a commonly used statistical method for determining the goodness of fit. This method can be used to analyze population data in which the members of the population fall into different categories.
chloroplast DNA (cpDNA) the genetic material found within a chloroplast.
chorionic villus sampling a method for obtaining cellular material from a fetus for the purpose of genetic testing.
chromatid following chromosomal replication in eukaryotes, the two copies that remain attached to each other in the form of sister chromatids.
chromatin the association between DNA and proteins that is found within chromosomes.
chromatin immunoprecipitation (ChIP) a method for determining whether proteins bind to a particular region of DNA. This method analyzes DNA-protein interactions as they occur in the chromatin of living cells.
chromatin remodeling a change in chromatin structure that alters the degree of compaction and/or the spacing and histone composition of nucleosomes.
chromatography a method of separating different macromolecules and small molecules based on their chemical and physical properties. A sample is dissolved in a liquid solvent and exposed to some type of matrix, such as a gel, a column containing beads, or a thin strip of paper.
chromocenter the central point where polytene chromosomes aggregate.
chromomere a dark band within a polytene chromosome.
chromosome the structures within living cells that contain the genetic material. Genes are physically located within the structure of chromosomes. Biochemically, chromosomes contain a very long segment of DNA, which is the genetic material, and proteins, which are bound to the DNA and provide it with an organized structure.
chromosome map a chart that depicts the linear arrangement of genes along a chromosome.
chromosome painting the use of probes to identify particular regions of chromosomes. The probes are usually assigned a computer-generated color.
chromosome territory in the cell nucleus, each chromosome occupies a nonoverlapping region called a chromosome territory.
chromosome theory of inheritance a theory of Sutton and Boveri that the inheritance patterns of traits can be explained by the transmission patterns of chromosomes during gametogenesis and fertilization.
chromosome walking a common method used in positional cloning in which a mapped gene or RFLP marker provides a starting point to molecularly "walk" toward a gene of interest via overlapping clones.
***cis*-acting element** a sequence of DNA, such as a regulatory element, that exerts a *cis*-effect.
***cis*-effect** an effect on gene expression due to genetic sequences that are within the same chromosome and often are immediately adjacent to the gene of interest.
cistron the smallest genetic unit that produces a positive result in a complementation experiment. A cistron is equivalent to a gene.
clade a group of species consisting of all descendents of the group's most common ancestor.
cladistic approach a way to construct a phylogenetic tree, also called a cladogram, by considering the various possible pathways of evolution and then choosing the most plausible tree.
cladogenesis (Greek, *clados*, "branch") during evolution, a form of speciation that involves the division of a species into two or more species.
cladogram a phylogenetic tree that has been constructed using a cladistic approach.
cleavage furrow a constriction that causes the division of two animal cells during cytokinesis.
clonal an adjective to describe a clone. For example, a clonal population of cells is a group of cells that are derived from the same cell.
clone the general meaning of this term is to make many copies of something. In genetics, this term has several meanings: (1) a single cell that has divided to produce a colony of genetically identical cells; (2) an individual that has been produced from a somatic cell of another individual, such as the sheep Dolly; (3) many copies of a DNA fragment that are propagated within a vector or produced by PCR.
closed complex the complex between transcription factors, RNA polymerase, and a promoter before the DNA has denatured to form an open complex.
closed conformation a tightly packed conformation of chromatin that cannot be transcribed.
cluster analysis the analysis of microarray data to determine if certain groups (i.e., clusters) of genes are expressed under the same conditions.
cM an abbreviation for centiMorgans; also see *map unit*.

coding strand the strand in DNA that is not used as a template for mRNA synthesis.

codominance a pattern of inheritance in which two alleles are both expressed in the heterozygous condition. For example, a person with the genotype I^AI^B has the blood type AB and expresses both surface antigens A and B.

codon a sequence of three nucleotides in mRNA that functions in translation. A start codon, which usually specifies methionine, initiates translation, and a stop codon terminates translation. The other codons specify the amino acids within a polypeptide sequence according to the genetic code.

codon bias in a given species, the phenomenon in which certain codons are used more frequently than others.

coefficient of inbreeding (F) see *inbreeding coefficient.*

cohesin a protein complex that facilitates the alignment of sister chromatids.

colinearity the correspondence between the sequence of codons in the DNA coding strand and the amino acid sequence of a polypeptide.

colony hybridization a technique in which a probe is used to identify bacterial colonies that contain a hybrid vector with a gene of interest.

combinatorial control the phenomenon common in eukaryotes in which the combination of many factors determines the expression of any given gene.

common ancestor someone who is an ancestor to both of an individual's parents.

comparative genomic hybridization (CGH) a hybridization technique to determine if cells (e.g., cancer cells) have changes in chromosome structure, such as deletions or duplications.

comparative genomics uses information from genome projects to understand the genetic variation between different populations and evolutionary relationships among different species.

competence factors proteins that are needed for bacterial cells to become naturally transformed by extracellular DNA.

competence-stimulating peptide (CSP) a peptide secreted by certain species of bacteria that allow them to become competent for transformation.

competent cells cells that can be transformed by extracellular DNA.

complementary describes sequences in two DNA strands that match each other according to the AT/GC rule. For example, if one strand has the sequence of ATGGCGGATTT, then the complementary strand must be TACCGCCTAAA.

complementary DNA (cDNA) DNA that is made from an RNA template by the action of reverse transcriptase.

complementation a phenomenon in which the presence of two different mutant alleles in the same organism produces a wild-type phenotype. It usually happens because the two mutations are in different genes, so the organism carries one copy of each mutant allele and one copy of each wild-type allele.

complementation test an experimental procedure in which the goal is to determine if two different mutations that affect the same trait are in the same gene or in two different genes.

complete nondisjunction during meiosis or mitosis, when all of the chromosomes fail to disjoin and remain in one of the two daughter cells.

complete transposable element see *autonomous transposable element.*

complex traits characteristics that are determined by several genes and are significantly influenced by environmental factors.

composite transposon a transposon that contains additional genes, such as antibiotic resistance genes, that are not necessary for transposition per se.

computer data file a file (a collection of information) stored by a computer.

computer program a series of operations that can analyze data in a defined way.

concordance in genetics, the degree to which pairs of individuals (e.g., identical twins or fraternal twins) exhibit the same trait.

condensation a change in chromatin structure to become more compact.

condense refers to chromosomes forming a more compact structure.

condensin a protein complex that plays a role in the condensation of interphase chromosomes to become metaphase chromosomes.

conditional alleles alleles in which the phenotypic expression depends on the environmental conditions. An example is temperature-sensitive alleles, which affect the phenotype only at a particular temperature.

conditional lethal allele an allele that is lethal, but only under certain environmental conditions.

conditional mutant a mutant whose phenotype depends on the environmental conditions, such as a temperature-sensitive mutant.

conglomerate a population composed of members of an original population plus new members that have migrated from another population.

conjugation a form of genetic transfer between bacteria that involves direct physical interaction between two bacterial cells. One bacterium acts as donor and transfers genetic material to a recipient cell.

conjugation bridge a connection between two bacterial cells that provides a passageway for DNA during conjugation.

conjugative plasmid a plasmid that can be transferred to a recipient cell during conjugation.

consensus sequence the most commonly occurring bases within a sequence element.

conservative model an incorrect model in which both strands of parental DNA remain together following DNA replication.

conservative transposition see *simple transposition.*

constitutive exon an exon that is always found in mRNA following splicing.

constitutive gene a gene that is not regulated and has essentially constant levels of expression over time.

constitutive heterochromatin regions of chromosomes that are always heterochromatic and are permanently transcriptionally inactive.

contig a series of clones that contain overlapping pieces of chromosomal DNA.

control element see *regulatory sequence or element.*

core enzyme the subunits of an enzyme that are needed for catalytic activity, as in the core enzyme of RNA polymerase.

corepressor a small effector molecule that binds to a repressor protein, thereby causing the repressor protein to bind to DNA and inhibit transcription.

core promoter a DNA sequence that is absolutely necessary for transcription to take place. It provides the binding site for general transcription factors and RNA polymerase.

correlation coefficient (*r*) a statistic with a value that ranges between −1 and 1. It describes how two factors vary relative to each other.

cosmid a vector that is a hybrid between a plasmid vector and phage λ. Cosmid DNA can replicate in a cell like a plasmid or be packaged into a protein coat like a phage. Cosmid vectors can accept fragments of DNA that are typically tens of thousands of base pairs in length.

C_0t curve a plot of C/C_0 versus C_0t.

cotransduction the phenomenon in which bacterial transduction transfers a piece of DNA carrying two closely linked genes.

cotransformation the phenomenon in which bacterial transformation transfers a piece of DNA carrying two closely linked genes.

cotranslational events that occur during translation.

covariance a statistic that describes the degree of variation between two variables within a group.

cpDNA an abbreviation for chloroplast DNA.

CpG island a group of CG sequences that may be clustered near a promoter region of a gene. The methylation of the cytosine bases usually inhibits transcription.

CREB protein (cAMP response element-binding protein) a regulatory transcription factor that becomes activated in response to specific cell-signaling molecules that cause the synthesis of cAMP.

cross a mating between two distinct individuals. An analysis of their offspring may be conducted to understand how traits are passed from parent to offspring.

cross-fertilization same meaning as *cross.* It requires that the male and female gametes come from separate individuals.

crossing over a physical exchange of chromosome pieces that most commonly occurs during prophase of meiosis I.

C-terminus see *carboxyl terminus.*

cycle threshold method (C_t method) in quantitative PCR, a method of determining the starting amount of DNA based on a threshold level at which the accumulation of fluorescence is significantly greater than the background level.

cyclic AMP (cAMP) in bacteria, a small effector molecule that binds to CAP (catabolite activator protein). In eukaryotes, cAMP functions as a second messenger in a variety of intracellular signaling pathways; in some cases, it binds to transcription factors such as the CREB protein.

cyclin a type of protein that plays a role in the regulation of the eukaryotic cell cycle.

cyclin-dependent protein kinases (CDKs) enzymes that are regulated by cyclins and can phosphorylate other cellular proteins by covalently attaching a phosphate group.

cytogeneticist a scientist who studies chromosomes under the microscope.

cytogenetic mapping the mapping of genes or genetic sequences using microscopy.

cytogenetics the field of genetics that involves the microscopic examination of chromosomes.

cytokinesis the division of a single cell into two cells. The two nuclei produced in M phase are segregated into separate daughter cells during cytokinesis.

cytological mapping see *cytogenetic mapping.*

cytoplasmic inheritance (also known as *extranuclear inheritance*) the inheritance of genetic material that is not found within the cell nucleus.

cytosine a pyrimidine base found in DNA and RNA. It base-pairs with guanine in DNA.

D

Dam see *DNA adenine methyltransferase.*

Darwinian fitness the relative likelihood that a genotype will survive and contribute to the gene pool of the next generation as compared with other genotypes.

database a computer storage facility that stores many data files such as those containing genetic sequences.

founder effect changes in allele frequencies that occur when a small group of individuals separates from a larger population and establishes a colony in a new location.
fraction (1) following centrifugation, a portion of the liquid contained within a centrifuge tube; (2) following column chromatography, a portion of the liquid that has been eluted from a column.
frameshift mutation a mutation that involves the addition or deletion of nucleotides not in a multiple of three and thereby shifts the reading frame of the codon sequence downstream from the mutation.
frequency distribution a graph that describes the numbers of individuals that are found in each of several phenotypic categories.
functional genomics the study of gene function at the genome level. It involves the study of many genes simultaneously.
functional protein microarray a type of protein microarray that monitors a particular kind of protein function, such as the ability to bind a specific drug.

G

G an abbreviation for guanine.
gain-of-function mutation a mutation that causes a gene to be expressed in an additional place where it is not normally expressed or during a stage of development when it is not normally expressed.
gamete a reproductive cell (usually haploid) that can unite with another reproductive cell to create a zygote. Sperm and egg cells are types of gametes.
gametogenesis the production of gametes (e.g., sperm or egg cells).
gametophyte the haploid generation of plants.
gap gene one category of segmentation genes.
G bands the chromosomal banding pattern that is observed when the chromosomes have been treated with the chemical dye Giemsa.
gel electrophoresis a method that combines chromatography and electrophoresis to separate molecules and macromolecules. Samples are loaded into wells at one end of the gel, and an electric field is applied across the gel that causes charged molecules to migrate from one side of the gel to the other.
gel mobility shift assay see *gel retardation assay.*
gel retardation assay a technique for studying protein-DNA interactions in which the binding of protein to a DNA fragment retards it mobility during gel electrophoresis.
gene a unit of heredity that may influence the outcome of an organism's traits. At the molecular level, a gene contains the information to make a functional product, either RNA or protein.
gene addition the addition of a cloned gene into a site in a chromosome of a living cell.
gene amplification an increase in the copy number of a gene.
gene chip see *DNA microarray.*
gene cloning the production of many copies of a gene using molecular methods such as PCR or the introduction of a gene into a vector that replicates in a host cell.
gene conversion the phenomenon in which one allele is converted to another due to genetic recombination and DNA repair.
gene dosage effect the phenomenon when the number of copies of a gene affects the phenotypic expression of a trait.
gene duplication an increase in the copy number of a gene. Can lead to the evolution of gene families.
gene expression the process in which the information within a gene is accessed, first to synthesize RNA and usually proteins, and eventually to affect the phenotype of the organism.
gene family two or more different genes within a single species that are homologous to each other because they were derived from the same ancestral gene.
gene flow changes in allele frequencies due to migration.
gene interaction when two or more different genes influence the outcome of a single trait.
gene knockin a type of gene addition in which a gene of interest has been added to a particular site in the mouse genome.
gene knockout when both copies of a normal gene have been replaced by an inactive mutant gene.
gene modifier effect when the allele of one gene modifies the phenotypic effect of the allele of a different gene.
gene mutation a relatively small mutation that affects only a single gene.
gene pool the totality of all genes within a particular population.
generalized transduction a form of transduction in which any piece of the bacterial chromosomal DNA can be incorporated into a phage.
general lineage concept a species concept which states that each species is a population of an independently evolving lineage.
general transcription factor (GTF) one of several proteins that are necessary for basal transcription at the core promoter.
gene rearrangement a rearrangement in segments of a gene, as occurs in antibody precursor genes.
gene redundancy the phenomenon in which an inactive gene is compensated for by another gene with a similar function.
gene regulation the phenomenon in which the level of gene expression can vary under different conditions.
gene replacement the swapping of a cloned gene made experimentally with a normal chromosomal gene found in a living cell.
gene therapy the introduction of cloned genes into living cells in an attempt to cure or alleviate disease.
genetically modified organism (GMO) an organism that has received genetic material via recombinant DNA technology.
genetic approach in research, the study of mutant genes that have abnormal function. By studying mutant genes, researchers may better understand normal genes and normal biological processes.
genetic code the correspondence between a codon (i.e., a sequence of three bases in an mRNA molecule) and the functional role that the codon plays during translation. Each codon specifies a particular amino acid or the end of translation.
genetic cross a mating between two individuals and the analysis of their offspring in an attempt to understand how traits are passed from parent to offspring.
genetic drift random changes in allele frequencies due to sampling error.
genetic linkage see *linkage.*
genetic linkage map see *genetic map.*
genetic map a chart that describes the relative locations of genes or other DNA segments along a chromosome.
genetic mapping any method used to determine the linear order of genes as they are linked to each other along the same chromosome. This term is also used to describe the use of genetic crosses to determine the linear order of genes. See also *linkage mapping.*
genetic marker any genetic sequence that is used to mark a specific location on a chromosome.
genetic mosaic see *mosaicism.*
genetic polymorphism when two or more alleles occur in population; each allele is found at a frequency of 1% or higher.
genetic recombination (1) the process in which chromosomes are broken and then rejoined to form a novel genetic combination. (2) the process in which alleles are assorted and passed to offspring in combinations that are different from the parents.
genetics the study of heredity.
genetic screening the use of testing methods at the population level to determine if individuals are heterozygous carriers for or have a genetic disease.
genetic testing the analysis of individuals with regard to their genes or gene products. In many cases, the goal is to determine if an individual carries a mutant gene.
genetic transfer the physical transfer of genetic material from one bacterial cell to another.
genetic variation genetic differences among members of the same species or among different species.
genome all of the chromosomes and DNA sequences that an organism or species can possess.
genome database a database that focuses on the genetic sequences and characteristics of a single species.
genome maintenance refers to cellular mechanisms that either prevent mutations from occurring and/or prevent mutant cells from surviving or dividing.
genome sequencing projects research endeavors that have the ultimate goal of determining the sequence of DNA bases of the entire genome of a given species.
genomic clone a clone made from the digestion and cloning of chromosomal DNA.
genomic imprinting a pattern of inheritance that involves a change in a single gene or chromosome during gamete formation. Depending on whether the modification occurs during spermatogenesis or oogenesis, imprinting governs whether an offspring will express a gene that has been inherited from its mother or father.
genomic library a DNA library made from chromosomal DNA fragments.
genomics the molecular analysis of the entire genome of a species.
genotype the genetic composition of an individual, especially in terms of the alleles for particular genes.
genotype-environment association when certain genotypes are preferentially found in particular environments.
genotype-environment interaction when the environmental effects on phenotype differ according to genotype.
genotype frequency the number of individuals with a particular genotype in a population divided by the total number of individuals in the population.
germ cells the gametes (i.e., sperm and egg cells).
germ line a lineage of cells that gives rise to gametes.
germ-line mutation a mutation in a cell of the germ line.

GloFish genetically modified aquarium fish that glow due to the introduction of genes that encode fluorescent proteins.
glucocorticoid receptor a type of steroid receptor that functions as a regulatory transcription factor.
goodness of fit the degree to which the observed data and expected data are similar to each other. If the observed and predicted data are very similar, the goodness of fit is high.
gradualism an evolutionary hypothesis suggesting that each new species evolves continuously over long spans of time. The principal idea is that large phenotypic differences that cause the divergence of species are due to the accumulation of many small genetic changes.
grande normal (large-sized) yeast colonies.
grooves in DNA, the indentations where the atoms of the bases are in contact with the surrounding water. In B DNA, there is a smaller minor groove and a larger major groove.
group I intron a type of intron found in self-splicing RNA that uses free guanosine in its splicing mechanism.
group II intron a type of intron found in self-splicing RNA that uses an adenine nucleotide within the intron itself in its splicing mechanism.
growth factors protein factors that influence cell division.
guanine a purine base found in DNA and RNA. It base-pairs with cytosine in DNA.
guide RNA in trypanosome RNA editing, an RNA molecule that directs the addition of uracil residues into the mRNA.
gyrase see *DNA gyrase.*

H

haplodiploid a species, such as certain bees, in which one sex is haploid (e.g., male) and the other sex is diploid (e.g., female).
haploid describes the phenomenon that gametes contain half the genetic material found in somatic cells. For a species that is diploid, a haploid gamete contains a single set of chromosomes.
haploinsufficiency the phenomenon in which a person has only a single functional copy of a gene, and that single functional copy does not produce a normal phenotype. Shows a dominant pattern of inheritance.
haplotype the linkage of particular alleles or molecular markers along a single chromosome.
Hardy-Weinberg equation $p^2 + 2pq + q^2 = 1$.
Hardy-Weinberg equilibrium the phenomenon by which under certain conditions, allele frequencies are maintained in a stable condition and genotypes can be predicted according to the Hardy-Weinberg equation.
helicase see *DNA helicase.*
helix–loop–helix domain a domain found in transcription factors that enables them to dimerize.
helix-turn-helix motif a structure found in transcription factor proteins that promotes binding to the major groove of DNA.
hemizygous describes the single copy of an X-linked gene in the male. A male mammal is said to be hemizygous for X-linked genes.
heritability the amount of phenotypic variation within a particular group of individuals that is due to genetic factors.
heterochromatin highly compacted DNA. It is usually transcriptionally inactive.
heterochronic mutation a mutation that alters the timing of expression of a gene and thereby alters the outcome of cell fates.
heterodimer when two polypeptides encoded by different genes bind to each other to form a dimer.
heteroduplex a double-stranded region of DNA that contains one or more base mismatches.
heterogametic sex in species with two types of sex chromosomes, the heterogametic sex produces two types of gametes. For example, in mammals, the male is the heterogametic sex, because a sperm can contain either an X or a Y chromosome.
heterogamous describes a species that produces two morphologically different types of gametes (i.e., sperm and eggs).
heterogeneity see *locus heterogeneity.*
heterogeneous nuclear RNA (hnRNA) same as pre-mRNA.
heterokaryon a cell produced from cell fusion that contains two separate nuclei.
heteroplasmy when a cell contains variation in a particular type of organelle. For example, a plant cell could contain some chloroplasts that make chlorophyll and other chloroplasts that do not.
heterosis the phenomenon in which hybrids display traits superior to either corresponding parental strain. Heterosis is usually different from overdominance, because the hybrid may be heterozygous for many genes, not just a single gene, and because the superior phenotype may be due to the masking of deleterious recessive alleles.
heterozygote an individual who is heterozygous.
heterozygote advantage a pattern of inheritance in which a heterozygote has a higher Darwinian fitness compared with either of the corresponding homozygotes.
heterozygous describes a diploid individual that has different copies (i.e., two different alleles) of the same gene.
***Hfr* strain** (for **high frequency of recombination**) a bacterial strain in which an F factor has become integrated into the bacterial chromosome. During conjugation, an *Hfr* strain can transfer segments of the bacterial chromosome.
hierarchical shotgun sequencing a genome sequencing strategy in which small DNA fragments are mapped prior to DNA sequencing.
highly repetitive sequences sequences that are found tens of thousands or even millions of times throughout the genome.
high stringency refers to highly selective hybridization conditions that promote the binding of DNA or RNA fragments that are perfect or almost perfect matches.
histone acetyltransferase an enzyme that attaches acetyl groups to the amino terminal tails of histone proteins.
histone code hypothesis the hypothesis that the pattern of histone modification acts much like a language or code in specifying alterations in chromatin structure.
histone deacetylase an enzyme that removes acetyl groups from the amino terminal tails of histone proteins.
histones a group of proteins involved in forming the nucleosome structure of eukaryotic chromatin.
hnRNA an abbreviation for heterogeneous nuclear RNA.
holandric gene a gene on the Y chromosome.
Holliday junction a site where an unresolved crossover has occurred between two homologous chromosomes.
Holliday model a model to explain the molecular mechanism of homologous recombination.
holoenzyme an enzyme containing all of its subunits, as in the holoenzyme of RNA polymerase that has σ factor along with the core enzyme.
homeobox a 180-bp consensus sequence found in homeotic genes.
homeodomain the protein domain encoded by the homeobox. The homeodomain promotes the binding of the protein to the DNA.
homeologous describes the homologous chromosomes from closely related species.
homeotic an adjective that was originally used to describe mutants in which one body part is replaced by another.
homeotic gene a gene that functions in governing the developmental fate of a particular region of the body.
homoallelic two or more alleles in different organisms that are due to mutations at exactly the same base within a gene.
homodimer when two polypeptides encoded by the same gene bind to each other to form a dimer.
homogametic sex in species with two types of sex chromosomes, the homogametic sex produces only one type of gamete. For example, in mammals, the female is the homogametic sex, because an egg can only contain an X chromosome.
homologous in the case of genes, this term describes two genes that are derived from the same ancestral gene. Homologous genes have similar DNA sequences. In the case of chromosomes, the two homologs of a chromosome pair are said to be homologous to each other.
homologous recombination the exchange of DNA segments between homologous chromosomes.
homologous recombination repair (HRR) also called homology-directed repair, occurs when the DNA strands from a sister chromatid are used to repair a lesion in the other sister chromatid.
homolog one of the chromosomes in a pair of homologous chromosomes.
homology structures that are similar to each other because they evolved from a common ancestor.
homology-base modeling the modeling of a three-dimensional structure of a molecule based on its homology to another molecule whose structure is already known.
homozygous describes a diploid individual that has two identical alleles of a particular gene.
horizontal gene transfer the transfer of genes from one individual to another individual that is not its offspring.
host cell a cell that is infected with a virus or bacterium.
host range the spectrum of host species that a virus or other pathogen can infect.
hot spots sites within a gene that are more likely to be mutated than other locations.
housekeeping gene a gene that encodes a protein required in most cells of a multicellular organism.
***Hox* complexes** a group of several *Hox* genes located in a particular chromosomal region.
***Hox* genes** mammalian genes that play a role in development. They are homologous to homeotic genes found in *Drosophila.*
Human Genome Project a worldwide collaborative project that provided a detailed map of the human genome and obtained a complete DNA sequence of the human genome.

mitotic nondisjunction an event in which chromosomes do not segregate equally during mitosis.
mitotic recombination crossing over that occurs during mitosis.
mitotic spindle apparatus (also known as the *mitotic spindle*) the structure that organizes and separates the chromosomes during M phase of the eukaryotic cell cycle.
model organism an organism studied by many scientists so that researchers can more easily compare their results and begin to unravel the properties of a given species.
moderately repetitive sequences sequences that are found a few hundred to several thousand times in the genome.
molecular clock the phenomenon by which the rate of neutral mutations can be used as a tool to measure evolutionary time.
molecular evolution the molecular changes in the genetic material that underlie the phenotypic changes associated with evolution.
molecular genetics an examination of DNA structure and function at the molecular level.
molecular marker a segment of DNA that is found at a specific site in the genome and has properties that enable it to be uniquely recognized using molecular tools such as gel electrophoresis.
molecular paleontology the analysis of DNA sequences from extinct species.
molecular pharming a recombinant technology that involves the production of medically important proteins in the mammary glands of livestock.
molecular profiling methods that enable researchers to understand the molecular changes that occur in diseases such as cancer.
molecular trait a trait that is observed at the molecular level. An example is the amount of a given protein in a cell.
monoallelic expression in the case of imprinting, refers to the phenomenon that only one of the two alleles of a given gene is transcriptionally expressed.
monohybrid an individual produced from a monohybrid cross.
monohybrid cross a cross in which an experimenter is following the outcome of only a single trait.
monomorphic a term used to describe a gene that is found as only one allele in a population.
monophyletic group see *clade.*
monoploid an organism with a single set of chromosomes within its somatic cells.
monosomic a diploid cell that is missing a chromosome (i.e., $2n - 1$).
monozygotic twins twins that are genetically identical because they were formed from the same sperm and egg.
morph a form or phenotype in a population. For example, red eyes and white eyes are different eye color morphs.
morphogen a molecule that conveys positional information and promotes developmental changes.
morphological trait a trait that affects the morphology (physical form) of an organism. An example is eye color.
mosaicism when the cells of part of an organism differ genetically from the rest of the organism.
motif the name given to a domain or amino acid sequence that functions in a similar manner in many different proteins.
M phase a general name given to nuclear division that can apply to mitosis or meiosis. It is divided into prophase, prometaphase, metaphase, anaphase, and telophase.
mRNA see *messenger RNA.*
mtDNA an abbreviation for mitochondrial DNA.
MTOC see *microtubule-organizing center.*
multinomial expansion equation an equation to solve genetic problems involving three or more types of unordered events.
multiple alleles when the same gene exists in two or more alleles within a population.
multiple sequence alignment an alignment of two or more genetic sequences based on their homology to each other.
multiplication method a method for solving independent assortment problems in which the probabilities of the outcome for each gene are multiplied together.
multipotent a type of stem cell that can differentiate into several different types of cells.
mutable site a site in a chromosome that tends to break at a fairly high rate due to the presence of a transposable element.
mutagen an agent that causes alterations in the structure of DNA.
mutant alleles alleles that have been created by altering a wild-type allele by mutation.
mutation a permanent change in the genetic material that can be passed from cell to cell or from parent to offspring.
mutation frequency the number of mutant genes divided by the total number of genes within the population.
mutation rate the likelihood that a gene will be altered by a new mutation.
myogenic bHLH protein a type of transcription factor involved in muscle cell differentiation.

N

n an abbreviation that designates the number of chromosomes in a set. In humans, $n = 23$, and a diploid cell has $2n = 46$ chromosomes.
narrow-sense heritability heritability that takes into account only those genetic factors that are additive.
natural selection refers to the process whereby differential fitness acts on the gene pool. When a mutation creates a new allele that is beneficial, the allele may become prevalent within future generations because the individuals possessing this allele are more likely to reproduce and pass the beneficial allele to their offspring.
natural transformation a natural process of transformation that occurs in certain strains of bacteria.
negative control transcriptional regulation by repressor proteins.
negative frequency-dependent selection a pattern of natural selection in which the fitness of a genotype decreases when its frequency becomes higher.
neutral mutation a mutation that has no detectable effect on protein function or no detectable effect on the survival of the organism.
neutral theory of evolution the theory that most genetic variation observed in natural populations is due to the accumulation of neutral mutations.
next-generation sequencing technologies newer DNA sequencing technologies that are more rapid and inexpensive.
nick translation the phenomenon in which DNA polymerase uses its 5′ to 3′ exonuclease activity to remove a region of DNA and, at the same time, replaces it with new DNA.
nitrogen mustard an alkylating agent that can cause mutations in DNA.
nitrous acid a type of chemical mutagen that deaminates bases, thereby changing amino groups to keto groups.
nonallelic homologous recombination recombination that occurs at sites within chromosomes due to the occurrence of repetitive sequences.
nonautonomous element a transposable element that lacks a gene such as transposase or reverse transcriptase that is necessary for transposition.
noncoding strand the strand of DNA within a structural gene that is complementary to the mRNA. The noncoding strand is used as a template to make mRNA.
noncomplementation the phenomenon in which two mutant alleles in the same organism do not produce a wild-type phenotype.
non-Darwinian evolution see *neutral theory of evolution.*
nondisjunction event in which chromosomes do not segregate properly during mitosis or meiosis.
nonessential genes genes that are not absolutely required for survival, although they are likely to be beneficial to the organism.
nonhomologous end-joining (NHEJ) protein a protein that joins the ends of DNA fragments that are not homologous. This occurs during site-specific recombination of immunoglobulin genes.
nonhomologous recombination the exchange of DNA between nonhomologous segments of chromosomes or plasmids.
non-LTR retrotransposon a type of retrotransposon that does not have long terminal repeats.
nonneutral mutation a mutation that affects the phenotype of the organism and can be acted on by natural selection.
nonparental see *recombinant.*
nonparental ditype (NPD) an ascus that contain cells that all have a nonparental combination of alleles.
nonrecombinant in a testcross, refers to a phenotype or arrangement of alleles on a chromosome that is not found in the parental generation.
nonsense codon a stop codon.
nonsense mutation a mutation that involves a change from a sense codon to a stop codon.
nontemplate strand a strand of DNA that is not used as a template during transcription.
nonviral-like retroelement a type of retroelement in which the sequence does not resemble a modern virus.
normal distribution a distribution for a large sample in which the trait of interest varies in a symmetrical way around an average value.
norm of reaction the effects of environmental variation on an individual's traits.
Northern blotting a technique used to detect a specific RNA within a mixture of many RNA molecules.
N-terminus see *amino terminus.*
nuclear genes genes that are located on chromosomes found in the cell nucleus of eukaryotic cells.
nuclear lamina a collection of fibers that line the inner nuclear membrane.
nuclear matrix (or **nuclear scaffold**) a group of proteins that anchor the loops found in eukaryotic chromosomes.
nucleic acid RNA or DNA. A macromolecule that is composed of repeating nucleotide units.
nucleoid a darkly staining region that contains the genetic material of mitochondria, chloroplasts, or bacteria.
nucleolus a region within the nucleus of eukaryotic cells where the assembly of ribosomal subunits occurs.
nucleoprotein a complex of DNA (or RNA) and protein.

nucleoside structure in which a base is attached to a sugar, but no phosphate is attached to the sugar.
nucleosome the repeating structural unit within eukaryotic chromatin. It is composed of double-stranded DNA wrapped around an octamer of histone proteins.
nucleosome-free region (NFR) a region within a chromosome where nucleosomes are not found.
nucleotide the repeating structural unit of nucleic acids, composed of a sugar, phosphate, and base.
nucleotide excision repair (NER) a DNA repair system in which several nucleotides in the damaged strand are removed from the DNA and the undamaged strand is used as a template to resynthesize a normal strand.
nucleus a membrane-bound organelle in eukaryotic cells where the linear sets of chromosomes are found.
null hypothesis a hypothesis that assumes there is no real difference between the observed and expected values.

O

ocher a stop codon with the sequence UGA.
octad a group of eight fungal spores contained within an ascus.
Okazaki fragments short segments of DNA that are synthesized in the lagging strand during DNA replication.
oligonucleotide a short strand of DNA, typically a few or a few dozen nucleotides in length.
oncogene a mutant gene that promotes cancer.
one-gene/one-enzyme hypothesis the idea, which later needed to be expanded, that one gene encodes one enzyme.
oogenesis the production of egg cells.
opal a stop codon with the sequence UAA.
open complex the region of separation of two DNA strands produced by RNA polymerase during transcription.
open conformation a loosely packed chromatin structure that is capable of transcription.
open reading frame (ORF) a genetic sequence that does not contain stop codons.
operator (or **operator site**) a sequence of nucleotides in bacterial DNA that provides a binding site for a genetic regulatory protein.
operon an arrangement in DNA in which two or more structural genes are found within a regulatory unit that is under the transcriptional control of a single promoter.
ordered octad an ascus composed of eight cells whose order depends on crossing over during meiosis.
ordered tetrad an ascus composed of four cells whose order depends on crossing over during meiosis.
ORF see *open reading frame.*
organelle a large specialized structure within a cell, which is often surrounded by a single or double membrane.
organism level when the level of observation or experimentation involves a whole organism.
organizing center in plants, a region of the meristem that ensures the proper organization of the meristem and preserves the correct number of actively dividing stem cells.
orientation-independent refers to certain types of genetic regulatory elements that can function in the forward or reverse direction. Certain enhancers are orientation independent.
origin of replication a nucleotide sequence that functions as an initiation site for the assembly of several proteins required for DNA replication.
origin of transfer the location on an F factor or within the chromosome of an *Hfr* strain that is the initiation site for the transfer of DNA from one bacterium to another during conjugation.
origin recognition complex (ORC) a complex of six proteins found in eukaryotes that is necessary to initiate DNA replication.
ortholog homologous genes in different species that were derived from the same ancestral gene.
outbreeding mating between genetically unrelated individuals.
outgroup in cladistics, a species or group of species that is most closely related to the ingroup.
ovary (1) in plants, the structure in which the ovules develop; (2) in animals, the structure that produces egg cells and female hormones.
overdominance an inheritance pattern in which a heterozygote is more vigorous than either of the corresponding homozygotes.
ovule the structure in higher plants where the female gametophyte (i.e., embryo sac) is produced.
ovum a female gamete, also known as an egg cell.
oxidative DNA damage changes in DNA structure that are caused by reactive oxygen species (ROS).
oxidative stress an imbalance between the production of reactive oxygen species (ROS) and an organism's ability to break them down.

P

p an abbreviation for the short arm of a chromosome.
pachytene the third stage of prophase of meiosis I.
pair-rule gene one category of segmentation genes.
palindromic when a sequence is the same in the forward and reverse direction.
pangenesis an incorrect hypothesis of heredity. It suggested that hereditary traits could be modified depending on the lifestyle of the individual. For example, it was believed that a person who practiced a particular skill would produce offspring that would be better at that skill.
paracentric inversion an inversion in which the centromere is found outside of the inverted region.
paralogs homologous genes within a single species that constitute a gene family.
parapatric speciation (Greek, *para,* "beside"; Latin, *patria,* "homeland") a form of speciation that occurs when members of a species are only partially separated or when a species is very sedentary.
parasegments transient subdivisions that occur in the *Drosophila* embryo prior to the formation of segments.
parental in a testcross, refers to a phenotype or arrangement of alleles on a chromosome that is the same as one or both members of the parental generation.
parental ditype (PD) an ascus that contains four spores with the parental combinations of alleles.
parental generation in a genetic cross, the first generation in the experiment. In Mendel's studies, the parental generation was true-breeding with regard to particular traits.
parental strand in DNA replication, the DNA strand that is used as a template.
parthenogenesis the formation of an individual from an unfertilized egg.
particulate theory of inheritance a theory proposed by Mendel. It states that traits are inherited as discrete units that remain unchanged as they are passed from parent to offspring.
P1 artificial chromosome (PAC) an artificial chromosome developed from P1 bacteriophage chromosomes.
paternal leakage the phenomenon in which species where maternal inheritance is generally observed, the male parent may, on rare occasions, provide mitochondria or chloroplasts to the zygote.
pattern the spatial arrangement of different regions of the body. At the cellular level, the body pattern is due to the arrangement of cells and their specialization.
pattern recognition in bioinformatics, this term refers to a program that recognizes a pattern of symbols.
PCR see *polymerase chain reaction.*
pedigree analysis a genetic analysis using information contained within family trees. In this approach, the aim is to determine the type of inheritance pattern that a gene follows.
pellet a collection of particles found at the bottom of a centrifuge tube.
peptide bond a covalent bond formed between the carboxyl group in one amino acid in a polypeptide chain and the amino group in the next amino acid.
peptidyl site (P site) a site on the ribosome that carries a tRNA along with a polypeptide chain.
peptidyl transfer the step during the elongation stage of translation in which the polypeptide is removed from the tRNA in the P site and transferred to the amino acid at the A site.
peptidyltransferase a complex that functions during translation to catalyze the formation of a peptide bond between the amino acid in the A site of the ribosome and the growing polypeptide chain.
pericentric inversion an inversion in which the centromere is located within the inverted region of the chromosome.
peripheral zone in plants, an area in the meristem that contains dividing cells that eventually differentiate into plant structures.
personalized medicine the application of genetic or molecular data in the treatment of disease.
petites mutant strains of yeast that form small colonies due to defects in mitochondrial function.
PFGE see *pulsed-field gel electrophoresis.*
PGD see *preimplantation genetic diagnosis.*
P generation the parental generation in a genetic cross.
phage see *bacteriophage.*
phage λ a bacteriophage that infects *E. coli.*
pharming see *molecular pharming.*
phenotype the observable traits of an organism.
phenylketonuria (PKU) a human genetic disorder arising from a defect in phenylalanine hydroxylase.
phosphodiester linkage in a DNA or RNA strand, a linkage in which a phosphate group connects two sugar molecules together.
photolyase an enzyme found in yeast and plants that can repair thymine dimers by splitting the dimers, which returns the DNA to its original condition.
photoreactivation a type of DNA repair of thymine dimers via photolyase that requires light.
phylogenetic tree a diagram that describes the evolutionary relationships among different species.
phylogeny the sequence of events involved in the evolutionary development of a species or group of species.
physical mapping the mapping of genes or other genetic sequences using DNA cloning methods.
physiological trait a trait that affects a cellular or body function. An example is the rate of glucose metabolism.

RFLP mapping the mapping of a gene or other genetic sequence relative to the known locations of RFLPs within a genome.
R group the side chain of an amino acid.
rho (ρ) a protein that is involved in transcriptional termination for certain bacterial genes.
rho-dependent termination transcriptional termination that requires the function of the rho protein.
rho-independent termination transcription termination that does not require the rho protein. It is also known as intrinsic termination.
ribonucleic acid (RNA) a nucleic acid that is composed of ribonucleotides. In living cells, RNA is synthesized via the transcription of DNA.
ribose the sugar found in RNA.
ribosome a large macromolecular structure that acts as the catalytic site for polypeptide synthesis. The ribosome allows the mRNA and tRNAs to be positioned correctly as the polypeptide is made.
ribosome-binding site a sequence in bacterial mRNA that is needed to bind to the ribosome and initiate translation.
ribozyme an RNA molecule with enzymatic activity.
right-left axis in bilateral animals, this axis determines the two sides of the body relative to the anteroposterior axis.
R loop experimentally, a DNA loop that is formed because RNA is displacing it from its complementary DNA strand.
RNA see *ribonucleic acid.*
RNA editing the process in which a change occurs in the nucleotide sequence of an RNA molecule that involves additions or deletions of particular bases, or a conversion of one type of base to a different type.
RNA-induced silencing complex (RISC) the complex that mediates RNA interference.
RNA interference the phenomenon that double-stranded RNA targets complementary RNAs within the cell for silencing or degradation.
RNA polymerase an enzyme that synthesizes a strand of RNA using a DNA strand as a template.
RNA primer a short strand of RNA, made by DNA primase, that is used to elongate a strand of DNA during DNA replication.
RNase an enzyme that cuts the sugar-phosphate backbone in RNA.
RNaseP a bacterial enzyme that is an endonuclease and cuts precursor tRNA molecules. RNaseP is a ribozyme, which means that its catalytic ability is due to the action of RNA.
RNA splicing the process in which pieces of RNA are removed and the remaining pieces are covalently attached to each other.
Robertsonian translocation the structure produced when two telocentric chromosomes fuse at their short arms.
root meristem an actively dividing group of cells that gives rise to root structures.
RT-PCR see *reverse transcriptase PCR* or *real-time PCR.*

S

satellite DNA in a density centrifugation experiment, a peak of DNA that is separated from the majority of the chromosomal DNA. It is usually composed of highly repetitive sequences.
SBS see *sequencing by synthesis.*
scaffold a collection of proteins that holds the DNA in place and gives chromosomes their characteristic shapes.
scaffold-attachment region (SAR) a site in the chromosomal DNA that is anchored to the nuclear matrix or scaffold.
SCE see *sister chromatid exchange.*
science a way of knowing about our natural world. The science of genetics allows us to understand how the expression of genes produces the traits of an organism.
scientific method a basis for conducting science. It is a process that scientists typically follow so they may reach verifiable conclusions about the world in which they live.
scintillation counting a technique that permits a researcher to count the number of radioactive emissions from a sample containing a population of radioisotopes.
search by content in bioinformatics, an approach to predict the location of a gene because the nucleotide content of a particular region differs significantly (due to codon bias) from a random distribution.
search by signal in bioinformatics, an approach that relies on known sequences such as promoters, start and stop codons, and splice sites to help predict whether or not a DNA sequence contains a gene.
secondary structure a regular repeating pattern of molecular structure, such as the DNA double helix or the α helix and β sheet found in proteins.
second-division segregation (SDS) in an ordered octad, a 2:2:2:2 or 2:4:2 arrangement of spores that occurs because two alleles do not segregate until the second meiotic division is completed.
sedimentation coefficient a measure of centrifugation that is normally expressed in Svedberg units (S). A sedimentation coefficient has the units of seconds: $1\ S = 1 \times 10^{-13}$ seconds.
segmental duplication a small segment of a chromosome that has a tandem duplication.
segmentation gene in animals, a gene, whose encoded product is involved in the development of body segments.
segment polarity gene one category of segmentation genes.
segments anatomical subdivisions that occur during the development of species such as *Drosophila.*
segregate when two things are placed in separate locations. For example, homologous chromosomes segregate into different gametes.
selectable marker a gene that provides a selectable phenotype in a cloning experiment. Many selectable markers are genes that confer antibiotic resistance.
selection coefficient one minus the fitness value.
selectionists scientists who oppose the neutral theory of evolution.
selection limit the phenomenon in which several generations of artificial selection results in a plateau where artificial selection is no longer effective.
selective breeding programs and procedures designed to modify the phenotypes in economically important species of plants and animals.
selenocysteine an amino acid that may be incorporated into polypeptides during translation.
self-fertilization fertilization that involves the union of male and female gametes derived from the same parent.
selfish DNA hypothesis the idea that transposable elements exist because they possess characteristics that allow them to multiply within the host cell DNA and inhabit the host without offering any selective advantage.
self-splicing refers to RNA molecules that can remove their own introns without the aid of other proteins or RNA.
semiconservative model the correct model for DNA replication in which the newly made double-stranded DNA contains one parental strand and one daughter strand.
semiconservative replication the net result of DNA replication in which the DNA contains one original strand and one newly made strand.
semilethal alleles lethal alleles that kill some individuals but not all.
semisterility when an individual has a lowered fertility.
sense codon a codon that encodes a specific amino acid.
sense strand the strand of DNA within a structural gene that has the same sequence as mRNA except that T is found in the DNA instead of U.
sequence complexity the number of times a particular base sequence appears throughout the genome of a given species.
sequence element in genetics, a sequence with a specialized function.
sequence recognition in bioinformatics, this term refers to a program that recognizes particular sequence elements.
sequence-tagged site (STS) a short segment of DNA, usually between 100 and 400 bp long, the base sequence of which is found to be unique within the entire genome. Sequence-tagged sites are identified by PCR.
sequencing see *DNA sequencing.*
sequencing by synthesis (SBS) a next-generation form of DNA sequencing in which the synthesis of DNA is directly monitored to deduce the base sequence.
sequencing ladder a series of bands on a gel that can be followed in order (e.g., from the bottom of the gel to the top of the gel) to determine the base sequence of DNA.
sex chromosomes a pair of chromosomes (e.g., X and Y in mammals) that determines sex in a species.
sex determination the factor(s) that determine whether an organism develops into a male or female.
sex-influenced inheritance an inheritance pattern in which an allele is dominant in one sex but recessive in the opposite sex. In humans, pattern baldness is an example of a sex-influenced trait.
sex-limited inheritance an inheritance pattern in which a trait is found in only one of the two sexes. An example would be beard development in men.
sex-limited traits traits that occur in only one of the two sexes.
sex linkage the phenomenon that certain genes are found on one of the two types of sex chromosomes but not both.
sex-linked gene a gene that is located on one of the sex chromosomes.
sex pilus (pl. **pili**) a structure on the surface of bacterial cells that acts as an attachment site to promote the binding of bacteria to each other.
sexual dimorphism species in which the males and females are morphologically distinct.
sexual reproduction the process whereby parents make gametes (e.g., sperm and egg) that fuse with each other in the process of fertilization to begin the life of a new organism.
sexual selection natural selection that acts to promote characteristics that give individuals a greater chance of reproducing.
shared derived character a trait shared by a group of organisms but not by a distant common ancestor.
Shine-Dalgarno sequence a sequence in bacterial mRNAs that functions as a ribosomal binding site.

shoot meristem an actively dividing group of cells that gives rise to shoot structures.
short-interfering RNAs (siRNAs) small RNA molecules that silence the expression of specific mRNAs via RNA interference.
short tandem repeat sequences (STRs) short DNA sequences that are repeated many times in a row. Often found in centromeric and telomeric regions.
shotgun sequencing a genome sequencing strategy in which DNA fragments to be sequenced are randomly generated from larger DNA fragments.
shuttle vector a cloning vector that can propagate in two or more different species, such as *E. coli* and yeast.
side chain in an amino acid, the chemical structure that is attached to the carbon atom (i.e., the α carbon) that is located between the amino group and carboxyl group.
sigma factor a transcription factor that recognizes bacterial promoter sequences and facilitates the binding of RNA polymerase to the promoter.
silencer a DNA sequence that functions as a regulatory element. The binding of a regulatory transcription factor to the silencer decreases the level of transcription.
silent mutation a mutation that does not alter the amino acid sequence of the encoded polypeptide even though the nucleotide sequence has changed.
similarity with regard to DNA, refers to a comparison of DNA sequences that have regions where the bases match up. Similarity may be due to homology.
simple Mendelian inheritance an inheritance pattern involving a simple, dominant/recessive relationship that produces observed ratios in the offspring that readily obey Mendel's laws.
simple sequence repeat (SSR) a short base sequence found within a chromosome that is repeated many times in a row.
simple translocation when one piece of a chromosome becomes attached to a different chromosome.
simple transposition a cut-and-paste mechanism for transposition in which a transposable element is removed from one site and then inserted into another.
SINEs in mammals, short interspersed elements that are less than 500 bp in length.
single-factor cross see *monohybrid cross.*
single-nucleotide polymorphism a genetic polymorphism within a population in which two alleles of the gene differ by a single nucleotide.
single-stranded binding protein a protein that binds to both of the single strands of DNA during DNA replication and prevents them from re-forming a double helix.
siRNA see *short interfering RNAs.*
sister chromatid exchange (SCE) the phenomenon in which crossing over occurs between sister chromatids, thereby exchanging identical genetic material.
sister chromatids pairs of replicated chromosomes that are attached to each other at the centromere. Sister chromatids are genetically identical.
site-directed mutagenesis a technique that enables scientists to change the sequence of cloned DNA segments.
site-specific recombination when two different DNA segments break and rejoin with each other at a specific site. This occurs during the integration of certain viruses into the host chromosome and during the rearrangement of immunoglobulin genes.
SMC proteins proteins that use energy from ATP to catalyze changes in chromosome structure.
snRNP a complex containing small nuclear RNA and a set of proteins, which are components of the spliceosome.
somatic cell any cell of the body except for germ-line cells that give rise to gametes.
somatic mutation a mutation in a somatic cell.
sorting signal an amino acid sequence or posttranslational modification that directs a protein to the correct region of the cell.
SOS response a response to extreme environmental stress in which bacteria replicate their DNA using DNA polymerases that are likely to make mistakes.
Southern blotting a technique used to detect the presence of a particular genetic sequence within a mixture of many chromosomal DNA fragments.
speciation the process by which new species are formed via evolution.
species a group of organisms that maintains a distinctive set of attributes in nature.
species concepts different approaches for distinguishing species.
spectrophotometer a device used by researchers to determine how much radiation at various wavelengths a sample can absorb.
spermatids immature sperm cells produced from spermatogenesis.
spermatogenesis the production of sperm cells.
sperm cell a male gamete. Sperm are small and usually travel relatively far distances to reach the female gamete.
spindle see *mitotic spindle apparatus.*
spindle pole during cell division in eukaryotes, one of two sites in the cell where microtubules originate.
spliceosome a multisubunit complex that functions in the splicing of eukaryotic pre-mRNA.
splicing see *RNA splicing.*
splicing factor a protein that regulates the process of RNA splicing.
spontaneous mutation a change in DNA structure that results from random abnormalities in biological processes.
spores haploid cells that are produced by certain species such as fungi (i.e., yeast and molds).
sporophyte the diploid generation of plants.
SR protein a type of splicing factor.
SSR see *simple sequence repeat.*
stabilizing selection natural selection that favors individuals with an intermediate phenotype.
stamen the structure found in the flower of higher plants that produces the male gametophyte (i.e., pollen).
standard deviation a statistic that is computed as the square root of the variance.
start codon a 3-base sequence in mRNA that initiates translation. It is usually 5′-AUG-3′ and encodes methionine.
stem cell a cell that has the capacity to divide and to differentiate into one or more specific cell types.
steroid receptor a category of transcription factors that respond to steroid hormones. An example is the glucocorticoid receptor.
stigma the structure in flowering plants on which the pollen land and the pollen tube starts to grow so that sperm cells can reach the egg cells.
stop codon a 3-base sequence in mRNA that signals the end of translation of a polypeptide. The three stop codons are 5′–UAA–3′, 5′–UAG–3′, and 5′–UGA–3′.
strain a variety that continues to produce the same characteristic after several generations.
strand in DNA or RNA, nucleotides covalently linked together to form a long, linear polymer.
stripe-specific enhancer in *Drosophila*, a regulatory region that controls the expression of a gene so that it occurs only in a particular parasegment during early embryonic development.
STRs see *short tandem repeat sequences.*
structural gene a gene that encodes the amino acid sequence within a particular polypeptide or protein.
STS see *sequence-tagged site.*
subcloning the procedure of making smaller DNA clones from a larger one.
submetacentric describes a chromosome in which the centromere is slightly off center.
subspecies a population within a species that has some distinct characteristics that differ from other members of the species.
subunit this term may have multiple meanings. In a protein, each subunit is a single polypeptide.
sum rule the probability that one of two or more mutually exclusive events will occur is equal to the sum of their individual probabilities.
supercoiling see *DNA supercoiling.*
supergroups a relatively recent way for evolutionary biologists to subdivide the eukaryotic domain.
supernatant following centrifugation, the fluid that is found above the pellet.
suppressor (or **suppressor mutation**) a mutation at a second site that suppresses the phenotypic effects of another mutation.
SWI/SNF family a group of related proteins that catalyze chromatin remodeling.
sympatric speciation (Greek, *sym*, "together"; Latin, *patria*, "homeland") a form of speciation that occurs when members of a species diverge while occupying the same habitat within the same range.
synapomorphy see *shared derived character.*
synapsis the event in which homologous chromosomes recognize each other and then align themselves along their entire lengths.
synaptonemal complex a complex of proteins that promote the interconnection between homologous chromosomes during meiosis.
synonymous codons two different codons that specify the same amino acid.
synteny group a group of genes that are found in the same order on the chromosomes of different species.
systematists biologists who study the evolutionary relationships among different species.

T

T an abbreviation for thymine.
tandem array (or **tandem repeat**) a short nucleotide sequence that is repeated many times in a row.
tandem mass spectrometry the sequential use of two mass spectrometers. It can be used to determine the sequence of amino acids in a polypeptide.
***Taq* polymerase** a thermostable form of DNA polymerase used in PCR experiments.
target-site primed reverse transcription (TPRT) a mechanism of DNA synthesis that occurs during the movement of certain types of retrotransposons.
TATA box a sequence found within eukaryotic core promoters that determines the starting site for transcription. The TATA box is recognized by a TATA-binding protein, which is a component of TFIID.
tautomer the forms of certain small molecules, such as bases, which can spontaneously interconvert between chemically similar forms.
tautomeric shift a change in chemical structure such as an alternation between the keto and enol forms of the bases that are found in DNA.

T DNA a segment of DNA found within a Ti plasmid that is transferred from a bacterium to infected plant cells. The T DNA from the Ti plasmid becomes integrated into the chromosomal DNA of the plant cell by recombination.
TE see *transposable element.*
telocentric describes a chromosome with its centromere at one end.
telomerase the enzyme that recognizes telomeric sequences at the ends of eukaryotic chromosomes and synthesizes additional numbers of telomeric repeat sequences.
telomerase reverse transcriptase (TERT) the enzyme within telomerase that uses RNA as a template to make DNA.
telomeres specialized DNA sequences found at the ends of linear eukaryotic chromosomes.
telophase the fifth stage of M phase. The chromosomes have reached their respective poles and decondense.
temperate phage a bacteriophage that usually exists in the lysogenic cycle.
temperature-sensitive allele an allele in which the resulting phenotype depends on the environmental temperature.
temperature-sensitive (ts) lethal allele an allele that is lethal at a certain environmental temperature.
temperature-sensitive (ts) mutant a mutant that has a normal phenotype at a permissive temperature, but a different phenotype, such as failure to grow, at the nonpermissive temperature.
template DNA a strand of DNA that is used to synthesize a complementary strand of DNA or RNA.
template strand see *template DNA.*
terminal deficiency see *terminal deletion.*
terminal deletion when a segment is lost from the end of a linear chromosome.
termination (1) in transcription, the release of the newly made RNA transcript and RNA polymerase from the DNA; (2) in translation, the release of the polypeptide and the last tRNA and the disassembly of the ribosomal subunits and mRNA.
termination codon see *stop codon.*
termination sequences (ter sequences) in *E. coli*, a pair of sequences in the chromosome that bind a protein known as the termination utilization substance (Tus), which stops the movement of the replication forks.
terminator a sequence within a gene that signals the end of transcription.
TERT see *telomerase reverse transcriptase.*
tertiary structure the three-dimensional structure of a macromolecule, such as the tertiary structure of a polypeptide.
testcross an experimental cross between a recessive individual and an individual whose genotype the experimenter wishes to determine.
tetrad (1) the association among four sister chromatids during meiosis; (2) a group of four fungal spores contained within an ascus.
tetraploid having four sets of chromosomes (i.e., $4n$).
tetratype (T) an ascus that has two parental cells and two nonparental cells.
TFIID a type of general transcription factor in eukaryotes that is needed for RNA polymerase II function. It binds to the TATA box and recruits RNA polymerase II to the core promoter.
thermocycler a device that automates the timing of temperature changes in each cycle of a PCR experiment.
30-nm fiber the association of nucleosomes to form a more compact structure that is 30 nm in diameter.
thymine a pyrimidine base found in DNA. It base-pairs with adenine in DNA.
thymine dimer a DNA lesion involving a covalent linkage between two adjacent thymine bases in a DNA strand.
Ti plasmid a tumor-inducing plasmid found in *Agrobacterium tumefaciens.* It is responsible for promoting tumor formation after a plant has been infected.
tissue-specific gene a gene that is highly regulated and is expressed in a particular cell type.
TNRE see *trinucleotide repeat expansion.*
topoisomerase an enzyme that alters the degree of supercoiling in DNA.
topoisomers DNA conformations that differ with regard to supercoiling.
totipotent a cell that possesses the genetic potential to produce an entire individual. A somatic plant cell or a fertilized egg is totipotent.
TPRT see *target-site primed reverse transcription.*
traffic signal see *sorting signal.*
trait any characteristic that an organism displays. Morphological traits affect the appearance of an organism. Physiological traits affect the ability of an organism to function. A third category of traits are those that affect an organism's behavior (behavioral traits).
***trans*-acting factor** a regulatory protein that binds to a regulatory element in the DNA and exerts a *trans* effect.
transcription the process of synthesizing RNA from a DNA template.
transcriptional start site the site in a gene where transcription begins.
transcription-coupled DNA repair a form of DNA repair that is initiated when RNA polymerase cannot transcribe over a damaged region.
transcription factors a broad category of proteins that influence the ability of RNA polymerase to transcribe DNA into RNA.
transcription-repair coupling factor (TRCF) a protein that recognizes when RNA polymerase is stalled over a damaged region of DNA and recruits DNA repair enzymes to fix the damaged site.
transduction a form of genetic transfer between bacterial cells in which a bacteriophage transfers bacterial DNA from one bacterium to another.
***trans*-effect** an effect on gene expression that occurs even though two DNA segments are not physically adjacent to each other. *Trans*-effects are mediated through diffusible genetic regulatory proteins.
transfection (1) when a viral vector is introduced into a bacterial cell; (2) the introduction of any type of recombinant DNA into a eukaryotic cell.
transfer RNA (tRNA) a type of RNA used in translation that carries an amino acid. The anticodon in tRNA is complementary to a codon in the mRNA.
transformation (1) when a plasmid vector or segment of chromosomal DNA is introduced into a bacterial cell; (2) when a normal cell is converted into a malignant cell.
transgene a gene from one species that is introduced into another species.
transgenic organism an organism that has DNA from another organism incorporated into its genome via recombinant DNA techniques.
transition a point mutation involving a change of a pyrimidine to another pyrimidine (e.g., C to T) or a purine to another purine (e.g., A to G).
translation the synthesis of a polypeptide using the codon information within mRNA.
translational regulatory protein a protein that regulates translation.
translational repressor a protein that binds to mRNA and inhibits its ability to be translated.
translesion synthesis (TLS) the synthesis of DNA over a template strand that harbors some type of DNA damage. This occurs via lesion-replicating polymerases.
translocation (1) when one segment of a chromosome breaks off and becomes attached to a different chromosome; (2) when a ribosome moves from one codon in an mRNA to the next codon.
translocation cross the structure that is formed when the chromosomes of a reciprocal translocation attempt to synapse during meiosis. This structure contains two normal (nontranslocated chromosomes) and two translocated chromosomes. A total of eight chromatids are found within the cross.
transposable element (TE) a small genetic element that can move to multiple locations within the chromosomal DNA.
transposase the enzyme that catalyzes the movement of transposable elements.
transposition the phenomenon of transposon movement.
transposon see *transposable element.*
transposon tagging a technique for cloning genes in which a transposon inserts into a gene and inactivates it. The transposon-tagged gene is then cloned using a complementary transposon as a probe to identify the gene.
transversion a point mutation in which a purine is interchanged with a pyrimidine, or vice versa.
trihybrid cross a cross in which an experimenter follows the outcome of three different traits.
trinucleotide repeat expansion (TNRE) a type of mutation that involves an increase in the number of tandemly repeated trinucleotide sequences.
triplex DNA a double-stranded DNA that has a third strand wound around it to form a triple-stranded structure.
triploid an organism or cell that contains three sets of chromosomes.
trisomic a diploid cell with one extra chromosome (i.e., $2n + 1$).
tRNA see *transfer RNA.*
***trp* repressor** a protein that binds to the operator site of the *trp* operon and inhibits transcription.
true-breeding line a strain of a particular species that continues to produce the same trait after several generations of self-fertilization (in plants) or inbreeding.
tumor-suppressor gene a gene that functions to inhibit cancerous growth.
two-dimensional gel electrophoresis a technique to separate proteins that involves isoelectric focusing in the first dimension and SDS-gel electrophoresis in the second dimension.
two-factor cross see *dihybrid cross.*

U

U an abbreviation for uracil.
unbalanced translocation a translocation in which a cell has too much genetic material compared with a normal cell.
unipotent a type of stem cell that can differentiate into only a single type of cell.
universal in genetics, the phenomenon that nearly all organisms use the same genetic code with just a few exceptions.

unordered octad an ascus composed of eight unordered cells.
unordered tetrad an ascus composed of four unordered cells.
up promoter mutation a mutation in a promoter that increases the rate of transcription.
up regulation genetic regulation that leads to an increase in gene expression.
uracil a pyrimidine base found in RNA.
UTR an abbreviation for the untranslated region of mRNA.
U-tube a U-shaped tube that has a filter at bottom of the U. The pore size of the filter allows the passage of small molecules (e.g., DNA molecules) from one side of the tube to the other, but restricts the passage of bacterial cells.

V

variance the sum of the squared deviations from the mean divided by the degrees of freedom.
variants individuals of the same species that exhibit different traits. An example is tall and dwarf pea plants.
V(D)J recombination site-specific recombination that occurs within immunoglobulin genes.
vector a small segment of DNA that is used as a carrier of another segment of DNA. Vectors are used in DNA cloning experiments.
vertical evolution the phenomenon that species evolve from preexisting species by the accumulation of gene mutations and by changes in chromosome structure and number. Vertical evolution involves genetic changes in a series of ancestors that form a lineage.
vertical gene transfer the transfer of genetic material from parents to offspring or from mother cell to daughter cell.
viral genome the genetic material of a virus.
viral-like retroelements retroelements that are evolutionarily related to known retroviruses.
viral vector a vector used in gene cloning that is derived from a naturally occurring virus.
virulent phage a phage that follows the lytic cycle.
virus a small infectious particle that contains nucleic acid as its genetic material, surrounded by a capsid of proteins. Some viruses also have an envelope consisting of a membrane embedded with spike proteins.
VNTRs also called minisatellites; segments of DNA that are located in several places in a genome and have a variable number of tandem repeats. The pattern of VNTRs was originally used in DNA fingerprinting.

W

Western blotting a technique used to detect a specific protein among a mixture of proteins.
whole-genome shotgun sequencing a genome sequencing strategy that bypasses the mapping step. The whole genome is subjected to shotgun sequencing.
wild-type allele an allele that is fairly prevalent in a natural population, generally greater than 1% of the population. For polymorphic genes, there may be more than one wild-type allele.
wobble base the first base (from the 5′ end) in an anticodon. This term suggests that the first base in the anticodon can wobble a bit to recognize the third base in the mRNA.
wobble rules rules that govern the binding specificity between the third base in a codon and the first base in an anticodon.

X

X chromosomal controlling element (Xce) a region adjacent to Xic that influences the choice of the active X chromosome during the process of X inactivation.
X inactivation a process in which mammals equalize the expression of X-linked genes by randomly turning off one X chromosome in the somatic cells of females.
X-inactivation center (Xic) a site on the X chromosome that appears to play a critical role in X inactivation.
X-linked genes (alleles) genes (or alleles of genes) that are physically located within the X chromosome.
X-linked inheritance an inheritance pattern in certain species that involves genes that are located only on the X chromosome.
X-linked recessive an allele or trait in which the gene is found on the X chromosome and the allele is recessive relative to a corresponding dominant allele.

Y

YAC see *yeast artificial chromosome.*
yeast artificial chromosome (YAC) a cloning vector propagated in yeast that can reliably contain very large insert fragments of DNA.
Y-linked genes (alleles) genes (or alleles of genes) that are located only on the Y chromosome.

Z

Z DNA a left-handed DNA double helix that is found occasionally in the DNA of living cells.
zygote a cell formed from the union of a sperm and egg.
zygotene the second stage of prophase of meiosis I.
zygotic gene a gene that is expressed after fertilization.

CREDITS

Photographs

Chapter 1
Opener: © Photo courtesy of the College of Veterinary Medicine, Texas A&M University/Corbis; 1.2(left): © Roslin Institute; 1.3a(right): © Advanced Cell Technology, Inc., Worcester, Massachusetts; 1.3b: © Photo taken by Flaminia Catteruccia, Jason Benton and Andrea Crisanti, and assembled by www.luciariccidesign.com.; 1.4: © Biophoto Associates/Photo Researchers; 1.5: © CNRI/Science Photo Library/Photo Researchers; 1.8: © Wildlife GmbH/Alamy; 1.10: © March of Dimes Birth Defects Foundation; 1.9a(top right): © Joseph Sohm; ChromoSohm Inc./Corbis; 1.9b: © Paul Edmondson/Photodisc Red/Getty Images RF; 1.13a: © George Musil/Visuals Unlimited; 1.13b: © SciMAT/Photo Researchers; 1.13c: © Steve Hopkin/ardea.com; 1.13d: © Brad Mogen/Visuals Unlimited; 1.13e: © Mark Smith/Photo Researchers; 1.13f: © J-M. Labat/Photo Researchers; 1.13g: © Wally Eberhart/Visuals Unlimited.

Chapter 2
Opener: © Chris Martin Bahr/SPL /Photo Researchers; 2.1: © SPL/Photo Researchers; 2.2b: © Nigel Cattlin/Photo Researchers.

Chapter 3
Opener: © Photomicrographs by Dr. Conly L. Rieder, Wadsworth Center, Albany, New York 12201-0509; 3.2b: © Burger/Photo Researchers; 3.2c: © Leonard Lessin/Peter Arnold; 3.6a(left): © Leonard Lessin/Peter Arnold; 3.6a(right): © Biophoto Associates/Photo Researchers; 3.8a-f: © Photomicrographs by Dr. Conly L. Rieder, Wadsworth Center, Albany, New York 12201-0509; 3.9a(1): © Dr. David M. Phillips/Visuals Unlimited ; 3.9b: © Ed Reschke; 3.11: © Diter von Wettstein.

Chapter 4
Opener: © Robert Calentine/Visuals Unlimited; 4.1: © Blickwinkel/Alamy; 4.5b: © Bob Shanley/The Palm Beach Post; 4.6a(1): © Peter Weimann/Animals Animals; 4.6a(2): © Tom Walker/Visuals Unlimited; 4.6b: © Sally Haugen/Virginia Schuett, www.pkunews.org; 4.6c: © Phototake Inc./Alamy; 4.7a-b: © Stan Flegler/Visuals Unlimited; 4.10: © Alan & Sandy Carey/Photo Researchers; 4.9a(top left): © Zig Leszcynski/Animals, Animals; 4.9b: © John T. Fowler/Alamy; 4.9c: © P. Wegner/Peter Arnold; 4.9d: © Gary Randall/Visuals Unlimited; 4.13a: © AP Images; 4.15a-c: National Parks Service, Adams National Historical Park; 4.15d: © Bettmann/Corbis ; 4.17 a-b: © Robert Maier/Animals, Animals; 4.18a: © Jane Burton/naturepl.com.

Chapter 5
Opener: © John Mendenhall, Institute for Cellular and Molecular Biology, University of Texas at Austin; 5.3a-b: © Courtesy of I. Solovei, University of Munich (LMU).; 5.3b: © Tim Davis /Photo Researchers; 5.6: © G.W. Willis, MD/Visuals Unlimited ; TA 5.1: Ronald G. Davidson, Harold M. Nitowsky, and Barton Childs. "Demonstration of Two Populations of Cells in the Human Female Heterozygous for Glucose-6-Phosphate Dehydrogenase Variants." *PNAS*. 50 (1963) f. 2, p. 484. Courtesy Harold M. Nitowsky. Reproduced with author permission.; 5.9: © Courtesy of Dr. Argiris Efstratiadis; 5.14a: Reproduced with permission from *The Journal of Cell Biology*, 1977, 72:687-694. Copyright 1977 The Rockefeller University Press.; 5.14b: Gibbs, SP., Mak, R., Ng, R., & Slankis, T. "The Chloroplast Nucleoid In Ochromonas Danica. II. Evidence For An Increase In Plastid DNA During Greening." *J Cell Sci*. 1974 Dec; 16(3):579-91. Fig. 1. By permission of the Company of Biologists Limited.

Chapter 6
Opener: © Educational Images Ltd/Custom Medical Stock Photo.

Chapter 7
Opener: © David Scharf/Peter Arnold; 7.4b: © Dr. L. Caro/Science Photo Library/Photo Researchers; 7.16a-b: © Carolina Biological Supply/Phototake.

Chapter 8
Opener: © BSIP/Phototake; 8.1a(center): © Michael Abbey/Photo Researchers; 8.1a(left): © Scott Camazine /Photo Researchers; 8.1a(right): © Carlos R Carvalho/Universidade Federal de Viçosa.; 8.1c: © C.N.R.I./Phototake; 8.4a: © Biophoto Assocates/Science Source/Photo Researchers; 8.4b: © Jeff Noneley.; 8.13b: © Paul Benke/University of Miami School of Medicine.; 8.13c: © Will Hart/PhotoEdit; 8.18a-b(top): © A. B. Sheldon; 8.19b: © David M. Phillips/Visuals Unlimited; 8.20a(left): © James Steinberg/Photo Researchers; 8.20b(bottom): © Biophoto Associates/Science Source/Photo Researchers; 8.20b(top): © Biophoto Associates/Science Source/Photo Researchers; 8.26: Robinson, T.J. & Harley, E.H. "Absence of geographic chromosomal variation in the roan and sable antelope and the cytogenetics of a naturally occurring hybrid." *Cytogenet Cell Genet*. 1995; 71(4): 363-9. Permission granted by S. Karger AG, Basel.

Chapter 9
Opener: © Ken Eward /Photo Researchers; 9.3: © Omikron/Photo Researchers; 9.11b: Pictorial Parade/Getty Images; 9.12a: From *The Double Helix*. By James D. Watson, 1968, Atheneum Press, NY. By permission of Cold Spring Harbor Laboratory Archives.; 9.15a© Barrington Brown/Photo Researchers; 9.15b: © Hulton|Archive by Getty Images; 9.17: © Laguna Design/Photo Researchers; 9.23b: © Alfred Pasieka/Photo Researchers.

Chapter 10
Opener: © Dr. Gopal Murti/Visuals Unlimited ; 10.3: © American Society for Microbiology. Reproduced with permission.; 10.10b: © Simpson's Nature Photography; 10.10c: © William Leonard; 10.16a-b: This article was published in *Cell*. 12(1). Thoma , F. & Koller, T. "Influence of histone H1 on chromatin structure." p. 103. f. 2a&c, Copyright Elsevier, 1977. Reproduced with permission.; TA 10.15: © Reprinted by permission from Macmillan Publishers Ltd. *Nature*. Subunit structure of chromatin. Markus Noll. 251: 5472, 249-251. 1974.; 10.17: Jerome Rattner/University of Calgary; 10.18b-c: Nickerson et al. "The nuclear matrix revealed by eluting chromatin from a cross-linked nucleus." *PNAS*. 94: 4446-4450. Figure 2a &b. © 1997 National Academy of Sciences, U.S.A.; 10.19a-b: Reprinted by permission from Macmillan Publishers Ltd. *Nature Reviews/Genetics*. Chromosome territories, nuclear architecture and gene regulation in mammalian cells. Cremer ,T. & Cremer, C. 2: 4, 292-301, 2001.; 10.21a(11nm): © Olins and Olins/Biological Photo Service; 10.21a(2nm): © Dr. Gopal Murit/Visuals Unlimited; 10.21b(30nm): Jerome Rattner/University of Calgary; 10.21c(300nm): This article was published in *Cell*. Nov; 12(3). Paulson, J.R. & Laemmli, U.K. "The structure of histone-depleted metaphase chromosomes." 817-2, 8. F. 5. Copyright Elsevier, 1977. Reprinted with permission.; 10.21d(1400nm & 700nm), 10.22a: © Peter Engelhardt/Department of Virology, Haartman Institute.; 10.22b: © Dr. Donald Fawcett/Visuals Unlimited.

Chapter 11
Opener: © Clive Freeman, The Royal Institution/Photo Researchers; p. 274: Meselson M, Stahl, F. "The Replication of DNA in Escherichia Coli." *PNAS*. Vol. 44, 1958. f. 4a, p. 673. Courtesy of M. Meselson.; 11.4: From *Cold Spring Harbor Symposia of Quantitative Biology*, 28, p. 43 (1963). Copyright holder is Cold Spring Habour Laboratory Press.; 11.8b: © Reprinted by permission from Macmillan Publishers Ltd. *The Embo Journal*. "Crystal structures of open and closed forms of binary and ternary complexes of the large fragment of Thermus aquaticus DNA polymerase I: structural basis for nucleotide incorporation." Ying Li et al. 17: 24, 7514-7525, 1998. ; 11.21: From *Journal of Molecular Biology*. Mar 14; 32(2). Huberman JA. Riggs AD. "Links On the mechanism of DNA replication in mammalian chromosomes." 327-41. Copyright Elsevier, 1968. Reprinted with permission.

Chapter 12
Opener: From Patrick Cramer, David A. Bushnell, Roger D. Kornberg. "Structural Basis of Transcription: RNA Polymerase II at 2.8 Ångstrom Resolution." *Science*, Vol. 292:5523, 1863-1876, June 8, 2001. Reprinted with permission of AAAS.; 12.12a(left): From Seth Darst, Bacterial RNA polymerase. *Current Opinion in Structural Biology*. Reprinted with permission of the author.; 12.12a: From Patrick Cramer, David A. Bushnell, Roger D. Kornberg. "Structural Basis of Transcription: RNA Polymerase II at 2.8 Ångstrom Resolution." *Science*, Vol. 292:5523, 1863-1876, June 8, 2001. Figure 1. Image courtesy of David Bushnell.; p. 316: Tilghman

et al. "Intervening sequence of DNA identified in the structural protion of a mouse β-glovin gene." *PNAS*. 1978, Vol. 75. f. 2, p. 727. This image is in the public domain.

Chapter 13
Opener: © Tom Pantages; 13.15a: © E. Kiseleva and Donald Fawcett/Visuals Unlimited; 13.15b: © Tom Pantages; 13.21: Miller , O. L. *Scientific American*. Vol. 228:3, 1973. p.35. Image courtesy O.L. Miller.

Chapter 14
Figure: 14.10b: © Mitchell Lewis, University Of Pennsylvania Medical Center/SPL/Photo Researchers.

Chapter 15
Opener: Zif268 zinc fingers bound to DNA (Pavletich and Pabo, 1991). Courtesy Song Tan, Penn State University, <http://www.bmb.psu.edu/faculty/tan/lab>.

Chapter 16
Opener: © Robert Brooker; 16.1a(left): © Phototake Inc./Alamy; 16.1a(right): © Phototake Inc./Alamy; 16.3a-b: Aulner et al. The AT-Hook Protein D1 Is Essential for Drosophila melanogaster Development and Is Implicated in Position-Effect Variegation. *Molecular and Cellular Biology*, February 2002, p. 1218-1232, F. 7. Vol. 22, No. 4. © 2002 © American Society for Microbiology. Image courtesy of Emmanual Kas.; 16.5: © Scott Aitken, www.scottpix.com; 16.22: © Dr. Kenneth Greer/Visuals Unlimited.

Chapter 17
Opener: © Matt Meadows/Peter Arnold; p. 460: Reprinted by permission from Macmillan Publishers Ltd. *Nature*. New Giemsa method for the differential staining of sister chromatids. Perry P. & Wolff S. 251:5471, 156-158, 1974.; 17.4: © Dr. John D. Cunningham/Visuals Unlimited.

Chapter 18
Opener: © Argus Fotoarchive/Peter Arnold.

Chapter 19
Opener: © Najlah Feanny/Saba/Corbis; 19.2: © M. Greenlar/Image Works; 19.3: R. L. Brinster and R. E. Hammer, School of Veterinary Medicine, University of Pennsylvania.; 19.5: Glofish © www.glofish.com ; 19.6: © Alan Handyside, Wellcome Images; 19.10: © Photo courtesy of the College of Veterinary Medicine, Texas A&M University/Corbis; 19.14: © Jack Bostrack/Visuals Unlimited; 19.16: Courtesy Monsanto.; 19.17: © Richard T. Nowitz/Phototake; 19.18: © Bill Barksdale/AGStockUSA.

Chapter 20
Opener: © Dr. Peter Lansdorp/Visuals Unlimited; 20.3: From Reid, T., Baldini, A., Rand, T.C., and Ward, D.C. "Simultaneous visualization of seven different DNA probes by in situ hybridization using combinatorial fluorescence and digital imaging microscopy. *PNAS*. 89: 4.1388-92. 1992. Courtesy Thomas Reid.; 20.16a: From *Pulsed Field Gel Electrophoresis, A practical guide*. Bruce Birren and Eric Lai. Figure .1, p. 109. Copyright Elsevier 1993. Reproduced with permission.; 20.16b: From Peter D. Butler and E. Richard Moxon. A physical map of the genome of Haemophilus influenzae type b. *J Gen Microbiol*. 1990 Dec; 136 (Pt 12):2333-42. Reprinted by permission of the Society for General Microbiology.

Chapter 21
Opener: © Alfred Pasieka/Photo Researchers; p 580: Courtesy Joseph DeRisi.; 21.5b: © Medical School. University of Newcastle upon Tyne/Simon Fraser/Photo Researchers.

Chapter 22
Opener: © Corbis RF; 22.1(bottom): © Mitch Reardon/Photo Researchers; 22.1(middle): © Hiroya Minakuchi /Seapics.com; 22.1(top): © Dr. P. Marazzi/Photo Researchers; 22.4: © Hulton-Deutsch Collection/Corbis; 22.8: © CNRI/SPL/Photo Researchers; 22.11: © Reprinted by permission from Macmillan Publishers Ltd. *Nature*. Transforming activity of DNA of chemically transformed and normal cells. Authors: Geoffrey M. Cooper, S. Okenquist & L. Silverman. 284: 5755, 418-421, 1980.; 22.20a-b: Courtesy Dr. Ruhong Li, Molecular and Cell Biology, University of California at Berkeley. Permission granted by The Duesberg Laboratory. Appeared in *Science*, Vol. 297, Number 5581, Issue 26, Jul 2002, p. 544. Author: Jean Marx, Title: "Debate Surges Over the Origins of Genomic Defects in Cancer."; 22.21a: Reprinted by permission from Macmillan Publishers Ltd. *Nature*. "Distinct types of Diffuse large B-cell lymphoma Identified by Gene Expression Profiling." Ash Alizadeh et al. 403:6769, 503-511, 2000. Image courtesy Ash Alizadeh.

Chapter 23
Opener: N.A. Callow/NHPA/PhotoShot; 23.3: Courtesy of E. B. Lewis., California Institute of Technology.; 23.8b-c: Christiane Nusslein-Vol.hard, *Development*, Supplement 1, 1991. © The Company of Biologists Limited.; 23.11(top): Christiane Nusslein-Vol.hard, *Development*, Supplement 1, 1991. © The Company of Biologists Limited.; 23.11(bottom three): Jim Langeland, Steve Paddock and Sean Carroll/University of Wisconsin - Madison; 23.13a: © Juergen Berger/Photo Researchers; 23.13b: F. R. Turner, Indiana University/Visuals Unlimited; p. 650: (both): Horvitz, H.R. & Sulston, J. (1980) "Isolation and genetic characterization of cell lineage mutants of the nematode Caenorhabditis elegans." *Genetics*. 96, 435-454. fig. 1a&b (1980).Courtesy Dr. Horvitz.; 23.18: © Photodisc/Object Series/Vol. 50/Getty Images RF; 23.21: © Jeremy Burgess/Photo Researchers; 23.24a-c: Elliott Meyerowitz and John Bowman. *Development*. Vol. 112:1-20, 1991. Courtesy of Elliott Meyerowitz.; p. 666: Courtesy Stephen Small/New York University.

Chapter 24
Opener: © PhotoDisc/Getty Images RF; 24.1a: © Miguel Castro/Photo Researchers; 24.1b: © D.Parer & E.Parer-Cook/ardea.com; 24.2: © Geoff Oxford; 24.7b: © Corbis RF; 24.13a: © OSF/Animals Animals; 24.15: © Paul Harcourt Davies/SPL/Photo Researchers, Inc; 24.16(right): © Gerald & Buff Corsi/Visuals Unlimited; 24.21: © Leonard Lessin/Peter Arnold; 24.22: © Mikael Karlsson/Alamy.

Chapter 25
Opener: PhotoDisc/Vol. 74/Getty Images RF; 25.1a: From Albert and Blakeslee, Corn and Man, *Journal of Heredity*. 1914, Vol. 5, pg. 51. Oxford University Press. This image is in public domain.; 25.3a(bottom): © Nigel Cattlin/Visuals Unlimited; 25.3a(top): © Grant Heilman Photography/Alamy; 25.12(mustard): © Inga Spence/Visuals Unlimited; 25.12(kohlrabi): © Nigel Cattlin/Photo Researchers ; 25.12(kale): © Valerie Giles/Photo Researchers; 25.12(broccoli, sprouts, cabbage, cauliflower): © Michael P. Gadomski/Photo Researchers; 25.11(bottom left): © Philip Gould/Corbis; 25.11(bottom right): © Juniors Bildarchiv/Alamy; 25.11(top left): © Henry Ausloos/Animals Animals; 25.11(top right): © Roger Tidman/Corbis.

Chapter 26
Opener: © LWA/Stephen Welstead/Blend Images/Corbis RF; 25.1a(left): © Mark Smith/Photo Researchers; 25.1a(Right): © Pascal Goetgheluck/ardea.com; 26.1b(left): © Rod Planck/Photo Researchers; 26.1b(right): © Ron Austing/Photo Researchers; 26.23(right): © Gerald and Buff Corsi/Visuals Unlimited; 26.3(left): © Paul & Joyce Berquist/Animals Animals; 26.20a(left): Courtesy Ed Laufer; 26.20b-c: Courtesy of Dr. J.M. Hurle. From *Development*. 1999 Dec; 126(23):5515-22. Reproduced with permission.; 26.22a-b: © Prof. Walter J. Gehring, University of Basel.

Appendix
1a: © Michael Gabridge/Visuals Unlimited; 1b: © Fred Hossler/Visuals Unlimited; 8: © Richard Wehr/custom Medical Stock Photo.

Line Art

Chapter 1
1.1 a,b: Courtesy of the U.S. Department of Energy Genome Programs. http://genomics.energy.gov

Chapter 9
9.18: Illustration, Irving Geis. Rights owned by Howard Hughes Medical Institute. Not to be reproduced without permission.

Chapter 12
12.12: Reprinted from "Bacterial RNA Polymerase" by Darst, S.A. Current Opinion in Structural Biology, vol. 11 (2) p. 157. Copyright 2001 with permission from Elsevier.

Chapter 14
14.10: From Crystal Structure of the Lactose Operon Repressor and its Complexes with DNA and Inducer by M. Lewis et al., Science, vol. 271, pp.1247-1254, March 1, 1996. Reprinted with permission from AAAS.

Chapter 20
20.18: From Whole-genome random sequencing and assembly of Haemophilus influenzae Rd by Fleischmann et al. Science 28 July 1995, vol. 269 pp. 496-512. Reprinted with permission of AAAS.

Chapter 22
22.21: Reprinted by permission from Macmillan Publishers LT: Nature "Distinct types of diffuse large B-cell lymphoma identified by gene expression profiling" by Alizadeh et al., February 2000, vol. 403, pp. 503-511, Copyright 2000.

Chapter 26
26.9: Reproduced with permission of FASEB from Olsen, G.J., and Woese, C.R. (1993) Ribosomal RNA: a key to phylogeny. FASEB Journal 7: 113-123. Permission conveyed through Copyright Clearance Center, Inc.

INDEX

Page numbers in *italics* indicate figures or tables. Page numbers preceded by A indicate material in the appendix.

A

D

F

G

H

I

J

K

L

M

N

O

P

Q

R

S

T